2021 IEEE 16th International Conference on Nano/Micro Engineered and Molecular Systems (NEMS 2021)

Xiamen, China
25 – 29 April 2021

Pages 1-946

IEEE Catalog Number: CFP21NME-POD
ISBN: 978-1-6654-3008-1

**Copyright © 2021 by the Institute of Electrical and Electronics Engineers, Inc.
All Rights Reserved**

Copyright and Reprint Permissions: Abstracting is permitted with credit to the source. Libraries are permitted to photocopy beyond the limit of U.S. copyright law for private use of patrons those articles in this volume that carry a code at the bottom of the first page, provided the per-copy fee indicated in the code is paid through Copyright Clearance Center, 222 Rosewood Drive, Danvers, MA 01923.

For other copying, reprint or republication permission, write to IEEE Copyrights Manager, IEEE Service Center, 445 Hoes Lane, Piscataway, NJ 08854. All rights reserved.

*** This is a print representation of what appears in the IEEE Digital Library. Some format issues inherent in the e-media version may also appear in this print version.

IEEE Catalog Number: CFP21NME-POD
ISBN (Print-On-Demand):   978-1-6654-3008-1
ISBN (Online):   978-1-6654-1941-3
ISSN: 2474-3747

Additional Copies of This Publication Are Available From:

Curran Associates, Inc
57 Morehouse Lane
Red Hook, NY 12571 USA
Phone: (845) 758-0400
Fax: (845) 758-2633
E-mail: curran@proceedings.com
Web: www.proceedings.com

2021 IEEE 16th International Conference on Nano/Micro Engineered and Molecular Systems (NEMS 2021)

Xiamen, China
25 – 29 April 2021

Pages 1-946

IEEE Catalog Number: CFP21NME-POD
ISBN: 978-1-6654-3008-1

Table of Contents

PL1: Plenary Talk I

Bioinspired Super-wettability System and Beyond   1
——Quantum-confined Superfluid: Energy Conversion, Chemical Reaction and Biological
Information Transfer
Lei JIANG

PL2: Plenary Talk II

Atomically Precise Chemical, Physical, Electronic, and Spin Contacts   3
Paul S. WEISS

PL3: Plenary Talk III

Pushing Laser Precision Engineering from Micro-scale to Nano-scale: Progress, Challenges and   5
Opportunities
Minghui HONG

PL4: Plenary Talk IV

From Supramolecular to Adaptive Chemistry – Contributions to Nanoscience and Technology   7
Jean-Marie LEHN

PL5: Plenary Talk V

Semiconductor Nanowires for Optoelectronics Applications   9
Chennupati JAGADISH

PL6: Plenary Talk VI

Nanomaterials of High-Surface Energy for Efficient Electrochemical Energy Conversion and   11
Storage
Shi-Gang SUN

PL7: Plenary Talk VII

Molecules in Motion: From Biology to Chemistry   13
Jean-Pierre SAUVAGE

PL8: Plenary Talk VIII

Nano-calorimetry: A New Tool for Materials Development   15
Joost J. VLASSAK

PL9: Plenary Talk IX

Transformative Impact of Printable Solar Cells for Meeting Next-Generation Energy Demands   17
Alex JEN

PL10: Plenary Talk X

Van der Waals Heterostructures   19
Kostya NOVOSELOV

CH1: Chihming Ho Award Session

Integrated Electrical Test Vehicle Co-designed with Microfluidics for Evaluating the Performance of   21
Embedded Cooling
Yuxin Ye, Nan Zhang, Lihang Yu, Bo Cong, Ruiwen Liu, Yanmei Kong and Binbin Jiao

Frequency Characteristics of Microfluidic Single Gate Oscillator   N/A
Zhou Zhou, Manman Xu, Hai Wang, Kunpeng Zhang, Daoheng Sun

Minimal-Invasive Levodopa Sensing Based on Differential Amperometry Microneedle Electrodes Decorated with Spike-like Au Nanoparticles   26
Hangxu Ren, Xiyu Mao, Shanshan Zhang, Yu Cai, Lu Fang, Xuesong Ye, Bo Liang

Silicon Nanowire Array Sensor for Highly Sensitive and Selective Detection of Nerve Agent Simulant Vapor via surface hydroxyl groups   30
Xingqi Liu, Hongpeng Zhang, Zhiping Huang, Xuanlin Yang, Shixing Chen, Yuelin Wang, Tie Li, Zhenxing Cheng

Magnetic Force Enabled Plant Seed Levitation to Simulate Microgravity Environment on A Microfluidic Chip   N/A
Jing Du, Lin Zeng, Zitong Yu, Sihui Chen, Xi Chen, Yi Zhang, Hui Yang

WK1: Wen H Ko Memorial Session

MEMS/SEMI Ecosystem in China --2021   N/A
Lucy HUANG

Forever Scaling—Pushing the Limit of Miniscule Sensors and Actuators   N/A
Zenghui WANG

Forever Pioneer- In Memory of Prof. Wen Ko   N/A
Tie LI

BC1: Best Conference Paper Session

Chip-level-microassembly Comb-drive XYZ-microstage with Large Displacements and Low Crosstalk   43
Gaopeng Xue, Masaya Toda, Xinghui Li, Takahito Ono

Effect of the Different Substrates and the Film Thickness on the Surface Roughness of Step Structure   47
Chenying Wang, Jiangtao Pu, Lei Li, Weixuan Jing, Yijun Zhang, Yaxin Zhang, Feng Han, Ming Liu, Wei Ren, Zhuangde Jiang

Efficient Infrared–Thermal-Electric Conversion with Textured Dielectric Film   51
Yunqian He, Yuelin Wang, Tie Li

Inertial Focusing Chip Based on Superposed Secondary Flows   55
Jianguo Feng, Qi Wang, Yuanming Ma, Honglong Chang, Gaobin Xu

A Sensorised Forcep Based on Piezoresistive Force Sensor for Robotic-assisted Minimally Invasive Surgery   60
Cheng Hou, Jiangjun Geng, Yuyang Sun, Tao Chen, Fengxia Wang, Hongliang Ren, Xiuli Zuo, Yanqing Li, Huicong Liu, Lining Sun

In situ Laser-assisted Micromachining of Environmentally-responsive Hydrogel Films   64
Hongjie Jiang, Manuel Ochoa

Three-dimensional Graphene FETs for pH Detection   68
Tao Deng, Yuning Li, Zhaohao Zhang, Yang Zhang, Zewen Liu

IS1: Advanced Micro/nano Fabrication, Materials and Applications

Surface and Bulk Micro/Nanomachining of Polymers   N/A
Junshan LIU

Flexible Healthcare Sensors   N/A
Xuewen WANG

Hexagonal Boron Nitride Phononic Crystal Waveguides for Classical and Quantum Signal Transduction   N/A
Yanan WANG

Micro-Fluidic Heat Dissipation Technology Based on Multi-elements Compatible MEMS Process   N/A
Yunna SUN

Accurate Three-dimensional Physical Simulations of Micro/Nano Fabrication Processes   N/A
Zaifa ZHOU

Micro-system based Multimodality Imaging System for Quantitative Molecular Imaging and Precision Health   N/A
Zhen QIU

IS2: Thermal management in MicroNano Systems

Thermal Transport of Mechanically Deformed 2D Materials *Baoxing XU*	N/A
Thermal Transport Properties of Metals and Metallic Nanostructures *Hua BAO*	N/A
Rational Design of Conductive Polymers for Flexible Thermoelectric Device *Hui LI*	N/A
Layered Two-Dimensional Materials Showing Two Extremes of Heat Transport *Hyejin JANG*	N/A
Layer Dependent Thermal Transport Properties of 2D WSe2 *Xian (Annie) ZHANG*	N/A
Temperature-Microstructure-Properties Relationships During Al-Sn-Al Thermo-Compression Wafer Bonding *Zhiyuan ZHU*	N/A

IS3: Resonant Micro/Nanoelectromechanical Systems

High Resolution Micro-sensors Based On Coupled Resonators/Modes *Chun Zhao*	N/A
Multimode Resonant Micromechanical Systems in Liquid for Biophysical Studies of Cancer Cells *Hao Jia*	N/A
Sensing with Serial or Paralell Micromechanical Resonators *Honglong Chang*	N/A
Atomically-Thin MoS_2 Nanoelectromechanical Resonators *Rui Yang*	N/A
Nonlinear MEMS: fundamentals and applications *Xueyong Wei*	N/A
Ultra-Wide Bandgap β-Ga2O₃ Resonant Nanoelectromechanical Systems (NEMS) *Xuqian Zheng*	N/A

IS4: Self-powered Micro/Nano Systems

Tribotronics for Active Mechanosensation and Self-Powered Microsystems *Chi ZHANG*	N/A
Pulsed Triboelectric Nanogenerator and Self-Powered Sensing System *Gang CHENG*	N/A
High Efficient Vibration Energy Harvesting for Self-powered Machine Monitoring Application *Jian HE*	N/A
Nanogenerators for Self-powered Wearable Physical Monitoring Systems *Wei TANG*	N/A
Triboelectric polymers and electrostatic power source *Xiangyu CHEN*	N/A
Self-Powered Triboelectric Pressure Sensors *Zhen WEN*	N/A

IS5: Bioelectronic Devices and Systems

Development of Self-powered Sensors towards Machine-learning Enabled Smart Systems *Chengkuo LEE*	N/A
Interactive Neuromorphic Synaptic Devices and Systems *Qijun SUN*	N/A
Minimally Invasive Devices for Biomedical Applications *Xi XIE*	N/A

Flexible bioelectronic devices for monitoring and regulating physiological activities *Xian HUANG*	N/A
Shape-adaptable Biomedical Devices *Xuemin DU*	N/A
Self-Powered Medical Electronics *Zhou Li*	N/A

IS6: Non-linear Micro/Nanoelectromechanical Systems

Nonlinear Interaction in Coupled Cantilevers and Its Sensing Applications *Cao XIA*	N/A
Insight of Sensing Mechanisms for Conjugated Polymer Based Gas Sensors *Jian SONG*	N/A
Ultrasensitive mass sensing utilizing nonlinear mode localization of MEMS curved beams with distributed electrodes *Jian ZHAO*	N/A
Nonlinear Broadband Energy Harvesting and Applications *Junyi CAO*	N/A
Dynamic-Based Micro and Nano Devices and Phenomena *Mohammad I. Younis*	N/A
Nonlinear dynamics of coupled MEMS resonators and its application *Ronghua HUAN*	N/A

S1: Micro/Nano/Molecular Fabrication I

Opportunities for Single-Molecule Electronics: A Ten-Year Perspective *Wenjing HONG*	138
Synchronized Surface Modification of TiO2 Composite Nanofiber Through Core-Shell Electrospinning *Hao Peng, Guoyi Kang, Jiaxin Jiang, Zungui Shao, Juan Liu, Yifang Liu, gaofeng zheng*	140
Precursor-Based ZnO Nano Inks for Printed Electronics *Wenxin Liu, Hao Dong, Di Wang, Xing Chen*	144
Efficient Manufacturing of Microdome Array for Advanced Electronic and Optical Devices *Zehua Xiang, Haobin Wang, Hang Guo, Ji Wan, Chen Xu, Liming Miao, Mengdi Han, Haixia Zhang*	148
Solvent-free Nanofabrication Based on Ice-assisted Electron-beam Lithography *Yu Hong, Ding Zhao, Min Qiu*	N/A
Ice-Assisted Electron-Beam Lithography for 3D Nanofabrication *Ding Zhao, Yu Hong, Min Qiu*	N/A

S2: Energy Conversion and Storage I

Conformal Bioelectronic Interfaces *Xiaodong CHEN*	156
Broadband Piezoelectric Harvester Utilizing Three-wafer Bonding for Self-powered Wireless Sensor System *Xin Chen, Daqiao Tong, Jian Zhang, Lingxiao Gao, Fayang Wang, Hongzhi Pan, Liang Cao, Xiaojing Mu*	158
Design and Fabrication of a Metal-silicon Actuator with Large Displacement and Low Voltage *Ying Bu, Jin Xie, Kaiquan Li, Zhuang Xiong, Bing Tang, Qi Tao, Jun Dai*	N/A
Flexible Single Electrode Triboelectric Nanogenerator *Yuchi Liu, Rami Ghannam, Xiaosheng Zhang*	N/A
Development of Closed-Loop Soft Actuators with Self-Sensing Capability for Advanced Wearable Assistive Robotics *Hwajoong Kim, Jaehong Lee*	N/A
Wearable Triboelectric Nanogenerator based on the Conductive Textile with Polyaniline Grafted *Yun Tian, Chuanhong Zhou, Zejue Yu, Zhangwei Liu, Chenyang Xing, Bo Meng, Zhengchun Peng*	N/A

S3: Micro/Nano Pores and Channels

Organic Molecular Sieve Membranes   170
Zhongyi JIANG

Regulatory Mechanism of Ionic/Molecular Transport Behaviors in Nanoscale Channels   N/A
Yaqi Hou, Xu Hou

Fabrication of Nanoslits with <111> Etching TSWE Method   174
Hao Hong, Li Ye, Ke Li, Pasqualina M Sarro, Guoqi Zhang, Zewen Liu

Confined Colloids Gating System for Tunable Fluid Transport   N/A
Zhizhi Sheng, Mengchuang Zhang, Jing Liu, Xu Hou

Characterization of ITO-SiNx Nanopores for Single-Biomolecular Sensing   180
Xin Zhu, Chaoming Gu, Xiaojie Li, Zhi Ye, Zhen Cao, Yang Liu

Mechanical Energy Harvester for Smart Shared Bicycle Application   184
Junlong Huang, Chen Bao, Anxin Luo, Fei Wang

S4: MEMS/NEMS I

Triboelectric Nanogenerator for Powering Body-implantable Electronics   189
Sang-Woo KIM

Analysis of Vibration Modes in TPoS Disk Resonators   N/A
Feilong Li, Cheng Tu, Jiannan Chen, XIAOSHENG ZHANG

A Resonant Differential Pressure Sensor Based on Bulk Silicon Technology   193
Chao Cheng, Yadong Li, Yulan Lu, Chao Xiang, Junbo Wang, Deyong Chen, Jian Chen

An Ultralow-Ripple Polarization Voltage Generator based on High- Voltage Bandgap Reference for MEMS Gyroscopes   197
Hua Chen, Zhen Meng, Xingcheng Zhang, Ke Liu

Electrochemical Micromachining of Copper Workpiece with Ultrashort Voltage Pulses   N/A
Zhen Ma, Lianhuan Han, Hantao Xu, Bingqian Du, Yang Wang, Dongping Zhan

Development of thin glass-based biconvex microlens via thermal expansion   203
Yusufu Aishan

Lubricated Non-immunogenic Neural Probes for Lifelong use brain-machine interfaces   N/A
Yeontaek Lee, Hyogeun Shin, Il-Joo Cho, Jungmok Seo

S5: Micro/Nano Fluidic I

Nanofluidic devices for Biosensing and Energy Conversion   209
Xinghua XIA

Inner Surface Design of Microfluidic channels for Microscale Flow Control   N/A
Shuli Wang, Xu Hou

Carbon Nanotube-Based Ionic Diodes: Design and Mechanism   N/A
Ran Peng, Yueyue Pan, Biwu Liu, Zhi Li, Peng Pan, Shuailong Zhang, Zhen Qin, Aaron R. Wheeler, Shirley Tang, Xinyu Liu

Research on a MEMS Detonated Device with Built-in Safety and Arming Device   215
Kexin Wang, Tengjiang Hu, Yulong Zhao, Wei Ren

Molecular Dynamics Simulation on Directional Wetting Behavior of a Nano Droplet on Titanium Oxides Surface   N/A
Chenhua Liu, Xiangmeng Li, Xijing Zhu

A Human Cornea-on-a-Chip for the Study of Epithelial Wound Healing by Extracellular Vesicles   N/A
Zitong Yu, Rui Hao, Jing Du, Xi Chen, Yi Zhang, Wei Li, Zhongze Gu, Hui Yang

Flow-through Electroporation Using Silver-PDMS Based 3D Sidewall Microelectrodes   228
Yao Cai, Yunyi Huang, Guangyu Qin, Duli Yu, Xiaoxing Xing

PS1: Poster Session I

Self-driven photoelectrochemical UV-visible photodetectors using ZrO_2@TiO_2 core-shell nanorod arrays modified with single-walled carbon nanotubes   232
Zhen Wang, Renrong Zheng, Na Wang, Haisheng San

Electronic Skin Based on Electrospun PVDF Nanofibers for Slipping Detection — 236
Lingke Yu, Wenchang Tu, Shanshan You, Lingling Liu, Daoheng Sun

A Flexible Hybridized Electromagnetic-Triboelectric Nanogenerator and its application for 3D Trajectory Sensing — N/A
Ji Wan, Haobin Wang, Liming Miao, Hang Guo, Chen Xu, Hai Xia (Alice) Zhang

A Simple Bandwidth Broadening Method of Terahertz Metamaterial Absorber by Partially Removing The Dielectric Layer — 242
Jia Xu, Jiahao Miao, Yi Liu, Yuan Tian, Xiaomei Yu

Tough Self-Healing Polymer Encapsulated Stretchable Conductive Fibers for Systematical Integration of Wearable Electronics — N/A
Chaebeen Kwon, Taeyoon Lee

Miniaturized Piezoresistive Sensors with Wide Working Range for Wearable Human Machine Interfaces — N/A
Hongcheng Xu, Ling Duan, Libo Gao

Water Drop Detection Contained in Lubricant Oil for Submersible Pump Fault Diagnosis — 251
Fumikazu Mizutani, Michio Shimimura, Akinori Mori, Yoshikazu Yoshida, Yoshihiro Hosokawa, Tadao Matsunaga, Sang-Seok Lee

Lateral characteristic calibration of an atomic force microscope using a Cr atomic deposition grating — 255

Junjie Wu, Xiaoyu Cai, Yuan Li, Yunxia Fu, Jiasi Wei

A Miniaturized PM2.5 Concentration Monitoring System Based on QCM Sensor and Optimized Structure Virtual Impactor — N/A
Yong Wang, Dongyang Chen, Qian Zhang, Yinshen Wang, Jin Xie

Preparation of Porous SnO2 and its Ethanol Detection Performance in MEMS Structure Gas Sensor — N/A
Xiao Zhang

High Electrochemical Activity of Oxygen-Doped Graphene Sheets Embedded Carbon Film — N/A
Yuanyuan Cao, Kaikai Sun, Haohua Zhong, Dongfeng Diao

Fast-scan MOEMS Mirror for HD Laser Projection Applications — 265
Huijun Yu, Wenjiang Shen, Peng Zhou

Modeling and Optimization of VLF Piezoelectric Antenna Towards Lower Frequency — 271
Huiliang Liu, Yao Chu, Weiguo Hou, Yinqiao Li, Yuan Yao, Yaxing Cai

Electrically Induced Wire-forming 3D Printing Technology of Flexible Liquid Metal — N/A
Xinpeng Wang, Liangtao Li, Yang Wang, Liang Hu

Amorphous carbon film based weak pressure sensor with ultrathin sensitive structure — N/A
Qi Zhang, Yulong Zhao

Bacterial Outer Membrane Vesicles Presenting Programmed Death 1 for Improved Cancer Immunotherapy via Immune Activation and Checkpoint Inhibition — N/A
Yao Li

Micro-fabricated alkali vapor cells for atomic spin gyroscope study — 282
Yao Chen, mingzhi Yu, Yintao Ma, Guoxi Luo, Zhuangde Jiang, Yu Bai, Libo Zhao

Dopamine modified nanopore for ultra-trace level detection — N/A
weijun li, xu hou

Nano/Micro Wearable Photo-sensor System Based on Novel Carbon Film — N/A
xi zhang, Dongfeng Diao

Electron-Induced Perpendicular Graphene Sheets Embedded Porous Carbon Film for Flexible Touch Sensors — N/A
Sicheng Chen, Lei Yang

Graphene Oxide Nanosheet Assisted Porous Chitosan Sponge for Hemostatic Applications — 293
Yuanyuan YANG, Yajing Shen

High-Performance Flexible Supercapacitors Based on Nitrogen-Doped Graphene Fiber Electrodes — N/A
Feng Han, Weixuan Jing, Bian Tian, Qijing Lin, Chenying Wang, Libo Zhao, Ping Yang, Zhuangde Jiang

Design and Simulation of a Wide-Bandwidth CMUTs Array with Dual-Mixed radii and Multi Operating Modes 299
Zichen Liu, Libo Zhao, Yihe Zhao, Jie Li, Zhikang Li, Yu Bai, Ping Yang, Qijing Lin, Zixuan Li, Jiawei Yuan, Zhuangde Jiang

Transparent, Anti-Freezing Hydrogels for Ultrasensitive Temperature and Strain Sensor Based on A Thin-Film Structure 303
Zixuan Wu, Haojun Ding, Yaoming Wei, Kai Tao, Jin Wu

Uniformly arrayed carbon nanofibers down to several nanometers N/A
Jufeng Deng, Chong Liu, Madou Marc

Self-Powered Transparent Stretchable 3D Motion Sensor N/A
Hang Guo, Haobin Wang, Zehua Xiang, Chen Xu, Wan Ji, Liming Miao, Jiayin Li, Mengdi Han, Haixia Zhang

ENHANCING SENSITIVITY USING ELECTROSTATIC SPRING IN COUPLING MODE-LOCALIZED MEMS ACCELEROMETER 311
Zheng Wang, XingYin Xiong, KunFeng Wang, WuHao Yang, ZhiTian Li, XuDong Zou

Self-powered Delta-Parallel-based Interface for Diversified Control Applications for 2/3D Control in Virtual/Real Space N/A
Cheng Hou, Jiangjun Geng, Tao Chen, Huicong Liu, Lining Sun

Microchip based Gigahertz Acoustic Streaming induced Cellular Internalization of Gold Nanorods N/A
Shan He, Xiaoyu Wu, Wenjun Li, Wei Pang, Xuexin Duan, Yanyan Wang

Direct-write 3D printed self-supporting flexible microstructure devices 319
cuiwen hu, haifeng chen, deyun mo, haishan lian, yinhuan yang, xiaojun chen

Biomimetic Curvatures Controlled Fluid Overflow System N/A
Zhichao Dong

A Flexible Pain Sensor Based on PDMS-AgNWs N/A
Chen Xu, Liming Miao, Haobin Wang, Hang Guo, Zehua Xiang, Haixia Zhang

Liquid gating membranes for multiphase separation N/A
Shijie Yu

A bionic fish with 4 IPMC pectoral fins N/A
zhao chun, Wang Yanjie, Tang Gangqiang, wang jiale, zhu denglin, Luo Minzhou

Ohmic Contact Characteristics of Silicon Carbide-based MEMS Devices 334
Chen Wu

Facile fabrication of NiO wrinkle micro/nanostructures and the application of enzyme-free glucose sensors 339
Shulan Jiang, Yong Tan

Hard protective transparent coatings for flexible electronics N/A
Oleksiy Penkov, Mahdi Khadom

High-efficient Electrochemical Betavoltaic Cell Using ZrO2 Modified TiO2 Nanorod Arrays 345
Renrong Zheng, Zhen Wang, Na Wang, Zan Ding, Haisheng San

Design and Fabrication of PDMS Thin Film Electromagnetic Actuator N/A
Hongguang Lu

Elastomers with ultrafast low temperature self-healing rate and low adhesion to ice N/A
Yizhi Zhuo

Spoof Plasmon Surfaces for Terahertz Sensing N/A
Yi Huang, Shuncong Zhong, Tingling Lin, Yujie Zhong

A Flexible Pressure/Flow Sensor Based on Graphene Piezoresistivity N/A
Zihao Dong, Yonggang Jiang, Deyuan Zhang

Hypoxia/pH dual-responsive nitroimidazole-modified chitosan/rose bengal derivative nanoparticles for enhanced photodynamic anticancer therapy N/A
xudong li, Yu Gao

Research on Key Technology of Novel RF MEMS Switch Designing and Fabricating 358
Yun Qi, Yuanyuan Yang, Wenjiang Shen

Design and Analysis of Dual Substrate RF MEMS Capacitive Switch with Low Actuation Voltage N/A
Yuanyuan Yang, Yun Qi, Wenjiang Shen

A pulse diagnosis instrument based on digital microfluidics system to imitate TCM — N/A
Shouju Yao, Hang Jin, Bin Qiu, Wenchang Tu, yike zhou, Shanshan You, Tianhao Wu, lingling liu, gonghan he, Daoheng Sun

Exploring the Therapeutic Effect of Sonodynamic Therapy in Cancer Cells under Different Ultrasonic Parameters and Treatment Conditions — N/A
Yilin Zheng, Jun Wang, Yu Gao

Stretchable and Flexible Fiber Electrode for Measuring Neural Signal with Au Nanoparticle Conductive Networks — N/A
Chihyeong Won, Taeyoon Lee

Broadband linear-to-circular polarization converter based on ultrathin metal nano-grating — 372
Yuanyi Fan, Ran Zhang, Ze Liu, Chuanlong Guan, Jinkui Chu

Skin Inspired Humidity and Pressure Sensor with Wrinkle-on-Sponge Structure — N/A
Liming Miao, Ji Wan, Chen Xu, Hang Guo, Haobin Wang, Zehua Xiang, Jiayin Li, Mengdi Han, Haixia Zhang

Probing zeta potential of glass in electrolyte solutions by colloidal probe technique — 378
Zhijian Liu, Jiaming Gao, Tianze Wu, Sen Wu, Zixiao Fan, Ziyi Yuan, Yongxin Song, Xinxiang Pan

High-Resolution Measurement Method Based on Array FOV and Coded Apertures — 382
Li Zhang, Fei Xing

Different Effects of Mass and Damping on Performance of Vibration and Wind Energy Harvesters — 386
Xiaokang Yang, Xuefeng He, Zhengguo Shang, Hui Huang, Chunlong Li, Yufei Liu

Nanocomposites Modified Platinum Wire Electrode for Detection of 17β-Estradiol Utilizing the Conformational Changes of Aptamers — N/A
Fanli Kong, Jinping Luo, Tao Ming, Shuai Sun, Yu Xing, Yan Cheng, Shihong Xu, Hongyan Jin, Xinxia Cai

In-situ deposited ion-imprinted polymers for electrochemical detection of trace cadmium in water — 392
Shuyu Xiao, Jingfang Hu, Jiyang Wang, Yu Song, Yansheng Li, Guowei Gao, Lei Qin

Fabrication of Three-dimensional Si based Electrodes for Integrated Li Ion Microbatteries — N/A
Chuang Yue

Electrochemical betavoltaic cell using black TiO_2 nanotube arrays modified with single-walled carbon nanotubes — 399
Na Wang, Renrong Zheng, Zhen Wang, Jiang Chen, Haisheng San

An Acoustic Microrobot Control System with Vision Feedback — N/A
Ying Wei, Hui Shen, Cong Zhao, Huan Ou and Xiaolong Lu

Electrochemical Paper Aptasensor Based on Biotin-Streptavidin System for Label Free Detection of 17β-Estradiol — N/A
Tao Ming, Jinping Luo, Juntao Liu, Shuai Sun, Yu Xing, Fanli Kong, Yan Cheng, Hongyan Jin, Xinxia Cai

Temperature Compensation for MOEMS Micromirror with Piezoresistive Angle Sensor — 407
Lei Qian, Huijun Yu, Jie Hu, Wenjiang Shen

Nanocrystalline Composed SnO_2 Inverse Opal for Highly Sensitive Acetone Gas Sensor at ppb-level — 412
Feng Jiang, Wen Zhang, Yu Wei

Study of the light emission from Eu^{3+} doped nanoporous organosilicate films — 417
JinMing Zhang, Jing Zhang, Yanrong Wang, Md. Rasadujjaman, Shuhua Wei, Jiang Yan

A Highly Sensitive Origami-paper-based Aptasensor For Detection Of Programmed Cell Death Protein 1 (PD-1) — N/A
Shuai Sun, Tao Ming, Jinping Luo, Juntao Liu, Yu Xing, Fanli Kong, Xinxia Cai

Investigation on Electrostatically Actuated Micromachined Gyroscope — N/A
junduo wang, Wenjiang Shen

HIGHLY SENSITIVITY RESONANT BI-DIRECTIONAL MAGNETIC FIELD MICRO-SENSOR — 425
Nouha ALCHEIKH, Sofiane Mbarek, Mohammad Younis

Microelectrode Array for Electrophysiology Detection of Amygdala in Free-moving Mice — N/A
Penghui Fan, Yilin Song, Yuchuan Dai, Yiding Wang, Botao Lu, Enhui He, Xinxia Cai

Photoluminescent Tough Gel Sheathed Suture for Near Infrared Bioimaging — N/A
Zhenwei Ma, Fan Yang, Dongling Ma, Jianyu Li

A Flexible Mechanical Composite Micro-Grating Tailored by One-Dimensional Ordered Wrinkle Patterns
Yi-hang Xin, Kai-Ming Hu, Xiu-Yuan Li, Er-Qi Tu, Wen-Ming Zhang
433

Vibration Characteristics of Piezoelectric Timoshenko Nanobeams in Viscoelastic Medium
Dong Wu, Dapeng Zhang, Mingwei Liu, Yongjun Lei
437

AC & DC Magnetic Field Sensor Based on Flexible Piezoelectric Polymer
Weilong Han, Jian Wen, Yifei Wu, Bo Niu, Yi Wu, Mingzhe Rong
443

Large Area Transient Films and Devices by Photonic Sintering of Two Dimensional Materials
Wenxing Huo, Zi'an Zhang, Zilun Wang, Ziyue Wu, Jiameng Li, Xian Huang
N/A

Fast reconstruction of Raman image based on the multi-channel imaging system and kernel function
Xian-Guang Fan, Long Liu, Ting Nie, Yi-Xin Lin, Ying-Jie Xu, Jian He, Xin Wang
N/A

Magnetically Induced Micropillar Arrays for Ultra-Sensitive Flexible Sensor with Wireless Recharging System
Libo Gao
N/A

A Capacitively Transduced Bulk Acoustic Wave MEMS Resonator with Low Bias Voltages
Zeji Chen, Qianqian Jia, Wenli Liu, Yinfang Zhu, Quan Yuan, Jinling Yang, Fuhua Yang
455

Detection of Single Protein Molecules Using MoS_2 Nanopores of Various Sizes
Chaoming Gu, Zhoubin Yu, Xiaojie Li, Xin Zhu, Zhen Cao, Zhi Ye, Chuanghong Jin, Yang Liu
459

Humidity Sensor Based on Thin-Film Piezoelectric-on-Substrate Resonator and MoS_2 for Multifunctional Applications
Hanyong Dong, Jintao Pang, Dongsheng Li, Qian Zhang, Jin Xie
463

Irreversible Electrowetting on Petal-mimetic Nanotextured Dielectric for Formation of Shape Controllable Polymeric Lenses
Huifen Wei, Xiangmeng Li
N/A

In-situ synthesized liquid metal microgels
Dawei Wang, Chennan Lu, Xiaohong Wang, Wei Rao
469

Design and Development of High Precision Magnetic Encoder Based on TMR MEMS Device
Zhang Bo, Jiang Yong, Li Yang, Chen Xiaoli, Wen Xiaolong
474

Flexible electrical stimulation device with Chitosan-Vaseline® dressing that accelerates wound healing in diabetes and corresponding stimulation optimization based on a precision layered skin model
Menglu Li, Xiaofeng Wang, Jikui Luo, Wei-qiang Tan, Xiaozhi Wang
N/A

Achieving ultrahigh energy density for triboelectric nanogenerator
Jingjing Fu, Yunlong Zi
N/A

Optical micro/nanofibre embedded soft film enables multifunctional flow sensing in microfluidic chips
Zhang Zhang, Jing Pan, Yao Tang, Yue Xu, Lei Zhang, Yuan Gong, Limin Tong
N/A

Dynamical Reversible Electrowetting with low voltage on the Dimethicone Infused Carbon Nanotube Array in Air
Lei Zhou, Miao Wang, Xu Hou
N/A

Trace mercury detection by ruthenium-based MOFs modified microelectrode
Chenyu Xiong, Yuhao Xu, Chao Bian, Ri Wang, Yong Xie, Mingjie Han, Shanhong Xia
N/A

Frequency Matching System of MEMS Gyroscope Based on Fuzzy Control Strategy
yifang Liu, Ruimin Liu, Gang Fu, Xin Yan, JunYu Chen, Gaofeng Zheng
491

High-yield and Large-size suspended graphene device Fabrication technique and Edge Burr Effect Analysis
Ying Liu, Qin Wang, Yong Zhang, Jing Qiu, Guan-jun Liu
N/A

Atomic study on the deform mechanism of CuTa/Cu and CuTa/Ta nanolaminates
Xiao Wang, Mengjie Wang, Min Liu, Yue Liu, Weidong Wang
498

Wireless Implantable Phototherapy Device for Oral Inflammation Repair
Shuang Huang, Cheng Yang, Ziqi Liu, Xinshuo Huang, Hui-Jiuan Chen
503

Development of Multi-functional Biocompatible Photopolymer
N/A

Seokyoung Bang, Dongha Tahk, Noo Li Jeon, Hong Nam Kim

Multiparametric Flexible Sensor Arrays system for in situ Immediate Diagnosis of Glossitis   509
Ziqi Liu, Keer Li, Shuang Huang, Xinshuo Huang, Lingfei Zhou, Tiancheng Sun, Hui-Jiuan Chen

Contribution discrimination of auxetic cantilever for increased piezoelectric output in vibration   513
energy harvesting
Lu Wang

Research on Embedded 3D Printing for Magnetic Soft Robots   518
Wei Zhang, Jiachen Li, Huicong Liu, Guoqing Jin

Research on Coupling Mechanism of Wireless Power Supply Equipment for Film Pressure   524
Wireless Sensing node of Water-lubricated bearing
Yifan Zhao, Nan Wang, Fan Jiang

Investigation of electrostatic-piezoelectric hybrid vibrational power generators with different   528
frequency broadening schemes
Yongqi Cao, Dezhi Nie, Jian Zhang, Yiwei Wang, Ronggang He, Zhe Zhao, Bowen Ji, Jianbing Xie, Kai Tao

Bio-inspired Flexible, Dual-modulation Synaptic Transistors Towards Artificial Visual Memory   N/A
Systems
Fuqin Sun, Qifeng Lu, Ting Zhang

Kinetic investigation into fast supramolecular processes and chemical reactions through   N/A
microfluidic NMR
Xinchang Wang, Liulin Yang, Xiaoyu Cao, Zhongqun Tian

Liquid Metal Nanoparticles Decorated with Graphene oxide for Enhanced Peroxidase-Like Activity   N/A
Xiaohong Wang, Dawei Wang, Chennan Lu, Wei Rao

Size-Dependent Particle Separating in Curved Microfluidic Chip   539
Yun Tian, Yongkang Wang, Yunfei Chen

Tailoring surface morphology and crystal quality characteristics in InGaAs/GaAs quantum well   N/A
structure by inserting high-temperature layer
Quhui Wang, Haizhu Wang, Xiaohui Ma

3D all-metal nano-cavity coupled metamaterial for refractive index sensing   546
kenan zhang, peimin lu, wengang wu, jia zhu, guanzhou lin, yun huang

Humidity Insensitive Nanogenerator Based on Natural Nanofibrils   N/A
Yanyuan Ba, Xiaowen Li, Xiaosheng Zhang

Design and Simulation of a Double-Ended Arrangement Infrared Thermopile   N/A
Yihao Guan, Cheng Lei

Nodding Duck Structured Hybrid Triboelectric-Electromagnetic Nanogenerator towards Ocean   N/A
wave energy harvesting and Self-Powered Monitoring
Liqiang Liu, Xiya Yang, Qunwei Tang

A Facile and Green Microwave Hydrothermal Method for Fabricating g-C3N4 Nanosheets with   N/A
Improved Hydrogen Evolution Performance
Hongmei Chen, Yanyun Fan, Zheng Fan, Danfeng Cui, Chenyang Xue, Wendong Zhang

One-step vapor deposition of fluorinated polycationic coating to fabricate antifouling and   N/A
anti-infective textile against drug-resistant
Ruixiang Zhao

Atomic Layer Deposition of Al2O3 on 2D MoS2 for Enhancing the Recovery rate of Gas Sensor   N/A
Sungjoo Wi, Inkyu Sohn, Youngjun Kim, Seungmin Jung, Hyungjun Kim

Nonlinear dynamic of coupling mechanical modes demonstrated by phononic frequency coms   N/A
Jiangkun Sun

Deep spatial frequency shift enabled chip-based sub-wavelength- resolution imaging   N/A
Mingwei Tang, Xiaowei Liu, Qing Yang, Xu Liu

A Low-Cost Digital Lithography System Supporting Visual Focusing and its Application on Optical   563
Fiber
Luming Wang, Zhimin Zhang, Ningning Luo, Haifeng Xiao, Long Ma, Qingwang Meng

Au-enabled nanostructured sapphire optical fiber sensor for in-situ gas sensing in harsh   N/A
environments

Kai Liu, Paul Ohodnicki, Jeffrey Wuenschell, Subhabrata Bera, Renhong Tang, Lin Li, Yi Liu, Tinghan Liu, Han Wu, Zhihua Huang, Henry Du

Research on the resolution of submicron lithographic projection system based on DMD for optical fiber end face 571
Qingwang Meng, Zhimin Zhang, Ningning Luo, Luming Wang, Menghan Xiong

Gold Nanoparticles Modified Sensitivity-Enhanced Uncooled Near-infrared Detector N/A
Xiaoyu Wu, Quanquan Guo, Wencheng Li, Chenyang Yu, Yanyan Wang

Research on Nonlinear Shrinkage Characteristics of Small Modulus Plastic Gear Based on Moldflow N/A
Xiansong He, Wangqing Wu, Yihua Lei

CS1: 学术好莱坞讲座

学术好莱坞讲座：现代科研论文的构思与写作 N/A
Fengyu LI

IS7: IS: Micro Nano and Molecular Systems for Diagnostics and Therapeutics

Fusion Application of Medical Sensors and Internet of Medical Things N/A
Jinhong GUO

Semiconducting Molecular Probes for Ultrasensitive Afterglow Imaging and Early Diagnosis N/A
Kanyi PU

Cell-based biosensing technology and system N/A
Ning HU

Evanescent wave fluorescent biosensor for the detection of microRNA N/A
Xiaohong ZHOU

3D Printing Magnetic Soft Milirobots by Recirculating Vat Polymerization for Point-of-Care Diagnostics N/A
Yi ZHANG

Active sensing and hybridized sensors towards tactile intelligence N/A
Bo MENG

IS8: IS: Micro-/Nanofluidic Devices and Systems

Patient-derived organoids analyzed on a superhydrophobic microwell array for predicting drug response of lung cancer patients within a week N/A
Peng LIU

Nanowire-integrated microfluidic devices to identify urinary microRNA groups for cancer detection N/A
Takao YASUI

Flexible filter based liquid biopsy N/A
Wei WANG

A Fully Integrated and Automated Microfluidic System for Rapid Testing of Respiratory Viruses N/A
Liangbin PAN

Bioinspired Surface Microfluidics for Digital PCR and Digital ELISA N/A
Tianzhun WU

Multi-dimensional manipulation of Solid/Liquid interaction N/A
Xu DENG

IS9: IS: Multidisciplinary Frontier: Advanced Chemistry Materials for MEMS/NEMS

Fabricating fluorescent glass Derived from Mesoporous Powders N/A
Lianjun WANG

Controllable liquid transfer for making high-performance thin-film devices N/A
Huan LIU

Photo-crosslinkable, insulating silk fibroin for bioelectronics with enhanced cell affinity N/A
Jie JU

Unconventional Nanotemplate-based Technique and Its Emerging Applications N/A

Liaoyong WEN

Paintable hydrogels based on silane chemistry
Xi YAO — N/A

Electrodepostion filling method of TSV for 3D integration
Yan WANG — N/A

IS10: IS: Multidisciplinary Frontier: Micro/Nano Energy and Smart Electronics

2D MEMS metamaterial modulators for THz Communication
Xiuhan LI — N/A

Fiber Structured Flexible Pressure Sensor and Electronic Skin
Zhaoling LI — N/A

Self-Powered Sensing Techniques Towards Intelligent Human-machine Interaction
Huicong LIU — N/A

Surface Plasmon Polaritons Enhanced 2D-Material Photodetector for Infrared Application
Wen HUANG — N/A

Regulation of Strong and Weak Bonds in Lithium Metal Batteries to Enhance Their Performance
Yong ZHAO — N/A

Achieving Ultrahigh Output Energy Density of Triboelectric Nanogenerator in High-Pressure Gas Environment
Yunlong ZI — N/A

IS11: IS: Electronic and Optoelectronic Fibers and Textiles

Tuning the Viscosity for a Multi-material Multi-functional Fiber
Chong Hou — N/A

Nanoparticles Enabled Drawing of Ultra-Long Metal Nanowires
Xiaochun Li — N/A

Electronic Multi-material Fiber Sensors and Energy Harvesters
Fabien Sorin — N/A

Advanced Functional Semiconductor Fibers and Textiles
Lei Wei — N/A

Computational Design of novel two-dimentional semiconducting silicon carbides for optoelectric applications
Liujiang Zhou — N/A

Nano-/Micro-enabled Multimaterial Fibers for Wearable and Biosensing Applications
Xiaoting Jia — N/A

IS12: IS: RF Micro/Nano Components and Systems

Investigation on Energy Damping Mechanisms and Q-enhancement Strategies in Piezoelectric-on-Silicon MEMS Resonators
Cheng TU — N/A

MEMS Resonators with High Q and Good Temperature Stability
Guoqiang WU — N/A

Novel RF MEMS devices using AlGaNGaN heterostructure
Haoshen ZHU — N/A

Mechanically-Flexible RF MEMS Resonators Enabled by FlexMEMS Technology
Menglun ZHANG — N/A

Chip-scale AlN Thin Film Transducers and RF Wake-up Receivers
Tao WU — N/A

SOI-MEMS based RF micromachined switches for wireless applications
Yong Zhu — N/A

BS1: Best Student Paper Session

Microstructural-PVDF Dielectric Layer Based High-Resolution Flexible Capacitive Pressure Sensor — 647

Zebang Luo, Jing Chen, Zhengfang Zhu, Lin Li, Yi Su, Lei Wang, Hui Li

Electrohydrodynamically Printed Multicolor Perovskite Image Sensor Array 652
Qilu Wang, Guannan Zhang, Hanyuan Zhang, Yongqing Duan, YongAn Huang

A Novel Piezoelectric Micromachined Ultrasonic Transducer with Adjustable Broad and Flat 656
Frequency Band
Lei Wang, Wei Zhu, Zhipeng Wu, Wenjuan Liu, Chengliang Sun

Stackable Triboelectric Nanogenerators for Self-powered Marine Monitoring Buoy 660
Jiale Dong, Hao Wang, Xiu Xiao, Taili Du, Yunpeng Zhao, Zhongqi Fan, minyi xu

Modulating the Electrical Transport Characteristics of a Metal-Semiconductor-Metal Structure by 665
Local Strain Gradient
Youchao Huang, Shenhui Ma, Jiaona Zhang, Chao Xie, Min Zhang

A high-frequency narrow-band filtering mechanism based on auto-parametric internal resonance 670
Rong Wang, Cao Xia, Dong F. Wang, Takahito Ono

Calibrate Silicon Nanowires Field Effect Transistor Sensor with its Photoresponse 676
Yi Yang, Yuelin Wang, Shixing Chen, Tie Li

Flexible and Transparent Ultraviolet Photodetector Enabled by Metal Doping ZnO Nanorods Based 680
on Mica Substrate
Yao huang, Hainan Zhang, Guanhua Dun, Yuhua Li, Renrong Liang, Yi Yang, Dan Xie, He Tian, Tian-ling Ren

S6: MEMS/NEMS II

Liquid Metal Nano Electronics 685
Jing LIU

Design and Manufacture of a High Precision MEMS Flexible Force Sensor 687
Huimin Zhang, Yanlu Feng, Wei Zhang

Linear stiffness tuning of MEMS triangular capacitors 691
Yiming Jin, Zhipeng Ma, Yixuan Guo, Tengfei Zhang, Xudong Zheng, Zhonghe Jin

Power Handling Capability Enhanced RF MEMS Switch Using Modified-Width Cantilevers Structure 695

Yulong Zhang, Zhuhao Gong, Huiliang Liu, Zewen Liu

Research on Motion Characteristics for Latching Mechanism of MEMS Safety and Arming Device 699
under Dual Environmental Forces
Shenghong Lei, Yun Cao, Weirong Nie, Zhanwen Xi, Na Xu, Weixiang Qiu

Modeling and Optimization of ScAlN-based MEMS Mirror with Large Static Two-axis Tilting Angle 704
Changhe Sun, Bolun Li, Wenqu Su, Yufei Liu, Yaming Wu

S7: Nanophotonics and Nanoscale Imaging

Wavelength conversion through plasmon-coupled surface states 709
Mona JARRAHI

Photonic Nanojet Produced by A Microfluidic Channel for Biofluid Monitoring 711
Guoqiang Gu, Pengcheng Zhang, Hui Yang

Fluorescence Enhancement Utilizing Dielectric Microbeads with Semi-open Microwells 715
Pengcheng Zhang, Bing Yan, Guoqiang Gu, Zitong Yu, Xi Chen, Zengbo Wang, Hui Yang

Spectrometer-free Refractometric Sensing Using Image Recognition on Centimeter-scale Gradient N/A
Nanostructures
Siyi Min, Shijie Li, Zhouyang Zhu, Yu Liu, Chuwei Liang, Jingxuan Cai, Fei Han, Yuyan Li, Xing Cheng,
Wen-Di Li

Strongly Anisotropic Monolayer InSe polarized light detector N/A
Xusheng Wang, xi zhang

Self-powered textile-based tactile sensors inspired by human-skin for multifunctional sensing of N/A
wearables and robots
Changyong Cao

Compare of SNOM, PTIR, PiFM and PFIR N/A
Hai-Long Wang, En-Ming You, Song-Yuan Ding, Zhong-Qun Tian

Full-spectrum optoelectronics based on bandgap-graded materials
Zongyin Yang

N/A

S8: Energy Conversion and Storage II

Flexible Bionic Intelligent Perception System
Jianhua ZHANG

729

Fabrication of laser scribed graphene stretchable supercapacitor by laser-assisted transfer printing strategy
Guangyuan Xu, Yanan Chen, Fangshuai Chen, Yanfang Meng, Yinji Ma, Xue Feng

731

Smart Power Management Microsystem for Distributed Renewable Energy Harvesting
Linhao Feng, Kai Huang, Mingxin Liu, Hongsheng Zhong, Xiaosheng Zhang

737

Card-based Hybrid Piezo-Triboelectric Nanogenerator for Simultaneously Harvesting Sliding Mechanical Energy
Danliang Wen, Peng Huang, Yanyuan Ba, Xiaosheng Zhang

N/A

Self-Powered Electrochemical Interfaces for Material and Energy Conversion
Shuyan Gao, Zhong Lin Wang

N/A

A Micro Device Array for Real-time Monitoring Cardiac Contraction
Li Wang, Jun Chen, Weiguang Su, Anqing Li, Pengbo Liu, Chonghai Xu

N/A

Wind-driven self-powered wireless environmental sensors for Internet of Things at long distance
di liu, baodong chen, jie an, chengyu li, jiajia shao, wei tang, chi zhang, zhonglin wang

N/A

Design of Self-powered Environment Monitoring Sensor Based on TEG and TENG
Jianhao Liu, Changxin Liu, Cong Zhao, Huaan Li, Guanghao Qu, Zhuofan Mao, Zhenghui Zhou

749

S9: Soft & Flexibe Devices and Applications I

Printed flexible/stretchable electronics and applications
Zheng Cui

754

Optical micro/nanofiber enabled compact tactile sensor for hardness discrimination
Yao Tang, Haitao Liu, Jing Pan, Zhang Zhang, Yue Xu, Lei Zhang, Limin Tong

N/A

A Flexible Triboelectric-Electromagnetic Hybridized Nanogenerator
Haitao Deng, Zhiyong Wang, Yanyuan Ba, Xinran Zhang, Danliang Wen, Xiaosheng Zhang

N/A

A Multifunctional ultrasensitive hybrid optical skin
Jing Pan, Lei Zhang, Limin Tong

N/A

An Integrated Stretchable Sensing Patch for Simultaneously Monitoring of Physiological and Biochemical Parameters
Wenhao Pan, Shiyi Xu, Xiyu Mao, Tianyu Li, Qingpeng Cao, Xuesong Ye, Bo Liang

762

A Textile Tactile Sensor for Dual-mode Proximity/Pressure Perception in Smart Robotics
Qinhua Guo, Weiguan Zhang, Zhengchun Peng

N/A

Modeling and Simulation of Flexible Vector Shear Flow Sensor Based on COMSOL Multiphysics
Guochang Liu, Wenping Cao, Haoyu Tan, Renxin Wang, Wendong Zhang

770

A Self-powered retractable device based on triboelectric nanogenerator for sensing of joint and spinal bending/stretching
Chengyu Li, Di Liu, Chaoqun Xu, Ziming Wang, Sheng Shu, Zhuoran Sun, Wei Tang, Zhonglin Wang

N/A

S10: Micro/Nano Fluidic II

Microfluidic Liquid Metal for Stretchable Biomed-electronics
Xingyu JIANG

777

Bioinspired Universal Flexible Elastomer-Based Microchannels
Baiyi Chen, Rongrong Zhang, Hexuan Fu, Xu Hou

N/A

Frequency Research of Microfluidic Wear Debris Detection Chip Based on Inductive Wheatstone Bridge
Yucai Xie, Hongpeng Zhang, Haotian Shi, Yuwei Zhang

780

Liquid Metal-Based Microfluidic for Sperm Thermotaxis.
Yimo Yan, Ran Liu, Boxuan Zhang

N/A

Digital Microfluidics for Efficient and Accurate Molecular Profiling of Single Circulating Tumor Cells — N/A
Qingyu Ruan, Jian Yang, Fenxiang Zou, Lingling Wu, Zhi Zhu, Chaoyong Yang

Bubble formation in nanopores — N/A
Alberto Giacomello, Roland Roth

S11: Nanoscale Robotics, Assembly, and Automation

Small-scale Wireless Robots for Medical Applications — 792
Metin SITTI

Order and Information in the Phases of Spinning Micro-disks — N/A
Wendong Wang, Gaurav Gardi, Paolo Malgaretti, Vimal Kishore, Metin Sitti

Single cell manipulation with acoustically powered microrobotic platforms — N/A
Xiaolong Lu, Ying Wei, Kangdong Zhao, Hui Shen, Wenjuan Liu

Simulation of the Shape-directed AC Driven Defective Micromotors — 798
Rencheng Zhuang, Dekai Zhou, Xiaocong Chang, Longqiu Li

3D printing of magnetically actuated miniature soft robots — 804
Zhongbao Wang, Yigen Wu, Dezhi Wu, Zhenyin Hai

Anisotropic spreading of droplets on striped electrodes — 809
Wei Wang, Yanbo Xie, Antoine Riaud

Low-cost Micro Search-Coil Magnetic Sensor with Self Calibration for the Internet of Things — 813
Hadi Tavakkoli, Izhar, XU ZHAO, Yi-Kuen Lee

S12: Micro/Nano Sensors and Actuators I

Nano size-effect enhanced sensitivity of gas detection — 818
Xinxin LI

Scalable Synthesis of SnO_2 Nanosheet Arrays on Chips for Ultralow Concentration NO_2 Detection — 820
Gaoqiang Niu, Changhui Zhao, Fei Wang

Thermalvoltage Measurement and Manipulation in Single-molecule Device — N/A
Junyang Liu, Hang Chen, Ping Zhou, Wenqiang Cao, Wenjing Hong

A Tungsten-Rhenium Thin Film Thermocouples Sensor Based on AlN Transition Layer — 826
Bian Tian, Bingfei Zhang, Zhongkai Zhang, Zhaojun Lliu, Jiangjiang Liu, Gong Cheng, Qijing Lin, Chen Wu,
Peng Shi, Zhuangde Jiang

Thin Film Antioxidative Coating for High-temperature Thin Film Sensors Made of Polymer-derived
Ceramics for Harsh Environment — N/A
Xiaochuan Pan, Zaifu Cui, Zhenyin Hai, Guochun Chen, Daoheng Sun

A New Method for Characterization of Single Cell Using System Identification — 832
Shuang Ma, Wenxue Wang, Lianqing Liu, Yuechao Wang, Tianlu Wang

Reversible Immunoaffinity Interface Enables Dynamic Manipulation of Trapping Force for
Accumulated Capture and Efficient Release of Circulating Rare Cells — N/A
Xiaofeng Chen, Lingling Wu, Chaoyong Yang

PS2: Poster Session II

Design and Optimization of Glass Frit Package Structure for Micro Pressure Switch — 840
Lingyun Wang, Daner Chen, Heng Xiong, yifang Liu

Damage profile model of nanostructure fabricated by Focused Helium Ion Beam — 845
Chenglong Liu, Qi Li, Qianhuang Chen, Yan Xing

Investigation of Broading Modulus Range of Soft Probes by Single Beam Acoustic Tweezer — 849
Huiyao Shi, Jialin Shi, Peng Yu, Lianqing Liu

Vertical Organic Synapse Expandable to 3D Crossbar Array — N/A
Yongsuk Choi, Seyong Oh, Chuan Qian, Jin-Hong Park, Jeong Ho Cho

Active-Powering Wearable Iontronic Tactile Sensing for Human Physiological Monitoring — N/A
Hongyan Sun, Yu Chang, Tingrui Pan

Negative Differential Resistance Effect in Graphene Nanoribbons Heterojunction — N/A
li cheng, Yu Zhu, Qingfeng Gong, Wenli Zhou

3D Printed Kenics Static Micromixer
Kunpeng Zhang, Gonghan He, Daoheng Sun   N/A

Viscoelastic deformation and process optimization of polymer microfluidic chip for rapid in-mold bonding
Ylhua Lei, wangqing Wu, Xiansong He   N/A

Factors Influencing Resolution of Optical Fiber End Face Processing in Digital Lithography
Menghan Xiong, Ningning Luo, Zhimin Zhang, Qingwang Meng   864

Study on friction heat generation mechanism by molecular dynamics
Changsheng He, Wangqing Wu   N/A

Through Glass Vias by Wet-etching Process in 49% HF Solution Using an AZ4620 Enhanced Cr/Au Mask
Guanghui Ding, Binghe Ma, Yuchao Yan, Weizheng Yuan, Jinjun Deng, Jian Luo   872

DNA-directed nanofabrication of ultra-scaled high-performance electronics
yahong chen, mengyu zhao, Zhi Zhu, Wei Sun   N/A

The Design of Love Wave Immunosensor for Real-time Detection of Cancer Biomarker from Human Saliva
Junyu Zhang, Tao Zhang, Yuantao Chen, Xiaojing Zhang, Hao Wan, Ping Wang   N/A

Measurement of Cardiomyocytes Motion Based on Surface Patterned Polydimethylsiloxane Membrane
Si Tang, Jialin Shi, Huiyao Shi, Peng Yu, Lianqing Liu   N/A

An Adaptive Octree Level Set Simulation Method of the Wet Etching Process for the Fabrication of Micro Structure on Sapphire Crystal
Ye Chen, Jin Qian, Xinyan Guo, Yan Xing   882

Surface Acoustic Wave (SAW) Devices Based on ScAlN/AlN/Si Layered Structures with Large Figure of Merit
Yan Liu, Binghui Lin, Yang Zou, Yao Cai, Wenjuan Liu, Chengliang Sun   N/A

A Flexible AgNPs-PDMS Substrate to Produce Ultrasensitive SERS Detection
Guanzhou Lin, Kenan Zhang, Yun Huang, Shengxiao Jin, Tian Kang, Yusa Chen, Liye Li, Peimin Lu, Wengang Wu   889

Fabrication and Properties of Two-Dimensional InSe Top-Gate Transistors and Their Applications in Logic Circuits
Wei Li, Xiaozhi Wang   N/A

Voxelated Meniscus-confined Electrodeposition of 3D Metallic Microstructures
Yutao Wang, Yuanliu Chen   N/A

A clogging rate prediction model based on porous microarray membranes for liquid biopsies
Yinghao He, Yaoping Liu, Wei Wang, Yufeng Jin   898

Biomimetic Electrode–Electrolyte Design for Efficient Electrocatalytic Nitrogen Fixation under Ambient Conditions
Yang Liu, Xinyi Zhang, Panagiotis Tsiakaras, Pei Kang Shen   N/A

Micro-droplet of Particulate Suspension Generated by a Pneumatic Ejection System
Weijie Bao, Shengnan Sun, Zhihai Wang, Yaohong Wang   904

Tweezers for Micro-droplet Transfer Based on Hydrophobic Non-parallel Plate Structure
Jiaqiang Wang, Liguo Chen, Xiongheng Bian, Zaichen Wang   909

Design and Preparation of a High Resolution Accelerometer Based on Graphene
Xiaodong Zhao, Yanlu Feng, Wei Zhang, Jianhui Liao   915

Nanorobotic Manipulation inside Scanning Electron Microscope for the Electrical and Mechanical Characterization of ZnO nanowires
Mei Liu, Aristide Djoulde, Quan Yang, Weilin Su, Lingli Kong, Jinjun Rao, Jinbo Chen, Zhiming Wang   919

Nano Mechanics Markers for Accessing the Effects of Ultraviolet(UVC) Disinfection
Yuxuan Xue, Ning Xi, Kaicheng Huang, Ye Ma   N/A

Multi-pulse triboelectric nanogenerator based on micro-gap corona discharge for enhancement of output performance
Ru Wang, Juan Cui, Yong-Qiu Zheng   N/A

Control over Electrical Property of a-InGaZnO Thin Film Transistors using coupled self-assembled molecular layer as Copper Diffusion Barrier.
Seungmin Lee, Minkyu Lee, Taeyoon Lee
N/A

Laterally-excited bulk-wave resonators (XBARs) with embedded electrodes in 149.5° Z-cut LiNbO₃
931

Xiyu Gu, Jieyu Liu, Yao Cai, Yan Liu, Chao Gao, Zhiwei Wen, Shishang Guo, Chengliang Sun

Research on Integrated Reliable Micro-High Explosive Train Model
Bo He, Wenzhong Lou, Hengzhen Feng, Yuecen Zhao, Yi Sun
935

Spatial and Temperature Control of Exciton Emission by Ferroelectric in Monolayer TMDs
bo wen, xi zhang, dongfeng diao
N/A

THz Hybrid Graphene &Metal Patches Metasurface for RCS Reduction and Digital Coding
Baolong Wang, Jiayi Yang, Shuangshuang Liu, Wei Xu, Guobin Chen, Di Feng, Lingjie Jia, Yan Meng, Xiuhan Li
941

A Flexible Circuit Fabricated by Tuning the Wettability of Liquid Metal
Chengjun Zhang, Qing Yang, Jingzhou Zhang, Jiale Yong, Xun Hou, Feng Chen
N/A

A method for automatic counting and labeling of cells stained with microporous membrane
Jiangcheng Cao, Tingting Hun, Wenbo Zhou, Zheng Liu, Wei Wang, Yufeng Jin
949

Design of a Controllable Push-triggered Microfluidic Chip for Vitrification Reagent Loading/unloading
Guangyi Cai, Boshi Jiang, Jiaxin Zhu, Fenglin Liu, Tianzhun Wu
953

BMP-2 Immobilized Lubricant-Infused Surface Coating for Orthopedic Implants with Anti-bacterial, Anti-inflammatory, and Osteogenic Functionalities
Jaegyu Park, Yeontaek Lee, Kijun Park, Jungmok Seo
N/A

Direct and All-dry Microfabrication of Ultramicroelectrode Based on Cold Atmospheric Microplasma Jet
Ye Xi, Longchun Wang, Zhejun Guo, Bin Yang, Jingquan Liu
959

Nanopore Surface Charge Sensing with Ion-Step Method
Jing Yang, Songyue Chen
963

Research on complex surface patterned conformal manufacturing system based on point cloud theory
Kaihan Yao, Junchuan Gao, Gonghan He, Zhenyin Hai, Daoheng Sun
N/A

A coplanar-electrode direct-current triboelectric nanogenerator with facile fabrication and stable output
Guoqiang xu, Yunlong Zi
N/A

The Mechanism and Realization of Using Ultrasonic Atomic Force Microscopy to Measure Subsurface Defects of Ultra-precision Components
Yuyang Wang, Chengjian Wu, Jinyan Tang, Yuanliu Chen
N/A

Breathable Graphene-based Hydrogel Strain Sensors
Xingchi Liu, Hengchang Bi, Xing Wu
N/A

Engineering Application of Micro Surface Acoustic Wave Sensing System in Temperature Detection of Copper BusBars in Transformer Cabinet of Electric Locomotive
Tinghan Liu
N/A

Immune-camouflage Coating for Implantable Medical Devices, Bioprinting, and Microfluidics
Kijun Park, Yejin Jo, Inwoo Kim, Soyeon Kim, Jungmok Seo
N/A

Slippery coated Implantable flexible microelectrode array (fMEA) for High-Performance Neural Interface
MD ESHRAT E ALAHI, Tian Zhou, Sara Khademi, Hao Wang, Tianzhun Wu
980

Light Extraction Efficiency Investigation of AlGaN-Based Deep Ultraviolet Light-Emitting Diodes on Nano-Patterned Sapphire Substrate
Wan Hui, Lei Yu, Lan Shuyu, Gong Liyan, Gui Chengqun, Zhou Shengjun
N/A

Equivalent Electrical Model and Experimental Analysis of a Novel Needle-ring Atmospheric Pressure Plasma Microjets Array
Lingju Xia, Junfeng Yang, Li Wen
986

A Wearable Health Monitoring System Self-powered by Human-motion Energy Harvester
990

Chen Bao, Anxin Luo, Yi Zhuang, Junlong Huang, Fei Wang

Improved Reservoir Computing by Carbon Nanotube Network with Polyoxometalate Decoration   994
Shuo Wu, Wenli Zhou, Kaiqiang Wen, Chengzhu Li, Qingfeng Gong

2D Microscopy of Magnetic Field Using Atomic Vapor Cell   998
Liu Chen

Atomistic resolved signals of amino acids for peptide sequencing by tunneling current analysis   N/A
Tommaso Civitarese, Giuseppe Zollo

Single-Molecule Electronics: Recent Advances and Perspectives   N/A
Junyang Liu, Wenjing Hong

Tunability of Optoelectronic Oscillator Based on SiO2 Optical Waveguide Ring Resonator   N/A
Zerong Jia, Yongqiu Zheng, Jiamin Chen, Liyun Wu, Chen Chen, Chenyang Xue

Lithographic Properties of Amorphous Solid Water   N/A
Shan Wu, Ding Zhao, Min Qiu

Chemical bonding of functional groups in Self-Assembled Monolayer (SAM) for Cu diffusion barrier   N/A
Minkyu Lee, Taeyoon Lee

Functionalized Nanochannels for Constructing Ultrasensitive Electrochemical Sensor   N/A
Wenrui Ma, Shunbo Li, Yi Xu

The Opposite Anisotropic piezoresistive effect ReS2   N/A
CHUNHUA AN, Jing Liu

Structural Design and Simulation of Ionic Transport and Mass Selection Process in MEMS   N/A
Time-of-Flight Mass Spectrometry Chip
Zongjia Cai, Ming Hao, Tianyuan Qi, Cong Wang, Kun Liu, Renchao Dou, Donghui Meng, Lichen Sun, Rongxin
Yan

Mechanisms of branch tip fusion in meshwork patterns   1018
Shan Guo, Mingzhu Sun, Xin Zhao

Leaf-like Self-assembled MXene/ZnOEP Hybrid Network for High-sensitive Temperature Sensing   1023
in Electronic Skin
Shijie Wan, Weiguan Zhang, Chenyang Xing, Zhengchun Peng

Energy Localization of Lamb Wave Resonator Using Dispersion Engineering   N/A
Yusi Zhu, Lidong Du, Zhan Zhao, Zhen Fang

Room temperature linear magnetoresistance in vertically aligned graphene nanocrystalline network   N/A
film
dong ding, xingze dai, chao wang, dongfeng diao

Preparation of Flame-retardant Lithium-ion Battery Separator by Coaxial Electrospinning   1035
Fangqin Shao, Guoyi Kang, Huatan Chen, Zungui Shao, Xiang Wang, Wengwang Li, Gaofeng Zheng

A Numerical Study on the Effects of Mechanical Properties of Red Blood Cells on Rheology in   N/A
Narrow Microchannels
De-Yun Liu, Peng Jing, Xiaolong Wang, Qiaodong Wei, Shenghong Zhang, Xiaobo Gong

Flexible Transparent Conductive Network Based on Liquid Film Rupture Self-assembly Method   N/A
Xinran Zhang, Haitao Deng, Danliang Wen, Yanyuan Ba, Haixia Zhang, Xiaosheng Zhang

An In2O3 Nanotubes based Gas Sensor Array combined with Machine Learning Algorithms for   1042
Trimethylamine Detection
Wenjie Ren, Changhui Zhao, Yingming Liu, Fei Wang

Study of touch mode MEMS capacitance vacuum gauge with circular diaphragm   N/A
xiaodong han, Gang Li, Yongjian Feng, Detian Li

PMUT-Based Air-Coupled Imaging And Surface Stain Detection   N/A
sheng sun, jianyuan wang, yuan ning, menglun zhang

3D printed water-soluble UV photopolymer for flexible sensor with sacrificial scaffolds and   N/A
indirected molding process
Chunjiang Wang, Wanhao Niu, Qihang Song, Xiangyu Mi, Jianxu Shi, Xiaoming Chen

Ultrasensitive Silicon Nanowire Field-Effect Biosensors Enabled by Functionalized with Gold   N/A
Nanoparticles and Aptamers
Qianhui Wei, Jianglan Yang, Han Xiao, Kuo Men, Qingzhu Zhang, Jing Zhang, Feng Wei

Self-Assembly of Electrospun Polymer Nanofibers: A General Phenomenon Generating Cylinder-Patterned Nanofibrous Structures
Tianhao Wu, Daoheng Sun

N/A

The Fabrication Of Three-layer Silicon Stacked Antenna
Yang Lei Cao, Jian Zhu, Fang Hou, Min Huang, Zong Lei Jiao

1057

Enhanced trapping stiffness based on SERS embedded microcavity
Yanhong Wang, Zhihui Li, Jingzhi Wu, Hengze Yang

N/A

Self-powered electro-tactile system for virtual tactile experiences
Yuxiang Shi

N/A

Influence of the Particle Size of Glass Powder on Sintering Characteristics in TGV Packaging
yifang Liu, Hongbo Sang, Zhenxiang Bu, Yulong Zhang, Guangen Gao, Lingyun Wang

1065

A self-powered system driven by random human walking energy for wearable healthcare applications
Fan Wang, Xiangyu Chen

N/A

An Electromagnetic-Piezoelectric-Triboelectric Hybridized Energy Harvester Towards Blue Energy
Yunfei Li, Tianyi Tang, Manjuan Huang, Xin Ma, Zhaohui Chen, Peijuan Cui, Huicong Liu, Lining Sun

1070

Internal Resonant Oscillation in Coupled Resonators for High-resolution Mass Sensing with A Wider Coupling Range
Nanxing Li, Cao Xia, Guowen Zheng, Xu Du, Dong F. Wang

1074

An Immersed Boundary Method for Mass Transfer through Porous Biomembranes under Large Deformations
Xiaolong Wang, Peng Jing, Qiaodong Wei, Shenghong Zhang, Xiaobo Gong

N/A

Elastic straining of free-standing monolayer graphene
Ke Cao

N/A

Multi-order Nonlinearities and Resulting Coherent Oscillations of the States in Quantum Dot-Nanomechanical Resonator Hybrid System
xinhe wang, Lin Cong, Guangwei Deng, Kaili Jiang, Xiaoyang Lin, Weisheng Zhao

1084

Investigation the Minimum Measurement Points for Calibration a High Precision NTC Thermistors in Cryogenic Field
Gong Xun, Xiaohe Tang, Yanjie Li, Minmin You, Zude Lin, Jingquan Liu

1088

Improve the Efficiency of Sidewall Reflector-type Optical Switch by Combining Inductively Coupled Plasma and Focused-ion-beam Etching
haoran xu, Jin Xie, Kaiquan Li, Zhuang Xiong, Weikang Dong, Kometani Reo, Jun Dai

N/A

A Triboelectric Nanogenerator with Gear Transmission for harvesting elastic potential energy
Xiaobo Lin, Lanxin Yang, Junfeng Zhong, Zhaoming Deng, Chenyang Xing, Bo Meng, Zhengchun Peng

N/A

A Facile Low-Cost Wireless Self-Powered Footwear System for Monitoring Plantar Pressure
huayi huang, Ziya Wang, Zhihao Zhu, Waner Lin, Zhengchun Peng

1096

Investigation of the Reliability of the Interconnection between Metal Electrode and Silicon Anchor in Silicon-on-Glass Process
MENGXIA LIU, Xianshan Dong, Jian Cui, Qiancheng Zhao

1102

A bistable criterion for the V-beam mechanism
Min Liu, Weidong Wang, Xiao Wang, D SY, Yingmin Zhu, Zimin Huo, Yijia Du

1106

Self-adaptive Microjet Array Cooling for RF High Power GaN Integration on Silicon
Miao Yu

1110

Wearable Wireless Sensing System with Ultrasensitive and Self-Cleaning Pressure Sensor for Electronic Skin
Xuan Li, Meng Wang, Weidong Wang, Libo Gao

N/A

Effect of deposition pressure on the tribological properties of Ti/WS2 composite films deposited by magnetron sputtering
Jun Ye

N/A

High-temperature Thin Film Temperature Sensor Made of Polymer-derived Ceramics for Harsh Environment
Zaifu Cui, Guochun Chen, Xiaochuan Pan, Zhenyin Hai, Daoheng Sun

N/A

A Contactless Switch for Cell Sorting by Area cooling

1126

Yigang Shen, Yaxiaer Yalikun, Yo Tanaka

Light-confined Plasmonic Probe for Photoelectric Characterization by Atomic Force Microscopy
Yaoping Hou, Chengfu Ma, Wenting Wang, Yuhang Chen
N/A

Alumina composite coatings electrical insulating properties on Ni-based superalloy
chao wu, Daoheng Sun
N/A

A multi-parameter flexible sensor for detection of water-quality and aquatic animal activities
Zhihong Wang, Zheng Gong, QIpei He, Deyuan Zhang, Yonggang Jiang
N/A

Real-time Monitoring and Analysis of Jet Behaviors in Electrohydrodynamic Direct-Writing
Yifang Liu, Junyu Chen, Jiaxin Jiang, Guoyi Kang, Huatan Chen, Jianyi Zheng, Gaofeng Zheng
1136

Molecular Insights into Distinct Detection Properties of α-Hemolysin, MspA, CsgG, and Aerolysin Nanopore Sensors
Wanqi Zhou
N/A

Large-scale Uniformly Hybrid Micro-nano Structure Wetting Solid Substrate for Surface-enhanced Raman Spectroscopy
Fanhong Chen, Qi Qi, Yupeng Zhao, Shaoxun Zhang, Yongmin Zhao, Anjie Ming, Changhui Mao
1142

Antibacterial polymeric films with killing and antifouling properties synthesized by initiated chemical vapor deposition
Qing Song
N/A

Electroosmotic flow in wild-type and mutated CgsG nanopore
Giovanni Di Muccio, Blasco Morozzo della Rocca, Mauro Chinappi
N/A

Surface Free Energy Characterization of Soft Materials through Computational Experiments
Francesco Maria Bellussi, Annalisa Cardellini, Lorenzo Chiavarini, Pietro Asinari, Matteo Fasano
N/A

An improved Adaptive Periodical Segment Matrix Algorithm for Signal Denoising in Real-Time ECG Sensing
Xinggu Liu, Liang He, Jinhua Li, Zhuqing Wang
N/A

Data Analysis with Machine Learning For High-Accuracy Multi-target Gas Sensor Array
Qihong Ning, Chun Huang, Liang He, Jinhua Li, Zhuqing Wang
N/A

Buckled Structure Formation on the Surface of Stretchable Conductive Fibers and Application for Hydrogen Sensor
Kukro Yoon, Taeyoon Lee
N/A

A Tunable Quasi-Zero Stiffness Mechanism for Thermal Compensation of a MEMS Gravimeter
Xiaopeng Zhang, Xueyong Wei, Yang Gao, Minghui Zhao, Yonghong Qi, Libo Zhao, Zhuangde Jiang
1157

Effect of a hydrophobic gas on the evaporation of water in confinement
Antonio Tinti, Gaia Camisasca, Alberto Giacomello
N/A

Electroosmotic Flux in Uncharged Solid-State Nanopores: An Atomistic Simulation Study
Matteo Baldelli, Sébastien Balme, Mauro Chinappi
N/A

The realization of ZrOxNy temperature sensors with good sensitivity and stability in the temperature range above 150K
Yanjie LI, Minmin YOU, Gong Xun, Xiuyan LI, Zude LIN, Jingquan LIU
1165

NEMS Sensors Based on Suspended Graphene
Xuge Fan, Frank Niklaus
1169

On-chip micro/nano devices for energy conversion and storage
Lin Xu, Liqiang Mai
N/A

A miniature infrared emitter with ultra-high emissivity
Dongsheng Shu, Jifang Tao, Yan Li, Xinye Fan
1175

Ultrathin elastic shape sensor used for endoscope shape reconstruction
Leixin Meng, Yuan Zhuang, Liqiang Wang, Qing Yang
N/A

CS2: 学术格莱美讲座

学术格莱美讲座：如何与编辑沟通？
Yan LI
N/A

IS13: IS: Multidisciplinary Frontier: Soft Rubbery Electronics and Stretchable Integrated Systems

E-skins with superhigh pressure resolution and tough interfaces
Chuanfei GUO — N/A

Soft Rubbery Electronics
Cunjiang YU — N/A

Three-dimensional soft electronic systems for biomedicine
Mengdi HAN — N/A

Design of artificial synapses and sensorimotor neurons
Wentao XU — N/A

Rational Design of Dielectric Materials for Flexible Capacitive Pressure Sensing Coatings
Zhuo LI — N/A

Mechanics-driven designs of soft network materials and their applications in stretchable integrated devices
Yihui ZHANG — N/A

IS14: IS: Multidisciplinary Frontier: Nanoplasmonics and Biosensors

Surface Plasmoinc Enhanced Exciton-Polariton Effect in Semiconductor Nanowires
Qing ZHANG — N/A

Optical Superoscillation for Label-free Subdiffraction Bioimaging
Guanghui Yuan — N/A

Implantable biochip for management of inflammation in lung cancer model
Guozhen Liu — N/A

In-Vitro Diagnostic Assays Enabled by Plasmon-Enhanced Photoacoustic Detection
Meng Lu — N/A

2D nanomaterials enhanced surface plasmon resonance for sensing applications
Shuwen Zeng — N/A

Hybrid metal/dielectric nanosystems and applications
Yali Sun — N/A

IS15: IS: Flexible and Wearable Microsystems for Sensing and Actuation

Trigger-Detachable Hydrogel Adhesives for Bioelectronic Interfaces
Ji LIU — N/A

Materials and Devices Designs for Bioelectronic Medicines
Jie ZHAO — N/A

Biodegradable materials for electronic medicine and biosensors
Lan YIN — N/A

Microscale Optoelectronic Devices for Biological Modulation and Sensing
Xing SHENG — N/A

Hybridized and Coupled Nanogenerators
Ya YANG — N/A

Deformable Structures for Flexible Smart Materials
Zunfeng LIU — N/A

IS16: IS: Advanced Scanning Probe Technology and Applications

Markerless fabrication of FinFET devices by Thermal Scanning Probe Lithography
Armin W. Knoll — N/A

Advanced scanning probes for nanofabrication and nanomeasurement
Huan Hu — N/A

Scanning Probe-Assisted Nanowire Circuitry
Pablo Ares — N/A

Advanced Lithography for Opto-Electronic Nanochips — N/A

Xiaorui Zheng

2D materials patterning by oxidation scanning probe lithography
Yu Kyoung Ryu   N/A

Spatially Confined Surface Reactions and Nanopatterning by Cantilever-free Polymer Tip Arrays
Zhuang Xie   N/A

IS17: IS: Advanced Microsystems for Biomedical Applications

Engineering and measuring systemic multi-organ interactions
Yi-Chin TOH   N/A

Intelligent Drug Delivery System Based on Microneedle Technology
Jongho PARK   N/A

Mechanosensing of biological samples with silicon-based microsystems
Mehmet Cagatay TARHAN   N/A

Single-molecule analysis of bio-molecules and its application
Rikiya WATANABE   N/A

Advanced biomedical microsystems for single-cell analysis
Soo Hyeon Kim   N/A

Surface Enhanced Raman Scattering Sensors for Diseases Detection
Tianxun GONG   N/A

Advanced microfluidics and PCR technology combined for point-of-care diagnostics
Sisi LI   N/A

IS18: IS: Multidisciplinary Frontier: Advanced Micro/Nano Electronics Plus (Bio-, Photo-, Energy-)

Angstrom-Porous MOF Materials for Efficient Ion Adsorption and Sieving
Huacheng ZHANG   N/A

Surface Free Energy Characterization of Soft Materials through Computational Experiments
Matteo FASANO   N/A

Soliton microcombs: integrated photonics powering metrology
Qifan YANG   N/A

Synthetic Embryology: Merging Stem Cells and Mechanical Microsystems to Forge
an Embryo-Free Future for Human Embryology
Yue SHAO   N/A

Controlled Synthesis and Devices Applications of 2D Crystal Arrays
Yu ZHOU   N/A

Active Ionic Artificial Skin: Ion Transport Mechanisms and Prototype Design
Zicai ZHU   N/A

S13: MEMS/NEMS III

Applications of Flexible Inorganic Thin-film Devices in Bioelectronics
Yuan LIN   1248

An On-Chip Inductive-Capacitive Sensor for the Detection of Wear Debris and Air Bubbles in
Hydraulic Oil
Haotian Shi, Hongpeng Zhang, Wei Li, Zhiwei Xu, Laihao Ma, Yucai Xie, Dian Huo   1250

AIN Contour Mode Resonators with Half Circle Shaped Reflectors
Zhifang Luo, shuai shao, Tao Wu   1255

Modeling of magnetic sensor based on BAW magnetoelectric coupling micro-heterostructure
Si Chen, Wanchun Ren, Junru Li, Chunrui Peng, Yang Gao   1259

The 3D Capacitance Modeling of Non-parallel Plates Based on Conformal Mapping
Yue Feng, Zilong Zhou, Wenlong Wang, Zehong Rao, Yanhui Han   1264

A Mode-localized Mass Sensor with Different Order Modes Coupling Induced by Asymmetric
Structures
Jiahao Song, Jian Zhao, Ming Lyu, Pengbo Liu, Heng Zhong, Xianze Zheng, Yinghai Tang   N/A

A Breathing Mode Dual-Ring Resonator with High Quality Factor Operating at Atmosphere Pressure
Wen Chen, Wenhan Jia, Guoqiang Wu — N/A

S14: Soft & Flexibe Devices and Applications II

Versatile E-Printing for Flexible Electronics
Yongan HUANG — 1277

A High-Resolution Self-powered Flexible Pressure Sensor Matrix Based on ZnO Nanowires
Qifeng Lu, Fuqin Sun, Lili Li, Yuanyuan Bai, Zihao Wang, Ting Zhang — 1279

Design and Preparation of a Microfluidic System for Stretching of Cells in Topographic Microstructures
Yingning He, Yuqian Yang, Yexin Gu, Tianjiao Mao, Yue Yu, Jiandong Ding — 1283

Stochastic analysis of the elctrothermal microactuator with fabrication error
Haotian Liu, Hao Chen, Yun Cao, Weirong Nie, Zhanwen Xi, Shenghong Lei — 1287

Dynamic Simulation of Nanogenerator Based on Finite Element Model Coupled with Lumped Parameter Elements
Yao Chu, Ruixing Han, Huiliang Liu, Xiongying Ye, Fei Tang — 1291

Direct Ink Writing of Soft Microscale Structures Using Pure Polydimethylsiloxane
Huyue Chen, Qifan Ding, Wen-Ming Zhang, Lei Shao — N/A

All-printed Flexible Tactile Sensors with High Sensitivity and Large Detection Range
Qifan Ding, Huyue Chen, Wen-Ming Zhang, Lei Shao — 1297

S15: Micro/Nano Sensors and Actuators II

Heterogeneous Integration for RF application
Jian ZHU — 1301

Chitosan/graphene oxide/MoS2/AuNPs modified electrochemical sensor for trace mercury detection
Ri Wang, Chenyu Xiong, Yong Xie, Mingjie Han, Yuhao Xu, Chao Bian, Shanhong Xia — 1303

Visual Chemical Detection Mechanism by a Liquid Gating System with Dipole-Induced Interfacial Molecular Reconfiguration
Yi Fan, Xu Hou — N/A

Quantitative Glucose Measurement on a Synthetic Paper Test Strip
Weijin Guo, Jonas Hansson, Wouter van der Wijngaart — 1310

Organic Field-effect Transistors Based Gas Sensors for Volatile Toluene Detection
Tengfei Guo, Yuelin Wang, Jian Song, Tie Li — N/A

P-dopant Enhanced Organic Thin Film Transistors based Gas Sensors for Volatile Benzenes Detection
Meng Liu, Yuelin Wang, Jian Song, Tie Li — N/A

Coupling Mechanism of the Micro Mass-spring System and Electrothermal Actuator and its application in optical switches
Changlei Feng, Jin Xie, Kaiquan Li, Zhuang Xiong, Bin Tang, Jun Dai — N/A

Improved Design of Polymer Micromachined Transmission for Flapping Wing Nano Air Vehicle
RASHMI KANT, Daisuke Ishihara, Ryotaro Suetsugu, Sunao Murakami, Prakasha Ramegowda — 1320

S16: Micro/Nano/Molecular Fabrication II

Atomic-scale Manufacturing based on TEM
Litao SUN — 1326

Research on Melt Electrowriting TPU Hydrophobic Microfiber Mesh for Directional Water Transport
Zungui Shao, Jiaxin Jiang, Junyu Chen, Huatan Chen, Guoyi Kang, Gaofeng Zheng — 1328

Object Manipulation with Freestanding Magnetic Microfibers Fabricated by FDM 3D Printing
Qing Lu, Ki-Young Song, Yue Feng — 1332

Electrochemical Nanoimprint Lithography
Hantao Xu, Lianhuan Han, Bingqian Du, Qinghui Meng, Yang Wang, Zhen Ma, Zhong-Qun Tian, Zhao-Wu Tian, Dongping Zhan — 1337

Adaptive Wavefront Shaping for Direct Laser Ablation Application   N/A
Chong Kuong Ng, Fan Zhang, Peng Tan, Yuanliu Chen

Simulation and Experimental Study on Single Pulse Ablation by a Femtosecond laser   N/A
Fan Zhang, Peng Tan, Chong Kuong Ng, Yuanliu Chen

Study on the Minimum Size of Molecule by Employing the Single-Molecule Plasmonic Optical Trapping Method   N/A
Yang Yang, Biao-Feng Zeng, Chun-An Huo, Jia Shi, Wenjing Hong, Zhong-Qun Tian

S17: M/NEMS IV

Soft Micro-robots and Micro-actuators: Materials, Design, Control and Applications   1347
Lining SUN

Ultrasensitive Mass Sensing Utilizing High-Order Mode Localization of Clamped-Clamped Microbeams with Distributed Electrodes   N/A
Ming Lyu, Jian Zhao, Pengbo Liu, Jiahao Song, Heng Zhong, Xianze Zheng, Yinghai Tang

Influence of Mass Loading Effect on Radiation Quality Factor of BAW Magnetoelectric Antenna   1353
Junru Li, Chunrui Peng, Yang Gao, Wanchun Ren, Xuefeng He

Optical Measurement of the Dynamic Response of an Electrothermal Microactuator   1358
Chen Hao, Wang Xingjie, Xi Zhanwen, Wang Jiong, Nie Weirong

A Wearable Strain Sensor Based on Fiber-structured PU/MXene/CNT Composite with Ultra-high Sensitivity and Broad Sensing Range   1362
Guoxi Luo, Qiankun Zhang, Yunyun Luo, Ke Chen, Wenke Zhou, Libo Zhao, Zhuangde Jiang

Nonlinear threshold mass sensor using the snap-through phenomenon of a clamped–clamped micromachined arch beam   N/A
Heng Zhong, Jian Zhao, Ming Lyu, Jiahao Song, Pengbo Liu, Xianze Zheng, Yu Huang

Feasibility Study of Wearable Muscle Disorder Diagnosing Based on Piezoelectric Micromachined Ultrasonic Transducer   1370
Mengjiao Qu, Hong Ding, Xuying Chen, Dengfei Yang, Dongsheng Li, Ke Zhu, Jin Xie

Multifunctional Cardiomyocyte-Based Biosensor for Electrophysiology-Mechanical Beating-Growth Viability Monitoring   1374
Dongxin Xu, Jiaru Fang, Ning Hu

S18: Soft & Flexibe Devices and Applications III

Developing biocompatible and implantable flexible pressure sensors for health monitoring applications   1378
Guozhen SHEN

An Electrical Double Layer-based Iontronic Tactile sensor for Detection of Biological Ionic Liquid   N/A
Yulu Liu, Shuyi Huang, Menglu Li, Xiangyu Zeng, Wei Li, Xiaozhi Wang

Strain-Engineered Bistable Clamped Thin Films: From Macroscale to Nanoscale Fabrication   N/A
Guangchao Wan, Ziao Tian, Borui Xu, Yongfeng Mei, Zi Chen

Highly Sensitive and Flexible Tactile Sensor Based on the Fabrication of Porous Graphene/Silicone Rubber Composites   1384
Zhijian Chen, Yancheng Wang, Deqing Mei, Jie Jin

Uniformly distributed self-filling micro-strips for high-performance pressure-sensitive sensor   1390
Zhiping Chai, Xingxing Ke, Han Chen, Jiaqi Zhu, Chuanfei Guo, Zhigang Wu

A Sensitive Flexible Strain Sensor via Anisotropy Microstructured Sensitized Surface Resistive Change for Human Motion Monitoring   1394
Wenjie Fei, Shuo Zhang, Zhigang Wu

A surface and interior material identification technology based on dual-mode sensor   1398
Zhuhui Yin, Ning Li, Weifeng Chen, Yue Jiang, Kaiyang Lan, Jiabin Peng, Zhengchun Peng

S19: MEMS/NEMS V

MEMS-based platforms for characterization of advanced functional materials   1403
Liviu NICU

Design and Fabrication of A latching Silicon-based MEMS Switch *Tongtong Cao, Tengjiang Hu, Yulong Zhao*	1405
MULTI-MODAL RESONANCE MEASUREMENT CAPABLE OF DISCERNING AND VISUALIZING EFFECTS DUE TO EXCITATION SCHEMES IN MEMS RESONATOR *Bo Xu, Jiankai Zhu, Tongqiao Miao, Jing Li, Qingyang Deng, Song Wu, Ting Wen, Fei Wang, Xiaoping Hu, Xuezhong Wu, Dingbang Xiao, Zenghui Wang*	1410
Nanostructure Vanadium-doped Zinc Oxide Film Sensor Endowed with Enhanced Piezoelectric Response *Wei Gao, Yu Zhang, Binghe Ma, Jian Luo, Jinjun Deng, Weizheng Yuan*	1415
Humidity Sensor with High Resolution and Fast Response Based on AlN Cantilever with Two Groups of Segmented Electrodes *Dongsheng Li, Hanyong Dong, Zihao Xie, Mengjiao Qu, Qian Zhang, Jin Xie*	1419
Design and Fabrication of Integrated Piezoelectric Micropump with Vortex enhancement Optimization *Boshen Liang, Louis Paquet, Yongbin Jeong, Paul Heremans, David Cheyns*	1423
3D-Printed Sugar Scaffold for High-Precision and Highly Sensitive Active and Passive Wearable Sensors *Dong Hae Ho, Panuk Hong, Joong Tark Han, Sang - Youn Kim, S. Joon Kwon, Jeong Ho Cho*	N/A

S20: Nanomaterials I

Structure-Controlled Synthesis of Single-Walled Carbon Nanotubes *Yan LI*	1430
A Novel Low-Temperature Post-Curing Transfer Method Of Graphene Wrinkling Surface For Strain Engineering *Kai-Ming Hu*	1432
Synthesis of Porous Co3O4-ZnO and its Performance for Sensitive Ethanol Detection *Xiao Zhang, Yaohua Xu, Feng Wei, Anjie Ming*	1436
Simultaneous Sensing of Refractive Index and Temperature Using a Symmetry-breaking Silicon Metasurface with Multiple Fano Peaks *Luo Zhao, Guoguo Kang, Junyi Wang, Haiyang Li, Cheng Zhang*	1441
Dynamic Curvature Nanochannel-Based Membrane with Anomalous Ionic Transport Behaviors *Miao Wang, Xu Hou*	N/A
Switchable PDT by a selfassembled FRET quenching nanoparticle *Ziying Li, Yu Gao*	N/A
Enhanced Ion Sensing Stability with Nanotextured Biosensors *Yanfang Wang, Yuanjing Lin*	1451

PS3: Poster Session III

Photonic chip-based ultrafast Raman soliton source *Zhao Li, Qingyang Du, Chaopeng Wang, Jinhai Zou, Juejun Hu, Zhengqian Luo*	N/A
Responsive DNA Hydrogel Facilitates Isolation and Retrieval of Circulating Tumor Cells *SHUGUANG XUAN, HONGTAO FENG, YULIN ZHOU, YUQING HUANG, Yan Chen*	N/A
Electron-beam Grayscale Lithography Using Solid Anisole *Rui Zheng, Ding Zhao, Min Qiu*	N/A
Velocity Random Walk Modelling of a Silicon MEMS Resonant Accelerometer based on Non-AGC Control Loop *Dong Li, QianCheng Zhao, Jian Cui*	1461
A Piezoresistive Pressure Microsensor Based on Simplified Fabrication Processes *QingGang Meng, Yulan Lu, Junbo Wang, Deyong Chen, Jian Chen, Bo Xie*	1465
Fabrication and Verification of a Novel Low-g MEMS Inertial Switch *Min Liu, Weidong Wang, Xiao Wang, D SY, Yingmin Zhu, Yijia Du*	N/A
Sensitive Acetone Gas Sensors based on MOF-derived Carbon Nanoparticles-decorated Mesoporous Fe2O3 Nanorods on MEMS *Li-Yuan Zhu, Kai-Ping Yuan, Xue-Yan Wu, Tao-Tao Wu, Hong-Liang Lu*	N/A

Hollow MXene Sphere-based Flexible E-skin for Multiplex Detection of Applied Force *Xue-Feng Zhao, Xiao-Hong Wen, Meng-Yang Liu, Hong-Liang Lu*	N/A
Triboelectric nanogenerator powered electrowetting-on-dielectric actuators *Dongyue Jiang, minyi xu, Jie Tan, Yutao Wang, Penghao Tian*	N/A
Mass fabrication of 4H-SiC high temperature pressure sensors by femtosecond laser etching *Lukang Wang, You Zhao, Yulong Zhao, Yu Yang, Bo Li, Taobo Gong*	1478
3D Hierarchical Nanoarchitecture AuNPs/MXene@PAMAM based Biosensor for cTnT Detection in Human Serum *Xin Liu, Yong Qiu, Hao Wan, Liujing Zhuang, Ping Wang*	1482
Insight of Volatile Benzenes Sensing Mechanisms for Conjugated Polymer Based Gas Sensors *Jian Song, Tengfei Guo, Meng Liu, Yuelin Wang, Tie Li*	N/A
Multidimensional characterization of optogenetically engineered cells based on an integrated platform *Jia Yang, Lipeng Zu, Wenxue Wang, Ning Xi, Lianqing Liu*	N/A
Design and Optimization of Microfluidic System Based on Energy Minimization *yike zhou, Daoheng Sun*	N/A
Effect of Scan Line Spacing on Laser-Reduced Graphene Oxide Based Temperature Sensing *Rui Chen, Wei Zhou, Chiqian Xiao, Tao Luo*	1493
High current density electron wind forces in metallic graphene nanoribbons *Ji Zhang, Tarek Ragab, Cemal Basaran*	N/A
A TENG-pressure combo-sensor for accurate material identification *Weifeng Chen, Ning Li, Zhuhui Yin, Yue Jiang, Jiabin Peng, Zhengchun Peng*	N/A
Investigation of Flexible Sweat Sensor for Sodium-Ion Concentration with a Combination of Two Sensing Mechanisms *Jiufu Zheng, Xuankai Xu, Yunzhi Hua, Xiaojin Zhao, Wei Xu*	1502
Long - line transmission design of ocean multi - parameter sensor data based on CR600 *Xuan Wang, Yongqiu Zheng, Juan Cui, Haoling Zhang*	N/A
Sensitive Wide-Frequency-Response Acoustic Sensor Based on Evanescent Field Excited in Semi-buried Optical Waveguide Ring Resonator *Yongqiu Zheng, chen chen, Jiamin Chen, Liyun Wu, Yuan Han, Chenyang Xue*	N/A
Fabrication of monodisperse magnetic nanorods for improving hyperthermia efficacy *Shan Zhao, Nanjing Hao, John X.J. Zhang, P.Jack Hoopes, Fridon Shubitidze, Zi Chen*	N/A
Pyro-Electrospinning for the Fabrication of Nanofiber Membrane-Embedded 3D Devices *Bin Qiu, feng xu, Shouju Yao, Wenchang Tu, yike zhou, Shanshan You, lingling liu, Tianhao Wu, Daoheng Sun*	N/A
Photothermal gel microvalve applied to precise flow control of microfluidic chip *Kehan Chen, Jingwen Pan, Minghao Xu, Shuai Wang, wenqiang zhang*	N/A
AlN Hybrid-Coupled Resonator With Phononic Crystal Reflector *Kangfu Liu, Yuxi Wang, Tao Wu*	1517
Multilayered Electret Generator for Self-powered Wireless Data Transmission *Zeyuan Cao, Shiwen Wang, Zibo Wu, Rong Ding, Xiongying Ye*	N/A
Optimization of S1 Lamb wave resonator with Al0.8Sc0.2N *shuai shao, Zhifang Luo, Tao Wu*	1523
Structure design and parameter exploration of a new type of plasma anemometer *xianlong Liu, Daoheng Sun*	N/A
Preparation of nanofiber membrane with controllable pore size and diameter based on auxiliary counter electrode *lingling liu, Dangheng Sun*	N/A
Optimizing Design, fabrication and calibration of thin film heat flux gauge on ITO/In2O3 thermopile *Xin Li*	N/A
Super-lightweight flexible wireless optogenetic stimulation device *Rui Luo, Ziqi Mei, Zheng You, Dahai Ren*	1535

Research on the Bouncing-Ball Based Triboelectric Nanogenerator for Self-powered Vibration Frequency Monitoring 1539
Xusheng Zuo, Taili Du, Fangyang Dong, Shunqi Li, Yongjiu Zou, Junhao Zhao, Peng Zhang, Yuewen Zhang, Peiting Sun, Minyi Xu

Surface modification of 3D printed microfluidic chip N/A
Shanshan You, Kunpeng Zhang, Shouju Yao, lingling liu, Tianhao Wu, yike zhou

Design of Piezoelectric Micro-Actuators Design Based on LiNbO$_3$ Thin Film 1545
Yushuai Liu, Zhiyuan Gao, Kangfu Liu, Tao Wu

Design of a Miniature Flat Plate-type Propellant actuator for MEMS Safe and Arm Devices N/A
Yaoxiong Wang, Jinhong Huang, Bin Tang, Chongfei Zhang, Jun Dai

Patterning multilayer alginate/PCL scaffolds for culturing myocardial tissue N/A
feng xu, qiang gao, bin lin, daohen sun

Electrospun Polyimide Nanofiber Separators for Lithium-ion Batteries 1554
Wenwang Li, Bangzhou Che, Jinghua Lin, Sinan Fu, Jiaxin Jiang, Gaofeng Zheng, Xiang Wang

Design and Research on High Overload Resistance of a Micro-mirror Structure N/A
Qiwei Wang, Jin Xie, Zhuang Xiong, Bing Tang, Jun Dai

Simulation and Analysis of Nano Robotic Manipulators Inside SEM 1560
Ziliang He, Lue ZHANG, Zhan Yang

PEDOT:PSS-based Nanopore Electrochemical Transistor Sensors for Particle Recognition N/A
Lin Li, Feng Zhou, Qiannan Xue, Xuexin Duan

Soft Robotic Manipulation System Capable of Stiffness Variation and Dexterous Operation for Safe Human-Machine Interactions N/A
Changyong Cao

Ultra-light Metamaterial for Sound Absorption Based on Miura-ori Sandwich Structure N/A
Yixin Wang, Jingwen Guo, Xingru Chen, Yi Fang, Xin Zhang, Hongyu Yu

High-temperature Thin Film Heat Flux Sensor Made of Polymer-derived Ceramics for Harsh Environment N/A
Guochun Chen, Zaifu Cui, Xiaochuan Pan, Zhenyin Hai, Daoheng Sun

CFD Analysis of Mixing Process in a Cross-shaped Micromixer N/A
Shuai Yuan, Bingyan Jiang, Tao Peng, Mingyong Zhou

High-detectivity Infrared Detector Based on Dual-layer Thermopile 1574
zhaohui yang, Gaobin Xu, Shirong Chen, Jianguo Feng, Yuanming Ma, Xing Chen

A High-Precision Mode Matching Method for Rate-Integrating Honeycomb Disk Resonator Gyroscope 1579
Sheng Yu, Xuezhong Wu, Xiang Xi, Jiangkun Sun, Qingsong Li, Dingbang Xiao, Yongmeng Zhang

Direct Electron-beam Patterning of Monolayer MoS2 with Water Ice N/A
Guangnan Yao, Ding Zhao, Yu Hong, Min Qiu

Acoustofluidic micromixer with bubble induced ultrasonic microstreaming N/A
PENG Tao

An Optical Microphone Based on Fabry-Perot Etalon Stability Structure N/A
Liyun Wu, Yongqiu Zheng, Jiamin Chen, Chen Chen, Xiaoqiang Hua, Chenyang Xue

Capacitive Stretchable Strain Sensors Based on Wavy-Structured Metal Electrodes 1589
Siqi YU, Hongyu Yu

Fabrication of Soft Magnetic Microstructures for Modulation of the Magnetic Field Distribution on the Micrometer Scale N/A
Fengshan Shen, Yan Yu, Yuexuan Li, Hongtao Feng, Yan Chen

Soft Interface Design for Electrokinetic Energy Conversion N/A
Jian Zhang, Xu Hou

An Underwater Material Recognition Device Based On The Seebeck Effect N/A
Nanxi Chen, Baichuan Shan, Jianhao Liu, Yuhang Fan, Kaiyuan Zhao, Mengze Li, Changxin Liu

Patterned Diphenylalanine Nanotubes Regulate the Behavior of Hippocampal Neurons N/A
Lipeng Zu, Huiyao Shi, Jia Yang, Yuanyuan Fu, Wenxue Wang, Ning Xi, Lianqing Liu

Temperature Compensation for MEMS Mass Flow Sensors Based on Back Propagation Neural Network 1601
Yan Wang, Shijin Xiao, Jifang Tao

Pressure Induced Transition from Wrinkling to Period-Doubling Instability in Flexible Tactile Sensors 1605
Xiuyuan Li, Kai-Ming Hu, Yi-hang Xin, Erqi Tu, Wenming Zhang

Improvement on the Uniformity of Deep Reactive Ion Etch for Electrically Isolated Samples N/A
Xiao Hu, Zhihan Zhen, Qiyu Huang

Design of a novel closed-loop electrostatic voltage sensor based on weakly coupled resonators 1611
zilong wang, Zhengwei Wu, Xiangming Liu, Yahao Gao, Simin Peng, Ren Ren, Fengjie Zheng, Yao Lv, Jun Liu, Hucheng Lei, Zhouwei Zhang, Wei Zhang, Jiachen Li, Chunrong Peng

Effect of the pitch on electroforming process of the microchannels N/A
Yanzhuo Dong

The Tunable Deformation of Microfluidic Strain Sensor Based on Auxetic Metamaterial 1617
Linna Mao, Dengji Guo, Sirong Huang, Taisong Pan

Effect of the substrate size on accuracy of in-situ stress measurements N/A
Jun Qiang, Bingyan Jiang

Measurement of Comb Frequency and Spacing Stability in Phononic Frequency Comb 1623
Qiqi Yang, Liu Xu, Ronghua Huan, Zhuangde Jiang, Adarsh Ganesan, Xueyong Wei

Inductive Effect of Nanofiber Deposition on Groove Structure on Cell Culture N/A
Wenchang Tu, yike zhou, Bin Qiu, Shanshan You, lingling liu, Tianhao Wu

Patterning and immobilization of silver Nanowires for flexible electronics by using Microwave 1629
sara khademi, Kiyumars Jalili, Hao Wang, MD ESHRAT E ALAHI, Tianzhun Wu

Rapid Fabrication of Flexible Strain Sensor for Plants Growth Monitoring N/A
Yicong Zhao, Jing Niu, Wenxing Huo, Xian Huang

Temperature and pressure dual-parameter sensing based on Fiber Bragg Grating 1635
Na Zhao, Qijing Lin, Liangquan Zhu, Kun Yao, Bian Tian, Libo Zhao, Zhuangde Jiang

Achieving High Energy-Storable Polythiourea Dielectric Materials in Molecular Design through Tuning H-Bonds N/A
Yang Feng, Guanghao Qu, Liuqing Yang, Huan Niu, Shengtao Li

Optical analysis of perovskite light-emitting diodes with nanostructured emissive layer via electrical simulation-assisted dipole location N/A
liyang chen, Zhuofei GAN, Dehu CUI, Jingxuan CAI, Wendi LI

Ultrathin elastic shape sensor used for endoscope shape reconstruction N/A
Leixin Meng, Yuan Zhuang, Liqiang Wang, Qing Yang

Microfluidics for Single-cell Multi-omics Analysis N/A
Xing Xu, Chaoyong Yang

A Biosensor Kit Based on Cu-MOF for Simultaneous Detection the Cortisol and Cyfra21-1 in Human Saliva N/A
Xinyi Wang, Tao Zhang, Shuqi Zhou, Liubing Kong, Wencheng Lin, Hao Wan, Ping Wang

Laser Induced Graphene Patterns on a Thin Polyimide Film via a cooling plate 1650
Qixiang Chen, Dezhi Wu, Zhiwen Chen, Hang Yu, Qinnan Chen, Daoheng Sun

A Robotic Dynamic Tactile Sensing System based on Electronic Skin 1655
Jishen Dai, Yu Xie, Dezhi Wu, Chen Songyue, Ting Fu, Wei Zhou

Characterization of an Asymmetrical Capacitive MEMS Tilt Sensor 1660
Yang Gao, Xiaopeng Zhang, Yonghong Qi, Maeda Ryutaro, Zhuangde Jiang, Xueyong Wei

A Reflective Color Filter Based on ITO-Ba0.5Sr0.5TiO3-ITO Nanofilms Capitalizing on a Black Layer 1664
Rui Wang, Jinying Zhang, Bingnan Wang, Xinye Wang, Defang Li, Jingyi Chen, Chenyu Guo, Xin Wang, Zhuo Li, Suhui Yang

High-performance rubbery electronics enabled by elastomeric composite materials N/A
Kyoseung Sim

A flexible rope-like structure sensor based on triboelectric nanognerator for multifunctional sensing N/A

Cong Zhao, Anaeli Elibariki Mtui, Xiangyu Liu, Kun Jiang, Jianye Wang, Chuan Wang, Minyi Xu

Giant Optical Activity in Achiral Plasmonic Au Nanocones Embedded into Alumina Nanohole Arrays — N/A

Yuyi Feng

Ferroelectric Nanocrack-based Nanoelectromechanical (NEM) Switches for Memory and Complementary Logic — N/A
Yaodong Guan, Zhe Guo, Qiang Luo, Jeongmin Hong, Long You

Conical Helmholtz Resonator-Based Triboelectric Nanogenerator for Harvesting of Acoustic energy — 1676
HongYong Yu, Taili Du, Hongfa Zhao, Qiqi zhang, Ling Liu, yue Huang, Anaeli Elibariki Mtui, Xiu Xiao, Minyi Xu

3D Cell Electrical Impedance Biosensor for Real-Time Drug Permeability Gradient Effect Monitoring — N/A
Yong Qiu, Xin Liu, Hao Wan, Ping Wang, Liujing Zhuang

Triboelectric Nanogenerator for Wind Energy Harvesting and Speed Sensing — N/A
Yan Wang, Chuanqing Zhu, Mengwei Wu, Hao Wang, Xiu Xiao, Guochang Wang, Jianchun Mi, Minyi Xu

Flexible Triboelectric Nanogenerator for Flow Control — N/A
Chuanqing Zhu, Xiangyi Wang, Mengwei Wu, Jialin Zhang, Chenxing Jia, Guochang Wang, Jianchun Mi, Minyi Xu

Design and fabrication of a multimode optical fiber surface plasmon resonance urea biosensor — 1686
Jiayin Li, Mengdi Han, Ji Wan, Haixia Zhang

A Resonant Differential Pressure Microsensor With a Stress Isolation Layer — 1691
Yadong Li, Chao Cheng, Yulan Lu, Jian Chen, Deyong Chen, Junbo Wang

A Novel Dual-channel Helmholtz Resonance Acoustic Energy Converter Based on Friction Nanogenerator — 1695
Ling Liu, Hongfa Zhao, Zhenhui Lian, Hongyong Yu, Qiqi Zhang, Wenxiang Li, Minyi Xu, Xiu Xiao

Fabrication of Multi-oriented Composite Nanofibrous Membrane by Electrospinning — 1700
Xiang Wang, Yongfu Xu, Jinghua Lin, Sinan Fu, Jiaxin Jiang, Gaofeng Zheng, Wenwang Li

A Breathable and Flexible Epidermal Glucose Sensor — N/A
Hailong Chen, Zhihua Pu, Dachao Li

Design of Low Sidelobe Level Milimeter-wave Planar Slotted Array Antenna Fed by Ridge Waveguide — N/A
Yan Cao, Yiming Tang

Modularized Hydrogel-Gate Field-Effect-Transistor Biosensors — N/A
Xiaochuan Dai

Bandwidth Expansion In Acoustic Filters based on AlN-on-Silicon Resonators — N/A
Jiannan Chen, Cheng Tu, Feilong Li, XIAOSHENG ZHANG

An Electromagnetic Actuator with Nonlinear Planar Micro-Spring for extended output range — N/A
Xuhan Dai

A High DC Current Output Salt Battery Hybrid and Enhanced by TENG — 1714
shuangshuang liu, Jiayi Yang, Zihao Niu, Yan Meng, Wei Xu, Di Feng, Baolong Wang, Meiqi Wang, Xiuhan Li

The Influence of Electrode on Elastic Constant C33D Extraction of Scandium-doped Aluminum Nitride Thin Film by Thickness-extensional Mode FBAR — 1718
hu xia

One-step preparation of underwater superoleophobic aluminum-coated copper mesh by pulsed laser cladding for enhanced oil-water separation — N/A
Junjin Lai, Rui Zhou

A Resonant High Pressure Sensor Based on Dual Cavities Design — 1725
Jie Yu, Yulan Lu, Deyong Chen, Junbo Wang, Bo Xie

MICROFLUIDIC DEVICE FOR ISOLATING CIRCULATING FETAL CELLS — N/A
Huimin Zhang, Chaoyong Yang

Numerical analysis of laser ablated structural size effect on enhanced anti-icing property of TC4 surface — 1732
Zhekun Chen, Rui Zhou, Yuhang Lin, Yi Zhu, Huangping Yan

Facile access to solvent-dependent luminescent carbon nanoparticles by laser ablation of activated carbon powders *Zhibin Chen, Rui Zhou, Longfan Li, Jingqin Cui*	1737
Fabrication of Superhydrophobic Surface by Inkjet Printing of Nano Silver Seeds on Porous Paper *Xinghao Zhang, Jin Xie, YU LIU*	N/A
A MEMS Voice Coil Motor with a 3D solenoid coil *Jiamian Sun, Zhi Tao, Haiwang Li, Kaiyun Zhu, Donghui Wang, Tiantong Xu, Hanxiao Wu*	1745
Tribovoltaic and Tribo-thermoelectric coupling effect on metal-semiconductor interface *Zhi Zhang, Chi Zhang*	N/A
Full printed flexible pressure sensor based on microcapsule controllable structure and composite dielectrics *Xiangyou Meng, Jing Zhao, Yaqin Pan, ziyun Han, Lixin Mo*	N/A
Nanoscale Triboelectrification Gated Transistor *Tianzhao Bu, Chi Zhang*	N/A
Modulated flexible artificial synaptic transistor based on graphene and ionic gel *Di Feng*	N/A
Preparation and research of ITO thin film strain gauge *Tao Yang, daoheng sun*	N/A
Screen Printing-Based Wearable Multi-Sensing Double-Chain Thermoelectric Generator *Tao Feng, Danliang Wen, Haitao Deng, Xiaosheng Zhang*	N/A
Optical Fiber Waveguiding Soft Photoactuators Exhibiting Giant Reversible Shape Change *Yongcheng He, Haohua Liang, Jiulin Gan, Zhongmin Yang, Meihua Chen*	N/A
Study of strain sensing behavior of self-powered stretchable mechanoluminescent optical fiber *Haohua Liang, Yongcheng He, Meihua Chen, Jiulin Gan, Zhongmin Yang*	N/A
The large piezoelectricity and high-power density of a 3D-printed multilayer copolymer in a rugby the ball-structured mechanical energy harvester *Xiaoting Yuan, Zhanmiao Li, Zhonghui Yu, Shuxiang Dong*	N/A
Tuning and Visualizing Motional Signal Transduction Efficiency in Atomically-Thin Nanoelectromechanical Resonant Structures *Jiankai Zhu, Jing Li, Xu Bo, Song Wu, Fei Xiao, Yachun Liang, Ting Wen, Fei Wang, Zenghui Wang*	N/A
Particle-Filled PHPS Silazane-Based Insulation Coating with High-temperature Electrical Insulating Properties on Ni-based Superalloy Substrates *Ji'an Lin, Qinnan Chen, Zaifu Cui, Zhenyin Hai, Daoheng Sun*	N/A
Research on torque calculation model of MR fluid-based micro-brake in thermal environment *Yan Zhang, Ying Liu, Jun Dai*	N/A
A Low Power MEMS microheater for MOS Gas Sensor Applications *Ziwei Lian*	N/A
Electron-beam-induced 3D Direct-writing of a Nanoelectrode Array for Intracellular Electrophysiological Recording *Wenqi ZHANG, Lixin Dong*	N/A
Electrically-responsive colloidal particle assembly in ionic surfactant solutions *Minqi Yang, Lisha Luo, Haiyang Fu, Hongjie Yin, Zhibin Yan, Mingliang Jin, Huicheng Feng, Guofu Zhou, Lingling Shui*	N/A

CS3: 学术奥斯卡讲座

学术奥斯卡讲座：如何做好学术报告？ *Haixia ZHANG*	N/A

IS19: IS: Multidisciplinary Frontier: Micro/Nano-Engineered Materials for Sensing and Applications

Flexible Human-Machine Interacting Sensors *Yanchao MAO*	N/A

Soft Robotic Manipulation System Capable of Stiffness Variation and Dexterous Operation for Safe Human-Machine Interactions
Changyong CAO   N/A

Borophene-Based Two-Dimensional Materials and Devices
Guoan TAI   N/A

Functional photonic crystals with superwettability
Jingxia WANG   N/A

Two-Dimensional Transition Metal Carbides and Nitrides for Terahertz Absorption Technology
Xu XIAO   N/A

Room Temperature Gas Sensors Based on Transition Metal Dichalcogenides Heterostructures
Zhi YANG   N/A

IS20: IS: Multidisciplinary Frontier: Functional Micro-nano Structures, Techniques and Applications

Nanoforests and nanoforest-based micro sensors
Haiyang MAO   N/A

Bioinspired Catalytic Materials Based on Metal-Sulfur Clusters
Jian LIU   N/A

Mass transport in atomic scale confinements
Sheng HU   N/A

Quality Heterostructure from Two-Dimensional Materials
Yang CAO   N/A

Fiber integrated multifunctional sensing and super-resolution imaging
Qing YANG   N/A

The applications of vanadium dioxide thin films on multi-functional sensor
Min GAO   N/A

S21: MEMS/NEMS VI

High Accuracy Resonant Accelerometers Based on QMEMS
Yulong ZHAO   1800

INFRARED THERMAL IMAGING SENSOR USING TRANSFERRED RUPHEN-BASED TEMPERATURE SENSITIVE FILM
Jialiang LI, Jun Hu, Yufei Zhai, Song Li, Yiming Yuan, Min Wang   N/A

Hybrid Frequency-time Domain Analysis of Bulk Acoustic Wave Circulator
yuan jing, zhang bingbin, Wu Chengfeng, gao yang   1804

A Piezoelectric Micromachined Ultrasonic Array with Frequency-selectable Tubes
Wei Zhu, Lei Wang, Zhipeng Wu, Wenjuan Liu, Chengliang Sun   N/A

A Micromachined Electrochemical Angular Accelerometer Based on Interdigital Electrodes
Tian Liang, Junbo Wang, Deyong Chen, Jian Chen, Bowen Liu, Chao Xu, Wenjie Qi, Xu She, Mingwei Chen, Anxiang Zhong, Yumo Duan   1813

Research on the Stability Threshold and Subharmonic Oscillation in a Parametric Excitation MEMS Resonator
Kuo Lu, Kai Wu, Qingsong Li, Xin Zhou, Yongmeng Zhang, Ming Zhuo, Xuezhong Wu, Dingbang Xiao   1817

Preparation and Detection of Micro-nano Electrode Array Based on Chipmunk Hypothalamus in Specific Brain Regions
Yiding Wang, Shengwei Xu, Chao Yang, Penghui Fan, Botao Lu, Gucheng Yang, Yinghui Li, Xinxia Cai   N/A

Large-Swing Low-Distortion High Voltage Amplifier Design for MEMS Gyroscopes
Hua Chen, Ke Liu, Zhen Meng   1825

S22: Bio Inspired Multiscale Interfaces

Bio-inspired multiscale adhesive interfacial materials
Shutao WANG   1830

Nanokits for Single Cell Analysis   1832

Dechen JIANG

SWCNTs/PEDOT:PSS Modified Microelectrode Array For Detection Of The Neuronal Electrophysiological Activity Of Cortical Neurons Under Glutamate Stimulation *Shihong Xu, Yu Deng, Enhui He, Shenwei Xu, Longzhe Sha, Kui Zhang, Yiling Song, Jinping Luo, Qi Xu, Xinxia Cai*	N/A
MWCNT/PEDOT:PSS Nanocomposites-modified Microelectrode Array For Recording Neural Activity Of hiPSCs-derived Mature Neurons To Different Concentrations Of K+ And Glu *Enhui He, You Zhou, Shihong Xu, Kui Zhang, Shengwei Xu, Yilin Song, Jinping Luo, Wanwan Zhu, Qi Xu, Xinxia Cai*	N/A
Bioinspired Liquid Gating Membrane-based Catheter with Anticoagulation and Positionally Drug Release Properties *Chunyan Wang, Xu Hou*	N/A
Ultra-thin flexible neuro probe utilizing biodegradable collagen microneedle *Dong Huang, Junshu Li, Zhongyan Wang, Zhitong Zhang, Qining Wang, Zhihong Li*	1839
Fabrication of polymer/metal composite micro/nano array structures and their applications in biological interfaces and actuators *Hongxu Chen*	1843

S23: Nanomaterials II

Design of Sub-nanometre Pores for Precise Separation *Jian JIN*	1847
Self-powered Flexible Supercapacitor for Human Motion Monitoring *Zheng Fan, Hongmei Chen, Yanyun Fan, Danfeng Cui, Shubin Yan, Chenyang Xue*	N/A
Optoelectronic Synapse using MoS2 van der Waals Homojunction with Interfacial Passivation for Fast Potentiation *Yizhen Ke, Lin Lin, Tianxun Gong, XIAOSHENG ZHANG, Wen Huang*	N/A
Bionic Tactile Sensor based on Triboelectric Nanogenerator for Motion Perception *Siyuan Wang, Peng Xu, Xinyu Wang, Hao Wang, Changxin Liu, Liguo Song, Guangming Xie, minyi xu*	1853
Triboelectric Nanogenerator Based on Silver Nanowires *Fengru Fan*	N/A
Functionalization and application of inorganic porous nanofibers *Yunqian Dai*	N/A

S24: Nature-inspired Energy Conversion and Harvesting: from Fundamentals to Applicaitons

NEMS-enabled Innovations at Interfaces for Water-Energy Nexus *Zuankai WANG*	1861
Application and Prospect of MEMS Technology to Geophysics *Heting Hong, Lvchao Ni, Hongchun Sun*	1863
Design and Analysis of MEMS Biaxial Coupled Resonance Accelerometer *Huimin Zhang, Yating Zhang, Wei Zhang*	1867
Development of Annular-Shaped Bernoulli Gripper for Contactless Gripping of Large-Size Silicon Wafer *Shihang Wang, Yancheng Wang, Deqing Mei, Songqiao Dai*	1871
Ultrasound-mediated Tough Tissue Adhesives Using Nanoparticles *Shuaibing Jiang, Zhenwei Ma, Tony Jin, Edmond Lam, Audrey Moores, Jianyu Li*	N/A
A Self-Powered Angle Sensor at Nanoradian-Resolution for Robotic Arms and Personalized Medicare *Ziming Wang, Jie An, Jinhui Nie, Jianjun Luo, Jiajia Shao, Tao Jiang, Baodong Chen, Wei Tang, Zhong Lin Wang*	N/A
Dielectrophoretic microdevice with planar electrodes for concentrating microparticles *Salini Krishnan, Alia Mohammmed Shaker Alblooshi, Fadi Alnaimat, Bobby Mathew*	N/A

S25: MEMS/NEMS VII

Semiconductor Biosensors for Global Health in the Era of AIoT   1885
Yikuen LEE

Frequency Output, A Multi-modes Vacuum Gauge with Highly Output linearity Based on   1888
Electrostatic Nonlinearity
wang chengxiang, Yunbing Kuang, Wu Yulie, Hou Zhanqiang, Zhang Yongmeng, Wu xuezhong,
Dingbang Xiao

High-resolution noncontact electrostatic voltage meter based on microsensor chip   1893
Xiaolong Wen, Pengfei Yang, Bo Zhang, Chunrong Peng

Preparation and Characterization of Ni-AAO Composite Scaffold Getter with Induction Heating   1897
Lujiang Liu, Lingyun Wang, Zhenxiang Bu, Kuixi Wu, Wenlong Lv, Dandan Gu

S26: Soft & Flexibe Devices and Applications IV

Bio-Inspired Ionic Skin for Theranostics   1901
Peiyi WU

Ultra-flexible and highly transparent hydrogel-based triboelectric nanogenerator for physiological   1903
signal monitoring
Jiahao Yu, Zhensheng Chen, Haozhe Zeng, Bowen Ji, Zixuan Wu, Jin Wu, Yunjia Li, Jianbing Xie,
Honglong Chang, Kai Tao

A Flexible Tactile Sensor for Three-dimensional Force Detection Based on Piezoelectric Sensing   1908
Yunyun Luo, Libo Zhao, Guoxi Luo, Zhikang Li, Ping Yang, Qiankun Zhang, Qijing Lin, Zhuangde Jiang,
Hongyan Wang, Yongshun Wu

Spontaneous breath analysis and healthcare assessment enabled by triboelectric nanogenerator   N/A
Yuanjie Su, Mingliang Yao, Guangzhong Xie, Huiling Tai, Yadong Jiang

S27: Soft & Flexibe Devices and Applications V

The Reinvention of silk   1914
Hu TAO

A Multifunctional Soft Robotic Finger based on Nano Temperature-Pressure Sensor for Material   N/A
Recognition
Xiaoshuang Zhang, Wentuo Yang, Cheng Zhou, Xueyou Sun, Mengying Xie, Xuexin Duan

Readout of Neural Activity in Mammalian Peripheral Olfactory System via Microelectrode arrays   N/A
Liujing Zhuang, Yan Duan, Suhao Wang, Qunchen Yuan, Jizhou Song, Ping Wang

A Three-electrode Multi-module Sensor for Accurate Bodily-kinesthetic Monitoring   N/A
Haobin Wang, Yu Song, Hang Guo, Ji Wan, Liming Miao, Chen Xu, Zhongyang Ren, Xuexian Chen,
Haixia Zhang

S28: Micro & Nanotechnologies for Biomedicine

Biomolecular Needling System for Medicals   1921
Beomjoon KIM

Physical Forces Influence the Self-organization of the Formation of Leader Cells During Collective   1923
Cell Migration
Mengyun Pan, Yongliang Yang, Lianqing Liu

TENG Enhanced and Self-Powered Sensors for Biomedical Signal Detection   N/A
Jieyu Dai, Zhou Li

April 25-29 , 2021   Xiamen, China

WELCOME

Welcome to the 16th IEEE International Conference on Nano/Micro Engineered & Molecular Systems (IEEE-NEMS 2021)!

The IEEE NEMS Conference series originated in 2005, is a premier conference series sponsored by the IEEE Nanotechnology Council. This Conference brings together annually the international MEMS community consisting of top players in academia and industry by providing them with the latest results on every aspect of M/NEMS, nanotechnology, and molecular technology. Previous 15 conferences were held in Zhuhai (2006), Bangkok (2007), Hainan Island (2008), Shenzhen (2009), Xiamen (2010), Kaohsiung (2011), Kyoto (2012), Suzhou (2013) , Hawaii (2014), Xi'an (2015), Matsushima Bay and Sendai (2016), Los Angeles (2017), Singapore (2018), Bangkok(2019). Despite the unprecedented challenges due to the COVID-19 pandemic, NEMS2020 was held online virtually.

At 2021, based on the improving situation of COVID19 at mainland China and fully support of Xiamen University, the committee decided to have NEMS2021 at Xiamen for attendees in mainland China, meanwhile provide the online broadcast to participants who NOT in mainland China. The conference has also enabled us to feature more attendees than other years. We hope that you will enjoy the presentations by our accomplished both on onsite and online.

We would like to express our sincerest gratitude to all the authors who submitted papers. Their high quality work serves as the foundation for the success of this conference. A total of 515 accepted paper out of 730 submissions were carefully selected by 90 experts comprising the Technical Program Committee (TPC).

The conference arranges presentation of accepted papers in parallel sessions with 10 plenary and 30 keynote and 128 invited talks 180 oral presentations, and 350 posters including late news. In addition, the TPC collectively nominated, based on quality, paper submissions as finalists for the Best Conference Paper Award, the Best Student Paper Award, the Best Conference Poster Award, and CM HO Best Paper Award in Micro/Nano Fluidics. These awards aim to recognize excellence amongst work and will be announced in a special ceremony at conference banquet.

More important, this year is the 100 anniversary of Xiamen University, we will celebrate it with a special memorial session of Prof. Wen H Ko, the pioneer and founder of Micro-Nano transducers field, the excellent alumni of Xiamen University, the first WHK Scholarship will be awarded too.

We gratefully acknowledge the all sponsors and benefactors for their contributions to this conference. The strong support from Xiamen University is highly appreciated.

In closing, we hope you enjoy the technical presentations, online networking, exhibition, and all the interactive features of the onsite/online platforms of the IEEE NEMS 2021 Conference!

General Chair
Haixia(Alice) ZHANG
Peking University

General Co-Chair
Daoheng SUN
Xiamen University

TPC Chair
Xu HOU
Xiamen University

ORGANIZING COMMITTEE

International Advisory Committee

Chih-Ming HO	Univ. of California
Masayoshi ESASHI	Tohoku Univ.
Meyya MEYYAPPAN	NASA AMES
Nicolas F. de ROOIJ	Univ. of Neuchatel
Toshio FUKUDA	Beijing Institute of Technology/Meijo Univ.

Steering Committee

Ning XI	Univ. of Hong Kong
Haixia (Alice) ZHANG	Peking Univ.
Daoheng SUN	Xiamen Univ.
Gwo-Bin LEE	National Tsing Hua Univ.
Osamu TABATA	Kyoto Univ.
Wen J. LI	City Univ. of Hong Kong
William C. TANG	Univ. of California at Irvine
Shuji TANAKA	Tohoku Univ.
Yu-Chong TAI	California Institute of Technology

IEEE-NEMS 2021 Committee

Honorary Chair:
Zhongqun TIAN   Xiamen University

General Chair:
Haixia (Alice) ZHANG   Peking University

General co-Chair:
Daoheng SUN   Xiamen University

Program Chair:
Xu HOU   Xiamen University

Program co-Chairs:
Jungmok SEO	Yonsei University
Jianyu LI	McGill University
Paolo MALGARETTI	Helmholtz Institute Erlangen-Nürnberg for Renewable Energy
Xian ZHANG	Stevens Institute of Technology

April 25-29, 2021   Xiamen, China

Zhigang CHEN   University of Southern Queensland

Invited Session Chair:
Xiaosheng ZHANG   Univ. of Electronic Science & Technology of China

Organizing Chair:
Wei ZHOU   Xiamen University

Award Committee Chair:
Wenjung LI   City University of HongKong

Sponsor Chair:
Liang HOU   Xiamen University

Industrial Chair:
Shenglin MA   Xiamen University

Promotion Chair:
Lingyun WANG   Xiamen University

Publication Chair:
Zhidong WANG   Chiba Institute of Technology
Xu HOU   Xiamen University

Technical Program Committee

Alberto GIACOMELLO	Sapienza Università di Roma
Arend van der ZANDE	UIUC
Azeemuddin SYED	International Institute of Information Technology Hyderabad
Baiyi CHEN	Xiamen University
Beomjoon KIM	The University of Tokyo
Binghao WANG	The University of Tokyo
Bo MENG	Shenzhen University
Boyang ZHANG	McMaster University
Cheng TU	University of Electronic Science and Technology of China
Daewoo KIM	Yonsei University
Dingbang XIAO	National University of Defense Technology
Frank NIKLAUS	KTH Royal Institure of Technology
Ho Cheung SHUM	University of Hong Kong
Hojeong JEON	Korea Institute of Science and Technology
Hong Nam KIM	Korea Institute of Science and Technology
Huacheng ZHANG	RMIT
Huicong LIU	Soochow University
Hyojin LEE	Korea Institute of Science and Technology
Jaehong LEE	DGIST
Jiajia CHEN	Xiamen University
Jianjun WANG	Institute of Chemistry Chinese Academy of Sciences
Jie JU	Henan University
Jinhye BAE	University of California San Diego
Kaichen XU	Zhejiang University

Mahmoud ALMASRI	University of Missouri
Massimo MASTRANGELI	Delft University of Technology
Matteo FASANO	Politecnico di Torino
Mauro CHINAPPI	Università di Roma Tor Vergata
Mengchuang ZHANG	Northwestern Polytechnical University
Mengdi HAN	Peking University
Miao WANG	Xiamen University
Min ZHANG	Peking University
Philippe BASSET	Univ Gustave Eiffel
Qihan LIU	University of Pittsburgh
Qijun SUN	Beijing Institute of Nanoenergy and Nanosystems, CAS
Qing YANG	Zhejiang University, China
Quan YUAN	Institute of Semiconductors, CAS
Rusen YANG	Xidian University
Seongjun PARK	KAIST
Shuai WANG	Xiamen University
Shuli WANG	Xiamen University
Shuo GAO	Beihang University
Stefano PALAGI	Scuola Superiore Sant'Anna
Tae-Eun PARK	Ulsan National Institute of Science and Technology
Tianxun GONG	University of Electronic Science and Technology of China
Tuhin SANTRA	Indian Institute of Technology Madras
Wei ZHOU	Xiamen University
Wendong WANG	Shanghai Jiao Tong University
Wenhui WANG	Tsinghua University
Wibool PIYAWATTANAMETHA	KMTIL
Xinge YU	City University of Hong Kong
Xipeng LI	Xiamen university
Xu XIAO	University of Electronic Science and Technology of China
Yaqi HOU	Xiamen University
Yi LI	SUSTech
Yinghui WANG	Institute of Microelectronics of Chinese Academy of Sciences
Yoshikazu HIRAI	Kyoto University
Yu ZHOU	Central South University
Yuan GAO	University of Illinois
Yuanjing LIN	Southern University of Science and Technology
Yuanzhi TAN	Xiamen University
Yuerui LU	Australian National University
Yunlong ZI	The Chinese University of Hong Kong
Zhizhi SHENG	Suzhou Institute of Nano-Tech and Nano-Bionics, CAS
Zhuoqing YANG	Shanghai Jiao Tong University
Zhuqing WANG	Sichuan University
Zongyou YIN	The Australian National University

April 25-29 , 2021   Xiamen, China

CONTESTS INFORMATION

The IEEE-NEMS Conference Series annually selects winners to receive the Best Conference Paper Award and the Best Student Paper Award. Starting from NEMS 2011, the Steering Committee has decided to also grant a Best Conference Poster Award annually at the conference. In addition, we are very grateful to a group of private donors in contributing financially to the conference to set up a CM HO Best Paper Award in Micro/Nano Fluidics, in honor of Prof. C.-M. Ho, of the University of California, Los Angeles. For any of these awards, factors to be considered in determining the winners are the significance of the new findings/applications, technical merits, originality, potential impact on the field, and clarity of presentation.

Best Conference Award Committee

Co-Chairs: Yong QIN(onsite), Wen Jung LI(online)

Best Conference Paper Award Committee

Zheng CUI, Weizheng YUAN, Yuan LIN, Chaoyong YANG

CM HO Best Paper Award
in Micro/Nano Fluidics Committee

Zheng CUI, Weizheng YUAN, Yuan LIN, Chaoyong YANG

Best Student Paper Award Committee

Zheng CUI, Weizheng YUAN, Yuan LIN, Jian CHEN

Best Conference Poster Award Committee

Zhengchun PENG, Xiongying YE, Wenbo DING, Mengdi HAN,
Xueyong WEI, Tao WU, Zenghui WANG, Jian CHEN

The 16th IEEE International Conference on Nano/Micro Engineered & Molecular Systems

BENEFACTORS

We gratefully acknowledge, at the time of printing, the financial contributions from the following:

CONFERENCE SPONSORS

BENEFACTORS

Microsystems & Nanoengineering

Evatec China Ltd.

Guangdong Sygole Intelligent Technology Co.,Ltd

BMF Precision Tech Inc.

TAN KAN KEE INNOVATION LABORATORY

Nanomaterials (MDPI)

Ningbo ABAX Sensing Electronic Technology Co. Ltd.

Xiamen High-End MEMS Technology Co., Ltd.

Suzhou Research Materials Microtech Co.,Ltd

XI'AN LEADMEMS SCI&TECH CO.,LTD.

Jiangsu Hinovaic Technology Co. Ltd.

ICEASY TECHNOLOGY CO.,LTD.

BEST PAPER AWARD BENEFACTOR

Microsystems & Nanoengineering/Springer Nature

April 25-29 , 2021   Xiamen, China

IEEE-NEMS 2021 Program

Location：Science & Art Center, Xiamen University（厦门大学思明校区，科学与艺术中心）

Sunday, 25 April

	Venue: Pre-function Area, 2rd Floor
13:00-21:00	Registration
18:30-20:30	Reception
	Venue: Exhibition Hall (1F)
13:00-21:00	Exhibition Set Up

Day1 - Monday, 26 April

	Main Venue: Concert Hall (2F)	
08:30 - 08:45	Opening Ceremony Chair: Haixia ZHANG	
08:45 - 09:30	Plenary Speaker Ia：Lei JIANG Chair: Zhongqun TIAN	
09:30 - 10:15	Plenary Speaker Ib：Paul WEISS Chair: Xu HOU	
10:15 - 10:45	Coffee Break & Group Photo	
10:45 - 12:45	Session Ia: Micro/Nano/Molecular Fabrication-I Keynote Speaker: Wenjing HONG Chair: Sheng HU 445,455,515,578,655	
10:45 - 12:45	Venue: Multi-function Hall (1F)	Venue: Conference Room No.01 (1F)
	Chihming Ho Award Chair: Yong QIN 60,99,263,511,544	Invited Session Ia Advanced Micro/nano Fabrication, Materials and Applications Chair: Zhuoqing YANG
	Venue: Conference Room No.04 (2F)	Venue: Conference Room No.05 (2F)
	Invited Session Ib Thermal management in MicroNano Systems Chair: Xian (Annie) ZHANG	Invited Session Ic Resonant Micro/Nanoelectromechanical Systems Chair: Zenghui WANG
	Venue: Conference Room No.07 (3F)	Venue: Sunshine Hall (3F)
	Session Ib Energy Conversion and Storage-I Keynote Speaker: Xiaodong CHEN Chair: Huan LIU 241,261,358,364,484	Session Ic Micro/Nano Pores and Channels Keynote Speaker: Zhongyi JIANG Chair: Xinghua XIA 93,127,161,340,411
12:45 - 13:30	Location: Qinye Canteen 3rd Floor	
	Lunch Break	
13:30 - 15:30	Main Venue: Concert Hall (2F)	
	Best Conference Paper Chair: Yong QIN 111,112,150,182,298,380,587	
	Venue: Multi-function Hall (1F)	Venue: Conference Room No.01 (1F)
	Wen H Ko Memorial Session Plenary Speaker Ic: Minghui HONG	Invited Session Id Self-powered Micro/Nano Systems Chair: Wei TANG
	Venue: Conference Room No.04 (2F)	Venue: Conference Room No.05 (2F)
	Invited Session Ie Bioelectronic Devices and Systems Chair: Zhou LI & Mengdi HAN	Invited Session If Non-linear Micro/Nanoelectromechanical Systems Chair: Dong F. WANG & Xueyong WEI
	Venue: Conference Room No.07 (3F)	Venue: Sunshine Hall (3F)
	Session Id MEMS/NEMS-I Keynote Speaker: Sang-Woo KIM Chair: Ting ZHANG 652,674,684,698,471,386	Session Ie Micro/Nano Fluidic-I Keynote Speaker: Xinghua XIA Chair: Zhongyi JIANG 64,361,114,552,591,456
	Main Venue: Concert Hall (2F)	
15:30 - 16:30	Plenary Speaker Id (Nobel Laureate)：Jean-Marie LEHN Chair: Yunbao JIANG	
	Venue: Exhibition Hall (1F)	
08:30 - 16:30	Exhibition	
16:30 - 18:30	Coffee Break & Poster Session I Accept Posters from No.4 to No.320 except 256 & 314	
17:00 - 18:00	中文讲座：学术好莱坞-现代科研论文的构思与写作，主讲人：李风煜教授（暨南大学）	
18:30	Adjourn for the Day	
	Location: Yifu Building	
19:00- 21:00	VIP Dinner (By Invited)	

The 16th IEEE International Conference on Nano/Micro Engineered & Molecular Systems

Day2 - Tuesday, 27 April			
	Main Venue: Concert Hall (2F)		
08:30 - 09:15	**Plenary Speaker IIa：Chennupati JAGADISH** Chair: Haixia ZHANG		
09:15 - 10:00	**Plenary Speaker IIb：Shigang SUN** Chair: Yan LI		
10:00 - 10:15	Coffee Break		
10:15 - 12:15	**Session IIa: MEMS/NEMS-II** Keynote Speaker: Jing LIU Chair: Dezhi WU 372,539,440,266,482,144		
10:15 - 12:15	Venue: Multi-function Hall (1F)		Venue: Conference Room No.01 (1F)
	Best Student Paper Chair: Yong QIN 2,149,335,407,448,460,469,664		**Invited Session IIa** **Micro Nano and Molecular Systems for Diagnostics and Therapeutics** Chair: Yi ZHANG
	Venue: Conference Room No.04 (2F)		Venue: Conference Room No.05 (2F)
	Invited Session IIb **Micro-/Nanofluidic Devices and Systems** Chair: Yan XU & Wenhui WANG		**Invited Session IIc** **Multidisciplinary Frontier: Advanced Chemistry Materials for MEMS/NEMS** Chair: Xi YAO
	Venue: Conference Room No.07 (3F)		Venue: Sunshine Hall (3F)
	Session IIb **Nanophotonics and Nanoscale Imaging** Keynote Speaker: Mona JARRAHI Chair: Xu XIAO 508,660,669,695,599,438,177		**Session IIc** **Energy Conversion and Storage-II** Keynote Speaker: Jianhua ZHANG Chair: Honglong CHANG 473,625,267,310,384,98,621
12:15 - 13:30	Location: Qinye Canteen 3rd Floor		
	Lunch Break		
13:30 - 15:30	Main Venue: Concert Hall (2F)		
	Session IId **Soft&Flexibe Devices and Applications-I** Keynote Speaker: Zheng CUI Chair: Wenli ZHOU 253,269,322,382,397,651,102		
	Venue: Multi-function Hall (1F)		Venue: Conference Room No.01 (1F)
	Session IIe **Micro/Nano Fluidic-II** Keynote Speaker: Xingyu JIANG Chair: Yan LI 613,371,336,697,132		**Invited Session IId** **Multidisciplinary Frontier: Micro/Nano Energy and Smart Electronics** Chair: Yong ZHAO & Xiuhan LI
	Venue: Conference Room No.04 (2F)		Venue: Conference Room No.05 (2F)
	Invited Session IIe **Electronic and Optoelectronic Fibers and Textiles** Chair: Fabien Sorin & Lei WEI		**Invited Session IIf** **RF Micro/Nano Components and Systems** Chair: Joshua E.-Y. LEE & Cheng TU
	Venue: Conference Room No.07 (3F)		Venue: Sunshine Hall (3F)
	Session IIf **Nanoscale Robotics, Assembly, and Automation** Keynote Speaker: Metin SITTI Chair: Xi XIE 159,200,417,293,251,368		**Session IIg** **Micro/Nano Sensors and Actuators-I** Keynote Speaker: Xinxin LI Chair: Xueyong WEI 631,403,461,488,529,622
15:30 - 16:30	Main Venue: Concert Hall (2F)		
	Plenary Speaker IIc (Nobel Laureate) : Jean Pierre SAUVAGE Chair: Xiaoyu CAO		
	Venue: Exhibition Hall (1F)		
08:30 - 16:30	Exhibition		
16:30 - 18:30	**Coffee Break & Poster Session II** Accept Posters from No.323 to No.526 include 256 & 314		
16:30 - 17:00	中文讲座：学术格莱美讲座--如何与编辑沟通？ 主讲人：李彦教授（北京大学，ACS Nano副主编）		
17:00 - 18:30	中文讲座：顶刊交流会 Top Journal Session		
18:30	Adjourn for the Day		

April 25-29 , 2021   Xiamen, China

Day3 - Wednesday, 28 April		
	Main Venue: Concert Hall (2F)	
08:30 - 09:15	**Plenary Speaker IIIa: Joost VLASSAK** Chair: Xu HOU	
09:15 - 10:00	**Plenary Speaker IIIb: Alex JEN** Chair: Xiaosheng ZHANG	
10:00 - 10:15	Coffee Break	
10:15 - 12:15	**Session IIIa** **MEMS/NEMS-III** Keynote Speaker: Yuan LIN Chair: Xiaosheng ZHANG 30,583,257,321,229,270	
10:15 - 12:15	Venue: Multi-function Hall (1F)	Venue: Conference Room No.01 (1F)
	Session IIIb **Soft&Flexibe Devices and Applications-II** Keynote Speaker: Yongan HUANG Chair: Libo GAO 129,134,211,146,218,219	**Invited Session IIIa** **Multidisciplinary Frontier: Soft Rubbery Electronics and Stretchable** Integrated Systems Chair: Cunjiang YU & Wei LAN
	Venue: Conference Room No.04 (2F)	Venue: Conference Room No.05 (2F)
	Invited Session IIIb **Multidisciplinary Frontier: Nanoplasmonics and Biosensors** Chair: Xiangwei ZHAO	**Invited Session IIIc** **Flexible and Wearable Microsystems for Sensing and Actuation** Chair: Zhengchun PENG & Xinge YU
	Venue: Conference Room No.07 (3F)	Venue: Sunshine Hall (3F)
	Session IIIc **Micro/Nano Sensors and Actuators-II** Keynote Speaker: Jian ZHU Chair: Tie LI 278,306,362,493,509,555,436	**Session IIId** **Micro/Nano/Molecular Fabrication-II** Keynote Speaker: Litao SUN Chair: Zenghui WANG 435,57,275,395,426,348
12:15 - 13:30	Location: Qinye Canteen 3rd Floor	
	Lunch Break	
13:30 - 15:30	Main Venue: Concert Hall (2F)	
	Session IIIe **M/NEMS-IV** Keynote Speaker: Lining SUN Chair: Hongzhong LIU 226,106,48,53,232,239,562	
	Venue: Multi-function Hall (1F)	Venue: Conference Room No.01 (1F)
	Session IIIf **Soft&Flexibe Devices and Applications-III** Keynote Speaker: Guozhen SHEN Chair: Rui ZHOU 430,576,533,561,563,565	**Invited Session IIId** **Advanced Scanning Probe Technology and Applications** Chair: Huan HU & Yu Kyoung RYU
	Venue: Conference Room No.04 (2F)	Venue: Conference Room No.05 (2F)
	Invited Session IIIe **Advanced Microsystems for Biomedical Applications** Chair: Soo Hyeon KIM	**Invited Session IIIf** **Multidisciplinary Frontier: Advanced Micro/Nano Electronics Plus (Bio-, Photo-, Energy-)** Chair: Yu ZHOU & Yue SHAO
	Venue: Conference Room No.07 (3F)	Venue: Sunshine Hall (3F)
	Session IIIg **MEMS/NEMS-V** Keynote Speaker: Liviu NICU Chair: Zewen LIU 203,105,116,173,538,418	**Session IIIh** **Nanomaterials-I** Keynote Speaker: Yan LI Chair: Jingxia WANG 13,11,34,284,186,676
15:30 - 16:30	Main Venue: Concert Hall (2F)	
	Plenary Speaker IIIc (Nobel Laureate) : Kostya NOVOSELOV Chair: Nanfeng ZHENG	
08:30 - 16:30	Venue: Exhibition Hall (1F)	
	Exhibition	
16:30 - 18:30	**Coffee Break & Poster Session III** Accept Posters from No.527 to No.720	
16:30 - 17:30	Industrial Panel	
17:30 - 18:30	中文讲座：学术奥斯卡讲座--如何做好学术报告? 主讲人：张海霞教授（北京大学）	
18:30	Adjourn for the Day	
19:00- 21:00	Location: Qinye Canteen 3rd Floor	
	Banquet & Award Ceremony	

The 16th IEEE International Conference on Nano/Micro Engineered & Molecular Systems

	Day4 - Thursday, 29 April	
	Venue: Multi-function Hall (1F)	**Venue: Conference Room No.01 (1F)**
8:30- 10:30	**Session IVa** **MEMS/NEMS-VI** Keynote Speaker: Yulong ZHAO Chair: Zewen LIU 242,311,334,637,338,223,163	**Session IVb** **Bio-inspired multiscale Interfaces** Keynote Speaker: Shutao WANG Keynote Speaker: Dechen JIANG Chair: Peiyi WU 199,209,391,560,23
	Venue: Conference Room No.04 (2F)	**Venue: Conference Room No.05 (2F)**
	Invited Session IVa **Multidisciplinary Frontier: Micro/Nano-Engineered** **Materials for Sensing and Applications** Chair: Xu XIAO & Tianxun GONG	**Session IVc** **Nanomaterials-II** Keynote Speaker: Jian JIN Chair: Jie JU 295,458,510,550,466
	Venue: Conference Room No.07 (3F)	
	Session IVd **Nature-inspired energy conversion and harvesting: from** **fundamentals to applicaitons** Keynote Speaker: Zuankai WANG Chair: Haisheng SAN 101,443,532,376,92,600	
10:45- 11:00	Coffee Break	
	Venue: Multi-function Hall (1F)	**Venue: Conference Room No.01 (1F)**
11:00 - 12:00	**Session IVe** **MEMS/NEMS-VII** Keynote Speaker: Yikuen LEE Chair: Songyue CHEN 396,646,365	**Session IVf** **Soft&Flexibe Devices and Applications-IV** Keynote Speaker: Peiyi WU Chair: Shutao WANG 51,90,28
	Venue: Conference Room No.04 (2F)	**Venue: Conference Room No.05 (2F)**
	Invited Session IVb **Multidisciplinary Frontier: Functional Micro-nano** **Structures, Techniques and Applications** Chair: Yang CAO	**Session IVg** **Soft&Flexibe Devices and Applications-V** Keynote Speaker: Hu TAO Chair: Guozhen LIU 491,598,115
	Venue: Conference Room No.07 (3F)	**Venue: Conference Room No.05 (2F)**
	Session IVh **Micro&Nanotechnologies for biomedicine** Keynote Speaker: Beomjoon KIM Chair: Guoan TAI 513,497	
	Location: Qinye Canteen 3rd Floor	
12:00 - 14:00	Farewell Lunch & Adjourn of Conference	
14:00	Lab Tour (Optional)	

The 16th IEEE International Conference on Nano/Micro Engineered & Molecular Systems

Plenary Speaker

Bioinspired Super-wettability System and Beyond
——Quantum-confined Superfluid: Energy Conversion, Chemical Reaction and Biological Information Transfer

Lei Jiang

Techinical Institute of Chemistry, Chinese Academy of Sciences, Beijing 100190, P.R. China

ABSTRACT

A new concept of "quantum-confined superfluid (QSF)" has been proposed for ultrafast ions and molecules transmission in biological ion channels, which are in a quantum way of single molecular or ionic chain with a certain number of molecules or ions. The biomimetic systems also exhibit QSF phenomena, such as ultrafast ions transport in artificial ion channels, and ultrahigh water flux in artificial water channels. The introduction of QSF concept in the fields of energy, chemistry and biology may create significant impact. As a challenge to the traditional theory, the concept of QSF will open up a new field of quantum ionics, promote the development and application of energy conversion material system, subvert the understanding of neural signal transmission in neuroscience and brain science, and promote the development of interfacial catalytic chemistry theory, and open up a new way for the future development of chemistry, chemical engineering and synthetic biology.

BIOGRAPHY

Lei Jiang is a Professor at the Technical Institute of Physics and Chemistry, Chinese Academy of Sciences. He is an academician of the Chinese Academy of Sciences, Academy of Sciences for the Developing World, and National Academy of Engineering, USA. His scientific contribution is learning from nature, discovering and establishing super-wettability system: from fundamental understanding to innovative applications. He has made a series of achievements, continuously leading the development of bioinspired super-wettability field in the world, and won many important international awards. Recent research interest includes introduction of quantum-confined superfluid into super-wettability system, and its applications in energy conversion, chemical reaction and biological information transfer.

978-1-6654-3008-1/21 $31.00 © 2021 IEEE

978-1-6654-3008-1/21 $31.00 © 2021 IEEE

April 25-29 , 2021   Xiamen, China

Atomically Precise Chemical, Physical, Electronic, and Spin Contacts

Paul S. Weiss

California NanoSystems Institute and Departments of Chemistry & Biochemistry, Bioengineering, and
Materials Science & Engineering, University of California, Los Angeles (UCLA), Los Angeles, California 90095, USA

Website: www.nano.ucla.edu

ABSTRACT

Two seemingly conflicting trends in nanoscience and nanotechnology are our increasing ability to reach the limits of atomically precise structures and our growing understanding of the importance of heterogeneity in the structure and function of molecules and nanoscale assemblies. By having developed the "eyes" to see, to record spectra, and to measure function at the nanoscale, we have been able to fabricate structures with precision as well as to understand the important and intrinsic heterogeneity of function found in these assemblies.

The physical, electronic, mechanical, and chemical connections that materials make to one another and to the outside world are critical and are intertwined in terms of their function. Just as the properties and applications of conventional semiconductor devices depend on these contacts, so do nanomaterials, many nanoscale measurements, and devices of the future. We discuss the important roles that these contacts can play in preserving key transport and other properties.

Initial nanoscale connections and measurements guide the path to future opportunities and challenges ahead. Band alignment and minimally disruptive connections are both targets and can be characterized in both experiment and theory. Chiral assemblies can control the spin properties and thus transport at interfaces. I discuss our initial forays into these areas in a number of materials systems.

BIOGRAPHY

Paul S. Weiss holds a UC Presidential Chair and is a distinguished professor of chemistry & biochemistry, bioengineering, and materials science & engineering at UCLA. His interdisciplinary research group includes chemists, physicists, biologists, materials scientists, mathematicians, bioengineers, electrical and mechanical engineers, computer scientists, clinicians, and physician scientists. They focus on the ultimate limits of miniaturization, exploring the atomic-scale chemical, physical, optical, mechanical, and electronic properties of surfaces, interfaces, supramolecular, and biomolecular assemblies. They develop new techniques to expand the applicability and chemical specificity of scanning probe microscopies. They apply these and other tools to study self- and directed assembly, and molecular and nanoscale devices. They advance nanofabrication down to ever smaller scales and greater chemical specificity to operate and to test functional molecular assemblies, and to connect to the chemical and biological worlds in neuroscience, gene editing, cancer immunotherapy, tissue engineering, and the microbiome. He has written over 400 publications, holds over 30 patents, and has given over 800 invited, plenary, keynote, and named lectures.

Prof. Weiss has been awarded a National Science Foundation (NSF) Presidential Young Investigator Award, a Sloan Fellowship, the American Chemical Society (ACS) Nobel Laureate Signature Award for Graduate Education in Chemistry, a Guggenheim Fellowship, a NSF Creativity Award, the ACS

Award in Colloid and Surface Chemistry, the ACS Southern California Section Tolman Medal, the ACS Patterson-Crane Award in Chemical Information, and the IEEE Nanotechnology Pioneer Award, among others. He is a fellow of the: American Association for the Advancement of Science, American Physical Society, American Vacuum Society, ACS, American Academy of Arts and Sciences, American Institute for Medical & Biological Engineering, Canadian Academy of Engineering, IEEE, Materials Research Society, and an honorary fellow of the Chinese Chemical Society and Chemical Research Society of India. He has been a visiting professor at the University of Washington and Kyoto University, and a distinguished visiting professor at the Kavli Nanoscience Institute and the Joint Center for Artificial Photosynthesis at Caltech. He is a visiting scholar at the Wyss Institute for Biologically Inspired Engineering at Harvard University. He held the Institut National de la Recherche Scientifique (INRS) Chaire d'excellence Jacques-Beaulieu and was a Fulbright Specialist for the Czech Republic. He is the founding editor-in-chief of *ACS Nano*.

978-1-6654-3008-1/21 $31.00 © 2021 IEEE

978-1-6654-3008-1/21 $31.00 © 2021 IEEE   4

The 16th IEEE International Conference on Nano/Micro Engineered & Molecular Systems

Pushing Laser Precision Engineering from Micro-scale to Nano-scale:

Progress, Challenges and Opportunities

Minghui HONG
Department of Electrical and Computer Engineering, National University of Singapore, Singapore 117508

Website: https://online.ece.nus.edu.sg/staff/web.asp?id=eleHMH

ABSTRACT

Laser precision engineering in micro/nano-scales has unique advantages as a non-contact process with flexible setup and high speed processing in ambient air. It is an important advanced manufacturing means for high quality micro/nano-structures' fabrication and related surface processing, especially to create new functional MEM/NEM devices and structures. In the past decades, we have witnessed its extensive applications from academic researches to production lines. Combined with other advanced processing tools, laser precision engineering's resolution has been pushed down to ~ 25 nm in laboratories, much smaller than the optical diffraction limit, which will soon play a much more important role in next-generation nano-manufacturing. In this talk, the research progress on dynamic laser-matter interactions will be reviewed. How to achieve small heat affected zone is the first key challenge in high quality laser precision engineering to push its resolution from micro-scale to nano-scale. To apply ultrafast femtosecond laser induced photo-thermal and photo-chemical effects is a must to solve this issue. The second key challenge is how to ensure high enough processing speed to meet various industrial needs as one beam laser processing at high resolution could not achieve this mission. Parallel laser beam processing becomes a versatile solution to fit the both requirements on high resolution and high speed at the same time. The third challenge is how to carry out the laser nano-structuring in far field as the near field effect requires our tiny optics working very close to sample surfaces, which confines the near field laser nanofabrication to be only suitable for a few super-smooth surface samples. Our recent research shows that hybrid fs laser processing in far field and in ambient air is a novel method to make ~15 nm features directly on Si surfaces (Fig. 1). Though we still need to understand clearly the complicated physics behind this unique experimental result, further fine tuning of our experimental setup and laser processing parameters is highly possible to bring our laser nano-fabrication resolution down to ~10 nm.

BIOGRAPHY

Prof. Minghui Hong specializes in laser microprocessing & nanofabrication, optical engineering and laser applications. He has co-authored 15 book chapters, 42 patents granted, and 450+ scientific papers in *Nature, Chemical Reviews, Nature Nanotechnology, Advanced Materials, Advances in Optics and Photonics, Nano Letters, Light: Science and Applications, ACS Nano, Science Advances, Nature Communications, and Laser & Photonics Reviews* etc. and 80+ plenary/keynote/invited talks in international conferences. He is a member of organizing committees for Laser Precision Micromachining International Conference (2001~2021), International Symposium of Functional Materials (2005, 2007 and 2014), Chair of International Workshop of Plasmonics and Applications in Nanotechnologies (2006), Chair of Conference on Laser Ablation (2009) and Chair of Asia-Pacific Near-field Optics Conference (2013 and 2019). Prof. Hong is invited to serve as an Editor of *Light: Science and Applications, Engineering, Science China G, Laser Micro/nanoengineering*, and Executive Editor-in-chief of *Opto-Electronic Advances*. Prof. Hong is Fellow of Academy of Engineering, Singapore (FSEng), Fellow of *Optical Society of America (OSA)*, Fellow of *International Society for Optics and Photonics (SPIE)*, and Fellow and Vice President of *International Academy of Photonics and Laser Engineering (IAPLE)* and Fellow and Vice President of *Institution of Engineers, Singapore (IES)*. Prof. Hong is currently a Full Professor and the Director of *Advanced Research and Technology Innovation Centre (ARTIC)*, Director of *Optical Science and Engineering Centre (OSEC)*, Department of Electrical and Computer Engineering, National University of Singapore. He is also a founder of *Phaos Technology Pte. Ltd.*

978-1-6654-3008-1/21 $31.00 © 2021 IEEE

From Supramolecular to Adaptive Chemistry – Contributions to Nanoscience and Technology

Jean-Marie Lehn

ISIS, University of Strasbourg; Strasbourg, France

Supramolecular chemistry is actively exploring systems undergoing self-organization, i.e. systems capable of spontaneously generating well-defined functional supramolecular architectures by self-assembly from their components, on the basis of the molecular information stored in the covalent framework of the components and read out at the supramolecular level through specific interactional algorithms, thus behaving as programmed chemical systems.

The implementation of molecular information controlled, "programmed" and functional systems allows the spontaneous but controlled generation of well-defined, functional molecular and supramolecular architectures of nanometric size through self-organization by design . It represents a means of performing programmed engineering and processing of functional nanostructures and offers attractive perspectives to nanoscience and nanotechnology.

Supramolecular entities as well as molecules containing reversible bonds are able to undergo a continuous change in constitution by reorganization and exchange of building blocks. This capability allows for self-organisation with selection and defines a Constitutional Dynamic Chemistry (CDC) on both the molecular and supramolecular levels. CDC introduces a paradigm shift with respect to constitutionally static chemistry. It takes advantage of dynamic constitutional diversity to enable variation and selection and thus allow for adaptation leading to the emergence of an adaptive chemistry towards a chemistry of complex systems.

- Lehn, J.-M., Supramolecular Chemistry: Concepts and Perspectives, VCH Weinheim, 1995.

- Lehn, J.-M., Dynamic combinatorial chemistry and virtual combinatorial libraries, Chem. Eur. J., 1999, 5, 2455.

- Lehn, J.-M., Programmed chemical systems : Multiple subprograms and multiple processing/expression of molecular information, Chem. Eur. J., 2000, 6, 2097.

- Lehn, J.-M., Toward complex matter: Supramolecular chemistry and self-organization, Proc. Natl. Acad. Sci. USA, 2002, 99, 4763.

- Lehn, J.-M., Toward self-organization and complex matter, Science, 2002, 295, 2400.

- Lehn, J.-M., From supramolecular chemistry towards constitutional dynamic chemistry and adaptive chemistry, Chem. Soc. Rev., 2007, 36, 151.

- Lehn, J.-M., in Constitutional Dynamic Chemistry, ed. M. Barboiu, Topics Curr. Chem, 2012, 322, 1-32.

- Lehn, J.-M., Perspectives in Chemistry – Steps towards Complex Matter, Angew. Chem. Int. Ed., 2013, 52, 2836-2850.

- Lehn, J.-M., Perspectives in Chemistry – Aspects of Adaptive Chemistry and Materials, Angew. Chem. Int. Ed., 2015, 54, 3276-3289.

BIOGRAPHY

Jean-Marie LEHN was born in Rosheim, France in 1939. In 1970 he became Professor of Chemistry at the Université Louis Pasteur in Strasbourg and from 1979 to 2010 he was Professor at the Collège de France in Paris. He is presently Professor at the University of Strasbourg Institute for Advanced Study (USIAS). He shared the Nobel Prize in Chemistry in 1987 for his studies on the chemical basis of "molecular recognition" (i.e. the way in which a receptor molecule recognizes and selectively binds a substrate), which also plays a fundamental role in biological processes. Over the years his work led him to the definition of a new field of chemistry, which he has proposed calling "supramolecular chemistry" as it deals with the complex entities formed by the association of two or more chemical species held together by non-covalent intermolecular forces, whereas molecular chemistry concerns the entities constructed from atoms linked by covalent bonds. Subsequently, the area developed into the chemistry of "self-organization" processes and more recently towards "adaptive chemistry", dynamic networks and complex systems. Author of over 1000 scientific publications, Lehn is a member of many academies and institutions. He has received numerous international honours and awards.

978-1-6654-3008-1/21 $31.00 © 2021 IEEE   8

April 25-29 , 2021   Xiamen, China

Plenary Speaker

Semiconductor Nanowires for Optoelectronics Applications

Chennupati Jagadish

ARC Centre of Excellence on Meta-Optical Systems, Research School of Physics, Australian National University, Canberra, ACT 2601, Australia

Website : https://physics.anu.edu.au/contact/people/profile.php?ID=106

ABSTRACT

Semiconductors have played an important role in the development of information and communications technology, solar cells, solid state lighting. Nanowires are considered as building blocks for the next generation electronics and optoelectronics. In this talk, I will introduce the importance of nanowires and their potential applications and discuss about how these nanowires can be synthesized and how the shape, size and composition of the nanowires influence their structural and optical properties. I will present results on axial and radial heterostructures and how one can engineer the optical properties to obtain high performance lasers, THz detectors, solar cells and to engineer neuronal networks. Future prospects of the semiconductor nanowires will be discussed.

BIOGRAPHY

Professor Jagadish is a Distinguished Professor and Head of Semiconductor Optoelectronics and Nanotechnology Group in the Research School of Physics, Australian National University. He has served as *Vice-President and Secretary Physical Sciences of the Australian Academy of Science* during 2012-2016. He is currently serving as Past President of IEEE Photonics Society, Past President of Australian Materials Research Society. Prof. Jagadish is the Editor-in-Chief of Applied Physics Reviews, Editor of 3 book series and serves on editorial boards of 19 other journals. He has published more than 950 research papers (680 journal papers), holds 5 US patents, co-authored a book, co-edited 15 books and edited 12 conference proceedings and 18 special issues of Journals. He has won the 2000 IEEE Millennium Medal and received Distinguished Lecturer awards from IEEE NTC, IEEE LEOS and IEEE EDS. He is a Foreign Member of US National Academy of Engineering, Fellow of the Australian Academy of Science, Australian Academy of Technological Sciences and Engineering, The World Academy of Sciences, US National Academy of Inventors, Indian National Science Academy, Indian National Academy of Engineering, Indian Academy of Sciences, National Academy of Sciences of India, IEEE, APS, MRS, OSA, AVS, ECS, SPIE, AAAS, FEMA, APAM, IoP (UK), IET (UK), IoN (UK) and the AIP. He received many awards including IEEE Pioneer Award in Nanotechnology, IEEE Photonics Society Engineering Achievement Award, OSA Nick Holonyak Jr Award, Welker Award, IUMRS Somiya Award, UNESCO medal for his contributions to the development of nanoscience and nanotechnologies and Lyle medal from Australian Academy of Science for his contributions to Physics, Beattie Steel Medal from Australian Optical Society and IEEE Education Award from Electron Devices Society. He has received Australia's highest civilian honor, AC, Companion of the Order of Australia, from the Australian Government.

978-1-6654-3008-1/21 $31.00 © 2021 IEEE

978-1-6654-3008-1/21 $31.00 © 2021 IEEE

Nanomaterials of High-Surface Energy for Efficient Electrochemical Energy Conversion and Storage

Shi-Gang Sun

State Key Laboratory of Physical Chemistry of Solid Surfaces, College of Chemistry and Chemical Engineering, Xiamen University, Xiamen 361005, China

Website : https://pcoss.xmu.edu.cn/en/info/1084/1082.htm; http://www.sungroup.ac/index.asp

ABSTRACT

Electrochemical energy conversion and storage paly a major role in mobile electronic devices, electric vehicles, renewable energy development and utilization, and are also widely applied in information technology and NEMS. The fast development of above application scenarios is raising increasing demands on electrochemical energy conversion and storage, including higher energy density, higher power density, higher safety, longer lifespan, lower cost and wider window of operation temperature. To fulfill this requirement and to promote the reduction of carbon emission, the key issue is to develop/exploit high-efficient electrocatalysts for fuel cells and high-performant electrode materials for batteries. The nanomaterials of high-surface energy possesses an open surface structure, and are appropriate high-efficient electrocatalysts with high-density of catalytic active sites and high-performant electrode materials with high-density of transportation channels. We have focused, in recent years, on structural design and controlled synthesis of nanomaterials of high-surface energy for both fuel cell catalysts and lithium batterie electrodes, the main progresses will be reported at the presentation.

References:

1. N. Tian et al. *Science*, 2007, 316, 732.
2. N. F. Yu, et al. *Angew. Chem. Int.* Ed. 2014, 53, 5097.
3. J. Xiao et al. *J. Am. Chem. Soc.* 2013, 135, 18754.
4. Z.Y. Zhou et al. *Chem. Soc. Rev.* 2011, 40, 4167. 5.
G. Z. Wei et al. *Adv. Mater.* 2010, 22, 4364.
6. F. Fu et al. *J. Mater. Chem. A*, 2013, 1, 3860.
7. T. Sheng et al. *Acc. Chem. Res.* 2016, 49, 2569.
8. Y.C. Wang et al. *ChemElectroChem.* 2018, 5, 1914.
9. S.J, Zhang, et al. *Small*, 2018, 14, 1801054.
10. X.C. Liu, et al. *Nano Energy*, 2018, 50, 685.
11. N.F. Yu, et al. *ACS Catal.* 2019, 9, 3144.
12. C. Xiao, et al. *Joule*, 2020, 4, 2562.

BIOGRAPHY

Shi-Gang Sun obtained his Bachelor of Science from Xiamen University, China, in 1982, Doctorat d'Etat (Docteur ès Sciences Physiques) from Université Pierre et Marie Curie (Paris VI), France, in 1986. After one-year post-doctoral research in the Laboratoire d'Electrochimie Interfaciale du CNRS, France, he returned to China by the end of 1987, and served as associate professor and later full professor in 1991 at the Department of Chemistry of Xiamen University till now. His main research interests of Prof. Sun include Electrocatalysis, Electrochemical SurfaceScience, Spectroelectrochemistry, Nanomaterials and Chemical power sources.

Prof. Sun is Academician of Chinese Academy of Sciences, fellow of International Society of Electrochemistry (ISE) and fellow of Royal Society of Chemistry (RSC). He has been awarded the "Brian Conway Prize" from International Society of Electrochemistry (ISE), "Distinguished Contribution Award" from the Chinese Society of Electrochemistry, "Le prix Franco-Chinois 2014-2015" jointly from Société Chimique de France (SCF) and Chinese Chemical Society (CCS), and the State Natural Science Award (2nd Degree) of China, "Achievement award" from Internatioanl Automotive Lithium Battery Associetion, and "Achievement award in spectroscopy" from Chinese Chemical Society and Chines Optics Society. He is now editorial board member of Journal of Electroanalytical Chemistry, Functional Materials Letters, ACS Energy Letters, Electrochemical Energy Review, National Science Review, Applied Chemistry and Journal of Solid State Electrochemistry, and is serving as associate editor to Electrochimica Acta, Spectral Analysis and Spectroscopy, Chinese Journal of Chemical Education, Acta Chimica Sinica, and editor-in-chief of the Journal of Electrochemistry.

Molecules in Motion: From Biology to Chemistry

Jean-Pierre Sauvage

Institut de Science et d'Ingénierie Supramoléculaires (ISIS), University of Strasbourg, F-67000 Strasbourg, France

ABSTRACT

An important family of molecules is that of interlocking or threaded rings named **catenanes** and **rotaxanes** respectively. The simplest catenane, a [2] catenane, consists of two interlocking rings. Rotaxanes consist of rings threaded by acyclic fragments (axes). These compounds played an important role in the emergence of the field named "**molecular machines**". This field has experienced a spectacular development, in relation to molecular devices at the nanometric level or as mimics of biological motors. In biology, motor proteins are omnipresent and crucial in a large variety of processes essential to life (ATPase, a rotary motor, being particularly impressive). Numerous examples of artificial molecular machines are based on simple or complex rotaxanes or catenanes. Non-interlocking ring compounds have also been used. In particular, light-driven rotary motors have been created by the team of Feringa. Finally, potential applications and future developments of this active research area will be mentioned.

BIOGRAPHY

Since the beginning of the 80s, Sauvage and his group have been interested in various fields including : (i) coordination photochemistry and solar energy conversion, (ii) CO_2 electrocatalytic reduction, (iii) chemical topology: catenanes, knots and rotaxanes, (iv) multifunctional ruthenium and iridium complexes for light-induced charge separation, (v) multifunctional porphyrins as models of the photosynthetic reaction centre as well as (vi) molecular switches and molecular machine prototypes such as a "swinging catenane", "muscles" or "compressors".
Sauvage received many awards.

The most recent ones are listed below :

2014	"Grand Prix de la Fondation de la Maison de la Chimie"
2016	Nobel Prize in Chemistry with J. Fraser Stoddart and Ben Feringa
	Honorary Fellow of the Royal Society of Chemistry (U-K)
	Foreign Member of the Russian Academy of Sciences
2017	Grand Officier de l'Ordre National du Mérite
	Honorary Doctorate, University of Athens (Greece)
2018	Honorary Doctorate, Uppsala University (Sweden)
2019	Honorary Doctorate, Universitat Politècnica de València (Spain)
	Foreign Associate of the U.S. National Academy of Sciences
2020	The Order of the Rising Sun, Gold and Silver Star, Japan

978-1-6654-3008-1/21 $31.00 © 2021 IEEE

Plenary Speaker

Nano-calorimetry: A New Tool for Materials Development

Joost J. Vlassak

Harvard School of Engineering and Applied Sciences, Harvard University, Cambridge, MA 02138, USA

Website : https://vlassakgroup.seas.harvard.edu/

ABSTRACT

Calorimetry has long been used to study chemical reactions and phase transitions in materials. The technique finds its origin in the mid-18th century when Scottish physician Joseph Black discovered the notion of latent heat and Lavoisier developed an ice calorimeter to measure the amount of heat given off during combustion of carbon or respiration of living organisms. Since then calorimetry has developed into a sophisticated technique indispensable in chemistry and materials science.

In this talk, I will show how the same technique can be used in the context of the Materials Genome Initiative. We use micromachining to fabricate arrays of calorimetric sensors that can perform measurements on samples as thin as a few nanometers at rates varying from isothermal to 10^5 K/s. The sensor arrays are ideally suited to explore complex materials systems using a combinatorial approach based on sputter-deposited thin-film composition spreads, especially when guided by computational materials science modeling. The methodology is illustrated for high-temperature shape memory alloys. Because of its large dynamic range, nanocalorimetry is also ideal for evaluating the kinetics of solid-state and solid-gas reactions. As an example, we use nano-calorimetry to evaluate the kinetics of solid-state reactions in Zr/B and Zr/B$_4$C multilayers and demonstrate that ultra-high temperature ceramics such as ZrB$_2$ and ZrB$_2$/ZrC alloys can be synthesized at moderate temperatures. Calorimetry traces show that the formation reactions typically proceed in two distinct steps: inter-diffusion and amorphization, followed by crystallization. First-principles calculations provide insight in the amorphization processes in the reactive multilayers and confirm the relatively low activation energies associated with these processes. Finally, I will present results on the non-Arrhenius crystallization kinetics of Cu$_{50}$Zr$_{50}$ metallic glass. Time permitting, a new isothermal device with applications in biophysics will be briefly discussed.

BIOGRAPHY

Joost J. Vlassak is currently the Abbott and James Lawrence Professor of Materials Engineering at Harvard University. Dr. Vlassak received his PhD degree in Materials Science from Stanford University. After working for several companies in Silicon Valley, Dr. Vlassak joined the School of Engineering and Applied Sciences at Harvard University in 2000. Dr. Vlassak is active in the field of the thermo-mechanical behavior of materials. He has developed experimental methods to characterize phase transitions and solid-state reactions in thin films, plastic deformation of coatings, elastic anisotropy in indentation, and fracture of thin films. Current experimental research projects focus on fracture of hydrogels, the use of hydrogels as ionic conductors in 3D printed devices, combinatorial calorimetry and calorimetry sensors for metabolic rate measurements in biological systems. Dr. Vlassak is co-author of more than 150 peer-reviewed scientific publications and several book chapters, and holds 7 US patents.

978-1-6654-3008-1/21 $31.00 © 2021 IEEE

April 25-29 , 2021   Xiamen, China

Transformative Impact of Printable Solar Cells for Meeting Next-Generation Energy Demands

Alex Jen
Department of Materials Science & Engineering and [2]Department of Chemistry, City University of Hong Kong, Kowloon, HK SAR

ABSTRACT

To complement silicon PV and make PV ubiquitous through a sustainable subsidy-free market model, it is crucial to develop low cost and high efficiency printable solar cells. Among different printable solar cell technologies, organic solar cells (OSCs) and perovskite solar cells (PVKSCs) are two candidates that have immense potential. They have inherently easier manufacturability, where low-temperature solution processing *via* high-throughput printing techniques like spray coating, inkjet printing, screen printing, blade coating, and slot die roll-to-roll (R2R) coating) can be used for manufacturing at scale. As a result, the capex requirements are significantly lower which will favor a low levelized-cost of electricity.

OSCs and PVKSCs also offer versatility in form factor for realizing flexible, semi-transparent, and color-tunable solar cells. These variants are pivotal for the success of building integrated photovoltaics (BIPVs) and niche market products (portable and wearable devices useful in fields like electronics, aviation, military, and medicine). Advancements in terms of non-fullerene acceptors (for OSCs) and multi-junction device architectures (for PVKSCs) have recently escalated the progression in power conversion efficiency (PCE). Record PCEs in lab scale devices have reached 19% and 25.5% for OSCs and PVKSCs, respectively, and efforts for scaling up and demonstrating large scale modules have begun through close cooperation between academia and industry. In this talk, I will provide an overview of progress and challenges for printable solar cells (OSCs and PVKSCs) toward commercialization and highlight our integral contributions over the last decade in this field.

BIOGRAPHY

Professor Alex Jen is the Lee Shau Kee Chair Professor of Materials Science and Chair Professor of Chemistry and Materials Science of the City University of Hong Kong. He also served as the Provost for CityU during 2016-2020. He is a distinguished researcher with more than 900 publications, 65,000 citations, and an H-index of 129. He is also a co-inventor for 63 patents and invention disclosures. His interdisciplinary research covers organic/hybrid functional materials and devices for photonics, energy, sensors, and nanomedicine.

For his pioneering contributions in organic photonics and electronics, Professor Jen was elected as an Academician by both the European Academy of Sciences and the Washington State Academy of Sciences. He is also a Fellow of several professional societies including AAAS, MRS, ACS, PMSE, OSA, SPIE. He was named by the Times Higher Education (THE) as one of the "Top 10 university researchers in Perovskite Solar Cell Research" from 2014 to 2017. In addition, he was recognized by Thomson Reuters as one of the "World's Most Influential Scientific Minds of 2015 and as a "Highly Cited researcher" in materials science from 2014-2019. He has also demonstrated strong capability in technology transfer, and was the founder of the Institute of Advanced Materials for Energy (i-AME) of the University of Washington.

He has also been appointed as the Changjiang Endowed Chair by the Ministry of Education, 1000 Talent Professor (Zhejiang Univ.), and as the World Class University Professor by the Korea Univ., and as the Distinguished Chair Professor by the National Taiwan University.

978-1-6654-3008-1/21 $31.00 © 2021 IEEE

978-1-6654-3008-1/21 $31.00 © 2021 IEEE

Van der Waals Heterostructures

Kostya S. Novoselov

Department of Materials Science and Engineering, National University of Singapore, Singapore, 117575, Singapore

ABSTRACT

One of the most important "property" of graphene is that it has opened a floodgate of experiments on many other 2D atomic crystals: BN, $NbSe_2$, TaS_2, MoS_2, *etc*. The resulting pool of 2D crystals is huge, and they cover a massive range of properties: from the most insulating to the most conductive, from the strongest to the softest.

If 2D materials provide a large range of different properties, sandwich structures made up of 2, 3, 4 … different layers of such materials can offer even greater scope. Since these 2D-based heterostructures can be tailored with atomic precision and individual layers of very different character can be combined together, - the properties of these structures can be tuned to study novel physical phenomena or to fit an enormous range of possible applications, with the functionality of heterostructure stacks is "embedded" in their design.

Already now such materials bring to life a number of exciting applications. In my lecture I will review some of them.

BIOGRAPHY

Prof Sir Konstantin 'Kostya' Novoselov FRS was born in Russia in August 1974. He has both British and Russian citizenship. He is best known for isolating graphene at The University of Manchester in 2004, and is an expert in condensed matter physics, mesoscopic physics and nanotechnology. Every year since 2014 Kostya Novoselov is included in the list of the most highly cited researchers in the world. He was awarded the Nobel Prize for Physics in 2010 for his achievements with graphene. Kostya holds positions of a Tan Chin Tuan Centennial Professor at the National University of Singapore. He is also part time Langworthy Professor of Physics and the Royal Society Research Professor at The University of Manchester.

He graduated from the Moscow Institute of Physics and Technology, and undertook his PhD studies at the University of Nijmegen in the Netherlands before moving to The University of Manchester in 2001. Later Professor Novoselov joint the National University of Singapore in 2019. Professor Novoselov has published more than 400 peer-reviewed research papers. He was awarded with numerous prizes, including Nicholas Kurti Prize (2007), International Union of Pure and Applied Science Prize (2008), MIT Technology Review young innovator (2008), Europhysics Prize (2008), Bragg Lecture Prize from the Union of Crystallography (2011), the Kohn Award Lecture (2012), Leverhulme Medal from the Royal Society (2013), Onsager medal (2014), Carbon medal (2016), Dalton medal (2016), Otto Warburg Prize (2019) among many others. He was knighted in the 2012 New Year Honours. 2020.

978-1-6654-3008-1/21 $31.00 © 2021 IEEE

Proceedings of the 16th Annual IEEE International
Conference on Nano/Micro Engineered and Molecular Systems
April 25-29, 2021

Integrated Electrical Test Vehicle Co-designed with Microfluidics for Evaluating the Performance of Embedded Cooling

Yuxin Ye[1,2], Student Member, IEEE, Nan Zhang[1,3], Lihang Yu[1,2], Bo Cong[1,2], Ruiwen Liu[1],
Yanmei Kong[1] *, Binbin Jiao[1] *, Member, IEEE

Abstract—From the dawn of the Information Age thermal management technology has played a key role in the continuing miniaturization, performance improvements, and higher reliability of electronic systems. The key advantage of embedded, versus conventional cooling is a reduction in the number of thermal interfaces between the heat source and the heat sink. Considerable research efforts have been devoted to improving the heat transfer efficiency through embedded cooling. However, there is a lack of a universal test platform to characterize the thermal performance of the embedded structure. In this work, an integrated electrical test vehicle co-designed with embedded cooling is proposed for evaluating the cooling performance of the embedded microfluidic structure and the overall dimension of the vehicle is under 30 $\times$ 15 $\times$ 2 mm^3. It provides a tool for analyzing the performance of embedded cooling and proposes a method for further study of embedded cooling mechanism.

I. INTRODUCTION

As the feature size decreases and integration degree increases, integrated circuits face a serious thermal challenge. Chip performance has improved due to increasing gate density as well as fabrication of multiple cores on a single chip. However, the inability to follow the Dennard scaling for supply voltage led to high chip heat flux dissipation densities, especially during the last decade [1-3]. Attention will be devoted to current thermal management requirements, driven by nano-electronics, which confront packaging engineers with the simultaneous "triple threat" of high-power, "hotspots" and 3D integration [4].

For efficient heat transfer, the net thermal resistance and heat spreading can be reduced by circulating the coolant through microchannels etched into the backside of the chip (also termed as embedded or direct chip backside microchannel cooling) [5]. It fulfills the requirements for low thermal resistance, low pressure drop, and excellent heat transfer coefficient in the field of applied engineering. Compared to the conventional thermal management approaches, embedded cooling can reduce the system level thermal resistance by up to 8x [6].

Considerable research efforts have been worked on microfluidic structures [7, 8], manifold structure [9, 10],

phase transformation mechanism [11] to optimize the performance of embedded cooling. However, even in state-of-the-art approaches, the knowledge of the thermal and electrical co-design in embedded cooling is still very limited, and there is a lack of effective analysis tools to evaluate the cooling performance of the embedded structure.

In this paper, an integrated electrical test vehicle co-designed with microfluidics is proposed to evaluate the thermal behavior of embedded cooling. The test vehicle is composed of a thermal test chip with heating unit array and PCB module. Embedded microchannel structure is fabricated on the thermal test chip substrate, and a closed electrical and flow test loop is formed through the assembly of the test chip and PCB module. The experiment results indicate that the test vehicle can evaluate the cooling performance of embedded structure and analyze the optimal working conditions. Moreover, it provides a useful analysis tool for further study of the heat transfer mechanism of embedded cooling.

II. EXPERIMENTAL SET-UP

A. Concept of Integrated Electrical Test Vehicle

Conventional backside attached microfluidic heat sinks are thermally joined to the chip through a thermal interface material (TIM) which combined with the thickness of substrate and heat sink base, results in heat spreading and high thermal resistance. In contrast, embedded cooling could effectively remove the interfacial thermal resistance (ITR) by directly fabricating the microfluidic structure in the substrate and shorten the heat transfer path between the hotspot and the coolant [12].

In this work, an integrated electrical test vehicle co-designed with microfluidics was proposed. The test vehicle is composed of thermal test chip and PCB module. The embedded microfluidic structure is etched on the substrate of the thermal test chip, and the cooling liquid can be directly introduced into the bottom of the chip to exchange heat with hot spots in the thermal test chip, as shown in Fig.1. Deep reactive ion etching process (DRIE) can realize the manufacture of different kinds of microfluidic structures, such as microchannels, micropillars, microgrooves, etc.

---

*Resrach supported by Thermal Characteristics of Thermos-sensitive Chip in Detection and Heat Dissipation Technology Foundation. (NO. E0GY14X001)

[1] Institute of Microelectronics of the Chinese Academy of Science, Beijing, 100029, China.

[2] University of Chinese Academy of Science, Beijing, 100049, China.

[3] School of Physics and Microelectronics, Zhengzhou University, Zhengzhou, 450001, China.

*Binbin Jiao and Yanmei Kong. Authors are with the Institute of Microelectronics of the Chinese Academy of Science, Beijing, 100029, China (corresponding author to provide phone: 010-82995931; E-mail: jiaobinb@ime.ac.cn &kongyanmei@ime.ac.cn).

The main function of the PCB module consists of two parts: (1) supplying power to the thermal test chip and reading the signal; (2) introducing the coolant into the embedded structure. Fig.2 plots its working principle and assembly method. The thermal test chip is assembled on the top layer of PCB. The bonding region are designed on the bottom of thermal test chip and top surface of PCB for filling solder to ensure the soldering strength. In addition, the power supply and signal readout circuits of the test chip are also designed on the top layer of PCB, and the electrical connection between chip and PCB is established through the wire bonding process.

In the process of chip design, the coolant is usually introduced directly into the embedded structure at the bottom of the chip to optimize the size of the embedded cooling chip. Therefore, the coolant delivery channels are machined on the bottom layer of the PCB model. Here, appropriate substrate material can be selected according to the test requirements. Visible polymethyl methacrylate substrate (PMMA) materials can be used to observe boiling phenomena during embedded cooling, on the other hand conventional PCB materials can be selected for reliable electronical expansibility. A closed flow channel is formed by the assembly of the top PCB and the bottom substrate, which connect the coolant I/O interface and the thermal test chip, as shown in Fig.2(b). The glue bonding method is adapted to effectively resist the leakage of coolant. Moreover, two negative temperature coefficient (NTC) thermistors are designed in the upstream and downstream channels to accurately measure the changes in coolant temperature during the cooling process, as shown in Fig.2(c).

The integrated thermal and electrical co-design test vehicle with microfluidics structure is an open platform with multi-variability and testability. The overall size of the multi-dimensional test vehicle is less than $30 \times 15 \times 2$ mm^3, meeting the rapidly increasing optimization requirements for the size, weight, power, and cost (SWaP-C) of electronics

B. Design and Fabrication of thermal test chip

The proposed of thermal test chip is designed to reproduce the thermal behavior of most vertical power devices [13]. In this work, the thermal test chip with multi-heating units is designed and fabricated, as shown in Fig.3(a). The overall size of thermal test chip is $6 \times 7 \times 0.4$ mm^3, which divided into a 4×5 array of 1×1 mm^2 square units in the central cross of the region. Each unit contains multiple parallel silicon resistance strips, which can simulate the heating behavior of local hotspots, the dimensions of the chip are shown in Table.1. Compared with the Pt micro-heater, the deposit silicon resistance strips in parallel has higher temperature stability and excellent process compatibility, which can be used as a low-cost micro-heat source solution.

The flexible connection of the heating array units can be realized by wire bonding, and each unit can be used as a hot spot to be combined according to the needs of the test. The heat transfer effect of hot spots with different sizes, shapes, and relative positions in the process of embedded cooling

can be accurately simulated. It accords with the thermal behavior of electronic chip with single hotspot and multi-hotspot.

TABLE 1.   DIMENSIONS OF THERMAL TEST CHIP

Dimension	Description	Value
w_m	Width of microchannel	70 μm
p_m	Pitch of microchannel	150 μm
h_m	Heigth of microchannel	300 μm
W_c	Width of TTC	7 mm
L_c	Length of TTC	6 mm
D_c	Depth of TTC	400 μm
N	Number of Heating Units	20 pcs

In this work, multiple parallel microchannels is designed as embedded microfluidic structure, which located below the heating cell array, as shown in Fig.3(b). Table1 illustrates the detail dimensions of the structure. Compared with other microfluidic structures, microchannel has better laminar flow characteristics, which ensures the thermal stability of the coolant during the heat exchange process. The above structure is processed by DRIE process on the substrate of thermal test chip, scanning electron microscopy (SEM) imagine in Fig.3(b) shows the actual morphology of the chip. A part of the region around the structure is reserved as a bonding ring to facilitate the bonding between the PCB and the chip.

C. Flow loop

The flow loop is constructed to measure the thermo-hydraulic characteristics of test vehicle, as shown in Fig.4. Deionized water was used as coolant and the temperature is controlled at 20 ℃. Infusion pump (Y-600, XYHY) can provide steady volumetric flow rates up to 1 L/min with maximum system pressures of 550 kPa. NTC thermistors (NTHS0402, VISHAY) and Differential pressure transducer (DPG409, OMEGA) are used to monitor temperature and pressure drops of coolant. Utilized low-ripple power supply and programming system to power hotspots. An IR camera is placed above the test vehicle and focused on the heat spot to probe the temperature distribution. Notably, the IR temperature measurement equipment was calibrated before experiments.

III. THEORETICAL ANALYSIS

This section describes simplified numerical models to predict the thermal resistance composition of above test vehicle, as shown in Fig.1. Employing these models is an effective way to obtain preliminary heat transfer data due to sophisticated fluid dynamic features and cost silicon processing techniques in experimental research.

The input electrical power P of the hotspot is obtained from the excitation voltage U and measured hotspot resistance R_h.

$$P = U^2 / R_h \qquad (1)$$

Considering the radiation and convective heat transfer between the hotspot and ambient, the actual heat flux should be expressed as:

$$q_{trans} = P/A_h - q_{loss} = \dot{m} g \int_{T_{c,in}}^{T_{c,out}} C_p(T)dt \quad (2)$$

Where C_p is the saturated specific heat of single-phase water, A_h is defined as the surface area of heat transfer and q_{loss} describes the heat flux lost by the heat exchange between hotspot and ambient. So q_{trans} is the net power transmitted to the coolant fluid.

In the process of evaluating cooling performance of the embedded microchannel structure, the vehicle total thermal resistance can effectively describe the energy conversion efficiency in heat transfer process, the total thermal resistance R_{total} can be described as:

$$R_{total} = (T_{h,max} - T_{c,in})/q_{trans} \quad (3)$$

Where $T_{h,max}$ is the maximum temperature of the hotspot, $T_{c,in}$ is the inlet temperature of coolant. In this work, the deionized water is selected as coolant, and the inlet temperature is controlled at 20 ℃ by reservoir. The NTC thermistor of test vehicle can also monitor the inlet temperature.

According to [14], the thermal resistance of embedded microchannel structure consists of three parts: conduction thermal resistance (R_{cond}), convection thermal resistance (R_{conv}), and coolant advection thermal resistance (R_{adv}). Where R_{cond} is the conduction thermal resistance in the solid substrate, which is calculated as[5]:

$$R_{cond} = t/KA_h \quad (4)$$

Where t is the thickness between the hotspot and microchannel structure. and K is defined as the thermal conductivity of the Si substrate, due to the temperature range of the heater is small in this experiment, K is regarded as 147 W/mK.

R_{adv} is the advection thermal resistance that indicates the coolant heating up as it absorbs heat [14], which is calculated as:

$$R_{adv} = (T_{c,out} - T_{c,in})/q_{trans} \quad (5)$$

Where $T_{c,in}$ is the inlet temperature of the coolant, and $T_{c,out}$ is the outlet temperature.

R_{conv} is the convection thermal resistance, which describes the convective heat transfer process between Si substrate and coolant, can be calculated as:

$$R_{conv} = R_{total} - R_{cond} - R_{adv} \quad (6)$$

IV. RESULT AND DISCUSSION

In this section, the cooling performance of the integrated thermal test vehicle with embedded structure was evaluated. The experiment focused on the thermal resistance composition of the test vehicle，and the main factors affecting the cooling of the embedded microfluidic structure .

Four heating units in the thermal test chip are connected by wire bonding as hotspots to change the local heat flux density of the hotspots, as shown in Fig.5. The microchannel structure is etched on the bottom of the thermal test chip as embedded microfluidic structure, Section Ⅱ(B) has illustrated the detail parameters.

Fig.6 plots the relationship between the heat flux density and the maximum temperature of the hotspot at different flow rates. As the heat flux increases, the maximum hotspot temperature rises linearly and this results are in good consistent with the previous research [5]. Under the volume flow rate of 125 mL/min, the maximum temperature of the hotspot from 0 to 800 W/cm² rise about 70 ℃. Moreover, this phenomenon becomes more obvious as the coolant flow rate decreases. It can be predicted that as the flow rate continues to decrease or the heat flux continues to increase, the chip enters a two-phase boiling state. Although boiling heat transfer has higher heat exchange capacity, the introduction of gas-liquid mixing state can easily cause instability in the cooling process. Therefore, in the current thermal design requirements, it is preferred to maintain the single-phase cooling state in the heat transfer process.

According to the definition of thermal resistance, the thermal resistance of the test vehicle during the cooling process can be calculated, as shown in Fig.7. In the single-phase cooling state, the total thermal resistance decreases with the increase of heat flux, and finally tends to a stable constant. At the same time, the variation trend of thermal resistance is not significantly affected by the volume flow of coolant. The total thermal resistance of the thermal test vehicle reaches the minimum value of 0.104 K/W when the heat flux is 800 W/cm² and volume flow reaches 125 mL/min. With the increase of flow rate, the total thermal resistance is easier to reach the equilibrium state.

The composition of the thermal resistance of the test vehicle is described in Fig.8 and Fig.9. For the embedded microchannel structure, convective thermal resistance dominates the total thermal resistance. At a coolant flow rate of 125 mL/min and heat flux of 800 W/cm², the heat resistance of convective heat transfer accounted for 81% of the total thermal resistance. Therefore, the cooling performance of embedded structure can be effectively improved by optimizing the embedded microfluidic structure to reduce the convective thermal resistance.

V. CONCLUSION

In this paper, an integrated electrical test vehicle co-designed with microfluidics is proposed for evaluating the thermal performance of embedded cooling. The thermal test chip and PCB module structure are the main components of the platform, where embedded microfluidic structure is fabricated on the thermal test chip substrate, and the closed electrical and flow loop is processed on the PCB module. In the experiment, the composition of the thermal resistance and the main factors affecting the cooling performance is analyzed. Experimental data shows that the test vehicle can accurately characterize the thermal behavior of embedded cooling. And it provides a tool for analyzing the performance of embedded cooling and proposes a method for further study of embedded cooling mechanism.

978-1-6654-3008-1/21 $31.00 © 2021 IEEE

ACKNOWLEDGMENT

The project was supported by Thermal Characteristics of Thermos-sensitive Chip in Detection and Heat Dissipation Technology Foundation. (NO. E0GY14X001). Thanks for the afford of the product-thermal test chip and technical support by Suzhou Rich Sensor Science & Technology Co., Ltd

REFERENCES

[1] C.S. Sharma, M.K. Tiwari, S. Zimmermann, T. Brunschwiler, G. Schlottig, B. Michel, D. Poulikakos, Energy efficient hotspot-targeted embedded liquid cooling of electronics, Applied Energy, 138 (2015) 414-422.

[2] C.S. Sharma, G. Schlottig, T. Brunschwiler, M.K. Tiwari, B. Michel, D. Poulikakos, A novel method of energy efficient hotspot-targeted embedded liquid cooling for electronics: An experimental study, Int J Heat Mass Tran, 88 (2015) 684-694.

[3] F.L.R. Avram Bar-Cohen, and David C. Deisenroth, Challenges and Opportunities in Gen3 Embedded Cooling with High-Quality Microgap Flow, (2018).

[4] J.G.F. Avram Bar-Cohen Joseph J. Maurer, DARPA's Intra/Interchip Enhanced Cooling (ICECool) Program, in: CS MANTECH Conference, New Orleans, Louisiana, USA, 2013.

[5] K.W. Jung, C.R. Kharangate, H. Lee, J. Palko, F. Zhou, M. Asheghi, E.M. Dede, K.E. Goodson, Embedded cooling with 3D manifold for vehicle power electronics application: Single-phase thermal-fluid performance, Int J Heat Mass Tran, 130 (2019) 1108-1119.

[6] J. Ditri, M.K. McNulty, S. Igoe, Embedded Microfluidic Cooling of High Heat Flux Electronic Components, P Ieee Les Eastm, (2014).

[7] D. Lorenzini, C. Green, T.E. Sarvey, X.C. Zhang, Y.C. Hu, A.G. Fedorov, M.S. Bakir, Y. Joshi, Embedded single phase microfluidic thermal management for non-uniform heating and hotspots using microgaps with variable pin fin clustering, Int J Heat Mass Tran, 103 (2016) 1359-1370.

[8] S. Feng, Y. Yan, H. Li, Z. He, L. Zhang, Temperature Uniformity Enhancement and Flow Characteristics of Embedded Gradient Distribution Micro Pin Fin Arrays Using Dielectric Coolant for Direct Intra-Chip Cooling, Int J Heat Mass Tran, 156 (2020).

[9] R. van Erp, R. Soleimanzadeh, L. Nela, G. Kampitsis, E. Matioli, Co-designing electronics with microfluidics for more sustainable cooling, Nature, 585(7824) (2020) 211-216.

[10] W. Escher, T. Brunschwiler, B. Michel, D. Poulikakos, Experimental Investigation of an Ultrathin Manifold Microchannel Heat Sink for Liquid-Cooled Chips, J Heat Trans-T Asme, 132(8) (2010).

[11] I. Mudawar, Recent Advances in High-Flux, Two-Phase Thermal Management, J Therm Sci Eng Appl, 5(2) (2013).

[12] Z. He, Y. Yan, Z. Zhang, Thermal management and temperature uniformity enhancement of electronic devices by micro heat sinks: A review, Energy, 216 (2021).

[13] X. Jordà, X. Perpiñà, M. Vellvehi, F. Madrid, D. Flores, S. Hidalgo, J. Millán, Low-cost and versatile thermal test chip for power assemblies assessment and thermometric calibration purposes, Applied Thermal Engineering, 31(10) (2011) 1664-1672.

[14] S. Adera, D.S. Antao, R. Raj, E.N. Wang, Hotspot Thermal Management via Thin-Film Evaporation-Part II: Modeling, Ieee T Comp Pack Man, 8(1) (2018) 99-112.

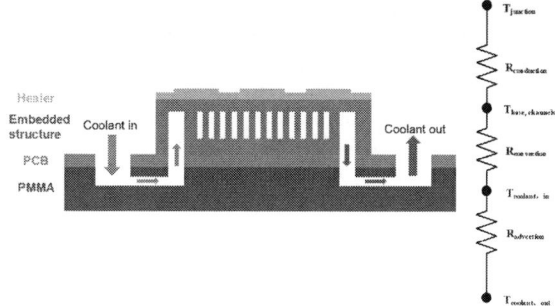

Figure 1. Concept of integrated electrical test vehicle and the composition of thermal resistance

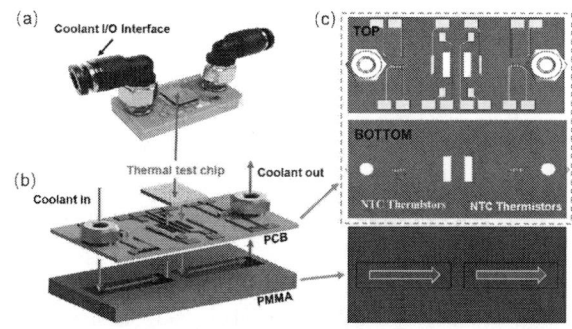

Figure 2. Scheme of. (a) Photograph of the integrated thermal test vehicle with liquid inlet and outlet interface;(b) Exploded illustration of integrated test vehicle with embedded cooling. (c) Overview of the 2 layers PCB model.

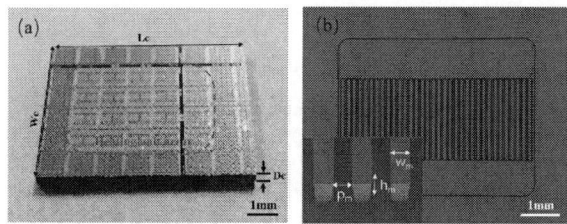

Figure 3. Thermal test chip details: (a) Top view of a TTC with a size of $6 \times 7 \times 0.4$ mm^3 and a 4×5 heating unit array in the center region; (b) SEM of the embedded microfluidic structure.

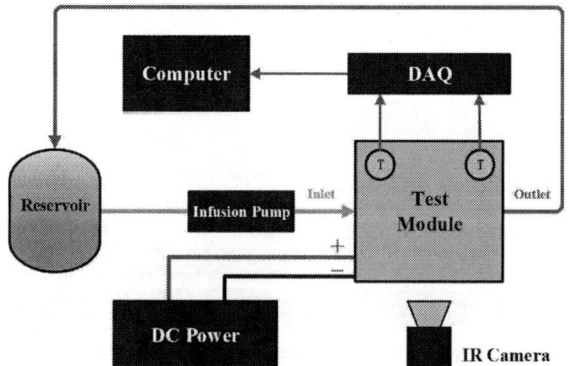

Figure 4. Schematic of the experimental flow loop

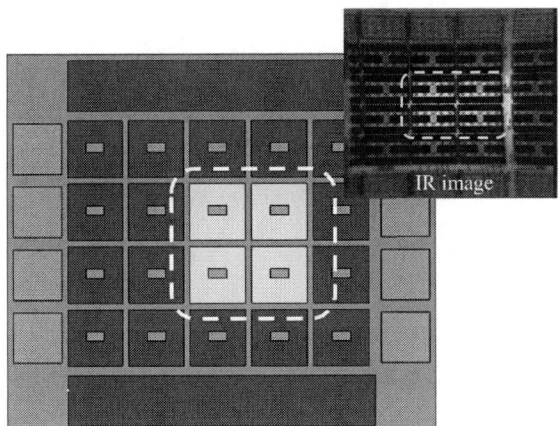

Figure 5. Hotspot definition in thermal test chip

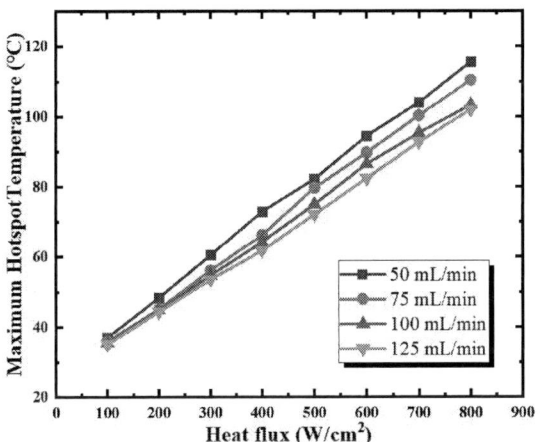

Fig.6 The relationship between the heat flux and maximum temperature of hotspot at different flow rates

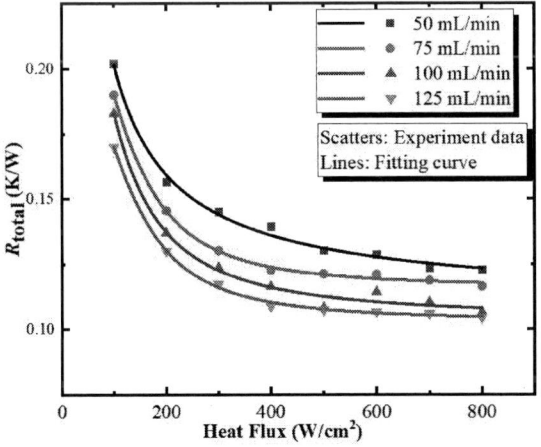

Figure 7. The relationship between the heat flux and total thermal resistance at different flow rates.

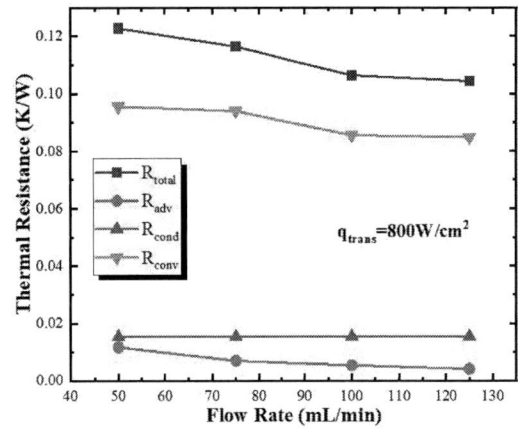

Figure 8. The relationship between thermal resistance composition and flow rate at the condition of heat flux density of 800 W/cm^2

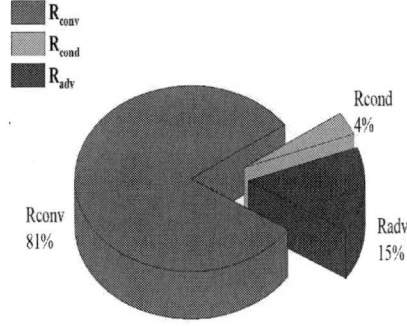

Figure 9. Composition of thermal resistance at 800 W/cm^2 and 125 ml/min.

Proceedings of the 16th Annual IEEE International
Conference on Nano/Micro Engineered and Molecular Systems
April 25-29, 2021

Minimal-Invasive Levodopa Sensing Based on Differential Amperometry Microneedle Electrodes Decorated with Spike-like Au Nanoparticles

Hangxu Ren, Xiyu Mao, Shanshan Zhang, Yu Cai, Shiyi Xu, Lu Fang, Xuesong Ye and
Bo Liang*, *Member, IEEE*

Abstract— Levodopa is the most commonly used drug for Parkinson's disease treatment since its inception. The concentration of levodopa must stay in a narrow range to achieve a better curative result which relies on the optimization of dosage. To adjust the dosage objectively and timely, it is of great significance to monitor the dynamic concentration of levodopa in Parkinson's patients. However, the high concentration of interferents coexist with levodopa can bring great inaccuracy to the detection. To address this issue, a novel high-accuracy levodopa biosensor based on differential amperometry microneedle electrodes is proposed. A pure levodopa signal can be extracted by two working electrodes with and without enzyme modification respectively. The result shows that this biosensor realize the dynamic detection of levodopa with the linear range of 0-20 μM, the sensitivity of 12.618 nA μM^{-1} cm^{-2}, and the detection limit of 0.18 μM (S/N=3). The system has good anti-interference performance to detect levodopa in high concentration of uric acid, glucose, and ascorbic acid with a variation of less than 5% (n = 3).

I. INTRODUCTION

Parkinson's disease (PD) is one of the most commonly neurodegenerative movement diseases in the world [1,2]. which severely affects the of life quality patients [3]. According to the research, there are 3 million PD patients in China, and it is being projected to reach around 5 million by 2030 [4]. Levodopa is the most effective drug for Parkinson treatment since 1960s[5], it can cross the blood-brain barrier and effectively improve the level of dopamine in the brain [6]. In addition, the concentration of levodopa in plasma directly contributes to the curative result, to avoid potential side effects caused by an abnormal high or low level of levodopa, such as dyskinesias, psychosis, orthostatic hypotension etc. [7], the concentration of levodopa should be controlled within a narrow range of therapeutic window. Therefore, it is of great significance to monitor the dynamic concentration of levodopa in patients who use levodopa.

Researchers have done great work toward the detection of levodopa based on various principles, such as spectrophotometry [8], capillary electrophoresis [9], fluorescence spectroscopy [10], high-performance liquid chromatography [11], and electrochemistry [12]. Due to its low-cost, high-sensitive, good linearity and a viable method for real-time sampling, electrochemical techniques are widely used on the monitoring of levodopa [13–17].

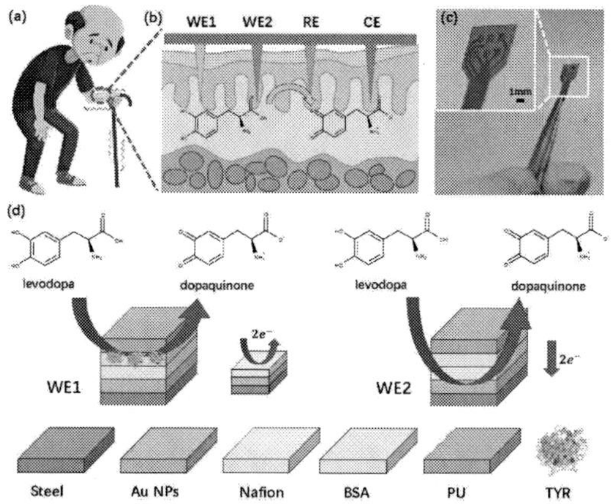

Figure 1. (a) A Parkinson's disease patient with minimal-invasive levodopa biosensor on his left arm. (b) The levodopa biosensor base on microneedle electrodes attached to epidermis. (c) Photograph of microneedle electrodes, and a zoom in photo. (d) Schematic of levodopa biosensor based on differential amperometry. From bottom to the top: the substrate was stainless steel, Au nanoparticle, Nafion membrane, enzyme layer and PU membrane. (Upper left) WE1 was modified with TYR and BSA. (Upper right) WE2 was modified with BSA.

However, there are many other electrochemically active substances in serum coexist with levodopa, which can bring deviation to the outcome while measuring the real-sample. Previous studies adapted a flexible non-invasive wearable sensor to measure levodopa in sweat [18], but it is a problem to obtain enough sweat on elder PD patients for real-time detection. Meanwhile, a dual-mode microneedle sensing platform for continuous minimally invasive electrochemical detection of levodopa using both square wave voltammetry and chrono-amperometry was reported [19]. It can record two different types of levodopa's signal on two electrodes at the same time offers a built-in redundancy. This kind of microneedle biosensor has a relatively larger size which is not convenient to wear during long time monitoring. Besides, its anti-interference ability and sensitivity still need to be improved. It is possible to conclude that accuracy and selectivity are two main factors need to consider when building a levodopa biosensor.

To achieve accurate and interference-resistant detection, this paper proposed a novel high-accuracy levodopa biosensor based on differential microneedle electrodes. It mainly consists of two working electrodes coated with four

This work was supported Zhejiang University K.P.Chao's High Technology Development Foundation; The Natural Science Foundation of Zhejiang Province (LQ20F010011, LY18H180006).

Hangxu Ren, Xiyu Mao, Shanshan Zhang, Yu Cai, Shiyi Xu, Xuesong Ye and Bo Liang are with Biosensor National Special Laboratory, Key Laboratory of Biomedical Engineering of Ministry of Education, College

of Biomedical Engineering andInstrument Science, Zhejiang University, Hangzhou, 310027, PR China (phone: 0571-87952756; e-mail: boliang1986@zju.edu.cn).

Lu Fang, is with College of Automation, Hangzhou Dianzi University, Hangzhou, 310018, PR China.

978-1-6654-3008-1/21 $31.00 © 2021 IEEE

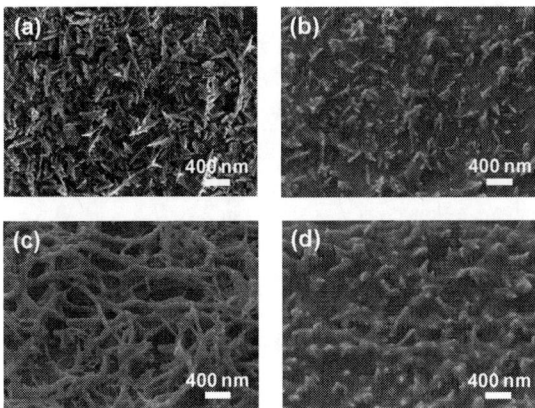

Figure 2. SEM images of microneedle electrode in different state of modification. (a) Spike-like Au nanoparticles, (b) modified with Nafion, (c) modified with poly-aniline, and (d) modified with tyrosinase and BSA after crosslinking.

layers as Fig. 1 (d) shows: metal catalytic layer, Nafion layer, enzyme layer, and biocompatible permeable membrane. One working electrode modified with tyrosinase which can convert levodopa into dopaquinone is WE1, the other modified with a blank control protein is WE2, so levodopa could be oxidized by Au. The Nafion layer is designed to isolate the weak current generated by the reaction that tyrosinase catalyzes levodopa to dopaquinone. Therefore, the signal detected by WE1 is only generated by other electrochemically active substances other than levodopa, and the signal detected by WE2 was generated by all the electrochemically active interferents and levodopa. Herein, the current difference between WE2 and WE1 is the signal generated by levodopa, this principle is denoted as differential amperometry.

II. EXPERIMENTAL

A. Materials and Reagents

Tyrosinase (TYR, 1190 U/mg) were purchased from Solarbio. Fetal bovine serum (FBS) was provided by CellMax (China). PBS solution (0.01 M, pH = 7.4) was purchased from Sangon Biotech (Shanghai, China). Bovine serum albumin (BSA), ascorbic acid (AA), uric acid (UA), glucose, and Chloroauric acid were purchased from Aladdin Co., Ltd. (Shanghai, China). Polyurethane (PU) was purchased from Sigma-Aldrich. Aluminum oxide polishing powder (d=0.05 μm) was purchased from HWRK Chem, Other reagents were purchased from Sinopharm Chemical Reagent Co., Ltd. (China) and were used without further modification.

B. Precondition of Microneedles

Microneedles were ultrasonically polished and cleaned in 80 mg/mL Aluminum oxide, deionized water (DI) for 10 min, respectively. Then, electrochemical cleaning process was done by cyclic voltammetry (-0.4V~0.6V, 10 cycles, scan rate 50 mV/s) until there are no significant redox peak. Finally, microneedles were risen with DI and dry at room temperature.

C. Modification of Levodopa Biosensors

To enhance the ability of anti-interference, the microneedle electrodes were modified four times to form four different layers. Metal catalytic layer: the microneedle electrode was immersed in 10 mM chloroauric acid at a constant voltage of 0 V for 300 s, Fig. 2 showed a spike-like Au nanoparticles (Au NPs) evenly covered the surface of microneedle, the diameter of spike Au NPs was about 200 nm. Nafion layer: 1% Nafion solution diluted in ethanol was applied to the electrode surface and allowed to dry in air form a thin film. Enzyme layer: for a better immobilization of tyrosinase, a poly-aniline film was first electrodeposition on the electrodes by galvanostatic with 0.1 mA for 100 s in 0.4 M aniline/1 M HCl solution, using Ag/AgCl as reference electrode and Pt wire as counter electrode. After electrodeposition, the electrode was immersed in deionized water for 30 min and then dried. Afterwards, WE1 was immersed in a solution of 5 mg/mL tyrosinase and 40 mg/mL BSA, and WE2 was immersed in a solution of 45 mg/mL BSA. Both electrodes were stored at 4°C for 24 h. After adsorption, the electrodes were rinsed in deionized water. Finally, both working electrodes were placed in a sealed bottle and 20 μL of 25% glutaraldehyde solution was added dropwise and placed in an incubator at 37 °C for 30 min.

Biocompatible polymer permeable membrane layer: To improve the linearity and stability of the microneedle electrodes, PU was selected to modify the outermost layer of the electrodes. 3% (w/w) PU solution was prepared by dissolved PU in 2% dimethyl formamide (DMF) and 98% tetrahydrofuran (THF). A copper wire with an aperture of 2 mm was dipped into the PU solution and the microneedle electrodes were inserted into the wire and pulled out once and then dried in air.

Both WE1 and WE2 were immersed in 10x PBS and stored in 4°C when not in use.

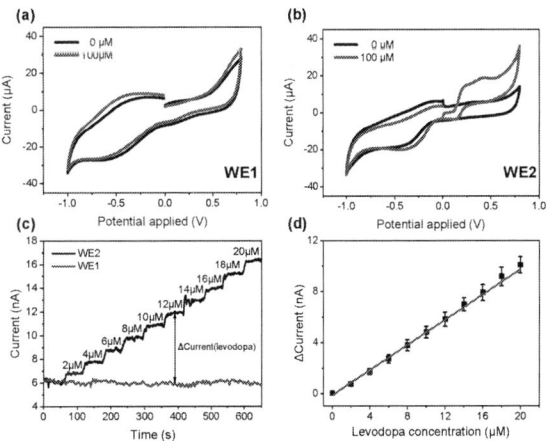

Figure 3. (a) CV curve of levodopa dissolved in PBS using the tyrosinase modified microneedle. (b) CV curve of levodopa dissolved in PBS using the blank enzyme modified microneedle. (c) Time-dependent current response of different levodopa concentrations. (d) The corresponding calibration curve (n=3).

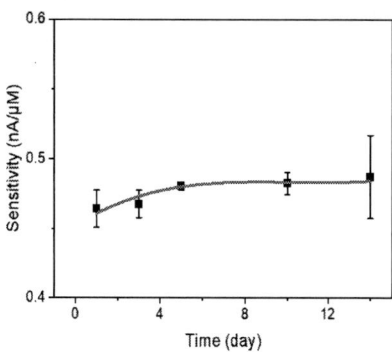

Figure 4. Sensitivity curve of WE2 stored for two weeks.

III. RESULTS AND DISSCUSSION

A. Performance of Levodopa Microneedle Biosensor

Due to the difference of modified enzyme layer, the electrochemical properties of the two microneedle electrodes for levodopa were different. By cyclic voltammetry measurement with the potential range of -1 V to 0.8 V at 100 mV/s in 0.01 M PBS solution (pH = 7.2), as shown in Fig.3(a)(b), it was obvious that WE2 modified with blank enzyme had a significant oxidation peak around 0.3 V, while WE1 modified with tyrosinase had no oxidation peak. This was because the tyrosinase modified on the WE1 had catalyzed levodopa into dopaquinone, and the Nafion film resisted the current to reach the metal catalytic layer. Fig3. (c) showed that two kinds of microneedle electrodes were used to detect the dynamic concentration change of levodopa (0-20 μM) in 0.01 M PBS solution under the applied potential of 0.3 V. About 2 μM levodopa was added every 60 s, and the response time was less than 10 s. A key point was to make sure the immersed length of microneedles stay the same. Waited around 200 s until the baseline current of WE1 and WE2 became steady and approximately the same value, then started recording signals. The current of WE1 remained unchanged while levodopa increased which validated that the WE1 only recorded the signal except levodopa. The difference between current response of the two microneedle electrodes to levodopa concentration was extracted, as shown in Fig.3(d), and the regression equation was y = 0.469x - 0.137. The modified length of the two microneedles was 7 mm, while the diameter of the microneedles was 0.16 mm so the sensitivity of the sensor to levodopa was 12.6 nA μM^{-1} cm^{-2}, with R^2 = 0.995, the limit of detection was 0.18 μM (S/N=3).

The two microneedle electrodes were immersed in 0.01 M PBS solution (pH = 7.2) and stored in refrigerator at 4°C for two weeks. There was no obvious current response in the concentration range of 0-20 μM levodopa for WE1 modified with tyrosinase. The sensitivity of WE2 modified with blank enzyme to levodopa was shown in Fig. 4. After two weeks of storage, the sensitivity remained 96.12% of the initial value (n = 3).

B. Characterization of Anti-interference Ability

The most difficult point in the detection of levodopa lies in the presence of high concentration of interfering substances, such as ascorbic acid, glucose, uric acid, which have certain electrochemical activity. With the presence of external potential, current signal will also be generated to interfere with the detection. Two kinds of microneedle electrodes were used to detect levodopa in the presence of high concentration of interferents, and the results can be seen in Fig.5. In this experiment, two kinds of microneedle electrodes had current response to the mixed solution of levodopa and different interferents at 0.3V applied potential. Because of the different modification of the two kinds of microneedle electrodes in the enzyme layer, WE1 could only detect the electrical signal of the interferents for the tyrosinase, while WE2 could detect the mixed signal of the interferents and levodopa, and the signal of levodopa could be extracted by subtracting the two signals. In the four groups of experiments, the concentration of levodopa was 10 μM, according to the previous calculation, the theoretical value of current response should be 4.69 nA. The error between the current value obtained by calculation and the theoretical value was less than 5%, which showed that the detection result of the sensor was reliable and effective.

C. In-vitro Evaluation

One of the difficulties in the detection of levodopa is the interference caused by sampling environment. To verify the effectiveness of levodopa biosensor, in vitro tests were carried out by detecting levodopa in serum. WE1 and WE2 were immersed in serum with 0 μM, 2 μM, 4 μM, 6 μM and 10 μM of levodopa respectively. The length of the microneedle immersed in serum was about 3 mm. Chronoamperometry response was recorded at 0.3V vs Ag/AgCl as shown in Fig.6(a), the response of WE1 remained unchanged, while the current of WE2 was linearly correlated with the concentration of levodopa. Take the

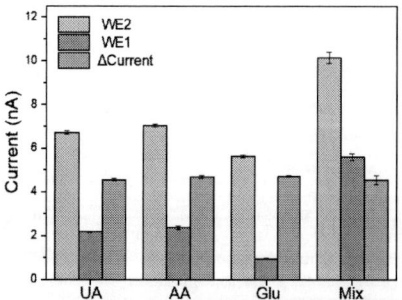

Figure 5. Detection of levodopa in the presence of various interferents. Concentration of levodopa in each group was 10μM, uric acid and ascorbic acid was 50μM, and glucose was 200μM, in Mix group the above three kinds of interferents were mixed with levodopa.

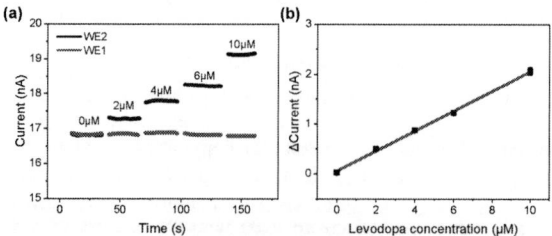

Figure 6. (a) Chronoamperometry response of WE1 and WE2 in the different concentration of levodopa (b) The correlation curve of extracted difference of current between WE2 and WE1 (n=3).

differential current between WE2 and WE1 to draw a graph as Fig.6(b). The regression equation was $y = 0.2 x - 0.023$, and the $R^2 = 0.998$. The sensitivity of the sensor to levodopa was 13 nA μM^{-1} cm^{-2}, which was similar to the previous experimental results, with a relative standard deviation of 3.1% (n = 3).

IV. CONCLUSION

In this paper, a high-accuracy levodopa biosensor based on differential amperometry microneedle electrodes was proposed, realizing a fast, real-time monitoring. Compared with previous reports, this novel minimal-invasive levodopa biosensor adapted an opposite strategy that indirectly detected the signal of levodopa by measuring the differential current between two working electrodes, one modified with tyrosinase and the other without it. Nafion effectively separated the current generated by levodopa oxidation reaction catalytic by tyrosinase, but it would reduce the sensitivity of at the same time. Nanotechnology for Au nanoparticle modification was applied to counter the decrease of sensitivity. Overall, the two working electrodes system, with a low detection limit, good reproductivity and sensitivity, provides another way towards Parkinson patients' daily health management. Further studies will focus on the optimization of the performance of microneedle electrodes and the development of an integrated wearable minimal-invasive levodopa monitoring system.

REFERENCES

[1] T. Pringsheim, N. Jette, A. Frolkis, and T. D. L. Steeves, "The prevalence of Parkinson's disease: A systematic review and meta-analysis," *Mov. Disord.*, vol. 29, no. 13, pp. 1583–1590, 2014.

[2] D. Berg, R. B. Postuma, C. H. Adler, and B. R. Bloem, "CME MDS Research Criteria for Prodromal Parkinson's Disease Key Definition Features of Prodromal PD," vol. 30, no. 12, pp. 1600–1609, 2015,.

[3] E. Ray Dorsey *et al.*, "Global, regional, and national burden of Parkinson's disease, 1990–2016: a systematic analysis for the Global Burden of Disease Study 2016," *Lancet Neurol.*, vol. 17, no. 11, pp. 939–953, 2018.

[4] G. Li *et al.*, "Parkinson's disease in China: A forty-year growing track of bedside work," *Transl. Neurodegener.*, vol. 8, no. 1, pp. 1–9, 2019.

[5] P. A. Lewitt, "Levodopa for the Treatment of Parkinson's Disease," *N. Engl. J. Med.*, vol. 359, no. 23, pp. 2468–2476, 2008.

[6] R. A. Hawkins, A. Mokashi, and I. A. Simpson, "An active transport system in the blood-brain barrier may reduce levodopa availability," *Exp. Neurol.*, vol. 195, no. 1, pp. 267–271, 2005.

[7] R. B. Postuma *et al.*, "MDS clinical diagnostic criteria for Parkinson's disease," *Mov. Disord.*, vol. 30, no. 12, pp. 1591–1601, 2015.

[8] M. Chamsaz, A. Safavi, and J. Fadaee, "Simultaneous kinetic-spectrophotometric determination of carbidopa, levodopa and methyldopa in the presence of citrate with the aid of multivariate calibration and artificial neural networks," *Anal. Chim. Acta*, vol. 603, no. 2, pp. 140–146, 2007.

[9] L. Zhang, G. Chen, Q. Hu, and Y. Fang, "Separation and determination of levodopa and carbidopa in composite tablets by capillary zone electrophoresis with amperometric detection," *Anal. Chim. Acta*, vol. 431, no. 2, pp. 287–292, 2001.

[10] T. Madrakian, A. Afkhami, and M. Mohammadnejad, "Simultaneous spectrofluorimetric determination of levodopa and propranolol in urine using feed-forward neural networks assisted by principal component analysis," *Talanta*, vol. 78, no. 3, pp. 1051–1055, 2009.

[11] D. Nyholm, H. Lennernäs, C. Gomes–Trolin, and S.-M. Aquilonius, "Levodopa Pharmacokinetics and Motor Performance During Activities of Daily Living in Patients With Parkinson's Disease on Individual Drug Combinations," *Clin. Neuropharmacol.*, vol. 25, p. 89, 2002.

[12] P. Prabhu, R. Suresh Babu, and S. Sriman Narayanan, "Amperometric Determination of L-Dopa by Nickel Hexacyanoferrate Film Modified Gold Nanoparticle Graphite Composite Electrode," *Sens. Actuators, B*, vol. 156, p. 606, 2011.

[13] B. Brunetti, G. Valdés-Ramírez, I. Litvan, and J. Wang, "A disposable electrochemical biosensor for l-DOPA determination in undiluted human serum," *Electrochem. commun.*, vol. 48, pp. 28–31, 2014.

[14] S. Shahrokhian and E. Asadian, "Electrochemical Determination of L-Dopa in the Presence of Ascorbic Acid on the Surface of the Glassy Carbon Electrode Modified by a Bilayer of Multi-Walled Carbon Nanotube and Poly-Pyrrole Doped with Tiron," *J. Electroanal. Chem.*, vol. 636, p. 40, 2009.

[15] A. Babaei and M. Babazadeh, "A Selective Simultaneous Determination of Levodopa and Serotonin Using a Glassy Carbon Electrode Modified with Multiwalled Carbon Nanotube/Chitosan Composite," *Electroanalysis*, vol. 23, p. 1726, 2011.

[16] H. Y. Yue *et al.*, "Selective Determination of L-Dopa in the Presence of Ascorbic Acid and Uric Acid Using a 3D Graphene Foam," *J. Solid State Electrochem.*, vol. 22, p. 3527, 2018.

[17] C. Yu *et al.*, "Differential coulometry based on dual screen-printed strips for high accuracy levodopa determination towards Parkinson's disease management," *J. Pharm. Biomed. Anal.*, vol. 190, p. 113498, 2020.

[18] L. C. Tai *et al.*, "Wearable Sweat Band for Noninvasive Levodopa Monitoring," *Nano Lett.*, vol. 19, no. 9, pp. 6346–6351, 2019.

[19] K. Y. Goud *et al.*, "Wearable Electrochemical Microneedle Sensor for Continuous Monitoring of Levodopa: Toward Parkinson Management," *ACS Sensors*, vol. 4, no. 8, pp. 2196–2204, 2019.

Proceedings of the 16th Annual IEEE International
Conference on Nano/Micro Engineered and Molecular Systems
April 25-29, 2021

Silicon Nanowire Array Sensor for Highly Sensitive and Selective Detection of Nerve Agent Simulant Vapor via Surface Hydroxyl Groups

Xingqi Liu[1,2], Hongpeng Zhang[1], Zhiping Huang[1],

Xuanlin Yang[1], Shixing Chen[2], Yuelin Wang[2], Tie Li[2*], Zhenxing Cheng[1*]

Abstract—Silicon nanowire array with high density of hydroxyl groups is presented here to detect vapor of methylphosphonate (DMMP), a simulant of the high toxic nerve agent. The whole fabrication process of silicon nanowire array sensor device was highly compatible with CMOS technology. It was found that the current in nanowires array treated with plasma exhibited a large and reversible change when exposed to DMMP vapor. The silicon nanowire array sensor can easily and rapidly detect DMMP vapor with a detection limit down to 0.2 ppm, the highest sensitivity reported to date. These initial promising results provide opportunities for manufacture of ultra-sensitive and low-cost nanowire sensors for DMMP vapor detection.

Keywords—silicon nanowire array; DMMP; sensitivity;

I. INTRODUCTION

The nerve agents such as sarin (GB) are one of the highest toxic chemicals (CWAs) and hold infamous status from their swift lethality and ease of manufacture. It is well-known that only 1 mg sarin can cause death within minutes of exposure. Especially after the Tokyo subway sarin gas attack (1995) and Syria (2013, 2017) wars, the detection of nerve agents has attracted extensive attention[1]. In order to timely preventing chemical weapon attacks, it will be very helpful to develop ultrasensitive and portable devices for sarin detection.

However, considering the high toxicity of sarin gas, a suitable surrogate is needed for security reason in laboratory. methylphosphonate (DMMP) is a good choice for its lower toxicity and similar P=O part structure with nerve agent sarin. Molecule structures of sarin and DMMP are shown in Fig. 1.

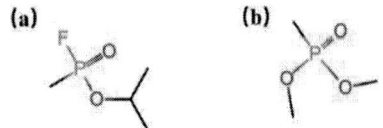

Figure 1. Molecule Structures of (a) sarin and (b) DMMP.

Currently, there exists various analytical techniques for the analysis of DMMP, such as gas chromatography, Raman spectrometry, and ion mobility spectrometry method[2]. But all of these methods exhibit limitations such as expensive, bulky and time-consuming. Nowadays, the recent developments in silicon nanowires provided unprecedented opportunities for the manufacture of ultra-sensitive and portable nanowire sensors[3]. Because of their unique large surface-to-volume ratio, silicon nanowires showed their ultra-sensitivity and fast response to the surrounding environment.

Since the interactions with target molecules occur on the silicon nanowire surface, the surface condition of the silicon nanowire is expected to have a strong impact on the sensor performance. Here, the silicon nanowire array device was treated with oxygen plasma to investigate the effect of the surface hydroxyl groups on sensor performance toward sarin simulant vapor. The silicon nanowire array sensor device was fabricated by a novel size-reduction design, which was highly compatible with CMOS technology[4]. After coating the silicon nanowire surface with high density of hydroxyl groups, the silicon nanowire array showed ultra-sensitive and selective response to DMMP vapor.

II. EXPERIMENTS

A. Fabrication of silicon nanowire array sensor

The fabrication of silicon nanowire array sensor device was highly compatible with CMOS technology, which providing a promising opportunity for ultrasensitive and portable sensors and its future large-scale manufacture. The fabrication of silicon nanowire array started from a 4-inch (111) oriented silicon-on-insulator (SOI) wafer. Standard photolithography was firstly applied to control the shape and size of windows. Then, the Si_3N_4 mask and the silicon underneath was dry etched with reactive ion etching process, forming adjacent rectangular etched window cavities. After anisotropic wet etching process in 40 wt% KOH solution, the silicon wall structure can be formed between adjacent etched window cavities with a width of about 350–400 nm. After the self-limiting oxidation process, inverted-triangle

---

This research was support by National Key Research and Development Program of China under No. 2017YFA0207100, 2017YFB0405403, and National Natural Science Foundation of China under No 31900937.

Xingqi Liu, Hongpeng Zhang, Zhiping Huang, Xuanlin Yang and Zhenxing Cheng are with Department of Chemical Defense, Institute of NBC Defense, PLA Army, Beijing 102205, China

Shixing Chen, Yuelin Wang and Tie Li are with the Science and Technology on Microsystem Laboratory, Shanghai Institute of Microsystem and Information Technology, Chinese Academy of Sciences, Shanghai 200050, China.
[1]*Zhenxing Cheng (e-mail: chengzx2018@sina.com)
[2]*Tie Li (email: tli@mail.sim.ac.cn);

978-1-6654-3008-1/21 $31.00 © 2021 IEEE

shape silicon nanowires were formed on the top center of walls. Subsequently, the oxidization layer was removed and the suspended silicon nanowires was released in buffered oxide etch solution.

Based on the novel size-reduction design, we can generate silicon nanowire array sensor devices on wafer-level. As shown in Fig. 2a, 3000 silicon nanowire array devices are successfully fabricated on one a 4-inch wafer, making it attractive for developing ultrasensitive and portable sensors and its future large-scale manufacture. Fig. 2b shows the tiny size and rectangle shape of one silicon nanowire array device. As shown in Fig. 2c, every silicon nanowire array device contains hundreds of same size parallel silicon nanowires. Fig. 2d shows every single silicon nanowire locates at the edge of adjacent etched cavities.

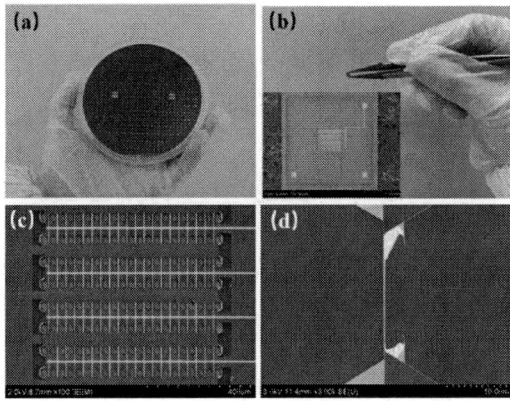

Figure 2. Photographs and SEM images of the silicon nanowire array device (a) The physical distribution of the 3000 silicon nanowire array devices on a 4-inch SOI wafer and (b) The photo and the SEM image of a silicon nanowire array device and (c) The SEM image of the silicon nanowires region and (d) The SEM image of a single silicon nanowire.

B. Sensor testing system

The setup for DMMP vapor flow and electrical measurement instruments are schematically shown in Fig. 3. DMMP vapor flow system consisted of three mass flow controllers and a four on/off valve connected with Teflon tube lines. Nitrogen gas was used to carry and dilute the DMMP vapor. And a 4-way valve was used to switch between nitrogen and the DMMP vapor. In order to eliminate the non-specific adsorption of analytes in the circuit before contacting the sensor surface, all the tubes and joints were used high-grade Teflon. During the test, the current of the silicon nanowires array was recorded by a Keithley 4200 semiconductor while vapor flow was switched between N_2 and DMMP vapor at intervals of a few seconds.

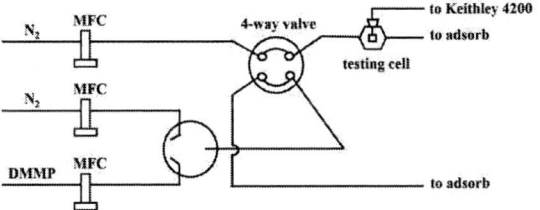

Figure 3. Schematic drawing of DMMP vapor flow and electrical measurement.

C. Surface Modification

As shown in Fig. 4, The one-step surface functionalization of silicon nanowires was plasma treatment by a surface microwave plasma cleaner. The plasma treatments process was carried (40 kHZ, 200 W) for 5 minutes to get enriched hydroxyl groups on the surface.

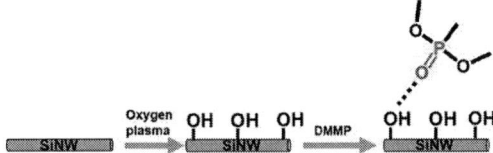

Figure 4. Graphic illustration of one-step surface functionalization of silicon nanowires and molecular sensing interaction.

III. RESULT AND DISCUSSION

For characterization of the hydroxyl terminated nanowire surface, surface studies measurements such as contact angle measurement and X-ray photoelectron spectroscopy (XPS) were firstly carried at room temperature. Considering the extremely tiny size of silicon nanowire array surface, XPS and contact angle measurements were performed on the same wafer, which was applied to fabricate the silicon nanowire array device.

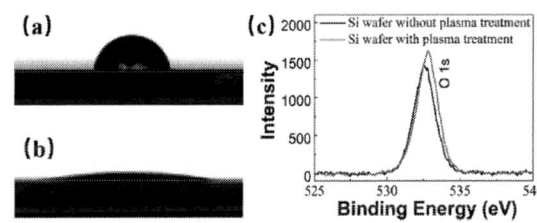

Figure 5. Contact angle measurements and XPS spectroscopy on SOI wafer (a) Contact angle measurements image of SOI wafer before plasma treatment and (b) Contact angle measurement image of SOI wafer after plasma treatment and (c) The O(1s) peak of XPS spectra of the SOI wafer before and after plasma treatment.

The contact angle measurement results by the drop shape method are shown in Fig. 5. After the SOI wafer was treated with plasma, the contact angle dramatically decreased. Given the hydrophilicity of hydroxyl groups, it indicates that treating silicon nanowire array with plasma can significantly enrich hydroxyl groups on surface.

XPS was also carried out on the same SOI wafer to further confirm the assumption, results were shown in Fig. 5c. the strong peak at the binding energy around 535 eV can be attributed to the typical O(1s) peak, which is from oxygen species on silicon surface. The intensity of the O(1s) peak increases with plasma treatment, suggesting that plasma treatment indeed enrich more oxygen species on silicon surface. And this could be explained that the plasma treatment provides more adsorption and binding sites to oxygen species (O^{2-}, O^-) adsorbed onto the surface[5].

Those surface measurements including contact angle measurements and XPS suggesting that the main role of the

978-1-6654-3008-1/21 $31.00 © 2021 IEEE

oxygen plasma treatment would be the removal of organic contaminants on the silicon nanowire surfaces. Furthermore, it could provide more adsorption and binding sites to hydroxyl groups and lead to stronger responses.

This phenomenon possibly provides opportunities for detection of DMMP vapor. Now that the plasma treatment indeed enriched surface hydroxyl groups on silicon nanowires. And it has been reported that the strength of hydroxyl bonds between hydroxyl groups of silicon surface with DMMP molecule is moderate and may be suitable to apply to repeatable sensitive materials[6][7]. Therefore, enriched hydroxyl groups on silicon nanowire array surface by plasma treatment could exist similar effects. The role of the silicon nanowires is proposed to important for both high surface-to-volume ratio and the hydroxyl terminated nanowire surface, which could connect to P=O part of DMMP. By functionalizing silicon nanowire surface with high density of hydroxyl groups, sensitive chemiresistive response to DMMP vapor would be highly possible to achieve.

All electrical measurements for silicon nanowire array sensor were carried on a Keithley 4200 semiconductor. The I-V curve measurements were firstly performed on both non-functionalized and functionalized silicon nanowire array sensors to characterization of enriched hydroxyl groups. As shown in Fig. 6a, the I-V curves are linear at ultralow voltage region indicates that the metal electrodes make ohmic contacts to the silicon nanowires[8]. And the current values slightly increase after modification indicates that the hydroxyl groups in the surface is equivalent to apply a negative gate voltage to silicon nanowires and further cause hole carrier to accumulate the nanowires surface.

Fig. 6b shows a repeatable response of the silicon nanowire array sensor device to DMMP vapor. The current signal increases rapidly when expose to DMMP vapor. The change in current signal begins immediately upon exposure of the silicon nanowire array device to the DMMP vapor, and stabilizes at a new value over a period of few seconds. As shown in Fig. 6c, the silicon nanowire array showed highly sensitive and selective sensor performance, with a detection limit down to 0.2 ppm. And current response of the sensor is clearly related to the concentration of DMMP vapor.

To evaluate the selectivity of silicon nanowire array sensor to DMMP vapor, we compared the responses to several interfering volatile organic compounds (VOCs). Certain volume of methylbenzene, dichloromethane, n-hexane and DMMP liquid were injected into Teflon gas bags to generate vapor at a concentration of 1 ppm. Then VOCs and DMMP vapor was transferred into glass syringe barrels. And the current was monitored in real-time while silicon nanowire array device was removed from ambient air into VOCs vapor of syringe barrel.

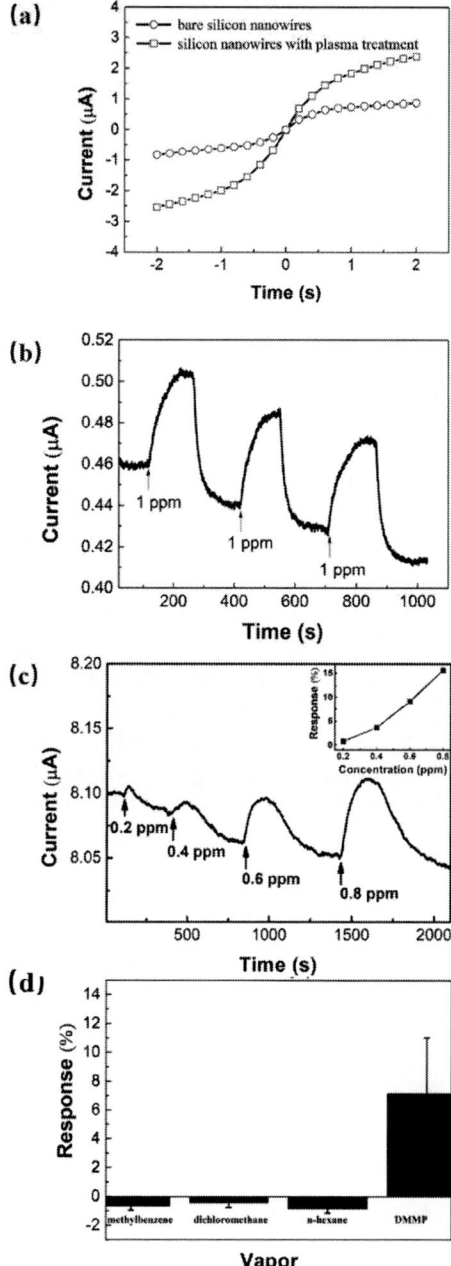

Figure 6. (a) The I–V curve of bare silicon nanowire array and silicon nanowire array treated with plasma and (b) The repeatable sensor response to DMMP vapor at a concentration of 1 ppm and (c) The sensor response to DMMP vapor at a series of concentrations and (d) Column plots of sensitivite response to methylbenzene, dichloromethane, n-hexane and DMMP vapor at a concentration of 1 ppm.

As shown in Fig. 6d, the silicon nanowire array response to DMMP vapor is unique from other interfering VOCs. This unique current increase of silicon nanowires is attributed to enriched hydroxyl groups on surface of silicon nanowires by oxygen plasma treatment. Because the P=O group act as hydrogen-bond acceptor [9], DMMP molecule can selectively interact with enriched hydroxyl groups of silicon nanowire surface.

IV. CONCLUSION

In conclusion, we demonstrated a simply way for fabrication and surface functionalization of silicon nanowire array to ultrasensitive detection of DMMP vapor. The fabrication of silicon nanowire array sensor device was highly compatible with CMOS technology. It was found that treating silicon nanowire array with plasma treatment can significantly enriched hydroxyl groups on the surface. After this one-step surface modification, the silicon nanowire array showed highly sensitive and selective sensor performance, with a detection limit down to 0.2 ppm. XPS and contact angle measurements indicated that sensitivity may be caused by hydrogen bonds between the hydroxyl terminated nanowire surface and the P=O part group of DMMP molecule. Due to both the CMOS fabrication compatibility and a relatively simple one-step surface functionalization, the silicon nanowire array sensor may provide us a promising opportunity for developing ultrasensitive and portable sensors and its future large-scale manufacture.

REFERENCES

[1]   Y. Seto, *On-site detection of chemical warfare agents.* INC, 2020.

[2]   Y. J. Jang, K. Kim, O. G. Tsay, D. A. Atwood, and D. G. Churchill, "Update 1 of: Destruction and detection of chemical warfare agents," *Chem. Rev.*, vol. 115, no. 24, pp. PR1–PR76, 2015.

[3]   X. Yang, A. Gao, Y. Wang, and T. Li, "Wafer-level and highly controllable fabricated silicon nanowire transistor arrays on (111) silicon-on-insulator (SOI) wafers for highly sensitive detection in liquid and gaseous environments," *Nano Res.*, vol. 11, no. 3, pp. 1520–1529, 2018.

[4]   Z. Lu, Y. Wang, and T. Li, "Novel Design and Fabrication of Silicon Nanowire Array on (111) Soi," *2019 20th Int. Conf. Solid-State Sensors, Actuators Microsystems Eurosensors XXXIII, TRANSDUCERS 2019 EUROSENSORS XXXIII*, vol. 1, no. 111, pp. 1712–1715, 2019.

[5]   S. Kim, C. Delker, P. Chen, C. Zhou, S. Ju, and D. B. Janes, "Oxygen plasma exposure effects on indium oxide nanowire transistors," *Nanotechnology*, vol. 21, no. 14, 2010.

[6]   L. Bertilsson, K. Potje-kamloth, H. Liess, D.- Neubiberg, I. Engquist, and B. Liedberg, "Adsorption of Dimethyl Methylphosphonate on Self-Assembled Alkanethiolate Monolayers," vol. 5647, no. 97, pp. 1260–1269, 1998.

[7]   S. M. Kanan and C. P. Tripp, "An infrared study of adsorbed organophosphonates on silica: A prefiltering strategy for the detection of nerve agents on metal oxide sensors," *Langmuir*, vol. 17, no. 7, pp. 2213–2218, 2001.

[8]   Z. Zhang *et al.*, "Quantitative analysis of current-voltage characteristics of semiconducting nanowires: Decoupling of contact effects," *Adv. Funct. Mater.*, vol. 17, no. 14, pp. 2478–2489, 2007.

[9]   J. Quenneville, R. S. Taylor, and A. C. T. Van Duin, "Reactive molecular dynamics studies of DMMP adsorption and reactivity on amorphous silica surfaces," *J. Phys. Chem. C*, vol. 114, no. 44, pp. 18894–18902, 2010.

Gap in pagination due to unavailable papers.

Pages 34-42

Proceedings of the 16th Annual IEEE International
Conference on Nano/Micro Engineered and Molecular Systems
April 25-29, 2021

Chip-level-microassembly Comb-drive XYZ-microstage with Large Displacements and Low Crosstalk

Gaopeng Xue[1,2,*], Masaya Toda[2], Xinghui Li[1], and Takahito Ono[2], *Member, IEEE*

Abstract— This paper presents a chip-level-microassembly comb-drive XYZ-microstage with large displacements and low crosstalk for the applications of scanning force microscope at cryogenic environment. The three-dimensional comb-drive XYZ-microstage, with no affection to the thermal variation, was accurately and orderly constructed with three components of a comb-drive XY-microstage for in-plane actuation, two comb-drive Z-actuators for out-of-plane actuation, and a base substrate using a chip-level-microassembly technology. This configuration can overcome the out-of-plane stroke-space limitation of conventional monolithic-wafer-based XYZ-microstages, and the crosstalk movements resulting from the coupling connection between in-plane and out-of-plane actuation units can be avoided. Additionally, we further conducted two aspects of designing the decoupling-motion structure and constraining the capacitance-decoupling crosstalk, to achieve low-crosstalk movements in the in-plane actuation unit. The folded-flexure springs with high stiffness were adopted to enhance the lateral stability of movable combs and improve the range of achievable strokes. Finally, the fabricated comb-drive XYZ-microstage, as a promising three-dimensional scanner, was capable of providing large displacements of 28.3 μm into +X direction, 20.9 μm into –X direction, 5.8 μm into +Y direction, 22.1 μm into –Y direction, and 50.5 μm into Z direction, respectively.

I. INTRODUCTION

Scanning force microscopy (SFM) techniques, including magnetic force microscopy (MFM), atomic force microscopy (AFM), friction force microscopy (FFM), and electrostatic force microscopy (EFM), as essential means of characterization, have been widely used in material surface science and biomedicine engineering [1]. Recently, combined with AFM detection technology, magnetic resonance force microscopy (MRFM) with an ultra-sensitive small-magnet-based cantilever was utilized to measure the densities of radicals with a non-invasive method on a nanoscale [2], which could break through the sensitivity constraint of the conventional MRI only with a millimeter-to-submillimeter resolution. Usually, magnetic resonance force measurements should be performed in a cryogenic environment [3], to reduce the thermomechanical noise and improve the cantilever-resonance sensitivity.

Microscanners for MRFM cryogenic measurements require large strokes and low crosstalk with small affections to thermal variation. Electrostatic actuation with comb-drive configuration would be the most applicable method in terms of large displacements at low temperatures and high flexibility for system integration [4]. However, because of conventional vertical microfabrication technologies, it was difficult to fabricate a three-dimensional (3D) XYZ-microstage from a monolithic wafer [5, 6]. Generally speaking, to construct a comb-drive XYZ-microstage capable of providing large displacements and low-crosstalk motion in multiple directions, an independent out-of-plane actuation unit needs to be configured, and also the decoupling connection between in-plane and out-of-plane actuation units is required to constrain the crosstalk motion.

In our previous research work [7–9], we proposed a chip-level-microassembly approach to establish a 3D comb-drive XYZ-microstage, and the out-of-plane stroke was considerably increased by adopting two large-displacement Z-actuators to actuate the XY-microstage independently. This configuration can overcome the out-of-plane stroke-space limitation of conventional XYZ-microstages fabricated from monolithic wafers [5, 6], and the crosstalk movements resulting from the coupling connection of multiple actuation units can be avoided. However, to integrate the assembled comb-drive XYZ-microstage into our MRFM measurement system feasibly, the displacements of the in-plane actuation unit, crosstalk movements at high actuation voltages, and specimen attachment of the scanning XYZ-microstage remain to be optimized.

In this study, we systematically developed a chip-level-microassembly approach to fabricate a comb-drive XYZ-microstage with large displacements and low-crosstalk movements in three directions for SFM. We quantitatively analyzed the support-spring structures with high stiffness, to enhance the lateral stability of the movable combs and improve the range of achievable strokes. Furthermore, two aspects of designing the decoupling-motion structure and constraining the capacitance-decoupling crosstalk were implemented to achieve low-crosstalk movements in the in-plane X- and Y-directions. Finally, the assembled comb-drive XYZ-microstage exhibited quite large displacements and low-crosstalk movements along three directions.

---

[1]Tsinghua Shenzhen International Graduate School, Tsinghua University, Tsinghua Campus, the University Town, Shenzhen 518055, China.

[2]Graduate School of Engineering, Tohoku University, 6-6-01 Aramaki-Aza-Aoba, Aoba-ku, Sendai 980-8579, Japan.

*Contacting Author: Gaopeng Xue is with Tsinghua Shenzhen International Graduate School, Tsinghua University, Tsinghua Campus, the University Town, Shenzhen 518055, China (phone number: +86-0755-

26032544; fax number: +86-0755-26036356; email address: xue.gaopeng@sz.tsinghua.edu.cn).

The project was supported by the following foundations: National Natural Science Foundation of China (Grant No. 52005291), Guangdong Basic and Applied Basic Research Foundation (Grant No. 2019A1515110373), and Shenzhen Science and Technology Program (Grant No. RCBS20200714114957381).

978-1-6654-3008-1/21 $31.00 © 2021 IEEE

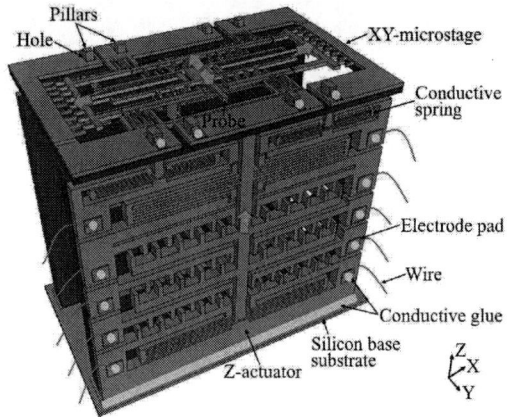

Fig. 1. Schematic of the proposed chip-level-microassembly comb-drive XYZ-microstage.

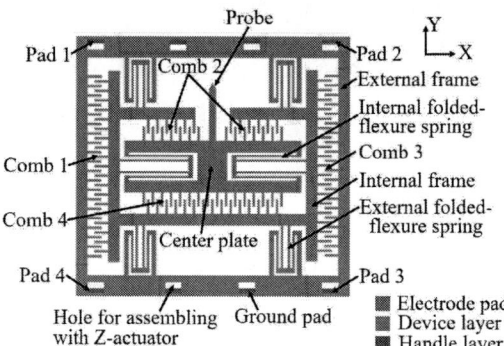

Fig. 2. Schematic of the decoupling-motion comb-drive XY-microstage.

II. DESIGN OF MICROASSEMBLY XYZ-MICROSTAGE

A. Chip-level-microassembly Concept

Fig. 1 shows a schematic of the chip-level-microassembly comb-drive XYZ-microstage, which consists of three components of a comb-drive XY-microstage for in-plane actuation, two comb-drive Z-actuators for out-of-plane actuation, and a bottom silicon base substrate. In the microassembly configuration, the two Z-actuators were mounted onto the grooves of the silicon base substrate in the vertical direction, and the XY-microstage was mounted on the two Z-actuators in the horizontal direction by inserting the pillars into the holes. We introduced soft mechanical springs, also called conductive springs, to realize the electrical interconnections between the actuation units of the two Z-actuators and XY-microstage. Conductive glue was utilized to connect the electrode pads electrically and fasten the assembled 3D structure. All the wires used to apply actuation voltages to the actuation units of XY-microstage and Z-actuators could be bonded onto the substrates of two Z-actuators together, which could effectively avoid the influence of wire-bonding process of the XY-microstage on the flexible structures of the support springs of two Z-actuators. Owing to this novel chip-level-microassembly technique, the coupling motion between the in-plane and out-of-plane actuation units could be eliminated completely. Note that the design and parameter of the comb-drive Z-actuator and silicon base substrate, referenced from our previously published article [9], will not be shown in this research work in detail.

B. Design of Comb-drive XY-microstage

As shown in Fig. 2, the in-plane comb-drive XY-microstage fabricated from a silicon-on-insulator (SOI) wafer is composed of two frames: an external stationary frame and an internal movable frame. The center plate is supported by one internal folded-flexure spring connected to the internal frame, and the internal frame is supported by two external folded-flexure springs connected to the external frame. A probe is outstretched from the center plate to an independent space, to achieve the attachment of the measurement sample and the integration of the MRFM measurement setup easily. Thus, the stationary electrode of

Comb 2 in the internal frame is separated into two parts by the probe. The external frame contains the stationary electrodes of Combs 1 and 3 to actuate the entire internal frame along the X-direction, and the internal frame contains the stationary electrodes of Combs 2 and 4 to actuate the center plate along the Y-direction. Based on this independent actuation mechanism between the external and internal frames, the center plate with the outstretched probe could be actuated with less crosstalk in the X- and Y-directions. Because all the electrode pads (pads 1, 2, 3, 4, and ground pads) are distributed on the external frame and comb electrodes (movable electrodes of combs 1 and 3 and stationary electrodes of combs 2, 4) are formed in the internal frame, some gaps are required to provide electrical insulation between the external and internal frames. There are eight holes near the electrode pads in the external stationary frame. Among these eight holes, four holes located at the corners are used to insert the pillars only for electrical interconnection, and four holes located in the center are not only for electrical connection but also for assembling with the pillars in Z-actuators. The size of the hole in the handle layer is larger than that in the device layer to avoid the connection between the handle layer of the XY-microstage and the pillars of the Z-actuator after assembly.

C. Capacitance Coupling Analysis in XY-microstage

Although this novel motion mechanism with independent actuation between external and internal frames could theoretically achieve less crosstalk in the X- and Y-directions, the crosstalk problem still presents clearly at a high actuation-voltage range, as reported in our previous work [9]. Because the electrode pads of four actuation units are distributed in the external frame, the coupling from actuation voltage signals is another key factor resulting in motion crosstalk. As shown in Fig. 3 (a), when the actuation voltage V_1 is applied to the electrodes of comb-1, the charge quantity Q_1 between the device and handle layers in the external frame is collected based on the equation $Q_1 = C_1 V_1$, where C_1 denotes the corresponding parallel-plate capacitance in the external frame. Charge Q_1 in the handle layer is mainly gathered near the parallel-plate electrode of capacitance C_1, and a certain percentage of charge $Q_2 = Q_1 \Delta_1$ drifts to the parallel-plate electrode of capacitance C_2. Thereafter, the charge Q_2 in the device layer is mainly gathered in the parallel-plate electrode of capacitance C_2, and a certain percentage of charge $Q_3 = Q_2 \Delta_2$ drifts to the

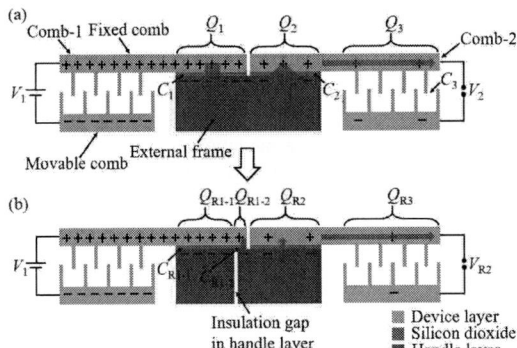

Fig. 3. Capacitance coupling analysis between different actuation units of XY-microstage: (a) no insulation gap in handle layer and (b) adding insulation gap in handle layer.

Fig. 4. Two structural layouts of folded-flexure spring applied in the comb-drive XY-microstage: (a) fixing the external four levers and (b) fixing the internal four levers.

comb electrode of comb-2. Finally, the voltage $V_2 = Q_3/C_3 = C_1V_1\Delta_1\Delta_2/C_3$ to comb-2, induced by the actuation voltage V_1 through capacitance coupling of C_1 and C_2, results in crosstalk motion of comb-2.

To constrain the undesired induced voltage V_2 from the capacitance coupling effect between the device and handle layers, insulation gaps are added to the handle layer, and the parallel-plate capacitance C_1 is divided into capacitances C_{R1-1} and C_{R1-2}, as shown in Fig. 3 (b). Only small capacitance C_{R1-2} plays the role of coupling capacitance, transporting the charge to the comb electrode of comb-2. The induced voltage to comb-2 is represented by $V_{R2} = C_{R1-2}V_1\Delta_1\Delta_2/C_3$. Therefore, the undesired induced voltage V_{R2} becomes small when the parallel-plate capacitance C_{R1-2} is small. The proper parallel distance of insulation gaps in the device and handle layers should be considered to avoid fragile structures of the external frame. Notably, the capacitance coupling effect from insulation gaps in the device and handle layers could be neglected owing to the small area of the parallel-plate and the large insulation gap.

D. Stiffness Enhancement of Support springs

According to lateral-instability analysis of the comb-drive configuration [10], a support spring with high stiffness is important for overcoming the lateral sticking of the combs and improving the range of achievable strokes. Normally, the support spring with folded-flexure configuration is widely applied in the comb-drive actuators, owing to its symmetric layout and robust structure. Fig. 4 (a) shows the conventional configuration of the folded-flexure spring, of which the external four levers are fixed on the anchors. We adopted this conventional spring applied in the comb-drive XY-microstage to support the entire internal frame. Notes that two of this conventional spring are always required in an actuation system, to guarantee the support balance of the movable structures. On the other hands, we further proposed a new configuration of the folded-flexure spring, where four internal levers are fixed on the anchors, as shown in Fig. 4 (b). Thus, the movable parts of comb-drive actuation unit could be supported by this newly-configured folded-flexure spring at four corners, which could construct a more stable and robust structure. In the design of the comb-drive XY-microstage, only one newly-configured folded-flexure spring is utilized to sufficiently support the center plate with the outstretched probe, as shown in Fig. 2.

The stiffness ratio R, derived from $R = mk_x/mk_y = l^2/w^2$, is unrelated to the number m of support springs and only depends on the length l and width w of the support spring levers, where k_x and k_y denote the spring constants in the lateral and stroke directions, respectively. Compared with the serpentine springs used ($l = 3,200$ µm, $w = 20$ µm for the X-axis and $l = 3,200$ µm, $w = 20$ µm for the Y-axis) in our previous research [9], the currently adopted folded-flexure springs ($l = 3,730$ µm, $w = 17$ µm for the X-axis and $l = 3,400$ µm, $w = 15$ µm for the Y-axis) possess a higher stiffness ratio (1.88 times along the X-axis and 2.01 times along the Y-axis), implying that a larger range of achievable strokes is more likely to be achieved.

III. MICROFABRICATION AND MICROASSEMBLY

Fig. 5 (a) showed the microfabrication results of the comb-drive XY-microstage with the locally magnified images, fabricated from an SOI wafer (200–1–400 µm) with dimensions of 20×20 mm². The used microfabrication technologies were mainly involved with the sputtering and lift-off of electrode pads, deep reactive ion etching (RIE) of Si-based device and handle layers, backside-alignment photolithography, wet-etching of SiO_2 layer, YAG-laser cutting of support beams, supercritical CO_2 drier, etc. Then, we introduced a base block of stainless steel as a support to guide the microassembly process precisely and orderly. The two Z-actuators with wire bonding were inserted into the two parallel grooves of the Si-base substrate and kept vertically by two flexible Mo-clamping plates (Fig. 5 (b)). A manipulator connected with a thin-Cu tray was controlled to transfer the XY-microstage, guaranteeing that the holes in

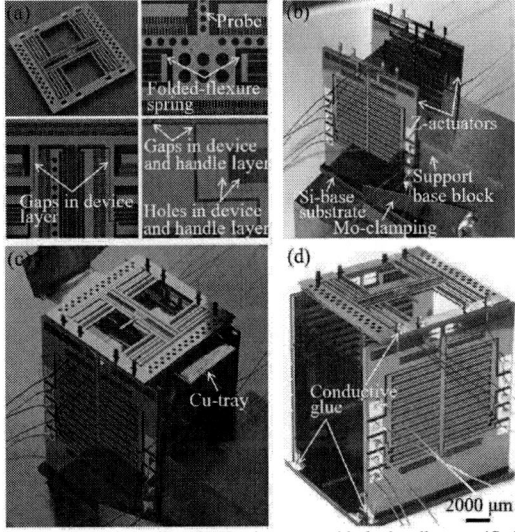

Fig. 5. Microfabrication of XY-microstage with the locally magnified images and microassembly of comb-drive XYZ-microstage.

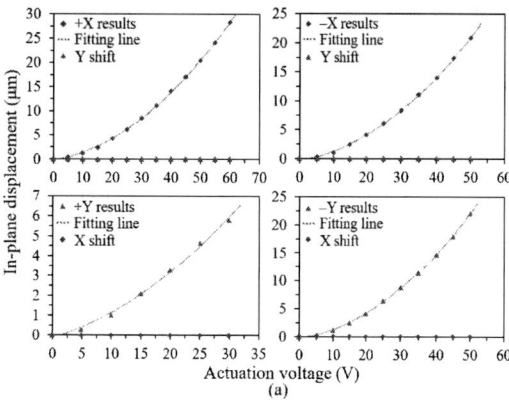

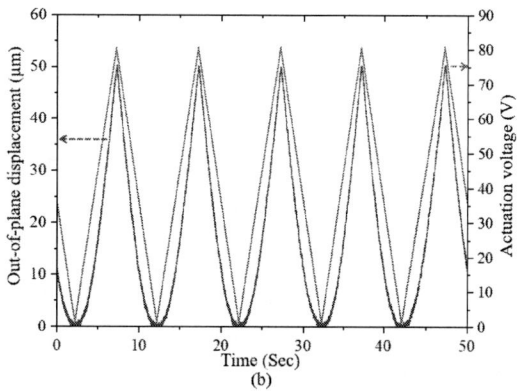

Fig. 6. Measured actuation performances of the fabricated XYZ-

the XY-microstage matched the pillars in Z-actuators completely (Fig. 5 (c)). After coating the conductive glue to realize the electrical interconnections between the electrode pads and solidifying the glue to fasten the connecting parts, the assembled 3D structure was released from the guide block and clamping plates (Fig. 5 (d)).

IV. EVALUATIONS AND DISCUSSIONS

To evaluate the actuation performance of the assembled XYZ-microstage, a DC power supply and a function generator with an amplifier were utilized to provide actuation voltages for the in-plane and out-of-plane actuation units, respectively. Fig. 6 (a) shows the in-plane displacements measured by an optical microscope with a digital camera (KEYENCE VHX 100), and Fig. 6 (b) presents the relationship between the out-of-plane displacement and triangle actuation voltages with time flow as measured by a KEYENCE laser focus displacement meter connected with a digital Pico-oscilloscope. The assembled comb-drive XYZ-microstage could provide quite large displacements of 28.3 µm @ 60 V along the +X-axis, 20.9 µm @ 50 V along the −X-axis, 5.8 µm @ 30 V along the +Y-axis, 22.1 µm @ 50 V along the −Y-axis, and 50.5 µm @ 80.9 V along the Z-axis. The coupling properties of in-plane cross-axis actuations depicted in Fig. 6 (a) were obtained using an optical microscope with a digital camera (KEYENCE VHX 100), verifying that the optimized structure of the in-plane actuation unit of XY-microstage provides excellent low-crosstalk movements. However, Fig.

6 (a) reveals that the stroke ranges of the obtained in-plane displacements in the ±X- and ±Y-directions are asymmetrical. The possible reasons for this characteristic are that the structural dimension deviations of the microfabrication process in terms of the comb-finger gap, support-spring lever, etc., have huge effects on the actuation performances of the comb-drive actuators.

Finally, the optimized support-spring structures with high stiffness could considerably enhance the lateral stability and improve the range of achievable strokes, and the proposed scheme with a decoupling-motion structural design and constrained capacitance-decoupling crosstalk could effectively provide low-crosstalk movements in the in-plane X- and Y-directions. Consequently, it was demonstrated that the assembled comb-drive XYZ-microstage with large displacements and low-crosstalk movements is a promising 3D scanning stage for SFM applications, including MRFM, in cryogenic environments.

V. CONCLUSIONS

This paper systematically and comprehensively presents a chip-level-microassembly comb-drive XYZ-microstage, which is composed of two comb-drive Z-actuators, a comb-drive XY-microstage, and a bottom base substrate, for 3D SFM scanning with large displacements and low crosstalk. In the future, we expect that the proposed chip-level-microassembly approach will be applicable to other multi-degree-of-freedom actuations with the integration of different driving modes.

ACKNOWLEDGMENT

This work was supported by Micro/Nanomachining Research Education Center (MNC) and Micro System Integration Center (µSIC) of Tohoku University, Japan.

REFERENCES

[1] G. Binnig, C. F. Quate, and Ch. Gerber, "Atomic force microscope," *Phys. Rev. Lett.*, vol. 56, pp. 930–933, 1986.

[2] J. A. Sidles, "Noninductive detection of single-proton magnetic resonance," *Appl. Phys. Lett.*, vol. 58, pp. 2854–2856, 1991.

[3] C. L. Degen, M. Poggio, H. J. Mamin, C. T. Rettner, and D. Rugar, "Nanoscale magnetic resonance imaging," *PNAS*, vol. 106, pp. 1313–1317, 2009.

[4] G. Xue, M. Toda, and T. Ono, "Comb-drive XYZ-microstage based on assembling technology for low temperature measurement systems," in *Proc. ICEP-IAAC*, 2015, pp. 83–88.

[5] Y. Ando, "Development of three dimensional electrostatic stages for scanning probe microscope," *Sens. Actuat. A-Phys.*, vol. 114, pp. 285–29, 2004.

[6] X. Liu, K. Kim, and Y. Sun, "A MEMS stage for 3-axis nanopositioning," *J. Micromech. Microeng.*, vol. 17, pp. 1796–1802, 2007.

[7] G. Xue, M. Toda, and T. Ono, "Assembled comb-drive XYZ-microstage with large displacements for low temperature measurement systems," in *Proc. IEEE MEMS*, 2015, pp. 14–17.

[8] G. Xue, M. Toda, and T. Ono, "Chip-level microassembly of XYZ-microstage with large displacements," *IEEJ Trans. Sensor. Micromachin.*, vol. 135, pp. 236–237, 2015.

[9] G. Xue, M. Toda, and T. Ono, "Comb-drive XYZ-microstage with large displacements based on chip-level microassembly," *J. Microelectromech. Syst.*, vol. 25, pp. 989–998, 2016.

[10] R. Legtenberg, A. W. Groeneveld, and M. Elwenspoek, "Comb-drive actuators for large displacements," *J. Micromech. Microeng.*, vol. 6, pp. 320–329, 1996.

978-1-6654-3008-1/21 $31.00 © 2021 IEEE

Proceedings of the 16th Annual IEEE International
Conference on Nano/Micro Engineered and Molecular Systems
April 25-29, 2021

Effect of the Different Substrates and the Film Thickness on the Surface Roughness of Step Structure

Chenying Wang[#], Jiangtao Pu[#], Lei Li, Weixuan Jing, Yijun Zhang,
Yaxin Zhang*, Feng Han, Ming Liu, Wei Ren and Zhuangde Jiang

Abstract— The surface roughness was important for the nanostructure, especially for the geometry standard in the future. The Al_2O_3 film was grown on Si and Si_3N_4 substrate by plasma-assisted atomic layer deposition (ALD) technique, and the nanostep structure was obtained on the Al_2O_3 thin film. The roughness of all kinds of surface (Si and Si_3N_4 substrates, Al_2O_3 films with different thickness and the step structures with different height) was researched by atomic force microscopy (AFM). Analysis of the surface roughness revealed that the roughness of the Al_2O_3 step was related to the film thickness and the type of substrate. The surface roughness of the lower surface of the step structure is larger than that of the substrate. Both the Si substrate and the Si_3N_4 substrate have the same rules in the rate of change of surface roughness, so the change of film roughness may be related to the initial roughness of the substrate or the substrate material. No matter the surface roughness of film or the surface roughness of step structure, the Si substrate was superior to that on the Si_3N_4 substrate. Because the surface energy of Si is smaller than that of Si_3N_4. The surface roughness can be further reduced by optimizing the process parameters or reducing the surface energy of the substrate through other processes.

I. INTRODUCTION

The nanostep structure is the basis of micro-nano machines. The height of step structure plays a crucial role for calibrating the measuring instruments. As the dimensions of a structure become smaller, the relative value of the surface roughness becomes larger. To some extent, the height of step structure is affected by surface roughness. The surface roughness is an important factor in evaluating the quality of step structure. For miniaturized functional structure and microelectromechanical (MEMS) device, the surface roughness can determine the performance and the stability via affect electrical properties of interconnects such as electrical capacity, electronic conductivity, surface energy, critical area, peak electric field, surface tension, sheet resistance, etc. [1]. Therefore, the effect of the surface roughness of the step structure on the measurement of the functional structure and the reliability of the MEMS devices cannot be ignored.

There are many studies focused on the effects of surface roughness on the performance of various structures. Z. Q. Chen et al. [2], A. B. Yu et al. [3], A. Albina et al. [4], O. Rezvanian et al. [5] analyzed the relationship between the

surface roughness and the performance of capacitive switches. The results demonstrated that the up- and done-states capacitance decreased as the roughness increased. The surface roughness also has a great influence on the thermal conductivity of various semi-conductive nanowires. J. S. Heron et al. [6], L. Liu et al. [7] probes that the surface roughness can reduce the thermal conductivity of crystalline Si nanowires. N. M. Pierre et al. [8] also probes that thermal conductivity reduced in rough Ge and GaAs nanowires. T. Wang et al. [9] has come to the conclusion that the rate of thermal bonding was reduced with the increase of surface roughness. In fact, the surface roughness is closely related to the properties of the materials and the machining process.

ALD is gaining attention as a thin film deposition method that can deposit uniform and conformal films on complex three-dimensional topographies. V. Miikkulainen et al. [10], S. Li et al. [11], Q. Ma et al. [12] studied the factors affecting the structure, morphology and performance of various thin films by ALD. J. P. He et al. [13], used ALD technique to grow Al_2O_3 films with nominal thicknesses of 400, 300 and 200 nm on silicon and soda lime glass substrates. H. S. Bahari et al. [14] reported the surface analysis of Cu substrate coated with ALD Al_2O_3 and its corrosion protection in NaCl solution, and found that the Al_2O_3 coating can greatly improve the corrosion resistance of Cu. K. J. Blakeney et al. [15] explored the new class of ALD precursor which provides a new way for thin film growth of difficult elements and materials by ALD. And in our previous work [16], we prepared a nanoscale step height structure by ALD and wet etching techniques. Obviously, ALD can precisely control the thickness of the thin film during the growth. It is potential method for the fabrication of nanoscale structures.

In this paper, we research the effect of different substrates and the film thickness on the surface roughness of step structure.

II. FABRICATION

The manufacturing process is divided into three main processes which are growth of the thin film, photolithograph and wet etching. The specific production process is shown in Fig. 1.

Resrach supported by the National Key R&D Program of China (Grant Nos. 2018YFF0212301), the National Natural Science Foundation of China (Grant Nos. 91748207), and State Key Laboratory of Robotics and Systems (HIT) (SKLRS-2020-KF-06).

[1]Collaborative Innovation Center of High-End Manufacturing Equipment, Xi'an Jiaotong University, Xi'an, China.

[2]School of Mechanical Engineering, Xi'an Jiaotong University, Xi'an, China

[3]School of Electronics and Information, Xi'an Jiaotong University, Xi'an, China.

[#] Chenying Wang and Jiangtao Pu contributed equally to this work.

*Contacting Author: Yaxin Zhang is with the School of Mechanical Engineering, Xi'an Jiaotong University, Xi'an, China. (phone: 82668616; e-mail: zhangyaxin@stu.xjtu.edu.cn).

978-1-6654-3008-1/21 $31.00 © 2021 IEEE

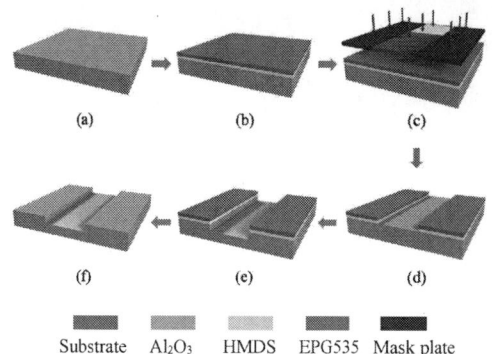

Figure 1. The production process (a) growth Al₂O₃ thin film on substrate. (b) coating and pre-bake. (c) exposuring. (d) development. (e) wet etching. (f) clean the sample.

In process (a). There are many film growth techniques, such as ALD, metal-organic chemical vapor deposition (MOCVD), magnetron sputtering (MS) and molecular beam epitaxy (MBE) etc. In general, MS technique is used to grow thicker film. MOCVD technique is used to grow organic compound. However, ALD and MBE technique are adept at growing thin films with smaller thickness. But MBE technique requires a good lattice match between the substrate and the film. ALD technique as the most apt technique for growing thin films with good quality was chosen to gain 10 nm and 20 nm Al₂O₃ thin films using trimethylaluminium (TMA) and water-vapor [17]. High quality Al₂O₃ thin films with 10 nm and 20 nm thickness were achieved in the case of 100 deposition cycles and 200 deposition cycles respectively.

In process (b). As enhancer, HMDS can increase the adhesion between the Al₂O₃ thin films and the photoresist. In order to volatilize organic solvents, the sample needs to be heated for 5 minutes at 95 ℃ after spin-coating HMDS and photoresist.

In process (c). Using the mask plate to UV-expose the part of the photoresist that needs to be etched.

In process (d). The 5 ‰ NaOH solution was chosen to remove the exposed photoresist. The development time was determined by the NaOH solution concentration and the thickness of the photoresist.

In process (e). The step structure was obtained after removing unwanted Al₂O₃ films by wet etching. The 70 % H₃PO₄ solution was chosen as the etching solution and heated to 60 ℃ in a water bath. The etching time was determined by the thickness of the film. The Al₂O₃ films of 10 nm and 20 nm thickness need to be etched 1 minute and 2 minutes respectively. The upper and lower surfaces of the step structure were made of the saved Al₂O₃ and the substrate.

In process (f). Removing the photoresist on the surface of the sample and cleaning the sample.

III. RESULTS AND DISCUSSION

The roughness of all kinds of surface (Si and Si₃N₄ substrates, Al₂O₃ films with different thickness and the step structures with different height) were measured by atomic force microscopy (AFM) in tapping mode (TM), on the square areas 5×5 μm². AFM images are shown in Fig. 2. The specific data are shown in TABLE 1 in detail.

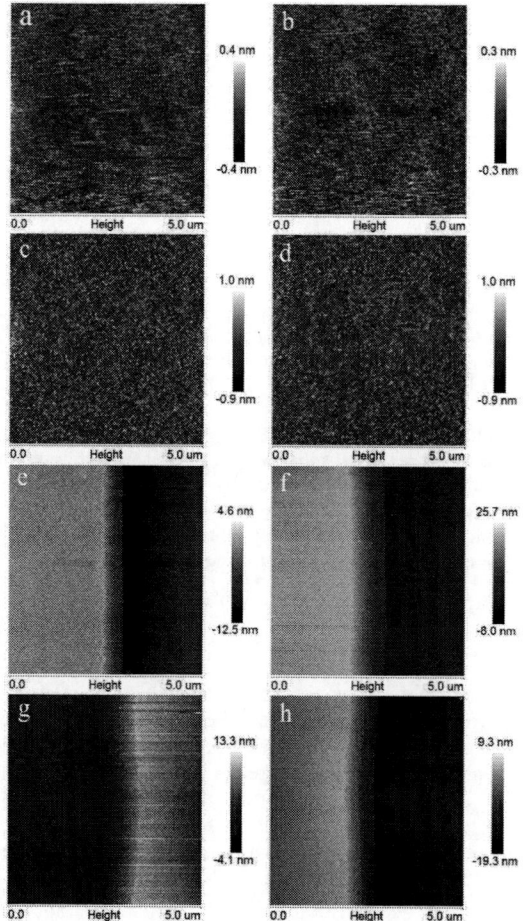

Figure 2. AFM images of all kinds of samples (a-b) The 10 nm and 20 nm Al₂O₃ thin films on the Si substrate. (c-d) The 10 nm and 20 nm Al₂O₃ thin films on the Si₃N₄ substrate. (e-f) The 10 nm and 20 nm step structures on the Si substrate. (g-h) The 10 nm and 20 nm step structures on the Si₃N₄ substrate.

TABLE I. THE SURFACE ROUGHNESS

	Bare (nm)	10 nm thin film (nm)	20 nm thin film (nm)
Si	0.138	0.079	0.069
Si₃N₄	0.310	0.221	0.201
	Bare (nm)	10 nm step (the lower surface) (nm)	20 nm step (the lower surface) (nm)
Si	0.138	0.155	0.204
Si₃N₄	0.310	0.322	0.539

(Ra)

For the Al_2O_3 thin film, the larger thickness of the thin film, the smaller the roughness of the film surface. There may be the following reason: in the process of growing thin film, the arrangement of atoms and the contact of atom layers are poetic. In other words, the atoms are not arranged tightly. With the increase of the film thickness, the imperfection was offset, the surface appearance of the substrate and the surface appearance of the Al_2O_3 thin film are complementary. So the surface roughness of the thin film is smaller than that of the substrate. In addition, the rate of film growth, temperature and pressure are the factors that affect the quality of the thin film. When the surface roughness of the step is greater than or equal to 10 % of the step height, the nanostep is meaningless, which means the height of the nanostep has a minimum value.

For the step structure, because of the upper surface protected by photoresist, the roughness of the upper surface and the surface roughness of the thin film equal in value. The surface roughness of the lower surface of the step structure is larger than that of the substrate. There may be the following reasons: (1) In the process of wet etching, the thin film was not corroded completely, portion of Al_2O_3 remained on the substrate; (2) In the process of wet etching and cleaning, the surface roughness is increased due to the expansion of original imperfection and the introduction of new imperfection; (3) Corrosion solution and analytical reagent is not clean enough, introduced new impurities lead to the increase of the surface roughness; (4) In the process of corrosion, the following chemical reactions will occur:

$$Al^{3+}+PO_4^{3-} \rightarrow AlPO_4. \tag{1}$$

The $AlPO_4$ is gel which is easy to attach to the surface of the substrate. This phenomenon will prevent the process of reaction. As shown in Fig. 3, the X-ray diffraction (XRD) results Fig. 3 is the X-ray diffraction (XRD) measurement result on the lower surface of the step. The peaks marked with red asterisks corresponds to the $AlPO_4$, which confirms that the $AlPO_4$ exists on the lower surface of the step.

As shown in Fig. 4, from the change in the slope of the two dashed lines, it can be seen that for the Al_2O_3 film grows on the substrate, as the film thickness increases, the rate of decrease in roughness slows down. And the change of the slope of the solid line show that for the step structure, as the height of the step increases, the roughness of the lower surface of the step increases at a faster rate.

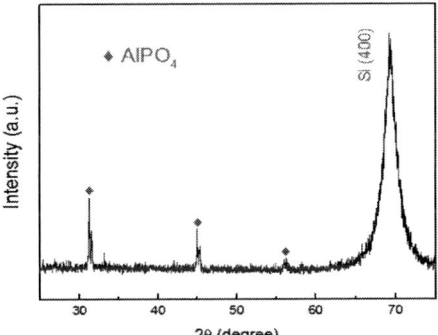

Figure 3. The XRD result on the lower surface of the step confirms that the $AlPO_4$ exists.

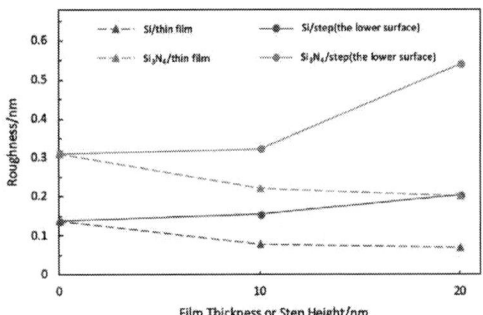

Figure 4. The relationship between film thickness/step height and surface roughness.

With the film thickness from 0 to 10 nm, the rate of surface roughness reduction of the Al_2O_3 films grown on the two substrates and increase of the roughness of the lower surface of the step is very close. It shows that the change of the film roughness may have nothing to do with the substrate material, only the roughness of the substrate. With the film thickness from 10 nm to 20 nm, the roughness of the Al_2O_3 film grown on both substrates gradually stabilizes, probably because of the limitation of the initial roughness of the substrate. The roughness of the Al_2O_3 film has been reduced to the limit. As the thickness of the film increases, the roughness will not change significantly. However, the roughness of the lower surface of the step on the Si_3N_4 substrate increases faster than that of the step on the Si substrate. It may be impurities are randomly introduced during the experiment, which requires new experiments to investigate. It is also possible that the change in surface roughness is indeed related to the substrate material.

If the change in surface roughness is related to the base material, we found that no matter the surface roughness of film or the surface roughness of step structure, the Si substrate was superior to that on the Si_3N_4 substrate. This phenomenon can be explained based on the interaction between different substrates and films. The bonding energy of Si-N bond (101.9 eV) is larger than that of Si-Si bond (99.5 eV) [18], so the surface energy of Si_3N_4 substrate is larger than that of Si substrate, which will establish more molecular adsorption and atomic adsorption on the Si_3N_4 substrate surface to form a stable state in the process of ALD. This results in the surface roughness of the Si_3N_4 substrate becoming larger. And this phenomenon leads to the lower surface of the Si_3N_4 step structure become rougher directly. These results show that we should choose substrate with small surface energy for the purpose of reducing the surface roughness. As for whether the change in surface roughness is really related to the substrate material, the most ideal is to select different substrates with the same initial roughness and choose more different material substrates to measure with more film thickness. This will be our next work.

IV. CONCLUSION

In this paper, the effects of different substrates and the thickness of the Al_2O_3 film obtained by ALD on the surface roughness of the step structures are studied. The results show that there is a negative correlation between the thickness of the Al_2O_3 film and the surface roughness, thus limiting the minimum height of the nanostep fabricated by ALD and wet etching.

In addition, the results show that the roughness of the upper surface of the step and the surface of the film equal in value, while the roughness of the lower surface is larger than that of the substrate. It is also found that the change of film roughness may be related to the initial roughness of the substrate or the substrate material. If it is related to the substrate material, since the surface energy of Si substrate is smaller than that of Si_3N_4 substrate, the Si substrate was superior to Si_3N_4 substrate regardless of the surface roughness of the film or the surface roughness of step structure.

Therefore, analysis of the results revealed that the surface roughness of the step structure can be reduced by the following methods: (1) Optimizing the process parameters of the production process (adjust the deposition rate, provide with stable pressure and temperature etc.) to obtain the perfect thin film; (2) Through annealing process to release the external stress, smaller stresses can lead to smaller strains, and then, surface energy will be reduced; (3) Doping, in which atoms of some other element are added to a crystal to obtain a more stable chemical properties and physical properties via reducing the surface energy. [19]

REFERENCES

[1] R. M. Patrikar, C. Y. Dong, W. J. Zhuang, "Modelling interconnects with surface roughness," in *Microelectronics Journal*, vol. 33, 2006, pp. 929-934.

[2] Z. Q. Chen, W. C. Tian, X. T. Zhang, "Effect of surface asperities on the capacitances of capacitive RF MEMS switches," in *J. Micromech.Microeng*, vol. 27, 2017, pp. 034002.

[3] A. B. Yu, A. Q. Liu, Q. X. Zhang, H. M. Hosseini, "Effects of surface roughness on electromagnetic characteristics of capacitive switches," in *J. Micromech.Microeng*, vol. 16, 2016, pp. 2157-2166.

[4] A. Albina, P. L. Taberna, J. P. Cambronne, P. Simon, E. Flahaut, T. Lebey, "Impact of the surface roughness on the electrical capacitance," in *Microelectronics Journal*, vol. 37, 2006, pp. 752-758.

[5] O. Rezvanian, M. A. Zikry, C. Brown, J. Krim, "Surface roughness,asperity contact and gold RF MEMS switch behavior," in *J. Micromech. Microeng*, vol. 17, 2007, pp. 2006-2015.

[6] J. S. Heron, T. Fournier, N. Mingo, O. Bourgeois, "Mesoscopic Size Effects on the Thermal Conductance of Silicon Nanowire," in *Nano Letters*, vol. 9, 2009, pp. 1861-1865.

[7] L. Liu, X. Chen, "Effect of surface roughness on thermal conductivity of silicon nanowires," in *Journal of Applied Physics*, 2010, 107. 033501.

[8] P. N. Martin, Z. Aksamija, E. Pop, U. Ravaioli, "Reduced Thermal Conductivity in Nanoengineered Rough Ge and GaAs Nanowires," in *Nano Lett*, vol. 10, 2010, pp. 1120-1124.

[9] T. Wang, J. Wu, T. Chen, F. Li, T. C. Zuo, S. B. Liu, "Surface roughness analysis and thermal bonding of microfluidic chips fabricated by CD/DVD manufacturing technology," in *Microsyst Technol*, vol. 23, 2017, pp. 1405-1409.

[10] V. Miikkulainen, M. Leskelae, M. Ritala, R. L. Puurunen, "Crystallinity of inorganic films grown by atomic layer deposition: Overview and general trends", in *Journal of Applied Physics*, vol. 44, 2013.

[11] S. Li, Y. Zhang, D. W. Yang, W. Yang, X. B. Chen, H. L. Zhao, J. Hou, P. Z. Yang, "Structure and optical properties of HfO_2 films on Si (100) substrates prepared by ALD at different temperatures", in *Physica B-condensed Matter*, vol. 584, 2020.

[12] Q. Ma, H. M. Zheng, Y. Shao, B. Zhu, W. J. Liu, S. J. Ding, D. W. Zhang, "Atomic-Layer-Deposition of Indium Oxide Nano-films for Thin-Film Transistors", in *Nanoscale Research Letters*, vol. 13, 2018, pp. 4.

[13] J. P. He, Y. G. Zhang, W. D. Shen, X. Liu, P. F. Gu, "Optical Properties of Al_2O_3 Thin Film Fabricated by Atomic Layer Deposition", in *Acta Optica Sinica*, vol. 30, 2010, pp. 277-282.

[14] H. S. Bahari, H Savaloni, "Surface analysis of Cu coated with ALD Al_2O_3 and its corrosion protection enhancement in NaCl solution: EIS and polarization", in *Materials Research Express*, vol. 6, 2019, pp. 086570.

[15] K. J. Blakeney, P. D. Martin, C. H. Winter, "Aluminum dihydride complexes and their unexpected application in atomic layer deposition of titanium carbonitride films", in *Dalton Transactions*, vol. 47, 2018, pp. 10897-10905.

[16] C. Y. Wang, S. M. Yang, W. X. Jing, W. Ren, Q. J. Lin, Y. J. Zhang, Z. D. Jiang, "Fabrication of Nanoscale Step Height Structure Using Atomic Layer Deposition Combined with Wet Etching," in *Chinese Journal of Mechanical Engineering*, vol. 29, 2016, pp. 91-97.

[17] D. Shah, D. I. Patel, T. Roychowdhury, D. Jacobsen, J. Erickson, M. R. Linford, "Optical function of atomic layer deposited alumina (0.5-41.0 nm) from 191 to 1688 nm by spectroscopic ellipsometry with brief literature review", in *Surface Science Spectra*, vol. 26, 2019.

[18] Y. Y. Liu, C. Du, K. Cao, R. Chen, X. L. Xu, J. Huang, B. Shan, "Influences of Deposition Power and Annealing Process on Al_2O_3 Film Deposited by PE-ALD", in *Semiconductor Technology*, vol. 43, 2018, pp. 610-15.

[19] W. L. Ding, J. Xu, J. Y. Q. Li, Y. Piao, P. Gao, X. L. Deng, C. Dong, "Characterization of silicon nitride films prepared by MW-ECR magnetron sputtering," in *Acta Physica Sinica*, vol. 55, 2006, pp. 1364-1368.

Proceedings of the 16th Annual IEEE International
Conference on Nano/Micro Engineered and Molecular Systems
April 25-29, 2021

Efficient Infrared–Thermal-Electric Conversion with Textured Dielectric Film

Yunqian He, Yuelin Wang, *Senior Member, IEEE*, and Tie Li

Abstract— Converting more infrared energy into heat while reducing heat loss from dielectric film to silicon substrate is very attractive for infrared–thermal-electric sensors. Thin FLat Dielectric (FLDI) film is known to possess low thermal conductivity, which is a conventional candidate for controlling heat loss. Meanwhile, high infrared absorption nanomaterials are adopted to compensate for the relatively low absorptivity of the FLDI film. However, this combination method may cause large weight stress, fabrication incompatibility, and high cost. In this paper, a new TExtured DIelectric (TEDI) film is proposed to simultaneously improve the heat conduction and infrared absorption properties in a simple, effective, and CMOS-compatible way. The fabricated thermopile platform with TEDI film can achieve about 55% enhancement in output voltage response and responsivity as well as about 47% in detectivity compared to the controlled thermopile with FLDI film. By this thermopile platform, the demonstrated heat conduction of the TEDI film can be reduced by about 28% while the tested infrared absorption can be increased by about 12%. This work may provide a simple and effective method toward engineering light-weight and thermally insulating FLDI films into efficient infrared–thermal-electric materials for achieving high-performance infrared or thermal sensors.

I. INTRODUCTION

Cost-effective MEMS infrared–thermal-electric sensors that have merits of low cost, low power, miniaturization, and low-frequency $(1/f)$ noise have been widely used in non-contact infrared detectors [1-3], NDIR sensing systems [4-6], infrared imagers [7-9], various gas/heat flow sensors [10-12]. The performance of the MEMS sensors was greatly influenced by the heat conduction and infrared absorption properties of the infrared absorber with CMOS-compatible FLDI film.

Reducing the heat conduction of the FLDI film can increase the utilization rate of the heat energy to reduce the power consumption of the thermal sensor [13] and enhance the responsivity and detectivity of the infrared sensor [2]. Isolating the dielectric film from the substrate is a greatly effective method to reduce the heat conduction loss to the substrate. The closed-film [14], suspended-film [13], and cantilever-beam structures [15] have been widely adopted in the infrared or thermal sensor. Thinning dielectric film or decreasing beam number or lowering width/ length rate of the beam or reducing active area are conventional ways to

This work was supported by the National Key Research and Development Program of China under Grant 2019YFB2005702 and 2018YFB2003001.
Yunqian He is with the Science and Technology on Microsystem Laboratory, Shanghai Institute of Microsystem and Information Technology, Chinese Academy of Sciences, Shanghai 200050, China, and

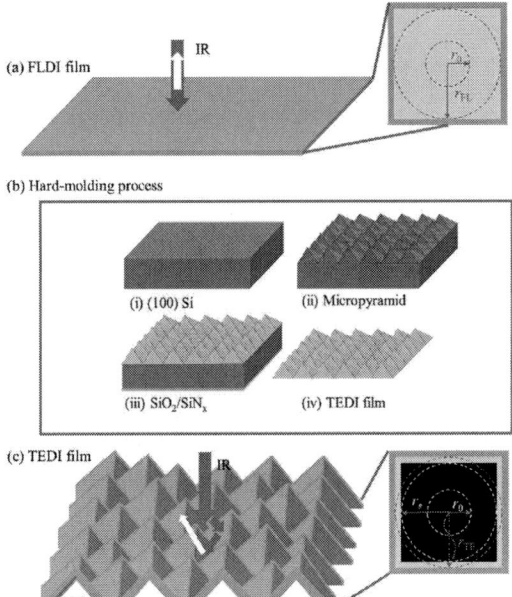

Fig. 1 Schematic of the (a) FLDI film, (b) hard-molding process, and (d) TEDI film.

control heat conduction loss and improve utilization rate of the heat energy. The performance of infrared sensors can be limited by the relatively low infrared absorptivity of the FLDI film [3]. Enhancing the infrared absorption property of the dielectric film is another effective method to improve the infrared sensor by increasing the converted infrared-heat energy. Nanomaterials, such as metal black (Au) [16], SU-8 [17], and carbon black [18], are combined on the surface of the FLDI film to obtain high infrared absorption. Advanced metamaterials [19, 20] are also introduced to control infrared absorption. However, this combination may result in large weight stress, fabrication incompatibility, and high cost.

Here, we present a new TEDI film with advantages of more infrared absorption and less heat conduction loss to enhance the performance of the infrared–thermal-electric sensors. The heat conduction and infrared absorption of the conventional FLDI film can be improved by texturing of the FLDI film. The heat conduction of the formed TEDI film

also with the University of Chinese Academy of Sciences, Beijing 100049, China. (e-mail: hyq1993@mail.sim.ac.cn).
Yuelin Wang and Tie Li are with the Science and Technology on Microsystem Laboratory, Shanghai Institute of Microsystem and Information Technology, Chinese Academy of Sciences, Shanghai 200050, China (e-mail: ylwang@mail.sim.ac.cn, tli@mail.sim.ac.cn).

978-1-6654-3008-1/21 $31.00 © 2021 IEEE

can be reduced by equivalent heat conduction extension as well as the infrared absorption can be increased by multiple reflection absorption of incident light. These improvements can be demonstrated by MEMS thermopile platforms.

II. DESIGN AND FABRICATION

A. TEDI film

The conventional FLDI film and the proposed TEDI film as well as the hard-molding process are shown in Fig. 1. The FLDI film (Fig. 1 (a)) is textured by a simple hard-molding process (Fig. 1 (b)), which mainly consists of making a random micropyramid silicon mold in a solution composed of 2 wt. % KOH, 5 vol. % IPA, and DI, forming dielectric films by thermal oxidation or LPCVD, and demolding for suspending the TEDI film (Fig. 1 (c)). Since the designed TEDI film is formed by the hard-molding process, using a micropyramid silicon as the mold. The TEDI film with area radius r_s will hold the angle characteristic (54.7°) between the (111) and (100) planes of the single-crystal silicon.

Thus, the equivalent conduction distance r_{TE} of the TEDI film is $r_{TE} = r_{FL} + 0.7r_s$. It is assumed that the thermal conduction loss through the air and the radiation are negligible and all heat is lateral through the dielectric film to the silicon substrate. According to the 2-D theoretical heat conduction model, the thermal conductance G of the dielectric film can be expressed as [21]:

$$G = \frac{2\pi\lambda t}{\ln(r/r_0)} \quad (1)$$

where λ is the thermal conductivity of the dielectric film, t is the film thickness, and r_0 and r are the radius of the heat source area and the heat conduction distance of the dielectric film, respectively.

Compared to the FLDI film, the thermal conductance increment ΔG of the designed TEDI film is a negative value as the radius r_{TE} is $0.7r_s$ larger than the r_{FL} and the $\ln(x)$ is an increase function. In other words, the thermal conductance of the conventional FLDI film can be reduced by $|\Delta G|$ through texturing of the FLDI film. Meanwhile, the infrared absorption of the TEDI film can be enhanced by multiple reflection absorption of the incident light on the surface of the textured structure, as shown in Fig. 1 (c).

B. Thermopile platform

In order to demonstrate the improvement in the infrared absorption and thermal conductance of the proposed TEDI film, both MEMS thermopile platforms with the controlled FLDI film and the experimental TEDI film are designed, as shown in Figs. 2 (a) and (b). The thermopile is mainly composed of 92 pairs of P+/N+poly-Si thermocouples and a 1.1 mm×1.1 mm infrared absorber. The area of the TEDI film is designed as 1.0 mm×1.0 mm. According to (1), the theoretical reduction $|\Delta G|$ in the thermal conductance of the TEDI film is about 30%. The reduction can be indirectly demonstrated by the inversely proportional relationship (3) between the voltage response V and thermal conductance G of the thermopile platform [22]. Besides, the infrared absorption improvement $\Delta\eta$ can be directly obtained by

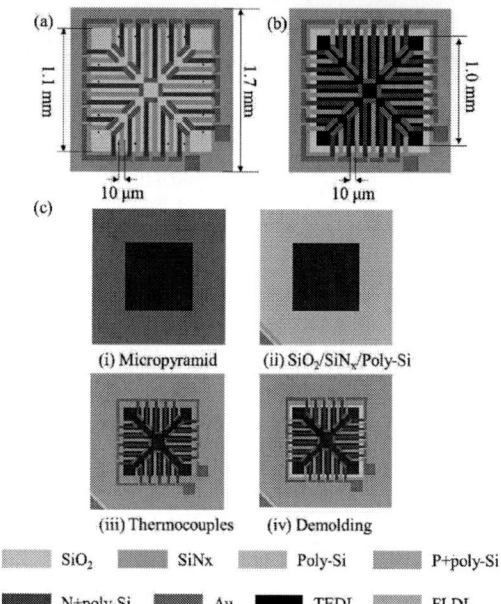

Fig. 2 Schematic of the (a) FLDI film-based thermopile, (b) TEDI film-based thermopile, and (c) fabrication process of the (b).

testing the infrared absorptivity η of the FLDI film and the TEDI film using FT-IR system.

$$V = N\Delta\alpha_{12} A_0 \phi_0 \frac{\eta}{G} \quad (2)$$

where N is the thermocouple number, $\Delta\alpha_{12}$ is the Seebeck coefficient difference between thermoelectric materials, A_0 is the infrared absorber area, ϕ_0 is the infrared radiation power density.

The fabrication process of the thermopile with TEDI film mainly includes: making a random micropyramid silicon mold, forming TEDI SiO₂/SiNₓ/poly-Si films, patterning 92 pairs of P+/N+poly-Si thermocouples, and demolding for releasing the infrared absorber with TEDI film. The detailed steps are shown in Fig. 2 (c).

(i-ii) The silicon mold was first made by wet etching. And then the 0.35 μm SiO₂ was thermally grown and the 1 μm SiNₓ and 0.6 μm poly-Si were deposited by LPCVD, as shown in Fig. 1 (b).

(iii) The poly-Si film was first heavily doped boron with doses of 9×10^{15} cm⁻² and phosphorus ion with doses of 8×10^{15} cm⁻². And then the 92 P+/N+poly-Si thermocouples were patterned by RIE of the doped region. Finally, a layer of TiW/Au was patterned.

(iv) The frontside of the substrate was first protected, and then the MEMS thermopile platform with TEDI film was released by backside etching of the random micropyramid silicon mold.

III. RESULTS AND DISCUSSION

The fabricated thermopile stages with FLDI and TEDI films for heat conduction reduction is shown in Figs. 3 (b)

978-1-6654-3008-1/21 $31.00 © 2021 IEEE

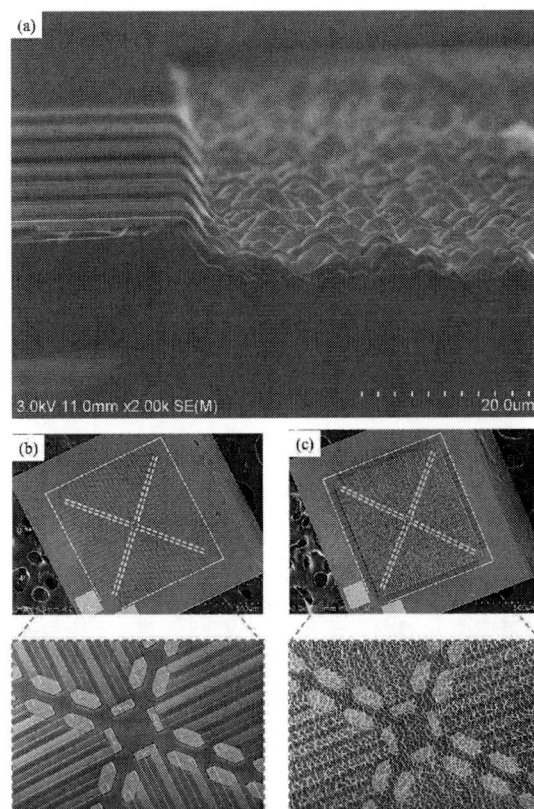

Fig. 3 SEM images of the formed TEDI film (a) and the fabricated MEMS thermopiles with (b) FLDI and (c) TEDI films.

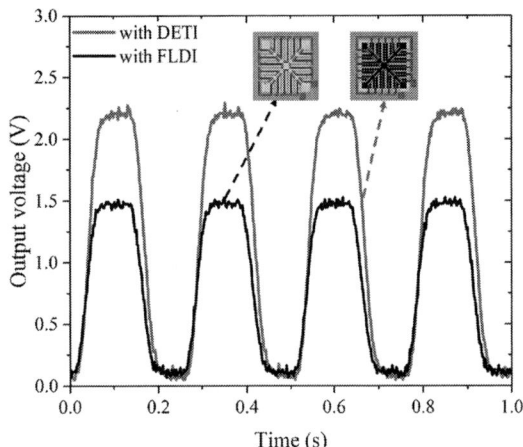

Fig. 4 Output voltage tested by blackbody system with 9 W·m⁻² at 500 K.

TABLE I
PERFORMANCE OF THE THERMOPILE

Thermopile types	FLDI film-based	TEDI film-based	Increment
Absorber area A (mm^2)	1.21	1.21	
Average absorptivity η (1)	0.57	0.64	12%
Resistance R (kΩ)	376.0	415.4	10%
Output voltage V (V)	1.33	2.07	55%
Responsivity Rv (V·W^{-1})	98.66	153.11	55%
Detectivity $D*$ (10^8 cm·Hz$^{1/2}$·W^{-1})	1.38	2.03	47%

and (c). The 92 P+/N+poly-Si thermocouples were well patterned on the surface of the controlled FLDI film and the designed TEDI film, as shown in Fig. 3 (b). The infrared absorber of the TEDI film-based thermopile was black in visible light, as shown in illustration of Fig. 3 (b). The output voltage of the thermopile was tested by blackbody system with 9 W·m⁻² at 500 K. The voltage response is shown in Fig. 4. The thermopiles with FLDI and TEDI films can achieve output voltage of about 1.33 V and 2.07 V, respectively. The voltage response of the TEDI film-based thermopile was enhanced by about 55% significantly. The other parameters of the fabricated MEMS thermopiles with FLDI and TEDI films are shown in Table I. The TEDI film-based thermopile has achieved high responsivity of 153.11 V·W⁻¹ and high detectivity of 2.03×10^8 cm·Hz$^{1/2}$·W⁻¹. The responsivity and detectivity were proportional to the output voltage response and the detectivity was also inverse to square root resistance [21]. Thus, the improvement (55%) in the responsivity was the same as that of the output voltage response while the improvement (45%) in the detectivity is slightly lower mainly due to the increase in the resistance R on the TEDI film.

The formed SiO$_2$/SiN$_x$ TEDI film is shown in Fig. 3 (a). The conventional FLDI film was textured and suspended. The infrared absorption (2.5 μm-14 μm) of the formed FLDI and TEDI films is shown in Fig. 5. The infrared

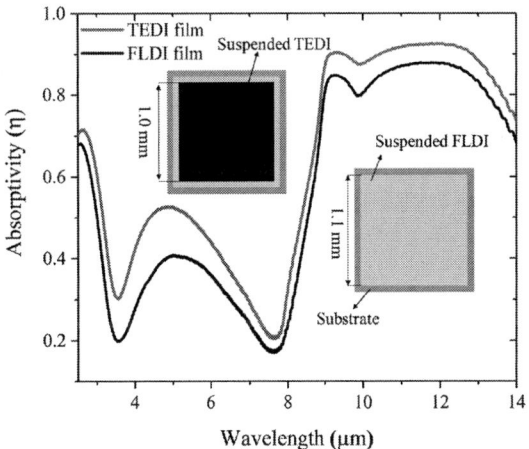

Fig. 5 Infrared absorption of the formed FLDI and TEDI films.

absorption of the formed TEDI film was higher than that of the controlled FLDI film, which may be mainly in virtue of

978-1-6654-3008-1/21 $31.00 © 2021 IEEE

the multiple reflection absorption of incident light on the textured surface. The tested infrared absorption η of the TEDI film can be enhanced by about $\Delta\eta \approx 12\%$. According to the energy conversions of the infrared-heat and the heat-voltage [22], the output voltage response was proportional to the infrared absorption η as well as inverse to the thermal conductance G. Refer to (2), the thermal conductance G of the designed thermopile with TEDI film can be reduced by about $|\Delta G| = 1-(1+\Delta\eta)/(1+\Delta V) \approx 28\%$, this value was in good agreement with the theoretical reduction (30%).

IV. CONCLUSION

Texturing dielectric film is proposed to improve the heat conduction and infrared absorption of the FLDI film in a simple, effective, and CMOS-compatible way in this work. The TEDI film-based thermopile platform with about 47%-55% performance enhancement is designed and fabricated to demonstrate that the infrared absorption property of the TEDI film can be increased by about 12% while the heat conduction can be reduced by about 28% simultaneously. These results indicate the feasibility and potential of using this TEDI film to develop high-performance infrared or thermal sensors.

REFERENCES

[1] A. Graf, M. Arndt, M. Sauer, and G. Gerlach, "Review of micromachined thermopiles for infrared detection," *Measurement Science and Technology*, vol. 18, pp. R59-R75, Jul, 2007.

[2] D. H. Xu, Y. L. Wang, B. Xiong, and T. Li, "MEMS-based thermoelectric infrared sensors: A review," *Frontiers of Mechanical Engineering*, vol. 12, pp. 557-566, Dec, 2017.

[3] D. H. Xu, B. Xiong, and Y. L. Wang, "Micromachined Thermopile IR Detector Module With High Performance," *Ieee Photonics Technology Letters*, vol. 23, pp. 149-151, Feb 1, 2011.

[4] T. A. Vincent, and J. W. Gardner, "A low cost MEMS based NDIR system for the monitoring of carbon dioxide in breath analysis at ppm levels.," *Sensors and Actuators B-Chemical*, vol. 236, pp. 954-964, Nov 29, 2016.

[5] F. F. de Hoyos-Vazquez, M. C. Carreno-de Leon, E. O. Serrano-Nunez, N. Flores-Alamo, and M. J. S. Rios, "Development of a novel non-dispersive infrared multi sensor for measurement of gases in sediments," *Sensors and Actuators B-Chemical*, vol. 288, pp. 486-492, Jun 1, 2019.

[6] T. V. Dinh, I. Y. Choi, Y. S. Son, and J. C. Kim, "A review on non-dispersive infrared gas sensors: Improvement of sensor detection limit and interference correction," *Sensors and Actuators B-Chemical*, vol. 231, pp. 529-538, Aug, 2016.

[7] D. Popa, S. Z. Ali, R. Hoppe, Y. Dai, and F. Udrea, "Smart CMOS mid-infrared sensor array," *Optics Letters*, vol. 44, pp. 4111-4114, Sep 1, 2019.

[8] Y. K. X. Zhang, and B. Yang, "Traffic Flow Detection Using Thermopile Array Sensor," *Ieee Sensors Journal*, vol. 20, pp. 5155-5164, May 15, 2020.

[9] A. Schaufelbuhl, N. Schneeberger, U. Munch, M. Waeliti, O. Paul, O. Brand, H. Baltes, C. Menolofi, Q. T. Huang, E. Doering, and M. Loefe, "Uncooled low-cost thermal imager based on micromachined CMOS integrated sensor array," *Journal of Microelectromechanical Systems*, vol. 10, pp. 503-510, Dec, 2001.

[10] W. Xu, X. Y. Wang, X. Zhao, and Y. K. Lee, "Two-Dimensional CMOS MEMS Thermal Flow Sensor With High Sensitivity and Improved Accuracy," *Journal of Microelectromechanical Systems*, vol. 29, pp. 248-254, Apr, 2020.

[11] C. B. Wen, J. T. Hong, F. Ru, Y. M. Li, and S. Quan, "A Novel Memristor-Based Gas Cumulative Flow Sensor," *Ieee Transactions on Industrial Electronics*, vol. 66, pp. 9531-9538, Dec, 2019.

[12] W. Tian, Y. Wang, H. Zhou, Y. L. Wang, and T. Li, "Micromachined Thermopile Based High Heat Flux Sensor," *Journal of Microelectromechanical Systems*, vol. 29, pp. 36-42, Feb, 2020.

[13] R. G. Spruit, J. T. van Omme, M. K. Ghatkesar, and H. H. P. Garza, "A Review on Development and Optimization of Microheaters for High-Temperature In Situ Studies," *Journal of Microelectromechanical Systems*, vol. 26, pp. 1165-1182, Dec, 2017.

[14] C. N. Chen, "Temperature Error Analysis and Parameter Extraction of an 8-14-mu m Thermopile With a Wavelength-Independent Absorber for Tympanic Thermometer," *Ieee Sensors Journal*, vol. 11, pp. 2310-2317, Oct, 2011.

[15] D. H. Xu, B. Xiong, and Y. L. Wang, "Design, fabrication and characterization of a front-etched micromachined thermopile for IR detection," *Journal of Micromechanics and Microengineering*, vol. 20, pp. 115004, Nov, 2010.

[16] C. N. Chen, and W. C. Huang, "A CMOS-MEMS Thermopile With Low Thermal Conductance and a Near-Perfect Emissivity in the 8-14-mu m Wavelength Range," *Ieee Electron Device Letters*, vol. 32, pp. 96-98, Jan, 2011.

[17] S. Ashraf, C. G. Mattsson, and G. Thungstrom, "Fabrication and Characterization of a SU-8 Epoxy Membrane-Based Thermopile Detector With an Integrated Multilayered Absorber Structure for the Mid-IR Region," *Ieee Sensors Journal*, vol. 19, pp. 4000-4007, Jun 1, 2019.

[18] A. De Luca, M. T. Cole, R. H. Hopper, S. Boual, J. H. Warner, A. R. Robertson, S. Z. Ali, F. Udrea, J. W. Gardner, and W. I. Milne, "Enhanced spectroscopic gas sensors using in-situ grown carbon nanotubes," *Applied Physics Letters*, vol. 106, pp. 194101, May 11, 2015.

[19] S. Ogawa, J. Komoda, K. Masuda, and M. Kimata, "Wavelength selective wideband uncooled infrared sensor using a two-dimensional plasmonic absorber," *Optical Engineering*, vol. 52, pp. 127104, Dec, 2013.

[20] S. Ogawa, K. Okada, N. Fukushima, and M. Kimata, "Wavelength selective uncooled infrared sensor by plasmonics," *Applied Physics Letters*, vol. 100, pp. 021111, Jan 9, 2012.

[21] I. H. Choi, and K. D. Wise, "A Silicon-Thermopile-Based Infrared Sensing Array for Use in Automated Manufacturing," *Ieee Transactions on Electron Devices*, vol. 33, pp. 72-79, Jan, 1986.

[22] C. Escriba, E. Campo, D. Esteve, and J. Y. Fourniols, "Complete analytical modeling and analysis of micromachined thermoelectric uncooled IR sensors," *Sensors and Actuators a-Physical*, vol. 120, pp. 267-276, Apr 29, 2005.

Proceedings of the 16th Annual IEEE International
Conference on Nano/Micro Engineered and Molecular Systems
April 25-29, 2021

Inertial Focusing Chip Based on Superposed Secondary Flows

Jianguo Feng, Qi Wang, Yuanming Ma, Honglong Chang*, *Member, IEEE* and Gaobin Xu*

Abstract—**Inertial focusing has been widely used in particle/cell separation and flow cytometry. In this work we presented an improved inertial focusing chip, consisting of an annular channel with obstacles distributed on the inner wall. Two types of secondary flows were generated and then superposed, thus enhanced the impact on particles. The simulation and experimental results showed that the enhanced secondary flow allowed precise focusing of particles and improved the focusing efficiency. The developed chip achieved a focusing width of ≈11.3 μm for 10.7 μm particles with a flow rate of ≈300 μL/min. This inertial focusing chip could be used for high-sensitive optical detection, especially for flow cytometry.**

I. INTRODUCTION

Particles/cells manipulation such as focusing [1-4], ordering [5, 6], separation [7-10] and filtration [11, 12] has been drawing extensive attention in last decade due to its wide range of applications, such as enrichment and isolation of target cells [13-15], sample pretreatment, water purification and micro flow cytometry [16, 17]. Generally, technologies for particles manipulation can be classified into active and passive methods based on the source of the manipulating forces. Active methods including dielectrophoresis [18], magnetophoresis [15], and acoustophoresis [19] use different external force fields, while passive technologies mainly rely on the channel geometry, such as deterministic lateral displacement [20] and inertial microfluidics [21, 22]. An active system can manipulate target particles in real time and achieve a precise position control. However, these methods always need complicated and expensive fabrication processes and set-ups. In contrast, passive technologies are always simple and robust and have a higher throughput.

Inertial microfluidics [23] is one of the most popular sheathless and passive methods for particles manipulation. Generally, it is usually accomplished based on inertial migration and secondary flows. Inertial migration of particles across streamlines was first described by Segre and Silberberg [24]. They observed that randomly suspending particles at the inlet of a straight pipe migrated laterally to an annulus (Fig. 1A), with a radius of 0.6 times of the pipe radius after a long enough distance. This effect was also called tubular pinch effect. Later, many groups focused on this phenomenon to attempt to explain the underlying mechanism. Theoretical and experimental analyses indicate that there are two main lift forces that dominate behind this phenomenon: 1) the shear-gradient lift force F_S, due to the curvature of the fluid velocity profile, which directs particles away from the channel center, and 2) the wall-induced lift force F_W, due to interaction between particles and channel wall, which directs the particles away from the wall. Under the action of these two main inertial lift forces and other forces (such as Magnus force and Saffman force), particles move across the streamlines and eventually reach the equilibrium positions. Different channel geometries have different focusing positions. For example, in square straight channels (aspect ratio ≈1), particles focus to four equilibrium positions near the center of each channel wall (Fig. 1B), while there are two focusing positions, located at the center of the long walls in a rectangular channel with a low aspect ratio of 0.5 (Fig. 1C).

Secondary flow is another important impact on particles manipulation. It can be generated into two types of channels: 1) curved channels [25] and 2) contraction-expansion array (CEA) channels [26]. Secondary flow in a curved channel is induced by the mismatch of the fluid momentum between fluid in center and the near-wall regions. The fluid near the channel center has lager inertia than that close to the wall. Therefore, it flows outward and induces the fluid near the channel wall flows inward based on the law of conservation of energy, eventually generating two symmetric and counter-rotating vortices, called Dean flow (Fig. 2A). In addition to Dean flow, fluid in a straight channel with CEA structures can also induce secondary flow, called geometry-induced secondary flow. Fig. 2B shows a straight channel

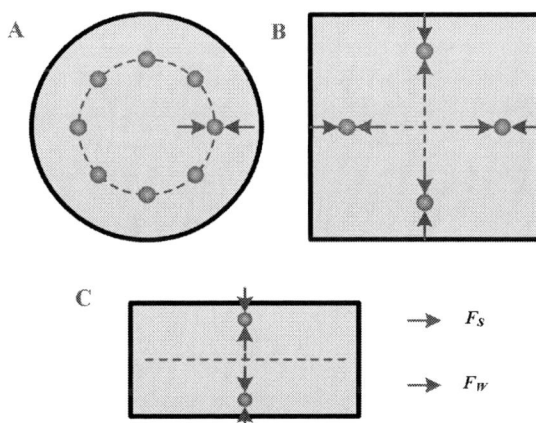

Fig. 1 Equilibrium positions based on shear-gradient lift force (F_S) and wall-induced lift force (F_W) in different channels: (A) circular, (B) square and (C) rectangular.

---

*This work was supported by following foundations: National Key Research and Development Program (2020YFB2008900), Key Research and Development Program of Anhui (1804a09020018) and University Collaboration and Innovation Project of Anhui (GXXT-2019-030).

Jianguo Feng, Qi Wang, Yuanming Ma, and Gaobin Xu are with the School of Electronic Science and Applied Physics, Hefei University of

Technology, Hefei, Anhui 230009, People's Republic of China. (phone: +86-15991752755; e-mail: fengjg@hfut.edu.cn; gbxu@hfut.edu.cn).

Honglong Chang is with the School of Mechanical Engineering, Northwestern Polytechnical University, Xi'an, Shaanxi 710072, People's Republic of China. (phone: +86-13488356886; e-mail: changhl@nwpu.edu.cn).

978-1-6654-3008-1/21 $31.00 © 2021 IEEE

with a series of pillars on one side of the channel wall, where local curved fluids can be generated beside the pillar regions because of the change of the channel cross-section. This kind of curved fluids can induce secondary flows like fluid inside the curved channels. Therefore, this geometry-induced secondary flow is also called Dean-like flow. Secondary flow can induce additional drag force that can be used to modify the inertial migration of particles.

A number of excellent works were reported for cells and particles focusing based on the inertial migration and secondary flows. Generally, they utilized different channel geometries, such as spiral or double spiral channels [27], curved channels [28], serpentine channels [29] and CEA channels [30]. Typically, Wu presented an 8-loop spiral microfluidic sorter based on Dean flow [27]. It can separate polymorphonuclear leukocytes and mononuclear leukocytes from diluted human blood (1−2% hematocrit) with high efficiency (>80%). However, dozens of centimeters' channel length led to high driving pressure. Chung reported an inertial focusing method that utilized a series of stepped channels to create localized geometry-induced secondary flow for particles focusing [31]. Although it could achieve single-stream particle focusing with high-throughput, the main disadvantage was the small channel dimensions (84 µm × 41.5 µm) and long channel length (≈6 cm) that could result in high driving pressure and a higher potential of the blockage in the channel. Zhang presented a microfluidic channel with asymmetrically patterned triangular contraction-expansion cavity arrays [32]. They utilized the geometry-induced secondary flow to focus the particles into a single stream.

It can be seen from above mentioned passive technologies, that only one kind of secondary flow, *i.e.* Dean flow or geometry-induced secondary flow is used to modify the track of particles inside the channels. This results in long channel length which can prolong the action time of secondary flow on particles, and small channel dimensions that can increase the secondary flow velocity (under the same driving pressure). However, long channel length and small channel dimensions have the disadvantages in higher driving pressures of particle suspensions and higher

potentials of clogging in the channels. Meantime, the equilibrium positions are near the channel wall, which can generate strong background noise when utilized in flow cytometry.

In our pervious paper, we have proposed an inertial focusing chip based on enhanced secondary flow [33]. However, the simulation results gave the wrong direction of chip design due to the improper selection of the channel cross sections. In this paper, we presented an optimized inertial focusing method that can generate Dean flow and geometry-induced secondary flow simultaneously. Simulation results showed that these two secondary flows were superposed, and thus totally enhanced. Based on this mechanism, we designed and fabricated the inertial focusing chips, and utilized fluorescent particles to test the focusing performance. The inertial chips achieved a focusing width of ≈11.3 µm, indicating a quite narrow distribution compared with particles diameter, which manifested the enhanced secondary flow improved the focusing efficiency and achieved precise focusing of particles. This presented inertial focusing chips could be used for high-sensitive optical detection especially for flow cytometry.

II. MATERIALS AND METHODS

A. Numerical simulations

In order to unveil the focusing mechanism behind the phenomenon and predict equilibrium positions of the particles, theoretical simulation was taken to evaluate secondary flows in different channel geometries by COMSOL Multiphysics 5.6 (Burlington, MA). We utilized laminar incompressible flow model to calculate the flow field with finite Reynolds numbers (Re) and applied a no slip boundary condition onto the channel walls. The flow rate at the inlet was 500 µL/min while the pressure at the outlet was set to zero.

B. Sample preparation

Rainbow fluorescent polystyrene particles (RFP-100-2, Spherotech, Inc.) were diluted in deionized water, obtaining a homogeneous suspension with the concentration of ≈ 4 × 10^5 particles/mL. The particles diameter is 10.7 µm with the density of 1.05 g/mL. Every particle contains a mixture of fluorophores that enable the Rainbow particles to be excited at any wavelength from 365 to 650 nm.

C. Experimental setup

A home-made pump system was applied to precisely and stably drive the particle suspension at a specific pressure. The inertial focusing chip was clamped on the stage of an inverted fluorescent microscope (Eclipse Ti-S, Nikon). The wavelength of the exciting light is 488 nm. A CCD camera (DS-Fi2, Nikon) and NIS-Elements software (Nikon) were used to record the experimental data and images.

III. RESULTS AND DISCUSSION

A. Numerical simulations of secondary flows

As we discussed in **Introduction**, secondary flows can be induced in CEA channels based on localized curved fluids near the obstacle regions. However, due to the fluid inertia, two curved fluids can generate secondary flows, as

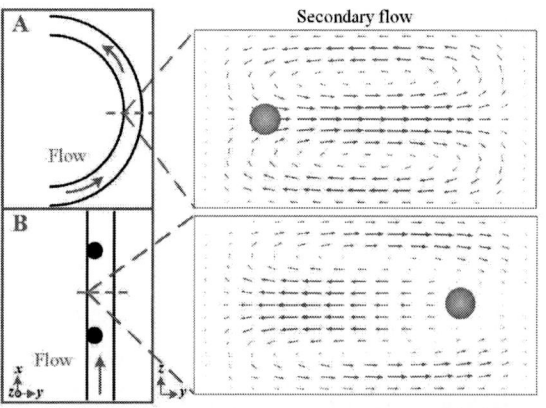

Fig. 2 Two types of secondary flows and their theoretical equilibrium positions (light purple circles): (A) Dean flow in a curved channel and (B) geometry-induced secondary flow in a straight channel with pillars on the wall. The red arrows represent the direction of secondary flow vortices.

978-1-6654-3008-1/21 $31.00 © 2021 IEEE

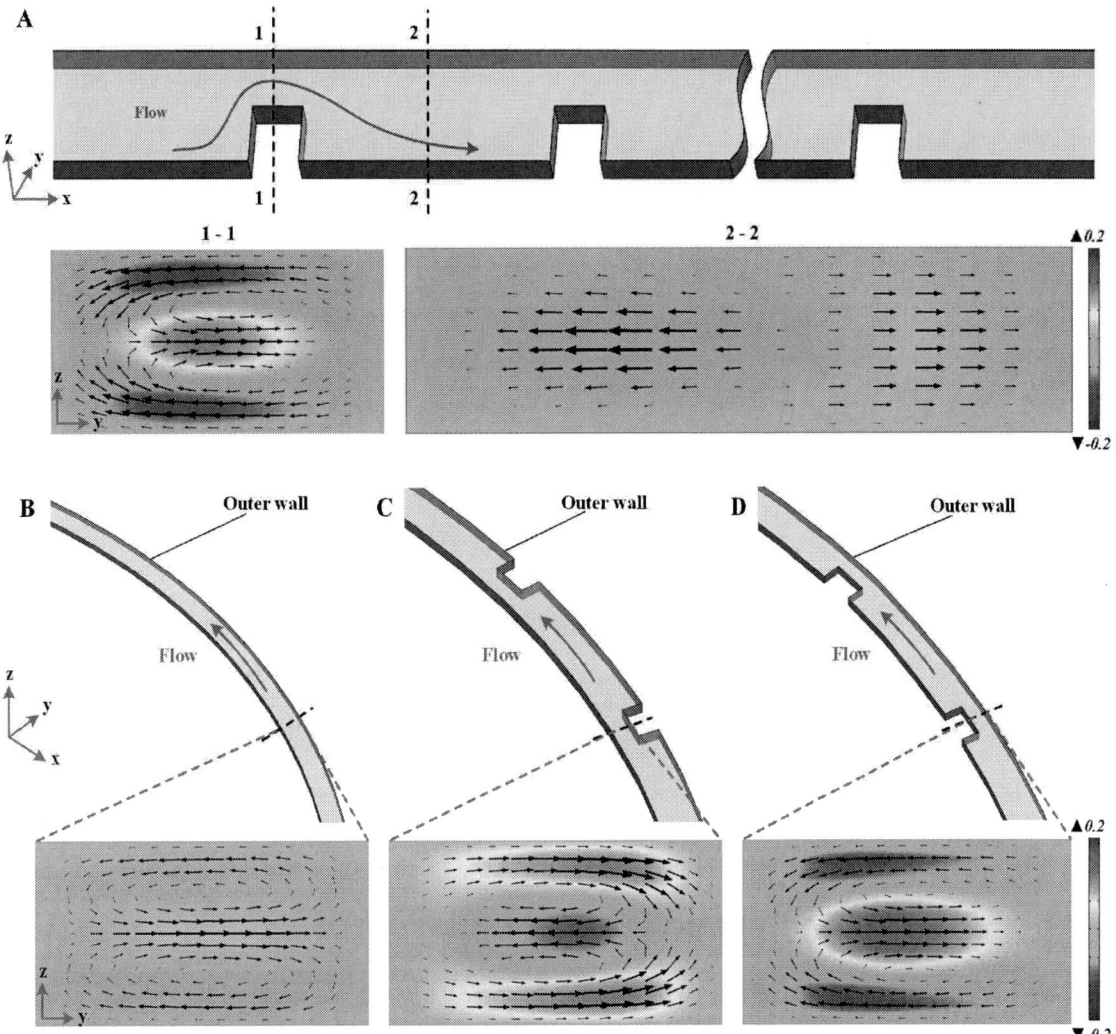

Fig. 3 Four types of channels and their theoretical simulation results: (A) straight channel with steps, (B) annular channel, (C) annular channel with steps at the outer wall, (D) annular channel with steps at the inner wall. The color contours represent the magnitude of secondary flow velocity, and the black arrows show the direction of secondary flow vortices. The legend ranges from -0.2 m/s to 0.2 m/s.

shown in Fig. 3A. One is in contraction area where main fluid flows in from expansion channel (cross-section 1), while another is in expansion area where the fluid flows out from contraction channel (cross-section 2). These two curved fluids have different curvature radii, velocities and adverse directions. Secondary flow in cross-section 1 has a much higher velocity than in cross-section 2 because of the higher flow velocity in contraction channel, contributing to a dominated impact on particle migration. Thus, we expected to improve and then utilize the strong secondary flow (cross-section 1) and suppress the influence of the weak secondary flow (cross-section 2) concurrently. Therefore, three types of channels were designed as shown in Fig. 3, where B is a part of annular channel; C is the annular channel with a series of steps at the outer wall, and D is the same annular channel with steps at the inner wall. It is clear that channel C and D is the combination of channel A and B. The channel B has a cross-section of 100 μm × 50 μm while it is

200 μm × 50 μm in combined channel C and D. The total length of the channel, intervals between adjacent steps and number of steps are same in four channels. The radius of annular channel is 5 mm. All the other settings in COMSOL are identical. We calculated the secondary flows, obtaining the velocity vector plots in channel cross-section. The color contours represent the magnitude of the secondary flow velocity, and the black arrows show the direction of secondary flow vortices.

As can be seen in Fig. 3, secondary flow vortices in channel A, B and D have same direction with centrifugal force, while they are opposite in channel C. Meanwhile, the secondary flow velocities are calculated as: $V_A \approx 0.16$ m/s, $V_B \approx 0.03$ m/s, $V_C \approx 0.14$ m/s and $V_D \approx 0.20$ m/s, respectively. Channel D has a larger velocity than the sum of channel A and B, indicating an enhanced secondary flow (improved >25%) due to the superposition of two types of

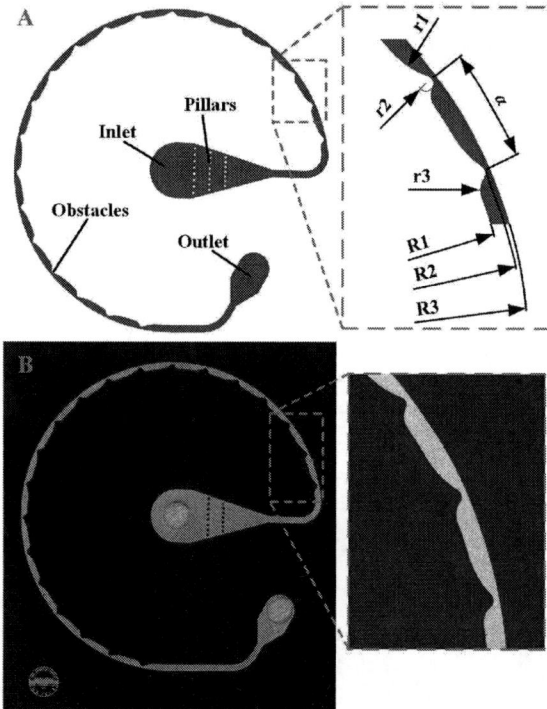

Fig. 4 The presented inertial focusing chip. (A) Chip design and a close-up of a part of channel with designed parameters. (B) The microscope image of the fabricated chip and a close-up of obstacles.

secondary flows. In the meantime, the strength of secondary flow in cross-section 2 decreases because of the opposite directions of two secondary flows, which will have a much weaker impact on particles focusing. As a result, channel D is the optimal chip structure for particle focusing.

B. Chip design and fabrication

Based on the simulation results, we designed the inertial focusing chip as shown in Fig. 4A. The focusing channel consisted of three quarters of an annular channel, obstacles at the inner wall and pillars at the inlet to prevent the channel clogging by large particles or debris. A close-up of a part of channel with designed parameters was also shown in Fig. 4A. The designed parameters of the channel are R1 = 4900 μm, R2 = 5050 μm, and R3 = 5100 μm, respectively. Instead of the angular steps, 20 smooth obstacles with the dimensions of 1000 μm (r1), 80 μm (r2) and 200 μm (r3) are placed on the inner wall with the space of α = 13° (≈1.2 mm). The total length of the channel is ≈ 25 mm, including 1 mm straight channel near the outlet for interrogation and detection.

The inertial focusing chips were fabricated on a Si substrate, with a diameter of ≈100 mm and a thickness of ≈500 μm. We first transferred the chip pattern onto the substrate using ultraviolet lithography with a positive photoresist (PR). Then, the Si substrate was etched using inductively coupled plasma etching, forming the microchannels with a depth of ≈50 μm. After the removal of PR, the Si wafer was bonded with a 500 μm glass wafer using anodic bonding. Finally, it was diced using a diamond dicing saw. The fabricated chip and a close-up of obstacles inside the annular channel were shown in Fig. 4B.

C. Inertial focusing of fluorescent particles

Rainbow fluorescent polystyrene particles with diameter of 10.7 μm were employed to test the focusing performance of the presented inertial chips. We utilized a home-made pump to drive the particles suspension with a constant pressure of 80 kPa. The experimental results of particles focusing are shown in Fig. 5. We first recorded the fluorescent streak images at different locations inside the channel as shown in Fig. 5A. Experimental results agree well with numerical simulations. Due to the enhanced the secondary flow and inertial migration, particles suffer the transverse motion and move to equilibrium position quickly. Even at the position of 16th obstacle, the focusing streak is narrow enough.

We then recorded the fluorescent streak at the interrogation area, which is located at the end of the channel (Fig. 5A). We performed curve fitting using a Gaussian distribution function with a custom written MATLAB (MathWorks, MA, USA) script to determine the full width at half maximum (FWHM) and the focusing position. The result is shown in Fig. 5B and C. The red dashed line represents channel walls, while the blue arrow is the direction of image processing. The focusing width (FWHM) was finally evaluated as ≈11.3 μm, indicating a quite narrow distribution compared with particles diameter. The center of the fluorescent streak was in ≈63 μm distance from the inner wall. At this condition, the particles suspension flow rate was ≈300 μL/min and the Re was ≈40. Moreover, the system throughput was ≈2000 particles/s.

We will then perform experiments using straight channel, annular channel, annular channel with obstacles on the outer

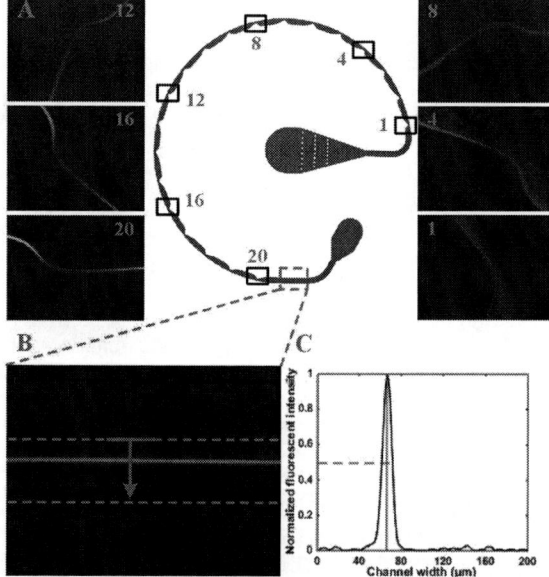

Fig. 5 Experimental results of particles focusing using the developed inertial chips. (A) The images of fluorescent streak at different obstacles indicated by black rectangles in channel structure. The read dashed rectangle at the end of the channel represents the interrogation area. (B) The image of fluorescent streak at interrogation area. The red dashed line represents channel walls and the blue arrow means the direction of image processing. (C) The fluorescence amplitude profile performed by a MATLAB script. The focusing width is evaluated as ≈11.3 μm.

978-1-6654-3008-1/21 $31.00 © 2021 IEEE

wall, and annular channel with obstacles on the inner wall to compare their focusing performance. Moreover, different driving pressure will be applied to test the focusing efficiency. Particles and cells with different diameters will also be used to verify the performance of our chips.

IV. CONCLUSION

In summary, we developed an inertial microfluidic chip for high-efficiency particles focusing. The chip combined an annular channel with obstacles distributed on the inner wall. Two types of secondary flows: Dean flow and geometry-induced secondary flow can be generated simultaneously and superposed inside the presented chips. The enhanced secondary flow improved the particles focusing efficiency. The particles focusing experimental results showed that our chip achieved a narrow focusing width of ≈ 11.3 μm for 10.7 μm particles and a high throughput of ≈ 2000 particles/s with a flow rate of ≈ 300 μL/min. The inertial focusing chip could be used in high-sensitive flow cytometry.

REFERENCES

[1] C. Liu, G. Hu, X. Jiang, and J. Sun, "Inertial focusing of spherical particles in rectangular microchannels over a wide range of Reynolds numbers," *Lab Chip,* vol. 15, no. 4, pp. 1168-77, 2015.

[2] X. Wang, M. Zandi, C. C. Ho, N. Kaval, and I. Papautsky, "Single stream inertial focusing in a straight microchannel," *Lab Chip,* vol. 15, no. 8, pp. 1812-21, 2015.

[3] J. Kim, J. Lee, C. Wu, S. Nam, D. Di Carlo, and W. Lee, "Inertial focusing in non-rectangular cross-section microchannels and manipulation of accessible focusing positions," *Lab Chip,* vol. 16, no. 6, pp. 992-1001, 2016.

[4] M. Lu *et al.*, "Microfluidic hydrodynamic focusing for synthesis of nanomaterials," *Nano Today,* vol. 11, no. 6, pp. 778-792, 2016.

[5] S. C. Hur, H. T. Tse, and D. Di Carlo, "Sheathless inertial cell ordering for extreme throughput flow cytometry," *Lab Chip,* vol. 10, no. 3, pp. 274-80, 2010.

[6] E. W. Kemna, R. M. Schoeman, F. Wolbers, I. Vermes, D. A. Weitz, and A. van den Berg, "High-yield cell ordering and deterministic cell-in-droplet encapsulation using Dean flow in a curved microchannel," *Lab Chip,* vol. 12, no. 16, pp. 2881-7, 2012.

[7] M. G. Lee, J. H. Shin, S. Choi, and J.-K. Park, "Enhanced blood plasma separation by modulation of inertial lift force," *Sens and Actuators B: Chem,* vol. 190, pp. 311-317, 2014.

[8] J. Zhang *et al.*, "A novel viscoelastic-based ferrofluid for continuous sheathless microfluidic separation of nonmagnetic microparticles," *Lab Chip,* vol. 16, no. 20, pp. 3947-3956, 2016.

[9] X. Zhang, Z. Zhu, N. Xiang, F. Long, and Z. Ni, "Automated microfluidic instrument for label-free and high-throughput cell separation," *Anal Chem,* vol. 90, no. 6, pp. 4212-4220, 2018.

[10] Y. Wang, J. Wang, X. Wu, Z. Jiang, and W. Wang, "Dielectrophoretic separation of microalgae cells in ballast water in a microfluidic chip," *Electrophoresis,* vol. 40, no. 6, pp. 969-978, 2019.

[11] A. A. S. Bhagat, S. S. Kuntaegowdanahalli, and I. Papautsky, "Inertial microfluidics for continuous particle filtration and extraction," *Microfluid and Nanofluid,* vol. 7, no. 2, pp. 217-226, 2008.

[12] A. J. Mach and D. Di Carlo, "Continuous scalable blood filtration device using inertial microfluidics," *Biotechnol Bioeng,* vol. 107, no. 2, pp. 302-11, 2010.

[13] H. Kim, S. Lee, J. H. Lee, and J. Kim, "Integration of a microfluidic chip with a size-based cell bandpass filter for reliable isolation of single cells," *Lab Chip,* vol. 15, no. 21, pp. 4128-32, 2015.

[14] J. F. Swennenhuis *et al.*, "Self-seeding microwell chip for the isolation and characterization of single cells," *Lab Chip,* vol. 15, no. 14, pp. 3039-46, 2015.

[15] J. Kim, H. Cho, S. I. Han, and K. H. Han, "Single-Cell Isolation of Circulating Tumor Cells from Whole Blood by Lateral Magnetophoretic Microseparation and Microfluidic Dispensing," *Analy Chem,* vol. 88, no. 9, pp. 4857-4863, 2016.

[16] M. E. Piyasena and S. W. Graves, "The intersection of flow cytometry with microfluidics and microfabrication," *Lab Chip,* vol. 14, no. 6, pp. 1044-59, 2014.

[17] J. Joslin *et al.*, "A fully automated high-throughput flow cytometry screening system enabling phenotypic drug discovery," *SLAS Discovery,* vol. 23, no. 7, pp. 697-707, 2018.

[18] D. Kim, M. Sonker, and A. Ros, "Dielectrophoresis: From Molecular to Micrometer-Scale Analytes," *Analy Chem,* vol. 91, no. 1, pp. 277-295, 2019.

[19] S. H. Kim, M. Antfolk, M. Kobayashi, S. Kaneda, T. Laurell, and T. Fujii, "Highly efficient single cell arraying by integrating acoustophoretic cell pre-concentration and dielectrophoretic cell trapping," *Lab Chip,* vol. 15, no. 22, pp. 4356-63, 2015.

[20] H. Chen, J. Sun, E. Wolvetang, and J. Cooper-White, "High-throughput, deterministic single cell trapping and long-term clonal cell culture in microfluidic devices," *Lab Chip,* vol. 15, no. 4, pp. 1072-83, 2015.

[21] S. S. Nathamgari *et al.*, "Isolating single cells in a neurosphere assay using inertial microfluidics," *Lab Chip,* vol. 15, no. 24, pp. 4591-7, 2015.

[22] J. Zhang *et al.*, "Fundamentals and applications of inertial microfluidics: a review," *Lab Chip,* vol. 16, no. 1, pp. 10-34, 2016.

[23] D. Di Carlo, "Inertial microfluidics," *Lab Chip,* vol. 9, no. 21, pp. 3038-46, 2009.

[24] G. SegrÉ and A. Silberberg, "Radial Particle Displacements in Poiseuille Flow of Suspensions," *Nature,* vol. 189, no. 4760, pp. 209-210, 1961.

[25] N. Xiang *et al.*, "High-throughput inertial particle focusing in a curved microchannel: Insights into the flow-rate regulation mechanism and process model," *Biomicrofluid,* vol. 7, no. 4, pp. 44116, 2013.

[26] D. Yuan *et al.*, "Continuous plasma extraction under viscoelastic fluid in a straight channel with asymmetrical expansion-contraction cavity arrays," *Lab Chip,* vol. 16, no. 20, pp. 3919-3928, 2016.

[27] L. Wu, G. Guan, H. W. Hou, A. A. Bhagat, and J. Han, "Separation of leukocytes from blood using spiral channel with trapezoid cross-section," *Anal Chem,* vol. 84, no. 21, pp. 9324-31, 2012.

[28] J. Sun *et al.*, "Size-based hydrodynamic rare tumor cell separation in curved microfluidic channels," *Biomicrofluid,* vol. 7, no. 1, pp. 11802, 2013.

[29] J. Zhang, W. H. Li, M. Li, G. Alici, and N. T. Nguyen, "Particle inertial focusing and its mechanism in a serpentine microchannel," (in English), *Microfluid and Nanofluid,* vol. 17, no. 2, pp. 305-316, 2014.

[30] M. G. Lee, S. Choi, and J. K. Park, "Inertial separation in a contraction-expansion array microchannel," *J Chromatogr A,* vol. 1218, no. 27, pp. 4138-43, 2011.

[31] A. J. Chung, D. R. Gossett, and D. Di Carlo, "Three dimensional, sheathless, and high-throughput microparticle inertial focusing through geometry-induced secondary flows," *Small,* vol. 9, no. 5, pp. 685-90, 2013.

[32] J. Zhang, M. Li, W. H. Li, and G. Alici, "Inertial focusing in a straight channel with asymmetrical expansion-contraction cavity arrays using two secondary flows," *J Micromech and Microeng,* vol. 23, no. 8, pp. 085023, 2013.

[33] J. Feng, W. Xun, and H. Chang, "Tiny inertial focusing chip based on enhanced secondary flow," 20th International Conference on Miniaturized Systems for Chemistry and Life Sciences, Dublin, Ireland, 9-13 October, 2016.

Proceedings of the 16th Annual IEEE International
Conference on Nano/Micro Engineered and Molecular Systems
April 25-29, 2021

A Sensorised Forcep Based on Piezoresistive Force Sensor for Robotic-assisted Minimally Invasive Surgery*

Cheng Hou[1], Jiangjun Geng[1], Yuyang Sun[1], Tao Chen[1], Member, IEEE, Fengxia Wang[1], Member, IEEE, Hongling Ren[2], Member, IEEE, Xiuli Zuo[3], Yanqing Li[3], Huicong Liu[1*], Member, IEEE and Lining Sun[1], Member, IEEE

Abstract— This paper reports a sensorized forcep with a minimized force sensing chip to facilitate Robotic-assisted Minimally Invasive Surgery (RMIS). A piezoresistive triaxial force sensor chip (2 mm × 2 mm) is developed and integrated in the grasping head of a continuum robot to provide additional tactile to the RMIS. Biocompatible hemisphere cap enhances the sensor's capability of triaxial force detection. A 3-dimensioanl (3D) force test is performed on the sensorized forcep. This simple strategy of configuration and sensing makes it possible in miniaturization of the forceps (outside diameter is less than 4 mm) for RMIS in the strictest operating space.

I. INTRODUCTION

Robotic-assisted minimally invasive surgery (RMIS) has introduced significant advantages such as smaller trauma, less postoperative pain, less damage to surrounding tissues, quicker recovery, and reduction in treatment cost. [1-4]. As the robots are comprised of a master-slave control system with a haptic mater device, they can provide desired accurate motions with the ability to scale the motions and cancel the hand tremor of the surgeon. But, it involves some limitations. RMIS mainly relies on the skills and experience of surgeons to perform the operation, with some high risk of mis-operation, eg. excessive grasping and traction force [4-8]. The lack of force sensing capability limits the surgeons' skill, since the surgeon judge the feedback force of the operation based on visual information and clinic experiences during RMIS [9-10]. The excessive force lead to additional damages to fragile tissues and organs in body, thereby increasing the risk of accidents. Accordingly, surgical robots with force sensing capability are to perform surgical tasks more accurately and security.

Recently, many research groups have conducted into the force sensing system of RMIS to overcome this problem [1-7]. However, the difficulty of sensor system integration still hinders its practical application in surgical instruments. [3]. And any locations of the surgical instrument except gripper are not suitable candidates due to inertia, friction, backlash, gravity and tension of cables actuating joints. The location of gripper without the aforementioned disturbances can provide accurate force since the direct contact with tissues. Thus, the excessive force of grasping and traction

can be prevented during operations.

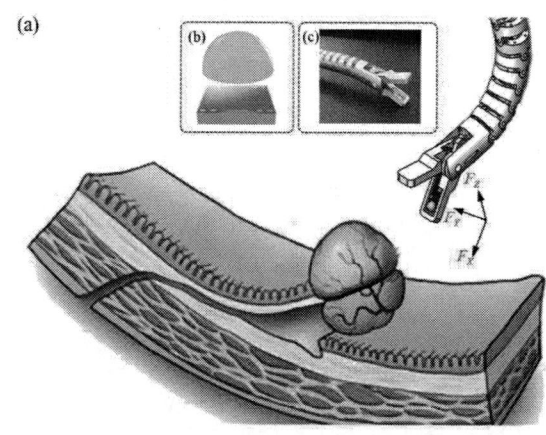

Figure 1. *(a) Overview of the sensorized forcep perceive three-dimensional forces during* Robotic assisted *minimally invasive surgery (RMIS). (b) The force sensor chip and PDMS cap. (c) The sensor integrated with the continuum robot.*

Recently, numerous studies have been conducted to integrate force sensors at the tips of surgical instruments. King et.al [1] integrated a piezoresistive tactile sensor into a forcep's tip. It can measure direct contact forces or grasping forces, but without the pulling forces. Okuda et.al [2] developed a 3-axises pressure sensors integrated at the surgical instrument and measured the grip force during laparoscopic surgery in pigs. However, the larger volume of this sensor restricts its application in such as Natural orifice transluminal endoscopic surgery (NOTES) which with more limited operative space. In addition, the sensors exposed to tissue/organ require relevant biocompatibility considerations. Kim et.al [3] developed a novel sensorized surgical forceps with 5-DOF force/torque (F/T) sensing capability based on capacitive. Two 3-DOF sensors integrated into the proximal region on upper and lower jaw to convert to 5-DOF F/T. However, the complex mechanical structure and assembly cannot confirm the consistency of grippers, and the smaller gripper require a higher level of processing, which also limits its application in natural endoscopic surgery such as minimally invasive

---

* This work is funded by the National Key R&D Program of China (2018YFB1307700); Postgraduate Research & Practice Innovation Program of Jiangsu Province.

[1]Cheng Hou, Jiangjun Geng, Yuyang Sun, Tao Chen, Fengxia Wang, Huicong Liu, Lining Sun are with the School of Mechanical and Electric Engineering, Jiangsu Provincial Key Laboratory of Advanced Robotics, Suzhou 215123, China

[2]Hongliang Ren is with the National University of Singapore Suzhou Research Institute, Suzhou 215125, China

[3]Department of Gastroenterology, Qilu Hospital, Cheeloo College of Medicine, Shandong University, Jinan, Shandong, 250012, China

Corresponding author: Huicong Liu (e-mail: hcliu078@suda.edu.cn).

978-1-6654-3008-1/21 $31.00 © 2021 IEEE

surgery of digestive endoscopy.

In this paper, a novel sensorized surgical forcep featuring one miniature (2 mm * 2 mm * 0.4mm) three-axis force sensor to the grasping head is presented. To design the miniature sensor with large range, orthogonal membrane arrangements of piezoresistance-sensing and PDMS cap are applied. One hemisphere biocompatible PDMS cap enhances the sensor's range and makes it capable of triaxial force detection. PDMS is filled in the slot and encapsulates the sensor and flexible print circuit board (FPCB), avoiding the exposure of the sensor and FPCB in the tissue and ensuring biocompatibility. In addition, this simple strategy of configuration and sensing makes it possible in miniaturization of the forceps (the outside diameter is less than 4 mm) for RMIS in the strictest operating space. Mass MEMS tape-out can reduce production costs and reduce surgical expenses. To calibrate the three force elements and evaluate the performance of the sensor, the forceps are installed in a surgical instrument. Through integration with a continuum robot, the grasp sensing capability of the manufactured prototype is evaluated.

II. EXPERIMENTAL METHODS

A. Fabrication of the Sensor Chip

Figure 2a shows the detailed sequence for the fabrication process flow of the piezoresistive sensor chip. A double side polished, n-type, (100) oriented silicon-on-insulator (SOI) wafer was used as a starting material. The thickness of device layer, substrate layer and buried silicon dioxide layer are 5 μm, 395 μm and 1 μm respectively. First, 3000 Å SiO_2 layer was grown on both side of the SOI wafer by the thermal oxidation process. Next, to pattern the piezoresistors, photolithography was conducted on front side of the wafer. Then P-type implantation of dosage 5×10^{14} ion cm^{-2} and energy of 20 keV is performed (tilt 7°and twist 0°). The size of the piezoresistor is 110 μm × 10 μm. Under the circumstance of rapid thermal annealing(RTA) at 900 degrees Celsius for 45 seconds, the dopants are activated to form the lightly piezoresistors. After thermal annealing, the depth of the joint is approximately 0.7 μm . Next, an extra SiO_2 layer of Plasma Enhanced Chemical Vapor Deposition (PECVD) 3000 Å for passivation purpose. After via open through RIE, the last P-type ion is implanted with a dosage of 2×10^{15} ion cm^{-2} and an energy of 35 keV (tilt 7°and twist 0°) followed the same RTA step to form a good ohmic contact. After that, sputtering and patterning aluminum layer with the thickness of 700 nm. Then metallization (annealing at 400 °C for 20 min) was performed for interconnection between aluminum and piezoresistors. After metallization, PECVD 1 μm silicon nitride film to compensate the compressive stress in the SiO_2 layer, also for the surface electrodes protection. Next, backside deep reactive ion etching (DRIE) is conducted to release the diaphragm structure up to the BOX layer. Eventually, buried SiO_2 layer was removed by the RIE process.

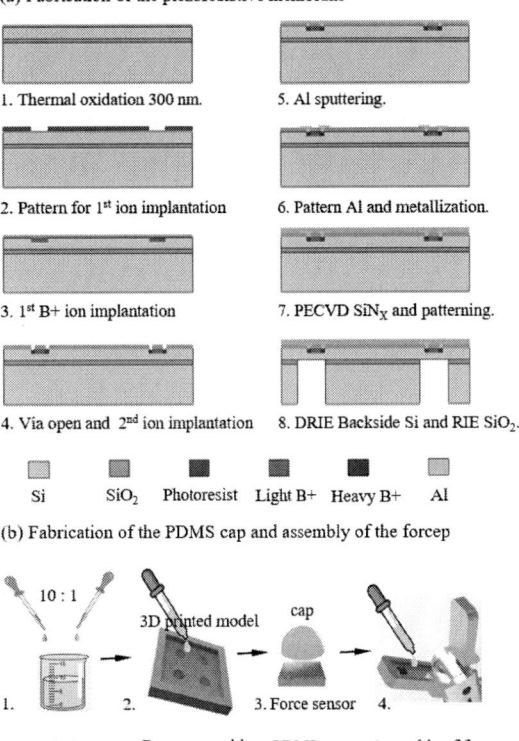

(a) Fabrication of the piezoresistive membrane

1. Thermal oxidation 300 nm.
2. Pattern for 1st ion implantation
3. 1st B+ ion implantation
4. Via open and 2nd ion implantation
5. Al sputtering.
6. Pattern Al and metallization.
7. PECVD SiN_x and patterning.
8. DRIE Backside Si and RIE SiO_2.

Si SiO_2 Photoresist   Light B+   Heavy B+   Al

(b) Fabrication of the PDMS cap and assembly of the forcep

1. PDMS   2. Reverse mould   3. Force sensor PDMS cap   4. Assembly of forcep

Figure 2. (a) Schematic illustration of the fabrication of the piezoresistive membrane. stetp 1: Thermal oxidation 300 nm SiO_2. Step 2: Pattern for 1st ion implantation. Step 3: 1st B+ ion implantation. Step 4: Via open and 2nd ion implantation. Step 5: Sputtering 700 nm Al. Step 6: Pattern Al and metallization. Step 7: PECVD SiN_x and patterning. Step 8: DRIE backside Si and RIE SiO_2. (b) Fabrication of the PDMS cap and assembly of the sensorized forcep. Step 1: Proportioning PDMS. Step 2: Pouring PDMS and high temperature heating to achieve inverted mold. Step 3-4: Assembly of the sensorized forcep.

B. Fabrication of the PDMS Cap

As shown in Figure 2b, in the fabrication process of the hemisphere PDMS cap, a three-dimensional (3D) printed mold with hemisphere structure is printed using a high-precision 3D printer (KONICA MINOLTA Form2). The PDMS (Sylgard184, Dow Corning Corp) was poured into the mold and stored in a vacuum box for 10 minutes to eliminate bubbles, and subsequently placed on a hot plate to heat it within 3 hours at 80 °C. In the process of the fabrication, the mixture ratio between the PDMS polymer base and its polymerization agent was 10:1.

C. Assembly of the Sensorized Surgical Forcep

After the fabrication of the sensor chip and PDMS cap, the fabricated sensor chip was first bonded and wired to an outer FPCB. One end of the FPCB bonded sensor is embedded and bonded to the rectangular slot (the height of the slot is 1 mm) of the lower forcep, as shown in Figure 2b. Then, the hemisphere PDMS cap was fixed at the center of the surface of the sensor chip. In the next step, the PDMS was poured into the rectangular slot until PDMS fills the slot, correspondingly the sensor chip was completely covered. Finally, storing and heating the PDMS with the same steps as before. A fully assembled sensorized forcep can be seen in Figure 2b.

III. EXPERIMENTAL RESULTS

The schematic drawing of the force sensor is illustrated in Figure 3a, while the optical microscope photo of a whole piezoresistive sensor chip is shown in Figure 3b. The sensor chip is in square shape with size of 2 mm × 2 mm ×0.4mm.

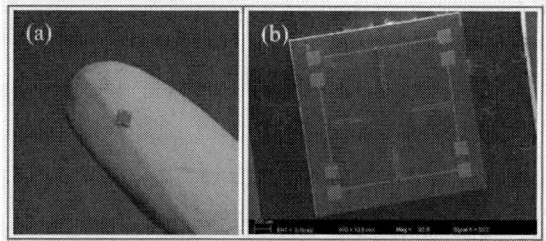

Figure 3. (a) The processed piezoresistor sensor chip on finger. (b) The optical microscope photo of a whole sensor chip

A. Sensing Characterizations of the Force Sensor

Figure 4a shows the four membrane of the piezoresistive sensor chip. In our experiment, the partial voltage method is used to measure the change in each membrane. As can be seen in Figure 4b, the four resistance piezoresistors were ion implanted and annealed. When the excitation voltage vary from -1V to 1V (Keithley 4200 A-SCS), the resistance value of the four piezoresistors are constant 3400 ohms, which has a good consistency.

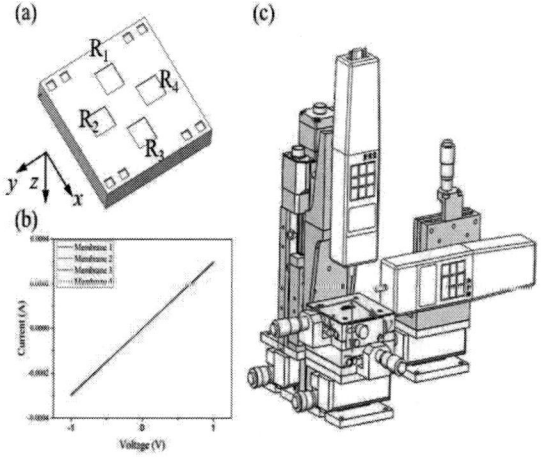

Figure 4 (a) Schematic illustration of four membrane. (b) The resistance of four membrane after fabrication. (c) Experimental setup for evaluating the responses of the sensorized forcep after assembly.

The experimental setup to evaluate the response of the force sensor to normal and lateral forces is illustrated in Figure 4c. The assembly sensorized lower forcep was fixed

on three-axis platform which can realize linear movement along the x-axis, y-axis, and z-axis with a resolution of 2 um. A digital force gauge (AIKOH RZ-2) is used to measure the normal force. The force in the lateral direction was measured by another digital force gauge (AIKOH RZ-2).

The signals as corresponding to the force on the x-, y-, and z-axis are defined as S_x, S_y, and S_z, respectively. Each corresponding signal can be calculated as follows:

$$S_z = -\frac{1}{4}\left(\frac{\Delta R_1}{R_1} + \frac{\Delta R_2}{R_2} + \frac{\Delta R_3}{R_3} + \frac{\Delta R_4}{R_4}\right) \quad (1)$$

$$S_x = \frac{\Delta R_1}{R_1} - \frac{\Delta R_3}{R_3} \quad (2)$$

$$S_y = \frac{\Delta R_2}{R_2} - \frac{\Delta R_4}{R_4} \quad (3)$$

Appling one normal force to the PDMS cap, all four membrane will generate deformation and stress under the action of the PDMS cap. Thus, the applied normal force can be detected through the average of the resistance change of the four membrane (equation (1)). Theoretically, all four piezoresistive membrane should have the same amount of deformation and stress. When a lateral force is applied to the PDMS cap, there is a torque around the center of the PDMS cap. This torque causes the PDMS rotate slightly. Consequently, one membrane will have a smaller deformation, while the other one is larger. Hence, the signal correspond to the lateral force are the difference in the resistance change of the two membrane according to the equation of (2-3).

The responses of the sensorized forcep integrated with force sensor to the normal and lateral forces are shown in Figure 5. As shown in Figure 5a, the change rate of resistances increase continuously with the increase of the normal force with a good linear relationship. Meanwhile, the change rate of the four piezoresistors has good consistency. When the force was applied with lateral force in the direction of x-axis, the difference between R_1 and R_3 increases with the increase of the applied force, while the difference between R_2 and R_4 is basically unchanged, as depicted in Figure 5b. As shown in Figure 5c, when the force was applied with lateral force in the direction of y-axis, the difference between R_2 and R_4 increases with the increase of the applied force, while the difference between R_1 and R_3 is basically unchanged. It can be seen that the signals corresponding to an axis was dominantly determined by the direction of the applied force. The relationships between outputs R_X, R_Y and R_Z and the forces applied in direction of each axis are shown in Figure 6 (a), (b) and (c), respectively. Three outputs from the resistance changes of four bridges were used to measure the normal force and the lateral forces. The results show that when force in a direction was applied, the output corresponding to that direction was larger than output in other directions. Moreover, the applied forces and the outputs are linear relationships. Crosstalk still exists in each direction, which is mainly due to the differences in the process of the piezoresistive sensor chip and the fabrication accuracy.

B. Results of the Sensorized Forcep

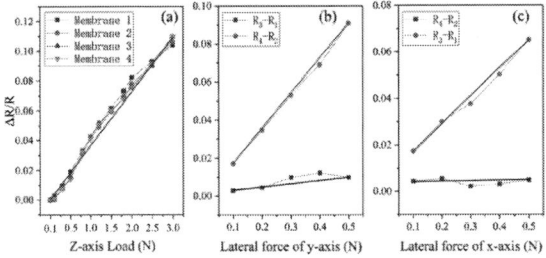

Figure 5. The responses of the sensorized forcep integrated with force sensor to (a) normal force, (b) lateral force in x-axis, and (c) lateral force in y-axis.

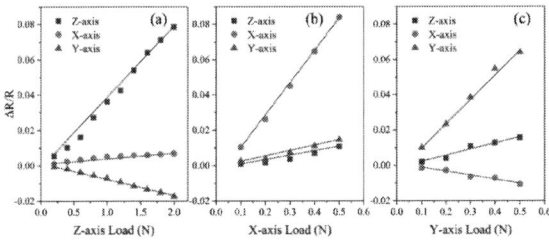

Figure 6 The relationship between applied forces and the outputs of the developed sensor.

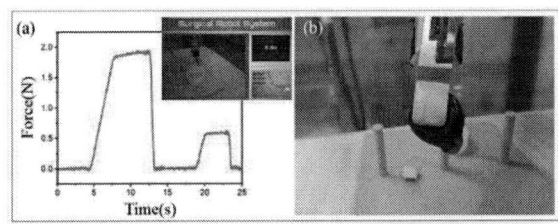

Figure 7. Figure (a) Sensor data in the process of peg transform task (b) Blueberry clipping experiment

We further assemble the sensorized forcep and continuum robot to the master-slave surgical robot system. Then a peg transfer task was carried out in which a ring near the trocar entry could be picked up by the manipulator and transferred to farther pegs in other configurations. In the process, the sensor record the interaction force , which will be transferred to the operator by visual feedback and auditory feedback as shown in the box of Figure 7a. The sensor data can be seen in Figure 7a. When the operator received the threshold alarm, the operator will consciously reduce the clamping force immediately.

Also, the master-slave and force feedback platform were used to complete the action of clapping blueberries. Due to the soft quality of blueberries, they were easy to be pinched by forceps in the absence of force feedback which could be seen in Figure 7b. The addition of force feedback made the operator have a certain sense of the force applied and reduced the risk of mis-operation to a certain extent.

IV. CONCLUSION

In this paper, we proposed a novel sensorized surgical forceps three-axis force sensor to the grasping head. The force sensor based on piezoresistive and membrane structure is presented. One hemisphere biocompatible PDMS cap enhances the sensor's measuring range and makes it capable of triaxial force detection. To avoid the exposure of the sensor and circuit in the tissue and ensuring biocompatibility, the PDMS is filled in the slot and encapsulates the sensor and circuit. In addition, this simple strategy of configuration and sensing makes it possible in miniaturization of the forceps (the outside diameter is less than 4 mm) for RMIS in the strictest operating space. To calibrate the three force elements and evaluate the performance of the sensor, the forceps are installed in a homemade surgical instrument. Through integration with a continuum robot, the 3-DOF sensing capability of the manufactured prototype is evaluated. The force sense information has effectively improved the safety of the clamping process, lateral forces will be included in future work. Through the vitro experiments, its effectiveness will be verified, and the quality of surgery will be improved.

ACKNOWLEDGMENT

This work is funded by the National Key R&D Program of China (2018YFB1307700); Postgraduate Research & Practice Innovation Program of Jiangsu Province.

REFERENCES

[1] C. H. King, M. O. Culjat, M. L. Franco, C. E. Lewis, E. P. Dutson, W. S. Grundfest, J. W. Bisley, "Tactile Feedback Induces Reduced Grasping Force in Robot-Assisted Surgery", IEEE Transactions on Haptics, vol. 2, pp. 103-110, 2009.

[2] Y. Okuda, A. Nakai, T. Sato, M. Kurata, I. Shimoyama, T. Oda, N. Ohkohci, "New device with force sensors for laparoscopic liver resection – investigation of grip force and histological damage", Minimally Invasive Therapy and Allied Technologies, 2020.

[3] U. Kim, Y. B. Kim, J. So, D. Y. Seok, H. R. Choi, "Sensorized Surgical Forceps for Robotic-Assisted Minimally Invasive Surgery", IEEE Transactions on Industrial Electronics, vol. 65, pp. 9604-9613, 2018.

[4] U. Kim, Y. B. Kim, D.-H. Lee, H. R. Choi, and D.-Y. Seok, "A Novel Six-Axis Force/Torque Sensor for Robotic Applications," IEEE/ASME Trans. Mechatronics, vol. 22, no. 3, pp. 1381–1391, 2016.

[5] U. Seibold and K. Bernhard, "Prototype of Instrument for Minimally Invasive Surgery with 6-Axis Force Sensing Capability," no. April, pp. 498–503, 2005.

[6] J. Peirs et al., "A micro optical force sensor for force feedback during minimally invasive robotic surgery," vol. 115, pp. 447–455, 2004.

[7] U. Kim, D. H. Lee, H. Moon, J. C. Koo, and H. R. Choi, "Design and realization of grasper-integrated force sensor for minimally invasive robotic surgery," IEEE Int. Conf. Intell. Robot. Syst., no. Iros, pp. 4321–4326, 2014.

[8] N. M. Bandari, "Hybrid piezoresistive-optical tactile sensor for simultaneous measurement of tissue stiffness and detection of tissue discontinuity in robot-assisted minimally invasive surgery," vol. 22, no. 7, 2019.

[9] L. Li, B. Yu, C. Yang, P. Vagdargi, R. A. Srivatsan, and H. Choset, "Development of an Inexpensive Tri-axial Force Sensor for Minimally Invasive Surgery," 2017.

[10] C. Dücső et al., "3D force sensors for laparoscopic surgery tool," Microsyst. Technol., vol. 24, no. 1, pp. 519–525, 2017.

Proceedings of the 16th Annual IEEE International
Conference on Nano/Micro Engineered and Molecular Systems
April 25-29, 2021

In situ Laser-assisted Micromachining of Environmentally-responsive Hydrogel Films

Hongjie Jiang[1]*, member, IEEE, Manuel Ochoa[2]

Abstract— Traditional micromachining techniques, such as lithography, etching, and micro-molding, to pattern hydrogel thin films, are time consuming, expensive, and do not scale well to large production. In this paper, we demonstrate a direct laser patterning of environmental responsive hydrogel films, allowing low cost, fast, and scalable fabrication for in situ mass production. We characterized and analyzed a series of transient features of the laser-engineered hydrogel, including the ablated width, sidewall quality, and resolution as a function of laser beam parameters and hydrogel thermal properties at different hydration states (from wet to fully dried). The optimal feature quality is achieved in hydrogels dried for 1-2 hours (40−60% weight loss), thus identifying a temporal window for a rapid end-to-end fabrication.

I. INTRODUCTION

Environmentally responsive hydrogels exhibit volumetric and geometric transitions in response to a variety of environmental stimuli such as temperature, light, pH, glucose, and magnetic/electrical signal [1]–[3]. Their sensing and actuating capabilities for bio-inspired applications can be easily tuned or enhanced by incorporating functional monomers or organic/inorganic agents/particles into the gel network [4]–[6]. Furthermore, the response time of such hydrogel-incorporated materials can be significantly improved by patterning the functional hydrogel into micro- or nanoscale. As a result, environmentally-responsive hydrogels can be engineered to address various biomedical applications such as biosensors, diagnostic imaging, and drug delivery [6]–[9].

Traditional micromachining techniques for generating micro- and nano-patterns on polymeric materials include micro-molding, dry-etching and photolithography [10]–[12]; however, these approaches suffer from various shortcomings. For example, micro-molding or soft lithography is a rapid and low-cost manufacturing technique, but its scalability and massive-fabrication capabilities require designing and manufacturing high-precision, expensive master molds. Such imprinting methods do not offer any means of in situ hydrogel modification (e.g., for individually tailoring a device for the end user immediately before use) after integration of multiple device components/layers. Photolithography or dry-etching usually suffers from low throughput, limited by the need for cleanroom facilities and multi-step processes that also increase production costs [13]. Moreover, since

photolithography typically requires processing with ultraviolet (UV) light, it is not always suitable for the fabrication of composite hydrogels as UV may either compromise the properties of the embedded particles or be blocked/scattered by those particles thus affecting the hydrogel polymerization. Recently, 3D printing/bioprinting technologies such as inkjet printing and micro-extrusion have been applied for the fabrication of 3D tissue constructs made of naturally biocompatible hydrogels to serve as tissue scaffolds. Although these approaches can be used to produce a well-defined 3D architecture with cells, the high printing quality requires a set of carefully tailored characteristics of polymeric inks, including quick gelation or specific levels of viscosity (a lower one preferred for inkjet while a higher one suitable for micro-extrusion), thus confining their general availability to only gelatin- or collagen-based materials. Additional limitations could exist in the clogging of the nozzle, resulting in higher costs due to frequently required manual maintenance [14].

A more suitable approach is the use of a scalable, standardized, rapid prototyping technique, namely laser engraving, which enables a diverse set of applications where controlled, localized heating is essential (e.g., cutting, welding, machining). Due to its high lateral resolution, low heat input, and high flexibility, a laser beam is capable of

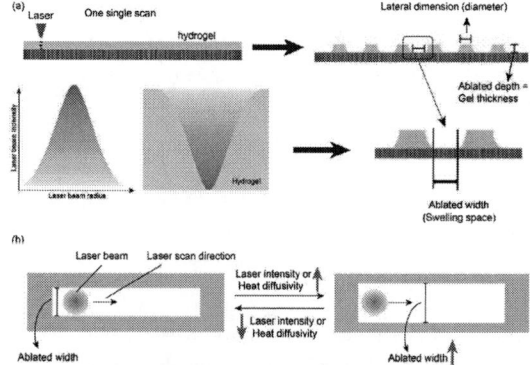

Fig. 1: Conceptual illustration of laser thermally etching hydrogel on an adhesion promoting substrate (a) to create a quadrilateral pyramid patterned hydrogel due to the Gaussian laser beam profile and (b) to achieve different ablated width regulated by laser beam intensity or by hydrogel heat diffusivity.

photochemically (at a laser beam wavelength smaller than

---

[1] Shien-Ming Wu School of Intelligent Engineering, South China University of Technology, Guangzhou 511442 China. (corresponding author, phone: 020-81182122; e-mail: jiang1029@ scut.edu.cn).

[2] DBA Ochoa Consulting, Pico Rivera, CA 90660 USA (e-mail: manny@manuelochoa.com).

978-1-6654-3008-1/21 $31.00 © 2021 IEEE

UV) or photothermally (at a laser beam wavelength from near infrared to mid infrared) ablating a wide range of materials such as metals, semiconductors, ceramics, or polymers [15], [16]. In this work, a CO_2 laser system (laser beam wavelength of 10.6 μm) is utilized to directly define micro features on an environmentally sensitive hydrogel film without damaging its integrity and sensitive properties. Compared to those abovementioned micro-machining techniques, laser micro-machining is more cost-effective and allows greater (but flexible) throughput and is thus ideal for scalable, in situ mass-manufacturing of micro-patterned hydrogels for complex biomedical micro-systems.

II. PRINCIPLE AND OPERATION

A. Principle and Theory

When a laser beam irradiates a hydrogel surface, the deposited thermal energy heats the hydrogel following the Gaussian distribution of the laser beam intensity [17], generating two thermally etched regions on hydrogels: the directly-affected region (where the laser beam is absorbed), and the surrounding area (heated by conduction of residual thermal energy from the first region). The combination of these two types of heating can be used to control the laser machining profile of a hydrogel, Fig. 1a.

The micro-features of the laser-engineered hydrogels depend on the laser beam profile and the time-varying thermal properties of the hydrogel. To increase the ablated width, one can increase the laser beam intensity (higher power or lower scanning speed) or the heat diffusivity of the hydrogel; the converse is also true, Fig. 1b. The heat diffusivity of the hydrogel depends on its moisture contents, which determines the thermal conductivity and heat capacity of the hydrogel [18]; decreasing the former weakens the heat propagation and concurrently reduces the ablated width, whereas reducing the latter speeds up the gel evaporation and concomitantly expands the ablated width. Therefore, there exists an optimum gel drying time (and moisture content) for creating the largest ablated width in the gel matrix.

B. Fabrication and Experiments

The environmental responsive hydrogel, chose in this wok, is poly methacrylic acid -co- acrylamide (mAA-co-AAm) gel, whose fabrication is detailed in the report [19]. The hydrogel was casted and processed directly in an adhesion-promoting substrate (GelBond® PAG Film, Lonza) to 0.5 mm, 1 mm, 1.5 mm, and 2 mm thickness. After one hour of polymerization, the hydrogel was left at room temperature for drying over 24 hours. Within the 24 hours, at an interval of one hour to four hours drying, a CO_2 laser system (10.6 μm, PLS6MW, Universal Laser Systems) was applied to conduct the micro-machining on the hydrogel, monitoring the parameters including the ablated width, sidewall quality and resolution. At the same time, the weight of the hydrogel at each interval was also hourly monitored. Finally, both the laser work and the weighing of the completely dried hydrogel after 24 hours are also performed as an extreme case of the change of the thermal properties of the hydrogel.

III. RESULTS AND DISCUSSION

Fig. 2 presents photographs of the top view of laser-micromachined poly mAA-co-AAm hydrogels at 1- or 2-hours drying time, where a single laser ablation on hydrogel generated a thermally etched microchannel. The bright areas within the microchannels indicate complete removal of hydrogel, whereas the surrounding shadows indicate partial removal. Non-uniformities of channel width along each trace can be attributed to overlap of the laser beam from one trace to the next; hence, when designing the manufacturing process, it is important to tune the laser parameters or hydrogel heat diffusivity (via its moisture content) to minimize the width of the partially-removed regions.

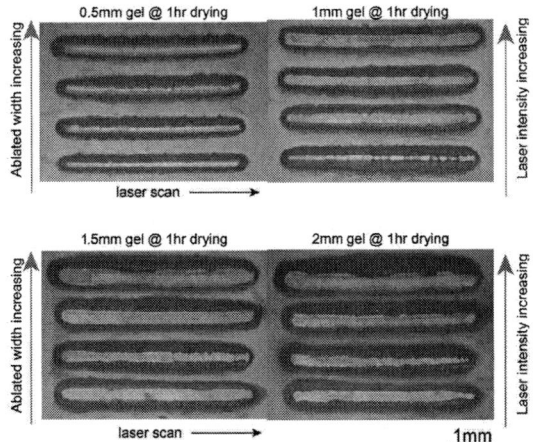

Fig. 2: Photographs of the 5 mm laser single ablation on hydrogels of 0.5 to 2 mm thick at 1~2 hours drying.

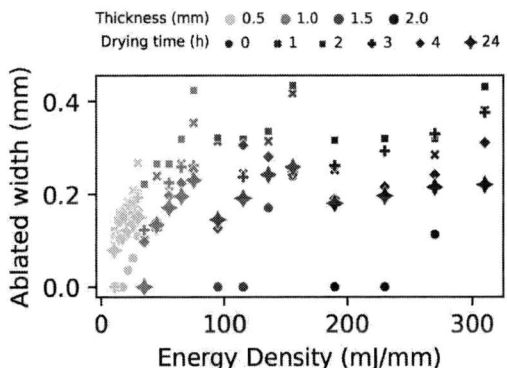

Fig. 3: The laser-ablated width as a combined function of laser beam intensity and fabrication timing, showing the laser micromachining able to generate a 30 μm to 440 μm wide thermal-ablation on hydrogels at a large range of 0.5 mm to 2 mm thick.

Fig. 3 presents a quantitative analysis of laser ablation in hydrogels, indicating a proportional relationship between the ablated width (mm) and the linear energy density (mJ/mm, defined as the laser beam power (W) divided by laser beam scan rate (mm/ms)). As the linear energy density increases, either increasing the laser power or decreasing the scanning rate (more heat accumulation), the ablated width increases; the converse is also true. This mechanism applies for hydrogels of 0.5 mm to 2 mm in thickness at a wide range of hydration states, i.e., hydrogel that was dried in ambient conditions for various durations between 0 h and 24 h. The results show that for hydrogels of thickness 0.5 mm to 2 mm, the laser micromachining enabled a tunable ablated width spanning a large range of 30 µm to 440 µm, Fig. 3. In order to achieve the largest ablated width with no thermal damage to the substrate, the 0.5 mm hydrogel required a linear energy density of 30 mJ/mm, which was further increased to 75 mJ/mm, 155 mJ/mm, and 310 mJ/mm by increasing the hydrogel thickness to 1 mm, 1.5 mm and 2 mm, individually. The proportional relationship between ablated width and laser intensity can be easily understood by the increase of hydrogel thickness, thus requiring more thermal energy-induced hydrogel evaporation.

As mentioned previously, the ablated width is also regulated by the time-dependent heat diffusivity of hydrogel. As shown in Fig. 4a, the maximum scalability occurred at 1 hour or 2 hours drying with a value of ablated width from 0.1 mm to 0.44 mm, which decreased to that of 0.1 mm to 0.3 mm at 4 hours drying and to the smallest one of 0.09 mm to 0.26 mm at 24 hours drying. Furthermore, Fig. 4b indicates the existence of an optimal processing window for laser micromachining hydrogels in terms of obtaining the largest ablated width within the least required drying time, which were 1 hour drying for hydrogel of thickness 0.5 mm or 2 hours drying for hydrogel of thickness 1 mm to 2 mm. Measurements of the heat diffusivity of the hydrogel show a hydrogel thickness-dependent evaporation dynamic; thinner hydrogels result in faster drying rates. Within the first two hours of drying, 0.5 mm thick hydrogel lost 60% of its water at a rate of 30 w%/hr, which reduced to 26 w%/hr, 22 w%/hr

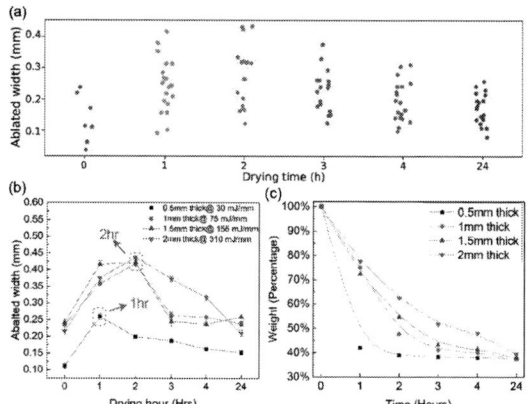

Fig. 4: (a-b) Achievable ablated widths for various drying times; the maximum width is achieved with 2 hours of drying. (c) The drying kinetics of hydrogel showing the weight lost as an exponential function of the reciprocal of the time.

and 18 w%/hr for hydrogels of 1 mm, 1.5 mm and 2 mm thickness, consequently resulting in 52%, 45%, and 37% weight loss, respectively, Fig. 4c. By relating the evaporation rate to the ablated width, it is reasonable to conclude that the optimal process window to achieve both the largest scalability and features of the ablated width is at 1~2 hours drying where the hydrogel loses 40%~60% of its water, depending on the initial thickness of the hydrogel.

The laser micro-machining quality was evaluated in terms of sidewall quality (S.Q), which can be defined as the ratio of ablated width to laser affected width (the sum of completely and partially removed hydrogel), Fig. 5a. The result was illustrated in Fig. 5b; for the hydrogels of 0.5 mm to 2 mm thickness, laser engraving can create a maximum S.Q of 0.45~0.55 mm/mm on average via 2~3 hours drying, 1.5~1.7 times greater than that of 0.3 mm/mm after 24 hours drying (fully dried hydrogel). Moreover, the S.Q can be further tuned by adjusting the linear energy density. For example, by reducing the linear energy density from 310 mJ/mm to 188 mJ/mm on 2 mm thick hydrogel, one can increase the S.Q from 0.51 mm/mm to 0.65 mm/mm.

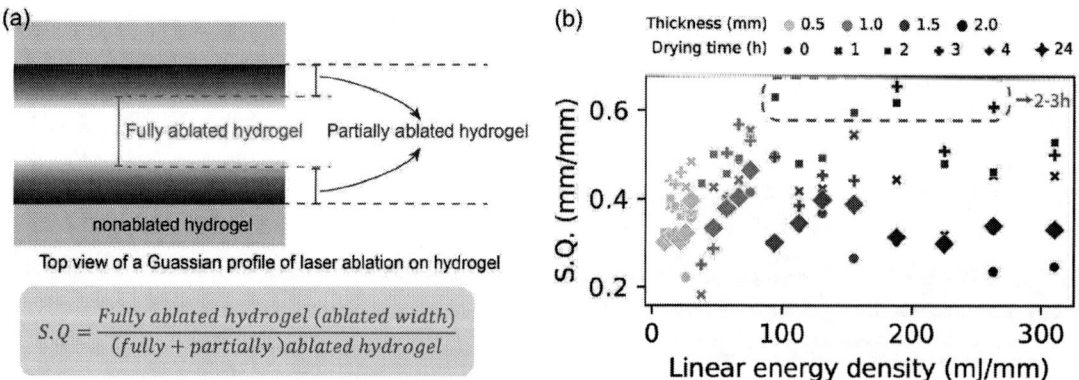

Fig. 5: (a) Demonstration of the S.Q of laser micro-machined hydrogels; (b) the result showing the largest S.Q of 0.61~0.65 mm/mm occurring at 2~3 hours drying for hydrogels of 1 mm to 2 mm thick.

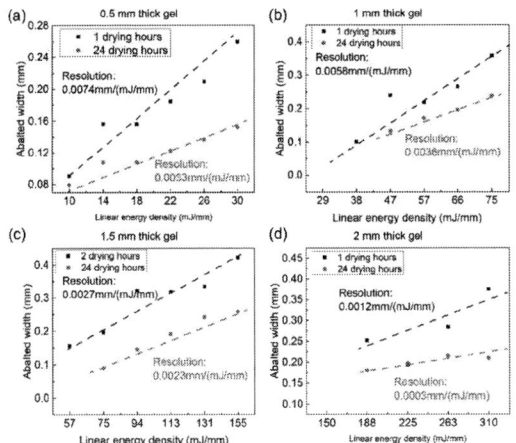

Fig. 6: Resolution analysis showing the comparison of resolution for laser micromachining hydrogels after 1~2 hour drying to that after 24 hours drying.

To demonstrate why it is important to apply the laser-machining on moisturized hydrogel instead of fully dried one, the resolution of laser-micromachining hydrogel was investigated in terms of the ratio of ablated width and linear energy density. In this regard, the resolution curvatures were compared at process times of 1~2 hours drying and of 24 hours drying. Fig. 6 presents the results. For 0.5 mm, 1 mm and 2 mm thick hydrogels, their maximum resolutions appeared at 1 hour drying, which were 0.0074, 0.0058 and 0.0012 mm/(mJ/mm) and 2.22, 1.59 and 4 times of 0.0033, 0.0036 and 0.0003 mm/(mJ/mm) at 24 hours drying, respectively; meanwhile, for 1.5 mm thick hydrogel, the maximum resolutions are 0.0027 mm/(mJ/mm) at 2 hours drying, which are 1.2 times of 0.0023 mm/(mJ/mm) at 24 hours drying. This comparison confirmed that laser engraving on hydrogels offers a rapid and temporally tunable micromachining process.

IV. CONCLUSION

We have developed a low-cost, rapid and tunable micromachining method, which uses the laser engraving system to micro-pattern the hydrogel in situ via tuning the laser beam profile and selecting an appropriate process timing. The protocol was demonstrated via laser machining hydrogels of four different thicknesses (0.5, 1, 1.5 and 2 mm) at different dehydration state (0-24 hours drying at room temperature) under a wide range of linear energy densities (10-300 mJ/mm) and within a single laser beam scan (5 mm ablation length). The results show that at 1~2 hours drying (the hydrogel loses 40%~60% weight) after the polymerization of hydrogels, a set of the best qualities, including the maximum scalability of the ablated width of 0.1 mm to 0.44 mm, the S.Q of 0.65 mm/mm and the resolution of 0.0074 mm/(mJ/mm), were achieved.

ACKNOWLEDGMENT

The authors would like to thank the help of Prof. Babak

Ziaie from Purdue university and Dr. Manuel Ochoa from DBA Ochoa Consulting for their technical support. This work is supported by H. Jiang's start-up fund from South China University of Technology.

REFERENCES

[1] R. A. Siegel, "Stimuli sensitive polymers and self regulated drug delivery systems: A very partial review," *J. Control. Release*, vol. 190, pp. 337–351, 2014.

[2] T. Garg, S. Singh, and A. Goyal, "Stimuli-sensitive hydrogels: an excellent carrier for drug and cell delivery," *Crit. Revews Ther. Drug Carr. Syst.*, vol. 30, no. 5, pp. 369–409, 2013.

[3] R. Masteiková, Z. Chalupová, Z. Sklubalová, and C. Z. Masteikova R, "Stimuli-sensitive hydrogels in controlled and sustained drug delivery.," *Medicina (Kaunas).*, vol. 39 Suppl 2, no. 2, pp. 19–24, 2003.

[4] J. H. Park, A. Kim, H. Jiang, S. H. Song, J. Zhou, and B. Ziaie, "A Wireless Chemical Sensing Scheme using Ultrasonic Imaging of Silica-Particle-Embedded Hydrogels (Silicagel)," *Sensors Actuators B Chem.*, vol. 259, pp. 552–559, 2018.

[5] H. Jiang *et al.*, "A Wireless Implantable Strain Sensing Scheme Using Ultrasound Imaging of Highly Stretchable Zinc Oxide/Poly Dimethylacrylamide Nanocomposite Hydrogel," *ACS Appl. Bio Mater.*, vol. 3, no. 7, pp. 4012–4024, 2020.

[6] J. H. Holtz and S. A. Asher, "Polymerized colloidal crystal hydrogel films as intelligent chemical sensing materials.," *Nature*, vol. 389, pp. 829–832, 1997.

[7] X. Zhao *et al.*, "Active scaffolds for on-demand drug and cell delivery," *Proc. Natl. Acad. Sci. U. S. A.*, vol. 108, no. 1, pp. 67–72, 2011.

[8] R. Bashir, J. Z. Hilt, O. Elibol, A. Gupta, and N. A. Peppas, "Micromechanical cantilever as an ultrasensitive pH microsensor," *Appl. Phys. Lett.*, vol. 81, no. 16, pp. 3091–3093, 2002.

[9] A. Baldi, Y. Gu, P. E. Loftness, R. A. Siegel, and B. Ziaie, "A hydrogel-actuated environmentally sensitive microvalve for active flow control," *J. Microelectromechanical Syst.*, vol. 12, no. 5, pp. 613–621, 2003.

[10] B. Ziaie, A. Baldi, M. Lei, Y. Gu, and R. A. Siegel, "Hard and soft micromachining for BioMEMS: Review of techniques and examples of applications in microfluidics and drug delivery," *Adv. Drug Deliv. Rev.*, vol. 56, no. 2, pp. 145–172, 2004.

[11] M. D. Tang, A. P. Golden, and J. Tien, "Molding of Three-Dimensional Microstructures of Gels," *J. Am. Chem. Soc.*, vol. 125, no. 43, pp. 12988–12989, 2003.

[12] M. Lei, Y. D. Gu, A. Baldi, R. A. Siegel, and B. Ziaie, "Soft mold-dry etch: A novel hydrogel patterning technique for biomedical applications," *Proc. 26th Annu. Int. Conf. Ieee Eng. Med. Biol. Soc. Vols 1-7*, vol. 26, pp. 1983-1986 5459, 2004.

[13] Z. Ding, A. Salim, and B. Ziaie, "Squeeze-film hydrogel deposition and dry micropatterning.," *Anal. Chem.*, vol. 82, no. 8, pp. 3377–82, 2010.

[14] F. Yanagawa, S. Sugiura, and T. Kanamori, "Hydrogel microfabrication technology toward three dimensional tissue engineering," *Regen. Ther.*, vol. 3, pp. 45–57, 2016.

[15] S. Mishra and V. Yadava, "Laser Beam MicroMachining (LBMM) - A review," *Opt. Lasers Eng.*, vol. 73, pp. 89–122, 2015.

[16] A. S. Holmes, "Laser fabrication and assembly processes for MEMS," *Proc. SPIE*, vol. 4274, pp. 297–306, 2001.

[17] D. Yuan and S. Das, "Experimental and theoretical analysis of direct-write laser micromachining of polymethyl methacrylate by CO2 laser ablation," *J. Appl. Phys.*, vol. 101, no. 2, p. 024901, 2007.

[18] X. Xie, D. Li, T.-H. Tsai, J. Liu, P. V. Braun, and D. G. Cahill, "Thermal Conductivity, Heat Capacity, and Elastic Constants of Water-Soluble Polymers and Polymer Blends," *Macromolecules*, vol. 49, no. 3, pp. 972–978, 2016.

[19] H. Jiang, M. Ochoa, F. Waimin, R. Rahim, and B. Ziaie, "A pH-regulated drug delivery dermal patch for targeting infected regions in chronic wounds," *Lab Chip*, vol. 19, pp. 2265–2274, 2019.

Proceedings of the 16th Annual IEEE International
Conference on Nano/Micro Engineered and Molecular Systems
April 25-29, 2021

Three-dimensional Graphene FETs for pH Detection

Tao Deng[#], Zhaohao Zhang, Yang Zhang, Yuning Li and Zewen Liu

Abstract—Three-dimensional (3D) graphene field-effect transistors (GFETs) with a microtubular structure were reported in this paper. The 3D GFET can be used to determine the pH value of analytes by directly measuring the Dirac point voltage. Experimental results demonstrated that the 3D GFET sensor has a sensitivity of 228 mV/pH and high linearity (R^2 = 0.9796). Considering that graphene can be functionalized by a lot of molecules, we expected to further develop 3D GFET sensors for more sensitive and specific biological and chemical detections.

I. INTRODUCTION

In recent years, the application of nanomaterials-based devices in the electrical detection of biological and chemical species has harvested a lot of attention in the fields of genomics, clinical diagnosis, due to its advantages of label-free and rapid detection, miniaturization, low cost and high sensitivity [1-4]. Graphene is an ideal nanomaterial to develop chemical and biological sensors because of its remarkable properties, such as ultrahigh carrier mobility, high electronic conductivity and ultra-large specific area [5-6]. The reported graphene sensors generally adopt graphene field-effect transistor (GFET) structures, especially liquid-gate GFET structures for liquid analytes [7-9]. When the biological or chemical species adsorb on the surface of the graphene, they will donate electrons to graphene or accept electrons from graphene, leading to conductance changes of graphene, thus, GFET can be used as biological or chemical sensors. To date, various GFET sensors have been proposed to detect pH [10-11], heavy metal ions [12], DNA and protein molecules [13-14], as well as COVID-19 causative virus [15]. However, most of the GFET sensors have a planar or two-dimensional (2D) architecture, due to the limitation of 2D lithography technology. In order to detect liquid analytes, the 2D GFET sensors have to integrate extra assistant facilities such as micro fluidic channels, cavities and pumps [9-10], which not only increases the complexity and cost, but also decreases the stability and accuracy.

Here, three-dimensional (3D) GFETs with a microtubular architecture were demonstrated, which offered a natural micro fluidic channel that could automatically suck the target analyte into the hollow microtube core to detect. Clear pH-dependent conductance characteristics were observed with 3D GFETs, a sensitivity of 228 mV/pH and high linearity (R = 0.9796) were obtained. Because of its high-sensitivity, simplicity, miniaturization, low cost and disposability, the 3D GFET shows great potential to be developed to a high-performance sensing platform for chemical and biological species.

II. EXPERIMENTS

A. Fabrication of 3D GFETS

A novel self-rolled-up approach was applied for fabricating the 3D GFETs [16]. Firstly, a 50 nm-thick Al sacrificial layer was sputtered on the silicon (Si) substrate, followed by compressive and tensile strained SiN_x layers were deposited partially on the Al layer. Then, a chromium/gold (Cr/Au) gate electrode with thicknesses of 10 and 50 nm, respectively, were sputtered on top of the SiN_x layers. After that, we deposited a silicon dioxide (SiO_2) dielectric layer with a thickness of 30 nm, to form a buried-gate structure. Subsequently, a single-layer CVD-grown graphene (purchased from ACS Material, LLC) was transferred onto the dielectric layer and patterned using the oxygen plasma etching. Fifthly, drain and source electrodes (Cr/Au, 10 nm/50 nm) were deposited on top of the graphene layer by the electron beam evaporation (EBE), and the 2D buried-gate GFET was completed, as displayed in Fig. 1a. Then, the Al sacrificial layer was selectively etched using a $FeCl_3$ solution, and the 2D GFET started to roll up driven by the strained SiN_x layers. Seventhly, the 3D GFET was completed (Fig. 1b). More information about the manufacturing processes could be found in our previous work [17]. The buried-gate 2D GFETs and rolled-up 3D GFETs were characterized by the Raman spectroscopy, as shown in Section III.

B. Electrical Testing and pH Detection

The transfer and output characteristic curves of the 3D GFETs were tested by using a probe station (Cascade Microtech., Summit 12000) and a semiconductor parameter analyzer (Keysight Tech., B1500A). The liquid analytes with different pH values were obtained by mixing a phosphate buffer solution at pH 7.4, a HCl solution and a NaOH solution. To detect the pH values, the liquid analytes were injected at one end of the 3D GFET microtube by using a tiny glass tube. The Dirac point voltage (V_{Dirac}) was defined as the buried-gate voltage (V_g) where the minimum drain-source current (I_{ds}) was obtained. We used the V_{Dirac} to determine the analyte's pH value. All of the experiments were conducted under the conditions of room temperature and atmospheric environment.

III. RESULTS AND DISCUSSIONS

A. Morphologies and Raman Spectra of 2D buried-gate GFETs and 3D rolled-up GFETs

The morphology of the 2D buried-gate GFET is shown in Fig. 1a, where five parallel devices were fabricated on the

---

T. Deng, Y. Zhang, Y. Li, are with the School of Electronic and Information Engineering, Beijing Jiaotong University, Beijing 100044, China (corresponding author (T. Deng) phone: 86-10-51683626; e-mail: dengtao@bjtu.edu.cn).

Z. Zhang is with the Institute of Microelectronics of Chinese Academy of Sciences (IMECAS), 100029 Beijing, China.

Z. Liu is with the School of Integrated Circuits, Tsinghua University, Beijing, 100084, China (e-mail: liuzw@tsinghua.edu.cn).

978-1-6654-3008-1/21 $31.00 © 2021 IEEE

same strained SiN_x layers. The dimensions (width × length) of the graphene conductive channel are 200 µm × 14 µm, and the drain and source electrodes have the same width of 116 µm. However, the width of the buried gate electrode is only 6 µm. By selectively etching the sacrificial layer, the strained SiN_x layers drive the 2D GFETs rolling up to form 3D GFETs, as displayed in Fig. 1b. It can be obviously seen that the five 3D GFETs shows pretty good uniformity. The zoomed-in micrograph shown in Fig. 1c indicates that the graphene conductive channel, source, drain and gate electrodes of a single 3D GFET are well-constructed. The diameter of the 3D rolled-up GFET microtube is around 58 µm, which can be easily controlled by adjusting the parameters of the SiN_x strained layers [17].

with each other tightly, forming an overlapping part. The 3D GFET's rolled-up winding number can be altered easily by varying the widths of the 2D GFET, as demonstrated in the previous work [17-18]. Fig. 2c displays a 6 × 3 array of the microtubes, including 90 devices. Since the manufacturing process of the 3D GFETs and the existing IC/MEMS processes are compatible , large-scale 3D devices can be produced with low cost. This feature is of importance for disposable and parallel sensing applications.

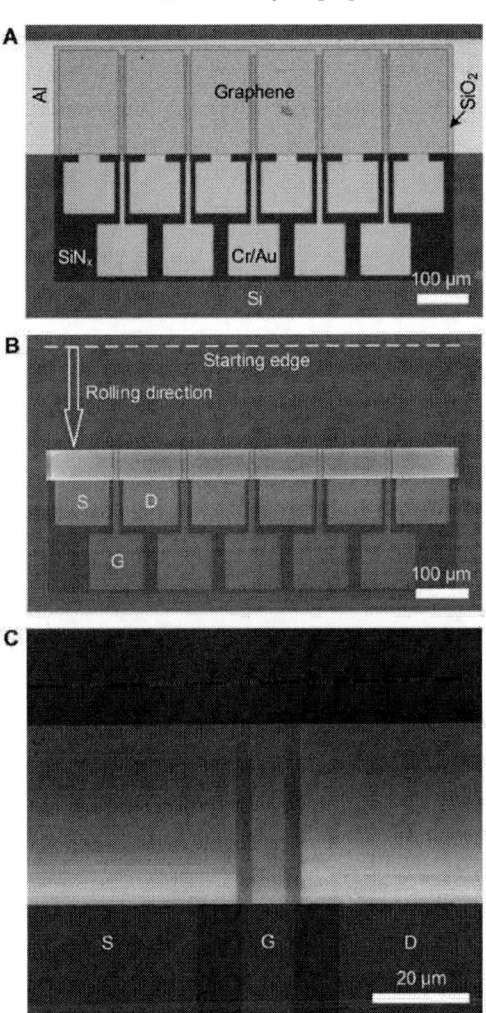

Figure 1. Mirographs of the 2D buried-gate GFETs and 3D rolled-up GFETs. (a) Optical micrograph of the 5 parallel 2D GFETs. (b) SEM micrograph of the 3D GFETs microtube after the rolled-up process. (c) Zoomed-in SEM mcirograph of the graphene channel of a single 3D rolled-up GFET.

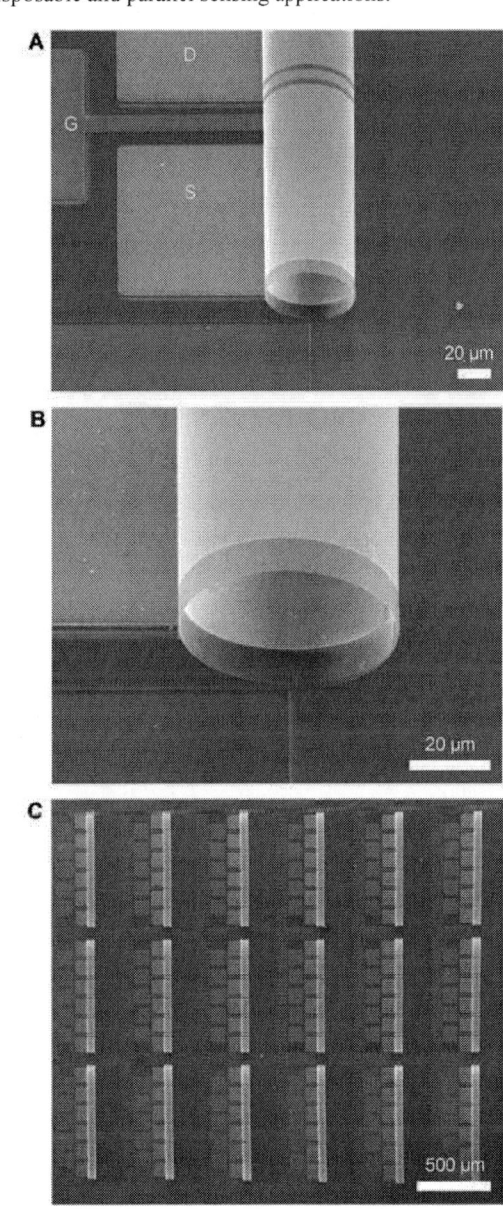

Figure 2. SEM mirographs of the 3D GFETs. (a) and (b) Side view and enlarged view of a 3D GFET. (c) 3D GFET microtubes array.

Fig. 2a shows the side view of the 3D rolled-up GFET and the hollow circular structure is obviously seen. The zoomed-in view in Fig. 2b illustrates that a complete circle is conformed, where the outside and inside walls contact

Figure 3 displays the Raman spectra of 2D and 3D GFETs. For the 2D structure, G band peak for the graphene locates at 1584.6 cm^{-1}, and 2D band peak locates at 2681.2 cm^{-1}. For the 3D structure, G band peak slightly shift to

978-1-6654-3008-1/21 $31.00 © 2021 IEEE

1585.5 cm^{-1}, while 2D band peak shift to 2679.8 cm^{-1}. These results verify that graphene exist in both structures. Moreover, the slight shifts of the two band peaks in the 3D graphene might be related to the strain originated from the unique rolled-up architecture [19, 20].

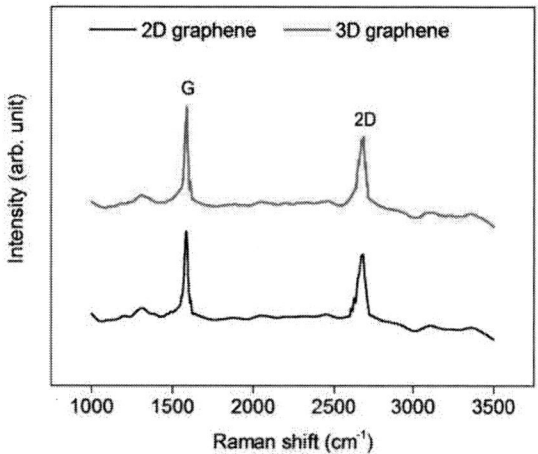

Figure 3.   Raman spectra for the graphene in 2D and 3D GFETs.

B. Electrical Properties of 3D GFETs

The transfer characteristics for the 3D rolled-up GFETs is displayed in Fig. 4a. The source-drain current (I_{ds}) firstly decreases and then increases as the buried gate voltage (V_{gs}) rises from -1 to 6 V, demonstrating a typical ambipolar behavior. The Dirac point voltage (V_{Dirac}) locates at 4.5 V, indicating a p-doped graphene conductive channel. In the reported back-gate 2D GFETs, a back gate voltage of tens to a hundred volt is always required to modulate the graphene conductance [21]. However, in our buried-gate 3D GFET a gate voltage of only several volts is required. This advantage is of importance in low voltage and energy-saving applications.

The output characteristics of the 3D rolled-up GFET are displayed in Fig. 4b. Under different V_{gs} varying from -1 to 3.5 V, the output I_{ds} increase linearly with the rising of the V_{ds}. This phenomenon indicates that the contacts between the graphene layer and the source/drain electrodes are ohmic contacts. At a certain value of bias voltage, the I_{ds} changes significantly with the V_{gs}, which indicates the 3D GFET has an excellent gate-control property.

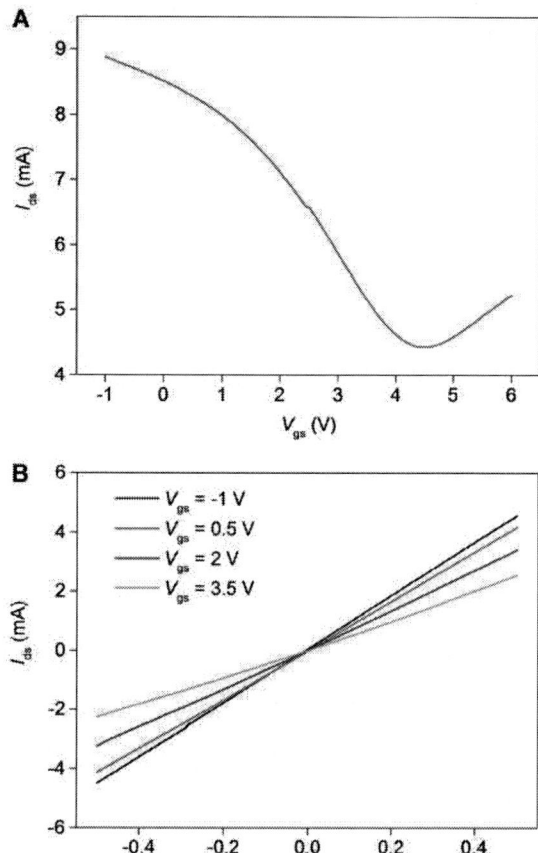

Figure 4.   Electrical characteristics of the 3D GFET. (a) Transfer characteristic curve. (b) Output characteristic curves.

C. pH Detection with 3D GFETs

Using the experimental set-up shown in Fig. 5a, the relationship between the 3D GFET's transfer characteristics and the pH values of analytes were investigated. Fig. 5b displays the I_{ds}-V_{gs} curves of a 3D GFET when it is filled with analytes with different pH values in the range of 5 to 9. With the increase in pH value, the Dirac point voltage (V_{Dirac}) of the 3D GFET shifted to the right. The phenomenon indicates the 3D GFET is capable to measure analyte's pH value via detecting the electrical properties.

Fig. 5c shows the value of the V_{Dirac} of the 3D rolled-up GFET against the pH value of analyte, indicating a linear relationship between the conductance of graphene and pH value over the range from 5 to 9. Similar phenomena have been observed in 2D liquid-gate GFETs [10] and two-terminal graphene devices [11] for pH sensing, which have been attributed to that the hydroxyl (OH$^-$) group's high concentration in the analyte enhanced the single-layer graphene's conductivity. The 3D GFET sensor demonstrates a sensitivity of 228 mV/pH and a high linearity ($R^2 = 0.9796$). This confirms the functionality of the 3D GFET as a high-performance sensing platform. Considering that 2D GFETs have been widely investigated as heavy mental ion sensors,

DNA sensors, protein sensors as well as virus sensors, the 3D GFETs would show great potentials in such fields.

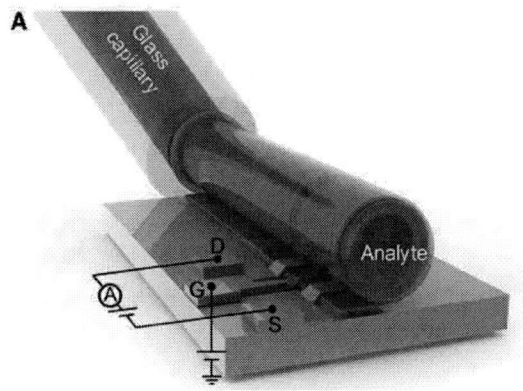

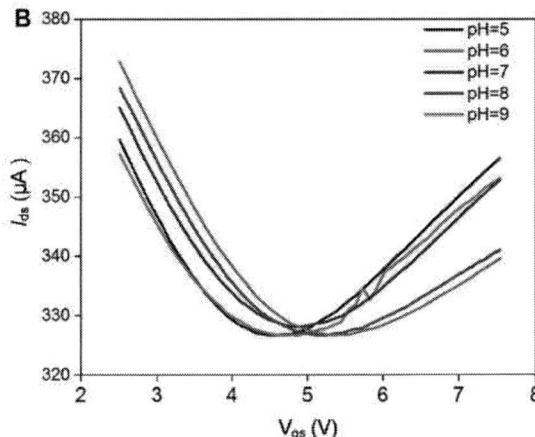

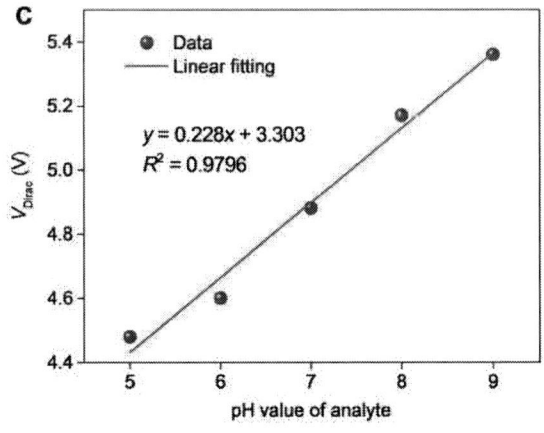

Figure 5. pH detection using the 3D GFET. (a) Schematic graph for the pH detecting experiment. (b) Transfer characteristic curves of the 3D GFET for various pH values in the range of 5-9. (c) The Dirac point voltage (V_{Dirac}) as a function of the analyte's pH value.

IV. CONCLUSION

A batch of 3D GFETs with microtubular architecture were manufactured using a technique of self-rolled-up. After the rolled-up process, the 3D graphene maintained the intrinsic properties of 2D graphene pretty well. We

demonstrated the pH value detection to show the functionality of the 3D GFET sensors. Analyte could be automatically sucked into the 3D GFET due to the large capillary force, and the analyte's pH value was determined by directly measuring of the Dirac point voltage. A sensitivity of 228 mV/pH and high linearity ($R^2 = 0.9796$) were obtained. As the 3D GFET has the characteristics of high sensitivity, simplicity, low cost, miniaturization and disposability, it has high potential to be used as high-performance sensing platforms for chemical and biological species.

ACKNOWLEDGMENT

The research was supported by the Beijing Natural Science Foundation (No. 4202062).

REFERENCES

[1] R. S. Andre, R. C. Sanfelice, A. Pavinatto, L. H. C. Mattoso, and D. S. Correa, "Hybrid nanomaterials designed for volatile organic compounds sensors: a review," *Materials & Design*, vol. 156, pp. 154–166, Oct. 2018.

[2] L. Syedmoradi, A. Ahmadi, M. L. Norton, and K. Omidfar, "A review on nanomaterial-based field effect transistor technology for biomarker detection," *Microchimica Acta*, vol. 186, no. 11, 739, Nov. 2019.

[3] V. Georgakilas, J. N. Tiwari, K. C. Kemp, J. A. Perman, A. B. Bourlinos, K. S. Kim, and R. Zboril, "Noncovalent functionalization of graphene and graphene oxide for energy materials, biosensing, catalytic, and biomedical applications," *Chemical Reviews*, vol. 116, no. 9, pp. 5464–5519, May 2016.

[4] W. J. Yin, J. Y. Sun, Y. Zhang, Y. Zhang, S. S. Li, M. Q. Zhu, H. Hong, Y. T. Ba, and T. Deng, "A novel three-dimensional Ag nanoparticles/rGO microtubular field effect transistor sensor for NO₂ detections," *Nanotechnology*, vol. 32, no. 2, 025304, Jan. 2021.

[5] D. R. Cooper, B. D'Anjou, N. Ghattamaneni, B. Harack, M. Hilke, A. Horth, N. Majlis, M. Massicotte, L. Vandsburger, E. Whiteway, and V. Yu, "Experimental review of graphene," *ISRN Condensed Matter Physics*, vol. 2012, pp. 1–56, 2014.

[6] A. K. Geim, and K. S. Novoselov, "The rise of graphene," *Nature Materials*, vol. 6, no. 3, pp. 183–191, Mar. 2007.

[7] N. Gao, T. Gao, X. Yang, X. C. Dai, W. Zhou, A. Q. Zhang, and C. M. Lieber, "Specific detection of biomolecules in physiological solutions using graphene transistor biosensors," *Proceedings of the National Academy of Sciences of the United States of America*, vol. 113, no. 51, pp. 14633–14638, Dec. 2016.

[8] R. Stine, S. P. Mulvaney, J. T. Robinson, C. R. Tamanaha, and P. E. Sheehan, "Fabrication, optimization, and use of graphene field effect sensors," *Analytical Chemistry*, vol. 85, no. 2, pp. 509–521, Jan. 2013.

[9] B. R. Goldsmith, L. Locascio, Y. N. Gao, M. Lerner, A. Walker, J. Lerner, J. Kyaw, A. shue, S. Afsahi, D. Pan, J. Nokes and F. Barron, "Digital biosensing by foundry-fabricated graphene sensors," *Scientific Reports*, vol. 9, 434, Jan. 2019.

[10] Y. Ohno, K. Maehashi, Y. Yamashiro, and K. Matsumoto, "Electrolyte-gated graphene field-effect transistors for detecting pH and protein adsorption," *Nano Letters*, vol. 9, no. 9, pp. 3318–3322, Sep. 2019.

[11] C. Y. Lee, K. F. Lei, S. W. Tsai, and N. M. Tsang, "Development of graphene-based sensors on paper substrate for the measurement of pH value of analyte," *BioChip Journal*, vol. 10, no. 3, pp. 182–188, Sep. 2016.

[12] K. Chen, G. Lu, J. Chang, S. Mao, K. Yu, S. Cui, and J. Chen, "Hg(ii) ion detection using thermally reduced graphene oxide decorated with functionalized gold nanoparticles," *Analytical Chemistry*, vol. 84, no. 9, pp. 4057–4062, May 2012.

[13] C. Zheng, L. Huang, H. Zhang, Z. Y. Sun, Z. Y. Zhang, and G. J. Zhang, "Fabrication of ultrasensitive field-effect transistor DNA biosensors by a directional transfer technique based on CVD-grown

graphene," *ACS Applied Materials and Interfaces*, vol. 7, no. 31, pp. 16953–16959, Aug. 2015.

[14] R. Hajian, S. Balderston, T. Tran, T. deBoer, J. Etienne, M. Sandhu, N. A. Wauford, J. Y. Chung, J. Nokes, M. Athaiya, J. Paredes, R. Peytavi, B. Goldsmith, N. Murthy, I. M. Conboy and K. Aran, "Detection of unamplified target genes via CRISPR-Cas9 immobilized on a graphene field-effect transistor," *Nature Biomedical Engineering*, vol. 3, no. 6, pp. 427–437, Jun. 2019.

[15] G. Seo, G. Lee, M. J. Kim, S. H. Baek, M. Choi, K. B. Ku, C. S. Lee, S. Jun, D. Park, H. G. Kim, S. J. Kim, J. O Lee, B. T. Kim, E. C. Park, and S. I. Kim, "Rapid detection of COVID-19 causative virus (SARS-CoV-2) in human nasopharyngeal swab specimens using field-effect transistor-based biosensor," *ACS Nano*, vol. 14, no. 4, pp. 5135–5142, Sep. 2020.

[16] J. Rogers, Y. G. Huang, O. G. Schmidt, and D. H. Gracias, "Origami MEMS and NEMS," *MRS Bulletin*, vol. 41, no. 2, pp. 123–129, Feb. 2016.

[17] T. Deng, Z. H. Zhang, Y. X. Liu, Y. X. Wang, F. Su, S. S. Li, Y. Zhang, H. Li, H. J. Chen, Z. R. Zhao, Y. Li, and Z. W. Liu, "Three-dimensional graphene field-effect transistors as high-performance photodetectors," *Nano Letters*, vol. 19, no. 3, pp. 1494–1503, Mar. 2019.

[18] S. S. Li, W. J. Yin, Y. N. Li, J. Y. Sun, M. Q. Zhu, Z. W. Liu, and T. Deng, "High sensitivity ultraviolet detection based on three-dimensional graphene field effect transistors decorated with TiO_2 NPs," *Nanoscale*, vol. 11, no. 31, pp.14912–14920, Aug. 2019.

[19] F. Ding, H. X. Ji, Y. H. Chen, A. Herklotz, K. Do̎rr, Y. F. Mei, A. Rastelli, and O. G. Schmidt, "Stretchable graphene: a close look at fundamental parameters through biaxial straining," *Nano Letters*, vol. 10, no. 9, pp.3453–3458, Sep. 2010.

[20] G. M. Mao, Q. Wang, Z. E. Chai, H. Liu, K. Liu and X. M. Ren, "Realization of uniaxially strained, rolled-up monolayer CVD graphene on a Si platform via heteroepitaxial InGaAs/GaAs bilayers," *RSC Advances*, vol. 7, no. 24, pp. 14481–14486, 2017.

[21] C. H. Liu, Y. C. Chang, T. B. Norris, and Z. H. Zhong, "Graphene photodetectors with ultra-broadband and high responsivity at room temperature," *Nature Nanotechnology*, vol. 9, no. 4, pp. 273–278, Apr. 2014.

Gap in pagination due to unavailable papers.

Pages 73-137

Keynote Speaker

Opportunities for Single-Molecule Electronics: A Ten-Year Perspective

Wenjing Hong

College of Chemistry and Chemical Engineering, State Key Laboratory of Physical Chemistry of Solid Surfaces
Xiamen University, 361005 Xiamen, China

Website : pilab.xmu.edu.cn

ABSTRACT

With the development and scaling downward of electronic devices, the dimension of electronic device has shrunk down to nanometer scale. Molecular electronics is an important subject of nanoelectronics, which will utilize rich varieties of organic and inorganic molecular materials as the functioning electronic component to realize the electronic behaviors as conventional electronic devices.

"More Moore": The original goal of molecular electronics is to build functional molecules from atomic level, and then to build electronic devices and circuits from the individual molecules. Towards this goal, quantum interference effect in the charge transport through single-molecule junction enables potential higher switching ratio and even lower subthreshold swing than traditional transistors when leads to the single-molecule transistors. However, the stability and the integration of multiple devices on chip remained as a major challenge for the fabrication of "molecular circuits".

"Beyond Moore": In recent years, we extended the application of single-molecule techniques to the quantitative analysis of the molecular physical-chemical process by counting the number of molecules. n this talk, I will share our recent effort towards quantitative analysis of the reaction rate, adsorption free energy, isomerization, the movement manipulation of molecules, and even the assembly of molecules at the scale of single or several molecules.

As perspective, recent advances of molecular electronics will lead to their device application on the field of sensors and thermoelectric devices in the next ten years.

BIOGRAPHY

Wenjing Hong is currently "Minjiang" Chair Professor of Chemical Engineering and Vice Dean in College of Chemistry and Chemical Engineering, Xiamen University. He is also the Assistant to Director (Prof. Zhong-qun Tian) in Tan Kah Kee Innovation Laboratory. Prof. Hong received his bachelor degree from Xiamen University, master degree from Tsinghua University, and Ph.D. degree from University of Bern, Switzerland. After his postdoctoral research at the University of Bern, Switzerland, he joined Xiamen University as a full professor in 2015. Prof. Hong's research mainly focused on single-molecule electronics, and the application of artificial intelligence for chemical engineering. 80+ peer-reviewed papers are published in top journals of chemistry and nanotechnology, including Nature Materials, 13 papers in Science Advances/Chem/Matter/Nature Communications, 17 papers in J. Am. Chem. Soc./Angew. Chem. Int. Ed., and 5 invited reviews in Chem. Soc. Rev., Acc. Chem. Res., Adv. Mater., etc. Besides the publications, the scientific instrument for single-molecule electronic characterization developped by his group has been used in 10+ research labs all over the world.

Prof. Hong was supported by the National Science Foundation for Excellent Young Scholars Program in 2017, he has undertaken serveral research projects such as the National Key Research and Development Program of China and the National Natural Science Foundation of China. In 2019, he was awarded the Chinese Chemical Society Award for Outstanding Young Chemist. In 2020, he won the second prize of Fok Ying Tung Education Foundation Young Faculty Award and the Fujian Province Youth Science and Technology Award.

Proceedings of the 16th Annual IEEE International
Conference on Nano/Micro Engineered and Molecular Systems
April 25-29, 2021

Synchronized Surface Modification of TiO_2 Composite Nanofiber Through Core-Shell Electrospinning

Hao Peng, Guoyi Kang, Jiaxin Jiang, Zungui Shao, Juan Liu, Yifang Liu, and Gaofeng Zheng*

Abstract— An online surface modification process based on core-shell electrospinning was provided to fabricate functional composite nanofibers. The polyethylene oxide (PEO) solution and titanium dioxide (TiO_2) suspension were provided through core and shell channel of spinneret respectively to obtain PEO/TiO_2 composite nanofibers. The PEO solution was stretched and solidified into nanofibers, and the TiO_2 nanoparticles were embedded into the PEO nanofibers. This method provided a good way to increase the distribution ratio of functional nanoparticles on the nanofiber surface. X-ray diffraction and Fourier transform infrared spectroscopy were used to confirm the materials component of composite nanofibers. The titanium mass content increased to 3.81 wt% by the core-shell electrospinning. The PEO/TiO_2 nanofibers enhanced the ultraviolet resistance furcation. This work demonstrates the possibility and validity of using coaxial electrospinning to modify the fiber surface at the same time of electrospinning process, it gives further insight into preparing composite nanofiber materials.

I. INTRODUCTION

In the past decades, electrospinning has attracted a great deal of attention due to the easy-control manufacturing of membrane with the small feature size, and extremely large specific surface area [1-3]. What's more, it is an efficient and economical way to fabricate composite nanofiber from organic materials [4, 5] and inorganic materials [6-8]. The inorganic nanoparticles were introduced to enhance the nanofibrous function of mechanical [9], physical [10], chemical [11-13], catalytic [2, 14]. However, the agglomeration behavior causes the obstacle that nanoparticles can't be distributed uniformly in the nanofiber so that hinder the high performance. A novel method for the efficient production of functional nanofibers is urgently required.

Hybrid electrospinning is one of the simple ways to obtain composite nanomaterials. Sultan Karagoz et al [15] fabricated PCL/TiO_2-Ag nanofibers by one-step electrospinning as a multifunctional material for the applications of surface-enhanced Raman scattering (SERS), photocatalysis and antibacterial. Liang et al [12] fabricated the electrospun PANI/MUCNT fibers on carbon paper as electrodes and obtained an excellent electrochemical performance. However, the problem of agglomeration still hinders the enhancement of the loading modified particles. Several post-processing methods have been developed for the preparation of composite nanofibers. Yu et al [16] used one-step in situ polymerization method to co-dope TiO_2,

sulfosalicylic (SSA), sodium dodecyl benzene sulfonate (SDBS) with polyaniline (PANI). Lu et al [17] combined electrospinning and hydrothermal reaction to obtain brushed-shaped PI/ZnO and (PI/Ag)/ZnO nanofiber. Lu et al [10] prepared electrospun C/Ni nanofiber film electrodes by reducing nickel acetate to nickel via low-temperature calcination. A good doping distribution can be obtained by indirect methods, but the excellent properties of electrospun fibers may destroyed during the complex process, so as to waste the raw materials. Thus, it is necessary to explore a simple and efficient process to realize the modification of electrospun fibers.

In this paper, the PEO/TiO_2 composite nanofiber membrane was prepared by coaxial electrospinning for UV protection. The TiO_2 nanoparticles, which was delivered through the core channel of spinneret distributed uniformly on the surface of the PEO nanofibers. The surface distribution ratio can be increased so that to enhance the functional performance. Thanks to the absorption of UV light and homogeneous three-dimensional distribution of TiO_2 nanoparticles, the PEO/TiO_2 electrospun nanofiber film has excellent shielding effect on UV light. This work providesd a novel strategy for the high efficiency production of functional nanofibers.

II. EXPERIMENTAL DETAILS

A. Materials

Polyethylene oxide (PEO) with an average molecular weight of 300,000 g/mol was purchased from Changchun Earth Fine Chemical Co., LTD. Nanosized titanium dioxide (analytical grade, 100 nm) was purchased from Aladdin China. Ethanol (C_2H_5OH, $\geq$ 99.7%) was purchased from Sinopharm Chemical Reagent Co., LTD. All the solvents were used without further purification.

B. Electrospinning process

The preparation process of nanofibers is illustrated in Fig. 1. Polyethylene oxide (PEO) solution with concentration of 8 wt% were dissolved in the mixture solvent of deionized water and ethanol (v: v=1:3). The blended solution was stirred at 60°C over 8 hours and used as the core materials. The TiO_2 nanoparticles were dissolved in the ethanol as shell materials and the mass concentration of TiO_2 nanoparticles were 0.5 wt%, 1 wt%, 2 wt% and 3 wt%, respectively. The TiO_2 dispersions were ultrasound-treated for 30 mins. As described in Fig. 1, the two kind solutions were delivered through the core and shell channel

---

*Resrach supported by the National Natural Science Foundation of China (No. 61772441), Science and Technology Planning Project of Fujian Province (No 2020H6003), Xiamen Municipal Science and Technology Project (No. 3502Z20193015), Natural Science Foundation of Guangdong Province (No. 2018A030313522), Science and Technology Planning Project of Shenzhen Municipality in China (No. JCYJ20180306173000073).

Hao Peng, Guoyi Kang, Jiaxin Jiang, Zungui Shao, Juan Liu, Yifang Liu and Gaofeng Zheng are with the Department of Instrumental and Electrical Engineering, Xiamen University, Xiamen 361102, China and the Shenzhen Research Institute of Xiamen University, Shenzhen 518000, China (corresponding author: Gaofeng Zheng, phone: +86-592-2194957; fax: +86-592-2182221; e-mail: zheng_gf@xmu.edu.cn).

978-1-6654-3008-1/21 $31.00 © 2021 IEEE

of spinneret respectively to obtain PEO/TiO$_2$ composite nanofibers.

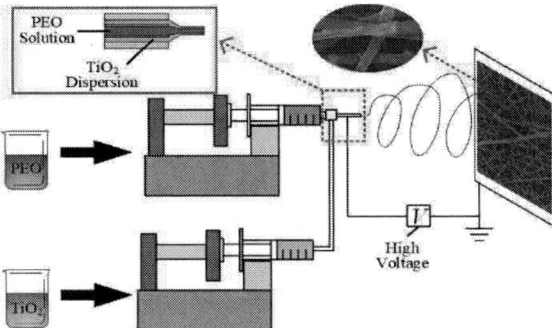

Figure 1. The schemiatic diagram of synchronized surface modification

As shown in TABLE I, six experiment groups were designed to achieve composite nanofiber with different components. These experiments were operated in the electrospinning equipment (NLM-0001, Xiamen Narai Technology Co. LTD, Xiamen, China). The coaxial spinneret comprised of core and shell channel, of which the diameter was 0.4 mm and 1.0 mm, respectively. The polymer solution was delivered to the inner channel of coaxial spinneret by a syringe pump, while the TiO$_2$ dispersions were delivered to the outer channel by another syringe pump. The flow rate was 300 μL/h and 100 μL/h for the inner and outer solution, respectively.

TABLE I.   COMPONENTS OF THE NANOFIBERS

Sample Name	Core	Shell
P	8 wt% PEO	--
P-T^1	8 wt% PEO+1 wt% TiO$_2$ mixed solutions	--
P/ T^1	8 wt% PEO	1 wt% TiO$_2$ dispersions
P/ T^2	8 wt% PEO	2 wt% TiO$_2$ dispersions
P/ T^3	8 wt% PEO	3 wt% TiO$_2$ dispersions
P-T^1/T^1	8 wt% PEO+1 wt% TiO$_2$ mixed solutions	1 wt% TiO$_2$ dispersions
P-T^1/T^2	8 wt% PEO+1 wt% TiO$_2$ mixed solutions	2 wt% TiO$_2$ dispersions
P-T^1/T^3	8 wt% PEO+1 wt% TiO$_2$ mixed solutions	3 wt% TiO$_2$ dispersions

The substrate was placed under the needle tip, and the distance between the tip and the substrate was 15 cm. The nanofibers were uniformly collected on the substrate. The electrospinning voltage was adjusted in the range of 10-15 kV to drive the jet, which was provided by a high voltage power supply. Then, the electrospinning composite membranes were vacuum-dried for 4 hours at 40°C to remove the residual solvent. Composite nanofibrous membrane were collected for 60 mins. All the experiments were carried out at ambient conditions.

C. Characterization

The surface morphologies of the samples were observed by a field emission scanning electron microscope (SEM,

Supar 55 Sapphire, Carl Zeiss Co. LTD., Germany). The chemical composition was determined by using Energy Dispersive Spectrometry (EDS, Oxford X-Max^N-80, Germany) for all the samples. The phase analysis of the nanofibers was taken by X-ray diffraction (XRD, XRD-7000, Shimadzu Co. LTD., Japan) at 2θ = 10-100°.

D. Anti-ultraviolet performance tests

Anti-ultraviolet performance of all the samples were evaluated by the electrospinning equipment, as shown in Fig. 2. All the samples were placed on a transparent substrate. The ultraviolet light was kept by an UV light (SHENYU V2), which can irradiate with a wavelength of 300-400 nm. The ultraviolet intensity was quantified by an ultraviolet photometer (TENMARS TM-213), which was 400 mm away from the UV light. The processed samples were taken between the UV light and the ultraviolet photometer while the distance between samples and the UV light was 180 mm. The anti-ultraviolet performance can be detected through the transmitted UV light.

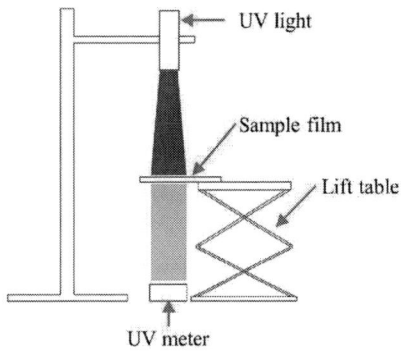

Figure 2. The anti-ultraviolet performance tests equipment

III. RESULTS AND DISCUSSION

A. Morphological characterization

The composite nanofibers were shown in Fig. 3. It can be seen that the fiber surface of PEO sample without adding TiO$_2$ is smooth. With the addition of TiO$_2$ nanoparticles, the surface of the fiber became rough, and TiO$_2$ nanoparticles with rice-granular distribution were observed on the fiber. Compared with the P-T nanofiber sample, the agglomeration of TiO$_2$ nanoparticles on the P/T nanofiber samples prepared by coaxial electrospinning was less obvious, the distribution of TiO$_2$ nanoparticles is more uniform, indicating that coaxial electrospinning was a simple and efficient method to reduce the agglomeration and to increase the content of TiO$_2$ nanoparticles. Hence, coaxial electrospinning is a promising method for preparing composite nanofiber materials.

Energy-dispersive spectroscopy (EDS) mapping was performed on P/T^3 samples to determine the presence of TiO$_2$ nanoparticle in the nanofibers, as shown in Fig. 4. The distribution of Ti elements was marked in red through the image. The green points indicating the distribution of carbon (C) elements matched with the PEO nanofiber image, while both the PEO and TiO$_2$ contain oxygen(O) functional groups, which is reflected in the distribution diagram of O elements (blue). The core-shell structures prepared by coaxial electrospinning contained uniformly dispersed TiO_2 nanoparticles.

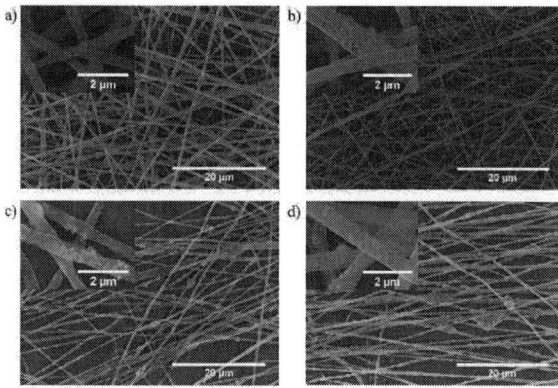

Figure 3.   Typical SEM images of (a) P, (b) P-T^1, (c) P/ T^3, (d) P-T^1/T^3 nanofiber membranes

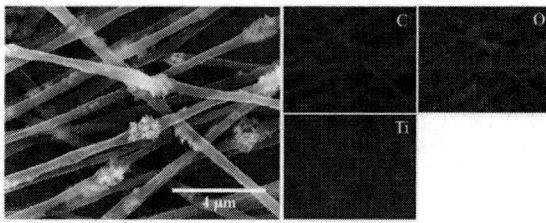

Figure 4.   C (green), O (blue), and Ti (red) EDS mapping of PEO/TiO$_2$ nanofiber membranes

Energy-dispersive spectroscopy (EDS) was performed to determine the presence of TiO_2 nanoparticles in the nanofibers. The results were showed in TABLE II. The content of Ti element was increased to 3.81 wt% (P-T^1/T^3) by coaxial electrospinning, which was three times than that in the sample (P-T^1) by hybrid electrospinning. Among the samples made by coaxial electrospinning, with the increasing of concentration of TiO_2 dispersion, the content of Ti element in the membranes increased. The results of elemental composition were agreed with the SEM and EDS mapping results.

TABLE II.   ELEMENTAL COMPOSITIONS OF ELECTROSPUN NANOCOMPOSITE MEMBRANES

Sample	Elemental composition as determined from EDS (weight %)		
	C	O	Ti
P	57.86	42.14	-
P-T^1	55.08	43.66	1.26
P/ T^1	55.86	42.71	1.43
P/ T^2	55.15	43.03	1.82
P/ T^3	54.01	43.73	2.26
P-T^1/T^1	53.81	43.10	3.09
P-T^1/T^2	54.85	41.80	3.35
P-T^1/T^3	51.89	44.30	3.81

XRD analysis of the P/T^3 nanofiber membranes was shown in Fig. 5. The characteristic peaks of PEO and TiO_2 nanoparticles can be clearly seen in the figure, of which the two distinctive peaks highlighted between $2\theta = 15 \sim 25°$

belong to amorphous PEO [18] and the diffraction arrange of these two characteristic peaks were $2\theta = 19.1°$ and $23.4°$ respectively. It can be identified that the main characteristic peaks and the corresponding lattice planes of rutile phase of TiO_2 nanoparticles were assigned at $27.5°$ (110), $36.2°$ (101), $41.2°$ (111), $44.1°$ (210), $54.4°$ (211), $56.7°$ (220), $62.8°$ (002) and $69.1°$ (301) [19].

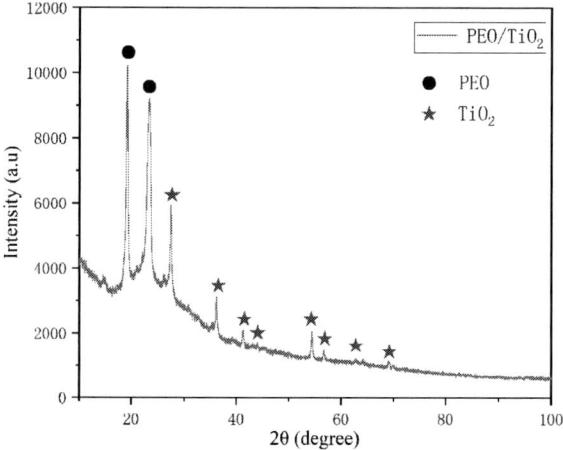

Figure 5.   XRD patterns of PEO/TiO$_2$ Nanofiber

B.   Anti-ultraviolet performance tests

The capacity of UV resistance for different membranes was quantified by the UV meter. The results were shown in Fig. 6. With the effect of TiO_2 nanoparticles, the UV resistance of the membranes had been improved to varying degrees. The UV intensity received by the UV meter in the absence of shade was 3600 $\mu W/cm^2$. It can be clearly seen that the UV resistance of electrospinning samples improved significantly with the addition of TiO_2 nanoparticles. Combined with the results of EDS analysis, the enhancement of TiO_2 content promotes the improvement of UV shielding performance. The PEO, PEO-TiO_2, PEO/TiO_2 and PEO-TiO_2/TiO_2 nanofiber were electrospun for 60 mins. Nearly 10% of UV rays passed through pure PEO fiber sample, meanwhile, this value decreased to 1.6% and 1.1% for P/T^3 sample and P^1/T^3 sample respectively. Compared to the UV intensity (134 $\mu W/cm^2$) through the hybrid electrospinning sample (P-T^1), the coaxial electrospinning samples (P/T^3 and P-T^1/T^3) exhibited lower UV transmittance (61 $\mu W/cm^2$ and 40 $\mu W/cm^2$). Combined with the SEM image and EDS analysis, we believed that the coaxial samples exhibit excellent UV resistance through the following two points: I) the enhancement of TiO_2 nanoparticle content in fibers; II) the distribution of TiO_2 nanoparticles on the surface of nanofibers.

978-1-6654-3008-1/21 $31.00 © 2021 IEEE

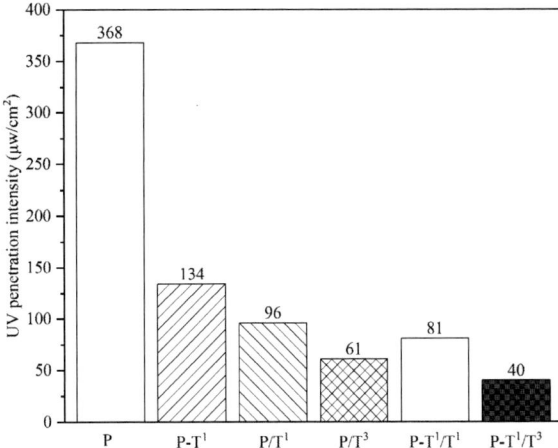

Figure 6. The UV resistance of P, P-T^1, P/T^1, P/T^3, P-T^1/T^1, P-T^1/T^3 membranes electrospun for 60 mins.

IV. CONCLUSIONS

In this study, PEO/TiO$_2$ nanofibers were successfully prepared by core-shell electrospinning, which has excellent UV resistance by adding TiO$_2$ nanoparticles. The problem of nanoparticle agglomeration can be overcome by coaxial electrospinning to gain special rough fiber morphology. With the content increasing and uniform distribution of TiO$_2$ nanoparticles, the nanofiber membranes have superior UV resistance compared to the membranes prepared by conventional hybrid electrospinning. Hence, the PEO/TiO$_2$ nanofibers made by core-shell electrospinning could be useful in many industrial applications.

ACKNOWLEDGMENT

This research was financially supported by the National Natural Science Foundation of China (No. 61772441), Science and Technology Planning Project of Fujian Province (No 2020H6003), Xiamen Municipal Science and Technology Project (No. 3502Z20193015), Natural Science Foundation of Guangdong Province (No. 2018A030313522), Science and Technology Planning Project of Shenzhen Municipality in China (No. JCYJ20180306173000073).

REFERENCES

[1] Z.-M. Huang, Y. Z. Zhang, M. Kotaki, and S. Ramakrishna, "A review on polymer nanofibers by electrospinning and their applications in nanocomposites," Composites Science and Technology, vol. 63, no. 15, pp. 2223-2253, 2003, doi: 10.1016/s0266-3538(03)00178-7.

[2] S. Zhu and L. Nie, "Progress in fabrication of one-dimensional catalytic materials by electrospinning technology," Journal of Industrial and Engineering Chemistry, vol. 93, pp. 28-56, 2021, doi: 10.1016/j.jiec.2020.09.016.

[3] V. Varkey, E. Tomlal Jose, and U. S. Sajeev, "Electrospinning technique for the fabrication of poly(styrene-co-methyl methacrylate) nanofibers and the effect of fiber diameter on UV–Visible absorption and thermal properties," Materials Today: Proceedings, vol. 33, pp. 2077-2081, 2020, doi: 10.1016/j.matpr.2020.01.591.

[4] D. P. Bhattarai, M. H. Kim, H. Park, W. H. Park, B. S. Kim, and C. S. Kim, "Coaxially fabricated polylactic acid electrospun nanofibrous scaffold for sequential release of tauroursodeoxycholic acid and bone morphogenic protein2 to stimulate angiogenesis and bone regeneration," Chemical Engineering Journal, vol. 389, 2020, doi: 10.1016/j.cej.2019.123470.

[5] N. O. San Keskin, A. Celebioglu, O. F. Sarioglu, T. Uyar, and T. Tekinay, "Encapsulation of living bacteria in electrospun cyclodextrin ultrathin fibers for bioremediation of heavy metals and reactive dye from wastewater," Colloids Surf B Biointerfaces, vol. 161, pp. 169-176, Jan 1 2018, doi: 10.1016/j.colsurfb.2017.10.047.

[6] J. Wang, S. Yao, Y. Ma, and W. Liu, "Electrode polarity effects in electrospinning organic/inorganic hybrid nanofibers," Ceramics International, vol. 47, no. 3, pp. 4352-4356, 2021, doi: 10.1016/j.ceramint.2020.09.285.

[7] J. G. Lee, O. S. Jeon, J.-H. Myung, and Y. G. Shul, "One-step fabrication of surface-decorated inorganic nanowires via single-nozzle electrospinning," Ceramics International, vol. 44, no. 10, pp. 11858-11861, 2018, doi: 10.1016/j.ceramint.2018.03.280.

[8] C. F. Armer, J. S. Yeoh, X. Li, and A. Lowe, "Electrospun vanadium-based oxides as electrode materials," Journal of Power Sources, vol. 395, pp. 414-429, 2018, doi: 10.1016/j.jpowsour.2018.05.076.

[9] C. Chen, Y. He, G. Xiao, Y. Wu, Z. He, and F. Zhong, "Zirconia doped in carbon fiber by electrospinning method and improve the mechanical properties and corrosion resistance of epoxy," Progress in Organic Coatings, vol. 125, pp. 420-431, 2018, doi: 10.1016/j.porgcoat.2018.09.027.

[10] J. Lu et al., "Super flexible electrospun carbon/nickel nanofibrous film electrode for supercapacitors," Journal of Alloys and Compounds, vol. 774, pp. 593-600, 2019, doi: 10.1016/j.jallcom.2018.09.383.

[11] Z. Wei, Y. Ren, H. Zhao, M. Wang, and H. Tang, "Controllable preparation and synergistically improved catalytic performance of TiC/C hybrid nanofibers via electrospinning for the oxygen reduction reaction," Ceramics International, vol. 46, no. 16, pp. 25313-25319, 2020, doi: 10.1016/j.ceramint.2020.06.325.

[12] J. Liang, S. Su, X. Fang, D. Wang, and S. Xu, "Electrospun fibrous electrodes with tunable microstructure made of polyaniline/multi-walled carbon nanotube suspension for all-solid-state supercapacitors," Materials Science and Engineering: B, vol. 211, pp. 61-66, 2016, doi: 10.1016/j.mseb.2016.04.014.

[13] X. Li, Y. Chen, H. Huang, Y.-W. Mai, and L. Zhou, "Electrospun carbon-based nanostructured electrodes for advanced energy storage – A review," Energy Storage Materials, vol. 5, pp. 58-92, 2016, doi: 10.1016/j.ensm.2016.06.002.

[14] K. Huang et al., "Catalytic behavior of electrospinning synthesized La0.75Sr0.25MnO3 nanofibers in the oxidation of CO and CH4," Chemical Engineering Journal, vol. 244, pp. 27-32, 2014, doi: 10.1016/j.cej.2014.01.056.

[15] S. Karagoz et al., "Synthesis of Ag and TiO$_2$ modified polycaprolactone electrospun nanofibers (PCL/TiO$_2$-Ag NFs) as a multifunctional material for SERS, photocatalysis and antibacterial applications," Ecotoxicol Environ Saf, vol. 188, p. 109856, Jan 30 2020, doi: 10.1016/j.ecoenv.2019.109856.

[16] J. Yu et al., "Cotton fabric finished by PANI/TiO$_2$ with multifunctions of conductivity, anti-ultraviolet and photocatalysis activity," Applied Surface Science, vol. 470, pp. 84-90, 2019, doi: 10.1016/j.apsusc.2018.11.112.

[17] F. Lu, J. Wang, Z. Chang, and J. Zeng, "Uniform deposition of Ag nanoparticles on ZnO nanorod arrays grown on polyimide/Ag nanofibers by electrospinning, hydrothermal, and photoreduction processes," Materials & Design, vol. 181, 2019, doi: 10.1016/j.matdes.2019.108069.

[18] C. Lu et al., "Thermal conductivity of electrospinning chain-aligned polyethylene oxide (PEO)," Polymer, vol. 115, pp. 52-59, 2017, doi: 10.1016/j.polymer.2017.02.024.

[19] S.-I. Oh, J.-C. Kim, M. A. Dar, and D.-W. Kim, "Synthesis and characterization of uniform hollow TiO$_2$ nanofibers using electrospun fibrous cellulosic templates for lithium-ion battery electrodes," Journal of Alloys and Compounds, vol. 800, pp. 483-489, 2019, doi: 10.1016/j.jallcom.2019.06.048.

Proceedings of the 16th Annual IEEE International
Conference on Nano/Micro Engineered and Molecular Systems
April 25-29, 2021

Precursor-Based ZnO Nano Inks for Printed Electronics*

Wenxin Liu, *Student Member, IEEE*, Hao Dong, Di Wang, Xing Chen

Abstract— **Printed electronics is a burgeoning additive manufacturing technique, which enabled extensive applications in micro/nano fabrication and flexible electronics. As a common printing material, metal oxide is employed in a variety of sensors and logic devices. However, synthesis of metal oxide ink with minimized particle size, uniform dispersion and appropriate rheological properties remains a challenge. In this study, we proposed two precursor-based inks for the printing of zinc oxide in nanoscale. The inks were tested using inkjet and dispensing printing method, respectively. The printed ZnO films showed porous nanostructures, which accommodates to the demands in sensor applications. Moreover, this study is also a demonstration of direct printing of metal oxide materials using precursor-based ink systems.**

I. INTRODUCTION

As a burgeoning additive manufacturing technique, printed electronics is simple, low-cost and eco-friendly, so that enabled substantial applications in micro/nano fabrication and flexible electronics. Nowadays, dispensing and inkjet printing are two extensively used methods for printed electronics, and their principles are described in Fig. 1. Dispensing printing makes a continuous deposition of materials by using air pressure to squeeze the ink from the container out of the needle. In contrast, inkjet printing is a "discrete" material deposition method. With a slight deformation of the piezocrystal in the inkjet head, the droplets were jetted, so as the to form the printed pattern jointly. On the other hand, as a common printed electronic material, metal oxide is a kind of inorganic semiconductor that utilized in sensors and logic device fabrication.

Recent studies of metal oxide printing intensively focused on the optimization of two types of ink system, including the direct dispersion of metal oxide nanoparticles [1], and the soluble precursors [2]. However, synthesis of nano ink with minimized particle size, uniform dispersion and appropriate rheological properties remains a challenge. Problems such as agglomeration of nano particles and nozzle clogging still occur during printing process. To figure out these issues, precursor-based nano ink is a promising method due to the small and homodispersed particles in molecular scale. Electrostatic repulsion exists between the particles so that the ink clogging could be prevented [3].

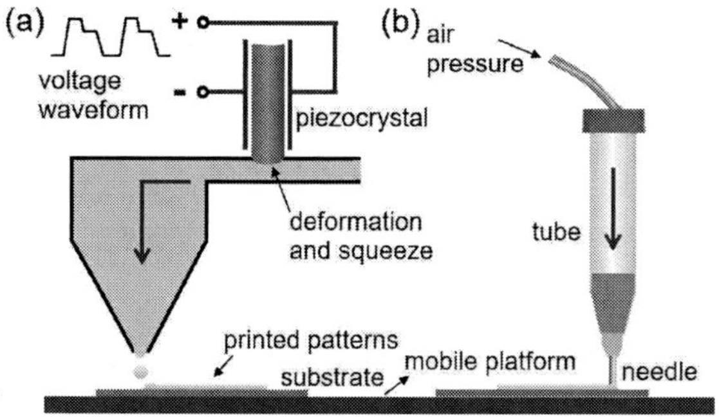

Fig. 1. Working principles of (a) inkjet printing and (b) dispensing printing.

However, physical properties of the precursor inks (e.g., viscosity, surface tension and volatility) should be further optimized to ensure the printability and avoid the quick gelation during printing.

In the current work, we proposed two precursor-based inks for the printing of zinc oxide in nanoscale. The ink system based on sol-gel process were printed on rigid substrates like Si and SiO_2 wafers, glass and ceramics, which could resist high temperatures in post-treatments. The other ink system based on solution process required a lower post-heating temperature, enabling the fabrication in flexible substrates with poor thermal stability. Both ink systems could be applied in dispensing printing, while the sol-gel precursor was further tested in inkjet printing with a higher precision. The printed ZnO films using both inks showed porous nanostructures, which were compatible with a gas or humidity sensing layer, and accommodated to the demands of sensor applications.

II. MATERIALS AND METHODS

A. Preparation of precursor inks

Two different precursor systems of ZnO were used for ink preparation. For the inks based on the sol-gel transition, zinc acetate dihydrate ($Zn(AC)_2{\cdot}2H_2O$, ZAD) was dissolved in isopropanol (C_3H_8O, IPA) with a concentration of 0.25 mol L^{-1}. Monoethanolamine (C_2H_7NO, MEA) was then

---

*Research supported by a Zhejiang Provincial Natural Science Foundation of China (No. LQ21F010003 to H.D), a China Postdoctoral Science Foundation (No. 2020M681952 to H.D) and Youth Science Fund Project of Zhejiang Lab (No. 2020MC0AA04 to H.D).

W. Liu is with the Key Laboratory of Biomedical Engineering of Ministry of Education, Zhejiang University, Hangzhou 310027, China, and Research Center for Intelligent Sensing, Zhejiang Lab, Hangzhou 311100, China. (e-mail: liuwx@zju.edu.cn).

H. Dong is with the Research Center for Intelligent Sensing, Zhejiang Lab, Hangzhou 311100, China, and Key Laboratory of Biomedical

Engineering of Ministry of Education, Zhejiang University, Hangzhou 310027, China. (corresponding author, phone: 0571-56390671; e-mail: cnhaodong@zhejianglab.com).

D. Wang is with the Research Center for Intelligent Sensing, Zhejiang Lab, Hangzhou 311100, China. (e-mail: diwang@zhejianglab.com).

X. Chen is with the Key Laboratory of Biomedical Engineering of Ministry of Education, Zhejiang University, Hangzhou 310027, China. (corresponding author, phone: 0571-87951183; e-mail: cnxingchen@zju.edu.cn).

978-1-6654-3008-1/21 $31.00 © 2021 IEEE

Fig. 2. The synthesis and preparation of (a) ink 1-1/1-2 and (b) ink 2.

added dropwise as a stabilizer while stirring. The prepared sol was aged for 24 hours and was observed to be transparent as DI water. The formation of colloid was indicated while the Tyndall Effect was shown. For dispensing printing, we employed glycerin (GLY) in a 1:3 volume ratio to increase the viscosity and surface tension of the sol. Meanwhile, the high boiling point of glycerin prevented the quick evaporation of the ink, which would lead to the blocking of the needle. This ink was marked as ink 1-1. For inkjet printing, the sol was mixed with ethylene glycol (EG) in a 1:3 volume ratio. A fluorocarbon surfactant (FC-4430, 3M) of 0.05 wt% was added to lower its surface tension and thus make it more printable. This ink was marked as ink 1-2.

The other precursor system was based on solution process. Zinc nitrate hexahydrate ($Zn(NO_3)_2 \cdot 6H_2O$, ZN) was dissolved in DI water with a concentration of 1.0 mol L^{-1}. 10 mL of 2.5 mol L^{-1} sodium hydroxide (NaOH) was then added dropwise to 15 mL of this solution under vigorous stirring. The mixed solution became turbid and was centrifuged, followed by the washing with DI water for 5 times to obtain the precipitate. 50 mL of 6.6 mol L^{-1} ammonia (NH_3, aq) was employed to dissolve the hydrated precipitate. After 6 hours of stirring, the supernatant was reserved to form a precursor solution with a concentration of 0.10 mol L^{-1} Zn. This solution was mixed with glycerin in a 1:1 volume ratio to form an ink marked as ink 2 for dispensing printing. The synthesis and preparation of the inks are illustrated in Fig. 2.

B. Dispensing and inkjet printing

A microelectronic printing platform (Mifang, Prtronic Scientific 2) with a dispensing needle of 34G (60 um) and a 16-nozzle print head was utilized for dispensing printing and inkjet printing, respectively. Silicon wafers and silicon dioxide wafers were employed as substrates for ink 1-1 and ink 1-2, while polyimide was for ink 2. Whereafter, the printed samples were dried on a hotplate (IKA, RCT basic IKAMAG®) at a temperature of 120 °C. Finally, the printed films of ink 1-1 and ink 1-2 were annealed at 500 °C, and the printed films of ink 2 were annealed at 280 °C.

C. Characterization

Jetting images were captured by the integrated camera of the printing platform. The contact angles of the inks on different substrates were measured by a contact angle meter

(Kino, SL150E) at room temperature of 25 °C. In addition, we calculated the surface tension of the ink using Image J-Fiji with Pendent Drop method. The pH and viscosity of the ink were measured by a pH meter (Mettler Toledo, Seven Compact S210) and a viscosity meter (IKA, Rotavisc lo-vi) at 25°C respectively. The chemical compositions were characterized by Fourier transform infrared spectroscopy (FTIR, Shimadzu, IrAffinity-1). The surface morphology of the films was determined by scanning electron microscopy (SEM, Hitachi, SU-8010).

III. RESULTS AND DISCUSSION

A. Ink Characterizations

Table. 1 illustrated the physicochemical properties of the ZnO nano ink. For inkjet printing, we further evaluated the printability of ink 1-2. Prior studies have defined sorts of dimensionless constants to characterize the behavior of ink drops, among which Reynolds number (Re), Weber number (We) and Z number are the most widely used ones [4]:

$$Re = \frac{\rho v d}{\eta} \qquad (1)$$

$$We = \frac{\rho v^2 d}{\gamma} \qquad (2)$$

$$Z = \frac{Re}{\sqrt{We}} = \frac{\sqrt{\gamma \rho d}}{\eta} \qquad (3)$$

In Eq. (1) and (2), ρ, v, η, and γ respectively represent the density, velocity, viscosity and surface tension of the ink drop, and d is the characteristic length, which is approximately equal to the nozzle diameter (20 um for Prtronic Scientific 2). By relating Re and We, Z number has given a more comprehensive description of the droplet behavior. From a prior simulation by Reis et al. [5], Z number was ranged in 1~10 for stable droplets formation. Besides, sufficient energy is needed for the ink drops to overcome the surface tension at the nozzle. Therefore, the minimum value of velocity was defined [6], which was also a critical condition of Weber number (We > 4). Finally, the impact of a drop on a substrate was taken into consideration. A drop could reach the substrate without splashing When $We^{1/2} \cdot Re^{1/4} < 50$ [7]. Fig. 3(a) showed a printability map with the restrictions mentioned above shown on it [4].

In addition to the parameters listed in Table. 1, we estimated the droplet velocity to be 3 m/s from the distance of two consecutive droplets, as illustrated in Fig. 3(c). The We, Re and Z numbers of ink 1-2 were 5.50, 4.86 and 2.07, respectively, which is marked with a red dot in the printable area in Fig. 3(a). By adjusting the voltage and waveform, spherical droplets of ink 1-2 without satellite drops were generated, as shown in Fig. 3(b).

TABLE I PHYSICOCHEMICAL PROPERTIES OF THE ZnO INKS

	[ZnO]ₗₙₖ (mol L^{-1})	Density (g cm^{-3})	pH	Viscosity (mPa S)	Surface tension (mN m^{-1})
Ink 1-1	0.0625	1.27	6.24	442	49.0
Ink 1-2	0.0625	1.07	7.17	13.2	35.1
Ink 2	0.0500	1.14	11.7	14.9	49.2

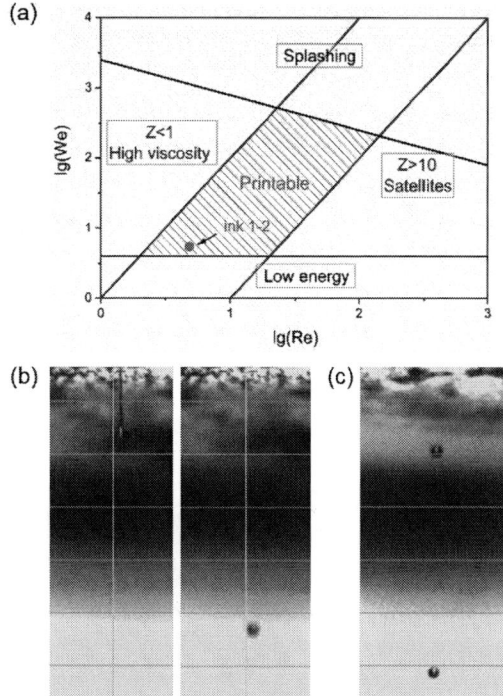

Fig. 3. The printability of ink 1-2 and the observed ink droplet morphology. (a) The printability map with a calculated point for ink 1-2; (b) A single ejected droplet leaving the nozzle; (c) Two continuous droplets ejected at a frequency of 7500 Hz.

Fig. 4 demonstrated the contact angles of the three inks on different substrates. After kept in ambient air for 1 hour, the sessile drops of ink 1-1 showed the best stability, while the contact angles of ink 1-2 and ink 2 emerged variations of more than 10%. This could be caused by the difference in volatility of solvents. Since the boiling point of glycerol (290 °C) is higher than that of ethylene glycol (197 °C), ink 1-2 showed a stronger volatility. Though glycerol was employed as a co-solvent, ammonia was the main factor of the instability of ink 2. These results showed that the instant post-treatments were needed for ink 1-2 and ink 2, but not necessary for ink 1-1. Therefore, the complexity of the procedure could be reduced in mass production when using ink 1-1.

B. Reactions Principles

The two precursor ZnO nano inks required different post-treatment temperatures for their different chemical processes. The inks based on the sol-gel process (ink 1-1 and ink 1-2) employed a metal-organic salt as a precursor, and the transition from sol to gel served as an intermediate step in obtaining the solid material of zinc oxide [8]. During the preparation and aging process of the solution, zinc acetate dehydrate hydrolyzed and polymerized. The zinc hydroxide molecules poly-condensed together and formed a final product of HO–(Zn–O–Zn)$_n$–OH [9]. The above reactions can be expressed as:

$$2Zn(AC)_2+H_2O \rightleftharpoons Zn(OH)(AC)+HAC+Zn^{2+}+2AC^- \quad (4)$$

$$Zn(OH)(AC)+H_2O \rightleftharpoons Zn(OH)_2+HAC \quad (5)$$

$$2HO-Zn-OH \rightarrow HO-Zn-O-Zn-OH+H_2O \quad (6)$$

In addition, MEA promoted the hydrolysis for its alkalinity and meanwhile formed a complex with part of the Zn^{2+}. Therefore, sol was stabilized with the two competing reactions.

After printing, the solvents evaporated, and the poly-condensation continued when the substrate undergoing pre-heating. Finally, the nanoscale ZnO was formed during annealing. This synthesis route is simple and cost effective [10]. However, despite a single report of decomposing the polymer at 200 °C [11], the annealing temperatures were usually higher than 400 °C for the complete decomposition of the polymer and thorough elimination of the organic components.

On the contrary, the ink based on solution chemistry was prepared through a carbon-free route. By adding NaOH to the inorganic salt precursor, zinc hydroxide was precipitated. The centrifuging and washing steps removed most of the ions. After dissolving in ammonia, an aqueous ammine-hydroxo zinc complex was formed. For the rapid decomposition kinetics of this complex and the absence of organics, high-purity ZnO could be produced at 150 °C ~ 300 °C [12, 13]. The low annealing temperature enables ink 2 to applying on flexible substrates like polyimide. The reactions are expressed as:

$$Zn(NO_3)_2+2NaOH \rightarrow Zn(OH)_2+2NaNO_3 \quad (7)$$

$$Zn(OH)_2+NH_3 \rightarrow Zn(OH)_2(NH_3)_x \rightarrow ZnO+xNH_3+H_2O \quad (8)$$

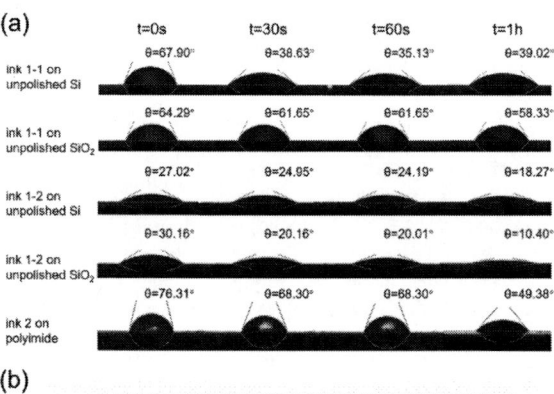

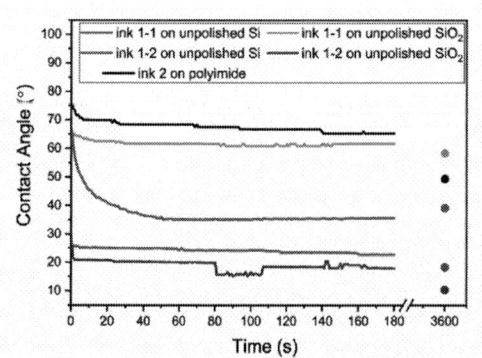

Fig. 4. (a) Contact angles of the ZnO precursor inks on different substrates after 0 s, 30 s, 60 s, and 1h; (b) Contact angles of the ZnO precursor inks on different substrates as a function of time in an hour.

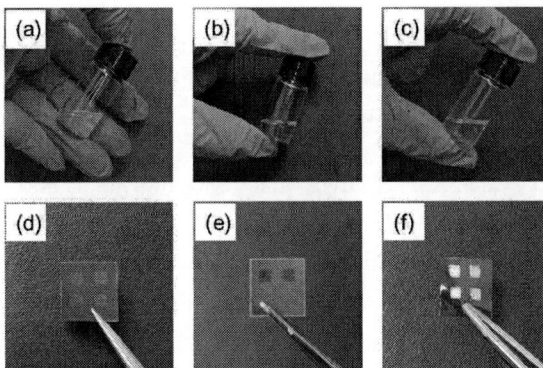

Fig. 5. The prepared inks and the printed films after post-treatments. All substrates are 1cm×1cm in size. (a) ink 1-1; (b) ink 1-2; (c) ink 2; (d) printed films of ink 1-1 on an unpolished SiO$_2$ wafer; (e) printed films of ink 1-2 on an unpolished Si wafer; (f) printed films of ink 2 on polyimide.

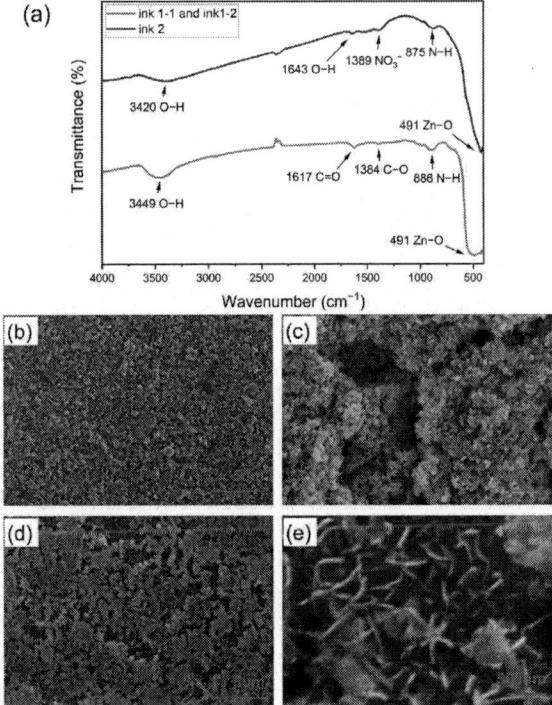

Fig. 6. Characterizations of the printed films. (a) the FTIR of the ZnO derived from the two different precursors; (b) and (c) the SEM images of the ZnO film derived from ink 1-1; (d) and (e) the SEM images of the ZnO film derived from ink 2.

C. Characterizations of Printed films

The prepared inks and printed films were shown in Fig. 5. Fig. 6 (a) illustrated the FTIR spectrums of ZnO derived from the inks based on sol-gel process (ink 1-1 and ink 1-2) and solution process (ink 2). Both spectrums showed an obvious peak ranged in 400~500 cm^{-1}, which confirmed the formation of ZnO at 500 °C for ink 1-1/1-2, and at 280 °C for ink 2 [14].

SEM images in Fig. 6 (b-d) showed that the two synthesis routes led to different morphologies of the printed

samples. The printed film of ink 1-1 was composed of spherical shaped ZnO nanoparticles with a diameter ranged in 20~30 nm. However, films printed by ink 2 showed a nanostructure like "dendrites" [9] with a few short nanowires radiating from one point. The diameters of the nanowires ranged in 40~80 nm. In addition, both films showed porous nanostructures with large specific surface areas, which is crucial in sensor fabrication.

IV. CONCLUSIONS

In this study, two novel precursor-based inks were proposed for the printing of zinc oxide in nanoscale. Both rigid and flexible substrates were tested with inkjet and dispensing printing methods. Furthermore, the inks based on sol-gel process are facile to synthesize, but require post treatments at 500 °C. The ink based on solution process can obtain ZnO at a lower temperature of 280 °C, although the synthesis is more complicated. Since precursor-based inks are employed in the synthesis of various metal oxides, our ink systems serve as a demonstration for the direct printing of metal oxides.

REFERENCES

[1] J. S. Gebauer, V. Mackert, S. Ognjanovic, and M. Winterer, "Tailoring metal oxide nanoparticle dispersions for inkjet printing," *J Colloid Interface Sci.*, vol. 526, pp. 400-409, Sep. 2018.

[2] M. Rieu et al., "Fully inkjet printed SnO$_2$ gas sensor on plastic substrate," *Sens. Actuator B-Chem.*, vol. 236, pp. 1091-1097, Nov. 2016.

[3] Z. Zhu et al., "Functional Metal Oxide Ink Systems for Drop-on-Demand Printed Thin-Film Transistors," *Langmuir*, vol. 36, pp. 8655-8667, Aug. 2020.

[4] B. Derby, "Inkjet Printing of Functional and Structural Materials: Fluid Property Requirements, Feature Stability, and Resolution," *Ann. Rev. Mater. Res.*, vol. 40, pp. 395-414, Jun.2010.

[5] N. Reis and B. Derby, "Ink Jet Deposition of Ceramic Suspensions: Modeling and Experiments of Droplet Formation," in *MRS OPL*, San Francisco, 2000, pp. 117-122.

[6] C. D. Paul et al., "Ink-jet printing of polymer light-emitting devices," in *Proc.SPIE*, San Diego, 2002, pp. 59-67.

[7] S. C. D. a. H. M. G., "An experimental investigation of fluid flow resulting from the impact of a water drop with an unyielding dry surface," *Proc. R. Soc. Lond. A*, vol.373, pp. 419-441, Jan. 1981,.

[8] X. Wang, M. Ahmad, and H. Sun, "Three-Dimensional ZnO Hierarchical Nanostructures: Solution Phase Synthesis and Applications," *Materials (Basel)*, vol. 10, pp.1304, Nov. 2017.

[9] H. Bahadur, A. K. Srivastava, R. K. Sharma, and S. Chandra, "Morphologies of sol-gel derived thin films of ZnO using different precursor materials and their nanostructures," *Nanoscale Res. Lett.*, vol. 2, pp. 469-475, Oct. 2007.

[10] O. Kassem, M. Saadaoui, M. Rieu, and J.-P. Viricelle, "A novel approach to a fully inkjet printed SnO$_2$-based gas sensor on a flexible foil," *J. Mater. Chem. C*, vol. 7, pp. 12343-12353, 2019.

[11] Y. Sun, J. H. Seo, C. J. Takacs, J. Seifter, and A. J. Heeger, "Inverted polymer solar cells integrated with a low-temperature-annealed sol-gel-derived ZnO Film as an electron transport layer," *Adv Mater*, vol. 23, pp. 1679-1683, Apr. 2011.

[12] S. T. Meyers, J. T. Anderson, C. M. Hung, J. Thompson, J. F. Wager, and D. A. Keszler, "Aqueous Inorganic Inks for Low-Temperature Fabrication of ZnO TFTs," *J. Am. Chem. Soc.*, vol. 130, pp. 17603-17609, Dec. 2008.

[13] Q. Xu et al., "Flexible Self-Powered ZnO Film UV Sensor with a High Response," *ACS Appl. Mater. Interfaces*, vol. 11, pp. 26127-26133, Jul. 2019.

[14] A. R. Nimbalkar and M. G. Patil, "Synthesis of ZnO thin film by sol-gel spin coating technique for H$_2$S gas sensing application," *Physica B*, vol. 527, pp. 7-15, Dec. 2017.

Proceedings of the 16th Annual IEEE International
Conference on Nano/Micro Engineered and Molecular Systems
April 25-29, 2021

Efficient Manufacturing of Microdome Array for Advanced Electronic and Optical Devices†

Zehua Xiang[1], Haobin Wang[1], Hang Guo[1,2], Ji Wan[1], Chen Xu[1,2], Liming Miao[1], Mengdi Han[3,*] and Haixia Zhang[1,2,*], *Member, IEEE*

Abstract— Aiming at improving the precision and decreasing roughness of micro-manufacturing, this work proposes a new method that combines non-uniform graphic compensation for template preparation and demolding with viscous fluid composed of mixtures of two types of polydimethylsiloxane (PDMS), Sylgard 184 and SE 1700. Notably, this method can widely apply to a bounty of micro-manufacturing methods, such as laser direct writing (LDW), 3D printing, and lithography. This paper exploits LDW to fabricate smooth convex microdome array (MA) on PDMS, and demonstrates the applications of such microstructures in electronics and optics. Specifically, the multifunctional microstructure can serve as a pressure sensor with high sensitivity and broad range, a strain senor for biaxial strain sensing and an anti-reflection film to improve photoelectric conversion efficiency of solar cells.

Keywords—laser direct writing; graphic compensation; viscous fluid; microdome array

I. INTRODUCTION

Micro-manufacturing is important for fabricating microstructures widely used in optical systems [1], electronic devices [2-4] and many other areas. Researchers have made extensive progresses on micro-manufacturing, such as lithography [5], ultraviolet (UV) imprint technology [6], laser direct writing [7]. Compared with other approaches, laser direct writing (LDW) is simple and easy to process. However, the fast processing feature usually leads to large roughness of the structure. Besides, the laser spots will expand out, making the processed samples not the same as the designed patterns.

In order to enhance the processing accuracy of LDW as well as reduce the roughness, many methods have been proposed such as wet etching assisting LDW [8], voxel modulation LDW [9]. However, these improved methods only apply to LDW.

In this paper, we innovatively create a new method including non-uniform graphic compensation for concave microdome array (MA) template and viscous fluid [10] demolding for smoothing the surface to fabricate the desired smooth convex MA which has many interesting applications. Superior to the methods mentioned above, our method can be widely applied in many manufacturing methods to make up for processing defects.

As for applications, we present three applications of MA structure. First, a pressure sensor with interlocked MA demonstrates excellent sensing performance such as a wide detection range and high linearity. Second, an MA-based strain sensor has the capability to detect the direction and magnitude of strains in two directions respectively. Third, an MA-based polydimethylsiloxane (PDMS) film can serve as an anti-reflection film to reduce the loss of incident light on the substrate surface, thus improving light utilization and photoelectric conversion efficiency.

II. EXPERIMENTAL METHODS

Fabrication of rough template with concave MA: The rough template is fabricated by a single-step programmable LDW process. It is synthesized by irradiation of Poly (methyl methacrylate) (PMMA) board with a CO_2 infrared laser (Vollerun Laser Technology Co., Ltd., 10.6 µm in wavelength). The power and scan rate are specifically decided by the material categories and thickness of the mold.

Fabrication of smooth convex MA on PDMS: The matrix materials compose mixtures of two types of PDMS, Sylgard 184 and SE 1700 (Dow Corning Co., USA) with a weight ratio of 3:2. Both materials are mixed with their curing agents in a 10:1 ratio by weight before commingling, followed by degassing for 30 minutes. The mixture is dropped on the template and heated on the hot plate at 75℃ for 3h to solidify. At last, a smooth PDMS convex MA can be peeled off from the template.

Fabrication of pressure sensor: The PDMS films with convex MA are spray-coated with ethanol solution of silver nanowire (Ag-NW) repeatedly until the resistivity of Ag-NW layer reaches the demand. Then, the pressure sensor is prepared by interlocking two MA PDMS films with the microdome surfaces facing each other. Conductive wires are bonded with the MA PDMS film by silver paste on the opposite edge of the sensor. Similarly, we prepare the pressure sensor with flat structure and single MA structure.

Fabrication of strain sensor: Firstly, flat PDMS film and PDMS films with different microdome spacing are prepared.

---

†This work was supported by National Key R&D Project from Minister of Science and Technology, China (2016YFA0202701, 2018YFA0108100) and the National Natural Science Foundation of China (Grant No. 61674004).

[1]National Key Laboratory of Science and Technology on Micro/Nano Fabrication, Institute of Microelectronics, Peking University, China.

[2]Academy for Advanced Interdisciplinary Studies, Peking University, China.

[3]Department of Biomedical Engineering, College of Future Technology, Peking University, China.

*Corresponding author. E-mail address: hmd@pku.edu.cn (M. Han); zhang-alice@pku.edu.cn (H. Zhang).

978-1-6654-3008-1/21 $31.00 © 2021 IEEE

Then, four conductive wires are bonded on four sides of the PDMS film after spray-coating Ag-NW.

Fabrication of anti-reflection film: Flat PDMS film and convex MA PDMS film with different microdome spacing are prepared using the method mentioned above. The concave MA on PDMS film is fabricated by demolding from convex MA PDMS film. The convex PDMS film is treated by Oxygen plasma to facilitate the later peeling.

III. RESULTS AND DISCUSSION

A. Design and process of smooth MA on PDMS

To get the smooth convex MA on PDMS, firstly, the rough template with a closely arranged concave MA is fabricated by a single-step programmable LDW process. For the structure with small size, fast LDW makes it attenuated by heat in the depth direction, leading to a microdome-shaped geometry. Notably, due to the fact that the circular spots processed by LDW will expand out and the transverse expansion is larger than longitudinal expansion, the programmed structure is not the same as the structure of the actual template. To solve the problem, we design the programmed template based on non-uniform graphic compensation. Therefore, the programmed structure is an array of ellipses with greater transverse length than vertical length rather than an array of circles (Fig. 1a <i>). Here, the distance between adjacent ellipses will be further discussed in the fourth paragraph of this section.

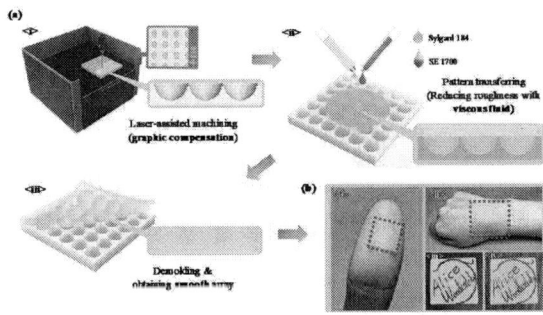

Figure 1. (a) Schematic diagram of the new method combining non-uniform compensation and viscous fluid, (b) optical images of the prepared PDMS film with MA.

In addition to enhancing the precision of LDW by graphical compensation, we also reduce the roughness by viscous fluid to decrease penetration, thus overcoming the manufacturing defects, as shown in Fig. 1a <ii>. We cast the mixture of Sylgard 184 and SE1700 PDMS at a weight ratio of 3:2 on the rough template. SE 1700 contains approximately 20 wt% fumed silica. Fumed silica nanoparticles consist of branched, chain-like, 3D secondary particles that cluster into tertiary particles which act as a rheological modifier that imparts an appropriate elastic modulus and yield stress. Therefore, the viscous PDMS will not penetrate into the rough parts of the template due to decreased liquidity. Compared with the template, the roughness of the convex MA on PDMS is significantly reduced. Generally speaking, the precision and roughness of LDW are improved by means of graphic compensation and viscous fluid.

Fig. 1b shows the optical images of PDMS film with MA. Benefiting from its high flexibility, it can be attached to our skin as a sensor for information sensing (Fig. 1b <i-ii>). Besides, the high light transmission feature makes it a proper optical device (Fig. 1b <iii-iv>).

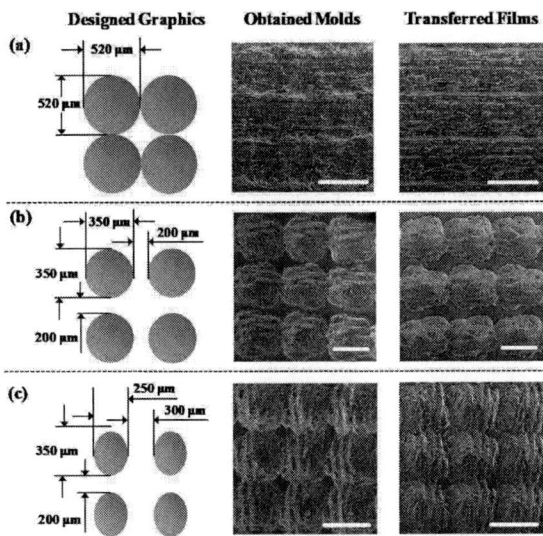

Figure 2. Comparison of three design methods and correspoding SEM photographs of obtained templates and transferred films: (a) non-compensation, (b) uniform compensation, (c) non-uniform compensation. Scale bars in (a)-(c) are 500 μm.

In Fig. 2, we compare the processing results by non-compensation, non-uniform compensation and uniform compensation. Evidently, by non-uniform compensation, the precision and resolution are effectively enhanced, which can be seen from the SEM photograph in Fig. 2c.

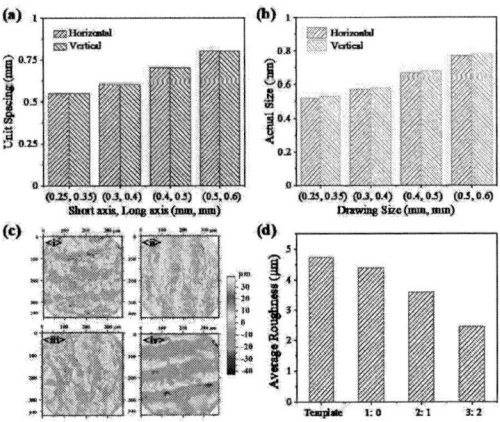

Figure 3. (a) Different non-uniform compensation for different microdome size, (b) the relationship between programmed size and obtained size of microdome, (c) roughness distribution and (d) average roughness of the template and transferred film with different weigh ratio of Sylgard and SE 1700.

Besides, we also study the different non-uniform compensation applied to microdome structure of different sizes, as shown in Fig. 3a. Additionally, Fig. 3b demonstrates the degree of image distortion in different sizes. The smaller size leads to a larger distortion due to the limited precision. Fig. 3c shows the roughness distribution of the template and the PDMS complementary mode structure with different weight ratios of Sylgard 184 and SE1700. We measure and average the roughness of each pixel, as shown in Fig. 3d. The Sylgard 184 completely transfers the roughness of template because of its strong penetration. With the increase of the weight ratio of SE1700, the roughness of the molded structure gradually decreases. The roughness is only 2.4 µm at the ratio of 3:2, half of the template. However, if we further increase the percentage of SE1700, the mixture PDMS will be too viscous to get the intact MA structure.

B. MA based pressure sensor

The pressure sensor in our work is formed by two parts. Both the top and bottom parts are Ag-NW coated PDMS with convex MA. Applying normal pressure will cause a deformation of the Ag-NW-PDMS structures, leading to an increase in the contact area between the top and the bottom Ag NW-PDMS structures. Such an increase in the contact area would form more conductive paths.

In order to evaluate the effect of the surface microdome on the sensing performance of the sensor, four different pressure sensors based on planar PDMS without surface microstructure, rough single MA type, rough and smooth interlocked MA type are fabricated.

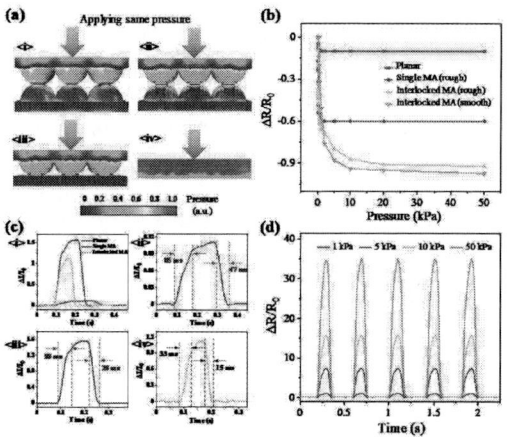

Figure 4. (a) COMSOL mechanical simulation and (b) sensing performance of four different pressure sensors, (c) current response of three types under a pressure of 1 kPa, (d) current response of smooth interlocked MA sensor under different applied pressures.

Fig. 4a shows the COMSOL mechanical simulation of the four types of sensors. The results indicate that interlocked MA bears larger pressure than single MA and planar one. As for the rough and smooth interlocked MA, the former has a larger pressure leading to its high sensitivity, while the latter demonstrates a more uniform increase of contact area, thus making its pressure response more linear. The experimental results agree well with simulation. Compared with planar type and single MA type, the interlocked MA sensors show a larger pressure detection range with more than 50kPa. Besides, due to the mixture of

SE1700, the PDMS with smooth interlocked MA has a higher Young's modulus, thus possessing a higher upper limit of detection than the rough one. In addition, the decreased roughness makes its resistance response highly linear to the applied normal pressure within 1 kPa (Fig. 4b). In Fig. 4c, we also test the response time of the different type sensors under a pressure of 1 kPa. It can be seen that the interlocked MA type sensor has the fastest response time of 33 ms, nearly three times faster than the planar type and twice than the single MA type.

According to Fig. 4d, the current through the sensor with smooth interlocked MA can respond repeatedly to the different applied pressures. Since the pressure is applied by periodic oscillations of Modal Shaker, the response current signal is a quasi-sine wave with different pulse widths, from which the high responsivity of the device can be seen.

C. MA based biaxial strain sensor

In this work, we proposes a MA based strain senor for biaxial strain sensing. As shown in Fig. 5a, the device is formed by a Ag-NW-PDMS film with convex MA on it. Applying strain can induce stress concentrations at microdome spacing along the force direction, leading to microcracks at corresponding regions of the Ag-NW film because of the convex MA structure. In particular, microcracks are only generated at the spacing of adjacent microdome along the x direction under the strain in the same direction. As a result, the conductive path in the x direction is destroyed, while remaining intact along the microdome in the y direction. Therefore, this strain sensor can distinguish two directions of applied force. In Fig. 5b, a 10% strain is applied in the x direction and the resistance in the x direction is greatly changed while the y direction is basically unchanged for the device with MA structure. For the planar strain sensor, there is not an excessive local stress, resulting in the similar resistance change in the two directions.

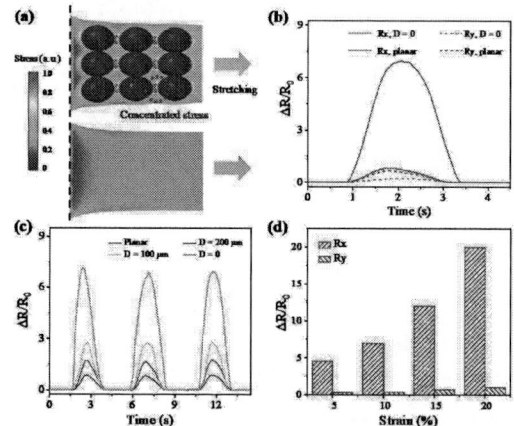

Figure 5. (a) COMSOL mechanical simulation of Ag-NW-PDMS film with MA versus planar Ag-NW-PDMS film, (b) resistance change of Ag-NW-PDMS film with MA versus planar Ag NWs-PDMS film in two directions at a 10% strain in the x direction, (c) resistance change of Ag-NW-PDMS film with MA of different spacing and planar Ag-NW-PDMS film at a 10% strain in the x direction, (d) different resistance change in two directions of closely arranged MA film at a strain of 5%, 10%, 15%, 20% respectively in the x direction.

Fig. 5c demonstrates the high sensitivity of the strain sensor with closely packed MA structure by comparing it with planar type and loosely arranged MA structures with the microdome spacing of 100 μm and 200 μm, respectively. We use the letter 'D' as an alternative of microdome spacing.

There is much to admire about the MA structure. For example, it can serve as a skeleton, enhancing the robustness of the device. With the MA structure, the sensor can tolerate a strain of 20%, as shown in Fig. 5d. The change of resistance in the x direction increases as the strain in the same direction rises from 5% to 20%. Similarly, the resistance in the y direction hardly changes.

D. MA based anti-reflection film

For solar cells, improving the efficiency of energy conversion is very important and the MA based PDMS film proposed in this work can play a role in it. By laminating a planar PDMS film on glass, the optical transmittance is slightly increased over a wide wavelength between 300 nm to 800 nm and the reflectance is evidently decreased. Fig. 6a illustrates the optical principle of the MA based PDMS film to decrease reflection and increase transmission of the incident light, which is mainly attributed to the step refractive index distribution of the constituent materials ($n_{air}/n_{PDMS}/n_{glass}$ = 1/1.43/1.52).

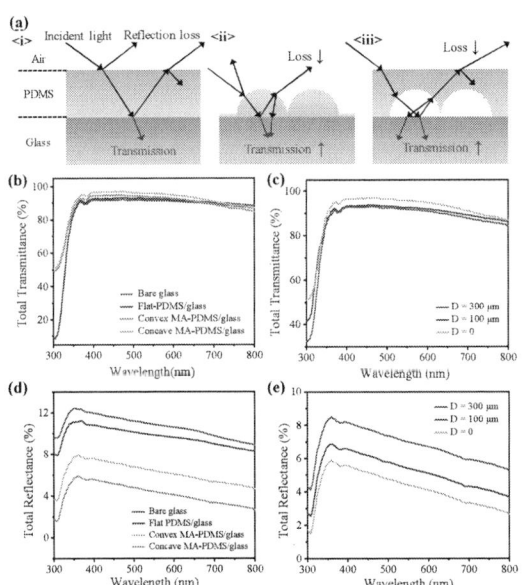

Figure 6. (a) Schematic diagram of reflection decrease and transmission increase, (b) total transmittance of four strutures and (c) concave MA struture with different microdome spacing, (d) total reflectance of four strutures and (e) concave MA struture with different microdome spacing.

Besides, compared with planar PDMS film, the MA on the PDMS can extend the optical path owing to the reflection and transmission between the concave MA and convex MA structures, which can effectively enhance the transmission of incident light by reducing their surface reflection. As shown in Fig. 6b and Fig. 6d, the concave MA-PDMS/glass has the best effect among three for reflection decrease (from 12.4% to 5.6%) and transmission increase (from 91.2% to 95.6%). In addition, we make a comparison of concave MA-

PDMS with different microdome spacing. It can be seen that the closely arranged type works much better due to the larger effective areas.

IV. CONCLUSION

In summary, we present an ingenious method combining graphic compensation and demolding with viscous fluid to improve the precision and reduce roughness of micro-manufacturing. Here, we specifically study the improved method applied in LDW. The smooth MA on PDMS made by this method offers unique features for a variety of functional devices, including pressure sensors with high sensitivity and broad range, strain sensors with capabilities in biaxial strain sensing, and anti-reflection films. In particular, the smooth MA based pressure sensor shows high linearity to applied pressure below 1 kPa and a wide detection range with more than 50 kPa. The MA based strain sensor can simultaneously detect the direction and magnitude of strains in two directions respectively. The MA based anti-reflection film can increase the optical transmittance and decrease reflectance over a wide wavelength between 300 nm to 800 nm. Furthermore, this idea can be widely applied in a great number of manufacturing methods. For example, the graphic compensation can also be used in lithography for picture designing; the viscous fluid can improve the roughness of 3D printing. In addition to higher light transmission efficiency in the optical field, the microdome array can simulate natural compound eyes and other structures for more diversified applications

REFERENCES

[1] Z. Ren, Q. Zheng, H. Zhang, et al., "Wearable and self-cleaning hybrid energy harvesting system based on micro/nanostructured haze film," *Nano Energy*, vol. 67, 104243, 2020.

[2] H. Wang, Y. Song, H. Zhang, et al., "A three-electrode multi-module sensor for accurate bodily-kinesthetic monitoring," *Nano Energy*, vol. 68, 104316, 2020.

[3] H. Guo, J. Wan, H. Zhang, et al., "Self-powered multifunctional electronic skin for a smart anti-counterfeiting signature system," *ACS Appl. Mater. Interfaces*, vol. 12, pp. 22357-22364, 2020.

[4] H. Wang, M. Han, Y. Song, and H. Zhang, "Design, Manufacturing and Applications of Wearable Triboelectric Nanogenerators," *Nano Energy*, vol. 81, 105627, 2021.

[5] K. Zhang, Y. H. Jung, Z. Ma, et al., "Origami silicon optoelectronics for hemispherical electronic eye systems," *Nat. Commun.*, vol. 8, pp. 1-8, 2017.

[6] J. Chen, J. Cheng, S. Chen, et al., "Precision UV imprinting system for parallel fabrication of large-area micro-lens arrays on non-planar surfaces," *Precis. Eng.*, vol. 44, pp. 70-74, 2016.

[7] Y. Yang, Y. Song, et al., "A laser-engraved wearable sensor for sensitive detection of uric acid and tyrosine in sweat," *Nat. Biotechnol.*, vol. 38, pp. 217-224, 2020.

[8] Z. Deng, F. Chen, X. Hou, et al., "Dragonfly-eye-inspired artificial compound eyes with sophisticated imaging," *Adv. Funct. Mater.*, vol. 26, pp. 1995-2001, 2016.

[9] D. Wu, J. Wang, H. Sun, et al., "Bioinspired fabrication of high-quality 3d artificial compound eyes by voxel-modulation femtosecond laser writing for distortion-free wide-field-of-view imaging," *Adv. Opt. Mater.*, vol. 2, pp. 751-758, 2014.

[10] Z. Wang, X. Yuan, S. Dong, et al., "3D-printed flexible, Ag-coated PNN-PZT ceramic-polymer grid-composite for electromechanical energy conversion," *Nano Energy*, vol. 73, 104737, 2020.

Gap in pagination due to unavailable papers.

Pages 152-155

Conformal Bioelectronic Interfaces

Xiaodong Chen

Innovative Centre for Flexible Devices (iFLEX), School of Materials Science and Engineering, Nanyang Technological University, Singapore

Website: https://personal.ntu.edu.sg/chenxd

ABSTRACT

Smart sensors not only enrich daily lives by providing enhanced smart functions, but also provide health information by monitoring body conditions. For example, patchable sensors have the potential to better interface with human skin, thus improving the sensitivity of detection of health indicators. However, the crucial aspects toward the advancement of such sensors rely on the development of novel mechanically durable materials, which allow maintaining the function under the deformed states. In this talk, I will present our latest progress on building conformal bioelectronic interfaces, manufacturing conformal sensors and their integration of individual devices into systems.

BIOGRAPHY

Dr. Xiaodong Chen is the President's Chair Professor in Materials Science and Engineering, Professor of Chemistry and Medicine (by courtesy) at Nanyang Technological University, Singapore (NTU). He is the Director of Innovative Centre for Flexible Devices (iFLEX) at NTU and the Director of Max Planck – NTU Joint Lab for Artificial Senses.

He received his B.S. degree (Honors) in chemistry from Fuzhou University (China) in 1999, M.S. degree (Honors) in physical chemistry from the Chinese Academy of Sciences in 2002, and Ph.D. degree (Summa Cum Laude) in biochemistry from University of Muenster (Germany) in 2006. After his postdoctoral fellow working at Northwestern University (USA), he started his independent research career as Nanyang Assistant Professor at Nanyang Technological University since 2009. He was promoted to Associate Professor with tenure in Sept 2013, then Full Professor in Sept 2016. He was appointed as the President's Chair Professor in Materials Science and Engineering in April 2019. His research interests include mechano-materials and devices, integrated nano-bio interface, and cyber-human interfaces.

He has been elected as a Fellow of the Academy of Engineering Singapore and a Fellow of Royal Society of Chemistry in UK. In addition, he was recognized by multiple prestigious awards and honors including Singapore NRF Investigatorship, *Small* Young Innovator Award, Singapore NRF Fellowship, Nanyang Research Award, *Lubrizol* Young Materials Science Investigator Award, Mitsui Chemicals-SNIC Industry Award in Materials and Nano-chemistry, and Friedrich Wilhelm Bessel Research Award from Alexander von Humboldt Foundation.

This page intentionally left blank.

Proceedings of the 16th Annual IEEE International
Conference on Nano/Micro Engineered and Molecular Systems
April 25-29, 2021

Broadband Piezoelectric Harvester Utilizing Three-wafer Bonding for Self-powered Wireless Sensor System

Xin Chen, Daqiao Tong, Jian Zhang, Lingxiao Gao, Fayang Wang, Hongzhi Pan, Liang Cao, and
Xiaojing Mu*

Abstract— The development prospect of the piezoelectric harvesters mostly subject to its operation bandwidth. This paper reports a piezoelectric energy harvester with the material of the AlN with a broadband of 37 Hz (743 Hz-780 Hz) through the simple method of mechanical limiter. The broadband harvester was packaged by a novel method of wafer-level vacuum packaging which employs two times of eutectic Al-Ge bonding to bond the layers of device wafer, the top wafer and the bottom wafer. And the peak-peak open-circuit output voltage is about 10 V, and an output density is 342.87 μW/cm² with the size of 0.6×1 cm². Furthermore, the reported energy harvester is hopeful applied in powering wireless sensor nodes with a wideband vibration frequencies.

I. INTRODUCTION

For the past few years, the rapid development of micro-nanofabricate technology has promoted the process of the miniaturization and low-power sensors. However, the growth of battery technology is far behind the evolution of sensor fabrication. Additionally, the piezoelectric harvesters can convert the vibration energy into usable electricity, and possess the merits of small and potable, and that have attracted the attention of many research institutions. Also it have been widely used in the territory of information-based weapons and equipment, ocean buoy, environmental health monitoring, et al. Whereas, the operation bandwidth of piezoelectric harvesters are usually limited to a narrow band due to the structure of the device, and the output power of the that one usually declines obviously when the input frequency diverges from the resonant frequency, as depicted in the Fig. 1a-b. Theses defect limits its efficiency of electro-mechanical transformation in the applied situation. Thus, many groups have done much efforts on broadening the vibrational frequencies and increasing the output power of the piezoelectric harvesters[1,2]. They have designed the array cantilever beams with different masses or designed the spring beam to connected the beams et al[3,4], these strategies always can broaden the bandwidth of the harvesters. In itself, these methods made the micromachining process more difficult and reduced reliability of the harvesters[5-7]. Hence, there is a demand for some piezoelectric harvesters which can overcome the limitation and harvest vibrational energy for a wideband frequency to applicate in the environment.

This reported manuscript presents an aluminum nitride (AlN) based piezoelectric energy harvester with a broadband of 37 Hz (743 Hz-780 Hz)) through a simple method of mechanical limiter. The peak-peak open-circuit output voltage of the harvester is about 10 V, and the peak output density is 342.87 μW/cm² with the size of 0.6×1 cm². In addition, the self-powered wireless temperature monitoring system is realized. The reported energy harvester is a promising candidate in applicating of powering wireless sensor nodes on the environment possessing the broadband vibration frequencies.

II. EXPERIMENTAL METHODS

A. Theoretical model for mechanical limiting of the wideband harvester with a mass

Based on the mechanical limiter and broaden frequency structure, an equivalent working model of the structure is proposed. According to the lumped parameter model, the piezoelectric nanogenerators coupling is equivalent to electrical viscous damping. The lumped parameter method is used to calculate the mechanical properties of the piezoelectric cantilever beam, and then the output voltage and output power of the structure are calculated through a stress distribution and a piezoelectric equation. Electrical performance, mechanical limiter and frequency broaden structure can be simplified into a lumped parameter model of spring-mass-damping.

Use theoretical knowledge of vibration mechanics:

$$m\ddot{u}_m(t) + c\dot{u}_m(t) + ku_m(t) = ma \tag{1}$$

$$\omega_n = \sqrt{k/m} \tag{2}$$

k is the sum of the spring constants of multiple cantilever beams of the common mass block; m is the total mass of the system, m is equal to the sum of the mass, m_0 is the mass block and the equivalent mass m_b^* represent the cantilever beam; u_m is the displacement of m, and it is the same as the lateral displacement of mass m_0; c is the total damping coefficient of the structure, which is equivalent to a linear viscous damping, also the sum of mechanical damping and electrical damping. From the above formula (1)-(2), the common mass model have not a significant broaden on the frequency band, but it can increase the natural frequency and amplitude output. And when this structure introduces a

---

[1]Key Laboratory of Optoelectronic Technology & Systems Ministry of Education, International R & D center of Micro-nano Systems and New Materials Technology, Chongqing University, Chongqing, China.
[2]State Key Laboratory of Mechanical Transmission, Chongqing University, Chongqing, China.

*Contacting Author: Xiaojing Mu is with Key Laboratory of Optoelectronic Technology & Systems Ministry of Education, International R & D center of Micro-nano Systems and New Materials Technology, Chongqing University, Chongqing, China. (phone:+86 15902396712; e-mail: mxjacj@cqu.edu.cn).

978-1-6654-3008-1/21 $31.00 © 2021 IEEE

mechanical limiter, the range of the frequency bandwidth will be increased. The simplified mechanical model of this theory is shown in the Fig. 2a-b. Obviously, the design of mechanical limiter can broaden the bandwidth and enhance the output performance of the piezoelectric harvester.

B. Electrical model -Output voltage

Based on the theory of Euler–Bernoulli's beam, the strain distribution in the piezoelectric layer is expressed as:

$$S_1(x,t) = -y_p u_b''(x,t) \qquad (3)$$

y_p indicates the position of the middle plane of the piezoelectric layer relative to the neutral plane, and the subscript "l" indicates that the strain is along the length of the cantilever beam. The stress of the cantilever beam is $T=E_p S_1$, and the maximum stress T_{max} occurs at the root of the fixed end at $x=0$:

$$T_{max} = E_p S_1(x,t)\big|_{x=0} \qquad (4)$$

E_p is the equivalent Young's modulus of the piezoelectric beam, which is generally expressed by the reciprocal of the elastic compliance constant s_{11}. The stress distribution along the width is ignored here. In order to avoid fatigue fracture of the piezoelectric beam, the allowable stress of the cantilever beam must be less than the mechanical strength during the device design process.

The electric displacement vector D in the piezoelectric layer can be expressed as:

$$D = d_{31}E_p S_1 = -y_p d_{31} E_p u_b''(x,t) \qquad (5)$$

Where d_{31} is the piezoelectric constant. The charge Q on the electrode obtained by integration is:

$$Q = w\int_0^l D\,dx \qquad (6)$$

For the piezoelectric single crystal, the upper and lower electrodes of the piezoelectric material form a capacitance $C_p=\varepsilon lw/h_p$, and ε is the dielectric constant. The voltage V between the upper and lower electrodes is:

$$
\begin{aligned}
V &= \frac{Q}{C_p} = \frac{h_p}{\varepsilon l}\int_0^l D\,dx \\
&= -\frac{d_{31}E_p}{\varepsilon}\frac{a_0}{\omega_n^2}\frac{1}{(1-\Omega^2)+i2\zeta\Omega}\frac{6y_p h_p(l+l_m)}{4l^3+6l^2 l_m+3ll_m^2}e^{i\omega t}
\end{aligned}
\qquad (7)
$$

When resonance occurs, $\omega=\omega_n$, $\Omega=1$, and the output voltage amplitude reaches the maximum value:

The above formula shows that the maximum output voltage amplitude V_{max} is directly proportional to acceleration amplitude a_0, piezoelectric layer thickness h_p, Young's modulus E_p, piezoelectric constant d_{31}, and inversely proportional to dielectric constant ε, damping ratio ζ, and square of resonance frequency ω_n.

C. Fabrication process

The fabrication of the wideband piezoelectric harvester is shown in the Fig. 3a. The hole device fabrication is divided into three parts: the device wafer layer, the top cap wafer layer and the bottom cap wafer layer.

For the device wafer layer, the fabrication starts with the depositing a seed layer of AlN (0.02 μm), a layer of Mo (0.2 μm), another two layers of AlN(1.2 μm) and Mo (0.2 μm) and the SiO₂ layer (0.2 μm) fabricated by the method of plasma. A flat wafer surface is fabricated which is critical for Al/Ge bonding and then SiO₂ (2 μm) is added, follows by a chemical mechanical polishing. Finally, top release etch is regulated.

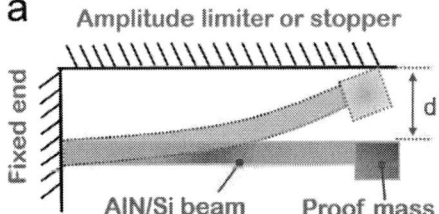

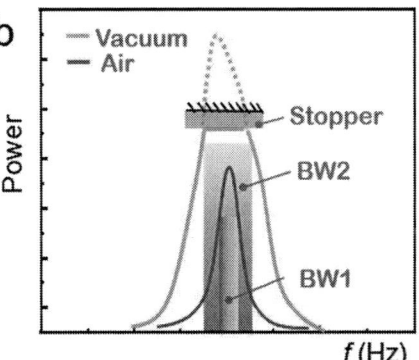

Fig. 2. (a) The equivalent operation model of the wideband harvester. (b) The principle of broad operation frequency of the harvester.

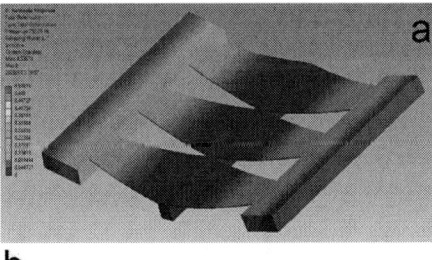

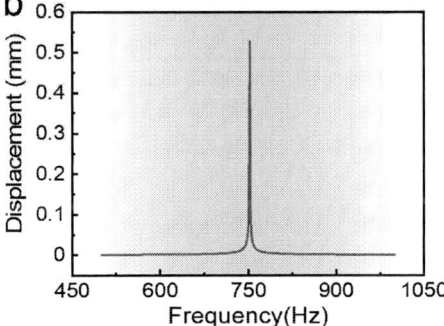

Fig. 1. (a) 3D illustration of the resonant model for the six cantilevers. (b) The resonant frequency of the piezoelectric nanogenerator.

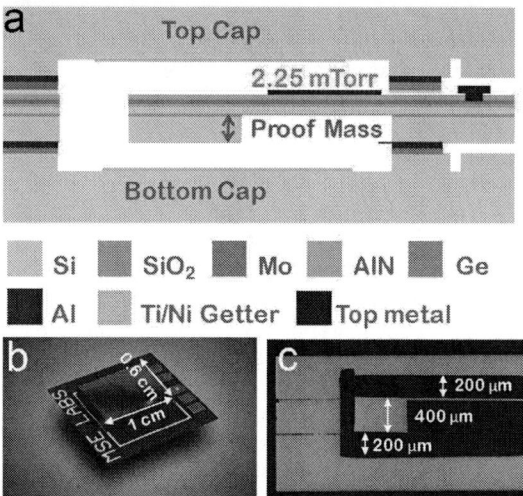

Fig. 3. (a) The structure of the wideband harvester. (b) The picture of the wideband nanogenerator. (c) The SEM picture for the cross section of the harvester.

For the top cap wafer, Ge and SiO$_2$ are deposited on a new wafer ahead of wafer which is cut down to 550 μm. Then, the layers of SiO$_2$, Ge and Si (the thickness is about 2.4 μm) were etched, then the layer of Si was deeply etched to create the structure of the cavity for the device. The device and top cap wafer are bonded through the way of the bonding of Al-Ge. The first few steps of bottom cap wafer fabrication steps are very similar to that of top cap fabrication procedures. Additionally, the Ti getter is added via a shadow mask. Then the bottom cap wafer is vacuum bonded to the prementioned top cap wafer and device wafer, also creating a vacuum cavity for the beams to vibrate with no air damping effect.

Then wafer-level vacuum packaged by a two-step three-wafer Al-Ge bonding: the first is top cap wafer bonding to the device wafer, then followed by bottom cap wafer bonding. And the picture of the wideband piezoelectric harvester is shown in the Fig. 3b. The SEM of the device section is diagramed in the Fig. 3c.

III. RESULTS AND DISCUSSIONS

A. Electrical test results

The output electricity of the wideband piezoelectric harvester have been measured through the vibration system which consists of an actuating vibration table, a power amplifier, a signal generator and a charge amplifier. An accelerometer was assembled onto the shaker together with the piezoelectric harvester, which provided the accelerated velocity of the shaker to the harvester. And the output electric of the harvester was tested at various frequency and acceleration via adjusting power amplifier and signal generator .

Fig.4 depicts the output electric of the designed wideband piezoelectric harvester. And the peak-peak open-circuit voltage response to the different frequencies at the acceleration from 0.3g to 2g is shown in Fig. 4a. The peak-peak voltage of the harvester reaches 5 V at the low acceleration of 0.3g. While the mechanical limits seems to

be unpolished at a small acceleration. In other hands, the response of the voltage generated essentially shows the sensitivity of measured harvester. While increasing the value of acceleration, the output voltage of the harvester is increased, and the function of mechanical limiter come into light. And the wideband characteristics appear at the acceleration of 1g, the output voltage reaches 10 V and the operation frequency bandwidth is 4 Hz. As the acceleration ranges from 1.5g to 2g, the bandwidth increases whereas the output voltage suffers a slight fall due to the amplitude of vibration of the beams exceeds the depth of the cavity of the capping wafers. And thus, the target of broaden the working frequency bandwidth is achieved through the mechanical limiter. Under the acceleration of 2g, the peak-peak open-circuit voltage of 9.56 V is obtained. Additionally, the band width of the broadband piezoelectric nanogenerator is 37 Hz (743 Hz-780 Hz).

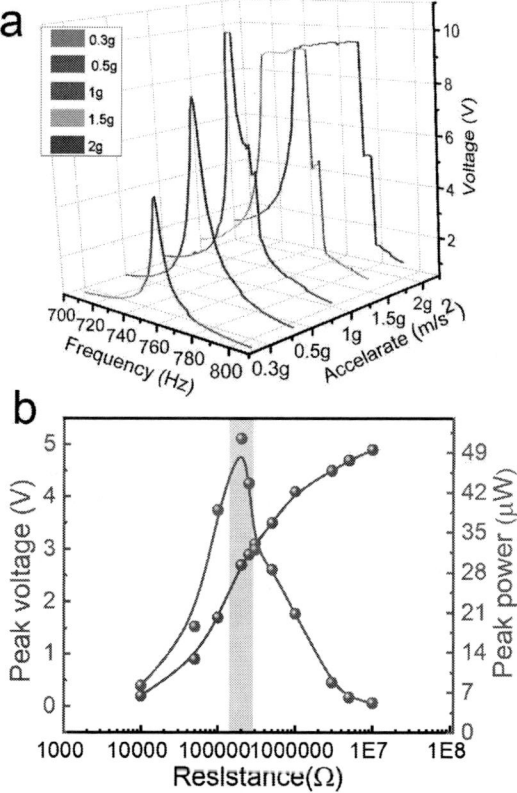

Fig. 4. (a) The output voltage of the wideband piezoelectric nanogenerator at different frequency on the acceleration ranges from 0.3g to 2g. (b) The output power of the broadband piezoelectric nanogenerator at the acceleration of 2g.

B. Application

Subsequently, the output electricity was investigated in the follow-up experiments and the self-powered wireless temperature system is realized. As the Fig. 5a displayed, the self-powered system is composed by an operation wideband harvester, a temperature sensor featured with low energy consumption (DS18B20), a wireless transmitter

(LRF215A), a wireless reception (LRF215U), a temperature monitoring system and the buck-boost energy management circuit (LTC3106). The power calculated to supply for sending a temperature sensor data is 0.85 mJ, and according to the numeration, the voltage of the battery (5.5 mAh) to set voltage of 3.4 V, which is enough for the temperature sensor transmitting. And the timing circuit was designed and set 10 s to send a data. As Fig. 5 diagramed, after the power management unit provides the wireless temperature sensor unit for a wireless communication, its stored energy would be consumed, resulting in a decrease in the output voltage of the power management unit, which is then supplemented by the power captured by the energy harvesting unit, and the output voltage of the power management unit is restored to set value. The results of the above experiments indicated that the wideband piezoelectric harvester possess the obious output performance.

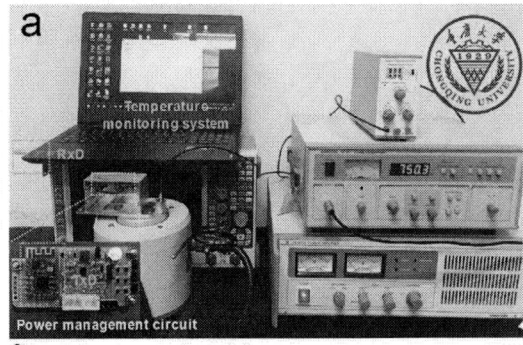

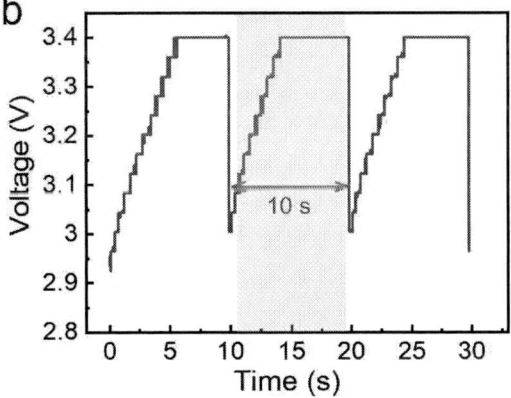

Fig. 5. (a) The picture of the self-powered wireless temperature monitoring system. (b) The charging and discharging curve of the wireless transmission.

IV. CONCLUSION

In summary, an aluminum nitride (AlN) based piezoelectric energy harvester with a broadband of 37 Hz (743 Hz-780 Hz) through the method of mechanical limit possessing the advantages of high reliability and simple micro-fabricate process is presented. The wideband piezoelectric harvester was micro-fabricated with the materials of AlN, and packaged by a novel wafer-level vacuum packaging strategy which employs two times of eutectic Al-Ge bonding. Results show that the peak-peak open-circuit output voltage is about 10 V, and an output density is 342.87 $\mu W/cm^2$ with the size of 0.6×1 cm^2. Furthermore, the self-powered wireless temperature monitoring system is designed to verify the applicability of the wideband piezoelectric harvester. And the reported harvester possessing the broadband vibration frequencies is a promising candidate in the application of powering wireless sensor nodes in an industrial environment.

ACKNOWLEDGMENT

This work was supported by the National Key Research and Development Program of China (No. 2019YFB2004800), the General Program of National Natural Science Foundation of China (NSFC, No.52075061), the Fundamental Research Funds for the Central Universities (No. 2019CDCGGD320), Chongqing University Scientific Research Reserve Top Talent Cultivation Program. The authors also gratefully acknowledge financial support from project graduate scientific research and innovation foundation of College of Optoelectronic Engineering, Chongqing University, Chongqing, China (Grant No. GDYKC202004).

REFERENCES

[1] H. Liu, C. Lee, T. Kobayashi, C. J. Tay, C. Quan. "Piezoelectric mems-based wideband energy harvesting systems using a frequency-up-conversion cantilever stopper". Sensor Actuat. A-Phys, vol. 186, 2012, pp.242-248.

[2] J. Q. Liu, H. B. Fang, Z. Y. Xu. "A MEMS-based piezoelectric power generator array for vibration energy harvesting". Microelectronics, vol. 5, 2008, pp.802-806.

[3] N. Wang, C. Sun, L. Y. Siow, H. Ji, Y. Gu. "AlN wideband energy harvesters with wafer-level vacuum packaging utilizing three-wafer bonding". IEEE International Conference on Micro Electro Mechanical Systems. IEEE, 2017.

[4] C. Sun, X. Mu, L. Y. Siow, W. M. Tsang, H. Ji, H. Chang, Q. Zhang, Y. Gu, and D.-L. Kwong, "A Miniaturization Strategy for Harvesting Vibration Energy Utilizing Helmholtz Resonance and Vortex Shedding Effect", IEEE Electron Device Lett., vol. 35, 2014, pp. 271-273.

[5] H. Liu, C. J. Tay, C. Quan, T. Kobayashi, and C. Lee, "Piezoelectric MEMS Energy Harvester for Low Frequency Vibrations with Wideband Operation Range and Steadily Increased Output Power", IEEE J. Microelectromech. Syst., vol. 20, 2011, pp. 1131-1142.

[6] W. Al-Ashtari., M. Hunstig, Hemsel T. "Enhanced energy harvesting using multiple piezoelectric elements: theory and experiments". Sensor Actuat. A-Phys. pp:138-146, 2013.

[7] N. Kawasaki, "Parametric study of thermal and chemical nonequilibrium nozzle flow," M.S. thesis, Dept. Electron. Eng., Osaka Univ., Osaka, Japan, 1993.

Gap in pagination due to unavailable papers.

Pages 162-169

Organic Molecular Sieve Membranes

Zhongyi Jiang

School of Chemical Engineering and Technology, Tianjin University, Tianjin, 300072, P.R. China

Website : http://www.jiang-lab.com/

ABSTRACT

Molecular separations that enable selective transport of target molecules from gas and liquid molecular mixtures, represent the most energy sensitive and significant demands. A number of emerging microporous organic materials have displayed the great potential as building blocks of molecular separation membranes, which not only integrate the rigid, engineered pore structures and desirable stability of inorganic molecular sieve membranes, but also exhibit high degree of freedom to create chemically-rich combinations/sequences. In this talk, we propose the concept of organic molecular sieve membranes (OMSMs) with the focus on the precise construction of membrane structures and efficient intensification of membrane processes. The platform chemistries, designing principles, assembly methods for the precise construction of OMSMs are elaborated. Conventional mass transport mechanisms are analyzed based on the interactions between OMSMs and penetrate(s). Particularly, 'STEM' guidelines of OMSMs are highlighted to guide the precise construction of OMSM structures and efficient intensification of OMSM processes. Emerging mass transport mechanisms are elucidated inspired by the phenomena and principles of the mass transport processes in biological realm. The representative applications of OMSMs in gas and liquid molecular mixture separations are highlighted.

BIOGRAPHY

Zhongyi Jiang is a Professor at School of Chemical Engineering and Technology of Tianjin University. He obtained a PhD degree from Tianjin University in 1994. He was a visiting scholar of University of Minnesota with Prof. Edward Cussler in 1997 and California Institute of Technology with Prof. David Tirrell in 2009. He is the winner of the National Science Fund for Distinguished Young Scholars in China, and a fellow of the Royal Society of Chemistry. He is an editorial board membrane of Journal of Membrane Science, Separation and Purification Technology, Green Chemical Engineering, etc. His research interests include biomimetic and bioinspired membranes and membrane processes, biocatalysis, and photocatalysis. Until now, he has co-authored over 560 peer-reviewed scientific papers, and the total citation times is over 22,000 with a h-index of 79.

Gap in pagination due to unavailable paper.

Pages 171-173

Proceedings of the 16th Annual IEEE International
Conference on Nano/Micro Engineered and Molecular Systems
April 25-29, 2021

Fabrication of Nanoslits with <111> Etching TSWE Method

Hao Hong, Li Ye, Ke Li, Pasqualina M. Sarro, Guoqi Zhang and Zewen Liu*

Abstract— In this paper, we report a modified three step anisotropic wet etching (TSWE) method to fabricate solid-state silicon nanoslits. The slit-opening process is performed by <111> crystal plane etching. The etching rate of the <111> crystal plane is reasonably slow as it is only 1/45 of the <100> etching rate, thus allowing and therefore good slits-opening controllability. By slowly etching the <111> crystal plane, the over-etching was effectively reduced. Perfectly rectangular nanoslits with different dimensions were successfully obtained. The smallest achieved feature size of the nanoslit is 8.3 nm.

I. INTRODUCTION

Solid-state nanopores/nanoslits have been used to study the fundamentals of ionic and polymer transport through nanofluidic channels [1], to develop detection tools for studying nucleic acids, proteins, and small molecules [2], and to characterize DNA nanostructure [3] over the past few years. More than that, various other applications for nanopores/nanoslits have also been demonstrated, such as gas separation [4], nanostencil lithography [5] and so on [6-10]. In principle, solid-state silicon is more robust, amenable to a larger range of operating conditions, compatible with semiconductor technology, so it is easier to integrate into electronics and embedded systems.

Common preparations of nanoslits mainly include dry and wet etching. Dry etching such as FIB and FEB could process sub-10nm nanoslits in a short time [11-14]. However, the high cost associated with it, has severely restricted the possibility of large-scale fabrication of nanoslits. Furthermore, the fabrication of nanoslits is difficult by FIB/FEM due to their serial manufacturing characteristics.

It has been demonstrated that the TSWE method, which is based on conventional semiconductor processing and MEMS wet etching techniques, could fabricate hollowed nanostructure effectively. By reducing the etching rate and improving the sensitivity of monitoring slit-opening events, good slits-opening controllability could be realized, thus nanoslits with different dimensions could be obtained. Feedback-based chemical etching method, as a simple and cost-effective strategy, has been successfully used to manufacture single crystal silicon (SCS) nanopores. Current feedback mechanism is an optional method, which monitors the pore-opening event by measuring the time

dependence of the electrical current across the silicon chip [15-16]. A bias voltage should be applied across the chip, and a measuring instrument with high sensitivity is needed to detect the electrical current in the liquid. Both the two methods could improve the sensitivity of monitoring slit-opening events and obtained extremely small feature size nanopores/nanoslits. Chen et al. use a color-feedback mechanism based on phenolphthalein mixed with KOH to turn red to monitor the slit-opening event, obtained the smallest nanoslit with the feature size of 5 nm [17]. Ye et al. detected the slit-opening event by a current feedback method and also got a small feature size nanoslit down to 3.8 nm [18]. In order to further reduce the over-etching, in this paper, the TSWE method is used to perform slow etching from the <111> crystal plane, thanks to the lower etching rate, and a nanoslit with feature size down to 8.3 nm is obtained.

II. EXPERIMENTS

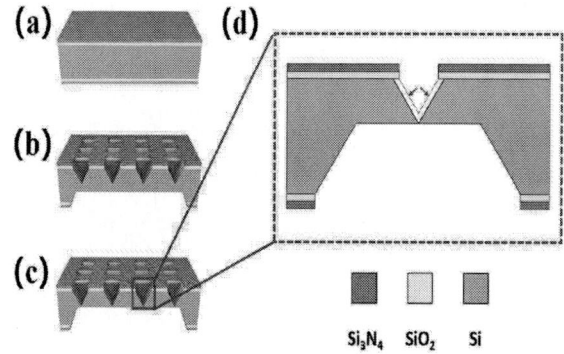

Fig. 1. Schematic illustration of the main process steps of the TSWE method: a) double-sided SiO_2 and Si_3N_4 deposited; b) the front cavity array and the back side etching window formed; c) the back-side thinning of silicon, d) etching model of the <111> planes.

The main process steps of the TSWE method are illustrated in Fig. 1. Single crystal silicon wafers with diameters of 4 inches and thickness of 300 μm were used as original substrates, all the wafers are double-polished, N-

---

*Research supported by Beijing Innovation Center for Future Chips, Beijing National Research Center for Information and The National Key R&D Program (2019YFA0707002).

H. Hong is with the Department of Microelectronics, Delft University of Technology, 2628 CD Delft, The Netherlands, and also with the Institute of Microelectronics, Tsinghua University, Beijing, China (e-mail: honghao@tsinghua.edu.cn).

L. Ye was with the Institute of Microelectronics, Tsinghua University, Beijing, China (e-mail: yel17@tsinghua.org.cn).

K. Li is with School of Electronic and Information Engineering, Beijing Jiaotong University, Beijing, China (e-mail: 19125019@bjtu.edu.cn).

P. M. Sarro and G. Zhang are the Department of Microelectronics, Delft University of Technology, 2628 CD Delft, The Netherlands (e-mail: P.M.Sarro@tudelft.nl; G.Q.Zhang@tudelft.nl).

Z. Liu is with the Institute of Microelectronics, Tsinghua University, Beijing, China (phone:+86 135-0123-8562; Fax:86-10-62771130; e-mail: liuzw@tsinghua.edu.cn).

978-1-6654-3008-1/21 $31.00 © 2021 IEEE

type. The concentration of KOH etching solution in each step was 33 wt.%.

At first, SiO_2 and Si_3N_4 layers were deposited on both sides of the silicon wafer grown by thermal oxidation and low-pressure chemical vapor deposition to be used as a stress buffer and mask layer respectively (refer to Fig. 1a). Then 33 wt.% KOH was used to form the front etching cavity array and the back side etching window as shown in Fig. 1b. Next, the back side thinning of silicon was conducted by the same KOH solution until it remaining about 2 μm. Finally, slow rate wet etching was in progress through the <111> crystal plane was performed until the nanoslits were opened. The slow etching rate could be explained as that two adjacent <100> crystal planes peel off each other, each silicon atom needs to break two chemical bonds. The density of dangling bonds is large, and it has higher surface energy and lower activation energy. For the <111> crystal plane, the separation of two adjacent <111> crystal planes require a chemical bond to be cut. So the <111> crystal plane has a low density of dangling bonds and low surface energy. And each silicon atom in the <111> plane is connected by three chemical bonds, so that it has a higher activation energy. As the etching rate increases with the increase of the dangling bond density, the dangling bond density of the <111> crystal planes is lower than for the <100> crystal planes. Thus the etching rate of the <111> crystal plane is reasonably slow as it is only 1/45 of the <100> etching rate at room temperature. Thanks to the lower etching rate, the over-etching was reduced effectively.

III.   RESULTS AND DISCUSSIONS

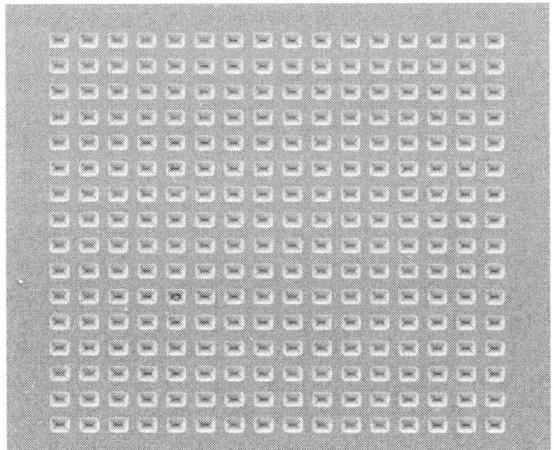

Fig. 2. SEM micrograph of the 16 × 16 etching cavity array.

After the first step of the wet etching, Fig. 2 exhibits a scanning electron microscope (SEM) micrograph of 16 × 16 etching cavity array, which could demonstrate that the front side etching cavity has been fabricated successfully over a large area. The size of each cavity is about 8 × 4 μm. Fig. 3 also shows a back side etched window, with a 0.5×0.5 mm^2 exposed area. From this enlarged figure, it could be seen that some nanoslits have been opened, which proves the proposed method could fabricate nanoslits effectively.

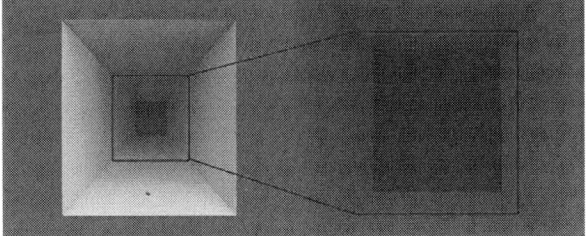

Fig. 3. SEM micrograph of the back etching window.

Furthermore, Fig. 4 gives a close-up of the obtained nanoslits. The shape of the nanoslits is rectangular, which is in accordance with the designed mask, and the distance between the upper and lower nanoslits is 8.035 μm. In addition, there still have some nanoslits which are not strictly rectangular. It could be explained by inhomogeneous etching, both the slit-opening time node of each side and the over-etching after the slit-opening event are different. Better controllability of uniformity in the wet etching process is necessary.

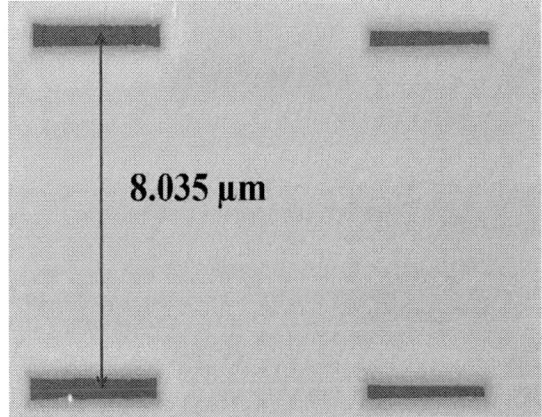

Fig. 4. SEM micrograph of the fabricated nanoslits array.

The best result is shown in Fig. 5, where the smallest feature size obtained 8.3 nm, is shown. Due to the slow rate etching performed from the <111> crystal plane, the over-etching is effectively reduced, and the sizes of the obtained nanoslits are also reduced in a controllable and reproducible way.

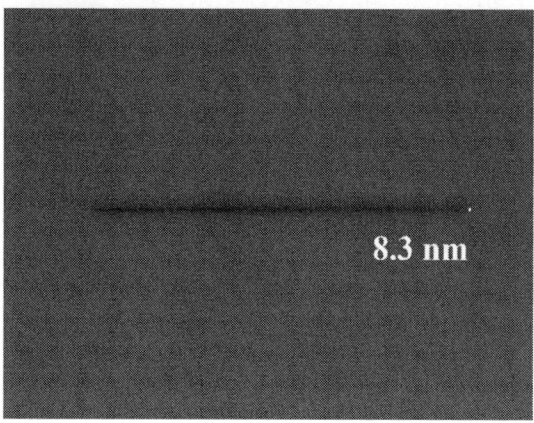

Fig. 5. The SEM micrographs of individual nanoslit fabricated by the TSWE method, with a feature size of 8.3 nm.

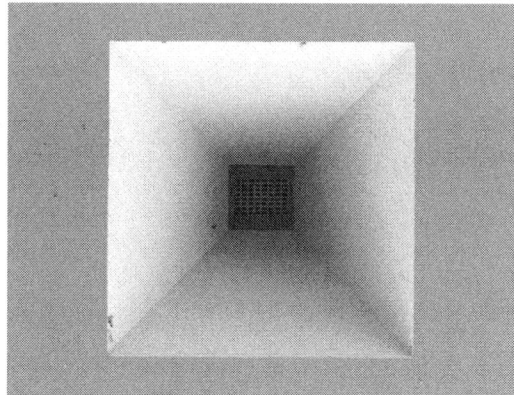

Fig. 6. SEM micrographs of fully opened nanoslit array.

In order to study the morphology and sizes of the completely opened nanoslits, Fig. 6 exhibits the SEM micrograph of the fully etched nanoslits. From the back etching window, an 8 × 8 nanoslit array has been obtained. all the nanoslits are rectangular as the front etching cavity is shown in Fig. 2. It could be demonstrated that all the nanoslits are opened. The next part will show the detail of the fully opened nanoslit.

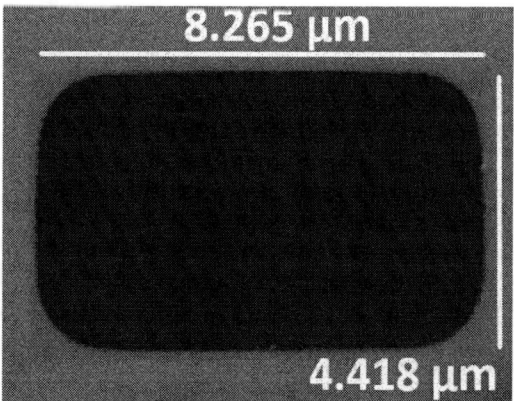

Fig. 7. SEM micrographs of a fully opened individual nanoslit.

Finally, after enough over-etching, the nanoslit with small dimension was fully opened, Fig. 7 shows a completely opened nanoslit, with the size of the nanoslit is 8.265 μm in length and 4.418 μm in width, which is corresponded with the front cavity.

IV. CONCLUSION

A modified TSWE method to fabricate solid-state silicon nanoslits was reported in this paper. The slit-opening process is performed by <111> crystal plane etching. Thanks to the reasonably slow <111> crystal plane etching rate compared with the <100> etching rate, the over-etching was reduced effectively and the good controllability of the slit-opening was realized. Perfectly rectangular nanoslits with different dimensions were successfully obtained. The smallest achieved feature size of the nanoslit is 8.3 nm. Duo the inhomogeneous etching, there still have some nanoslits which are not strictly rectangular. The proposed method provides a controllable and reproducible way obtain various nanostructure.

ACKNOWLEDGMENT

This research was supported by Beijing Innovation Center for Future Chips, Beijing National Research Center for Information and The National Key R&D Program (2019YFA0707002).

REFERENCES

[1] Briggs, K. et, "DNA translocations through nanopores under nanoscale preconfinement," *Nano Lett.* 18,660-668, 2018.

[2] Yusko, E. C. et al, "Real-time shape approximation and fingerprinting of single proteins using a nanopore," *Nat. Nanotechnol.* 12, 360-367, 2017.

[3] Karau, P. & Tabard-cossa, V, "Capture and translocation characteristics of short branched DNA labels in solid-state nanopores," *ACS Sens.* 3, 1308-1315, 2018.

[4] P. Wang, W. Li, C. Du, X. Zheng, X. Sun, Y. Yan, and J. Zhang, "CO_2/N_2 separation via multilayer nanoslit graphene oxide membranes: Molecular dynamics simulation study", *Computational Materials Science*, 140 , pp. 284-289, 2017.

[5] T. Deng, J. Chen, M. Li, Y. Wang, C. Zhao, Z. Zhang, and Z. Liu, "Controllable Shrinking of Inverted-pyramid Silicon Nanopore Arrays by Dry-oxygen Oxidation", *Nanotechnology*, 24, 50, 6, 2013.

[6] K. L. Wu, Z. X. Chen and X. F. Li, "Real gas transport through nanopores of varying cross-section type and shape in shale gas reservoirs,"*Chem. Eng. J*, 281, 813-825, 2015.

[7] D. Rodriguez-Larrea and H. Bayley, "Multistep protein unfolding during nanopore translocation," *Nat. Nanotechnol*, 8, 288-295, 2013.

[8] S. Benner, R. Chen, N. Wilson, R. Abu-Shumays, N. Hurt, K. R. Lieberman, D. W. Deamerm, W. B. Dunbar and M. Akeson, "Sequence-specific detection of individual DNA polymerase complexes in real time using a nanopore," *Nat. Nanotechnol*, 2, 718-724, 2007.

[9] J. Nivala, D. B. Marks and M. Akeson, "Unfoldase-mediated protein translocation through an α-hemolysin nanopore," *Nat. Biotechnol*, 31, 247-250, 2013.

[10] J. Nivala, L. Mulroney, G. Li, J. Schreiber and M. Akeson, "Discrimination among protein variants using an unfoldase-coupled nanopore," *ACS Nano*, 8, 12365-12375, 2014.

[11] J. Shim, J. A. Rivera and R. Bashir, "Electron beam induced local crystallization of HfO2 nanopores for biosensing application," *Nanoscale*, 5, 10887-10893, 2013.

[12] S. W. Kowalczyk, D. B. Wells, A. Aksimentiev and C. Dekker, "Slowing down DNA translocation through a nanopore in lithium chloride," *Nano Lett*, 12, 1038-1044, 2012.

[13] I. Yanagi, T. Ishida, K. Fujisaki and K-ichi Takeda, "Fabrication of 3 nm-thick Si3N4 membranes for solid-state nanopore using the poly-Si sacrificial layer process," *Sci. Rep*, 5, 14656, 2015.

[14] G. F. Schneider, S. W. Kowalczyk, V. E. Calado, G. Pandraud, H. W. Zandbergen, L. M. K. Vandersypen and C. Dekker, "DNA translocation through graphene nanopores," *Nano Lett*, 10, 3163-3167, 2010.

[15] D. Pedone, M. Langecker, G. Abstreiter and U. Rant, "A nanoparticles and DNA molecules in a femtoliter compartment: confined diffusion and narrow escape," *Nano Lett*, 11, 1561-1567, 2011.

[16] S. R. Park, H. Peng and X. S. Ling, "Fabrication of nanopores in silicon chips using feedback chemical etching," *Small*, 3, 116-119, 2007.

[17] Q. Chen, Y. Wang, T. Deng and Z. Liu, "Fabrication of nanopores and nanoslits with feature sizes down to 5 nm by wet etching method," *Nanotechnology*, 29, 085301, 2018.

[18] L. Ye, H. Hong and Z. Liu, "Fabrication of Single-Crystal Silicon Nanoslits with Feature Sizes Down to 4 nm and High Length-Width Ratios," *2020 IEEE 33rd International Conference on Micro Electro Mechanical Systems (MEMS)*, Vancouver, BC, Canada, pp. 279-282, 2020.

Gap in pagination due to unavailable paper.

Pages 178-179

Proceedings of the 16th Annual IEEE International
Conference on Nano/Micro Engineered and Molecular Systems
April 25-29, 2021

Characterization of ITO-SiN$_x$ Nanopores for Single-Biomolecular Sensing

Xin Zhu, Chaoming Gu, Xiaojie Li, Zhi Ye, Zhen Cao, and Yang Liu[*]

Abstract— **Nanopores incorporating electrical gates made of ITO material have highly-desired potential of electrically modulating the translocation process of biomolecules. In this work, we report our experimental study of such devices under the condition of no electrical gate biases being applied. This baseline characterization is to demonstrate their suitability for translocation sensing. The nanopore devices incorporating ITO gate electrodes are fabricated by TEM drilling, yielding~10 nm pore sizes. dsDNA translocation experiments, as well as noise characterization, are conducted using these devices and compared with bare SiNx nanopore control devices. We particularly examine the individual signals and statistics of the folding and knotting configurations of translocating dsDNAs. It is demonstrated that, without applying gate biases, the incorporated ITO layers only have moderate effects on the dsDNA translocation process, suggesting their general suitability for translocation sensing.**

INTRODUCTION

Solid-state nanopores have been extensively studied for their great potential of detecting biomolecules at the single-molecular level[1,2]. It is highly desired to integrate electrical functions at the front-end of nanopores to extend detection capabilities by embedding gates or FETs in nanopores[3]. The choice of proper active layer materials is crucial for such devices. Metals such as gold have been used as gate materials but are not compatible with the usual TEM nanopore fabrication process. Focus ion beams (FIB) is needed for such devices, resulting in rather large nanopore diameter (~100 nm), making detecting single biomolecule difficult. Paik *et al.* used nanopore array to modulate the translocation rate of DNAs, the modulation effect is examined by PCR analysis, but no single bio-molecule translocation events were detected[4]. TiN is another option that is compatible with the TEM process. Harrer *et al.* fabricated such devices with complex process achieving ~10 nm pore diameters, but they still did not demonstrate translocation events[5].

Indium tin oxide (ITO) is a gate material extensively used in TFTs owing to its transparent, cheap, and non-toxic property[6]. More importantly, ITO is compatible with the common nanopore TEM drilling method. Our previous work has demonstrated that nanopores incorporating ITO gate electrodes are capable of modulating the passage of dsDNAs with different folding or knotting configuration states by applying gate biases[7]. In this study, we complement that

work by investigating their translocation and noise characteristics under the condition of no gate biasing. These results are compared to those of bare SiN$_x$ control nanopores. With such a comparison, we aim to identify whether the additional ITO layer itself hinders the translocation process and the general suitability of ITO-gated nanopores for translocation sensing.

DEVICE FABRICATION AND CHARACTERIZATION

The main fabrication process of ITO-SiN$_x$ nanopores includes sputtering ~20 nm ITO on a ~10 nm free-standing SiN$_x$ membrane and the following TEM drilling process, as shown in Fig. 1. The low-stress SiN$_x$ membrane used in this report is purchased from Norcada. ITO thin-film is deposited through magnetron sputtering (Sputter JSS-450-1). 400°C N$_2$ annealing is conducted to fully release the stress right after the sputtering process. We find devices are prone to be broken without the annealing process.

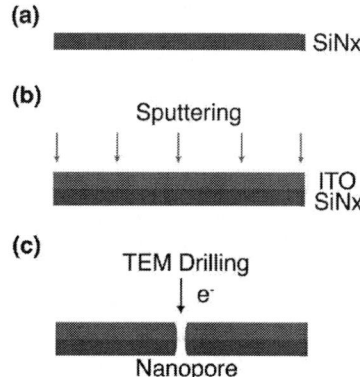

Fig. 1 Fabrication process of ITO-SiN$_x$ device. (a, b) ~20 nm ITO is deposited through sputtering on ~10 nm low-stress SiN$_x$ membrane (c) nanopore is drilled through TEM

We use TEM (FEI Tecnai G2-F20) to drill a nanopore on the SiNx-ITO double-layer structure. During the drilling process, ITO is removed gradually at first and then SiN$_x$ is exposed as shown in Fig. 2(a). The initial nanopore is formed in ~ 5 min and expand rapidly. TEM figure of a ~5 nm ITO-SiN$_x$ device is shown in Fig. 2(b). Clear crystalline ITO and amorphous SiN$_x$ can be identified in the TEM figure. The

---

*This work was supported by National Natural Science Foundation of China (61774132, 61574126).
Xin Zhu, Chaoming Gu, Xiaojie Li, Ye Zhi, Zhen Cao, and Yang Liu are with College of Information Science and Electronic Engineering, Zhejiang University, Hangzhou 310027, China.

*Contacting Author: Yang Liu is with College of Information Science and Electronic Engineering, Zhejiang University, Hangzhou 310027, China (email: yliu137@gmail.com).

978-1-6654-3008-1/21 $31.00 © 2021 IEEE

FFT of Fig. 2(b) further indicates the co-exist of ITO and SiNx membrane.

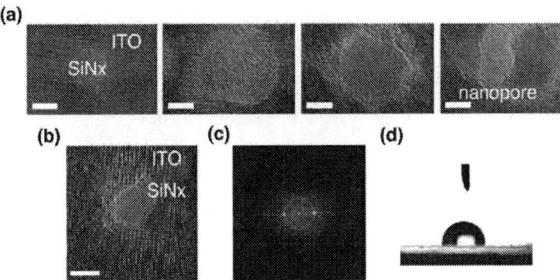

Fig. 2 Characterization of the ITO-SiNx devices. (a) Serious TEM pictures of SiNx-ITO membrane during the drilling process, the time interval is ~10 s, scale bar: 10 nm. (b) Typical ITO-SiNx nanopore device, scale bar: 5 nm. (c) FFT of nanopore region in (b). (d) hydrophilic characterization of ITO-SiNx device

The diameter of the final device can be well controlled by adjusting drilling time. To avoid further changing the size and shape of nanopores, we will take a snap-shot of devices reaching the desired size and stop further electron irradiation to the devices. It's noteworthy that as the nanopore size has a significant effect on the translocation experiment, we usually control the final diameter of nanopores to be ~10 nm[8]. The nanopores are thus ~30 nm in length and ~10 nm in diameter.

The ITO-SiNx devices exhibit hydrophobic property as shown in Fig. 2(d). Air plasma treatment is therefore adopted to fully prime the devices just before the translocation experiment. After the plasma process, we immediately package the PCB-Chip assembly with two PTFE fluidic cells using two custom-made silicone O-rings. This process ensures the formation of a giga-ohm seal. Ethanol is firstly introduced into the cells to initially wet the device. It is then gradually replaced by buffered saline solution. In this study, a 2 M LiCl solution with 10 mM Tris-EDTA buffer (pH=8) is used. Trans-membrane current is recorded and amplified by a patch-clamp amplifier (Axopatch 200B) and digitalized by a DAQ system (Axon Digidata 1550). The translocation events are identified and analyzed from the traces using Translyzer[9].

TRANSLOCATION EXPERIMENT AND DISCUSSIONS

We conduct translocation experiments of two ~10 nm ITO-SiNx devices and one ~10 nm SiNx control device. The measurement set-up follows previous work[7,8,10] and is shown in Fig. 3. Analytes used for all devices are 5 nM λDNAs (double-strand, 48.5 kbp). The persistence length of λDNAs at high saline solution such as 2 M in our experiment is ~50 nm. Folding and knots configurations of DNAs are expected during translocation[11,12].

Ionic current traces of one ITO-SiNx device and one bare SiNx device are shown in Fig. 4(a). The trans-membrane voltage for both devices is 500 mV. The open-pore ionic current for both cases is consistent with the model, indicating fully wetting of devices. We can observe that for the ITO-SiNx case, the current trace is lower than the bare SiNx case, which is attributed to the higher effective thickness of cylindrical nanopore in the ITO-SiNx case. More importantly, higher

baseline fluctuation is observed in the ITO-SiNx case. Further PSD analysis as shown in Fig. 4(b) indicates higher 1/f noise of ITO-SiNx devices than bare SiNx control. Additional ITO layer might introduce surface traps which finally deteriorate the Signal-Noise-Ratio (SNR) of devices[13]. Nevertheless, clear translocation events can be identified in both cases.

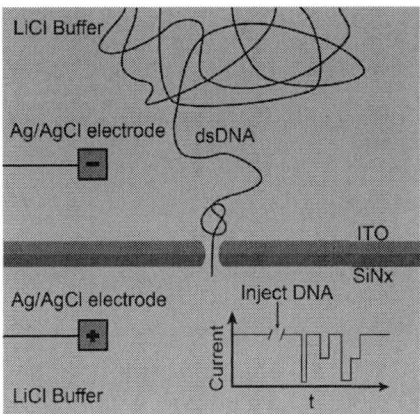

Fig. 3 Schematic of ds-λDNA translocation experiment set-up. The nanopore devices are immersed in suffered LiCl solution. Two Ag/AgCl electrodes are used to apply trans-membrane voltage.

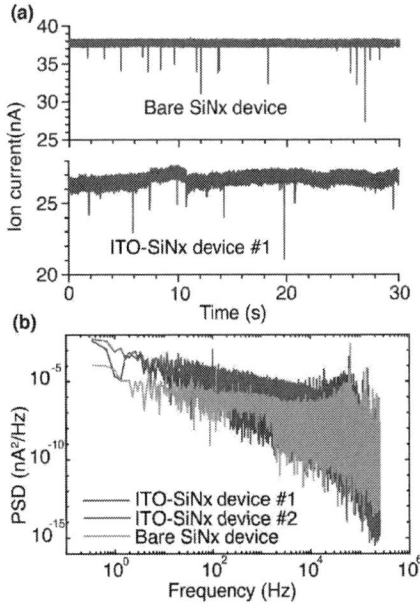

Fig. 4 (a) Current trace of ITO-SiNx device and bare-SiNx device. (b) PSD analysis of two ITO-SiNx devices and one SiNx device, the time interval is 3 s. Trans-membrane voltage for all devices is 500 mV

The events statistics of one ITO-SiNx device and one bare SiNx device are shown in Fig. 5. Total 395 translocation events are extracted from 18 min trace for bare SiNx device and 128 events in 6 min for ITO-SiNx case. We can see that the additional ITO layer has a moderate effect on the translocation rate. Plotting maximum blockage versus dwell time of events, we can recognize clusters of events from the scatter plot of both devices (Fig. 5(a, d)). Linear, folding, and knots are attributed to 1st, 2nd, and 3rd clusters of events respectively as shown in Fig. 5(b, e). Classical "type 1" "type 21" and "type 213" configurations[5] are identified from Fig. 5(a, d). We can observe that the events distribution pattern and individual event configuration are similar for two devices, while the specific ratio of folding/knots to linear events is of significant difference.

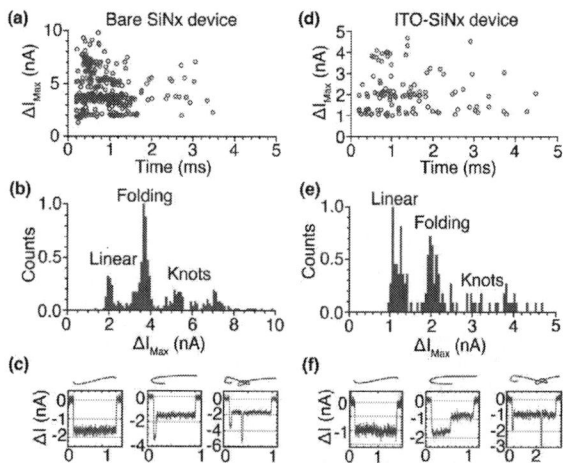

Fig. 5 DNA configuration analysis of ITO-SiNx device and bare SiNx device. (a)Scatter plot of translocation events under 500 mV transmembrane for bare SiNx device, current blockage used here is maximum descend of each event (n = 395 events, t = 18 min). (b) DNA configuration distribution in (a). (c) typical linear, folding and knots configuration and corresponding morphology in (b) (d-f) similar analysis of ITO-SiNx device (n = 128 events, t = 6 min). Trans-membrane voltage is 500 mV for both devices

We conduct ECD (event charge deficit) analysis to further demonstrate the discrepancies between the two sorts of devices. ECD was first proposed by Li. J et al.[14]. ECD is usually regarded as a constant value under the same bias, regardless of diverse configuration within DNA while traversing nanopore[14,15]. Events of three devices including two ITO-SiNx devices and one SiNx device all follow a hyperbolic pattern in their dwell time vs average blockage level relation, as shown in Fig. 6(a). This finding is consistent with the ECD model. We can also observe that a small portion of events (black circle) deviate away from the hyperbolic fitting (green dash). Those events might be attributed to the interaction between biomolecule and devices[2]. Such interactions might prolong the translocation process. Interestingly, we find ITO-SiNx devices are likely to occupy the center of the fitting curve while SiNx control dominates the upper-half. This finding indicates a lower portion of events are complex configurations for ITO-SiNx double-layer structure, which is the opposite for bare SiNx control[12]. The configuration distribution variation between two sorts of devices is consistent with previous configuration analysis. It's noteworthy that although two types of devices both obey the ECD law, the dwell time and specific ECD value vary from device to device as shown in Fig.6 (b-c).

Such differences might originate from effective thickness variation and moderate interaction between ITO and DNAs.

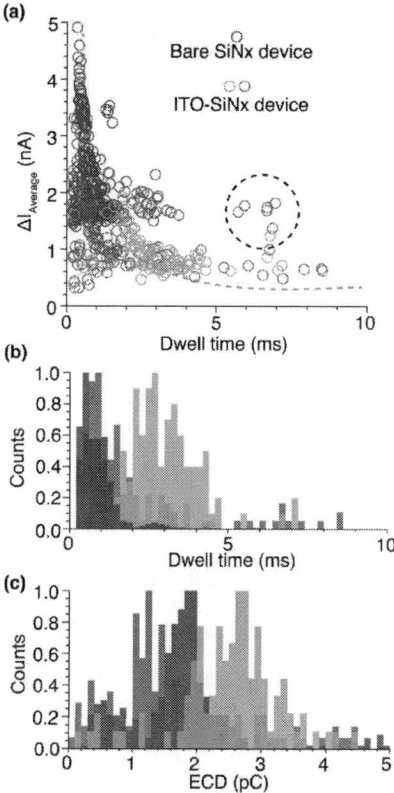

Fig. 6 ECD analysis of ITO-SiNx and bare SiNx devices. (a) Scatter plot of translocation events extracted from two ITO-SiNx devices and one SiNx device, current blockage used here is the average value of each event (blue circle: n = 395 events; red circle: n = 128 events; yellow circle: n=104 events). (b)&(c) Corresponding dwell time distribution and ECD distribution of (a). The trans-membrane voltage is 500 mV for all devices

SUMMARY

We fabricate and characterize ITO-SiNx double-layer devices, and conduct translocation experiments for ITO-SiNx devices and bare SiNx control. Our control experiment and further analysis indicate that the additional layer of ITO does not hinder the dsDNA translocation process under the condition of no gate biasing, except for its influence on dsDNA configuration distribution. Our findings show ITO is a great alternative for further monolithic integration of gates or FETs at the front end of the nanopores.

ACKNOWLEDGMENT

We acknowledgment Xiaowei Wang from the center of microscopy, Zhejiang University for the help of TEM drilling and characterization. The Sputtering process is done at ZJU Micro-Nano Fabrication Center.

REFERENCES

[1] D. Branton, D. W. Deamer, A. Marziali, "The potential and challenges of nanopore sequencing," Nat Biotechnol, vol. 26, no. 10, pp. 1146–1153, Oct. 2008.

[2] S. Garaj, W. Hubbard, A. Reina, J. Kong, D. Branton, and J. A. Golovchenko, "Graphene as a subnanometre trans-electrode membrane," Nature, vol. 467, no. 7312, pp. 190–193, Aug. 2010.

[3] L. Xue, H. Yamazaki, R. Ren, M. Wanunu, A. P. Ivanov, and J. B. Edel, "Solid-state nanopore sensors," Nature Reviews Materials, pp. 1–21, Sep. 2020.

[4] K.-H. Paik, Y. Liu, V. Tabard-Cossa, M. J. Waugh, D. E. Huber, J. Provine, R. T. Howe, R. W. Dutton, and R. W. Davis, "Control of DNA capture by nanofluidic transistors.," ACS Nano, vol. 6, no. 8, pp. 6767–6775, Aug. 2012.

[5] S. Harrer, P. S. Waggoner, B. Luan, A. Afzali-Ardakani, D. L. Goldfarb, H. Peng, G. Martyna, S. M. Rossnagel, and G. A. Stolovitzky, "Electrochemical protection of thin film electrodes in solid state nanopores," Nanotechnology, vol. 22, no. 27, pp. 275304–7, May 2011.

[6] A. Jiang, Y. Yuan, N. Liu, "Transparent Capacitive-Type Fingerprint Sensing Based on Zinc Oxide Thin-Film Transistors," IEEE Electron Device Lett., vol. 40, no. 3, pp. 403–406, Feb. 2019.

[7] X. Zhu, X. Wang, Z. Cao, "Nanopores incorporating ITO electrodes for electrical gating of DNA at different folding states," IEEE IEDM, 26.1, Dec.2017

[8] X. Zhu, Li X, Gu C, "Experimental study of excessively-long translocation time of single DNA through sub-5 nanometer solid-state nanopores," IOP Conference Series: Earth and Environmental Science, vol. 632, no. 5, 052072, Oct. 2020.

[9] C. Plesa and C. Dekker, "Data analysis methods for solid-state nanopores," Nanotechnology, vol. 26, no. 8, pp. 1–8, Jan. 2015.

[10] C. Gu, Z. Yu, X. Li, "Experimental study of protein translocation through MoS2 nanopores," Appl. Phys. Lett., pp. 1–6, Nov. 2019.

[11] C. Plesa, D. Verschueren, S. Pud, "Direct observation of DNA knots using a solid-state nanopore," Nature Nanotechnology, pp. 1–6, Aug. 2016.

[12] R. K. Sharma, I. Agrawal, L. Dai, P. S. Doyle, and S. Garaj, "Complex DNA knots detected with a nanopore sensor," Nat Commun, pp. 1–9, Sep. 2019.

[13] R. M. M. Smeets, U. F. Keyser, N. H. Dekker, and C. Dekker, "Noise in solid-state nanopores," Proc. Natl. Acad. Sci. U.S.A., vol. 105, no. 2, pp. 417–421, Jan. 2008.

[14] J. Li, M. Gershow, D. Stein, E. Brandin, and J. A. Golovchenko, "DNA molecules and configurations in a solid-state nanopore microscope," Nat Mater, vol. 2, no. 9, pp. 611–615, Sep. 2003.

[15] M. Mihovilovic, N. Hagerty, and D. Stein, "Statistics of DNA Capture by a Solid-State Nanopore," Phys. Rev. Lett., vol. 110, no. 2, pp. 028102–5, Jan. 2013.

Proceedings of the 16th Annual IEEE International
Conference on Nano/Micro Engineered and Molecular Systems
April 25-29, 2021

Mechanical Energy Harvester for Smart Shared Bicycle Application

Junlong Huang, Chen Bao, Anxin Luo and Fei Wang*, *Senior Member, IEEE*

Abstract—**The human body-related movements contain ultra-low frequency characteristics, and energy harvesters at ultra-low frequency have received extensive attention in energy harvesting. Bicycles are one of the common transportation modes in cities, but the energy supply for devices such as smart locks on the shared bicycles is a challenging task. By combining ultra-low frequency energy harvesters with the motion of riding a bicycle, milliwatt-level energy output can be continuously generated. And through attaching micro-sensors, the shared bicycles can become a self-supplied mobile sensing node.**

I. INTRODUCTION

With the introduction of low-carbon mobility initiatives, shared bicycles have been put into use in significant numbers in Chinese cities [1]. However, the smart locks with wireless transceiver systems used for shared bicycles require sufficient power to work properly and the positioning devices on some other bicycles also require energy supply [2]. The energy supply of existing commercial shared bicycle is mainly provided by chemical batteries, solar panels and small generators like bicycle hubs [3]. However, chemical batteries are difficult to replace and faced environmental pollution problem. Solar panels are susceptible to weather leading to unstable in energy collection efficiency. Therefore, the use of mechanical energy into electrical energy has become a possible technological trend, such as bicycle hubs or other mechanical energy harvesters. Yang et al. [4] proposed an energy harvesting structure using the conversion of up and down vibration into rotation when the bicycle through a speed bump. Wang et al. [5] proposed a energy harvester based magneto-fluid that using low-frequency magneto-fluid vibrations caused by changes in acceleration during braking, acceleration or steering to harvest energy. Nevertheless, these existing methods are not suitable for harvesting energy on relatively gentle roads in cities.

Another option is to attach micro energy harvesters inside these smart locks devices. Some teams have already designed MEMS energy harvesters with high bandwidth, high energy density and stable operation in harsh environments [6–8]. However, these MEMS devices require high resonance frequencies work environment, making it difficult to apply directly in bicycle riding.

Based on the previously published structure of the ultra-low frequency energy harvester [9] and the test method of

All Author are with the School of Microelectronics, Southern University of Science and Technology, Shenzhen 518055, China.

*Corresponding Author: Fei Wang is with the School of Microelectronics, Southern University of Science and Technology, Shenzhen 518055, China. (email: wangf@sustech.edu.cn)

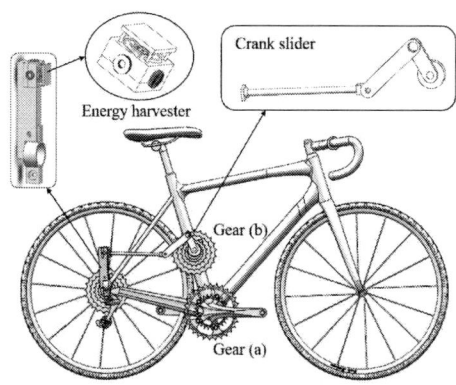

Fig. 1. The schematic diagram of energy harvesting system.

crank slider in low frequency energy harvesting application [10, 11], we propose a new energy harvesting system applicable to bicycle structures. This system harvests energy during the normal cycling, not only for powering the smart lock, but also for other sensors on the bicycle, enhancing cycling safety and enabling the bicycle to act as a mobile sensing node to collect data information widely in the city. Moreover, it can also be applied to other devices that perform energy harvesting through pressing [12–14].

II. DEVICE DESIGN

The entire structure of the bicycle equipped with the energy harvesting system is shown in Fig. 1. We added a new chain to the bicycle's pedal pivot to drive gear (b). At the same time, an energy harvesting device is mounted at the rear of the bicycle by a fixing assembly. The rotation of gear (b) drives the crank slider mechanism to press the energy harvester periodically. Through the crank-slider, the energy harvester can be adjusted to work at perfection by adjusting the rotation speed through the teeth number of gears.

We abstractly modeled the crank slider as shown in Fig. 2. By theoretical derivation, we derived the expressions for the end velocity of the crank slider as a function of the rotational speed and the length of the slider rods, where:

$$v = \omega r \cos(\omega t) + \frac{\omega r^2 \sin(2\omega t)}{2\sqrt{l^2 - r^2 \cos^2(\omega t)}}$$

Due to the higher speed is beneficial to increase the output power [9]. We choose different ratios between l and r and plot the maximum velocity for the end of crank rod as shown in

978-1-6654-3008-1/21 $31.00 © 2021 IEEE

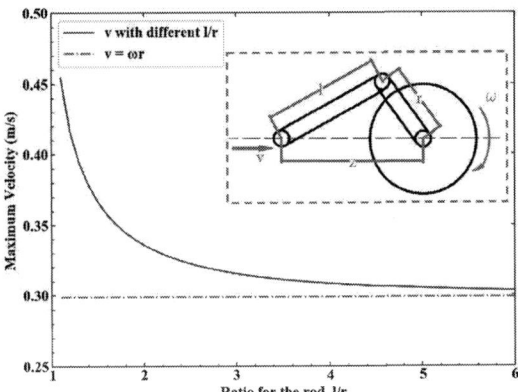

Fig. 2. Crank slider model with relevant parameter and the maximum speed at the end of the crank slider for different lengths l/r of the slider rod. ($r = 0.08\,\text{m}$, and, $\omega = 3.74\,\text{rad/s}$).

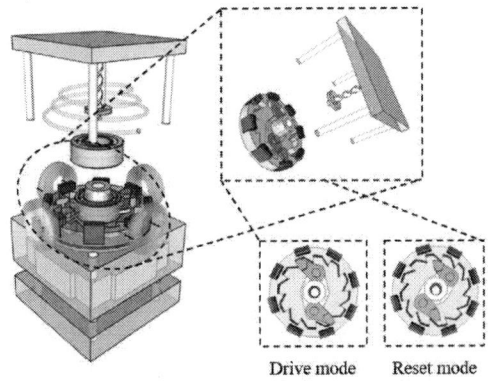

Fig. 3. Schematic diagram of the harvesting device in the previous study [9] with drive and reset modes.

Fig. 2. The maximum velocity gradually decreases to a constant value (ωr) as l/r increases. This shows that we can get a high speed by increasing r appropriately. Also, the model facilitates us to adjust the length of the slider bar to match with different sizes of bicycles.

The crank-slider is used to convert the rotational motion into a linear motion. The main device of the whole system is the ultra-low frequency energy harvester based on an inertial rotor [9]. The linear motion causes the threaded rod to drive the rotor to rotate with magnets fixed on it, thus generating a periodically varying magnetic flux to produce an induced current in the coils. When the crank slider is reset, the rotor will continue to rotate with inertia due to the ratchet pawl structure. The overall assembly is shown in Fig. 3.

III. EXPERIMENTS AND RESULTS

According to the schematic, the actual diagram of our assembled bicycle with the energy harvester installed is shown

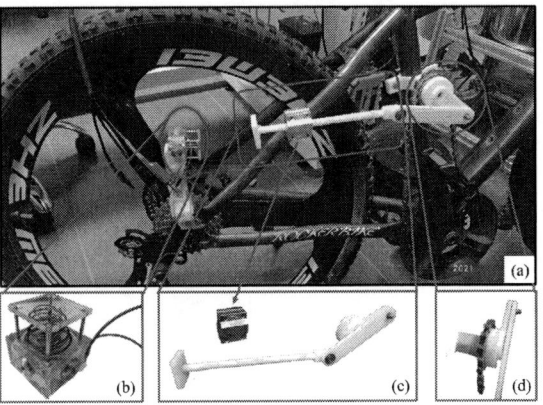

Fig. 4. The diagram of proposed energy harvesting system: (a) The bicycle energy harvester system, (b) Energy harvester, (c) Crank slider mechanism, (d) Spindle with gears.

in Fig. 4. In the diagram, in addition to the energy harvester part, we have marked the spindle with gears connected to the tooth gear carried by the bicycle turning pedals. The spindle and the drive rod form a crank slider mechanism. The drive rod is also connected to the push rod via a bearing, and the push rod presses the energy harvester through the limiter which ensures that the push rod moves in the horizontal direction.

In order to reduce the mass of the device and to ensure the mechanical strength, we replaced the previous metal devices with devices obtained through 3D printing using resin as the material. Some dimensions and parameters of the crank slider mechanism and energy harvester are shown in Table I.

Then, we connected the output pins of the energy harvester to the Keithley 6514 electrostatic meter. The voltage waveform and values can be recorded after opening the Labview acquisition program for voltage signals on the PC and setting the voltage range sensitivity and sampling rate parameters.

During the experiment, we drove the entire bicycle energy harvesting system through turning the bike pedals by hand. It should be noticed that for measuring the rotational speed, we used the unit of rpm (revolutions per minute) which is one of

TABLE I
DESIGN PARAMETERS OF THE BICYCLE ENERGY HARVESTER.

Description	Values	Units
Number of teeth on chainwheel	34	-
Number of teeth on spindle	15	-
Length of spindle rod	54	mm
Length of drive rod	105	mm
Length of push rod	205	mm
Resistance of energy harvester	86.4	Ω
Size of energy harvester	$35 \times 35 \times 32$	mm^3

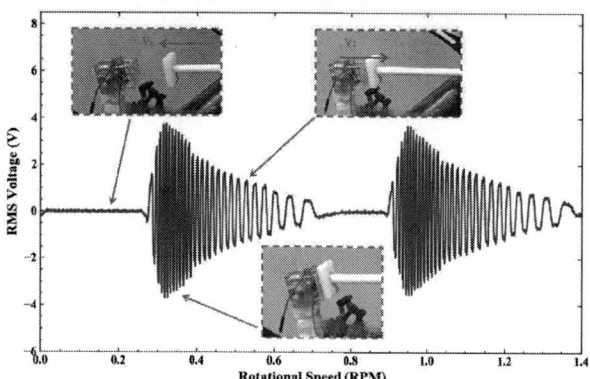

Fig. 5. The energy harvesting process of the push rod contacting and pressing the energy harvester.

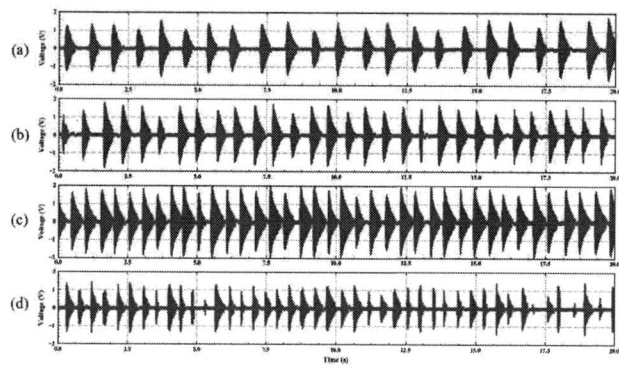

Fig. 7. The the output voltage waveform with $86.4\,\Omega$ resistor in different rotational speed: (a) 30 rpm, (b) 40 rpm, (c) 50 rpm, (d) 60 rpm.

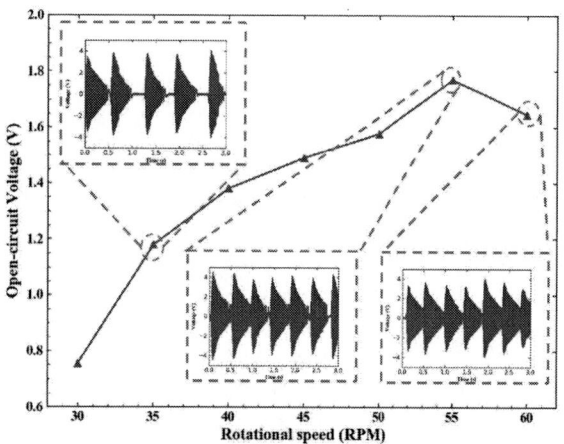

Fig. 6. The output voltage and voltage waves at the open circuit condition.

the key parameter to measure the cycling process.

When the bicycle turns the pedal to drive the chainwheel at 50 rpm, the rotational speed of the spindle is 113 rpm. According to the calculation, and the end of the push rod can reach $0.73\,\mathrm{m/s}$. With the rotation of the spindle, the crank slider drives the push rod contacting, pressing and releasing the energy harvester, and the working schematic is shown in the Fig. 5.

In the study of optimal pedal speeds for bicycle riding, Coast et al. [15] selected trained bicycle racers and found that the appropriate pedal speed for riding was between 60 and 80 rpm. Since untrained people experience more resistance at high pedal frequencies, we set the pedal speed to 30 to 60 rpm for testing in order to meet the daily riding of the shared bicycle rather than for racing.

We used the sound come from the metronome as a signal to turn the bicycle pedal by hand, and the measured Root-Mean-

Square output voltage at the open circuit condition is shown in Fig. 6.

From the figure, we could see that the open circuit voltage gradually increases with the increase of pedal speed at first. This is because the increase in pedal speed leads to an increase in the frequency at which the energy harvester is pressed. When the pedal speed reaches 55 rpm, the equivalent pressing frequency reaches $2.1\,\mathrm{Hz}$. At such a high pressing frequency, the ratchet wheel which carrying the magnet inside the energy harvester spins at high speed for an increased time, so the output voltage increases accordingly.

However, when the pedal speed increases from 55 to 60 rpm, the open circuit voltage drops slightly. From the voltage waveform figure, we could see that when the pedal speed is too high, the excessive pressing frequency causes the twist lever in the energy harvester to retract quickly. This will additionally increase the frictional damping between the pawl and ratchet, resulting in a rapid decay of the maximum output voltage after each press of the energy harvester, thus causing the overall RMS output voltage to drop. We can further increase the output voltage of the system in the high pedal speeds situation by selecting the suitable materials of ratchet and pawls to reduce the frictional damping inside the energy harvester. It is also possible to change the ratio of the rod lengths in the crank slider to reduce the pressing speed so that it will increase the pedal speed at which peak output is reached.

The total internal resistance of the energy harvester with four coils fixed is $86.4\,\Omega$, so we connect the output of the energy harvester to a resistor box adjusted to the same resistance value and the output voltage waveform obtained is shown in Fig. 7. It is obvious from the figure that as the rotation speed increases, the number of presses on the energy harvester per unit time increases. When the rotation speed is too large, the peak voltage per cycle is affected by the additional drag brought by the same as in the case of an open circuit.

978-1-6654-3008-1/21 $31.00 © 2021 IEEE

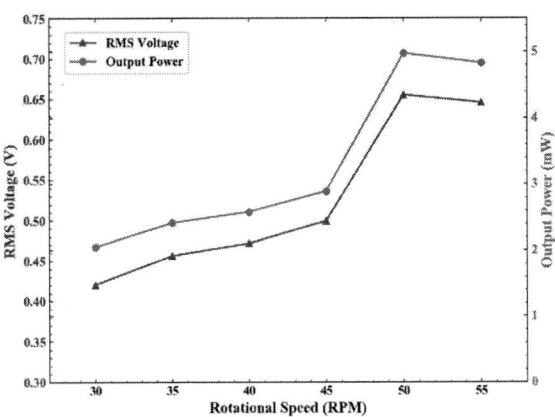

Fig. 8. The maximum output power and output voltage of the bicycle energy harvester with the increase of rotational speed.

The results of our measured RMS output voltage and maximum output power with matching resistors connected are shown in Fig. 8. The maximum output power of 4.97 mW appears at a rotation speed of 50 rpm, which is a very suitable speed for daily bike sharing rides.

IV. CONCLUSION

We have successfully used a crank slider mechanism to convert the rotation of bicycle pedals into linear horizontal motion and drive ultra-low frequency inertial rotor energy harvesting to form a bicycle energy harvesting system. The power output of 2.05 mW to 4.97 mW level was successfully collected in the range of daily riding speed from 30 rpm to 55 rpm of pedal rotation speed. The highest output power of 4.97 mW can be reached at 50 rpm. When the pedal rotation speed is greater than 55 rpm, the output power and voltage will drop. In the future, we can try to improve this decreasing trend by reducing the frictional resistance of the pawl and ratchet inside the collector and adjusting the ratio of the rod length in the crank slider.

Theoretically, the milliwatts of power obtained by the bicycle energy harvesting system could intermittently power several sensors. And it can be further applied to the shared bicycle to make it a mobile sensing node in the city. At the same time, this way of converting rotation into linear motion and thus harvesting low frequency energy can also provide a new idea for bicycle and other low frequency energy harvesting region.

ACKNOWLEDGMENT

This work was financially supported in part by the Shenzhen Science and Technology Innovation Committee under Grant JCYJ20200109105838951, in part by Guangdong Natural Science Funds under Grant 2018A050506001 and in part by "Climbing Program" Special Funds under Grant pdjh2021c0078.

REFERENCES

[1] S. Cai, X. Long, L. Li, H. Liang, Q. Wang, and X. Ding, "Determinants of intention and behavior of low carbon commuting through bicycle-sharing in china," *Journal of Cleaner Production*, vol. 212, pp. 602 – 609, 2019.

[2] B. Op het Veld, D. Hohlfeld, and V. Pop, "Harvesting mechanical energy for ambient intelligent devices," *Information Systems Frontiers*, vol. 11, no. 1, pp. 7–18, 2009.

[3] V. Kumar and D. P. Verma, "Design and fabrication of planetary drive magnet pedal power hub-dynamo," *Advances in Materials Science and Engineering: An International Journal (MSEJ)*, vol. 2, no. 3, 2015.

[4] Y. Yang, Y. Pian, and Q. Liu, "Design of energy harvester using rotating motion rectifier and its application on bicycle," *Energy*, vol. 179, pp. 222–231, 2019.

[5] S. Wang, Y. Liu, D. Li, and Z. Zhang, "A ferrofluid-based planar vibration energy harvester for smart lock of shared bicycle," *International Journal of Applied Electromagnetics and Mechanics*, vol. 61, no. 2, pp. 293–300, 2019.

[6] Y. Zhang, T. Wang, A. Luo, Y. Hu, X. Li, and F. Wang, "Micro electrostatic energy harvester with both broad bandwidth and high normalized power density," *Applied Energy*, vol. 212, pp. 362–371, 2018.

[7] A. Luo, Y. Zhang, X. Guo, Y. Lu, C. Lee, and F. Wang, "Optimization of mems vibration energy harvester with perforated electrode," *Journal of Microelectromechanical Systems*, pp. 1–10, 2021.

[8] A. Luo, Y. Xu, Y. Zhang, M. Zhang, X. Zhang, Y. Lu, and F. Wang, "Spray-coated electret materials with enhanced stability in a harsh environment for an mems energy harvesting device," *Microsystems and Nanoengineering*, vol. 7, no. 1, p. 15, 2021.

[9] A. Luo, Y. Zhang, X. Dai, Y. Wang, W. Xu, Y. Lu, M. Wang, K. Fan, and F. Wang, "An inertial rotary energy harvester for vibrations at ultra-low frequency with high energy conversion efficiency," *Applied Energy*, vol. 279, p. 115762, 2020.

[10] K. Fan, H. Qu, Y. Wu, T. Wen, and F. Wang, "Design and development of a rotational energy harvester for ultralow frequency vibrations and irregular human motions," *Renewable Energy*, vol. 156, pp. 1028–1039, 2020.

[11] K. Fan, P. Xia, Y. Zhang, H. Qu, G. Liang, F. Wang, and L. Zuo, "Achieving high electric outputs from low-frequency motions through a double-string-spun rotor," *Mechanical Systems and Signal Processing*, vol. 155, p. 107648, 2021.

[12] Q. Tan, K. Fan, K. Tao, L. Zhao, and M. Cai, "A two-degree-of-freedom string-driven rotor for efficient energy harvesting from ultra-low frequency excitations," *Energy*, vol. 196, p. 117107, 2020.

[13] K. Fan, J. Liu, M. Cai, M. Zhang, T. Qiu, and L. Tang, "Exploiting ultralow-frequency energy via vibration-to-

rotation conversion of a rope-spun rotor," *Energy Conversion and Management*, vol. 225, p. 113433, 2020.

[14] K. Fan, M. Cai, F. Wang, L. Tang, J. Liang, Y. Wu, H. Qu, and Q. Tan, "A string-suspended and driven rotor for efficient ultra-low frequency mechanical energy harvesting," *Energy Conversion and Management*, vol. 198, p. 111820, 2019.

[15] J. Coast, R. Cox, and H. Welch, "Optimal pedalling rate in prolonged bouts of cycle ergometry," *Medicine and science in sports and exercise*, vol. 18, no. 2, p. 225230, April 1986.

April 25-29 , 2021   Xiamen, China

Triboelectric Nanogenerator for Powering Body-implantable Electronics

Sang-Woo Kim

School of Advanced Materials Science and Engineering, SKKU Advanced Institute of Nanotechnology (SAINT), Samsung Advanced Institute for Health Sciences & Technology (SAIHST), Sungkyunkwan University (SKKU), Suwon 16419, Republic of Korea

Website : http://nesel.skku.edu/

ABSTRACT

In this presentation, I firstly introduce the fundamentals and possible device applications of TENGs, including their basic operation modes. Then the different improvement parameters will be discussed. As the first main topic, I will report transcutaneous ultrasound energy harvesting using triboelectric technology. Implantable medical devices (IMDs) are designed to perform or augment the functions of existing organs by using monitoring, measuring, processing units, and the actuation control. Conventional IMDs are powered with primary batteries that require frequent surgeries for maintenance and replacement. Therefore, IMDs require a new reliable and safe powering system to avoid the need for frequent surgeries. Recently my group demonstrated that ultrasound was used to deliver mechanical energy through skin and liquids and demonstrated that a thin inplantable vibrating triboelectric nanogenerator (TENG) is able to effectively harvest it. Ultrasound TENG (US-TENG) was triggered with an applied 20-kHz ultrasound at 3 W/cm2 reaching 9.71 $V_{\text{(root mean square [RMS])}}$ and 427 μA_{RMS}. The measured output current was enhanced two orders of magnitude compared with conventional TENGs, with a similar level of surface charge density, triggered in low-frequency mechanical environments. Interestingly, to experimentally simulate clinical conditions closer to human in the laboratory, we inserted US-TENG under porcine tissue, showing that it fully charged a rechargeable Li-ion battery having a capacity of 0.7 mAh. As the second topic, I will deal with our very recent demonstration of a commercial coin battery-sized high-performance inertia-driven TENG (I-TENG) based on body motion and gravity. In a preclinical test, we demonstrate that the encapsulated device successfully harvested energy using real-time output voltage data monitored via a Bluetooth low-energy information-transmitting system. Details will be presented in the conference site.

BIOGRAPHY

Dr. Sang-Woo Kim is currently an SKKU Distinguished Professor (SKKU Fellow) at Sungkyunkwan University (SKKU) and Director of the BK21 FOUR SKKU MSE Program. He received a Ph.D. in Electronic Science and Engineering from Kyoto University in 2004 and completed his postdoctoral research at University of Cambridge. From 2015 to 2016, he was a visiting professor at Georgia Institute of Technology as a sabbatical stay. His recent research interest is belonging to triboelectric/piezoelectric nanogenerators for powering wearable and body-implatable electronics, self-powered sensors and neurostimulation, and 2D materials including graphene, h-BN, and TMDs.. Prof. Kim has published over 270 research papers including Science, Nature Journals, etc (h-index of 69) and presented more than 100 Plenary, Keynote, Invited talks in international/domestic conferences.

Prof. Kim served as Chairman of the 4th NGPT (Nanogenerators and Piezotronics) conference at SKKU in 2018 and Director of SAMSUNG-SKKU Graphene/2D Research Center. He is currently serving as Head of the MSE graduate program at SKKU, an Associate Editor of *Nano Energy*, and an Executive Board Member of *Advanced Electronic Materials*. He received the Award of the Ministry of Science and Technology (Science Technology Researcher Award for this month, 2020), the Award for Excellent Basic Research (2019), the Research Award of the National Academy of Engineering of Korea (2018), The Republic of Korea President's Award for Scientific Excellence (2015), etc.

978-1-6654-3008-1/21 $31.00 © 2021 IEEE

Gap in pagination due to unavailable paper.

Pages 190-192

Proceedings of the 16th Annual IEEE International
Conference on Nano/Micro Engineered and Molecular Systems
April 25-29, 2021

A Resonant Differential Pressure Sensor Based on Bulk Silicon Technology

Chao Cheng, Yadong Li, Yulan Lu, Chao Xiang, Jian Chen, Junbo Wang*, *Member, IEEE* and
Deyong Chen*, *Member, IEEE*

Abstract—A resonant differential pressure sensor based on bulk silicon technology was put forward in this study. SOI-MEMS technology was employed to fabricate sensor die. A SOI wafer for differential pressure sensing and a GOS (glass on silicon) wafer as a cap were anodically bonded to achieve vacuum packaging for resonators. Finite element simulation was conducted, confirming that the intrinsic resonant frequencies of resonators changed with the variation of differential pressure linearly. Open-loop test results showed the resonators demonstrated a high-Q value of 22000, and differential pressure experiment results indicated that the sensitivity of sensor for differential pressure measurement was characterized as 52.75 Hz/kPa.

I. INTRODUCTION

Resonant differential pressure sensors have been widely used in petrochemical, industry, metallurgy, aerospace and other fields due to the excellent stability, quasi-digital output, high resolution, and high precision advantage [1-2]. In the past literature reports, the resonant differential pressure sensor basically adopts thin-film vacuum packaging process, such as Yokogawa Electric Corporation developed a resonant differential pressure sensor where selective epitaxial growth process and sacrificial layer releasing were used to achieve vacuum packaging [3-4]. In addition, the university of Wisconsin and Honeywell adopted polysilicon deposition process and sacrificial layer etching to achieve vacuum packaging [5-7]. However, the above-mentioned thin-film vacuum packaging method is complicated and difficult to implement.

Based on the problems in the above reports, a resonant differential pressure sensor was proposed where bulk-silicon technology was used to achieve vacuum packaging for resonators. The newly fabricated sensor possessed a high Q-value of 22000 and the sensitivity of this microsensor for differential pressure measurement was quantified as 52.75 Hz/kPa.

II. SENSOR DESIGN AND FEM SIMULATION

Fig. 1(a) shows the structure of the designed sensor, which includes two parts namely the SOI wafer and the GOS wafer, which functioned as a sensitive element and vacuum packaging cap. The sensitive element is composed of the pressure sensitive diaphragm I formed on handle layer and three H-shaped resonators formed on device layer. Central,

side resonators are located at the middle and edge area of the diaphragm while the compensated resonator is positioned on the frame area of the sensor respectively. These three

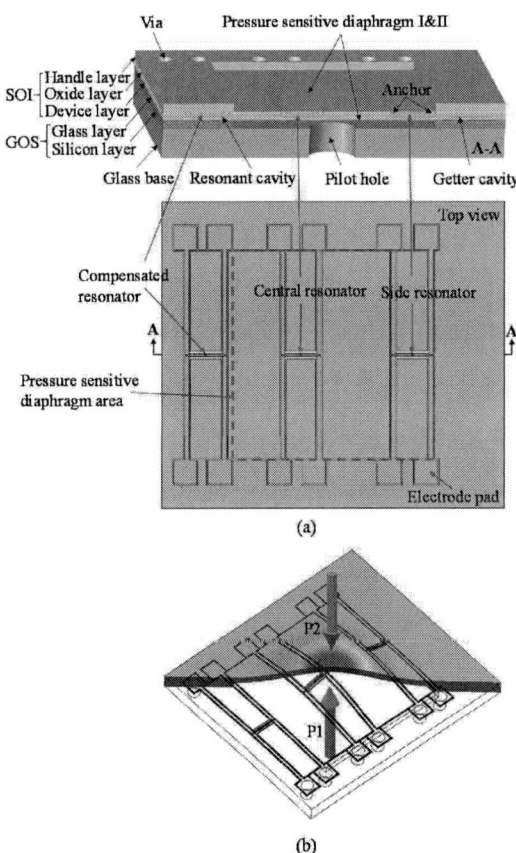

Figure 1. (a) Schematic of proposed resonant differential pressure sensor, composed of a SOI wafer (a device layer with resonators, an oxide layer with anchor structure and a handle layer with pressure sensitive diaphragm and via) as a sensitive element and a GOS wafer (a silicon layer with pressure sensitive diaphragm and a glass layer with resonant cavity and getter cavity) for vacuum packaging. (b) The differential pressure under measurement bends the diaphragms, inducing different stress of resonators, leading to the resonant frequency increased or decreased.

This study was supported by the National Key R&D Program of China (Grant No. 2018YFB2002302).

Chao Cheng, Yadong Li, Yulan Lu, Chao Xiang, Jian Chen, Junbo Wang and Deyong Chen are with the State Key Laboratory of Transducer Technology, Aerospace Information Research Institute, Chinese Academy of Sciences, Beijing 100190, China, and also with the School of Electronic,

Electrical and Communication Engineering, University of Chinese Academy of Sciences, Beijing 100049, China (e-mail: chengchao18@mails.ucas.ac.cn).

*Corresponding author: Junbo Wang (tel: +86-010-588-87191, e-mail: jbwang@mail.ie.ac.cn).

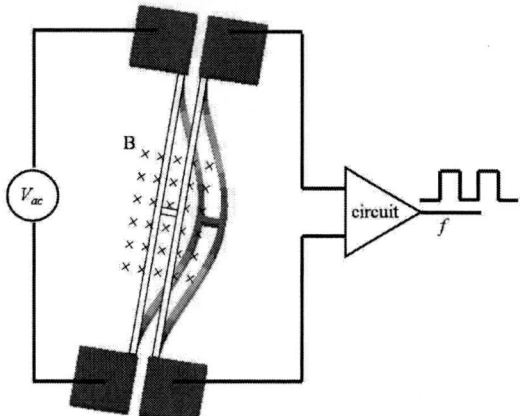

Figure 2. The working principle of resonator: H-shaped resonator works in a constant magnetic field (B), motivated by AC signal, and a signal with intrinsic resonant frequency can be detected through closed-loop circuit.

resonators are attached to pressure sensitive diaphragm through the anchor structure in the buried oxide layer. The differential output increases the sensitivity of sensor for measurement of differential pressure compared to single resonator output. The frequency output of the compensated resonator will be used for the subsequent static pressure compensation. The GOS wafer is composed of a glass layer with three resonant cavities and a getter cavity and a single crystal silicon layer with pressure sensitive diaphragm II. The SOI wafer is anodically bonded with GOS wafer, which make resonators operate in vacuum environment and a glass base with pilot hole is bonded with the fabricated sensor die eventually. The differential pressure (P1-P2) causes deformation of the diaphragms, and the stress is transferred to resonators through anchors, leading to the drift of resonant frequency of the resonators as Fig. 1(b). Furthermore, the differential pressure can be calculated through detecting changes of the frequency.

The proposed sensor is based on electromagnetic excitation and electromagnetic detection. Fig. 2 shows the working principle of resonator, which is formed by an excitation beam, a detection beam and one connecting block.

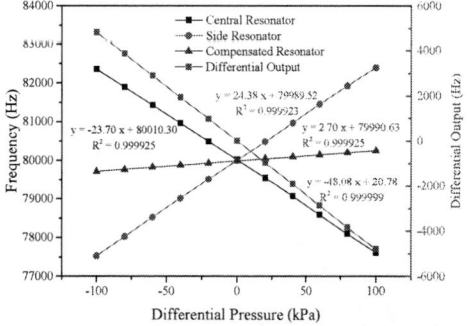

Figure 3. The simulation results of differential pressure performance of the proposed sensor, which indicated the variation of frequencies and the differential output of central and side resonators along with differential pressure.

The excitation beam with AC signal will generate Lorentz force in the constant vertical magnetic field (B), which will drive the detection beam to move, thus the electromagnetic induced voltage is generated at both ends of the detection beam due to the Faraday law of electromagnetic induction.

Fig. 3 shows the simulation results of the differential pressure performance of the proposed sensor, which indicated that the frequency response of central and side resonators changed in the opposite direction with differential pressure sensitivities of -23.70 Hz/kPa and 24.38 Hz/kPa and the sensitivity of differential output of -48.08 Hz/kPa. In addition, the compensated resonator was not very sensitive to differential pressure with a differential pressure sensitivity of 2.70 Hz/kPa, which will be used to implement static pressure compensation.

III. FABRICATION

A. Fabrication Process

The MEMS-fabrication was adopted to manufacture the designed sensor as shown in Fig. 4, mainly including the fabrication of SOI wafer, GOS wafer and SOI-GOS wafer.

First of all, a 4-inch SOI wafer was cleaned (see Fig. 4(a)). Next, photolithography was utilized to pattern the device layer of SOI to form the resonators and the oxide layer under the resonators was removed through HF (see Fig. 4(b)).

As for the fabrication process of GOS wafer, a GOS wafer was formed by anodic bonding of a silicon wafer and

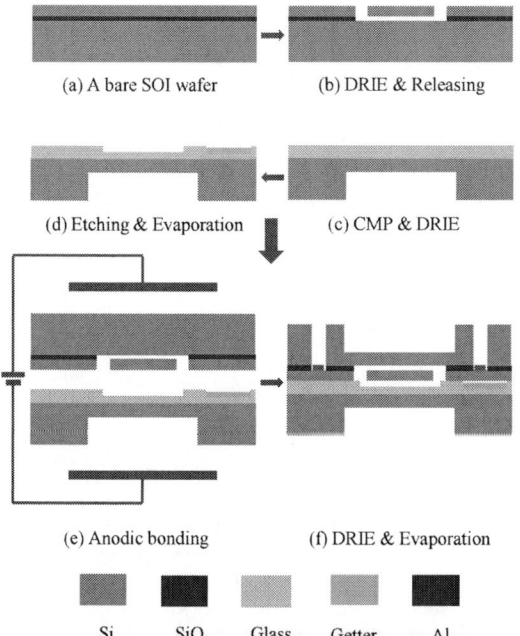

Figure 4. The fabrication processes of resonant differential pressure sensor. (a) a bare SOI wafer; (b) DRIE to form resonators and release the resonators; (c) CMP and DRIE of GOS wafer; (d) resonant cavities and getter cavity were formed, followed by evaporating getter; (e) anodic bonding between SOI wafer and GOS wafer to form vacuum packaging for resonators; (f) DRIE to form diaphragm and vias on the handle layer, and evaporated Al electrode.

978-1-6654-3008-1/21 $31.00 © 2021 IEEE

a glass wafer firstly. Then, CMP was performed on the glass surface of GOS wafer and DRIE was performed on the silicon surface of GOS wafer to form pressure sensitive diaphragm I (see Fig. 4(c)). Next, gaseous HF was utilized to etch the polished glass surface of GOS to form getter cavities and the resonant cavities (see Fig. 4(d)).

In the last part of SOI-GOS wafer, the SOI wafer was anodically bonded with the fabricated GOS wafer where vacuum packaging was realized (see Fig. 4(e)). The bonding voltage as a key parameter in the process of anodic bonding was carefully considered, which can refer to following pull-in part. Eventually, DRIE was implemented to form the pressure sensitive diaphragm II and vias, and evaporated Al electrode (see Fig. 4(f)).

B. Pull-in

The pull-in effect may occur in the fabrication process of anodic bonding between SOI wafer and GOS wafer, causing the resonators physical contact with the glass layer of GOS, leading to the resonators' invalidation. The bonding voltage of anodic bonding will generate electrostatic force, which can account for the possible pull-in between the resonators and glass layer. Therefore, the bonding voltage need to be considered seriously. The formula of pull-in voltage is

$$V_{pull-in} = \sqrt{\frac{8kd^3}{27\varepsilon_0 S}} \qquad (1)$$

The d is the vertical distance between the resonator and glass layer of GOS, namely the depth of resonant cavity (10 μm). The ε_0 and S is dielectric constant in vacuum and the surface area of resonator, respectively. The k is the equivalent spring constant of double-ended fixed beam, expressed as following.

$$k = \frac{32Ebh^3}{L^3} \qquad (2)$$

Where the E, b, h, L are the Young's modulus, the width, the height and the length of resonator, respectively. The pull-in voltage was quantified as 2297 V through theoretical calculation. In addition, FEM simulation based on COMSOL was conducted to simulate pull-in effect. Fig. 5 shows the simulation result, which indicated pull-in will

occur when the bonding voltage reaches 2610 V. The FEM simulation was basically consistent with the theoretical calculation, which indicated the bonding voltage of 420 V in the actual anodic bonding was safe and reliable.

IV. RESULTS

Fig. 6(a) shows the view of fabricated device after dicing with a device size of 10.2 mm × 11.8 mm. The open-loop measurement for resonators was conducted with results

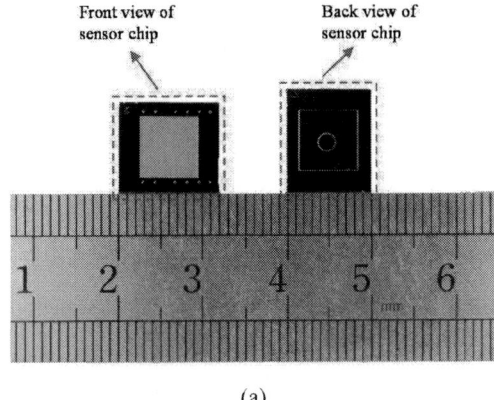

(a)

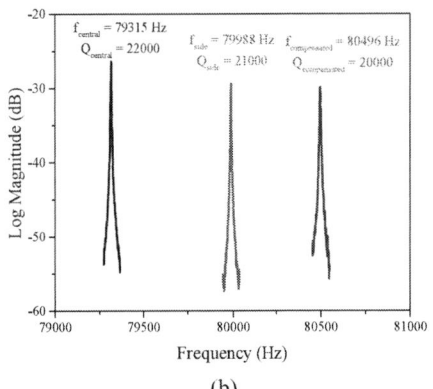

(b)

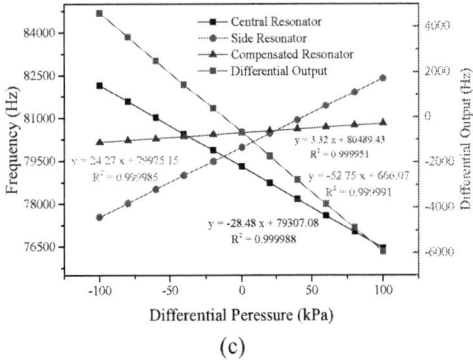

(c)

Figure 6. (a) The view of fabricateded sensor chip. (b) The open Loop results, implying the vacuum packaging for the resonators. (c) Frequency resposes of three resonators with varied differential pressure at the temperature of 25°C and static pressure of 110 kPa.

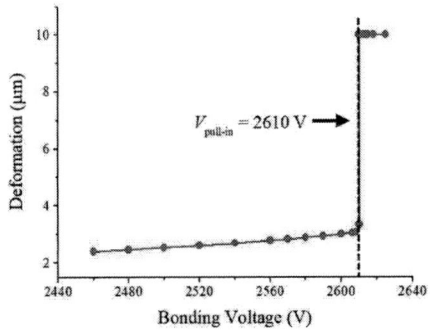

Figure 5. The simulated variation relationship between the deformation of resonator under the electrostatic force and the bonding voltage.

978-1-6654-3008-1/21 $31.00 © 2021 IEEE   195

shown in Fig. 6(b). Under the temperature of 25°C and differential pressure of 0 kPa, the central resonator located an intrinsic frequency of 79315 Hz and a Q value of 22000, the side resonator located an intrinsic frequency of 79988 Hz and a Q value of 21000, and the compensated resonator located an intrinsic frequency of 80496 Hz and a Q value of 20000, respectively, which revealed the effectiveness of vacuum packaging. The closed-loop experiment of developed sensor was conducted to characterize the differential pressure performance as shown in Fig. 6(c). Test results produced the differential pressure sensitivity and correlation coefficient of -28.48 Hz/kPa, 0.999988 for the central resonator, 24.27 Hz/kPa, 0.999985 for the side resonator and 3.32 Hz/kPa, 0.999951 for the compensated resonator respectively at the temperature of 25°C and static-pressure of 110 kPa. Furthermore, the differential pressure sensitivity of the sensor was characterized 52.75 Hz/kPa. The results were roughly consistent with simulation results, confirming the feasibility of the developed microsensor.

V. CONCLUSION

A resonant differential pressure sensor based on bulk-silicon technology was proposed in this paper. Bulk-silicon technology was used to fabricated device. Open-loop test indicated three resonators demonstrated a Q-value higher than 20000. Closed-loop test characterized the differential pressure performance of sensor with a differential pressure sensitivity of 52.75 Hz/kPa. These results confirm the feasibility of developed resonant differential pressure sensor, which can implement effective measurement of differential pressure.

ACKNOWLEDGMENT

Special thanks should give to my supervisor, professor Wang and professor Chen. They instructed the experiments carried out in this paper.

REFERENCES

[1] H. Hu, L.Q. ZHONG, and Q. ZHOU, "Actuality and development of the differential pressure sensor technology," Machine Tool & Hydraulics, vol. 41, no. 11, pp. 187-190, Nov. 2013.

[2] T. Saigusa, and H. Kawayama, "Intelligent differential pressure transmitter using micro-resonators," Proceedings of the 1992 International Conference on Industrial Electronics, Control, Instrumentation, and Automation. IEEE, 1992.

[3] K. Ikeda, H. Kuwayama, T. Kobayashi, T. Watanabe, T. Nishikawa, T. Yoshida, K. J. S. Harada, "Silicon pressure sensor integrates resonant strain gauge on diaphragm," Sensors and Actuators A: Physical, vol. 21, no. 1-3, pp. 146-150, 1990.

[4] K. Ikeda, H. Kuwayama, T. Kobayashi, T. Watanabe, T. Nishikawa, T. Yoshida, K. J. S. Harada, "Three-dimensional micromachining of silicon pressure sensor integrating resonant strain gauge on diaphragm," Sensors and Actuators A: Physical, vol. 23, no. 1-3, pp. 1007-1010, 1990.

[5] H. Guckel, J. J. Sniegowski, T. R. Christenson, F. J. S. Raissi, "The application of fine-grained, tensile polysilicon to mechanicaly resonant transducers," Sensors and Actuators A: Physical, vol. 21, no. 1-3, pp. 346-351, 1990.

[6] H. Guckel, C. Rypstat, M. Nesnidal, J. D. Zook, D. W. Burns, and D. K. Arch, "Polysilicon resonant microbeam technology for high performance sensor applications," Technical Digest IEEE Solid-State Sensor and Actuator Workshop. IEEE, 1992.

[7] D. W. Burns, J. D. Zook, R. D. Horning, W. R. Herb, H. J. S. Guckel, "Sealed-cavity resonant microbeam pressure sensor," Sensors and Actuators A: Physical, vol. 48, no. 3, pp. 179-186, 1995.

Proceedings of the 16th Annual IEEE International
Conference on Nano/Micro Engineered and Molecular Systems
April 25-29, 2021

An Ultralow-Ripple Polarization Voltage Generator based on High-Voltage Bandgap Reference for MEMS Gyroscopes

Hua Chen*, *member, IEEE*, Zhen Meng*, Xingcheng Zhang, and Ke Liu

Abstract—*This paper proposes an ultralow-ripple direct-current (DC) bias voltage generator for the polarization requirement of high-performance MEMS gyroscopes. The generator is based on a high-voltage (HV) bandgap reference whose power supply is provided by an integrated Dickson-type DC-DC converter. The DC-DC converter can output a 25 V source with a 400 μA load current. In addition, an HV buffer is connected behind the HV bandgap reference for isolation and better driving. With a 10 pF load capacitance exhibited by the MEMS gyroscope, the post-simulated output voltage of the generator is 20 V with a ripple of 5 mV$_{P-P}$ and a settling time of 13.5 μs. If the load capacitance is increased to 50 pF, the ripple is reduced to 1 mV$_{P-P}$ with a settling time of 16.7 μs.*

Keywords— Charge pump, low ripple, bandgap, DC bias, MEMS gyroscope.

I. INTRODUCTION

Microelectromechanical system (MEMS) gyroscopes have achieved great success in low- and medium-precision applications such as automotive electronics and consumer electronics [1]. With the performance improvement and the low price, they are expected to be popularized in high-end applications such as personal navigation, global navigation satellite system, inertial navigation, internet of things, ubiquitous sensors, and big data [2]. In recent years, MEMS gyroscope performance has gradually approached the navigation level [3, 4]. In addition to the device level optimization, it also needs to optimize the peripheral circuit. Because the capacitive MEMS gyroscope uses electrostatic excitation and capacitance detection, it needs to apply a polarization voltage on the mass [3, 4]. Low-ripple polarization voltage is important because it can effectively reduce the spur and noise in the drive-mode and sense-mode signals. On the other hand, configurable polarization voltage is vital for the mode-matching gyro. If the ripple of the voltage is small, the accuracy of the mode-matching is high [3]. Therefore, how to design a high precision polarization voltage on-chip has become an important research topic.

Generally, on-chip high voltage (HV) generation is based on the inductor-less Dickson charge pump [5] wherein the diode responsible for charge transfer can be realized by the MOS transistor [5-10] or the Schottky diode [11]. In this work, a diode made in a p-type island, which is superimposed on an n-type buried layer, is used because the turn-on voltage is lower and the parasitic is smaller. Besides, compared with the MOS-based structure, this circuit is simpler. In terms of reducing the ripple of the charge pump output, literature [8] used a phase-shift clock technique to reduce the ripple by 3 times; literature [9] adopted parallel

dual-branch charge-transfer switches and a current-controlled oscillator; literature [11] employed dual Dickson charge pumps; literature [10] utilized a low dropout regulator; literature [12] introduced a separated dynamic clock voltage scaling technique. This work attempts to reduce the ripple from a perspective of system architecture rather than module optimization. When applying the technique of literature [10], we found that the power PMOS transistor enters the deep linear region because the source voltage (charge pump output) is much higher than the gate voltage. Hence, this method cannot effectively suppress the ripple. Because of this, we propose a new scheme that is based on an HV bandgap reference to achieve an ultralow ripple. The highlight of this scheme is that the HV bandgap reference decouples the final output from the charge pump output, and outputs an adjustable voltage by tuning the reference current or the output resistor.

II. CIRCUIT DESIGN

The capacitive MEMS gyroscope and its peripheral circuit are shown in Fig. 1. The figure shows the drive mode and the sense mode is similar. In the drive mode, the MEMS gyro mass vibrates along the horizontal direction. A 20 V DC bias voltage is applied to the mass as a polarization voltage. A low-voltage alternating current (AC) voltage is connected to the input electrode for electrostatic excitation. The output electrode is linked to the external readout circuit for capacitance detection.

The block diagram of the polarization voltage generator is shown in Fig. 2. The DC-DC converter converts the 5 V power supply into 25 V which is used for the HV bandgap reference and the HV buffer. The HV bandgap reference generates a 20V reference voltage that does not change with

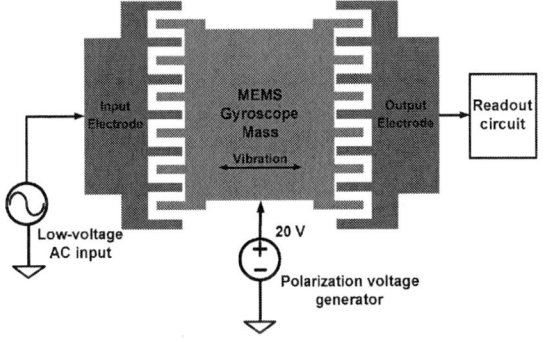

Figure 1. The Capacitive MEMS gyroscope and its interface circuit.

All authors are with Smart Sensing Center, Institute of Microelectronics Chinese Academy of Sciences (CAS), 3 Beitucheng West Road, Beijing, 100029, China.

*Contacting Author: Zhen Meng and Hua Chen are both with Institute of Microelectronics CAS; (phone: +86-010-82995709; e-mail: mengzhen@ime.ac.cn; chenhua111@mails.ucas.ac.cn).

978-1-6654-3008-1/21 $31.00 © 2021 IEEE

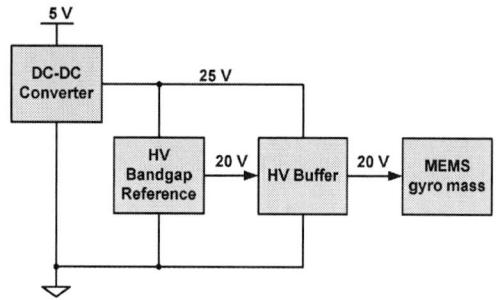

Figure 2. The proposed high-voltage (HV) polarization voltage generator.

temperature. The voltage is then buffered by the HV buffer and acts on the gyro mass. The HV bandgap reference and the HV buffer maintain a high power-supply-rejection-ratio (PSRR) to suppress the ripple of the 25 V power supply transmission to the outputs of these two blocks.

The block diagram of the DC-DC converter is shown in Fig. 3. The converter works in a skipping mode. When the 25 V output voltage reaches its final value, the logic control circuit periodically turns on or off the ring oscillator to regulate the charge pump output. Based on the HHGrace 0.35 μm Bipolar-CMOS-DMOS (BCD) process, the high-efficiency DC-DC converter with a 400 μA load current was designed. As shown in Fig. 3, the charge pump, the ring oscillator, and the bandgap reference are the three key modules. For the charge pump, a 6-stage diode-based Dickson structure [5, 11] was adopted. Herein, the pump capacitance is 4 pF, the pump frequency is 100 MHz, and the pump amplitude is 5 V. Besides, the diode used in the charge pump is fabricated in an isolated p-type island and its forward conduction voltage is 0.6 V. For the ring oscillator, a 5-stage current-hungry inverters ring [13] was chosen to achieve an oscillation frequency of 100 MHz. The phase noise at 10 kHz offset is -23 dBc/Hz, and the one at 1 MHz offset is -83 dBc/Hz. For the bandgap, supplied by a 5 V power, it outputs a 1.25 V reference voltage for the comparator.

The proposed HV bandgap reference is shown in Fig. 4. The popular Banba architecture [14] was adopted to generate a temperature-insensitive reference current (I_4), which flows through the output resistor R3. If the temperature coefficient of the R3 is zero, then the output voltage V_{ref} is not sensitive to the temperature. Besides, by tuning the R3 value, a configurable V_{ref} is achieved. Note that the R3 consists of two series resistors with opposite temperature coefficients.

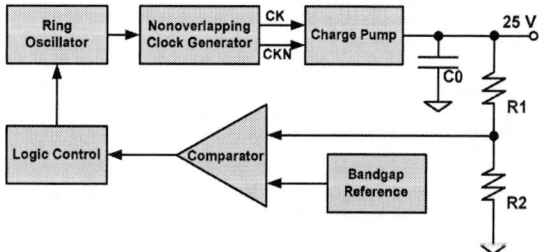

Figure 3. Block diagram of the high-efficiency DC-DC converter.

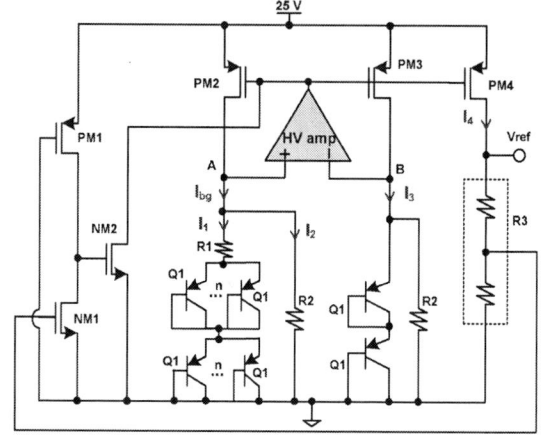

Figure 4. The proposed HV bandgap reference.

The HV p-n-p transistors were stacked to provide a suitable common-mode voltage for the HV amplifier which let point A and point B voltages be equal. Because the size of the HV MOS transistors of PM2 and PM3 is equal, the I_{bg} and the I_3 are identical. The tap voltage extracted from the R3 drives the startup circuit which is made by the HV transistors of PM1, NM1, and NM2.

The startup circuit closes down when the bandgap reference is powered up and operates stably. At this time, the bandgap reference current I_{bg} is:

$$I_{bg} = I_1 + I_2 = \frac{2\ln(n)*V_T}{R_1} + \frac{2V_{EB1}}{R_2} = \frac{2}{R_2}\left(\frac{R_2\ln(n)}{R_1}V_T + V_{EB1}\right)$$

Here, n is the number of the parallel HV p-n-p transistors of the left branch, V_T is the thermoelectric potential of the p-n junction of the HV p-n-p transistor, about 26 mV at 27°C, and V_{EB1} is the forward turn-on voltage of the HV p-n-p transistor. By adjusting the R2, the reference current is configurable. Let the size of the PM4 be m times that of the PM2, the HV reference voltage is:

$$V_{ref} = I_4 * R_3 = mI_{bg} * R_3 = \frac{2mR_3}{R_2}\left(\frac{R_2\ln(n)}{R_1}V_T + V_{EB1}\right)$$

By tuning the values of the m, the R2, or the R3, the V_{ref} is also tunable.

To obtain a 20V reference voltage with zero temperature coefficient at room temperature, the following design steps were adopted. Firstly, the temperature characteristic of the junction voltage V_{EB1} was simulated, and the coefficient of V_T was calculated to be 21. Therefore, the bandgap current I_{bg} has a zero temperature-coefficient at 27°C. Secondly, let n=7, then R2/R1=10.8. Thirdly, set I_{bg}=10 μA, then R2=125 kΩ, and then R1=11.5 kΩ. Fourthly, since the reference voltage V_{ref} is 20 V and the R3 cannot be too large, a high current I_4 was chosen. Hence, let m=5, then the R3 value was figured out to be 400 kΩ.

978-1-6654-3008-1/21 $31.00 © 2021 IEEE

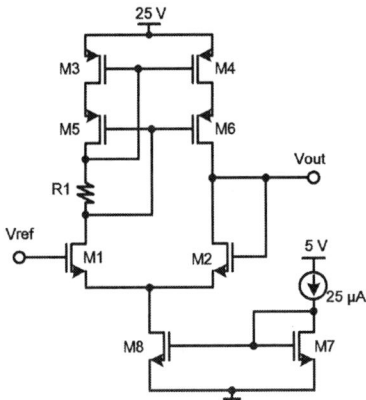

Figure 5. The presented HV buffer circuit.

The presented HV buffer is shown in Fig. 5. The power supply is 25 V and the input voltage V_{ref} is 20 V, so the HV NMOS transistors were chosen for the input differential pair named M1 and M2. To achieve a high PSRR, a self-biased cascode structure consisting of four HV PMOS transistors and one high-poly resistor was employed to realize the current mirror load. The self-biased structure can achieve a wide output swing. Since the source terminal voltage of the M1~M2 is high, the M7 and M8 still use the HV NMOS transistor. However, the gate of this type of transistor does not need to withstand high voltage, but only the drain. To realize the buffer function, the output voltage is directly fed back to the gate of M2. Thanks to the small output resistance, the buffer can drive a large external load.

III. EXPERIMENT RESULT

The ultralow-ripple HV DC bias generator was designed by the HHGrace 0.35 μm BCD process, and the tape-out top layout is shown in Fig. 6. The chip area with IO pads is 1860 μm * 960 μm. In the DC-DC converter module, two separate 5 V supplies were employed to prevent the large signal of the ring oscillator from polluting the sensitive bandgap. Besides, the two supplies were connected to a large of on-chip decoupling capacitors. The output voltage of the HV buffer was led out by a customized high-voltage IO.

The total layout parasitics were extracted and the full-chip post-simulation was done. For the DC-DC converter, the output voltage and its ripple are shown in Fig. 7. From the transient waveform, it can be seen that the output voltage is

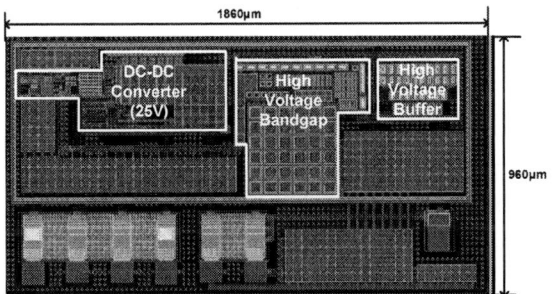

Figure 6. The tape-out layout of the proposed HV polarization voltage generator

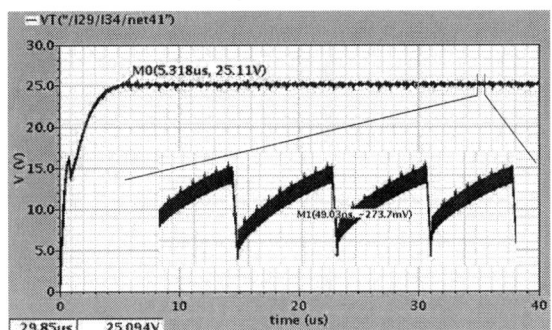

Figure 7. The post-simulated output voltage of the DC-DC converter, and the inset shows the ripple voltage.

25 V with a ripple of 273.7 mV$_{p-p}$. The settling time is 5.3 μs. Hence, the DC-DC converter can provide a stable HV source and sufficient load current for the HV bandgap reference and the HV buffer. If the output voltage is directly used as the DC bias for the MEMS gyro, the ripple level is too large.

The output voltage and its ripple of the HV buffer are shown in Fig. 8. Note that the load capacitance exhibited by the MEMS gyro was modeled as 10 pF. The output voltage V_{out} is 20 V with a ripple of 5 mV$_{p-p}$ and the settling time is 13.5 μs. The ripple is much smaller, compared with the one of the DC-DC converter. On the other hand, the setting time is increased by about 8 μs, which is due to the start-up process of the HV bandgap reference. Besides, the frequency of the ripple shown in the inset comes from the pump clock frequency in the DC-DC converter.

If the load capacitance presented by the gyro was modeled as 50 pF, the output voltage and its ripple of the HV buffer are shown in Fig. 9. The output is 20 V with a ripple of 1 mV$_{p-p}$. It can be seen that the larger the load capacitance is, the smaller the output voltage ripple is. This is because the load capacitance and the parasitic resistance form a passive low-pass filter. The larger the load capacitance is, the lower the upper cut-off frequency will be, and the easier it is to suppress the ripple. In addition, the settling time is increased to 16.7 s. This is because the larger the load capacitance is, the longer the charging and discharging time is.

For the HV polarization voltage generator, the total power dissipation is 7 mW under a supply of 5 V. The summary of

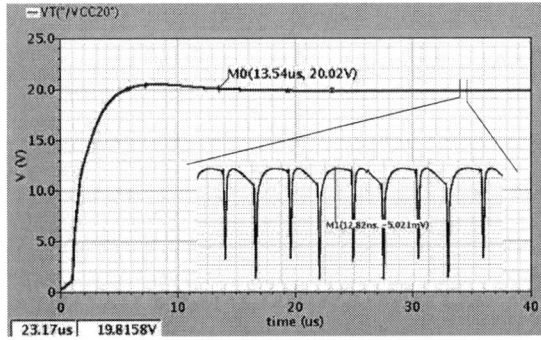

Figure 8. The post-simulated output voltage of the HV buffer at 10 pF load, and the inset shows the ripple voltage.

978-1-6654-3008-1/21 $31.00 © 2021 IEEE

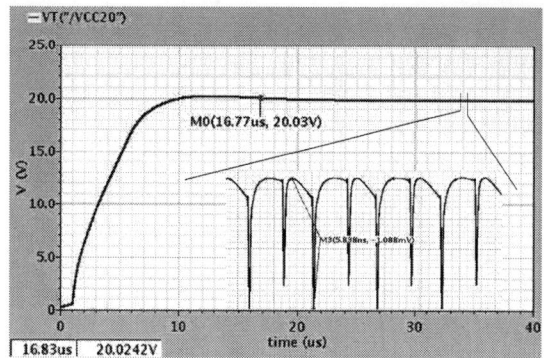

Figure 9. The post-simulated output voltage of the HV buffer at 50 pF load, and the inset shows the ripple.

post-simulation results and the comparison with other related works are shown in Table I. Compared with other works, the proposed DC bias generator with an HV bandgap reference shows an ultralow ripple and moderate power consumption.

IV. CONCLUSION

For the high-precision DC bias requirement of MEMS gyroscopes, this paper presents an ultralow-ripple polarization voltage generator which is based on an integrated DC-DC converter and an HV bandgap reference. The DC-DC converter outputs a 25 V power supply with a 400 μA load current capability for the HV bandgap reference and the HV buffer. With a 10 pF load capacitance, the post-simulated output voltage of the generator is 20 V with a ripple of 5 mV$_{p-p}$. The settling time is 13.5 μs. If the load capacitance is increased to 50 pF, the ripple is reduced to 1 mV$_{p-p}$ with a little longer settling time of 16.7 μs. Subsequent research will focus on the DC-DC converter optimization, such as using the current-controlled oscillator, the separated dynamic clock voltage scaling technique, etc. The proposed ultralow-ripple HV generator can be used in other sensors with strict requirements for the polarization voltage.

ACKNOWLEDGMENT

Hua Chen would like to thank Jianzhong Zhao and Zhi Li for their valuable discussions and assist help in the use of LDMOS transistors and ESD design.

REFERENCES

[1] J. Marek, "MEMS for automotive and consumer electronics," in *2010 IEEE International Solid-State Circuits Conference - (ISSCC),* 7-11 Feb. 2010 2010, pp. 9-17.

[2] E. T. Benser, "Trends in inertial sensors and applications," in *2015 IEEE International Symposium on Inertial Sensors and Systems (ISISS) Proceedings,* 23-26 March 2015 2015, pp. 1-4.

[3] Q. S. Li *et al.,* "0.04 degree-per-hour MEMS disk resonator gyroscope with high-quality factor (510 k) and long decaying time constant (74.9 s)," *Microsystems & Nanoengineering,* vol. 4, Nov 2018, Art no. 32.

[4] Y. Zhao *et al.,* "A Sub-0.1 degrees/h Bias-Instability Split-Mode MEMS Gyroscope With CMOS Readout Circuit," *IEEE Journal of Solid-State Circuits,* vol. 53, no. 9, pp. 2636-2650, Sep 2018.

[5] J. F. Dickson, "On-chip high-voltage generation in MNOS integrated circuits using an improved voltage multiplier technique," *IEEE Journal of Solid-State Circuits,* vol. 11, no. 3, pp. 374-378, 1976.

[6] J. T. Wu and K. L. Chang, "MOS charge pumps for low-voltage operation," *IEEE Journal of Solid-State Circuits,* vol. 33, no. 4, pp. 592-597, Apr 1998.

[7] A. H. Alameh and F. Nabki, "A 0.13-um CMOS Dynamically Reconfigurable Charge Pump for Electrostatic MEMS Actuation," *IEEE Transactions on Very Large Scale Integration (VLSI) Systems,* vol. 25, no. 4, pp. 1261-1270, 2017.

[8] Z. Luo, M. Ker, W. Cheng, and T. Yen, "Regulated Charge Pump With New Clocking Scheme for Smoothing the Charging Current in Low Voltage CMOS Process," *IEEE Transactions on Circuits and Systems I: Regular Papers,* vol. 64, no. 3, pp. 528-536, 2017.

[9] B. Rumberg, D. W. Graham, and M. M. Navidi, "A Regulated Charge Pump for Tunneling Floating-Gate Transistors," *IEEE Transactions on Circuits and Systems I: Regular Papers,* vol. 64, no. 3, pp. 516-527, 2017.

[10] J. Gao, T. Gu, K. Nie, Z. Gao, and J. Xu, "A Low-Ripple Charge Pump With Novel Compensator for Transient-Response Improvement in CMOS Image Sensors," *IEEE Transactions on Circuits and Systems II: Express Briefs,* pp. 1-1, 2020.

[11] M. Marx, S. Rombach, S. Nessler, D. De Dorigo, and Y. Manoli, "A 141-uW High-Voltage MEMS Gyroscope Drive Interface Circuit Based on Flying Capacitors," *IEEE Journal of Solid-State Circuits,* vol. 54, no. 2, pp. 511-523, 2019.

[12] C. C. Huang, F. Liu, Q. Q. Wang, and Z. L. Huo, "A Small Ripple Program Voltage Generator Without High-Voltage Regulator for 3D NAND Flash," *IEEE Transactions on Circuits and Systems Ii-Express Briefs,* vol. 67, no. 6, pp. 1049-1053, Jun 2020.

[13] H. Hwang, B. Jo, S. Park, S. Kim, C. Jeong, and J. Moon, "A 13.56 MHz CMOS ring oscillator for wireless power transfer receiver system," in *TENCON 2014 - 2014 IEEE Region 10 Conference,* 22-25 Oct. 2014 2014, pp. 1-4.

[14] H. Banba *et al.,* "A CMOS bandgap reference circuit with sub-1-V operation," *IEEE Journal of Solid-State Circuits,* vol. 34, no. 5, pp. 670-674, 1999.

TABLE I. POST-SIMULATION RESULTS OF THE HV POLARIZATION VOLTAGE GENERATOR AND COMPARISON WITH OTHER RELATED WORKS

References	[8]	[9]	[10] *	[11]	[12]	This work *	
Process	0.18 μm CMOS	0.35 μm CMOS	0.11 μm CMOS	0.35 μm CMOS	0.18 μm CMOS	0.35 μm BCD	
VDD (V)	3.3	2.5	3.3	3.3	1.8	5	
V$_{out}$ (V)	10.5	7.5~16	4	6~22.4	15.5~23	20	
# of stages	3	6	2	8	15	6	
Frequency (MHz)	NA	1~20	25	0.1~2.6	10~20	100	
C$_{pump}$ (pF)	50	18	1	NA	8	4	
C$_{load}$ (pF)	100	80	300	65	50	10	50
Max ripple (mV$_{p-p}$)	283	18	1	46	2	5	1
Power (mW)	11.55	NA	1.48	0.14	NA	7	

*Simulation results

978-1-6654-3008-1/21 $31.00 © 2021 IEEE

Gap in pagination due to unavailable paper.

Pages 201-202

Proceedings of the 16th Annual IEEE International
Conference on Nano/Micro Engineered and Molecular Systems
April 25-29, 2021

Development of thin glass-based biconvex microlens fabrication technique via thermal expansion

Yusufu Aishan, Yaxiaer Yalikun, and Yo Tanaka

Abstract—We report a simple and efficient glass microlens fabrication technique by thermally inflating the microcavities on the thin glass slide. Due to the joint effect of high-temperature and vacuum pressure, the microcavities are gradually formed into hollow, transparent, and biconvex microlens. Based on this method, we fabricated a wide range of biconvex glass microlens that diameters from 30 μm to 1 mm on the thickness of 100 μm glass slide at the single and arrays of distributions. We evaluated the microlens morphologies, optical imaging quality, and magnification power.

I. INTRODUCTION

Tremendous progress has been made in microlens fabrication techniques since Antonie van Leeuwenhoek constructed the first microscope with a small optical objective lens in the 17th century[1] and started a new era for microbiological studies by enabled to observe small targets in more details. Since then owing to the microlens compactness and compatibility advantages, it gradually becomes a key element in many applications. In particular, along with the constant development of microfabrication and micro-electromechanical system (MEMS) technologies, it provided more advancements in microlens fabrication techniques and promoted the various miniaturized optics integration with microfluidics. Up to the present, dozens of microlens fabrication methods are introduced in the past as micromachining[2], thermal reflow[3], laser direct writing[4] but the microlens fabricated in those methods have some common shortages as low efficiency, transparency, thermal and mechanical instabilities. Conventionally, the microlens is fabricated from various materials such as liquid[5], plastic[6], polydimethylsiloxane (PDMS)[7]. But none of these alternative materials can compete with glass, because glass has better transparency at most wavelengths, better mechanical and thermal stabilities, and better durability in extreme environments. It is also commonly admitted that fabricating microlenses using glass materials has more limitations and difficulties because it usually required complicated production processes, involved expensive equipment and at low production efficiency. Therefore, a practical and productive glass microlens fabrication technique is needed in the field.

---

*This Research supported by Grant-in-Aid for Scientific Research on Innovative Areas (No. 19H05338), RIKEN Junior Research Associate Program, and the Future Young Researchers Support Project of Osaka University, Japan.

Y. A. Author, is with the graduate School of Frontier Biosciences, Osaka University, 1-3 Yamadaoka, Suita, Osaka 565-0871, Japan

Y. Y. Author, is with the graduate School of Nara Institute of Science and Technology, 8916-5 Takayamacho, Ikoma, Nara 630-0192, Japan

Y. T Author, is with the center for Biosystems Dynamics Research (BDR), RIKEN, 1-3 Yamadaoka, Suita, Osaka 565-0871, Japan
E-mail: yo.tanaka@riken.jp. TEL& FAX: +81-6-6105-5132.

II. THEORY AND METHODOLOGY

This fabrication concept was originated from the conventional glass blowing technique that manipulating the glass shape in high-temperature conditions by blowing air into the softened glass chamber. Similarly, this microlens fabrication method uses the thermal expansion principle to expand microcavities between two thin glass slides to form a biconvex microlens-like dome structure, as shown in Fig.1 (a) – (d) which are the main procedures in this proposed microlens fabrication technique.

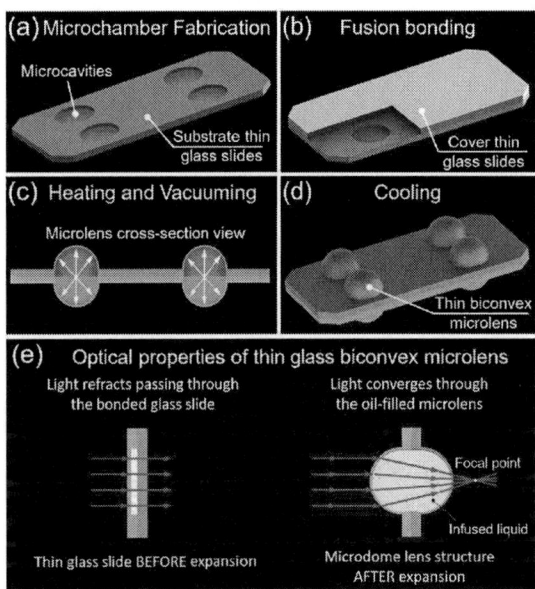

Fig.1. The main procedures of the thin glass microlens fabrication and its optical feature illustrations. (a) A thin glass slide that has microcavities (via wet-etching process) is prepared that used as the substrate glass slide. (b) The substrate and the cover thin glass slides are pre-bonded together. (c) The dome structural deformation occurred due to the thermal expansion of air trapped in the microcavities and vacuum environment of the furnace chamber. (d) After cooling, the microlens is permanently formed. (e) The optical imaging features comparison before and after the microlens formed and filled with liquid. Images are not to scale.

Two thin glass slides are needed which are preferably made from the same glass material to guarantee a similar expansion and shrinking ratio[8] during the heating and cooling procedures. Before the fabrication process, The glass slides were annealed at 520 °C for about 5 h long to prevent cracking and to relieve their internal stresses. The entire microlens fabrication procedures can be briefly summarized as the following steps: first, as in Fig.1(a), a wet etching technique is applied for obtaining microcavities on the substrate glass slide at the desired depths and sizes on one thin glass slide that serving as the substrate. Second, Fig.1 (b) after immersing the substrate and the cover thin

978-1-6654-3008-1/21 $31.00 © 2021 IEEE   203

glass in piranha solution for about 30 minutes to make them highly hydrophilic and easy to bond. Then, those two slides are temporarily bonded together by pressing on them and keeps the microcavities in the middle and ready for the fusion bonding. Third, Fig.1(c) under the joint effect of the air volume thermal expansion and the vacuum pressure of the furnace, at the high-temperature condition the thin glass viscosity decreases[9], the air trapped inside the microcavities gradually expands and pushing the thin glass slides in opposite directions. Eventually biconvex, thin, hollow spherical microlens structures are formed. Fourth, Fig.1 (d), via steps of cooling processes a hard, hollow biconvex microlens structure is formed on the bonded thin glass slide[10]. This method enabled fabricating single or arrays of microlens at different sizes simultaneously.

III. EXPERIMENTAL

3.1. Materials and chemicals

Borosilicate thin glass slides obtained from Matsunami Glass (Osaka, Japan) were used for the fabrication and the size was 30 × 70 ×0.1 mm. Mineral oil (SMR-100, Miyazaki, Japan) was introduced into the expanded hollow chamber of the microlens using a syringe pump (Chemyx Fusion-400; Stafford, TX). A fluorescents microscope (IX-71; Olympus, Tokyo, Japan) with an objective 1.6 × lens (NA, 0.08) with excitation/emission wavelengths of 480/520 nm was used. The microchannel observation images were captured with commercial interface software (cellSens, Olympus).

3.2. Substrate thin glass slide preparation

Previous to the microlens fabrication, the thin glass slides were annealed between 520 °C to 540 °C for 5 hours, for the purpose of preventing the glass slides from cracking under the high-temperature condition and to completely relieve their residual internal stresses, became more stress-resistant. Then two metal membrane layers Chromium (Cr) and gold (Au) were sputter deposited (EIS-220, Elionix, Tokyo, Japan) on one of the annealed glass slides respectively for 45 min and 30 min to get a total metal layer between 130 nm to 160 nm thickness. Next, the photoresist (OFPR) was spin-coated on top of the metal layer at an average thickness of 1 µm. The glass slide was exposed to UV light (DL-1000, NanoSystem Solutions, Inc., Uruma, Japan) to get micropatterns distribution of microlens. After that, the conventional wet-etching technique was applied to obtain microcavities on the substrate thin glass slide and the etching speed in 25 wt% HF was between 7 to 9 µm/min. Finally, the deposited metal layers and photoresist layer were washed off with suitable reagents, to leave only the thin glass slide with wet-etched microcavities as the substrate glass slide.

3.3. Thin microlens fabrication

Both cover and substrate thin glass slides were immersed in piranha solution (sulfuric acid: hydrogen peroxide = 3:1) for 30 min to remove organic dust or particles. This cleaning made them highly hydrophilic. The two thin glass slides were laid one over the other while keeping the microcavities between them and they were pre-bonded by gently pressing on the top slide until no extra air remained between them except for the air trapped in the microcavities. Then, the pre-bonded glass slide was put on an aluminum oxide plate (80 × 80 × 5mm) and placed into a programmable vacuum furnace (KDF-900GL; Denken, Kyoto, Japan) for fusion bonding. The reason for using the aluminum plate was to prevent the glass slide fusion bond to the furnace floor during the process. The specific temperature conditions and durations for the microlens formation is listed in Table.1:

TABLE I.   THERMAL PROCESSES OF MICROLENS FABRICATION

Steps	Temperature (°C)	Duration (h)	Stage
1	25 to 250	2	
2	250 to 360	1	Fusion bonding
3	360 to 450	1	
4	450 to 600	3	
5	Maintain at 600 & Vacuum	5	Heating and vacuuming
6	600 to 500	3	Cooling
7	500 to Room temperature	Natural cooling	

IV. RESULTS AND DISCUSSION

4.1. Various microlens fabrication

By this fabrication technique, we made a variety of microlenses diameters ranged from 30 µm to 1 mm as shown in Fig.2. The thicknesses of each thin glass slides was 100 µm and the etching depth on the substrate glass slide was 25µm on average.

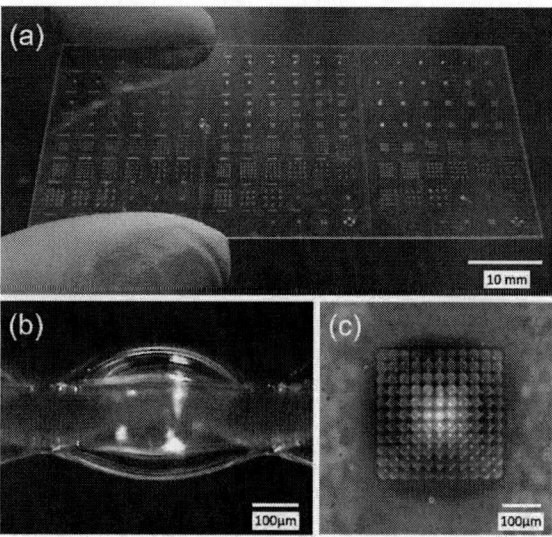

Fig. 2. Illustrations of the fabricated microlens at various sizes and distributions based on the proposed method. (a) A whole plate of fabricated microlens at single and arrays of distribution. (b) Side view of a microlens that diameter at 500 µm. (c) A microlens array diameter at 30 µm.

The microlens formation mainly depends on the glass slide thickness, the substrate etching depths, and the thermal processing condition. The surface uniformity, diameters, and expanded heights are important indicators in evaluating the fabricated microlens capability. The surface roughness and the curvature quality are crucial in determining the microlens optical performance. Fig.3 showed some of the microlens surface morphological assessment results that obtained by using the stylus profiler (The DektakXT, MA) which has a measuring resolution of 0.1 μm.

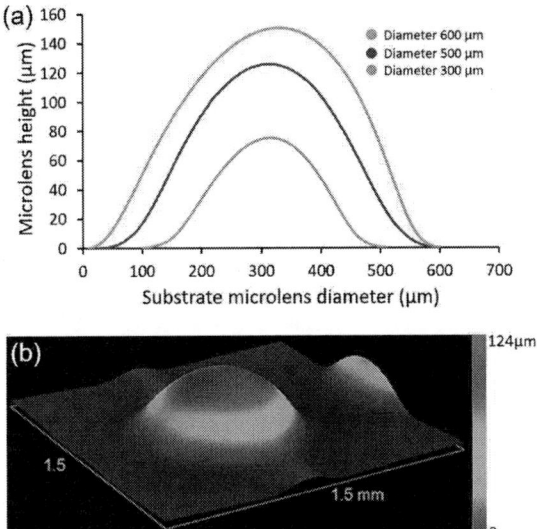

Fig. 3. Morphological evaluation results of comparing different diameters of microlenses under the same thermal condition. (a) The green line indicates the diameter of 300 μm microlens. The purple line indicates the microlens diameter at 500 μm. The blue line indicates the microlens diameter at 600 μm. All substrate glass thickness at 100 μm and etching depth 25μm on average. (b) 3D surface profiler image of the purple line microlens.

Based on the evaluation results as in Fig.3, it can be confirmed that this thermal expanding technique is feasible for fabricating various sizes of thin glass microlenses. Fabricated microlens diameters can vary from few micrometers to the millimeter range and can be flexibly distributed a single or arrays of order on the glass slide. Besides, this entire fabrication procedure is free from further polishing or any extra processes. Also, fabricating more extreme and accurate microlens is highly foreseeable by optimizing the photolithography and wet-etching process during the fabrication.

4.2. Microlens optical characterization

The optical properties of the microlens as focal length, magnification power are very important and those parameters are determined by the expansion heights, diameters, and surface curvatures[11]. Magnification power is related to the focal position of the microlens itself which includes the object distance, the image distance, and the

focal length[12]. Therefore the distance calibration is important in optical imaging. To evaluate the imaging properties of the thin glass microlens, we observed a glass micro ruler plate from various distances using a measuring microscope with an objective lens of 10 × lens (NA, 0.28), as the image in Fig.4 (a). First, a 500 μm diameter oil-filled microlens was fixed to an adjustable jig holder and positioned over the micro ruler plate, as the image in Fig.4 (b). Then, we observed the micro ruler image changes when the jig was shifted vertically up and down above the micro ruler to alter the microlens focus level to obtain different magnified images, as shown in Fig.4 (c) – (f).

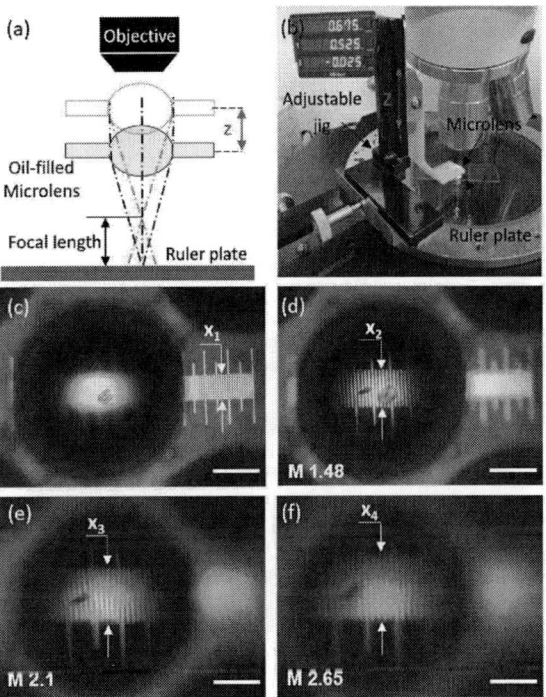

Fig.4: The evaluation of fabricated microlens imaging feature by changing the distance between microlens with the target plate. (a) The concept image of the experiment. By altering the distance between the oil-filled microlens and the micro ruler plate as the jig holder was lifted up or down within the distance Z. (b) Photo of the actual setup showing the oil-filled microlens attached to the jig holder placed under the microscope. From (c) to (f) are the images that the microlens placed at different distances to the micro ruler plate. The scale bar is 200 μm.

As Fig.4 (c) showed, first, we directly focused on the micro ruler plate without the oil-filled microlens by lowering the microscope 103 μm which produced the primary scale of the ruler plate. Then, we calibrated the microscope focus to the ruler plate through the microlens by lifting the microlens 185 μm from afar, a clear image was produced at a magnification of 1.43, as in Fig.4 (d). Next, the image (e) was obtained by lifting the microlens for 248 μm and the magnification was 2.1. Repeatedly, we kept lifting the jig for another 410 μm distance and 2.65 of magnification was achieved in the system, as in Fig.4 (f). But the image started

978-1-6654-3008-1/21 $31.00 © 2021 IEEE

blurring and we calibrated the focus by lowering the microscope. These results confirmed that the oil-filled glass microlens possesses higher optical magnification and stability advantages compared to the conventional liquid microlens[13].

4.3. Applications of the thin microlens with microfluidics

A 500 μm oil-filled thin glass microlens was integrated with a microfluidic glass chip that thickness at 200 μm. The microchannel width 130 μm and 50 μm depth. We tested the fabricated microlens magnification power by observing the fluorescent microbeads flowing at a constant speed in the microchannel. The observed fluorescent microbeads diameter was 10 μm. The overall experiment setup and observing results are illustrated in the below Fig.5.

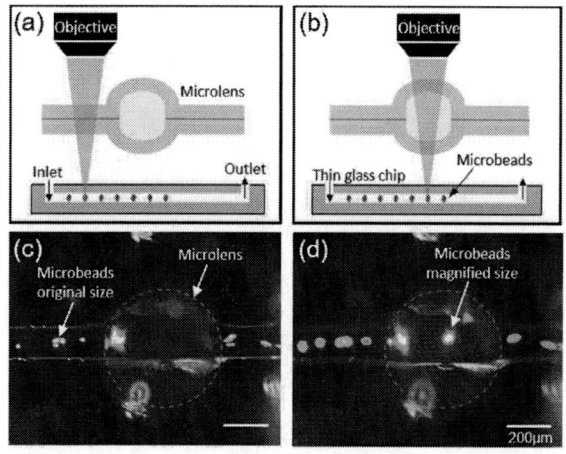

Fig.5: The oil-filled thin glass microlens on-chip integration demonstration. (a) and (b) are observing fluorescent microbeads with or without microlens. (c) and (d) are the actual results observing microbeads with or without microlens.

As shown in Fig.5(b), the magnification was obvious when the fluorescent microbeads flowed past the microlens. The results proved that this microlens has high potentials in integrating with microfluidic glass chip for various applications.

V. CONCLUSION

We evaluated the optical features of the fabricated microlens and demonstrated a simple integration with a microfluidic system. The followings can be summarized; First, from the fabrication method perspective, unlike other microlens expansion techniques that have used plastics, PDMS or SU-8 materials, we used glass which has advantages of high transparency, low surface flow resistance, and relatively smooth morphologies. Additionally, unlike conventional heterogeneous anodic bonding, as silicon with glass bonding[14] or metal with glass bonding[15], we used homogeneous glass material bonding which reduced the glass cracking and increased the fabrication efficiency.

Second, from the application perspective, this new type of microstructures can provide more options for on-chip fast focusing, sensitive detection[16], and accurate imaging applications[17, 18]. Also, it can contribute to improving the efficiency of entire systems: for example, in constructing high-frequency magnetic resonance devices which required thinner and stronger hollow walls. Furthermore, this thermally expanded biconvex shape is preferred for developing the on-chip continuous 3D cell culturing devices because dome curvature increases the cell spheroid cell formation than culturing cells on the conventional flat surface.

REFERENCES

[1] M. Karamanou, "Anton van Leeuwenhoek (1632-1723): Father of micromorphology and discoverer of spermatozoa" *Revista Argentina de Microbiologia.* vol. 42, no. 4, pp. 311–314, 2010.

[2] C. R. King, "Out-of-plane refractive microlens fabricated by surface micromachining" *IEEE Photonics Technology Letters*, vol. 8, no. 10, pp. 1349–1351, 1996.

[3] H. Yang, "High fill-factor microlens array mold insert fabrication using a thermal reflow process" *Journal of Micromechanics and Microengineering*, vol. 14, no. 8, pp. 1197–1204, 2004.

[4] F. Chen, "Maskless fabrication of concave microlens arrays on silica glasses by a femtosecond-laser-enhanced local wet etching method" *Optics Express*, vol. 18, no. 19, pp. 20334-20343, 2010.

[5] X. Zeng and H. Jiang, "Liquid tunable microlenses based on MEMS techniques" *Journal of Physics D: Applied Physics*, vol. 46, no. 32, pp. 323001, 2013.

[6] H. Ottevaere, "Comparing glass and plastic refractive microlenses fabricated with different technologies" *Journal of Optics A: Pure and Applied Optics*, vol. 8, no. 7, pp. S407-S429, 2006.

[7] T.-K. Shih, "Fabrication of PDMS (polydimethylsiloxane) microlens and diffuser using replica molding" *Microelectronic Engineering*, vol. 83, no. 11–12, pp. 2499–2503, 2006.

[8] S. Nomura and K. Hayashi, "The Impact of Thermal Shrinkage of Glass Carriers on Achieving Fine Pitch Wiring Through Fan-Out WLP/PLP Process" in *Proceedings - Electronic Components and Technology Conference*, vol. 2018, pp. 979–984, 2018.

[9] D. Senkal, "Demonstration of 1 Million Q-Factor on Microglassblown Wineglass Resonators With Out-of-Plane Electrostatic Transduction" *Journal of Microelectromechanical Systems*, vol. 24, no. 1, pp. 29–73, 2015.

[10] Y. Aishan, "Thin glass micro-dome structure based microlens fabricated by accurate thermal expansion of microcavities" *Applied Physics Letters*, vol. 115, pp. 263501, 2019.

[11] T. Zhou, "Investigation on the viscoelasticity of optical glass in ultraprecision lens molding process" *Journal of Materials Processing Technology*, vol. 209, no. 9, pp. 4484–4489, 2009.

[12] L. Li, "Displaceable and focus-tunable electrowetting optofluidic lens" *Optics Express*, vol. 26, no. 20, pp. 25839–25848, 2018.

[13] B. Jin, "Adaptive liquid lens driven by elastomer actuator" *Optical Engineering*, vol. 55, no. 1, pp. 017107, 2016.

[14] M. M. R. Howlader, "Hybrid plasma bonding for void-free strong bonded interface of silicon/glass at 200°C" *Talanta*, vol. 82, no. 2, pp. 508–515, 2010.

[15] D. Briand, "Bonding properties of metals anodically bonded to glass" *Sensors and Actuators A Physical*, vol. 114, no. 2, pp. 543-549, 2004.

[16] Y. Aishan, "Pneumatically actuated thin glass microlens for on-chip multi-magnification observations" *Actuators*, vol. 9, no. 73, 2020.

[17] Y. Aishan, "Accurate rotation of ultra-Thin glass chamber for single-cell multidirectional observation" *Applied Physics Express*, vol. 13, no. 2, pp. 026502, 2020.

[18] Y. Yuan, "Fabrication of ultra-thin glass sheet by weight-controlled load-assisted precise thermal stretching" *Sensors and Actuators A: Physical*, vol. 321, no. 15, pp. 112604, 2021

Gap in pagination due to unavailable paper.

Pages 207-208

Nanofluidic devices for Biosensing and Energy Conversion

Xing-Hua Xia

State Key Lab of Analytical Chemistry for Life Science, School of Chemistry and Chemical Engineering, Nanjing University, Nanjing 210023, China

Website : https://xxh.nju.edu.cn/

ABSTRACT

The ionic transport properties of nanochannels with size comparable to the electric double layer thickness are dominantly determined by the surface properties and can be regulated by surface recognition reaction and external fields such as light and heat. Thus, understanding of influence of surface properties on the mass transport behavior in nanochannel will help us to exploit the interesting application of the nanofluidic device in biosensing and energy conversion.

This talk will introduce how the surface properties regulate the mass transport behavior through the nanochannels for biosensing and energy conversion. Our results show that the surface charges play a determining role when the channel size is approximately close to the electric double layer thickness. As the nanochannel surface is negatively charged, anions will be accumulated in front of the channel, while cations can easily pass through. Based on these unique phenomena, biomolecules carrying different charges can be easily separated and label-free DNA assays are developed. In addition, rapid protein concentration, efficient florescence labeling and purification can be achieved on a micro/nanofluidic devices. At the end, we will show the integrate of plasmonics to nanochannels for the possibility of light-electricity conversion on a nanofluidics devices. Our results demonstrate the micro/nano-fluidic devices are promising platform for bioanalysis and energy conversion.

BIOGRAPHY

Xing-Hua Xia is currently a Professor at the School of Chemistry and Chemical Engineering, Nanjing University, China. He received the M.Sc. degree in chemistry from Xiamen University, China, in 1989 and the Ph.D. degree of chemistry in 1996 from the University of Bonn, Germany. During the following five years, he worked at the Universität der Bundeswehr München, at the Fritz-Haber-Institute of the Max-Planck-Society in Berlin, Germany and the Debye Institute of the University of Utrecht, The Netherlands, performing postdoctoral studies.

Since 2001, he took the Professor position at the School of Chemistry and Chemical Engineering, Nanjing University, China. He was awarded as the National Science Fund for Outstanding Young Scholars (2001) and Changjiang Chair Professor from the MOE of China (2009). His research activities were awarded the first class prize of natural science researches from the Ministry of Education of China (2011). He is now an editor of Journal of Electroanalytical Chemistry and member in editorial boards of Scientific Reports, Analytical Chemistry, Talanta and a few Chinese Chemical Journals. The current research activities of his group are focused on interfacial behaviors of biomolecules and biosensors, plasmon enhanced spectroscopies and electrochemistry, bioinspired catalytic materials chemistry, biorecognition and reaction kinetics of bioevents within confined nanospace, *e.g.*, mass transport properties in nanofluidics. He has published more than 370 peer review papers.

Gap in pagination due to unavailable papers.

Pages 210-214

Proceedings of the 16th Annual IEEE International
Conference on Nano/Micro Engineered and Molecular Systems
April 25-29, 2021

Research on a MEMS Detonated Device with Built-in Safety and Arming Device

Kexin Wang[1], Tengjing Hu[1,*], Yulong Zhao[1] and Wei Ren[1,2]

Abstract— **The MEMS detonated device is the new generation of pyrotechnics, the safety of which influence the whole munition system, so the device is only controlled by single electro signal is unreliable. In order to solve the difficulties, a MEMS detonated device with built-in safety and arming device is proposed in this paper. The device is composed of the detonator, the safety and arming device (SAD), the cover plate and the PCB. The detonator is a NiCr alloy bridge foil with Al/CuO energetic composite film. The SAD is set on the top of the detonator to add a mechanical insurance. With 11V driven voltages of electro-thermal actuators, the SAD can realize the bidirectional large-displacement movement of the barrier, which can reciprocally switch the device condition between safety condition and arming condition. The SAD has a Silicon/Ni enhanced barrier. In safety condition, it can block the flame which the detonator under 64V voltage stimulating generate. And the flame can go through barrier in arming condition.**

I. INTRODUCTION

The detonated device is the forepart of fuze in munitions to control the detonated occasion, so its performance seriously affects the safety and reliability of weapon system. Confined by fabrication technology, traditional detonated devices can meet the basic functional requirements, but can't achieve the further development of intelligent and miniaturization in weapon system for its bulky size. To overcome the difficulties, the conception of micro detonated device based on MEMS technology is initiated. Through the fabrication of Micro-Nano processing, the MEMS detonated device can integrate the SAD and detonator with microscale structure.

The detonator can transfer stimulating electro energy into other energy forms, usually included explosive [1], laser [2] or high-speed flyer [3]. If the detonator is merely controlled by electric signal is unsafety, it needs the Safety and Arming Device (SAD) to add a mechanical insurance. The SAD is usually placed between detonator and initiating explosive, and the basic principle of it is interrupting the energy transited by a movable barrier in safety condition [4]. Divided by driving principle, the actuation methods of SAD are mainly included inertial force [5], electro-thermal [6] and pyrotechnics [7].

To improve the safety and reliability of the detonator, a MEMS detonated device with built-in SAD is proposed in

---

1 State Key Laboratory for Manufacturing System Engineering, Xi'an Jiaotong University, Xi'an, China.
2 Science and Technology on Applied Physical Chemistry Laboratory, Shaanxi Applied Physical Chemistry Research Institute, Xi'an, China
*Contacting Author: TengJing Hu is with the State Key Laboratory for Manufacturing System Engineering, Xi'an Jiaotong University, Xi'an, China. (email: htj047@xjtu.edu.cn)

this paper. The detonated device is mainly composed of the detonator and the SAD. The detonator is a Ni-Cr alloy bridge foil with an Al/CuO energetic composite film covered, which can generate flame under electro signal stimulation. The SAD is fabricated by SOI wafer, and the barrier is driven by the V-shape electro-thermal actuators in device layer. The handle layer has a frame under the silicon barrier to assemble Ni enhanced barrier. The MEMS detonated device can realize the condition switch function and safety and arming function.

II. MODELING

A. The Overall Structure of the MEMS detonated device

The overall size of the MEMS detonated device with built-in SAD is $17.8 \times 13.4 \times 2.5 mm^3$, which is composed of the detonator, cover plate, SAD and PCB, as shown in Fig. 1. All of the components are assembled in line. The main

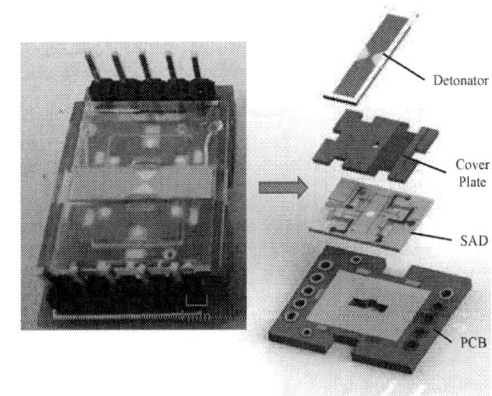

Figure 1. The basic structure of of the MEMS detonated device.

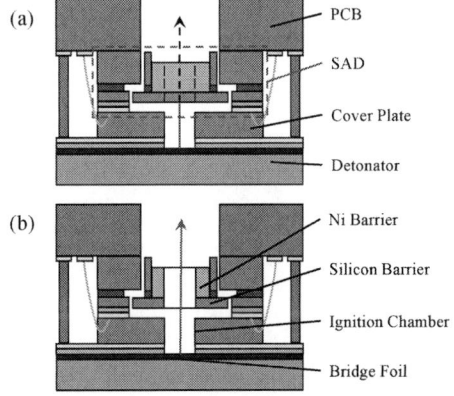

Figure 2. The basic principle of of the MEMS detonated device. (a) The safety condition. (b) The arming condition.

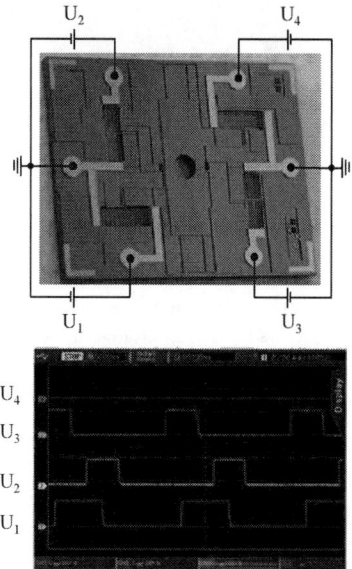

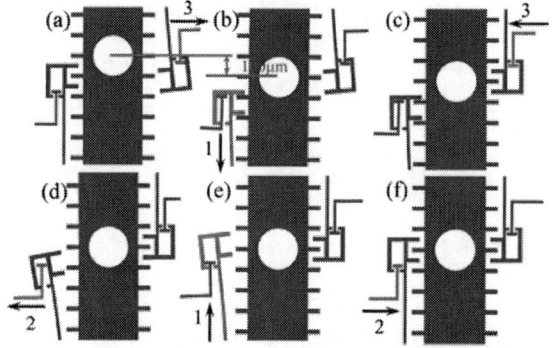

Figure 3. The basic structure of of the SAD.

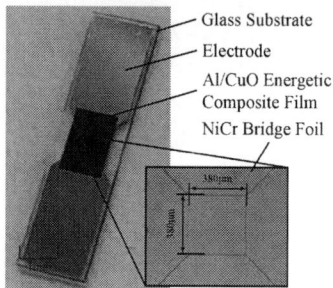

Figure 5. The basic structure of of the detonator.

$$k_s \frac{d^2 T(x)}{dx^2} + J^2 \rho = \frac{S}{h} \frac{T(x) - T_r}{R_T} \qquad (1)$$

With the thermal boundary conditions:

$$T(0) = T(L) = T_r \qquad (2)$$

Here, k_s is the thermal conductivity of silicon. J represents the density of the electrical current, $J = V\cos\theta/\rho L$. V is the voltage that applied on the actuator. ρ is the electrical resistivity of silicon. R_T is the thermal resistivity between the bottom of the structure and the surface of the substrate. T_r is the reference temperature, which equals to the room temperature. The thermal expansion generated by the V-shape electro-thermal actuator can be expressed as equation (3):

$$d = \frac{L}{2}\left(\sqrt{\frac{(1 + \alpha(T - T_r))^2}{\cos^2 \theta} - 1} - \tan \theta\right) \qquad (3)$$

Here, α is the thermal expansion coefficient of silicon. T is the average temperature of the beam. Based on previous research, the V-shape beam can generate at least 10μm displacement when the applied voltage is 11V. But for the thermal expansion coefficient of silicon is too low, the actuators need a soft lever mechanism to enlarge the deformation, and the proportion of it is 1:10. The pawls designed on the output of the soft lever mechanism can generate 100μm displacement under 11V voltage applied.

The designed displacement of barrier is 1mm, which is too large for electro-thermal actuators. The pawl and barrier compose a rack mechanism. It can realize the large displacement by the accumulation of multiple small step. The SAD has four V-shape electro-thermal actuators to control the movement of four pawls independently. The four groups electro signal is shown in Fig. 3 and the operation process is shown in Fig. 4. Every small step movement can be divided into six sub-steps. Two pawls are used to driven the barrier and the other two can control the engage between the driven pawls and barrier. Once condition switch needs 10 step. Due to the pawls symmetrically placed on the two side of barrier, the SAD can realize the bidirectional large-displacement movement of the barrier.

Figure 4. The operation process of the SAD. (a) Right disengage. (b) Pulling. (c) Right engage. (d) Left disengage. (e) Aligning. (f) Right engage.

functional components are the detonator and SAD. The PCB can connect the circuit of detonator and SAD, the hole in the center of which can filled in initiating explosive. The function of cover plate is to restrict the movement of SAD and connect the detonator and SAD by an ignition chamber designed in the center of it.

The basic principle of the MEMS detonated device is shown in Fig. 2. When the device in safety condition, the barrier covers the ignition chamber, the flame generated by detonation signal will be blocked. Otherwise, the bridge foil, the ignition chamber and the hole of barrier are kept in a line in arming condition, and the flame will go through the barrier and trigger the initiating explosive.

B. The Basic Principle of the SAD

The structure of the SAD is shown in Fig. 3. The barrier is driven by V-shape electro-thermal actuators. The basic unit of actuators is V-shape silicon beam. According to the previous research [6], the temperature distribution of the beam can be obtained by the following equations:

TABLE I. BASIC PARAMETER OF THE DETONATOR

Parameter	Value
Side length of the bridge	$380\mu m$
Layer thickness of NiCr alloy	$800nm$
Single layer thickness of Al	$25nm$
Single layer thickness of Cu	$50nm$
Proportion of Al and Cu	1:2
Total thickness of energetic film	$2.25\mu m$

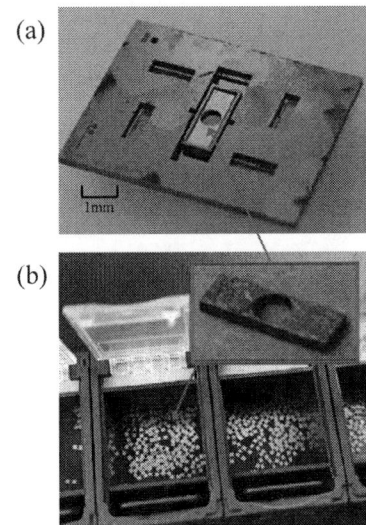

Figure 6. The assembling of Ni enhanced barrier. (a) The backside of SAD. (b) The Ni enhanced barrier.

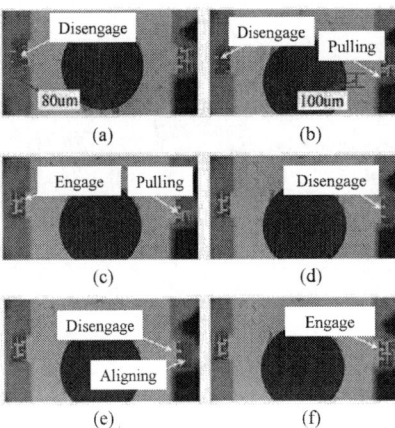

Figure 7. The test result of operation process. (a) Right disengage. (b) Pulling. (c) Right engage. (d) Left disengage. (e) Aligning. (f) Right engage.

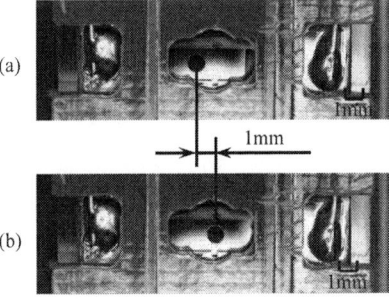

Figure 8. The test result of condition switch function. (a) Safety condition. (b) Arming condition.

C. The Basic Principle of the Detonator

The structure of the detonator is shown in Fig. 5. The detonator is a nickel-chromium alloy bridge foil based on glass substrate, with an Al/CuO energetic composite film covered. The main parameter is shown in Tab. I. The diameter of ignition chamber in cover plate is 500μm, so the bridge foil is a square bridge with 380μm side length to match it. Stimulated by the surge voltage applied on the electrode, the bridge foil will change into high temperature plasma. Through thermal conduction and condensation of plasma, the high temperature will catalytic the Al/CuO energetic composite film react self-diffusion thermite reaction and generate flame. The energetic film can raise the energy output of the detonator.

III. FABRICATION

The SOI wafer (50μm device layer, 3μm buried layer, 400μm handle layer) is used to make movable structure in the SAD by a standard process. The Ni enhanced barrier is fabricated by femtosecond laser and assembled on the frame of handle layer by epoxy glue as show in Fig. 6. Compared with LIGA and electroplating process, the femtosecond laser can fabricate more materials and make the SAD can applied in more detonated environment. The fabrication of the cover plate is carried out on the 4-inch 300μm thick double side polished wafer by a standard

process and the PCB is also fabricated by standard process. All the film is sputtered on the 500μm thick glass substrate to make the detonator. After the independent fabrication, the four parts will be fixed by a fixture and connected by epoxy glue to keep the bridge foil, ignition chamber and barrier in aligning.

IV. TEST AND DISCUSSION

A. The Test of Condition Switch Function

Based on the driven principle of SAD, the barrier need four group electro signal applied in specific sequence. The voltage of electro signal will directly affect the displacement of electro-thermal actuators. The relationship of the actuator displacement and drive voltage already has numerous studies in previous research [6]. According to the result of test, the displacement of pawl is near 100μm in 11V, which is equal the step distance of barrier. The barrier driven by 10 cycle of electro signal can reach to 1mm large-displacement movement. The test of the operation process is show in Fig. 7 and result of the condition switch function test is shown in Fig. 8. In arming position, the bridge foil, ignition chamber and the hole of barrier are kept in aligning to wait Ignition signal. In safety position, the ignition chamber is all covered by barrier. Once switch time is no more than 2s.

978-1-6654-3008-1/21 $31.00 © 2021 IEEE 217

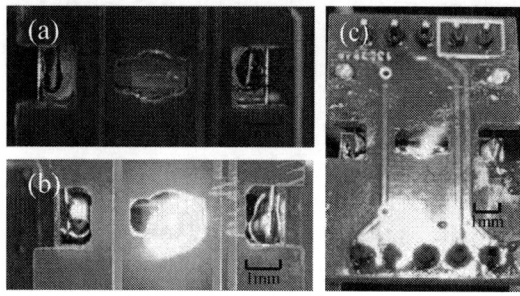

Figure 9. The test result of safety and arming function. (a) Safety condition. (b) Arming condition. (c) The oblique view of arming condition.

TABLE II.   SPECIFIC PARAMETERS OF THE ELECTROSTATIC DISCHARGE TEST

Parameter	Value
Charging voltage	$50\pm0.5kV$
Capacitance	$500\pm25pF$
Discharge resistance	$5\pm0.25k\Omega$
Discharge loop inductance	$<5\mu H$

B. The Test of Safety and Arming Function

The test of safety and arming function is designed in two groups, one of them is stimulating the detonator in safety position and the other one is in arming position. The result of the test is shown in Fig. 9. To verify the safety of device, the voltage of electro signal is high to 64V. In safety position, the light of detonator can be seen from the gap of SAD, and the barrier still keep intact after the pound of flame. In arming position, the flame of detonator can go through the barrier smoothly and has a certain height.

C. The Environmental Adaptability Test of the MEMS Detonated Device

The MEMS detonated device aims to be applied in weapon system, so it must have the ability to adapt the complicated environment of munitions, which include high and low temperature cycle test, vibration test and electrostatic discharge test. In high and low temperature cycle test, the environmental temperature of device has 10 cycles from -40°C to 50°C with 5°C/min temperature change rate, and every cycle last 4h, the temperature curve of which is shown in Fig. 10. The vibration test is carried out in X, Y, Z direction, for 5 minutes each. The random vibration spectrum can be described as: 20Hz-80Hz +3dB/oct, 80Hz-350Hz 0.2g²/Hz, and 350Hz-2000Hz -3dB/oct, shown in Fig. 11. The specific parameters of the electrostatic discharge test are shown in Tab. II. After the three test, the structure of all the devices are keep intact and have integral function same as before.

V. CONCLUSION

The principle, fabrication and function test of a MEMS detonated device with built-in safety and arming device are presented in this paper. In order to improve the safety

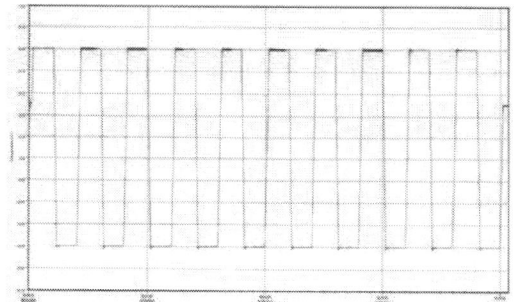

Figure 10. The curve of temperature cycle tese.

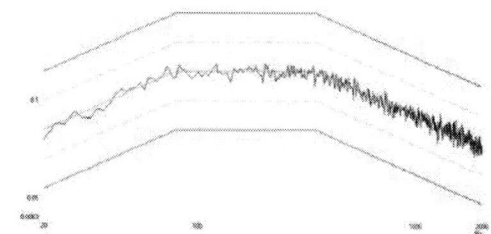

Figure 11. The curve of vibration signal.

of detonator, the SAD is set on the top of the detonator to form an integrated device. The device is composed of the detonator, the SAD, the cover plate and the PCB. The principle of the MEMS detonated device is reasonable and the fabrication of it is reliable. After test, the device can make 1mm bidirectional movement under 11V voltage driven, and it can disrupt the energy transit in safety position. The device has the condition switch function and safety and arming function. The total size of the MEMS detonated device is 17.8×13.4×2.5mm³. If composed with initiating explosive, it will become a high integrated pyrotechnic. So the device has great potential in the application of weapon miniaturization. However, the size of the SAD is much bigger then the detonator. The further study should focus on how to enhance the degree of miniaturization and integration of the device.

ACKNOWLEDGMENT

This work is supported by the National Key R&D Program of China (2017YFB1102900) and the Fundamental Research Funds for the Central Universities (xzy012019004).

REFERENCES

[1] K. Zhang, C. Rossi, M. Petrantoni, and N. Mauran, "A Nano Initiator Realized by Integrating Al/CuO-Based Nanoenergetic Materials With a Au/Pt/Cr Microheater," *Journal of Microelectromechanical Systems*, vol. 17, pp. 832-836, 2008.

[2] K. R. Cochran, L. Fan and D. L. Devoe, "Moving reflector type micro optical switch for high-power transfer in a MEMS-based safety and arming system," *Journal of Micromechanics & Microengineering*, vol. 14, pp. 138-146, 2004.

[3] C. Xu et al., "An electro-explosively actuated mini-flyer launcher," *Sensors and Actuators A: Physical*, vol. 292, pp. 17–23, Jun. 2019.

[4] W. H. Maurer, G. H. Soto, and D. R. Hollingsworth, "Method for utilizing a MEMS safe arm device for microdetonation," U.S. Patent 7 007 606 B1, Mar. 7, 2006.

[5] J.-H. Jeong et al., "Miniature mechanical safety and arming device with runaway escapement arming delay mechanism for artillery fuze," Sensors and Actuators A: Physical, vol. 279, pp. 518–524, Aug. 2018.

[6] T. Hu, K. Fang, Z. Zhang, X. Jiang, and Y. Zhao, "The Research on MEMS S&A Device with Metal-Silicon Composite Structure," Journal of Microelectromechanical Systems, vol. 28, pp. 1088-1099, 2019.

[7] H. Pezous et al., "Integration of a MEMS based safe arm and fire device," Sensors and Actuators A: Physical, Phys., vol. 159, pp. 157–167, May 2010.

Gap in pagination due to unavailable papers.

Pages 220-227

Proceedings of the 16th Annual IEEE International
Conference on Nano/Micro Engineered and Molecular Systems
April 25-29, 2021

Flow-through Electroporation Using Silver-PDMS Based 3D Sidewall Microelectrodes*

Yao Cai[†], Yunyi Huang[†], Guangyu Qin, Duli Yu, and Xiaoxing Xing

Abstract— **Microfluidic electroporators integrated with 3D microelectrodes and working with flow-through manner facilitate high throughput transfection and benefit integration of back-end processing module for the electroporated cells. Here we demonstrate for the first time the flow-through electroporation using 3D sidewall microelectrodes made of silver-PDMS (AgPDMS). Such 3D AgPDMS structure, as a result of low-cost and simple casting process, greatly simplifies the fabrication as compared to existing electroporators incorporating 3D electrodes. Meanwhile, it allows flexible control over the height of the electrode with smooth sidewall profile, which in turn projects rather uniform electric field through deep channel. Delivery of the membrane-impermeable dye of propidium iodide achieves efficiency and viability both at ~80% for Hela cells, and 79% efficiency with 93% viability for A549 cells. We also show the device capability for plasmid DNA transfection on hard-to-transfect Hela cells. Further, we demonstrate intracellular delivery of nanometer-sized quantum dots (QDs). We believe that the innovative device is a useful addition to the microfluidic electroporation toolbox. It holds great potential as a powerful tool for low-cost and high throughput gene transfection as well as engineered nanoparticles delivery for biological applications.**

I. INTRODUCTION

Intracellular delivery of exogenous materials (i.e., DNA, RNA, protein and nanoparticles) is extensively involved in molecular biology study as well as clinical practice such as gene therapy and regenerative medicine [1-4]. Among the various delivery methods, electroporation as a vector-free technique, enables effective transfection for a wide range of cell types with simple configuration and low toxicity, and thus has drawn increasing attention during the past decades. Commercialized instrumentations perform electroporation in batch mode with not only high voltage (>1kV) but also high consumption of costly transfection material. In comparison, microfluidic electroporation platforms offer improved transfection efficiency and survival rate with precisely controlled microenvironment, which in turn significantly save the expensive reagent and reduce the activation voltage.

Flow-through electroporation device has drawn increasing interest because it leverages cell transfection in

*Research supported by the National Natural Science Foundation of China (Grant No. 61804007), Fundamental Research Funds for the Central Universities (Grant No. XK1802-4 and No. buctylkjcx06). the Research Funds from Beijing Advanced Innovation Center for Soft Matter Science and Engineering (Grant No. BAIC201607).
†authors contribute equally.
All authors are with the College of Information Science and Technology, The Beijing University of Chemical Technology, Beijing, China.
Duli Yu is also with the Beijing Advanced Innovation Center for Soft Matter Science and Engineering
Contacting Author: Xiaoxing Xing, phone: +86 (010) 64436719, email: xxing@mail.buct.edu.cn).

high throughput and facilitates post-processing of the activated cells [1]. Coplanar metal electrodes patterned on chamber floor or metal wires embedded in channel sidewalls are popular electrode configurations with simple fabrication [2][3]. However, these designs pose electric field non-uniformity either due to the rapid field decay from the planar electrode surface or the intensified field near the electrode edges [4][5]. In contrast, electroplating of thick metal produces smooth and vertical sidewall profile and thus projects uniform field for electroporation, yet at the expenses of fabrication cost and complexity [4][6]. Alternatively, researchers present flow-through electroporation device fabricated by low-cost injection molding of various conducting materials such as electrolyte and liquid metal [4][5]. Nevertheless, challenging control of the electrode-fluid interface leads to compromised reproducibility of such device [7].

3D microelectrodes made of Polydimethylsiloxane (PDMS) mixed with conducting particles (e.g., micro/nano particles made of silver, cooper, carbon, etc) benefit from simple and reliable PDMS casting process with accurately patterned features. Recent works further simplify such casting process to one-step for the flow channel and electrode molding, and exploit the molded device for dielectrophoretic cell sorting [8][9]. Here, we further broaden the utility of the simply replica-molded 3D AgPDMS structure to serve as vertical sidewall electrodes for flow-through electroporation. To the best of our knowledge, this is the first attempt of conducting-PDMS electrodes for electroporation. As illustrated in Figure 1, the device has a structural layer entirely made of AgPDMS, sandwiched between a PDMS cap layer and a glass substrate. The structural layer consists of a pair of 3D AgPDMS electrodes that also define the flow channel between their facing sidewalls. The channel at the central

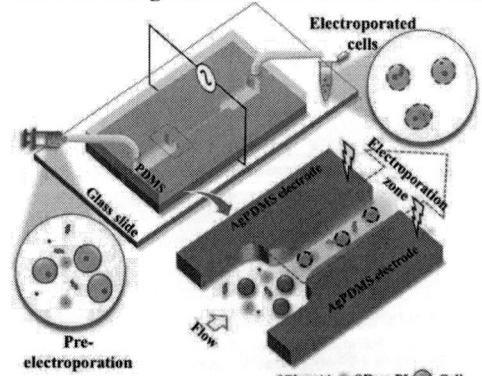

Figure 1. Schematic illustration of the flow-through electroporation chip. Zoom-in view presents the AgPDMS layer with long constriction of electroporation zone sandwiched between 3D AgPDMS electrodes featuring uniform sidewall profile.

978-1-6654-3008-1/21 $31.00 © 2021 IEEE

electroporation zone emerges with 50 μm constricted width for minimized voltage to induce electroporation as cells flowing through. We demonstrate fabrication of such device with vertical sidewall featuring uniform profile and easily turned depth up to 100 μm, which facilitates projection of uniform field strength in deep channel. We verify the device capability for electroporation with efficient delivery of the cell-impermeable dye, propidium iodide (PI), respectively to Hela and A549 cells. Then we perform preliminary test using the device for transfection of green fluorescence protein (GFP) plasmid on Hela cells. We also demonstrate our device for effective intracellular delivery of quantum dots (QDs) of ~25 nm, which potentially facilitate a wide range of biological applications such as cell imaging and diagnostics.

II. MATERIALS AND METHODS

A. Device Fabrication

Figure 2(a) depicts the fabrication process from a cross-sectional view incorporating the sidewall electrode pair and the flow channel in between. The process was detailed in previous publications [8][9], except that here only one layer of SU8 photoresist (2050, MicroChem Corp, MA, US) was lithographically patterned on the mold for casting AgPDMS electrodes with uniform sidewall profile (step i). Next was the casting of the entire device structure, including the flow sidewall and the electrodes, in only one-step molding of AgPDMS. The AgPDMS composite, a paste-like matter made by thoroughly mixing of 85 *wt*% of silver microparticles and 15 *wt*% of PDMS gel, was filled into the SU8 mold. The filled mold was pressed against a piece of copy paper and forced to slide on the paper to clear the excessive AgPDMS waste, and then the filled AgPDMS was cured at 70 °C for 3 hours (step ii). Next a PDMS cap layer was cured on top of the mold, crosslinking with the AgPDMS layer underneath (step iii). The PDMS-AgPDMS structure was then peeled off from the mold, punched with inlet/outlet ports, and glued onto a glass slide pre-coated with a thin layer of semi-cured PDMS as the adhesive layer (step iv). The outward sidewalls of the AgPDMS electrodes were connected to copper wires with silver epoxy for connection with the voltage source.

B. Cells and Reagents

Human cervical cancer cells (Hela) and human lung adenocarcinoma cells (A549) were cultured in the Dulbecco's Modified Eagle Medium supplemented with 10% fetal bovine serum and 3% Penicillin/Streptomycin, and maintained in a humidified incubator at 37 °C with 5% of CO_2. The membrane-impermeable dye PI of 50 μg/ml (Sigma, MO, US) was applied for intracellular delivery test, being companied with the calcein-AM of 2 μg/ml (Aladdin, Shanghai, China), a membrane-permeable dye for viable cell stain. Cell transfection was tested using a pAcGFP1-C3 plasmid at 40 μg/ml, which encoded green fluorescence protein (GFP) (Takara Bio, CA, USA). CdSe QDs (Mesolight, Suzhou, China) were applied at 160 nM for delivery. These delivery materials were respectively prepared in electroporation buffer containing 8 mM Na_2HPO_4, 2 mM KH_2PO_4, and 250 mM sucrose (pH = 7.4).

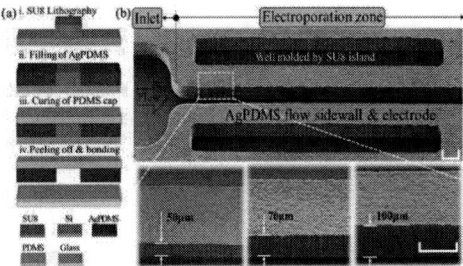

Figure 2. (a) Process flow of the AgPDMS casting process. (b) SEM image of the device from the upstream wide channel to the long-constricted channel in the electroporation zone. Inset images: AgPDMS sidewall electrode molded with various height. Scale bar: 100 μm.

C. Experimental

Prior to the experiment, cells were harvested, washed by centrifuge in the electroporation buffer free of delivery materials (100×g, 5min), and then resuspended in the same buffer at cell density of ~5e5 cells/ml. As the experiment started, the cell suspension was infused into the device inlet with the flow rate controlled by a syringe pump (Longer, Baoding, China), and the sample from the outlet was collected into a 1.5ml tube. Alternating current signal facilitating bubble suppression was exploited here, with a sine voltage delivered from a function generator (RIGOL, Beijing, China). An inverted fluorescence microscope (ECLIPSE TI2-U, Nikon, Japan) equipped with CMOS camera (ORCA-Flash4.0, Hamamatsu, Japan) was used for monitoring and examination of the delivery results. For the PI delivery, the electroporated cells were further incubated for 30min before being checked on glass slide. For pAcGFP1-C3 plasmid transfection and QDs delivery, the collected samples were centrifuged to be resuspended in fresh culture medium, and subsequently seeded into a 24-well plate. GFP expression and QDs were examined 24 h after the electroporation experiment, respectively by the fluorescence microscope and confocal microscope (SP8, Leica, Germany). All electroporation experiments were run in parallel with control cell groups incubated with delivery materials whereas in absence of electric activation.

III. RESULTS AND DISCUSSION

A. Fabricated Device

Figure 2(b) depicts the scanning electron microscopy (SEM) images of the fabricated device. The SEM image partially reveals the flow channel with 400 μm width at upstream, which subsequently evolves into a long-constricted electroporation zone at the midstream with 50 μm width and 24 mm length. The sidewalls of the flow channel, serving also as the poration electrodes, exhibit rather smooth and uniform profile. The wells within the bulk AgPDMS are molded by SU8 islands repeatedly distributed in the large-area mold cavities casting the bulk electrodes, which assist substantial filling of AgPDMS. The three inset images show the fabricated devices with various channel depths (i.e., the electrode thickness) up to 100 μm, as a result of simple adjustment for the SU8 layer thickness on the mold during the spin coating process. Such AgPDMS electrode thickness exceeds that of the existing thick metal (less than 50 μm) electrodes made by

978-1-6654-3008-1/21 $31.00 © 2021 IEEE   229

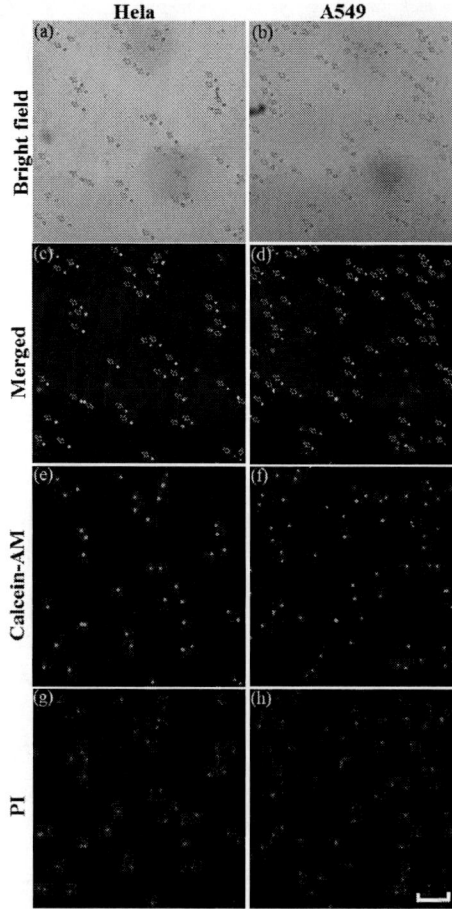

Figure 3. Micrographs of Hela and A549 cells after electroporation under (a-b) bright field, (c-d) merged views from fluorescence filters of (e-f) calcein-AM and (g-h) PI. Electric field intensity: 700 V/cm for Hela cells and 600 V/cm for A549 cells. Frequency: 100 kHz. Residence time: 120 ms. Scale bar: 200 μm.

electroplating [6], despite of being produced by the more time-saving and cost-effective casting process. Such deep electrodes potentially allow higher flow rate for enhanced throughput, and meanwhile maintain the electric field uniformity for electroporation throughout the channel depth by the uniform profile.

B. Electroporation Efficiency and Cell Viability

We first assessed the electroporation efficiency by PI delivery and meanwhile monitored the cell viability by calcein-AM that probed the cell enzymatic activity. Figure 3 shows the representative image groups taken for Hela and A549 cells for the presence of Calcein-AM and PI. Each merged image presented a majority of yellow-colored cells (pointed by arrows) as a result of the merged green- and red- fluorescence, which corresponded to the successfully electroporated cells, with PI intake (red) while remaining viable (green). To find the optimal field intensities for the two cell types, we evaluated the electroporation efficiency and viability against the field intensity from a 100 kHz sine voltage. Cells got continuously exposed to the electric field for ~120 ms when traveling through the electroporation zone, at 30 μl/min flow rate through a 50 μm deep device.

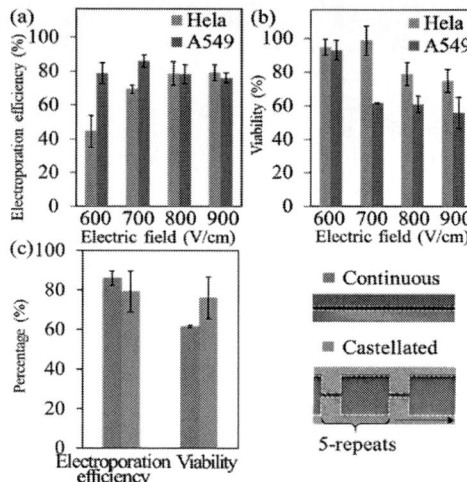

Figure 4. (a) Electroporation efficiency and (b) viability versus electric field intensity for Hela and A549 cells. (c) Electroporation efficiency and viability for the device with continuous-long constriction and the device with castellated sidewall, where A549 cells get exposed to 700 V/cm field respectively with residence time of 120 ms and 5× 24ms. Frequency: 100 kHz. Error bar: mean ± s.d..

The electroporation efficiency referred to the percentage of the electroporated cells against the viable cells, and the viability was the fractionation of the viable cells with respect to the entire cell amount. As illustrated in Figure 4(a-b), Hela cells exhibited increased electroporation efficiency approaching 80% with intensified electric field up to 900 V/cm, where the viability had moderate decline to 75%. In contrast, A549 cells presented acceptably high electroporation efficiency of 79% and high viability of 93% at low field intensity of 600 V/cm. The efficiency peaked (86%) at further increased electric field of 700 V/cm, whereas with rapid decline of cell viability of ~60%.

Previous work reported that long residence time for cells exposed to high field in the electroporation zone potentially induced hazardous effect such as Joule heating, which could be alleviated by pulsed voltage [10]. Here we also involved a design with castellated sidewall that gave rise to five repeats of alternating wide (400 μm) and narrow (50 μm) sections within the channel, as illustrated in the lower-panel SEM image in Figure 4(c). Cells traveling through this device experienced gated electric field being effectively high for electroporation only in the narrow sections. We tested this device for PI delivery to A549 cells with 700 V/cm, under which the long constriction device exhibited low viability of 60%. Figure 4(c) reveals that the castellated device, by segmenting 120 ms-long cell residence time to 5×24 ms, maintained 15% higher viability at the expense of relatively small efficiency drop of 6% compared to the device with long-constriction electroporation zone. Within the field intensity range (600~900 V/cm) covered in our test, our device achieved higher electroporation efficiency and comparable viability for PI delivery to Hela cells, as well as both improved efficiency and viability for A549 cells, as compared to the existing works [11-13]. We believe that the device performance could be further improved with optimization regarding the frequency and residence time.

978-1-6654-3008-1/21 $31.00 © 2021 IEEE

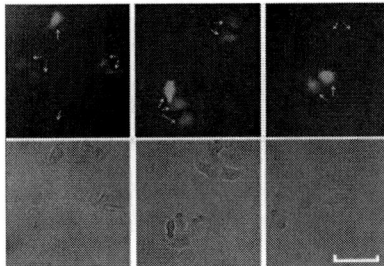

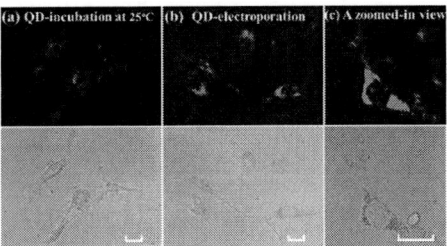

Figure 5. Fluorescence (upper panel) and bright field (lower panel) micrographs of Hela cells observed 24 h after the GFP transfection experiment. Electric field intensity: 1000 V/cm. Frequency: 100 kHz. Residence time: 180 ms. Scale bar: 100 μm

Figure 6. Fluorescence (upper panel) and bright-field (lower panel) confocal microscopic images of Hela cells delivered with QDs via (a) incubation and (b-c) electroporation. Electric field intensity: 700 V/cm. Frequency: 100 kHz. Scale bar: 25μm.

C. Plasmid Transfection and QDs Delivery

We performed preliminary test on the device for DNA transfection using the 50 μm deep device with long-constriction electroporation zone. pAcGFP1-C3 plasmid were transfected to Hela cells under 600 to 1000 V/cm field intensity, and GFP expression was examined 24 h after the transfection experiments. By gradually increasing the electric field intensity, we found that the transfection emerged to be noticeable under microscope at 1000V/cm (not optimized), as illustrated in Figure 5. Further optimizations are undergoing regarding the frequency, residence time, and the plasmid concentration for enhanced transfection efficiency.

To examine the device ability to deliver a range of materials other than PI and plasmid DNA, we tested intracellular delivery of ~25 nm sized CdSe QDs into Hela cells using the long-constriction device with 50 μm depth. QDs are fluorescent nanoparticles with extraordinary brightness and long-term photostability compared to organic dyes, and thus have wide range of applications such as imaging for cytosolic proteins and diagnostics. Figure 6(b-c) exhibit successful delivery of QDs at 700 V/cm, where the spread cells at 24 h after electroporation show rather uniform and bright cytosol fluorescence excluding the nuclear. In comparison, the control cells incubated with the QDs emerge with much weaker brightness from the scattered fluorescence spots (Figure 6a), as a result of endocytosis-based delivery.

IV. CONCLUSIONS AND FUTURE DIRECTIONS

In conclusion, we present the first electroporation device featuring 3D sidewall electrodes made of AgPDMS for flow-through electroporation. The 3D AgPDMS electrode is produced by one-step casting process much simpler and lower-cost than the state of art electroporation devices. The accurately patterned AgPDMS sidewall electrodes have flexibly turned sidewall height and smooth sidewall profile that project uniform electric field throughout the entire channel depth for more homogeneous electroporation performance. Cellular uptake of PI via electroporation using the device has been demonstrated with improved efficiency and/or viability compared to the previous works. Plasmid transfection encoding GFP is also successfully achieved at a preliminary step with a room for further optimization regarding the electroporation conditions. The capability of QDs delivery further suggests

the device utilities for delivery of wider range of nanomaterials and thus broadened potential in biological applications. In our ongoing work, we configure the 3D AgPDMS electrodes as highly parallel interdigitated digits for ultra-high throughput electroporation.

ACKNOWLEDGEMENT

Y. C. thanks Dr. Dandan Sui from the Analysis Center for the kind training the confocal microscopy.

REFERENCES

[1] Z. Wei, X. Li, D. Zhao, H. Yan, Z. Hu, Z. Liang, Z. Li, "Flow-through cell electroporation microchip integrating dielectrophoretic viable cell sorting," *Anal. Chem.*, vol. 86, no. 20, pp. 10215-10222, 2014.

[2] A. Adamo, A. Arione, A. Sharei, K.F. Jensen, "Flow-through comb electroporation device for delivery of macromolecules," *Anal. Chem.*, vol 85, no. 3, pp. 1637-1641, 2013.

[3] A. Chang, X. Liu, H. Tian, L. Hua, Z. Yang, S. Wang, "Microfluidic Electroporation Coupling Pulses of Nanoseconds and Milliseconds to Facilitate Rapid Uptake and Enhanced Expression of DNA in Cell Therapy," *Sci. Rep.*, vol. 10, no. 1, 2020.

[4] Y. Luo, L. Yobas, "Flow-through electroporation of mammalian cells in decoupled flow streams using microcapillaries," *Biomicrofluidics*, vol. 8, no. 5, 2014.

[5] C. Han, X. He, J. Wang, L. Gao, G. Yang, D. Li, S. Wang, X. Chen, Z. Peng, "A low-cost smartphone controlled portable system with accurately confined on-chip 3D electrodes for flow-through cell electroporation," *Bioelectrochemistry*, vol. 134, pp. 107486, 2020.

[6] H. Lu, M. Schmidt, K. Jensen, " A microfluidic electroporation device for cell lysis," *Lab Chip*, vol. 5, no. 1, pp. 23-29, 2005.

[7] S. Puttaswamy, P. Xue, Y. Kang, Y. Ai, "Simple and low cost integration of highly conductive three-dimensional electrodes in microfluidic devices," *Biomed. Microdevices*, vol. 17, no. 1, pp. 4, 2015.

[8] X. Nie, Z. Liang, Y. Lu, Y. Cai, C. Zhang, D. Yu, X. Xing, "Bidirectional Cell Sliding on Active Tracks for High Throughput Dielectrophoretic Cell Sorting in Continuous-Flow," *Proc. of MEMS 2020*, pp. 1018-1021, 2020.

[9] X. Nie, Z. Zhang, C. Han, D. Yu, X. Xing, "Dielectrophoretic Cell Separation using Conducting Silver PDMS Microelectrodes Featuring Non-Uniform Sidewall Profile," *MEMS 2019*, pp. 39-42, 2019.

[10] H. Wang, C. Lu, "Microfluidic electroporation for delivery of small molecules and genes into cells using a common DC power supply," *Biotechnol. Bioeng.*, vol. 100, no. 3, pp. 579-586, 2008.

[11] H. He, D. Chang, Y. Lee, "Using a micro electroporation chip to determine the optimal physical parameters in the uptake of biomolecules in HeLa cells," *Bioelectrochemistry*, vol. 70, pp. 363-368, 2007.

[12] J. Valley, S. Neale, H. Hsu, A. Ohta, A. Jamshidi, M. Wu, " Parallel single-cell light-induced electroporation and dielectrophoretic manipulation," *Lab Chip*, vol. 9, no. 12, pp. 1714-1720, 2009.

[13] M. Kim, T. Kim, Y. Cho, " Cell electroporation chip using multiple electric field zones in a single channel," *Appl. Phys. Lett.*, vol. 101, no. 22, 2012.

Proceedings of the 16th Annual IEEE International
Conference on Nano/Micro Engineered and Molecular Systems
April 25-29, 2021

Self-driven Photoelectrochemical UV-visible Photodetectors Using ZrO$_2$@TiO$_2$ Core-shell Nanorod Arrays Modified with Single-walled Carbon Nanotubes

Zhen Wang[1,2], Renrong Zheng[1,2], Na Wang[1,2], and Haisheng San[1,2,*]

Abstract— We report a self-driven photoelectrochemical (PEC) photodetector (PD) using Ar-annealed ZrO$_2$@TiO$_2$ core-shell nanorod arrays (TNRAs) modified with single-walled carbon nanotubes (SWCNTs) to enhance the photoelectric performance and extend the spectral range from UV to visible light. It is demonstrated that the PECPD based on SWCNTs/ZrO$_2$@TNRAs structure has higher responsivity (155.56 mA/W and 1.75 mA/W) and faster response time (80 ms) than pure TNRAs-based PECPD under ultraviolet and visible irradiation at zero bias. The performance enhancement can be associated to the fact that ZrO$_2$@TiO$_2$ core-shell structure can effectively suppress the recombination of photoexcited electrons and I$_3^-$ in electrolyte and the SWCNTs loaded on the TNRAs can act as hole acceptors, providing sufficient active reactions sites for redox species.

Keywords—TiO$_2$ nanorod arrays; core-shell structure; carbon nanotubes; photoelectrochemical photodetectors

I. INTRODUCTION

In the past decades, high-performance photodetectors (PDs) with excellent optical responsivity, good stability and superior sensitivity have attracted considerable attention. It is noticeable that the PDs based on photoelectrochemical (PEC) mechanism can be driven and generate the photoexcited-current without external bias. The self-powered characteristic is suitable for the sensing applications in remote and inaccessible locations [1-3]. The ultraviolet photodetector (UVPD) that effectively utilize UV energy to drive the sensing element is a typical self-powered sensor for the UV optical sensing. However, most of the current PECPDs often suffer from low sensitivity as well as responsivity [4, 5]. One-dimensional TiO$_2$ nanorod arrays (1-D TNRAs) with ordered and oriented nanostructures are considered to be the excellent candidates for the fabrication of highly sensitive, self-driven PECPDs due to its wide bandgap, high-aspect-ratio and high electron mobility [6, 7].

Generally, pure TNRAs often exhibit a low UV responsivity and high interface resistance due to the lack of built-in potential for separating electron-hole pairs (EHPs). The effective strategy is to composite TNRAs with other semiconductors to form the core-shell structures, e.g., MgO@TNRAs core-shell structure [8, 9], which have been widely applied in photovoltaic devices owing to their high separation efficiency of EHPs through the differences in bandgap. In addition, rapid transmission of carriers in interface can also increase the photoresponsivity. It is

reported that low-dimensional carbon materials can provide efficient pathways for carrier transport, which is beneficial to the improvement of interface resistance. Yang et al. used TNRAs modified with a layer of double-walled carbon nanotubes (DWCNTs) as the photoanodes of dye-sensitive solar cells, the device efficiency can reach 10.24 % [10].

In this work, we report a self-driven UV-visible PECPD using ZrO$_2$@TiO$_2$ core-shell structure modified with single-walled carbon nanotubes to enhance the performance of PECPDs.

II. MATERIALS AND METHODS

A. Fabrication of devices based on nanostructure

Well-aligned TNRAs were grown on FTO/glass (2 mm thickness, 7 Ω square) by a one-step solvothermal method. The hydrothermal synthesis was conducted at 150 °C for 20 h in a sealed autoclave. After cooled to room temperature, the samples were taken out and rinsed extensively with deionized water and alcohol for several times. Next, the TRNAs samples were thermally annealed at 450 °C in argon for better crystallinity (denoted as Ar-TNRAs). Next, the Ar-TNRAs were composited with ZrO$_2$ through chemical bathing deposition to form a ZrO$_2$@Ar-TNRAs core-shell structure. Finally, the uniform SWCNTs were modified on the surface of ZrO$_2$@Ar-TNRAs. Experimentally, the SWCNTs/ZrO$_2$@Ar-TNRAs/FTO/glass structure was used as photoanode and the Pt sheet was acted as counter electrode. The polyiodide (I$^-$/I$_3^-$) electrolyte consist of 0.5 mol/L of LiI, 0.25 mol/L of I$_2$, 0.5 mol/L of 4-tert-butylpyridine, 0.6 mol/L of 1,3-dimethylimidazolium and, which was injected into the micro-gap between the photoanode and counter electrode, and two metal clamps were attached to the two electrodes, respectively. Next, the assembled structure was put into a plastic mould and buried by the transparent resin glue. The packaged devices were transferred subsequently to the oven for thermocuring at 80

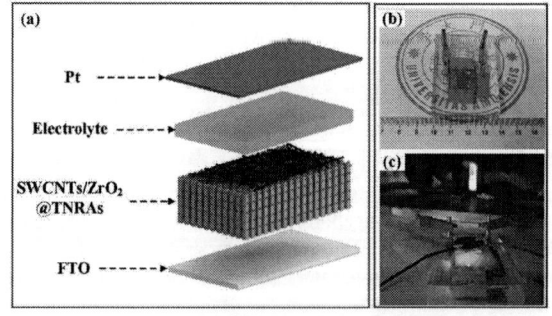

Figure 1. (a) Schematic diagram of FTO/SWCNTs/ZrO$_2$@Ar-TNRAs/ electrolyte/Pt structure; (b) photographs of the packaged device and (c) the device under testing.

[1]Pen-Tung Sah Institute of Micro-Nano Science and Technology, Xiamen University, Xiamen 361005, Fujian, China.

[2]Shenzhen Research Institute of Xiamen University, Shenzhen 518000, Guangdong, China.

*Corresponding Author: Haisheng San is with the Xiamen University, Xiamen, China (phone: +86-592-2181340; E-mail: sanhs@xmu.edu.cn).

978-1-6654-3008-1/21 $31.00 © 2021 IEEE   232

°C for 2 hours. The schematic diagram and the actual photographs of the devices are shown in Fig. 1. The effective area of devices is ~ 3 cm^2. For a comparative study, the PECPDs based on Ar-TNRAs and ZrO$_2$@Ar-TNRAs structures were also prepared.

B. Material characterizations and device measurements

The surface morphologies of the samples were investigated using field emission scanning electron microscopy (FESEM, ZEISS Sigma HD microscope). The crystal structures of samples were characterized by X-ray diffraction analysis (XRD, Rigaku Ultima IV) using Cu-Kα radiation at 50 kV in a 2θ range from 20° to 80°. The Raman scattering was performed using Raman spectrometer (Renishaw in via, UK) with a laser wavelength of 532 nm. The UV-visible (UV-vis) spectrophotometer (Varian, Cary 5000 with Integrating Sphere Attachment) was used to measure the UV-visible absorption spectra in the wavelength range of 200 ~ 700 nm. The photoresponse properties of the devices were measured using the electrochemical workstation (Chenhua CHI660E, China) in a Faraday cage.

III. RESULTS AND DISCUSSION

Fig. 2(a) and Fig. 2(b) respectively show the typical top-view FESEM images of the Ar-TNRAs and ZrO$_2$@Ar-TNRAs structures. It is found that the well-aligned and high-density TiO$_2$ nanorods with cubic-columnar shape are uniformly grown on the FTO/glass substrate. Fig. 2(c) shows the cross-sectional view of the Ar-TNRAs. After 20 h of growth, the free-standing TNRAs are nearly perpendicular to the FTO/glass substrate with the average length of 3 μm and diameter of 100 nm. Fig. 2(d) exhibits the top-view FESEM images of the Ar-TNRAs modified with SWCNTs, it can be clearly observed that the flexible SWCNTs are twined around each other and anchored on the surface of TNRAs, forming a large-range conductive network extended in all directions and providing abundant effective pathways for the rapid transport of carriers.

Fig. 3(a) exhibits the XRD spectra of Ar-TNRAs, ZrO$_2$@Ar-TNRAs and SWCNTs/ZrO$_2$@Ar-TNRAs grown on FTO/glass substrates, respectively. The diffraction peaks at 36.1°, 41.3°, 62.8° and 69.9° are well corresponded to (101), (111), (002) and (112) crystal planes of the tetragonal rutile TiO$_2$ [JCPDS no. 65-0192], respectively [8]. Owing to the low amount of ZrO$_2$ and SWCNTs loaded on TNRAs, the diffraction peaks of ZrO$_2$ and SWCNTs are not observed in the XRD pattern. Fig. 3(b) shows a comparison of the Raman spectra for all samples. It can be seen that the Raman spectrum of the Ar-TNRAs sample has three typical characteristic bands at ~609 cm^{-1}, ~443 cm^{-1} and ~241 cm^{-1}, which are completely consistent with the A$_{1g}$, E$_g$ and multiphoton vibration modes of rutile TiO$_2$. The spectrum of SWCNTs/ZrO$_2$@Ar-TNRAs structure mainly show four different characteristic bands at ~153 cm^{-1} (radial breathing mode), ~1341 cm^{-1} (the 1D band), ~1592 cm^{-1} (the G band) and ~2674 cm^{-1} (the 2D band) [11]. Furthermore, the location and intensity of these characteristic bands indicate that the SWCNTs modified on the surface of samples have obvious metallic properties

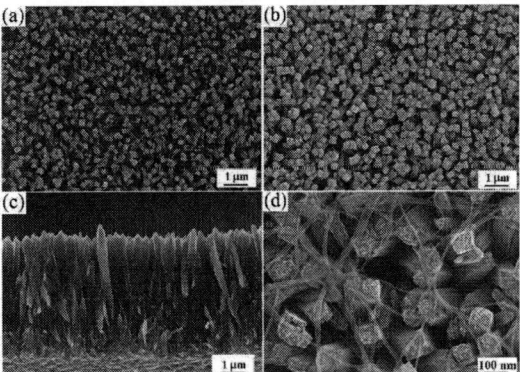

Figure 2. (a) Typical top-view FESEM images of Ar-TNRAs and (b) ZrO$_2$@Ar-TNRAs; (c) cross-view FESEM image of Ar-TRNAs; (d) typical top-view FESEM image of SWCNTs/ZrO$_2$@Ar-TNRAs.

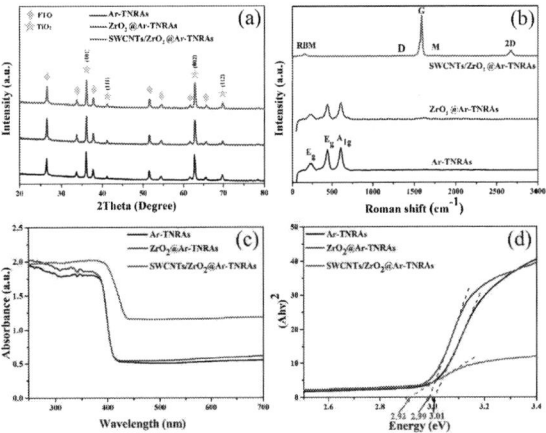

Figure 3. (a) XRD spectra and (b) Raman spectra of Ar-TNRAs, ZrO$_2$@Ar-TNRAs and SWCNTs/ZrO$_2$@Ar-TNRAs samples; (c) UV-vis light absorbance spectra and (d) the calculated bandgap values of Ar-TNRAs, ZrO$_2$@Ar-TNRAs and SWCNTs/ZrO$_2$@Ar-TNRAs samples.

(m-SWCNTs) with few defect states, which is beneficial to the improvement of photoelectric performance [12]. Fig. 3(c) shows the UV-visble absorption spectra of Ar-TNRAs, ZrO$_2$@Ar-TNRAs and SWCNTs/ZrO$_2$@Ar-TNRAs samples. It can be found that all samples exhibit the very high absorbance in the UV region, but the SWCNTs/ZrO$_2$@Ar-TNRAs sample show a slightly red-shift in the UV absorption edge and an enhanced absorption in visible region, which can be attributed to the surface plasmon resonance effect (SPR) induced by m-SWCNTs [13]. Fig. 3(d) shows the specific bandgap (Eg) of all samples, which can be calculated by the Kubelka-Munk function from the absorption edges of spectra. It can be seen that the bandgap of ZrO$_2$@Ar-TNRAs (about 2.99 eV) is slightly lower than that of Ar-TNRAs (about 3.01 eV), which may be related to the increasement of surface charge. With the loading of SWCNTs on samples, the bandgap is narrowed to $E_g \approx 2.92$ eV. The SPR exciton absorption of SWCNTs should be responsible for the bandgap narrowing, and similar results can also be found in the MWCNTs/ZnO structures [12].

978-1-6654-3008-1/21 $31.00 © 2021 IEEE

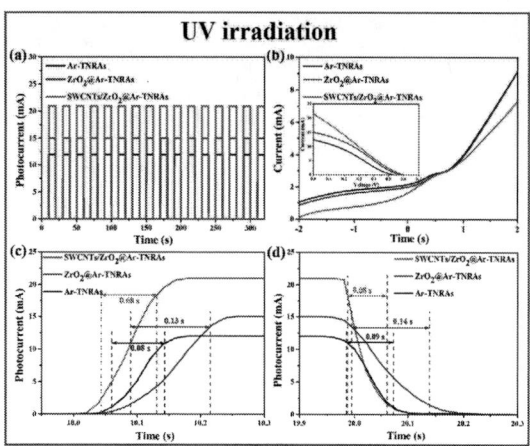

Figure 4. Photoelectrical characteristics of PECPDs based on Ar-TNRAs, ZrO$_2$@Ar-TNRAs and SWCNTs/ZrO$_2$@Ar-TNRAs structures under 45 mW/cm^2 UV irradiance. (a) Cyclic I-t photoresponse and (b) I-V curve; (c) enlarged rising and (d) decaying edges of photocurrent response.

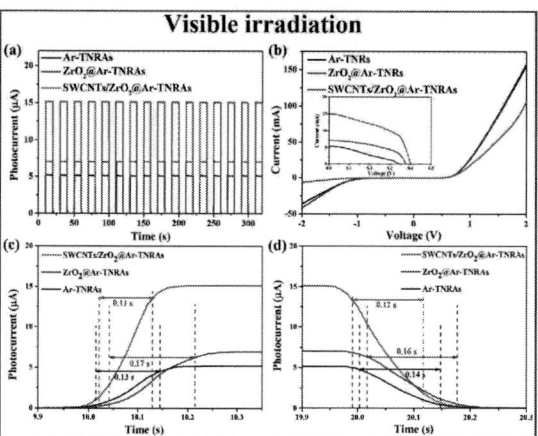

Figure 5. Photoelectrical characteristics of PECPDs based on Ar-TNRAs, ZrO$_2$@Ar-TNRAs and SWCNTs/ZrO$_2$@Ar-TNRAs structures under 2.86 mW/cm^2 of visible light. (a) Cyclic I-t photoresponse and (b) I-V curve; (c) Enlarged rising and (d) decaying edges of photocurrent response.

Fig. 4(a) shows a comparison of time-dependent cyclic photoresponse of PECPDs based on Ar-TNRAs, ZrO$_2$@Ar-TNRAs and SWCNTs/ZrO$_2$@Ar-TNRAs samples under 45 mW/cm^2 of UV irradiation (365 nm) from FTO side with a 10 s of on/off internal at zero bias. It can be seen that all devices keep good repeatability and stability. In comparison with the photocurrent of 12.02 mA in device based on Ar-TNRAs structure, the photocurrent of device based on the ZrO$_2$@Ar-TNRAs structure is increased up to 15.08 mA, indicating that the ZrO$_2$/TiO$_2$ heterojunction structure play a key role in enhancing the photoresponse of device. With the loading of SWCNTs on the surface of ZrO$_2$@Ar-TNRAs structure, it can be found that the photorecurrent is further increased up to 21.43 mA. Meanwhile, the responsivity and the on/off current ratio reaches 155.56 mA/W and 1.59×10^5 under UV irradiation, respectively. The great improvement in performance should be attributed to the formation of super large-range conductive network of SWCNTs, resulting in the improvement of photoconduction and thus the increasement of carrier separation efficiency. The SWCNTs not only provide more efficient pathways for the transmission of photogenerated holes, but also supply sufficient active reaction sites for redox species, which greatly enhances the photoresponsivity of the device. The photovoltaic effects of the PECPDs based on Ar-TNRAs, ZrO$_2$@Ar-TNRAs and SWCNTs/ZrO$_2$@Ar-TNRAs structure were investigated through measuring the current-voltage (I-V) characteristics under 45 mW/cm^2 of UV irradiation (365 nm), as shown in Fig. 4(b). It can be found that all of devices exhibit standard diode I-V characteristics. Compared with these devices based on Ar-TNRAs and ZrO$_2$@Ar-TNRAs structures, the device based on SWCNTs/ZrO$_2$@Ar-TNRAs structure exhibits higher open-circuit voltage (V_{oc} = ~ 0.60 V) and short-circuit current (I_{sc} = ~ 21.44 mA). Fig. 4(c) and Fig. 4(d) show the enlarged rising and decaying edges of the photoresponse of the devices based on Ar-TNRAs, ZrO$_2$@Ar-TNRAs and SWCNTs/ZrO$_2$@Ar-TNRAs structures under 45 mW/cm^2 of UV irradiance, respectively. It is noted that the device based on SWCNTs/ZrO$_2$@

Ar-TNRAs structures show a faster rise time and decay time (~80 ms) than other devices, which can be attributed to the fact that the SWCNTs/ZrO$_2$@Ar-TNRAs structure can effectively promote the rapid transmission and separation of photogenerated carriers, and the cyclic regeneration of I$^-$/I$_3^-$ redox couples in the electrolyte can be also accelerated. This implies that the SWCNTs/ZrO$_2$@Ar-TNRAs structure is more preferable to self-powered PECPDs.

Fig. 5(a) shows a comparison of time-dependent cyclic photoresponse of PECPDs based on Ar-TNRAs, ZrO$_2$@Ar-TNRAs and SWCNTs/ZrO$_2$@Ar-TNRAs structures under 2.86 mW/cm^2 of visible irradiation (465 nm) from FTO side with a 10 s of on/off internal at zero bias. It can be seen that all devices also keep good repeatability and stability. Compared with the Ar-TNRAs and ZrO$_2$@Ar-TNRAs structures, the both visible photocurrent and photoresponse of PECPDs based on the SWCNTs/ZrO$_2$@ Ar-TNRAs structures are significantly enhanced (~15.11 μA and 1.75 mA/W), which can be attributed to the surface plasmon resonance effect (SPR) induced by m-SWCNTs. The SWCNTs can not only increase the surface charge density of nanostructure, but also absorb visible light to excite hot electrons, which can be directly transferred to the conduction band of TiO$_2$ by tunnelling effect. Fig. 5(b) displays the (I-V) characteristics of all devices under 2.86 mW/cm^2 of visible irradiation. Compared with Ar-TNRAs and ZrO$_2$@Ar-TNRAs structures, the PECPDs based on SWCNTs/ZrO$_2$@ Ar-TNRAs structure also exhibit higher V_{oc} (~0.40 V) and I_{sc} (~15.12 μA). Fig. 5(c) and Fig. 5(d) show the enlarged rising and decaying edges of the photoresponse of the PECPDs based on Ar-TNRAs, ZrO$_2$@Ar-TNRAs and SWCNTs/ZrO$_2$@Ar-TNRAs structures, respectively. The PECPDs based on SWCNTs/ZrO$_2$@Ar-TNRAs structure show the fastest response (~110 ms). This indicates that the SWCNTs/ZrO$_2$@Ar-TNRAs composite structure has a good application prospect in the field of wide-band photoelectric response.

978-1-6654-3008-1/21 $31.00 © 2021 IEEE

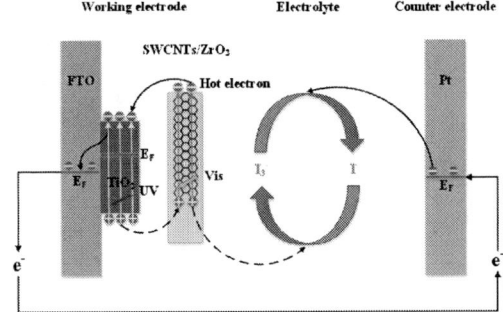

Figure 6. Schematic diagram of photovoltaic enhancement mechanism based on FTO/SWCNTs/ZrO₂@Ar-TNRAs/electrolyte/Pt structure devices.

Fig. 6 shows the schematic diagram of photovoltaic enhancement mechanisms of PECPDs based on FTO/SWCNTs/ZrO$_2$@Ar-TNRAs/electrolyte/Pt structure. When the UV incident light is irradiated on the surface nanostructure from FTO/glass, the electrons on the valence band (VB) of TiO$_2$ are excited to the conduction band (CB), and some excited electrons will diffuse into the FTO electrode and then move to the Pt counter electrode through the external circuit. With the catalytic effect of Pt, the I$_3^-$ in the electrolyte will undergo a reduction reaction with electrons at the interface of the electrolyte/Pt. Considering that the TNRAs cannot be completely covered by the modified ZrO$_2$ particles, the photogenerated holes in TiO$_2$ will migrate to the interface of photoanode/electrolyte and undergo an oxidation reaction with I$^-$ in the electrolyte. With the continuous UV or visible irradiation, the entire reaction process can be carried out in a loop continuously. For the radiation of visible light, the SPR effect excites the hot electrons from m-SWCNTs, which can be directly transferred to the CB of TiO$_2$ by tunnelling effect, contributing to the generation of photocurrent.

IV. CONCLUSION

In summary, we report a self-powered PECPD based on SWCNTs/ZrO$_2$@Ar-TNRAs core-shell structure. Free-standing and high-crystallized Ar-TNRAs were fabricated on FTO/glass through one-step solvothermal method and following thermal treatment in argon, and then the ZrO$_2$ and SWCNTs were composited successively on the surface of Ar-TNRAs, which was confirmed by the measurements of SEM and Roman. The introduction of SWCNTs can significantly enhance the absorption of UV and visible light due to the SPR effect, which was verified by the measurements in UV-visible light absorption. The photoelectric performance of devices based on SWCNTs/ZrO$_2$@Ar-TNRAs structure was effectively enhanced due to the increasement of separation efficiency of photogenerated EHPs in ZrO$_2$/TiO$_2$ junction as well as the inhibition for the recombination of photo-induced electrons with I$_3^-$ in electrolyte. This work provides the multiple strategies to enhance the photoelectric performance of PECPDs, which are considered to be very promising for the practical applications of self-driven UV-visible PECPDs.

ACKNOWLEDGMENT

This work was supported by the National Natural Science Foundation of China (Grant No. 61574117), the Natural Science Foundation of Guangdong Province (Grant No. 2018B030311002).

REFERENCES

[1] J. Zhou, L. Chen, Y. Wang, Y. He, X. Pan, E. Xie, "An overview on emerging photoelectrochemical self-powered ultraviolet photodetectors," *Nanoscale*, Vol. 8(1), pp. 50-73, 2016.

[2] L. Sang, M. Liao, Y. Koide, "High-temperature ultraviolet detection based on InGaN Schottky photodiodes," *Applied Physics Letters*, Vol. 99(3), pp. 031115-1, 2011.

[3] W. Tian, Y. Wang, L. Chen, L. Li, "Self-Powered Nanoscale Photodetectors," *Small*, Vol. 1701848, pp. 1-22, 2017.

[4] C. Chen, B. F. Hu, Z. Wang, "Face-to-face intercrossed ZnO nanorod arrays with extensive NR-NR homojunctions for a highly sensitive and self-powered ultraviolet photodetector," *Nano Energy*, Vol. 65, pp. 104042, 2019.

[5] C. Cao, C. Hu, X. Wang, "UV sensor based on TiO$_2$ nanorod arrays on FTO thin film," *Sensors and Actuators B: Chemical*, Vol. 156(1), pp. 114-119, 2011.

[6] Z. Wang, S. Ran, B. Liu, D. Chen, G. Shen, "Multilayer TiO$_2$ nanorod cloth/nanorod array electrode for dye-sensitized solar cells and self-powered UV detectors," *Nanoscale*, Vol. 4(11), pp. 3350-8, 2013.

[7] X. Li, C. Gao, H. Duan, Nanocrystalline TiO$_2$ film based photoelectrochemical cell as self-powered UV-photodetector. *Nano Energy*, Vol. 1(4), pp. 640-645, 2012.

[8] Y. Gao, J. Xu, S. Shi, H. Dong, Y. Cheng, C. Wei, X. Zhang, S. Yin, L. Li, "TiO$_2$ nanorod arrays based self-powered UV photodetector: heterojunction with NiO nanoflakes and enhanced UV photoresponse," *ACS Appl Mater Interfaces*, Vol. 10(13), pp. 11269-11279, 2018.

[9] S. Ni, F. Guo, D. Wang, "Effect of MgO surface modification on the TiO$_2$ nanowires electrode for self-powered UV photodetectors," *ACS Sustainable Chemistry & Engineering*, Vol. 6(6), pp. 7265-7272, 2018.

[10] L. Yang, W. F. Leung, "Electrospun TiO$_2$ nanorods with carbon nanotubes for efficient electron collection in dye-sensitized solar cells," *Advanced Materials*, Vol. 25(12), pp. 1792-1795, 2013.

[11] K. C. Silva, P. Corio, J. J. Santos, "Characterization of the chemical interaction between single-walled carbon nanotubes and titanium dioxide nanoparticles by thermogravimetric analyses and resonance Raman spectroscopy," *Vibrational Spectroscopy*, Vol. 86, pp. 103-108, 2016.

[12] Y. R. Park, N. Liu, C. J. Lee, "Photoluminescence enhancement from hybrid structures of metallic single-walled carbon nanotube/ ZnO films," *Current Applied Physics*, Vol. 13(9), pp. 2026-2032, 2013.

[13] M. Kim, K. J. Wan, "Purification of aromatic hydrocarbons using Ag-multiwall carbon nanotube-ZnO nanocomposites with high performance," *J. Ind. Eng. Chem*, Vol. 47, pp. 94-101, 2017.

Proceedings of the 16th Annual IEEE International
Conference on Nano/Micro Engineered and Molecular Systems
April 25-29, 2021

Electronic Skin Based on Electrospun PVDF Nanofibers for Slipping Detection*

Lingke Yu, Wenchang Tu, Lingling Liu, Shanshan You, Daoheng Sun*

Abstract—The capability to detect slipping is supplementary to functional electronic skin and electrospun Poly (Vinylidene Fluoride) (PVDF) nanofiber attracts attention from both research and industry areas due to its outstanding piezoelectricity. Herein, we designed and fabricated an electronic skin for slipping sensing based on electrospun PVDF nanofiber mat. The PVDF nanofiber mat fluctuates as an object contacts and slips upon it. Piezoelectric test shows the open circuit voltage consists of wave-like signal during the slipping, differently from the pressing motion. Moreover, the amplitude of the open circuit voltage of PVDF nanofiber mat increases with the slipping speed.

I. INTRODUCTION

With the increasing demand for robots and prosthesis from areas like industry, health caregiving and recovery, electronic skin mimicking human skin becomes a key element and receives more and more attentions[1, 2]. Additional to capabilities of bending, stretching, pressing and temperature sensing, slipping detection is essential as well for electronic skin. Within the last decades, researches have been carried out to realize this function with sensors based on mechanisms such as piezoresistance[3], capacitance[4, 5] and piezoelectricity[6]. Among them, piezoelectric sensor especially made of polymer such as Poly (vinylide fluoride) (PVDF) outstands for its rapid response, wide frequency range and flexibility.

Electrospinning is a method for fast producing piezoelectric PVDF nanofiber or fiber mat. Its in-situ mechanical stretching and electric poling induce the formation and orientation of β phase crystal with strong dipole moment[7, 8]. Due to the high porosity of electrospun PVDF fiber mat, its Young's modulus is decreased compared to bulk PVDF, making it more flexible. Therefore, it can be used as candidate material for electronic skin to sense slippage.

Herein, we designed and fabricated a flexible piezoelectric sensor for slipping detection based on porous electrospun PVDF nanofiber mat. The fiber mat is

*Resrach supported by National Natural Science Foundation of China (U2005214, 51475398) and Key Technology Projects of Xiamen (3502Z20191020).

Daoheng Sun is with Department of Mechanical and Electrical Engineering, Xiamen University, China (corresponding author, e-mail: sundh@xmu.edu.cn).

Lingke Yu is with Department of Mechanical and Electrical Engineering, Xiamen University, China (847889047@qq.com).

Wenchang Tu is with Department of Mechanical and Electrical Engineering, Xiamen University, China (tuwenchang@stu.xmu.edu.cn).

Lingling Liu is with Department of Mechanical and Electrical Engineering, Xiamen University, China (youssdyx@163.com).

Shanshan You is with Department of Mechanical and Electrical Engineering, Xiamen University, (19920201151452@stu.xmu.edu.cn).

sandwiched between copper electrodes and response to the mechanical deformation caused by slippage. The unique feature of output voltage under slipping is compared to that of pressing for demonstration of its feasibility. And the influence of slippage speed and contact force on piezoelectric response has also been tested and discussed.

II. EXPERIMENTAL

A. Electrospinning of PVDF nanofiber mat

PVDF powder purchased from Sigma-Aldrich with molecular weight of 543000g/mol is dissolved in binary solvent DMF and acetone. They are fully mixed on magnetic stirrer for 24 hours, forming viscous solution with concentration of 16wt%. Then it is pumped from a 2.5mL syringe by a precision pump (Harvard 11 Pico Plus) to a metallic needle with round nozzle (inner diameter 250μm). A PVDF nanofiber mat was electrospun with high DC voltage as shown in Fig 1a and the parameters used during the process is listed in Table 1.

B. Package of the electronic skin

Scanning electron microscope (SEM, LEO 1530) was used to observe the surface morphology of eletrospun PVDF nanofibers. Then the PVDF nanofiber mat was sandwiched between two PI (Polyimide) membranes with 4×4 copper electrode array. Finally the flexible sensing unit is glued to a prosthetic hand made of wood, as shown in Fig 1b and c.

C. Piezoelectric measurement

Then the packaged electronic skin was placed on a flat acrylic board upon an X-Y precise motion stage of Newport XMS50 to test its piezoelectric output under the condition of slippage. It was also placed on a modal shaker (Sinocera JZK-5) for the pressing test as comparison, as depicted in Fig 1d and Fig 1e. Meanwhile the open circuit voltage of PVDF nanofiber mat during mechanical deformation was measured with an oscilloscope (Tektronix TDS 2014B).

TABLE I.   PARAMETERS FOR ELECTROSPINNING OF PVDF NANOFIBER MAT

Parameters	Value
PVDF solution concentration	16%
Solvent volume ratio (DMF:Acetone)	5:5
Applied voltage	10kV
Distance between nozzle and collector	10cm
Supply rate of solution	200μL/hr

978-1-6654-3008-1/21 $31.00 © 2021 IEEE

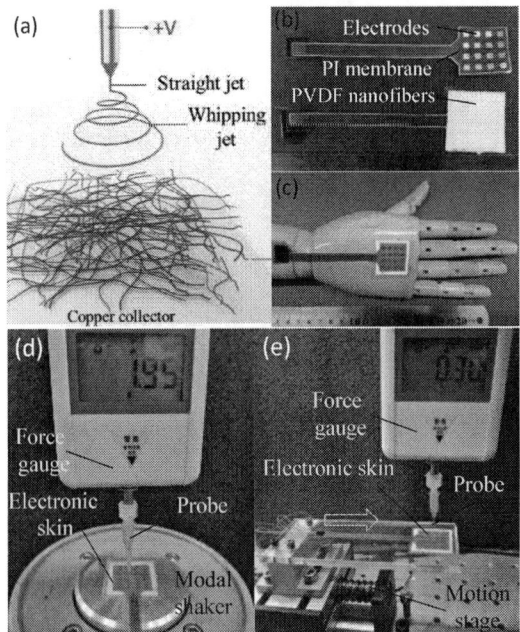

Figure 1. Fabrication and test of the electronic skin: (a) schematics of setup for electrospinning; (b) electrospun PVDF nanofiber mat and PI substrate with electrodes; (c) packaged electronic skin on prosthetic hand; (d) electronic skin on modal shaker for presing test; (e) electronic skin on motion stage for slippage test

III. RESULTS AND DISCCUSIONS

A. Morphology

The electrospun PVDF nanofiber mat possesses good porosity as shown in Fig 2a. The software ImageJ was used for fiber diameter and mat thickness measurement. About fifty nanofibers was randomly selected and the results show the mat has mean fiber diameter about 0.546μm. Most of the fibers are within the range from 200 to 800μm (Fig 2c). And thickness of the mat is about 348μm, as measured from Fig 2b.

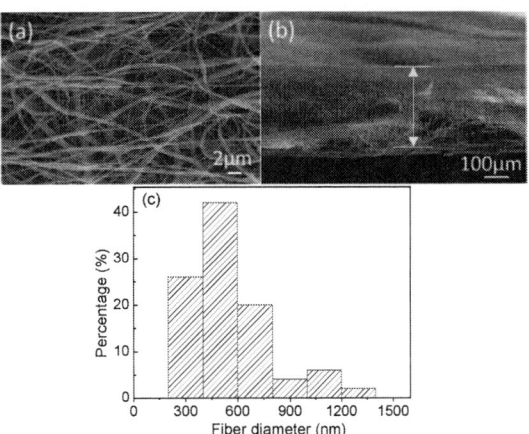

Figure 2. Electrospun PVDF nanofiber mat morphology : (a) SEM of PVDF nanofiber mat; (b) cross section of the fiber mat; (c) statistics of PVDF fiber diameter

B. Feature of piezoelectric signal

The mechanism of slipping detection is shown in Fig 3a. Once an object (such as fingertip or a plastic rod) starts to slip on surface of the sensing unit, the contacted part of PVDF nanofiber mat is compressed due to its porous structure. Suppose the fingertip keeps moving in one direction, the PVDF nanofiber mat endures fluctuation-like dynamic strain continuously, thus outputting wave-like piezoelectric voltage until the slipping stops, as illustrated in Fig 3a.

In our experiment, the open circuit voltage of PVDF nanofiber mat under the conditions of press and slip is measured for comparison. When the electronic skin is pressed, its open circuit voltage consists of two peaks corresponding to "press" and "release", as illustrated in Fig 3b. Differently, when it is slipped, multiple peaks occurred in the measured open circuit voltage (red curve of Fig 3b). The difference of open circuit voltage signals can be used for recognizing the motion type.

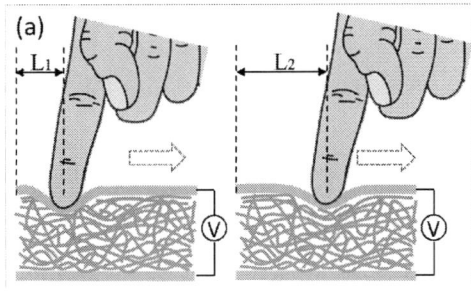

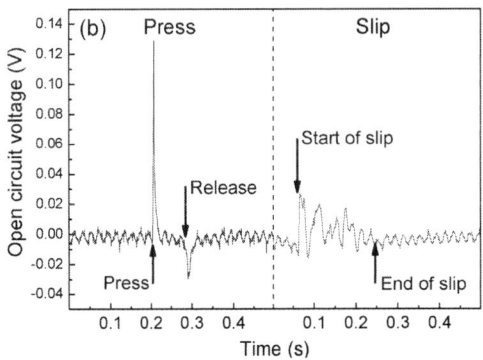

Figure 3. Mechanism of sensing slip and open circuit voltage of electronic skin: (a) schematics of PVDF nanofiber mat fluctuation caused by slip; (b) open circuit voltage of PVDF nanofiber mat under pressing and slipping

According to the electrical and mechanical coupling equation[9] for piezoelectric PVDF sensor, the relation between electric displacement and the strain can be expressed as:

$$V_{oc} = Q/C_{PVDF} = \left(\int d_{33} E_{33} A \dot{\varepsilon_{33}} dt \right) \Big/ C_{PVDF} \qquad (1)$$

in which V_{oc} refers to open circuit voltage, Q for charges generated of piezoelectricity, C_{PVDF} for capacitance of PVDF fiber mat between electrode, d_{33} for piezoelectric coefficient normal to fiber mat, E_{33} for Young's modulus of PVDF fiber mat at its thickness direction, A for contact area between electrode and PVDF fiber mat, $\dot{\varepsilon}_{33}$ for strain rate applied to fiber mat. It can be seen from the above equation that the output open circuit voltage is decided by the slippage speed and contact force, supposing other parameters kept constant.

C. Influence of slippage speed

As the above analysis mentioned, the open circuit voltage of electronic skin increases with faster slipping speed, as shown in Fig 4a. Increasing the slipping speed from 0.1mm/s to 10 mm/s, the mean open circuit voltage ascends from 16.4mV to 86.9mV (Fig 4b). This is caused by the faster fluctuation of PVDF nanofiber mat, resulting in higher strain rate and less piezoelectric charge leakage through PVDF nanofiber mat[10]. So it can be used for detecting the slipping speed.

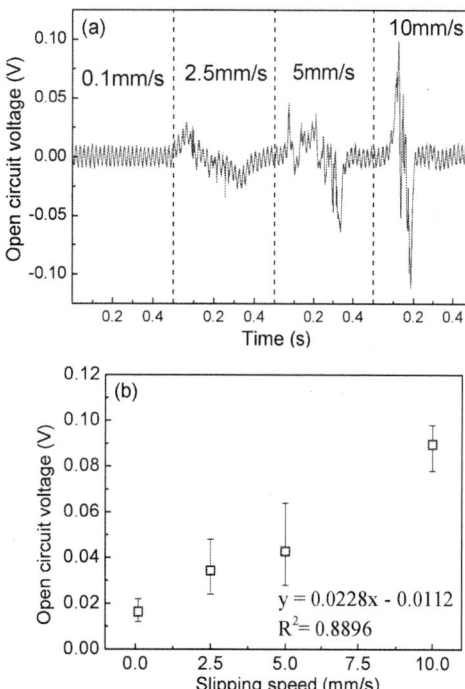

Figure 4. Open circuit voltage of electronic skin: (a) output signal with different slipping speed; (b) mean open circuit voltage with different slipping speed.

After linear fitting the mean open circuit voltage in Fig4b, it is obtained the fitting equation as:

$$y = 0.0228x - 0.0112 \qquad (2)$$

$$R^2 = 0.8896 \qquad (3)$$

The sensitivity to slippage speed of this electronic skin is 0.0228V/mms⁻¹ for the range of 0~10mm/s.

D. Influence of contact force

Furthermore, the contact force normal to the electronic skin has influence on its piezoelectric output. As shown in Fig 5, when the speed of motion stage is kept at 2.5mm/s,

the open circuit voltage signal becomes stronger with increasing the contact force from 0.2N to 1.4N, as shown in Fig 5a. The mean open circuit voltages corresponding to each contact force are 11.8, 19.9, 25.2, 32.9mV. And we also calculated the fitting equation from the mean open circuit voltage in Fig 5b, expressed as:

$$y = 0.0068x + 0.0054 \qquad (4)$$

$$R^2 = 0.9938 \qquad (5)$$

Therefore the sensitivity to contact force of this electronic skin is 0.0068V/N within the range from 0.2 to 1.4N.

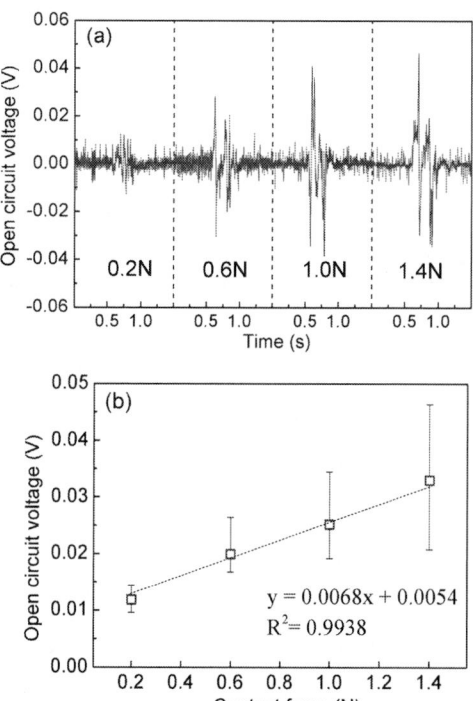

Figure 5. Open circuit voltage of electronic skin: (a) output signal with different contact force; (b) mean open circuit voltage with different contact force.

IV. CONCLUSION

An electronic skin based on electrospun porous PVDF nanofiber mat has been fabricated and demonstrated in this paper. The PVDF nanofiber mat as sensing element is sandwiched between electrodes. Due to the elasticity caused by porosity of PVDF fiber mat, it responds to slippage with wave-like deformation. And its open circuit voltage is in accordance with the mechanical deformation, differentiating from that of pressing. The piezoelectric response of this electronic skin increases with slipping speed and contact force, exhibiting sensitivity about

$0.0228V/mms^{-1}$ and $0.0068V/N$, within the range of $0{\sim}10mm/s$ and $0.2 {\sim}1.4N$ respectively. However, there are still plenty room of improvement, and our future work would be focused on the stability of the output voltage and durability test.

ACKNOWLEDGMENT

The authors would like to thank Mr Gonghan He and Miss Yipeng Cai from Department of Electrical and Mehancial Engineering of Xiamen University for their help in SEM work.

REFERENCES

[1] A. Chortos, J. Liu, and Z. Bao, "Pursuing prosthetic electronic skin," *Nat Mater,* vol. advance online publication, 07/04/online 2016.

[2] A. Chortos and Z. Bao, "Skin-inspired electronic devices," *Materials Today,* vol. 17, pp. 321-331, 2014/09/01/ 2014.

[3] R. A. Romeo, C. M. Oddo, M. C. Carrozza, E. Guglielmelli, and L. Zollo, "Slippage Detection with Piezoresistive Tactile Sensors," *Sensors (Basel, Switzerland),* vol. 17, p. 1844, 2017.

[4] S. Asano, M. Muroyama, T. Nakayama, Y. Hata, Y. Nonomura, and S. Tanaka, "3-Axis Fully-Integrated Capacitive Tactile Sensor with Flip-Bonded CMOS on LTCC Interposer," *Sensors (Basel, Switzerland),* vol. 17, p. 2451, 2017.

[5] S. J. Kim, S. G. Baek, H. Moon, H. R. Choi, and J. C. Koo, "Development of a capacitive slip sensor using internal air gap," *Microsystem Technologies,* vol. 24, pp. 4471-4476, 2018/11/01 2018.

[6] Y. Xin, H. Tian, C. Guo, X. Li, H. Sun, P. Wang, *et al.*, "PVDF tactile sensors for detecting contact force and slip: A review," *Ferroelectrics,* vol. 504, pp. 31-45, 2016/11/13 2016.

[7] G. Zhu, Z. Zeng, L. Zhang, and X. Yan, "Piezoelectricity in β-phase PVDF crystals: A molecular simulation study," *Computational Materials Science,* vol. 44, pp. 224-229, 2008/12/01/ 2008.

[8] X. Cai, T. Lei, D. Sun, and L. Lin, "A critical analysis of the α, β and γ phases in poly (vinylidene fluoride) using FTIR," *RSC Advances,* vol. 7, pp. 15382-15389, 2017.

[9] J. Sirohi and I. Chopra, "Fundamental understanding of piezoelectric strain sensors," *Journal of intelligent material systems and structures,* vol. 11, pp. 246-257, 2000.

[10] C. Chang, V. H. Tran, J. Wang, Y.-K. Fuh, and L. Lin, "Direct-write piezoelectric polymeric nanogenerator with high energy conversion efficiency," *Nano letters,* vol. 10, pp. 726-731, 2010.

Gap in pagination due to unavailable paper.

Pages 240-241

Proceedings of the 16th Annual IEEE International
Conference on Nano/Micro Engineered and Molecular Systems
April 25-29, 2021

A Simple Bandwidth Broadening Method of Terahertz Metamaterial Absorber by Partially Removing The Dielectric Layer

Jia Xu, Jiahao Miao, Yi Liu, Yuan Tian and Xiaomei Yu[*]

Abstract—In this paper, we demonstrate a simple bandwidth broadening method of terahertz (THz) metamaterial absorber (MA) by partially removing the dielectric layer of the MA. Two THz MAs with the dielectric layer fully covered and partially removed in the first and third quadrants were designed, while the top resonators of the MAs are in the same multiplexed cross form. The simulated FWHM (full width at half maximum) increases from 1.12 THz to 1.93 THz and the measured FWHM increases from 1.86 THz to 2.71 THz as the dielectric layer is partially removed in the first and third quadrants. The bandwidth broadening is caused by merging the blue-shifted resonant peaks corresponding to the resonances in the first and third quadrants with the resonant peaks corresponding to the resonances in the second and fourth quadrants. This design provides valuable guidance in designing broadband THz MAs.

I. INTRODUCTION

Metamaterial absorber (MA) is an important terahertz (THz) element in many applications, such as thermal energy harvesting, thermo-photovoltaic energy conversion, thermal imaging, and emissivity control, etc. A sandwich structure, consisting of arrays of subwavelength metallic pattern on a dielectric spacer, backed by a metallic ground plane, is widely applied in realizing the MAs. Since the first THz MA was reported by T. Hu et al. [1], MAs worked in THz region have been attracted considerable attentions due to the paucity of suitable materials in nature in this frequency range. However, most of the previous THz MAs have common drawbacks of absorption bandwidth due to the absorption mechanism being based on electromagnetic resonance, which usually work at limited frequency ranges and not suitable for many applications. To obtain a broadband absorption, two methods by merging several close absorption peaks into a wider absorption bandwidth have generally been proposed. The first one is to nest the multiplexed resonators with different sizes horizontally in a planar structure [2-4], however, the bandwidth is relatively narrow. Rongzhou Gong et al. demonstrated a polarization-insensitive broadband terahertz metamaterial absorber based on coplanar multi-squares films with a bandwidth of 0.8 THz corresponding to the absorption greater than 90% [2]. Another method to broaden the bandwidth is to use multilayer structures [5-7], but the fabrication process of such structures is relatively complex and expensive. In 2018, Siyu Tan et al. proposed an alternative design for broadband metamaterial absorption by incorporating two interlaced fishnet layers and achieved a broadband absorber with an FWHM (full width at half maximum) of about 0.99 THz [7]. In this work, we demonstrate a simple approach to broaden the bandwidth of THz MA by partially removing the dielectric layer with the top

multiplexed resonators unchanged. The FWHM was broadened by 45.7% experimentally.

II. DESIGN AND SIMULATIONS

A MA with multiplexed resonators horizontally in a planar structure was first designed, and the schematic diagram of the MA is illustrated in Fig. 1. The MA employs a typical metal/dielectric layer/metal structure with a 500 nm aluminum (Al) ground plane at the bottom, two dielectric layers in the middle and a 400 nm Al resonance layer on top. The resonance layer contains four cross resonators multiplexed in the four quadrants with different dimensions, and the optimized parameters of the resonators are shown in Table 1. The four multiplexed resonators offset from the center by 10 μm separately with a period of 40μm. The dielectric layers consist of a top 600 nm SiN_x dielectric layer and a 1.8 μm polyimide (PI) dielectric layer. The SiN_x and PI were selected as the middle dielectric layers due to their favorable EM properties in THz region.

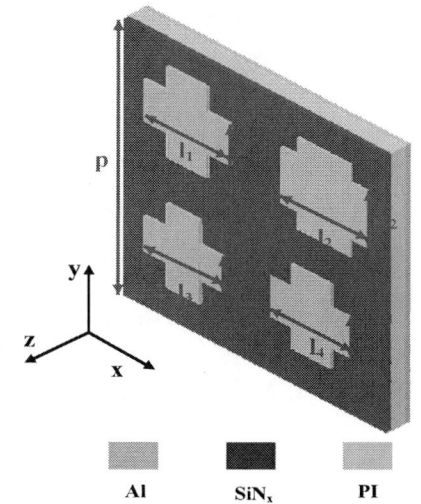

Figure 1. Schematic diagram of the multiplexed resonators MA.

TABLE I.   DESIGN PARAMETERS OF RESONATORS

Parameters	p	l_1	l_2	l_3	l_4	w_1	w_2	w_3	w_4
Value (μm)	40	13	13	12	12	6	8	5	6

To evaluate the absorption performance of the MA, a finite-difference time-domain simulation was carried out.

All Authors are with the National Key Laboratory of Science and Technology on Micro/Nano Fabrication, Institute of Microelectronics, Peking University, Beijing, China.

*Contacting Author: Xiaomei Yu (email: yuxm@pku.edu.cn).

978-1-6654-3008-1/21 $31.00 © 2021 IEEE

During the simulation, the incident light is perpendicular to the metamaterial surface and the electric component is along x axis. The simulated absorption spectrum of the MA is plotted in Fig. 2(a), where a resonant peak is clearly observed at 6.27 THz with the maximum absorption of 99%. The FWHM is calculated to be 1.12 THz with a Q factor of 5.60. The MAs with the cross resonators in the first and the third quadrants and the second and fourth quadrants were simulated separately, and it can be seen from Fig. 2(a) that four peaks corresponding to four cross resonators merge to form the resonant peak at 6.27 THz since they are close to each other sufficiently. This absorption mechanism is also confirmed by the power loss density at 6.27 THz in Fig. 2(b). All the four cross resonators show the obvious power loss at the resonance frequency.

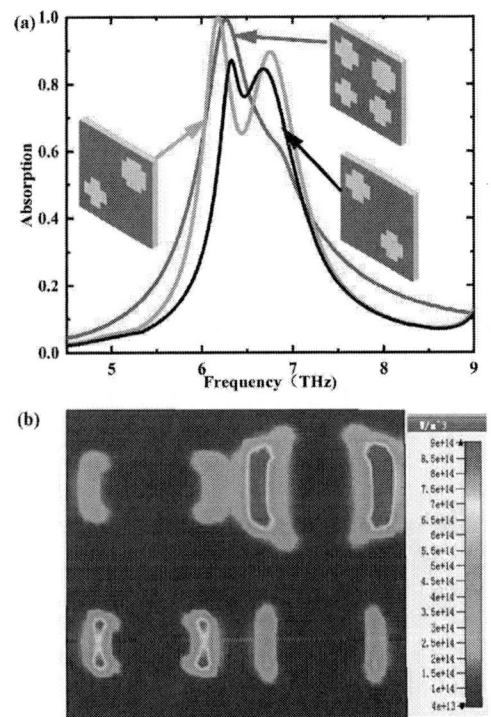

Figure 2. (a) Simulated absorption spectrum of the MA with the dielectric layer fully covered. (b) Power loss density at the resonant peak of the MA with four resonators.

To broaden the bandwidth of the above MA further, the SiN_x dielectric layer was removed outside the upper cross resonators in the first and third quadrants, while the resonance layer kept unchanged, as shown in Fig. 3. The simulated absorption spectrum of the broadband MA is shown in Fig. 4. Three resonant peaks occur at 6.41 THz, 6.91 THz and 7.63 THz with the maximum absorptions of 92%, 84% and 85% respectively. The three resonant peaks together merge to a broadband absorption. The central frequency is taken at 7.14 THz, and the FWHM of the broadband is calculated to be 1.93 THz with a Q factor of 3.70. To analyze the reason for the bandwidth broadening by removing the SiN_x dielectric layer outside the upper cross resonators in the first and third quadrants, the simulated absorption spectra with the cross resonators only kept in the first and third quadrants and only

kept in the second and fourth quadrants are also illustrated in Fig. 4. Compared to the MA with the dielectric layer fully covered, partially removing the dielectric layer outside the upper cross resonators in the first and third quadrants results in a blue shift of the resonance peaks. However, the removing has little effect on the absorption performance for the two cross resonators in the second and fourth quadrants. According to the equivalent circuit model [8], the resonance peaks blue-shifting are caused by the reduction of equivalent capacitance due to the dielectric layer area reducing. So, the bandwidth broadening is caused by merging the blue-shifted resonant peaks corresponding to the resonators in the first and third quadrants with the resonant peaks corresponding to the resonators in the second and fourth quadrants.

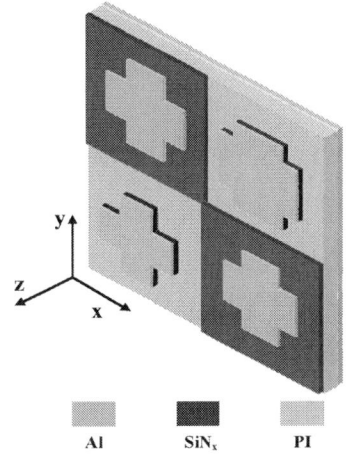

Figure 3. Schematic diagram of the broadband MA with the SiNx dielectric layer removed outside the upper cross resonators in the first and third quadrants.

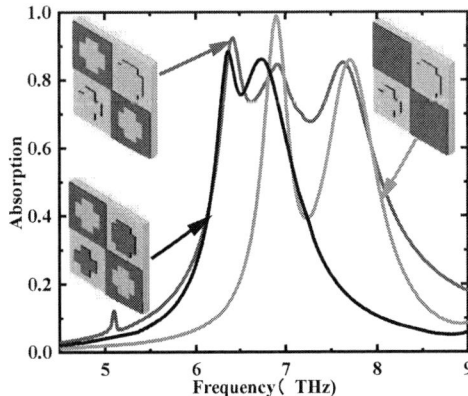

Figure 4. Simulated absorption spectra of the broadband MA.

The power loss density distribution at the three resonance frequencies of 6.41 THz, 6.91 THz and 7.63 THz for the broadband MA are also simulated, as illustrated in Fig. 5(a)-5(c). The resonant peak at 6.41 THz is induced by the resonance of the cross resonators in the second and fourth quadrants, which plays the dominant role in energy dissipation. The resonance of the cross resonators in the first and fourth quadrants does mainly contribute to the absorption at 6.91 THz. At the resonance frequency of 7.63 THz, the energy loss is mainly caused by the cross resonator in the third quadrant.

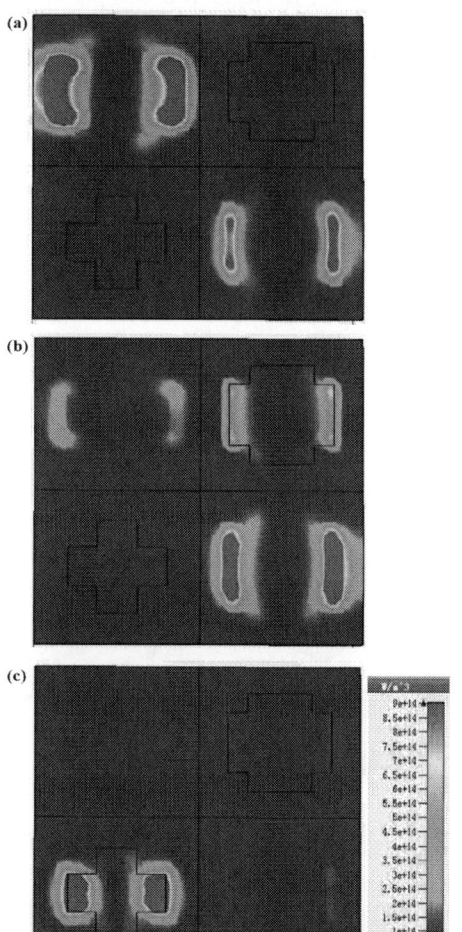

Figure 5. Power loss density in the dielectric layer of the broadband MA at (a) 6.41 THz, (b) 6.91 THz and (c) 7.63 THz.

III. FABRICATION AND EXPERIMENTAL RESULTS

The detailed fabrication process of the THz MAs is shown in Fig. 6. Firstly, a 500 nm Al film was sputtered on the Si substrate, which serves as the ground plane of the THz MA. Then a 1.8 μm thick PI layer was spin coated onto the Al layer. Next, a 600 nm thick SiN_x layer was deposited on the PI layer by using plasma enhanced chemical vapor deposition (PECVD) technique. Another 400 nm-thick Al film was sputtered and defined as the top resonance layer by using wet etching. Finally, the SiN_x layer outside the upper cross resonators in the first and third quadrants was removed by using reactive ion etching. Figure 7 shows the SEM images of the fabricated THz MAs and their corresponding single unit cell.

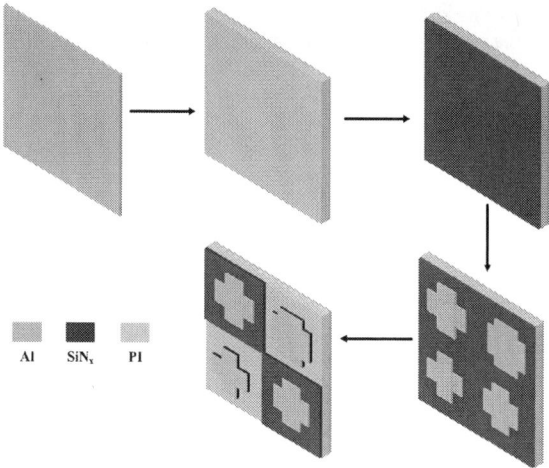

Figure 6. Schematic of the fabrication process of the THz MAs.

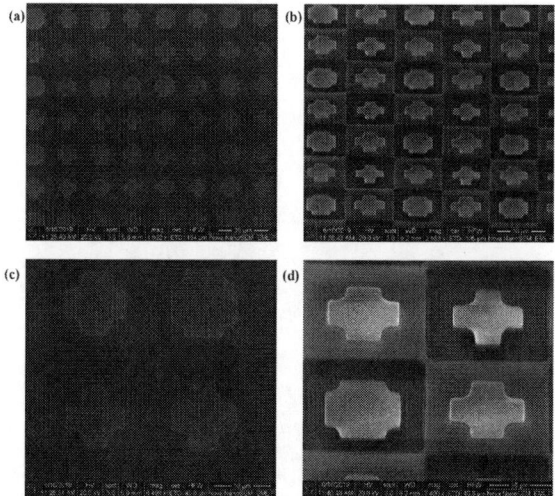

Figure 7. SEM images of (a, b) the two THz MAs and (c, d) their corresponding single unit cell.

The absorption performance of the fabricated MAs was characterized by a Fourier-transform infrared (FTIR) spectrometer. Absorption spectra were obtained by $A = 1 - R$ since the transmission through the samples is nearly zero owing to the continuous Al ground plane. The measured absorption spectra of the two THz MAs are shown in Fig. 8. The MA with the SiN_x dielectric layer fully covered displays a single resonant peak at 6.34 THz with a maximum absorption of 90%, and has an FWHM of 1.86 THz between 5.41 THz and 7.27 THz. The Q factor was calculated to be 3.41. The MA with the SiN_x dielectric layer partially removed outside the upper cross resonators in the first and third quadrants has three main resonant peaks at 6.50 THz, 7.15 THz and 8.08 THz with the maximum absorption of 83%, 75% and 76% respectively. The FWHM of 2.71 THz (from 5.80 THz to 8.51 THz) was obtained, which is 1.46 times as wide as the FWHM of the MA with the SiN_x dielectric layer fully covered, and the Q factor was calculated to be 2.64.

978-1-6654-3008-1/21 $31.00 © 2021 IEEE   244

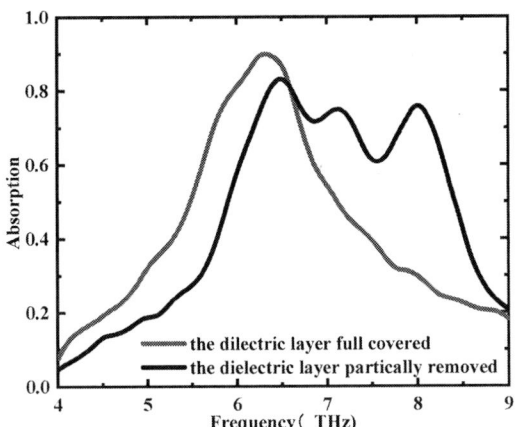

Figure 8.   Measured absorption spectra of the designed MAs.

Both the experimental and simulated results show that partially removing the dielectric layer is an effective method to broaden the bandwidth of the multiplexed resonators MA, but the bandwidth of the experimental results is a little bit wider than the simulated result. We attribute the variation to the fabrication errors, including the accurate size controlling of the patterns, thickness errors and surface roughness.

IV. CONCLUSION

In summary, we have designed and experimentally demonstrated a simple bandwidth broadening method of THz MA by partially removing the dielectric layer of the planar structure MA, while the top multiplexed cross resonators are kept unchanged. The simulated FWHM is increased by 72.3% as the dielectric layer is partially removed in the first and third quadrants, while measured FWHM is increased by 35.0% at the same time. This design method provides valuable guidance in designing the broadband THz MAs.

ACKNOWLEDGMENT

The work was funded by the National Natural Science Foundation of China (grants No. 61935001).

REFERENCES

[1] H. Tao, N. I. Landy, C. M. Bingham, X. Zhang, R. D. Averitt, and W. J. Padilla, "A metamaterial absorber for the terahertz regime: design, fabrication and characterization," *Opt. Express*, vol. 16, no. 10, pp. 7181-7188, 2008.

[2] Y. Cheng, Y. Nie, and R. Gong, "A polarization-insensitive and omnidirectional broadband terahertz metamaterial absorber based on coplanar multi-squares films," *Optics & Laser Technology*, vol. 48, pp. 415-421, 2013.

[3] Y. Wen, W. Ma, J. Bailey, G. Matmon, and X. Yu, "Broadband terahertz metamaterial absorber based on asymmetric resonators with perfect absorption," *IEEE Trans. on Terahertz Science and Technology*, vol. 5, no. 3, pp. 406-411, 2015.

[4] D. Jia, J. Xu, and X. Yu, "Ultra-broadband terahertz absorption using bi-metasurfaces based multiplexed resonances," *Opt. Express*, vol. 26, no. 20, pp. 26227-26234, 2018.

[5] J. Wang, Y. Deng, Y. Wu, S. Lai, and W. Gu, "A single-resonant-structure and optically transparent broadband THz metamaterial absorber," *Journal of Infrared, Millimeter, and Terahertz Waves*, vol. 40, no. 6, pp. 648-656, 2019.

[6] W. Pan, X. Yu, J. Zhang, and W. Zeng, "A novel design of broadband terahertz metamaterial absorber based on nested circle rings," *IEEE Pho. Tech. Letters*, vol. 28, no. 21, pp. 2335-2338, 2016.

[7] S. Tan, F. Yan, N. Xu, J. Zheng, W. Wang, and W. Zhang, "Broadband terahertz metamaterial absorber with two interlaced fishnet layers," *AIP Advances*, vol. 8, no. 2, p. 025020, 2018.

[8] G. Duan, J. Schalch, X. Zhao, A. Li, C. Chen, R. D. Averitt, and X. Zhang, "A survey of theoretical models for terahertz electromagnetic metamaterial absorbers," *Sensors and Actuators A: Physical*, 2019.

Gap in pagination due to unavailable papers.

Pages 246-250

Proceedings of the 16th Annual IEEE International
Conference on Nano/Micro Engineered and Molecular Systems
April 25-29, 2021

Water Drop Detection Contained in Lubricant Oil for Submersible Pump Fault Diagnosis

Fumikazu Mizutani[1,2], Michio Shimomura[1,2], Akinori Mori[1], Yoshikazu Yoshida[2],
Yoshihiro Hosokawa[1], Tadao Matsunaga[1], and Sang-Seok Lee[*1]
[1]Graduate School of Engineering, Tottori University, Tottori, Japan
[2]Sakuragawa Pump MFG Co., Ltd., Osaka, Japan
[*]sslee@tottori-u.ac.jp, *Senior Member, IEEE*

Abstract— In this paper, we propose a method to detect microscaled water drop contained in lubricant oil for a submersible pump fault diagnosis. The real-time deterioration detection of the submersible pump is quite important to guarantee working continuously during pumping water. However, it is rarely reported on proper real-time monitoring tool to predict the submersible pump fault. Because the submersible pump is always using in water, detection method to predict fault is limited. Moreover, vibration, solid housing and 24 hours continuous working of pump are also affected fault detection method selection. However, our detection method uses a simple optical measurement setup consisted of LEDs and photodiodes, by which real-time detection of submersible pump fault is realized. The most common damage of submersible pump is water inflow into oil room due to mechanical seal degradation. The oil room is filled up with lubricant oil to support motor axis rotation. We detect microscaled water drops flowed into oil room by using optical measurement setup and predict the submersible pump fault. We performed experiment to verify the method. As a result, we confirmed that the effectiveness of the proposed optical method for the detection of water drop contained in lubricant oil. Furthermore, we found that the photodiode output voltage changes according to power law with change of inflow water quantity.

Keywords—submersible pump; fault diagnosis; optical measurement setup; power law

I. INTRODUCTION

Since submersible pump was invented by Armais Arutunoff in 1928 [1], it had been applied to various industries, construction fields and natural disaster area mainly to extract water or oil. Basically, to use submersible pump, the whole device is submerged in water or oil that to be pumped. Moreover, to maintain a constant fluid level or to lower fluid level, in many cases, submersible pumps are continuously working for 24 hours. Especially, from the viewpoint of water environment control, submersible pump has much important role. Therefore, to predict fault of the submersible pump and maintain its performance in advance are very crucial. However, it is rarely reported on proper real-time monitoring apparatus to predict the submersible pump fault. Although some pump manufacturers attempted to monitor the submersible pump status, there was no scientific verification about monitoring method and exact monitoring is not achieved yet. To diagnose fault of the

submersible pump, a study used sound that came from a bolt hitting of the pump by a test hammer was reported [2]. However, its method cannot be real-time monitoring and it is doubtful in terms of reliability. If we consider the importance of submersible pump in the field, the real-time monitoring of the pump status is very important issue and the development of real-time monitoring system should be required.

On the other hand, there are few options for installing real-time monitoring tool in submersible pump in terms of inside structure of the pump. Furthermore, the submersible pump is submerged in fluid that to be pumped and is working for 24 hours, it is vibrated severely, and its housing is quite solid. These factors give restriction to select of real-time monitoring method for predicting submersible pump fault.

To predict the submersible pump fault, we are developing sensor system that can be overcome above mentioned restrictions. The sign for pump fault can be abnormal vibration, abnormal sound, pumping performance degradation. Here, we focus on detection of contaminations in oil room in the pump, which is directly related to pumping performance degradation. Actually, it is the most common damage of the submersible pump. If the mechanical seal enveloping the rotating axis for impeller is deteriorated, then water inflow into the oil room. The oil room is filled up with lubricant oil. Contaminants such as water and small particles increasing means that the deterioration of mechanical seal becomes worse. It causes pumping performance degradation and consequently the submersible pump is out of order. The developing sensor system is aiming at detection of water drop as one of contaminants in the oil room. By detecting water drop flowed into oil room, we can predict the maintenance timing of the mechanical seal.

In this study, we propose a simple optical method to detect water drop flowed into oil room of a submersible pump. By monitoring the quantity (volume) of water drop contained in lubricant oil, the maintenance timing of mechanical seal of the pump can be easily predicted. Proposed optical method is quite simple that utilizes a set of LED and photodiode (PD). We use the red LED as a light source, whose wavelength has the feature that transmittance is different between water and oil. The photodiode detects the transmitted LED light intensity differences depending on

978-1-6654-3008-1/21 $31.00 © 2021 IEEE   251

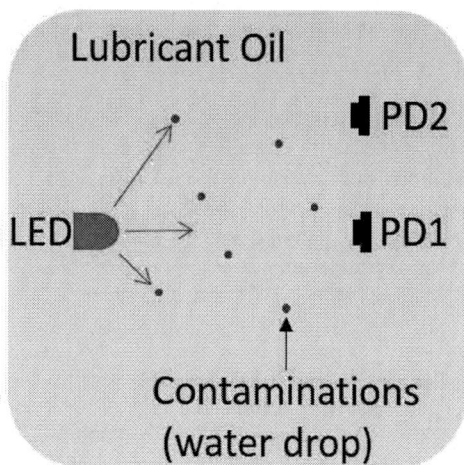

Fig. 1. Schematic view of working principle of proposed optical measurement setup. Water drops contained in lubricant oil is detected by transmitted or scattered LED light intensity change measured with photodiode (PD).

water drop volume.

II. MEASUREMENT PRINCIPLE AND OPTICAL MEASUREMENT SETUP

A. Measurement Principle

To detect the water drop contained in lubricant oil of the oil room, we propose a simple optical measurement setup consisting of a LED and a PD. Previously, for water resources, we had proposed a simple optical measurement setup to detect turbidity and chlorophyll *a* concentration simultaneously [3,4]. Proposed optical measurement setup is suitable for detecting small contaminations in the fluid. Furthermore, implementation is quite simple and apparatus is small sized. Therefore, it can be applicable to contaminations detection in the lubricant oil as well. In Fig. 1, schematic view of working principle of proposed optical measurement setup is shown. The emitted LED light is reflected by water drops in lubricant oil. The traveled LED light intensity through the lubricant oil and water drops mixture decreases and measured by photodiode (PD). We can expect that if water drops become more, the traveled LED light intensity becomes less. The PD output voltage can be easily transmitted to wireless communication module installed outside pump, which makes us to achieve easily the real-time monitoring system for the submersible pump.

B. Optical Measurement Setup

To verify the usefulness of the proposed optical measurement setup for the detection of water drops contained in lubricant oil, we fabricated a measurement system. The oil room in the submersible pump is modelized as the measurement system. In Fig. 2, schematic view of the fabricated measurement system is shown. Inside measurement system, usual lubricant oil is filled up and its volume (2L) is similar to a real commercial submersible pump (UEX series [5] manufactured by Sakuragawa pump MFG, Co., Ltd.). To model the real commercial pump, we stirred the lubricant oil with stirrer. In the vessel of measurement system, four posts are prepared as shown in

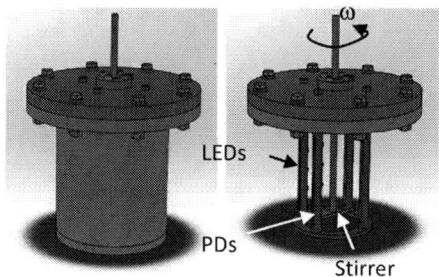

Fig. 2. Schematic views of fabricated measurement setup to verify proposed water drop detection method using LEDs and photodiodes. Left figure shows total view and right figure shows inside the measurement setup.

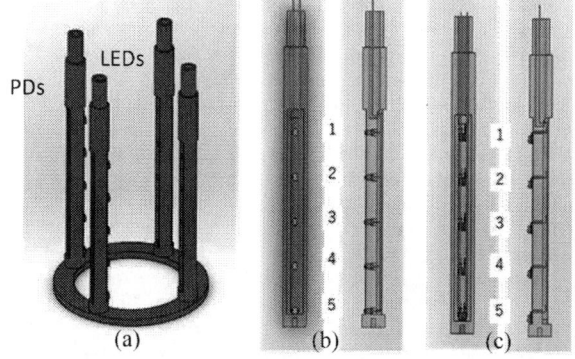

Fig. 3. Schematic views of four posts prepared inside the measurement system; (a) four posts are prepared for optical and electrical measurements. (b) LEDs are located vertically at one post. Front (left) and cross-sectional views (right) are represented. (c) PDs are located vertically at another post confronted the LEDs post. Front (left) and cross-sectional views (right) are represented.

Fig. 2 and Fig. 3(a). Two posts and another two posts are prepared for the optical measurements and the electrical measurements, respectively. In this paper, we only focus on optical measurement. The LEDs are located vertically at one post for optical measurement (see Fig. 3(b)). On the other hand, the PDs are located vertically at another post confronted the LEDs post (see Fig. 3(c)). These posts are fixed firmly both to keep optical axis, and not to affect by vibration of the measurement system and rotation of the lubricant oil. To investigate the location dependency of the light source (LED), we prepared five LEDs and PDs, respectively as shown in Fig. 3(b) and 3(c).

Circuit for measurement is shown in Fig. 4. The LED light intensity is obtained by measuring the voltage of resistance R2 in Fig. 4. Absorption coefficient of water for visible wave is biggest in red region [6]. Therefore, we used a red LED as light source that has the wavelength of 624 nm. The response wavelength of PD is 300 nm to 1000 nm.

III. EXPERIMENT RESULTS AND DISCUSSION

We performed optical measurement to verify the working principle. In the experiment, water drop volume was changed as 10 ml, 20 ml, 40 ml, 80 ml, 120 ml, 160 ml and 200 ml, and rotation speed of stirrer was 1500 rpm. In

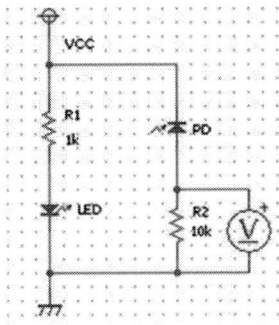

Fig. 4. Circuit used in measurement. The LED intensity is obtained by measuring the voltage of the resistance R2. The resistance R2 is 10 kΩ.

Fig. 5 to 7, the PD1, PD3 and PD5 output voltages are shown when we used the LED5, respectively. The LED and PD numbers are represented in Fig. 3(b) and (c), respectively.

Though the water drops volumes used in the experiment were small and cannot be distinguish with naked eyes, we could confirm the PD output voltage decrease as we expected.

It means that proposed optical measurement setup can be applicable to water drop detection flowed into oil room of the submersible pump. Furthermore, we found that the PD output voltages decrease according to power law. The exponents are -0.292, -0.324 and -0.443 for the LED5-PD1, LED5-PD3 and LED5-PD5 cases, respectively, as shown in Fig. 5 to 7. The PD output voltages decrease rate is bigger when the distance between the LED and PD is longer. Moreover, we could confirm that the PD output voltage value becomes bigger when the distance between the LED and PD is shorter.

Though the lubricant oil was stirred by stirrer in the experiment, we could achieve reasonable results that means proposed optical measurement setup can be also applicable to rotating and vibrating apparatus.

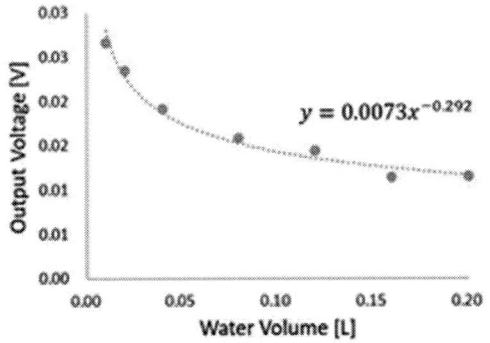

Fig. 5. The PD1 output voltages versus water drops volume contained in the lubricant oil when the LED5 was used and the stirrer rotation is 1500 rpm.

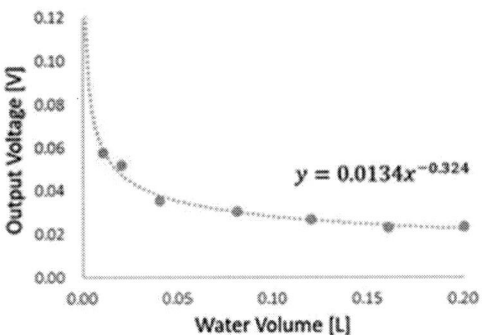

Fig. 6. The PD3 output voltages versus water drops volume contained in the lubricant oil when the LED5 was used and the stirrer rotation is 1500 rpm.

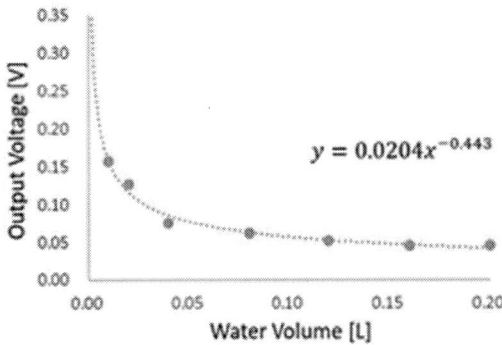

Fig. 7. The PD5 output voltages versus water drops volume contained in the lubricant oil when the LED5 was used and the stirrer rotation is 1500 rpm.

IV. CONCLUSION

We proposed an optical measurement method to detect water drop contained in lubricant oil for a submersible pump fault diagnosis. The optical measurement method consists of a set of LED and PD. In order to verify usefulness of the proposed method, we fabricated measurement system modeling oil room of the submersible pump. We performed the experiment to detect water drops contained in the lubricant oil. As a result, we successfully distinguish water drop volume change by the PD output voltage change. Moreover, we found that the PD output voltage decreases according to power law by increasing water drop volume.

We conclude that the proposed optical measurement method is suitable for detection of submersible pump fault. We will connect sensor system to a communication module to establish for the real-time monitoring. Moreover, the proposed method will be also applied to detect other contaminants in lubricant oil.

ACKNOWLEDGMENT

The authors would like to thank Mr. D. Taketoshi for technical advice and support for fabrication of the measurement system.

REFERENCES

[1] D. Everett, "Submergible Pump," The Encyclopedia of Oklahoma History and Culture, https://www.okhistory.org/publications/enc/entry.php?entry=SU001.G.

[2] N. Yabuki, K. Ueta and J. Kotani, "Experiment of sound monitoring of bolts and submersible pumps and development of distributed sound database system", Journal of Applied Mechanics, vol. 7, pp. 1159-1166, 2004.

[3] R. Komiyama, T. Kageyama, M. Miura, H. Miyashita and S.-S. Lee, "Turbidity monitoring of lake water by trasmittance measurement with a simple optical setup", Proc. Of IEEE Sensors 2015, pp. 1-4, doi: 10.1109/ICSENS.2015.7370456, 2015.

[4] R. Isoyama, M. Taie, T. Kageyama, M. Miura, A. Maeda, A. Mori and S.-S. Lee, "A feasibility study on the simutaneous sensing of turbidity and chlorophyll a concentration using a simple optical measurement setup", Micromachines, vol. 8, 112, doi:10.3390/mi8040112, 2017.

[5] http://sakuragawa.co.jp/en/itemseries

[6] R. Pope and E. Fry, "Absorption spectrum (380–700 nm) of pure water. II. Integrating cavity measurements", Applied Optics, vol. 36, pp. 8710-8723, 1997.

Proceedings of the 16th Annual IEEE International
Conference on Nano/Micro Engineered and Molecular Systems
April 25-29, 2021

Lateral characteristic calibration of an atomic force microscope using a Cr atomic deposition grating

Junjie Wu, Xiaoyu Cai, Yuan Li, Yunxia Fu and Jiasi Wei*

Abstract— **This paper presents the lateral characteristic calibration of an atomic force microscope (AFM) using a Cr atomic deposition grating. A Cr atomic deposition grating with a pitch of 212.8 nm was used to calibrate the lateral properties of the AFM. The pitch of the Cr atomic deposition grating can be directly traced to the standing laser wavelength. The coefficient of the AFM has been corrected according to the calibration results. Then, a 3-μm grating was measured before and after calibration to verify the effectiveness of the calibration. The uncertainty of the measurement result was analyzed.**

I. INTRODUCTION

High precision measurement of key parameters has become a hot and difficult issue in the field of nanotechnology recently. Atomic force microscope (AFM) plays a vital role in the measurement and characterization of microstructures because of its high resolution and ability of three-dimensional measurement [1-3]. The most important dimensions in 3D contours normally include line edge roughness, line width roughness, sidewall roughness, and sidewall angle. These parameters have a strong effect on the device performance in general [4]. So the accuracy of AFM in the measurement is of particular importance.

For numerous AFMs, the micro displacement is provided by the piezo-transducers (PZT) with a voltage applied to it [5]. The PZT suffers from the non-linear and creep behaviors, these will introduce measurement errors. Hence, it is necessary to be calibrated. Nano dimensional standard is widely used in the calibration of AFMs, including the lateral and vertical characteristics. However, the standard should also be calibrated by the professional metrology institution to ensure traceability. In this paper, a Cr atomic deposition grating was introduced to calibrate the lateral characteristic of AFM. The dimension of the grating can be traced to the laser wavelength directly.

II. EXPERIMENTAL ARRANGEMENT

A. The AFM to be calibrated

The NaniteAFM from Nanosurf AG has been employed in the calibration. The scan head has a maximum scan range of 110 μm × 110 μm × 22 μm. The typical *z*-measurement noise can be 350 pm and 90 pm in the static and dynamic mode, respectively. Since the scan head is driven by a piezo-transducers (PZT) in all the *x*, *y* and *z* directions, it is necessary to be calibrated periodically to ensure accuracy.

The AFM is composed of a scan head, an analog controller and a digital controller, as shown in Fig. 1. A laser diode is used to generate a laser beam, which is then focused on the backside of a cantilever and received by a quadrant detector. A tip is attached to the end of a microfabricated cantilever in order to measure the atomic-range forces between the tip and the sample surface. The other end of the cantilever is fixed to the PZT with a special fixture. The PZT provides micro displacement when measuring. The signals from the quadrant detector is then captured and processed by the AFM controller and transmitted to PC for further processing.

B. The Cr atomic deposition grating

Laser-focused atomic deposition is a new technique for nano fabrication, in which, atoms are deposited onto a substrate through a near-resonant laser standing wave and formed into parallel lines onto a substrate. The pitch fabricated by this technique is directly traced to the standing laser wavelength [6]. The grating standard sample is fabricated using laser-focused Cr atomic deposition method. The principle is based on the radiation pressure of a standing laser wave field with the particular spatial distribution to control atomic motion and create grating structures. Fig.2 is the schematic diagram of laser-focused atomic deposition based on standing laser wave. The period of the grating is 212.8 nm, which is half the wavelength of the laser.

Cr is an ideal element for atomic deposition because of its low surface diffusion and high chemical stability [7]. Fig.3 shows the fabricated Cr laser atomic deposition grating by Tongji University. Although the Cr laser deposition grating has self-traceability, the dimension of the grating is also verified by the nano measuring machine equipped with a metrological AFM. The measurement result of the nano

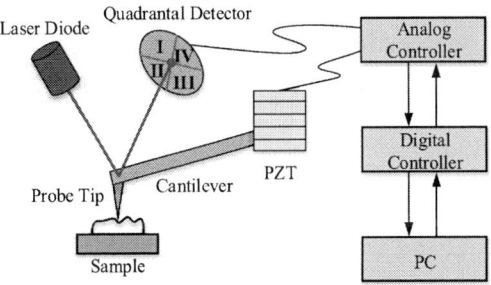

Figure 1. Schematic diagram of the AFM system.

---

* This study has been sponsored by Shanghai Sailing Program (No. 19YF1441700), Science and Technology Plan of Shanghai Administration for Market Regulation (No. 2019-02), and the National Key Research and Development Project (No. 2019YFB2004900).

Jiasi Wei is with the Shanghai Institute of Measurement and Testing Technology, No. 1500, Zhangheng Rd., Shanghai, China (corresponding author to provide phone: 021-38839800-33353; fax: 021-50798605; e-mail: weijs@ simt.com.cn).

978-1-6654-3008-1/21 $31.00 © 2021 IEEE

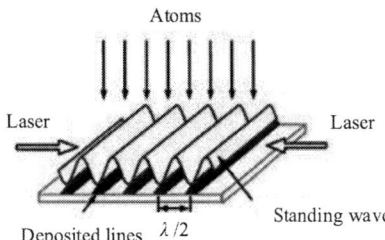

Figure 2. Schematic diagram of laser-focused atomic deposition.

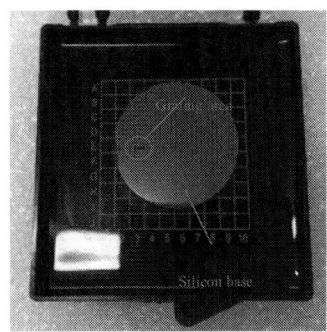

Figure 3. Picture of the Cr laser atomic deposition grating.

measuring machine can be directly traced to the definition of meter. The verified value of the fabricated Cr atomic deposition grating is 212.7 nm with a standard deviation of 0.02 nm.

III. CALIBRATION AND RESULTS EVALUATION

For calibration of scales in the lateral direction, the 1D line gratings or 2D arrays of known average pitch is used generally. For calibration of the vertical scale, a known step height or a groove is usually applied. This paper only discusses the lateral calibration of AFMs using a special transfer standard, the Cr laser atomic deposition grating. Because of the grating is one dimensional, it is necessary to calibrate the AFM in the x and y direction, respectively.

A. The orthogonal angel search

The pitch of 1D grating is the lateral distance between the same phase points of two adjacent structures. If the scanning direction is not orthogonal to the grid direction, a cosine error will occur, which will make the measurement result too large. Therefore, it is very important to select the orthogonal scan angle when measuring. In order to obtain the orthogonal scan angle, the double-angle deflection method is used to calculate the orthogonal angle. Fig. 4 shows the process of orthogonal angle extraction in 1D grating measurement.

Firstly, place the grating on the sample table, and a base line is set along the x axis. Then, rotate the grating with an angle of α_1 and scan the grating along the x axis. Calculate the pitch P_1. Continue to rotate another angle of α_2-α_1, and take the second scan. Calculate the pitch P_2. Using "(1)" to calculate the orthogonal angle, and rotate the sample back in an angle of α_2-φ. Thus, the orthogonal scan direction is obtained. Repeat operation can be taken to achieve a better accuracy of the angle.

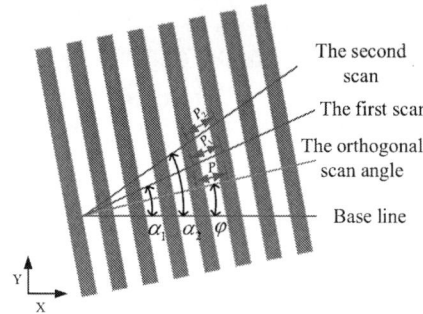

Figure 4. The orthogonal angle extraction.

$$\varphi=\arctan[(P_1*\cos\alpha_1-P_2*\cos\alpha_2)/(P_1*\sin\alpha_1-P_2*\sin\alpha_2)] \quad (1)$$

B. Calibration of the AFM

After searching the orthogonal scan angle, measurement of the Cr laser atomic deposition grating was conducted. Before the first measurement, the original calibration coefficient of the AFM was deleted through the software interface. In this condition, the AFM is in an uncalibrated state. Then, ten times measurements were taken along the orthogonal direction of the grating. The fast Fourier transform (FFT) method is applied to evaluate the grating. The FFT method calculates the grid pitch by measuring the spatial frequency of the sample surface topography information in the frequency domain. This method can reduce the influence of the ripple fluctuation and the contour signal glitch.

The average value of ten times measurement is 176.4 nm. Since the output of the AFM has good linearity, the calibration coefficient can be defined using a dimensionless coefficient. The coefficient is calculated as dividing the standard value by the measured value of the Cr laser atomic deposition grating. According to the calculating result, the calibrated coefficient was set as 1.206. To verify the effectiveness of the calibration, a second measurement of the Cr laser atomic deposition grating was conducted. Fig.5 shows the 3D reconstructed image and the 2D single line profile of the grating. The average

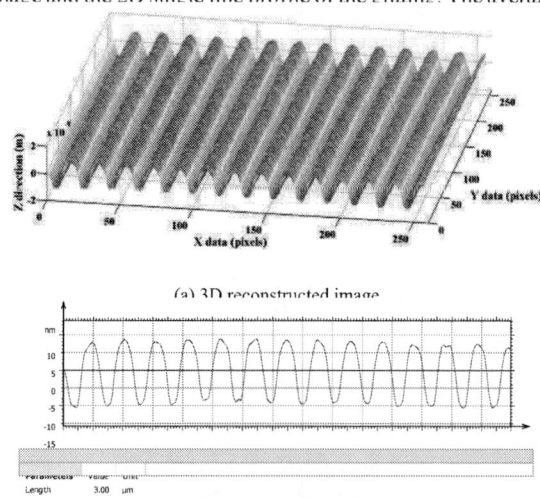

(a) 3D reconstructed image

(b) 2D single line profile

Figure 5. The 3D reconstructed image and 2D single line profile.

value of ten times measurement is 212.6 nm, with a standard deviation of 0.2 nm.

C. Measurement of the grating sample

After calibration of the AFM, the measurement of a 3-μm grating was conducted to verify the accuracy of the calibration in the micron scale. The 3-μm grating was calibrated using the nano measuring machine equipped with a metrological AFM. The calibrated value of the grating is 2999.7±3 nm ($k=2$). In the measurement, the AFM was working in the tap mode. The probe tip model of Tap190Al-G was chosen in the measurement. The scan range was set as 20 μm × 20 μm and the sampling rate was set as 256 points per line.

The grating sample was separately measured with the AFM before and after calibration for comparison. In the experiment, the orthogonal angel was searched firstly. Then ten times repeated measurement was taken in the same position of the grating sample. The mean value of the grating measured before calibration is 2609.3 nm, with a standard deviation of 2.91 nm. While the mean value measured after calibration is 2998.6 nm, with a standard deviation of 2.86 nm. Fig.6 shows the 3D reconstructed image and the 2D single line profile of the grating measured after calibration.

By comparing the measurement results before and after calibration, the validity of the calibration was verified. The Cr laser atomic deposition grating can be used to calibrate the lateral characteristic of AFM accurately.

D. Measurement uncertainty analysis

AFM is a type of scanning probe microscope (SPM). According to the calibration specification of the SPMs, the measurement uncertainty of AFM can be calculated as follows:

- The measurement model. The measurement result of AFM can be described as:

$$l_i = l_s + \Delta l_r + \Delta l_e + \Delta l_d + \Delta l_p \qquad (2)$$

In "(2)", l_i is the measurement result of AFM, l_s is the calibration result of the measured sample, Δl_r is the measurement error from the measurement repeatability, Δl_e is

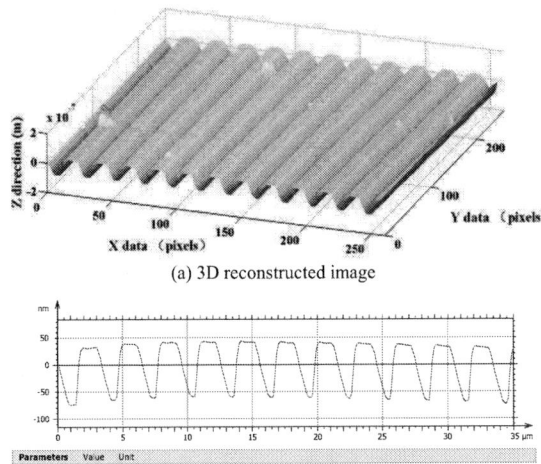

(a) 3D reconstructed image

(b) 2D single line profile

Figure 6. The 3D reconstructed image and 2D single line profile.

the measurement error from the thermal deformation of the measured sample, Δl_d is the error from the image tilt correction, and Δl_p is the error from the orthogonality of x and y axis.

- The combined variance. Because each input is independent of each other, the combined variance can be described as:

$$u_c^2(l_i) = u^2(l_s) + u^2(\Delta l_r) + u^2(\Delta l_e) + u^2(\Delta l_d) + u^2(\Delta l_p) \qquad (3)$$

In "(3)", $u(l_s)$ is the standard uncertainty from the sample, $u(\Delta l_r)$ is the standard uncertainty from the measurement repeatability, $u(\Delta l_e)$ is the standard uncertainty from the thermal deformation of the measured sample, $u(\Delta l_d)$ is the standard uncertainty from the image tilt correction, and $u(\Delta l_p)$ is the standard uncertainty from the orthogonality of x and y axis. The sensitivity coefficient of each item is 1.

- The combined standard uncertainty. The combined standard uncertainty can be described as:

$$u_c(l_i) = \sqrt{u^2(l_s) + u^2(\Delta l_r) + u^2(\Delta l_e) + u^2(\Delta l_d) + u^2(\Delta l_p)} \qquad (4)$$

- The extended uncertainty. The extended uncertainty can be described as:

$$U = k * u_c(l_i), \; k=2 \qquad (5)$$

During the measurement, the environment temperature is controlled in the range of 20±0.5℃. The uncertainty of the calibrated sample is 3 nm and k is 2, so the $u(l_s)$ is calculated as 1.5 nm. The standard deviation of ten times measurement is 2.86 nm, so the $u(\Delta l_r)$ is 1.43 nm. Considering the small temperature fluctuations and the low temperature coefficient of the sample material, the $u(\Delta l_e)$ can be neglected. The tile of the sample is controlled to less than 5°, so the $u(\Delta l_d)$ is less than 0.12 nm. The $u(\Delta l_p)$ is calculated as 0.24 nm. Finally, the $u_c(l_i)$ is calculated as 2.09 nm. And the U is calculated as 4.18 nm ($k=2$).

IV. CONCLUSIONS

Micro- and nano scale measurement is a complicated operation, the environment noise, temperature, air disturbance should be considered. The surface properties, temperature coefficient and surface contamination may also affect the measurement result. Therefore calibration of the measurement instrument is necessary to ensure traceability. This paper presents the application of a Cr atomic deposition grating in the lateral characteristic calibration of an AFM. The effectiveness of the calibration was verified by measuring a 3-μm grating sample. Because of the special preparation process, the pitch of the Cr atomic deposition grating can be directly traced to the standing laser wavelength. Thus, the traceability chain can be shortened if the Cr atomic deposition grating is used in the calibration.

ACKNOWLEDGMENT

Special thanks are given to the research team of Prof. Tongbao Li in Tongji University for providing the Cr laser atomic deposition grating.

REFERENCES

[1] A. Boccaccio, L. Lamberti, M. Papi, M. D. Spirito and C. Pappalettere. "Effect of AFM probe geometry on visco-hyperelastic characterization of soft materials," *Nanotechnology*, vol. 26, pp. 1–15, July 2015.

[2] A. Cordes, B. Bunday and Eric Cottrell. "Sidewall slope sensitivity of critical dimension atomic force microscopy," *Journal of Micro/Nano lithography Mems & Moems*, vol. 11, pp. 011011, February 2012.

[3] G. Valdrè, D. Moro and G. Ulian. "Scanning probe microscopy with vertically oriented cantilevers made easy," *Measurement Science and Technology*, vol. 23, pp. 126103, June 2012.

[4] R. Zhang, S. Wu, L. Liu, N. Lu, X. Fu, S. Gao, et al. "A CD probe with a tailored cantilever for 3D-AFM measurement," *Measurement Science and Technology*, vol. 29, pp. 125011, November 2018.

[5] G. Dai, F. Pohlenz, M. Xu, L. Koenders, H. Danzebrink and G. Wilkening. "Accurate and traceable measurement of nano- and microstructures," *Measurement Science and Technology*, vol. 17, pp. 545-552, January 2006.

[6] L. Lei, Y. Li, X. Deng, G. Fan, X. Cai, X. Cheng, et al. "Laser-focused Cr atomic deposition pitch standard as a reference standard," *Sensors and Actuators A: Physical*, vol. 222, pp. 184-193, February 2015.

[7] M. Ma, T. Li, W. Wu, Y. Xiao, P. Zhang, and W. Gong. "Laser-Focused Atomic Deposition for Nanascale Grating," *Chinese Physics Letters*, vol. 28, pp. 073202, July 2011.

Gap in pagination due to unavailable papers.

Pages 259-264

Proceedings of the 16th Annual IEEE International
Conference on Nano/Micro Engineered and Molecular Systems
April 25-29, 2021

Fast-scan MOEMS Mirror for HD Laser Projection Applications

Huijun Yu, Peng Zhou, Wenjiang Shen*, *Member, IEEE*

Abstract—MOEMS mirror has many applications, such as machine vision, laser projection and so on. In this paper, an electromagnetically-driven biaxial (2D) scanning mirror and its application for HD laser projection system have been presented. A single metal coil, which is deposited on the surface of the exterior frame, is used for generating two pairs of torques and driving the mirror to scan along both axes. Finite Element Method (FEM) is used for designing and optimizing the mechanical properties of the mirror. The mechanical deflection angles of the mirror are measured to be ±8.5° along slow axis with driving current of 100 mA at 60 Hz, and ±12.5° along fast axis with driving current of 100 mA at resonant frequency around 37 kHz. Moreover, four-terminal piezoresistive sensors are integrated in the torsional beams of the mirror. The deflection of the mirror along two axes can be monitored through these sensors and the sensitivity of the piezoresistor is 3 mV/V@1deg. Finally, the mirrors are used for laser display application, where the image 720P （1280x720） resolution at 60 Hz frame rate is demonstrated.

I. INTRODUCTION

THE benefits of miniaturization through Microelectromechanical Systems (MEMS) can be seen in many applications, especially in optics and photonics[1-5]. Recently, there has been increasing interest in micromirror and a host of researches have been done[4, 6, 7]. Hah employed comb-drive actuators to actuate the micromirror[8]. It completes a mechanical scan angle of 7.2° with a high voltage of 65 V. Jung also utilized comb-drive actuators to complete micromirror, which can scan by two axes[7]. In the piezoelectric micromirror devices, piezoelectric materials are used for actuation. Zhong presented the piezoelectric unimorph applied to the micromirror[6], it can reach a mechanical rotation angle of 4.88°. Xie presented a micromirror, which is thermally actuated by poly-silicon heaters[9]. Arda illustrated a high performance 2-D micromirror actuated through the effect of an electromagnetic torque, which is generated by the current carrying coil [10], it has full optical scan angles of 65° and 53° for slow (60 Hz sawtooth) and fast (21.3 kHz sinusoid) scan directions, respectively. But these performance parameters cannot meet the performance requirements of HD laser projection for MOEMS mirror. If the resolution of laser scanning projection needs to reach above 720P, the fast

scanning frequency of the MOEMS mirror must reach above 27 kHz.

Based on the requirements, this paper proposed a 1mm diameter two-dimensional MOEMS scanning mirror driven by electromagnetic method, can offer the high frequency and the driving moment of the large angle required. Also, four-terminal piezoresistive elements are integrated in the beams of the micromirror. Piezoresistive effect has been widely utilized in micro-mechanical sensors. The strain alters the resistivity of the piezoresistive element when the flexible structure deforms. The piezoresistive element is an essential part of the micromirror. With the feedback signals from the four-terminal piezoresistor, the micromirror can complete the closed-loop control.

The working principle of the 2D scanning micromirror is explained in section II. Calculations and simulations for mechanical structure and four-terminal piezoresistive element are also summed up in this section. Section III depicts the fabrication process of the 2-D scanning MEMS mirror. Section IV describes the measurement results of this micromirror and its application in laser projection display. Finally, section V concludes the main remarks for this paper.

II. PRINCIPLES AND DESIGN

A. Working Principle

It is shown in Fig.1, it demonstrates the working principle of the MOEMS mirror. Through the Lorentz's force which is generated by the outer magnetic field and the electric coil. The force makes the mirror deflect.

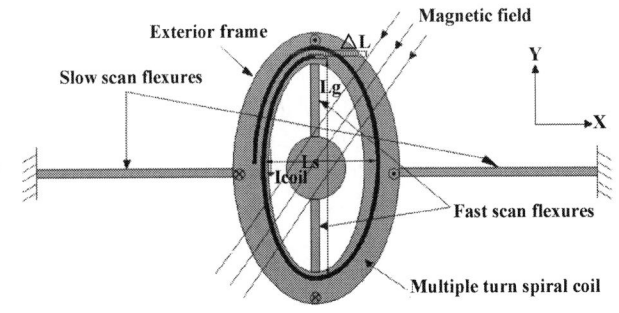

Fig. 1. Schematic of the 2-D micromirror. I_{coil} is the coil current, L_s and L_g are the radius of the elliptical coil, ΔL is the pitch.

The Lorentz's force can be calculated by the following equation[11]

$$F = \oint_L idl \times B \qquad (1)$$

where F is the Lorentz's force, i is the current flowing through the coil and B is the strength of the outer magnetic field. The force induced torque is applied normal to the external magnetic field. As a result, the micromirror would scan around the axis. The torque also can be calculated by the following equation

---

This research was funded by the National Key Research and Development Program of China (Grant No. 2018YFF01010901 and No. 2016YFB0401903).

Huijun Yu is with School of Nano-Tech and Nano-Bionics, University of Science and Technology of China, Hefei 230026, China.(phone: +86 512 62872688; e-mail: hjyu2012@sinano.ac.cn).

Peng Zhou and Wenjiang shen are with Key Lab of Nanodevices and Applications, Suzhou Institute of Nano-tech and Nano-bionics, Chinese Academy of Sciences, Suzhou 215123, China.(phone: +86 512 62872688; e-mail: wjshen2011@sinano.ac.cn).

978-1-6654-3008-1/21 $31.00 © 2021 IEEE

$$T = iBcos\theta \sum_{m=0}^{N-1} \pi \cdot \left(\frac{L_s}{2} + m \cdot \Delta L\right) \cdot \left(\frac{L_g}{2} + m \cdot \Delta L\right) \quad (2)$$

where i the current, B is the magnetic field strength, θ is the angle between the outer magnetic field and the coil, N is the number of the spiral coil turns, L_s and L_g are the radius of the elliptical coil, and ΔL is the pitch.

B. Mechanical Design of Micromirror

For laser projection display application, the characteristics of the 2D scanning micromirror determine the display format. The resolution of the projection image can be calculated by these equations.

$$N = \frac{\theta_{opt}D}{a\lambda} = \frac{4\theta_{mech}D}{a\lambda} \quad (3)$$

$$f_s = \frac{F_r N}{2} \quad (4)$$

where D is the diameter of the inner mirror, a is the shape factor of the mirror (in general a=1), λ is the wavelength of the incident laser beam, θ_{opt} and θ_{mech} are the optical scan angle and the mechanical scan angle, respectively. F_r is the refresh rate or the frequency of the slow scan for the display, f_s is the frequency of the fast scan. Through these equations of the image resolution, it can be known that $\theta_{mech}D$ is the important factor.

A series of models and equations are applied to explain the theoretical model of the 2D scanning micromirror. Firstly, the MEMS mirror has two parts, the exterior frame and the inner mirror, connected together with the fast scan flexures. The exterior frame is linked to the substrate with the slow scan flexures. A lumped model which describes the motion of the micromirror is presented in Fig. 2. In this figure, K_{fast} is the spring constant of the fast scan flexures and c_2 is the damping coefficient of the inner mirror. These parameters form a mass-damping-spring system. Similarly for the integral part, including the inner mirror and exterior frame, the K_{slow} and c_1 form another mass-damping-spring system. The two systems form a second-order system which analyses the motion of the micromirror. θ_1 and θ_2 in the graph of the lumped model are the mechanical rotation angles along the slow scan flexures and fast scan flexures, respectively.

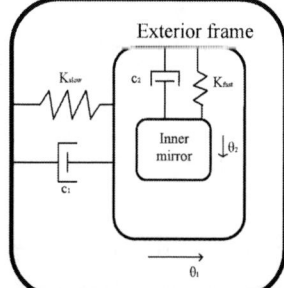

Fig. 2. The lumped model of the micromirror.

The mass-damping-spring system, which is second-order vibration system, can explain the motion of the micromirror. The second-order equation is

$$T = I_m\ddot{\theta} + c\dot{\theta} + K\theta \quad (5)$$

where T is the driving torque which is produced by the Lorentz's force, I_m is the inertia of the integral part. I_m can be calculated by equation (6) as below[12]

$$I = \iiint_V \rho r^2 dV \quad (6)$$

where V is the volume of the micromirror, ρ is the material density of the micromirror and r is the distance between the integral point and the torsion beam. The spring constant K, can be calculated based on the following equation [12]

$$K = \frac{2K_{ab}ab^3G}{L_f} \quad (7)$$

where a(width), b(height) and L_f (length)are the parameters of the scan flexure, G is torsional modulus(Pa) and K_{ab} is given by[13]

$$K_{ab} = \begin{cases} 5.33 - 3.36\frac{b}{a}\left(1 - \frac{b^4}{12a^4}\right), a > b \\ 5.33 - 3.36\frac{a}{b}\left(1 - \frac{a^4}{12b^4}\right), b > a \end{cases} \quad (8)$$

The resonant frequency of the micromirror can be attained while I and K are known.

$$f_n = \frac{\omega_n}{2\pi} = \frac{1}{2\pi} \cdot \sqrt{\frac{K}{I}} \quad (9)$$

When the frequency of the driving signal is f, the scan angle θ can be calculated by the following equations[14],

$$\theta = \frac{\theta_o}{\sqrt{\left[1 - \left(\frac{f}{f_n}\right)^2\right] + (2\zeta\frac{f}{f_n})^2}} \quad (10)$$

$$\zeta = \frac{c}{2\sqrt{IK}} \quad (11)$$

$$\theta_o = \frac{T}{K} \quad (12)$$

where θ_o is the static rotation angle, which is the theoretical rotation angle under the actuation of a static torque, T is the driving torque, ζ is air damping ratio, f_n is the resonant frequency, c is the damping coefficient. When the frequency of the driving signal is equal to the resonant frequency of the micromirror, the scan angle of the resonant state can be obtained by[14]

$$\theta_r = \theta_o \cdot Q \quad (13)$$

$$Q = \frac{1}{2\zeta} \quad (14)$$

where θ_r is the scan angle of the resonant state, Q is the quality factor.

Based on these analysis results, we design dimensions of the mirror. The thickness of the mirror is 40 μm and the diameter of the inner mirror is 1mm. The length and the width of the fast scan flexure are 640 μm and 150 μm, respectively. The length and the width of the slow scan flexure are 1800 μm and 40 μm.

C. Piezoresistive Sensor for Micromirror

Fig. 3(a) shows the schematic diagram of the four-terminal piezoresistive element. The shear stress makes the piezoresistive element distort and changes the resistivity of the piezoresistive element. When the current is flowing into the two terminals of the element, we can detect the voltage signals in the other two terminals and the output voltage value changes as the resistivity changes. The output voltage can be expressed by the following equation[15]:

$$V_{34} = \frac{W}{L}V_{12}\frac{\rho_6}{\rho_0} = \frac{W}{L}V_{12}\pi_{44}T_6 \qquad (15)$$

where W(width) and L(length) are the size of the piezoresistive element. V_{12} and V_{34} are the voltage of the different two terminals. π_{44} is the shear piezoresistance coefficient and T_6 is the shear stress which is proportional to the rotation angle of the mirror.

In this work, four-terminal piezoresistive element is integrated in the micromirror device to monitor the rotation angle of the micromirror. With the output of the piezoresistive sensor, the close loop control for the micromirror can be achieved.

We design two types of piezoresistive element, as shown in Fig. 3(b). One is the traditional type, where the four terminals are connected together through the solid sensor zone. The other is Newmann four-terminal piezoresistive element[15], which has a center insulation island in the sensor zone. Compared with the traditional four-terminal piezoresistive device, the Newmann four-terminal piezoresistive element enhances the potential gradients at the output contacts, which has a better sensitivity[15]. The piezoresistive element is placed in the root of the torsional beam as it is shown in Fig. 3(c) and (d).

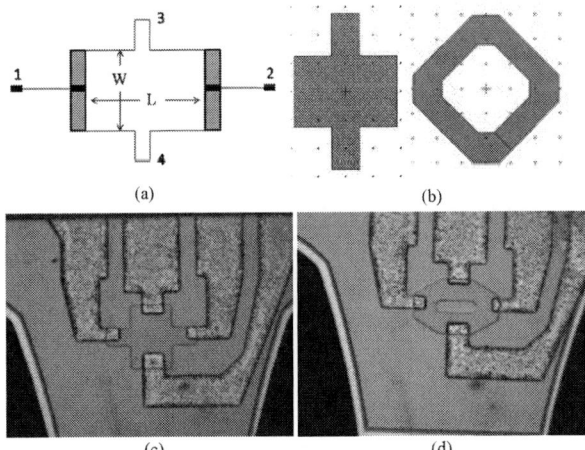

Fig.3. (a) The schematic diagram of the piezoresistive element (b) the four-terminal piezoresistive element designs. (c) traditional four-terminal piezoresistive element . (d) Newmann four-terminal piezoresistive element.

III. FABRICATION

The device fabrication process is shown in Fig. 4(a) to Fig. 4(h). The fabrication starts with a silicon-on-insulator (SOI) wafer. The top silicon of the SOI wafer will form the mechanical structures of the micromirror. The suspended

micromirror structures will be released after etching away the underneath handle silicon layer and buried oxide layer (BOX). Fig. 5 shows the picture of the packaged MOEMS mirror with external magnets.

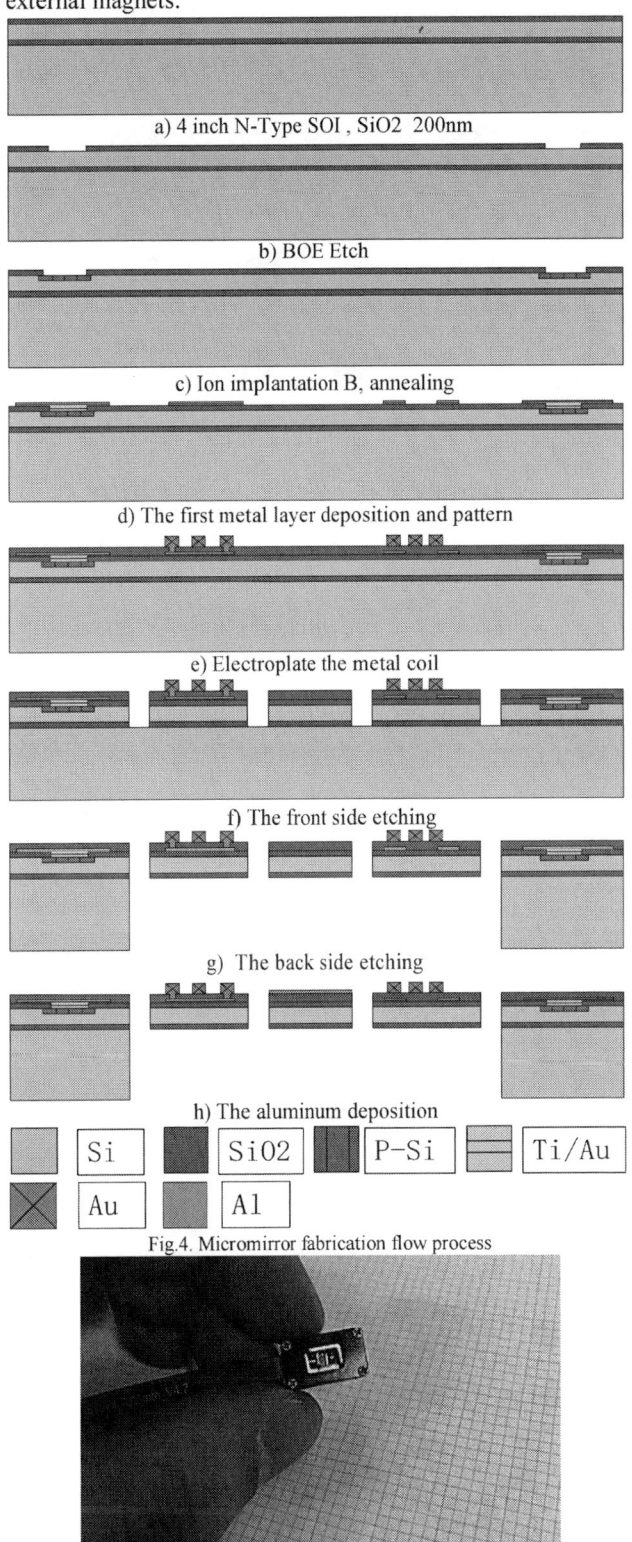

a) 4 inch N-Type SOI , SiO2 200nm

b) BOE Etch

c) Ion implantation B, annealing

d) The first metal layer deposition and pattern

e) Electroplate the metal coil

f) The front side etching

g) The back side etching

h) The aluminum deposition

Fig.4. Micromirror fabrication flow process

Fig.5. The package die assembly of the MEMS mirror

IV. RESULTS AND DISCUSSIONS

In the section, we will present the measurement results and discuss the characteristics of the micromirror. We will also demonstrate its application for laser image projection display.

A. The Measurement Results and Discussions

The schematic diagram of the measure system is described in Fig.6. The micromirror is driven by signals generated from signal generator. The laser deflects from mirror surface and the light position can be precisely detected in a 1.5mm×1.5mm position sensitive detector (PSD)[16].

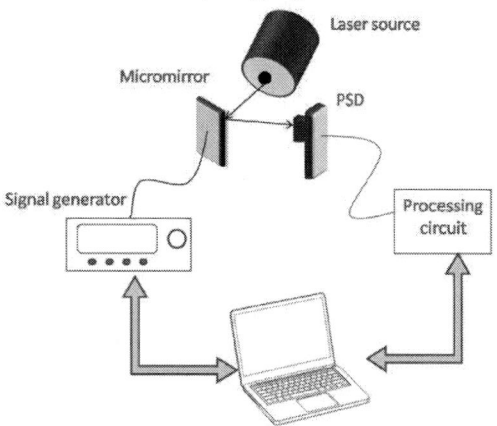

Fig.6. The schematic diagram of the measure system

We first measured the mechanical deflection angle α along the slow scan flexures. The frequency of the sinusoidal driving signal of the slow scan flexures is 60 Hz. As the driving current value (valid values) is adjusted from 10 mA to 100 mA, the mechanical deflection angle α is also increasing. As shown in Fig. 7, while the drive current is 100 mA, the slow mechanical scan angle is about ±8.5°. Similarly, when the driving signal is around 37 kHz, the inner mirror scans along the fast scan flexures. The fast mechanical scan angle is about ±12.5° when the drive current is 100 mA.

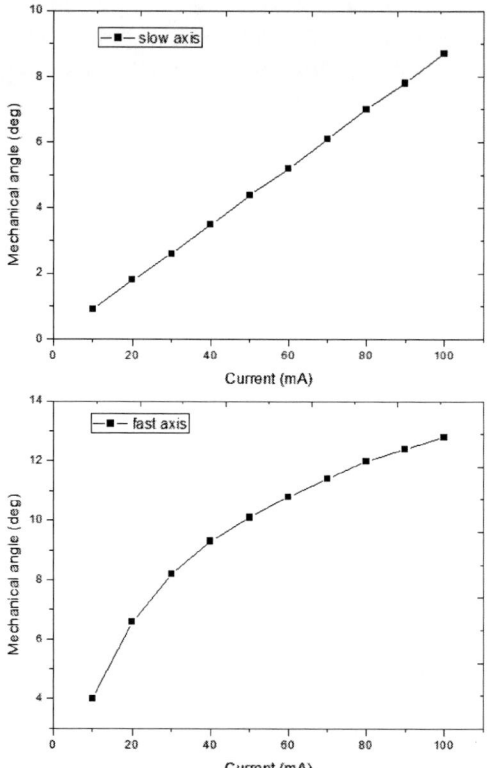

Fig.7. The relationship between the mechanical rotation angle and the superimposed drive current.

B. The Measurement Results of the Four-Terminal Piezoresistive Element

The performance of the piezoresistive element is also tested. Fig.8 presents the sensor signal as a function of the mirror rotation angle. For the traditional four-terminal piezoresistive element, the sensitivity is 2.6 mV/V@1deg. The sensitivity of the Newmann piezoresistive element which has the insulation island in the center is 3 mV/V@1deg. The supply voltage is 1 V.

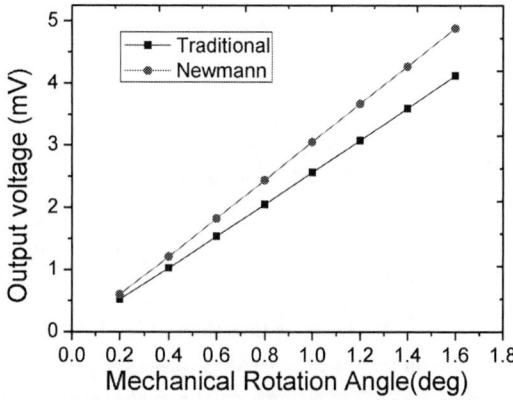

Fig.8. Sensor signals as the function of the mirror rotation angle of the four-terminal piezoresistive element.

C. Laser Projection Image

The 2D scanning micromirror is used to demonstrate image display[17]. A 2-D image on the screen can be projected by the combination of micromirror 2-D scanning and RGB lasers' modulation.

Our current display system can create the images with 1280×720 pixels resolution at 60 Hz frame rate, as shown in Fig. 9. Image quality still needs to be improved, such as to improve the resolution by increasing the resonant frequency of the fast scan and to compensate the image distortion by using the ramp wave as the driving signal of the slow scan.

Fig.9. The RGB images created by laser projection system.

V. CONCLUSION

In this paper, a 2D scanning micromirror with electromagnetic actuation has been designed, fabricated and tested. The single driving coil is used for achieving the scan of the micromirror in two directions by mechanical de-coupling of driving signals with different frequency components. The fabricated device can provide large scanning angles in both directions. For slow scan, ±8.5° of mechanical rotation angle is achieved with driving current 100 mA. For fast scan, ±12.5° of scanning is demonstrated with driving current 100 mA at resonant frequency around 37 kHz. The integrated four-terminal piezoresistive sensor can monitor the scan angles of micromirror, and the sensitivity is 3mV/V@1deg for the Newmann sensor with the insulation island in the center. By using the 2-D scanning of the micromirror, we demonstrate the laser image projection with 720P (1280x720) resolution at 60 Hz refreshing frame rate.

ACKNOWLEDGMENT

The authors would like to thank the SINANO's Nano Fabrication Facility (NFF) for helping in the fabrication process of the micromirror and the computer center for helping in the simulation of the micromirror design.

REFERENCES

[1] O. G. Solgaard, A.A.; Howe, R.T.; Lee, L.P.; Peter, Y.-A.; Zappe, H.;, "Optical MEMS: From Micromirrors to Complex Systems," *Journal of Microelectromechanical Systems*, vol. 23, no.3, pp. 517-538, June 2014.

[2] P. R. Patterson, H. Dooyoung, N. Hung, H. Toshiyoshi, C. Ru-min, and M. C. Wu, "A scanning micromirror with angular comb drive actuation," *Micro Electro Mechanical Systems, 2002. The Fifteenth IEEE International Conference on*, pp. 544 - 547, 24-24 Jan. 2002.

[3] Z. Lixia, M. Last, V. Milanovic, J. M. Kahn, and K. S. J. Pister, "Two-axis scanning mirror for free-space optical communication between UAVs," *Optical MEMS, 2003 IEEE/LEOS International Conference on*. pp. 157-158, 18-21 Aug. 2003.

[4] D. Raboud, T. Barras, F. Lo Conte, L. Fabre, L. Kilcher, F. Kechana, *et al.*, "MEMS based color-VGA micro-projector system," *Procedia Engineering*, vol. 5, pp. 260-263, 2010.

[5] I. H. P. S.W. Nam, J.A. Jeon,Jiwoo Nam,J. Lee,J.H. Park,J. Yang,T. Ebisuzaki,Yoshiya Kawasaki,Yoshiyuki Takizawa,S. Wada, and JEM-EUSO Collaboration, "A New LIDAR Method using MEMS Micromirror Array for the JEM-EUSO mission," *PROCEEDINGS OF THE 31st ICRC, ŁOD´Z 2009*, pp. 1-4, July 2009.

[6] J. Zhong, X. Xiong, Z. Yao, J. Hu, and P. Patra, "Design and Optimization of Piezoelectric Dual-Mode Micro-Mirror," *Springer Netherlands* pp. 411-416, 2010/01/01 2010.

[7] D. Jung, D. Kallweit, T. Sandner, H. Conrad, H. Schenk, and H. Lakner, "Fabrication of 3D comb drive microscanners by mechanically induced permanent displacement," vol. 7208, pp. 72080A-72080A-11, 2009.

[8] D. Hah, P. R. Patterson, H. D. Nguyen, H. Toshiyoshi, and M. C. Wu, "Theory and experiments of angular vertical comb-drive actuators for scanning micromirrors," *Selected Topics in Quantum Electronics, IEEE Journal of*. vol. 10, pp. 505 - 513, May-June 2004.

[9] H. Xie, Y. Pan, and G. K. Fedder, "Endoscopic optical coherence tomographic imaging with a CMOS-MEMS micromirror," *Sensors and Actuators A: Physical*, vol. 103, pp. 237-241, 1/15/ 2003.

[10] H. U. Arda D. Yalcinkaya, "Two-Axis Electromagnetic Microscanner for High Resolution Displays," *Journal of Microelectromechanical Systems*, vol. 15, no.4, pp. 786-794, Aug 2006.

[11] M. N. O. Sadiku, "Elements of Electromagnetics , 5rd ed," *Oxford. U.K.: Oxford Univ. Press*, 2001.

[12] H. Urey, "Torsional MEMS scanner design for high-resolution display systems," *Proc. SPIE*, vol. 4773, pp. 27-37, July 2002.

[13] H. Urey, C. Kan, and W. O. Davis, "Vibration mode frequency formulae for micromechanical scanners," *Journal of Micromechanics and Microengineering*, vol. 15, pp. 1713-1721, 2005.

[14] J. M. KRODKIEWSKI, "MECHANICAL VIBRATION," *THE UNIVERSITY OF MELBOURNE Department of Mechanical and Manufacturing Engineering*, 2008.

[15] M. Doelle, D. Mager, P. Ruther, and O. Paul, "Geometry optimization for planar piezoresistive stress sensors based on the pseudo-Hall effect," *Sensors and Actuators A: Physical*, vol. 127, pp. 261-269, 3/13/ 2006.

[16] I. C. Khan, J.; Ben-Mrad, R.; He, S.; Schertzer, M. J., "Performance of an electrostatic actuated micromirror in a vacuum and non-vacuum packaging,," *IECON 2012 - 38th Annual Conference on IEEE Industrial Electronics Society*, pp. 3987-3990, Oct. 2012.

[17] M. P. H. David W. Wine, Lorne Jenkins, Hakan Urey, Thor D. Osborn, "Performance of a Biaxial MEMS-Based Scanner for Microdisplay Applications," *Conf. on MOEMS and Miniaturized Systems, SPIE*, vol. 4178, pp. 186-196, 2000.

978-1-6654-3008-1/21 $31.00 © 2021 IEEE

Proceedings of the 16th Annual IEEE International
Conference on Nano/Micro Engineered and Molecular Systems
April 25-29, 2021

Modeling and Optimization of VLF Piezoelectric Antenna Towards Lower Frequency*

Huiliang Liu, Yao Chu, Weiguo Hou, Yinqiao Li, Yuan Yao, and Yaxing Cai

Abstract —This work marks the first performance analysis of structure optimization of piezoelectric disc antenna and its impact on the characteristics and performance. Compared with the state-of-art technology, three distinctive advances have achieved: (a) first demonstration of central perforation on piezoelectric disc to further reduce the resonance frequency of the acoustically driven antenna; (b) simulation results and discussion of the characteristic and performance of the proposed design; and (c) an outlook and feasibility analysis of spaceborne low-frequency telecommunication with acoustically driven antenna on satellites.

I. INTRODUCTION

Along with the rapid development of communication technology in the past decades, high-speed wireless communication has become ubiquitous in urban area. We have been used to having convenient means of communication around us. However, there is still a vast region on earth, even if with the most advanced technology, out of implementation of wireless communication. That is the expanse underwater or underground in the application scenarios for marine, mining, etc.[1]. For example, wireless communication with underwater objects has always been a major problem in the field of emergency communication. It is hard to carry out efficient search and rescue measures in vehicle disaster or industrial accident, since the electromagnetic penetration is relatively poor in water, rock and other media for the conventional frequency band with shorter wavelength. It is also difficult to maintain low attenuation during long-distance propagation. In contrast, the longer wavelength of very low frequency (VLF, 3 kHz to 30 kHz) and ultra-low frequency (ULF, 0.3 kHz to 3 kHz) radiation allows it to travel tens of meters through ground and water and thousands of kilometers beyond the horizon through the air. However, the promotion of VLF application was slow during the last few decades. One of the biggest challenges of VLF technology is the efficiency of antenna. With the conventional electrical antenna, the long wavelength of VLF calls for an enormous physical size that stretches for miles to achieve high efficiency. The miniaturization of VLF transmitter generally requires additional matching network to mitigate the loss, which can weigh hundreds of pounds and therefore limit their flexible application in practice.

*Research partially supported by the China Postdoctoral Science Foundation (2020M670361).

H. Liu, W. Hou, Y. Li, Y. Yao, and Y. Cai are with the Institute of Telecommunication and Navigation Satellite, China Academy of Space Technology, Beijing, 100094, CHINA.

Y. Chu is with the Department of Precision Instrument, Tsinghua University, Beijing, 100084, CHINA (e-mail: chuyao@tsinghua.edu.cn).

In the last decade, piezoelectric material was introduced to develop compact RF antenna [2]. By means of the piezoelectric effect, which converts mechanical stress to a buildup of electrical charge, the Q-factor of the system can be greatly improved without the external matching network [3]. The design philosophy shifts from electrical to piezoelectric antennas utilizing strain-driven currents at mechanical resonance frequency, which offers a promising solution for dramatical advance on the performance over electrical antennas [4]. Previously, researchers have demonstrated that the lead zirconate titanate (PZT) antenna is 6000 times more efficient than an electrical antenna with comparable size [5].

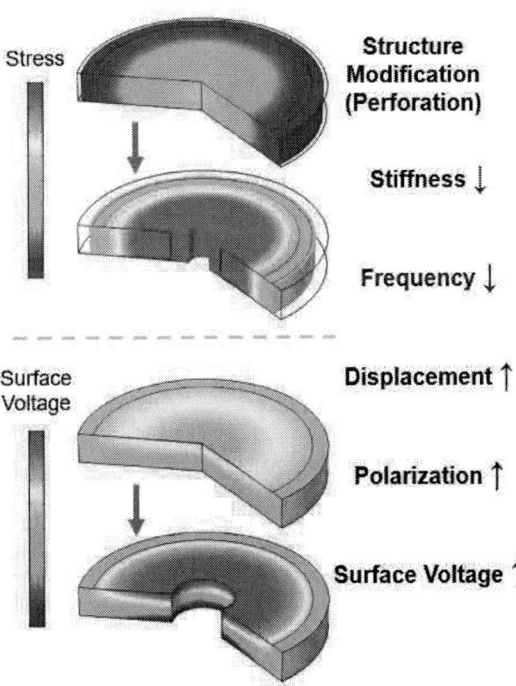

Figure 1. Schematic of the structure modification philosophy on the acoustically driven antenna to improve the performance. The central perforation is implemented on the piezoelectric disc to reduce the stiffness and decrease the resonance frequency, which could result in the augment of displacement, surface voltage, along with the polarization of the piezoelectric material.

The proposed structure is illustrated in Figure 1. Compared with the conventional acoustically driven antenna, an additional perforation is introduced on the piezoelectric disc at the center. The expected improvement includes the reduction of stiffness and decrease of the

978-1-6654-3008-1/21 $31.00 © 2021 IEEE

resonance frequency. It also results in the augment of displacement, surface voltage, along with the polarization of the piezoelectric material. In this work, the simulation results of the characteristics and performance are demonstrated based on the proposed design. An outlook and the feasibility analysis is conducted on spaceborne low-frequency telecommunication with acoustically driven antenna on satellites.

II. DEVICE DESIGN

Acceleration/deceleration of charges can generate far-field radiation and it can be depicted by Larmor formula [6]. The radiated magnetic field at a distance R from the accelerated charge can be expressed as:

$$|B| = \frac{\mu_0 q}{4\pi} \frac{a}{cR} \sin \theta \qquad (1)$$

where μ_0 is the permeability of free space, q is the amount of charge, a is the charge acceleration, c is the light speed, and θ is the angle between the acceleration vector and the observation vector.

This work concentrates on the analysis of structural modification impact on the characteristics and performance of piezoelectric disc antenna. The proposed structure includes a PZT disc with diameter of $8\,cm$, thickness of $1\,cm$ and patterned silver electrodes along the disc circumference of $0.5\,cm$, as shown in Figure 2. The variable r_hole represents the radius of central perforation. The dilation-mode vibration of the resonator is actuated by applying a sinusoidal voltage on the metalized edges of the disc. The equivalent electrical circuit model for the proposed design at mechanical resonance is also shown in Figure 2. In the model, R_m and R_{rad} represent the equivalent resistance corresponding to the energy dissipation through mechanical loss and radiation, respectively [5].

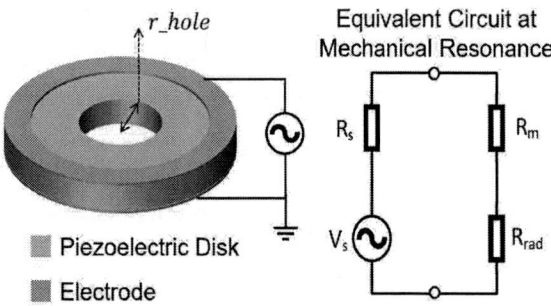

Figure 2. Schematic of the proposed design and the equivalent circuit at mechanical resonance of the acoustically driven antenna. The three-dimensional view of the PZT disc with diameter of $8\,cm$, thickness of $1\,cm$ and patterned silver electrodes along the disc circumference of $0.5\,cm$. The radius of central perforation is parameterized as r_hole. Equivalent electrical circuit model for the proposed design at mechanical resonance.

III. SIMULATION RESULTS

The modeling and simulation were carried out in COMSOL Multiphysics with MEMS toolbox. The

eigenfrequency simulations were conducted firstly to determine the resonance frequency of the proposed structures. Then, the frequency-domain simulations were conducted to figure out the characteristics including induced stress, the displacement of the surface, internal polarization, surface voltage, and terminal impedance. Finally, the radiated magnetic field can be calculated through Equation (1) with the internal polarization which reflects the surface charge density of the PZT disc.

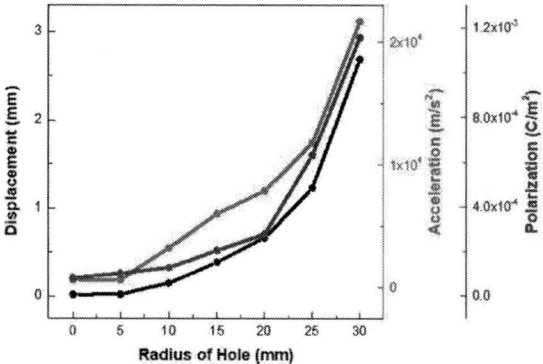

Figure 3. Characteristics of the proposed design, including the displacement, the acceleration, and the polarization of the PZT disc in terms of the hole radius.

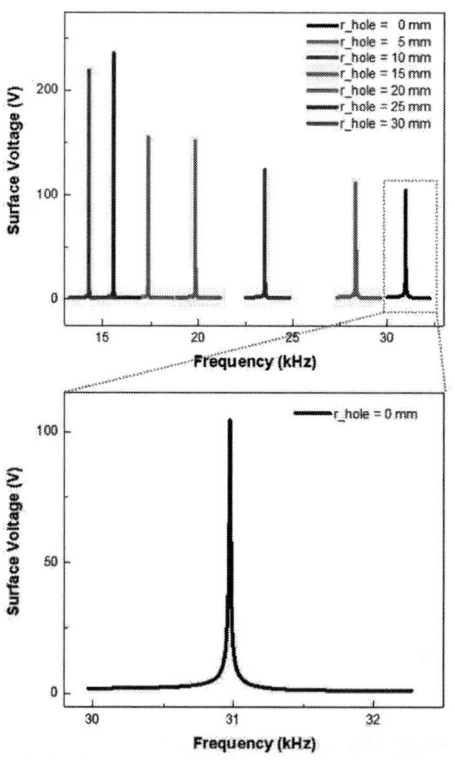

Figure 4. Simulation results of the surface voltage with a driven voltage of $1\,V$ with the hole radius of $0, 5, 10, 15, 20, 25$, and $30\,mm$ (upper) and a zoom-in vision of the results with hole radius of $0\,mm$.

Figure 3 illustrates the characteristics of the proposed structure including the maximum displacement of surface, acceleration and the internal polarization. As the radius of hole increases, the stiffness of structure decreases along with a considerable augment of the induced stress. Figure 4 illustrates the simulation results of the surface voltage with a driven voltage of $1\,V$ with the hole radius of $0, 5, 10, 15, 20, 25$, and $30\,mm$. The surface voltage decreases as the hole gets smaller. The trend has a deviation with the maximum radius of $30\,mm$. In this case, the proportion of unmetalized region is smaller than the silver electrode area. The uneven distribution and edge effect cause a small reduction of the maximum surface voltage compared with the result at radius of $25\,mm$.

decreases as the hole becomes larger. Further modification can be implemented to fit the motional resistance to the resistance of the driving source by changing the width of the electrode.

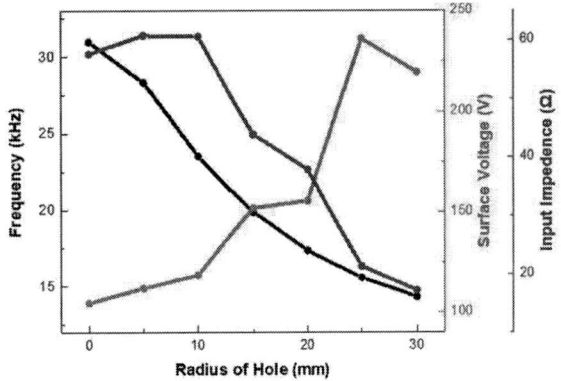

Figure 6. Characteristics of the proposed design, including the resonance frequency, the surface voltage, and the input impedance in terms of the hole radius.

Figure 6 summarized the peak value of properties mentioned in Figure 4 and Figure 5, in terms of the hole radius. Figure 7 shows the reduction of both the simulated magnetic field radiation at $100\,m$ with $1\,W$ input power and the mass of the proposed device, as the hole gets smaller. The charge amount q, in equation (1), is the surface integral of charge density (polarization). The acceleration a equals the product of the dipole distance and the squared resonant frequency. The decreasing resonant frequency dominates the calculated value of the induced magnetic field.

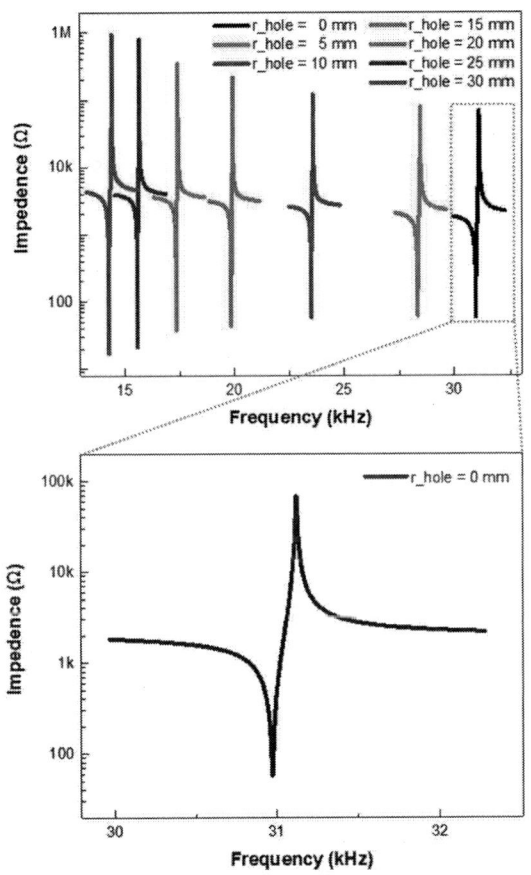

Figure 5. Simulation results of the impedance at the input terminals of the PZT disc with the hole radius of $0, 5, 10, 15, 20, 25$, and $30\,mm$ (upper) and a zoom-in vision of the results with hole radius of $0\,mm$.

Figure 5 illustrates the simulation results of the impedance at the input terminals of the PZT disc. Upon excitation, the actuated acoustic wave is reflected by the boundaries which induced a standing acoustic wave. The resonance frequency of the dilation mode decreases from $30.97\,kHz$ to $14.29\,kHz$ kHz as the radius of hole increases from $0\,mm$ to $30\,mm$. The motional resistance is around $60\,\Omega$ when the radius is smaller than $15\,mm$ and

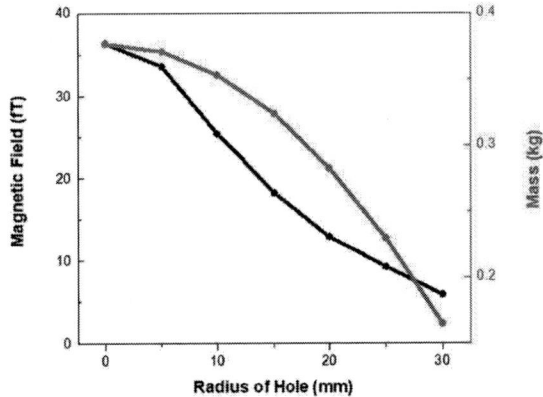

Figure 7. The simulated magnetic field radiation at $100\,m$ with $1\,W$ input power and the mass of the proposed device in terms of hole radius.

IV. ANALYSIS AND DISCUSSIONS

Figure 8 illustrates the schematic of the spaceborne low-frequency telecommunication application with acoustically driven antenna on satellites. It's not a novel research topic in the aerospace industrial field. The exploration of VLF communication system on aerobat or spacecraft can be traced back to 1980s [7, 8]. The paradigm of the spaceborne VLF antenna was first focused on the tether style. The tether primarily acts as the thrust, which utilizes electromotive force to induce the electric potential along the tether system to boost the orbit, during the interaction with the geomagnetic field [9]. A side effect of tether current flow is the production of electromagnetic waves at low frequency into the ionosphere, which comes from the electron transfer from the tether back into the plasma. This feature offers a promising solution of VLF antenna in space, along with several in-orbit trials (TSS-1, SEDS-2, ProSEDS, etc.) [10]. However, the tether length, ranging from $200\,m$ to $25\,km$, causes the extraordinary complexity of the system and poses the challenge of deployment and implementation.

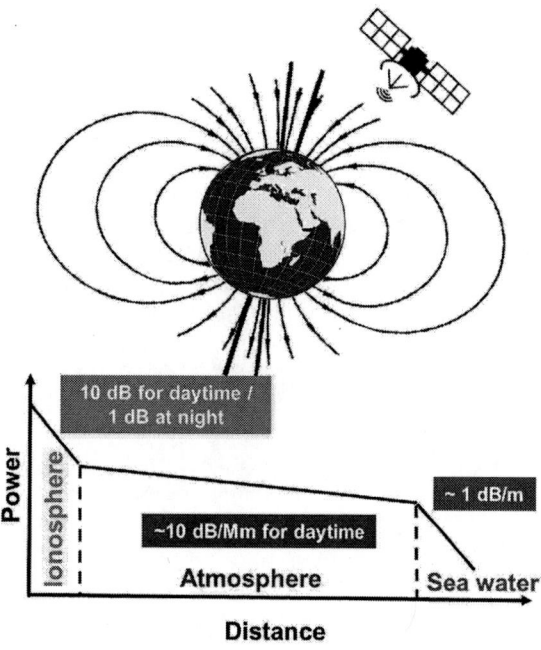

Figure 8. The schematic of the spaceborne low-frequency communication application with acoustically driven antenna on satellites (upper) and the schematic of link budget analysis at $1\,kHz$.

Figure 8 illustrates the schematic of link budget at $1\,kHz$. The electromagnetic wave at $1\,kHz$ could penetrate the ionosphere with attenuation of $10\,dB$ for daytime and $1\,dB$ at night [11]. The whistler wave front normals within 20 degrees of a field line will be coupled and transported, which offers an effective path towards the mid and high latitudes [12]. The attenuation rate for propagation in the earth – ionosphere waveguide is relatively lower at VLF and ULF. In the worst-case scenario, the attenuation is $10\,dB/Mm$ 10 dB/Mm at $1\,kHz$ [13]. The attenuation rate in sea water is *c.a.* $1\,dB/m$ at $1\,kHz$. To sum up, the link

attenuation is around $40\,dB$ from the low earth orbit satellite to an object under 20-meter sea water by day. The driven power is not the major limitation of the low frequency transmitter, since the satellite platform generally offers kilo-watt level power for payload at $50\,V$ or $100\,V$. The direction of future efforts should be focused on the improvement of efficiency, implementation at ULF band and the configuration of piezoelectric disc arrays.

V. CONCLUSION

To sum up, we have proposed a modification process to guide the acoustically driven antenna towards lower frequency. The performance analysis of structure modification impact on the characteristics and performance of piezoelectric disc antenna has been demonstrated. The PZT disc with central perforation is simulated using finite element modeling to summarize the trend of properties along with the change of physical parameter. Finally, the outlook and feasibility analysis of spaceborne low-frequency telecommunication with acoustically driven antenna on satellites are conducted to figure out the challenges and direction of efforts.

ACKNOWLEDGMENT

Y. Chu thanks Shuimu Tsinghua Scholar Program for the financial support during postdoctoral research.

REFERENCES

[1] M.A. Kemp, M. Franzi, A. Haase, E. Jongewaard, M.T. Whittaker, M. Kirkpatrick, et al., A high Q piezoelectric resonator as a portable VLF transmitter, *Nature Communications*, 10(2019) 1715.

[2] T. Nan, H. Lin, Y. Gao, A. Matyushov, G. Yu, H. Chen, et al., Acoustically actuated ultra-compact NEMS magnetoelectric antennas, *Nature Communications*, 8(2017) 296.

[3] J.P. Domann, G.P. Carman, Strain powered antennas, *Journal of Applied Physics*, 121(2017).

[4] A.E. Hassanien, M. Breen, M.-H. Li, S. Gong, A theoretical study of acoustically driven antennas, *Journal of Applied Physics*, 127(2020) 014903.

[5] A.E. Hassanien, M. Breen, M.-H. Li, S. Gong, Acoustically driven electromagnetic radiating elements, *Scientific Reports*, 10(2020) 17006.

[6] Griffiths, DavidJ, Introduction to electrodynamics /-3rd ed: Pearson Education Asia Limited; 2006.

[7] M.S. Wheeler, G.R. Beach, VLF Communication system, U.S. Patents1984.

[8] P. Bannister, J. Harrison, C. Rupp, R. King, M. Cosmo, E. Lorenzini, et al., Orbiting transmitter and antenna for spaceborne communications at ELF/VLF to submerged submarines, *rpsa*, (1993).

[9] P.M. Banks, Review of electrodynamic tethers for space plasma science, *Journal of Spacecraft and Rockets*, 26(1989) 234-9.

[10] M. Dobrowolny, E. Melchioni, Electrodynamic aspects of the first tethered satellite mission, *Journal of Geophysical Research: Space Physics*, 98(1993) 13761-78.

[11] A.S. Jursa, Handbook of Geophysics and the Space Environment. 4th edition (Final), ; Air Force Geophysics Lab., Hanscom AFB, MA (USA)1985, p. Medium: X; Size: Pages: 1065.

[12] P.D. Cotton, A.J. Smith, T.G. Wolf, W.L. Poulsen, D.L. Carpenter, The propagation of mixed polarization VLF (f ⩽5 kHz) radio waves in the Antarctic Earth-ionosphere waveguide, *Radio Science*, 27(1992) 593-610.

[13] L. Barclay, Propagation of radiowaves: Institution of Engineering and Technology; 2003.

Gap in pagination due to unavailable papers.

Pages 275-281

Proceedings of the 16th Annual IEEE International
Conference on Nano/Micro Engineered and Molecular Systems
April 25-29, 2021

Micro-fabricated alkali vapor cells for atomic spin gyroscope study

Yao Chen, Mingzhi Yu, Yintao Ma, Guoxi Luo, Zhuangde Jiang*, Yu Bai, Libo Zhao

Abstract— The combination of MEMS technology with atomic spin devices could fabricate chip scale atomic sensors. In most of the atomic devices based on alkali metals, the key component is an alkali vapor cell. Here we describe an alkali vapor cell fabrication technic for atomic spin gyroscope study. In the alkali vapor cell, alkali metal is filled in an oxygen and water free glovebox and noble gas isotopes such as ^{129}Xe and ^{131}Xe are also filled. Compared to an atomic magnetometer, in an atomic spin gyroscope, the isotope enriched noble gases are typically very expensive. We need to do recycling of these expensive gases after filling in the anode bonding chamber or else the gases will be wasted. Due to the large volume of the anode bonding chamber utilized in the fabrication process, a recycling platform is built based on liquid nitrogen. The relaxation time is a very important parameter of an alkali vapor cell. We developed a platform with magnetic field shielding system, non-magnetic field heating and single beam atomic spin detection subsystem. The spin relaxation time of the noble gases is tested in this study.

I. INTRODUCTION

Due to its high sensitivity, atomic gyroscope finds a wide range of application in fundamental physics study and applied instruments. The application includes rotation sensing[1, 2], testing physics beyond the standard model: anomalous spin forces detection[3, 4], low field NMR detection in a micro-fluid chip[5], etc. For nuclear magnetic resonance gyroscope (NMRG) which is a kind of atomic spin gyroscope is promising for fabricating with MEMS technology. The size, cost of NMRG could be reduced a lot with the combination of MEMS technology. It is believed that NMRG could be the next generation miniature gyroscopes.

The key component of atomic spin gyroscopes is an alkali vapor cell in which isotope enriched gases such as ^{129}Xe and ^{131}Xe, alkali metal which is very active to water and oxygen, nitrogen gas for quenching are filled. The gases with nuclear spins could be utilized for rotation sensing for their spin angular momentum could directed to one direction in the inertial space. Thus, atomic spin gyroscope could be used for inertial navigation and it is a kind of quantum gyroscope. There are two methods to fabricate an atomic magnetometer. One is the traditional machining technology in which the alkali vapor cell is made of glass and the glass vapor cell is handled through a torch. The

other method is the micro-machining method in which the vapor cell is fabricated through glass-silicon anode bonding[6]. It is obvious that the bonding technology can fabricate smaller vapor cell and the cost is lower. The combination of atomic gyroscope with micro-machining technology could fabricate chip-scale atomic gyroscopes. With the merits of small size and low costs, chip-scale atomic gyroscope may find a wide range of application in the industry, including unmanned aerial vehicle, self-driving cars, etc.

Several articles had been focused on studying the fabrication of MEMS alkali vapor cell for atomic magnetometers[6-9]. However, very few articles had been focus on alkali vapor cell fabrication for atomic spin gyroscope study[10]. It is believed that there is one reason that make such a few research institutes doing MEMS atomic spin gyroscope study. It is the fabrication cost of the vapor cell. The silicon-glass bonding chamber is typically very large. The alkali vapor cell is quite small. The volume maybe around 1mm³. When we filled the alkali vapor cell with ^{129}Xe or ^{131}Xe gas, we need to fill the full chamber whose volume is larger than 10L with the gases. There would be a waste of a lot of expensive noble gas enriched gases. Thus, we need to recycle the gases after the bonding process. Here we have developed a method based on liquid nitrogen to do the recycling. Testing the alkali vapor cells are also very important. In this paper, we measured the spin precession of noble gases and spin relaxation of noble gases are tested.

II. CURRENT RESULTS

A. Alkali Vapor Cell Fabrication

The fabrication of the Xe MEMS cell is mainly composed of alkali filling and expensive gas recycling. We etched several rectangular and cylinder holes in the wafer which owns a thickness of 2mm. The plasma ion etching is utilized to do such deep etching. We first etch the silicon from one side and then the other side is etched. Both sides are etched for a long time and about 1mm thick wafer is etched. In the vapor cell we fabricated, we also designed a very tiny tunnel to let alkali metal move from the reservoir to the working chamber. The rectangular chamber is the storage chamber which is utilized to store alkali metal and the cylindrical chamber is the working chamber in which the

---

* This work is supported by Open Research Projects of Zhejiang Lab under grant number 2019MB0AB02, China Postdoctoral Science Foundation under grant number 2020M683462, Jiangsu Province Youth Foundation under grant number BK20200244 and Jiangsu Province Foreign Expert Program under grant number BX2020032.

Yao Chen, Mingzhi Yu, Yintao Ma, Guoxi Luo, Zhuangde Jiang and Libo Zhao are all with the State Key Laboratory for Manufacturing Systems Engineering, the International Joint Laboratory for Micro/Nano Manufacturing and Measurement Technologies, Overseas Expertise Introduction Center for Micro/Nano Manufacturing and Nano

Measurement Technologies Discipline Innovation, Xi'an Jiaotong University (Yantai) Research Institute for Intelligent Sensing Technology and System, School of Mechanical Engineering, Xi'an Jiaotong University, Xi'an 710049, China (e-mail: yaochen@xjtu.edu.cn; 18369904909@163.com; 1064212021@qq.com; luoguoxi@xjtu.edu.cn; zdjiang@xjtu.edu.cn; libozhao@xjtu.edu.cn). (*Corresponding author: Zhuangde Jiang)

Y. Bai is with Suzhou Research institute, Xi'an Jiaotong University, Suzhou, Jiangsu 215123, China (e-mail: ybai@xjtu.edu.cn).

pumping light is directed. Figure. 4 shows the alkali vapor cell we fabricated and we can see the main structure of the vapor cell. The glass is attached to the wafer through anode bonding.

The filling of the alkali metal is directly. We construct a glovebox to do the alkali metal filling. We fix our bonding chamber into a glove box in which the gases are purified to reduce the concentration of oxygen and water. The concentration of water and oxygen is under 0.01ppm after the purifying. Due to the low level of oxidation gases in our system, we can directly open the Cs or other alkali metal and let them expose to the glovebox atmosphere. The commercially available glass tube filled with alkali metal Cs is directly open in the glass box and then a small droplet of Cs is filled in the cylindrical silicon cell before anode bonding. The bonding temperature in our experiment is around 479K.

For gas filling, we have developed a gas filling station and vacuum generation station. We pump out the gases in the bonding chamber and then 2000Pa of Xe noble gases and 1 atm nitrogen gas are filled. Molecular pump is utilized in our experiment and the pumping speed is around 100L/s. The vacuum can easily get to 10^{-3} Pa in our experiment and we think it is enough because all the gases has been purified in our glove box. There is very little contamination of the bonding chamber. We covered the other glass after the filling of gas we need. Forces is added to the top of the glass for good contact of silicon and glass. We did a last step of anode bonding to seal the vapor cell.

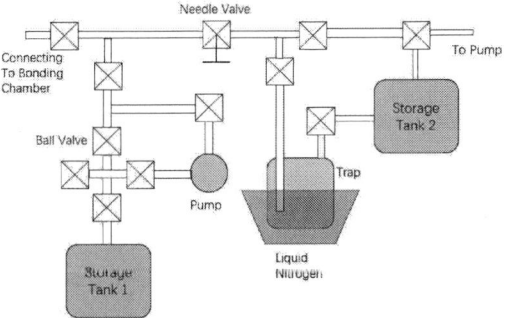

Figure. 1 The principal of gas recycling.

As we have illustrated, gas recycling is very important for atomic spin gyroscope vapor cell fabrication. After the last step of anode bonding, we pump out the residual gas in the bonding chamber which is enriched several noble gas isotopes into a large tank 1 showed in Figure. 1. Here an oil free scroll pump is utilized. Several ball valves are utilized for gas handling and vacuum isolation. We chose ball valve for their low leak parameters and high-pressure withstanding. Then the store gases are ready for recycling. A needle valve is utilized to handle the gases gradually into the Trap. Liquid nitrogen is filled in the trap dewar. Due to large molecular mass, ^{129}Xe and ^{131}Xe will condensate on the surface of the trap. Other gases such as nitrogen and helium will not condensate if there is low pressure of gases in the trap. There is a buffer storage tank 2 utilized in the system before the pump for the recycling. We gradually pumped out the gases in the liquid nitrogen trap when the noble gases are trapped by the low temperature. Xe noble gases will absorbed by the trap as we pump out other gases. After the pump out of

nitrogen gas and helium gas, we let the dewar to normal temperature and absorbed Xe gas will release from the surface. A gas bottle is utilized and at the same time, we let the bottle stay in liquid nitrogen, finally Xe gases will again condensed into the bottle. The condensed gases could be used for the next time vapor cell fabrication. Since we have known the percentage of ^{129}Xe in the bottle, we only need to handle one isotope to do a new mixing of the noble gas. Note that in order to know the concentration of ^{129}Xe and ^{131}Xe, we need to do a test with mass spectral meter. Figure. 1 shows the basic principle of gas recycling.

Figure. 2 The setup of the gas recycling system.

Figure. 2 shows the setup of the recycling system. Most of the valves utilized in this experiment are ball valves and several needle valves are utilized to handle the gases with flow control. Most of the connections are KF flanges for it is strong enough for gas pressure of below 5 Atms. Several gas bottles are fixed in this system. A baking skill is utilized to heat the system as we do the pumping. The outgassing of the surface will let the vacuum and gas handling pipes be more cleaner. The liquid nitrogen trap is a home made one and the volume of the trap is around 2 Liter.

B. Experimental Setup for Xe Signal Measurement

After the cell fabrication, we do a relaxation time measurement of the noble gases. Figure. 3 shows the platform for the relaxation time measurement. This platform could also be utilized to do magnetic field measurement. The details of the experiment could be found in this reference[11]. The laser utilized is a TOPTICA DFB laser. The center wavelength is 894nm which is the D1 line of Cs atoms. A reference Cs cell with no buffer gases is utilized to monitor the wavelength of the laser. The vapor cell is fixed at the center of the magnetic field shield. The shied composed of 3 layer of μ-metal and one layer of MzZn ferrite. The inner diameter of the ferrite is 10cm. After degaussing, the residual magnetic field at the center of the shield is smaller than 3nT. A set of 3 axis coil is utilized to generate magnetic field. We added holding magnetic field along the pumping light direction to let noble gases nuclear precess around it. The laser light is directed into the magnetic field though a fiber. In order to reduce the magnetic field produced by the fiber end, we let the fiber core go through a homemade fiber ferrule whose main parts is ceramic. This end of the fiber mainly composed of non-magnetic material. Finally, the laser light is circularly polarized and goes through the vapor cell with two sides.

Figure. 3 The platform for nuclear spin relaxation testing.

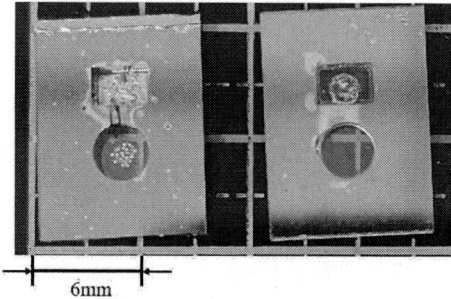

6mm

Figure. 4 The fabricated alkali vapor cell.

The vapor cell is heated with a 3Watt laser with wavelength of 1550nm. The temperature of the small vapor cell could be easily heated to around 403K which is enough for the Cs metal to produce enough alkali density. Thermal isolation is carefully done to hold the heat produced by the laser. The reason why we use laser for heating is because no magnetic field can be produced by the laser heating. After the laser passing through the vapor cell, a photo diode is utilized to do the detection. The absorption of the light is monitored to know the precession of nuclear spins. Note that if we directly monitor the nuclear spin precession through laser light absorption, the signal is very weak. Thus, a modulation technology is utilized. Around 150nT of modulation magnetic field is added to the perpendicular direction of the holding magnetic field. The modulation frequency is around 1000Hz. There will be a 1000Hz signal after the laser light pass through the vapor cell. We demodulated the absorption signal at 1000Hz then the laser could detect the nuclear spin magnetic field perpendicular to the holding magnetic field by the alkali atoms. The Glass-Silicon Bonding technology is utilized to fabricate an alkali vapor cells. We directly filled a small droplet Cs metal into a rectangular reservoir showed in Figure. 4. The vapor cells are filled with 2KPa natural abundance Xe gas and 1 Atm nitrogen gas. Finally, the last glass plate is bonded with the

silicon base. The diameter of the vapor cell hole in this experiment is 3mm and the thickness of the silicon is 2mm. There are several small passes between the reservoir and the working cavity of the vapor cell. The evaporated alkali metal could go pass the tunnel to the working cavity. The metal with gold color is Cs metal.

C. Nuclear Spin Precession Measurement

After the filling process, we need to know if we have filled noble gases in the vapor cell. As we have talked, the apparatus utilized in this paper is similar to the apparatus in this reference[11]. We tuned the temperature of the vapor cell to a point when the optical depth (OD) reached 2. The holding magnetic field along the pumping laser direction is set to 220nT. After about 1min's pumping, we suddenly change the magnetic field in the direction perpendicular to the holding field for several seconds and then change this field back again to its original value. Nuclear spins of ^{129}Xe and ^{131}Xe will both rotate around the holding magnetic field. They will produce variation magnetic field in the rotation plane which is perpendicular to the pumping laser. This field could be sensed and enhanced by Fermi Contact interaction by alkali metal Cs. Cs atoms could be utilized to form a single beam absorption magnetometer to test this variation field. The nuclear spins will rotate freely around the holding magnetic field and we call this rotation signal the free induction decay signal(FID). Figure. 5 shows the FID signal of the Xeon noble gases. Note that in Xeon gas, the natural abundance of ^{129}Xe is 26.44% and the natural abundance of ^{131}Xe is 21.18%. We could see both of the two isotope's signals. From Figure. 5 we can see the clear precession signal of ^{129}Xe. The gyro-magnetic ratio of ^{129}Xe is 0.0117Hz/nT and we can calculate the precession frequency is about 2.6Hz. The precession signal of ^{131}Xe is not so clear and there seems to be something there. We can see the deviation of the FID signal of ^{129}Xe from an exponential decay. The fluctuation signal seems to be the ^{131}Xe precession signal.

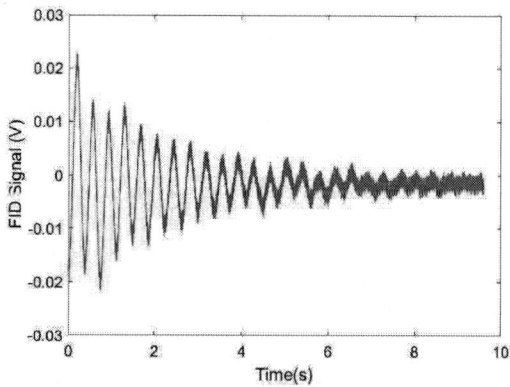

Figure. 5 The Free Induction Decay (FID) signal of the noble gases.

In order to see the ^{131}Xe precession signal, we make a power spectrum of the recorded time domain signal. From Figure. 6 we can see the clear peak of ^{129}Xe precession signal and the peak is at around 2.5Hz. This is agree with the calculated number. There seems to be 3 small peaks

978-1-6654-3008-1/21 $31.00 © 2021 IEEE

within the frequency range of 0.5 and 1Hz and we believe that this is the signal from ^{131}Xe atoms. As we know that the ^{131}Xe atoms own a nuclear quadrupole moment and with a nuclear spin of $I=3/2$. Theoretically, we will see 3 peaks.

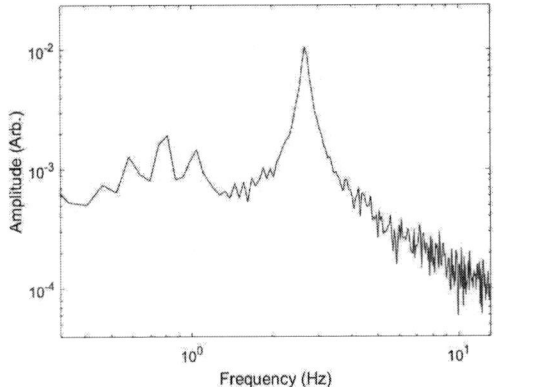

Figure. 6 The frequency spectrum of the precession signal of noble gases.

We note that further work should be done to improve the alkali vapor cells. The relaxation of ^{129}Xe and ^{131}Xe in a silicon-glass vapor cell should be carried out. New surface coating other than Cs-H film should be developed to prolong the relaxation time of ^{131}Xe atoms.

III. CONCLUSION

In conclusion, we have developed a method for atomic spin gyroscope alkali vapor cell fabrication. The vapor cell fabrication is based on MEMS technology. Since expensive ^{129}Xe and ^{131}Xe gases are needed to be filled in the vapor cell, we have designed a method based on liquid nitrogen trap to recycle the wasted gases for the next time gas filling. Alkali metal is directly filled in the vapor cell in a glove box. Since the vapor cell only owns two sides, we developed a single beam absorption atomic magnetometer based on Cs to do the detection of nuclear spin precession. The FID signal of ^{129}Xe and ^{131}Xe is recorded. Further improvement of the measurement should be done to measure the relaxation time of Xe isotopes.

ACKNOWLEDGMENT

This work is supported by Open Research Projects of Zhejiang Lab under grant number 2019MB0AB02, China Postdoctoral Science Foundation under grant number 2020M683462, Jiangsu Province Youth Foundation under grant number BK20200244 and Jiangsu Province Foreign Expert Program under grant number BX2020032.

REFERENCES

[1] Y. Chen et al., "Spin exchange broadening of magnetic resonance lines in a high-sensitivity rotating K-Rb-21Ne co-magnetometer," *Scientific reports*, vol. 6, p. 36547, 2016.

[2] T. Kornack, R. Ghosh, and M. V. Romalis, "Nuclear spin gyroscope based on an atomic comagnetometer," *Physical review letters*, vol. 95, no. 23, p. 230801, 2005.

[3] W. Ji et al., "New Experimental Limits on Exotic Spin-Spin-Velocity-Dependent Interactions by Using SmCo5 Spin

Sources," *Physical review letters*, vol. 121, no. 26, p. 261803, 2018.

[4] M. Bulatowicz et al., "Laboratory Search for a Long-Range T-Odd, P-Odd Interaction from Axionlike Particles Using Dual-Species Nuclear Magnetic Resonance with Polarized Xe 129 and Xe 131 Gas," vol. 111, no. 10, p. 102001, 2013.

[5] R. Jiménez-Martinez et al., "Optical hyperpolarization and NMR detection of 129Xe on a microfluidic chip," *Nature communications*, vol. 5, p. 3908, 2014.

[6] S. Knappe et al., "A chip-scale atomic clock based on 87Rb with improved frequency stability," *Optics express*, vol. 13, no. 4, pp. 1249-1253, 2005.

[7] L.-A. Liew, S. Knappe, J. Moreland, H. Robinson, L. Hollberg, and J. Kitching, "Microfabricated alkali atom vapor cells," *Applied Physics Letters*, vol. 84, no. 14, pp. 2694-2696, 2004.

[8] R. Han, Z. You, F. Zhang, H. Xue, and Y. Ruan, "Microfabricated vapor cells with reflective sidewalls for chip scale atomic sensors," *Micromachines*, vol. 9, no. 4, p. 175, 2018.

[9] J. Zhang et al., "Integration of the SERF Magnetometer and the Mz Magnetometer Using Micro-Fabricated Alkali Vapor Cell," in *2018 IEEE 68th Electronic Components and Technology Conference (ECTC)*, 2018, pp. 2007-2012: IEEE.

[10] G. Buchs et al., "Nuclear spin decoherence time in MEMS atomic vapor cells for applications in quantum technologies," in *AIP Conference Proceedings*, 2018, vol. 1936, no. 1, p. 020011: AIP Publishing LLC.

[11] Y. Chen et al., "A single beam Cs-Ne SERF magnetometer with differential laser power noise suppression method," 2021.

Gap in pagination due to unavailable papers.

Pages 286-292

Proceedings of the 16th Annual IEEE International
Conference on Nano/Micro Engineered and Molecular Systems
April 25-29, 2021

Graphene Oxide Nanosheet Assisted Porous Chitosan Sponge for Hemostatic Applications

Yuanyuan Yang, *Student Member, IEEE*, Yajing Shen*, *Senior Member, IEEE*

Abstract— **Porous sponges have received great attention in the fields of hemostatic applications, which has been required in clinical occasions over the past decades. In this work, we develop the graphene oxide nanosheet assisted chitosan sponge with simple freeze-drying process. Since chitosan owns good clotting capability, antibacterial property as well as biodegradability, the porous sponge could be applied in the hemostasis as well as wound healing. By adding the graphene oxide, the mechanical strength could be further enhanced for better wound patching. Such sponge owns high porosity exhibiting good water absorption ability. Besides, additional drugs could be loaded into sponge exhibiting time-dependent releasing performance. Biodegradability is also found in porous sponge. This work offers a simple but efficient approach for the fabrication of porous sponge, exhibiting great potential for hemostatic treatment.**

INTRODUCTION

Excessive hemorrhage is one of the leading causes of death nowadays during the surgery. It is important to reducing the blood loss in time during the emergency medical treatments. For instance, when suffering from the severe skin loss or even the nonhealing wounds, it is needed to immediately cover the wound surface with the dressing to protect the skin from blood and proteins loss. Furthermore, it could prevent the bacterial infection and further tissue damage.[1] Thus, the biocompatible materials with effective hemostatic property are urgently required. Several materials have been developed for hemostasis including collagen, gelatin, chitosan and so on [2-4]. Among them, chitosan owning good clotting capability, biocompatibility as well as biodegradability [5-9]. It shows great potential for use in drug delivery devices, allowing sustained delivery of various drugs [10, 11]. It also owns hemostatic properties, which makes it a potential candidate in the wound management area.[12, 13]

However, the hemostatic property of chitosan for extensive bleeding wounds is still limited. The increasing utilization of chitosan-based materials require further modification of their property. Recently, the porous materials have attracted great attentions which are widely applied in the areas such as medicine, tissue engineering, oil–water separation and so on [14-18]. The porous structure could help the storage of the liquid with the inner hole. Thus, the hemostatic performance of chitosan could be enhanced by developing porous chitosan material. Even so, low mechanical properties of chitosan still restrict its performance as sponge when be used in hemostatic applications. To solve this problem, nanofillers could be added at low loading rate. The mechanical property of chitosan solution could be enhanced when the nanofillers are dispersed homogeneously [19]. As one of nanofillers, graphene materials owns honeycomb and one-atom-thick structure, exhibiting outstanding electronic, mechanical and thermal properties.[20] Due to the benefits of graphene, graphene-based polymers are widely researched.[21]

In this work, we develop a graphene oxide (GO) assisted chitosan-based sponge with simple fabrication process. Due to the adding of GO, the resulted sponge exhibits high porosity and good mechanical property. The water absorption rate is also highly enhanced. What's more, drugs could be loaded into the porous sponge for specific treatment. According to the experiment, the chitosan/GO-polydopamine (PDA) sponge present good drug releasing performance. Also, the sponge exhibits high biodegradability which could be used for in situ biodegradable wound dressing. Such porous sponge offers an efficient while low-cost and simple fabrication approach for hemostatic materials as well as its future applications.

EXPERIMENT SECTION

Porous chitosan/GO-PDA sponge is prepared by simple fabrication process as shown in Fig. 1. At the beginning, GO-PDA solution is prepared by adding 7 mg DA powder into 15 ml 2 mg/ml GO solution. After that, the mixture is stirred, and the PH value is adjusted to 8.5 using Tris. And then, the resulted solution is stirred in 45 ℃ for 12 hours. In the meanwhile, by adding chitosan power in 2% aqueous acetic acid solution, 10 mg/ml chitosan solution is obtained after 1 hour sonication. After that, GO-PDA solution is slowly added to chitosan solution and mixed for 2 hours (the weight of chitosan to GO is 500:1, 100:1, 50:1, 25:1 and 17:1 separately). Then the resulted solution is poured into the mold and frozen for 1 more hour. The frozen mixture is finally under freeze-drying for 15 hours to get the porous sponge. To further characterize the inner structure, the chitosan/GO sponge is characterized by a field emission scanning electron microscopy (FEI Quanta 450 FESEM).

Y. Yang is with the Department of Biomedical Engineering, City University of Hong Kong, Kowloon, Hong Kong 999077, China

Y. Shen is with the Department of Biomedical Engineering, City University of Hong Kong, Kowloon, Hong Kong 999077, and also with the City University of Hong Kong Shenzhen Research Institute, Shen Zhen 518000, China (e-mail: yajishen@cityu.edu.hk).

This work was supported by ShenZhen (China) Basic Research Project (Grant No. JCYJ20170413140519030) and CityU grant 7005086, 7005209, 9680182. (Corresponding author: Yajing Shen.)

978-1-6654-3008-1/21 $31.00 © 2021 IEEE

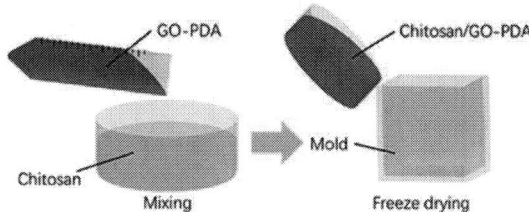

Figure 1. Fabrication process of sponge.

The percent of porosity of the sponge is characterized by the ethanol method. At first, a 1.5 mL tube is filled with absolute ethanol and the weight of it is recorded as w_1. Then a certain volume of ethanol is poured out and the sponge with the weight of w_s is immersed into the tube. Then the tube is placed in the vacuum chamber ensuring that the ethanol can fill the inter pore canal of the sponge for full immersing. Then the tube is added with the ethanol to full itself again. In this case, the total weight is recorded as w_2. Finally, the sponge is withdrawn from the tube with certain amount of ethanol trapped in their pores. Then the weight of ethanol left in the tube is recorded as w_3. With such procedures, the porosity of the sponge θ can be acquired according to the equation:

$$\theta = \frac{w_2 - w_s - w_3}{w_1 - w_3} \times 100\% \qquad (1)$$

For the water absorption analysis, sponge samples are first placed into a vacuum chamber for drying. After that, the sponge samples are immersed into distilled water. After soaking for 1 minute at room temperature, the sponge samples are removed from distilled water. The water absorption rate A_b is then calculated according to the equation:

$$A_b = \frac{w_t}{w_0} \times 100\% \qquad (2)$$

where w_0 is the weight of sponge sample on dried state and w_t is the weight of the sponge sample after immersed into water.

To prepare drug-loaded sponge, fluorescein sodium powder (FL) is added to GO solution and the mixture is stirred for 12 h. The unbound FL particles in the supernatant is then removed from the mixture solution by centrifugation at 9500rpm for 1 h. This process is repeated for 3 times for the full removal of the unbound FL. The GO–FL composite is then dispersed in distilled water and stirred for 1 more h. By repeating the fabrication step of chitosan-GO-PDA sponge as mentioned before, the drug-loaded porous sponge is obtained. To test the drug release performance, sponge is immersed in distilled water and the liquid samples are collected according to time. After that, UV-Visible spectroscopy is carried out on dispersions of sponge by a Perkin Elmer Lambda 900 spectrometer.

RESULTS AND DISCUSSION

To investigate the GO-PDA' effect to porous sponge, a series of chitosan-GO-PDA sponges are prepared with different chitosan to GO-PDA weight ratio r (r = 500, 100, 50, 25 and 17). The morphologies of the sponge samples are characterized using digital camera and SEM. As shown in SEM image of Fig. 2, the sponge owns high porosity. By adjusting the concentration of chitosan-GO-PDA solution before freeze-drying, the density of sponges after expansion differs even less to 0.019 mg/cm³. Fig. 3 gives the compressive strength of chitosan-GO-PDA sponge with increasing loading strain. Strengthen by the GO nanosheet, the porous sponge exhibits good mechanical strength. This enhancement is resulted from the strong interfacial interactions between the GO nanosheets and the chitosan matrix. Firstly, the GO carboxyl groups and the chitosan amino groups owns electrostatic interactions between each other. Besides, the oxygen-containing groups of the GO nanosheets and the chitosan (i.e. hydroxyl, carboxyl and amino groups) can also form strong H-bonding interactions. Therefore, the high interfacial adhesion established between GO nanosheet and chitosan ensures the combination of them and the stress applied to the sponge can be transferred through the GO nanosheets, enabling the sponge sample high strength.

Figure 2. Digital images of sponges with different weight ratio r (chitosan: GO-PDA) and the SEM image of sponge.

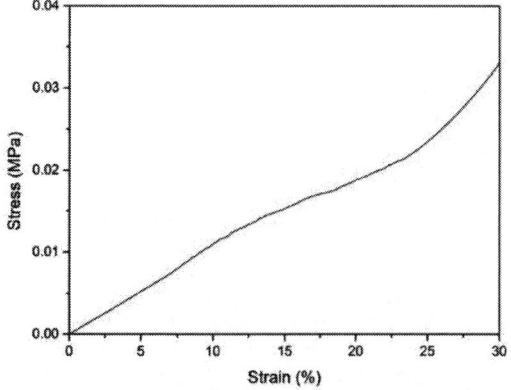

Figure 3. Mechanical performance of sponge.

The relationship between the porosity and the GO-PDA weight ratio (r=500, 100, 50, 25 and 17) is demonstrated in Fig. 4. As exhibited, the chitosan-GO-PDA sponge exhibits high porosity, which would increase with higher GO-PDA weight ratio that the sponge sample group with r=17 owns highest porosity about 97%. The water absorption rate of different sponge samples is shown as Fig. 5. The results indicate that the presence of GO-PDA has an influence on the water absorption properties. As exhibited, the water absorption rate would decrease with higher GO-PDA concentration and sponge sample group with r=500 owns highest water absorption rate up to 9000%. The reason is that the more GO nanosheet will create more extensive crosslinkages and then harden the porous composite. Therefore, the polymer chains are stretched to restrict the gel expansion. Nevertheless, all the chitosan-GO-PDA sponges exhibit extraordinarily high water absorption abilities which are more than 4000%. Such capability ensures the diffusion of blood to the surface and interior regions of the porous sponge for the high adsorption abilities of blood.

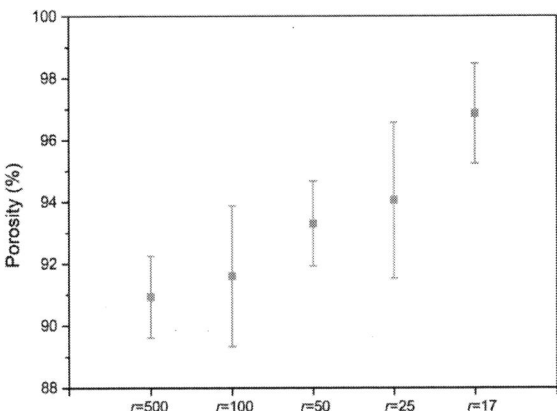

Figure 4. Porosity of different chitosan-GO-PDA sponge samples (r=500, 100, 50, 25 and 17).

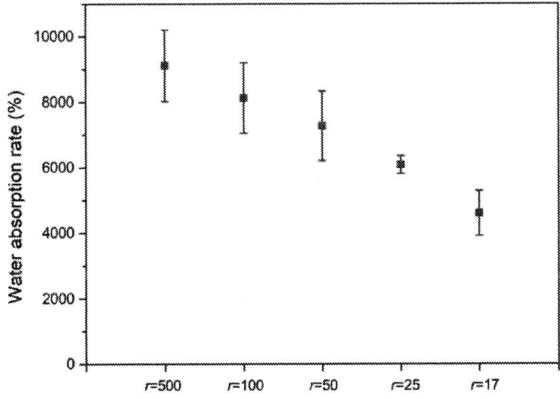

Figure 5. Water absorption rate of different chitosan-GO-PDA sponge samples (r=500, 100, 50, 25 and 17).

To apply the porous sponge into hemostatic applications as well as other wound treatment, specific drug could be loaded as well. Herein, the drug release performance of such material is also calibrated. Fluorescein sodium (FL) is used as the drug here to analyze the release rate because it could be well dispended in liquid solution. It has distinct absorption peaks between 450 to500 nm. Also, its molecular weight is 376 g mol^{-1}, which is similar to several drugs. To load the FL into chitosan/GO sponge, FL power is added during the fabrication process. Since the release rate and is important to the function of sponge for drug treatment, it is then analyzed using UV-Vis spectroscopy. Fig. 6 shows the cumulative quantity (%) of drug release in a PBS solution over 120 mins. The release profiles for the samples containing different loading FL ratios are different. Here, two sample groups are set with FL concentration equal to 4.8 wt % and 9.1 wt % of the sponge with the same chitosan-GO-PDA content in the sponge. As exhibited, the sample containing a 4.8 wt % FL loading reaches a drug level plateau after 30 mins with 2.35% of the drug is released. The sample containing a 9.1 wt % loading also reach the plateau in 30 mins, 2.96% of the drug in the sample is released. Such results show the possibility of acting as a rapid drug delivery system. The relationship between the time and drug release amount is resulted from the drug concentration gradient in the solution. To be specific, the drug concentration in the release medium is small at the beginning that the gradient is high. Therefore, the diffusion of drug from the sponge to the release medium is fast. As the drug concentration in the release medium increases, the gradient turns to be smaller and the diffusion rate of the decreases. By changing the the drug loading, the drug release rate can be altered for specific application.

Biodegradation of sponge is vital in a wound healing process, which affected cell vitality, growth as well as the host response. In this work, in vitro biodegradation study of the chitosan-GO-PDA sponge is conducted in dish as Fig.7 and Fig.8. Fig. 7 shows the macroscopic observation biodegradation of sponge's absorption interface. During the experiment, the sponge is immersed into water film. After 6 h, the sponge in absorption interface turns to liquid state due to degradation. Besides, the biodegradation of whole sponge is conducted. The whole

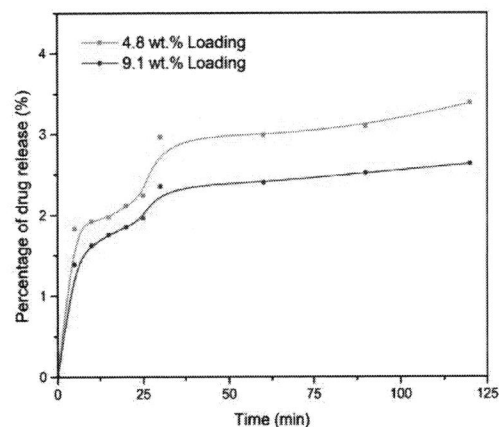

Figure 6. Drug release performance of sponge.

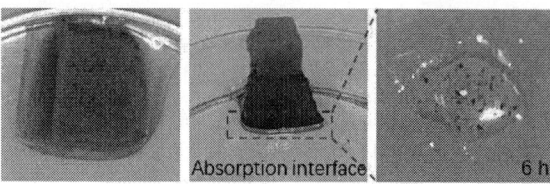

Figure 7. Biodegradation of absorption interface.

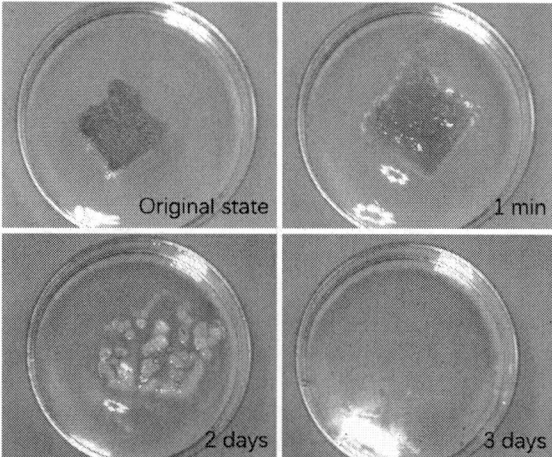

Figure 8. Biodegradation of whole sponge.

sponge is immersed. As shown in Fig. 8, the sponge swells and finally turn to liquid state after 3 days. The high degradation ability of sponge shows its potential to be used as an in situ biodegradable wound dressing.

CONCLUSION

In this work, we fabricate GO assisted chitosan-based sponge with simple fabrication process. Due to the adding of GO, the resulted sponge exhibits high porosity and good mechanical property. The water absorption rate is also studied which is up to 9000%. What's more, drugs releasing performance is tested using FL be loaded sponge. With such time-dependent drug releasing ability, such sponge could be used for specific treatment. Biodegradation test is also conducted, which shows good biodegradability both in absorption interface and whole sponge. Such porous sponge with simple fabrication process while high porosity, good mechanical strength and drug releasing performance is promising for the hemostatic Applications.

REFERENCES

[1] E. Denkbaş, unknown, r. B, E. Öztürk, Özdem, unknown, et al., "Norfloxacin-loaded chitosan sponges as wound dressing material," Journal of biomaterials applications, vol. 18, pp. 291-303, 2004.

[2] X. Jiang, Y. Wang, D. Fan, C. Zhu, L. Liu, and Z. Duan, "A novel human-like collagen hemostatic sponge with uniform morphology, good biodegradability and biocompatibility," Journal of Biomaterials Applications, vol. 31, pp. 1099-1107.

[3] M. S. Kim, K. D. Hong, K. W. Shin, S. H. Kim, S. H. Kim, M. S. Lee, et al., "Preparation of porcine small intestinal submucosa sponge and their application as a wound dressing in full-thickness skin defect of rat," vol. 36, pp. 0-60.

[4] R. W. Hutchinson, K. George, D. Johns, L. Craven, and P. Shnoda, "Hemostatic efficacy and tissue reaction of oxidized regenerated cellulose hemostats," Cellulose, vol. 20, pp. 537-545, 2012.

[5] R. Justin and B. Chen, "Characterisation and drug release performance of biodegradable chitosan–graphene oxide nanocomposites," Carbohydrate Polymers, vol. 103, pp. 70-80.

[6] S. B. Rao and C. P. Sharma, "Use of chitosan as a biomaterial: Studies on its safety and hemostatic potential," Journal of Biomedical Materials Research, vol. 34, pp. 21-28, 1997.

[7] A. Busilacchi, A. Gigante, M. Mattioli-Belmonte, S. Manzotti, and R. A. A. Muzzarelli, "Chitosan stabilizes platelet growth factors and modulates stem cell differentiation toward tissue regeneration," Carbohydrate Polymers, vol. 98, pp. 665-676.

[8] H. L. Lai, A. Abu'Khalil, and D. Q. Craig, "The preparation and characterisation of drug-loaded alginate and chitosan sponges," International journal of pharmaceutics, vol. 251, pp. 175-181, 2003.

[9] Y.-J. Seol, J.-Y. Lee, Y.-J. Park, Y.-M. Lee, I.-C. Rhyu, S.-J. Lee, et al., "Chitosan sponges as tissue engineering scaffolds for bone formation," Biotechnology letters, vol. 26, pp. 1037-1041, 2004.

[10] J. Akbuga, "The effect of the physicochemical properties of a drug on its release from chitosonium malate matrix tablets," International journal of pharmaceutics, vol. 100, pp. 257-261, 1993.

[11] H. Tozaki, J. Komoike, C. Tada, T. Maruyama, A. Terabe, T. Suzuki, et al., "Chitosan capsules for colon-specific drug delivery: improvement of insulin absorption from the rat colon," Journal of pharmaceutical sciences, vol. 86, pp. 1016-1021, 1997.

[12] T. Anilkumar, J. Muhamed, A. Jose, A. Jyothi, P. Mohanan, and L. K. Krishnan, "Advantages of hyaluronic acid as a component of fibrin sheet for care of acute wound," Biologicals, vol. 39, pp. 81-88, 2011.

[13] B. Anisha, D. Sankar, A. Mohandas, K. Chennazhi, S. V. Nair, and R. Jayakumar, "Chitosan–hyaluronan/nano chondroitin sulfate ternary composite sponges for medical use," Carbohydrate polymers, vol. 92, pp. 1470-1476, 2013.

[14] K. Quan, G. Li, D. Luan, Q. Yuan, L. Tao, and X. Wang, "Black hemostatic sponge based on facile prepared cross-linked graphene," Colloids and Surfaces B: Biointerfaces, vol. 132, pp. 27-33, 2015.

[15] H. Wang, E. Wang, Z. Liu, D. Gao, R. Yuan, L. Sun, et al., "A novel carbon nanotubes reinforced superhydrophobic and superoleophilic polyurethane sponge for selective oil–water separation through a chemical fabrication," Journal of Materials Chemistry A, vol. 3, pp. 266-273, 2015.

[16] S. Dinescu, M. Ionita, A. M. Pandele, B. Galateanu, H. Iovu, A. Ardelean, et al., "In vitro cytocompatibility evaluation of chitosan/graphene oxide 3D scaffold composites designed for bone tissue engineering," Bio-medical materials and engineering, vol. 24, pp. 2249-2256, 2014.

[17] J. Ge, L.-A. Shi, Y.-C. Wang, H.-Y. Zhao, H.-B. Yao, Y.-B. Zhu, et al., "Joule-heated graphene-wrapped sponge enables fast clean-up of viscous crude-oil spill," Nature nanotechnology, vol. 12, p. 434, 2017.

[18] L. Fan, C. Luo, M. Sun, X. Li, and H. Qiu, "Highly selective adsorption of lead ions by water-dispersible magnetic chitosan/graphene oxide composites," Colloids and Surfaces B: Biointerfaces, vol. 103, pp. 523-529, 2013.

[19] X. Yang, Y. Tu, L. Li, S. Shang, and X.-m. Tao, "Well-dispersed chitosan/graphene oxide nanocomposites," ACS applied materials & interfaces, vol. 2, pp. 1707-1713, 2010.

[20] R. R. Nair, P. Blake, A. N. Grigorenko, K. S. Novoselov, T. J. Booth, T. Stauber, et al., "Fine structure constant defines visual transparency of graphene," Science, vol. 320, pp. 1308-1308, 2008.

[21] T. Figueroa, S. Carmona, S. Guajardo, J. Borges, C. Aguayo, and K. Fernández, "Synthesis and characterization of graphene oxide chitosan aerogels reinforced with flavan-3-ols as hemostatic agents," Colloids and Surfaces B: Biointerfaces, vol. 197, p. 111398, 2020.

978-1-6654-3008-1/21 $31.00 © 2021 IEEE

Gap in pagination due to unavailable paper.

Pages 297-298

Proceedings of the 16th Annual IEEE International
Conference on Nano/Micro Engineered and Molecular Systems
April 25-29, 2021

Design and Simulation of a Wide-Bandwidth CMUTs Array with Dual-Mixed radii and Multi Operating Modes

Zichen Liu, Libo Zhao*, Yihe Zhao, Jie Li, Zhikang Li, Yu Bai, Ping Yang, Qijing Lin, Zixuan Li, Jiawei Yuan, Zhuangde Jiang

Abstract— A capacitive micromachined ultrasonic transducers (CMUTs) array with dual-mixed radii is designed to improve bandwidth and realize multi-modes operating for immersed ultrasonic applications. The two-size (TS) CMUT array is composed of two sub-arrays with different membrane radii, but the same cell arrangement. The two sub-arrays are separately connected to external power sources to achieve separate control of each array. An electromechanical-acoustic coupling simulation model of the TS-CMUTs array is established by finite element method to investigate their impedance and conductance properties under different DC voltages in the frequency domain. The results indicate that the designed CMUTs can operate with a significant improved frequency bandwidth ranging from 9.75 MHz to 13.35 MHz by adjusting the DC bias voltage. Compared with traditional CMUTs arrays, the proposed CMUTs array can contribute to an increase of up to 270% in the bandwidth.

I. INTRODUCTION

Ultrasonic transducers show promising prospects in various fields, such as ultrasonic imaging and therapy, industrial inspection, chemical production, marine terrain exploration, and military sonar [1], because of their advantages of high transmission frequency, good directivity, and strong penetration. Indexes such as frequency and bandwidth are the key performance parameters of ultrasound, which have always been the important contents in ultrasound research. Compared with traditional piezoelectric ultrasonic transducers, capacitive micromachined ultrasonic transducer (CMUT) technology has been emerged as a high potential candidate for their widespread applications in the fields of ultrasonic imaging, non-destructive testing, and flow measurement. CMUTs have many advantages, such as low cost, wide bandwidth, and high electromechanical coupling coefficient [1-2]. CMUT cell is mainly composed of vibrating membrane, cavity, pillars, and other main structures. The size of the vibrating membrane and the thickness of the cavity will affect the working frequency and

bandwidth of CMUT cell. Bandwidth and center frequency are regarded as important performance-related properties of CMUTs. A wide bandwidth can achieve a higher detection resolution [2]. Multi-frequency ultrasound transducers could be used at several fields such as co-registered image-guided high-intensity ultrasound therapy [3].

Previous studies have made significant efforts to analyze how to increase the bandwidth of CMUTs and make CMUTs work at multiple frequencies. Generally, the bandwidth of CMUTs can be improved by changing the structural parameters such as their cavity depth or thickness, radius of vibrating membranes [4-5]. Besides, adopting a proper array arrangement can not only increase the bandwidth [6], but also realize multi-frequency operation [3]. However, the array structures designed in previous works can only achieve a single performance improvement.

In this manuscript, a wide-bandwidth CMUTs array with dual-mixed radii is designed and analyzed by finite element method (FEM) simulations. To simplify the calculation, the CMUT unit structure is simplified to a finite element model containing only the vibrating membrane and the cavity. The results demonstrate that the proposed array design can achieve the multi-frequency operation by adjusting the DC bias voltage and can significantly improve the bandwidth up to 270% in comparison with traditional CMUTs arrays.

II. MODELING AND METHOD

A. Structure and Principle

The CMUT cell is a typically electrostatic-capacitive structure, which is composed of the top vibrating membrane and the bottom silicon substrate with a vacuum cavity to improve the energy conversion efficiency. The main structure of the CMUT cell is shown in Fig. 1. From bottom to top, there are the bottom electrode, the bottom substrate, the insulating layer, the pillar, the cavity, and the vibrating membrane. The vibrating membrane is composed of the top

---

This work was supported in part by the National Natural Science Foundation of China (Grant Nos. 51875449, 51890884, U1909221), and Jiangsu Province Foreign Expert Program (Grant No. BX2020032).

ZC. Liu, LB. Zhao, J. Li, ZK. Li, P. Yang, QJ. Lin, ZX. Li, JW. Yuan, ZD. Jiang are with State Key Laboratory for Manufacturing Systems Engineering, the International Joint Laboratory for Micro/Nano Manufacturing and Measurement Technologies, Overseas Expertise Introduction Center for Micro/Nano Manufacturing and Nano Measurement Technologies Discipline Innovation, Xi'an Jiaotong University (Yantai) Research Institute for Intelligent Sensing Technology and System, and School of Mechanical Engineering, Xi'an Jiaotong University, Xi'an, Shaanxi 710049, China (e-mail: mhlzc1996@stu.xjtu.edu.cn; libozhao@mail.xjtu.edu.cn; xjlijie@stu.xjtu.edu.cn; zhikangli@xjtu.edu.cn; summer86517@126.com;

xjjingmi@163.com; 915605320@qq.com; 3203987404@qq.com; zdjiang@xjtu.edu.cn). (*Corresponding author: Libo Zhao)

YH. Zhao is with State Key Laboratory for Manufacturing Systems Engineering, the International Joint Laboratory for Micro/Nano Manufacturing and Measurement Technologies, 111 Innovation and Intelligence Base for Micro/Nano Manufacturing and Nano Measurement Technologies, Xi'an Jiaotong University (Yantai) Research Institute for Intelligent Sensing Technology and System, School of Mechanical Engineering, Xi'an Jiaotong University, Xi'an, Shaanxi 710049, China, and Integrated Circuits Laboratory, Institute of Microengineering, School of Engineering, École polytechnique fédérale de Lausanne, Neuchâtel 2000, Switzerland (e-mail: johnzhaoyihe@stu.xjtu.edu.cn).

Y. Bai is with Suzhou Research institute, Xi'an Jiaotong University, Suzhou, Jiangsu 215123, China (e-mail: ybai@xjtu.edu.cn).

978-1-6654-3008-1/21 $31.00 © 2021 IEEE   299

electrode and silicon membrane [1]. The CMUTs array can not only operate in the receiving mode, but also in the emission mode.

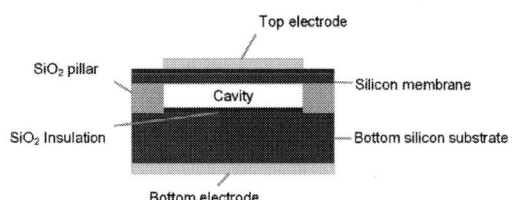

Figure 1. The cross-section schematic of one circular CMUT cell.

When it operates in the receiving mode, a direct current (DC) bias voltage needs to be applied between the top and bottom electrodes, the vibrating membrane is pulled to the substrate by an electrostatic force to arrive at a balanced location. The top membrane generates forced vibration under the co-excitation of the electrostatic forces and the ultrasonic signals to create the displacement and capacitance changes, which can generate the current signals [1].

As for the emission mode, the CMUTs should operate under a co-excitation of a DC bias voltage and an alternating current (AC) voltage to generate the ultrasonic waves with a specific frequency [1].

The natural resonant frequency [7] of a circular CMUTs cell without external loads is given by:

$$f_0 = \frac{10.21}{2\pi R^2}\sqrt{\frac{D}{\rho h}} \qquad (1)$$

where, R, ρ, h are the radius, density and thickness of the top vibrating membrane, respectively. D is the flexural stiffness, whose expression is:

$$D = \frac{Eh^3}{12(1-v^2)} \qquad (2)$$

where, E is the Young modulus, and v is the Poisson ratio.

The collapse voltage [7] of the circular CMUTs cell is given by:

$$V_{pi} = \frac{5.369d_0}{R^2}\sqrt{\frac{Dd_0}{\varepsilon_0}} \qquad (3)$$

where, d_0 is the distance between the top and bottom electrodes, ε_0 is the vacuum dielectric constant. Two high-frequency circular CMUT cells with different radii are designed for high-bandwidth immersion applications .

B. Modeling of Dual-Mixed radii CMUTs Array

The parameters of two different circular CMUTs cells are shown in Table I. For convenience of fabrication process, all the cells are designed with the unified membrane thickness and the gap height.

TABLE I.   PARAMETERS OF THE TWO KINDS OF CMUT CELLS

Parameter	Designs	
	Type-A	*Type-B*
Top membrane radius, R (μm)	18	20
Top membranet thicness, t (μm)	2	2
Cavity height, g (μm)	0.2	0.2
Insulation SiO₂ thicness, t_0 (μm)	0.05	0.05
Collapse voltage, V_{pi} (V)	191	154

The two-size (TS) CMUTs array is designed and composed of Type-A and Type-B sub-arrays with separated electrodes, and the center-to-center distance between each cell is set to be 20μm, as shown in Fig. 2. The DC bias voltage can be separately applied through an external voltage divider circuit with one total DC voltage. To investigate their impedance and conductance characteristics, two three-domain coupling models of TS-CMUTs and traditional CMUTs array are established by COMOSOL software, as shown in Fig. 3. A 4×4 square array with two membrane sizes is modeled. The simulation parameters are shown in Table II.

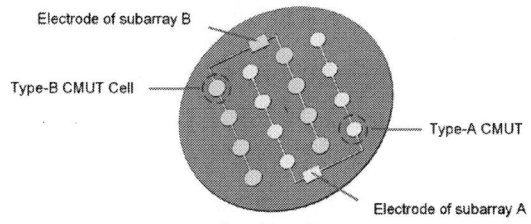

Figure 2. The design of the TS-CMUTs array.

TABLE II.   PARAMETERS OF THE SIMULATION MODEL

Parameter	Values
Si, Young modulus, E (Pa)	169E9
Si, Poisson ratio, v	0.28
Vacuum dielectric constant, ε_0 (pF/μm)	8.85E-6
Si, relative permittivity, ε_1	11 7
Si, density, ρ_1 (kg/m³)	2332
Water, density, ρ_2 (kg/m³)	1000
Water, sound velocity, c (m/s)	1500

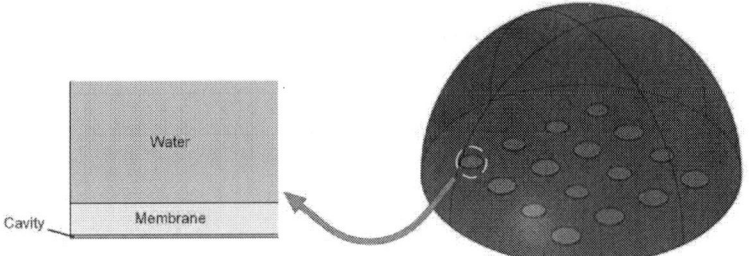

Figure 3. Simulation model of TS-CMUTs by FEM in COMSOL software.

III. RESULT AND DISCUSSION

When different bias voltages are applied to TS-CMUTs by different methods, it can operate in different modes. For comparison, the 4×4 CMUTs array containing only Type-A cells and the 4×4 CMUTs array containing only Type-B cells are simulated when applied with the maximum bias voltage (80% of the collapse voltage), which are 152V and 122V, respectively. The parameters and results are shown in Table III. For comparison, Table III also contains four loading methods of the DC bias voltage of the TS-CMUTs. Simultaneously, the resonance frequency, center frequency, -3dB bandwidth and fractional bandwidth are analyzed in the corresponding operating mode.

a)

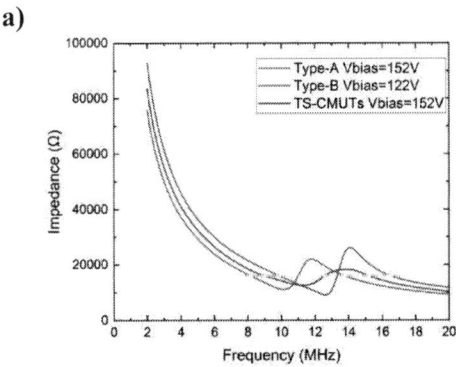

b)

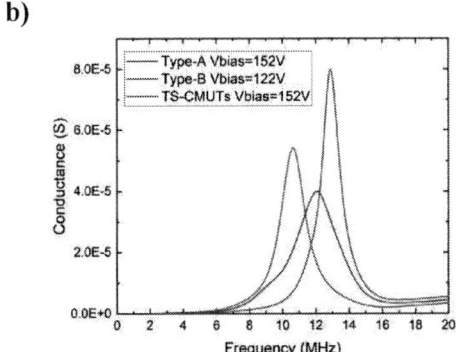

Figure 4. The electrical characterization of 4×4 array with Type-A, 4×4 array with Type B, and 4×4 array with TS-CMUTs loaded with maximum bias voltage. a) The impedance. b) The conductance

A. TS-CMUTs Operating in the Normal Mode

When the electrodes of the two subarrays of TS-CMUTs are applied with the same voltage at the same time, it operates in the normal mode. When the maximum bias voltage (152V) is applied to TS-CMUTs (Table III), the impedance and conductance of TS-CMUTs are compared with the traditional arrays, as shown in Fig. 4. It is observed that compared with the traditional CMUTs array, the bandwidth of TS-CMUTs operating in the normal mode is improved by 60.8% (Table III). However, the conductance amplitude of TS-CMUTs is slightly reduced.

B. TS-CMUTs Operating in the Wideband Mode

When the DC bias voltages applied to the subarrays of Type-A and Type-B are set to be 95 V and 122 V (Table III), respectively, TS-CMUTs have the largest bandwidth of 4.25 MHz and the largest fractional bandwidth of 37.69%. Compared with traditional arrays, the bandwidth is increased by 270% as shown in Fig. 5. In addition, fractional bandwidth allows to analyze and compare the performance of CMUTs operating at different frequencies, which is defined as the bandwidth divided by the center frequency. The fractional bandwidth of TS-CMUTs operating in wide bandwidth mode reaches 37.7%, which is also a significant improvement. The underly mechanism is based on multiple modals fusion, which is found effective in improving the bandwidth [6].

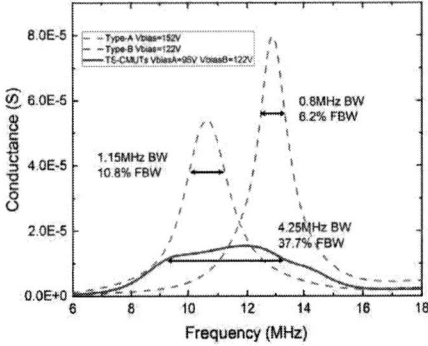

Figure 5. The -3dB bandwidths and fractional bandwidths of 4×4 array with Type-A, 4×4 array with Type B, 4×4 array with TS-CMUTs operated in wide bandwidth mode.

TABLE III   PARAMETERS AND RESULTS OF THE SIMULATION MODEL

Parameter		Designs					
		TS-CMUTs Array				Traditional Array	
		Normal	*Wide band*	*High frequency*	*Low frequency*	*Type-A**	*Type-B**
DC Bias V*otage*, V_{DC} (V)	Sub-aray A, V_{DCa} (V)	152	95	152	0	152	122
	Sub-aray B, V_{DCb} (V)	152	122	0	122		
AC v*ltage*, V_{AC} (V$_{pp}$)		0.5	0.5	0.5	0.5	0.5	0.5
Resonant frequency, f_r (MHz)		12.10	11.90	13.35	9.75	12.90	10.65
Center frequency, f_c (MHz)		12.01	11.28	13.23	10.13	12.9	10.63
Bandwidth, *BW* (MHz)		1.85	4.25	1.75	2.45	0.80	1.15
Fractional bandwidth, *FBW* (%)		15.32	37.69	13.23	24.20	6.20	10.82

C. TS-CMUTs Operating in the Multi-frequency Modes

Fig. 6 shows the resonant frequencies of the TS-CMUTs under different DC bias voltages. When the DC bias voltages applied to the subarrays of Type-A and Type-B are set to be 0 V and 122 V (Table III), respectively, indicating only sub-array B is operating, the resonance frequency of TS-CMUTs is 9.75MHz. Similarly, when only sub-array A is operating, the resonance frequency of TS-CMUTs reaches 13.35MHz. By adjusting the DC bias voltage, the TS-CMUTs can operate at different resonance frequencies ranging from 9.75 MHz to 13.35MHz.

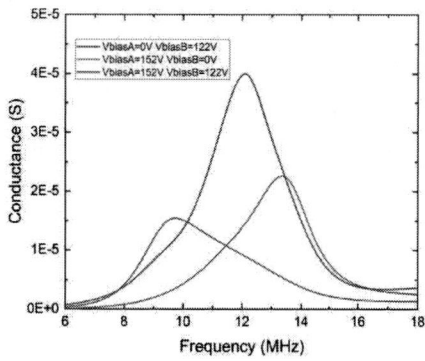

Figure 6. The conductance of 4×4 array with TS-CMUTs loaded with different bias voltages at different frequencies.

IV. CONCLUSION

We proposed a wide-bandwidth CMUTs array with dual-mixed radii and multi operating modes for immersed ultrasonic applications. The simulation results show that the TS-CMUTs array has the higher -3dB bandwidth and fractional bandwidth in the wide-bandwidth mode. The multi-frequency operation could be achieved by adjusting the DC bias voltages applied to the two subarrays. However,

the TS-CMUTs would reduce the conductance. The further research will focus on improving this phenomenon by adjusting the distance among the cells in the array.

REFERENCES

[1] Brenner. K, et al. " Advances in Capacitive Micromachined Ultrasonic Transducers." Micromachines, vol. 10, no. 2, pp.152-179, 2019.

[2] Öralkan, et al. "Capacitive Micromachined Ultrasonic Transducers: Next-Generation Arrays for Acoustic Imaging?" IEEE Transactions on Ultrasonics Ferroelectrics & Frequency Control, vol. 49, pp.1596-1609, 2002.

[3] M. Maadi, and R. J. Zemp. "Modelling of large-scale multi-frequency CMUT arrays with circular membranes." IEEE Ultrasonics Symposium, pp. 1-4, 2016.

[4] M. N. Senlik, et al. "Bandwidth, power and noise considerations in airborne cMUTs." IEEE International Ultrasonics Symposium Proceedings, pp. 438-441, 2009.

[5] M. N. Senlik, S. Olcum, and A. Atalar. "Improved performance of cMUT with nonuniform membranes." IEEE Ultrasonics Symposium, pp.597-600, 2005.

[6] C. Bayram, et al. "Bandwidth improvement in a cMUT array with mixed sized elements." IEEE Ultrasonics Symposium, pp. 1956-1959, 2005.

[7] Z. K. Li, et al. "Resonant frequency analysis on an electrostatically actuated microplate under uniform hydrostatic pressure," J. Phy. D: Appl Phys, vol. 46, no. 195108, 2013.

Proceedings of the 16th Annual IEEE International
Conference on Nano/Micro Engineered and Molecular Systems
April 25-29, 2021

Transparent, Anti-Freezing Hydrogels for Ultrasensitive Temperature and Strain Sensor Based on A Thin-Film Structure

Zixuan Wu, Haojun Ding, Yaoming Wei, Kai Tao, Jin Wu*

Abstract— A thin film sandwich structure (TFSS) was designed to improve the wearing comfort, sensing performance, immunity to humidity, and miniaturization of the hydrogel sensor synchronously. The TFSS sensor was fabricated using the layer-by-layer spin-coating technology with a thickness of only 12.15 μm for the LiBr-percolated hydrogel layer. Benefiting from the ultrathin device structure, the low thermal capacity of LiBr solution, new capacitance measurement mode and ionic transport manner, the TFSS thermistor displays unprecedented sensitivity of 24.54%/°C, ultrafast response time (0.19 s), recovery time (0.08 s), high resolution (0.8 °C) and a broad detection range (-30~96 °C). The sensitivity of our temperature sensor is one order of magnitude higher than those of state-of-the-art stretchable temperature sensors. In addition to temperature, the tensile strain can also be selectively detected with distinguishable signals, endowing the bimodal sensor with multifunctional sensing capability.

I. INTRODUCTION

Stretchable electronics have attracted increasing attention due to their prospective applications for wearable devices, physiological signals monitoring, energy harvesting, prosthetic skin, and implantable medical electronics [1, 2]. Especially in the emerging era of the internet of things (IoT), multifunctional detection with high performance in an integrated device is imperative [3]. The stretchable sensors adhered to human skin can precisely detect physiological signals with wearing comfort [4]. Among the various variables for human health and activity monitoring, temperature and strain sensors can be used to characterize a number of important physiological states [5, 6]. For example, temperature fluctuations can indicate fever, chills, and inflammation [7]. The strain sensors worn in different parts can record signals such as body movements, pulse, heartbeat, expressions, etc [8]. Currently, two kinds of strategies were used for the preparation of stretchable temperature and strain sensors: combination of sensitive nanomaterials with insulating stretchable elastomers, and the use of intrinsically stretchable sensitive materials. For example, Lee and coworkers fabricated a stretchable temperature sensor with a tunable thermal index by embedding silver nanowires and graphene film inside a polydimethylsiloxane (PDMS) matrix [9]. Wu and coworkers found that the polyacrylamide (PAM)/κ-carrageenan double network (DN) hydrogel was sensitive to both temperature and strain as well as featured the ultrastretchability (1200% tensile strain) [1, 10, 11].

However, the nanomaterial itself does not have stretchability, and the synthesis process is complicated and expensive. Intrinsically stretchable sensitive materials such as ionic gels, hydrogels, etc. have problems such as evaporation, leaking, and instability. Moreover, sensors made of these materials are generally bulky, which does not meet the trends of miniaturization and integration. At the same time, it is often difficult to distinguish different sensing parameters synchronously when using the single sensitive material to measure multiple parameters. Therefore, it is urgent to address the issues of miniaturization and multifunctional detection of the stretchable sensitive materials.

Herein, we propose a stretchable temperature and strain sensor based on a thin film sandwich structure (TFSS), which can detect temperature and strain synchronously in capacitance mode and conductance mode. The TFSS sensor possesses anti-freezing and anti-drying abilities due to the salt percolation treatment and encapsulation. The sensor displayed a high thermal sensitivity (24.54%/°C) in capacitance mode, which is nearly 10 times that (2.59%/°C) in conductance mode. Importantly, the TFSS endows the sensor with fast response speed and wearing comfort, enabling it to apply to various wearable applications, such as real-time monitoring body temperature, motions, respiratory, physical state, etc.

II. EXPERIMENTAL METHODS

A. Preparation of the solution of PAM/Carrageenan

1.5 g of carrageenan, 7.5 g of acrylamide, 0.375 g of 2-hydroxy-4'-(2-hydroxyethoxy)-2-methylpropiophenone, 0.09 g of potassium chloride (KCl), and 4.5 mg of N,N-methylenebis(acrylamide) (MBA) were dissolved in 41 mL of deionized water. The solution was stirred at 95 °C for 4 h for complete mixing. The homogenous solution was degassed and then kept stirring at 75 °C for the subsequent step.

B. Fabrication of the TFSS Sensor

First, Sylgard 184 and cross-linking agent were mixed at 10:1 in weight and then spin-coated on the hexamethyl-disilazane (HMDS)-treated glass flake at 500 rpm, and then cured at 70 °C for 2 h to form PDMS layer. The PDMS layer became hydrophilic after oxygen plasma treatment. Then, the hydrogel precursor solution was spin-coated on the PDMS at 500 rpm. Subsequently, the hydrogel layer was polymerized in 50wt% LiBr aqueous solution by storing at

---

Zixuan Wu, Haojun Ding, Yaoming Wei and Jin Wu are with the State Key Laboratory of Optoelectronic Materials and Technologies and the Guangdong Province Key Laboratory of Display Material and Technology, School of Electronics and Information Technology, Sun Yat-sen University, Guangzhou 510275, China.

Kai Tao is with The Ministry of Education Key Laboratory of Micro and Nano Systems for Aerospace, Northwestern Polytechnical University, Xi'an, 710072, PR China
*Contacting Author: Jin Wu; (e-mail: wujin8@mail.sysu.edu.cn)

978-1-6654-3008-1/21 $31.00 © 2021 IEEE

6 °C for 1 h and UV exposure for 30 min, followed by immersing in LiBr solution for another 30 min. The residual solution was removed from the surface of the sample and edges of the hydrogel were cut. The silver pastes were coated on the two ends of the hydrogel to fabricate the electrodes. The sample was encapsulated with a top layer of PDMS by spin-coating process. Finally, the TFSS was peeled off from the glass flake and the residual HMDS was washed with ethanol and water.

III. Results and Discussion

The fabrication process of the TFSS sensor is shown in Figure 1a, which is described in detail in experimental methods section. Notice that the PDMS layer at the bottom needed to be warmed up to 75 °C before spin-coating the hydrogel, ensuring that the hydrogel layer spread evenly during the spin-coating process. Since the pre-shaped hydrogel layer was too thin to dehydrate, the polymerization must be carried out in LiBr aqueous solution until the hydrogel layer was endowed with anti-drying ability. The schematic diagram of the device structure is shown in Figure 1b. Since both PDMS and hydrogel were soft materials, the device exhibited good flexibility. The silver paste electrodes were thicker than the hydrogel layer and partially covered the edges of the hydrogel to ensure the conformal contact between the electrode and the hydrogel. Thanks to the spin-coating process, the bulk hydrogel was miniaturized into a thin-film sensor, and its light transmittance was greatly improved. The transmission of visible light through TFSS surpassed 90% due to both the high transparency of PDMS and that of the ultrathin hydrogel layer with the thickness of only 12.15 μm (Figure 1c-1e).

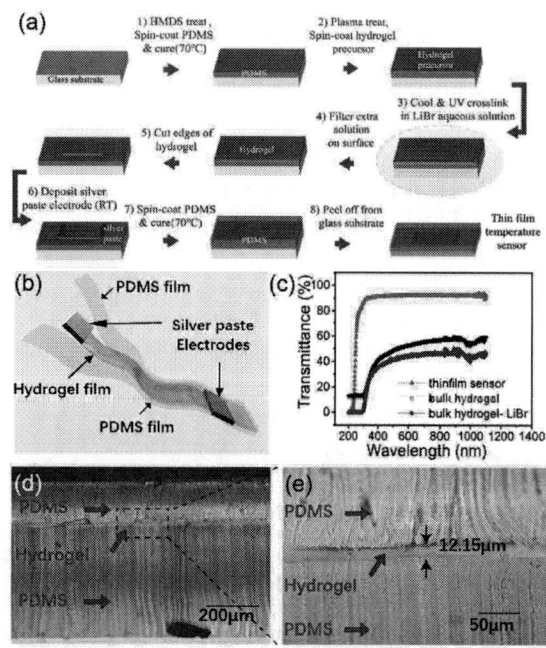

Figure 1. (a) Scheme of preparation of the TFSS sensor. (b) Scheme showing the sandwich structure of TFSS sensor. (c) Transmittance spectra of TFSS, pristine and 50wt% LiBr bulk hydrogels, respectively. (d-e) Optical images of the cross-sectional profile of TFSS, where the hydrogel is 12.15 μm.

As an ionically conducting hydrogel, the ions transport inner the hydrogel was easily affected by temperature. The thermal sensing performance of the TFSS was assessed by recording the capacitance and conductance variations of the sensor when the temperature changed. The response is defined as normalized capacitance and conductance variation ($\Delta C/C_0$ and $\Delta G/G_0$), in which the ΔC or ΔG is the capacitance/conductance variation relative to the original value C_0 or G_0 at 22.1 °C. As shown in Figure 2, in both capacitance and conductance mode, the TFSS displayed a high sensitivity to temperature within a wide range of temperature (-30 to 95.3°C). Especially when the temperature was above room temperature, the sensitivity was higher. The TFSS displayed an ultrahigh sensitivity (24.54%/°C) to temperature above 55°C in capacitance mode, which was nearly 10 times that in conductance mode (2.59%/°C). The TFSS broke through the low detecting limit of ordinary hydrogels. Furthermore, it can operate normally at -30°C. The advantage of normal operation at subzero temperature was attributed to the LiBr percolation. The ionic hydration of LiBr in hydrogel broke the connection between water molecules to prevent the hydrogel from forming icy crystals below 0°C. As such low temperature, the TFSS still displayed the high sensitivity of 0.93%/°C and 1.15%/°C in capacitance mode and conductance mode, respectively, suggesting the TFSS was capable of working in extreme conditions.

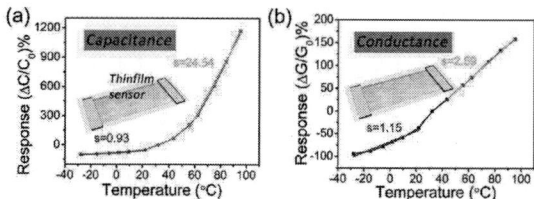

Figure 2. (a) and (b) Relative capacitance and conductance responses as a function of temperature, respectively, for the TFSS. Inset: schematic of TFSS.

The mechanism of thermal sensing in conductance mode is easily understood that the mobility of ions increases with elevated temperature. However, the mechanism in capacitance mode has not yet been revealed. To specify the impact of each part of TFSS on thermal sensing, a structure of parallel plate capacitors was hypothesized. As shown in Figure 3a, the TFSS was composed of three capacitors: C_P (dielectric layer: PDMS), C_H (dielectric layer: hydrogel), and C_W (dielectric layer: water) in the horizontal direction as expressed in (1) and (2). At low temperature, the water was locked in the hydrogel, where the C_W was inexistent. As temperature elevated, both the PDMS and hydrogel expanded, leading to increase the depth (d) and enlarge the sectional area (A). These variations of geometric parameters change the value of capacitance. Since the thickness (th) of capacitors film was extremely thin compared with the depth and width (one-hundredth to one-thousandth), the variations of depth and width can be ignored. Therefore, the geometric effect is simplified as the change of thickness of the thin film. The increased thickness of PDMS and hydrogel were responsible for the increment of capacitance. Besides, the water molecules evaporated with the increased temperature and then condensed in the interfaces of PDMS-hydrogel, gradually forming water layers. What's more, the relative dielectric constant (ε_W=78.5, 25°C) of water was higher than PDMS [12] (ε_p =3) and LiBr aqueous solution [13] ($\varepsilon_H = \varepsilon_{LiBr} \leq 40$), which further explained the increased normalized capacitance as expressed in (3).

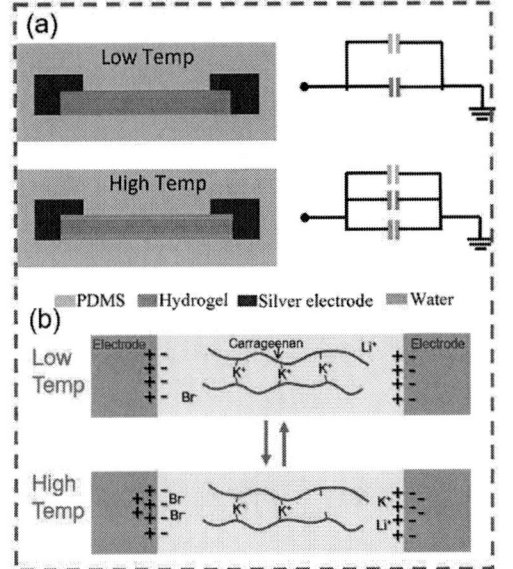

Figure 3. (a) Schematic illustrating the structure of TFSS and the equivalent circuit diagram of the temperature sensor (left) at low (above) and high (below) temperatures, respectively. (b) Schematic illustrating the temperature-dependent adsorption of the ions at the hydrogel-electrode interface.

$$C_{total} = C_p + C_H + C_W \tag{1}$$

$$C_{total} = \frac{\varepsilon_P \varepsilon_0 A_P}{d_P} + \frac{\varepsilon_H \varepsilon_0 A_H}{d_H} + \frac{\varepsilon_W \varepsilon_0 A_W}{d_W} \tag{2}$$

$$Response(\%) = \frac{\Delta C}{C_0} \times 100\% = \frac{\varepsilon_P \cdot th_{P^*} + \varepsilon_H \cdot th_{H^*} + \varepsilon_W \cdot th_{W^*}}{\varepsilon_P \cdot th_P + \varepsilon_H \cdot th_H} \times 100\% \tag{3}$$

In addition to the geometry and phase transition, the ions accumulation induced by the temperature at the interface of electrode-hydrogel was also responsible for the changed capacitance. As temperature elevated, the sol-gel transition occurred, resulting in K+ release from the double helix structure of κ-carrageenan [14]. The released K+ migrated to the double layer at the interface. Besides, the Li+ and Br- ions in the hydrogel were easily adsorbed to the double layer with increased temperature [15]. The adsorption effect accounts for the increased capacitance that the increased concentration of ions accumulated at the electrode.

The miniaturization of the temperature sensor endowed it with superior wearing comfort, which was in line with the needs of practical applications. The bulk hydrogel-based sensors often show sluggish response speed when detecting temperature, which impeded their applications in real-time monitoring of physiological signals such as breath. In this case, the TFSS sensor showed an optimized solution for hydrogel-based sensors. For example, the TFSS was capable of real-time monitoring the respiratory and body temperature. To record respiratory signals, the TFSS was conformally attached to the upper lip. As shown in Figure 4, the dynamic responses recorded the respiratory due to the temperature variation induced by exhalation and inhalation. Since the temperature of the body was higher than the

surrounding environment, the exhaled gas heated up the TFSS, resulting in the increased capacitance. On the contrary, the inhaled gas cooled the TFSS and led to the decreased capacitance. The recorded data can also reflect the respiratory rate and intensity. With the increased intensity of activities, the respiratory rate increased from 18 times/min (sitting) to 30 times/min (rope skipping) and 43 times/min (fast running), respectively. The accurate detection of respiratory rate was attributed to the short response (0.19s) and recovery (0.08s) time (Figure 4c). Fundamentally, the thin film structure speeded up the heat transfer. At the same time, the increased intensity of activity made the body warmer, leading to a higher temperature of exhaled gas. Accordingly, the response was increased from 2.7% to 140% and 210.5%, respectively.

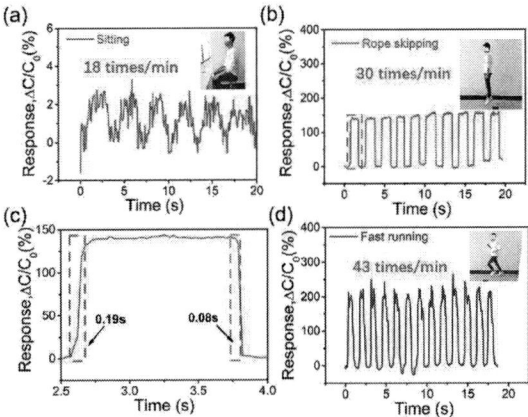

Figure 4. (a-d) Dynamic capacitance responses to different activities: (a) sitting, (b) rope skipping, and (d) fast running. The short response time (0.19 s) and recovery time (0.08 s) were extracted from the dynamic response curve in (c).

In addition to temperature, the TFSS was able to detect strain because the mechanical deformation changed the resistance of the hydrogel. The ultrathin and stretchable structure was conducive to be conformally attached on different parts of human body and stretching together with skin, accurately detecting tiny physiological signals. For instance, the TFSS attached on the throat to monitoring swallowing (Figure 5a), and attached on the finger to monitor the finger bending (Figure 5b).

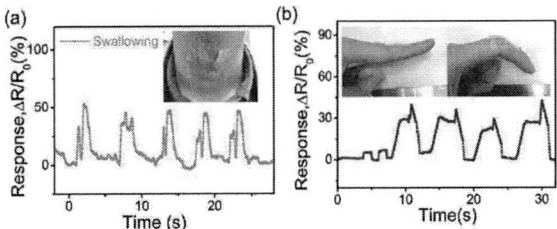

Figure 5. (a) Dynamic response to repeated swallowing. (b) Dynamic response to repeated bending and relaxation of the forefinger.

Previous multifunctional sensors can detect multiple signals, which is similar to skin. However, it is still a challenge to distinguishably detect strain and temperature simultaneously. To solve this problem, we proposed a new method to distinguish strain and temperature signals by utilizing the capacitance and conductance modes at the same time. The TFSS showed ultrahigh thermal sensitivity in capacitance mode. Compared with the thermal sensitivity, the sensitivity to strain (gauge factor < 1) was negligible. The compound plot of response towards temperature and strain in capacitance mode was shown in Figure 6a, which enabled us to acquire the temperature and ignore the impact of strain. Since the sensitivities of strain and temperature in conductance mode were similar, both stimuli must be considered. The temperature obtained from capacitance mode can be used to calculate the value of strain in the compound plot in conductance mode as shown in Figure 6b.

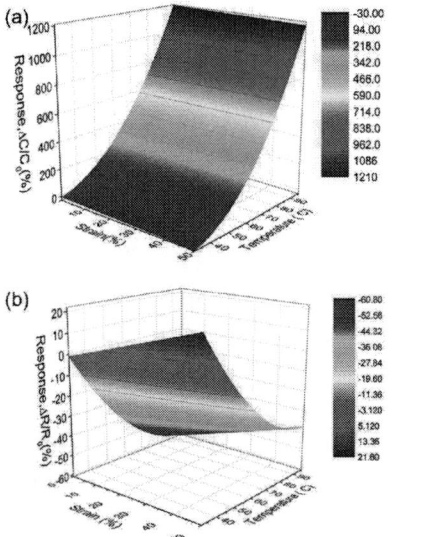

Figure 6. Compound plots of responses of TFSS towards temperature and strain stimuli in capacitance (a) and resistance (b) modes, which were calculated with a linear superposition of temperature and strain responses.

IV. CONCLUSION

A dual-functional temperature and strain sensor based on a thin film sandwich structure was fabricated by spin-coating process. The TFSS sensor showed high transparency (>90%) due to the ultrathin hydrogel (12.15μm) and intrinsically transparent PDMS layers. With the LiBr percolation, the frost resistance of the TFSS sensor was highly improved and it can operate at an extremely low temperature as -30 °C. Meanwhile, the encapsulation of PDMS inhibited the water evaporation of hydrogel, endowing the sensor with anti-drying ability and stability. A horizontal plate capacitance model and temperature-dependent ions accumulation in the electrode-hydrogel interface were accounted for the capacitance variation. The thermal sensitivity in capacitance mode (24.54%/°C) was much higher than that in conductance mode (2.59%/°C). Moreover, the dual-functional detection was achieved by recording data in capacitance mode and conductance mode simultaneously by utilizing the sensitivity differences in a different modes. Benefiting from the miniaturization and high performance of the TFSS sensor, it can be conformally attached on human body to real-time monitor physiological signals with wearing comfort.

ACKNOWLEDGMENT

J.W. acknowledges financial supports from the National Natural Science Foundation of China (61801525), the Guangdong Basic and Applied Basic Research Foundation (2020A1515010693), the Guangdong Natural Science Funds Grant (2018A030313400), the Science and Technology Program of Guangzhou (201904010456), and the Fundamental Research Funds for the Central Universities, Sun Yat-sen University (19lgpy84).

REFERENCES

[1] Z. Wu, X. Yang, and J. Wu, "Conductive hydrogel- and organohydrogel-based stretchable sensors," *ACS Appl. Mater. Interfaces,* vol. 13, pp. 2128-2144, 2021.

[2] K. Tao et al., "Origami-inspired electret-based triboelectric generator for biomechanical and ocean wave energy harvesting," *Nano Energy,* vol. 67, p. 104197, 2020.

[3] K. Xu, Y. Lu, and K. Takei, "Multifunctional skin-inspired flexible sensor systems for wearable electronics," *Adv. Mater. Technol.,* vol. 4, p. 1800628, 2019.

[4] X. Wu et al., "A wearable, self-adhesive, long-lastingly moist and healable epidermal sensor assembled from conductive MXene nanocomposites," *J. Mater. Chem. C,* vol. 8, pp. 1788-1795, 2020.

[5] F. Lu, Y. Wang, C. Wang, S. Kuga, Y. Huang, and M. Wu, "Two-dimensional nanocellulose-enhanced high-strength, self-adhesive, and strain-sensitive poly(acrylic acid) hydrogels fabricated by a radical-induced strategy for a skin sensor," *ACS Sustain. Chem. Eng.,* vol. 8, pp. 3427-3436, 2020.

[6] J. Chen et al., "Multifunctional conductive hydrogel/thermochromic elastomer hybrid fibers with a core–shell segmental configuration for wearable strain and temperature sensors," *ACS Appl. Mater. Interfaces,* vol. 12, pp. 7565-7574, 2020.

[7] G. Ge et al., "Muscle-inspired self-healing hydrogels for strain and temperature sensor," *ACS Nano,* vol. 14, pp. 218-228, 2020.

[8] L. Wang et al., "Tough, adhesive, self-healable, and transparent ionically conductive zwitterionic nanocomposite hydrogels as skin strain sensors," *ACS Appl. Mater. Interfaces,* vol. 11, pp. 3506-3515, 2019.

[9] C. Yan, J. Wang, and P. S. Lee, "Stretchable graphene thermistor with tunable thermal index," *ACS Nano,* vol. 9, pp. 2130-2137, 2015.

[10] J. Wu et al., "Highly stretchable and transparent thermistor based on self-healing double network hydrogel," *ACS Appl. Mater. Interfaces,* vol. 10, pp. 19097-19105, 2018.

[11] J. Wu et al., "Ultrastretchable and stable strain sensors based on antifreezing and self-healing ionic organohydrogels for human motion monitoring," *ACS Appl. Mater. Interfaces,* vol. 11, pp. 9405-9414, 2019.

[12] S. R. A. Ruth, L. Beker, H. Tran, V. R. Feig, N. Matsuhisa, and Z. Bao, "Rational design of capacitive pressure sensors based on pyramidal microstructures for specialized monitoring of biosignals," *Adv. Funct. Mater.,* vol. 30, p. 1903100, 2020.

[13] A. Levy, D. Andelman, and H. Orland, "Dielectric constant of ionic solutions: a field-theory approach," *Phys. Rev. Lett.,* vol. 108, p. 227801, 2012.

[14] J. Wu et al., "Stretchable, stable, and room-temperature gas sensors based on self-healing and transparent organohydrogels," *ACS Appl. Mater. Interfaces,* vol. 12, pp. 52070-52081, 2020.

[15] H. Ota et al., "Highly deformable liquid-state heterojunction sensors," *Nat. Commun.,* vol. 5, p. 5032, 2014.

Gap in pagination due to unavailable papers.

Pages 307-310

Proceedings of the 16th Annual IEEE International
Conference on Nano/Micro Engineered and Molecular Systems
April 25-29, 2021

ENHANCING SENSITIVITY USING ELECTROSTATIC SPRING IN COUPLING MODE-LOCALIZED MEMS ACCELEROMETER

Z Wang[1,2], X Y Xiong[2], K F Wang[2, 3], W H Yang[2], Z T Li[2], and X D Zou[1,2,3]

[1]QiLu Research Institute, Aerospace Information Research Institute Chinese Academy of Sciences, Jinan, CHINA

[2] The State Key Laboratory of Transducer Technology, Aerospace Information Research Institute Chinese Academy of Sciences, Beijing, CHINA and

[3] School of Electronic, Electrical and Communication Engineering, University of Chinese Academy of Sciences, Beijing, CHINA

Abstract

This paper reports a novel designed electrostatic spring structure that use to significantly enhance the sensitivity of mode-localized resonant accelerometer (ML-RXL) without increasing the device size. Moreover, the sensitivity can be flexibly adjusted by the bias voltage. Finite element method (FEM) simulation results presented demonstrate that the sensitivity of ML-RXL is increased from 59.322 AR/g to 96.034 AR/g with a linear operating range of 0.3g by applying the bias voltage of 30V.

KEYWORDS

Mode localization, Amplitude ratio, MEMS accelerometer, Electrostatic Spring, Sensitivity.

Introduction

Comparing with the conventional frequency shift resonant sensors, mode-localized MEMS devices can yield 2-3 orders of sensitivity enhancement by measuring the amplitude ratio (AR) [1][2] and show the remarkable common mode rejection capabilities [3][4] even in the environment with changeable ambient temperature or pressure. The coupling stiffness generated by using electrostatic [5][6] or mechanical structure [7][8] between resonators is the key parameter to the sensitivity of mode-localized MEMS devices. Generally, enhancing the sensitivity of mode-localized MEMS devices is achieved with a particularly designed weak coupling stiffness between the resonators. However, this method causes the reduction of frequency-split between two coupled modes, thus results in a reduction in bandwidth, which brings a huge challenge for the design of the interface circuit. Besides, recent studies show that using a high degree of freedom (DOF) resonator system can also enhance the sensitivity [9], but higher DOF means larger device size.

This work proposes a new method of using electrostatic spring to enhance the sensitivity of mode-localized MEMS devices without weakening the coupling stiffness or increasing the DOF of the resonator system. Moreover, the sensitivity can be flexibly adjusted by tuning the bias voltage and the noise caused by the electrostatic voltage can be ignored when using a low-noise voltage source [10]. Finite-element method (FEM) simulation results are presented to verify the enhancement of the sensitivity.

Theory Analysis

The structure schematic diagram of ML-RXL is shown in figure 1 (a). Two identical clamped-clamped resonators are coupled by the micro-lever mechanical structure, which is wide enough to cause the weak coupling stiffness between resonators[8]. The electrostatic spring structures, as shown in figure 1(b), form a pair of capacitance with the proof mass, which generates a pair of cancelled electrostatic forces and doubles the softening effect of electrostatic stiffness. The mode shape of the resonator system is shown in figure 1(c).

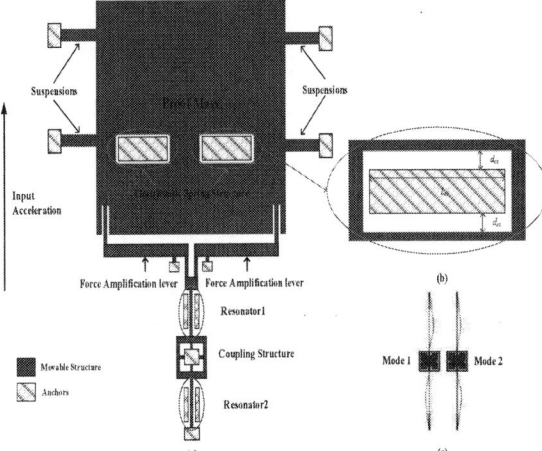

Figure 1: (a) The structure schematic diagram of ML-RXL; (b) The detail of electrostatic spring structure;(c) COMSOL simulation results of the vibration mode shape of the coupling resonators.

The mass-damper-spring lumped parameter model of the ML-RXL sensing device is shown in figure 2. Two identically designed resonators (with masses $m_1 = m_2 = m$ and stiffness $k_1 = k_2 = k$) are coupled by the micro-lever mechanical structure (k_c). The electrostatic spring softening effect induced by the parallel plate structure can effectively reduce the axial stiffness of the proof mass system, which leads to a stronger axial force applied on the resonator system that is amplified by the micro-lever structure. The axial force changes its stiffness with respect to the other resonator will lead to the mode-localization effect between the two resonators. When the displacement of the proof mass is much smaller than the gap between the proof mass and electrostatic

978-1-6654-3008-1/21 $31.00 © 2021 IEEE   311

spring structures, the sensitivity (S_{AR}) of ML-RXL can be approximated as [10][11]

$$S_{AR} = \frac{\partial AR}{\partial A_{cc}} \approx S_0 \frac{1}{1 - k_{es}/k_{susp}}$$

Where A_{cc} is the input acceleration, S_0 is the sensitivity of ML-RXL without bias voltage applied on electrostatic spring structures, k_{es} is the electrostatic spring, k_{susp} is the effective mechanical stiffness on proof mass in the axial direction.

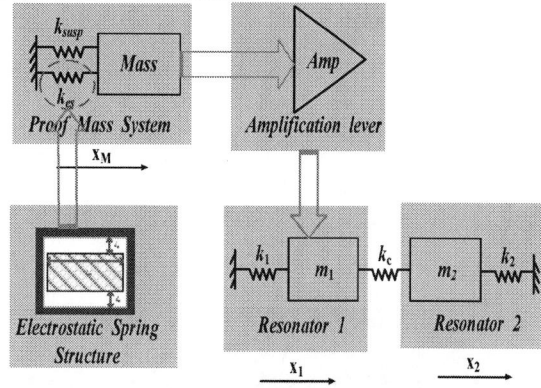

Figure 2: Mass-damper-spring lumped parameter model of the ML-RXL sensing device.

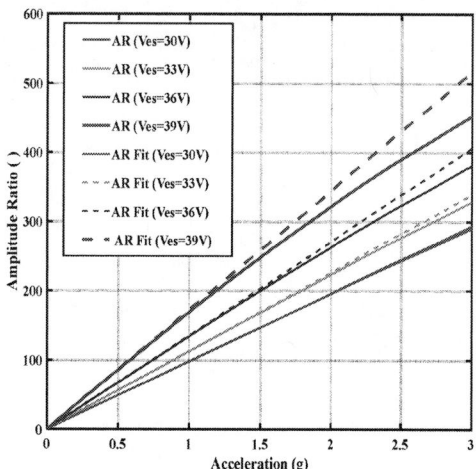

Figure 3: Theoretical results of the amplitude ratio (AR) of ML-RXL as a function of the input acceleration varying from 0g to 3g with different bias voltage varying from 30V to 39V with the step of 3V.

Hence the sensitivity can be improved with the increasing bias voltage, as shown in figure 3. However, as the bias voltage increasing, the displacement of the proof mass increases while the gap between the proof mass and electrostatic spring structures decreases, which results in non-linearity of sensitivity and the reduction of the linear operating range, as shown in figure 4. The nonlinearity of the sensitivity is less than 0.1% within the input acceleration corresponding to the blue curve segments, the

nonlinearity of the sensitivity is less than 0.5% and larger than 0.1% within the input acceleration corresponding to the red segments, the nonlinearity of the sensitivity is less than 1% and larger than 0.5% within the input acceleration corresponding the green segments. Therefore, it is necessary to make a trade-off between the sensitivity and the linear operation range.

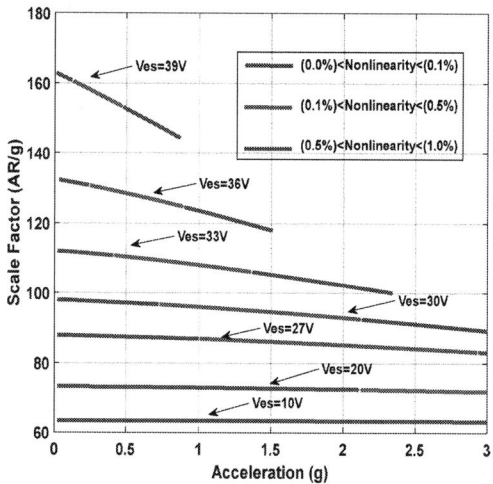

Figure 4: Theoretical results of the sensitivity of ML-RXL as a function of the input acceleration with different bias voltage applying on the electrostatic spring structures.

Simulation And Results

Silicon ML-RXL is placed in an air-filled domain to simulate the electrostatic stiffness softening effect (using COMSOL 5.4), as shown in figure 5. The Simulation Parameter of ML-RXL is summarized in table 1. Only the electric potential of the electrostatic spring structures is 30V and the electric potential of other components is 0V. FEM simulation results of the coupled modes resonant frequencies and AR are shown in figure 6 and figure 7. The linear range is defined as the sensitivity nonlinearity less than 0.1% to achieve better linearity. The sensitivity of AR can be enhanced from 59.322 AR/g to 96.034 AR/g by applying bias voltage (V_{es}=30V) on the electrostatic spring structures when working in the linear operating range.

Table 1: The Design Parameter of ML-RXL

Parameters	Value
Total size	5.2mm×4.2mm
Proof mass	1.50mg
Device thickness	40um
Lever amplification ratio	22.39
d_{es}	3um
Length of tine beam	400um
Width of tine beam	7um
k_c/k	4.5×10^{-5}
Scale factor	59.322AR/g

978-1-6654-3008-1/21 $31.00 © 2021 IEEE   312

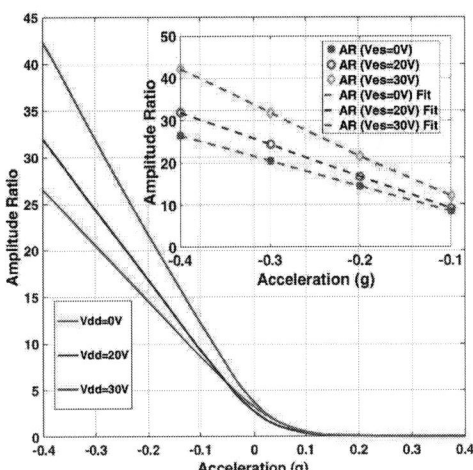

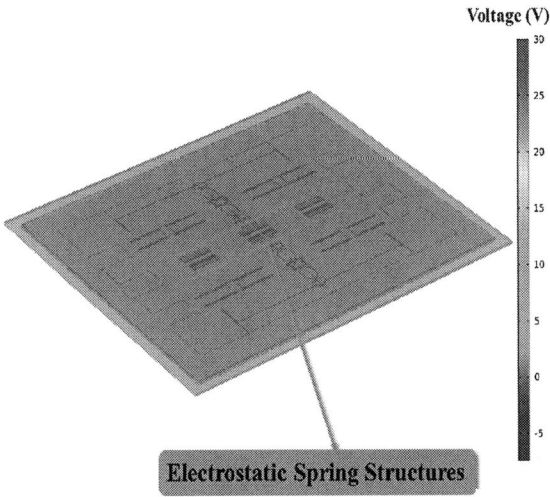

Figure 5: FEM simulation model and electric potential distribution of ML-RXL with bias voltage (30V) applied on the electrostatic spring structures.

Figure 7: FEM simulation results of the amplitude ratio (AR) as a function of the input acceleration varying from -0.4g to +0.4g for different bias voltage varying from 0V to 30V. Inset shows the linear region with the input acceleration varying from -0.4g to -0.1g.

Conclusion

This paper presents a novel designed electrostatic spring structures to realize greater axial force acted on the resonator system for mode-localized resonant sensor. FEM simulation results demonstrate that using electrostatic spring structures can significantly enhance the sensitivity. More in-depth study for device fabrication and experiments with interface circuits is the subject of our future work.

Acknowledgment

This work was supported by the National Key Research and Development Program of China (Grant No. 2018YFB2002300).

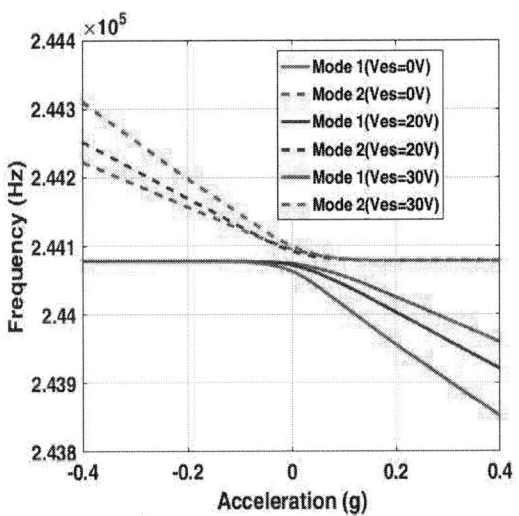

Figure 6: FEM simulation results of the coupled modes (solid line indicates mode 1 and the dashed line indicates mode 2) resonant frequencies shift as a function of the input acceleration varying from -0.4g to +0.4g for different bias voltage varying from 0V to 30V with the step of 10 V.

References:

[1] Zhao C, Wood G S, Xie J, et al. A force sensor based on three weakly coupled resonators with ultrahigh sensitivity[J]. Sensors and Actuators A: Physical, 2015, 232: 151-162.

[2] Zhang H, Li B, Yuan W, Kraft M, Chang H. An acceleration sensing method based on the mode localization of weakly coupled resonators[J]. Journal of microelectromechanical systems, 2016, 25(2): 286-296.

[3] Thiruvenkatanathan P, Yan J, Seshia A A. Differential amplification of structural perturbations in weakly coupled MEMS resonators[J]. IEEE transactions on ultrasonics, ferroelectrics, and frequency control, 2010, 57(3): 690-697.

[4] Zhang H, Zhong J, Yuan W, Yang J, Chang H. Ambient pressure drift rejection of mode-localized resonant sensors[C]//2017 IEEE 30th International

Conference on Micro Electro Mechanical Systems (MEMS). IEEE, 2017: 1095-1098.

[5] Thiruvenkatanathan P, Yan J, Seshia A A. Ultrasensitive mode-localized micromechanical electrometer[C]//2010 IEEE International Frequency Control Symposium. IEEE, 2010: 91-96.

[6] Wood G S, Zhao C, Pu S H, et al. An Investigation of Structural Dimension Variation in Electrostatically Coupled MEMS Resonator Pairs Using Mode Localization[J]. IEEE Sensors Journal, 2016, 16(24).

[7] Pandit M, Zhao C, Sobreviela G, et al. A mode-localized MEMS accelerometer with 7μg bias stability[C]//2018 IEEE Micro Electro Mechanical Systems (MEMS). IEEE, 2018: 968-971.

[8] Wang Z, Xiong X Y, Li Z T, et al. Enhancing Parametric Sensitivity Using Micro-Lever Coupler in Mechanical Coupling Mode-Localized Mems Accelerometer[C]//2019 20th International Conference on Solid-State Sensors, Actuators and Microsystems & Eurosensors XXXIII (TRANSDUCERS & EUROSENSORS XXXIII). IEEE, 2019: 1846-1849.

[9] Kang H, Yang J, Chang H. A mode-localized accelerometer based on four degree-of-freedom weakly coupled resonators[C]//2018 IEEE Micro Electro Mechanical Systems (MEMS). IEEE, 2018: 960-963.

[10] Xiong X Y, Zou X, Wang Z, et al. Using Electrostatic Spring Softening Effect To Enhance Sensitivity Of Mems Resonant Accelerometers[J]. IEEE Sensors Journal, 2020.

[11] Zotov S A, Simon B R, Trusov A A, et al. High quality factor resonant MEMS accelerometer with continuous thermal compensation[J]. IEEE Sensors Journal, 2015, 15(9): 5045-5052.

Gap in pagination due to unavailable papers.

Pages 315-318

Proceedings of the 16th Annual IEEE International
Conference on Nano/Micro Engineered and Molecular Systems
April 25-29, 2021

Direct-write 3D printed self-supporting flexible microstructure devices*

Cuiwen Hu, Haifeng Chen, Deyun Mo, Haishan Lian, Yinhuan Yang, Xiaojun Chen

Abstract— Conventional 3D printing methods often require supporting structures to maintain the structure of the suspended parts in the manufacture of metamaterials, micro-nano arrays, micro-channels. The remaining uncured resin in the microstructure and the removal of the sacrificial layer material by post-processing are a key challenge for structural molding. Therefore, we propose a free-form-based direct writing 3D printing technology for manufacturing flexible microstructures without self-supporting materials. We proved that the 3D printed microstructure can withstand the bending moment it generates to prevent the collapse of the flexible microstructure. This method provides a new way for the manufacture of flexible microstructure devices such as soft robots, flexible sensing, biological tissues, micro-nano fluidics, and metamaterials.

Keyword—Direct-write 3D printing, self-supporting, flexible microstructure devices

I. INTRODUCTION

Self-supporting flexible micro-nano structures (such as interconnected pores, ultra-slender flow channels, membrane-cavity, honeycomb, etc.) are typical features of macro devices such as soft robots, flexible sensors, biological tissues, micro-nano fluidics, and metamaterials. In particular, the self-supporting flexible membrane-cavity and honeycomb microstructure unit provides large geometric deformation, distributed drive and sensing under the action of external force[1]. The movable membrane-cavity pump/valve, ultra-slender micro-channels structures in microfluidic control realize the functions of micro-scale fluid restriction/guidance, transportation, logic control, and distributed drive[2-4]. However, this type of microstructure manufacturing is still limited to laboratories. In the manufacturing process, it is often necessary to consider the matching of self-supporting properties and flexible functions, which inevitably poses new challenges to structural materials and manufacturing processes.

Soft lithography and additive manufacturing are the most promising methods for manufacturing this type of structure-device. For PDMS materials, Quake[5] et al. fabricated an on-chip gas/liquid driven microvalve array by using soft lithography. There are many problems in the production of micropumps/valves with this method, such as the inability to produce true 3D structures, the large-area bonding rate and consistency are difficult to control, the deformation and

collapse of the movable membrane, and insufficient stiffness[6]. In view of the reconciliation of manufacturing accuracy and efficiency, addit8ive manufacturing technology has more potential for application in the manufacture of flexible micro-nano-scale structures and devices. It is expected to be a supplement or even an alternative to soft lithography technology to enhance the manufacturability of flexible microstructure devices[7]. Several 3D printing methods (such as SLA[8], jetting MJM[9]) have been used to manufacture micro-nano structure devices. However, there may be uncured residual resin or sacrificial support materials temporarily used to form hollow structures during microstructure manufacturing. Based on the needs of post-processing, the degree of automation and molding quality of flexible microstructure device manufacturing are affected[10]. In addition, the microstructure devices made of the materials used in these two 3D printing technologies are low in flexibility and cannot be integrated with existing structures and sensor devices, which limits the development of such devices in the fields of soft robots, flexible skin, and flexible electronic sensing.

Direct-write 3D printing technology provides a free-form method for the manufacture of flexible microstructures. The filaments can be deposited on the substrate to manufacture flexible microstructure devices[11]. The challenge for printing inks to directly manufacture hollow structures is that the mechanical strength of the uncured polymer is insufficient to resist the creep of the printed structure[12][13]. This usually requires the use of sacrificial support materials[14]. One possible solution is to print part of the wall structure in advance and glue it on the top with a flat plate or glass[15]. To create a hollow structure without supporting material, it is necessary to ensure that the maximum stress in the self-supporting structure is less than the yield strength of the ink after extrusion.

Therefore, direct-write 3D printing technology based on free prototyping is employed to manufacture flexible microstructures without the need for post-processing. One-step manufacturing of self-supporting microstructures is achieved by selecting inks with appropriate yield strength and controlling the outline of the printed hollow structure. Different from the traditional 3D printing idea of printing and curing at the same time, the viscoelastic ink directly prints the hollow structure in an air atmosphere without any

---

*Resrach supported by the National Science Foundation of China (No. 52005239), Guangdong Basic and Applied Basic Research Foundation (No. 2019A1515110637, No.2020A1515010165), Guangdong Province Youth Innovation Talent Project (No. 2019KQNCX076), Science and Technology Planning Project of Zhanjiang (No. 2020A01044, NO. 2020B01363, No. 2018A02009).

Corresponding author: Xiaojun Chen is with School of Mechanical and Electronic Engineering, Lingnan Normal University, Zhanjiang, China (phone: +86-18030250016; e-mail: chxj@lingnan.edu.cn).

support structure. During the entire printing process, the printed self-supporting microstructures are quickly cross-linked and cured in an air atmosphere, so that they have sufficient mechanical strength to support subsequent printing of the superstructure. The rapid manufacturing of unsupported flexible microstructures is demonstrated, such as hollow triangular structures, trapezoidal structures, and circular ring structures. This method provides a new way for the manufacture of flexible microstructure devices such as soft robots, flexible sensing, biological tissues, micro-nano fluidics, and metamaterials.

II. MATERIALS AND METHODS

Materials

Flexible transparent silicone (Loctite, SI 595) was purchased from Henkel Loctite (China) Co., Ltd. The PMMA boards and glass slides were used as the carrier platform for the printed parts, and was purchased from Deyao Plastic Material Factory. Liquid metal (Galinstan, the mass fractions of Ga, In and Sn are 62.5%, 21.5% and 16% respectively) were purchased from Beijing Dream-ink Technology Co., Ltd.

Self-supporting microstructure printing

3D printing machines (EFL-BP6601) are used to manufacture self-supporting microstructures and devices. It consists of the X-Y-Z platform to move the prints in three directions. The Y moving axis is located on the platform, and the temperature heating plate is installed (0-200 degrees adjustable). A detachable print nozzle is installed on the Z axis. The printing material is regulated by air pressure control using a pressure controller. The microstructure printing process is similar to the conventional 3D printing process. Import the designed print model into the slicing software for slicing, and then set the printing parameters in the printing program such as extrusion pressure, platform movement speed, material extrusion speed, first layer height, etc., and then start printing to complete the entire printing process.

Characterization

The printing process is photographed and recorded by the mobile phone. The printing line width is measured by an optical microscope.

III. RESULTS AND DISCUSSION

The schematic diagram of direct writing 3D printed self-supporting flexible microstructures is shown in Figure 1. We manufactured inverted V-shaped structures, inclined wall structures, circular ring structures, and triangular structures by using elastic silicone printing materials. The single line width resolution of the structure is an important criterion for measuring the accuracy of 3D printing.

We investigated the relationship between the printing layer height (the distance between the nozzle and the printing substrate), the extrusion pressure, the printing speed, and the material placement time and the line width diameter, as shown in Figure 2. As shown in Figure 2a, the larger the print layer height setting, the smaller the line width; when the layer height is set to 0.3mm, the line width is about 280 microns. However, when the layer height exceeds 0.4mm, the printed straight layer spirals, so the line width becomes larger. The extrusion speed of the printing material is also a key factor affecting the line width. Obviously, as the extrusion pressure increases, the printed line width also increases. When the extrusion pressure is set to less than 25 PSI, the printed structure appears intermittent and cannot form a whole structure. Therefore, under the condition of ensuring that the printing line width is small enough, we choose the extrusion pressure to be between 27-30PSI.

We also investigated the relationship between nozzle movement speed and line width. As shown in Figure 2c, as the nozzle moving speed increases, the printed line width gradually decreases. However, when the nozzle speed is greater than 5mm/s, the phenomenon of wire breakage is likely to occur, and the quality of the printed wire is poor. When the printing speed is 2-4mm/s, the printing line width

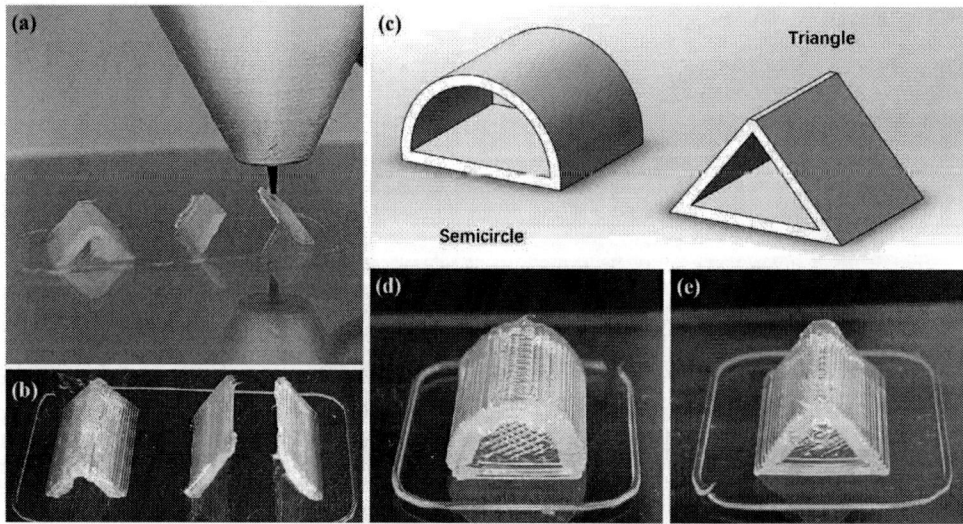

Fig. 1. (a) Schematic of self-supporting 3D printed microstructure. (b-e) Self-supporting microstructure.

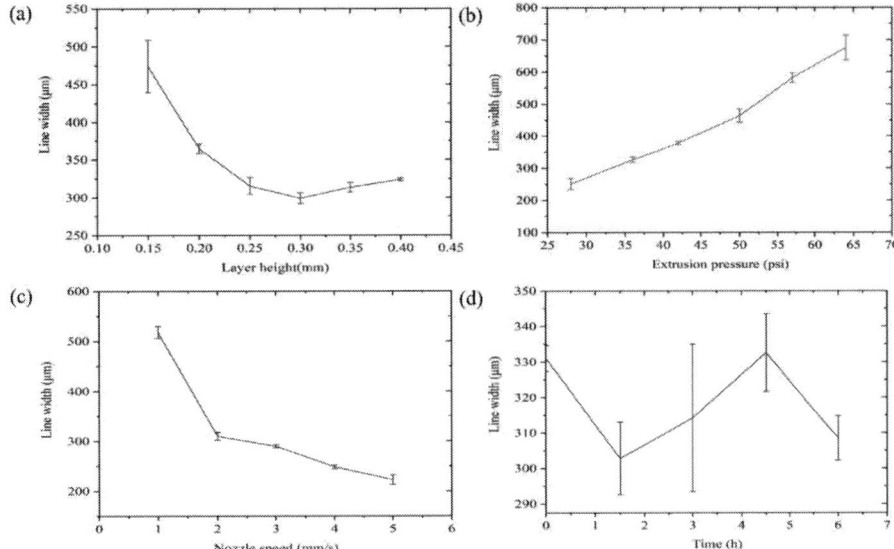

Fig. 2. The relationship between printing layer height, extrusion pressure, moving speed and material placement time and printing line width.

is stable between 250-300 microns. Since the selected printing material can be cured in a room temperature environment, the placement time of the material in the printing needle has an impact on the printing effect. We placed the printing material in the needle tube for 6 hours, and discarded the part of the needle in contact with the air each time. Then in the printing experiment, the measured printing line width floated in the range of 300-330 microns.

The suspended microstructure can be formed without supporting structure, which is closely related to the properties of the material. Transparent silica gel was used as 3D printing material, and it is quickly cured when exposed to the air without ultraviolet radiation or heating. It exhibits high elongation and adhesion, with a Young's modulus of 189.7 kPa. When printing a self-supporting structure or a hollow structure, the mechanical strength of the uncured printing material can fully resist the creep of the flexible structure is a key factor. In order to demonstrate the feasibility of printing self-supporting structures, we printed cuboids with different inclination angles (Figure 3).

When the structure angle is 90 degrees, the extruded filaments are stacked layer by layer in the Z-axis direction. When the structural angles are 75°, 60°, 45°, and 30°, the extruded filaments are stacked layer by layer in the Z direction, and stacked in the X direction to form a structural angle. However, when the structure angle is less than 30°, it

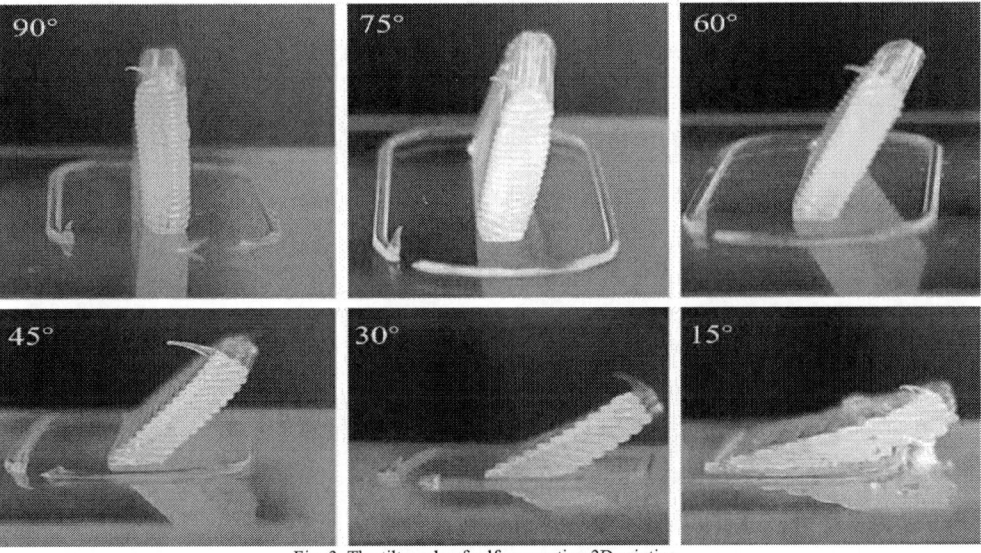

Fig. 3. The tilt angle of self-supporting 3D printing.

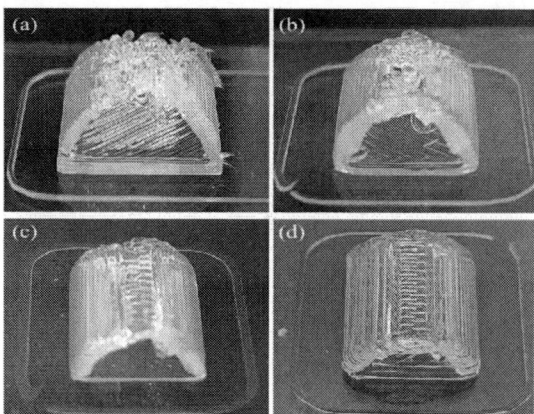

Fig. 4. The influence of arc diameter on 3D printing self-supporting microstructure.

is easy to collapse at the root of the structure (the part with the greatest stress). The main reason is that the yield stress of the printing material is not enough to resist the bending stress of the printing structure at the root.

When printing the arc-shaped structure, the influence of different arc diameters on the forming structure was investigated (Figure 4). We investigated semi-circular arc structures with inner diameters of 8.8mm, 5.8mm, 5.4mm, 4.5mm, and wall thicknesses of 0.6mm. As shown in Figures 4a and 4b that the arc-shaped structure can be formed, but the forming quality on the top of the microstructure is poor. The main reason is that there is no support structure to provide support for the printing filament connection when printing a large arc radius, which causes the filament to collapse. When the radius of the arc decreases, the printing filaments are tightly connected and supported by the viscoelasticity of the printing material. Therefore, it can print out a structure with better forming quality, as shown in Figure 4d.

The smaller the arc diameter, the better the quality of the structure. It is easier to construct micro-channels, micro-pillar-cavities and other microstructures. We fabricated Y-shaped microchannels with internal diameters of 0.8mm and 0.6mm, respectively (Figure 5). We inject water into the inlet of the microchannel, and then observe the outflow of water at the outlet. The experiment demonstrates that the inlet and outlet of the microchannel chip are interconnected.

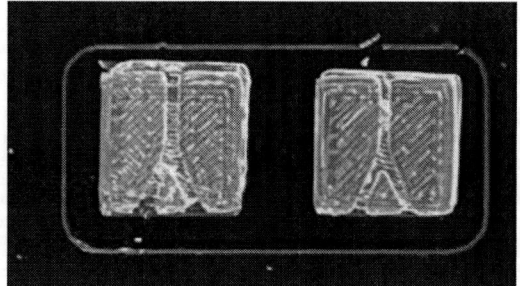

Fig. 5. Self-supporting 3D printed Y-shaped micro channel.

Fig.6. Liquid metal elastomer structure

In addition, we also printed a 6*6 array of liquid metal elastomer structure (Figure 6). We test the elasticity of the force exerted by the mesh structure (Figure 7). It shows that the mesh structure has excellent elasticity and will not break or deform under the action of tension. When pressure is applied, the liquid metal flowing inside the elastomer structure is connected, which will expand its application in flexible electronics and sensing in the future.

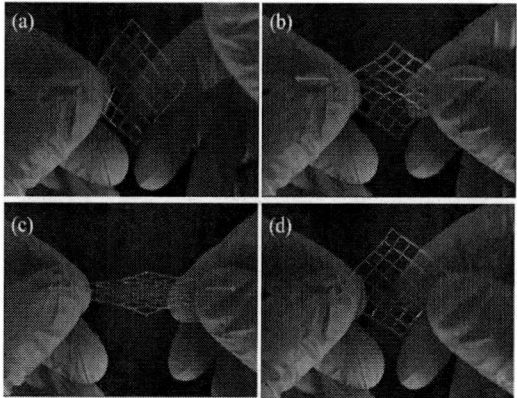

Fig. 7. Tensile test of self-supporting 3D printed silicone material.

IV. CONCLUSION

In summary, we report a direct-write 3D printing method curable at room temperature, which can be used to fabricate flexible self-supporting microstructures without the need for supporting materials. Due to the high viscoelasticity and room temperature curable characteristics of silicone material, it provides structural support and maintains the structure shape deposited during the printing process. Therefore, flexible self-supporting structures such as inclined structures, semi-circular arc structures, and microchannel structures can be manufactured. The composite material prepared by doping liquid metal with silica gel is also printable, and this direct writing method greatly broadens the range of printing materials. The article demonstrates the feasibility of direct-write 3D printing in the manufacture of flexible self-supporting microstructures. In the future, this method has a very broad application prospect in the manufacture of flexible electronics, soft robots, microfluidic electronics and other devices.

REFERENCES

[1] G. O. Young, "Synthetic structure of industrial plastics (Book style with paper title and editor)," in *Plastics*, 2nd ed. vol. 3, J. Peters, Ed. New York: McGraw-Hill, 1964, pp. 15–64.

[2] W.-K. Chen, *Linear Networks and Systems* (Book style). Belmont, CA: Wadsworth, 1993, pp. 123–135.

[3] H. Poor, *An Introduction to Signal Detection and Estimation.* New York: Springer-Verlag, 1985, ch. 4.

[4] B. Smith, "An approach to graphs of linear forms (Unpublished work style)," unpublished.

[5] E. H. Miller, "A note on reflector arrays (Periodical style—Accepted for publication)," *IEEE Trans. Antennas Propagat.*, to be published.

[6] J. Wang, "Fundamentals of erbium-doped fiber amplifiers arrays (Periodical style—Submitted for publication)," *IEEE J. Quantum Electron.*, submitted for publication.

[7] C. J. Kaufman, Rocky Mountain Research Lab., Boulder, CO, private communication, May 1995.

[8] Y. Yorozu, M. Hirano, K. Oka, and Y. Tagawa, "Electron spectroscopy studies on magneto-optical media and plastic substrate interfaces(Translation Journals style)," *IEEE Transl. J. Magn.Jpn.*, vol. 2, Aug. 1987, pp. 740–741 [*Dig. 9th Annu. Conf. Magnetics* Japan, 1982, p. 301].

[9] M. Young, *The Techincal Writers Handbook.* Mill Valley, CA: University Science, 1989.

[10] J. U. Duncombe, "Infrared navigation—Part I: An assessment of feasibility (Periodical style)," *IEEE Trans. Electron Devices*, vol. ED-11, pp. 34–39, Jan. 1959.

[11] S. Chen, B. Mulgrew, and P. M. Grant, "A clustering technique for digital communications channel equalization using radial basis function networks," *IEEE Trans. Neural Networks*, vol. 4, pp. 570–578, July 1993.

[12] R. W. Lucky, "Automatic equalization for digital communication," *Bell Syst. Tech. J.*, vol. 44, no. 4, pp. 547–588, Apr. 1965.

[13] S. P. Bingulac, "On the compatibility of adaptive controllers (Published Conference Proceedings style)," in *Proc. 4th Annu. Allerton Conf. Circuits and Systems Theory*, New York, 1994, pp. 8–16.

[14] G. R. Faulhaber, "Design of service systems with priority reservation," in *Conf. Rec. 1995 IEEE Int. Conf. Communications,* pp. 3–8.

[15] W. D. Doyle, "Magnetization reversal in films with biaxial anisotropy," in *1987 Proc. INTERMAG Conf.*, pp. 2.2-1–2.2-6.

[16] G. W. Juette and L. E. Zeffanella, "Radio noise currents n short sections on bundle conductors (Presented Conference Paper style)," presented at the IEEE Summer power Meeting, Dallas, TX, June 22–27, 1990, Paper 90 SM 690-0 PWRS.

[17] J. G. Kreifeldt, "An analysis of surface-detected EMG as an amplitude-modulated noise," presented at the 1989 Int. Conf. Medicine and Biological Engineering, Chicago, IL.

[18] J. Williams, "Narrow-band analyzer (Thesis or Dissertation style)," Ph.D. dissertation, Dept. Elect. Eng., Harvard Univ., Cambridge, MA, 1993.

[19] N. Kawasaki, "Parametric study of thermal and chemical nonequilibrium nozzle flow," M.S. thesis, Dept. Electron. Eng., Osaka Univ., Osaka, Japan, 1993.

[20] J. P. Wilkinson, "Nonlinear resonant circuit devices (Patent style)," U.S. Patent 3 624 12, July 16, 1990.

Gap in pagination due to unavailable papers.

Pages 324-333

Ohmic Contact Characteristics of Silicon Carbide-based MEMS Devices

Chen Wu, Xudong Fang*, Zhihong Feng*, Qiang Kang, Yuanjie Lv，Yuefei Yan，Zhuangde Jiang

Abstract—The stability and reliability of electrical ohmic contacts are the key to the stable operation of silicon carbide power electronic devices in extreme environments. Firstly, the current research status and common problems of SiC ohmic contacts has been analyzed. Secondly, a carrier model for the ohmic contact between metals and SiC based on the tunneling effect is established. By increasing the doping concentration of the substrate SiC, the Schottky barrier width can be reduced, which is more conducive to the formation of ohmic contacts. Thirdly, 200nm thick Ni metal layer was deposited on 4H-SiC substrate with a doping concentration of NA=2e19cm-3. Finally, the effect of high-temperature thermal annealing temperature on metal ohmic contact is studied and results show that Ni can form ohmic contact with SiC when the annealing temperature exceeds 800°C. Scanning electron microscope and atomic force microscope are used to observe the contact surface. After high-temperature thermal annealing, the electrical properties are guaranteed, but metals indicate defects such as cluster precipitation. Therefore, it is possible to prevent the contact layer metal from being damaged by adding other covering metals.

I. INTRODUCTION

Silicon (Si) microelectromechanical systems have achieved important success in the application of commercial sensors. However, with the development of MEMS technology, the limitations of Si devices have become more and more significant, especially for applications under extreme conditions such as high temperature, high frequency, and high power[1]. In the high temperature area, the Si band gap is small (1.12eV), and intrinsic excitation occurs at room temperature. As the operating temperature increases, the concentration of electron-hole pairs in the depletion layer increases, causing the p-n junction of Si-based materials to be severely degraded at operating temperatures above 150°C [2]. In addition, when the temperature of the working environment reaches more than 600°C, the elastic coefficient of the Si material itself will be seriously attenuated. Therefore, the operating temperature of Si devices can only be 200℃. Therefore, in order to meet the monitoring requirements of equipment operating conditions in high temperature and harsh environments, such

as real-time monitoring of temperature, pressure and acceleration, the third generation of wide-bandgap semiconductors has emerged in recent years[1]. As shown in Tab. 1, the third-generation wide-bandgap semiconductor represented by silicon carbide (SiC) has significant mechanical, chemical and electrical advantages such as high carrier migration rate, high hardness, and acid and alkali corrosion resistance[1]. These advantages make it widely used in microelectronics and micromechanical systems. The SiC material itself can withstand temperatures exceeding 2300°C without significant material performance degradation, but so far, the using temperature of SiC-based devices has been reported as high as 800°C[3]. The deterioration of the metal-semiconductor contact performance results in the failure of the key problem of long-term work under high temperature [4]. Therefore, the long-term operation of SiC devices under high temperature conditions must solve the problem of high temperature stability of ohmic contacts.

Table.1. Performance comparison of SiC and other common semiconductors.

Performance parameter	Si	GaAs	3C-SiC	4H-SiC	6H-SiC	GaN
Crystal structure						
Band gap(eV)	1.12	1.43	2.3	3.3	2.9	3.39
Melting point(K)	1690	1510	3100	3100	3100	1227
Electron migration rate($cm^2/v\cdot s$)	1400	8500	1000	900	400	<400
Breakdown electric field strength($/10^5 V\cdot m^{-1}$)	2.5	3	40	40	40	20
Thermal conductivity($W/c\cdot K$)	1.5	0.5	4.9	4.9	4.9	1.5
Saturation drift rate($cm\cdot s^{-1}$)	1.0	2	2.5	2.7	2.5	2.5

There are more than 250 different crystal types of SiC that have been discovered[1]. However, only three crystal types are commonly used, including α-SiC represented by

---

Xudong Fang is with the School of Mechanical Engineering, Xi'an Jiaotong University, 710049, Xi'an, China. (Corresponding author, phone: 86+13689257568; e-mail: dongfangshuo30@xjtu.edu.cn)

Zhihong Feng is with National Key Laboratory of Application Specific Integrated Circuit, Hebei Semiconductor Research Institute, Shijiazhuang 050051, China. (Corresponding author, phone: 0311-87091855; e-mail: ga917vv@163.com.)

Chen Wu is with the School of Mechanical Engineering, Xi'an Jiaotong University, 710049, Xi'an, China. (e-mail: chen_wu69@stu.xjtu.edu.cn)

Qiang Kang is with the School of Mechanical Engineering, Xi'an Jiaotong University, 710049, Xi'an, China. (e-mail: kangxjtu@stu.xjtu.edu.cn)

Yuanjie Lv is with National Key Laboratory of Application Specific Integrated Circuit, Hebei Semiconductor Research Institute, Shijiazhuang 050051, China. (e-mail: yuanjielv@163.com.)

Yuefei Yan is with Xidian University, Key Laboratory of Electronic Equipment Structure Design, Ministry of Education, No.2 South Taibai Road, Xi'an 710071, China. (e-mail: yfyan530@163.com.)

Zhuangde Jiang is with the School of Mechanical Engineering, Xi'an Jiaotong University, 710049, Xi'an, China. (e-mail: zdjiang@xjtu.edu.cn)

3C-SiC and β-SiC represented by 4H and 6H-SiC. Existing SiC ohmic contact research mainly focuses on hexagonal 4H and 6H-SiC[5]-[12]. This is because 3C-SiC can only be epitaxial on other substrates and cannot be used as a bulk material. Existing research mainly focuses on the annealing parameters of SiC ohmic contacts formed by specific metal combinations, high temperature diffusion characterization of deposited metals, contact resistance values, etc.[13]-[15]. There are also some common problems that remain to be resolved, including poor process repeatability, unclear ohmic contact formation mechanism, and single contact evaluation index.

In this paper, firstly, the ohmic contact formation mechanism based on the work function of metals and semiconductors is described. Secondly, the standard process flow for preparing the ohmic contact of the conventional SiC pressure sensor is proposed. Finally, the ohmic contact of the Ni-based SiC pressure sensor with ohmic characteristics is prepared according to the process flow. This work has reference value for studying the ohmic contact of SiC-based electronic devices with low resistance and high temperature stability.

II. MECHANISM MODEL

Metal-semiconductor contacts include two types, ohmic contacts and rectifying Schottky contacts. Among them, ohmic contact means that the current-voltage relationship (IV curve) between the metal electrodes is linear regardless of whether a forward voltage or a reverse voltage is applied, and the resistance between the metal and the semiconductor can be ignored compared with the resistance of the device, and the Schottky characteristic is the rectification characteristic, and there is only output current under one direction voltage bias[18]-[19]. The Schottky barrier height ϕ_B between the metal and semiconductor determines whether the electrical characteristics is ohmic contact or Schottky contact, and is defined as:

$$q\phi_B = q(\phi_m - \chi_s) \quad (1)$$

Where the Schottky barrier height ϕ_B satisfies the relationship represented by Equ.(1), and $q\phi_B$ equals the difference between the metal work function $q\phi_m$ and the semiconductor electron affinity $q\chi_s$, and its physical meaning is the energy required for electrons to enter the semiconductor from the metal.

As shown in Fig.1, if there is a metal satisfy that $\phi_m < \phi_s$, electrons in the metal will be transferred to the semiconductor, thereby reducing the energy level at the metal/semiconductor interface. At this time, if the voltage is applied, electrons can flow freely in two directions, and the contact appears as an ohmic contact. However, since the 4H-SiC material has an electron affinity of $3.1eV$[1], almost no metal has a work function that can satisfy the $\phi_m < \phi_s$ condition. That's why metal deposition on semiconductors often appears as Schottky contacts. Therefore, in the absence of a metal that matches the work function of SiC, it is difficult to reduce the height of the Schottky barrier. At this time, the tunneling flow of electrons can be achieved by reducing the width of the barrier, then ohmic contact will be formed between the SiC and metals.

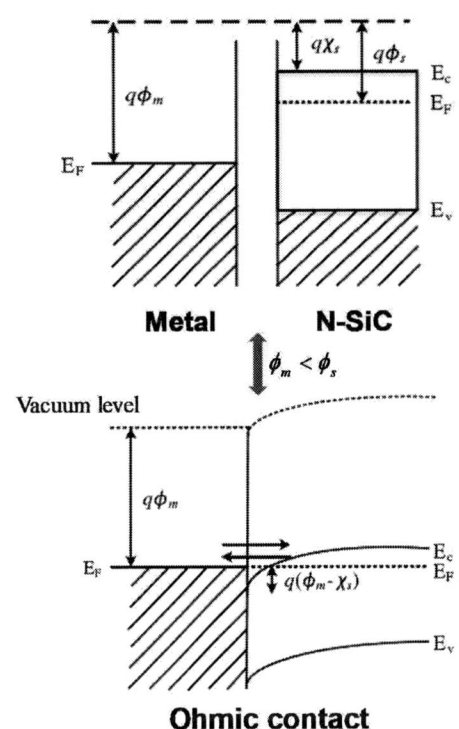

Figure 1. Energy band diagram before and after contact between metal and n-type semiconductor: ($\phi_m < \phi_s$)

Through the carrier transport mechanism of metals and semiconductors, we can find a good way to reduce the width of the barrier. The characteristic energy E_{00} of the following formula (2) can be used to explain and classify the carrier transport mechanism of the metal-SiC ohmic contact[18]-[19]:

$$E_{00} = (h/4\pi) \times \sqrt{N/(m^*\varepsilon_s)} \quad (2)$$

Among them, h is the Planck constant, m^* is the effective carrier mass, and N is the substrate doping concentration, and ε_s is the dielectric constant of the SiC semiconductor. By comparing the relationship between the characteristic energy and the electron energy, the mechanism model of the metal-SiC contact can be judged. When the doping concentration N is high, the characteristic energy will be greater than the electron energy kT. At this time, through the field emission near the fermi level, the tunneling of carriers will be realized. The ohmic contact mechanism model established above shows that increasing the substrate doping concentration can achieve the ohmic contact characteristics between SiC and metal.

978-1-6654-3008-1/21 $31.00 © 2021 IEEE   335

III. FABRICATION PROCESS

Taking the ohmic contact research of SiC-based MEMS pressure sensor as an example, the research results of typical SiC-based ohmic contact are summarized, and the existing common problems are analyzed. In the sensor signal output structure, we hope that the contact point exhibits typical ohmic contact characteristics, which can effectively reduce the voltage drop at the contact point, thereby reducing the power consumption of the sensor[14]. Fig. 2 is a schematic diagram of a typical MEMS high temperature pressure sensor. Pressure acts on the diaphragm, causing the piezoresistive bars to produce stress and strain. ohmic contact electrodes located on the diaphragm connect the piezoresistive bars to form a Wheatstone bridge to realize electrical signal connection and external output.

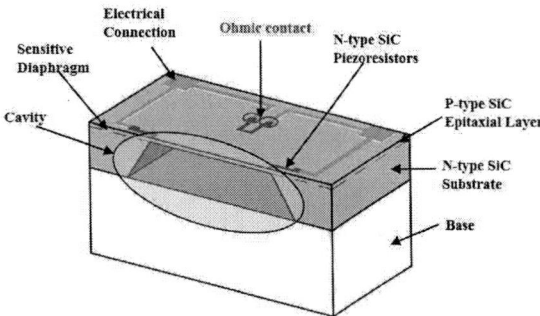

Figure 2. Schematic diagram of piezoresistive pressure sensor chip structure ohmic contact electrode.

The typical fabrication process of ohmic contact is shown in Fig.3. It starts with a 2 inch 4H-SiC wafer with thickness of 350μm as shown in Fig.3. In step (i), the aluminum is implanted into the SiC single crystal substrate by competitive epitaxy to form a P-type doped layer. And then another N-type doped layer is grown on the top of P-type layer so that P-N junction can form after step(ii). It is worth noting that the epitaxial growth sequence of P-type and N-type can be changed. In the process of device manufacturing, when P-type ohmic contacts is needed, the N-type can be grown at first and finally the P-type is grown on the upper layer. The other subsequent processes remain unchanged. Then, the surface layer of N-doped SiC is etched to form a pressure sensitive element. After the step (iii), a layer of silicon dioxide needs to be deposited on the entire surface of the wafer as an electrical insulation isolation layer. In step (v), contact windows are formed between the metal layers and SiC. Magnetron sputtering or chemical vapor deposition equipment is used to prepare the metal layer required for ohmic contact. Finally, through a high-temperature thermal annealing step (vii), the metal and SiC atoms are all activated, and ohmic contact is formed by forming alloy phase or a change in the high-temperature surface state of the material interface. High temperature thermal annealing is a key step in the formation of contact properties, and the determination of annealing parameters, time and atmosphere requires a lot of experiments to explore and verify.

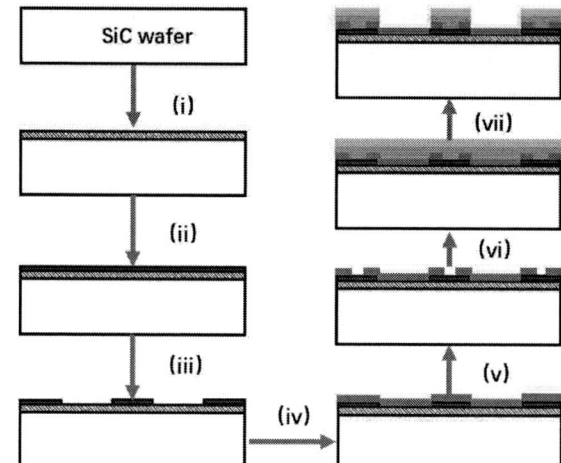

Figure 3. A fabrication process of ohmic contact.

IV. TESTING AND ANALYSIS

A variety of different metals including Ti, Ni, Pt, W, Al, etc. are currently used as SiC ohmic contacts[4]–[11]. The reason why SiC ohmic contact has become a difficult point is that no matter which of the metal types, metal layer thickness, annealing temperature, annealing time and other process parameters changes, it will cause different contact characteristics. Therefore, according to the standard preparation process of ohmic contact proposed in this study, we prepared a Ni metal layer with a thickness of 200nm on a 4H-SiC substrate with a doping concentration of $NA=2e19cm^{-3}$, and rapid thermal annealing was carried out at three different temperatures of 800℃, 1000℃ and 1050℃ for 120s to study the contact characteristics of Ni and SiC at different annealing temperatures.

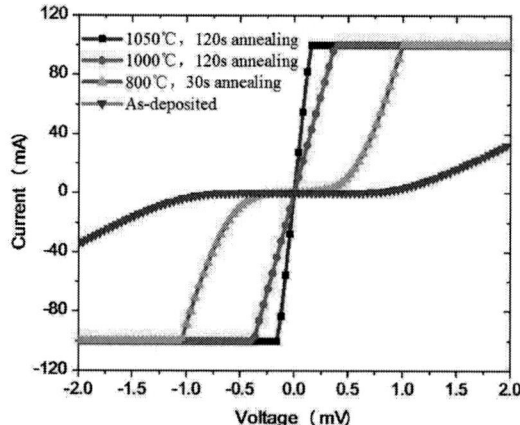

Figure 4. Ohmic contact I-V characteristics of Ni and SiC.

The semiconductor analyzer was used to apply a source voltage of -2V~2V to the electrodes to measure the current-voltage IV characteristics between the electrodes. The result is shown in Fig. 4. It is Schottky contact as deposited, after low temperature annealing (<800℃), the IV contact performance is still Schottky contact. Many previous

978-1-6654-3008-1/21 $31.00 © 2021 IEEE

experiments have proved that only when the rapid annealing temperature reach 1000 ℃, Ni can form a stable ohmic characteristic with the substrate SiC. when the temperature reaches 1000°C, the microscopic change of the interface causes the electrical properties to change from Schottky contact to ohmic contact. This shows that high-temperature annealing changes the metal-semiconductor interface properties, and within a certain range, the increase in annealing temperature can promote the conversion of the system from Schottky contact to ohmic contact.

Scanning electron microscope(SEM) and atomic force microscope (AFM) are used to observe the surface morphology and roughness of Ni ohmic contact in the as-deposited state and after annealing, as shown in Fig. 5 and Fig.6. Obviously, the surface of the as-deposited metal is smooth and uniform without obvious defects. After high-temperature annealing, granular clusters begin to appear on the metal surface. When working in a high-temperature environment, we speculate that such clusters will further expansion and may eventually lead to tearing and wrinkling of the metal layer, which may be one of the important factors for the ohmic contacts failure at high temperatures. Although the electrical performance is guaranteed after thermal annealing, it still causes certain defects and damage to the metal material itself. Therefore, in the future research, we propose a scheme of deposition multiple layers of metal to cover and protect the contact layer and reduce the damage to the contact surface caused by high temperature annealing.

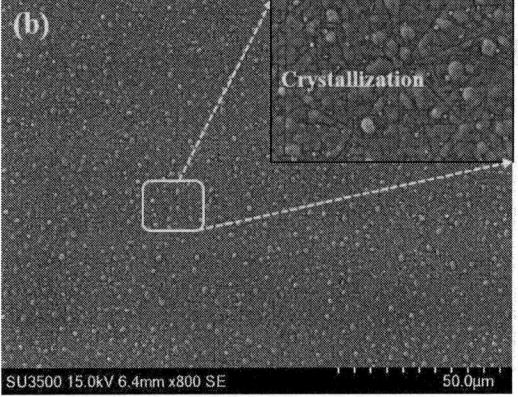

Figure 5. Topography of Ni contact electrode.
(a) As deposited; (b) After high temperature annealing.

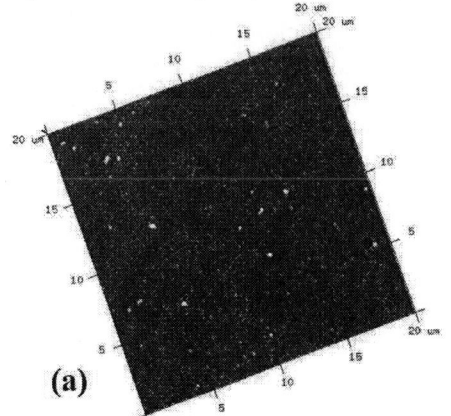

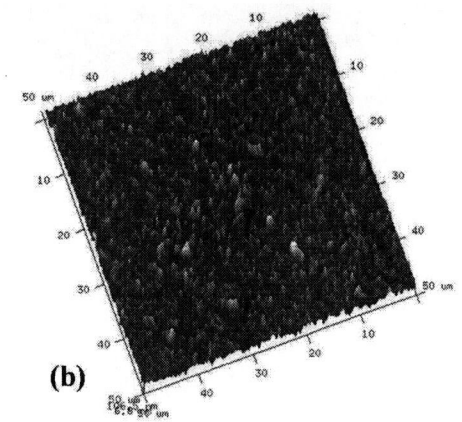

Figure 6. Surface roughness of Ni contact electrode.
(a) As deposited; (b) After high temperature annealing.

V. CONCLUSION

A conclusion section is not required. Ohmic contact is widely needed in many microelectronics and MEMS devices. It has been widely studied in recent years, especially in high temperature components. However, there are still some challenges, such as the lack of physical models, low process repeatability, and poor high temperature stability, and so on. Based on the above challenges, this work systematically expounds the physical model of tunneling to form SiC ohmic contact, and increasing the doping concentration can make SiC ohmic contacts easier to form. Taking the SiC pressure sensor as an example, an ohmic contact sample with Ni as the contact metal was prepared, and through continuous attempts at annealing parameters, experimental results show that Ni may form ohmic contact with high doping concentration 4H-SiC when temperature exceeds 800°C, moreover, annealing parameters of nitrogen, 1000°C, 120s is finally determined to form stable Ni metal ohmic contact with N type 4H-SiC. However, the SEM observation results show that the metal layer appears clusters and precipitation phenomena after annealing. Therefore, it is possible to prevent the contact layer metal from being damaged by adding other covering metals, such as Pt, Au, etc.

ACKNOWLEDGMENT

The authors would like to thank the support from National Natural Science Foundation of China (No. 51703180), China Postdoctoral Science Foundation (No.2017M610634), the Fundamental Research Funds for the Central Universities (No. xpt012020006, xjj2017024), Shaanxi Postdoctoral Science Foundation (No.2017BSHEDZZ73), National Key Research & Development (R&D) Program of China (Grant No. 2016YFB0501600), 111 Program (No. B12016) and the Recruitment Program of Global Experts (Grant No. WQ2017610445).

REFERENCES

[1] Hoang-Phuong Phan, Dzung Viet Dao, Koichi Nakamura, Sima Dimitrijev, and Nam-Trung Nguyen, Journal of Microelectromechanical Systems ,2015.

[2] Jin, S., X. A. Fu, and M. Mehregany. "Ohmic contacts on n-type polycrystalline silicon carbide with Ti/TaSi2/Pt." TRANSDUCERS 2009-2009 International Solid-State Sensors, Actuators and Microsystems Conference. IEEE, 2009.

[3] Okojie R S , Lukco D , Nguyen V , et al. 4H-SiC Piezoresistive Pressure Sensors at 800 °C With Observed Sensitivity Recovery. IEEE Electron Device Letters, 2015, pp.174-176.

[4] Larger R , Frechette L G . "Very high temperature (800°C) ohmic contact of Au/Ni2Si on n-type polycrystalline silicon carbide aged in air." Solid-state Sensors, Actuators & Microsystems Conference IEEE, 2011.

[5] Barda B , Petr Machá, Marie Hubiková. "Ti and Ti/Sb ohmic contacts on n-type 6H–SiC." Microelectronic Engineering. 2008, pp. 2022-2024.

[6] Hallin C , Yakimova R , B. Pécz, et al. "Improved Ni ohmic contact on n-type 4H-SiC." Journal of Electronic Materials. 1997, pp.119-122.

[7] Guo, Hui , Y. Zhang , and Microelectronic School Xidian University, Xi. "Fabrication of Ti-Al Ohmic Contacts to N-type 6H-SiC with P+ Ion Implantation." The Sixth International Workshop on Junction Technology, 2006.

[8] Cole M W , Joshi P C , Hubbard C , et al. "Thermal stability and performance reliability of Pt/Ti/WSi/Ni ohmic contacts to n-SiC for high temperature and pulsed power device applications." Journal of Applied Physics. 2002.

[9] Kuchuk A V , Guziewicz M , Ratajczak R , et al. "Thermal degradation of Au/Ni2Si/n-SiC ohmic contacts under different conditions." Materials Science & Engineering: B (Advanced Functional Solid-State Materials), 2009, pp.38-41.

[10] Yu H , Zhang X , Shen H , et al. "Thermal stability of Ni/Ti/Al ohmic contacts to p-type 4H-SiC." JOURNAL OF APPLIED PHYSICS, 2015, pp. 1179-317.

[11] Basak, D. , and S. Mahanty . "Ti/Ni/Ti/Au ohmic contact to n-type 6H-SiC." Materials Science & Engineering B (Solid-State Materials for, Advanced Technology) . 2003, pp.177-180.

[12] Laariedh F , Lazar M , Cremillieu P , et al. "The role of nickel and titanium in the formation of ohmic contacts on p-type 4H–SiC." Semiconductor Science and Technology, 2013.

[13] Li, Yan Liang, et al. "Extremely Thermal Stable Ni/W/TaSi2/Pt Simultaneous Ohmic Contacts to N-Type and P-Type 4H-SiC." Materials Science Forum. Vol. 924. Trans Tech Publications Ltd, 2018.

[14] Xudong F, Chen W, et al. A 350°C piezoresistive n-type 4H-SiC pressure sensor for hydraulic and pneumatic pressure tests. Journal of Micromechanics and Microengineering. 2020.

[15] Delucca, J. M, and S. E. Mohney . "Approaches to High Temperature Contacts to Silicon Carbide." MRS Proceedings, 1996, pp.137.

[16] Ervin M H , Jones K A , Lee U , et al. "Approach to optimizing n-SiC Ohmic contacts by replacing the original contacts with a second metal." Journal of Vacuum Science & Technology B (Microelectronics and, Nanometer Structures), 2006, pp.1185-1189.

[17] Okojie R S , Lukco D . "Simultaneous ohmic contacts to n-type 4H-SiC by phase segregation annealing of co-sputtered Pt-Ti." Journal of Applied Physics, 2016, pp. 215-301.

[18] M S Shur. Physics of Semiconductor Devices, 1969.

[19] SHUR M, RUMYANTSEV S, LEVINSHTEIN M. SiC Materials and Devices. WORLD SCIENTIFIC, 2008

Proceedings of the 16th Annual IEEE International
Conference on Nano/Micro Engineered and Molecular Systems
April 25-29, 2021

Facile fabrication of NiO wrinkle micro/nanostructures and the application of enzyme-free glucose sensors

Shulan Jiang[1*], Yong Tan[2]

Abstract—Enzyme-free glucose sensors are of great significance for monitoring patients' blood glucose, but the fabrication of high performance sensors is still challenging. In this study, we have fabricated NiO wrinkle micro/nanostructures through UV/Ozone treatment of uncured PDMS and then sputtering of NiO/Cu films. The obtained NiO wrinkle structures were demonstrated as flexible enzyme-free glucose sensors and the performances have been evaluated. The glucose sensor shows good electrochemical performance, and the sensitivity is about 655.8 $\mu A \cdot mmol^{-1} \cdot cm^{-2}$. The proposed strategy of integration of active materials on flexible wrinkle structures has potential in improving flexible sensor performance.

I. INTRODUCTION

Diabetes mellitus is a serious disease that has endangered our health because of the increased level of glucose concentration in blood [1]. The monitor of glucose concentration is of great significance. Researchers are pursuing low-cost and user-friendly glucose monitoring sensors using various types of transducers such as thermal, optical, electrochemical, and acoustic, etc [2]. Conventional electrochemical glucose biosensors entail the use of enzymes such as glucose oxidase or glucose hydrogenase, and the working mechanism of the sensors is based on the enzymatic reaction [3]. The enzyme-based glucose sensors still present some shortcomings arisen from the intrinsic nature of enzyme molecules such as rigorous operating conditions, complicated immobilization techniques, a certain chemical instability and high cost [4-5]. So enzyme-free glucose sensors have attracted great attentions for the rapid electron transfer and reducing environmental interferences [6].

Precious metals and their alloys exhibit high catalytic activity towards electro-oxidation of glucose, while the high costs will restrict the real practical applications. Other alternatives based on metals like Ni and Cu and metal oxides such as CuO, Co_3O_4 and NiO have been studied as electrocatalysts sensors of glucose [7]. Particularly, nickel and nickel oxide electrodes are good candidates as electrocatalysts for enzyme-free glucose sensor because of their good stability and reproducibility [8]. Ni-based nanomaterials show high catalytic activity for glucose due to the redox couple $Ni(OH)_2/NiOOH$ in alkaline medium while glucose directly oxidizes to glucolactone [9]. The NiO nanofilm sputtered on micro/nanostructures with large surface area will increase the electrocatalytic active sites thus enhance the sensor performance.

In this study, we have proposed the fabrication of enzyme-free glucose sensor by UV/Ozone treatment of uncured PDMS, then curing of PDMS and sputtering of NiO/Cu nanofilms. The morphologies of the NiO wrinkle structures were characterized and the electrochemical performances were evaluated. The integration of NiO nanofilm on the wrinkle structures is beneficial for the glucose sensors.

II. EXPERIMENTAL SECTION

The PDMS (Sylgard 184, Dow and Corning, USA) precursor was prepared by mixing base and crosslinking agent in ratio of 10:1. After degassing in a vacuum kettle, the precursor mixtures were spin coated onto the glass substrate ($12 \times 12 \times 1$ mm^3) for 60 s at 1000 rpm. Then the samples were standing for 15 min to avoid the flow of the PDMS precursor on the substrate. The uncured PDMS samples were then exposed to 184.9 nm and 253.7 nm UV source (PSDP UV 8T, Novascan, America) for 6 min, (shown in Figure 1). Next, the treated samples were cured by heating them on hotplate for 60 min with various curing temperatures of 90°C. The self-wrinkle structures were fabricated by thermal polymerization and cross-linking of UV/ozone treated uncured PDMS samples, and then peeled off the PDMS samples. Cu nanofilm was firstly sputtered on the self-wrinkle structures as conductive layer. Then NiO nanofilm was integrated onto the self-wrinkle structures for enzyme free glucose sensing. The nanofilms were all sputtered by magnetron sputtering coating system (TPR-450). The bare PDMS was also sputtered with NiO/Cu nanofilms for comparison. The morphologies of the wrinkle surfaces were observed by atomic force microscope (AFM) and optical three-dimensional (3D) surface profiler (SuperView W1).

The wrinkle structures integrated with NiO/Cu films were used as working electrode for enzyme-free glucose sensor, with Ag/AgCl and Pt plate as reference electrode and counter electrode, respectively. The electrochemical performances were evaluated by Cyclic Voltammetry (CV) and Constant Potential Amperometry (CPA) techniques using electrochemical workstation (CHI660).

---

*Research supported by the National Science Foundation of China (No. 51775458).

Dr. Shulan Jiang is now with School of Mechanical Engineering and Electric Information, China University of Geosciences, Wuhan, 430074 China. (Corresponding author, 86-027-67883273; e-mail: jiangshulan@cug.edu.cn).

Yong Tan was with School of Mechanical Engineering, Southwest Jiaotong University, Chengdu, 610031, China. (e-mail: 805889724@qq.com).

978-1-6654-3008-1/21 $31.00 © 2021 IEEE

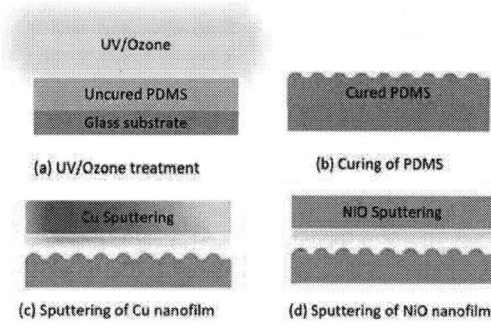

Figure 1. Typical fabrication procedure of wrinkle structures

III. RESULTS AND DISCUSSION

When uncured PDMS was exposed to UV light and oxidized by the UV/ozone system which could continuously generate atomic oxygen and ozone, a thin stiff SiO_x layer formed due to the photo-oxidation reaction [10]. Then, the PDMS sample was cured by heating on a hotplate and it would shrink because the cross-linking of the monomer and the reducing of the total volume during curing. So the self-wrinkle structures were formed by the interface mismatch between bottom cured PDMS and top stiff SiO_x layer to generate surface self-wrinkle patterns autonomously.

The effect of NiO nanofilm deposition time on the morphologies of wrinkle structures was discussed and the results are shown in Figure 2. When sputtered for 5 min and 10 min, NiO nanofilms on self-wrinkle PDMS substrate show the integration of labyrinth wrinkles and self-wrinkles of the substrate. The results show difference morphologies when sputtered for 20 min, which exhibit three kinds (labyrinth wrinkles, hump wrinkles and self-wrinkles) of wrinkles nesting. We have found the integration of NiO nanofilm on bare PDMS substrate show various patterns (as shown in the three-dimensional profiles). When sputtered for 5 min and 10 min, the patterns show labyrinth wrinkles. The patterns show the nesting of labyrinth wrinkles and hump wrinkles when sputtered for 20 min. It's suggested that the labyrinth and hump wrinkles were formed by the compressive stress generated by the cooling process of NiO/Cu nanofilm-substrate system to room temperature.

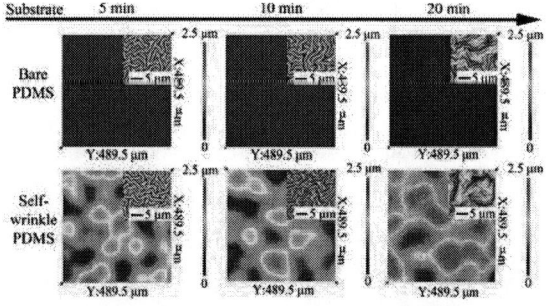

Figure 2. The 3D profiles and AFM images of the wrinkle structures sputtered for 5 min, 10 min and 20 min, respectively.

The relationship between sputtering time and amplitude was shown in Figure 3. The amplitudes of the NiO wrinkle structures formed on bare PDMS substrate and self-wrinkle PDMS substrate all increase with sputtering time. For the NiO wrinkle structures formed on bare PDMS substrate, the amplitude increases from 26.28 nm to 58.19 nm. While for the NiO wrinkle structures formed on self-wrinkle PDMS substrate, the amplitude increases from 22.71 nm to 67.43 nm.

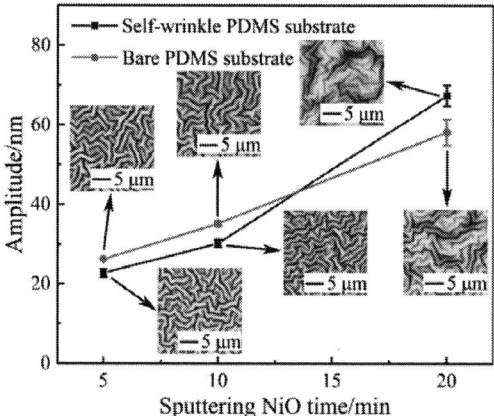

Figure 3. The relationsheep between amplitude and sputtering time. The insets are the AFM images of the wrinkle surface.

The electrochemical performances were evaluated by cyclic voltammetry technique, and the results are shown in Figure 4. The samples are tested in 0.1 M NaOH at different scan rates. The CV curves show increased cyclic current as the increase of scan rates. The oxidation peak current density increases from 0.1 mA cm^{-2} to 0.75 mA cm^{-2}, the reduction peak current density increased from 0.08 mA cm^{-2} to 0.6 mA cm^{-2}. It shows the reaction is a surface controlled process.

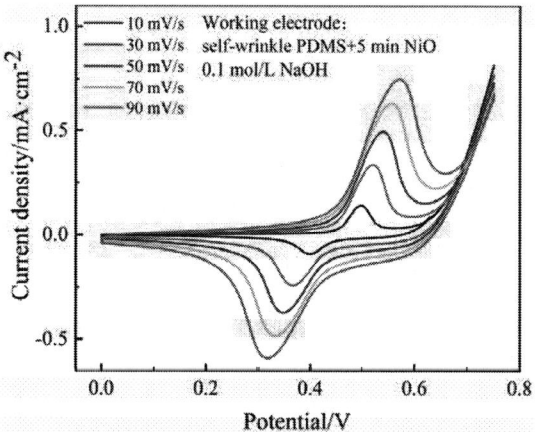

Figure 4. CV curves of the electrode with NiO nanofilm sputtered self-wrinkle PDMS substrate at different scan rates.

The CV curves of the glucose sensors testing in 0.1 M NaOH with different glucose concentrations were shown in Figure 5(a). With the increases of glucose concentration, the Ni^{2+} were oxidized to Ni^{3+} and the peak current also increases. The peak voltage was between 0.3 V~0.6 V, which is the same with the literature reported of 0.55 V [11].

978-1-6654-3008-1/21 $31.00 © 2021 IEEE   340

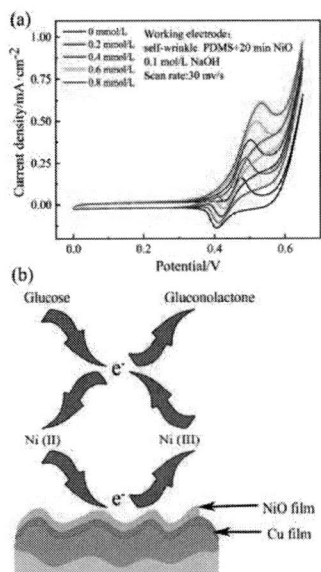

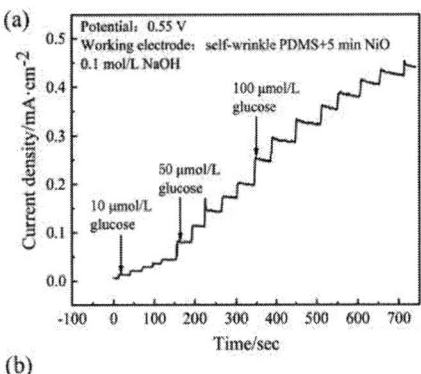

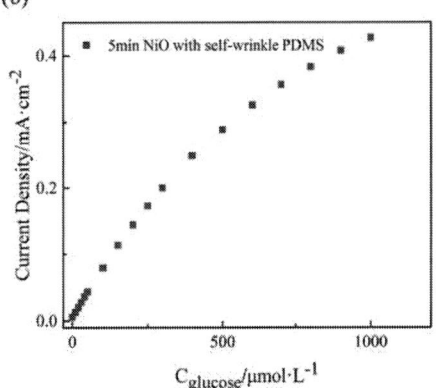

Figure 5. (a) CV curves of the electrode with NiO nanofilm sputtered self-wrinkle PDMS substrate at different glucose concentrations. (b) The reaction mechanism of the NiO film on wrinkle structures with glucose.

The reaction mechanism of NiO to glucose occurs according to the following reactions [12].

$$NiO + H_2O \rightarrow NiO \cdot H_2O \rightarrow Ni(OH)_2 \quad (1)$$

$$Ni(OH)_2 + OH^- \leftrightarrow NiOOH + H_2O + e^- \quad (2)$$

The NiO surface will converts into the hydrated α-$Ni(OH)_2$ and anhydrous β-$Ni(OH)_2$ forms when immersion in an alkaline electrolyte firstly according to Eq.1 [13]. Then the $Ni(OH)_2$ species will convert to their respective oxyhydroxide species (γ-NiOOH and β-NiOOH) when the electrode is cycled between ca. 0 and 1 V (c.f., Eq. (2)).

Figure 6(a) displays the amperometric current-time curve obtained with the self-wrinkle PDMS structure   n optimized sensing conditions (i.e., 0.1 mol L^{-1} NaOH and +0.55 V) upon successive additions of glucose. The electrode response (i.e., less than 5 s to achieve 95% of the steady-state) was almost instantaneous, which showed good performance. The fast response was attributed to the high number of nickel electrocatalytic centers available on the NiO/wrinkle electrode surface to react with glucose. Even for glucose concentrations as small as 10 μmol L^{-1}, a good sensor response was obtained. Figure 6(b) shows the calibration curve for current density vs. concentration of glucose for the electrode sputtering with 5 min of NiO nanofilm. The sensitivity of the wrinkle structures based electrode was 655.8 μA·$mmol^{-1}$·cm^{-2}. We have compared the performance of the electrode, which was sputtered with Cu film as conductive layer and NiO nanofilm on flat PDMS substrate. The results show that the sensitivity is 512.7 μA·$mmol^{-1}$·cm^{-2}, which is not as good as that of the NiO nanofilm on self-wrinkle structures.

Figure 6. (a) Amperometric response of the working electrode to successive additions of a certain solution of glucose in 0.1 mol/L NaOH at the voltage of 0.55 V. (b) Calibration curve for current density vs. concentration of glucose for the electrode sputtering with 5 min of NiO nanofilm.

IV. CONCLUSION

NiO/PDMS wrinkle structures were fabricated by UV/Ozone treatment of uncured PDMS and then sputtering of NiO/Cu nanofilms. The electrochemical performance showed advantages compared with the NiO nanofilm sputtered on the flat PDMS surface. The sensitivity of the NiO nanofilm/PDMS wrinkle structures is about 655.8 μA·$mmol^{-1}$·cm^{-2}. The proposed method is low-cost and facile and the obtained structures have potential for flexible glucose sensors.

The discussion of the thickness of NiO nanofilm to the sensor performance will be studied in the future. And the stress formed on the substrate when sputtering of metal films will be considered for further study.

ACKNOWLEDGMENT

This work was supported by the National Science Foundation of China (No. 51775458).

REFERENCES

[1] B. Insight, "American Diabetes Association". June 2013, 21 (2013) 31.

[2] D.W. Hwang, S. Lee, M. Seo, T.D. Chung, "Recent advances in electrochemical non-enzymatic glucose sensors-a review". *Analytica Chimica Acta*, 1033 (2018) pp: 1-34.

[3] J. Wang, "Electrochemical glucose biosensors". *Chemical Reviews,* 108 (2008) pp. 814-825.

[4] K.E. Toghill, R.G. Compton, "Electrochemical non-enzymatic glucose sensors: a perspective and an evaluation". *International Journal of Electrochemical Science,* 5 (2010), pp: 1246-1301.

[5] X.H. Kang, Z.B. Mai, X.Y. Zou, P.X. Cai, J.Y. Mo, "A sensitive nonenzymatic glucose sensor in alkaline media with a copper nanocluster/multiwall carbon nanotube-modified glassy carbon electrode". *Analytical Biochemistry,* 363 (2007), pp: 143-150.

[6] S. Shahrokhian, E. Khaki Sanati, H. Hosseini, "Advanced on-site glucose sensing platform based on a new architecture of free-standing hollow Cu(OH)$_2$ nanotubes decorated with CoNi-LDH nanosheets on graphite screen-printed electrode". *Nanoscale,* 11 (2019) pp: 12655-12671.

[7] Y. Mu, D. Jia, Y. He, Y. Miao, H.L. Wu, "Nano nickel oxide modified non-enzymatic glucose sensors with enhanced sensitivity through an electrochemical process strategy at high potential". *Biosensors and Bioelectronics,* 26 (2011), pp: 2948-2952.

[8] A. Sun, J. Zheng, Q. Sheng, "A highly sensitive non-enzymatic glucose sensor based on nickel and multi-walled carbon nanotubes nanohybrid films fabricated by one-step-co-electrodeposition in ionic liquids". *Electrochimica Acta,* 65 (2012), pp: 64-69.

[9] H. Tian, M. Jia, M. Zhang, J. Hu, "Nonenzymatic glucose sensor based on nickel ion implanted-modified indium tin oxide electrode". *Electrochimica Acta,* 96 (2013), pp: 285-290.

[10] E. Yilgor, O. Kaymakci, M. Isik, S. Bilgin, I. Yilgor, "Effect of UV/ozone irradiation on the surface properties of electrospun webs and films prepared from polydimethylsiloxane-urea copolymers". *Applied Surface Science,* 258(10) (2012) pp: 4246-4253.

[11] Ghanbari, K. and Z. Babaei, "Fabrication and characterization of non-enzymatic glucose sensor based on ternary NiO/CuO/polyaniline nanocomposite". *Analytical Biochemistry,* 2016, 498, pp: 37-46.

[12] F.J. Garcia-Garcia, P. Salazar, F. Yubero, A.R. González-Elipe, Non-enzymatic glucose electrochemical sensor made of porous NiO thin films prepared by reactive magnetron sputtering at oblique angles. *Electrochimica Acta,* 201 (2016) pp: 38-44.

[13] H. Bode, K. Dehmelt, J. Witte, "Zur kenntnis der nickelhydroxidelektrode. II. Über die oxydationsprodukte von nickel(II)-hydroxiden". *Zeitschrift Fur Anorganische Und Allgeneine Chemie.* 366 (1969) pp: 1-21.

Gap in pagination due to unavailable paper.

Pages 343-344

Proceedings of the 16th Annual IEEE International
Conference on Nano/Micro Engineered and Molecular Systems
April 25-29, 2021

High-efficient Electrochemical Betavoltaic Cell Using ZrO₂ Modified TiO₂ Nanorod Arrays

Renrong Zheng[1,2], Zhen Wang[1,2], Na Wang[1,2], Zan Ding[1], and Haisheng San[1,2,*]

Abstract-This paper presents a novel and high-efficient EBC based on free-standing TNRAs sensitized with ZrO₂. A sandwich type electrochemical betavoltaic cell (EBC) was assembled using a 10 mCi of ^{63}Ni/Ni planar sheet as the counter electrode, ZrO₂/Ar-TNRAs/FTO as working electrode filled by acetonitrile electrolyte containing I⁻/I₃⁻ ions. The EBCs demonstrate an effective betavoltaic effect with open-circuit voltage of 0.276 V, short-circuit current density of 1.041 μA·cm⁻² and the energy conversion efficiency of 9.27%. The experimental results indicate that the interface electrochemical reaction can effectively enhance betavoltaic effect.

I. INTRODUCTION

In recent years, the Nano-Electro-Mechanical-System (NEMS) take on a rapid development and application, but some special applications in remote and inaccessible locations are still limited due to the lack of high-efficient and long-life micropower sources.[1] Radioisotope cells, which involve the generation of power by coupling a beta source to an energy conversion structure, are an ideal choice for NEMS power supply due to its characteristics of miniaturization, high energy density, long-life and insensitivity to environment and temperature [2-4]. For converting radioactive decay energy into electrical energy, it is believed that the direct energy conversion mechanism based on betavoltaic effect is a promising technique for powering the microsystems [5,6]. Betavoltaic cells can converter the energy of beta particles emitted from radioisotopes to electricity when the beta particles interact with a semiconductor *p-n* junction to create electron-hole pairs (EHPs) that are drawn off as current [3]. Presently, the main problem of betavoltaic cells is the low energy conversion efficiency (ECE) (≤5%) [7], which limits their practical application. For this reason, Kwon et.al. demonstrated a water-based nuclear cell using ^{90}Sr/^{90}Y radiation source for the generation of electricity [8]. Su et al. developed a dye-sensitized betavoltaic battery (DSBS) using C-14 beta-radiation source for the first time [9]. These nuclear cells based on electrochemical mechanism provide a new strategy in the development of betavoltaic cells.

In this work, we first demonstrate a novel and high-efficient electrochemical betavoltaic cell (EBC) using TiO₂ nanorod arrays grown on FTO/glass (TNRAs/ FTO/glass) as anodes, a ^{63}Ni/Ni sheet as counter electrode

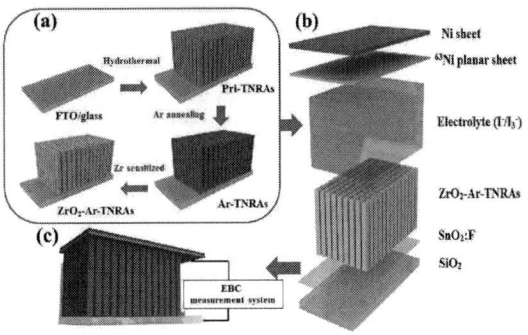

Figure 1. Schematic illustration of fabrication processes of EBC based on ZrO₂/Ar-TNRAs. (a) Growth of TNRAs and thermal treatment; (b) 3-D constitution and (c) 3-D assembled diagrams of EBC.

and acetonitrile electrolyte containing I⁻/I₃⁻ ions. TNRAs, fabricated by direct hydrothermal growth on FTO substrate, which are an excellent candidate for EBCs due to their wide bandgap. large surface area, excellent chemical stability, and cost-effective fabrication. By the modification of ZrO₂ on TNRAs, the EBCs can achieve a significant increase of ECE when compared to other results reported.

II. MATERIALS AND METHODS

A. Synthesis of materials and fabrication of devices

Fig. 1 shows the fabrication and assembly processes of the EBC based on Ni/^{63}Ni/electrolyte/ZrO₂-Ar-TNRAs/FTO /glass structure. The TNRAs was synthesis on F-doped SnO₂ (SnO₂:F) thin film on glass substrate (FTO/glass) using a one-step hydrothermal method followed by a thermal annealing treatment, as shown in Fig. 1(a). The commercial electroconductive glass FTO (FTO/glass, 20 mm×20 mm×2.2 mm, ≤14 Ω/ square, Nanotech, China) were ultrasonic cleaning in isopropanol, ethanol, and acetone (1:1:1) mixed solution for 1 h. For hydrothermal process, the FTO was put into a 30 mL of Teflon-lined stainless-steel autoclave with a mixed solution consisting of 0.4 mL Tetrabutyl titanate (C₁₆H₃₆O₄Ti), 12 mL Hydrochloric acid (HCl, 36 wt%) and 12 mL deionized water at 150°C for 20 h. The prepared pristine TNRAs (Pri-TNRAs) sample were cleaned using deionized water and absolute ethanol, and then the Pri-TNRAs samples were annealed in different atmosphere (Air, Ar or 5%H₂/Ar) at 450 °C for 1 h to obtain Air-TNRAs, Ar-TNRAs and H₂-TNRAs samples (Fig. 2(a)).

As-prepared Ar-TNRAs were immersed into a mixed solution with 50 mL of 0.1 M Zirconyl nitrate [ZrO(NO₃)₂] and 1 mL of polyethylene glycol by ultrasonic treatment for

---

[1]Pen-Tung Sah Institute of Micro-Nano Science and Technology, Xiamen University, Xiamen 361005, Fujian, China.

[2]Shenzhen Research Institute of Xiamen University, Shenzhen 518000, Guangdong, China.

*Corresponding Author: Haisheng San is with the Xiamen University, Xiamen, China (phone: +86-592-2181340; E-mail: sanhs@xmu.edu.cn).

978-1-6654-3008-1/21 $31.00 © 2021 IEEE

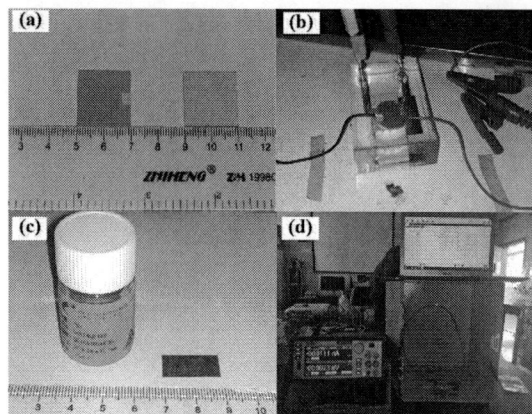

Figure 2. (a) A comparison of Ar-TNRAs/FTO/glass and FTO/glass; (b) Photographs of UV detector based on Ni/electrolyte/ZrO₂ -Ar-TNRAs/ FTO structure, (c) ^{63}Ni sheet and (d) EBC test system.

30 min. Next, the ammonia solution ($NH_3 \cdot H_2O$) was dropwise-added into the reaction solution to adjust the pH to 9. The duration of immersion was 25 min, and then the samples were heated in air at 50°C for 24 h. The alkalescency feature of $NH_3 \cdot H_2O$ is hardly to corrode the glass. The obtained samples with the sensitization of ZrO_2 on the Ar-TNRAs is marked as ZrO_2-Ar-TNRAs.

To assembly the EBCs, a 10 mCi of Nickel-63 (^{63}Ni) sheet (^{63}Ni/Ni, 10 × 20 mm², HTA, China, Fig. 2(c)) is used as β-radioisotopes source. The ^{63}Ni/Ni sheet was assembled on the surface of TNRAs/FTO/glass, and the organic electrolyte was filled into the micro-gap between the anode and counter electrode to form the Ni/^{63}Ni/electrolyte/ TNRAs/FTO sandwich structure. The organic electrolyte using acetonitrile as the solvent consist of 0.5 M of LiI, 0.25 M of I_2, 0.5 M of 4-tert-butylpyridine and 0.6 M of 1,2-Dimethyl-3-propylimidazo-lium iodide. Fig. 1(b) shows a schematic diagram of 3-D configuration of the EBC device. By mechanically clamping the anode and counter electrode together, an excellent electrical connect was achieved between TNRAs nanostructures and ^{63}Ni/Ni sheet. A Ni sheets (10 mm × 20 mm × 0.3 mm) was also assembled on the top surface of the TNRAs samples to form Ni/electrolyte/TNRAs/FTO devices for a compared study. To avoid the leakage of organic electrolyte, the devices were sealed by epoxy resin pouring sealant. The active area of the EBC is 10 mm×15 mm (Fig. 1(c)).

B. Materials characterizations and device measurements

X-ray diffraction (XRD) analysis were performed on X-ray diffractometer (Rigaku Ultima IV, Japan) by using Cu Kα radiation at 40 kV and 40 mA (λ = 1.5416 Å). The morphology and composition of the fabricated samples were investigated using a scanning electron microscopy (SEM, ZEISS microscope, Germany). The UV response measurements were conducted at zero bias and under the 2.90 mW·cm⁻² of UV radiation with a 10 s interval of on/off cycle. The UV response characteristics of the fabricated devices were characterized using an electrochemical workstation (Chenhua CHI660E, China) under room temperature (Fig. 2(b)). I-V measurements for the EBCs

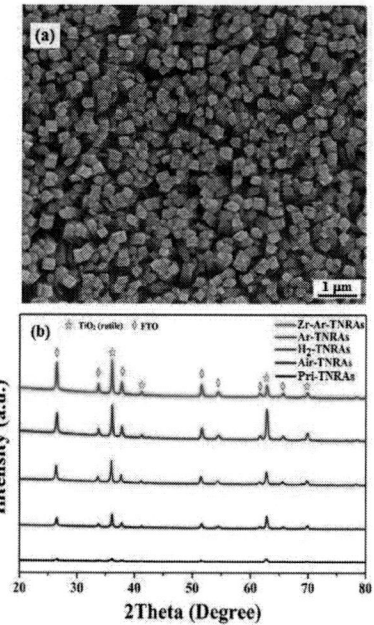

Figure 3. (a) top-view and partial enlarged FESEM images of and (b) XRD spectra of TiO₂ nanorod arrays (TNRAs) respectively.

were performed in a Faraday cage by using a Keithley Model 2450 source meter (Fig. 2(d)).

III. RESULTS AND DISCUSSION

The top-view SEM image of the TNRAs was shown in Fig. 3(a). The ZrO₂-Ar-TNRAs exhibits well-aligned and ultrahigh-density of TiO₂ nanorod arrays with better vertical orientation to FTO substrate. The TiO₂ nanorods have a tetragonal morphology with 1.5 ~ 1.6 μm in length and 180 ~ 240 nm in diameter. XRD patterns of the ZrO₂-Ar-TNRAs, Ar-TNRAs, H₂-TNRAs and Pri-TNRAs are shown in Fig. 3(b). The 2θ signals can be indexed to the tetragonal rutile phase of TiO₂ structure (JCPDS#71-0650). Furthermore, the main (100) diffraction peak of the rutile phase TiO₂ was observed in all samples, and the (101) diffraction peaks of the anatase phase TiO₂ was not observed, meaning that TRNAs samples were pure rutile phase structure. On account of the content of ZrO₂ is too less to be detected by X-ray. All samples still retain the pure rutile phase through different atmosphere annealing process, but the intensity of Ar-TRNAs diffraction peaks has a slightly weakening, which may be ascribed to the lattice strain and slightly shrink of crystallite size due to the formation of oxygen vacancies (O_V) and Ti^{3+} defects.

A cost-effective approach to investigate and optimize the performance of EBCs is to measure the devices by means of UV irradiation instead of beta irradiation. The device performance was investigated through time-dependent photoresponse measurements using 2.90 mW·cm⁻² of UV irradiation on the FTO/glass window (Fig.4(a)). The photocurrent can be periodically generated with a 10 s interval of on/off cycle of UV radiation, indicating a good stability in output performance. The

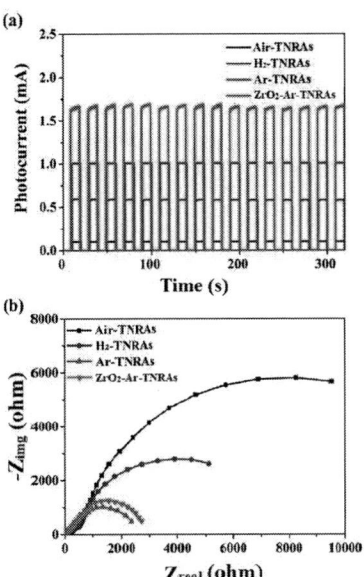

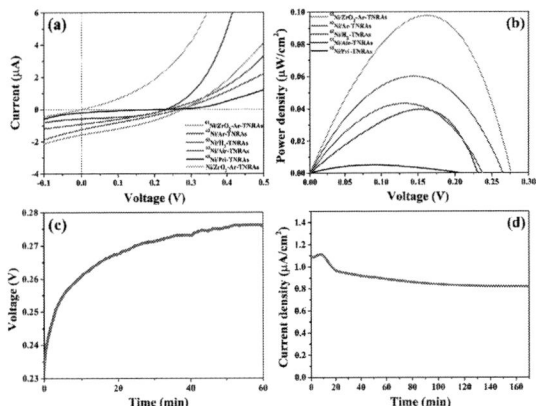

Figure 5. (a) I-V and (b) P-V characteristics of EBC based on Ni/[63]Ni /electrolyte/TNRAs/FTO structure and Ni/electrolyte/ZrO_2-Ar- TNRAs /FTO structure; Long time (c) V-T and (d) J-T characteristics of EBC based on Ni/[63]Ni/electrolyte/ZrO_2-Ar-TNRAs/FTO.

Figure 4. (a) A comparison of time-dependent photoresponse properties and (c) Nyquist plots of all devices.

device based on ZrO_2-Ar-TNRAs structure has a highest photocurrent of ~ 1.60 mA than that of the others. Electrical properties of TRNAs are characterized by electrochemical impedance spectroscopy (EIS). Fig. 4(b) depict the Nyquist plots of all devices at zero-bias and in dark conditions. The Ar-TNRAs structure exhibits the lowest impedance of photoanode/electrolyte interface due to the existence of Ti^{3+} and O_V defects beneficial for the suppression of EHPs recombination and thus the increase of the performance of EBC. [10-11] It is demonstrated that a low extent of ZrO_2 sensitized on TNRAs has little effect on the interface impedance. [12, 13]

Table I shows the measured results of all electrical parameters of EBCs. When compared with the EBC based on Ar-TNRAs structure (ECE = 5.70%, P_{max}= 60.1 nW·cm⁻², the EBC based on ZrO_2-Ar-TNRAs structure exhibits a highest ECE of 9.27% with P_{max}= 97.7 nW·cm⁻². Furthermore, it needs to note the radiation damage of

acetonitrile electrolyte induced by the beta radiation [14]. Decomposition of acetonitrile may occur by the beta radiation, resulting in intramolecular reduction due to the generation of carbon-centered [15]. Therefore, the EBCs used in our research were performed a stability and reliability evaluation, as shown in Fig. 5(c)-(d). The V_{oc} increase from 0.23 V in initial stage and then trend to reach a stable value in 0.275 V after long time of work. At the same time, the J_{sc} has a slight decrease in initial stage and then trend to reach a stable value at ~ 0.832 μA·cm⁻² after 160 min test.

A schematic diagram of working mechanisms of EBC based on ZrO_2-Ar-TNRAs nanostructure is depicted in Fig. 6. The band gap of ZrO_2 is about 5 eV, which is much larger than that of rutile TiO_2 (3.2 eV), and when beta particles are radiated onto ZrO_2/TNRAs, beta-excited EHPs are generated in both ZrO_2 and TiO_2. The potential of ZrO_2 conductive band (CB) is more negative than that of TiO_2, which will form a matching type-II energy level structure at the TiO_2/ZrO_2 interface [16]. Some of the beta-excited electrons will migrate from the CB of ZrO_2/TNRAs to the FTO glass and finally move to the Ni counter electrode through external circuit. Such CB structure will inhibit the

TABLE I. ELECTRICAL PARAMETERS OF EBC BASED ON DIFFERENT WORKING ELECTRODES

Beta Source:Anode structures of EBCs	Electrical Parameters of EBCs					
	Open-circuit voltage V_{oc} (V)	Short-circuit current I_{sc} (μA)	Short-circuit current density J_{sc} (μA·cm²)	Max power density P_{max} (nW·cm²)	Filling factor FF	Total efficiency η (%)
Ni:ZrO_2-Ar-TNRAs/FTO	0	0	0	/	/	/
[63]Ni:Pri-TNRAs/FTO	0.208	0.198	0.132	50.0	0.182	0.47
[63]Ni:Air-TNRAs/FTO	0.230	0.573	0.382	39.7	0.452	3.77
[63]Ni:H_2-TNRAs/FTO	0.237	0.898	0.599	43.3	0.305	4.11
[63]Ni:Ar-TNRAs/FTO	0.265	1.239	0.826	60.1	0.275	5.70
[63]Ni:ZrO_2-Ar-TNRAs/FTO	0.276	1.562	1.041	97.7	0.340	9.27

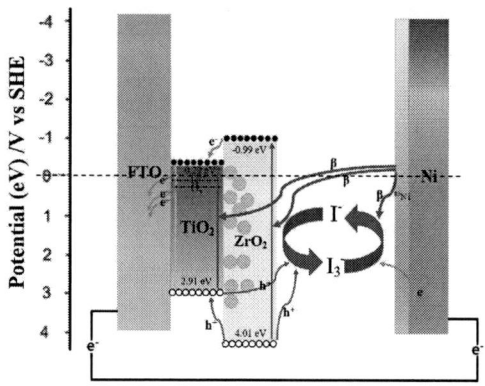

Figure 6. Schematic diagram of working mechanisms of EBCs based on ZrO$_2$-Ar-TNRAs nanostructure.

recombination of beta-excited electrons with I$_3^-$ in the electrolyte, which is beneficial to the increase of beta-current of EBC. Meanwhile, the potential of ZrO$_2$ valence band (VB) is more positive than that of TiO$_2$, which will inhibit the beta-excited holes in TiO$_2$ to move to ZrO$_2$/electrolyte interface and react with I$^-$ in the electrolyte. Thanking to the fact that the TNRAs cannot be completely covered by the ZrO$_2$ particles, some beta-excited holes can migrate the interface and undergo an oxidation reaction with I$^-$ to form I$_3^-$ in the electrolyte. These oxidized redox mediators (I$_3^-$) diffuse to the surface of the Ni/^{63}Ni/ electrolyte counter electrode interface to capture the beta-excited electrons accumulated in the electrode to form I$^-$, thereby resulting in the regeneration of redox mediators and the achievement of a complete charge transfer cycle.

IV. CONCLUSION

In conclusion, we report a novel and high-efficient EBC based on free-standing TNRAs sensitized with ZrO$_2$. The TNRAs were grown on FTO conductive glass using hydrothermal method, which were sensitized with ZrO$_2$ by precipitation method. A sandwich type EBC was assembled using a 10 mCi of ^{63}Ni/Ni planar sheet as the counter electrode, ZrO$_2$-Ar-TNRAs/FTO as working electrode filled by acetonitrile electrolyte containing I$^-$/I$_3^-$ ions. The EBCs demonstrate an effective betavoltaic effect with open-circuit voltage (V_{oc}) of 0.276 V, short-circuit current density (J_{sc}) of 1.041 µA·cm^{-2}, max power density (P_{max}) of 97.7 nW·cm^{-2} and ECE of 9.27%.

ACKNOWLEDGMENT

This work was supported by the National Natural Science Foundation of China (Grant No. 61574117), the Natural Science Foundation of Guangdong Province (Grant No. 2018B030311002).

REFERENCES

[1] C. Chen, J. Chen, Z. Wang, J. Zhang, and H. San, "Free-standing ZnO nanorod arrays modified with single-walled carbon nanotubes for betavoltaics and photovoltaics", *J. Mater. Sci. Technol.*, vol.54, pp.48-57, 2020.

[2] P. Rappaport, "The Electron-voltaic effect in p-n junctions induced by Beta-particle bombardment", *Phys. Rev.*, vol.93, pp.246-247, 1954.

[3] L. Olsen, "Betavoltaic energy conversion", *Energ. Convers. Manage.*, vol.13, pp.117-127, 1973.

[4] J. Zheng, H. Dang, X. Feng, P.-H. Chien, and Y.-Y. Hu, "Li-ion transport in a representative ceramic–polymer–plasticizer composite electrolyte: Li$_7$La$_3$Zr$_2$O$_{12}$–polyethylene oxide–tetraethylene glycol dimethylether", *J. Mater. Chem. A*, vol.5, pp.18457-18463, 2017.

[5] C. L. Weaver, R. J. Schott, M. A. Prelas, D. A. Wisniewski, J. B. Rothenberger, and E. D. Lukosi, "Radiation resistant PIDECα cell using photon intermediate direct energy conversion and a ^{210}Po source", *Appl. Radiat. Isot.*, vol.132, pp.110-115, 2018.

[6] M. Lu, G.-g. Zhang, K. Fu, G.-h. Yu, D. Su, and J.-f. Hu, "Gallium nitride schottky betavoltaic nuclear batteries", *Energ. Convers. Manage.*, vol.52 pp.1955-1958, 2011.

[7] C. Chen, N. Wang, and H. San, "Electro-chemically reduced graphene oxide on well-aligned titanium dioxide nanotube arrays for betavoltaic enhancement", *ACS Appl. Mater. & Inter.*, vol.8, pp.24638-24644, 2016.

[8] B. H. Kim, and J. W. Kwon, "Plasmon-assisted radiolytic energy conversion in aqueous solutions", *Sci. Rep.-UK*, vol. 4, pp. 5249-5257, 2014.

[9] Y. Hwang, Y. H. Park, H. S. Kim, D. H. Kim, S. Ali, S. Sorcar, M. C. Flores, M. R. Hoffmann, and S.-I. In, "C-14 powered dye-sensitized betavoltaic cells", *Chem. Commun.*, vol.56, pp.7080-7083, 2020.

[10] Y. Ma, N. Wang, J. Chen, C. Chen, H. San, and J. Chen, "Betavoltaic enhancement using defect-engineered TiO$_2$ nanotube arrays through electrochemical reduction in organic electrolytes", *ACS Appl. Mater. Inter.*, vol.10, pp.22174-22181, 2018.

[11] N. Wang, Y. Ma, J. Chen, C. Chen, H. San, "Defect-induced betavoltaic enhancement in black titania nanotube arrays", *Nanoscale*, vol.10, pp.13028-13036, 2018.

[12] Kim, S. Suh, M.J. Choi, and Y. S. Kang, "Fabrication of SrTiO$_3$-TiO$_2$ heterojunction photoanode with enlarged pore diameter for dyesensitized solar cells", *J. Mater. Chem. A*, vol.1, pp.11820-11827, 2013.

[13] H. Sun, L. Pan, X. Piao, and Z. Sun, "Long afterglow SrAl$_2$O$_4$:Eu,Dy phosphors for CdS quantum dot-sensitized solar cells with enhanced photovoltaic performance", *J. Mater. Chem. A*, vol.1, pp.6388-6392, 2013.

[14] Y. Liu, X. Tang, M. Liu, Z. Xu, Z. Zhang, and M. Fang, "Effect of 10 MeV electron irradiation on dye-sensitized solar cells", *Radiat. Eff. Defect. S.*, vol.172, pp. 342-353, 2017.

[15] M. A. Rauf, and S. S. Ashraf, "Radiation induced degradation of dyes-An overview", *J. Hazard. Mater.*, vol.166, pp.6-16, 2009.

[16] R. S. Selinsky, Q. Ding, M. S. Faber, J. C. Wright, and S. Jin, "Quantum dot nanoscale heterostructures for solar energy conversion", *Chem. Soc. Rev.*, vol.42 pp.2963-2985, 2013.

Gap in pagination due to unavailable papers.

Pages 349-357

Proceedings of the 16th Annual IEEE International
Conference on Nano/Micro Engineered and Molecular Systems
April 25-29, 2021

Research on Key Technology of Novel RF MEMS Switch Designing and Fabricating

Yun Qi, Yuanyuan Yang, Wenjiang Shen*

Abstract— RF MEMS switches attract much attention in RF MEMS devices. Compared with FET and PIN switches, RF MEMS switches have much lower power consumption, lower insertion loss and higher isolation at high frequency, and can be widely used in satellite, radar, communication systems and wireless systems. This paper proposes one contact switch with single-ended cantilever beam and three kinds of capacitive switch with flexural hinge structure. The key technologies of structure design, simulation, fabrication process and test analysis are studied. COMSOL and ANSYS HFSS are used as the tools for analysis, in order to optimize the switch structure. This paper introduces the layout design and fabrication process. A method of constructing the top electrode and coplanar waveguide simultaneously is adopted to simplify the fabrication process. According to the test results, the actuation voltage of contact switches and capacitive switches are 43.5 V and 4.6~15.0 V. At the operating frequency of 4.5 GHz, the isolation of contact switches and three kinds of capacitive switches are -32.5 dB, -17.3 dB, -21.5 dB and -8.8 dB, the insertion loss are -0.47 dB, -1.4 dB, -2.0 dB and -1.2 dB respectively.

Keywords—RF MEMS switch; actuation voltage; insertion loss; isolation

I. INTRODUCTION

Radio Frequency Micro-electro-mechanical System (RF MEMS) is one of the important application fields of MEMS technology. In order to improve the performance of Radio Frequency system, the micro-machining technology is applied to manufacture passive devices in microwave communication to achieve high integration inside the chip [1,2]. At present, the RF MEMS switch is one of the most important elementary units in RF MEMS devices [3], which controls the On\Off of the RF signal through the movement of the micro-mechanical structure. Compared with the traditional FET and PIN switch, the passive device RF MEMS switch eliminates the influence of P-N and metal-semiconductor junctions, which has the advantages [4] of almost zero power consumption, low insertion loss, high isolation, low cost, small size and good linearity, and

*The resrach was supported by the National Key Research and Development Program of China (Grant No.2018YFF01010901 and No.2016YFB0401903)

Yun Qi is with the Nano Science and Technology Institute, University of Science and Technology of China, Suzhou 215123, China. (phone: +86 512 628 76188; e-mail: yunqi@mail.ustc.edu.cn)

Yuanyuan Yang is with the School of Nano-Tech and Nano-Bionics, University of Science and Technology of China, Hefei 230026, China (phone: +86 512 628 76188; e-mail: yang2o@mail.ustc.edu.cn)

Wenjiang Shen is with Key Lab of Nanodevices and Applications, Suzhou Institute of Nano-Tech and Nano-Bionics, Chinese Academy of Sciences, Suzhou 215123, China. (phone: +86 512 628 76188; e-mail: wjshen2011@Sinano.ac.cn)

is widely used in phased radar, aerospace satellite, wireless communication and other fields.

Recently, RF MEMS switches have been the hot spot of research by well-known companies and universities, such as Raytheon [5], Motorola, Samsung, University of California [6], Hastpa University of Turkey [7], Indian Institute of Technology [8-10], Southeast University [11], Tsinghua University [12], Peking University. In current research, two different types of RF MEMS switches, series and shunt switches have been developed. The typical series switch consists of a thin metal cantilever bridge fixed at one end over the signal transmission line of the coplanar waveguide (CPW), while the shunt switch consists of a metal membrane which is located above the signal line and the both ends are fixed to the ground line. The shunt capacitive switch is suitable for higher operating frequency and lower actuation voltage, but its lifetime and reliability are limited by the dielectric layer [13].

Generally, there are many serious problems in fabrication process flow and performance of the switches studied by current researches. To produce the suspended top electrode and reduce the actuation voltage of the switch, the process flow is usually complex and failure may happen in actual application. Thus it is necessary to develop the fabrication process and physical design of the switches. In this paper, we proposed one series contact switch with single-ended cantilever beam and three kinds of shunt capacitive switch with flexure hinge structure to achieve high-performance switches with high operating frequency, low actuation voltage, low insertion loss and high isolation.

II. MATERIALS AND METHODS

A. Materials

High resistancte silicon is selected as substrate for its relatively high dielectric constant, high resistivity and low cost. Au is selected as the material for CPW transmission lines due to the chemical stability and low impedance. In addition, Au is also selected as the material for the top electrode for its low modulus of elasticity, high yield strength and conductivity.

B. Design

As for the design of RF MEMS switch, the structure dimensions of CPW should be taken into consideration firstly. In order to achieve the travelling wave match, the characteristic impedance of CPW lines should be 50 Ω which avoids unnecessary insertion loss [14]. Ansoft HFSS is used to analyze the characteristic impedance of CPW lines. Table 1 summarizes characteristics of CPW lines and the simulation result is 49.43 Ω which has a margin of error less than 5%.

978-1-6654-3008-1/21 $31.00 © 2021 IEEE

TABLE I.   THE CHARACTERISTICS OF CPW LINES

Symbol	Description	Value
W	width of the signal line	120 μm
G	space between the signal and ground line	80 μm
h	thickness of transmission line	2 μm
H	thickness of substrate	500 μm

Most commonly electrostatic force actuation method is used for the actuation of the MEMS devices. When a DC voltage is applied to the driving electrode, the top electrode will move downward due to the electrostatic force until it makes contact with the bottom structure. According to the statics model of switch, the actuation voltage [15] V_p is given by

$$V_p = \sqrt{\frac{8kg_0^3}{27\varepsilon_0 A}} \qquad (1)$$

Where g_0 is air gap, k is the spring constant, ε_0 is the permittivity of vacuum and A is the overlap area between electrodes. For series contact switch, the spring constant of cantilever beam is given by

$$K = \frac{2Ewt^3}{3l^3} \qquad (2)$$

Where w, t, and l are width, thickness, and length of the cantilever beam, respectively. E is the Young's modulus. Equation (2) indicates that reducing the thickness and increasing the length of beam can effectively reduce the actuation voltage. Thus the overall structure diagram of the designed series contact switch and the structure diagram of the cantilever beam are shown in Figure 1. The calculated spring constant of the beam is 5.34 N/m, and the air gap (g_0) is 2 μm. Then the calculated actuation voltage is 7.53 V.

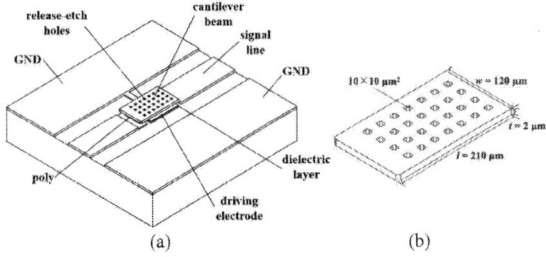

Figure 1.   (a) The overall structure diagram of the designed series contact switch and (b) the structure diagram of the cantilever beam

As for the shunt capacitive switch with flexural hinge structure, the spring constant [16] of a single beam is given by

$$k_{1,2\ldots,s} = \frac{Ewt^3}{l^3} \qquad (3)$$

Where w, t, and l are width, thickness, and length of the single beam, respectively. And s is the number of single beams in flexure hinge structure. Then the spring constant of the single-end flexural hinge could be expressed by

$$\frac{1}{K_m} = \frac{1}{k_1} + \frac{1}{k_2} + \frac{1}{k_3} + \cdots + \frac{1}{k_s} \qquad (4)$$

Then the spring constant of the membrane could be concluded as following

$$K_{total} = nK_m \qquad (5)$$

Where n is the number of flexural hinges which connect the membrane and ground line. The above equations indicate that reducing the thickness or increasing the length of the single beam, decreasing the number of flexural hinges or increasing the flexural degree of the single beam can effectively reduce the actuation voltage. Therefore, we proposed three kinds of capacitive switches with different flexural hinge structure as shown in Figure 2. The calculated spring constant of the U-type, Curve-1 type and Curve-2 type are 8.26 N/m, 1.82 N/m and 1.12 N/m, respectively. Then the calculated actuation voltage are 13.81 V, 5.34 V and 5.02 V, respectively.

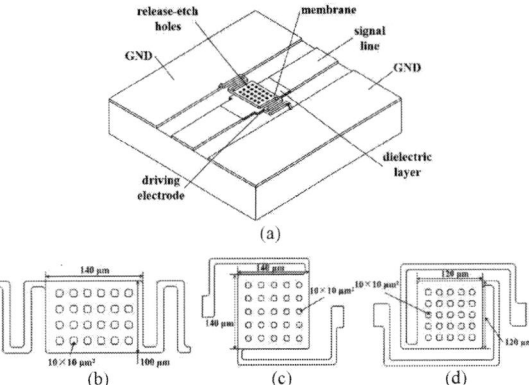

Figure 2.   (a) The overall structure diagram of the designed shunt capacitive switch. Different flexural hinge structure of (b) the U-type, (c) the Curve-1 type and (d) the Curve-2 type

C. Simulation

In this paper, the MEMS module in COMSOL Multiphysics was applied to simulate the switch to obtain the actuation voltage simulation results and the displacement structure diagram of the upper electrode. Figure 3 shows the vertical displacement of the top electrode under different applied voltages. From the simulation results, it can be seen that the actuation voltage of the contact switch, the U-type, Curve-1 type and Curve-2 type capacitive switch are 10.10 V, 8.69V, 4.10V and 2.44 V, respectively.

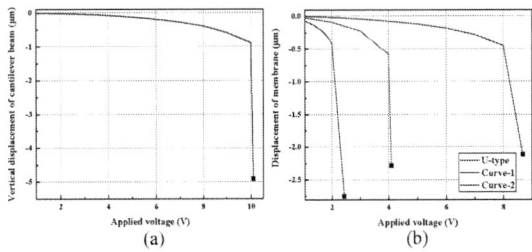

Figure 3.   The vertical displacement of the top electrode under different applied voltages of (a) contact switches and (b) capacitive switches

Further, the stress state of switch under actuation voltage is simulated by COMSOL which is shown in Figure 4. According to the stress state, the maximum stresses are 23.7 MPa, 18.8 MPa, 18.5 Mpa and 8.38 MPa respectively which are far less than the yield strength of gold (32 MPa). It indicates that the top electrode will not deform or warp under the actuation voltage, thus the designed structure is reasonable.

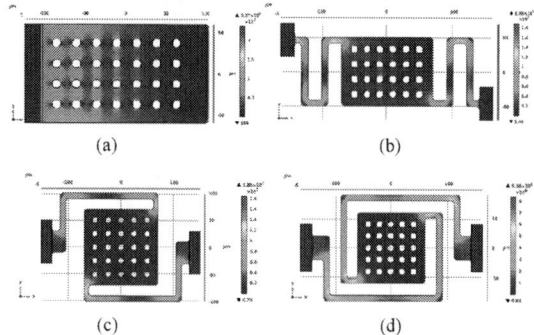

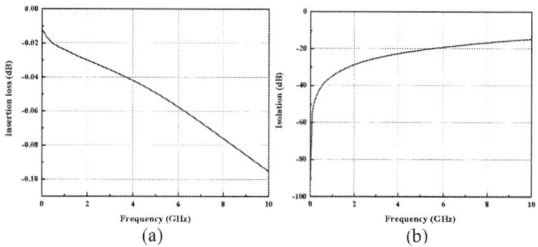

Figure 4. The stress state under actuation voltage of (a) contact switches and capacitive switches in (b) the U-type, (c) the Curve-1 type and (d) the Curve-2 type

In this paper, the RF performance simulations of MEMS switch are carried out by Ansoft HFSS software, and the simulation results of contact switches are shown in Figure 5.

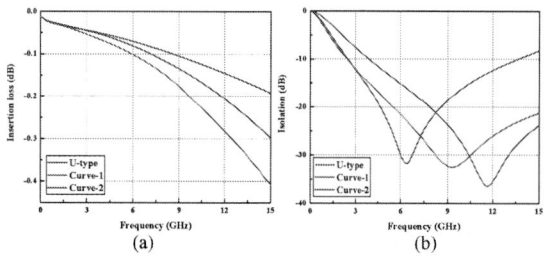

Figure 5. The simulated (a) insertion loss and (b) isolation of the contact switches

It can be seen from Figure 5 that the insertion loss is better than -0.05 dB at 5 GHz, and the isolation is better than -20 dB at 5 GHz. With the increase of operating frequency, the RF performance of the contact switches becomes worse.

In addition, the simulation results of capacitive switches are shown in Figure 6.

Figure 6. The simulated (a) insertion loss and (b) isolation of the capacitive switches

It can be seen from the Figure 6 that the insertion loss of the three types are better than -0.1 dB at 5 GHz, and the isolation of the U-type, Curve-1 type and Curve-2 type are better than -12 dB, -18 dB and -20 dB at 5 GHz. The RF performance of the capacitive switches become worse with

the increase of operating frequency at a certain range. Moreover, the capacitive switches are more suitable for higher operating frequency.

III. FABRICATION

A. Fabrication process flow

The fabrication process of the contact MEMS switch can be compatible with the process of the shunt capacitive MEMS switch. Figure 7 presents the fabrication process flow of the switch. The switch was fabricated on the high-resistivity silicon substrate with the thickness of 500 μm and the dielectric constant of 11.9. The SiO_2 layer with a thickness of 500 nm was formed by plasma enhanced chemical vapor deposition (PECVD) process to insulate the switch from the substrate. 400 nm thickness of Si_3N_4 was deposited to form the bumps which only exist in contact switch. Then, 250 nm thickness of Au was deposited and patterned to define DC bias electrode. The Si_3N_4 which acts as a dielectric layer with the thickness of 200 nm was deposited using PECVD. Next, 2 μm thickness of polyimide was spin-coated and cured to create a sacrificial layer. 2 μm Au was electroplated to form the top metallic electrode and the CPW transmission lines simultaneously. Finally, the sacrificial layer was removed by using oxygen plasma to eliminate stiction between the electrodes.

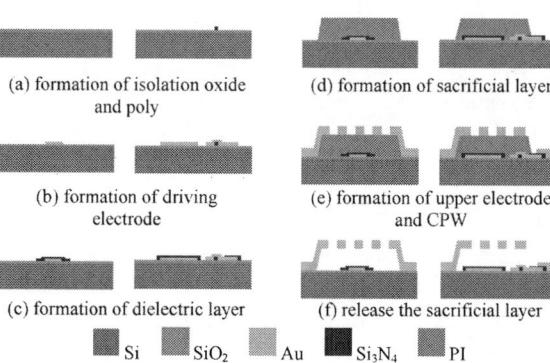

Figure 7. Fabrication process flow of the RF MEMS switches

B. Results of process flow

The contact switches with single-ended cantilever beam and three kinds of shunt capacitive switches were successfully prepared. The scanning electron microscope (SEM) images of the fabricated switches are shown in Figure 8. Furthermore, the surface profile curves and 3D surface topography of the top electrodes on the switches are shown in Figure 9.

It can be seen from the Figure 8 and Figure 9 that the surface flatness and structure dimensions of the fabricated contact switches are ideal compared with the design models, which indicates the results of process flow basically meet the design requirements. Also, the three kinds of capacitave switch meet the design requirements. However, the cantilever beam of contact switches warped upward, this is due to the residual thermal stress during the releasing process of polyimide. Warped beam is harmful for the mechanical performance because it leads to larger actuation voltage in practical measurement.

978-1-6654-3008-1/21 $31.00 © 2021 IEEE   360

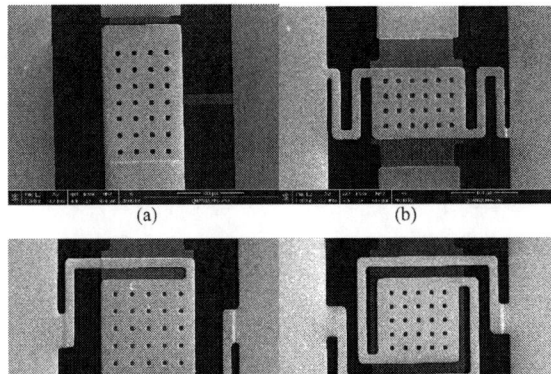

Figure 8. The scanning electron microscope (SEM) images of the fabricated (a) contact switches and capacitive switches in (b) the U-type, (c) the Curve-1 type and (d) the Curve-2 type

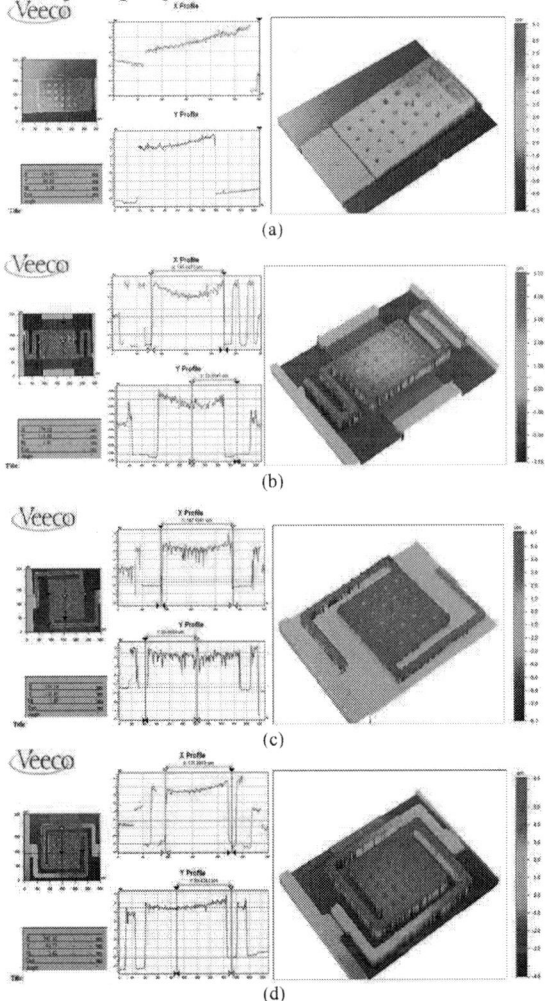

Figure 9. The surface profile curves and 3D surface topography of the top electrode in (a) contact switches and capacitive switches of (b) the U-type, (c) the Curve-1 type and (d) the Curve-2 type

IV. RESULTS

A. Mechanical properties

In order to test the actuation voltage of the contact switches, the Agilent B1500A semiconductor device analyzer and Istatv probe station were adopted. A scanning voltage of 0~100 V with a step length of 0.5 V was applied between the driving electrode and the top electrode of the switch, and a DC bias voltage of 1.5 V was applied to both ends of the signal transmission line to measure the current variation of the circuit. Figure 10 shows the measured I-V characteristic curve of the switch which indicates the cantilever beam was pulled down when the applied voltage achieved 43.5 V.

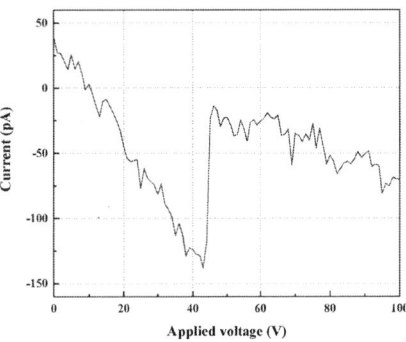

Figure 10. I-V characteristic measured of the contact switches

On the other hand, the actuation voltage of the capacitive switches was tested by applying a voltage of 0~40 V with a step length of 0.2 V between electrodes. Figure 11 shows the measured C-V characteristic curve of the switch with a measurement frequency of 1 MHz. When the applied voltage reaches a certain value, the capacitance will increase suddenly and then turn to be constant, indicating that the top electrode moved downward and contacted with the bottom structure due to the actuation voltage. Thus the measured actuation voltage values of the U-type, Curve-1 and Curve-2 capacitive switches are 15.0 V, 14.2 V and 4.6 V respectively.

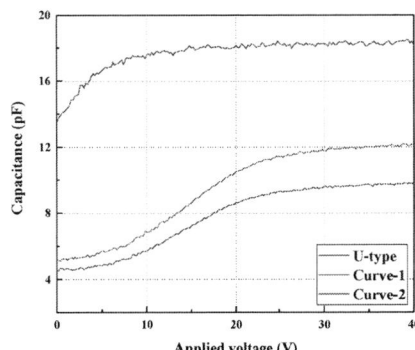

Figure 11. C-V characteristic measured of the capacitive switches

However, the measured actuation voltage values of both contact and capacitive switches are slightly larger than the simulation values. This are mainly caused by the residual stress during the release process of the polyimide, which leads to a minor upward warp of the top electrode and then the measured actuation voltage values tends to be larger.

B. Radio-frequency performance

The Agilent E5071C network analyzer, Istatv probe station and Keysight E36106B programmable DC power supply were employed to test the radio-frequency performance of the switches. The DC power supply was connected to the top electrode and the driving electrode, and at the meantime the two ports of the network analyzer were connected to the input and output of the coplanar waveguide transmission line. Then the scattering-parameters of the switches were measured in the frequency range of 0–40 GHz by applying a scanning voltage of 0~100 V with a step length of 1 V. Figure 12 shows the measurement and simulation results for the contact switches in the OFF state and in the ON state. According to the results, the measured isolation is -32.5 dB at 4.5 GHz which shows a good agreement with the simulation results, while the measured insertion loss is -0.47 dB at 4.5 GHz which is worse than the simulation results. This is mainly because of the high contact impedance introduced by the test system and the residual carbide at the bottom after the releasing process of polyimide.

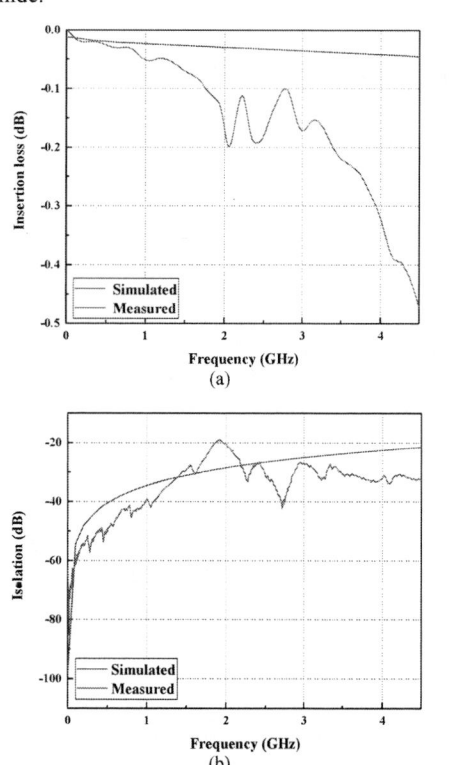

Figure 12. Measurement and simulation results for the contact switches (a) in the ON state and (b) in the OFF state

Figure 13 shows the measurement results for the capacitive switches in the ON state and in the OFF state. The measured insertion loss of the U-type, Curve-1 and Curve-2 capacitive switch is -1.4 dB、-2.0 dB and -1.2 dB

at 4.5 GHz, respectively. It is obviously that the measurement results tend to be worse than the simulation results, this is due to the high contact impedance introduced by the test system and the signal interference during from the top electrode to the signal transmission line. In addition, the measured isolation of the U-type, Curve-1 and Curve-2 capacitive switch is -17.3 dB、-21.5 dB and -8.8 dB at 4.5 GHz respectively which shows a good agreement with the simulation results.

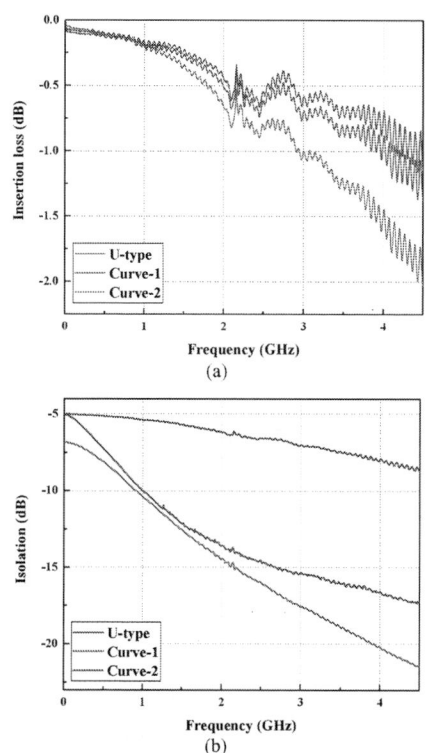

Figure 13. Measurement results for the capacitive switches (a) in the ON state and (b) in the OFF state

V. CONCLUSION

This paper presents the design, fabrication, and testing of the contact switches with single-ended cantilever beam and three kinds of shunt capacitive switches. The contact switches exhibit low actuation voltage (43.5 V), very high isolation (-32.5 dB) and low insertion loss (-0.47 dB) at 4.5 GHz. Further, according to the static model and simulation results of switch, a extremely low actuation voltage (4.6 V) RF MEMS switch is designed and optimized by comparing three different capacitive switches. The measured isolation and insertion loss of the capacitive switch is -21.5 dB and -1.2 dB at 4.5 GHz respectively which are in good agreement with the simulation results. Due to the small size (2.1 mm×2.1 mm), low actuation voltage and excellent RF performance, the proposed RF MEMS switch is an excellent candidate for satellite, radar systems, electronic countermeasures and wireless communication.

978-1-6654-3008-1/21 $31.00 © 2021 IEEE

ACKNOWLEDGMENT

The research was supported by the National Key Research and Development Program of China (Grant No. 2016YFB0402003 and No.2016YFB0401903). The authors are grateful to SINANO's Nano Fabrication Facility (NFF) for their support in the fabrication process of RF MEMS switches and related tests.

REFERENCES

[1] Yao J J. RF MEMS from a device perspective[J]. J.micromech.microeng, 2000, 10(4):960-131706704.

[2] Sharma A K , Gupta N . Microelectromechanical System (MEMS) Switches for Radio Frequency Applications - A Review[J]. Sensors and Transducers, 2013, 148(1):11-21.

[3] Zhu Jian. RF MEMS devices design, fabrication, and application[M]. Beijing: National Defense Industry Press, 2012:1-8.

[4] Rebeiz G M, Muldavin J B . RF MEMS switches and switch circuits[J]. Microwave Magazine IEEE, 2001, 2(4):59-71.

[5] Yao Z J , Chen S . Micromachined low-loss microwave switches[J]. Journal of Microelectromechanical Systems, 1999, 8(2):129-134.

[6] Yang H H , Zareie H , Rebeiz G M . A High Power Stress-Gradient Resilient RF MEMS Capacitive Switch[J]. Journal of Microelectromechanical Systems, 2015, 24(3):599-607.

[7] Demirel K, Yazgan E , Demir I , et al. A folded leg Ka-band RF MEMS shunt switch with amorphous silicon (a-Si) sacrificial layer[J]. Microsystem Technologies, 2016, 23(5):1-10.

[8] Jaiswal A , Dey S , Abegaonkar M P , et al. High isolation RF MEMS SPDT switch for 60 GHz ISM band antenna routing applications[C]// 2016 Asia-Pacific Microwave Conference (APMC). IEEE, 2016.

[9] Shekhar S, Vinoy K J, Anathasuresh G K. Low-voltage high-reliability MEMS switch for millimeter wave 5G applications[J]. Journal of Micromechanics and Microengineering. 2018.

[10] Angira M, Bansal D, Kumar P, et al. A novel capacitive RF-MEMS switch for multi-frequency operation[J]. Superlattices and microstructures, 2019.

[11] Sun J , Li Z , Zhu J , et al. Design of DC-contact RF MEMS switch with temperature stability[J]. AIP Advances, 2015, 5(4).

[12] Li M , Zhao J , You Z , et al. Design and fabrication of a low insertion loss capacitive RF MEMS switch with novel micro-structures for actuation[J]. Solid State Electronics, 2017.

[13] Chee J , Kami R , Fisher T S , et al. DC-65 GHz characterization of nanocrystalline diamond leaky film for reliable RF MEMS switches[C]// European Microwave Conference. IEEE, 2005.

[14] Rebeiz G M , Muldavin J B , Schoenlinner B , et al. RF MEMS: Theory, Design, and Technology[M]. 2004. Rebeiz G M, Muldavin J B , Schoenlinner B , et al. RF MEMS: Theory, Design, and Technology[M]. 2004.

[15] Peroulis D, Pacheco S P , Sarabandi K , et al.Electromechanical considerations in developing low-voltage RF MEMS switches[J]. Microwave Theory & Techniques IEEE Transactions on, 2003, 51(1):259-270.

[16] Reza A H , Saeed K . Design and simulation of a novel RF MEMS shunt capacitive switch with a unique spring for Ka-band application[J]. Microsystem Technologies, 2018:1-10.

Gap in pagination due to unavailable papers.

Pages 364-371

Proceedings of the 16th Annual IEEE International
Conference on Nano/Micro Engineered and Molecular Systems
April 25-29, 2021

Broadband linear-to-circular polarization converter based on ultrathin metal nano-grating *

Yuanyi Fan, Ran Zhang, Ze Liu, Chuanlong Guan, and Jinkui Chu*

Abstract—A metamaterial based on ultrathin metal nano-grating is proposed to realize the linear-to-circular (LTC) polarization conversion in the visible light band. Meanwhile, a perfect quarter-wave plate of the desired operating wavelength can also be obtained using the subwavelength metal nano-grating's inverse polarizing effect. In this paper, we first optimized metal nano-grating's size parameters through the finite difference time domain (FDTD) method, and then an LTC polarization converter based on ultrathin nano-grating featured with 12 nm thickness ($<\lambda$/40), 180 nm period, 120 nm linewidth was fabricated on the target substrate by nanoimprint lithography (NIL), metal thermal evaporation (MTE) and metal nanostructures transfer (MNT) processes. The simulation and experimental results demonstrate that the LTC polarization converter based on ultrathin metal nano-grating has the advantages of simple structure and excellent performance. The ultrathin metal nano-grating structure also makes it easy to fabricate with large area, low cost, and high efficiency.

I. INTRODUCTION

Polarization is a basic physical parameter of electromagnetic wave. The polarization modulation can enable us applied light in the fields of polarization manipulation, optical sensing, imaging, and communication. Traditional polarization modulators are made of bulk birefringent materials, such as solid crystal or liquid crystal. However, the large size of traditional birefringent materials limits the polarization modulators' miniaturization and integration. Metamaterial is a kind of artificial two-dimensional material with subwavelength size and spacing periodic array, which shows unusual electromagnetic parameters for an incident plane wave. It can control the phase, amplitude, polarization, and lead to many interesting electromagnetic phenomena, such as holography [1], [2], perfect absorption [3], [4], abnormal reflection, and refraction [5]. Therefore, metamaterials make it possible to miniaturize and integrate optical devices. As a basic optical device, polarization converter has attracted many scholars to realize it through metamaterials. The polarization converter based on dielectric metamaterial [6], [7] has high transmittance because it has no ohmic loss. However, micro-nano structures based on dielectric metamaterials usually have a large aspect ratio due to their low birefringence. The combination of metal metal-nanostripe and dielectric [8] can effectively increase the birefringence of metamaterials, but

the fabrication process is relatively complex. Additionally, metamaterials based on metal nanostructures [9],[10] usually need to precisely control the parameters of the cell to achieve aπ/2 phase difference between TM and TE polarization at the desired operating wavelength. The precise parameters control also brings difficulties to the production. Through the above analysis, the main problem of current metamaterial-based polarization converters is that the working bandwidth is narrow or the manufacturing process is more complicated.

In this paper, a metamaterial based on ultra-thin metal nano-grating is proposed and fabricated to realize the LTC polarization conversion in the visible light band. On the one hand, as a self-complementary metamaterial [11],[12], ultra-thin metal nano-grating can achieve π/2 phase difference in a wide band. On the other hand, the ultrathin metal nano-grating structure makes it easy to fabricate with large area, low cost, and high efficiency. Besides, a perfect quarter-wave plate of the desired operating wavelength can also be obtained using the subwavelength metal nano-grating's inverse polarizing effect [13],[14].

II. THEORY AND ANALYSIS

We first analyze the LTC polarization converter theoretically. The incident linearly polarized light propagating along the +z-axis direction can be expressed as:

$$\mathbf{E}_i(\mathbf{r},t) = \begin{bmatrix} I_x \\ I_y \end{bmatrix} e^{i(kz-wt)} \quad (1)$$

where I_x and I_y represent complex amplitudes component along the x-axis and y-axis, k represents the wave vector, ω is the frequency. The Jones vector $\begin{bmatrix} I_x \\ I_y \end{bmatrix}$ determines the polarization state and intensity of polarized light. The Jones vectors of linearly polarized light with polarization angle θ, left- and right-circularly polarized light can be written as $\begin{bmatrix} \cos\theta \\ \sin\theta \end{bmatrix}$, $\frac{1}{\sqrt{2}}\begin{bmatrix} 1 \\ i \end{bmatrix}$ and $\frac{1}{\sqrt{2}}\begin{bmatrix} 1 \\ -i \end{bmatrix}$. The transmission polarized light can be written as:

$$\mathbf{E}_t(\mathbf{r},t) = \begin{bmatrix} T_x \\ T_y \end{bmatrix} e^{i(kz-wt)} . \quad (2)$$

*********This work was supported in part by the National Natural Science Foundation of China under Grant 51675076, in part by the Science Fund for Creative Research Groups under Grant 51621064, and in part by the Fundamental Research Funds for the Central Universities under Grant DUT20LAB303. (Corresponding author: Jinkui Chu.)
Yuanyi Fan, Ran Zhang, Ze Liu, Chuanlong Guan and Jinkui Chu are with the School of Mechanical Engineering, Dalian University of

Technology, Dalian 116024, China, and also with the Key Laboratory for Micro/Nano Technology and System of Liaoning Province, Dalian University of Technology, Dalian 116024, China (e-mail: fanyuanyi@mail.dlut.edu.cn; zhangr@dlut.edu.cn; liuze@mail.dlut.edu.cn; guan357@mail.dlut.edu.cn; chujk@dlut.edu.cn).

978-1-6654-3008-1/21 $31.00 © 2021 IEEE

The incident and transmitted polarized light can be correlated by the polarization converter's Jones matrix, which can be expressed as:

$$\begin{bmatrix} T_x \\ T_y \end{bmatrix} = \begin{bmatrix} T_{xx} & T_{xy} \\ T_{yx} & T_{yy} \end{bmatrix} \cdot \begin{bmatrix} I_x \\ I_y \end{bmatrix}. \quad (3)$$

For the device without linear polarization conversion effect, the transmitted polarized light can be written as:

$$\begin{bmatrix} T_x \\ T_y \end{bmatrix} = \begin{bmatrix} T_{xx} I_x \\ T_{yy} I_y \end{bmatrix}. \quad (4)$$

From equation (4) and the Jones vectors of linearly and right- and left-circularly polarized waves, it can be determined that for a high-performance LTC polarization converter with the incident linearly polarized light $\begin{bmatrix} \cos\theta \\ \sin\theta \end{bmatrix}$, the phase difference between T_{xx} and T_{yy} must be equal to $n\pi/2$, where n represents an odd number. Further, the condition $\dfrac{T_{xx}}{\sin\theta} = \dfrac{T_{yy}}{\cos\theta}$ must be satisfied. Also, when the transmittance is equal ($T_{xx}=T_{yy}$) with the phase difference $n\pi/2$ (n represents an odd number), the LCT polarization converter behaves as a perfect quarter-wave plate.

III. DESIGN AND SIMULATION

To meet the above-mentioned conditions to achieve high-performance broadband LTC polarization conversion, and make the fabrication process simple, a self-complementary metamaterial based on ultrathin metal nano-grating is proposed. The structure schematic of the proposed metal nano-grating LTC polarization converter is shown in Fig. 1. The substrate is fused silica, the middle dielectric layer is UV-curable resist, and the top layer is aluminum metal nano-grating. The performance of the LTC polarization converter is simulated by the finite difference time domain (FDTD) method. To improve optimization efficiency, metal nano-grating can be simulated by a two-dimensional cross-section. In the two-dimensional simulation, the X direction is periodic boundary condition, and the Z direction is absorbing boundary condition. Linearly polarized light propagates straight along the +z-axis direction. The background refractive index is assumed as 1, the electric permittivity of the UV-curable resist layer is 1.6 in the simulation wave band, and the optical constant of aluminum metal nano-grating layer and fused silica substrate in the visible light regime is downloaded from the URL https://refractiveindex.info/.

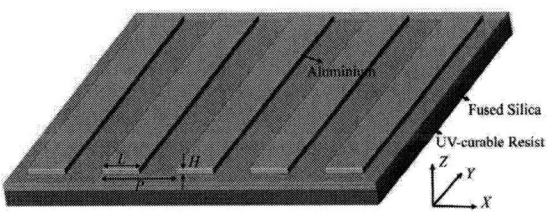

Fig. 1. The structure schematic of the proposed LTC polarization converter.

We design an LTC polarization converter that works in the visible light region (about 380 nm ~760 nm). To avoid diffraction, the period P of the metal nanowire grid is less than half wavelength (fixed at 180 nm period in this paper). The metal nano-grating with a duty cycle of 0.5 is a self-complementary metamaterial. Compared with previous proposals of polarization converters, the advantage of LTC polarization converters based on self-complementary metamaterials is the rigorous constancy of the phase difference between TE and TM polarized light, which is just about $\pi/2$ in a wider operating wavelength range. Besides, this advantage can be maintained even under oblique incidence. According to the assumption of self-complementary metamaterials, the component is an infinitely thin perfect electrical conductor [11]. In the real manufacturing process, it is difficult to meet the assumption. Therefore, we study the phase difference and transmittance of TM and TE polarized light by changing the linewidth L and thickness H of metal nano-grating in the simulation process.

We first studied the influence of linewidth variation on the transmittance and phase difference of TM and TE polarized light through numerical simulation. Fig. 2(a) and 2(b) show the transmittance and phase difference of TM and TE polarized light with metal nano-grating P=180 nm, H=12 nm, and L=60, 90, 120, 150 nm, respectively. As shown, with the increase of linewidth, the transmittance of TM and TE polarized light decreases, and the phase difference increases. The metal nano-grating can achieve $\pi/2$ phase difference in a wide band with P=180 nm, H=12 nm, and L=120 nm. Then, we studied the influence of nano-grating thickness variation on the transmittance and phase difference of TM and TE polarized light. Fig. 2(c) and 2(d) show the transmittance and phase difference of TM and TE polarized light with metal nano-grating P=180 nm, L=120 nm, and H=8, 10, 12, 14 nm, respectively. With the increase of thickness, the transmittance difference between TM and TE polarized light increases. For the linear polarizer, we pursue a larger transmittance difference. For polarization converter, we expect TM and TE polarized light to have larger transmittance and smaller transmittance difference. The change of the H has little effect on the phase difference shown in Fig. 2(d). Therefore, the LTC polarization converter based on metal nano-grating has a large tolerance of manufacturing error.

Subwavelength metal nano-grating is usually used as linear polarizers with the TM transmittance exceeding the TE transmittance. However, the subwavelength metal nano-grating also has a reverse polarization effect with the TE transmittance exceeding the TM transmittance. Fig. 2(d) shows that the change of the H has little effect on the phase difference. Therefore, by adjusting the nano-grating's L, the transmittance of TM and TE polarized light can be made equal while keeping the phase difference equal to $\pi/2$. Then, we can get a perfect quarter-wave plate at the desired operating wavelength. As shown in Fig. 2(d), the transmittance of TM and TE polarized light are equal at λ=575 nm, and the phase difference is $\pi/2$, which is a nearly perfect quarter-wave plate with the metal nano-grating P=180 nm, H=12 nm, and L=120 nm.

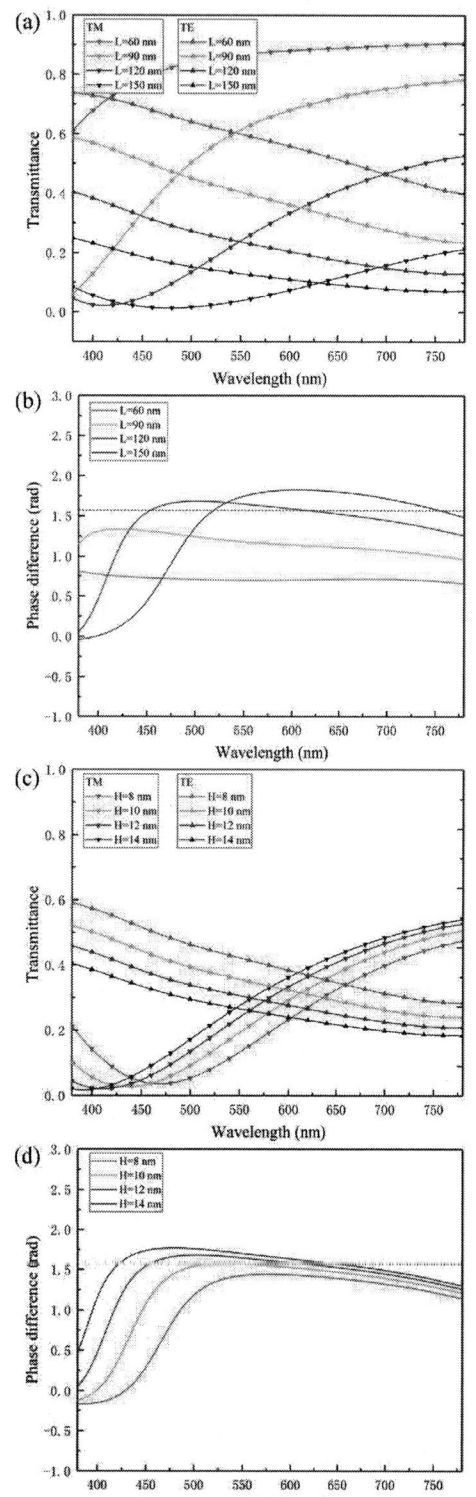

Fig. 2. The transmittance and phase difference of TM and TE polarized light (a,b) with metal nano-grating P=180 nm, H=12 nm and L=60, 90, 120, 150 nm, (c,d) with metal nano-grating P=180 nm, L=120 nm and H=8, 10, 12, 14 nm.

IV. RESULTS AND DISCUSSION

To further verify the performance of our proposed LTC polarization converter, we fabricated a sample featured with the metal nano-grating P=180 nm, H=12 nm, L=120 nm and tested its transmittance and phase difference of TM and TE polarized light. The ultrathin metal nano-grating structure also makes it easy to fabricate with large area, low cost, and high efficiency. The fabrication process schematic illustration is shown in Fig. 3. Firstly, the IPS (Intermediate Polymer Sheet) soft template was fabricated by nanoimprint lithography (Fig. 3 (a)); then the aluminum layer was deposited on the surface of the IPS soft template by metal thermal evaporation process (Fig. 3(b)); finally, the metal nano-grating on the surface of the IPS was transferred to the silica substrate by UV curing metal nanostructure transfer process, and the silica substrate was pre-coated with UV-curable resist (Fig. 3(c，d)). The detailed fabricating process can refer to our published paper [15].

In this process, the feature size of the original nickel-metal mold is featured with 180 nm period, 60 nm linewidth, and 100 nm height. The feature size of the obtained IPS soft template is completely complementary to the original nickel-metal mold. Besides, to facilitate the transfer of the metal nano-grating, the height of the original nickel template nano-grating should be greater than the thickness of the thermally evaporated metal layer. As shown, Fig. 4(a) shows the surface SEM image of the IPS soft template. The period and linewidth of the nano-grating are 180 nm and 120 nm, respectively. Fig. 4(b) shows the surface SEM image of the IPS soft template after aluminum deposition. Fig. 4(c) shows the silica surface SEM image after transferring the metal nano-grating. Aluminum metal nano-grating cross-sectional image is shown in Fig. 4(d). The thickness of the aluminum nano-grating is 12 nm. The aluminum nano-grating's cross-sectional morphology is not a perfect rectangle, which is related to the deformation of the IPS soft template during the NIL process and the deposition directionality of the thermally evaporated metal aluminum.

We tested the performance of the LTC polarization converter by a spectrometer. During the testing process, we first tested the TM and TE polarization transmittance of the LTC polarization converter. Then the LTC polarization converter was combined with a commercial linear polarizer. According to the principle of coherent light interference, the phase difference is calculated. Fig. 5 shows the transmittance and phase difference of TM and TE polarized

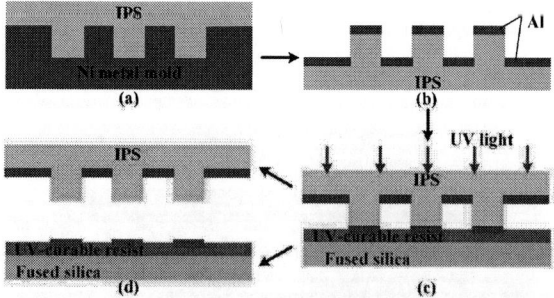

Fig. 3. The fabrication process schematic illustration of the LTC polarization converter. (a) Thermal NIL process. (b) Metal thermal evaporation process. (c) Flexible contact UV-curable process. (d) Transfer of metal nano-grating.

light of the LTC polarization converter. It can be found that the simulation results of transmittance and phase difference of TM and TE polarized light are slightly different from the test. The possible reasons include the change of geometric and optical parameters of the aluminum nano-grating. The cross-sectional morphology of aluminum nano-grating is not a perfect rectangle, which is related to the deformation of the IPS soft template during the NIL process and the deposition directionality of the thermally evaporated metal aluminum. Also, impurities in the thermal evaporation process and later natural oxidation will further affect the optical parameters of the metal aluminum.

V. CONCLUSION

In conclusion, we presented an ultrathin metal nano-grating metamaterial to realize the linear-to-circular (LTC) polarization conversion in the visible light band. The simulation and experimental results demonstrate that the LTC polarization converter has the advantages of simple structure and excellent performance. The ultrathin metal nano-grating structure also makes it easy fabricated with large area, low cost, and high efficiency. In addition，the change of the metal nano-grating's thickness has little effect on the phase difference. Therefore, by adjusting the nano-grating's linewidth, the transmittance of TM and TE

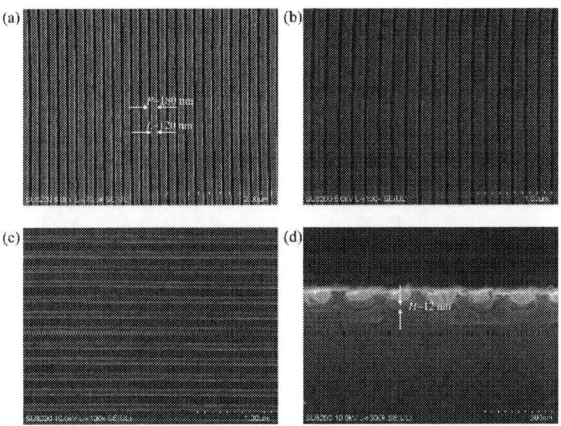

Fig. 4. SEM surface images of (a) IPS soft template, (b) IPS soft template after aluminum deposition, (c) silica substrate after metal nano-grating transfer. SEM cross-sectional image of (d) silica substrate after metal nano-grating transfer. P=180 nm, L=120 nm and H=12 nm.

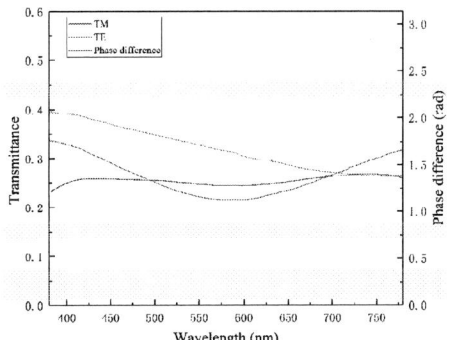

Fig. 5. The transmittance and phase difference of TM and TE polarized light of the fabricated LTC polarization converter.

polarized light can be made equal while keeping the phase difference equal to $\pi/2$. Then, a perfect quarter-wave plate can be designed at the desired operating wavelength.

However, simulation results of LTC's transmittance and phase difference of TM and TE polarized light are slightly different from the test. The reason may be due to the manufacturing error and the optical parameters of the metal aluminum obtained by thermal evaporation are not completely consistent with those optical parameters used in the simulation process. In our future work, we will optimize the process parameters of nanoimprint lithography and metal thermal evaporation to obtain better cross-section morphology of the nano-grating. Furthermore, the optical parameters of the thermally evaporated aluminum will be measured and used for optimizing nano-grating geometric parameters.

REFERENCES

[1] J. Guo et al., "Polarization multiplexing for double images display," Opto-Electronic Adv., vol. 2, no. 7, pp. 18002901–18002906, 2019.

[2] X. Xie et al., "Plasmonic Metasurfaces for Simultaneous Thermal Infrared Invisibility and Holographic Illusion," Adv. Funct. Mater., vol. 28, no. 14, pp. 1–6, 2018.

[3] N. Bai, F. Zhong, J. Shen, H. Fan, and X. Sun, "A thermal-insensitive ultra-broadband metamaterial absorber," J. Phys. D. Appl. Phys., vol. 54, no. 9, p. 095101, Mar. 2021.

[4] J. Zhou et al., "Cross-Shaped Titanium Resonators based Metasurface for Ultra-Broadband Solar Absorption," IEEE Photonics J., vol. 13, no. 1, 2021.

[5] N. Yu et al., "Light propagation with phase discontinuities: Generalized laws of reflection and refraction," Science (80-.)., vol. 334, no. 6054, pp. 333–337, 2011.

[6] S. Wang et al., "Arbitrary polarization conversion dichroism metasurfaces for all-in-one full Poincaré sphere polarizers," Light Sci. Appl., vol. 10, no. 1, 2021.

[7] J. Yang and T. Lan, "High-efficiency, broadband, and wide-angle all-dielectric quarter wave plate based on anisotropic electric and magnetic dipole resonances: publisher's note," Appl. Opt., vol. 58, no. 5, p. 1297, 2019.

[8] M. Nyman, S. Maurya, M. Kaivola, and A. Shevchenko, "Optical wave retarder based on metal-nanostripe metamaterial," Opt. Lett., vol. 44, no. 12, p. 3102, 2019.

[9] J. J. Cadusch, T. D. James, and A. Roberts, "Experimental demonstration of a wave plate utilizing localized plasmonic resonances in nanoapertures," Opt. Express, vol. 21, no. 23, p. 28450, 2013.

[10] X. T. Zhou, R. C. Jin, J. Wang, J. Q. Li, and Z. G. Dong, "All-metal metasurface polarization converter in visible region with an in-band function," Appl. Phys. Express, vol. 12, no. 9, 2019.

[11] J. D. Baena, J. P. Del Risco, A. P. Slobozhanyuk, S. B. Glybovski, and P. A. Belov, "Self-complementary metasurfaces for linear-to-circular polarization conversion," Phys. Rev. B - Condens. Matter Mater. Phys., vol. 92, no. 24, pp. 1–9, 2015.

[12] L. Maiolo et al., "Quarter-wave plate metasurfaces on electromagnetically thin polyimide substrates," Appl. Phys. Lett., vol. 115, no. 24, 2019.

[13] G. G. Kang, I. Vartiainen, B. F. Bai, H. Tuovinen, and J. Turunen, "Inverse polarizing effect of subwavelength metallic gratings in deep ultraviolet band," Appl. Phys. Lett., vol. 99, no. 7, pp. 10–13, 2011.

[14] G. Kang, J. Dong, I. Vartiainen, P. Paakkonen, and J. Turunen, "Investigation of inverse polarization transmission through subwavelength metallic gratings in deep ultraviolet band," Opt. Commun., vol. 311, pp. 33–37, 2013.

[15] Y. Fan, R. Zhang, Z. Liu, D. Huang, and J. Chu, "Direct metallic nanostructures transfer by flexible contact UV-curable nano-imprint lithography," Appl. Phys. Express, vol. 12, no. 9, 2019.

Gap in pagination due to unavailable paper.

Pages 376-377

Proceedings of the 16th Annual IEEE International
Conference on Nano/Micro Engineered and Molecular Systems
April 25-29, 2021

Probing zeta potential of glass in electrolyte solutions by colloidal probe technique

Zhijian Liu, Jiaming Gao, Tianze Wu, Sen Wu, Zixiao Fan, Ziyi Yuan, Yongxin Song, and *Xinxiang Pan*

Abstract— On the microfluidic chip, the zeta potential is an important parameter dominating the direction and magnitude of the electroosmotic flow, electrophoresis and other electrokinetic transport phenomenon. However, the traditional methods for zeta potential measurement mainly work based on the indirect methods, which would introduce many errors. In this paper, a novel direct method based on AFM colloidal probe technique was developed to measure the zeta potential. Firstly, to simplify the theoretical model, the approaching speed of the colloidal probe, at which the hydrodynamic force could be neglected, was determined. Then by fitting the experimental data with the theoretical model, the zeta potential of glass in electrolyte solutions was directly calculated.

Index Terms— zeta potential, electrolyte solutions, atomic force microscope, colloidal probe

I. INTRODUCTION

On the microfluidic chip, the zeta potential is an important parameter dominating the direction and magnitude of the electroosmotic flow, electrophoresis and other electrokinetic transport phenomenon [1]. So during the related application, the measurement of the zeta potential is very crucial. The traditional methods for zeta potential measurement mainly include the current monitoring method [2] and the streaming potential method [3]. Unfortunately, these methods would work based on an external electric or flow field that would influence the zeta potential of the solid-liquid interface [4]. On the other hand, the atomic force microscope (AFM) has developed from a traditional three-dimensional nano topography observation equipment to a weak-interaction forces measurement. More importantly, the measurement of the interaction forces by AFM could be carried out in air, liquid or vacuum condition. Here, the AFM was used to measure the forces that are mainly induced by the interaction of the electric double layers (EDL) between the glass spheres immersed in electrolyte solutions. By fitting the measured forces with the theoretical model, the zeta potential was further calculated.

II. THE PRINCIPLE OF THE MEASUREMENT

As figure 1 shown, when a glass sphere approaches to another one anchored on a substrate in an electrolyte solution, the glass sphere would be affected by the hydrodynamic force (F_H), Derjaguin-Landau-Verwey-Overbeek (DLVO, (F_{DLVO})) forces and other short range forces (F'). Specially, when the separation (D) between the glass spheres is about one to several Debye lengths, the short range forces could be neglected [5]. In this case, the interaction forces between the glass spheres only include the hydrodynamic force and the DLVO force. Moreover, when the glass sphere approaches to another one at a very slow speed, the hydrodynamic force could also be neglected [6]. Therefore, the forces acted on the glass sphere could be further simplified to the DLVO forces.

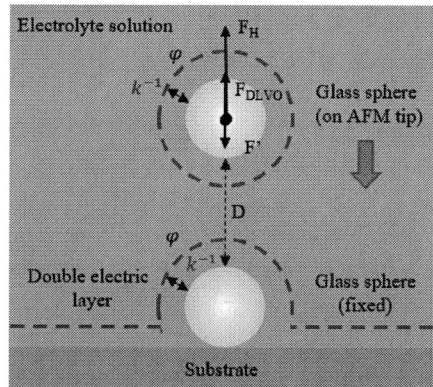

Figure 1. The principle of the measurement

It is well known that the DLVO forces only includes the Van der Waals force and the EDL interaction force [7]. The theoretical model of the DLVO has been developed well and could be found in the following:

This work was supported by National Key Research and Development Program of China (2017YFC1404603); National Natural Science Foundation of China (51909019, 51479020); Fundamental Research Funds for the Central Universities (3132019330, 3132019336). (*Corresponding author: Zhijian Liu*).

Z. Liu is with the College of Marine Engineering, Dalian Maritime University, Dalian, 116026, China (phone: 0411-84723190; e-mail: liuzhijian@dlmu.edu.cn).

J. Gao is with the College of Marine Engineering, Dalian Maritime University, Dalian, 116026, China (e-mail: gaojiaming@dlmu.edu.cn).

T. Wu is with the College of Marine Engineering, Dalian Maritime University, Dalian, 116026, China (e-mail: wutianze@dlmu.edu.cn).

S. Wu is with the College of Marine Engineering, Dalian Maritime University, Dalian, 116026, China (e-mail: dlmuwusen@163.com).

Z. Fan is with the College of Marine Engineering, Dalian Maritime University, Dalian, 116026, China (e-mail: fanzixiao@dlmu.edu.cn).

Z. Yuan is with the College of Marine Engineering, Dalian Maritime University, Dalian, 116026, China (e-mail: 1329953441@qq.com).

Y. Song is with the College of Marine Engineering, Dalian Maritime University, Dalian, 116026, China (e-mail: yongxin@dlmu.edu.cn).

X. Pan is with the College of Marine Engineering, Dalian Maritime University and the College of Electronics, Dalian, 116026, China and Information Engineering, Guangdong Ocean University, Zhanjiang 524088, China. (email: dmupanxx@gmail.com).

$$F = \frac{4\pi R \varepsilon k \varphi^2}{e^{+kD} - e^{-kD}}(1 - e^{-kD}) - \frac{AR}{6D^2} \qquad (1)$$

where R is the radius of the glass sphere (here, the sizes of the two glass spheres are same), ε is the dielectric constant, k^{-1} is the Debye length, φ is the zeta potential, A is the Hamaker constant and D is the separation distance between the glass spheres. So when the relationship between the interaction forces (F) and the separation distance (D) in the process of approaching is measured by AFM, the zeta potential of the glass sphere could be directly calculated.

III. EXPERIMENTS

The preparation of a AFM colloidal probe was performed by referring the previous paper [8]. Briefly, a glass microsphere, 20 μm in diameter, was glued on the tip of a triangular cantilever using UV-curable adhesive. Then, the AFM colloidal probe was placed in a clean chamber for 24 hours to solidify. The microscope image of the fabricated probe is as shown in figure 2. Referencing the previous literatures [9], [10], the deflection sensitivity, together with the spring constant, of the fabricated probe were calibrated.

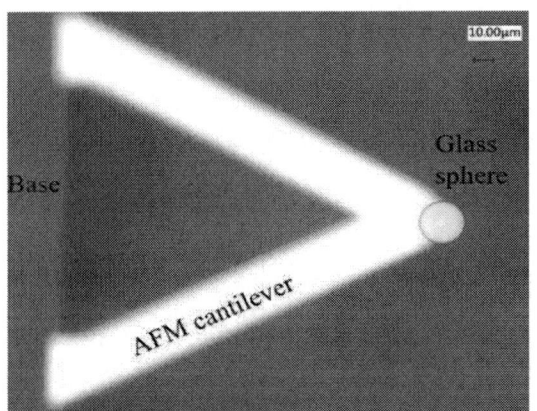

Figure 2. The microscope image of the fabricated AFM colloidal probe

In the experiments, three electrolyte solutions (pure water, 0.1 mM and 1 mM NaCl) were employed. Here, the pure water was prepared by a pure water generation system and its resistance is as large as 18.2 MΩ.cm. Then, the 0.1 mM and 1 mM NaCl solutions were prepared with dissolution NaCl powder into the pure water with a certain weight ratio. In the experiments, all solutions above were prepared freshly and then, degassed in a vacuum box for at least 30 mins.

As figure 3 shown, the experiments were performed by an AFM which was operated under contact mode in liquid. During the measurement, the glass spheres were firstly placed in a home-made fluid cell and one of them (far away from other spheres) was selected as sample. Then one kind of electrolyte solution was introduced into the fluid cell carefully by a pipette. Then, The AFM was operated to make the colloidal probe approach to and retract from the selected glass sphere with a set speed. Meanwhile, by a four-quadrant photodetector in AFM system, the cantilever vertical deflection of the colloidal probe was continuous measured. The deflection profile (the relationship between the vertical

deflection of the probe and the distance in Z axis) can be obtained in this way. Then, using an AFM post-processing software, the deflection profile could be converted to the force profile (the relationship between the interaction force and separation) with the deflection sensitivity and spring constant of the probe. All the measurements above were carried out at room temperature (about 25 °C).

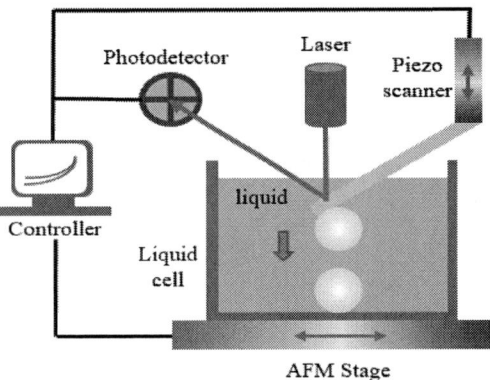

Figure 3. The experimental system of the measurement

IV. RESULTS AND DISCUSSIONS

A. Determination of approaching speed

As indicated in the Part II, during approaching, the speed of the colloidal probe should be low enough so that the hydrodynamic force could be neglected. In order to determine the value of the speed at which the hydrodynamic force could be neglected, the force profiles were measured when changing the speed of the probe in a large range. The force profiles of the colloidal probe during both approaching (filled dots) and retracting (hollow dots) with different speeds are shown in figure 4. It is obvious that the forces during retracting are lower than that during approaching when the speed are 3 and 5μm/s. These differences are mainly induced by the hydrodynamic force. However, when the speed is as low as 0.4μm/s, the forces during retracting are nearly same to that during approaching. Therefore, the approaching speed of colloidal probe should be no more than 0.4μm/s when using the zeta potential measurements. Actually, the speed of the colloidal probe was set 0.3μm/s in the measurement.

B. Determination of zeta potential

Figure 5 shows the typical force profiles when the colloidal probe approached the microsphere on substrate in three different electrolyte solutions. It is obvious that the forces for all cases increase when the separation deceases. Specially, when the solution selected is pure water (data marked in red in figure 5), the force appears (larger than 0.01nN) when the separation between spheres is about 350 nm. It should be noted that in pure water, the thickness of EDL is no more than 200 nm [11]. So, it is acceptable that the separation at which two EDLs of spheres begin to interact with each other is below 400nm. Consequently, the results here agree mainly with the theoretical analysis. In addition, by increasing the concentration, the interaction force appeared at a smaller separation, about 180 nm for 0.1

978-1-6654-3008-1/21 $31.00 ©2021 IEEE

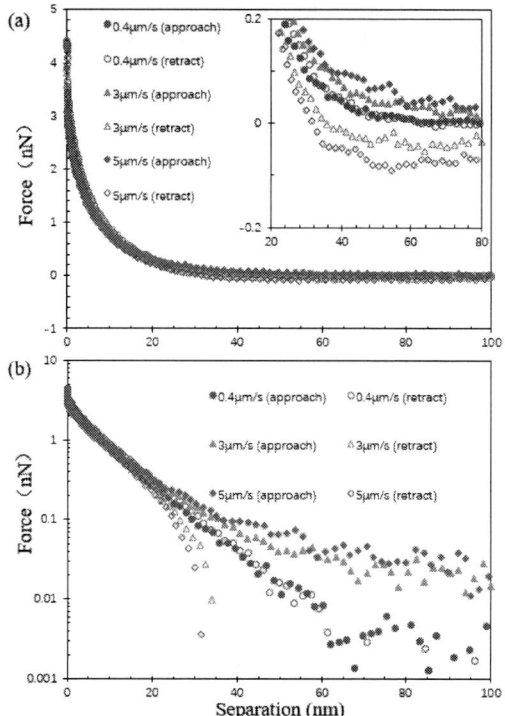

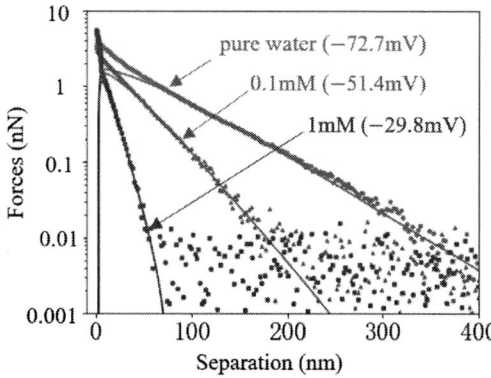

Figure 4. The force profiles of the colloidal probe during both approaching (filled dots) and retracting (hollow dots) with the speed of 5µm/s (red diamond), 3µm/s (green triangle) and 0.4µm/s (blue circle). To clarify the difference between the data with different speed, both regular (a) and logarithmic (b) plots are shown.

Figure 5. the measured interaction profiles between the colloidal probe and a glass microsphere in different electrolyte solutions.

mM and 60 nm for 1 mM NaCl. Therefore, the separation at which the repulsive force appears should decrease while increase the concentration as the EDL thickness decreases. The fitting results (with equation 1) are shown as solid lines in figure 5. The zeta potential of the glass sphere decreases obviously with the ionic concentration. More specifically, when the ionic concentration increases from pure water to 0.1mM and 1 mM NaCl, the zeta potential measured decreases from -72.7 mV to -51.4 mV and -29.8 mV.

What's more, it is obvious that the solid line cannot fit the experimental data well while the separation is below one Debye length. Actually, this point was also found by previous researchers [12]. It could be attributed to that the interaction forces between spheres in the short separation are very complex and many other short-range forces (hydration, solvation, surface roughness and so on) may be included. In this case, the DLVO theory is inapplicable.

C. Comparison the measured zeta potential with others

The comparison between the zeta potential measured in this work and reported in previous paper are shown in Table 1. Generally speaking, the zeta potential measured in this work is a little higher than that measured by traditional streaming potential in the literature [13]. The differences may be caused by different measuring principles. Even so, the zeta potential measured by AFM colloidal probe is a more direct method and it could be speculated that the results would be more reliable.

TABLE I.   THE COMPARISON OF ZETA POTENTIALS

Solutions	Zeta potential (mV)	
	Value in Ref.[13]	*Value in this study*
Pure water	-62	-72.7
0.1 mM NaCl	-29	-51.4
1 mM NaCl	-22	-29.8

V.   CONCLUSION

The zeta potential of glass sphere in electrolyte solutions was directly measured by AFM colloidal probe in this work. The results prove that the AFM colloidal probe is a very reliable technique to measure the weak interaction forces in liquid environment. In the future, this technique would be developed to study the zeta potential for some special interfaces (such as soft matter or biological samples) which would be very difficult to measure using traditional methods.

REFERENCES

[1] M.R. Hossan, D. Dutta, N. Islam, P. Dutta, "Review: Electric Field Driven Pumping in Microfluidic Device," Electrophoresis 39(5-6) (2018) 702-731.

[2] B. Wang, R.D. Oleschu, J.H. Horton, "Chemical Force Titrations of Amine- and Sulfonic Acid-Modified Poly(dimethylsiloxane)," Langmuir 21 (2005) 1290-1298.

[3] J. Jachowicz, M. D. Berthiaume, "Heterocoagulation of silicon emulsions on keratin fibers," Journal of Colloid and Interface Science 133(1989)118-134.

[4] F. Lu, J. Yang, D.Y. Kwok, "Flow Field Effect on Electric Double Layer during Streaming Potential Measurements," J. Phys. Chem. B 108 (2004) 14970-14975.

[5] Y. Yang, K.M. Mayer, J.H. Hafner, "Quantitative Membrane Electrostatics with the Atomic Force Microscope," Biophys J 92(6) (2007) 1966-1974.

[6] F.J. Montes Ruiz-Cabello, G. Trefalt, P. Maroni, M. Borkovec, "Electric Double-layer Potentials and Surface Regulation Properties Measured by Colloidal-probe Atomic Force Microscopy," Phys Rev E Stat Nonlin Soft Matter Phys 90(1) (2014) 1-10.

[7] M. Leivers, J.M. Seddon, M. Declercq, E. Robles, P. Luckham, "Measurement of Forces between Supported Cationic Bilayers by Colloid Probe Atomic Force Microscopy: Electrolyte Concentration and Composition," Langmuir 35(3) (2019) 729-738.

[8] W.A. Ducker, T.J. Senden, "Measurement of Forces in Liquids Using a Force Microscope," Langmuir 8 (1992) 1831-1836.

[9] H.-J. Butt, B. Cappella, M. Kappl, "Force Measurements with the Atomic Force Microscope: Technique, Interpretation and Applications," Surface Science Reports 59(1-6) (2005) 1-152.

[10] G. Han, H.-S. Ahn, "Calibration of Effective Spring Constants of Colloidal Probes Using Reference Cantilever Method," Colloids and Surfaces A: Physicochemical and Engineering Aspects 489 (2016) 86-94.

[11] H.-J. Butt, G. K., M. Kappl, "Physics and Chemistry of Interfaces," 2nd edn, Wiley-VCH, Weinheim (2006).

[12] N. Eom, D.F. Parsons, V.S.J. Craig, "Roughness in Surface Force Measurements: Extension of DLVO Theory to Describe the Forces between Hafnia Surfaces," J Phys Chem B 121(26) (2017) 6442-6453.

[13] Y. Gu, D. Li, "The ζ-Potential of Glass Surface in Contact with Aqueous Solutions," Journal of Colloid and Interface Science 226(2) (2000) 328-339.

Proceedings of the 16th Annual IEEE International
Conference on Nano/Micro Engineered and Molecular Systems
April 25-29, 2021

High-Resolution Measurement Method Based on Array FOV and Coded Apertures

Li Zhang, and Fei Xing*

Absrtact -It is difficult to realize large field of view (FOV) and high-resolution at the same time. How to achieve high-resolution measurement under the premise of large FOV imaging is an important research issue. Aiming at the problem of single point target recognition, especially the angle measurement of single light source, this paper proposes a high-resolution measurement method with array FOV and coded apertures. A mask is added in front of the image sensor. The coded apertures on the mask divide the large FOV into several sub fields of view (sub-FOV). Different sub-FOV are time-sharing multiplexed on the image sensor. Apertures encoded by position information are arranged on the sub-FOV. When a single light source irradiates the apertures, the diffraction spot is imaged on the image sensor. The centroid method can accurately determine the centroid of light spot to submicron scale or even nanoscale. Analyzing the coded information contained in the imaging can realize the recognition of sub-FOV, and then calculate the incident angle of light to realize the target recognition. A mask is designed to realize the method. Through experiments, this mask can achieve a large FOV of 120°×120° with an angle measurement accuracy of 0.003°(3 σ).

I. INTRODUCTION

The contradiction between large field of view (FOV) and high- resolution is a classic contradiction in instrument design [1]. Large FOV requires short focal length, and high-resolution requires long focal length, which is difficult to achieve in a system at the same time. The existing instruments only have large FOV, such as fisheye camera [2], or only have high-resolution, such as various microscopes. Dai Qionghai's team [3] has developed a large-scale instrument, which uses 35 cameras to splice and collect the large FOV light after passing through the complex optical system, so as to achieve micron resolution imaging within the centimeter level FOV. However, in many cases, we do not need to analyze the details of the object, but focus on an important attribute or feature of the object. In this regard, we can use this thinking to study more professional measurement methods and achieve higher performance instruments. Single light source is a common measurement object. For example, the sun sensor can measure the attitude of the satellite by measuring the incident angle of the sun [4]. AFM can measure the nanoscale shape by measuring the angle of the reflected laser beam [5]. In the development of a sun sensor, Wei Mingsong[6] used the technology of image sensor multiplexing to realize the measurement of large FOV with

high utilization rate of the instrument. However, the mask has two types of apertures，which named attitude update apertures and positioning apertures. The imaging characteristics of the two kinds of apertures are similar, which is not conducive to rapid recognition and solution.

Based on the above ideas, this paper proposes and implements a general method for measuring the beam angle of a single light source, and describes how this method can achieve large FOV and high-resolution at the same time. The basic principle of this method is to measure the incident angle of the light beam of the perceived object. On the basis of ensuring long focal length, large FOV is realized by array FOV arrangement and image sensor multiplexing. According to the coding information on the FOV of different arrays, the information of which sub fields of view (sub-FOV) is collected by the image sensor is identified. Then calculate the incident angle of the object, and finally realize the object recognition. In this paper, an optical mask is designed and combined with other hardware circuits, such as image sensor, to realize the above measurement method. An experiment is designed to verify the feasibility of this method. The measurement method can achieve large FOV and high-resolution, compact instrument, high resource utilization. And this measurement method has important practical value in the context of single target perception, light angle measurement, vehicle formation and so on.

II. BACIS PRINCIPLES

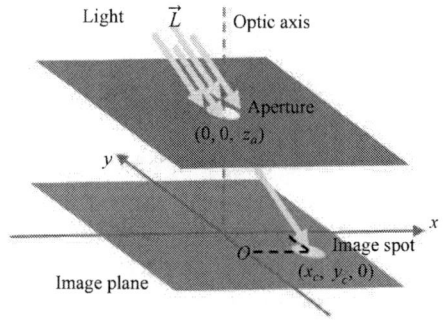

Figure 1. Schematic diagram of pinhole imaging and angle measurement.

The principle of this method is shown in Fig. 1.

---

1 Department of Precision Instrument, Tsinghua University, Beijing 100084, China.
2 Beijing Innovation Center for Future Chips, Tsinghua University, Beijing, 100084, China.
3 State Key Laboratory of Precision Measurement Technology and

Instrument, Tsinghua University, Beijing 100084, China
* Contacting Author: Fei Xing is with Department of Precision Instrument, Tsinghua University, Beijing 100084, China. (e-mail: xingfei@mail.tsinghua.edu.cn)

According to the principle of pinhole imaging, the light from the light source propagates to the small aperture and forms an image spot on the image plane. The angle between the line from the centroid of the image spot to the center of the aperture and the optical axis is the incident angle of the light.

Taking the projection of the aperture center on the image sensor as the coordinate origin, the O-xyz coordinate system is established. Assuming the coordinate of the aperture center is $(0, 0, z_a)$, the coordinates of the spot centroid formed by the beam passing through the aperture are $(x_c, y_c, 0)$, then the incident direction $\vec{L}$ of light can be expressed by two included angles, such as (1) and (2).

$$L_x = \arctan\left(\frac{x_c}{z_a}\right) \qquad (1)$$

$$L_y = \arctan\left(\frac{y_c}{z_a}\right) \qquad (2)$$

Where: L_x is the angle between the incident light of the perceived object and the z-axis in the O-xz plane of the system coordinate system, L_y is the angle between the incident light of the perceived object and the z-axis in the O-yz plane of the system coordinate system. After calculating the direction vector $\vec{L}$ of incoming and outgoing beams, the target can be perceived.

The determination of image spot centroid is different from object imaging, which requires accurate detail measurement but has the limitation of diffraction limit. And the resolution is generally in the micron scale. As an abstract feature of image pattern, the centroid of image spot can be calculated from the information of the whole two-dimensional image pattern, so it is not limited by the diffraction limit and can achieve nanoscale resolution. There are many methods to identify the image spot centroid, and the centroid method is commonly used. The accuracy of 0.05 pixel [7] can be achieved by choosing the way that accords with the imaging features and designing the centroid recognition algorithm reasonably. For micron scale pixels, the resolution of centering is nanoscale.

III. METHOD

In the case of nanoscale centering resolution of the image spot centroid, the angle resolution is mainly complemented by the focal length. The longer the focal length is, the higher the angle resolution is, as shown in Fig. 2.

In order to achieve high resolution, long focal length is needed. The FOV of a single aperture is limited by the focal length and the size of the image sensor. In the case of long focal length, the FOV of a single aperture is very small, as shown in Fig. 2(a) and Fig. 2(b). However, we can achieve a large FOV through the arrangement of multiple apertures, as shown in the Fig. 2(c). The relative positions of different apertures and image sensors are different, and their FOVs are also different. Multiple sub-FOV are seamlessly spliced together to form a large FOV.

The sub-FOV corresponding to each aperture is long focal length. That is to say, the image spot angle measurement using each aperture is high resolution, which can achieve high accuracy. The apertures are arranged on the two-dimensional plane, and the two-dimensional FOV is divided into two-dimensional sub fields to form the array FOV.

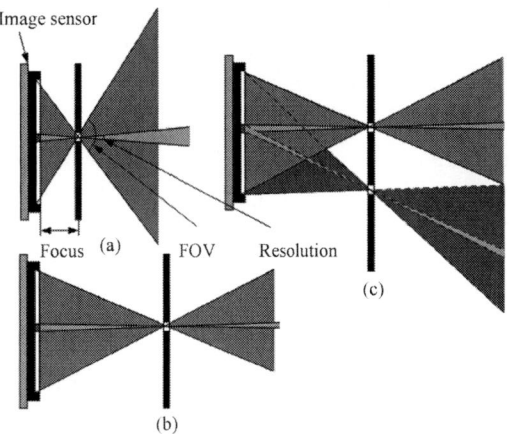

Figure 2. The contradiction between large FOV and high-resolution. (a) Large FOV requires short focal length, which leads to low resolution; (b) high resolution requires long focal length, which leads to small FOV; (c) multiple apertures are arranged to achieve large FOV and high-resolution at the same time.

For a single target to be detected, the aperture will form an image spot on the image sensor. From the second II, it can be seen that the vector formed by the line between the centroid of the imaging spot and the aperture is the direction of light incidence. When multiple apertures can be imaged on the image sensor, the incident vector of light can be accurately calculated only when the corresponding relationship between the aperture and the imaging spot is clear. The imaging spots formed by different apertures are some diffraction spots with similar characteristics, so it is difficult to identify every aperture through the information of a single image spot. Therefore, it is necessary to code the arrangement of apertures. Under the illumination of parallel beam or nearly parallel beam, the image spots formed by the apertures have the same coding information. According to the coding information of the centroid position of one image spot and its surrounding image spots, the matching relationship between the image spot and the aperture can be determined. And then the incident angle of the beam relative to the image sensor coordinate system can be calculated. Finally, the high-resolution recognition of a single target is realized.

The size of the sensitive surface of the image sensor is taken as the distribution range of apertures in a sub-FOV. In this area, the FOV-positioning-triangle, as shown in Fig. 3, is arranged. In the coordinate system of the plane where the aperture is located, the abscissa distance between the center position coordinates of aperture 1 and the center position coordinates of aperture 2 is the X-direction location coding, and the ordinate distance between the center position coordinates of aperture 2 and the center position coordinates of aperture 3 is the Y-direction location coding. In order to reduce the random error, we should arrange as many apertures as possible in the sub-FOV. Increase the difference between the abscissa of aperture 2 and aperture 3, so that the apertures evenly cover the whole sub-FOV.

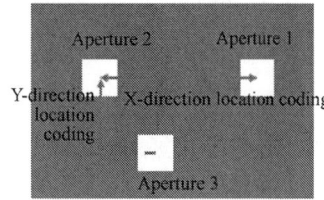

Figure 3. Aperture coding method.

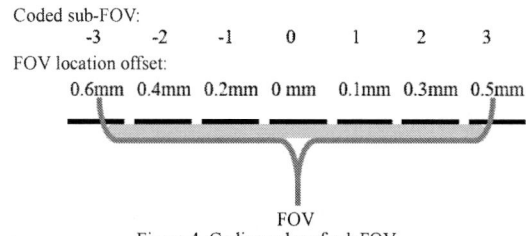

Figure 4. Coding rules of sub FOV.

The coding rules of aperture are shown in the Fig. 4. The X-direction location coding and Y-direction location coding in the FOV-positioning-triangle of different sub-FOV have different values. By identifying the values of the X-direction and Y-direction location coding and comparing them with the coding rules, we can determine which sub-FOV the orientation triangle is located in. Furthermore, we can determine which sub-FOV the three holes in the FOV-positioning-triangle and the three imaging spots belong to. For holes that are not in the same sub-FOV, false-FOV-positioning-triangles will also be formed. These false-FOV-positioning-triangles are not helpful to determine which sub-FOV that the aperture belongs to, but can determine the boundaries of different sub-FOV. According to the known aperture arrangement and the distance between the aperture and the sub-FOV boundary, the one-to-one correspondence between the aperture and the image spot can be realized. After the corresponding relationship between aperture and image spot is determined, the incident vector of light can be calculated accurately.

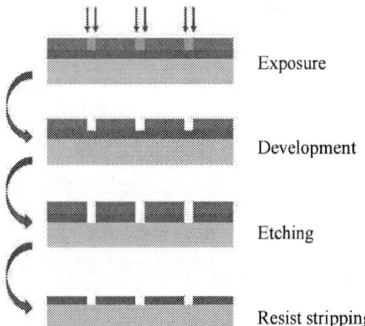

Figure 5. The manufacturing technology of coding aperture.

The coded apertures are produced by photolithography, as shown in the Fig. 5. The photoresist is coated on the chromium plated quartz substrate, and the laser or electron beam is used to expose according to the designed array FOV and coded apertures. After the photoresist at the exposure is removed, the chromium layer is exposed and removed by etching. After resisting stripping, the required array FOV and coded apertures can be obtained.

The physical mask is shown in Fig. 6. When receiving the light from a certain angle, the image collected by the image sensor below is as shown in the Fig. 7.

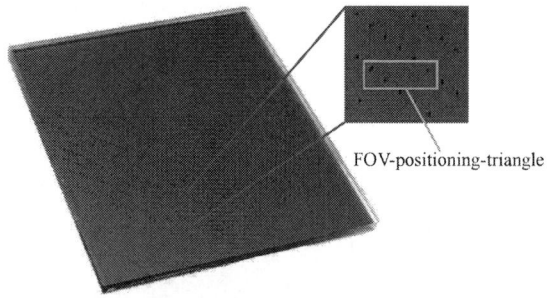

Figure 6. Physical picture of mask.

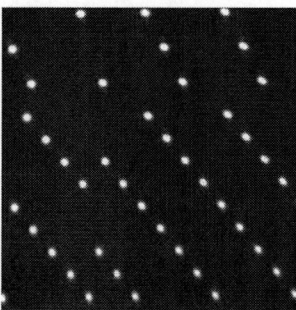

Figure 7. Encoded image collected by image sensor.

In order to identify the corresponding relationship between the collected image spots and the apertures on the mask, this paper designs the recognition algorithm. Firstly, the FOV-positioning-triangle in the acquired image is identified, then the junction point of the sub-FOV is identified, and then the corresponding relationship between the acquisition area and the mask area is identified. Finally, the matching between the small aperture and the image spot in the corresponding area is realized.

IV. EXPERIMENT

In this paper, an experimental system is built, as show in Fig. 8. The mask is placed on the image sensor and fixed on the high-accuracy three-axis turntable. The turntable drives the mask and the image sensor to move together. The light from the light source irradiates the mask plate after passing through the collimator, and the image sensor collects the image of the imaging spots. The image is transmitted to the computer through the serial port, and the incident angle of light is calculated through the processing of recognition algorithm. Taking the set value of the turntable as the true value, the measurement error is obtained by comparing the calculated value with the true value.

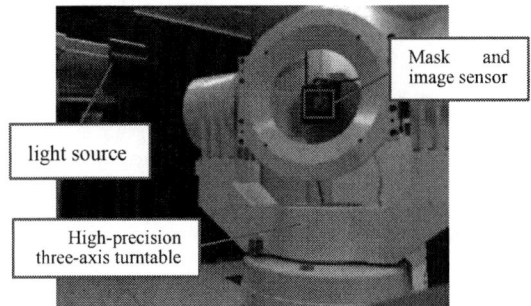

Figure 8. Experimental device.

Firstly, the two test curves selected in this paper are: Y-axis of turntable keeps 0 ° unchanged, X-axis turns from - 60 ° to 60 ° with interval of 0.5 ° sampling points (first test curves); X-axis of turntable keeps 0 ° unchanged, Y-axis turns from - 60 ° to 60 ° with interval of 0.5 ° sampling points (second test curves). The collected image is processed. The angle measurement error in X direction and Y direction are shown in the Fig. 9.

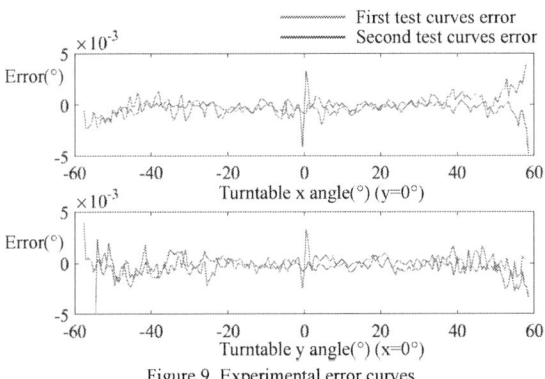

Figure 9. Experimental error curves.

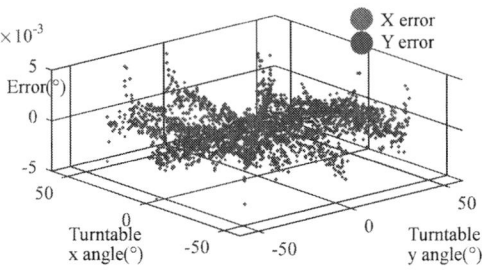

Figure 10. Experimental error scatter-plot; Blue: Angle measurement error in X direction; Red: Angle measurement error in Y direction

Then, 8 test curves are selected in the whole FOV, which are 3 horizontal lines, 3 vertical lines and 2 diagonal lines. The horizontal line of the test is that the central-axis of the turntable remains unchanged and the outer-axis of the turntable changes from - 60 ° to 60 °; the vertical line of the test is that the outer-axis of the turntable remains unchanged and the central-axis of the turntable changes

from - 60 ° to 60 °; the oblique line of the test is that the central-axis and the outer-axis of the turntable rotate at the same time. Each curve is separated by 0.5 ° to collect data, and the collected image is processed and compared with the set value of the turntable. Finally, it can be seen that the angle measurement accuracy of 0.003 ° can be achieved on the premise of ensuring 120 ° × 120 ° large FOV.

V. CONCLUSION

This paper presents a method to measure both large FOV and high-resolution. This method is realized by array FOV and coded imaging. The long focal distance ensures the high-resolution of the FOV, but reduces the field angle. Through the arrangement of apertures, different FOV is composed of array FOV, and it is spliced into a large FOV. The array FOV is projected on the same image sensor, which is difficult to recognize. In this paper, we use coded apertures to distinguish different sub-FOV. This paper introduces the principle of this method and makes a mask which uses the principle to measure. Through experiments, the system can achieve the accuracy of 0.003 ° in light incidence angle measurement with the large FOV of 120 °×120 °. This paper prove the feasibility of this measurement method. The method has practical value in the fields of single target recognition and ray angle measurement.

REFERENCES

[1] AM. Lohmann, RG. Dorsch, D. Mendlovic, Z. Zalevsky, C. Ferreira, "Space-bandwidth product of optical signals and systems," *JOURNAL OF THE OPTICAL SOCIETY OF AMERICA A-OPTICS IMAGE SCIENCE AND VISION*, vol.13, pp.470-473, 1996.

[2] J. Kannala, and SS. Brandt, "A generic camera model and calibration method for conventional, wide-angle, and fish-eye lenses," *IEEE TRANSACTIONS ON PATTERN ANALYSIS AND MACHINE INTELLIGENCE*, vol.28, pp.1335-1340, 2006.

[3] Jingtao fan, Jinli Suo, Jiamin Wu, Hao Xie, Yibing Shen, Feng Chen, Guijin Wang, Liangcai Cao, Guofan jin, Quansheng He, Tianfu Li, Guoming Luan, Lingjie Kong, Zhenrong Zheng, and Qionghai Dai, "Video-rate imaging of biological dynamics at centimetre scale and micrometre resolution," *NATURE PHOTONICS*, vol. 13, pp.809, 2019.

[4] P. Ortega, G. Lopez-Rodriguez, J. Ricart, M. Dominguez, LM. Castaner, JM. Quero, CL. Tarrida, J. Garcia, M. Reina, A. Gras, M. Angulo, "A Miniaturized Two Axis Sun Sensor for Attitude Control of Nano-Satellites," *IEEE SENSORS JOURNAL*, vol.10, pp.1623-1632, 2010.

[5] J. Kwon, J. Hong, YS. Kin, DY. Lee, K. Lee, SM. Lee, and SI. Park, "Atomic force microscope with improved scan accuracy, scan speed, and optical vision," *REVIEW OF SCIENTIFIC INSTRUMENTS*, vol.74, pp.4378-4383, 2003.

[6] MinSong Wei, Fei Xing, Zheng You, and Geng Wang, "Multiplexing image detector method for digital sun sensors with arc-second class accuracy and large FOV," *OPTICS EXPRESS*, vol.22, pp.23094-23107, 2014.

[7] HaiYang Zhan, Fei Xing, Li Zhang, "Analysis of optical measurement resolution limit for close-to-atomic scale manufacturing," *Acta Physica Sinica*, 2020.

Proceedings of the 16th Annual IEEE International
Conference on Nano/Micro Engineered and Molecular Systems
April 25-29, 2021

Different Effects of Mass and Damping on Performance of Vibration and Wind Energy Harvesters*

Xiaokang Yang, Xuefeng He, Zhengguo Shang, Hui Huang, Chunlong Li, and Yufei Liu

Abstract—Piezoelectric energy harvesters which scavenge energy from ambient vibrations and winds attract increasing attention as the power sources of wireless sensor nodes. For these two types of harvesters, the optimal parameters are much different, but few references discussed this type of differences, which will be evaluated by deliberately designed experiments in this work. By taking the harvesters with the same piezoelectric composite beam configuration but different equivalent masses as examples, the differences were analyzed theoretically and experimentally. The equivalent masses are modified by changing the masses of the proof mass or the bluff body. In vibration energy harvesting, as we know, heavier mass is preferred for the lower resonant frequency and the higher electrical output. Decreasing the mechanical damping ratio is helpful to enlarge the output power at resonance but adversely narrows the frequency bandwidth. On the contrary, in wind energy harvesting, theoretical results show that smaller mechanical damping ratio is preferred and the experimental results show that lighter mass is preferred for the higher output power and the lower cut-in speed of galloping. The conclusions are valuable to developing efficient small- or micro-scale WEHs.

Keywords—energy harvesting; vibration; wind-induced vibration; piezoelectricity; galloping

I. INTRODUCTION

More and more wireless sensor nodes have been integrated into the world for structural health monitoring, environmental monitoring, and security[1,2]. These nodes are commonly powered by small batteries which require expensive and time-consuming maintenance such as replacing or recharging[3]. Small- or micro-scale kinetic energy harvesters (KEHs) are promising power sources of wireless sensor nodes, which can transfer ambient kinetic energy into electricity and may be used to replace or charge small batteries[4]. Vibrations and winds are ubiquitous kinetic energies in environments, as a result, vibration energy harvesters (VEHs) and wind energy harvesters (WEHs) attract most attention in recent years. Compared with electromagnetic or electrostatic KEHs, piezoelectric ones possess the advantages of structural simplicity and high power density.

Piezoelectric composite cantilever with a proof mass or bluff body at the free end, as shown in Figure , can be used

to scavenge not only vibration energy but also wind energy, and even to scavenge them at the same time [5-9]. The proof mass at the free end of the cantilever for vibration energy scavenging also functions as the bluff body for wind energy scavenging. As we know, the equivalent mass and mechanical damping of KEHs strongly affect the output performance. Literature has systematically studied the effects of mass and mechanical damping on the performance of VEHs. But there are few references which systematically discussed the differences of the effects of mass and mechanical damping on VEHs and WEHs. In this work, to analyze these effects, the piezoelectric vibration energy harvesters (PVEHs) and the piezoelectric wind energy harvesters (PWEHs) with the same piezoelectric composite beam configuration but different proof masses or bluff bodies will be theoretically and experimentally analyzed.

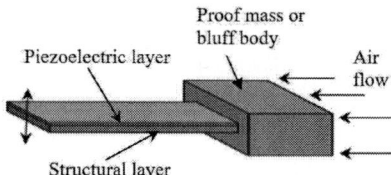

Figure 1. Schematic of a cantilevered piezoelectric energy harvester for vibrations and winds.

II. THEORETICAL ANALYSIS

The lumped-parameter and distribute-parameter models of PVEHs have been proposed and used to understand the characteristics and to optimize the device structure[4,5]. When the base excitation frequency ω matches the natural frequency, the harvester produces the output power given by[4]

$$P = \frac{m\xi_e A^2}{4\omega(\xi_m+\xi_e)^2},\tag{1}$$

where m is the equivalent mass, ξ_m and ξ_e are the mechanical and electrical damping ratios, respectively, and A is the acceleration magnitude of base excitations. The effects of mass and damping on VEHs have been discussed systematically in literature, suggesting that heavy tip mass

---

* This work was financially supported by the R&D project of State Grid Corporation of China under Grant 5700-202036164A-0-0-00. *(Corresponding author: X. F. He.)*

X. K. Yang, X. F. He, Z. G. Shang, and Y. F. Liu are with Key Laboratory of Optoelectronic Technology and Systems of the Education Ministry of China, Chongqing University, Chongqing 400044, China (e-mail: yangxiaokang@haust.edu.cn, hexuefeng@cqu.edu.cn; zhengry@cqu.edu.cn; yufei.liu@cqu.edu.cn).

H. Huang and C. L. Li are with Electric Power Intelligent Sensing Technology and Application State Grid Corporation Joint Laboratory, Global Energy Interconnection Research Institute Co., Ltd., Beijing 102209, China (e-mail: huanghui@geiri.sgcc.com.cn; lichunlong@geiri.sgcc.com.cn).

X. K. Yang is also with School of Mechatronics Engineering, Henan University of Science and Technology, Luoyang 471000, China (e-mail: yangxiaokang@haust.edu.cn).

978-1-6654-3008-1/21 $31.00 © 2021 IEEE

should be used to increase the power density[4,5]. As a result, high density materials such as tungsten have been used to make the proof mass to increase m. But the selection of ξ_m is more complex because decreasing ξ_m may increase the output power at resonance, but adversely decreases the frequency bandwidth of VEHs. Therefore, the smaller mechanical damping is not always the better and there is a trade-off between the maximum output power and the bandwidth.

Vortex-induced vibration (VIV), galloping or flutter have been utilized by PWEHs to scavenge wind energy[9-15]. Among them, galloping-based PWEHs have the advantages of both structural simplicity and high electrical output. WEHs can produce relatively high electrical outputs only when the wind speed is higher than a specified value, the cut-in speed. In the following, the effects of mass and mechanical damping ratio on wind energy scavenging are analyzed by using the galloping-based PWEHs as the examples. For a galloping PWEH with the configuration given in Fig. 1, when air flows from the free to the clamped ends of the cantilever with the speed of U, the equations of a galloping PWEH are given by [7-9]

$$m\ddot{y} + \left(c - \frac{1}{2}\rho ULDA_1\right)\dot{y} + ky + \frac{1}{2}\rho U^2 LD\left[A_3\left(\frac{\dot{y}}{U}\right)^3 - A_5\left(\frac{\dot{y}}{U}\right)^5 + A_7\left(\frac{\dot{y}}{U}\right)^7\right] + \Theta V = 0 \qquad (2)$$

$$C_p\dot{V} + \frac{V}{R} - \Theta\dot{y} = 0 \qquad (3)$$

where the overdot represents a derivative with respect to time, y is the bluff body displacement in the transverse flow direction, L and D are length and height of the bluff body, respectively, ρ is the density of air, V is the voltage across the resistor R connected to the harvester, m, c, k, Θ and C_p are the equivalent mass, the mechanical damping coefficient, the stiffness, the electromechanical coupling coefficient and the capacitance of the harvester, respectively, and A_1, A_3, A_5 and A_7 are the transverse force coefficients determined by the shape of the bluff bodies. The mechanical damping coefficient c is related to the mechanical damping ratio ξ_m by $c = 2\xi_m\sqrt{mk}$. As the turbulence of the small wind tunnel in our laboratory is about 6.5%, A_1, A_3, A_5 and A_7 are set as 2.83, 90.75, -4248.3 and 30146.43, respectively[16].

The effects of the mass and mechanical damping ratios on PWEHs were analyzed by solving Eqs. (2) and (3). The same piezoelectric cantilever configuration but different bluff bodies were used in the simulations, and in the experiments in Section III. The parameters completely determined by the piezoelectric cantilever were set as fixed values, that is $k = 10.173N/m$, $D = 4mm$, $L = 10mm$, $\xi_m = 0.03$, $\rho = 1.25kg/m^3$, $C_p = 1.328 \times 10^{-10}F$ and $\Theta = 2.009 \times 10^{-6}C/m$.

First, the effects of the mass were simulated. With the increase of the wind speed, the simulated output powers across the optimal resistances for the cases with different equivalent masses of $m = 0.5m_0$, $m = m_0$, $m = 2m_0$ and $m = 5m_0$, where $m_0 = 2.631 \times 10^{-10}kg$, were worked out, as shown in Figure . The cut-in speed of galloping

decreases with decreasing equivalent mass. But, for a specified wind speed with galloping, the output power of the harvester with a lighter bluff body is lower. Therefore, the lighter bluff body is not always the better and there is a trade-off between the cut-in speed and the output power.

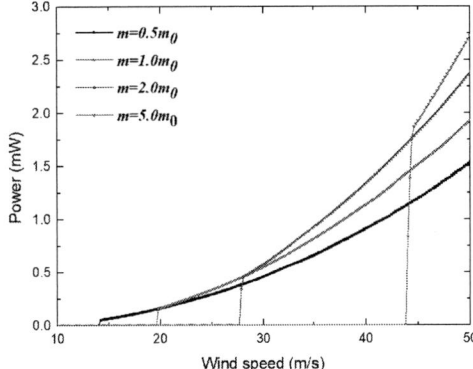

Figure 2. Simulated RMS voltage versus wind speed for devices with different mass.

Then, the effects of the mechanical damping ratio were simulated by setting $\xi_m = 0.02$, $\xi_m = 0.03$, $\xi_m = 0.04$ and $\xi_m = 0.06$, respectively. In the simulations, the equivalent mass was fixed as $m = m_0$. The output powers across the optimal resistance were worked out and plotted in Figure . The simulated cut-in speed of galloping decreases and the output power increases when the mechanical damping ratio increases. Therefore, for wind energy harvesting, smaller damping ratio is always preferred. There is no doubt for this conclusion, and it is almost unnecessary to be verified by experiments.

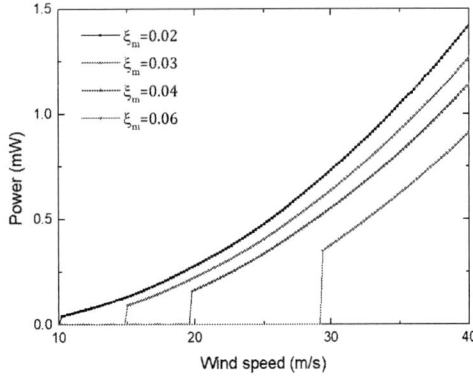

Figure 3. Simulated RMS voltage versus wind speed for devices with different mechanical damping.

In summary, for vibration energy scavenging, heavier proof mass is preferred for the higher output power and the lower resonant frequency. But smaller mechanical damping ratio is not always the better and there is a trade-off between the maximum output power and the frequency bandwidth. For wind energy scavenging, smaller mechanical damping ratio is preferred for the smaller cut-in speed and the higher output power. When the bluff body mass decreases, the cut-in speed decreases but the output power at galloping also decreases. Therefore, according to the simulation results, lighter bluff body is not always the better and there is a trade-off between the cut-in speed and the output power. In the following, the effects of the mass will be further studied by experiments.

III. EXPERIMENTAL RESULTS AND DISCUSSIONS

Experiments were deliberately designed to evaluate the different effects of the mass on PVEHs and PWEHs. Two harvester prototypes (Dev 1 and Dev 2) were fabricated, as shown in Figure . Their cantilevers are almost the same, with the free length of 8 mm and width of 4 mm. The cantilevers are composed of a 92-µm-thick polyethylene terephthalate (PET) and a 52-µm-thick polyvinylidene fluoride (PVDF). The dimensions of the cuboid proof masses or bluff bodies are the same of 4mm×4mm×10mm. The proof masses or bluff bodies of Dev 1 and Dev 2 are made of polymethyl methacrylate (PMMA) and foam, respectively, as a result Dev 1 is much heavier than Dev 2. The equivalent mass of Dev 2 is approximately equal to m_0, that is $m = m_0$, and the value of m_0 may be found in Section II.

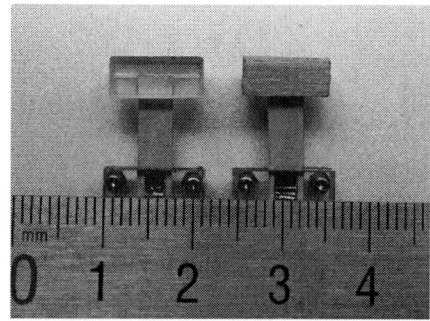

Figure 4. Photograph of harvester prototypes.

The vibration energy scavenging performance of the prototypes were characterized on an electromagnetic shaker. Under harmonic base excitations with the acceleration amplitude of 0.2 g (where g = 9.81 m·s⁻²), the root mean square (RMS) voltages across the optimal resistances were measured, as shown in Figure . The experimental optimal resistances of Dev 1 and Dev 2 are 39 MΩ and 15 MΩ, respectively. When the excitation frequency increases, and the experimental maximum RMS voltages are 15.83 and 2.79 V at 32.2 and 98.8 Hz, respectively. The maximum output powers at resonance of Dev 1 and Dev 2 are 6.43 and 0.52 µW, respectively. The frequencies of vibrations in natural environments are generally low and researchers always aim to decrease the resonant frequencies of PVEHs to match that of the ambient vibrations in most cases[4,17,18]. Therefore, it can be concluded that, for vibration energy scavenging, the device with heavier proof mass (Dev 1) is much better than the other (Dev2) for the higher output power and lower resonant frequency. The conclusion is the same with the theoretical result and has been arrived by many researchers.

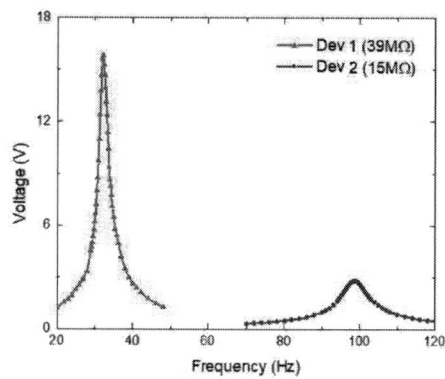

Figure 5. Experimental RMS voltage across optimal resistances versus base excitation frequency.

The prototypes were then tested in a small wind tunnel to evaluate the wind energy scavenging performances. With the increase of the wind speed, the RMS voltage across the optimal resistances were measured, as shown in Figure . The cut-in speed of galloping for the device with the lighter bluff body (Dev 2) is much lower and the electrical output is much higher than the other (Dev1). For example, when wind speed is 13.2 m/s, the RMS voltages across the optimal loads of Dev 1 and Dev 2 are about 3.74 and 27.39 V, with the output power of about 0.36 and 50.01 µW, respectively. Therefore, the experiments show that, for PWEHs, the device with lighter bluff body (Dev 2) is better than the other (Dev1) for the higher output power and lower cut-in speed.

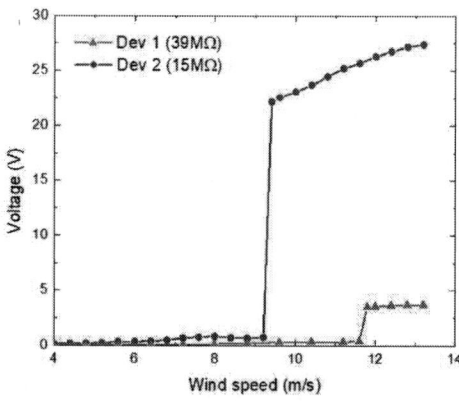

Figure 6. RMS voltage across optimal resistances versus wind speed.

It should be noted that, for wind energy scavenging, there is an opposite tendency of the theoretical and experimental relationships between the bluff body mass of the harvester and the electrical output in the galloping wind speed range. The numerical simulations show that, in the galloping ranges, the output power of the device with the lighter bluff body is lower than the other, as shown in Figure . But the experimental results show that, in galloping ranges, the output power of the device with the lighter bluff body is much higher than the other, as shown in Figure . The main reason is that the mechanical damping ratios of the devices with the heavier and lighter bluff bodies in the simulations are set as the same value of 0.03. In fact, the mechanical damping ratio changes with the mass. But most references set a constant mechanical damping ratio in the optimization of PWEHs. Above results in this work clearly demonstrate that the simple treatment of mechanical damping by neglecting its dependency on the mass is improper and may cause wrong predictions. More work needs to be conducted to find more proper methods to set the mechanical damping in the optimization of PWEHs.

IV. CONCLUSION

The effects of the mass and mechanical damping on VEHs and WEHs are different. Taking the harvesters with the same piezoelectric composite beam configuration but different masses, which are modified by changing the mass of the bluff bodies or proof mass, these effects were theoretically analyzed, and the effect of the mass was experimentally. The theoretical analysis show that heavier proof mass is preferred for a PVEH and smaller mechanical damping ratio is preferred for a PWEH. The smaller mechanical damping ratio is not always the better for a PVEH and there is a trade-off between the maximum output power at resonance and the frequency bandwidth. By setting the mechanical damping ratio as a constant value, the simulation results show that the lighter proof mass is not always the better for a PWEH and there is a trade-off between the output power and cut-in speed. But the experiments show that the lighter proof mass is always the better for a PWEH for the higher output power and the lower cut-in speed of galloping. According to the simulations, at a specified wind speed with galloping, the output power of the device with lighter bluff body is lower, which is opposite to the experimental results. The main reason is the improper treatment of mechanical damping. For the optimization of PWEHs, more proper treatment methods for the mechanical damping need to studied in the future.

REFERENCES

[1] Y. L. Zi and Z. L. Wang, "Nanogenerators: An emerging technology towards nanoenergy," *APL Mater.*, Vol. 5, No. 7, 2017.

[2] Y. Chen, N. Chiotellis, L.-X. Chuo, C. Pfeiffer, Y. Shi, R. G. Dreslinski, A. Grbic, T. Mudge, D. D. Wentzloff, D. Blaauw, and H. S. Kim , "Energy-Autonomous Wireless Communication for Millimeter-Scale Internet-of-Things Sensor," *IEEE J. Sel. Area. Comm.*, vol. 34, no. 12, pp. 3962-3977, 2016.

[3] Z. Yan and A. Abdelkefi, "Nonlinear characterization of concurrent energy harvesting from galloping and base excitations," *Nonlinear Dynam.*, vol. 77, no. 4, pp. 1171–1189, 2014.

[4] S. Roundy, P. K. Wright, and J. Rabaey, "A Study of Low Level Vibrations as a Power Source for Wireless Sensor Nodes," *Comput. Commun.*, vol. 26, no. 11, pp. 1131–1144, 2003.

[5] A. Erturk and D.J. Inman, "An experimentally validated bimorph cantilever model for piezoelectric energy harvesting from base excitations," *Smart Mater. Struct.*, vol.18, no. 2, 2009.

[6] X. F. He, Z. G. Shang, Y. Q. Cheng, and Y. Zhu, "A micromachined low-frequency piezoelectric harvester for vibration and wind energy scavenging," *J.Micromech. Microeng.*, vol. 23, no. 12, 2013.

[7] A. Bibo and M. F. Daqaq, "On the optimal performance and universal design curves of galloping energy harvesters," *Appl. Phys. Lett.*, vol. 104, no. 2, 2014.

[8] Y. Yang, L. Zhao, and L. Tang, "Comparative study of tip cross-sections for efficient galloping energy harvesting," *Appl. Phys. Lett.*, vol. 102, no. 6, 2013.

[9] X. F. He, X. K. Yang, and S. L. Jiang, "Enhancement of wind energy harvesting by interaction between vortex-induced vibration and galloping," *Appl. Phys. Lett.*, vol. 112, no. 3, 2018.

[10] A. Abdelkefi, "Aeroelastic energy harvesting: A review," *Int. J. Eng. Sci.*, vol. 100, pp. 112–135, 2016.

[11] A. B. Rostami and M. Armandei, "Renewable energy harvesting by vortex-induced motions: Review and benchmarking of technologies," *Renew. Sust. Energ. Rev.*, vol. 70, pp. 193–214, 2017.

[12] M. F. Daqaq, A. Bibo, I. Akhtar, A. H. Alhadidi, M. Panyam, B. Caldwell, and J. Noel, "Micropower Generation Using Cross-Flow Instabilities: A Review of the Literature and Its Implications," *J. Vib. Acoust.*, vol. 141, no. 3, 2019.

[13] H. L. Dai, A. Abdelkefi, and L. Wang, "Theoretical modeling and nonlinear analysis of piezoelectric energy harvesting from vortex-induced vibrations," *J. Intel. Mat. Syst. Str.*, vol. 25, no. 14, pp. 1861–1874, 2014.

[14] J. L. Wang, C. Y. Zhang, S. H. Gu, K. Yang, H. Li, Y. Y. Lai, and D. Yurchenko, "Enhancement of low-speed piezoelectric wind energy harvesting by bluff body shapes: Spindle-like and butterfly-like cross-sections," *Aerosp. Sci. Technol.* vol.103, no. 10, 2020.

[15] M. Bryant, E. Wolff, and E. Garcia, "Aeroelastic flutter energy harvester design: the sensitivity of the driving instability to system parameters," *Smart Mater. Struct.*, vol. 20, no.12, 2011.

[16] X. Yang, X. He, J. Li, and S. Jiang, "Modeling and verification of piezoelectric wind energy harvesters enhanced by interaction between vortex-induced vibration and galloping," *Smart Mater. Struct.*, vol. 28, no. 11, 2019.

[17] K. Q. Fan, Y. W. Zhang, H. Y. Liu, M. L. Cai, and Q. X. Tan, "A nonlinear two-degree-of-freedom electromagnetic energy harvester for ultra-low frequency vibrations and human body motions," *Renew. Energy*, vol. 138, pp. 292-302, 2019.

[18] H. J. Zhang, S. L. Jiang, and X. F. He, "Impact-based piezoelectric energy harvester for multidimensional, low-level, broadband, and low-frequency vibrations," *Appl. Phys. Lett.*, vol. 110, no. 22, 2017.

Gap in pagination due to unavailable paper.

Pages 390-391

Proceedings of the 16th Annual IEEE International
Conference on Nano/Micro Engineered and Molecular Systems
April 25-29, 2021

In-situ deposited ion-imprinted polymers for electrochemical detection of trace cadmium in water

Shuyu Xiao[1,2,4] Jingfang Hu[1,2,3,4*] Jiyang Wang[1,2] Yu Song[1,2,3,4*] Yansheng Li[1,2,3,4] Guowei Gao[1,2,3*] Lei Qin[2*]

Abstract—**Two kinds of in-situ deposited ion imprinted polymers (IIPs) using electropolymerization method were presented for electrochemical detection of trace cadmium in water, which are ion imprinted polypyrrole/reduce graphene oxide modified glassy carbon electrode (IIPpy/rGO/GCE) and ion imprinted polyo-phenylenediamine/reduced graphene oxide modified glassy carbon electrode (IIPoPD/rGO/GCE).Electrochemical analysis of cadmium on the two modified electrodes was investigated by square wave anodic stripping voltammetry (SWASV) under the optimized conditions. The experiment results showed that IIPpy/rGO/GCE had wider linear range (1-100μg/L) and IIPoPD/rGO/GCE possessed lower detection limit (0.13 μg/L). The two kinds of modified electrodes are both green, simple, reused and compatible for trace cadmium detection in water with highly selectivity.**

I. INTRODUCTION

Cadmium is a highly toxic heavy metal pollutant, which can accumulate in human kidneys and destroy their filtration function, and may cause health problems such as vomiting, abdominal pain, fracture, lung cancer and so on [1].As early as 1974, the United Nations Environment Programme and the International Commission on Heavy Metals in Labor and Health identified cadmium as a key pollutant, and the International Agency for Research on Cancer listed cadmium (II) as a class 1 carcinogen [2].So it is of great practical significance to detect trace amount of cadmium.

Traditional optical detection methods such as ultraviolet spectrophotometry[3],atomic absorption spectrophotometry [4], inductively coupled plasma mass spectrometry [5] and fluorescence method [6] can obtain accurate detection results for cadmium ions, but their measurement instruments are large in volume and time-consuming, which is not conducive to field and on-line application. Electrochemical analysis [7] is based on the principle that the electrode and the solution to be tested form a simple chemical battery, and then according to the electrical or electrochemical properties of the substance to be tested, Some of the electrical parameters, such as potential and current, can be

quantitatively analyzed by establishing relationship between the change value of the parameter and the concentration or content of the substance to be measured. Compared with the traditional detection methods, the electrochemical method has the advantages of being easy to miniaturization, high sensitivity and low cost. Many literatures have reported that inorganic materials, organic materials and biomaterials, etc have been used as electrode modified materials for cadmium electrochemical detection [8-10]. Among these various modifiers, IIPs is one of the organic materials with highly selectivity and stability for specific recognition of target heavy metal ions.

The superiority of IIPs is their reliable selectivity for the capture of target ion [11]. IIPs take the target ions as the template and polymerize with specific functional monomers under the action of crosslinkers. After polymerization, the template ions are eluted to obtain a large number of three-dimensional pores that are completely consistent with the spatial size and shape structure of the template ions. These sites can specifically rebind the target ions in the presence of similar ions. Due to most IIPs are non conductiv organic polymers, IIPs always were ground to powder and assembled into carbon paste electrode for electrochemical sensor development [12-13]. This assembled modication method is not only complicate and time consuming, but also difficult to have a good uniformity between sensor and sensor, which, however, resulted limited application. In this paper, we found the conducting polymers of PPy and PoPD can be used for IIP preparation, because they both have – NH_2 group which could have complexation with metal-ions during polymerization. So, herein electrochemical polymerization method, which is thought simple, controllable to have uniform film thickness by adjusting the electrodeposition parameters of time, potential, and current [14], was used for in-situ preparation IIPs on electrode surface using o-phenylenediamine (oPD) and pyrrole (py) as functional monomer repectively, and cadmium as template. In order to improve the electron trasfer rate, reduced graphene oxide (rGO) was modified on electrode surface prior to IIPs electropolymerization.

The aim of the present work was to introduce the two simple and electrochemical sensors based on in-situ deposited IIPs modified electrodes for trace cadmium detection and comparative study was investigated on their preparation, performances and real application.

II. EXPERIMENT N

Both experiments used a three-electrode system, with glassy carbon electrode as working electrode, platinum electrode as counter electrode and silver/silver chloride

[1]School of Automation, Beijing Information Science & Technology University, Beijing 100101, China
[2]Beijing key Laboratory of sensor, Beijing Information Science & Technology University, Beijing 100101, China
[3]Key Laboratory of Modern Measurement and Control Technology, Ministry of Education, Beijing Information Science and Technology University, Beijing 100192, China
[4]State Key Laboratories of Transducer Technology, Shanghai Institute of Microsystems and Information Technology, Chinese Academy of Sciences, Shanghai 200050, China.

978-1-6654-3008-1/21 $31.00 © 2021 IEEE

electrode as reference electrode. In order to ensure the accuracy of the experiment process, the electrode should be pretreated before the experiment, and the glassy carbon electrode and platinum electrode should be polished by Al_2O_3. The imprinted film was in situ polymerized on the electrode surface by electro-polymerization.

Preparation of this novel electrochemical sensor includes two steps: the first step is to reduce graphene oxide to reduced graphene oxide by cyclic voltammetry (CV) and to deposit on the surface of a glass carbon electrode with a deposition time of 600 seconds. The second step is the in situ polymerization of cadmium ion imprinted membrane. The polished electrode is put into the Pre-prepared solution for electro-polymerization. Finally, electrochemical peroxidation was selected to remove the template ions.

Ion imprinted membrane with py as functional monomer is prepared as follows: 0.1mol/L py, 250mg/L Cd(II) and 0.1mol/L LiClO4 are added into the aqueous solution, stirring to obtain a homogeneous mixed solution, and then CV scanning (-0.2 to 0.8 V, scanning rate of 50 mV/s) applied to rGO/GCE, for 10 cycles in the mixed solution, and finally, in0.1mol/L NaOH solution, the template Cd(II) ions were removed 10 times with CV(-0.4 to +1V) running at a scanning rate of 50mV/s . The method for preparing the ion imprinted polymers with oPD as functional monomer is as follows: rGO/GCE is soaked in 0.1MHAC-NaAc buffer (pH=5.2), containing 0.04mol/L oPD and 0.01mol/L cadmium (II), and is subjected to electro-polymerization for 20 times in the mixed solution by CV(0.0 to +0.8V). At last, in 0.1mol/L HCl solution, the template ions were removed by electrochemical oxidation at 1.5V for 15min.

III. CHARACTERIZATION OF THE IMPRINTED COMPOSITES

Scanning electron microscope (SEM) has super high resolution, which can observe the three-dimensional morphology of the surface of the sample at nanometer scale, and has a large observation range. Although the transmission electron microscope (TEM) shows a two-dimensional image, it can see the structure and morphology of the inner layer. Therefore, these two instruments are used to characterize the surface morphology and internal structure of the composites, as shown in Fig.1 and Fig.2. Fig.1a and d show the SEM images of graphene deposited by these two sensors, the typical folding shape of graphene can be clearly observed by the figure, indicating that electrodeposition of graphene oxide on the electrode surface can be successfully achieved by electrochemical reduction method [15]. Fig.1b and e show the SEM images of the two sensors after in situ polymerization of ionic imprinted membranes. It is obvious that a dense and uniform polymer film is formed on the surface of reduced graphene oxide. Through this film, the wrinkled graphene materials are still observed, which indicates that the corresponding ionic imprinted polymers are successfully polymerized on the surface of reduced graphene oxide; Fig.1c and f show the SEM images of the two sensors after elution of template ions. It can be seen that the surface of the eluted polymer is rougher and sparse compared with the surface of the polymer before elution. This indicates that the template Cd(II) ion has been successfully eluted and generated

corresponding binding sites. Fig.2 is the TEM scan result, the same effect can be observed as Fig. 1.

To further verify whether electrochemical elution can successfully elute template cadmium ions in imprinted membranes, we used energy dispersive spectrometer (EDS) to analyze the elements of the samples before and after elution. Take the IIPoPD/rGO/GCE sensor as an example, as shown in Fig.3. Fig.3A is a EDS diagram of the pre-elution imprinted complex. It can be seen that the pre-elution imprinted complex contains a large number of carbon elements (C), oxygen elements (O), a small amount of nitrogen elements (N), and cadmium characteristic peaks.Fig.3B is the elemental analysis EDS image of the imprinted composite after elution. Compared with before elution, there are still a large number of carbon, oxygen and nitrogen elements, but there is no characteristic peak of cadmium in the imprinted complex after elution. This indicates that the electrochemical method adopted in this experiment successfully eluted the template ions.

IV. PERFORMANCE TEST OF SENSOR

Under the optimized conditions, Cd(II) calibration curves of two sensors (IIPoPD/rGO/GCE and IIPpy/rGO/GCE) were obtained by SWASV. As shown in Fig.4, the green solid line shows the Cd(II) calibration curve of IIPoPD/rGO/GCE, there is a good linear relationship between SWASV stripping peak values of and Cd (II) at 1-50μg/L, and the linear fitting curve is I=0.21846c+7.97271(I is the peak current of oxidation peak, and c is the Cd(II) concentration) .The correlation coefficient is 0.99899. According to S/N=3, the detection limit is 0.13ng/mL, and the sensitivity is 0.22μA/μg/L. The electrodeposited rGO is helpful to improve the sensitivity. The red solid line shows the Cd(II) calibration curve of IIPpy/rGO/GCE, it has a good linear relationship between SWASV stripping peak values and Cd (II) at 1-100μg/L, the linear equation is I=0.14292c+6.88719, the linear correlation coefficient is 0.992, and the detection limit is 0.26 ng/mL.

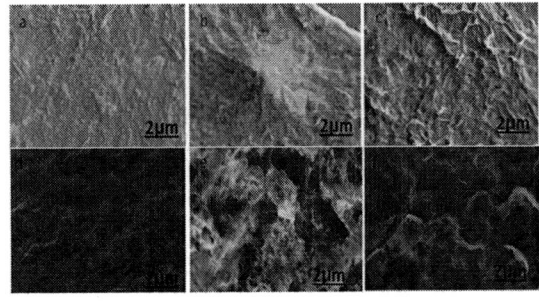

Figure 1.   a) SEM image of rGO/GCE (Py as functional monomer) b) SEM image of IIPpy/rGO/GCE before template removal c) SEM image of IIPpy/rGO/GCE after template removal d) SEM image of rGO/GCE (functional monomer with oPD) e) SEM image of IIPoPD/rGO/GCE before template removal f) SEM image of IIPoPD/rGO/GCE after template removal.

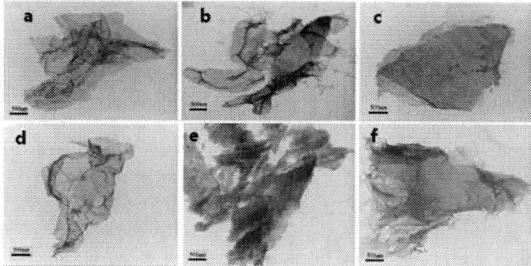

Figure 2. a) TEM image of rGO/GCE (Py as functional monomer) b) TEM image of IIPpy/rGO/GCE before template removal c) TEM image of IIPpy/rGO/GCE after template removal d) TEM image of rGO/GCE (functional monomer with oPD) e) TEM image of IIPoPD/rGO/GCE before template removal f) TEM image of IIPoPD/rGO/GCE after template removal

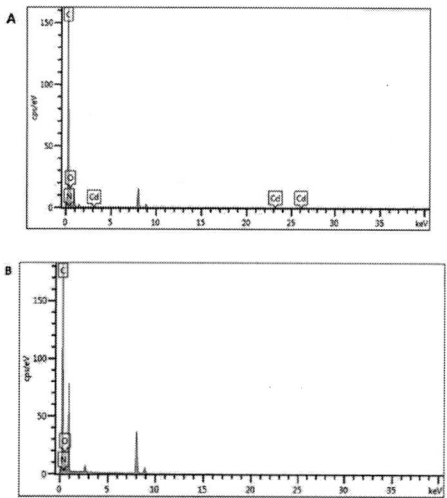

Figure 3. (A) EDS spectrum of Cd(II)- IIPoPD/rGO/GCE; (B) EDS spectrum of IIPoPD/rGO/GCE

Generally, the limit value of cadmium content in all kinds of water quality in China is 5 or 10μg/L, the two ion imprinted electrochemical sensors we have studied can measure the cadmium content in the commonly used water quality in China and judge whether it exceeds the standard.

Table 1 compares the performance of the two sensors prepared in this paper with other modified electrodes. As can be seen from the table, the sensor prepared by this experiment has a better linear range and lower detection limit than the Cd (II) electrochemical sensor in other literatures.

V. SELECTIVITY, REPEATABILITY AND STABILITY OF THE SENSOR

In the actual detection process of water samples, due to the existence of many interference ions, it is easy to interfere with the detection of electrochemical sensors, so we have done the following experiments to test the

performance of the sensors. According to IIPoPD/rGO/GCE sensor, Mg(II), Hg(II), Zn(II), Fe(II), Cu(II), Ni(II) and Mn(II) metal ions with the concentration of 200μg/L were added into 10μg/L Cd(II) standard solution for experiment. The experimental results show that the overall detection error is less than 5% even after adding 20 times of interference ions. For the IIPpy/rGO/GCE sensor, 10 times the concentration of Mg(II), Zn(II), Cr(VI), Mn(II) and Cr(III) were added into the standard solution of Cd(II) at a concentration of 10μg/L. The results showed that the overall detection deviation was still less than 5%, which indicated that the two sensors studied in this paper had excellent selectivity, which was mainly due to the introduction of ion imprinting technology.

Fig.5 shows the repeatability of IIPoPD /rGO/GCE and IIPpy /rGO/GCE sensors. Five sensors prepared under the same conditions were tested, and the experimental results showed good repeatability. The stability of the two sensors was then tested every 1 hour, after many tests, it was found that the response current of the two sensors to Cd (II) remained above 95% of the initial calibration value, mainly because the ion imprinted sensitive film prepared by electro-polymerization was stable and not easily damaged. It can be preserved for a long time.

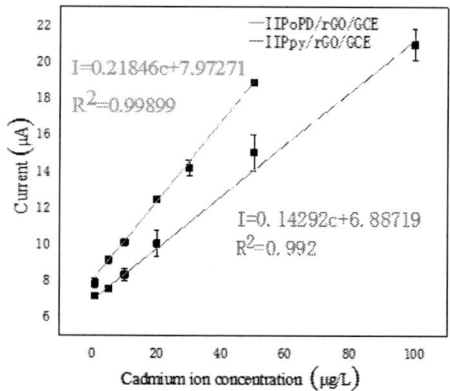

Figure 4. The SWASV response of different concentrations of Cd(II) at IIPpy/rGO/GCE and IIPoPD/rGO/GCE

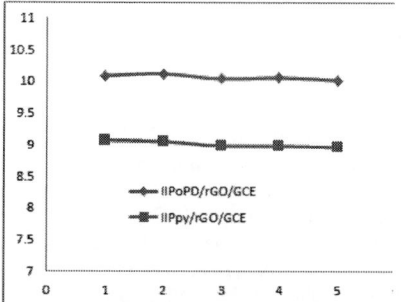

Figure 5. the repeatability of IIPoPD /rGO/GCE and IIPpy /rGO/GCE sensors

VI. Detection of Actual Water Samples

In order to investigate the ability of the sensor to detect trace cadmium ions in actual water samples and to determine whether it has good practical application ability, we will collect actual water samples outside the laboratory for testing the two sensors. 1000 ml of water samples were collected in three different places in Beijing and stored in polyethylene bottles and brought back to the laboratory. After the analysis of some water samples by atomic absorption spectrometry, the cadmium content in the three water samples was less than 1μg/L, because of the special geographical location and strict water quality management in Beijing, it can be considered that there is no cadmium ion in the actual water samples in Beijing. Hence, Cd(II) of the known concentration was introduced into the water sample for analysis and determination for testing needs. The test results of the two sensors are shown in Table 2 and Table 3, respectively.

According to the table, The recovery rate of the IIPoPD/rGO/GCE sensor in the actual water sample is between 94%~106.4%, and the recovery rate of the IIPpy/rGO/GCE sensor in the actual water sample is between 93.29 and 108.36%, all of which have a good recovery range. Therefore, the two sensors can be applied to the actual detection of real water samples in the field.

TABLE I. COMPARISON OF PREPARED ELECTRODES WITH OTHER REPORTED CADMIUM ION MODIFIED ELECTRODES

Modified Electrode	Linear Range (μg/L)	LOD (μg/L)	Reference
Modified carbon nanotube electrode with bismuth film	2–100	0.7	[16]
Glass Carbon Electrode Modified by Tin Film	10–110	1.1	[17]
Bi powder modified carbon paste electrode	10–100	1.2	[18]
IIPpy/rGO/GCE	1-100	0.26	This Work
IIPoPD /rGO/GCE	1-50	0.13	This Work

TABLE II. CD (II) CONCENTRATION DETECTION IN REAL WATER SAMPLES (IIPoPD/RGO/GCE SENSOR)

Real Water Samples	Cd(II) Concentration Added (μg/L)	Detection Results (μg/L)	Recovery (%)
Water sample 1	1	0.94	94
Water sample 2	5	5.32	106.4
Water sample 3	20	19.68	98.4

TABLE III. CD (II) CONCENTRATION DETECTION IN REAL WATER SAMPLES (IIPY/RGO/GCE SENSOR)

Real Water Samples	Cd(II) Concentration Added (μg/L)	Detection Results (μg/L)	Recovery (%)
Water sample 1	10	9.71	97.1
Water sample 2	50	54.18	108.36
Water sample 3	100	93.29	93.29

ACKNOWLEDGMENT

We acknowledge financial supports from National Natural Science Foundation of China (No. 61601037, No. 61701475) and Open foundation of State Key Laboratories of Transducer Technology (Chinese Academy of Sciences) (No. SKT1902)

REFERENCES

[1] T. Alizadeh, "A carbon paste electrode impregnated with cd2+ imprinted polymer as a new and high selective electrochemical sensor for determination of ultra-trace cd2+ in water samples". Journal of Electroanalytical Chemistry. vol. 657, pp. 98-106,2011

[2] S. Magnus, "Distribution and concentration of cadmium in human kidney". Academic Press,vol.39,pp.7,1986

[3] T. G. Levitskaia, "Direct Spectrophotometric Analysis of Cr(VI) Using a Liquid Waveguide Capillary Cell", Applied Spectroscopy, vol.62, pp.9,2008

[4] J. Komárek, "Determination of palladium and platinum by electrothermal atomic absorption spectrometry after deposition on a graphite tube", Spectrochimica Acta Part B: Atomic Spectroscopy, vol .54 pp.739-743,Jun,1999

[5] D.C Lambkin, "The problem of arsenic interference in the analysis of soils for cadmium by inductively coupled plasma-optical emission spectrometry", Science of the Total Environment, vol.256,Jun,2000

[6] L.Wang, "Preparation and application of a novel core/shell organic nanoparticle as a fluorescence probe in the selective determination of Cr(VI)", Spectrochimica acta. Part Λ, Molecular and biomolecular spectroscopy, vol.62,Nov,2005

[7] K.A.Yao, Instrument analysis, Nanjing University Press,2017

[8] L.Jing, "Nafion‐graphene nanocomposite film as enhanced sensing platform for ultrasensitive determination of cadmium,"ElectrochemistryCommunications,vol.11,pp.1085-1088, Mar, 2009

[9] N. Shahbazi, "Probe for sensitive direct determination of sulphide ions based on gold nanoparticles," IET Nanobiotechnology, vol.12,Dem,2018

[10] .Jiang , "An ionophore‐Nafion modified bismuth electrode for the analysis of cadmium(II),"Electrochemistry Communications, vol.12,pp.202-205,2010

[11] J. Fu, "Current status and challenges of ion imprinting", Journal of Materials Chemistry A,vol,3,pp,13598-13627,May,2015

[12] A.Hamid, "Determination of cadmium(II) using carbon paste electrode modified with a Cd-ion imprinted polymer",Microchimica Acta,vol,178,Jul,2012

[13] H,Y.Mou, "Research Progress of Ion Imprinting Polymer", Chemical progress,vol,30,pp,2467-2480,Nov,2011

[14] Y.Y.Peng," Recent Innovations of Molecularly Imprinted Electrochemical Sensors Based on Electropolymerization Technique", Current Analytical Chemistry ,vol.11,Sep,2015

[15] K.Chen, "Three-dimensional porous graphene-based composite materials: electrochemical synthesis and application", Journal of Materials Chemistry ,vol.39,pp. 20968-20976,2012

[16] G.H. Hwang, "Determination of trace metals by anodic stripping voltammetry using a bismuth-modified carbon nanotube electrode" , Talanta, vol,2,pp, 301-308.,2008,

[17] W.W. Zhu, "Simultaneous determination of chromium(III) and cadmium(II) by differential pulse anodic stripping voltammetry on a stannum film electrode" ,Talanta, , vol,5,pp,1733-1737,2007

[18] S.B. Hočevar, "Novel electrode for electrochemical stripping analysis based on carbon paste modified with bismuth powder",Electrochimica Acta, vol,4,pp,706-710, 2005,

Gap in pagination due to unavailable paper.

Pages 397-398

Proceedings of the 16th Annual IEEE International
Conference on Nano/Micro Engineered and Molecular Systems
April 25-29, 2021

Electrochemical betavoltaic cell using black TiO₂ nanotube arrays modified with single-walled carbon nanotubes

Na Wang[1,2], Renrong Zheng[1,2], Zhen Wang[1,2], Jiang Chen[1,2], and Haisheng San[1,2,*]

Abstract – **This paper reports a design and fabrication of electrochemical-betavoltaic cell (EBC), which consists of black TiO₂ nanotube arrays (BTNTAs) as work electrode, and polyiodine (I^-/I_3^-) solution as electrolyte, and a 10 mCi of ^{63}Ni/Ni (10×20 mm²) sheet as counter electrode as well as beta source. By modifying single-wall carbon-nanotubes (SWCNTs) on the surface of BTNTAs, the EBC exhibits an energy conversion efficiency of ~5.12% with V_{oc} = 80.0 mV, J_{sc} = 1.59 µA/cm², and P_{max} = 54 nW/cm². The enhancement of betavoltaic effect can be attributed to the large surface area of SWCNTs/BTNTAs with high-density active sites and excellent electric conductivity.**

I. INTRODUCTION

In the recent year, a rapid development of wireless sensing microsystems and information technology have significantly increased the interest for independent, sustainable, and maintenance-free cells. Betavoltaic cells, which can directly convert the energy of beta particles emitting from a radioisotope source into electricity, are the ideal choices as a micropower system due to their characteristics of miniaturization, high energy density, long lifetime and insensitivity to environment and temperature [1-3].

In 1953 Paul Rappaport reported first betavoltaic cell based on a ^{90}Sr-^{90}Y radioisotope source, which demonstrated an energy conversion efficiency (ECE) of 0.2 % [4]. From then on, the betavoltaic cells have attracted considerable researchers to investigate betavoltaic effect in different materials, such as gallium arsenide (GaAs) [5], aluminium gallium arsenide (AlGaAs) [6], silicon carbide (SiC) [7], gallium nitride (GaN) [8] and indium gallium phosphide(InGaP) [9]. Sun's work has demonstrated that the ECE of betavoltaic cell using three-dimensional (3-D) porous silicon *p-n* diodes was around 10 times larger than that using 2-D planar one [10]. Up to now, there are few reports on the betavoltaic cells using 3D nanostructures. TiO₂ nanotube arrays (TNTAs) fabricated by direct electrochemical anodization of titanium (Ti) are suggested to be preferred candidate for betavoltaic cells due to their wide bandgap, high specific surface area, excellent chemical stability, and low-cost preparation [11]. However, the high interface impedance between TNTAs and electrodes results in a high recombination of electron-hole pairs (EHPs) and thus a low ECE [12, 13].

In this study an electrochemical betavoltaic cell (EBC) using defect-engineered black TiO₂ nanotube arrays modified with single-wall carbon-nanotubes SWCNTs to enhance the ECE of betavoltaic cell. The verification and optimization and of beta-anode materials were performed through using ultraviolet (UV) photo-response measurements for the electrochemical UV photodetectors (EUVPDs) with similar device structure with EBC.

II. EXPERIMENTAL SECTION

A. Preparation of the SWCNTs@BTNTAs

Free-stand TNTAs were prepared on titanium (Ti) sheets (20 µm thickness, 99.99% purity) using a standard double-electrode electrochemical anodization method in a electrochemical reaction system, where the Ti sheet was used as working electrode and the Pt sheet as counter electrode. The electrolyte solution consists of 97 vol% ethylene glycol (EG), 0.5 wt% Ammonium fluoride (NH₄F) and 3.0 vol% deionized water. The electrochemical anodization was conducted at room temperature by using a stable voltage (50 V) for 45 min. the prepared TNTAs samples were annealed in high-purity argon at 500 °C for 2 hours, and the as-prepared samples were denoted as BTNTAs. As a comparison, the anodized samples prepared using air-anneal at the same temperature were denoted as TNTAs. Next, the SWCNTs were ultrasonically dissolved in alcohol solution and then magnetically stirred for 30 min to form uniform SWCNTs solution. Subsequently, the as-synthesized solution of 0.05 ml was added onto the surface of BTNTAs, and then was heated in air at 60 °C for 10 min to achieve a good adhesion of SWCNTs on BTNTAs. The as-prepared samples were denoted as SWCNTs@BTNTAs.

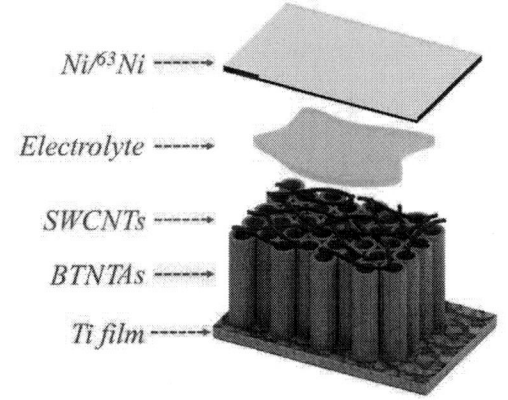

Figure 1. Schematic 3-D decomposed diagram of EBC based on SWCNTs@BTNTAs structure.

---

[1]Pen-Tung Sah Institute of Micro-Nano Science and Technology, Xiamen University, Xiamen 361005, Fujian, China.

[2]Shenzhen Research Institute of Xiamen University, Shenzhen 518000, Guangdong, China.

*Corresponding Author: Haisheng San is with the Xiamen University, Xiamen, China (phone: +86-592-2181340; E-mail: sanhs@xmu.edu.cn).

978-1-6654-3008-1/21 $31.00 © 2021 IEEE

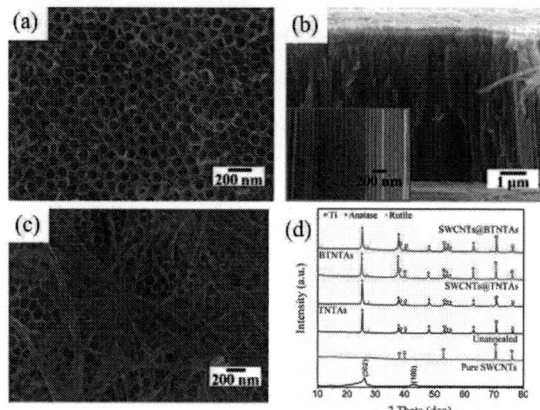

Figure 2. (a) Typical top-view and (b) side-view FESEM images of BTNTAs; (c) top-view FESEM image of the BTNTAs composited with 0.1%@SWCNTs and (d) the X-ray diffraction pattern.

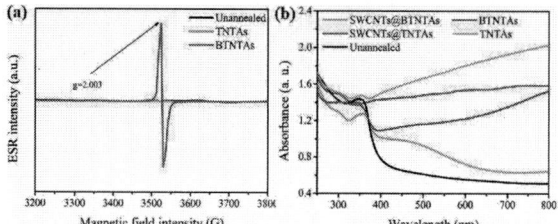

Figure 3 (a) ESR spectra and (b) UV-Vis light absorbance spectra.

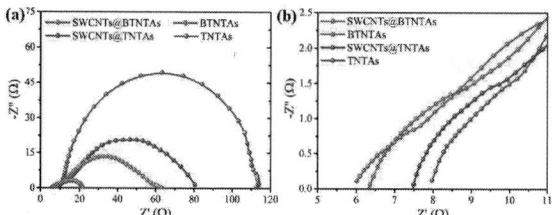

Figure 4. (a) Nyquist plots of TNTAs, BTNTAs, SWCNTs@TNTAs and SWCNTs@BTNTAs structures and (b) the enlarged Nyquist plots in the high-frequency region.

B. Assembly of the EBCs

A sandwich-type of EBC was packaged using the SWCNTs@BTNTAs/Ti structure as anode, a ^{63}Ni sheet as cathode, and polyiodine (I^-/I_3^-) as electrolyte. The schematic 3-D decomposited diagram of EBC structure is shown in Fig. 1. The effective active area of EBC were approximately 1.5 cm^2. As a comparison, the EBCs based TNTAs, BTNTAs, SWCNTs@TNTAs were also prepared using same processes.

III. RESULTS AND DISCUSSION

The typical top-view FESEM image of the BTNTAs is exhibited in Fig. 2(a). The BTNTAs are composed of highly ordered and tightly arranged nanotubes with the average pore diameter of approximately 100 nm and the wall thickness of about 10 nm. It has been confirmed that the morphology and microstructure size of the BTNTAs did not change obviously before and after thermal treatment in different atmospheres. Fig. 2(b) exhibits the cross-sectional view of the BTNTAs. The length of free-standing BTNTAs grown perpendicular to the Ti substrate are about 10 μm. The partial-enlarged image (see inset in Fig. 2(b)) shows the densely packed and aligned nanotube arrays with clear nanotube-nanotube (NT-NT) contact interface. It is considered that the vertically aligned nanotubes create a 3-D conductive matrix [14]. The NT-NT contact interface form a NT-NT homojunction beneficial for the separation of EHPs [14, 15]. Fig. 2(c) exhibits the top-view FESEM image of the BTNTAs modified with SWCNTs. It can be clearly observed that the 1-D SWCNTs wind each other and anchor on the upper surface of BTNTAs. The SWCNTs conductive network provide abundant active sites, which are beneficial for rapid transferring and reaction of redox species [16]. Fig. 2(d) exhibits the X-ray diffraction patterns for unannealed TNTAs, TNTAs, BTNTAs, SWCNT@TNTAs and SWCNT@BTNTAs samples. The XRD pattern of pristine TNTAs suggests the amorphous nature of unannealed TNTAs. The TNTAs samples show a strong (101) diffraction peak for anatase phase at $2\theta = 25.2°$. In contrast, the BTNTAs samples shows two strong anatase diffraction peak at $2\theta_{(110)} = 27.2°$ and $2\theta_{(101)} = 36.9°$. The crystallized TiO$_2$ can improve the charge carrier transport due to the increase of carrier mobility. Furthermore, the annealed TNTAs also show a diffraction peak for rutile phase at $2\theta_{(110)} = 27.4°$ with extremely weak intensity. It has been reported that the reduced anneal can facilitate the transition of anatase phase to rutile phase[17], and it is suggested that the mixed anatase/rutile phase may make a homojunction beneficial for the separation of EHPs and the decrease of EHPs recombination[18].

Fig. 3(a) exhibits the ESR spectra of unannealed TNTAs, TNTAs, and BTNTAs samples. The presence of oxygen vacancy (OVs) defects are confirmed in BTNTAs sample with a distinct ESR signals at $g = 2.003$ when compared with the unannealed and TNTAs samples. Thermal treatment in an oxygen-deficient atmosphere (e.g. inert gases) can lead oxygen molecules to be released from the TiO$_2$ and generate a large number of OVs defects. However, in an oxygen-rich environment (e.g. air), it is difficult to generate OVs. The OVs defects can form a shallow donor level (namely the electron traps) directly below conduction band (CB).

Fig. 3(b) exhibits the UV-visible (UV-Vis) absorption spectrum for all samples. All samples show strong light absorption characteristics in the UV region as a result of the wideband gap of TiO$_2$. In addition, compared with the BTNTAs sample and SWCNTs-modified sample, the unannealed and TNTAs samples exhibits a weaker absorption in the visible region. On account of the effects of OVs and Ti^{3+} defects, Ar annealed BTNTAs sample exhibits about 2.5 times higher absorbance in visible region than TNTAs and unannealed samples. Furthermore, the absorption in visible region can be further improved by the SWCNTs modification, and the visible absorption enhances with the increase of wavelength in the visible region. The surface plasmon resonance effect (SPR) induced by SWCNTs should be responsible for the visible absorption. Similar results also can be found in the graphene/ZnO [19] and SWCNTs/ZnO structures [20].

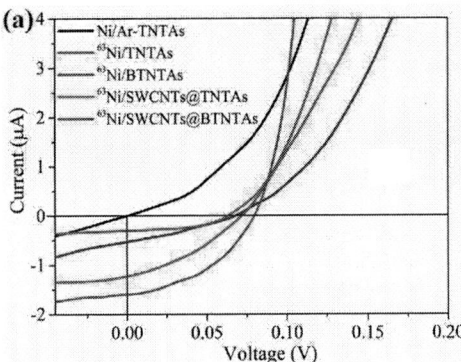

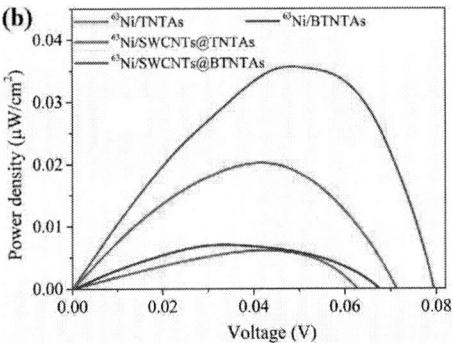

Figure 5. (a) *I-V* and (b) *P-V* characteristics of EBCs based on TNTAs, BTNTAs, SWCNTs@TNTAs and SWCNTs@BTNTAs structures.

Fig. 4(a) and Fig. 4(b) respectively show the electrochemical impedance spectra of TNTAs/Ti, BTNTAs/Ti, SWCNTs@TNTAs/Ti and SWCNTs@BTNTAs/Ti structures using I^-/I_3^- electrolyte in the low-frequency region and high-frequency region. The semicircle curves under low frequency measurement corresponds to the diffusion impedance (R_{diff}) at the Ti/TNTAs/electrolyte interface, and the semicircle curves under high frequency measurement corresponds to the charge transfer impedance R_{ct} and double-layer capacitance $\mathrm{CPE_1}$ at the counter electrode/electrolyte interface. The curvature radius of the two semicircles are directly proportional to the corresponding resistance value. It can be seen from the Fig.4 that the semicircular radio of the BTNTAs/Ti and SWCNTs@BTNTAs/Ti structures are smaller than that based on TNTAs/Ti structure. This indicates that Ar-annealed treatment can bring about low charge transfer resistance in interface due to the increase in conductivity of structure material. In addition, the SWCNTs@BTNTAs/Ti structures has lowest radio in all structures, which means the smallest R_{ct} and R_{diff}. This implies that the modification of SWCNTs on BTNTAs can enhance the interface conductivity as well as charge separation efficiency, thereby effectively decrease the interface impedance.

Fig. 5(a) and (b) shows the current-voltage (*I-V*) and power-voltage (*P-V*) characteristics of the EBCs with 10 mCi beta source, respectively. All of the betavoltaic cells exhibit betavoltaic effect with stable short-circuit current (I_{sc}) and open-circuit voltage (V_{oc}). It can be found from

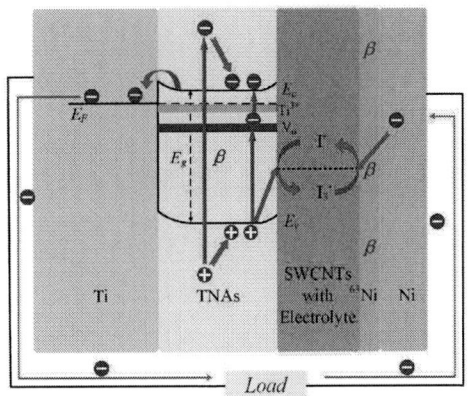

Figure 6. Schematic diagrams of energy band and operating principle of EBC based on Ni/⁶³Ni/electrolyte/SWCNTs@BTNTAs/Ti structure.

Table 1 that the EBC based on SWCNTs@BTNTAs structure has maximum ECE of 5.12% with V_{oc} = 80 mV, J_{sc} = 1.06 μA/cm², and P_{max} = 54 nW/cm². In comparison with BTNTAs structure, an around 3.9 times of increase in P_{max} could be attributed to SWCNTs modified on the defect-engineered BTNTAs, which not only decrease the interface resistance between anode and electrolyte, but also act as hole acceptors for accumulating beta-generated holes and providing sufficient active reaction sites for redox species.

Fig. 6 shows the schematic diagrams of energy band and operating principle of EBC based on the Ni/⁶³Ni/electrolyte/SWCNTs@BTNTAs/Ti structure. It can be seen that a complete cycle of charge transport process is achieved. With the driving force caused by the potential difference between the redox potential of I^-/I_3^- and the Fermi energy-level of working electrodes, the beta-excited electrons transfer to the Ti substrate from the CB of the TiO₂ and then pass through the external circuit to the Ni counter electrode. The beta-excited holes in VB of TiO₂ will transfer to SWCNTs network and undergo an oxidation reaction with I^- anions in the electrolyte, as shown in following equation:

$$3I^- + 2h^+ \rightarrow I_3^- \qquad (1)$$

The oxidized redox mediators (I_3^-) diffuse to the ⁶³Ni/Ni counter electrode where they carry out a reduction reaction with the electrons in counter electrode to regenerate the reduced redox mediators (I^-), as shown in following equation:

$$I_3^- + 2e^- \rightarrow 3I^- \qquad (2)$$

It can be considered that the oxidation reaction of I^- to I_3^- on the SWCNTs@BTNTAs/Ti anode can be enhanced through the improvement of electrical conductivity of anode material and the increase of charge transport rate due to more reaction sites induced by the three-dimensional nanostructure. Under the irradiation of the beta source, the reaction process can continuously circulate and output a steady flow of current.

978-1-6654-3008-1/21 $31.00 © 2021 IEEE

TABLE I. ELECTRICAL PARAMETERS OF EBC BASED ON DIFFERENT WORKING ELECTRODES

Beta source:Anode structures of EBCs	Electrical Parameters of Electrochemical Betavoltaic Cells					
	Open-circuit voltage V_{oc} (mV)	Short-circuit current I_{sc} (μA)	Short-circuit current density J_{sc} ($\mu A/cm^2$)	Max power density P_{max} (nW/cm^2)	Filling factor FF	Total efficiency η (%)
Ni:BTNTAs/Ti	0	0	0	/	/	/
[63]Ni:TNTAs/Ti	63.3	0.30	0.20	9.48	0.499	0.90
[63]Ni:BTNTAs/Ti	68.4	0.53	0.35	11.01	0.304	1.05
[63]Ni:SWCNTs@TNTAs/Ti	72.2	1.22	0.81	31.30	0.434	2.97
[63]Ni:SWCNTs@BTNTAs/Ti	80.0	1.59	1.06	54.00	0.425	5.12

IV. CONCLUSION

In summary, we reports a design and fabrication of electrochemical betavoltaic cells, which consist of black TiO$_2$ nanotube arrays as work electrode, and polyiodine (I$^-$/I$_3^-$) solution as electrolyte, and a 10 mCi of [63]Ni/Ni sheet as counter electrode as well as beta source. By I–V measurements, the EBC based on BTNTAs modified with SWCNTs has maximum ECE of ~5.12% with V_{oc} = 80 mV, J_{sc} = 1.06 $\mu A/cm^2$, and P_{max} = 54.0 nW/cm^2. The performance enhancement is attributed to the fact that SWCNTs modification on the defect-engineered BTNTAs not only decrease the interface resistance between anode and electrolyte, but also act as hole acceptors for accumulating beta generated holes and providing sufficient active reactions sites for redox species.

ACKNOWLEDGMENT

This work was supported by the National Natural Science Foundation of China (Grant No. 61574117), the Natural Science Foundation of Guangdong Province (Grant No. 2018B030311002).

REFERENCES

[1] L. C. Olsen, P. Cabauy, and B. J. Elkind, "Betavoltaic power sources," *Physics Today,* vol. 65, no. 12, pp. 35-38, 2012.

[2] S. T. Revankar, "Advances in Betavoltaic Power Sources," *Journal of Energy and Power Sources,* vol. 1, no. 6, pp. 321-329, 2014.

[3] C. C. A. B. E *et al.,* "Free-standing ZnO nanorod arrays modified with single-walled carbon nanotubes for betavoltaics and photovoltaics," *Journal of Materials Science & Technology,* vol. 54, pp. 48-57, 2020.

[4] Rappaport and P., "The Electron-Voltaic Effect in p-n Junctions Induced by Beta-Particle Bombardment," *Physical Review,* vol. 93, no. 1, pp. 246-247, 1954.

[5] A. Waris, Y. Kusumawati, A. S. Alfarobi, I. K. Aji, and K. Basar, "Preliminary design of betavoltaic battery using Co-60 and Pm-147 with GaAs substrate," in *International Conference on Theoretical & Applied Physics,* 2016.

[6] H. Chen, L. Jiang, and X. Chen, "Design optimization of GaAs betavoltaic batteries," *Journal of Physics D Applied Physics,* vol. 44, no. 21, p. 215303, 2011.

[7] A. A. Svintsov, A. A. Krasnov, M. A. Polikarpov, A. Y. Polyakov, and E. B. Yakimov, "Betavoltaic battery performance: Comparison of modeling and experiment," *Applied Radiation & Isotopes,* vol. 137, p. 184, 2018.

[8] C. Munson, Q. Gaimard, K. Merghem, S. Sundaram, and A. Ougazzaden, "Modeling, Design, Fabrication and Experimentation of a GaN-based, 63Ni Betavoltaic Battery," *Journal of Physics D Applied Physics,* vol. 51, no. 3, 2017.

[9] S. Butera, M. D. C. Whitaker, A. B. Krysa, and A. M. Barnett, "Investigation of a temperature tolerant InGaP (GaInP) converter layer for a 63Ni betavoltaic cell," *Journal of Physics D Applied Physics,* vol. 50, no. 34, 2017.

[10] W. Sun, N. P. Kherani, K. D. Hirschman, L. L. Gadeken, and P. M. Fauchet, "A Three-Dimensional Porous Silicon p-n Diode for Betavoltaics and Photovoltaics," *Advanced Materials,* vol. 17, no. 10, pp. p.1230-1233, 2005.

[11] Q. Zhang, R. Chen, H. San, G. Liu, and K. Wang, "Betavoltaic effect in titanium dioxide nanotube arrays under build-in potential difference," *Journal of Power Sources,* vol. 282, pp. 529-533, 2015.

[12] X. Chen, L. Liu, P. Y. Yu, and S. S. Mao, "Increasing Solar Absorption for Photocatalysis with Black Hydrogenated Titanium Dioxide Nanocrystals," *Science,* vol. 331, no. 6018, pp. 746-750, 2011.

[13] C. Chen, N. Wang, P. Zhou, H. San, K. Wang, and X. Chen, "Electrochemically Reduced Graphene Oxide on Well-Aligned Titanium Dioxide Nanotube Arrays for Betavoltaic Enhancement," *Acs Applied Materials & Interfaces,* p. 24638, 2016.

[14] Grimes and A. Craig, "Synthesis and application of highly ordered arrays of TiO$_2$ nanotubes," *Journal of Materials Chemistry,* vol. 17, no. 15, pp. 1451-1457, 2007.

[15] G. Liu, N. Hoivik, X. Wang, S. Lu, K. Wang, and H. Jakobsen, "Photoconductive, free-standing crystallized TiO2 nanotube membranes," *Electrochimica Acta,* vol. 93, pp. 80-86, 2013, doi: 10.1016/j.electacta.2013.01.116.

[16] X. Pan, K. Zhu, G. Ren, N. Islam, J. Warzywoda, and Z. Fan, "Electrocatalytic properties of a vertically oriented graphene film and its application as a catalytic counter electrode for dye-sensitized solar cells," *Journal of Materials Chemistry A,* vol. 2, no. 32, pp. 12746-12753, 2014.

[17] D. A. H. Hanaor and C. C. Sorrell, "Review of the anatase to rutile phase transformation," *Journal of Materials Science,* vol. 46, no. 4, pp. 855-874, 2011.

[18] X. Yu *et al.,* "Rutile Nanorod/Anatase Nanowire Junction Array as Both Sensor and Power Supplier for High-Performance, Self-Powered, Wireless UV Photodetector," *Small,* vol. 12, no. 20, pp. 2759-2767, 2016.

[19] P. S. Chandrasekhar and V. K. Komarala, "Graphene/ZnO nanocomposite as an electron transport layer for perovskite solar cells; the effect of graphene concentration on photovoltaic performance," *Rsc Advances,* vol. 7, no. 46, pp. 28610-28615, 2017.

[20] M. Kim and W. K. Jo, "Purification of aromatic hydrocarbons using Ag–multiwall carbon nanotube–ZnO nanocomposites with high performance," *Journal of Industrial & Engineering Chemistry,* p. S1226086X16304579, 2016.

Gap in pagination due to unavailable papers.

Pages 403-406

Proceedings of the 16th Annual IEEE International
Conference on Nano/Micro Engineered and Molecular Systems
April 25-29, 2021

Temperature Compensation for MOEMS Micromirror with Piezoresistive Angle Sensor

Lei Qian, Huijun Yu, Jie Hu, Wenjiang Shen*

Abstract—Assuring MOEMS micromirror work with well-controlled deflection angle in varying environmental condition is of vital importance in practical application. To solve this problem, we develop an electro-magnetically driven micromirror with piezoresistive angle sensor, which is conjunct with platinum resistance temperature detector. The proposed n-type piezoresistive angle sensor is placed on the torsion beam of micromirror to monitor the rotation angle. The test results show high sensitivity of 10.86 mV/° with 3.3V power supply, but it exhibits instability at different temperature. Then the platinum resistance temperature sensor is integrated as an accurate solution for local temperature compensation. It shows linear response with slope of 1.207 Ω/°C to temperature excitation. The device is calibrated and tested in environmental chamber to apply temperature variation over a range from -20 °C to 75 °C. After compensation, the mechanical angle error of micromirror is 0.943° compared to uncompensated one of 3.21°.

I. INTRODUCTION

Micro-optical-electro-mechanical system (MOEMS) micromirror, as a fundamental element, is playing vital role applications from image acquisition systems for example like endoscopes, to highly accurate scanning systems like laser radar and scanning spectrometers. MOEMS micromirror can integrate mechanical structures, optical components and circuits on a single silicon substrate while behaving accurately and reliably. In addition to that, it has many outstanding features including distinct deflection angle, but consumes respectively low energy. Small size and lower fabrication cost also make it easy to achieve mass production. In practical applications, the dynamic characteristics of micromirror such as precision of mechanical amplitude and speed of response are supposed to be considered in the design and the scheme of control.

For the reason that optical beam need to be controlled as exact as possible, assuring stable resonant oscillation with well commanded amplitude under varying environmental conditions is very necessary. Therefore, simple open-loop

---

* The research was supported by the National Key Research and Development Program of China (Grant No.2018YFF01010901 and No.2016YFB0401903).

Lei Qian and Huijun Yu is with School of Nano-Tech and Nano-Bionics, University of Science and Technology of China, Hefei 230026, China (phone: +86 512 628 76188;e-mail:lei621@mail.ustc.edu).

Jie Hu is with Nano Science and Technology Institute, University of Science and Technology of China, Suzhou 215123, China.(phone: +86 512 628 76188;e-mail: hj1549@mail.ustc.edu.).

Wenjian Shen is with Key Lab of Nanodevices and Applications, Suzhou Institute of Nano-Tech and Nano-Bionics, Chinese Academy of Sciences, Suzhou 215123, China (phone: +86 512 628 76188; e-mail: wjshen2011@sinano.ac.cn).

---

control can hardly meet the accuracy requirement of MOEMS micromirror deflection angle[1]. So it is a natural and creative idea to integrate an angle sensor inside the MOEMS devices so as to make the system more robust. There are many ways to obtain the position feedback signal, among which mainstream methods are capacitive sensing[2], piezoelectric sensing [3]and piezoresistive sensing[4]. Piezoresistive sensor is one of the most successful commercial MOEMS device that has a large field of applications such like industrial control, consumer electronics and aerospace, which tends to be used in measuring strain and stress because of its excellent performance in measurement precision and manufacture process[5]. Combined the advantages of the other two sensing ways, piezoresistance pressure sensor can easily be integrated on the micromirror[4] and output clear signal that generates from the strain of torsion beam. However, piezoresistance is so sensitive to temperature variation so it can hardly provide credible position information when it comes to thermal disturbance[6].

In response to this problem, this contribution presents a new MOEMS micromirror with two sensors for strain and temperature to realize closed loop control system. The proposed temperature sensor is made up by platinum that has linear response to temperature excitations and it can be used to achieve temperature compensation of piezoresistive sensor. On the other hand, piezoresistive angle sensor that configures as 4 Wheatstone bridges is placed on torsion beam to detect the mechanical angle of micromirror according to the strain of torsion beam. The control logic unit would adjust the driving signal after receiving the feedback information.

II. DESIGN

A. Micromirror with piezoresistive angle sensor

Electromagnetically driven micromirror is used in this study. As shown in fig.1 mirror connects with two torsion beam and metal driving coil is plated on it. Sinusoidal signal is applied to form driving current while permanent magnetic is placed beneath the mirror to provide sufficient magnetic flux. When actuating current flows through the coil, Lorentz force generates on coil and actuates micromirror move. The angle of micromirror depends on the frequency and amplitude of driving signal. With the rotation of micromirror, torsion beam deforms and stress occurs on the piezoresistance[7].

As a bulk material is under stress and strain occurs, its resistance would change (ΔR) which is a function of geometry effects and resistivity variation ($\Delta \rho$). For semiconductor like silicon and germanium, resistivity changes with strain are extraordinarily more obvious than

978-1-6654-3008-1/21 $31.00 © 2021 IEEE   407

geometry effects due to the change in band structure. This phenomenon is called piezoresistive effect. According to piezoresistive theory, relative change of resistivity is proportional to stress applied and can be described by

$$\frac{\Delta \rho}{\rho} = \pi T \qquad (1)$$

Where T is stress and ρ is stress-free resistivity. π represents piezoresistance coefficient whose value depends on the direction of electric field and stress. π_{11}, also called longitudinal piezoresistance coefficient, is applied if stress and[8] electric field are parallel ,while π_{12}, also called transverse piezoresistance coefficient, is suitable if stress is vertical to direction of electric field. The values of these are shown in table 1. Sheer piezoresistance coefficient π_{44} is so small that π_{11} and π_{12} are the most expected.

In this paper, n- type piezoresistive angle sensor made by ion implantation is integrated on torsion beam of micromirror to detect the motion position. The angle sensor employs a configuration of full wheatstone bridge with two longitudinal and two transverse piezoresistors to increase sensitivity. When impurity concentration of piezoresistor is less than $10^{18}/cm^3$[9], the sensitivity is the highest despite being susceptible to temperature varying. So the doping dose is determined to be $6.5\times10^{14}/cm^2$. Practically, π_{12} and π_{11} have to be transformed from silicon crystallographic coordinate system to the coordinate system of <100> plane, which is crystal orientation of substrate, by direction cosines, where θ is the orientation of the piezoresistors on the substrate[10]:

$$\pi'_{11} = \pi_{11} - 2(\pi_{11} - \pi_{12} - \pi_{44})\sin^2\theta\cos^2\theta \qquad (2)$$

$$\pi'_{12} = \pi_{11} + 2(\pi_{11} - \pi_{12} - \pi_{44})\sin^2\theta\cos^2\theta \qquad (3)$$

$$\pi'_{44} = (\pi_{11} - \pi_{12} - \pi_{44})\sin 2\alpha\cos 2a \qquad (4)$$

When θ is 90°, π_{11} and π_{12} all reach their maximum value for n-type silicon that means the highest sensitivity and the calculated value is shown in Table 1.

TABLE I.   TABLE1 VALUE OF PIEZORESISTANCE COEFFCIENT [8]

Material	Piezoresistance coefficient ($10^{-11}Pa^{-1}$)		
	π'_{11}	π'_{12}	π'_{44}
n-type silicon	-102.2	+53.7	-13.6

The full wheatstone bridge output will be

$$V_o = \frac{V_i}{4}\left(\frac{\Delta R_2}{R_2} - \frac{\Delta R_3}{R_3} + \frac{\Delta R_4}{R_4} - \frac{\Delta R_1}{R_1}\right) \qquad (5)$$

Where Vi is source voltage, $\Delta R_i/R_i$ (i=1, 2, 3, 4) are piezoresistor relative changes.

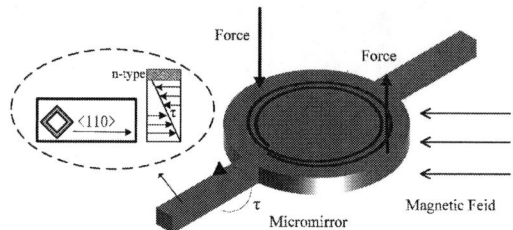

Figure1: basic structure of micromirror with piezoresistive angle sensor

B. Temperature Sensor

Thin film resistance temperature detector has many advantages such as high sensitivity, fast thermal response and wide range of temperature measurement[11]. Among the various material, platinum(Pt) is usually the preferred metal for its great precision and high stability, which allows it work in harsh environment[12]. In addition, obvious temperature coefficient of resistance (TCR) is a vital reason for choosing platinum. TCR (α) is the key index to determine the performance of temperature sensor, which refers to the sensitivity of resistance to temperature changes and it can be described by[13]

$$R(T) = R(T_0)[1 + \alpha(T - T_0)] \qquad (6)$$

$$\alpha = \frac{R(T) - R(T_0)}{R(T_0)}(T - T_0) \qquad (7)$$

Where R (T0) is the resistance at temperature T0, and R(T) is the resistance at temperature T. Further, $\alpha(R)$ is a more commonly used quantity that indicates the ratio of resistance variation to the temperature change, which is given as

$$\alpha(R) = \alpha R(T_0) = \frac{R(T) - R(T_0)}{(T - T_0)} \qquad (8)$$

The resistance of platinum thin film is up to its size of shape, which is generally designed to be polygonal-shape or spiral type. Polygonal shape is chosen because the spiral one is easy to exert the influence of inductance results from current change. Temperature sensor resistance is calculated as

$$R = \rho\frac{l}{a \times t} \qquad (9)$$

The TCR of bulk platinum at 25°C is 39.2×10^{-4}/K and the resistivity is 10.6×10^{-8}. However, in almost all cases, thin film can hardly reach this level, because metal resistivity depends on different scattering mechanism, such as phonon scattering, impurity scattering, interface scattering and so on, among which only phonon scattering is affected by thermal. In order to achieve the properties of bulk material, the thickness of membrane tends to be ten to twenty times of electron men free path and the effects of other scattering are decreased. Based on the analyses above, the temperature sensor we employed is 30um in width, 27.5mm in length and 300nm in thickness with resistance of approximately 324Ω, and theoretically, $\alpha(R)$ is about 1.27Ω/°C.

C. Modeling

To realize closed loop to the micromirror, the proposed temperature sensor is used to detect the disturbance of external thermal disturbance. The piezoresistive sensor is used to monitor the angle of micromirror. Then the control unit will adjust the actuating signal to keep deflection angle of the micromirror stable. Firstly, when the temperature variations interfere with the piezoresistvie sensor, the temperature sensor will pick up the changes in temperature and produce potential changes. Then the FPGA control unit will receive a set of digital signal that is corresponding to temperature variation. Secondly, calibrated angle data will be found in look-up-table according to the temperature data. On the other hand, the angle sensor keeps measuring the motion of micromirror and feed back to control chip. Finally, by comparing the measured angle data and calibrated data, error signal is acquired and the control unit adjusts driving signal according to it. The process of adjustment adopts stepping technique which means that as long as the error exists, the control unit keeps adjusting the driving signal by a constant step size.

III. FABRICATION

The test device is composed by a micromirror, piezoresistive angle sensor and platinum temperature sensor. The whole microfabrication process is carried out in micro/nanofabrication facility and based on MEMS/NEMS manufacture technology. These microstructures are implemented by using both-sides polished 6 inches SOI wafer with a device layer of 55 um and a handle layer of 600um in thickness, which are insulated by a 1 um-thick buried oxide layer. The microfabrication process flow is shown as in Fig.2.

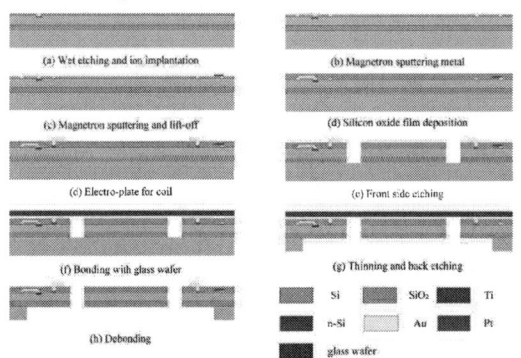

Figure 2. Microfabrication process flow

The wafers are initially put in oxidation furnace to form a silicon dioxide film on the surface that acts as mask during the doping process. After photoresist rotated-coated on the wafer and exposure, the graphics of ion implantation windows are displayed, which are also the shape of four-terminal piezoresistive sensors. Then 7:1 BOE etched through the oxide layer to open the doping windows.

Phosphorus ion is injected into top silicon layer and the dose of ion implantation is 6.5×10^{14}/cm2. After annealing activation, square resistance is tested out to be 104Ω by four pointprobe sheet resistivity measurement. In purpose to lead out piezoresistive sensor signal, Magnetron sputtering and reactive ion etching (RIE) are used to realized Au and Ti film that used as signal wire and electrode. RTP annealing at 450 °C for 10min is followed to create good ohmic contact while minimizing Schottky barrier between the metal layer and underlying doped silicon.

In use of probe station and semiconductor parameter analyze, the results of current-voltage sweep measurement exhibit good linearity, which confirms a good ohmic contact. Then magnetron sputtering and lift-off process are carried out to get temperature sensor with 300nm platinum film which takes 20nm Titanium as adhesive layer. Silica membrane is deposited by plasma enhanced chemical vapor (PECVD) for insulation and also for protecting temperature sensors. Drive coil of micromirror was electroplated and the coil material is gold for its outstanding electrical conductivity. Next, shape of micromirror is formed by using deep reactive ion etching (DRIE). After that, the SOI is stuck to a glass wafer by temporary bonding. In order to reach the required thickness of the device, the wafer gets thinned to and polished. Finally, the back structure was revealed by lithography and released by DRIE and glass wafer is debonded marking the end of all process. Fig.3 is the actual graphic of angle sensor and temperature sensor.

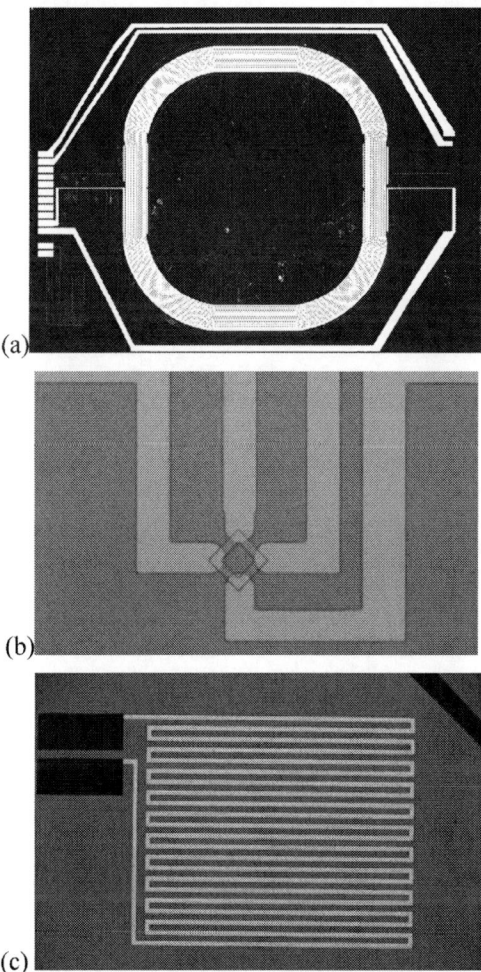

Figure 3. (a) full view of MEMS chip (b) Piezoresistive angle sensor and (c) platinum resistance temperature detector

IV. TEST SETUP AND RESULTS

A. Test setup

The preliminary experimental test is to verify the performance of the proposed platinum temperature sensor. In this experiment, the fabricated chip is attached on probe station that can change the temperature, and the measurement of the temperature sensor is conducted by semiconductor parameter analyze. Calibrated data acquisition and closed-loop control test are carried out in environmental chamber that control the thermal load, to apply temperature variation over a range from -20°C to 75°C. The temperature range is chosen to simulate the surrounding changes that micromirror may exposed to, for example, during the night and day temperature variations. 3.3V voltage bias is supplied by a source meter to piezoresistive angle sensor and its output is measured by scopes. In addition to that, function signal generator provides the micromirror with a specific frequency and amplitude driving signal. The micromirror reflects a light beam from laser source to form a scanning light, which can be used to calculate the optical angle of the micromirror by triangle method. Fig.4 shows the test setup employed in this study. The calculation formula is

$$\theta = 2\beta = 2\arctan(\frac{l}{2s}). \tag{10}$$

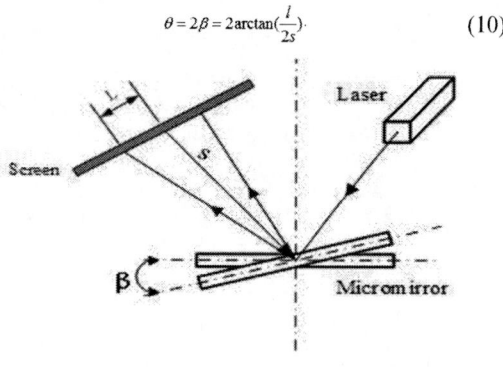

Figure 4. Schematic mechanical angle measuring

B. Results

The TCR (temperature coefficient of resistance) of the platinum temperature was first measured by applying a thermal load. Temperature varied from 25 °C to 100 °C .As shown in Fig.5, initial resistance of temperature sensor is 486.81 Ω. It exhibited a remarkable linear relationship between its resistance and the temperature change and the measured value of α(R) is 1.207Ω/°C, which showed that the proposed temperature sensor succeeded in capturing precisely the local changes in temperature. Need to add that, the measured value of initial resistance and TCR is different to calculated value for the reason that properties of thin film can hardly completely agree with the theoretical one due to experimental condition limitation. Similarly, what we can get in Fig.6 is the ability of the piezoresistive angle sensor to detect the position of micromirror and the influence of thermal disturbance to the sensitivity of pirzoresistive angle sensor. It is obvious that with the increase of deflection angle, the output of piezoresistive sensor increased and the sensitivity at room temperature was 10.86mV/°C. However, because piezoresistance coefficient varies with temperature

change, the angle sensor output also exhibited a consistent trend that the sensitivity decreased to 9.18 mV/°C at 75°C, but increased to 12.34mV/°C at -20°C. The capability of the novel compensation system to reduce the thermal impact on the stability of the angle to micromirror was also investigated in this section. Preliminarily, the output of angle sensor at rated was calibrated by applying thermal load in an environmental chamber. Fig.7 (a) reports the output of the piezoresistance Bridge at rated angle for 28.07° of micromirror. Fig.7 (b) shows both compensated and uncompensated angle at different temperature and it can be clearly seen that the uncompensated angle of micro-mirror can lead to very large errors, while the maximum deviation can reach up to 11.4°. For the compensated series, the absolute error dropped to only 3.4%.

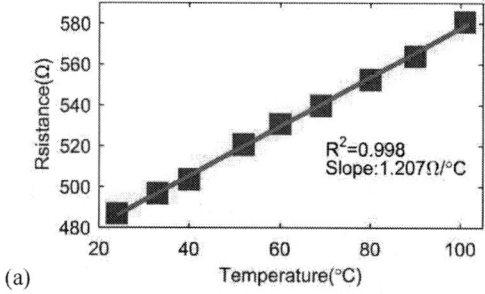

(a)

Figure 5.Typical results for temperature versus resistance change of platinum temperature.

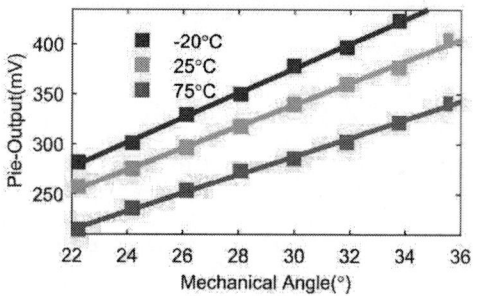

Figure 6. Measured output of angle sensor at different mechanical angle .

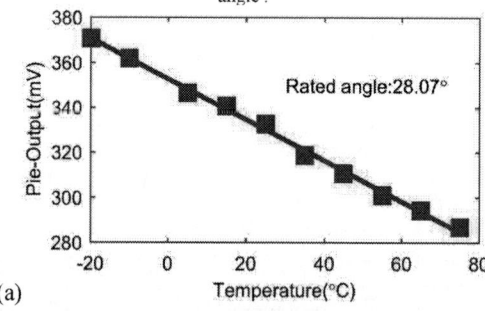

(a)

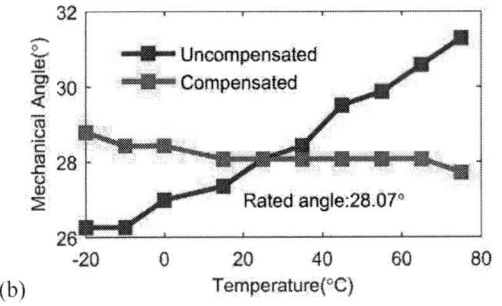

(b)

Figure 7. (a) output of angle sensor at rated mechanical angle and (b) Measured mechanical angle of micromirror before and after temperature compensation.

V. CONCLUSION

In this paper, a novel temperature compensation system for piezoresistive angle sensor on MOEMS micromirror has been presented. The system takes advantage of a platinum resistance temperature sensor to detect the local temperature changes in close to the piezoresistive angle bridge. A device with temperature sensor and piezoresistor bridge was microfabricated to test the performance of the compensator and angle-sensor. The experiment results showed that the compensation system is helpful to precise control of micromirror. The suggested temperature sensor was able to capture the temperature variation and the output signal of angle sensor could also reflect the motion of micromirror.

This compensation method is beneficial for the application of micromirror in changing environment especially in use of laser projection or LIDAR, since the two sensors provide advantage to closed-loop control of micromirror. In addition to that, sensing elements can easily be integrated into fabrication process of MEMS at low cost.

ACKNOWLEDGMENT

The research was supported by the National Key Research and Development Program of China (Grant No. 2018YFF01010901 and No.2016YFB0401903). The authors are grateful to SINANO's Nano Fabrication Facility (NFF) for their support in the fabrication process of micromirror and related tests.

REFERENCES

[1] A. Tortschanoff *et al.*, "Position encoding and phase control of resonant MOEMS mirrors," *Sensors and Actuators A: Physical,* vol. 162, no. 2, pp. 235-240, 2010.

[2] A. A. Kuijpers *et al.*, "Towards embedded control for resonant scanning MEMS micromirror," (in English), *Proceedings of the Eurosensors Xxiii Conference,* vol. 1, no. 1, pp. 1307-+, 2009.

[3] K. Meinel *et al.*, "2D Scanning Micromirror with Large Scan Angle and Monolithically Integrated Angle Sensors Based on Piezoelectric Thin Film Aluminum Nitride," *Sensors,* vol. 20, no. 22, 2020.

[4] M. Sasaki, M. Tabata, T. Haga, and K. Hane, "Piezoresistive rotation angle sensor integrated in micromirror," (in English), *Japanese Journal of Applied Physics Part 1-Regular Papers Brief Communications & Review Papers,* vol. 45, no. 4b, pp. 3789-3793, Apr 2006.

[5] A. A. Barlian, W. T. Park, J. R. Mallon, Jr., A. J. Rastegar, and B. L. Pruitt, "Review: Semiconductor Piezoresistance for Microsystems," *Proc IEEE Inst Electr Electron Eng,* vol. 97, no. 3, pp. 513-552, 2009.

[6] K. Matsuda, K. Suzuki, K. Yamamura, and Y. Kanda, "Nonlinear Piezoresistance Effects in Silicon," (in English), *Journal of Applied Physics,* vol. 73, no. 4, pp. 1838-1847, Feb 15 1993.

[7] H. Yu, P. Zhou, K. Wang, Y. Huang, and W. Shen, "Optimization of MOEMS Projection Module Performance with Enhanced Piezoresistive Sensitivity," *Micromachines (Basel),* vol. 11, no. 7, Jun 30 2020.

[8] C. S. Smith, "Piezoresistance Effect in Germanium and Silicon," (in English), *Physical Review,* vol. 94, no. 1, pp. 42-49, 1954.

[9] Y. Kanda, "A Graphical Representation of the Piezoresistance Coefficient in Silicon," *IEEE TRANSACTIONS ON ELECTRON DEVICES,* pp. 64-70, 1982.

[10] J. C. Suhling and R. C. Jaeger, "Silicon Piezoresistive Stress Sensors and Their Application in Electronic Packaging," (in English), *Ieee Sensors Journal,* vol. 1, no. 1, pp. 14-30, Jun 2001.

[11] L. Y. a. E. M. Jonathan T. W. Kuo "Micromachined Thermal Flow Sensors-A Review." 2012.

[12] F. Lacy, "Using Nanometer Platinum Films as Temperature Sensors (Constraints From Experimental, Mathematical, and Finite-Element Analysis)," (in English), *Ieee Sensors Journal,* vol. 9, no. 9, pp. 1111-1117, Sep 2009.

[13] R. M. Tiggelaar, R. G. R. Sanders, A. W. Groenland, and J. G. E. Gardeniers, "Stability of thin platinum films implemented in high-temperature microdevices," (in English), *Sensors and Actuators a-Physical,* vol. 152, no. 1, pp. 39-47, May 21 2009.

Proceedings of the 16th Annual IEEE International
Conference on Nano/Micro Engineered and Molecular Systems
April 25-29, 2021

Nanocrystalline Composed SnO$_2$ Inverse Opal for Highly Sensitive Acetone Gas Sensor at ppb-level*

First A. Feng Jiang, Second B. Wen Zhang*, and Third C. Yu Wei

Abstract— The exhaled acetone concentration of diabetic patients is much higher than that of normal persons, which can be used for early warning, large-scale screening, and adjuvant treatment for diabetes. Therefore, a non-invasive exhaled acetone sensor with a low detection limit, high sensitivity, and high reliability is widely studied. Their sensing performance largely depends on the chemical composition of sensing materials, surface modification, the microstructure of the sensing layer. In this paper, Tin oxide (SnO$_2$) nanocrystalline composed inverse opal structure were successfully prepared, using polystyrene microspheres as templates, as a sensing material for an acetone gas sensor. Compared with pure SnO$_2$ prepared by the hydrothermal method, inverse opal SnO$_2$ has a much higher response at a low concentration of 500ppb-10ppm, Up to 5.6 response at 1ppm(240°C), which is already suitable for diabetes screening without doping noble metal. The improvement of gas sensitivity is mainly due to the small size of the SnO$_2$ nanoparticles and the pore structure composed of the nanocrystalline. The relatively high response, fast response and recovery speed, and good stability make it a promising gas sensor material for screening and monitoring of diabetes.

I. INTRODUCTION

Diabetes is a long-term chronic metabolic disorder caused by insufficient insulin secretion or inadequate use of insulin. World Health Organization (WHO) report reveals diabetes among the top 10 causes of death and disability worldwide. According to the survey and forecast of the 9th edition of "IDF global diabetes survey" released by the International Diabetes Federation (IDF), China is the country with the largest number of diabetes patients, reaching 116.4 million, of which 35.5 million are over 65 years old, and it is expected to grow to 54.3 million by 2030. More than 65 million people with diabetes in China have not yet been diagnosed, and these people have no obvious symptoms of diabetes. If they are not detected early and intervened in time, the risk of diabetes-related complications will increase, thus greatly adding the medical expenses related to diabetes.

At present, two kinds of blood glucose detection methods are widely used in the world: venous blood detection by the biochemical analyzer and fingertip blood detection by blood glucose analyzer. Biochemical analyzer detection of venous blood can directly and accurately reflect the glucose level in the blood circulation system, which is the gold standard for measuring blood glucose concentration[1]. However, due to the complex measurement process and expensive equipment, it is only suitable for hospitals; blood glucose meter measurement requires multiple blood sampling every day, which brings psychological pressure and risk of infection to patients. The above two methods are traumatic to the human body, and can not achieve continuous monitoring of blood glucose, which may miss the peak value of blood glucose changes.

In recent years, the minimally invasive detection method has been developed, which uses microneedle to puncture the skin to continuously obtain the tissue fluid of the human body, and then calculates the glucose concentration in the blood by measuring the glucose concentration in the tissue fluid. Although the method of minimally invasive can reduce the direct damage to the human body and monitor the blood glucose in real-time for some time, it has high requirements for microneedles and implanted materials and still has some risk of infection.

In 1969, Tassopoulos' group proposed that the content of ketone in the blood is positively correlated with exhaled acetone, so the blood glucose value can be calculated by detecting the content of exhaled acetone, which can be used for early warning, large-scale screening, and adjuvant treatment for diabetes[2]. As a component of exhaled breath, the concentration of acetone in healthy people is about 300-800 ppb, while the concentration in diabetic patients is more than 1.8 ppm, some even as high as 12 ppm[3]. Therefore, the development of a simple, portable, low-cost, and miniaturized acetone sensor is of great significance for the screening and monitoring of diabetes. This detection method has the advantages of non-invasive, non-destructive, no side effects, and can be extended to a variety of cancer, respiratory disease screening, and further combined with emerging technologies such as smart wear, big data cloud service, to provide important case-based data for smart medicine.

It is reported in the literature that many substances have high sensitivity to acetone, and the detection limit reaches ppb level, which is expected to be used for the detection of acetone in exhaled breath. Common materials include various metal oxides, such as WO$_3$, ZnO[4], SnO$_2$[5], TiO$_2$, In$_2$O$_3$, Co$_3$O$_4$, Fe$_3$O$_4$, etc. In addition, many perovskite oxide materials, such as LaFeO$_3$, SmFeO$_3$, NdFeO$_3$, also have high sensitivity to acetone. The composition, morphology, surface modification, doping metal catalyst, and other modifications of the material have further improved the sensitivity of the material. Generally, sensing materials are made into special structures, such as nanowires,

*Resrach supported by Natural Science Basic Research Plan in Shaanxi Province of China (Program No. 2020JM-027)..

F. A. Feng Jiang is with School of Chemistry, Xi'an Jiaotong University, Xi'an, Shaanxi 710049, China (e-mail: jiangfeng4015@stu.xjtu.edu.cn).

S. B. Wen Zhang, is School of Chemistry, Xi'an Jiaotong University, Xi'an, Shaanxi 710049, China (corresponding author phone: 86-029-82663914; e-mail: zhangwen@mail.xjtu.edu.cn).

T. C. Yu Wei is with School of Chemistry, Xi'an Jiaotong University, Xi'an, Shaanxi 710049, China (e-mail: weiyuix@stu.xjtu.edu.cn).

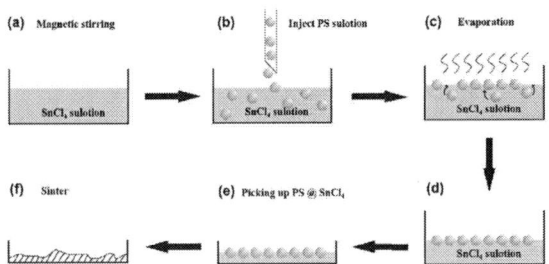

Fig.1 Inverse opal SnO₂ preparation flow chart.

nanosheets[1], nanotubes[6], hollow[4, 7] and core-shell structures[8], to increase the specific surface area and improve the active sites. The oxide semiconductor materials and graphene are compounded uniformly to improve the specific surface area and enhance the electron transport ability[9]. In addition, the construction of hetero-junction is also a useful way to improving the detection sensitivity[10]. The hetero-junction interface provides channels for electron transformation, which is conducive to the adsorption of oxygen and the redox reaction of acetone. Precious metals such as platinum, gold, and palladium can also be used for surface modification[8, 11]. Noble metals change the concentration of free electrons and holes in the material through the surface modification, mainly has been ascribed to Fermi-level control or spill-over effect and catalytic oxidation, so as to change the conductivity of the material and achieve the response to the target gas.

SnO₂ is a typical n-type wide bandgap semiconductor (Eg = 3.6 eV at 300 K). It is widely used as a gas sensing material because of its high conductivity and good stability. This kind of metal oxide semiconductor gas sensor material has the advantages of simple preparation, good repeatability, low cost, and good stability, so it has the most application prospect. In this paper, SnO₂ with inverse opal structure was prepared by a simple method. Using polystyrene microspheres as templates, inverse opal SnO₂ nanoparticles were formed through simple chemical adsorption and sintering. Compared with the pure SnO₂ sample prepared by a simple hydrothermal method, the prepared inverse opal structure shows better gas sensitivity to acetone and absorbs more acetone molecules through the generated pore channels and enlarged active sites.

II. EXPERIMENTAL

A. Synthesis of SnO₂

Tin (IV) chloride pentahydrate ($SnCl_4 \cdot 5H_2O$), ethanol (C_2H_5OH), ammonium hydroxide ($NH_3 \cdot H_2O$) were purchased from Sinopharm Chemical Reagent Co. Ltd. of China. polystyrene (PS) spheres (diameter: 800 nm) were purchased from Rigor science Co., Ltd. of China. All of the chemicals were analytical grade and used without further purification. Deionized water was used for all of the experiments.

1.8g (5 mmol) $SnCl_4 \cdot 5H_2O$ was dispersed uniformly in a mixture of 20 mL ethanol and 25 mL deionized water. 6 ml ammonium hydroxide was added slowly to the solution, stirring while dripping, until a paste-like mixture was obtained and the mixture was transferred to a 100 mL Teflon-lined stainless steel autoclave and held at 180℃ for 24 h. Then, the product was washed three times with deionized water and ethanol absolute successively and dried at 100℃ for 6 h. After calcining at 500℃ for 1 h, yellowish SnO₂ powder was obtained.

B. Synthesis of inverse opal SnO₂

1.1g (3 mmol) $SnCl_4 \cdot 5H_2O$ was dissolved in 7 mL ethanol and 10 mL deionized water under magnetic stirring for 30 min. Slowly added 300μl of PS microsphere dispersion liquid and held at 50℃. When the PS spheres floated on the liquid surface, transferred the film to a crucible. Then, the spheres were dried in air for 1 h at 60℃ and heat-treated at 500℃ for 2 h (1℃/min) to decompose the PS template and form the inverse opal SnO₂ (Fig. 1).

C. Characterization

The compositions and crystal structure of the SnO₂ and inverse opal SnO₂ were characterized with Bruker D8 Advance Polycrystal X-ray diffraction (XRD). The microstructures of the SnO₂ and inverse opal SnO₂ were characterized by scanning electron microscopy (SEM, Zeiss GEMINI 500) at an operating voltage of 5 kV. The specific internal microstructures of the samples were investigated by field emission transmission electron microscope (FE-TEM, JEOL JEM-F200). The gas sensing performances were investigated by a gas sensing measurement instrument (WS-30A, Wei Sheng Electronics Co., Ltd., China).

D. Gas sensing test

After grinding the prepared product, it is dispersed in deionized water to form a uniform paste. The paste is evenly coated on the surface of the ceramic tube substrate to ensure that the surface was uniformly covered, the surface of the ceramic tube has a pair of gold electrodes connected with platinum wires. The Ni-Cr alloy coil was placed in the ceramic tube as a heater to adjust the working temperature by controlling the voltage. The sensors were fixed on a testing board in 18 L glass chamber of the WS-30A system. By injecting different stoichiometric acetone solvents onto the evaporation platform in the glass chamber, different volume fractions of acetone gas can be obtained. The

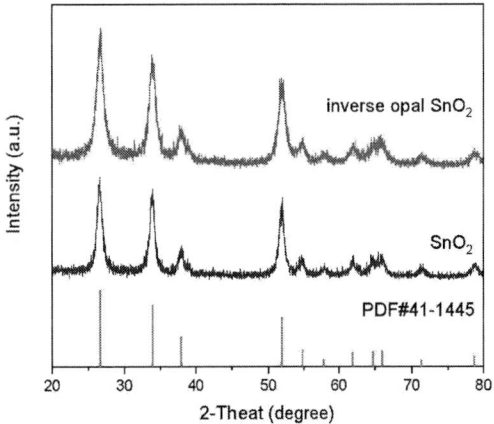

Fig.2 XRD patterns of SnO₂ and Inverse opal SnO₂ nanoparticles.

ambient temperature was 20°C, and the ambient relative humidity was 15%. For a typical n-type semiconductor gas sensor, the response value (S) is defined as S = Ra / Rg (where Ra and Rg are the resistance of the sensor in air and test gas respectively). Meanwhile, the response and recovery time are defined as the time needed for the gas sensor to reach 90% of the total response change in target gas and the air, respectively[12].

III. RESULT AND DISCUSSION

A. Characterization of inverse opal SnO2

The X-ray diffraction (XRD) patterns the crystal structure of the prepared SnO_2 nanoparticles and inverse opal SnO_2 nanoparticles. As shown in Fig. 2, most of the diffraction peaks of the product can be well indexed by tetragonal rutile SnO_2 (PDF#41-1445). It can be seen that due to their nanocrystalline nature, significant peak broadening was observed in both samples, and after the formation of the inverse opal structure, the diffraction peaks at 26.6°, 33.9°and 51.8° showed wider peak. It shows that the inverse opal SnO_2 crystal particles are smaller. According to the Debye-Scherrer formula, the grain size of SnO_2 can be calculated to be 10.5nm, and the grain size of inverse opal SnO_2 is 7.1nm.

Fig. 3 shows the morphological characteristics of the SnO_2 nanoparticles and inverse opal SnO_2. From Fig.3a, It can be seen that the surface of the macrostructure after aggregation of SnO_2 nanoparticles prepared by the hydrothermal method is smooth. Fig.3b shows the microstructure after dispersion. The size of dispersed SnO_2 nanoparticles is smaller and very smooth. Fig. 3c and Fig. 3d show the morphological characteristics of inverse opal SnO_2. It can be seen that by the PS microsphere template, primary pore structures of different sizes are formed on the surface of the particles (Fig. 3c). The microstructures have formed trenches and pore channels (Fig. 3d), which increase the number of active sites and enhance the gas-sensitive characteristics while laying the foundation for rapid response and recovery.

The TEM analysis results of pure SnO_2 nanoparticles and inverse opal SnO_2 nanoparticles were present in Fig.4. As shown in Fig. 4a, the diameter of the SnO_2 nanoparticles prepared by the hydrothermal method is about 10 nm. The intergranular lattice spacing is 0.335nm and 0.263nm, corresponding to the (110) and (101) crystal planes of tetragonal rutile SnO_2 (Fig. 4b-c), respectively. Fig. 4d shows an image of inverse opal SnO_2 nanoparticles. Its diameter is about 5nm, which is significantly smaller than pure SnO_2 nanoparticles. The intercrystalline lattice spacing is 0.339nm and 0.268nm, corresponding to the (110) and (101) crystal planes of tetragonal rutile SnO_2 (Fig. 4e-f), respectively. Therefore, in this work, the reaction with acetone gas mainly occurred at the (110) and (101) crystal facet of the SnO_2 nanoparticles. Among the various crystal planes of SnO_2, the (110) and (101) crystal planes are mainly exposed on the surface due to their lower surface energy, which reduces the binding energy and adsorption energy of acetone molecules, making it easier to adsorb acetone molecules to the surface of SnO_2[1].

B. Gas sensing properties

Using 100 ppm acetone as the test analyte, the optimal working temperature of inverse opal SnO_2 nanoparticles was determined. As shown in Fig. 5a, the response of the inverse opal SnO_2 sensor increases as the temperature increases from 80°C to 240°C, and then decreases when the temperature exceeds 240°C. The maximum response of inverse opal SnO_2 to 100 ppm acetone is 53.1, which is higher than the response of SnO_2 (40.3). In addition, even at a temperature of 100°C, the response of the inverse opal SnO_2 sensor reached 7.2, which indicates that the material can also respond to acetone at low temperatures. Fig. 5b shows the response of the inverse opal SnO_2 sensor under different humidity. Different saturated salt solutions provide a series of the relative humidity (RH) of 11%-95%,

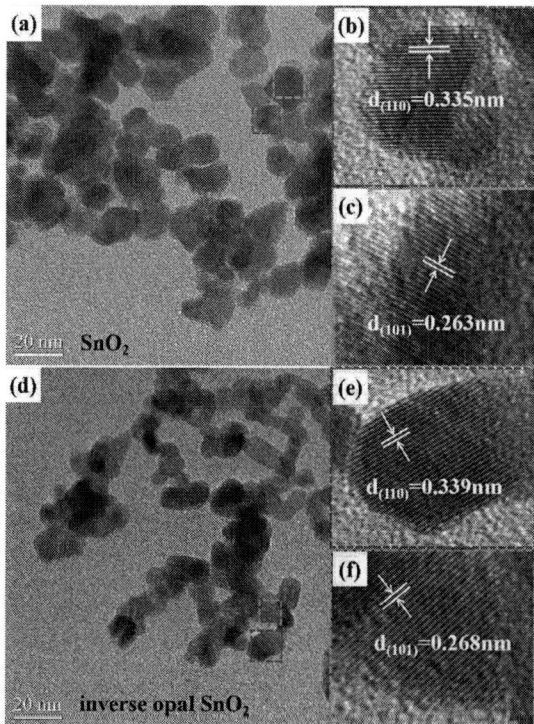

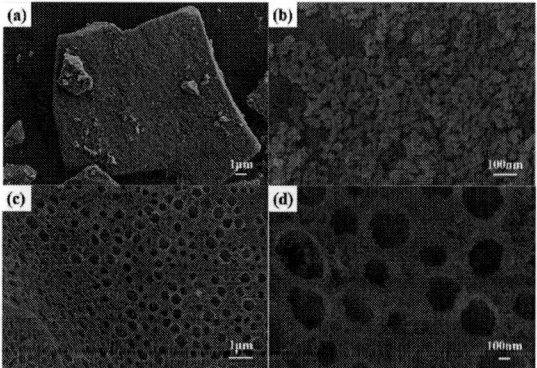

Fig. 3 SEM (a,b) of SnO_2 nanoparticles; SEM (c,d) of inverse opal SnO_2 nanoparticles.

Fig. 4 TEM (a-c) of SnO_2 microspheres; TEM (d-f) of inverse opal SnO_2 nanoparticles.

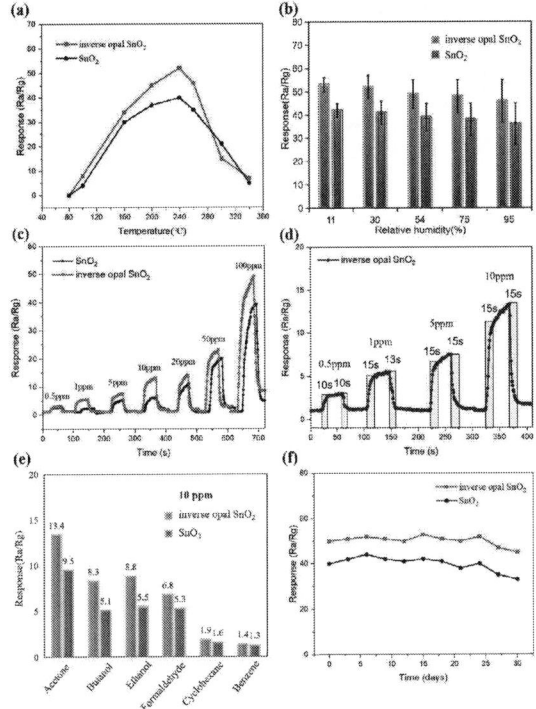

Fig. 5 The gas-sensitive properties of inverse opal SnO_2 nanoparticles. Gas response of sensors (a) at various working temperatures and (b) at various relative humidity. (c) Sensor response to different concentrations of acetone at 240°C. (d) Response-recovery curve at 1-10ppm. Inverse opal SnO_2 sensor (e) selectivity to acetone and (f) working stability.

LiCl(11% RH), MgCl₂(30% RH), Mg(NO₃)₂ (54% RH), NaCl (75% RH), KNO₃ (95% RH), respectively. It can be seen that with the increase of humidity, the influence of humidity on the gas sensitivity of inverse opal SnO_2 becomes greater. The error range of the response value under low humidity is only 3%, but at 95%RH, the error value is close to 10%. Therefore, in the subsequent research work, some treatment should be done to minimize the humidity of target gas before testing in order to improve the response of the gas-sensitive material.

Fig. 5c shows the response and recovery curve of the sensor in the range of 500 ppb to 100 ppm at 240°C (Fig. 5a). When acetone was injected into the test equipment, the response started to increase and then stabilized. When the acetone diffusion disappears, the response drops to the baseline level. The magnitude of the response change increases with the increase in concentration. SnO_2 also has a more obvious response at low concentrations. It can be seen that in the range of 500ppb to 100ppm, the response of inverse opal SnO_2 is significantly improved compared to pure SnO_2, especially at 1ppm and 10ppm, there is almost a three-fold increase in response. In particular, the response reached 5 at 1 ppm, which indicates that the detection limit of the inverse opal SnO_2 is low. Inverse opal SnO_2 is significantly improved compared to pure SnO_2, especially at 1ppm and 10ppm, there is almost a three-fold increase in response. In particular, the response reached 5 at 1 ppm,

which indicates that the detection limit of the inverse opal SnO_2 is low.

The response and recovery time from 500ppb to 10ppm are shown in Fig. 5d. The response time and recovery time from 500ppb to 10ppm are shown in Fig. 5b. It can be seen that with the increase of acetone gas concentration, the response time and recovery time do not increase, and remain at about 15s, indicating that the SnO_2 nanoparticles have a relatively fast response and recovery, as well as stability under different concentrations of acetone gas. These properties demonstrate its potential as a gas-sensitive material for diabetes screening and monitoring.

In practical applications, selectivity is an important indicator of gas detection by gas sensors. Fig. 5e shows the response of two samples to 10 ppm of acetone, n-butanol, ethanol, formaldehyde, cyclohexane, and benzene at 240°C. Compared with traditional SnO_2, inverse opal SnO_2 has a higher response to all target gases and a better response to acetone than other target gases, that the sensor has good selectivity to acetone. At the same time, because different gas molecules have different sizes, we can cover the surface of the sensor with a layer of material that acts as a filter. By adjusting the pore size of the material, the surface of the sensor can only be exposed to the target gas, which enhances the selectivity of the gas sensor.

Besides, as shown in Fig. 5f, to study the stability, the inverse opal SnO_2 sensor was tested ten times in 30 days at 100 ppm acetone and 240°C, and the response only slightly decreased after 25 days. Indicating that the inverse opal SnO_2 sensor has excellent stability and can work stably for a long time.

TABLE I. COMPARISON OF GAS SENSING PROPERTIES OF VARIOUS MATERIALS TOWARD ACETONE

sample	Operating temperature (°C)	Acetone concentration (ppm)	Response (Ra/Rg)	Ref.
Iron oxide foam	300	5	1.65	[13]
α-Fe₂O₃/ multi-walled carbon nanotube	220	1	2.05	[14]
pod-like SnO_2	280	1	2	[15]
SnO₂/ZnSnO₃	290	5	5	[16]
Au-NiO@SnO₂	300	1	4.5	[11]
Co₃O₄/SnO₂	220	1	1.8	[17]
inverse opal SnO_2	240	1	5.6	This work

Table I shows the response comparison of other reported materials to acetone gas sensing. It can be seen that our undoped pure inverse opal structure SnO_2 shows a higher gas response and a lower working temperature, so in subsequent explorations, an appropriate amount of precious metals and metal oxides can be doped to form the compound can better improve the gas sensing characteristics.

C. Sensing mechanism

SnO_2 nanoparticles belong to a typical n-type semiconductor. There are a large number of free electrons on the surface. When the SnO_2 nanoparticle gas sensor is exposed to the air at 150°C-300°C, oxygen molecules from the air are adsorbed on the surface and combine with the electrons on the surface of the gas-sensitive material to generate oxygen ions, as in

$$O_2 \text{ (ads)} + 2e^- \rightarrow 2O^- \text{ (ads)} \qquad (1)$$

When SnO_2 nanoparticles come into contact with reducing acetone gas, the absorbed oxygen ions will react with the acetone molecules, and the electrons will be released back to the surface of the material. The increase of free electrons will inevitably lead to an increase of the conductivity of the material, so the resistance of the gas-sensitive material decreases in the acetone atmosphere. This process can be expressed by Eq. (2).

$$CH_3COCH_3(gas) + 8O^-(ads) \rightarrow 3CO_2(gas) + 3H_2O(gas) + 8e^- \quad (2)$$

Inverse opal SnO_2 nanoparticles are smaller in size and have a porous structure, which enriches active sites, accelerates the speed of electron transmission, is more conducive to the transfer, diffusion and adsorption of acetone molecules, and can consume more acetone molecules. Therefore, SnO_2 with inverse opal structure responds faster.

IV. CONCLUSION

The inverse opal SnO_2 nanoparticles were successfully synthesized by a simple and easy synthesis method. The microstructure of the irregular pores generated and the smaller size of nanoparticles was verified by SEM and TEM. The formed grooves and increased active sites make the inverse opal SnO_2 nanoparticles exhibit good sensing performance for detecting acetone at low concentrations (1-10ppm). Due to its high response at 1ppm (S=5.6), relatively fast response recovery time (15s), and excellent cycle stability, inverse opal SnO_2 nanoparticles are expected to become a gas-sensitive sensor material for screening and monitoring diabetes.

ACKNOWLEDGMENT

This work is supported by the Natural Science Basic Research Plan in Shaanxi Province of China (Program No. 2020JM-027).

REFERENCES

1. Kim, K., et al., *Catalyst-free Highly Sensitive SnO₂ Nanosheet Gas Sensors for Parts per Billion-Level Detection of Acetone.* ACS Applied Materials & Interfaces, 2020. **12**(46): p. 51637-51644.

2. Tassopoulos, C.N., D. Barnett, and T.R. Fraser, *Breath-acetone and Blood-sugar Measurements in Diabetes.* Lancet, 1969. **1**(7609): p. 1282-+.

3. Park, S., *Acetone gas detection using TiO₂ nanoparticles functionalized In₂O₃ nanowires for diagnosis of diabetes.* Journal of Alloys and Compounds, 2017. **696**: p. 655-662.

4. Rao, J., et al., *Construction of Hollow and Mesoporous ZnO Microsphere: A Facile Synthesis and Sensing Property.* ACS Applied Materials & Interfaces, 2012. **4**(10): p. 5346-5352.

5. Li, L., H. Lin, and F. Qu, *Synthesis of mesoporous SnO₂ nanomaterials with selective gas-sensing properties.* Journal of Sol-Gel Science and Technology, 2013. **67**(3): p. 545-555.

6. Zhu, S.M., et al., *Biotemplate fabrication of SnO₂ nanotubular materials by a sonochemical method for gas sensors.* Journal of Nanoparticle Research, 2010. **12**(4): p. 1389-1400.

7. Li, J., et al., *Facile Synthesis and Acetone Sensing Performance of Hierarchical SnO₂ Hollow Microspheres with Controllable Size and Shell Thickness.* Industrial & Engineering Chemistry Research, 2016. **55**(12): p. 3588-3595.

8. Li, X., et al., *Design of Au@ZnO Yolk–Shell Nanospheres with Enhanced Gas Sensing Properties.* ACS Applied Materials & Interfaces, 2014. **6**(21): p. 18661-18667.

9. Kalidoss, R., et al., *Comparative Study on the Preparation and Gas Sensing Properties of Reduced Graphene Oxide/SnO₂ Binary Nanocomposite for Detection of Acetone in Exhaled Breath.* Analytical Chemistry, 2019. **91**(8): p. 5116-5124.

10. Yuan, K., et al., *Fabrication of a Micro-Electromechanical System-Based Acetone Gas Sensor Using CeO₂ Nanodot-Decorated WO₃ Nanowires.* ACS Applied Materials & Interfaces, 2020. **12**(12): p. 14095-14104.

11. Wang, X., et al., *SnO₂ core-shell hollow microspheres co-modification with Au and NiO nanoparticles for acetone gas sensing.* Powder Technology, 2020. **364**: p. 159-166.

12. Jin, W.X., et al., *Hydrothermal synthesis of monodisperse porous cube, cake and spheroid-like α-Fe₂O₃ particles and their high gas-sensing properties.* Sensors and Actuators B: Chemical, 2015. **220**: p. 243-254.

13. Han, D. and M. Zhao, *Facile and simple synthesis of novel iron oxide foam and used as acetone gas sensor with sub-ppm level.* Journal of Alloys and Compounds, 2020. **815**: p. 152406.

14. Jia, X., et al., *Preparation and enhanced acetone sensing properties of flower-like α-Fe₂O₃/multi-walled carbon nanotube nanocomposites.* Sensors and Actuators B: Chemical, 2019. **300**: p. 127012.

15. Yu, H., et al., *Fabricating pod-like SnO₂ hierarchical micro-nanostructures for enhanced acetone gas detection.* Materials Science in Semiconductor Processing, 2021. **121**: p. 105451.

16. Cheng, P., et al., *SnO₂/ZnSnO₃ Double-shelled Hollow Microspheres Based High-performance Acetone Gas Sensor.* Sensors and Actuators B: Chemical, 2020: p. 129212.

17. Xu, Y., et al., *Highly sensitive and selective electronic sensor based on Co catalyzed SnO₂ nanospheres for acetone detection.* Sensors and Actuators B: Chemical, 2020. **304**: p. 127237.

Proceedings of the 16th Annual IEEE International
Conference on Nano/Micro Engineered and Molecular Systems
April 25-29, 2021

Study of the light emission from Eu^{3+} doped nanoporous organosilicate films*

Jinming Zhang, Jing Zhang, Yanrong Wang, Md. Rasadujjaman, Shuhua Wei, and Jiang Yan

Abstract— Rare earth doped Nanoporous organosilicate are promising for luminesce materials application because of their unique structural and large internal surface area. In this paper, we have systematically investigated the photoluminescence characteristics of the silica based materials. In this study, the Eu^{3+} were assembled into the pores of nanoporous organosilicate glass (OSG) by sol-gel method. In addition there are obvious red luminescence peak of the films. However the doping of Eu^{3+} can change the porosity of the films. The photoluminescence characters of different Eu^{3+} ions doped OSG films have been studied. And all the prepared films have showed a stronger emission of Eu^{3+} at 615nm. The optimal content of Eu^{3+} has been confirmed according to the sample with strongest emission.

I. INTRODUCTION

Strong luminescence from silicon based light emitters is detected at room temperature. This discovery is very attractive. Because this indicates potential applications in Si-based optoelectronic devices, especially blue and UV luminescence devices, which means these devices are easy to manufacture or integrate with CMOS. So luminescent silica based materials attracted enormous research attention in recent years[1]. There are many kinds ways to make luminescent silica based materials. Studies have shown that sol-gel method is an appropriate method for the preparation of new luminescent nano-materials[2-4]. Porous organosilicon glass films have the advantages of high hydrophobicity, large internal surface area and uniform micropores, and are an excellent rare earth ion carrier material[5,6]. Therefore Luminescent silica based materials prepared by sol-gel technology have broad application prospects in various kinds of fields such as optics, luminescent solar collectors, intelligent window photochromic panels, environmental and biological impurities sensors, tunable lasers, active waveguides, linear and nonlinear optical materials, and biomarkers[7-10].

To improve the luminescence intensity of silica-based materials, many kinds of chemical elements are added into the silica-based materials. One kind of the doped elements is

*Resrach supported by the National Natural Science Foundation of China (Grant No. 61874002), Natural Science Foundation of Beijing, China (4182021) and Scientific Research Startup Foundation of North China University of Technology.

All authors are with the School of information North China University of Technology, Beijing, China.

Jinming Zhang(e-mail: 18811558924@163.com).

Jing Zhang(e-mail:zhangj@ncut.edu.cn).

Yanrong Wang(corresponding author phone: 86-15652626140; e-mail: wangyanrong@ncut.edu.cn).

Md. Rasadujjaman(e-mail: rasadphy@duet.ac.bd).

Shuhua Wei(e-mail: weishuhua@ncut.edu.cn).

Jiang Yan(e-main: jiangy@ncut.edu.cn).

rare earth element. This kind of chemical element can provide the luminous center. It is well known that the most common valence state of rare earth ions is the trivalent state. Different trivalent rare earths (such as Sm^{3+}, Ce^{3+}, Tb^{3+}, Er^{3+}, and Yb^{3+}) are doped into silica gel, and different sensitizers (such as Al ion) are added to obtain the luminescence performance of strong emission, which has been widely reported[11-14]. Recently reports on the luminescence of europium trivalent (Eu^{3+}) doped in silica keep increasing[15,16]. These results show that the rare earth alone cannot emit light in the silica matrix without any energy transfer activator[17]. Therefore, it is of great significance to study photoluminescence of porous OSG films doped only with rare earth ions.

In this work, we report the synthesis and properties of Eu^{3+} doped porous OSG films. The porosity dependence, the concentration dependence and the antenna effect of Eu^{3+} doped OSG films are analyzed respectively.

II. EXPERIMENT

The nanoporous OSG films were prepared by acid-catalyzed hydrolysis and polycondensation of 1,2-bis(trimethoxysilyl)ethane(BTMSE) and Methyltrimethoxysilane (MTMS) with deionized water. Tetrahydrofuran (THF) was added to produce a homogeneous solution. Brij 30 (C_{12} $H_{25}(OCH_2OCH_2)_4OH$ with molar mass 362 g/mole) was used as a surfactant to obtain porous structure by a self-assembly process[18]. The molar ratio of BTMSE:MTMS:H2O was 0.47:0.53:3.087, and a small amount of a 0.0053M HCl was added as a catalyst. The prepared solution was stirred continue for 10~12h at 60 ℃, and then the gel was stand for 24h. Different content of $Eu(NO_3)_3 \cdot 6H_2O$ was added into the prepared precursor mixture and remarked as #1, #2, #3, and #4, as shown in Table 1.

The films were deposited on 150mm diameter silicon wafers (1－10 Ω •cm) at the rotation speed of 4500 rpm and cured on the hot plate at 150℃, 40 min to remove solvents (soft bake), and final curing at 400℃, 20 min remove NO3- and to obtain Eu ions-doped OSG films (hard bake).

The Eu^{3+} doped OSG films were treated by O_2 plasma in ICP-8000, then soaking them in 2% v/v 3-aminopropyltriethoxysilane solution for 1.5h. To measure the chemical components of the samples, Fourier Transform Infrared Spectroscopy (FTIR, JASCO FT/IR-6300) was used. The measurement settings are 64 times scan, 4 cm^{-1} resolution and 4000-400 cm^{-1} wavenumber range of. The luminescence of the film was measured by photoluminescence (PL) on JASCO FP-8300. The Xe lamp was used as a light source, and the excitation and emission spectra range in wavelength is from 200nm to 750nm, and the slit width is 5nm. The porosities of OSG films were characterized by using heptane adsorption isothermsmeasurement on ellipsometry(EP) based apparatus at 21℃.

III. RESULT AND DISCUSSION

Fig. 1 shows the FT-IR spectra of the pristine OSG films with a porosity of 20%. Curve a and Curve b in Fig. 1 represents the FTIR spectra of Soft Bake film and Hard Bake film, respectively. The band range from 3700 to 3000 cm^{-1} in curve a(Fig. 1) can be attributed to the Si-OH vibration, the band range from 3000 to $2750cm^{-1}$ and 1400 cm-1 can be attributed to the CHx group, the absorption peak at $1750cm^{-1}$ can be attributed to the silanol groups O-H absorption, the band range from 1280 to $1270cm^{-1}$ can be attributed to the terminal $Si-CH_3$ bonds that are important components of OSG films, the range from 1300 to $1000cm^{-1}$ corresponds to the Si-O-Si skeleton, these were the characteristic absorption peaks of porous OSG matrix[19]. The result shows that Hard Bake removes amounts of CHx group and all of the Si-OH, whose existence will increase the adsorption of water molecules. This can effectively reduce the PL quenching of water molecules[20].

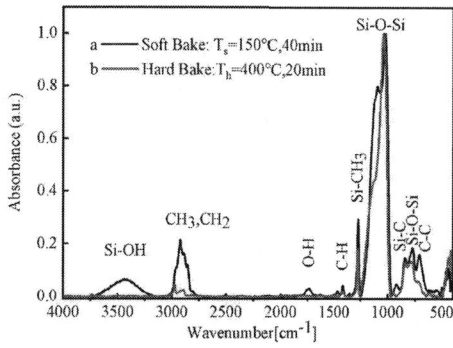

Fig. 1. FT-IR spectra of the pristine OSG films with a porosity of 20% Soft Bake (curve a) and Hard Bake (curve b).

Table 1 shows the EP results of Eu^{3+} doped OSG films. The results indicates that the porosity of the films decrease with the increase of Eu^{3+} content. It is inferred that for the low content of Eu^{3+}(e.g. sample #2), only small amout of Eu^{3+} was deposited on the wall of the pall. As for the high content of Eu^{3+}(e.g. sample 4), Eu^{3+} first fills small pores. Therefore, the pores closed very fast at the neck because of the filling, leading to that some pores became unaccessible. So that for sample #4, RI doesn't change much, but the change of full porosity is significant. Without Eu^{3+} doping, the pores are completely open. For sample #2, small amount of Eu^{3+} deposited on the neck of the pore. For sample #3, more Eu^{3+} was deposited on pore neck. For sample #4, many pores closed very fast at the opening and then the porosity as shown in Fig. 2. reduced. It indicates that doping Eu^{3+} ions make pore size smaller and pore filling is becoming diffusion limited.

Table 1. Effect of Eu^{3+} content on EP

Sample No.	Effect of Eu^{3+} content on EP			
	Eu^{3+} content (wt%)	Thickness (nm)	Refractive index(RI)	Porosity (%)
#1	0	478.0	1.325	25
#2	1.7	493.6	1.329	25
#3	12.1	485.2	1.333	21
#4	25.8	465.7	1.320	13

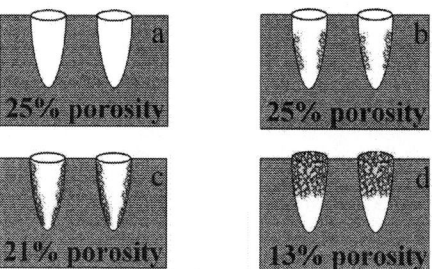

Fig. 2. The possible model of porosity with different concentration of Eu^{3+} ions. (a) Eu^{3+} ions content is 0.0 wt%; (b) Eu^{3+} ions content is 1.7 wt%; (c) Eu^{3+} ions content is 12.1 wt%; (d) Eu^{3+} ions content is 25.8 wt%.

Fig 3 Shows the PL excitation spectra of sample #3. The excitation spectra of the sample were obtained by monitoring the emission wavelength of 615nm. This peak was the strongest emission wavelength in the sample, which are attributed to the f–f transitions within the $Eu^{3+}4f^6$ electron configuration[21]. This excitation spectra contains of a broad band and a narrow peak. The broad band at about 230nm is due to the charge-transfer band of Si-O-Si deficient centers-Eu^{3+}[22]. Fig 4 shows the emission spectra of films with different Eu^{3+} content. All of these emission spectrum showed the narrow peaks assigned to the f–f transitions of the Eu^{3+}, which is due to the shielding of the 4f orbitals by the outer $5s^2$ and $5p^6$ orbitals. There are three emission peaks including 590nm, 615nm and 689nm of all the samples. The emission band at 590nm could be attributed to the typical $^5D_0$-$^7F_1$ transition of Eu^{3+}. The strongest emission band appeared near 615nm could be attributed to the typical $^5D_0$-$^7F_2$ transition of Eu^{3+}. The emission band at 689nm could be attributed to the $^5D_0$-$^7F_4$ transition of Eu^{3+}. The emission peak at 615nm belongs to the electric dipole transition, and the fluorescence intensity is greater than the magnetic dipole transition at 590nm. It indicates that europium ion is in the inversion position of symmetry center and is strongly dependent on the local symmetry environment[23]. The strongest red emission occurs when Eu^{3+} content is 12.1wt%. High concentration with Eu^{3+} doping will close the hole of the film leading to luminescence intensity quenching.

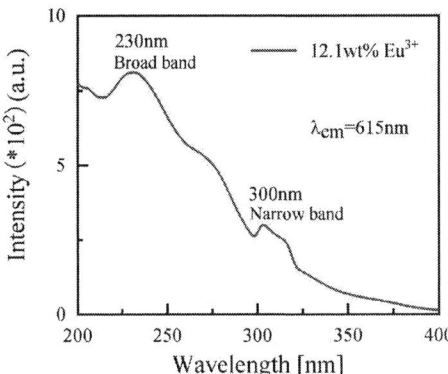

Fig 3 Excitation spectra of 12.1wt% Eu³⁺ doped OSG films.

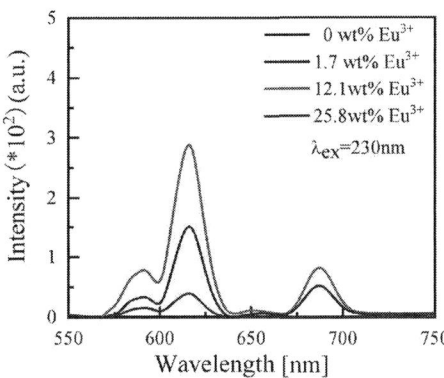

Fig. 4 Emission spectra of different content of Eu³⁺ doped OSG films.

APTES treatment of 12.1wt% Eu³⁺ doped OSG films increase the energy transfer efficiency by the antenna effect as shown in Fig.5. Fig 5 (b) shows the FTIR spectrum of APTES treatment of Eu³⁺ doped OSG films the bands at around 900-1250 cm⁻¹ originating from the $Si - O - Si$ asymmetric/symmetric stretching, which shift towards higher binding energy/wavenumber due to the increase of CHx groups and this suggests the formation of more $Si-O-Si$ cage type structures[24]. Fig4 (c) shows that the band at 615 cm⁻¹ are attributed to the Short-wave level- $\equiv$ C-H deformation vibration. The increased spectral intensity by the treatment with the antenna molecules indicates that supply the small absorption coefficient of rare earth ions in the UV-visible region, and the non-radiation transition enhances the characteristic emission of Eu³⁺ at 615nm Fig 4 (d).

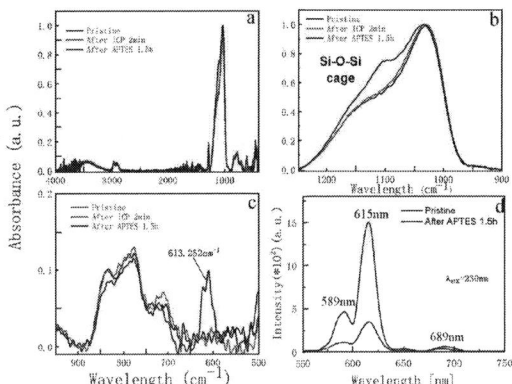

Fig. 5. FTIR spectrum of APTES treatment of 12.1wt% Eu³⁺ ions doped OSG film (a) and in large of 1250-950cm⁻¹ (b), 950-500cm⁻¹ (c). PL emission spectrum of APTES treatment of 12.1wt% Eu³⁺ ions doped OSG films (d).

IV. CONCLUSION

The luminescence film can be prepared by sol-gel method with doped Eu³⁺. Comparing the PL result of the films with different Eu³⁺, it can be found that the film with Eu³⁺ of 12.1wt% has the strongest luminescent intensity at 615nm. The doped Eu³⁺ ions replaces Si-CH₃ of the film and forms the Si-Eu band, which makes pore size smaller and pore filling becoming diffusion limited. In addition, high concentration doping leads to PL quenching. Finally the antenna effect without radiative transition enhances the characteristic emission of Eu³⁺.

ACKNOWLEDGMENT

Resrach supported by the National Natural Science Foundation of China (Grant No. 61874002), Natural Science Foundation of Beijing, China (4182021) and Scientific Research Startup Foundation of North China University of Technology.

REFERENCES

[1] Zhi-Yong W, Ke-Xin L and Xiao-Tang R 2010 Mechanism and enhancement of photoluminescence from silicon nanocrystals implanted in SiO₂ matrix Chinese Phys. B191–5

[2] Miyata H, Fukushima Y, Okamoto K, Takahashi M, Watanabe M, Kubo W, Komoto A, Kitamura S, Kanno Y and Kuroda K 2011 Remarkable birefringence in a TiO₂-SiO₂ composite film with an aligned mesoporous structure J. Am. Chem. Soc. 13313539–44

[3] De la Cruz J, Palomino Merino R, Trejo-García P, Espinosa J E, Aceves Torres R, Moreno-Barbosa E, Gervacio-Arciniega J J and Soto E 2019 Luminescent properties of a hybrid SiO₂-PMMA matrix doped with terbium Opt. Mater. (Amst). 8742–7

[4] Liu J, Wang X, Xuan T, Li H and Sun Z 2014 Photoluminescence and thermal stability of Mn²⁺ co-doped SrSi₂O₂N₂:Eu²⁺green phosphor synthesized by sol-gel method J. Alloys Compd. 593128–31

[5] Dutt A, Matsumoto Y, Godavarthi S, Santana-Rodríguez G, Santoyo-Salazar J and Escobosa A 2014 White bright luminescence at room temperature from TEOS-based thin films via catalytic chemical vapor deposition Mater. Lett. 131295–7

[6] Rani N and Ahlawat R 2019 Role of Ceria Nanocrystals on Morphology and Luminescence of Eu³⁺ doped SiO₂ nanopowder J. Lumin. 208135–44

[7] Nayef U M, Hussein H T and Abdul Hussien A M 2018 Study of photoluminescence quenching in porous silicon layers that using for chemical solvents vapor sensor Optik (Stuttg). 1721134–9

[8] Van Der Voort P, Esquivel D, De Canck E, et al. Periodic mesoporous organosilicas: from simple to complex bridges; a comprehensive overview of functions, morphologies and applications[J]. Chemical Society Reviews, 2013, 42(9): 3913-3955.

[9] Wen D and Shi J 2013 A novel narrow-line red emitting $Na_2Y_2B_2O_7$:Ce^{3+},Tb^{3+},Eu^{3+} phosphor with high efficiency activated by terbium chain for near-UV white LEDs Dalt. Trans. 42 16621–9

[10] Wirnsberger G, Yang P, Scott B J, Chmelka B F and Stucky G D 2001 Mesostructured materials for optical applications: From low-k dielectrics to sensors and lasers Spectrochim. Acta-Part A Mol. Biomol. Spectrosc. 572049–60

[11] Jin L, Li D, Xiang L, Wang F, Yang D and Que D 2013 Energy transfer from luminescent centers to Er^{3+} in erbium-doped silicon-rich oxide films Nanoscale Res. Lett. 81–6

[12] Sun J, Zhang X S, Yuan L L, Feng Z J, Ling Z and Li L 2014 White emission from Tm^{3+}/Tb^{3+}/Eu^{3+} co-doped fluoride zirconate under ultraviolet excitation Chinese Phys. B 232–6

[13] Pawlik N, Szpikowska-Sroka B, Sołtys M and Pisarski W A 2016 Optical properties of silica sol-gel materials singly- and doubly-doped with Eu^{3+} and Gd^{3+} ions J. Rare Earths 34 786–95

[14] Wu-Chang D, Yan L, Yun Z, Jian-Chuan G, Yu-Hua Z, Bu-Wen C, Jin-Zhong Y and Qi-Ming W 2009 A comparison of silicon oxide and nitride as host matrices on the photoluminescence from Er^{3+} ions Chinese Phys. B 183044–8

[15] Zhang Q, Sheng Y, Zheng K, Qin X, Ma P and Zou H 2015 Novel organic-inorganic amorphous photoactive hybrid films with rare earth (Eu^{3+}, Tb^{3+}) covalently embedded into silicon-oxygen network via sol-gel process Mater. Res. Bull. 70379–84

[16] Zhang W long, Liu Y, Yu H and Dong X ting 2019 Eu and Tb co-doped porous SiO_2 aerogel composite and its luminescent properties J. Photochem. Photobiol. A Chem. 37947–53

[17] Khan A F, Yadav R, Singh S, Dutta V and Chawla S 2010 Eu^{3+} doped silica xerogel luminescent layer having antireflection and spectrum modifying properties suitable for solar cell applications Mater. Res. Bull. 451562–6

[18] Whitesides G M and Grzybowski B 2002 Self-assembly at all scales Science (80-). 2952418–21

[19] Nenashev R, Wang Y, Liu C, Kotova N, Vorotilov K, Zhang J, Wei S, Seregin D, Vishnevskiy A, Leu J (Jim) and Baklanov M R 2017 Effect of Bridging and Terminal Alkyl Groups on Structural and Mechanical Properties of Porous Organosilicate Films ECS J. Solid State Sci. Technol. 6 N182–8

[20] Tagaya M, Ikoma T, Yoshioka T, Motozuka S, Minami F and Tanaka J 2011 Efficient synthesis of Eu(III)-containing nanoporous silicas Mater. Lett. 652287–90

[21] Chen W, Sammynaiken R and Huang Y 2000 Photoluminescence and photostimulated luminescence of Tb^{3+} and Eu^{3+} in zeolite-Y J. Appl. Phys. 881424–31

[22] Lu Q, Wang Z, Wang P and Li J 2010 Structure and luminescence properties of Eu^{3+}-doped cubic mesoporous silica thin films Nanoscale Res. Lett. 5761–8

[23] Binnemans K, Lenaerts P, Driesen K and Go C 2004 A luminescent tris (2-thenoyltrifluoroacetonato) europium (III) complex covalently linked to a 1 , 10-phenanthroline- functionalised sol–gel glass 191–5

[24] Redzheb M A 2018 Synthesis and Characterization of Mesoporous Organosilica Films for Low-k Dielectric Application 1–177

Gap in pagination due to unavailable papers.

Pages 421-424

Proceedings of the 16th Annual IEEE International
Conference on Nano/Micro Engineered and Molecular Systems
April 25-29, 2021

Highly Sensitive Resonant Bi-Directional Magnetic Field Micro-Sensor *

N. Alcheikh, S. Ben Mbarek, M. I. Younis, *Member, IEEE*

Abstract— This paper demonstrates a micro-machined ultra-sensitive bi-directional magnetic sensor based on the Lorentz-force principle. The magnetometer relies on detecting the resonant frequency of electrothermally actuated straight micro-resonator operated near the buckling point. Optical sensing is utilized to capture the frequency shift while the device is operated at ambient pressure and temperature. The micro-sensor shows a high normalized sensitivity (S) of 9.48 $(mA.T)^{-1}$ compared to the literature and good linearity of 0.12 % in wide range of magnetic field. Also the sensor consumes low power around 0.2 mW. These features encourage for low cost applications.

I. INTRODUCTION

Recently, Micro-electro-mechanical systems (MEMS) magnetometers based on the Lorentz-force principle have been attracted further attention for various applications, such as telecommunications, electronic compasses, electronic compasses, inertial navigation systems, and biomedical [1-4]. MEMS magnetometers are simple in design with low power consumption, low cost, and highly sensitive [5]. The operating principle of most resonating micro-sensors relies on the frequency variation of the heated structures due to magnetic field (frequency shift) [6-8]. Resonating Lorentz-force micro-structures have the advantages of high accuracy, high outstanding stability, low power consumption, high sensitivity, large output signal, and immunity to noise.

Various sensing methods have been used for micro-resonators based magnetic sensors including piezoelectric, piezoresistive, capacitive, and optical sensing [9-12]. Despite device imperfections from the intrinsic losses, optical sensing is more robust technique, requires simple read–out electronic circuits, and has higher immunity to electromagnetic interference. Also, optical sensing does not require operation in vacuum [9].

In [13-14], the buckling point of heated in-plane clamped-clamped straight micro-beam has been used to realize ultra-sensitive gas and pressure sensors. The devices are based on measuring the resonance frequency shift when varying the air pressure (gas type and concentrations). Also, based on the same principle, we showed the possibility to demonstrate bi-directional magnetic Lorentz-force [15].

*Research supported by KAUST Foundation.

The authors are with the Physical Science and Engineering Division, King Abdullah University of Science and Technology, Thuwal 23955-6900, Saudi Arabia (e-mails: nouha.alcheikh@kaust.edu.sa, sofiane.benmbarek@kaust.edu.sa, and mohammad.younis@kaust.edu.sa.

Corresponding author: Mohammad I. Younis; phone: +966-2-808-0597.

In this work, we demonstrate a wide-range ultra-sensitive bi-directional magnetometer. The device is based on operating an electrothermally heated straight micro-beam near buckling, which offers significant frequency shift leading to high sensitivity.

II. DEVICE DESCRIPTION AND MEASUREMENT METHOD

A schematic of the Lorentz-force magnetometer is shown in Fig. 1. The tested micro-beams are fabricated from silicon-on-insulator (SOI) wafer (SOIMUMPs process) by MEMSCAP. They are clamped from both sides at the anchors. They have lengths L (600 µm and 800 µm), thickness b (25 µm), and width h (2 µm). The air-gap(g) between the micro-beams and the electrodes is 8 µm, Fig. 1. As shown, a DC electrothermal voltage source V_{Th} is used to induce a current I_{Th} passing through the micro-resonator. This heats it up by Joule's heating, and hence under z-axis magnetic field (B), Lorentz-force (F_L) is generated normal to the micro-beam in y-axis. Depending on the direction of B, the micro-beam can be buckled down (+y-axis) or up (-y-axis), which results in downward or upward shift in the resonant frequency.

To measure the resonant frequency shift of the micro-beams, we use a laser Doppler vibrometer from Polytec (MSA 500) with a white noise signal [16]. The micro-beams are driving electrostatically by an AC harmonic voltage and a DC static bias of 10 V. The sensor is tested under atmospheric pressure and room temperature. We should mention that, the proposed sensing technique face some challenges, such as environmental temperature variation. However, the proposed micro-sensor is simple in design with low cost fabrication.

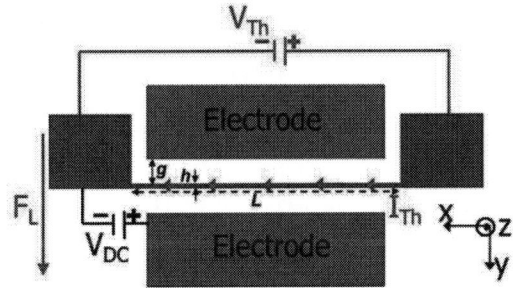

Figure 1. *Schematic of the proposed magentic sensor based on in-plane clamped-clamped straight micro-beam electrothermally actuated.*

III. RESULTS AND DISCUSSION

Figure 2 shows the measured variation of the resonant frequency of a 600 µm micro-beam length versus bias current (I_{Th}) for various values of bi-directional z-axis

978-1-6654-3008-1/21 $31.00 © 2021 IEEE   425

magnetic fields (B). As shown in Fig. 2, when B= 0 mT, the frequency decreases initially toward zero when increases I_{Th}, which increases the compressive force, until buckling where the frequency drops to a minimum value. Note that because the micro-beam is initially biased at V_{DC} = 10 V, the frequency does not drop to zero at buckling (perturbed bifurcation).

To measure the frequency response of the device (FM), a permanent magnet is used under the resonators to generate B, Fig. 1. The magnetic field was varied from 70 mT to 140 mT, in positive and negative directions, by moving the permanent magnet nearer or farther the device. As shown in Fig. 2, adding B in the positive z direction causes upward shift in the frequency, whereas B in the negative z direction causes an opposite effect. This indicates bi-directionality of the sensor. Note that at the buckling point I_{Th}=1.6 mA, the frequency shifts due to B is maximum.

The sensitivity of the micro-senor S (1/T) is defined as the relative change in the resonant frequency ($\Delta f/f_0$) versus the input magnetic field (∂B) [17]. The frequency shift (Δf) is expressed as ($f-f_0$), where f_0 and f are the frequency of the micro-resonator at 0 mT and with B, respectively. Figure 3 shows $\Delta f/f_0$ versus B at I_{Th}= 1.6 mA (around the buckling point). As indicated from the slope, the sensor has high sensitivity of S=2.33/T. In addition to high sensitivity, the sensor shows high linearity of 0.5 %.

Figure 4 shows the measurement results for another case study of 800 μm micro-beam. As indicated in Fig. 4a, the buckling point is around I_{Th}=0.27 mA. Figure 4b shows sensitivity of S=2.56/T for (+B). To evaluate the performance of both studied beams, we normalize the sensitivity with respect to I_{Th}. Accordingly, for the 600 μm micro-beam length, S=1.456 $(mA.T)^{-1}$, and for the 800 μm micro-beam length, S=9.48 $(mA.T)^{-1}$. Both results indicate much higher sensitivity compared to the state of the art.

In addition to having high sensitivity, low power consumption and good resolution are also important factors. At I_{Th}=0.27 mA (bucking point), the magnetometer presents a low power consumption around 0.2 mW due to electrothermal actuation. Hence, based on simple design (straight micro-beam), high sensitivity is presented for a low current. However, the device power can be improved after vacuum packaging.

In our previous work, we showed that for 800 μm micro-beam length, the sensitivity highly depends on the temperature variation where the Thermal Coefficient of Frequency TCF is found to be around −0.006/°C [15]. Moreover, it suspects that by doping the devices with high concentrations of an n-type dopant, the device TCF can be highly reduced [18].

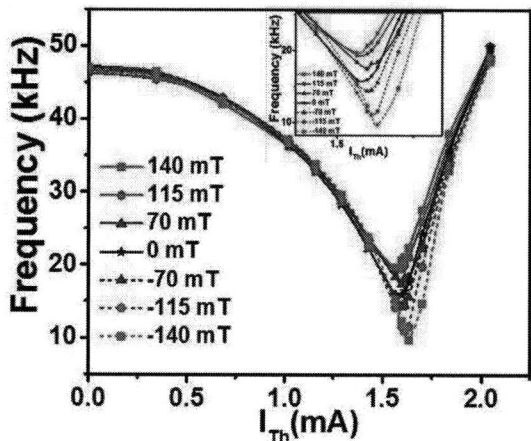

Figure 2. *Measured resonant frequency variation of 600 μm micro-beam length versus bias current (I_{Th}) for various values of bi-directional z-axis magnetic fields (B). Inset shows the enlarged view of resonant frequency variation around buckling.*

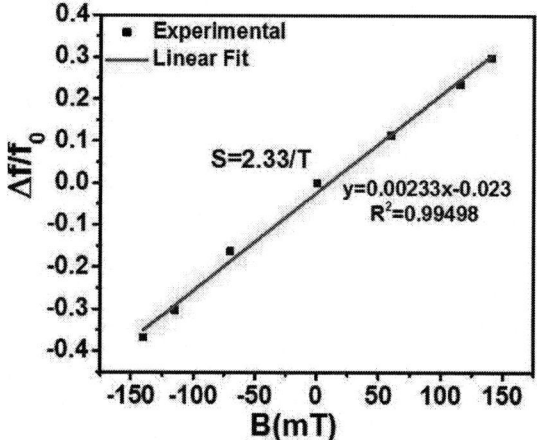

Figure 3. *Measured the relative change in the resonant frequency ($\Delta f/f_0$) with B for I_{Th}= 1.6 mA (around buckling point). f_0 represent the resonant frequency of the beam at 0 mT. The sensitivity of the micro-sensor (S) denotes the linear coefficient of the variation of $\Delta f/f_0$ versus B.*

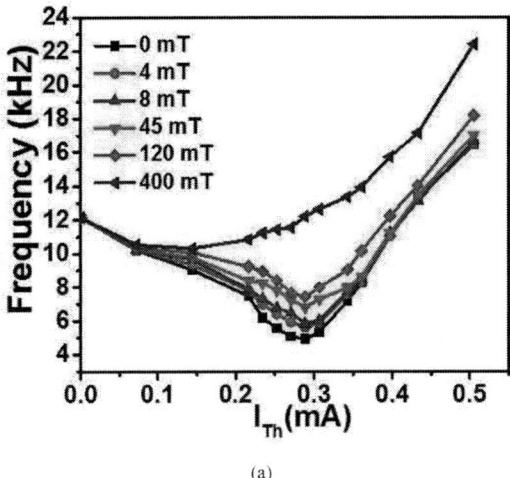

(a)

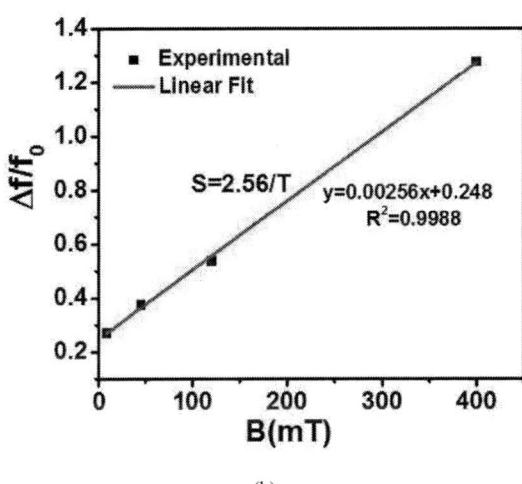

(b)

Figure 4. *(a) Resonant frequency variation results of 800 μm micro-beam length versus I_{Th} for various values B in z-axis. (b)The relative change in the resonant frequency ($\Delta f/f_0$) versus B at I_{Th}= 0.27 mA (around buckling point).*

TABLE I.   THE PERFORMANCE COMPARISONS OF THE PROPOSED MEMS Z-AXIS MAGNETIC FIELD SENSORS WITH FM-LORENTZ FORCE BASED MAGNETOMETERS.

Reference	Surface (mm^2)	Power (mW)	Sensitivity (mA.T)$^{-1}$	Non-linearity (%)
[8]	4.8	7.5e-5	160e-6	0.4
[17]	0.48	40	33.9e-6	0.1
[19]	0.308	10	2.8	0.5
This work	0.012	3.7	1.456	0.5
This work	0.016	0.2	9.48	0.2

Table. 1 shows the performance comparisons of the proposed micro-sensor with MEMS magnetometers based on frequency-modulation-Lorentz force (FM). As seen from Table 1, the device sensitivity in this work is significantly higher compared to other reported FM magnetometers. Moreover, with small dimensions, the device has low power consumption with good linearity for a wide magnetic field range of 140 mT. However, the dissipation power can be improved after vacuum packaging.

IV. CONCLUSION

We experimentally showed an ultra-sensitive bi-directional Lorentz-force magnetometer based on operating a heated straight beam resonator around the buckling point. The resonant frequency variation of the micro-beam under magnetic fields is measured. The results indicate high sensitivity compared to the state-of-the-art Lorentz-force magnetometers. The proposed sensor has the advantages of simplicity, low cost, low consumption, good linearity, and scalability. These attractive features pave the way for deeper investigations on the implementation of such proposed magnetometers for even lower range magnetic fields in the μT. Next phases of this research are planned to test the devices at low pressure for ultra-low consumption and low field detection.

ACKNOWLEDGMENT

We acknowledge financial support from King Abdullah University of Science and Technology.

REFERENCES

[1] A. Herrera-May, J. Soler-Balcazar, H. Vázquez-Leal, J. Martínez-Castillo, M. Vigueras-Zuñiga, and L. Aguilera-Cortés, "Recent advances of MEMS resonators for lorentz force based magnetic field sensors. Design, Applications and Challenges," *Sensors*, vol. 16, pp. 1359, 2016

[2] M. Olga, N. Dudchenko, and A. Dudchenko, "Doxorubicin magnetic conjugate targeting upon intravenous injection into mice: High gradient magnetic field inhibits the clearance of nanoparticles from the blood," *Journal of magnetism and magnetic materials*, vol. 293, pp. 473-482, 2005.

[3] A. L. Herrera-May, L. A. Aguilera-Cortés, P. J.García-Ramírez, P.J., E. and Manjarrez, " Resonant magnetic field sensors based on MEMS technology," *Sensors*, vol. 9, pp.7785-7813, 2009.

[4] M. Díaz-Michelena, "Small magnetic sensors for space applications," *Sensors*, vol. 9, pp. 2271-88, 2009.

[5] J. M. Allen, Taylor & Francis: Boca Raton, FL, USA, 2005.

[6] B. Bahreyni, C. Shafai C, "A resonant micromachined magnetic field sensor," *IEEE Sensors Journal*, vol. 7, pp. 1326-34, 2007.

[7] V. Kumar , S. Pourkamali, "Lorentz force MEMS magnetometer with frequency modulated output," *IEEE International Conference on Micro Electro Mechanical Systems (MEMS)* , Jan 24, pp. 589-592, 2016.

[8] H. Liang, S. Liu, B. Xiong, "In-Plane-Sense Magnetometer Based on Torsional MEMS With Vertically-Interlaced Combs via Self-Alignment Technique," IEEE Electron Device Letters, vol. 6, pp. 900-3, 2020.

[9] A.L. Herrera-May, A. L. Aguilera-Cortés, P. J. García-Ramírez, E. Manjarrez E, "Resonant magnetic field sensors based on MEMS technology," *Sensors*, vol. 10, pp. 7785-813, 2009.

[10] B. Park, M. Li, S. Liyanage, C. Shafai, " Lorentz force based resonant MEMS magnetic-field sensor with optical readout," *Sensors and Actuators A: Physical*, vol. 241,pp.12-18, 2016.

[11] G. Wu , D. Xu, B. Xiong, D. Feng, Y. Wang, "Resonant magnetic field sensor with capacitive driving and electromagnetic induction sensing," *IEEE electron device letters*, vol. 34, pp. 459-61,2013.

[12] A. L. Herrera-May, M. Lara-Castro, F. López-Huerta, P. Gkotsis, J. P. Raskin, E. Figueras, "A MEMS-based magnetic field sensor with simple resonant structure and linear electrical response, " *Microelectronic Engineering*, vol. 142, pp. 12-21, 2015.

[13] A. Z. Hajjaj, N. Alcheikh N, M. A. Hafiz, S. Ilyas, M. I. Younis, "A scalable pressure sensor based on an electrothermally and electrostatically operated resonator," *Applied Physics Letters*, vol. 111, pp. 223503, 2017.

[14] A. Z. Hajjaj, N. Jaber, N. Alcheikh , M . I. Younis, "A Resonant Gas Sensor Based on Multimode Excitation of a Buckled Microbeam," *IEEE Sensors Journal*, vol. 20, pp. 1778-85, 2019.

[15] N. Alcheikh, M. I. Younis, "Resonator-Based Bidirectional Lorentz Force Magnetic Sensor," IEEE Electron Device Letters, vol. 42, pp. 406 - 409, 2021.

[16] [Online].Polytech: http://www.polytec.com/us/.

[17] W. Zhang W, J.E. Lee, , "Frequency-based magnetic field sensing using Lorentz force axial strain modulation in a double-ended tuning fork. ," *Sensors and Actuators A: Physical*, vol. 211, pp. 1451-52, 2014.

[18] A. Hajjam, A. Logan, S. Pourkamali, "Doping-induced temperature compensation of thermally actuated high-frequency silicon micromechanical resonators," *Journal of microelectromechanical system*, vol. 21, pp. 681–687, Jun. 2012.

[19] G. Laghi, S. Dellea, A. Longoni, P. Minotti, A. Tocchio, S. Zerbini, G. Langfelder," G. Torsional MEMS magnetometer operated off-resonance for in-plane magnetic field detection," *Sensors and Actuators A: Physical*, vol. 229, pp. 218-26, 2015.

Gap in pagination due to unavailable papers.

Pages 429-432

Proceedings of the 16th Annual IEEE International
Conference on Nano/Micro Engineered and Molecular Systems
April 25-29, 2021

A Flexible Mechanical Composite Micro-Grating Tailored By One-Dimensional Ordered Wrinkle Patterns

Yi-Hang Xin, Kai-Ming Hu*, Xiu-Yuan Li,Er-Qi Tu and Wen-Ming Zhang[1]

Abstract— Optical grating devices based on periodic micro and nanostructures are widely employed to precisely manipulate light propagation, which are significant for light sensors, optical data storage, information technologies and automatic driving. However, the parameters of rigid periodic structures are difficult to tune after manufacturing, which seriously limits their capacity of in-situ light manipulation. Here, a novel mechanically tuned flexible micro-grating is proposed for optical encryption, which is composed of two one-dimensional (1D) ordered wrinkle patterns. 1D ordered wrinkles in the anthracene-contained copolymer (PAN)/poly (dimethylsiloxane) (PDMS) system are prepared by selective exposure under a strip mask. By changing the angle between two gratings, the diffraction pattern of single point laser transforms from a 1D pattern to an array.

I. INTRODUCTION

With the advantages of light weight, low cost and simple operation, periodically tunable diffraction grating is a promising choice for dynamic modulation of beam directivity [1]. Owing to their adjustability,gratings based on wrinkle are applied to multi-directional stress/strain measure [2-4]，light collection[5] and characterization method of the wrinkled surface morphology[6,7].

The flexible material has a low modulus so it can withstand large deformation, which provides the micro-grating tailored by wrinkle patterns with dynamic adjustment ability. The period and amplitude of this kind of grating can be regulated. Mechanical stretching could tune the grating period [8-10]. Rectangular and sinusoidal phase gratings are orthogonally coupled to fabricate a mechanically tunable composite grating [11,12].The intensity of grating diffraction pattern can be changed by adjusting the amplitude [13]. Takei and co-workers change the amplitude of surface wrinkles by controlling the cylindrical oil bubble in the PDMS interlayer [14]. In addition ,gratings based on wrinkle are able to be erased.Through the photothermal effect of near-infrared radiation [15-17],the control of electrical signals [18], or the distance from the heat source [19], we can achieve the purpose of dynamic switching of wrinkle gratings.

Compared with one-dimensional(1D) grating, the control of two-dimensional(2D) grating is more difficult.2D diffraction images can be obtained when a point laser goes through single grating with 2D ordered wrinkle patterns[20, 21]. However, it is difficult to regulate the pattern style of the multiple exposure 2D gratings after manufacturing, and it is unable to control multiple parameters at the same time. If it is used for optical encryption,the dynamic encryption

capabilities are limited. Compared to single grating with 2D wrinkle patterns, the fabrication of single 1D grating device in flexible composite grating system is easier, and the assembly and adjustment of these gratings can be more flexible,which can obtain more diverse results.

In this paper, we propose a flexible composite micro-grating to dynamically tune the diffraction behavior of light for optical encryption. This grating system can dynamically regulate the 2D diffraction pattern by tuning the relative position between the two micro-gratings tailored by 1D ordered wrinkle patterns through mechanical methods.

II. RESEARCH METHOD

A. A Flexible Mechanical Composite Micro-Grating

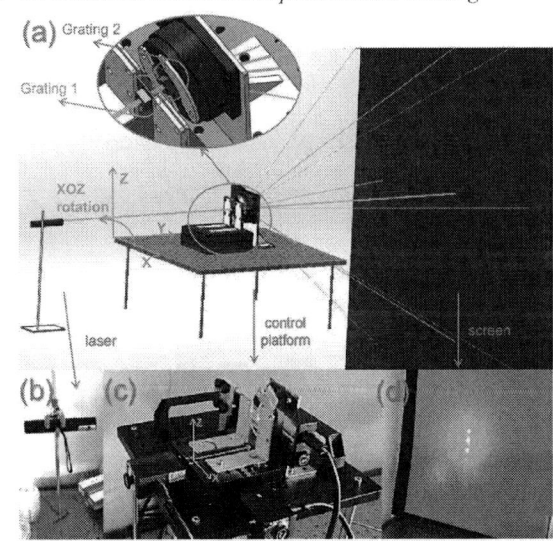

Figure 1: Mechanical composite micro-grating system, where the central line direction of the two gratings is defined as y axis, the horizontal plane is xoy plane, and the vertical direction is z direction: (a) overall schematic diagram of mechanical composite micro-grating system,(b) laser source,(c) control platform with two 1D wrinkle micro-gratings,(d) screen.

As shown in Figure 1, mechanical composite wrinkle micro-grating system consists of a laser source, two 1D wrinkle micro-gratings, a clamping device and a screen. The laser source and the screen are respectively arranged on each sides of the clamping adjustment device, and the clamping

All authors are with State Key Laboratory of Mechanical System and Vibration, School of Mechanical Engineering, Shanghai Jiaotong University, 200240, Shanghai, China

(corresponding author Kai-Ming Hu, Tel.: +86-18321842038, e-mail: hukaiming@sjtu.edu.cn; corresponding author Wen-Ming Zhang, Tel.: +86-2154744990, email: wenmingz@sjtu.edu.cn).

978-1-6654-3008-1/21 $31.00 © 2021 IEEE

adjustment device holds the two edges of the two gratings respectively. The clamping adjustment device can regulate the geometric parameters of a single grating by mechanical stretching in the x direction, and the maximum frequency is 5 Hz. The clamping adjustment device can also regulate the relative position relationship of the two gratings by translation or rotation. Grating 1 can be translated in x and y directions and Grating 2 can be rotated along y axis. The translation accuracy is 1 μm and the rotation accuracy is 0.005° in the xoz plane. The diffraction phenomenon of combined grating can be dynamically controlled by the above methods, so as to achieve the purpose of optical encryption.

B. Micro-Grating Tailored By 1D Ordered Wrinkle Patterns

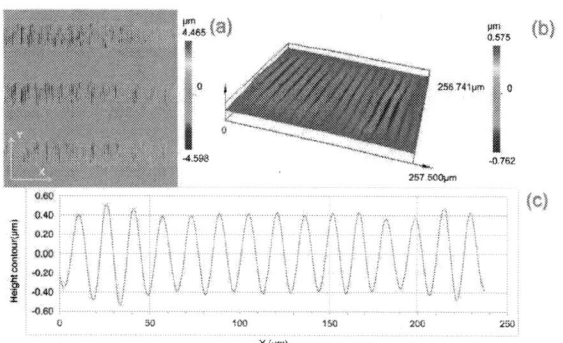

Figure 2: Three-dimensional (3D) laser scanning confocal microscope (LSCM) images of the sample with the width of the wrinkled and wrinkle-free belts about 1:1: (a) 2D image of the height of 1D ordered wrinkle patterns,(b) 3D image of the height topography,(c) the height topological curves of orange line marked in Figure 2(a).

The grating with 1D ordered wrinkle patterns in the bilayer system is prepared by spin-coating of PAN on PDMS elastomer. 6% PAN solution is first spin-coated on a PDMS substrate with the rotation speed of 4000 rpm and then irradiated by UV light under a 200-μm-strip photomask for 27 min, resulting in the relatively stiff areas on the surface film.After the sample is heated to 110 ° and then cooled at room temperature ,the exposed areas will be triggered 1D ordered wrinkles spontaneously. The 1D ordered wrinkle formation of PDMS-PAN is due to the mismatch of modulus and thermal expansion coefficient between the hard skin layer and the soft substrate, which is caused by the photocrosslinking of the top layer of PDMS-PAN. (Figure 2a-b).

Selective exposure using a mask with 1D stripe morphology will trigger 1D ordered wrinkle pattern. The 1D ordered wrinkle patterns are characterized by LSCM, which shows a sinusoidal topography with wavelength 15.5 μm and amplitude 400 nm (Figure 2c). Due to the soft boundary condition of the unexposed region and the yield stress distribution of the surface film caused by selective UV exposure, the thermal stress can be released along the y

direction, but not along the x direction, resulting in $\sigma_x > \sigma_c > \sigma_y$, where σ_c is the critical stress value causing wrinkles (Figure 2a). Therefore, unidirectional wrinkles are triggered along x direction (Figure 2b).

By controlling the heating time and the temperature, the width ratio of wrinkled and wrinkle-free belt can be controlled.Short heating time or low heating temperature will lead to a narrow wrinkled belts and no obvious diffraction phenomenon.On the contrary, with overexposure, the stress boundary will differ from the photomask pattern. As a result, the wrinkles will not completely be 1D sinusoidal shape perpendicular to the boundary and the collimation wil be poor. The samples with the width ratio of the wrinkled and wrinkle-free belts about 1:1 are nice experimental objects for 1D gratings (Fig. 2a).

III. RESULTS AND DISCUSSION

A. diffraction theory

According to the law of diffraction, the grating equation of wrinkled bilayer systems under the applied tensile strain can be expressed as

$$\lambda_{wrinkle}sin\theta_m = m\lambda \qquad (1)$$

where $\lambda_{wrinkle}$ are the wavelength of wrinkle patterns; m and θ_m denote the diffraction order and the angle of the m th diffraction order, respectively; λ means the wavelength of the incident light.

In the experiment, a laser emitter with the wavelength of 532 nm is used, and the average wavelength of 1D diffraction grating is 15.5 μm. The calculated first order diffraction angle is $sin\theta_1 = \frac{m\lambda}{\lambda_{wrinkle}} = 0.03432$.

B. Diffraction grating system experimental results

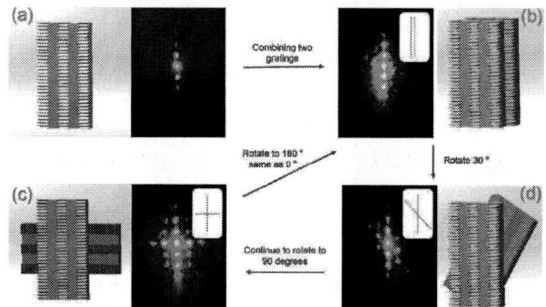

Figure.3: The change of diffraction array caused by adjusting the angle of double-layer 1D grating. Blue and orange lines are used to represent the array principal axis caused by the direction of single grating. (a)Diffraction photos and grating schematic diagram of single 1D grating, (b) Diffraction photos and grating schematic diagram of composite micro-gratings.The angle between two 1D gratings is 0°. (c) Diffraction photos and grating schematic diagram of composite micro-gratings.The angle between two 1D gratings is 30 ° (d) Diffraction photos and grating schematic diagram of composite micro-gratings. The angle between two 1D gratings is 90 °

When the diffraction angle is small, by approximating $sin\theta_m = \frac{x}{l}$, the following formula can be obtained as

$$\lambda_{wrinkle} \frac{x}{l} = m\lambda \qquad (2)$$

where x represents the distance between adjacent diffraction spots on the same axis, and l represents the distance between the grating and the screen.

For a 1D grating, the diffraction point of 1D wrinkle patterns aligns in a line along x-axis (Figure 3a). The experimental results show that the diffraction angle of 1D grating is: $sin\theta' = \frac{x}{l} = \frac{5.5mm}{160mm} = 0.03437$.

Compared with the theoretical value of diffraction angle calculated above, it can be seen that the 1D ordered wrinkles used in the experiment can be approximately regarded as a diffraction grating with grating constant 15.48 µm, which is highly consistent with the measured sinusoidal wrinkles wavelength 15.50µm.

For mechanical composite micro-gratings, the diffraction pattern is the same as one of the 1D grating when the included angle between two gratings is 0° (Figure 3b).When the included angle changed from 0° to 90°, the diffraction pattern evolved from 1D line to 2D parallelogram array (Figure 3c-d). The direction of the two diagonal lines of the parallelogram array is the same as the direction of the sine wrinkle array. The brightness distribution of the parallelogram array gradually decreases from the central light spot to the light spot on the two diagonal lines,and the light spot around has the lowest brightness. In the process of rotation, the distance between adjacent spots on the same axis remains unchanged.

C. Diffraction result analysis

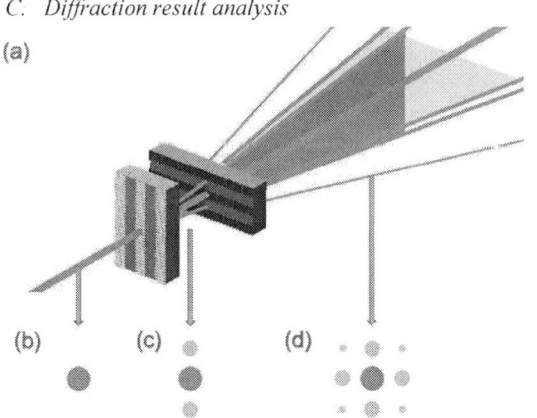

Figure 4: Schematic diagram of diffraction principle.(a)The beam goes through two layers of gratings and diffracts into 2D array (Only the first order diffraction is considered),(b)Schematic diagram of laser source,(c)Schematic diagram of Grating 1 diffraction image,(d)Schematic diagram of Grating 1 and Grating 2 diffraction images.

The reason why the diffraction pattern after passing through the composite grating becomes a parallelogram is that twice diffractions emerge when the laser passes through

composite micro-gratings (Figure 4a) .

According to the law of diffraction, [22], when a point laser goes through Grating 1, the light diffracts into 1D line along x axis; when the 1st diffraction light goes through Grating 2, as a new light source ,1D diffraction light diffracts into 2D parallelogram array (Figure 4c-d).

Therefore, the direction of the two diagonal lines of the parallelogram array is the same as the direction of the principal axis of the two gratings (the x direction in the 1D ordered wrinkle sample in Fig. 2a). During the process of mechanical rotation, the force provided by clamping device does not cause the deformation of Grating 2, thus the wrinkle morphology of Grating 2 remains unchanged, and the diffraction spot spacing along the x direction of Grating 2 does not change as well.

IV. CONCLUSION

A flexible mechanical composite micro-grating system is proposed. In order to solve difficulty of dynamically adjusting the 2D wrinkle grating, a mechanical composite tuning method is introduced. Compared with the traditional method of multiple selection area exposure, the flexible mechanical composite micro-grating system has the advantages of simple fabrication process, combination with high degree of freedom and rich combination results. The diffraction pattern of the light source can be dynamically regulated and encrypted by several control parameters. It is helpful to improve the key strength in the process of optical encryption.

ACKNOWLEDGMENT

The authors gratefully acknowledge the supports by the National Science Foundation for Distinguished Young Scholars (11625208), Major Program (12032015) and Young Scientists of China (11802173), the Program of Shanghai Academic/Technology Research Leader (19XD1421600).

REFERENCES

[1] T. Ma, H. Liang, G. Chen, B. Poon, H. Jiang, and H. Yu, "Micro-strain sensing using wrinkled stiff thin films on soft substrates as tunable optical grating," Optics Express, vol. 21, no. 10, pp. 11994-12001, 2013/05/20 2013.

[2] J. T. Yang et al., "Fabrication of micron and submicron gratings by using plasma treatment on the curved polydimethylsiloxane surfaces," (in English), Optical Materials, Article vol. 72, pp. 241-246, Oct 2017.

[3] J.-t. Yang, J. Tang, Y.-b. Wang, H. Guo, and J. Liu, "Tunable grating for stress measurement," Optics and Precision Engineering, vol. 26, no. 7, pp. 1596-603, July 2018.

[4] H. Guo et al., "Vectorial strain gauge method using single flexible orthogonal polydimethylsiloxane gratings," (in English), Scientific Reports, Article vol. 6, p. 10, Mar 2016, Art. no. 23606.

[5] S. Schauer et al., "Disordered diffraction gratings tailored by shape-memory based wrinkling and their application to photovoltaics," Optical Materials Express, vol. 8, no. 1, pp. 184-198, 2018/01/01 2018.

[6] T. Ohzono and H. Monobe, "Morphological Transformation of a Liquid Micropattern on Dynamically Tunable Microwrinkles," Langmuir, vol. 26, no. 9, pp. 6127-6132, May 4 2010.

[7] S. Peng, W. Lil, and J. Zhang, "Diffraction-Pattern Based on Spontaneous Wrinkled Thin Films," Materials Transactions, vol. 58, no. 1, pp. 1-5, 2017 2017.

[8] J. S. Liu et al., "Transfer printing via a PAA sacrificial layer for wrinkle-free PDMS metallization," (in English), Journal of Materials Science-Materials in Electronics, Article vol. 31, no. 3, pp. 2347-2352, Feb 2020.

[9] D. Yin et al., "Stability Improved Stretchable Metallic Gratings With Tunable Grating Period in Submicron Scale," Journal of Lightwave Technology, vol. 33, no. 15, pp. 3327-3331, Aug 1 2015.

[10] L. Xu, X. Wang, Y. Kim, T. C. Shyu, J. Lyu, and N. A. Kotov, "Kirigami Nanocomposites as Wide-Angle Diffraction Gratings," ACS Nano, vol. 10, no. 6, pp. 6156-62, Jun 28 2016.

[11] Y. Meng, S. Zhang, K. Wu, J. Li, and L. Li, "Mechanically Tunable Bilayer Composite Grating for Unique Light Manipulation and Information Storage," Advanced Optical Materials, vol. 7, no. 2, 2019.

[12] Y. Meng, X. Gong, Y. Huang, and L. Li, "Mechanically tunable opacity effect in transparent bilayer film: Accurate interpretation and rational applications," Applied Materials Today, vol. 16, pp. 474-481, 2019.

[13] C. Harrison, C. M. Stafford, W. H. Zhang, and A. Karim, "Sinusoidal phase grating created by a tunably buckled surface," (in English), Applied Physics Letters, Article vol. 85, no. 18, pp. 4016-4018, Nov 2004.

[14] A. Takei, H. Fujita, and Ieee, "Wrinkle Meets MEMS: Tunable Grating and Hydrophobic Surface," in 26th Ieee International Conference on Micro Electro Mechanical Systems(Proceedings IEEE Micro Electro Mechanical Systems, 2013, pp. 568-571.

[15] H. H. Hou, J. Yin, and X. S. Jiang, "Smart Patterned Surface with Dynamic Wrinkles," (in English), Accounts of Chemical Research, Review vol. 52, no. 4, pp. 1025-1035, Apr 2019.

[16] F. Li, H. H. Hou, J. Yin, and X. S. Jiang, "Near-infrared light-responsive dynamic wrinkle patterns," (in English), Science Advances, Article vol. 4, no. 4, p. 8, Apr 2018, Art. no. eaar5762.

[17] L. Zhang et al., "Realizing Dynamic Diffraction Gratings Based on Light-Direct Writing of Responsive 2D Ordered Patterns," Acs Materials Letters, vol. 2, no. 9, pp. 1135-1141, Sep 8 2020.

[18] I. T. Lin, Y. S. Choi, C. Wojcik, T. S. Wang, S. Kar-Narayan, and S. K. Smoukov, "Electro-responsive surfaces with controllable wrinkling patterns for switchable light reflection-diffusion-grating devices," (in English), Materials Today, vol. 41, pp. 51-61, Dec 2020.

[19] Y. Wang, Y. Zhai, A. Villada, S. N. David, X. B. Yin, and J. L. Xiao, "Programmable localized wrinkling of thin films on shape memory polymers with application in nonuniform optical gratings," (in English), Applied Physics Letters, Article vol. 112, no. 25, p. 5, Jun 2018, Art. no. 251603.

[20] L. Zhou, K. Hu, W. Zhang, G. Meng, J. Yin, and X. Jiang, "Regulating surface wrinkles using light," National Science Review, vol. 7, no. 7, pp. 1247-1257, 2020.

[21] H. Park et al., "Multidirectional Wrinkle Patterns Programmed by Sequential Uniaxial Strain with Conformal yet Nontraceable Masks," Macromolecular Rapid Communications, vol. 38, no. 19, Oct 2017, Art. no. 1700311.

[22] K. M. Hu et al., "Delamination - Free Functional Graphene Surface by Multiscale, Conformal Wrinkling," Advanced Functional Materials, vol. 30, no. 34, 2020.

978-1-6654-3008-1/21 $31.00 © 2021 IEEE

Proceedings of the 16th Annual IEEE International
Conference on Nano/Micro Engineered and Molecular Systems
April 25-29, 2021

Vibration Characteristics of Piezoelectric Timoshenko Nanobeam in Viscoelastic Medium*

Dong Wu[1,2], Dapeng Zhang[1,2], Mingwei Liu[1,2], Yongjun Lei[1,2]

1 College of Aerospace Science and Engineering, National University of Defense Technology, Changsha, 410073, China

2 Hunan Key Laboratory of Intelligent Planning and Simulation for Aerospace Missions, Changsha, Hunan 410073, China

Contact author: zhangdapenghit@126.com

Abstract—The mechanical behavior of piezoelectric components has a marked impact on the performance of nano electromechanical systems (NEMS). Consequently, it is of great importance to investigate the mechanical behavior of piezoelectric nanomaterials. This work takes the nanobeam placed in a visco-Pasternak medium as the research object. The vibration governing equations and the corresponding boundary conditions of the system are obtained based on the nonlocal Timoshenko theory and the Hamiltonian principle, considering the effect of the piezoelectric effect and the flexoelectric effect. The solutions of the governing equations are obtained by utilizing the transfer function method (TFM). On this basis, the analysis of the effects, including the nonlocal effect, the flexoelectric effect, and the viscoelastic medium, on the nanobeam's vibration characteristics is carried out by analyzing the numerical result. Theoretical guidance for the design and analysis of the piezoelectric nano components can be offered through this paper.

Key word— nonlocal elastic theory, piezoelectric effect, flexoelectric effect, Timoshenko beam model, viscoelastic medium

I. INTRODUCTION

Piezoelectric energy harvesting device[1], with the characteristics of high energy density and simple working mode[2], has great application potential in the power supply of nano components as it can collect the mechanical energy generated by vibration in the environment and convert it into electric energy which can be used by nano components. The investigation of the flexoelectric effect is a fundamental problem involved in the design, analysis, and application of piezoelectric energy harvesting devices, which is of great significance in the engineering and theoretical field.

The flexoelectric effect considers that the electric polarization intensity of non-uniform strain dielectric materials is affected by the electric field, strain, and strain gradient. At the nanometer scale, the strain gradient of piezoelectric nano components is so large that the flexoelectric effect cannot be ignored. The main methods to study the mechanical behavior of piezoelectric nanomaterials include experimental measurement, molecular dynamics simulation, and continuum mechanics theory. Both experimental

*Resrach supported by National Natural Science Foundation of China (11902348); Natural Science Foundation of Hunan Province, China (2020JJ5650); Science Project of the National University of Defense Technology (ZK20-27).

measurement and molecular dynamics simulation results show that the mechanical properties of piezoelectric nanomaterials have obvious scale effects on the nanoscale[3,4], and classical continuum mechanics is difficult to accurately describe the mechanical properties of piezoelectric nanomaterials because the scale effects of materials are ignored[5]. The nonlocal theory[6] made up for the shortcomings of continuum mechanics in terms of scale effect and has been extensively applied in describing the mechanical behavior of nanomaterials. Some scholars[7-9] believe that to accurately predict the mechanical properties of piezoelectric nanomaterials, the nonlocal elasticity theory is necessary to be combined with the strain gradient theory. Lei et al.[10,11] researched the dynamic response of a nanobeam which is under external viscoelastic damping on the basis of the nonlocal elasticity and the Euler-Bernoulli/Timoshenko beam models. Yang et al. [12] studied the effect law of the surface effects on the piezoelectric Euler-Bernoulli nanobeam's bending deformation. Yan et al. [13] employing the nonlocal elasticity to analyze the surface effect's impact law on the Euler-Bernoulli nanobeam's buckling behavior and dynamic response characteristics. In practical engineering applications, nano components are mostly placed in a viscoelastic medium, whose equivalent mechanical models are beams/plates placed in a viscoelastic medium. However, there are relatively few studies that comprehensively consider the scale effect, the flexoelectric effect, and the viscoelastic medium factors.

A free vibration theoretical analysis model of a nanobeam placed in a visco-Pasternak foundation is proposed in this research. By applying the nonlocal Timoshenko beam model and Hamilton's principle, the governing equations and the corresponding boundary conditions are established. Moreover, the solutions of the governing equations are obtained by employing TFM. Finally, the nonlocal parameter, flexoelectric coefficients, and the damping coefficient's effect law on the system's natural frequency are carefully discussed. Theoretical guidance for piezoelectric nano components' design and analysis is provided in this study.

II. GOVERNING EQUATIONS AND BOUNDARY CONDITIONS

A nanobeam, having height, length, and width are h, L, and b, respectively, resting on a viscoelastic medium, which is modeled through the visco-Pasternak medium[14] is shown in Fig.1. The cartesian coordinate system was established with the left end of the neutral axis of the nanobeam as the origin,

978-1-6654-3008-1/21 $31.00 © 2021 IEEE   437

the x-axis points to the right along the neutral axis, and the z-axis point to the transverse vibration direction of the nanobeam, orthogonal to the x-axis at the same time. According to the continuum theory of nanometer materials, considering the nanobeam's scale effect and the coupling between polarization and strain gradient but ignoring the effect of high-order polarization gradient is ignored.

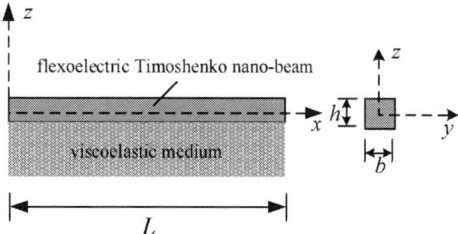

Figure 1. The geometry of a flexoelectric nanobeam supported by a viscoelastic medium

Considering the nanobeam's scale effect by employing the nonlocal elasticity theory, the constitutive equation can be expressed as[15]

$$(1-(e_0a)^2\nabla^2)\boldsymbol{\sigma}_{ij}=\boldsymbol{c}_{ijkl}\boldsymbol{\varepsilon}_{kl}+\boldsymbol{d}_{ijk}\boldsymbol{P}_k \quad (1)$$

$$(1-(e_0a)^2\nabla^2)\boldsymbol{\tau}_{ijk}=\boldsymbol{f}_{ijkl}\boldsymbol{P}_l \quad (2)$$

$$\boldsymbol{E}_i=\boldsymbol{a}_{ij}\boldsymbol{P}_j+\boldsymbol{d}_{jki}\boldsymbol{\varepsilon}_{jk}+\boldsymbol{f}_{jkli}\boldsymbol{\varepsilon}_{jk,l} \quad (3)$$

where e_0a and $\boldsymbol{a}_{ij}$ represent the nonlocal parameter and the reciprocal of the dielectric constant, respectively; $\boldsymbol{\sigma}_{ij}$ is the nonlocal stress tensor; $\boldsymbol{c}_{ijkl}$ is the elastic coefficient of the piezoelectric material; $\boldsymbol{\varepsilon}_{ij}$ is the classical strain tensor; $\boldsymbol{P}$ is the polarization tensor; $\boldsymbol{d}_{ijk}$ is the piezoelectric coefficients of the piezoelectric material; $\boldsymbol{\tau}_{ijk}$ is the nonlocal stress tensor; $\boldsymbol{f}_{ijkl}$ is the flexoelectric coefficients of the piezoelectric material; $\boldsymbol{E}$ is the electric field tensor, respectively; $\boldsymbol{f}_{ijkl}$, $\boldsymbol{c}_{ijkl}$, and $\boldsymbol{d}_{ijk}$ respectively represent the flexoelectric coefficients, the piezoelectric coefficients, and the elastic coefficient of the material of the piezoelectric nanobeam.

A Timoshenko beam model is used to model the piezoelectric beam, therefore, the nonzero elements of the nonlocal stress tensor $\boldsymbol{\sigma}_{ij}$, the nonlocal higher-order stress tensor $\boldsymbol{\tau}_{ijk}$, and the electric field $\boldsymbol{E}$ can be written as

$$(1-(e_0a)^2\nabla^2)\boldsymbol{\sigma}_{xx}=c_{11}\boldsymbol{\varepsilon}_{xx}+d_{31}\boldsymbol{P}_z \quad (4)$$

$$(1-(e_0a)^2\nabla^2)\boldsymbol{\sigma}_{xz}=kc_{44}\boldsymbol{\gamma}_{xz} \quad (5)$$

$$(1-(e_0a)^2\nabla^2)\boldsymbol{\tau}_{xxz}=f_{3113}\boldsymbol{P}_z \quad (6)$$

$$(1-(e_0a)^2\nabla^2)\boldsymbol{\tau}_{xzx}=f_{3131}\boldsymbol{P}_z \quad (7)$$

$$\boldsymbol{E}_z=a_{33}\boldsymbol{P}_z+d_{31}\boldsymbol{\varepsilon}_{xx}+f_{3113}\boldsymbol{\varepsilon}_{xx,z}+f_{3131}\boldsymbol{\gamma}_{xz,x} \quad (8)$$

where k is the shear correction coefficient and $k=\dfrac{5}{6}$ is widely accepted for a beam with a rectangular section.

Assuming that the polarization direction of the material along to the z-axis. What's more, in order to simplify the model, only the z component of the electric field is considered, hence the electric field is given by

$$E_z=-\partial\Phi/\partial z \quad (9)$$

In terms of the Timoshenko beam model theory, noticing the axial compression load, the displacement field of the beam can be written as

$$u=-z\varphi(x,t), v=0, w=w(x,t) \quad (10)$$

where u is the displacement component parallel to the x-axis, v is the displacement component along to the y-axis, and w represents the displacement component parallel to the z-axis, φ is the rotation of the cross-section relative to the z-axis.

It is assumed that there is no free potential in the nanobeam, which can be obtained by Gauss's law

$$\epsilon_0 E_{z,z}+P_{z,z}=0 \quad (11)$$

where $\epsilon_0=8.85\times10^{-12}$ is the permittivity constant of the vacuum or the air.

According to the Timoshenko beam model theory, the nonzero elements of the strain gradient and the strain can be shown as

$$\varepsilon_{xx}=-z\frac{\partial\varphi}{\partial x}, \varepsilon_{xx,z}=-\frac{\partial\varphi}{\partial x}$$

$$\gamma_{xz}=-\varphi+\frac{\partial w}{\partial x}, \gamma_{xz,x}=-\frac{\partial\varphi}{\partial x}+\frac{\partial^2 w}{\partial x^2} \quad (12)$$

Substituting Eq.(9) and (12) into Eq.(11), taking Eq.(8) into consideration, consequently, the z-axis components of the electric potential, the electric field intensity, and the polarization intensity can be obtained respectively. Thus, the expressions of rotation angle φ and deflection w for nonlocal stress and nonlocal higher-order stress can be shown as

$$(1-(e_0a)^2\nabla^2)\sigma_{xx}=\frac{d_{31}f_{3113}}{a_{33}}\frac{\partial\varphi}{\partial x}-\frac{d_{31}V}{a_{33}h}+$$
$$\frac{d_{31}f_{3131}}{a_{33}}\left(\frac{\partial\varphi}{\partial x}-\frac{\partial^2 w}{\partial x^2}\right)+\left(\frac{\epsilon_0 d_{31}^2}{1+\epsilon_0\alpha_{33}}-c_{11}\right)z\frac{\partial\varphi}{\partial x} \quad (13)$$

$$(1-(e_0a)^2\nabla^2)\sigma_{xz}=kc_{44}\left(-\varphi+\frac{\partial w}{\partial x}\right) \quad (14)$$

$$(1-(e_0a)^2\nabla^2)\tau_{xxz}=\frac{\epsilon_0 d_{31}f_{3113}}{1+\epsilon_0\alpha_{33}}z\frac{\partial\varphi}{\partial x}+$$
$$\frac{f_{3131}f_{3113}}{a_{33}}\left(\frac{\partial\varphi}{\partial x}-\frac{\partial^2 w}{\partial x^2}\right)+\frac{f_{3113}^2}{a_{33}}\frac{\partial\varphi}{\partial x}-\frac{f_{3113}V}{a_{33}h} \quad (15)$$

$$(1-(e_0a)^2\nabla^2)\tau_{xzx}=\frac{\epsilon_0 d_{31}f_{3131}}{1+\epsilon_0\alpha_{33}}z\frac{\partial\varphi}{\partial x}+$$
$$\frac{f_{3113}f_{3131}}{a_{33}}\frac{\partial\varphi}{\partial x}-\frac{f_{3131}V}{a_{33}h}+\frac{f_{3131}^2}{a_{33}}\left(\frac{\partial\varphi}{\partial x}-\frac{\partial^2 w}{\partial x^2}\right) \quad (16)$$

According to the Hamiltonian's principle, which can be shown as

$$\int_0^t (\delta W+\delta\Pi_k-\delta U)dt=0 \quad (17)$$

where W is the work done by the external force, Π_k is the kinetic energy of the nanobeam, and U is the strain energy. Integrate the first-order variation of W, Π_k, and U over [0, T]

$$\int_0^t \delta W\,dt = \int_0^t \left(\bar{N}\delta w + \bar{M}\delta\varphi \right)\Big|_0^L dt - \int_0^t \int_0^L b N_Q \delta w\,dx\,dt \tag{18}$$

$$\int_0^t \delta \Pi_k\,dt = \int_0^t \int_0^L -\rho\left(A\frac{\partial^2 w}{\partial t^2}\delta w + I\frac{\partial^2 \varphi}{\partial t^2}\delta\varphi \right)dx\,dt \tag{19}$$

$$\int_0^t \delta U\,dt = \int_0^t \int_V \left(\sigma_{xx}\delta\varepsilon_{xx} + \sigma_{xz}\delta\gamma_{xz} \right)dV\,dt + \int_0^t \int_V \left(\tau_{xzz}\delta\varepsilon_{xx,z} + \tau_{zxx}\delta\gamma_{xz,x} \right)dV\,dt \tag{20}$$

where $\bar{N}$ is the applied transverse force of the nanobeam, $\bar{M}$ is the applied bending moments of the nanobeam, $A = bh$ is the cross-sectional area of the nanobeam, $I = \int_A z^2\,dA$ is the moment of inertia of the nanobeam. N_Q is the force of the viscoelastic medium on the nanobeam. For the visco-Pasternak viscoelastic medium model, considering the scale effect, N_Q can be expressed as

$$(1 - (e_0 a)^2 \nabla^2) N_Q = k_w w - k_G \nabla^2 w + c_t \partial w/\partial t \tag{21}$$

where k_w is the shear elastic modulus of the visco-Pasternak medium, k_G is the Winkler elastic modulus of the medium, and c_t is the damping coefficient of the medium.

Substituting Eq. (19)-(21) into Eq.(18) and take the arbitrariness of $\delta\varphi$ and δw on $x \in [0, L]$ into consideration at the same time, the vibration governing equations of the system can be written as

$$\frac{\partial Q}{\partial x} - \frac{\partial^2 M^\gamma}{\partial x^2} - b N_Q = \rho A \frac{\partial^2 w}{\partial t^2} \tag{22}$$

$$Q - \frac{\partial M}{\partial x} - \frac{\partial M^\gamma}{\partial x} = \rho I \frac{\partial^2 \varphi}{\partial t^2} \tag{23}$$

What's more, the related boundary conditions can be displayed as

$$\varphi = \text{const or } \boldsymbol{M} + \boldsymbol{M}^\gamma = -\bar{\boldsymbol{M}} \tag{24}$$

$$w = \text{const or } \boldsymbol{Q} - \frac{\partial \boldsymbol{M}^\gamma}{\partial x} = \bar{\boldsymbol{N}} \tag{25}$$

$$\frac{\partial w}{\partial x} = \text{const or } \boldsymbol{M}^\gamma = 0 \tag{26}$$

where $\boldsymbol{M}$ is the bending moment of the nanobeam, $\boldsymbol{Q}$ is the shear force of the nanobeam and $\boldsymbol{M}^\gamma$ is the torsional moment of the nanobeam. They are derived as

$$\boldsymbol{M} = \int_A (\sigma_{xx}z + \tau_{xxz})\,dA, \boldsymbol{Q} = \int_A \sigma_{xz}\,dA, \boldsymbol{M}^\gamma = \int_A \tau_{zxx}\,dA \tag{27}$$

In order to facilitate calculation and analysis, dimensionless parameters are introduced, they are given by

$$\bar{x} = \frac{x}{L}, \bar{w} = \frac{w}{L}, \eta = \frac{h}{L}, \bar{c}_{11} = c_{11}L, \bar{c}_{44} = c_{44}L, \bar{k}_w = \frac{k_w L^2}{\bar{c}_{44}}$$

$$\bar{k}_G = \frac{k_G}{\bar{c}_{44}}, \tau = \frac{t}{L}\sqrt{\frac{\bar{c}_{44}}{\rho L}}, \alpha = \frac{e_0 a}{L}, \bar{c}_t = \frac{c_t L}{\sqrt{\rho \bar{c}_{44}L}} \tag{28}$$

Substituting Eq.(28) into Eq.(22)-(26), the dimensionless form of the free vibration governing equations are shown as

$$E_1 \frac{\partial^4 \bar{w}}{\partial \bar{x}^4} + E_2 \frac{\partial^2 \bar{w}}{\partial \bar{x}^2} - E_3 \frac{\partial^3 \bar{\varphi}}{\partial \bar{x}^3} - E_2 \frac{\partial \bar{\varphi}}{\partial \bar{x}} - \left(1 - \alpha^2 \frac{\partial^2}{\partial \bar{x}^2}\right)\left(\bar{k}_w \bar{w} - \bar{k}_G \frac{\partial^2 \bar{w}}{\partial \bar{x}^2} + \bar{c}_t \frac{\partial \bar{w}}{\partial \tau}\right) \tag{29}$$

$$= \left(1 - \alpha^2 \frac{\partial^2}{\partial \bar{x}^2}\right)\eta \frac{\partial^2 \bar{w}}{\partial \tau^2}$$

$$E_3 \frac{\partial^3 \bar{w}}{\partial \bar{x}^3} + E_2 \frac{\partial \bar{w}}{\partial \bar{x}} - E_4 \frac{\partial^2 \bar{\varphi}}{\partial \bar{x}^2} - E_2 \bar{\varphi} \tag{30}$$

$$= \left(1 - \alpha^2 \frac{\partial^2}{\partial \bar{x}^2}\right)\frac{\eta^3}{12}\frac{\partial^2 \bar{\varphi}}{\partial \tau^2}$$

where

$$E_1 = \frac{f_{3131}^2 \eta}{a_{33}\bar{c}_{44}L}, \quad E_2 = k\eta$$

$$E_3 = \left(\frac{f_{3113}f_{3131}}{a_{33}\bar{c}_{44}L} + \frac{f_{3113}^2}{\bar{c}_{44}a_{33}L}\right)\eta$$

$$E_4 = \left[\frac{\epsilon_0 d_{31}^2 L}{\bar{c}_{44}(1 + \varepsilon_0 a_{33})} - \frac{\bar{c}_{11}}{\bar{c}_{44}}\right]\frac{\eta^3}{12} \tag{31}$$

$$+ \frac{(f_{3131} + f_{3113})^2 \eta}{\bar{c}_{44}a_{33}L}$$

For free vibration, the applied transverse force and the applied bending moment are ignored, hence the boundary conditions can be rewritten as

$$E_1 \frac{\partial^3 \bar{w}}{\partial \bar{x}^3} + E_2 \frac{\partial \bar{w}}{\partial \bar{x}} - E_3 \frac{\partial^2 \bar{\varphi}}{\partial \bar{x}^2} - E_2\bar{\varphi} = 0$$
$$\text{or } \bar{w} = \text{const} \tag{32}$$

$$E_4 \frac{\partial \bar{\varphi}}{\partial \bar{x}} - E_3 \frac{\partial^2 \bar{w}}{\partial \bar{x}^2} = 0 \text{ or } \bar{\varphi} = \text{const} \tag{33}$$

$$E_3 \frac{\partial \bar{\varphi}}{\partial \bar{x}} - E_1 \frac{\partial^2 \bar{w}}{\partial \bar{x}^2} = 0 \text{ or } \frac{\partial \bar{w}}{\partial \bar{x}} = \text{const} \tag{34}$$

III. SOLUTION OF THE GOVERNING EQUATIONS

Employing TFM to obtain the solution of the dimensionless governing equations shown in Eq. (29) and (30). According to TFM, the state vector $\boldsymbol{\eta}(\bar{x}, \Omega)$ is defined as

$$\boldsymbol{\eta}(\bar{x}, \Omega) = \left[\bar{W}, \frac{d\bar{W}}{d\bar{x}}, \frac{d^2\bar{W}}{d\bar{x}^2}, \frac{d^3\bar{W}}{d\bar{x}^3}, \Phi, \frac{d\Phi}{d\bar{x}}\right]^T \tag{35}$$

Using Eq.(35), rewrite Eq.(29) and (30) as

$$\frac{\mathrm{d}\boldsymbol{\eta}(\overline{x},\Omega)}{\mathrm{d}\overline{x}} = \boldsymbol{F}(\Omega)\boldsymbol{\eta}(\overline{x}) \tag{36}$$

where

$$\boldsymbol{F}(\Omega) = \begin{bmatrix} 0 & 1 & 0 & 0 & 0 & 0 \\ 0 & 0 & 1 & 0 & 0 & 0 \\ 0 & 0 & 0 & 1 & 0 & 0 \\ -\dfrac{A_2}{B_1} & 0 & -\dfrac{B_2}{B_1} & 0 & 0 & \dfrac{B_3}{B_1} \\ 0 & 0 & 0 & 0 & 0 & 1 \\ 0 & \dfrac{E_4}{A_4} & 0 & \dfrac{E_3}{A_4} & \dfrac{A_5}{A_4} & 0 \end{bmatrix} \tag{37}$$

where

$$A_1 = E_1 - \overline{k}_G \alpha^2, \quad A_2 = \mathrm{i}\,\overline{c}_t \Omega + \overline{k}_w - \Omega^2 \eta,$$
$$A_3 = \alpha^2 \Omega^2 \eta - \mathrm{i}\,\overline{c}_t \Omega \alpha^2 - \overline{k}_w \alpha^2 - \overline{k}_G - E_2,$$
$$A_4 = E_4 + \alpha^2 \Omega^2 \frac{\eta^3}{12}, \quad A_5 = \Omega^2 \frac{\eta^3}{12} - E_2, \tag{38}$$
$$B_1 = A_1 - \frac{E_3^2}{A_4}, \quad B_2 = A_3 - \frac{E_3 E_2}{A_4}, \quad B_3 = E_4 + \frac{E_3 A_5}{A_4}$$

The solutions of Eq. (36) can be determined as

$$\boldsymbol{\eta}(\overline{x},\Omega) = e^{\boldsymbol{F}(\Omega)\overline{x}} \boldsymbol{\eta}(0,\Omega) \tag{39}$$

Similarly, rewrite the dimensionless boundary conditions as

$$\boldsymbol{M}(\Omega)\boldsymbol{\eta}(0) + \boldsymbol{N}(\Omega)\boldsymbol{\eta}(1) = 0 \tag{40}$$

where $\boldsymbol{M}(\Omega)$ and $\boldsymbol{N}(\Omega)$ are dimensionless boundary condition selection matrix, varying with the boundary conditions. Substituting Eq.(40) into Eq. (39), it can be obtained that

$$[M(\Omega) + N(\Omega)e^{F(\Omega)}]\eta(0,\Omega) = 0 \tag{41}$$

The characteristic equation of Eq. (41) is determined as

$$\det[\boldsymbol{M}(\Omega) + N(\Omega)e^{\boldsymbol{F}(\Omega)}] = 0 \tag{42}$$

Utilizing numerical methods can obtain the solutions of Eq. (42).

IV. NUMERICAL RESULTS AND DISCUSSION

Firstly, verification of the proposed beam model and the solving method of the governing equations are presented by comparing with the results obtained in the literature [16]. On this basis, the effects of the nonlocal parameter, the flexoelectric coefficient, and the damping coefficient of the medium on the free vibration characteristics of the piezoelectric nanobeam are systematically investigated. The related parameters of the piezoelectric nanobeam used in this section are set as follows[16-18]: the length L, height h and width b of the nanobeam are 20nm, 2nm, and 2nm, respectively; the elastic modulus c_{11} and c_{44} of the nanobeam materials are 131GPa and 42.9GPa, respectively; the transverse flexoelectric coefficients f_{3113} and tangential flexoelectric coefficients f_{3131} are 5V and 4V, respectively; the mass density ρ, reciprocal of the dielectric constant a_{33} and piezoelectric coefficient d_{31} are 6020 kg/m^3, 0.79×10^8 V·m/C and 1.87×10^8 V/m, respectively; the Winkler elasticity modulus, the shear modulus, and the damping coefficient of the visco-Pasternak medium are 0.1GPa/nm, 0.25GPa/nm, and 0.1 MPa·ns/nm, respectively.

A. Model Validation

Ref.[16] investigated a micro-scale piezoelectric beam, considering the surface effect and the flexoelectric effect. However, the scale effect of the micro-scale piezoelectric beam and the viscoelasticity of the medium are ignored. In this section, to verify the theoretical model and the solving method presented in this paper, the viscoelastic medium is firstly ignored and the first-order dimensionless natural frequencies calculated by the newly proposed model in this paper are compared with those obtained by Ref.[16].

In Tab.1, some numerical results are presented and compared for various slenderness ratios of the nanobeam. It can be seen from Tab. 1 that the relative error between the solutions of the governing equation proposed in this paper and those obtained in Ref.[16] is less than 1%, which verifies the correctness of the solution proposed in this paper.

B. Effect of the nonlocal parameter α

Fig.2 demonstrates the real part ratios of the dimensionless natural frequencies vary with different nonlocal parameters, where "C" means clamped boundary condition, "F" represents freedom boundary condition, and "S-S' is simply supported boundary condition. Fig.2 shows a phenomenon that all the real part ratios of the dimensionless natural frequencies decrease with the increase of nonlocal parameters α, and the reduction amplitude increases with the increase of the frequency order and the connection stiffness, indicating that the real parts of the natural frequencies decrease. The real part ratios of the dimensionless natural frequency of the nanobeam resting in the viscoelastic medium decrease less than that of the nanobeam without a viscoelastic medium because the viscoelastic medium can strengthen the nanobeam. In conclusion, increasing the structural stiffness will lead to weakness of the nonlocal effect.

TABLE. 1 The comparison between the solutions obtained in this paper and those obtained in Ref.[16]

	L/h					
	6	8	10	16	20	30
Present	0.45147	0.34583	0.27950	0.17668	0.14172	0.094732
Ref.[16]	0.45163	0.34596	0.27960	0.17675	0.14177	0.094767
Relative Error/%	0.03543	0.03758	0.0358	0.03960	0..03527	0.03693

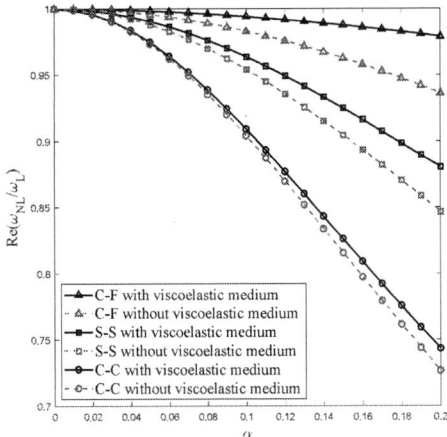

Figure 2. Effect of nonlocal parameter α on the real part of the dimensionless natural frequency of the nanobeam

C. Effect of flexoelectric coefficients

Numerical results are shown in this section to discuss the effect of the transverse flexoelectric coefficient f_{3113} and the tangential flexoelectric coefficient f_{3131} on the real parts of the dimensionless natural frequencies of a simply supported nanobeam. Results are obtained by setting nonlocal parameter $\alpha = 0.2$.

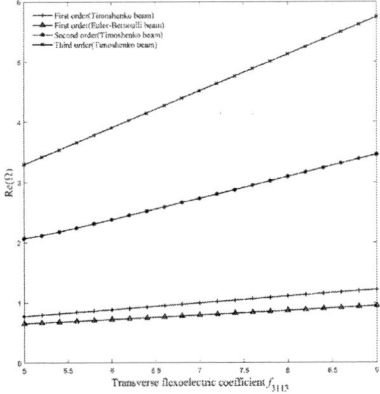

Figure 3. The real parts of the dimensionless natural frequencies vary with the transverse flexoelectric coefficient f_{3113}

Fig.3 shows the real parts of the dimensionless natural frequencies vary with transverse flexoelectric coefficients f_{3113}. The real parts of the dimensionless natural frequencies increase with the increase of the transverse flexoelectric coefficient f_{3113} linearly, and the increased amplitude becomes larger at large frequency order, indicating that the nanobeam's effective stiffness is strengthened by the transverse flexoelectric coefficient f_{3113}. Compare between the dimensionless natural frequencies of the Timoshenko nanobeam and the Euler-Bernoulli nanobeam, the real parts of the former are larger than the latter. Furthermore, the impact effect of the transverse flexoelectric coefficient f_{3113} on the Timoshenko nanobeam is greater than that of the

Euler-Bernoulli nanobeam, as considering the torsion effect reduces the stiffness of the nanobeam.

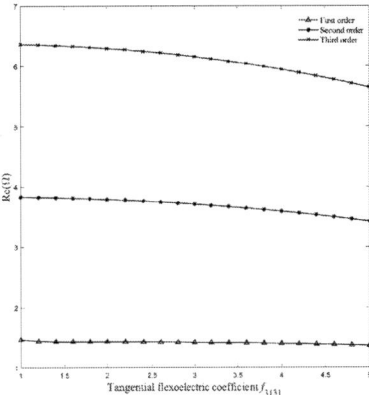

Figure 4. The real parts of the nanobeam's dimensionless natural frequencies vary with the tangential flexoelectric coefficient f_{3131}

Fig. 4 shows the real parts of the nanobeam's dimensionless natural frequencies vary with the transverse flexoelectric coefficients f_{3131}. As shown in Fig.4 that the real parts of the dimensionless natural frequencies decrease with the increase of the tangential flexoelectric coefficient f_{3131}, indicating that the tangential flexoelectric coefficient f_{3131} weakens the structural stiffness of flexoelectric nanobeams.

D. Effect of the damping coefficient

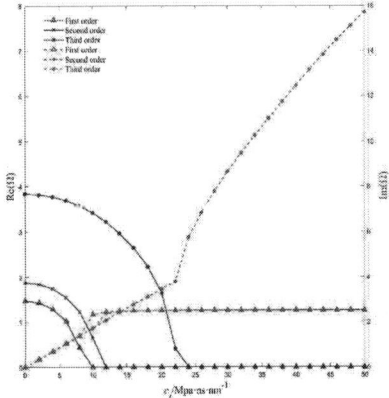

Figure 5. The nanobeam's dimensionless natural frequencies vary with the damping coefficient c_t

Fig. 5 displays the nanobeam's dimensionless natural frequencies at different damping coefficients of the visco-Pasternak medium. As shown in Fig. 5 that while the damping coefficient c_t surpasses a certain damping coefficient, which is different at imparity case and named as the critical damping coefficient c_{t_cri}, the real parts of the dimensionless natural frequencies decrease to zero, indicating that reciprocating vibration no longer occurs; at the same time, the growth rate of the imaginary part experienced a sharp rise to a gradual decline. While $c_t < c_{t\_cri}$, the real parts of the dimensionless natural frequencies decline with the damping coefficient c_t increases and the decline rate gradually increases; however, the imaginary parts of the dimensionless natural frequencies grow with the damping coefficient c_t increase linearly.

V. CONCLUSION

The free vibration characteristics of the piezoelectric nanobeam placed in the visco-Pasternak medium were analyzed in this paper. The governing equations and the related boundary conditions of the system were obtained by employing the nonlocal Timoshenko model and Hamilton's principle. Applying TFM to obtain the numerical solutions of the governing equations. Results were presented to verify the novel model and theory compared with the reference. The impact of the nonlocal parameter, flexoelectric coefficients, and the damping coefficient on the dimensionless natural frequency of the nanobeam are systematically discussed. Results are as follow:

(1) The nonlocal parameter weakens the effective structural stiffness of the piezoelectric nanobeam.

(2) The transverse flexoelectric coefficient strengthens the nanobeam, and the impact effect on the Timoshenko nanobeam is greater than that on the Euler-Bernoulli nanobeam. While the nanobeam's stiffness is weakened by the tangential flexoelectric coefficient.

(3) The imaginary parts of the nanobeam's dimensionless natural frequencies rise as the damping coefficient grows, however, the real parts decrease as the damping coefficient rises.

REFERENCES

[1] N. Zhao, S. Zhang, F.-R. Yu, et al. "Exploiting interference for energy harvesting: A survey, research issues and challenges," *IEEE Access*, vol. 5, pp. 10403-10421, 2017.

[2] C.-L. Chen, Z.-Q. Li, X. Liang, et al. "Electromechanical Coupling Model and Performance Analysis of the Unimorph Cantilever Beam-based Flexoelectric Energy Harvester," *Chinese Journal of Solid Mechanics*, vol. 41, no.2, pp. 159-169, 2020.

[3] C.-Q. Chen, Y. Shi, Y.-S. Zhang, et al. "Size dependence of Young's modulus in ZnO nanowires," *Physical Review Letters*, vol.96, no. 7, pp. 1-4, 2006.

[4] R. Agrawal, B. Peng, E. E. Gdoutos, et al. "Elasticity size effects in ZnO nanowires-a combined experimental-computational approach," *Nano letters*, vol. 8, no. 11, pp. 3668-3674, 2008.

[5] Z.-B. Shen. "Vibration characteristics of micro/nano mass sensor via nonlocal elasticitytheory," Ph.D. dissertation, National University of Defense Technology, 2012.

[6] P. Duhem. "Le potentiel thermodynamique et la pression hydrostatique," *Annales Scientifiques De L École Normale Supérieure*, vol. 10, no. 3, pp. 183-230, 1893.

[7] Y.-S. Li, W.-J. Feng, Z.-Y. Cai. "Bending and free vibration of functionally graded piezoelectric beam based on modified strain gradient theory," *Composite Structures*, vol. 115, pp. 41-50, 2014.

[8] F. Ebrahimi, M. R. Barati, A. Dabbagh. "A nonlocal strain gradient theory for wave propagation analysis in temperature-dependent inhomogeneous nanoplates," *International Journal of Engineering Science*, vol. 107, pp. 169-182, 2016.

[9] D.-P. Zhang, Y.-J. Lei, S. Adhikari. "Flexoelectric effect on vibration responses of piezoelectric nanobeams embedded in viscoelastic medium based on nonlocal elasticity theory," *Acta Mechanica*, vol. 229, no.6, pp. 2379-2392, 2018.

[10] Y. Lei, S. Adhikari, M. I. Friswell. "Vibration of nonlocal Kelvin–Voigt viscoelastic damped Timoshenko beams," *International Journal of Engineering Science*, vol. 66-67, pp. 1-13, 2013.

[11] Y. Lei, T. Murmu T, S. Adhikari, et al. "Dynamic characteristics of damped viscoelastic nonlocal Euler–Bernoulli beams," *European Journal of Mechanics. A: Solids*, vol. 42, pp.125-136, 2013.

[12] F. Yang. "Effect of surface energy on the bending of piezoelectric nanowires," *Electronic Components and Materials*, vol. 35, no. 8, pp. 46-49, 2016.

[13] Z. Yan, L.-Y. Jiang. "The vibrational and buckling behaviors of piezoelectric nanobeams with surface effects," *Nanotechnology*, vol.22, no. 24, pp. 245703, 2011.

[14] A. M. Zenkour, M. Sobhy. "Nonlocal piezo-hygrothermal analysis for vibration characteristics of a piezoelectric Kelvin–Voigt viscoelastic nanoplate embedded in a viscoelastic medium," *Acta Mechanica*, vol. 229, no. 1, pp. 3-19. 2018.

[15] M. S. Majdoub, P. Sharma, T. Cagin. "Dramatic enhancement in energy harvesting for a narrow range of dimensions in piezoelectric nanostructures," *Physical Review B: Condensed Matter and Materials Physics*, vol. 78, no. 12, pp. 1-4, 2009.

[16] Y. M. Yue, K. Y. Xu, T. Chen. "A micro scale Timoshenko beam model for piezoelectricity with flexoelectricity and surface effects," *Composite Structures*, vol. 136, pp. 278-286, 2016.

[17] Z. Yan, L.-Y. Jiang. "Size-dependent bending and vibration behaviour of piezoelectric nanobeams due to flexoelectricity," *Journal of Physics D: Applied Physics*, vol. 46, no. 35, pp.1-7, 2013.

[18] X.-J. Li, Y. Luo. "Flexoelectric effect on vibration of piezoelectric microbeams based on a modified couple stress theory," *Shock and Vibration*, vol.2017, pp. 1-7, 2017.

Proceedings of the 16th Annual IEEE International
Conference on Nano/Micro Engineered and Molecular Systems
April 25-29, 2021

AC & DC Magnetic Field Sensor Based on Flexible Piezoelectric Polymer

Weilong Han, Jian Wen, Yifei Wu*, Bo Niu, Yi Wu, and Mingzhe Rong

Abstract— Using flexible polymer to make passive magnetic field sensor is a feasible way. Flexible piezoelectric polymer combined with permanent magnets were used to build a cantilever beam structure to measure AC and DC magnetic fields. In magnetic field, the piezoelectric material vibrates in different modes under the drive of a permanent magnet. The magnetic field measurement mechanism was studied and a finite element simulation model of the piezoelectric cantilever beam was built. The sensor successfully detected AC magnetic field (sine-wave, 50Hz, 0-20 Oe) and DC magnetic field (0-50 Oe). Under the AC and DC magnetic field, the voltage outputs with the sensitivity of the sensor are 0-22 V with 1.1V/Oe and 0-1 V with 20 mV/Oe, respectively. Compared with existing technologies, this paper provides a room-temperature sensor that measures both AC and DC magnet field through a simple and low-costed way.

I. INTRODUCTION

Magnetic sensing technology is one of the important research fields in sensors. New magnetic sensors using smart materials can be used in a variety of applications, including current-carrying wire status monitoring in smart grid applications [1-3], magnetic flux detection of interlocking nails in bone fracture surgery [4], and magnetic energy collection and magnetic-flux detection/conversion of magnetic energy-harvest-powered wireless sensing systems [5] and so on. Traditional magnetic sensors include giant magnetoresistance sensor [6], superconducting quantum interface devices(SQUIDS) [7] and sensors based on the Hall effect. Although traditional sensors have advantages in sensitivity, the instruments and costs are expensive and some cannot work at room temperature. Using new functional materials such as piezoelectric and magnetostrictive materials to measure magnetic field can solve the above problems.

Previously, researchers made a variety of structures to respond to magnetic fields such as piezoelectric cantilever beam and magnetostrictive-piezoelectric-magnetostrictive laminated composites. Yu et al detected the frequency and amplitude of an AC magnetic field at room temperature using ZnO piezoelectric nanowires(NWs) with cantilever structure [8]. Youshimura et al reported a prototype of the AC magnetic field energy harvester using a PZT piezoelectric bimorph cantilever and a neodymium magnet [9]. The resonance frequency was adjusted around 50 Hz in order to obtain maximum output power. Hung et al demonstrated a miniature magnetic-force-based, three-axis, AC magnetic sensor with piezoelectric/vibrational energy-

harvesting functions [10]. All the above designs used cantilever structure to respond to AC magnetic field, and used piezoelectric ceramics as signal output layer. They did not consider DC magnetic field measurements because when there is only DC magnetic field, the magnet cannot vibrate, and the piezoelectric has no continuous output.

For the magnetoelectric composites, the DC bias is required when measuring AC magnetic field which causes high energy consumption. Guo et al fabricated flexible magnetoelectric laminated materials using magnetostrictive material Metglas and PVDF for the application to the AC current measurement [11]. Under the magnetic field (sine-wave, 1000 Hz, 0-9 Oe), the voltage output with the sensitivity of the sensor are 0-100 mV with 3.916-10.679 mV/Oe. Because of its high resonance frequency, the magnetoelectric laminated material will produce obvious output under high frequency AC magnetic field which is not suitable for low frequency magnetic field measurement.

In this paper, an AC and DC magnetic field sensor based on flexible piezoelectric polymer PVDF cantilever and a magnet was demonstrated. In magnetic field, the piezoelectric material vibrates in different modes under the drive of a permanent magnet, so the strength and direction of the magnetic field can be obtained by measuring the piezoelectric output voltage. The sensor can measure the low frequency AC magnetic field passively. The measurement of DC magnetic field needs additional low frequency AC signal. The magnetic field measurement mechanism was studied and a finite element simulation model of the piezoelectric cantilever beam was built. The sensor successfully detected AC magnetic field (sine-wave, 50Hz, 0-20 Oe) and DC magnetic field (0-50 Oe). Under the AC and DC magnetic field, the voltage output with the sensitivity of the sensor are 0-22 V with 1.1V/Oe and 0-1 V with 20 mV/Oe, respectively. In comparison to existing technologies, the sensor can measure both AC and DC magnetic field and work passively with higher sensitivity.

II. STRUCTURE AND SIMULATION

A. Structure and Working Principle

Fig. 1(a) shows the structure of the device which mainly includes magnet, flexible piezoelectric polymer, extraction electrodes and support base. The device parameters are as follows: the piezoelectric PVDF size is 25×16 mm^2, and the thickness is 28 μm. The permanent magnet has a diameter of φ8mm, a thickness of 4mm and a weight of 3g.

This work was supported by the science and technology project of State Grid Corporation of China (Grant No. 5226SX1800FA).

W. L. Han (phone: +86 13720521272; e-mail: hwl120306@ stu.xjtu.edu.cn), J. Wen, Y. F. Wu, Y. Wu, M. Z. Rong are with the State

Key Laboratory of Electrical Insulation and Power Equipment, Xi'an Jiaotong University, Xi'an 710049, P.R.C.

B. Niu is with the Shaanxi Electric Power Science Research Institute. (*Corresponding author: yifei208@xjtu.edu.cn).

978-1-6654-3008-1/21 $31.00 © 2021 IEEE

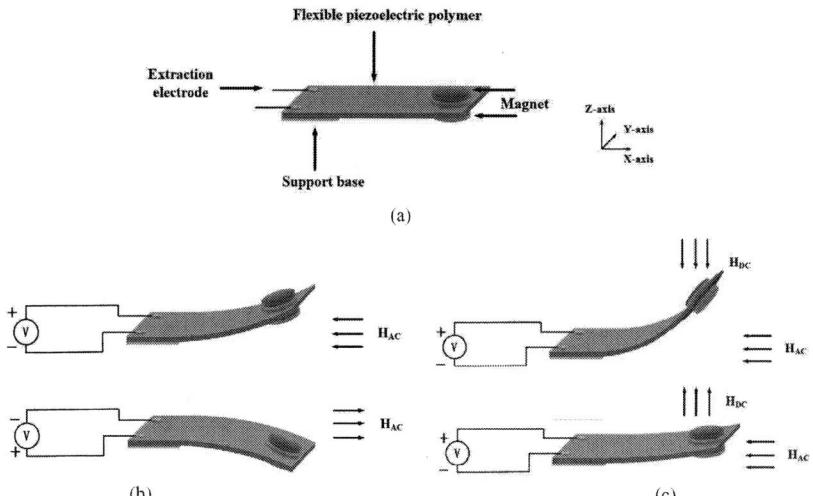

Figure 1. (a) The structure and components of the device. (b) Measuring X-axis AC magnetic field. (c) Measuring Z-axis DC magnetic field.

In Fig. 1(b), the magnet placed in the AC magnetic field is coupled with the AC magnetic field to vibrate, driving the piezoelectric cantilever beam structure to vibrate to generate electrical signals, thus realizing the conversion of energy from magnetic energy to mechanical to electrical energy. When the magnet is in the X-axis AC magnetic field H_{ac}, it will vibrate along the Z-axis. The stronger the AC magnetic field, the greater the vibration amplitude of cantilever beam and the greater the voltage output of piezoelectric. In Fig. 1(c), adding Z-axis DC magnetic field H_{dc} on the basis of AC magnetic field H_{ac}, the amplitude will be increased or decreased. When the Z-axis H_{dc} is in the same direction with magnet field, the output is weakened (shown in Fig. 1(c) below). Otherwise the output is enhanced. The piezoelectric output is different from that in the case of only AC magnetic field, so the voltage variation can be used to characterize the DC magnetic field.

The magnetic field in X-axis will exert magnetic torque M on the cantilever beam. The magnetic torque M can be calculated by the following equation: [12]

$$M = \boldsymbol{m} \times \boldsymbol{B} \approx (\boldsymbol{B_r} \cdot V) \times \boldsymbol{B} \qquad (1)$$

Where B is the measured magnetic field, V is the volume of the magnet, B_r is the residual flux density of magnet.

For the cantilever beam structure in this paper, the voltage induced on the PVDF can be calculated by the following equation: [12]

$$U = -\frac{3Md_{31}}{2Wh\varepsilon_{33}^T} = k_c M \qquad (2)$$

Where M is magnetic torque applied on the cantilever, d_{31} the piezoelectric constant, ε_{33}^T the dielectric permittivity under constant stress, W the width of the beam, h the thickness of the beam. When the device is determined, k_c is a constant. From (1) and (2), there will be the following equation:

$$U = k_c M = k_c (\boldsymbol{B_r} \cdot V) \times \boldsymbol{B} \qquad (3)$$

Under AC magnetic field H_{ac}, the piezoelectric output voltage is U_1 and the voltage is U_2 when adding DC magnetic field H_{dc}. B_1 the magnetic induction corresponding to H_{ac}. B_2 the magnetic induction corresponding to H_{dc}. Then the voltage variation caused by DC magnetic field H_{dc} can be calculated as follows:

$$\begin{aligned} \Delta U &= U_2 - U_1 \\ &= k_c (\boldsymbol{B_r} \cdot V) \times (\boldsymbol{B_2} + \boldsymbol{B_1}) - k_c (\boldsymbol{B_r} \cdot V) \times \boldsymbol{B_1} \\ &= k_c (\boldsymbol{B_r} \cdot V) \times \boldsymbol{B_2} \end{aligned} \qquad (4)$$

From (4), k_c, B_r, and V are constants, so the variation of voltage is linear with the intensity of DC magnetic field.

B. Simulations

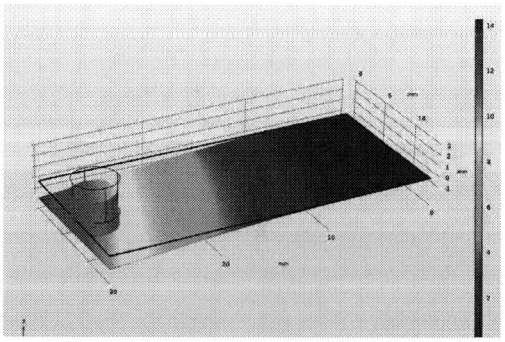

Figure 2. The piezoelectric cantilever beam model

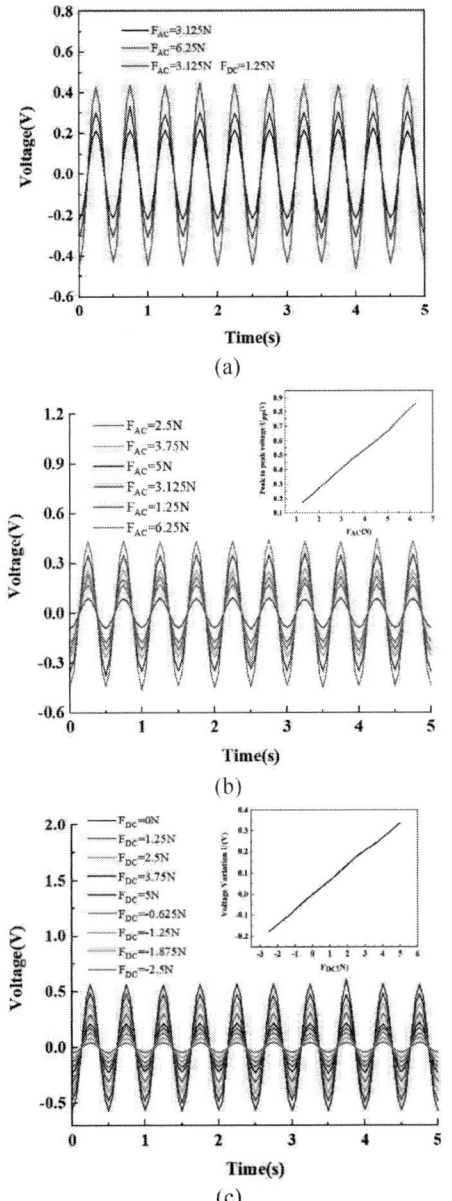

Figure 3. (a) Piezoelectric output under F_{AC} and output under the combined action of F_{AC} and F_{DC}. (b) Influence of different sinusoidal forces F_{AC} on piezoelectric output. (c) Influence of different F_{DC} on piezoelectric output

The piezoelectric cantilever beam model was established in COMSOL. Fig. 2 shows the structure of the model mainly including PVDF film and cylindrical magnet. A 2Hz sinusoidal force F_{AC} in Z direction is applied to the magnet to simulate the effect of X-axis AC magnetic field. A sinusoidal force F_{DC} in Z direction is applied to simulate the effect of Z-axis DC magnetic field. The change of piezoelectric output voltage can reflect the applied force and the change of magnetic field.

Fig. 3(a) shows under a sinusoidal force F_{AC} with a peak value of 3.125 N, the piezoelectric output voltage is an approximate sine wave with a peak value of 0.216 V. When the peak value of sinusoidal force becomes 6.25 N, the voltage peak value increases to 0.432 V. When the 1.25 N Z-axis force F_{DC} is added on the basis of F_{AC}, the output voltage increases to 0.305V. Fig. 3(b) shows piezoelectric output voltage waveforms under different sinusoidal forces F_{AC} changing from 1.25 N to 6.25 N. The inset shows that the peak to peak voltage increases linearly with the increase of the peak value of the sinusoidal force. Fig. 3(c) shows piezoelectric output voltage waveforms under different Z-axis forces F_{DC} changing from -2.5 N to 5 N. The inset shows that the voltage amplitude increases linearly with the increase of the peak value of F_{DC}.

III. RESULTS AND DISCUSSION

Then the proposed magnetic sensor was tested in 50 Hz AC magnetic field and DC magnetic field. The measured magnetic field is provided by the three-dimensional Helmholtz coil, the voltage regulator provides 50Hz AC current, and the DC source provides DC current. The intensity of magnetic field is measured by Gauss meter. The piezoelectric output signal is detected by an oscilloscope.

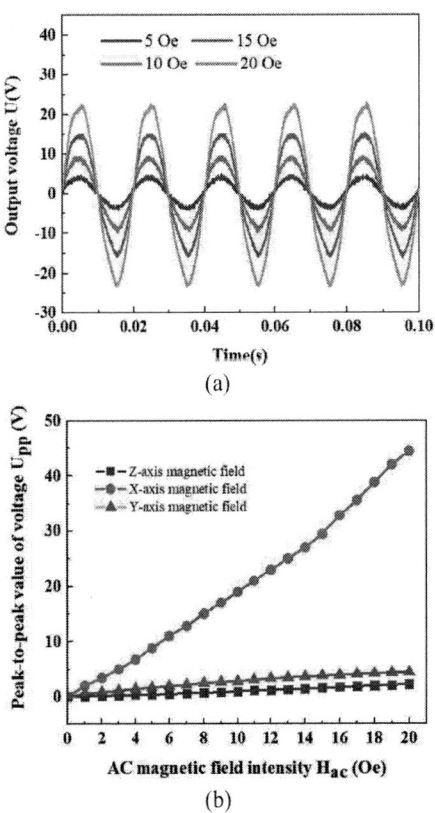

Figure 4. (a) Piezoelectric output voltage under 50Hz AC magnetic field. (b) Peak to peak value of piezoelectric output voltage variation with AC magnetic field.

978-1-6654-3008-1/21 $31.00 © 2021 IEEE

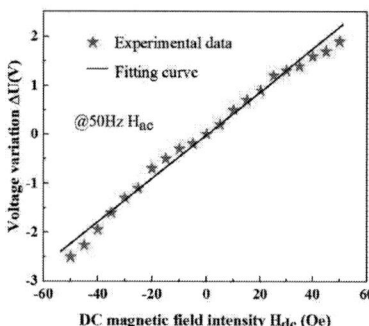

Figure 5. Peak to peak value of piezoelectric output voltage variation with DC magnetic field.

TABLE I. COMPARISON OF THIS WORK AND OTHER SENSORS

Table Head	Comparison Item			
	Measured magnetic field	*Output voltage*	*Sensitivity*	*Sensor volume*
Yu et al [8]	AC (0-2.4 Oe)	0-10 mV	0-8mV/Oe	1600 mm³
Hung et al [10]	AC(0.2-3.2 Oe)	1.13-26.15 mV	8.79 mV/Oe	8000 mm³
Guo et al [11]	AC(0-9 Oe)	0-100 mV	3.916-10.679 mV/Oe	4.5 mm³
This work	AC (0-20 Oe) DC (-50-50 Oe)	AC(0-22 V) DC(0-1 V)	AC(1.1V/Oe) DC (20 mV/Oe)	212.2 mm³

When different intensity AC magnetic fields were applied, the peak-to-peak value of the piezoelectric output voltage signal was recorded, and the corresponding curve of the two was obtained. When measuring DC magnetic field, an X-axis AC magnetic field must be applied in advance so that the piezoelectric device outputs an AC voltage signal. Subsequently, the measured DC magnetic field was applied, the peak-to-peak value of the AC signal changes, and the relationship curve between the DC magnetic field and the change value of the voltage signal was obtained.

50Hz, 0-20 Oe AC magnetic field was measured. Fig. 4(a) shows the piezoelectric output voltage waveforms under different intensity AC magnetic fields. The output voltage is approximately 50Hz sine wave. As the measured magnetic field increases from 5 Oe to 20 Oe, the voltage amplitude increases from 4V to 22V. Fig. 4(b) shows the relationship between the peak-to-peak value of the piezoelectric output voltages and the AC magnetic field strength. The AC magnetic field with different axial directions has different force on the magnet. As the X-axis AC magnetic field strength increases from 0 to 20 Oe, the peak-to-peak output voltage linearly increases from 0 V to 44 V. As the Y-axis AC magnetic field strength increases from 0 to 20 Oe, the peak-to-peak output voltage linearly increases from 0 V to 4.35 V. As the Z-axis AC magnetic field strength increases from 0 to 20 Oe, the peak-to-peak output voltage linearly increases from 0 V to 2.5 V. This shows that the sensitivity to the magnetic field in X direction is the highest. In other words, when the magnetic field intensity is determined, the output voltage is the highest with X-axis direction magnetic field.

0-50 Oe bidirectional DC magnetic field was also measured. Fig. 5 shows the relationship between the variation value of the piezoelectric output voltages and the DC magnetic field strength. When the DC magnetic field strength is -50 Oe, 0 and 50 Oe respectively, the voltage value is -2.5 V, 0, and 1.9 V. The experimental results show that the piezoelectric output signal has a good linear relationship with AC magnetic field and DC magnetic field strength.

The sensitivity of magnetic sensor in this paper and other representative piezoelectric cantilever and laminated composites magnetic sensors are summarized in TABLE I for comparison. In the work of Hung et al, under the x-axis magnetic field (sine-wave, 100 Hz, 0.2-3.2 Oe) and the z-axis magnetic field (sine-wave, 142 Hz, 0.2-3.2 Oe), the voltage output with the sensitivity of the sensor are 1.13-26.15 mV with 8.79 mV/Oe and 1.31-8.92 mV with 2.63 mV/Oe, respectively. In the work of Guo et al, under the magnetic field (sine-wave, 1000 Hz, 0-9 Oe), the voltage output with the sensitivity of the sensor are 0-100 mV with 3.916-10.679 mV/Oe. In the case of little difference in sensor volume, the output voltage and magnetoelectric conversion coefficient of the sensor in this paper is larger. The function of measuring DC magnetic field is also unique.

IV. CONCLUSION

In this paper, an AC and DC magnetic field sensor based on flexible piezoelectric polymer PVDF cantilever and a magnet was demonstrated. It can passively measure power frequency magnetic field and measure DC magnetic field with low-power consumption. The magnet placed in the AC magnetic field is coupled with the AC magnetic field to vibrate, driving the piezoelectric cantilever beam structure to vibrate to generate electrical signals, thus realizing the conversion of energy from magnetic energy to mechanical to electrical energy. A finite element simulation model of the piezoelectric cantilever beam was built, and the magnetic field measurement mechanism was studied. The result of the numerical calculations indicates that the output has a good linearity with the magnetic field strength. The sensor detected AC magnetic field (sine-wave, 50Hz, 0-20 Oe) and DC magnetic field (0-50 Oe). Under the AC and DC magnetic field, the voltage output with the sensitivity of the sensor are 0-22 V with 1.1V/Oe and 0-1 V with 20 mV/Oe, respectively. The dependence of the measured output voltage on the intensity of magnetic field agree well with the simulation results. The sensor can also distinguish the direction of the AC magnetic field and has potential applications for wireless sensing systems and environmental monitoring.

978-1-6654-3008-1/21 $31.00 © 2021 IEEE

REFERENCES

[1] C. Reig, M. D. Cubells-Beltran, and D. R. Munoz, "Magnetic field sensors based on giant magnetoresistance (GMR) technology: applications in electrical current sensing," *Sensors*, vol. 9, pp. 7919-7942, 2009.

[2] Y. Ouyang, J. L. He, J. Hu, and S. X. Wang, "A current sensor based on the giant magnetoresistance effect: design and potential smart grid applications," *Sensors*, vol. 12, pp. 15520-15541, 2012.

[3] P. C. Yeh, T. K. Chung, C. H. Lai, and C. M. Wang, "A magnetic-piezoelectric smart material-structure utilizing magnetic force interaction to optimize the sensitivity of current sensing," *Applied Physics a-Materials Science & Processing*, vol. 122, 2016.

[4] T. H. Wong, T. K. Chung, T. W. Liu, H. J. Chu, W. Y. Hsu, P. C. Yeh, C. C. Chen, M. S. Lee, and Y. S. Yang, "Electromagnetic/magnetic-coupled targeting system for screw-hole locating in intramedullary interlocking-nail Surgery," *IEEE Sensors Journal*, vol. 14, pp. 4402-4410, 2014.

[5] T. K. Chung, P. C. Yeh, H. Lee, C. M. Lin, C. Y. Tseng, W. T. Lo, C. M. Wang, W. C. Wang, C. J. Tu, P. Y. Tasi, and J. W. Chang, "An attachable electromagnetic energy harvester driven wireless sensing system demonstrating milling-processes and cutter-wear/breakage-condition monitoring," *Sensors*, vol. 16, 2016.

[6] M. N. Baibich, J. M. Broto, A. Fert, F. Nguyen Van Dau, F. Etienne, G. Creuzet, A. Friederich and J. Chazelas, "Giant magnetoresistance of (001)Fe/(001)Cr magnetic superlattices," *Phys. Rev. Lett.* vol. 61, 1988.

[7] Weinstock H 1996 *SQUID Sensors: Fundamentals, Fabrication, and Applications* (New York: Springer) 329

[8] A. Yu, M. Song, Y. Zhang, J. Z. Kou, J. Y. Zhai and Z. L. Wang, "A self-powered AC magnetic sensor based on piezoelectric nanogenerator," *Nanotechnology*, vol. 25, 2014.

[9] T. Yoshimura, K. Izumi, Y. Ueno, T. Minami, S. Murakami and N. Fujimura, "Piezoelectric energy harvesting from AC current-carrying wire," *Japanese Journal of Applied Physics*, vol. 58, 2019.

[10] C. F. Hung, P. C. Yeh and T. K. Chung, "A miniature magnetic-force-based three-axis AC magnetic sensor with piezoelectric/vibrational energy-harvesting functions," *Sensors*, vol. 17, 2017.

[11] X. R. Guo, X. J. Yu, G. F. Lou, "Preparation process and magnetoelectric properties of flexible Metglas/PVDF magnetoelectric material," *Advanced Technology of Electrical Engineering and Energy*, vol. 38, no. 6, 2019.

[12] Z. Xing, K. Xu, "Investigation of low frequency giant magnetoelectric torque effect," *Sens. Actuators, A* vol. 189, pp. 182-186, 2013.

Gap in pagination due to unavailable papers.

Pages 448-454

Proceedings of the 16th Annual IEEE International
Conference on Nano/Micro Engineered and Molecular Systems
April 25-29, 2021

A Capacitively Transduced Bulk Acoustic Wave MEMS Resonator with Low Bias Voltages

Zeji Chen, Qianqian Jia, Wenli Liu, Yinfang Zhu, Quan Yuan, Jinling Yang, and Fuhua Yang

Abstract— This work presents a high-Q capacitively transduced bulk acoustic wave (BAW) mode resonator which can be driven into vibrations with low bias voltages, distinguished from conventional BAW devices which are excited via tens of Volts. Comprehensive analysis was provided to have an insight into the transduction mechanism of the resonator. The structural design was optimized in terms of the enlarged transduction area and minimized energy dissipation. A simple and reliable silicon on insulator (SOI)-based fabrication process was exploited. The resonator with promising Q values over 50000 can be driven with a low bias voltage of 2 V. Besides, the nonlinearity was also investigated for stable operation of the devices. The proposed resonator could have great potential applications in advanced monolithic RF front end transceivers.

I. INTRODUCTION

Future wireless communication systems feature miniaturization, multi-mode, reconfigurability, and higher compactness [1]. The oscillators for frequency references play a vital role. In recent years, radio frequency micro-electro-mechanical system (RF-MEMS) resonators with IC compatibility, small size, low power consumption, and superior reliability have emerged as a core device for future RF front end transceivers [2]. Combining MEMS resonators with CMOS sustaining circuits, monolithic MEMS oscillators can be implemented to replace conventional off-chip quartz crystals [3, 4]. Among various vibrating modes, the bulk acoustic wave (BAW) modes have been demonstrated to yield preferable $f \times Q$ products for their low dissipations and high stiffness [5]. Besides, the resonance frequencies of BAW modes are determined by the lateral dimensions, which achieves multi-frequency functionality on a single chip.

However, driving and sensing of such high-stiffness resonators require considerably high DC bias voltages (V_P), which poses a bottleneck against their practical applications [6, 7]. In [8], a 12 MHz length extensional mode resonator with Q of 1.8×10^5 needed an extremely high V_P up to 100 V. In [9], a vacuum encapsulated Lamé mode resonator with excellent Q value of 1.48×10^6 was reported, while the required V_P was 50 V. Some efforts have been made to reduce bias voltages. A 17.6 MHz Lamé mode resonator driven by 2.5 V was implemented based on a high gain-bandwidth trans-impedance amplifier. Nevertheless, the Q of 8000 in vacuum was not high enough [10].

Simultaneously reducing bias voltages and maintaining high Q values are prerequisites to explore IC integrable MEMS oscillators. In this work, a silicon on insulator (SOI)-based BAW mode resonator with both high Q values and low bias voltages was developed. Systematical optimizations in terms of structural design as well as fabrication process have been accomplished.

II. RESONATOR DESIGN

A novel Lamé mode resonator was designed. The resonance frequency was determined by the square length L:

$$f_0 = \frac{1}{\sqrt{2}L}\sqrt{\frac{E}{2\rho(1+\upsilon)}}, \qquad (1)$$

where E, ρ, and υ relate to the Young's modulus, density, and Poisson's ratio, respectively. The resonance frequency of the proposed resonator is 51 MHz. The capacitive transduction which is apt to achieve preferable Q values was adopted. With an AC signal v_i and a DC bias voltage V_P applied to the driving electrode and resonator respectively, the resonator was driven into vibration. A time-harmonic motional current, i_{out}, was generated at the sensing electrode. In order to reduce bias voltages, it is critical to express i_{out} explicitly and take the corresponding optimization strategies.

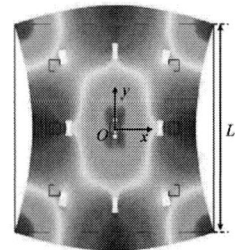

Figure 1. The displacement distributions of the Lamé mode.

The electrostatic force F_e takes the form as:

$$F_e = \left(\frac{\partial C}{\partial u}\right)V_P V_i = \frac{\varepsilon_0 L_{ele}h}{d_0^2}V_P V_i, \qquad (2)$$

where $\left(\frac{\partial C}{\partial u}\right)$ is the electrode-resonator overlap capacitance variation per unit displacement [11], ε_0, d_0, h, and V_i denote the permittivity, spacing gap, resonator thickness, and the

---

This work was supported by the National Natural Science Foundation of China (61734007, 61874116, and 61804150), and the Youth Innovation Promotion Association of CAS (29E07RQC03).

The authors are with the Institute of Semiconductors, Chinese Academy of Sciences (CAS), Beijing 100083, China, also with the Center of Materials Science and Optoelectronics Engineering, University of Chinese

Academy of Sciences, Beijing 100049, China. Z.J. Chen, Q.Q. Jia, W.L. Liu, Y.F. Zhu, Q. Yuan, and J.L. Yang are also with the State Key Laboratory of Transducer Technology, Shanghai, 200050, China. (e-mail: chenzeji@semi.ac.cn; jiaqianqian@semi.ac.cn; liuwenli@semi.ac.cn; yfzhu@semi.ac.cn; yuanquan@semi.ac.cn; jlyang@semi.ac.cn; fhyang@semi.ac.cn.)

978-1-6654-3008-1/21 $31.00 © 2021 IEEE

amplitude of the AC signal, respectively, L_{ele} refers to the electrode length overlapping the resonator. The mode shape is given in Fig. 1, which can be described by the following displacement distribution equations [12]:

$$X_{\text{mode}}(x,y) = D\sin\left(\frac{\pi}{L}x\right)\cos\left(\frac{\pi}{L}y\right),$$
$$Y_{\text{mode}}(x,y) = -D\cos\left(\frac{\pi}{L}x\right)\sin\left(\frac{\pi}{L}y\right), \tag{3}$$

where D is the vibrational amplitude. The origin is set at the square center. Supposing that the sensing electrode is perpendicular to the x direction, only the displacements along the x direction contribute to i_{out}. The displacement of a certain point at the edge can be expressed as:

$$dX\left(\frac{L}{2}, y\right) = Q\frac{dF_e}{k_{eff}\left(\frac{L}{2}, y\right)}, \tag{4}$$

where $k_{eff}\left(\frac{L}{2}, y\right)$ refers to the equivalent stiffness at this point, taking the form as [13]:

$$k_{eff}\left(\frac{L}{2}, y\right) = \omega_0^2 \frac{\rho h \int_{-\frac{L}{2}}^{\frac{L}{2}}\int_{-\frac{L}{2}}^{\frac{L}{2}}\left[\left(X_{\text{mode}}(x',y')\right)^2 + \left(Y_{\text{mode}}(x',y')\right)^2\right]dx'dy'}{\left(X_{\text{mode}}\left(\frac{L}{2}, y\right)\right)^2}, \tag{5}$$

where ω_0 is the angular resonance frequency. The output motional current at this point can be expressed using the following equation [13]:

$$di_{out}(y) = \omega_0 V_P \frac{\varepsilon_0 h}{d_0^2} X\left(\frac{L}{2}, y\right) dy, \tag{6}$$

where ω_0 is the angular resonance frequency. The relationship between the displacements of two points at the edge can be written as [13]:

$$\frac{dX\left(\frac{L}{2}, y_1\right)}{dX\left(\frac{L}{2}, y_2\right)} = \frac{X\left(\frac{L}{2}, y_1\right)}{X\left(\frac{L}{2}, y_2\right)} = \frac{X_{\text{mode}}\left(\frac{L}{2}, y_1\right)}{X_{\text{mode}}\left(\frac{L}{2}, y_2\right)}, \tag{7}$$

Substituting (7) into (6), and taking the integral from $-\frac{L_{ele}}{2}$ to $\frac{L_{ele}}{2}$, the total output current within the sensing electrode can be attained:

$$i_{out} = \omega_0 V_P \frac{\varepsilon_0 h}{d_0^2} X\left(\frac{L}{2}, 0\right) \int_{-\frac{L_{ele}}{2}}^{\frac{L_{ele}}{2}} \frac{X_{\text{mode}}\left(\frac{L}{2}, y\right)}{X_{\text{mode}}\left(\frac{L}{2}, 0\right)} dy. \tag{8}$$

Combining (4), (5), (7) and (8) yields:

$$i_{out} = M\left(\int_{-\frac{L_{ele}}{2}}^{\frac{L_{ele}}{2}} X_{\text{mode}}\left(\frac{L}{2}, y\right) dy\right)^2. \tag{9}$$

Where M is a constant:

$$M = \frac{Q V_P^2 V_i \frac{\varepsilon_0^2 h}{\rho \omega_0 d_0^4}}{\int_{-\frac{L}{2}}^{\frac{L}{2}}\int_{-\frac{L}{2}}^{\frac{L}{2}}\left[\left(X_{\text{mode}}(x',y')\right)^2 + \left(Y_{\text{mode}}(x',y')\right)^2\right]dx'dy'}. \tag{10}$$

The routings for reducing bias voltages can be inferred from (9 - 10). The electrodes should be well configured to enlarge the transduction area. The energy dissipations

should be suppressed for higher Q values. Additionally, the spacing gap plays a more important role for its fourth power dependence, which is expected to be sufficiently shrunk.

A. Enlarged transduction area

Four electrodes parallel to the resonator sidewalls were adopted to enlarge the transduction area. According to the mode shape given in Fig. 1, the adjacent edges are vibrating out of phase, thus, the two opposite electrodes were connected for driving, while the other twos for sensing.

B. Quality factor optimization

For high Q values, according to:

$$\frac{1}{Q} = \sum \frac{1}{Q_i}, \tag{11}$$

where i denotes a certain type of loss source, and the total Q is determined by various loss mechanisms. The thermoelastic damping (TED), anchor loss, and air damping are major sources of energy dissipations for MEMS resonators [14].

The TED arises from the stress induced thermal gradients formed along different vibrating regions, the irreversible heat flow causes the energy dissipation. For Lamé mode, during the vibrations, the resonator volume is conserved, therefore, the TED is insignificant.

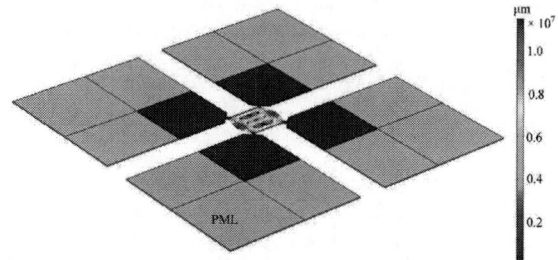

Figure 2. The simulation model for anchor loss of the described resonator.

To minimize the anchor loss, the supporting beams were located at the square corners, corresponding to the nodal regions. It should be noted that the dimensions of supporting beams are associated with the anchor loss [15]. To have an optimal supporting beam, as shown in Fig. 2, a simulation model was established to characterize the Q_{anchor} values under different supporting beam dimensions. The perfectly matched layers (PML) are employed to absorb the elastic waves transmitting in the semi-infinite substrate [16]. The PML thickness is set as the resonance wavelength. In this way, a complex-valued eigenfrequency ω_r can be obtained so that the Q_{anchor} can be calculated via:

$$Q_{anchor} = \frac{\text{Re}(\omega_r)}{2\,\text{Im}(\omega_r)}. \tag{12}$$

Utilizing the parametric sweep method, the optimal supporting beam design can be obtained.

In air, it has been verified that the squeezed film damping (SFD) is significant for Lamé mode resonators [17]. When the resonator is vibrating, the air within the spacing gap is moved in and out alternatively, thus, a varied pressure field is generated, which induces a force on the sidewall, a portion of energy is thus dissipated. For proposed Lamé mode resonators with nano-scale spacing gaps, the SFD could contribute most to the energy loss.

III. FABRICATION PROCESS

To implement nano-scale spacing gaps for greatly reduced bias voltages, a simple and reliable SOI fabrication process was employed [18]. The resonators as well as electrical routings were patterned by inductively coupled plasma (ICP) dry etch. The handling layer was grounded to suppress the feedthrough signals. The nano-scale spacing gap was defined using thermal oxidation. Subsequently, the heavily doped LPCVD polysilicon is deposited and patterned to form the electrodes. Then, the lift-off process was utilized to produce the electrode pads. Finally, the device was released using the hydrofluoric acid (HF) solution. The scanning electron microscope (SEM) photograph of the fabricated resonator is given in Fig. 3.

Figure 3. The SEM picture of the fabricated resonator.

IV. RESULTS AND DISCUSSIONS

The resonator was firstly tested on a RF probe station at atmosphere. A high precision U8032A voltage source was employed to provide DC bias voltages. An E5071C network analyzer was utilized to generate an AC driving signal with RF power of 0 dBm and measure the S_{21} transmission curves.

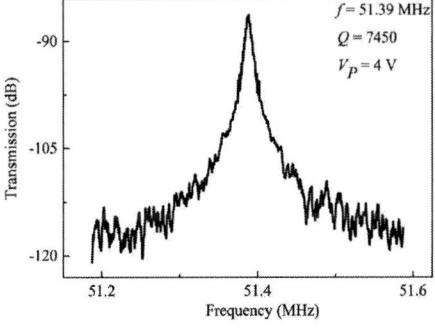

Figure 4. The frequency response of the Lamé mode resonator in air.

Fig. 4 shows the frequency response of the Lamé resonator with resonant frequency of 51.39 MHz and Q of 7450 in air. A significant resonance peak was attained with low bias voltages of 4 V, which is a preferable performance

among reported Lamé mode resonators in air. A high signal-to-noise ratio over 30 dB was obtained. No significant distortions of the transmission spectrum were observed, which verifies that the grounded substrate can effectively suppress the feedthrough signal. The Q is much lower than the calculated one, which can be attributed to severe SFD, consistent with the tendency that SFD dominates the Q values of Lamé mode resonators in air.

The resonator was then measured on a Lakeshore CRX-4K probe station. With a high vacuum of 8×10^{-5} bar, the SFD is sufficiently suppressed. As shown in Fig. 5, a substantially enhanced Q value of 52400 was achieved, corresponding to the $f \times Q$ product of 2.69×10^{12}. Moreover, the significantly improved Q leads to a further reduced V_P of 2 V, while the signal-to-noise ratio maintained around 30 dB. Compared with the reported BAW resonators vibrating at lower frequencies [19, 20], the proposed device with much higher stiffness exhibited comparable Q values and dramatically reduced bias voltages.

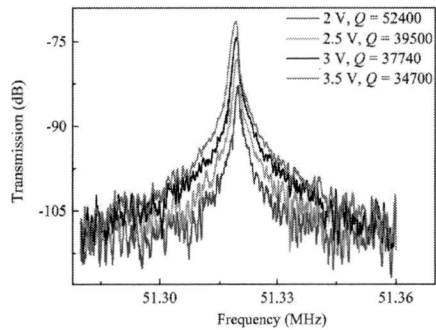

Figure 5. The frequency responses under different bias voltages.

Since the resonator can be excited with a very low voltage, a moderate V_P increase can substantially improve the vibrational amplitude. Hence, it is essential to characterize the frequency responses of the proposed resonator under different bias voltages and determine the stable working conditions of the devices. The frequency responses under higher bias voltages were also characterized and plotted in Fig. 5, wherein the AC signal was remained as 0 dBm and V_P was increased by a step of 0.5 V. The frequency shifts are due to the tuning of electrical stiffness [11]. With the increasing V_P, the curves were tilted into higher frequencies, which infers that the resonator has been driven into the nonlinear regime. The dynamic equation for a nonlinear system takes the form as [21]:

$$m\ddot{x} + \zeta \dot{x} + k_1 x + k_2 x^2 + k_3 x^3 = F_e, \quad (13)$$

where m and ζ denote the equivalent mass and damping, k_1, k_2, and k_3 are the linear, 2nd and 3rd effective stiffness, respectively. The critical amplitude of nonlinearity can be expressed as [22]:

$$x_c = \frac{4\sqrt{2}}{\sqrt{9\sqrt{3}Q|\kappa_{non}|}}, \quad (14)$$

where κ_{non} is a coefficient associated with the high-order stiffness [22]:

978-1-6654-3008-1/21 $31.00 © 2021 IEEE   457

$$\kappa_{non} = \frac{3}{8}\frac{k_3}{k_1} - \frac{5}{12}\left(\frac{k_2}{k_1}\right)^2.\qquad(15)$$

According to (14-15), the nonlinearity is generally observed in resonators with low linear stiffness and high Q values. For high-stiffness BAW mode resonators, it suffers less from nonlinearity than low-stiffness flexural modes. However, the nonlinearity can still occur in BAW mode resonators with high Q values. It can be indicated from Fig. 6 that for stable operation, the applied AC signal and V_P of the proposed resonator should not exceed 0 dBm and 2.5 V.

V. Conclusion

In this work, a high-performance capacitively transduced Lamé mode resonator was demonstrated. A reduced voltage down to 2 V and high Q value over 5×10^4 were simultaneously implemented based on the optimized structural designs and fabrication process. The described methodologies can be adapted to various high-stiffness BAW resonators, enabling high-end IC integrable RF devices in future wireless communications.

References

[1] J. Thompson et al., "5G wireless communication systems: prospects and challenges [Guest Editorial]," IEEE Commun. Mag., 2014, vol. 52, no. 2, pp. 62-64. DOI: 10.1109/MCOM.2014.6815889

[2] C. T.-C. Nguyen, "MEMS technology for timing and frequency control," IEEE Trans. Ultrason. Ferroelectr. Freq. Control, 2007, vol. 54, no. 2, pp. 251-270. DOI: 10.1109/freq.2005.1573895

[3] G. Wu, J. Xu, E. J. Ng, and W. Chen, "MEMS Resonators for Frequency Reference and Timing Applications," J. Microelectromech. Syst., 2020, vol. 29, no. 5, pp. 1137 - 1166. DOI: 10.1109/JMEMS.2020.3020787

[4] J. Van Beek and R. Puers, "A review of MEMS oscillators for frequency reference and timing applications," J. Micromech. Microeng., 2011, vol. 22, no. 1, p. 013001.

[5] R. Abdolvand, B. Bahreyni, J. Lee, and F. Nabki, "Micromachined resonators: A review," Micromachines, 2016, vol. 7, no. 9, p. 160. DOI: 10.3390/mi7090160

[6] J. E. Lee, J. Yan, and A. A. Seshia, "Low loss HF band SOI wine glass bulk mode capacitive square-plate resonator," J. Micromech. Microeng., 2009, vol. 19, no. 7, p. 074003. DOI: 10.1088/0960-1317/19/7/074003

[7] J. E. Lee, J. Yan, and A. A. Seshia, "Study of lateral mode SOI-MEMS resonators for reduced anchor loss," J. Micromech. Microeng., 2011, vol. 21, no. 4, p. 0450DOI: 10. DOI: 10.1088/0960-1317/21/4/045010

[8] T. Mattila et al., "A 12 MHz micromechanical bulk acoustic mode oscillator," Sens. Actuator A-Phys., 2002, vol. 101, no. 1-2, pp. 1-9. DOI: 10.1016/s0924-4247(02)00204-2

[9] G. Wu, D. Xu, B. Xiong, Y. Wang, Y. Wang, and Y. Ma, "Wafer-level vacuum packaging for MEMS resonators using glass frit bonding," Journal of microelectromechanical systems, 2012, vol. 21, no. 6, pp. 1484-1491. DOI: 10.1109/JMEMS.2012.2211572

[10] T. T. Chen, J. C. Huang, Y. C. Peng, C. H. Chu, and S. S. Li, "A 17.6-MHz 2.5V ultra-low polarization voltage MEMS oscillator using an innovative high gain-bandwidth fully differential trans-impedance voltage amplifier," in 2013 IEEE 26th International Conference on Micro Electro Mechanical Systems (MEMS 2013) 2013. DOI: 10.1109/MEMSYS.2013.6474349

[11] J. Wang, Z. Ren, and C.-C. Nguyen, "1.156-GHz self-aligned vibrating micromechanical disk resonator," IEEE Trans. Ultrason. Ferroelectr. Freq. Control, 2004, vol. 51, no. 12, pp. 1607-1628. DOI: 10.1109/tuffc.2004.1386679

[12] M. Ziaei-Moayyed, D. Elata, E. Quévy, and R. Howe, "Differential internal dielectric transduction of a Lamé-mode resonator," J. Micromech. Microeng., 2010, vol. 20, no. 11, p. 115036. DOI: 10.1088/0960-1317/20/11/115036

[13] M. Akgul, L. Wu, Z. Ren, and C. T.-C. Nguyen, "A negative-capacitance equivalent circuit model for parallel-plate capacitive-gap-transduced micromechanical resonators," IEEE Trans. Ultrason. Ferroelectr. Freq. Control, 2014, vol. 61, no. 5, pp. 849-869. DOI: 10.1109/tuffc.2014.2976

[14] M. Weinberg, R. Candler, S. Chandorkar, J. Varsanik, T. Kenny, and A. Duwel, "Energy loss in MEMS resonators and the impact on inertial and RF devices," in 2009 International Solid-State Sensors, Actuators and Microsystems Conference (TRANSDUCERS 2009), 2009, pp. 688-695: IEEE. DOI: 10.1109/SENSOR.2009.5285418

[15] L. Khine and M. Palaniapan, "High-Q bulk-mode SOI square resonators with straight-beam anchors," J. Micromech. Microeng., 2008, vol. 19, no. 1, p. 015017. DOI: 10.1088/0960-1317/19/1/015017

[16] D. S. Bindel and S. Govindjee, "Elastic PMLs for resonator anchor loss simulation," International Journal for Numerical Methods in Engineering, 2005, vol. 64, no. 6, pp. 789-818. DOI: 10.1002/nme.1394

[17] M. S. Hajhashemi, A. Rasouli, and B. Bahreyni, "Performance optimization of high order RF microresonators in the presence of squeezed film damping," Sens. Actuator A-Phys., 2014, vol. 216, pp. 266-276. DOI: 10.1016/j.sna.2014.05.014

[18] T. Wang, Z. Chen, Q. Jia, Q. Yuan, J. Yang, and F. Yang, "A Novel High Q Lamé-Mode Bulk Resonator with Low Bias Voltage," Micromachines, 2020, vol. 11, no. 8, p. 737. DOI: 10.3390/mi11080737

[19] V. Kaajakari, T. Mattila, A. Oja, J. Kiihamaki, and H. Seppa, "Square-extensional mode single-crystal silicon micromechanical resonator for low-phase-noise oscillator applications," IEEE Electron Device Lett., 2004, vol. 25, no. 4, pp. 173-175. DOI: 10.1109/led.2004.824840

[20] O. Holmgren, K. Kokkonen, V. Kaajakari, A. Oja, and J. V. Knuuttila, "Direct optical measurement of the Q values of RF-MEMS resonators," in IEEE Ultrasonics Symposium, 2005., 2005, vol. 4, pp. 2112-2115: IEEE. DOI: 10.1109/ultsym.2005.1603298

[21] L. Shao, T. Niu, and M. Palaniapan, "Nonlinearities in a high-Q SOI Lamé-mode bulk resonator," J. Micromech. Microeng., 2009, vol. 19, no. 7, p. 075002. DOI: 10.1088/0960-1317/19/7/075002

[22] V. Kaajakari, T. Mattila, A. Oja, and H. Seppa, "Nonlinear limits for single-crystal silicon microresonators," J. Microelectromech. Syst., 2004, vol. 13, no. 5, pp. 715-724. DOI: 10.1109/jmems.2004.835771

Proceedings of the 16th Annual IEEE International
Conference on Nano/Micro Engineered and Molecular Systems
April 25-29, 2021

Detection of Single Protein Molecules Using MoS₂ Nanopores of Various Sizes*

Chaoming Gu, Zhoubin Yu, Xiaojie Li, Xin Zhu, Zhen Cao, Zhi Ye, Chuanghong Jin and Yang Liu

Abstract— MoS₂ nanopores have recently been demonstrated to be capable of detecting translocation of proteins. In this work, we further investigate the effect of MoS₂ nanopore size on the translocation process. The MoS₂ nanopores are fabricated using TEM drilling that yields nanopore diameters of ~10 nm and ~20 nm. We use these nanopores to detect bovine serum albumin (BSA) proteins and analyze the translocation signal statistics as well as the shape of individual signals. It is found that larger MoS₂ nanopores can reduce the threshold voltage for translocation and increase the translocation event rate. Translocation at reduced bias voltages may prolong individual translocation process, which can be beneficial for improving the sensing resolution.

I. INTRODUCTION

In recent decades, solid state nanopores are playing an important role in the field of detecting single biomolecules such as DNAs and proteins, showing benefits of better stability, more flexible pore size control and more integration and modification capabilities over biological nanopores [1]. Using the common sensing mechanism of ionic current blockade caused by the biomolecules translocation process, solid state nanopores are now able to reveal detailed structure and configuration of DNAs [2]–[6] and protein molecules [7]–[9]. However, the channel lengths of the usual SiN$_x$ nanopores (10~30 nm) still restrict the detection resolution and makes it hard to sense the minuscule details of biomolecules such as the DNA bases. MoS₂ nanopores are a type of more recently studied solid state nanopores that are based on two-dimensional substrates. The monolayer thickness of MoS₂ is ~6.5 Å, which translates into higher spatial resolution in the ionic current blockade signals. Furthermore, compared to another 2-D materials graphene, MoS₂ nanopores have weaker interaction with biomolecules, which makes the biomolecules easier to go through the nanopore. The DNA detection using MoS₂ nanopores has been broadly studied, such intrinsic advantages of MoS₂ nanopores have been experimentally validated[10]–[12]. As for protein detection using MoS₂ nanopores, a recent simulation work has indicated its feasibility[13]. Our previous work experimentally demonstrated that the MoS₂ nanopores can indeed detect individual protein molecules [14], showing that the protein translocation is easier compared to similar experiments conducted by graphene nanopores[15], [16].

In this work, we further investigate the translocation of BSA protein molecules through MoS₂ nanopores with various pore diameters. The aim is to understand what role the pore size plays in the translocation process and its implications in protein sensing applications.

II. FABRICATION

The fabrication of MoS₂ nanopores includes three steps: SiN$_x$ substrate pore drilling, MoS₂ mechanical exfoliation and transferring, and MoS₂ nanopore drilling.

The substrate is a silicon round chip and in the center is a free-standing rectangular window with ~15 nm thickness SiN$_x$ membrane on it (Norcada Inc.). To make sure the MoS₂ membrane transferred later is free-standing too, we use focused ion beam (FIB) (FEI Quanta 3D FEG with a gallium ion beam) to drill a large pore with 1 μm in diameter on the center of the window.

MoS₂ membranes are acquired by mechanical exfoliation. First, using tapes to exfoliate a thin MoS₂ from the bulk. Then, still using tapes to exfoliate the thin MoS₂ we already got repeatedly to make it thinner. A PDMS stamp is utilized to transfer the MoS₂ pieces from the tape. By observing them under the optical microscope, we can distinguish which MoS₂ piece has less layers by comparing the color of it against the background. The lower contrast means thinner thickness[17]. After choosing the qualified MoS₂ piece and identifying its location on the PDMS stamp, the MoS₂ membrane then can be transferred onto the FIB pore by three-dimensional transferring platform Fig 1 (a), (c) show the FIB pore covered by the transferred MoS₂ membranes. The thickness of them are both around 3 nm (~4 layers). Further confirmation of MoS₂ thickness can be conducted by other measurements like Raman spectroscopy method[14], [18].

The MoS₂ nanopores are drilled by transmission electronic microscope (TEM) (FEI Tecnai G^2 F20). By adjusting the related parameters like electron dose, we can control the nanopore diameters to meet our demands. Here we drill two different nanopores. The large pore has ~20 nm in diameter (Fig. 1 (b)) and the small pore with ~10 nm in diameter (Fig. 1 (d)). Under the TEM we can see that the shape of the pore is irregular round. So far, the nanopore shape has no significant bearing on our later protein detection experiments.

---

*Research supported by National Natural Science Foundation of China (61774132).

Chaoming Gu, Xiaojie Li, Xin Zhu, Zhen Cao, Zhi Ye, and Yang Liu are with College of Information Science and Electronic Engineering, Zhejiang University, Hangzhou 310027, China (e-mail: yliu137@gmail.com).

Zhoubin Yu and Chuanghong Jin are with State Key Laboratory of Silicon Materials, School of Materials Science and Engineering, Zhejiang University, Hangzhou 310027, China.

978-1-6654-3008-1/21 $31.00 © 2021 IEEE   459

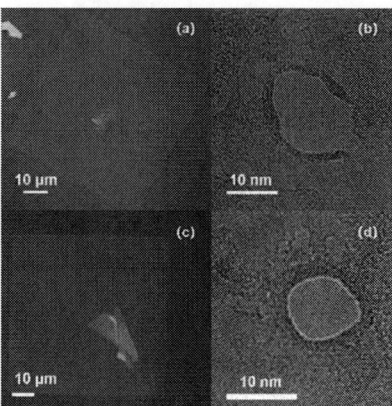

Fig. 2 (a), (c) Optical microscopic images of the FIB-drilled pore covered with an exfoliated MoS2 thin membrane. The color contrast between the MoS$_2$ membrane and the window is low, which means the MoS$_2$ membrane is very thin. (b), (d) TEM images of the nanopore drilled by TEM on the MoS2 membrane with pore diameter as ~20 nm (large pore) and ~10 nm (small pore) respectively.

III. RESULTS AND DISCUSSIONS

Bovine serum albumin (BSA) protein molecules are tested in our experiments. The native BSA molecule is an ellipse approximately with ~14 nm long axis and ~4 nm short axis[7]. Its molecular wight is about 66 kDa. BSA has an isoelectric point in pH 5.1~5.5[19], hence when it is in the alkaline solution it will carry negative charges.

Fig. 2 illustrates the experiment setup. 45 nM BSA (BBI Life Science) is dissolved in 1 M KCl solution buffered at pH 8 using tris-EDTA to let BSA molecules be negatively charged. The MoS$_2$ nanopore device divides the reservoir into two parts, the Cis part and the Trans part. BSA molecules start translocation from Cis (negative voltage) driven by the bias voltage, go through the nanopore, and then reach Trans (positive voltage). The PDMS seal ensures that MoS$_2$ nanopore is the only channel that BSA molecules can pass through. The blockade ionic current caused by BSA translocation will be amplified and recorded by patch clamp (Axon 200B) and analyzed later.

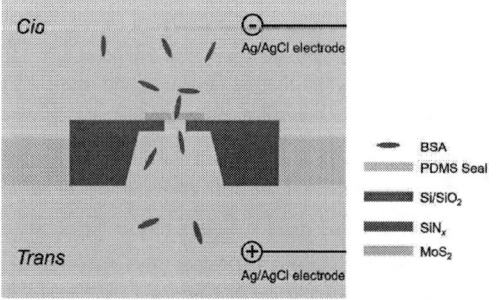

Fig. 1 The schematic of the measurement setup of BSA translocation

Two MoS$_2$ nanopore devices are tested: device A with a large nanopore (~20 nm in diameter) and device B with a small nanopore (~10 nm in diameter). Both devices are under the same experimental conditions. In the experiments we find that the BSA translocation signals begin appearing at 200 mV in device A, while in device B, the signals don't appear until the bias voltage is increased to 400 mV. This result indicates that larger pore can decrease the threshold voltage of BSA translocation. The larger pore may lower the barrier around the nanopore entry and increase the chance for BSA molecules to be captured from the bulk part of the solution. Therefore, the molecules only need small driven force (lower bias voltage) to get close to the nanopore and go through it. The lower threshold voltage has positive effects on the durability of the MoS$_2$ nanopores. It can reduce the possibility of current leakage and MoS$_2$ membrane break for its very thin thickness. Lower threshold voltage also avoids changing the shape and the size of the pore itself, excluding the potential influences from irrelevant variables.

Fig. 3 shows the BSA translocation traces through MoS$_2$ nanopores with various pore sizes. Fig. 3 (a) is the trace of device A at 400 mV and (b) is that of device B at 400mV. For two different MoS$_2$ nanopore traces at the same bias voltage, the fluctuation level of the baselines is similar, and the noise level of both devices is close, too, around 2 nA, which indicates that the pore size is not the main factor responsive to the instability and the noise of the baseline. The signal noise ratio (SNR) also has no obvious distinction between two various pore sizes. While there is a significant variance in events rate. In the same testing interval e.g. 2.5 s, the amount of the BSA translocation signals through the device A is 26, 6.5 times more than the device B, 4 events in all, which means that larger pore sizes can facilitate the BSA translocation. Easier translocation allows us to collect more signals in a short period so that the signals statistical analysis in the next step can be more effective.

The dwell time of a signal is an important parameter. For the single biomolecule detection, the longer dwell time can demonstrate more detail information of a molecule, including its structure, configuration etc. In Fig. 4, the events distribution of dwell time of device A and device B is illustrated with Gaussian fitting. For both devices, the major part of events has a very short dwell time (< 1ms), and some of them may reflect the situation that BSA molecules just appear at the entrance or go in the pore partially and move back, but not pass through the nanopore[20]. While in the longer dwell time area (>1ms), it's clear that the events of device B, the small pore, are rare. At 400 mV bias voltage, there are barely any events of device B distributed in the area the dwell time longer than 10 ms (see the inset). In the contrast, most of the events longer than 10 ms are from the device A. Fig. 4 shows the events distribution of the device A at three different bias voltages: 200 mV, 300 mV and 400 mV. We notice that even at a lower voltage like 200 mV, there is still a number of longer signals (>10 ms) of large pore, while for the small pore, the threshold voltage is not even reached. Fig. 4 also signifies that the bias voltage doesn't play a role in the dwell time distribution. This result shows that a larger pore size can prolong the translocation

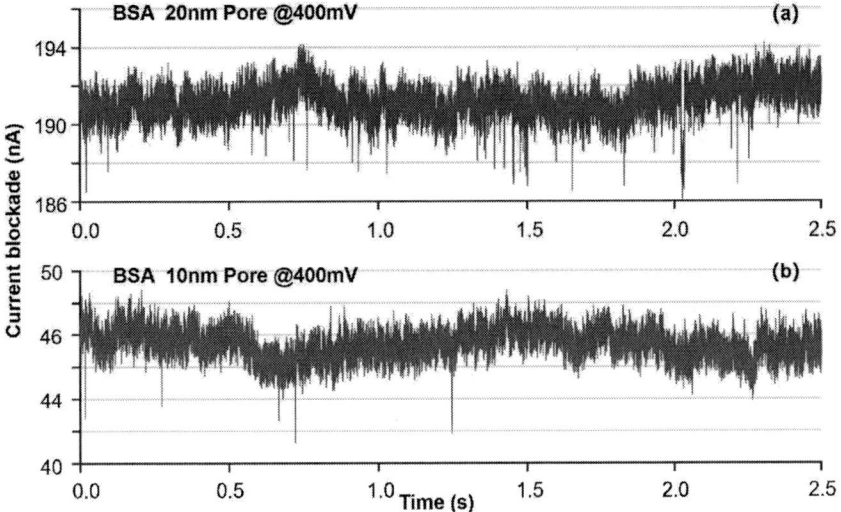

Fig. 3 (a), (b) Two BSA translocation traces in 2.5 s through the large and small MoS$_2$ nanopores under 400 mV respectively.

time of BSA molecules. One possible explanation is that larger pores give enough room for the BSA molecules to translocate through the nanopore in different angles and therefore increase the possibility for BSA to interact with the MoS$_2$ nanopores, which may slow the velocity of the translocation.

Longer dwell time, combined with higher resolution provided by MoS$_2$ nanopores, exhibits more details in a single translocation event to us. Fig. 5 shows two single events of device A at 400 mV bias voltage. Both of them have dwell time longer than 10 ms. Fig. 5 (a) is a typical kind of event, with a current blockade around 3 nA and a dwell time around 13 ms. The blockade part is flat. Statistically, this kind of events have 2~3 nA blockade. Fig. 5 (b) shows a different kind of event with an extra level in the amplitude. The first level has 2 nA blockade, and the second level in the middle has the same blockade current as

the first level but a much shorter dwell time(~2 ms). This may due to that two BSA molecules translocate through the MoS$_2$ nanopore concurrently with different velocity. This kind of event, like Fig. 5 (b), doesn't show up frequently and the total amount is few. But they only appear in the large pore, because it is not easy for two BSA molecules to pass through the small pore with the narrow space and the stronger confinement.

IV. CONCLUSION

In this work, we use MoS$_2$ nanopores with two pore sizes to detect single BSA molecules. The experiment results show that larger pores can effectively decrease the translocation threshold voltage, which can in turn prolong the lifetime of the MoS$_2$ nanopore device and improve its stability. The larger pores also facilitate the translocation process and increase the translocation dwell time. This is beneficial for obtaining higher resolution in the translocation detection. It is observed that some translocation signals exhibit characteristics of two BSA molecules passing through the nanopores at the same time.

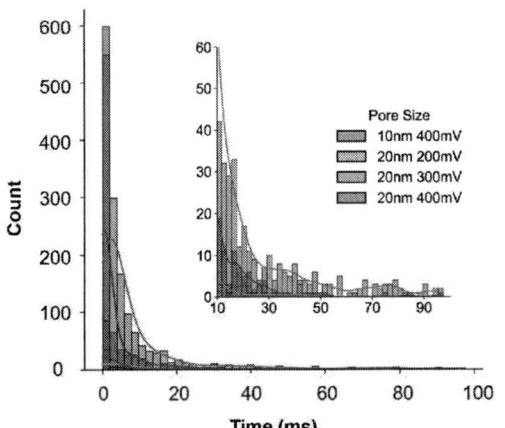

Fig. 4 The statistical distribution of the translocation dwell time of the small (~10 nm) and the large (~20 nm) MoS$_2$ nanopore under different voltages with Gaussian fitting. Inset: the distribution of the dwell time > 10 ms.

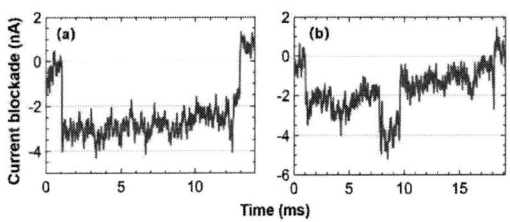

Fig. 5 (a) A typical long translocation signal under 400 mV and (b) A long signal with two levels in the amplitude under 400 mV from device A.

REFERENCES

[1] C. Dekker, "Solid-state nanopores," *Nature Nanotechnology*, vol. 2, no. 4, Art. no. 4, Apr. 2007.

[2] C. Plesa *et al.*, "Direct observation of DNA knots using a solid-state nanopore," *Nature Nanotech*, vol. 11, no. 12, pp. 1093–1097, Dec. 2016.

[3] R. Kumar Sharma, I. Agrawal, L. Dai, P. S. Doyle, and S. Garaj, "Complex DNA knots detected with a nanopore sensor," *Nat Commun*, vol. 10, no. 1, p. 4473, Dec. 2019.

[4] J. Li, M. Gershow, D. Stein, E. Brandin, and J. A. Golovchenko, "DNA molecules and configurations in a solid-state nanopore microscope," *Nature Materials*, vol. 2, no. 9, Art. no. 9, Sep. 2003.

[5] S.-C. Liu, Q. Li, Y.-L. Ying, and Y.-T. Long, "Detection of structured single-strand DNA via solid-state nanopore," *ELECTROPHORESIS*, vol. 40, no. 16–17, pp. 2112–2116, 2019.

[6] N. A. W. Bell, M. Muthukumar, and U. F. Keyser, "Translocation frequency of double-stranded DNA through a solid-state nanopore," *Phys. Rev. E*, vol. 93, no. 2, p. 022401, Feb. 2016.

[7] D. Fologea, B. Ledden, D. S. McNabb, and J. Li, "Electrical characterization of protein molecules by a solid-state nanopore," *Appl. Phys. Lett.*, vol. 91, no. 5, p. 053901, Jul. 2007.

[8] D. S. Talaga and J. Li, "Single-Molecule Protein Unfolding in Solid State Nanopores," *J. Am. Chem. Soc.*, vol. 131, no. 26, pp. 9287–9297, Jul. 2009.

[9] L. Restrepo-Pérez, S. John, A. Aksimentiev, C. Joo, and C. Dekker, "SDS-assisted protein transport through solid-state nanopores," *Nanoscale*, vol. 9, no. 32, pp. 11685–11693, 2017.

[10] J. Feng *et al.*, "Identification of single nucleotides in MoS2 nanopores," *Nature Nanotech*, vol. 10, no. 12, pp. 1070–1076, Dec. 2015.

[11] K. Liu, J. Feng, A. Kis, and A. Radenovic, "Atomically Thin Molybdenum Disulfide Nanopores with High Sensitivity for DNA Translocation," *ACS Nano*, vol. 8, no. 3, pp. 2504–2511, Mar. 2014.

[12] K. Liu *et al.*, "Geometrical Effect in 2D Nanopores," *Nano Lett.*, vol. 17, no. 7, pp. 4223–4230, Jul. 2017.

[13] H. Chen, L. Li, T. Zhang, Z. Qiao, J. Tang, and J. Zhou, "Protein Translocation through a MoS2 Nanopore:A Molecular Dynamics Study," *J. Phys. Chem. C*, vol. 122, no. 4, pp. 2070–2080, Feb. 2018.

[14] C. Gu *et al.*, "Experimental study of protein translocation through MoS2 nanopores," *Appl. Phys. Lett.*, vol. 115, no. 22, p. 223702, Nov. 2019.

[15] Y. P. Shan *et al.*, "Surface modification of graphene nanopores for protein translocation," *Nanotechnology*, vol. 24, no. 49, p. 495102, Dec. 2013.

[16] G. Goyal, Y. B. Lee, A. Darvish, C. W. Ahn, and M. J. Kim, "Hydrophilic and size-controlled graphene nanopores for protein detection," *Nanotechnology*, vol. 27, no. 49, p. 495301, Dec. 2016.

[17] M. M. Benameur, B. Radisavljevic, J. S. Héron, S. Sahoo, H. Berger, and A. Kis, "Visibility of dichalcogenide nanolayers," *Nanotechnology*, vol. 22, no. 12, p. 125706, Mar. 2011.

[18] H. Li *et al.*, "From Bulk to Monolayer MoS2: Evolution of Raman Scattering," *Advanced Functional Materials*, vol. 22, no. 7, pp. 1385–1390, 2012.

[19] T. Peters, "Serum Albumin," in *Advances in Protein Chemistry*, vol. 37, C. B. Anfinsen, J. T. Edsall, and F. M. Richards, Eds. Academic Press, 1985, pp. 161–245.

[20] M. Zwolak and M. Di Ventra, "Physical approaches to DNA sequencing and detection," *Rev. Mod. Phys.*, vol. 80, no. 1, pp. 141–165, Jan. 2008.

Proceedings of the 16th Annual IEEE International
Conference on Nano/Micro Engineered and Molecular Systems
April 25-29, 2021

Humidity Sensor Based on Thin-Film Piezoelectric-on-Substrate Resonator and MoS$_2$ for Multifunctional Applications

Hanyong Dong, Jintao Pang, Dongsheng Li, Qian Zhang and Jin Xie*, *Member, IEEE*

Abstract— This work proposed a thin-film piezoelectric-on-substrate (TPoS) resonator based humidity sensor with Molybdenum disulfide (MoS$_2$) film as sensing material. The small size resonator provided advantages of high stability and high resolution and the MoS$_2$ contributed high sensitivity. The proposed humidity sensor performed with sensitivity of 434 Hz/%RH, high stability of below 0.367%, fast response time of 1.8 s and recovery time of 2.8 s. The sensor is potential for multifunctional applications for its good properties. We demonstrated the sensor's performance in human breath analysis and detection of finger's moving in a noncontact way as examples.

Keywords— Relative humidity sensor, piezoelectric-on-substrate resonator, Molybdenum disulfide, breath detection, noncontact detection.

I. INTRODUCTION

Humidity sensor is a traditional type sensor with long development history. It was previously used in agriculture or meteorology to monitor the environment's humidity in a large scale. As development of MEMS technology and the growing demand for humidity sensor applied in more precise and dynamic conditions, the humidity sensor is required for miniaturization and high performance [1]. For example, using humidity sensor to detect human breath process is a typical method for noncontact respiration monitoring, while many illnesses and discomfort lead to an abnormal respiratory rate and depth [2, 3]. And applying humidity sensor to detect humidity level around human skin is another important application, which can provide various physiological information which reflect human's health status. [4]. Monitoring the humidity change around the fingertip can realize noncontact detection of finger position and motion, which is potential for gesture recognition [5].

A MEMS humidity sensor typically consists of transducing platform and sensing material. Comparing to traditional capacitive humidity sensors [6-8], which face problems of instability and high impedance, piezoelectric resonator sensors have advantages of high stability, high resolution and CMOS compatibility [9]. Among resonators, the TPoS resonator is an outstanding transducing platform with advantages of small size, high quality factor and low motional impedance [10, 11]. As for sensing material, MoS$_2$ is a popular two-dimensional nanomaterial with special superiority in humidity sensing including large surface-to-

volume ratio, low interlayer force and stronger combination with metal interface [12]. In this paper, we proposed a humidity sensor based on TPoS resonator and MoS$_2$ film. The humidity sensing performance of the sensor was investigated and its applications in human breath analysis and finger motion detection were studied.

II. EXPERIMENTAL METHOD

A. Fabrication of TPoS resonator

The proposed humidity sensor is constructed by a TPoS resonator and a MoS$_2$ film, as demonstrated in Figure 1. The resonator comprises a AlN thin film as piezoelectric layer with a pair of Al electrodes on the top and a silicon substrate in the bottom. The size of the device is 189 μm × 189 μm and the thickness of electrodes, AlN layer and silicon layer are 0.2 μm, 0.5 μm and 10 μm respectively.

Figure 1. Schematic of humidity sensor.

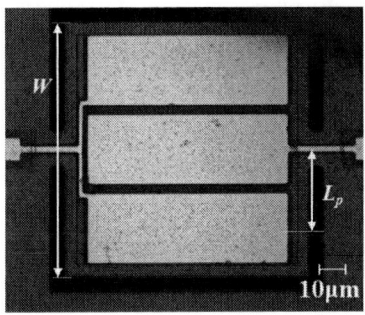

Figure 2. Electron microscopy photo of the uncoated resonator.

---

*This work is supported by the National Natural Science Foundation of China (51875521), the Zhejiang Provincial Natural Science Foundation of China (LZ19E050002) and the Science Fund for Creative Research Groups of National Natural Science Foundation of China (51821093).

All of authors are with the State Key Laboratory of Fluid Power and Mechatronic Systems, Zhejiang University, Hangzhou 310027, People's Republic of China.
Corresponding author: *Jin Xie, E-mail: xiejin@zju.edu.cn.

978-1-6654-3008-1/21 $31.00 © 2021 IEEE 463

A commercial MEMS process (PiezoMUMPs, MEMSCAP) was adopted to fabricate the resonator. The AlN piezoelectric layer was deposited over a doped silicon-on-insulator (SOI) wafer with 10 μm Si layer by reactive sputtering and patterned by wet-etched. Next the metal electrodes layer consisting of 20 nm chrome and 1 μm aluminum was deposited and patterned. Then the Si layer was etched until the buried oxide by deep reactive ion etching (DRIE) to define the device. Finally, a DRIE was used from the backside to release the device. Figure 2 shows the electron microscopy photo of the uncoated resonator and its resonant frequency of the resonator can be estimated as

$$f = \frac{1}{2(W/n)}\sqrt{\frac{E_{eff}}{\rho_{eff}}} = \frac{1}{2L_p}v_{a,eff} \qquad (1)$$

where W is the width of the resonator, E_{eff} is the effective young's modulus of the device along the width orientation, ρ_{eff} is the effective density of the device, and n is the number of electrode fingers. Besides, the equation can be expressed as a function of finger pitch L_p and effective acoustic velocity along the width $v_{a,eff}$ [10].

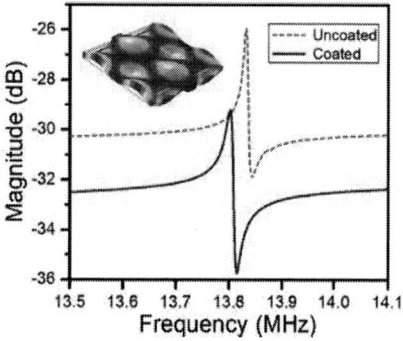

Figure 3.   S21 curves of uncoated and coated resonator.

B. Preparation of humidity sensor

MoS₂ was chosen as the humidity sensing material for its excellent properties. Firstly, the MoS₂ powder was treated by ultrasonic exfoliating to prepare MoS₂ dispersions. Then the MoS₂ film was formed by drop-casting from a 0.5 μL drop of 0.2 mg/ml MoS₂ dispersions as the sensing layer. Figure 3 shows S21 curves of the device measured before

and after coating and the inset shows the resonance mode shape simulated in COMSOL. The sensor is operated in 13.8 MHz. When the relative humidity varies, the sensing layer would adsorb or desorb water molecule and leads to the resonant frequency shifts.

C. Experimental setup

The experimental setup to generate flows in different humidity degree is shown in Figure 4. The ratios of the wet air and dry air were controlled by adjusting the tunable valve, from this the relative humidity in the sealed metallic chamber was controlled and calibrated by a commercial hygrometer. The sensor was placed in a metallic chamber to avoid disturbance. A network analyzer was used for measurement of the sensor's resonant frequency and a LABVIEW program was used to real-time record the sensor response. All experiments were carried out at 23.5 ± 0.5°C.

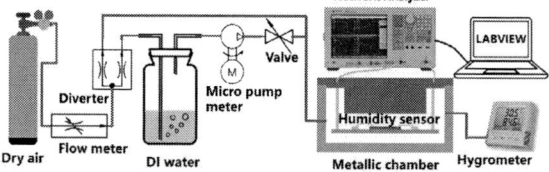

Figure 4.   Schematic of experimental setup for humidity sensor.

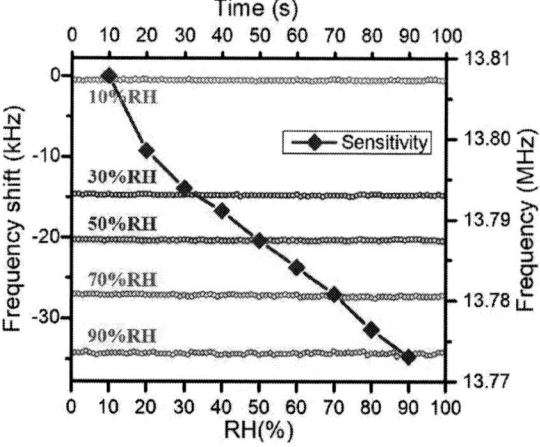

Figure 5.   Sensitivity and short term stability of the sensor.

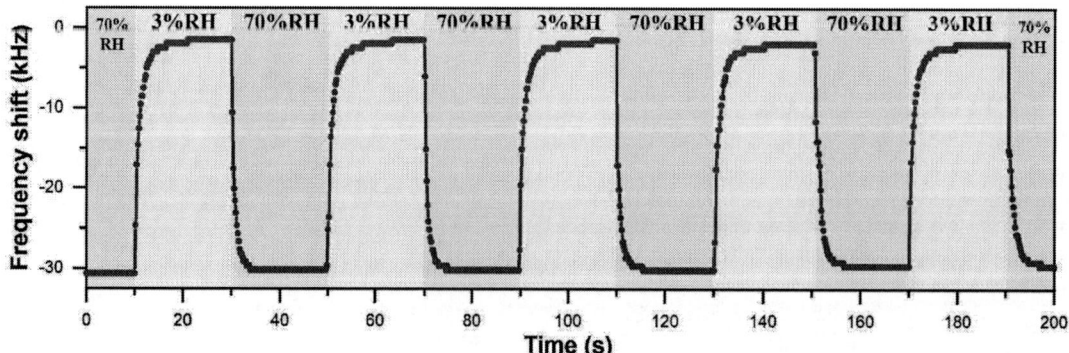

Figure 6.   Short term repeatability of the sensor when the humidity was circularly tuned between 70%RH and 3%RH.

978-1-6654-3008-1/21 $31.00 © 2021 IEEE

III. EXPERIMENTAL RESULT

A. Characteristics of the proposed sensor

As displayed in the Figure 5, the relative humidity was adjusted from 10%RH to 90%RH with 10%RH as the sample interval and the resonant frequency shifts of the sensor were record by a network analyzer for 3 times and averaged. The sensor performed a 34.7 kHz frequency shift from 10%RH to 90%RH and an averaged sensitivity of 434 Hz/%RH. In addition, to test the short-term stability of the sensor, the relative humidity was kept unchanged in 100 s and the output frequency of the sensor was recorded with an interval of 1 s. The short-term stability was represented as variation which can be expressed as

$$Var = \frac{f_{RMSD}}{\Delta F} \times 100\% \qquad (2)$$

where f_{RMSD} was the Root-Mean-Square Deviation (RMSD) of the recorded frequencies in 100 s at a certain relative humidity and ΔF was the full-range frequency shift. As a result, the short-term stability of the sensor at 10%RH, 30%RH, 50%RH, 70%RH, 90%RH were 0.153%, 0.184%, 0.219%, 0.329% and 0.367% respectively. The variations got larger when humidity increased due to the more drastic water molecule exchanging process between sensor and environment.

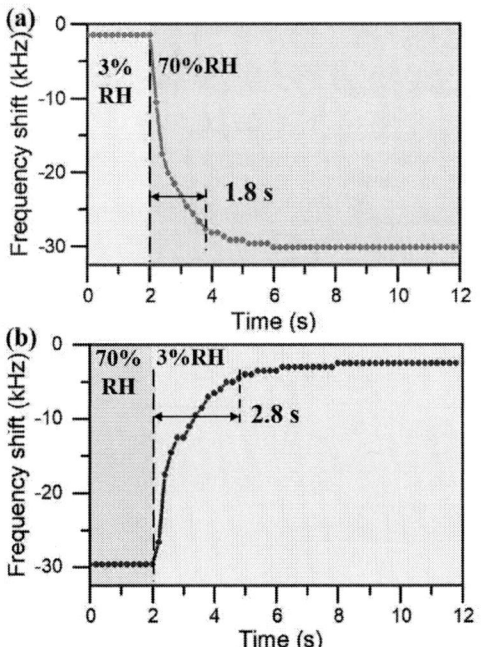

Figure 7. Response and recovery time of the proposed sensor.

The sensor performed a good short-term repeatability as shown in Figure 6. The relative humidity was circularly adjusted at 3%RH and 70%RH and the resonant frequency shifts of the sensor was recorded with an interval of 0.2 s. Besides, the response and recovery time, which were defined as the time that the frequency shift reaches 90% of the total frequency shift value, can be read out from the curve as

shown in Figure 7. The sensor performed a short response time of 1.8 s and recovery time of 2.8 s.

B. Applications of the proposed sensor

Humidity sensor has several applications in our daily life. The proposed humidity sensor was competent in practical applications due to its good performance. In this work, the capability of the sensor applied in breath analysis and noncontact detection of finger moving was verified.

Human breath is one typically activity along with humidity change, the outside humidity increases when human exhale and decreases when inhale. The sensor was installed 5 cm below the nose for detection and the output resonant frequency shift was recorded. As shown in Figure 8, three types of breath were tested and the response of the sensor was recorded. The breath rates were about 5 times/min, 12 times/min and 80 times/min in deep, normal and fast breath respectively. And the graph has shown that when breath speeded up, the amplitude of the sensor response would decrease. That is because in a more dynamic condition, the sensor is more difficult to reach saturation. Thus this application requires for a high sensitive and fast-response sensor. In this work, the proposed sensor can clearly distinguish the exhale and inhale process in all conditions.

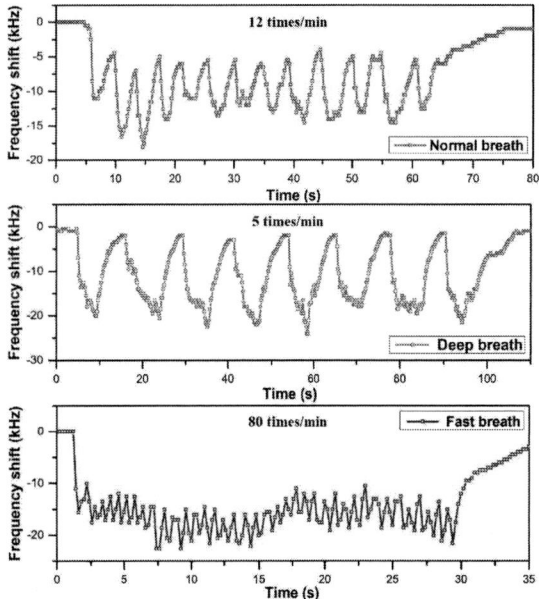

Figure 8. Application in breath detection and the sensor response to 3 types human breath (a) Normal breath (12 times/min); (b) Deep breath (5 times/min); (c) Fast breath (80 times/min)..

Noncontact finger motion detection is another application for humidity sensor. Because of the continue sweat evaporation, the relative humidity around human's skin would rise. Particularly, this phenomenon would intensify at the fingertip and generate a sharp humidity gradient filed. Thus, we can use a humidity sensor to detect the position or movement of the finger in a noncontact way.

Figure 9(a) shows that the sensor can measure the distance of a bare finger. Besides, a comparative measurement to a finger with a rubber glove verified that the response is principally caused by the moisture around the fingertip and slightly caused by temperature change. The influence of temperature can be reduced by adding a reference bare resonator in future work. Figure 9(b) shows a real-time response of a finger vertically moving from 20 mm to 1 mm to the sensor up and down. The moving speed of the finger can be reflected from the magnitude of the sensor response. In future work, more information of finger motion can be detected by applying an array of sensor. The proposed sensor can clearly distinguish the finger motion and this method has potential to be used as an assistance in gesture recognition or noncontact human-machine interaction.

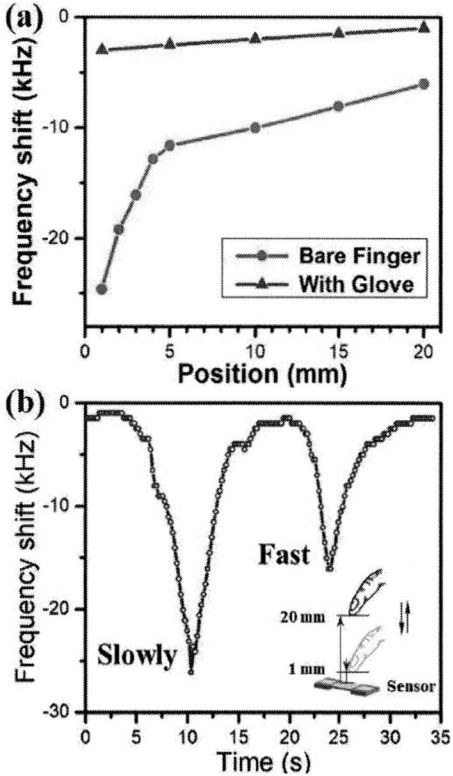

Figure 9. Application in noncontact detection. (a) measurement of finger's positions; (b) measurement of finger's vertical moving.

IV. CONCLUSION

Benefiting from the high operation frequency and stability of TPoS resonator and high properties of MoS$_2$, the proposed novel relative humidity sensor showed a high sensitivity of 434 Hz/%RH, high stability with variation below 0.367%, fast response time of 1.8 s and recovery time of 2.8 s. The sensor has an excellent performance in human breath analysis and noncontact finger motion detection. Clear detections of up to 80 time/min breath and finger moving within 20 mm were achieved, indicating the proposed sensor have potential for human activities

monitoring, assistance of medical diagnosis and noncontact gesture recognition.

REFERENCES

[1] C.-Y. Lee and G.-B. Lee, "Humidity Sensors: A Review," Sensor Letters, vol. 3, no. 1, pp. 1-15, 2005.

[2] F. Q. Al-Khalidi, R. Saatchi, D. Burke, H. Elphick, and S. Tan, "Respiration rate monitoring methods: a review," Pediatr Pulmonol, vol. 46, no. 6, pp. 523-9, Jun 2011.

[3] J. Dai et al., "Ultrafast Response Polyelectrolyte Humidity Sensor for Respiration Monitoring," ACS Appl Mater Interfaces, vol. 11, no. 6, pp. 6483-6490, Feb 13 2019.

[4] J. Wu et al., "Carbon Nanocoil-Based Fast-Response and Flexible Humidity Sensor for Multifunctional Applications," ACS Appl Mater Interfaces, vol. 11, no. 4, pp. 4242-4251, Jan 30 2019.

[5] N. Li et al., "A fully inkjet-printed transparent humidity sensor based on a Ti3C2/Ag hybrid for touchless sensing of finger motion," Nanoscale, vol. 11, no. 44, pp. 21522-21531, Nov 28 2019.

[6] J. Zhao et al., "Highly Sensitive MoS2 Humidity Sensors Array for Noncontact Sensation," Adv Mater, vol. 29, no. 34, Sep 2017.

[7] Y. Feng et al., "TaS2 nanosheet-based ultrafast response and flexible humidity sensor for multifunctional applications," Journal of Materials Chemistry C, vol. 7, no. 30, pp. 9284-9292, 2019.

[8] X. Yu, X. Chen, X. Ding, X. Yu, X. Zhao, and X. Chen, "Facile fabrication of flower-like MoS2/nanodiamond nanocomposite toward high-performance humidity detection," Sensors and Actuators B: Chemical, vol. 317, 2020.

[9] S. Tadigadapa and K. Mateti, "Piezoelectric MEMS sensors: state-of-the-art and perspectives," Measurement Science and Technology, vol. 20, no. 9, 2009.

[10] R. Abdolvand, H. M. Lavasani, G. K. Ho, and F. Ayazi, "Thin-Film Piezoelectric-on-Silicon Resonators for High-Frequency Reference Oscillator Applications," Ieee Transactions on Ultrasonics Ferroelectrics And Frequency Control, vol. 55, no. 12, pp. 2596-2606, Dec 2008.

[11] B. P. Harrington and R. Abdolvand, "In-plane acoustic reflectors for reducing effective anchor loss in lateral–extensional MEMS resonators," Journal of Micromechanics and Microengineering, vol. 21, no. 8, 2011.

[12] S.-L. Zhang, H.-H. Choi, H.-Y. Yue, and W.-C. Yang, "Controlled exfoliation of molybdenum disulfide for developing thin film humidity sensor," Current Applied Physics, vol. 14, no. 3, pp. 264-268, 2014.

Gap in pagination due to unavailable paper.

Pages 467-468

Proceedings of the 16th Annual IEEE International
Conference on Nano/Micro Engineered and Molecular Systems
April 25-29, 2021

In-situ synthesized liquid metal microgels

Dawei Wang, [ab†] Chennan Lu, [ab†] Xiaohong Wang, [ab] and Wei Rao [ab*]

Abstract— Microgels have attained extensive attention due to their unique physical and chemical properties, the combination with the same flexible emerging material—liquid metal to fabricate hybrid microgels may exhibit complementary advantages and may expand its applications. In this work, we proposed a kind of *in-situ* cross-linking liquid metal microgels, which can be used as fully flexible theranostics, showing encouraging potential in various biomedical applications. As a preliminary attempt, we fabricated Ga-alginate microgel (denoted as ALGa microgel) by a facile method, and further loaded with anti-cancer drugs (DOX·HCl). Herein, the gallium is not only used as a crosslinking agent, but also a source of various functions. The preliminary studies demonstrated that the developed ALGa microgels maintained a good morphology, uniform size, high stability, excellent biocompatibility and may integrate functions of drug loading, stimulus-response and imaging, thereby making it possible to achieve synergistic treatment of chemotherapy and hyperthermia to eliminate cancer cells. Overall, this pattern provides the possibility for the convenient preparation of multifunctional liquid metal microgels.

I. INTRODUCTION

Microgel as a special class of colloids with an intramolecular cross-linked structure, show a series of performance advantages, such as, high porosity, tunable architecture, adjustable dimensions, modifiability, mechanical strength, permeability and responsive nature to different environmental stimuli, etc., making them potential candidates for drug delivery, antifouling coatings, catalysis, tissue engineering, water purification and responsive materials [1-6]. The synthesis of microgels requires certain technical processes to form an ordered structure with specific functions. However, current commonly used methods, such as emulsion polymerisation and precipitation polymerisation, have limited efficiency or difficuly for purification [2]. In addition, the microgels prepared by these methods have limited functionality and difficult to match complex application requirements. Therefore, we try to change the strategy to control the release of the cross-linking agent, so that the polymerisation occurs in a specific area inside the composite microgels. As an alternative idea, the gallium-based liquid metals (LMs) with unique amorphous properties (superb flexibility, excellent fluidity, shape transformability, and low viscosity) may have good compliance to *in-situ* fabricate hybrid microgels.

The soft nature of the LMs make them possible to synthesize micro/nano-scaled particles by convenient preparation methods, such as ultrasonication, moulding,

fluidic jetting, shear-mixing and microfluidic flow-focusing [7]. During the preparation of liquid metal particles, the thin gallium oxide "skin" (akin to the protective aluminum oxide coating on metallic aluminum [8, 9]) can be rapidly ruptured, thereby releasing gallium ions because of the strong reactivity when exposing to oxygen and water. Therefore, offering the possibility of *in-situ* cross-linking with some polysaccharides (e.g., alginate, gellan gum, carboxymethyl cellulose, hyaluronic acid and quaternized chitosan) [10], some proteins (e.g., soy protein, whey protein) and other cross-linkable polymers (Scheme. 1). This kind of liquid metal microgels can be used as fully flexible theranostics which may exhibit complementary advantages of microgel and liquid metal, showing encouraging potential in various biomedical applications, such as molecular imaging, drug carriers, biomedical devices and cancer therapy. On one hand, the microgels can serve as a protective barrier to hinder internal liquid metal droplets from coalescence and oxidation, enhancing stability and improving biosafety. On the other hand, the internal liquid metal may provide diverse featured properties (such as high electrical conductivity, thermal conductivity, electromagnetic properties, facile functionalization accessibility, catalytic properties and photothermal/photodynamic effect) [7, 11-15], enriching the functionality of the hybrid microgels. As a preliminary attempt, we fabricated Ga-alginate microgel (denoted as ALGa microgel) by a facile method, and further loaded with anti-cancer drugs (DOX·HCl). The basic features of ALGa microgels have been characterized, and the preliminary functionalities were also analyzed.

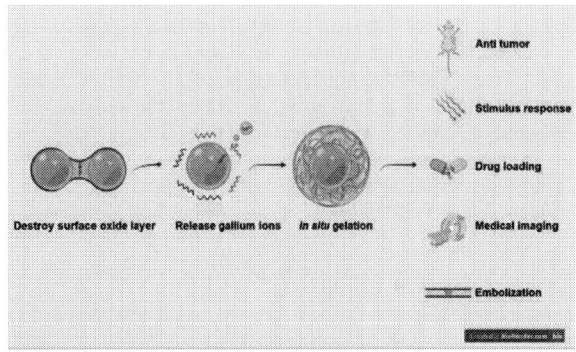

Scheme. 1. Schematic illustration of *in-situ* synthesized liquid metal microgels, and potential biomedical applications.

This work was financially supported by the National Key R&D Program of China (2018YFC1705106) and the National Natural Science Foundation of China (No. 51706236 & No. 51890893).

a Technical Institute of Physics and Chemistry, Chinese Academy of Sciences, Beijing 100190, China

b School of Future Technology, University of Chinese Academy of Sciences, Beijing 100049, China

† The authors contribute equally

* Corresponding author: Wei Rao, Phone: +86-10-82543719 Email: weirao@mail.ipc.ac.cn

978-1-6654-3008-1/21 $31.00 © 2021 IEEE

II. Experimental Procedures

A. Experimental Materials

Gallium (Ga, purity 99.99 %) was provided by the Aluminum Co. Ltd of China. Sodium alginate was obtained from Sinopharm Chemical Reagent Co. Ltd (Shanghai, China). All other chemicals were analytic grade and acquired from Sinopharm Chemical Reagent Co. Ltd (Shanghai, China).

B. Preparation of ALGa Microgels

In brief, little gallium metals were placed in the sodium alginate aqueous solution (2 wt %) and the *in-situ* cross-linked liquid metal microgels (denoted as ALGa microgels) were synthesized by ultrasonic method (Ultrasonic Disruptor, Emerson, power 70 %, 20 min, ice-water bath). Afterward, the ALGa microgels were washed with DI water multiple times by centrifugation (4000 rpm, 10 min), and then freeze-dried for further characterization.

C. Characterizations of ALGa Microgels

The scanning electron microscope (SEM, Quanta FEG 250, FEI, America) was used for morphological characterization of the lyophilized ALGa microgels at voltage of 10 kV. The SEM-EDX (SEM combined with energy dispersive X-ray spectroscopy) was used to analyze the elemental composition and the corresponding element mapping. Besides, the Raman spectral of ALGa microgels and alginate were detected with a micro-confocal laser Raman spectrometer (inVia-Reflex, Renishaw, UK). An aqueous solution of ALGa microgels was used for evaluating the size distribution and zeta potential with nanoparticle size potentiometer (Zetasizer nanoseries, Malvern Instruments Ltd. Malvern, UK).

D. Loading of drugs in ALGa Microgels

Drug-loaded ALGa microgels were prepared relying on the water swelling properties of dry microgels [16]. In brief, the weighted lyophilized ALGa microgels (~10 mg) were mixed with DOX·HCl (0.50 mg/mL, 2 mL of aqueous solution), then shook using an oscillator (200 rpm, 37 °C, 48 h). After centrifugation and washing with DI water, the sample was freeze-dried for further use. The remaining concentration of DOX·HCl in the mixture solution (the washing liquid and the supernatant of centrifugation) was detected using ultra-micro UV-vis spectrophotometer (Thermo Scientific, NanoDrop One, America) with 479 nm wavelength. Drug encapsulation efficiency (E_{de}) and drug loading capacity (C_{dl}) of the ALGa microgels were calculated by Equation (1, 2) [17]. More than three parallel experiments were conducted.

$$C_{dl} = \frac{M_{drug}}{M_{dlm}} \times 100\% \tag{1}$$

$$E_{de} = \frac{M_{drug}}{M_{df}} \times 100\% \tag{2}$$

where E_{de}, C_{dl} are the drug encapsulation efficiency and the drug loading capacity, M_{dlm} is the weight of drug-loaded microgels, M_{de} is the weight of feeding drugs, M_{drug} is the weight of loaded drugs in the microgels.

E. NIR Laser Photothermal Characteristics and Magnetic Heating Property of ALGa Microgels

For exploration of the photothermal characteristics, the lyophilized ALGa microgels were irradiated with an 808 nm NIR laser under the power of 1.0 W/cm^2 for about 60 s, and the temperature changes within the laser irradiation range were monitored using infrared camera (SC620, FLIR, USA). In addition, the temperature response of ALGa microgels under alterative magnetic fields (≈150 KHz) was also monitored to investigate the inductive heating property.

F. X-ray Visibility of ALGa Microgels

The X-ray visibility of ALGa microgels with different concentrations (25, 50, 100, 200, 500, 1000 ppm, aqueous solution) were assessed using small animal micro CT (Pingsheng Medical Technology (Kunshan) Co., Ltd., SuperNova, China) with scanning voltage of 60 kV and current of 500 μA.

G. Biocompatibility of ALGa Microgels

Mouse embryonic fibroblasts (3T3 cells) and human melanoma cells (C8161 cells) were utilized to evaluate the *in vitro* cytotoxicity of ALGa microgels. The cells were cultured with standard medium, 89 % basal medium (DMEM (Dulbecco's Modified Eagle's Medium, Gibco) for 3T3 cells, DMEM (high glucose, Gibco) for C8161 cells) supplemented with 10 % FBS (fetal bovine serum, Biological Industries) and 1 % PS (Penicillin-Streptomycin Solution, Corning). The cells suspension in the culture medium with a density of 5×10^4 cells/mL were seeded in a 96-well plate (100 μL/well) and incubated in a humidified 5% CO$_2$ incubator at 37 °C for 24 hours . The culture solution was removed after cells adhering to the plate, and then replaced with 100 μL new culture medium containing various concentrations of ALGa microgels (0.1, 1.0, 10.0, and 100.0 mg/mL). Subsequently, after co-cultivated in humidified 5% CO$_2$ incubator at 37 °C for 24 or 48 hours , 10 μL of CCK-8 (Cell Counting Kit-8, Dojindo) with 100 μL of fresh medium were added to each well for staining (incubation at 37 °C in 5 % CO$_2$ for 2 hours) Finnally, the optical densities (OD) were measured using a multi-function microplate reader (Varioskan Flash, Thermo Scientific, Massachusetts, USA) at 450 nm and calculated cell viability by Equation (3).At least three parallel experiments were performed on each cell line during each time period.

$$C_v = \frac{OD_{eg} - OD_b}{OD_{cg} - OD_b} \tag{3}$$

The wells filled with only medium were regarded as background and cells that were cultured with no ALGa microgels were considered as control groups (defined as 0 and 100% cell viability). Where OD_{eg} is the OD of the experimental group, OD_{cg} is the OD of the control group, and OD_b is the OD of background.

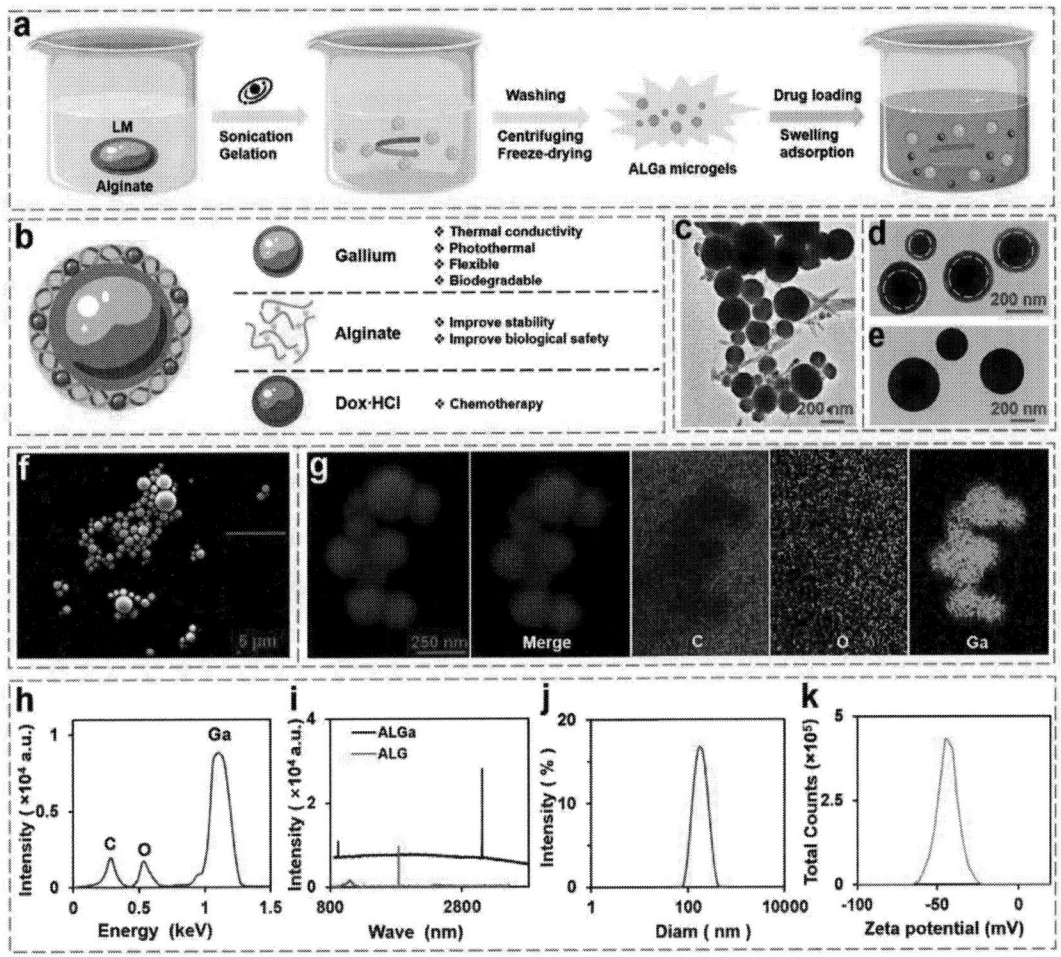

Fig. 1. Schematic illustration of both fabrication (a) and microstructure (b) of ALGa microgels. Characterization of ALGa microgels: The TEM images of ALGa microgels (c,d), and Ga nanoparticles (e); (f) The SEM image of the ALGa microgels; (g) The element mapping of C, O and Ga; (h) The EDS spectra of ALGa microgels; (i) The Raman spectroscopy of ALGa microgels and alginate; The size (j) and zeta potential (k) of ALGa microgels.

III. RESULTS AND DISCUSSIONS

In this work, we have successfully synthesized *in-situ* cross-linked liquid metal microgels (denoted as ALGa microgels), which can be served as multifunctional theranostics. During the ultrasonic dispersion process (Fig. 1a), the liquid metal particle cleavage to the nanoscale and the oxide film on the surface would rupture; Once the chemically active gallium expose and gallium ions release, the cross-linking of alginic acid and polyvalent metal ions can quickly self-assemble a gel on the liquid metal nanoparticle interface, thereby *in-situ* forming microgels, enhancing stability and improving biosafety [10]. After that, the commonly clinically used anticancer drugs DOX·HCl were loaded into the ALGa microgels using the swelling adsorption method due to the strong electrostatic interaction between the positively charged DOX·HCl and the negatively charged carboxyl group in alginate (Fig. 1a and b) [16, 18].

Next, the basic characterizations of ALGa microgels were performed. The typical TEM images of ALGa

microgels (Fig. 1c and d) showed that the gel shell (tens of nanometers in thickness) formed on the liquid metal nanoparticle interface compared with the bare Ga nanoparticle surface (Fig. 1e). And the typical spherical shape and smooth surface of ALGa microgels can be clearly observed from Fig. 1f. Besides, the SEM-EDS mapping and spectra were employed to analyze the elemental composition of ALGa microgels, as can be seen from Fig. 1g and h, clearly demonstrated that the main components of the particles are Ga while C and O from the alginate-gel mainly concentrate on the surface. Meanwhile, the Raman spectroscopy of ALGa microgels also indicates the success of *in-situ* synthesized liquid metal microgels (Fig. 1i). Afterward, the size distribution and zeta potential of ALGa microgels were tested by a nanoparticle size potentiometer. The normal distribution trend of size was observed with average value of 164.4 ± 15.4 nm (Fig. 1j), and the zeta potential reflects high stability with an average value of -44.2 ± 6.2 mV (Fig. 1k). In short, the developed ALGa microgels as a hydrogel composite material maintained a

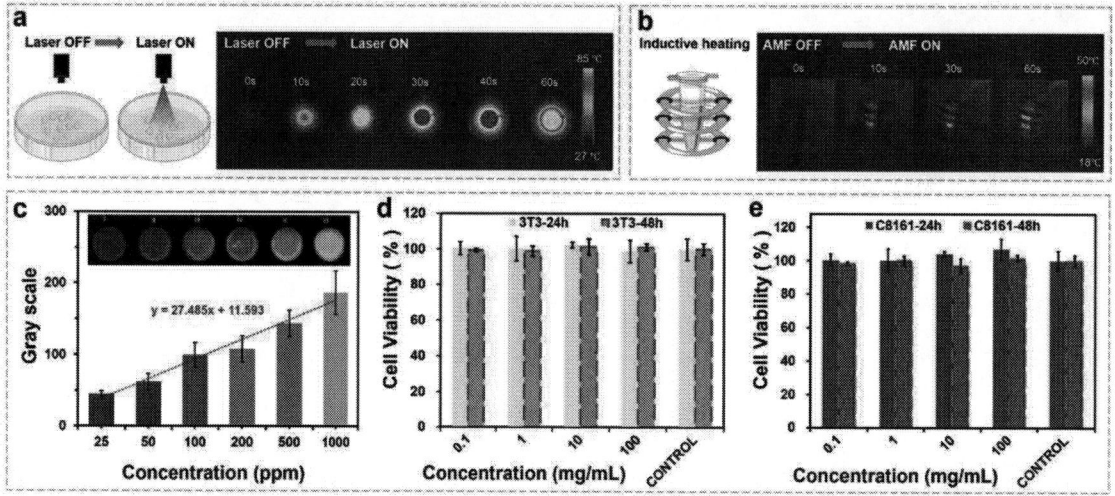

Fig. 2. (a) The photothermal susceptibility of ALGa microgels under NIR laser irradiation; (b) The inductive heating property of ALGa microgels under alterative magnetic field; (c) The CT images of ALGa microgels with varied content (sample i-viwith 25, 50, 100, 200, 500, 1000 ppm), and corresponding cray scale; The cell viability of 3T3 cells (d) and C8161 cells (e) after co-cultured with various ALGa microgels (0.1, 1.0, 10.0, and 100.0 mg/mL) for 24 h and 48 h.

good morphology, uniform size, high stability and may beneficial for subsequent drug loading.

As a kind of hybrid microgels, liquid metal microgels may exhibit complementary advantages of microgel and liquid metal. Herein, the preliminary functionalities of drug loading, stimulus-response and imaging were analyzed. Primarily, anticancer drug DOX·HCl was successfully loaded using the aforementioned method, drug encapsulation efficiency (E_{de}) and the drug loading capacity (C_{dl}) of ALGa microgels reached 46.67 ± 7.57 %, 4.63 ± 0.69 %, respectively, indicating that the ALGa microgels have nice drug loading capacity. Besides, the prepared ALGa microgels maintain superior photothermal conversion efficiency (Fig. 2a) as some previous advancements have substantiated, showing great potential for cancer photothermal therapy [12, 13, 19]. Meanwhile, ALGa microgels also possess excellent magnetic heating property due to its inherent metallic nature (Fig. 2b), which enables their unique application as non-magnetic agents in magnetic hyperthermia [20, 21]. Moreover, the ALGa microgels also retain the high-density property of liquid metal, making them radiopaque under X-ray/CT, which fetch up for the insufficient imaging abilities of the pure polymeric microgels to meet the current clinical diagnostic requirements. As shown in Fig. 2c, the CT brightness of ALGa microgels increased with the increase of concentration, and the corresponding gray scales allow a more intuitive observation. Furthermore, the *in vitro* cytotoxicity of ALGa microgels were tested through cells co-culture experiments using mouse embryonic fibroblasts (3T3 cells) and human melanoma cells (C8161 cells). As shown in Fig. 2d and e, the cytotoxicity of each experimental group can be intuitively judged by the cell viabilities. The cells viabilities of ALGa microgels after co-incubated for 24 h and 48 h were approximately 100 %, and no significant differences were found between the control groups with the experimental groups, indicating that these microgels do not

affect cell proliferation. These findings imply that the biocompatible ALGa microgels can be served as multifunctional theranostic agent, thereby making it possible to achieve synergistic treatment of chemotherapy and hyperthermia to eliminate cancer cells. Therefore, it could be of interest in biomedical applications, but further research is needed.

IV. CONCLUSION

In this work, we proposed a kind of *in-situ* synthesized liquid metal microgels, which may exhibit complementary advantages of microgels and liquid metals. Namely, such hybrid microgel can be served as fully flexible theranostics, showing encouraging potential in various biomedical applications, such as molecular imaging, drug carriers, biomedical devices and cancer therapy. As an attempt, we fabricated Ga-alginate microgel (denoted as ALGa microgel) by a facile method, and further loaded with anti-cancer drugs (DOX·HCl). The preliminary studies demonstrated that the biocompatible ALGa microgels maintained a good morphology, uniform size, high stability and may integrate functions of drug loading, stimulus-response and imaging, thereby making it possible to achieve synergistic treatment of chemotherapy and hyperthermia to eliminate cancer cells. In summary, this is an exploratory work for the future development of multifunctional liquid metal microgels, which may receive attention in biomedical applications.

REFERENCES

[1] F. Scheffold, "Pathways and challenges towards a complete characterization of microgels," *Nat Commun,* vol. 11, no. 1, p. 4315, Sep 4 2020.

[2] B. R. Saunders, N. Laajam, E. Daly, S. Teow, X. Hu, and R. Stepto, "Microgels: From responsive polymer colloids to biomaterials," *Adv Colloid Interface Sci,* vol. 147-148, pp. 251-62, Mar-Jun 2009.

[3] G. Agrawal and R. Agrawal, "Stimuli-Responsive Microgels and Microgel-Based Systems: Advances in the Exploitation of Microgel

Colloidal Properties and Their Interfacial Activity," *Polymers (Basel),* vol. 10, no. 4, Apr 9 2018.

[4] S. Seiffert, "Small but Smart: Sensitive Microgel Capsules," *Angewandte Chemie International Edition,* vol. 52, no. 44, pp. 11462-11468, 2013.

[5] D. Klinger and K. Landfester, "Stimuli-responsive microgels for the loading and release of functional compounds: Fundamental concepts and applications," *Polymer,* vol. 53, no. 23, pp. 5209-5231, 2012.

[6] C. W. Pester, A. Konradi, B. Varnholt, P. van Rijn, and A. Böker, "Responsive Macroscopic Materials From Self-Assembled Cross-Linked SiO2-PNIPAAm Core/Shell Structures," *Advanced Functional Materials,* vol. 22, no. 8, pp. 1724-1731, 2012.

[7] M. Zhang, S. Yao, W. Rao, and J. Liu, "Transformable soft liquid metal micro/nanomaterials," *Materials Science and Engineering: R: Reports,* vol. 138, pp. 1-35, 2019.

[8] R. C. Chiechi, E. A. Weiss, M. D. Dickey, and G. M. Whitesides, "Eutectic gallium-indium (EGaIn): a moldable liquid metal for electrical characterization of self-assembled monolayers," *Angew Chem Int Ed Engl,* vol. 47, no. 1, pp. 142-4, 2008.

[9] R. J. Larsen, M. D. Dickey, G. M. Whitesides, and D. A. Weitz, "Viscoelastic properties of oxide-coated liquid metals," *Journal of Rheology,* vol. 53, no. 6, pp. 1305-1326, 2009.

[10] X. Li *et al.*, "Liquid Metal Droplets Wrapped with Polysaccharide Microgel as Biocompatible Aqueous Ink for Flexible Conductive Devices," *Advanced Functional Materials,* vol. 28, no. 39, 2018.

[11] J. Yan, Y. Lu, G. Chen, M. Yang, and Z. Gu, "Advances in liquid metals for biomedical applications," *Chem Soc Rev,* vol. 47, no. 8, pp. 2518-2533, Apr 23 2018.

[12] X. Sun *et al.*, "Shape tunable gallium nanorods mediated tumor enhanced ablation through near-infrared photothermal therapy," *Nanoscale,* vol. 11, no. 6, pp. 2655-2667, 2019.

[13] S. A. Chechetka, Y. Yu, X. Zhen, M. Pramanik, K. Pu, and E. Miyako, "Light-driven liquid metal nanotransformers for biomedical theranostics," *Nat Commun,* vol. 8, p. 15432, May 31 2017.

[14] Q. Wu *et al.*, "Dual-Functional Supernanoparticles with Microwave Dynamic Therapy and Microwave Thermal Therapy," *Nano Letters,* vol. 19, no. 8, pp. 5277-5286, 2019/08/14 2019.

[15] X. Sun *et al.*, "Liquid Metal Microparticles Phase Change Medicated Mechanical Destruction for Enhanced Tumor Cryoablation and Dual - Mode Imaging," *Advanced Functional Materials,* 2020.

[16] D. Wang, Q. Wu, R. Guo, C. Lu, M. Niu, and W. Rao, "Magnetic Liquid Metal Loaded Nano-in-Micro Spheres as Fully Flexible Theranostics for SMART Embolization," *Nanoscale,* 10.1039/D1NR01268A 2021.

[17] Q. Wang, S. Liu, F. Yang, L. Gan, X. Yang, and Y. Yang, "Magnetic alginate microspheres detected by MRI fabricated using microfluidic technique and release behavior of encapsulated dual drugs," *Int J Nanomedicine,* vol. 12, pp. 4335-4347, 2017.

[18] J. Liu, Y. Zhang, C. Wang, R. Xu, Z. Chen, and N. Gu, "Magnetically Sensitive Alginate-Templated Polyelectrolyte Multilayer Microcapsules for Controlled Release of Doxorubicin," (in English), *Journal of Physical Chemistry C,* vol. 114, no. 17, pp. 7673-7679, May 6 2010.

[19] Y. Lu *et al.*, "Enhanced Endosomal Escape by Light-Fueled Liquid-Metal Transformer," *Nano Lett,* vol. 17, no. 4, pp. 2138-2145, Apr 12 2017.

[20] X. Wang *et al.*, "Printed Conformable Liquid Metal e - Skin - Enabled Spatiotemporally Controlled Bioelectromagnetics for Wireless Multisite Tumor Therapy," *Advanced Functional Materials,* vol. 29, no. 51, 2019.

[21] D. Wang *et al.*, "Non-Magnetic Injectable Implant for Magnetic Field-Driven Thermochemotherapy and Dual Stimuli-Responsive Drug Delivery: Transformable Liquid Metal Hybrid Platform for Cancer Theranostics," *Small,* vol. 15, no. 16, p. e1900511, Apr 2019.

Proceedings of the 16th Annual IEEE International
Conference on Nano/Micro Engineered and Molecular Systems
April 25-29, 2021

Design and Development of High Precision Magnetic Encoder Based on TMR MEMS Device

Bo Zhang，Yong Jiang，Yang Li，Xiaoli Chen，Xiaolong Wen

(Beijing Engineering Research Center of Detection and Application for Weak Magnetic Field，School of Mathematics and Physics，University of Science and Technology Beijing，Beijing 100083，China)

Abstract—**The MEMS magneto-sensitive chip based on the TMR principle is used as the sensitive element of the magnetic encoder. In terms of hardware development, after optimizing the design of the magnetic circuit, developing a precise rotating structure, designing a stabilized power supply circuit and a low-noise signal conditioning circuit, and digitally converting the analog voltage signal with high signal-to-noise ratio through analog-to-digital conversion and then entering the microprocessor. A high-precision magnetoelectric rotary encoder was designed and developed, and an automatic acquisition system was used to collect and store the encoder data. In terms of program design, a data processing, correction, and calibration algorithm was designed and written for the original output data of sensitive components, and MATLAB software was used to simulate, calculate and verify the angle data calculated by the encoder for further verification. The developed encoder has an angle measurement accuracy under actual working conditions. The encoder is installed on a high-precision time grid turntable system for testing, and the encoder's testing accuracy has reached a good level of 0.0469°.**

I. INTRODUCTION

Servo system is a general term for equipment that enables the system terminal executive structure to realize dimensional actions according to control instructions, such as displacement, rotational speed and torque[1]. It determines the precision, control speed and stability of automation machinery, so it is the core of industrial automation equipment. The encoder, as the front-end rotational position detection device of the servo control system, is the core component that determines the control accuracy of the servo system.

Magnetic encoder is characterized by non-contact signal induction, which has high resolution, high frequency response, long service life and high reliability, and can be applied to a variety of harsh use environments. It is more and more widely used in dozens of industrial technical fields and automotive electronics, artificial intelligence, robots, automatic detection and high-end equipment and industries [2].

Magnetic encoder has many principles to realize detection. How to further improve the measurement accuracy and realize high-efficiency mass production for sensitive element chips with different physical principles is one of the important issues for industrialization of this kind of sensor. Accord to different physical effects, magnetic encoder can be divided into Hall type[3], anisotropic magnetoresistance resistance [4], giant magnetoresistance resistance[5] and tunnel magnetoresistance resistance.Compared with other magnetic induction technologies, TMR technology has higher sensitivity and signal-to-noise ratio, lower power consumption, and is less affected by temperature, which is more suitable for developing high-end encoders with high accuracy and reliability.

In this paper, a magnetic sensor based on TMR principle is selected, through the design of structure, magnetic circuit and circuit, a data processing, correction and calibration algorithm is designed for its output signal under rotating magnetic field, which improves the accuracy of angle measurement by encoder.

II. PRINCIPLE OVERVIEW

TMR magneto-sensitive elements have the advantages that other types of magneto-sensitive elements do not have, and have excellent performances such as higher sensitivity, higher signal-to-noise ratio, greater magnetoresistance effect, and less influence by temperature, thus showing extremely high application value[6].

TMR refers to the effect that the tunneling resistance changes with the relative direction of ferromagnetic materials on both sides in ferromagnetic-insulator thin film-ferromagnetic structure materials.

The magnetization direction of the free layer can change with the change of the external magnetic field[7]. By using this effect, the magnetic sensor chip is placed at a certain position away from the surface of the magnetic drum. When the magnetic drum rotates along with the rotating shaft, the magnetic field generated by the magnetic drum changes. and the magnetization direction of the free layer changes with the rotating magnetic field, resulting in a change in the included angle between the magnetization direction of the free layer and the magnetization direction of the fixed layer,

978-1-6654-3008-1/21 $31.00 © 2021 IEEE   474

thus changing the magnetoresistance. By this method, the rotation angle of the magnetic drum can be measured [8].

III. SYSTEM DESIGN

A. Overall hardware design of encoder

In this paper, the magnetic MEMS device based on TMR effect developed by Tokyo Dengikagaku Kogyo K.K(TDK) is selected as the magnetic sensor of magnetic encoder. The main control module consists of STM32F373CCT6 single chip microcomputer of STMicroelectronics and related supporting circuits, TPS54302DDC power supply DC-DC of Texas instruments is selected as the power supply module. To further reduce the power ripple, SGM2036 chip is used to design LDO power chip for secondary voltage stabilization, and SN65HVD75 chip is used as RS-485 Modbus-RTU module to communicate with the upper computer.

Two Wheatstone bridges are designed in the magnetic MEMS device package. The basic principle is to use four TMR tunnel magnetoresistance modules R1, R2, R3 and R4, and a square circuit is composed of four resistors. when no current passes through the bridge, the bridge is in a balanced state. at this time:

$$R1 \cdot R2 = R3 \cdot R4. \qquad (1)$$

At this time, the required resistance value of TMR magnetoresistive module can be obtained and converted into voltage signal for output [9].

Fig. 1 is a schematic diagram of the internal Whiston bridge circuit of the magnetic MEMS device. In which white and gray arrows indicate the magnetic field direction of the fixed layer. It can be seen from the schematic diagram that the two Wheatstone bridges can measure the voltage values of SIN and COS at the included angle θ between the magnetic MEMS device and the measured magnetic field [10].

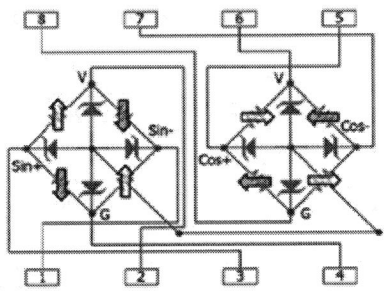

Figure 1. Whiston bridge circuit inside magnetic sensor chip

The internal structure of the magnetic encoder is shown in Fig. 2. The magnetic drum magnetized in radial direction is embedded with the rotating shaft to be measured. The magnetic MEMS device and the matching driving circuit module are welded on the circuit board [11]. The rotating

shaft drives the magnetic drum to rotate, thus forming a magnetic field signal with sine distribution changing with the angle above the magnetic MEMS device [12]. At the same time, the magnetic MEMS device detects the changed magnetic field and outputs the corresponding two-way sine and cosine voltage signals through the Wheatstone bridge [13].

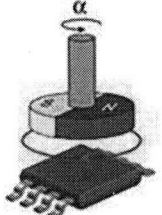 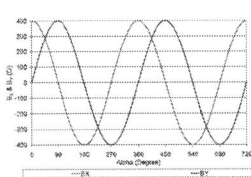

Figure 2. Schematic diagram of working principle of magnetic encoder

B. Design of MCU module and peripheral circuit

The microcontroller and peripheral circuits constitute a micro-control unit(Microcontroller Unit, MCU), which is used to collect and process the original data of magnetic sensors. In order to improve the detection accuracy of the magnetic encoder, the microprocessor with high bit ADC is selected as much as possible. At the same time, considering that some fields need encoder to measure the rotation angle at high speed, the microprocessor is required to have higher main frequency.

According to the above requirements, the STM32F373CCT6 single chip microcomputer developed by SGS-THOMSON Microelectronics is finally selected, which adopts the ARM®32-bit Cortex®-M4 CPU with FPU, with the main frequency of 72MHz, DSP instruction set and up to 84 high-speed I/O ports (GPIO) [14].

Modbus RTU based on serial communication is a compact communication mode which adopts binary data representation, with CRC cyclic redundancy check after each data frame format [12], and uses serial data transmission to communicate. RS-485 protocol is a serial communication standard, which defines the voltage, impedance, signal level and other parameters in the circuit, but does not define the software protocol, so it is convenient and flexible to use and easy to operate [15]. In this project, RS-485 serial communication mode is adopted for communication hardware circuit [14], SN65HVD75 chip is adopted for level conversion, Modbus-RTU protocol is adopted for communication protocol.

IV. SOFTWARE DESIGN

A. Algorithm architecture design

The magnetic sensitive element is designed with double Wheatstone bridges inside, one bridge is a sine bridge and the other is a cosine bridge, that is, when the magnetic field above the magnetic sensitive element rotates and causes the magnetic field to change, the output voltage signals of the two bridges differ in phase by 90°, and the angle of magnetic field rotation can be calculated by finding the arctangent function (arctan) corresponding to the sine and cosine values.

978-1-6654-3008-1/21 $31.00 © 2021 IEEE

In the framework based on the above principle, in order to further improve the encoder angle measurement accuracy, it is proposed to calibrate, process and angle solve the TMR magnetic sensitive element raw output data waveform by the following steps.

(1)Standardization of the original output data waveform: Due to the influence of mechanical assembly errors on the magnetic sensitive element, the magnetic drum and the rotating shaft, it is impossible to achieve complete parallelism between the magnetic sensitive element and the magnetic drum, which will result in the voltage signal induced by the magnetic sensitive element during the rotation of the magnetic drum not being a strictly standard sine and cosine curve, resulting in deviations in angle resolution, which will affect the measurement accuracy of the encoder. For these reasons, the output data waveform of the magnetic sensitive element is first corrected to a standard sine and cosine curve.

(2)Data waveform offset correction: Due to the process, the Wheatstone bridge cannot be prepared as a completely physically symmetrical circuit, i.e., the voltage output is not zero at zero magnetic field, resulting in an output sine cosine voltage curve that is not symmetrical about the zero point, thus causing inconvenience to subsequent signal processing. An algorithm is needed to correct the offset of the output signal of the magnetically sensitive element.

(3)Data waveform normalization: Due to the preparation process, the sensitivity of the TMR magnetic sensitive devices on the two bridges cannot be identical, i.e., the output voltage amplitudes of the two Wheatstone bridges cannot be exactly equal under the same magnetic field environment, so the normalization algorithm is needed to equalize the amplitudes of both signal waveforms to "1" to provide standard data for the subsequent solution angle.

(4)Correction of phase difference of sine and cosine data waveforms: Due to the device.

(5)Angle solving of the corrected data waveform: The corrected data waveform is solved using the arctangent function (arctan) algorithm to derive the drum rotation angle value.

B. Algorithm principles and data processing

After collecting the raw data output from the sensitive element, curve fitting of the raw data waveform is performed. It can be seen that the raw data waveform is a graph of two waveforms close to the trigonometric function, but compared with the standard sine and cosine function, the raw data waveform is obviously deviated in terms of phase, amplitude symmetry and period symmetry, so the fitted raw data waveform needs to be normalized to sine and cosine.The original data waveform is first corrected in the amplitude offset of the detected voltage to make it

symmetric about the X-axis. This process is shown in (2)-(5),

$$c_1 = \frac{\mathrm{Max}[V\cos(\theta)] + \mathrm{Min}[V\cos(\theta)]}{2} \quad (2)$$

$$c_2 = \frac{\mathrm{Max}[V\sin(\theta)] + \mathrm{Min}[V\sin(\theta)]}{2} \quad (3)$$

$$V'\cos(\theta) = V\cos(\theta) - C_1 \quad (4)$$

$$V'\sin(\theta) = V\sin(\theta) - C_2 \quad (5)$$

After the correction by this step, the data waveform can be transformed as shown in Fig. 3.

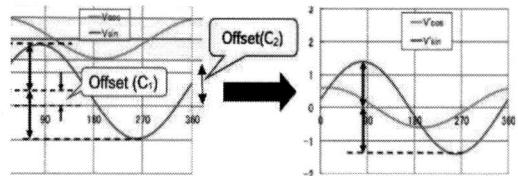

Figure 3. Schematic diagram of the amplitude shift correction of the original data waveform

The second step is to simultaneously normalize the ordinate voltage amplitudes of the two groups of sine and cosine waveforms of the original data, and reduce the amplitudes of different waveforms to approximately unit 1 in equal proportion. The process is shown in (6) - (9).

$$A_1 = \frac{\mathrm{Max}[V\cos(\theta)] - \mathrm{Min}[V\cos(\theta)]}{2} \quad (6)$$

$$A_2 = \frac{\mathrm{Max}[V\sin(\theta)] - \mathrm{Min}[V\sin(\theta)]}{2} \quad (7)$$

$$V''\cos(\theta) = \frac{1}{A_1} \cdot V'\cos(\theta) \quad (8)$$

$$V''\sin(\theta) = \frac{1}{A_2} \cdot V'\sin(\theta) \quad (9)$$

After this step is corrected, the waveform amplitudes of the two groups of original data can be corrected in equal proportion, as shown in Fig. 4.

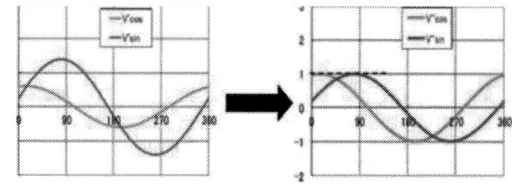

Figure 4. Modification diagram of waveform amplitude of original data

A theta angle varying in the range of 0° to 360° should have a standard set of sine and cosine waveforms with a period of 360°, an amplitude of unit 1, and a phase difference of 90° between its SIN and COS values (vertical coordinates) and the corresponding angle (horizontal coordinates).At this point, the vertical coordinates of the two functions of the waveform in the same coordinate system for the arctan (tangent) value, you can get the corresponding horizontal coordinate value, that is, the corresponding angle value.For the output data waveform of the magnetic encoder in this project, the phase difference between the two sine and cosine waveforms is corrected to 90° to perform the angle solution.0°-360° range change of theta angle, its SIN and COS vector relationship as shown in Fig. 5, and due to the magnetic sensing element preparation process and mechanical assembly errors and other reasons, resulting in the sine and cosine bridge output of the two sets of original data sine and cosine waveform phase difference is not the standard 90 °, at this time, assuming that the phase difference between $V'' \cos(\theta)$ and $V'' \sin(\theta)$ function waveform after the previous step correction is α, and then according to the principle of vector superposition According to the principle of vector superposition, the phase difference α is added to and subtracted from the horizontal coordinates of the two waveforms, i.e., the waveforms are shifted to the left and right by α degrees, and the two new function waveforms are added and subtracted to obtain two new function waveforms $V_{C+S}(\theta)$ and $V_{C-S}(\theta)$. At this time, the phase difference of these two function waveforms can be corrected to the standard 90°.

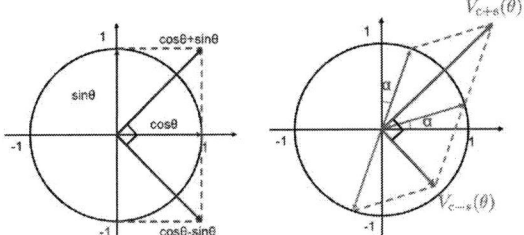

Figure 5. Schematic diagram of the vector form of phase difference correction

The process of phase difference correction using vector addition and subtraction is shown in (10) and (11).

$$V_{C+S}(\theta) = V\cos(\theta+\alpha) + V\sin(\theta-\alpha) \quad (10)$$

$$V_{C-S}(\theta) = V\cos(\theta+\alpha) - V\sin(\theta-\alpha) \quad (11)$$

However, for $V_{C+S}(\theta)$ and $V_{C-S}(\theta)$, the phase difference between the two new function waveforms is 90°, but because it is obtained by adding and subtracting the $V'' \cos(\theta)$ and $V'' \sin(\theta)$ function waveform vectors in the previous step, the amplitude is no longer the standard "1" due to the superposition of the vectors, so it is necessary to use the amplitude correction algorithm to correct the amplitude of the two function waveforms again, the correction algorithm formula is similar to the above steps, as shown in (12)-(15)

$$B_1 = \frac{\text{Max}[V_{C+S}(\theta)] - \text{Min}[V_{C+S}(\theta)]}{2} \quad (12)$$

$$B_1 = \frac{\text{Max}[V_{C-S}(\theta)] - \text{Min}[V_{C-S}(\theta)]}{2} \quad (13)$$

$$V'_{C+S}(\theta) = \frac{1}{B1} V_{C+S}(\theta) \quad (14)$$

$$V'_{C-S}(\theta) = \frac{1}{B2} V_{C-S}(\theta) \quad (15)$$

Fig. 6 shows the vector form of data waveform phase difference correction and amplitude correction. It can be seen that the new function waveforms $V'_{C+S}(\theta)$ and $V'_{C-S}(\theta)$ generated by the original data waveform after multiple corrections at this time have a phase difference of 90° and an amplitude of "1", which are already basically standard sine and cosine functions, and the arctan algorithm can be used to find the forward and backward tangent values to find the angle value corresponding to each acquisition point of the original data waveform.

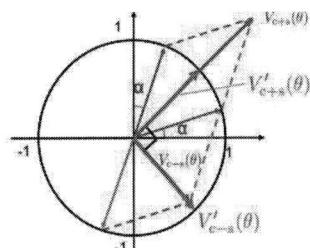

Figure 6. Schematic diagram of the vector form of phase difference correction and amplitude correction

The arctan values of the $V'_{C+S}(\theta)$ and $V'_{C-S}(\theta)$ function waveforms can be calculated directly using (16), which is the angular value θ_{cal} obtained by processing, correcting, and solving the raw data collected by the magnetic encoder.

$$\theta_{cal} = \arctan\left(\frac{V'_{C+S}(\theta)}{V'_{C-S}(\theta)}\right) \quad (16)$$

Since the two sets of function waveforms are vectorized and subtracted during the phase difference correction, the function waveform has a 45° offset in the direction of the horizontal coordinate, so when solving for the value θ_{cal}, 45° needs to be subtracted, i.e., as shown in (17),

$$\theta_{cal} = \arctan\left(\frac{V'_{C+S}(\theta)}{V'_{C-S}(\theta)}\right) - 45° \quad (17)$$

(18)-(21) are used to further process and calibrate the solved angle values θ_{cal}.

$$\text{AngleError} = \theta_{cal} - \theta_{act} \quad (18)$$

$$D = \frac{\text{Max}[\text{AngleError}] - \text{Min}[\text{AngleError}]}{2} \quad (19)$$

$$\theta_{cal}(\theta_{cal} > \theta_{act}) = \theta_{cal}(\theta_{cal} > \theta_{act}) - D \quad (20)$$

$$\theta_{cal}(\theta_{cal} < \theta_{act}) = \theta_{cal}(\theta_{cal} < \theta_{act}) + D \quad (21)$$

After the processing and calibration of the above steps, the error between the solved angle value θ'_{cal} and the actual angle value θ_{act} will be reduced significantly, and if necessary, the above processing and calibration steps can be repeated to further reduce the error.

V. DATA ACQUISITION

The magnetic encoder studied in this paper is automatically data collected using a fully automated encoder calibration turntable system designed and developed in our laboratory. The system allows setting the acquisition speed, number of turns and acquisition interval, and collecting different numbers of raw data points (i.e. the output voltage value of the magnetic sensitive element sine cosine bridge) for the magnetic encoder by setting different parameters.

As shown in Fig. 7, the assembled encoder is mounted on the high-precision rotary table, and the rotary table rotation angle is controlled by the upper computer control system, and the voltage signals of the two sine and cosine bridges output from the encoder are collected and stored at the same time.

978-1-6654-3008-1/21 $31.00 © 2021 IEEE

Figure 7. Magnetic encoder and data acquisition experimental platform

The acquisition system is set up to collect 100 points evenly during one rotation of the encoder shaft from 0 to 360°, i.e., a set of voltage signals and standard angle values of the rotary table are collected at an interval of 3.6°and stored in the flash memory of the microcontroller.

A. MATLAB Simulation and Inspection

The raw data of the magnetic encoder magnetic sensing chip is analyzed and processed using MATLAB's ability to quickly analyze and process large amounts of data, and the data correction algorithm studied in this paper is used to perform several simulations with different parameters to ensure and this algorithm can achieve the desired goal.

As can be seen from （20）, after the simulation test, the calculated Angle value and the actual Angle value are almost equal. At the same time, the relation formula between the calculated Angle value θ_cal and the actual Angle value θ_act is obtained through Matlab linear fitting

$$\theta_{cal} = 1.00000563 * \theta_{act} + 0.02326594 \qquad （22）$$

From the above, it can be seen that the relationship between the two is a straight line with a slope of 1.00000563 and an offset of 0.02326594, which can approximate θ _cal ≈ θ _act, i.e. calculate the Angle value ≈ the actual Angle value.This formula fully shows that the magnetic encoder data processing and correction algorithm designed in this paper can achieve the goal more ideal.

B. Encoder turntable test

In order to further verify the measurement accuracy of the encoder in actual working conditions, the collected data are burned into the flash storage area of the encoder microcontroller, and the Angle is calculated and output in real time through the internal algorithm of the microcontroller.

In the first method, the 100 groups of data collected were fitted directly through Step (5) in the algorithm architecture design without algorithm modification, and the Angle value was calculated by arctangent trigonometric function, which was compared with the standard Angle of the turntable to analyze the error.

100 groups of data from the second way, according to the algorithm in the design of architecture (1) (2) (3) (4) (5) steps are cosine standardization, normalization of deviation correction, phase difference correction, processing, etc., finally passed the arctangent trigonometric functions, it is concluded that the encoder real-time output value point of

view, and compared with standard turntable Angle, analyzing the error.

The high-precision time-grid turntable of the laboratory is selected as the standard turntable for the test, and the Angle positioning accuracy of the turntable can reach 1 arc second, that is, about 0.0003°.Install the magnetic encoder on the rotating shaft of the turntable and ensure the installation concentricity.In the process of rotation, the output solution Angle of the magnetic encoder is tested, and the output Angle of the encoder and the standard Angle of the turntable are recorded in real time, and the error of the output Angle of the encoder is analyzed.

Fig. 8 shows the developed encoder, and Fig. 9 shows the physical drawings of magnetic encoder and high-precision time-gate turntable system during the test.

Figure 8. Magnetic encoder

Figure 9. Physical diagram of magnetic encoder and high-precision time grid turntable system during test

After the experimental test, the detailed data of the solution Angle and standard Angle obtained by the two methods before and after the test are shown in Table 1.

Table 1 Values of calculated angles and standard angles before and after correction

Corrected angle (°)	Uncorrected angle (°)	Standard angle (°)
−0. 046899101	0. 5	0. 000
29. 95622641	30. 24623729	30. 001
60. 04422916	59. 89813419	60. 003
89. 98865088	89. 73130561	90. 000
120. 0397437	119. 9993795	120. 001
149. 9887801	150. 276908	150. 000
179. 9880458	180. 1086395	180. 003
209. 9940365	209. 6714314	210. 000
239. 9954999	239. 5058863	240. 002
269. 9738471	269. 7256388	270. 001
300. 0200934	300. 1274426	300. 002
330. 024742	330. 3764632	330. 001

Corrected angle (°)	Uncorrected angle (°)	Standard angle (°)
360. 0123232	360. 3965464	360. 002

In order to more visually display the improvement effect of the output accuracy of the magnetic encoder modified by the algorithm, the distribution of the calculation Angle error obtained by the two methods was placed in the same coordinate system, and the comparison of the calculation Angle error of the magnetic encoder with or without the algorithm correction is shown in Fig. 10.

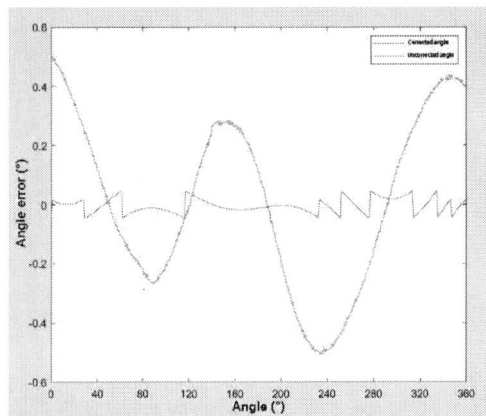

Figure 10. Comparison of Angle errors in solution with
or without algorithm correction

It can be seen from the above content that after algorithm processing and modification, the magnetic encoder has achieved a better measurement effect and its positioning accuracy is better than 0.0469°, which has obvious technical advantages compared with the existing conventional encoder products in related fields.

VI. CONCLUSION

This article research and design a kind of based on TMR magnetoelectric type magnetic susceptibility chips rotary encoder, aiming at its raw output data to write a data processing, correction and calibration algorithm, using MATLAB software for the simulation and test, and further on the grid when high precision turntable system has carried on the actual test, finally the output of the encoder accuracy reached 0.0469 °, compared to the conventional encoder product positioning accuracy of 0.3 ° in this field, this paper developed at the encoder.The work in this paper provides a new design idea and research method for the encoder field, which helps to promote the research progress of high-end encoder products in China to a certain extent.

VII. REFERENCES

[1] W. Wang, B. Xie, Z. Zuo and H. Fan, "Adaptive Backstepping Control of Uncertain Gear Transmission Servosystems With Asymmetric Dead-Zone Nonlinearity," IEEE Transactions on Industrial Electronics, vol. 66; no. 5; pp. 3752-3762, 2019.

[2] M. Spanghero, G. Magni, E. Boselli, M. Piombino, F. Mason and G. Cozzi, "Prediction of metabolisable energy content of commercial total mixed rations (TMR) for lactating dairy cows based on gas production measured into two TMR fractions," Anim. Feed Sci. Tech., vol. 226, pp. 65-70, 2017.

[3] P. BalasubramanianD.L. Maskell and N.E. Mastorakis, "Asynchronous early output majority voter and a relative-timed asynchronous TMR implementation," Microelectron. Reliab., vol. 114, pp. 113781, 2020.

[4] W. Deng and J. Yao, "Asymptotic Tracking Control of Mechanical Servosystems with Mismatched Uncertainties," IEEE/ASME Transactions on Mechatronics, pp. 1, 2020.

[5] Schüthe T, Albounyan A, Riemschneider K R. Two-Dimensional Characterization and Simplified Simulation Procedure for Tunnel Magnetoresistive Angle Sensors[C]//2019 IEEE Sensors Applications Symposium (SAS). IEEE, 2019: 1-6.

[6] S. HanG. Kong and S. Choi, "A Detection Scheme With TMR Estimation Based on Multi-Layer Perceptrons for Bit Patterned Media Recording," IEEE Transactions on Magnetics IEEE Transactions on Magnetics, vol. 55; 55, no. 7; 7, pp. 1-4, 2019.

[7] Bhaskarrao N K, Anoop C S, Dutta P K. Oscillator-less Direct-Digital Front-End Realizing Ratiometric Linearization Schemes for TMR-Based Angle Sensor[J]. IEEE Transactions on Instrumentation and Measurement, 2019, 69(06): 3005-3014.

[8] Y. Liu, W. Jose, L. Zhu, M. Dovek and P. Wang, "Effects of mechanical stress on the resistance of TMR devices," J. Magn. Magn. Mater., vol. 310, no. 2, Part 3, pp. e924-e926, 2007.

[9] G. Basile, S. Mottard, D. Tremblay and E. Boghossian, "Targeted Muscle Reinnervation (TMR) in Extremity Malignant Tumors Requiring Amputation," Operative Techniques in Orthopaedics, vol. 30, no. 2, pp. 100803, 2020.

[10] Shen Z H, Min H. Combination method of DC-DC converter and LDO to improve efficiency and load regulation[J]. Electronics Letters, 2011, 47(10): 615-617.

[11] Y.K. Zheng, et al., "Side Shielded TMR Reader With Track-Width-Reduction Scheme," IEEE Transactions on Magnetics IEEE Transactions on Magnetics, vol. 42; 42, no. 10; 10, pp. 2303-2305, 2006.

[12] W. Miao, X. Liu, K.H. Lam and P.W.T. Pong, "DC-Arcing Detection by Noise Measurement With Magnetic Sensing by TMR Sensors," IEEE Transactions on Magnetics IEEE Transactions on Magnetics, vol. 54; 54, no. 11; 11, pp. 1-5, 2018.

[13] N. WongpanitP. O. Å. Persson and S. Tungasmita, "Etching behaviors of tunneling magnetoresistive (TMR) materials by ion beam etching system," Materials Today: Proceedings, vol. 5, no. 7, Part 1, pp. 15186-15191, 2018.

[14] N.T.H. Nguyen, et al., "Reconfiguration Control Networks for FPGA-based TMR systems with modular error recovery," Microprocess. Microsy., vol. 60, pp. 86-95, 2018.

[15] M.C. Collivignarelli, F. Castagnola, M. Sordi and G. Bertanza, "Treatment of sewage sludge in a thermophilic membrane reactor (TMR) with alternate aeration cycles," J. Environ. Manage., vol. 162, pp. 132-138, 2015.

Gap in pagination due to unavailable papers.

Pages 480-490

Proceedings of the 16th Annual IEEE International
Conference on Nano/Micro Engineered and Molecular Systems
April 25-29, 2021

Frequency Matching System of MEMS Gyroscope Based on Fuzzy Control Strategy

Yifang Liu[1], Ruimin Liu[1], Gang Fu[1], Xin Yan[2] , Junyu Chen[1] and Gaofeng Zheng[1*]

Abstract—Due to the deviation between the actual value and the theoretical value of the structure size of the micromachined gyroscope, the difference of material characteristics and the change of environmental factors, the resonant frequencies of the driving mode and the detection mode of the micromachined gyroscope are mismatched, which will greatly reduce the sensitivity of angular velocity detection. Based on the characteristics of the parameter uncertainty and strong nonlinearity of the gyroscope modal frequency matching system, a fuzzy control strategy was designed to optimize the modal frequency matching process. The Simulink simulation model of the mode matching fuzzy control was researched. The simulation results showed that the phase difference between the two mode output signals was 90 ° and the response time of the whole fuzzy closed-loop control system was less than 0.4 s. Based on FPGA, the fuzzy controller of mode matching was realized, and the frequency difference was controlled within 5 Hz.

BACKGROUND

With the development of MEMS (Micro-Electro-Mechanical Systems) technology, MEMS gyroscopes with small size, low cost, and high precision have appeared. In 1991, the first micromachined gyroscope was born. In the nearly 30 years, the MEMS gyroscopes have developed greatly. On the one hand, the resolution sensitivity and bias stability of the MEMS gyroscopes had been improved by adjusting the structural design and processing technology, and the Brownian noise had been reduced [1-4]. Different process methods had been used to combine the MEMS gyroscope mechanical structure and signal control circuit on the chip [5-8], which improved the miniaturization and digital integration of the system.

However, there are many disturbing factors such as the deviation between the actual value of the structure size and the designed value, the difference in material characteristics, and the drift of ambient temperature and pressure. These factors will change the vibration frequency of the resonator of the micromachined vibratory gyroscope, which makes the frequencies of both driving mode and detection mode mismatch. The reduced sensitivity of the MEMS gyroscopes becomes a bottleneck for its high-precision applications. Therefore, it is important to study the decoupling of the internal signals and the frequency matching control for improving the measurement accuracy and sensitivity of MEMS gyroscopes.

In many cases, the driving mode and detection mode of micromachined gyroscopes were controlled independently. The driving mode was controlled to drive the gyro to vibrate normally and the detection mode was controlled to measure the input angular velocity. There had been a relatively mature scheme for closed-loop control of drive mode. However, the closed-loop control system for detecting mode was complicate, which required the modal matching while improving the stability and increasing the bandwidth. For performing modal frequency matching control by electrostatic force feedback method, Chunhua used fuzzy algorithm and neural network algorithm to design a controller through Labview and FPGA (Field Programmable Gate Array) to establish the relationship between the input voltage and the difference between the two modal resonance frequencies. It needed to collect experimental data in different parameters in advance, and to look up the input voltage according to the frequency difference in the modal matching process [9, 10]. The relationship between frequency difference and input voltage could also be obtained indirectly by the stiffness, damping, amplitude or phase of the gyro. Prikhodko[11] and Yangcheng[12] established the relationship between the amplitude or phase of the output signal at the detection end and the input voltage. Ezekwe[13] input an oscillating signal at the detection end to analyze the matching degree between both the amplitude and phase and the drive signal. Chang[14] introduced drivers at the detection end to analyze the response signal and adjusted the frequency through structures such as phase-locked loop (PLL). Prikhodko[15] analyzed the completion progress of modal matching based on the jitter output signal of the quadrature electrode.

The structural parameters of the MEMS capacitive gyroscopes are relatively complex. It is difficult to study the control system based on the mathematical model built from the internal structure only. In addition, the control system would be disturbed easily by external pressure and ambient temperature. Therefore, higher requirements are put forward for the frequency matching system control of the drive mode and detection mode of gyroscope, which can maintain the stability, accuracy and timeliness of the system while disturbed by the both internal and external factors. The parameters of the gyroscope modal frequency matching system are uncertain. They change with strong nonlinear characteristics. So in this paper, a fuzzy control strategy was designed to optimize the process of modal frequency matching.

THE PRINCIPLE OF MODAL MATCHING OF MEMS CAPACITIVE GYROSCOPE

When the MEMS capacitive gyroscope detected the angular velocity, it could be equivalent to the mass-spring-damping system, which is shown in Fig. 1. The driving direction is perpendicular to the detection

1- Department of Instrumental and Electrical Engineering, Xiamen University, Xiamen, CHINA.
2- Aviation Key Laboratory of Science and Technology on Inertia, FACRI, Xi'an, CHINA.
*Contacting Author: (phone: +86-592-2194957 ; email: zheng_gf@xmu.edu.cn)

978-1-6654-3008-1/21 $31.00 © 2021 IEEE

direction. In other words, the phase difference is 90°. The gyroscope vibrates linearly along the driving direction. When it feels the angular velocity perpendicular to the vibration plane, the gyroscope will vibrate linearly in this direction due to Coriolis force. Therefore, the input angular velocity can be measured by the voltage.

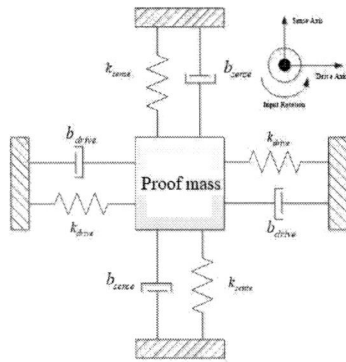

Fig. 1. Schematic diagram of mass spring damping system

The driving end of gyroscope output a current signal under the force F(t)=Fsin($\omega_{drive}t$). After the 90° phase shift, a voltage signal with amplitude of about 1V was obtained through a transimpedance amplifier:

$$V_{drive2}(t) = -A_x \omega_{drive} R_f V_P \frac{C_{x0}}{d_0} \cos\omega_{drive} t \quad (1)$$

Where A_x was the amplitude in the driving direction, $\omega_{drive=\sqrt{k_{drive}/m_{drive}}}$ was the natural angular frequency of the driving mode, R_f was the I-V(current-voltage) conversion resistance of the transimpedance amplifier, V_P was a fixed DC voltage applied to the fixed plate when driving, $\frac{C_{x0}}{d_0}$ was the solution result of the comb tooth structure in the driving direction. The amplitude and phase of the output voltage signal were extracted in the driving direction for further processing.

Then the signal in the detection direction was extracted. There were manufacturing process errors, which caused the detection mode of the gyroscope to be affected by the orthogonal error force before the angular velocity was input. And there was an upward movement speed in the movable comb teeth of the comb-tooth capacitor. Therefore, the detection mode output a weak current signal, which was transformed into a weak output voltage by cross capacity amplifier:

$$V_{sense1}(t) = \int \frac{I_{senseout}}{C_m} dt = \int V_P v_y \frac{C_{y0}}{d_1}$$

$$= \frac{C_{y0} V_P A_y A_x \omega_{drive}^2 m_{drive} \sin\theta}{C_m d_1} \cos(\omega_{drive}t + \delta_2) \quad (2)$$

Where C_m was the value of the cross capacity, $\frac{C_{y0}}{d_1}$ was the solution result of the comb tooth structure in detection direction, A_y was the amplitude in detection direction, m_{drive} was the effective mass of driving mode, θ was the angle between the actual driving direction and designed driving direction of the gyroscope, δ_2 was the offset phase:

$$\delta_2 = \arctan\frac{b_{sense}\omega_{drive}}{k_{sense} - m_{sense}\omega_{drive}^2} = \arctan\frac{2\xi(\omega_{drive}/\omega_{sense})^2}{1 - \frac{\omega_{drive}^2}{\omega_{sense}^2}} \quad (3)$$

Demodulate driving signal V_{drive2} and detection signal V_{sense1} through phase detector, and extract the phase difference information between driving signal and detection signal. After processed by the multiplier and the second-order low-pass filter of the phase detector, an error signal containing phase difference information was obtained:

$$V_{error}(t) = -\frac{m_{drive}R_f C_{x0} C_{y0} V_P^2 A_x^2 A_y \omega_{drive}^3 \sin\theta}{C_m d_0 d_1} \cos\delta_2 \quad (4)$$

The purpose of modal matching control of the gyro was to achieve $\omega_{drive} = \omega_{sense}$. It could find $2\xi(\omega_{drive}/\omega_{sense})^2 / 1 - \frac{\omega_{drive}^2}{\omega_{sense}^2} \to \infty$ and $\delta_2 = 90°$ by formula (3). And formula (4) showed that when it achieved modal matching, $V_{error}(t) = 0$. The controller of the closed-loop feedback was set by the error signal containing phase difference information. The error signal was fed back to the gyro through the controller output DC offset signal. Change the resonant angular frequency of the detection mode through a DC offset signal combining formula (5).

$$\omega_s = \sqrt{\frac{k_{eff}}{m_{sense}}} = \sqrt{\frac{k_{sense} + KV^2}{m_{sense}}}, K < 0 \quad (5)$$

When the error signal was adjusted to zero, the gyro achieved modal matching.

THE SIMULATION OF FUZZY CONTROLLER FOR MODE MATCHING

The principle block diagram of the modal matching fuzzy controller of the micromechanical gyroscope was shown in Fig. 2.

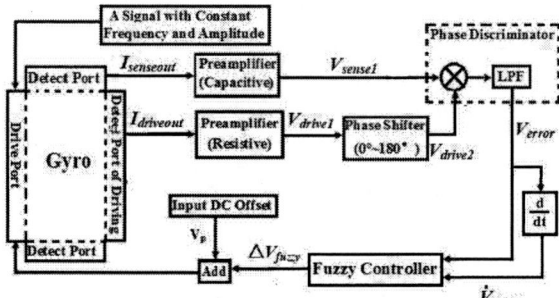

Fig. 2 The principle block diagram of the modal matching fuzzy controller

The phase relationship between the quadrature error signal and the detection signal of the drive end was used to analyze the modal matching of the system. The modal matching system was divided into three parts. The first part extracted two modal weak current signals $I_{driveout}$ and $I_{senseout}$. Then got V_{drive1} by transimpedance amplification and got V_{sense1} by transcapacity amplification. The second part demodulated V_{drive2} and V_{sense1} to obtain an error signal containing the phase difference between the two signals. The third part calculated the relationship between the error signal and the input DC offset signal of the gyroscope by the fuzzy controller. In order to improve the response speed and stability of the modal matching controller, the error signal V_{error} and the error signal change rate $\dot{V}_{error}$ were used as the input of the fuzzy controller. According to the fuzzy control strategy, the increment $\triangle V_{fuzzy}$ to the input DC offset was calculated and fed back to the gyro.

The Simulink simulation block diagram of the modal matching fuzzy control system was shown in Fig. 3. According to the frequency magnitude of the two modes, the simulation step was set to 10^{-6}, simulation time was set to 10 s.

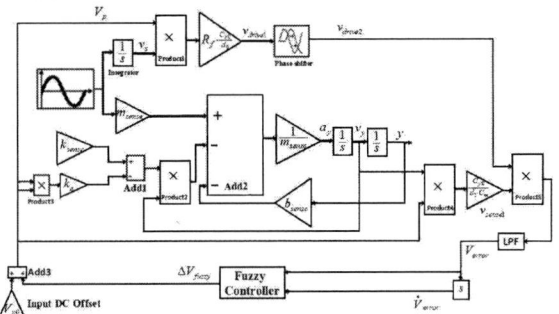

Fig. 3. Simulink simulation block diagram of modal matching fuzzy control

In the debugging process, the debugging result was judged by the input error signal of the fuzzy controller V_{error} and the output DC offset incremental signal $\triangle V_{fuzzy}$. The two signals in the middle of the debugging and after the debugging were shown in Fig. 4 and Fig. 5, respectively. It could be seen that the response time of the entire closed-loop system was less than 0.4s when using fuzzy control to achieve modal matching simulation. The oscillation of the input error signal was eliminated, and the DC offset output signal also reached a stable value smoothly after the fuzzy control.

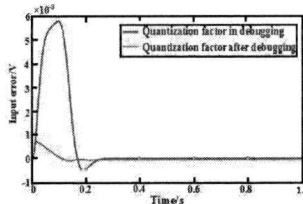

Fig. 4. The comparison of input error signal between mid-stage and finished debugging of quantization factor

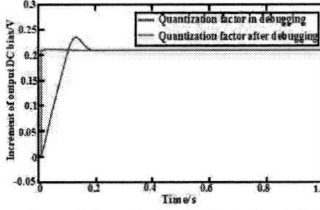

Fig. 5 The comparison of the output DC offset signals between mid-stage and finished debugging of quantization factor

After the simulation, the quantization factor of V_{error} was 340, the quantization factor of $\dot{V}_{error}$ was 0.001, and the scale factor of $\triangle V_{fuzzy}$ was 0.1. The method of fuzzy quantification of certainty was designed as the "and" method to "take the smallest". The "contain" method was used to "take the smallest", the "merge" method was used to "take the bigger" and the "or" method was used to "take the bigger". The final membership degree value was

calculated through the quantification of the certainty degree, and the center of gravity method was selected for defuzzification. The outputs of the two modes of the fuzzy closed-loop control system with the adjusted parameters were shown in Fig. 6, respectively. In the process of gyro modal frequency matching, the output of the detection mode of fuzzy control system always increased steadily. After reaching stability, the phase difference between the two modes reached 90°, which showed that the system had achieved mode matching.

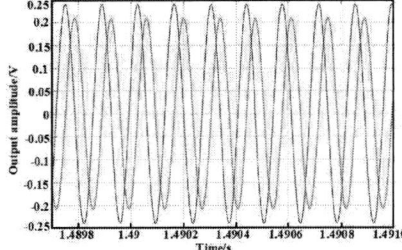

Fig. 6 Partially zoom in waveform when stable: drive modal (blue) and detection modal (pink)

DESIGN AND EXPERIMENT OF FREQUENCY MATCHING CONTROLLER

(1) The hardware structure of the modal matching controller

The internal structure of the modal matching fuzzy controller was shown in Fig. 7. There was a gyro signal generation analog module inside the FPGA before the phase detection module and the control algorithm. The signal analog module generated a drive signal and a detection signal as the input of the phase discrimination module to build a closed loop control system. The register generated two sine signals DDS1 and DDS2 with the same frequency. The DDS2 signal was phase-shifted by the band-pass filter and analyzed by the DDS1 signal. Then the phase detector output a signal to the fuzzy control algorithm module, and generated a feedback signal to change the center frequency of the band-pass filter to make it close to the frequency of the DDS1 signal. In practice, the XC6SLX9 chip of the Spartan6 series was chosen to design the modal matching fuzzy controller.

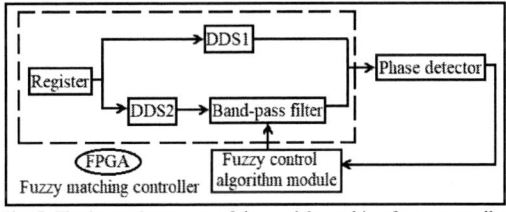

Fig. 7 The internal structure of the modal matching fuzzy controller

In order to simplify the system and save resources, convert the sine signals of the drive end and the detection end into square wave signals, and then input them to the phase detection module of FPGA. The phase detection algorithm process was as follow: first filter the two channels of signals, and then do two counts: one is pha_cyc, which counts the high level signal of the channels; the other is pha_err, which is the number of high levels of the phase difference signal that perform the exclusive OR operation on the two channels. Finally, the phase difference between the two signals $\triangle\varphi$ was obtained by formula (6):

$$\triangle\varphi = 360° \times \frac{pha\_err}{pha\_cyc} \qquad (6)$$

(2) Design of fuzzy control algorithm

The fuzzy controller for modal frequency matching adopted dual input and single output, the inputs were V_{error} and $\dot{V}_{error}$, and the output $\triangle V_{fuzzy}$ of the controller was used to change the resonance frequency of the gyro detection mode. When fuzzing the membership function of V_{error}, we divided it into five fuzzy sets {NB,NS,ZO,PS,PB}, and set the scope of its universe to [-64,64]. The membership function of the input $\dot{V}_{error}$ was divided into three fuzzy sets{N,Z,P},and the scope of its universe was set to [-32,32].The membership function of the output $\triangle V_{fuzzy}$ was divided into five fuzzy sets {NB,NS,ZO,PS,PB}, and the scope of the universe was set to [-64,64].

According to the two inputs, the fuzzy rule library was formulated as shown in Table 1. When the deviation increased in the negative direction and the deviation was also negative, select the maximum output in the positive direction, and vice versa. When the deviation increased in the negative direction and the deviation was not positive, choose the maximum output in the positive direction, and vice versa. When the deviation changed to 0 and the deviation increased slightly in the negative direction, the positive output was selected as a smaller output, and vice versa.

Table 1 Fuzzy rule of modal matching controller

$\dot{V}_{error}\backslash V_{error}$	NB	NS	ZO	PS	PB
N	PB	PB	ZO	NS	NS
Z	PB	PS	ZO	NS	NB
P	PS	PS	ZO	NB	NB

The quantification method of certainty was designed as a small method, and the defuzzification method adopted the weight method. Calculate the results according to the random point theory and use Table 2 for statistics.

Table 2 The theoretical calculation results of the input and output of fuzzy controller

$\dot{V}_{error}\backslash V_{error}$	-35	-15	10	36
-66	64	48	47	32
-35	58	38	32	32
-20	32	22	18	20
10	-10	-12	-14	-13
38	-32	-32	-36	-54
68	-32	-41	-50	-64

(3) Experiment and verification of fuzzy controller

When conducting experimental research on the modal matching fuzzy controller, the phase detector module detected the phase difference signal between the band-pass filter phase-shifted DDS signal and the sine signal generated by the DDS. The phase difference signal compared with the control target value. The calculated deviation and the rate of change of the deviation were used as the input of the control algorithm. The DDS signal passed through a band-pass filter with the DDS signal frequency as the center frequency and produced a phase shift value as the control target. The output of the control algorithm was the address increment that stored the tap coefficients of the band-pass filter with different center frequencies. The address stored the coefficients corresponding to the center frequency of the band-pass filter from low to high. If the center frequency of the initial band-pass filter was greater than the DDS frequency, its corresponding coefficient was placed in the highest address in the ROM. If it was lower than the DDS frequency, its corresponding coefficient was placed in the lowest address in the ROM.

The modal matching fuzzy control algorithm was used for debugging. Set the initial value of the center frequency of the band-pass filter to 3381 Hz, and the frequency of the sinusoidal signal generated by the DDS module was 3051 Hz.

The address increments output by the fuzzy controller were sequentially searched to the center frequency of the band-pass filter corresponding to each address. The frequency following situation with DDS signal frequency could be got in Fig. 8. It was showed that the center frequency of the band-pass filter gradually decreased during the control process. There was no oscillation. And the final oscillation amplitude was also small. According to the partially amplified waveform, it was shown that when reaching stability, the frequency of the band-pass filter was changed in the range of 3051Hz~3056Hz, and the frequency error of the analog modal matching was within 5 Hz.

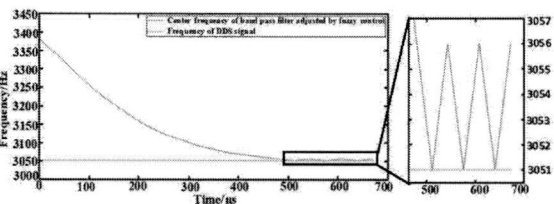

Fig. 8 Center frequency and DDS frequency

CONCLUSION

The modal matching process of the micromachined capacitive gyroscope was theoretically analyzed and a modal matching fuzzy control system was constructed, which included the extraction of two modal output signals, current and voltage conversion, amplification and filtering, etc. The second part was signal demodulation. It was used to extract the phase difference of the two modal output signals. The third part was to use the controller to calculate the DC signal based on the phase difference signal and fed it back to the gyro. The simulation result showed that the phase difference of the two modal output signals after control was 90°, which verified the realization of the fuzzy controller for gyro modal matching. The response speed and stability of the gyro system were improved by optimizing the fuzzy control rules, the quantization factor of the fuzzy input and the scale factor of the fuzzy output. The modal matching fuzzy controller was designed based on FPGA, and the deviation change input was designed as 3 fuzzy sets. The frequency difference during modal matching was controlled within 5 Hz, which had high control accuracy and stability.

ACKNOWLEDGMENT

This research was financially supported by the Aviation Science funds (Aviation Key Laboratory of Science and Technology on Inertia) (No. 20150868002, 20180868001), Science and Technology Planning Project of Shenzhen Municipality in China (JCYJ20180306173000073), Natural Science Foundation of Guangdong Province (2018A030313522), and Xiamen Municipal Science and Technology Projects (3502Z2019015).

REFERENCES

[1] F. Ayazi, K. Najafi, "Design and fabrication of high-performance polysilicon vibrating ring gyroscope", Proceedings MEMS 98: The Eleventh Annual International Workshop on IEEE, Heidelberg, 25-29 Jan. 1998, pp. 621-626.

[2] F. Ayazi, K.Najafi, "A HARPSS polysilicon vibrating ring gyroscope", Journal of Microelectromechanical Systems, Vol.10, pp. 169-179, Jun. 2001.

[3] G. He, K.Najafi, "A single-crystal silicon vibrating ring gyroscope", Micro Electro Mechanical Systems: The Fifteenth IEEE International Conference, Las Vegas, 24-24 Jan. 2002, pp. 718-721.

[4] M. Lutz, W. Golderer, J. Gerstenmeier, et al, "A precision yaw rate sensor in silicon micromachining", Proceedings of International Solid State Sensors and Actuators Conference, IEEE, Chicago, 19-19 June 1997, pp. 847-850.

[5] C.G. Ju., X. Li. , J.J. Zou, Q. Wei, B. Zhou, R. Zhang ,"An Auto-Tuning Continuous-Time Bandpass Sigma-Delta Modulator with Signal Observation for MEMS Gyroscope Readout Systems", Sensors, Vol.20, pp.1-14, Apr. 2020.

[6] R.S. Lv, Q. Fu, L. Yin, Y. Gao, W. Bai, W.B. Zhang, Y.F. Zhang, W.P. Chen, X.W. Liu, "An Interface ASIC for MEMS Vibratory Gyroscopes with Nonlinear Driving Control", Micromachines, Vol.10, pp.1-15, Apr. 2019.

[7] D.Z. Xia, Y.W. Hu, L. Kong, C.Y. Chang, "Design of a Digitalized Microgyroscope System Using Sigma Delta Modulation Technology", IEEE Sensors Journal, Vol.15, pp.3793-3806, Jul. 2015.

[8] B.R. Johnson, E. Cabuz, H. B. French, R.Supino,"Development of a MEMS gyroscope for northfinding applications" , IEEE/ION Position, Location and Navigation Symposium, Indian Wells, 4-6 May 2010, pp168-170.

[9] C.H. He, Q.C. Zhao, D.C. Liu, L.G. Dong, Z.C. Yang, G.Z. Yan, "An automatic real-time mode-matching MEMS gyroscope with fuzzy and neural network control", The 17th International Conference on Solid-State Sensors, Actuators and Microsystems IEEE, Barcelona, 16-20 June 2013, pp.54-57.

[10] C.H.He, Q.C.Zhao, D.C. Liu, L.G. Dong, Z.C.Yang, G.Z.Yan, "A MEMS Vibratory Gyroscope With Real-Time Mode-Matching and Robust Control for the Sense Mode", IEEE Sensors Journal, Vol.15, pp.2069-2077, Apr. 2015.

[11] I.P. Prikhodko, S. Nadig, J.A. Gregory, A.C. William, W.J. Michael, "Half-a-month stable 0.2 degree-per-hour mode-matched MEMS gyroscope", IEEE International Symposium on Inertial Sensors and Systems (INERTIAL), Kauai , 27-30 March 2017, pp.1-4.

[12] C.Yang, H.S. Li, L. Xu, K.P. Zhu, "Automatic mode-matching technology for silicon MEMS gyroscope based on low-frequency modulation signal", Journal of Chinese Inertial Technology, Vol.24, pp.542-547, Aug. 2016.

[13] C.D. Ezekwe, B.E. Boser, "A Mode-Matching Closed-Loop Vibratory Gyroscope Readout Interface With a 0.004 /s/ Noise Floor Over a 50 Hz Band", IEEE Journal of Solid-State Circuits, Vol.43, pp.3039-3048, Aug. 2009.

[14] B.S. Chang, W.T. Sung, J.G. Lee, K.Y. Lee, SK. Sung, "Automatic mode matching control loop design and its application to the mode matched MEMS gyroscope", IEEE International Conference on Vehicular Electronics and Safety Beijing, 13-15 Dec. 2007,pp.1-6.

[15] I.P. Prikhodko, J.A. Gregory, W.A. Clark, J.A. Geen, M.W. Judy, C.H. Ahn, T.W. Kenny, "Mode-matched MEMS Coriolis vibratory gyroscopes: Myth or reality?", IEEE/ION Position, Location and Navigation Symposium , Savannah, 11-14 April 2016,pp.1-4.

Gap in pagination due to unavailable paper.

Pages 496-497

Atomic study on the deform mechanism of CuTa/Cu and CuTa/Ta nanolaminates

Xiao Wang, Mengjie Wang, Min Liu, Yue Liu and Weidong Wang, Senior *Member, IEEE*

Abstract— **The purpose of this paper is clarifying the deform mechanism of CuTa/Cu and CuTa/Ta nanolaminates during nanoindentation test. Firstly, two kinds of amorphous and nanocrystalline interface model were built. Additionally, the Cu layer thickness in CuTa/Cu are λ between 3 Å and 12 Å. The Ta layer thickness in CuTa/Ta are μ between 3 Å and 12 Å. Secondly, nanoindentation test were introduced to get the elastic modulus and hardness of nanolaminates. As the results claimed that different deform mechanisms play the dominant role during the indentation load and unload. More exactly, when the Cu layer is less than 7 Å the Cu layer dominate the deform mechanism, the elastic modulus and harness increase with increasing λ. As λ increasing much more, the CuTa layer will dominate the deform mechanism. In the contrast, the elastic modulus and hardness will decrease with increasing λ. Similarly as CuTa/Cu model, CuTa/Ta model has an obviously change in the dominant role of deform mechanism. In the end, the influence of penetrate rate was discussed.**

I. INTRODUCTION

Cu-Ta alloy plays an important role in representation of immiscible alloy systems. Because of its excellent thermal and mechanical properties, it has broad application prospects in the fields of electronics, aerospace, medicine and health. As early as 1989, Subramanlan *et al.* [1] determined the phase diagram of the Cu-Ta system through experimental means, which opened the prelude to the study of the Cu-Ta alloy system. In the following decades, many scholars conducted a large number of studies on the preparation method, microstructure and mechanical properties of the Cu-Ta system through experiments and simulations. Darling *et al.* [2-10] have been committed to the preparation and application of Cu-Ta alloys, and have made many achievements in atomic simulation research. They proposed a method for preparing Cu-Ta alloy by mechanical alloying, and obtained samples with more excellent thermodynamic properties through continuous optimization of the process. In addition, the use of magnetron sputtering to prepare Cu-Ta films is also a research hot topic and an important preparation method [11-17]. Zeng *et al.* [11] measured the elastic modulus and hardness of the Cu-Ta film prepared by sputtering through the nanoindentation experiment. They proposed the law of the change of the modulus and hardness with Ta content. Wang *et al.* [12] found that when the sputtering time exceeds one minute, the obtained Cu-Ta film has good antibacterial ability and is expected to be widely used in surgical materials in the future.

Although the preparation technology of Cu-Ta alloy is constantly improving, more research is needed to obtain an alloy with stable structure and excellent performance. It has attracted many people's attention in recent years that preparing multilayer film to improve the mechanical properties of alloy [18-25]. Carpenter *et al.* [18] prepared CuNi multilayer film by sputtering method. From the results of nanoindentation experiments, the effectiveness of the multilayer film structure in improving the hardness of the CuNi system is verified. Zeng *et al.* [19] successfully prepared Bulk Cu/Ta nanolamellar multilayers through cross accumulative roll bonding, which has excellent strength and thermal stability. Gu *et al.* [20] sputtered CuTa/Cu nanolaminate structure, verified the size effect by nanoindentation test, and the hardness and modulus of this structure have been greatly improved.

In this paper, atomic simulations were used to study the indentation mechanical behavior of the CuTa/X nanolaminate structure, and the influence of different nanocrystalline layer thicknesses was also discussed.

II. EXPERIMENT

A. Physical modeling

In order to explore the deform mechanism of CuTa/Cu and CuTa/Ta nanolaminates, the simulant nanoindentation test was executed by molecular dynamics (MD). The two models was built as follow, CuTa/Cu model is formed with CuTa amorphous layer and Cu nanocrystalline layer. The Cu layer thickness is λ vary from 3 Å to 12 Å and the CuTa layer thickness keep a constant of 12 Å. They were combined to a single composite layer as we called it one cycle. Thus, the total model formed with 5 cycles. Similarly as CuTa/Cu layer, CuTa/Ta model formed with 5 cycles. Every cycle built in array as the Fig. 1(b) shows. The two models in bulk volume are 120 Å × 120 Å × z_depth (range from 75 Å to 120 Å). Additionally, the Ta concentration in CuTa amorphous layer is 50 at% initially.

B. Methodology of atomistic simulations

The indentation tests were used by MD method using the Large-scale Atomistic/Molecular Massively Parallel Simulation (LAMMPS) [26]. The embedded atom models are used to describe the force on Cu-Cu and Ta-Ta respectively. The angular-dependent interatomic potential (ADP) [27] was set to describe the function of Cu-Ta. In addition, the force

---

Weidong Wang is with the School of Mechano-Electronic Engineering, and the Research Center of Micro-nano Systems, Xidian University, Xi'an 710071, China. (corresponding author, e-mail: wangwd@ mail.xidian.edu.cn).

Xiao Wang and Min Liu are Ph.D. candidate students in the School of Mechano-Electronic Engineering, Xidian University, Xi'an 710071, China. (e-mails: wangxiao9626@outlook.com)

Yue Liu is Ph.D. candidate student in the School of Aeronautics, Northwestern Polytechnical University, Xi'an 710072, China. (emails: lxyxy@mail.nwpu.edu.cn)

Mengjie Wang is graduate student in the School of Mechano-Electronic Engineering, Xidian University, Xi'an 710071, China. (e-mails: mengjie.W@stu.xidian.edu.cn)

978-1-6654-3008-1/21 $31.00 © 2021 IEEE

between diamond tip and CuTa/x system were described in LJ model. The time step was set to 1 fs. At the start of simulation, energy minimization and relaxation were carried out to make the system reach the equilibrium state with lowest energy. During the indentation test, the model boundary was set as periodic on the x direction and y direction, and shrink on the z direction.

From above model, ten samples were used for nanoindentation test, which include λ and μ equal 3, 5, 7, 9 and 12 Å respectively. One diamond sphere was introduced to simulate the indenter. And the diameter of tip is 40 Å. To simulate the experiment process, the load rate is 0.2 Å/Ps, the unload rate is faster than load rate one time. The load time is 80 Ps and the unload time is 40 Ps. After test, the elastic modulus and hardness can be obtained from the Depth-Force curve. The exactly way to calculate the parameters we used is O&P method [28].

Specifically, it can be obtained by the following equations. Firstly, the unloading profile can be fitting with Eq. 1.

$$P_u = a(h - h_r)^m \qquad (1)$$

Where a is the fitting parameter, h_r is the residual depth, and the index m is the shape parameter of the indenter. Secondly, the elastic modulus can be calculated by Eq. 2.

$$\frac{1}{E_r} = \frac{1-v_s^2}{E_s} + \frac{1-v_i^2}{E_i} \qquad (2)$$

Where E_r is the reduced elastic modulus, E_s is the elastic modulus of the material to be tested, E_i is the elastic modulus of the indenter, and V_s and V_i are the Poisson's ratio of the material to be tested and the Poisson's ratio of the indenter, respectively.

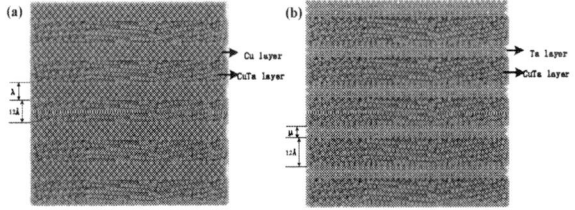

Figure 1. (Color online) physical model of different CuTa/x interface. (a) The model of CuTa/Ta interface, the Cu layer depth is λ range from 3 Å to 12 Å, the CuTa amorphous layer depth is 12 Å. (b) The model of CuTa/Ta interface, the Ta layer depth is μ range from 3 Å to 12 Å, the CuTa amorphous layer depth is 12 Å. The Cu atoms are shown in golden color, green color represents the Ta atoms.

III. RESULTS

A. Nanoindentation test

Fig. 2 shows the Depth-Force curve of the nanoindentation test performed on the CuTa/Cu (3Å) model. In the initial stage, when the indenter is not in contact with the surface of the substrate, due to the existence of van der Waals force, the top of the indenter is attracted downward by the atoms on the surface of the substrate. After the indenter moved 0.6 Å from the initial position, it began to touch the surface of the substrate. The force on the top of the indenter also becomes a positive value, that is, a repulsive force opposite to the indentation

direction is generated. As the indentation depth increases, the load also increases until it reaches the maximum indentation depth of 15 Å. And then unloading stage begin. After the indenter left the surface of the substrate due to plastic deformation, the residual depth of the substrate was about 12 Å. The elastic modulus and hardness of the sample can be obtained through unloading stage.

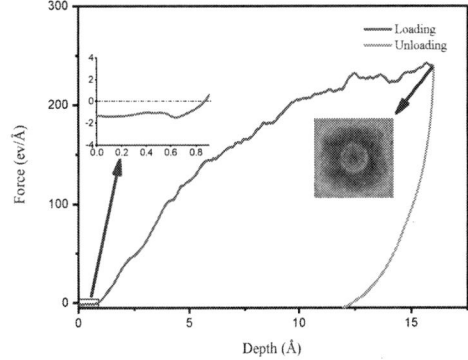

Figure 2. (Color online) Depth-Force curve, using a spherical diamond indenter to press into the substrate at a speed of 0.2 Å/Ps and leave the substrate at a speed of 0.4 Å/Ps. The red curve represents the loading process, and the green curve represents the unloading process.

Nanoindentation tests were perefomed on CuTa/X models with different layer thicknesses and the Depth-Force curves are shown in Fig. 3. A diamond indenter were used with the base of the indenter at a distance of 15 Å. The thickness of the model differs from that originally envisaged due to the movement of the atoms on the surface of the model during the energy minimization process. As a result, the energy minimization process reduces the z-directional length of the entire simulated system. The actual distance between the indenter and the model surface will therefore be less than 15 Å. For the reliability of the experimental results, the indentation depth corresponding to the moment when the contact force is greater than 0 is taken as the contact orgin point. From then on, the indentation depth increases at a constant rate until a predetermined time step is reached. Unloading stage then begins and the depth at which the indenter completely leaves the surface is taken as the residual depth. Since the elastic modulus and hardness are obtained from the unloading curve, the experimental results are reliable.The obtained modulus and hardness are show in Tab. 1. The results from inedntation simulation tests have a great agreement with the results of Gu et al.[20].

TABLE I. ELASIC MODULUS & HARDNESS

Layer thickness (Å)	Physical Model			
	CuTa/Cu		CuTa/Ta	
-	Er (GPa)	HM (GPa)	Er (GPa)	HM (GPa)
3	131.15	9.18	170.10	9.48
5	141.42	9.84	215.61	12.81
7	161.99	8.44	237.85	13.59
9	152.64	8.94	215.97	8.79
12	134.53	6.39	192.93	11.73

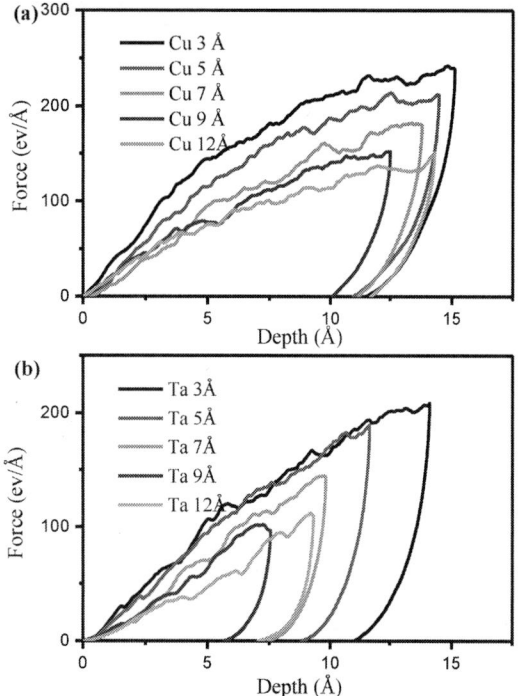

Figure 3. (Color online) Load displacement curves of nanoindentation for models with different Cu/Ta layer thicknesses. (a) CuTa/Cu (λ= 3, 5, 7, 9, 12 Å); (b) CuTa/Ta (μ= 3, 5, 7, 9, 12 Å).

As the Tab.1 lists, the hardness of CuTa/Cu model increases with increasing Cu layer thickness, which is agreement with the mixture rule. As for CuTa/Ta model, the hardness increases firstly and then drops when the Ta layer comes to 7 Å. Four snapshots were collected during the unloading stage for a comprehensive in the hardness evolution.

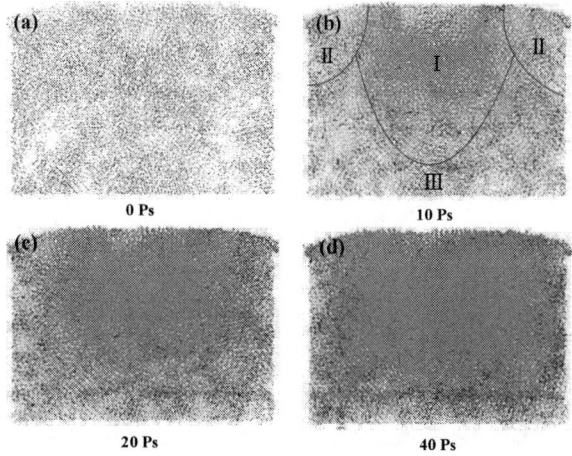

Figure 4. (Color online) Displacement vectors of the atoms during unloading, with the arrow pointing from the initial position of the atom to the position after unloading is completed. (a)-(d) correspond to the distribution of the displacement vectors of the atoms from the beginning of unloading to 40 Ps of unloading. (a) 0 Ps; (b) 10 Ps; (c) 20 Ps; (d) 40 Ps.

In Fig. 4, the motion of the atoms inside the block structure during the unloading phase is shown with the help of Ovito

software [29]. At an initial t = 10 Ps, the larger values of the displacement vectors of the atoms are mainly concentrated in region I. As the indenter continues to move upwards, the atoms in region I stabilize and the values of the displacement vectors no longer change significantly. The atoms in region II start to move towards the center of the pressed region (spherical crater). When the unloading time reaches 20 Ps, the overall atomic distribution stabilizes, except for a small convergence of the atoms in region III towards the center of the spherical crater.

B. Hall-Petch (H-P) Effect

It is clear from Tab. 1 that the elastic modulus increases and then decreases with the thickness of the nanocrystalline layer for both CuTa/Cu and CuTa/Ta. The inflection point is 7 nm. At less than 7 nm, the elastic modulus increases with the increasing thickness of the nanocrystalline layer. Above 7 nm, the elastic modulus decreases as the thickness of the nanocrystalline layer increases. This phenomenon is very similar to the relationship between grain size and strength described by the Hall-Petch relationship. In Fig. 5(b), the elastic modulus versus nanocrystalline layer thickness is fitted, which is consistent with the Hall-Petch relationship and the inverse Hall-Petch relationship perfectly.

As for the hardness, the CuTa/Cu model decreases with increasing Cu layer thickness, which basically conforms to the mixing rule. As we know, the hardness of Cu is relatively low and the addition of Cu layers to the CuTa amorphous nano-layers will inevitably reduce the hardness according to the mixing rule, which is clearly confirmed by the experimental results. For the CuTa/Ta model, the hardness decreases rather than increases at a layer thickness of 9 nm. The severe plasticity deformation and defects may be the mainly two reasons for that, which can be confirmed from Fig .3(b). The residual indentation depth decrease to about 5 Å when the Ta layer thickness is 9 Å. Combining the state of energy minimization of CuTa/Ta (9 Å), it is clear that there are many voids at the contact surface of amorphous and nanocrystalline layers.

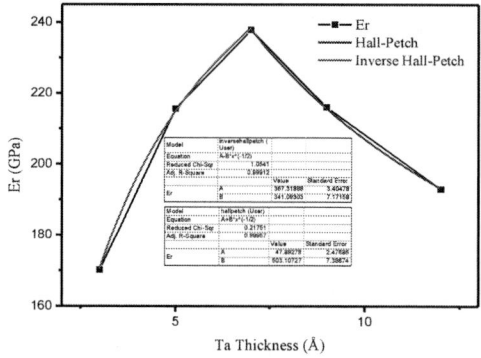

Figure 5. (Color online) Variation of elastc modulus with the Ta layer thickness satisfy the Hall-Petch relationship and Inverse Hall-Petch relationship.

C. Penetrate rate

At the end, the effect of different indentation rates on the hardness and modulus of the model CuTa/Cu (3 Å) was discussed. As the indentation rate increases, the hardness and modulus increase essentially linearly. The results obtained in the interval of 5-20 m/s are more plausible and closer to the experimental values, which is also in agreement with the results of Lu *et al.* [21] As describe in Fig. 6(b), the origin region is reliable when we simulate the nanoindentation test. Beyond this interval, the hardness and modulus increase linearly and the results obtained are not consistent with the actual results. Additionally, if the penetrate rate is less than 5 m/s, the simulation system will consume much time and space.

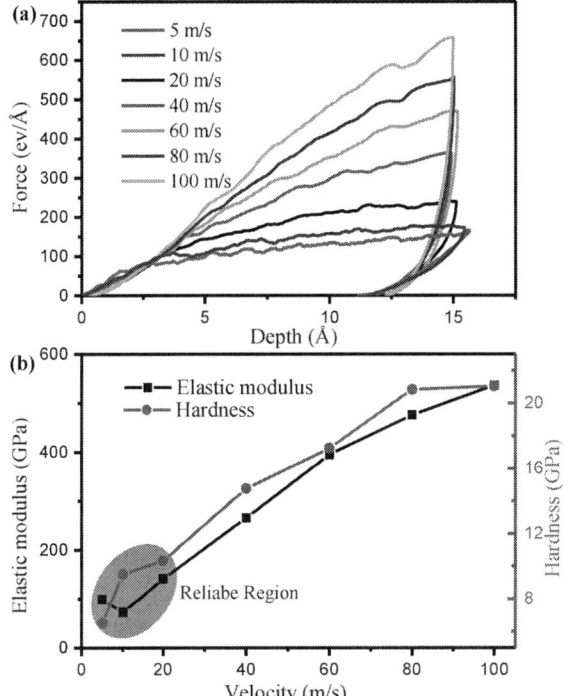

Figure 6. (Color online) Results of nanoindentation experiments at different indentation rates. (a) Depth-Force curves at different indentation rates; (b) Elastic modulus and hardness at different indentation rates.

IV. CONCLUSION

In this article CuTa/x laminates with different thicknesses of nanocrystalline layers were studied. The hardness and elastic modulus of the different models were statistically determined by nanoindentation experiments. The elastic modulus for the CuTa/Cu model was found to strictly satisfy the Hall-Petch relationship and the inverse Hall-Petch relationship, specifically the relationship between the elastic modulus and the thickness of the nanocrystalline layer bounded by 7 nm satisfies Eq. 3 and 4, respectively. For the CuTa/Ta model, the elastic modulus is also bounded by 7 nm, with the modulus increasing with thickness for values less than this and decreasing with thickness for values greater than this. The hardness of both models basically fluctuates around the mean value and is not sensitive to the thickness of the nanocrystalline layer. In addition the displacement vectors of atoms during nanoindentation were visualized using Ovito and it was found that the motion of atoms during indentation showed distinct regional properties. Finally, the effect of

different indentation rates on the hardness and modulus of the model were investigated, and found that the values of hardness and modulus are closer to the experimental values in the velocity interval of 5-20 m/s. Increase of the penetrate rate will result in a linear increase in modulus and hardness. Therefore, when performing simulated nanoindentation experiments, controlling the indentation rate is an important factor in ensuring correct and reliable experimental results.

ACKNOWLEDGMENT

These authors would like to acknowledge the financial support of the project from the National Key Laboratory of Science and Technology on Vacuum Technology and Physics (No. HTKJ2019KL510007), the National Natural Science Foundation of China (No. 42004157), and the Key Research and Development Program of Shaanxi (Program No.2020GY-252).

REFERENCES

[1] Subrananlan PR, Laughlin DE. The Cu-Ta (copper-tantalum) system, Bulletin of Alloy Phase Diagrams, 1989,10(6):652-686.

[2] Darling KA, Huskins EL, Schuster BE, Wei Q, Kecskes LJ. Mechanical properties of a high strength Cu-Ta composite at elevated temperature. Materials Science and Engineering a-Structural Materials Properties Microstructure and Processing. 2015,638:322-8.

[3] Darling KA, Rajagopalan M, Komarasamy M, Bhatia MA, Hornbuckle BC, Mishra RS, et al. Extreme creep resistance in a microstructurally stable nanocrystalline alloy. Nature. 2016;537(7620):378-+.

[4] Darling KA, Roberts AJ, Mishin Y, Mathaudhu SN, Kecskes LJ. Grain size stabilization of nanocrystalline copper at high temperatures by alloying with tantalum. Journal of Alloys and Compounds. 2013,573:142-50.

[5] Darling KA, Srinivasan S, Koju RK, Hornbuckle BC, Smeltzer J, Mishin Y, et al. Stress-driven grain refinement in a microstructurally stable nanocrystalline binary alloy. Scripta Materialia. 2021,191:185-90.

[6] Darling KA, Tschopp MA, Guduru RK, Yin WH, Wei Q, Kecskes LJ. Microstructure and mechanical properties of bulk nanostructured Cu–Ta alloys consolidated by equal channel angular extrusion. Acta Materialia. 2014,76:168-85.

[7] Kale C, Turnage S, Garg P, Adlakha I, Srinivasan S, Hornbuckle BC, et al. Thermo-mechanical strengthening mechanisms in a stable nanocrystalline binary alloy - A combined experimental and modeling study. Materials & Design. 2019,163.

[8] Kale C, Srinivasan S, Hornbuckle BC, Koju RK, Darling K, Mishin Y, et al. An experimental and modeling investigation of tensile creep resistance of a stable nanocrystalline alloy. Acta Materialia. 2020,199:141-54.

[9] Hornbuckle BC, Kale C, Srinivasan S, Luckenbaugh TL, Solanki KN, Darling KA. Revealing cryogenic mechanical behavior and mechanisms in a microstructurally-stable, immiscible nanocrystalline alloy. Scripta Materialia. 2019,160:33-8.

[10] Bhatia MA, Rajagopalan M, Darling KA, Tschopp MA, Solanki KN. The role of Ta on twinnability in nanocrystalline Cu–Ta alloys. Materials Research Letters. 2016,5(1):48-54.

[11] Zeng F, Gao Y, Li L, Li DM, Pan F. Elastic modulus and hardness of Cu–Ta amorphous films. Journal of Alloys and Compounds. 2005,389(1-2):75-9.

[12] Wang S, Zhu W, Yu P, Wang X, He T, Tan G, et al. Antibacterial nanostructured copper coatings deposited on tantalum by magnetron sputtering. Mater Technol. 2015,30(B2):B120-B5.

[13] Müller CM, Sologubenko AS, Gerstl SSA, Spolenak R. On spinodal decomposition in Cu–34at.% Ta thin films – An atom probe tomography and transmission electron microscopy study. Acta Materialia. 2015,89:181-92.

[14] Powers M, Derby B, Shaw A, Raeker E, Misra A. Microstructural characterization of phase-separated co-deposited Cu-Ta immiscible alloy thin films. Journal of Materials Research. 2020;35(12):1531-42.

[15] Qin W, Fu L, Xie T, Zhu J, Yang W, Li D, et al. Abnormal hardness behavior of Cu-Ta films prepared by magnetron sputtering. Journal of Alloys and Compounds. 2017,708:1033-7.

[16] Raeker E, Powers M, Misra A. Mechanical performance of co-deposited immiscible Cu-Ta thin films. Scientific Reports. 2020,10(1).

[17] Tian W, Dai JY, Zhang LJ, Chang YQ, Bao MD, Yu SW. Microstructure and properties of nanocrystalline Cu-Ta thin films prepared by direct current magnetron sputtering. Surface Engineering. 2020:1-9.

[18] Carpenter JS, Misra A, Anderson PM. Achieving maximum hardness in semi-coherent multilayer thin films with unequal layer thickness. Acta Materialia. 2012,60(6-7):2625-36.

[19] Zeng LF, Gao R, Fang QF, Wang XP, Xie ZM, Miao S, et al. High strength and thermal stability of bulk Cu/Ta nanolamellar multilayers fabricated by cross accumulative roll bonding. Acta Materialia. 2016,110:341-51.

[20] Gu C, Wang F, Huang P, Xu KW, Lu TJ. Structure-dependent size effects in CuTa/Cu nanolaminates. Materials Science and Engineering: A. 2016,658:381-8.

[21] Lu L, Huang C, Pi WL, Xiang HG, Gao FS, Fu T, et al. Molecular dynamics simulation of effects of interface imperfections and modulation periods on Cu/Ta multilayers. Computational Materials Science. 2018,143:63-70.

[22] Muller CM, Sologubenko AS, Gerstl SSA, Suess MJ, Courty D, Spolenak R. Nanoscale Cu/Ta multilayer deposition by co-sputtering on a rotating substrate. Empirical model and experiment. Surf Coat Technol. 2016,302:284-92.

[23] Wang F, Huang P, Xu M, Lu TJ, Xu KW. Shear banding deformation in Cu/Ta nano-multilayers. Materials Science and Engineering a-Structural Materials Properties Microstructure and Processing. 2011;528(24):7290-4.

[24] Wei MZ, Cao ZH, Shi J, Pan GJ, Xu LJ, Meng XK. Evolution of interfacial structures and creep behavior of Cu/Ta multilayers at room temperature. Materials Science and Engineering a-Structural Materials Properties Microstructure and Processing. 2015,646:163-8.

[25] Zhou Q, Li JJ, Wang F, Huang P, Xu KW, Lu TJ. Strain rate sensitivity of Cu/Ta multilayered films: Comparison between grain boundary and heterophase interface. Scripta Materialia. 2016,111:123-6.

[26] Plimpton S. Fast Parallel Algorithms for Short-Range Molecular Dynamics. J Comp Phys,1995, 117:1-19.

[27] Purja Pun GP, Darling KA, Kecskes LJ, Mishin Y. Angular-dependent interatomic potential for the Cu–Ta system and its application to structural stability of nano-crystalline alloys. Acta Materialia. 2015,100:377-91.

[28] Oliver WC, Pharr GM. An improved technique for determining hardness and elastic modulus using load and displacement sensing indentation experiments. Journal of Materials Research, 1992,7(6), 1564-1583.

[29] Stukowski A. Visualization and analysis of atomistic simulation data with OVITO – the Open Visualization Tool Modelling Simul. Mater. Sci. Eng. 18 (2010), 015012.

Proceedings of the 16th Annual IEEE International
Conference on Nano/Micro Engineered and Molecular Systems
April 25-29, 2021

Wireless Implantable Phototherapy Device for Oral Inflammation Repair*

Shuang Huang, Cheng Yang, Ziqi Liu, Xinshuo Huang, Hui-Jiuan Chen

Abstract— LED light source is widely used in stomatology treatment and cosmetology. However, since the large size and insufficient intensity of the wireless light source, it is difficult to implant the phototherapy device into the mouth for long-term treatment. This paper proposed to fabricate a wireless implantable micro phototherapy device for oral inflammation repair. AlGaAs light-emitting diodes were fabricated according to the wavelength and optical power density of inflammatory tissue repair. The LED chip is used as the treatment light source, whose chip and LED are organically integrated to ensure the miniaturization of the device and to increase the overall illumination area and optical power density as much as possible. Based on the magnetic resonance coupled wireless power supply mode, a near-field wireless power supply system is designed to maintain the normal operation of the implanted device with coupling parameters such as coupling mode, harmonic mode. Vibration frequency, coupling coefficient, coil design details such as quality factor, etc. were optimized to ensure sufficient energy supply to the device. In this paper, the optical, electrical and temperature performance of the whole system is characterized. Specific in vivo long-term clinical trials will be conducted later on the basis of these studies.

Keywords—Wireless implant device; phototherapy; near infrared light emitting diode; oral inflammation.

I. INTRODUCTION

One of the main goals of stomatology is to repair the inflammation of oral tissue and recover the tissue inflammation caused by trauma, disease and infection. Oral inflammation and wound injury diseases need to be treated with appropriate methods. So far, mechanical treatment is the mainstream method of periodontal inflammatory diseases.[1] However, traditional mechanical treatment alone may not be able to completely eliminate bacteria or achieve the best wound healing effect. Therefore, nowadays, chemical therapy is also used to assist mechanical therapy, such as the use of antibiotics[2], so as to achieve the goal of completely eradicating bacteria and promoting wound healing and tissue regeneration after debridement. This kind of mechanical therapy combined with chemotherapy has been used in the treatment of aggressive or severe advanced periodontitis. In addition to the above treatment methods, the use of laser and light-emitting diode for periodontal

Hui-Jiuan Chen is with the School of electronic and Information Engineering, Sun Yat-Sen University, Guangzhou 510006, China (Corresponding author, e-mail: chenhuix5@mail.sysu.edu.cn). Other authors are all with the same affiliate.

phototherapy has also attracted extensive attention in the periodontal academia, which has shown a variety of beneficial effects in clinical manifestations, treatment process and treatment results[3]. Compared with mechanical therapy and chemotherapy, phototherapy is milder and does not cause mechanical damage or other side effects.[4-6] Nowadays, the combination of phototherapy with mechanical therapy or chemotherapy is becoming more and more popular. This topic is to discuss the treatment methods to promote the repair of alveolar tissue in the specific period after surgical wound and before artificial root implantation.

So far, LED chips have been widely used in stomatology, but the implanted optoelectronic devices are still in the stage of scientific experiments. The reason for this is that there are many difficulties in achieving compatibility between precision electronic instruments and the fragile biological tissue environment: 1. the use of biocompatible materials to prepare miniaturization phototherapy devices, 2. wireless miniaturized devices in power to meet the treatment of oral tissue, 3. short-term phototherapy will not cause oral tissue damage. Therefore, we show the treatment methods of alveolar tissue repair by using wireless implanted optoelectronic devices in the specific period after the operation wound and before the artificial root implantation.

II. FABRICATION OF WIRELESS IMPLANTABLE PHOTOTHERAPY DEVICE

After tooth extraction, due to the damage of the tissue, the blood oozing from the wound will flow into and fill the alveolar. After the coagulation factor plays a role, the fibrin in the blood will interweave with each other, and gather the cells and proteins in the blood to form a blood clot (thrombus). Here, to accelerate the regeneration and healing of bone tissue during blood coagulation, the general process of wireless implantable phototherapy is shown in **Fig. 1**.

The near-infrared light with 808 nm wavelength has a significant effect in promoting the repair of alveolar wound tissue. Among them, the radiation dose, optical power density and other parameters will directly affect the of treatment, so the LED chip as the treatment light source is especially important. To reduce the economic and time cost of device processing, we will use the organic metal chemical vapor deposition method for the epitaxial growth of AlGaAs material[7], and use silicon as the substrate material. The manufacturing process of AlGaAs near infrared LED chip is shown in **Fig. 2**. The overall size of the chip is about 320 µm × 320 µm, the thickness is about 175 ± 25 µm, the size of the

978-1-6654-3008-1/21 $31.00 © 2021 IEEE

light emitting area is about 285 μm, and the size of the pad is about 105 μm. the N electrode and p electrode are made of gold. The micron size led device is suitable for millimeter size implantable devices.

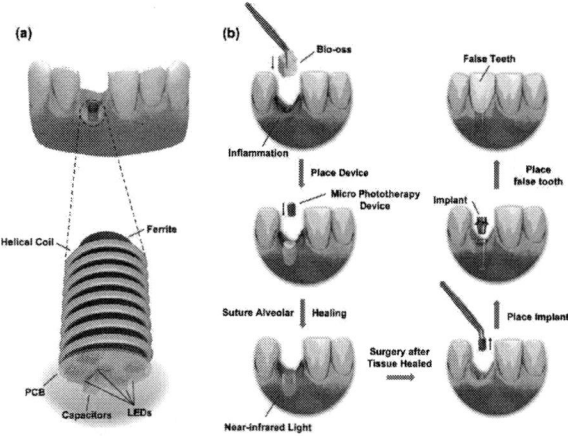

Figure 1. Phototherapy after oral tissue damage. (a) Schematic diagram of tissue damage treated by wireless implantable phototherapy device. (b) The procedure of phototherapy for the treatment of tissue damage caused by tooth extraction.

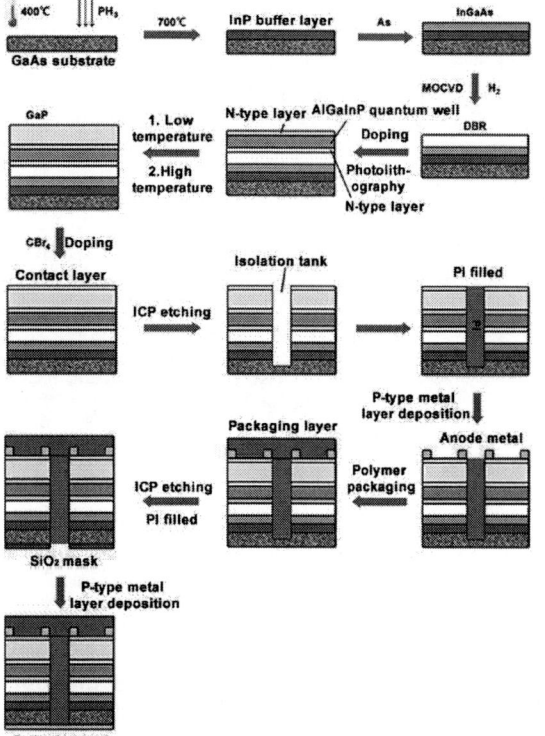

Figure 2. The process of AlGaAs near infrared LED chip.

III. CURRENT RESULTS

In this paper, the classical resonant circuit is synthesized, and the resonant circuit of the implantable device is simulated and compared. Three parameters are fixed in the simulation: inductance L is 432 μH, power frequency f is 700 kHz, and coupling coefficient K is 0.52. The simulation results are shown in **Figs. 3a and 3b**. When the circuit mode is the fourth, the coupling coefficient K is 0.52, and the power frequency f is 700 kHz. The output power of the load is calculated by simulation with the change curve of coil parameters, as shown in **Fig. 3c**. The L1 / 2 / R parameter of the coil with inductance of 400 ~ 500 μH can reach above 0.85. The receiver implantable device integrated with LED and wireless receiving coil is shown in **Fig. 4 (left)**. The optical diagram of the circuit board at the receiving end is shown in **Fig. 4 (right)**.

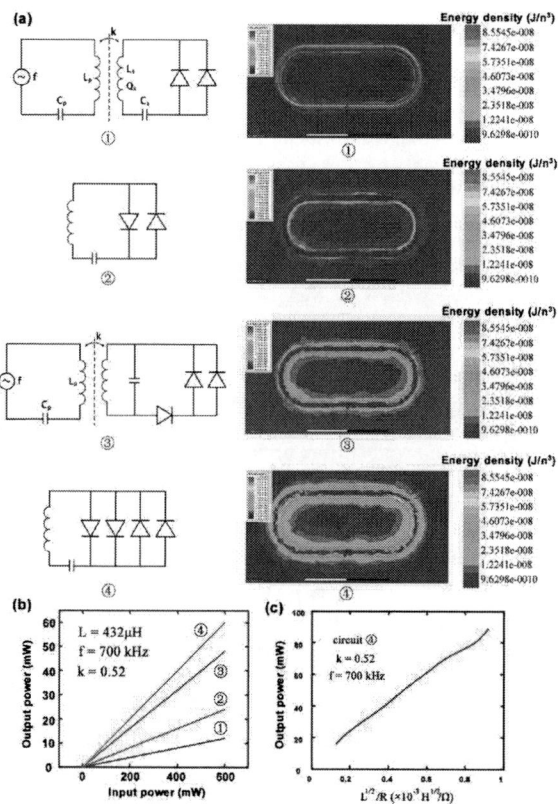

Figure 3. (a) Four kinds of circuits and their simulation diagrams of flux density. (b) Four kinds of simulation results of output power-input power curve under different circuit forms. (c) The output power curve of load with coil parameters is calculated by simulation, the coupling coefficient K is 0.52 and the power frequency f is 700 kHz.

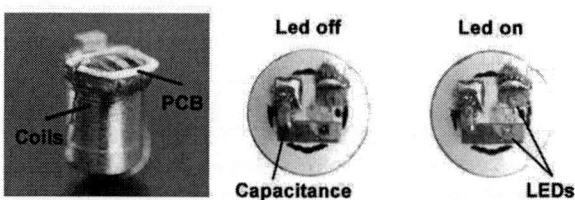

Figure 4. Physical picture of wireless implantable phototherapy device. The winding specifications of the magnetic rod coil are as follows: the diameter of the magnetic rod is 2 mm, the length is 4 mm, the enameled wire with wire diameter of 0.1mm is used for winding, the total outer

diameter of the final coil is 3.3mm, the number of coils is 350, and the inductance is 453 μH.

After the integration of the devices, we carried out in vitro wireless phototherapy simulation. **Fig. 5** shows the model device used to test the optical power and temperature in vitro. Here, the test equipment and probe need to be replaced to measure other parameters. For the convenience of testing, we can measure the distance D between two coils to simply characterize the coupling degree between devices. The working frequency f of the system can be characterized by the return loss curve of the device tested by the network analyzer. In addition, we use signal generator as AC voltage source to supply power for the system in the actual test process. The power supply is a square wave with 50% duty cycle, and the frequency f and voltage V are adjustable. V is also one of the parameters that affect the input power and output power density of the system. To sum up, we can test the relationship between frequency f, voltage V and output power density, as well as the influence of distance D on transmission efficiency. The test results are shown in **Figs. 6a-e**. Similarly, the thermal characteristics of the transmitting coil may also affect the normal operation of the system. The test results of thermal characteristics are shown in **Fig. 6f**.

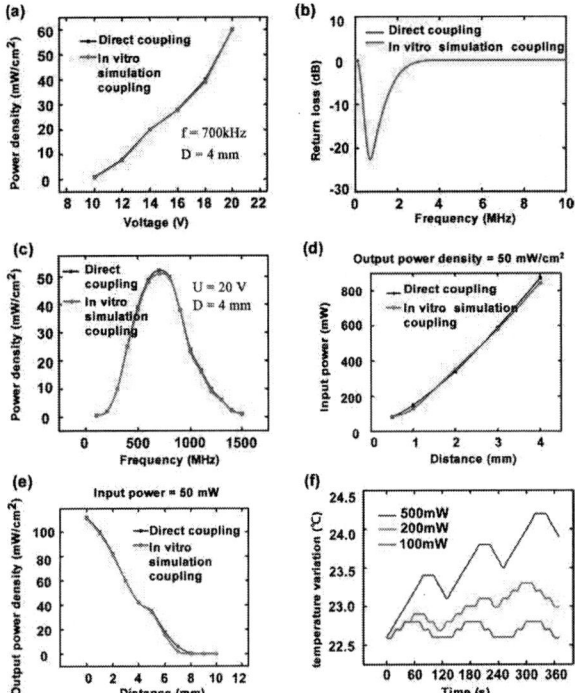

Figure 6. In vitro measurement of electrical, optical, RF and thermal properties of devices. (a) Energy power density - voltage curve of devices. (b) Return loss - frequency curve of devices. (c) Energy power density frequency curve of devices. (d) Input power - distance curve required by device at fixed output power density. (e) Output power - density distance curve of device at fixed input power. (f) Temperature characteristic curve of transmitting coil. In the test of the temperature characteristics of the device, we use 100 milliwatts, 200 milliwatts and 500 milliwatts input power drivers to light up, and turn off the power after 60 s to let the device cool down naturally.

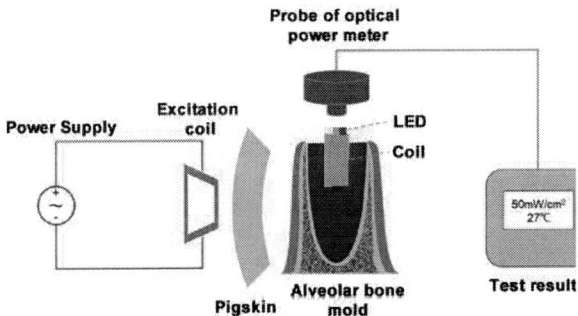

Figure 5. Schematic diagram of in vitro simulation test device. In the actual measurement process, we can adjust the voltage value and frequency of the external power supply, as well as the distance between the external excitation coil and the alveolar bone mold, and test the electrical, optical and thermal properties of the device through the terminal power meter, thermometer or network analyzer. Pig skin is used to simulate the skin tissue outside the mouth, while the alveolar bone mold is filled with blood and loaded with hydroxyapatite to simulate the actual treatment scene.

In addition to the numerical test of the temperature change of the device, we also took thermal imaging photos of the transmitter and the receiver in the in vitro experiment. The test results are shown in **Fig. 7** and **Fig. 8**. From the beginning of power on, the thermal images and their changes of receiver and transmitter from 0s to 30 s, 60 s, 90 s and 120 s were captured. It can be seen from the figure that the temperature of the implantable device rises slowly within 120 s, but the temperature rises slowly when it reaches about 27 °C. It should be considered that the environmental temperature has an impact on it. The temperature rise of the emitter is also stable within $1 \sim 2$ °C, which will not affect the normal operation of the device.

From the in vitro test of the wireless oral implantable micro phototherapy device, its overall performance shows a relatively stable trend. It not only ensures that enough light power can cure the damaged tissue, but also ensures that the device itself can work normally within 2 minutes of the work cycle not burning the gum tissue. At the same time, the device heating and electromagnetic wave propagation itself will not cause harm to human body.

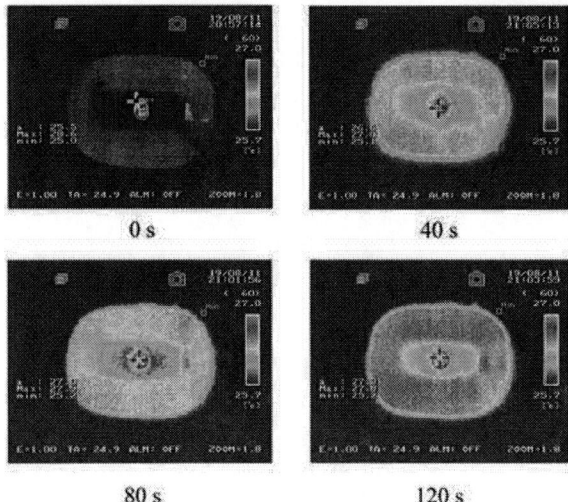

Figure 7. Thermography of transmitting coil. The input power drivers of 100 milliwatts, 200 milliwatts and 500 milliwatts are used to light up, and the power supply is cut off after 60 s to let the device cool down naturally.

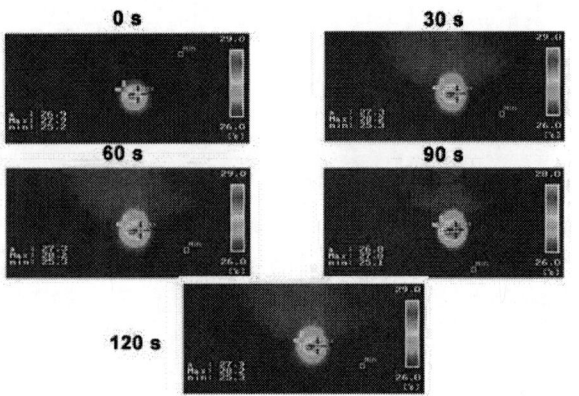

Figure 8. Thermogram of receiving coil. The input power drivers of 100 milliwatts, 200 milliwatts and 500 milliwatts are used to light up, and the power supply is cut off after 60 s to let the device cool down naturally.

IV. SUMMARY

In this work, a wireless implantable micro phototherapy device using for oral inflammation repair was proposed. To ensure that the device can work normally in the actual application and achieve the purpose of treating the damaged alveolar tissue, we designed, prepared and optimized of some parameters. Especially, AlGaAs light-emitting diodes was fabricated, and the miniaturized LED chip was used as the therapeutic light source. And the whole device in vitro, including light, electricity, radio frequency, temperature and so on were tested. How to balance the accuracy and real-time of electronic devices is the key problem of future implantable biomedical electronic devices with the complexity of diagnosis, treatment and nursing, human function and its practical needs in the design and development. This work provided a hardware basis for optical treatment of oral inflammation and a basis for verification of clinical experiments of advanced optical treatment in the future.

ACKNOWLEDGMENT

The authors would like to acknowledge financial support from the National Natural Science Foundation of China (61901535, 31900954), Guangdong Province Key Area R&D Program (Grant No.2018B030332001), Science and Technology Program of Guangzhou, China (Grant No. 201907010038) and Guangdong Basic and Applied Basic Research Foundation (Grant No. 2019A1515012087). The authors wish to thank 100 Talents Program of Sun Yat-Sen University (76120-18841213).

REFERENCES

[1] D. Matthews, "Conclusive support for mechanical nonsurgical pocket therapy in the treatment of periodontal disease. How effective is mechanical nonsurgical pocket therapy?," *Evidence-based dentistry,* vol. 6, no. 3, pp. 68-9, 2005 2005.

[2] R. Mutschelknauss, "Drug therapy for periodontal diseases. Possibilities and limits," *Zahnarztliche Praxis,* vol. 24, no. 14, pp. 376-7, 1973-Jul-20 1973.

[3] C. D. Eduardo *et al.*, "Laser phototherapy in the treatment of periodontal disease. A review," (in English), *Lasers in Medical Science,* Review vol. 25, no. 6, pp. 781-792, Nov 2010.

[4] C. B. Rosa *et al.*, "Effect of the laser and light-emitting diode (LED) phototherapy on midpalatal suture bone formation after rapid maxilla expansion: a Raman spectroscopy analysis," *Lasers in Medical Science,* vol. 29, no. 3, pp. 859-867, May 2014.

[5] H.-z. Cui, Z.-j. Li, and X.-h. Fan, "The development of light emitting diode therapy in biology and medicine," *Laser Technology,* vol. 30, no. 6, pp. 638-56, Dec. 2006.

[6] G. B. Altshuler and V. V. Tuchin, "Light emitting toothbrush for oral phototherapy," Patent US 07223270, 2007. Available: <Go to ISI>://BIOSIS:PREV200700413048.

[7] K. Streubel, N. Linder, R. Wirth, and A. Jaeger, "High brightness AlGaInP light-emitting diodes," (in English), *Ieee Journal of Selected Topics in Quantum Electronics,* Article vol. 8, no. 2, pp. 321-332, Mar-Apr 2002, Art. no. Pii s1077-260x(02)03774-7.

Gap in pagination due to unavailable paper.

Pages 507-508

Proceedings of the 16th Annual IEEE International
Conference on Nano/Micro Engineered and Molecular Systems
April 25-29, 2021

Multiparametric Flexible Sensor Arrays System for in situ Immediate Diagnosis of Glossitis*

Ziqi Liu, Keer Li, Shuang Huang, Xinshuo Huang, Lingfei Zhou, Tiancheng Sun and Hui-Jiuan Chen

Abstract-This paper presents the first multiparametric flexible sensor arrays system for in situ immediate diagnosis of glossitis, which may furnish clues for certain systemic diseases. The system includes a multiparametric flexible sensor array and accompanying signal processing circuitry. The flexible sensor arrays are obtained by facile micro-fabrication and chemical modifications. The sensors are affixed to the tongue mucosa for simultaneous monitoring the signals of sodium, potassium, pH, and temperature variations, which are transmitted to a smart terminal via the circuitry. In vitro tests have demonstrated excellent response linearity, sensitivity, selectivity, and reproducibility of this sensor arrays system. The signal processing circuitry provides reliable support for integrated testing of multiparametric sensors. Simulating the tongue state with agar, the system significantly detected changes in marker levels from the exudate on the agar surface. The multiparametric flexible sensor arrays system is expected to tackle the "black box" issues in several clinical diagnostics in the future.

I. INTRODUCTION

Glossitis generally refers to chronic, non-specific inflammation of the tongue. The state of the tongue can provide clues for some systemic diseases in the clinic[1]. The association of diseases including systemic psoriasis[2], diabetes[3], protein-calorie malnutrition[4], and celiac disease[5] with glossitis has been extensively studied. However, the diagnosis of glossitis still relies on biopsy, blood tests, and microscopic examination of saliva or mucosal smears[6, 7], which substantially burdens the time and cost for both physicians and patients, as well as the early detection and treatment of these diseases.

The local physiological responses related to inflammation are mainly manifested in telangiectasis, slow blood flow, and partial components exudation into tissues. These processes are predisposed to cause changes in several physiological parameters, providing the possibility of a rapid diagnosis of glossitis. Typically, the absolute concentration of sodium in oral cavity is positively correlated with the degree of inflammation, while potassium exhibit relative levels that are more than three times higher than in serum[8]. Besides, both increased pH and temperature are common responses to infective inflammation.

*Research supported by the National Natural Science Foundation of China (61901535, 31900954).
Hui-Jiuan Chen is with the School of electronic and Information Engineering, Sun Yat-Sen University, Guangzhou 510006, China (Corresponding author, e-mail: chenhuix5@mail.sysu.edu.cn). Other authors are all with the same affiliate.

Herein, we propose a multiparametric flexible sensor arrays system, which provides a basis for the immediate diagnosis of glossitis. The system can be manufactured and employed easily and quickly. The flexible sensor arrays are affixed to the tongue mucosa and monitor variations in sodium, potassium, pH, and temperature in real-time. The signals derived from the sensors are transmitted to a smart terminal via a signal processing circuitry. The multiparametric sensors have shown excellent performance in in vitro electrochemical testing. Benefited from the development of the dedicated signal processing circuitry, the system allows the simultaneous monitoring and output of data on several physiological indicators. Furthermore, the testing results in the environment in which the tongue state is simulated provides a convincing basis for the clinical application of the system. We believe that the multiparametric flexible sensor arrays system is a valuable strategy to address the diagnosis of glossitis as well as other diseases.

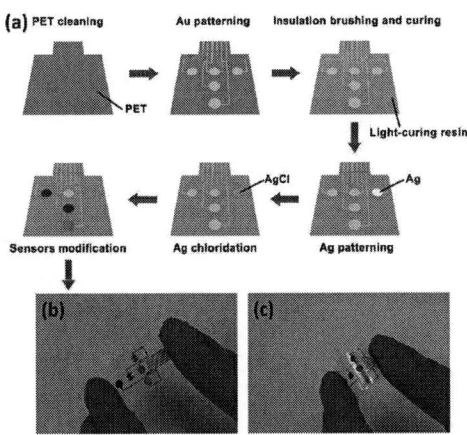

Figure 1. (a) Fabrication process of the multiparametric flexible sensor array. (b) and (c) Photographs of the multiparametric flexible sensor array after modification.

II. EXPERIMENTAL SECTION

A. Reagents and Apparatus

All of the following reagents were purchased from Sigma Aldrich: sodium ionophore X, sodium tetrakis[3,5-bis(trifluoromethyl)phenyl] borate (Na-TFPB), valinomycin, sodium tetraphenylborate (NaTPB), bis(2-ethylhexyl) sebacate (DOS), polyvinyl chloride (PVC), polyaniline emeraldine, tetrahydrofuran (THF), cyclohexanone (CYC), dimethyl sulfoxide (DMSO), 3,4-ethylenedioxythiophene (EDOT), poly(sodium 4-styrenesulfonate) (NaPSS), and polyvinyl butyral resin BUTVAR B-98 (PVB). The graphene dispersion was

978-1-6654-3008-1/21 $31.00 © 2021 IEEE

obtained from XFNANO, and the agarose was received from BBI CO., LTD. All reagents were not additionally treated before use.

B. Fabrication of flexible electrode arrays

The multiparametric flexible sensor arrays are fabricated as shown in Fig. 1a. Briefly, the flexible electrode arrays on PET are obtained by 20 nm Cr/60 nm Au deposited via magnetron sputtering with a metal mask. Light-curing resins are brushed on the non-electrode areas and cured under UV light for 1 min to eliminate signal disturbances from the conductive path. To prepare Ag/AgCl reference electrode, 200 nm Ag was deposited in the reference electrode area, followed by 1 min reaction in 6 M HNO_3 solution and 0.1 M $FeCl_3$ solution, respectively.

C. Preparation of sensors

The remaining electrodes were chemically modified individually to prepare the corresponding sensors. Before modification of the sodium, potassium and pH sensors, PEDOT:PSS was deposited on the electrode surface by electropolymerization in a solution containing 0.01 M EDOT and 0.1 M NaPSS.

The sodium and potassium sensors were obtained by drop-casting 10 μl THF solution containing 0.15 wt% Na ionophore X, 0.08 wt% Na-TFPB, 4.81 wt% PVC, and 9.55 wt% DOS or 4 μl CYC solution containing 0.46 wt% valinomycin, 0.12 wt% NATPB, 7.56 wt% PVC, and 14.96 wt% DOS, respectively. To enhance the reliability of the preparation, the sensors were dried overnight before use after drop-casting.

The preparation of the pH sensor included, first, 50 mg polyaniline emeraldine was dissolved in 20 mL DMSO, followed by thorough stirring before drop-casting 5 μL of this solution to the electrode and dried at room temperature. The electrode was then placed in the vacuum chamber with 2 mL of HCl and kept for 6 h. The color of the electrode surface gradually changed from dark blue to dark green. Finally, the electrode was dried with nitrogen to obtain the pH sensor. 4 μL graphene dispersion was drop-casted onto the electrode surface and dried to prepare the temperature sensor.

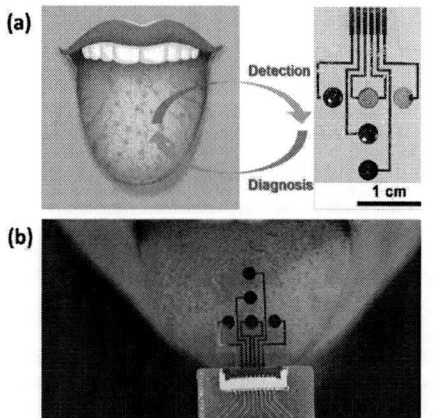

Figure 2. (a) Photograph of a flattened multiparametric flexible sensor array and schematic of an application scenario. (b) Photograph of a multiparametric flexible sensor array on a subject's tongue.

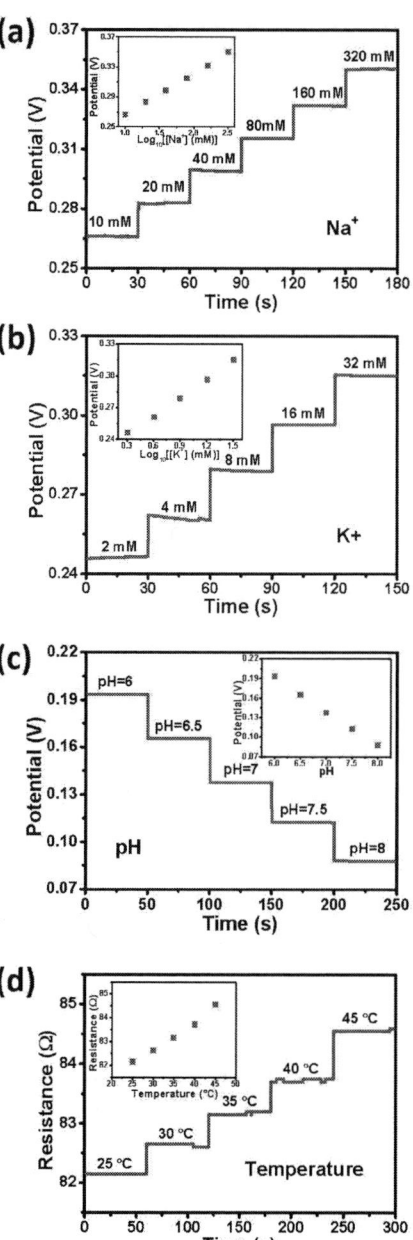

Figure 3. The open-circuit potential responses of the sodium (a), potassium (b) and pH (c) sensors in NaCl, KCl and PBS (pH = 6-8) solutions. (d) The resistance response of the temperature sensor to temperature changes (25–45 °C) in PBS.

III. RESULT AND DISCUSSION

A multiparametric flexible sensor array fabricated by the process described above is shown in Figs. 1b and 1c. Figs. 2a and 2b depict that the array has a small absolute-size (~1.75 cm*2.75 cm), which is appropriate for affixing to the tongue. The sensors of this system have shown remarkable performance in vitro tests. As shown in Figs. 3a, 3b and 3c, the sodium, potassium and pH sensors exhibited linear open-circuit potential responses with concentrations of 10-160 mM ($R^2 > 0.999$), 2-32 mM ($R^2 >$

0.998) and pH = 6-8 ($R^2 > 0.998$), respectively. As presented in Fig. 3d, the temperature sensor shows a linear sensitivity of 0.12 $\Omega/^{\circ}C$ from 25 to 45 $^{\circ}C$ ($R^2 > 0.98$). Resulting from the average tenfold concentration change sensitivity is 55.6 mV for sodium, 57.5 mV for potassium, and 52.9 mV for hydrogen ion, these sensors also showed a near-Nerstian behavior (the theoretical sensitivity of the ISE-based sensors according to the Nerstian equation should be 59).

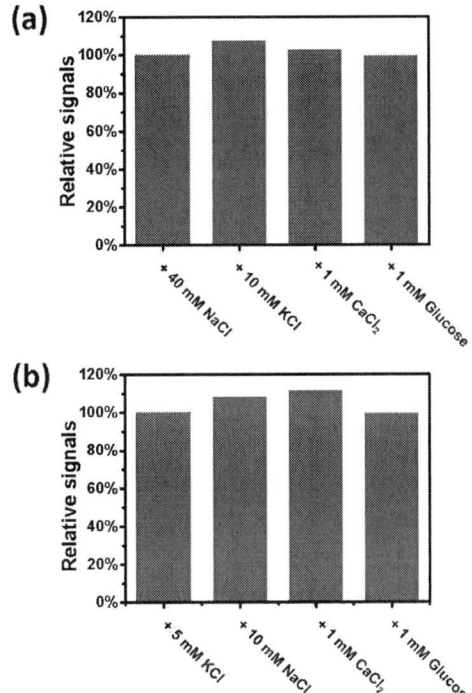

Figure 4. The selectivity study of sodium (a) and potassium (b) sensors.

The tongue is enclosed in the mouth, which is a complex environment. Various substances in saliva and other fluids may interfere with the response signal of the sensors. Therefore, it is necessary to evaluate the selectivity of the sensor with substances commonly found in saliva including sodium, potassium, calcium and glucose. For sodium or potassium sensor, the sensing target substance was added first and then the interfering substances were added to compare their relative response signals. As displayed in Figs. 6a and 6b, the results show that the sodium and potassium sensors exhibited excellent response stability for interfering substances, indicating that they are capable of withstanding with disturbances in oral environment.

As a further step, the reproducibility of these three sensors was verified (Figs. 4a, 4b, and 4c), which are free from interference with each other in the integration test (Fig. 4d). The significance of characterizing the reproducibility of the sensor is to illustrate the reliability of the sensor during preparation and application.

Apart from the evaluation of the sensor performance, the accuracy of the signal processing circuit is also of significance for the signal output of the whole system. As

illustrated in Fig. 5a, the circuit module was developed based on a voltage follower, differential amplifier, and 1Hz low-pass filter, and the results of the signal transmission studies showed extremely high accuracy ($R^2 = 0.9998$) (Fig. 5b). These results strongly contribute to the successful application of the multiparametric flexible sensor arrays system.

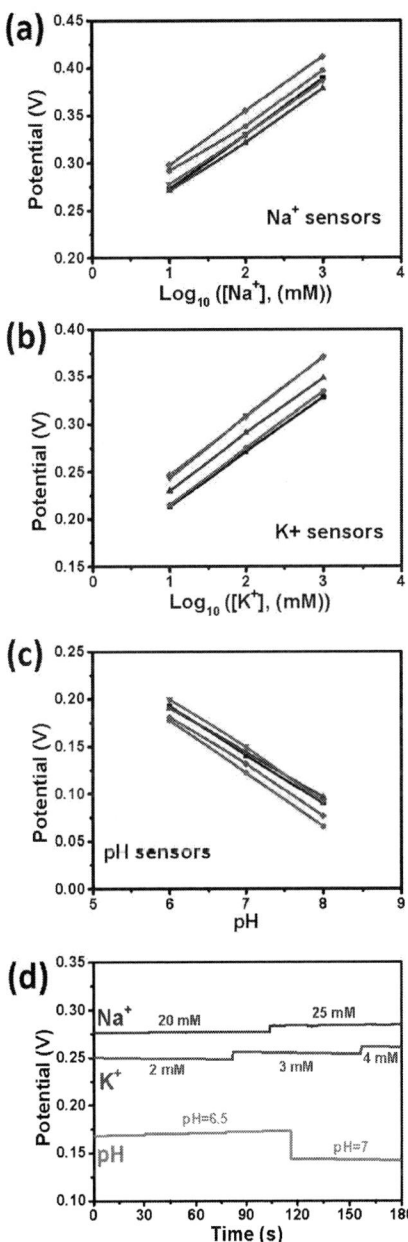

Figure 5. The reproducibility of the sodium (a), potassium (b) and pH (c) sensors (five samples for each type of sensor). (d) Integration and interference studies of the sensor array.

978-1-6654-3008-1/21 $31.00 © 2021 IEEE

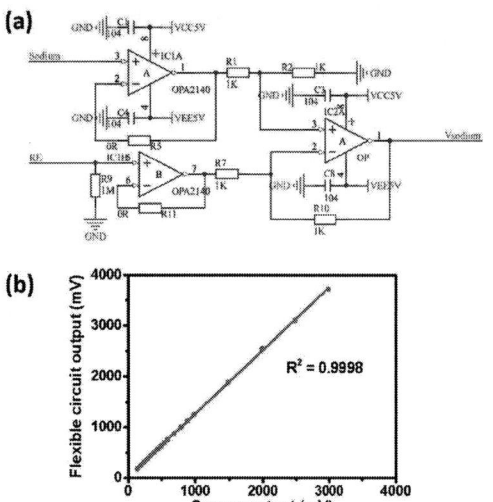

Figure 6. (a) Schematic diagram of signal-conditioning circuit. (b) Signal transmission studies based on circuit module.

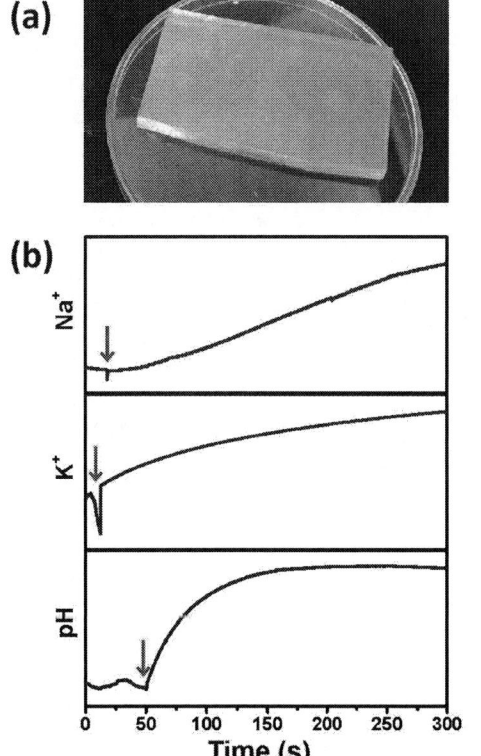

Figure 7. (a) Photograph of the agarose for tongue simulation. (b) The relative response signal changes of sodium, potassium and pH sensors after dropping the marker solution on top of the agarose.

To further evaluate the potential of the multiparametric flexible sensor arrays system for clinical applications, agarose was used to simulate the tongue, largely attributed to their similar surface properties and mechanical properties (Fig. 7a). A flexible sensor array was inserted

into the under-surface of the agarose to simulate its affixation to the tongue in a clinical scenario. 100uL of the marker solution was dropped on top of the agarose and the trend of the response signal from the under-surface of the sensor was observed, the results are illustrated in Fig. 7b. Considering that the dropping solution preferred to diffuse throughout the agarose, resulting in a dilution of the marker concentration, a high concentration of the marker solution was chosen to eliminate this effect. Once the solution was dropped, the response signals of the sodium, potassium and pH sensors all demonstrated a certain tendency of continuous response and eventual stabilization, suggesting that these sensors are capable of receiving changes in in the levels of markers from the in vivo surface in a simulated clinical environment. For sodium and potassium sensors, the response signals were not fully released within 300 s after the solution dropped, primarily because of the response latency, which depends on both the rate of diffusion and the response speed of the sensors. Nevertheless, since the increase or decrease in the levels of markers from glossitis are generally relative, these sensors still have the ability to detect rapidly in practical applications.

IV. CONCLUSION

In this work, the multiparametric flexible sensor arrays system was developed for in situ immediate diagnosis of glossitis. When inflammation occurs, it typically contributes to an increase in pH as well as temperature resulting from the accumulation of bacteria and leukocytes. Increased vascular permeability induces variability in the ratio of sodium and potassium concentrations. Therefore, monitoring these indicators is of great significance for the diagnosis of glossitis. The multiparametric flexible sensor arrays system has been fully evaluated in different environments, providing a reliable basis for subsequent clinical applications.

REFERENCES

[1] J. A. Byrd, A. J. Bruce, and R. S. Rogers, "Glossitis and other tongue disorders," *Dermatologic Clinics,* vol. 21, no. 1, pp. 123-134, 2003/01/01/ 2003.

[2] W. G. de Campos, C. V. Esteves, L. G. Fernandes, C. Domaneschi, and C. A. L. Júnior, "Treatment of symptomatic benign migratory glossitis: a systematic review," *Clinical Oral Investigations,* vol. 22, no. 7, pp. 2487-2493, 2018/09/01 2018.

[3] G. P. Wysocki and T. D. Daley, "Benign migratory glossitis in patients with juvenile diabetes," *Oral Surgery, Oral Medicine, Oral Pathology,* vol. 63, no. 1, pp. 68-70, 1987/01/01/ 1987.

[4] T. Bøhmer and M. Mowé, "The association between atrophic glossitis and protein-calorie malnutrition in old age," *Age and Ageing,* vol. 29, no. 1, pp. 47-50, 2000.

[5] L. Pastore, L. Lo Muzio, and R. Serpico, "Atrophic glossitis leading to the diagnosis of celiac disease," (in eng), *N Engl J Med,* vol. 356, no. 24, p. 2547, Jun 14 2007.

[6] A. F. Sharabi and R. Winters, "Glossitis," in *StatPearls*Treasure Island (FL): StatPearls Publishing Copyright © 2020, StatPearls Publishing LLC., 2020.

[7] M. Erriu, F. Canargiu, G. Orrù, V. Garau, and C. Montaldo, "Idiopathic atrophic glossitis as the only clinical sign for celiac disease diagnosis: a case report," *Journal of Medical Case Reports,* vol. 6, no. 1, p. 185, 2012/07/04 2012.

[8] R. S. Kaslick *et al.,* "Quantitative Analysis of Sodium, Potassium and Calcium in Gingival Fluid from Gingiva in Varying Degrees of Inflammation," *Journal of Periodontology,* vol. 41, no. 2, pp. 93-97, 1970.

Proceedings of the 16th Annual IEEE International
Conference on Nano/Micro Engineered and Molecular Systems
April 25-29, 2021

Contribution discrimination of auxetic cantilever for increased piezoelectric output in vibration energy harvesting

Lu Wang[1,2], Libo Zhao[1,2,*], Zhuangde Jiang[1,2], Xiang Li[2], Ping Yang[1,2], Yu Bai[3], Guoxi Luo[1,2], Qijing Lin[1,2], and Maeda Ryutaro[1,2]

Abstract—Piezoelectric vibration energy harvester (PVEH) with auxetic layer has the advantage of increased voltage output and decreased resonant frequency. To give contribution discrimination of auxetic cantilever for increased piezoelectric output in vibration energy harvesting, this paper summarizes three influencing factors including neutral layer shift, negative Poisson's ratio, and damping loss factor. Three kinds of PVEH with different substrate structures which including solid plate, pores plate, and auxetic plate were designed and fabricated. Their piezoelectric voltage and power curves are simulated in FEA and tested in the experiment. Three contributing factors were identified and calculated for increased piezoelectric output in vibration energy harvesting by auxetic cantilever including neutral layer shift, negative Poisson's ratio, and damping loss factor. This research gives a constructive reference for the design of high output and low frequency PVEH.

Keywords—Auxetic layer, neutral layer shift, negative Poisson's ratio, Resonant frequency, PVEH.

I. INTRODUCTION

Vibration energy harvesting concerns with converting ambient vibration energy to electrical energy for power supply of small electronic devices. This promising approach as a sustainable self-power source leaves out the trouble of routine replacing the battery. A typical application of this technology such as self-powered wireless sensor nodes [1] in a distributed network can be used for state monitoring of machine, transportation, and human [2]. The vibration energy exists in the form of specific frequency such as modal frequency or operating frequency of the machine. Cantilever beam with tip proof mass is a traditional inertia device utilized in vibration energy harvesting via piezoelectric mechanism. The piezoelectric layer is often bonded at the surface of substrate layer of the beam in unimorph or bimorph style. The piezoelectric voltage is mainly proportional to the strain. Because resonant state can cause more strain in piezoelectric layer. the resonant frequency of vibration energy harvester should be design follow that specific frequency. Increase the strain in the piezoelectric layer by mechanical design is hotly pursued in optimal energy harvesting power output. Basic approach is to increase proof mass and increase the thickness of the substrate layer [3]. However, the volume and mass of

[1]State Key Laboratory for Manufacturing Systems Engineering, International Joint Laboratory for Micro/Nano Manufacturing and Measurement Technologies, Overseas Expertise Introduction Center for Micro/Nano Manufacturing and Nano Measurement Technologies Discipline Innovation, Xi'an Jiaotong University (Yantai) Research Institute for Intelligent Sensing Technology and System, and School of Mechanical Engineering, Xi'an Jiaotong University, Xi'an 710049, China
[2]School of Mechanical Engineering, Xi'an Jiaotong University, Xi'an 710049, China
[3]Suzhou Research institute, Xi'an Jiaotong University, Suzhou, Jiangsu 215123, China
*Corresponding author: libozhao@mail.xjtu.edu.cn

piezoelectric vibration energy harvester (PVEH) are also increased and more strain energy is wasted in the substrate layer.

Auxetic metamaterials and structures with negative Poisson's ratio expand (contract) transversely under uniaxial tension (compression) [4]. This unusual behavior exhibits a larger strain in area contrast to conventional materials and structures [5]. Both d31 and d32 contribute to the piezoelectric output, it is reported to have a large enhancement in energy harvesting with auxetic metamaterials and structures [6]. William J.G [7] reported the peak electric power produced by an auxetic piezoelectric energy harvester is 14.4 times that of an plain harvester. However, the main contribution of the gain is the low elastic modulus of the auxetic substrate layer and induce large strain in the piezoelectric layer under the same lateral stress. Qiang Li [8] presented a piezoelectric bimorph with auxetic behaviors can increase the output power of vibration energy harvesting, which is 2.76 times that of a conventional bimorph in COMSOL simulation. Eghbali [9] designed two different auxetic structure as boosters and pasted on a conventional cantilever beam to increase the strain of the piezoelectric layer. Adding this intermediate booster in the non-resonant low frequency range can increase the extracted power by 3.9 and 7.0 times, respectively, but the resonance frequency has increased. Eghbali [10] proposed an auxetic lattice resonator supported by an acoustic rectangular tube to increase the efficiency of acoustic energy harvesting by 10 times.

Except for energy harvesting, auxetic metamaterials and structures also contribute to the ultrasensitive strain sensors. He [11] demonstrated a stretchable resistive strain sensor, which self-assembled a conductive network on a polydimethylsiloxane monolithic substrate with a sine wave and auxiliary binary structure, and a bare substrate. Compared with its traditional counterpart, the sensitivity is greatly improved by 22 times. Zhang [12] used polyurethane foam to expand and contact with polytetrafluoroethylene (PTFE) to manufacture a contact type triboelectric self-powered strain sensor.

There are some advanced processes to fabricate auxetic metamaterials and structures such as 3D printing and MEMS technology. By 3D printing, Li [13] designed mechanical metamaterials primary structural frame attached with a square array of piezoelectric cantilevers to achieve vibration isolation and energy harvesting. By MEMS technology, Umino [14] made a metamaterial SU8 elastic layer by photolithography on one polyvinylidene fluoride (PVDF) film. Ichige [15] coated and patterned the SU-8 elastic layer to form the metamaterial structure. Tsukamoto [16] used the three-dimensional lithography method of a vibration energy harvester to produce a flexible three-dimensional grid structure with periodic holes to reduce the resonance frequency and increase the output power.

978-1-6654-3008-1/21 $31.00 © 2021 IEEE   513

Despite auxetic metamaterials and structures shown good performance in energy harvesting and ultrasensitive sensing, the multifactor mechanism should be discussed. In order to give contribution discrimination of auxetic cantilever for increased piezoelectric output in vibration energy harvesting, this paper summarizes three influencing factors including neutral layer shift, negative Poisson's ratio, and damping loss factor. Three kinds of PVEH with different substrate structures which including solid plate, pores plate, and auxetic plate were designed and fabricated. Their piezoelectric voltage and power curves are simulated in FEA and tested in experiment. This research gives a constructive reference for the design of high output and low frequency PVEH.

II. PRINCIPLE AND MODELING

A. The principal of PVEH with auxetic structure

PVEH with auxetic structure proposed in this work is shown in Fig. 1. A unimorph cantilever consists of polylactic acid (PLA) substrate layer, epoxy bonding layer, and PVDF layer. The cantilever is fixed on the base by a fixture and add a proof mass at the tip. As a novel design, the auxetic structure has occupied half thickness of the substrate layer to increase the piezoelectric output and decrease the resonant frequency.

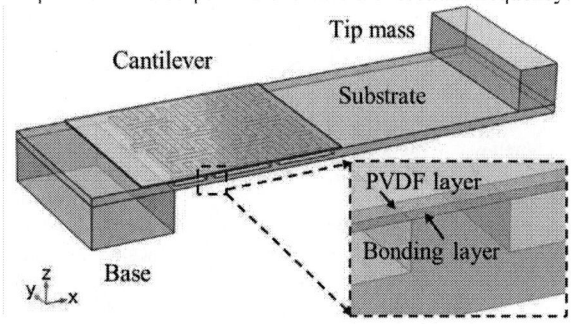

Fig. 1. Schematic of PVEH with auxetic structure.

When the base is excited, the tip mass will vibrate driven by inertia force, and cantilever will have bending deformation and produces strain at the root surface particular. The piezoelectric layer will generate electric charge in alternating tension and compression. Schematic of shift neutral layer contributes to the increased piezoelectric output in vibration energy harvesting is shown in Fig. 2. There is no strain in the neutral layer, and the further away from the neutral layer, the greater the strain. When the top half part of substrate layer is transformed into auxetic layer, the neutral layer will shift away from the piezoelectric layer because the elastic modulus of auxetic layer (E_A) is lower than that of substrate layer (E_S). The stiffness and resonant frequency of PVEH will decrease due to the decrease of total elastic modulus of the cantilever. The piezoelectric output will increase due to the increase of strain in piezoelectric layer.

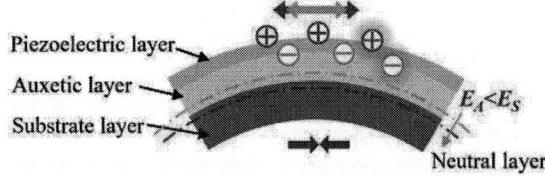

Fig. 2. Schematic of shift neutral layer contributes to the increased piezoelectric output in vibration energy harvesting.

Traditional piezoelectric layer is operated as d31 mode in cantilever configuration. Due to the positive Poisson's ratio of normal material and structure, the d32 mode will generate the opposite piezoelectric charge with the d31 mode. Thanks to the negative Poisson's ratio of auxetic structure, the piezoelectric performance can be enhanced by the d32 mode and d31 mode together. Fig. 3 shown the schematic of d32 mode and d31 mode both contributes to the increased piezoelectric output in vibration energy harvesting by negative Poisson's ratio. The open-circuit voltage of the piezoelectric layer is:

$$U_{OC} = \left(\frac{d_{31}}{\varepsilon_{33}} \overline{\sigma_{11}} + \frac{d_{32}}{\varepsilon_{33}} \overline{\sigma_{22}} \right) t_p \tag{1}$$

$$\overline{\sigma_{22}} = -\mu \overline{\sigma_{11}} \tag{2}$$

where ε_{33} and t_p are the permittivity and thickness of the piezoelectric layer, respectively; $\overline{\sigma_{22}}$ and $\overline{\sigma_{11}}$ are the transverse (y-axis) and average longitudinal (x-axis) stress in the piezoelectric layer, respectively.

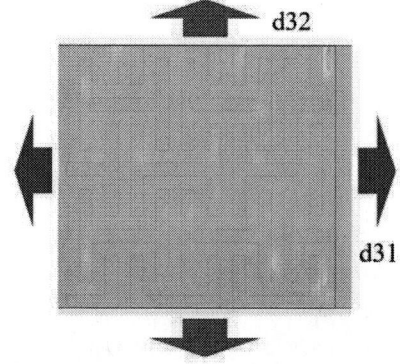

Fig. 3. Schematic of d32 mode and d31 mode both contributes to the increased piezoelectric output in vibration energy harvesting by negative Poisson's ratio.

In order to distinguish the contributions, three kinds of PVEH with different substrate structures were designed and shown in Fig. 4, which including PVEH1 with solid plate, PVEH2 with pores plate and PVEH3 with auxetic plate. Pores plate has positive Poisson's ratio, but auxetic plate has negative Poisson's ratio, and they have almost the same elastic modulus, which is lower than solid plate.

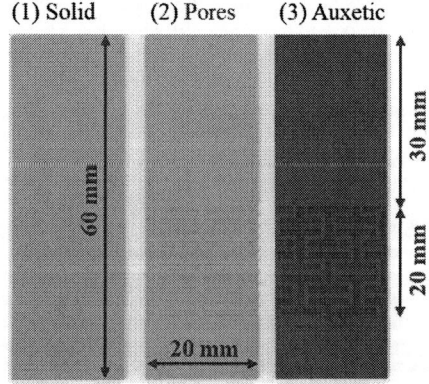

Fig. 4. Different substrates of PVEH, (1) Solid, (2) Pores, (3) Auxetic.

B. FEA Model and Simulation

Electromechanical coupling analysis of PVEH is deploied in COMSOL Multiphysics software by finite element analysis (FEA). 3D geometric and material parameters of PVEH are shown in Table I. The thickness of Ag electrode layer and epoxy resin bonding layer are neglected in FEA modeling. The flexibility matrix of PVDF is:

3.781e-010[1/...	-1.482e-010[1...	-1.724e-010[1...	0[1/Pa]	0[1/Pa]	0[1/Pa]
-1.482e-010[1...	3.781e-10[1/Pa]	-1.724e-010[1...	0[1/Pa]	0[1/Pa]	0[1/Pa]
-1.724e-010[1...	-1.724e-010[1...	1.092e-9[1/Pa]	0[1/Pa]	0[1/Pa]	0[1/Pa]
0[1/Pa]	0[1/Pa]	0[1/Pa]	1.428e-09[1/Pa]	0[1/Pa]	0[1/Pa]
0[1/Pa]	0[1/Pa]	0[1/Pa]	0[1/Pa]	1.11e-09[1/Pa]	0[1/Pa]
0[1/Pa]	0[1/Pa]	0[1/Pa]	0[1/Pa]	0[1/Pa]	1.11e-09[1/Pa]

The coupling matrix of PVDF is:

0[C/N]	0[C/N]	0[C/N]	0[C/N]	0[C/N]	0[C/N]
0[C/N]	0[C/N]	0[C/N]	0[C/N]	0[C/N]	0[C/N]
1.358e-011[C/...	1.476e-012[C/...	-3.38e-011[C/N]	0[C/N]	0[C/N]	0[C/N]

TABLE I.   DESIGN PARAMETERS OF PVEHS

Symbol	Parameters	Value
w_s	Substrate width	20 mm
w_a	Auxetic line width	1 mm
l_s	Substrate length	60 mm
l_c	Cantilever length	50 mm
l_m	Mass length	5 mm
t_s	Substrate thickness	1.0 mm
t_a	Auxetic thickness	0.5 mm
t_p	PVDF thickness	0.11 mm
t_m	Mass thickness	5 mm
R_{load}	Load resistance	100 MΩ
Acc	Excitation acceleration	0.25 g (2.45 m/s²)
n_s	Damping loss factor	0.04
ρ_{PLA}	PLA density	800 kg/m³
ρ_{Cu}	Cu density	8300 kg/m³
ρ_{PVDF}	PVDF density	1780 kg/m³
E_{PLA}	PLA elastic modulus	3.2 GPa
μ_{PLA}	PLA Poisson's ratio	0.3

For the geometric mesh grid, the active cantilever beam is thinner and the inactive structure is thicker. In the field of structural mechanics physics, the foundation of PVEH is fixed. The damping loss coefficient of the structure is set to 0.04. The direction of the piezoelectric axis of the piezoelectric layer is along the Z axis. PVEH applies the vibration acceleration of an object load along the Z axis with an amplitude of 0.25g.

In the circuit physical field, load resistance of 100 MΩ is applied with top and bottom surface of PVDF for open-circuit state. The voltage U at load resistor is used as the output voltage of PVEH. The FEA simulation data is export and plot in MATLAB. The simulation voltage-frequency response curves of different PVEHs are shown in Fig. 5.

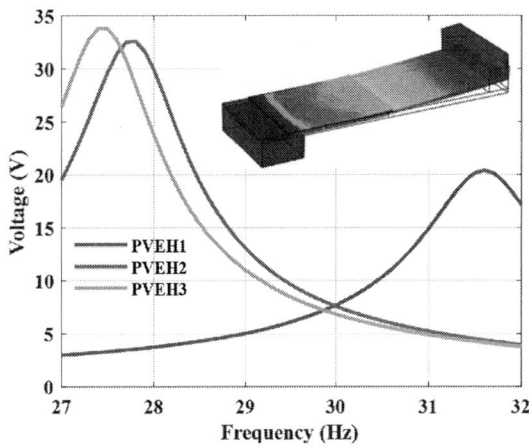

Fig. 5.   Simulation voltage-frequency response curves of different PVEHs.

The simulation results of PVEHs with different substrates are mainly compared in first vibration mode. PVEH3 has lower resonant frequency of 27.4 Hz than PVEH1 resonant frequency of 31.6 Hz. As for open-circuit voltage, PVEH3 is also 70% higher than PVEH1. Interestingly, PVEH2 also has low resonant frequency of 27.7 Hz and high open circuit voltage of 32.7 V, which is close with those of PVEH3. This indicates low elastic modulus in pores and auxetic substrate induced neutral layer shift contributes more to the decreased resonant frequency and increased piezoelectric output in vibration energy harvesting. As for benefits of negative Poisson's ratio, a positive $\overline{\sigma}_{22}$ can help to increase the U_{OC}, but d_{32} is close to one-tenth of d_{31} for PVDF. Therefore, there is little increase in U_{OC} compare PVEH3 with auxetic substrate to PVEH2 with pores substrate.

III. RESULTS AND DISCUSSION

A. Fabrication

PLA substrate layers of solid, pores, and auxetic plates are fabricated by 3D printing (Anycubic 4max pro) with accuracy of 0.05 mm. Prototypes photos are shown in Fig. 6. PVDF layer is coated by silver electrodes on both sides, and bonded with substrate layers by epoxy resin. Nickel fabric is used to lead out electrode wire. Cantilever is clamped on the exciter (JZK-10) by a fixture. Open-circuit voltage is measured by an oscilloscope (MSOX3052A) which has the probe resistance of 100 MΩ.

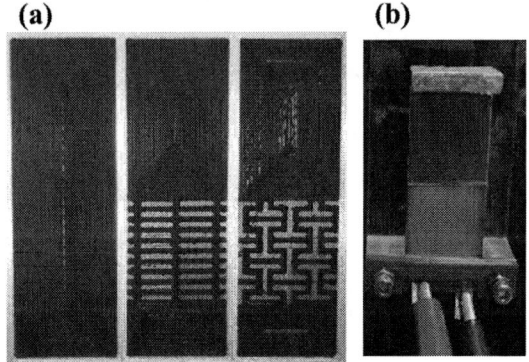

Fig. 6.   Prototypes photos. (a) 3D printing substrate layers of solid, pores, and auxetic plates. (b) PVEH prototypes.

B. Experimental results

Sweep excitation from 20-35 Hz in 0.25 g acceleration to find resonant frequency of each PVEHs. Constant resonant frequency excitation with varied load resistances to find optimal load resistance. The root mean square (RMS) power of load resistor $P = 0.5U^2/R_{load}$ is used as the RMS electric output power of PVEH. Experimental curves of different PVEHs are shown in Fig. 6, which include voltage-frequency response curves, peak voltage and RMS power curves with varied load resistances of three PVEHs. Performance comparison between PVEHs with different substrate layers are shown in Table II, which including maximum open-circuit peak voltage at resonant frequency and maximum RMS power at optimal load resistance.

The experimental results are not as good as the simulation results, but the trend is consistent. It shows PVEH3 has a peak voltage of 18.4 V which is 64% increased than that of PVEH1 with 11.23 V. The RMS power of PVEH3 with 5.182 μW is 2.56 times of that of PVEH1 with 2.024 μW at the same optimal load resistance of 50 MΩ. The peak power of PVEH2 is 2.876 μW, which is not as high as that of PVEH3. PVEH3 has the lowest damping loss factor with 0.0334 and PVEH1 has the highest damping loss factor with 0.0459.

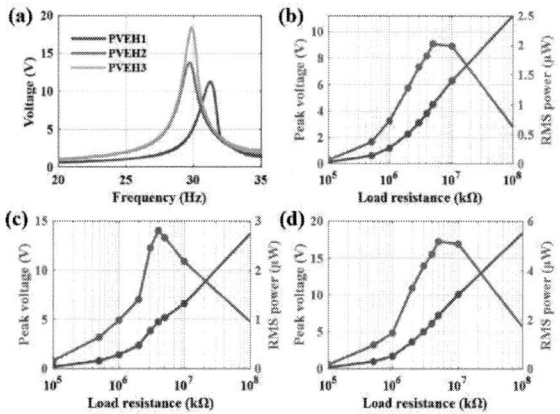

Fig. 7. Experimental curves of different PVEHs. (a) Voltage-frequency response curves of three PVEHs. Peak voltage and RMS power curves with varied load resistances of (b) PVEH1, (c) PVEH2, (d) PVEH3.

TABLE II.   PERFORMANCE COMPARISON BETWEEN PVEHs WITH DIFFERENT SUBSTRATE LAYERS.

Performance	PVEH vibration type		
	PVEH1	PVEH2	PVEH3
Substrate layer	Solid	Pores	Auxetic
Resonant Frequency (Hz)	31.3	29.7	29.9
Peak voltage (V)	11.23	13.77	18.43
RMS power (μW)	2.024	2.876	5.182
Optimal load resistance (MΩ)	50	40	50
Loss factor	0.0459	0.0421	0.0334

C. Contribution discrimination

Three contributing factors were identified for increased piezoelectric output in vibration energy harvesting by auxetic cantilever including neutral layer shift, negative Poisson's ratio, and damping loss factor. Neutral layer shifts away from the piezoelectric layer because of the low elastic modulus in pores layer and auxetic layer. The resonant frequency of PVEH will decrease due to the decrease of total elastic modulus of cantilever and the piezoelectric output will increase due to the increase of strain in piezoelectric layer. The negative Poisson's ratio of auxetic structure can enhance the piezoelectric performance with d32 and d31 mode together. Damping loss factor has an inverse proportional relationship with piezoelectric voltage output in resonant frequency.

Here, the raising effects of these three factors on the open-circuit voltage are quantitatively discussed. Because damping loss factor contributes 9.0% increased peak voltage by comparison between PVEH2 and PVEH1, the neutral layer shift contributes 12.5% increased peak voltage by dividing damping loss factor contribution in totally 22.6% increased. PVEH3 has peak voltage of 64% totally increased than that of PVEH1. It is found that damping loss factor contributes 33.4% increased peak voltage by comparison between PVEH3 and PVEH1. By dividing neutral layer shift contribution of 12.5% increased and damping loss factor contribution of 22.6% increased in 64% totally increased, negative Poisson's ratio contributes 18.9% increased.

IV. CONCLUSION

We have presented the principle of neutral layer shift and negative Poisson's ratio in PVEH with auxetic layer. Three kinds of PVEH with different substrate structures which including solid plate, pores plate and auxetic plate were designed and fabricated. Their piezoelectric voltage and power curves are simulated in FEA and tested in experiment. Experiment result analysis shows 64% totally increased piezoelectric voltage output compared between PVEH3 and PVEH1, where neutral layer shift contribution of 12.5% increased and damping loss factor contribution of 22.6% increased, and negative Poisson's ratio contributes 18.9% increased.

ACKNOWLEDGMENT

This work was supported in part by the National Natural Science Foundation of China (Grant Nos. 51890884, 91748207), Jiangsu Province Foreign Expert Program (Grant No. BX2020032), and the Recruitment Program of Global Experts (Grant No. WQ2017610445).

REFERENCES

[1] L. Wang et al., "System level design of wireless sensor node powered by piezoelectric vibration energy harvesting," Sensors and Actuators A: Physical, vol. 310, 2020, doi: 10.1016/j.sna.2020.112039.

[2] L. Wang et al., "Self-sustained autonomous wireless sensing based on a hybridized TENG and PEG vibration mechanism," Nano Energy, 2020, doi: 10.1016/j.nanoen.2020.105555.

[3] L. Wang et al., "A packaged piezoelectric vibration energy harvester with high power and broadband characteristics," Sensors and Actuators A: Physical, vol. 295, pp. 629-636, 2019, doi: 10.1016/j.sna.2019.06.034.

[4] X. Ren, R. Das, P. Tran, T. D. Ngo, and Y. M. Xie, "Auxetic metamaterials and structures: a review," Smart Materials and Structures, vol. 27, no. 2, 2018, doi: 10.1088/1361-665X/aaa61c.

[5] P. U. Kelkar, H. S. Kim, K. H. Cho, J. Y. Kwak, C. Y. Kang, and H. C. Song, "Cellular Auxetic Structures for Mechanical Metamaterials: A Review," Sensors, vol. 20, no. 11, Jun 1 2020, doi: 10.3390/s20113132.

[6] T. Fey et al., "Mechanical and electrical strain response of a piezoelectric auxetic PZT lattice structure," Smart Materials and Structures, vol. 25, no. 1, 2016, doi: 10.1088/0964-1726/25/1/015017.

[7] W. J. G. Ferguson, Y. Kuang, K. E. Evans, C. W. Smith, and M. Zhu, "Auxetic structure for increased power output of strain vibration energy harvester," Sensors and Actuators A: Physical, vol. 282, pp. 90-96, 2018, doi: 10.1016/j.sna.2018.09.019.

[8] Q. Li, Y. Kuang, and M. Zhu, "Auxetic piezoelectric energy harvesters for increased electric power output," *AIP Advances,* vol. 7, no. 1, 2017, doi: 10.1063/1.4974310.

[9] P. Eghbali, D. Younesian, and S. Farhangdoust, "Enhancement of piezoelectric vibration energy harvesting with auxetic boosters," *International Journal of Energy Research,* vol. 44, no. 2, pp. 1179-1190, 2019, doi: 10.1002/er.5010.

[10] P. Eghbali, D. Younesian, and S. Farhangdoust, "Enhancement of the low-frequency acoustic energy harvesting with auxetic resonators," *Applied Energy,* vol. 270, 2020, doi: 10.1016/j.apenergy.2020.115217.

[11] H. Yu *et al.,* "Two-Sided Topological Architecture on a Monolithic Flexible Substrate for Ultrasensitive Strain Sensors," *ACS Appl Mater Interfaces,* vol. 11, no. 46, pp. 43543-43552, Nov 20 2019, doi: 10.1021/acsami.9b14476.

[12] S. L. Zhang, Y.-C. Lai, X. He, R. Liu, Y. Zi, and Z. L. Wang, "Auxetic Foam-Based Contact-Mode Triboelectric Nanogenerator with Highly Sensitive Self-Powered Strain Sensing Capabilities to Monitor Human Body Movement," *Advanced Functional Materials,* vol. 27, no. 25, 2017, doi: 10.1002/adfm.201606695.

[13] Y. Li, E. Baker, T. Reissman, C. Sun, and W. K. Liu, "Design of mechanical metamaterials for simultaneous vibration isolation and energy harvesting," *Applied Physics Letters,* vol. 111, no. 25, 2017, doi: 10.1063/1.5008674.

[14] T. T. Y. Umino, S. Shiomi, "Development of vibration energy harvester with 2D mechanical metamaterial structure," *PowerMEMS 2017,* 2017.

[15] N. K. R. Ichige, Y. Umino1, T, "Size Optimization of Metamaterial Structure for Elastic Layer of a Piezoelectric Vibration Energy Harvester," *PowerMEMS 2019,* C 2019.

[16] T. Tsukamoto, Y. Umino, K. Hashikura, S. Shiomi, K. Yamada, and T. Suzuki, "A Polymer-based Piezoelectric Vibration Energy Harvester with a 3D Meshed-Core Structure," *J Vis Exp,* no. 144, Feb 20 2019, doi: 10.3791/59067.

Research on Embedded 3D Printing for Magnetic Soft Robots

Wei Zhang,   Jiachen Li,   Huicong Liu,   Guoqing Jin*

Abstract— Traditional 3D printing technology is difficult to fabricate soft robots with complex structures and functions. However, embedded 3D printing technology has the ability to fabricate complex structures, including sensors, actuators, control system and power system integrated with soft robots. In this paper, based on the developed embedded direct inkjet 3D printing system, the magnetic slurry is printed inside Ecoflex materials to fabricate magnetically driven soft robots. Firstly, Comsol was used to study the rheological properties and viscosity properties of printing materials. Secondly, the optimal printing parameters are determined by the material ratio, mechanical stretching and 3D printing parameter experiments; Finally, the magnetic inchworm and gecko soft robots were printed by the determined optimal parameter, and bionic experiments were performed to verify the feasibility of integrated manufacturing of embedded 3D printing soft robots.

Keywords— *embedded 3D printing, soft robot, process parameter optimization, magnetic driven*

I. INTRODUCTION*

In recent years, soft robots with high flexibility and safety have been widely used in industrial production, exploration of complex environments, medical rehabilitation, etc [1-3].

Currently, the manufacturing methods of soft robots mainly focus on mold pouring [4], manual manufacturing [5] and external 3D printing [6]. As the design of the structure and function of soft robot becomes more and more complex, traditional manufacturing methods could not meet the requirements of the intelligent. Embedded 3D printing technology can fabricate complex internal geometric structures, and functional new materials can also be freely formed in the supporting solution [7-8], which provides technical support for the fabrication of intelligent software robots.

In recent years, some related researches of embedded 3D printing have been published. Whitesides research group of Harvard University created the internal circuit of entirely soft, autonomous octopus by embedded 3D printing technology [9]. Wood RJ research group of Harvard University designed sheet-type flexible strain sensors and sensing gloves by embedded 3D printing to verify sensor performance [10]. Truby, RL. Designed with a soft gripper with functional materials embedded based on the technology of embedded 3D printing[11]. Hinton TJ of Carnegie Mellon

University demonstrated the freeform reversible embedding of PDMS material, which broadened the range of materials that can be applied to additive manufacturing [12]. Using embedded 3D printing technology, the Dvir of Technion-Israel University in Israel has successfully printed a retractable heart [13]. A. Lee of Carnegie-Mellon University went a step further and printed heart tissue that can pump blood normally by embedded 3D bioprinting [14].

However, researchers mainly focused on new materials and biological properties, there are few studies on the optimization of process parameters in the printing process. This paper focuses on the optimization of the process parameters of embedded 3D printing magnetic slurry, printing the paste mixed with magnetic powder in two-component silicone to fabricate high quality intelligent soft robots.

First, magnetic slurry of different proportions were prepared for studying the rheological properties of the two-component silicone Ecoflex00-30. Secondly, some parameters are considered, such as the fluid pressure in the needle, the flow rate and the movement of the needle in the support material, to simulate suitable printing parameters. Then, the combination of printing parameters is determined through the experiment of material ratio, printing shape, printing thickness and filling density. Finally, the magnetic drive inchworm and gecko soft robot were printed by the optimal combination of parameters, and several tests were carried out to verify the feasibility of integrated manufacturing of magnetic soft robots by embedded 3D printing.

II. MATERIALS

Embedded 3D printing requires support materials to fix the geometric position of the uncured functional paste. As shown in Fig.1(a), the support material needs to behave as fluid when the needle moves during the printing process, which meets the Bingham equation, and its mathematical expression is:

$$\tau = \tau_s + \eta_1 \times \dot{\gamma} == \eta \frac{dv}{dr} + \tau_s \qquad |\tau| > \tau_s$$
$$\dot{\gamma} = 0 \qquad\qquad\qquad |\tau| \le \tau_s \qquad (1)$$

Based on the equation above, the flow of the viscoplastic matrix around the cylindrical needle is quantified to simulate embedded 3D printing processing.

The curve of storage modulus and loss modulus after Ecoflex00-30 mixed for 10min and 30min was tested by a rotational rheometer (Kinexus) in Fig.1(b-d). The mixed material is shear-thinning, but the viscosity is too low to support itself during 3D printing.

The magnetic slurry is made of magnetic particles NdFeB and Ecoflex00-30, mixed and degassed by a planetary centrifugal mixer (Thinky), and can be used for research after the finished product is stable. The whole process takes about 20 minutes.

---

* Research supported by National Natural Science Foundation of China (61773274).

Wei Zhang is with Soochow University, Robotics and Microsystem Research Center, Suzhou,CHINA (e-mail: sven_zhang@sz.acmlife.org)

Jiachen Li is with Soochow University, Robotics and Microsystem Research Center, Suzhou, CHINA (e-mail: ljjcc3107@gmail.com)

Huicong Liu is with Soochow University, Robotics and Microsystem Research Center, Suzhou, CHINA (e-mail: hcliu078@suda.edu.cn)

Guoqing Jin is with Soochow University, Robotics and Microsystem Research Center, Suzhou, CHINA (e-mail: gqjin@suda.edu.cn)

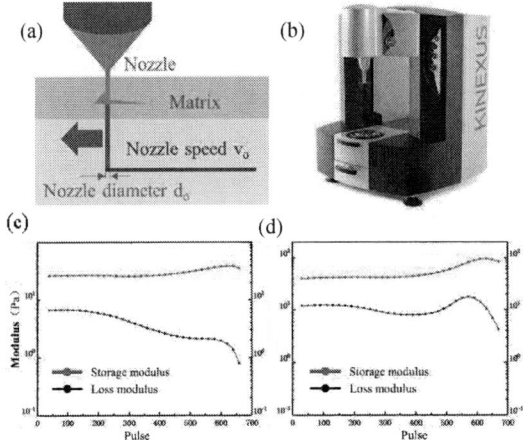

Figure 1. *(a)* embedding printing model; *(b)* rheometer;

(c) rheological analysis; *(d)* PDMS viscosity change

III. SIMULATION

Ecoflex 00-30 is a non-Newtonian fluid, so Comsol Multiphysics (Computational Fluid Dynamics) is used for fluid simulation.

In part A, internal flow field of different nozzle types was simulated. First, the simulation of internal flow field of Ecoflex for two different types of nozzles has been done, and the functional differences of the two nozzles were compared. Straight needle is selected to reduce the damage to the supporting fluid. Second, perform more detailed simulation for the embedded straight nozzle, and provide pre-research data for subsequent embedded 3D printing parameter experiments.

In part B, needle embedded movement was simulated, the fluid parameters take the median value by Grosskopf [8], which are marked in Tab.1. In order to obtain the appropriate range of shear thinning of the support material, Comsol was used to calculate the embedded printing process, and simulate the dynamic flow velocity and shear rate of the adjacent fluid.

TABLE I. RHEOLOGICAL PARAMETERS CONDITIONS MATRIX MATERIAL

$m(pa*s)$	n	τ_y (N/m2)	mp (s)	$\dot{\gamma}_{ref}$ (1/s)	$\rho(kg/m3)$
10	0.9	20	5	0.1	1100

A. Simulation of Pressure and Flow Rate

The air pressure P is used as a parameterized simulation scan to simulate the pressure and flow rate U of the needle. Fig.2(a-b) is the simulation diagram of the pressure and flow rate when the inner diameter of the needle is 0.44mm.

The extrusion pressure P directly affects the final flow rate U. The flow rate U can be accurately simulated by inputting the simulated pressure P, which provides accurate data for appropriate process combination in the following research.

B. Simulation of Embedded Movement

Embedded printing process was calculated and simulated by Comsol to obtain the appropriate range of shear thinning

of the support material. As shown in Fig.2(c-d), the dynamic flow velocity and shear rate of the nearby fluid are simulated by inserting the needle into the Bingham fluid.

When the moving speed v0 of needle, the range of high flow rate and shear rate (red area) expands, the maximum shear rate increases, but the maximum flow rate remains almost unchanged.

Fluid flows in the area when the shear stress exceeds the yield stress. For embedded printing that requires support, suitable scope of work should be controlled in double needle diameter range.

IV. PARAMETER EXPERIMENTS

Parameter optimization experiments were utilized to implement embedded 3D printing. Firstly, the magnetic size and mechanical strength of materials with different mixing ratios were tested. Secondly, embedded 3D printing experiment was carried out to analyze the influence of shape, thickness and filling density on magnetism under the same area. Finally, the best combination of printing parameters was determined to manufacture magnetic driven software robots.

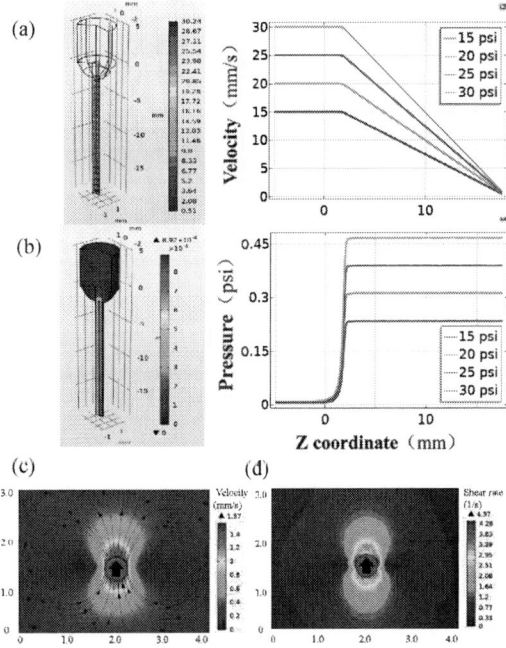

Figure 2. Simulation diagram *(a)* flow rate;

(b) pressure; *(c)* flow rate; *(d)* shear rate

A. Material Ratio

Mix NdFeB particles with different mass fraction ratios with Ecoflex to test the maximum magnetic force. As shown in Fig.3(a), different proportions of magnetic slurry were placed in petri dishes ,which diameter are 20 mm, ensure that the liquid level is flush with the petri dish, and the thickness of the sheet is the same. Subsequently, the petri dishes were placed in a -15psi negative pressure for curing, and magnetic tests were carried out after 24 hours.

The experimental device is shown in Fig. 3(b), magnetic force is provided by the suction cup electromagnet (P25/11), which can absorb 5kg copper plates at DC24V and 0.15A. At the same time, a high-precision electronic tension meter (AIGU ZP-10) was used to test the maximum shedding force of 9 different ratios of flakes. Lubricated linear slide module of the stepper motor is used to achieve a uniform speed of 30mm/min. The peak data before shedding was measured multiple times, and the results were summarized in a scatter box plot in Fig.3(c).

It can be observed that the maximum shedding force of the flakes increases as the proportion of magnetic powder particles increases. The ratio of 2:1 is the extreme point of the median growth rate, and the overall graph shows a linear upward trend. However, the excessively high proportion of particles leads to increased viscosity, which makes it difficult to extrude and is unsuitable for 3D printing. Therefore, it is necessary to consider the relationship between viscosity and magnetic force.

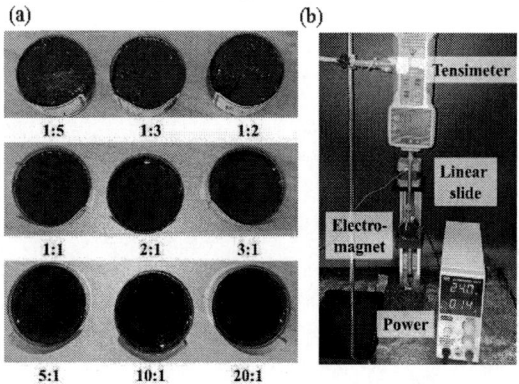

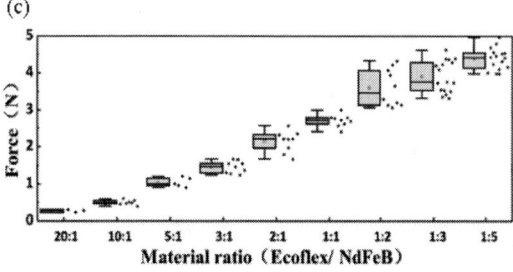

Figure 3. *(a)* material ratio; *(b-c)* magnetic test experiment

B. Tensile Test

Tensile experiments were performed on the mixed slurry of NdFeB particles and Ecoflex in different proportions to test the mechanical properties after curing. The tensile specimens, which length L=150mm, were manufactured by casting, and solidified for 24h under negative pressure. Then the test was performed by electric tensile testing machine (ZQ-990), stretching at the speed of 60mm/min, as shown in Fig.4(a-c). Finally, the results of tensile strength-time and tensile strength-strain are arranged in Fig.4(d-e), and the results that were too similar were combined.

The slurry has almost no fluidity and cannot be molded and cast when particles exceed 50%wt.

The maximum tensile increment of the fully cured Ecoflex00-30 is approximately 200% of the original size and nominal strain will gradually reduce by increasing the proportion of particles. The maximum stretching increment

is reduced to 100% of the original size when the particles reach 50%wt. When the mechanical strength of the two materials is too different, the material will fall off when stretched or contracted, which will reduce the service life. Summarizing the two tests, choose the optimal material ratio of 2:1.

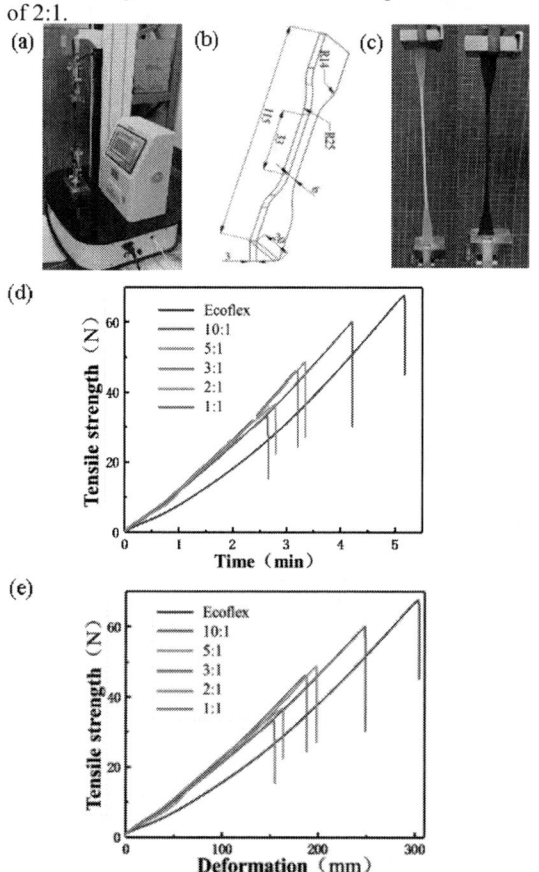

Figure 4. *(a)* tensile test; *(b)* tensile specimen size; *(c)* tensile specimen;

(d) strength-time;*(e)* strength- deformation diagram

C. Experiment of Embedded 3D Printing

The Bingham property of the support material keeps the printed paste in a fixed position, and even some complex patterns can be printed (Fig.5(a)).Based on the previous simulation results, orthogonal experiments are carried out from the three parameters of shape, thickness and packing density. As shown in Fig.5(b), embedded 3D printing in a mold filled with silicone, and the maximum magnetic force is tested after curing.

Printing parameters in Fig.5(c): (1) Shape: circles, squares, triangles and pentagrams of the same area. (2) Density:20%, 40%, 60%, 80% and 100%. (3) Thickness: 0.5mm, 1mm, 1.5mm, 2mm and 2.5mm. The maximum shedding force experiment was performed on 100 groups of subjects, and the data was summarized in a scatter box plot in Fig.5(d-f).

In general, circle has the best magnetic properties, followed by square and triangle, star is the worst under a single factor. The filling density η and the magnetism are approximately proportional to the linear relationship.There is almost no magnetic force when $\eta<60\%$. The magnetic increases with the increase of the thickness t, and when the thickness$t>1mm$, the growth rate of maximum magnetic force gradually decrease.

However, excessive filling density and thickness (η=100%, t=2.5mm) will cause diffusion of the magnetic paste. Spreading beyond the preset area is unacceptable in high-precision embedded 3D printing. So the best parameter combination is: circle, η=80% 、 t=2mm. Under this condition, the maximum magnetic force provided by a circular area (r=5mm) is about 1N.

V. Soft Robot Experiments

We use self-built gantry DIW embedded 3D printer for soft robot printing experiments. Ecoflex 00-30 was used as supporting material and the printing material uses magnetic paste (2:1). All subsequent experimental products were printed with the following parameters.

A. Inchworm Soft Robot

Embedded 3D printing has unique advantages of fabricating complex internal geometric structures integrated with soft robots. The designed inchworm soft robot is based on the Yeoh model [15] with CAD software SolidWorks. As shown in Fig.6 (a-b), the length, width and height are 56mm, 16mm and 8mm, and the magnetic functional area is embedded in the two sides of soft robot. As shown in Fig.6(c), it is the crawling process of the soft robot with imitated inchworm magnetic drive on aluminum plates and glass.

The experiment used five different materials of the same thickness (aluminum plate, PC, paper, PVC and glass) to perform a vertical plane crawling experiment. Same number of rubidium magnets were used on the back to attract the head and tail parts (shown in Fig.6(d)). Due to roughness of surface, the maximum distance, height and success rate are different. The average data is summarized in Tab.2 and the surface roughness Ra is measured by the handheld roughness meter (TR200).

As the roughness of surface increases, the crawling amplitude of the inchworm gradually becomes smaller, and can be arched at an angle of about 75° on the smooth glass plane. However, the silicone will be adsorbed on the smooth surface, and reduce the rate of successful vertical crawling success. On the rougher PVC board, the crawling success rate reaches 85%, but the distance of a single vertical crawl is 66.5% of the smooth surface, and the crawling speed is about 4.2mm/s.

Meanwhile, five slope crawling experiments were carried out on a rougher PVC board to verify the feasibility of magnetic drive inchworm crawling on different slopes (Fig.6(e)), which providing effective evidence for embedded 3D printing soft robots.

TABLE II. INCHWORM VERTICAL CRAWLING EXPERIMENT

Material	Ra(μm)	Distance /step(mm)	Height h(mm)	Height h(mm)
Glass	0.006	15.8	19.1	25%
PC	1.6	14.1	17.9	40%
AL	3.2	13.7	16.2	55%
Paper	7.4	11.3	14.9	70%
PVC	25	10.5	13.6	85%

B. Gecko Soft Robot

The green gecko soft robot was cast by a mold and the magnetic functional areas were printed in four 'palms' positions. Afterwards, a gait experiment was carried out on the printed gecko. Gecko crawls vertically in five ways: 1 moves; 3 moves; 1, 3 moves; 1, 2 moves; 1, 3 moves, and summary the distance and success rate 20 times in Fig.7.

The success rate of single-foot crawling is over 80%, and the distance is about 60% of between two feet. While crawling with two feet on the same side is prone to rollover, and easy to fail. In general, the alternating movement of the front and rear feet makes the whole body in an 'S' shape, which is a relatively successful and natural crawling gait.

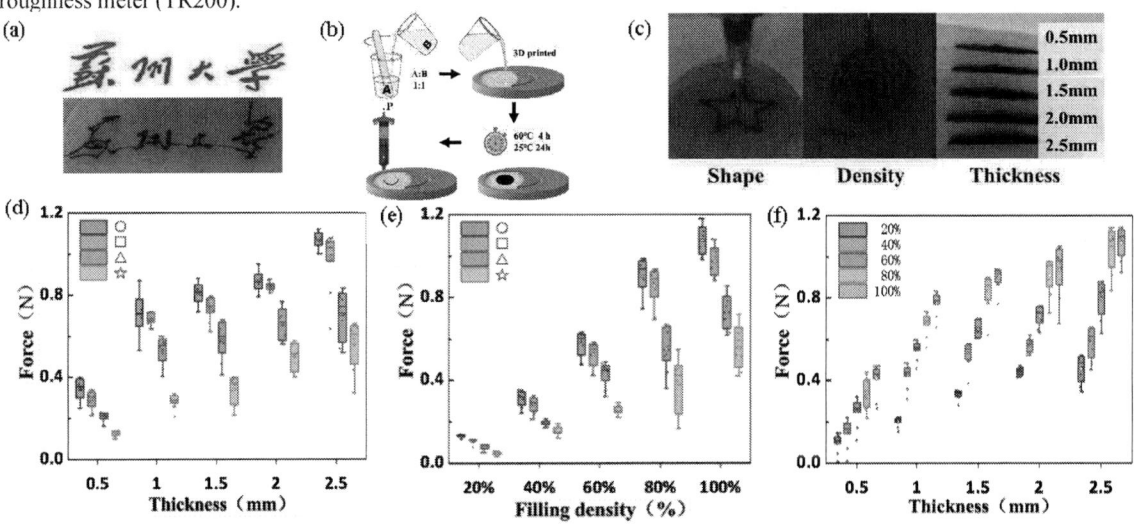

Figure 5. (a) complex graphics printing; (b) embedded printing process; (c) parameter experiment;

(d) shape-thickness;(e) shape-filling density; (f) filling density-thickness

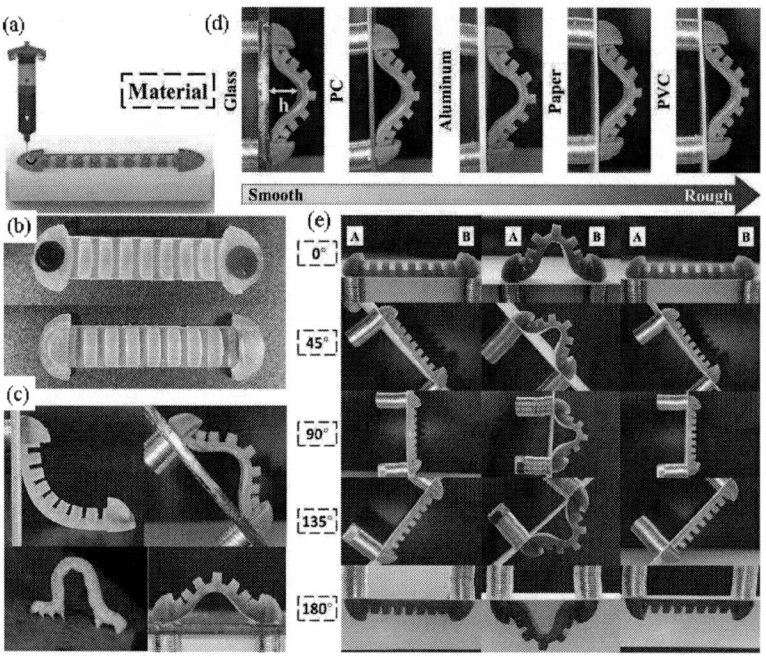

Figure 6. *(a)* printing process;*(b)* embedded printing inchworm;*(c)* crawling diagram;

(d) vertical crawling of different materials with the same thickness;*(f)* crawling at different angles

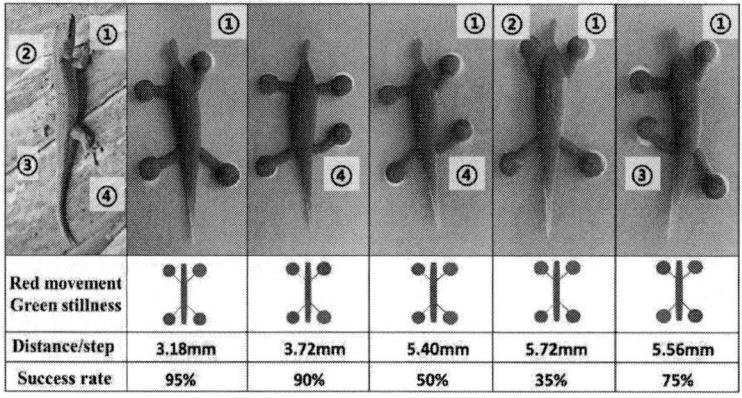

Red movement Green stillness					
Distance/step	3.18mm	3.72mm	5.40mm	5.72mm	5.56mm
Success rate	95%	90%	50%	35%	75%

Figure 7. Gait experiment of gecko like soft robot

VI. CONCLUSION

Based on the built embedded 3D printing system, this paper focused on the optimal parameter combination of embedded 3D printing magnetic functional materials through materials, simulations and experiments. Main conclusions are as follows:

- The relationship between pressure P and flow rate U is simulated. Simultaneously, shear rate of the needle moving in Bingham fluid was simulated, and the optimal range of the critical needle moving speed was obtained.
- Magnetic force and mechanical strength of different proportions of magnetic materials were tested to determine suitable material parameters. Secondly, the embedded 3D printing experiment was carried out and the best parameter combination is: circle, η=80%、

t=2mm.
- Finally, inchworm and gecko soft robots were fabricated by the selected optimal parameters. In the future, this research can be applied to the cleaning of smooth surfaces or pipe crawling. All printed soft robots showed high quality, verifying the feasibility of integrated fabricating of soft robots with embedded 3D printing technology.

ACKNOWLEDGMENT

Research supported by National Natural Science Foundation of China (61773274).

REFERENCES

[1] Rus D , Tolley M T . Design, fabrication and control of soft robots[J]. Nature, 2015, 521(7553):467-75.

[2] Xiaoxia, Le, Wei, et al. Recent Progress in Biomimetic Anisotropic Hydrogel Actuators[J]. Advanced Science, 2019.

[3] Bingbing Hu, Guoqing Jin,et al. Design and Fabrication of a Multi-actuator Soft Robot Inspired by Young Tiger Beetle.[J]ROBOT,2018,40(05):626-633.

[4] Gong Z , Xie Z , Yang X , et al. Design, fabrication and kinematic modeling of a 3D-motion soft robotic arm[C]// IEEE International Conference on Robotics & Biomimetics. IEEE, 2016.

[5] Chen, Xueya, Liang,et al. Pneumatically Actuated Soft Robotic Arm for Adaptable Grasping[J]. Journal of Solid Mechanics, 2018(5):608-622.

[6] Research on 3D Printing Silicone Soft Materials for Soft Robots [J]. China Mechanical Engineering, 2020,31(05):603-609+629.

[7] Wang J , Liu Y , Fan Z , et al. Ink-based 3D printing technologies for graphene-based materials: a review[J]. Advanced Composites and Hybrid Materials, 2019.

[8] Grosskopf, Abigail, K,et al. Viscoplastic Matrix Materials for Embedded 3D Printing[J]. Acs Applied Materials & Interfaces, 2018.

[9] Fitzgerald, Daniel J, Whitesides,et al. An integrated design and fabrication strategy for entirely soft, autonomous robots[J]. Nature, 2016.

[10] Muth J T , Vogt D M , Truby R L , et al. Embedded 3D Printing of Strain Sensors within Highly Stretchable Elastomers[J]. Advanced Materials, 2014, 26(36):6307-12.

[11] Soft Somatosensitive Actuators via Embedded 3D Printing[J]. Advanced Materials, 2018, 30(15).

[12] Hinton T J , Hudson A , Pusch K , et al. 3D Printing PDMS Elastomer in a Hydrophilic Support Bath via Freeform Reversible Embedding[J]. ACS Biomaterials ence and Engineering, 2016:1781-1786.

[13] Noor N , Shapira A , Edri R , et al. 3D Printing of Personalized Thick and Perfusable Cardiac Patches and Hearts[J]. Advanced Science, 2019.

[14] Lee A, Hudson A R , Shiwarski D J , et al. 3D bioprinting of collagen to rebuild components of the human heart[J]. ence (New York, N.Y.), 2019,365(6452):482-487.

[15] Wu P , Jiangbei W , Yanqiong F . The Structure, Design, and Closed-Loop Motion Control of a Differential Drive Soft Robot[J]. Soft Robotics, 2017, 5(1):71

Proceedings of the 16th Annual IEEE International
Conference on Nano/Micro Engineered and Molecular Systems
April 25-29, 2021

Research on Coupling Mechanism of Wireless Power Supply Equipment for Film Pressure Wireless Sensing node of Water-lubricated bearing

Yifan Zhao, Nan Wang*, Fan Jiang

Abstract—The wireless sensing node used to monitor the film pressure of water-lubricated bearing is installed on shaft and runs at high speed with shaft, it is time-consuming to replace the battery which has finite lifetime. Therefore, the continuous power supply for wireless sensing node is difficult to be solved. In this paper, the non-contact radio energy transmission method is proposed. Compared with the electrically conductive slip ring, the wear and spark discharge can be avoided by this method. The nested coil structure is adopted as the coupling mechanism of inductively coupled power transfer device, and the LCC-S compensation topology is utilized to compensate the primary and secondary circuit. The theoretical model and physical model of inductive power transmission coupling compensation system are established, and the primary and secondary coils are co-simulated by Maxwell and Simplorer software. Finally, the simulation results are compared with the theoretical ones, and the results show that the error between simulation and theory results is less than 5%, which also verifies the feasibility and accuracy of this method.

Key Words—Wireless sensing node, water-lubricated bearing, the non-contact radio energy transmission, non-contact radio energy transmission, coupling mechanism, LCC-S compensation topology.

I. INTRODUCTION

Water film pressure is an important parameter of water-lubricated bearings, and the important characteristics including the water film thickness, the bearing capacity and lubrication mode can be obtained through in-depth study of water film pressure [1]. Fig. 1 shown the water-lubricated bearing test-rig. In order to supply the power to wireless sensing node rotating with shaft for water film pressure, the common methods including battery power supply, and generator power supply are proposed. Because of the finite life, and the time-consuming replacement of battery, the battery is unable to used for the continuously monitoring of wireless sensor node installed on shaft rotates at high-speed. The conductive slip ring is used for power transmission in the method of generator power supply, and which has frictional wear problems, and will lead to unsafe accidents such as spark discharge. Therefore, the wireless power supply method is proposed, and the wireless power transfer technology is used in this paper.

[1]Shaanxi Key Laboratory of Industrial Automation, Hanzhong, Shaanxi, China.
[2]School of Mechanical Engineering, Shaanxi University of Technology, Hanzhong, Shaanxi, China.
*Contacting Author: Nan Wang is the associate professor of the School of Mechanical Engineering, Shaanxi University of Technology; No.1 East Ring Road, Hantai District, Hanzhong, Shaanxi, China email: heroyoyu@126.com).

Wireless power transfer mainly uses space invisible medium to complete. It can be divided into microwave radiation type, magnetic coupling resonance type and electromagnetic induction type. Microwave radiation type is to convert electric energy into microwave, using antenna to achieve directional long distance wireless power supply. However, microwave transmission is susceptible to the influence of medium, which limits its application occasions. It is generally applied in special fields such as aerospace [2]. Microwave radiation type is the use of the near field resonance concept of strong coupling, the resonance frequency of the basic principle is the same between two objects can achieve efficient energy exchange, can implement several meters within the scope of the wireless power supply [3,4]. Electromagnetic induction radio energy transmission uses electromagnetic induction to realize the wireless transmission of electric energy, suitable for a variety of occasions, so it is widely used.

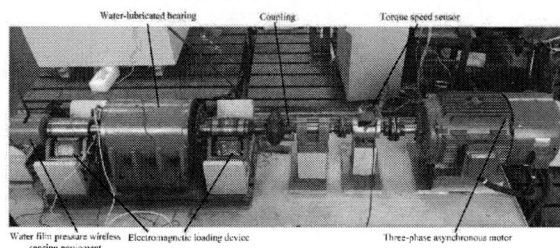

Figure 1. Water-lubricated bearing test-rig.

II. SYSTEM STRUCTURE AND THEORETICAL ANALYSIS

A. System Structure

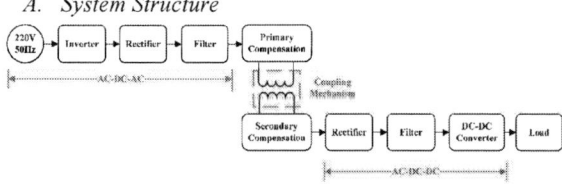

Figure 2. The overall system structure.

The difficulty of continuous monitoring of water film pressure wireless sensing node can be solved by wireless power supply technology. The overall system structure is shown in Fig. 2 . The electrical energy transformation part, compensation part, electromagnetic coupling part and load part are included in it.

Electromagnetic coupling mechanism can be equivalent to a loosely coupled transformer, the air gap between the primary and secondary coil which result in the leakage

978-1-6654-3008-1/21 $31.00 © 2021 IEEE   524

inductance of the larger. In order to make the secondary coil produce enough of the induced electromotive force, the primary coil needs to be completed prior to power AC - DC - AC conversion, but the high frequency AC electric flux into the primary coil will cause the reactive power of the system is too high, affect the power factor of system. Therefore, it is necessary to increase the resonance compensation topology in the primary and secondary coils to improve the power factor of the system. After the inductive voltage compensation of the secondary side coil, it is connected to the load after the conversion of AC-DC-DC.

B. Circuit Analysis

The LCC-S compensation topology based on mutual inductance model is shown in Fig. 3. U_1 is the output voltage of full-bridge inverter.

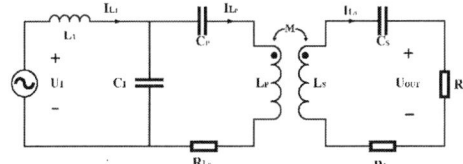

Figure 3.   Equivalent circuit of LCC-S topology.

Fundamental AC impedance analysis was used for theoretical analysis. Assuming that the input of the full-bridge inverter circuit is U_{in}, so the effective value of the output fundamental voltage U_1 can be written as[5]

$$U_1 = \frac{2\sqrt{2}}{\pi} U_{in} \qquad (1)$$

Fundamental AC impedance analysis was used for theoretical analysis. When the compensation circuit resonates, the following relationship should be satisfied

$$\begin{cases} \omega L_1 = \dfrac{1}{\omega C_1} \\[2mm] \omega L_p = \dfrac{1}{\omega C_p} + \dfrac{1}{\omega C_1} \\[2mm] \omega L_s = \dfrac{1}{\omega C_s} \end{cases} \qquad (2)$$

The configuration of LCC-S compensation topology parameters can be obtained by (2)

$$\omega = \frac{1}{\sqrt{L_1 C_1}} = \frac{1}{\sqrt{(L_P - L_1)C_P}} = \frac{1}{\sqrt{L_s C_s}} \ , \quad \omega = 2\pi f \qquad (3)$$

The primary side impedance Z_1 and secondary side impedance Z_2 are given by

$$\begin{cases} Z_1 = R_{L1} + j\omega L_1 + \dfrac{1}{j\omega C_1} / /(\dfrac{1}{j\omega C_p} + j\omega L_p + R_{Lp}) \\[3mm] Z_2 = R_l + R_{Ls} + j\omega L_s + \dfrac{1}{j\omega C_s} \end{cases} \qquad (4)$$

The equivalence relationship is established as follows

$$\begin{cases} j\omega M \dot{I}_{Lp} - Z_2 \dot{I}_{Ls} = 0 \\[2mm] j\omega M \dot{I}_{Ls} + Z_M \dot{I}_{Lp} = 0 \end{cases} \qquad (5)$$

The reflection of the secondary side coil to the primary side coil can be obtained

$$Z_M = \frac{\omega^2 M^2}{Z_2} \qquad (6)$$

Combined with (1), Z_1, Z_2, Z_M can be deduced as

$$\begin{cases} Z_1 = R_{L1} + \dfrac{\omega^2 L_1^2}{R_{LP} + Z_M} \\[3mm] Z_2 = R_l + R_{Ls} \\[3mm] Z_M = \dfrac{\omega^2 M^2}{R_l + R_{Ls}} \end{cases} \qquad (7)$$

The LCC-S topology input current $\dot{I}_1$ can be given by

$$\dot{I}_1 = \frac{\dot{U}_1}{Z_1} \qquad (8)$$

The primary coil current $\dot{I}_{Lp}$ and secondary coil current $\dot{I}_{Ls}$ are calculated by

$$\dot{I}_{Lp} = \frac{\dfrac{1}{j\omega C_1} \dot{I}_1}{\dfrac{1}{j\omega C_p} + j\omega L_p + Z_M + R_{Lp} + \dfrac{1}{j\omega C_1}} = \frac{-j\omega L_1 \dot{U}_1}{(Z_M + R_{Lp})Z_1} \qquad (9)$$

$$\dot{I}_{Ls} = \frac{j\omega M \dot{I}_{Lp}}{Z_2} = \frac{\omega^2 M L_1 \dot{U}_1}{Z_2(Z_M + R_{Lp})Z_1} \qquad (10)$$

Furthermore, the LCC-S compensation output voltage $\dot{U}_{out}$ can be calculated as

$$\dot{U}_{out} = \dot{I}_L R_l = \frac{\omega^2 M L_1 R_l \dot{U}_1}{Z_2(Z_M + R_{LP})Z_1} \qquad (11)$$

Ignore the coil internal resistance, the $\dot{U}_{out}$ can be expressed as

$$\dot{U}_{out} = \frac{M}{L_1} \cdot \dot{U}_1 \qquad (12)$$

It can be concluded from (12) that the LCC-S compensation topology is load-independent when the coil resistance is ignored, and the output voltage can be adjusted through L_1.

III. Simulation Analysis of Coupling Mechanism

As the core of the wireless power supply system, the coupling mechanism completes the energy conversion between electricity and magnetism. Aiming at the problem that the water film pressure wireless sensing node rotates with high speed along the shaft, in order to avoid the wire intertwist, the coupling mechanism requires the secondary coil to rotate coaxially and primary coil to be fixed on the base.

A. Simplified Model Building

The coupling mechanism can be divided into plane type and spiral nested type according to the coil structure. Due to the fixed distance between the spindle and the base, too many turns of the planar coil will contact with the base and cause wear and tear, thus leading to the failure of the whole power supply system. However, the increase of turns of the single layer coil of the nested structure is not restricted by the radial distance, so the nested coil structure is selected as the coupling mechanism. The parameters of coupling mechanism model are shown in Table I. The size of the shaft diameter of the test stand and the rotating working environment are taken into account.

TABLE I. PARAMETERS OF COUPLING MECHANISM MODEL

Type	Value
Primary-side coil size	52mm
Secondary-side coil size	57mm
Air gap	5mm

The simplified model of the coupling mechanism is shown in Fig. 4. The external and internal coils of the nested structure act as primary and secondary coils respectively, the external supporting structure and the details of coil winding are ignored.

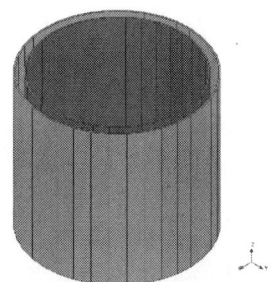

Figure 4. Simplified model of coupling mechanism

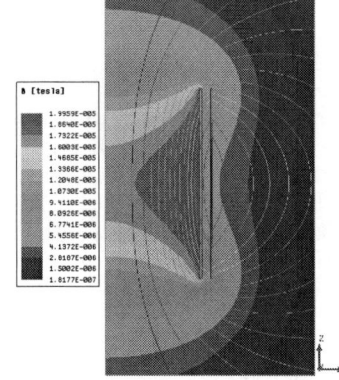

Figure 5. Field intensity distribution in 2D static

Through 2D static field simulation with Maxwell software, the magnetic field intensity and magnetic induction line distribution in the RZ coordinate system are shown in Fig. 5. Where the color is darker and the magnetic induction lines are denser, the magnetic field intensity is greater. As can be seen from the figure, the field intensity is concentrated on the edge and inside of the coil.

B. Influence of Coil Turns

Maxwell software was used to conduct parameterized scanning of the established simplified model and the results were shown at Fig. 6 and Fig. 7.

In Fig. 6, primary coil self-induction, secondary coil self-induction and mutual induction are represented by L_P, L_s and M respectively. The results show that the inductance of the simplified model is proportional to the number of turns.

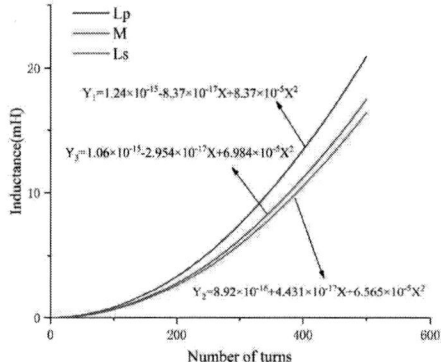

Figure 6. Inductance varies with number of turns

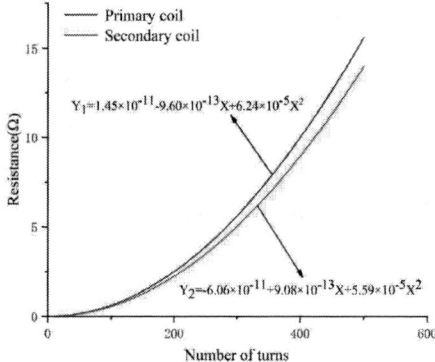

Figure 7. Resistance varies with number of turns

By observing the change of coil resistance with the number of turns in the simplified model in Fig. 7, it can be seen that the coil internal resistance of the simplified model is also proportional to the number of turns, and the primary coil internal resistance is higher than the secondary coil due to the larger size of the primary coil.

The results shown in Fig. 6 and Fig. 7 are consistent with the theory, jointly verifying the accuracy of the simulation of the simplified model. The number of turns of the original secondary coil is set as 100:100 according to the actual situation.

C. Co-Simulation

In order to ensure the accuracy of the results, Maxwell software and Simplorer software were used to co-simulate the coupling mechanism. The circuit was built in Simpliorer, and the simplified model of the coupling mechanism under transient field was imported into it. According to (1), the input voltage is a square wave voltage with a RMS of 279V, and the voltage frequency is set to 85kHz to ensure that the receiving coil senses sufficient induced emf. Meanwhile, the compensation topology parameters of LCC-S are calculated by (2), and the specific simulation parameters are shown in Table II.

TABLE II.   PARAMETERS OF COMPENSATION TOPOLOGY

Parameter	Value	Parameter	Value
U_{in}	$310\,V$	f	$85\,kHz$
L_p	$839.38\,\mu H$	L_s	$700.44\,\mu H$
M	$658.68\,\mu H$	L_1	$646.26\,\mu H$
C_1	$0.0054\,\mu F$	C_p	$0.0307\,\mu F$
C_s	$0.0054\,\mu F$	R_{Lp}	$0.624\,\Omega$
R_{Ls}	$0.559\,\Omega$	R_L	$8\,\Omega$

Set the configured parameters in the Simplorer software as shown in Fig.8.

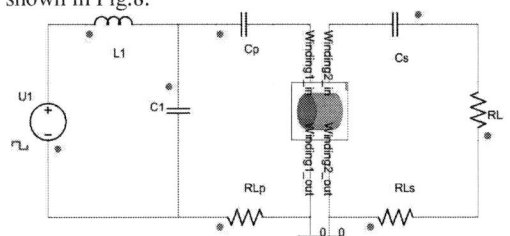

Figure 8.   Simplorer simulation circuit

Then, the field intensity distribution of the simplified model for observing the transient field under the co-simulation is shown in Fig. 9. The simulation results shown the electromagnetic induction intensity is mainly concentrated in the inner and edge parts of the coupling mechanism, which is the same as the 2D static field simulation results in Fig. 5.

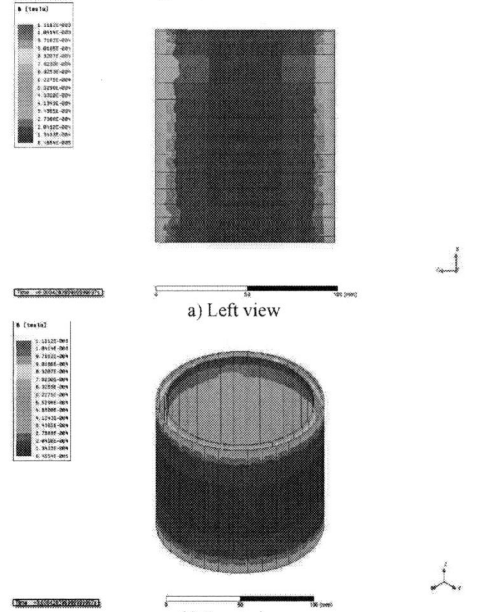

a) Left view

b) Front view

Figure 9.   Field strength distribution of coupling mechanism

The simulation results of voltage waveform in Simplorer software are shown in Fig. 10 and Fig. 11. The peak value of the output voltage is about 395V, less than 5% different from the theoretical value. It can be seen that after LCC-S compensation, the coupling mechanism can meet

the demand of 24V power supply of wireless sensor nodes under subsequent circuit processing.

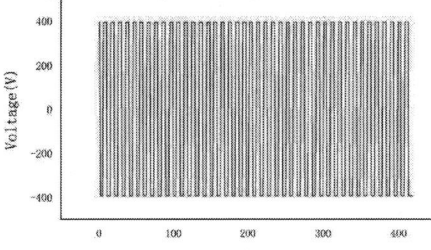

Figure 10. Input voltage waveform

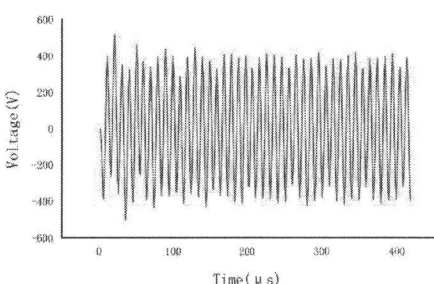

Figure 11. Output voltage waveform

IV. CONCLUSION

This paper in the view of the water film pressure wireless sensor node of water-lubricated bearing is difficult to supply power under rotating condition, put forward the wireless power supply solutions and establish the simplified model of electromagnetic coupling mechanism. The simplified model of the coupling mechanism was simulated in Maxwell and Simplorer, through LCC-S compensation topology, the magnetic field distribution and circuit input and output waveforms were obtained, which verified the feasibility of the proposed method.

ACKNOWLEDGMENT

This work supported by National Natural Science Foundation of China (51605269), Support Program for Young Outstanding Talents in Universities and Colleges of Shaanxi Province, China (SLGQD1802).

REFERENCES

[1] Nan Wang, Litao Yang, Yingxuan liang, Peng Wang. Non-contact Electromagnetic Loading Monitoring System for Water-lubricated Bearings[J]. China Mechanical Engineering, 2019, 30(24): 3004-3009.

[2] Matsumoto H. Research on solar power satellites and microwave power transmission in Japan[J]. IEEE microwave magazine, 2002, 3(4): 36-45.

[3] Karalis A, Joannopoulos J D, Soljacic M. Efficient wireless non-radiative mid-range energy transfer[C]. Proceedings of the 2006 AIP Industrial Physics Forum, San Francisco, USA, 2006: 34-48.

[4] Kurs A, Karalis A, Moffatt R, et al. Wireless power transfer via strongly coupled magnetic resonances[J]. Science, 2007, 317(5834): 83-86.

[5] Hu A P. Selected resonant converters for IPT power supplies[D], New Zealand: Department of Electrical and Electronic Engineering, University of Auckland, 2001.

Proceedings of the 16th Annual IEEE International
Conference on Nano/Micro Engineered and Molecular Systems
April 25-29, 2021

Investigation of electrostatic-piezoelectric hybrid vibrational power generators with different frequency broadening schemes

Yongqi Cao, Dezhi Nie, Jian Zhang, Yiwei Wang, Ronggang He, Zhe Zhao, Jin Wu, Bowen Ji, Jianbing Xie, and Kai Tao

Abstract— This paper proposes an electrostatic-piezoelectric hybrid vibrational power generator with different frequency broadening schemes. Both the nonlinear frequency broadening mechanisms and the synergized effect of the electrostatic-piezoelectric hybrid structures are investigated. On the one hand, we adopt the curved fixture structure, which has a 25% increase in bandwidth compared with the ordinary stopper structure. On the other hand, by integrating the electrostatic structure, the half-power bandwidth of the piezoelectric cantilever beam is 16Hz to 20Hz, and the peak power is 3.6mW, the half-power bandwidth of the electrostatic structure is 14.5Hz to 19.5Hz, and the peak power is 2.2mW. This means that under the same space utilization, the performance is improved by 60%. In this paper, the hybrid generator's structure and performance are optimized, and finally the response bandwidth and performance are improved. In general, the device designed in this paper has advantages such as larger bandwidth and better performance.

I. INTRODUCTION

Over the past few decades, tremendous advances have been made in microelectronic systems, with devices becoming smaller and requiring less energy [1]. However, limited by service life and energy density of the traditional batteries, the power supply scheme of these systems is still challenging. So people hope to design some devices for recovering the energy in the environment to power these microelectronic systems autonomously. Meanwhile, these lowpower electronic devices pose a challenge to cheap, flexible, portable, and sustainable energy resources [2–6]. Therefore, many researchers are devoted to the demand of these microelectronic systems, among which vibration energy collection has become a research hotspot [7-16].

The energy conversion forms based on environmental vibration are electrostatic, electromagnetic, piezoelectric and composite [11, 17 21]. The small piezoelectric cantilever beam type linear narrow-band generator proposed by Bai et al. [22] can acquire the average power of 50μW and 20μW, respectively, when placed on the arm and top of the human body. However, linear narrow-band generators cannot adapt to changing environmental vibrations at any time and cannot provide a stable output. So, Shahruz et al.

[23] proposed an array piezoelectric broadband generator. The different natural frequencies of the array cantilever beams increase the bandwidth of the generator. Xue et al. [24] also proposed a piezoelectric bimorph cantilever beam array broadband generator, which uses different resonance frequencies caused by different wafer thicknesses to increase the bandwidth of the generator. But the average power of the array generator is low. Leland et al. [25] proposed a clamped beam adjustable resonant generator with axial compression preload. This generator can change the beam stiffness to change the natural frequency and increase the bandwidth. But the adjustable resonant generator cannot respond quickly to changes in excitation. In addition to the above two types of generators that achieve broadband, Li et al. [26] proposed a hardened extrusion-type nonlinear generator that can recover energy at a lower vibration level. Weiqun Liu et al. [27] proposed a nonlinear generator with curved surface fixtures, effectively increasing bandwidth and power.

In general, there are many methods to implement broadband generators, but the nonlinear generator is a more effective solution due to its large bandwidth when carrying out energy recovery. So, this paper proposes a hybrid nonlinear generator with vibration energy harvesting. On the one hand, we adopt the curved fixture structure to increase the bandwidth. On the other hand, electrostatic [28-33] generator structure is integrated with the piezoelectric generator structure to make the power of the nonlinear generator have a better increase.

II. THE CONCEPTION OF DEVICE

A. Device structure and process

This paper designs a hybrid nonlinear generator that combines two power generation methods. The device's structure part comprises piezoelectric and electrostatic power generation structures. As depicted in Figure 1, combined with the energy conversion circuit based on the LTC3588-1 chip, this hybrid nonlinear generator integrates the electrostatic and piezoelectric power generation structures relying on the cantilever structure.

Among them, the piezoelectric structure uses a MEMS piezoelectric vibration energy harvester design scheme

---

*Research supported by National Natural Science Foundation of China Grant (No. 51705429).

Yongqi Cao, Dezhi Nie, Jian Zhang, Yiwei Wang, Ronggang He, Zhe Zhao, Bowen Ji, Jianbing Xie, and Kai Tao are with Ministry of Education Key Laboratory of Micro and Nano Systems for Aerospace, School of Mechanical Engineering, Northwestern Polytechnical University, Xi'an 710072, China (corresponding author to provide phone: +86-029-88460434; fax: +86-029-88460434; e-mail: taokai@nwpu.edu.cn).

Bowen Ji is also with Unmanned System Research Institute, Northwestern Polytechnical University, Xi'an 710072, China (e-mail: bwji@nwpu.edu.cn).

Jin Wu is with School of Electronics and Information Technology, Sun Yat-sen University, Guangzhou 510275, China (e-mail: wujin8@mail.sysu.edu.cn).

Yongqi Cao and Dezhi Nie contributed equally.

based on piezoelectric thick film. To enhance the electrical performance, the piezoelectric structure adopts the bonding and thinning technology based on the intermediate layer to fabricate the film with excellent piezoelectric properties. At the same time, the mass block attached to the free end of the cantilever beam further improves the output performance.

In the piezoelectric structure, the PZT piezoelectric material is used as the piezoelectric functional layer, the piezoelectric cantilever beam is used as the main structure of the energy harvester, and a mass block is added at the free end of the beam. The electrostatic structure is composed of two electrodes. One electrode is made of Cu material attached to the top of the package shell. The other electrode is the FEP electret pre-charged film attached to the Cu material layer. The piezoelectric cantilever beam is used as support layer.

The circuit mainly consists of LTC3588-1, which integrates a full wave bridge rectifier with low loss and a buck converter with high efficiency. The package is a 3D printed shell made of resin material. The entire device collects vibration energy in the environment by piezoelectric and electrostatic power generations and then converts into a well-regulated output to power application microcontrollers, sensors, data converters, and wireless transmission components.

The piezoelectric power generation in this paper is mainly based on piezoelectric materials' positive piezoelectric effect. In the actual working process, the device's piezoelectric layer constantly repeats the charging and discharging process under the influence of the alternating forces provided by the external vibration environment, so as to realize the conversion of mechanical energy to electrical energy.

The working principle of electrostatic power generation is to convert external mechanical disturbances into capacitance changes of variable capacitors under a constant bias voltage, which causes the charge flow between the two plates, thereby converting mechanical energy into electrical energy under external excitation.

Figure 2 shows the movement of the two internal power generation structures when the device is subjected to vibration. When subjected to external vibration, the free end mass of the beam causes the cantilever beam to vibrate up and down, and the piezoelectric crystal produces a piezoelectric effect. The change in the distance between the FEP electrode layer of the electrostatic structure and the other electrode on the top of the shell produces an electrostatic effect to complete the work process. The vibration of the cantilever beam causes the two power generation structures of piezoelectric and electrostatic to start working at the same time. The power generation of the two power generation structures is based on the movement of the piezoelectric cantilever beam, which simplifies the complexity of the device, and the process of movement of two structures driven by a single drive structure is realized.

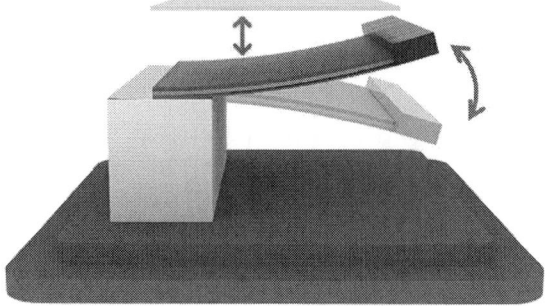

Figure 2. Schematic diagram of power generation structure movement.

B. Devices and circuits in kind

In this paper, two kinds of power generation structures are integrated and encapsulated in the 3D printing shell together with the circuit. Figure 3 shows the appearance of the device after packaging and the actual circuit used in this paper. As shown in the figure, this paper integrates the two power-generation structures and circuits in a small-sized package shell. Under ensuring its good performance, the size of the device is reduced as much as possible, so that the device can adapt to more applications.

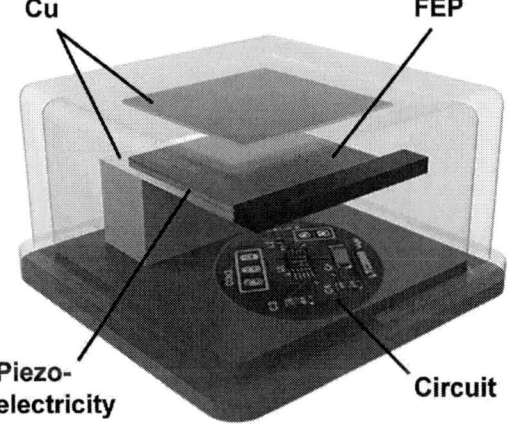

Figure 1. Schematic diagram of device structure concept.

Figure 3. Schematic diagram of power generation structure movement.

III. EXPERIMENTAL DESIGN

First of all, this paper designs two fixtures, one is an ordinary linear fixture, the other is a curved fixture. The design will test the respective characteristics of the two fixtures in the experiment for comparative analysis. In addition, on the basis of these two fixtures, stoppers will be added at the middle and the end of the ordinary linear fixtures, to explore the nonlinear effects of piezoelectric cantilever beams under three different mechanical structures. When the cantilever beam is deformed by external excitation, the cantilever beam will come into contact with the fixture, the effective length of the beam will be shortened, and the rigidity of the system will increase, thereby introducing nonlinear effects.

Figure 4 shows the overall comparison of the non-linear effects of the cantilever beam caused by the use of the curved fixture and the setting of stoppers at the middle and the end of the cantilever on the basis of the ordinary linear fixture. The two cases of the ordinary linear fixture setting stoppers are as shown in figure 4 (a) and (b), the stoppers are respectively set at the middle and the end of the cantilever beam. When the cantilever beam is deformed due to vibration, the cantilever beam contacts the stopper. At this time, the effective length of the cantilever beam will change greatly, but this change is fixed. Figure 4 (C) shows a curved fixture, which makes the effective length of the cantilever beam have different changes in one job. So the nonlinear effect curves of the three structures are also significantly different.

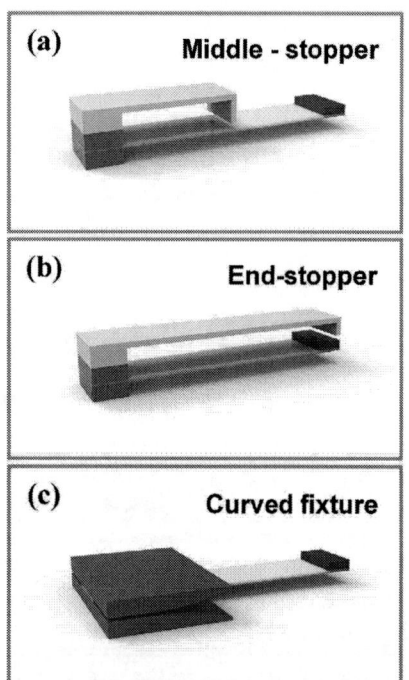

Figure 4. Three non-linear test structures.

IV. TESTING RESULTS

First, this paper explores the influence of the ordinary linear fixture and the curved fixture on the performance of piezoelectric power generation structures. As shown in figure 5, when the external load is 10kΩ and the excitation amplitude increases from 1Vpp to 6Vpp, the non-linear effect exhibited by the curved fixture makes the frequency band continue to expand, which shows the good non-linear effect of the curved fixture. It can be seen that the bandwidth of the curved fixture is 25 Hz to 33 Hz under 4Vpp, while the bandwidth of the ordinary linear fixture is 22 Hz to 27 Hz. The curved fixture bandwidth under 5Vpp is 26 Hz to 35 Hz, and the bandwidth of the ordinary linear fixture is 22 Hz to 27 Hz. At 4Vpp and 5Vpp, the curved fixture has expanded bandwidth compared to the ordinary linear fixture. Compared with the ordinary linear fixture, the curved fixture's bandwidth increased by 60% and 80% respectively. At this time, the power of the curved fixture is 112% and 120% of the ordinary linear fixture.

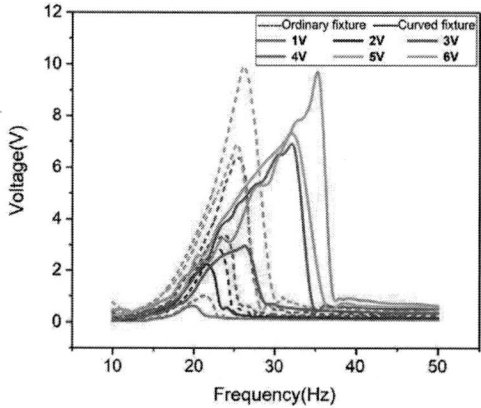

Figure 5. Common fixture and curved fixture test comparison.

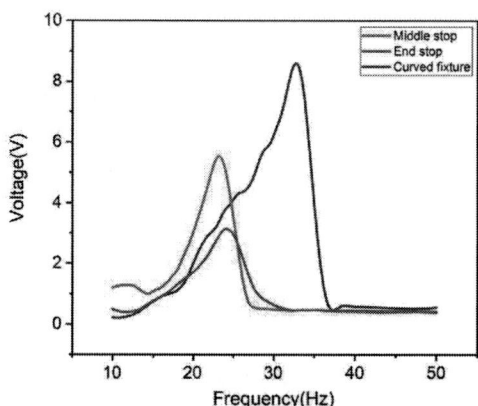

Figure 6. Three structural non-linear tests.

On the basis of the ordinary linear fixture, the experiment set stoppers at the middle and the end of the cantilever. The nonlinear effects of the cantilever caused by these two structures and using the curved fixture are tested in Figure 6. It can be seen in the figure that although the non-linearity of placing the stopper at the end may not be very obvious due to small acceleration, it can be seen that the nonlinear effect of placing the stopper at the middle and using the curved fixture is more obvious. Obviously, it can be seen that the bandwidth and performance of using the curved fixture have been greatly improved compared to the ordinary linear fixture of setting stoppers. Using the curved fixture, the performance of the device is improved by nearly 2 times, and the bandwidth expansion is nearly 125% of the setting stoppers.

After exploring the advantages of the curved fixture, the experiment adds an electrostatic structure on this basis. And we conduct the nonlinear effect experiment by forward and reverse sweep experiments to explore the nonlinear effect after adding the electrostatic structure. The result is shown in Figure 7. After adding the electrostatic structure, the half-power bandwidth of the piezoelectric cantilever beam is 16 Hz to 20 Hz, and the half-power bandwidth of the electrostatic structure is 14.5 to 19.5 Hz. At this time, the electrostatic external load is 10kΩ and the peak power is 3.6mW, while the piezoelectric external load is 2kΩ and the peak power is 2.2mW. It can be seen that although the addition of the electrostatic structure does not expand the frequency bandwidth, it improves the performance of the generator. This means that under the same space utilization, the performance is improved by 60%.

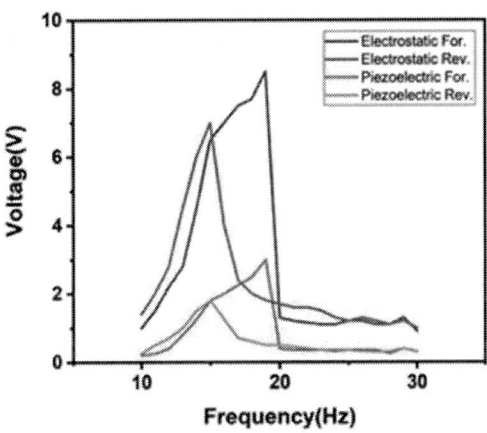

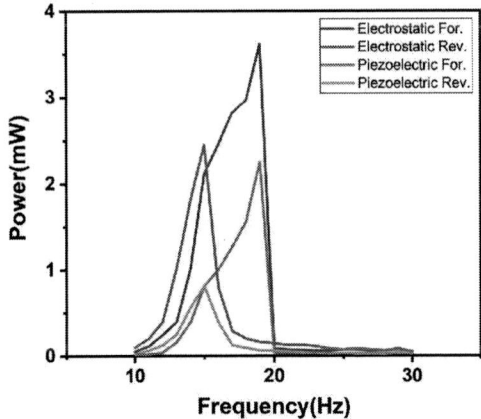

Figure 7. Non-linear curve with electrostatic structure added.

V. CONCLUSION

This paper proposes a hybrid nonlinear generator with two power generation structures: piezoelectric and electrostatic. The influence of the ordinary linear fixture and the curved fixture on the performance of the generator

is explored, and it is verified that compared with the ordinary linear fixture, the curved fixture has excellent bandwidth expansion capabilities and can improve the performance. While exploring the nonlinear effect of the piezoelectric cantilever beam, we integrate the electrostatic power generation structure into the piezoelectric structure, and the nonlinear experiment of piezoelectric and electrostatic addition is carried out. After adding the electrostatic structure, the half-power bandwidth of the piezoelectric cantilever beam was 16Hz to 20Hz and the peak power is 3.6mW. The half-power bandwidth of the electrostatic structure is 14.5Hz to 19.5Hz, and the peak power is 2.2mW. Although the addition of electrostatic structure does not expand the frequency bandwidth, it improves the performance of the generator. This means that under the same space utilization, the performance is improved by 60%. Finally, we optimize the structure and performance of the hybrid generator and have achieved the improvement of the bandwidth and performance. In general, the device designed in this paper has the advantages of larger bandwidth, better performance and smaller size, and can be used in many vibration environments with small spaces.

ACKNOWLEDGMENT

This research is supported by National Natural Science Foundation of China Grant (No. 51705429), the Fundamental Research Funds for the Central Universities, Guangdong Natural Science Funds Grant (2018A030313400), Science, Technology and Innovation Commission of Shenzhen Municipality.

REFERENCES

[1] A. Pirisi, M. Mussetta, G. Gruosso, and R. E. Zich, "Bio-inspired optimization techniques for wireless energy transfer," *Proceedings of the International Conference on Electromagnetics in Advanced Applications (ICEAA)*, Sydney, Australia, pp. 731-732, 2010.

[2] H. Liu, J. Zhong, C. Lee, S.-W. Lee, and L. Lin, "A comprehensive review on piezoelectric energy harvesting technology: materials, mechanisms, and applications," *Applied Physics Reviews, vol. 5, no. 4*, 2018.

[3] Y. J. Li, K. Tao, B. George, and Z. C. Tan, "Harvesting Vibration Energy: Technologies and Challenges," *IEEE Industrial Electronics Magazine*, 2020.

[4] J. L. Wang, L. H. Tang, L. Y. Zhao, G. B. Hu, R. J. Song, and K. Xu, "Equivalent circuit representation of a vortex‐induced vibration‐based energy harvester using a semi‐empirical lumped parameter approach," *International Journal of Energy Research, vol. 44, no. 6*, pp. 4516-4528, 2020.

[5] L. C. Zhao, H. X. Zou, G. Yan, F. R. Liu, T. Tan, K. X. Wei, and W. M. Zhang, "Magnetic coupling and flextensional amplification mechanisms for high-robustness ambient wind energy harvesting," *Energy Conversion and Management*, vol. 201, pp. 112166, 2019.

[6] H. L. Cao, Y.J. Zhang, Z. Q. Han, X. L. Shao, J. Y. Gao, K. Huan, Y. B. Shi, J. Tang, C. Shen, and J. Liu, "Pole-zero temperature compensation circuit design and experiment for dual-mass MEMS gyroscope bandwidth expansion," *IEEE-Asme Transactions on Mechatronics*, vol. 24, no. 2, pp. 677–688, 2019.

[7] K. Tao, H. P. Yi, L. H. Tang, J. Wu, P. H. Wang, N. Wang, L. X. Hu, Y. Q. Fu, J. M. Miao, and H. L. Chang, "Piezoelectric ZnO thin films for 2DOF MEMS vibrational energy harvesting," *Surface & Coatings Technology*, vol. 359, pp. 289–295, 2019.

[8] C. Shearwood, and R. Yates, "Development of an electromagnetic microgenerator," *Electronics Letters*, vol. 33, no. 22, pp. 1883-1884, 1997.

[9] S. Meninger et al. "Vibration-to-electric energy conversion," *IEEE Transactions on Very Large Scale Integration (VISI) Systems*, vol. 9, no. 1, pp. 64-74, 2001.

[10] H. B. Fang, J. Q. Liu, Z. Y. Xu, L. Dong, L. Wang, D. Chen, B. C. Cai, and Y. Liu, "Fabrication and performance of MEMS-based piezoelectric power generator for vibration energy harvesting," *Microelectronics Journal*, vol. 37, no. 11, pp. 1280-1284, 2006.

[11] L. Wang, and F. G. Yuan, "Vibration energy harvesting by magnetostrictive material," *Smart Materials and Structures*, vol. 17, no. 4, 2008.

[12] K. Tao, J. Wu, L. H. Tang, X. Xia, S. W. Lye, J. M. Miao and X. Hu, "A novel two-degree-of-freedom MEMS electromagnetic vibration energy harvester,"*Journal of Micromechanics and Microengineering*, vol. 26, no. 2-3, pp. 035020, 2016.

[13] J. L. Wang, G. B. Hu, Z. Su, G. P. Li, W. Zhao, L. H. Tang, and L. Y. Zhao, "A cross-coupled dual-beam for multi-directional energy harvesting from vortex induced vibrations," *Smart Materials and Structures*, vol. 28, no. 12, 2019.

[14] X. B. Shan, H. L. Li, Y. C. Yang, J. Feng, Y. C. Wang, and T. Xie, "Enhancing the performance of an underwater piezoelectric energy harvester based on flow-induced vibration," *Energy*, vol. 172, pp. 134–140, 2019.

[15] Y. Wu, J. Qiu, S. Zhou, H. Ji, Y. Chen, and S. Li, "A piezoelectric spring pendulum oscillator used for multidirectional and ultra-low frequency vibration energy harvesting," *Appl. Energy*, vol. 231, pp. 600–614, 2018.

[16] J. Chen, G. Zhu, W. Yang, Q. Jing, P. Bai, Y. Yang, T. C. Hou, and Z. L. Wang, "Harmonic-resonator-based triboelectric nanogenerator as a sustainable power source and a sedf-powered active vibration sensor," *Advanced Materials*, vol. 25, no. 42, pp. 6094-6099, 2013.

[17] W. L. Lu and Y. M. Hwang, "Analysis of a vibration-induced micro-generator with a helical micro-spring and induction coil," *Microelectronics Reliability*, vol. 52, no. 1, pp. 262-270, 2012.

[18] H. Si, J. L. Dong, L. Chen, L. Z. Sun, X. D. Zhang, and M. T. Gao, "Study of the ambient vibration energy harvesting based on piezoelectric effect," *International Journal of Nanoscience*, vol. 14, no. 1/2, pp. 14600171-14600177, 2014.

[19] J. K. Huang, R. C. O'Handley, and D. Bono , "New high-sensitivity hybrid magnetostrictive/ electroactive magnetic field sensors," *Smart Structures and Materials 2003: Smart Sensor Technology and Measurement Systems*, San Diego, CA, USA, vol. 5050, pp. 229-237, 2003.

[20] M. A. Karami and D. J. Inman, "Nonlinear hybrid energy harvesting utilizing a piezo -magneto-elastic spring," *Active and Passive Smart Structures and Integrated Systems 2010, Pts 1 and 2*, vol. 7643, no. 2, pp. 379-380, 2010.

[21] V. R. Challa, M. G. Prasad, and F. T. Fisher, "A coupled piezoelectric-electromagnetic energy harvesting technique for achieving increased power output through damping matching," *Smart Materials & Structures*, vol. 18, no. 9, pp. 7566-7579, 2009.

[22] Y. Bai, P. Tofel, Z. Hadas, J. Smilek, P. Losak, P. Skarvada, and R. Macku, "Investigation of a cantilever structured piezoelectric energy harvester used for wearable devices with random vibration input," *Mechanical Systems and Signal Processing*, no. 106, pp. 303-318, 2018

[23] S. M. Shahruz, "Design of mechanical band-pass filters for energy scavenging," *Journal of sound and vibration*, vol. 292, no. 3-5, pp. 987-998, 2006.

[24] H. A. Xue, Y. T. Hu, and Q. M. Wang, "Broadband piezoelectric energy harvesting devices using multiple bimorphs with different operating frequencies," *IEEE transactions on ultrasonics, ferroelectrics, and frequency control*, vol. 55, no. 9, pp. 2104-2108, 2008.

[25] E. S. Leland and P. K. Wright, "Resonance tuning of piezoelectric vibration energy scavenging generators using compressive axial preload," *Smart Materials and Structures*, vol. 15, no. 5, pp. 1413-1420, 2006.

[26] H. T. Li, Z. Yang, J. Zu, and W. Y. Qin, "Numerical and experimental study of a compressive-mode energy harvester under random excitations," *Smart Materials and Structures*, vol. 26, no. 3, 2017.

[27] W. Q. Liu, C. Z. Liu, B. Y. Ren, Q. Zhu, G. D. Hu, and W. Q. Yang, "Bandwidth increasing mechanism by introducing a curve fixture to the cantilever generator," *Applied Physics Letters*, vol. 109, no. 4, 2016.

[28] K. Tao, H.P. Yi, Y. Yang, L. H. Tang, Z. S. Yang, J. Wu, H. L. Chang, and W. Z. Yuan, "Miura-origami-inspired electret/triboelectric power generator for wearable energy harvesting with water-proof capability," *Microsystems & Nanoengineering*, vol. 6, no, 1, pp. 1-11, 2020.

[29] X. Q. Zhang, P. Pondrom, G. M. Sessler, and X. C. Ma, "Ferroelectret nanogenerator with large transverse piezoelectric activity," *Nano Energy*, vol. 50, pp. 52-61, 2018.

[30] K. Tao, L. H. Tang, J. Wu, S. W. Lye, H. L. Chang, and J. M. Miao,"Investigation of multimodal electret-based MEMS energy harvester with impact-induced nonlinearity," *Journal of Microelectromechanical Systems*, vol. 27, no. 2, pp. 276–288, 2018.

[31] Y. L. Zhang, T. Y. Wang, A. X. Luo, Y. S. Hu, X. Li, and F. Wang, "Micro electrostatic energy harvester with both broad bandwidth and high normalized power density," *Applied Energy*, vol. 212, pp. 362-371, 2018.

[32] K. Tao, H. P. Yi, Y. Yang, H. L.Chang, J. Wu, L. H. Tang, Z. S. Yang, N. Wang, L. X. Hu, Y. Q. Fu, J. M. Miao, and W. Z. Yuan, "Origami-inspired electret-based triboelectric generator for biomechanical and ocean wave energy harvesting," *Nano Energy*, vol. 67, pp.104197, 2020.

[33] K. Tao, S. W. Lye, L. H. Tang, J. M. Miao, and X. Hu, "Out-of-plane electret-based MEMS energy harvester with the combined nonlinear effect from electrostatic force and a mechanical elastic stopper," *Journal of Micromechanics and Microengineering*, vol. 10, pp. 104014, 2015.

Gap in pagination due to unavailable papers.

Pages 533-538

Proceedings of the 16th Annual IEEE International
Conference on Nano/Micro Engineered and Molecular Systems
April 25-29, 2021

Size-Dependent Particle Separating in Curved Microfluidic Chip*

Yun Tian, Yongkang Wang and Yunfei Chen*

Abstract— **A passive microfluidic chip was designed for separating microparticles of different sizes. Microparticles with different sizes are separated from different outlets through spiral microchannels as the particles of various sizes are subjected to different inertial forces in the curved microchannel. Based on the design rules of the inertial microfluidic chip, some critical features of the microfluidic chip including curvature of the microchannel and flow rate are optimized by finite element method (FEM) analysis. The optimized microfluidic chips are fabricated by a soft lithography process. The experimental results revealed that as the flow rate increased to 400 μL/min, particles began to inertial focus, thus obtaining relatively good separating results. Increased the flow rate further, however, the particle inertial focus diverged, which is not conducive to the separating of different particles.**

I. INTRODUCTION

Worldwide, cancer has become one of the major public health problems that seriously threaten human life and health. The World Health Organization (WTO) report shows [1] that one in six people die of cancer each year all over the world and about 9.6 million people worldwide died of cancer in 2018. As many as 90% of cancer-related deaths are related to cancer metastasis [2]. If cancer can be diagnosed and treated early, the cancer mortality rate can be greatly reduced. Cancer metastasis occurs through cancer cells released from the primary tumor or the metastatic site. The released cancer cells circulate in the body through the lymphatic system or peripheral blood, and then invade remote tissues and form new metastases [3]. The cancer cells that circulate in the human body and can form new metastases are called circulating tumor cells (CTCs), which are clinically regarded as important indicators for early cancer diagnosis [4].

Microfluidic technology has entered people's vision in recent decades. This technology uses micrometer-level channels to achieve precise control of fluids. It has many advantages including a fast-response process, less sample solution consumption, low cost and so on. Now it has been applied to cell separating [5], capture [6], concentration [7], and other occasions. At present, microfluidic chips are mainly used for separating CTCs in two ways. One is active separating, including dielectrophoresis separating [8], magnetic separating [9], acoustic separating [10], and so on; the other is passive separating, including size-based filtering separating [11], deterministic lateral displacement separating [12], inertial force separating [13], and so on. Active separating has the advantages of high stability and accuracy but has the disadvantages of low throughput,

complex operation, and high cost. Compared with active separating, passive separating can handle higher throughput without the need for external field generators, which is more affordable and can keep cells active. The separation of circulating tumor cells from the blood is still very challenging as the concentration of circulating tumor cells is extremely low (~109 cells/mL) [14]. This study proposed a passive separating method based on the principle of inertial force, and then fabricated a spiral microchannel for separating microparticles of different sizes. Some critical features of the microfluidic chip including curvature of the microchannel and flow rate are optimized by finite element method (FEM) analysis using COMSOL Multiphysics 5.3a. Polystyrene microspheres of different diameters are employed to represent blood cells and CTCs for the FEM simulation. With the advantages of easy operation, high reliability, and no requirement of a sheath flow, the spiral microchannel has great potential for contributing to cell sorting applications.

II. THEORY AND MECHANISM

The particle inertial focusing effect in the direct microchannel is caused by the lateral migration of particles in the microchannel. This unique lateral migration effect mainly results from the interaction of two forces: the lift induced by the velocity shear gradient (F_{SL}) pushes the particles in the fluid away from the flow channel center, and the wall induced lift (F_{WL}) pushes the particles away from the channel walls [15]. The narrow space between particles and the channel wall will cause the unbalanced forces around particles, which generates lift to keep particles away from the wall [16]. The shear gradient lift (F_{SL}) results from the curvature of the parabolic velocity profile. The particle velocity at both sides of the vertical direction is different, so the shear gradient is also different, resulting in a pressure difference between the top and bottom of the particle. Due to the pressure difference, the shear gradient lift force is exerted on the particles, pushing the particles to the channel wall (until the wall surface induces a lift balance). Asmolov proposed an analytical expression of the net lift force exerted on rigid particles in the Poiseuille flow ($\frac{a}{D_h} \ll 1$), which is as follows [17]:

$$F_L = \frac{\rho U_m^2 a_p^4}{D_h^2} f_L(Re_c, X_p) \tag{1}$$

Where U_m is the maximum fluid velocity, which can be estimated by twice the average characteristic velocity ($2U_f$), f_L is the lift coefficient which depends on the particle

*Resrach supported by the National Key Research and Development Program of China (No. 2018YFB1105400).

All authors are with the Jiangsu Key Laboratory for Design and Manufacture of Micro-Nano Biomedical Instruments, School of Mechanical Engineering, Southeast University, Nanjing, China.

*Contacting Author: Yunfei Chen is with the Jiangsu Key Laboratory for Design and Manufacture of Micro-Nano Biomedical Instruments, School of Mechanical Engineering, Southeast University, Nanjing, China (email: yunfeichen@seu.edu.cn).

978-1-6654-3008-1/21 $31.00 © 2021 IEEE   539

position within the channel (X_p) and the channel Reynolds number (Re_c), a_p is the particle size, D_h is the hydraulic diameter.

In the parabolic velocity profile of Poiseuille flow in a microchannel, the fluid near the channel centerline has a higher velocity than the fluid near the wall. When fluid flows through the curved channel, the imbalance of the centrifugal force and the radial pressure gradient causes the center line fluid to flow outward. In order to meet mass conservation in the closed channel, the fluid near the outer wall will flow back along the upper and lower bottom surfaces of the channel. As a result, two vortices with opposite rotation directions are generated in the vertical main flow direction. This phenomenon is called Dean flow or secondary flow [18]. Because of Dean flow in the cross-section, the particles in the radial direction will be affected by the Dean drag force besides the inertial lift in the curved channel. The F_D expression of the Dean drag force is as follows [19,20]:

$$F_D \propto \frac{\rho U_m^2 a_p D_h}{R} \qquad (2)$$

III. MATERIALS AND METHODS

A. Chip design and simulations

The microfluidic chip consists of a 5-loop Archimedean spiral microchannel with one inlet and three outlets (Fig.1a). A liquid inlet groove with a diameter of 2 mm and three outlet grooves with a diameter of 1 mm are designed for fluids in and out, respectively. The cross-sectional size of the microfluidic chip is 150 μm in width and 40 μm in height while the distance between adjacent channels is 500 μm.

To verify that the designed spiral microchannel has the capability of separating CTCs in the blood, FEM is employed to simulate the particle behavior in the microchannel. The FEM simulation was performed on COMSOL Multiphysics 5.3a software. According to the above description, the lift force and the Dean drag force acting on particles in the microchannel are both functions of the particle diameter, so the particle diameter is a very important parameter. For simplicity, blood cells are represented by 7 μm particles while CTCs are represented by 15 μm particles. We released 200 7 μm and 15 μm particles respectively and set the flow rate to 400 μL/min at the entrance.

B. Fabrication

The microfluidic device was fabricated using a standard soft lithography process. First, a layer of negative photoresist (SU-8 2050, Microchem) was spin-coated on the silicon substrate, and then through the process of ultraviolet light exposure and development, the male mold of the spiral microchannel was obtained. The photoresist height determines the microchannel height, which can be controlled by the spin coating speed. After that, the PDMS polymer and the crosslinked polymer (Sylgard 184, Dow Corning) were mixed at a ratio of 10:1, and the mixed PDMS was poured on the silicon substrate male mold, and vacuum treatment was performed to remove the bubbles in the PDMS. Cured in an oven at 80°C for 1.5 hours. After that, the PDMS was separated from the male mold to obtain the PDMS with the

required spiral microchannel structure, and holes were punched at the corresponding position. We put the PDMS and the glass slides in a plasma cleaner (PDC-002, Harric Plasma) for 3 minutes, took them out after 3 minutes and quickly bonded them, put them in an oven after bonding, and kept warm at 120°C for 3 hours for bonding strength.

C. Sample preparation

In the experiment, polystyrene microspheres of 7 μm and 15 μm are used to mimic blood cells and CTCs, respectively. Polystyrene microspheres are more regular and more rigid than cells, which can better characterize the sorting performance of spiral microchannels. The particle diameters used to prepare the solution were 7 μm (ABT-8-0700, Biotyscience) and 15 μm (ABT-8-1500, Biotyscience) particles. Prepared 7 μm particle solution, we took 50 μL of 7 μm polystyrene microsphere solution with a concentration of 5% w/v and mixed with 6mL deionized water to obtain a 7 μm particle concentration of about 3.25×10^6/mL. Prepared 15μm particle solution, we took 10 μL 15 μm polystyrene microsphere solution with a concentration of 5% w/v and mixed with 10 mL deionized water to obtain a 15 μm particle concentration of about 4.25×10^4/mL. The prepared solution was shaken sufficiently to ensure a uniform distribution of the particles in the solution.

D. Experimental set-up

The experiment platform mainly included a syringe pump (Legato100, KDScientific), optical microscope (IX7, Olympus), CCD camera (Exi Blue, QImaging), and computer. The syringe pump was connected to the inlet end of the microfluidic chip through a catheter, the microfluidic chip was placed on the stage of the optical microscope, and the outlet led to three conduits connected to the collection tube. The real-time picture in the spiral microchannel during the whole experiment was recorded by a high-speed camera, which can be viewed on the computer through the Image-Pro Express software and taken. The taken pictures can be post-processed by the software ImageJ.

IV. RESULTS AND DISCUSSION

A. Simulation results

According to the previous section, the particle trajectory diagram in the spiral microchannel can be obtained. The distribution of the two kinds of particles at the outlets is shown in Fig. 1b and 1c. The blue particles represent 7 μm particles, and the red particles represent 15 μm particles. Fig. 1d shows the number of particles flowing out of the left outlet, Fig. 1e shows the number of particles flowing out of the middle outlet. The final statistical result was that 200 7 μm particles and 21 15 μm particles flowed out from the left outlet, 178 15 μm particles flowed out from the middle outlet, and 1 15 μm particle flowed out from the right outlet, which revealed that the separating rate of the designed spiral microchannel is 89%, and the separating purity reaches 100%.

978-1-6654-3008-1/21 $31.00 © 2021 IEEE

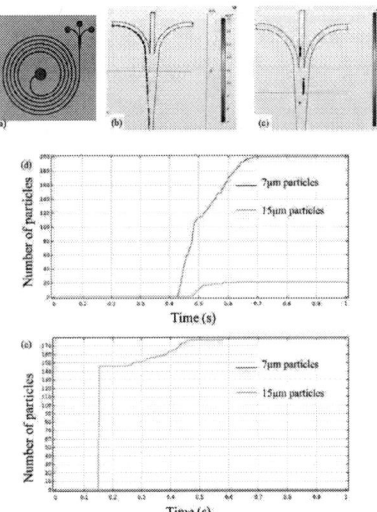

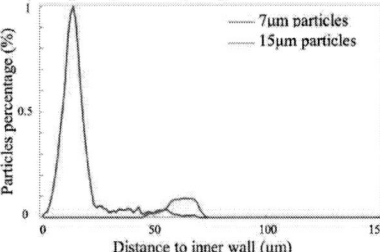

Figure 3. Particles distribution in the width direction of the microchannel.

microchannel has the function of screening 7 μm and 15 μm particles. At the same time, it can be found that the range of 7 μm particle distribution is relatively large, which may interfere with the purity of 15 μm particles.

Figure 1. (a) Schematics of a microfluidic chip with one inlet and three outlets. (b) 7 μm particles at exit; (c) 15 μm particles at exit; (d) The number of particles flowed from inner outlet; (e) The number of particles flowed from a middle outlet.

B. Separation of binary polystyrene particles

Install a syringe with a configured 7 μm particle solution on the syringe pump. According to the simulation results, set the flow rate to 400 μL/min. Observed after the fluid flow reached a stable state. Fig. 2a and 2b show 7 μm particles inertial-focused situation in microchannel while 2c and 2d show the situation of 15 μm particles. This phenomenon is consistent with the simulation results. Fig. 2b and 2d are 100 real-time photos superimposed to get pictures.

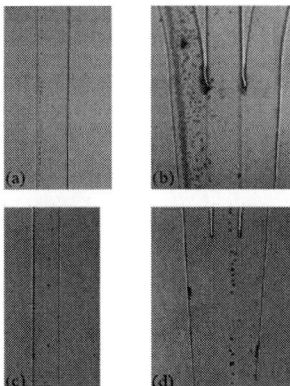

Figure 2. (a) 7 μm particles at microchannel; (b) 7 μm particles at exit; (c) 15 μm particles at microchannel; (d) 15 μm particles at microchannel.

By taking multiple photos in a row, we can obtain the particle's location information. Fig. 3 depicts the inertial-focused position of different particles in the microchannel, indicating that both 7 μm and 15 μm particles present an inertial focusing state. Obviously, most 7 μm particles located at the area near the inner wall, but 15 μm particles located at the microchannel centerline area. 7 μm particles are mainly focused in the interval of 10-20 μm from the inner wall, and 15 μm particles are mainly focused in the interval of 50-70 μm from the inner wall. Therefore, the

According to the above solution preparation ratio, we prepared 3 mL 7 μm solution and 5 mL 15 μm solution. In order to simulate the ratio of CTCs to blood cells in the human body, the 15 μm particles concentration in the prepared solution should be much lower than the 7 μm particles concentration, so the two solutions concentration after mixing was approximate to the 7 μm concentration, 3.25×10^6/mL. The flow rate was set to 400 μL/min, 100 photos were taken continuously and vertically superimposed with a high-speed camera. The distribution of the mixed particles after inertially focused in the microchannel is shown in Fig. 4. 15 μm particles are marked by yellow circles, which is consistent with the simulation results that the particles of different sizes are well separated.

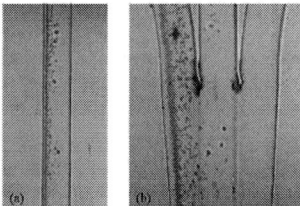

Figure 4. The distribution of mixed solution particles in the microchannel.

Fig. 5 presents the distribution of the mixed particles and the pure particles after being inertially focused under the same experimental conditions. It can be seen from Fig. 4 and Fig. 5 that when the two particles were mixed, the inertial focus position of the two particles changed due to the interaction between the particles. The 7 μm particles were mainly focused in the interval of 15-25 μm from the inner wall, and the 15 μm particles were mainly focused in the interval of 40-50 μm. After the two particles were mixed, the focusing interval between the two particles was closer, but it was still large enough to realize separating.

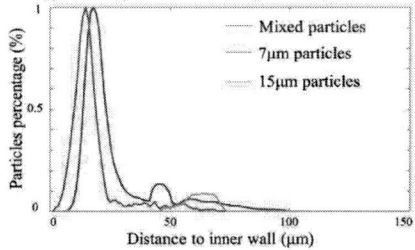

Figure 5. Distribution of mixed particles in the microchannel along the width direction.

In addition to the particle diameter, the flow velocity is also an important factor that affects the particle inertial focus position. The 7 μm particle solution and the mixed solution were configured in the above manner, and the prepared 7 μm particle solution had a concentration of 3.25×10^6/mL, and the mixed solution had a concentration of 4.28×10^6/mL. For the 7 μm particle solution, we set the syringe pump flow rate to 300 μL/min, collected the particle distribution information in the flow channel after the fluid flows stably. Then we increased the flow rate to 400 μL/min and 500 μL/min in the same way, and recorded the distribution of particles in the flow channel respectively. We calculated the position information of 7 μm particles at three flow rates and plot them on Fig 6. As the flow velocity increases, the focus position of 7 μm particles shifted to the microchannel center, and then showed a slight divergence trend. 7 μm particles were focused in the interval of 10-20 μm from the inner wall of the flow channel.

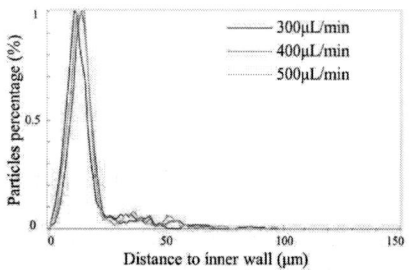

Figure 6.   7 μm particles distribution at different flow rates.

For the mixed particle solution, the flow rate was set to 300 μL/min, 400 μL/min, 500 μL/min, and 600 μL/min, and the particle distribution in the flow channel was recorded respectively. Fig. 7 shows the particle distribution in the flow channel. The peak near the inner wall was the 7 μm particles focus position, and the peak near the center was the 15 μm particles focus position. With the increase of the flow rate, the 7 μm focus position migrated to the inner wall, and the 15 μm focus position migrated to the inner wall of the flow channel. The increase of the flow rate caused the 7 μm particles to have a tendency to diverge, which harms an adverse effect on the purity of the collected 15 μm particles.

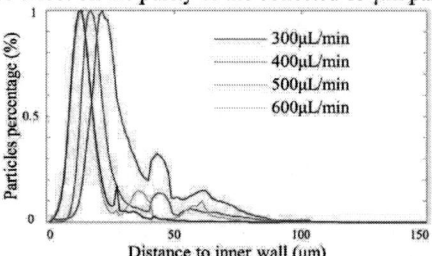

Figure 7.   Distribution of mixed particles at different flow rates.

V. CONCLUSION

This paper mainly studied some critical influencing factors of the microchannel with a rectangular cross-section of 150×40μm in the inertial separation of 7 μm particles and 15 μm particles. The comparison of simulation and experiment showed that the experimental results were consistent with the simulation results, and the structure of the spiral microchannel had a screening effect between 7 μm and 15 μm particles. While ensuring the realization of the function, the influence of the flow velocity on the particle inertial focusing and screening effect was explored. If the flow rate is too small, the inertial focusing interval of 7 μm particles and 15 μm particles will be too close, which is not conducive to the screening of the two kinds of particles and can not reach the maximum separation speed. Excessive flow rate leads to a divergent tendency of 7 μm particles, which will reduce the purity of the collected 15 μm particles and bring additional interference to subsequent operations. The optimal flow rate should be guaranteed to be around 400 μL/min.

ACKNOWLEDGMENT

This work was funded by the National Key Research and Development Program of China (No. 2018YFB1105400).

REFERENCES

[1] C. P. Wild, E. Weiderpass, et al, *World Cancer Report: Cancer Research for Cancer Prevention.*   Lyon, France: International Agency for Research on Cancer, 2020.

[2] K. Pantel, C. Alix-Panabières, "Circulating tumour cells in cancer patients: challenges and perspectives," *Trends in Molecular Medicine*, vol. 16, pp. 398-406, Sep. 2010.

[3] R. A. Harouaka, M. Nisic, et al, "Circulating Tumor Cell Enrichment Based on Physical Properties," *Journal of Laboratory Automation*, vol. 18, pp. 455-468, Jul. 2013.

[4] T. A. Burinaru, M. Avram, et al, "Detection of Circulating Tumor Cells Using Microfluidics," *ACS Combinatorial Science*, vol. 20, pp. 107-126, Jan. 2018.

[5] Y. Gou, S. Zhang, et al, "Sheathless Inertial Focusing Chip Combining a Spiral Channel with Periodic Expansion Structures for Efficient and Stable Particle sorting," *Analytical Chemistry*, vol. 92, pp. 1833-1841, Dec. 2020.

[6] J. Che, V. Yu, et al, "Classification of large circulating tumor cells isolated with ultra-high throughput microfluidic Vortex technology." *Oncotarget*, vol. 7, pp. 12748-12760, Jan. 2016.

[7] X. J. Zhang, Z. X. Zhu, et al, "Automated microfluidic instrument for label-free and high-throughput cell separation," *Analytical Chemistry*, vol. 90, pp. 4212-4220, Mar. 2018.

[8] M. Sun, J. Xu, et al, "Creating a capture zone in microfluidic flow greatly enhances the throughput and efficiency of cancer detection," *Biomaterials*, vol. 197, pp. 161-170, Mar. 2019.

[9] L. Kermanshah, M. Poudineh, et al, "Dynamic CTC phenotypes in metastatic prostate cancer models visualized using magnetic ranking cytometry," *Lab Chip*, vol. 18, pp. 2055-2064, Jun. 2018.

[10] P. Li, Z. M. Mao, et al, "Acoustic separation of circulating tumor cells," *Proceedings of the National Academy of Sciences of the United States of America*, vol. 112, pp. 4970-4975, Apr. 2015.

[11] S. J. Hao, Y. Wan, et al, "Size-based separation methods of circulating tumor cells," *Advanced Drug Delivery Reviews*, vol. 125, pp. 3-20, Feb. 2018.

[12] T S. H. Tran, B D. Ho, et al, "Open channel deterministic lateral displacement for particle and cell sorting," *Lab Chip*, vol. 17, pp. 3592-3600, Sep. 2017.

[13] M. Dhar, J. N. Lam, et al, "Functional profiling of circulating tumor cells with an integrated vortex capture and single-cell protease activity assay," *Proceedings of the National Academy of Sciences*, vol. 115, pp. 9986-9991, Oct. 2018.

[14] J. M. Jackson, M. A. Witek, et al, "Materials and microfluidics: enabling the efficient isolation and analysis of circulating tumour cells," *Chemical Society Reviews*, vol. 46, pp. 4245-4280, Jun. 2017.

[15] H. Amini, W. Lee, et al. "Inertial microfluidic physics," *Lab Chip*, vol. 14, pp. 2739-2761, May 2014.

[16] J. Zhou, I. Papautsky, "Fundamentals of inertial focusing in microchannels," *Lab Chip*, vol. 13, pp. 1121-1132, Jan. 2012.

[17] E. S. Asmolov. "The inertial lift on a spherical particle in a plane Poiseuille flow at large channel Reynolds number," *Journal of Fluid Mechanics*, vol. 381, pp. 63-87, Feb. 1999.

[18] A. Shiriny, M. Bayareh, "Inertial focusing of CTCs in a novel spiral microchannel," *Chemical Engineering Science*, vol. 229, pp. 116102, Jan. 2021.

[19] D. Di Carlo, D. Irimia, et al, "Continuous inertial focusing, ordering, and separation of particles in microchannels," *Proceedings of the National Academy of Sciences*, vol. 104, pp. 18892-18897, Dec. 2018.

[20] N. Xiang, H. Yi, et al, "High-throughput inertial particle focusing in a curved microchannel: Insights into the flow-rate regulation mechanism and process model," *Biomicrofluidics*, vol. 7, pp. 054101, Sep. 2013.

Gap in pagination due to unavailable paper.

Pages 544-545

Proceedings of the 16th Annual IEEE International
Conference on Nano/Micro Engineered and Molecular Systems
April 25-29, 2021

3D all-metal nano-cavity coupled metamaterial for refractive index sensing

Kenan Zhang[1,2], Guanzhou Lin[2], Jia Zhu[2], Yun Huang[2], Peimin Lu[1, *] and Wengang Wu[2,3, *]

Abstract— **Three-dimensional (3D) all-metal metamaterial with subwavelength structures are widely used to excite local surface plasma polaritons through the coupling of electromagnetic fields with metallic electron plasmas, because of its diversity and flexibility. Herein, a novel fabrication process combined with electron-beam-lithography and electroplating processes, has been developed to produce an 3D all-gold metamaterial. The metamaterial consists of underlying gold film and golden mushroom array, producing a metal–insulator–metal structure in which the dielectric layer can be filled with air or other liquid to be tested. After filling with different liquids, there are obvious changes in the reflection spectrum, and the sensitivity could reach 480.91 nm per refraction index unit, so it has potential in the field of real-time refractive index detection.**

I. INTRODUCTION

Metamaterials refer to a type of composite that is artificially designed to have special properties that are not found in natural materials. According to different application scenarios, metamaterials can be divided into acoustic, thermal, intelligent and electromagnetic metamaterials. Electromagnetic metamaterials are mainly used to modulate electromagnetic waves, which are usually periodic arrays composed of subwavelength resonant metal structures. By reasonably designing the morphology and materials of these subwavelength structures, various parameters such as polarization[1], phase[2], orbital angular momentum[3], spectrum[4] of electromagnetic wave can be regulated.

All-metal metamaterials with subwavelength structure are a kind of electromagnetic metamaterials, they are often used to stimulate localized surface plasmons polaritons (SPPs), which are based on the interaction between electromagnetic fields and free electrons in the metal surface or subwavelength metal nanostructures[5]. The plasmon system can confine light to less than the diffraction limit of light, so it has always been a research focus[6-7]. This interaction leads to an increase in the optical near-field on the surface of the nanostructure.

The optical properties of subwavelength metal structures often depend on their shape, structure and composition[8-10], like one-dimensional gratings[11,12], two-dimensional holes[13], three-dimensional (3D) vertically stacked circular arrays of nanoparticles and so on. while

three-dimensional (3D) structures have been widely studied due to their multiple structural parameters and unique optical properties. For example, Mao et al have fabricated 3D helical plasmonic nanostructures to rotate light polarization[14], Zhu et al make a 3D cavity-coupled metamaterial with different shapes to have different structure color for biosensor[15].

In addition, three-dimensional metal–insulator–metal (MIM) structures are often used to coupling SPPs excited by the upper and lower metals because it limits the electromagnetic field energy to a space smaller than the diffraction limit $(\lambda/2n)^3$, where λ is the wavelength of incident light and n is the refractive index of the surrounding environment[16]. This limitation leads to an accompanying field enhancement, which is essential for plasmonic, light is confined in the cavity for a long time to increase the photo lifetime and enhance the interaction of light matter[17-18]. As a result, a slight adjustment of the dielectric layer will produce a significant change in the spectrum. Although the nano-cavity with sub-wavelength volume has high field enhancement and quality factor, the cavity is often filled with solids and it is inconvenient to change the intermediate dielectric layer.

In this paper, a 3D all-metallic nano-cavity coupled metamaterial has been developed. This metamaterial is composed of a lower layer of gold film and golden mushroom array, forming a MIM structure. A nano-cavity is formed under each golden mushroom. When the height of the nano-cavity is less than or equal to the decay length of SPPs, the light can be confined to the dielectric layer and the interaction between light and matter can be enhanced. In addition, the medium layer in the MIM structure is air, which is easily filled by other liquids or gases to be tested. Since the dielectric layer is only a few hundred nanometers, this structure can excite cavity-coupled plasmonic modes in the visible light and near-infrared range. In general, it exhibits a distinct reflection spectrum changes with the different liquids filled in the nanocavity. The sensitivity can reach 480.91 nm per refraction index unit (RIU) which provides great potential for real-time refractive index detection. We choose Au as the material here, because the technology of electroplating Au is mature, electroplating stability and repeatability is good; On the other hand, Au has good

Research supported by the National Natural Science Foundation of China under Grant No. 61974004 and 61931018.
[1]College of Physics and Information Engineering, Fuzhou University, Fujian 350108, P.R. China.
[2]National Key Laboratory of Science and Technology on Micro/Nano Fabrication, Institute of Microelectronics, Peking University, Beijing 100871, P.R. China.

[3]Frontiers Science Center for Nano-optoelectronics, Peking University, Beijing 100871, China.
*Contacting Author: Peimin Lu (email: lpm@fzu.edu.cn), Wengang Wu (email: wuwg@pku.edu.cn).

978-1-6654-3008-1/21 $31.00 © 2021 IEEE

biocompatibility, it can be further used in other operations such as biological modification and biosensor.

II. FABRICATION-METHOD

The metamaterial structure can be obtained using electron-beam-lithography (EBL) and electroplating processes. As shown in the figure 1(a), step 1, Ti/Au with a thickness of 5/50 nm is deposited on a glass by electron beam evaporation. This metal film serves as both a reflective layer and a seed layer to prepare for subsequent electroplating. Step 2 (figure 1(b)), a layer of positive photoresist polymethyl methacrylate (PMMA) is coated on the gold film. The rotate speed and the concentration of the solution of PMMA both determine the thickness of PMMA, and the thickness of PMMA determine the height of the nano-cavity. Next, a serious of circular holes with a diameter of 200 nm and a period of 1600 nm were defined by EBL (figure 1(c)). Step 3 (figure 1(d)), set the electroplating voltage to a rectangular wave with a 20% duty cycle (period 100 ms, positive level 20 ms, negative level 80 ms), the maximum voltage is 10 V, and the current is less than 1 mA/cm^2. Step 4 (figure 1(e)), connect the platinum sheet to the positive electrode of the electroplating source, the sample to the negative electrode, and put them in the electroplating solution to start electroplating for 5 min. As shown in Figure 1, the gold pillars will continue to grow out of the holes in the photoresist to form a mushroom array. Step 5 (figure 1(f)), put the wafer in acetone for 2 hours to remove the photoresist. Finally, the air nano-cavity array is made.

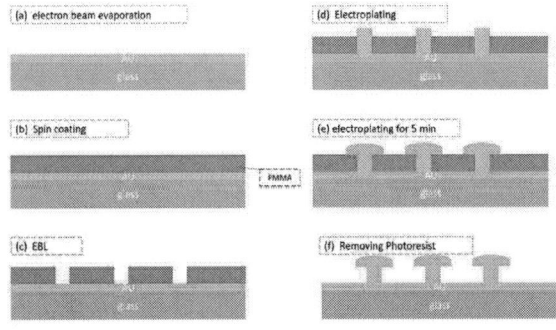

Figure 1. Schematic fabrication process for the mushroom array, the height of the nano-cavity depends on the thickness of the PMMA.

As shown in the figure 2, the scanning-electron-microscopy (SEM) pictures of the mushroom under different current conditions (left 5 mA/cm^2, right 1 mA/cm^2) when electroplating time is 5 minutes. The magnitude of the current determines the roughness of the nano mushroom. We can see that the roughness of the mushroom array in the right image has been significantly improved. In order to get a better nano-cavity, we choose 1 mA/cm^2 for processing.

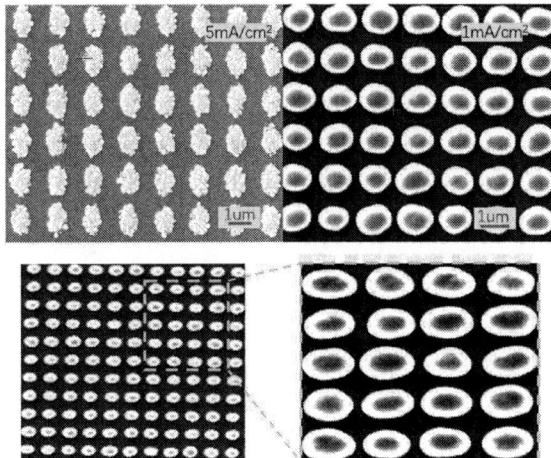

Figure 2. SEM images of the structures after electroplating under different current conditions when electroplating time is 5 minutes.

Figure 3 shows the SEM images of the mushroom at different electroplating times when the current is 1mA/cm^2. When the electroplating is starting, mushrooms begin to grow out of the circular hole. After 5 minutes on the electricity, the mushrooms gradually become larger; After eight or ten minutes of electroplating, the mushrooms would stick together so that light could not enter the nano-cavity, so we chose five minutes as electroplating time.

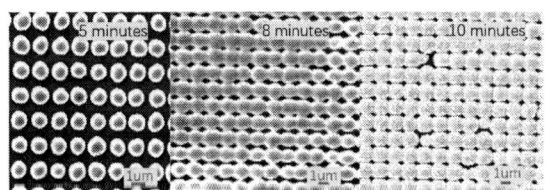

Figure 3. SEM images of the structures after electroplating under different electroplating times when the current is 1mA/cm^2.

In order to characterize the height of the nano-cavity, atomic force microscope (AFM) was used to characterize the surface morphology after EBL. As shown in the figure 4, the hole depth is 289 nm, which will be filled with gold pillars later. Therefore, the hole depth is the same as the height of the nano-microcavity.

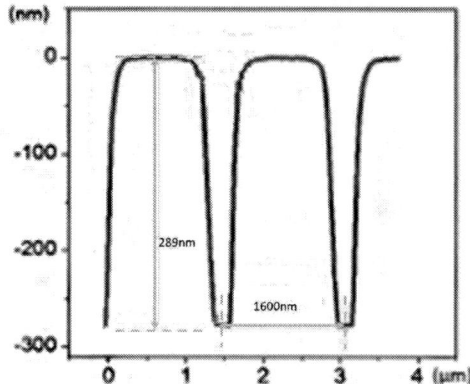

Figure 4.   Characterization of the surface morphology after EBL.

III.   SERSING MEASUREMENT

The figure 5 shows the diagram of spectrum measurement. The incident light is emitted from above to excite the SPPs mode on the Au surface. The electromagnetic waves are reflected back and forth in the nano-cavity under the mushroom, and the electromagnetic energy is confined in the nano-cavity for a long time. As a result, light waves of the corresponding wavelength are coupled into the cavity and then continue to attenuate, so strong absorption is observed at the corresponding wavelength.

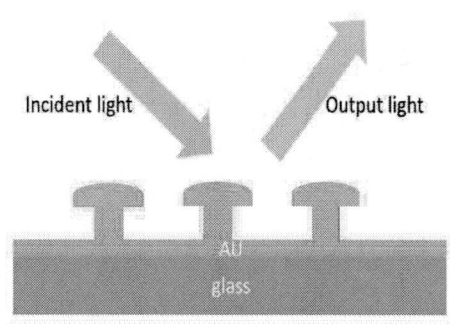

Figure 5.   The diagram of spectrum measurement.

Figure 6 shows the color of the metamaterial under the light microscope as the nanocavity is filled with air and different liquids (water, acetone, ethanol, IPA (iso-Propyl alcohol)). We can see that when the dielectric layer is air, the metamaterial appears rose red, and when the nanocavity is filled with different liquids, the metamaterial appears sky blue.

The experimental reflectance spectra were collected under normal incident light using a microscope-mounted IdeaOptics visible spectrophotometer. As shown in figure 7(a), when the structure is immersed in the liquid with different refractive index, the spectrum will have obvious

changes. With the increase of refractive index ranging from 1.331 to 1.375, the position of the peak reflectivity will be red-shifted. Figure 7(b) shows dip and peak position as a function of refraction index and the sensitivity could reach 480.91 nm/RIU. This nanocavity coupling structure provides a new opportunity for sensitivity and real-time refractive index sensing.

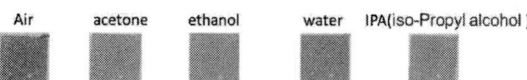

Figure 6.   color changes of the structures immersed in different mediums.

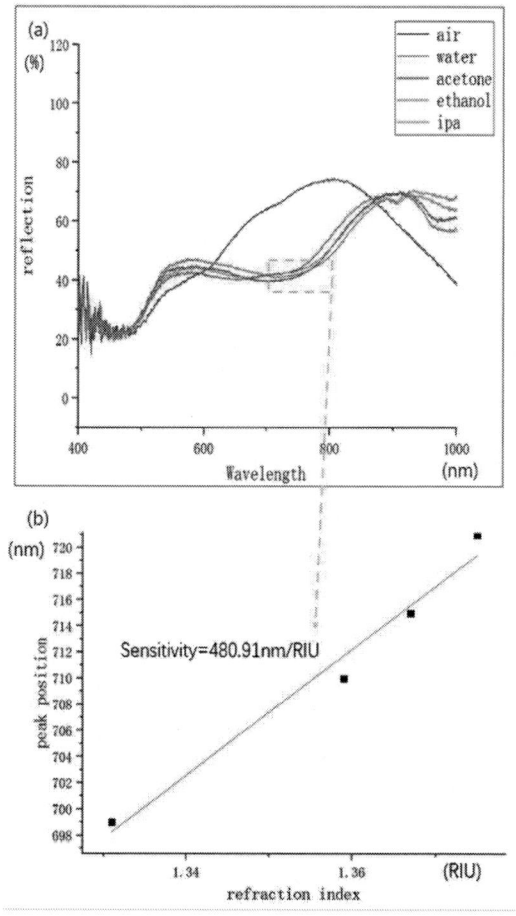

Figure 7.   Spectral shift (a) of the structures immersed in different mediums, (b) Dip and peak position as a function of refraction index.

IV.   CONCLUSION

In general, we combine EBL and electroplating to propose a processing method for making 3D all-metallic nano-cavity coupled metamaterial. The dielectric layer is an air nanocavity, where electromagnetic energy is localized and the coupling strength is continuously enhanced. The experiment results of the structure also support this theory. Since the nano-cavity can be filled with other liquids, the analyte can also enter the nano-cavity, which can be used for high-sensitivity detection. The structures have sensitive spectrum changes through different liquids in the nano-cavity. The sensitivity can reach 480.91 nm/RIU, and the structure is composed entirely of Au, which has great potential for real-time visual biochemical sensing.

ACKNOWLEDGMENT

This work is supported by the National Natural Science Foundation of China under Grant No. 61974004 and 61931018.

REFERENCES

[1] Gansel J K, Thiel M, Rill M S, et al. Gold helix photonic metamaterial as broadband circular polarizer. Science, 2009, 325(5947): 1513-1515.

[2] Khorasaninejad M, Chen W T, Devlin R C, et al. Metalenses at visible wavelengths: Diffraction-limited focusing and subwavelength resolution imaging. Science, 2016, 352(6290): 1190-1194.

[3] Karimi E, Schulz S A, De Leon I, et al. Generating optical orbital angular momentum at visible wavelengths using a plasmonic metasurface. Light: Science & Applications, 2014, 3(5): e167-e167.

[4] Chen Y, Duan X, Matuschek M, et al. Dynamic color displays using stepwise cavity resonators. Nano letters, 2017, 17(9): 5555-5560.

[5] N. J. Halas, S. Lal, W. S. Chang, S. Link and P. Nordlander, Chem. Rev., 2011, 111, 3913–3961.

[6] J. Zheng, W. Yang, J. Wang, J. Zhu, L. Qian, and Z. Yang, "An ultranarrow SPR linewidth in the UV region for plasmonic sensing," Nanoscale, vol. 11, no. 9, pp. 4061-4066, Mar 7 2019.

[7] A. Kristensen et al., "Plasmonic colour generation," Nature Reviews Materials, vol. 2, no. 1, 2016.

[8] J. R. Fan, J. Zhu, W. G. Wu and Y. Huang, Small, 2017, 13, 1601710. 3.

[9] R. M. Bakker, Y. F. Yu, R. Paniagua-Dominguez, B. Luk'yanchuk and A. I. Kuznetsov, Nano Lett., 2017, 17, 3458–3464. 4.

[10] Z. Mu, H. Yu, M. Zhang, A. Wu, G. Qi, P. K. Chu, Z. An, Z. Di and X. Wang, Nano Lett., 2017, 17, 1552–1558.

[11] L. Duempelmann, A. Luu-Dinh, B. Gallinet and L. Novotny, ACS Photonics, 2015, 3, 190–196.

[12] Y. Takashima, M. Haraguchi and Y. Naoi, Sens. Actuators, B, 2018, 255, 1711–1715.

[13] T. D. James, P. Mulvaney and A. Roberts, Nano Lett., 2016, 16, 3817–3823.

[14] Y. Mao, Y. Zheng, C. Li, L. Guo, Y. Pan, R. Zhu, J. Xu, W. Zhang and W. Wu, Adv. Mater., 2017, 29, 1606482.

[15] Zhu, Jia, et al. "Three-dimensional cavity-coupled metamaterials for plasmonic color and real-time colorimetric biosensors." Nanoscale 12(2020).

[16] S. A. Maier, Plasmonics: fundamentals and applications, Springer, 2007.

[17] Y. Nagasaki, M. Suzuki and J. Takahara, Nano Lett., 2017, 17, 7500–7506.

[18] V. J. Sorger, R. F. Oulton, J. Yao, G. Bartal and X. Zhang, Nano Lett., 2009, 9, 3489–3493.

Gap in pagination due to unavailable papers.

Pages 550-562

A Low-Cost Digital Lithography System Supporting Visual Focusing and its Application on Optical Fiber*

Luming Wang, Zhimin Zhang, Ningning Luo*, Haifeng Xiao, Long Ma and Qingwang Meng

Abstract—We here present a DMD based low-cost digital lithography system, and report the results of microstructure fabrication experiments on both glass slide substrates and single mode optical fiber end faces. The system possesses the ability to rapidly fabricate microstructures in sub-micron scale, and the characteristic of fabricating on unconventional substrates such as optical fiber end faces. Through the processing procedure of "visible light – chromatic aberration and optical path difference correction – ultraviolet light", the function of visual focusing was supported, without sacrificing part of the photoresist film. With above-mentioned features, the system is promising to accelerate the processing and construction of experimental sub-micron scale microstructure prototypes, especially micro-element prototypes on fiber end faces.

I. BACKGROUND

A microstructure fabrication method with the feature of low-cost and the ability of field processing would help to promote the experimental research of MEMS and micro devices. Fabrication methods such as Electron Beam Lithography [1], Focused Ion Beam Lithography (FIBL) [2-3] and Nanoimprint [4] take the advantages of high accuracy and high contrast. Taking the FIBL as an example, this method leaving scratches by focusing and accelerating the ion beam, bombarding the surface of substrates. A FIBL equipment can usually focus the ion beam to a few nanometers to achieve nanoscale microstructures processing. These methods have long been extensively used in the processing of both integrated circuit editing, nano-scale microstructures fabrication, and other micro operations in nanoscale. Yet, the above methods require sophisticated processing techniques and expensive equipments. Besides, for communication optical fiber devices working in the C-band, sub-micron scale feature size is sufficient to generate optical diffraction modulations in a high efficiency. Therefore, for optical fiber devices, it is necessary to find a balance between equipment costs and fabrication accuracy.

Relying on the development of MEMS, Digital Micromirror Device (DMD) has rapidly risen and became popular. DMD (we use DLP7000, Texas Instrument)

*The work is supported by National Natural Science Foundation of China (61704070, 61464008) and Natural Science Foundation of Jiangxi Province (20202BAB202012).

All authors are with the Key Laboratory of Opto-Electronic Information Science and Technology of Jiangxi Province, and Key Laboratory of Nondestructive Test (Ministry of Education), Nanchang Hangkong University, Nanchang, China.

Contacting Author: Ningning Luo is with the Key Laboratory of Opto-Electronic Information Science and Technology of Jiangxi Province, and Key Laboratory of Nondestructive Test (Ministry of Education), Nanchang Hangkong University, Nanchang, China (phone:+86-13879177625; email:ningningluo2002@126.com).

integrated 1024×768 micromirrors on the area of $14.008 \times 10.506 mm^2$, each micromirror has a size of $13.68 \times 13.68 \mu m^2$. By controlling the deflection of micromirrors between ± 12 degrees according to the digital signal, it can selectively reflect the spatial light field, and thus possesses the ability to modulate light intensity distribution. Based on the advantages of fast, flexible and controllable modulation of light fields, it has widely attracted the interest of researchers to DMD based digital lithography [5-6]. Without the need for physical masks required in conventional lithography method [7], maskless lithography is more suitable for microstructure prototypes fabrication. Although limited by the size of micromirrors and the optical properties of projection objective lens, the theoretical fabrication accuracy limit is sub-micron [8] (the specific diffraction limit will be discussed below), in some cases, it is sufficient.

Researchers have reported experiment results of microstructures fabrication on common substrates [9] and fiber end faces [10] by DMD based lithography method, having achieved 1.5μm and 2.2μm feature size, respectively. The results reported in the above papers illustrate the operability of fabricating microstructures by using DMD based digital photolithography method, especially on the optical fiber end faces. However, due to various factors, including but not limited to the diffraction limit mentioned above, it is troublesome for researchers to increase the fabrication accuracy to the sub-micron scale, especially for fiber end faces as substrate. For the experiment system of the latter report, a novel method was proposed. Researchers welded a small section of Coreless Silica Fiber (CLF) right after a Single Mode Fiber (SMF), the emitted beam from SMF was expanded in the CLF. The length of the CLF is accurately calculated so that the emitted beam could cover the entire end face of the CLF, thereby increasing the available area for fiber element fabrication. It is an effective method in the case of insufficient fabrication accuracy, but the processing of SMF and CLF based hybrid fiber is highly complicated, and direct processing on bare fiber is needed. Furthermore, in paper [10], a microscope has been used as projection element, the costs of the exposure equipment are increased.

In this report, we discuss a DMD based digital lithography system that supports visual focusing for rapid field fabrication of microstructures, without the use of microscope, to reduce the costs significantly. Through the direct photoresist spin-coating method on fiber end faces we proposed, photolithography could be directly processed on the end faces of optical fiber patch cords. In the functional verification experiment, the fabrication of 2-Dimentional microstructures with the minimum feature size of approximately 0.5μm was achieved. As the application target of the method we discussed in this paper, the experiment of processing microstructures on optical fiber end faces were also achieved, which exhibited the adaptability and practicality of this system for different types of substrates.

II. NUMERICAL ANALYSIS

Before setting up the experimental system, we carried out the theoretical calculations on the fabrication accuracy of the digital photolithography system, and designed digital masks to project micro-patterns with specific physical dimensions. We here considering an important factor that cause a effect on the accuracy of photolithography: the Numerical Aperture (NA) of the projection objective lens. The following calculations are in the category of undulatory optics. For measuring the optical diffraction limit, with NA as a parameter, the Rayleigh criterion formula used was,

$$d = 0.61\lambda / NA \tag{1}$$

Where d is the optical diffraction limit of the digital photolithography system; λ is the wavelength of the exposure light source, in the experiment we proposed, an i-line UV LED with λ=365nm was selected. Substituting the objective lens parameter selected in this experiment (magnification m=40x, NA=0.65) into (1), the diffraction limit caused by the objective lens was obtained as d≈342.54nm.

For the DMD we use, DLP7000, the size of a single digital micromirror is 13.68×13.68μm^2, thus we abstract it as pixel size p=13.68μm. Without considering the optical deviation of the projection part, it could be simply calculated that the actual size of a single pixel after projection is p_s=p/m=342nm. Factor p_s is very close but still smaller than the diffraction limit d, thus it will be difficult to clearly distinguish each pixel after the reduced projection. Moreover, the optical deviation of the projection part has to be considered. Therefore, the minimum feature size of the digital masks we made should be at least 2 pixels, to ensure a clear divide between lines. In the experiment we proposed, we first focused on the fabrication resolution verification, to evaluate the ability for fabricating microstructures on optical fiber end faces by digital photolithography method.

III. CURRENT EXPERIMENT SETUP

The working flow of the DMD based rapid digital photolithography method was summarized in Fig. 1. After careful cleaning, substrate was uniformly coated with a layer of photoresist by a spin coating device, and different photoresist film thickness for various microstructures were obtained through different rotation speeds. Soft-bake process was taken in a hot plate (for flat substrates such as slides) or a hot wind oven (for stereoscopic substrates such as fiber tips), to evaporate part of solvent in the photoresist. Then the substrate was exposed in the exposure device. The exposed substrate was developed in developer solution for a few seconds, and microstructures could be observed after careful rinsing. Conventional physical mask is needless, since the ability of DMD to modulate spatial light field based on electrical signals. Therefore, a single fabrication

processing could be completed within minutes based on the system we proposed.

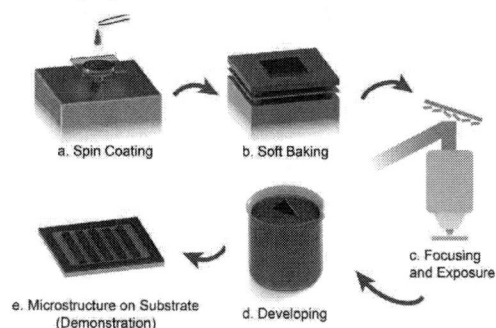

Figure 1. Schematic workflow for the microfabrication process.

Photo of the exposure device we proposed was shown in Fig. 2. In the exposure device, an ultraviolet (UV) LED emits a UV light beam. Its light intensity distribution was modulated by the DMD reflective spatial light modulator, and then projected onto the substrate through a 40× objective lens. Due to the short working distance of the objective lens with high magnification, precise focusing is needed before exposure. But UV irradiation while focusing will result in a failure of a part of photoresist, which is not feasible on substrates with tiny available area such as fiber end faces. Thus, for the visualization of focusing process, a visible light beam in red was introduced, and combined with the UV light beam to a single illuminating light beam through a 50:50 splitter prism. During the process of focusing, only visible light source was used. The visible light field reflected from the substrate will be introduced into the tube lens of the CCD camera by another 50:50 splitter prism, to observe the image of substrate surface. The UV light source is used only during the exposure process. According to different microstructures, the exposure process may take 1-3s.

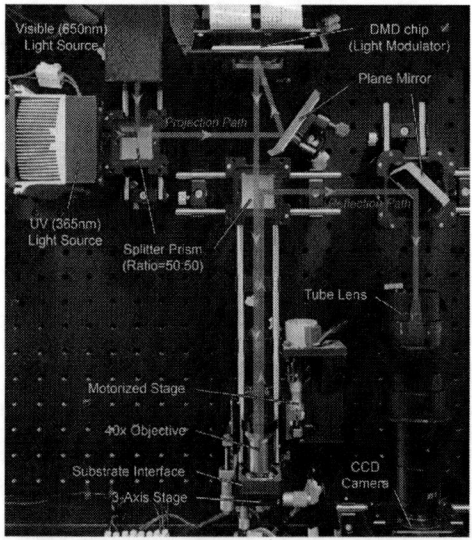

Figure 2. Top-viewed photograph of the DMD based digital lithography exposure device. Colored markings demonstrate that only approximate position.

The application on optical fibers of this method will offer the ability to rapidly fabricate micro elements on optical fiber end faces. As a preliminary step, here we report a photoresist spin-coating device dedicated to fiber end faces, as shown in Fig. 3(a), which allows direct spin-coating of photoresist on the end faces of optical fiber patch cords. Although only the fiber core (including cladding) with a diameter of 62.5μm needs to be spin-coated with photoresist, we directly use fiber patch cord equipped with a ceramic ferrule. The entire surface of the fiber ferrule end face with a diameter of about 2.5mm will be completely spin-coated with photoresist, in order to reduce the influence of surface tension on the flatness of photoresist film, and avoid the edge effect of spin-coating (the edge of photoresist film will be significantly thicker than the center) on the fiber end face. Furthermore, in order to lock the optical fiber into the exposure device, an optical fiber patch cord clamping structure was made, photo was shown in Fig. 3(b).

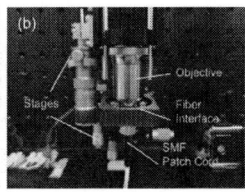

Figure 3. (a) Photograph of the spin-coating turntable with a optical fiber patch cord installed. (b) Photograph of the fiber interface of the exposure device.

IV. CURRENT RESULTS

A. The experiment of microstructures fabrication on glass slide substrates

In this part, an i-line positive type photoresist, KMP-C7600 was used for fabrication, and a thin photoresist film was spin-coated on the substrate by a spin-coating device. Soft bake was taken at 90°C for 60s. We use sub-micron scale binary optical diffraction microstructures, which will generate optical diffraction effect under visible light irradiation, to test the fabrication ability of the exposure device. Digital masks of linear gratings and Dammann grating with different feature sizes were applied to DMD chip respectively. The visible light source was first turned on, and the position of adjust stage was adjusted to image the digital mask on the substrate clearly. Prior to switching the light source, stage position must be precisely adjusted to compensate for the chromatic aberration and optical path difference caused by different wavelengths in the objective lens. Afterwards, the UV light source was turned on to expose for about 2s. A 4‰ NaOH solution was used as the developer, and the development process takes about 3s. After carefully washing away residual photoresist, binary microstructures we need couldbe formed on the surface of the substrate.

Linear gratings and Dammann grating were fabricated by the experimental system we proposed, and the minimum feature size is approximately 0.5μm. Fig. 4(a)-(c) shows the photomicrographs of the above microstructures under a microscope. The optical diffraction modulation effects under a dot green laser irradiation of above microstructures were shown in Fig. 4(d)-(f).

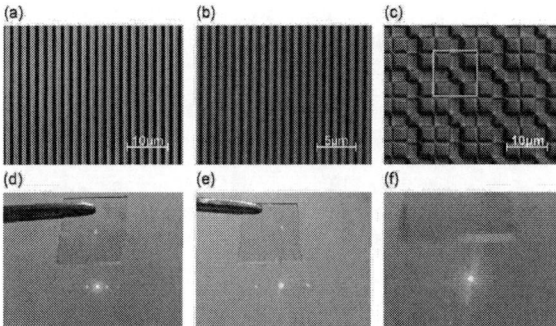

Figure 4. Photomicrographs of microstructures fabricated on glass slide substrates and optical diffraction effects under a beam of green laser irradiation modulated by the microstructures. (a) and (d), linear grating with a period of about 2μm. (b) and (e), linear grating with a period of about 1μm. (c) and (f), Dammann grating with a beam splitting ratio of 5x5, and the minimum feature size is about 700nm. In sub-graph (c), the yellow rectangle indicate one standard period of the Dammann grating.

In order to further evaluate the fabrication result, an optical profiler was used to measure the surface morphology, to evaluate the parameters such as straightness and contrast of lines. Limited by the measurement capability of the profiler we used, only micron-scale microstructures could be effectively calculated for depth information. Therefore, we measured the linear grating with a period of 2μm (8pixels), and observed its surface morphology under the profiler, as shown in Fig.5.

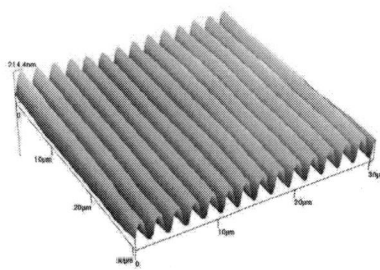

Figure 5. The surface morphology of the linear grating with a period of 2μm, fabricated by the photolithography system we proposed. The photo was taken with an optical profiler under a 50× interference objective lens.

Through the experiment above, the microstructure fabrication function of the DMD based photolithography system has been verified to be effective. However, the microscopic measurement results of microstructures we fabricated are inconsistent with the theoretical calculation result we discussed in the second part of this paper. With a width of 2pixels in the digital mask, the line was actually produced with a size of 0.5μm. It could be calculated that the actual reduction magnification factor is about 54.7×.

B. The experiment of microstructures fabrication on SMF end faces

The above-mentioned experiment on glass slide substrates has illustrated that, the DMD based photolithography system we proposed possesses the fabrication accuracy of 0.5μm. For a SMF core area with a diameter of about 9μm, a microstructure contains more than 10 periods will be able to modulate optical diffraction

978-1-6654-3008-1/21 $31.00 © 2021 IEEE

effects effectively. Therefore, we try to fabricate functional optical microstructures directly on the end faces of SMF patch cords.

The experiment setup and reagent selection in this part were kept the same with the above experiment. The device shown in Fig. 3(a) was used to spin-coat a layer of photoresist on the end faces of the SMF patch cords. It is necessary to note that, because of the area of fiber end face is tiny, the photoresist we dropped will inevitably flow to the side of the ceramic ferrule and remain on the surface. If the soft-bake was processed directly, an adhesion force will be generated between the photoresist and the surface of ceramic ferrule, farther, the diameter of ceramic ferrule will be changed significantly. When we try to lock the fiber into the interface of exposure device next, the residual photoresist will greatly affect the fiber fixing accuracy, and even cause the fiber ferrule to be too thick to be inserted into the fiber interface. Therefore, before soft-backing, the side of the fiber ceramic ferrule must be carefully cleaned with moist cotton wipes.

After spin-coating and carefully cleaning, the entire fiber patch cord was placed in a hot-wind oven for soft-baking at 100°C for 60s. The exposure and developing process were also consistent with the experiment performed on glass slide substrates. Photomicrographs of binary microstructures on the SMF end faces were shown in Fig. 6(a) and (c). During observation, the fiber patch cords were fixed on the microscope by a 3D printed holder. When a green laser was coupled into the unprocessed facet of the optical fiber, the laser beam was emitted from the processed facet, and then optical diffraction effects modulated by the microstructures on fiber core could be observed, as shown in Fig. 6(b) and (d).

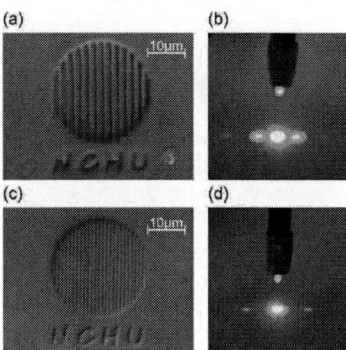

Figure 6. Photomicrographs of the microstructures fabricated on SMF end faces and its optical diffraction effects. In the photomircrographs, the fiber core with a diameter of 9μm is located in the central of microstructures' circular area. The letters "NCHU" are not involved in light modulation. (a) and (b), linear grating with a period of about 2μm. (c) and (d), linear grating with a period of about 1μm.

V. CONCLUSION

A DMD based photolithography method and system, with the function of visualize observation was proposed in this paper. By using the 40× objective lens as the reduced projection element, through experimental verification, the fabrication accuracy of 0.5μm was achieved. Furthermore, the surface morphology of microstructures was measured,

and, by fabricating linear gratings, Dammann gratings as beam splitting elements, the optical diffraction modulation results of these microstructures were evaluated.

In order to apply the DMD based photolithography method to the fabrication of optical fiber tip elements, a direct photoresist spin-coating device for optical fiber end faces and a microstructure fabrication working flow dedicated for optical fiber end faces were proposed. The sub-micron scale fabrication accuracy and the specially designed fabrication working flow possess the ability to fabricate elements directly on the end faces of optical fiber patch cords, without the preparation of hybrid fiber of the method proposed in paper [10] that is difficult to process. Although the fabrication accuracy is still relatively limited, it is sufficient for fiber tip elements fabrication such as beam spilitters and other functional optical microstructures [11-12]. The fabrication method, system and working flow are relatively more convenient and suitable for the wide applications of fiber tip devices in fiber optic systems.

ACKNOWLEDGMENT

The work is supported by National Natural Science Foundation of China (61704070, 61464008) and Natural Science Foundation of Jiangxi Province (20202BAB202012).

REFERENCES

[1] Y. Lin, Y. Zou, Y. Mo, J. Guo, and R. G. Lindquist, "E-beam patterned gold nanodot arrays on optical fiber tips for localized surface plasmon resonance biochemical sensing," in *Sensors (Switzerland)*, vol. 10, no. 10, pp. 9397 – 9406, 2010.

[2] R. S. Rodrigues Ribeiro, P. Dahal, A. Guerreiro, P. A. S. Jorge, and J. Viegas, "Fabrication of Fresnel plates on optical fibres by FIB milling for optical trapping, manipulation and detection of single cells," in *Scientific Reports*, vol. 7, no. 1, 2017.

[3] K. Sloyan, H. Melkonyan, and M. S. Dahlem, "Fabrication of 2D and 3D Photonic Structures using Focused Ion Beam," in *2020 IEEE Photonics Conf. IPC 2020 - Proc.*, pp. 4469 – 4480, 2020.

[4] A. Koshelev et al., "High refractive index Fresnel lens on a fiber fabricated by nanoimprint lithography for immersion applications," in *Opt Lett*, vol. 41, no. 15, pp. 3423-6, 2016.

[5] K. F. Chan, Z. Feng, R. Yang, A. Ishikawa, and W. Mei, "High-resolution maskless lithography," in *J. Microlithogr. Microfabr. Microsystems*, vol. 2, no. 4, pp. 331 – 339, 2003.

[6] H. Martinsson, T. Sandstrom, A. Bleeker, and J. D. Hintersteiner, "Current status of optical maskless lithography," in *J. Microlithogr. Microfabr. Microsystem*s, vol. 4, no. 1, pp. 1 – 15, 2005.

[7] Z. Zhang, Y. Gao, N. Luo, K. Zhong, and Z. Liu, "Multi-direction digital moving mask method for fabricating continuous microstructures," in *Optica Applicata*, vol. 45, no. 1, 2015.

[8] M. Kang, C. Han, and H. Jeon, "Submicrometer-scale pattern generation via maskless digital photolithography," in *Optica*, vol. 7, no. 12, p. 1788, 2020.

[9] Q. Zheng, J. Zhou, Q. Chen, L. Lei, K. Wen, and Y. Hu, "Rapid Prototyping of a Dammann Grating in DMD-Based Maskless Lithography," in *IEEE Photonics Journal*, vol. 11, no. 6, pp. 1-10, 2019.

[10] J. B. Kim and K. H. Jeong, "Batch fabrication of functional optical elements on a fiber facet using DMD based maskless lithography," in *Opt Express*, vol. 25, no. 14, pp. 16854-16859, 2017.

[11] R. S. Rodrigues Ribeiro, P. Dahal, A. Guerreiro, P. Jorge, and J. Viegas, "Optical fibers as beam shapers: from Gaussian beams to optical vortices," in *Opt. Lett.*, vol. 41, no. 10, p. 2137, 2016.

[12] M. Zaboub, A. Guessoum, N. E. Demagh, and A. Guermat, "Fabrication of polymer microlenses on single mode optical fibers for light coupling," in *Opt. Commun.*, vol. 366, pp. 122 – 126, 2016.

Gap in pagination due to unavailable paper.

Pages 567-570

Proceedings of the 16th Annual IEEE International
Conference on Nano/Micro Engineered and Molecular Systems
April 25-29, 2021

Research on the resolution of submicron lithographic projection system based on DMD for optical fiber end face

Qingwang Meng, Zhimin Zhang, Ningning Luo*, Luming Wang, and Menghan Xiong

Abstract— We report a direct and facile method to study the limit resolution of optical fiber microlithography, which is based on DMD (digital micromirror device) the fiber end face of submicron uv lithography system, with 365 nm uv light, supplemented by 620-630 nm red light for focusing the reference light source, using the Reduced projection objective lens will focus the DMD modulation of the ultraviolet light on the spin of the photoresist fiber end face, realize the optical fiber end surface microstructure (single exposure precision 900 nanometers, single exposure area of more than hundred microns) preparation.By changing the magnification of the objective and Number of pixels in a single cycle of digital mask, the influence of the Objectives with different magnification and the pixel point on the precision and integrity of the microstructure was analyzed, and the lithography experiment was carried out to verify and get the optimal solution between the two factors.

I. INTRODUCTION

With the continuous innovation and development of the research field of optical fiber system, the corresponding scientific research results begin to be widely used in the world.In the field of micro-nano structure application, such as fiber optic tweezers[1], fiber probe is widely used in single-cell scale biological online observation and microscopic operation[2-3].Up to now, the processing methods of optical fiber devices include scanning electron beam lithography[5-6], focused ion beam lithography[7] and femtosecond laser direct writing[8]. In addition, DMD maskless numberword lithography with strong programmable, convenient and flexible mask switching and high processing efficiency can be used to fabricate optical fiber devices[9].However, based on the DMD maskless uv light on the application of optical fiber end face the theory and technology in many aspects there are still many problems unsolved, such as about 365 nm uv light on the photoresist surface to achieve some kind of linear grating cycle line width, linear grating mask generated by digital micromirror single wide number of pixels and the relationship between the reduction ratio lens imaging, can get a variety of preparation plan.and through the optical simulation and on the glass substrates, optical fiber end face a large number of lithography experiment validated the contrast,The optimal combination of DMD pixel and zoom

All authors are with the Key Laboratory of Opto-Electronic Information Science and Technology of Jiangxi Province, and Key Laboratory of Nondestructive Test (Ministry of Education), Nanchang Hangkong University, Nanchang, China.

*Contacting Author: Ningning Luo (phone:+86-13879177625; email:ningningluo2002@126.com).

objective in the preparation of submicron linear grating is obtained.

II. RESEARCH PRINCIPLE AND PROCEDURE

As shown in Fig. 1, the overall system schematic diagram of this study on ultraviolet lithography of optical fiber end face microstructure is listed. It is not difficult to see from the diagram that the optical path of the whole system is very simple and clear.First of all, the 365 nm uv light is emitted by the uv LED through a beam splitter 1 first, and participate in the digital mask initially combined beams focusing red light LED, after 1 reflex reflector light next to the surface of the dmd(The DMD used in this system consisted of 1920 by 1080 micromirror units, with a single micromirror size of 7.56μm, an interval of 1μm between adjacent micromirrors, and a micromirror inclination of ±12°), and then through the Numbers on the DMD mask for spatial light modulation, this mask to carry through the reduction of the objective information of uv light focused on the base of spin positive photoresist (glass or optical fiber end face), so as to complete the whole process of lithography.And the whole focusing process can be observed and adjusted in real time by CCD.

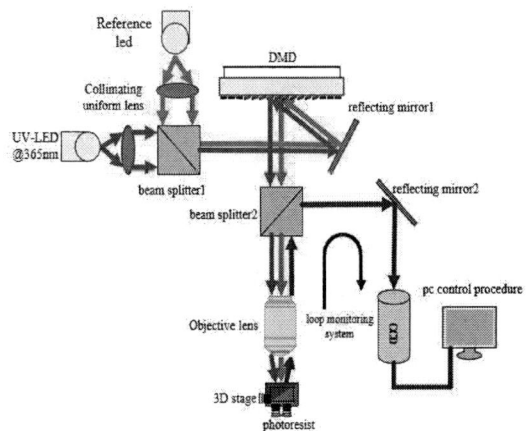

Figure 1. System optical path diagram.

Next, we built the entire system optical path, as shown in Fig. 2. The whole experiment process was carried out in the yellow light region without ultraviolet light.In addition, the whole laboratory adopts ultra-clean seal treatment, and the optical path components used in the experiment are purchased from optical device companies that meet the national standard,The mobile platform used in the experiment is a three-axis flexible displacement platform produced by Thorlab Company in the United States, with the minimum resolution of 500nm, which can ensure the

978-1-6654-3008-1/21 $31.00 © 2021 IEEE   571

position adjustment of DMD more accurately and It makes the experiment more rigorous and reliable.

Figure 2.   The whole experimental light path object and light path trajectory diagram.

Because the uv LED light show is gaussian distribution and the divergence Angle of emergent light is larger, on the one hand lead to ultraviolet light utilization rate is not high, on the other hand the parallel light irradiation to the DMD surface can lead to DMD mask serious distortion of ultraviolet light modulation distortion, in order to solve this problem, the Fig. 3 (a) of the collimating lens structure, can limit the LED light divergence Angle to between 3 ° and 13 °, effectively alleviate the ultraviolet light collimation.In order to demonstrate this effect directly, TracePro software was used to simulate the trajectory of LED light emitted after passing through the collimating lens in Fig. 3 (b-c). As can be seen from the ray tracing diagram, there is a big difference in the ray emission Angle between Fig. 3(b) without the collimating lens and Fig.3(c) with the collimating lens.

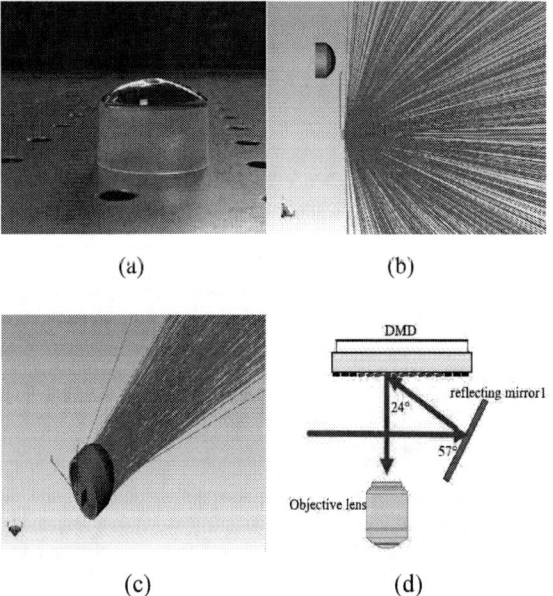

Figure 3.   Preliminary shaping of divergence Angle of UV LED and TracePro ray tracing(a-b-c) and The incident Angle of ultraviolet light passing through the mirror(d).

Due to the digital micromirror device (DMD) of a single pixel unit work, micro lens unit 12 ° inclination of plus or minus, in order to ensure that the digital micromirror device with mask information in the space of the modulation light vertical irradiation in the spin of the positive photoresist surface of basement (glass or optical fiber end face), which react with photoresist, obtained by geometric calculation incident light irradiation in the mirror 1 when the incident Angle for 33 °, as shown in Fig. 3 (d).

III.   ANALYSIS OF LIGHT INTENSITY DISTRIBUTION ON PIXEL PLAN

Lithography system based on DMD is the use of single photon absorption effect of the positive photoresist exposure to realize the design processing, processing of resolution determined by the scope of single photon absorption effect, can be through a single pixel focus projection light field intensity distribution and the threshold effect of photoresist, to control the scope of the single photon absorption effect.Due to the high magnification of the projection objective lens used in the experiment, the vector diffraction theory is needed to calculate the intensity distribution of the projection light field of a single DMD pixel at the focus (i.e., the single-photon point spread function).Considering that a single pixel of DMD is a rectangular element, the diffraction pattern of a single pixel can be regarded as Fraunhofer moment hole diffraction, and its model schematic diagram is shown in Fig. 4.

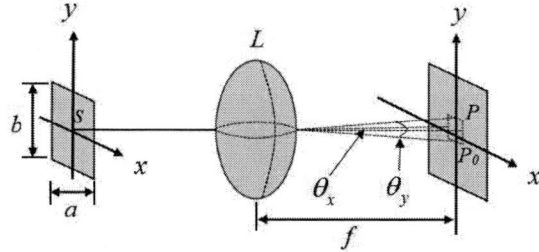

Figure 4.   Imageing principle of single pixel in focal plane.

The equivalent image size a and b of a single DMD pixel unit at the entry pupil of the objective lens are determined by the focal length f and the numerical aperture NA of the objective lens. Its center is taken as the origin of coordinates, and the complex amplitude of point P on the observation screen in the focal plane is

$$E(x, y) = A\left(\frac{\sin \alpha}{a}\right)\left(\frac{\sin \beta}{\beta}\right)\exp\left[ik\left(\frac{x^2 + y^2}{2f}\right)\right] \quad (1)$$

Where, α and β are optical coordinates, respectively

$$\alpha = \frac{k\sin(\theta_x)a}{2} = \frac{\pi}{\lambda}a\sin(\theta_x) = \frac{\pi a}{\lambda f}x = a'x, \quad (2)$$

978-1-6654-3008-1/21 $31.00 © 2021 IEEE   572

$$\beta = \frac{k \sin(\theta_y) b}{2} = \frac{\pi}{\lambda} b \sin(\theta_x) = \frac{\pi b}{\lambda f} y = \beta' y. \quad (3)$$

It can be seen that the complex amplitude of point P is related to its two coordinates, where A is the normalized amplitude value and F is the focal length of the objective lens.By substituting Equations (2) and (3) into Equation (1), the complex amplitude distribution formula of a single DMD pixel at point P in the focal plane can be obtained. Considering the relationship between the light intensity and the complex amplitude, the light intensity distribution can be obtained

$$I(x, y) = |E(x, y)|^2. \quad (4)$$

When multiple pixels of DMD chip work at the same time, the geometric projection relationship of the central coordinates of a single pixel on the image plane can be considered. The light intensity of a point on the focal plane is the sum of the complex amplitudes of each pixel at that point, i.e

$$E(x, y) = \sum_{m=-M/2}^{M/2} \sum_{n=-N/2}^{N/2} A \left[\frac{(\sin\alpha')(x-md)}{\alpha'(x-md)} \frac{(\sin\beta')(y-nd)}{\beta'(y-nd)} \right] \times$$

$$\exp\left\{ ik \left[\frac{(x-md)^2 + (y-nd)^2}{2f} \right] \right\}, \quad (5)$$

Where, M and N represent the total number of pixels in x and y axes respectively, M and N correspond to the position information of DMD pixels on the object plane respectively, the M on x axis and the N on y axis, and d is the displacement difference of adjacent DMD pixels in the image plane.Therefore, based on the superposition theory of the projected light intensity of the above pixels, we can obtain the light intensity distribution of different pixels on the focal plane.

In addition, we also simulated and analyzed the intensity distribution of the photomask of the ultraviolet light source used in this experiment under the three different reduction magnification objective lenses, as shown in Fig. 5(a-c).It is not difficult to see from Fig. 5 that, with the gradual increase of the objective lens reduction magnification, the digital mask light field in the base part presents a Gaussian distribution and the light field intensity per unit area keeps increasing, but the consequent result is that the mask light field area keeps decreasing in multiple.

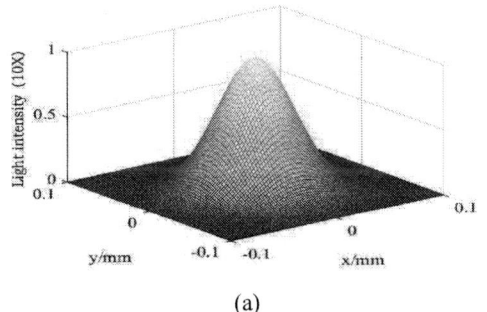

(a)

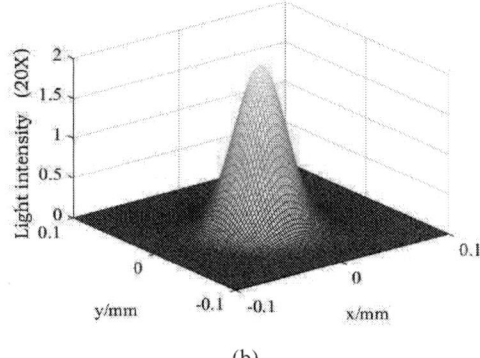

(b)

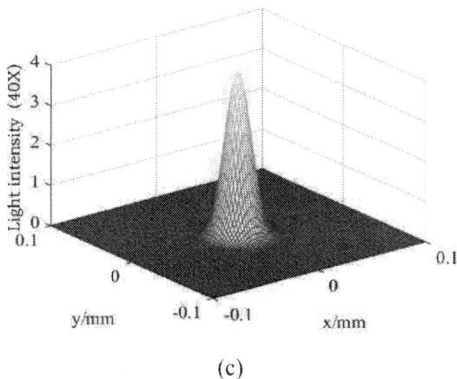

(c)

Figure 5. A simulation of the intensity of UV light modulated by a DMD digital mask focused on the photoresist surface through an objective lens (a-10x, Na =0.25, b-20x, Na =0.4, c-40x, Na =0.65).

Through the preliminary analysis of the imaging effect of the digital mask through the three kinds of zoom zoom objective lens, the relationship between the intensity of the projection light field and the exposure area under the three states is grasped on the whole, and it is not difficult to see that the exposure area and the exposure intensity are inversely proportional and nonlinear

IV. EXPERIMENTAL RESULTS AND ANALYSIS

On the basis of this experiment, a large number of lithography experiments were carried out. With other experimental conditions unchanged, the experiments were divided into three parts.The first part is to set the single-period pixel points of the raster mask to two, and the reduction of the objective lens to 10X.The second part is to set the single period pixel points of the raster mask to 4, and the reduction of the objective lens to 20X.The third part is to set the single-period pixel points of the raster mask to 8, and the reduction ratio of the objective lens to 40X.As shown in Fig. 6, the experimental results of the three experiments on glass substrates are listed (the observation and collection of the whole experimental results were all observed under an olibus 100X optical microscope and Linear grating linewidth 920nm).

Contrast experiment results (a2, b2, c2), integrity first a2 microstructure compared with b2, c2 is more bad, carved out of the grating microstructure contour line is not obvious and serious distortion, the depth of the microstructure is not enough, but you couldn't see preliminary b2, c2 differences, then we use the b2, c2 after partial enlarged view, you can see the c3 line width integrity and line depth must be better than b3, evidenced by b3, c3 structure judging color depth, the color of c3 is in b3 to dark, illustrate the depth of the c3 lithography is better than b3,The contours of the lines are less distorted.

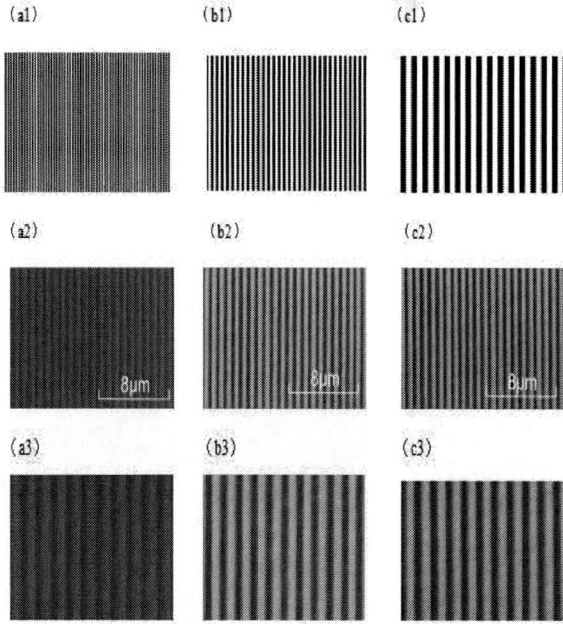

Figure 6. Digital mask for three kinds of pixels(a1- 2 pixels ,b1- 4 pixels,c1- 8 pixels),Comparison of experimental results of three experimental schemes (a2, b2, c2) on glass substrates and a partially enlarged view(a3, b3, c3).

Next, on the basis of previous experiments, we used a single-mode fiber with UV positive photoresist uniformly rotated on the end face in advance to replace the glass substrate, and kept other experimental conditions unchanged.

The basic process of photoresist experiment mainly includes substrate surface treatment, gluing, pre-drying, exposure, post-drying, developing, cleaning residual glue (not necessary), etc. but because the glass substrate used in the previous experiment is different from the optical fiber end face material used now, the photoresist has different adhesion to the two substrates.To solve this problem, we changed the pre-baking time and pre-baking temperature to compensate for this difference.

TABLE I. GLASS AND OPTICAL FIBER EXPERIMENTAL PARAMETERS

experiment parameter	experimental procedure		
	Gluing(r/s)	Pre-drying(T°)	Exposure(s)
glass	1000	90	2
Optical fiber	900	100	3

In table I details the use of glass substrate and the optical fiber end face when lithography key parameters of experimental basement for grating (well before the glue machine speed, drying time, exposure time) the quantitative control of the dependent variable, such as the experimental data are in after hundreds of thousands of times of repeated experiments, summed up the experimental steps of the optimal solution, any inaccurate experiment parameter Settings will result in huge experimental results deviation even experiment fails, the line is fuzzy linear grating microstructure distortion, uneven line width, thus affecting the experimental data acquisition.

In determining the fiber microstructure lithography experiment required above several key parameters, we will special fixture fixture replacing fiber base, under existing experimental conditions, the experiment is divided into three groups, three groups of the only independent variable is essence shrinks the equipment parameters of the object lens, then the first set of objective (10 x, Na = 0.25), the second objective (20 x, Na = 0.4), the third group objective (40 x, Na = 0.65).Then a large number of experiments were carried out to verify the three groups of experiments successively, and the experimental parameters were adjusted slightly in time according to the experimental results. After integrating a large number of experimental results, the three groups of experimental results as shown in Fig. 7 were finally obtained.

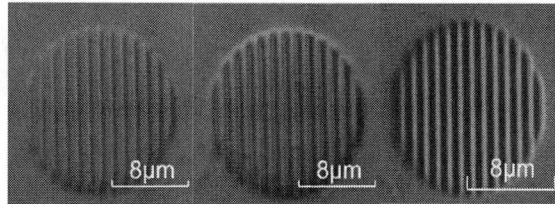

a. (10X, 2 pixels)   b. (20X, 4 pixels)   c. (40X, 8 pixels)

Figure 7. Comparison of experimental results of three experimental schemes (a2-a3, b2-b3, c2-c3) on Optical fiber end face.

In Fig. 7, comparing the three groups of experiment results, first under 10 x object lens, the optical fiber end face of the linear grating lines with limited depth and line width is not uniform, microstructure distortion is serious, but in the second group used 20 x objective experimental results in the figure, while the microstructure outline clear, line width is more uniform, but in the 100 x optical microscope can be observed clearly to the whole surface has obvious thin film interference fringe grating microstructure, the preparation of linear grating during the line is still not enough depth, short of using the standard, the final results from the third set of experiments compared with the former two experiments uniform depth of the first line and the line width is the most best,The distortion of the microstructure is small, which preliminarily indicates that the experimental results of these three experimental schemes can prove that under the same experimental conditions, when the digital lithography technology is used to prepare the submicron microstructure of a certain fixed specification, it is easier to obtain better experimental results by using high power objective lens.

V. CONCLUSION

A DMD based optical fiber end face submicron lithography system is studied to explore the best method to fabricate the optical fiber end face (first on the glass substrate) with a certain resolution. We in the preparation of 900 - nm line width of the linear grating as an example, proposed three kinds of implementation scheme, from the end of the experimental results, both in the glass substrate and the optical fiber end face, the third kind of experimental results of the experiment scheme is optimal, which proves that the multiple objective and processing in DMD lithography system structure and that there was a linear relationship between the objective lens ratio is higher, the preparation of the microstructure of line width uniformity, the better, the lower the structure distortion degree. However, the problem is that the higher the objective lens multiple, the smaller the area of the microstructure prepared at one time. Therefore, it is hoped that this work can provide a reference for the field of digital lithography.

ACKNOWLEDGMENT

The work is supported by National Natural Science Foundation of China (61704070, 61464008) and Natural Science Foundation of Jiangxi Province (20202BAB202012) and Postgraduate Innovation Special Fund of Nanchang Hangkong University (YC2020-059) .

REFERENCES

[1] Optical Tweezers to Replace 'Bots [J]. Nuts & volts,2018,39(1):9-9.
[2] C.Liberale,P.Minzioni,F.Bragheri,F.DE Angelis,E.Di Fabrizio,and I. Cristiani,"Miniaturized all fiber probe for three-dimensional optical trapping and manipulation," Nat. Photonic 1(12), 723-727(2007).
[3] Qiu, Jianrong, Shen Yi, Shangguan, Ziwei, "ALL-fiber probe for optical coherence tomography with an extented depth of focus by a high-efficient fiber-based fliter," Optics Communications: A Journal Devoted to the Rapid Publication of Short Contribution in the Field of Optics and Interaction of Light with Matter.413:276-2822018.
[4] Yu Zhang, Li Zhao, Yunhao Chen, Zhihai Liu, Yaxun Zhang, Enming Zhao, Jun Yang, Li bo Yuan. "Single optical tweezers based on elliptical core fiber," Optics Communications,2016.
[5] Manfrinato V R, Zhang L H, Su D, " Resolution limits of electron-beam lithography toward the atomic scale," Nano Letters,13(4):1555-1558,2013.
[6] Park J H,Steingart D A, Kodambaka S, " Electrochemical electron beam lithography: Write, read, and erase metallic nanocrystals on demand," Science Advances,3(7):e1700234,2017.
[7] Schroder T, Trusheim M E, Walsh M, " Scalable focused ion beam creation of nearly lifetime-limited single quantum emitters in diamond nanostructures," Nature Commu nications,3(7):e1700234,2017.
[8] Zhao Y Y, Zhang Y L, Zheng M L, " Three-dimensional Luneburg lens at optical frequencies," Laser Photonics Reviews,10(4): 665-672, 2016.
[9] Jae-Beom Kim, Ki-Hun Jeong, "Batch fabrication of functional optical elements on a fiber facet using DMD based maskless lithography," Opt. Express,25(14), 16854-16859(2017).

Gap in pagination due to unavailable papers.

Pages 576-646

Proceedings of the 16th Annual IEEE International Conference on Nano/Micro Engineered and Molecular Systems
April 25-29, 2021

Microstructural-PVDF Dielectric Layer Based High-Resolution Flexible Capacitive Pressure Sensor

Zebang Luo[1], Jing Chen[1], Zhengfang Zhu[1], Lin Li[1], Yi Su[1], Lei Wang[1, 2], Hui Li[1, 2, *]

1. Shenzhen Institutes of Advanced Technology, Chinese Academy of Sciences, Shenzhen 518055, Guangdong, China

2. CAS Key Laboratory for Health Informatics, Shenzhen Institutes of Advanced Technology, Chinese Academy of Sciences, Shenzhen 518055, Guangdong, China

Email: hui.li1@siat.ac.cn

Abstract—Flexible capacitive pressure sensors are gaining increasing attention due to their excellent performance to sensing the external contact and pressure. Here, we report a high-resolution (10.3 dpi) flexible capacitive pressure sensor matrix with Microstructure-PVDF film and transferable electrode array (MPTEA) for pressure mapping. The sensor presents high linear sensitivity under 20 kPa (7.6 MPa^{-1}), fast response time (<100 ms), and excellent flexibility which can wrap on the finger very well. Moreover, the capacitance of each pixel in the MPTEA exhibited obviously change when get pressing, and different complex pressure patterns could be successfully located by it. The sensor has huge potential in various application, including robots, touchpad, human-machine interfaces and real-time pressure mapping.

Keywords—flexible sensor, microstructure-PVDF, electrode array, pressure mapping

I. INTRODUCTION

Flexible pressure sensors, which can imitate skin to sense external stimulus [1-3] through converting pressure stimulation into electrical signals [4], have excited great research interests. With the development of flexible pressure sensors based on different physical conversion mechanisms, such as piezo-resistive [5-7], piezo-capacitive [8-12], and piezo-electric [13-15], which has a great potential in motion detection [16, 17], soft robots [18, 19], prosthetic [20] and diagnostics [20, 21]. Recently, the ability of flexible pressure sensor to realize pressure mapping is a interesting topic for these applications [22]. Compared to piezoelectric and piezoresistive sensors, the flexible capacitive pressure sensor can be easier prepared for detecting the pressure distribution [23, 24]. Additionally, flexible capacitive pressure sensor has edge effect caused by fringe capacitance, which can sense the proximity of object and have great potential in human-machine interfaces, touch screen, etc.

At present, many different flexible pressure sensor matrices have been explored for detecting the surface pressure distribution [12, 25-27]. Kulkarni et al. reported a 3 × 3 pixelated sensor array based on AgNWs electrode array and PDMS dielectric layer, and the width and pitch of the electrode are each 6mm, which the resolution is 2.5 dpi (by calculation) [25]. Lee et al. prepared a 5 × 5 pixelated flexible pressure sensor array with an area of 33mm × 33mm based on ITO electrode array and porous PDMS layer, and the resolution is 3.8 dpi [26]. Shen et al. fabricated 5 × 5 pixelated flexible capacitive pressure sensor matrix by CNC@XNBR@Ag electrodes and PU dielectric layer, and the resolution is 2.1 dpi [27]. Shuai et al. prepared a flexible capacitive pressure sensor array based on Ag electrodes and unstructured PDMS dielectric layer, which the resolution is 4.23 dpi [12].Although these matrices can detect the surface pressure distribution at a certain extent, but the large spacing between electrodes leads to low resolution, which will limit them to reflect the contours of the surface objects. It is important to reduce the distance between the electrodes to improve the resolution of flexible capacitive pressure sensor matrix.

Here, we improved the resolution of flexible capacitive pressure sensor matrix via optimized electrodes array fabrication method, and the resolution is as high as 10.3 dpi. The optimized electrodes array fabrication method was consisted of PDMS-paper-based mask and vacuum-filtration, which can generally reduce the pitch of electrodes array. In addition, we fabricated a Microstructure-PVDF (MP) stable film for dielectric layer, which can enhance the sensitive and stability of the sensor. Moreover, the PEDOT-MWCNTs/MP/PEDOT-MWCNTs sandwich-structured sensor possesses impressive compressibility and flexibility. The flexible 8 × 8 pixelated flexible pressure capacitive sensor array based on MP film with Transferable Electrode Array (MPTEA) can map different 3D-printed letters, and excellent pressure sensitivity and long-term durability shows extensive potential in pressure mapping.

II. DESIGN AND FABRICATION

Fig. 1 shows the preparation process of MPTEA. First, prepared a clean glass and silicon wafer. 2g PVDF (Sigma-Aldrich, Inc.) was dissolved in 8g NMP to get PVDF solution (20 wt%) and spined PVDF solution on the glass with 500 rpm for 30 s, then placed on hot plate at 40°C to get a 20 μm PVDF thick film. Using photolithography to prepare microstructured wafer mold. The mixture of PDMS (Sylgard 184, Dow Corning) elastomer to cross-linker (10:1) was prepared by centrifugal stirring and diluted with n-hexane (2:1 n-hexane to PDMS). Then, spined diluted PDMS mixture solution on the microstructured mold and the glass with 20 um PVDF film to fabricate the microstructured layer and PVDF/PDMS layer.

978-1-6654-3008-1/21 $31.00 © 2021 IEEE

After PDMS cured, peeled off micropillars and PVDF/PDMS layer from microstructured mold and glass. Plasma treatment on the surface of two peeled films for 3 mins. The surface of treated films are contacted and combined by covalent bonds to form a Microstructure-PVDF (MP) film. Next, mixed 0.03 g MWCNTs (Aladdin, Inc.), 1.5 ml PEDOT:PSS (1.5 wt%, Heraeus, Inc) and 50 ml ethanol up for 1 min. The PEDOT:PSS/MWCNTs/ethanol solution was dispersed by intelligent ultrasonic processor at ice bath 1 hour. After sonication process, the filter paper (101, BKMAM, Inc) was used to filtered the dispersed PEDOT:PSS/MWCNTs/ethanol solution to get electrode solution. High conductive and flexible PEDOT:PSS/MWCNTs electrodes array was prepared by vacuum filtration of electrode solution with PVDF membrane and PET filtration mask. Then, spined the mixture Ecoflex-0030 on the surface of the MP film with 2500 rpm for1 min to get adhesive layer. While Ecoflex-0030 semi-cured, put filter paper with the patterned electrode on semi-cured Ecoflex-0030. After Ecoflex-0030 cured at room temperature, removed the PVDF membrane to get the electrode array remaining on the dielectric layer. The same method was used to keep the electrode array on the other side. Integrate the electrode array and dielectric layer into a whole by transferring electrode from filter membrane to dielectric layer via Ecoflex-0030, which can greatly enhance the stability of the sensor.

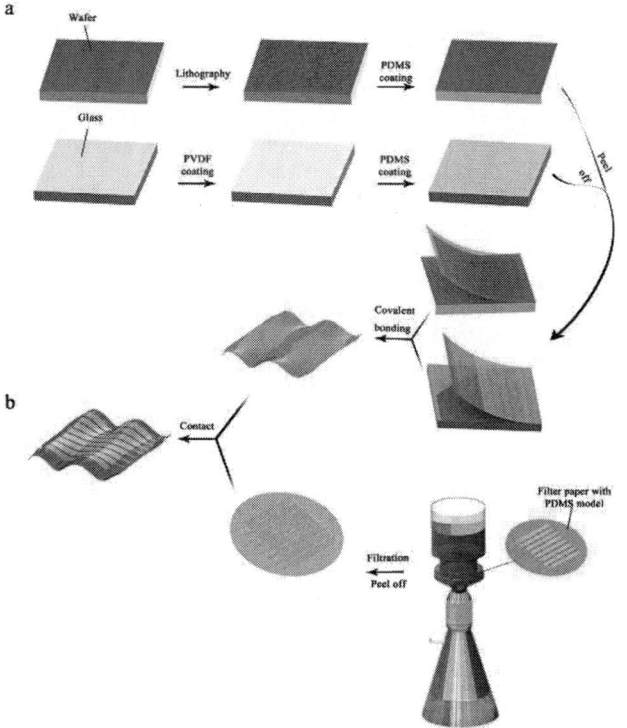

Fig. 1: *Device fabrication. a) The preparation of the Microstructure-PVDF film. b) Preparation of electrode and the integration of flexible capacitive pressure sensor array.*

The schematic of MPTEA was shown in Fig. 2. The MPTEA is consisted of two PEDOT:PSS-MWCNTs electrodes array and a Microstructure-PVDF layer between them. The SEM image of pillars microstructure as shown in Fig. 2b, which can improve the compressibility of the dielectric layer to enhance sensor's sensitivity and shorten the response time to external pressure. Fig. 2b is the SEM image of PEDOTS:PSS-MWCNTs electrode. Here, the cross-linked PEDOT:PSS and MWCNTs structure can significance improve the conductive of electrode which the conductivity is high to 259 S/cm. Meanwhile, this structure make the electrode with adequate flexible to support the sensor can be twisted and bent as shown in Fig. 2d and 2e.

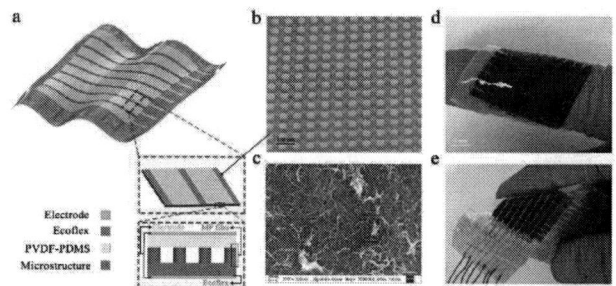

Fig. 2: *Schematic of the cross-type MPTEA. a) Schematic of an 8 × 8 cross-type flexible pressure sensor. Insert: Partial enlarged oblique and cross-sectional view of the structural design of the top and bottom. b) The SEM image of the pillars microstructure of MPTEA. c) SEM image of the PEDPT:PSS/MWCNTs electrode. d) Photographs of a fabricated 8 × 8 cross-type flexible sensor matrix on the finger with good flexibility. e) The picture show that the wired sensor matrix twisted well.*

III. RESULTS AND DISCUSSION

The capacitance of the sensor can be describe by Equation 1.

$$C = \varepsilon_0 \varepsilon_{MP} \left(\frac{A}{d} \right) \tag{1}$$

where ε_{MP} and ε_0 refer to the relative dielectric permittivity of the MP film and dielectric constant of vacuum, respectively. A is the sensor area and d is the vertical distance between two electrodes. To quantitatively evaluate the performance of the pressure sensor with Microstructure-PVDF dielectric layer, the change of relative capacitance ($\Delta C/C_0$) while linearly increasing pressure was measured as shown in Fig. 3a and 3b. The sensitivity of the flexible capacitive pressure sensor S is typically defined as Equation 2.

$$S = \delta(\Delta C/C_0)/\delta p \tag{2}$$

, where C_0 and ΔC are the initial capacitance of the sensor and the capacitance change with the applied pressure of the sensor, respectively, and p represents the applied pressure. As shown in Fig. 3a, a sensitivity of 7.6 MPa^{-1} was obtained of the sensor with Microstructure-PVDF dielectric layer in the lower pressure range (under 20 kPa). This value is considerably higher than other structured pressure sensor. Fig. 3b shows the relative capacitance change of sensor with different dielectric layer structure under 20 kPa, it is obviously that MPTEA presents excellent linearity and higher sensitivity.

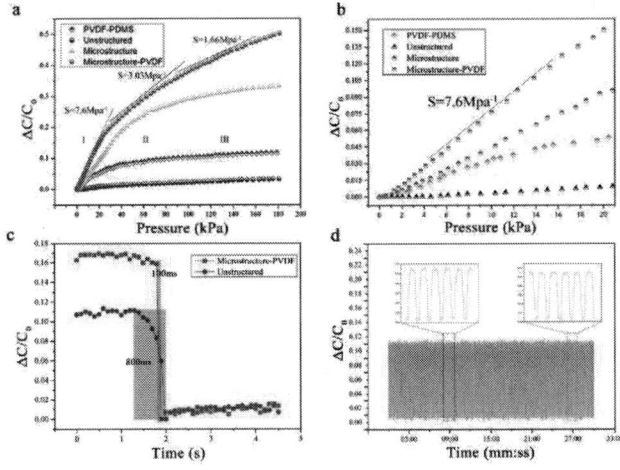

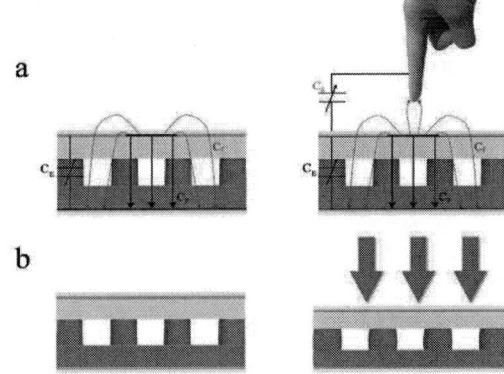

Fig. 4: a) Finger approaching the top electrode of the sensor. The finger reduces the capacitance between the electrodes (C_E) by coupling itself with the projected field (C_g). b) MPTEA before and under application of pressure.

Fig. 3: a) Relative capacitance change of flexible capacitive pressure sensor under a wide range of linearly increasing applied pressures, up to pressure in excess of 180kpa. The pressure region have be divided into 3 regions (0-20 kPa, 20-90 kPa and 90-180 kPa) and the maximum sensitivity is 7.6MPa⁻¹ at range I (0-20 kPa). as shown in graph. The high-pressure ranges (II and III) show that the sensitivities were 3.03 MPa⁻¹ and 1.66 MPa⁻¹ respectively. b) The relative capacitance change of the sensor with different dielectric layer to increasing pressure from 0 to 20 kPa. c) The reaction time of sensor to external pressure released. d) More than 500 cycles testing of flexible pressure sensor with Microstructure-PVDF film.

The micropillars structured layer has more deformable surfaces than unstructured PDMS layer. Here, increased surface of micropillars allows the MPTEA to be reversibly and rapidly released the elastic energy of the deformation caused by external pressure, which can generally reduce the influence of PDMS viscoelastic behavior. Owing to this, the relaxation time of the sensor with microcylinder structured dielectric layer is much shorter than the sensor with unstructured layer as shown in Fig. 3c, which are 100ms and 800ms separately. Meanwhile, the Microstructure-PVDF dielectric layer make the sensor have great toughness, and it can still work after more than 500 cycle tests at 30 minutes as shown in Fig. 3d.

The capacitive pressure sensor will produce a field that extends beyond the surface of sensor encapsulated with PDMS when it works, and the capacitance between tow electrode C_E is composed of the plate capacitance C_P and the fringe capacitance C_f as shown in Fig. 4a (left). The grounded conductors will affect the external electric field while it approaches to device acts as the third electrode, which capacitively couples to the top electrode of the flexible capacitive sensor, which represented by the variable capacitor C_g in Fig. 4a (right). The present of coupling capacitance C_g will reduce the fringe capacitance between electrode C_f which result the change of capacitance between two electrodes C_E. According to equation 1, the deformation changed of the dielectric layer during the sensor under pressure causes a corresponding change in the capacitance as shown in Fig 4b. And the capacitance of sensor will also change while the finger approaching the sensor. because of fringe capacitance C_f.

The contact sensing performances of the MPTEA encapsulated with PDMS were studied by forming a proximity and direct contact between grounded conductor (human finger) and the top surface of the sensor as shown in Fig. 5. Excellent stability and repeatability were obtained as shown in Fig. 5a, which indicates that the device can efficiently detect electric field changes between top electrode and nearby objects. Since human are a grounded conducting medium, while the finger is approaching the sensor's electrode, which definitely disturbs the external electric field. While finger contacting MPTEA, C_E decreases as C_f is reduced by forming coupling capacitance C_g with top electrode and finger. While C_p remain unchanged because no geometry deformation is introduced. The distance between finger and top electrode also will change coupling capacitance C_g. While the finger approaching, C_g increased and C_E decreased. Meanwhile, distinguishable changes in capacitance of the approach sensors are observed of contact and approach sensing modes as shown in Fig. 5b, indicating a good sensitivity and stability of touching performance.

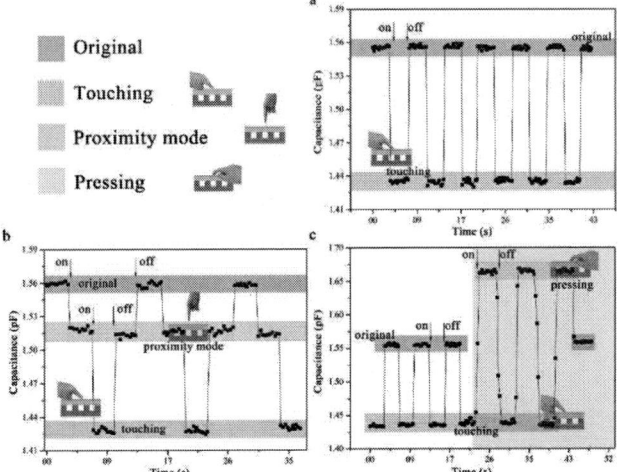

Fig. 5: The contact performance of MPTEA. a) The change of capacitance before and after finger touching. b) Capacitance changes of the device under finger proximity mode and touching mode. (Proximity mode: Place the finger about 1cm from the device.) c) The capacitance change of MPTEA in touching mode and pressing mode.

By pressing the device with finger, the capacitance changes significantly as shown in Fig 5c. The contact of finger and MPTEA leads to capacitance decreasing as shown in Fig 5d. Physical pressing generally induces an increase of C_E since the distance between tow electrodes is reduced, C_p increased. The increase of capacitance induced by pressing is clearly larger than decrease by finger contact, demonstrating that the potential of MPTEA to detect both touch and pressing of human finger.

The 3D-printed H-, I- and heart-shaped molds were designed with specific size of the pixels of MPTEA to establish the ability of MPTEA to measure the pressure distribution of the surface objects, as shown in Fig. 6a. Because of resin materials, the 3D printed molds are extremely light, and the weight of H-, i- and heart-shaped molds are 1.68, 1.88, and 1.54g respectively. The corresponding distribution and the relative capacitance changes of MPTEA's pixel to different shaped molds as shown in Fig. 6b-d. With the performance of high sensitivity and low detection limit, MPTEA can easily measure the surface pressure distribution while different 3D-printed molds were placed on it as shown in Fig. 6b and c. In order to better reflect the resolution of MPTEA and the degree of response of each pixel to complex objects, the 3D printed heart-shaped mold with the correspond size to each pixel was placed on MPEA. The corresponding distribution and the relative capacitance change of each pixel are shown in Fig. 6d. Complex 3D-printed heart-shaped mold can be mapped by MPEA clearly, and the correspond of each pixel in contact with the object changes almost the same.

For the objects with different shapes, the output signals of MPTEA could well map the different objects which placed on it. The results show that MPTEA has high sensitivity and high resolution, which performs like our skin to map external pressures, and has the potential application in the wearable electronics and human-machine interactive operations.

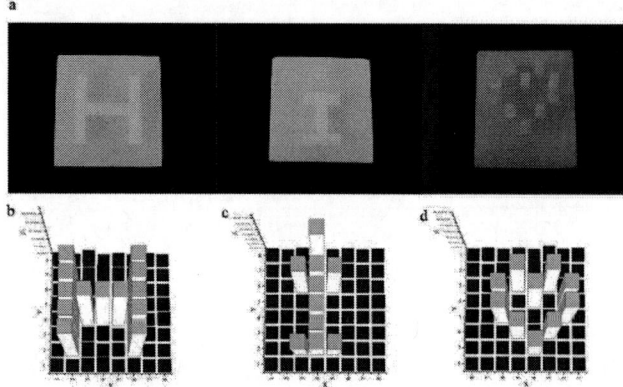

Fig. 6: *a) The 3D-printed letters H, i and heart-shaped objects. b-d) were the corresponding distributions of the normalized capacitance change with different shaped molds on the MPTEA.*

IV. CONCLUSION

In this paper, the high-performance MPTEA was developed. A simple way to prepare transferable electrode array for top and bottom electrode of capacitive pressure sensor is greatly improve the resolution of flexible capacitive pressure sensor

matrix. The resolution of sensor matrix is high to 10.3 dpi, which can map different shaped objects such as the letter of H- and i-shaped. The sensor presented excellent durability which can still work normally after more than 500 cycle tests under 15 kPa high pressure. This high-resolution sensor matrix has characteristics similar to human skin, which can response to external stimuli and map the surface pressure distribution. Meanwhile, high stability and reliability of the sensor shows great potential in biomedical applications.

ACKNOWLEDGMENT

This research was supported by National Key Research and Development Program of China (2019YFC1711701), National Natural Science Foundation of China (61803364, U1913216, U1713219), and Shenzhen Fundamental Research Project (JCYJ20180302145549896).

REFERENCES

[1] Asghar, W., et al., Piezocapacitive Flexible E - Skin Pressure Sensors Having Magnetically Grown Microstructures. Advanced Materials Technologies, 2020. 5(2): p. 1900934.

[2] Ma, Z., et al., Advanced electronic skin devices for healthcare applications. J Mater Chem B, 2019. 7(2): p. 173-197.

[3] Seminara, L., et al., Towards integrating intelligence in electronic skin. Mechatronics, 2016. 34: p. 84-94.

[4] Pang, C., et al., A flexible and highly sensitive strain-gauge sensor using reversible interlocking of nanofibres. Nat Mater, 2012. 11(9): p. 795-801.

[5] Ding, Y., et al., Flexible and Compressible PEDOT:PSS@Melamine Conductive Sponge Prepared via One-Step Dip Coating as Piezoresistive Pressure Sensor for Human Motion Detection. ACS Appl Mater Interfaces, 2018. 10(18): p. 16077-16086.

[6] Lee, S., et al., A transparent bending-insensitive pressure sensor. Nat Nanotechnol, 2016. 11(5): p. 472-8.

[7] Shi, J., et al., Multiscale Hierarchical Design of a Flexible Piezoresistive Pressure Sensor with High Sensitivity and Wide Linearity Range. Small, 2018. 14(27): p. e1800819.

[8] He, Z., et al., Capacitive Pressure Sensor with High Sensitivity and Fast Response to Dynamic Interaction Based on Graphene and Porous Nylon Networks. ACS Appl Mater Interfaces, 2018. 10(15): p. 12816-12823.

[9] Kwon, D., et al., Highly Sensitive, Flexible, and Wearable Pressure Sensor Based on a Giant Piezocapacitive Effect of Three-Dimensional Microporous Elastomeric Dielectric Layer. ACS Appl Mater Interfaces, 2016. 8(26): p. 16922-31.

[10] Ruth, S.R.A., et al., Rational Design of Capacitive Pressure Sensors Based on Pyramidal Microstructures for Specialized Monitoring of Biosignals. Advanced Functional Materials, 2019.

[11] Wan, S., et al., Graphene oxide as high-performance dielectric materials for capacitive pressure sensors. Carbon, 2017. 114: p. 209-216.

[12] Shuai, X., et al., Highly Sensitive Flexible Pressure Sensor Based on Silver Nanowires-Embedded Polydimethylsiloxane Electrode with Microarray Structure. ACS Appl Mater Interfaces, 2017. 9(31): p. 26314-26324.

[13] Huang, T., et al., Phase-Separation-Induced PVDF/Graphene Coating on Fabrics toward Flexible Piezoelectric Sensors. ACS Appl Mater Interfaces, 2018. 10(36): p. 30732-30740.

[14] Liu, M., et al., Large-Area All-Textile Pressure Sensors for Monitoring Human Motion and Physiological Signals. Adv. Mater. Weinheim, 2017. 29(41).

[15] Yan, C., et al., Epidermis-Inspired Ultrathin 3D Cellular Sensor Array for Self-Powered Biomedical Monitoring. ACS Appl Mater Interfaces, 2018. 10(48): p. 41070-41075.

[16] Lee, W., S.H. Hong, and H.W. Oh, Characterization of Elastic Polymer-Based Smart Insole and a Simple Foot Plantar Pressure Visualization Method Using 16 Electrodes. Sensors (Basel), 2018. 19(1).

[17] Sengupta, D., Y. Pei, and A.G.P. Kottapalli, Ultralightweight and 3D Squeezable Graphene-Polydimethylsiloxane Composite Foams as Piezoresistive Sensors. ACS Appl Mater Interfaces, 2019. 11(38): p. 35201-35211.

[18] He, Q., et al., Electrically controlled liquid crystal elastomer-based soft tubular actuator with multimodal actuation. Sci Adv, 2019. 5(10): p. eaax5746.

[19] Yan, X., et al., Carbon fibre based flexible piezoresistive composites to empower inherent sensing capabilities for soft actuators. Soft Matter, 2019. 15(40): p. 8001-8011.

[20] Choi, S., et al., Recent Advances in Flexible and Stretchable Bio-Electronic Devices Integrated with Nanomaterials. Adv. Mater. Weinheim, 2016. 28(22): p. 4203-18.

[21] Khan, Y., et al., Monitoring of Vital Signs with Flexible and Wearable Medical Devices. Adv. Mater. Weinheim, 2016. 28(22): p. 4373-95.

[22] Li, J., et al., Recent progress in flexible pressure sensor arrays: from design to applications. Journal of Materials Chemistry C, 2018. 6(44): p. 11878-11892.

[23] Chen, S., et al., Matrix-Addressed Flexible Capacitive Pressure Sensor With Suppressed Crosstalk for Artificial Electronic Skin. IEEE Transactions on Electron Devices, 2020. 67(7): p. 2940-2944.

[24] Pyo, S., J. Choi, and J. Kim, Flexible, Transparent, Sensitive, and Crosstalk-Free Capacitive Tactile Sensor Array Based on Graphene Electrodes and Air Dielectric. Advanced Electronic Materials, 2018. 4(1).

[25] Kulkarni, M.R., et al., Transparent Flexible Multifunctional Nanostructured Architectures for Non-optical Readout, Proximity, and Pressure Sensing. ACS Appl Mater Interfaces, 2017. 9(17): p. 15015-15021.

[26] Lee, B.-Y., et al., Low-cost flexible pressure sensor based on dielectric elastomer film with micro-pores. Sensors and Actuators A: Physical, 2016. 240: p. 103-109.

[27] Gao, Z., et al., A Self-Healable Bifunctional Electronic Skin. ACS Appl Mater Interfaces, 2020. 12(21): p. 24339-24347.

Proceedings of the 16th Annual IEEE International
Conference on Nano/Micro Engineered and Molecular Systems
April 25-29, 2021

Electrohydrodynamically Printed Multicolor Perovskite Image Sensor Array

Qilu Wang[1#], Guannan Zhang[1#], Hanyuan Zhang[1], Yongqing Duan[1]*, *Member, IEEE* and YongAn Huang[2], *Member, IEEE*

Abstract— Direct multicolor perovskite image sensors have broad application prospects in electronic eyes, wearable sensors, optical communications and so on. However, high-resolution perovskite multicolor image sensors with simple and cost-effective preparation methods are rarely reported at present. Here, we used electrohydrodynamic (EHD) printing to realize high-resolution perovskite multicolor image sensor based on pixelized bandgap perovskite array to identify color, which successfully realized the clear mapping of the light source signal and color recognition. Through the optimization of printing and annealing process to achieve controllably fabrication of ~1 μm patterned films; The potential of higher-resolution devices has been proven by designing a device of 2 μm channel length and 30 μm channel width, which exhibited R and D* values of 1.625 A/W and ~4.451×10^11 Jones. This work paves the way for the realization of high-resolution multicolor perovskite image sensors.

I. INTRODUCTION

The future development of image sensors tends to color recognition and high-resolution, which will play an important role in the fields of electronic eyes, wearable sensor equipment and optical communication. In order to realize color recognition, there are four main approaches including:1) using color filter array or prisms, which is the most widely used in multicolor image sensor,[1-4] 2) choosing narrowband absorption materials as absorber,[5] 3) using optical interference or plasmonic effect on chosen wavelength,[6] 4) controlling surface-charge recombination or self-trapped states to realize narrowband response.[7-11] However, most multicolor image sensors based on these strategies require integrated additional equipment, which increases the cost of sensors, limits the pixel density and flexible application scenarios[12-14] and loses other band information. Combination of gradient bandgap materials and integration of these light-sensitive materials arrays in the same multicolor image sensor is an ideal way to achieve color recognition.

Hybrid organic–inorganic perovskite exactly meets the needs of light-sensitive materials for multicolor image sensors. It has a series of excellent photoelectric properties, such as tunable bandgap ability, strong light collection ability, high carrier mobility and small exciton binding energy, which rivaling inorganic materials such as silicon or InGaAs and novel two-dimensional materials such as MoS_2 and graphene.[15-17] In recent years, patterned perovskite

image sensors have been widely studied. For example, Lee et al. introduced a multi-channel image sensor array with a hydrophobic-hydrophilic spin coating method, with an optimal resolution of 6350 dpi.[18] However, only one material can be used in the spin-coating process, so spin-coated perovskite arrays are impossible to identify color which are only applied to monochromatic perovskite imaging sensor. Jie Xue et al. introduced a multicolor perovskite image sensor whose patterning method is placing different solutions at each location of the array non-automatically. They used this method to construct 10×10 $CsPbX_3$ arrays on glass substrate. [10] By tuning the thickness and bandgap of $CsPbX_3$, they realized a narrow band response of red, green and blue based on surface-charge recombination. But the large thickness of ~20 μm caused by this color recognition method and the large pixels of 1cm×1cm caused by the patterning method seriously restricted high-resolution and flexible application scenarios. At present, perovskite sensors with high resolution and color recognition are rarely reported. The key challenge of multicolor image perovskite sensors is the appropriate color recognition method and efficient preparation method, which can realize high-resolution array manufacturing of a series of different bandgap perovskite materials. The common perovskite preparation methods such as spin coating, evaporation, doctor-blade coating and freeze-drying can not meet the above requirements. In the previous work, we applied electrohydrodynamic printing to full-color display, through optimizing the EHD printing process, a high-resolution dot matrix of 5 μm is achieved, providing a new idea for meeting this challenge. [19]

In this paper, we used EHD printing to prepare perovskite film arrays with different bandgaps to realize high-resolution multicolor image sensor. EHD printing has unique advantages in high-resolution patterning and wide compatibility of solution viscosity.[20] Through the optimization of printing and annealing process, stable printing-on-demand mode and high-resolution patterning can be realized. Meanwhile, the inherent defects of perovskite, such as rich pinholes and small grain size, can be eliminated to realize perovskite film morphology controllably fabrication. And we realized array manufacturing of a series of different band gap perovskite materials whose absorption cutoff wavelength can evolve from 440 nm to 740 nm, and then the color can be distinguished by using the strong band edge absorption and

[1]State Key Laboratory of Digital Manufacturing Equipment and Technology, Huazhong University of Science and Technology, Wuhan 430074, China.
[2]Guangdong Sygole Intelligent Technology Co.,Ltd, Dongguan 523000, China

[#]These authors contributed equally to this work.
*Contacting Author: Yongqing Duan is with State Key Laboratory of Digital Manufacture Equipment and Technology, Huazhong University of Science and Technology, Wuhan 430074, China. (Phone/fax: +86-27-8754-3072; e-mail: duanyongqing@hust.edu.cn).

978-1-6654-3008-1/21 $31.00 © 2021 IEEE

cutoff edge difference value. Furthermore, perovskite film array was applied in an imaging sensor, from which the clear mapping of the light source signal and color recognition were successfully obtained.

II. EXPERIMENTAL SECTION

2.1. Solution Configuration

Figure 1a showed the schematic diagram of the device structure and its fabrication process. The sensor was composed of column electrode, row electrode, perovskite pixels and insulation layer at the intersection of row electrode and column electrode, and perovskite pixels were prepared by EHD printing. A precondition for enabling stable drop-on-demand printing is preparing perovskite inks with good stability and reliable processability in air. At the same time, to achieve different bandgap of perovskite absorber subpixels, we first anticipated the bandgap variable inks in the EHD printing process. The precursor inks for EHD printing were prepared by adding methylammonium halide and lead halide into ionic liquid solvent. MAAc was chosen here for two reasons: one is solution stability and facile fabrication of perovskite in ambient air; for the other one, high viscosity is beneficial to improve printability and crystallinity[21] in Figure 1b. It can be observed that the absorption cutoff wavelength could be effectively tuned in the range from 440 nm to 740 nm, showing apparently strong band edge absorption, and making it possible to realize color recognition. Furthermore, we can adjust the cut-off wavelength on the scale of several nanometers by carefully adjusting the proportion of halogen elements, and then we can achieve more accurate color recognition.

The 0.58 mol/L perovskite inks were prepared by mixing MAX (X=Cl, Br, I) and PbX_2 (X=Cl, Br, I) in MAAc at 80 °C for 2 h with constant stirring. The molar ratio for PbI_2:MACl was 1:1. Lead bromide ($PbBr_2$, 99.9%), lead iodide (PbI_2, 99.9%), Lead chloride ($PbCl_2$, 99.9%), methylammonium iodide (MAI, 99.9%), methylammonium bromide (MABr, 99.9%), methylammonium chloride (MACl, 99.9%), methylammonium acetate (MAAc) were all purchased from Xi'an polymer light technology corp. All reagents were used without further purification.

MAPbBr$_2$Cl, MAPbI$_{0.87}$Br$_{2.13}$ and MAPbI$_{2.4}$Br$_{0.6}$ films. (b)The microscope figure of electrodes. (c) The image of observation microscope of printing process.

2.2. The Printing Process

The perovskite ink was supplied to the glass nozzle whose diameter at the tip is 10~55 μm. The printing process was realized by applying the high voltage between the nozzle and the substrate at a height of about 40~160 μm. During printing, the substrate was heated at 80 °C. After printing, the substrate with perovskite patterns were annealed at 100 °C for 10 min to complete crystallization. All printing and annealing process were done in ambient conditions. A high-resolution EHD printer (EHD Jet-H professional type, Guangdong Sygole Intelligent Technology Co. Ltd.) was used to print inks into the pixels.

2.3. Fabrication of The Device

We followed these procedures to fabricate the structure: The substrate was cleaned with acetone, isopropyl alcohol, ethanol, and deionized water for 10 min each in sequence and dried with nitrogen. Electrodes were prepared through lithography technique (MA6 SUSS), vacuum evaporating, and lift-off process. Cr/Au column electrodes with thicknesses of 10 nm/80 nm were achieved by vacuum evaporating. Then, PMMA photoresist was spin-coated onto the above substrate at 3000 rpm for 40 s and baked at 80 °C for 10 min as the insulating layer. Through exposure and development, the PMMA film was wiped off except every intersection region. Then Au electrodes were deposited by vacuum evaporating again using a mask as row electrode. The overall structure of the electrodes was displayed in Figure 1c. Perovskite pixels were prepared as above elaborated, as shown in Figure 1d.

III. RESULTS AND DISCUSSION

3.1. Optimization of the Perovskite Printing Resolution

Apart from the aforementioned ink design, control of EHD printing process parameters is vital for stable drop-on-demand printing and high-resolution patterning. A pulse voltage was utilized in EHD printing, where the peak voltage was chosen to induce a jetting mode for a short duration and the baseline voltage was chosen to make sure that no jetting appeared. With the increase of peak voltage and the decrease of pulse frequency, the average dot diameter increased as a higher peak voltage lead to more ink pulled down and higher pulse frequency corresponded to smaller jetting time and smaller droplet size. As we know, small nozzle diameter helps to receive a small droplet size. A nozzle with diameter of 10 μm was used to print dot array and line array. Dot matrix and lines array could be efficiently created as shown in Figure 2a-c, a perovskite dot array with minimal diameter of ~1 μm was fabricated as shown in Figure 2c. High-quality perovskite films with scant pinholes were obtained on a hot substrate of 80℃ and post-anneal of 100℃, which were a better choice for sensors, as shown in Figure 2d.

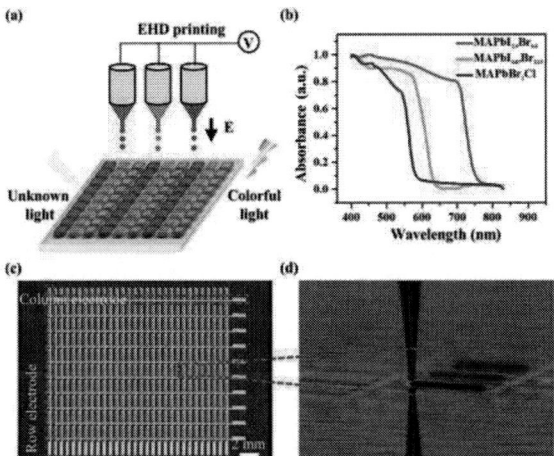

Figure 1. (a) Schematic illustration of the EHD printing fabrication process of multicolor sensor array. (b) Normalized absorption spectra of

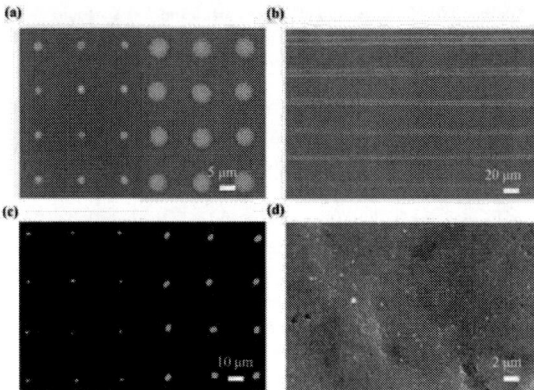

Figure 2. (a) Microphotograph of the EHD printed perovskite microarrays with MAPbX₃ dot diameter of 2 and 6 μm, respectively. (b) Microphotograph of the EHD printed perovskite lines with width of 5~10 μm, respectively. (c) The EHD printed high-resolution dot matrix with MAPbX₃ dot diameter of 1 and 3 μm, respectively. (d) SEM image of MAPbI$_{2.4}$Br$_{0.6}$ films formed at 80 ℃ substrate temperature.

3.2. Photovoltaic Characterization

The final multicolor image sensor utilized photoconductive structure (Au–perovskite–Au structure) and the energy band diagram were shown in Figure 3a and 3b. The device composed of pixels (10×10×3 subpixels) was designed and fabricated on the glass substrate, with the perovskite films of different spectral responses on subpixel 1~3 (MAPbBr$_2$Cl, MAPbI$_{0.87}$Br$_{2.13}$ and MAPbI$_{2.4}$Br$_{0.6}$). Specifically, our method can be used to make higher-resolution devices, and a sensor with 2 μm channel length was designed and prepared as shown in Figure 3c. I–V characteristics of sensor with MAPbI$_{2.4}$Br$_{0.6}$ film with 2 μm channel length were shown under dark and illumination by using a green laser diode with the increasing light intensity. The sensor of 2 μm channel length and 30 μm channel width exhibited the best R and D* values of 1.625 A/W and ~4.451×10^{11} Jones (light intensity=1 mW/cm²), respectively. Furthermore, it could be reasonably speculated that a resolution of 726 dpi could be achieved when the electrode width and the channel length are both 2 μm, and the channel width was 30 μm. It would be higher if we further decreased the line width and pixel interval.

To illustrate the color recognition method and potential clearly, sensors based on MAPbI$_{2.4}$Br$_{0.6}$, MAPbI$_{0.87}$Br$_{2.13}$, and MAPbBr$_2$Cl were characterized under light source with wavelength of 470, 530, and 680 nm used as blue, green, and red light source, respectively. Normalized current of sensors based on different materials of the absorber under blue, green, and red illuminations were shown in Figure 3d. As expected, the sensor based on MAPbBr$_2$Cl only responded to the blue light, the sensor based on MAPbI$_{0.87}$Br$_{2.13}$ responded to both blue and green light, while the sensor based on MAPbI$_{2.4}$Br$_{0.6}$ responded to three light sources, which were consistent with the designed absorption spectra of the perovskite films at corresponding positions.

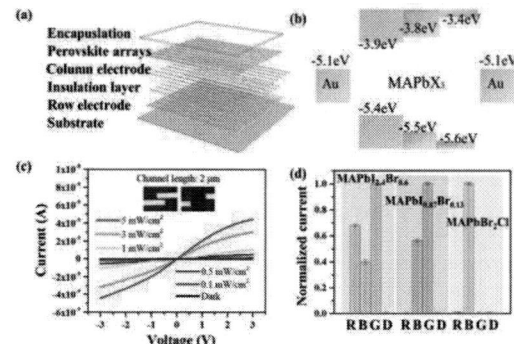

Figure 3. (a) Layer by layer structure of perovskite image sensor array. (b) Energy band illustration of the device. (c) I–V characteristics of sensor with MAPbI$_{2.4}$Br$_{0.6}$ film and 2μm channel length under dark and different intensities of green light. (d) Normalized current of sensors based on MAPbI$_{2.4}$Br$_{0.6}$, MAPbI$_{0.87}$Br$_{2.13}$, and MAPbBr$_2$Cl of the absorber under blue, green, and red illuminations.

3.3. Image Application

To present device applications, the image sensor based on MAPbI$_{2.4}$Br$_{0.6}$, MAPbI$_{0.87}$Br$_{2.13}$, and MAPbBr$_2$Cl was integrated on the same substrate as one device, which successfully captured an image patterned through a shadow-mask and identified the set light source color. As shown in Figure 4a, a pre-designed shadow mask was placed between the light source and the device to map a figure. Normalized current mapping result using the image sensor array after patterned illumination through a shadow-mask was displayed in Figure 4b. The pixels exposed to illumination displayed higher current values, while others occluded by the designed mask were roughly equal to the dark current in the voltage. Clear line patterns could be showed in the normalized current mapping results, which indicated the excellent imaging performance of multicolor image sensor. It is worth noting that due to the difference in the responsiveness of different materials to light, the follow-up work can adopt a similar strategy to electroluminescent devices. The poorly photosensitive materials can be prepared in a larger area for better application in full-color imaging. We predict that it has great potential in electronic eyes, wearable sensors and optical communication applications. However, the size of pixels and photovoltage performance should be further improved for the future optoelectronics devices.

IV. CONCLUSION

This work introduces a facile, fast, and high-resolution patterning method of inorganic–organic hybrid perovskite thin films. The developed process can be applied to array preparation of multiple perovskite materials on the same substrate. By optimizing the EHD printing process, we successfully developed high-resolution (~1 μm) perovskite films array and multicolor image sensor. The potential of higher-resolution devices has been proven by designing a device of 2 μm channel length, which exhibited R and D* values of 1.625 A/W and ~4.451×10^{11} Jones. The EHD

978-1-6654-3008-1/21 $31.00 © 2021 IEEE

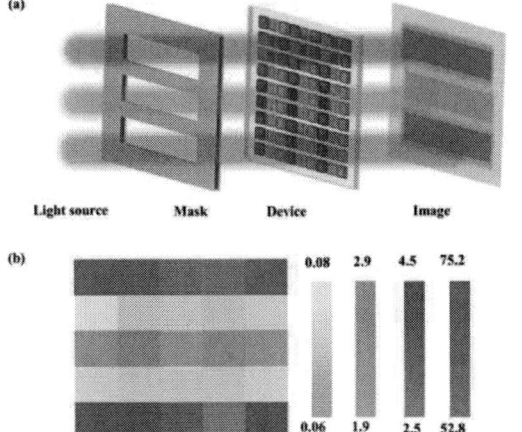

Figure 4. (a) Schematic illustration of the image sensor array to detect multipoint light distribution. (b) Normalized current mapping result using the image sensor array after patterned illumination through a shadow-mask, The unit of value is nA.

printing process has a high potential to become a versatile tool for fabrication of high-resolution multiplexed perovskite optoelectronic device array, such as image sensors and light emitting diodes, compatible with large-scale manufacturing application. The current work provides an important guideline for the patterning of perovskite films for the next-generation multicolor image sensor.

ACKNOWLEDGMENT

This work was financially supported by the National Key Research and Development Program of China (2018YFA0703200), the National Natural Science Foundation of China (52075209, 11932009) and the Program for Guangdong Introducing Innovative and Enterpreneurial (2017ZT07G331). The authors would also like to thank Flexible Electronics Manufacturing Laboratory in Comprehensive Experiment Center for advanced manufacturing and equipment technology.

REFERENCES

[1] C. Liu, Q. Zhang, D. Wang, G. Zhao, X. Cai, L. Li, H. Ding, K. Zhang, H. Wang, D. Kong, L. Yin, L. Liu, G. Zou, L. Zhao, and X. Sheng, "High Performance, Biocompatible Dielectric Thin-Film Optical Filters Integrated with Flexible Substrates and Microscale Optoelectronic Devices," Advanced Optical Materials, vol. 6, no. 15, 2018.

[2] W. L. Tsai, C. Y. Chen, Y. T. Wen, L. Yang, Y. L. Cheng, and H. W. Lin, "Band Tunable Microcavity Perovskite Artificial Human Photoreceptors," Adv Mater, vol. 31, no. 24, pp. e1900231, Jun, 2019.

[3] P. Wang, S. Liu, W. Luo, H. Fang, F. Gong, N. Guo, Z. G. Chen, J. Zou, Y. Huang, X. Zhou, J. Wang, X. Chen, W. Lu, F. Xiu, and W. Hu, "Arrayed Van Der Waals Broadband Detectors for Dual-Band Detection," Adv Mater, vol. 29, no. 16, Apr, 2017.

[4] M. Dandin, P. Abshire, and E. Smela, "Optical filtering technologies for integrated fluorescence sensors," Lab Chip, vol. 7, no. 8, pp. 955-77, Aug, 2007.

[5] A. Sobhani, M. W. Knight, Y. Wang, B. Zheng, N. S. King, L. V. Brown, Z. Fang, P. Nordlander, and N. J. Halas, "Narrowband photodetection in the near-infrared with a plasmon-induced hot electron device," Nat Commun, vol. 4, pp. 1643, 2013.

[6] B. Y. Zheng, Y. Wang, P. Nordlander, and N. J. Halas, "Color-selective and CMOS-compatible photodetection based on aluminum plasmonics," Adv Mater, vol. 26, no. 36, pp. 6318-23, Sep, 2014.

[7] Y. Fang, Q. Dong, Y. Shao, Y. Yuan, and J. Huang, "Highly narrowband perovskite single-crystal sensors enabled by surface-charge recombination," Nature Photonics, vol. 9, no. 10, pp. 679-686, 2015.

[8] K. S. Cho, K. Heo, C. W. Baik, J. Y. Choi, H. Jeong, S. Hwang, and S. Y. Lee, "Color-selective photodetection from intermediate colloidal quantum dots buried in amorphous-oxide semiconductors," Nat Commun, vol. 8, no. 1, pp. 840, Oct 10, 2017.

[9] S. Lefler, R. Vizel, E. Yeor, E. Granot, O. Heifler, M. Kwiat, V. Krivitsky, M. Weil, Y. E. Yaish, and F. Patolsky, "Multicolor Spectral-Specific Silicon Nanodetectors based on Molecularly Embedded Nanowires," Nano Lett, vol. 18, no. 1, pp. 190-201, Jan 10, 2018.

[10] J. Xue, Z. Zhu, X. Xu, Y. Gu, S. Wang, L. Xu, Y. Zou, J. Song, H. Zeng, and Q. Chen, "Narrowband Perovskite Sensor-Based Image Array for Potential Application in Artificial Vision," Nano Lett, vol. 18, no. 12, pp. 7628-7634, Dec 12, 2018.

[11] M. E. Cryer, and J. E. Halpert, "300 nm Spectral Resolution in the Mid-Infrared with Robust, High Responsivity Flexible Colloidal Quantum Dot Devices at Room Temperature," ACS Photonics, vol. 5, no. 8, pp. 3009-3015, 2018.

[12] H. Park, Y. Dan, K. Seo, Y. J. Yu, P. K. Duane, M. Wober, and K. B. Crozier, "Filter-free image sensor pixels comprising silicon nanowires with selective color absorption," Nano Lett, vol. 14, no. 4, pp. 1804-9, 2014.

[13] D. H. Auston, K. P. Cheung, and P. R. Smith, "Picosecond photoconducting Hertzian dipoles," Applied Physics Letters, vol. 45, no. 3, pp. 284-286, 1984.

[14] B. P. Carrow, and J. F. Hartwig, "Distinguishing between pathways for transmetalation in Suzuki-Miyaura reactions," J Am Chem Soc, vol. 133, no. 7, pp. 2116-9, Feb 23, 2011.

[15] H. C. Ko, M. P. Stoykovich, J. Song, V. Malyarchuk, W. M. Choi, C. J. Yu, J. B. Geddes, 3rd, J. Xiao, S. Wang, Y. Huang, and J. A. Rogers, "A hemispherical electronic eye camera based on compressible silicon optoelectronics," Nature, vol. 454, no. 7205, pp. 748-53, Aug 7, 2008.

[16] C. Choi, M. K. Choi, S. Liu, M. S. Kim, O. K. Park, C. Im, J. Kim, X. Qin, G. J. Lee, K. W. Cho, M. Kim, E. Joh, J. Lee, D. Son, S. H. Kwon, N. L. Jeon, Y. M. Song, N. Lu, and D. H. Kim, "Human eye-inspired soft optoelectronic device using high-density MoS2-graphene curved image sensor array," Nat Commun, vol. 8, no. 1, pp. 1664, Nov 21, 2017.

[17] W. Lee, Y. Liu, Y. Lee, B. K. Sharma, S. M. Shinde, S. D. Kim, K. Nan, Z. Yan, M. Han, Y. Huang, Y. Zhang, J. H. Ahn, and J. A. Rogers, "Two-dimensional materials in functional three-dimensional architectures with applications in photodetection and imaging," Nat Commun, vol. 9, no. 1, pp. 1417, Apr 12, 2018.

[18] W. Lee, J. Lee, H. Yun, J. Kim, J. Park, C. Choi, D. C. Kim, H. Seo, H. Lee, J. W. Yu, W. B. Lee, and D. H. Kim, "High-Resolution Spin-on-Patterning of Perovskite Thin Films for a Multiplexed Image Sensor Array," Adv Mater, vol. 29, no. 40, Oct, 2017.

[19] M. Zhu, Y. Duan, N. Liu, H. Li, J. Li, P. Du, Z. Tan, G. Niu, L. Gao, Y. Huang, Z. Yin, and J. Tang, "Electrohydrodynamically Printed High‐Resolution Full‐Color Hybrid Perovskites," Advanced Functional Materials, vol. 29, no. 35, 2019.

[20] J. U. Park, M. Hardy, S. J. Kang, K. Barton, K. Adair, D. K. Mukhopadhyay, C. Y. Lee, M. S. Strano, A. G. Alleyne, J. G. Georgiadis, P. M. Ferreira, and J. A. Rogers, "High-resolution electrohydrodynamic jet printing," Nat Mater, vol. 6, no. 10, pp. 782-9, Oct, 2007.

[21] L. Chao, Y. Xia, B. Li, G. Xing, Y. Chen, and W. Huang, "Room-Temperature Molten Salt for Facile Fabrication of Efficient and Stable Perovskite Solar Cells in Ambient Air," Chem, vol. 5, no. 4, pp. 995-1006, 2019.

Proceedings of the 16th Annual IEEE International
Conference on Nano/Micro Engineered and Molecular Systems
April 25-29, 2021

A Novel Piezoelectric Micromachined Ultrasonic Transducer with Adjustable Broad and Flat Frequency Band

Lei Wang, Wei Zhu, Zhipeng Wu, Wenjuan Liu, Chengliang Sun

Abstract— **This paper presented a modified piezoelectric micromachined ultrasonic transducer (pMUT) with an adjustable large and flat frequency band in liquid-coupled operation, which is named as Cr-pMUT. A cavity in the center of the pMUT and released holes are formed by the same fabrication step, the cavity size and relative position between cavity and pMUT (wall thickness) determine the width and flatness of the frequency band. In this work, an equivalent circuit model of the Cr-pMUT is introduced to express the crosstalk behavior which is reason of the broadband performance. By theoretically analyzing and parameters optimizing, a broad and flat frequency band can be achieved, and it provides a simple method to tune the bandwidth. Compared to the r-pMUT and traditional pMUT, the Cr-pMUT achieve wider -3dB frequency band by 6% and 537% respectively with the same diaphragm size in FC-84.**

I. INTRODUCTION

Nowadays, micromachined ultrasonic transducers (MUTs) play an important role in different fields such as medical imaging [1], fingerprint scanning [2] and range-finding [3]. Capacitive MUTs (cMUTs) and piezoelectric MUTs (pMUTs) are two most typical devices classified by the working principle. Compared to cMUTs, pMUTs do not need small gap and high DC bias between electrodes, which can be driven by low AC voltage, thereby reducing the circuit complexity. However, the pMUTs are limited by the low bandwidth [4] for normal liquid-coupled devices.

In recent studies, A. Hajati et al. used an array of pMUTs with different size to improve the bandwidth which exceeded 55% at the center frequency of 5MHz in water [5]. Same behavior has also been observed in pMUT arrays reported in [6]. For a single pMUT, a ring-shaped pMUT has been presented to extend the bandwidth up to 160% in FC-84 [7]. However, the center velocity fluctuates drastically in frequency band, which affects the stability of the pMUTs working in chirp waveforms.

In this work, a modified Cr-pMUT structure with a cavity in the center of the pMUT is presented. In contrast to previous traditional pMUT and r-pMUT cell, the Cr-pMUT has wider bandwidth and flatter center pressure-frequency response in the band. In this study, an equivalent circuit model of the Cr-pMUT is established to explain the phenomenon of the flat band in frequency response profile. Since the interference acoustic wave generated in cavity, the radiation resistance keeps a constant without fluctuation. By optimizing the size of the cavity and wall thickness, a broader and flatter frequency band of the Cr-pMUT can be

realized in FC-84 compared with traditional pMUT and r-pMUT reported in [7].

II. WORKING PRINCIPLE AND MODEL ESTABLISHMENT

A. Structure Design

The structure of the Cr-pMUT element consists of an annular diaphragm and a cavity on the center of the diaphragm, as shown in Fig. 1, which can be realized by cavity SOI (CSOI) wafer. Aluminum nitride (AlN) and Silicon (Si) are adopted for piezoelectric thin film and structure layer materials in this study respectively, and the top electrode with the coverage of 55% of the inner cavity is applied to the piezoelectric diaphragm. The inner cavity works in vacuum and the outer cavity works in medium (FC-84). The structure parameters and material properties of the Cr-pMUT are presented in Table I and Table II.

TABLE I.

SUMMARY OF THE STRUCTURE PARAMETERS OF THE Cr-PMUT CELLS

Symbol	Dimension		
	Description	*Value*	*Unit*
h_0	Cavity depth	340	μm
h_1	Structure layer thickness	1	μm
h_2	Piezoelectric layer thickness	1	μm
h_3	Electrode layer thickness	0.1	μm
r_0	Top electrode center radius	300	μm
r_1	Cavity radius	140	μm
t	Wall thickness	60	μm

TABLE II.

SUMMARY OF THE MATERIAL PROPERTIES OF THE Cr-PMUT CELLS

Symbol	Material Properties of AlN		
	Description	*Value*	*Unit*
Y_1	Young's modulus of AlN	340	GPa
v_1	Poisson's ratio of AlN	0.3	/
ρ_1	Density of AlN	3300	kg/m³
d_{31}	Piezoelectric coefficient of AlN	2.2	pm/V
Symbol	Material Properties of Si		
	Description	*Value*	*Unit*

*Resrach supported by the Hubei Provincial Major Program of Technological Innovation (Grant No. 2019AAA052).

Lei Wang, Wei Zhu, Zhipeng Wu, Wenjuan Liu and Chengliang Sun are both with the Institute of Technological Sciences, Wuhan University, Wuhan, 430072 China.

Corresponding author: Wenjuan Liu (e-mail: lwjwhu@whu.edu.cn), Chengliang Sun (phone: +86-027-6877-6588; e-mail: sunc@whu.edu.cn).

Y_2	Young's modulus of Si	170	GPa
v_2	Poisson's ratio of Si	0.28	/
ρ_2	Density of Si	2329	kg/m³
Symbol	**Material Properties of FC-84**		
	Description	*Value*	*Unit*
ρ_3	Density of FC-84	1730	kg/m³
c	The speed of sound in FC-84	543	m/s

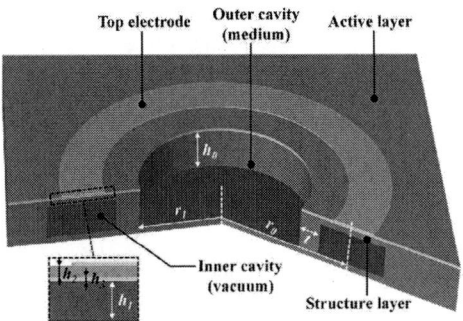

Fig. 1. Schematics of Cr-pMUT structure.

B. Equivalent Circuit Model of the Cr-pMUT

An equivalent circuit model is built to analyze the crosstalk behavior in the medium due to the size of the Cr-pMUT is smaller than the acoustic wavelength, as shown in Fig. 2. V_{in} is the input voltage of the transducer, Z_{pm} is the mechanical impedance of the circular diaphragm, Z_a is the self-radiation impedance, Z_{cross} is the mutual-radiation impedance of the r-pMUT reported in [7]. $Z_{cross1} + Z_{cross2}$ is the mutual-radiation impedance of the Cr-pMUT, where Z_{cross1} is caused by the ring-shaped geometry and Z_{cross2} is caused by the cavity.

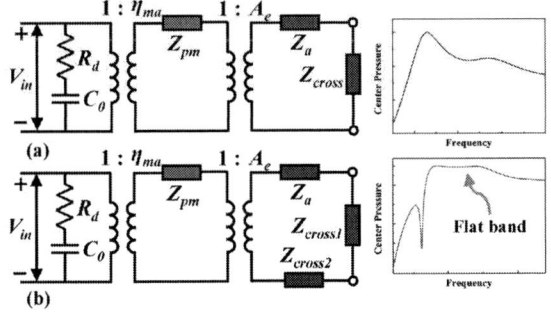

Fig. 2. Equivalent circuit model and pressure-response schematics. (a) r-pMUT condition. (b) Cr-pMUT condition.

The equivalent circuit model for the r-pMUT reported in [7] with fluctuation in frequency band is presented in Fig. 2(a). By comparison, the Cr-pMUT with additional crosstalk effect caused by cavity is presented in Fig. 2(b), which can flatten the frequency band.

To verify the phenomenon shown in Fig. 2, radiation impedance amplitude, radiation resistance R_{cross}, and radiation reactance X_{cross} based on the equivalent circuit model is presented in Fig. 3.

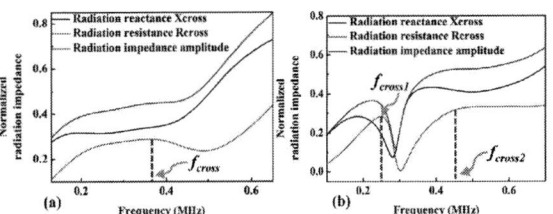

Fig. 3. Frequency response of normalized radiation impedance of the crosstalk effect. (a) r-pMUT condition. (b) Cr-pMUT condition.

The average velocity of the r-pMUT and Cr-pMUT can be presented [4]:

$$|v_{real}| = \frac{\eta V_{in}/A_{eff}}{\sqrt{(R_m+\frac{R_{free}}{A_{eff}^2})^2+(X_m+\frac{X_{free}}{A_{eff}^2})^2+\left(\frac{R_{cross}^2}{A_{eff}^4}+\frac{X_{cross}^2}{A_{eff}^4}\right)+2I_{cross}}} \quad (1)$$

Where, R_m and X_m are mechanical resistance and reactance; R_{free} and X_{free} are radiation resistance and reactance without crosstalk; R_{cross} and X_{cross} are radiation resistance and reactance with crosstalk; A_{eff} is the effective vibration area of the Cr-pMUT.

I_{cross} is expressed as follows [4]:

$$I_{cross} = \left(R_m + \frac{R_{free}}{A_{eff}^2}\right)\frac{R_{cross}}{A_{eff}^2} + \left(X_m + \frac{X_{free}}{A_{eff}^2}\right)\frac{X_{cross}}{A_{eff}^2} \quad (2)$$

As shown in Fig. 3, the crosstalk behavior can be introduced by the mutual radiation impedance. The value of the fluctuation in (1) is mainly determined by I_{cross} dominated by R_{cross}. So, the average velocity is suppressed in peak of the radiation resistance curve. And the average velocity of the area around the peak will increase due to the low radiation resistance. Specifically, there are two peak frequencies in frequency-radiation response of the Cr-pMUT as shown in Fig. 3(b) compared to the single peak response of the r-pMUT as shown in Fig. 3(a). The first peak response is contributed by the cavity while the second peak response is contributed by the ring-shaped geometry. A flat transition area around the second peak can be formed as shown in the curve of the radiation resistance. Therefore, a broad and flat band can be formed within the pressure-frequency response. An FEA model is established using COMSOL Multiphysics to study the characteristics of the Cr-pMUT, as shown in Fig. 4. A perfect match layer (PML) is used to avoid sound reflection.

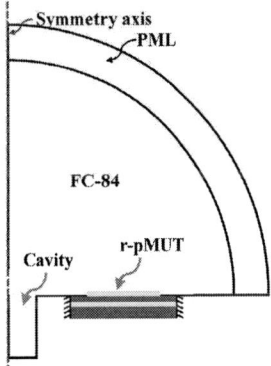

Fig. 4. FEA model of the Cr-pMUT including cavity, r-pMUT and medium (FC-84).

978-1-6654-3008-1/21 $31.00 © 2021 IEEE

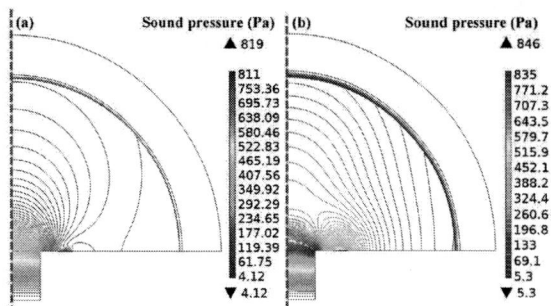

Fig. 5. Sound pressure contour distribution of the Cr-pMUT. (a) At the f_{cross1}; (b) At the f_{cross2}.

Fig. 5 shows the simulation results of the Cr-pMUT. Fig. 5 (a) and Fig. 5 (b) show the sound pressure contour distribution at the f_{cross1} and f_{cross2}, respectively. At the f_{cross1}, the cavity works as a radiator. And, at the f_{cross2}, the pressure at the open end of the cavity trends to be zero, which can prove the validity of the equivalent circuit model.

III. GEOMETRY OPTIMIZATION

A. Cavity Depth Optimization

In order to study the influence of cavity size on the bandwidth of the Cr-pMUT, a systematic simulation has been studied. In order to make the figures clearer so as to make it feasible for reader to analyze the trend of the curves, no more than five curves are placed in one figure. In this case, the thickness of the AlN and Si layer are both determined as 1μm. As illustrated in Fig. 6, there are twenty Cr-pMUTs with different cavity depth when the cavity radius and wall thickness are fixed as 100 μm and 150 μm, respectively. It is clear that f_{cross2} goes down with the increase of cavity depth. The bandwidth is much narrow with h_0 = 20-300 μm, which cannot be used as a broadband device, as shown in Fig. 6(a)-(c). The bandwidth is improved when h_0 is greater than 300 μm, especially when h_0 equals to 340 μm, as shown in Fig. 6(d). Therefore, the cavity depth h_0 plays an important role in the formation of the broadband, which is the first parameter that needs to be determined of the Cr-pMUT.

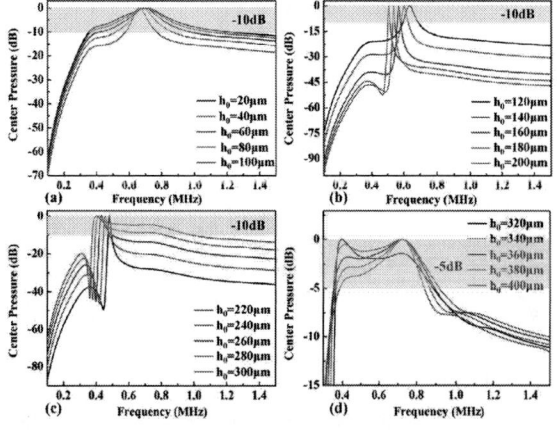

Fig. 6. Frequency response of center pressure with different cavity depth. (a) Center pressure curve with h_0=10-100 μm. (b) Center pressure curve with h_0=120-200 μm. (c) Center pressure curve with h_0=220-300 μm. (d) Center pressure curve with h_0=320-400 μm.

B. Cavity Radius Optimization

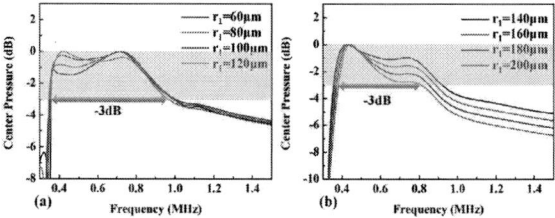

Fig. 7. Frequency response of center pressure with different cavity radius. (a) Center pressure curve with r_1=60-120 μm. (b) Center pressure curve with r_1=140-200 μm.

The variation of the bandwidth obtained by changing r_1 can be seen in Fig. 7. The bandwidth goes up with the increase of r_1 from 60 μm to 80 μm when cavity depth and wall thickness are fixed as 340 μm and 150 μm respectively, since the f_{cross} is close to the natural frequency (f_{free}) of the ring-shape pMUT. And then the bandwidth begins to reduce with the increase of r_1 from 120 μm to 200 μm as f_{cross} is far from f_{free}, thus the two resonances are not sufficiently matched to have a pronounced effect. In addition, the reduced sound speed in FC-84 lowers the acoustic resonance frequency, such it occurs at r_1=100 μm and r_0=300 μm. It is clear that cavity radius affects the size of bandwidth of the Cr-pMUT.

C. Wall Thickness Optimization

In this part, the effect of the wall thickness on the bandwidth of the Cr-pMUT is analyzed. As shown in Fig. 8(a), the bandwidth is no more than 200 kHz when t is less than 70 μm. However, the bandwidth increases to more than 600 kHz and remains stable while the value of the t is between 80 μm and 170 μm as shown in Fig. 8(b) and Fig. 8(c). In addition, the bandwidth gradually decreases as t continues to increase as shown in Fig. 8(d). This phenomenon is mainly determined by the relative position of f_{cross} and f_{free}. Therefore, a flat bandwidth can be obtained by adjusting the cavity radius when the cavity depth is fixed as 340 μm and the wall thickness is within the range which bandwidth is stable especially when t equals to 130 μm.

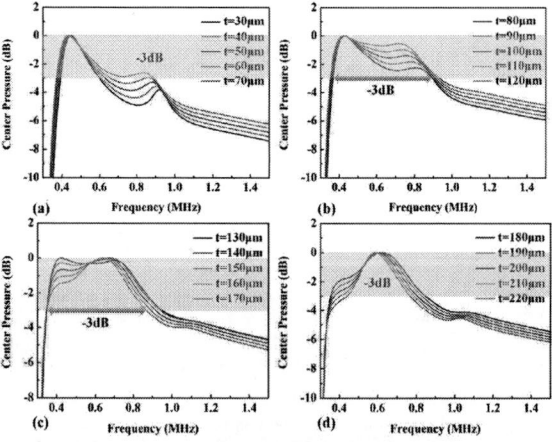

Fig. 8. Frequency response of center pressure with different wall thickness. (a) Center pressure curve with t=30-70 μm. (b) Center pressure curve with t=80-120 μm. (c) Center pressure curve with t=130-170 μm. (d) Center pressure curve with t=180-220 μm.

IV. BANDWIDTH COMPARISON

In this section, the bandwidth comparison between the Cr-pMUT, r-pMUT reported in [7] and traditional pMUT is analyzed with different top electrode center radius when cavity depth and wall thickness are fixed as 340 μm and 130 μm. As shown in Fig. 9, the traditional pMUT and r-pMUT have a sharp center pressure curve with smaller bandwidth. The largest fractional bandwidth, defined as the -3dB frequency range divided by the resonance frequency ($BW_f = \Delta f / f_0$), is 158% of the r-PMUT with $r_0 = 300$ μm and 26.2% of the traditional pMUT, as shown in Fig. 9(a).

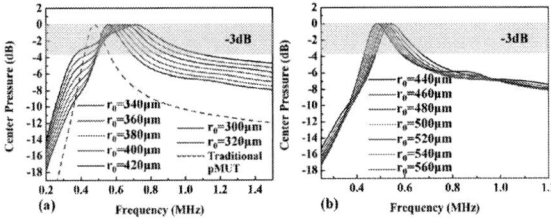

Fig. 9. Frequency response of center pressure of the r-pMUT and traditional pMUT. (a) Center pressure curve with r_0=300-420 μm (The dotted line indicates the traditional pMUT). (b) Center pressure curve with r_0=440-560 μm.

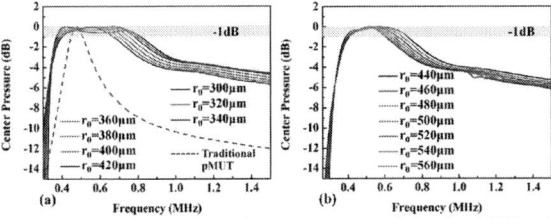

Fig. 10. Frequency response of center pressure of the Cr-pMUT and traditional pMUT. (a) Center pressure curve with r_0=300-420 μm (The dotted line indicates the traditional pMUT). (b) Center pressure curve with r_0=440-560 μm.

As expected, the Cr-pMUT exhibits a wider and flatter frequency band by fixing t and h_0 as 130 μm and 340 μm, respectively. As shown in Fig. 10, the center pressure of the in-band ripples is within 1dB, indicating a flat frequency band. In addition, the biggest -3dB bandwidth of the Cr-pMUT is more than 167% which is 6% and 537% higher than the r-pMUT and traditional pMUT, respectively. And BW_f varies between 93% and 167% for the smallest and largest Cr-pMUT simulation respectively when the range of r_0 is between 300 μm and 560 μm, where the biggest bandwidth occurs with r_0 equaling to 300 μm. These results show that the Cr-pMUT has great potential as a wide and flat adjustable frequency band device.

V. CONCLUSION

In this work, a modified pMUT, which named as Cr-pMUT, is presented. And theory analysis models and FEA methods are applied to study the Cr-PMUT. Both the theory results and the simulation results indicate that the broadband phenomenon arises from the resonance of the cavity in the center of the pMUT and ring-shaped geometry. An extra acoustic impedance is added to the pMUT cell through the additional crosstalk effect. An equivalent circuit model is established for the Cr-pMUT. The optimal parameters for three kinds of cavity size of the Cr-pMUT are simulated and investigated by using COMSOL software. A -3dB bandwidth of 167% is achieved for the Cr-pMUT in FC-84, which is increased by 6% and 537% compared to the traditional pMUT and r-pMUT. The resonant cavity can act as a radiator in place of extra pMUT cell. Therefore, the Cr-pMUT has great potential as a wide and flat adjustable frequency band device.

ACKNOWLEDGMENT

This work was supported by the Hubei Provincial Major Program of Technological Innovation (Grant No. 2019AAA052).

REFERENCES

[1] Yang Y, Tian H, Wang Y F, et al. An ultra-high element density pMUT array with low crosstalk for 3-D medical imaging[J]. Sensors, 2013, 13(8): 9624-9634.

[2] Akhbari S, Sammoura F, Eovino B, et al. Bimorph piezoelectric micromachined ultrasonic transducers[J]. Journal of Microelectromechanical Systems, 2016, 25(2): 326-336.

[3] Przybyla R J, Shelton S E, Guedes A, et al. In-air rangefinding with an aln piezoelectric micromachined ultrasound transducer[J]. IEEE Sensors Journal, 2011, 11(11): 2690-2697.

[4] Xu T, Zhao L, Jiang Z, et al. An analytical equivalent circuit model for optimization design of a broadband piezoelectric micromachined ultrasonic transducer with an annular diaphragm[J]. IEEE transactions on ultrasonics, ferroelectrics, and frequency control, 2019, 66(11): 1760-1776.

[5] Hajati A, Latev D, Gardner D, et al. Three-dimensional micro electromechanical system piezoelectric ultrasound transducer[J]. Applied Physics Letters, 2012, 101(25): 253101.

[6] Suzuki K, Nakayama Y, Shimizu N, et al. Study on Wide-band Piezoelectric Micro-machined Ultrasound Transducers (pMUT) by Combined Resonance Frequencies and Controlling Poling Directions[C]//2018 IEEE International Ultrasonics Symposium (IUS). IEEE, 2018: 1-3.

[7] Eovino B E, Akhbari S, Lin L. Broadband ring-shaped PMUTS based on an acoustically induced resonance[C]//2017 IEEE 30th International Conference on Micro Electro Mechanical Systems (MEMS). IEEE, 2017: 1184-1187.

Proceedings of the 16th Annual IEEE International
Conference on Nano/Micro Engineered and Molecular Systems
April 25-29, 2021

Stackable Triboelectric Nanogenerators for Self-Powered Marine Monitoring Buoy*

Jiale Dong, Hao Wang, Xiu Xiao, Taili Du, Yunpeng Zhao, Zhongqi Fan and Minyi Xu*

Abstract— Powering marine distributed nodes (e.g. monitoring buoys) has always been challenging under the harsh conditions in oceans. Triboelectric nanogenerator (TENG), on account of its robustness and cost, provide a novel solution to this problem. In this study, a stackable TENG (S-TENG) consisting of polytetrafluoroethylene pellets, an acrylic plates coated with aluminum film and a 3D-printed brace, has been proposed to harvest wave energy efficiently. Experiments show that each S-TENG unit produces a maximum peak power of 25.22mW while a power density of $34.65W/m^3$ is achieved. The output characteristic of multiple S-TENG units parallel connected is studied. It turns out that the output performance yields linear increase with the unit number without having to use rectifier bridge. This feature indicates that self-powered marine monitoring buoy is achievable by S-TENG integration.

I. INTRODUCTION

With seaway transportation carrying out most of world trade and the boosting ocean exploration, marine monitoring is more and more a critical issue [1-4]. However, much of the ocean (though it covers 70% of the planet's surface) has not been covered by the monitoring systems [2, 5, 6]. Marine buoys, with functions including collecting oceanographic data, transmitting real-time signals and detecting fishery resources, are essential in-place marine monitoring nodes [7]. The battery-powered marine buoys are troubled with sustainability limitations [8]. Therefore, developing the self-powered marine buoys is more preferable. Wave, as a high density energy resource, is of great accessibility for monitoring buoys [6, 9-11]. At present, wave energy is usually harvested with the electromagnetic generators [11-13]. However, they are not always the best choice in the case of a floating monitoring buoy on observation of its volume and weight.

Triboelectric nanogenerator (TENG), was invented as an effective technology to harvest random mechanical vibrations [4, 14-22]. Compared with EMG, TENG performs better within low frequency range, which is practically the frequency of most waves [4, 20, 23-33]. Xu et al. have developed a kind of tower-like TENG to harvest wave energy from arbitrary direction [30]. Zhang et al. developed a self-powered intelligent buoy system through a multilayered TENG [29]. Nevertheless, the influences from the hydrodynamic parameters has not been considered.

Therefore, a study on buoy-integrated wave energy TENG is very relevant to realizing self-powered marine monitoring.

Based on our previous work, a stackable triboelectric nanogenerator (S-TENG) is proposed to power the marine monitoring buoy. The S-TENG is designed to efficiently harvest wave energy in real ocean environment. Each TENG layer is of two acrylic plates attached by aluminum electrodes and PTFE pellets in-between. Each unit is composed of 10 TENG layers to take advantage of its easy stackable features. Experiments shows that for one S-TENG unit (10 layer), the maximum peak power can reach 25.22mW while the power density can reach $34.65W/m^3$. Furthermore, the output characteristic of multi S-TENG units is studied and analyzed.

II. RESULTS AND DISCUSSION

2.1 Forming the S-TENG

A self-powered buoy integrated with wave energy harvesting S-TENG units is shown in Fig. 1a. As a marine monitoring node, the buoy carries the corresponding monitoring apparatus. The wave energy converting module, (as shown in Fig. 1b) consists of cylindrical acrylic sealed case and 7 parallel connected S-TENG units (each unit has 10 layers). The acrylic case itself is anti-corrosive and it has been totally sealed to prevent the internal structure from corrosion. Fig. 1c is the photograph of the sealed device, the height and diameter of the cylindrical case is 40cm and 33cm, respectively. The 7 S-TENG units inside are hexagonal arranged and mechanically fixed to make the electrical output in-phase. The height of each S-TENG unit is 20cm and the diameter is 10cm. The freestanding structure is designed and optimized based on our previous work [30] and the schematic diagram is shown in Fig. 1d. The PTFE pellets rolls freely in the 2-D space while the PTFE pellets make frictions with the two aluminum electrodes attached on the acrylic plate. The gap between the two plates is just large enough to ensure free rolling (the diameter of the PTFE pellet is 10.5mm and the height of the 3D printed brace is 10.6mm).

The S-TENG's working principle has been depicted in Fig. 1e. The PTFE pellets rolls freely between the two Aluminum electrodes under the excitation of waves, and they will become negatively charged after contacting with

---

*Resarch supported by National Science Foundation of China

Jiale Dong is with Dalian Maritime University, Dalian, China (email: daviddongjiale@163.com)

Hao Wang is with Dalian Maritime University, Dalian, China (email: hao8901@dlmu.edu.cn)

Xiu Xiao is with Dalian Maritime University, Dalian, China (email: xiaoxiu@dlmu.edu.cn)

Taili Du is with Dalian Maritime University, Dalian, China (email: dutaili@dlmu.edu.cn)

Yunpeng Zhao is with Dalian University of Technology, Dalian, China (email: Ypzhao@dlut.edu.cn)

Zhongqi Fan is with Dalian University of Technology, Dalian, China (email: zhongqifan@mail.dlut.edu.cn)

*Corresponding author: Minyi Xu is with Dalian Maritime University, Dalian, China (phone: +86 18941134769; e-mail: xuminyi@dlmu.edu.cn).

the Al electrodes. When PTFE pellets moves back and forth positive charges are induced on the electrodes. Therefore, an induction current is generated in the circuit connecting the electrodes.

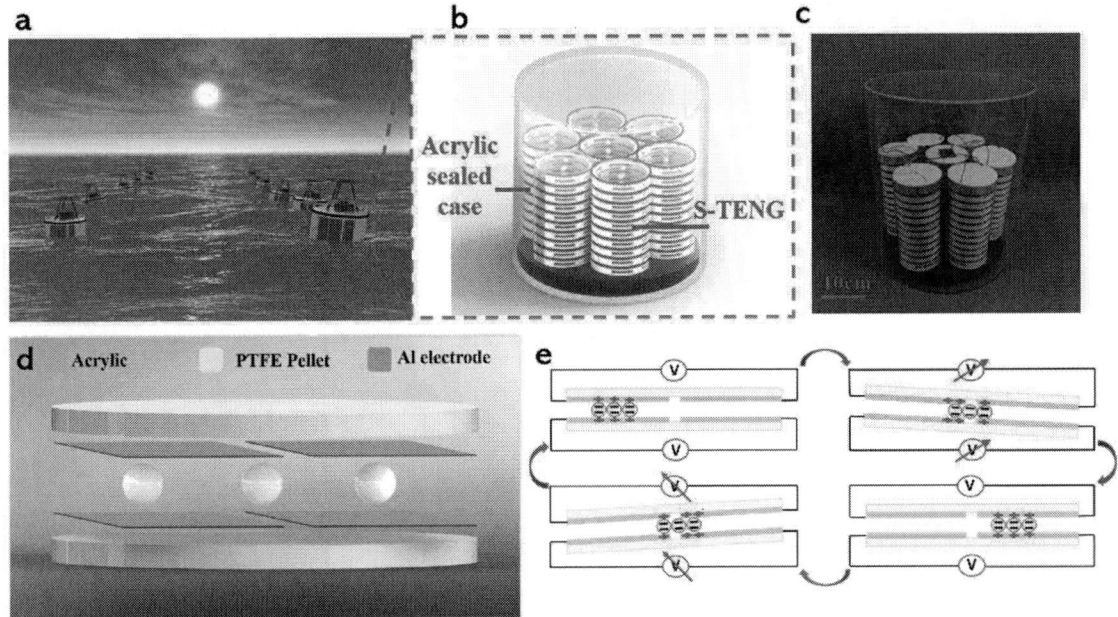

Figure 1. The buoy design and working principle. (a) Schematic illustrations of the self-powered buoy . (b) Zoom-in of wave energy converting module . (c) The fabricated self-powered buoy. (d) The structure of one S-TENG layer. (e) The working principle of one S-TENG layer.

2.2 Output performance of one S-TENG Unit

The electrical outputs of one S-TENG unit have been systematically investigated on a wave environment simulation platform. As depicted in Fig. 2a, a linear motor has been adapted to provide accurate sway motion at different frequencies (0.4Hz < f < 2Hz) and amplitudes (50mm < A < 130mm). To simulate ocean waves, the frequency is set to 0.4Hz, 1.2HZ and 2Hz. As shown in Fig. 2b, as the frequency f increases (while the sway amplitude A is set to 130mm), the short-circuit current I_{sc} increases before reaching its peak value of 11.94μA at 2Hz. On the other hand, the transferred charge Q_{sc} increases when f changes from 0.4Hz to 1.2Hz and reaches its peak value of 0.70μC at 1.2Hz (as shown in Fig. 2c). Then the value of Q_{sc} decreases a little when f increases to 2Hz. At the frequency of 0.4Hz, the input energy is not large enough to drive all the PTFE pellets roll freely across the two electrodes, thus the transferred charge Q_{sc} is quite small. The influence of the motion amplitude is shown in Fig. 2d and 2e. Under the same frequency (2Hz), it have observed that I_{sc} increases gradually when the amplitude A increases from 50mm to 130mm, and achieves the peak value (12.28μA) when the amplitude A is 130mm. Larger wave amplitude means more input energy, therefore larger I_{sc} is produced by S-TENG unit. Q_{sc} remains mostly unchanged as after the PTFE pellets are fully activated and charged. The outputs of a single S-TENG unit under different resistance are shown in Fig.2f. The peak power reaches 25.22mW while the maximum power density gets as high as 34.65W/m^3.

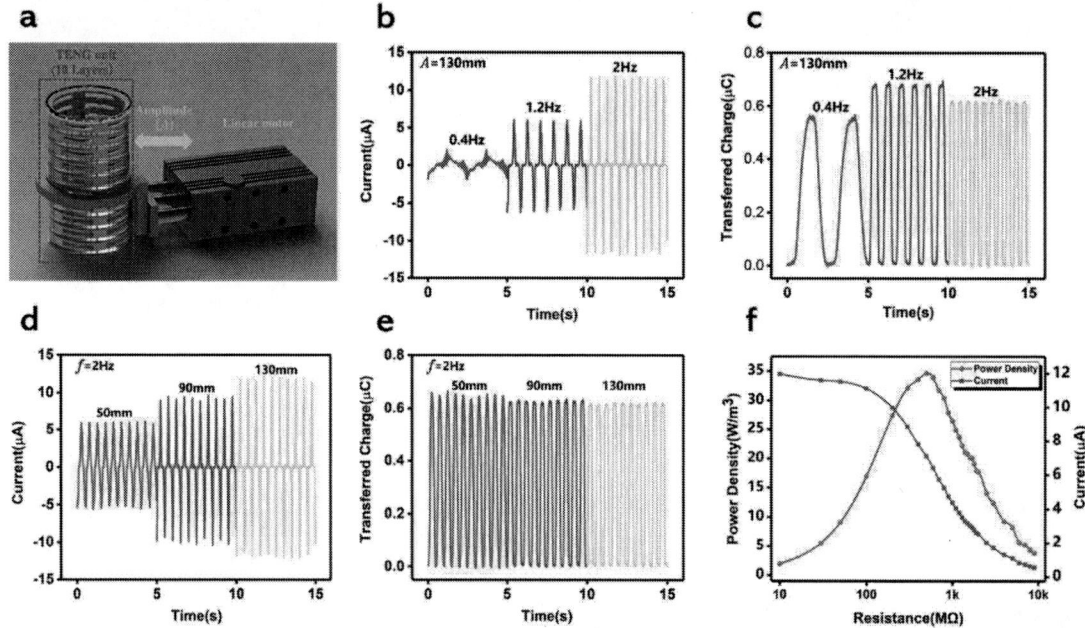

Figure 2. Electrical outputs of a single S-TENG unit with wave frequencies and amplitudes. (a) Diagram of one S-TENG unit driven sinusoidally by the linear motor. (b) The short-circuit current I_{sc} under various frequencies at the amplitude 130mm (c) The transferred charge Q_{sc} under various frequencies at the amplitude 130mm. (d) The short-circuit current I_{sc} under various amplitudes at the same frequency 2Hz and the (e) The transferred charge Q_{sc} under various amplitudes at the same frequency 2Hz. (f) The output current and power density with the circuit resistance.

2.3 Outputs of multiple S-TENG units

Previous studies have shown that parallel connecting more TENG units can promote electrical output. But due to the phase difference of AC signal of each TENG unit, a rectifier bridge is necessary, which causes energy loss and increases circuit complexity. The S-TENG's electrodes are arranged to the same direction to make the output AC in-phase.

Here a sway motion simulation is performed to verify the output characteristics of multiple parallel connected TENGs, as shown in Fig.3a. A flatbed cart is connected to the linear motor generating the sway motion. 7 S-TENG units (with 70 layers in total) have been fixed onto the pedestals on the flatbed cart through screw. All of the 70 S-TENG layers have parallel connected without the rectifier bridge. The frequency of the driving linear motor is set to 1Hz to simulate ocean wave and the amplitude is set to 130mm.

As the parallel connected S-TENG layers increases (Fig.3b), the corresponding output I_{sc} boosts from 5.44μA (10 layers), 14.78μA (30 layers), 23.99μA (50 layers) to finally 35.98μA (70 layers). The relationship between the output transferred charge Q_{sc} and the number of the parallel connected S-TENG units is shown in Fig.3c. The average output transferred charge Q_{sc} yields (similar) increasing trend with the number of the S-TENG unit. The short-circuit current I_{sc} and the transferred charge Q_{sc} yield linear correlation with the number of parallel connected S-TENG units (as shown in Fig.3d). It's believed that the output will keep linear increase with greater number of S-TENG units as the output from each unit can be kept in-phase with each other by arranging S-TENG's electrodes to the same direction. This characteristic of S-TENG provides a feasible way to promote its total power output when integrated into a buoy to achieve a self-powered marine monitoring node.

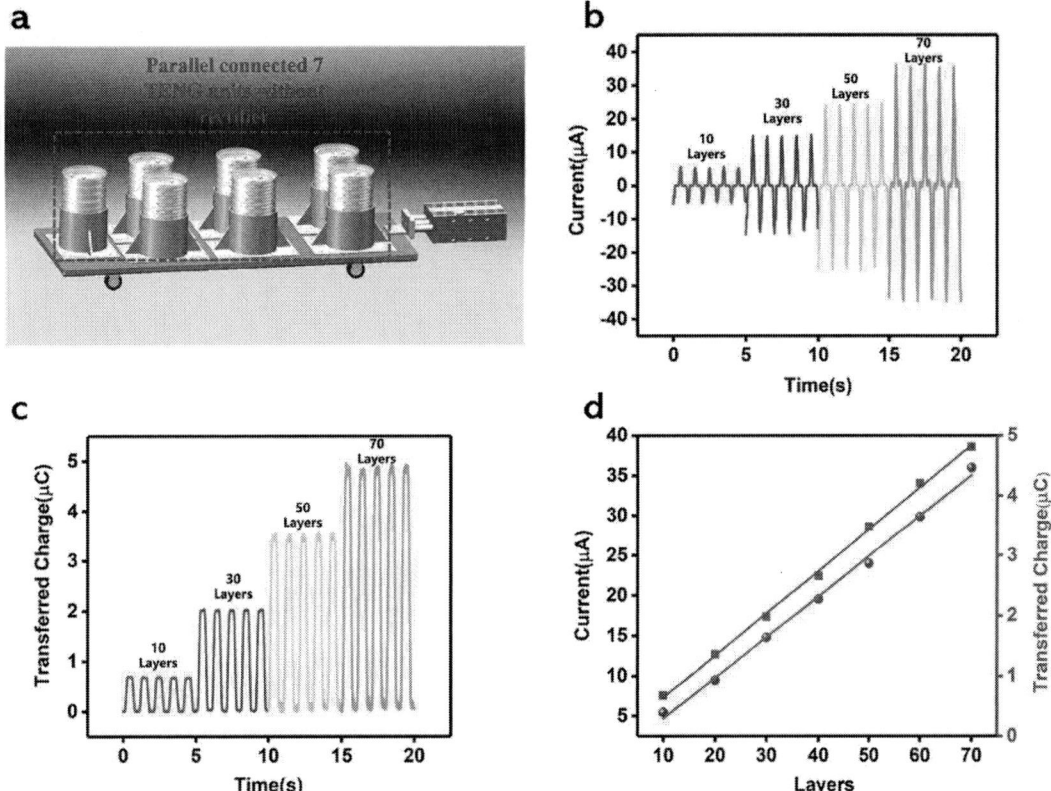

Figure 3. Output characteristics of 7 S-TENG units parallel connected. (a) The wave sway motion simulating platform. The output short-circuit current I_{sc} (b) and transferred charge Q_{sc}(c) with the number of S-TENG units increasing. (d) The variation of output current I_{sc} and transferred charge Q_{sc} with respct to the number of S-TENG units.

III. CONCLUSION

A stackable triboelectric nanogenerator (S-TENG) to be integrated with a buoy has been investigated systematically. This compact S-TENG consists of two acrylic plates with aluminum electrodes attached to their surfaces and PTFE pellets in-between the plates, which can be stacked into parallel connection very easily. In the experiments, each S-TENG unit (consisting of 10 layers) could produce a maximum power of 25.22mW with a power density of 34.65W/m³. It is also found that the output increases linearly with the layer number of S-TENG. This means that by simply integrating more S-TENG units, the required power for a self-powered buoy could possibly be reached.

REFERENCES

[1] Liu, K.; Liu, Y.; Yang, Z.; Li, M.; Guo, Z.; Guo, Y.; Hong, F.; Yang, X.; He, Y.; Feng, Y., Oceansense. ACM SIGMOBILE Mobile Computing and Communications Review 2010, 14 (2).

[2] Kroger, S.; Law, R. J., Sensing the sea. Trends Biotechnol 2005, 23 (5), 250-6.

[3] Alverson, K.; Baker, D. J., Taking the pulse of the oceans. Science 2006, 314 (5806), 1657.

[4] Wang, Z. L.; Jiang, T.; Xu, L., Toward the blue energy dream by triboelectric nanogenerator networks. Nano Energy 2017, 39, 9-23.

[5] Xu, G.; Shen, W.; Wang, X., Applications of wireless sensor networks in marine environment monitoring: a survey. Sensors 2014, 14 (9), 16932-54.

[6] Schiermeier, Q.; Tollefson, J.; Scully, T., Electricity without carbon. Nature 2008, 454, 816-823.

[7] Albaladejo, C.; Soto, F.; Torres, R.; Sanchez, P.; Lopez, J. A., A low-cost sensor buoy system for monitoring shallow marine environments. Sensors 2012, 12 (7), 9613-34.

[8] Seah, W. K. G.; Eu, Z. A.; Tan, H.-P., Wireless sensor networks powered by ambient energy harvesting (WSN-HEAP) - Survey and challenges. In 2009 1st International Conference on Wireless Communication, Vehicular Technology, Information Theory and Aerospace & Electronics Systems Technology, 2009; pp 1-5.

[9] Scruggs, J.; Jacob, P., Harvesting Ocean Wave Energy. Science 2009, 323 (5918), 1176-1178.

[10] Wang, Z. L., Catch wave power in floating nets. Nature 2017, 542 (7640), 159-160.

[11] Tollefson, J., Power from the oceans: Blue energy. Nature 2014, 508 (7496), 302-304.

[12] Westwood, A. J. R., Ocean power: Wave and tidal energy review. 2004, 5 (5), 50-55.

[13] Falcão, A. F. d. O., Wave energy utilization: A review of the technologies. Renewable Sustainable Energy Rev. 2010, 14 (3), 899-918.

[14] Wang, Z. L., Triboelectric nanogenerators as new energy technology for self-powered systems and as active mechanical and chemical sensors. ACS nano 2013, 7 (11), 9533.

[15] Wang, Z. L., Triboelectric nanogenerators as new energy technology and self-powered sensors - principles, problems and perspectives. Faraday Discuss. 2014, 176, 447-58.

[16] Wang, S.; Lin, L.; Wang, Z. L., Triboelectric nanogenerators as self-powered active sensors. Nano Energy 2015, 11, 436-462.

[17] Wang, Z. L., On Maxwell's displacement current for energy and sensors: the origin of nanogenerators. Mater. Today 2017, 20 (2), 74-82.

[18] Wang, Z. L., Nanogenerators, self-powered systems, blue energy, piezotronics and piezo-phototronics – A recall on the original thoughts for coining these fields. Nano Energy 2018, 54, 477-483.

[19] Luo, J.; Wang, Z. L., Recent advances in triboelectric nanogenerator based self-charging power systems. Energy Storage Materials 2019.

[20] Mariello, M.; Guido, F.; Mastronardi, V. M.; Todaro, M. T.; Desmaële, D.; De Vittorio, M., Nanogenerators for harvesting mechanical energy conveyed by liquids. Nano Energy 2019, 57, 141-156.

[21] Wu, C.; Wang, A. C.; Ding, W.; Guo, H.; Wang, Z. L., Triboelectric Nanogenerator: A Foundation of the Energy for the New Era. Advanced Energy Materials 2019, 9 (1), 1802906.

[22] Rodrigues, C.; Nunes, D.; Clemente, D.; Mathias, N.; Correia, J.; Santos, P.; Taveira-Pinto, F.; Morais, T.; Pereira, A.; Ventura, J., Emerging Triboelectric Nanogenerators for Ocean Wave Energy Harvesting: State of the Art and Future Perspectives. Energy & Environmental Science 2020.

[23] Niu, S.; Zhou, Y. S.; Wang, S.; Liu, Y.; Lin, L.; Bando, Y.; Wang, Z. L., Simulation method for optimizing the performance of an integrated triboelectric nanogenerator energy harvesting system. Nano Energy 2014, 8, 150-156.

[24] Wang, X.; Niu, S.; Yin, Y.; Yi, F.; You, Z.; Wang, Z. L., Triboelectric Nanogenerator Based on Fully Enclosed Rolling Spherical Structure for Harvesting Low-Frequency Water Wave Energy. Adv. Energy Mater. 2015, 5 (24), 1501467.

[25] Shi, Q.; Wang, H.; Wu, H.; Lee, C., Self-powered triboelectric nanogenerator buoy ball for applications ranging from environment monitoring to water wave energy farm. Nano Energy 2017, 40, 203-213.

[26] Xiao, T. X.; Jiang, T.; Zhu, J. X.; Liang, X.; Xu, L.; Shao, J. J.; Zhang, C. L.; Wang, J.; Wang, Z. L., Silicone-Based Triboelectric

[27] Xu, L.; Jiang, T.; Lin, P.; Shao, J. J.; He, C.; Zhong, W.; Chen, X. Y.; Wang, Z. L., Coupled Triboelectric Nanogenerator Networks for Efficient Water Wave Energy Harvesting. ACS nano 2018, 12 (2), 1849-1858.

[28] Liang, X.; Jiang, T.; Liu, G.; Xiao, T.; Xu, L.; Li, W.; Xi, F.; Zhang, C.; Wang, Z. L., Triboelectric Nanogenerator Networks Integrated with Power Management Module for Water Wave Energy Harvesting. Advanced Functional Materials 2019.

[29] Xi, F.; Pang, Y.; Liu, G.; Wang, S.; Li, W.; Zhang, C.; Wang, Z. L., Self-powered intelligent buoy system by water wave energy for sustainable and autonomous wireless sensing and data transmission. Nano Energy 2019, 61, 1-9.

[30] Xu, M.; Zhao, T.; Wang, C.; Zhang, S. L.; Li, Z.; Pan, X.; Wang, Z. L., High Power Density Tower-like Triboelectric Nanogenerator for Harvesting Arbitrary Directional Water Wave Energy. ACS nano 2019.

[31] Yang, X.; Xu, L.; Lin, P.; Zhong, W.; Bai, Y.; Luo, J.; Chen, J.; Wang, Z. L., Macroscopic self-assembly network of encapsulated high-performance triboelectric nanogenerators for water wave energy harvesting. Nano Energy 2019, 60, 404-412.

[32] Zhang, X.; Yu, M.; Ma, Z.; Ouyang, H.; Zou, Y.; Zhang, S. L.; Niu, H.; Pan, X.; Xu, M.; Li, Z.; Wang, Z. L., Self-Powered Distributed Water Level Sensors Based on Liquid–Solid Triboelectric Nanogenerators for Ship Draft Detecting. Advanced Functional Materials 2019.

[33] Chen, X.; Gao, L.; Chen, J.; Lu, S.; Zhou, H.; Wang, T.; Wang, A.; Zhang, Z.; Guo, S.; Mu, X.; Wang, Z. L.; Yang, Y., A chaotic pendulum triboelectric-electromagnetic hybridized nanogenerator for wave energy scavenging and self-powered wireless sensing system. Nano Energy 2020, 69.

Nanogenerator for Water Wave Energy Harvesting. ACS applied materials & interfaces 2018, 10 (4), 3616-3623.

Proceedings of the 16th Annual IEEE International
Conference on Nano/Micro Engineered and Molecular Systems
April 25-29, 2021

Modulating the Electrical Transport Characteristics of a Metal-Semiconductor-Metal Structure by Local Strain Gradient

Youchao Huang, Shenhui Ma, Jiaona Zhang, Chao Xie and Min Zhang*, *Member, IEEE*

Abstract— Metal-semiconductor-Metal (M-S-M) structure is widely used in electronic devices, which plays a key role in device performance. Here, we propose a simple and effective way to modulate the contact characteristics of aluminum (Al)-amorphous indium gallium zinc oxide (a-IGZO)-molybdenum (Mo) structure by applying local strain at the interface between Al layer and a-IGZO layer. Four types of *I-V* curves including forward rectifying, backward rectifying, linear and approximately symmetric are achieved under different pressures. Due to the nanoscale strain gradient induced by inhomogeneous local strain, flexoelectric potential is generated at the interface of the Al and the a-IGZO, which leads to the energy band bending. Therefore, the barrier height modulation can be realized. Meanwhile, the charges induced by flexoelectric effect in the a-IGZO layer can also modulate the barrier height at the a-IGZO/Mo interface. This work shows great potential to enhance the performance of electronic devices and it is promising for novel electromechanical applications.

I. INTRODUCTION

Electronic devices with two electrode contacts generally consist of a metal-semiconductor-metal (M-S-M) structure [1-3]. In general, the contact of metal and semiconductor forms Ohmic contact or Schottky contact, which mainly depends on the material property and nature of interface. As a result, the electrical transport characteristics (current-voltage curves, i.e. *I-V* curves) of typical M-S-M structures containing two metal-semiconductor contacts can be classified into four types: linear, forward rectifying, backward rectifying as well as symmetric curves. The performance and reliability of devices based on M-S-M structure are strongly affected by the effective barrier height and width of the metal-semiconductor contacts [2]. Therefore, great efforts have been devoted to changing materials and optimizing interface properties in convention so as to improve device performance [4, 5]. However, these methods are complicated and hard to effectively control the contact property.

Strain engineering is an effective way to regulate the properties of materials and optimize the device performance by inherent or external strain. The piezoelectric effect and flexoelectric effect can be utilized to modulate the contact characteristic of M-S-M structure by strain. For non-centrosymmetric materials such as the wurtzite structure ZnO, GaN and InN, strong piezoelectric effect can occur by

applying a strain to tune charge carrier transport at the contact [6-8]. J. He et al. showed a piezoelectric gated diode with a Ti-ZnO nanowire-Ti structure, which can work like a CMOS device to tune the local contact by applying strain without applying gate voltage [6]. Similarly, Y. Hu et al. proposed that the local contact could be tuned step-by-step from Schottky contact to Ohmic contact or vice versa by controlling the coupling among mechanical, photonic, and electrical properties of ZnO nanowires [7]. Although these methods can modulate the contact characteristics, they are difficult to be applied widely due to the material limit, that is, wurtzite structure nanowire.

Unlike piezoelectric effect only occurring in non-centrosymmetric crystal, flexoelectricity can be generated by inhomogeneous strain in all dielectric materials and some semiconductors [9-11]. Flexoelectric effect is one of the electromechanical effects, which shows a linear response between polarization and strain gradient. Recently, D. Lee et al. showed the unique coupling between electronic transport and the mechanical strain gradient in a dielectric epitaxial thin film [12]. In the device with Pt-HoMnO$_3$-Pt structure, they observed three types of *I-V* curves under different strain gradients. L. Wang et al. demonstrated a strong flexoelectric effect in centrosymmetric semiconductors such as Si, TiO$_2$ and Nb-SrTiO$_3$ [10]. By using an atomic force microscopy (AFM) tip to apply stress on semiconductors, the barrier height between metal and semiconductor could be modulated and hence the electrical transport was changed. However, these methods to obtain giant strain gradient are complicated due to the severe film growth conditions or precise manipulation of AFM tip.

Herein, we demonstrate a simple method to modulate the contact characteristics of aluminum (Al)-amorphous indium gallium zinc oxide (a-IGZO)-molybdenum (Mo) structure based on the flexoelectric effect. Because of the nanoscale surface roughness interaction, a large strain gradient will occur when applying a strain to the interface between the Al and the a-IGZO layer. The flexoelectric polarization charges generated by the strain gradient can be used to modulate the barrier of both interfaces and further enable the M-S-M structure with various electrical transport characteristics. This exploitation of strain gradient can provide a route to improve the performance of electronic devices and develop new electromechanical applications.

---

*Research was supported by Shenzhen Science and Technology Innovation Grant JCYJ20180507181702150, Guangdong Science and Technology Plan 2019A05051001, and National Nature Science Foundation of China 62074008.

Youchao Huang, Shenhui Ma, Chao Xie, and Min Zhang are with the School of Electronics and Computer Engineering, Peking University, Shenzhen 5180555, China.

Jiaona Zhang is with the School of Electronics and Computer Engineering, The Hong Kong University of Science and Technology, Hong Kong, China.

*Corresponding Author: Min Zhang (phone: 86-0755-26032482; zhangm@ece.pku.edu.cn).

II. EXPERIMENTS

Fig. 1(a) is the separated M-S-M structure, which is composed of upper and bottom parts. The upper part of 90 nm-thick Al was sputtered on a polyethylene naphthalene (PEN) substrate with an area of 0.5 cm × 2 cm as the top metal electrode. The bottom part was formed by 100 nm-thick Mo as the bottom metal electrode and a-IGZO as the semiconductor layer. Mo and a-IGZO were sequentially sputtered on a 2-inch clean glass substrate. The flow rates of argon (Ar) and oxygen (O$_2$) during the growth of a-IGZO thin films are 47 sccm and 3 sccm, respectively. After fabrication, a piece of copper foil was attached on Mo for testing.

During the measurement, the upper part was adhered on the load and the bottom part was fixed to a horizontal stable base. The force was applied on the M-S-M structure by an electrodynamic force equipment (ZHIQU Precision Instruments ZQ-990B), as shown in Fig. 1(b). And the *I-V* curves were measured by a semiconductor characterization system (Keithley 4200-SCS). Additionally, the fabricated a-IGZO film was characterized by X-ray diffraction (XRD) and AFM.

III. RESULT AND DISCUSSION

Different forces (4.3 N, 7.3 N, 30.0 N and 80.0 N) were applied on the fabricated M-S-M structure with 550 nm-thick a-IGZO layer. And four types of representative *I-V* curves have been obtained including the forward rectifying, approximately symmetric, backward rectifying, and linear curves, as shown in Fig. 2(a)-(d). Theoretically, the *I-V* curves of our M-S-M structure should keep linear according to the work function of materials. However, when it was applied with force of 4.3 N or 30.0 N, the Schottky characteristic can be observed with the opposite conduction direction. The approximately symmetric curve indicates that there exist back-to-back Schottky diodes in the M-S-M structure. According to the back-to-back Schottky image force barrier model, ln(I) is linear with $V^{1/4}$ [13]. And the curve in Fig. 2(e) verifies this assumption. Fig. 2(d) indicates that the metal-semiconductor contacts in the M-S-M structure are the Ohmic contacts. For comparison, we also fabricated a M-S-M structure by successively depositing Mo, a-IGZO and Al layers on a 2-inch glass substrate with the same parameters. As shown in Fig. 2(f), the integrated M-S-M structure shows the Ohmic contact characteristic, which is in consistence with the theory as expected. That is, the contact characteristic is dependent on the pressure

applied on the separated structure.

Fig. 3 (a) and (b) illustrate the relationship between the contact characteristic of the M-S-M structure and applied force ranging from 1.0 N to 51.0 N. When the applied force is increased from 1.0 N to 6.2 N, the curves show the forward rectifying and the resistance gradually decreases. However, when the applied force is increased from 7.3 N to 51.0 N, the curves are approximately symmetric firstly, and then they become backward rectifying. The resistance does not reduce as expected due to the increase in the effective area induced by force. Moreover, the curve become linear when the applied force further increases to 80.0 N as shown in Fig. 2(d). These phenomena are similar to the modulation of local contact by piezoelectric effect in some extent.

In addition, the *I-V* curves of M-S-M structures with different thicknesses of a-IGZO film were measured. We observed the similar phenomena that the electrical transport characteristics of M-S-M structure can be modulated continuously with increased force. Fig. 3(c) shows the *I-V* curves of different M-S-M structures under the force of 1.0 N. The current gradually decreases when the thickness of a-IGZO layer increases from 40 nm to 550 nm, which can be attributed to the variation of barrier height and resistance of a-IGZO layer.

The aforementioned phenomena of four types of *I-V* curves in our experiment results cannot be explained by the piezoelectric effect. Firstly, there is no obvious XRD peak in the 2θ range of 10-90° indicating that the a-IGZO film is not the crystal but the amorphous morphology, as shown in

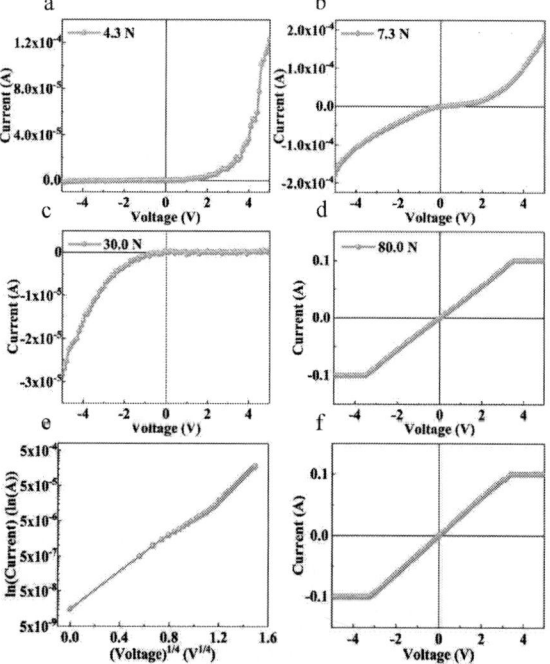

Figure 2. *I-V* curves in M-S-M structure with 550 nm-thick a-IGZO. (a) Forward rectifying curve under 4.3 N, (b) approximately symmetric curve under 7.3 N, (c) backward rectifying under 30.0 N, and (d) linear curve under 80.0 N. (e) The relationship of logarithm of current with $V^{1/4}$. (f) The integrated M-S-M structure by successive deposition shows the Ohmic characteristic.

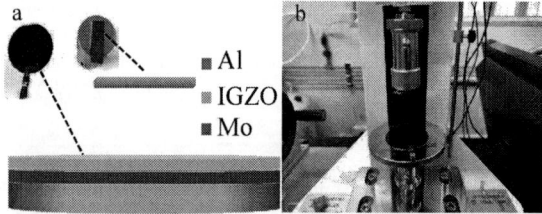

Figure 1. (a) The diagram of the separated M-S-M structure. (The inset is the photograph of the separated M-S-M structure with bottom part (left) and upper part adhered on the load (right)). (b) The test equipment for applying force.

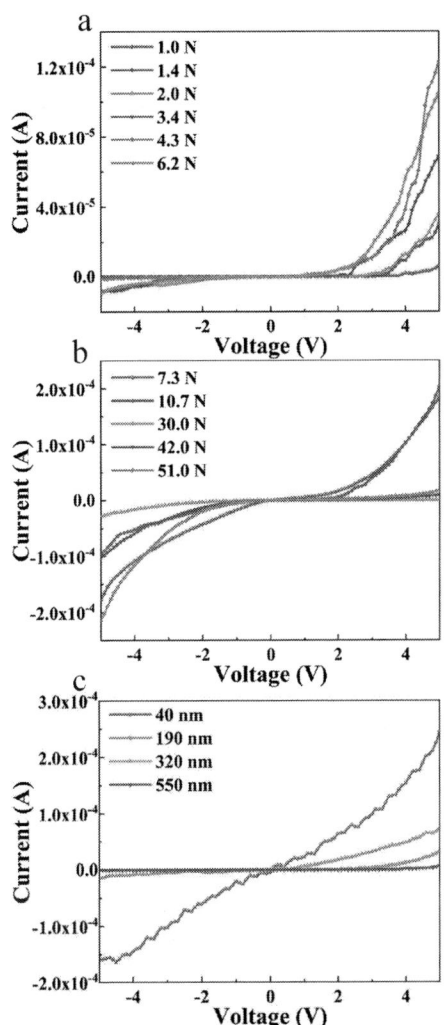

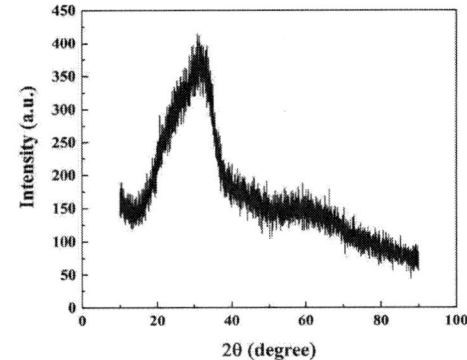

Figure 4. XRD pattern of the a-IGZO layer on a glass substrate.

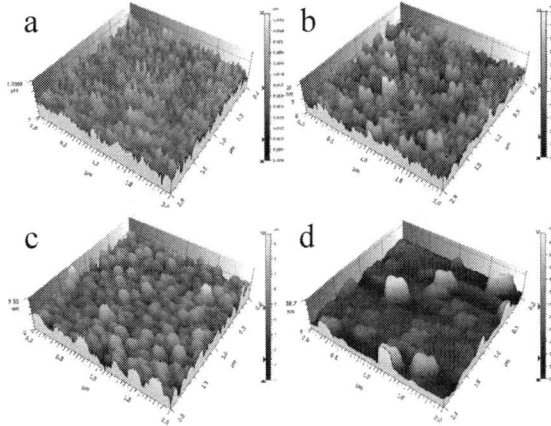

Figure 5. 3D AFM images of surface morphology of the a-IGZO layers with different thicknesses. (a) 40 nm-thick a-IGZO layer with RMS roughness of 0.722 nm, (b) 320 nm-thick a-IGZO layer with RMS roughness of 1.32 nm, (c) 550 nm-thick a-IGZO layer with RMS roughness of 1.26 nm, and (d) 320 nm-thick a-IGZO layer with RMS roughness of 7.77 nm in another region.

Figure 3. *I-V* curves of M-S-M structure with 550 nm-thick a-IGZO under (a) small and (b) large forces. (c) *I-V* curves of M-S-M structure with different a-IGZO layer thicknesses under 1.0 N force.

Fig. 4. All oxide piezoelectric materials such as $Pb(Ti,Zr)O_3$ and ZnO, are crystals [14]. So, the a-IGZO is not piezoelectric material. Secondly, four types of electrical characteristics appear, which cannot be observed by piezoelectric effect because the barrier height at the interface only increases or decreases monotonically under the pressure.

On the other hand, the flexoelectric effect can be used to explain these phenomena qualitatively. The flexoelectric potential on the surface of a-IGZO is induced by the large strain gradient at nanoscale asperities due to the flexoelectric effect. The direction and magnitude of flexoelectric potential are dependent on the applied force, asperity size, local topography, and material properties (e.g., crystallographic orientation) [15]. As shown in Fig. 5(a)-(c), the surface morphology and root mean square (RMS) roughness of a-IGZO film varies with different a-IGZO film thicknesses. Therefore, the M-S-M structures with various thicknesses of

a-IGZO layer have shown four types of *I-V* curves under the same pressure as illustrated in Fig. 3(c). Moreover, the RMS roughness of different regions of a-IGZO layer has shown large variation, as shown in Fig. 5(b) and (d), which is inevitably introduced due to the film deposition process. Hence, it is possible to observe different types of curves from a same M-S-M structure (Fig. 3(a) and (b)) by modulating the force and effective contact area.

The modulation of contact characteristic by the flexoelectric potential will be further discussed here. The schematic diagram of the M-S-M structure is shown in Fig. 6(a). Based on the work functions of Al, Mo and n-type semiconductor a-IGZO (4.2 eV, 4.5 eV, 4.9 eV, respectively), the energy band profile is illustrated in Fig. 6(b), which is theoretically Ohmic contact. Due to strain gradient, the flexoelectric polarization field is induced in the inner a-IGZO film leading to the polarization charges and potential on the a-IGZO surface. When the negative flexoelectric polarization charges are induced at the Al/a-IGZO interface by the polarization field, the positive flexoelectric polarization charges will be induced at the a-IGZO/Mo interface. Therefore, a negative flexoelectric

978-1-6654-3008-1/21 $31.00 © 2021 IEEE

potential is generated at the top surface of a-IGZO. When the flexoelectric field is strong enough, a large number of negative polarization charges will lead to the band bending upward, which changes the distribution of free carriers in the a-IGZO layer and finally results in the variation of contact characteristic from Ohmic to Schottky at the interface. Meanwhile, the positive charges will make the band bend downward at the a-IGZO/Mo interface (Fig. 6(c)) corresponding to the forward rectifying curve. In contrast, if the flexoelectric polarization field is weak, the M-S-M contact is still Ohmic, but the band profile changes slightly (Fig. 6(d)). When the positive and negative polarization charges accumulate at the Al/a-IGZO interface and a-IGZO/Mo interface, respectively, a positive flexoelectric potential is generated at the top surface of a-IGZO. Similarly, if there are enough negative polarization charges, the contact characteristic will be altered from Ohmic to Schottky (Fig. 6(e)) corresponding to the backward rectifying curve. Otherwise, the M-S-M structure remains as the Ohmic contact. Under the conditions above, the polarization charges play a dominant role to control the band, while the flexoelectricity affecting the Al is relatively weak. In reality, the large strain gradient in semiconductor also affects the band structure in metal, which has been observed in recent studies [10, 16]. As for the back-to-back Schottky contacts as shown in Fig. 2(b), this effect should be considered. Originally, the polarization field in the a-IGZO layer leads to the accumulation of positive and negative charges at Al/a-IGZO and a-IGZO/Mo interfaces, respectively. At the same time, the potential behaving as a gate voltage reduces the Fermi-level energy of Al, which results in a Schottky contact, as illustrated in Fig. 6(f) [17].

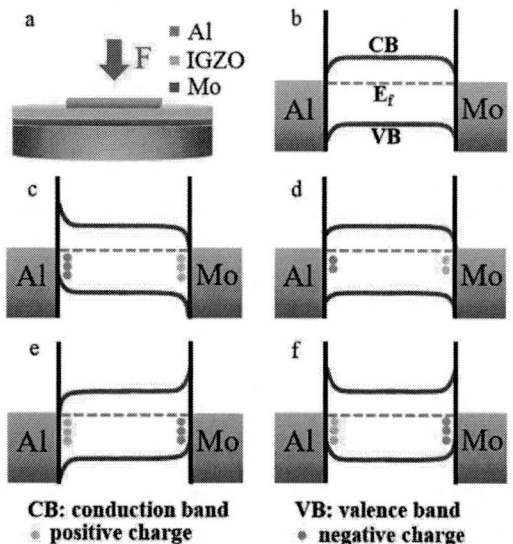

CB: conduction band
· positive charge

VB: valence band
· negative charge

Figure 6. The physical mechanism of modulating the electrical transport characteristic on the M-S-M structure by flexoelectric effect. (a) The schematic diagram of the M-S-M structure. (b) The band diagram of the M-S-M structure without the flexoelectric effect. (c) The band diagram for forward rectifying curve. (d) The band diagram for liner curve. (e) The band diagram for backward rectifying curve. (f) The band diagram for approximately symmetric curve.

IV. CONCLUSION

In this paper, the modulation of electrical transport characteristics on M-S-M structures is demonstrated by applying strain on the semiconductor surface. The mechanism of modulation mainly comes from two steps: (1) Local strain on the rough surface results in the large strain gradient. (2) The flexoelectric polarization charges induced by strain gradient modulates the barrier height at the Al/a-IGZO and a-IGZO/Mo interfaces, which enables the M-S-M structure with four types of contact characteristics under different forces. Due to universal nature of flexoelectricity, we can control the carrier transport in M-S-M structures by mechanical strain instead of depending on the property of materials conventionally. This work provides a pathway to design and fabricate flexoelectricity-modulated electronic devices.

REFERENCES

[1] S. Song, Y. Sim, S.-Y. Kim, J. H. Kim, I. Oh, W. Na, D. H. Lee, J. Wang, S. Yan, Y. Liu, J. Kwak, J.-H. Chen, H. Cheong, J.-W. Yoo, Z. Lee, and S.-Y. Kwon, "Wafer-scale production of patterned transition metal ditelluride layers for two-dimensional metal–semiconductor contacts at the Schottky–Mott limit," Nature Electronics, vol. 3, no. 4, pp. 207-215, 2020.

[2] Y. Liu, J. Guo, E. Zhu, L. Liao, S. J. Lee, M. Ding, I. Shakir, V. Gambin, Y. Huang, and X. Duan, "Approaching the Schottky-Mott limit in van der Waals metal-semiconductor junctions," Nature, vol. 557, no. 7707, pp. 696-700, 2018.

[3] J. Feng, C. Gong, H. Gao, W. Wen, Y. Gong, X. Jiang, B. Zhang, Y. Wu, Y. Wu, H. Fu, L. Jiang, and X. Zhang, "Single-crystalline layered metal-halide perovskite nanowires for ultrasensitive photodetectors," Nature Electronics, vol. 1, no. 7, pp. 404-410, 2018.

[4] C. Y. Huang, T. J. Lin, and P. C. Liao, "High-performance metal-semiconductor-metal ZnSnO UV photodetector via controlling the nanocluster size," Nanotechnology, vol. 31, no. 49, p. 495203, 2020.

[5] D. Y. Guo, Z. P. Wu, Y. H. An, X. C. Guo, X. L. Chu, C. L. Sun, L. H. Li, P. G. Li, and W. H. Tang, "Oxygen vacancy tuned Ohmic-Schottky conversion for enhanced performance in β-Ga2O3 solar-blind ultraviolet photodetectors," Applied Physics Letters, vol. 105, no. 2, p. 023507, 2014.

[6] J. H. He, C. L. Hsin, J. Liu, L. J. Chen, and Z. L. Wang, "Piezoelectric Gated Diode of a Single ZnO Nanowire," Advanced Materials, vol. 19, no. 6, pp. 781-784, 2007.

[7] Y. Hu, Y. Chang, P. Fei, R. L. Snyder, and Z. L. Wang, "Designing the electric transport characteristics of ZnO micro/nanowire devices by coupling piezoelectric and photoexcitation effects," ACS nano, vol. 4, no. 2, pp. 1234-1240, 2010.

[8] Y. Peng, M. Que, H. E. Lee, R. Bao, X. Wang, J. Lu, Z. Yuan, X. Li, J. Tao, and J. Sun, "Achieving high-resolution pressure mapping via flexible GaN/ZnO nanowire LEDs array by piezo-phototronic effect," Nano Energy, vol. 58, pp. 633-640, 2019.

[9] P. Zubko, G. Catalan, and A. K. Tagantsev, "Flexoelectric Effect in Solids," Annual Review of Materials Research, vol. 43, no. 1, pp. 387-421, 2013.

[10] L. Wang, S. Liu, X. Feng, C. Zhang, L. Zhu, J. Zhai, Y. Qin, and Z. L. Wang, "Flexoelectronics of centrosymmetric semiconductors," Nature Nanotechnology, vol. 15, no. 8, pp. 661-667, 2020.

[11] J. Narvaez, F. Vasquez-Sancho, and G. Catalan, "Enhanced flexoelectric-like response in oxide semiconductors," Nature, vol. 538, no. 7624, pp. 219-221, 2016.

[12] D. Lee, S. M. Yang, J. G. Yoon, and T. W. Noh, "Flexoelectric rectification of charge transport in strain-graded dielectrics," Nano Letters, vol. 12, no. 12, pp. 6436-40, 2012.

[13] C. Y. Nam, D. Tham, and J. E. Fischer, "Disorder effects in focused-ion-beam-deposited Pt contacts on GaN nanowires," Nano Letters, vol. 5, no. 10, pp. 2029-2033, 2005.

[14] H. Liu, J. Zhong, C. Lee, S.-W. Lee, and L. Lin, "A comprehensive review on piezoelectric energy harvesting technology: Materials,

mechanisms, and applications," *Applied Physics Reviews,* vol. 5, no. 4, p. 041306, 2018.

[15] C. A. Mizzi, A. Y. W. Lin, and L. D. Marks, "Does Flexoelectricity Drive Triboelectricity?," *Physical Review Letters,* vol. 123, no. 11, p. 116103, 2019.

[16] P. C. Lou, W. P. Beyermann, and S. Kumar, "Experimental evidence of hidden spin polarization in silicon by using strain gradient," 2020, unpublished.

[17] L. Liang, Q. Chen, J. Lu, W. Talsma, J. Shan, G. R. Blake, T. T. M. Palstra, and J. Ye, "Inducing ferromagnetism and Kondo effect in platinum by paramagnetic ionic gating." *Science Advance,* vol. 4, no. 4, p. eaar2030, 2018.

Proceedings of the 16th Annual IEEE International
Conference on Nano/Micro Engineered and Molecular Systems
April 25-29, 2021

A high-frequency narrow-band filtering mechanism based on auto-parametric internal resonance

Rong Wang, Cao Xia, Dong F. Wang, *Member, IEEE* and Takahito Ono

Abstract—**This paper proposes a new high-frequency narrow-band filtering mechanism using the internal resonance of an inverted T-shaped resonant structure. The proposed filtering mechanism is theoretically analyzed by combining the multi-scale method with the numerical iterative algorithm. For potential narrow-band filtering application, the influence laws of parameters related to the inverted T-shaped resonant structure on the internal resonance response characteristics are systematically studied. Two possible filtering systems integrated respectively with MEMS Switch and Double Frequency Circuit are further considered and discussed. The above two filtering systems have advantages of a good amplitude stability, a clear edge distribution in the passband frequency domain, and a strong suppression ability outside the passband frequency domain.**

Key words: narrow-band filtering, internal resonance, filtering characteristics, filtering systems, systematical study.

I. INTRODUCTION

MEMS technology is a frontier research area produced by the cross fusion of microelectronic technology and mechanical and optical fields. MEMS device has the characteristics of miniaturization, diversification and integration, suitable for a variety of applications. In recent years, the nonlinear behavior of MEMS device has been widely research [1-4], nonlinear dynamics characteristics of micro-nano resonators are important contents of the MEMS research, and how to make use of these nonlinear dynamic characteristics to solve practical problems in life has become an interesting research hot spot [5]. Many teams have conducted the application study of MEMS devices' nonlinear behavior, and some achievements have been made in the fields of mass sensing [6-10], logic devices [11], actuators [12], friction characterization [13], energy harvester [14-18], atomic force microscope [19], etc.

With the development and breakthrough of electronic technology and communication technology, the precision of filter signal processing is required to be higher and higher. The nonlinear internal resonance has shown its potential value in other fields, only when the frequency of the excitation signal satisfies certain internal resonance condition, can the energy transfer between modals occur,

this opens up a promising new research direction for the development of new filters.

Different from most mechanical filters that adopt direct linear resonance to filter [20-23], this paper proposes a novel MEMS band-pass filter based on the nonlinear internal resonance characteristic of the inverted T-shaped resonant beam structure. Furthermore, we systematically study the influence of the parameters related to the inverted t-type resonant beam on the filter characteristics of the resonator.

II. STRUCTURE DESIGN AND BASIC WORKING PRINCIPLE

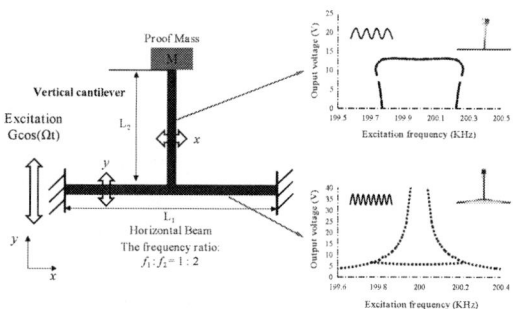

Figure 1. The inverted T-shaped resonant structure. The horizontal beam L_1 vibrates along the Y-axis driven by the external signal to be filtered. The vertical cantilever L_2 will vibrate along the Y-axis with the horizontal beam. Only when there is a frequency signal satisfying specific resonance condition in the filtering signal, will L_2 be parameterized excited by L_1 to generate X-axis vibration and output the corresponding electrical signal through the piezoelectric layer, which frequency is half of the effective signal in the signal to be filtered.

The inverted T-shaped resonant beam designed in this paper for narrowband filtering is shown in Fig. 1. The structure is mainly composed of horizontal beam and vertical cantilever through mechanical coupling, their natural frequencies are close to the high-order and low-order modal frequencies of the structure respectively, and the frequency ratio is close to an integer 2:1.

R. Wang is with Micro Engineering and Micro Systems Laboratory (JML), School of Mechanical and Aerospace Engineering, Jilin University, Changchun, 130022 China (e-mail: wangrong19@mails.jlu.edu.cn).

C. Xia is with Micro Engineering and Micro Systems Laboratory (JML), School of Mechanical and Aerospace Engineering, Jilin University, Changchun, 130022 China (e-mail: xiacao19@mails.jlu.edu.cn).

D. F. Wang is with Micro Engineering and Micro Systems Laboratory (JML), School of Mechanical and Aerospace Engineering, Jilin University, Changchun, 130022 China. He is also with Research Center for Ubiquitous

MEMS and Micro Engineering, National Institute of Advanced Industrial Science and Technology (AIST), Tsukuba, 305-8564 Japan (principal corresponding author: +86-(0)431-8509-4698; fax: +86-(0)431-8509-4698; e-mail: dongfwang@jlu.edu.cn).

T. Ono is with Department of Mechanical Systems and Design, Tohoku University, Sendai, 980-8579 Japan (e-mail: ono@nme.mech.tohoku.ac.jp).

III. EXPERIMENTAL VERIFICATION

A. Experimental set-up and measurements

As shown in Fig. 2, a macroscopic experimental platform is designed for preliminary experimental verification. The experimental device is fixed on the top of a vibrator, and the vibrator drove the inverted T-shaped resonant structure in the vertical direction. The beams in the experimental resonator structure are mainly made of copper, and piezoelectric films are attached to these beams, which output vibrations from both horizontal beam and vertical cantilever.

B. Experimental results

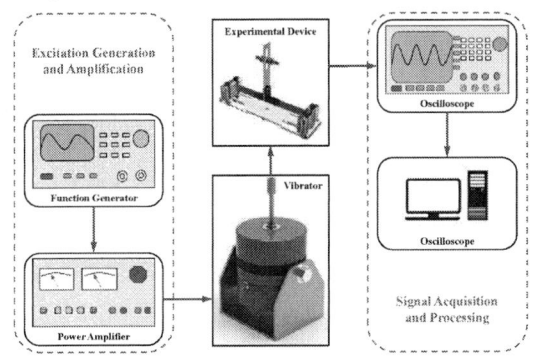

Figure 2. Schematic diagram of macroscopic experimental device

In the experiment, the inverted T - shaped resonant structure is excited with sweep frequency mode. When internal resonance occurs, the vertical beam has an obvious horizontal vibration with a vibration frequency that is half of the frequency of the excitation signal. Fig. 3 shows the output voltage of the piezoelectric film of the vertical beam at different excitation frequencies.

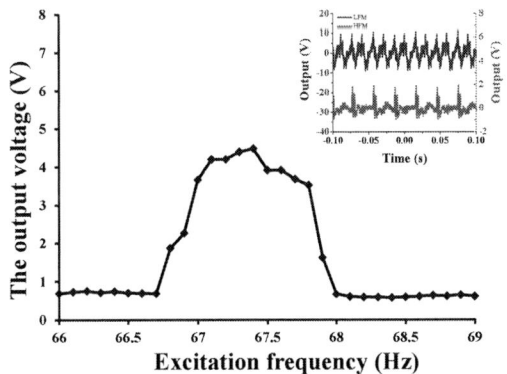

Figure 3. The low frequency modal output characteristic curve of invert resonant beam structure. The upper right subgraph is Transient response time domain diagram of low and high frequency modals.

Through experimental research, we can find that when the frequency of the excitation signal is close to the natural frequency of the horizontal beam (high frequency modal), internal resonance occurs, part of the vibration energy is transferred to the vertical beam (low frequency modal),

causing the vertical beam to vibrate at half the frequency of the excitation signal.

By selecting a suitable frequency in the passband frequency domain as the excitation signal frequency and changing the driving voltage to change the amplitude of the excitation signal, we can obtain the relation between the low frequency modal output and the amplitude of the excitation signal, the experimental data and relationship curve are shown in TABLE 1 and Fig. 4.

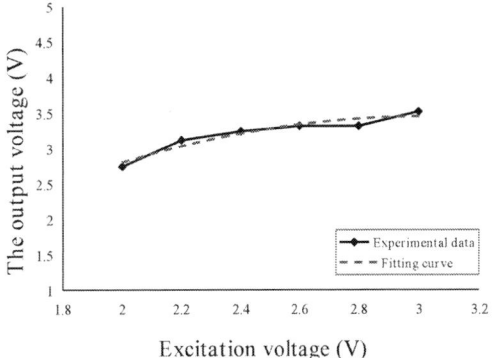

Figure 4. The relationship between the output amplitude of low frequency modal and the amplitude of excitation signal.

Figure 5. THE EXPERIMENTAL DATA

Excitation Voltage(V)	The results of multiple experiments					
	1	2	3	4	V_1	V_2
2	3.6	2.9	3.1	3	3.15	2.75
2.2	3.8	3.4	3.5	3.4	3.525	3.125
2.4	4.0	3.6	3.6	3.4	3.65	3.25
2.6	4.0	3.7	3.8	3.4	3.725	3.325
2.8	3.9	3.7	3.7	3.6	3.725	3.325
3	4.1	3.8	3.9	3.9	3.925	3.525

IV. THEORETICAL DERIVATION

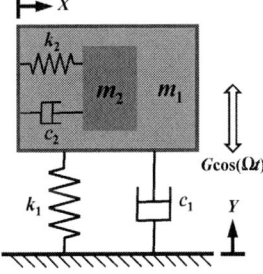

Figure 6. Simplified mass-spring damping modal of the inverted T-shaped resonant structure.

It is found that the amplitude change of excitation signal will also affect the filter pass-band bandwidth which is the frequency domain range of internal resonance response. Therefore, from the theoretical and simulation level, we systematically research the influence of excitation signal amplitude and parameters related to the resonator structure on the internal resonance characteristic of the structure, especially focusing on the low frequency modal output, which is associated with the filter application potential of the inverted T-shaped resonant structure.

A. Theoretical solution process

The simplified mass-spring-damping model of the inverted T-shaped resonant structure is shown in Fig. 5, which can be described as the following dimensionless modal coupled motion equations [24-25]:

$$\ddot{y} + 2\varepsilon\xi_1\dot{y} + \omega_1^2 y - \varepsilon k\mu(\dot{x}^2 + x\ddot{x}) = \varepsilon G\cos\Omega t \quad (1)$$

$$\ddot{x} + 2\varepsilon\xi_2\dot{x} + \omega_2^2 x - \varepsilon ky\ddot{x} = 0 \quad (2)$$

Where y, x represent Longitudinal high frequency modal amplitude and transverse low frequency modal amplitude : ξ_1, ξ_2 are the coefficients associated with high frequency modal damping and low frequency modal damping; ω_1, ω_2 represent high-order natural frequency and low-order natural frequency; k, μ, G, Ω represent modal coupling coefficient, nonlinear coefficient, excitation signal amplitude and excitation frequency respectively; ε is a small perturbation parameter.

L_1 can be regarded as a fixed beam with two ends, and its natural frequency can be determined by the equation (3) [26]:

$$\omega_1^2 = \frac{48EI}{(\frac{48m_l}{\pi^4}+m_z)l^3} \quad (3)$$

L_2 can be regarded as a cantilever, and its natural frequency can be determined by the equation (4) [26]:

$$\omega_2^2 = \frac{3EI}{(0.24m_l+m_z)l^3} \quad (4)$$

The multi-scale method [27-30] can be used to approximate this nonlinear dimensionless modal coupled motion equation of invert resonant beam, and finally, we can obtain the following simplified system of differential equations (5.1-5.4):

$$\frac{1}{2}k\mu\omega_2^2\alpha_x^2\cos\varphi_1 - \omega_1\alpha_y\sigma - \frac{1}{2}G\cos\varphi_2 = 0 \quad (5.1)$$

$$\frac{1}{2}k\mu\omega_2^2\alpha_x^2\sin\varphi_1 + \xi_1\omega_1\alpha_y + \frac{1}{2}G\sin\varphi_2 = 0 \quad (5.2)$$

$$\frac{1}{4}k\omega_1^2\alpha_x\alpha_y\cos\varphi_1 - \frac{1}{2}\omega_2\alpha_x(\sigma + \sigma_1) = 0 \quad (5.3)$$

$$\frac{1}{4}k\omega_1^2\alpha_x\alpha_y\sin\varphi_1 - \xi_2\omega_2\alpha_r = 0 \quad (5.4)$$

The approximate solution of the final periodic solution can be expressed as follows:

$$x = \alpha_x \cos\left(\frac{\Omega t+\varphi_1+\varphi_2}{2}\right) + O(\varepsilon) \quad (6.1)$$

$$y = \alpha_y \cos(\Omega t + \varphi_2) + O(\varepsilon) \quad (6.2)$$

Where, α_x is the low frequency modal response amplitude, α_y is the high frequency modal response amplitude, and φ_1, φ_2 are the reference phase.

B. Numerical simulation results

Based on the multiscale method, the simplified partial differential equations of the modal coupling motion equations are achieved. Using numerical iterative algorithm and MATLAB, we can carry out the numerical simulation analysis. Specific variable Settings and parameter Settings are shown in TABLE II and TABLE III.

As shown in Fig. 6, the results of theoretical study and simulation study are consistent with the experimental phenomenon. In addition, we also found that the frequency domain edge of the response of the vertical beam has jump and mutation phenomenon, and the amplitude outside the frequency domain of the response is almost zero, these are beneficial to realize quite efficient band-pass filtering.

TABLE I.   VARIABLE SETTINGS

The variable name	MATLAB program variable Settings	
	Partial differential equations	MATLAB
Low frequency modal amplitude	α_x	a1
High frequency modal amplitude	α_y	a2
Phase 1	φ_1	c1
Phase 2	φ_2	c2

TABLE II.   PARAMETER SETTINGS

The Parameter name	MATLAB program parameter Settings	
	Partial differential equations	MATLAB
High-order modal frequency	ω_1	w1
Low-order modal frequency	ω_2	w2
High frequency modal damping	ξ_1	s1
Low frequency modal damping	ξ_2	s2
Modal coupling coefficient	k	k
Nonlinear coefficient	μ	u
Excitation signal amplitude	G	G

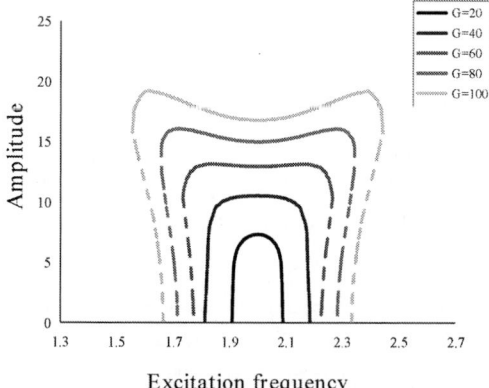

Figure 7. Low frequency modal response characteristics of invert T-shaped resonant structure.

According to Fig. 6, the amplitude of excitation signal will affect the response amplitude and frequency domain bandwidth of the vertical beam (low frequency modal of the inverted T-shaped resonant structure), which may affect its performance in band-pass filtering.

Therefore, in the application of high frequency narrowband filtering with inverted T-shaped resonant beam structure, the amplitude of the filtering signal should be kept constant as far as possible to obtain a stable filter band. Other methods to overcome this problem can be studied in the future.

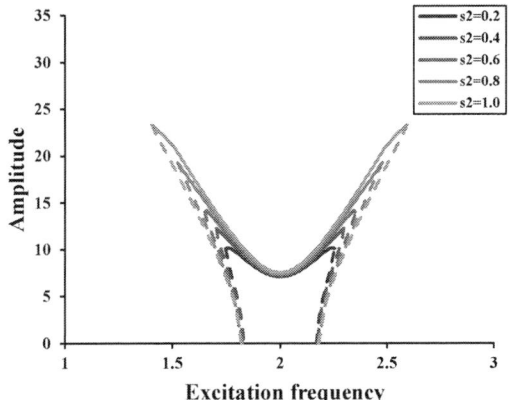

Figure 8. Low frequency modal outputs with different low frequency modal dampers

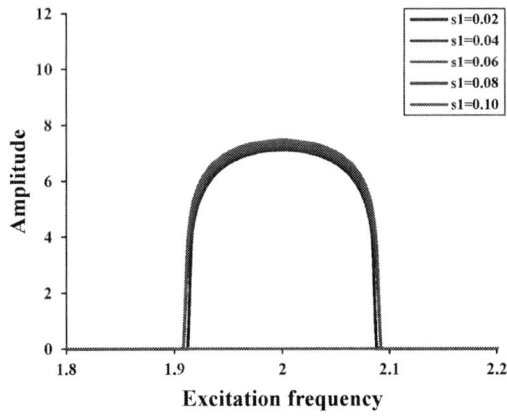

Figure 9. Low frequency modal outputs with different high frequency modal dampers

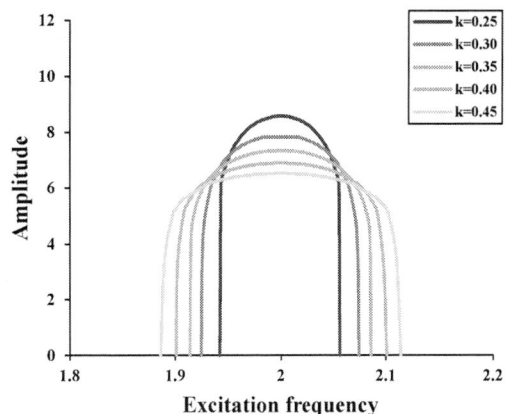

Figure 10. Low frequency modal outputs with different modal coupling coefficients

As shown in Fig. 7, 8 and 9, we systematically studied the relationship between the low frequency modal output characteristics and the parameters related to the inverted T-shaped resonant structure, including the low frequency modal damping, the high frequency modal damping and the modal coupling coefficient.

Through simulation study, we found that:

(1) The damping of the low frequency modal (vertical cantilever) has a significant effect on the output characteristics of the low frequency modal of the resonator, small damping changes can cause the output characteristics of the resonator to deviate significantly from those required.

(2) The influence of high frequency modal damping (horizontal beam) on the output characteristic of low frequency modal of resonator is not very obvious, damping change just can cause a slight change in the output amplitude of the resonator. Moreover, the high frequency modal damping seems to be more related to the threshold excitation amplitude of the internal resonance response.

(3) The change of the modal coupling coefficient has obvious effect on the low frequency modal output characteristics of the resonator. However, with the increase of modal coupling coefficient, the passband bandwidth increases, while the output amplitude of low frequency modal decreases, so the coupling strength of coupling mode should be designed reasonably to meet the filtering requirements.

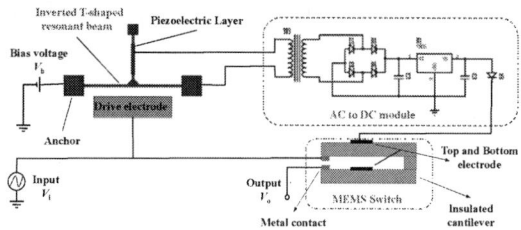

Figure 11. Filtering system integrated with MEMS switches. In this system, the low frequency modal signal is input into the AC to DC module, which generates a voltage stabilized electrical signal to control the on or off of MEMS switch, this filtering system is suitable for single frequency signal filtering.

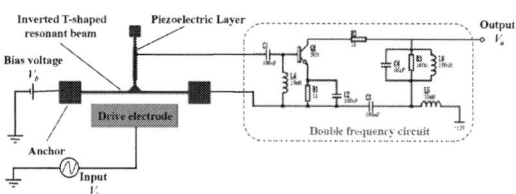

Figure 12. Filtering system integrated with Double Frequency Circuit. In this system, when the effective electrical signal passes through the inverted T-shaped resonant structure, the vertical cantilever (low frequency modal) outputs the electrical signal with half the original frequency, and then the signal is input to the frequency doubling circuit for frequency doubling processing, finally the effective filtering signal we need can be obtained.

V. DISCUSSION

It should be noted that the low frequency vibration output of the inverted T-shaped resonant structure is half of the frequency of the excitation signal, that is to say, after the filter signal passes through the resonator, the frequency of the effective signal has been changed to half of the original one. In addition to the special filter which needs one half octave, it is also necessary to design a suitable post-processing module to realize the conventional high frequency narrowband filtering. As shown in Fig. 10 and 11, two possible filtering systems combined with MEMS switches and double frequency circuits are tentatively proposed, and their feasibility remains to be researched in the future.

VI. CONCLUSION

In conclusion, this paper proposes a novel high-frequency narrow-band filtering mechanism based on the nonlinear internal resonance characteristic of the inverted T-shaped resonant structure. Firstly, from the experimental level, it is preliminarily verified that the internal resonance characteristics of inverted t-shaped resonant beam structure have the potential to be applied to high-frequency narrowband filtering. Secondly, we systematically study the influence of the parameters related to the inverted T-shaped resonant structure on the filter characteristics of the resonator, which is instructive to design a reasonable resonator structure for obtaining better filtering function.

ACKNOWLEDGMENT

This work was supported by National Natural Science Foundation of China (NSFC, Grant No. 51975250 & Grant No. 51675229). This work was also supported by Free Exploration Key Project of Jilin Natural Science Foundation (NSFJ, Grant No. 2020122366JC), Scientific Research Foundation for Leading Professor Program of Jilin University (Grant No. 419080500171 & No. 419080500246), as well as Graduate Innovation Fund of Jilin University (Grant No. 101832020CX101).

REFERENCES

[1] B. Jeong, H. Cho, H. Keum, S. Kim, D. M. Mcfarland, L.A. Bergman, W.P. King, and A.F. Vakakis, "Complex nonlinear dynamics in the limit of weak coupling of a system of microcantilevers connected by a geometrically nonlinear tunable nanomembrane," *Nanotechnology*, vol. 25, no. 46, pp. 465501, Oct. 2014.

[2] I. S. Arroyo, and H. D. Zanette, "Duffing revisited: phase-shift control and internal resonance in self-sustained oscillators," *The European Physical Journal B*, vol. 89, no. 1, pp. 12, Jan. 2016.

[3] S. M. Pourkiaee, S. E. Khadem, M. Shahgholi, and S. Bab, "Nonlinear modal interactions and bifurcations of a piezoelectric nanoresonator with three-to-one internal resonances incorporating surface effects and van der Waals dissipation forces," *Nonlinear Dynamics*, vol. 88, no. 3, pp. 1785-1816, May. 2017.

[4] C. Chen, D. H. Zanette, D. A. Czaplewski, S. Shaw, and D. López, "Direct observation of coherent energy transfer in nonlinear micromechanical oscillators," *Nature Communications*, vol. 8, pp. 15523, May. 2017.

[5] S. Ilyas, K. N. Chappanda, and M. I. Younis, "Exploiting nonlinearities of micro-machined resonators for filtering applications," *Applied Physics Letters*, vol. 110, no. 25, pp. 253508, Jun. 2017.

[6] J. Chaste, A. Eichler, J. Moser, G. Ceballos, R. Rurali, and A. Bachtold, "A nanomechanical mass sensor with yoctogram resolution," *Nature nanotechnology*, vol. 7, no. 5, pp. 300-303, May. 2012.

[7] G. Zheng, X. Du, C. Xia, S. Wan, X. Wang, and D.F. Wang, "Oscillation in coupled resonator systems: Part IV - study on 1:3 internal resonance applicable to sensor devices of high sensitivity," in 20th Design, Test, Integration & Packaging of MEMS and MOEMS, Roma, Italy, 2018, pp. 141–144.

[8] X. Du, D. F. Wang, C. Xia, S. Isao, and R. Maeda, "Internal resonance phenomena in coupled ductile cantilevers with triple frequency ratio–part I: experimental observations," IEEE Sensors Journal, vol. 19, pp. 5475–5483, Jul. 2019.

[9] X. Du, D. F. Wang, C. Xia, I. Shimoyama, and R. Maeda, "Internal resonance phenomena in coupled ductile cantilevers with triple frequency ratio-part II: a mass sensitivity amplification schemes," IEEE Sensors Journal, vol. 19, pp. 5484–5492, Jul. 2019.

[10] C. Xia, D. F. Wang, T. Ono, T. Itoh, and R. Maeda, "A mass multi-warning scheme based on one-to-three internal resonance," Mechanical Systems and Signal Processing, vol. 142, 106784, Mar. 2020.

[11] M. A. A. Hafiz, L. Kosuru, and M. I. Younis, "Microelectromechanical reprogrammable logic device," *Nature Communications*, vol. 7, pp. 11137, Mar. 2016.

[12] A. M. Fennimore, T. D. Yuzvinsky, W. Han, M. S. Fuhrer, J. Cumings, and A. Zettl, "Rotational actuators based on carbon nanotubes," *Nature*, vol. 424, no. 6947, pp. 408-410, Jul. 2003.

[13] D. F. Wang, C. Xia, X. Du, G. Zheng, X. Liu, and G. Liu, "Synchronized Cu-cantilever structure for kinetic friction characterization." IEEE Sensors Journal, vol. 18, no. 18, pp. 7375-7382, Sep. 2018.

[14] Q. Shi, T. Wang, and C. Lee, "MEMS based broadband piezoelectric ultrasonic energy harvester (PUEH) for enabling self-powered implantable biomedical devices," *Scientific reports*, vol. 6, pp. 24946, Apr. 2016.

[15] Y. Jia, S. Du, E. Arroyo, and A.A. Seshia, "Autoparametric resonance in a piezoelectric MEMS vibration energy harvester," in

Proc. 31st IEEE International Conference on Micro Electro Mechanical Systems, Belfast, 2018, pp. 226-229.

[16] J. Xu, and J. Tang, "Multi-directional energy harvesting by piezoelectric cantilever-pendulum with internal resonance," *Applied Physics Letters*, vol. 107, no. 21, pp. 213902, Nov. 2015.

[17] L. Chen, W. Jiang, M. Panyam, and M. F. Daqaq, "A broadband internally resonant vibratory energy harvester," *Journal of Vibration and Acoustics*, vol. 138, no. 6, pp. 061007,Dec. 2016.

[18] L. Xiong, L. Tang, and B. R. Mace, "A comprehensive study of 2:1 internal-resonance-based piezoelectric vibration energy harvesting," *Nonlinear Dynamics*, vol. 91, no. 3, pp. 1817-1834, Feb. 2018.

[19] Randi. Potekin, S. Dharmasena, H. Keum, X. Jiang, J. Lee, "Multi-frequency Atomic Force Microscopy Based on Enhanced Internal Resonance of an Inner-Paddled Cantilever," *Sensors and Actuators A: Physical*, vol. 273, pp. 206-220, Apr. 2018.

[20] L. Lin, R.T. Howe, and A.P. Pisano, "Microelectromechanical filters for signal processing," *Journal of Microelectromechanical systems*, vol. 7, no. 3, pp. 286-294, Sep. 1998.

[21] F. D. Bannon, J. R. Clark, and C. T. C. Nguyen, "High-Q HF microelectromechanical filters," *IEEE Journal of solid-state circuits*, vol. 35, no. 4, pp. 512-526, Apr. 2000.

[22] Y. Yu, A. C. Wong, and T. C. Nguyen, "VHF free-free beam high-Q micromechanical resonators," *Journal of Microelectromechanical Systems*, vol. 9, no. 3, pp. 347-360, Sep. 2000.

[23] J. F. Rhoads, S. W. Shaw, K. L. Turner, and R. Baskaran, "Tunable microelectromechanical filters that exploit parametric resonance," *Journal of Vibration and Acoustics*, vol. 127, no. 5, pp. 423-430, Oct. 2005.

[24] Y. Jia, and A. A. Seshia, "An auto-parametrically excited vibration energy harvester," *Sensors and Actuators A: Physical*, vol. 220, pp. 69-75, Dec. 2014.

[25] Y. Jia, J. Yan, K. Soga, and A. A. Seshia, "Parametrically excited MEMS vibration energy harvesters with design approaches to overcome the initiation threshold amplitude," *Journal of Micromechanics and Microengineering*, vol. 23, no. 11, pp. 114007, Nov. 2013.

[26] M. Cartmell, "Introduction to linear, parametric, and nonlinear vibrations," *Journal of Engineering Mechanics*, vol. 119, no. 3, pp. 642, Mar. 1993.

[27] B. Jeong, C. Pettit, S. Dharmasena, H. Keum, J. Lee, J. Kim, S. Kim, D. M. Mcfarland, L. A. Bergman, and A. F. Vakakis, "Utilizing intentional internal resonance to achieve multi-harmonic atomic force microscopy," *Nanotechnology*, vol. 27, no. 12, pp. 125501, Mar. 2016.

[28] D. H. Zanette, "Effects of noise on the internal resonance of a nonlinear oscillator," *Scientific Reports*, vol. 8, no. 1, pp. 5976, Apr. 2018.

[29] M. M. Kamel, "Bifurcation analysis of a nonlinear coupled pitch–roll ship," *Mathematics and Computers in Simulation*, vol. 73, no. 5, pp. 300-308, Jan. 2007.

[30] A. Sarrafan, B. Behraad, and G. Farid, "Development and characterization of an H-shaped microresonator exhibiting 2: 1 internal resonance," *Journal of Microelectromechanical Systems*, vol. 26, no. 5, pp. 993-1001, Oct. 2017.

Proceedings of the 16th Annual IEEE International
Conference on Nano/Micro Engineered and Molecular Systems
April 25-29, 2021

Calibrate Silicon Nanowires Field Effect Transistor Sensor with its Photoresponse

Yi Yang, Yuelin Wang, Shixing Chen*, and Tie Li*

Abstract— Silicon nanowires (SiNWs) field effect transistor (FET) sensor has been considered to be one of the most promising candidates in the low cost detection of biological and chemical molecules. However, because of silicon nanowires' size distribution, the responses to the same sample of different SiNWs FET sensors show obvious fluctuation. Such problem brings great obstacles to used SiNWs FET sensors in high-precision detection. Here, we proposed a new strategy to calibrate SiNWs FET sensors by using its photoresponse. The foundation of this method is the similar relationship between SiNWs FET sensor response to the target and its own response to the illumination. The validity of this method were confirmed through the results of experiments that calibrated SiNWs FET sensor can accurately get the concentration results of the target in serum samples from patients with breast cancer. This work represents an important step toward the using of SiNWs FET sensor for accurate quantitative biochemical sensing and is expected to open up exciting opportunities for widely clinical application in the future.

Index Terms—Silicon nanowires (SiNWs), calibration, photoresponse, target response

I. INTRODUCTION

It is a huge demand for the high-precision detection of biochemical molecules, such as tumor biomarkers[1, 2] and volatile organic compounds (VOCs)[3]. Thus, a variety of biochemical sensors play an irreplaceable role during recognizing the important targets because of their advantages in miniaturization[4], sensitivity[5], cost[6] and response speed[7].

A considerable of biochemical sensors are based on the principle of field effect transistor (FET), including graphene FET sensor[8], carbon nanotubes FET sensor[9], zero dimensional quantum dot FET sensor[10], one dimensional SiNWs FET sensor[11], two dimensional nanoribbons FET sensor[12] and a large quantity of thin film FET sensors[13]. Compared with the SiNWs FET sensor, almost all those FET sensors are not compatible with standard complementary mental oxide semiconductor (CMOS) process. Thus SiNWs FET sensors are the most suitable for detection of biological and chemical molecules for its almost perfect ability to be batch manufactured. However, due to silicon nanowires'

size distribution, several SiNWs FET sensors have different response to the same target and there is no useful method to calibrate SiNWs FET sensor[14-19].

In this paper, we proposed an innovative strategy in which SiNWs FET sensors' response to the illumination can be used to calibrate its own response to the target, as shown in Figure 1a. Due to the non-contact characteristics of photoresponse measurement, this calibration method has advantages of being pollution-free to SiNWs FET sensors. The rationality of this strategy is the relationship between SiNWs FET sensors' responses to the target and illumination. Experiments were designed to prove the validity of this new approach to the calibration of SiNWs FET sensors. Moreover, we utilized SiNWs FET sensor, calibrated by its own photoresponse, to detect breast cancer marker and the detection error was small.

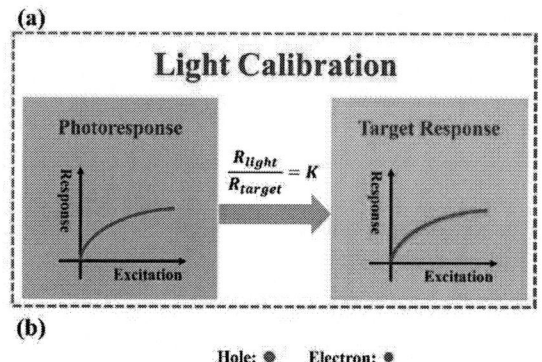

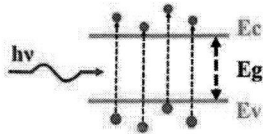

Figure. 1. (a) The diagram of calibration of SiNWs FET sensor's target response using its own photoresponse. (b) The schematic of photo generated carriers.

II. MODELING AND THEORETICAL ANALYSIS

SiNWs FET sensor is extremely sensitive to the illumination and the fundamental principle is that the electron in valence band absorbed the energy of an incident

* This work was supported by the National Key Research and Development Program of China under No.2017YFB0405403, 2017YFA0207100, and National Natural Science Foundation of China under No 31900937 (Corresponding author: Shixing Chen, Tie Li.).
Yi Yang is with the Science and Technology on Micro-system Laboratory, Shanghai Institute of Microsystem and Information Technology, Chinese Academy of Sciences, Shanghai 200050, China, and

also with the University of Chinese Academy of Sciences (UCAS), Beijing 100190, China. (e-mail: hanshu@mail.sim.ac.cn)
Yuelin Wang, Shixing Chen and Tie Li are with the Science and Technology on Microsystem Laboratory, Shanghai Institute of Microsystem and Information Technology, Chinese Academy of Sciences, Shanghai 200050, China. (e-mail: sxchen@mail.sim.ac.cn, tli@mail.sim.ac.cn)

978-1-6654-3008-1/21 $31.00 © 2021 IEEE

photon and was excited into conduction band, leaving unpaired hole in valence band. The schematic of photo generated carriers is shown in Figure. 1b.

The photocurrent in silicon nanowire can be given by[20, 21]:

$$I_{light} = nqAv = q\alpha\beta\tau_0\tau_t \cdot \frac{P_t}{1 + \left(\frac{P_t}{P_0}\right)^k} \qquad (1)$$

where q is the elementary charge, α is the absorption coefficient of silicon nanowire, β is quantum yield or the number of electron hole pairs excited by every single photon, τ_0 is the lifetime of carrier when the interface traps were fully occupied, τ_t is the lifetime of carrier at time t, k is the attenuation factor of carrier's lifetime, P_0 is the incident optical power when the interface traps were fully occupied, and P_t is the incident optical power at time t. The formula 1 can be simplified appropriately into[22-25]:

$$I_{light} = c \cdot P^k \quad k\epsilon(0,1) \qquad (2)$$

where c is a simplified coefficient.

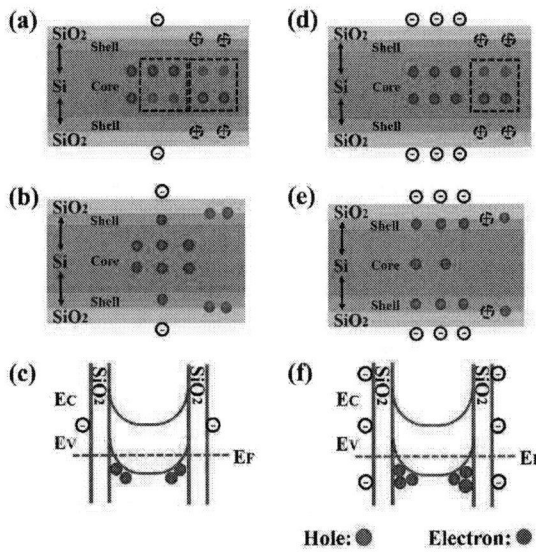

Hole: ●   Electron: ●

Figure. 2. (a) The schematic of generated electron hole pairs in silicon nanowire during the process of photoresponse. (b) The schematic of the separation of electrons and holes during the process of photoresponse. (c) The band diagram of SiNWs FET sensors under illumination. (d) The schematic of generated electron hole pairs in silicon nanowire during the process of target response. (e) The schematic of the separation of electrons and holes during the process of target response. (f) The band diagram of SiNWs FET sensors in target detection.

The process of SiNWs FET sensors' photoresponse is depicted in Figure. 2a, Figure. 2b and Figure. 2c. Electron hole pairs are generated inside of silicon nanowire, which is shown in Figure. 2a. And in Figure. 2b, carriers captured by interface traps and attracted by surface charge, which plays a very important role in SiNWs FET sensors' photoresponse. Because of the separation of electrons and holes, "core and shell" structure appeared inside silicon nanowire. Such "core and shell" structure leads to the increasing of carriers' lifetime and larger photo current.

Holes are accumulated and the layer of charge is formed on the surface of silicon nanowire. This charge layer changes the surface potential of silicon nanowire, which is equal to applying a gate voltage to SiNWs FET sensor. Because of the existence of negative charge on surface of silicon nanowire, valence band and conduction band bends upward at the interface. The band diagram of SiNWs FET sensor under illumination is shown in Figure. 2c.

The "core and shell" structure performs a significant role as well as in target response. Thus, the target response of SiNWs FET sensor is similar to its own photoresponse on the basic principle. Difference between target response and photoresponse is that target response has more fixed charge on top of oxide layer and photoresponse has larger amount of electron hole pairs inside silicon nanowire. The process of SiNWs FET sensors response to target and corresponding band diagram is shown in Figure. 2d, Figure. 2e and Figure. 2F. Based on similar electron transport process, we proposed that the formulas of SiNWs FET sensors response to target and response to illumination have the similar form and some quantitative relationship can be built up between them.

III. Experiments and Result Discussion

In order to verify the correctness of above suppose, ingeniously designed experiments were carried out. Firstly, we modified the surface of SiNWs FET sensors, which is necessary for specific detection of the target. Secondly, gradually increasing light was applied to the surface of SiNWs FET sensor and the output current was recorded by computer. Finally, we tested the target response of SiNWs FET sensor, the signal rising with concentration of the target increased, and the output current was also recorded. The results of these experiments displayed in Figure. 3a. The curve of SiNWs FET sensor's response to target is almost coincident with its own photoresponse. Such coincidence is a strong evidence that the target response of SiNWs FET sensor has similar form to the photoresponse and can be given by:

$$I_{target} = m \cdot C^n \quad n\epsilon(0,1) \qquad (3)$$

where m is a simplified coefficient, C is the concentration of target, and n is an attenuation factor of carrier's lifetime in target response. We also used this formula to fit the experimental data of SiNWs FET sensor's response to the target and comparison between the fitting results of photoresponse and target response was carried. These two fitting results are listed in Table 1 and the fitting error of target response is almost the same size as the fitting error of photoresponse, which demonstrates the rationality of this formula's application in SiNWs FET sensor's target response.

Table 1. Fitting results of photoresponse and target response

$I_{light} = c \cdot P^k$			$I_{target} = m \cdot C^n$		
	Value	Error		Value	Error
c	1.36438	0.09327	m	0.28797	0.00788
k	0.33342	0.01572	n	0.35826	0.0194

We repeated above experiments using different SiNWs FET sensors for many times and recorded every devices' responses to targets and illumination respectively. Based on recorded data, the value of photoresponse and target response for SiNWs FET sensors can be calculated out and listed in Table 2. We also computed the ratio of every SiNWs FET sensor's photoresponse to its own target response, and put all these devices' ratio in a bar chart. The number of devices in each cell is counted and corresponding normalized probability of each cell is calculated. Such normalized probability follows normal distribution, which is shown in Figure. 3b. The mathematical expectation of this normal distribution is 27.8 and the most appropriate value of the ratio of sensor's photoresponse to its own target response is 27.8. Conclusion can be described by the following formula:

$$\frac{R_{l1}}{R_{t1}} = \frac{R_{l2}}{R_{t2}} = \cdots = \frac{R_{ln}}{R_{tn}} = K \qquad (4)$$

where K is a constant, R_l is the photoresponse of SiNWs FET sensor, R_t is the target response of SiNWs FET sensor.

Table 2. The value of the ratio of SiNWs FET sensor's photoresponse and target response.

Number	Photoresponse	Target Response	Ratio
1	13.86	0.501	27.67
2	18.84	0.661	28.5
3	35.956	1.273	28.24
4	15.12	0.538	28.104
5	19.36	0.731	26.48
6	14.33	0.5301	27.03
7	13.01	0.44	29.56
8	11.71	0.487	24.04
9	10.059	0.34	29.41
10	10.28	0.36	28.56
11	9.64	0.328	29.39
12	13.02	0.51	25.53
13	8.54	0.29	29.45
14	16.40	0.603	27.2

Therefore, if the experimental conditions remain exactly the same, the value of SiNWs FET sensor's photoresponse is proportional to the value of device's target response. Thus, photoresponse of SiNWs FET sensor can be a powerful tool to characterize sensor's ability of signal conversion and normalize its own target response.

Based on above theoretical analysis and experimental results, we proposed a method of calibration of SiNWs FET sensor's target response using its own photoresponse. The specific calibration procedure was as follows: firstly, we recorded the response signals of several SiNWs FET sensors to the same target and the curve of every SiNWs FET sensor's photoresponse; secondly, calculated the value of each sensor's photoresponse, normalized its own target response and made the standard curve of target response; thirdly, used the formula 3 to fit the standard curve of target response and obtain the fitting parameters; finally, tested the photoresponse of a new SiNWs FET sensor, and used this photoresponse to normalize sensor's target response, and then substituted the normalized target response into the

fitting result to calculate the real concentration of target.

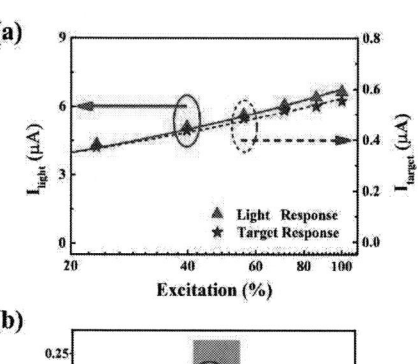

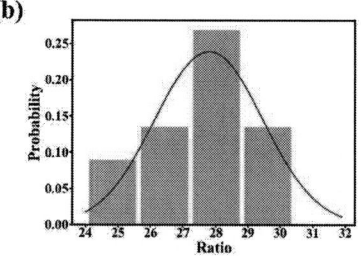

Figure. 3. (a) The curves of SiNWs FET sensor's photoresponse and target response and the fitting results of these two kinds of response. (b) The bar chart of the ratio of SiNWs FET sensor's photoresponse to target response and its normalized normal distribution curve.

Follow above calibration process, we designed experiments to detect carcinoembryonic antigen (CEA), a tumor marker of breast cancer, in serum solution of patients with breast cancer using SiNWs FET sensors. Three SiNWs FET sensors were randomly selected to detect standard sample, which was prepared by adding a fixed amount of target into 100 times diluted normal human serum and then diluted into different concentration. These three SiNWs FET sensors' responses to the standard target were recorded and shown in Figure. 4a. Those three responses were normalized by their own photoresponse and the curves of normalized target responses are shown in Figure. 4b. Then we obtained the standard curve of SiNWs FET sensor's target response, which is shown in Figure. 4c. The data were fitted with formula 3 and the result of data fitting is given by:

$$I_{target} = 0.006434 \cdot C^{0.2954} \qquad (5)$$

Finally, we randomly selected a SiNWs FET sensor and tested its photoresponse and target response to detect CEA in diluted serum from patient with breast cancer. The value of relative variation of current was calculated from the output current of this SiNWs FET sensor and substituted into formula 5. The concentration of CEA in this sample can be calculated out. A comparison between the test data of our SiNWs FET sensor and the clinical data was shown in Figure. 4d.

Using this calibration method, SiNWs FET sensor can detect the target quantitatively, which is a pivotal step forward for real usage of SiNWs FET sensors. The result of SiNWs FET sensor's detection is 368 ng/mL, and the test result given by clinical methods is 355 ng/mL. Detection error is so small that demonstrates our calibration method is useful which can make our SiNWs FET sensor's quantitative detection much better than before.

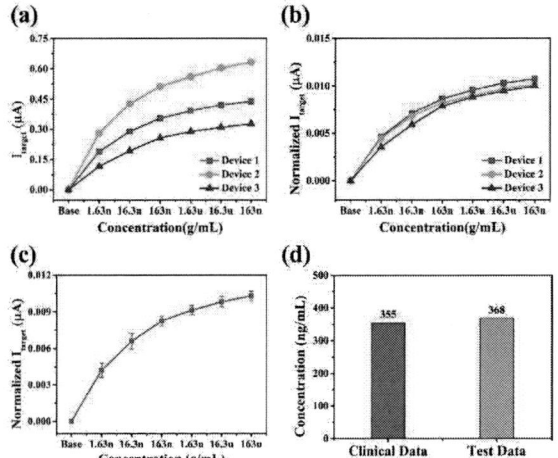

Figure. 4. (a) Three SiNWs FET sensors' relative variation of output current to standard target. (b) Three SiNWs FET sensors' normalized relative variation of output current to standard target. (c) The standard curve of SiNWs FET sensor's response to standard target. (d) The result of a comparison between the test data of our SiNWs FET sensor and the clinical data.

IV. CONCLUSION

In this paper, we proposed that SiNWs FET sensor's photoresponse can be used to calibrate its own target response. In order to verify the validity of this calibration method, we firstly put forward that the target response and the photoresponse of SiNWs FET sensor have similar response formula from the perspective of theoretical model. The results of designed experiments proved the existence of such similarity in qualitative and quantitative respect respectively. Successful application of this calibration method in the detection of breast cancer marker in real test environment and very small detection error indicated the feasibility and efficiency of this calibration method using in large-scale early screening of biomarkers. Such efficient calibration method promoted an excellent solution to the problem of SiNWs FET sensor's quantitative detection, which will promote the wide application in the detection of biological and chemical molecules in the future.

REFERENCES

[1] D. F. Hayes, V. R. Zurawski, Jr., and D. W. Kufe, "Comparison of circulating CA15-3 and carcinoembryonic antigen levels in patients with breast cancer," *J Clin Oncol*, vol. 4, no. 10, pp. 1542-50, Oct 1986.

[2] B. Liu *et al.*, "Tumor-suppressing roles of miR-214 and miR-218 in breast cancer," *Oncol Rep*, vol. 35, no. 6, pp. 3178-84, Jun 2016.

[3] A. Mirzaei, S. G. Leonardi, and G. Neri, "Detection of hazardous volatile organic compounds (VOCs) by metal oxide nanostructures-based gas sensors: A review," *Ceramics International*, vol. 42, no. 14, pp. 15119-15141, 2016.

[4] A. R. Gao *et al.*, "Enhanced Sensing of Nucleic Acids with Silicon Nanowire Field Effect Transistor Biosensors," *Nano Letters*, vol. 12, no. 10, pp. 5262-5268, Oct 2012.

[5] J. Li, S. Pud, M. Petrychuk, A. Offenhausser, and S. Vitusevich, "Sensitivity Enhancement of Si Nanowire Field Effect Transistor Biosensors Using Single Trap Phenomena," *Nano Letters*, vol. 14, no. 6, pp. 3504-3509, Jun 2014.

[6] A. Agarwal, K. Buddharaju, I. K. Lao, N. Singh, N. Balasubramanian, and D. L. Kwong, "Silicon nanowire sensor array

using top-down CMOS technology," *Sensors and Actuators a-Physical*, vol. 145, pp. 207-213, Jul-Aug 2008.

[7] A. Kim *et al.*, "Ultrasensitive, label-free, and real-time immunodetection using silicon field-effect transistors," *Applied Physics Letters*, vol. 91, no. 10, Sep 3 2007.

[8] V. Singh, D. Joung, L. Zhai, S. Das, S. I. Khondaker, and S. Seal, "Graphene based materials: Past, present and future," *Progress in Materials Science*, vol. 56, no. 8, pp. 1178-1271, 2011.

[9] R. H. Baughman, A. A. Zakhidov, and W. A. de Heer, "Carbon nanotubes--the route toward applications," *Science*, vol. 297, no. 5582, pp. 787-92, Aug 2 2002.

[10] C. A. Mirkin, R. L. Letsinger, R. C. Mucic, and J. J. Storhoff, "A DNA-based method for rationally assembling nanoparticles into macroscopic materials," *Nature*, vol. 382, no. 6592, pp. 607-9, Aug 15 1996.

[11] G. F. Zheng, X. P. A. Gao, and C. M. Lieber, "Frequency Domain Detection of Biomolecules Using Silicon Nanowire Biosensors," *Nano Letters*, vol. 10, no. 8, pp. 3179-3183, Aug 2010.

[12] Z. Zeng *et al.*, "Single-layer semiconducting nanosheets: high-yield preparation and device fabrication," *Angew Chem Int Ed Engl*, vol. 50, no. 47, pp. 11093-7, Nov 18 2011.

[13] D. J. Lipomi *et al.*, "Skin-like pressure and strain sensors based on transparent elastic films of carbon nanotubes," *Nat Nanotechnol*, vol. 6, no. 12, pp. 788-92, Oct 23 2011.

[14] N. Elfstrom, R. Juhasz, I. Sychugov, T. Engfeldt, A. E. Karlstrom, and J. Linnros, "Surface charge sensitivity of silicon nanowires: Size dependence," *Nano Letters*, vol. 7, no. 9, pp. 2608-2612, Sep 2007.

[15] J. S. Jie, W. J. Zhang, K. Q. Peng, G. D. Yuan, C. S. Lee, and S. T. Lee, "Surface-Dominated Transport Properties of Silicon Nanowires," *Advanced Functional Materials*, vol. 18, no. 20, pp. 3251-3257, Oct 23 2008.

[16] Y. Kutovyi *et al.*, "Liquid-Gated Two-Layer Silicon Nanowire FETs: Evidence of Controlling Single-Trap Dynamic Processes," *Nano Letters*, vol. 18, no. 11, pp. 7305-7313, Nov 2018.

[17] M. P. Lu, E. Vire, and L. Montes, "Ionic screening effect on low-frequency drain current fluctuations in liquid-gated nanowire FETs," *Nanotechnology*, vol. 26, no. 49, Dec 11 2015.

[18] Y. Paska, T. Stelzner, S. Christiansen, and H. Haick, "Enhanced Sensing of Nonpolar Volatile Organic Compounds by Silicon Nanowire Field Effect Transistors," *Acs Nano*, vol. 5, no. 7, pp. 5620-5626, Jul 2011.

[19] G. D. Yuan *et al.*, "Tunable electrical properties of silicon nanowires via surface-ambient chemistry," *ACS Nano*, vol. 4, no. 6, pp. 3045-52, Jun 22 2010.

[20] C. Soci *et al.*, "ZnO nanowire UV photodetectors with high internal gain," *Nano Letters*, vol. 7, no. 4, pp. 1003-1009, Apr 2007.

[21] T. Yoshino, S. Yokoyama, and T. Fujii, "Effect of light irradiation on native oxidation of silicon surface," *Japanese Journal of Applied Physics Part 1-Regular Papers Short Notes & Review Papers*, vol. 40, no. 4a, pp. 2223-2224, Apr 2001.

[22] M. T. Ahmadi, H. H. Lau, R. Ismail, and V. K. Arora, "Current-voltage characteristics of a silicon nanowire transistor," *Microelectronics Journal*, vol. 40, no. 3, pp. 547-549, Mar 2009.

[23] H. H. Fang and W. D. Hu, "Photogating in Low Dimensional Photodetectors," *Advanced Science*, vol. 4, no. 12, Dec 2017.

[24] N. Guo *et al.*, "Anomalous and Highly Efficient InAs Nanowire Phototransistors Based on Majority Carrier Transport at Room Temperature," *Advanced Materials*, vol. 26, no. 48, pp. 8203-8209, Dec 23 2014.

[25] H. Kind, H. Q. Yan, B. Messer, M. Law, and P. D. Yang, "Nanowire ultraviolet photodetectors and optical switches," *Advanced Materials*, vol. 14, no. 2, pp. 158-+, Jan 16 2002.

Proceedings of the 16th Annual IEEE International
Conference on Nano/Micro Engineered and Molecular Systems
April 25-29, 2021

Flexible and Transparent Ultraviolet Photodetector Enabled by Metal Doping ZnO Nanorods Based on Mica Substrate

Yao Huang[&], Hainan Zhang[&], Guanhua Dun[&], Yuhua Li, Renrong Liang, Yi Yang, Dan Xie, He Tian and Tian-Ling Ren, Senior Member, IEEE

Abstract—Flexible ultraviolet (UV) photodetectors based on Ga-doped and Fe-doped ZnO Nanorods (NRs) were successfully fabricated respectively on mica substrate with superior transparency. In this work, $Zn_{0.987}Ga_{0.013}O$ NRs and $Zn_{0.984}Fe_{0.016}O$ NRs were synthesized using a simple and low-cost hydrothermal method. Meanwhile, the effects of doping elements on the morphology, structure and photoelectric properties of ZnO NRs were systematically studied. The doping of Ga element enhances the UV responsivity of the device reaching 71.44 A/W. While the doping of Fe element is favorable to increase the response speed by 3 times and reduce the dark current to 10^{-10} A. After 100 cycles of bending, the photodetection performance of the device remains almost unchanged, which indicates the device has a good potential for wearable applications.

Index Terms—Flexible ultraviolet photodetector, ZnO nanorods, Ga-doped, Fe-doped, Mica substrate.

I. INTRODUCTION

Nowadays, the ultraviolet photodetector plays an important role in many applications from industry to scientific research, benefiting from its advantages of high detection accuracy and good invisibility. The application of UV photodetector in bending, folding, stretching, or other conditions reveals the increasing demands for flexible devices. Meanwhile, the wearable UV detector has been the new hot spot. The mica substrate not only possesses the good property of flexibility and transparency, but also has a unique advantage of high temperature resistant up to 600 ℃, which is hard to achieve by other flexible substrates. Therefore, the mica substrate has wide potential for flexible UV photodetector [1][2] .

In most cases, zinc oxide (ZnO) has the direct band-gap of 3.37 eV and larger exciton binding energy up to 60 meV [3]. The wide band gap of ZnO corresponds to the ultraviolet band, indicating that it is an extremely competitive material for ultraviolet photodetector. ZnO NRs also have the huge

---

*This work was supported by National Key R&D Program (2016YFA0200400), National Natural Science Foundation (61874065, U20A20168 ,51861145202,11604011), and National Basic Research of China(2015CB352101).

Y.Huang, H. Zhang, G. Dun, R. Liang, D. Xie, H. Tian, Y. Yang and T. Ren are with the Institute of Microelectronics & Beijing National Research Center for Information Science and Technology (BNRist), Tsinghua University, Beijing 100084, China. T. Ren are also with Center for Flexible Electronics Technology, Tsinghua University, Beijing 100084, China.

Y. Li are with the School of Electronic and Information Engineering, Beijing Jiaotong University, Beijing 100044, China.

&These authors contributed equally to this work. Correspondence and requests for materials should be addressed to Tian-Ling Ren (email: RenTL@tsinghua.edu.cn).

---

advantages of high mechanical stability, low cost, simple growth method, non-toxicity, good biocompatibility and safe biochemical properties [4].

To further improve the photoelectric properties of ZnO, the doping is an effective strategy, which can effectively enhance the carrier concentration in the material, increase the conductivity, change the light absorption characteristics, thereby improving the photoelectric performance of the device. R.R.Prabhakar et al. [5] synthesized Mn-doped ZnO NRs by chemical vapor deposition method, and researched the changes of the photophysical properties as the result of Mn doping. H.H.Zhang et al. [6], M.K.Gupta et al. [7] prepared Na-doped and K-doped ZnO NRs and applied them to UV photodetectors.

In this study, Ga and Fe element were chosen for the doping elements. Ga element were incorporated into the ZnO crystal lattice to replace Zn atoms, and acted as n-type donors, resulting an increased electrical conductivity. In addition, the atomic radius (0.062 nm) and ionic radius (0.126 nm) of Ga are closest to those of Zn (0.074 nm, 0.131 nm), and the Ga-O bond length (0.192 nm) is almost equal to the Zn-O bond (0.197 nm) [8], So Ga element is a good choice as for doping. Besides, through doping diluted transition metals, such as Fe [9], Ni [10], Mn [11], Co [12], into the ZnO lattice, the photoelectric characteristics of ZnO can also be changed and RT ferromagnetism can be realized at the same time [13]. Therefore, Fe-doped ZnO is a promising material integrated with semiconductor, optoelectronic and magnetic properties, making it beneficial to spintronic devices and fast photodetector. In this work [14][15], $Zn_{0.987}Ga_{0.013}O$ NRs and $Zn_{0.984}Fe_{0.016}O$ NRs were successfully prepared by a low-cost hydrothermal method [16][17], and ultraviolet photodetectors were also fabricated on flexible mica substrates. The photoelectric characteristic and flexible performance of the ZnO UV photodetector were systematically studied. The doping of Ga or Fe element not only enhanced optoelectric performance of ZnO NRs, the device fabricated in this work also exhibits superior fatigue resistance to bending and transparency, which has great application potential in wearable electronic devices.

II. EXPERIMENTAL DETAILS

A. Device fabrication

All chemicals were of analytical grade and were used without further purification. Ga-doped and Fe-doped ZnO NRs were synthesized on the mica substrate by a low-cost hydrothermal method. The fabrication of the device fabrication contains three steps. Firstly, the mica substrates were cleaned by ultrasonic cleaning in acetone, sequentially washing with isopropanol, alcohol and drying by nitrogen gas. Secondly, the 60 nm thick ZnO seed layer was obtained on a mica substrate through magnetron sputtering with a pristine ZnO target (99.999% purity). After that, the ZnO seed layer was annealed at 300 °C for 40 minutes. Then, 10/50 nm thick layers of Ti/Au were deposited through photolithography, sputtering and exfoliating to form the interdigital electrode contacts. Finally, Ga-doped and Fe-doped ZnO NRs were synthesized on the ZnO seed layer using hydrothermal method. The precursor solution of Ga-doped ZnO NRs was prepared by separately adding 1.1751 g of Zinc nitrate hexahydrate (Zn (NO$_3$)$_2$ • 6H$_2$O), 0.0128 g Gallium nitrate hydrate (Ga (NO$_3$)$_3$ • xH$_2$O) and 0.5608 g of hexamethylenetetramine (C$_6$H$_{12}$N$_4$, HMTA) in 100 mL of deionized water under a magnetic stirrer, corresponding to a concentration of 39.5, 0.5 and 40 mM, respectively. Zinc chloride hexahydrate (ZnCl$_2$ • 6H$_2$O) (39.5 mM), iron (II) chloride tetrahydrate (FeCl$_2$ • 4H$_2$O) (0.5 mM) and HMTA (40 mM) were used as precursors for Fe-doped ZnO NRs. Then the treated mica substrates respectively immersed in the precursor solution were placed in an oven heated at 90 °C. In the end, the resulting samples were washed with isopropanol and deionized water successively.

B. Characterization

The field-emission scanning electron microscopy (FE-SEM, Carl Zeiss Microscopy, Merlin) was used to observe the morphology of the prepared ZnO NRs. The chemical composition of the doped ZnO was characterized by the X-ray Photoelectron Spectroscopy techniques (XPS,250XI). X-ray diffraction (XRD, Bruker D8) was used to identify the crystal structure and crystallographic properties of the material. The optical properties were analyzed by photoluminescence spectra (PL, LABRAM HR) [18]. In this experiment, a Cascade Summit probe station was used to fix the device, which was connected to a semiconductor parameter performance analyzer (Agilent B1500A). The UV photoelectric characteristics were measured using an ultraviolet light source with the wavelength of 365 nm at different illumination intensities.

III. RESULTS AND DISCUSSION

Fig. 1(a) (b) (c) show the top-view FESEM images of undoped, Ga-doped and Fe-doped ZnO NRs, in low magnification and high magnification (inset images), respectively. Three samples exhibited neatly arranged hexagonal pillar arrays with diameters of approximately 80, 50 and 30 nm, respectively. From Fig. 1 (d) (e) (f), it also can be observed that the lengths of undoped and Ga-doped ZnO NRs were both about 700 nm with aligned growth orientation, but the Fe-doped ZnO NRs were relatively messy with smaller length of 500 nm.

For the purpose of verifying the substitutions of the Ga or Fe ions in the ZnO NRs, XPS characterization was performed. Fig. 2(a) shows Zn 2p spectrum of undoped, Ga-doped and Fe-doped ZnO NRs, respectively, which divided into Zn 2p$_{3/2}$ and Zn 2p$_{1/2}$. The O 1s spectra of the three samples is shown in Fig. 2(b). The O 1s spectrum in the undoped ZnO NRs has obvious asymmetric characteristics, indicating the presence of multi-component

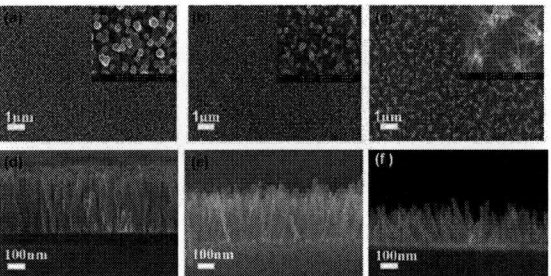

Figure 1. (a)(c)(d) top-view FESEM images of undoped, Ga-doped and Fe-doped ZnO nanorods respectively, (d)(e)(f) cross-view FESEM images of three samples respectively.

oxygen. The peaks in the spectrum are the sum of the three peaks located at 530.5 eV, 531.5 eV and 532.7 eV. The first peak on the low binding energy (530.5 eV) corresponds to the O^{2-} ion in the ZnO NRs. The intermediate binding energy (531.5 eV) peak is originated from oxygen defects, and its intensity represents the concentration of oxygen vacancies. The highest binding energy peak usually corresponds to the oxygen molecules absorbed on the crystal surface. The symmetry of the O 1s spectrum of the doped sample is improved, indicating that the Ga and Fe atoms enter the ZnO lattice to make up for the oxygen vacancies in the sample, thereby reducing defects. Fig. 2 (c)(d) show Ga 2p spectrum of Ga-doped ZnO NRs and Fe 2p spectrum of Fe-doped ZnO NRs, respectively. Ga 2p was divided into Ga 2p$_{3/2}$ and Ga 2p$_{1/2}$. It can be seen from the spectrum that the binding energies at the peak positions are 1118.75 eV and 1145.60 eV, respectively, and the bimodal spacing is 26.8 eV, indicating that Ga element exists as Ga^{3+}. In addition, the binding energies of 710.50 eV and 723.95 eV correspond to the peak positions of Fe 2p$_{3/2}$ and Fe 2p$_{1/2}$, respectively, and the bimodal spacing is 13.40 eV. Because the bimodal spacings of Fe, Fe^{2+}, and Fe^{3+} are 13.1 eV, 13.4 eV, and 13.6 eV, respectively, the iron element exists as Fe^{2+} in the zinc oxide lattice. The element content of the prepared doped ZnO NRs were measured by Energy Dispersive Spectrometer (EDS), which revealed that the percentage of Ga or Fe replacing zinc atoms is 1.3 at% and 1.6 at%,

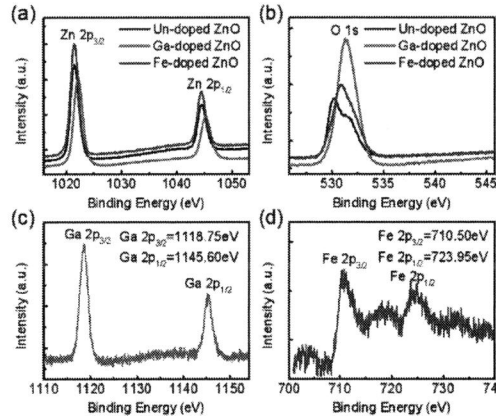

Figure 2. XPS of (a) Zn 2p of undoped, Ga-doped and Fe-doped ZnO nanorods, and (b) O 1s spectrum of three samples, (c) Ga 2p spectra of Ga-doped ZnO nanorods and (d) Fe 2p spectrum of Fe-doped ZnO nanorods.

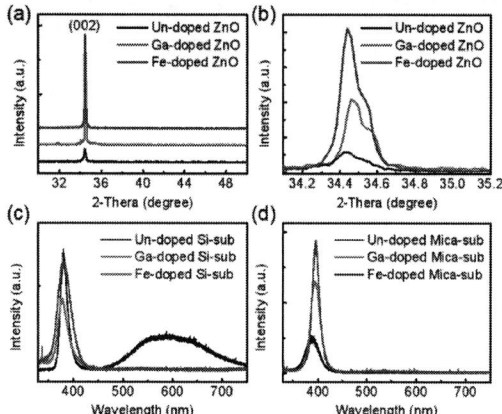

Figure 3. (a) (b) XRD patterns of undoped, Ga-doped and Fe-doped ZnO nanorods. (c) photoluminescence (PL) spectrum of three samoles, in which the black line displays undoped ZnO, red line shows Ga-doped ZnO and blue line is Fe-doped ZnO on silicon substrate. (d) PL of three samoles on mica substrate.

respectively.The doped compounds are $Zn_{0.987}Ga_{0.013}O$ and $Zn_{0.984}Fe_{0.016}O$, basically confirming to the expected doping ratio.

In this experiment, the crystal orientation of ZnO NRs was determined by XRD analyses (Fig. 3(a)). The apparent (002) peak at 34.5° indicated that all fabricated NRs exhibited the hexagonal wurtzite structure. The three samples all grown in a c-axis orientation with excellent single crystal characteristic. These consequences reveal that there are no other impure phases in doping ZnO, such as Ga_2O_3, FeO or Fe_3O_4. In Fig. 3(b), we observed that the peak position of Ga-doped ZnO was shifted towards higher 2θ. It indicated that the Ga^{3+} ions with a radius of 0.062 nm replaced the Zn^{2+} with a radius of 0.074 nm in the ZnO NRs crystal matrix, resulting in the decrease of the lattice size. The result of Fe-doped ZnO was opposite to Ga-doped ZnO, because of the larger radius of Fe^{2+} (0.076 nm).

Fig. 3(c) and (d) show the photoluminescence (PL) spectra of three samples (undoped, Ga-doped and Fe-doped ZnO NRs) on silicon substrate and mica substrate, respectively. The samples exhibit two typical luminescence behaviors, namely ultraviolet (UV) emission and deep-level emission. It is obvious that all types of ZnO NRs exhibit two emission behavior. The emission peak originated from the band-to-band recombination of ZnO is centered at approximately 385 nm. This peak belongs to the UV-band. Relatively, the peak located in the visible-band usually comes from deep-level emissions in result of the oxygen vacancies and zinc interstitials. The deep-level emission intensity of ZnO NRs on the mica substrate is much lower than that of the silicon substrate, indicating that the former haves fewer energy level defects. It is clear that the doped ZnO NRs emit a lower broadband signal in the visible band. This phenomenon is due to the addition of Fe and Ga elements, which reduces the oxygen vacancies.

In our work, high-performance flexible photodetectors based on undoped, Ga-doped and Fe-doped ZnO NRs were prepared on mica substrate of good flexibility and

transparent. The schematic diagram and optical image of the photodetector are shown in Fig. 4(a) and (b), respectively.

To investigate the UV detection performance, we exposed the device to a 365 nm ultraviolet irradiation with the light power density from 0.172 to 18.9 mW/cm^2. The device was fixed at a certain position from the light source. The illumination intensity of UV light can be adjusted by the UV light source generator. Fig. 4(c) (d)and(e) display the current-voltage (I-V) curves of the photodetector based on undoped, Ga-doped and Fe-doped ZnO NRs in the dark and various power ultraviolet irradiation, respectively. In the dark, oxygen molecules are easily attracted to the external surface of ZnO NRs and trap more free electrons from the conduction band of ZnO. The phenomenon causes the loss of electrons on the surface of the ZnO and reducing the carrier concentration in the sample, thereby the device has a smaller dark current. Under UV illumination, photo-induced carriers are generated in ZnO NRs. Under the surface electric field, the holes migrate from the inside to the surface of ZnO, and combine with oxygen ions. The oxygen molecules are released from the surface again. The surface reactions can be described as follow:

$$O_2(g) + 2e^- \rightarrow O_2^{2-}(ad) \qquad (1)$$

$$O_2^{2-}(ad) + 2h^+ \rightarrow O_2(g) \qquad (2)$$

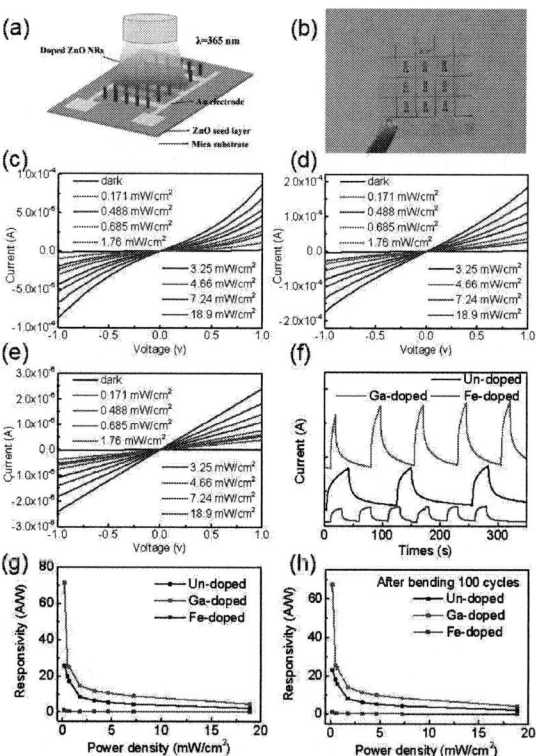

Figure 4.(a) Schematic diagram of the doped ZnO nanorods UV photodetector, (b) optical image of the prepared device, (c) (d) (e) the measured I-V curves of undoped, Ga-doped and Fe-doped ZnO UV photodetector under dark and various UV light power irradiation, respectively, (f) the I-T response of the devices, (g) the responsivity(R) of pristine device, (h) the responsivity(R) of device after 100 cycles bending, respectively.

978-1-6654-3008-1/21 $31.00 © 2021 IEEE

To evaluate the photoelectric characteristic of the device, light responsivity, which is an important parameter, was defined as the following formula:

$$R(A/W) = \frac{I_{ph}}{P_\lambda} = \frac{I_p - I_d}{D \cdot S}$$

I_{ph} is the photocurrent, where I_p and I_d are the currents under ultraviolet light and darkness, respectively. P_λ is the UV light power, and its value is equal to the product of D and S. With the introduction of Ga doping, the UV responsivity reached 71.44 A/W under a 1 V bias, which were improved by approximately 3 times compared with undoped ZnO NRs. Ga doping not only enhances the electron density in the device, but also decreases the lateral depth of the depletion layer. In attention, more oxygen molecules can be attracted to the external surface of Ga-doped ZnO NRs, promoting the transfer of electrons to the surface. As a result, the photoresponse of Ga-doped device is significantly enhanced. The responsivity of the Fe-doped device is relatively small (0.96 A/W under a 1 V bias). However, its application in the field of low-power devices is of great significance. Meanwhile, the current-time (I-T) response of the photodetector was also tested and the results were shown in Fig. 5(f). It's worth noting that the response times of Fe-doped device reduced 3 times.

The UV responsivity of these devices are much larger than commercial photodetectors based on SiC or GaN (0.2 A/W), and are competitive to the results in other reports. The device prepared by Y.Li et al. [19] had a maximum responsivity of 1.22 A/W under a 5V bias. H.I et al. [20] used a catalyst-free gas-solid mechanism to synthesize well-arranged nanorods on Si (100) with a responsivity of 0.22 A/W. Therefore, compared with other research results, the devices prepared in our work had huge advantage.

Furthermore, the UV photodetector on mica substrate also exhibits some favorable properties due to the doping of Ga or Fe element. It can be obviously observed that after 100 cycles bending under 7mm bending radii of curvature, the photoresp. once of the three devices (undoped, Ga-doped and Fe-doped ZnO NRs respectively) have almost unchanged (Fig. 4(h)), indicating that the flexible UV photodetectors have good fatigue resistance to bending. In addition, benefit from the good flexible properties of mica substrate, the largest bending loading goes far more than the above. Moreover, because of fewer optical absorption and reflection in the visible band, the mica substrate used in this work has excellent transparency (Fig. 4(b)), which enables the device to be applied for the transparent integrated optoelectronics.

IV. CONCLUSION

In summary, Ga-doped and Fe-doped ZnO NRs were successfully synthesized using a low-cost hydrothermal method, and the percentage of each element replacing zinc atoms are 1.3 at% and 1.6 at%, respectively. The XPS results suggested that Ga^{3+} and Fe^{2+} were incorporated into the ZnO lattice, resulting a lower deep-level emission peak in the visible band from PL spectra. Moreover, ultraviolet photodetectors based on undoped, Ga-doped and Fe-doped

ZnO nanorods were fabricated respectively on the mica substrate. Due to the superior flexible and bendable properties of mica substrate, the photodetector can be fitted to surfaces of different shapes for various applications. In addition, the doping of Ga element can help to obtain a large responsivity to 71.44 A/W, and the doping of Fe element is favorable to increase the response speed by 3 times and reduce dark current. By optimizing the doping condition and the device fabrication process, metal-doped ZnO NRs have great potential for future UV photodetectors with high performance, low cost and flexibility.

REFERENCES

[1] Xia, Younan, Peidong Yang, Yugang Sun, Yiying Wu, Brian Mayers, Byron Gates, Yadong Yin, Franklin Kim, and Haoquan Yan, "One-Dimensional Nanostructures: Synthesis, Characterization, and Applications," Advanced Materials, 15 (5): 353–89, Mar. 2003.

[2] Yi, G. C., Wang, C., & Park, W. I., "ZnO nanorods: synthesis, characterization and applications," Semiconductor Science and Technology, 20(4), S22, Mar. 2005.

[3] Lu, F., Cai, W., & Zhang, Y., "ZnO hierarchical micro/nanoarchitectures: solvothermal synthesis and structurally enhanced photocatalytic performance," Advanced Functional Materials, 18(7), 1047-1056, Apr. 2008.

[4] Zhou J, Xu NS, Wang ZL, "Dissolving behavior and stability of ZnO wires in biofluids: a study on biodegradability and biocompatibility of ZnO nanostructures," Advanced Materials, 18(18):2432-5, Sep. 2006.

[5] Prabhakar, R.R., Mathews, N., Jinesh, K.B., Karthik, K.R.G., Pramana, S.S., Varghese, B., Sow, C.H. and Mhaisalkar, S., "Efficient multispectral photodetection using Mn doped ZnO nanowires. Journal of Materials Chemistry," 22(19), pp.9678-9683, May. 2012.

[6] Zhang, H.H., Pan, X.H., Li, Y., Ye, Z.Z., Lu, B., Chen, W., Huang, J.Y., Ding, P., Chen, S.S., He, H.P. and Lu, J.G., "The role of band alignment in p-type conductivity of Na-doped ZnMgO: Polar versus non-polar," Applied Physics Letters, 104(11), p.112106, Mar. 2014.

[7] Gupta, M.K., Sinha, N. and Kumar, B., "p-type K-doped ZnO nanorods for optoelectronic applications," Journal of Applied Physics, 109(8), p.083532, Apr. 2011.

[8] Jin, B.J., Bae, S., Lee, S.Y. and Im, S., "Effects of native defects on optical and electrical properties of ZnO prepared by pulsed laser deposition," Materials Science and Engineering: B, 71(1-3), pp.301-305, Feb. 2000.

[9] Khayatian, A., Asgari, V., Ramazani, A., Akhtarianfar, S.F., Kashi, M.A. and Safa, S., "Diameter-controlled synthesis of ZnO nanorods on Fe-doped ZnO seed layer and enhanced photodetection performance," Materials Research Bulletin, 94, pp.77-84, Oct. 2017.

[10] Chu, Y.L., Ji, L.W., Hsiao, Y.J., Lu, H.Y., Young, S.J., Tang, I.T., Chu, T.T. and Chen, X.J., "Fabrication and Characterization of Ni-Doped ZnO Nanorod Arrays for UV Photodetector Application," Journal of The Electrochemical Society, 167(6), p.067506, Mar. 2020.

[11] Chey, C.O., Liu, X., Alnoor, H., Nur, O. and Willander, M., Fast piezoresistive sensor and UV photodetector based on Mn-doped ZnO nanorods. physica status solidi (RRL)–Rapid Research Letters, 9(1), pp.87-91, Jan. 2015.

[12] Liu, C.W., Chang, S.J., Hsiao, C.H., Lo, K.Y., Kao, T.H., Wang, B.C., Young, S.J., Tsai, K.S. and Wu, S.L., "Noise properties of low-temperature-grown Co-doped ZnO nanorods as ultraviolet photodetectors," IEEE Journal of Selected Topics in Quantum Electronics, 20(6), pp.89-95, Mar. 2014.

[13] Wu, J.J., Liu, S.C. and Yang, M.H., "Room-temperature ferromagnetism in well-aligned Zn 1− x Co x O nanorods," Applied Physics Letters, 85(6), pp.1027-1029, Aug. 2004.

[14] Ueda, K., Tabata, H. and Kawai, T., "Magnetic and electric properties of transition-metal-doped ZnO films," Applied Physics Letters, 79(7), pp.988-990, Aug. 2001.

978-1-6654-3008-1/21 $31.00 © 2021 IEEE

[15] Sharma, P., Gupta, A., Rao, K.V., Owens, F.J., Sharma, R., Ahuja, R., Guillen, J.O., Johansson, B. and Gehring, G.A., "Ferromagnetism above room temperature in bulk and transparent thin films of Mn-doped ZnO," Nature materials, 2(10), pp.673-677, Oct. 2003.

[16] Zou, L., Xiang, X., Wei, M., Li, F. and Evans, D.G., "Single-crystalline ZnGa2O4 spinel phosphor via a single-source inorganic precursor route," Inorganic chemistry, 47(4), pp.1361-1369, Feb. 2008.

[17] Ajmal, H.M.S., Khan, F., Nam, K., Kim, H.Y. and Kim, S.D., "Ultraviolet Photodetection Based on High-Performance Co-Plus-Ni Doped ZnO Nanorods Grown by Hydrothermal Method on Transparent Plastic Substrate," Nanomaterials, 10(6), p.1225, Jun. 2020.

[18] Zhang, H., Zhao, Y., Geng, X., Huang, Y., Li, Y., Liu, H., Liu, Y., Li, Y., Wang, X., Tian, H. and Liang, R., "Au Nanoparticles-Decorated Surface Plasmon Enhanced ZnO Nanorods Ultraviolet Photodetector on Flexible Transparent Mica Substrate," IEEE Journal of the Electron Devices Society, 7, pp.196-202, Dec. 2018.

[19] Yingying, L., Chuanwei, C., Xiang, D., Junshan, G. and Haiqian, Z., "Facile fabrication of UV photodetectors based on ZnO nanorod networks across trenched electrodes," Journal of Semiconductors, 30(6), p.063004, Jun. 2009.

[20] Abdulgafour, H.I., Hassan, Z., Ahmed, N.M. and Yam, F.K., "Comparative study of ultraviolet detectors based on ZnO nanostructures grown on different substrates," Journal of Applied Physics, 112(7), p.074510, Oct. 2012.

April 25-29 , 2021   Xiamen, China

Keynote Speaker

Liquid Metal Nano Electronics

Jing Liu

Department of Biomedical Engineering, School of Medicine, Tsinghua University, Beijing 100084, China
& Technical Institute of Physics and Chemistry, Chinese Academy of Sciences, Beijing 100190, China

Website : www.liquidmetallab.com

ABSTRACT

Conventional micro/nano electronics are generally complex, time, material and energy consuming to make. From an alternative, the recently initiated direct printing of electronics and then making of functional devices via the liquid metal ink and its allied material and machines. Such basic way also opens big potential towards straightforward fabrication of ever smaller circuits, say the nano electronics and the MEMS and NEMS thus involved. This talk is dedicated to present an overview and perspective on the fundamentals and technological endeavors of the new horizontal liquid metal printed nano electronics. The challenges thus raised and the opportunities offered from the nanoscience will be clarified. Particularly, a group of key strategies through nanotechnologies to innovate and enhance various liquid metal electronic inks, printing machines as well as new generation transformable nano electronics will be illustrated. The fundamental mechanisms lying behind will be interpreted. Further, several representative desktop printing methods developed in the lab will be taken as examples to illustrate their pervasive adaptability in quickly making nano electronics and future ending user NEMS devices. Typical utilization of high performance liquid metal printed nano electronics in diverse situations such as tiny soft device, biomedical health care system with high sensitivity, smart home appliances, and advanced printed MEMS/NEMS etc. will be discussed. Prospects within the area spanning from revolutionizing material, machine to new conceptual practices etc. will be given.

BIOGRAPHY

Jing Liu is a jointly appointed Professor of Tsinghua University (THU) and Technical Institute of Physics and Chemistry, Chinese Academy of Sciences. He received his double bachelor's degrees (B.E. in Engineering & B.S. in Physics) in 1992, and Ph.D. (in Engineering and Biomedicine) in 1996, all from THU. Dr. Liu had ever been a Post-Doctorial Research Associate at Purdue University and Senior Visiting Scholar at Massachusetts Institute of Technology. He works intensively at the interdisciplinary areas among liquid metal, biomedical engineering and thermal science and has published 16 popular books on the cutting edge frontiers, 21 invited book chapters and more than 500 peer reviewed journal articles (over 30 selected as cover or back cover story), 200 issued invention patents. Dr. Liu pioneered a group of very fundamental discoveries and technological breakthrough on liquid metals which were frequently featured over the world. Many of his inventions have been widely used in industry and by the society. He is a recipient of numerous awards like: The William Begell Medal, 2015 R&D 100 Award Finalist etc.

978-1-6654-3008-1/21 $31.00 © 2021 IEEE

978-1-6654-3008-1/21 $31.00 © 2021 IEEE

Proceedings of the 16th Annual IEEE International
Conference on Nano/Micro Engineered and Molecular Systems
April 25-29, 2021

Design and Manufacture of a High Precision MEMS Flexible Force Sensor*

Huimin Zhang, Yanlu Feng, Wei Zhang

Abstract— **Flexible electronics is the research hotspot of micro devices nowadays. In this paper, a new type of flexible force sensor is proposed which adopts a cantilever beam structure, and four piezoresistors are arranged at the position of maximum stress. The size and distribution position of the resistors are determined by mechanical calculation and simulation. The flexible force sensor adopts MEMS etching technology to etch a cantilever beam on the monocrystalline silicon slice. By thinning the silicon substrate and filling the flexible material Parylene on the surface, the flexible three-dimensional structure of silicon-based micro-nano structure is realized through Parylene.**

I. INTRODUCTION

Force sensors are widely used. The traditional force sensors are generally based on semiconductor rigid materials, but because of their poor flexibility, it is difficult to adapt to the application scene and demand of the rapid development of science and technology. Compared with the traditional force sensor, the flexible force sensor has the characteristics of light and thin, which can measure the force of complex curved surface, small volume and so on. In principle, flexible force sensors can be divided into piezoresistive type, piezoelectric type and capacitive type. Their working principle is that changes in force will change the resistance, capacitance and voltage of the sensor[1].

Flexible electronics as a new type of electronic technology, with its unique flexibility, ductility and high efficiency, low cost manufacturing process, in information, energy, medical, national defense and other fields has a wide range of application prospects, such as flexible electronic display, organic light-emitting diode OLED, printing RFID, thin film solar panels, etc[2]. At present, there are two main preparation strategies for flexible electronic devices: one is to directly use flexible functional materials to achieve flexible electronic devices. The other is to transfer thinned silicon-based electronic components to a flexible substrate by means of transfer printing. For the first method, flexible electronic devices directly applied by flexible organic semiconductor materials have good flexibility, but the low mobility and unlinearity are still two important factors limiting their performance. Compared with organic semiconductor materials, rigid semiconductor materials

represented by silicon not only have higher mobility, but also have formed a set of mature preparation process. For example, Professor John A. Rogers et al. realized thin and flexible electronic devices by transferring silicon structures to flexible substrates using elastomers[3][4].

In this paper, a flexible force sensor based on Parylene is proposed, and its size is $1 \times 1mm^2$ and simulation electrical sensitivity is 0.03V/N.

II. PRINCIPLE AND DESIGN

The force transducer consists of a proof mass and a cantilever beam. When a cantilever beam is subjected to a force, it will bend and deform. Because the thickness of the cantilever beam must satisfy the minimum deflection theory as shown in (1), and also needs to satisfy the elastic limit of silicon.

$$\omega < \frac{h}{5} \qquad (1)$$

$$\sigma = 8 \times 10^7 Pa$$

Where ω is the maximum deflection of a cantilever beam under stress, h is the thickness of the cantilever beam. Therefore, the cantilever beam is designed to be 50 μm thick. Stress and strain distribution diagram can be obtained by mechanical simulation analysis with ANSYS software as shown in Fig.1. In order to obtain the maximum output, the piezoresistor should be arranged in the area with high stress as far as possible.

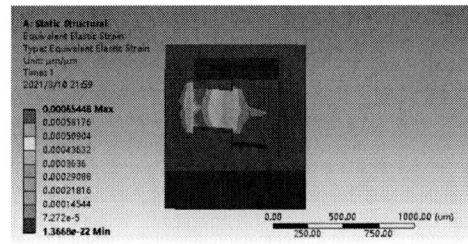

Figure.1. Stress distribution diagram of cantilever beam

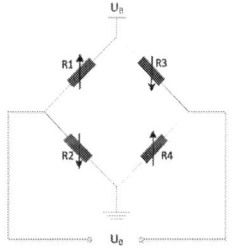

Figure.2. Circuit diagram

Huimin Zhang is with College of Enigeering, Peking University, Beijing 100871, China(1901213621@pku.edu.cn)

Yanlu Feng is with Beijing Institute of Collaborative Innovation, Beijing 100871, China(1194355789@qq.com)

Contacting Author: Wei Zhang is with School of Electronics Engineering and Computer Science, Peking University, Beijing 100871, China(weizhang@pku.edu.cn)

978-1-6654-3008-1/21 $31.00 © 2021 IEEE   687

The MEMS flexible force sensor, as a mechanical quantity sensitive device, generally adopts the form of Wheatstone bridge shown in Fig.2. In order to improve the full-scale output, reduce the zero-point temperature drift and improve linearity, the internal design of the sensor adopt a full-bridge circuit and a constant voltage source to supply power. When the cantilever beam is subjected to external load and produces strain, the resistivity changes, and the resistance made of its material also changes. The relation between piezoresistor and stress can be expressed as (2):

$$\frac{\Delta R}{R} = \pi_x \sigma_x + \pi_y \sigma_y \tag{2}$$

Where π_x and π_y are the longitudinal and transverse piezoresistive coefficients of silicon materials. σ_x and σ_y are the longitudinal and transverse stresses on the piezoresistor. For the P-type injection region, π_x and π_y are equal in size and opposite in sign, so we can get (3) and (4):

$$\frac{\Delta R}{R} = \frac{1}{2}\pi_{44}|\sigma_x - \sigma_y| \tag{3}$$

$$R = \frac{R_0}{W}\int_0^L (\pi_x\sigma_x + \pi_y\sigma_y)dx = \frac{0.5\pi_{44}R_0}{W}\int_0^L (\sigma_x - \sigma_y)dx \tag{4}$$

Therefore, as long as the longitudinal and transverse stress difference on the resistor is maximized, the maximum rate of change of the resistor can be caused and the maximum voltage output can be obtained on the bridge.

The fixed end stress of the cantilever beam is extracted through secondary development of ANSYS, and the distribution of the local area with equal stress difference can be obtained by analyzing the joint stress data, and then the piezoresistor layout position can be obtained in Fig.3. The relationship between the output voltage and the input voltage is shown as (5)[5].

$$V_{OUT} = \frac{\pi_{44}(B-A)}{4L + \pi_{44}A + \pi_{44}B}V_B \tag{5}$$

Where A and B are integrals of the stress difference along piezoresistive R_1 and R_2. Next, it is necessary to extract the transverse and longitudinal stress difference of each node on the surface of the cantilever beam, then calculate the integral value, and select the four equivalent regions with the largest integral value of the stress difference to arrange the piezoresistor. Of the four, two are arranged laterally and two vertically, and the full bridge circuit can be used to obtain the maximum output.

The resistance of the piezoresistor depends on the uniformity of the doping. Design should consider: (a)Doping consistency; (b)The bridge is designed as a closed loop; (c)Doping type and doping concentration. In order to make the zero output small and the sensitivity temperature characteristic good, the surface impurity concentration is required to be in $10^{18}\sim 10^{21}/cm^3$. Because the hole mobility decreases with the increase of temperature, the membrane resistance increases with the increase of temperature. The higher the average impurity concentration, the smaller the change of mobility with the concentration, and the smaller the temperature coefficient of the membrane resistance. So this paper chooses $R_s=250\Omega$. The

relation between piezoresistor and sheet resistance satisfies the following equation (6) and (7).

$$R = R_S \frac{L}{b} \tag{6}$$

$$P = \frac{I^2 R}{WL} = \frac{I^2 R_S \frac{L}{W}}{WL} = \frac{I^2 R_S}{W^2} \tag{7}$$

Where P is the maximum power consumption per unit area; R_S is the sheet resistance; R is piezoresistor; W is the width of the resistor; L is the length of the resistor. Because the end of the resistor is close to the pins, the resistor value decreases, which can be corrected by formula (8):

$$R = R_S\left(\frac{L_1}{W_1} + \frac{L_2}{W_2} + 2K_1 + nK_2\right) \tag{8}$$

Where L_1 and L_2 are the lengths of the resistance; K_1 is is the end correction factor; K_2 is the angle correction factor; n is the number of corners.

In addition, because the piezoresistive effect of turning is negative, it is necessary to widen the turning section to reduce the resistance. So resistor change and resistor value satisfy the following relationship:

$$\frac{\Delta R}{R} = \frac{\Delta R_1 - \Delta R_2}{R_1 + R_2} = \frac{(R_1 - R_2)\pi\sigma}{R_1 + R_2} = \frac{1 - \frac{L_2}{L_1}\frac{W_1}{W_2}}{1 + \frac{L_2}{L_1}\frac{W_1}{W_2}}\pi\sigma \tag{9}$$

The substrate of the sensor is N-type silicon, and the piezoresistor is P-type silicon. When the piezoresistor is placed along [011] on the (100) crystal plane of P-type silicon.

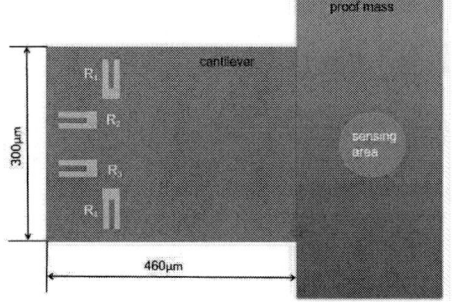

Figure.3. Diagram of resistance distribution on cantilever beam.

III. FABRICATION

The processing process of the sensor is shown in Fig. 4. (a) Prepare a silicon wafer. (b) Oxidation to form a SiO2 layer, which acts as a buffer and protection. (c) Photolithography, followed by the removal of photoresist after boron ion implantation on the patterned silicon substrate beneath the SiO2 layer. (d) A second photolithography and re-doping of boron ions at both ends of the resistor to form an ohmic contact region, which reduces the contact resistance between the metal lead and the resistor. (e) RIE after photolithography, which requires a good contact interface, a surface concentration that meets ohmic contact requirements, and no natural oxide layer on the contact surface. (f) The contact electrode is formed by sputtering the metal in the ohmic contact zone at both ends of the resistor strip. First, a layer of metal Ti is sputtered to

978-1-6654-3008-1/21 $31.00 © 2021 IEEE   688

enhance the adhesion, and then Au is sputtered. The thickness of Ti is approximately 200-500Å and that of Au is approximately 2000-5000Å. (g) Photolithography. (h) Parylene is grown on the surface of the device to cover a metal electrode. (i) Thinning silicon wafers on the back.

Silvaco software is used to simulate two injection, including concentrated boron injection into the contact area of the piezoresistor and light boron injection into the body of the piezoresistor, a fire discharge, and LPCVD silicon oxide process as shown in Fig.5. And the process layout of the flexible force sensor is shown in Fig.6.

In order to accurately take the measurement signal, this paper adopts the method of multi-point measurement, that is, the design of 3 × 3 array force sensing structure. By predefined isolation slot and polymer retention deposition, the microstructures are coated, and then the flexible structure of high-performance silicon-based sensor is achieved through polymer graphics and substrate release. In this way, the overall flexibility of the array is achieved while maintaining the performance of silicon-based sensor parts. The flexible machining technology is not only used in the processing of silicon-based devices, but also used in the array of flexible force sensors as substrate interconnection[6][7].

In the flexible processing of sensor array, Parylene material is used as the flexible substrate and silicon-based devices with good electrical properties are used as the functional structure. The flexibility of silicon-based devices is realized by array of silicon-based devices and filling the array gap with Parylene. First, the silicon-based sensor is discretized through the microgroove, and then Parylene is filled into the microgroove, and the interconnection of the discretized units can be realized through the hole of Parylene and metallization. Finally, the excess silicon under the groove is removed by deep etching on the back, making the entire array a flexible structure connected by Parylene (band interconnection). In this way, the good electrical properties of silicon-based devices are retained, and the damage of device structure and the influence of sensor performance are avoided by bending. The flexible packaging technology is shown in Fig.7.

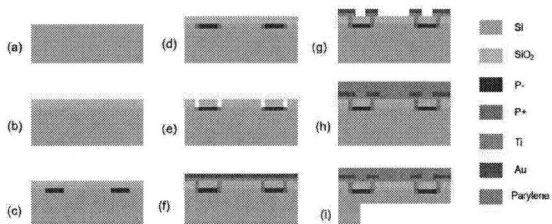

Figure.4. Processing technology of flexible force sensor.

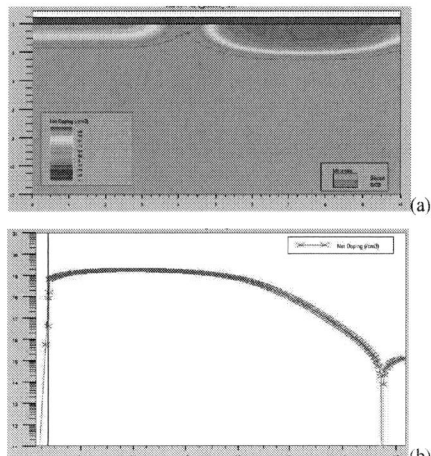

Figure.5.(a)Simulation of junction depth and sheet resistance (b)Concentration distribution in the light boron region.

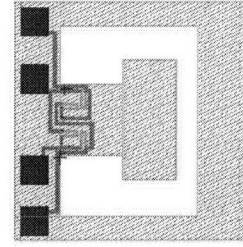

Figure.6. The process layout of the flexible force sensor.

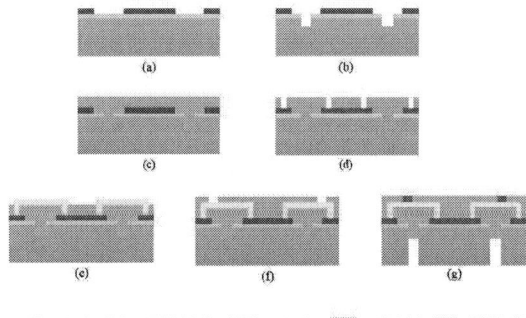

Figure.7 Packaging technology of flexible force sensor array.

IV. .SIMULATION AND RESULTS

When the sensor is attached to the surface of the object to be measured, the proof mass of sensor captures the force. The constant voltage U_B of Wheatstone circuit is 3V, the other end of the piezoresistors are grounded, using COMSOL to simulate voltage output. Then the voltage output distribution and the variation curve of input force with output voltage are shown in Fig.8. It can be seen from Fig.8(b) that the voltage output has a linear relationship with the input force value, and the sensitivity S_0 is 0.03V/N.

978-1-6654-3008-1/21 $31.00 © 2021 IEEE

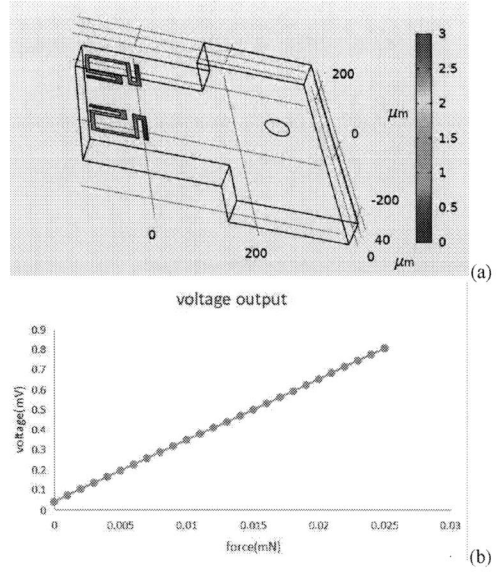

voltage output

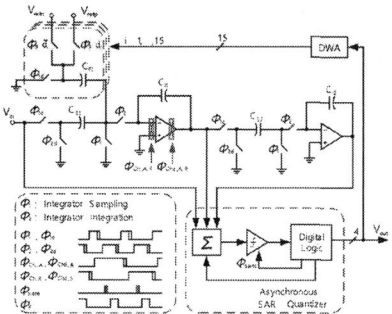

Figure.11. Modulator circuit structure of $\sum \Delta ADC$.

Figure.8. (a)The voltage output distribution. (b)The variation curve of input force with output voltage.

The overall technical flow of the special chip for signal acquisition and processing is shown in Fig.9, which is realized by the top-down design method. The front-end detection circuit is connected with the bracelet sensor to detect the resistance change of the sensor and convert the signal into voltage output. Low noise is the key to the design of this circuit. Chopper Stabilization is used in the design of front-end detection circuit to reduce the influence of low frequency non-ideal factors such as circuit noise and offset as shown in Fig.10. In the sensing chip, multiple channels $\sum \Delta ADC$ are needed to collect the signals of various sensors.

It has the advantages of high precision and low power consumption when quantifying low frequency signal, which is suitable for the application of this project. The circuit structure of ADC modulator is shown in Fig.11. Chopper Stabilization technology is used in the integrator to further reduce op amp noise.

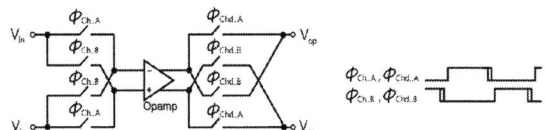

Figure.9. Front-end detection circuit using Chopper Stabilization technology

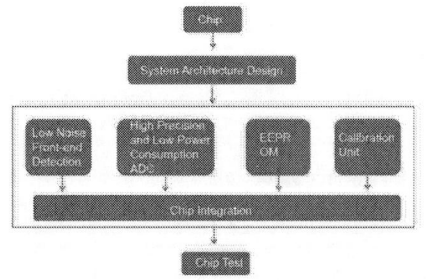

Figure.10. General technical flow chart.

V. CONCLUSION

In this paper, a flexible force sensor is designed, of which the cantilever structure can meet the needs of high range and high sensitivity measurement. The structure and layout of the cantilever beam and the piezoresistor are optimized by finite element simulation to obtain the required detection range and sensitivity. Then considering the inverse piezoresistive effect at the corner of the sheet resistance, the resistors are modified in size. The process of Parylene based flexible force sensor is introduced in detail, and the technical route of three dimensional flexible packaging technology is briefly described. Finally, the voltage output of the flexible force sensor is simulated, which can get a good linear relationship between the force and the voltage, and the circuit structure matching with the flexible force sensor array is designed.

ACKNOWLEDGMENT

I would like to thank my supervisor professor Zhang for his guidance on my academic and scientific research. Thanks to the students in the research process for mutual discussion.

REFERENCES

[1] Zhang Yanhong, Yang Chen, Zhang Zhaohua, et al. A novel pressure microsensor with 30-μm-thick diaphragm and meander-shaped piezoresistors partially distributed on high-Stress bulk silicon region[J]. IEEE Sensors Journal, 2007, 12(7): 1742-1746.

[2] Huang Y, Fan X, Chen S-C, Zhao N, Adv. Funct. Mater, 2019, 29, 1808509.

[3] K. J. Lee，M. J. Motala，M. A. Meitl，et al.，Large-Area，Selective Transfer of Microstructured Silicon: A Printing- Based Approach to High-Performance Thin-Film Transistors Supported on Flexible Substrates, Advanced Materials, 2005, 17(19):2332–2336.

[4] E. Menard, R. G. Nuzzo, J. A. Rogers, Bendable single crystal silicon thin film transistors formed by printing on plastic substrates，Applied Physics Letters, 2005, 86(9):257-206.

[5] Han Ruirui, Zhang Zhaohua, Ren Tiantian, et al. Simulation Method for the Sensitivity and Linearity of Piezoresistive Pressure Sensors[J]. MEMS and Sensors, 1671-4776(2012)02-0096-06.

[6] A. Dhar, J. C. Loach, P. J. Barton, J. T. Larsen, A.W.P. Poon, Low-background temperature sensors fabricated on parylene substrates, Journal of Instrumentation, 2015, 10, P12002.

[7] Bing Han, Peng Wang, Huichao Jin, Zhishan Hou, Xue Bai, Wettability and surface energy of parylene F deposited on PDMS, Physics Letters A, Volume 384, Issue 25, 7 September 2020, 126628.

Proceedings of the 16th Annual IEEE International
Conference on Nano/Micro Engineered and Molecular Systems
April 25-29, 2021

Linear stiffness tuning of MEMS triangular capacitors *

Yiming Jin, Zhipeng Ma, Yixuan Guo, Tengfei Zhang, Xudong Zheng and Zhonghe Jin

Abstract— **Based on the proposed MEMS triangular capacitors previously, we performed a further investigation on the stiffness tuning characteristics of a spring-softening triangular capacitor including the fringe effect and effective stiffness nonlinearity and the effects on the performance of a MEMS accelerometer. Both the numerical and experimental results demonstrate a linear yet reduced stiffness tuning ability of the triangular capacitor as a result of fringe effect. With applying a very large tuning voltage (e.g. 15 V), the linearity of the stiffness tuning is significantly degraded. With the stiffness tuning of the triangular capacitor, the sensitivity of the MEMS accelerometer is improved by 20%, and the corresponding Allan bias instability is slight decreased as expected.**

I. INTRODUCTION

MEMS inertia sensors have achieved a great success owing to its lost cost, weight and power consumption, and high reliability and precision of acceleration and angular rate measurements [1-4]. Both micromachined capacitive accelerometers and gyroscopes demand an electrostatic stiffness tuning which enables compensating the fabrication tolerance, matching the resonance modes and extending the operation modes [5-7]. We have proposed the MEMS triangular capacitors for the stiffness tuning of the inertia sensors previously [8-9], which are deployed with two parallel plates (rectangular and triangular shapes respectively). The lateral displacement of the movable plate produces a change of capacitance between the triangular capacitors, resulting into an effective linear stiffness in theory. Limited by the design rules and fabrication processes, the triangular capacitor with low ratio between overlapped width and capacitive gap cannot be simply regarded as ideal parallel-plate capacitors, requiring a further investigation of the characteristics related to stiffness tuning including the fringe effect, the nonlinearity of stiffness tuning and even the cross-sensitivity [10-12]. In this paper, we built both the analytical and numerical models of the spring-softening triangular capacitor and investigated its effective negative stiffness and nonlinearity with and without considering the fringe effect. To verify the predictions, we designed and fabricated a MEMS accelerometer with deployed triangular capacitor, and characterized both the effective stiffness tuning ability and the effects on the sensitivity and bias instability of acceleration measurement. In this paper, the structures of triangular capacitor and the designed and fabricated accelerometer deployed with the triangular capacitor are first illustrated. Then, the analytical and numerical models of the triangular capacitor are developed and their predictions are compared with the experimental

results. Finally, the performance of the accelerometer with tunable stiffness is characterized in terms of sensitivity and Allan bias instability.

II. TRIANGULAR CAPACITOR AND ACCELEROMETER

A. Triangular capacitor

The main structure of the triangular capacitor with linear spring softening is illustrated in Fig. 1(a). The movable electrode is a rectangular plate which can be suspended and constrained by the folded beams, allowing the lateral displacement in x axis. The other fixed electrode is composed of an array of triangular plates, with a predefined gap away from the movable plate. As a result, with applying a DC bias voltage between the electrodes, the induced lateral electrostatic forces are cancelled out when the movable plate is aligned with equal overlapped area over the triangular shapes, while the effective stiffness still exists in x axis.

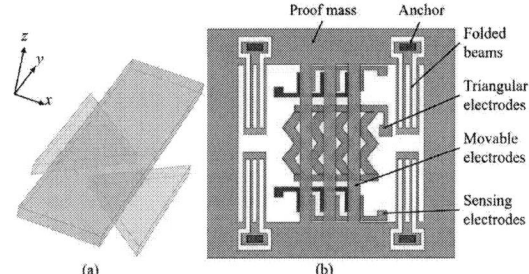

Fig. 1 Schematic view of (a) the triangular capacitor and (b) the MEMS accelerometer deployed with the triangular capacitor.

B. MEMS accelerometer

A MEMS accelerometer with an array of triangular capacitors is designed and shown in Fig. 1(b). The proof mass attached with the movable rectangular plates are suspended by four folded beams which have good decoupling between x and y axes. The sensing and stiffness tuning electrodes are deployed on another layer with a pre-defined gap. The sensing electrodes consists of two groups of differential rectangular electrodes excited by two opposite AC voltages with a carrier frequency of 156.25 kHz. The stiffness tuning electrodes are an array of integrated triangles, forming triangular capacitors together with the movable rectangular plates. The accelerometer with triangular capacitors is fabricated based on a standard silicon-on-glass (SOG) process, in which the sensing and stiffness tuning

---

*Research supported by Zhejiang Provincial Natural Science Foundation of China under Grant LQ19F040011.

All authors are with the Micro-Satellite Research Center of Zhejiang University, Hangzhou, China (corresponding to Z. Ma: 0571-8795-2991; e-mail: mazhipeng@zju.edu.cn).

978-1-6654-3008-1/21 $31.00 © 2021 IEEE

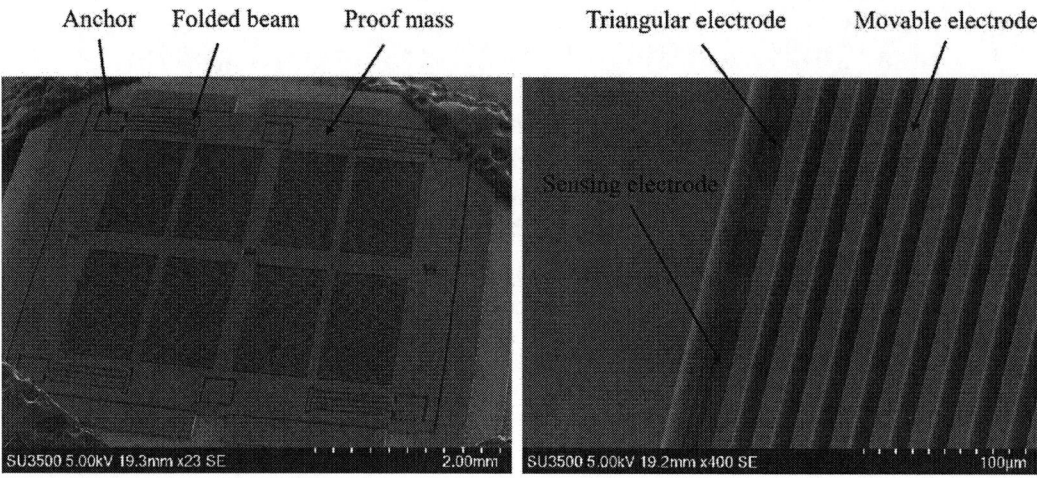

Fig. 2 The SEM images of fabricated MEMS accelerometer.

electrodes are patterned on a Ti/Au layer pre-coated on the glass wafer, the capacitive gap and the electrical contact are formed by patterning the silicon wafer before the anodic bonding between the silicon and glass wafers, and the movable silicon structure are released by patterning top surface of the remaining silicon wafer after thinning the silicon wafer to a thickness of about 60 μm. The micromachined accelerometer is shown in Fig. 2. The fabricated accelerometer is mounted onto a customed PCB board with CV pickoff circuit, A/D and D/A converter and FPGA chip. The demodulation and low-pass filtration of the displacement signal is implemented in FPGA chip. The tuning voltage is generated by the FPGA and applied to the triangular electrodes through the D/A converter and the amplification circuit.

III. LINEAR STIFFNESS TUNING

As mentioned above, the triangular capacitor is regarded as a pair of parallel plates without considering the fringe effect. The effective stiffness of the triangular capacitor can be simply derived by differentiating the electrostatic energy twice and represented by Eq. (3):

$$U = \frac{\varepsilon N \tan\theta (x_0^2 + x^2)}{g} V_{app}^2 \qquad (1)$$

$$F = \frac{\partial U}{\partial x} = \frac{2\varepsilon N \tan\theta x}{g} V_{app}^2 \qquad (2)$$

$$k_e = \frac{\partial F}{\partial x} = \frac{2\varepsilon N \tan\theta}{g} V_{app}^2 \qquad (3)$$

where U, F and k_e are the electrostatic energy, electrostatic force and the effective stiffness of the triangular capacitor, respectively, ε is the permittivity, θ, g and N are the ratio between overlapped length and width, the capacitive gap, and the number of triangular capacitors, respectively, x_0 and x are the initial overlapped width and the displacement, respectively.

The resulted effective stiffness is opposite to the mechanical stiffness. The ratio of the overlapped width and the capacitive gap of the triangular capacitor is relatively small, in this case, about 2. The influence of the fringe effect on the effective stiffness of the triangular capacitor cannot be neglected anymore. Based on finite element method (FEM), the electrical field across the triangular capacitor is numerically calculated, indicating innegligible fringe effect, as shown in Fig. 3. FEM analysis is carried out using Comsol 5.5, in which a three-dimensional (3D) modeling is built with grounding the proof mass, applying the bias tuning voltage to the triangular electrodes, and fixing the anchors. To analyze the effect of the tuning voltage and capacitive gap, the FEM simulations are performed with applying both the electrostatic and mechanics physics. The effective stiffness is derived by sweeping the applied force on the proof mass while recording the corresponding displacement.

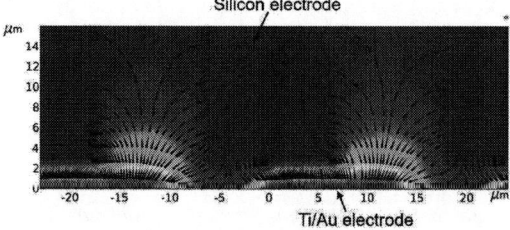

Fig. 3 Electrical field across the triangular capacitors.

To verify both predictions, the fabricated MEMS accelerometer are characterized with applying different tuning voltages to the triangular capacitors. The physical parameters of the fabricated accelerometer are calibrated by the frequency sweeping and the pull-in experiments as shown in Table 1.

A. Stiffness tuning ability

As revealed in Eq. (3), the effective stiffness of the triangular capacitor is proportional to the square of the tuning voltage without consideration of the fringe effect. The square relationship between the effective stiffness and the tuning voltage is still valid for the case taking into account of the fringe effect, as indicated by the numerical and experimental results in Fig. (4). However, the effective stiffness estimated by the numerical method is about 70% of the calculated stiffness using Eq. (3). This is attributed to the fringe effect of the electrical field across the triangular capacitor. The measured effective stiffness of the triangular capacitor is close to that estimated by the numerical method. Since the capacitive gap between the triangular capacitor is affected by the fabrication tolerance, the measured effective stiffness might be slightly deviated from the estimated one using FEM.

Figure (5) shows the effect of the capacitive gap on the effective stiffness of the triangular capacitor. Both predictions indicate that the effective stiffness is inversely proportional to the capacitor gap.

Table 1 Physical parameters of the fabricated accelerometer.

Description	Parameter	Value
Mass	m	2.04×10^{-6} [kg]
Mechanical stiffness	k	22.67 [N/m]
Frequency	f	530.6 [Hz]
Capacitive gap	g	3 [um]
Triangular pair	N	1452
Triangular shape	$\tan\theta$	2.5
Sensing gain	k_{xc}	3.6×10^{-8}
CV gain	k_{cv}	10^{12}

B. Nonlinearity of stiffness tuning

The nonlinearity of the effective stiffness arising from the triangular capacitors is evaluated by both the simulation and the measurement. As shown in Fig. (6), as the increase of the tuning voltage, the nonlinearity of the effective stiffness slightly increases. The measured nonlinearity of the effective stiffness is much higher than that estimated from the simulation taking into account of the fringe effect. The input acceleration during the simulation is limited to ± 0.1 g due to the limitation of FEM solver while the input acceleration of the measurement varies within ± 1 g. When the tuning voltage is larger than 13 V, the measured nonlinearity is dramatically increased from about 0.6% to more than 1.4%, probably due to the cross-coupling effect arising from the out-of-plane electrostatic force.

IV. ACCELEROMETER PERFORMANCE

The scale factor of the test accelerometer is calibrated by using the earth gravity that can be varied by rotating the index table, as shown in Fig. 7. The calibrated scale factors

with applying different tuning voltages shown in Fig. 8, is increased when the tuning voltage increases. With applying a tuning voltage of 15 V to the triangular capacitor, the scale factor of the accelerometer is increased by about 20% when compared with the scale factor of the accelerometer without stiffness tuning. The Allan variance analysis is performed for the accelerometer with different tuning voltages under the zero-acceleration input. The measurements last for over 12 hours with a sampling rate of 10 Hz and the best 1-hour data with least fluctuation is used for the Allan variance analysis. The results of Fig. 9 indicate that the accelerometer with tuning voltage of triangular electrode can improve the bias instability.

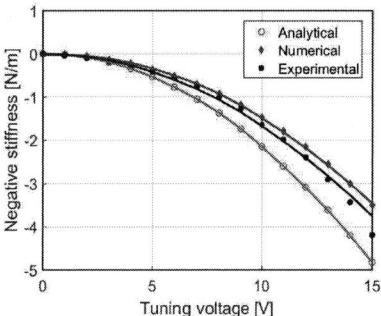

Fig. 4 Effective negative stiffness as a function of tuning voltage.

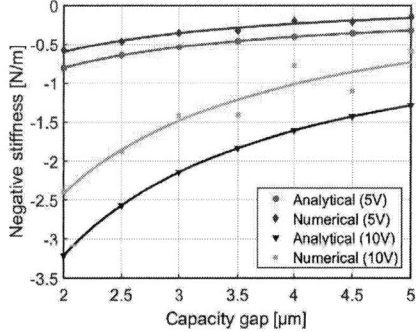

Fig. 5 Effective negative stiffness as a function of capacitive gap.

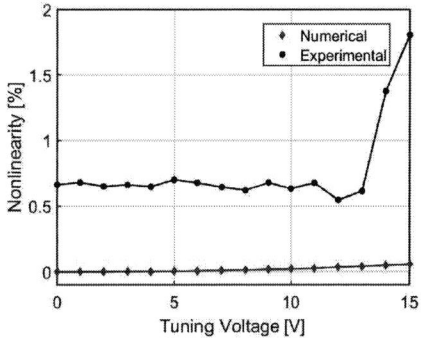

Fig. 6 Nonlinearity of effective stiffness as a function of tuning voltage.

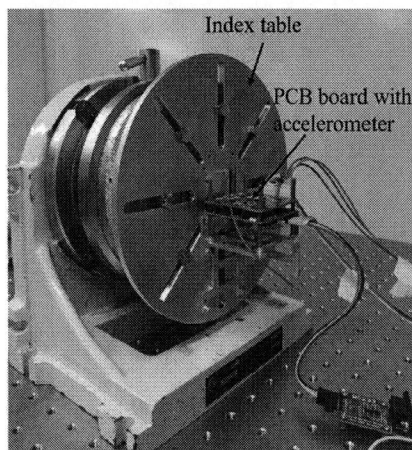

Fig. 7 Experimental setup of the MEMS accelerometer.

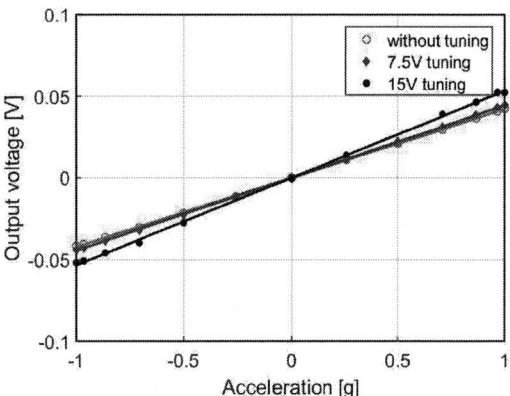

Fig. 8 Scale factors of the MEMS accelerometer with different tuning voltages.

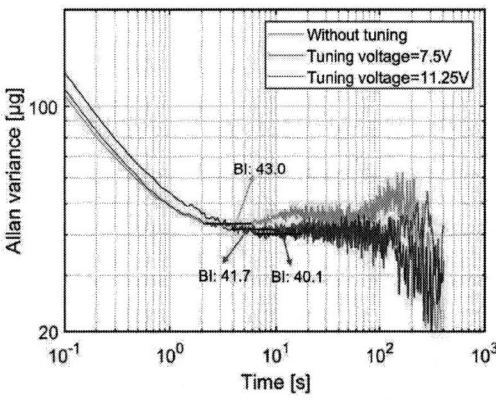

Fig. 9 Allan variance analysis of the MEMS accelerometer with different tuning voltages.

V. CONCLUSION

An investigation of the stiffness tuning characteristics of the spring-softening triangular capacitor is presented in this paper. The FEM model is built to take into account of the fringe effect of the triangular capacitor, and the numerical results indicate that the triangular capacitor has a linear yet reduced stiffness tuning ability. A MEMS accelerometer deployed with the triangular capacitor is designed and fabricated based on a standard SOG process. The linearity and the reduction of the stiffness tuning are verified by the experimental results. However, the linearity is significantly degraded when apply a large tuning voltage (e.g. 15 V) across the triangular capacitor. With the linear stiffness tuning by the triangular capacitors, the sensitivity of the accelerometer is improved by up to 20%, and the Allan bias instability is slightly reduced.

REFERENCES

[1] R. Mukhiya et al., "MEMS accelerometer-driven fuel-control system for automobile applications," in *Proc. SPIE Int. Soc. Opt. Eng.*, Melbourne, VIC, Australia, 2008, vol. 7268, p. 72680Q.

[2] D. Hollocher et al., "A very low cost, 3-axis, MEMS accelerometer for consumer applications," in *Proc. IEEE Sens.*, Christchurch, New zealand, 2009, pp. 953–957.

[3] K. H. Kim, J. S. Ko, Y. H. Cho, K. Lee, B. M. Kwak, and K. Park, "A skew-symmetric cantilever accelerometer for automotive airbag applications," *Sens. Actuators A Phys.*, vol. 50, no. 1–2. pp. 121–126, 1995.

[4] J. A. Geen, S. J. Sherman, J. F. Chang, and S. R. Lewis, "Single-chip surface micromachined integrated gyroscope with 50°/h allan deviation," in *IEEE J. Solid State Circuits*, 2002, vol. 37, no. 12, pp. 1860–1866.

[5] Z. X. Hu, B. J. Gallacher, J. S. Burdess, S. R. Bowles, and H. T. D. Grigg, "A systematic approach for precision electrostatic mode tuning of a MEMS gyroscope," *J. Micromech. Microeng.*, vol. 24, no. 12, 2014.

[6] Y. Guo, Z. Ma, T. Zhang, X. Zheng, and Z. Jin, "A stiffness-tunable MEMS accelerometer," *J. Micromech. Microeng.*, vol. 31, no. 2, pp. 1–10, 2021.

[7] S. Sung, W. T. Sung, C. Kim, S. Yun, and Y. J. Lee, "On the mode-matched control of MEMS vibratory gyroscope via phase-domain analysis and design," *IEEE ASME Trans. Mechatron.*, vol. 14, no. 4, pp. 446–455, 2009.

[8] H. Wu, X. Zheng, Y. Lin, Z. Ma, and Z. Jin, "Linear Parametric Amplification /Attenuation Without Spring Hardening /Softening Effect in MEMS Gyroscopes," in *Proc. IEEE Int. Conf. Micro Electro Mech. Syst. MEMS*, Vancouver, Canada, 2020, vol. 2020-Janua, pp. 757–760.

[9] X. Zheng, H. Wu, Y. Lin, Z. Ma, and Z. Jin, "Linear parametric amplification/attenuation for MEMS vibratory gyroscopes based on triangular area-varying capacitors," *J. Micromech. Microeng.*, vol. 30, no. 4, 2020.

[10] K. Y. Park, C. W. Lee, H. S. Jang, Y. Oh, and B. Ha, "Capacitive type surface-micromachined silicon accelerometer with stiffness tuning capability," *Sens. Actuator A-Phys.*, vol. 73, no. 1–2, pp. 109–116, 1999.

[11] L. Dong, Y. Pan, J. Lou, and J. Bao, "Study of the influence of fringe edge on MEMS capacitive accelerometers self-calibration," *Microsyst. Technol.*, vol. 21, no. 6, pp. 1179–1186, 2015.

[12] S. G. Adams, F. M. Bertsch, K. A. Shaw, and N. C. MacDonald, "Independent tuning of linear and nonlinear stiffness coefficients," *J. Microelectromech. Syst.*, vol. 7, no. 2, pp. 172–180, 1998.

[13] K. B. Lee, L. Lin, and Y. H. Cho, "A closed-form approach for frequency tunable comb resonators with curved finger contour," *Sens. Actuator A-Phys.*, vol. 141, no. 2, pp. 523–529, 2008.

Proceedings of the 16th Annual IEEE International
Conference on Nano/Micro Engineered and Molecular Systems
April 25-29, 2021

Power Handling Capability Enhanced RF MEMS Switch Using Modified-Width Cantilevers Structure*

Yulong Zhang, Zhuhao Gong, Huiliang Liu, and Zewen Liu

Abstract—A power handling capability enhanced radio frequency micro-electro-mechanical system (RF MEMS) switch using modified-width cantilevers structure is presented. Among the cantilevers of the switch, the inner cantilevers width is larger than the outer one, which results in different forces and resistances on different contacts when they are actuated. The proposed design can suppress the skin-effect-led uneven current distribution between the inner and outer contacts, which is beneficial to enhance the power handling capability. A thermo-electro-mechanical model is built on the basis of finite element analysis (FEA) to analyse the devices. The simulations indicate the effects of the design. The switch is fabricated and tested. The results demonstrate that the proposed structure can bear 2 W RF continuous wave (10GHz) for 2 hours under prolonged ON-state condition, which is higher than the switch with equi-width cantilever structure.

Keywords—RF MEMS, switch, power handling capability, improvement, modified-width cantilevers.

I. INTRODUCTION

Radio frequency micro-electro-mechanical system (RF MEMS) switches are attracting more and more attentions for their superior RF performances, including low loss, high isolation, high linearity [1]. However, improvement of the power handling capability is still an important task for widespread applications of the DC-contact RF MEMS switch.

Previous studies show several methods to enhance the power handling capability of the DC-contact RF MEMS switch: special materials, such as transition metals and their alloys [2]; new structures, such as push-pull lever, stoppers [3-4]; more contacts or more parallel switches [5]. The use of special materials can improve the ability of contact temperature tolerance. However, the high resistivities of transition metals and alloys can bring in more loss, which is harmful for RF performance. The introduction of new structures, such as push-pull lever, stoppers, always results in more RF signal mismatch, which needs careful design. The method of more contacts or more parallel switches can improve power handling capability, but it is limited by skin effect of RF signal, which means the effect of more contacts and switches will be decreased. Meanwhile, some previous

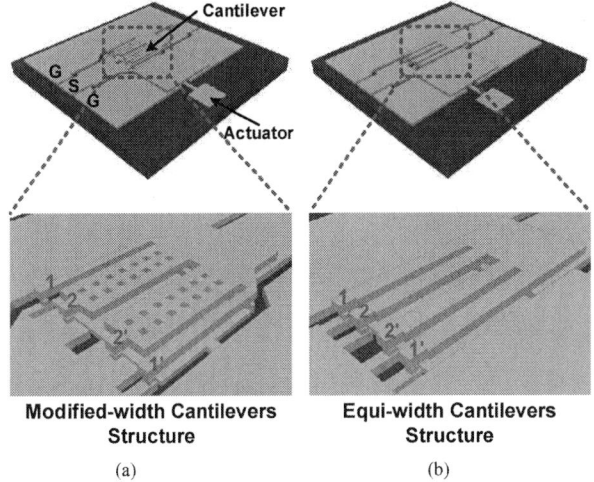

Figure 1. Schematic views of the RF MEMS switches: (a) Overview and close-view of the switch with modified-width cantilevers structure; (b) Overview and close-view of the switch with equi-width cantilevers structure.

studies combine more than one method to improve the power performance, such as the use of Ru alloys and stoppers [4], new cantilever and multi-contact [6], multi-switch and special material [7].

In this work, we present a novel power handling capability enhanced RF MEMS switch using modified-width cantilevers structure. A developed thermo-electro-mechanical model and power handling analysis indicate that the design can suppress the uneven current distribution led by skin-effect and the theoretical power handling capability can be improved. The simulations are conducted by using finite element analysis (FEA) softwares, including CoventorWare [8], High Frequency Structure Simulator (HFSS), and ePhysics. The device is fabricated by using surface manufacturing process. The test results show that the proposed structure can improve the power handling performance. Meanwhile, a switch with equi-width cantilever structure is also simulated, fabricated and tested as a comparison.

II. MODELLING AND SIMULATION

The proposed RF MEMS switch using modified-width cantilevers structure and switch using equi-width cantilevers structure are illustrated in Fig. 1(a) and (b), respectively. The widths of inner and outer cantilevers of modified-width cantilevers switch (MWS) are 58μm and 12μm; and the widths of the four cantilevers of equi-width cantilevers switch (EWS) are 20μm. There are four contacts

---

*The research is supported by National Key R&D Program of China No. 2018YFB2002801.

Y. Zhang and Z. Liu are with the Institute of Microelectronics, Tsinghua University, Beijing, 100084, China.

Z. Gong is with the Fujian Electronics & Information Co., Fujian, 350005, China..

H. Liu is with the Institute of Telecommunication and Navigation Satellite, China Academy of Space Technology, Beijing, 100094, CHINA.

The corresponding author is Z. Liu. (e-mail: liuzw@tsinghua.edu.cn)

978-1-6654-3008-1/21 $31.00 © 2021 IEEE

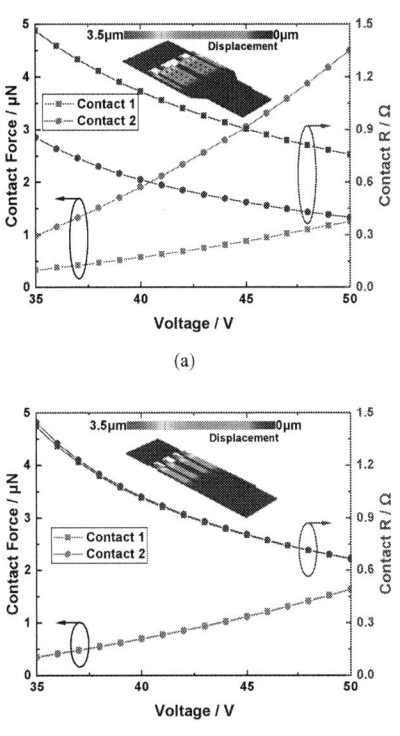

(a)

(b)

Figure 2. Simulated contact force and calculated contact resistance of the switches versus driving voltage: (a) Switch with modified-width cantilevers structure; (b) Switch with equi-width cantilevers structure.

in the switches, in which the outer contacts are denoted as 1 and 1', while the inner contacts are denoted as 2 and 2'.

When RF signal passes through the EWS, the currents distributed on the outer contacts (1 and 1') are higher than the inner contacts (2 and 2'), because of the skin effect, which is harmful for power handling capability improvement. The proposed switch MWS can diminish the uneven current distribution effect. Owing to the wider width, the inner cantilevers can share more current than the outer narrower cantilevers. Meanwhile, the inner cantilevers bear higher electrostatic force than outer cantilever, which results in a relative smaller contact resistance on inner contacts than that of outer contacts.

The contact forces (F_c) and resistances (R_c) are studied by CoventorWare. The simulated pull-in voltages of the two switches are both about 30V, and the contact forces simulation is conducted from 35V to 50V, as shown in Fig. 2. The contact resistance is calculated by using (1), in which ρ is resistivity of gold, H is hardness of gold [9].

$$R_c = \frac{\rho}{2}\sqrt{\frac{\pi H}{F_c}} \qquad (1)$$

For EWS (Fig. 2b), the contact forces and resistances are same among inner and outer contacts. However, the relationship changes for MWS (Fig. 2a). The inner contact forces of MWS are higher than outer one, so the contact resistances of inner contacts are smaller than that of outer

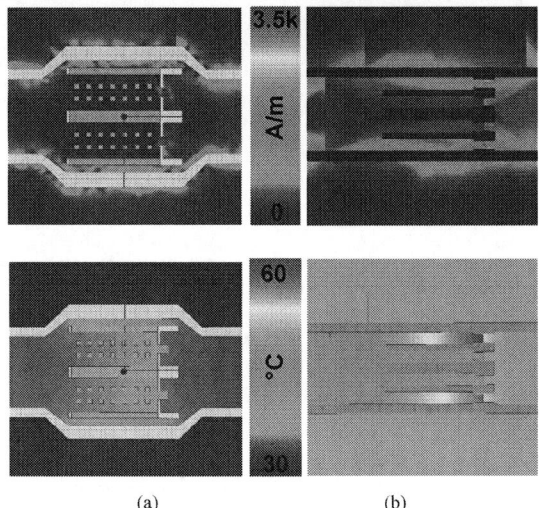

(a)   (b)

Figure 3. Simulated current density field using HFSS and static temperature field using ePhysics with 1W RF signal passing through: (a) Current density field (upper left) and static temperature field (bottom left) of the modified-width cantilevers structure; (b) Current density field (upper right) and static temperature field (bottom right) of the equi-width cantilevers structure.

contacts. To ensure a robust working condition, the actuation voltage is set to be 45V for the following electromagnetic (EM) simulation and thermal simulation.

Based on the CoventorWare simulation results, the EM simulation and thermal simulation are conducted using ANSYS (HFSS and ePhysics), as shown in Fig. 3. The ANSYS simulations are under conduction of 1W 10 GHz RF signal passing through the switches. The current density field and static temperature field of MWS are shown in Fig. 3a, while the current density field and static temperature field of EWS are shown in Fig. 3b. It is easy to draw a conclusion that the current density and temperature of MWS are both lower than that of EWS, which indicate that the MWS can hold a higher RF Power.

To obtain theoretical power handling capabilities (P_{th}) of the two switches, the temperature of contacts (T_c) are calculated with (2), in which P is RF power passing through, k is the temperature improvement factor, b is the environment temperature, L is Lorenz's constant, Z_0 is characteristic impedance of the transmission line.

$$T_c = \sqrt{\frac{PR_c^2}{16LZ_0} + \left(kP + b\right)^2} \qquad (2)$$

Equation (2) is derived with the help of Holm's contact theory [9], Fourier's heat conduction law and Ohm's law. The calculated T_c results are plotted in Fig. 4. When T_c is higher than a threshold which is related to the melting point, the probability of micro-melting on contact will increase sharply. The threshold of gold micro-melting is set to be 170°C, third of melting point of gold. Then the P_{th} of MWS and EWS are 2.3W and 1.4W, respectively.

978-1-6654-3008-1/21 $31.00 © 2021 IEEE

TABLE 1. SIMULATED AND CALCULATED PERFORMANCES OF THE RF MEMS SWITCHES WITH MODIFIED-WIDTH AND EQUI-WIDTH CANTILEVERS STRUCTURES.

Switches	Actuation Voltage	F_c(μN)		R_c(Ω)		J_c*(kA/m)		α**	$T_{cantilever}$***(°C)		P_{th}(W)
		Con1	Con2	Con1	Con2	Con1	Con2		1	2	
Modified-Width Cantilevers Structure	45V	0.882	3.06	0.906	0.487	5.31	5.26	0.990	42.4	36.9	2.3
Equi-Width Cantilevers Structure		1.12	1.11	0.803	0.806	8.82	3.29	0.373	56.4	45.8	1.4

* J_c is the simulated RF currents passing through the contact;

** α $(\alpha=J_{c2}/J_{c1})$ is the calculated current uneven factor;

*** $T_{cantilever}$ is the simulated highest temperature of the cantilevers (contacts not included) with 1W RF wave (10GHz) passing through.

TABLE 2. MEASURED POWER HANDLING CAPABILITY RESULTS OF THE SWITCHES WITH MODIFIED-WIDTH AND EQUI-WIDTH CANTILEVERS STRUCTURES AT 10GHz.

Switches	Actuation Voltage	DC current × Cycles	CW RF Power × Time
Modified-Width Cantilevers Structure	45V	>250mA × 10^5	2.0W × 2h
Equi-Width Cantilevers Structure	45V	>250mA × 10^5	1.5W × 2h

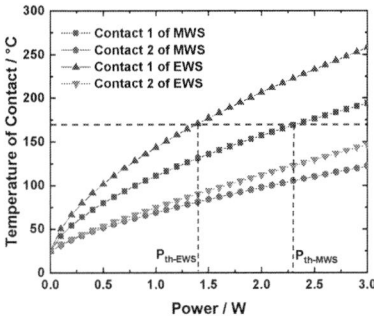

Figure 4. Calculated temperature of contacts in different switches versus different RF power with theoretical power (P_{th}) marked.

The thermo-electro-mechanical model for DC-contact RF MEMS switch power handling capability analysis is well established. The parameters used in the thermo-electro-mechanical model are listed in Table 1. The results of the model indicate that the MWS design suppresses the uneven current distribution well, which can improve the power handling capability of the device.

III. FABRICATION AND MEASUREMENT

The switches are fabricated using surface manufacturing process, in which the Au electroplating and polyimide (PI) sacrificial layer releasing are the key processes in the fabrication. The scanning electron microscope (SEM) photographs of MWS and EWS are shown in Fig. 5, with overviews of the switches, zoom-in views of the cantilevers, and detailed view of the contacts.

The pull-in and lift-off voltages of the switches are tested to be 42V and 21V, respectively, which are higher than the design, due to the process variation. The S-parameters of the MWS are tested with the help of probe station (Cascade Summit 12000M) and Vector Network Analyzer (VNA) Agilent E8683C. The tested S-parameters

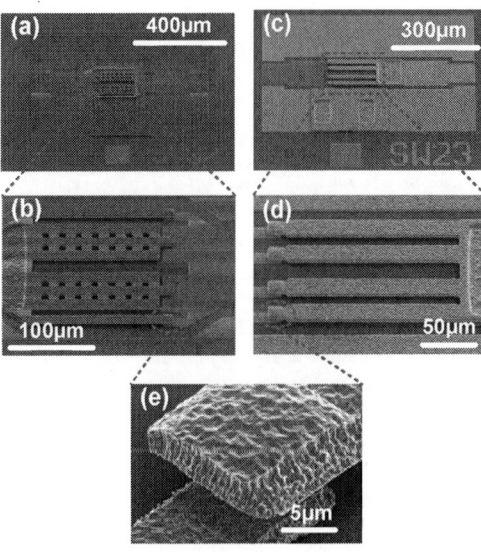

Figure 5. SEM photographs of the switches: (a)(b) Overview of the whole switch and zoom-in view of the modified-width cantilevers; (c)(d) Overview of the whole switch and zoom-in view of the equi-width cantilevers; (e) Detailed view of the Au-Au contact of the switches.

are plotted in Fig. 6, with fitted results using Advanced Design Systems (ADS). The insertion loss (IL) and return loss (RL) of MWS are about -1dB and -15dB @10GHz, respectively, at ON-state. The Isolation of MWS is -15dB @10GHz at OFF-state.

The power handling capability test is also carried out to verify the design of MWS, and the results are listed in Table 2. Both of the MWS and EWS can bear 250mA DC current more than 10^5 cycles under cold-switching condition. But the RF power handling performance of the two switches are different. The MWS can bear 2.0W CW RF (10GHz) signal more than 2h, while the EWS can only bear 1.5W × 2h, which is lower than MWS. The tested results agree well with the design and simulation.

978-1-6654-3008-1/21 $31.00 © 2021 IEEE

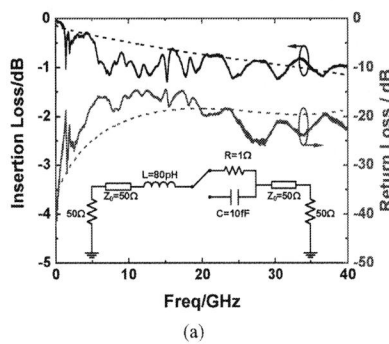

(a)

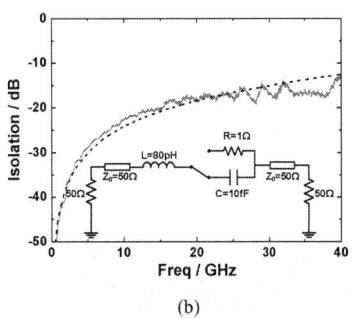

(b)

Figure 6. Measured S-Parameters and fitted results using ADS of the switch with modified-width structure: (a) Insertion Loss and Return Loss at ON-state; (b) Isolation at OFF-state.

IV. CONCLUSION

Design, Modeling, Fabrication and test of a DC-contact RF MEMS switch using modified-width cantilevers structure are conducted in this paper. A thermo-electro-mechanical model for DC-contact RF MEMS switch power handling capability analysis is established. The simulation and test results demonstrate that the modified-width cantilevers structure design can suppress the uneven current distribution effectively, which is beneficial for power handling capability enhancement. This method can be used together with other means to further improve the power performance of the device.

ACKNOWLEDGMENT

The authors gratefully acknowledge the technical support of SiMEMS Micro/nano System Co., Ltd, Suzhou, Jiangsu Province, China. And the devices were fabricated in Nano Fabrication Facility at the Suzhou Institute of Nano-tech and Nano-bionics, China.

REFERENCES

[1] G. M. Rebeiz, *RF MEMS Theory, Design, and Technology.* New Jersey: J. Wiley & Sons, 2003.

[2] P. M. Zavracky, N. E. McGruer, R. H. Morrison, and D. Potter. "Microswitches and microrelays with a view toward microwave applications", *Int J RF and Microwave Comp Aid Eng*, vol. 9, pp. 338-347, 1999.

[3] Y. Zhu, L. Han, L. Wang, J. Tang and Q. Huang, "A Novel Three-State RF MEMS Switch for Ultrabroadband (DC-40 GHz)

Applications," in *IEEE Electron Device Letters*, vol. 34, no. 8, pp. 1062-1064, Aug. 2013.

[4] C. D. Patel and G. M. Rebeiz, "A High-Reliability High-Linearity High-Power RF MEMS Metal-Contact Switch for DC–40-GHz Applications," in *IEEE Transactions on Microwave Theory and Techniques*, vol. 60, no. 10, pp. 3096-3112, Oct. 2012.

[5] R. Stefanini, M. Chatras, P. Blondy and G. M. Rebeiz, "Miniature MEMS Switches for RF Applications," *Journal of Microelectromechanical Systems*, vol. 20, no. 6, pp. 1324-1335, Dec. 2011.

[6] Y. Zhang, Z. Gong, H. Liu and Z. Liu, "Distributed Multicontact RF MEMS Switch for Power Handling Capability Improvement," *2019 20th International Conference on Solid-State Sensors, Actuators and Microsystems & Eurosensors XXXIII (TRANSDUCERS & EUROSENSORS XXXIII)*, Berlin, Germany, 2019, pp. 869-872.

[7] (2021) Menlo Micro. [Online] https://www.menlomicro.com/

[8] (2021) CoventorWare. [Online] https://www.coventor.com/

[9] R. Holm, *Electrical Contacts - Theory and Application*, 4th ed. New Jer-sey: Springer, 1967.

Proceedings of the 16th Annual IEEE International
Conference on Nano/Micro Engineered and Molecular Systems
April 25-29, 2021

Research on Motion Characteristics for Latching Mechanism of MEMS Safety and Arming Device under Dual Environmental Forces

Shenghong Lei[1], Yun Cao[1]*, Weirong Nie[1], Zhanwen Xi[1], Na Xu[2], Weixiang Qiu[1]

[1]School of Mechanical Engineering, Nanjing University of Science and Technology, China

[2] Hubei Aerospace Flight Vehicle Institute, China
Corresponding author: Yun Cao, E-mail: caoyun0620@sina.com

Abstract—To improve the matching design of safety and arming device parallel to the projectile axis with micro detonation sequence and plastic deformation under setback force, a MEMS safety and arming device with a latching mechanism placed perpendicularly to the projectile axis has been proposed. The proposed device is more beneficial to increase the force area of mechanism and to reduce the energy loss of micro detonation. The motion of the latching mechanism under dual environmental forces is theoretically modeled, and the criteria of reliable latching are given. Furthermore, dynamic finite element simulations are carried out to study the motion characteristics under dual environmental forces. The rerults in simulations indicate that the latching mechanism can achieve latching reliably under the rotary speed of 7000-18000 rpm with no plastic deformation.

Keywords—MEMS, safety and arming device, dual environmental forces, latching mechanism

I. INTRODUCTION

With the development of ammunition technology, the fuze has been endowed with more functions, and higher requirements have been put forwards for the volume distribution and full utilization of each part of the fuze. Hence, the research on the miniaturization of the fuze safety system has become a trend. Microelectromechanical System (MEMS) has been developed rapidly in the military field due to its advantages and has become one of the key technologies to realize miniaturization and function expansion of fuze. Since Charles H. Robinson proposed the MEMS safety and arming device (SAD) [1], the design and optimization of the MEMS SAD have made great development [2-6]. An important application direction of MEMS technology in fuze is the MEMS SAD.

In the previous research, most of the latching mechanisms are placed along the projectile axis. There are some problems that plastic deformation occurred in the mechanism due to setback force and the matching design of SAD with micro detonation sequence. And most studies in the motion characteristics analysis for MEMS SAD have only focused on the fluence of a single environmental force. However, MEMS SAD is subject to more than one environmental force in the realistic ballistic environment. To solve the problems above, a MEMS SAD with a latching

mechanism placed perpendicularly to the amunition axis has been proposed here. And the dual environmental forces are used to analyze the motion characteristics for the SAD with latching mechanism in the ballistic environment.

II. DEVICE DESIGN

We put forwards a MEMS SAD placed perpendicularly to the projectile axis in our work. The structure has shown in FIG.1. We introduce the MEMS SAD designed by us briefly here to make the article complete. The MEMS SAD has consisted of setback mechanism, centrifugal mechanism, slider, micro spring, frame, and latching mechanism. Among them, the latching mechanism plays an extremely pivotal role in ensuring the ammunition reliable arming by dual environmental forces produced in the internal and external ballistic. The latching mechanism is composed of latching beam, locking head, positioning rigid block. The designed mechanism will meet its requirements that latching reliably and no plastic deformation of the mechanism under the rotary speed of 7000-18000 rpm corresponding to setback acceleration of 5000-16000 g.

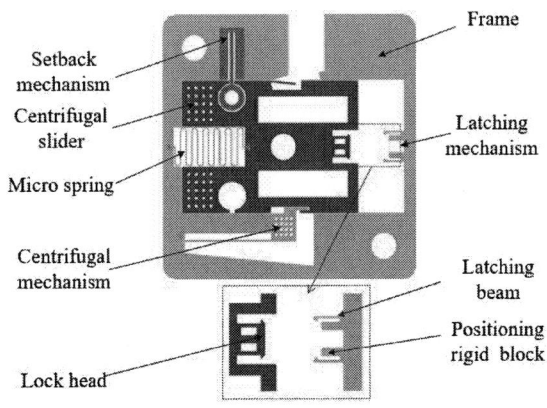

FIG. 1 MEMS SAD designed by us

III. THEORETICAL ANALYSIS

It is a prerequisite for reliable latching that the lock head can fully catch the latching beams. Assuming that the spring-tension remains constant when the latching beam is

978-1-6654-3008-1/21 $31.00 © 2021 IEEE   699

opened, the dynamic equation before reliable latching can be expressed as,

$$m\frac{d^2X}{dt^2} = \begin{cases} ma_c(t) - k\Delta X - \mu m(a_s(t)+g) & X \le x1 \\ ma_c(t) - \mu m(a_s(t)+g) - k\Delta X & \\ \quad -2\cdot F_N(sin\alpha - \mu cos\alpha) & x1 < X \le x2 \end{cases} \quad (1)$$

Herein,

$$\begin{cases} a_s(t) = \begin{cases} A_s \cdot \sin(\pi/T_s)t & t \le T_s \\ 0 & t > T_s \end{cases} \\ a_c(t) = (\pi/30)^2 \cdot n(t)^2 \cdot (r0 + \Delta X) \\ n(t) = \begin{cases} n_{\max}[1 - \cos(\pi \bullet t / T_s)] & t \le T_s \\ n_{\max} & t > T_s \end{cases} \end{cases} \quad (2)$$

Wherein, X is the displacement of the lock head, m is the mass of the slider and micro spring, k is the modulus of elasticity of micro spring, ΔX is the axial displacement of the lock head from static to entry of the lock seat completely, F_N is the extrusion pressure between the latching beam and lock head when the lock head opens the latching beam, α is the angle between the side of the lock head and its central axis, μ is the coefficient of friction. In addition, $a_s(t)$ is the setback acceleration with time, A_s and T_s is the pulse width and the amplitude of it, $a_c(t)$ is the centrifugal acceleration with time. $n(t)$ is the rotary speed with time, n_{max} is the maximum value of rotary speed. The key parameters of the mechanism as shown in FIG. 2. r_0 is the initial rotation radius of the slider. $x1$ is the displacement of the lock head in contact with the latching beam, and $x2$ is the displacement of the lock head just into the latching beam.

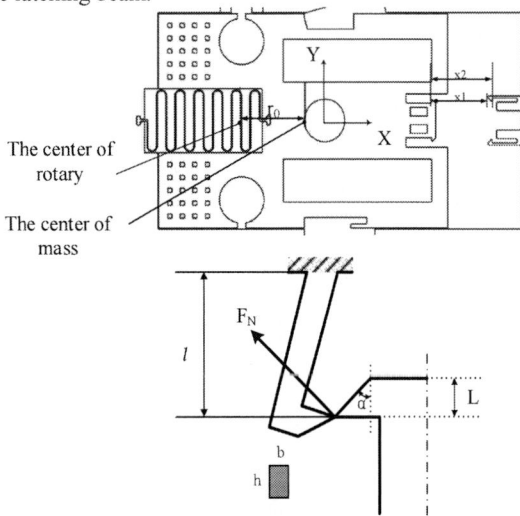

FIG. 2 Key parameters and force analysis of the mechanism

The latching beam could be considered as an elastic beam in the process of latching. The force analysis between the latching beam and lock head is shown in FIG. 2. The deflection of the latching beam is analyzed by the beam deflection curve equation. The deflection is calculated by

$W = (-Fx^2)/6EI(3l - x)$. When $x = l$, the deflection value at the position of l is $W = (-Fl^2)/3EI$. Moreover, the F is $F = F_N \cos\alpha - \mu \cdot F_N \sin\alpha$.

The relation between axial displacement X and deflection is $W = X\tan\alpha$. And the moment of inertia for the latching beam is expressed as $I = (hb^3)/12$.

According to the formula above, we can get the expression of F_N is that

$$F_N = (Ehb^3 tan\alpha)\cdot X / (4l^3(cos\alpha - \mu sin\alpha)) \quad (3)$$

When $X=L$, the lock head just enters the critical state of the lock seat, and the maximum value of F_N is obtained.

$$F_{Nmax} = (Ehb^3 tan\alpha)\cdot L / (4l^3(cos\alpha - \mu sin\alpha)) \quad (4)$$

Wherein, h is the thickness of the latching beam, b is the width of the latching beam, L and l is the length of lock head and latching beam respectively.

Combining the equation above, the motion of the slider at this critical state is that:

$$\frac{d^2X}{dt^2} = \frac{ma_c(t) - k\Delta X}{m} - \frac{(LEhb^3 tan\alpha)(\sin\alpha + \mu\cos\alpha)}{(2ml^3(cos\alpha - \mu sin\alpha))} \quad (5)$$

The formula above can be used to determine whether the latching mechanism is latching reliably or not. In another word, if $dX/dt > 0$, it means that the latching mechanism can latch reliably, if not, the latching mechanism can't.

IV. DYNAMIC SIMULATION ANALYSIS

To better and clearer understand the working process and the motion characteristics of the latching mechanism under dual environmental forces, the finite element model of the MEMS SAD is built as shown in FIG. 3. The finite element method model is simulated by ABAQUS explicit dynamic analysis software.

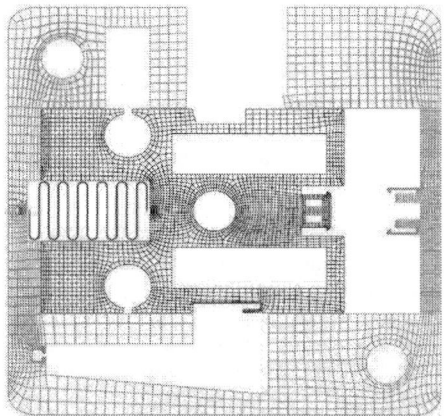

FIG. 3 FEM model of the MEMS SAD

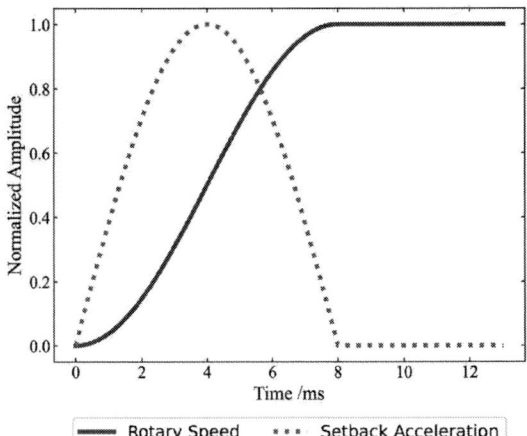

FIG. 4 The pulse width and normalized amplitude of load

The material parameters setting during the simulation are shown in Table I. Besides, the pulse width and amplitude of load are shown in FIG. 4.

TABLE I. THE MATERIAL PARAMETERS SET

Mass Density (tonne/mm³)	Young's Modulus (MPa)	Poisson's Ratio	Yield Stress (MPa)
8.91e-9	1.8e5	0.3	750

To explore the performance of the latching mechanism under dual environmental forces, the mechanism is simulated by loading the load condition 1-12 as shown in Table II.

TABLE II. THE LOAD CONDITION SET

Load Condition	Setback Acceleration (g)	Pulse Width (ms)	Rotary Speed (rpm)
1	5000	8	7000
2	6000	8	8000
3	7000	8	9000
4	8000	8	10000
5	9000	8	11000
6	10000	8	12000
7	11000	8	13000
8	12000	8	14000
9	13000	8	15000
10	14000	8	16000
11	15000	8	17000
12	16000	8	18000

FIG. 5 displays the displacement result of the latching mechanism by loading load condition 1. FIG. 5(a) shows that the latching mechanism cannot be locked when the

setback and centrifugal acceleration act together. The latching mechanism moves rapidly towards to arming position when the action of the setback force ends. The displacement of the lock head reaches 3.005 mm at 8.905ms to complete the first locking. Due to the existence of a rigid block for positioning in the lock seat, the lock head and the rigid block of the lock seat have collision and rebound briefly. After the first rebound, the lock head continues to move to the lock seat under the action of centrifugal acceleration, then the lock head completes the second locking at 10.18ms. After that, the lock head and the rigid block for positioning continue to collide and rebound many times as shown in FIG. 5(b). However, the lock head never breaks away from the constraint of the locking beam during the rebound process, and finally, the lock head stays in close contact with the lock seat. It suggests that the latching mechanism realizes stable locking.

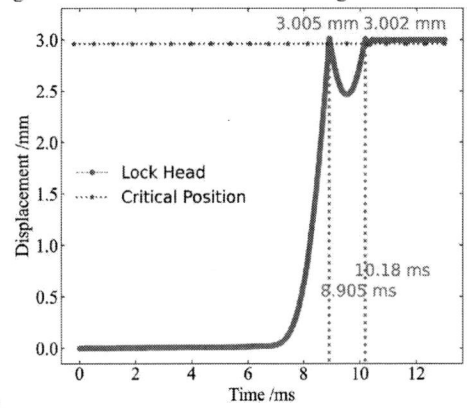

(a).

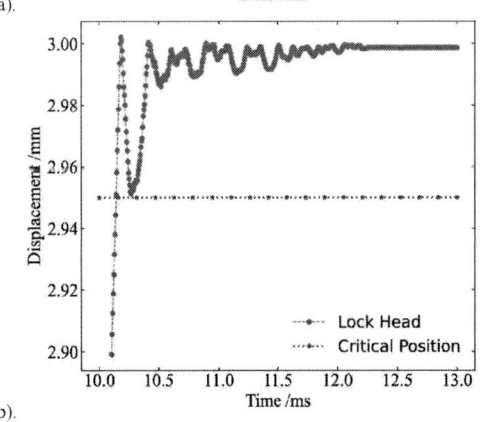

(b).

FIG. 5 Displacement result under load condition 1. (a) displacement of the whole process(b) displacement after reliable latching

FIG. 6 shows that the maximum stress for the MEMS SAD is located at the micro spring, not the latching mechanism when the setback and centrifugal acceleration are applied together. Moreover, the maximum stress is much lower than the yield stress of the material.

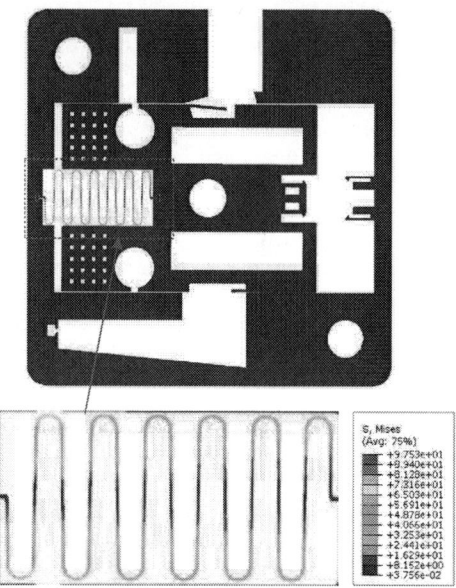

FIG. 6 Contour of maximum stress for SAD

FIG. 7(a) tells us that the maximum stress for the latching mechanism is located at the root of the latching beam under the action of centrifugal force when the lock head is opening the latching beam. FIG. 7(b) presents the maximum stress is located between the lock head and rigid block when they have collision briefly. Besides, the values of maximum stress are less than the yield stress of the material. The results, as shown in FIG. 7, indicate that there is no plastic deformation in the latching mechanism at 7000rpm.

Furthermore, we take the rotary speed 7000 rpm and setback acceleration 5000g as an initial value, 1000 rpm and 1000g as a step length to explore the performance of the micro SAD module at higher setback acceleration and rotary speed.

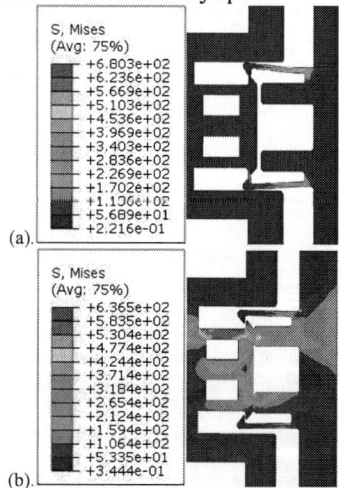

FIG. 7 Contour of maximum stress for a latching mechanism. (a). the maximum stress of the latching beam (b). maximum stress between the lock head and positioning rigid block

The displacement results of the mechanism at higher setback acceleration and rotary speed are shown in FIG. 8. It suggests that the latching mechanism can latch reliably finally, and the motion mode for them is similar to that at load condition 1, no more details here.

FIG. 9(a) and FIG. 9(b) display the maximum stress values for dangerous sections of the latching mechanism. On the one hand, all of the maximum stress is less than the yield stress of the material as seen from FIG. 9. In another word, there is no plastic deformation in the latching mechanism at higher rotary speed.

All the results above indicate that the mechanism designed in this paper performs well at the rotary speed of 7000-18000 rpm corresponding to the setback force of 5000-16000 g. It means that the latching mechanism designed meets its requirements.

V. CONCLUSION

In this paper, a MEMS safety and arming device with a latching mechanism has been proposed. Under dual environmental forces, the motion of the latching mechanism is theoretically modeled, and the criteria of reliable latching are given. To study the motion characteristics under dual environmental forces, the dynamic finite element simulations are carried out. The results in simulations indicate that the latching mechanism can achieve latching reliably under the rotary speed of 7000-18000 rpm with no plastic deformation.

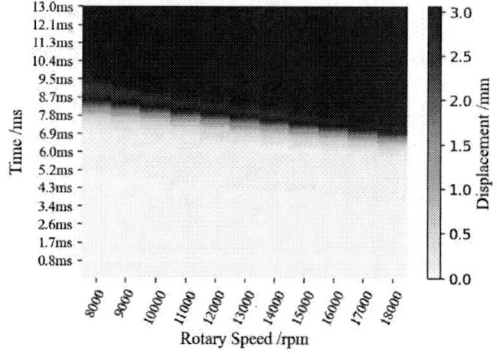

FIG. 8 Displacement results of the mechanism at the higher rotary speed

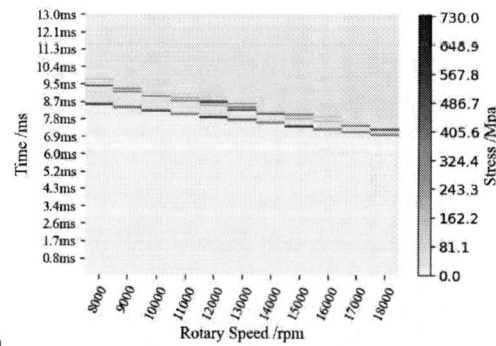

(a).

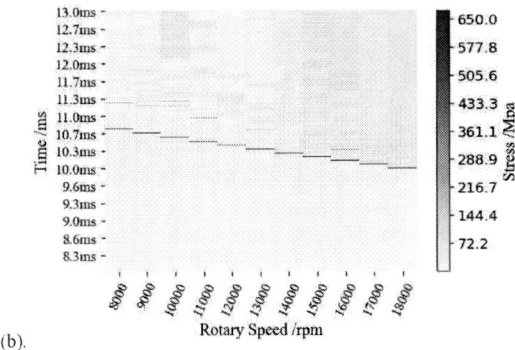

(b).

FIG. 9 Maximum stress of the latching mechanism. (a). the maximum stress of the latching beam. (b). maximum stress between the lock head and rigid positioning block

VI. NEXT

The MEMS SAD has been fabricated, as shown in FIG. 10. A further test will be tested to verify the validity of the latching mechanism.

FIG. 10 The fabricated MEMS SAD

ACKNOWLEDGMENT

This work is supported by the National Natural Science Foundation of China (No.51805268) and the Fundamental Research Funds for the Central Universities (No.30920021101).

REFERENCES

[1] C. H. Robinson, R. H. Wood, T. Q. Hoang, and U. S. O. A. The, "Miniature MEMS-based electro-mechanical safety and arming device," *U.S. Patent 6964231B1, 2005-11-15 2005.*

[2] C. H. Robinson and R. H. Wood, "Ultra-miniature electro-mechanical safety and arming device," *U.S. Patent 8276515B1,* 2012.

[3] F. Chenyang, L. Xiaojie and W. Falin, "MEMS slide block locking mechanism based on elastic supporting structure," in *2012 International Conference on Manipulation, Manufacturing and Measurement on the Nanoscale,* 2012, pp. 254-258.

[4] L. W. F. Y. Wang Fufu, "Parametric Research of MEMS Safety and Arming System," in *IEEE-NEMS 2013,* 2013.

[5] J. O. Seok, J. Jeong, J. Eom, S. S. Lee, C. J. Lee, S. M. Ryu, and J. S. Oh, "Ball driven type MEMS SAD for artillery fuse," *Journal of micromechanics and microengineering,* vol. 27, p. 15032, 2016.

[6] J. Jeong, J. Eom, S. S. Lee, D. W. Lim, Y. I. Jang, K. W. Seo, S. S. Choi, C. J. Lee, and J. S. Oh, "Miniature mechanical safety and arming device with runaway escapement arming delay mechanism for artillery fuze," *Sensors and Actuators A: Physical,* vol. 279, pp. 518-524, 2018.

Proceedings of the 16th Annual IEEE International
Conference on Nano/Micro Engineered and Molecular Systems
April 25-29, 2021

Modeling and Optimization of ScAlN-based MEMS Mirror with Large Static Two-axis Tilting Angle

Changhe Sun[1,2,3,*], Bolun Li[2], Wenqu, Su[2], Yufei Liu[3], and Yaming Wu[4], *Member, IEEE*

Abstract— A particularly efficient piezoelectric MEMS mirror with the effective area of 10×10 mm2 for optical scanning applications has been developed using $Sc_xAl_{1-x}N$ material as the actuator with Sc content up to x = 0.43. The architecture of this novel MEMS mirror is comprised of a large-size reflective micromirror plate and four cloverleaf-shaped $Sc_xAl_{1-x}N$ micro-cantilever actuators. After optimization of the structural parameters, the simulated static tilting angle of $Sc_{0.41}Al_{0.59}N$ MEMS mirror reaches to a peak value of ±20.5 °@120 V_{DC} with good linearity and two-axis operation, which is much larger than previous results. In addition, the dependence of the static tilting angle on Sc concentration has also been investigated theoretically.

Keywords— Scanning mirror; ScAlN; Static tilting angle; Piezoelectric actuation; Rotation transformation

I. INTRODUCTION

Microelectromechanical systems (MEMS) based scanning mirrors have been proven as an indispensable technology in laser radar, optical communication, optical switches, high-resolution displays and biomedical imaging [1]. The main advantages of MEMS scanning mirrors over the conventional large-scale mechanisms include: miniaturized size, low-power consumption, easy integration and cost efficiency. Typically, a reliable MEMS scanning mirror, for instance used in LIDAR imaging, is desired to provide a specified tilting angle and retain its position at arbitrary moment according to the applied voltage. Moreover, the ideal scanning mirror should have good mechanical linearity, high-accuracy control and a pupil diameter of less than 10 mm. Till now, there are a variety of actuation methods applied for MEMS scanning mirrors, mostly based on electrostatic, electromagnetic, electrothermal and piezoelectric actuation principles [2-5]. Though the electrostatic, electromagnetic and electrothermal actuators can exhibit some unique characteristics for certain applications,

some common challenges confronted largely prevent to achieve both large 2-Degrees-of-Freedom (2-DoF) tilting angles and static non-resonant operation. Instead, numerous impressive piezoelectric actuation MEMS mirrors reported in the previous literature have demonstrated to achieve resonant/static actuation mode, linear control, fast response, and low power consumption characteristics [1, 5-7].

A piezoelectric micromirror is less affected by air damping, thermal or electromagnetic interference than those driven by electrostatic, thermal or electromagnetic force. There are a wide variety of designs for piezoelectric MEMS mirrors, with most of them deploying a large mechanical rotation angle and a low operation voltage level [6, 7]. Baran et al developed a 1.4-mm-wide PZT thin-film actuation scanning mirror with the combination of mechanical amplification, exhibiting an optical scan angle of 38.5° at 24 V with the resonant frequency of 40 kHz [8]. Chen et al presents a MEMS thin-PZT cantilever driven micro-lens actuator capable of delivering a large out-of-plane displacement of 145um at 22 V driving voltage, with a resonant frequency of 2 KHz [9]. Such superior architectures are almost entirely based on PZT piezoelectric material. while the challengeable patterning process and poor compatibility with the mainstream CMOS and/or MEMS process greatly limits its wide use in the MEMS area.

Table 1. Comparison between ScAlN, AlN and PZT

Performance	ScAlN	AlN	PZT
Material Category	Non-ferroelectric		Ferroelectric
$e_{31,f}$ [C/m²]	~3.16	~1.1	~21
ε_r	~16.7	~10	~1300
$e_{31,f}^2 / \varepsilon_0 \varepsilon_r$	0.598	0.121	0.339
V_{DC} [V]	~200	~200	~2
Linearity	Linear		hysteresis
Directionality	Bidirectional		Unidirectional
Long-term Stability	Stable		Degrading
CMOS compatibility	Yes		No

AlN-based piezoelectric actuation strategy for a MEMS mirror application has been presented with a modified silicon-on-insulator-based MEMS process, demonstrating full CMOS compatibility, excellent linear control and various mirror movement modes [1]. However, both dynamic and static mirror operation modes of this device have the tilting angles of

This work was supported by the open fund of Cooperative Innovation Center of Unconventional Oil and Gas (Ministry of Education & Hubei Province), Yangtze University under UOG2020-03, the grant project of Science and Technology Research Program from Educational Committee of Hubei Province under Q20201309, and Science and Technology Development Funds of Yangtze University.

[1]Cooperative Innovation Center of Unconventional Oil and Gas (Ministry of Education & Hubei Province), Yangtze University, Wuhan 430100, China

[2]National Demonstration Center for Experimental Electrotechnics and Electronics Education, Yangtze University, Jingzhou 434023, China

[3]Key Laboratory of Optoelectronic Technology & Systems (Chongqing University), Ministry of Education, Chongqing, 400044, China

[4]State Key Laboratory of Transducer Technology, Shanghai Institute of Microsystem and Information Technology, Chinese Academy of Sciences, Shanghai 200050, China

*Contacting author: Changhe Sun is with the Electronics & Information School, Yangtze University, China (E-mail: chhesun@163.com)

only about 0.2°/V and 0.005°/V, respectively, which is attributed to low piezoelectric coefficients of AlN material. Based on Table 1, ScAlN as a significantly higher performance piezoelectric material than AlN and PZT, can be used to deliver large force enabling a mechanical tilting angle of ±14° at 150 V_{DC}, in addition to the aforementioned inherent superiorities in AlN-based micromirrors. Although only a handful of ScAlN-based micromirrors have been created with the improvement of piezoelectric coefficient $e_{31,f}$ [10, 11], up to now the most pressing research needed to significantly promote MEMS scanning mirror technology lies primarily in pushing novel advanced designs to gain their full potential. Therefore, aiming at further improving theoretical performance of the piezoelectric MEMS mirror, a novel ScAlN-based MEMS actuation mechanism is developed here for mirror applications, theoretically offering a maximum two-axis tilting angle of ±20.5° at 120 V DC voltage.

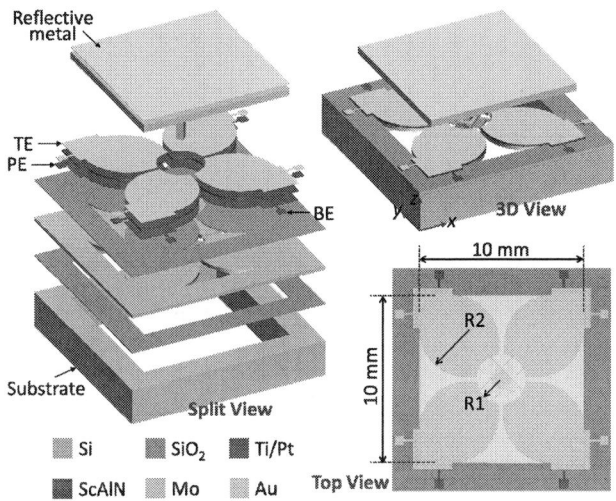

Fig. 1. Structure configuration of ScAlN MEMS mirror.

II. ELECTROMECHANICAL DESIGN

Due to the abovementioned factors, a particularly efficient piezoelectric MEMS scanning mirror with the effective area of 10×10 mm^2 has been designed for investigating the static electromechanical performance using $Sc_xAl_{1-x}N$ material as the actuator with Sc content up to $x = 0.43$. As illustrated in Fig. 1, the architecture of the MEMS mirror is mainly comprised of a reflective micromirror plate and four cloverleaf-shaped $Sc_xAl_{1-x}N$-based piezoelectric unimorph microcantilevers (PE) laminated with a top electrode (TE) and bottom electrode (BE), which are connected to a silicon-on-insulator (SOI) substrate having a 10-µm-thick device layer. The micromirror plate is coated with Au thin film to increase the reflective coefficient of the reflector. To explore a much larger mechanical tilting angle than 15° within 150 V DC voltage, four S-shaped meandering springs are adopted to connect the PEs and micromirror plate. Particularly, such spring design can avoid high stress concentrations and leverage its rotation

transformation to enlarge the two-axis deflection of the micromirror. Both the insulating SiO$_2$ layers and $Sc_xAl_{1-x}N$ film have the thickness of 1 µm.

When the PE actuators connected with two opposite sides of the mirror are applied DC voltages of the opposite polarities, the mirror will be rotated by some torque produced by two opposite deflects under the electrostatic forces. Fig.2 illustrates the mechanism of this newly developed static $Sc_xAl_{1-x}N$ MEMS scanning mirror. It should be noted that two opposite driving voltages are applied on opposite actuator pairs for tilting the micromirror along one axis (x or y), while for arbitrary tilting motions all actuators need be deployed.

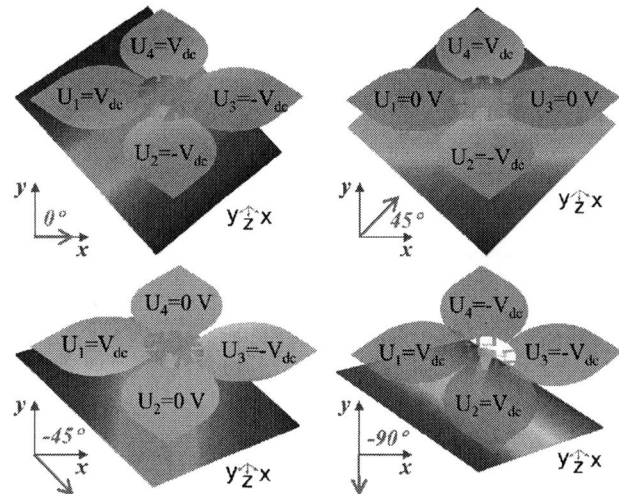

Fig. 2. Static simulation analysis of mirror with different voltage excitations.

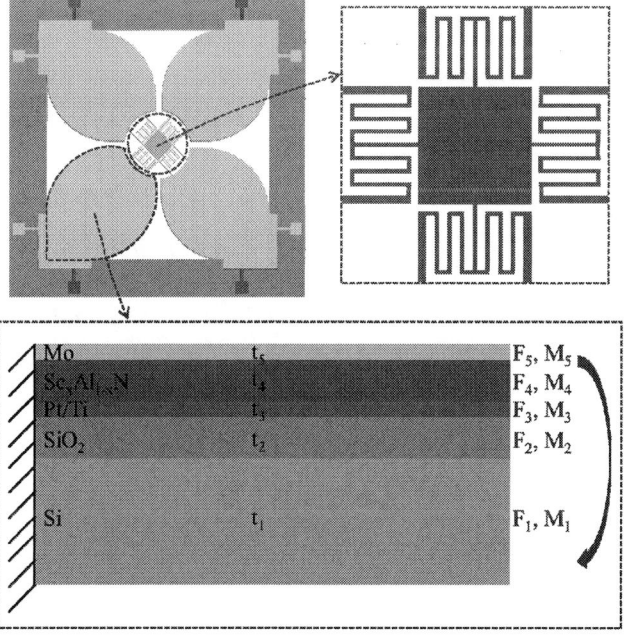

Fig. 3. Simplified structure of the PE actuator part in the MEMS mirror.

III. ANALYTICAL MODELING

In order to pursuit the best behavior of the $Sc_xAl_{1-x}N$ MEMS scanning mirror, it is very essential to theoretically analyze the $Sc_xAl_{1-x}N$ actuator part and optimize its structural parameters. Therefore, an analytical model based on thin plate theory is developed to better understand the device. In this model, the static deflection characteristics of the PE actuator part is investigated by approximately considering it is equivalent to a multi-layer cantilever with a fixed boundary condition at one end and roller boundary condition at the S-shaped meandering spring connection end, as shown in Fig. 3.

Here it is assumed both the residual stress and shear effects can be ignored to simplify the mathematical modeling. According to the equilibrium conditions of the axial forces and moments acting on each layer, the total force F_{tot} and moment M_{tot} can be expressed as:

$$F_{tot} = \sum_{i=1}^{5} F_i = 0 \tag{1}$$

$$M_{tot} = \sum_{i=1}^{5} M_i = \sum_{j=1}^{5} F_j \left(\sum_{i=1}^{j-1} t_i + \frac{t_j}{2} \right) \tag{2}$$

The curvature $1/r$ of the i th-layer in the actuator part can be approximately written as: $1/r = M_i/E_iI_i$, where E_iI_i is the flexural rigidity. Note that the total thick of the multilayer actuator part is much small so that the curvature of each layer can be assumed to be same. Thus, the curvature $1/r$ is estimated by the following equation:

$$\frac{1}{r} = \frac{1}{\sum_{i=1}^{5} E_iI_i} \left[\frac{t_1}{2} \quad t_1 + \frac{t_2}{2} \quad \sum_{i=1}^{2} t_i + \frac{t_3}{2} \quad \sum_{i=1}^{3} t_i + \frac{t_4}{2} \quad \sum_{i=1}^{4} t_i + \frac{t_5}{2} \right] \mathbf{F}$$

$$= \frac{1}{\sum_{i=1}^{5} E_iI_i} \mathbf{DF} \tag{3}$$

where $\mathbf{F} = \begin{bmatrix} F_1 & F_2 & F_3 & F_4 & F_5 \end{bmatrix}^T$ is the column vector of axial forces. By equating the strain at the interface between the i th layer and $(i+1)$ th layer, we can get:

$$d_{31}E_i + \frac{F_i}{A_iE_i} - \frac{t_i}{2r} = d_{31}E_{i+1} + \frac{F_{i+1}}{A_{i+1}E_{i+1}} - \frac{t_{i+1}}{2r} \tag{4}$$

where d_{31} is the piezoelectric coefficient of the $Sc_xAl_{1-x}N$ layer, which is associated with the Sc content x, can be approximately written by the following fitting equation [9]:

$$d_{31}(x) = -2.87 + 0.862x + 5.35x^2 - 2.89x^3 \tag{5}$$

Based on Eq. (3) and Eq. (4), we can yield the final expression of $1/r$ without the forces. The tip deflection z_p of the actuator part with a length l_p can be expressed as:

$$z_p = \frac{l_p^2}{2r} \tag{6}$$

Fig. 4 shows the free body diagram of the single PE actuator with two fixed boundaries, meandering springs as the rotation transformer (RT) and a mirror connector. Consider the angle θ_p at the tip deflection of the PE actuator, where

$\theta_p = M_A l_p / \sum_{i=1}^{5} E_iI_i$, the rotation angle of the mirror connector can be estimated by the following relationship:

$$\theta_m = \arcsin \left[\frac{z_p + l_s \sin(\theta_p)}{l_m} \right] \tag{7}$$

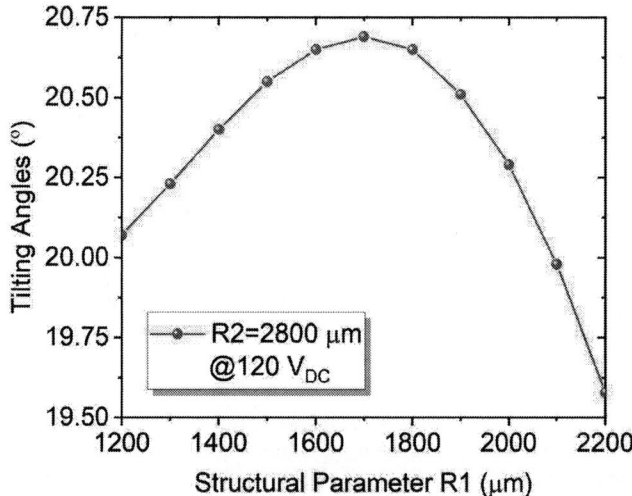

Fig. 4. Free body diagram of the single PE actuator

IV. RESULTS AND DISCUSSION

To quantitatively analyze the electromechanical performance of the proposed mirror mechanism, the structural parameters including R1 and R2 (in Fig. 1) of the actuators is optimized through the finite-element-modeling (FEM) simulations using commercial software COMSOL. To reduce the simulation load, all boundaries of four cloverleaf-shaped actuators connected to the SOI substrate are assumed to be clamped. The simulated tilting angle of the $Sc_{0.41}Al_{0.59}N$-based MEMS mirror reaches to a peak value of 20.5°@120 V_{DC} at moderate-size structure parameters R1 and R2, as shown in Fig. 5 and Fig. 6.

Fig. 5. Simulated tilting angles of the scanning mirror with the varying structural parameter R1.

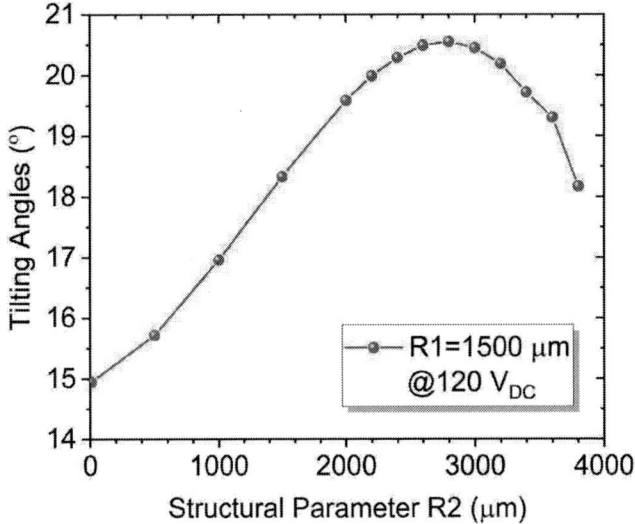

Fig. 6. Simulated tilting angles of the scanning mirror with the varying structural parameter R2.

In the case of static voltage actuation, the tilting angle of the constant mirror structure will linearly increase with applied DC voltage, providing a reasonably high slope of 0.173°/V in Fig. 7, which is much larger than previously reported results, especially when compared with AlN-based scanning mirrors. The outstanding rotation performance may be mainly attributed to the optimal structure design of the cloverleaf-shaped PE actuator and the adoption of the rotation transformer to further enlarge the tilting angle obtained by piezoelectric actuation.

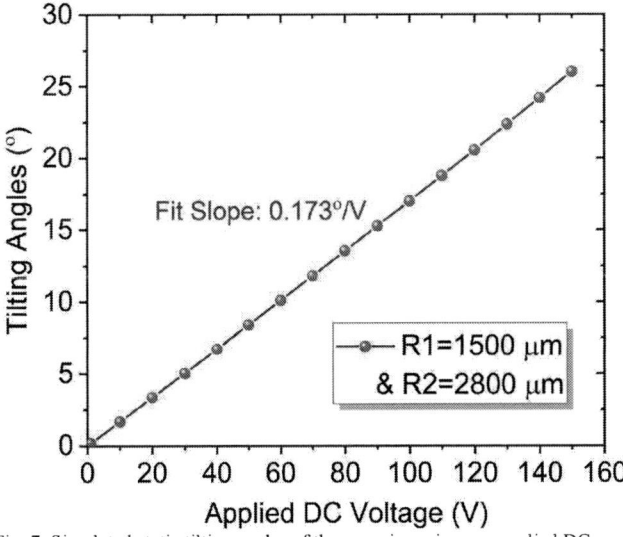

Fig. 7. Simulated static tilting angles of the scanning mirror vs applied DC voltages.

In addition, since the Sc concentration x in the $Sc_xAl_{1-x}N$ film closely associated to its polarization and piezoelectricity will largely determine the deflection response of the scanning mirror, according to Eq. (4) – Eq. (6), a series of FEM simulations with different Sc concentrations have also been carried out by combining experimentally obtained and theoretically computed piezoelectric properties in Ref. [5, 10,

11]. Fig. 8 demonstrates that there is positive correlation between the mechanical tilting angle and the Sc concentration but with a remarkable nonlinearity, which can be attributed to nonlinear piezoelectricity and Young's modulus of the $Sc_xAl_{1-x}N$ film with increasing Sc concentration [10]. The simulated data also indicates the static mechanical tilting angle of $Sc_{0.43}Al_{0.57}N$-based MEMS mirror is about 3.5 times larger than that of AlN-based same architecture and also much larger than previously reported ScAlN-based micromirrors.

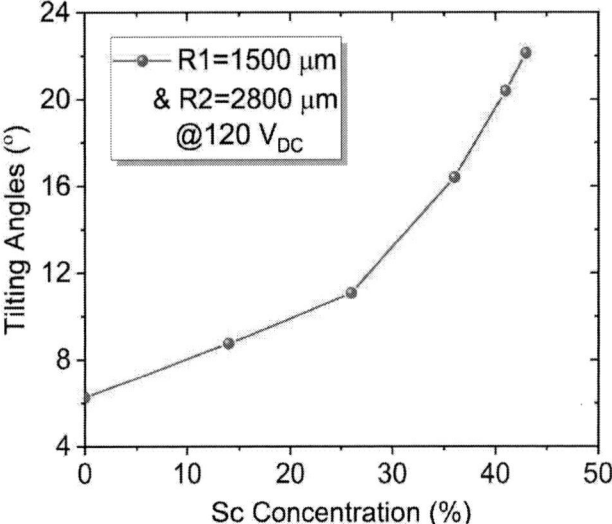

Fig. 8. Simulated static tilting angles of the scanning mirror with different Sc concentrations.

V. CONCLUSIONS

In conclusion, a novel $Sc_xAl_{1-x}N$ based piezoelectric MEMS scanning mirror with an effective area of 10×10 mm^2 has been developed. The architecture is comprised of four cloverleaf-shaped $Sc_xAl_{1-x}N$ micro-cantilever actuators and a large-size reflective micromirror plate. The theoretical model of this novel structure has been established, demonstrating that the adoption of an appropriate rotation transformer design can remarkably enlarge the tilting angle. After optimization of the structural parameters, the simulated static tilting angle of $Sc_{0.41}Al_{0.59}N$ based MEMS mirror reaches to a peak value of $\pm20.5°@120$ V$_{DC}$, especially offering good linearity and two-axis operation, which is much larger than previous results. Besides, the dependence of the static tilting angle on Sc concentration has also been investigated. Although the static mechanical behavior of the proposed $Sc_xAl_{1-x}N$ device can be comparable to dynamic response behavior of some PZT-based scanning mirrors, there are still some challenges to overcome: three-dimensional wafer-level microfabrication process, high-efficiency and long-term stable ScAlN film with Sc content of over 41%, and much higher-performance actuator structure design, which will be focused in the later works.

REFERENCES

978-1-6654-3008-1/21 $31.00 © 2021 IEEE

[1] J. Shao, Q. Li, C. Feng, et al. "AlN based piezoelectric micromirror," *Optics Letters*, 2018, 43(5): 987-990.

[2] C. -L. Hung, H. Y. –H. Lai, T. W. Lin, et al. "An electrostatically driven 2D micro-scanning mirror with capacitive sensing for projection display," *Sensors and Actuators A: Physical*, 2015, 222:122-129.

[3] Y. Zhou, Q. Wen, Z. Wen, et al. "An electromagnetic scanning mirror integrated with blazed grating and angle sensor for near infrared micro spectrometer." *Journal of Micromechanics & Microengineering*, 2017, 27(12): 125009.

[4] K. Jia, S. Pal, H. Xie. "An Electrothermal Tip–Tilt–Piston Micromirror Based on Folded Dual S-Shaped Bimorphs," *Journal of Microelectromechanical Systems*, 2009, 18(5): 1004-1015.

[5] S. Gu-Stoppel, T. Lisec, S. Fichtner, et al. "AlScN based MEMS quasi-static mirror matrix with large tilting angle and high linearity," *Sensors and Actuators A: Physical*, 2020, 312, 112107.

[6] S. –H. Chen, A. Michael, C. Y. Kwok. "Design and Modeling of Piezoelectrically Driven Micro-Actuator With Large Out-of-Plane and Low Driving Voltage for Micro-Optics," *Journal of Microelectromechanical Systems*, 2019, 28(5): 919-932.

[7] K. H. Koh, T. Kobayashi, C. Lee. "Investigation of piezoelectric driven MEMS mirrors based on single and double S-shaped PZT actuator for 2-D scanning application," *Sensors and Actuators A: Physical*, 2012, 184: 149-159.

[8] U. Baran, D. Brown, S. Holmstrom, et al. "Resonant PZT MEMS Scanner for High-Resolution Displays," *Journal of Microelectromechanical Systems*, 2012, 21(6):1303-1310.

[9] S. H. Chen, A. Michael, C. Y. Kwok. "Design and Modeling of Piezoelectrically Driven Micro-Actuator With Large Out-of-Plane and Low Driving Voltage for Micro-Optics," *Journal of Microelectromechanical Systems*, 2019, 28(5):919-932.

[10] M. Akiyama, K. Umeda, A. Honda, et al. "Influence of scandium concentration on power generation figure of merit of scandium aluminum nitride thin films,". *Applied Physics Letters*, 2013, 102: 021915.

[11] M. A. Caro, S. Zhang, M. Ylilammi, et al. "Erratum: Piezoelectric coefficients and spontaneous polarization of ScAlN (2015 J. Phys. Condens. Matter 27 245901)," *Journal of Physics Condensed Matter*, 2015, 27: 245901.

The 16th IEEE International Conference on Nano/Micro Engineered & Molecular Systems

Wavelength conversion through plasmon-coupled surface states

Mona Jarrahi

Electrical and Computer Engineering Department, University of California Los Angeles, 90095, CA, USA

Website : https://www.seas.ucla.edu/~mjarrahi/mjarrahi.html

ABSTRACT

Surface states generally degrade semiconductor device performance by raising the charge injection barrier height, introducing localized trap states, inducing surface leakage current, and altering the electric potential. Therefore, there has been an endless effort to use various surface passivation treatments to suppress the undesirable impacts of the surface states. We show that the giant built-in electric field created by the surface states can be harnessed to enable passive wavelength conversion without utilizing any nonlinear optical phenomena. Photo-excited surface plasmons are coupled to the surface states to generate an electron gas, which is routed to a nanoantenna array through the giant electric field created by the surface states. The induced current on the nanoantennas, which contains mixing product of different optical frequency components, generates radiation at the beat frequencies of the incident photons. We utilize the unprecedented functionalities of plasmon-coupled surface states to demonstrate passive wavelength conversion of nanojoule optical pulses at a 1550 nm center wavelength to terahertz regime with record-high efficiencies that exceed nonlinear optical methods by 4-orders of magnitude. The presented scheme can be used for optical wavelength conversion to different parts of the electromagnetic spectrum ranging from microwave to infrared regimes by using appropriate optical beat frequencies.

BIOGRAPHY

Mona Jarrahi received her B.S. degree in Electrical Engineering from Sharif University of Technology in 2000 and her M.S. and Ph.D. degrees in Electrical Engineering from Stanford University in 2003 and 2007. She served as a Postdoctoral Scholar at the University of California Berkeley from 2007 to 2008. After serving as an Assistant Professor at the University of Michigan Ann Arbor, she joined the University of California Los Angeles in 2013 where she is currently a Professor of Electrical Engineering and the Director of the Terahertz Electronics Laboratory. Prof. Jarrahi has made significant contributions to the development of ultrafast electronic and optoelectronic devices and integrated systems for terahertz, infrared, and millimeter-wave sensing, imaging, computing, and communication systems by utilizing novel materials, nanostructures, and quantum well structures as well as innovative plasmonic and optical concepts. The outcomes of her research have appeared in 200 publications and 160 keynote/plenary/invited talks and have received a significant amount of attention from scientific news outlets including Huffington Post, Popular Mechanics, EE Times, IEEE Spectrum, Optics & Photonics News Magazine, Laser Focus world, and Photonics Spectra Magazine. Her scientific achievements have been recognized by several international and national prestigious awards including the Presidential Early Career Award for Scientists and Engineers (PECASE); Friedrich Wilhelm Bessel Research Award from Alexander von Humboldt Foundation; Moore Inventor

Fellowship from the Gordon and Betty Moore Foundation; Kavli Fellowship by the USA National Academy of Sciences (NAS), Grainger Foundation Frontiers of Engineering Award from the USA National Academy of Engineering (NAE); Breakthrough Award from Popular Mechanics Magazine; Research Award from Okawa Foundation; Early Career Award in Nanotechnology from the IEEE Nanotechnology Council; Outstanding Young Engineer Award from the IEEE Microwave Theory and Techniques Society; Booker Fellowship from the USA National Committee of the International Union of Radio Science; Lot Shafai Mid-Career Distinguished Achievement Award from the IEEE Antennas and Propagation Society; Early Career Award from the USA National Science Foundation (NSF); Young Investigator Awards from the USA Office of Naval Research (ONR), the Army Research Office (ARO), and the Defense Advanced Research Projects Agency (DARPA); the Elizabeth C. Crosby Research Award from the University of Michigan; and the Distinguished Alumni Award from Sharif University of Technology. Prof. Jarrahi is a Fellow of IEEE, OSA, SPIE, and IOP societies and has served as a distinguished lecturer of IEEE, traveling lecturer of OSA, and visiting lecturer of SPIE societies.

978-1-6654-3008-1/21 $31.00 © 2021 IEEE

978-1-6654-3008-1/21 $31.00 © 2021 IEEE

Proceedings of the 16th Annual IEEE International
Conference on Nano/Micro Engineered and Molecular Systems
April 25-29, 2021

Photonic Nanojet Produced by A Microfluidic Channel for Biofluid Monitoring*

Guoqiang Gu, Pengcheng Zhang, and Hui Yang*

Abstract—Owing to the properties like narrow beam waist and high intensity focusing, photonic nanojet (PNJ) has attracted sustainable attention in many fields, such as single nanoparticle or molecular detection, fluorescent or Raman scattering enhancement, subwavelength nanopatterning, super-resolution imaging, etc. The characteristics of PNJ are mainly affected by the material compositions and morphology features of the investigated microstructures. In this paper, we present a microfluidic channel structure to produce PNJs for biofluid monitoring. The proposed scheme is numerically studied by using the finite element method. Through the change of PNJ parameters, such as the maximum light intensity, working distance and full width at half-maximum, the refractive indices of the injected biofluids can be determined.

I. BACKGROUND

Biofluid monitoring plays an important role in the field of life science. The physical properties of biofluid such as refractive index, density, viscosity, conductivity and opacity can provide abundant information for monitoring [1]. Refractive index is a very interesting property as it contains a lot of important formation on such as blood glucose level and label-free molecular interaction [2, 3]. Optical measurement on the refractive index has attracted wide attentions owing to its reliability, fast response and high sensitivity [4]. Photonic nanojet (PNJ), a scattered beam of light with a high-intensity main lobe and very narrow waist that can be generated upon illumination on a transparent dielectric symmetric microstructure. Therefore, PNJ is a good candidate for monitoring the refractive index of liquid [5, 6], as the field distributions and shape features of the PNJ are mainly determined by the geometrical morphologies and material properties of the microstructures [7-11]. The experimental characterization of PNJ can be performed using the method proposed in Ref. [12].

Here, we design a microfluidic channel structure consisted of a glass tube and filled with biofluids. The microfluidic structure is designed with geometry specific for the generation of PNJ. The characteristics of the PNJ are theoretically investigated by numerical simulation using finite element method (FEM). The parameters of the PNJ, i.e. maximum light intensity I_{max}, working distance (WD) and full-width at half maximum (FWHM) changing with the refractive indices of different biofluids filled in the microfluidic channel, are analyzed. The formed two

different PNJ modes based on the inflection point of the first light refraction interface potentially provide more ways for probing. Our research will facilitate the study of integrating PNJ phenomenon with microfluidic devices as well as the exploitation of new applications based on such integrated system.

II. RESEARCH APPROACH

The schematic of the research model is shown in Fig.1(a). The microfluidic channel made of glass consists of one hollow microtubule in the middle and two cuboids on both sides. Biofluids can be injected through the inlet channel of the cuboid on the left, filling the hollow microtubule and flowing out of the cuboid on the right. The illumination light is incident on the top of the microfluidic

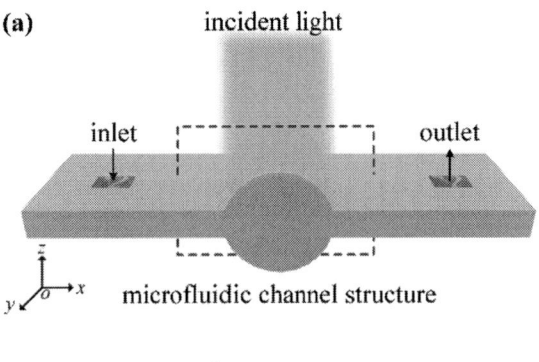

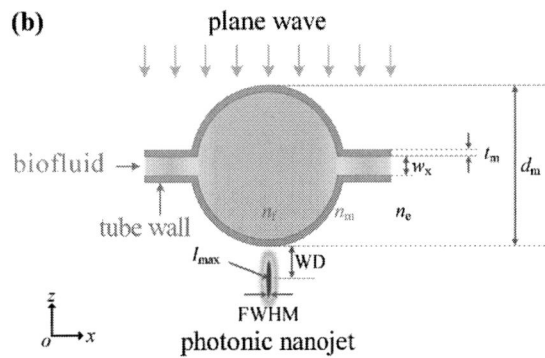

Fig. 1. Schematic diagrams of plane-wave-illuminated microfluidic channel system. (a) Three-dimensional (3D) structural model and (b) two-dimensional (2D) cross-section showing the simulation model.

---

*Research supported by the Guangdong Basic and Applied Basic Research Foundation (2019A1515011242), Key-Area Research and Development Program of Guangdong Province (2019B020226004), National Natural Science Foundation of China (61805271, 62074155), and Shenzhen Science and Technology Innovation Commission (JCYJ20170818154035069, KCXFZ202002011008124).

All authors are with the Laboratory of Biomedical Microsystems and Nano Devices, Bionic Sensing and Intelligence Center, Institute of Biomedical and Health Engineering, Shenzhen Institutes of Advanced Technology, Chinese Academy of Sciences, 518055 Shenzhen, China (phone: +86-755-8639-2675; e-mail: hui.yang@siat.ac.cn).

978-1-6654-3008-1/21 $31.00 © 2021 IEEE

channel system. The model is assumed infinite long along y-axis direction.

Fig. 1(b) shows the cross-section of the region marked by the dotted box in Fig. 1(a), corresponding to the 2D model used in the following numerical studies. All the results are obtained from the 2D FEM-based full-wave simulation. The unpolarized plane wave with monochromatic wavelength (λ) and unit light intensity (I) is propagating from the top boundary to the bottom boundary of the computational domain along negative z-axis direction. n_b, n_m and n_e are the refractive indices of the filled-biofluid, the microfluidic channel and the environmental medium, respectively. d_m is the diameter of the middle cylinder, t_m is the tube wall thickness of the microfluidic channel, w_x is the width of the

inlet along x-axis direction. Three main PNJ parameters are discussed in this work: I_{max} is the maximum light intensity in the axial profile, WD is the distance from the vertex at the bottom of the middle cylinder to the point of I_{max}, FWHM is the beam waist of the central diffraction lobe along x-axis direction. The origin of the coordinate axis is defined at the center of the microfluidic channel system.

III. RESULTS AND DISCUSSION

Fig. 2(a) shows the light intensity distribution of a water flow filled microfluidic channel that is illuminated by a plane wave. The wavelength of the monochromatic wave is $\lambda = 532$ nm, and the intensity is $I = 1$. The refractive indices of the glass tube, filled water and environmental medium are $n_m = 1.45$, $n_w = 1.33$ and $n_e = 1$, respectively. Here, the geometric features of the investigated model are $d_m = 5$ μm, $t_m = 0.2$ μm, $w_x = 0.6$ μm, respectively. A typical PNJ is formed on the rear side of the middle microcylinder. The intensity distribution along the line between the point $x = 0$ μm, $z = -2.5$ μm and the point $x = 0$ μm, $z = -6.5$ μm, which is marked with red dashed line in Fig. 2(a), is shown in Fig. 2(b). The maximum light intensity I_{max} is located at the point of $x = 0$ μm, $z = -3.55$ μm. The value of I_{max} is more than ten times higher than the intensity of the incident plane wave. The working distance WD equals to 1.97λ, which means the focal point position is beyond the decay length of the evanescent wave ($\lambda/2\pi$ [7]). At the I_{max} point along x-axis direction, the intensity profile is drawn in Fig. 2(c). By fitting the distribution curve with a Gaussian approximation, the obtained FWHM is ~0.51λ. It is within the subwavelength scale and close to the classical diffraction limit.

A distinct advantage of the microfluidic channel structure presented in this work is the sensitivity of PNJ parameters to the filled fluidic media. Fig. 3 shows the correlation between PNJ parameters and the refractive indices of different biofluids. The dielectric properties of the

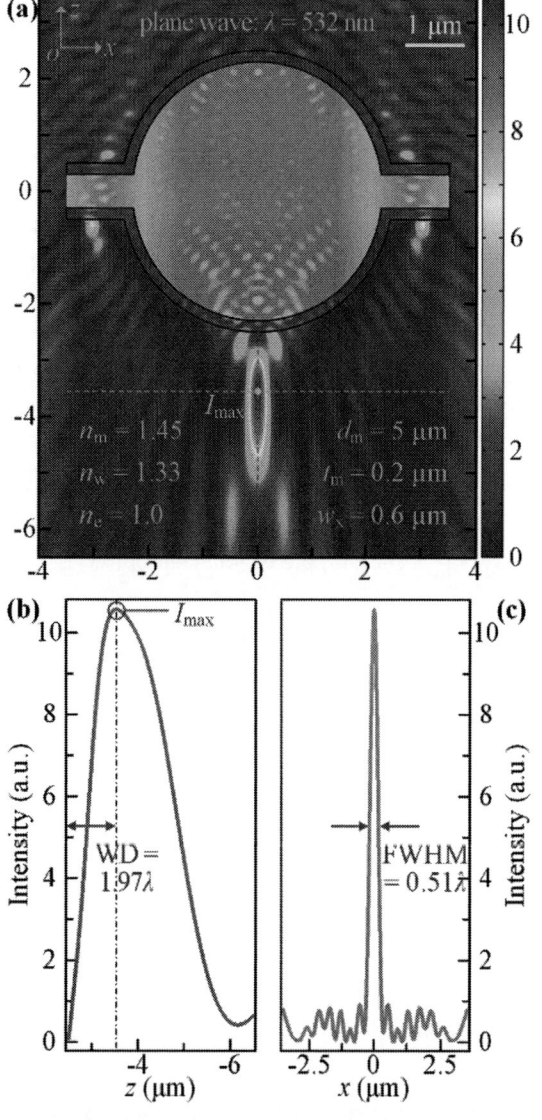

Fig. 2. (a) FEM simulation results for a typical example of PNJ generated by illuminating a microfluidic channel system with $n_m = 1.45$, $n_w = 1.33$, $n_e = 1$, $d_m = 5$ μm, $t_m = 0.2$ μm, $w_x = 0.6$ μm, and $\lambda = 532$ nm, $I = 1$. (b) Light intensity profile along z-axis (from $z = -2.5$ μm to $z = -6.5$ μm). (c) Lateral light intensity at the I_{max} point along x-axis direction.

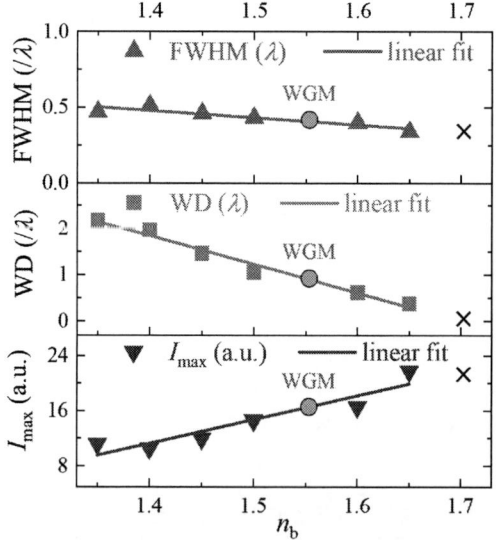

Fig. 3. PNJ parameters (I_{max}, WD and FWHM) for the injected biofluids with different refractive indices. The gray circles and black crosses represent the possible WGMs and the near-field focusing (not PNJ).

microfluidic channel and the surrounding medium, as well as the dimensional parameters of the microchannel keep the same as the parameters given in Fig. 2(a). As the refractive indices of the biofluids to be measured increase from 1.35 to 1.65 with step size of 0.05, the I_{max} shows an increasing trend, while the WD and FWHM both have a decreasing trend. All the lateral beamwidths are equal to or below the diffraction limit. The cases of $n_b = 1.55$ and $n_b = 1.70$, which are marked with circle and cross symbol in Fig. 3, represent the special situation of the possible resonant whispering gallery modes (WGMs) and not the normally defined PNJ, respectively.

As shown in Fig. 4(a), the distribution of the optical field around the interface between the filled biofluid and the tube is similar to the resonant WGMs [13]. Due to the generation of WGMs, the photon is confined within the microfluidic channel and travels back and forth, leading to the I_{max} point moving toward the inside of the microfluidic channel. In transverse electric (TE) mode, we calculated the resonant modes with radial quantum number $n = 1$, angular quantum number $l = 27$ [14]. Fig. 4(b) and Fig. 4(c) are the mode field distributions in the equator and polar directions. The resonant wavelength of the fundamental mode for the microfluidic channel cavity is $\lambda \approx 531.32$ nm, which is very close to the illumination wavelength. Because the refractive index difference between the filled biofluid of $n_b = 1.55$ and the glass tube of $n_m = 1.45$ is very small, the ability of microfluidic channel to confine the photons is weak and thus the weak light fields existed in the periphery of the cavity. When the refractive index of the filling biofluid further increases to 1.70, which is marked with the cross symbol in Fig. 3, the I_{max} point returns to the interior of the microfluidic channel. As can be seen from Fig. 4(d), the propagation light

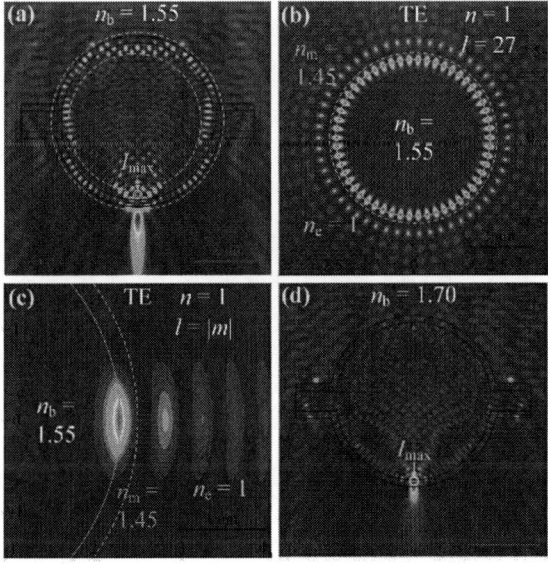

Fig. 4. (a) optical field distribution of the microfluidic channel with the refractive index of filled biofluid is $n_b = 1.55$. The green lines marked field distribution is the possible existed resonant WGMs. The TE mode field distributions of the microfluidic channel cavity along (b) equal direction with $n = 1$, $l = 27$ and (c) polar direction with $n = 1$, $l = 27$, $l = |m|$. m is the azimuthal quantum number. (d) optical field distribution of the microfluidic channel with the refractive index of filled biofluid is $n_b = 1.70$.

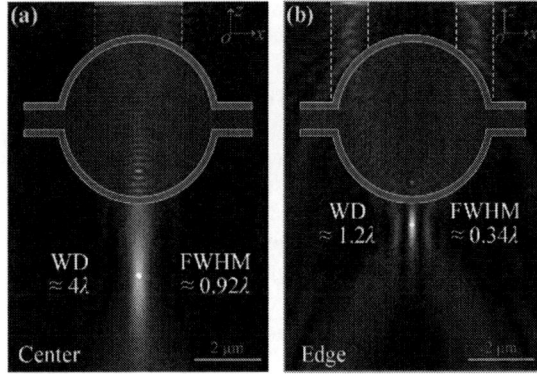

Fig. 5. Two different types of PNJs with (a) long WD, wide FWHM and (b) short WD, narrow FWHM used for biofluids monitoring that is generated from the middle region and two edge regions illuminated by the plane wave.

is focused within the microfluidic channel system and there exists no PNJ.

There is a universal rule for plane-wave-illuminated circular-structured dielectric microparticles: when the two edge regions of the first light refraction interface are illuminated, a tight focus with small FWHM and short WD is formed; on the other hand, when the middle region of the first light refraction interface is illuminated, a loose focus with large FWHM and long WD is formed. The inflection point, i.e. the boundary point between the edge and the middle regions, can be solved by a specific mathematical and physical method [15]. Based on the knowledge of the inflection point, we calculate the optical field distributions of the microfluidic channel structure under two different irradiation conditions. Since the circular boundary part of the microfluidic channel structure plays a major role in the formation of PNJ, we set the irradiation light to only illuminate the circular boundary part of the designed structure. The coordinates of the calculated two inflection points are (-1.36 μm, 2.1 μm) and (1.36 μm, 2.1 μm), respectively. Fig. 5(a) and Fig. 5(b) show the light field distributions of the abovementioned two cases. It is clear to see that the PNJs for these two cases are different. When the middle region is illuminated, the formed PNJ has a long WD (~ 4λ) and a wide FWHM (~ 0.92λ) without any secondary diffraction side lobes. While the two edge regions are illuminated, the produced PNJ has one central main lobe and two secondary side lobes. The WD and FWHM of the main lobe is ~ 1.2λ and ~ 0.34λ, respectively. Obviously, such two different types of PNJs will potentially provide more approaches for the monitoring biofluids.

IV. CONCLUSION

In summary, we designed and demonstrated a microfluidic channel structure for the generation of PNJs. The model is preliminarily investigated by numerical calculation with FEM-based simulation and theoretical analysis with the method of mathematical physics. The formed PNJ has a beam waist close to the diffraction limit, a maximum light intensity an order of magnitude higher than the intensity of the illumination source and a propagation distance far beyond the evanescent field region without significant diffraction. Through the variation of the PNJ parameters, the injected biofluids into the microfluidic channel with different refractive indices can be effectively perceived and monitored. The WGMs appear when the injected biofluid causes the captured photon to satisfy the resonant condition in the microfluidic channel cavity. The PNJs of different light field structures from the division of middle and edge irradiated regions based on the inflection point indicate more approaches for refractive index detection. Compared with other monitoring and detection methods, this approach provides advantages such as straightforwardness, high throughput and high sensitivity, and the possibility of multiplex detection. The proposed technique here will not only provide new approaches to integrate the PNJs into microfluidic devices, but also open new opportunities for developing new PNJ-based techniques for microfluidic biosensing applications.

REFERENCES

[1] Z. Xu, K. Han, I. Khan, X. Wang, and G. L. Liu, "Liquid refractive index sensing independent of opacity using an optofluidic diffraction sensor," *Opt. Lett.*, vol. 39, no. 20, pp. 6082-6085, Oct. 2014.

[2] R. O. Esenaliev, K. V. Larin, I. V. Larina, and M. Motamedi, "Noninvasive monitoring of glucose concentration with optical coherence tomography," *Opt. Lett.*, vol. 26, no. 13, pp. 992-994, Jul. 2001.

[3] D. J. Bornhop, J. C. Latham, A. Kussrow, D. A. Markov, R. D. Jones, and H. S. Srensen, "Free-solution label-free molecular interactions studied by back-scattering interferometry," *Science*, vol. 317, no. 5845, pp. 1732-1736, Sep. 2007.

[4] Q. Wang and W.-M. Zhao, "Optical methods of antibiotic residues detections: A comprehensive review," *Sens. Actuators B Chem.*, vol. 269, pp. 238-256, Sep. 2018.

[5] M. P. Sentis, F. R. Onofri, and F. Lamadie, "Photonic jet reconstruction for particle refractive index measurement by digital in-line holography," *Opt. Express*, vol. 25, no. 2, pp. 867-873, Jan. 2017.

[6] H. Yang and M. A. Gijs, "Micro-optics for microfluidic analytical applications," *Chem. Soc. Rev.*, vol. 47, no. 4, pp. 1391-1458, Jan. 2018.

[7] G. Gu, J. Song, H. Liang, M. Zhao, Y. Chen, and J. Qu, "Overstepping the upper refractive index limit to form ultra-narrow photonic nanojets", *Sci. Rep.*, vol. 7, no. 1, pp. 5635, Jul. 2017.

[8] G. Gu, J. Song, M. Chen, X. Peng, H. Liang, and J. Qu, "Single nanoparticle detection using a photonic nanojet", *Nanoscale*, vol. 10, no. 29, pp. 14182-14189, Jul. 2018.

[9] A. Heifetz, S.-C. Kong, A. V. Sahakian, A. Taflove, and V. Backman, "Photonic nanojets", *J. Comput. Theor. Nanos.*, vol. 6, no. 9, pp. 1979-1992, Sep. 2009.

[10] B. S. Luk'yanchuk, R. Paniagua-Domínguez, I. Minin, O. Minin, and Z. Wang, "Refractive index less than two: Photonic nanojets yesterday today and tomorrow", *Opt. Mater. Express*, vol. 7, no. 6, pp. 1820-1847, Jun. 2017.

[11] J. Zhu and L. L. Goddard, "All-dielectric concentration of electromagnetic fields at the nanoscale: the role of photonic nanojets", *Nanoscale Adv.*, vol. 1, no. 12, pp. 4615-4643, Nov. 2019.

[12] P. Ferrand, J. Wenger, A. Devilez, M. Pianta, B. Stout, N. Bonod, E. Popov, and H. Rigneault, "Direct imaging of photonic nanojets", *Opt. Express*, vol. 16, no. 10, pp. 6930-6940, May. 2008.

[13] G. C. Righini, Y. Dumeige, P. Féron, M. Ferrari, G. Nunzi Conti, D. Ristic, and S. Soria, "Whispering gallery mode microresonators: fundamentals and applications", *Riv. Nuovo Cimento Soc. Ital. Fis.* vol. 34, no. 7, pp. 435-488, 2011.

[14] C. C. Lam, P. T. Leung, and K. Young, "Explicit asymptotic formulas for the positions widths and strengths of resonances in Mie scattering", *J. Opt. Soc. Amer. B Opt. Phys.*, vol. 9, no. 9, pp. 1585-1592, Sep. 1992.

[15] G. Gu, P. Zhang, S. Chen, Y. Zhang, and H. Yang, "Inflection point: a new perspective on photonic nanojets", *arXiv:2012.09569*, Dec. 2020.

Proceedings of the 16th Annual IEEE International
Conference on Nano/Micro Engineered and Molecular Systems
April 25-29, 2021

Fluorescence Enhancement Utilizing Dielectric Microbeads with Semi-open Microwells

Pengcheng Zhang[1], Bing Yan[2], Guoqiang Gu[1], Zitong Yu[1], Xi Chen[1], Zengbo Wang[2] and Hui Yang[1]*

Abstract—Dielectric microbeads can converge light into a narrow beam with high intensity which allows for enhancing the fluorescent signals. However, a critical challenge for the experimental realization is introducing the fluorescent samples into this extremely small beam area. Here, we design and fabricate dielectric microbeads with semi-open microwells in which localized converging light beam of high intensity can be generated. We show that fluorescent microspheres can be efficiently loaded into the semi-open microwells and thus simultaneously illuminated by the converging beam generated by the micro-optical structures without any further manipulations. Pronounced fluorescent enhancement of around 9 folds can be obtained in comparison with the fluorescent microspheres on glass substrates, stemming from the excellent convergent effect of this micro-optical structure.

I. INTRODUCTION

Enhancing the signal of chromophores (i.e., fluorescent molecules or quantum dots) is a highly desirable goal for the practical applications in optical detection and imaging. The common strategy includes the enhancement of the local excitation intensity, the emission rate, or the radiation collection efficiency [1, 2]. These can be generally realized by properly tailoring the electromagnetic environment, via using plasmon coupling structures, photonic crystals, and dielectric microbeads, with demonstrated fluorescent enhancement up to several orders of magnitude [3-9]. Among these approaches, dielectric microbeads are considered as a direct and cost-effective route to enhance the fluorescence signal without requiring expensive nanofabrication facilities or complex near-field configurations. Upon illumination, dielectric microbeads could modulate the wavefront of the incident electromagnetic field, which is converged into a tiny beam of high intensity with subwavelength dimensions along the three directions of space, providing a universal and simple optical structure for fluorescent enhancement applications [10, 11]. However, the converging area is in general generated in an open space with extremely small effective area, making it technically difficult to precisely introduce the fluorescent samples into, hindering their usage in real applications. In this contribution, we fabricate dielectric microbeads with semi-open microwells to efficiently introduce the fluorescent samples into the converging area and simultaneously to enhance their fluorescent signals. The incident light beam can be modulated by this micro-optical

structure and is confined in the semi-open microwells with high field intensity that is significantly exceeding the incident light. Fluorescent microspheres can be passively trapped inside the semi-open microwells during the evaporation process and further be enhanced by the high-intensity converging light up to about one order of magnitude.

II. METHODS

1. Fabrication of dielectric microbeads with semi-open microwells

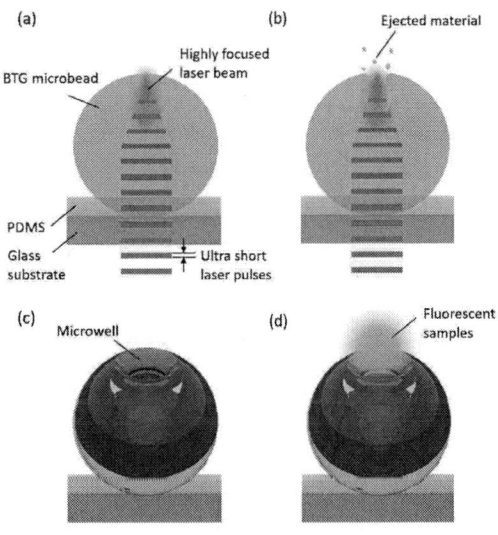

Fig. 1. (a) Barium titanate glass (BTG) microbead is illuminated from the bottom by femtosecond laser and a highly focused laser beam is generated. (b) BTG are melted and materials are removed from the foci. (c) A semi-open microwell is formed. (d) Signals of the fluorescent samples in the microwell can be enhanced.

The fabrication process is schematically illustrated in Fig. 1. Dielectric microbeads composed of barium titanate glass (BTG) with diameter ~80 μm and refractive index of 1.9 are used in our experiment. The BTG microbeads are coated on the glass substrate with a layer of uncured polydimethylsiloxane (PDMS). After the PDMS is cured, unattached BTG microbeads are discarded, leaving a single layer of BTG microbeads on the glass substrate. Femtosecond laser is utilized to fabricate semi-open microwells on the BTG microbeads. Upon exposing to a

---

[1]Laboratory of Biomedical Microsystems and Nano Devices, Bionic Sensing and Intelligence Center, Institute of Biomedical and Health Engineering, Shenzhen Institutes of Advanced Technology, Chinese Academy of Science, Shenzhen, China.

[2]School of Computer Science and Electronic Engineering, Bangor University, Dean Street, Bangor, Gwynedd LL57 1UT, UK.

*Contacting author: Hui Yang is with the Laboratory of Biomedical Microsystems and Nano Devices, Bionic Sensing and Intelligence Center, Institute of Biomedical and Health Engineering, Shenzhen Institutes of Advanced Technology, Chinese Academy of Science, 518055 Shenzhen, China (phone: +86-755-8639-2675; e-mail: hui.yang@siat.ac.cn).

978-1-6654-3008-1/21 $31.00 © 2021 IEEE

femtosecond laser from the bottom of the glass substrate, a focused high-intensity beam is generated near the top surface inside the BTG microbeads. Due to the heating effect generated by the highly focused laser beam, the illuminated BTG material is melted and ejected from the BTG microbeads, resulting a single semi-open microwell on each BTG microbead.

2. *Introduce of fluorescent microspheres in the microwell by droplet drying self-assembly*

The chip with BTG glass microspheres is cleaned twice with 2-proponal, rinsed with Milli-Q water and dried with nitrogen. Subsequently, it is treated with air plasma for 1 minutes to increase its hydrophilicity. Then 10uL of the fluorescent microspheres was dropped on the chip. Due to the hydrophilicity of the surface, the droplet spreads out over the whole surface area which benefits the evaporation of the liquid. During the droplet drying process, small amounts of the liquid that is trapped in the microwell evaporates slowly, resulting in capillary forces which draw fluorescent microspheres close by into the microwell. Then the chip is completely dried with nitrogen for further characterization.

3. *Fluorescence measurement under light microscope*

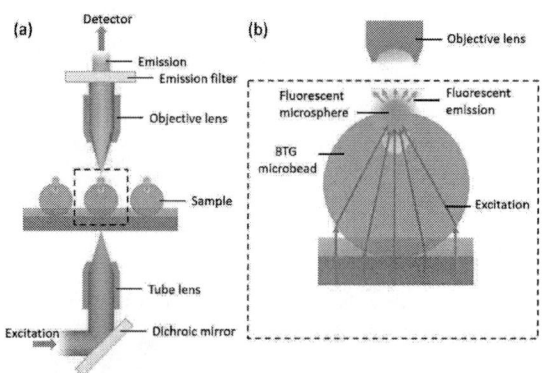

Fig. 2. (a) Optical configuration for recording the signal of the fluorescent microspheres and (b) the enlarged sketch showing the optical path in the black dashed box of (a).

The chip with the fluorescent microspheres is characterized under a light microscope (ZEISS Observer 7) with the light path shown in Fig. 2. The chip is illuminated by a power linearized LED light source (ZEISS Colibri 7) as the fluorescent excitation light. The emission fluorescent signal is collected by a 10× objective with NA of 0.25 (ZEISS Objective N-Achroplan) or a 40× objective with NA of 0.55 (ZEISS Objective LD A-Plan). The objective is focused on the focal plane of the fluorescent microspheres or the outer contour of the BTG glass spheres.

III. Results and Discussion

The prepared BTG microbeads with semi-open microwells are characterized under the light microscope and the scanning electron microscope (SEM, ZEISS Gemini 500), as shown in Fig. 3. The dielectric microspheres with semi-open microwells exhibit a dark circular area on the center of the top, arising from the scattering effect of the walls surrounding the microwell, Fig. 3 (a). The SEM images show the detailed view of the semi-open microwells,

Fig. 3 (b) and (c). As a seen from the SEM images, a larger outer contour is on the top of the dielectric microsphere and then the height of the contour gradually decreases and eventually narrows into a semi-open microwell with a diameter of ~6μm. The interior of the semi-open microwell is observed as an ellipsoidal cavity with a maximum cross-sectional profile of ~13 μm and a height of ~20 μm, as illustrated in Fig. 3 (d).

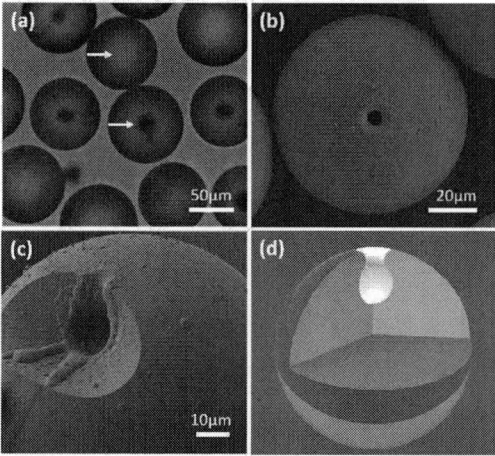

Fig. 3. (a) Optical microscope image and (b) (c) SEM images of the BTG microbeads with semi-open microwell. (d) Sketch illustration of the microbead with semi-open microwell.

Simulation based on finite element method (FEM) shows the ability of this micro-optical structure to converge light inside the semi-open microwell, as presented in Fig. 4. The light intensity inside the semi-open microwell exhibits pronounced enhancement compared with that of the incident light.

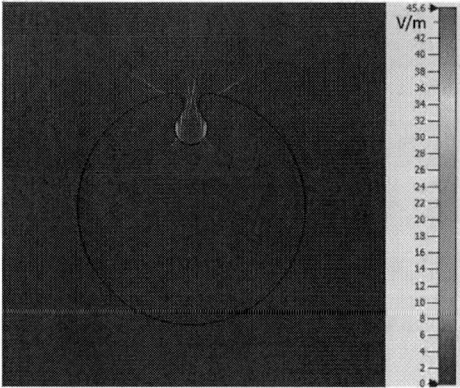

Fig. 4. Simulations based on finite element method (FEM) showing the ability of this micro-optical structure (diameter 80 μm) to converge light inside the microwell.

This micro-optical structure allows to load fluorescent samples directly into the semi-open microwells and to enhance their fluorescent signals. Since for the fluorescent microspheres, the intensity of the emission light has roughly a linear relationship with the intensity of the excitation light, that is, the confined light in the semi-open microwell. Thus the fluorescent microspheres can be utilized as indicators to report the intensity of the confined light in the microwells. It is hypothesized that the fluorescent microspheres loaded in the microwells are exposed and excited by the confined light generated by the micro-optical structure with high intensity, thus exhibiting a higher intensity on their emission light, in comparison with the fluorescent microspheres outside the microwells, i.e., on the glass substrate. Fluorescent microspheres of different sizes (1μm, 5μm, 10μm and 15μm in diameter, respectively) are loaded into the microwells. Self-assembly is used to load the fluorescent microspheres into the microwells. In the self-assembly process, the geometry of the semi-open microwell benefits a slower evaporation of the suspension containing the fluorescent microspheres, resulting in capillary forces which draw fluorescent microspheres close by into the microwells. It is observed that microwells can be filled at a yield approaching 70% after 5 loading steps (each assembly process is seen as one step) and the loading efficiency depended on the size of the fluorescent microspheres and the number of loading steps. Unfilled microwells can be further eliminated by additional loading steps with a more dilute suspension. However, multiple fluorescent microspheres loaded in one single microwell or located beneath the dielectric microbeads would bring interference on determination of the fluorescent enhancement effect in the following experiments. In order to eliminate this interference, it is necessary to load only one microsphere in each single microwell, which can be achieved by employing a dilute suspension and performing more loading steps. Here, samples with single microsphere loading are utilized in the following experiments.

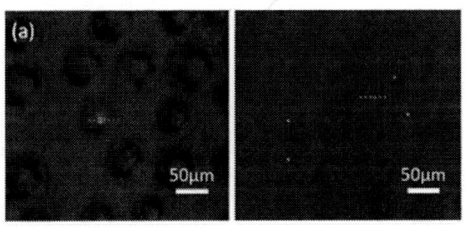

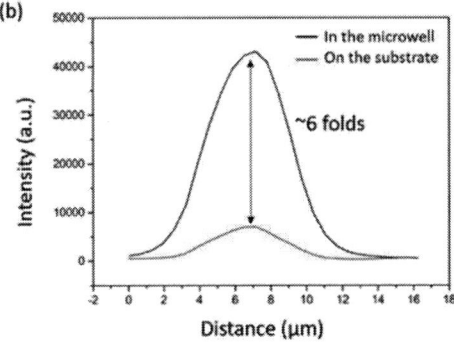

Fig. 5. (a) Optical images of the fluorescent microspheres (diameter 5 μm) in the semi-open microwell (left) and on the substrate (right). (b) The intensity profile of the corresponding fluorescent microspheres.

After loading, the samples are completely dried with nitrogen and their fluorescent signal are collected in the air ambient under the optical configuration as shown in Fig. 2. Their arithmetic mean emission intensity (I_{emi}^{*}) is the calculated and obtained. As a comparison, the arithmetic mean emission intensity of the microspheres on the glass

substrate (I_{emi}) is also collected in the identical parameters (excitation power and the exposure time) under the same optical configuration as the control. It should be noticed that, only the fluorescent microspheres loaded at the expected position are included in our statistics. The electrostatic force may cause some fluorescent microspheres to randomly adhere to the inner surface of the microwell, especially for the spheres of smaller size with relative large spatial freedom (1μm and 5μm). This leaves them at a higher position in the longitudinal direction or an off-center position, which can be distinguished by reading their focal length or measuring their geometric position to the center, and are excluded in our statistics. As expected, microspheres inside the microwells exhibit higher emission intensity than that outside the microwells, Fig. 5. About 6 folds of enhancement can be obtained from the fluorescent microspheres with diameter of 5 μm, Fig. 5(b). The enhancement effect is attributed to the light focusing effect of the microwells due to the generation of confined light with higher intensity that significantly exceeds that of the illuminating light.

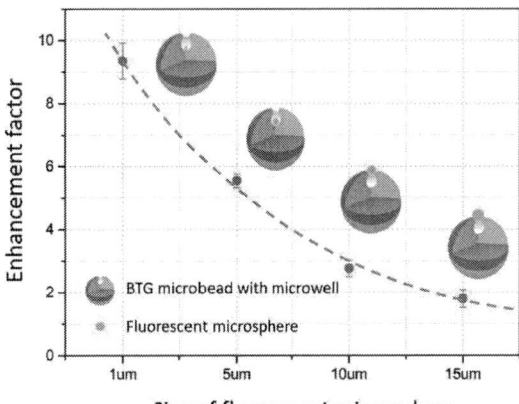

Fig. 6. The enhancement factor v.s. the size of the fluorescent beads used in the experiment. The sketch inside shows the relative position between the fluorescent microspheres and the microwells.

To quantify the enhancement effect of this optical microstructure, we define fluorescence enhancement factor η_F as

$$\eta_F = \frac{I_{emi}^{*}}{I_{emi}}$$

where I_{emi}^{*} represents the arithmetic mean emission intensity of the fluorescent microspheres in the microwell and I_{emi} represents the arithmetic mean emission intensity of the fluorescent microspheres on the glass substrate. Fluorescence enhancement factor of different sizes are measured respectively. The fluorescence enhancement factor η_F versus their sizes are shown in Fig. 6. It is shown that, about one order (9.5 folds) of η_F is obtained on the microspheres with diameter of 1μm. with the size of the fluorescent microspheres decreasing, the η_F increases. The relationship between the η_F and size of the fluorescent microspheres arises from the longitudinal position of the fluorescent microspheres in the semi-open microwells.

Due to the ellipsoidal geometry structure of the microwells, these fluorescent microspheres exhibit different

978-1-6654-3008-1/21 $31.00 © 2021 IEEE

height distributions along the longitudinal direction, inline images in Fig. 6. These semi-open microwells provide decent area for the accommodation of fluorescent samples as well as for enhancing their fluorescent signals. Our study offers a versatile platform to enhance the signal of fluorescent samples via dielectric microbeads, which has great potential in general for biosensing applications.

IV. CONCLUSION

In summary, dielectric microbeads with semi-open microwells are fabricated and their ability to enhance the fluorescent signals is demonstrated. Fluorescent microspheres can be passively trapped inside the semi-open microwells, which allows the efficient introducing of the fluorescent microspheres inside the light converging area. The light converging ability of this micro-optical structure is simulated based on finite element method. We show that the incident light can be confined and converged in the semi-open microwells with high field intensity and signals of the fluorescent microspheres inside the semi-open microwells can be enhanced up to one order of magnitude. It is observed that fluorescence enhancement factor is related to the size of the fluorescence microspheres. Fluorescent microspheres of smaller size exhibit higher fluorescence enhancement factor due to the position in the longitudinal direction. This superiorities of the micro-optical structure on the light converging ability possess great potentials in biosensing applications.

REFERENCES

[1] D. Gerard, A. Devilez, H. Aouani, B. Stout, N. Bonod, J. Wenger, et al., "Efficient excitation and collection of single-molecule fluorescence close to a dielectric microsphere," Journal of the Optical Society of America B-Optical Physics, vol. 26, pp. 1473-78, Jul 2009.

[2] D. Gerard, J. Wenger, A. Devilez, D. Gachet, B. Stout, N. Bonod, et al., "Strong electromagnetic confinement near dielectric microspheres to enhance single-molecule fluorescence," Optical Express, vol. 16, pp. 15297-303, Sep 15 2008.

[3] M. Saboktakin, X. Ye, U. K. Chettiar, N. Engheta, C. B. Murray, and C. R. Kagan, "Plasmonic enhancement of nanophosphor upconversion luminescence in Au nanohole arrays," ACS Nano, vol. 7, pp. 7186-92, Aug 27 2013.

[4] W. Zhang and H. Lei, "Fluorescence enhancement based on cooperative effects of a photonic nanojet and plasmon resonance," Nanoscale, vol. 12, pp. 6596-6602, Mar 28 2020.

[5] N. Ganesh, W. Zhang, P. C. Mathias, E. Chow, J. A. Soares, V. Malyarchuk, et al., "Enhanced fluorescence emission from quantum dots on a photonic crystal surface," Nature Nanotechnology, vol. 2, pp. 515-20, Aug 2007.

[6] D. Y. Li, D. L. Zhou, W. Xu, X. Chen, G. C. Pan, X. Y. Zhou, et al., "Plasmonic photonic crystals induced two-order fluorescence enhancement of blue perovskite nanocrystals and its application for high-performance flexible ultraviolet photodetectors," Advanced Functional Materials, vol. 28, pp. 1804429, Oct 10 2018.

[7] L. L. Liang, D. B. L. Teh, Dinh, W. D. Chen, Q. S. Chen, Y. M. Wu, et al., "Upconversion amplification through dielectric superlensing modulation," Nature Communications, vol. 10, pp. 1391, Mar 27 2019.

[8] H. Yang, M. Cornaglia, and M. A. Gijs, "Photonic nanojet array for fast detection of single nanoparticles in a flow," Nano Letter, vol. 15, pp. 1730-5, Mar 11 2015.

[9] H. Aouani, P. Schon, S. Brasselet, H. Rigneault, and J. Wenger, "Two-photon fluorescence correlation spectroscopy with high count rates and low background using dielectric microspheres," Biomedical Optics Express, vol. 1, pp. 1075-83, Nov 1 2010.

[10] A. Heifetz, S. C. Kong, A. V. Sahakian, A. Taflove, and V. Backman, "Photonic nanojets," Journal of Computational and Theoretical Nanoscience, vol. 6, pp. 1979-92, Sep 1 2009.

[11] I. V. Minin, O. V. Minin, and Y. E. Geints, "Localized EM and photonic jets from non-spherical and non-symmetrical dielectric mesoscale objects: Brief review," Annalen der Physik, vol. 527, pp. 491-497, 2015.

Gap in pagination due to unavailable papers.

Pages 719-728

Flexible Bionic Intelligent Perception System

Jian-Hua Zhang

Key Laboratory of Advanced Display and System Application, Ministry of Education, Shanghai University,
Shanghai 200072, China

ABSTRACT

To construct an intelligent perception system with the same function as the organism is promising in the application of personal health surveillance, man-machine interaction, and artificial electronic skin. The perception function of the organism includes both the acquisition of the external information through sensing organs and the cognitive competence by means of learning, judgment, and reasoning of the nervous system. Thus, an intelligent perception system mainly depends on two basic units: flexible sensor and artificial synapse. However, the device structure and fabrication process of flexible sensor hampers not only the further improvement of its sensing properties, including the sensitivity and linearity range, but also the massive large-area manufacture in the form of high-density arrays. In terms of artificial synapse, the search for new material systems and device structures to simulate synaptic properties remains challenges.

This talk will introduce flexible piezoresistive pressure sensors with highly sensitivity and ultrabroad linear response range based on biomimetically textured porous materials and pollen-shaped hierarchical structure. On the other hand, taking advantage of high-density manufacturing and signal amplification of thin film transistor (TFT), we developed the active electrode array based on TFT to realize the high resolution perception of human EMG signals. Meanwhile, kinds of synaptic transistors based on TFT with high migration rate were designed and prepared to simulate the memory property of the human brain. These works provide the innovative application of TFT in the field of intelligent perception, which are promising for the application of medical physiological signal, man-machine interaction, etc.

BIOGRAPHY

Jian-Hua Zhang is currently a Professor in the Key Laboratory of Advanced Display and System Application, Ministry of Education, Shanghai University, China. She received the Ph.D. degree of engineering in 1999 the Shanghai University, China. During the following two years, she worked at the City University of Hong Kong in Hongkong and Heriot-Watt University, UK, performing postdoctoral studies.

Since 2005, she took the Professor position in Key Laboratory of Advanced Display and System Application, Ministry of Education, Shanghai University, China. She was awarded as the National Science Fund for Outstanding Young Scholars (2018). Her research activities were awarded the first prize of technical invention in Shanghai (2014), the first prize of science and technology progress in Shanghai twice (2016, 2020). She is now an editor of Journal of Bionic Engineering and Advance In Manufacturing, and chairman in China OLED technical committee in Society for Information Display (SID). Her group endeavors on the development of semiconductor manufacturing and packaging process and its applications in flexible intelligent display. Prof. Zhang is active in the field of heterogeneous integration of integrated circuit (IC), display technology and sensor.

978-1-6654-3008-1/21 $31.00 © 2021 IEEE   730

Proceedings of the 16th Annual IEEE International
Conference on Nano/Micro Engineered and Molecular Systems
April 25-29, 2021

Fabrication of laser scribed graphene stretchable supercapacitor by laser-assisted transfer printing strategy

Guangyuan Xu[a, b], *IEEE Member*, Yanan Chen[c], Fangshuai Chen[c], Yanfang Meng[a, b], Yinji Ma[a, b], and Xue Feng[a, b] *

[a]*AML, Department of Engineering Mechanics, Tsinghua University, Beijing 100084, China.*
[b]*Center for Flexible Electronics Technology, Tsinghua University, Beijing 100084, China.*
[c]*Department of Material Science and Engineering, Tianjin University, Tianjin 100072, China*

An extensive range of nanocarbon, such as 0D fullerene, 1D carbon nanotube, 2D graphene, and 3D carbon aerogels, has attracted significant attention. Graphene, being an outstanding electrode material has already been evidenced in its performance for fabricating supercapacitors. Direct laser writing was reported to be an innovative methodology that enabled patterning graphene electrodes within a single-step. Laser scribed graphene (LSG) is produced on polyimide (PI) substrate, which is not stretchable. Most researchers report transferring LSG from PI onto the stretchable substrate to fabricate stretchable LSG sensors and supercapacitors. Transfer printing is a critical procedure for LSG embedding on a stretchable substrate. But no one has indeed investigated the transfer printing of LSG. In this work, we report a laser-assisted method to improve the success rate of transfer printing. The resulting LSG/PDMS is utilized in fabricating the in-plane stretchable supercapacitor. Its current density was improved from 0.02 to 0.6 mA/cm^2 with the rising scan rate from 10 to 200 mV/s. The highest capacitance was calculated as 18 mF/cm^2 at the consistent current density of 0.02 mA/cm^2. Also, the LSG supercapacitor showed excellent cycling stability after 1000 cycles. The laser-assisted strategy has permitted the successful promises for developing LSG stretchable wearable electronics.

Keywords: laser scribed graphene; laser-assisted transfer printing; stretchable; supercapacitor

I. INTRODUCTION

Polyimide (PI) is the most common substrate for laser scribed graphene (LSG) patterning. PI is a very good flexible thin-film substrate, and is usually used as the substrate in flexible electronic technology. But the electronics on PI substrates can only serve as flexible electronics due to PI cannot be stretched [1]. Polydimethylsiloxane (PDMS) is an ideal material for stretchable substrates, and its modulus is

close to that of human skin. Therefore, developing the strategy to transfer print electronics onto PDMS is important for making stretchable electronics integrating with human. The discovery of LSG patterned from polymeric substrate in 2014 has awakened much attention in recent years [2]. Direct laser writing of LSG patterns on PI has produced great promises for efficient fabrication of carbon-based electronics [3, 4]. An extensive range of applications, including: sensors [5], supercapacitor [6, 7], catalysis [8], adsorbent [9], and triboelectric nanogenerators [10] have been well studied.

As far as we see, there is little research which has been conducted on the transfer printing of LSG from PI onto PDMS. The strategy of transfer printing is critical to determine the quality of LSG electronics on stretchable PDMS. In this work, we discover a laser-assisted strategy on the bottom backing side of PI that can result in a preliminary crack in the middle of LSG. The laser scribing can highly improve the quality of LSG electronics transferred onto PDMS. This laser-assisted transfer approach is simple, but facile and efficient for fabricating high-quality stretchable LSG/PDMS supercapacitors. We report this efficient laser-assisted transfer printing methodology of LSG electronics for the first time in the paper. The transfer strategy will enable the successful fabrication of LSG stretchable and wearable electronics to facilitate our human life.

II. METHODOLOGY

The laser used for LSG patterning and bottom backing side treatment was performed by a CO_2 Universal Laser Systems (VLS2.30). The wavelength of CO_2 laser is 10.6 μm. The various parameters are adjusted between the speed (0-28 cm/s), the power (0-35 W), also the pulses per inch, PPI (0-1000). The optimized parameters of patterning LSG electrodes are power at 3.85 W, speed at 0.28 cm s^{-1}, pulses per inch (PPI) at 1000, and Z-distance at 2 mm. The parameters of laser-assisted transfer printing strategy are power at 8.75 W, speed at 28 cm/s, and PPI at 1000, as well as Z-distance at 2 mm.

---

*Contacting Author: Xue Feng is with the Department of Mechanical Engineering, Tsinghua University, Beijing, China (phone:+86-10-6278-1465; fax:+86-10-6278-1465; e-mail: fengxue@tsinghua.edu.cn)

978-1-6654-3008-1/21 $31.00 © 2021 IEEE

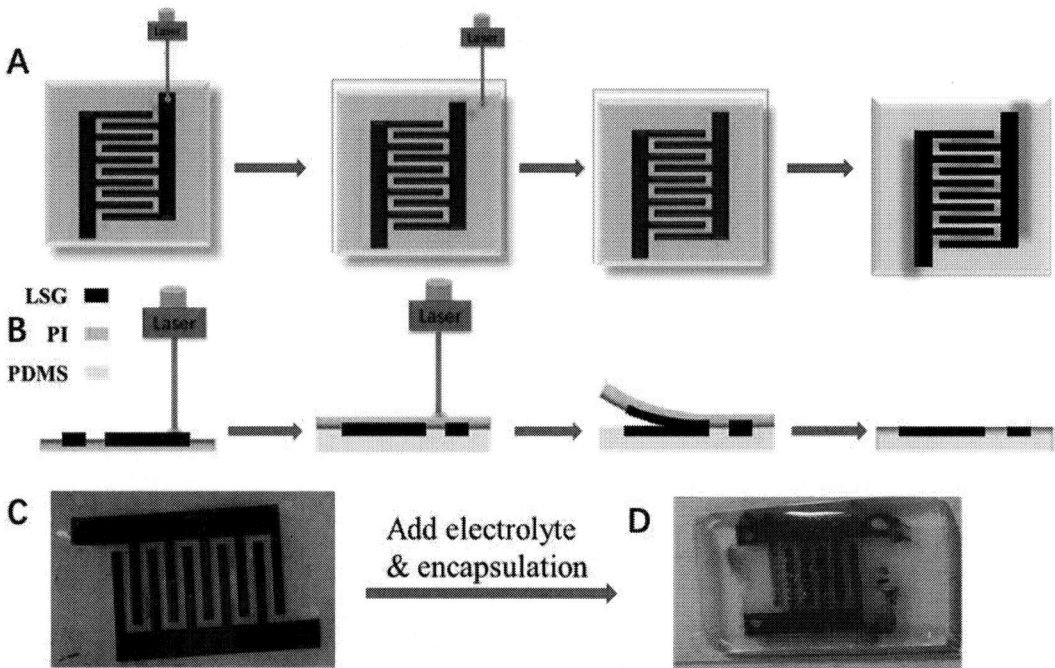

Figure 1. The schematic representation of the fabrication of LSG stretchable supercapacitor by a laser-assisted method. (A) The schematic vertical view of transfer process of LSG device generated from PI onto stretchable PDMS substrate. (B) The sectional view of transfer printing process of LSG device carbonized from PI onto PDMS. Detailed demonstration (from left to right): laser scribing of LSG supercapacitor electrodes, cast PDMS and decent laser treatment on the back side of PI, separate the PDMS and PI layers, transfer LSG electrode embedded into PDMS. (C) The photograph of LSG electrode embedding on PDMS. (D) The real photograph of a stretchable LSG supercapacitor fabricated by laser-assisted strategy after the treatment of bonding the lead wire, adding gel electrolyte, and package of LSG/PDMS stretchable supercapacitor.

III RESULTS AND DISCUSSION

The fabrication process of the stretchable LSG interdigital supercapacitor is shown in Figure 1. Multiple LSG electrodes are engraved by laser with a chosen pattern on PI. The intensity of the laser is controlled by adjusting the parameters such as power, speed, and PPI of the CO_2 laser source. The laser can engrave the surface and carbonize PI converting to LSG, but not cut through the bottom backing sheet. And, a layer of PDMS is spin casted on the LSG electrodes on the top fronting side of PI. The PDMS layer is warmed up to 70 °C and is solidified for 2 hours. Then, a laser scribing with low power intensity (Condition: power at 8.75 W, speed at 28 cm/s, and PPI at 1000) is conducted on the bottom backing side of PI. A transfer process could easily peel off the LSG electrodes embedded in PDMS from PI. Through the assistance of laser treatment, the LSG interdigital supercapacitor electrodes would be transferred onto the stretchable PDMS substrate with a high quality. The transfer printing of LSG/PDMS is highly better compared to the direct transfer without laser treatment on the bottom backing sheet of PI.

A subsequent step is to attach the common conductive coper tape on the top fronting side of LSG/PDMS electrodes. And then, the PVA/H_2SO_4 gel electrolyte is uniformly casted on the surface of LSG/PDMS supercapacitor electrodes. After waiting for 12 hours until the electrolyte is completely immersed into the microstructure of LSG, a layer of Ecoflex polymer is spin coated on the top of electrolyte, which serves as the package of LSG/PDMS stretchable electrodes. The in-plane stretchable LSG supercapacitor is assembled entirely and ready for mechanical and electrochemical measurements.

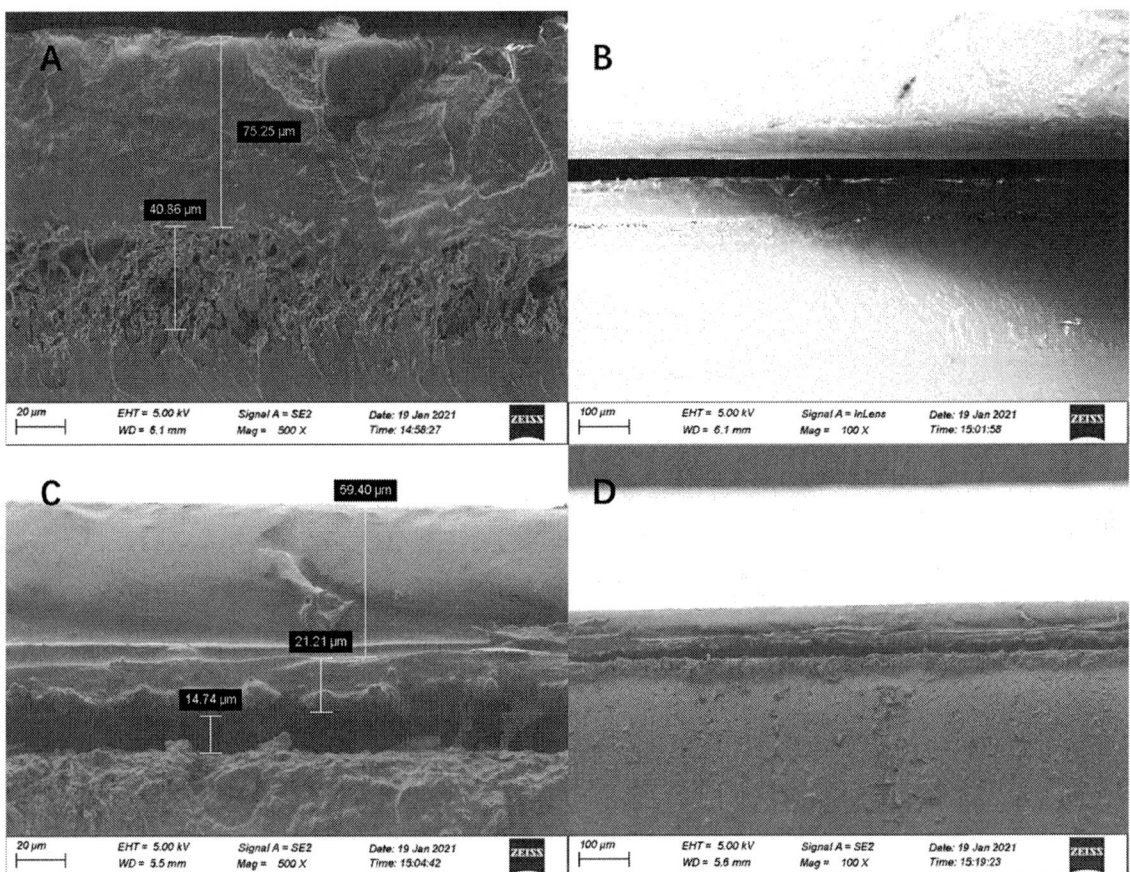

Figure 2. The cross-sectional SEM image compares the transfer printing of LSG electrodes without (A&B) and with (C&D) the laser treatment on the bottom backing side of PI

Figure 2 insulates an investigation of the cross-sectional structure of LSG during the transfer printing. Figure 2A shows a cross-sectional scanning electron microscopy (SEM) image by directly transferring the LSG without the laser scribing on PI backing side. Figure 2B indicates the SEM image of all the cross sections of PDMS-LSG-PI interface of same sample as Figure 1A. The LSG layer was approximately 40 μm in middle with amorphous porous structure. We did not find the crack structure initiation in the sandwich system of PDMS-LSG-PI. Therefore, it will become difficult to fracture the LSG embedded in PDMS when peel off it from PI. This transfer process is in accordance with the least energy loss principle. The interfacial interaction energy of LSG-PI is higher than that of LSG-PDMS interface. So that the fracture will initiate on the LSG-PDMS during the peel-off, which means the transfer of LSG/PDMS will not likely to happen.

On the contrary, Figure 2C demonstrates a SEM image by laser-assisted transfer strategy of LSG with the laser scribing on bottom backing side of PI. Figure 2D points to the SEM image of the entire cross section of PDMS-LSG-PI interface of same electrode as Figure 2C. We could figure out an obvious crack of 15 μm width in the middle of

LSG, which is resulted by the laser scribing on the bottom backing side of PI. The laser scribing is a photo thermal reaction process, and can introduce energy into PDMS-LSG-PI. PDMS is a stretchable polymer and performs elastic deformation when receiving energy. But PI is not a stretchable plastic and performs plastic deformation when receives energy. The different deformation of PDMS and PI after receiving energy leads to the crack initiation in the middle of LSG. And then the transfer of LSG/PDMS is easy because the fracture has already appeared in LSG. According to the least energy loss principle, LSG can be easily transferred from PI to PDMS by a simple peel-off procedure

978-1-6654-3008-1/21 $31.00 © 2021 IEEE

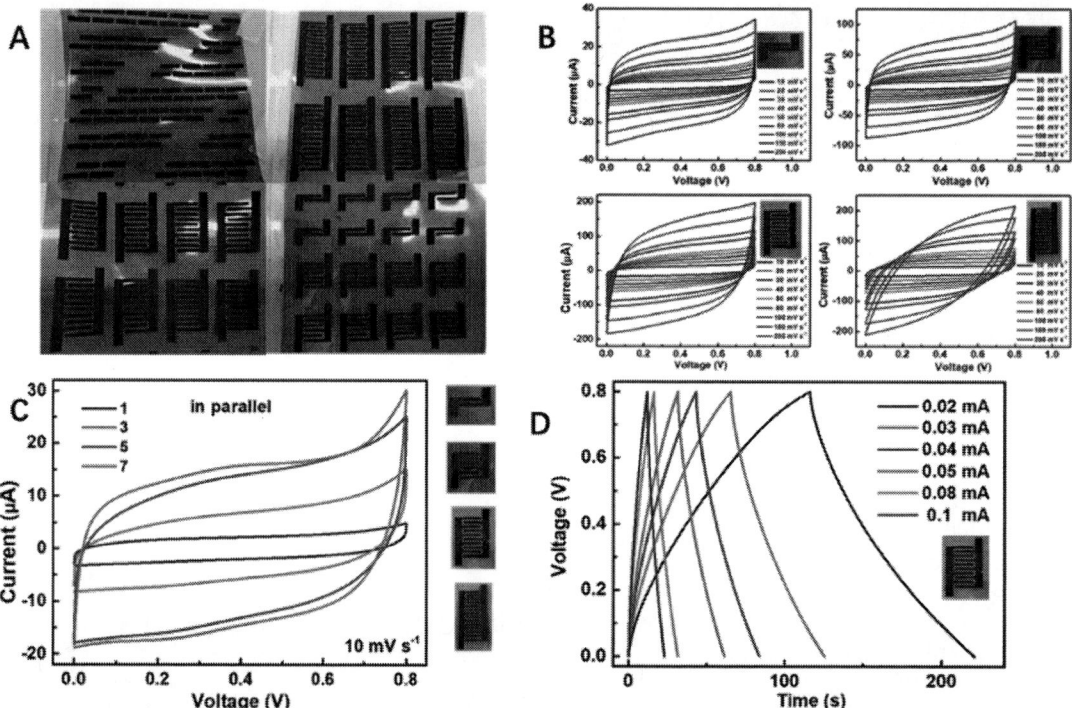

Figure 3. The electrochemical characterization of LSG fabricated on PI as the function of supercapacitors in PVA/H_2SO_4 electrolyte. (A) Photograph of laser scribing LSG electrodes with linear and interdigital in-plane geometries. (B) CV curves of LSG as the function of supercapacitors achieved at scan rates of 10 to 200 mV/s of diverse interdigital geometries. (C) Comparison of CV results obtained at LSG supercapacitors with different interdigital geometries (1, 3, 5, 7 electrodes on each collector). (D) GCD profiles of LSG performed as the function of MSC obtained at the current of 0.02 to 0.1 mA/cm.

Electrochemical characterizations of the LSG/PI in-plane symmetric supercapacitors are completed with a PVA/H_2SO_4 gel electrolyte. Figure 3A shows the real photos of different LSG symmetric interdigital supercapacitor electrodes on PI. They include one, three, five, and seven electrodes on each collector, respectively. Figure 3B presents the cyclic voltammetry (CV) results of the fabricated LSG/PI interdigital supercapacitors with varied geometries at different scan rates from 10 to 200 mV/s. The voltage ranges between 0 and 0.8 V, almost displaying rectangular shape, which indicates an effective double-layer capacitive behavior of LSG/PI supercapacitors. The highest capacitance was calculated as 18 mF/cm^2 at the consistent current density of 0.02 mA/cm^2.

It is clear that the areal capacitance of electrodes increases with the rise of LSG thickness. In Figure 3C, the CV results from different geometries interdigital electrodes

are compared. This curve shows the capacitance of LSG/PI supercapacitors increases with interfacial area increases. Figure. 3D demonstrates the galvanostatic charge and discharge (GCD) results of the LSG/PI device with five electrodes on anode is achieved at the current density from 0.02 mA/cm^2 to 0.1 mA/cm^2. The longer discharge time can be detected for the device with smaller current density. The GCD curves of the device is nearly equicrural triangle, which indicates the oxidation and reduction behavior does not likely to happen on the interface.

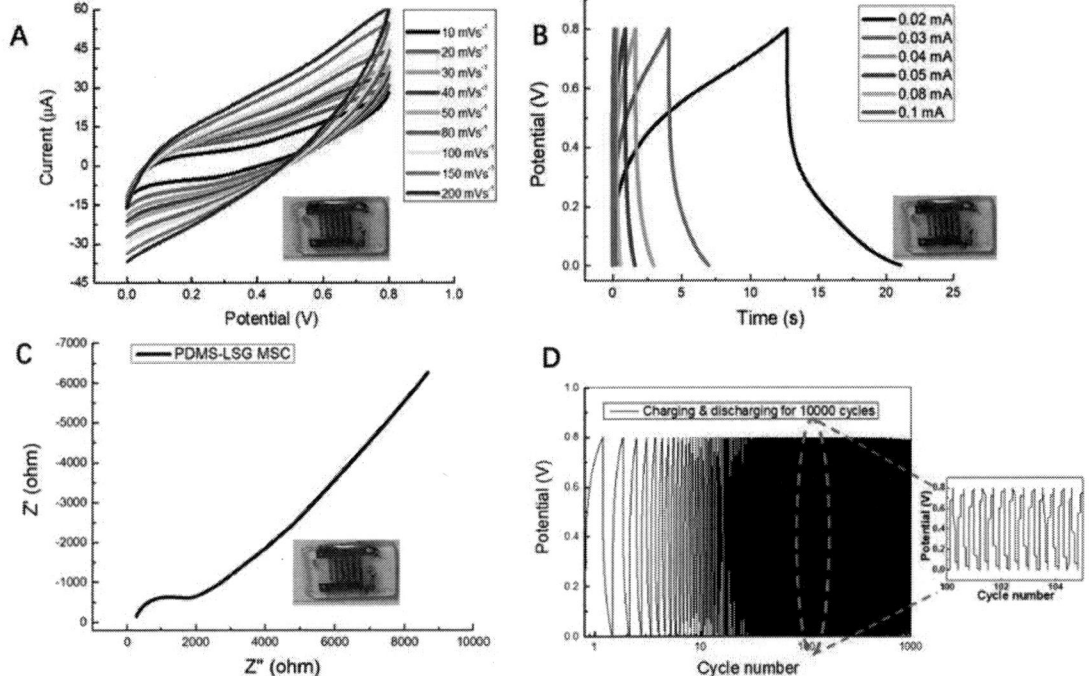

Figure 4. The electrochemical performance of LSG interdigital electrode after transferring onto PDMS as the function of stretchable supercapacitor in PVA/H_2SO_4 electrolyte. (A) CV profile of LSG-PDMS as the function of supercapacitor with scan rate of 10 to 200 mV/s. (B) GCD curve of LSG-PDMS obtained at current of 0.02 to 0.1 mA/cm^2. (C) EIS profile of LSG-PDMS ranging from 0.1 Hz to 10000 Hz. (D) Cycling stability of LSG-PDMS as the function of supercapacitor for 1000 cycles, measured at current of 0.1 mA/cm^2.

After the fabrication of LSG/PDMS stretchable supercapacitors, the device with three electrodes on each collector was selected for electrochemical measurements. The electrochemical performance of the LSG/PDMS in-plane stretchable supercapacitors are measured using PVA/H_2SO_4 gel electrolyte. Figure 4A illustrates the CV curves of the resultant LSG/PDMS interdigital supercapacitors at changed scan rates from 10-200 mV/s The voltage varieties between 0 and 0.8 V, exhibiting nearly rectangular shape, which designates an effective double-layer capacitive behavior of LSG/PDMS device.

Figure. 4B reveals the GCD curve of the stretchable LSG/PDMS device at the current density ranging from 0.02 mA/cm^2 to 0.1 mA/cm^2. Figure 4C presents the electrical impedance spectroscopy (EIS) result observed on the LSG/PDMS supercapacitor, and an obvious semi-cycle resistance can be observed, indicating the presence of internal resistance of LSG/PDMS device. In addition, the LSG/PDMS stretchable supercapacitor exhibits excellent cycling stability during charging and discharging for 10000 cycles. The capacitance of LSG/PDMS supercapacitor keeps stable after charging and discharging 1000 times (Figure 4D).

IV. CONCLUSION

The procedure of laser scribing on the bottom backing side of PI is demonstrated to be an effective strategy for transfer printing LSG from PI to PDMS. The transfer quality of LSG/PDMS electrode is highly better than that without the laser treatment on the backing side of PI. The interdigital stretchable LSG/PDMS supercapacitor is thus fabricated and its electrochemical properties are measured through CV, GND, EIS and cycling stability test. The laser-assisted transfer printing approach permits the successful promises for developing LSG based stretchable and wearable electronics.

ACKNOWLEDGMENT

Dr. Guangyuan Xu would like to acknowledges the Department of Engineering Mechanics, and the Center for Flexible Electronics Technology at Tsinghua University. Also, Dr. Guangyuan Xu would like to thank the support from "Shuimu Tsinghua Scholar" fellowship and China "Postdoc exchange plan of talents introduce" program.

REFERENCES

1. Wang L, Yu X, Wang D, Zhao X, Yang D, urRehman S, Chen C, Zhou H, Dang G: **High modulus and high strength ultra-thin polyimide films with hot-stretch induced molecular orientation**. *Materials Chemistry and Physics* 2013, **139**(2):968-974.

2. Lin J, Peng Z, Liu Y, Ruiz-Zepeda F, Ye R, Samuel ELG, Yacaman MJ, Yakobson BI, Tour JM: **Laser-induced porous graphene films from commercial polymers**. *Nature communications* 2014, **5**:5714.

3. Chyan Y, Ye R, Li Y, Singh SP, Arnusch CJ, Tour JM: **Laser-Induced Graphene by Multiple Lasing: Toward Electronics on Cloth, Paper, and Food**. *ACS Nano* 2018.

4. Ye R, James DK, Tour JM: **Laser-Induced Graphene**. *Accounts of Chemical Research* 2018.

5. Rahimi R, Ochoa M, Yu W, Ziaie B: **Highly Stretchable and Sensitive Unidirectional Strain Sensor via Laser Carbonization**. *ACS Applied Materials & Interfaces* 2015, **7**(8):4463-4470.

6. Peng Z, Lin J, Ye R, Samuel ELG, Tour JM: **Flexible and Stackable Laser-Induced Graphene Supercapacitors**. *ACS Applied Materials & Interfaces* 2015, **7**(5):3414-3419.

7. Peng Z, Ye R, Mann JA, Zakhidov D, Li Y, Smalley PR, Lin J, Tour JM: **Flexible Boron-Doped Laser-Induced Graphene Microsupercapacitors**. *ACS Nano* 2015, **9**(6):5868-5875.

8. Ren M, Zhang J, Tour JM: **Laser-induced graphene synthesis of Co_3O_4 in graphene for oxygen electrocatalysis and metal-air batteries**. *Carbon* 2018, **139**:880-887.

9. Rathinam K, Singh SP, Li Y, Kasher R, Tour JM, Arnusch CJ: **Polyimide derived laser-induced graphene as adsorbent for cationic and anionic dyes**. *Carbon* 2017, **124**:515-524.

10. Stanford MG, Li JT, Chyan Y, Wang Z, Wang W, Tour JM: **Laser-Induced Graphene Triboelectric Nanogenerators**. *ACS Nano* 2019, **13**(6):7166-7174.

Proceedings of the 16th Annual IEEE International
Conference on Nano/Micro Engineered and Molecular Systems
April 25-29, 2021

Smart Power Management Microsystem
for Distributed Renewable Energy Harvesting

Linhao Feng[1,#], Kai Huang[2,#], Mingxin Liu[2], Hongsheng Zhong[2,*], Xiaosheng Zhang[2,*]

Abstract— **In this work, we proposed a smart power management microsystem for distributed renewable energy harvesting which could combine various renewable energy, especially solar energy and wind energy, and convert multiple input to stable single output, and demonstrated the feasibility of the system by building the model of photovoltaic cell and wind turbine and simulating the performance of maintaining stable power output in changing environment. For verifying the practicability of the smart power management microsystem, we designed a double input parallel boost circuit based on STM32 and proved its ability to manage the output optimally.**

I. INTRODUCTION

The greenhouse effect caused by burning fossil fuels have greatly threatened the survival and development of human society, so it's extremely urgent for environment to be carbon neutral. Therefore, the utilization of renewable energy, especially solar energy and wind energy, as an important move to be carbon neutral has attracted intensive interests. Several papers focusing on the multiple renewable energy management, which is based on the complementary characteristics of wind and solar generation and determines the scale of solar photovoltaic panels, wind turbines and batteries according to the load demand and irradiation, wind speed and other factors, have been reported [1,2].

In a traditional electrical grid, electrical power is generated by large and centralized power plants, however it's unaccommodated for relatively massive and dispersed renewable energy. To implement the optimal utilization of renewable energy, it's necessary to design a smart distributed energy generation system. In previous study, a distributed model predictive control(DMPC) [3] method was proposed to manage the energy generation. We design the energy control microsystem based on DMPC, as shown in Fig. 1, which is divided in energy management optimizing the power distribution for the lowest cost and the most stable output under the premise of meeting the requirements in safety, and power regulation tracking the output determined by the energy management.

Most of the distributed hybrid energy generation control systems only concentrated on the simulation results and lacked of the design and application of the actual system. Otherwise, the management and maintenance of distributed generation system was also less involved. Herein, we only

[1]POWERCHINA SICHUAN ELECTRIC POWER ENGINEERING CO.,LTD.
[2]University of Electronic Science and Technology of China, School of Electronic and Engineering
[#]These authors contribute equally to this work.
[*]Corresponding Author: hszhong@uestc.edu.cn, zhangxs@uestc.edu.cn

focus on the solar energy and wind energy, which are the most widely used clean new energy sources, as the main input energy, and utilize the complementary characteristics of wind energy and solar energy in time and season to maximize the utilization of renewable resources. Hence, we designed an intelligent power management.

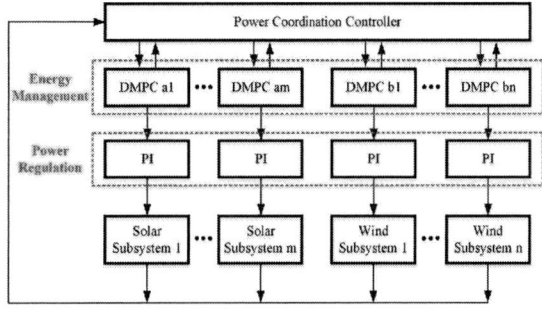

Fig. 1. Schematic diagram of the designed distributed energy controller based on DMPC.

Microsystem for distributed renewable energy harvesting to predict, manage and control power distribution for the off-grid system is shown in Fig. 2. The whole system consists of four parts which are electric energy generation, conversion, control, energy storage and consumption. Clean energy including wind energy, solar energy, fuel cell, biomass energy and so on as the input will be transferred to DC-DC boost and converted to standard 48 V, it is convenient to manage the whole system. Energy storage is also contained in the system, it will store or release the electricity according to the power generation and consumption to maintain the stable output. The electric control section based on DMPC is the emphasis of our research and we design a prototype, a double input parallel boost circuit based on STM32, and proved its ability to manage the output optimally which demonstrated the feasibility of the presented microsystem.

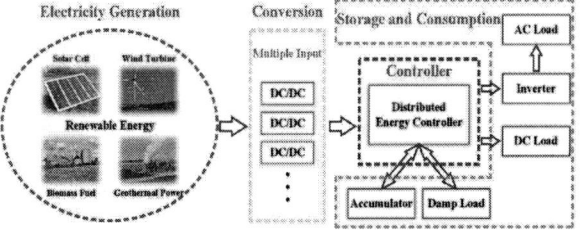

Fig. 2. Schematic of smart power management microsystem for distributed renewable energy harvesting.

978-1-6654-3008-1/21 $31.00 © 2021 IEEE

II. SIMULATION AND EXPERIMENT

A. Solar Cell and Wind Turbine Model

Among all the so-called renewable energy, solar and wind power have already been widely used and concerned for its amplitude development foreground. For the industry application of the solar and wind energy, a series of researches concentrating on the model of photovoltaic solar cell and wind turbine were implemented [4-7]. Photovoltaic solar cell can be simplified to an equivalent circuit in Fig. 3 and an equation (1).

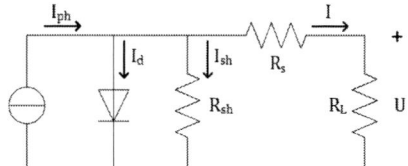

Fig. 3. Simplified-equivalent circuit of photovoltaic cell.

$$I = I_{ph} - I_o \left\{ \exp\left[\frac{q(U+IR_S)}{AKT}\right] - 1 \right\} - \frac{U+IR_S}{R_{sh}} \qquad (1)$$

Using the photovoltaic solar cell parameter which manufacturers provided, we constructed our model with Matlab/Simulink. The simulation results are shown in Fig. 4(a-d).

Meanwhile, wind turbine model is constructed with the equation (2) [7]. Where P_w is the power extracted from the wind, ρ is the air density, C_p is the power coefficient, λ is the tip speed ratio which is the ratio between blade tip speed v_t and wind speed at hub height upstream the rotor v, β is the pitch angle of rotor blades, A_r is the area covered by the rotor. For details, please refer to [7].

$$P_w = \frac{1}{2}\rho A_r C_p(\lambda, \beta)v^3 \qquad (2)$$

The simulation results is shown in Fig. 4(e,f). Based on the model, a principal of maximum power point tracking(MPPT) algorithm to maintain the optimized power point(MPP) is proposed as follows.

B. MPPT

In accordance with the simulation results above, both photovoltaic and wind power generation have the unique maximum power point(MPP). In order to decrease the power wastage and enhance economic performance, MPPT algorithm is indispensable. In the section we mainly discuss the photovoltaic power generation system. In recent years, numerous methods to achieve MPPT have been reported [8], MPPT techniques for photovoltaic system is to maximize the power output by tracking the MPP influenced by temperature and irradiance. Among all the strategies, perturb and observe maximum power point tracking method(P&O) is the most extensively used method due to its low-cost implementation. Please refer to [8] for details, in brief, the theory of P&O is similar with dichotomy: if the operating voltage is perturbed in a defined direction, and the output power increased, it means that the operating point is moving towards the MMP, and the voltage will be perturbed in the same direction further, otherwise it is moving away from MMP.

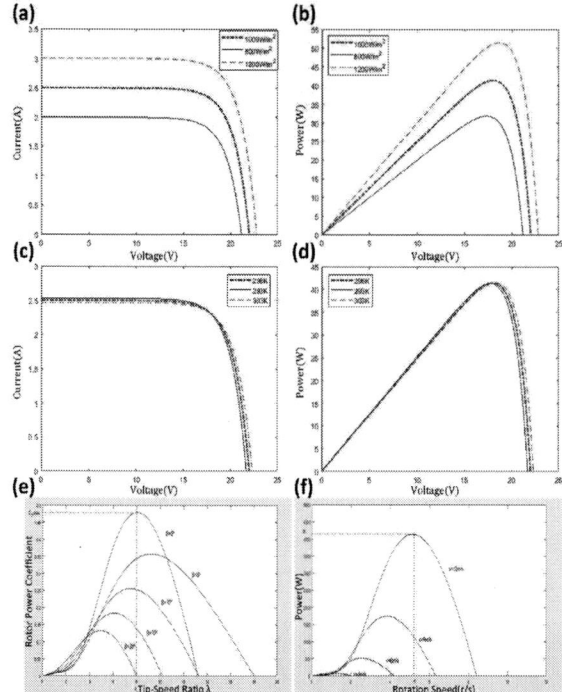

Fig. 4. (a) and (d) show the simulation results of the solar cell in changing environment. (e) and (f) present the effect of tip-speed ratio and rotation speed to the output of wind turbine.

Traditional P&O has a drawback that, at steady state, operating point oscillation often happened around MMP causing the output instable. Based on P&O, we proposed an advanced perturb and observe method combined with dichotomy(Advanced P&O), its characteristics are as follow:

1) Variable-step: to shorten the tracking time, before the sampling point approach MPP, a larger step size is used to track. Relatively, when the sampling point is approaching the MPP in a certain range, it will implement a smaller step.

2) Dichotomy: the operating point will be continuously perturbed in the determined direction until the perturbed output power is less, and then, the step will be shortened by half.

3) Threshold: it is almost impossible to achieve the perfect condition, P(V+ △ V)-P(V)=0, causing the system operating point oscillation. Therefore, setting a proper threshold E is necessary, when |P(V+ △ V)-P(V)|<E, the operating point is recognized as MPP.

On the basis of the algorithm above, we simulated the system under the environment of Matlab/Simulink and the operating voltage control was implemented by boost circuit. Setting the temperature at 298 K and initial irradiation 1000 W/m², at 0.5s, the irradiation was reduced to 800 W/m². Fig. 5 shows the simulation results that system output will oscillate for a while and achieve a new MPP in changed condition immediately. In conclusion, with the MPPT technique, renewable energy depending on the changeable nature environment is easier to carry out the optimal utilization.

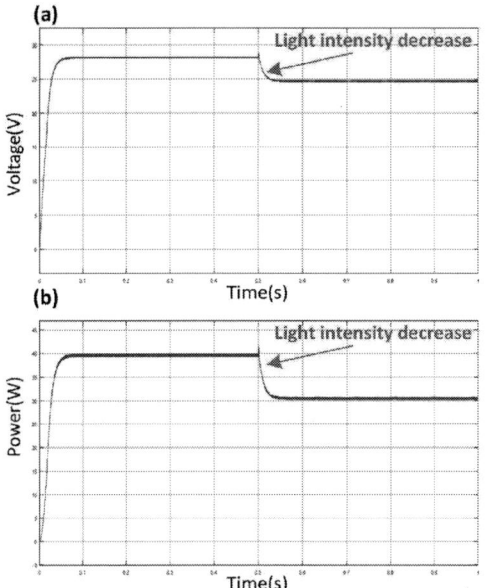

Fig. 5. The output of solar cell maintained in MPP in different conditions.

C. DMPC

In the section, we discussed DMPC, the principle of our designed system. Firstly, the power demands of end-users and the maximum power output under different condition are variable in practice. To meeting the total power demand anytime, battery banks were introduced in distributed system, equation (3) presented the current energy storage [9].

$$Q_{BA}(t) = Q_{BA}(t-1) + \eta_{BA}P_{BA}\Delta t \qquad (3)$$

Where Q_{BA} is the current energy storage, η_{BA} is the coefficient of charge/discharge efficiency, P_{BA} is power of charge/discharge, and Δt is the time of charge/discharge.

We adopt the technique for order preference by similarity to an ideal solution(TOPSIS) as the principle of the proposed DMPC algorithm. Herein, generating cost $f_{01}(P_w^*, P_s^*)$ and output smoothness $f_{02}(P_w^*, P_s^*)$ are considered as the main factor of system optimization scheduling.

$$f_{01}(P_w^*, P_s^*) = \arg\min(G_1 + G_2) \qquad (4)$$

$$f_{02}(P_w^*, P_s^*) = \arg\min(F_1 + F_2) \qquad (5)$$

Where G_1, G_2 is the generating cost of photovoltaic and wind energy, and F_1, F_2 is the power fluctuation of photovoltaic and wind energy. Concretely, they are shown in equation (6).

$$\begin{cases} G_1 = C_s \times \sum_{i=1}^{m} P_{si} + \sum_{i=1}^{m} N_{si}(B_s + 20 \times M_s) \\ G_2 = C_w \times \sum_{j=1}^{n} P_{wj} + \sum_{j=1}^{n} N_{wj}(B_w + 20 \times M_w) \\ F_1 = \sum_{i=1}^{m} \left\| \frac{P_{si}^* - P_{si}}{P_{si}} \right\|^2 \\ F_2 = \sum_{j=1}^{n} \left\| \frac{P_{wj}^* - P_{wj}}{P_{wj}} \right\|^2 \end{cases} \qquad (6)$$

Where m, n present the amount of photovoltaic and wind power subsystems, C_s, C_w are the generating cost per kilowatt hour of photovoltaic and wind power subsystems,

$P_{si}, P_{wj}, P_{si}^*, P_{wj}^*$ are the original and optimized power output, N_{si}, N_{wj} are the amount of module each subsystem, B_s, B_w are the installation cost, M_s, M_w are the maintenance cost, we assume the service life is 20 years. Meanwhile, the constrain is presented as follow.

$$\begin{cases} P_{load} = \sum_{i=1}^{m} P_{si}^* + \sum_{j=1}^{n} P_{wj}^* + P_b \\ 0 \le P_{si} \le P_{sim} \\ 0 \le P_{wj} \le P_{wjm} \\ N_{si} \ge 1 \\ N_{wj} \ge 1 \end{cases} \qquad (7)$$

P_b is the battery charge/discharge power, when the output of system can't meet the demand of end-user, $P_b > 0$, otherwise, $P_b < 0$. P_{sim}, P_{wjm} are the maximum power. We assume the amount of photovoltaic and wind power subsystems are both 2, M_s equates 10 and M_w equates 1. Other parameters were provided from local manufacturers.

With the TOPSIS algorithm, finally, we obtain the optimal output value in different condition. Specific scheduling process is as follow. Firstly, the power demand is collected from the load and is sent to energy management for analysis and the power output of each subsystems is also collected. And then, the distributed energy controller calculates the maximum output power of each subsystem in different conditions. In next step, the energy management part calculates the optimal output value of each subsystems with the DMPC based on TOPSIS. Finally, the power regulation part obtains the optimal output value and controls the output of each subsystems.

In the Matlab/Simulink environment, we carry out our simulation researching the effect of irradiation and load demand to system. The simulation results are shown in Fig. 6(e,f). (e) revealed the effect of irradiation under the same load demand 1400 KW, irradiation (c) varied evenly between 80-120 KW/m². Relatively, (f) is the simulation result under the invariable irradiation (d) and changeable load demand (b), all the simulations were assumed at 298 K and the wind speed (a) varied irregularly between 7-10 m/s. In conclusion, the power output is related with the environment change, and with the constructed DMPC algorithm, we maintained the smooth output and also took economic costs into account, which is significant for the industrialization of distributed renewable energy utilization.

D. Prototype

With the concept of smart power management micro-system for distributed renewable energy harvesting, we designed a double input parallel boost circuit based on STM32 and proved its ability to manage the output optimally. In the system, a DC power was assumed as the photovoltaic cell and wind turbine power input, and with the parallel boost circuit boosting the voltage to 48V and STM32 controlling the circuit, the output was regulated.

Fig. 7(a) shows the construct of the designed circuit, we adopt the double loop control strategy, output voltage sampled by divider resistance is sent to UC3842 for duty cycle control which deciding the output voltage. At the same time, current signal is also sent to UC3842 realizing the current inner loop control and overcurrent protection.

978-1-6654-3008-1/21 $31.00 © 2021 IEEE

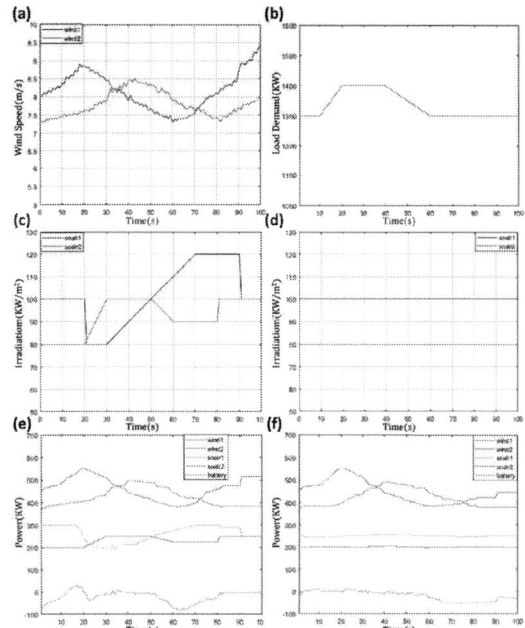

Fig. 6. Simulation results in different irradiation and load demand.

Herein, a digital potentiometer regulated by STM32 is inserted to adjust the duty circle of the two boost modules as shown in Fig. 7(b). Firstly, we set the current ratio $K=I_1:I_2$, and if the actual current ratio K^* monitored by STM32 didn't equal K, the digital potentiometer will redistribute the resistance of the two modules until $K^*=K$. System output was stabilized at 48 V, therefore, we can realize the power regulation by controlling the distribution of current. Finally, the measurement was implemented as shown in Fig. 7(c-e), in (c) and (d), we set K=0.5 and successfully maintained the standard 48 V output stabilized at a relatively steady level. After calculation, the system efficiency is above 90% and ripple factor is below 2%, which prove the good performance of the system. In (e), we discussed its ability to dominate the current distribution, with the same input voltage 20V and current 2.63A, setting different K, the system can still meet the design requirements of distributing current in any ratio.

III. CONCLUSION

This new design smart power management microsystem for distributed renewable energy harvesting is presented and the simulation model of solar cell and wind turbine is constructed. We improved and simulated an algorithm based on P&O to track the MPP of the solar cell and discussed DMPC, the principle of our system. In the end, we verified the feasibility of the proposed microsystem by the designed smart power management microsystem.

ACKNOWLEDGMENT

This work is supported by NSFC (No. 62074029, No. 61804023), the Key R&D Program of Sichuan Province (No. 2020ZHCG0038), and the Fundamental Research Funds for the Central Universities (No. ZYGX2019Z002).

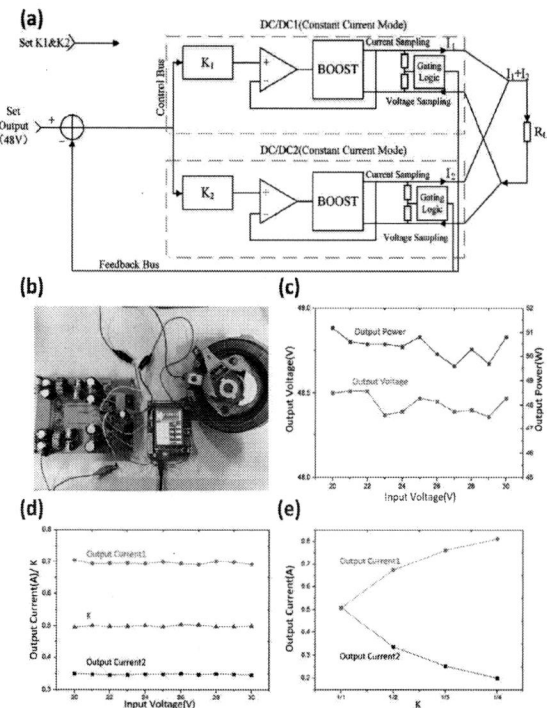

Fig. 7. (a,b) show schematic diagram and photograph of the small designed microsystem. (c) demonstrates the output was stabilized. (d,e) prove that the system can control the output current in different current ratio K.

REFERENCES

[1] C. Abbey and G. Joos, "Energy Management Strategies for Optimization of Energy Storage in Wind Power Hybrid System," *2005 IEEE 36th Power Electronics Specialists Conference*, Recife, 2005, pp. 2066-2072.

[2] M. Cirrincione et al., "Intelligent energy management system," *2009 7th IEEE International Conference on Industrial Informatics*, Cardiff, Wales, 2009, pp. 232-237.

[3] W. Qi, J. Liu and P. D. Christofides, "Distributed Supervisory Predictive Control of Distributed Wind and Solar Energy Systems," in *IEEE Transactions on Control Systems Technology*, vol. 21, no. 2, March 2013, pp. 504-512.

[4] S. Weixiang, C. F. Hoong, W. Peng, L. P. Chiang and K. S. Yang, "Development of a mathematical model for solar module in photovoltaic systems," *2011 6th IEEE Conference on Industrial Electronics and Applications*, Beijing, China, 2011, pp. 2056-2061.

[5] Caisheng Wang and Hashem Nehrir, "Power management of a stand-alone wind/photovoltaic/fuel-cell energy system," *2008 IEEE Power and Energy Society General Meeting - Conversion and Delivery of Electrical Energy in the 21st Century*, Pittsburgh, USA, 2008, pp. 1-1.

[6] H. Tian, F. Mancilla-David, K. Ellis, E. Muljadi, and P. Jenkins, "A cell-to-module-to-array detailed model for photovoltaic panels," *Solar Energy*, vol. 86, no. 9, 2012, pp. 2695-2706.

[7] J. G. Slootweg, S. W. H. de Haan, H. Polinder and W. L. Kling, "General model for representing variable speed wind turbines in power system dynamics simulations," in *IEEE Transactions on Power Systems*, vol. 18, no. 1, Feb. 2003, pp. 144-151.

[8] N. Femia, G. Petrone, G. Spagnuolo and M. Vitelli, "Optimization of perturb and observe maximum power point tracking method," in *IEEE Transactions on Power Electronics*, vol. 20, no. 4, 2005, pp. 963-973.

[9] P. Kou, D. Liang, F. Gao and L. Gao, "Coordinated Predictive Control of DFIG-Based Wind-Battery Hybrid Systems: Using Non-Gaussian Wind Power Predictive Distributions," in *IEEE Transactions on Energy Conversion*, vol. 30, no. 2, June 2015, pp. 681-695.

Gap in pagination due to unavailable papers.

Pages 741-748

Proceedings of the 16th Annual IEEE International
Conference on Nano/Micro Engineered and Molecular Systems
April 25-29, 2021

Design of Self-powered Environment Monitoring Sensor Based on TEG and TENG

Jianhao Liu, Changxin Liu, Cong Zhao, Huaan Li, Guanghao Qu, Zhuofan Mao, Zhenghui Zhou

Abstract— This paper designs a self-powered environmental monitoring sensor device based on Thermoelectric Power Generation(TEG) and Triboelectric Nanogenerator(TENG). The device is composed of an energy supply unit and an environmental monitoring sensor unit. The energy supply unit is composed of a TEG unit and the thin film flapping TENG unit. The environmental monitoring sensor unit monitors the wind speed and direction through the voltage generated by the thin film flapping TENG, and the voltage of the rain energy TENG represents the magnitude of rainfall. The thin film flapping type TENG in the energy supply part can achieve an output voltage of 244V and an output current of 35uA at a wind speed of 12m/s. The TEG unit can obtain an output voltage of 4.27V and an output current of 328mA at the hot end of 110°C. In the experiment, compared with a single unit, the charging voltage is increased by 93.3% and 28.9% in the process of charging the capacitor based on the high-current TEG and high-voltage TENG composite device. The composite energy harvesting device can successfully drive the circuit management unit and realize the self-supply of the environmental monitoring sensor unit.

Keywords—Thermoelectric Power Generation (TEG), Triboelectric Nanogenerator (TENG), self-powered, sensor

I. INTRODUCTION

In today's increasingly urgent energy crisis, more and more people pay attention to energy collection. The current technological development has greatly reduced the energy consumption of electronic devices and improved their work efficiency. Converting the ubiquitous mechanical energy in the surrounding environment into electrical energy to drive electronic devices has become an effective way to solve the energy supply problem of electronic devices[1]. Among them, the development of wireless sensor network technology has a very important impact on the progress of environmental detection technology. Because wireless networks have the advantages of low cost, high real-time performance, and no pollution, this technology is widely used in environmental monitoring[2]. Collecting and transmitting information through the wireless network saves a lot of human resources and improves the performance of the environmental detection system[3]. However, how to provide suitable power for these wireless sensor network units has become an important issue facing researchers. Generally speaking, the power consumption of micro devices is very low, so it is of great significance and practical value to develop a power system that can collect energy from

the surrounding environment to power electronic devices[4]. This paper proposes a self-powered environmental monitoring sensor based on multi-energy complementary of Thermoelectric Power Generation (TEG) and Triboelectric Nanogenerator (TENG). It mainly consists of two parts: an energy supply unit and a sensing unit. The energy supply part is based on two energy harvesting devices based on TEG and TENG. The two devices are coupled to realize the self-supply of environmental monitoring sensors by collecting solar and wind energy in the environment into electrical energy. The sensing part mainly monitors the wind speed, wind direction and rainfall in the environment. The sensing part is realized through TENG. Among them, the wind speed and direction are realized by the wind-induced vibration film flapping Triboelectric Nanogenerator, The amount of rain is realized by the rain energy Triboelectric Nanogenerator. The magnitude of TENG voltage can characterize wind speed and rainfall. The wind direction is realized by four perpendicular wind-induced vibration film flapping type TENGs on the lower chassis that can rotate.

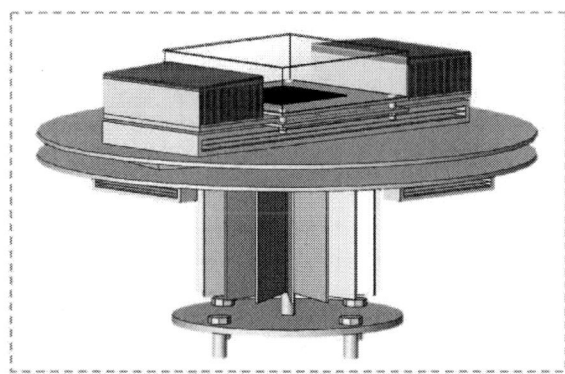

Fig.1. The model diagram based on TEG and TENG self-powered environmental monitoring sensors.

The model diagram of the self-powered environmental monitoring sensor device is shown in Figure 1. It can not only collect energy from the surrounding environment in real time to supply power to the environmental monitoring sensor, but also has the advantages of simple structure, low construction cost, and high working reliability. It has good practicality Value and development prospects.

---

*Resrach supported by the Natural Science Foundation of Liaoning Province, the National Key R & D Program and the Fundamental Research Funds for the Central Universities.

*Contacting Author: Changxin Liu. İs with College of marine engineering,

Dalian Maritime University, China. (phone:15842478808;email:liu_changxin @dlmu.edu.cn).

All authors are with Onshore Engine Room Laboratory, College of Marine Engineering, Dalian Maritime University, No.1,Linghai Road, Dalian 116026, China.

978-1-6654-3008-1/21 $31.00 © 2021 IEEE

II. RESULTS AND DISCUSSION

A. A. Working Principle

Triboelectricity is a common physical phenomenon. It is an electrification effect caused by contact, that is, the process of friction between one material and another material is charged[5]. The triboelectric nanogenerator (TENG) was first invented by Zhonglin Wang's research team in 2012. TENG is a new type of power generation technology that is excellent at collecting energy in the environment and converting it into electrical energy. Its working principle is based on the coupling of frictional electrification effect and electrostatic induction effect[6]. It has the advantages of simple structure, light weight, wide selection of materials, low cost, and high efficiency output even under low-frequency mechanical motion. It is considered to be a potential solution for building self-powered systems and has been applied in the fields of the Internet of Things, sensor networks, robotics and artificial intelligence[7].

The working principle of the triboelectric generator (TENG) is based on the coupling of triboelectric effect and electrostatic induction, the working principle diagram of the wind-induced vibration film flapping TENG is shown in Figure 2a.In the initial state, the dielectric film is in contact with the lower copper electrode under the action of gravity. Because the dielectric PTFE is relatively electronegative, positive and negative triboelectric charges are generated on the surface of the copper electrode and the membrane (Figure 2a i). Under the action of wind, the membrane separates from the lower copper electrode and vibrates upward. At this stage, triboelectrons flow from the upper copper electrode to the lower copper electrode through an external circuit, thereby generating a transient current (Figure 2a ii). When the film is in contact with the upper copper electrode, all positive charges are displayed on this electrode (Figure 2a iii). Subsequently, the reverse movement of the membrane causes the electrons to travel backwards through the external circuit (Figure 2a iv). Under the action of wind, the entire power generation cycle continues to circulate, and the mechanical energy of the wind is successfully converted into electrical energy[8].

The working mechanism of the rain energy TENG is shown in Figure 2b. Raindrops falling from the air will generate triboelectric charges, and then the surface of the droplets will be charged (Figure 2b i). When the positively charged water droplets gradually approach the surface of the PTFE film (Figure 2b ii), a positive potential difference is generated between the copper electrode and the ground electrode. Electrons will flow from the ground electrode to the copper electrode to shield this potential difference, until the electrons completely shield the potential difference (Figure 2b iii) and reach a balanced state, resulting in a positive pulse current. When the water droplets leave the PTFE membrane (Figure 2b iv), a negative potential difference is formed between the copper electrode and the ground electrode. Electrons will flow from the copper electrode to the ground electrode, resulting in a negative current pulse. So if water flows continuously into and out of the surface of the device, then a continuous power output can be obtained. The amount of rainfall is proportional to the output voltage of the rain energy TENG, so the voltage can be used to characterize the amount of rainfall [9].

When there is a temperature difference between the two ends of different thermoelectric materials, the characteristics of the thermoelectric material will generate an electromotive force at the two ends of the material to form a current and generate a thermoelectric effect, which is called thermoelectric power generation. Thermoelectric power generation is mainly based on the principle of the Seebeck effect. As shown in Figure 2c, P-type and N-type are conductors of two different materials. In the loop formed by them, If there is a temperature difference between the two contact points, an electromotive force will be generated in the loop formed by P and N. The above phenomenon was first discovered by the German physicist Seebeck, so people call it the Seebeck effect. Its appearance has laid a solid theoretical foundation for the research of thermoelectric power generation technology[10-14].

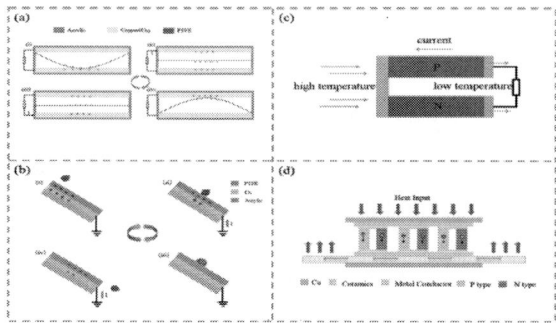

Fig.2. (a) The working principle of the thin film flapping TENG. (b) The working mechanism of the rain energy TENG. (c) Schematic diagram of Seebeck effect. (d) Schematic diagram of Working principle of the TEG

The solar TEG unit is composed of a solar heat collection material on the top, a thermoelectric power generating sheet between them, and a micro heat pipe array and heat dissipation fins on the bottom. The solar heat absorption film on the top can absorb the heat of the solar energy and act as a heat source to heat up the hot end of the TEG unit. As shown in Figure 2d, the micro heat pipe array at the bottom is composed of an external thin metal plate and a small amount of liquid working fluid inside. Each micro heat pipe array is composed of multiple independently operating micro heat pipes. Each micro heat pipe has a micro fin structure inside. The working fluid inside is heated and vaporized, and flows inside the micro heat pipe array to the condensing end. The vaporized working fluid exchanges heat with the outside at the condensing end and then liquefies again and then flows back to the heated end. Such repetition can take away the heat from the cold end of the thermoelectric power generation sheet and ensure a certain temperature difference between the two ends of the TEG. The thermoelectric power generation sheet is composed of 128 pairs of PN sections, and its electrical connection mode is series, and the thermal connection mode is parallel. Based on the Seebeck effect, under the action of the temperature difference between the two ends of the thermoelectric power sheet, the negatively charged free electrons of the N-type semiconductor material and the positively charged holes of the P-type semiconductor material transition from the high temperature end to the low temperature end, thereby forming a potential difference, becoming thermoelectromotive force.

B. Output Performance of the Device

In order to explore the working performance of the energy supply part, the working performance of TEG unit and TENG unit were tested. First, an experimental study was carried out on the heating characteristics of the hot end of the TEG unit. The solar heat-absorbing coating has good heat-collecting characteristics under the sun. In order to facilitate the experimental research, we use a 100° constant temperature electric heating device to simulate the solar heat source to experiment on the hot end of the TEG device, and obtain the curve of TEG working voltage and working current with time, as shown in Figure 3(a),3(b) . As the hot end heats up, the output voltage and output current of TEG gradually increase. The maximum output voltage of TEG can reach 4.27V, and the maximum output current can reach 328mA. The temperature difference between the hot and cold end is an important factor that affects the performance of the TEG unit. Therefore, change the wind speed of the external environment and explore the influence of the wind speed on the TEG performance. As the wind speed increases from 2m/s to 16m/s, the TEG output voltage increases from 3.25V to 4.27V, and the output current increases from 250mA to 328mA, as shown in Figure 3(c). This is because as the wind speed increases, the heat dissipation capacity of the micro heat pipe array at the cold end of the TEG increases. The temperature of the cold end drops from 67°C to 53°C, and the temperature difference between the cold and hot ends increases from 43°C to 57°C, as shown in Figure 3(d) .

The experimental data is shown in Figure 4 (c). The results show that when the wind speed increases from 3.39m/s to 12m/s, the output voltage of TENG increases first and then decreases with the increase of wind speed, while the output current of TENG increases linearly with the increase of wind speed. The reason is that as the wind speed increases, the contact frequency between the PTFE film of TENG and the upper and lower copper plates increases, and the charge accumulated on the surface of the Cu film increases, so the working voltage of TENG shows a rising trend. As the wind speed exceeds a certain critical value, the beating frequency of TENG's PTFE membrane increases, resulting in insufficient contact between TENG's PTFE membrane and Cu membrane, and thus the contact area decreases. Therefore, the output voltage of TENG shows a downward trend when it exceeds a certain wind speed. At the same time, the increase in wind speed leads to an increase in the beating frequency of the PTFE membrane, which in turn leads to an increase in the electron transfer speed between the upper and lower Cu membranes, and the number of electrons flowing in a unit time increases, so the output current of TENG shows an upward trend. In addition, when TENG works as a power source, its performance is related to the load resistance. As shown in Figure 4(d), changing the load resistance of TENG, its output current decreases with the increase of wind speed, and the output power first rises and then decreases with the increase of wind speed. At 5MΩ, its output power reaches its maximum. The output power is now 0.79mW, and the output current is 25.1uA.

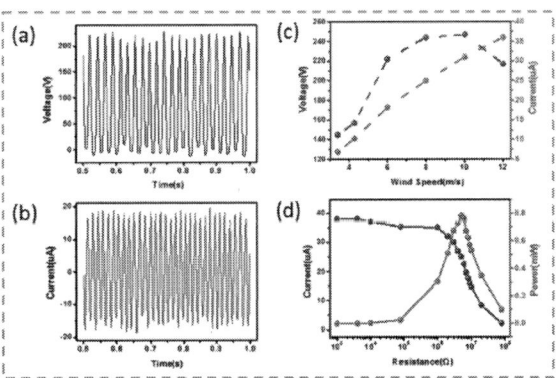

Figure 4. (a) Output voltage of TENG (b) Out current of the TENG. (c) Dependence of output performance of the TENG on the wind speed. (d) Dependence of output current and power of the TENG on external resistances.

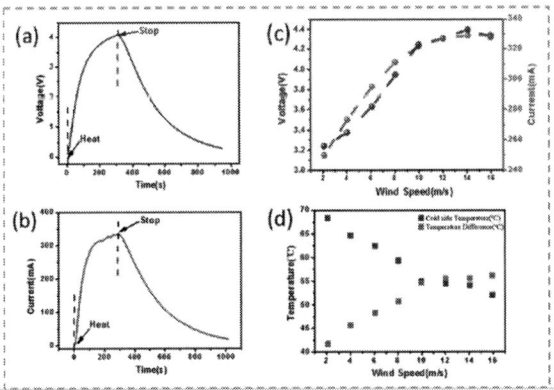

Figure 3. (a) Solar TEG output voltage (b) Solar TEG output current. (c) Dependence of output performance of the TEG on the wind speed. (d) The dependence of the temperature difference of the TEG on the wind speed.

In the wind-induced vibration film flapping TENG unit, as shown in Figure 4 (a), the output current of TENG is 35uA, and as shown in Figure 4 (b), its output voltage can reach up to 244V. Through experimental research, we know that different wind speeds have a certain influence on the output voltage and output current of TENG, and wind speed has a significant influence on the output performance of TENG.

In order to explore the performance of the composite power generation device in practical applications, the TEG-TENG composite energy harvesting device proposed in this research is used to charge a 47uF capacitor. After the AC output of the TENG unit is processed by a full-wave rectifier bridge, it is converted into DC power. parallel output with TEG unit, A Schottky diode is used to isolate the two power generating units, as shown in Figure 5(a). For example, Figure 5(b) shows the charging process of composite energy harvesting. First, when the hot end is gradually heating up, TEG is used to charge the capacitor. After the TEG hot end

978-1-6654-3008-1/21 $31.00 © 2021 IEEE

starts to heat for 240s, the capacitor is charged to 4V due to the low voltage output of TEG. Next, use the rectified TENG to continue charging the same capacitor. After 200s, the capacitor can be increased to 17V, and the charging voltage is increased by 325%. It shows that the composite power generation device can well compensate the low voltage limit of the TEG unit. Figure 5(c) shows that TEG alone is used to charge a 47uF capacitor. After 125s, the capacitor is charged to 3V. TENG alone can charge the capacitor to 4.5V after 125s. For composite power generation devices, it can be charged after 125s. The capacitor is charged to 5.8V. Compared with the output voltage performance of the separate TEG unit and the TENG unit, the charging performance of the composite energy harvesting device is improved by 93.3% and 28.9%, respectively.

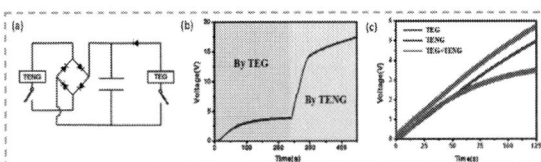

Figure.5. (a) Power management unit used in hybrid energy converter charging the capacitor. (b) The measured voltage curve of a 47uF capacitor charged by solar TEG, Wind TENG. (c) The measured voltage curve of a 47uF capacitor charged by solar TEG, Wind TENG, hybrid energy converter, separately.

For the sensing part, the wind speed and direction are realized by the wind-induced vibration film flapping Triboelectric Nanogenerator. The relationship between the wind speed, the film flapping frequency and the output voltage is shown in Figure 6(a), 6(b). It can be seen from the figure that within a certain range, the wind speed is positively correlated with the beat frequency of the film and the output voltage of TENG, so the output voltage can be used to characterize the wind speed. The wind direction can be achieved by rotating the four vertical TENGs on the chassis.

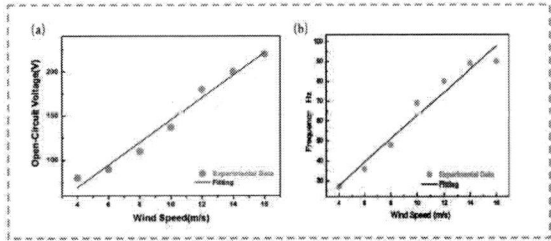

Fig.6. (a) The open circuit voltage of TENG under different wind speeds. (b) The tapping frequency of TENG films under different wind speeds.

The monitoring of rainfall is mainly realized through the rain energy Triboelectric Nanogenerator. In order to illustrate the performance of rain energy TENG in collecting rain energy, the water tap is used to provide water flow to drive rain energy TENG work. The flow rate of the water

flow is set at 60 ml/s, and the distance between the device and the tap outlet is set at 20 cm. Measure the output voltage and output current to characterize the output performance of the device. As shown in Figure 7(a) and (b), the peak voltage of the device can reach 3.5V, and the maximum current can reach 10μA. In addition, the influence of the water flow rate on the performance of the device was also investigated. The output voltage showed an increasing trend when the flow rate increased from 0ml/s to 120ml/s, and then remained almost unchanged, as shown in Figure 7(c).

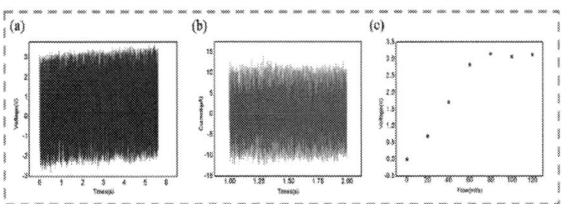

Fig.7. (a) The open circuit voltage of rain energy TENG. (b) The short circuit current of TENG. (c) Variation of the voltage of rain energy TENG with fluid flow.

III. SUMMARY AND PERSPECTIVES

In short, based on the Seebeck effect, contact electrification and electrostatic induction effects, solar energy and wind energy in the natural environment can be collected at the same time or separately. At 110°C at the hot end, the thermoelectric power generation unit can generate a maximum of 4.27V, 328mA of output electrical energy, and at a wind speed of 16m/s, the wind energy TENG unit can generate 244V, 35uA of electrical energy. Self-powered environmental monitoring sensor based on multi-energy complementary of Thermoelectric Power Generation (TEG) and Triboelectric Nanogenerator (TENG) has been systematically studied. By collecting solar and wind energy in the surrounding environment, the energy in the environment is converted into electrical energy to supply power to the sensing unit. For the composite power generation unit, the 47uF capacitor can be charged to 5.8V after 125s. Compared with the output voltage performance of the separate TEG unit and the TENG unit, the charging performance of the composite energy harvesting device is improved by 93.3% and 28.9%, respectively. This power supply method provides a potential solution for environmentally-friendly off-grid power supplies. At the same time, using TENG for environmental monitoring and sensing is also an innovation.

ACKNOWLEDGMENT

We gratefully acknowledge financial support of the Natural Science Foundation of Liaoning Province, Grant/Award No. 20161063; the National Key R & D Program, Grant/Award No. 2017YFC14046; the Fundamental Research Funds for the Central Universities, Grant/Award Nos. 3132018255, 3132019330.

REFERENCES

[1] J. Y. Hwang, J. G. An, A. Aziz, J. H. Kim, J. S. Song. "Interworking Models of Smart City with Heterogeneous Internet of Things Standards," IEEE Communications Magazine. vol. 57, pp.74-79, 2019.

[2] M. Dong, R. R.Huang, Y. Zheng, S. J. Zhao, J. Chen. "Discussion Development of Internet of Things and Wisdom Agriculture, " in Modern Agricultural Science and Technology. Vol. 14, pp.338-340, 2016.

[3] Z. L. Wang, "Entropy theory of distributed energy for internet of things" in Nano Energy. vol 58, pp.669-672, 2019.

[4] X. Cao, Y. Jie, N. Wang, Z. L. Wang. " Triboelectric nanogenerators driven self-powered electrochemical processes for energy and environmental science" in Advanced Energy Materials, 2016.

[5] S. Wang, L. Lin, Z. L. Wang. "Triboelectric nanogenerators as self-powered active sensors " in Nano Energy, 2015.

[6] Z. L. Wang, L. Lin, J. Chen, S. M. Niu, Y. L. Yu. Triboelectric Nanogenerator. Beijing Science and Technology Press, 2017.

[7] W. Ding, A. C. Wang, C. Wu, H. Guo, Z. L. Wang. "Human-machine interfacing enabled by triboelectic nanogenerators and triboelectrics" In Adwanced Materials Technologies,2018.

[8] Y Wang, J Wang, X Xiao, S. Wang, M. Y. Xu. "Multi-functional wind barrier based on triboelectric nanogenerator for power generation, self-powered wind speed sensing and highly efficient windshield" in Nano Energy, vol 73, 2020.

[9] Q. J. Liang, Research on Water Flow Energy Harvesting and Vibration Sensor Based on Frictional Electric Effect, University of Science and Technology Beijing, 2017.

[10] C. X. Liu. Experimental Study of Thermoelectric Power Generation System and its Application, Dalian University of Technology, 2015

[11] C. X. Liu, W. X. Ye, H. A. Li, J. H. Liu, C. Zhao, Z. F. Mao "Experimental study on cascade utilization of ship's waste heat based on TEG‐ORC combined cycle" in International Journal of Energy Research, 2020.

[12] C. X. Liu, X. X. Pan, X. F. Zheng, Y. Y. Yan, W. Z. Li. "An experimental study of a novel prototype for two-stage thermoelectric generator from vehicle exhaust" in Journal of the Energy Institute. vol 89. pp. 271-281, 2015.

[13] W. X. Ye, C. X. Liu, J. H. Liu, C. Zhao, T. L. Dai, K. F. Ma. "Experimental study on TEG-ORC combined cycle for cascade recovery of multiple waste heat from ships" in Journal of Xi'an Jiaotong University, vol 8. pp. 1-9, 2020.

[14] C. X. Liu, W. X. Ye, J. H. Liu, G. P. Lv, T. Q. Zhao, J. M. Dong. "TEG-ORC combined cycle performance for cascade utilization of multiple waste heat from ships" in Journal of Engineering Science. vol 1 pp.1-10, 2020.

The 16th IEEE International Conference on Nano/Micro Engineered & Molecular Systems

Printed flexible/stretchable electronics and applications

Zheng Cui

Suzhou Institute of Nanotech and Nanobionics, Chinese Academy of Sciences, P.R. China

Website : www.perc-sinano.com

ABSTRACT

Lithography-based patterning and subtractive pattern transfer have been the dominant fabrication technologies for the fabrication of either integrated circuit chips or electronic circuit boards. In the past decade, printing fabrication has risen to prominence because of its additive nature. Printing enables patterning in large area and at low cost, can be of high throughput and less pollutant to environment, and most interestingly, independent of substrate properties. For the fabrication of a flexible or stretchable electronic system, conventional lithographic patterning has to rely on a rigid carrier substrate to perform the fabrication processes, printing can directly construct circuit patterns on a flexible or stretchable substrate. The rapid development in solution forms of various functional electronic materials has made printing as a preferred technology for the fabrication of flexible or stretchable electronics.

The author's group has been engaged in the development of printed flexible and stretchable electronics technologies since 2009 when the Printable Electronics Research Center (PERC) was setup. In the past 10 years, printable conductors, semiconductors and light emitting inks have been developed. Thin-film transistors, light emitting devices, sensors and solar cells have been made by inkjet printing, screen printing or an innovative hybrid printing technique. The printed flexible/stretchable electronics have found applications in touch panels, RFID tags, wearable sensors and smart textiles as shown in Figure 1.

BIOGRAPHY

Professor Zheng Cui had his PhD in 1988 and worked in the UK for 20 years, first at Cambridge University and then at Rutherford Appleton Laboratory in the field of micro and nanofabrication technologies. In 2009, he returned to China and setup the Printable Electronics Research Center (PERC). The Center has established research themes from electronic ink formulation to printing process development with applications including printed solar cells, printed thin-film transistors, printed organic light emission, and printed flexible/stretchable/wearable electronics. He has authored and coauthored over 280 technical papers, 7 books on the subject of micro and nanofarication technologies and 4 books on the subject of printed electronics. Some of the developed technologies at PERC have been transferred to industry.

978-1-6654-3008-1/21 $31.00 © 2021 IEEE

Gap in pagination due to unavailable papers.

Pages 755-761

Proceedings of the 16th Annual IEEE International
Conference on Nano/Micro Engineered and Molecular Systems
April 25-29, 2021

An Integrated Stretchable Sensing Patch for Simultaneously Monitoring of Physiological and Biochemical Parameters

Wenhao Pan, Shiyi Xu, Xiyu Mao, Tianyu Li, *Qingpeng Cao, Xuesong Ye, and* Bo Liang[*], *Member, IEEE*

Abstract—Application of wearable sensors for sports monitoring provides real-time insight into an athlete's physical condition. At present, most wearable sensing devices measure only a small number of physical or electrophysiological parameters, but neglect the rich chemical information available from biomarkers in human body fluid. This work developed a multi-parameter monitoring device based on flexible patch, which integrated a variety of sensors, and is capable of continuous, real-time and synchronous monitoring of physiological parameters (e.g. ECG, PPG) and biochemical parameters (e.g. sodium, potassium, glucose in sweat). The flexible patch was designed with an island-bridge structure, making itself miniaturized and stretchable. In addition, an Android application was developed to receive the Bluetooth low energy packets of the monitoring device. And the monitoring device was tested in signal acquisition of all sensors. Finally, on-body experiments were carried out, to successfully verify the function of multi-parameter monitoring. The results show that our multi-parameter monitoring device provided in this work is suitable for sports scenes, which allows the feedback of body status through the comprehensive evaluation of individuals, and has a good application prospect for athletes to monitor physiological load during training.

I. INTRODUCTION

Real-time monitoring athletes' physical state during training can not only provide real-time feedback on the athletes' physical conditions, but also help rationalize training plans and evaluate training results. Wearable monitoring devices currently on the market that support sports monitoring mainly measure physical or electrophysiological parameters such as step count, heart rate and pulse rate. These devices ignore the richer biochemical information provided by the large number of biomarkers in human body fluids. Biochemical parameters can be measured by conventional means (clinical laboratories or large field devices)[1], but this results in a certain lag. The collection of physiological and biochemical data is time-sensitive and if the data is not collected in time, the data obtained will be inaccurate.

Sweat is an easily collected body fluid, rich in biomarkers that reflect a person's physiological condition[2]. With the development of flexible materials and flexible electronics, wearable sensors have been rapidly developed through innovations in materials and structures[3, 4], which have led to a strong development in sweat in-situ detection

technology. In recent years, there is a large number of literature in the field of biochemical parameter monitoring at home and abroad, including in the laboratory of our team, demonstrating that epidermal biosensors that fit to the skin can continuously measure biochemical markers such as electrolytes and metabolites in sweat[5-11]. The transition from blood analysis to in situ sweat analysis through sweat sensors could provide a non-invasive and potentially dynamic approach to health assessment. Exercise produces large amounts of sweat, which largely solves the problem of sweat sampling and therefore sweat sensors are uniquely suited to monitor body parameters during exercise. Combining biochemical detection of body surface sweat with basic physiological parameters, continuous, real-time and simultaneous monitoring of multiple body parameters can provide a more comprehensive assessment of an athlete's physical condition for the purpose of scientific training.

In this work, we developed a multi-parameter monitoring device based on flexible patch for human body that allows continuous, real-time, simultaneous monitoring of basic physiological parameters (e.g. ECG and pulse waves) and biochemical parameters (sodium, potassium, glucose) in sweat. When used in exercise monitoring, electrocardiogram measurements can help monitor heart health and function. The periodicity of the PPG signal corresponds to the cardiac rhythm, and thus, Heart rate can be estimated using the PPG signal[12]. Excessive loss of potassium and sodium in sweat could result in dehydration hyponatremia, hypokalemia or muscle cramps[13]; sweat glucose is reported to be metabolically related to blood glucose[14].

The flexible patch device designed and manufactured in this work included an electronic circuit patch and sensor patches, as shown in Fig.1. The electronic circuit patch was based on a thin FPCB and was modularized using an "island-bridge" structure, with the control and communication module as the "central island", the power supply module and the individual detection modules as "peripheral islands" and the traces between the control and communication module and the other modules as the "bridge" (Fig.1c-d). ECG sensor patch (Fig.1a), printed ECG electrodes on the front side, electrodeposited copper foil contacts on the back side, and electrical connections on both sides through copper holes (Fig.1f). The circular patch was cut into two

---

This work was supported Zhejiang University K.P.Chao's High Technology Development Foundation; The Natural Science Foundation of Zhejiang Province (LQ20F010011, LY18H180006)

Wenhao Pan, Shiyi Xu, Xiyu Mao, Qingpeng Cao, Xuesong Ye and Bo Liang are with Biosensor National Special Laboratory, Key Laboratory of

Biomedical Engineering of Ministry of Education, College of Biomedical Engineering and Instrument Science, Zhejiang University, Hangzhou, 310027, PR China (e-mail: boliang1986@zju.edu.cn)

978-1-6654-3008-1/21 $31.00 © 2021 IEEE

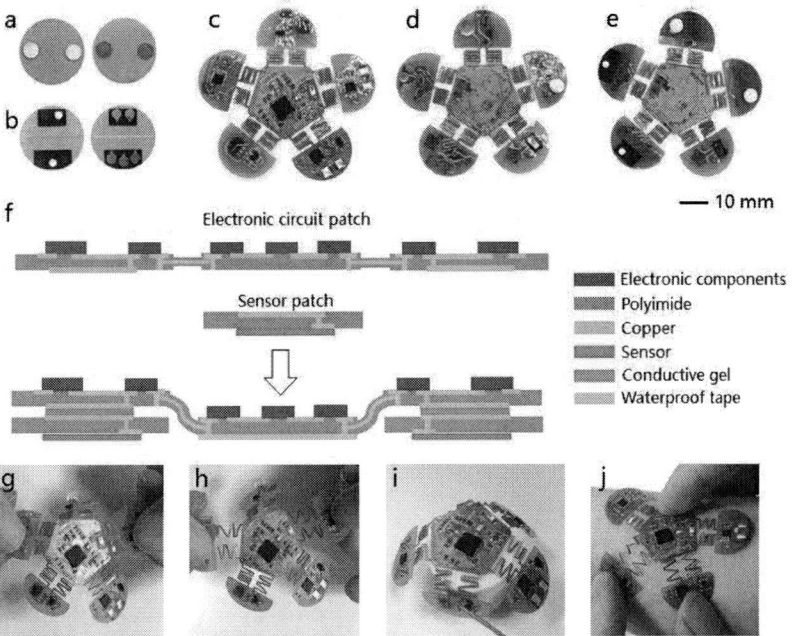

Figure 1 Flexible patches. (a) Photograph of an ECG sensor patch, with Ag/AgCl printed ECG electrodes on the front (left) and copper foil contacts on the back (right); (b) Photograph of a sweat sensor patch, with printed electrode arrays on the front (left) and copper foil contacts on the back (right); (c, d) Photograph of the front and back sides of an electronic circuit patch respectively; (e) The photograph after the sensor patch and the electronic circuit patch were assembled; (f) Fabrication process of the patch; (g-j): Demonstration of the stretchability of flexible patches.

semicircles along the diameter and placed on two 'islands' to ensure the collection distance of the ECG electrodes (Fig.1e). Sweat sensor patch (Fig.1b), printed electrochemical electrode array on the front and electrodeposited copper foil contacts on the back and electrical connections on both sides through copper holes (Fig.1f). The circular patch was cut into two semicircles along the diameter, which were ion sensor and glucose sensor part. The construction of the serpentine wires allowed the whole patch to be well stretched and to adapt to different curves of the skin area (Fig.1g-j).

II. METHOD

A. Electronic circuit system design

Aiming at the low power consumption and miniaturization requirements of the multi-parameter wearable monitoring device, integrated electronic components were selected for the circuit design and reducing the device size, as shown in Fig2. A lithium polymer battery (3.7 V, 45 mAh) was used as the power source, and the battery voltage was converted to a 3.3 V output through a low dropout regulator (LDO) to power the entire device. In the 'active mode', the device consumed, on average, 1.45 mA from a 3.7 V supply.

Control and communication module: At the core of the FPCB was a Bluetooth low energy System on Chip (BLE SoC) Texas Instruments (TI) CC2541. By using the built-in 12-bit Analog to Digital Converter (ADC) and hardware interface (such as I2C and UART), we processed the signals from the sensor and front end circuit. And the data was transmitted to a mobile terminal equipment as BLE packets through the built-in Bluetooth transceiver. A Johanson Technology 2.4 GHz chip antenna (2450AT18A100) and a Johanson Technology chip Balun-BPF (2450BM15A0002) were used for wireless transmission.

Amperometric detection front end for glucose: The sweat glucose detection front end circuit was realized by TI LMP91000 analog front end (AFE) chip. While the AFE applied a constant potential (−0.1 V) between the reference and the working electrodes, it also measured the current from the working electrode and outputted it as an analog voltage value by using the linear transimpedance amplifier (TIA). The output was digitized by the ADC integrated into the controller. The AFE chip was programmable through the I2C interface driven by the controller.

Potentiometric detection front end for ion: The sweat ion detection front end circuit was designed by the potentiometric method to measure the open circuit potential (OCP) of the working electrode and the reference electrode. A voltage follower circuit with an operational amplifier (AD8606 from Analog Devices Inc.) was used for measuring the potential signal from the electrodes. The output of the voltage follower circuit was followed by an RC low-pass filter, to minimize the noise and interference in the measurements. The output was digitized by the ADC integrated into the controller.

978-1-6654-3008-1/21 $31.00 © 2021 IEEE

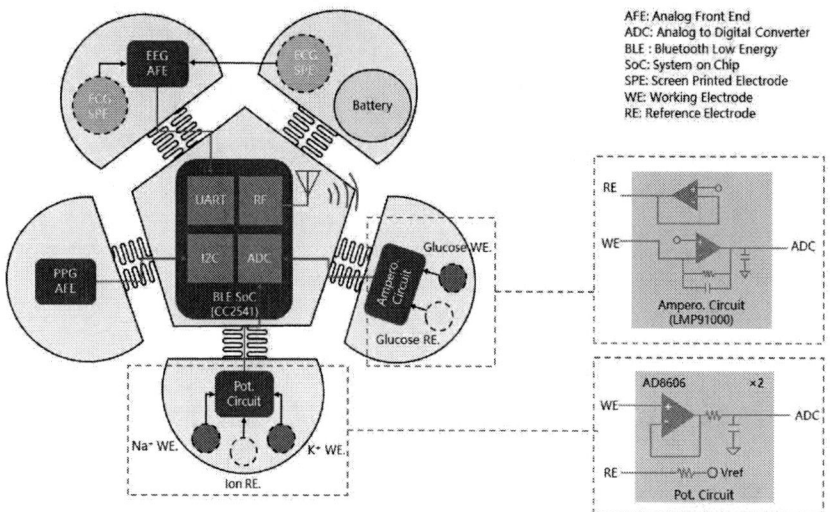

AFE: Analog Front End
ADC: Analog to Digital Converter
BLE : Bluetooth Low Energy
SoC: System on Chip
SPE: Screen Printed Electrode
WE: Working Electrode
RE: Reference Electrode

Figure 2 System design of the sensor patch. The ECG signal is obtained using an AFE and read by the SoC via UART interface. Meanwhile, the PPG signal is obtained using another AFE and read by the SoC via I2C interface. The potential signal from the ion sensor is detected by a potentiometric circuit and the current signal from the glucose sensor is detected by an amperometric circuit and they are converted to analogue voltage signals and output to the ADC of the SoC. All data are then processed and transmitted using the BLE SoC to a mobile phone.

ECG detection front end: The BMD101 SoC from Neurosky was used for biopotential measurements to record the electrocardiogram signals from the fabricated electrocardiogram electrodes.

PPG detection front end: The MAX30102 chip from Maxim Integrated was used to design the front-end for pulse wave signal acquisition based on the photoplethysmography (PPG). The MAX30102 includes internal LEDs, photodetectors, optical elements, and low-noise electronics with ambient light rejection. The chip is programmable through the I2C interface driven by the controller.

B. Sensor preparation

In this work, screen printing technology was used to prepare carbon electrodes and Ag/AgCl electrodes with low cost, simple process and mass production. Use carbon electrodes to make ion-selective electrodes and enzyme-based electrochemical bioelectrodes, which were used for the detection of body surface sweat electrolytes (sodium ions, potassium ions) and metabolites (glucose) respectively. And carbon electrodes for glucose sensor were doped with Prussian Blue to reduce the oxidation peak potential. The Ag/AgCl electrodes were used as reference electrodes of sweat sensor and electrophysiological electrodes of ECG sensor.

Screen-printing of the electrodes: The first layer of silver/silver chloride ink was printed, followed by carbon ink, and finally the blue insulator layer. The cure condition of every layer was at 90 °C for 1 min. The glucose sensor included a working electrode and a reference electrode. The ion sensor included a sodium working electrode, a potassium working electrode and a common reference electrode. All sweat electrodes were round shapes with a 2.5 mm diameter, while electrophysiological electrodes were 4.0 mm.

Preparation of the glucose sensor: The method of layer-by-layer drop-coating to form a film was used to modify the polymer film on the surface of the working electrode to fix glucose oxidase (GOx) to prepare a glucose sensor. The

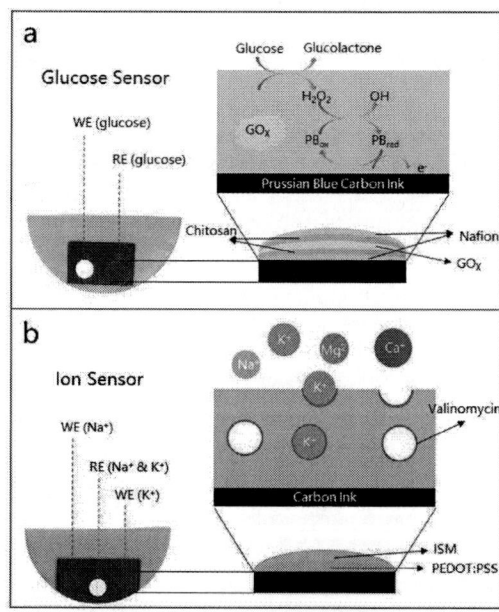

Figure 3 Photograph and schematic of electrochemical sensors. (a) Photograph of glucose sensor and schematic of electrode modification, along with the corresponding recognition and transduction events. (b) Photograph of ion sensor and schematic of electrode modification, along with the corresponding recognition and transduction events (taking potassium as an example).

materials included GOx solution (10 mg/mL containing 40 mg/ml bovine serum albumin stabilizer), chitosan solution (5 % chitosan and 3 % acetic acid) and Nafion solution. The drop casting sequence was 2 μL Nafion solution, 3 μL chitosan solution, 2 μL GOx solution, 3 μL chitosan solution, and 2 μL Nafion solution. Each drop needed to be dried for 30 minutes under ambient conditions before proceeding to the next step. The sensor was stored at 4 °C when not in use.

Preparation of the K⁺/Na⁺ selective sensor: Different ion selective electrodes modified different ion selective membranes (ISM) on the surface of carbon electrodes. Firstly, A PEDOT: PSS polymer film was electroplated on the working electrodes, which can effectively reduce the potential drift. The plating current was 14 μA, while the plating time was 700s. Thereafter, 4 μL ISM solutions was dripped on the working electrode, and the 2 μL reference electrode solution was dripped on the reference electrode five times. The drying time was about 5 minutes for each 2 μL drop, and the electrode was dried at 4 °C over 12 hours until the solvent completely evaporated. K⁺ ISM included valinomycin, and Na⁺ ISM included sodium ionophore X.

Preparation of the electrocardiogram sensor: The electrocardiogram printed electrodes were covered with conductive hydrogel adhesives. The patch was then affixed to a medical-grade adhesive sheet required for applying to human skin.

III. RESULTS

A. Performance test of sweat sensor

Combination of sweat detection circuitry and sweat sensor on a flexible patch in a standard solution. Time-current curves for the glucose electrode and open-circuit potential response curves for the sodium/potassium ion electrode were recorded. Applied a constant potential of -0.1 V between the glucose working and reference electrodes. The data was recorded via Android phone APP and saved to the local end and the test results are shown in Fig.4.

Fig.4a shows the amperometric response of the glucose biosensor to increasing glucose concentrations in the physiological range of 0–1.0 mM. It is evidenced from this figure that the biosensor responds linearly to the lactate concentrations in this range with a sensitivity of 1.5 μA/mM Fig.4b-c show the potentiometric response of the potassium and sodium sensors to increasing concentrations in the physiological range of 1–32 mM and 4-128 mM respectively. The sensitivity is 64.2 mV per decade of concentration for K⁺, and 72.5 mV per decade of concentration for Na⁺. But there were fluctuations and burrs in the signal. This is the fact that the ion detection circuit does not use an integrated analogue front-end module, but rather a self-designed and built potential detection circuit, and the circuit filtering needs to be further enhanced. The open circuit potential response curve of the ion sensor has a peak wave before each step, because of the large sample size of the ion solution added in each drop, ranging from tens to hundreds of microlitres, resulting in a large signal response caused by a sudden increase in the concentration of the solution near the electrode. As the time increases, the solution concentration

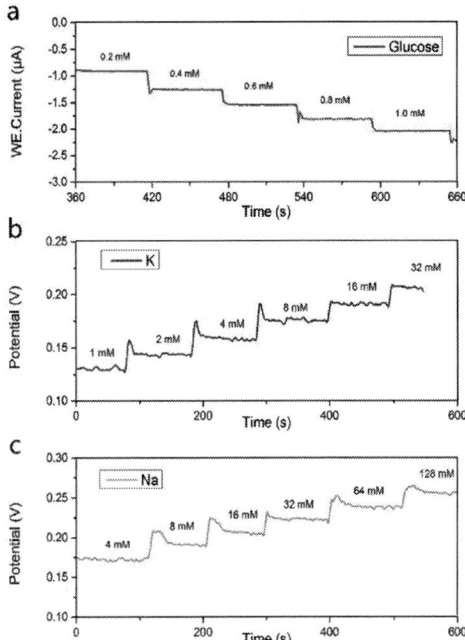

Figure 4 (a) Amperometric response of the glucose biosensor to increasing lactate concentration from 0 to 1.0 mM, with 0.2 mM increments in PBS (pH 7). Applied voltage = - 0.1 V versus Ag/AgCl. (b) Potentiometric response of the potassium sensor to 1, 2, 4, 8, 16, 32 mM KCl in DI water. (c) Potentiometric response of the sodium sensor to 4, 8, 16, 32, 64, 128 mM NaCl in DI water.

becomes homogeneous and the response curve gradually became smoother.

C. ECG signal acquisition test

The patch was mounted under the subject's lower left rib cage and the subject's single-lead ECG signals were recorded before and after exercise, as shown in Fig.5. Fig.5a, 5c show the raw ECG amplification signal from the BMD101 packet, with the amplitude concentrated between

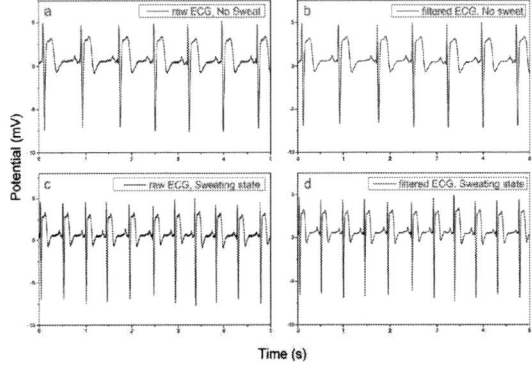

Figure 5 ECG signals before cycling (no sweat state) (a) and during cycling (sweating state) (c), as well as their smoothed filtered signals (b, d)

5 mV and -10 mV, and the waveforms are clear, with characteristic ECG bands such as the QRS band. Fig.5b, 5d used a ten-point twice smoothing filter to eliminate the burr and make the waveform smoother. It can be seen that the ECG detection module in this work had reliable ECG acquisition performance and the prepared ECG sensor patch was less susceptible to interference of sweat on ECG signal acquisition.

D. Pulse wave signal acquisition test

The raw pulse waveform data stored on the phone was exported and plotted as a corresponding signal waveform, as shown in the Fig.6a, which shows the presence of a large amount of noise. Fig.6b shows the waveform after a sliding mean filter with a window size of 10, which removes most of the high frequency noise and maintains the original curve characteristics. This filtering algorithm is simple and can be ported to a microcontroller program for fast real-time processing.

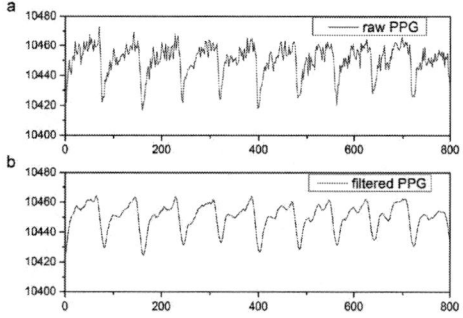

Figure 6 PPG raw signal (a) and filtered signal (b).

E. On-body test

The on-body test offers two wearing options: on the forearm or under the lower left rib cage, as shown in the Fig.7. In this work, the data recording for the on-body experiments was done under the lower left rib cage because the single-lead ECG system of the device is close to the chest to collect a better ECG signal, and because sweating is evident in the under-chest area, making it easy to complete the sweat detection. Before fixing the patch, the skin needs to be wiped with an alcohol pad in order to avoid contamination of the sweat or electrodes with impurities such as oil on the skin surface. When fixing the flexible patch the serpentine lead can be stretched appropriately to complete the best wearing effect. An appropriate amount of conductive hydrogel was added to the electrodes of the ECG

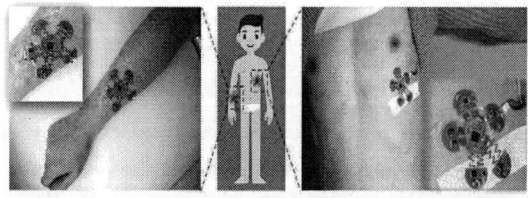

Figure 7 Photograph of the flexible patch mounted on a person's right forearm (left) and lower left rib cage (right). The sensor side faces down toward the skin, and the component side faces up.

sensor patch to improve the adhesion and conductivity between the ECG electrodes and the skin. The sweat sensor patch was underlain by a layer of fabric absorbent cotton.

In order to obtain sweat continuously and ensure the continuity of sweat detection, indoor-cycling was chosen for testing in this paper. As a continuous aerobic exercise, indoor-cycling can induce changes in heart rate and also induce a large amount of sweat, which can be used to verify the fluctuation of glucose, sodium and potassium in sweat. The cycling intensity was based on a base level of 15 km/h. Subjects were asked to stay on the bike for a total of 40 minutes, including a 5 minute warm-up phase at 50 % intensity (7.5 km/h), a 25 minute phase at 100 % intensity (15 km/h) and a 10 minute rest phase (Fig.8a). In order to ensure that the sweat production was sufficient to reach a measurable level, the sweat sensors data was only recorded from the 15th minute onwards. As the measurements were taken on the chest, the heart rate was extracted from the ECG signal and not from the PPG signal.

After warming up, the subject's heart rate quickly rose from around 80 bpm at rest to over 100; as the intensity of exercise increased, the heart rate continued to rise, varying between 110 and 140 bpm, with large fluctuations possibly due to slight displacement of the electrode contact points caused by body shaking or other exercise disturbances; after stopping exercise, the subject's After stopping the exercise,

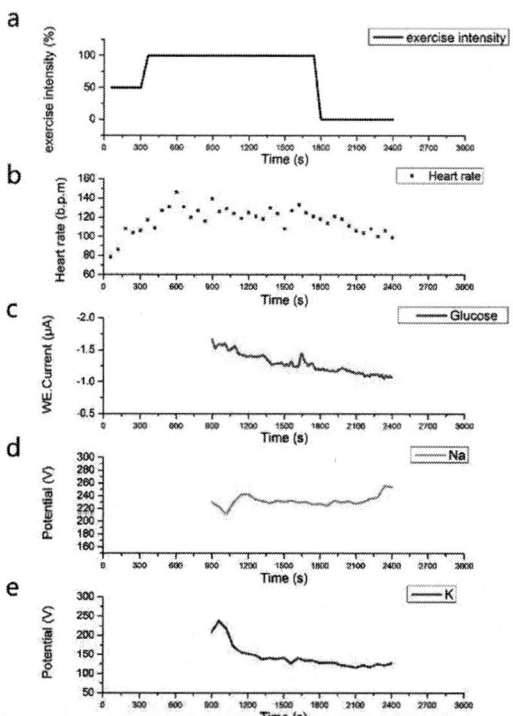

Figure 8 Real-time on-body evaluation of the flexible patch showing the exercise intensity (a), heart rate (b), and glucose level (c), potassium level (d), sodium level (e) in sweat. As the measurement was taken on the chest, the heart rate was extracted from the ECG signal and not from the PPG signal.

the subject's body remained stationary and the heart rate began to drop steadily to around 100 bpm (Fig.8b). The subject was fasting at the time of the test and their sweat glucose level was about a few tenths of their blood glucose concentration. As the exercise progressed, the glucose concentration gradually decreased to a more stable level (Fig.8c). This is probably because the body begins to consume blood glucose rapidly as exercise progresses, causing a change in the ratio of glucose in the sweat ducts to glucose in the surrounding tissue fluid or blood, and thus a change in the amount of glucose secreted into the sweat. Whether this process occurs before or after heavy sweating needs to be investigated further. Sodium ion concentrations are generally on a slow upward trend, indicating a continued increase in sweat production (Fig.8d). As the rate of sodium reabsorption in the sweat ducts is not related to the rate of sweating, sodium can be an important parameter in assessing the rate of sweating. A decreasing trend in potassium ion concentration indicates that the body is losing potassium ion (Fig.8e). Physiologically, a low potassium concentration is a sign of dehydration and can alert athletes to rehydrate when the potassium concentration is low.

IV. CONCLUSION

In summary, we have proposed a multi-parameter monitoring device based on the flexible patch, which can simultaneously monitor the basic physiological parameters of the human body (e.g. electrocardiogram, pulse wave), and biochemical parameters in sweat (e.g. sodium, potassium and glucose). The device integrated biosensors and physical sensors, was powered by a lithium battery and communicated with mobile terminal equipment (smart mobile phone) through Bluetooth low energy technology to realize real-time transmission of various parameters. Due to a flexible, modular, integrated and conformal design method, the whole device was stretchable, miniaturized and had low power consumption. This work has achieved good results in signal conduction, signal conditioning, data processing, wireless transmission, system integration, etc. Such a combined biosensor and physical sensor application is a reference for the development of multimodal sensing platforms and has strong application prospects in exercise monitoring. Further study will be focused on optimizing the sensors structure and improving anti-jamming ability.

REFERENCES

[1] A. J. Tudos, G. A. J. Besselink, and R. B. M. Schasfoort, "Trends in miniaturized total analysis systems for point-of-care testing in clinical chemistry," *Lab on a Chip,* vol. 1, no. 2, pp. 83-95, 2001.

[2] Z. Sonner *et al.*, "The microfluidics of the eccrine sweat gland, including biomarker partitioning, transport, and biosensing implications," *Biomicrofluidics,* vol. 9, no. 3, May 2015.

[3] J. A. Rogers, T. Someya, and Y. G. Huang, "Materials and Mechanics for Stretchable Electronics," *Science,* vol. 327, no. 5973, pp. 1603-1607, Mar 26 2010.

[4] C. F. Wang, C. H. Wang, Z. L. Huang, and S. Xu, "Materials and Structures toward Soft Electronics," *Advanced Materials,* vol. 30, no. 50, Dec 13 2018.

[5] S. Emaminejad *et al.*, "Autonomous sweat extraction and analysis applied to cystic fibrosis and glucose monitoring using a fully integrated wearable platform," *Proceedings of the National Academy of Sciences of the United States of America,* vol. 114, no. 18, pp. 4625-4630, May 2 2017.

[6] A. J. Bandodkar *et al.*, "Epidermal tattoo potentiometric sodium sensors with wireless signal transduction for continuous non-invasive sweat monitoring," *Biosensors & Bioelectronics,* vol. 54, pp. 603-609, Apr 15 2014.

[7] W. Gao *et al.*, "Fully integrated wearable sensor arrays for multiplexed in situ perspiration analysis," *Nature,* vol. 529, no. 7587, pp. 509-+, Jan 28 2016.

[8] T. Guinovart, A. J. Bandodkar, J. R. Windmiller, F. J. Andrade, and J. Wang, "A potentiometric tattoo sensor for monitoring ammonium in sweat," *Analyst,* vol. 138, no. 22, pp. 7031-7038, 2013.

[9] Q. P. Cao *et al.*, "A Smartwatch Integrated with a Paper-based Microfluidic Patch for Sweat Electrolytes Monitoring," *Electroanalysis,* Nov 4 2020.

[10] J. Kim *et al.*, "Simultaneous Monitoring of Sweat and Interstitial Fluid Using a Single Wearable Biosensor Platform," *Advanced Science,* vol. 5, no. 10, Oct 2018.

[11] S. Imani *et al.*, "A wearable chemical-electrophysiological hybrid biosensing system for real-time health and fitness monitoring," *Nature Communications,* vol. 7, no. 1, May 2016.

[12] Z. L. Zhang, Z. Y. Pi, and B. Y. Liu, "TROIKA: A General Framework for Heart Rate Monitoring Using Wrist-Type Photoplethysmographic Signals During Intensive Physical Exercise," *IEEE Transactions on Biomedical Engineering,* vol. 62, no. 2, pp. 522-531, Feb 2015.

[13] D. B. Speedy, T. D. Noakes, and C. Schneider, "Exercise-associated hyponatremia: a review," *Emerg Med (Fremantle),* vol. 13, no. 1, pp. 17-27, Mar 2001.

[14] M. S. Talary, F. Dewarrat, D. Huber, and A. Caduff, "In vivo life sign application of dielectric spectroscopy and non-invasive glucose monitoring," *Journal of Non-Crystalline Solids,* vol. 353, no. 47-51, pp. 4515-4517, Dec 1 2007.

Gap in pagination due to unavailable paper.

Pages 768-769

Proceedings of the 16th Annual IEEE International
Conference on Nano/Micro Engineered and Molecular Systems
April 25-29, 2021

Modeling and Simulation of Flexible Vector Shear Flow Sensor Based on COMSOL Multiphysics

Guochang Liu, Wenping Cao, Haoyu Tan, Renxin Wang*, and Wendong Zhang*, *Senior Member, IEEE*

Abstract— Three-dimensional MEMS devices, which provide a mechanical response with high strain deformation, have been widely used in flexible electronics due to their excellent properties. In this paper, we introduce a kind of stretchable flexible buckling crossbeam-cilia Vector shear flow sensor. The transient model of shear flow signal simulation was established through COMSOL Multiphysics soft, a multi-physics coupling simulation software. The voltage signal output ratio of the sensor X channel and Y channel is as high as 0.44%, so it has good directivity, and the velocity sensitivity is as high as 6.4×10^{-3} Vms^2/kg. It has laid a foundation for the manufacture and application of sensors and opened up a new road for ocean exploration.

I. INTRODUCTION

In the past few decades, three-dimensional MEMS structures have attracted more and more researchers' attention. Scalable electronics represent a fundamentally different and even more challenging technology with an interesting, unique ability to bend and conform to complex curved surfaces. The method based on the control of loading-path trajectories is a common method for fabricating three-dimensional structures and is also a hot research direction in the field of flexible electronics (1-10). The great potential of controllable buckling has been realized in small-scale soft materials and has been widely used in biomedical devices (11-15), energy storage (16-18), optical devices (19,20), and other fields. However, no application of this structure in water flow detection has been reported.

In this paper, we propose a design concept to simulate the performance of three-dimensional structures in shear flow. First, a three-dimensional crossbeam-cilia structure was assembled using a biaxial preloaded flexible substrate and a two-dimensional precursor of silica. Then, the buckling vector shear flow sensor structure is used to detect the flow signal detection research. It is challenging to study the transient and amplitude uncertainty of the ocean current signal. In the simulation, the COMSOL Multiphysics transient study provides a powerful solution for this situation (21,22).

II. TECHNICAL BACKGROUND

A. Working principle

The shear probe comprises cilia, buckling crossbeam, central block, four ends Pad, and piezoresistors, as shown in

Figure 1 and Figure 2(b). The sensor is attached to the stretched flexible substrate, then release substrate to make four ends of the beam inward specified distance to achieve the crossbeam's buckling, as shown in Figure 2.

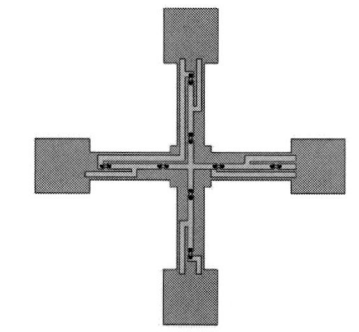

Figure 1. Diagram of piezoresistors on crossbeam.

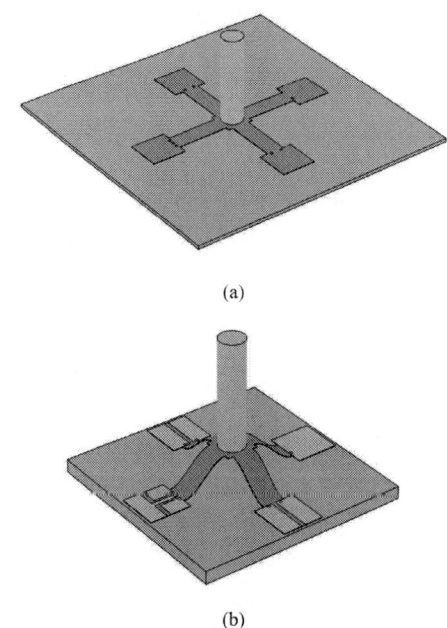

(a)

(b)

Figure 2. (a) 2D crossbeam structure on the stretched flexible substrate; (b) buckling crossbeam structure on the released flexible substrate.

Resrach supported by the National Natural Science Foundation of China as National Major Scientific Instruments Development Project (Grant No. 61927807), National Natural Science Foundation of China (Grant No. 51875535), The Fund for Shanxi '1331 Project' Key Subject Construction and Innovation Special Zone Project.

All authors are with State Key Laboratory of Testing Technology, North University of China, Taiyuan 030051, China.

*Contacting Authors: Renxin Wang is with State Key Laboratory of Testing Technology, North University of China, Taiyuan 030051, China. (phone: +86-3922131; email: wangrenxin@nuc.edu.cn); Wendong Zhang is with State Key Laboratory of Testing Technology, North University of China, Taiyuan 030051, China. (phone: +86-3923640; email: wdzhang@nuc.edu.cn).

978-1-6654-3008-1/21 $31.00 © 2021 IEEE

The flow signal acts on the cilia and then passes to the crossbeam, causing the stress change of the crossbeam, which in turn causes the resistance value of the piezoresistors to change. The output X and Y voltage realize the signal vector detection through each Wheatstone bridge, which is shown in Figure 3. When the sensor receives the flow signal, the output signal of the X channel is:

$$E_{Ax} = \frac{(R_1 + \Delta R_1)(R_3 + \Delta R_3) - (R_2 + \Delta R_2)(R_4 + \Delta R_4)}{(R_1 + \Delta R_1 + R_2 - \Delta R_2)(R_3 + \Delta R_3 + R_4 - \Delta R_4)} E_{in} \quad (1)$$

Where E_{Ax} and E_{in} are the input and output voltages, respectively; And ΔR is the resistance change of the piezoresistor.

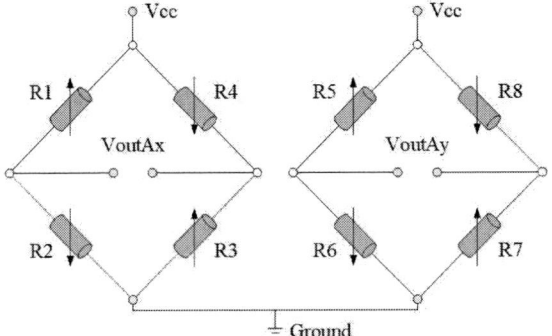

Figure 3. Wheatstone Bridge.

When the water flow acts on the sensor, the force exerted on the cilia is:

$$F = \frac{1}{2} \rho U^2 A \sin 2\alpha \quad (2)$$

Where U is the instantaneous velocity of the fluid, α is the angle of the fluid's direction and the sensor's cilia, ρ is the density of water, and A is the cilia's cross-sectional area. E_{rms} is defined as root-mean-square voltage meaning $E_{Ax} = \sqrt{2} E_{rms}$. The output voltage of the sensor has a linear relationship with the output voltage, so:

$$C = \frac{E_{rms}}{F} \quad (3)$$

Where we define C as the coefficient and $S_f = CA/2$ as the velocity sensitivity. Finally, we can calculate the X channel velocity sensitivity of the sensor by the following formula:

$$S_f = \frac{E_{Ax}}{\sqrt{2} \rho U^2 \sin 2\alpha} \quad (4)$$

The same velocity sensitivity of the Y channel can be found.

B. Simulation equations

The Solid mechanics interface is a branch of structural mechanics based on solving the equation of motion and solid materials' constitutive model. It calculates the results of displacement, stress, and strain. In this study, we added the structural mechanic's description of inward displacement into the structure, which is following the prestressing state of the buckling crossbeam this study. The steady-state form of solid mechanics is:

$$0 = \nabla \cdot (FS)^T + \mathbf{F}v \quad (5)$$

$$F = I + \nabla \mathbf{u} \quad (6)$$

Where F is the displacement gradient, $\mathbf{u}$ is the structural displacement vector, S is the stress. For linear elastic material:

$$S = S_{ad} + J_i F_{inel}^{-1} (\mathbf{C} : \varepsilon_{el}) F_{inel}^{-T} \quad (7)$$

$$\varepsilon_{el} = \frac{1}{2}(F_{el}^T F_{el} - I) \quad (8)$$

$$F_{el} = F F_{inel}^{-1} \quad (9)$$

$$S_{ad} = S_0 + S_{ext} + S_q \quad (10)$$

$$\varepsilon = \frac{1}{2}\left[(\nabla \mathbf{u})^T + \nabla \mathbf{u} + (\nabla \mathbf{u})^T \nabla \mathbf{u} \right] \quad (11)$$

$$\mathbf{C} = \mathbf{C}(E, v) \quad (12)$$

The sensor receives the flow signal in the time-dependent water. The transient laminar flow interface can effectively represent water impacting the cilia at different periods with different velocities. The general transient form based on the Navier-Stokes equation is:

$$\rho \frac{\partial \mathbf{u2}}{\partial t^2} + \rho(\mathbf{u2} \cdot \nabla)\mathbf{u2} = \nabla \cdot \left[-\rho \mathbf{I} + \mathbf{K} \right] + \mathbf{F} \quad (13)$$

$$\rho \nabla \cdot \mathbf{u2} = 0 \quad (14)$$

$$\mathbf{K} = \mu(\nabla \mathbf{u2} + (\nabla \mathbf{u2})^T) \quad (15)$$

Where ρ is fluid's density, $\mathbf{u2}$ is velocity vector, $\mathbf{K}$ is viscous stress tensor, $\mathbf{F}$ is volume force vector.

Stress on the beam of the sensor causes the conductivity to change of the piezoresistors. The change in the conductivity is caused by the shift in the material energy band structure. It is also caused by the relative instability of carrier mobility and number density. We introduce the electric currents-single layer shell interface to describe this physical phenomenon. Among the time-domain related research types, the dynamic-state considering the piezoresistive of single-layer shell structures is used:

$$\nabla_T \cdot (d_s \mathbf{J}) = d_s Q_{j,v} \quad (16)$$

$$\mathbf{J} = (\sigma + \varepsilon_0 \varepsilon_r \frac{\partial}{\partial t})\mathbf{E} + \mathbf{J_e} \quad (17)$$

$$\mathbf{E} = -\nabla_T V \quad (18)$$

Where, $\mathbf{J}$ is Current density, Q_j is an external current source, $\mathbf{J_e}$ is an externally generated current density, d_s is shell thickness.

III. METHODOLOGY

A. Geometry and Properties of materials

The crossbeam, central block, and four-end Pad are cuboids made of SiO_2, while the cilia are cylinders made of Si. Specific structural parameters are shown in Table 1. Material's properties we need are shown in Table 2. The piezoresistors are a single-shell structure of p-Silicon, and its relative dielectric constant and number density are 4.5 and 1.32×10^{19} $1/cm^3$, respectively. The fluid domain is water with a sound velocity of 1480 m/s and a density of 1000 kg/m^3.

Table 1. Specific parameters of sensitive structures.

Structure name	Parameter (μm)
Length of the crossbeam	300
Width of the crossbeam	100
Thickness of the crossbeam	3
Length of the Pad's side	200
Length of the central block's side	150
Height of the cilia	700
Radius of the cilia	60
Length of the piezoresistor	30
Width of the piezoresistor	10
Thickness of the piezoresistor	0.34

Table 2. Material properties.

Properties	Si	SiO2
Young's modulus (GPa)	130	70
Poisson's ratio	0.27	0.2
Density (kg/m³)	2329	2200

B. Meshing

The mesh precision of the finite element model greatly affects the accuracy of simulation, but the structure's size also limits the size of the finite element mesh. Crossbeam is a structure with a large length and width but very thin thickness, so to form a high-quality mesh, the mesh size must be reduced to fit the thickness. A mapping method is used to set the length and width as the appropriate boundary mesh and then sweep in the thickness direction. This is important for we focus on the change of stress on the beam, so it is more necessary to refine it. Relatively speaking, the cilia only play the role of transferring the force to the beam. We do not require that much grid for the cilia, and we use a sparse grid to reduce the amount of computation. The piezoresistive material uses a single layer shell structure with thickness made dense by mapping mesh. The fluid domain is a conventional hydrodynamic mesh. The overall meshing and quality histogram are shown in Figure 4.

The grid skewness of sensors and fluid domains set in this way is almost all greater than 0.4, among which the grid skewness on the beam that we pay the most attention to is all 1, which meets the precision requirements of the physical field.

C. Physics interfaces

The shear flow sensor transmits the cilia's vibration caused by the flow information to the crossbeam, which outputs the voltage signal through the Wheatstone bridge.

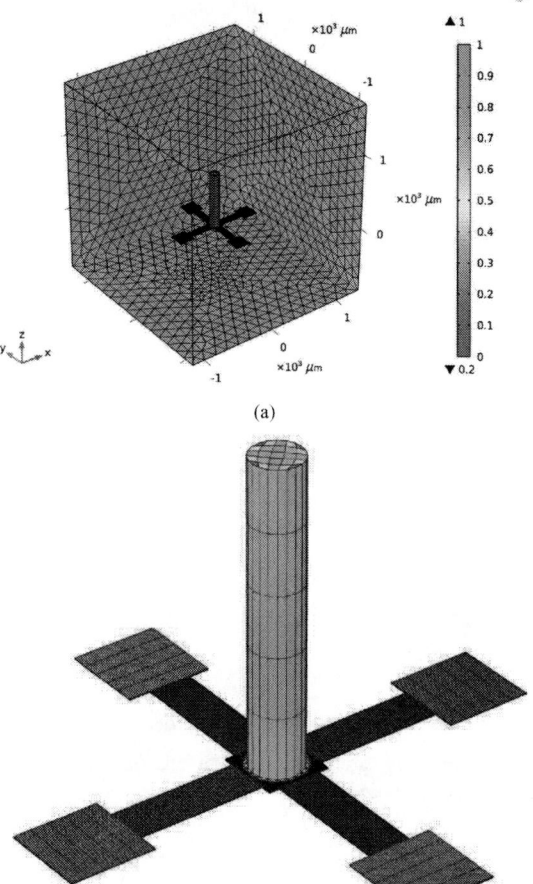

(a)

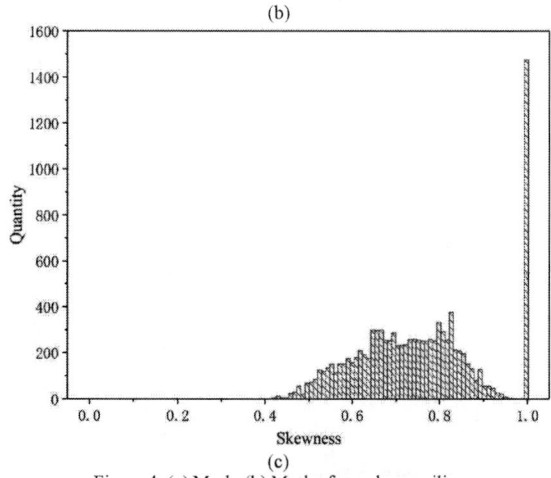

(b)

(c)

Figure 4. (a) Mesh; (b) Mesh of crossbeam-cilia; (c) Mesh quality histogram.

This process involves several physical phenomena. Therefore, in CMOSOL Multiphysics 5.5, three physical field modules are selected to simulate these physical effects: solid mechanics module, fluid mechanics module, and AC/DC module.

The key point of this structure design is to realize the buckling of the crossbeam structure. In the crossbeam buckling process, we simulated the release process of different proportions of prestretched substrates by progressively forcing the four-end Pad from 0 to 20um to the center. It should be noted that due to the large deformation of 2D precursor films during warping, geometric nonlinearity must be considered.

The hydrodynamic physical field calibrated the velocity sensitivity. The boundary of the fluid domain is provided with inlet, outlet, and open boundary, respectively. The inlet is the boundary of the net inflow of fluid. The velocity of the inlet can be specified, which is set to 1m/s. The exit boundary condition is the boundary with the net outflow from the water, and the speed or pressure can be specified. Here, it is set as the exit with the pressure of 0Pa. An open boundary is a boundary where fluid can flow freely in and out of a fluid domain, thus simulating a vast water body. The interface of "Electric Currents-Single Layer Shell" is introduced, and a 5V DC voltage source is added to supply power for the Wheatstone bridge.

IV. RESULTS AND DISCUSSIONS

Compared with the traditional hydrophone and flow sensor, a shear flow sensor with prestressed and prestrained island-bridge structure is proposed in this paper. The first step to study the stress and pre-strain sensor is the crossbeam's buckling, and the fluid shock is applied based on the warpage of the first step. As shown in Figure 5(a), the maximum stress tensor is reached at the extreme ends of the beam. The stress tensors at both ends are -5.7×10^8 N/m^2 and 5.5×10^8 N/m^2, respectively, as shown in Figure 5(b).

At a flow rate of 1m/s, the cilia were subjected to a pressure of 545Pa, as shown in Figure 6. As shown in Figure 7, the water flow will create a vortex after passing the cilia, which will affect the sensor's stress change. Due to water vortices' influence, the most extreme ends of the beam still have the greatest strain variation (3.1×10^6 and -1.7×10^6 N/m^2, respectively) where the piezoresistors are placed, and we can see that in Figure 8. In the fluid domain, the transient study of the flow impulse with flow velocity from 0 to 1 m/s and then a sinusoidal curve with a base value of 1 m/s and an amplitude of 0.1 m/s is carried out in the X channel. For a 5V DC power supply to Wheatstone Bridge, the bridge's output voltage changes with the flow rate, as shown in Figure 9. It can be seen that the voltage variation of the Wheatstone bridge in the X channel changes with the flow velocity. When the flow rate is 1m/s, the output voltage difference between the two ends of the Wheatstone bridge is 9mV, obtained from the velocity sensitivity formula. The velocity sensitivity of the sensor constructed in this paper can reach $S_f = 6.4 \times 10^{-3}$ Vms^2/kg. Simultaneously, the output voltage in the Y channel is 0.04 mV, the voltage signal output ratio of the sensor X channel and Y channel is as high as 0.44%. The vector shear flow sensor shows good directivity in this paper.

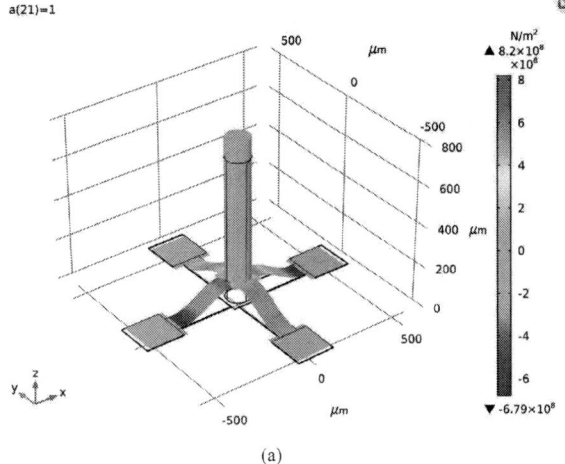

(a)

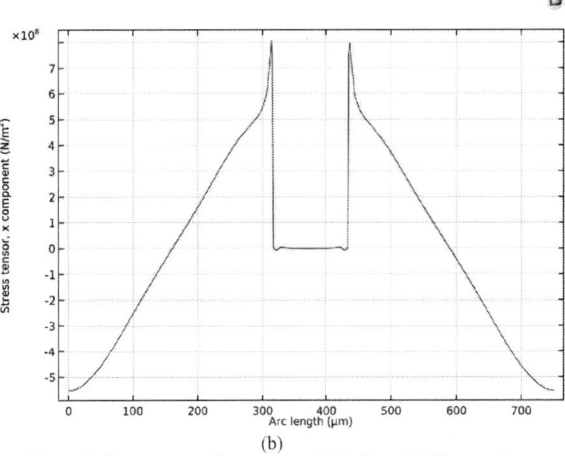

(b)

Figure 5. Stress tensor, X component (a) Surface, (b) Line graph

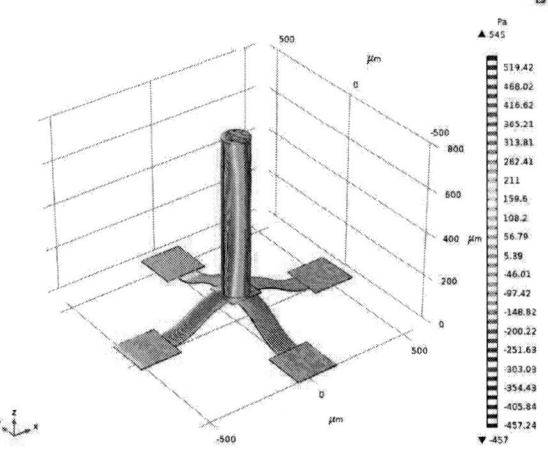

Figure 6. Contour of pressure.

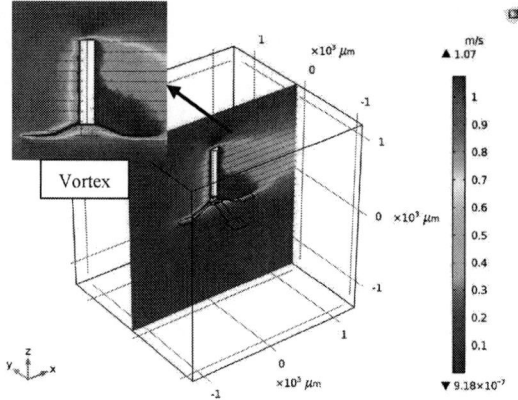

Figure 7. Slice: Velocity magnitude and direction.

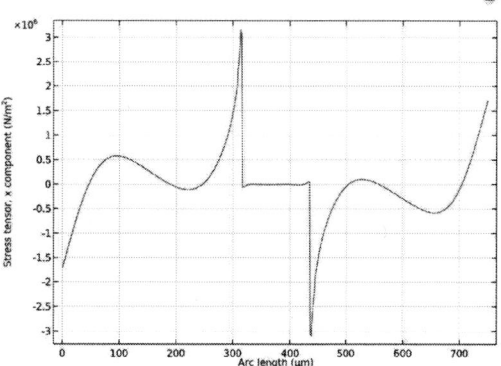

Figure 8. Strain variation of buckling beam.

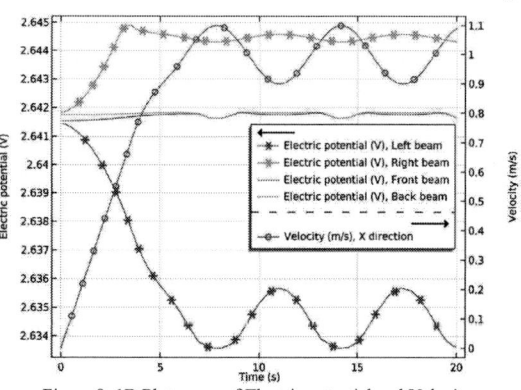

Figure 9. 1D Plot group of Electric potential and Velocity.

V. CONCLUSION

In this paper, a kind of flexible high sensitivity vector shear flow sensor is designed, and its detection mechanism is introduced. It is a great extension of the research on shear flow detection of island-bridge structures. Simulation analysis shows that the sensor has an excellent performance in two-dimensional vector shear flow detection, and the sensitivity up to 6.4×10^{-3} Vms^2/kg. Compared with the PNS shear flow probe, the detection dimension is increased, and the sensitivity is greatly improved. The research in this paper has laid a theoretical foundation for the production of the new shear flow sensor, but it is only in the design and simulation stage. In the near future, it is expected to be applied to underwater robots and turbulent subsurface buoy and provides new opportunities for ocean turbulence observation.

REFERENCES

[1] S. Xu, "Assembly of micro/nanomaterials into complex, three-dimensional architectures by compressive buckling". Science 347, 154-159, 2015.

[2] X. Guo, "Two- and three-dimensional folding of thin film single-crystalline silicon for photovoltaic power applications." Proc. Natl. Acad. Sci. U.S.A. 106, 20149-20154, 2009.

[3] . C. Py, "Capillary origami: Spontaneous wrapping of a droplet with an elastic sheet." Phys. Rev. Lett. 98, 156103, 2007.

[4] V. Y. Prinz, "A new technique for fabricating three-dimensional micro and nanostructures of various shapes." Nanotechnology 12, 399-402, 2001.

[5] T. H. Ware, M. E. McConney, J. J. Wie, V. P. Tondiglia, T. J. White, "Voxelated liquid crystal elastomers." Science 347, 982-984, 2015.

[6] C. Baek, P. M. Reis, "Rigidity of hemispherical elastic gridshells under point load indentation." J. Mech. Phys. Solids 124, 411-426, 2019.

[7] C. Baek, A. O. "Sageman-Furnas, M. K. Jawed, P. M. Reis, Form finding in elastic gridshells." Proc. Natl. Acad. Sci. U.S.A. 115, 75-80, 2018.

[8] J. Marthelot, F. L. Jimenez, A. Lee, J. W. Hutchinson, P. M. Reis, "Buckling of a pressurized hemispherical shell subjected to a probing force." J. Appl. Mech. 84, 121005, 2017.

[9] R. Al-Rashed, F. López Jiménez, J. Marthelot, P. M. Reis, "Buckling patterns in biaxially pre-stretched bilayer shells: Wrinkles, creases, folds and fracture-like ridges." Soft Matter 13, 7969-7978, 2017.

[10] J. Cui, F. R. Poblete, Y. Zhu, "Origami/kirigami-guided morphing of composite sheets." Adv. Funct. Mater. 28, 1802768, 2018.

[11] X. Dai, W. Zhou, T. Gao, J. Liu, C. M. Lieber, "Three-dimensional mapping and regulation of action potential propagation in nanoelectronics-innervated tissues." Nat. Nanotechnol. 11, 776–782, 2016.

[12] T. G. Leong, "Tetherless thermobiochemically actuated microgrippers." Proc. Natl. Acad. Sci. U.S.A. 106, 703-708, 2009.

[13] V. Magdanz, S. Sanchez, O. G. Schmidt, "Development of a sperm-flagella driven micro-bio-robot." Adv. Mater. 25, 6581–6588, 2013.

[14] Z. Yan, "Three-dimensional mesostructures as high-temperature growth templates, electronic cellular scaffolds, and self-propelled microrobots." Proc. Natl. Acad. Sci. U.S.A. 114, E9455-E9464, 2017.

[15] S. Yao, P. Swetha, Y. Zhu, "Nanomaterial-enabled wearable sensors for healthcare." Adv. Healthc. Mater. 7, 1700889, 2018.

[16] H. Ning, "Holographic patterning of high-performance on-chip 3D lithium-ion microbatteries." Proc. Natl. Acad. Sci. U.S.A. 112, 6573-6578, 2015.

[17] J. H. Pikul, H. Gang Zhang, J. Cho, P. V. Braun, W. P. King, "High-power lithium ion microbatteries from interdigitated three-dimensional bicontinuous nanoporous electrodes." Nat. Commun. 4, 1732, 2013.

[18] H. Zhang, X. Yu, P. V. Braun, "Three-dimensional bicontinuous ultrafast-charge and discharge bulk battery electrodes." Nat. Nanotechnol. 6, 277–281, 2011.

[19] P. V. Braun, "Materials chemistry in 3D templates for functional photonics." Chem. Mater. 26, 277–286, 2014.

[20] Z. Liu, "Nano-kirigami with giant optical chirality." Sci. Adv. 4, eaat4436, 2018.

[21] L Zhong, "COMSOL Multiphysics Simulation of Ultrasonic Energy in cleaning tanks." COMSOL conference, 2015.

[22] Giwercman B, Jensen E T, N Høiby, "An Introduction to COMSOL Multiphysics Version 5.1". Antimicrobial Agents & Chemotherapy, 35(5):1008-1010, 2015.

Gap in pagination due to unavailable paper.

Pages 775-776

Microfluidic Liquid Metal for Stretchable Biomed-electronics

Xingyu JIANG

Southern University of Science and Technology, Shenzhen, China

Website : http://www.jiangxingyu.com

ABSTRACT

Our interest in making stretchable bioelectronics stems from our work in microfluidic chips that allow many biomedical functions, such as diagnosis, screening/synthesis of therapeutics, as well as tissue engineering. Integrating liquid metal and elastic polymer-based microfluidic chips into stretchable electronic circuits, we find that these stretchable circuits can dramatically expand the capability of micro-/nano-scale structures in a variety of biomedical applications, such as biomedical sensing, tissue engineering, drug discovery and biological computing. For example, a fully stretchable blood oxygen sensor and a fully stretchable sweat detection device can be prepared by using liquid metal elastic polymer microfluidic devices. These new materials can also be used to prepare healable electronic blood vessels and vascular stents that allow gene therapy on targeted live tissues.

BIOGRAPHY

Xingyu Jiang is a Chair Professor at the Southern University of Science and Technology, Shenzhen, China. He obtained his BS at the University of Chicago (1999) and PhD at Harvard University (Chemistry, 2004). In 2005, he joined the National Center for NanoScience and Technology. He moved to the Southern University of Science and Technology in 2018. His research interests include microfluidics and nanomedicine. He was awarded the National Science Foundation of China's Distinguished Young Scholars Award, the Human Frontier Science Program Young Investigator Award, and Tencent ExplorePrize. He is a Fellow of the Royal Society of Chemistry (UK) and American Institute of Medical and Biological Engineering. He has published over 300 peer-reviewed papers.

Gap in pagination due to unavailable paper.

Pages 778-779

Frequency Research of Microfluidic Wear Debris Detection Chip Based on Inductive Wheatstone Bridge

Yucai Xie, Hongpeng Zhang, Haotian Shi, and Yuwei Zhang

Abstract— A microfluidic wear debris detection chip based on an inductive Wheatstone bridge is proposed. The chip consists of two double-wire solenoid coils, a precision resistor, and a potentiometer to form a Wheatstone bridge structure. Two double-wire solenoid coils are used as detection unit and reference unit respectively. The condition of the bridge balance and the fluctuation voltage when the bridge is unbalanced is analyzed theoretically. The relationship between fluctuating voltage and excitation frequency was explored through experiments, and the 150μm iron particles in hydraulic oil were detected. The research provides technical support for the modularization and portability of the microfluidic wear debris sensor, which is of great significance for the fault diagnosis and prevention of the hydraulic system damages.

I. PREFACE

Highly contaminated hydraulic oil causes vibration in the hydraulic system leading to the failure of the system. The contamination is as a results of wear debris generation due to contact of two to frictional surfaces. During normal operation, the allowable concentration and size of debris particle is usually 10-20 μm, but abnormal wear occurs when the concentration of the wear debris increases from 50-100 μm, which can lead to hydraulic system failure [1-2].

The recent conventional oil detection technology includes laboratory detection method and particle counting method, laboratory detection method includes analytic ferrographic method and spectrometer method. Particle counting methods include acoustic detection, capacitance detection, inductance detection and optical detection. Particle counting method generate pulses when the particles pass through the detection device. The number of the pulse generated indicates the number of the particle and the amplitude indicates the size of the particles, this method can distinguish the properties of the particle and the analysis can reveal the health condition of the hydraulic system [3-4]. Acoustic detection [5] and optical detection methods [6] have higher accuracy. However, according to their principles, they are susceptible to environmental noise and

oil transmittance. In large mechanical systems, environmental noise is large and oil transmittance is poor, so these two methods have certain limitations in oil detection. The capacitance detection method [7] has a high detection accuracy, but it is susceptible to the influence of water droplets. By contrast, the inductance detection method is not affected by water droplets and bubbles, and it is not limited by the light transmittance of oil, so it is more suitable for the detection of metal particles in the oil.

Fu [8] designed a three-coil inductive wear debris sensor. It uses the external amplifier and filter circuit to process the output signal, and collects data through the data acquisition module and LabVIEW, which gets rid of the shackles of the impedance analyzer. However, this structure should ensure that the structure of the excitation coil at both ends is exactly the same, otherwise, the magnetic field of the detecting coil in the middle cannot be completely offset, which will affect the detection accuracy. Fan[9] designed a double-coil sensor based on the principle of inductance balance, with one coil serving as the detection coil and the other as the reference coil. Du[10] proposed the way of microfluidic inductance detection chip and made a microfluidic device based on an inductance sensor. In addition, the coil external amplifier circuit was used to amplified the signal and the LC circuit[11-14] was proposed to improve the detection accuracy. Zhang [15-18] improved the detection accuracy and accomplished the detection of a variety of pollutants by optimizing the structure of the chip and adding high magnetic conductivity materials.

This paper describes a low-cost microfluidic wear debris detection chip based on an inductive Wheatstone bridge. The chip is excited by a waveform generator and the signal was collected by data acquisition card. This design was able to detect 150μm iron particles in hydraulic oil.

II. CHIP DESIGN AND THEORETICAL ANALYSIS

A. Chip Design

The fabrication process of the chip is as follows; first, double-wire winded into a 200 turns coil. The bridge

---

*Resrach supported by the Natural Science Foundation of China (51679022), the Fundamental Research Funds for the Central Universities(3132019034) and the Technology Innovation Foundation of Dalian(2019J12GX023).

Yucai Xie received the B.E. degree in measurement and control technology and instrumentation from Liaoning University of Technology, Jinzhou, China, in 2019. He is currently pursuing the M.D. degree with the Mechanical and Electrical Integration, Dalian Maritime University, Dalian, China. His research interests include oil detection technology and microfluidic technology.

Hongpeng Zhang (Corresponding author) received Ph．D．from Dalian Maritime University in 2005．He is now a professor and

Ph．D．supervisor in Dalian Maritime University．His main research interests include marine engineering，mechatronics and microfluidics.

Haotian Shi received his B.Eng. degree in Dalian Maritime University in 2017．Now he is a Ph. D. student in Dalian Maritime University．His main research interests includes marine engineering, mechatronics and microfluidic technology.

YuWei Zhang received the B.E. degree in energy and environmental engineering from Dalian Ocean University, Sichuan, China, in 2019. She is currently pursuing the M.D. degree with the Mechanical and Electrical Integration, Dalian Maritime University, Dalian, China. Her research interests include oil detection technology and microfluidic technology.

consists of two coils, 300 Ω RJ711 precise resistance, and 2 k Ω potentiometer. 500μm copper wire was inserted through the coil to form the channel, then it was fixed on the glass substrate to form the mold. Secondly, the PDMS and curing agent were mixed in the ratio of 10:1, air bubbles in the mixture were removed in an air vacuum for 40 minutes. The mixture was poured into the mold and dried in a drying oven. After the mixture was completely solidified, the mold was removed from the oven and the copper wire was extracted. Finally, holes were cut at both ends of the chip to serve as oil inlet and outlet. After the above steps, the fabrication of the chip was completed

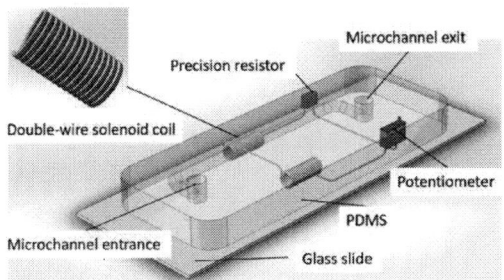

Fig. 1 Microfluidic wear debris particle detection chip

B. Theoretical Analysis

The detection area of the chip is the straight channel area consisting of double-wire solenoid coil.

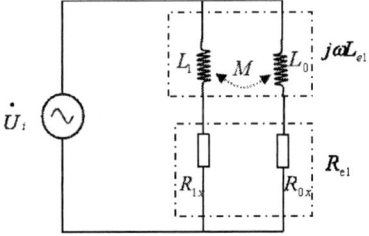

Fig. 2 Schematic diagram of two-wire solenoid coil

In addition to self-inductance L_1 and L_2, produced by to coils there is mutual inductance M between the coils. For two identical solenoid coils, $L_1 = L_2 = L$. Under high-frequency AC excitation, the equivalent inductance L_e of the double-wire solenoid coil is:

$$L_e = \frac{L + M}{2} . \qquad (1)$$

On the other hand, according to previous research results [19-20], when particle pass though, the coil resistance changes as follows:

$$\Delta R = -\frac{8\pi^2 \mu_0 f N^2}{W^2 + d_1^2} \operatorname{Im}(K_p) . \qquad (2)$$

The inductance change of the coil is:

$$\Delta L = \frac{4\pi\mu_0 N^2}{W^2 + d_1^2} \operatorname{Re}(K_p) . \qquad (3)$$

Where, μ_0 is the vacuum permeability, N is the number of coil turns, d_1 is the diameter of the coil (equal to the diameter of the micro-channel), W is the axial width of the coil, and K_p is the complex susceptibility coefficient, which is related to the size of metal particles, relative permeability, electrical conductivity and the excitation frequency of the power supply and its given by

$$K_P = \sqrt{-j\omega\mu_r\mu_0\sigma} , \qquad (4)$$

ω is the angular frequency that excites the alternating current; μ_r is the relative permeability of metal particles; σ is the electrical conductivity of the metal particle.

According to the above formula, the equivalent inductance change of the double coil caused by metal particle can be written:

$$\Delta L_e = \frac{\Delta L + \Delta M}{2} = \frac{\frac{4\pi\mu_0 N^2}{W^2 + d^2} \operatorname{Re}(K_p) + \Delta M}{2} . \qquad (5)$$

The mutual inductance is related to the structure of the coils, the material, and the relative permeability of the particle. Besides, since $R_1 = R_2 = R$, the equivalent resistance of the parallel coil is changed as follows:

$$\Delta R_e = \frac{\Delta R}{2} = -\frac{4\pi\mu_0 f N^2}{W^2 + d^2} \operatorname{Im}(K_p) . \qquad (6)$$

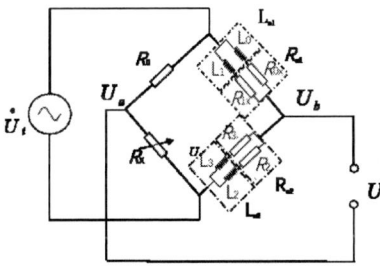

Fig. 3 Circuit diagram of inductive Wheatstone bridge

The schematic diagram of the chip circuit is shown in Figure 3.

$$U_a = U_i \frac{R_x}{R_x + R_0} , \qquad (7)$$

$$U_b = U_i \frac{R_{e2} + j\omega L_{e2}}{(R_{e1} + R_{e2}) + j\omega(L_{e2} + L_{e2})} . \qquad (8)$$

The valid value of U_b is:

$$|U_b| = U_i \sqrt{\frac{R_{e2}^2 + \omega^2 L_{e2}^2}{(R_{e1} + R_{e2})^2 + \omega^2 (L_{e1} + L_{e2})^2}} . \qquad (9)$$

The condition for the bridge to reach equilibrium is:

$$R_0(j\omega L_{e2} + R_{e2}) = R_x(j\omega L_{e1} + R_{e1}), \qquad (10)$$

Where ω is the angular frequency, namely $\omega = 2\pi f$, f is the excitation frequency of AC voltage, L_{e1} and L_{e2} are the basic equivalent inductance of the two coils, R_{e1} and R_{e2} are the basic equivalent resistance of the two coils respectively.

Without particle, the potentiometer bridge balance between the coils and resistance are the same, at this time the voltage difference between the two endpoints is 0V. When a particle goes through the detection area, the equivalent inductance of the coil and basic equivalent resistance change respectively to L_1 L_2 and R_1 R_2, according to the above formula, the voltage fluctuation is as follows:

$$U = U_i \left[\frac{R_{e2} + j\omega L_{e2}}{(R_{e1} + R_{e2}) + j\omega(L_{e1} + L_{e2})} - \frac{R_x}{R_0 + R_x} \right]. \quad (11)$$

Its amplitude is:

$$|U| = U_i \left[\sqrt{\frac{R_{e2}^2 + \omega^2 L_{e2}^2}{(R_{e1} + R_{e2})^2 + \omega^2(L_{e1} + L_{e2})^2}} - \frac{R_x}{R_0 + R_x} \right]. \quad (12)$$

III. DETECTION COIL SIMULATION

The magnetic field distribution of the coil was simulated using in COMSOL software. In the simulation model, the parameters were set as follows: the number of coil turns is 15, the material is defined as copper, plus a 4mm diameter sphere as the air domain, the physical field is added as the magnetic field, the coil domain is added for the analog coil, and the frequency domain is set as 2V and 2MHz. One group without particles and the other group with iron particles of 150μm added to the center of the coil. Figure 4 shows the cross-section of the coil magnetic field cloud image on the X-axis. Figure 4 (a) and (b) are magnetic field cloud images without an iron particle and with a 150μm iron particle respectively.

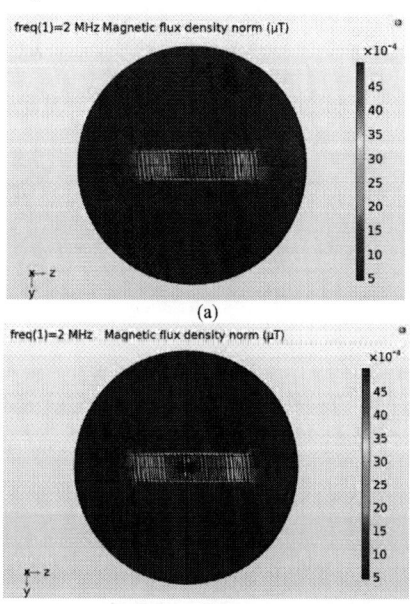

Fig. 4 The cross-section of the coil magnetic field cloud

As shown in Figure 4 (b), The iron particle is magnetized by a magnetic field, which enhances the magnetic field of the surrounding coil. At the same time, the eddy current effect caused by the magnetic field weakens the surrounding magnetic field. As a result, the magnetic field density mode changes as shown in the figure above. From the simulation diagram,

As shown in figure 4, it can be seen that the iron particle magnetic field produced by the magnetization effect is greater than the eddy current effect. Therefore, the size of particles can be distinguished according to the density of the magnetic field, and the chip circuits transforms the resistances and inductances signals to voltage signal which can be used to calculate the particle size.

Figure 5 shows the variation of magnetic field density norm along the Z-axis in two cases.

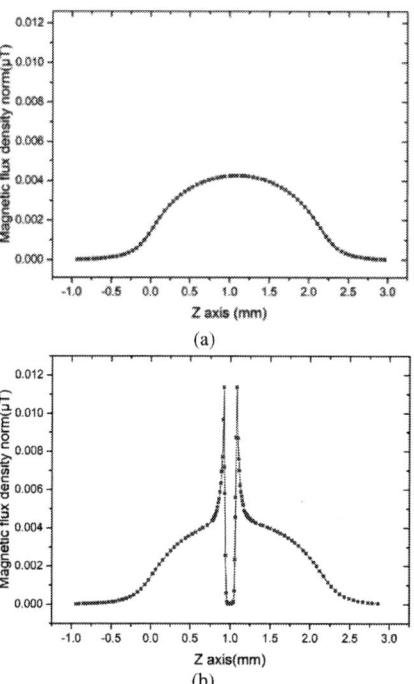

Fig. 5 The Curve of magnetic field density norm

Figure 5 (a) shows the variation of the magnetic field density norm along the Z-axis without particle, and Figure 5 (b) shows the variation of the magnetic field density norm along the Z-axis with 150μm iron particle. As shown in Figure 5. (a), the magnetic flux density in the middle segment of the double-wire solenoid coil is stronger as compare to both ends of the coil. The flux density of the middle segment is about 0.0049μT, and that of the two ends is about 0.0015μT. The presence of iron particle in the coil, changes the variation of the magnetic field density norm along the Z-axis to M-shaped pattern as shown in Figure 5 (b). The peak value of the magnetic field density norm on both sides of the particle is 0.0113μT, which is 2.3 times that of 0.0049μT without iron particles. The simulation results show that when the coil is excited at a high-frequency with alternating current, it generates magnetic field, and when ferromagnetic particles pass through the detection region it magnetized and increases in the density of the magnetic flux.

IV. TESTING PLATFORM CONSTRUCTION

The voltage detection system is shown in Figure 6 below. The detection system is composed of a detection chip, oscilloscope(Agilent Technologies MS071048), waveform generator(33600A Series Keysight), microscope (Moticam 2306), computer and data acquisition card (DAQ card) (NI USB_6211).

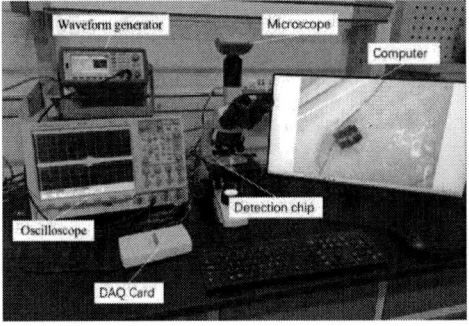

Fig. 6 Microfluidic oil detection system

V. EXPERIMENTAL ANALYSIS

Iron particles were selected to represent ferromagnetic particles in the experiment. Oil samples containing iron particles of a different sizes were prepared. The preparation process is as follows: firstly, different mesh sieves were used obtained iron particles of different size range, then an electronic balance was used in measuring 5mg of each particle size and mixed with 100mL of the hydraulic oil (L-Great Wall HM 46) by an oscillator (IKA S25), which was oscillated evenly for 30 minutes. Finally, 1mL of each mixture was used to carry out the experiment. In the experiment, a waveform generator was used to provide a sine wave signal with 10V for the testing chip. The oscilloscope was used to observe the basic voltage output of the chip in real-time. By adjusting the potentiometer the basic voltage value was reduced to minimum in order to obtain the minimum basic voltage value. The oil sample was injected into the detection chip, As shown in Figure 6, the particle in the detection coil was observed through the microscope. The output voltage signal was collected through the DAQ card.

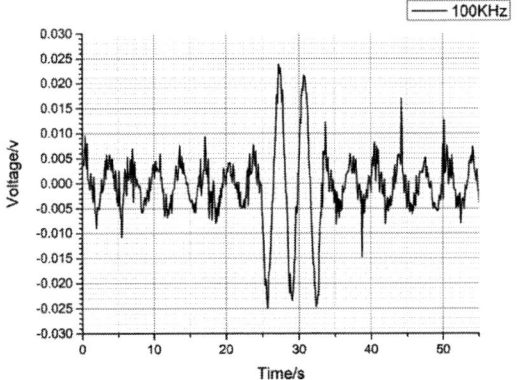

Fig. 7 Voltage signal of 338 μm iron particles.

From Equation 12, it is observed that the voltage output is related to the excitation frequency, therefore the frequency characteristics of the chip was explored by varying the frequency from 20-220khz, with the waveform generator generating 10v, while detecting 338μm iron particles.

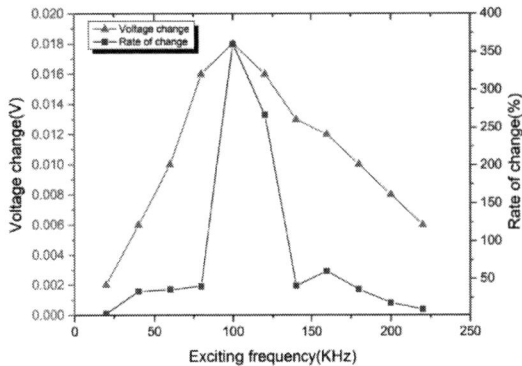

Fig. 8 Relationship between fluctuating voltage and frequency

As shown in Figure 8, the experimental results show that when the excitation frequency of the detection coil is 100KHz, the fluctuating voltage generated is the largest, which is 0.018V, and its voltage change rate is 360%. The reason is that when the coil is in an alternating magnetic field with a lower frequency, the magnetization effect produced by the particles passing by is weak, and the relative change rate is lower. When the coil is in an alternating magnetic field with a higher frequency, the particles produce a stronger magnetization effect. At the same time, the eddy current effect is also significantly enhanced, which offsets the effect of the original magnetic field enhancement, so the relative change rate is low.

When the coil is excited at 100KHz, the magnetization effect of the particle is stronger than the eddy current effect, and the magnetic field changes greatly, which significantly improves the detection of the particles. The signal generated using the optimum frequency while detecting 338μm iron is as shown in figure 7.

A sine wave signal with an amplitude of 10V and an excitation frequency of 100KHz was used to excite the detection chip, and the oil samples to be tested mixed with particles of different sizes are respectively injected into the detection chip. Among them, the first coil is the detection coil and the second coil is the reference coil. The output end of the detection chip was connected to AI1 and GND of the data acquisition card respectively. The sampling frequency of the LabVIEW data acquisition unit was set at 10Hz, the number of points to be sampled was set at 1000, the physical channel was selected as AI1, and the configuration of the input terminal was set as RES. Inject the oil mixed with iron particles of different sizes into the detection chip. When the particles pass through the coil, under the combined action of the eddy current effect and magnetization effect of the particles, the composite magnetic flux and impedance characteristics of the coil change. The mixed oil with an iron particle of different sizes was experimented and their fluctuation voltage was collected.

978-1-6654-3008-1/21 $31.00 © 2021 IEEE

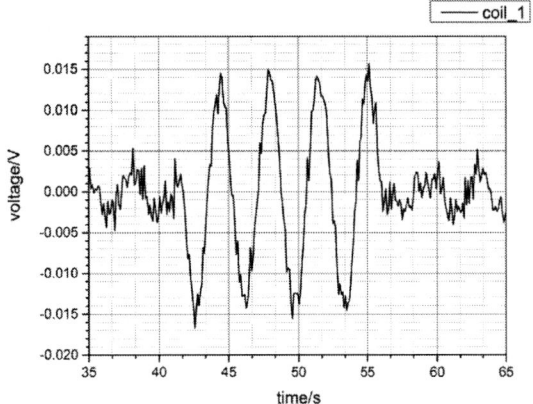

Fig. 9 Voltage signal of 310μm iron particles.

The voltage signal generated by the passage of 310μm iron particles is as shown in figure 9, where the time occupied by the fluctuating voltage is related to the speed of the particle.

At the optimum excitation frequency of 100Khz, the relationship between the particle size and the fluctuating voltage is shown in Figure 10.

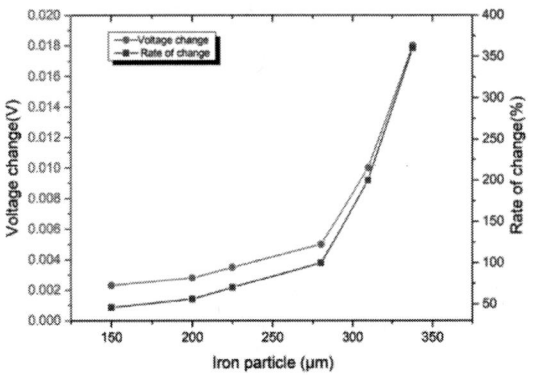

Fig. 10 Relationship between fluctuating voltage and particle size

The relationship curve between the fluctuating voltage and particle size is established from the experimental data, As shown in figure 10, the relationship between the particle size and the fluctuating voltage is nonlinear. Increase in size of the particle causes increase in the fluctuating voltage. When the particle size is less than 300μm, the fluctuation voltage changes slowly, but when the particle size is 300μm or above, there is a drastic change in the fluctuating voltage. Therefore, this curve can provide distinction of particle sizes.

VI. CONCLUSION

In this paper, an inductive microfluidic wear debris detection chip based on a Wheatstone bridge is introduced. The chip's Wheatstone bridge structure consists of two double-wire solenoid coils, a precision resistor, and a potentiometer. Firstly, the feasibility of using Wheatstone bridge to test the chip is analyzed theoretically and the changes in magnetic flux density, when the particle passes

through the detection coil was obtained through simulation. The chip was used to test the metal particles in the oil. The experimental results show that the optimal excitation frequency of the chip is 100kHz. Under the optimal excitation frequency, the minimum iron particles of about 150μm can be detected. This research method completely gets rid of the difficulties and biasness of impedance analyzer. Due to the portability of the chip it can be combined with an embedded system to provide a new detection method for modularization and portability. In the next study, it is necessary to apply physical and digital filtering to the chip to improve detection accuracy. At the same time, it is necessary to design an error processing circuit to improve the anti-interference ability of the data.

REFERENCES

[1] ZHANG H P, HUANG W, ZHANG Y D, et al. Design of the microfluidic chip of oil detection〔C〕. Applied Mechanics and Materials, 2012, 117: 517-520.

[2] DU L, ZHU X, HAN Y, et al. High throughput wear debris detection in lubricants using a resonance frequency division multiplexed sensor〔J〕. Tribology Letters, 2013, 51(3): 453-460.

[3] KUMA R M, MUKHE R JEE P S, MIS R A N M. Advancement and current status of wear debris analysis for machine condition monitoring: A review〔J〕. Industrial Lubrication and Tribology, 2013, 65 (1): 3-11.

[4] WU T, PENG Y, WU H, et al. Full-life dynamic identification of wear state based on on-line wear debris image features〔J〕. Mechanical Systems and Signal Processing, 2014, 42(1): 404-414.

[5] Zhang Jie,Drinkwater Bruce W.,Dwyer Joyce Rob S.. Monitoring of Lubricant Film Failure in a Ball Bearing Using Ultrasound[J]. Journal of Tribology,2006,128(3).

[6] Kwon O K, Kong H S, Han H G, et al. On-line measurement of contaminant level in lubricating oil: US 2000.

[7] ZENG L, ZHANG H P, LIU E C, LIU H X, CUI F Y. Study on the Rapid detection and counting of small Particles in Marine hydraulic Oil[J]. Machine tools and hydraulics, 2015,43 (17): 20-23+15.

[8] FU J J, ZHAN H Q, GU J. Detection Circuit Design of Three-coil Inductive Particle Sensor [J]. Instrument Technique and Sensor, 2012, (02): 5-7.

[9] FAN H B, ZHANG Y T, REN G Q, et al. Experimental study of an on-line monitoring sensor for wear particles in oil〔J〕. Tribology, 2010, 30 (4): 338-343.

[10] DU L, CARLETTA J, VEILLETTE R, et al. A magnetic Coulter counting device for wear debris detection in lubrication[C]// ASME 2009 International Mechanical Engineering Congress and Exposition. 2009.

[11] Du,L.;Carletta,J.E.;Veillette,R.J.;Zhe,J.InductiveCoulterCounting:D etectionandDifferentiationofMetalWearParticlesinLubricant.Smart Mater.Struct.2010,19,057001.

[12] Du,L.;Zhu,X L.;Han,Y.;Zhao,L. ImprovingSensitivityofanInductive PulseSensorforDetectionofMetallicWearDebrisinLubricantsUsingP arallelLCResonanceMethod.Measure.Sci.Technol.2013,24,075106.

[13] Du,L.;Zhe,J.InstrumentationCircuitryforanInductiveWearDebrisSen sor.InNewCircuitsandSystemsConference(NEWCAS).IEEE10thInt. 2012,501–504..

[14] Du,L.;Zhe,J.AHighThroughputInductivePulseSensorforOnlineOilD ebrisMonitoring.Tribol.Int.2011,44,175–179.

[15] Haotian Shi, Hongpeng Zhang*, Laihao Ma, Yuqing Sun, Haiquan Chen, Xupeng Zhao, Chenzhao Bai and Yufei Zhang, "Inductive-capacitive Coulter counting: detection and differentiation of multi-contaminants in hydraulic oil using a microfluidic sensor," IEEE Sensors Journal..

[16] Haotian Shi, Hongpeng Zhang, Changzhi Gu, Lin Zeng. A multi-parameter on-chip impedance sensor for the detection of particle contamination in hydraulic oil[J]. Sensors and Actuators A: Physical, 2019, 293:150-159.

[17] Chenzhao Bai, Hongpeng Zhang, Lin Zeng, Xupeng Zhao, and Zilei Yu. (2019). "High-Throughput Sensor to Detect Hydraulic Oil Contamination Based on Microfluidics." Ieee Sensors Journal 19(19): 8590-8596.

[18] Chenzhao Bai, Hongpeng Zhang *, Lin Zeng, Xupeng Zhao, Laihao Ma, Inductive Magnetic Nanoparticle Sensor based on Microfluidic Chip Oil Detection Technology, Micromachines 2020, 11, 183.

[19] [19]Xingming Zhang, Hongpeng Zhang, Yuqing Sun, Haiquan Chen, and Yindong Zhang. Research on the output characteristics of microfluidic inductive sensor[J]. Journal of Nanomaterials, 2014, 725246.(15).

[20] Yu Wu. Detection of foreign particles in lubrication oil with a microfluidic chip[J]. Industrial Lubrication and Tribology, 2018,70(8).

Gap in pagination due to unavailable papers.

Pages 786-791

The 16th IEEE International Conference on Nano/Micro Engineered & Molecular Systems

Small-scale Wireless Robots for Medical Applications

Metin Sitti

Director, Max Planck Institute for Intelligent Systems, Stuttgart, Germany
Professor, Department of Information Technology and Electrical Engineering, ETH
Zurich, Switzerland
Professor, Koc University, School of Medicine & College of Engineering, Istanbul, Turkey

URL: http://pi.is.mpg.de

ABSTRACT

Wireless small-scale medical robots have the potential to revolutionize healthcare, since they have the unique capability of non-invasive access and operation inside hard-to-reach and unprecedented small spaces inside the human body. However, due to miniaturization limitations on on-board actuation, powering, sensing, computing and communication, new methods need to be introduced in creating such tiny robots. First, cell-driven biohybrid microrobots are proposed, where a synthetic microrobot body is driven by attached biological microorganisms (bacteria and microalgae) or cells (macrophages). Such biohybrid microswimmers loaded with drugs are shown *in vitro* to deliver drugs actively and locally to cancerous tissues. Next, external magnetic fields are used to actuate and steer rolling microrobots, spinning helical microswimmers and soft-bodied magnetic millirobots. Inspired by soft-bodied small animals, soft millirobots are demonstrated to be able to have seven locomotion modalities to be able to navigate in complex environments, such as inside the human body. Also, a baby jellyfish-inspired soft swimmer is shown to realize many diverse functions by controlling the local fluidic flow around them towards medical use.

BIOGRAPHY

Professor Sitti has pioneered many research areas, including small-scale wireless medical robots, gecko-inspired microfiber adhesives, bioinspired miniature robot locomotion and physical intelligence. He was a professor in Department of Mechanical Engineering and Robotics Institute at Carnegie Mellon University in Pittsburgh, USA during 2002-2014 and a research scientist at University of California at Berkeley during 1999-2002. He is currently director at Max Planck Institute for Intelligent Systems in Stuttgart, Germany since 2014. As side academic appointments, he is a professor in Department of Information Technology and Electrical Engineering at ETH Zurich, Switzerland and professor in School of Medicine & College of Engineering at Koç University, Turkey. He received the BSc (1992) and MSc (1994) degrees in electrical and electronics engineering from Boğaziçi University, Turkey, and the PhD degree (1999) in electrical engineering from the University of Tokyo, Japan. He is an IEEE Fellow. He founded nanoGriptech Inc. in 2012 to commercialize his lab's gecko-inspired microfiber adhesive technology, branded as Setex®. He is an IEEE Fellow. He has published over 460 peer-reviewed articles, 2 books and over 15 patents.

Prof. Sitti received the Breakthrough of the Year Award in the Falling Walls World Science Summit (2020), ERC Advanced Grant (2019), Rahmi Koç Science Prize (2018), SPIE Nanoengineering Pioneer Award (2011) and NSF CAREER Award (2005). He has won over fifteen awards at major conferences, such as Best Paper Award in Robotics Science and Systems Conference (2019), IEEE/ASME Best Mechatronics Paper Award (2014) and Best Paper Award in the IEEE/RSJ Intelligent Robots and Systems Conference (2009, 1998). He is the editor-in-chief of *Progress in Biomedical Engineering* and *Journal of Micro-Bio Robotics* and associate editor in *Science Advances* and *Extreme Mechanics Letters* journals.

Gap in pagination due to unavailable papers.

Pages 793-797

Simulation of the Shape-directed AC Driven Defective Micromotors*

Rencheng Zhuang, Dekai Zhou, Xiaocong Chang, and Longqiu Li, *Member, IEEE*

Abstract— **Electric-driven micro/nanomotors (EDMNMs) have been widely researched because of their unique advantages, such as non-necessity of fuel, high mobility, and flexible controllability. However, most EDMNMs rely on asymmetric dielectric constants of the heterogeneous materials while ignoring the influence of the structure. In this study, we propose a novel type of defective micromotors (DMMs) with asymmetric structure that can be driven by the AC electric field. A theoretical model was established and the AC electrokinetic behavior of DMMs was studied theoretically. The propulsion mechanisms of DMMs can be attributed to self-dielectrophoresis and induced-charge electrophoresis at different frequency ranges. This approach provides a novel and universal approach to design highly controllable EDMNMs with asymmetric structure. The DMMs are expected to achieve spatially complex movement and can be widely applied in biomedicine, targeted delivery, micro/nanosensors, and micromechanical systems.**

I. INTRODUCTION

Micro/nanomotors are exquisite machines designed at micro/nanoscale, which can convert other forms of energy [1-5] into mechanical energy so as to produce motion. Considerable artificial micro/nanomotors have been used a great deal over recent years in multidisciplinary fields, such as in precision medicine [6], environmental remediation [7], and micro/nanoengineering fields [8]. Electric-driven micro/nanomotors (EDMNMs) have gained a wealth of attention because of their unique advantages such as non-necessity of fuel, high mobility, and flexible controllability. Due to these outstanding features, researches on the propulsion mechanisms [9], cluster behaviors [10], and applications [11] of EDMNMs have become one of the most attractive topics.

According to the different electric frequency ranges, there are two propulsion mechanisms for the EDMNMs under a uniform AC electric field. The propulsion mechanisms can be summarized as self-dielectrophoresis (sDEP) under high-frequency ranges and induced-charge electrophoresis (ICEP) under low-frequency ranges. The existing EDMNMs are mostly Janus structures composed of two different materials with a large difference in dielectric constants. Therefore, they are significantly restricted by the materials, and it is difficult for them to achieve complex moving behaviors under the sDEP propulsion mechanism [12]. In order to perform some more challenging and

complex tasks, independent of the substrate and precise control are both indispensable for EDMNMs. Lee et al. successfully achieved 3D helical trajectories control of EDMNMs under ICEP propulsion mechanism through changing the asymmetric metal patches on the surface of spherical particles [13]. By constructing the asymmetric shape of EDMNMs to achieve the complex motion has been proved to be an effective method theoretically [14-16]. However, the construction of these EDMNMs is still a significant challenge due to the difficulties in the fabrication of micro/nanomotors with sophisticated structures.

In this study, we propose a novel type of defective micromotors (DMMs) with asymmetric structure that can be driven by AC electric field. We have successfully fabricated DMMs with single conductive material through the glancing angle deposition technique and membrane electrodeposition method in the previous study. Here, we establish a theoretical model and theoretically study the AC electrokinetic behavior of the DMMs. The simulation results can be used to design EDMNMs with asymmetric structures and high mobility. The DMMs are expected to achieve spatially complex movement and can be widely applied in micro/nanosensors, environmental science and engineering, targeted delivery, and micromechanical systems.

II. MODELING AND SIMULATION

The DMMs are microtubes with asymmetric defects and composed of single conductive material (Fig. 1a). We established a global Cartesian coordinate system to better describe the DMMs. The y and x axes are parallel to the long and short axes of the DMMs, respectively, and the z-axis is normal to the polarization plane of the electric field. When the AC electric field is applied, the motion of the DMMs can be divided into two steps (Fig. 1a). First, the DMMs are subject to the electric-induced dipole torque and turn from horizontal state to vertical state, which is known as the electro-orientation behavior. Due to the polarizability along the y-axis of the DMMs is significantly higher than that along the x-axis, the electric-induced dipole torque tends to align the y-axis of the DMMs parallel to the electric field. Then, the DMMs exhibit the directional motion under the electric field. The moving behavior of the DMMs is highly dependent on the AC electric field frequency (ω). The frequency ranges are distinguished by the relationship

---

*This work was supported by the National Natural Science Foundation of China (51822503, 51875141, 51905135, 52005138 and U20A20297), Natural Science Foundation of Heilongjiang Province of China (LH2020E044), China Postdoctoral Science Foundation (2019M651275, 20200M6709006 and BX20190097), Project (HIT. NSRIF. 2020033) supported by Natural Scientific Research Innovation Foundation in Harbin

Institute of Technology, Heilongjiang Postdoctoral Science Foundation (LBH-Z19018 and LBH-Z20061).

Rencheng Zhuang, Dekai Zhou, Xiaocong Chang and Longqiu Li are with the State Key Laboratory of Robotics and System and Key Lab for Micro-systems and Micro-structures Manufacturing, Harbin Institute of Technology, Harbin 150001 China (phone: 86-18686853862; email: longqiuli@hit.edu.cn).

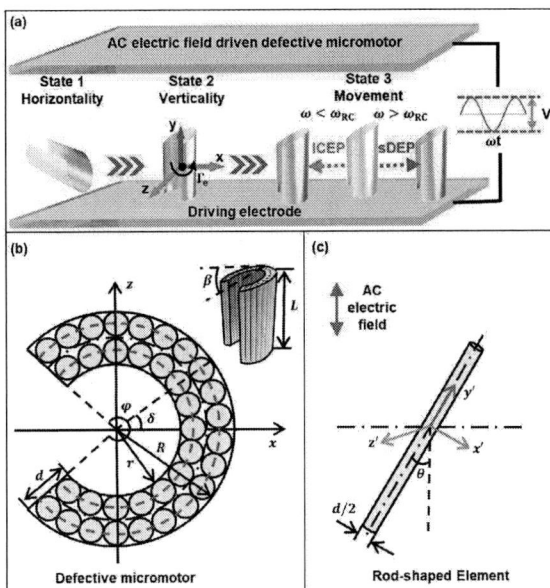

Figure 1. Schematic illustration of the moving behaviors and the theoretical model of the AC electric-driven DMMs. (a) Schematic of the electric cell and the moving behaviors of the DMMs under AC electric field. (b) The superposition of a finite number of the rod-shaped elements used to model the DMMs. The global coordinate system is aligned with its origin coincident with the center of the DMMs. (c) The local coordinate system establishes for the rod-shaped elements. The coordinate system is aligned with the y' component parallel to the long axis of the element.

between the applied electric field frequency and the characteristic frequency of the DMMs (ω_{RC}), which is the reciprocal of the characteristic time scale for the electro-diffusion of the ions in the solution. At high-frequency ranges ($\omega > \omega_{RC}$), the propulsion mechanism can be attributed to sDEP and the DMMs move with the defect-free side forward. Conversely, the propulsion mechanism is attributed to ICEP at low-frequency ranges ($\omega < \omega_{RC}$) and the DMMs move with the defect side forward.

The characteristic frequency of the DMMs can be approximately calculated as follows:

$$\omega_{RC} = \frac{2Dg(K)}{\lambda_D \bar{L}} \sim 107 \text{ kHz} \tag{1}$$

where $\lambda_D = \sqrt{\varepsilon_m D/\sigma_m}$ is the Debye length for deionized water, ε_m is the dielectric coefficient of the medium, D is the diffusion coefficient of the ionic species, σ_m is the conductivity of the medium, $\bar{L} = L - R\tan(\beta)$ is the average length of the DMMs, L is the maximum length along the y-axis of the DMMs, β is the oblique angle of the DMMs, R is the outer diameter of the DMMs, $K = 2d/\bar{L}$ is the aspect ratio of the DMMs, d is the wall thickness of the DMMs, and $g(K) = \ln(1/K)$ is the dimensionless function of the aspect ratio. In order to study the AC electrokinetic behavior of the DMMs, we have established a theoretical model (Fig. 1b). The global coordinate system is aligned with its origin coincident with the center of the DMMs. The superposition of a finite number of the rod-shaped elements is used to model the DMMs. For simplify, we assume that the DMMs are composed of double-layer rod-shaped elements (Fig. 1c). The local coordinate system is aligned

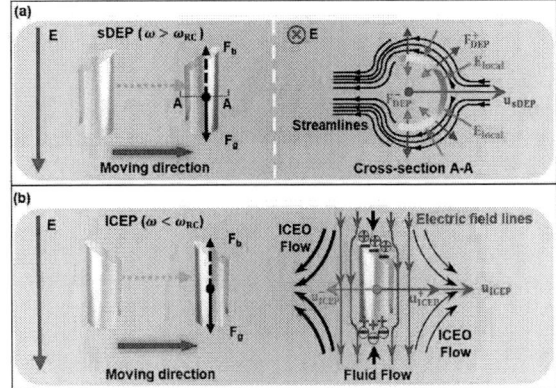

Figure 2. Schematic of the propulsion mechanisms of the DMMs under different frequency ranges. (a) Schematic of the sDEP propulsion mechanism at high electric field frequencies ($\omega > \omega_{RC}$). (b) Schematic of the ICEP propulsion mechanism at low electric field frequencies ($\omega < \omega_{RC}$).

with y' component parallel to the long axis of the rod-shaped element.

According to the established theoretical model, we can investigate the electro-orientation behavior of the DMMs using the electro-orientation model [17]. The electric-induced dipole torque on the DMMs is approximately equal to the sum of the torque on each rod-shaped element. Here, we assume that the DMMs are perfectly polarizable (there are no Faradaic currents at the electrolyte/DMMs interface). Furthermore, the DMMs can orientate freely under the applied AC electric field, although they are much denser than water. For the sake of universality, we assume that the direction of the electric field is not aligned with any of the axes of the rod-shaped element. For the AC electric field of the form $\boldsymbol{E}(t) = E_0 \text{Re}\left[\left(\sin\theta \mathbf{e}_{x'} + \cos\theta \mathbf{e}_{y'}\right)e^{i\omega t}\right]$, the time-averaged torque on the rod-shaped element can be calculated as follows [18]:

$$\boldsymbol{\tau}_i = \frac{1}{2}\text{Re}\left[\left(\boldsymbol{p}_{y'} + \boldsymbol{p}_{x'}\right) \times \left(\boldsymbol{E}_{y'} + \boldsymbol{E}_{x'}\right)^*\right] \tag{2}$$

where θ is the angle between the applied electric field and the y' axis of the rod-shaped element; $\boldsymbol{E}_{y'} = E_{0y'}e^{i\omega t}\mathbf{e}_{y'}$ and $\boldsymbol{E}_{x'} = E_{0x'}e^{i\omega t}\mathbf{e}_{x'}$ are the component of the AC electric field along the y' and x' axes of the element, respectively; $\mathbf{e}_{y'}$ and $\mathbf{e}_{x'}$ are the unit vector along the y' and x' axes of the element, respectively; and $\boldsymbol{p}_{y'}$ and $\boldsymbol{p}_{x'}$ are the electric-induced dipole along the y' and x' axes of the element, respectively. The induced dipole can be calculated as follows:

$$p_{y'} = \frac{\pi\varepsilon_m L'^3 E_{0y'}}{6} \frac{i\Omega_d}{1 + i\Omega_d \ln(1/K')} \tag{3}$$

$$p_{x'} = \frac{\pi\varepsilon_m L' d^2 E_{0x'}}{8} \frac{i\Omega_d - 1}{i\Omega_d + 1} \tag{4}$$

where $K' = d/2L'$ is the aspect ratio of the elements, L' is the length of the single element, $\Omega_d = C_{DL}\omega d/4\sigma_m$ is the nondimensional frequency, and C_{DL} is the capacitance of the double layer. Due to the dipole along the x' axis of the element is much smaller than that along the y' axis, the dipole along the x' axis can be negligible for simplicity. Therefore, the electric-induced dipole torque on the DMMs can be calculated as follows:

$$\Gamma_e = \sum_{i=1}^{N} \tau_i = \sum \frac{\pi \varepsilon_m L_i'^3}{24} \frac{\Omega_d^2 \ln(1/K\prime)}{1+\Omega_d^2 \ln^2(1/K\prime)} E_0^2 \sin 2\theta \mathbf{e}_z \quad (5)$$

where $N = [(4r + 2d) \cdot \varphi]/d$ is the number of the elements, r is the inner radius of the DMMs, φ is the central angle of the DMMs, and $\mathbf{e}_z$ is the unit vector normal to the polarization plane of the electric field.

After the y' axis of the DMMs is parallel to the direction of the applied electric field, the DMMs exhibit moving behaviors in the opposite direction under different frequencies. At high frequencies ($\omega > \omega_{RC}$), the DMMs are polarized and behave as a conductor in the AC electric field. In the direction along the y axis of the DMMs, the gravity (F_g) and buoyancy (F_b) are balanced. The applied electric field acts on the electric-induced dipole and provides the dielectrophoretic force (DEPF) for driving the DMMs (Fig. 2a). Due to the defective structure of the DMMs, the electric field gradient around the DMMs is asymmetric, which results in asymmetric DEPF. The DEPF acting on the DMMs can be divided into two radial forces, F_{DEP}^+ and F_{DEP}^-, which respectively provide positive and negative contributions to the motion of the DMMs. Because F_{DEP}^+ is greater than F_{DEP}^-, the DMMs move toward the defect-free side due to the net DEPF. The time-averaged net DEPF can be calculated as follows:

$$F_{DEP} = \sum_{i=1}^{n_1} F_{DEP,i}^+ - \sum_{j=1}^{n_2} F_{DEP,j}^+ \quad (6)$$

$$F_{DEP,i}^+ = \frac{\pi d^2 L_i'}{96} \varepsilon_m \mathrm{Re}[f_{CM}] \nabla |E|^2 \cos\delta_i \quad (7)$$

$$F_{DEP,j}^- = \frac{\pi d^2 L_j'}{96} \varepsilon_m \mathrm{Re}[f_{CM}] \nabla |E|^2 \cos\delta_j \quad (8)$$

where $n_1 = [(r + 0.75d) \cdot \varphi]/d$ and $n_2 = [(r + 0.25d) \cdot \varphi]/d$ are the halves of the outer-layer and inner-layer elements, respectively; $\delta_i = \pm i \cdot \varphi/2n_1$ and $\delta_j = \pm j \cdot \varphi/2n_2$ are the central angles of the outer-layer and inner-layer elements, respectively; $f_{CM} = (\varepsilon_r^* - \varepsilon_m^*)/[\varepsilon_m^* + A(\varepsilon_r^* - \varepsilon_m^*)]$ is the Clausius–Mossotti factor of the elements; $\varepsilon_r^* = \varepsilon_r - j\sigma_r/\omega$ and $\varepsilon_m^* = \varepsilon_m - j\sigma_m/\omega$ are the complex permittivity of the elements and medium, respectively; $A = d^2/4L'^2 [\ln(4L'/d) - 1]$ is the depolarization factor of the elements; ε_r is the dielectric coefficient of the elements; σ_r is the conductivity of the elements; and $j = \sqrt{-1}$.

At low frequencies ($\omega < \omega_{RC}$), the ions in the medium accumulate in the double-layer capacitor around the DMMs generating a dipolar induced charge cloud that directs the fluid flow (Fig. 2b). The fluid flow is known as the induced-charge electro-osmosis (ICEO) flow. Due to the defective structure of the DMMs, the electric double-layer around the DMMs is asymmetric, which results in the fluid flow velocity at the defect-free side is higher than that at the defective side. Therefore, the DMMs move with the defective side forward and the propulsion mechanism can be attributed to ICEP. The ICEP velocity of the DMMs subject to the slip velocity field can be calculated as follows [19]:

$$u_{ICEP} = u_{ICEP}^+ - u_{ICEP}^- \quad (9)$$

$$u_{ICEP}^+ = -\frac{\varepsilon_m}{3V_1 \eta} \int_{-\varphi/2}^{\varphi/2} (\mathbf{n} \cdot \mathbf{r}) E_s \psi R \overline{L} d\varphi \quad (10)$$

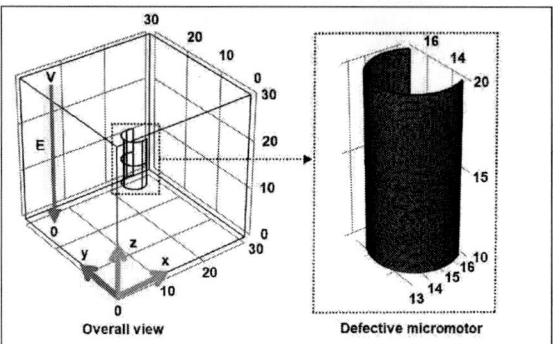

Figure 3. The numerical simulation model of the DMMs with deionized water as the environment and AC electric field as the driven source.

$$u_{ICEP}^- = -\frac{\varepsilon_m}{3V_2 \eta} \int_{-\varphi/2}^{\varphi/2} (\mathbf{n} \cdot \mathbf{r}) E_s \psi r \overline{L} d\varphi \quad (11)$$

where u_{ICEP}^+ is the velocity caused by the ICEO flow in the same direction as that of the motion of the DMMs; u_{ICEP}^- is the velocity caused by the ICEO flow opposite to the direction of the motion of the DMMs; $V_1 = n_1 d^2 \overline{L}/24$ and $V_2 = n_2 d^2 \overline{L}/24$ are the volumes of the outer-layer and inner-layer elements, respectively; E_s is the local electric field at the surface; ψ is the potential drop across the capacitor at the interface of the DMMs; η is the viscosity of the fluid; $\mathbf{n}$ is the direction vector normal to the DMMs surface; and $\mathbf{r}$ is the radial coordinate. The local electric field at the surface of the DMMs can be calculated as follows [20]:

$$E_s = (\mathbf{I} - \mathbf{nn}) \cdot \left[\frac{\mathbf{ee}}{(1-L_\parallel)} + \frac{(\mathbf{I} - \mathbf{ee})}{(1-L_\perp)}\right] \cdot E \quad (12)$$

where $\mathbf{e}$ is the directional vector orients along the y' axis of the rod-shaped elements; and $L_\parallel$ and $L_\perp$ are the polarization factors along the y' axis and x' axis of the elements, respectively. The polarization factors can be calculated as follows [21]:

$$L_\parallel = -\frac{K'^2}{(1-K'^2)} + \frac{K'^2}{2(1-K'^2)^{3/2}} \ln\left(\frac{1+(1-K'^2)^{1/2}}{1-(1-K'^2)^{1/2}}\right) \quad (13)$$

$$L_\perp = \frac{1}{2(1-K'^2)} - \frac{K'^2}{4(1-K'^2)^{3/2}} \ln\left(\frac{1+(1-K'^2)^{1/2}}{1-(1-K'^2)^{1/2}}\right) \quad (14)$$

The diffusion distribution of the ions around the surface of the elements can be calculated by relating the local electric double layer potential to the local charge density as follows [22]:

$$\nabla^2 \psi = -\frac{\rho}{\varepsilon_m} = -\frac{e}{\varepsilon_m}(z^+ c^+ - z^- c^-) \quad (15)$$

where ρ is the charge density, z is the charge number, and c is the ionic concentration. The ionic concentration can be calculated by the mass balance as follows [23]:

$$\frac{\partial c^\pm}{\partial t} + \nabla j^\pm = 0 \quad (16)$$

$$j^\pm = -D^\pm \left(\nabla c^\pm \mp \frac{z^\pm e}{k_B T} \nabla \psi\right) - c^\pm u \quad (17)$$

where k_B is the Boltzmann constant, T is the temperature, and $\boldsymbol{u}$ is the fluid velocity that can be calculated as follows [24]:

$$\nabla \cdot \boldsymbol{u} = \boldsymbol{0} \tag{18}$$

$$-\eta \nabla^2 \boldsymbol{u} + \nabla p = \boldsymbol{0} \tag{19}$$

Since analytic solutions of the electric field, transportation of ions, and local velocity field are difficult to be obtained. We used the finite element method to study the electrokinetic behavior of the DMMs. A finite element model used to quantitatively calculate the electric field gradient and the ICEO slip velocity around the DMMs under AC electric field was performed in COMSOL™ 5.4 with fully coupled electrostatics, transport of diluted species, and creeping flow modules. We investigated the effects that the dimension of the DMMs has on the electrokinetic behavior. A 3D cuboid geometry was used to model the electric cell with deionized water: 30 μm high, 30 μm long, and 30 μm wide. The DMM 10 μm long, 5 μm outer diameter, and 50 nm wall thickness was in the center of the cuboid geometry (Fig. 3). Mesh of the local medium part and the DMM were refined. The electrostatic equations were solved in the cuboid domain. The electrostatic boundary conditions were electric field in the z-direction, at the lower wall (z = 0) was grounded while the upper wall a voltage of 10 V was applied. The electric displacement was continuity on the DMM surfaces, and the insulating boundary conditions were applied on the other surfaces. The top and bottom surfaces (z = 0 and z = 30) were the driving electrodes while the other domain surfaces were of bulk electrolyte. The bulk concentration conditions were used for the ions on the domain surfaces and no penetration of the ion conditions were used on the DMM surfaces. For the hydrodynamic problem no-penetration and no-slip boundary conditions were taken at the DMM surfaces and the edge of the domain surfaces.

III. RESULTS AND DISCUSSION

Under AC electric field, the DMMs show moving behaviors in the opposite direction at different frequency ranges. This results from the different propulsion mechanisms-sDEP and ICEP. To investigate the moving behavior of the proposed DMMs, we simulated the electric field gradient at high frequencies and the flow field at low frequencies. From (7) and (10), we know that the AC electrokinetic behavior of DMMs is significantly dependent on the dimension of the DMMs. Therefore, we have investigated the effects that three key parameters have on the electrokinetic behavior of the DMMs. Here, we assume that the interface between the cylindrical surface of the DMMs with a radius of R and the medium is the outer layer. The interface between the cylindrical surface of the DMMs with a radius of r and the medium is the inner layer.

At high frequencies ($\omega > \omega_{RC}$), the propulsion mechanism of the DMMs can be attributed to sDEP, which results from the interaction between the applied electric field and the electric-induced dipole. Due to the asymmetry of the defective structure, the induced electric field gradient around the DMMs is asymmetric (Fig. 4a). The magnitude of the electric field gradient in the outer layer of the DMMs is higher than that in the inner layer. Therefore, the outer layer

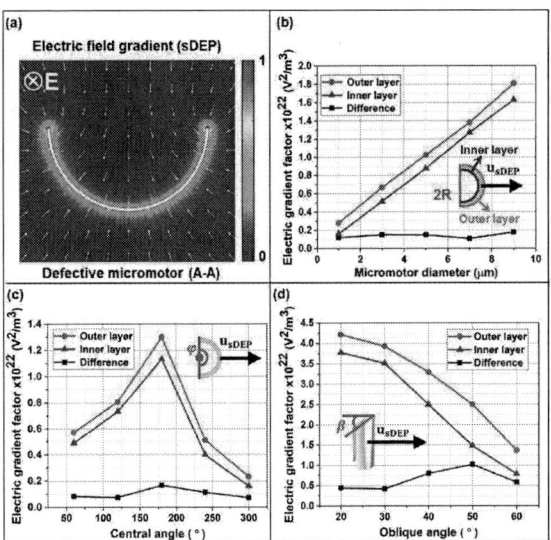

Figure 4. A simulation illustration of the electrokinetic behavior of the DMMs at high frequencies under a given voltage amplitude of 10V. (a) Numerical simulation results of the electric field strength gradients and electric field lines around the DMMs at 200kHz (normalized by the maximum of the electric field strength). The white arrows indicate the direction of the electric field gradient. (b) Outer diameter-dependence of the electric field gradient factor for the outer and inner layers of the DMMs and their differences. (c) Central angle-dependence of the electric field gradient factor for the outer and inner layers of the DMMs and their differences. (d) Oblique angle-dependence of the electric field gradient factor for the outer and inner layers of the DMMs and their differences.

of the DMMs is subject to a higher radially outward DEPF and the DMMs move with the defect-free side forward. We further studied the effects that the diameter, central angle, and oblique angle have on the electrokinetic behavior of the DMMs under high frequencies. First, we investigated the influence of the outer diameter on the electrokinetic behavior of the DMMs, the central angle is 180 degrees and the oblique angle is zero. In order to get the more accurate results of the electrokinetic behavior through the electric field gradient, we investigated the relationship between the diameter of the DMMs and the electric gradient factor $\nabla|E|^2$. Since the changes in diameter have a direct effect on the DEPF, we used two coefficients $\chi_{rw} = r + 0.75d$ and $\chi_{rn} = r + 0.25d$ to modulate the results of the electric gradient factor. The difference in the electric field gradient factor grows with the increased diameter of the DMMs (Fig. 4b). The growing trend has a power law as a function of the diameter, and the exponent of the function is less than one. We can further get the relationship between the diameter of the DMMs and the sDEP velocity under high frequencies, namely, $u_{sDEP} \propto R^P$ ($0 < P < 1$). Then, we investigated the relationship between the electric gradient factor and the central angle of the DMMs (Fig. 4c). In this condition, the diameter is 5 μm and the oblique angle is zero. According to the simulation results, the electric gradient factor difference reaches the peak value at a central angle of 180 degrees. This confirms that when the central angle is 180 degrees, the DMMs have significant mobility, which provides a valuable reference for the design of the DMMs. Finally, we studied the influence of the oblique angle on the moving behavior of the DMMs, the diameter is 5 μm and the central angle is 180 degrees (Fig. 4d). The simulation results show that the electric gradient factor difference reaches the peak value at the oblique angle of 50 degrees. Furthermore, it can be seen from the results that no matter how many degrees of the oblique angle is taken, the electric gradient factor is always higher than that there is no oblique angle. This proves that the oblique angle has a significant effect on increasing the mobility of the DMMs.

At low frequencies ($\omega < \omega_{RC}$), the propulsion mechanism of the DMMs can be attributed to ICEP, which results from the ICEO flow. Due to the defective structure of the DMMs, the ICEO flow around the DMMs is asymmetric. According to the numerical simulation results, the local slip velocity around the DMMs is radially outwards and the slip velocity on the outer layer is higher than that on the inner layer (Fig. 5a). Therefore, the DMMs move with the defective side forward under low-frequency ranges and the velocity is determined by the difference in the slip velocity between the outer layer and inner layer. We further investigated the effect that the diameter, central angle, and oblique angle have on the moving behavior of the DMMs. First, we studied the influence of the outer diameter on the ICEP velocity of the DMMs (Fig. 5b). In this condition, the central angle is 180 degrees and the oblique angle is zero. The ICEP velocity can be obtained through the difference in the ICEO slip velocity between the outer layer and inner layer of the DMMs. As with the simulation process of the relationship between the diameter of the DMMs and the electric gradient factor, here we use two coefficients $\varsigma_{rw} = R/(r + 0.75d)$ and $\varsigma_{rn} = R/(r + 0.25d)$ to modulate the ICEO slip velocity. The simulation results show that the ICEO slip velocity difference reaches the peak value when the diameter of the DMMs is 5 μm. This indicates that the DMMs with a diameter of 5 μm have a higher mobility under low frequencies. Then, we investigated the relationship

between the central angle of the DMMs and the ICEO slip velocity difference, the diameter of the DMMs is 5 μm and the oblique angle is zero (Fig. 5c). According to the simulation results, there are two peak values for the difference in the ICEO slip flow, which appear at the central angles of the DMMs are 120 degrees and 240 degrees, respectively. The reason behind this particular aberration is that the ICEO flow generated by the part of the DMMs exceeding 180 degrees will be offset with the part less than 180 degrees. Therefore, the difference in the ICEO slip flow has two peak values. Finally, we studied the influence of the oblique angle on the differences in the ICEO slip velocity (Fig. 5d). In this condition, the diameter is 5 μm and the central angle is 180 degrees. The simulation results show that the differences in the ICEO slip velocity decrease with the increasing oblique angle. This indicates that the smaller oblique angle can increase the mobility of the DMMs under the low-frequency electric field. Furthermore, it can be known from the simulation results that the oblique angle has a small effect on the moving behavior of the DMMs under low electric frequency ranges.

Based on the aforementioned numerical simulation results, it can be known that the proposed DMMs can exhibit motion behaviors under uniform AC electric field according to the propulsion mechanisms-sDEP and ICEP. Furthermore, by changing the frequency of the AC electric field, the direction of the DMMs can be easily controlled. According to the simulation results that the influences of the diameter, central angle, and oblique angle on the electrokinetic behavior of the DMMs, we can design the DMMs with higher mobility. Based on the fabrication method we proposed in the previous study, these key parameters for the DMMs can be easily modulated. The diameter can be controlled by changing the pore size of the membrane, the central angle can be controlled by changing the time of the electrodeposition, and the oblique angle can be controlled by changing the angle of the evaporation.

IV. CONCLUSION

In this work, we have proposed a novel defective micromotor (DMM) with an asymmetric structure, which can be driven by a uniform AC electric field. A theoretical model for the DMMs was established and the moving behavior of the DMMs under different frequency ranges have been theoretically proved. We investigated the electrokinetic behavior of the DMMs and the simulation results can guide us to design the DMMs with higher mobility. The DMMs eliminate the restriction of materials and are expected to achieve spatial movement in a complex environment. This innovative result will provide a new approach for designing electric-driven micro/nanomotors and can be widely applied in drug delivery, biomedicine, environmental remediation, micro/nanosensors, and micromechanical systems.

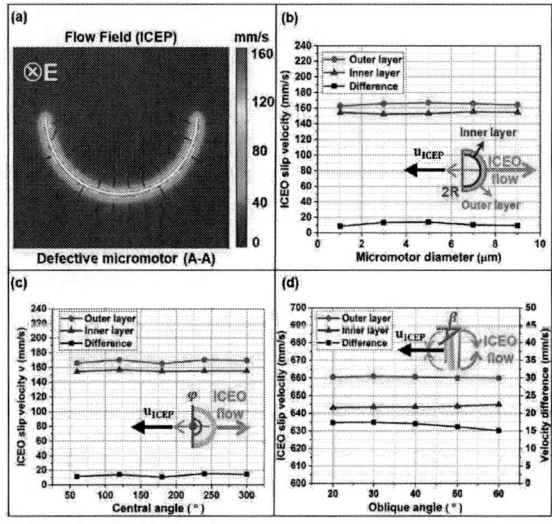

Figure 5. A simulation illustration of the electrokinetic behavior of the DMMs at low frequencies under a given voltage amplitude of 10V. (a) Numerical simulation results of the slip velocity around the DMMs and the direction of the flow field (red arrows) at 20kHz. (b) Outer diameter-dependence of the ICEO slip velocity around the outer and inner layers of the DMMs and their differences. (c) Central angle-dependence of the ICEO slip velocity around the outer and inner layers of the DMMs and their differences. (d) Oblique angle-dependence of the ICEO slip velocity around the outer and inner layers of the DMMs and their differences.

REFERENCES

[1] W. Wang, L. A. Castro, M. Hoyos, and T. E. Mallouk, "Autonomous motion of metallic microrods propelled by ultrasound," *ACS nano*, vol. 6, no. 7, pp. 6122-6132, May 2012.

[2] F. Ji, D. Zhou, G. Zhang, and L. Li, "Numerical Analysis of Visible Light Driven Gold/Ferric Oxide Nanomotors," *IEEE Transactions on Nanotechnology*, vol. 17, no. 4, pp. 692-696, Mar. 2018.

[3] A. Boymelgreen, G. Yossifon, and T. Miloh, "Propulsion of active colloids by self-induced field gradients," Langmuir, vol. 32, no. 37, pp. 9540-9547, 2016.

[4] F. M. Weinert and D. Braun, "Observation of slip flow in thermophoresis," *Phys. Rev. Lett.*, vol. 101, no. 16, p. 168301, Oct. 2008.

[5] P. Fischer and A. Ghosh, "Magnetically actuated propulsion at low Reynolds numbers: towards nanoscale control," *Nanoscale*, vol. 3, no. 2, pp. 557-563, Dec. 2011.

[6] P. L. Venugopalan, B. Esteban-Fernández de Ávila, M. Pal, A. Ghosh, and J. Wang, "Fantastic voyage of nanomotors into the cell," ACS nano, vol. 14, no. 8, pp. 9423-9439, 2020.

[7] M. Safdar, J. Simmchen, and J. Jänis, "Light-driven micro-and nanomotors for environmental remediation," Environmental Science: Nano, vol. 4, no. 8, pp. 1602-1616, 2017.

[8] X. Zhang, C. Chen, J. Wu, and H. Ju, "Bubble-propelled jellyfish-like micromotors for DNA sensing," ACS applied materials & interfaces, vol. 11, no. 14, pp. 13581-13588, 2019.

[9] T. B. Jones, "Basic theory of dielectrophoresis and electrorotation," IEEE Eng. Med. Biol. Mag., vol. 22, no. 6, pp. 33-42, 2003.

[10] J. Yan, M. Han, J. Zhang, C. Xu, E. Luijten, and S. Granick, "Reconfiguring active particles by electrostatic imbalance," Nature materials, vol. 15, no. 10, pp. 1095-1099, 2016.

[11] M. Z. Bazant and T. M. Squires, "Induced-charge electrokinetic phenomena," Current Opinion in Colloid & Interface Science, vol. 15, no. 3, pp. 203-213, 2010.

[12] W. Liu, Y. Ren, Y. Tao, H. Yan, C. Xiao, and Q. Wu, "Buoyancy-free janus microcylinders as mobile microelectrode arrays for continuous microfluidic biomolecule collection within a wide frequency range: A numerical simulation study," Micromachines, vol. 11, no. 3, p. 289, 2020.

[13] J. G. Lee, A. M. Brooks, W. A. Shelton, K. J. Bishop, and B. Bharti, "Directed propulsion of spherical particles along three dimensional helical trajectories," Nature communications, vol. 10, no. 1, pp. 1-8, 2019.

[14] T. M. Squires and M. Z. Bazant, "Breaking symmetries in induced-charge electro-osmosis and electrophoresis," J. Fluid Mech., vol. 560, p. 65, 2006.

[15] E. Yariv, "Induced-charge electrophoresis of nonspherical particles," Phys. Fluids, vol. 17, no. 5, p. 051702, 2005.

[16] A. M. Brooks, S. Sabrina, and K. J. Bishop, "Shape-directed dynamics of active colloids powered by induced-charge electrophoresis," Proceedings of the national academy of sciences, vol. 115, no. 6, pp. E1090-E1099, Jan. 2018.

[17] J. J. Arcenegui, P. García-Sánchez, H. Morgan, and A. Ramos, "Electro-orientation and electrorotation of metal nanowires," PhRvE, vol. 88, no. 6, p. 063018, 2013.

[18] J. J. Arcenegui, P. García-Sánchez, H. Morgan, and A. Ramos, "Electric-field-induced rotation of Brownian metal nanowires," PhRvE, vol. 88, no. 3, p. 033025, 2013.

[19] K. A. Rose, J. A. Meier, G. M. Dougherty, and J. G. Santiago, "Rotational electrophoresis of striped metallic microrods," PhRvE, vol. 75, no. 1, p. 011503, 2007.

[20] K. A. Rose, B. Hoffman, D. Saintillan, E. S. Shaqfeh, and J. G. Santiago, "Hydrodynamic interactions in metal rodlike-particle suspensions due to induced charge electroosmosis," PhRvE, vol. 79, no. 1, p. 011402, 2009.

[21] M. Fair and J. Anderson, "Electrophoresis of nonuniformly charged ellipsoidal particles," J. Colloid Interface Sci., vol. 127, no. 2, pp. 388-400, 1989.

[22] P. García-Sánchez and A. Ramos, "Electrorotation of a metal sphere immersed in an electrolyte of finite debye length," PhRvE, vol. 92, no. 5, p. 052313, 2015.

[23] O. D. Velev, S. Gangwal, and D. N. Petsev, "Particle-localized AC and DC manipulation and electrokinetics," Annual Reports Section "C"(Physical Chemistry), vol. 105, pp. 213-246, 2009.

[24] A. Ramos, P. García-Sánchez, and H. Morgan, "AC electrokinetics of conducting microparticles: A review," Current Opinion in Colloid & Interface Science, vol. 24, pp. 79-90, 2016.

Proceedings of the 16th Annual IEEE International
Conference on Nano/Micro Engineered and Molecular Systems
April 25-29, 2021

3D printing of magnetically actuated miniature soft robots

Zhongbao Wang, Yigen Wu, Dezhi Wu, and Zhenyin Hai

Abstract— Magnetically actuated soft robots possess great potential for applications in medical surgery, bioengineering, and micromanipulation due to its fast response, easiness to manipulate, and untethered control. However, it remains a challenge to fabricate flexible magnetically actuated soft robots by magnetic materials with a high viscosity ($\geq 10^4$ Pa.s). Therefore, we proposed a microscale Weissenberg effect-based 3D printing technique for fabrication of magnetically actuated soft robots via the programmable magnetization of high-viscosity magnetic polymer precursor prepared by mixing magnetic microparticles with elastic polymer. The magnetic microparticles in precursor solution can be reoriented by applying a magnetic field around nozzle during direct writing, which imparts the programmable sequences of magnetic polarities to printed structures. Meanwhile, we investigated the effect of main process parameters on the morphologies of printed structures. 3D shape transformations of a 2D patterned structure and the grasping experiment of a magnetically actuated cross shaped soft gripper verified the effectiveness of the programmable magnetization of high-viscosity flexible magnetic materials for the proposed 3D printing technique.

Keywords—soft robot, magnetic actuator, Weissenberg effect, 3D printing

I. INTRODUCTION

In recent years, soft robots have attracted abundant attention of researchers due to their unique intrinsic properties, such as lightweight, infinite degrees of freedom, and low-cost fabrication [1]-[3]. Many typical soft robots reported in previous literatures mainly imitated the shapes and functionalities of the soft-bodied natural creatures, such as octopus [4], caterpillar [5], earthworm [6], and starfish [7][8]. Great efforts will still be made to develop the micro-scale soft robots because of their significant potential for applications in rescue, minimally invasive surgery, targeted therapy and micromanipulation. The bulk structures of soft robots are mainly fabricated by the compliant materials with the Young's modulus of $10^4 \sim 10^9$ MPa. Thanks to the low resistance of these materials to pressure, the soft robots can squeeze through the gap smaller than its normal size and enter some narrow or sinuous spaces [9].

The main actuation methods of soft robots include fluidic elastomer actuation (hydraulic and pneumatic actuation), smart materials, chemical reaction, and magnetic actuation. Many remarkable soft robots [10]-[12] driven by pneumatic soft actuators (PSAs) have been reported owing to abundant air supply, low cost, lightweight and no pollution. However, the response speed of the PSAs is limited by the air supply

pressure, and the air pressure and flow rate change nonlinearly with time. Smart materials can serve as a section of the robot's body, so it is easy to realize the integrated design of the actuator and body of the robot. Generally, shape memory alloys (SMAs) and shape memory polymers (SMPs) [13]-[14] employed to imitate the muscle fibers of the soft-bodied animals can deform continuously by heating, but there are some restrictions such as heating and cooling hysteresis. Electroactive polymers (EAPs) include ionic EAPs and electric EAPs, while the poor coupling efficiencies of ionic EAPs [15] result in lower actuation speeds and a high voltage inducing an actuation strain is required for electric EAPs [16]-[19], which will bring design difficulties and safety problems. The chemical reaction actuation [20]-[21] easily achieves the miniaturization and integration of soft robots, but its chemical reaction process is not controllable.

Magnetic actuation is a non-contact energy transfer, which has the advantages of fast response, safe human-robot interaction, untethered control and low cost. Lately, some impressive magnetically actuated miniature soft robots [22]-[26] have been developed to meet demands for applications in medical surgery, bioengineering, and micromanipulation. Such robots were mostly fabricated by manually assembling microscale magnetic components via mixing magnetic microparticles with elastic polymer, and then a controllable magnetic field generated by energized windings or permanent magnets was used to actuate these robots [23]-[25] to achieve monotonous locomotion. However, these robots can only obtain simple magnetization profiles, which is not suitable for sophisticated motions. To obtain the desired magnetization profiles, 3D printing of the programmable ferromagnetic domains of flexible magnetic materials was proposed [27], by reorienting of magnetic microparticles in flexible matrix to achieve the programmable morphologies of flexible magnetic materials. Manufacturing methods based on different 3D printing techniques, such as direct ink writing (DIW) [27], ultraviolet lithography (UVL) [28], and direct laser writing (DLW) [29], have been successfully used to pattern magnetic microparticles in the flexible matrix to realize the programmed magnetization profiles of flexible magnetic materials.

Nevertheless, commercial 3D printing equipment could not be used directly to print homemade materials. Furthermore, it is difficult for 3D printing technique to base on microextrusion, such as direct ink writing, which use pressure to push a precursor solution to manipulate a high-

---

This work was supported by National Natural Science Foundation of China (52075464, 91648114) and Science and Technology Program of Shenzhen City (JCYJ20180306172700388).

Z. Wang is with Department of Mechanical and Electrical Engineering, Xiamen University, Xiamen, China (e-mail: zbao-wang@139.com).

D. Wu is with Department of Mechanical and Electrical Engineering, Xiamen University, Xiamen, China (corresponding author e-mail: wdz@xmu.edu.cn).

Y. Wu and Z. Hai are with Department of Mechanical and Electrical Engineering, Xiamen University, Xiamen, China.

978-1-6654-3008-1/21 $31.00 © 2021 IEEE

viscosity magnetic solution. This results in a long-time delay and clogging in printing due to high flow resistance, especially for the small nozzles. Consequently, we proposed a microscale Weissenberg effect-based 3D printing method for fabrication of magnetically actuated soft robots via the programmable magnetization of high-viscosity magnetic polymer precursor. The polymer precursor is filled with magnetic microparticles and the microparticles can be reoriented by applying a magnetic field around the nozzle during direct writing [27][30], which imparts the programmable sequences of magnetic polarities to the printed structures. Several micro-robots were fabricated by using the proposed 3D printing technique, and we demonstrated that these micro-robots can achieve fast reversible three-dimensional (3D) shape transformations under an actuation magnetic field.

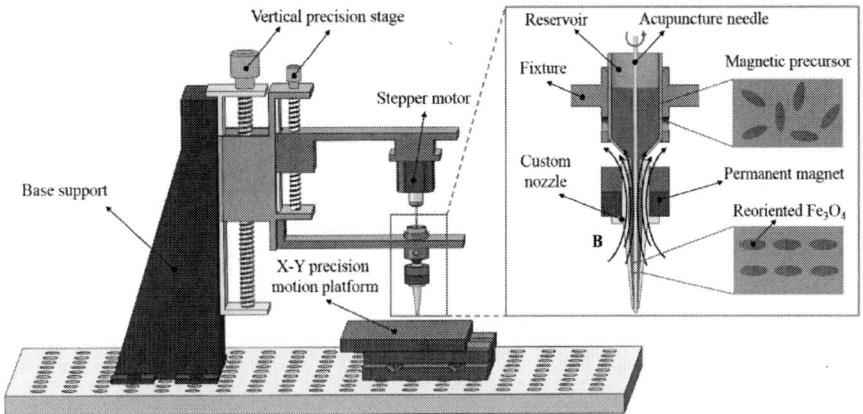

Fig. 1. Working principle schematic of the microscale Weissenberg effect-based 3D printing system for programmable magnetization.

II. RESULTS

A. The microscale Weissenberg effect-based 3D printing system

A novel microscale Weissenberg effect-based 3D printing system as depicted in Fig. 1 was built to realize the programmable magnetization of high-viscosity magnetic polymer precursors. The proposed 3D printing system was designed basing on our previous work [31], which utilized the Weissenberg effect to pump the high viscosity non-Newtonian precursor solution at the pico-liter level. The right side of Fig. 1 shows the local cutaway view containing the core components of this 3D printing system, in which the reorientation principle of magnetic microparticles is illustrated. A hollow cylindrical permanent magnet was installed around the customized nozzle to generate a directional magnetic field $\mathbf{B}\approx45$ mT, which induced magnetic microparticles in the magnetic precursor solution reoriented along $\mathbf{B}$ direction. Patterning sequences of magnetic polarities could be applied to the printed structures by programmatically controlling X-Y precision motion platform. In addition, by heating the collection device on the motion platform to accelerate the curing of deposited fiber, the 3D structure with a certain height can also be printed.

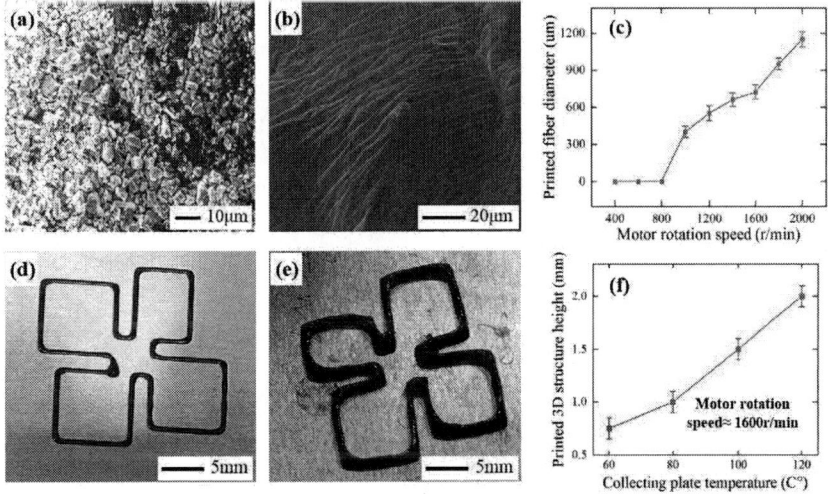

Fig. 2. Effect of main process parameters on the microscale Weissenberg effect-based 3D-printed structures. SEM images of Fe3O4 microparticles (a) and the printed fiber (b). (c) Effect of the rotation speed of stepper motor on the diameter of printed fibers. The printed 2D sample (d) and 3D sample (e). (f) Effect of the temperature of collection plate on the height of printed 3D samples.

978-1-6654-3008-1/21 $31.00 © 2021 IEEE

B. Effect of main process parameters on 3D-printed structures

The magnetic precursor solution for 3D printing was prepared by mixing ferrite microparticles with the superelastic silicone rubber matrix. Firstly, we mixed Part A and Part B of Ecoflex 00-30 in a 1:1 weight ratio, stirring with an electric stirrer at 1,600 rpm for 1 minute to prepare a flexible matrix. Then, 50 wt % Fe_3O_4 microparticles with an average size of 5μm (Fig. 2(a)) were dispersed in the flexible matrix prepared in the previous step, and stirring at 2,000 rpm for 3 minutes to prepare the flexible ferromagnetic composite. Finally, the prepared ferromagnetic composite was magnetized to saturation by a capacitor impulse magnetizer to obtain the magnetic precursor solution. To investigate the effect of main process parameters on the microscale Weissenberg effect-based 3D printing structures, we firstly studied the effect of the rotation speed of the stepper motor on the diameter of printed fibers (Fig. 2 (c)), and the experimental results showed that pumping precursor solution based on the Weissenberg effect couldn't be realized when the rotation speed of stepper motor was lower than 800 rpm. This was because the acupuncture needle with lower rotation speed produced less shear force on the magnetic precursor solution, which hardly overcame the intrinsic attraction and adhesion force of the magnetic precursor solution with high viscosity. The diameter of printed fibers increased with the rotation speed when the rotation speed was higher than 800 rpm, and Fig. 2(d) showed a 2D sample printed at 1500 rpm with a line width of about 400 microns. To achieve the printing of 3D structures, we heated the collection plate on the motion platform to accelerate the curing of the deposited fibers so as to realize the stacking of fibers layer by layer. As illustrated in Fig. 2(f), the height of printed 3D samples increased with the temperature of collection plate within a certain range. Due to the poor thermal conductivity of the polymer, the printed 3D structures had a wide bottom and narrow top, as showed in Fig. 2(e).

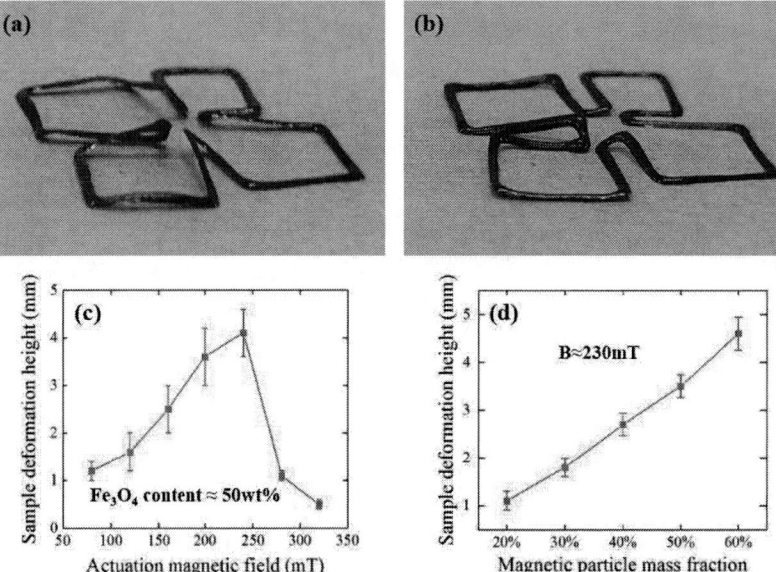

Fig. 3. Three-dimensional (3D) shape transformations of the printed two-dimensional (2D) patterned planar sample. 3D shape transformations of a 2D patterned planar structure under an actuation magnetic field (a) and the same field with reverse direction (b). Effect of the actuation magnetic field (c) and the magnetic particle mass fraction (d) on the deformation height of printed samples.

C. 3D shape transformations of the printed 2D patterned planar sample

To verify the effectiveness of the programmable magnetization for the proposed 3D printing technique, a two-dimensional patterned planar structure as illustrated in Fig. 2(d) was printed by using our method and would undergo three-dimensional shape transformation when it was placed in an actuation magnetic field perpendicular to this planar structure (Fig. 3(a)). Switching the direction of the actuation magnetic field, the deformation direction of this planar structure was also changed (Fig. 3(b)). In addition, we investigated the effect of the actuation magnetic field on the deformation height of this planar structure. As shown Fig. 3(c), the deformation height of this planar structure increased with the actuation magnetic field when the actuation magnetic field was lower than 240 mT, but the deformation height of the 2D structure sharply decreased when the field was higher than 240 mT. This is because that excessive actuation magnetic field overcame the intrinsic coercivity of the printed planar structure and magnetized it again, thus changing the original patterned magnetic polarities. Fig. 3(d) demonstrated that the deformation height of the 2D structure increased with the magnetic particle mass fraction.

D. Magnetically actuated cross shaped soft gripper

Finally, a cross shaped soft gripper was fabricated to demonstrated the effectiveness of the proposed microscale Weissenberg effect-based 3D printing technique. Fig. 4(a) revealed the design dimensions and magnetic polarity distribution of the cross shaped gripper, and the length of a single finger is 10 mm and the width is 1.5 mm. The motion path of X-Y precision motion platform is shown in Fig. 4(b) for printing this soft gripper. To demonstrate the grasping ability of the cross shaped gripper, a simple grab experiment was carried out, as shown in Fig. 4(c). The experimental results showed that the cross shaped gripper could grasp a 1.5 g object with size of 4×5×6 mm under an actuation magnetic field of 180 mT.

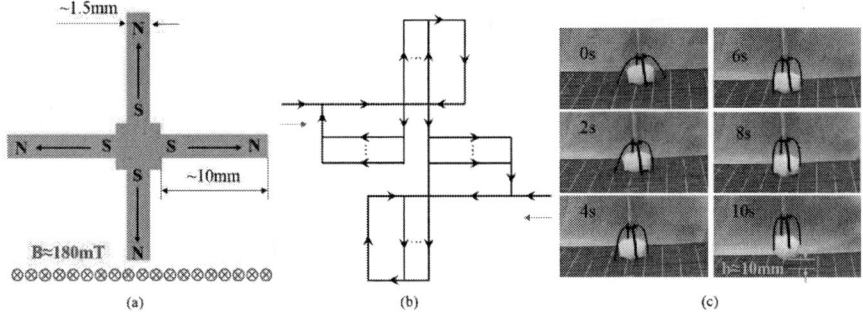

Fig. 4. Magnetically actuated cross shaped soft gripper. (a) The design dimensions and polarity distribution of the cross shaped gripper. (b) The motion path of X-Y precision motion platform. (c) Grasping test of the magnetically actuated soft gripper.

III. EXPERIMENTAL METHODS

A. Magnetically actuated cross shaped soft gripper

The preparation procedure of magnetic precursor solution is as follows: Firstly, we mixed Part A and Part B of Ecoflex 00-30 (Smooth-on Inc., America) in a 1:1 weight ratio, stirring with an electric stirrer (ZY3057, ART EXHIBITION, China) at 1,600 rpm for 1 minute to prepare a flexible matrix. Then, 50wt% Fe_3O_4 microparticles with an average size of 5 μm were dispersed in the flexible matrix prepared in the previous step, and stirring at 2,000 rpm for 3 minutes to prepare the flexible ferromagnetic composite. Finally, the prepared ferromagnetic composite was magnetized to saturation by a capacitor impulse magnetizer (MAG1240, YongGui, China) to obtain the magnetic precursor solution.

B. Experimental device operation

The acupuncture needle was rotated by a stepper motor (42BYGH23, HANPOSE, China). A CCD camera (UI-2250SE-C-HQ, China) was used to record the printing procedure. Magnetic microparticles and the printed fibers were characterized by the scanning electron microscope (SU-70, Hitachi, Japan).

IV. CONCLUSIONS

In this paper, we proposed a novel microscale Weissenberg effect-based 3D printing method for fabrication of magnetically actuated soft robots via the programmable magnetization of high-viscosity flexible magnetic materials. We studied the effect of the actuation magnetic field and the magnetic particles mass fraction on the deformation of printed samples under an actuation magnetic field. To verify the effectiveness of the programmable magnetization for the proposed 3D printing technique, a 2D planar structure was printed by using our method and can achieve 3D shape transformations under an actuation magnetic field. In addition, a magnetically actuated cross shaped soft gripper was fabricated by using the proposed 3D printing technique, and we demonstrated that the cross shaped gripper could grasp a 1.5 g object with size of 4×5×6 mm under an actuation magnetic field of 180 mT. In future work, we are committed to 3D printing of more complex, multi-functional magnetically actuated soft robots based on the proposed 3D printing technique.

ACKNOWLEDGMENT

This work was in part supported by National Natural Science Foundation of China (52075464, 91648114) and Science and Technology Program of Shenzhen City (JCYJ20180306172700388).

REFERENCES

[1] S. Chen, Y. Cao, M. Sarparast, H. Yuan, L. Dong, X. Tan, and C. Cao, "Soft Crawling Robots: Design, Actuation, and Locomotion", *Advanced Materials Technologies*, vol. 5, no. 2, pp. 1900837, 2020.

[2] R. F. Shepherd, F. Ilievski, W. Choi, S. A. Morin, A. A. Stokes, A. D. Mazzeo,X. Chen, M. Wang, and G. M. Whitesides, "Multigait soft robot", *Proceedings of the national academy of sciences,* vol. 108, no. 51, pp. 20400-20403, 2011.

[3] E. W. Hawkes, L. H. Blumenschein, and J. D. Greer, "A soft robot that navigates its environment through growth", *Science Robotics*, vol. 2, no. 8, 2017.

[4] C. Laschi, M. Cianchetti, B.Mazzolai , L. Margheri, M. Follador, and P. Dario, "Soft robot arm inspired by the octopus", *Advanced Robotics*, vol. 26, no. 7, pp. 709-727, 2012.

[5] H. Zeng, O. M. Wani, P. Wasylczyk, and A. Priimagi, "Light-driven, caterpillar-inspired miniature inching robot", *Macromolecular rapid communications,*vol. 39, no. 1, pp. 1700224, 2018.

[6] A. A. Calderón, J. C. Ugalde, L. Chang, Z. J. Cristóbal, and N. Pérez-Arancibia, "An earthworm-inspired soft robot with perceptive artificial skin", *Bioinspiration & biomimetics*, vol. 14, no. 5, pp. 056012, 2019.

[7] A.Poungrat and T. Maneewarn, "A starfish inspired robot with multi-directional tube feet locomotion", *Proc. IEEE Int. Conf. Robotics and Biomimetics (ROBIO)*, pp.712-717, 2017.

[8] W. L. Scott and D. A. Paley, "Geometric Gait Design for a Starfish - Inspired Robot Using a Planar Discrete Elastic Rod Model", *Advanced Intelligent Systems*, vol. 2, no. 6, pp. 1900186, 2020.

[9] D. Trivedi, C. D. Rahn, W. M. Kier, and D.W. Lan, "Soft robotics: Biological inspiration, state of the art, and future research", *Applied bionics and biomechanics*, vol. 5, no. 3, pp. 99-117, 2008.

[10] B. Mosadegh, P. Polygerinos, C. Keplinger, S. Wennstedt, R. F. Shepherd, U. Gupta, J. Shim, K. Bertoldi, C. J. Walsh, and G. M. Whitesides, "Pneumatic Networks for Soft Robotics that Actuate Rapidly", *Advanced Functional Materials*, vol. 24, no. 15, pp. 2163-2170, 2014.

[11] A. Rafsanjani, Y. Zhang, B. Liu, S. M. Rubinstein, and K. bertoldi, "Kirigami skins make a simple soft actuator crawl", *Science Robotics*, vol. 3, no. 15, pp. eaar7555, 2018.

[12] M. S. Verma, A. Ainla, D. Yang, D. Harburg, and G. M. Whitesides, "A soft tube-climbing robot", *Soft robotics*, vol. 5, no. 2, pp. 133-137, 2019.

[13] W. Wang , and S. H. Ahn, "Shape memory alloy-based soft gripper with variable stiffness for compliant and effective grasping", *Soft robotics*, vol. 4, no. 4, pp. 379-389, 2017.

[14] X. Huang, K. Kumar, and M. K. Jawed, "Chasing biomimetic locomotion speeds: Creating untethered soft robots with shape memory alloy actuators", *Science Robotics*, vol. 3, no. 25, pp. eaau7557, 2018.

[15] C. Wang, K. Sim, J. Chen, H. Kim, Z. Rao, Y. Li, W. Chen, J. Song, R. Verduzco, and C. Yu, "Soft ultrathin electronics innervated adaptive fully soft robots", *Advanced Materials*, vol. 30, no. 13, pp. 1706695, 2018.

[16] J. Cao, L. Qin, J. Liu , Q. Ren, C. C. Foo, H. Wang, H. P. Lee, and J. zhu, "Untethered soft robot capable of stable locomotion using soft electrostatic actuators", *Extreme Mechanics Letters*, vol. 21, pp. 9-16, 2018.

[17] X. Ji, X. Liu, V. Cacucciolo, M. Imboden, Y. Civet, A. E. Haitami, S. Cantin, Y. Perriard, and H. Shea, "An autonomous untethered fast soft robotic insect driven by low-voltage dielectric elastomer actuators", *Science Robotics*, vol. 4, no. 37, pp. eaaz6451, 2019.

[18] T. Li, Z. Zou, G. Mao, X. Yang, Y. Liang, C. Li, S. Qu, Z. Suo, and W. Yang, "Agile and resilient insect-scale robot", *Soft robotics*, vol. 6, no. 1, pp. 133-141, 2019.

[19] W. Sun, F. Liu, Z. Ma, C. Li, and J. Zhou, "Soft mobile robots driven by foldable dielectric elastomer actuators", *Journal of Applied Physics*, vol. 120, no. 8, pp. 084901, 2016.

[20] M. Wehner, R. L. Truby , D. J. Fitzgerald , B. Mosadegh, G. M. Whitesides, J. A. Lewis, and R. J. Wood, "An integrated design and fabrication strategy for entirely soft, autonomous robots", *Nature*, vol. 536, no. 7617, pp. 451-455, 2016.

[21] N. W. Bartlett, M. T. Tolley, J. T. B. Overvelde, J. C. Weaver, B. Mosadegh, K. Bertoldi, G.M. Whitesides, R. J. Wood, "A 3D-printed, functionally graded soft robot powered by combustion", *Science*, vol. 349, no. 6244, pp. 161-165, 2015.

[22] T. N. Do, H. Phan, T. Q. Nguyen, and V. Yon, "Miniature soft electromagnetic actuators for robotic applications", *Advanced Functional Materials*, vol. 28, no. 18, pp. 1800244, 2018.

[23] W. Hu, G. Z. Lum, M. Mastrangeli, and M. Sitti, "Small-scale soft-bodied robot with multimodal locomotion", *Nature*, vol. 554, no. 7690, pp. 81-85, 2018.

[24] S. Ijaz, H. Li , M. C. Hoang , C. Kim, D. Bang, E. Choi, and J. Park, "Magnetically actuated miniature walking soft robot based on chained magnetic microparticles-embedded elastomer", *Sensors and Actuators A: Physical*, vol. 301, pp. 111707, 2020.

[25] J. Wang, N. Jiao, X. Wang , D. Lin, S. Tung, and L. Liu, "An electromagnetic anglerfish-shaped millirobot with wireless power generation", *Biomedical microdevices*, vol. 21, no. 1, pp. 1-8, 2019.

[26] D. Hua, X. Liu, S. Sun, M. A. Sotelo, Z. Li, and W. Li, "A Magnetorheological Fluid Filled Soft Crawling Robot with Magnetic Actuation", *IEEE/ASME Transactions on Mechatronics*, 2020.

[27] Y. Kim, H. Yuk, R. Zhao, S. A. Chester, and X. Zhao, "Printing ferromagnetic domains for untethered fast-transforming soft materials", *Nature*, vol. 558, no. 7709, pp. 274-279, 2018.

[28] T. Xu, J. Zhang, M. Salehizadeh, O. Onaizah, and E. Diller, "Millimeter-scale flexible robots with programmable three-dimensional magnetization and motions", *Science Robotics*, vol. 4, no. 29, pp. Eaav4494, 2019.

[29] H. Deng, K. Sattari, Y. Xie , P. Liao, Z. Yan, and J. Lin, "Laser reprogramming magnetic anisotropy in soft composites for reconfigurable 3D shaping", *Nature Communications*, vol. 11, no. 1, pp. 6325, 2020.

[30] D. Kokkinis, M. Schaffner, and R.S. André, "Multimaterial magnetically assisted 3D printing of composite materials", *Nature Communications*, vol. 6, pp. 8643, 2015.

[31] Mei X, Chen Q, Wang S, "The microscale Weissenberg effect for high-viscosity solution pumping at the picoliter level", *Nanoscale*, vol.10, no.15, 2018.

Proceedings of the 16th Annual IEEE International
Conference on Nano/Micro Engineered and Molecular Systems
April 25-29, 2021

Anisotropic spreading of droplets on striped electrodes

Wei Wang, Yanbo Xie, and Antoine Riaud

Abstract — Anisotropic spreading of droplets on a heterogeneous solid surface is important for fundamental research and industrial applications. Compared to structurally and chemically heterogeneous surfaces, electrowetting could control the surface free-energy landscape of the solid-liquid interface dynamically with the applied voltage. In this work, we study the anisotropic spreading of droplets on striped electrodes (interdigitated electrodes) with voltages applied. The contact angles and contact radii of the contact area observed parallel and perpendicular to the striped electrodes are presented quantitatively. Our research may help the understanding of droplet spreading in heterogeneous electric fields and assist in the applications of electrowetting.

I. INTRODUCTION

Wetting occurs when a droplet encounters a solid surface and spreads to reach a mechanical equilibrium state [1]. The properties of wetting, usually characterized by the contact angle, present the chemical-physical interactions between liquid and solid surface and determine the spreading process and equilibrium morphology of droplets. It has demonstrated crucial importance in various applications, involving 3D printing [2], liquid manipulation [3], self-cleaning surfaces [4], and droplet-based electricity generators [5]. Hence, the capacity to control the wetting and spreading of droplets on solid surfaces has attracted interest from fundamental research and industrial applications, especially for heterogeneous surfaces.

Young [6] has proposed that the equilibrium contact angle is determined by the balance of interfacial tensions at the liquid-solid-vapor triple-phase contact line. Thermodynamically, the capability to control the equilibrium contact angle equals the ability to change the liquid-solid-vapor interfacial free energy [1]. In the last several decades, advances in nanomanufacturing technology provide an unprecedented ability to customize the free-energy landscape of solid surfaces by modifying the chemical distribution [7] and topography [8] of surfaces. In his pioneer work, Lippmann [9] showed that the electrostatic charge induced by an external electrical potential was able to decrease the solid-liquid interfacial tension. Irrespective of its great potential as a liquid handling technique (i.e. electrowetting [10]), the capability to dynamically control the spreading of droplets on solid surfaces provides a highly versatile approach for fundamental studies of the wettability of liquid on solid surfaces. For structurally [11] and

chemically [7] corrugated surfaces, experimental observations have demonstrated that the spreading of droplets in directions parallel and perpendicular to the stripes obeys the same scaling law to time, but the contact line moves faster along the parallel direction. The anisotropic spreading of droplets in different directions results in an elliptical contact area elongated parallel to the stripes [7, 11]. Similar results have also been reported in electrowetting works [12]. However, quantification of how such changes alter the spreading dynamics and morphology of droplets on a solid surface is still an active area of research.

In this work, we studied the effect of electrowetting on the anisotropic spreading of droplets on striped electrodes. The difference between the advancing contact angle, receding contact angle, and contact radius measured in the directions parallel and perpendicular to the stripes is found negligible for voltages in a certain range, beyond which the circular contact area transits to an elliptical shape. The droplet dynamics and anisotropic spreading during the impact and condensation on the striped electrodes are similar to those observed on structurally and chemically corrugated surfaces under the effect of electric fields.

II. EXPERIMENTAL SETUP

The experiments were carried out following the setup in our previous work [13]. A periodic array of interdigitated electrodes (50 µm finger width and spacing, 130 nm thick) was patterned on indium-tin-oxide glasses using standard lithography and wet etching. The interdigitated electrodes ensure the generation of a corrugated electric field on the solid surface and alter the solid-liquid interfacial free-energy landscape. This corrugated solid-liquid free-energy landscape is reminiscent of the structural and chemical stripes used in other studies of droplet wetting and spreading [7, 8, 11]. Cyanoethyl pullulan (408 nm thick) and polytetrafluoroethylene (60 nm thick) were spin-coated successively as dielectric and hydrophobic layers, respectively, to insulate the liquid from electrodes and reversibly control the contact angle.

In experiments, a 40 µL deionized water droplet was deposited slowly on the coplanar electrowetting platform

---

W. Wang is with MOE Key Laboratory of Material Physics and Chemistry under Extraordinary Conditions, School of Physical Science and Technology, Northwestern Polytechnical University, Xi'an, 710072, China (e-mail: wangv@nwpu.edu.cn).

Y. Xie is with MOE Key Laboratory of Material Physics and Chemistry under Extraordinary Conditions, School of Physical Science and

Technology, Northwestern Polytechnical University, Xi'an, 710072, China (corresponding author, e-mail: ybxie@nwpu.edu.cn).

A. Riaud is with State Key Laboratory of ASIC and System, School of Microelectronics, Fudan University, Shanghai 200433, China (corresponding authors, e-mail: antoine_riaud@fudan.edu.cn).

978-1-6654-3008-1/21 $31.00 © 2021 IEEE

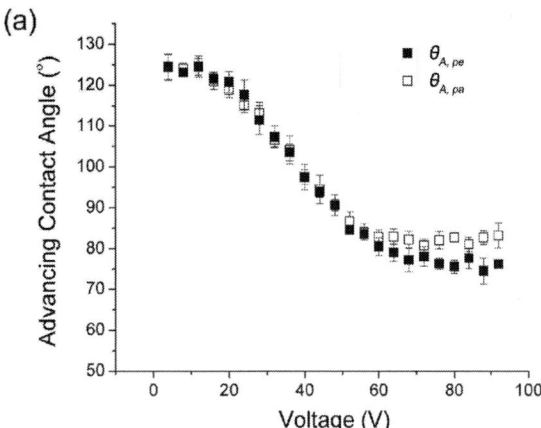

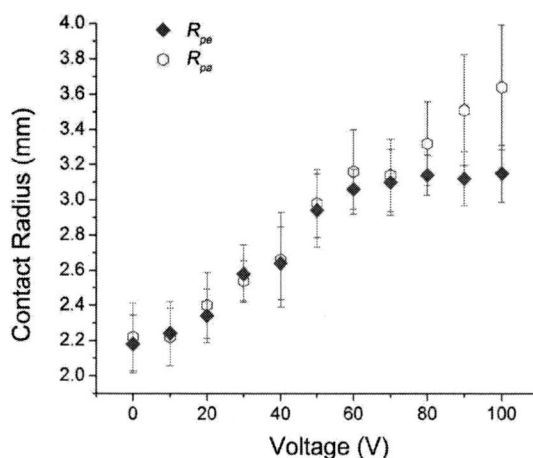

Fig. 2 Contact radii measured in directions parallel (R_{pa}) and perpendicular (R_{pe}) to the striped electrodes. Each test was repeated 5 times.

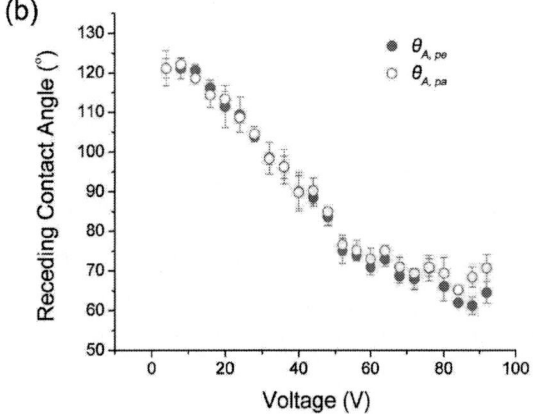

Fig. 1 Contact angles measured in directions parallel and perpendicular to the striped electrodes. (a) Advancing contact angles and (b) receding contact angles observed in directions parallel ($\theta_{A,pa}$ and $\theta_{R,pa}$) and perpendicular ($\theta_{A,pe}$ and $\theta_{R,pe}$) to the striped electrodes. Each test was repeated 5 times.

with its electrodes applied to a given voltage. After the droplet spread to reach an equilibrium, its contact radius was measured in two directions (parallel and perpendicular to the striped electrodes). A steel tubing is then inserted into the top of the droplet to replenish or extract liquid at 0.1 µL/min. The relatively low flow rate ensures a near mechanical equilibrium shape and negligible dynamic contact angle hysteresis of the droplet [14]. Due to the static contact angle hysteresis, the contact line remained pinned until the contact angle reached the advancing or receding contact angle [13, 15]. Side view images of the droplet are captured with a goniometer (DSA30, KRÜSS, Germany) in directions parallel and perpendicular to the striped electrodes for the advancing and receding contact angles. After each acquisition, the surface is cleaned, and the experiment is repeated with a different voltage.

III. RESULTS AND DISCUSSIONS

To clarify the spreading anisotropy of droplets on the surface, the contact angle and contact radius at different applied voltages were observed experimentally in two directions (parallel and perpendicular to the striped electrodes). The spreading of liquid on a solid surface is usually a result of the competition between the driving force and energy dissipation attributed mainly to the viscous resistance and molecular friction of moving contact lines [1]. The driving force is determined by the imbalance of surface tensions. In this work, it contains an electro-capillary force exerted by the electrowetting effect, which could be quantified by modifying the solid-liquid interfacial tension in Young's equation based on Lippmann's equation [10].

The advancing contact angles (θ_A) and receding contact angles (θ_R) are demonstrated in Fig. 1, with the subscripts *pa* and *pe* specifying whether the measurements were conducted in a direction parallel or perpendicular to the striped electrodes, respectively. Consistent with previous studies of electrowetting on interdigitated electrodes [12, 13], the contact angle decreases approximately in a quadratic relation with the applied voltage until the voltage increases to about 50 V_{DC} and the plots then grow into the contact angle saturation regime. The contact angles measured in two directions are approximately equal for applied voltage smaller than 60 V_{DC}, above which the difference emerges. A further increase of the applied voltage would enlarge the difference between the contact angles observed in two directions, but its value gets stable for voltages higher than about 80 V_{DC} due to the contact angle saturation effect.

Moreover, we also got the spreading of droplets in the directions parallel and perpendicular to the striped electrodes in experiments, which was characterized by the contact radii of droplets in two directions. Given that the

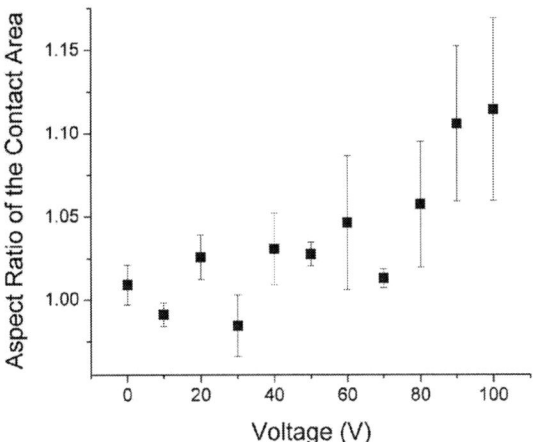

Fig. 3 Aspect ratio of the elliptical contact area. The droplet is elongated along the direction in which the striped electrodes extend. Each test was repeated 5 times.

electro-capillary force is proportional to the change of contact angles under the effect of electrowetting, the same value of contact angles measured in two directions for voltages smaller than 60 V_{DC} indicates equal electro-capillary forces exerted on the contact line in two directions. Thus, in this voltage range, the spreading of droplets in directions parallel and perpendicular to the striped electrodes obeys the same relation to the applied voltage. The contact radii in two directions are equal, indicating a circular contact area below 60 V_{DC}. However, the isotropic spreading of droplets terminates when continuing to increase the applied voltage. The divergence of contact angles in two directions induces the anisotropic spreading of droplets as a larger capillary force is exerted in the direction parallel to the striped electrodes than that in the perpendicular direction. Thus, similar to droplets on structurally corrugated surfaces [11], the spreading of liquid starts along the direction in which the striped electrodes extend. The motion of the contact line stops after a certain distance due to the combined action of contact angle hysteresis and surface tensions, but the contact angle keeps increasing under the effect of viscous and inertial forces. As the contact angle becomes large enough to its static advancing contact angle, the liquid quickly crosses a free-energy barrier formed by a pair of interdigitated electrodes. The process described above is repeated until the system gets stable, forming an elongated elliptical contact area. Similar phenomena are also seen and discussed in the work by Wang and Zhao [11] for droplet spreading on corrugated substrates under the effect of electrowetting.

For quantitative analysis, after the droplet was deposited on the platform with its electrodes applied to a given voltage and spread to reach an equilibrium shape, the contact radii were experimentally measured in the directions parallel and perpendicular to the striped electrodes. The experimental observations are shown in Fig. 3. Consistent with our

analysis based on the evolution of contact angles and the resulting spreading dynamics of droplets, the contact radii in two directions are found equal to each other, indicating a circular contact area at voltages smaller than 60 V_{DC}. The further increase of the voltage applied decreases the contact angle and exerts a stronger electro-capillary force on the contact line. This drags the droplet to extend along the direction parallel to the striped electrodes. However, as the applied voltage alters the solid-liquid interfacial tension, the resulting comb-like corrugated free-energy landscape on the solid-liquid interface behaves like virtual free-energy barriers of a comb-like shape that hinder the motion of droplets in the direction perpendicular to the striped electrodes. Thus, for applied voltage larger than 60 V_{DC}, the free-energy barriers become strong enough that the droplet mostly spreads along these barriers. In this condition, the anisotropic spreading of droplets elongates the contact area to approximately an elliptical shape. We further calculated the aspect ratio of the elliptical contact area and present the results in Fig. 3. The increase of aspect ratio is smaller than 5% when applied voltage increases from 0 to 60 V_{DC}, but grows fast for higher voltages as the droplet is elongated by the corrugated electric field. It has been proposed that on structurally corrugated substrates, the structures may also guide the liquid to spread along the grooves [11].

IV. CONCLUSION

To quantify the spreading anisotropy of droplets on striped electrodes, we report the experimental measurements of contact angles and contact radii in directions parallel and perpendicular to the electrodes. The results propose a gradual evolution of the solid-liquid contact area from circular to elliptical. This quantitative study provides a viable alternative for droplet elongation in practical applications.

ACKNOWLEDGMENT

We present our thanks to Kaidi Zhang and Qi Wang for their help in the fabrication and test of electrowetting chips.

REFERENCES

[1] D. Bonn, J. Eggers, J. Indekeu, J. Meunier, and E. Rolley, "Wetting and spreading." Review of Modern Physics, vol. 81, pp. 739-805, 2009.

[2] Y. Zhong, Z. Dong, C. Li, H. Du, N. X. Fang, L. Wu, and Y. Song, "Continuous 3D printing from one single droplet." Nature Communications, vol. 11, pp. 4685, 2020.

[3] R. Malinowski, I. P. Parkin, and G. Volpe, "Advances towards programmable droplet transport on solid surfaces and its applications." Chemical Society Reviews, vol. 49, pp. 7879-7892, 2020.

[4] Y. Lu, S. Sathasivam, J. Song, C. R. Crick, C. J. Carmalt, and I. P. Parkin, "Robust self-cleaning surfaces that function when exposed to either air or oil." Science, vol. 347, pp. 1132-1135, 2015.

[5] W. Xu, H. Zheng, Y. Liu, X. Zhou, C. Zhang, Y. Song, X. Deng, M. Leung, Z. Yang, R. X. Xu, Z. L. Wang, X. C. Zeng, and Z. Wang, "A droplet-based electricity generator with high instantaneous power density." Nature, vol. 578, pp. 392-396, 2020.

[6] T. Young, "An essay on the cohesion of fluids." Philosophical Transactions of the Royal Society of London Series I, vol. 95, pp. 65-87, 1805.

[7] V. G. Damie, and K. Rykaczewski, "Nano-striped chemically anisotropic surfaces have near isotropic wettability." Applied Physics Letters, vol. 110, pp. 171603, 2017.

[8] D. Wang, Q. Sun, M. J. Hokkanen, C. Zhang, F. -Y. Lin, Q. Liu, S. -P. Zhu, T. Zhou, Q. Chang, B. He, Q. Zhou, L. Chen, Z. Wang, R. H. A. Ras, and X. Deng, "Design of robust superhydrophobic surfaces." Nature, vol. 582, pp. 55-59, 2020.

[9] G. Lippmann, "Relations entre les phénomènes électriques et capillaires." Annales de Chimie et de Physique, vol. 5, pp. 494-549, 1875.

[10] F. Mugele, and J. C. Baret, "Electrowetting: From basics to applications." Journal of Physics: Condensed Matter, vol. 17, pp. R705-R774, 2005.

[11] Z. Wang, and Y. -P. Zhao, "Wetting and electrowetting on corrugated substrates." Physics of Fluids, vol. 29, pp. 067101, 2017.

[12] G. McHale, C. V. Brown, M. I. Newton, G. G. Wells, and N. Sampara, "Dielectrowetting driven spreading of droplets." Physical Review Letters, vol. 107, pp. 186101, 2011.

[13] W. Wang, Q. Wang, K. Zhang, X. Wang, A. Riaud, and J. Zhou, "On-demand contact line pinning during droplet evaporation." Sensors and Actuators B: Chemical, vol. 312, pp. 127983, 2020.

[14] V. S. Nikolayev, and D. A. Beysens, "Relaxation of nonspherical sessile drops towards equilibrium." Physical Review E, vol. 65, 046135, 2002.

[15] H. B. Eral, D. J. C. M. 't Mannetje, and J. M. Oh, "Contact angle hysteresis: a review of fundamentals and applications." Colloid and Polymer Science, vol. 291, pp. 247-260, 2013.

Proceedings of the 16th Annual IEEE International
Conference on Nano/Micro Engineered and Molecular Systems
April 25-29, 2021

Low-cost Micro Search-Coil Magnetic Sensor with Self Calibration for the Internet of Things

Hadi Tavakkoli, Izhar, Xu Zhao, Yi-Kuen Lee

Abstract— **In this paper, a low-cost search-coil magnetic field sensor (SCMS) with an integrated self-calibration is presented. The sensor consists of a pair of micro coils (a sensing coil and a coil for self-calibration). A theoretical model and FEM model were developed for SCMS's systematic design analysis with different key parameters. We found that the normalized SNR as a function of track width w_c has two operation regimes: noise dominant and sensitivity dominant and the existence of the critical track width (w_c) ranging from 50µm to 100 µm to achieve optimal SNR. The optimized micro SCMS, with the calculated noise equivalent magnetic induction of 250nT/Hz$^{0.5}$ and the dynamic range of ±26mT at 50 Hz with amplification gain of 100 and ±5 V power supply, is successfully fabricated and tested using a low-cost four-layer Printed Circuit Board (PCB). There is a good agreement among the theoretical predictions, FEM simulations, and experimental results. The low-cost micro SCMS is promising for energy monitoring of large-scale data centers in the era of the Internet of Things.**

I. INTRODUCTION

Various magnetic sensors have been implemented by different transduction principles, such as Hall devices, Giant Magneto-Resistance (GMR), MEMS Lorentz force, Anisotropic Magneto-Resistive (AMR), optical fiber, fluxgate, search coil and Superconducting Quantum Interference Device (SQUID) [1]. The Search coil magnetic sensors (SCMS) have the highest input dynamic range (10^{-15} ~ 1 Tesla) [1] and can be self-powered. In the literature, several SCMSs are reported [2] [3]. These SCMSs are usually realized by wrapping a long wire around the core to achieve a large number of turns [4]. Since these sensors are solenoid type, therefore they are bulky (e.g. 10cm long and 17mm diameter [2]) and difficult to employ in micromachined applications. Besides, they cannot be realized using CMOS MEMS fabrication technology.

The size of SCMS can be reduced by using a planar spiral inductor coil design [5]. Besides, low-cost batch fabrication technologies (PCB, CMOS/CMOS MEMS foundry processes) can be used to realize these sensors and produce significant cost savings. To date, very limited research has been reported on micro fabricated SCMS [6]. They presented micro-coil with the die size of

10.7mm×10.7mm and the coil size of 3.16mm×3.16mm, which can be used for a high magnetic field greater than 1 mT measurement. In this work, we conducted the theoretical modeling and FEM simulations for the design optimization of SCMS's key performance design parameters (sensitivity, noise, SNR, track width, thickness, spacing, outer and inner diameters).

Furthermore, we successfully designed and fabricated a low-cost PCB-based self-calibrated SCMS with a planar coil design. Moreover, an on-chip characterization setup was fabricated on the same chip to achieve the self-calibration function which is important for the Internet of Things application.

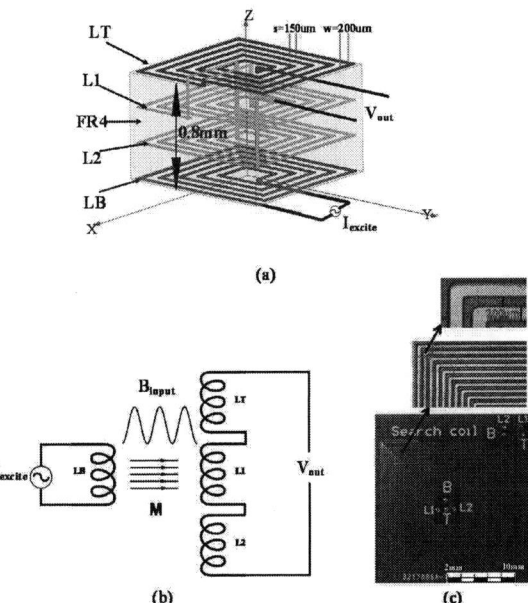

Figure 1.(a) The 3D schematic diagram for the Search Coil Magnetic Field Sensor (SCMS) with self-calibration, the inductor at the bottom layer (L_B) is used for excitation and the other three micro coils are employed for sensing. This structure is used for the self-calibration, (b)the equivalent circuit of the fabricated SCMS, (c)the fabricated SCMS using low-cost 4 layers PCB fabrication process .

Hadi Tavakkoli, Izhar, Xu Zhao are with the Department of Mechanical and Aerospace Engineering, The Hong Kong University of Science and Technology, Hong Kong SAR (e-mail: htavakkoli@connect.ust.hk).

*Corresponding author: Yi-Kuen Lee is with the Department of Mechanical and Aerospace Engineering, The Hong Kong University of

Science and Technology, Hong Kong SAR and with Guangzhou HKUST Fok Ying Tung Reseach Institute (e-mail: meyklee@ust.hk).

978-1-6654-3008-1/21 $31.00 © 2021 IEEE

II. DESIGN PROCEDURE

A. Working principle

The working principle of the 4-layer self-calibrated micro SCMS is based on the Faraday induction law that is demonstrated in Figure. 1. When an air core inductor experiences an alternating magnetic field, the voltage (V_{out}) will be induced on the inductor, which can be calculated by Eqn. (1) (for the case of constant coil area and uniform magnetic field):

$$V_{out} = -\frac{\partial \varphi}{\partial t} = -\frac{\partial (A_{eff}B)}{\partial t} = -A_{eff}\frac{\partial B}{\partial t} \qquad (1)$$

where φ is the magnetic flux, t is time, A_{eff} is an effective area that experiences the alternating magnetic field, and B is magnetic field density normal to the surface of the inductor. Although by adding a high permeability core, the sensitivity will increase, the sensor may suffer from the saturation field, hysteresis, and nonlinearity of the core. If the input magnetic field is in the form of sinusoidal waveform ($B=B_{max}Sin(2\pi ft)$, where f is frequency), the output voltage can be calculated by (2):

$$V_{out} = 2\pi f A_{eff} B_{max} Cos(2\pi ft) \qquad (2)$$

If the SCMS is realized by the solenoid-type coil, the effective area can be calculated by multiplying the single turn area by total number of turns approximately (correction factor is required when there is multilayer stacking [7]). The spiral inductors which are shown in Figure 1(a) can be approximated by concentric closed areas [8] and the effective area can be calculated by summing up the area of all turns.

The presented SCMS is realized using low-cost 4-layer PCB fabrication processes, the 3D model is illustrated in Figure 1(a), and it consists of four concentric stacked planar inductors. Each layer includes one planar inductor, and the bottom inductor (LB) is utilized to generate a magnetic field for on-chip calibration. The remaining three inductors (L_1, L_2 and L_T) are connected in an in-phase series configuration to measure the magnetic field. The electrical model is shown in Figure 1b, and the fabricated SCMS is displayed in Figure 1(c). If an exciting current (I_{excite}) is applied to the L_B, the induced voltage on the sensing coil can be calculated by Eqn. (3) [8].

$$V_{out} = -M\frac{dI_{excite}}{dt} \qquad (3)$$

where M is the mutual inductance between the exciting and sense coils. If the input current to the excite coil, and the input uniform magnetic field are both in sinusoidal waveform ($B=B_{max}Sin(2\pi ft)$ and $I_{excite}=i_{max}Sin(2\pi ft)$), the maximum output voltage ($V_{out,max}$) can be calculated and measured. Therefore, From the analysis of Eqns. (1) and (2) the equivalent maximum uniform magnetic field can be determined as follows:

$$B_{max} = \frac{Mi_{max}}{A_{eff}} \qquad (4)$$

Eqn. (4) can be used for the self-calibration structure. By applying a known exciting current and calculation of the mutual inductance between the excite coil and the sense coil, and the effective area, the output voltage at a certain uniform magnetic field can be calculated. The mutual inductance the excite coil, and the sense coil is calculated from FEM simulation. The input-output response can be obtained by substituting Eqn. (4) into Eqn. (1). The output voltage is measured at different input excitation currents at 1 kHz frequency. The calculation, simulation, and measurement results for normalized output voltage with respect to frequency are illustrated in Figure 2. The 3D FEM simulations were performed with the quasistatic model, using ANSYS Maxwell (Ansys Inc., Pennsylvania, USA),which is in the flat frequency band of the inductor. The self and mutual inductances are independent of frequency in quasistatic model [9].

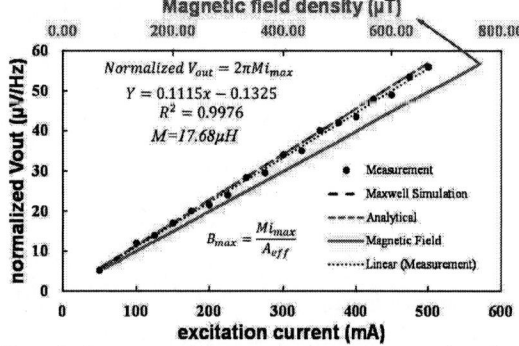

Figure 2. Comparison of the normalized sensor output voltage from the theory, numerical simulations and experiments of the fabricated sensor at Iexcite =50mA~500mA and fexcite = 1kHz. The induced output voltage is utilized to calculate the sensitivity of the sensor by the proposed on-chip characterization structure. The equivalent uniform magnetic field for the sensitivity calculation is obtained by the mutual inductance value.

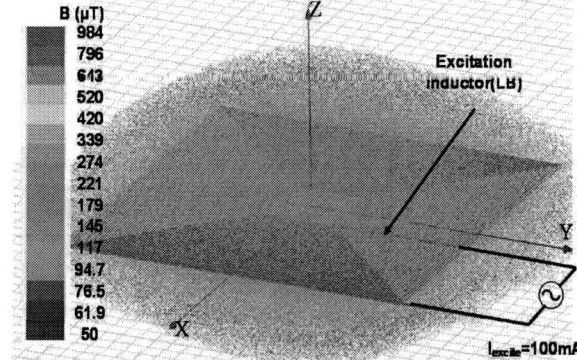

Figure 3. The predicted magnetic field density from the numerical simulations around the excitation coil at 100mA.

978-1-6654-3008-1/21 $31.00 © 2021 IEEE

The quasistatic solution is valid as along as the maximum physical dimensions (L) of the electrical components are much smaller than the wavelength. The wavelength at 50Hz is 6×10^6 meters, much larger than the length of the fabricated coil ($L< 1m$). In Figure 2 the excite frequency is constant (1 kHz), and the input current to the excite coil is swept from 50mA~500mA. The input-output response of the SCMS is calculated by Eqns. (1) ~ (4). The noise of the senor is measured by hp dynamic signal analyzer 35665 and the calculated noise equivalent magnetic induction of the fabricated SCMS on 4-layer PCB is 12.5nT/Hz $^{0.5}$ at 1 kHz.

The predicted magnetic field profile from the FEM model around the excite coil at 100mA current is depicted in Figure 3. Note that the generated magnetic field around the excite coil is independent of the excitation frequency for the low frequency applications, which meet the assumption of quasistatic model. The magnetic field around the excitation coil is non-uniform, and the equivalent uniform magnetic field that can generate the same output voltage on the sense coil is almost 136 μT.

B. Signal to Noise Ratio (SNR) calculation and optimization

At low frequencies, the inductor can be approximated by an ideal inductor, series resistance, and shunt capacitor on the isolating substrate[5]. If the spiral inductor is fabricated on a semiconductor substrate such as silicon, the eddy current loss has to be considered and the electrical model must include the substrate capacitance and the resistance [10]. The electrical model of the sensing coil at very low frequencies where the shunt capacitance can be overlooked is shown in Figure 4. The thermal noise which originates from the DC resistance of the coil has to be considered in the SNR calculation[11].

In the design of SCMS, sensitivity and the thermal noise are correlated. For the solenoid type, by increasing the number of turns, the sensitivity also increases. On the other hand, the resistance of the coil increases due to the length increment of the coil, which leads to an increase of thermal noise. The sensitivity enhancement is preferred, but the noise will degrade the sensor performance. Therefore, the

optimized dimensions should be calculated for maximum SNR. Furthermore, the size of the SCMS should be considered from application and cost perspective.

The spiral inductor utilized in this work can be described by track width, track spacing, inner and outer diameters. In the fixed outer diameter of the spiral inductor, by decreasing the track width and spacing, the number of turns increases. As a result, the effective area of the sensor increases. Therefore, the sensitivity will enhance according to (1). On the other hand, the length of the coil increases, and the track width decreases, which leads to an increase in the DC resistance of the inductor.

By increasing the track width, the thermal noise will decrease, and it tends to reach zero in an extreme case, but after a certain value the thermal noise reduction is not helpful for SNR improvement of the sensor, since the input-referred noise of the amplifier becomes the dominant noise source. Therefore, the input-referred noise of the amplifier must also be included in the noise calculation. The noise of the sensor can be calculated by (5):

$$V_n = \sqrt{thermal\ noise^2 + amplifier\ noise^2} \tag{5}$$

The normalized SNR_n is defined by (6):

$$SNR_n = \frac{normalized\ sensitivity}{V_n} = \frac{\frac{V_{out}}{B_{max} \times f}}{V_n} \tag{6}$$

By considering the readout amplifier noise, the critical track widths for maximum SNR can be calculated. w_{ci} ($i=1\sim3$) is calculated for common dimensions in the PBC fabrication technology, as shown in Figure 5. It can be seen that on the left side of the critical track width, SNR is enhanced by increasing the track width. This is due to a reduction in thermal noise generated by the coil resistance. On the right side of the critical width, the sensitivity decreases, and the dominant factor in the SNR changes from noise reduction to sensitivity reduction, and the SNR starts to drop as shown in Figure 5.

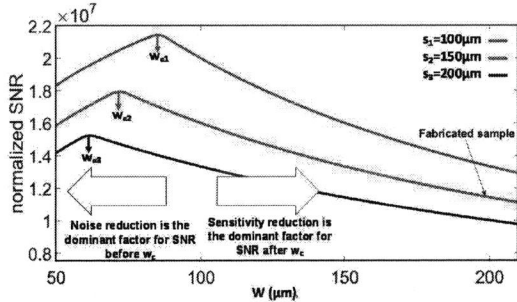

Figure 4. The calculated SNR as a function of the track width at different track spacing with an outer diameter of 2cm in the range of minimum dimension fabricated by PCB technology. By increasing the track width, the total noise will decrease, but after the critical dimension the dominant noise source is the amplifier noise, and the SNR starts to decrease due to the sensitivity reduction.

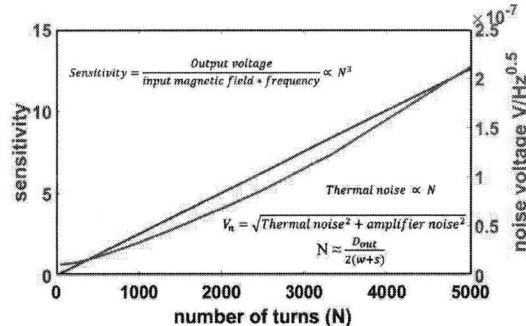

Figure 5. The Sensitivity and the noise voltage calculation of the SCMS. Considering amplifier noise as a function of the number of turns for the outer diameter (Dout) of 2cm, minimum feature size, and track thickness of 1um for 3 stacked spiral inductors.

To design an optimized SCMS, track spacing (s), track width (w), inner diameter (D_{in}), and outer diameter (D_{out}) of the spiral inductor should be optimized to achieve maximum SNR. The number of turns (N) is a function of coil parameters and can be calculated by (7):

$$N = \frac{D_{out} - D_{in} + 2s}{2(w + s)} \qquad (7)$$

The inner diameter and the track spacing are usually much smaller than the outer diameter of the spiral inductor. Therefore, the number of turns can be approximated by (8):

$$N \approx \frac{D_{out}}{2(w + s)} \qquad (8)$$

It is calculated that the number of turns (N), which includes all the spiral inductor dimensional parameters, is the dominant parameter in the SCMS design and the sensitivity and sensor noise are proportional to N^3 and N, respectively. This is depicted in Figure 6. Therefore, the track spacing which has no effect on the resistance of the inductor at low frequencies, should be set to the minimum value allowed by the fabrication limitations. Although decreasing the track spacing will increase the coupling capacitance between the tracks, the coupling capacitance can be ignored here, since the application of the proposed SCMS is for low frequencies.

Furthermore, the normalized SNR calculation is illustrated in Figure 7 (D_{out} = 2cm, and minimum feature size $\geq 1\mu$m). The dominant noise source changes around the critical point from the amplifier noise to the sensor's thermal noise. At the right side of the critical point (by increasing the number of turns), the DC resistance of the coil increases dramatically due to the track width reduction and the coil length increment. Therefore, by increasing the number of turns, the SNR starts to drop. On the left side of the critical point, the thermal noise of the coil can be ignored compared to the amplifier noise, since the DC resistance decreases substantially. Hence, the amplifier noise becomes the dominant noise source in this region.

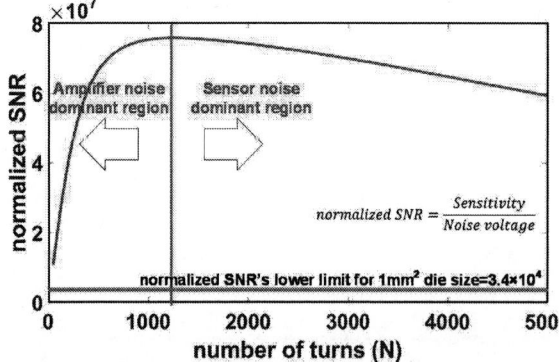

Figure 6. The SNR calculation of the SCMS with minimum feature size and track thickness of 1um and Dout=2cm. The dominant noise source changes after the critical dimension from the amplifier noise to the sensor noise. By utilizing microfabrication technology, a substantial improvement in the sensitivity and SNR can be achieved.

On the other hand, N is proportional to the outer diameter (D_{out}) of the coil at fixed w and s, so the sensor sensitivity and noise are proportional to the dominant length scale D_{out}^3 and D_{out}, respectively. In addition, by comparing Figure 5 and Figure 7, it can be concluded that by utilizing microfabrication technology, higher SNR can be achieved. Besides, the SCMS size can be reduced to 1mm², typical MEMS sensors' length scale. The sensitivity of our PCB-based SCMS can be easily improved based on a parametric analysis, optimization, and integration of the on-chip ASIC circuit by using CMOS MEMS technology. Furthermore, while other sensors use bulky setups for characterization, the design of self-calibrated SCMS will be useful for IoT applications, especially in large-scale data centers. The fabricated sensor's calculated noise equivalent magnetic induction with amplification gain of 100 and ±5 V power supply at 50 Hz is 250nT/Hz$^{0.5}$ with a dynamic range of ±26mT similar to the MLX91205 Hall sensor (Melexis Technologies, Belgium) can be used for smart homes and for monitoring energy use in data centers [12].

III. CONCLUSION

The theoretical modeling and FEM simulations of the SCMS were developed for systematic design analysis for spiral coil configuration, and a low-cost optimized SCMS with the size of 2.073cm×2.073cm was successfully fabricated using low-cost 4-layer PCB. The fabricated low-cost magnetometer on PCB can reach the sensitivity of commercially available Hall sensors at low frequencies. Although the fabricated magnetometer is relatively large compared to the commercially available CMOS Hall sensors, its merit is that the PCB realization of the sensor is much cheaper than the CMOS process. The SNR as a function of track width shows two operation regimes (noise dominant and sensitivity dominant) and the existence of the critical track width for maximum SNR. By decreasing the track width and the spacing, higher SNR can be achieved. CMOS MEMS technology will be used to develop SCMS integrated with on-chip electronics in near future works for energy monitoring of large-scale data centers in the era of the Internet of Things.

ACKNOWLEDGMENT
This work is partially supported by Hong Kong Research Grants Council (No. 16201220) and partially by Guangzhou Science and Technology Committee (No. 201807010045).

REFERENCES

[1] A. L. Herrera-May, L. A. Aguilera-Cortés, P. J. García-Ramírez, and E. Manjarrez, "Resonant magnetic field sensors based on MEMS technology," *Sensors*, vol. 9, no. 10, pp. 7785–7813, 2009.

[2] C. Coillot, J. Moutoussamy, R. Lebourgeois, S. Ruocco, and G. Chanteur, "Principle and performance of a dual-band search coil magnetometer: A new instrument to investigate fluctuating magnetic fields in space," *IEEE Sens. J.*, vol. 10, no. 2, pp. 255–260, 2010.

[3] F. Han, S. Harada, and I. Sasada, "Fluxgate and search coil hybrid: A low-noise wide-band magnetometer," *IEEE Trans. Magn.*, vol.

48, no. 11, pp. 3700–3703, 2012.

[4] A. Grosz, E. Paperno, S. Amrusi, and B. Zadov, "A three-axial search coil magnetometer optimized for small size, low power, and low frequencies," *IEEE Sens. J.*, vol. 11, no. 4, pp. 1088–1094, 2011.

[5] C. Massin, G. Boero, F. Vincent, J. Abenhaim, P. A. Besse, and R. S. Popovic, "High-Q factor RF planar microcoils for micro-scale NMR spectroscopy," *Sensors Actuators, A Phys.*, vol. 97–98, no. November 2001, pp. 280–288, 2002.

[6] T. M. A. B. T. Azmi and N. Bin Sulaiman, "Simulation Analysis of Micro Coils for the Fabrication of Micro-scaled Search Coil Magnetometer," *Proc. - 6th Int. Conf. Comput. Commun. Eng. Innov. Technol. to Serve Humanit. ICCCE 2016*, pp. 183–186, 2016.

[7] S. Tumanski, "Induction coil sensors - A review," *Meas. Sci. Technol.*, vol. 18, no. 3, 2007.

[8] H. Tavakkoli, E. Abbaspour-Sani, A. Khalilzadegan, A. M. Abazari, and G. Rezazadeh, "Mutual inductance calculation between two coaxial planar spiral coils with an arbitrary number of sides," *Microelectronics J.*, vol. 85, no. December 2018, pp. 98–108, 2019.

[9] C. R. Paul, *Inductance: Loop and Partial.* 2009.

[10] C. P. Yue, C. Ryu, J. Lau, T. H. Lee, and S. S. Wong, "Physical model for planar spiral inductors on silicon," *Tech. Dig. - Int. Electron Devices Meet.*, vol. 94305, pp. 155–158, 1996.

[11] A. Nourmohammadi, S. M. H. Feiz, and M. H. Asteraki, "Investigation of Noise Reduction and SNR Enhancement in Search Coil Magnetometers at Low Frequencies," *physics.ins-det*, no. 2, pp. 1–6, 2014.

[12] H. Shoukourian, T. Wilde, A. Auweter, and A. Bode, "Monitoring Power Data: A first step towards a unified energy efficiency evaluation toolset for HPC data centers," *Environ. Model. Softw.*, vol. 56, pp. 13–26, 2014.

978-1-6654-3008-1/21 $31.00 © 2021 IEEE

April 25-29 , 2021   Xiamen, China

Nano Size-effect Enhanced Sensitivity of Gas Detection

Xinxin Li

State Key Lab of Transducer Technology, Shanghai Institute of Microsystem and Information Technology,
Chinese Academy of Sciences, Shanghai 200050, China

ABSTRACT

Nano size-effect is the soul of nanoscience and technology, where the feature nano-scale dimension dominates the properties of the nanomaterial. Recently more and more functionalized nanomaterials are proposed and developed for biochemical sensing applications. Beyond the increase in specific surface area, nanostructure can exhibit more significant property change when the solid geometry is shrunk down to nano-scale. Along with the size decrease, the atom/molecule number at surface is comparable to that in the body and the former may be dominant in surface activity. The author's group have worked on the topic for many years and quite a lot of nano size-effects on sensitivity of biochemical nanosensors have been found experimentally and analytically revealed by using quantitative characterization of thermodynamic/kinetic parameters, which reflect the essential of phenomenal behavior of the sensing nanomaterial. The presentation will introduce a series of nanosensors developed in the group by using ZnO nanowires.

BIOGRAPHY

Prof. Xinxin Li received his BS degree from Tsinghua University, PhD degree from Fudan University. He ever worked in HKUST, Nanyang Technological University of Singapore, Tohoku University of Japan. In MEMS and transducer technology field, Prof. Xinxin Li is a senior scientist. He is now serving as a professor and Director of the State Key Lab of Transducers Technology, Shanghai Institute of Microsystem and Information Technology, Chinese Academy of Sciences. He received National Science Fund for Distinguished Youth Scholars in 2007.

He has authored/co-authored more than 300 SCI journal papers. He became the first IEEE MEMS TPC member from mainland China and totally served the committee for four years. From 2011 to 2019, he also served as the only International Steering Committee member from mainland China for the Transducers conference and he has authored/co-authored 72 papers there (including an invited presentation in 2009 at Denver, USA). He has invented more than 100 patents on MEMS and sensor technology, including more than 30 licensed to industrial field for volume production. He is now serving as associate editor or editorial member for the SCI journals of J Micromech Microeng, Scientific Reports, Microsyst & Nanoeng, and, Micromachines.

This page intentionally left blank.

Proceedings of the 16th Annual IEEE International
Conference on Nano/Micro Engineered and Molecular Systems
April 25-29, 2021

Scalable Synthesis of SnO_2 Nanosheet Arrays on Chips for Ultralow Concentration NO_2 Detection*

Gaoqiang Niu, Changhui Zhao, and Fei Wang*, Senior Member, IEEE

Abstract— **This work reports a chemiresistive type of NO_2 sensor based on patterned SnO_2 nanosheet arrays (NSAs), which are in-situ synthesized on MEMS chips via a scalable one-step homogeneous precipitation approach. Dry etching of SnO_2 NSAs is studied by using an inductively couple plasma (ICP) process. As-grown SnO_2 nanosheets demonstrate ultrathin thickness with cross-linked networks. The SnO_2 NSAs based sensor exhibits ultrasensitive (150.9, at 10 ppm) to NO_2 gas at 150 °C with a wide concentration range from 30 ppb to 50 ppm. Especially, this sensor also shows liner relationships between sensor response and NO_2 concentration within the range of 30-1000 ppb and 2-50 ppm, respectively, indicating a promising route for detection of ultralow concentration of NO_2. Our results provide a new strategy toward in-situ wafer-level fabrication of SnO_2 NSAs based micro gas sensing devices.**

I. INTRODUCTION

As a pungent, toxic, corrosive gas, NO_2 is a primary contributor to air pollution and acid rain. Especially, human health will be adversely affected by long-term inhalation of ultralow concentration of NO_2. Over the past decades, nanostructured metal oxide semiconductor (MOS) based chemiresistive gas sensors have been intensively studied for the purpose of promoting their NO_2 sensing performances (sensitivity, selectivity, response/recovery times, and stability) [1, 2]. Due to the corrosive nature of NO_2 gas, it is still a challenge to develop NO_2 sensor with high sensitivity and good stability. Recently, on-chip growth of SnO_2 nanosheet arrays (NSAs) has attracted more attention to realize ultrasensitive detection of trace gases, in some cases, even at sub-ppm levels [3-6]. In addition, it is also urgently needed to explore a facile and scalable approach to fabricate wafer-level sensing films based on nanoarrays.

In term of the wafer-level fabrication process of MEMS gas sensors, the patterning of the SnO_2 film on the MEMS chips is not mature. Previously, a lift-off technique has been proposed to pattern SnO_2 sol-gel films because of the relatively weak adhesion strength between SnO_2 films and SiO_2 sacrificial layer instead of Si_3N_4 substrate [7]. Considering the high adhesion of SnO_2 films on the substrate prepared by in-situ growth, sputtering or chemical-vapor deposition, the above patterning process is not applicable. Another problem of patterning SnO_2 layer is the incompatibility between lithography processes and the synthesis of SnO_2 film [8-10]. The spin coating of photoresists carried out in lithography process may destroy

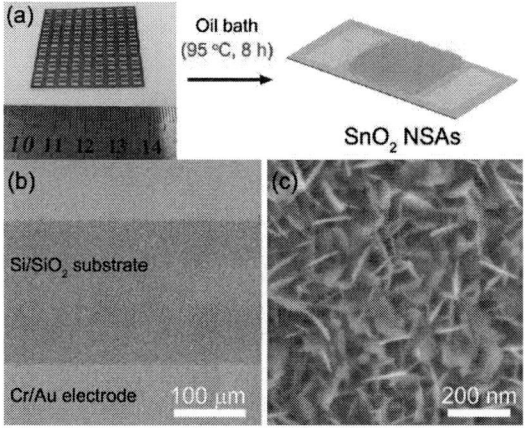

Fig. 1. (a) Digital image of MEMS chips and schematic illustration of the SnO_2 NSAs. (b) Low- and (c) high-magnification SEM images of as-prepared SnO_2 NSAs.

the uniformity and porous structure of SnO_2 films. Consequently, patterned fabrication process of SnO_2 NSAs on the MEMS chip is urgently needed.

In this work, we present a NO_2 sensor based on pure SnO_2 NSAs using a homogeneous precipitation method, which shows ultrasensitive to NO_2 gas with a detection limit down to 30 ppb. An inductively couple plasma (ICP) etching process is explored to pattern the SnO_2 NSAs, indicating the good compatibility of on-chip growth of SnO_2 NSAs with MEMS technology.

II. EXPERIMENT

A. Synthesis and Patterning of SnO_2 NSAs on Chips

2 μm SiO_2 was thermally grown on a 4-inch silicon wafer to prepare the insulation layer. Afterwards, Cr/Au (10/100 nm) were deposited and patterned with a lift-off process to fabricate the interdigital electrodes, as shown in Fig. 1a. The wafer was immersed into 100 mL deionized water containing urea (0.04 M) and $SnCl_2 \cdot 2H_2O$ (0.03 M). After maintaining at 95 °C for 8 h, the wafer surface was covered with SnO_2 NSAs.

Fig.2a shows the patterning fabrication process of SnO_2 NSAs. SnO_2 layer was patterned using standard lithograph techniques, and the remaining photoresist served as a mask to protect the patterned SnO_2 NSAs from the ICP etching.

This work was supported in part by The National Key Research and Development Program of China under Grant 2020YFB2008604, in part by the Shenzhen Science and Technology Innovation Committee under Grant JCYJ20170412154426330.

G. Niu and C. Zhao are with the School of Microelectronics, Southern University of Science and Technology, Shenzhen 518055, China.
*Contacting Author: F. Wang is with the School of Microelectronics, Southern University of Science and Technology, Shenzhen 518055, China (phone: 86-755-88018509; e-mail: wangf@sustech.edu.cn).

978-1-6654-3008-1/21 $31.00 © 2021 IEEE

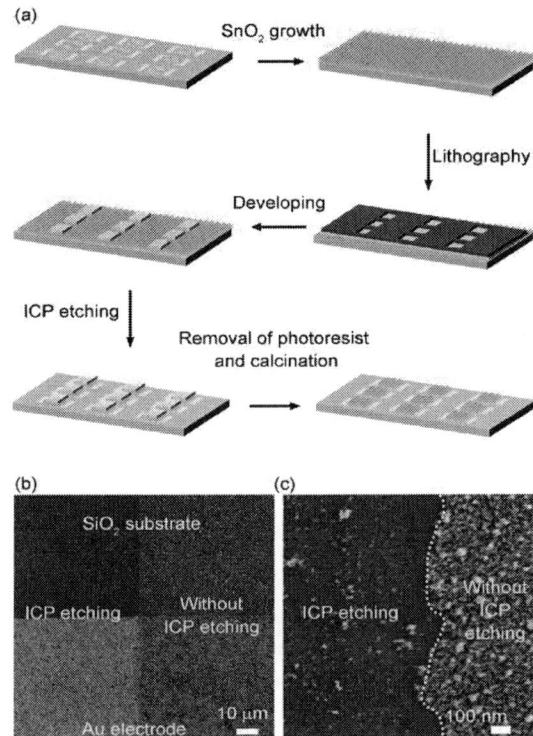

Fig.2. (a) The fabrication process of the patterned SnO$_2$ nanosheet arrays based sensor. (b) Low- and (c) high-magnification SEM images of patterned SnO$_2$ NSAs.

Then, the uncovered SnO$_2$ NSAs were etched by the etching gases (SF$_6$ and BCl$_3$). Remaining photoresist could be easily removed by acetone, and the wafer was annealed at 400 °C for 2 h in air to obtain SnO$_2$ NSAs. Finally, as-prepared wafer was diced into small chips for gas sensing measurements. The synthesized chips were characterized by field emission scanning electron microscope (FESEM, Gemini 300, Carl Zeiss).

B. Gas-Sensing Measurements

Fig. 3 shows digital image of the CGS-4TPs gas-sensing measurement system. When a certain volume of NO$_2$ (19900 ppm, in nitrogen) was injected into a 15 L chamber, the resistance of the sensor was recorded in real time. All sensors were measured on a hotplate with controlled operation temperature. The environmental relative humidity is about 24%.

III. RESULT AND DISCUSSION

A. Morphological Characteristics

Figs. 1b and 1c show the SEM images of the SnO$_2$ NSAs, revealing that the vertically-aligned ultrathin nanosheets (thickness <10 nm) can deposit directly onto the MEMS chips. The magnified image (Fig. 1c) confirms the surface morphology of the cross-linked networks and some semi-open spaces among these nanosheets. Such a structure provides a large portion of exposed surface area, which can

Fig. 3. CGS-4TPs gas-sensing measurement system, the inset shows the photo of test platform.

facilitate gas adsorption and promote the sensitivity to low concentration of target gas.

Figs. 2b and 2c display the SEM images of the SnO$_2$ NSAs with or without ICP etching. Obviously, it can be seen that SnO$_2$ NSAs can be etched from the SiO$_2$ substrate and the interdigital Au electrode. A detailed observation of the chip also demonstrates that the SnO$_2$ NSAs maintain the original morphology. This result indicates that adhesion between SnO$_2$ and substrate is strong enough to ensure that the NSAs will not be damaged.

B. NO$_2$ Sensing Properties

Transient resistance curve of SnO$_2$ NSAs based sensor to various concentrations of NO$_2$ at 150 °C is displayed in Fig. 4. Once the sensor is exposed to NO$_2$ gas, the resistance of the sensor increases rapidly and recovers to its initial value after exposing to air. With an increase of NO$_2$ concentration, the sensor resistance also increases monotonically, which means the increase of response.

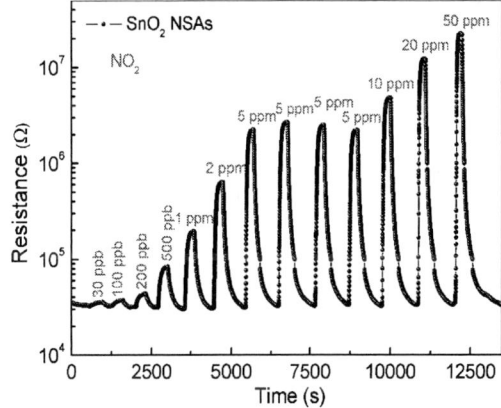

Fig. 4. Transient resistance curve of SnO$_2$ NSAs based sensor to various NO$_2$ concentrations at 150 °C

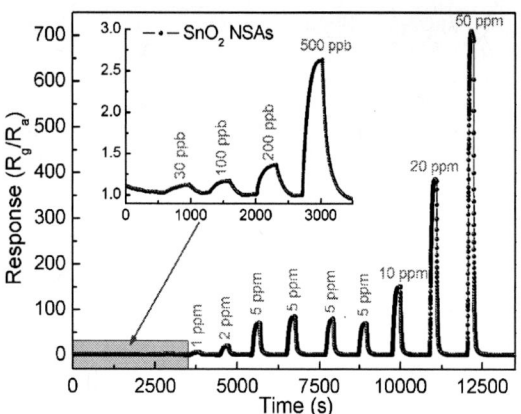

Fig. 5. Response curve of SnO$_2$ NSAs to different concentrations of NO$_2$ measured at 150 °C (inset shows the low concentration level)

The sensor response is defined as the ratio of sensor resistance in NO$_2$ (Rg) and in air (Ra). Fig. 5 indicates that the SnO$_2$ NSAs based sensor achieves high responses to NO$_2$ at 150 °C. The response value (Rg/Ra) is as high as 150.9 to 10 ppm NO$_2$. Especially, this sensor exhibits a remarkable response to ultralow concentration of NO$_2$, with an experimental limit of detection (LOD) of 30 ppb (inset of Fig. 5).

Reproducibility of SnO$_2$ NSAs based sensor is also investigated in Fig. 4 and Fig. 5. SnO$_2$ NSAs based sensor was repeatedly measured four times to 5 ppm NO$_2$. And transient resistance and response curves repeat periodically with NO$_2$ concentration, suggesting an excellent reproducibility of the SnO$_2$ NSAs based sensor.

Moreover, Fig. 6 further confirms the liner relationship between sensor response and NO$_2$ concentration within the range of 30-1000 ppb. Similarly, the SnO$_2$ NSAs also

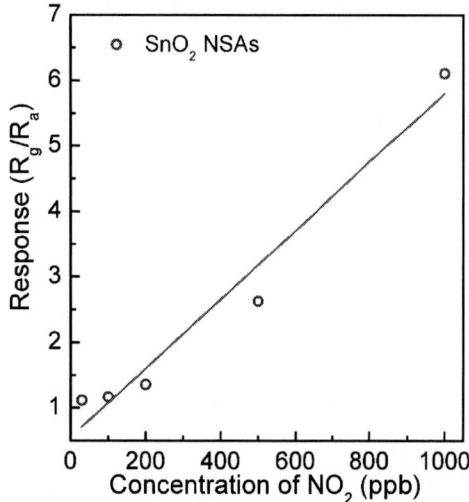

Fig. 6. Relationship between the SnO$_2$ NSAs sensor response and NO$_2$ concentration (30-1000 ppb).

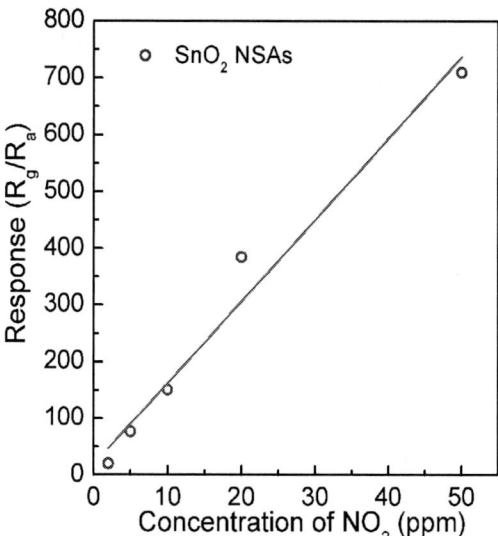

Fig. 7. Relationship between the SnO$_2$ NSAs sensor response and NO$_2$ concentration (2-50 ppm).

present an excellent liner relationship at high NO$_2$ concentrations (2-50 ppm), as shown in Fig. 7, which demonstrates the feasibility of the sensor for real application.

C. NO$_2$ Sensing Mechanism

As a typical n-type MOS based sensitive material, the NO$_2$ sensing mechanism of SnO$_2$ is generally explained as the reaction of NO$_2$ with electrons on the surface of SnO$_2$, resulting in the resistance increase of the sensor [11]. In air, oxygen molecules will adsorb on the surface of SnO$_2$ nanosheets, and capture free electrons from the conduction band of SnO$_2$ to form oxygen species. It is commonly accepted that the chemisorbed oxygen species (O$_2^-$, O$^-$, and O^{2-}) depend on the operation temperature. Herein, the operation temperature is 150 °C, and O$^-$ is the dominant species. At the same time, an electron depletion layer is generated on the surface of SnO$_2$. When the SnO$_2$ NSAs based sensor is exposed to NO$_2$ gas, the NO$_2$ serves as a stronger oxidant than oxygen and captures more electrons form SnO$_2$ nanosheets, forming to a thicker electron depletion layer. Consequently, the resistance of the sensor will further increase in NO$_2$ atmosphere. Due to the large exposed surface area of SnO$_2$ NSAs on the chip, the sensor enables to achieve high sensitivity to ultralow concentration of NO$_2$.

IV. CONCLUSION

A highly sensitive NO$_2$ sensor based on SnO$_2$ NSAs has been fabricated by combining one-step homogeneous precipitation method and MEMS technique. The as-grown SnO$_2$ NSAs can be patterned via lithography process and ICP dry etching. The robust of SnO$_2$ NSAs on the wafer displays an excellent compatibility with the MEMS fabrication technique, which is a promising approach for scalable fabrication of micro gas sensors. The SnO$_2$ NSAs based sensor exhibits ultrasensitive to NO$_2$ gas (150.9 to 10 ppm) and low detection limit (30 ppb). Especially, this chip also shows excellent reproducibility and liner relationships between sensor response and NO_2 concentration within the range of 30-1000 ppb and 2-50 ppm, respectively.

REFERENCES

[1] R. Kumar, O. Al-Dossary, G. Kumar, A. Umar, "Zinc oxide nanostructures for NO_2 gas–sensor applications: A review," *Nano-Micro Lett.*, vol. 7, 2015, pp. 97-120.

[2] A. Mishra, S. Basu, N.P. Shetti, K.R. Reddy, "Metal oxide nanohybrids-based low-temperature sensors for NO_2 detection: a short review," *J. Mater. Sci. Mater. Electron*, vol. 30, 2019, pp. 8160-8170.

[3] H. Gong, C. Zhao, F. Wang, "On-chip growth of SnO_2/ZnO core-shell nanosheet arrays for ethanol detection," *IEEE Elect. Dev. Lett.*, vol. 39, 2018, pp. 1065-1068.

[4] H. Gong, C. Zhao, G. Niu, W. Zhang, F. Wang, "Construction of 1D/2D α-Fe_2O_3/SnO_2 hybrid nanoarrays for sub-ppm acetone detection," *Research*, vol. 2020, 2020, pp. 11.

[5] C. Zhao, H. Gong, G. Niu, F. Wang, "Ultrasensitive SO_2 sensor for sub-ppm detection using Cu-doped SnO_2 nanosheet arrays directly grown on chip," *Sens. Actuators B: Chem.*, vol. 324, 2020, pp. 128745.

[6] G. Niu, C. Zhao, H. Gong, X. Leng, and F. Wang, "NiO nanoparticle-decorated SnO_2 nanosheets for ethanol sensing with enhanced moisture resistance," *Microsyst. Nanoeng.*, vol. 5, no. 21, 2019, pp. 1-8.

[7] B. Esfandyarpour, S. Mohajerzadeh, S. Famini, A. Khodadadi, and E. Asl Soleimani, "High sensitivity Pt-doped SnO_2 gas sensors fabricated using sol–gel solution on micromachined (100) Si substrates," *Sens. Actuators B: Chem.*, vol. 100, no. 1-2, 2004, pp. 190-194.

[8] L. Francioso, M. Russo, A. M. Taurino, and P. Siciliano, "Micrometric patterning process of sol–gel SnO_2, In_2O_3 and WO_3 thin film for gas sensing applications: Towards silicon technology integration," *Sens. Actuators B: Chem.*, vol. 119, no. 1, 2006, pp. 159-166.

[9] A. Ebrahimi, A. Pirouz, Y. Abdi, S. Azimi, and S. Mohajerzadeh, "Selective deposition of CuO/SnO_2 sol–gel on porous SiO_2 suitable for the fabrication of MEMS-based H_2S sensors," *Sens. Actuators B: Chem.*, vol. 173, 2012, pp. 802-810.

[10] L. Francioso, D. S. Presicce, A. M. Taurino, R. Rella, P. Siciliano, and A. Ficarella, "Automotive application of sol–gel TiO_2 thin film-based sensor for lambda measurement," *Sens.Actuators B: Chem.*, vol. 95, no. 1-3, 2003, pp. 66-72.

[11] J.H. Dang, N. Lee, A. Mirzaei, M.S. Choi, H.S. Choi, H. Park, H. Jeon, S.S. Kim, and H.W. Kim., "SnS-functionalized SnO_2 nanowires for low-temperature detection of NO_2 gas," *Mater. Charact.*, 2021, p. 110986.

Gap in pagination due to unavailable paper.

Pages 824-825

Proceedings of the 16th Annual IEEE International
Conference on Nano/Micro Engineered and Molecular Systems
April 25-29, 2021

A Tungsten-Rhenium Thin Film Thermocouples Sensor Based on AlN Transition Layer

TIAN Bian[1,*], ZHANG Bingfei[1], ZHANG Zhongkai[1], LIU Zhaojun[1], LIU Jiangjiang[1], CHENG Gong[1]

Qijing Lin[1], WU Chen[1], SHI Peng[1], JIANG Zhuangde[1]

1 Xi'an Jiaotong University, Xi'an 710049, China

* Email: t.b12@xjtu.edu.cn

Abstract-A tungsten-rhenium (W-Re) thin film thermocouples (TFTCs) temperature sensor for ultra-high temperature detection is proposed to solve the problem of turbine inlet temperature measurement. The TFTCs is composed of silicon carbide (SiC) substrate, aluminium nitride (AlN) transition layer and W-Re3/W-Re25 alloy thermoelectric layer with a stacking structure. The thermoelectric characteristics and thermomechanical stability of the designed multi-layer heterogeneous films are simulated with the finite element simulation software. Combined with magnetron sputtering technology, the designed TFTCs is prepared and tested through the self-built calibration platform. The result shows that the fabricated TFTCs has excellent linearity and repeatability and can meet the requirements of stable, reliable and rapid response high temperature measurement.

I. INTRODUCTION

With the development of aviation and aerospace technology, aero-engine is developing in the direction of high compression and high thrust-weight ratio. This situation will lead to higher operating temperatures on turbine blade surfaces, combustion chamber walls and other parts. The hot end surface will face a severe test, especially the temperature at the front of the turbine can reach 1600K~2000K [1-2]. Earlier studies have shown that creep, ablation, fatigue fracture and other failure forms of engine turbine blade are easy to occur under high temperature environment. Therefore, the high temperature measurement of aero-engine components has always been an important problem to be solved in the aviation field.

TFTCs with two-dimensional temperature sensitive structure has the advantages of small heat capacity, small volume and fast response [3-4]. Besides, its overall structure can be approximately regarded as a two-dimensional plane structure so that it is almost no interference in the flow field on the surface of the object to be measured, which can realize the in-situ measurement of engine turbine blade surface temperature. In view of the limitations of traditional high-temperature measurement methods, this paper designs and fabricates a new kind of ultrahigh-temperature tungsten-rhenium sensor with a transition layer. The structure of sensor is optimized by using finite element simulation technology, which can reduce the thermal stress as much as possible between the film layers and improves the thermal mechanical stability of the sensor. In this way, the purpose of stable working and accurate measurement in

high temperature environment of the sensor can be realized by theoretical direction.[1]

II. SENSOR DESGIN

TFTCs is a kind of temperature sensor based on Seebeck effect whose output voltage depends on the temperature difference between the hot and cold ends when the materials of thermo-electrodes are determined. The electromotive force (EMF) can be calculated by the following equation:

$$E_{AB} = S_{AB}(T_h - T_c) \qquad (1)$$

Where E_{AB} is the EMF of TFTCs, S_{AB} is the Seebeck coefficient of TFTCs, T_h is the temperature of the hot end, and T_c is the temperature of the cold end.

In order to ensure stable output under high temperature, the tungsten-rhenium alloy (W-Re3/W-Re25) with high melting point (＞3000°C) are selected as thermoelectric materials. The SiC with the high temperature resistance and great shock resistance is selected as the substrate material [5]. Meanwhile, a transition layer is necessary to reduce the thermal stress. The structure of the TFTCs is shown in Fig.1.

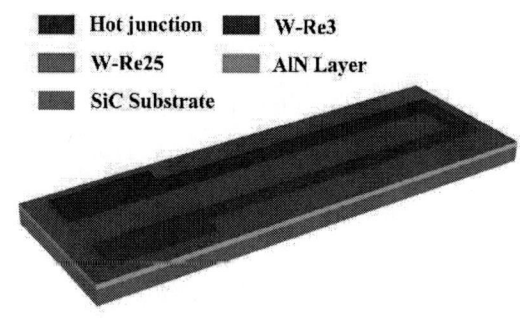

Figure 1. The structure of W-Re TFTCs

The thermoelectric electrodes with an overall length of 90mm and width of 3mm. In order to form a hot end, the thermoelectric electrodes overlap each other in a rectangular area of 4mm×27mm. And the cold end is two rectangular

---

This work is supported by National Key Research and Development Project (2019YFB2004501), National Natural Science Foundation of China (No.91748207), 111 Program (No.B12016), China Postdoctoral Science Foundation（2020M683461）, National Key Research and Development Project (2020YFB2009101).

978-1-6654-3008-1/21 $31.00 © 2021 IEEE   826

areas of 10 mm×18mm where the positive and negative wires are connected respectively to output thermoelectric potential signal.

III. SIMULATION

The finite element analysis (FEA) is used to analyze the thermoelectric properties and thermomechanical stability of W-Re TFTCs. We choose heat transfer module of COMSOL Multiphysics for the steady state thermoelectric simulation. For the analysis of thermoelectric properties, we simulate the thermoelectric output and distribution when hot end is heated from 400K to 1500K and cold end remains at 293K. The result of simulation is shown in Fig.2-3. The main physical parameters of W-Re TFTCs are listed in Table I.

TABLE I. The physical parameters of W-Re TFTCs

Physical parameters	WRe3	WRe25
Seebeck coefficient (μV/K)	5	23.5
Conductivity (S/m)	60000	35336
Thermal conductivity (W/m·K)	35	43
Heat capacity (J/Kg·K)	176	150
Density (kg/ m³)	19900	19700

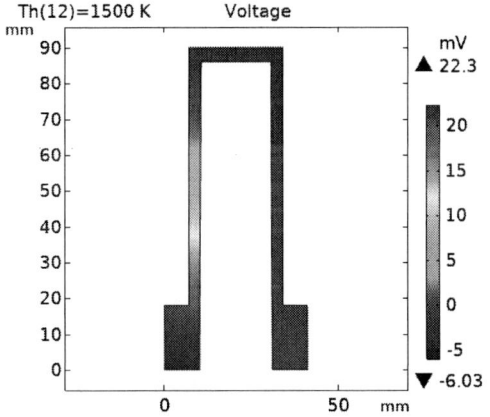

Figure 2. The thermoelectric distribution of W-Re TFTCs.

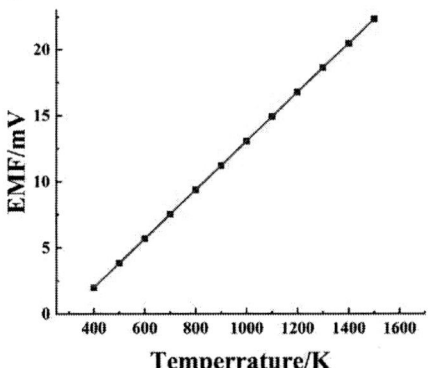

Figure 3. The simulation of thermoelectric output of W-Re TFTCs.

As shown in Fig.2, the EMF is 22.3mV when the temperature difference between two electrodes is 1206K. In Fig.3, the slope of the EMF output curve fitted by softare is 0.0185mV/K which is the same as the sensitivity of W-Re thermocouples (18.5μV/K).

According to the residual stress calculation model of progressively deposited coatings with simple plane geometry proposed by Tsui and Clyne, the thermal stress of the film can be calculated by the following equation [6]:

$$\sigma_{th} = \frac{E_{ef} \int_{T_r}^{T_D} (\alpha_s - \alpha_f) dT}{1 + 4(E_{ef}/E_{es})(h/H)} \quad (2)$$

Where s_{th} is the thermal stress between film and subsrate, E_{ef} and E_{es} are the effective Young's modulus of film and substrate respectively, and h, H, T_D, T_r, α_f, α_s mean thickness of film, thickness of substrate, service temperature, room temperature, thermal expansion coefficient of film and substrate material. Table II lists the required physical parameters for simulation.

The relationships between temperature, thickness of substrate, thin film and the maximum thermal stress are investigated respectively, shown in Fig.4.

TABLE II. The physical parameters of simulation.

Physical parameters	Materials			
	W-3Re	W-25Re	SiC	AlN
Elastic modulus/(GPa)	360	363	400	340
Poisson's ratio	0.28	0.28	0.17	0.25
Coefficient of thermal expansion/($10^{-6}K^{-1}$)	4.57	5.0	3.6	4.5
Thermo conductivity/(W/m·K)	174	140	83.6	20

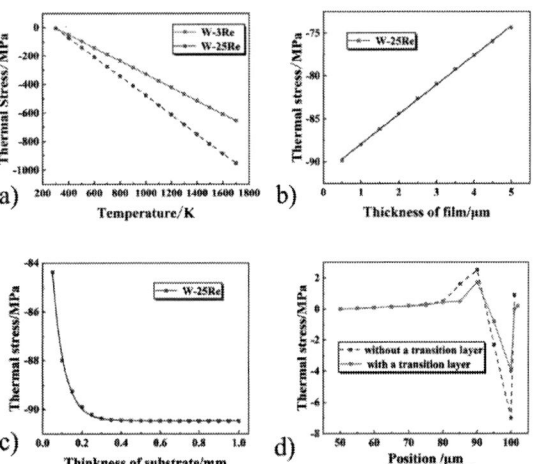

Figure 4. The thermal mechanical stability simulation of W-Re TFTCs.

The maximum thermal stress between the thin films shows an upward trend with the increase of working

978-1-6654-3008-1/21 $31.00 © 2021 IEEE

temperature and substrate thickness, and it decreases with the increase of film thickness. Especially, working temperature has a more dramatic impact than the other parameters. At the same time, the AlN transition layer can effectively reduce the shear stress between thermoelectric layer and the substrate.

IV. FABRICATION

The process flow for the TFTCs is shown in Fig.5. Briefly, AlN transition layer is deposited on the SiC substrate which has been cleaned by acetone and anhydrous ethanol with ultrasonic cleaning 10 minutes. Then photoresist is evenly spun on AlN transition layer and patterned for deposition of W-Re3 and W-Re25 electrodes. The Physical map of fabricated W-Re TFTCs is shown in Fig.6.

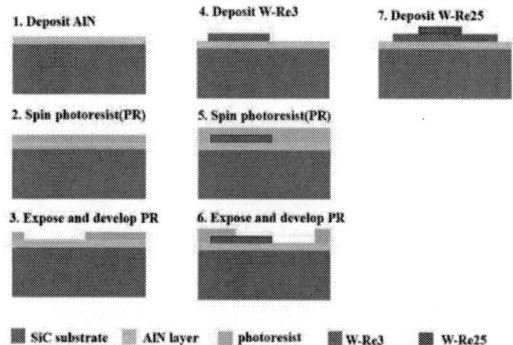

Figure 5.   The fabrication process of W-Re TFTCs.

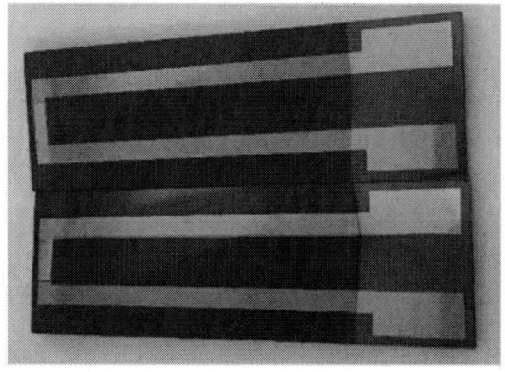

Figure 6.   The samples of fabricated W-Re TFTCs.

V. CALIBRATION

A static calibration system is designed to study the thermoelectric characteristics of fabricated TFTCs as shown in figure 7. The prepared W-Re TFTCs are tested in a muffle furnace (LHT0820 Germany Nabertherm). Two K-type thermocouples are respectively used to measure the temperature of the hot end and cold end. Moreover, the cold junctions of the thermocouples are cooled by circulating cold water to maintain a steep temperature gradient. The thermoelectric voltage of the W-Re TFTCs and K-type thermocouples are recorded with a data collector (LR8410-30 Japan HIOKI).

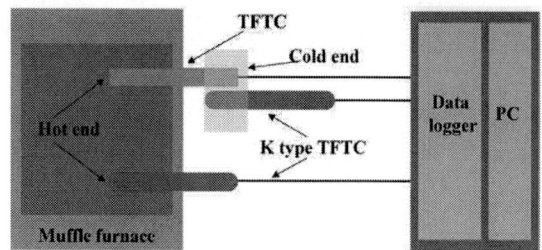

Figure 7.   The static test system of W-Re TFTCs.

The EMF can be observed by figure 8 under a rising temperature difference from 0°C to 1000°C in muffle furnace. The EMF fitting curve is shown in Fig. 8, which is given as

$$E = 0.034T - 21.74 \qquad (3)$$

Where T is the temperature difference between the hot end and cold end, and E is the EMF of W-Re TFTCs(mV). The equation shows the sensitivity of WRe3–WRe25 TFTCs is 34μV/K, which is much higher than 18μV/°C of traditional standard C-type thermocouples. In addition, the actual output curve exhibits excellent linearity whose fitting degree of linearity can reach 0.99553. [7]

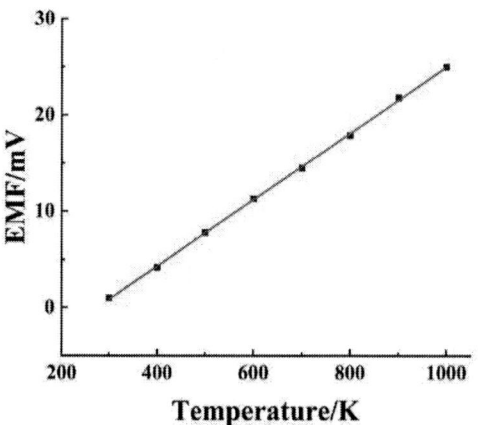

Figure 8.   The fitted curve of EMF of TFTCs

In order to assess the repeatability of prepared W-Re TFTCs, the repeatability error can be calculated by the following equation:

$$\sigma_R = \frac{3S_{max}}{Y_{FS}} \qquad (4)$$

Where σ_R is the repeatability error, S_{max} is the maximum standard deviation, and Y_{FS} is the full range output of W-Re TFTCs.

The fabricated TFTCs is tested for nine cycles over 5 hours in a muffle furnace in Fig.9. According to the data in table III, the repeatability error of W-Re TFTCs is only 0.74% in the range of 100°C-400°C which shows a great potential for practical applications.

978-1-6654-3008-1/21 $31.00 © 2021 IEEE

TABLE III.   The results of repeatability test

Temperature difference(°C)	1st (mV)	2nd (mV)	3rd (mV)	4th (mV)	5th (mV)	6th (mV)	7th (mV)	8th (mV)	9th (mV)
100	1.29	1.31	1.31	1.30	1.31	1.31	1.33	1.31	1.33
150	3.08	3.24	3.23	3.08	3.21	3.02	3.24	3.23	3.06
200	5.13	5.13	5.1	5.02	5.18	5.13	5.13	5.1	5.02
250	7.23	7.29	7.15	7.18	7.11	7.23	7.29	7.15	7.18
300	9.62	9.61	9.48	9.49	9.53	9.62	9.61	9.48	9.49
350	12.0	12.0	11.9	11.8	11.9	12.0	12.0	11.9	11.8
400	14.5	14.5	14.5	14.4	14.4	14.5	14.5	14.5	14.4

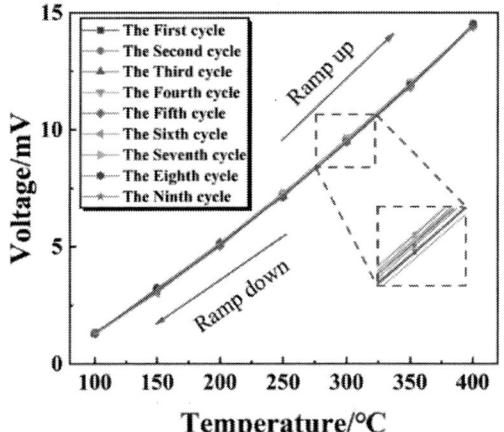

Figure 9.   The repeatability test of W-Re TFTCs

VI.   CONCLUSION

In this paper, a kind of W-Re TFTCs which has high accuracy and excellent thermomechanical stability is proposed and fabricated. This study makes some contributions to the field of high temperature measurement. The results of simulation show that the maximum thermal stress and shear stress between thermoelectric layer and the substrate can be effectively reduced by using the introduction of AlN transition layer, leading the stability of service in high temperature environment to get a significant improvement. The static test shows that the prepared W-Re TFTCs fabricated by magnetron sputtering technology shows great performances in sensitivity (34μV/K), excellent linearity and small repeatability error (0.74%). An issue that is not addressed in this study is the design and fabrication of the protective layer in order to seek the breakthrough of ultra high temperature applications

REFERENCE

[1] Tian B, Liu Z, Zhang Z, et al. Effect of film deposition rate on the thermoelectric output of tungsten-rhenium thin film thermocouples by DC magnetron sputtering[J]. Journal of Micromechanics and Microengineering, 2020, 30(6):065004.

[2] Bandara S, Gunapala S. Multiple waveband temperature sensor[C]. Proceedings of Directed Energy Test & Evaluation Conference. NASA, USA, 2006.

[3] YS Wang, XM Dong, W Liu et al. Research on Developments of High Temperature Testing Technology for Aero-Engine[J]. Measurement & Control Technology, 2017, 36(09):16.

[4] Zhang Z, Tian B, Yu Q, et al. Mechanical properties analysis and process optimization for tungsten-rhenium thin film thermocouples sensor[C]. IEEE SENSORS, 2017.

[5] Tian B, Liu Y, Zhang Z, et al. WRe26–In$_2$O$_3$ probe-type thin film thermocouples applied to high temperature measurement[J]. Review of Scientific Instruments, 2020, 91(7):074901.

[6] Tsui YC, Clyne TW. An analytical model for predicting residual stresses in progressively deposited coatings Part 2: Cylindrical geometry[J]. Thin Solid Films, 1997, 306(1): 23-33.

[7] Liu Z, Tian B, Fan X, et al. A temperature sensor based on flexible substrate with ultra-high sensitivity for low temperature measurement[J]. Sensors and Actuators A: Physical, 2020, 315:112341.

Gap in pagination due to unavailable paper.

Pages 830-831

Proceedings of the 16th Annual IEEE International
Conference on Nano/Micro Engineered and Molecular Systems
April 25-29, 2021

A New Method for Characterization of Single Cell Using System Identification

Shuang Ma[1,2,3], Wenxue Wang[*1,2], Yuechao Wang[1,2], Lianqing Liu[*1,2] and Tianlu Wang[4]

1 State Key Laboratory of Robotics, Shenyang Institute of Automation, Chinese Academy of Sciences, Shenyang, China;
2. Institutes for robotics and Intelligent Manufacturing, Chinese Academy of Sciences, Shenyang, China
3. University of Chinese Academy of Sciences, Beijing, China
4. Department of Radiotherapy, Cancer Hospital of China. Medical University, Liaoning Cancer Hospital and Institute, Shenyang, China
*Corresponding author:wangwenxue@sia.cn; lqliu@sia.cn

Abstract—**The mechanical properties of cells reflect the state of macro-organism. While many approaches are available for performing mechanical property investigations on cells, to characterize dynamic mechanical property of a single cell is challenging. We measure the mechanical properties of a single cell using an Atomic Force Microscopy and develop a novel method to characterize dynamic mechanical property of a single cell based on the theory of system identification. We evaluated the new method by comparing its performance with the other methods in classifying different kinds of cells. The results show that the method we proposed outperformed other methods by achieving average classification accuracy of more than 90%. This work provides a new attempt for characterization of a single cell system, which is of great significance for studying the dynamic changes of the mechanical properties of single cells and analyzing or controlling the state of cells.**

Keywords—single cell; dynamic mechanical property; Atomic Force Microscopy; system identification

I. INTRODUCTION

Cells are the fundamental units in living organisms which contain important biological information. The mechanical property of the single cell is tightly related to physiological and pathological activities of living organisms [1]. On the one hand, from molecules to cells, to the level of tissues, there will be changes in mechanical properties in almost every basic unit due to the influence of the pathophysiological environment [2] .On the other hand, cells interact with adjacent cells or extracellular matrix(ECM) in the surrounding microenvironment, generating mechanical forces transduction, thereby affecting the physiological behavior of cells. Taken the highly complex tumors as an example: It is commonly known that cancer cells are often softer than their normal counterpart. [3]. Correspondingly, the stiffness of the cancerous tissue and surrounding adjacent tissues will also change [4]. In a growing tumor, the tumor cells undergo epithelial-mesenchymal transition (EMT), while the increased cell density produces compressive mechanical stress that acts on the tumor cells to

promote cell proliferation [5]. Research on mechanical properties of cells in the pathophysiological process is of great significance to study the occurrence and development of cancer and to provide valuable reference for clinical treatment.

In the past few decades, many methods have developed to measure the mechanical properties of single cells. The appearance of Atomic Force Microscopy (AFM) provides a new way for single cell analysis with its unique advantages comparing with other methods such as optical tweezers stretching, magnetic bead rotation, shear fluid method, through-hole analysis, micro pipette method, etc. AFM can obtain unique and increasingly important information that is not available with other techniques, including morphology information, as well as mechanical information and other physical properties of cells and cell-substrate adhesions [6]. In the past few years, researchers have used AFM to measure the mechanical properties of cells. By measuring the mechanical properties of cells, such as stiffness and viscoelastic, it shows significant differences in the mechanical properties of normal cells and cancer cells [7]. It is proposed that the difference in actin tissue between tumor cells and normal cells may directly lead to changes in the cell mechanics of cancer cells [8]. The mechanical property of cells have been shown strong correlation with their collective migration and invasiveness, so it can be used as a biological index which can effectively indicate the pathophysiological state changes of cells[9].

Single-cell analysis based on AFM measurements provides new possibilities for label-free detection of cell state. Previous studies on the mechanical properties of cells mainly focused on the Young's modulus of cells [10]. Young's modulus, as the most basic material property of cell surface, can only reflect the stiffness of cell surface. In fact, the cell itself is a viscoelastic body[11]. Due to the fluidity of cytoplasm or other reasons, the viscoelastic properties of cells are always in dynamic change [12]. Research that regards viscoelasticity as a label-free biomarker for cancer is relatively lacking. Moreover，the analysis of a single cell based on the Young's modulus of a single time point cannot reflect cell states changing with time. Single cell analysis based on the viscoelastic parameters of continuous time curves overcomes these problems but involves too many parameters to directly reflect the dynamic changes of cells. The mathematical modeling of a single cell based on dynamic viscoelastic parameters can overcome the shortcomings mentioned above, which is helpful to understand

978-1-6654-3008-1/21 $31.00 © 2021 IEEE

the behavior of single cell and study the mechanism of some diseases. But due to the signal pathway inside the cell is very complex, it is very difficult to model the single cell mathematically.

To solve the above problems, this paper proposes a single-cell mechanical property modeling method based on system science. This method is used to model and analyze four different kinds of cell lines, and apply to the classification of different types of cells. Our method provides a novel sight to research on dynamical mechanical properties of single cells.

II. METHOD AND MATERIALS

A. Modeling of Dynamic Mechanical Property of a Single Cell

The view of system science is that the internal characteristics of the system need to be characterized by external input and output characteristics. As shown in Fig. 1, a single cell is regarded as a system. We apply mechanical stimulation to the cell and measure the mechanical response of the cell. Then the stack vector state space equation is established regarding the mechanical stimulus on a single cell as input and the mechanical properties of the single cell as output. In this paper, the Young's modulus and viscoelastic parameters are taken as the mechanical properties of the cells, according to which the state space equations are established respectively. The internal structure information of the system can be inferred according to the description of input and output.

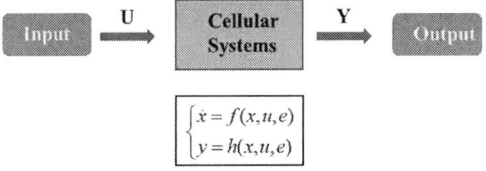

Fig. 1. Schematic diagram of cell dynamics behavior from the view of system science.

Among them, the state $x(t)$ describes the change of the internal state of the cell system and the variable y describes the output of the system under the action of input excitation.

In this paper, a single cell is regarded as a linear time invariant system approximately. Mathematically, the following state space equation is used to describe a single cell system, that is:

$$\begin{cases} \dot{x}(t) = Ax(t) + Bu(t) + Ke(t) \\ y(t) = Cx(t) + Du(t) + e(t) \end{cases} \quad (1)$$

Among them, A、B、C、D are coefficient matrixes, K is the disturbance matrix, $u(t)$ is the input of the system, $y(t)$ is the output of the system, $e(t)$ is the disturbance.

B. Modeling a Single Cell Based on Young's Modulus

The AFM is used to measure the Young's modulus of a single cell at equal intervals over a continuous period of time. Therefore, for each cell system, we get the corresponding n input and output pairs at n time points: $(u(t), y(t) : t = 1 \ldots n)$. In this part, we only consider the mechanical stimulation of AFM to the cell and assume that the mechanical stimulus is constant. We make input vector $u(t) = (0)_{1\times1}$ and the Young's modulus at the corresponding time point as the output, that

is $y_{YoungsModulus}(t) = (YoungsModulus(t))_{1\times1}$. In order to facilitate system identification, we interpolate the inputs and outputs of two adjacent time points respectively. So for each cell system, we get 17 corresponding input and output pairs $(u(t), y_{YoungsModulus}(t) : t = 1 \ldots 17)$ at 17 time points.

Suppose that the present moment is $k = \lceil 17/2 \rceil$, f is a future time and $f = \lceil 17/2 \rceil + 1, \lceil 17/2 \rceil + 2, \cdots, n$, p is a certain moment in the past and $p = 1, 2, \cdots \lceil 17/2 \rceil - 1$. The stack vector state space equations based on the Young's modulus of the past and future moments were obtained as shown below.

$$y_f = \Gamma_f x(k) + H_f u_f + G_f w_f + v_f \quad (2)$$
$$y_p = \Gamma_p x(k-p) + H_p u_p + G_p w_p + v_p \quad (3)$$

Thus, the Hankle matrix based on the Young's modulus is obtained, as shown below.

$$Y_f = \Gamma_f X_f + H_f U_f + G_f W_f + V_f \quad (4)$$
$$Y_P = \Gamma_p X_p + H_p U_p + G_p W_p + V_p \quad (5)$$

C. Modeling a single Cell Based on Viscoelastic Parameters

The viscoelastic parameters of a single cell were obtained at a fixed time point for a continuous period of time under the action of AFM. Similarly, in the process of system identification of single cell system based on viscoelastic parameters, we make input vector $u(t) = (0 \ 0 \ 0 \ 0 \ 0)_{1\times5}$. The viscoelastic parameters obtained at the corresponding time point are taken as the output, we make output vector $y_{Visco\_Para}(t) = (k_0(t) \ k_1(t) \ b_1(t) \ k_2(t) \ b_2(t))_{1\times5}$. By means of mean value interpolation for input and output of adjacent time points, we also get corresponding input and output pairs for each cell system at 17 time points: $(u(t), y_{Visco\_Para}(t) : t = 1 \ldots 17)$.

We make present moment $k = \lceil 17/2 \rceil$, f is a future time and $f = \lceil 17/2 \rceil + 1, \lceil 17/2 \rceil + 2, \cdots, n$, p is a certain moment in the past and $p = 1, 2, \cdots \lceil 17/2 \rceil - 1$. The stack vector state space equations based on the viscoelastic parameters of the past and future moments of the single cell system were obtained as shown below.

$$y_f = \Gamma_f x(k) + H_f u_f + G_f w_f + v_f \quad (6)$$
$$y_p = \Gamma_p x(k-p) + H_p u_p + G_p w_p + v_p \quad (7)$$

Thus, the Hankle matrix based on the viscoelastic parameters is obtained as shown below.

$$Y_f = \Gamma_f X_f + H_f U_f + G_f W_f + V_f \quad (8)$$
$$Y_P = \Gamma_p X_p + H_p U_p + G_p W_p + V_p \quad (9)$$

According to the Hankle matrixes based on the Young's modulus and the viscoelastic parameters we mentioned above, we can obtain the order and coefficient matrixes of the corresponding system according to the subspace identification method. After the system order and system parameters are given, we use the eigenvalues of coefficient matrix A to characterize the single cell. This method can be used to describe the dynamic changes of the mechanical properties (Young's modulus or viscoelastic parameters) of a single cell.

D. Cell Culture and Preparation

The cell lines used in this study were obtained from the Shenyang Pharmaceutical University. HEK-293 cells (human embryonic kidney cell line), MCF-7 cells (low invasive human breast cancer cell line), L929 cells (mouse fibroblast cell line), MDA-231 cells(high invasive breast cancer cell line) were cultured in RPMI-1640 medium (Hyclone, USA) containing 10% fetal bovine serum and 1% penicillin-streptomycin solution at 37° C(5% CO2).

E. Cell mechanical property measurement by AFM

In order to obtain the mechanical property of the cells, we use Bioscope Catalyst AFM(Bruker, America) (Fig.2A) to perform indentation experiments on living cells（HEK-293、MCF-7、L929、MDA-231). AFM is used in combination with an inverted microscope（Ti,Nikon,Japan）. The material of the probe is silicon nitride. The model of the probe is MLCT (Bruker, USA) and the elastic coefficient of the cantilever beam is 0.01 N / m (Bruker). The detection environment is a liquid culture medium and the detection mode is a contact mode. The cell lines used in our study are all adherent living cells.

Before performing cell detection, firstly, we use a probe to obtain a force curve in the blank area of the base to correct the deflection sensitivity of the cantilever beam. Then we use the thermal noise module of AFM to obtain the precise elastic coefficient of the cantilever beam. Next, we begin to measure the mechanical properties of cells. Under the guidance of an optical microscope, we find the target cell in the field of vision and control the AFM probe to move to the center of the target cell. Then the AFM probe is used to measure the elasticity and viscoelasticity of the cell in approach-stay-retract mode[13](Fig.2C).

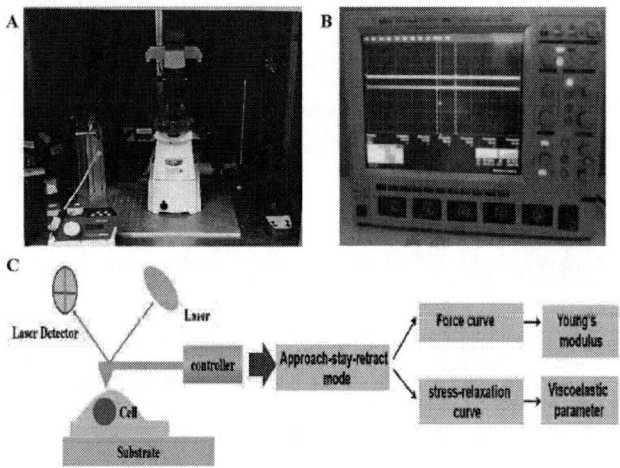

Fig.2 Experimental platform and method A: AFMB: Oscilloscope C: Simultaneous measurement of cell elasticity and viscoelasticity

First of all, the tip of the needle approaches the cell at a certain speed (4um/s). When the tip of the AFM probe is pressed on the sample, it leads to the rapid deformation of the cell and the rapid increase of the force on the cantilever beam. After the tip touches the cell, it continues to press down and the cell produces indentation under the action of the loading force of the probe. In the stress-relaxation stage, the piezoelectric (PTZ) actuator is kept at a constant depth. As the cell continues to deform, it results in a gradual decrease in the deflection of the cantilever beam. When the displacement of the cantilever beam reaches the set maximum value (1um), the tip stops pressing and stays on the cell for a period of time (6s). Then the probe retracts from the cell surface to the initial position in the vertical direction and the fallback time is set to 3s. In the process of approach-stay-retract, the force curve was recorded by AFM control software. The Young's modulus reflecting cell elasticity was extracted from the force curve. The stress-relaxation curve was recorded by the oscilloscope (LeCroy,America) in Fig.2B and the viscoelastic parameters of cell viscoelasticity were extracted from the stress-relaxation curve. A typical force curve obtained from the surface of MCF-7 cells is shown in Fig.3A. Because the probe stays on the cell surface for 3s, the end of the approximation curve does not coincide with the initial end of the regression curve. Fig. 3B is a typical stress-relaxation curve obtained on the surface of MCF-7 cells, which shows the continuous attenuation of the force during the relaxation process.

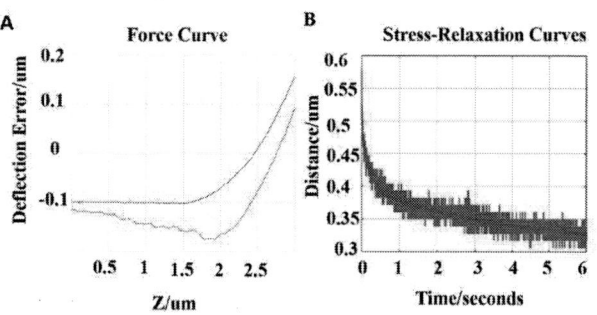

Fig.3 The force curve and stress-relaxation curve of the cell. (A) Force curve of the cell (B) stress-relaxation curve of the cell

Fifty cells of each type (HEK293, MCF-7, L929, MDA231) were selected for measurement. For each cell, 10 force curves and 10 stress-relaxation curves as shown in Fig.3A and Fig.3B were obtained at 0 h, 0.5 h, 1 h, 1.5 h, 2 h, 2.5 h, 3.5 h, 3 h, 3.5 h, 4 h, respectively. All of the force curves and stress-relaxation curves were obtained at different locations in the central area of the cell. In this study, we use the same cantilever beam to detect all the cells in the experiment thereby ensuring that the relevant parameters (loading speed, residence time, pressure displacement) in the AFM indentation experiment are consistent.

F. Data Processing

1)Young's modulus Calculating

As we use a tapered tip in the indentation experiment, we use the Sneddon model [14] to extract Young's modulus from the force curve:

$$F = \frac{2E\delta^2 \tan\theta}{\pi(1-v^2)} \quad (10)$$

In the formula, v is the Poisson's ratio of the cell (usually the cell is considered as an incompressible material, so $v=0.5$), F is the probe loading force, δ is the indentation depth, E is the

Young's modulus of the cell which is the half-open angle of the tapered tip. According to Hooke's law, we can get F:

$$F = kx \qquad (11)$$

k is the elastic coefficient of the cantilever beam and x is the deflection of the cantilever beam (which can be obtained directly from the force curve).

2) Viscoelastic Parameters Calculating

The improved Maxwell model proposed by Wang Bo [13] is used to extract the cell viscoelastic parameters according to the stress-relaxation curve recorded by the oscilloscope. In this model, a single cell is regarded as a system which can be written using a state-space equation as following:

$$\begin{cases} \dot{x}_i = -\dfrac{k_i}{b_i} x_i + \dfrac{k_i}{b_i} u, i = 1, 2, \cdots, n \\ y = -\sum_{i=1}^{n} k_i x_i + (k_0 + \sum_{i=1}^{n} k_i) u \end{cases} \qquad (12)$$

in which u and y are the system input and output, respectively, the state variable x_i represents the movement distance of the point between the spring and damper in the *ith* path, and k_i and b_i are the elastic and viscous parameters of the corresponding springs and dampers, respectively. The states x_i were closely related to the cell deformation. According to the order and parameters identification, the system is a two-order system, that is $n = 2$. The least square method is then used to identify five viscoelastic parameters k_0, k_1, b_1, k_2, b_2.

IV. RESULTS AND DISCUSSION

A. Comparison between identification curve and real value

1) Comparison of Young's modulus identification results and real values

According to the single cell modeling and system identification results based on Young's modulus，it is found that the mechanical stimulus-Young's modulus system is a second-order system.

Fig.4A shows a comparison between the identification results (blue curve) and the real Young's modulus (red curve) of a single cell system. The cell is randomly selected and the system regards mechanical stimulus as input and Young's modulus as output.

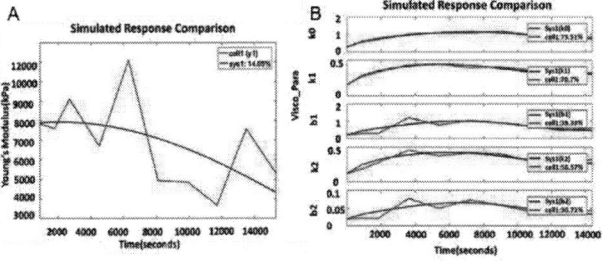

Fig.4 Fitting diagram of system identification results and real mechanical parameters of cells (A)Comparison between the system identification results based on mechanical stimulation input-Young's modulus output and the actual Young's modulus of cells (B) Comparison between the system identification results based on mechanical stimulation input-viscoelastic parameters output and the actual viscoelastic parameters of cells

2) Comparison of viscoelastic parameters identification results and real values

By using the n4sid method to identify the order and parameters of the system, it is found that the mechanical stimulation- viscoelastic parameter system is a third-order or fourth-order system. Fig.4B shows a comparison between the identification results (blue curve) and the real viscoelastic parameters (red curve) of a single cell system. The cell is randomly selected and the system is regarding mechanical stimulus as input and viscoelastic parameters as output.

By comparing Fig. 4A and Fig. 4B, we find that from the graph point of view, the fitness of identification curve of Fig. 4B to the real curve of cell is better than that of Fig. 4A ; from the fit value point of view, the fit value of cell system identification based on viscoelastic parameter is larger, which means that the identification result based on mechanical stimulation input- viscoelastic parameter output is closer to that of cell true state.

B. Processing of matrix A

1) Processing of matrix A based on Young's modulus

After the single cell system is identified based on the mechanical stimulation input - Young's modulus output, coefficient matrix A of the system is used to characterize the state of a single cell. The eigenvalues of the matrix A obtained from the parameter identification are calculated, and the eigenvalues are processed according to the following flow chart:

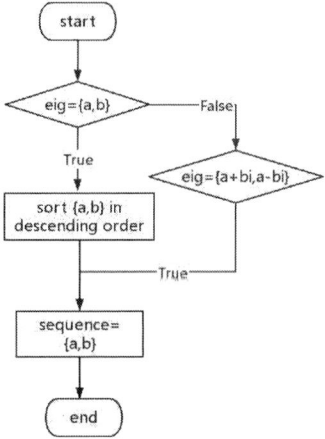

In this system, a real number sequence is used to represent a single cell. The dynamic change of Young's modulus of a single cell with time is included in the characterization.

2) Processing of matrix A based on viscoelastic parameters

After the single cell system is identified based on the mechanical stimulation input- viscoelastic parameter output, the eigenvalues of the matrix A are calculated. For different order conditions, the eigenvalues are processed as following flow chart:

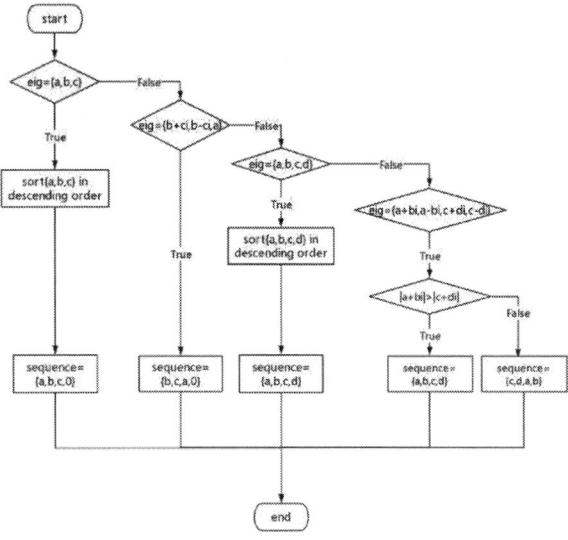

The single cell is characterized by the sequence above in the system, which includes the dynamic change information of viscoelasticity of single cell over time.

C. Performance comparison of cell classification results

1) Classification of three different kinds of cells

Three different kinds of cell lines (HEK293 ($N1 = 50$), L929 ($N2 = 50$), MCF-7 ($N3 = 50$)) are characterized based on the Young's modulus and viscoelastic parameters. Machine learning algorithms have been widely used in bioinformatics and computational biology. In this study, twenty-two classification functions of classification learner in MATLAB are used to classify different kinds of cells. Fig.5 shows the performance comparison of cell classification (L929、MCF7、HEK293)using machine learning algorithms based on Young's modulus and viscoelastic parameters. Fig.5A is the box chart which shows the classification accuracy of the twenty-two kinds of machine learning algorithms based on Young's modulus under the condition of nine single time point, continuous time curve and system identification. Fig.5B is the box chart which shows the classification accuracy of the twenty-two kinds of machine learning algorithms based on viscoelastic parameters under the condition of nine single time point, continuous time curve and system identification.

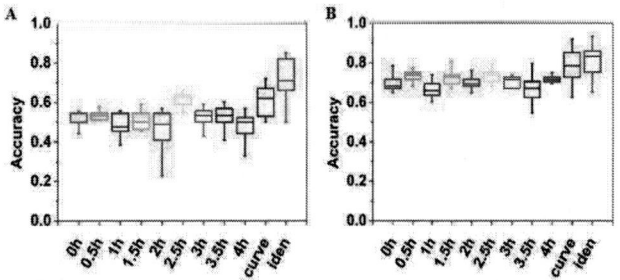

Fig.5 The performance comparison of cell classification (HEK293 ($N1 = 50$), L929 ($N2 = 50$), MCF-7 ($N3 = 50$)). (A)The classification accuracy box chart of the twenty-two kinds of machine learning algorithms based on Young's modulus under the condition of single time point, continuous time curve and system identification. (B)The

classification accuracy box chart of the twenty-two kinds of machine learning algorithms based on viscoelastic parameters under the condition of single time point, continuous time curve and system identification.

2) Classification of high and low invasive cells

We will carry out cell classification of low invasive breast cancer cell line MCF-7 ($N1 = 50$) and high invasive breast cancer cell line MDA-231 ($N2 = 50$) based on Young's modulus and viscoelastic parameters. Twenty-two classification functions in classification learner in MATLAB are used to classify different kinds of cells. Fig.6A is the box chart which shows the classification accuracy of the twenty-two kinds of machine learning algorithms based on Young's modulus under the condition of nine single time point, continuous time curve and system identification. Fig.6B is the box chart which shows the classification accuracy of the twenty-two kinds of machine learning algorithms based on viscoelastic parameters under the condition of nine single time point, continuous time curve and system identification.

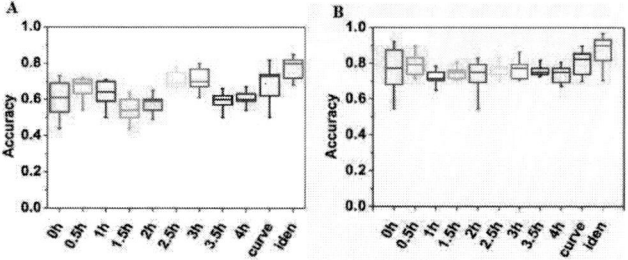

Fig.6 The performance comparison of cell classification (MDA231 ($N1 = 50$), MCF-7 ($N2 = 50$)). (A)The classification accuracy box chart of the twenty-two kinds of machine learning algorithms based on Young's modulus under the condition of single time point, continuous time curve and system identification. (B)The classification accuracy box chart of the twenty-two kinds of machine learning algorithms based on viscoelastic parameters under the condition of single time point, continuous time curve and system identification.

From Fig.5 and Fig.6, we found that in terms of single time point, continuous time curve or system identification methods, the number of methods with higher accuracy based on viscoelastic parameters classification is larger than that based on Young's modulus. We can conclude that the classification accuracy based on viscoelastic parameters is higher than that of Young's modulus, whether it is based on single time point, continuous time curve or system identification method. From Fig.5 and Fig.6, we can also conclude that in general, the accuracy of cells classification based on the mechanical parameters (viscoelastic parameters and Young's modulus) of continuous time curve is higher than that based on the mechanical parameters of single time point; the accuracy of cell classification based on the system identification results of mechanical parameters of a single cell is higher than that based on the mechanical parameters of continuous time curve. It is worth noting that the results based on mechanical stimulation input - viscoelastic parameter output identification have the highest classification accuracy and the best classification accuracy can reach more than 90%.

This classification result is not unexpected. Young's modulus of cells is only the basic material characteristics of cell surface. However, viscoelasticity of cells can reflect the flow of cytoplasm inside cells and is closely related to the physiological activities of cells. Therefore, the classification efficiency based on Young's modulus is lower than the classification efficiency based on viscoelastic parameters whether it is based on single time point or continuous time curve or system identification method.

Living cells are always in dynamic change. The mechanical property of continuous time points can reflect the dynamic characteristics of cells. Therefore, the classification accuracy based on the mechanical characteristics parameters of continuous time points is higher than that based on single time points. The method based on the system identification takes into account the dynamic changes of cells and describes single cell with eigenvalues. Compared with the continuous time curve method, the redundant information is removed, the dimension of variables is greatly reduced and the classification efficiency is improved.

III. CONCLUSION

In this study，a novel approach based on the system identification is proposed to characterize a single cell based on the mechanical property. By comparing with other methods, our proposed method based on the identification results of viscoelastic parameters displays a superior performance than the other algorithms in terms of the accuracy or vector dimension in single cell characterization.

ACKNOWLEDGE

This work is supported by the National Key R&D Program of China (Grant No. 2018YFB1304700), the National Natural Science Foundation of China (Grant Nos. U1908215, 61925307, 61903265, 91748212, U1613220, 91848201, U1813210, 61821005, 61927805)，the Instrument Developing Project of the Chinese Academy of Sciences (Grant No. YJKYYQ20180027), and the Key Research Program of Frontier Sciences, CAS (Grant No. QYZDB-SSW-JSC008),Liaoning Province Natural Science Foundation [20180550741], Shenyang Major Scientific Research Projects [19-112-4-090], Liaoning Province Key Area Joint Open Fund [2019-KF-01-01].

REFERENCE

[1] Marta Urbanska, Hector E. Muñoz, Josephine Shaw Bagnall, Oliver Otto,Scott R. Manalis, Dino Di Carlo,and Jochen Guck, "A comparison of microfluidic methods for highthroughput cell deformability measurements", Nat Methods. 2020 Jun; 17(6): 587–593.

[2] Jun-ichiro Jo, Jian-Qing Gao,Yasuhiko Tabata,"Biomaterial-based delivery systems of nucleic acid for regenerative research and regenerative therapy",Regen Ther. 2019 Dec 1; 11: 123–130.

[3] Sungmin Nam, Vivek Kumar Gupta, Hong-pyo Lee, Joanna Y. Lee, Katrina M. Wisdom, Sushama Varma, Eliott Marie Flaum, Ciara Davis, Robert B. West, and Ovijit Chaudhuri,"Cell cycle progression in confining microenvironments is regulated by a growth-responsive TRPV4-PI3K/Akt-p27Kip1 signaling axis",Sci Adv. 2019 Aug; 5(8): eaaw6171.

[4] Yong Liu, Chuanping Yang, Chengsong Cao, Qing Li, Xin Jin, and Hanping Shi, "Hsa_circ_RNA_0011780 Represses the Proliferation and Metastasis of Non-Small Cell Lung Cancer by Decreasing FBXW7 via Targeting miR-544a", Onco Targets Ther. 2020; 13: 745–755.

[5] Kazunori Iwasaki, Ryo Ninomiya, Toshitaka Shin, Takeo Nomura, Tooru Kajiwara, Naoki Hijiya, Masatsugu Moriyama, Hiromitsu Mimata, and Fumihiko Hamada,"Chronic hypoxia‐induced slug promotes invasive behavior of prostate cancer cells by activating expression of ephrin‐B1",Cancer Sci. 2018 Oct; 109(10): 3159–3170.

[6] Mohamed Yassine Amarouch, Jaouad El Hilaly, and Driss Mazouzi ,"AFM and FluidFM Technologies: Recent Applications in Molecular and Cellular Biology",Scanning. 2018; 2018: 7801274.

[7] Kaoru Uesugi, Fumiaki Shima, Ken Fukumoto,Ayami Hiura,Yoshinari Tsukamoto,3 Shigeru Miyagawa,Yoshiki Sawa, Takami Akagi, Mitsuru Akashi, and Keisuke Morishima,"Micro Vacuum Chuck and Tensile Test System for Bio-Mechanical Evaluation of 3D Tissue Constructed of Human Induced Pluripotent Stem Cell-Derived Cardiomyocytes (hiPS-CM)", Micromachines (Basel). 2019 Jul; 10(7): 487.

[8] Ailing Fu, Yixue Hou, Zhenyao Yu, Zizhen Zhao, and Zesheng Liu,"Healthy mitochondria inhibit the metastatic melanoma in lungs",Int J Biol Sci. 2019; 15(12): 2707–2718.

[9] María Anguiano, Data curation, Formal analysis, Investigation, Methodology, Software, Validation, Visualization, Writing – original draft,#1 Xabier Morales,"The use of mixed collagen-Matrigel matrices of increasing complexity recapitulates the biphasic role of cell adhesion in cancer cell migration: ECM sensing, remodeling and forces at the leading edge of cancer invasion",PLoS One. 2020; 15(1): e0220019.

[10] Ming Xiong, , Junying Li, Shuling Yang, Fansen Zeng, Yali Ji, Jiang Liu, Qiaoping Wu, Qingjun He, Ronglong Jiang, Fuyuan Zhou, "Influence of Gender and Reproductive Factors on Liver Fibrosis in Patients With Chronic Hepatitis B Infection,Clin Transl Gastroenterol. 2019 Oct; 10(10): e00085.

[11] Katarzyna Jasińska-Konior, Olga Wiecheć, Michał Sarna, Agnieszka Panek, Jan Swakoń, Marta Michalik, Krystyna Urbańska, and Martyna Elas,"Increased elasticity of melanoma cells after low-LET proton beam due to actin cytoskeleton rearrangements,Sci Rep. 2019; 9: 7008.

[12] Reiko Irifuku, Yuhki Yanase, Tomoko Kawaguchi, Kaori Ishii, Shunsuke Takahagi, and Michihiro Hide,"Impedance-Based Living Cell Analysis for Clinical Diagnosis of Type I Allergy",Sensors (Basel). 2017 Nov; 17(11): 2503.

[13] Bo Wang, Wenxue Wang, Yuechao Wang, Bin Liu and Lianqing Liu, "Dynamical Modeling and Analysis of Viscoelastic Properties of Single Cells",Micromachines 2017, 8, 171

[14] Mi Li1, Lianqing Liu, Ning Xi, Yuechao Wang,"Atomic force microscopy studies on cellular elastic and viscoelastic properties",Science China LIfe Science, 2018 Vol.61 No.1:57‐67

Gap in pagination due to unavailable paper.

Pages 838-839

Proceedings of the 16th Annual IEEE International
Conference on Nano/Micro Engineered and Molecular Systems
April 25-29, 2021

Design and Optimization of Glass Frit Package Structure for Micro Pressure Switch

Lingyun Wang[1], Daner Chen[2], Heng Xiong [3] and Yifang Liu[2*]

Abstract— A novel packaging structure is designed and the packaging process is optimized for micro pressure sensors with glass frit bonding technology in this paper. The packaging structure with inner-outer reference walls, metal lead break, and thermal stress release break is designed. The height of the reference walls can precisely control the thickness of glass frit layer. The technological parameters of screen printing across metal lead steps are also explored in this article originally. By optimizing the presintering atmosphere, peak temperature and holding time, the covering issue of frit on the metal lead surface is solved. The micro holes in the glass frit are effectively removed by optimizing the bonding temperature. The results of experiments show that when the peak temperature of presintering is 400 °C and the presintering atmosphere is 30 minutes in vacuum and in air respectively, the coverage rate of glass frit on the metal leads reaches 85%, which meets the packaging requirements. When the bonding temperature is 500 °C, the micro holes in the interlayer of glass frit are completely removed and the bonding strength of the pressure sensor is 12.52 MPa.

I. BACKGROUND

MEMS (Microelectromechanical systems) switches have gained attention as an alternative to conventional transistors due to their lower energy consumption and abrupt switching characteristics that provide for a very large resistance ratio between programmed states in the development of the Internet of Things (IoT) [1-4]. For more than twenty years, the study of the MEMS pressure switch has made some certain process, but there are still many difficulties in the development. The main problems exist in electrical contact and packaging. First, when the MEMS pressure works, the metal electrodes contact with each other, the quality of electrical contact seriously affects the service life and reliability of the switch. Secondly, the MEMS pressure switch can be electrically connected with the external circuit through metal leads. How to make a structure compatible with metal leads seal well and improve the bonding strength of leads is a difficulty that people pay attention to. The existence of the metal leads poses a great challenge to the encapsulating method. For example, aluminum lead which is commonly seen in MEMS is easy to form alloys with Si at high temperature (The aluminum-silicon melting point is 577 °C.). And the aluminum lead is also easy to react with SiO_2 to destroy the stability and reliability of the device [5,6]. Therefore, the compatibility between bonding method and metal lead

1- Department of Mechanical and Electrical Engineering, Xiamen University, Xiamen, CHINA.
2- Department of Instrumental and Electrical Engineering, Xiamen University, Xiamen, CHINA
3- Aviation Key Laboratory of Science and Technology on Inertial Technology, FACRI, Xi'an710065,China
*Contacting Author: (email: yfliu@xmu.edu.cn)

process should be considered when a packaging method is selected. Existing MEMS pressure switches are mostly encapsulated by Si-Si direct bonding [7] or silicon-glass anode bonding[8,9] These two bonding methods have high requirements for the packaging interface. For example, the roughness of the bonding surface is required to be less than 10Å. And the warpage of the silicon wafer is required to be less than 5μm [10], etc. The selection of electrode materials has also been limited. In Si-Si direct bonding and anodic bonding, metal leads are usually placed by trenching and embedding wire or through-hole interconnection. These two ways both require insulation for isolation, so the process is complex. Glass frit bonding is a wafer-level encapsulation technology with low melting point glass material as the interlayer, which can realize hermetic encapsulation[11, 12]. The advantages of glass frit bonding include the following: hermetic sealing, high accommodation and less selectivity of bonding surface materials (silicon, silicon oxide, silicon nitride, aluminum, or glass), high tolerance of roughness for the bond interface, and eliminating the need for the process to achieve electrical isolation between the feedthroughs and the sealing rings [13, 14]. At present, there are many examples of successfully sealing electrode leads with glass frit. For example, L. Hofmann et al. have used glass frit to encapsulate the Cu lead while preparing the gyroscope [15]. Also, the excellent bonding performance and flow characteristics of the glass frit can compensate for the gap left by the metal wire to form a strong and robust packaging structure. Therefore, this article intends to use the glass frit bonding technology with good interface compatibility to directly package the passive MEMS pressure switch with lead.

II. PACKAGING STRUCTURE DESIGN

The difficulty in the process of glass frit bonding is that the molten frit can extend or even pollute the MEMS device during bonding, which will lead to the failure of encapsulation. The line width of glass frit is uneven and seriously expands during the bonding of glass frit. Some scholars have studied the methods of solution to these two problems. The BTT (Barrier Trench Technology) proposed by Chen xiao et al. effectively controlled the area of the glass frit [16]. But there are drawbacks to this approach. Improper selection of the width and depth of the groove will cause the glass frit overflow or under-filling, which will affect the sealed performance of final encapsulation. For this reason, we designed a composite bonding structure with reference wall, outside micro groove and inside micro groove, which can precisely control the bonding thickness of glass frit and avoid the excessive glass frit to contaminate the MEMS structure [17].

978-1-6654-3008-1/21 $31.00 © 2021 IEEE

In view of this, the packaging structure with outer and inner reference wall with some breaks is designed in this article and is shown in Fig. 1. The inner reference wall is used to effectively isolate glass frit from both pressure-sensitive element and metal electrode so as to prevent them from being polluted. The outer reference wall has two breaks on each of the three edges where no electrode leads pass to give the excess glass frit a certain space for extension. So the excess glass frit won't overflow the reference wall due to extension. The stress concentration will be reduced at the same time. Both the inner and outer reference walls have an outlet for the passing metal lead. And a space of 200 μm is left on each side of the electrode lead to make up the alignment error. The height of the inner-outer reference wall is used to precisely control the thickness of the intermediate bonding layer of the glass frit. The glass frit accumulates at the break when bonding to ensure the sealing performance at the outlet of the lead. The distance between the inner and outer reference wall is 800 μm. The height and width of both reference walls are 8 μm and 100 μm respectively. The distance between the outer edge of the outer reference wall and the silicon wafer edge is 300 μm. The distance between the inner edge of the inner reference wall and the edge of the pressure sensitive film is 200 μm.

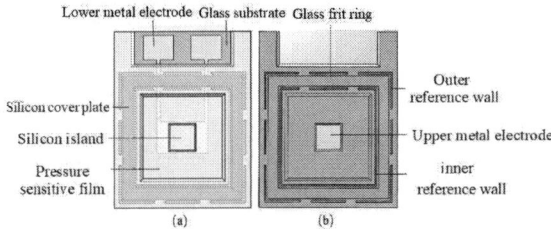

Fig. 1 The composite packaging structure with outer and inner reference wall (a) Overall structural perspective; (b) Silicon cover view

III. PACKAGING PROCESS AND RESULTS

The quality of glass frit packaging depends on the size and the number of holes in the glass frit layer after bonding, which is closely related to the presintering process and thermal compression bonding parameters. Before presintering, the glass frit is patterned onto a silicon substrate by screen printing. The 5643W glass frit produced by Koartan company is used in this paper. Its sealing temperature is 425~450 ℃, and the thermal expansion coefficient is 4~10×10^{-6}/°C.

A. Accurate Cross-step Screen Printing

The printing equipment used in the experiment is an MS-300F semi-automatic screen printing machine produced by Marabu. The screen printing includes sample preparation, substrate clamping, graphic alignment, frit processing and printing. Cross-step printing of glass frit is the difficulty of this experiment. The stencil mask with 325 mesh stainless screen is used in the experiment. The thickness of screen emulsion and the line width of stencil are 60 μm and 300 μm, respectively. And the printing speed is 30 mm/s. The cross-step screen printed glass frit on the

metal lead is shown in Fig.2. The thickness and line width of the metal lead are 1μm and 200μm, respectively. It can be seen that the screen printed glass frit can effectively cover metal lead. And there is no fault. And the line width of the glass frit remains the same.

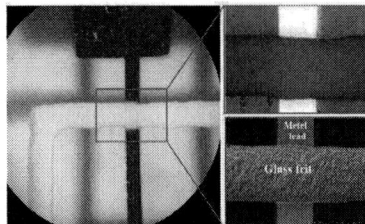

Fig. 2 The effect of glass frit cross-step printing

B. Presintering of Glass Frit on Composite Substrate

(1) Three-stage Bubble Removal Process

The glass substrate needs to be static for 10 minutes after screen printing, until the glass frit is smoothed in the air and the volatile solvent slowly evaporates at room temperature. After that, the presintering proceeds. The purpose of presintering is to completely remove water, organic solvent and organic binder from the glass frit, so as to thoroughly melt the glass powder and form a piece of smooth and compact glass [18].

In order to inhibit the generation of micro-nano holes in the glass frit layer, the initial temperature curve of the glass frit in the presintering experiment is shown in Fig.3. The organic solvent and organic binder are removed by holding them at 150°C and 300°C for 30 minutes, respectively. The liquid glass forms at an air atmosphere of 450°C for 30 minutes. Then another 30 minutes at a vacuum environment of 450°C will remove bubbles in the liquid glass. The last 60 minutes at an air atmosphere of 500°C will make the holes in the surface of the liquid glass flat. Finally, let the glass substrate stands still until its temperature slowly drops to room temperature.

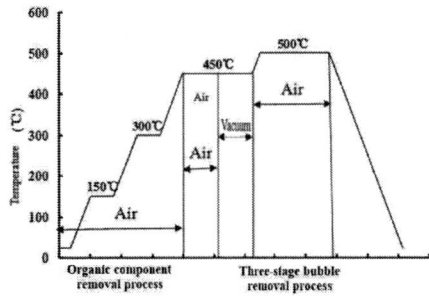

Fig.3 Presintering temperature curve of three-stage bubble removal process

As shown in Fig. 4, the sintered glass frit is transparent and its surface appears to be dense and smooth. However, due to the different wettability between frit-electrode and frit-substrate, the glass frit flows through the electrode lead and spreads along the length of the electrode lead, resulting in the frit coverage area is almost 0. Such a thin glass frit is not sufficient to form a good sealing when bonding.

Therefore, the presintering process is improved on the basis of the three-stage bubble removal process to obtain the glass frit ring with the required sealing thickness, especially focusing on the low coverage problem of glass frit on the metal lead.

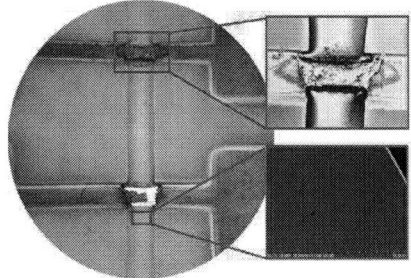

Fig. 4. The frit appearance of three-stage bubble removal presintering

(2) The Change of Presintering Atmosphere and Peak Temperature

The results of Kerry Cheung [19] showed that the melting of glass frit needs to be carried out in a convective atmosphere in which the oxygen helps to burn out the organic solvent to obtain clear glass. From this we can know that the fluidity of glass paste will be limited in vacuum. Therefore, three controlled tests are set up. They are processed only in vacuum, only in air, and in air followed by in vacuum, respectively. All of the peak temperatures are still 450℃. The backflow stage at 500℃ in the three-stage process is removed. The temperature of the substrate is immediately reduced right after vacuum reflow, while the other parameters remained unchanged.

In addition, many studies show that the peak temperature of presintering has the greatest influence on the results of presintering. If the peak temperature is too low, the glass frit can't be fully melted. It will result in an uneven surface so as to affect the bonding quality. If the peak temperature is too high, the metal electrode has the risk of eutectic melting. Combined with the previous research results in which the glass frit can't be completely melt less than 400℃, another group with the 400℃ for 60 minutes peak temperature is set. The parameters of the four control experiments are shown in Table 1.

Table 1. Parameters of presintering process of four control groups

Number	Peak temperature/℃	Holding time /min	Sintering atmosphere
1	450	60	Vacuum
2	450	60	Air
3	450	60	30min air + 30min vacuum
4	400	60	30min air + 30min vacuum

The experimental results are shown in Fig.5. In the first group, as shown in Figure 5(a), the glass frit covers about 50% of area on the metal lead and it has the tendency to shrink to both the sides of the metal lead along the width direction. At the same time, its texture is similar to that of the unsintered glass frit, with a black surface and large pits. It can be inferred that the vacuum environment is helpful to control the flowability of glass frit. But too long time in vacuum environment is not conducive to the backflow and filling of glass frit. Moreover, the organic solvent is not completely volatilized and remains on the surface, so it is necessary to control the action time of the vacuum environment.

In the second group, the surface of the frit is relatively smooth without black particles, but a lot of small bubbles are found in the frit layer, and the cover thickness of glass frit on the metal lead is obviously insufficient, as shown in Fig.5(b). According to the three-stage process, an air-only presintering environment can't exhaust the gas so as to remove all the bubbles, but the oxygen in the air is helpful to burn out the organic solvent and obtain the clear glass.

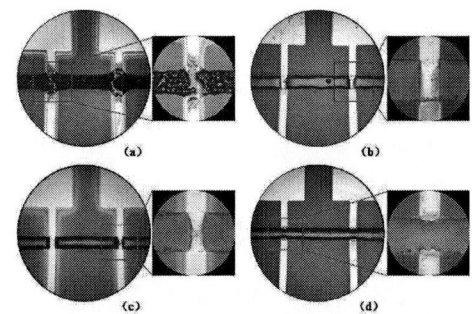

Fig. 5 Presintering effect of four control groups (a)First group, 450℃, vacuum-only; (b) Second group, 450℃, air-only; (c) Third group, 450℃, air and vacuum (d) Fourth group, 400℃, air and vacuum

On the basis of the second group, the substrate in third group is vacuumed for half an hour in the duration of peak temperature. The density of frit is greatly improved comparing with the previous group. There are no obvious bubbles in the glass frit, but only a small amount of frit remains on the metal lead, as shown in Fig. 5(c). The vacuum holding environment of 30 minutes improves the presintering effect. However, there is still a long way to go to get the glass frit ring with the target sealing thickness. Compared with the third group, the peak temperature is reduced in the fourth group, and the presintering effect is good, as shown in Fig.5(d). Although the glass paste on the metal lead slightly shrinks, the coverage area is about 85% of the area after screen printing, which is enough for packaging.

(3) Thermal Compression Bonding

Before bonding, the silicon cover plate with pressure sensitive structure and upper metal electrode is cleaned with deionized water. And the glass substrate with lower metal electrode and the presintered glass frit ring is cleaned too. After the silicon cover plate at the bottom and the glass substrate at the upper are put into the stainless steel hard plate for alignment, the thermal compression bonding is carried out on Awb04 bonding machine.

Bonding parameters mainly include bonding temperature, bonding pressure, cooling rate and heating rate. The heating rate has little effect on the glass frit bonding results and it's usually not considered. Based on the consideration of thermal stress release, the cooling rate should be as slow as possible. But it will lead to too long time and significantly reduced production efficiency, so the cooling rate of 5 °C / min is selected. When the temperature drops to 300 °C, the bonding procedure is ended and nitrogen is introduced for cooling.

The thickness of glass frit is further compressed under the condition of heating and pressure. The degree of compression is limited by the inner-outer reference walls on the silicon cover. In order to precisely control the distance between the upper and lower metal electrodes, the bonding pressure should ensure that the bonding thickness of glass frit is compressed to the same height as the reference wall. According to previous experiments, 10 atm can meet the bonding requirements.

The bonding temperature should ensure that the glass frit can be melt sufficiently. The recommended bonding temperature of 5643W glass frit is 425~450°C, so initially we set the temperature as 450°C when bonding. The surface morphology of the glass frit after bonding is shown in Fig. 6(a). The glass frit looks like gray ceramic, and the expansion of it is limited within the reference walls. There are many holes in the frit layer, and the large holes are mainly concentrated near the groove. Because the bonding temperature is not high enough, the glass frit cannot melt completely, resulting in the difficulty of removing bubbles, and the holes are formed during the cooling process.

To compare the influence of bonding temperature on bonding effect, we add two experiments in which the peak bonding temperature is 480°C and 500°C, respectively. And the other experimental parameters remain. The experimental results are shown in Fig. 6 (b) and 6(c). With the increase of bonding temperature, the number of holes in the glass frit layer decreases. There are no obvious holes in the layer of glass frit after bonding at 500°C.

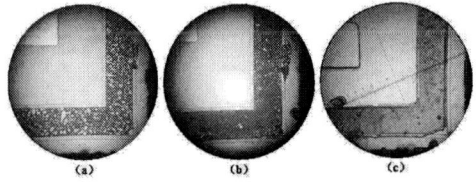

Fig. 6. Surface morphology of glass frit at different bonding temperatures (a)450°C; (b) 480°C; (c)500°C

The cross section of the bonding sheet is prepared and observed by scanning electron microscope, as shown in Fig. 7. In Fig. 7, (a), (b), (c) correspond to the bonding situations at 450°C, 480°C, 500°C respectively. It can be seen from Figure 7(a) that many small holes distribute in the glass frit layer at 450°C. When the bonding temperature rises to 480°C, we can find in Fig. 7(b) that the number of holes decreases, and there are large holes occasionally. When the bonding temperature is 500°C, we can find in Fig. 7(c) that the glass frit layer is dense without obvious holes, which is consistent with the results of the surface morphology. Therefore, the temperature of thermal compression bonding of the glass frit is set to be 500°C.

The universal electronic testing machine WDW-10 is used to conduct pull test on packaging devices at different bonding temperatures. When the bonding temperature is 450°C, the glass frit is not completely molten. The tensile strength is 4.98 MPa which is too low because there are many holes in the frit layer. When the bonding temperature is 480°C and 500°C, the tensile strength can both meet the target requirements (10MPa). The average value is 12.52 MPa when the bonding temperature is 500°C.

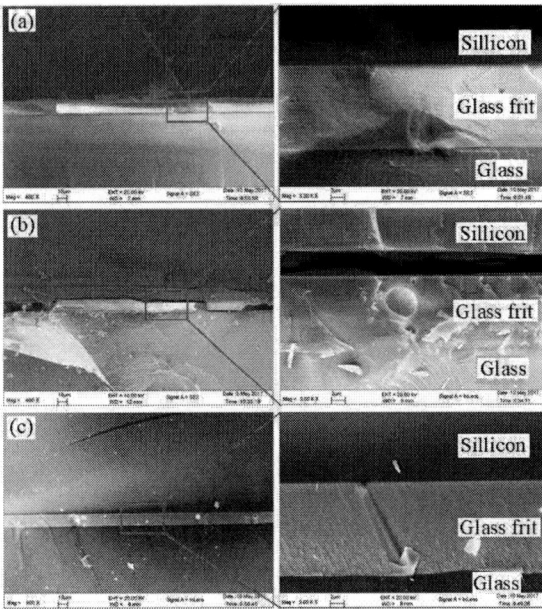

Fig. 7. SEM of cross section of glass frit at different bonding temperatures (a) 450°C; (b) 480°C; (c) 500°C

IV. CONCLUSION

In this paper, the packaging structure of passive MEMS pressure switch is designed on the basis of considering the coverage of glass frit on the metal lead and not polluting the sensitive structure. It mainly consists of the inner-outer reference walls, which have the design of the metal lead's compatible outlet and the break to release the thermal stress. By optimizing screen printing parameters, presintering peak temperature and presintering atmosphere, the wettability of glass frit on the metal lead surface is improved. And the coverage of glass frit on the metal lead surface is increased from 0% to 85%. When the bonding temperature reaches 500°C, the holes in the intermediate layer of glass frit are successfully removed. The bonding strength of the passive MEMS pressure switch reached 12.52MPa.

ACKNOWLEDGMENT

This research was financially supported by the Aviation Science funds (Aviation Key Laboratory of Science and Technology on Inertia) (No. 20150868002, 20180868001), Science and Technology Planning Project of Shenzhen Municipality in China (JCYJ20180306173000073), Natural Science Foundation of Guangdong Province

(2018A030313522), and Xiamen Municipal Science and Technology Projects (3502Z2019015).

REFERENCES

1. U. Sikder, G. Usai, TT. Yen, K. Horace-Herron, L. Hutin, TJK. Liu, "Back-End-of-Line Nano-Electro-Mechanical Switches for Reconfigurable Interconnects", IEEE Electron Device Letters, Vol.41, pp.625-628, Apr.2020.

2. V.Singh, V.Kumar, A.Saini, P.K.Khosla, S.Mishra, "Design and Development of the MEMS-Based High-g Acceleration Threshold Switch", Journal of Microelectromechanical System, Vol.30, pp.24-31, Feb.2021.

3. M.Liu, Y.M. Zhu, C.Wang, Y.Chen, Y.L. Wu, H.Zhang, Y.J. Du, W.D.Wang, "A Novel Low-g MEMS Bistable Inertial Switch With Self-Locking and Reverse-Unlocking Functions", Journal of Microelectromechanical System, Vol.29, pp.1493-1503, Dec.2020.

4. K. Kato, V. Stojanovic, T.J.K. Liu, "Non-volatile nano-electro-mechanical memory for energy-efficient data searching", IEEE Electron Device Lett., Vol. 37, pp. 31-34, Jan. 2016.

5. X.M Yang, "Ohmic Contact of High-current-density" Devices", Semiconductor Optoelectronics, Vol.23, pp. 274-27, May 2002.

6. X.U. Wei, Y.C.Wang, L.Luo, "Wafer level hermetic package of MEM S by glass solder at low temperature", Journal of Functional Materials and Devices, Vol.11, pp.343-346, Feb.2005.

7. C.S. Yu, M.H. Zhan, F.F.Hu, L.Y. Li, K.X. He, G.B. Xu, "Research and Design of a Novel Island Membrane Pressure", Micronanoelectronic Technology, Vol.52, pp. 446-451, Sep.2015.

8. R. Joyce, M. George, L. Bhanuprakash, D.K .Panwar, R.R. Bhatia, S. Varghese, J. Akhtar, "Investigation on the effects of low-temperature anodic bonding and its reliability for MEMS packaging using destructive and non-destructive techniques", Journal of Materials Science-materials in Electronics, Vol.29, pp. 217-231, Jan.2018.

9. R. Joyce, K. Singh, S. Varghese, J. Akhtar, "Stress reduction in silicon/oxidized silicon–Pyrex glass anodic bonding for MEMS device packaging: RF switches and pressure sensors", Journal of Materials Science-materials in Electronics, Vol.26, pp. 411-423, Jan.2015.

10. R. Tummala, V. Madisetti, "System on chip or system on package?" IEEE Des. Test. Comput. Vol.16, pp.48–56 Feb. 1999.

11. G.Q.Wu, D.H. Xu, X.Sun, B.Xiong, YL. Wang, "Wafer-Level Vacuum packaging for Microsystems Using Glass Frit bonding", IEEE Transactions on Components, Packaging and Manufacturing Technology, Vol.3, pp.1640-1646, Oct. 2013.

12. R.Knechtel, "Glass frit bonding: an universal technology for wafer level encapsulation and packaging", Microsystem Technologies, Vol.12, pp.63-68, Dec.2005.

13. G.Q. Wu, D.H. Xu, B. Xiong, Y.C. Wang, Y.L. Wang, Y.L. Ma, "Wafer-level vacuum packaging for MEMS Resonators using glass frit bonding", Journal of Microelectromechanical Systems, Vol.21, pp. 1484-1491, Dec.2012.

14. N. Lorenz, S. Millar, M. Desmulliez, DP. Hand, "Hermetic glass frit packaging in air and vacuum with localized laser joining", J. Micromech. Microeng., vol. 21, pp. 045-039, Apr. 2011.

15. L. Hofmann, S. Dempwolf, D. Reuter, R. Ecke, K.Gottfried, SE.Schulz, R. Knechtel, T. Gessner, "3D integration approaches for MEMS and CMOS sensors based on a Cu through-silicon-via technology and wafer level bonding", Conference on Smart Sensors, Actuators, and MEMS VII 1st SPIE Conference on Cyber-Physical Systems, Barcelona, pp. 951709-951709-12, May 04-06, 2015.

16. X. Chen, P. Yan, JJ. Tang, WG. Ning, GW. Xu, L. Luo, "Application of WLG with barrier trench structure in precision screen printing technology by glass frit", 11th International Conference on Electronic Packaging Technology & High Density Packaging, Xi'an, 16-19 Aug. 2010, pp.71-73.

17. Y.F. Liu, D.E Chen, L.W.Lin, G.F. Zheng, J.Y. Zheng, L.Y. Wang, D.H. Sun, "Glass frit bonding with controlled width and height using a two-step wet silicon etching procedure", Journal of Micromechanics and Microengineering, Vol.26, pp.1-8, Mar.2016.

18. W.T. Zhou, H.L. Bi, Z.H. Yu, XD. Wang, Y.J. Cheng, ZH. Xi, W.Wei, "Fabrication of the nanofluidic channels type leak assembly based on the glass frit sealing method", Journal of Vacuum Science & Technology, Vol.37, pp. 050603-1-050603-5, Sep. 2019 .

19. K.Cheung, *Die-level glass frit vacuum packaging for a micro-fuel processor system*, Massachusetts Institute of Technology, 2005.

Proceedings of the 16th Annual IEEE International
Conference on Nano/Micro Engineered and Molecular Systems
April 25-29, 2021

Damage profile model of nanostructure fabricated by Focused Helium Ion Beam

Chenglong Liu, Qi Li, Qianhuang Chen,and Yan Xing*

Abstract—Focused Helium Ion Beam （FHIB） can process nanostructures with high resolution, but the damage caused by FHIB cannot be ignored, therefore quantitative modeling and calculation are needed. The damage of He+ mainly includes the amorphization of crystal materials and the helium bubbles produced by helium accumulation in processing materials. In this paper, a helium bubble evolution model based on the reaction rate theory is proposed, which can simulate the evolution process of helium bubble in the substrate. Under a certain dose and energy input, the evolution of helium bubbles at different depths can be calculated, and the size characteristics and concentration of helium bubbles at different depths in the substrate after processing can be obtained. At the same time, two sets of experiments experiments were carried out by helium ion microscopy. In the first set of experiments，silicon substrate was injected with He+ with dose of 6.25 × 10^18 He+/cm^2 and injection energy of 10kev, 15kev and 35kev . In the second set of experiments，silicon substrate was injected with He+ at energy of 35keV and dose of 2.5×10^18 He+/cm^2 , 5×10^18 He+/cm^2, 6.25×10^18 He+/cm^2. Finally, the experimental results are compared with the simulation results.

I. INTRODUCTION

In recent years, the FHIB with its highly focused gas field ionization source and small beam-sample interaction volume, can image microstructures with high depth of field and high contrast. Therefore, it has been successfully applied to a variety of micro-nano manufacturing processes, such as Imaging of sub-nanostructures [1], milling of nanoholes, deposition of nanopillars, and repair of EVU photolithography masks [2] and so on. However, in the process of FHIB application, FHIB will cause significant damage to the substrate material, including the amorphization of the material [3,4], the nano-scale helium bubble and the surface expansion of the material, etc. The formation of helium bubbles will further lead to a series of helium brittle problems such as surface spalling and elongation, creep fracture time and significantly reduced fatigue life. The degree of brittleness depends on the size and density of helium bubbles. In this paper, the formation mechanism and evolution law of helium bubbles are studied, which will provide theoretical and experimental guidance for reducing substrate damage during focused helium ion beam processing.

When helium are introduced into solid materials by injection, because of its insolubility, helium atoms tend to be trapped in vacancies and impurity atoms. The migration process of a single helium atom in a solid is very complex. In the process of migration, helium atoms may be trapped by dislocations in crystal materials, or trapped in vacancy or

defect clusters , and finally form visible helium bubbles. In this paper, the evolution of helium bubbles in silicon substrate after helium ion implantation is simulated by using the reaction rate theory at multi-scale.

II. METHODOLOGY

2.1 Overview of the evolution of helium bubbles

The formation of helium bubbles includes four stages: injection, nucleation, growth, migration and coalescence. Migration and coalescence usually occurs when the material reaches high temperature (T > 0.5T_m, T_m is the melting temperature). Since the experiment was carried out at room temperature, the migration and coalescence mechanism of helium bubbles was not considered in our model. During the implantation process, helium ions will collide with the atoms of the substrate material nucleus, causing the crystalline material atoms to shift and produce Frenkel Pairs (FPs), and neutral helium atoms will be deposited in the gaps or replacement positions of the crystal lattice. During the deposition process, the nucleation process will also occur at the same time. The helium atoms and FPs in the substrate combine with each other during diffusion to form a stable helium bubble nucleus. With the increase of helium bubble concentration, it is more likely to absorb excess helium and vacancies in the existing helium bubbles than to form new helium bubbles on the substrate. After that, the growth process of helium bubbles is dominant, and the density of helium bubbles changes little for a long time. The flow chart of experiment and reaction rate model is shown in Fig. 1.

2.2 Nucleation model of helium bubbles

In the helium bubble nucleation model, the concentration changes of different defects are described by the rate theory equation. The reaction rate theory describes the diffusion of point defects and the interaction between them, and finally obtains the dynamic evolution of defects in one-dimensional space. The basic types of defects included in this model are self interstitial atoms (SIAs, I), vacancies (V), helium atoms (He) and complex clusters formed by binary reactions (I_n, V_n, He_n and $He_n V_m$, m and n are the number of defects in clusters)

$$\frac{\partial C_\theta}{\partial t} = G_\theta + D_\theta \nabla^2 C_\theta + \Sigma[k(\theta',\theta)C_\theta - k(\theta,\theta')C_\theta] - L_\theta \quad (1)$$

Where C_θ is the concentration of defect θ, G_θ is the yield, D_θ is the diffusion coefficient, and k is the reaction coefficient. Table 1 shows the types of defect reactions and their reaction coefficients and thermal emission coefficients.

All authors are with Jiangsu Key Laboratory for Design and Manufacture of Micro-Nano Biomedical Instruments, School of chanical Engineering, Southeast University, Nanjing 211189, China

*Contacting Author: Y. Xing is with the Jiangsu Key Laboratory for Design and Manufacture of Micro-Nano Biomedical Instruments, School of

Mechanical Engineering, Southeast University, Nanjing 211189, China (phone: +86 025-52098413; email: xingyan@seu.edu.cn)

978-1-6654-3008-1/21 $31.00 © 2021 IEEE

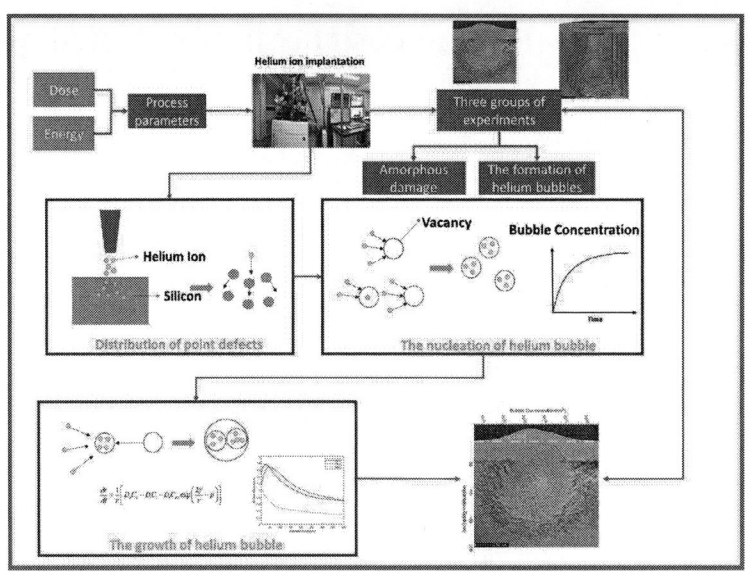

Fig. 1 Flow chart of experiment and reaction rate model

Helium bubbles are essentially stable clusters, which are generated by continuous reaction of point defects after injection. Due to the variety of defective reaction combinations involved in the reaction process, the model needs to be simplified. In this paper, the model is simplified by the following assumptions: (1) Ignoring the inherent defects of the material, all the defects in the substrate are produced by helium ion implantation. (2) Any complex defect cluster containing three helium atoms is considered as a critical nucleus. After the formation of the critical nucleus, as long as it absorbs any vacancy or helium atom, it is considered as forming a stable helium bubble nucleus. (3) Because the mobility of clusters is closely related to their size, and only a single point defect (such as vacancy, self interstitial atom and helium atom) is considered to migrate in a short time scale, the migration of complex defect clusters is ignored in the

model, and only a single point defect is considered to move. (4) After the formation of helium bubbles, they only interact with a single point defect, ignoring the interaction between clusters and helium bubbles.

We use SRIM to obtain the deposition concentration of defects in silicon. Combined with the helium ion flux, we obtain the deposition rates $G_{I/V/He}$ of defects at different depths.

2.3 Growth model of Helium bubble

The growth of helium bubble mainly includes three factors: (1) the absorption of vacancy and self interstitial atom; (2) the absorption of helium atom; (3) the pressure balance inside and outside the helium bubble. The change expression of average bubble size is as follows:

$$\frac{dr}{dt} = \frac{1}{r}\left[D_v C_v - D_i C_i - D_v C_{ev}\exp(\frac{2\gamma}{r} - p)\right] \quad (2)$$

Where D_v and D_i are the diffusion coefficients of vacancies and self-interstitial atoms, C_v and C_i are the concentrations of vacancies and self-interstitial atoms, C_{ev} is the equilibrium concentration of vacancies, r is the radius of the helium bubble, γ is the surface tension, and p is the gas pressure in the bubble calculated using a virial expansion of the helium equation of state up to the third order. Formation energy, migration energy and binding energy are key parameters in the reaction rate theory. The parameters used in this paper are carefully selected from published experimental values or from ab initio/MD calculations. The migration energies of SIA, V and He used in this paper are $E_I^m = 0.07$ ev[5], $E_v^m = 0.3$ ev[6], $E_{He}^m = 0.68$ ev[7], respectively. For the complex defects included in the reaction rate equation, the binding energies are $E_{He-v}^b = 3.13$ ev, $E_{He2-v}^b = 5.55$ ev, $E_{He2}^b = 0.048$ ev.

The above-mentioned series of reaction rate equations are solved by full implicit iterative scheme using MATLAB. Once the injection dose reaches the preset value, the calculation will be stopped and the data will be stored.

Table 1 Reaction types and reaction coefficients

Reaction type	Reaction coefficient
$I + V \Leftrightarrow 0$	k_{I+V}^+
$I + HeV \rightarrow He$	$k_{IIoV_n+I}^+$
$I + He_2V \rightarrow He_2$	$k_{He_2V+I}^+$
$V + He \Leftrightarrow HeV$	k_{He+V}^+, k_{HeV}^-
$V + HeV \Leftrightarrow HeV_2$	$k_{He+V_2}^+, k_{HeV_2}^-$
$V + He_2 \Leftrightarrow He_2V$	$k_{He_2+V}^+, k_{He_2V}^-$
$V + He_2V \Leftrightarrow He_2V_2$	$k_{He_2V+V}^+, k_{He_2V_2}^-$
$V + He_3 \rightarrow He_3V$	$k_{He_3V}^+$
$He + He \Leftrightarrow He_2$	$k_{He+He}^+, k_{He_2}^-$
$He + HeV \Leftrightarrow He_2$	$k_{He_2V+He}^+$
$He + He_2 \Leftrightarrow He_3$	$k_{He+He}^+, k_{He_2}^-$
$He + He_2V \Leftrightarrow He_3V$	$k_{He+He}^+, k_{He_2}^-$
$He + He_3 \rightarrow He_4$	$k_{He+He}^+, k_{He_2}^-$

III. RESULT AND DISCUSSION

In order to verify the accuracy of the model, we compare the simulation results with the experimental results. All experiments were performed at room temperature, using helium ion microscope (ZEISS ORION NanoFab) to inject helium ion beams into the silicon substrate (100) in a line scan manner.

In the first set of experiments, the dose was set to 6.25×10^{18}He$^+$/cm^2. Helium ions with energies of 10keV, 15keV and 35keV were implanted into the silicon substrate respectively. The TEM damage cross section of the implanted silicon substrate is shown in Fig. 2. It can be observed that at a low energy of 10 keV, the size of the helium bubble in the substrate is significantly larger than when the injection energy is 15 keV and 35 keV. There may be two reasons for this phenomenon: (1) Self-trap mechanism. (2) Some of the helium bubbles coalesced, resulting in the growth of helium bubbles.

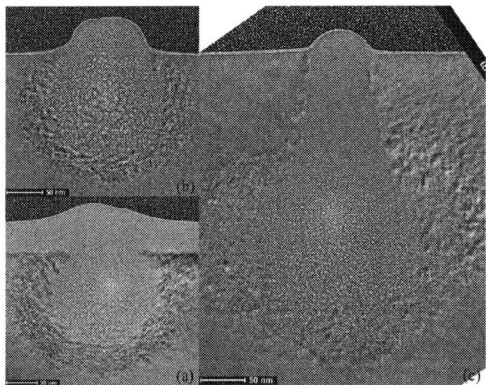

Fig .2 TEM images of cross-section silicon sample of damage profile with ion dose of 6.25×10^{18} /cm^2(a) 10 keV (b) 15 keV and (c) 15 keV

The results of the reaction rate equation under the same parameters are shown in Fig 3. It can be found that the size of helium bubbles increases first and then decreases with the energy of 15 keV and 30 keV in the reaction rate model, reaching the maximum radius of 1.67 nm and 1.75 nm at the depth of 27 nm and 50 nm respectively, and then reaching the steady state when the radius reaches about 0.7 nm. When the implant energy is 10keV, the radius of the helium bubble decreases from the initial value of 1.07nm with the increase of depth, and finally reaches the steady-state value of 0.3nm at 260nm. The variation trend of the concentration was similar to the average radius, showing a trend of first increasing and then decreasing. The helium bubble concentration reaches the maximum at 120nm, 300nm and 350nm, and then decreases continuously. Table 2 shows the statistical data of experiments and simulations of this parameter. Comparing the simulation results with the experimental results, we find that the average radius of helium bubble in the experiment deviates greatly from the reaction rate model at the energy of 10kev. The main reason for the deviation is that the self trap mechanism and helium bubble coalescence are not considered in the growth model.

When the energy is set to 35kev, the doses is 3.75×10^{18}He$^+$/cm^2, 5×10^{18}He$^+$/cm^2, 6.25×10^{18}He$^+$/cm^2, respectively. The calculated helium bubble size and concentration change curve is shown in Fig. 4. We found that

Table 2 Experimental and simulation results with dose of 6.25×10^{18}He$^+$ / cm2 and injection energy of 10keV, 15kev and 35kev respectively

Energy (kev)	Type	Average radius (nm)	Concentration (/m^3)
10	Experiment	1.90	2.69×10^{26}
10	Simulation	0.76	1.51×10^{26}
15	Experiment	1.57	4.43×10^{26}
15	Simulation	1.41	1.72×10^{26}
35	Experiment	1.08	1.10×10^{27}
35	Simulation	0.77	7.74×10^{26}

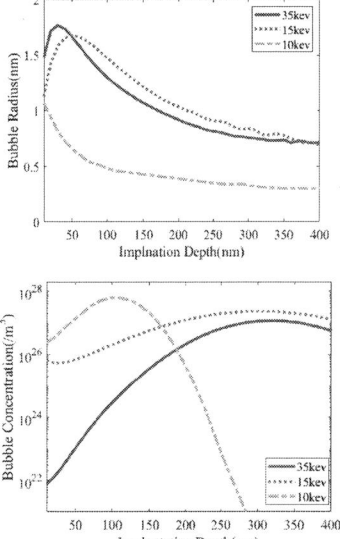

Fig. 3 The concentration and size of helium bubbles varied with depth at a dose of 6.25×10^{18}He$^+$/cm^2 with injected energies of 10, 15 and 35 keV

the size changes and concentration changes of the helium bubbles in the three cases are very similar. The average size of the helium bubble has a short-distance growth process value near the surface, and then continues to decrease and finally converges. The statistical data of the experiment and simulation results are shown in Table 3. It can be found that both the experimental and simulation results show that the average concentration and size of helium bubbles continue to increase with the increase of the dose. Because the injection energy is the same, the difference is proportional to the yield of the point defects substituted in the reaction rate model. High dose will cause more point defects at the corresponding

Table 3 Experimental and simulation results with energy of 35 keV and injection dose of 3.75×10^{18}He$^+$ / cm^2, 5×10^{18}He$^+$ / cm^2 and 6.25×10^{18}He$^+$ / cm^2 respectively

Dose (He$^+$ / cm^2)	Type	Average radius (nm)	Concentration (/m^3)
3.75×10^{18}	Experiment	1.09	6.92×10^{26}
3.75×10^{18}	Simulation	0.61	2.02×10^{26}
5×10^{18}	Experiment	1.15	7.06×10^{26}
5×10^{18}	Simulation	0.67	4.92×10^{26}
6.25×10^{18}	Experiment	1.08	1.10×10^{27}
6.25×10^{18}	Simulation	0.77	7.74×10^{26}

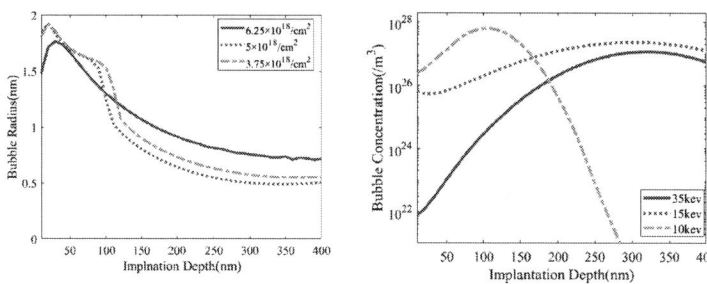

Fig. 4 The concentration and size of helium bubbles with energy of 35 keV and dose of $3.75 \times 10^{18} He^+ / cm^2$, $5 \times 10^{18} He^+/cm^2$ and $6.25 \times 10^{18} He^+/cm^2$

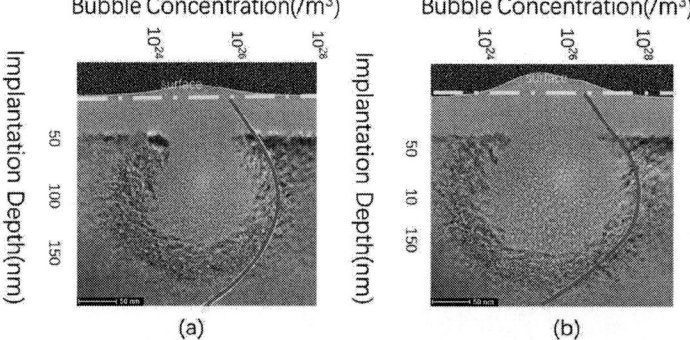

Fig. 5 Simulations and experimental results of Concentration helium bubbles at dose of 5×10^{18} /cm^2, 3.75×10^{18} /cm^2

depth of the substrate, which will lead to the nucleation and growth of helium bubbles more easily.

In Fig. 5, the helium bubble concentration curves of $5 \times 10^{18} He^+/cm^2$ and $6.25 \times 10^{18} He^+/cm^2$ at the energy of 10kev are plotted on the TEM images. It can be found that when the dose is $5 \times 10^{18} He^+/cm^2$, it is difficult to observe the presence of helium bubbles in the shallow layer of the TEM image.

IV. CONCLUSION

The accuracy of the model is verified by comparing the calculated results with the experimental results. The results show that: (1) In the case of low energy, the radius of the helium bubble is larger and the concentration is lower than high energy. (2) In the case of high energy, Increased dose will increase the average radius and concentration of helium bubbles (3) The concentration of helium bubble is an important factor to observe helium bubble. when the helium bubble concentration reaches a certain order of magnitude (about Cb >10^{26}/m^3), the existence of helium bubbles can be observed. (4) In the case of low energy, the calculation results deviate greatly from the experiments. The deviation of calculation results may be caused by self-trapping or coalesce.

ACKNOWLEDGMENTS

We acknowledge financial supports from the National Natural Science Foundation of China (No.51875104) and the National Key R&D Program of China (No. 2018YFB2002600).

REFERENCES

[1] D. Emmrich, A. Beyer, A. Nadzeyka, S. Bauerdick, J.C. Meyer, J. Kotakoski, A. Gölzhäuser, Nanopore fabrication and characterization by helium ion microscopy, Appl. Phys. Lett. 108 (2016).

[2] C.M. Gonzalez, W. Slingenbergh, R. Timilsina, J.-H. Noh, M.G. Stanford, B.B. Lewis, K.L. Klein, T. Liang, J.D. Fowlkes, P.D. Rack, Evaluation of mask repair strategies via focused electron, helium, and neon beam induced processing for EUV applications, in: Extrem. Ultrav. Lithogr. V, SPIE, 2014: p. 90480M.

[3] Shao T , Chen Q , Xing Y , et al. An Experiment Based Damage Profile Function for Focused Helium Ion Beam Process in Fabrication of Micro/Nano Structures[C] 2020 IEEE 33rd International Conference on Micro Electro Mechanical Systems (MEMS). IEEE, 2020.

[4] Chen Q , Shao T , Xing Y . An Experiment-Based Profile Function for the Calculation of Damage Distribution in Bulk Silicon Induced by a Helium Focused Ion Beam Process[J]. Sensors, 2020, 20(8):2306.

[5] Hallén A , Keskitalo N , Josyula L , et al. Migration energy for the silicon self-interstitial[J]. Journal of Applied Physics, 1999, 86(1):214-216.

[6] Caliste D , Pochet P . Vacancy-Assisted Diffusion in Silicon: A Three-Temperature-Regime Model[J]. Physical Review Letters, 2006, 97(13):135901.

[7] Pizzagalli L , Charaf-Eddin A . Migration of noble gas atoms in interaction with vacancies in silicon[J]. Semiconductor Science & Technology, 2015, 30(8):085022.

Proceedings of the 16th Annual IEEE International
Conference on Nano/Micro Engineered and Molecular Systems
April 25-29, 2021

Investigation of Broading Modulus Range of Soft Probes by Single-beam Acoustic tweezer

Huiyao Shi[1,2,3], Jialin Shi[1,2], Peng Yu[1,2], and Lianqin Liu[1,2]*, Member, IEEE

Abstract— Atomic force microscopy has been developed as a primary method to quantitatively measure the nanomechanical properties of materials and cells. The elastic modulus of the sample can be acquired by measuring the deformation of the probe and the indentation of the sample. The spring constant of the probes directly determines the resolution and range in the modulus measurement. When the material modulus span is large, the probe cannot achieve an effective indentation depth, resulting in a soft probe is not competent for a stiff sample. Here we simulated the feasibility of driving a soft probe together with the acoustic radiation force and the acoustic steaming force generated by a single-beam acoustic tweezer. The single-beam acoustic tweezer can generate a driving force of more than 10 nN on the cantilevers. This work provides a reference for wide-range mechanical properties measuring method from 1 kPa to 1 GPa using a single soft probe.

I. INTRODUCTION

Quantitative measuring mechanical properties in micro/nanoscale has received extensive attention in medical science[1] and advanced nanomaterials[2]. Atomic force microscope (AFM) has become an effective tool for nanomechanical measurement due to its sub-nanometer-level imaging resolution and pN-level mechanical resolution. The typical modulus measuring principle of AFM is pressing a tiny probe to the surface of the sample, and obtain a force curve by measuring the deflection of the cantilever. At the same time, the difference between the expansion of the piezo-stage and the deflection of the cantilever represents the deformation of the sample. Then the relationship between stress and strain on the sample is calculated by the contact mechanism models.

Therefore, in the process of mechanical properties measurement, we need to meet two conditions: firstly, the sample should be hard enough to allow the cantilever to produce measurable deformation. Then, the sample should be soft enough, so that the probe can be pressed into the sample within its linear strain range. This brings about the problem of stiffness matching between AFM probes and samples.

---

*This work was supported by the National Natural Science Foundation of China (Grants 61927805), Youth Program of National Natural Science Foundation of China (Grants 61903359).

[1]State Key Laboratory of Robotics, Shenyang Institute of Automation, Chinese Academy of Sciences, Shenyang, China.

[2]Institutes for Robotics and Intelligent Manufacturing, Chinese Academy of Sciences, Shenyang 110169, China.

[3]University of Chinese Academy of Sciences, Beijing, China.

* Lianqing Liu is with State Key Laboratory of Robotics, Shenyang Institute of Automation, Chinese Academy of Sciences, Shenyang, China (e-mail: lqliu@sia.cn).

Broadening the measuring range of AFM probes by additional force is a promising method. The magnetic force has been proved to broaden the range of a soft probe [3]. However, the magnetic field is a low energy conversion method and will cause a sharp temperature rise on the probes. A possible alternative is the sound field driving method. Ultrasonic transducers have high energy conversion efficiency. High-frequency ultrasonic transducers can generate a positive excitation force exceeding 10 nN through acoustic radiation force[4] and acoustic steaming force[5]. In this paper, we have calculated the acoustic radiation force and the acoustic steaming force when a probe is driven by the high-frequency ultrasonic transducer. These results indicated that a commercialized soft probe (k=0.01 N/m) driven by a single-beam acoustic tweezer has a broad measurement range from kPa to GPa. This work provided a theoretical basis and simulation verification for the design of an ultrasonic-driven AFM system.

II. NUMERICAL MODEL

A. Model system and computational domain

The transducer and AFM probe are immersed in water(Figure.1a), the probe is selected as a commercial triangle cantilever with a spring constant of 0.01 N/m. Neglecting the matching of fluid environment and acoustic transducer. We simplified the transducer into an arc structure inserted into the water. The diameter of transducer is D and the focus radius is R (Figure 1b).

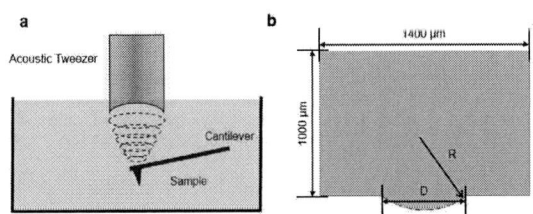

Figure 1. (a) End-view of single-beam acoustic tweezer actuated AFM, the transducer and cantilever is immersed in a sink. (b) The corresponding simulation area is about 1000 μm height and 1400 μm width. The transducer is set as an arc with a diameter of D and the focus center is R.

B. Model parameters

The simulation is set as a two-dimensional model, and the center frequency of the transducer is 100 MHz. The relevant parameter of the medium is listed in Table 1.

Table 1 Model parameters. The parameters are taken from the COMSOL Material Library and all given at temperature T=25°C.

978-1-6654-3008-1/21 $31.00 © 2021 IEEE

Water	Name	Value	Unit
Density	ρ_0	1000	kg · m⁻³
Viscosity	μ	2.98×10^{-3}	Pa · s
Absorption	α	0.025	1 · m⁻¹
Speed	c_0	1495	m · s⁻¹
Temperature	T_0	293.7	K
Frequency	f_0	100	MHz

C. Results

Acoustic actuated AFM system schematically illustrated in Figure 2a. The mechanical properties of samples can be measured by a tiny cantilever (Figure 2b). Typically, to ensure the accuracy of the modulus, the cantilever must work in a small linear deformation range, and ensure that the ratio of the deformation of the cantilever to the indentation of the sample is between 1/3 and 3. There is a modulus measuring range of each AFM probes, which brings about the problem of modulus matching in the characterization of mechanical properties by AFM (Figure 3c).

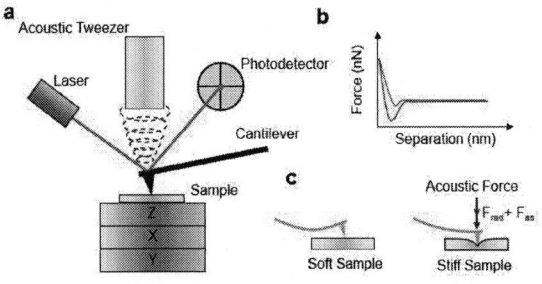

Figure 2. (a) Schematic of acoustic-actuated AFM system, samples are placed on a piezo-stage. Laser and acoustic are focused on the cantilever, and the laser is reflected onto a photodetector. (b) A typical force curve is exhibited, blue and red curves represent approaching and away from the samples. (c) A soft probe cannot indent into a stiff sample, while the acoustic force and help a soft probe pressed into the sample.

In order to improve the probe modulus measuring range, the single-beam acoustic tweezer is applied to increase the pressure on the samples. We simulated acoustic pressure under different diameters and focus centers of the transducer. As shown in figure 3a, acoustic pressure at the focus center mainly depends on the frequency and the amplitude of the transducer. With the increase of vibration velocity, acoustic pressure increase from 0 to 3 MPa. The larger size of the transducer also leads to a slightly increased pressure.

We measured the beam size of the focus center, the lateral resolution of the beam is about 18 μm. Acoustic will not only cause sound radiation force but also lead to a stable

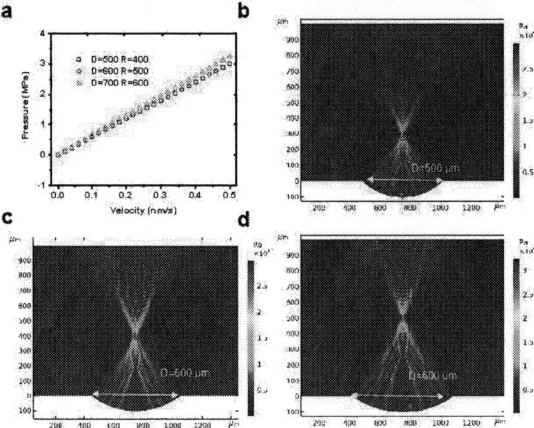

Figure 3. (a) Simulation of acoustic pressure in various velocities. With the increase of velocity, acoustic pressure increase from 0 to 3 MPa. Simulation of acoustic pressure field at the condition of the focus center and diameter of the transducer is 400 μm and 500 μm respectively (b); 500 μm and 600 μm respectively (c); 600 μm and 700 μm respectively, (d).

acoustic streaming field. When the pressure increases, the velocity of the acoustic streaming also increases and the maximum velocity point moves backward. The acoustic flow field presents a jet state (Figure.5a,c).

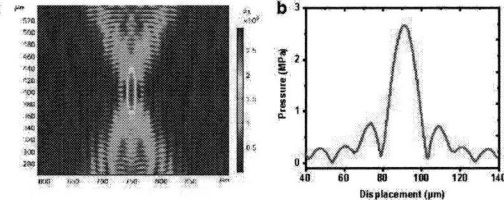

Figure 4. Focus center of single-beam acoustic tweezer. (b) The lateral resolution (-6dB beamwidth) is 18 μm.

From the simulation of the acoustic flow field, we can observe that acoustic streaming also can increase the force applied on a cantilever. The velocity will increase to 0.27 m/s when the pressure is over 3MPa.

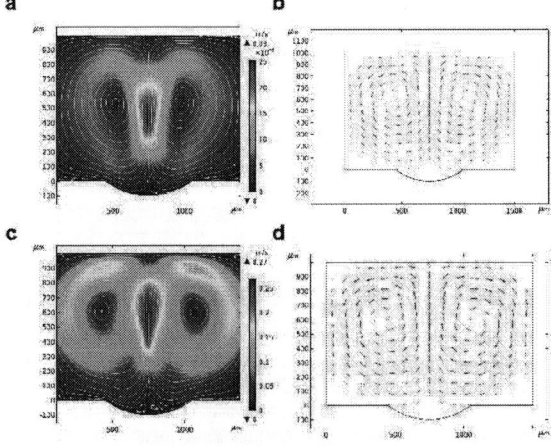

Figure 5. (a) Acoustic streaming field and streamline(b) when the transducers work at 0.1 nm/s. (c) Acoustic streaming field and streamline(d) when the transducers work at 0.5 nm/s.

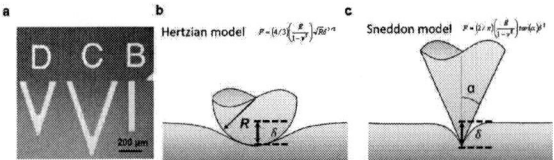

Figure 6. (a) Commercial AFM probe, the spring constant of MLCT-C is about 0.01 N/m. The theory used to compute modulus by (b) Hertzian and (c) Sneddon model.

Details of cantilever were characterized (Figure 6a), we stimulated the force applied by a commercial probe (MLCT-C). The spring constant of the cantilever is about 0.01 N/m. The modulus of the sample can be calculated by Hertzian model and Sneddon model (Figure 6bc).

III. MATH

From the above simulation, we can understand that the radiation force generated by acoustic pressure and the fluid drag force generated by acoustic streaming field will have a synergistic effect.

When a cantilever target was placed along the path of a focused acoustic wave beam, a time-averaged force can generate, which is known as the acoustic radiation pressure. At the focal plane, the amplitude of the acoustic pressure wave was assumed as P_{ac} and the pressure reflection coefficient can be defined as Γ.

$$U = \frac{P_{ac}^2}{2\rho c^2}\left(1+|\Gamma|^2\right)$$

Using the relation that the average intensity of the incident beam is given by $I_{ac} = P_{ac}^2/(2\rho c)$. Acoustic radiation pressure on the cantilever, F_{arf}, can be expressed in terms of the intensity as

$$F_{arf} = I_{ac}\frac{\left(1+|\Gamma|^2\right)}{c}$$

The total force applied to the cantilever can be found by integrating the radiation pressure, and hence it is proportional to the average power incident on the cantilever.

The intensity of acoustic power was integrated on the focus plane (3.4756×10^{-4} W), and the force applied on the cantilever was estimated as 232.5 nN. The deflection sensitivity and tip radius of MLCT-C are 20 nm/V and 20 nm. In order to ensure the accuracy of modulus measuring, the trigger voltage is below 1V. The force produced by the deformation of a cantilever is about 0.2 nN. Assuming that sample can be indented into 2 nm and the modulus is about 12 GPa.

For an incompressible and Newtonian fluid, an acoustic steaming motion was caused by viscous attenuation of the sound field. Reynolds stress difference of fluid element is $-\left(\partial\overline{\rho_0 u_i u_j}/\partial x_i\right)hS$.

The spatial variation of Reynolds stress is expressed as

$$F_j = \frac{-\partial\overline{\rho_0 u_i u_j}}{\partial x_i}$$

Navier-Stokes equation of steady acoustic streaming was expressed as:

$$\rho_0\left(\overline{u_i\partial u_j}/\partial x_i\right) = -\partial\overline{P}/\partial x_j + \eta\nabla^2\overline{u_j} + F_j$$

For an incompressible fluid, mass conservation law reads:

$$\partial\overline{u_i}/\partial x_i = 0$$

Generally, if the ultrasound wave propagated on the longitudinal scale much larger than its wavelength, a steady flow can arise from the ultrasound absorption, which is called Eckart steaming and the streaming motion can be described by the following equation.

$$\rho_0 u_i \cdot \frac{\partial u_j}{\partial x_i} = F_j - \frac{\partial p}{\partial x_j} + \mu\frac{\partial^2 u_j}{\partial x_i \partial x_i}$$

For a cantilever, the steaming force was estimated as

$$F_{asf} = \rho A u(x,0)\cdot u(x,0)$$

Considering the Reynolds number.

$$Re = \frac{\rho v d}{\mu}$$

μ is the dynamic viscosity of the fluid n water, μ =2.98$\times$ 10^{-3} Pa·s, ρ is the density of the fluid, ρ =1000 kg·m^{-3}.

The steaming force was estimated as 74 nN. Combing with the acoustic steaming force and acoustic radiation force, the force applied to the probe increased from 0.5 nN to 306 nN. The measurement limit of the probe is extended to 16GPa.

The results in this paper are the simulation and theoretical estimation of the mechanical characterization of the sample by the acoustic-driven AFM. Acoustic transducer can bring a large driving force on a probe by acoustic radiation force and acoustic streaming force. But the acoustic steaming may bring the vibration of AFM probe, eventually lead to inaccurate measurement. While in the simulation, we found that under lower acoustic pressure, the velocity of acoustic streaming will be greatly reduced. Our simulation provides partial theoretical verification for the development of an acoustic-driven AFM system. The advantage of acoustic actuated AFM is its board modulus measuring range. We may succeed in a modulus range from kPa to GPa by a soft probe (0.01 N/m). In biological system, tissues can soft as cells(kPa) and stiff as bones(GPa). When measuring multi-level tissue structure, changing probes will introduce lots of errors. Acoustic-driven AFM may provide new methods for the mechanical properties measurement in biological tissue level

IV. CONCLUSION

In summary, we presented acoustic actuated AFM to extend the modulus measuring range of the soft probe. The single-beam acoustic tweezer can increase the force applied on the cantilever by acoustic pressure and acoustic streaming. The results indicated there is about 300 nN will be applied to the cantilever. By this method, we may succeed in a modulus range from kPa to GPa by a soft probe.

ACKNOWLEDGMENT

Huiyao Shi thanks Daojin Lin and Wenjun Tan for assistance with the simulation and valuable discussion.

REFERENCES

[1] M. Li, N. Xi, Y. Wang, and L. Liu, "Atomic force microscopy for revealing micro/nanoscale mechanics in tumor metastasis: from single cells to microenvironmental cues," Acta Pharmacol. Sin., (2020).

[2] V. Basavalingappa, S. Bera, B. Xue, J. O Donnell, S. Guerin, P. Cazade, H. Yuan, E. U. Haq, C. Silien, K. Tao, L. J. W. Shimon, S. A. M. Tofail, D. Thompson, S. Kolusheva, R. Yang, Y. Cao, and E. Gazit, "Diphenylalanine-Derivative Peptide Assemblies with Increased Aromaticity Exhibit Metal-like Rigidity and High Piezoelectricity," ACS Nano, 14, pp. 7025-7037, (2020).

[3] X. Meng, H. Zhang, J. Song, X. Fan, L. Sun, and H. Xie, "Broad modulus range nanomechanical mapping by magnetic-drive soft probes," Nat. Commun., 8, (2017).

[4] C. Fei, Y. Li, B. Zhu, C. T. Chiu, Z. Chen, D. Li, Y. Yang, K. Kirk Shung, and Q. Zhou, "Contactless microparticle control via ultrahigh frequency needle type single beam acoustic tweezers," Appl. Phys. Lett., 109, p. 173509, (2016).

[5] R. B. H. Slama, B. Gilles, M. B. Chiekh, and J. C. Bera, "Characterization of focused-ultrasound-induced acoustic streaming," Exp. Therm. Fluid Sci., 101, pp. 37-47, (2019).

Gap in pagination due to unavailable papers.

Pages 853-863

Proceedings of the 16th Annual IEEE International
Conference on Nano/Micro Engineered and Molecular Systems
April 25-29, 2021

Factors Influencing Resolution of Optical Fiber End Face Processing in Digital Lithography

Menghan Xiong, Ningning Luo[*], Zhimin Zhang and Qingwang Meng

Abstract—We report a method based on digital micromirror processing on the end face of optical fiber, and successfully fabricated functional microdevices at the submicron level on the optical fiber end face, This method does not require a physical mask, is low cost and allows rapid prototyping. This system produces microdevices with resolution up to submicron. In this paper, the influence factors of the imaging resolution of the DMD (Digital Micromirror Device) processing system are also analyzed, and the results of theoretical simulation and experiment are combined, which are of great significance to improve the resolution of digital lithography.

I. INTRODUCTION

Optical fiber has become an important carrier of information transmission in modern communication technology because of its low loss, long distance transmission capacity， wide frequency band and large capacity of information transmission, free from electromagnetic interference and many other advantages. The functional optical components of the fiber end face provide new features for high-performance compact fiber systems, on the basis of maintaining the advantages of large capacity, no interference and easy networking of optical fiber communication. The processing methods of optical fiber end devices used in the market are mainly divided into the following: femtosecond laser direct writing, electron beam lithography and focused ion beam lithography. These processing methods have high resolution. However, these methods are not special for processing of optical fiber end components and the cost of a single equipment is more than RMB 2 million. There is no low-cost and fast special equipment for digital processing of optical fiber end components on the market.

In this paper, a digital lithography system is designed and built for processing direct integrated functional components of optical fiber end face. By using DMD optical fiber end face micromachining method, the digital mask displayed in real time can spatially modulate the ultraviolet light, the shape and size of optical fiber end face microstructure are controlled precisely. In order to meet the requirements of semiconductor processing, the resolution of the digital lithography system is an important parameter in the whole system, and the precision of the resolution of the system is the main factor restricting the machining size.

All authors are with the Key Laboratory of Opto-Electronic Information Science and Technology of Jiangxi Province, and Key Laboratory of Nondestructive Test (Ministry of Education), Nanchang Hangkong University, Nanchang, China.

*Contacting Author: Ningning Luo (phone:+86-13879177625; email:ningningluo2002@126.com).

Therefore, it is very important to analyze the influencing factors of system resolution.This paper simulates the lithography system, compares the experimental results, and analyzes and discusses the resolution factors in detail.

II. DIGITAL LITHOGRAPHY BASED ON DMD

The digital lithography system uses DMD to replace the mask in the traditional lithography technology. The experimental schematic diagram and system device are shown in Fig.1 and Fig.2. Digital fiber processing system consists of lighting system, projection system and imaging system. The lighting system consists of a UV LED (Ultraviolet Light-Emitting Diode) source(365 nm in center wavelength, and 120 mW power), a light collimator and a DMD(1024x 768 arrays of 13.68 x 13.68μm micrromirrors). The projection system includes a projection objective lens (40*, NA=0.65), a CCD (Charge Coupled Device) and a beam splitter. The imaging system mainly consists of base, control box and 3D displacement platform. The ultraviolet beam emitted by the light source through the plane light of the collimator enters the DMD at a fixed angle. The DMD lens modulates the digital mask designed by the computer. The beam reflected by the DMD passes through the beam splitter lens and enters the projection objective lens for imaging on the pupil behind. The three-dimensional displacement platform accurately controls the position of the substrate in three directions, so that the optical fiber coated with photoresist coating on the substrate is placed on the pupil of the projection objective lens. The light reflected from the end face of the fiber is amplified by the projection objective lens, and then divided by the beam splitter to the CCD image sensor to monitor the reflection pattern of the end face of the fiber in real time. The optical fibers of different sizes can be patterned by different magnification objective lenses.

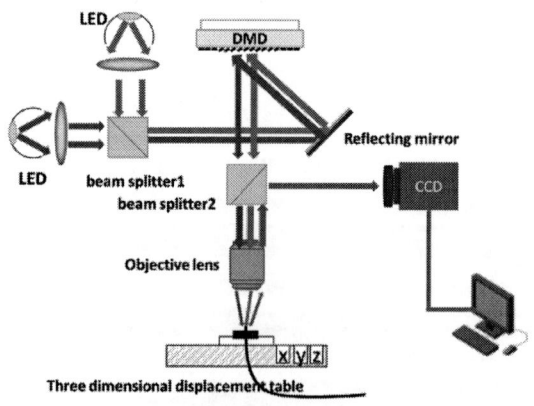

Figure 1. System optical path diagram.

978-1-6654-3008-1/21 $31.00 © 2021 IEEE 864

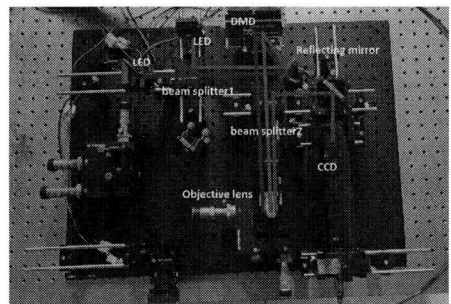

Figure 2. The whole experimental light path object and light path trajectory diagram.

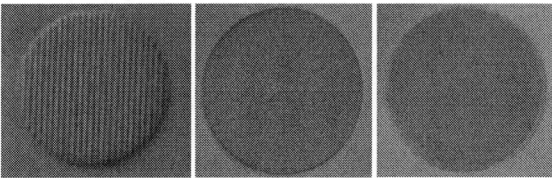

Figure 3. Microstructure of optical fiber end face

Fig. 3 shows some of the microstructures, such as linear and ring gratings, and orthogonal gratings, prepared by this system on the optical fiber end faces. The optical fiber processing system based on DMD can achieve spatial resolution of 700nm.

III. INFLUENCING FACTORS OF RESOLUTION

In the manufacture of optical fiber end devices, small size functional devices are expected, so in order to meet the requirements of device precision, system resolution is a very important factor. In the optical fiber end face digital lithography system, the main factors affecting the resolution of lithography technology are light source, DMD and projection objective lens.

A. The light source

DMD digital lithography generally uses an ultraviolet wavelength laser as the light source. If the wavelength is too long, the light absorption of DMD micromirror array will be large, and the life of DMD will be damaged. The modulated ultraviolet light needs to pass through the projection objective lens and then image at the end face of the optical fiber, so the wavelength of the light source has a great influence on the projection objective lens imaging. A projection objective has a diffraction limit

$$R = k\frac{\lambda}{NA}. \tag{1}$$

NA is the numerical aperture of the projection objective lens, λ is the exposure wavelength, and k is the process factor. It can be seen from the above formula that in order to improve the resolution and reduce the minimum diffraction value of the projection objective, the wavelength of the light source should be reduced. Therefore, the UV light source with the wavelength of 365nm is selected as the light source of the system. In digital lithography, the DMD modulated beam requires high uniformity of the light source. If the light

source for exposure has poor uniformity and collimation, it will lead to the phenomenon of different thickness and interval depth of the microstructural lines after exposure. A linear grating with a period of 4 pixels and a line width of 27.36 microns is simulated using a central light source. The simulation results are shown in Fig.3. It can be clearly seen from the figure that the edge of the grating structure is not exposed enough due to insufficient light intensity, while the center is overexposed due to excessive light intensity. Fig.4 and Fig.5 are the experimental results.

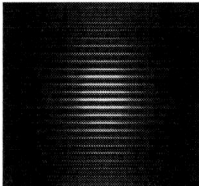

Figure 4. Simulation results of linear grating

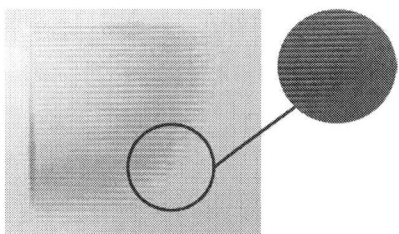

Figure 5. Experimental results of linear grating

B. DMD

The special structure of micromirror array of DMD determines that DMD is a mixed element of reflection and diffraction. DMD projection imaging depends on the surface of the micro mirror according to the angle of reflection, according to the law of reflection; moreover, because of the two-dimensional structure of DMD, the beam passing through DMD presents the diffraction properties of the shining grating. When all micromirrors are "open", DMD is similar to a two-dimensional diffraction grating with a period of d (typically 13.68μm) and a pore size of p (typically 12.68μm). Therefore, it is necessary to consider the effects of reflection and diffraction when analyzing the modulation law of DMD beam and coherent imaging law under coherent light irradiation.

The theoretical model of spatial light modulation characteristics of DMD is established. Assume that M(x, y) is a black and white with a contrast of 1, then the output function of the mask after passing through the DMD micromirror is,

$$M'(x,y) = M(x,y)\left[\sum_{m=0}^{M-1}\sum_{n=0}^{N-1}\delta(x - mT_x, y - nT_y) * rect\left(\frac{x}{W_x}, \frac{y}{W_y}\right)\right]$$

$$= \frac{1}{T_xT_y}M(x,y) * \left[rect\left(\frac{x}{W_x}, \frac{y}{W_y}\right) * comb\left(\frac{x}{T_x}, \frac{y}{T_y}\right)\right]$$

$$\tag{2}$$

W_x and W_y are unit widths of each micromirror respectively; T_x and T_y are pixel cycles in horizontal and vertical directions of the micromirror; M and N are the number of DMD micromirrors on X and Y axes respectively; $comb(\dfrac{x}{T_x},\dfrac{y}{T_y})$ is a two-dimensional dressing function.

rect is the rectangle function. The Fraunhofer diffraction of $M'(x,y)$ is the spectrum of the above equation, obtained by applying the Fourier transform of $M'(x,y)$

$$M'(f,g)=\sum_{m=0}^{M-1}\sum_{n=0}^{N-1}\frac{W_xW_y}{T_xT_y}*M\left(f-\frac{m}{T_x},g-\frac{n}{T_Y}\right)*sinc(fW_x)sinc(gW_y)$$
(3)

Ignoring the constant term independent of distribution, the spectrum distribution obtained from the Fourier transform of the sampled image of DMD is modulated by the *sinc* function, where:

$$\sin c(x)=\frac{\sin(\pi x)}{\pi x}.$$
(4)

The above equation shows that M'(f, g) is copied in large quantities and modulated by the sinc function, and a series of diffractive spots will be observed on the observation surface, each of which contains image $M'(f,g)$

A linear grating with a grating period of 7 pixels was designed and simulated, and the imaging results and profiles obtained were shown in Fig.6 and Fig.7

Figure 6. Simulation results of raster imaging

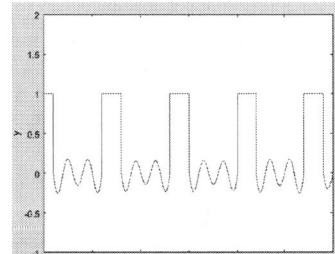

Figure 7. Raster imaging simulation plane results

C. Projection objective lens

The resolution of the projection objective is a very important parameter in the whole digital lithography system, and it is the main factor determining the resolution of the system. In order to achieve high resolution microstructure, projection objective lens with large zoom ratio is generally used. However, according to the formula given above, the minimum feature size that can be resolved by projection objective lens can be obtained,

$$R=k\frac{\lambda}{NA}.$$
(5)

It can be seen that in order to improve the resolution and reduce the minimum size that can be resolved by the projection objective lens, the methods of reducing the exposure wavelength and increasing the numerical aperture can be adopted. Moreover, the aperture of the projection objective lens also greatly determines the quality of the imaging results. The aperture of the projection objective lens is analyzed in detail here.

As shown in Fig. 8, set the focal length f, object distance p and image distance q of the lens. Under paraxial conditions, ignoring the absorption of the lens, the phase modulation effect of the lens is described by the complex amplitude transmittance $t(x,y)$, which can be written as,

$$t(x,y)=\frac{U_1'(x,y)}{U_1(x,y)}=\exp[-i\frac{k}{2f}(x^2+y^2)].$$
(6)

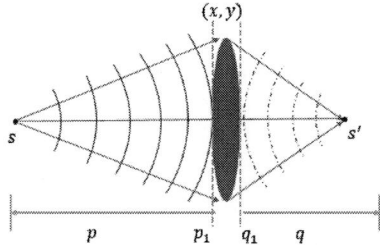

Figure 8. Schematic diagram of lens phase transformation

Considering the finite size of lens aperture, $p(x,y)$ is used to denote aperture function or pupil function

$$P(x,y)=\begin{cases}1,\text{within the aperture}\\0,\text{outside the aperture}\end{cases}.$$
(7)

The projection objective imaging is shown in Fig.9. f is the focal length of a positive lens, the distances from the input surface Σ_0 and output surface Σ_i to the thin lens are d_1 and d_2 respectively, The light field distribution in the plane of the object is $U_0(x_0,y_0)$, on the observation plane is $U_i(x,y)$. The object light field from the object surface reaches the lens plane through diffraction, if the Fresnel diffraction condition is satisfied, the complex amplitude is $U_1(x',y')$, after passing through the lens, the complex amplitude distribution of the light field becomes,

$$U_1'(x',y')=U_1(x',y')\exp(-ik\frac{x'^2+y'^2}{2f})P(x',y').$$
(8)

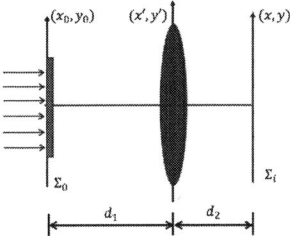

Figure 9. Schematic diagram of lens imaging coordinates

Where $P(x, y')$ is the lens aperture function. The light field reaches the observation surface through diffraction, according to Fresnel diffraction, its complex amplitude is distributed is,

$$U(x,y) = \frac{\exp(ikd_2)}{i\lambda d_2} \iint_{-\infty}^{+\infty} U_1'(x',y') \exp[ik\frac{(x-x')^2+(y+y')^2}{2d_2}]dx'dy' \quad (9)$$

The calculation of $U(x, y)$ can also be done using the Fourier algorithm to obtain the distribution of the light field on the observed surface

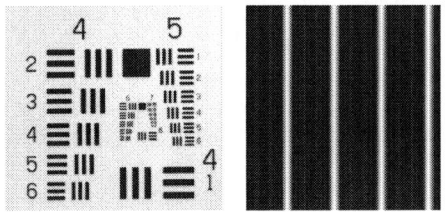

Figure 10. Resolution resolution plate and linear grating mask with DMD diffraction

The simulation of projection objective lens imaging is carried out by using resolution plate and linear grating mask with DMD diffraction. This is shown in Fig.10.

In order to observe the influence of different lens aperture on the imaging effect, the projection objective lens with different lens aperture was designed for simulation. Fig.11 shows projection objective lens imaging with lens apertures of 10mm, 20mm, and 80mm respectively.

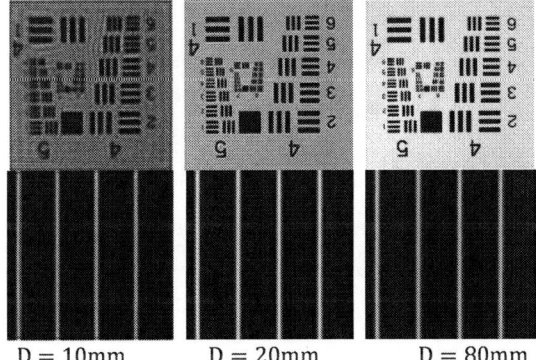

Figure 11. Simulation imaging under three lens apertures

IV. CONCLUSION

As can be seen from the results of simulation and experimental verification, excessive exposure leads to distortion of the central part structure of the grating obtained in the experiment, while the edge of the grating is not deep enough due to insufficient light intensity, and the farther away from the center, the shallower the depth, which is consistent with the theoretical analysis conclusion of the experiment. When the grating is used to simulate the diffraction of DMD, it can be clearly seen that while the DMD reflects the digital mask to achieve the purpose of beam modulation, it will produce the diffraction effect of the mask itself. The experiment using the binary grating results have obvious diffraction phenomenon, which is in accordance with the theory. In the projection analysis results, the larger the aperture of the projection objective lens, the greater the minimum resolution of the objective lens, and the aperture of the objective lens also greatly affects the experimental results, the larger the aperture, the clearer the projection objective lens imaging. Therefore, in the experiment of digital lithography, the light source with good uniformity should be selected, and the light source should be collimated by the collimation device. In optical fiber end-face imaging, a projection objective lens with large numerical aperture should be used to improve the fineness of the end-face microstructure.

V. DISCUSSION

In the experiment, it was found that part of the experimental imaging results deviated from the theoretical value when the two-dimensional linear grating was made on the optical fiber end face. When the optical fiber end face was focused, it was found that the self-reproducing image of the grating could be observed at different positions. After analyzing the phenomenon, it was determined that the self-imaging effect was the cause.

The principle of self-image effect is as follows: Under the irradiation of monochromatic parallel light, the periodic object in the light field behind it will have a period of $z = 2p^2 / \lambda$, and the recurrent image of the periodic object will clearly appear, where p is the period of the object, λ is the wavelength of incident light; in addition, a phase shift self image with π phase shift occurs at $\frac{1}{2}$ of each propagation cycle, and the autoimage positions corresponding to different wavelengths increase with the decrease of the wavelength, filling in the photolithographic dark areas created by other wavelengths, i.e., a weighted incoherent superposition.

According to the principle of Fourier optics, the complex amplitude distribution on the object plane can be regarded as the result of the linear combination of plane waves with different spatial frequencies, which can be expressed as follows,

$$E(x,y,0) = \iint U(f_x, f_y) \exp[i2\pi(xf_x + yf_y)]df_x df_y \quad (10)$$

$U(f_x, f_y)$ represents the Fourier spectrum of the field distribution in the aperture plane. According to the angular spectrum theory, when these plane waves with different spatial frequencies propagate from the object plane to the observation plane at a finite distance, a phase factor related to distance z and spatial frequencies (f_x, f_y) will be introduced into the corresponding angular spectrum. As shown below,

$$\exp(ikz\gamma) = \exp(ikz\sqrt{1 - \lambda^2 f_x^2 - \lambda^2 f_y^2}). \quad (11)$$

Therefore, the complex amplitude distribution on the observation plane at distance z can be expressed as a recombination of these plane waves after propagation, as shown below,

$$E(x, y, z) = \iint U(f_x, f_y)\exp(ikz\gamma)\exp[i2\pi(xf_x + yf_y)]df_x df_y$$

$$(12)$$

When a plane wave passes through a linear grating with a complex amplitude transmittance of $A_n = \frac{1}{p}\int_{-\frac{p}{2}}^{\frac{p}{2}} T(x)exp(-i2\pi\frac{nx}{p})dx$, p is the periodic constant of the grating, the field distribution at any distance can be simplified as,

$$E(x, y, z) = \sum_{-\infty}^{+\infty} A_n exp(-i\pi\lambda f_n^2 z)exp(i2\pi f_n x). \quad (13)$$

$f_n = n / p$ it can be seen from the equation that when the phase factor satisfies $\exp\left(-i\pi\lambda f_n^2 z_s\right) = 1$.

In this case, the optical amplitude distribution at $z_s = k * \frac{2p^2}{\lambda}\left(k = 0, 1, 2, 3 \ldots\right)$ can be expressed as,

$$E(x, z_S) = \sum_{-\infty}^{+\infty} A_n exp(i2\pi f_n x). \quad (14)$$

Each diffraction wave, including the fundamental diffraction wave, recovers its original complex amplitude at the same time, that is, the self-image of the grating appears. If the phase factor satisfies $\exp\left(-i\pi\lambda f_n^2 z_{SR}\right) = -1$. In this case, the optical amplitude distribution at $z_{SR} = \left(k - \frac{1}{2}\right) * \frac{2p^2}{\lambda}\left(k = 0, 1, 2, 3 \ldots\right)$ can be expressed as,

$$E(x, z_{SR}) = \sum_{-\infty}^{+\infty} A_n exp(i2\pi f_n x - \pi). \quad (15)$$

In the above equation, a π phase shift is introduced, and the light intensity distribution will be reversed by equivalent contrast, that is, the phase shift self-image of one-dimensional periodic micro-nano structure will be formed. When lithography is done with polychromatic light, the self-image positions corresponding to different wavelengths increase with the decrease of wavelength, it can fill in the photolithographic dark areas produced by other wavelengths, which is called weighted incoherent superposition.

In optical fiber end face processing, an idea of using self-image effect is proposed. Using different wavelengths of the light source, the imaging position of the self-image is different, can be in a position of superposition imaging. In theory, it can achieve the effect of periodic frequency doubling, improve the resolution of optical fiber end micromachining and greatly improve the precision of optical fiber end machining.

VI. SUMMARY

This paper presents a method for processing fiber end face functional devices, this method can quickly and cheaply integrate micro-optical devices on the optical fiber end face. The digital lithographic fiber system based on DMD has high spatial resolution. A variety of functional components are fabricated on different kinds of fibers, and the minimum resolution of the device is 700nm. At the same time, the theoretical and experimental analysis of the imaging resolution of the DMD lithography system is carried out in this paper, which plays a great role in improving the imaging resolution of the system and provides a new direction for the optical fiber end face integration processing method.

ACKNOWLEDGMENT

This work was supported by the Innovation Fund Designated for Graduate Students of Nanchang Hangkong University (YC2019048) and National Natural Science Foundation of China (61704070, 61464008) and Natural Science Foundation of Jiangxi Province (20202BAB202012).

REFERENCES

[1] H.-L Chien, et al."Maskless lithography based on oblique scanning of point array with digital distortion correction," *Optics and Lasers in Engineering* , 2021, p.136.

[2] T. Ampere, "Recent developments in nanofabrication using ion projection lithography，" *Weinheim an der Bergstrasse*, 2005,1(6).

[3] W.He, W.D.Zhang, F.Y.Meng, L.Q.Zhu, "Electron beam lithography inscribed varied-line-spacing and uniform integrated reflective plane grating fabricated through line-by-line method," *Optics and Lasers in Engineering*,2021,p. 138.

[4] N. Magotra, S. Divakar, Chengjie Tu, S. R. J. Brueck and Xiaolan Chen, "Digital image processing applied to imaging interferometric lithography,"*Conference Record of Thirty-Second Asilomar Conference on Signals, Systems and Computers (Cat. No.98CH36284)*, Pacific Grove, CA, USA, 1998, pp. 989-993 vol.2, doi: 10.1109/ACSSC.1998.751411.

[5] H. Yeh and K. Chen, "Development of a Digital-Convolution-Based Process Emulator for Three-Dimensional Microstructure Fabrication Using Electron-Beam Lithography," in *IEEE Transactions on Industrial Electronics*, vol. 56, no. 4, pp. 926-936, April 2009, doi: 10.1109/TIE.2008.2006030.

[6] R. Freytag, E. Böttcher, "Ring-shaped CO2 laser beams advance optical fiber processing," *Laser Focus World* , 2018,p. 54.

[7] R.S. Mahendran, L. Wang, "Fiber-optic sensor design for chemical process and environmental monitoring," *Optics and Lasers in Engineering,* vol.47, no.10, pp. 1069-1076, 2009.

[8] D.-H. Dinh, H. L. Chien, Y. C. Lee, "Maskless lithography based on digital micromirror device (DMD) and double sided microlens and spatial filter array," *Optics & Laser Technology*, vol.113, pp. 407-415, 2019.

Gap in pagination due to unavailable paper.

Pages 870-871

Proceedings of the 16th Annual IEEE International
Conference on Nano/Micro Engineered and Molecular Systems
April 25-29, 2021

Through Glass Vias by Wet-etching Process in 49% HF Solution Using an AZ4620 Enhanced Cr/Au Mask

Guanghui Ding, Binghe Ma*, Yuchao Yan, Weizheng Yuan, Jinjun Deng, and Jian Luo

Abstract— **Through glass vias on a high-quality borosilicate glass wafer (*i.e.*, BOROFLOAT® 33) of 500 μm in thickness were accomplished in 49% HF solution using an AZ4620 enhanced Cr/Au mask, without pin holes or defects on the surface. Here, the dehydrated AZ4620 photoresist film plays an important in etching process, which not only enhances the Cr/Au masks by filling micro cracks on them, but also resists the etchant as the first barrier for a longer etching time (*e.g.*, 150 min for 6 μm thick AZ4620 photoresist in 49% HF solution). No peeling off of AZ4620 photoresist film occurs due to its excellent adhesive properties to Au film. The etching rate of borosilicate glass wafer in depth, which is found to be significantly affected by the dimension of masks, decreases with increasing etching time due to the deposition of etching products. The undercut etching rate of glass wafers can be twice of that in depth, resulting in an inclination of the side wall of the vias (or cavities), which ranges from 45° to 50° and is found to be slightly depended on the quality of the masks, the adhesive strength between the interface, for example.**

I. INTRODUCTION

High-quality borosilicate glass wafers are widely used in the fabrication of MEMS devices due to their excellent physical and chemical properties. The glass wafers themselves can act as critical components of bio-MEMS or optical-MEMS devices because of their high transparency, ensuring a good observation on the activities in biochips or low optical signal loss during the transmission [1-2]. Meanwhile, their high electric insulation and similar coefficient of thermal expansion with that of silicon makes a great bonding ability to silicon wafers, acting as supporting substrate or protective cap during the fabrication of MEMS devices, such as physical sensors and actuators. More recently, backside interconnections are required in MEMS devices, on the one hand, to miniaturize their overall size and on the other hand, to reduce parasitic electrical signals for high precious measurements/actuations. Therefore, as one of the most popular techniques that are able to realize the backside interconnections for MEMS devices, through glass vias (TGV) technique has been widely used as signal paths in a variety of MEMS sensors/actuators [3-4].

Through vias on glass wafers can be achieved by using mechanical drilling, laser machining, sand blowing, wet etching and other methods [5]. Of which, mechanical drilling is able to machine deep vertical vias on a glass wafer. But for wafer level applications where hundreds of through vias are required, this method is apparently of low efficiency and may result in broken due to machining stress. Similar problems also exist in laser machining and sand blowing process and the machined surface of them can be very rough [6-7]. Wet etching process for through glass vias, however, has been widely studied for signal paths in fabrication of MEMS devices and their packages. If a high aspect ratio is not required, this method can obtain through vias on glass up to 1 mm in thickness, using optimized masks for long-time etching in 49% HF solution. In this process, the endurance of masks in high-concentration HF solution and their adhesive properties with glass wafers plays an important role in the quality of wet etching results. The most popular masks for wet etching of glass wafers in high concentration HF solution are Cr/Au films, together with photoresist layer for stronger endurances [8-9]. Such the combination has been proved as a good way to realize through vias on glass wafers, but with optimized parameters and special process.

In this paper, we investigated the wet etching process of borosilicate glass wafers in 49% HF solution, using a Cr/Au mask, together with a dehydrated AZ4620 photoresist film for enhancements. One the one hand, the parameters of masks, *e.g.*, the thickness of Cr/Au film are studied to show the best combination for wet etching process, finding an effective way to realize through glass vias with low costs. On the other hand, the size of wet etching window on the mask is of interest because such the etching process is isotropic, which means that the topography of through vias on glass wafers is determined by the relationship between the etching rate and undercutting rate. This is helpful for readers to get proper parameters during their research.

II. METHODS AND PROCESS

A 4-inch borosilicate glass wafer (BOROFLOAT® 33, Schott Co.) with a thickness of 500 μm was used in this study. The etching masks consisting of 50 nm Cr and 200 nm Au are sputtered on the top of the glass wafer, followed by a 10 μm thick AZ4620 photoresist (AZ Electronic Materials) film, which was spined using a coating machine. Although Cr film can increase the adhesive strength between Au film and glass wafers, its thickness should be smaller than 100 nm to avoid micro-cracks due to residual stress. An UV film was adhesive to the backside of the glass wafer to protect the backside surface of glass wafers from 49% HF solution.

The etching procedure is illustrated in Fig. 1. The glass wafer should be pre-cleaned in $H_2SO_4/H_2O_2 = 4:1$ solution for 10 minutes, then rinsed in de-ionized water for 120

---

*Resrach supported by the National Natural Science Foundation of China with grants no. 51775446 and 51735011.

All authors are with the School of Mechanical Engineering, Northwestern Polytechnical University in Xi'An City, China. (email: mabh@nwpu.edu.cn).

978-1-6654-3008-1/21 $31.00 © 2021 IEEE

seconds to remove most of the adhesives, see Fig. 1 (a). This is followed by drying the wafer with high-pressure Nitrogen gas and then on a 120 °C hot plate for 10 min. After that, 50 nm thick Cr and 200 nm thick Au film are sputtered on the top of the glass wafer successively before patterning the masks using standard photolithography process, see Fig. 1 (b) and (c). In this process, sputtering Au film is divided into two steps: the first step is sputtering 100 nm Au and wait for some time to cool down the glass wafer, ensuring the formation of micro-cracks, due to the release of thermal stress, and the second step is sputtering the rest 100 nm Au film to fill the micro-cracks on Au film, if they exist, to obtain good masks. Then an AZ4620 photoresist film of 12 μm in thickness is spin-coated on Cr/Au film with a rotation speed of 1300 rpm (round per minute) and patterned using the standard photolithography process. This is followed by 30-minute hard bake process at a temperature of 120 °C for dehydration and by doing this, the film able to resist 49% HF solution for at least 3 hours. After wet etching of Cr/Au film to form etching windows on the masks, the bottom of the glass wafers is protected using anti-acid UV film, which can be easily removed by exposure process in the end, see Fig. 1 (d). Put the glass wafer into 49% HF solution (under room temperature) and keep it still without stirring during the etching process, helping to obtain through vias on the glass wafer with perfect uniformities, see Fig. 1 (e). Finally, remove all the masks and protective UV films after wet etching process and re-clean the glass wafer for further process, see Fig. 1 (f).

film after 150-minute etching process in 49% HF solution is about 50 nm, indicating a good endurance for acting as a mask in this study. However, the AZ4620 photoresist film can't survive a minute and will be peeled off very soon if it is merely spin-coated on glass wafers or Cr film due to its poor adhesive properties to these substrates. Only shallow cavities on glass wafers, *e.g.*, less than 20 μm, can be accomplished using Cr + AZ4620 photoresist film, but with pin holes and micro-creeps on the surface because the defects and micro-cracks of Cr film, see Fig. 2. Thus, an Au film is required, on the one hand, to obtain a stronger adhesive of photoresist film to the substrate, and on the other hand, to resist HF solution during the etching process. With the help of Au film, no peeling off except for micro-cracks on AZ4620 photoresist film can be found after wet etching process, as shown in Fig. 3. Luckily, these micro-cracks won't cause obvious defects or micro-creeps on the surface of glass wafer since there is an Au film to prevent HF solution.

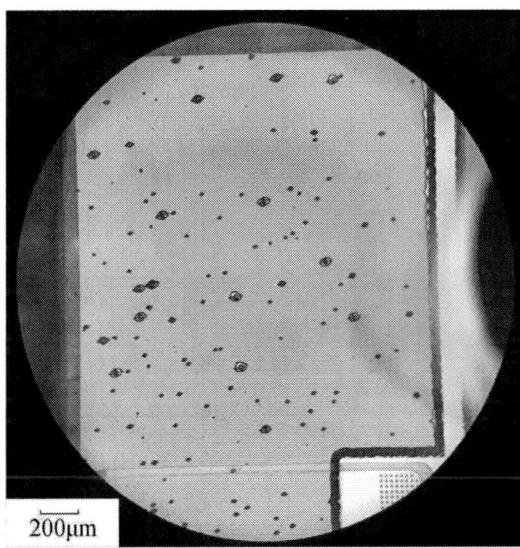

Figure 2. Pin holes on the glass wafer after 3 minutes etching in 49% HF solution using 100 nm Cr film and 5 μm dehydrated AZ4620 photoresist film for enhancement.

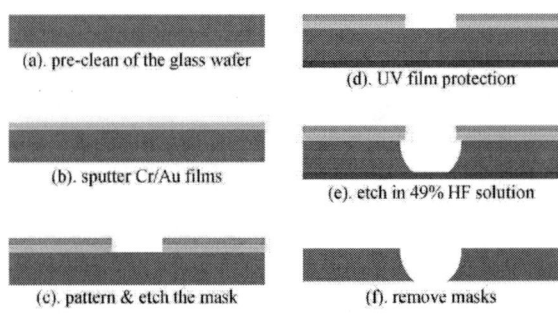

Figure 1. Experimental procedures of through vias on glass wafers using wet etching process in 49% HF solution.

(a). pre-clean of the glass wafer

(b). sputter Cr/Au films

(c). pattern & etch the mask

(d). UV film protection

(e). etch in 49% HF solution

(f). remove masks

A series of etching windows (circles) on the masks and the etching time of glass wafers in 49% HF solution were studied in order to understand the etching process detailly. The etching rate of glass in depth and lateral for different etching windows, and surface quality are of interest to be discussed in the following section. A conclusion will be given to shown optimized parameters for through glass vias etching in the end.

III. RESULTS AND DISCUSSION

A. Quality of Etching Masks

The etching rate of AZ4620 photoresist film (spin-coated on an Au film) in 49% HF solution was measured as low as 40 nm/min after 30-minute hard bake on a 120 °C hot plate. The surface roughness of AZ4620 photoresist

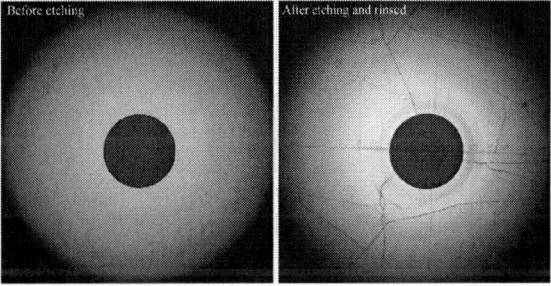

Figure 3. Micro-cracks on the dehydrated AZ4620 photoresist film after 10 minutes wet etching process in 49% HF solution. The Cr/Au film still remains in good conditions, acting as barriers to prevent pin holes or defects on the surface of glass wafers.

B. Etching Rate in Depth

Etching rate is a key parameter during wet etching process. Fig. 4 presents the cross-view profiles of through vias on the glass wafers in 49% HF solution using 50 μm, 300 μm, 500 μm, and 1000 μm etching windows (circles) on the mask under different etching time, which ranges from 10 to 150 minutes. Initial etching rate of glass wafers in depth remains similar, *e.g.*, 7 μm/min for etching windows greater than 300 μm. The etching rate in depth is also found to increase significantly when etching windows on the mask becomes greater for long-time etching process. According to our results, the average etching rate in depth increases from 2.3 μm/min for 50 μm etching window to 4.6 μm/min for 1000 μm etching window at an etching time of 100 minutes. Meanwhile, the etching rate decreases with etching time for all cases because of the deposition of etching products, which acts as a barrier to prevent further chemical reaction during such the etching process. We can also conclude that through glass vias of 500 μm can be accomplished in 150 min etching period using 49% HF solution when the etching windows are greater than 1000 μm, where the diameter of through vias is 1600 μm for lower surface (far away from the mask) and 3000 μm for upper surface (next to the mask), respectively. But for smaller etching windows, longer time is required for through glass vias.

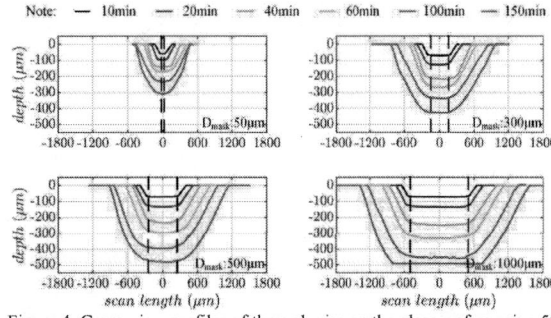

Figure 4. Cross-view profiles of through vias on the glass wafers using 50 μm, 300 μm, 500 μm, and 1000 μm etching windows (circles, indicated by black dashed lines) on the mask under different etching time.

C. Undercut Etching Rate

The undercut etching rate of glass wafers also decreases with increasing etching time, which can be as much as twice of etching rate in depth for some cases, as shown in Fig. 5. This results in an inclination of the side wall of the vias (or cavities) during the etching process and that's why the aspect ratio of wet etching of glass wafers is low. Such the inclination of the wall is of great benefit for backside electrical interconnections since it is much easier to deposit metal layers on that by using sputter process. The problem is that, however, the size of holes is difficult to be minimized, especially for thick glass wafers.

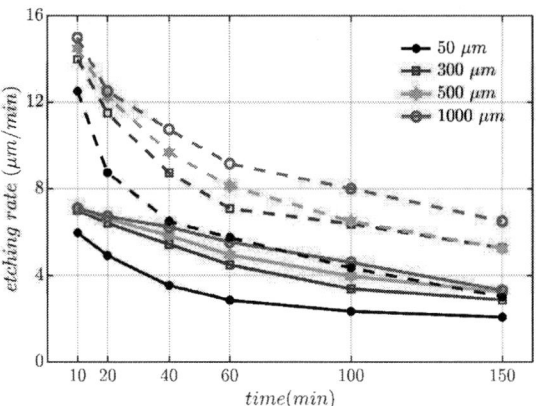

Figure 5. The etching rate in depth (solid lines) and undercut (dashed lines) for different size of etching window with increasing etching time.

Fig. 6 presents SEM (Scanning Electron Microscope) images of etching process of glass wafers in 49% HF solution with 500 μm etching window under different time. The uniformity of etching process remains good and the angle of side wall ranges from 45° to 55°, which doesn't change too much even if the etching time is much different. Also, it takes around 150 minutes to etch through the glass wafer with a single-side undercut length of 675 μm. Apparently, the wet etching process is stable and can be well repeated if the parameters are kept unchanged.

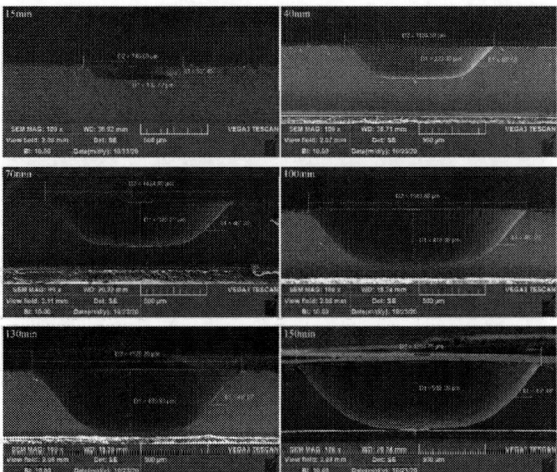

Figure 6. SEM images of wet etching process of glass wafers in 49% HF solution with 500 μm etching windows under different etching time.

D. Surface Quality

An example of SEM images is shown in Fig. 7, where we present an overall topography of deep cavities etched on glass wafers in 49% HF solution for 60 minutes. No pin holes and defects are observed on the surface of glass and the uniformity of cavities is good.

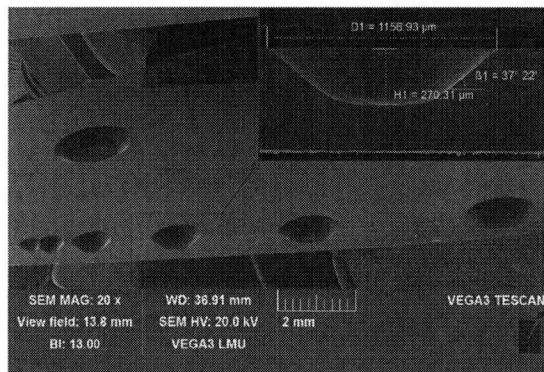

Figure 7. SEM images of etched glass wafers in 49% HF solution after 60 minutes. The enlarged view on the right top is obtained using 300 μm etching window.

IV. CONCLUSION

In this paper, we investigated wet etching process of borosilicate glass wafers in 49% HF solution using an AZ4620 photoresist film enhanced Cr/Au mask. Defects of micro-cracks on the mask are the main reasons to cause pin holes on the glass wafers. Thus, one can use two steps to solve this problem: the first step is to sputter Au film twice to reduce micro-cracks and obtain a high-quality metal mask, and the second step is to use a dehydrated AZ4620 photoresist film to enhance the endurance of the mask in high-concentration HF solution [10]. It is not necessary to increase the thickness of Au film, e.g., up to 1 μm, because the dehydrated photoresist layer is also able to resist HF solution for a quite long time. For example, the etching rate of dehydrated AZ4620 photoresist film in 49% HF solution can be as low as 40 nm/min, which is much lower than that of borosilicate glass wafers. This is good to low the cost of making masks.

The size of etching windows is also important for through glass vias since the etching products can be easily deposited on the bottom during the etching process. This would slow down etching rate of glass wafers in depth, resulting in small aspect ratio of through vias. Thus, the AZ4620 enhanced Cr/Au masks are good choices for wet etching of glass wafers to get through vias.

REFERENCES

[1]. M. Stjernström, J. Roeraade, "Method for fabrication of microfluidic systems in glass", *Journal of Micromechanics and Microengineering*, 1998, 8(1): 33.

[2]. J.W. Liu, Q.A. Huang, J.T. Shang and J.Y. Tang, "Micromachining of pyrex 7740 glass for micro-fluidic devices", *14th International Conference on Miniaturized Systems for Chemistry and Life Sciences, Groningen, The Netherlands*, 2010, 1907-1909.

[3]. G. H. Ding, B. H. Ma, J. J. Deng, W. Z. Yuan, and K. Liu, "Accurate Measurements of Wall Shear Stress on a Plate with Elliptic Leading Edge", *Sensors*, 2018, 18(8):2682.

[4]. J. Y. Jin, S. H. Yoo, B. W. Yoo and Y. K. Kim, "A wafer-level vacuum package using glass-reflowed silicon through-wafer interconnection for nano/micro devices" *Nanosci. Nanotechnology*, 2012, 5252–62.

[5]. P. Pawar, R. Ballav, and A. Kumar, "Machining Processes of Pyrex Glass: A Technological Review", *6th International & 27th All India Manufacturing Technology, Design and Research Conference* 2016.

[6]. T. Abe, X. H. Li and M. Esashi, "Endpoint detectable plating through femtosecond laser drilled glass wafers for electrical interconnections", *Sensors Actuators: A*, 2003, 108 234–8

[7]. E. Belloy, S. Thurre, E. Walckiers, A. Sayah and M. A. Gijs "The introduction of powder blasting for sensor and microsystem applications", *Sensors Actuators: A*, 2000, 84 330–7.

[8]. J. Y. Jin, S. Yoo, J. S. Bae, and Y. K. Kim. "Deep wet etching of borosilicate glass and fused silica with dehydrated AZ4330 and a Cr/Au mask", *Journal of Micromechanics & Microengineering*, 2014, 24(1):5003.

[9]. F. E. H. Tay, C. Iliescu, J. Jing, and J. M. Miao, "Defect-free wet etching through pyrex glass using Cr/Au mask", *Microsystem Technologies*, 2006, 12(10-11):935-939.

[10]. C. Iliescu, F. E. H. Tay and J. M. Miao, "Strategies in deep wet etching of Pyrex glass", *Sensors Actuators: A,* 2007, 133 395–400.

Gap in pagination due to unavailable papers.

Pages 876-881

Proceedings of the 16th Annual IEEE International
Conference on Nano/Micro Engineered and Molecular Systems
April 25-29, 2021

An adaptive octree level set simulation method of the wet etching process for the fabrication of micro structure on sapphire crystal

Ye Chen, Jin Qian, Xinyan Guo, Yan Xing*

Abstract— This paper introduces a MEMS wet etching process simulation platform based on adaptive octree level set method (LSM). Under the limited memory, the mesh adaptive technique for octree can solve the level set simulation problem of MEMS wet etching with large scale, high aspect ratio and fine structure. The octree grid is nonuniform, and the advanced interpolation technology can accurately obtain the signed distance value of the neighbor grid. An advanced interpolation technology is also proposed to solving this problem. Finally, LSM based on adaptive octree technology is applied to the simulation of C-plane sapphire wet etching process. The simulation results indicate that this method can improve the calculation accuracy, reduce the calculation memory and reduce the calculation time in the process of MEMS wet etching simulation.

I. INTRODUCTION

LSM is a method to predict the evolution of the interface. Its core idea is to use the high-dimensional methods to solve the low-dimensional problems. Compared with other methods of interface evolution, such as the Cellular Automata (CA) method, LSM can solve the complicated topological changes. In recent years, the LSM has been widely used in fields such as crystal growth, computational fluids, image processing, and process optimization.

Due to the miniaturization of MEMS devices, it is difficult to manufacture MEMS microstructures with traditional processing techniques. Therefore, processing techniques such as FIB, dry etching and wet etching are often used to manufacture MEMS microstructures. The wet etching morphology of the crystal is determined by many factors such as crystal material, etchant, etching temperature, etching time and mask shape. Only by selecting appropriate etching parameters can the etching profile desired by the user be obtained. Therefore, it is very important to select a simulation method that can accurately predict the etching profile. The LSM is an effective method for predicting the interface evolution process of MEMS wet etching.

The accuracy and effectiveness of MEMS wet etching process simulation based on the LSM has been proved by many scholars at home and abroad[1]. Compared with the standard level set method, the advanced LSM based on narrow-band method [2] and sparse field method [3] reduces the calculation time effectively by reducing the number of calculation points. However, few literatures can solve these problems, which include low computational accuracy and lack of memory on the large-scale array, high aspect ratio and high-precision MEMS structure wet etching simulation.

Therefore, this paper proposes a MEMS wet etching process simulation platform based on adaptive LSM, which can effectively improve the calculation accuracy, reduce the calculation memory and lower the running time[4].

II. METHODOLOGY

A. Level Set Method

LSM stipulates that the interface at any time is the set of points where the level set function(LSF) is equal to zero. Figure. 1 is a basic flow chart of the operation of the LSM. It can be seen that the LS method includes three basic steps: initialization, numerical calculation and output data. Among them, the most important part is the numerical calculation. The proper difference format and reinitialization process are very important to the accuracy and precision of the simulation results[6].

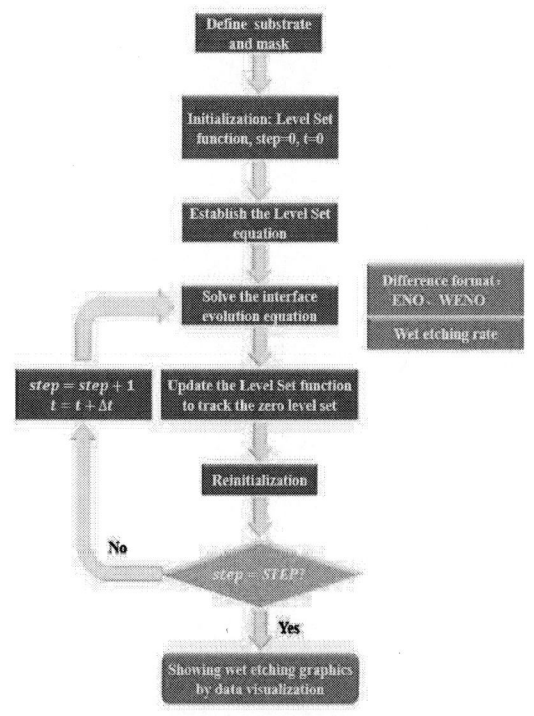

Figure 1. Standard LSM flow chart

---

*Resrach supported by the National Natural Science Foundation of China (No.51875104) and the National Key R&D Program of China (No. 2018YFB2002600).

All authors are with Jiangsu Key Laboratory for Design and Manufacture of Micro-Nano Biomedical Instruments, School of Mechanical Engineering, Southeast University, Nanjing 211189, China.

*Contacting Author: Y. Xing is with Jiangsu Key Laboratory for Design and Manufacture of Micro-Nano Biomedical Instruments, School of Mechanical Engineering, Southeast University, Nanjing 211189, China (phone: +86 025-52098413; email: xingyan@seu.edu.cn).

978-1-6654-3008-1/21 $31.00 © 2021 IEEE

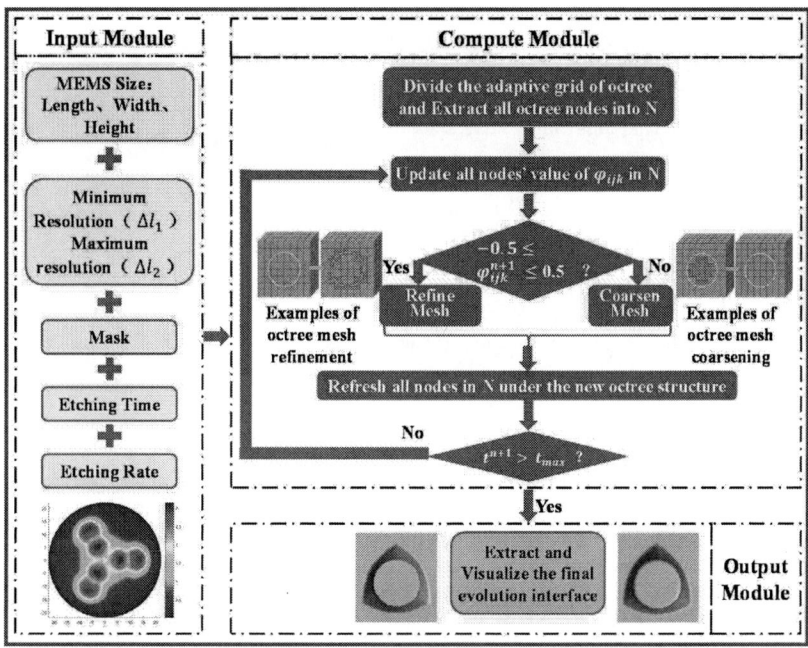

Figure 2. The flow chart of octree LS method

To ensure the characteristics of LSF, LSF is usually represented by a signed distance function. In other words, the absolute value of the function value of all points in the calculation domain of LSF is the shortest distance to the interface. It is stipulated that the value of the point inside the interface is negative, and the value of the point outside the interface is positive[8].

At any time, the signed distance value of the interface is zero, that is, $\phi(x^\rightarrow, t) = 0$. Therefore, the partial derivative of the signed distance function with respect to time and space can be used to describe the motion interface. The general partial differential equation of the LSF is shown in (1):

$$\frac{\partial \phi}{\partial t} + \vec{S} \cdot \nabla\phi + V_N |\nabla\phi| = b_k |\nabla\phi| \qquad (1)$$

In (1), $\vec{S}$ represents the velocity field imposed by the external environment, and V_N represents the velocity field along the normal direction of the moving interface, and b_k represents the velocity field dependent on the curvature. In the process of interface evolution, in order to ensure sufficient smoothness of the LSF, the modulus of the function gradient $|\nabla\phi|$ must always be guaranteed to be 1, and it is monotonous near the normal direction.

In the process of wet etching of crystals, the evolution of the moving interface is only related to the rate in the normal direction. Therefore, the level set equation can be simplified as (2):

$$\frac{\partial \phi}{\partial t} + V_N |\nabla\phi| = 0 \qquad (2)$$

Because the level set equation is a specific form of the Hamilton-Jacobi equation $\phi_t + H(\nabla\phi) = 0$, where $H(\nabla\phi) = V_N |\nabla\phi|$. Due to the characteristics of wet etching, we use the Lax-Friedichs (LF) format[5] to numerically solve

the Hamilton-Jacobi equation. Based on the central discrete format, applying (2), the LF format can be written as the basic form of (3):

$$\phi_{i,j,k}^{n+1} = \phi_{i,j,k}^{n} - \Delta t [H(\beta_x^+, \beta_y^+, \beta_z^+) - \alpha_F(\alpha_x \beta_x^- + \alpha_y \beta_y^- + \alpha_z \beta_z^-)] \qquad (3)$$

Among them, $\beta_l^{\pm} = \frac{1}{2}[(a_{i,j,k}^{+l} \pm a_{i,j,k}^{-l})]$, $a_{i,j,k}^{\pm l} = \frac{\pm\phi_{(i,j,k)\pm 1} \mp \phi_{i,j,k}}{\Delta l}$, $l = x, y, z$. α_F is the overall viscosity coefficient, $\alpha_x, \alpha_y, \alpha_z$ are the dissipation coefficients in the selected dimension respectively, and their value are as shown in (4):

$$\alpha_x = \max \left| \frac{\partial H}{\partial \phi_x} \right|, \alpha_y = \max \left| \frac{\partial H}{\partial \phi_y} \right|, \alpha_z = \max \left| \frac{\partial H}{\partial \phi_z} \right| \qquad (4)$$

When using finite differences to solve partial differential equations, CFL (Courant-Friedrichs-Lewy) conditions are used to limit the time step Δt to enhance its stability, expressed as (5):

$$\Delta t = \frac{0.5}{max(\alpha_x + \alpha_y + \alpha_z)} \qquad (5)$$

However, due to the existence of systematic errors, the LSF no longer satisfies the characteristics of the signed distance function after a few steps of numerical solution. Therefore, in order to ensure the accuracy of the solution, it is necessary to reinitialize the LSF every certain time step.

However, large-scale array, high aspect ratio and high-precision wet etching level set simulation of MEMS structures are prone to problems of low calculation accuracy and insufficient memory. Therefore, this paper proposes the octree LSM. And the core of this method is to keep the dense grid near the interface and the sparse grid far away from the interface during the evolution of the interface, so as to achieve the advantage of reducing the memory under the premise of ensuring the accuracy of the operation. Fig. 2 shows the flow chart of the octree LSM. The core technology of the octree LS method are grid adaptive technology and interpolation technology.

B. Octree structure grid

An octree is a tree-like data structure used to describe a three-dimensional space. Fig. 3 is the generation process of an octree, the left side is the division of the three-dimensional space area, and the right side is the corresponding node relationship. Each node of the octree represents a cube element. Each node may have no child nodes or eight child nodes. The volume of the cube element represented by these eight child nodes is equal to the volume of the parent node. A unit without a parent node is the root of the tree, a unit with a parent node but no child nodes is a leaf, and a unit with both parent and child nodes is the tree trunk. An important parameter of the octree is the depth of the node. The depth value can be used to indicate the number of times the cube unit represented by the node is divided. It is stipulated that the depth value of the root node is 0, the depth value of each child node is one more than the depth value of its parent node, and the depth values of the eight child nodes of the same parent node are the same. The octree structure is the basis of the octree grid, and each node of the octree can be regarded as an octree grid. The cube element represented by the number of root nodes is the delineated calculation space, which is the grid element with the largest side in the octree grid structure. The octree grid represented by the trunk node must be subdivided into eight sub-grid units, and the grid unit represented by the leaf node will not be subdivided again. The deeper the depth of the octree grid, the more times it is subdivided, and the side lengths of grid cells of the same depth are the same. If the side length of the root grid is one, the side length l_n of the grid with a depth of n can be expressed as (6):

$$l_n = \frac{l}{2^n} \qquad (6)$$

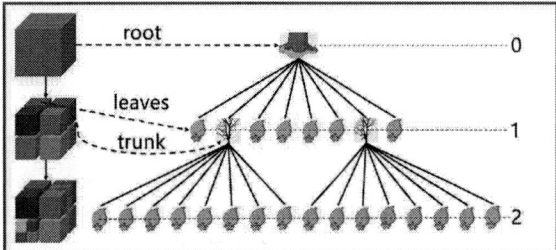

Figure 3. Octree grid structure diagram

C. Adaptive technology of octree structure grid

The core of the octree grid adaptive technology is to always keep the dense grid near the interface and the sparse grid far away from the interface during the evolution of the interface, so as to achieve the advantage of reducing memory while ensuring the accuracy of calculations. As shown in Fig. 4, the adaptive technology of the octree grid mainly includes

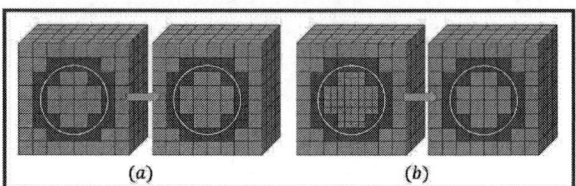

Figure 4. (a) Mesh refinement and (b) Mesh coarsening

two parts: grid refinement and grid coarsening. Among them, mesh refinement is the key to ensure the calculation accuracy of the LSM. Mesh coarsening is the key to saving memory in LSM. Fig. 4(a) refines the mesh on the interface to improve the accuracy of the numerical calculation. In Fig. 4(b), the mesh inside the interface is of little significance to the evolution of the interface, so it is coarsened.

Fig. 5(a) shows the adaptive technology during the initialization of the octree grid. The principle of initialization is to ensure that each grid contains only one material. Of course, the user can also determine the maximum and minimum dimensions of the grid according to the corresponding resolution requirements. This meshing technique makes the grids near the interface dense and the grids far away from the interface sparse. Fig. 5(b) shows the adaptive technology of the octree grid in the interface evolution process. The octree grid structure is constantly changing with the change of the interface, always ensuring the highest grid resolution at the interface, and the resolution far away from the interface is reduced.

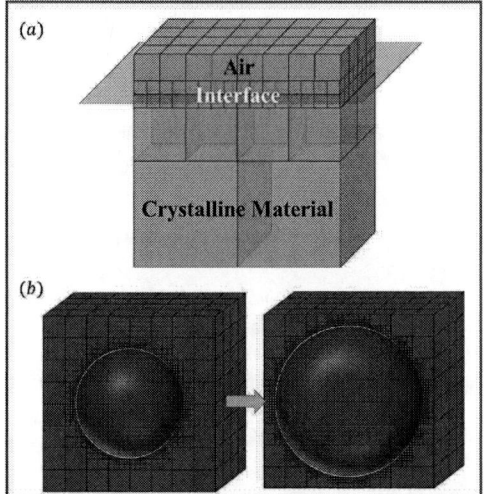

Figure 5. (a)The adaptive technique of octree mesh initialization and (b) The adaptive technology of octree mesh in the process of evolution

D. Interpolation technology of octree structure grid

Fig. 6(a) shows that there are three types of nodes in the octree structure grid. The first one, the most ideal normal nodes, connects 6 edges. The second one is T-shaped node, which may connect 4 or 5 edges. And the last one is boundary node, which may connect 3, 4 or 5 edges. For a normal node, we can directly obtain its neighbor nodes in three directions.

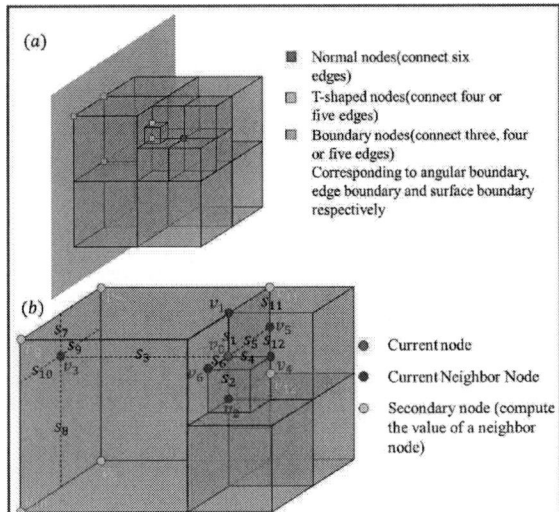

Figure 6. (a) Three types of octree structure grid nodes and (b) T-shaped nodes

For the boundary nodes and T-shaped nodes, we use linear interpolation to obtain the neighbor nodes. Fig. 6(b) is a T-shaped node connecting 4 edges. Its upper, lower, front, and right neighbor nodes can be obtained directly, the left neighbor node needs to be obtained through the trilinear interpolation method, and the rear neighbor node can be obtained through the linear interpolation method to obtain.

III. RESULT AND DISCUSSION

In order to verify the accuracy of the octree-based level set process simulation system, it was compared with the experimental results many times. In order to highlight the superiority of this algorithm, it is compared with the simulation of the standard level set algorithm. Fig. 7(a) shows the etching result of C-plane sapphire. The etching conditions are as follows: temperature (236℃), $H_2SO_4:H_3PO_4 = 3:1$, square mask with side length ($25\mu m$), time (180min). It can be seen from Fig. 7(a) that because the C-plane sapphire has three symmetry in the XOY plane, the characteristic structural surfaces of the groove are A, B, C, D, and E, respectively. Compare the simulated etch topography of the octree LS (Fig. 7(c)) and the etch topography simulated by the standard LSM (Fig. 7(b)) with the experimental topography (Fig. 7(a)), The results show the accuracy of octree LS to simulate square grooves.

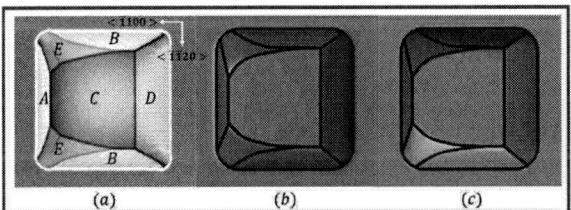

Figure 7. (a)Etched microstructure and (b) Simulated result by standard LSM and (c) Simulated result by octree LSM

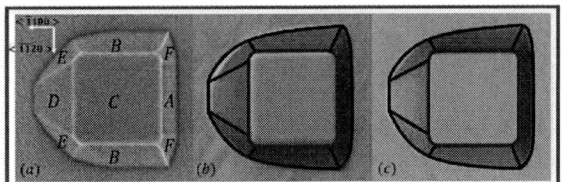

Figure 8. (a)Etched rectangle microstructure and (b) Simulated result by standard LSM and (c) Simulated result by octree LSM

Fig. 8(a) is the experimental result of the sapphire C-plane surface etched under the condition of the square mask. The etching conditions are as follows: temperature (236℃), $H_2SO_4:H_3PO_4 = 3:1$, square mask, time (180min). It can be seen from Fig. 8(a) that due to the three-symmetrical nature of wet etching of sapphire, the obvious symmetry can be found. The two edge directions of the boss are respectively $< 11\bar{2}0 >$ and $< 1\bar{1}00 >$, as shown in Fig. 8(a). It is easy to know that the characteristic structural surfaces of the boss are A, B, C, D, E, and F respectively. Fig. 8(b) is the processing condition simulated by the standard LSM, which has a high coincidence with the actual experiment, which verifies the accuracy of the standard LSM. Fig. 8(c) is a simulation diagram of the LS simulation system based on the octree. It is found that the overlap with the actual experimental diagram and the standard LSM simulation diagram is very high, which verifies the accuracy of the model and the comparison of the standard LSM's substitutability.

Fig. 9(a) is the experimental result of etching a complex composite structure. The etching conditions are as follows: temperature (236℃), $H_2SO_4:H_3PO_4 = 3:1$, combined mask, time (180min). It can be seen from Fig. 9(a) that due to the three-symmetric nature of wet etching of sapphire, the internal bosses show obvious three-symmetry. The two edge directions of the square groove are $< 11\bar{2}0 >$ and $< 1\bar{1}00 >$, as shown in Fig. 8(a). It is easy to know that the characteristic structural planes of the complex composite structure are A, B, C, D, E, and F respectively. By observing Fig. 9(b) and Fig. 9(c), we can see the accuracy of the octree LS method for simulating complex graphics.

Fig. 10 is a comparison of the running time and memory usage of the two methods. It can be seen that the octree LSM occupies less memory than the standard LSM under the same operation scale, and the octree LSM's running time is also lower than that of the standard LSM. And with the increase of the computing scale, these two advantages of the octree LSM continue to be highlighted. When the computing scale reaches 256*256*256, the computing time of the standard LSM is 4

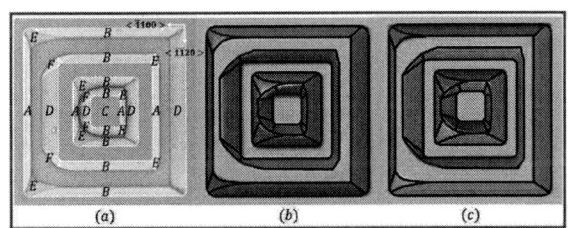

Figure 9. (a)Etched composed microstructure and (b) Simulated result by standard LSM and (c) Simulated result by octree LSM

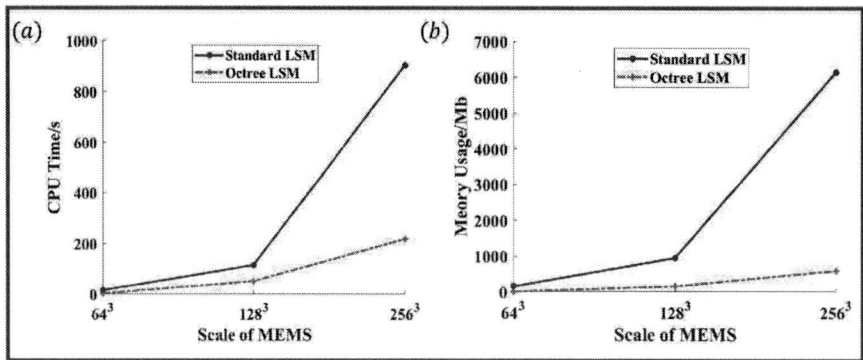
Figure 10. The comparison between the standard LSM and the octree LSM

times that of the octree LSM, and the memory of the standard LS The usage is 11.25 times that of octree LSM.

IV. CONCLUSION

In this paper, a MEMS wet etching process model based on Octree LSM is developed. Through three sapphire wet etching mask experiments, the results of the wet etching experiments and the simulation results of the MEMS wet etching process method based on the standard LSM are compared with the simulation results of the MEMS wet etching process method based on the octree LSM. The comparison verifies the accuracy. Meantime, it also verifies applicability of the process simulation method and its advantages compared to the standard LSM. The results show that: (a) Compared with the standard LSM, the accuracy of the octree LSM is proved; (b) The simulation speed of the octree LSM is faster than the simulation speed of the standard LSM under the same operation scale; (c) Under the same computing scale, the memory usage of the octree LSM is lower than that of the standard LSM; (d) During simulating large-scale arrays, high aspect ratios, and high-precision MEMS structures, the problems of low calculation accuracy and insufficient memory are prone to occur. The octree LSM can solve these problems successfully.

ACKNOWLEDGMENT

We acknowledge financial supports from the National Natural Science Foundation of China (No.51875104) and the National Key R&D Program of China (No. 2018YFB2002600).

REFERENCES

[1] Level set methods and dynamic implicit surfaces: By Stanly Osher and Ronald Fedkiw. Springer, New York. (2003). 273 pages.
[2] Jang G W , Kambampati S , Chung H , et al. Configuration optimization for thin structures using level set method[J]. Structural and Multidisciplinary Optimization, 2019. [3] C.S. Kim, R.G. Hobbs, A. Agarwal, Y. Yang, V.R. Manfrinato, M.P. Short, J. Li, K.K. Berggren, Focused-helium-ion-beam blow forming of nanostructures: Radiation damage and nanofabrication, Nanotechnology. 31 (2020).
[3] Saien, Soudeh; Moghaddam, Hamid Abrishami; Fathian, Mohsen (2017). A unified methodology based on sparse field level sets and boosting algorithms for false positives reduction in lung nodules detection. International Journal of Computer Assisted Radiology and Surgery.
[4] Jia-Cheng, Zai-Fa, Zhou, et al. Three-Dimensional Simulation of DRIE Process Based on the Narrow Band Level Set and Monte Carlo Method.[J]. Micromachines, 2018.
[5] OSHER S,FEDKIW R P. LEVEL SET METHODS AND DYNAMIC IMPLICIT SURFACES[M], SPRINGER VERLAG, 2003.
[6] J. Zhang, Y. Xing, M. A. Gosálvez, X. Qiu, X. Lin and C. Zhang, "Level Set Simulation of Surface Evolution in Anisotropic Wet Etching of Patterned Sapphire Subtrate," 2019 IEEE 32nd International Conference on Micro Electro Mechanical Systems (MEMS), Seoul, Korea (South), 2019, pp. 361-364, doi: 10.1109/MEMSYS.2019.8870817.
[7] Y. Xing, Z. Guo, G. Wu and M. A. Gosálvez, "Characterization of Orientation-Dependent Etching Properties and Surface Morphology of Sapphire Crystal in Wet Etching," 2019 20th International Conference on Solid-State Sensors, Actuators and Microsystems & Eurosensors XXXIII (TRANSDUCERS & EUROSENSORS XXXIII), Berlin, Germany, 2019, pp. 281-284, doi: 10.1109/TRANSDUCERS.2019.8808344.
[8] Yan Xing, Jin Qian, Miguel A. Gosálvez, Jie Zhang, Yunze Zhang, Simulation-based optimization of out-of-plane, variable-height, convoluted quartz micro needle arrays via single-step anisotropic wet etching, Microelectronic Engineering, Volume 231,2020,111375,ISSN 0167-9317,

Gap in pagination due to unavailable paper.

Pages 887-888

Proceedings of the 16th Annual IEEE International
Conference on Nano/Micro Engineered and Molecular Systems
April 25-29, 2021

A Flexible AgNPs-PDMS Substrate to Produce Ultrasensitive SERS Detection

Guanzhou Lin[1], Kenan Zhang[1,2], Yun Huang[1], Shengxiao Jin[1], Tian Kang[1], Yusa Chen[1,3], Liye Li[1], Peimin Lu[2] and Wengang Wu[1,4]*

Abstract— **In this paper, we proposed a simple Surface-Enhanced Raman Scattering (SERS) detection method for the possible pesticide residues on fruit surfaces. 3-aminopropyl triethoxysilane（APTES）is used to modify Ag nanoparticles (AgNPs) on polydimethylsiloxane (PDMS) membranes. The PDMS was attached to the fruit epidermis to adsorb the substance, and then the SERS test was performed to analyze whether the epidermis had pesticide residues. In the present study, the AgNPs were modified using the Tollens synthesis method while glucose was set to be the reducing agent, and the AgNPs were synthesized into a uniform size of 20-50nm. In this study, we fabricated a simple flexible SERS substructure, which achieved the detection limits of R6G and Thiram at 10^{-8} M. In the future, our work can be extended to biomedical applications, or focusing on the characteristic substances that may threaten the national defense and security, which has a broad application scenario.**

I. Introduction

Pesticide residues on the fruit surfaces have always been a problem that plagues the public. Traditional pesticide testing methods require time and relatively high cost [1]. For pesticide detection, the traditional methods mainly include mass spectrometer-chromatography, liquid chromatography, electrophoresis and other methods, but these methods often require complex equipment and operation process, which is not easy to rapid and low-cost detection [2-4]. As a mature detection technology, SERS has been widely used in biomedicine and substance detection. In general, SERS effects occur in precious metal nanoparticles (such as Au, Ag) or rough metal surfaces, and surface plasmon resonance occurs in the presence of light, resulting in the enhanced Raman signals of molecules that are close to or adsorbed on the surface [5]. SERS has high sensitivity and narrow band, which provides spectral fingerprint information, and it has anti-bleaching properties and has no damage to the sample [6].

PDMS is one of Bio-MEMS's most popular polymer materials, it has many advantages such as optical transparency, excellent elasticity, chemical inertness, gas permeability and biocompatibility. PDMS has many applications, it works relatively simple and has a lower manufacturing cost. In recent years, there is a great interest in polymer materials combined with nanoparticles[7],

especially in optical films, color conversion devices and SERS substrates[8]. The flexible substrate is a commonly used method for SERS detection, and the surface of the object is detected by depositing the nanoparticles on the bendable flexible material, which aims to the in situ detection [9].

Thiram is a dithiocarbamate fungicide. As a common pesticide, Thiram is mainly used to control plant diseases, and has an effect irritating to human skin and mucous membranes [10]. In recent years, many methods have been developed, including plasma absorption [11], colorimetric method [12], fluorescence analysis [13], electrochemical [14] and resonance Rayleigh scattering [15] to detect Thiram residues on vegetables and fruits. The previous test process is cumbersome and complicated, and takes a long time, and the detection is not so flexible to cope with the requirements of detection.

In this paper, we propose a method to detect Thiram on apple surface by a flexible SERS substrate which modified Ag nanoparticles on PDMS. The synthesized AgNPs are modified on the PDMS by APTES, and the modified PDMS is attached to the apple to analyze the pesticide. The method is simple and easy to operate, friendly to the environment. The strategy can achieve uniform distribution of AgNPs, which lead to the SERS signals with good consistency and sensitivity. The detection method used is quick, easy to operate, environmental friendly. Finally, we demonstrated that the SERS substrate can achieve the detection of Rhodamine 6G (R6G) and Thiram (pesticide) at a satisfying level (10^{-8}M), and also has good consistency and enhancement effect.

---

*Research supported by National Key Laboratory of Science and Technology on Micro/Nano Fabrication, Institute of Microelectronics, Peking University.

G.Z. Lin is with the National Key Laboratory of Science and Technology on Micro/Nano Fabrication, Institute of Microelectronics, Peking University, Beijing 100871 China (email: tk968810@pku.edu.cn);

K.N. Zhang is with the National Key Laboratory of Science and Technology on Micro/Nano Fabrication, Institute of Microelectronics, Peking University, Beijing 100871, China, and the College of Physics and Information Engineering, Fuzhou University, Fujian 350108, P.R. China;

Y. Huang, S.X. Jin, T. Kang, Y.S. Chen, L.Y. Li is with the National Key Laboratory of Science and Technology on Micro/Nano Fabrication, Institute of Microelectronics, Peking University, Beijing 100871, China ;

P.M. Lu is with College of Physics and Information Engineering, Fuzhou University, Fujian 350108, P.R. China;

W.G. Wu is with the National Key Laboratory of Science and Technology on Micro/Nano Fabrication, Institute of Microelectronics, Peking University, Beijing 100871, China and the Frontiers Science Center for Nano-optoelectronics, Peking University, Beijing 100871, China. (corresponding: +86-10-62757553, email: wuwg@pku.edu.cn)

978-1-6654-3008-1/21 $31.00 © 2021 IEEE

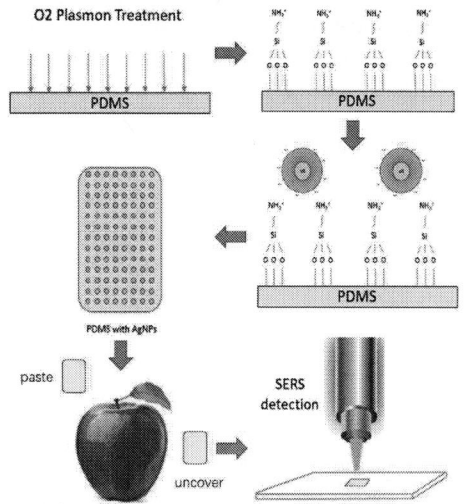

Figure 1: The main principles of the study. AgNPs synthesized by the modified Tollens method were modified on PDMS, and the substance was adhered to the surface of the apple, and then subjected to SERS detection.

II. MATERIALS AND METHODS

A. Materials

Silver nitrate (99.9%), R6G (C28H31N2O3Cl, 99%) and Thiram (C6H12N2S2, 99.9%) were purchased from Sigma-Aldrich (Darmstdt, Germany), Ammonia (25% w/w aqueous solution) and D-glucose were supplied by Beijing Chemical Works (Beijing, China), APTES (98%) was purchased from Aladdin, and SYLGARD 184 Silicone Elastomer Base and SYLGARD 184 Silicone Elastomer Curing Agent were purchased from Dow Corning Corporation (Midland, MI, USA). All the reagents used in this work were of analytical grade. Deionized water (Milli-Q purification system, Millipore Co., Bedford, MA, USA) was used for all experiments.

B. Characterizations

UV-visible spectra were obtained from Agilent Cary 8454 spectroscopy system (Agilent Technologies Inc., Santa Clara, CA, USA), the path length of the cell is 1 cm. The PDMS was treated by BD-20AC Laboratory Chrona Treater (Electro-Technic Products Inc., Chicago, IL, USA). Scanning electron microscope (SEM) images were obtained from FEI NanoSEM 430(FEI, Hillsboro, OR, USA). The Raman spectra were obtained from Renishaw inVia Reflex Raman Microscope and Spectrometer (Renishaw, Gloucestershire, UK). The 633nm laser and 50x objective is used in this research. The integration time of all spectra acquisition for each measurement was set to be 5s. The laser power was set at 10%, 1.7 mW. We selected five random spots from the same SERS substrate, which provided the spectra for the final analysis.

C. Preparation of Ag NPs

The nanoparticles prepared by the Tollens method [16]. We synthesized the AgNPs under the aqueous glucose solution of 0.2 mol, 0.5 mol, 1 mol, 2 mol. The AgNPs were obtained from the chemical reaction of the silver nitrate and ammonia, by adding the ammonia under the stirring, the silver ammonia complex solution is prepared. According to the former research, we chose the silver ammonia complex solution at 0.005 mol to add in the 0.2mol glucose every 30min, repeated 10-15 times and finally completed the synthesize of AgNPs.

D. Fabrication of the SERS substrates

The curing agent was added in a ratio of 10:1 with Slygard 184, placed in a petri dish, and baked in a 70 Celsius degrees oven for two hours to produce PDMS films [17]. The PDMS was cut into small squares of about 1 cm*1 cm and using plasma treatment to make the surface super-hydrophilic. The PDMS was then soaked into a 10% aqueous solution of APTES for 10 hours. The modified PDMS membrane was then rinsed with alcohol to remove the unbonded APTES molecules. Different concentrations of AgNPs solution were taken, and PDMS was immersed in the AgNPs solution for 10 hours to bind the particles with the modified APTES on PDMS, and then the AgNPs were modified on the flexible substrate PDMS.

III. RESULT AND DISCUSSION

A. Characterization of AgNPs

Figure 2 shows the UV-Vis spectra of the AgNPs, it can be observed that the absorption peak is at 414 nm, which indicates that AgNPs are approximately spherical and relatively homogeneous in a stable colloidal solution [17]. Figure 3 is a view of SEM observed to modify AgNPs in different concentrations of glucose on PDMS. As the glucose concentration increases, the size of the prepared AgNPs decreases, and the surface morphology is also affected. According to literature analysis, we chose 0.2M AgNPs solution [18].

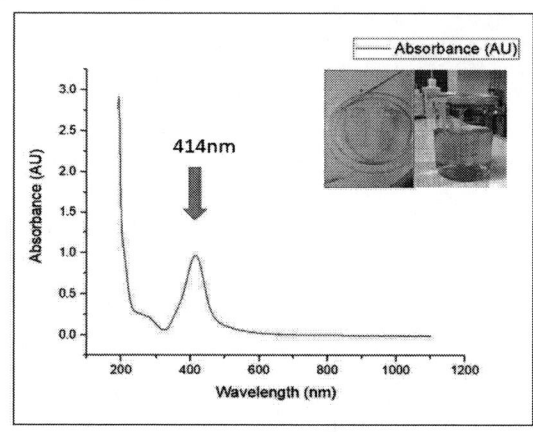

Figure 2: The UV-Vis spectrum of AgNPs shows that there is a single distinct absorption peak at 414 nm, indicating that the synthesized AgNPs are approximately spherical.

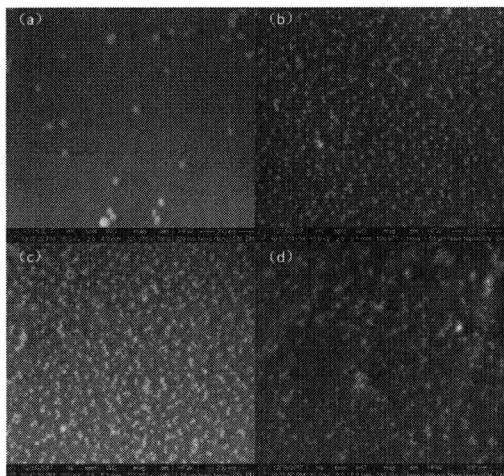

Figure 3: Particle morphology of different concentrations of AgNPs. (a) to (d) are 0.2, 0.5, 1.0, 2.0 M glucose-synthesized AgNPs, respectively.

B.   Detection mechanism

The SERS signal can be achieved when the analyte is close enough the the AgNPs., and the sensitivity will be risen. The AgNPs is easy to be oxidated, so we use glucose to provide the protection of the nanoparticles. Besides, during the modification process, the negatively charged glucose binds to the positively charged amino group of APTES, which is immobilized on the PDMS membrane by electronegativity, and the silver nanoparticles are encapsulated in the glucose protective shell. For the samples used for testing, Thiram was previously dissolved in ethanol and diluted with deionized water. At the time of testing, the glucose is dissolved and the silver nanoparticles are exposed to the substance. Then the substrate will produce the SERS signal of the organic substances on the apple skin.

C.   Sensitivity

SERS detection sensitivity is a key indicator. In this experiment, in order to verify the effect of the proposed APTES-modified PDMS, we selected different concentrations of R6G solution for SERS testing. It can be seen from the Figure 3(a) that at a concentration of 10^{-5}M the strong peak appears at 612 cm^{-1},774 cm^{-1}, 1127 cm^{-1}, 1185 cm^{-1}, 1310 cm^{-1}, 1360 cm^{-1}, 1509 cm^{-1}, 1573 cm^{-1}, which is same to the previous reports. At which peak positions, when the concentration is lowered to 10^{-8} M can still be discerned. After further investigation, the reproducibility of the SERS substrate has shown a good performance. The RSD(average relative standard deviation) of the spectra intensities at 1509 cm^{-} from the 5 random points of the substrate was 4.6%, which demonstrated the excellent performance of the substrates.

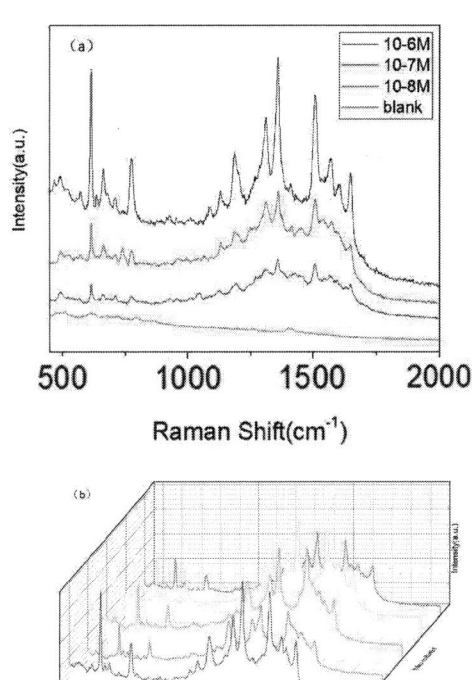

Figure 4: The SERS spectrum of R6G. From Fig 4(a), it can be seen that the characteristic peak of R6G is still clearly identifiable by 10^{-8} M. Fig 4(b) shows the spectrum of R6G at 10^{-6} M, 5 random points was collected and the RSD was 4.6% to prove the good detection performance of the prepared substrate.

Because of the high local electric field near the surface of the substrate, the enhanced intensity of the SERS signal is commonly much higher than the normal Raman scattering signal. The enhancement factor (EF) is an important parameter to describe the enhancement effect of the SERS substrate. The EF was calculated as follows [19]:

$$ EF = \frac{I_{SERS}N_{Bulk}}{I_{Raman}N_{Surface}} \quad (1) $$

where I_{SERS} is the intensity of the vibrational mode in the surface-enhanced spectrum of a given mode, I_{Raman} is the intensity of the same mode in the Raman spectrum, N_{Bulk} is the number of macromolecules used for bulk sample detection, and $N_{Surface}$ is digitally adsorbed on the SERS active substrate[20]. The calculated EF was about $2.36*10^5$. It can be seen that the substrate we prepared has good SERS enhancement properties.

D.   Application on Thiram

Thiram is a typical pesticide which is widely used in agriculture. In this research, the SERS substrate was used to complete the trace detection of Thiram. Figure 5 shows the different concentrations of Thiram SERS lines under the PDMS SERS substrate. The main Raman bands include 563 cm^{-1} attributed to υ(S-S), 1147 cm^{-1} corresponding to ρ (CH3) and υ(C-N), and 1383 cm^{-1} corresponding to δ_s(CH3) and υ(C-N), and 1511 cm^{-1} corresponding to υ(C-

N), δ(CH3), and ρ(CH3) [18]. It can be seen that the signal at the position of 1383cm^{-1} can be clearly discerned until 10^{-8} M, so the scheme provides a method for quickly detecting the surface of the apple surface of the Thiram residue, the precision of which reaches 10^{-8}M, which is much lower than The national standard (5 mg/kg, about 2×10^{-5} M) has this broad application prospect in the detection of pesticides on the surface of fruits.

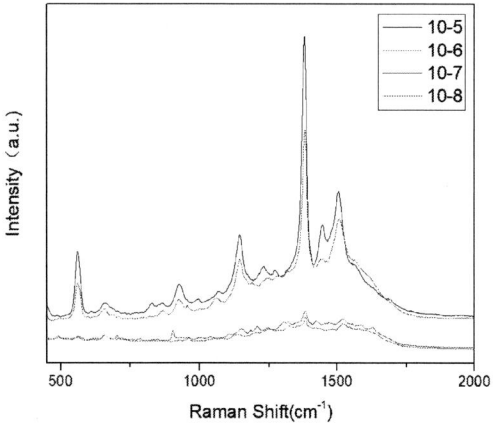

Figure 5: SERS lines of different concentrations of Thiram. Starting from 10^{-5} M, the most obvious characteristic peak at 1383 cm^{-1} is always observable. In this study, the detection limit reached 10^{-8} M.

IV. CONCLUSIONS

In summary, we proposed a simple and convenient method for testing the pesticide residues on apple surface. We modified AgNPs on a flexible substrate PDMS and completed the detection by the direct contact with apples. We demonstrated that this SERS flexible substrate can be used for ultrasensitive detection of R6G and Thiram (10^{-8} M). The detection limit of Thiram is lower than the maximum residue of fruits and vegetables required by national standards (5 mg/kg, equal to 2×10^{-5} M). The flexible SERS substrate will be extend to biomedical field and provide more significant applications.

ACKNOWLEDGMENT

This work is supported by the National Natural Science Foundation of China under Grant No. 61974004 and 61931018. We also thank the Electron Microscopy Laboratory of Peking University, and the School of Earth and Space Science of Peking University for the use of their equipment.

REFERENCES

[1] Seiber J N, Kleinschmidt L A. Contributions of pesticide residue chemistry to improving food and environmental safety: past and present accomplishments and future challenges[J]. Journal of agricultural and food chemistry, 2011, 59(14): 7536-7543.

[2] Liu T, Zhang C, Peng J et al. Residual Behaviors of Six Pesticides in Shiitake from Cultivation to Postharvest Drying Process and Risk Assessment. Journal of agricultural and food chemistry. 2016, 64(47),8977-8985.

[3] Ferreira J A, Ferreira J M, Talamini V, et al. Determination of pesticides in coconut (Cocos nucifera Linn.) water and pulp using

modified QuEChERS and LC–MS/MS. Food chemistry 213 (2016): 616-624.

[4] Uclés A, Lopez S H, Hernando M D, et al. Application of zirconium dioxide nanoparticle sorbent for the clean-up step in post-harvest pesticide residue analysis. Talanta 144 (2015): 51-61.

[5] Kneipp, K, Kneipp, H, Itzkan I, et al. Ultrasensitive chemical analysis by Raman spectroscopy. Chemical Reviews. 1999, 99, 2957−2976.

[6] Ando J, Fujita K, Smith N I, et al. Dynamic SERS imaging of cellular transport pathways with endocytosed gold nanoparticles, Nano Letters. 11 (2011): 5344 − 5348.

[7] Uhlenhaut D I, Smith P, and Caseri W, Color switching in gold—polysiloxane elastomeric nanocomposites, Advanced Materials, vol. 18, no.13, pp. 1653-1656, 2006.

[8] Ee H S and Agarwal R, Tunable metasurface and flat optical zoom lens on a stretchable substrate, Nano letters, vol. 16, no. 4, pp. 2818-2823, 2016.

[9] Shiohara A, Langer J, Polavarapu L, et al. Solution processed polydimethylsiloxane/gold nanostar flexible substrates for plasmonic sensing[J]. Nanoscale, 2014, 6(16): 9817-9823.

[10] Yuan C, Liu R, Wang S, et al. Single clusters of self-assembled silver nanoparticles for surface-enhanced Raman scattering sensing of a dithiocarbamate fungicide[J]. Journal of Materials Chemistry, 2011, 21(40): 16264-16270.

[11] Rastegarzadeh S, Abdali S. Colorimetric determination of thiram based on formation of gold nanoparticles using ascorbic acid[J]. Talanta, 2013, 104: 22-26.

[12] Rohit J V, Kailasa S K. Cyclen dithiocarbamate-functionalized silver nanoparticles as a probe for colorimetric sensing of thiram and paraquat pesticides via host–guest chemistry[J]. Journal of Nanoparticle Research, 2014, 16(11): 2585.

[13] Bhamore J R, Jha S, Mungara A K, et al. One-step green synthetic approach for the preparation of multicolor emitting copper nanoclusters and their applications in chemical species sensing and bioimaging[J]. Biosensors and Bioelectronics, 2016, 80: 243-248.

[14] Charoenkitamorn K, Chailapakul O, Siangproh W. Development of gold nanoparticles modified screen-printed carbon electrode for the analysis of thiram, disulfiram and their derivative in food using ultra-high performance liquid chromatography[J]. Talanta, 2015, 132: 416-423.

[15] Parham H, Pourreza N, Marahel F. Determination of thiram using gold nanoparticles and resonance Rayleigh scattering method[J]. Talanta, 2015, 141: 143-149.

[16] Y. Yin, Z.Y. Li, Z. Zhong, B. Gates, Y. Xia, and S.Venkateswaran, Synthesis and characterization of stable aqueous dispersions of silver nanoparticles through the Tollens process, Journal of Materials Chemistry, vol. 12, no. 3, pp. 522-527, 2002.

[17] Zhu J, Chen X Y, Fan I R, et al. A green chemistry synthesis of Ag nanoparticles and their concentrated distribution on PDMS elastomer film for more sensitive SERS detection, 2017 19th International Conference on Solid-State Sensors, Actuators and Microsystems (TRANSDUCERS)

[18] Zhu J, Lin G, Wu M, et al. Large-Scale Fabrication of Ultrasensitive and Uniform Surface-Enhanced Raman Scattering Substrates for the Trace Detection of Pesticides[J]. Nanomaterials, 2018, 8(7): 520.

[19] Severyukhina A N, Parakhonskiy B V, Prikhozhdenko E S, et al. Nanoplasmonic chitosan nanofibers as effective SERS substrate for detection of small molecules[J]. ACS applied materials & interfaces, 2015, 7(28): 15466-15473.

[20] Kumar P, Khosla R, Soni M, et al. A highly sensitive, flexible SERS sensor for malachite green detection based on Ag decorated microstructured PDMS substrate fabricated from Taro leaf as template, Sensors and Actuators B: Chemical, 2017

Gap in pagination due to unavailable papers.

Pages 893-897

Proceedings of the 16th Annual IEEE International
Conference on Nano/Micro Engineered and Molecular Systems
April 25-29, 2021

A clogging rate prediction model based on porous microarray membranes for liquid biopsies

Yinghao He, Yaoping Liu, Wei Wang and Yufeng Jin

Abstract-This paper presents a filter clogging model for the screening of rare tumour cells by microporous array filter membranes for liquid biopsies. In clinical practice, the ability to achieve high throughput and low clogging rates with microporous array membranes is one of the most important indicators of the reliability of liquid biopsy technology. However, the occurrence of clogging due to various complex factors such as pore size, number of pores and sample concentration makes it difficult to isolate target cells from the membrane. By considering the size and number of the tumour cells and the pore size of the filter membrane, this paper allows us to predict the trend of clogging rate after filter membrane filtration from the change in concentration. The reliability of the model was demonstrated through mathematical validation reasoning and experimental results, thus helping subsequent extensive filtration experiments on liquid biopsies to select more appropriate blood dilution concentrations for sample pre-treatment to ensure that the clogging rate is within a reasonable range, and ultimately improving the accuracy of capturing tumour markers and liquid biopsies.

I. INTRODUCTION

In liquid biopsy technology, the use of porous microarray membranes to screen blood for rare tumour cells can effectively aid in the early screening of cancer patients, detect the effectiveness of treatment, and reduce recurrence rates, especially in the early screening of humans for cancer has shown great promise [1]. At the same time, the scientific research on the clogging mechanism of cells in biomedical systems is a promising area, as it is both fundamental and practical significance, as well as complex [2]. However, there has been little discussion of the clogging that occurs during filtration of porous microarray membranes in liquid biopsies, and the severity of the clogging (shown in Fig. 1) affects the high throughput and purity of the results (both of which are critical indicators of the accuracy of early cancer screening).

The filtration based on cell size and dimensions of porous microarray membranes may become the preferred option for achieving early cancer screening in the future due to its ease of operation and low cost. Also, in terms of filtering the results of circulating tumor cells, porous microarray filter membranes are effective in achieving two key metrics, high throughput and high purity. However, circulating tumor cells inevitably clog when passing through small pores in porous microarray membranes, especially for clinical samples with large numbers of components.

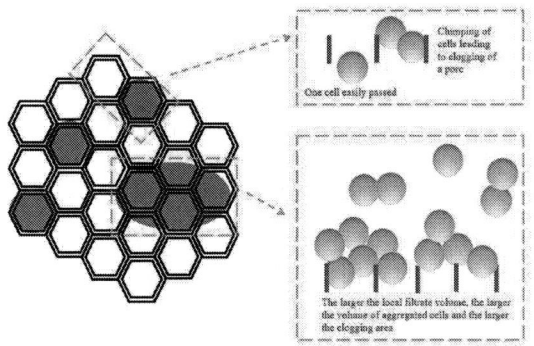

Fig. 1. A diagram of a partial clogging after filtration. As shown, the picture on the left indicates a small area of the porous microarray membrane, where each hexagonal pore represents a pore in the membrane, and the gray part indicates that the area has been clogged with clumps of cells. The two pictures on the right side with dashed lines indicate that one cell can easily pass through the pore (the cell diameter is smaller than the pore diameter), two cells clumped together may lead to clogging of the pore, and multiple pores clogged (due to the large concentration of cells at this location, the phenomenon of cell clumping tends to occur, resulting in a large area of clogging).

In the filtration process of circulating tumour cells, there are many factors that contribute to clogging. We mainly consider the size, dimensions and number of cells, the concentration and volume of the sample solution, and the pore size of the small pores in the filter membrane.

Previous paper proposed a model that considers only the several parameters of the filter membrane to estimate the clogging rate roughly [1]. And this paper proposes a mathematical model that can predict the trend of clogging rate, further addressing the effect of filtrate on clogging rate, by considering filtrate concentration and cell size, and combining various parameters of the membrane. The consistence between the theoretical model and the experimental results is also discussed. This model will be widely applied to the pretreatment of porous microarray membrane filtration experiments in liquid biopsy techniques.

II. METHODS AND EXPERIMENT

A. Clogging rate prediction model

In this model, we consider that each cell is randomly placed in the container with a uniform probability distribution in space before filtering, so that the initial position of the cells is an independent random variable. As

---

*Research supported by TSV 3D Integrated Micro/Nano system Lab and Shenzhen Science and Technology Innovation Committee. (Corresponding author: Yaoping Liu and Wei Wang.)

Yinghao He and Yufeng Jin are with Peking University ShenZhen Graduate School, Guangdong, 518055, China. (E-mail: 1801213284@pku.edu.cn, yfjin@pku.edu.cn).

Yaoping Liu and Wei Wang are with Institute of Microelectronics, Peking University, Beijing, 100872, China. (E-mail: lyping@pku.edu.cn, w.wang@pku.edu.cn).

978-1-6654-3008-1/21 $31.00 © 2021 IEEE

in the rest of the paper, N is the number of cells initially placed in the filtrate, V is the total volume of the filtrate initially, and Ω is a small region of the solution inside the container and its volume denoted by $|\Omega|$ (shown in Fig. 2), so that the probability that exactly i of the N cells happen to be clustered in Ω to begin with is

$$p(k=i)=C_N^i \; (|\Omega|/V)^i (1-|\Omega|/V)^{\,N-i} \qquad (1)$$

According to (1), the probability p of cell clustering depends on such four values: the total number of cells N, the assumed number of aggregated cells i, a small selected region Ω, and the total volume of the sample solution. In order to consider the critical condition for any one pore to clog, we define Ω as follows, i.e. $|\Omega| = hA$, where h is the diameter of the cell and A is the area of a pore, thus $V \gg |\Omega|$ and $N \gg i$, the asymptotic value $p(k=i)$ is

$$p(k=i)=\frac{1}{i\,!} \; (C|\Omega|)^i \; e^{-C|\Omega|} \qquad (2)$$

In the above equation, C is the initial concentration of the filtrate, C=N/V. (The above analysis of the approximate model is also valid for the single pore filtration model [3].)

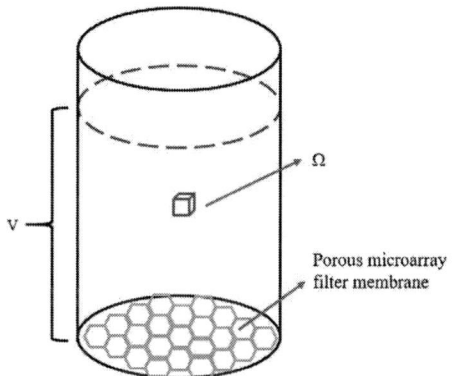

Fig. 2. The cylinder is the container in which the filtrate is loaded during filtration. V is the total volume of the sample solution at the start of solution filtration, with the highest liquid level as the blue dashed line. Ω is any part of the sample solution and the volume size is represented by the absolute value symbol in mathematics, i.e. $|\Omega|$. Subsequently, we define $|\Omega| = rA$ when estimating the critical state of clogging. It is worth noting that there is no restriction on the shape of Ω, regardless of how large the volume of $|\Omega|$ is, i.e., Ω is arbitrary in shape at any volume. The hexagonal grid at the bottom of the cylinder is a schematic representation of a porous microarray filter membrane.

Since the volume of each pore is constant, we assume that the critical state for a pore clogging is the simultaneous capture of k_0 cells, and that k_0 is proportional to the Ω volume divided by cell volume, i.e. $k_0 = \lambda|\Omega| / V_{cell}$. The λ related to cell-to-cell and cell-to-filter membrane and container interactions during filtration and $0 < \lambda < 1$. So when $i = k_0$, $p(k=k_0)$ is the probability of a critical clogging of a small pore. Therefore, the probability that a cluster of cells that would allow a small pore to be clogged is $p(k>k_0)$.

Considering that clustered cells of varying degrees that meet the critical condition can clog membrane pores to different degrees, the mathematical expectation of the number of membrane pores that can be clogged by filtrate of initial volume V and initial number of cells N is

$$M=\sum_{s=0}^{\infty} \frac{V}{(k_s/k_0)|\Omega|} \; p(k=k_s) \qquad (3)$$

In the above equation, the $k_S=k_0+1,2,3\ldots$. Since $0<C\Omega\ll 1$, the (3) can be approximated as

$$M= e^{-C|\Omega|} k_0 V \frac{1}{(k_0+1)\,!} \; (C)^{k_0}(\Omega)^{k_0\text{-}1} \qquad (4)$$

In this equation, the only unknown quantity is λ in k_0. We can calculate λ from any one of a set of experimental data, and then use (4) to predict the clogging rate at any filtrate concentration.

B. Porous microarray membranes and Cell preparation

The Parylene C porous microarray membranes used for the experiments were prepared based on our previously developed molding technique [4, 5, 6]. And we selected microporous arrays with micropore diameters of 25 µm and 15 µm with thicknesses of 10 µm. It is important to note that each pore in the porous microarray membrane has the same size and pore spacing. Each pore is a regular hexagon, but in the theoretical calculations we treat the area of each pore as an approximation to the area of a circle.

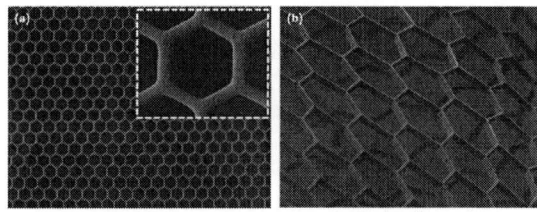

Fig. 3. SEM images of the filter membrane structure. The top view (a) and the side view (b).

A549 (Human lung cancer cells) was used for filtration validation in this work and the median diameter of this batch of A549 cells was 16 µm after measurement experiments. We divided the them into two sets of experiments for filtration through 25 µm and 15 µm microporous arrays, respectively.

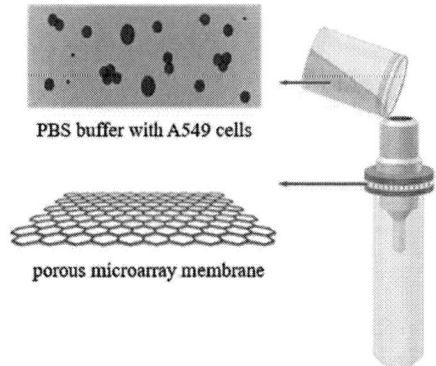

Fig. 4. Schematic illustration of the filtration unit. PBS buffer with A549 cells is poured into the fixture inlet and the cells are filtered through the membrane to complete the filtration.

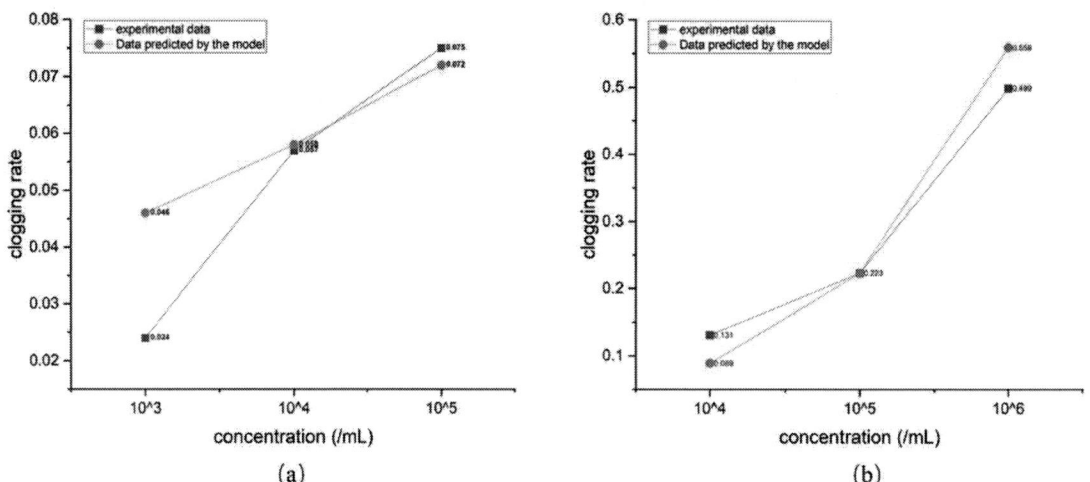

Fig. 5. The line graph of clogging rate at different concentrations. The (a) and (b), respectively, correspond to the results of the filtration of A549 cells through the 15 μm and 25 μm filter membranes at different concentrations, the data are also the same as in tables I and II.

TABLE I. Comparison of model prediction data and experimental data. (15 μm membrane filtration)

Concentration	Volume	λ	Model prediction value	Experiment date
10^5/mL	500 μL	0.83	0.0719	0.0752 (±0.002)
10^4/mL	5000 μL	0.83	0.0578	0.0566 (±0.004)
10^3/mL	50000 μL	0.83	0.0464	0.0239 (±0.004)

TABLE II. Comparison of model prediction data and experimental data. (25 μm membrane filtration)

Concentration	Volume	λ	Model prediction value	Experiment date
10^6/mL	500 μL	0.64	0.559	0.499 (±0.045)
10^5/mL	5000 μL	0.64	0.223	0.223 (±0.023)
10^4/mL	50000 μL	0.64	0.089	0.131 (±0.009)

III. RESULTS AND DISCUSSION

As can be seen from TABLE I, we selected the same number of cells (5×10^4) for each experiment. After several filtration experiments, three sets of experimental data were obtained: three different clogging rates were obtained by filtering cells with 15 μm porous microarray membranes at three different concentrations.

According to (4), with only the parameters λ in k_0 unknown, we can solve for λ to be approximately equal to 0.83 by using the "middle" set of data ($C=10^4$/mL, V=5000 μL) from the three sets of data obtained above. Based on the calculated $\lambda = 0.83$, we can then calculate the predicted values of the clogging rate at $C = 10^3$/mL and $C = 10^5$/mL by (4). The clogging rates obtained in the laboratory using filtration experiments and those predicted by the theoretical model have been plotted separately in (a) of Fig. 5.

As can be seen from TABLE II, we selected the same number of cells (5×10^5) for each experiment. And three different clogging rates were obtained by filtering cells with 25 μm porous microarray membranes at three different

concentrations. Based on the same approach, we can calculate the predicted values of the clogging rate of the theoretical model for each of the three different concentrations, where $\lambda=0.64$. The corresponding data has been plotted in (a) of Fig. 5.

Judging from the degree of agreement between the experimental data and the model prediction data, the mathematical model requires only one set of data to calculate the parameter λ to reasonably predict the trend of the effect of concentration change on the clogging rate and to explain the clogging mechanism of clustered cells.

Moreover, as can be seen from the results, the flow rate of cells during filtration must be different due to the different sizes of porous microarray filter membranes used in the two sets of experiments (15 μm diameter in one set and 25 μm in the other). It also affects the clumping between cells, and the collision of cells with the filter membrane and container, resulting in different criteria for clumping cells that cause clogging, so λ is not the same.

The clogging rates of 25 μm porous microarray membranes is more severe than that of 15 μm membranes

978-1-6654-3008-1/21 $31.00 © 2021 IEEE

for the same concentration (when C = 10^5/mL, using 25 μm membrane filtration, clogging rate is 0.223; using 15 μm membrane filtration, clogging rate is 0.075. And when C = 10^4/mL, using 25 μm membrane filtration, clogging rate is 0.131; using 15 μm membrane filtration, clogging rate is 0.057). The reason for the greater clogging rate with the large pore filtration than with the small pore filtration may be due to the difference in the number of cells selected. 25 μm membrane filtration uses 5×10^5 cells, whereas 15 μm membrane filtration uses 5×10^4 cells (a tenfold difference in number).

IV. CONCLUSION

This work presents a model for filter clogging of microporous array filter membranes for liquid biopsy screening of rare tumour cells. In this paper, by considering the size and number of tumour cells and the pore size of the filter membrane, we will be capable of predicting the variation of the clogging rate at different concentrations. The agreement between the values predicted by the theoretical model and the actual clogging rate demonstrates the reliability of the model. This work will help researchers to efficiently pre-treat diluted clinical samples in filtration experiments for liquid biopsies, reducing the occurrence of unnecessary clogging rates and achieving high throughput and purity filtration.

ACKNOWLEDGMENT

This work was financially supported by TSV 3D Integr ated Micro/Nano system Lab (ZDSYS201802061805105), the Shenzhen Science and Technology Innovation Commi ttee (JCYJ20190808155007550).

REFERENCES

[1] Y. Liu, et al., " Concentration Match Based Modulation for Clogging-Free Filtration Through Micropore Arrays" *20th International Conference on Solid-State Sensors, Actuators and Microsystems and Eurosensors XXXIII(TRANSDUCERS and EUROSENSORS)*, JUN. 23-27, 2019, Berlin, GERMANY,270-272

[2] Emilie Dressaire and Alban Sauretab, "Clogging-of-microfluidic-systems", *Soft Matter*, vol 13, pp. 37-48, 2017

[3] Goldsztein GH, "Volume of suspension that flows through a small orifice before it clogs", *Siam Journal on Applied Mathematics*, vol 66, pp. 228-236, 2005

[4] Y. Liu, et al., "Filtration membrane with ultra-high porosity and pore size controllability fabricated by Parylene C molding technique for targeted cell separation from bronchoalveolar lavage fluid (BALF)", *Proceeding of the 18th International Conference on Solid-State Sensors, Actuators and Microsystems (Transducers 2015)*, Jun. 21–25, 2015, Anchorage, USA, 1767–1769.

[5] Y. Liu, et al., "2.5-Dimensional Parylene C micropore array with a large area and a high porosity for high throughput particle and cell 'separation", *Microsystems & Nanoengineering*, 2018, 4: 13 (1–12).

[6] Li Tingyu and Liu Yaoping, et al., "A rapid liquid biopsy of lung cancer by separation and detection of exfoliated tumor cells from bronchoalveolar lavage fluid with a dual-layer "PERFECT" filter system", *THERANOSTICS*, vol 10, pp. 6517-6529, 2020

Gap in pagination due to unavailable paper.

Pages 902-903

Proceedings of the 16th Annual IEEE International
Conference on Nano/Micro Engineered and Molecular Systems
April 25-29, 2021

Micro-droplet of Particulate Suspension Generated by a Pneumatic Ejection System

Weijie Bao[1], Shengnan Sun[2], Zhihai Wang*[1], and Yaohong Wang[2]

Abstract— **Micro-droplet generation for solid particle suspensions is studied experimentally over a wide range of volume fraction of solid (η), using a home build pneumatic ejection system. Ejection is actuated by a solenoid valve, setting to "conduction" state for short period of time Δt. Pressurized gas of P_0 enters the reservoir, creating a pressure pulse P(t), forcing the liquid out via a tiny nozzle to produce a micro-droplet. Here, P(t) is measured by a high-speed sensor, and the ejection process is examined by high speed photography and image processing. Single droplet can be ejected for suspensions with η up to 33%. For η less than about 18%, the required pressure amplitude increases roughly linearly with η. With η increased above 18%, the demand for pressure amplitude is significantly higher than the linearly increasing trend. Especially as η is increased above 24%, the liquid band stretches much longer, and the break-up is delayed drastically.**

Keywords—micro-droplet, pneumatic, particulate suspension

I. INTRODUCTION

The technique of droplet on demand (DOD) was in the earlier years employed for inkjet printing [1]. Recently it has found more applications in the field of biomedicine [2], printed electronics [3][4], additive manufacturing etc. [5]. A lot of applications involve mixing solid particles in liquids. The generation of droplets with solid particles has potential applications in drug production [6]. Inkjet of conductive micro/nano particle suspensions is widely used in printed electronics [7]. Inkjet printing of ceramic particles is used in additive manufacturing [8].

Traditional DOD methods use either thermal or piezoelectric actuation. Previously, piezoelectric print-heads are popular in the experiments where micro-droplets with solid particles are ejected [9]. Recently, progress have been made in the region of non-standard micro-droplet ejection techniques [10]. The micro-droplet ejection based on electrohydrodynamic (EHD) method has great potential for the enhancement of printing resolution and for reducing the risk of nozzle clogging. There are also experimental studies involving the EHD ejection of particulate suspensions [7]. However, EHD ejection state is highly sensitive to many experimental factors, so a well-controlled operation is not easy [11].

Pneumatic ejection is also an example of non-standard ejection technique. A typical pneumatic ejector is schematically shown in Fig. 1. A reservoir is linked with a

high-pressure gas source via a high-speed solenoid valve, and is connected with the ambience via a venting tube. The solenoid valve is set to "conduction state" for a short period of time Δt, and pressurized gas of pressure P_0 rashes into the reservoir, then releases via the venting tube. This would generate a pressure pulse P(t) within the reservoir. The pressure pulse would force the liquid flowing out of the nozzle to form a liquid belt, which then breaks up to produce micro-droplets. The pneumatic micro-droplet ejector is simple and reliable to operate. It can also work with high viscosity and/or high temperature liquids [12][13]. Pneumatic micro-droplet ejection has been employed in the fields of electronic packaging [14], 3D printing [15][16], bio-medicine etc. [17][18].

Other than the application researches, some also focus on the ejection process itself. Pneumatic micro-droplet ejection is due to multi-phase flow driven by the pressure P(t), and also to instability associated with the surface tension [19]. Effects of relevant ejection parameters, including source pressure and actuation of the solenoid valve, have been studied experimentally [20]. This ejection process may be studied by solving a multi-phase flow problem, where the VOF (volume of fluid) method is taken to track the interface [21]. According to our literature survey, there is no experimental study on the pneumatic ejection of droplets containing solid particles. In this paper, a home-made pneumatic ejection system to employed to produce micro-droplet of particulate suspensions in a wide range of volume fraction of solid. The pressure waveform in the reservoir is measured. The ejection processes are examined by high-speed photography and by image processing methods. Relevant parameters such as droplet volume and break-up time are measured.

II. PNEUMATIC MICRO-DROPLET GENERATOR AND ITS WORKING PRINCIPLES

0Referring to our previous publication [22], the lab-build pneumatic micro-droplet ejection system is schematically shown in Fig. 1. The whole experimental setup includes a micro-droplet ejector and monitoring modules. The ejector includes a reservoir, a high-speed solenoid valve, and air path (mainly, the venting tube), the nozzle is attached to the bottom of the reservoir, and the pressure of the gas source is tuned by a pressure regulator; the visual monitoring module consists of a high-speed CMOS Camera and a LED, and a high-speed pressure sensor is used for measuring pressure in the reservoir. Both the ejector and the monitor are controlled

1 Faculty of Information, Beijing University of Technology, Beijing 100124, China

2 Center for Applied Mathematics, Tianjin University, Tianjin 300072, China

*Contacting Author: Zhihai Wang is with Faculty of Information, Beijing University of Technology, Beijing 100124, China (Phone:86-18810898336; E-mail: wangzhihai@bjut.edu.cn)

978-1-6654-3008-1/21 $31.00 © 2021 IEEE

by PC through a STM32 micro-controller. Pure Nitrogen gas is used in our experiments.

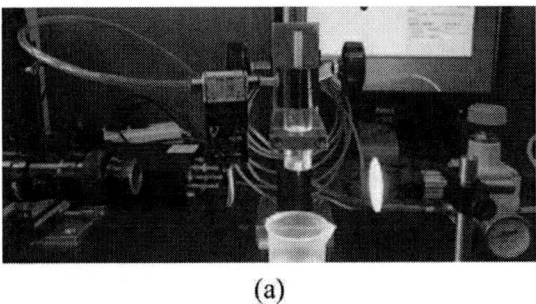

(a)

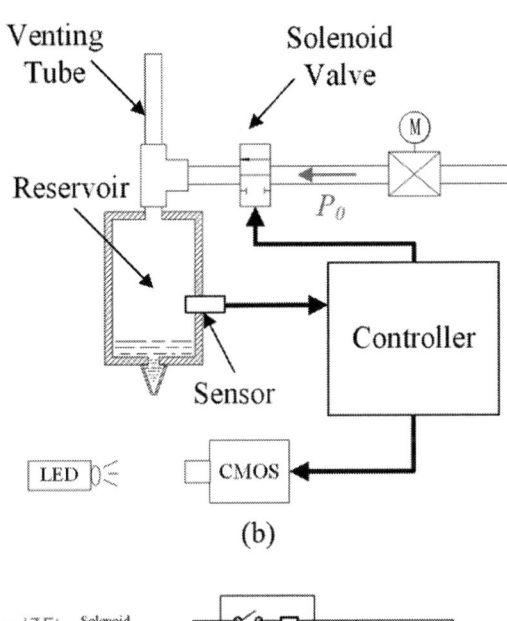

(b)

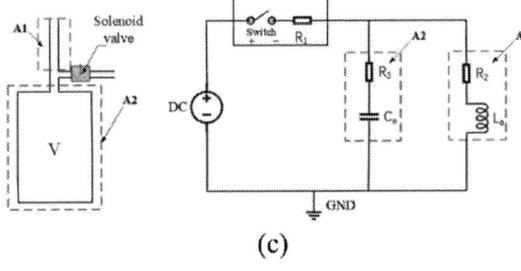

(c)

Fig. 1 (a) photograph of the lab-build pneumatic micro-droplet ejector, (b) Schematic representation of the setup. The ejector consist of a liquid reservoir, a nozzle, a pressure regulator, a solenoid valve, and a venting tube. The ejection process is studied by a high speed camera illuminated with LED. The gas pressure in the reservoir is measured via a high speed sensor. The operation of the system is realized by a controller. (c) An model based on electro-acoustic analogy is used for analyzing the pressure pulse generation in the reservoir.

A. Measuring Pressure Pulse P(t) in the Reservoir

As just introduced, P(t) produces the driving force for the ejection process. A high-speed pressure sensor (sampling frequency 1MHz, pressure range -20kPa ~ +40kPa) is employed for collecting the pressure pulse P(t) in the reservoir. An oscilloscope (sampling frequency 50kHz) takes the output signal from the pressure sensor and actuation signal generated by the micro-controller. The rising edge of this actuation signal is taken as triggering for the solenoid valve, and the width of the actuation signal set the period of the "conduction" state. The rising edge is naturally taken as the starting point $t = 0$ for other measurements. Starting from $t = 0$, pressure pulse is recorded as P(t) (0 - 100ms). After about 100ms, the pressure is rather small. Some typical pressure pulse measured in this experiment are shown in Fig. 2(a).

B. The Generation of Pressure Pulse P(t) in the Reservoir

The generation of pressure pulse in the reservoir has been discussed in our previous publication [22]. Specifically, as shown in Fig. 1(c), the voltage across the capacitor in the RLC circuit and the pressure in the reservoir follow the same equation. Therefore, an electro-acoustic analogy can be made. In this method, acoustic compliance, acoustic inertance, and acoustic resistance correspond to capacitance, inductance, and electric resistance respectively. The solenoid valve is equivalent to a switch and a series resistance. The gas source of constant pressure P_0 is equivalent to the constant voltage source. For the electric circuit, after the switch is turned on, the voltage source charges the capacitor. After the switch is turned off, the capacitor and the inductor form a LC oscillator, damped by the resistor. The pressure oscillation, well-known as Helmholtz oscillation [23], could be understood in the circuit theory as LC oscillation. The duration of the first positive pressure period equals to roughly Δt, plus quarter period of the Helmholtz oscillation. The amplitude of the pressure pulse is determined by P_0. The electro-acoustic analogy works well only for that P_0 is much lower than the ambient pressure. In this experiment, the source pressure P_0 ranges from 20kPa (far below the ambient pressure at 1atm) to 300kPa (significantly higher than the ambient pressure). Fig. 2(a) shows the pressure pulse waveforms P(t) for a set of P_0. As shown in Fig. 2(b), the amplitude of the first positive pressure cycle, referred as P_{MAX}, does not increase linearly with P_0. In addition, with the increase of P_0, the duration of the first positive pressure cycle increases, that is, P(t) enters negative pressure at a later time. And the negative pressure peak is also delayed. All those nonlinear effects cannot be explained by this electro-acoustic analogy model, as it is linear in nature.

C. Visual Monitoring of Ejection State

A high-brightness LED and a high-speed CMOS camera (mini AX100, Photron limited) are used to study the ejection process. Photo-taking is done within a back-lightening setup,

978-1-6654-3008-1/21 $31.00 © 2021 IEEE

where the nozzle is placed between the camera and the LED. The frame rate is 4000 fps, and the exposure time is 50μs.

The image processing methods has been described in our previous publication [22]. Firstly, a region of interest (ROI) is specified to speed up the image processing. In the second step, for grayscale images, binarization is done by using the Otsu algorithm. Due to refraction of the light, there is small bright region in the droplet. To better extract of the droplet information, in the 3rd step, an algorithm is employed to have the bright region refilled. In the final step, the droplets are marked after a connected region analysis. By image processing methods, the number of droplets, the geometrical sizes for each droplet can be extracted. To calculate the volume of the droplet from the image, the width of the droplet is measured versus vertical coordinate z. The volume of the droplet can be calculated by $V = \sum_z \pi [w(z)/2]^2$.

III. EXPERIMENTAL STUDIES ON DROPLET EJECTION FOR PARTICULATE SUSPENSIONS OF DIFFERENT VOLUME FRACTION OF PMMA

The ejection experiment is performed with aqueous glycerol solution with PMMA microspheres mixed in. The diameter of PMMA microspheres is 15μm. The concentration of glycerol in the solution is 69%. At this concentration, the density of liquid matches the density of PMMA. Therefore, sediment of solid microspheres during the ejection process can be ignored. The volume fraction of PMMA, referred as η in the rest of this paper, ranges from 0 to 33%, at a step of 3%. A 600μm internal diameter nozzle (Musashi engineering) is taken in this experiment. The source pressure P_0 is tuned by a pressure regulator (SMC Corporation) in the range of 20 - 350 kPa. The liquid level is hold steady during the experiment, so the hydrostatic pressure does not change. Meanwhile, the behavior of Helmholtz oscillation does not change as the volume V above the liquid is fixed.

A. Amplitude of actuation for particle suspensions at different volume fraction of solid

Usually, every time the solenoid valve is actuated, single droplet is expected. The number of droplets ejected depends mostly on the amplitude of the pressure pulse. Fig. 2(a) shows typical pressure pulse waveforms P(t) in the reservoir that can stably eject single droplet for a set of volume fraction of PMMA microspheres (η). The conduction time of solenoid valve is set to be Δt = 2ms, and the source pressures are also labeled in the figure. The relationship between P_{MAX} and P_0 is not linear, which has been described in the previous section. With η increased from 0 to 33%, Fig. 2(c) shows operating range and average value of the source pressure P_0 for the generation of single droplet. The inset panel shows those data for P_{MAX}. For η less than about 18%, the required P_{MAX} increases approximately linearly versus η. For η above 18%, the P_{MAX} needs to eject single droplet deviates from the linear trend significantly.

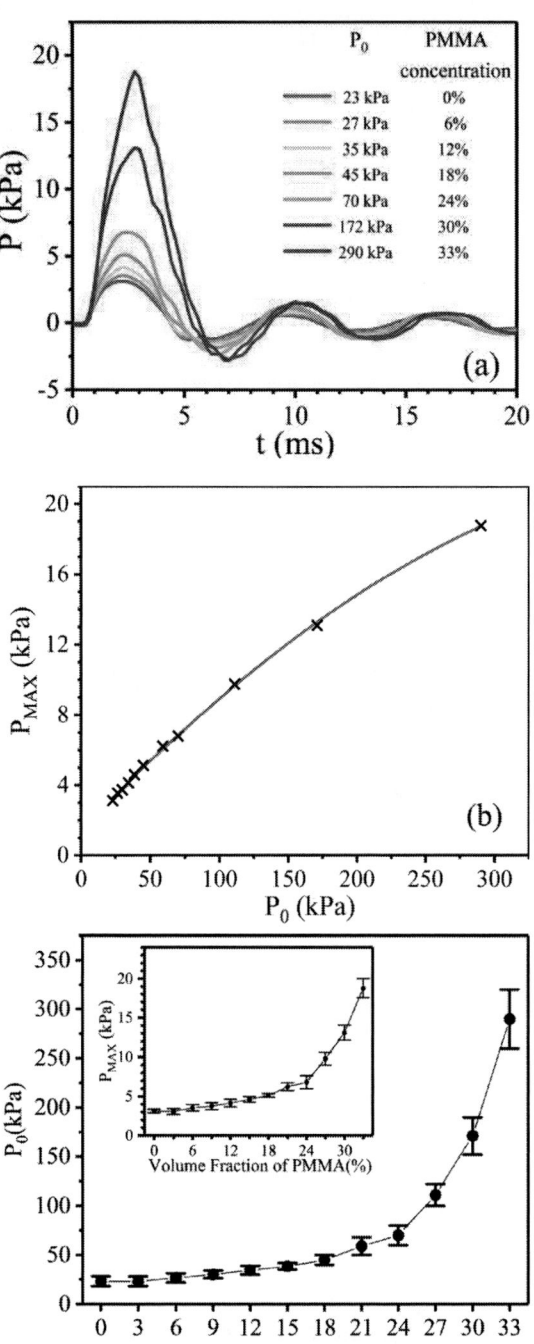

Fig. 2. (a) Typical pressure waveforms taken at several P_0, which can produce single droplet for PMMA suspensions of differentη. (b) Variation of P_{MAX} versus source pressure P_0. (c) Source pressure P_0 and pulse amplitude P_{MAX} (inset) required for ejecting single droplet, for PMMA suspensions with η ranging from 0 to 33%.

B. Ejection process for PMMA suspensions and aqueous Glycerol solution (PMMA free)

Fig. 3(a) shows the droplet ejection process for aqueous glycerol solution without PMMA microspheres, where P_0 = 25kPa, Δt = 2ms. The photos have been processed as previously described. The break-up occurs at around 7ms. Fig. 3(b) shows the droplet ejection process for suspension with 33% volume fraction of PMMA, where $P_0 = 300$kPa, $\Delta t = 2$ms. The break-up time is about 13ms, which is drastically delayed compared with the case of $\eta = 0\%$. Before the break-up, the liquid band at the nozzle is stretched significantly longer than that for aqueous glycerol solution without PMMA microspheres. Although not shown here, the phenomenon of liquid band stretching becomes significant only for η larger than 24%.

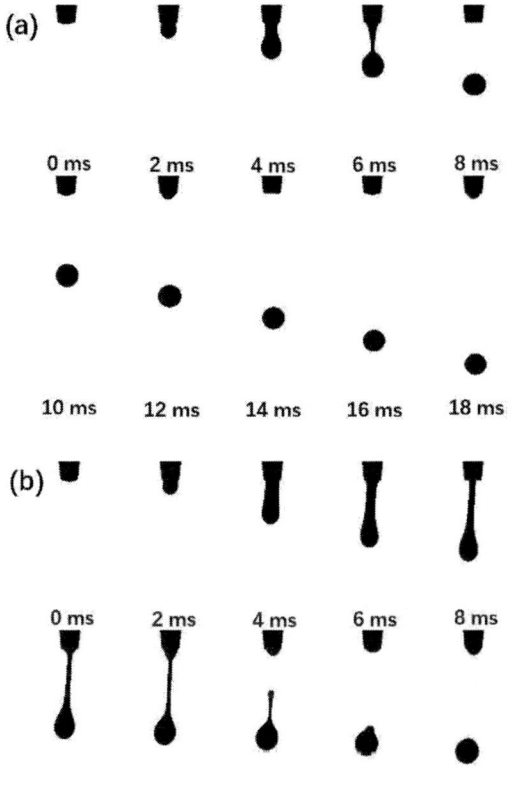

Fig. 3. Ejection process for liquid without PMMA micro-spheres (a), and for liquid of $\eta = 33\%$ (b). The source pressures for ejection are 25kPa and 300kPa respectively. Δt is set to 2ms. The ejection frequency is 10Hz.

C. Volume of micro-droplet and break-up time for particle suspensions at different η, high frequency ejection

Fig. 4(a) shows the volume of the micro-droplet for a set of volume fractions of PMMA (η). The droplet volume increases slightly with η. Moreover, when η is larger than 24%, the fluctuation of droplet volume becomes significant.

By using the high-speed photography, the break-up time (t_B) can be accurately measured. Fig. 4(b) shows the break-up time for several η. With η increased from zero, t_B tends to increase slightly, and then drastically increases for η higher than 21%. For suspension of the highest PMMA

concentration ($\eta = 33\%$), the break-up time can reach 12-14ms. This would inevitably limit the frequency of micro-droplet ejection. We have tried in the experiment 50Hz ejection frequency. It is shown that the stability and consistency of the ejection state are as good as those for aqueous glycerol solution ($\eta = 0\%$).

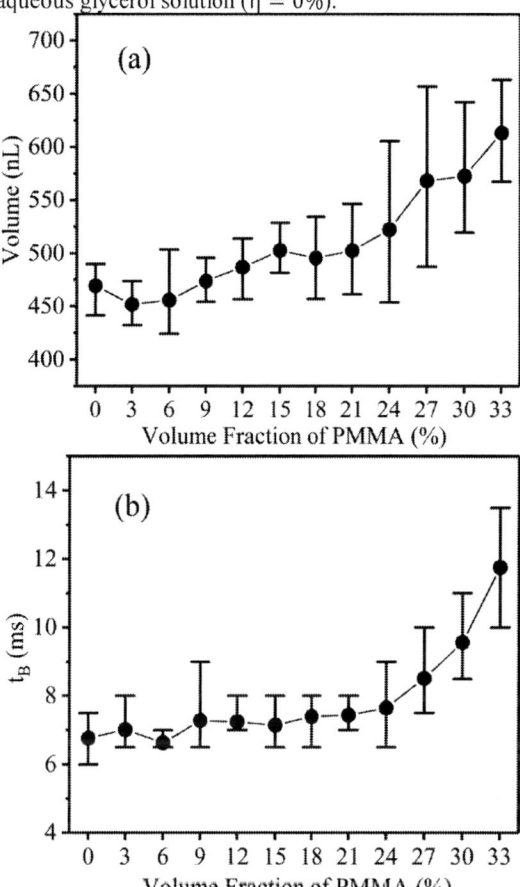

Fig. 4. Volume of micro-droplet (a) and break-up time (b) for a set of suspensions of different volume fraction of PMMA microsphere.

D. Discussions

As early as 1906, Einstein pointed out that for sufficiently low particle concentration, the effective viscosity coefficient of the fluid increases linearly versus η [24]. A large number of studies have shown that the effective viscosity coefficient increases sharply and deviates from the linear trend as η is enhanced above 20% [24]. In our experiment, the sharp increase of P_{MAX} required for single droplet ejection (Fig. 2(c)) may be attributed to the change of effective viscosity.

It was proposed that negative pressure contributes to the break-up of the liquid belt and the generation of droplets [20]. This is consistent with our previous experimental results with DI water. As shown in Fig. 2(a), negative pressure cycle is indeed delayed as P_0 is enhanced for higher η.

However, it can be seen from Fig. 3 and Fig. 4 that for the suspension with $\eta = 33\%$ the pressure pulse has entered the negative region at t = 6ms, but the break-up does not occur until t = 12-13ms. In addition, from Fig. 4(b), the break-up can occurred during the positive pressure period. All these results indicate that the delay of the negative pressure cycle is not sufficient to explain the significant delay of the break-up. It is well known viscous fluids are more resistant to flow instability. The stretching of liquid band and delay of break-up for high η suspensions are more likely due to the enhancement of effective viscosity. Further numerical simulations based on multiphase flow of effective viscosity may be able to interpret these experimental data more quantitatively. It is worth mentioning that the pressure waveform in the reservoir can be accurately measured. Therefore, the setting of inlet boundary conditions can be simple and reliable.

IV. CONCLUSIONS

The process of micro-droplet ejection for particulate suspension with different volume fraction of solid is studied experimentally, by using a pneumatic ejection system. Single droplet can be ejected for suspensions with η up to 33%. For η less than about 18%, the required pressure amplitude P_{MAX} increases roughly linearly with η. With η increased above 18%, the demand for pressure amplitude is significantly higher than the linearly increasing trend. Theoretical models predict that in the similar η range, the effective viscosity of the particulate suspension as a function of η deviates from the linear increasing trend significantly. Our experimental results are qualitatively consistent with the theoretical models. As η is increased, the volume of the droplet slightly increases. Especially as η is increased above 24%, the liquid at the nozzle tends to stretch to form much longer liquid band, while the break-up is delayed significantly. Those behaviors may also be due to the enhancement of effective viscosity. We speculate the long break-up time might limit the ejection frequency. It is shown that for particulate suspension of η = 30%, the ejection frequency can reach 50Hz, and the stability and consistency of the ejection are comparable to those for aqueous glycerol solution (without solid particle).

REFERENCES

[1] W. Zapka, Handbook of Industrial Inkjet Printing: A Full System Approach, Wiley-VCH, 2018.

[2] H. Gudapati, M. Dey, and I. Ozbolat, "A Comprehensive Review on Droplet-based Bioprinting: Past, Present and Future," Biomaterials, vol. 102, pp. 20-42, 2016.

[3] Z. Cui, Printed Electronics: Materials, Technologies and Applications, Wiley, 2016.

[4] N. C. Raut and K. Al-Shamery, " printing metals on flexible materials for plastic and paper electronics," J. Mater. Chem. C, vol. 6, pp. 1618-1641, 2018.

[5] M. Vaezi, H. Seitz, and S. Yang, "A review on 3D micro-additive manufacturing technologies," Int J Adv Manuf Technol, vol. 67, pp. 1721-754, 2013.

[6] A. J. Radcliffe, J. L. Hilden, Z. K. Nagy and G. V. Reklaitis, "Dropwise Additive Manufacturing of Pharmaceutical Products Using Particle Suspensions," Journal of Pharmaceutical Sciences Volume 108, Issue 2, pp. 914-928, 2019.

[7] M. W. Lee, S. An, , N. Y. Kim, J. H. Seo, J. Y. Huh and H. Y. Kim,. "Effects of pulsing frequency on characteristics of

electrohydrodynamic inkjet using micro-Al and nano-Ag particles," Experimental Thermal & Fluid ence vol. 46, pp. 103-110 , 2013.

[8] B. Derby, "Additive Manufacture of Ceramics Components by Inkjet Printing," Engineering, vol.1, issue 1, pp. 113-123, 2015.

[9] H. Yoo and C. Kim, "Generation of inkjet drop of particulate gel," Korea-Australia Rheology Journal,vol. 27, issue 3, pp. 189-196, 2015.

[10] O. A. Basaran, H. Gao, P. P. Bhat, "Nonstandard Inkjets," Annual Review of Fluid Mechanics, vol. 45, pp. 85-113, 2013.

[11] M. S. Onses, E. Sutanto, P. M. Ferreira, A. G. Alleyne, and J. A. Rogers, "Mechanisms, Capabilities, and Applications of High-Resolution Electrohydrodynamic Jet Printing," Small, vol. 11, pp. 4237-4266, 2015.

[12] S. A. Banitabaei and A. Amirfazli, "Pneumatic drop generator: Liquid pinch-off and velocity of single droplets," Colloids and Surfaces A: Physicochem. Eng. Aspects, vol. 505, pp. 204–213,2016.

[13] S. I. Moqadam, L. Madler and N. Ellendt, "A High Temperature Drop-On-Demand Droplet Generator for Metallic Melts," Micromachines, vol. 10, pp. 477, 2019.

[14] J. Luo, L. Qi, S. Zhong, J. Zhou, H. Li, "Printing solder droplets for micro devices packages using pneumatic drop-on-demand (DOD) technique," Journal of Materials Processing Technology, vol. 212, pp. 2066-2073, 2012.

[15] W. Cao and Y. Miyamoto, "Freeform fabrication of aluminum parts by direct deposition of molten aluminum," Journal of Materials Processing Technology, vol. 173, pp. 209-212, 2006.

[16] N. Lass, L. Riegger, R. Zengerle and P. Koltay, "Enhanced Liquid Metal Micro Droplet Generation by Pneumatic Actuation Based on the StarJet Method," Micromachines, vol. 4, pp. 49-66, 2013.

[17] I. H. Choi, H. Kim, S. Lee, S. Baek and J. Kim, "lug-in nanoliter pneumatic liquid dispenser with nozzle design flexibility," BIOMICROFLUIDICS, vol. 9, p. 064102, 2015.

[18] Z. Chen, X. Zhang, P. Chen, W. Li, K. Zhou, L. Shi, K. Liu and C. Liu, "3D Multi-Nozzle System with Dual Drives Highly Potential for 3D Complex Scaffolds with Multi-Biomaterials," International Journal of Precision Engineering and Manufacturing, vol. 18, issue 5, pp. 755-761, 2017.

[19] J. Eggers and E. Villermaux, "Physics of liquid jets," Rep. Prog. Phys. vol. 71, pp. 036601, 2008

[20] S. Cheng and S. Chandra. "A pneumatic droplet-on-demand generator," Experiments in fluids, vol.34, issue 6, pp. 755-762, 2003.

[21] J. Luo, L. Qi, J. Zhou, X. Hou and H. Li, "Modeling and characterization of metal droplets generation by using a pneumatic drop-on-demand generator," Journal of Materials Processing Technology, vol. 212, pp. 718-726, 2012.

[22] F.Wang, Y. W. Wang, W. J. Bao, H. Y. Zhang, J. G. Li and Z. H. Wang "Controlling Ejection State of a Pneumatic Micro-droplet Generator Through Machine Vision Methods," International Journal of Precision Engineering and Manufacturing, vol. 21, pp.633–640, 2020.

[23] C. M. Harris, Handbook of Acoustical Measurements and Noise Control, McGraw-Hill, New York, 1991.

[24] N. Petford, "Which effective viscosity," Mineralogical Magazine, Vol. 73, issue 2, pp. 167-191, 2009.

Proceedings of the 16th Annual IEEE International
Conference on Nano/Micro Engineered and Molecular Systems
April 25-29, 2021

Tweezers for Micro-droplet Transfer Based on Hydrophobic Non-parallel Plate Structure

JiaQiang Wang, Liguo Chen*, Xiongheng Bian, Zaichen Wang

***Abstract*-In the tweezers based on the hydrophobic non-parallel plate structure, the droplet can be controlled to move up and down by changing the distance between the two plates of the tweezers and the angle between the tweezers plate and the vertical direction, so that the liquid can be captured from the hydrophobic substrate and then release. In this paper, a micro-droplet transfer forceps based on a hydrophobic non-parallel plate structure is produced, which can precisely control the distance between the two plates of the forceps and the angle between the forceps plate and the vertical direction, so as to realize the manipulation of the droplets.**

Keywords-Droplet; Tweezers; Non-parallel plate; Hydrophobicity; Capillary force.

I. INTRODUCTION

The capture and release operations of droplets have received more and more attention. The existing droplet capture and release operation methods mainly adjust the adhesion of the surface contacting the droplets, that is, adjust the high adhesion capture droplets and adjust the low adhesion release droplets. There are two ways to adjust the adhesion of the surface in contact with the droplets: one is to make the surface respond to external stimuli, such as light [1], magnetic [2], electricity [3], and temperature[4], through the selection of surface materials, thereby changing Its adhesion. The other is to make a stretchable flexible surface and deform its surface microstructure by stretching to change its adhesion [5][6]. In addition, there are studies on mechanical modeling of droplets in non-parallel plates and analysis of the force and movement of the droplets [7][8]. We have also analyzed the force and movement of the droplets in the hydrophobic non-parallel plate, but due to the limitations of experimental equipment, a large number of experiments have not been carried out to verify, and no applied tweezers have been developed.

This paper introduces a micro-droplet transfer tweezers based on the hydrophobic non-parallel plate structure. Based on the theoretical basis of the force and movement analysis of the droplets in the hydrophobic non-parallel plate structure, an application-type tweezers platform is built, which can be used by a computer host It is controlled to meet the requirements of theoretical analysis and achieve the capture and release of droplets.

II. MECHANICAL MODEL ANALYSIS

Study the movement of the droplet in the tweezers. For the droplet in the tweezers, establish a mechanical model,

analyze the force of the droplet, and study and discuss what factors affect the movement of the droplet in the tweezers. Perform qualitative analysis through theory, and then conduct quantitative research through experiment.

Fig. 1 shows a schematic diagram of the state of the droplet in the hydrophobic non-parallel plate. The contact angle between the droplet and the hydrophobic surfaces of the two plates is an obtuse angle (actually about 120°).

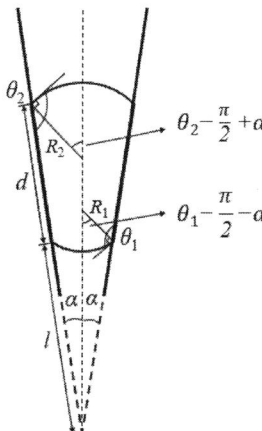

Fig. 1. Force model of droplets in hydrophobic nonparallel plate structure.

Where R_1 is the radius of curvature of the lower end of the droplet, R_2 is the radius of curvature of the upper end of the droplet, θ_1 is the contact angle of the lower end of the droplet in the hydrophobic non-parallel plate, θ_2 is the contact angle of the upper end of the droplet in the hydrophobic non-parallel plate, α Is the angle between the non-parallel plate and the vertical direction, d is the width of the droplet in the hydrophobic non-parallel plate, and l is the distance between the extended intersection of the lower end of the droplet and the non-parallel plate.

Make the following assumptions for the model in Fig. 1: (1) The surface of the droplet is spherical, and the shape of the droplet in the figure is an arc. (2) The gravity of the droplet is negligible. (3) The model is symmetric about the central axis.

Then the volume of the droplet in the non-parallel plate can be expressed as:

$$V \approx 2\pi d^2(d + 2l)\sin\alpha \qquad (1)$$

Robotics & Microsystem Center, Collaborative Innovation Center of Suzhou Nano Science and Technology, Soochow University, Suzhou 215123, China. Email: hbhuang@suda.edu.cn; chenliguo@suda.edu.cn

978-1-6654-3008-1/21 $31.00 © 2021 IEEE

From the geometric relationship in the figure, we can get:

$$l\sin\alpha = R_1\sin\left(\theta_1 - \frac{\pi}{2} - \alpha\right) = -R_1\cos(\theta_1 - \alpha) \quad (2)$$

$$(l+d)\sin\alpha = R_2\sin\left(\theta_2 - \frac{\pi}{2} + \alpha\right) = -R_2\cos(\theta_2 + \alpha) \quad (3)$$

So you can get:

$$R_1 = -\frac{l\sin\alpha}{\cos(\theta_1 - \alpha)} \quad (4)$$

$$R_2 = -\frac{(l+d)\sin\alpha}{\cos(\theta_2 + \alpha)} \quad (5)$$

Combining with Young-Laplace Eqn., suppose the atmospheric pressure is P_o and the pressure inside the upper and lower ends of the droplet are P_{i1} and P_{i2} respectively, then:

$$P_{i1} = \frac{2\gamma}{R_1} + P_o = -\frac{2\gamma\cos(\theta_1 - \alpha)}{l\sin\alpha} + P_o \quad (6)$$

$$P_{i2} = \frac{2\gamma}{R_2} + P_o = -\frac{2\gamma\cos(\theta_2 + \alpha)}{(l+d)\sin\alpha} + P_o \quad (7)$$

It can be seen from Eqns. (6) and (7) that there are two main controllable factors that affect the direction of movement of the droplet in the hydrophobic non-parallel plate: the distance between the two plates of the tweezers and the angle between the tweezers plate and the vertical direction.

First, analyze the influence of the distance between the two plates of the tweezers. Assuming that the width of the lower end of the droplet is L, it can be obtained from the geometric relationship in the figure, $L = 2l\sin\alpha$, then l increases with the increase of L.

When the droplet is in the equilibrium position, that is $P_{i1} = P_{i2}$, we can get:

$$\frac{\cos(\theta_1 - \alpha)}{l} = \frac{\cos(\theta_2 + \alpha)}{l+d} \quad (8)$$

For P_{i1} and P_{i2} to find the derivative with respect to l, we can get:

$$\dot{P}_{t1} = -\frac{2\gamma\cos(\theta_1 - \alpha)}{l^2\sin\alpha} = -\frac{2\gamma\cos(\theta_1 - \alpha)}{l\sin\alpha} \cdot \frac{1}{l} \quad (9)$$

$$\dot{P}_{t2} = -\frac{2\gamma\cos(\theta_2 + \alpha)}{(l+d)^2\sin\alpha} = -\frac{2\gamma\cos(\theta_2 + \alpha)}{(l+d)\sin\alpha} \cdot \frac{1}{l+d} \quad (10)$$

It can be seen from Eqns. (9) and (10) that, obviously, $\dot{P}_{t1} > \dot{P}_{t2}$, the rate of change of P_{i1} with the increase of l is greater than the rate of change of P_{i2} with the increase of l. After the droplet is in the equilibrium position, increase the distance L between the two plates of the tweezers, that is, increase l. It can be seen from Eqns. (6) and (7) that P_{i1} and P_{i2} decrease at the same time, but the decrease of P_{i1} is less than that of P_{i2}. It is much smaller, then $P_{i1} < P_{i2}$, so the droplet will move downward; in the same way, reduce the distance between the two plates of the tweezers, then $P_{i1} > P_{i2}$, so the droplet will move upward.

Then, analyze the influence of the angle between the tweezers plate and the vertical direction. θ_a and θ_r are the advancing angle and the receding angle between the droplet and the forceps plate, respectively. If the droplet moves downward, then, $\theta_1 = \theta_a$, $\theta_2 = \theta_r$; if the droplet moves upward, then, $\theta_1 = \theta_r$, $\theta_2 = \theta_a$. The change of the angle between the tweezers plate and the vertical direction is similar to the change of the distance between the two plates of the tweezers. When the angle between the tweezers plate and the vertical direction is increased, $P_{i1} < P_{i2}$, the drop will move downward; when the angle between the tweezers plate and the vertical direction is reduced, $P_{i1} > P_{i2}$, so the droplet will move upward.

Fig. 2a shows a schematic diagram of the capture process of droplets. The droplet capture process is divided into two steps: the first step is to set a proper initial distance between the tip of the tweezers and the angle between the plate and the vertical direction so that the droplet can enter the tweezers; the second step, the tweezers reach a proper height, By reducing the tip of the tweezers to a suitable distance, the droplet can be captured.

First, the first step of the analysis is to set a proper initial distance between the tip of the tweezers and the angle between the plate and the vertical direction, so that the droplets can enter the tweezers. The force analysis model is shown in Fig. 2b. The contact angles of the droplets with the hydrophobic surfaces of the two plates and the substrate are obtuse. Assuming that the droplet is circular, the model is symmetric about the central axis. Where L is the initial width of the tip of the forceps, α is the angle between the initial plate and the vertical direction, θ is the contact angle between the droplet and the two hydrophobic plates, ψ is the contact angle between the droplet and the hydrophobic substrate, and R is the liquid The radius of curvature of the drop.

From the geometric relationship in the figure, we can get:

$$L = 2R\sin\left(\theta - \frac{\pi}{2} + \alpha\right) = -2R\cos(\theta + \alpha) \quad (11)$$

It can be seen from experiments that during the downward movement of the tweezers, if the width of the tip of the tweezers is narrow, the droplet cannot enter the tweezers, but will be squeezed to one side of the tweezers. It is assumed here that the advancing angle between the droplet and the two hydrophobic plates is θ_a. If the droplet can enter the tweezers, it should be satisfied $\theta > \theta_a$. From the Eqn. (11), it can be seen that L increases with the increase of θ, then L must satisfy:

$$L > -2R\cos(\theta_a + \alpha) \quad (12)$$

According to this formula, the value ranges and critical conditions of L and α can be obtained.

Then, in the second step of analysis, the tweezers reach a suitable height, and the droplets can be captured by reducing the tip of the tweezers to a suitable distance. The force analysis model is shown in Figs. 2c and 2d. The contact angles of the droplets with the hydrophobic surfaces of the two plates and the substrate are all obtuse angles. Assuming that the droplet profile is a circular arc, the model is symmetrical about the central axis. Where L is the initial

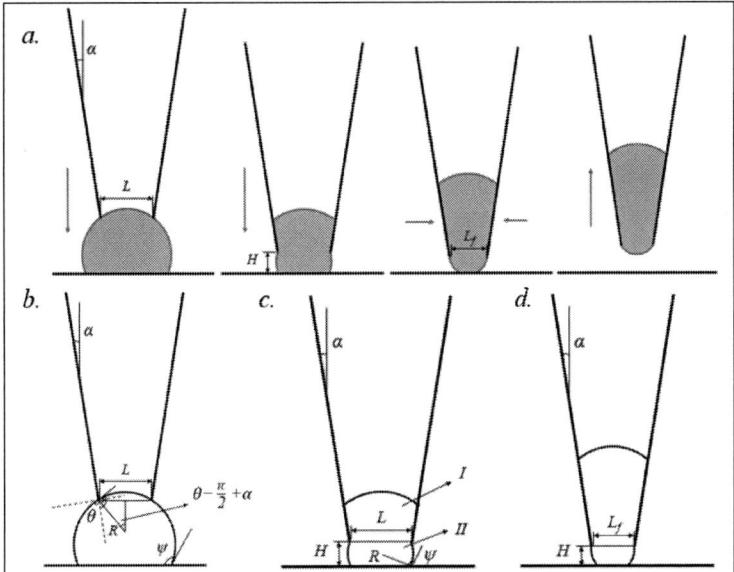

Fig. 2. Force and motion model of droplet capture process. (a) Schematic illustration of the droplet capture process. (b) Analysis of the initial tip distance of the tweezers and the angle between the plate and the vertical direction. (c) Analysis of the final height of the tweezers. (d) Analysis of the final distance of the tip of the tweezers.

width of the tip of the tweezers, α is the angle between the initial plate and the vertical direction, L_u is the width of the upper end of the drop in the tweezers, H is the final suitable height reached by the tweezers, and ψ is the contact angle of drop and the hydrophobic substrate, R is the radius of curvature of the drop, I is the part of the drop inside the tweezers, II is the part of the drop between the tweezers and the substrate, and L_f is the final distance between the tips of the tweezers.

It can be obtained from the geometric relationship in the figure:

$$H = 2R \sin\left(\psi - \frac{\pi}{2}\right) = -2R \cos \psi \qquad (13)$$

Combining with Young-Laplace Eqn., suppose the atmospheric pressure is P_o, then the internal pressure P_{II} of the droplet II between the tweezers and the substrate is:

$$P_{II} = \frac{2\gamma}{R} + P_o = -\frac{4\gamma \cos \psi}{H} + P_o \qquad (14)$$

The internal pressure P_I of the droplet I in the tweezers can be obtained according to the force analysis of the droplet in the hydrophobic non-parallel plate structure:

$$P_I = -\frac{4\gamma \cos(\theta_2 + \alpha)}{L_u} + P_o \qquad (15)$$

Therefore:

$$\Delta P = P_{II} - P_I \qquad (16)$$

When the tweezers move downwards, it will squeeze the droplets between the tweezers and the substrate, then H will decrease, it can be seen from Eqn. (14) that P_{II} will increase; while when the droplets move upwards, L_u will increase, and then P_I will decrease. Therefore, if ΔP will increase, the droplet will move upward and enter the tweezers continuously.

It can be seen from experiments that in the process of squeezing the tweezers to reduce the distance between the tips of the tweezers, if the height H of the tweezers is large, the droplets cannot completely enter the tweezers, but will leak to both sides. It is assumed that the advancing angle of the droplet and the hydrophobic substrate is ψ_a. If the droplet can completely enter the tweezers without leaking, $\psi < \psi_a$ should be satisfied. From the Eqn. (13), it can be seen that H increases with the increase of ψ, then H must satisfy :

$$H < -2R \cos \psi_a \approx -L \cos \psi_a \qquad (17)$$

And

$$H < \frac{L_u \cos \psi_a}{\cos(\theta_a + \alpha)} \qquad (18)$$

After the tweezers reach the appropriate height H, squeeze the tweezers to reach the appropriate final tip distance L_f. According to the force analysis of the droplet in the hydrophobic non-parallel plate structure, in this process, the Laplace pressure at the lower end of the droplet is lower than the upper end, then the Laplace of the droplet When the pressure is upward, the droplet moves upward, and then the tweezers move upward to achieve the capture and separation of the droplet from the substrate.

It can be seen from experiments that during this process, if the final tip distance L_f of the tweezers is large, the droplets cannot be completely captured and leave the substrate when the tweezers move upward. This is due to the adhesion between the droplets and the substrate and the liquid There is a hysteresis in the contact angle between the drop and the tweezers plate. To achieve complete capture of the droplet, the Laplace pressure of the droplet must be upward and overcome the hysteresis of the contact angle between the droplet and the tweezers plate. Let θ_a and θ_r be the advancing angle and retreating angle between the droplet and the tweezers plate respectively, if the droplet moves downward, then $\theta_1 = \theta_a$, $\theta_2 = \theta_r$; if the droplet moves upward, then, $\theta_1 = \theta_r$, $\theta_2 = \theta_a$. According to Eqns. (1) (8), the relationship between the width L of the lower end of the liquid drop in the tweezers and the angle α between the tweezers plate and the vertical direction can be obtained. According to this relationship, the value range and critical value of L_f that completely capture the droplet can be obtained.

The force analysis of the droplet release process is similar to the previous force analysis. Two cases can be considered: the first is to increase the distance between the two plates of the tweezers, then the Laplace pressure on the droplet will go down and the droplet will move towards; The second type is to increase the distance between the two plates of the tweezers and the angle between the two plates of the tweezers and the vertical direction at the same time. Both factors make the Laplace pressure of the droplet downward, and the droplet moves downward to realize the release.

First, consider the first case, we only increasing the distance between the two plates of the tweezers. It can be seen from experiments that the successful release of the droplets cannot be achieved, and the droplets will stick to the two plates of the tweezers and cannot fall. This is because the angle between the tweezers plate and the vertical direction is small, that is, the downward Laplace pressure of the droplet in the tweezers is small, and the hysteresis force caused by the hysteresis of the contact angle between the droplet and the tweezers plate cannot be overcome. At this time, to achieve the release of the droplets, the droplets can be brought into contact with the substrate first, and the release can be achieved by relying on the adhesion of the substrate.

Then, consider the second case, we increase the distance between the two plates of the tweezers and the angle between the two plates of the tweezers and the vertical direction at the same time, as shown in Fig. 3. Through experiments, this situation can achieve the successful release of droplets. Both factors make the Laplace pressure of the droplet downward, so that the droplet can overcome the hysteresis force caused by the hysteresis of the contact angle between the droplet and the tweezers plate. The droplet will first gather at the tip of the tweezers, and then release the droplet.

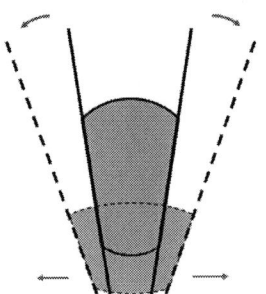

Fig. 3. Schematic illustration of the droplet release process.

It can be seen that in the process of droplet release, the main influencing factor is the angle between the tweezers plate and the vertical direction. Similarly to the previous analysis on L_f, the value range and critical value of α to achieve droplet release can be obtained according to the obtained relationship.

III. EXPERIMENT AND RESULT DISCUSSION

From the previous analysis, the main control requirements for tweezers to capture and release droplets are: (1) Control the distance between the two plates of the tweezers; (2) Control the angle between the tweezers plate and the vertical direction; (3) Control the tweezers The height from the base. Fig. 4 shows the experimental platform of tweezers based on theory. Fig. 4a shows the control requirements of the tweezers movement. Fig. 4b shows the 3D model diagram. The system consists of a linear motor controlling the distance between the two plates of the tweezers, and two rotating steps. The motor controls the angle between the two plates of the tweezers and the vertical direction, and a linear sliding table controls the height of the tweezers. Fig. 4c is a physical image of the tweezers. For better experimental results, two 0.1mm thick rectangular glass sheets are pasted on the tweezers. Due to the size limitation of the stepper motor, and the minimum angle between the two plates of the tweezers and the vertical direction should be 0°, the tweezers are designed into the L shape in the figure. By controlling the movement of the linear motor, the two tweezers board can be fitted together.

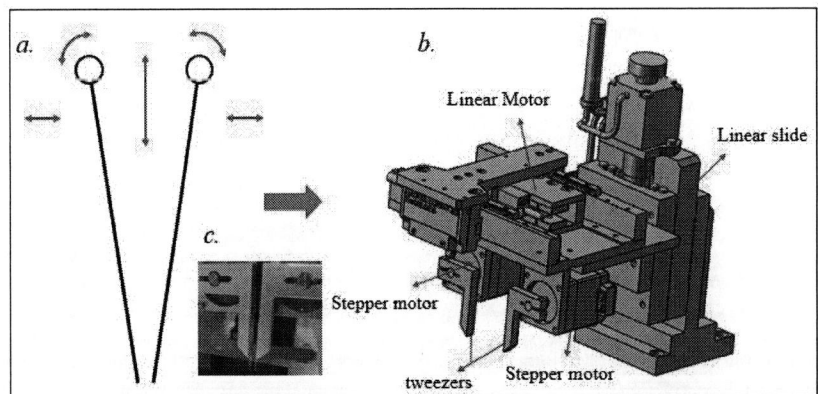

Fig. 4. Tweezers experimental platform. (a) Control requirements for the movement of tweezers. (b)3 D model diagram of tweezers system. (c) Physical drawing of the tweezers section.

Then we conduct experimental research. According to the theoretical analysis of the previous droplet capture process, the experimental research of the droplet capture process is mainly divided into three parts. Experiments are performed on each part, and the experimental results are analyzed and discussed to verify the previous theoretical analysis. The droplet size studied in this experiment is 4μl.

Investigate the initial distance L between the tip of the forceps and the angle α between the forceps plate and the vertical direction, as shown in Fig. 5a. During the downward movement of the tweezers, if the initial tip distance of the tweezers is narrow, the droplet cannot enter the tweezers, but will be squeezed to one side of the tweezers. Experiments were performed for different combinations of L and α to verify whether the droplets could enter the tweezers under each combination parameter, and the experimental results were obtained. In Fig. 5a, the abscissa is α and the ordinate is L. The droplet can enter the tweezers with a black "o", and the droplet cannot enter the tweezers with a red "x". The black broken line indicates that the droplet can enter the tweezers. Enter the critical curve of the tweezers. From the experimental results in the figure, the appropriate parameters can be obtained, α is 5°, and L is 0.7mm.

Investigate the tweezers are lowered to a suitable height H, as shown in Fig. 5b. In the process of squeezing the tweezers to reduce the width of the tip of the tweezers, if the height of the tweezers is large, the droplets cannot completely enter the tweezers, but will leak to both sides. Experiments were performed on different combinations of H and α to verify whether the droplets would leak from both

sides of the tweezers under each combination parameter, and the experimental results were obtained. In Fig. 5b, the abscissa is α and the ordinate is H. The droplets will not leak from both sides of the tweezers with black "o", and the droplets will leak from both sides of the tweezers with red "x", black The broken line represents the critical curve that the droplet will not leak from both sides of the tweezers under each parameter. From the experimental results in the figure, the appropriate parameters can be obtained, α is 5°, and H is 0.3mm.

Investigate the squeezing of the two plates of the forceps to the appropriate final tip distance L_f, as shown in Fig. 5c. In the process of squeezing the two plates of the tweezers, if the distance between the final tip of the tweezers is large, the droplet cannot overcome the adhesion of the substrate during the upward movement of the tweezers, and cannot be completely captured away from the substrate, but will stick to the substrate. Experiments are performed on different combinations of L_f and α to verify whether the droplets will be completely captured and leave the substrate under each combination parameter, and the experimental results are obtained. In Fig. 5c, the abscissa is α and the ordinate is L_f. The droplet can be completely captured and leave the substrate with a black "o", and the droplet cannot be completely captured and leave the substrate with a red "x", and the black broken line is in Under each parameter, the droplet can be completely captured and leave the critical curve of the substrate. From the experimental results in the figure, the appropriate parameters can be obtained, α is 5°, and L_f is 0.2 or 0.3mm. It can be known from the experiment that the smaller L_f, the better the capture effect, so it is more appropriate to choose 0.2mm for L_f here.

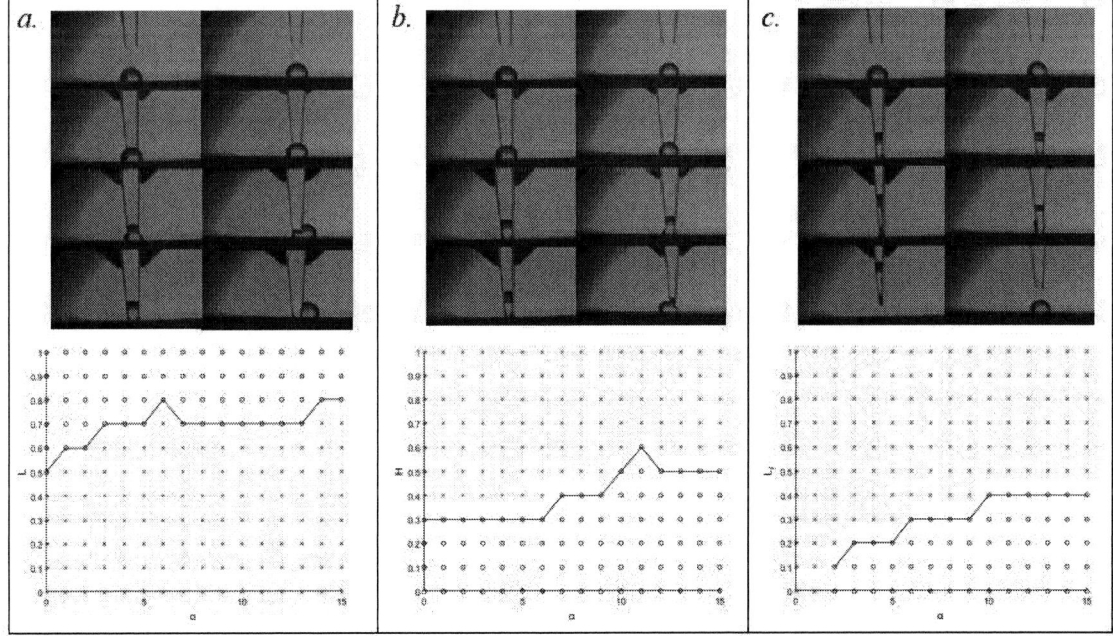

Fig. 5. Experiment of tweezers capture process. (a) Whether droplets can enter tweezers and experimental data curves. (b) Whether the droplet will leak from both sides of the tweezers and the experimental data curve. (c) Whether the droplet will be completely captured off the substrate and the experimental data curve.

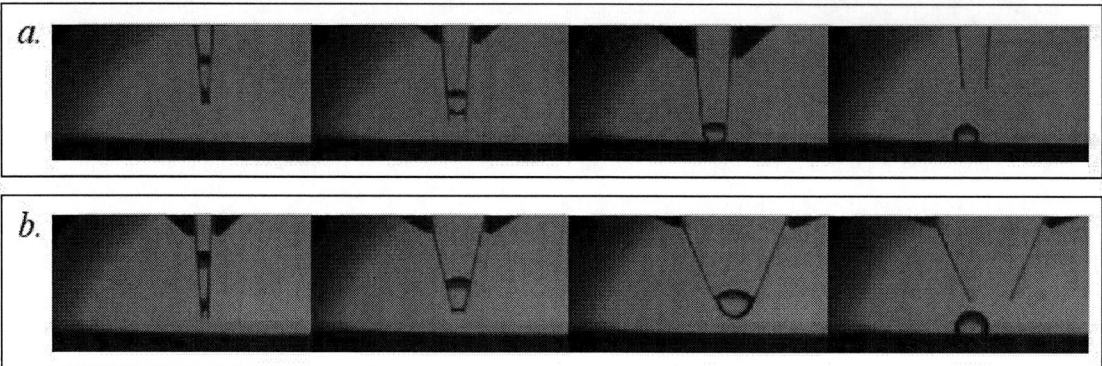

Fig. 6. Experiment of tweezers release process. (a) The process of contact and re-release of droplets with the substrate. (b) At the same time, the distance between the tweezers and the angle between the tweezers and the vertical direction is increased.

For the release process of the droplet, two cases are mainly studied: (1) Do not change the angle between the tweezers plate and the vertical direction, only increase the distance between the two plates of the tweezers, but make the droplet contact the substrate, as shown in Fig. 6a Shown. When the droplet is brought into contact with the substrate, the release of the droplet can be achieved by relying on the adhesion of the substrate. In this way, the influence of the angle between the tweezers plate and the vertical direction and the volume of the droplet is small. Obviously, this situation can only be that the substrate is hydrophobic or hydrophilic, that is, the hydrophobicity of the substrate is less than or equal to the hydrophobicity of the side of the tweezers, otherwise the adhesion of the substrate is small, and the release of droplets cannot be achieved. (2) Simultaneously increase the distance between the two plates of the tweezers and the angle between the tweezers plate and the vertical direction, as shown in Fig. 6b. In the process of increasing the angle between the tweezers plate and the vertical direction, the distance between the two plates of the tweezers also slowly increases. At this time, the droplets in the tweezers will first move downward and gather at the tip of the tweezers. The contact area of the plates becomes smaller, and then slowly increase the distance between the two plates of the tweezers to make the droplets fall. It can be known from the experiment that the critical value of the final angle α between the tweezers plate and the vertical direction is 18° in order to achieve the successful release of the droplets. If the final angle α between the tweezers plate and the vertical direction is small, the droplets will still stick to the sides of the tweezers.

IV. CONCLUSION

In this work, we first established a mechanical model to analyze the force movement of the droplet in the hydrophobic non-parallel plate. The analysis showed that when the distance between the two plates of the tweezers is increased or the vertical direction of the tweezers is increased When the angle is between, the droplet has a downward movement trend, and vice versa. The impact of three main factors on the droplet capture process is also analyzed, which are the distance between the initial tip of the tweezers and the angle between the tweezers plate and the vertical direction, the tweezers are lowered to a suitable height, and the two plates of the tweezers are squeezed to a

suitable final tip. Distance, the release process of the droplet is also analyzed. Finally, an experimental platform for microdroplet transfer tweezers based on the hydrophobic non-parallel plate structure was made, and the theoretical analysis was verified by experiments, and suitable parameters were obtained.

ACKNOWLEDGMENT

Authors acknowledge financial support from the Key University Science Research Project of Jiangsu Province (17KJA460008) and Natural Science Foundation of Jiangsu Province of China (BK20171215).

REFERENCES

[1] Li, Chao, et al. "In situ fully light-driven switching of superhydrophobic adhesion." *Advanced Functional Materials*, pp. 760-763, 2012.

[2] Yang, Chao, Lei Wu, and Gang Li. "Magnetically responsive superhydrophobic surface: in situ reversible switching of water droplet wettability and adhesion for droplet manipulation." *ACS applied materials & interfaces*, pp. 20150-20158.

[3] Xu, Lianyi, et al. "Electrochemically Tunable Cell Adsorption on a Transparent and Adhesion‐Switchable Superhydrophobic Polythiophene Film." *Macromolecular rapid communications*, pp. 1205-1210, 2015.

[4] Tokudome, Yasuaki, et al. "Switchable and reversible water adhesion on superhydrophobic titanate nanostructures fabricated on soft substrates: photopatternable wettability and thermomodulatable adhesivity." *Journal of Materials Chemistry A*, pp. 58-61, 2014.

[5] Wang, Zhiwei, et al. "Stretchable superlyophobic surfaces for nearly-lossless droplet transfer." *Sensors and Actuators B: Chemical*, pp. 649-654, 2017.

[6] Wang, Yongzhen, et al. "Smart superhydrophobic shape memory adhesive surface toward selective capture/release of microdroplets." *ACS applied materials & interfaces*, pp. 10988-10997, 2019.

[7] Luo, Cheng, Xin Heng, and Mingming Xiang. "Behavior of a liquid drop between two nonparallel plates." *Langmuir*, pp. 8373-8380, 2014.

[8] Ataei, Mohammadmehdi, et al. "Stability of a liquid bridge between nonparallel hydrophilic surfaces." *Journal of colloid and interface science*, pp. 207-217, 2017.

Proceedings of the 16th Annual IEEE International
Conference on Nano/Micro Engineered and Molecular Systems
April 25-29, 2021

Design and Preparation of a High Resolution Accelerometer Based on Graphene

Xiaodong Zhao[1], Yanlu Feng[2], Wei Zhang[1], Jianhui Liao[1]

Abstract— Because of the contribution to the development of modern science and technology, scanning tunneling microscopy (STM) effect has been recognized generally for the Nobel Physics Prize in 1986. However, due to the limitation of micromachining technology, tunnel spacing of the tunneling accelerometer cannot reach the order of 1nm. The tunnel effect is generated by loading a pull-down voltage on the deflection electrode. This method not only greatly increases the complexity of the process, but also reduces the consistency of the device, making it difficult to achieve high-accuracy measurement, which limits the increasingly widespread application of tunneling accelerometer. The paper takes this as a breakthrough point to research the design and preparation method of tunneling accelerometer based on graphene. At first, the overall structure of the accelerometer is studied through simulation calculations according to the requirements of the indicators. After that, graphene and cantilever are integrated, then the atomic tunnel junction is prepared accurately by using electro-combustion technology. Finally, tunnel current is detected in real time by controlling the closed-loop feedback circuit. And acceleration is obtained through feedback voltage. For the tunneling accelerometer, the research work will provide powerful support for the improvement of the performance and the expansion of application area.

I. INTRODUCTION

With the continuous maturity of silicon micromachining technology, silicon accelerometers occupy an increasingly important position in the sensor market. Miniaturization, intelligence, and integration have become the development direction of acceleration sensors, and their applications have gradually expanded to navigation, microgravity measurement, hydroacoustic measurement, and seismology.

Since the American scientists G. Bining and H. Rohrer won the Nobel Prize in Physics for successfully developing a scanning microscope based on the quantum tunneling effect in 1986, applications based on the quantum tunneling effect have become a research hotspot. Existing researches all use micro-machining technology to make the tunnel spacing, but due to process limitations, the tunnel spacing cannot reach 1nm (existing studies have shown that tunneling current can only be generated when the tunnel spacing is less than 1nm)[1-3]. The usual approach is to apply voltage to the deflection pull-down electrode and pull the tunnel tip spacing to 1nm until the tunnel effect is generated, this state is then regraded as the balance point of the device[4-5]. This method has the following disadvantages: (1) at present, the tunnel tip is

---

[1]School of Electronics Engineering and Computer Science, Peking University, Beijing, China.

[2]Institute of optics and electronics, Beijing Institute of Collaborative Innovation, Beijing, China.

Contacting Author: Xiaodong Zhao is with Institute of Microelectronics, Peking University. No.5 Yiheyuan Road Haidian District, Beijing, P.R.China (phone: 86-15011083487; email: zxdong@pku.edu.cn).

basically obtained by KOH etching. Due to the rapid etching of convex corner, the height and initial spacing of the tunnel tip are difficult to control, and the repeatability of the device is poor, which is not suitable for mass manufacturing; (2) there is a nonlinear exponential attenuation relationship between the tunnel current and the tunnel gap, so the feedback control system must be used to suppress various low-frequency noises in order to improve the accuracy of accelerometer. However, due to the limitation of micromachining technology, the initial tunnel spacing is difficult to reach 1nm[6-7]. Therefore, it is necessary to add an additional circuit control module to make the accelerometer reach the initial tunnel current state, which will inevitably increase the power consumption of the device, increase the noise source, and affect the test accuracy of the accelerometer.

MEMS accelerometers are the mainstream direction of future development. In order to reduce the process complexity and low-frequency noise of traditional tunnel accelerometers, this paper combines atomic-level controlled electrode preparation technology with MEMS technology, utilizing the super-sensitive characteristics of tunnel current and tunnel junction distance, develop a high-precision accelerometer.

II. PRINCIPLE

The magnitude of the tunnel current I is a measure of the degree of overlap of the electron wave functions, which is related to the distance between the tip and the sample and the average barrier height on the surface of the sample. It satisfies the following relationship:

$$I \propto kV_t e^{-\sqrt{\varphi}x}. \tag{1}$$

where, I is tunnel current, V_t is tunnel bias voltage, $\alpha = 10.25 / nm(ev)^{1/2}$, φ is tunnel junction barrier height, x is electrode spacing.

It can be seen from (1) that the tunnel current changes exponentially with the change of displacement, and the tunnel accelerometer detects the change of acceleration based on this principle.

III. STRUCTURE DESIGN AND PROCESSING OF ACCELEROMETER

Based on the above working principle and the inherent characteristics of graphene materials, the paper designs a new type of high-precision tunnel accelerometer. Fig. 1 is a schematic diagram of the structure, including electrodes, tunnel electrodes, springs, masses, tunnel junctions and graphene. After power-on, the tunnel junction (tunnel gap ≤ 1nm) made of graphene generates tunnel current. The tunnel current has an exponential relationship with the tunnel spacing. This state is regarded as the initial equilibrium state. When an

978-1-6654-3008-1/21 $31.00 © 2021 IEEE

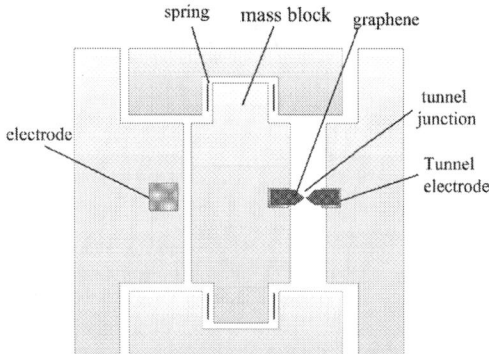

Figure 1. Structure diagram of accelerometer

external acceleration is applied, the mass shifts. As the tunnel spacing changes, the tunnel current changes accordingly, the initial equilibrium state is broken, the feedback voltage changes, and the feedback electrode pulls the mass back to the initial equilibrium state through electrostatic force. The feedback voltage in this process reflects the magnitude of the external acceleration in real time. The value of acceleration can be obtained by detecting the feedback voltage.

A. Simulation analysis

In order to meet the high-precision test requirements, the paper uses the full-scale 1mg and ng-level resolution as the design index, and uses ANSYS to perform simulation calculations. The initial distance between the electrodes is 1nm, and the mass displacement corresponding to the application of 1mg acceleration is analyzed. The tunnel current is calculated by (1). Furthermore, the tunnel current value corresponding to the ng-level resolution can be obtained. Table 1 lists the simulation calculation results.

TABLE I. CALCULATION RESULTS

Electrode Initial spacing	Simulated displacement (1mg)	Tunnel current (1mg)	Displacement (100ng)	Tunnel current (100ng)
1nm	8.5E-04nm	28.88pA	8.50E-08nm	28.35pA

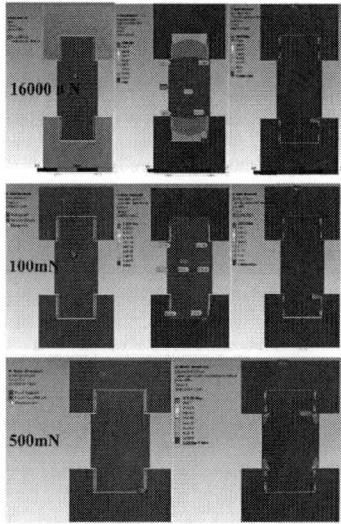

Figure 3. Simulation results

In order to ensure the sensitivity of the accelerometer in one direction in the lateral direction (X direction), further modal frequency is simulated. Fig. 2 shows the simulation results of the sixth-order mode. The third-order mode is X-direction translation, which is about 7.7kHz away from the second-order frequency and about 6.3kHz away from the fourth-order frequency.

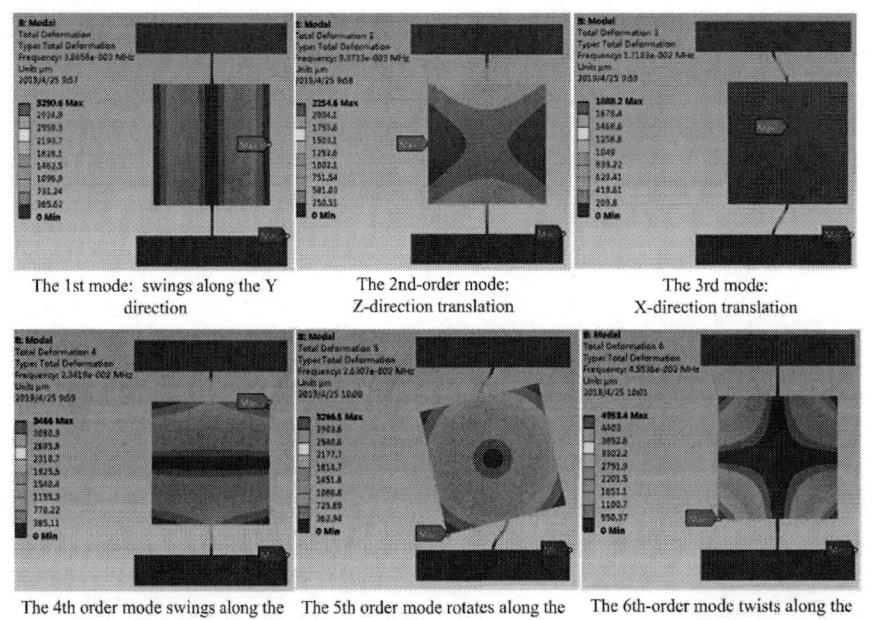

The 1st mode: swings along the Y direction	The 2nd-order mode: Z-direction translation	The 3rd mode: X-direction translation
The 4th order mode swings along the XY plane	The 5th order mode rotates along the XY plane	The 6th-order mode twists along the XY plane

Figure 2. Model simulation results

978-1-6654-3008-1/21 $31.00 © 2021 IEEE

In addition, the integration of graphene and the cantilever beam is achieved by external stress, different stresses are applied in the longitudinal direction of the accelerometer cantilever beam, and the overall stress change of the accelerometer is observed through simulation analysis. Fig. 3 shows the equivalent stress changes when a longitudinal force of 16000μN, 100mN, and 500mN is respectively applied to the mass. The simulation results show that as the longitudinal force increases, the equivalent stress increases from 199Mpa to 263Mpa. When a longitudinal force of 500mN is applied, the mass limit is 1μm, and the equivalent stress changes little, which shows that the structural design can meet the integration mode of graphene.

B. Processing and preparation

According to the simulation design, the layout is drawn, and the high-precision accelerometer is prepared by standard micro-machining technology. Fig. 4 shows the main process preparation flow. The specific steps are:

(1) Select 7740 glass as the substrate.

(2) The limit groove is lithographically etched on the back of the silicon wafer with a height of 1 μm to ensure that the mass block is in a free and movable state.

(3) Anode bonding, aligning and bonding the glass substrate and the back of the silicon wafer.

(4) Thin the silicon wafer KOH to the required thickness.

(5) Continue to etch the shallow groove at the marked position on the front of the silicon wafer, and sputter the metal to corrode to form the feedback electrode and the tunnel current emitter electrode.

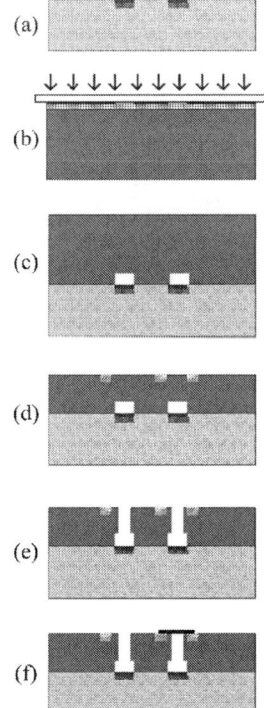

Figure 4. Process flow chart of preparing

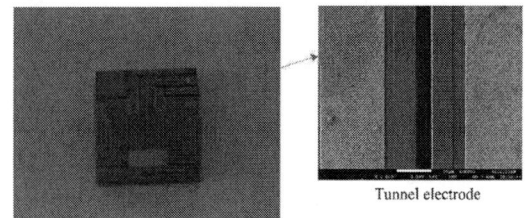

Figure 5. Accelerometer sample

(6) Deeply etch the front surface of the silicon wafer and etch through the silicon wafer to generate freely movable springs and masses.

The accelerometer is processed according to the above preparation process, and its sample is shown in Fig. 5.

C. Accurate preparation of molecular-scale tunnel junctions in graphene electrodes

The graphene is transferred on the processed accelerometer tunnel electrode, and the nanometer-level tunnel junction is prepared on the graphene by the feedback-controlled electro-combustion method. The process flow chart is shown in Fig. 6. First, the single-layer or few-layer graphene is obtained by mechanical peeling, and the graphene is transferred to the Si wafer by dry transfer. And then electrodes are vapor-deposited on both ends of the silicon wafer, and photolithography or electron beam exposure is performed on the graphene. A narrow strip is formed in the middle position, and the narrow strip of graphene is burned off by applying a voltage at both ends of the electrode to form a tunnel gap. The tunnel current and the applied voltage form a feedback control loop. The tunnel current directly reflects the size of the tunnel spacing. The magnitude of the applied voltage can be changed by measuring the current between the graphene tunnels in real time.

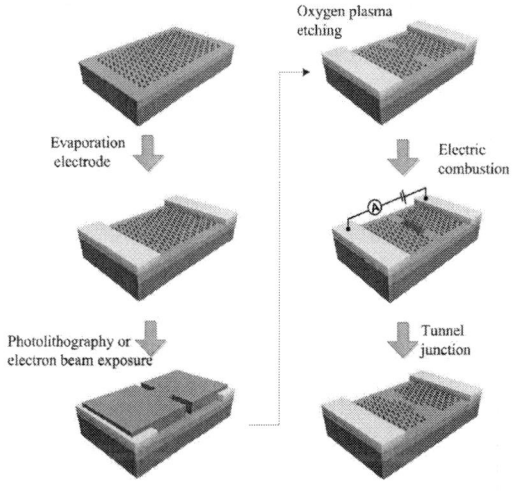

Figure 6. Schematic diagram of the preparation process of graphene tunnel junction

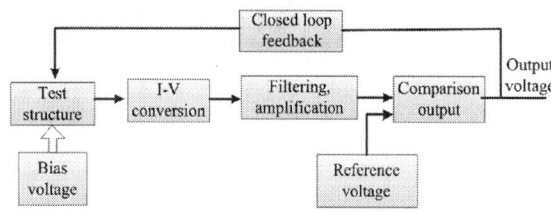

Figure 7. Feedback circuit design block diagram

D. Feedback circuit design

The accelerometer with high sensitivity and good stability is usually operated in a closed loop state, and the external acceleration is measured by the magnitude of the feedback control voltage. The method used in the thesis is to feed back the distance between the electrodes by the feedback circuit to keep the mass in the initial state, so that the feedback voltage can reflect the change of acceleration in real time. The system block diagram of the overall scheme is shown in Fig. 7.

In the Fig. 7, the voltage bias circuit provides the tunnel junction voltage for the tunnel accelerometer; the I-V conversion circuit realizes the conversion of the tunnel current to the voltage. And then the converted voltage is followed by filtering and differentially amplified with the reference voltage. The output voltage is fed back to the feedback electrode, as the driving voltage of the tunnel accelerometer. According to the structure parameters of the accelerometer, it is preliminarily estimated that the maximum electrostatic driving force of the mass is 10.73V.

IV. CONCLUSION

A high-precision tunnel accelerometer is developed using the precise preparation technology of single molecule and single atom measurement. Different from the traditional longitudinal accelerometer, it is based on the characteristics of graphene material and adopts a laterally sensitive method to sense the external micro acceleration. The mass is connected to the fixed surface through springs. The setting of four springs ensures that the mass moves in one dimension and the sensitivity improve. The tunnel junction is made of graphene, which avoids the limitation of traditional technology, and ensures the controllability of tunnel spacing. The accelerometer has the characteristics of high precision, low cost and small volume. It can not only be widely used in the automotive industry, but also has great application prospects in the military, such as improving the precision strike ability of weapons and equipment.

The new high-precision accelerometer proposed in the paper mainly completed the overall structure design, simulation and processing of the accelerometer in the early stage, and realized the transfer of graphene and the preparation of the tunnel junction. In the later stage, a feedback control circuit will be designed and optimized. Through real-time detection of the tunnel current, the mass will be maintained at the initial equilibrium position to ensure the high-precision characteristics of the accelerometer. In addition, it is necessary to build an accelerometer test platform, provide standard acceleration, and complete the test of the sensitivity and resolution of the accelerometer.

REFERENCES

[1] B.Boxenhorn and P.Greiff, "Monolithic Silicon Accelerometers," Sensors and Actuators, vol.21-23, pp.273-277,1990.

[2] .F.Rudolf, A.Jornod, J.Bergqvist and H.Leathold, "Precision Accelerometers with ug resolution," Sensors and Actuators, vol.21-23, pp.297-302, 1990.

[3] Y.Matsumoto and M.Esashi, "Integrated silicon capacitive accelerometer with PLL servo technique," Sensors and Actuators,vol.21-23, pp.209-217, 1990.

[4] LIU C H, KENNY T W. "A high-precision, wide-bandwidth micro-machined tunneling accelerometer," Journal of Microelectromechanical Systems, vol.10, pp.425-433, 2001.

[5] CHINGWEN Y, KHAL IL N. "A low-voltage bulk-silicon tunneling—based silicon microacceleromete," Technical Digest-International Electron Devices Meeting，1995, pp. 593-596.

[6] XUE Wei, WANG Jing, CUI Tian-hong. "Highly sensitive micromachined tunneling sensors," Optics and Precision Engineering, vol.12, pp.491-503, 2004.

[7] Dong Haifeng,HaoYilong, Jia Yubin, etal. "Fabrication and Characterization of Tunneling Current of Anodic Bonded Dry-Etched MEMS Tunneling Accelerometer," Chinese Journal of Semiconductors, vol.12, pp.58-63, 2004.

Proceedings of the 16th Annual IEEE International
Conference on Nano/Micro Engineered and Molecular Systems
April 25-29, 2021

Nanorobotic Manipulation inside Scanning Electron Microscope for the Electrical and Mechanical Characterization of ZnO nanowires

Mei Liu, Aristide Djoulde, Quan Yang, Weilin Su, Lingli Kong, Jinjun Rao, Jinbo Chen, and Zhiming Wang

Abstract— **Revealing novel material behavior at the nanoscale is hindered by the difficulties of conducting well-instrumented tests. Hence, the nanorobotic manipulation inside scanning electron microscope system is considered as an efficient method in nano-measurements. We investigate the electrical and mechanical properties of Zinc oxide (ZnO) nanowires by direct manipulation with the aid of a scanning electron microscope (SEM) compatible nanomanipulator system and atomic force microscope (AFM). We report the contact resistance of ZnO nanowire-Au/tungsten/ZnO nanowire junctions, real-time impedance analysis, the nanoindentation hardness and elastic modulus of ZnO nanowires. This research will provide a better understanding of the electrical and mechanical behavior of a single ZnO nanowire for future applications in nanoscale field effect-transistors and nanosensors.**

I. INTRODUCTION

Nanowires related research has become one of the hottest topics in nanoscience and nanotechnology[1]. As an important nanowire under active research, zinc oxide (ZnO) nanowire started to receive great attention due to its excellent functional properties such as strong binding energy, wide-bandgap, semiconductivity and biocompatibility[2, 3]. It is revered as the perfect building blocks for field-effect transistors (FETs), light-emitting diodes (LEDs), lasers diodes, biosensors, gas sensors and nanogenerators[4-6]. Moreover, ZnO nanowire applications rely strongly on the electrical and mechanical properties of ZnO nanowires[7].

ZnO-FETs consist of a single ZnO nanowire connected with two electrodes at two ends and placed on a flat substrate that serves as a gate electrode. However, high contact resistance between ZnO nanowire and electrodes limits the performance of ZnO-FETs. On the other hand, ZnO-nanogenerator applications search to combine piezoelectric and semiconductive properties of ZnO to harvest electrical energy by waste mechanical stress such as vibrations, compressions and relaxations from the strain energy induced in the ZnO. Therefore, it is important to study the mechanical behavior of ZnO nanowire since the energetic performance and durability of ZnO-nanogenerators depend on the elastic deformation that can be supported. Finally, impedance analysis of ZnO nanowires is necessary since it is a candidate material for highly sensitive gas sensors and biosensors for medical diagnosis and environmental pollution monitoring.

* Corresponding author: mliu@shu.edu.cn.
Mei Liu, Djoulde Aristide, Yang Quan, Weilin Su, Kong Lingli, Rao Jinjun, Chen Jinbo, and Wang Zhiming are with Shanghai Key Laboratory of Intelligent Manufacturing and Robotics, School of Mechatronic Engineering and Automation, Shanghai University, Shanghai, China.

Various experiments have been performed for the mechanical and electrical characterization of ZnO nanowires including bending and tensile loading modes and two point-probes electrical measurement respectively[8]. The above-mentioned mechanical testing methods were limited to the determination of elastic properties only, and were difficult to measure the applied force accurately. The results in some studies showed that the Young's modulus (20 to 800 GPa) was essentially independent from the diameter of ZnO nanowires, while in some studies diameter dependence was observed[9]. At present, electrical properties vary widely among experiments, and may reflect a combination of inherent contact resistance and extraneous factors such as contamination. ZnO nanowire was usually on a flat silicon substrate and a probe was moved downward to contact with the nanowire. The total resistance was measured, but the possibility of a ZnO nanowire bridging the electrode pair was highly uncertain and often caused probe tip damage and sometimes inadvertently severs the nanowire[10]. Few impedance studies and characterization via nanoindentation technique of ZnO nanowire have been found in the literature.

To date, scanning electron microscope (SEM) and atomic force microscope (AFM) have become the most popular equipments for nano operations. SEM has a large built-in vacuum chamber and real-time imaging capability, and can be equipped with a nanomanipulator system, AFM, to achieve a more powerful 'nano-laboratory' for nanorobotic manipulation[10-13]. Therefore, nanorobotic manipulation inside SEM is an efficient method to flexibly pick up a nanowire, adjust the gap and the contact length between a nanowire and the electrodes[14] as well as use SEM for the fast approach of the AFM cantilever before the nanoindentation test[15].

In this paper, we present an efficient method to evaluate the nanoindentation hardness (H) and elastic modulus (E) of ZnO nanowires using an SEM compatible AFM installed inside SEM. Current-voltage measurement using a two-point probe technique on ZnO nanowire is performed with the assistance of nanomanipulator system for fabricating ZnO nanowire-Au/tungsten/ZnO nanowire junctions. The contact resistance R_{ZA}, R_{ZT}, R_{ZZ} for ZnO-Au, ZnO-Tungsten, ZnO-ZnO nanowires junctions respectively are calculated from the experimental total resistance R_T. Real-time impedance analysis of ZnO nanowire is also described.

978-1-6654-3008-1/21 $31.00 © 2021 IEEE

919

II. ELECTRICAL CHARACTERIZATION EXPERIMENT

A. System configuration and Sample preparation

A nanomanipulation system (LifeForce TNI, Canada) with four nanomanipulators was used here within a SEM (SU3500) for the electrical characterization of ZnO nanowires (Fig.1. (a)). Initially, ZnO nanowires (purchased from Nanjing XFNANO Materials Tech Co., Ltd, China) powder were magnetically stirred in alcohol for 10 minutes and spin-coated on a clean silicon substrate. Before ZnO nanowires deposition, the Pt was coated on the silicon substrate via the magnetron sputtering. The morphologies of the ZnO nanowires were studied with SEM. The deposited film was fractured by means of cracking the silicon substrate using a diamond scriber, and the resulting straight free-standing nanowires cantilevers formed the ZnO nanowire forest (Fig.1. (b)). The experimental total resistance is measured with a source-measure instrument (Keithley 2280S system source meter) connected through its two channels. Because the scanning electron beam induces a charge on the sample surface, while recording the electrical current, the electron beam is switched off.

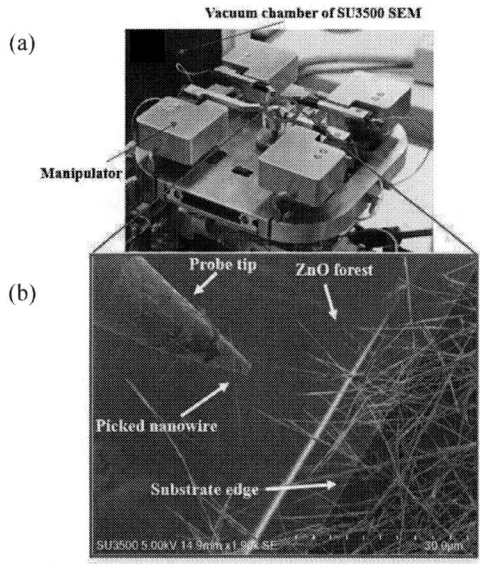

(a)

(b)

Figure 1. System configuration for the electrical characterization. (a) The four manipulators nanorobotic manipulation system embedded into SEM vacuum chamber.(b) SEM image of the sample.

B. Contact resistance calculation

Fig. 2 is the proposed schematics to measure the total resistance and evaluate the contact resistance of ZnO nanowire-Au/tungsten tip/ZnO nanowire junctions. The model considers a single ZnO nanowire contacted by two metal electrodes. To form the bridge with the nanowire, a ZnO nanowire is initially picked up from the forest by an electrode mounted on the manipulator assisted by EBID. During the manipulation, the relative position of the electrode and nanowire can be estimated by focussing on each object and the correspondent difference of the working distance. Once they touch each other, the electrostatic force at the metal will assist in establishing a rather stable connection. The picked nanowire is carried towards the

second electrode and simply placed on the surface of the metal by an arbitrary length. It should be noted that the electrodes are Au-coated AFM cantilevers and tungsten probes. Referring to Fig. 2., the total resistance R_T between two electrodes is as follows[16]:

$$R_T = \left(4\frac{\rho_{Nw}}{\pi}\right)\left(\frac{L}{D^2}\right) + \sum_n R_C + 2R_P \quad (1)$$

Where n is the index of the total number of contact resistances R_C at the junctions, L is the transmission length, whereas ρ_{Nw} and D are the resistivity and the diameter respectively of the nanowire. The resistances of the electrodes R_P were measured using the 4 wire resistance measurement method with the precise Keitley DMM7510 instrument. The Transfer length method (TLM) measurement is utilized to calculate the contact resistance and the corresponding resistivity of the nanowire.

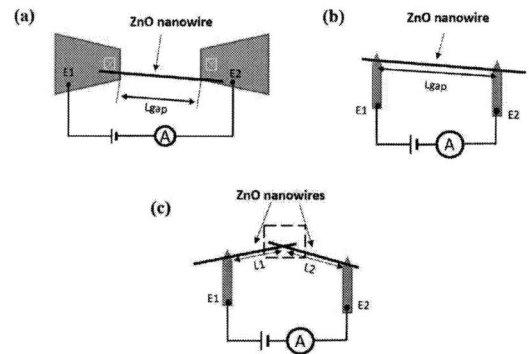

Figure 2. Schematics of the experimental setup used to measure I–V characteristics. (a) ZnO nanowire-Au interface. (b) ZnO nanowire-tungsten tip interface. (c) ZnO nanowire-ZnO nanowire interface.

C. Impedance Measurements

Measurements were performed with an Agilent Impedance Analyzer, using 10 mV, 100 mV, and 500 mV signal in the frequency range from 20 Hz to 1 MHz. The ZnO nanowire is bridged (with the same strategy as the one developed in the above section) between two tungsten electrodes connected to the impedance analyzer device. Real-time impedance analysis of ZnO nanowire is then perfomed.

III. NANOINDENTATION EXPERIMENT

The commercially available SEM compatible AFM (TNI EM-AFM, CA, Canada) with imaging and nanoindentation capabilities was used for the experiments. The force and displacement resolutions are better than 5 nN and 0.2 nm, respectively. For both imaging and nanoindentation, we have used Au-coated silicon piezoresistive cantilever (Hitachi High-Tech Science Corporation, K-U001354500, JAPAN) with its tip radius of 10nm. The force constant of the AFM cantilever is 40 N/m. The testing process is visually observed in real-time under the SEM feedback. The nanoindentation entails applying a load with an indenter tip of known geometry on ZnO nanowires, and recording the tip displacement. The load-displacement obtained from the instrument's software (which provides built-in calculation tools to analyze and display the data collected) allowed to calculate the nanoindentation hardness and elastic modulus.

The nanoindentation hardness H is defined as the maximum applied indentation load divided by the projected contact area of the indented. From the unloading load-displacement curve, hardness can be obtained as follows:

$$H = \frac{F_{max}}{A(h_c)} \quad (2)$$

Where F_{max} is the maximum recorded force, $A(h_C)$ is the indentation projected area of the contact surface between the indenter and the ZnO nanowire that is related to the contact depth h_C.

The commonly used equation in nanoindentation for calculation of Young's modulus is as follows:

$$\frac{1}{E_r} = \frac{1-v^2}{E} + \frac{1-v_i^2}{E_i} \quad (3)$$

Where E and E_i indicate the nanoindentation elastic modulus for the AFM tip and the ZnO nanowire respectively, whereas v and v_i are the respective Poisson's ratio of the indenter and the indented nanowire. For silicon (n-type), E_i=169 Gpa and v_i =0.27. E_r is the reduced contact modulus between the tip and the nanowire, and can be evaluated using the Sneddon Equation for any indenter geometry as follows:

$$E_r = \frac{\sqrt{\pi}}{2\beta} \cdot \frac{S}{\sqrt{A(h_c)}} \quad (4)$$

Where β is a constant that depends on the geometry of the indenter (β=1 for tetrahedral intender). S is the experimentally measured contact stiffness which is determined from the initial slope of the initial portion of the unloading curve.

It should be noted that (2) and (4) highly depend on the indentation projected area $A(h_c)$ which is calculated based on the indenter geometry. For instance, the Oliver and Pharr model determines the contact depth with the following equation:

$$A(h_c) = C_0 h_c^2 + C_1 h_c + C_2 h_c^{0.5} + C_3 h_c^{0.25} + C_4 h_c^{1/8} + C_5 h_c^{1/16} \quad (5)$$

For a perfect pyramidal indenter, the relationship is[17]:

$$A(h_c) = \pi \tan^2 \theta h_c^2 \quad (6)$$

Where h_c is calculated using the Oliver and Pharr model with $h_c = h_{max} - \epsilon.F_{max}/S$. ϵ is a geometric constant, which is 0.72 for a tetrahedral indenter.

IV. RESULTS AND DISCUSSION

A. Morphology characterization

Fig. 3 presents the SEM images and the diameter distribution of the ZnO nanowires used for this study. The network of the nanowires obtained shows nanowires with lengths of 1-50 μm and a diameter ranging from 50 to 250 nm (Fig. 3(b)). Only nanowires having an almost constant diameter which does not vary with length were targeted for the different characterizations as shown in the inset of Fig. 3(b).

(a)
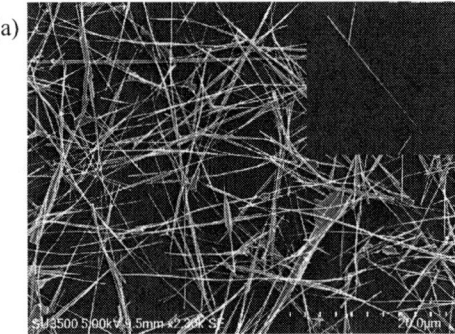

(b)
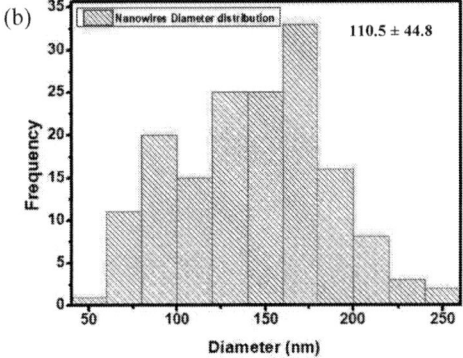

Figure 3. Nanowire morphologies: (a) SEM image of ZnO nanowires lying on a silicon substrate. The Inset shows a manified image of a single ZnO nanowire almost uniform in diameter.(b) diameter distribution of the dispersed ZnO nanowires measured by SEM.

B. Electrical measurements

Using the four moveable probes, several resistance measurements were performed on various nanowires. Prior to the measurements, a single ZnO nanowire is removed from its forest and bridged on two electrodes as shown in Fig. 4. Subsequently a current voltage sweep is carried out.

Fig. 5(a) shows the corresponding I-V curves plotted with over 2000 data points. As we can see, the current flowing in the region −5 v < V <+10 v indicates a schottky barrier formed at a junction between the nanowire and the electrodes. The equivalent electric model is a resistor (corresponding to the nanowire passed through) connected on both sides by two diodes (representing the two Schottky contacts) as shown in inset of Fig. 5(a). At least three measurements at the same positions were recorded and the average data were plotted to ensure the repeatability and well established electrical connection between the electrodes and the nanowires. It is possible that the slight change between the different measures may results from the fact that the contact area has changed.

Fig. 5(b) shows the total resistance R_T measurements obtained from the I–V characteristics as a function of the length over the square of the diameter of the nanowire. R_T of each curve was determined from the linear section at forward bias. A linear curve is fitted with the experimental data collected for R_T versus (L/D^2), yielding an intercept at $L = 0$ of $2R_c + 2R_P$.

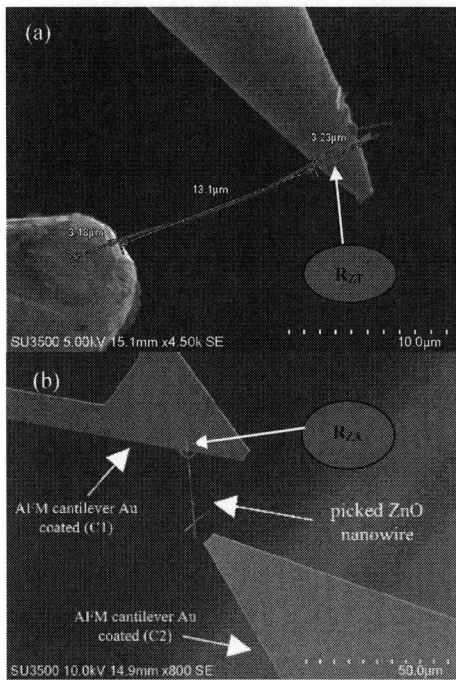

Figure 4. SEM images of a single ZnO nanowire in contact with two electrodes (a) ZnO-Tungsten contact resistance (R_{ZT}) measurements. (b) ZnO-AU contact resistance (R_{ZA}) measurements.

The measures of a total number of 2 nanowires whose diameters vary between 139 nm and 160 nm were listed in table I and II. The contact resistances R_{ZT} and R_{ZA} were calculated to be 29.4883 KΩ and 45.737 KΩ respectively. The nanowire resistivity was estimated to be between 1.5×10^{-3} and 2.2×10^{-3} $\Omega.cm$.

TABLE I. ZnO -Tungsten junction parameters determined from 2 point-probe measurement

Sample	Junction parameters					
	D (nm)	L (μm)	R_T (KΩ)	R_P (Ω)	R_{ZT} (KΩ)	ρ_W (Ω.cm)
1	139	15.15	80.87 ±9.8	0.002	29.48	2.2 x 10⁻³
		11.5	77.18 ±8.4			
		8.59	71.4 ±7.9			

TABLE II. ZnO -Au junction parameters determined from 2 point-probe measurement

Sample	Junction parameters					
	D (nm)	L (μm)	R_T (KΩ)	R_P (Ω)	R_{ZA} (KΩ)	ρ_W (Ω.cm)
2	137.5	3.42	95.093 ±6.3	1.2	45.737	1.5 x 10⁻³
		3.78	95.474 ±5.5			
		7.92	99.855 ±4.8			

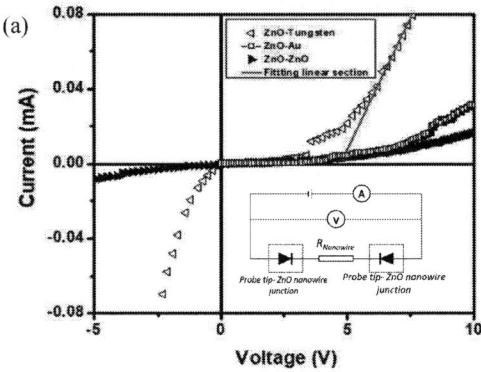

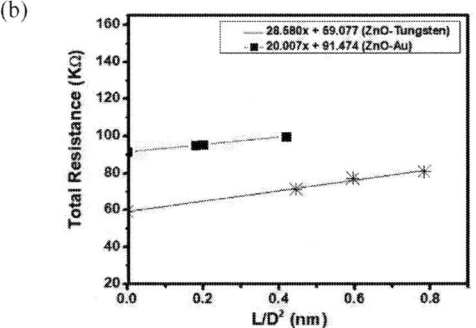

Figure 5. Two probe I-V characteristics of a single ZnO nanowire with Tungsten, Au and ZnO in a SEM vaccum room. The inset is the schematic of the back-to-back Schottky contacts formed among metals eletrodes and ZnO nanowire;. (b) Plot of R_T versus L/D^2.

With the known data of contact resistance of tungsten tip/ZnO and the resistivity of the nanowire, the contact resistance between two overlapping ZnO nanowires as illustrated in Fig. 6 were calculated. The junction resistance R_{ZZ} can be extracted from the measured total resistance between L_1+L_2 given by $R_T = R_{NW1} + R_{NW2} + R_{ZZ} + 2R_{ZT} + 2R_P$. The junction parameters were listed in table III. The analysis of our data leads to an average ZnO-ZnO junction resistance R_{ZZ} of 247.63 ± 5.1 KΩ.

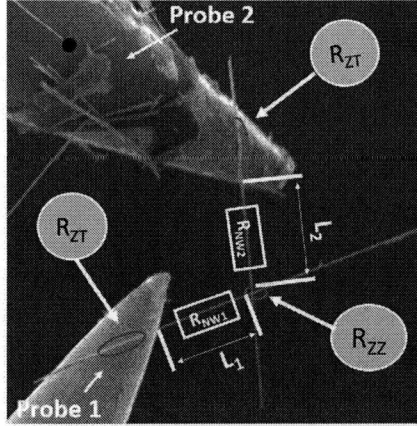

Figure 6. SEM image of two ZnO nanowires overlapped for the evaluation of the ZnO-ZnO contact resistance (R_{ZZ}) measurement where L_1 and L_2 are the inner lenghts.

TABLE III.   ZnO -ZnO JUNCTION PARAMETERS DETERMINED FROM 2 POINT-PROBE MEASUREMENT

Sample	Junction parameters					
	D_1 (nm)	D_2 (nm)	L_1 (μm)	L_2 (μm)	R_T (KΩ)	R_{ZZ}(KΩ)
3	134	150	6.01	5.59	322.864	247.63 ± 5.1

By comparing the contact resistances, we can notice that there is a strong dependence on the nature of the contact in the I–V characteristics. ZnO-ZnO junction offered larger (much higher) contact resistance/high blocking barrier than that offered by ZnO-Au and ZnO-tungsten junctions. A very good agreement was achieved[18-22] and the reliability of our experimental procedure is validated, enabling a meaningful estimation of the electrical properties of ZnO nanowires.

Fig. 7 is the impedance measurements of a single ZnO nanowire. A number of two nanowires with diameter $D{\sim}137$ nm, length $L_1{\sim}22.1$ μm and $L_2{\sim}18$ μm, bridging two tungsten tips were selected. The corresponding average impedances were listed in Table II. Prior to the measurements, the compensation of the combined test fixture of the impedance analyzer and nanomanipulator system was carried out since is considered as an electronic circuit. As shown in Fig. 7 the increase in frequency with the increase in the signal from 10 mV to 500 mV is associated with the increase in the impedance.

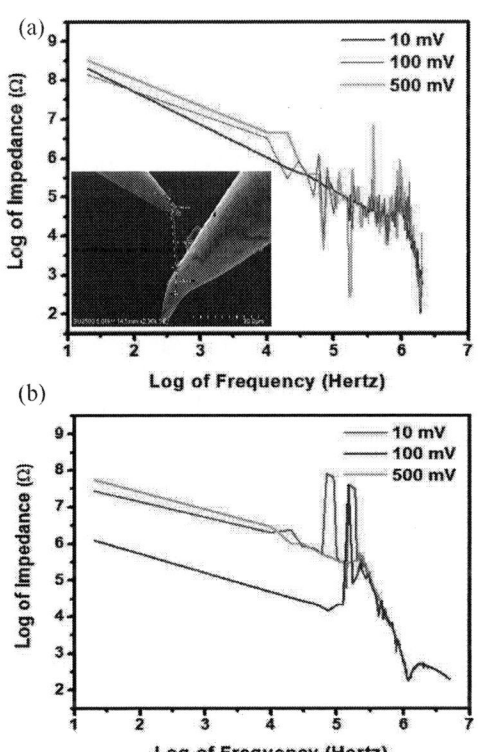

Figure 7.   Impedance measurements on two diferent ZnO nanowires bridging two tungsten tips.

TABLE IV.   EXTRACTED IMPEDANCE PARAMETERS OF A SINGLE ZnO NANOWIRE

Frequencies	Impedances range		
	10 mV	100 mV	500 mV
20 Hz	~200.8 MΩ	~207.8 MΩ	~207.8 MΩ
100 Hz	~74.5 kΩ	~79 kΩ	~79 kΩ
300 Hz	~35.1k Ω	~23 kΩ	~23 kΩ
1 MHz	~726.25 Ω	~806.5 Ω	~806.5 Ω

C. Nanoindentation

A drop of solution with ZnO nanowires was put on a clean silicon substrate. As shown Fig. 8(a) the AFM cantilever approached the ZnO nanowires before the indentation operation using the SEM feedback image to prevent the probe from touching the substrate and damaging the tip. A tetrahedral intender tip was used to locate and image a single nanowire and estimate the position of the indenter tip. Fig. 8(b) shows AFM image of ZnO nanowires obtained from AFM scanning contact mode with a scanning rate of 0.5 Hz and scan size of 10 μm x 10 μm. The nanoindentation were performed on various nanowires.

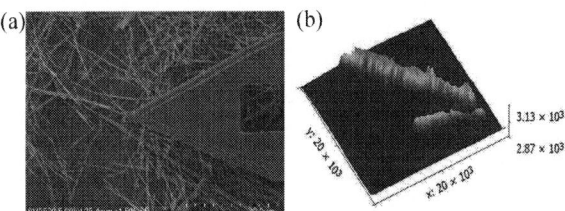

Figure 8.   (a) SEM image of an AFM cantilever approaching ZnO nanowires before the indentation operation. (b) 3D AFM image of ZnO nanowires.

Fig. 9 shows typical load-displacement curves obtained from the instrumented nanoindentation in which each plot represents a loading and unloading cycle. Prior to the nanoindentation on the nanowires, measurement was carried out on a flat clean silicon for the AFM cantilever spring (40 N/m) calibration. The measurements were made in a maximum the depth of 100-140 nm.

As shown in table V, the nanoindentation modulus (E) and Hardness (H) were estimated to be respectively in the range of 2.6-17.24 MPa and 55-79 MPa for ZnO nanowires in the range of 100-170 nm.

TABLE V.   THE CALCULATED NANOIDENTATION HARDNESS AND YOUNG'S MODULUS FOR ZnO NANOWIRES DIAMETERS RANGING FROM 100-170 NM

Sample	Hardness (MPa)		Young's modulus (MPa)	
	Min	Max	Min	Max
1	2.6	17.24	55	79

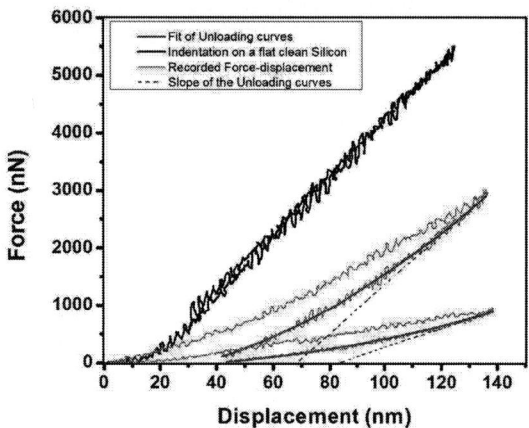

Figure 9. A typical load-displacement curves obtained by indenting on ZnO nanowires. The curves in red correspond to the highest and the lowest maximum load obtained at the end of the loading .

V. Conclusion

In summary, we have successfully characterized the electrical (contact resistance, resistivity, impedance) and mechanical (elastic modulus, hardness) properties of ZnO nanowires by using commercially available SEM compatible nanomanipulator system and atomic force microscope (AFM). Experimental results showed a contact resistance ZnO-ZnO junction of 247.63 KΩ much higher than that offered by ZnO-Au junction and ZnO-tungsten junction respectively 45.737 KΩ and 29.488 KΩ. The resistivity of the ZnO nanowires were measured to be between 1.5×10^{-3} and 2.2×10^{-3} Ω.cm. The AFM nanoindentation technique allowed measurements of nanoindentation modulus E and Hardness H of ZnO nanowires ranging from 100 to 170 nm. H is 9.92 ± 7.32 MPa while E is 67 ± 12 MPa. The results are in perfect correlation with those presented in the literature. Thus, nanorobotic manipulation inside SEM should be selected for convenient, efficient and easy characterization of nanomaterials; and can be expanded to other nanoscale devices characterization. This research could be beneficial for future nanowire applications in nanodevices and nanosensors. However, our results cannot completely exclude the influence of the temperature and the contamination (dust deposition) on the contact area between the nanowire and the electrodes which are perfect investigations for future work.

Acknowledgment

This project was supported by the National Natural Science Foundation of China (No. 51205245,61573236); The Joint Specialized Research Fund for the Doctoral Program of Higher Education; The Scientific Research Foundation for the Returned Overseas Chinese Scholars, State Education Ministry.

References

[1] S. Wang, Z. Shan, and H. Huang, "The Mechanical Properties of Nanowires," (in eng), *Adv Sci (Weinh),* vol. 4, no. 4, p. 1600332, Apr 2017.

[2] S. Bagga, J. Akhtar, and S. Mishra, "Synthesis and applications of ZnO nanowire: A review," *AIP Conference Proceedings,* vol. 1989, no. 1, p. 020004, 2018.

[3] Y. Zhang, M. K. Ram, E. K. Stefanakos, and D. Y. Goswami, "Synthesis, Characterization, and Applications of ZnO Nanowires," *Journal of Nanomaterials,* vol. 2012, p. 624520, 2012.

[4] J. H. Na, M. Kitamura, M. Arita, and Y. Arakawa, "Hybrid p-n junction light-emitting diodes based on sputtered ZnO and organic semiconductors," *Applied Physics Letters - APPL PHYS LETT,* vol. 95, pp. 3303-253303, 2009.

[5] W. Zhang, R. Zhu, V. Nguyen, and R. Yang, "Highly sensitive and flexible strain sensors based on vertical zinc oxide nanowire arrays," *Sensors and Actuators A: Physical,* vol. 205, pp. 164-169, 2014.

[6] S. Rafique, A. K. Kasi, Aminullah, J. K. Kasi, M. Bokhari, and S. Zafar, "Fabrication of Br doped ZnO nanosheets piezoelectric nanogenerator for pressure and position sensing applications," *Current Applied Physics,* vol. 21, pp. 72-79, 2021.

[7] D. Yang *et al.,* "Patterned growth of ZnO nanowires on flexible substrates for enhanced performance of flexible piezoelectric nanogenerators," *Applied Physics Letters,* vol. 110, no. 6, p. 063901, 2017.

[8] F. Xu, Q. Qin, A. Mishra, Y. Gu, and Y. Zhu, "Mechanical properties of ZnO nanowires under different loading modes," *Nano Research,* vol. 3, no. 4, pp. 271-280, 2010.

[9] X. Wang *et al.,* "Growth Conditions Control the Elastic and Electrical Properties of ZnO Nanowires," *Nano Letters,* vol. 15, no. 12, pp. 7886-7892, 2015.

[10] C. Ru *et al.,* "Automated four-point probe measurement of nanowires inside a scanning electron microscope," in *2010 IEEE International Conference on Automation Science and Engineering,* 2010, pp. 533-538.

[11] M. Liu *et al.,* "Interactive Manipulation of Nonconductive Microparticles in Scanning Electron Microscope by a Virtual Nano-hand Strategy," (in eng), *Micromachines (Basel),* vol. 10, no. 10, Oct 2 2019.

[12] C. Shi *et al.,* "Recent advances in nanorobotic manipulation inside scanning electron microscopes," *Microsystems & Nanoengineering,* vol. 2, no. 1, p. 16024, 2016.

[13] U. Mick, V. Eichhorn, T. Wortmann, C. Diederichs, and S. Fatikow, "Combined nanorobotic AFM/SEM system as novel toolbox for automated hybrid analysis and manipulation of nanoscale objects," in *2010 IEEE International Conference on Robotics and Automation,* 2010, pp. 4088-4093.

[14] N. Yu *et al.,* "Characterization of the Resistance and Force of a Carbon Nanotube/Metal Side Contact by Nanomanipulation," *Scanning,* vol. 2017, p. 5910734, 2017.

[15] C. Jiang, H. Lu, H. Zhang, Y. Shen, and Y. Lu, "Recent Advances on In Situ SEM Mechanical and Electrical Characterization of Low-Dimensional Nanomaterials," *Scanning,* vol. 2017, p. 1985149, 2017.

[16] J. Y. Park, J. Cho, and S. C. Jun, "Review of contact-resistance analysis in nano-material," *Journal of Mechanical Science and Technology,* vol. 32, no. 2, pp. 539-547, 2018.

[17] G. Guillonneau, G. Kermouche, J.-M. Bergheau, and J.-L. Loubet, "A new method to determine the true projected contact area using nanoindentation testing," *Comptes Rendus Mécanique,* vol. 343, no. 7, pp. 410-418, 2015.

[18] S. LeBlanc, S. Phadke, T. Kodama, A. Salleo, and K. E. Goodson, "Electrothermal phenomena in zinc oxide nanowires and contacts," *Applied Physics Letters,* vol. 100, no. 16, p. 163105, 2012.

[19] C. Y. Chen, P. H. Chang, K. T. Tsai, and J. H. He, "Electrical and optoelectronic characterization of a ZnO nanowire contacted by focused-ion-beam-deposited Pt," in *2010 3rd International Nanoelectronics Conference (INEC),* 2010, pp. 1177-1178.

[20] D. H. Weber *et al.,* "Determination of the specific resistance of individual freestanding ZnO nanowires with the low energy electron point source microscope," *Applied Physics Letters,* vol. 91, no. 25, p. 253126, 2007.

[21] X. Li, J. Qi, Q. Zhang, and Y. Zhang, "Fabrication and electrical property of individual ZnO nanowire based mesfet," *Procedia Engineering,* vol. 27, pp. 1471-1477, 2012.

[22] S. Das, J. Kar, and J.-M. Myoung, "Junction Properties and Applications of ZnO Single Nanowire Based Schottky Diode," 2011.

Gap in pagination due to unavailable papers.

Pages 925-930

Proceedings of the 16th Annual IEEE International
Conference on Nano/Micro Engineered and Molecular Systems
April 25-29, 2021

Laterally-excited bulk-wave resonators (XBARs) with embedded electrodes in 149.5° Z-cut LiNbO₃

Xiyu Gu, Jieyu Liu, Yao Cai, Yan Liu, Chao Gao, Zhiwei Wen, Shishang Guo, Chengliang Sun

Abstract— **In this paper, we present a new design of laterally-excited bulk wave resonators (XBARs) which contains narrow embedded interdigital electrodes (IDEs). The proposed XBAR has a high electromechanical coupling coefficient (K^2_{eff}) of 33.51% and the smooth impedance curve. With the assistance of 149.5° Z-cut lithium niobate (LiNbO₃) in XBARs, we obtain a high K^2_{eff} of 42.54%. We investigate the impact on K^2_{eff} and spurious modes of XBARs with different depths of electrodes (T_e). XBARs with the optimization of embedded electrodes show a great prospect in fabricating broad bandwidth filters.**

INTRODUCTION

The new 5G filters' (n77, n78 and n79) demands on high frequency (3-5 GHz) and very large bandwidth, are puzzling many experts. BAW filters, occupying a big RF filter market, exhibit a great potential in high adjustable frequency range (1.5~5 GHz). However, its low-moderate level of electro-mechanical coupling in AlN is considering [1].

Many researches have been devoted to LiNbO₃, due to its strong piezoelectric coefficients. Many experts explored the preparation method of thin mono-crystalline Lithium Niobate membrane. LiNbO₃ have been prepared by various methods, such as CVD [2], LPE [3], epitaxial growth by melting [4], ion-diffusion [5], and RF sputtering [6]. In 1998, M. Levy used ion-slicing technology to fabricate devices based on lithium niobate membranes [7]. This ion-slicing technology has opened the door of fabricating high frequency resonators. Michio Kadota and Takashi Ogami fabricated the first-order asymmetric (A1) Lamb wave resonator in sub-micrometer (395nm) thin LiNbO₃ membranes, which has high phase velocity, large electromechanical coupling coefficient (K^2_{eff}) and high frequency than 5GHz [8,9].

In 2018, V. Plessky proposed laterally-excited bulk-wave resonators (XBARs) based on 400-nm-thick membrane of LiNbO₃. This resonator has a high K^2_{eff} at above 4GHz and some spurious modes [10]. This resonator's structure is similar to the SAW resonator. With a pitch of a few microns, its IDEs is placed periodically on the top of LiNbO₃, and the pitch is much wider than the electrode width, as shown in Fig.1(a).

In this work, we present a new design of laterally-excited bulk-wave resonators (XBARs) which contains narrow embedded interdigital electrodes (IDEs). For the strongest

electric field part distributed in the membrane, the proposed XBAR eliminate many spurious modes to achieve the smooth impedance curve. Meantime, it has a high K^2_{eff} of 33.51%. From 0 to 25 GHz, we analyze several main horizontal shear vibration modes within the membrane. Thus, when the embedded depths of electrodes (T_e) are fixed, the K^2_{eff} is increased by rotating 149.5° from the Z-axis about the material X-axis. For 149.5° Z-cut LiNbO₃, we investigate the impact of different embedded depths of electrodes and the impact of different thicknesses of LiNbO₃.

DESIGN OF RESONATOR

The structure of a typical XBAR is showed in Fig. 1(a). Briefly, that structure is similar to the SAW resonator, whose IDEs are deposited on thin mono-crystalline LiNbO₃ membrane. But there are the most differences in the depth of LiNbO₃ membrane and the pitch of IDEs. In XBARs, the depth of LiNbO₃ membrane is less than 1μm and the pitch of IDEs is few microns. These characteristics make XBARs' success in high frequency. Fig.1(b) shows the curved electric field lines distribution in the traditional electrodes of alternating polarity. The electric field, which spreads in thin LiNbO₃ membrane, is not the strongest part. As the depth of thin LiNbO₃ membrane increasing, electric field strength is weakening gradually. But XBARs mainly use the e₂₄ and e₁₅ piezoelectric coefficient of LiNbO₃ to excite the horizontal displacements on the membrane surface.

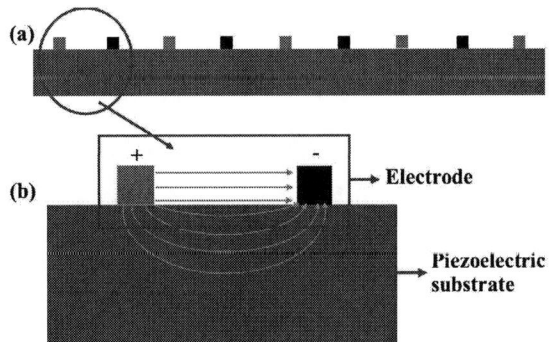

Fig. 1. Schematic diagram of the device structure of a traditional IDEs XBARs. (a) Cross-sectional diagram of the resonator. (b) Illustration of the curved electric field lines distribution.

For the new design of XBARs, the narrow IDEs are embedded in thin LiNbO₃ membrane, as shown in Fig.2(a). Fig.2(b) shows the electric field lines distribution excited by

---

*Resrach supported by National Key R&D Program of China (No. 2020YFB2008800).

X. Gu and S. Guo are with the School of Physics and Technology, Wuhan University, Wuhan, 430072 China.

J. Liu, Y. Cai, Y. Liu, C. Gao, and Z. Wen are with the Institute of Technological Sciences, Wuhan University, Wuhan, 430072 China.

Corresponding author: Shishang Guo (e-mail: gssyhx@whu.edu.cn), Chengliang Sun (phone: +86-027-6877-6588; email: sunc@whu.edu.cn).

978-1-6654-3008-1/21 $31.00 © 2021 IEEE

embedded IDEs. As the horizontal electric field is enhanced, the e_{24} and e_{15} piezoelectric coefficient of LiNbO$_3$ are used fully. The horizontal vibrations may be more intense. Under this situation, the energy of other spurious modes will reduce. Then, the spurious modes at resonance may be eliminated.

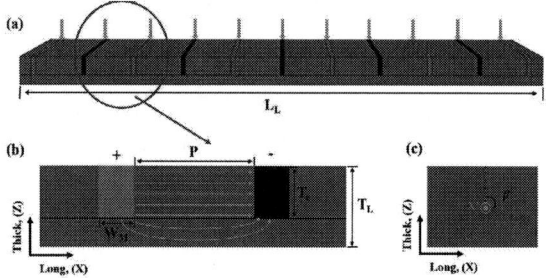

Fig. 2. (a) Schematic diagram of a new design with narrow embedded IDEs. (b) Cross-sectional diagram of the resonator and illustration of the curved electric field lines distribution. (c) Side view about β° Z-cut LiNbO$_3$ thin film.

SIMULATION AND ANALYSIS

In simulation, Z-cut LiNbO$_3$ and molybdenum (Mo) are adopted for piezoelectric thin film and electrode materials. Table I discloses XBARs parameters. We use 300 nm-thick LN membrane and 1 μm-wide Mo electrodes, spaced at P = 10 μm in simulated devices. Under this condition, the electrode pitch P is significantly larger than electrodes width or the LN membrane thickness, which is typical XBARs structure. When the T_L is stable, the embedded depths of electrodes (T_e) is equal to 2/3 T_L in this design. We obtain smooth impedance curve by using COMSOL software to simulate it, as shown in Fig.3. Fig.3 also shows the impedance curve of typical XBARs in the same parameters. Comparing between the two curves, we fine the spurious modes at resonance are eliminated for this design. In this paper, K_{eff}^2 is adopted as the integral effective coupling coefficient of XBAR device.

The IEEE standard definition is [11]

$$K_{eff}^2 = \frac{\pi}{2}\frac{f_s}{f_p} cot(\frac{\pi}{2}\frac{f_s}{f_p}).\qquad(1)$$

The simplified formula is

$$K_{eff}^2 = \frac{\pi^2}{4}\frac{f_s}{f_p}\frac{f_p-f_s}{f_p}\qquad(2)$$

Where f_s is the series resonance frequency and f_p is the parallel resonance frequency of the resonator (see in Fig. 3), and the formula (2) is adopted in this paper. There is a high K_{eff}^2 of 33.51% in this design. This result confirms the horizontal vibrations are more intense. Mechanical energy converted from electrical energy becomes bigger.

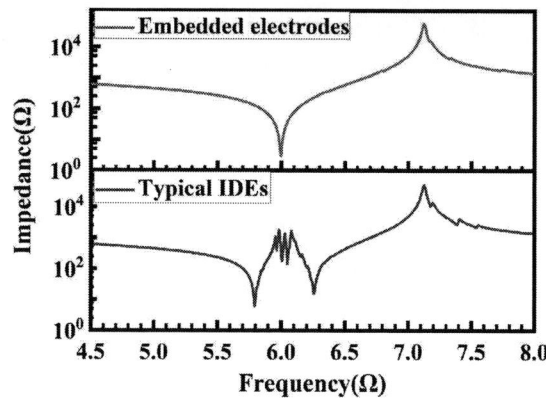

Fig. 3. Impedance curves of the typical XBARs and a new design with narrow embedded IDEs.

For investigating its potentiality in design of filters, we simulate the impedance curve from 0 to 25 GHz at A = 2/3. As shown in Fig.3, there are several main peaks in impedance curve, which represent different modes. After analyzing these mechanical vibrations in X-axis of the substrate, we find they belong to the anti-symmetric shear wave (A_0, A_1, A_3,...) and the vibration in x-y plane are the largest displacement in three planes. The modes of resonance at 3.065GHz and 3.385GHz are A_0 mode. The main vibration mode of resonance at 6GHz is A_0 mode. The horizontal A_3 harmonic is far to the right of Anti-Resonance. This consequence shows the anti-symmetric shear wave is leading. At the same time, the horizontal A_3 harmonic at 19.665GHz may be the demand for the future. For the distance between resonance mode and undesired modes is bigger than 2.5GHz, there is no problem to design filter. In term of eliminating spurious modes, it reveals the large potential of this design.

TABLE I. XBARs PARAMETERS

Symbol	Content
T_L	Thickness of LiNbO$_3$
L_L	Length of LiNbO$_3$ thin film
W_L	Width of LiNbO$_3$ thin film
L_M	Length of Molybdenum electrodes
W_M	Width of Molybdenum electrodes
N	Number of electrodes
P	Pitch
T_e	Molybdenum electrodes Embedded depth
A	The ratio of T_e to T_L

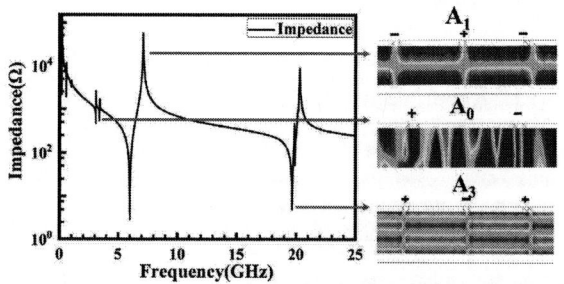

Fig. 4. Impedance curve of the new XBARs from 0 to 25 GHz, indicating main XBARs modes and spurious modes.

Rotated Z-cut LiNbO$_3$ substrates are studied in this work due to their commercial availability. In a rotated Z-cut LiNbO$_3$ with rotation β, the X-axis of the material is within the substrate and the normal vector (n) of the substrate is rotated −β from the Z-axis about the material X-axis, as shown in Fig.2(c). When β is equal to 149.5°, we obtain the 300 nm-thick 149.5° Z-cut LiNbO$_3$ membrane. Under this situation, the K_{eff}^2 of this design is improved to 42.54%. It reveals the large potential of Z-cut LiNbO$_3$. When the thickness of LN membrane is stable T_L=300nm, the effect of different embedded depths of electrodes is studied. We find the impedance curve is smoother, when A takes the value between 2/3 and 5/6, as shown in Fig.5. Such results indicate that some vertical components of electric filed are important.

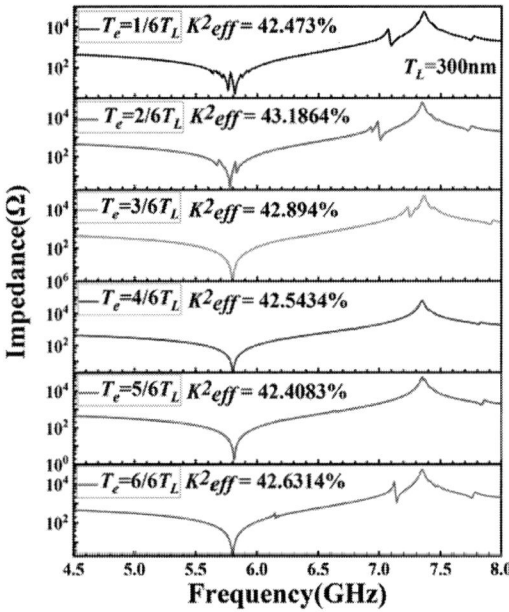

Fig. 5. Impedance curves of different A, when LiNbO$_3$ membrane thicknesses is 300nm.

When the embedded depths of electrodes (T_e) is equal to 2/3 T_L and other parameters is fixed, we change 100° Z-cut LiNbO$_3$ membrane thicknesses. Fig.6 shows this impact on K_{eff}^2 and spurious modes of XBARs. As the membrane thickness is thickening, A_0 mode is moving to resonance frequency. The resonance frequency and K_{eff}^2 also are affected, as shown in Fig.7. Simulations results show an almost 1/T_L relation with frequency with a small perturbation. It also shows influence of thickness on K_{eff}^2. In general, K_{eff}^2 is increasing with thickness thickening. So, it is critical for the membrane thickness to XBAR.

CONCLUSION

For fabricating the satisfying filters, resonators are invented by many experts. The new design of XBARs, containing narrow embedded IDEs in thin LiNbO$_3$ membrane, exhibits a high K_{eff}^2 and inconspicuous spurious

modes. The optimization with embedded electrodes can be adopted to improve the performance of XBARs.

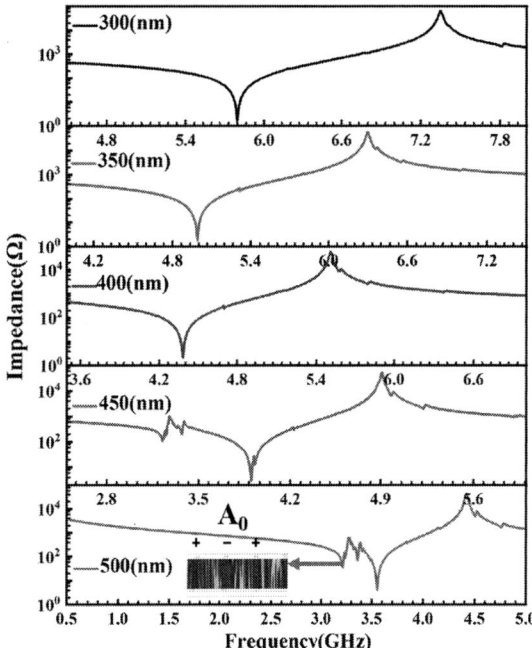

Fig. 6. Impedance curves of different LiNbO$_3$ membrane thicknesses in $T_e = 2/3\ T_L$.

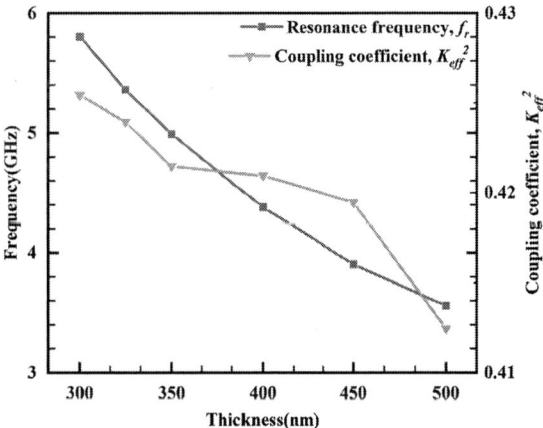

Fig. 7. Electromechanical coupling coefficient and resonance frequency vs different LiNbO$_3$ membrane thicknesses in $T_e = 2/3\ T_L$.

REFERENCES

[1] C. S. Lam, "A review of the timing and filtering technologies in smartphones," in 2016 IEEE International Frequency Control Symposium (IFCS), 2016, pp. 1–6.

[2] Curtis B J, Brunner H R. The growth of thin films of lithium niobate by chemical vapour de position[J]. Materials Research Bulletin, 1975, 10(6): 515-520.

[3] Baudrant A, Vial H, Daval J. Liquid phase epitaxy of LiNbO3 thin films for integrated optics[J]. Materials Research Bulletin, 1975, 10(12): 1373-1377.

978-1-6654-3008-1/21 $31.00 © 2021 IEEE

[4]   Miyazawa S, Sugii K, Uchida N. Optical waveguiding in LiNbO3 single‐crystal film grown by the EGM technique[J]. Journal of Applied Physics, 1975, 46(5): 2223-2228.

[5]   Kaminow I P, Carruthers J R. Optical waveguiding layers in LiNbO3 and LiTaO3[J]. Applied Physics Letters, 1973, 22(7): 326-328.

[6]   Takada S, Ohnishi M, Hayakawa H, et al. Optical waveguides of single‐crystal LiNbO3 film deposited by rf sputtering[J]. Applied Physics Letters, 1974, 24(10): 490-492.

[7]   M. Levy et al., "Fabrication of single-crystal lithium niobate films by crystal ion slicing", Applied Physics Letters 73, pp. 22932295, 1998.

[8]   M. Kadota et. al., "High-Frequency Lamb Wave Device Composed of MEMS Structure Using LiNbO$_3$ Thin Film and Air Gap", IEEE Transactions on Ultrasonics, Ferroelectrics, and Frequency Control, vol. 57, no. 11, 2010

[9]   Michio Kadota and Takashi Ogami, "5.4 GHz Lamb Wave Resonator on LiNbO$_3$ Thin Crystal Plate and Its Application." Japanese Journal of Applied Physics 50 (2011) 07HD11.

[10] V. Plessky, S. Yandrapalli, P.J. Turner, L.G. Villanueva, J. Koskela and R.B. Hammond, "5 GHz laterally-excited bulk-wave resonators (XBARs) based on thin platelets of lithium niobate." ELECTRONICS LETTERS, 24th January 2019 Vol. 55 No. 2 pp. 98–100.

[11] R. Aigner, in Proceedings of the IEEE Symposium on Acoustic Wave Devices for Future Mobile Communication Systems (IEEE, 2007), pp. 85–91.

Proceedings of the 16th Annual IEEE International
Conference on Nano/Micro Engineered and Molecular Systems
April 25-29, 2021

Research on Integrated Reliable Micro-High Explosive Train Model

Bo He*, *Student Member, IEEE*, Wen-Zhong Lou, Heng-Zhen Feng, *Member, IEEE*, Yue-Cen Zhao, Yi Sun

Abstract—In this paper, in order to make the micro-high explosive train more integrated and reliable, the motion state of flyer driven by two-point initiation copper azide (primary explosive) is studied, and a metal film bridge for two-point ignition is designed. The simulation tests of explosion model and metal bridge model are carried out by AUTODYN and COMSOL respectively. The results show that: 1) the appropriate two-point initiation distance can make the flyer obtain sufficient kinetic energy in a shorter acceleration chamber; 2) the parallel metal bridge structure has more intense temperature rise and higher initiation instantaneity.

I. INTRODUCTION

As the core function module of micro fuze, micro-high explosive train has the function of amplifying energy and initiating the main charge of ammunition [1-3]. But now, with less energetic agents in micro-high explosive train to improve system integration, the reliability is more difficult to guarantee。

Researchers mainly study the micro-primary explosive and micro-igniter. Compared with traditional energetic materials, nano energetic materials have the advantages of smaller particle size, larger specific surface area and more atoms on the surface, which can fully release its energy during explosion, higher energy density, faster energy release speed, better stability and safety [4-6]. Copper azide is an excellent primary explosive with high energy and environmental protection, but its sensitivity is high [7]. With the wide attention of micro-charge technology, the sensitivity of copper azide obtained by this technology is much lower than that of traditional copper azide [8-9]. This makes it a trend to use copper azide instead of non-environmental primary explosive. Semiconductor bridge and metal bridge are widely used as micro igniters due to their mass production and high process consistency [10-11]. both of which form plasma through material phase transformation and detonate the primary explosive by electric explosion [12-13]. Metal bridge has a simpler process than semiconductor bridge.

The materials and manufacturing processes of primary explosive and micro igniter have been studied in the existing research, but there are few reports on the reliable transmission of detonation energy in different modes of micro-high explosive train.

In this paper, we establish the model of micro-high explosive train, and then according to the simulation experiment of different initiation modes, we design and simulate two kinds of metal film bridges as micro point firearm according to the obtained two-point initiation

distance, so as to realize the integrated and reliable micro detonation sequence model.

II. MICRO-HIGH EXPLOSIVE TRAIN MODEL

The model of micro-high explosive train is shown in Fig.1. The energy transfer sequence is as follows: The electric current flows through the micro-igniter and the electric explosion occurs, igniting the micro-primary explosive. The detonation energy drives the flyer to shear fracture and accelerates fully in the accelerating chamber between the micro-primary explosive and the lead. The high kinetic energy flyer hits the lead to realize the ignition of the second. The secondary charge detonates the booster. Finally, the energy is input into the warhead to detonate the main charge.

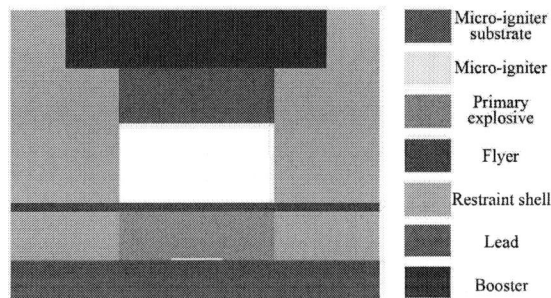

Figure 1. Micro-high explosive train.

III. FLYER DRIVEN BY PRIMARY EXPLOSIVE

The simulation model, modeling in AUTODYN software, of driving flyer is shown in Fig.2. The primary explosive is copper azide, which has excellent initiation performance and is very friendly to the environment. The charge density of copper azide is 2.29 g/cm^3 and the explosion velocity is 4.7 km/s. The JWL equation of copper azide is [14]

$$p = A\left(1 - \frac{\omega}{R_1 V}\right)e^{-R_1 V} + B\left(1 - \frac{\omega}{R_2 V}\right)e^{-R_2 V} + \frac{\omega E}{V} \ (1)$$

where p is the pressure of detonation product; V is the relative volume; E is the initial specific energy; other parameters are A =410 GPa, B =4.5 GPa, R_1 =4.90, R_2 =1.3, ω =0.3.

---

*Resrach supported by National Defense Basic Scientific Research Program of China (JCKY2019602B004).

Bo He, Wen-Zhong Lou, Heng-Zhen Feng, Yue-Cen Zhao, Yi Sun all are with the School of Mechatronical Engineering, Beijing Institute of

Technology, Beijing, China (corresponding author to provide e-mail: cqrhb0928@126.com).

978-1-6654-3008-1/21 $31.00 © 2021 IEEE

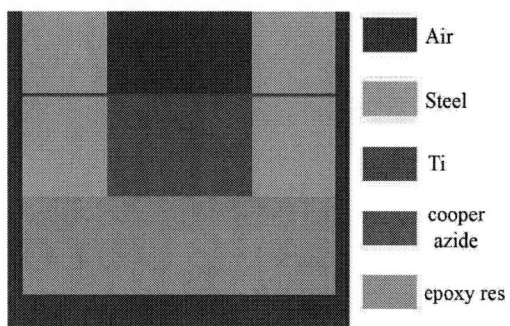

Figure 2. Simulation model of driving flyer.

The lead is impacted by a high kinetic energy flyer to produce detonation. The shock initiation criteria of most heterogeneous explosives is given by [15]

$$P^n \cdot \tau > K \qquad (2)$$

where P is the impact incident pressure, relating to the velocity and thickness of the flyer; τ is the pulse duration, which is related to the material and shape of the flyer; n and K is a constant, which are determined by experiments.

Eq.2 indicates that the kinetic energy and the shape of flyer are related to the reliability of initiation of the lead.

The size parameters of the simulation model are as follows: the size parameters of copper azide are 1×0.6 mm; the thickness of Ti flyer is 20 μm; the length of acceleration chamber is 500um. If the acceleration chamber is too long, the flyer will roll in the chamber. On the contrary, the flyer can not get enough kinetic energy because the acceleration chamber is too short. In order to ensure the reliability of the lead for ignition, in this paper, the effects of single-point ignition and two-point ignition on the motion characteristics of flyer are simulated. The pressure distributions under different initiation modes are shown in Fig.3.

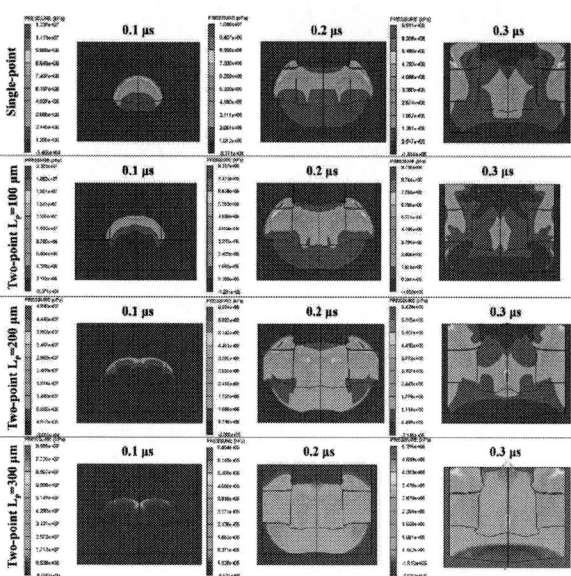

Figure 3. Pressure distributions under different initiation modes. (L_p is the distance between the initiation point of two points and the center of symmetry)

From the Fig.3, we observe that when two-point ignition at the same time, the flyer can obtain more sufficient kinetic energy in a shorter acceleration time, which can be explained by the Mach reflection caused by the oblique collision of two detonation waves when they meet [16], the stress concentration (especially in the center of symmetry, the stress concentration phenomenon is the most obvious), and the greater instantaneous acceleration obtained by the flyer. It is noted that the larger the L_p is, the higher the average velocity of the flyer will be, however, the easier the center of the flyer will be torn off. We monitored the motion state of flyer under different initiation modes, and the velocity (Fig.1(a), (c)) and the displacement (Fig.1(b), (d)) of flyer are shown in Fig.4.

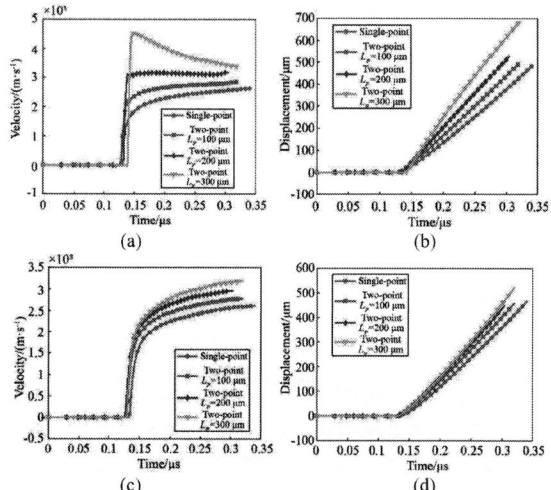

Figure 4. Flyer motion state under different initiation modes. (*(a) Velocity of flyer center; (b) Displacement of flyer center; (c) Velocity of 200 μm from the center of flyer center; (d) Displacement of 200 μm from the center of flyer center*)

From the Fig.4(a), we observe that the velocity change in the center of flyer is the most obvious. In different initiation modes, Fig.4 (c) shows the velocity change of 200um away from the center of flyer. When L_P =300 μm, the instantaneous velocity in the center of flyer is much higher than that in other positions, which may lead to the fracture of the center of flyer and make it impossible to initiate the next charge. Therefore, the distance between two points of initiation needs to be selected. The appropriate two points of initiation can shorten the acceleration chamber and improve the integration of micro detonation sequence.

IV. DESIGN & SIMULATION OF MICRO INITIATOR

We designed two "Butterfly" metal film bridges in parallel and in series with L_P =200 μm (i.e. the distance between two initiation points is 400 μm) as the micro igniter for initiating copper azide. The model of the micro igniter is shown in Fig.5. Fig.5(a) is the material structure of micro igniter. The substrate is single-crystal silicon; the middle is a layer of silicon dioxide as the insulation layer; the top layer is nickel which is not easy to oxidize in air. The metal film bridge can be batch prepared by surface silicon technology.

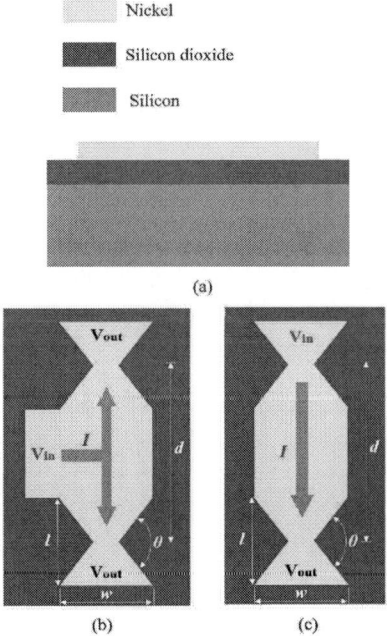

(a)

(b)   (c)

Figure 5. Metal film bridge structure. ((a) Material composition; (b) Parallel structure; (c) Series structure)

The resistance of a single "Butterfly" metal bridge region can be obtained by

$$ R = \frac{\rho}{h} \frac{l}{2w} \frac{\left[2w - l\cot\left(\theta/2\right)\right]}{\left[w - l\cot\left(\theta/2\right)\right]} \quad (3) $$

where ρ is the resistivity, and the conductivity of nickel at room temperature is 6.84 $\Omega \cdot \mu$m ; d =400 μm; w =200

μm; l =300 μm; θ =115° ; the thickness of the metal film bridge is 1 μm.

The Joule thermal finite element simulation of two kinds of metal film bridges is carried out by COMSOL software. By inputting equivalent constant pressure 3 V, the temperature distributions of two kinds of structures are shown in Fig.6. Fig.6(a) is the temperature nephogram of parallel structure at 3 μs, and the maximum temperature of bridge area has reached 4600 K. Fig.6(b) is the temperature nephogram of series structure at 6 μs, and the maximum temperature of bridge area is 3600 K.

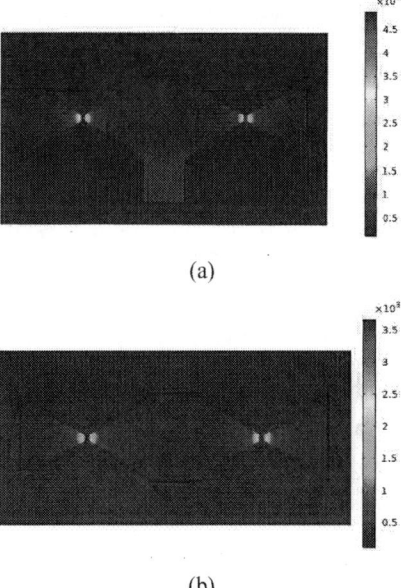

(a)

(b)

Figure 6. Temperature distributions of two structures ((a) Parallel structure at 3us; (b) Series structure at 6us)

Fig.7 shows the temperature change trend of the center of the bridge region of the two structures with time. When the parallel igniter and the series igniter continue to heat up above 3003 K (the vaporization temperature of nickel), electric explosion occurs. From the comparison of the two igniters' temperatures, it can be seen that under the same structural parameters, the temperature rise of the parallel structure is more intense, and it is used as a micro igniter with higher instantaneity.

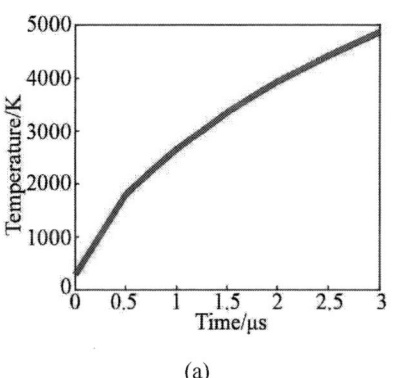

(a)

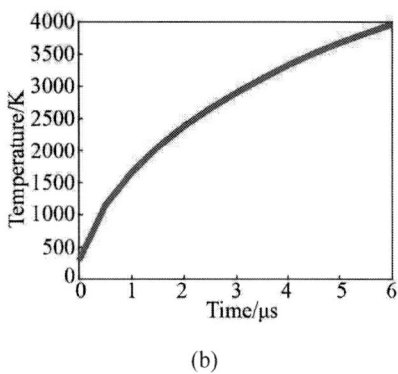

(b)

Figure 6. Temperature vs. time. (*(a) Parallel structure; (b) Series structure*)

V. CONCLUSION

In this paper, we analyze the initiation process of micro igniter (metal film bridge) - primary explosive (copper azide) - flyer (TI) in the micro detonation sequence. The simulation results show that the flyer can maintain a good shape and obtain sufficient velocity in a shorter acceleration chamber due to the Mach reflection in the appropriate distance of two-point initiation, and the micro-igniter used for two-point initiation is designed. The micro-high explosive train can be more integrated and reliable when two points are detonated at the same time. Whether the micro booster sequence can work reliably should be considered when simultaneous initiation cannot be guaranteed.

REFERENCES

[1] C. Rossi, and D. Esteve, "Micropyrotechnics, a new technology for making energetic microsystems: review and prospective - Review," Sensors and Actuators a-Physical, vol. 120, no. 2, pp. 297-310, May 2005.

[2] H. Pezous, C. Rossi, M. Sanchez, F. Mathieu, X. Dollat, S. Charlot, L. Salvagnac, and V. Conedera, "Integration of a MEMS based safe arm and fire device," Sensors and Actuators a-Physical, vol. 159, no. 2, pp. 157-167, May 2010.

[3] J. Lee, and T. Kim, "MEMS solid propellant thruster array with micro membrane igniter," Sensors and Actuators a-Physical, vol. 190, pp. 52-60, Feb 2013.

[4] K. Zhang, C. Rossi, P. Alphonse, C. Tenailleau, S. Cayez, and J.-Y. Chane-Ching, "Integrating Al with NiO nano honeycomb to realize an energetic material on silicon substrate," Applied Physics a-Materials Science & Processing, vol. 94, no. 4, pp. 957-962, Mar 2009.

[5] D. G. Xu, Y. Yang, H. Cheng, Y. Y. Li, and K. L. Zhang, "Integration of nano-Al with Co3O4 nanorods to realize high-exothermic core-shell nanoenergetic materials on a silicon substrate," Combustion and Flame, vol. 159, no. 6, pp. 2202-2209, Jun 2012.

[6] Y. Yang, D. Xu, and K. Zhang, "Effect of nanostructures on the exothermic reaction and ignition of Al/CuOx based energetic materials," Journal of Materials Science, vol. 47, no. 3, pp. 1296-1305, Feb 2012.

[7] M. Zhou, Z. Li, Z. Zhou, T. Zhang, B. Wu, L. Yang, and J. Zhang, "Antistatic Modification of Lead Styphnate and Lead Azide for Surfactant Applications," Propellants Explosives Pyrotechnics, vol. 38, no. 4, pp. 569-576, Aug 2013.

[8] F. Forohar, V. J. Bellitto, W. M. Koppes, and C. Whitaker, "The adsorption of hydrazoic acid on single-walled carbon nanotubes," Materials Chemistry and Physics, vol. 112, no. 2, pp. 427-431, Dec 2008.

[9] V. Pelletier, S. Bhattacharyya, I. Knoke, F. Forohar, M. Bichay, and Y. Gogotsi, "Copper Azide Confined Inside Templated Carbon Nanotubes," Advanced Functional Materials, vol. 20, no. 18, pp. 3168-3174, Sep 2010.

[10] R. Z. Xie, L. M. Li, L. Liu, X. M. Ren, and Y. Xue, "Study of Design and Performance of Micro Initiation Train," Binggong Xuebao/Acta Armamentarii, vol. 38, no. 3, pp. 460-465, Mar 2017.

[11] J. Wang, Y. Li, B. Zhou, and Z. Z. Gao, "Firing process and spectrum diagnosis of semiconductor bridge for high output energy micro-initiator," Sensors and Actuators a-Physical, vol. 270, pp. 108-117, Feb 2018.

[12] J. U. Kim, C. O. Park, M. I. Park, S. H. Kim, and J. B. Lee, "Characteristics of semiconductor bridge (SCB) plasma generated in a micro-electro-mechanical system (MEMS)," Physics Letters A, vol. 305, no. 6, pp. 413-418, Dec 2002.

[13] D. A. Benson, M. E. Larsen, A. M. Renlund, W. M. Trott, and R. W. Bickes, "Semiconductor bridge – a plasma generator for ignition of explosives," Journal of Applied Physics, vol. 62, no. 5, pp. 1622-1632, Sep 1987.

[14] G. Jian, Q. Zeng, J. Guo, B. Li, and M. Li, "Simulation of flyers driven by detonation of copper azide," Explosion and Shock Waves, vol. 36, no. 2, pp. 248-252, Mar 2016.

[15] H. S. Yadav, "Initiation of detonation in explosives by impact of projectiles," Defence Science Journal, vol. 56, no. 2, pp. 169-177, Apr 2006.

[16] C. S. Morawetz, "Potential-theory for regular and Mach reflection of a shock at a wedge," Communications on Pure and Applied Mathematics, vol. 47, no. 5, pp. 593-624, May 1994.

Gap in pagination due to unavailable paper.

Pages 939-940

Proceedings of the 16th Annual IEEE International Conference on Nano/Micro Engineered and Molecular Systems
April 25-29, 2021

THz Hybrid Graphene &Metal Patches Metasurface for RCS Reduction and Digital Coding

Baolong Wang, Jiayi Yang, Shuangshuang Liu, Wei Xu, Guobin Chen, Di Feng, Lingjie Jia,

Yan Meng, Xiuhan Li*

Abstract—Metasurface that shows excellent ability in controlling electromagnetic (*EM*) waves has attracted tremendous attention. Based on the polarization conversion and the Pancharatnam–Berry (*PB*) phase, a hybrid metasurface that consists of graphene and metal patches (*MGMPs*) is presented to realize reduction of the radar cross section (*RCS*) and the beam reconfiguration of terahertz waves. The unit cells of MGMPs are designed for the linearly polarized (*LP*), the left circularly polarized (*LCP*) and the right circularly polarized (*RCP*) waves. The MGMPs can realize the polarization conversion and produce the phase difference by using coding elements "0" and "1" in combination. The MGMPs are generated using random code, which lead to the redistribution of electromagnetic waves energy and achieve RCS reduction in a wide frequency band range and incident angle. In addition, polarization conversion ratio (*PCR*) of higher than 90% is achieved. The simulation results show that the MGMPs can achieve the RCS reduction in a wide frequency range from 0.35 THz to 1 THz (*96.3%*) at the normal incidence for THz waves. The super working bandwidth further indicates that the MGMPs have huge application potential in the field of RCS reduction.

Keywords: Metasurface; Pancharatnam-Berry phase; Radar Cross Section reduction; Terahertz waves; The Interface Effect

I. INTRODUCTION

The realization of EM waves control technology in a wide frequency band has always been the focus of people's exploration, including anomalous reflection [1-4], focusing antenna [5], RCS reduction [6], and programmable logic device [7]. There is no material with both negative electrical conductivity and dielectric constant in nature. Metamaterials are artificially designed subwavelength atoms, which have extraordinary performance in regulating electromagnetic wave beams. Many magical physical phenomena have been realized using metamaterials, e.g. negative refractive index [8], invisibility cloak, perfect imaging, and shifting phantom ability [9].

Metasurface is two-dimensional metamaterials. Compared with metamaterials, metasurface shows advantages of low loss, ultra-thin thickness, low cost, and conformal properties [10]. Many fascinating physical phenomena have been realized using metasurface, such as beam steering [11, 12], information transmission [13], phase shifters, harmonic control, filtering [14, 15], etc. The

Interface Effect (IE) is the basic theory for metasurface [15]. The IE is that different metasurface unit cells have different amplitude, phase, polarization and frequency spectrum adjustment effects for the incident EM waves. In 2014, the concept of digital coding metasurface was proposed by Cui et al. [16], that is, using digital coding elements "0" and "1" to represent two different metasurface unit cells, which have phase difference of 180° with respect to the same THz waves. Moreover, due to digital coding metasurface can achieve abnormal reflection, refraction [17, 18] and focusing of electromagnetic waves, the combination of coding metasurface and digital information theory can achieve electromagnetic beam waves regulation such as RCS reduction.

The polarization-insensitive unit cell based on phase control uses different metal patch to achieve different phase reflection. The MGMPs composed of the unit cells can change the propagation direction of EM waves to redistribute the reflected waves energy in different directions. In recent years, electric field sensitive materials have been used to form metasurface to achieve more flexible control abilities. However, there are relatively few researches on compositing unit cells combining graphene and metal patches or apertures, and the effective frequency band of other works are slightly narrower than the proposed unit cell [19, 20].

In this work, a hybrid MGMPs unit cell is presented. The MGMPs unit cell is based on polarization conversion, which can realize polarization conversion. The relative bandwidth has reached 96.3%, covering the frequency band from 0.35 THz to 1 THz. Moreover, the polarization conversion characteristics of the MGMPs are not sensitive to the incident angle changes. Our works have effectively achieved RCS reduction in the terahertz band.

II. UNIT CELL DESIGN

As depicted in Fig.1, the MGMPs unit cell is a six-layer E-shaped multilayer sandwich structure. The period of the MGMPs unit cell is p=150 μm. The bottom metal plane material is copper, and its thickness t=0.5 μm and a conductivity of 5×10^8 S/m. Both the upper dielectric substrate and the fifth layer dielectric substrate are made of PTFE with the loss tangent is tanδ = 0.0002 and the permittivity is ε = 2.1, and the upper dielectric substrate

---

*Xiuhan Li is corresponding author.

*Xiuhan Li works in Beijing Jiaotong University (phone: +86-10-5168-3981; fax: +86-10-5168-3682; e-mail: lixiuhan@bjtu.edu.cn).

Baolong Wang, Jiayi Yang, Shuangshuang Liu, Wei Xu, Guobin Chen, Di Feng, Lingjie Jia, Yan Meng are with Beijing Jiaotong University (e-mail: 19125050@bjtu.edu.cn, 20111006@bjtu.edu.cn).

thickness is h_1=7 μm. The geometrical parameters of the E-shaped metal patch are a=115.5 μm, b=12.5 μm and c=37.5 μm. The gap between the E-shaped metal patch and the small square is i=0.5 μm, and the third layer of the unit cell is made of silicon with the permittivity $\varepsilon_s = 11.9$, the conductivity of 0.00025 and the thickness $h_{si} = 0.2\ \mu m$. The fourth layer of the unit cell is graphene. After optimization, the graphene chemical potential is set to 0.8 eV, and the period of the graphene is p_1=50 μm. The thickness of the fifth layer of PTFE is 68.57 μm. (See Table 1 for specific the unit cell parameters.)

According to the polarization conversion theory, due to the structure asymmetry of the MGMPs unit cell, the reflection phase responses in the x-direction and the y-direction are different. By optimizing the MGMPs unit cell's geometric parameters and chemical potential of the graphene, stable broadband phase regulation can be realized.

The simulation tool was the commercial software CST Microwave Studio. The simulation process used unit cell boundary conditions and frequency domain solvers. Fig.2(a-d) show the amplitude and phase responses, PCR and phase difference respectively from 0.2 THz to 1.2 THz.

From Fig.2, we can know that under the incidence of LP terahertz waves, the reflected waves are mainly the cross-polarized waves. Simulation results show that the MGMPs unit cells have three resonance frequencies in the range of 0.2 THz – 1.2 THz, which are 0.35 THz, 0.6 THz and 0.95 THz. The broadband characteristics of the unit cells are attributed to the existence of multiple resonance modes. Fig.2(a) shows the cross-polarized waves and co-polarized waves reflection amplitude. Fig.2(b) shows the digital coding elements "0" and "1" reflection phase responses under THz waves normal incidence. Fig.2(c) shows that PCR greater than 90% has been achieved. Fig.2(d) shows that the reflection phase control characteristics of digital coding elements "0" and "1" are very stable. In order to optimize and compare whether the addition of graphene can improve the unit cell PCR characteristics, a simulation test was carried out.

As shown in Fig.3, case 1 is a unit cell embedded in graphene, while case 2 is not. Simulation results show that the PCR of case 1 is slightly higher than case 2, and case 1 can maintain a wide effective frequency region.

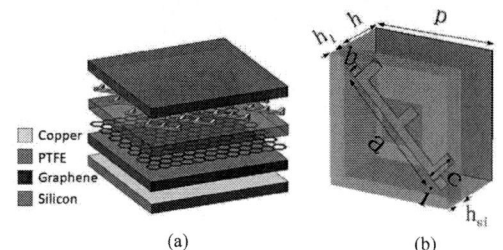

Fig. 1. The MGMPs unit cell. (a) Structural model, (b) schematic diagram of parameters.

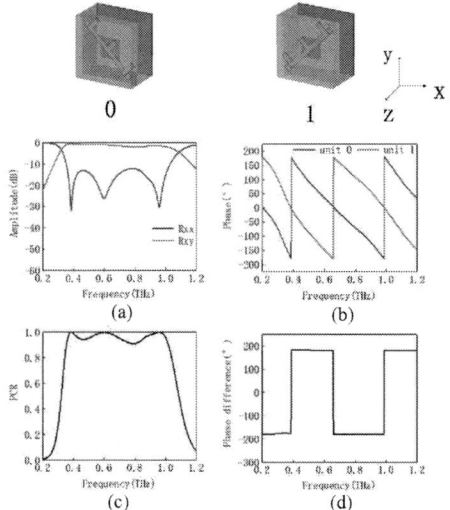

Fig. 2. The digital coding elements "0" and "1" built by the MGMPs unit cell. (a) Co-polarization reflection amplitude R_{xx} and cross-polarization reflection amplitude R_{xy}. (b) The reflection phase. (c) PCR of digital coding elements "0" and "1". (d) The phase difference between digital coding unit cells "0" and "1".

Table 1. Specific parameters of the unit cell

Parameter	Size(μm)	Parameter	Size(μm)
t	0.5	p	150
p_1	50	i	0.5
h_{si}	0.2	h_1	7
h	66.85	a	142.5
b	12.5	c	37.5

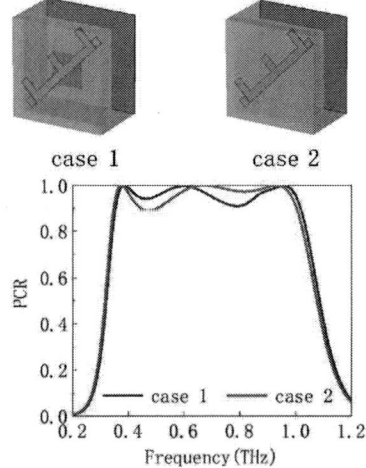

Fig. 3. PCR results without integrated graphene and integrated graphene. Case 1: unit cell embedded in graphene. Case 2: unit cell not embedded in graphene.

978-1-6654-3008-1/21 $31.00 © 2021 IEEE

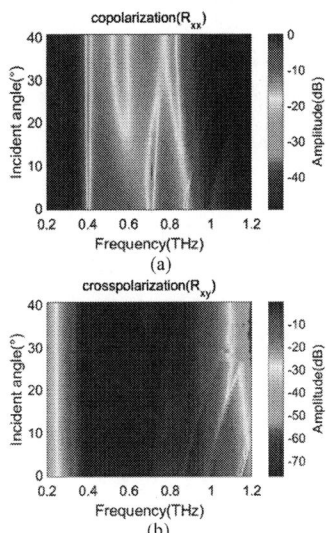

Fig. 4. The MGMPs unit cell's co-polarization reflection amplitude and cross-polarization reflection amplitude. (a) co-polarization reflection amplitude. (b) cross-polarization reflection amplitude.

Fig.4 shows that the relationship between the co-polarization and the cross-polarization reflection amplitude characteristic with the incident angle under the incidence of different frequencies terahertz waves. Fig.4(a) shows that when the LP waves incidence angle changes from 0° to 40°, the co-polarization reflective amplitude does not change much in a wide bandwidth. Similarly, as shown in Fig.4(b), it shows the cross-polarization reflective amplitude when the LP waves incidence angle changes from 0° to 40°. For different LP waves incidence angles, the MGMPs unit cell almost maintains a wide effective frequency region.

Fig.5 and Fig.6 show that the E-shaped metal patch and the bottom metal plane current distribution at three resonance frequencies respectively. Fig.5(a-c) show that E-shaped metal patch's current distribution at resonance frequencies under transverse electric (*TE*) waves normal incidence. Fig.5(d-f) show that the bottom metal plane's current distribution at resonance frequencies under TE waves normal incidence. Similarly, Fig.6(a-c) respectively show the E-shaped metal patch's current distribution at resonance frequencies under Transverse Magnetic (*TM*) terahertz waves normal incidence, and Fig.6(d-f) respectively show under TM terahertz waves normal incidence, the bottom metal plane's current distribution at resonance frequencies. According to Fig.5 and Fig.6, the current components of the E-shaped metal patch are not uniform, which causes the phase difference of the terahertz waves, and the polarization conversion is realized. At different resonance frequencies, the GMGPs unit cells produce different resonance modes, namely electric resonance and magnetic resonance. At 0.35 THz and 0.6 THz, magnetic resonance occurs, and at 0.95 THz, electric resonance occurs. The existence of multiple resonance modes leads to the effective wide bandwidth characteristics of the MGMPs unit cell.

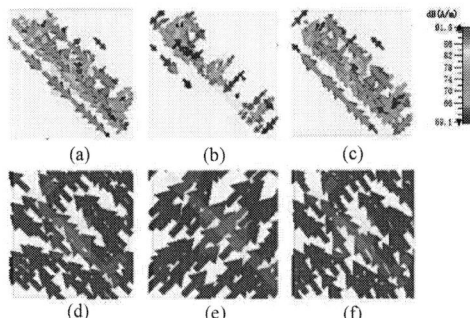

Fig. 5. Current distribution for the E-shaped metal patch, bottom metal plane under TE waves. E-shaped metal patch current distribution at (a) 0.35 THz, (b) 0.6 THz and (c) 0.95 THz, respectively. Bottom metal plane current distribution at (d) 0.35 THz, (e) 0.6 THz and (f) 0.95 THz, respectively.

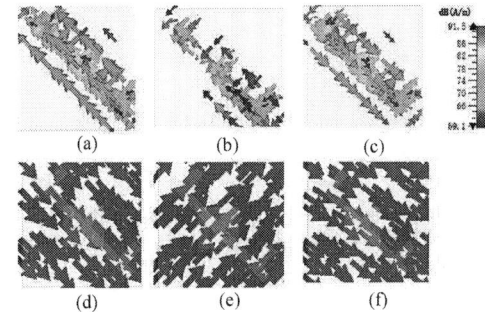

Fig. 6. Current distribution for the E-shaped metal patch, bottom metal plane under TM waves. E-shaped metal patch current distribution at (a) 0.35 THz, (b) 0.6 THz and (c) 0.95 THz, respectively. bottom metal plane current distribution at (d) 0.35 THz, (e) 0.6 THz and (f) 0.95 THz, respectively.

PB phase theory can design multi-bit coding for circularly polarized (*CP*) waves. In order to explore whether the proposed unit cells can realize multi-bit coding for CP waves, we designed four digital coding unit cells. Two digital coding elements with inverse phase information are needed to design 1-bit MGMPs for RCS reduction. Similarly, 2-bit MGMPs for RCS reduction need four digital coding elements with phase difference of 90°. The fixed phase difference is represented by 00, 01, 10, 11 respectively. Using PB phase theory, rotating the E-shaped metal patch at different angles to obtain different digital coding elements. In detail, this is achieved by the top metallic pattern's rotation angle α, as shown in Fig.7(a). Then, in order to obtain the four digital coding elements, the rotation angle α is changed from 22.5° to 157.5°, and the step size is 45°. Fig.7(b) describe the co-polarization and the cross-polarization reflection amplitude characteristic of the multi-bit MGMPs unit cells under the incidence of different frequencies LCP terahertz waves. Fig.7(c) shows the reflection phase characteristics of the multi-bit MGMPs unit cells. The results show that the multi-bit MGMPs unit cells have a stable phase regulation in a wide effective frequency region. Fig.7(d) shows the E-shaped metal patch patterns of the MGMPs unit cells.

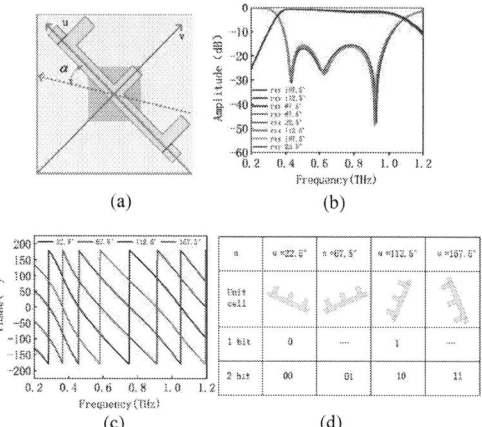

(a)　　　　　　　(b)

(c)　　　　　　　(d)

Fig. 7. Simulated performance of multi-bit MGMPs unit cells. (a) Rotation diagram of the unit cell's top metallic pattern. (b) The reflectivity curve of the MGMPs unit cells with rotation angles of 22.5°, 67.5°, 112.5°, 157.5°. r_{xy} represents the cross-polarization reflection amplitude, r_{xx} represents the co-polarization reflection amplitude. (c) The phase of the unit cells with rotation angles of 22.5°, 67.5°, 112.5°, 157.5°. (d) Schematic diagram of multi-bit MGMPs unit cells.

III. DESIGN AND SIMULATION OF CODING METASURFACES

The various far-field characteristics of metasurface are determined by the unit cells unique coding. The far-field mode scattering function based on the PB phase coding metasurface is [3]:

$$f(\theta,\varphi) = f_e(\theta,\varphi)\sum_{m=1}^{M}\sum_{n=1}^{N}\exp\{-i\{\varphi(m,n)+ \\ KD_x\sin\theta(m-1/2)\cos\varphi+KD_y(n-1/2)\cos\varphi\sin\varphi\}\}, \quad (1)$$

where θ is the elevation angle, φ is the azimuth angle, and $f_e(\theta,\varphi)$ is the unit cell pattern function. The directionality Dir (θ,φ) of the metasurface is given by:

$$Dir(\theta,\varphi) = \frac{4\pi|f(\theta,\varphi)^2|}{\int_0^{2\pi}|f(\theta,\varphi)^2|\sin\theta d\theta d\varphi}. \quad (2)$$

Due to polarization conversion and Generalized Snell's Law (GSL) [6], the RCS can be reduced by forming random sequences coding metasurface. The MGMPs unit cells can construct the metasurface with multiple scattering functions, Fig.8(a-d) show 1-bit and 2-bit MGMPs generated by random coding. Fig.9 shows that far-field of different coding metasurface. Due to phase response of the MGMPs are randomly distributed, the incident EM waves energy will be redirected to multiple directions.

The far-field scatter pattern is computed at three resonance frequencies for using MGMPs to achieve RCS reduction (Fig.10). The RCS reduction of 1-bit MGMPs is more than 5 dB at three resonance frequencies. Due to the phase random distribution of the MGMPs, conventional reflection waves are not dominant. The PEC phase is evenly distributed, causing conventional reflection waves are dominant. Therefore, the MGMPs have more diffused scattering than PEC.

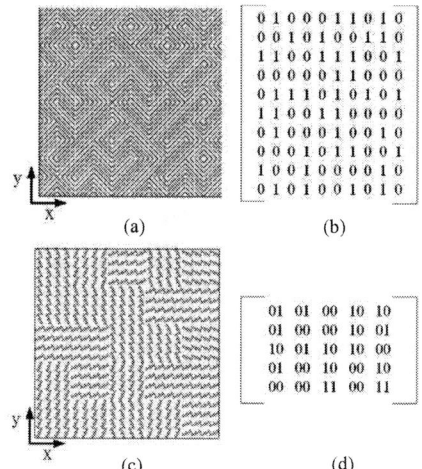

(a)　　　　　　　(b)

(c)　　　　　　　(d)

Fig. 8. The proposed coding metasurface and corresponding coding matrices. (a) and (b) 1-bit MGMPs, (c) and (d) 2-bit MGMPs.

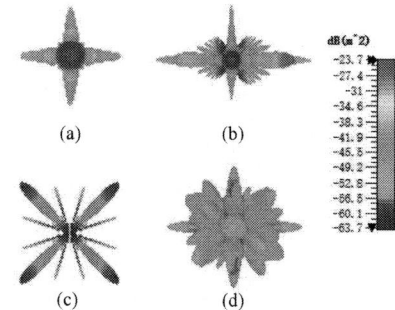

(a)　　　　　　　(b)

(c)　　　　　　　(d)

Fig. 9. Far-field diagrams of different coding matrices. (a) PEC. (b) 0101···/0101··· coding matrix. (c) 0101···/1010··· coding matrix. (d) The 1-bit MGMPs.

The scattering mode of the MGMPs varies with frequencies, it is due to the phase characteristics of the MGMPs is affected by frequencies. Fig.11(a) shows the RCS comparison of PEC and 1-bit MGMPs, and Fig.11(b) shows the RCS comparison of PEC and 2-bit MGMPs. One can see that for LCP and RCP waves, the RCS reduction is similar and in the 0.35 THz to 1 THz frequency band, RCS reduction more than 5 dB. Table 2 compares the proposed unit cells with other works. the MGMPs unit cells have an ultra-wideband effective frequency band (96.3%) compared with other works.

In order to research the MGMPs E-field distribution, the E-field distribution at 0.6 THz of PEC and 1-bit MGMPs are investigated. As shown in Fig.12, the 1-bit MGMPs have more diffused scattering than PEC. Fig.13 simulated the RCS of 1-bit MGMPs, 2-bit MGMPs and PEC under terahertz waves normal incidence at 0.6 THz. Fig.13 shows that the terahertz waves are redirected in multiple directions, 1-bit MGMPs and 2-bit MGMPs for RCS reduction can achieve a maximum RCS of -10dB.

978-1-6654-3008-1/21 $31.00 © 2021 IEEE

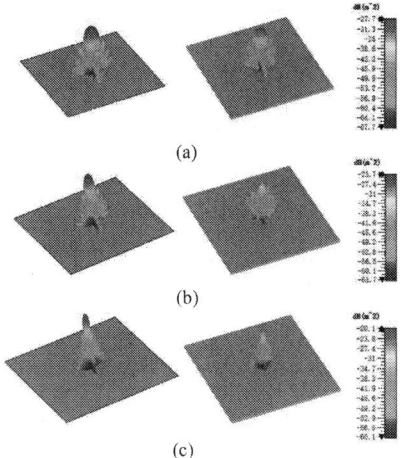

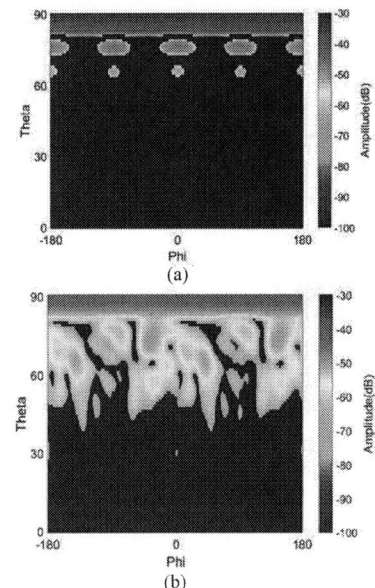

Fig. 10. PEC (left) and 1-bit MGMPs at (a) 0.38 THz (b) 0.6 THz (c) 0.96 THz 3D far-field scattering patterns.

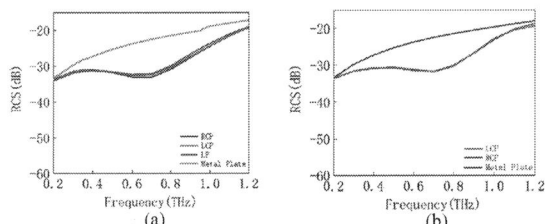

Fig. 11. (a) RCS of PEC and 1-bit MGMPs at LP, LCP, RCP normal incidence. (b) RCS of PEC and 2-bit MGMPs at LCP, RCP normal incidence.

Fig. 12. Electric field distribution at 0.6 THz (a) PEC (b) 1-bit MGMPs.

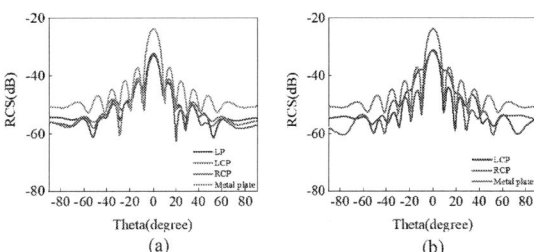

Fig. 13. (a) RCS for 1-bit MGMPs and PEC at 0.6 THz (b) RCS for 2-bit MGMPs and PEC at 0.6 THz.

Additionally, Fig.14(a) and Fig.14(b) show the near-field energy distribution of the 1-bit MGMPs and 2-bit MGMPs for RCS reduction at 0.1 THz and 0.6 THz frequency points. The simulation results show that in the effective frequency region of the unit cells, 1-bit MGMPs and 2-bit MGMPs have stronger near-field energy scattering, while in the ineffective frequency region of the unit cells, the EM waves energy is not redirected, at this time, the MGMPs function is equivalent to PEC.

Table 2. Comparison of the unit cell effective frequency band

Referencce	Frequency Band	Relative bandwidth of Unit Cell(%)
[2]	15-40GHz	90.9
[3]	0.8-1.6THz	66.7
[5]	9-18GHz	66.7
[7]	8-12GHz	40
[8]	0.34-0.49THz	36
[11]	16-33GHz	69.4
[12]	40-65GHz	47.6
[21]	1.0-2.0THz	66.7
Proposed design	0.35-1.0THz	96.3

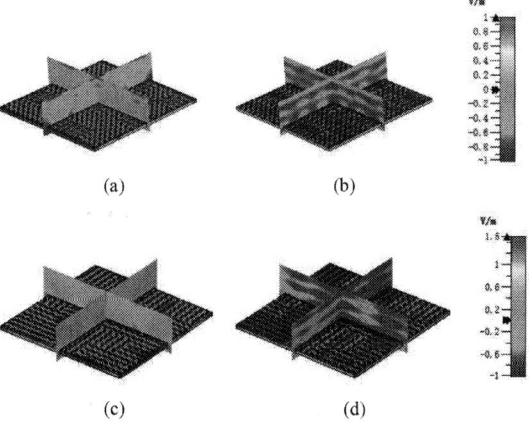

Fig. 14. The near-field of 1-bit MGMPs under LP waves normal incidence (a) at 0.1 THz (b) at 0.6 THz. (c) The near-field of 2-bit MGMPs under LCP waves normal incidence at 0.1 THz (d) at 0.6 THz.

IV. CONCLUSION

Based on polarization conversion and PB phase, the MGMPs have been designed to achieve RCS reduction and beam reconfiguration of terahertz waves. The unit cells of MGMPs are designed for LP, LCP and RCP waves. PCR of higher than 90% is achieved. The simulation result shows that the MGMPs can achieve RCS at the LP waves normal incidence. The relative bandwidth has reached 96.3%, covering the frequency band from 0.35 THz to 1 THz. The reduction of 5 dB further indicates that the MGMPs have huge application potential in the field of RCS reduction.

ACKNOWLEDGMENT

This work is supported by the Fundamental Research Funds for the Central Universities+2020JBZD011, National Natural Science Foundation of China (60706031 and 61574015), Beijing; National Science Foundation (4122058), National Key R&D Program of China (2016YFB1200203).

.

REFERENCES

[1] Akbari M, Samadi F, Sebak A, et al. "Superbroadband Diffuse Wave Scattering Based on Coding Metasurfaces: Polarization Conversion Metasurfaces," *IEEE Antennas and Propagation Magazine*, vol.61, no.2, pp. 40-52, 2019.

[2] Saifullah Y, Waqas A B, Yang G, et al. "4-Bit Optimized Coding Metasurface for Wideband RCS Reduction," *IEEE Access*, vol.7, pp. 122378-122386, 2019.

[3] Jiu-Sheng L, Jian-Quan Y. "Manipulation of Terahertz Wave Using Coding Pancharatnam–Berry Phase Metasurface," *IEEE Photonics Journal*, vol.10, no.5, pp. 1-12, 2018.

[4] Hosseininejad S E, Rouhi K, Neshat M, et al. "Digital Metasurface Based on Graphene: An Application to Beam Steering in Terahertz Plasmonic Antennas," *IEEE Transactions on Nanotechnology*, vol.1, no.1, pp. 99, 2019.

[5] Wang H, Li Y, Chen H, et al. "Multi-Beam Metasurface Antenna by Combining Phase Gradients and Coding Sequences," *IEEE Access*, vol. 7, pp. 62087-62094, 2019.

[6] Gao L H, Cheng Q, Yang J, et al. "Broadband diffusion of terahertz waves by multi-bit coding metasurfaces," *Light: Science & Applications*, vol.4, no.9, pp. e324, 2015.

[7] Cui T J, Qi M Q, Wan X, et al. "Coding metamaterials, digital metamaterials and programmable metamaterials," *Light Science & Applications*, vol.3, no.10, pp. 218, 2014.

[8] Liu Y, Zhao X. "Perfect Absorber Metamaterial for Designing Low-RCS Patch Antenna," *IEEE Antennas & Wireless Propagation Letters*, vol.13, pp. 1473-1476, 2014.

[9] FENG Rui, YI Jianjia. "Multi-beam generating lens antenna based on space transformation," *Proceedings of the 2017 National Antenna Conference. Chinese Institute of Electronics*, 2017.

[10] Paquay M, Iriarte J C, Ederra I, et al. "Thin AMC Structure for Radar Cross-Section Reduction," *IEEE Transactions on Antennas & Propagation*, vol.55, no.12, pp. 3630 3638, 2017.

[11] Tran M C, Pham V H, Ho T H, et al. "Broadband microwave coding metamaterial absorbers," *entific Reports*, vol.10, no.1, pp. 1810, 2020.

[12] Zhao Y, Huang C, Song Z, et al. "A Digital Metamaterial of Arbitrary Base Based on Voltage Tunable Liquid Crystal," *IEEE Access*, vol.7, pp. 79671-79676, 2019.

[13] Polat E O, Uzlu H B, Balci O, et al. "Graphene-Enabled Optoelectronics on Paper," *ACS Photonics*, vol.3, no.6, pp. 964-971, 2016.

[14] Xu, Bing–zheng, Gu C Q, Li Z, et al. "A novel structure for tunable terahertz absorber based on graphene," *Optics Express*, vol.21, no.20, pp. 23803, 2013.

[15] Xu W, Lv T, Guo H, et al. "A graphene-metamaterial hybrid structure for the design of reconfigurable low pass terahertz filters," *Microwave and Optical Technology Letters*, pp. 1-6, 2020.

[16] Liu S, Cui T J, Xu Q, et al. "Anisotropic coding metamaterials and their powerful manipulation of differently polarized terahertz

waves," *Light: Science & Applications*, vol.5, no.5, pp. e16076, 2016.

[17] Neto A H C, Guinea F, Peres N M R, et al. "The electronic properties of graphene and its bilayer," *Vacuum*, vol.83, no.1, pp. 1248–1252, 2009.

[18] Yatooshi T, Ishikawa A, Tsuruta K. "Terahertz wavefront control by tunable metasurface made of graphene ribbons," *Applied Physics Letters*, vol.107, no.5, pp. 788, 2015.

[19] Zhou Y, Cao X Y, Gao J, et al. "RCS reduction for grazing incidence based on coding metasurface," *Electronics Letters*, vol.53, no.20, pp. 1381-1383, 2017.

[20] Wenwei, Liu, Shuqi. "Realization of broadband cross polarization conversion in transmission mode in the terahertz region using a single-layer metasurface," *Optics Letters*, pp. 61831012-61871355, 2015.

[21] Zhou C, Li J S. "Polarization conversion metasurface in terahertz re gion," *Chinese Physics B*, vol.29, no.7, pp. 078706, 2020.

AUTHOR INDEX

Aishan, Yusufu 203
Alahi, Md Eshrat E 980, 1629
Alcheikh, N. 425
Bai, Yu 282, 299, 513
Bai, Zaiqiao 1084
Bao, Chen 184, 990
Bao, Weijie 904
Bian, Chao 1303
Bian, Tian 826
Bian, Xiongheng 909
Bingfei, Zhang 826
Bu, Zhenxiang 1065, 1897
Cai, Guangyi 953
Cai, Xiaoyu 255
Cai, Yao 228, 931
Cai, Yaxing 271
Cai, Yu 26
Cao, Jiangcheng 949
Cao, Liang 158
Cao, Qingpeng 762
Cao, Tongtong 1405
Cao, Wenping 770
Cao, Yongqi 528
Cao, Yun 699, 1287
Cao, Zhen 180, 459
Chai, Zhiping 1390
Chang, Honglong 55, 1903
Chang, Xiaocong 798
Che, Bangzhou 1554
Chen, Daner 840
Chen, Deyong 193, 1465, 1691, 1725, 1813
Chen, Fangshuai 731
Chen, Fanhong 1142
Chen, Guobin 941
Chen, Haifeng 319
Chen, Han 1390
Chen, Hao 1287, 1358
Chen, Hongxu 1843
Chen, Huatan 1035, 1136, 1328
Chen, Hua 197, 1825
Chen, Hui-Jiuan 503, 509
Chen, Huyue 1297
Chen, Jiang 399
Chen, Jian 193, 1465, 1691, 1813
Chen, Jinbo 919
Chen, Jingyi 1664
Chen, Jing 647
Chen, Junyu 491, 1136, 1328

Chen, Ke 1362
Chen, Liguo 909
Chen, Mingwei 1813
Chen, Qianhuang 845
Chen, Qinnan 1650
Chen, Qixiang 1650
Chen, Rui 1493
Chen, Shirong 1574
Chen, Shixing 30, 676
Chen, Si 1259
Chen, Songyue 963, 1655
Chen, Tao 60
Chen, Weifeng 1398
Chen, Wu 826
Chen, Xiaojun 319
Chen, Xiaoli 474
Chen, Xing 144, 1574
Chen, Xin 158
Chen, Xi 715
Chen, Xuying 1370
Chen, Yanan 731
Chen, Yao 282
Chen, Ye 882
Chen, Yunfei 539
Chen, Yusa 889
Chen, Zeji 455
Chen, Zhaohui 1070
Chen, Zhekun 1732
Chen, Zhensheng 1903
Chen, Zhibin 1737
Chen, Zhijian 1384
Chen, Zhiwen 1650
Cheng, Chao 193, 1691
Cheng, Zhenxing 30
Cheyns, David 1423
Chu, Jinkui 372
Chu, Yao 271, 1291
Cong, Bo 21
Cong, Lin 1084
Cui, Jian 1102, 1461
Cui, Jingqin 1737
Cui, Peijuan 1070
Dai, Jishen 1655
Dai, Songqiao 1871
Deng, Guang-Wei 1084
Deng, Jinjun 872, 1415
Deng, Qingyang 1410
Deng, Tao 68

Ding, Guanghui	872
Ding, Haojun	303
Ding, Hong	1370
Ding, Jiandong	1283
Ding, Qifan	1297
Ding, Zan	345
Djoulde, Aristide	919
Dong, Fangyang	1539
Dong, Hai-Feng	998
Dong, Hanyong	463, 1419
Dong, Hao	144
Dong, Jiale	660
Dong, Siyan	1106
Dong, Xianshan	1102
Du, Bingqian	1337
Du, Taili	660, 1539, 1676
Du, Xu	1074
Du, Yijia	1106
Duan, Yongqing	652
Duan, Yumo	1813
Dun, Guanhua	680
Fan, Xinye	1175
Fan, Xuge	1169
Fan, Yuanyi	372
Fan, Zhongqi	660
Fan, Zixiao	378
Fang, Hou	1057
Fang, Jiaru	1374
Fang, Lu	26
Fang, Xudong	334
Fei, Wenjie	1394
Feng, Di	941, 1714
Feng, Heng-Zhen	935
Feng, Jianguo	55, 1574
Feng, Linhao	737
Feng, Xue	731
Feng, Yanlu	687, 915
Feng, Yue	1264, 1332
Feng, Zhihong	334
Fu, Gang	491
Fu, Sinan	1554, 1700
Fu, Ting	1655
Fu, Yunxia	255
Ganesan, A.	1623
Gao, Chao	931
Gao, Guangen	1065
Gao, Guowei	392
Gao, Jiaming	378
Gao, Lingxiao	158
Gao, Wei	1415
Gao, Y.	1660
Gao, Yahao	1611

Gao, Yang	1157, 1259, 1353, 1804
Gao, Zhiyuan	1545
Geng, Jiangjun	60
Gong, Cheng	826
Gong, Qingfeng	994
Gong, Taobo	1478
Gong, Xun	1088, 1165
Gong, Zhuhao	695
Gu, Chaoming	180, 459
Gu, Dandan	1897
Gu, Guoqiang	711, 715
Gu, Xiyu	931
Gu, Yexin	1283
Guan, Chuanlong	372
Guo, Chenyu	1664
Guo, Chuan Fei	1390
Guo, Dengji	1617
Guo, Hang	148
Guo, Shan	1018
Guo, Shishang	931
Guo, Weijin	1310
Guo, Xinyan	882
Guo, Yixuan	691
Guo, Z. J.	959
Hai, Zhenyin	804
Han, Feng	47
Han, Lianhuan	1337
Han, Mengdi	148, 1686
Han, Mingjie	1303
Han, Ruixing	1291
Han, Weilong	443
Han, Yanhui	1264
Hansson, Jonas	1310
He, Bo	935
He, Ronggana	528
He, Xuefeng	386, 1353
He, Yinghao	898
He, Yingning	1283
He, Yunqian	51
He, Ziliang	1560
Heremans, Paul	1423
Hong, Hao	174
Hong, Heting	1863
Hosokawa, Yoshihiro	251
Hou, Cheng	60
Hou, Weiguo	271
Hou, Zhanqiang	1888
Hu, Cuiwen	319
Hu, Jie	407
Hu, Jingfang	392
Hu, Kai-Ming	433, 1432, 1605
Hu, Ning	1374

Hu, Tengjiang 1405
Hu, Tengjing 215
Hu, Xiaoping 1410
Hua, Yunzhi 1502
Huan, R. 1623
Huang, Dong 1839
Huang, Huayi 1096
Huang, Hui 386
Huang, Junlong 184, 990
Huang, Kai 737
Huang, Manjuan 1070
Huang, Min 1110
Huang, Shuang 503, 509
Huang, Sirong 1617
Huang, Xinshuo 503, 509
Huang, Yao 680
Huang, Yongan 652
Huang, Youchao 665
Huang, Yue 1676
Huang, Yunyi 228
Huang, Yun 546, 889
Huang, Zhiping 30
Hun, Tingting 949
Huo, Dian 1250
Huo, Zimin 1106
Ishihara, Daisuke 1320
Izhar 813
Jalili, Kiyumars 1629
Jeong, Yongbin 1423
Ji, Bowen 528, 1903
Jia, Lingjie 941
Jia, Qianqian 455
Jian, Zhu 1057
Jiang, A. Feng 412
Jiang, Boshi 953
Jiang, Fan 524
Jiang, Hongjie 64
Jiang, Jiaxin 140, 1136, 1328, 1554, 1700
Jiang, Kaili 1084
Jiang, Shulan 339
Jiang, Yong 474
Jiang, Yue 1398
Jiang, Z. 1623, 1660
Jiang, Zhuangde 47, 282, 299, 334, 513, 1157, 1362,
.... 1635, 1908
Jiangjiang, Liu 826
Jiao, Binbin 21
Jin, Chuanghong 459
Jin, Guoqing 518
Jin, Jie 1384
Jin, Shengxiao 889
Jin, Yiming 691

Jin, Yufeng 898, 949
Jin, Zhonghe 691
Jing, Weixuan 47
Kang, Guoguo 1441
Kang, Guoyi 140, 1035, 1136, 1328
Kang, Qiang 334
Kang, Tian 889
Ke, Xingxing 1390
Khademi, Sara 980, 1629
Kong, Lingli 919
Kong, Yanmei 21
Kuang, Yunbin 1888
Lee, Sang-Seok 251
Lee, Yi-Kuen 813
Lei, Cao Yang 1057
Lei, Hucheng 1611
Lei, Jiao Zong 1057
Lei, Shenghong 699, 1287
Lei, Yongjun 437
Li, Bolun 704
Li, Bo 1478
Li, Chengzhu 994
Li, Chunlong 386
Li, Defang 1664
Li, Dongsheng 463, 1370, 1419
Li, Dong 1461
Li, Haiwang 1745
Li, Haiyang 1441
Li, Huaan 749
Li, Hui 647
Li, Jiachen 518, 1611
Li, Jiayin 1686
Li, Jie 299
Li, Jing 1410
Li, Junru 1259, 1353
Li, Junshi 1839
Li, Keer 509
Li, Ke 174
Li, Lei 47
Li, Lili 1279
Li, Lin 647
Li, Liye 889
Li, Longfan 1737
Li, Longqiu 798
Li, Nanxing 1074
Li, Ning 1398
Li, Qingsong 1579, 1817
Li, Qi 845
Li, Shunqi 1539
Li, Tianyu 762
Li, Tie 30, 51, 676
Li, Wei 1250

Li, Wenwang 1035, 1328, 1554, 1700
Li, Wenxiang 1695
Li, Xiang 513
Li, Xiaojie 180, 459
Li, Xinghui 43
Li, Xiu-Yuan 433, 1432, 1605
Li, Xiuhan 941, 1714
Li, Xiuyan 1165
Li, Yadong 193, 1691
Li, Yang 474
Li, Yanjie 1088, 1165
Li, Yanqing 60
Li, Yansheng 392
Li, Yan 1175
Li, Yinqiao 271
Li, Yuan 255
Li, Yuhua 680
Li, Yunfei 1070
Li, Yuning 68
Li, Yunjia 1903
Li, Z T 311
Li, Zhihong 1839
Li, Zhikang 299, 1908
Li, Zhuo 1664
Li, Zixuan 299
Lian, Haishan 319
Lian, Zhenhui 1695
Liang, Boshen 1423
Liang, Bo 26, 762
Liang, Renrong 680
Liang, Tian 1813
Liang, Wenjie 1084
Liao, Jianhui 915
Lin, Guanzhou 546, 889
Lin, Jinghua 1554, 1700
Lin, Qijing 299, 513, 826, 1635, 1908
Lin, Waner 1096
Lin, Xiaoyang 1084
Lin, Yuanjing 1451
Lin, Yuhang 1732
Lin, Zude 1088, 1165
Liu, Bowen 1813
Liu, Changxin 749, 1853
Liu, Chenglong 845
Liu, Chen 998
Liu, Fenglin 953
Liu, Guochang 770
Liu, Haotian 1287
Liu, Huicong 60, 518, 1070
Liu, Huiliang 271, 695, 1291
Liu, J. Q. 959
Liu, Jianhao 749

Liu, Jieyu 931
Liu, Jingquan 1088, 1165
Liu, Juan 140
Liu, Jun 1611
Liu, Kangfu 1517, 1545
Liu, Ke 197, 1825
Liu, Lianqing 832, 1923
Liu, Lianqin 849
Liu, Lingling 236
Liu, Ling 1676, 1695
Liu, Lujiang 1897
Liu, Mei 919
Liu, Mengxia 1102
Liu, Mingwei 437
Liu, Mingxin 737
Liu, Ming 47
Liu, Min 498, 1106
Liu, Ruimin 491
Liu, Ruiwen 21
Liu, Shuangshuang 941, 1714
Liu, Wenjuan 656
Liu, Wenli 455
Liu, Wenxin 144
Liu, Xiangming 1611
Liu, Xingqi 30
Liu, Xin 1482
Liu, Yang 180, 459
Liu, Yan 931
Liu, Yaoping 898
Liu, Yifang 140, 491, 840, 1065, 1136
Liu, Yingming 1042
Liu, Yi 242
Liu, Yue 498
Liu, Yufei 386, 704
Liu, Yushuai 1545
Liu, Zewen 68, 174, 695
Liu, Ze 372
Liu, Zheng 949
Liu, Zhijian 378
Liu, Zichen 299
Liu, Ziqi 503, 509
Lou, Wen-Zhong 935
Lu, Chennan 469
Lu, Kuo 1817
Lu, Peimin 546, 889
Lu, Qifeng 1279
Lu, Qing 1332
Lu, Yulan 193, 1465, 1691, 1725
Luo, Anxin 184, 990
Luo, Guoxi 282, 513, 1362, 1908
Luo, Jian 872, 1415
Luo, Ningning 563, 571, 864

Luo, Rui	1535
Luo, Tao	1493
Luo, Yunyun	1362, 1908
Luo, Zebang	647
Luo, Zhifang	1255, 1523
Lv, Wenlong	1897
Lv, Yao	1611
Lv, Yuanjie	334
Ma, Binghe	872, 1415
Ma, Laihao	1250
Ma, Long	563
Ma, Shenhui	665
Ma, Shuang	832
Ma, Xin	1070
Ma, Yinji	731
Ma, Yintao	282
Ma, Yuanming	55, 1574
Ma, Zhen	1337
Ma, Zhipeng	691
Mao, Changhui	1142
Mao, Linna	1617
Mao, Tianjiao	1283
Mao, Xiyu	26, 762
Mao, Zhuofan	749
Matsunaga, Tadao	251
Mbarek, S. Ben	425
Mei, Deqing	1384, 1871
Mei, Ziqi	1535
Meng, Qinggang	1465
Meng, Qingwang	563, 571, 864
Meng, Yanfang	731
Meng, Yan	941, 1714
Meng, Zhen	197, 1825
Miao, Jiahao	242
Miao, Liming	148
Miao, Tongqiao	1410
Min, Huang	1057
Ming, Anjie	1142, 1436
Mizutani, Fumikazu	251
Mo, Deyun	319
Mori, Akinori	251
Mtui, Anaeli Elibariki	1676
Mu, Xiaojing	158
Murakami, Sunao	1320
Ni, Lvchao	1863
Nie, Dezhi	528
Nie, Weirong	699, 1287, 1358
Niklaus, Frank	1169
Niu, Bo	443
Niu, Gaoqiang	820
Niu, Zihao	1714
Ochoa, Manuel	64

Ono, Takahito	43, 670
Ouyang, Su	1718
Pan, Hongzhi	158
Pan, Mengyun	1923
Pan, Taisong	1617
Pan, Wenhao	762
Pan, Xinxiang	378
Pang, Jintao	463
Paquet, Louis	1423
Peng, Chun-Rui	1259
Peng, Chunrong	1611, 1893
Peng, Chunrui	1353
Peng, Hao	140
Peng, Jiabin	1398
Peng, Shi	826
Peng, Simin	1611
Peng, Zhengchun	1023, 1096, 1398
Pu, Jiangtao	47
Qi, Qi	1142
Qi, Wenjie	1813
Qi, Y	1660
Qi, Yonghong	1157
Qi, Yun	358
Qian, Jin	882
Qian, Lei	407
Qin, Guangyu	228
Qin, Lei	392
Qin, Lifeng	1718
Qiu, Weixiang	699
Qiu, Yong	1482
Qu, Guanghao	749
Qu, Mengjiao	1370, 1419
Ramegowda, Prakasha C.	1320
Rao, Jinjun	919
Rao, Wei	469
Rao, Zehong	1264
Rasadujjaman, Md.	417
Rashmikant	1320
Ren, Dahai	1535
Ren, Hangxu	26
Ren, Hongling	60
Ren, Ren	1611
Ren, Tian-Ling	680
Ren, Wan-Chun	1259
Ren, Wanchun	1353
Ren, Wei	47, 215
Ren, Wenjie	1042
Riaud, Antoine	809
Rong, Mingzhe	443
Ryutaro, M.	1660
Ryutaro, Maeda	513
San, Haisheng	232, 345, 399

Sang, Hongbo 1065
Sang, Jun-Jun 998
Sarro, Pasqualina M. 174
Shang, Zhengguo 386
Shao, Fangqin 1035
Shao, Lei 1297
Shao, Shuai 1255, 1523
Shao, Zungui 140, 1035, 1328
She, Xu 1813
Shen, Wenjiang 265, 358, 407
Shen, Yajing 293
Shen, Yigang 1126
Shi, Haotian 780, 1250
Shi, Huiyao 849
Shi, Jialin 849
Shimomura, Michio 251
Shu, Dongsheng 1175
Song, Ki-Young 1332
Song, Liguo 1853
Song, Yongxin 378
Song, Yu 392
Su, Weilin 919
Su, Wenqu 704
Su, Yi 647
Suetsugu, Ryotaro 1320
Sun, Changhe 704
Sun, Chengliang 656, 931
Sun, Daoheng 236, 1650
Sun, Fuqin 1279
Sun, Hongchun 1863
Sun, Jiamian 1745
Sun, Jiangkun 1579
Sun, Lining 60, 1070
Sun, Ming-Zhu 1018
Sun, Peiting 1539
Sun, Shengnan 904
Sun, Tiancheng 509
Sun, Yi 935
Sun, Yuyang 60
Tan, Haoyu 770
Tan, Yong 339
Tanaka, Yo 203, 1126
Tang, Fei 1291
Tang, Tianyi 1070
Tang, Xiaohe 1088
Tao, Jifang 1175, 1601
Tao, Kai 303, 528, 1903
Tao, Zhi 1745
Tavakkoli, Hadi 813
Tian, Bian 1635
Tian, He 680
Tian, Yuan 242

Tian, Yun 539
Tian, Zhao-Wu 1337
Tian, Zhong-Qun 1337
Tian, Zhou 980
Toda, Masaya 43
Tong, Daqiao 158
Tu, Er-Qi 433, 1432, 1605
Tu, Wenchang 236
Van Der Wijngaart, Wouter 1310
Wan, Hao 1482
Wan, Ji 148, 1686
Wan, Shijie 1023
Wang, Baolong 941, 1714
Wang, Bingnan 1664
Wang, Chengxiang 1888
Wang, Chenying 47
Wang, Dawei 469
Wang, Di 144
Wang, Dong F. 670, 1074
Wang, Donghui 1745
Wang, Fayang 158
Wang, Fei 184, 820, 990, 1042, 1410
Wang, Fengxia 60
Wang, Haobin 148
Wang, Hao 660, 980, 1629, 1853
Wang, Hongyan 1908
Wang, Jiaqiang 909
Wang, Jiong 1358
Wang, Jiyang 392
Wang, Junbo 193, 1465, 1691, 1725, 1813
Wang, Junyi 1441
Wang, K F 311
Wang, Kexin 215
Wang, L. C. 959
Wang, Lei 647, 656
Wang, Lingyun 840, 1065, 1897
Wang, Lukang 1478
Wang, Luming 563, 571
Wang, Lu 513
Wang, Meiqi 1714
Wang, Mengjie 498
Wang, Nan 524
Wang, Na 232, 345, 399
Wang, Ping 1482
Wang, Qilu 652
Wang, Qining 1839
Wang, Qi 55
Wang, Renxin 770
Wang, Ri 1303
Wang, Rong 670
Wang, Rui 1664
Wang, Shihang 1871

Wang, Siyuan 1853
Wang, Tianlu 832
Wang, Weidong 498, 1106
Wang, Wei 809, 898, 949
Wang, Wenlong 1264
Wang, Wenxue 832
Wang, Xiang 1035, 1328, 1554, 1700
Wang, Xiaohong 469
Wang, Xiao 498, 1106
Wang, Xinhe 1084
Wang, Xinjie 1358
Wang, Xinye 1664
Wang, Xinyu 1853
Wang, Xin 1664
Wang, Yancheng 1384, 1871
Wang, Yanfang 1451
Wang, Yang 1337
Wang, Yanrong 417
Wang, Yan 1601
Wang, Yaohong 904
Wang, Yiwei 528
Wang, Yongkang 539
Wang, Yuechao 832
Wang, Yuelin 30, 51, 676
Wang, Yuxi 1517
Wang, Zaichen 909
Wang, Zengbo 715
Wang, Zenghui 1410
Wang, Zhen 232, 345, 399
Wang, Zhihai 904
Wang, Zhiming 919
Wang, Zhongbao 804
Wang, Zhongyan 1839
Wang, Zihao 1279
Wang, Zilong 1611
Wang, Ziya 1096
Wang, Z 311
Wei, C. Yu 412
Wei, Feng 1436
Wei, Jiasi 255
Wei, Shuhua 417
Wei, X 1623, 1660
Wei, Xueyong 1157
Wei, Yaoming 303
Wen, Jian 443
Wen, Kaiqiang 994
Wen, Li 986
Wen, Ting 1410
Wen, Xiaolong 474, 1893
Wen, Zhiwei 931
Wu, Cheng-Feng 1804
Wu, Chen 334

Wu, Dezhi 804, 1650, 1655
Wu, Dong 437
Wu, Hanxiao 1745
Wu, Jin 303, 528, 1903
Wu, Junjie 255
Wu, Kai 1817
Wu, Kuixi 1897
Wu, Sen 378
Wu, Shuo 994
Wu, Song 1410
Wu, Tao 1255, 1517, 1523, 1545
Wu, Tianze 378
Wu, Tianzhun 953, 980, 1629
Wu, Wengang 546, 889
Wu, Xuezhong 1410, 1579, 1817, 1888
Wu, Yaming 704
Wu, Yifei 443
Wu, Yigen 804
Wu, Yi 443
Wu, Yongshun 1908
Wu, Yulie 1888
Wu, Zhengwei 1611
Wu, Zhigang 1390, 1394
Wu, Zhipeng 656
Wu, Zixuan 303, 1903
Xi, Xiang 1579
Xi, Y 959
Xi, Zhanwen 699, 1287, 1358
Xia, Cao 670, 1074
Xia, Hu 1718
Xia, Lingju 986
Xia, Shanhong 1303
Xiang, Chao 193
Xiang, Zehua 148
Xiao, Chiqian 1493
Xiao, Dingbang 1410, 1579, 1817, 1888
Xiao, Haifeng 563
Xiao, Shijin 1601
Xiao, Shuyu 392
Xiao, Xiu 660, 1676, 1695
Xie, Bo 1465, 1725
Xie, Chao 665
Xie, Dan 680
Xie, Guangming 1853
Xie, Jianbing 528, 1903
Xie, Jin 463, 1370, 1419
Xie, Yanbo 809
Xie, Yong 1303
Xie, Yucai 780, 1250
Xie, Yu 1655
Xie, Zihao 1419
Xin, Yi-Hang 433, 1432, 1605

Xing, Chenyang	1023
Xing, Fei	382
Xing, Xiaoxing	228
Xing, Yan	845, 882
Xiong, Chenyu	1303
Xiong, Heng	840
Xiong, Menghan	571, 864
Xiong, Xy	311
Xu, Bo	1410
Xu, Chao	1813
Xu, Chen	148
Xu, Dongxin	1374
Xu, Gaobin	55, 1574
Xu, Guangyuan	731
Xu, Hantao	1337
Xu, Jia	242
Xu, L.	1623
Xu, Minyi	660, 1539, 1676, 1695, 1853
Xu, Na	699
Xu, Peng	1853
Xu, Shiyi	26, 762
Xu, Tiantong	1745
Xu, Wei	941, 1502, 1714
Xu, Xuankai	1502
Xu, Yaohua	1436
Xu, Yongfu	1700
Xu, Yuhao	1303
Xu, Zhiwei	1250
Xue, Gaopeng	43
Yalikun, Yaxiaer	203, 1126
Yan, Bing	715
Yan, Huangping	1732
Yan, Jiang	417
Yan, Xin	491
Yan, Yuchao	872
Yan, Yuefei	334
Yang, B.	959
Yang, Cheng	503
Yang, Dengfei	1370
Yang, Fuhua	455
Yang, Hui	711, 715
Yang, Jiayi	941, 1714
Yang, Jing	963
Yang, Jinling	455
Yang, Junfeng	986
Yang, Pengfei	1893
Yang, Ping	299, 513, 1908
Yang, Q.	1623
Yang, Quan	919
Yang, Suhui	1664
Yang, W H	311
Yang, Xiaokang	386

Yang, Xuanlin	30
Yang, Yinhuan	319
Yang, Yi	676, 680
Yang, Yongliang	1923
Yang, Yuanyuan	293, 358
Yang, Yuqian	1283
Yang, Yu	1478
Yang, Zhan	1560
Yang, Zhaohui	1574
Yao, Kun	1635
Yao, Yuan	271
Ye, Li	174
Ye, Xiongying	1291
Ye, Xuesong	26, 762
Ye, Yuxin	21
Ye, Zhi	180, 459
Yin, Zhuhui	1398
Yoshida, Yoshikazu	251
You, Minmin	1088, 1165
You, Shanshan	236
You, Zheng	1535
Younis, M. I.	425
Yu, Duli	228
Yu, Hang	1650
Yu, Hongyong	1676, 1695
Yu, Hongyu	1589
Yu, Huijun	265, 407
Yu, Jiahao	1903
Yu, Jie	1725
Yu, Lihang	21
Yu, Lingke	236
Yu, Miao	1110
Yu, Mingzhi	282
Yu, Peng	849
Yu, Sheng	1579
Yu, Siqi	1589
Yu, Xiaomei	242
Yu, Yue	1283
Yu, Zhoubin	459
Yu, Zitong	715
Yuan, Jiawei	299
Yuan, Jing	1804
Yuan, Quan	455
Yuan, Weizheng	872, 1415
Yuan, Ziyi	378
Zeng, Haozhe	1903
Zhan, Dongping	1337
Zhang, B. Wen	412
Zhang, Bing-Bin	1804
Zhang, Bo	474, 1893
Zhang, Cheng	1441
Zhang, Dapeng	437

Zhang, Guannan.....................652
Zhang, Guoqi.....................174
Zhang, H......................1660
Zhang, Hainan.....................680
Zhang, Haixia.....................148, 1686
Zhang, Hanyuan.....................652
Zhang, Hao.....................1110
Zhang, Hongpeng.....................30, 780, 1250
Zhang, Hongze.....................1110
Zhang, Huimin.....................687, 1867
Zhang, Jian.....................158, 528
Zhang, Jiaona.....................665
Zhang, Jing.....................417
Zhang, Jinming.....................417
Zhang, Jinying.....................1664
Zhang, Kenan.....................546, 889
Zhang, Li.....................382
Zhang, Lue.....................1560
Zhang, Min.....................665
Zhang, Nan.....................21
Zhang, Pengcheng.....................711, 715
Zhang, Peng.....................1539
Zhang, Qiankun.....................1362, 1908
Zhang, Qian.....................463, 1419
Zhang, Qiqi.....................1676, 1695
Zhang, Ran.....................372
Zhang, Shanshan.....................26
Zhang, Shaoxun.....................1142
Zhang, Shuo.....................1394
Zhang, Tengfei.....................691
Zhang, Ting.....................1279
Zhang, Weiguan.....................1023
Zhang, Wei.....................518, 687, 915, 1611, 1867
Zhang, Wen-Ming.....................433, 1297, 1432, 1605
Zhang, Wendong.....................770
Zhang, X......................1660
Zhang, Xiaopeng.....................1157
Zhang, Xiaosheng.....................737
Zhang, Xiao.....................1436
Zhang, Xingcheng.....................197
Zhang, Yang.....................68
Zhang, Yating.....................1867
Zhang, Yaxin.....................47
Zhang, Yijun.....................47
Zhang, Yongmeng.....................1579, 1817, 1888
Zhang, Yuewen.....................1539
Zhang, Yulong.....................695, 1065
Zhang, Yuwei.....................780
Zhang, Yu.....................1415
Zhang, Zhaohao.....................68
Zhang, Zhimin.....................563, 571, 864
Zhang, Zhitong.....................1839

Zhang, Zhouwei.....................1611
Zhao, Changhui.....................820, 1042
Zhao, Cong.....................749
Zhao, Hongfa.....................1676, 1695
Zhao, Junhao.....................1539
Zhao, Libo.....................282, 299, 513, 1157, 1362, 1635, 1908
Zhao, Luo.....................1441
Zhao, Minghui.....................1157
Zhao, Na.....................1635
Zhao, Qiancheng.....................1102, 1461
Zhao, Weisheng.....................1084
Zhao, Wenyu.....................1096
Zhao, Xiaodong.....................915
Zhao, Xiaojin.....................1502
Zhao, Xin.....................1018
Zhao, Xu.....................813
Zhao, Yifan.....................524
Zhao, Yihe.....................299
Zhao, Yongmin.....................1142
Zhao, You.....................1478
Zhao, Yue-Cen.....................935
Zhao, Yulong.....................215, 1405, 1478
Zhao, Yunpeng.....................660
Zhao, Yupeng.....................1142
Zhao, Zhe.....................528
Zhaojun, Liu.....................826
Zheng, Fengjie.....................1611
Zheng, Gaofeng.....................140, 491, 1035, 1136, 1328, 1554, 1700
Zheng, Guowen.....................1074
Zheng, Jianyi.....................1136
Zheng, Jiufu.....................1502
Zheng, Renrong.....................232, 345, 399
Zheng, Xudong.....................691
Zhong, Anxiang.....................1813
Zhong, Hongsheng.....................737
Zhongkai, Zhang.....................826
Zhou, Dekai.....................798
Zhou, Lingfei.....................509
Zhou, Peng.....................265
Zhou, Rui.....................1732, 1737
Zhou, Wei.....................1493, 1655
Zhou, Wenbo.....................949
Zhou, Wenke.....................1362
Zhou, Wenli.....................994
Zhou, Xin.....................1817
Zhou, Zhenghui.....................749
Zhou, Zilong.....................1264
Zhu, Jiankai.....................1410
Zhu, Jian.....................1110
Zhu, Jiaqi.....................1390
Zhu, Jiaxin.....................953
Zhu, Jia.....................546

Zhu, Kaiyun 1745
Zhu, Ke 1370
Zhu, Liangquan 1635
Zhu, Wei 656
Zhu, Xin 180, 459
Zhu, Yinfang 455
Zhu, Yingmin 1106
Zhu, Yi 1732
Zhu, Zhengfang 647
Zhu, Zhihao 1096
Zhuang, Liujing 1482
Zhuang, Rencheng 798
Zhuang, Yi 990
Zhuangde, Jiang 826
Zhuo, Ming 1817
Zou, X D 311
Zou, Yongjiu 1539
Zuo, Xiuli 60
Zuo, Xusheng 1539

IEEE
445 Hoes Lane
Piscataway, NJ 08854-4141

ISBN 978-1-6654-3008-1

2021 IEEE 16th International Conference on Nano/Micro Engineered and Molecular Systems (NEMS 2021)

Xiamen, China
25 – 29 April 2021

Pages 947-1926

IEEE Catalog Number: CFP21NME-POD
ISBN: 978-1-6654-3008-1

2021 IEEE 16th International Conference on Nano/Micro Engineered and Molecular Systems (NEMS 2021)

Xiamen, China
25 – 29 April 2021

Pages 947-1926

IEEE Catalog Number: CFP21NME-POD
ISBN: 978-1-6654-3008-1

**Copyright © 2021 by the Institute of Electrical and Electronics Engineers, Inc.
All Rights Reserved**

Copyright and Reprint Permissions: Abstracting is permitted with credit to the source. Libraries are permitted to photocopy beyond the limit of U.S. copyright law for private use of patrons those articles in this volume that carry a code at the bottom of the first page, provided the per-copy fee indicated in the code is paid through Copyright Clearance Center, 222 Rosewood Drive, Danvers, MA 01923.

For other copying, reprint or republication permission, write to IEEE Copyrights Manager, IEEE Service Center, 445 Hoes Lane, Piscataway, NJ 08854. All rights reserved.

*** *This is a print representation of what appears in the IEEE Digital Library. Some format issues inherent in the e-media version may also appear in this print version.*

IEEE Catalog Number:	CFP21NME-POD
ISBN (Print-On-Demand):	978-1-6654-3008-1
ISBN (Online):	978-1-6654-1941-3
ISSN:	2474-3747

Additional Copies of This Publication Are Available From:

Curran Associates, Inc
57 Morehouse Lane
Red Hook, NY 12571 USA
Phone: (845) 758-0400
Fax: (845) 758-2633
E-mail: curran@proceedings.com
Web: www.proceedings.com

Table of Contents

PL1: Plenary Talk I

Bioinspired Super-wettability System and Beyond
——Quantum-confined Superfluid: Energy Conversion, Chemical Reaction and Biological
Information Transfer
Lei JIANG

1

PL2: Plenary Talk II

Atomically Precise Chemical, Physical, Electronic, and Spin Contacts
Paul S. WEISS

3

PL3: Plenary Talk III

Pushing Laser Precision Engineering from Micro-scale to Nano-scale: Progress, Challenges and
Opportunities
Minghui HONG

5

PL4: Plenary Talk IV

From Supramolecular to Adaptive Chemistry – Contributions to Nanoscience and Technology
Jean-Marie LEHN

7

PL5: Plenary Talk V

Semiconductor Nanowires for Optoelectronics Applications
Chennupati JAGADISH

9

PL6: Plenary Talk VI

Nanomaterials of High-Surface Energy for Efficient Electrochemical Energy Conversion and
Storage
Shi-Gang SUN

11

PL7: Plenary Talk VII

Molecules in Motion: From Biology to Chemistry
Jean-Pierre SAUVAGE

13

PL8: Plenary Talk VIII

Nano-calorimetry: A New Tool for Materials Development
Joost J. VLASSAK

15

PL9: Plenary Talk IX

Transformative Impact of Printable Solar Cells for Meeting Next-Generation Energy Demands
Alex JEN

17

PL10: Plenary Talk X

Van der Waals Heterostructures
Kostya NOVOSELOV

19

CH1: Chihming Ho Award Session

Integrated Electrical Test Vehicle Co-designed with Microfluidics for Evaluating the Performance of
Embedded Cooling
Yuxin Ye, Nan Zhang, Lihang Yu, Bo Cong, Ruiwen Liu, Yanmei Kong and Binbin Jiao

21

Frequency Characteristics of Microfluidic Single Gate Oscillator
Zhou Zhou, Manman Xu, Hai Wang, Kunpeng Zhang, Daoheng Sun

N/A

Minimal-Invasive Levodopa Sensing Based on Differential Amperometry Microneedle Electrodes Decorated with Spike-like Au Nanoparticles   26
Hangxu Ren, Xiyu Mao, Shanshan Zhang, Yu Cai, Lu Fang, Xuesong Ye, Bo Liang

Silicon Nanowire Array Sensor for Highly Sensitive and Selective Detection of Nerve Agent Simulant Vapor via surface hydroxyl groups   30
Xingqi Liu, Hongpeng Zhang, Zhiping Huang, Xuanlin Yang, Shixing Chen, Yuelin Wang, Tie Li, Zhenxing Cheng

Magnetic Force Enabled Plant Seed Levitation to Simulate Microgravity Environment on A Microfluidic Chip   N/A
Jing Du, Lin Zeng, Zitong Yu, Sihui Chen, Xi Chen, Yi Zhang, Hui Yang

WK1: Wen H Ko Memorial Session

MEMS/SEMI Ecosystem in China --2021   N/A
Lucy HUANG

Forever Scaling—Pushing the Limit of Miniscule Sensors and Actuators   N/A
Zenghui WANG

Forever Pioneer- In Memory of Prof. Wen Ko   N/A
Tie LI

BC1: Best Conference Paper Session

Chip-level-microassembly Comb-drive XYZ-microstage with Large Displacements and Low Crosstalk   43
Gaopeng Xue, Masaya Toda, Xinghui Li, Takahito Ono

Effect of the Different Substrates and the Film Thickness on the Surface Roughness of Step Structure   47
Chenying Wang, Jiangtao Pu, Lei Li, Weixuan Jing, Yijun Zhang, Yaxin Zhang, Feng Han, Ming Liu, Wei Ren, Zhuangde Jiang

Efficient Infrared–Thermal-Electric Conversion with Textured Dielectric Film   51
Yunqian He, Yuelin Wang, Tie Li

Inertial Focusing Chip Based on Superposed Secondary Flows   55
Jianguo Feng, Qi Wang, Yuanming Ma, Honglong Chang, Gaobin Xu

A Sensorised Forcep Based on Piezoresistive Force Sensor for Robotic-assisted Minimally Invasive Surgery   60
Cheng Hou, Jiangjun Geng, Yuyang Sun, Tao Chen, Fengxia Wang, Hongliang Ren, Xiuli Zuo, Yanqing Li, Huicong Liu, Lining Sun

In situ Laser-assisted Micromachining of Environmentally-responsive Hydrogel Films   64
Hongjie Jiang, Manuel Ochoa

Three-dimensional Graphene FETs for pH Detection   68
Tao Deng, Yuning Li, Zhaohao Zhang, Yang Zhang, Zewen Liu

IS1: Advanced Micro/nano Fabrication, Materials and Applications

Surface and Bulk Micro/Nanomachining of Polymers   N/A
Junshan LIU

Flexible Healthcare Sensors   N/A
Xuewen WANG

Hexagonal Boron Nitride Phononic Crystal Waveguides for Classical and Quantum Signal Transduction   N/A
Yanan WANG

Micro-Fluidic Heat Dissipation Technology Based on Multi-elements Compatible MEMS Process   N/A
Yunna SUN

Accurate Three-dimensional Physical Simulations of Micro/Nano Fabrication Processes   N/A
Zaifa ZHOU

Micro-system based Multimodality Imaging System for Quantitative Molecular Imaging and Precision Health   N/A
Zhen QIU

IS2: Thermal management in MicroNano Systems

Thermal Transport of Mechanically Deformed 2D Materials
Baoxing XU   N/A

Thermal Transport Properties of Metals and Metallic Nanostructures
Hua BAO   N/A

Rational Design of Conductive Polymers for Flexible Thermoelectric Device
Hui LI   N/A

Layered Two-Dimensional Materials Showing Two Extremes of Heat Transport
Hyejin JANG   N/A

Layer Dependent Thermal Transport Properties of 2D WSe2
Xian (Annie) ZHANG   N/A

Temperature-Microstructure-Properties Relationships During Al-Sn-Al Thermo-Compression Wafer   N/A
Bonding
Zhiyuan ZHU

IS3: Resonant Micro/Nanoelectromechanical Systems

High Resolution Micro-sensors Based On Coupled Resonators/Modes
Chun Zhao   N/A

Multimode Resonant Micromechanical Systems in Liquid for
Biophysical Studies of Cancer Cells
Hao Jia   N/A

Sensing with Serial or Paralell Micromechanical Resonators
Honglong Chang   N/A

Atomically-Thin MoS_2 Nanoelectromechanical Resonators
Rui Yang   N/A

Nonlinear MEMS: fundamentals and applications
Xueyong Wei   N/A

Ultra-Wide Bandgap β-Ga2O3 Resonant Nanoelectromechanical Systems (NEMS)
Xuqian Zheng   N/A

IS4: Self-powered Micro/Nano Systems

Tribotronics for Active Mechanosensation and Self-Powered Microsystems
Chi ZHANG   N/A

Pulsed Triboelectric Nanogenerator and Self-Powered Sensing System
Gang CHENG   N/A

High Efficient Vibration Energy Harvesting for Self-powered Machine Monitoring Application
Jian HE   N/A

Nanogenerators for Self-powered Wearable Physical Monitoring Systems
Wei TANG   N/A

Triboelectric polymers and electrostatic power source
Xiangyu CHEN   N/A

Self-Powered Triboelectric Pressure Sensors
Zhen WEN   N/A

IS5: Bioelectronic Devices and Systems

Development of Self-powered Sensors towards Machine-learning Enabled Smart Systems
Chengkuo LEE   N/A

Interactive Neuromorphic Synaptic Devices and Systems
Qijun SUN   N/A

Minimally Invasive Devices for Biomedical Applications
Xi XIE   N/A

Flexible bioelectronic devices for monitoring and regulating physiological activities Xian HUANG	N/A
Shape-adaptable Biomedical Devices Xuemin DU	N/A
Self-Powered Medical Electronics Zhou Li	N/A

IS6: Non-linear Micro/Nanoelectromechanical Systems

Nonlinear Interaction in Coupled Cantilevers and Its Sensing Applications Cao XIA	N/A
Insight of Sensing Mechanisms for Conjugated Polymer Based Gas Sensors Jian SONG	N/A
Ultrasensitive mass sensing utilizing nonlinear mode localization of MEMS curved beams with distributed electrodes Jian ZHAO	N/A
Nonlinear Broadband Energy Harvesting and Applications Junyi CAO	N/A
Dynamic-Based Micro and Nano Devices and Phenomena Mohammad I. Younis	N/A
Nonlinear dynamics of coupled MEMS resonators and its application Ronghua HUAN	N/A

S1: Micro/Nano/Molecular Fabrication I

Opportunities for Single-Molecule Electronics: A Ten-Year Perspective Wenjing HONG	138
Synchronized Surface Modification of TiO2 Composite Nanofiber Through Core-Shell Electrospinning Hao Peng, Guoyi Kang, Jiaxin Jiang, Zungui Shao, Juan Liu, Yifang Liu, gaofeng zheng	140
Precursor-Based ZnO Nano Inks for Printed Electronics Wenxin Liu, Hao Dong, Di Wang, Xing Chen	144
Efficient Manufacturing of Microdome Array for Advanced Electronic and Optical Devices Zehua Xiang, Haobin Wang, Hang Guo, Ji Wan, Chen Xu, Liming Miao, Mengdi Han, Haixia Zhang	148
Solvent-free Nanofabrication Based on Ice-assisted Electron-beam Lithography Yu Hong, Ding Zhao, Min Qiu	N/A
Ice-Assisted Electron-Beam Lithography for 3D Nanofabrication Ding Zhao, Yu Hong, Min Qiu	N/A

S2: Energy Conversion and Storage I

Conformal Bioelectronic Interfaces Xiaodong CHEN	156
Broadband Piezoelectric Harvester Utilizing Three-wafer Bonding for Self-powered Wireless Sensor System Xin Chen, Daqiao Tong, Jian Zhang, Lingxiao Gao, Fayang Wang, Hongzhi Pan, Liang Cao, Xiaojing Mu	158
Design and Fabrication of a Metal-silicon Actuator with Large Displacement and Low Voltage Ying Bu, Jin Xie, Kaiquan Li, Zhuang Xiong, Bing Tang, Qi Tao, Jun Dai	N/A
Flexible Single Electrode Triboelectric Nanogenerator Yuchi Liu, Rami Ghannam, Xiaosheng Zhang	N/A
Development of Closed-Loop Soft Actuators with Self-Sensing Capability for Advanced Wearable Assistive Robotics Hwajoong Kim, Jaehong Lee	N/A
Wearable Triboelectric Nanogenerator based on the Conductive Textile with Polyaniline Grafted Yun Tian, Chuanhong Zhou, Zejue Yu, Zhangwei Liu, Chenyang Xing, Bo Meng, Zhengchun Peng	N/A

S3: Micro/Nano Pores and Channels

Organic Molecular Sieve Membranes 170
Zhongyi JIANG

Regulatory Mechanism of Ionic/Molecular Transport Behaviors in Nanoscale Channels N/A
Yaqi Hou, Xu Hou

Fabrication of Nanoslits with <111> Etching TSWE Method 174
Hao Hong, Li Ye, Ke Li, Pasqualina M Sarro, Guoqi Zhang, Zewen Liu

Confined Colloids Gating System for Tunable Fluid Transport N/A
Zhizhi Sheng, Mengchuang Zhang, Jing Liu, Xu Hou

Characterization of ITO-SiNx Nanopores for Single-Biomolecular Sensing 180
Xin Zhu, Chaoming Gu, Xiaojie Li, Zhi Ye, Zhen Cao, Yang Liu

Mechanical Energy Harvester for Smart Shared Bicycle Application 184
Junlong Huang, Chen Bao, Anxin Luo, Fei Wang

S4: MEMS/NEMS I

Triboelectric Nanogenerator for Powering Body-implantable Electronics 189
Sang-Woo KIM

Analysis of Vibration Modes in TPoS Disk Resonators N/A
Feilong Li, Cheng Tu, Jiannan Chen, XIAOSHENG ZHANG

A Resonant Differential Pressure Sensor Based on Bulk Silicon Technology 193
Chao Cheng, Yadong Li, Yulan Lu, Chao Xiang, Junbo Wang, Deyong Chen, Jian Chen

An Ultralow-Ripple Polarization Voltage Generator based on High- Voltage Bandgap Reference for 197
MEMS Gyroscopes
Hua Chen, Zhen Meng, Xingcheng Zhang, Ke Liu

Electrochemical Micromachining of Copper Workpiece with Ultrashort Voltage Pulses N/A
Zhen Ma, Lianhuan Han, Hantao Xu, Bingqian Du, Yang Wang, Dongping Zhan

Development of thin glass-based biconvex microlens via thermal expansion 203
Yusufu Aishan

Lubricated Non-immunogenic Neural Probes for Lifelong use brain-machine interfaces N/A
Yeontaek Lee, Hyogeun Shin, Il-Joo Cho, Jungmok Seo

S5: Micro/Nano Fluidic I

Nanofluidic devices for Biosensing and Energy Conversion 209
Xinghua XIA

Inner Surface Design of Microfluidic channels for Microscale Flow Control N/A
Shuli Wang, Xu Hou

Carbon Nanotube-Based Ionic Diodes: Design and Mechanism N/A
Ran Peng, Yueyue Pan, Biwu Liu, Zhi Li, Peng Pan, Shuailong Zhang, Zhen Qin, Aaron R. Wheeler,
Shirley Tang, Xinyu Liu

Research on a MEMS Detonated Device with Built-in Safety and Arming Device 215
Kexin Wang, Tengjiang Hu, Yulong Zhao, Wei Ren

Molecular Dynamics Simulation on Directional Wetting Behavior of a Nano Droplet on Titanium N/A
Oxides Surface
Chenhua Liu, Xiangmeng Li, Xijing Zhu

A Human Cornea-on-a-Chip for the Study of Epithelial Wound Healing by Extracellular Vesicles N/A
Zitong Yu, Rui Hao, Jing Du, Xi Chen, Yi Zhang, Wei Li, Zhongze Gu, Hui Yang

Flow-through Electroporation Using Silver-PDMS Based 3D Sidewall Microelectrodes 228
Yao Cai, Yunyi Huang, Guangyu Qin, Duli Yu, Xiaoxing Xing

PS1: Poster Session I

Self-driven photoelectrochemical UV-visible photodetectors using $ZrO_2@TiO_2$ core-shell nanorod 232
arrays modified with single-walled carbon nanotubes
Zhen Wang, Renrong Zheng, Na Wang, Haisheng San

Electronic Skin Based on Electrospun PVDF Nanofibers for Slipping Detection *Lingke Yu, Wenchang Tu, Shanshan You, Lingling Liu, Daoheng Sun*	236
A Flexible Hybridized Electromagnetic-Triboelectric Nanogenerator and its application for 3D Trajectory Sensing *Ji Wan, Haobin Wang, Liming Miao, Hang Guo, Chen Xu, Hai Xia (Alice) Zhang*	N/A
A Simple Bandwidth Broadening Method of Terahertz Metamaterial Absorber by Partially Removing The Dielectric Layer *Jia Xu, Jiahao Miao, Yi Liu, Yuan Tian, Xiaomei Yu*	242
Tough Self-Healing Polymer Encapsulated Stretchable Conductive Fibers for Systematical Integration of Wearable Electronics *Chaebeen Kwon, Taeyoon Lee*	N/A
Miniaturized Piezoresistive Sensors with Wide Working Range for Wearable Human Machine Interfaces *Hongcheng Xu, Ling Duan, Libo Gao*	N/A
Water Drop Detection Contained in Lubricant Oil for Submersible Pump Fault Diagnosis *Fumikazu Mizutani, Michio Shimimura, Akinori Mori, Yoshikazu Yoshida, Yoshihiro Hosokawa, Tadao Matsunaga, Sang-Seok Lee*	251
Lateral characteristic calibration of an atomic force microscope using a Cr atomic deposition grating *Junjie Wu, Xiaoyu Cai, Yuan Li, Yunxia Fu, Jiasi Wei*	255
A Miniaturized PM2.5 Concentration Monitoring System Based on QCM Sensor and Optimized Structure Virtual Impactor *Yong Wang, Dongyang Chen, Qian Zhang, Yinshen Wang, Jin Xie*	N/A
Preparation of Porous SnO2 and its Ethanol Detection Performance in MEMS Structure Gas Sensor *Xiao Zhang*	N/A
High Electrochemical Activity of Oxygen-Doped Graphene Sheets Embedded Carbon Film *Yuanyuan Cao, Kaikai Sun, Haohua Zhong, Dongfeng Diao*	N/A
Fast-scan MOEMS Mirror for HD Laser Projection Applications *Huijun Yu, Wenjiang Shen, Peng Zhou*	265
Modeling and Optimization of VLF Piezoelectric Antenna Towards Lower Frequency *Huiliang Liu, Yao Chu, Weiguo Hou, Yinqiao Li, Yuan Yao, Yaxing Cai*	271
Electrically Induced Wire-forming 3D Printing Technology of Flexible Liquid Metal *Xinpeng Wang, Liangtao Li, Yang Wang, Liang Hu*	N/A
Amorphous carbon film based weak pressure sensor with ultrathin sensitive structure *Qi Zhang, Yulong Zhao*	N/A
Bacterial Outer Membrane Vesicles Presenting Programmed Death 1 for Improved Cancer Immunotherapy via Immune Activation and Checkpoint Inhibition *Yao Li*	N/A
Micro-fabricated alkali vapor cells for atomic spin gyroscope study *Yao Chen, mingzhi Yu, Yintao Ma, Guoxi Luo, Zhuangde Jiang, Yu Bai, Libo Zhao*	282
Dopamine modified nanopore for ultra-trace level detection *weijun li, xu hou*	N/A
Nano/Micro Wearable Photo-sensor System Based on Novel Carbon Film *xi zhang, Dongfeng Diao*	N/A
Electron-Induced Perpendicular Graphene Sheets Embedded Porous Carbon Film for Flexible Touch Sensors *Sicheng Chen, Lei Yang*	N/A
Graphene Oxide Nanosheet Assisted Porous Chitosan Sponge for Hemostatic Applications *Yuanyuan YANG, Yajing Shen*	293
High-Performance Flexible Supercapacitors Based on Nitrogen-Doped Graphene Fiber Electrodes *Feng Han, Weixuan Jing, Bian Tian, Qijing Lin, Chenying Wang, Libo Zhao, Ping Yang, Zhuangde Jiang*	N/A

Design and Simulation of a Wide-Bandwidth CMUTs Array with Dual-Mixed radii and Multi Operating Modes — 299
Zichen Liu, Libo Zhao, Yihe Zhao, Jie Li, Zhikang Li, Yu Bai, Ping Yang, Qijing Lin, Zixuan Li, Jiawei Yuan, Zhuangde Jiang

Transparent, Anti-Freezing Hydrogels for Ultrasensitive Temperature and Strain Sensor Based on A Thin-Film Structure — 303
Zixuan Wu, Haojun Ding, Yaoming Wei, Kai Tao, Jin Wu

Uniformly arrayed carbon nanofibers down to several nanometers — N/A
Jufeng Deng, Chong Liu, Madou Marc

Self-Powered Transparent Stretchable 3D Motion Sensor — N/A
Hang Guo, Haobin Wang, Zehua Xiang, Chen Xu, Wan Ji, Liming Miao, Jiayin Li, Mengdi Han, Haixia Zhang

ENHANCING SENSITIVITY USING ELECTROSTATIC SPRING IN COUPLING MODE-LOCALIZED MEMS ACCELEROMETER — 311
Zheng Wang, XingYin Xiong, KunFeng Wang, WuHao Yang, ZhiTian Li, XuDong Zou

Self-powered Delta-Parallel-based Interface for Diversified Control Applications for 2/3D Control in Virtual/Real Space — N/A
Cheng Hou, Jiangjun Geng, Tao Chen, Huicong Liu, Lining Sun

Microchip based Gigahertz Acoustic Streaming induced Cellular Internalization of Gold Nanorods — N/A
Shan He, Xiaoyu Wu, Wenjun Li, Wei Pang, Xuexin Duan, Yanyan Wang

Direct-write 3D printed self-supporting flexible microstructure devices — 319
cuiwen hu, haifeng chen, deyun mo, haishan lian, yinhuan yang, xiaojun chen

Biomimetic Curvatures Controlled Fluid Overflow System — N/A
Zhichao Dong

A Flexible Pain Sensor Based on PDMS-AgNWs — N/A
Chen Xu, Liming Miao, Haobin Wang, Hang Guo, Zehua Xiang, Haixia Zhang

Liquid gating membranes for multiphase separation — N/A
Shijie Yu

A bionic fish with 4 IPMC pectoral fins — N/A
zhao chun, Wang Yanjie, Tang Gangqiang, wang jiale, zhu denglin, Luo Minzhou

Ohmic Contact Characteristics of Silicon Carbide-based MEMS Devices — 334
Chen Wu

Facile fabrication of NiO wrinkle micro/nanostructures and the application of enzyme-free glucose sensors — 339
Shulan Jiang, Yong Tan

Hard protective transparent coatings for flexible electronics — N/A
Oleksiy Penkov, Mahdi Khadem

High-efficient Electrochemical Betavoltaic Cell Using ZrO2 Modified TiO2 Nanorod Arrays — 345
Renrong Zheng, Zhen Wang, Na Wang, Zan Ding, Haisheng San

Design and Fabrication of PDMS Thin Film Electromagnetic Actuator — N/A
Hongguang Lu

Elastomers with ultrafast low temperature self-healing rate and low adhesion to ice — N/A
Yizhi Zhuo

Spoof Plasmon Surfaces for Terahertz Sensing — N/A
Yi Huang, Shuncong Zhong, Tingling Lin, Yujie Zhong

A Flexible Pressure/Flow Sensor Based on Graphene Piezoresistivity — N/A
Zihao Dong, Yonggang Jiang, Deyuan Zhang

Hypoxia/pH dual-responsive nitroimidazole-modified chitosan/rose bengal derivative nanoparticles for enhanced photodynamic anticancer therapy — N/A
xudong li, Yu Gao

Research on Key Technology of Novel RF MEMS Switch Designing and Fabricating — 358
Yun Qi, Yuanyuan Yang, Wenjiang Shen

Design and Analysis of Dual Substrate RF MEMS Capacitive Switch with Low Actuation Voltage — N/A
Yuanyuan Yang, Yun Qi, Wenjiang Shen

A pulse diagnosis instrument based on digital microfluidics system to imitate TCM
Shouju Yao, Hang Jin, Bin Qiu, Wenchang Tu, yike zhou, Shanshan You, Tianhao Wu, lingling liu, gonghan he, Daoheng Sun

N/A

Exploring the Therapeutic Effect of Sonodynamic Therapy in Cancer Cells under Different Ultrasonic Parameters and Treatment Conditions
Yilin Zheng, Jun Wang, Yu Gao

N/A

Stretchable and Flexible Fiber Electrode for Measuring Neural Signal with Au Nanoparticle Conductive Networks
Chihyeong Won, Taeyoon Lee

N/A

Broadband linear-to-circular polarization converter based on ultrathin metal nano-grating
Yuanyi Fan, Ran Zhang, Ze Liu, Chuanlong Guan, Jinkui Chu

372

Skin Inspired Humidity and Pressure Sensor with Wrinkle-on-Sponge Structure
Liming Miao, Ji Wan, Chen Xu, Hang Guo, Haobin Wang, Zehua Xiang, Jiayin Li, Mengdi Han, Haixia Zhang

N/A

Probing zeta potential of glass in electrolyte solutions by colloidal probe technique
Zhijian Liu, Jiaming Gao, Tianze Wu, Sen Wu, Zixiao Fan, Ziyi Yuan, Yongxin Song, Xinxiang Pan

378

High-Resolution Measurement Method Based on Array FOV and Coded Apertures
Li Zhang, Fei Xing

382

Different Effects of Mass and Damping on Performance of Vibration and Wind Energy Harvesters
Xiaokang Yang, Xuefeng He, Zhengguo Shang, Hui Huang, Chunlong Li, Yufei Liu

386

Nanocomposites Modified Platinum Wire Electrode for Detection of 17β-Estradiol Utilizing the Conformational Changes of Aptamers
Fanli Kong, Jinping Luo, Tao Ming, Shuai Sun, Yu Xing, Yan Cheng, Shihong Xu, Hongyan Jin, Xinxia Cai

N/A

In-situ deposited ion-imprinted polymers for electrochemical detection of trace cadmium in water
Shuyu Xiao, Jingfang Hu, Jiyang Wang, Yu Song, Yansheng Li, Guowei Gao, Lei Qin

392

Fabrication of Three-dimensional Si based Electrodes for Integrated Li Ion Microbatteries
Chuang Yue

N/A

Electrochemical betavoltaic cell using black TiO2 nanotube arrays modified with single-walled carbon nanotubes
Na Wang, Renrong Zheng, Zhen Wang, Jiang Chen, Haisheng San

399

An Acoustic Microrobot Control System with Vision Feedback
Ying Wei, Hui Shen, Cong Zhao, Huan Ou and Xiaolong Lu

N/A

Electrochemical Paper Aptasensor Based on Biotin-Streptavidin System for Label Free Detection of 17β-Estradiol
Tao Ming, Jinping Luo, Juntao Liu, Shuai Sun, Yu Xing, Fanli Kong, Yan Cheng, Hongyan Jin, Xinxia Cai

N/A

Temperature Compensation for MOEMS Micromirror with Piezoresistive Angle Sensor
Lei Qian, Huijun Yu, Jie Hu, Wenjiang Shen

407

Nanocrystalline Composed SnO2 Inverse Opal for Highly Sensitive Acetone Gas Sensor at ppb-level
Feng Jiang, Wen Zhang, Yu Wei

412

Study of the light emission from Eu3+ doped nanoporous organosilicate films
JinMing Zhang, Jing Zhang, Yanrong Wang, Md. Rasadujjaman, Shuhua Wei, Jiang Yan

417

A Highly Sensitive Origami-paper-based Aptasensor For Detection Of Programmed Cell Death Protein 1 (PD-1)
Shuai Sun, Tao Ming, Jinping Luo, Juntao Liu, Yu Xing, Fanli Kong, Xinxia Cai

N/A

Investigation on Electrostatically Actuated Micromachined Gyroscope
junduo wang, Wenjiang Shen

N/A

HIGHLY SENSITIVITY RESONANT BI-DIRECTIONAL MAGNETIC FIELD MICRO-SENSOR
Nouha ALCHEIKH, Sofiane Mbarek, Mohammad Younis

425

Microelectrode Array for Electrophysiology Detection of Amygdala in Free-moving Mice
Penghui Fan, Yilin Song, Yuchuan Dai, Yiding Wang, Botao Lu, Enhui He, Xinxia Cai

N/A

Photoluminescent Tough Gel Sheathed Suture for Near Infrared Bioimaging
Zhenwei Ma, Fan Yang, Dongling Ma, Jianyu Li

N/A

A Flexible Mechanical Composite Micro-Grating Tailored by One-Dimensional Ordered Wrinkle Patterns 433
Yi-hang Xin, Kai-Ming Hu, Xiu-Yuan Li, Er-Qi Tu, Wen-Ming Zhang

Vibration Characteristics of Piezoelectric Timoshenko Nanobeams in Viscoelastic Medium 437
Dong Wu, Dapeng Zhang, Mingwei Liu, Yongjun Lei

AC & DC Magnetic Field Sensor Based on Flexible Piezoelectric Polymer 443
Weilong Han, Jian Wen, Yifei Wu, Bo Niu, Yi Wu, Mingzhe Rong

Large Area Transient Films and Devices by Photonic Sintering of Two Dimensional Materials N/A
Wenxing Huo, Zi'an Zhang, Zilun Wang, Ziyue Wu, Jiameng Li, Xian Huang

Fast reconstruction of Raman image based on the multi-channel imaging system and kernel function N/A
Xian-Guang Fan, Long Liu, Ting Nie, Yi-Xin Lin, Ying-Jie Xu, Jian He, Xin Wang

Magnetically Induced Micropillar Arrays for Ultra-Sensitive Flexible Sensor with Wireless Recharging System N/A
Libo Gao

A Capacitively Transduced Bulk Acoustic Wave MEMS Resonator with Low Bias Voltages 455
Zeji Chen, Qianqian Jia, Wenli Liu, Yinfang Zhu, Quan Yuan, Jinling Yang, Fuhua Yang

Detection of Single Protein Molecules Using MoS_2 Nanopores of Various Sizes 459
Chaoming Gu, Zhoubin Yu, Xiaojie Li, Xin Zhu, Zhen Cao, Zhi Ye, Chuanghong Jin, Yang Liu

Humidity Sensor Based on Thin-Film Piezoelectric-on-Substrate Resonator and MoS_2 for Multifunctional Applications 463
Hanyong Dong, Jintao Pang, Dongsheng Li, Qian Zhang, Jin Xie

Irreversible Electrowetting on Petal-mimetic Nanotextured Dielectric for Formation of Shape Controllable Polymeric Lenses N/A
Huifen Wei, Xiangmeng Li

In-situ synthesized liquid metal microgels 469
Dawei Wang, Chennan Lu, Xiaohong Wang, Wei Rao

Design and Development of High Precision Magnetic Encoder Based on TMR MEMS Device 474
Zhang Bo, Jiang Yong, Li Yang, Chen Xiaoli, Wen Xiaolong

Flexible electrical stimulation device with Chitosan-Vaseline® dressing that accelerates wound healing in diabetes and corresponding stimulation optimization based on a precision layered skin model N/A
Menglu Li, Xiaofeng Wang, Jikui Luo, Wei-qiang Tan, Xiaozhi Wang

Achieving ultrahigh energy density for triboelectric nanogenerator N/A
Jingjing Fu, Yunlong Zi

Optical micro/nanofibre embedded soft film enables multifunctional flow sensing in microfluidic chips N/A
Zhang Zhang, Jing Pan, Yao Tang, Yue Xu, Lei Zhang, Yuan Gong, Limin Tong

Dynamical Reversible Electrowetting with low voltage on the Dimethicone Infused Carbon Nanotube Array in Air N/A
Lei Zhou, Miao Wang, Xu Hou

Trace mercury detection by ruthenium-based MOFs modified microelectrode N/A
Chenyu Xiong, Yuhao Xu, Chao Bian, Ri Wang, Yong Xie, Mingjie Han, Shanhong Xia

Frequency Matching System of MEMS Gyroscope Based on Fuzzy Control Strategy 491
yifang Liu, Ruimin Liu, Gang Fu, Xin Yan, JunYu Chen, Gaofeng Zheng

High-yield and Large-size suspended graphene device Fabrication technique and Edge Burr Effect Analysis N/A
Ying Liu, Qin Wang, Yong Zhang, Jing Qiu, Guan-jun Liu

Atomic study on the deform mechanism of CuTa/Cu and CuTa/Ta nanolaminates 498
Xiao Wang, Mengjie Wang, Min Liu, Yue Liu, Weidong Wang

Wireless Implantable Phototherapy Device for Oral Inflammation Repair 503
Shuang Huang, Cheng Yang, Ziqi Liu, Xinshuo Huang, Hui-Jiuan Chen

Development of Multi-functional Biocompatible Photopolymer N/A

Seokyoung Bang, Dongha Tahk, Noo Li Jeon, Hong Nam Kim

Multiparametric Flexible Sensor Arrays system for in situ Immediate Diagnosis of Glossitis   509
Ziqi Liu, Keer Li, Shuang Huang, Xinshuo Huang, Lingfei Zhou, Tiancheng Sun, Hui-Jiuan Chen

Contribution discrimination of auxetic cantilever for increased piezoelectric output in vibration   513
energy harvesting
Lu Wang

Research on Embedded 3D Printing for Magnetic Soft Robots   518
Wei Zhang, Jiachen Li, Huicong Liu, Guoqing Jin

Research on Coupling Mechanism of Wireless Power Supply Equipment for Film Pressure   524
Wireless Sensing node of Water-lubricated bearing
Yifan Zhao, Nan Wang, Fan Jiang

Investigation of electrostatic-piezoelectric hybrid vibrational power generators with different   528
frequency broadening schemes
Yongqi Cao, Dezhi Nie, Jian Zhang, Yiwei Wang, Ronggang He, Zhe Zhao, Bowen Ji, Jianbing Xie, Kai Tao

Bio-inspired Flexible, Dual-modulation Synaptic Transistors Towards Artificial Visual Memory   N/A
Systems
Fuqin Sun, Qifeng Lu, Ting Zhang

Kinetic investigation into fast supramolecular processes and chemical reactions through   N/A
microfluidic NMR
Xinchang Wang, Liulin Yang, Xiaoyu Cao, Zhongqun Tian

Liquid Metal Nanoparticles Decorated with Graphene oxide for Enhanced Peroxidase-Like Activity   N/A
Xiaohong Wang, Dawei Wang, Chennan Lu, Wei Rao

Size-Dependent Particle Separating in Curved Microfluidic Chip   539
Yun Tian, Yongkang Wang, Yunfei Chen

Tailoring surface morphology and crystal quality characteristics in InGaAs/GaAs quantum well   N/A
structure by inserting high-temperature layer
Quhui Wang, Haizhu Wang, Xiaohui Ma

3D all-metal nano-cavity coupled metamaterial for refractive index sensing   546
kenan zhang, peimin lu, wengang wu, jia zhu, guanzhou lin, yun huang

Humidity Insensitive Nanogenerator Based on Natural Nanofibrils   N/A
Yanyuan Ba, Xiaowen Li, Xiaosheng Zhang

Design and Simulation of a Double-Ended Arrangement Infrared Thermopile   N/A
Yihao Guan, Cheng Lei

Nodding Duck Structured Hybrid Triboelectric-Electromagnetic Nanogenerator towards Ocean   N/A
wave energy harvesting and Self-Powered Monitoring
Liqiang Liu, Xiya Yang, Qunwei Tang

A Facile and Green Microwave Hydrothermal Method for Fabricating g-C3N4 Nanosheets with   N/A
Improved Hydrogen Evolution Performance
Hongmei Chen, Yanyun Fan, Zheng Fan, Danfeng Cui, Chenyang Xue, Wendong Zhang

One-step vapor deposition of fluorinated polycationic coating to fabricate antifouling and   N/A
anti-infective textile against drug-resistant
Ruixiang Zhao

Atomic Layer Deposition of Al2O3 on 2D MoS2 for Enhancing the Recovery rate of Gas Sensor   N/A
Sungjoo Wi, Inkyu Sohn, Youngjun Kim, Seungmin Jung, Hyungjun Kim

Nonlinear dynamic of coupling mechanical modes demonstrated by phononic frequency coms   N/A
Jiangkun Sun

Deep spatial frequency shift enabled chip-based sub-wavelength- resolution imaging   N/A
Mingwei Tang, Xiaowei Liu, Qing Yang, Xu Liu

A Low-Cost Digital Lithography System Supporting Visual Focusing and its Application on Optical   563
Fiber
Luming Wang, Zhimin Zhang, Ningning Luo, Haifeng Xiao, Long Ma, Qingwang Meng

Au-enabled nanostructured sapphire optical fiber sensor for in-situ gas sensing in harsh   N/A
environments

Kai Liu, Paul Ohodnicki, Jeffrey Wuenschell, Subhabrata Bera, Renhong Tang, Lin Li, Yi Liu, Tinghan Liu, Han Wu, Zhihua Huang, Henry Du

Research on the resolution of submicron lithographic projection system based on DMD for optical fiber end face *Qingwang Meng, Zhimin Zhang, Ningning Luo, Luming Wang, Menghan Xiong*	571
Gold Nanoparticles Modified Sensitivity-Enhanced Uncooled Near-infrared Detector *Xiaoyu Wu, Quanquan Guo, Wencheng Li, Chenyang Yu, Yanyan Wang*	N/A
Research on Nonlinear Shrinkage Characteristics of Small Modulus Plastic Gear Based on Moldflow *Xiansong He, Wangqing Wu, Yihua Lei*	N/A

CS1: 学术好莱坞讲座

学术好莱坞讲座：现代科研论文的构思与写作 *Fengyu LI*	N/A

IS7: IS: Micro Nano and Molecular Systems for Diagnostics and Therapeutics

Fusion Application of Medical Sensors and Internet of Medical Things *Jinhong GUO*	N/A
Semiconducting Molecular Probes for Ultrasensitive Afterglow Imaging and Early Diagnosis *Kanyi PU*	N/A
Cell-based biosensing technology and system *Ning HU*	N/A
Evanescent wave fluorescent biosensor for the detection of microRNA *Xiaohong ZHOU*	N/A
3D Printing Magnetic Soft Milirobots by Recirculating Vat Polymerization for Point-of-Care Diagnostics *Yi ZHANG*	N/A
Active sensing and hybridized sensors towards tactile intelligence *Bo MENG*	N/A

IS8: IS: Micro-/Nanofluidic Devices and Systems

Patient-derived organoids analyzed on a superhydrophobic microwell array for predicting drug response of lung cancer patients within a week *Peng LIU*	N/A
Nanowire-integrated microfluidic devices to identify urinary microRNA groups for cancer detection *Takao YASUI*	N/A
Flexible filter based liquid biopsy *Wei WANG*	N/A
A Fully Integrated and Automated Microfluidic System for Rapid Testing of Respiratory Viruses *Liangbin PAN*	N/A
Bioinspired Surface Microfluidics for Digital PCR and Digital ELISA *Tianzhun WU*	N/A
Multi-dimensional manipulation of Solid/Liquid interaction *Xu DENG*	N/A

IS9: IS: Multidisciplinary Frontier: Advanced Chemistry Materials for MEMS/NEMS

Fabricating fluorescent glass Derived from Mesoporous Powders *Lianjun WANG*	N/A
Controllable liquid transfer for making high-performance thin-film devices *Huan LIU*	N/A
Photo-crosslinkable, insulating silk fibroin for bioelectronics with enhanced cell affinity *Jie JU*	N/A
Unconventional Nanotemplate-based Technique and Its Emerging Applications	N/A

Liaoyong WEN

Paintable hydrogels based on silane chemistry *Xi YAO*	N/A
Electrodepostion filling method of TSV for 3D integration *Yan WANG*	N/A

IS10: IS: Multidisciplinary Frontier: Micro/Nano Energy and Smart Electronics

2D MEMS metamaterial modulators for THz Communication *Xiuhan LI*	N/A
Fiber Structured Flexible Pressure Sensor and Electronic Skin *Zhaoling LI*	N/A
Self-Powered Sensing Techniques Towards Intelligent Human-machine Interaction *Huicong LIU*	N/A
Surface Plasmon Polaritons Enhanced 2D-Material Photodetector for Infrared Application *Wen HUANG*	N/A
Regulation of Strong and Weak Bonds in Lithium Metal Batteries to Enhance Their Performance *Yong ZHAO*	N/A
Achieving Ultrahigh Output Energy Density of Triboelectric Nanogenerator in High-Pressure Gas Environment *Yunlong ZI*	N/A

IS11: IS: Electronic and Optoelectronic Fibers and Textiles

Tuning the Viscosity for a Multi-material Multi-functional Fiber *Chong Hou*	N/A
Nanoparticles Enabled Drawing of Ultra-Long Metal Nanowires *Xiaochun Li*	N/A
Electronic Multi-material Fiber Sensors and Energy Harvesters *Fabien Sorin*	N/A
Advanced Functional Semiconductor Fibers and Textiles *Lei Wei*	N/A
Computational Design of novel two-dimentional semiconducting silicon carbides for optoelectric applications *Liujiang Zhou*	N/A
Nano-/Micro-enabled Multimaterial Fibers for Wearable and Biosensing Applications *Xiaoting Jia*	N/A

IS12: IS: RF Micro/Nano Components and Systems

Investigation on Energy Damping Mechanisms and Q-enhancement Strategies in Piezoelectric-on-Silicon MEMS Resonators *Cheng TU*	N/A
MEMS Resonators with High Q and Good Temperature Stability *Guoqiang WU*	N/A
Novel RF MEMS devices using AlGaNGaN heterostructure *Haoshen ZHU*	N/A
Mechanically-Flexible RF MEMS Resonators Enabled by FlexMEMS Technology *Menglun ZHANG*	N/A
Chip-scale AlN Thin Film Transducers and RF Wake-up Receivers *Tao WU*	N/A
SOI-MEMS based RF micromachined switches for wireless applications *Yong Zhu*	N/A

BS1: Best Student Paper Session

Microstructural-PVDF Dielectric Layer Based High-Resolution Flexible Capacitive Pressure Sensor	647

Zebang Luo, Jing Chen, Zhengfang Zhu, Lin Li, Yi Su, Lei Wang, Hui Li

Electrohydrodynamically Printed Multicolor Perovskite Image Sensor Array   652
Qilu Wang, Guannan Zhang, Hanyuan Zhang, Yongqing Duan, YongAn Huang

A Novel Piezoelectric Micromachined Ultrasonic Transducer with Adjustable Broad and Flat   656
Frequency Band
Lei Wang, Wei Zhu, Zhipeng Wu, Wenjuan Liu, Chengliang Sun

Stackable Triboelectric Nanogenerators for Self-powered Marine Monitoring Buoy   660
Jiale Dong, Hao Wang, Xiu Xiao, Taili Du, Yunpeng Zhao, Zhongqi Fan, minyi xu

Modulating the Electrical Transport Characteristics of a Metal-Semiconductor-Metal Structure by   665
Local Strain Gradient
Youchao Huang, Shenhui Ma, Jiaona Zhang, Chao Xie, Min Zhang

A high-frequency narrow-band filtering mechanism based on auto-parametric internal resonance   670
Rong Wang, Cao Xia, Dong F. Wang, Takahito Ono

Calibrate Silicon Nanowires Field Effect Transistor Sensor with its Photoresponse   676
Yi Yang, Yuelin Wang, Shixing Chen, Tie Li

Flexible and Transparent Ultraviolet Photodetector Enabled by Metal Doping ZnO Nanorods Based   680
on Mica Substrate
Yao huang, Hainan Zhang, Guanhua Dun, Yuhua Li, Renrong Liang, Yi Yang, Dan Xie, He Tian, Tian-ling Ren

S6: MEMS/NEMS II

Liquid Metal Nano Electronics   685
Jing LIU

Design and Manufacture of a High Precision MEMS Flexible Force Sensor   687
Huimin Zhang, Yanlu Feng, Wei Zhang

Linear stiffness tuning of MEMS triangular capacitors   691
Yiming Jin, Zhipeng Ma, Yixuan Guo, Tengfei Zhang, Xudong Zheng, Zhonghe Jin

Power Handling Capability Enhanced RF MEMS Switch Using Modified-Width Cantilevers Structure   695

Yulong Zhang, Zhuhao Gong, Huiliang Liu, Zewen Liu

Research on Motion Characteristics for Latching Mechanism of MEMS Safety and Arming Device   699
under Dual Environmental Forces
Shenghong Lei, Yun Cao, Weirong Nie, Zhanwen Xi, Na Xu, Weixiang Qiu

Modeling and Optimization of ScAlN-based MEMS Mirror with Large Static Two-axis Tilting Angle   704
Changhe Sun, Bolun Li, Wenqu Su, Yufei Liu, Yaming Wu

S7: Nanophotonics and Nanoscale Imaging

Wavelength conversion through plasmon-coupled surface states   709
Mona JARRAHI

Photonic Nanojet Produced by A Microfluidic Channel for Biofluid Monitoring   711
Guoqiang Gu, Pengcheng Zhang, Hui Yang

Fluorescence Enhancement Utilizing Dielectric Microbeads with Semi-open Microwells   715
Pengcheng Zhang, Bing Yan, Guoqiang Gu, Zitong Yu, Xi Chen, Zengbo Wang, Hui Yang

Spectrometer-free Refractometric Sensing Using Image Recognition on Centimeter-scale Gradient   N/A
Nanostructures
Siyi Min, Shijie Li, Zhouyang Zhu, Yu Liu, Chuwei Liang, Jingxuan Cai, Fei Han, Yuyan Li, Xing Cheng,
Wen-Di Li

Strongly Anisotropic Monolayer InSe polarized light detector   N/A
Xusheng Wang, xi zhang

Self-powered textile-based tactile sensors inspired by human-skin for multifunctional sensing of   N/A
wearables and robots
Changyong Cao

Compare of SNOM, PTIR, PiFM and PFIR   N/A
Hai-Long Wang, En-Ming You, Song-Yuan Ding, Zhong-Qun Tian

Full-spectrum optoelectronics based on bandgap-graded materials
Zongyin Yang

N/A

S8: Energy Conversion and Storage II

Flexible Bionic Intelligent Perception System
Jianhua ZHANG

729

Fabrication of laser scribed graphene stretchable supercapacitor by laser-assisted transfer printing strategy
Guangyuan Xu, Yanan Chen, Fangshuai Chen, Yanfang Meng, Yinji Ma, Xue Feng

731

Smart Power Management Microsystem for Distributed Renewable Energy Harvesting
Linhao Feng, Kai Huang, Mingxin Liu, Hongsheng Zhong, Xiaosheng Zhang

737

Card-based Hybrid Piezo-Triboelectric Nanogenerator for Simultaneously Harvesting Sliding Mechanical Energy
Danliang Wen, Peng Huang, Yanyuan Ba, Xiaosheng Zhang

N/A

Self-Powered Electrochemical Interfaces for Material and Energy Conversion
Shuyan Gao, Zhong Lin Wang

N/A

A Micro Device Array for Real-time Monitoring Cardiac Contraction
Li Wang, Jun Chen, Weiguang Su, Anqing Li, Pengbo Liu, Chonghai Xu

N/A

Wind-driven self-powered wireless environmental sensors for Internet of Things at long distance
di liu, baodong chen, jie an, chengyu li, jiajia shao, wei tang, chi zhang, zhonglin wang

N/A

Design of Self-powered Environment Monitoring Sensor Based on TEG and TENG
Jianhao Liu, Changxin Liu, Cong Zhao, Huaan Li, Guanghao Qu, Zhuofan Mao, Zhenghui Zhou

749

S9: Soft & Flexibe Devices and Applications I

Printed flexible/stretchable electronics and applications
Zheng Cui

754

Optical micro/nanofiber enabled compact tactile sensor for hardness discrimination
Yao Tang, Haitao Liu, Jing Pan, Zhang Zhang, Yue Xu, Lei Zhang, Limin Tong

N/A

A Flexible Triboelectric-Electromagnetic Hybridized Nanogenerator
Haitao Deng, Zhiyong Wang, Yanyuan Ba, Xinran Zhang, Danliang Wen, Xiaosheng Zhang

N/A

A Multifunctional ultrasensitive hybrid optical skin
Jing Pan, Lei Zhang, Limin Tong

N/A

An Integrated Stretchable Sensing Patch for Simultaneously Monitoring of Physiological and Biochemical Parameters
Wenhao Pan, Shiyi Xu, Xiyu Mao, Tianyu Li, Qingpeng Cao, Xuesong Ye, Bo Liang

762

A Textile Tactile Sensor for Dual-mode Proximity/Pressure Perception in Smart Robotics
Qinhua Guo, Weiguan Zhang, Zhengchun Peng

N/A

Modeling and Simulation of Flexible Vector Shear Flow Sensor Based on COMSOL Multiphysics
Guochang Liu, Wenping Cao, Haoyu Tan, Renxin Wang, Wendong Zhang

770

A Self-powered retractable device based on triboelectric nanogenerator for sensing of joint and spinal bending/stretching
Chengyu Li, Di Liu, Chaoqun Xu, Ziming Wang, Sheng Shu, Zhuoran Sun, Wei Tang, Zhonglin Wang

N/A

S10: Micro/Nano Fluidic II

Microfluidic Liquid Metal for Stretchable Biomed-electronics
Xingyu JIANG

777

Bioinspired Universal Flexible Elastomer-Based Microchannels
Baiyi Chen, Rongrong Zhang, Hexuan Fu, Xu Hou

N/A

Frequency Research of Microfluidic Wear Debris Detection Chip Based on Inductive Wheatstone Bridge
Yucai Xie, Hongpeng Zhang, Haotian Shi, Yuwei Zhang

780

Liquid Metal-Based Microfluidic for Sperm Thermotaxis.
Yimo Yan, Ran Liu, Boxuan Zhang

N/A

Digital Microfluidics for Efficient and Accurate Molecular Profiling of Single Circulating Tumor Cells — N/A
Qingyu Ruan, Jian Yang, Fenxiang Zou, Lingling Wu, Zhi Zhu, Chaoyong Yang

Bubble formation in nanopores — N/A
Alberto Giacomello, Roland Roth

S11: Nanoscale Robotics, Assembly, and Automation

Small-scale Wireless Robots for Medical Applications — 792
Metin SITTI

Order and Information in the Phases of Spinning Micro-disks — N/A
Wendong Wang, Gaurav Gardi, Paolo Malgaretti, Vimal Kishore, Metin Sitti

Single cell manipulation with acoustically powered microrobotic platforms — N/A
Xiaolong Lu, Ying Wei, Kangdong Zhao, Hui Shen, Wenjuan Liu

Simulation of the Shape-directed AC Driven Defective Micromotors — 798
Rencheng Zhuang, Dekai Zhou, Xiaocong Chang, Longqiu Li

3D printing of magnetically actuated miniature soft robots — 804
Zhongbao Wang, Yigen Wu, Dezhi Wu, Zhenyin Hai

Anisotropic spreading of droplets on striped electrodes — 809
Wei Wang, Yanbo Xie, Antoine Riaud

Low-cost Micro Search-Coil Magnetic Sensor with Self Calibration for the Internet of Things — 813
Hadi Tavakkoli, Izhar, XU ZHAO, Yi-Kuen Lee

S12: Micro/Nano Sensors and Actuators I

Nano size-effect enhanced sensitivity of gas detection — 818
Xinxin LI

Scalable Synthesis of SnO_2 Nanosheet Arrays on Chips for Ultralow Concentration NO_2 Detection — 820
Gaoqiang Niu, Changhui Zhao, Fei Wang

Thermalvoltage Measurement and Manipulation in Single-molecule Device — N/A
Junyang Liu, Hang Chen, Ping Zhou, Wenqiang Cao, Wenjing Hong

A Tungsten-Rhenium Thin Film Thermocouples Sensor Based on AIN Transition Layer — 826
Bian Tian, Bingfei Zhang, Zhongkai Zhang, Zhaojun LIiu, Jiangjiang Liu, Gong Cheng, Qijing Lin, Chen Wu, Peng Shi, Zhuangde Jiang

Thin Film Antioxidative Coating for High-temperature Thin Film Sensors Made of Polymer-derived Ceramics for Harsh Environment — N/A
Xiaochuan Pan, Zaifu Cui, Zhenyin Hai, Guochun Chen, Daoheng Sun

A New Method for Characterization of Single Cell Using System Identification — 832
Shuang Ma, Wenxue Wang, Lianqing Liu, Yuechao Wang, Tianlu Wang

Reversible Immunoaffinity Interface Enables Dynamic Manipulation of Trapping Force for Accumulated Capture and Efficient Release of Circulating Rare Cells — N/A
Xiaofeng Chen, Lingling Wu, Chaoyong Yang

PS2: Poster Session II

Design and Optimization of Glass Frit Package Structure for Micro Pressure Switch — 840
Lingyun Wang, Daner Chen, Heng Xiong, yifang Liu

Damage profile model of nanostructure fabricated by Focused Helium Ion Beam — 845
Chenglong Liu, Qi Li, Qianhuang Chen, Yan Xing

Investigation of Broading Modulus Range of Soft Probes by Single Beam Acoustic Tweezer — 849
Huiyao Shi, Jialin Shi, Peng Yu, Lianqing Liu

Vertical Organic Synapse Expandable to 3D Crossbar Array — N/A
Yongsuk Choi, Seyong Oh, Chuan Qian, Jin-Hong Park, Jeong Ho Cho

Active-Powering Wearable Iontronic Tactile Sensing for Human Physiological Monitoring — N/A
Hongyan Sun, Yu Chang, Tingrui Pan

Negative Differential Resistance Effect in Graphene Nanoribbons Heterojunction — N/A
li cheng, Yu Zhu, Qingfeng Gong, Wenli Zhou

3D Printed Kenics Static Micromixer *Kunpeng Zhang, Gonghan He, Daoheng Sun*	N/A
Viscoelastic deformation and process optimization of polymer microfluidic chip for rapid in-mold bonding *Ylhua Lei, wangqing Wu, Xiansong He*	N/A
Factors Influencing Resolution of Optical Fiber End Face Processing in Digital Lithography *Menghan Xiong, Ningning Luo, Zhimin Zhang, Qingwang Meng*	864
Study on friction heat generation mechanism by molecular dynamics *Changsheng He, Wangqing Wu*	N/A
Through Glass Vias by Wet-etching Process in 49% HF Solution Using an AZ4620 Enhanced Cr/Au Mask *Guanghui Ding, Binghe Ma, Yuchao Yan, Weizheng Yuan, Jinjun Deng, Jian Luo*	872
DNA-directed nanofabrication of ultra-scaled high-performance electronics *yahong chen, mengyu zhao, Zhi Zhu, Wei Sun*	N/A
The Design of Love Wave Immunosensor for Real-time Detection of Cancer Biomarker from Human Saliva *Junyu Zhang, Tao Zhang, Yuantao Chen, Xiaojing Zhang, Hao Wan, Ping Wang*	N/A
Measurement of Cardiomyocytes Motion Based on Surface Patterned Polydimethylsiloxane Membrane *Si Tang, Jialin Shi, Huiyao Shi, Peng Yu, Lianqing Liu*	N/A
An Adaptive Octree Level Set Simulation Method of the Wet Etching Process for the Fabrication of Micro Structure on Sapphire Crystal *Ye Chen, Jin Qian, Xinyan Guo, Yan Xing*	882
Surface Acoustic Wave (SAW) Devices Based on ScAlN/AlN/Si Layered Structures with Large Figure of Merit *Yan Liu, Binghui Lin, Yang Zou, Yao Cai, Wenjuan Liu, Chengliang Sun*	N/A
A Flexible AgNPs-PDMS Substrate to Produce Ultrasensitive SERS Detection *Guanzhou Lin, Kenan Zhang, Yun Huang, Shengxiao Jin, Tian Kang, Yusa Chen, Liye Li, Peimin Lu, Wengang Wu*	889
Fabrication and Properties of Two-Dimensional InSe Top-Gate Transistors and Their Applications in Logic Circuits *Wei Li, Xiaozhi Wang*	N/A
Voxelated Meniscus-confined Electrodeposition of 3D Metallic Microstructures *Yutao Wang, Yuanliu Chen*	N/A
A clogging rate prediction model based on porous microarray membranes for liquid biopsies *Yinghao He, Yaoping Liu, Wei Wang, Yufeng Jin*	898
Biomimetic Electrode–Electrolyte Design for Efficient Electrocatalytic Nitrogen Fixation under Ambient Conditions *Yang Liu, Xinyi Zhang, Panagiotis Tsiakaras, Pei Kang Shen*	N/A
Micro-droplet of Particulate Suspension Generated by a Pneumatic Ejection System *Weijie Bao, Shengnan Sun, Zhihai Wang, Yaohong Wang*	904
Tweezers for Micro-droplet Transfer Based on Hydrophobic Non-parallel Plate Structure *Jiaqiang Wang, Liguo Chen, Xiongheng Bian, Zaichen Wang*	909
Design and Preparation of a High Resolution Accelerometer Based on Graphene *Xiaodong Zhao, Yanlu Feng, Wei Zhang, Jianhui Liao*	915
Nanorobotic Manipulation inside Scanning Electron Microscope for the Electrical and Mechanical Characterization of ZnO nanowires *Mei Liu, Aristide Djoulde, Quan Yang, Weilin Su, Lingli Kong, Jinjun Rao, Jinbo Chen, Zhiming Wang*	919
Nano Mechanics Markers for Accessing the Effects of Ultraviolet(UVC) Disinfection *Yuxuan Xue, Ning Xi, Kaicheng Huang, Ye Ma*	N/A
Multi-pulse triboelectric nanogenerator based on micro-gap corona discharge for enhancement of output performance *Ru Wang, Juan Cui, Yong-Qiu Zheng*	N/A

Control over Electrical Property of a-InGaZnO Thin Film Transistors using coupled self-assembled molecular layer as Copper Diffusion Barrier.
Seungmin Lee, Minkyu Lee, Taeyoon Lee

N/A

Laterally-excited bulk-wave resonators (XBARs) with embedded electrodes in 149.5° Z-cut LiNbO$_3$

931

Xiyu Gu, Jieyu Liu, Yao Cai, Yan Liu, Chao Gao, Zhiwei Wen, Shishang Guo, Chengliang Sun

Research on Integrated Reliable Micro-High Explosive Train Model
Bo He, Wenzhong Lou, Hengzhen Feng, Yuecen Zhao, Yi Sun

935

Spatial and Temperature Control of Exciton Emission by Ferroelectric in Monolayer TMDs
bo wen, xi zhang, dongfeng diao

N/A

THz Hybrid Graphene &Metal Patches Metasurface for RCS Reduction and Digital Coding
Baolong Wang, Jiayi Yang, Shuangshuang Liu, Wei Xu, Guobin Chen, Di Feng, Lingjie Jia, Yan Meng, Xiuhan Li

941

A Flexible Circuit Fabricated by Tuning the Wettability of Liquid Metal
Chengjun Zhang, Qing Yang, Jingzhou Zhang, Jiale Yong, Xun Hou, Feng Chen

N/A

A method for automatic counting and labeling of cells stained with microporous membrane
Jiangcheng Cao, Tingting Hun, Wenbo Zhou, Zheng Liu, Wei Wang, Yufeng Jin

949

Design of a Controllable Push-triggered Microfluidic Chip for Vitrification Reagent Loading/unloading
Guangyi Cai, Boshi Jiang, Jiaxin Zhu, Fenglin Liu, Tianzhun Wu

953

BMP-2 Immobilized Lubricant-Infused Surface Coating for Orthopedic Implants with Anti-bacterial, Anti-inflammatory, and Osteogenic Functionalities
Jaegyu Park, Yeontaek Lee, Kijun Park, Jungmok Seo

N/A

Direct and All-dry Microfabrication of Ultramicroelectrode Based on Cold Atmospheric Microplasma Jet
Ye Xi, Longchun Wang, Zhejun Guo, Bin Yang, Jingquan Liu

959

Nanopore Surface Charge Sensing with Ion-Step Method
Jing Yang, Songyue Chen

963

Research on complex surface patterned conformal manufacturing system based on point cloud theory
Kaihan Yao, Junchuan Gao, Gonghan He, Zhenyin Hai, Daoheng Sun

N/A

A coplanar-electrode direct-current triboelectric nanogenerator with facile fabrication and stable output
Guoqiang xu, Yunlong Zi

N/A

The Mechanism and Realization of Using Ultrasonic Atomic Force Microscopy to Measure Subsurface Defects of Ultra-precision Components
Yuyang Wang, Chengjian Wu, Jinyan Tang, Yuanliu Chen

N/A

Breathable Graphene-based Hydrogel Strain Sensors
Xingchi Liu, Hengchang Bi, Xing Wu

N/A

Engineering Application of Micro Surface Acoustic Wave Sensing System in Temperature Detection of Copper BusBars in Transformer Cabinet of Electric Locomotive
Tinghan Liu

N/A

Immune-camouflage Coating for Implantable Medical Devices, Bioprinting, and Microfluidics
Kijun Park, Yejin Jo, Inwoo Kim, Soyeon Kim, Jungmok Seo

N/A

Slippery coated Implantable flexible microelectrode array (fMEA) for High-Performance Neural Interface
MD ESHRAT E ALAHI, Tian Zhou, Sara Khademi, Hao Wang, Tianzhun Wu

980

Light Extraction Efficiency Investigation of AlGaN-Based Deep Ultraviolet Light-Emitting Diodes on Nano-Patterned Sapphire Substrate
Wan Hui, Lei Yu, Lan Shuyu, Gong Liyan, Gui Chengqun, Zhou Shengjun

N/A

Equivalent Electrical Model and Experimental Analysis of a Novel Needle-ring Atmospheric Pressure Plasma Microjets Array
Lingju Xia, Junfeng Yang, Li Wen

986

A Wearable Health Monitoring System Self-powered by Human-motion Energy Harvester

990

Chen Bao, Anxin Luo, Yi Zhuang, Junlong Huang, Fei Wang

Improved Reservoir Computing by Carbon Nanotube Network with Polyoxometalate Decoration   994
Shuo Wu, Wenli Zhou, Kaiqiang Wen, Chengzhu Li, Qingfeng Gong

2D Microscopy of Magnetic Field Using Atomic Vapor Cell   998
Liu Chen

Atomistic resolved signals of amino acids for peptide sequencing by tunneling current analysis   N/A
Tommaso Civitarese, Giuseppe Zollo

Single-Molecule Electronics: Recent Advances and Perspectives   N/A
Junyang Liu, Wenjing Hong

Tunability of Optoelectronic Oscillator Based on SiO2 Optical Waveguide Ring Resonator   N/A
Zerong Jia, Yongqiu Zheng, Jiamin Chen, Liyun Wu, Chen Chen, Chenyang Xue

Lithographic Properties of Amorphous Solid Water   N/A
Shan Wu, Ding Zhao, Min Qiu

Chemical bonding of functional groups in Self-Assembled Monolayer (SAM) for Cu diffusion barrier   N/A
Minkyu Lee, Taeyoon Lee

Functionalized Nanochannels for Constructing Ultrasensitive Electrochemical Sensor   N/A
Wenrui Ma, Shunbo Li, Yi Xu

The Opposite Anisotropic piezoresistive effect ReS2   N/A
CHUNHUA AN, Jing Liu

Structural Design and Simulation of Ionic Transport and Mass Selection Process in MEMS   N/A
Time-of-Flight Mass Spectrometry Chip
Zongjia Cai, Ming Hao, Tianyuan Qi, Cong Wang, Kun Liu, Renchao Dou, Donghui Meng, Lichen Sun, Rongxin
Yan

Mechanisms of branch tip fusion in meshwork patterns   1018
Shan Guo, Mingzhu Sun, Xin Zhao

Leaf-like Self-assembled MXene/ZnOEP Hybrid Network for High-sensitive Temperature Sensing   1023
in Electronic Skin
Shijie Wan, Weiguan Zhang, Chenyang Xing, Zhengchun Peng

Energy Localization of Lamb Wave Resonator Using Dispersion Engineering   N/A
Yusi Zhu, Lidong Du, Zhan Zhao, Zhen Fang

Room temperature linear magnetoresistance in vertically aligned graphene nanocrystalline network   N/A
film
dong ding, xingze dai, chao wang, dongfeng diao

Preparation of Flame-retardant Lithium-ion Battery Separator by Coaxial Electrospinning   1035
Fangqin Shao, Guoyi Kang, Huatan Chen, Zungui Shao, Xiang Wang, Wengwang Li, Gaofeng Zheng

A Numerical Study on the Effects of Mechanical Properties of Red Blood Cells on Rheology in   N/A
Narrow Microchannels
De-Yun Liu, Peng Jing, Xiaolong Wang, Qiaodong Wei, Shenghong Zhang, Xiaobo Gong

Flexible Transparent Conductive Network Based on Liquid Film Rupture Self-assembly Method   N/A
Xinran Zhang, Haitao Deng, Danliang Wen, Yanyuan Ba, Haixia Zhang, Xiaosheng Zhang

An In2O3 Nanotubes based Gas Sensor Array combined with Machine Learning Algorithms for   1042
Trimethylamine Detection
Wenjie Ren, Changhui Zhao, Yingming Liu, Fei Wang

Study of touch mode MEMS capacitance vacuum gauge with circular diaphragm   N/A
xiaodong han, Gang Li, Yongjian Feng, Detian Li

PMUT-Based Air-Coupled Imaging And Surface Stain Detection   N/A
sheng sun, jianyuan wang, yuan ning, menglun zhang

3D printed water-soluble UV photopolymer for flexible sensor with sacrificial scaffolds and   N/A
indirected molding process
Chunjiang Wang, Wanhao Niu, Qihang Song, Xiangyu Mi, Jianxu Shi, Xiaoming Chen

Ultrasensitive Silicon Nanowire Field-Effect Biosensors Enabled by Functionalized with Gold   N/A
Nanoparticles and Aptamers
Qianhui Wei, Jianglan Yang, Han Xiao, Kuo Men, Qingzhu Zhang, Jing Zhang, Feng Wei

Self-Assembly of Electrospun Polymer Nanofibers: A General Phenomenon Generating Cylinder-Patterned Nanofibrous Structures
Tianhao Wu, Daoheng Sun

N/A

The Fabrication Of Three-layer Silicon Stacked Antenna
Yang Lei Cao, Jian Zhu, Fang Hou, Min Huang, Zong Lei Jiao

1057

Enhanced trapping stiffness based on SERS embedded microcavity
Yanhong Wang, Zhihui Li, Jingzhi Wu, Hengze Yang

N/A

Self-powered electro-tactile system for virtual tactile experiences
Yuxiang Shi

N/A

Influence of the Particle Size of Glass Powder on Sintering Characteristics in TGV Packaging
yifang Liu, Hongbo Sang, Zhenxiang Bu, Yulong Zhang, Guangen Gao, Lingyun Wang

1065

A self-powered system driven by random human walking energy for wearable healthcare applications
Fan Wang, Xiangyu Chen

N/A

An Electromagnetic-Piezoelectric-Triboelectric Hybridized Energy Harvester Towards Blue Energy
Yunfei Li, Tianyi Tang, Manjuan Huang, Xin Ma, Zhaohui Chen, Peijuan Cui, Huicong Liu, Lining Sun

1070

Internal Resonant Oscillation in Coupled Resonators for High-resolution Mass Sensing with A Wider Coupling Range
Nanxing Li, Cao Xia, Guowen Zheng, Xu Du, Dong F. Wang

1074

An Immersed Boundary Method for Mass Transfer through Porous Biomembranes under Large Deformations
Xiaolong Wang, Peng Jing, Qiaodong Wei, Shenghong Zhang, Xiaobo Gong

N/A

Elastic straining of free-standing monolayer graphene
Ke Cao

N/A

Multi-order Nonlinearities and Resulting Coherent Oscillations of the States in Quantum Dot-Nanomechanical Resonator Hybrid System
xinhe wang, Lin Cong, Guangwei Deng, Kaili Jiang, Xiaoyang Lin, Weisheng Zhao

1084

Investigation the Minimum Measurement Points for Calibration a High Precision NTC Thermistors in Cryogenic Field
Gong Xun, Xiaohe Tang, Yanjie Li, Minmin You, Zude Lin, Jingquan Liu

1088

Improve the Efficiency of Sidewall Reflector-type Optical Switch by Combining Inductively Coupled Plasma and Focused-ion-beam Etching
haoran xu, Jin Xie, Kaiquan Li, Zhuang Xiong, Weikang Dong, Kometani Reo, Jun Dai

N/A

A Triboelectric Nanogenerator with Gear Transmission for harvesting elastic potential energy
Xiaobo Lin, Lanxin Yang, Junfeng Zhong, Zhaoming Deng, Chenyang Xing, Bo Meng, Zhengchun Peng

N/A

A Facile Low-Cost Wireless Self-Powered Footwear System for Monitoring Plantar Pressure
huayi huang, Ziya Wang, Zhihao Zhu, Waner Lin, Zhengchun Peng

1096

Investigation of the Reliability of the Interconnection between Metal Electrode and Silicon Anchor in Silicon-on-Glass Process
MENGXIA LIU, Xianshan Dong, Jian Cui, Qiancheng Zhao

1102

A bistable criterion for the V-beam mechanism
Min Liu, Weidong Wang, Xiao Wang, D SY, Yingmin Zhu, Zimin Huo, Yijia Du

1106

Self-adaptive Microjet Array Cooling for RF High Power GaN Integration on Silicon
Miao Yu

1110

Wearable Wireless Sensing System with Ultrasensitive and Self-Cleaning Pressure Sensor for Electronic Skin
Xuan Li, Meng Wang, Weidong Wang, Libo Gao

N/A

Effect of deposition pressure on the tribological properties of Ti/WS2 composite films deposited by magnetron sputtering
Jun Ye

N/A

High-temperature Thin Film Temperature Sensor Made of Polymer-derived Ceramics for Harsh Environment
Zaifu Cui, Guochun Chen, Xiaochuan Pan, Zhenyin Hai, Daoheng Sun

N/A

A Contactless Switch for Cell Sorting by Area cooling

1126

Yigang Shen, Yaxiaer Yalikun, Yo Tanaka

Light-confined Plasmonic Probe for Photoelectric Characterization by Atomic Force Microscopy *Yaoping Hou, Chengfu Ma, Wenting Wang, Yuhang Chen*	N/A
Alumina composite coatings electrical insulating properties on Ni-based superalloy *chao wu, Daoheng Sun*	N/A
A multi-parameter flexible sensor for detection of water-quality and aquatic animal activities *Zhihong Wang, Zheng Gong, QIpei He, Deyuan Zhang, Yonggang Jiang*	N/A
Real-time Monitoring and Analysis of Jet Behaviors in Electrohydrodynamic Direct-Writing *Yifang Liu, Junyu Chen, Jiaxin Jiang, Guoyi Kang, Huatan Chen, Jianyi Zheng, Gaofeng Zheng*	1136
Molecular Insights into Distinct Detection Properties of α-Hemolysin, MspA, CsgG, and Aerolysin Nanopore Sensors *Wanqi Zhou*	N/A
Large-scale Uniformly Hybrid Micro-nano Structure Wetting Solid Substrate for Surface-enhanced Raman Spectroscopy *Fanhong Chen, Qi Qi, Yupeng Zhao, Shaoxun Zhang, Yongmin Zhao, Anjie Ming, Changhui Mao*	1142
Antibacterial polymeric films with killing and antifouling properties synthesized by initiated chemical vapor deposition *Qing Song*	N/A
Electroosmotic flow in wild-type and mutated CgsG nanopore *Giovanni Di Muccio, Blasco Morozzo della Rocca, Mauro Chinappi*	N/A
Surface Free Energy Characterization of Soft Materials through Computational Experiments *Francesco Maria Bellussi, Annalisa Cardellini, Lorenzo Chiavarini, Pietro Asinari, Matteo Fasano*	N/A
An improved Adaptive Periodical Segment Matrix Algorithm for Signal Denoising in Real-Time ECG Sensing *Xinggu Liu, Liang He, Jinhua Li, Zhuqing Wang*	N/A
Data Analysis with Machine Learning For High-Accuracy Multi-target Gas Sensor Array *Qihong Ning, Chun Huang, Liang He, Jinhua Li, Zhuqing Wang*	N/A
Buckled Structure Formation on the Surface of Stretchable Conductive Fibers and Application for Hydrogen Sensor *Kukro Yoon, Taeyoon Lee*	N/A
A Tunable Quasi-Zero Stiffness Mechanism for Thermal Compensation of a MEMS Gravimeter *Xiaopeng Zhang, Xueyong Wei, Yang Gao, Minghui Zhao, Yonghong Qi, Libo Zhao, Zhuangde Jiang*	1157
Effect of a hydrophobic gas on the evaporation of water in confinement *Antonio Tinti, Gaia Camisasca, Alberto Giacomello*	N/A
Electroosmotic Flux in Uncharged Solid-State Nanopores: An Atomistic Simulation Study *Matteo Baldelli, Sébastien Balme, Mauro Chinappi*	N/A
The realization of ZrOxNy temperature sensors with good sensitivity and stability in the temperature range above 150K *Yanjie LI, Minmin YOU, Gong Xun, Xiuyan LI, Zude LIN, Jingquan LIU*	1165
NEMS Sensors Based on Suspended Graphene *Xuge Fan, Frank Niklaus*	1169
On-chip micro/nano devices for energy conversion and storage *Lin Xu, Liqiang Mai*	N/A
A miniature infrared emitter with ultra-high emissivity *Dongsheng Shu, Jifang Tao, Yan Li, Xinye Fan*	1175
Ultrathin elastic shape sensor used for endoscope shape reconstruction *Leixin Meng, Yuan Zhuang, Liqiang Wang, Qing Yang*	N/A

CS2: 学术格莱美讲座

学术格莱美讲座：如何与编辑沟通？ *Yan LI*	N/A

IS13: IS: Multidisciplinary Frontier: Soft Rubbery Electronics and Stretchable Integrated Systems

E-skins with superhigh pressure resolution and tough interfaces
Chuanfei GUO   N/A

Soft Rubbery Electronics
Cunjiang YU   N/A

Three-dimensional soft electronic systems for biomedicine
Mengdi HAN   N/A

Design of artificial synapses and sensorimotor neurons
Wentao XU   N/A

Rational Design of Dielectric Materials for Flexible Capacitive Pressure Sensing Coatings
Zhuo LI   N/A

Mechanics-driven designs of soft network materials and their applications in stretchable integrated   N/A
devices
Yihui ZHANG

IS14: IS: Multidisciplinary Frontier: Nanoplasmonics and Biosensors

Surface Plasmoinc Enhanced Exciton-Polariton Effect in Semiconductor Nanowires
Qing ZHANG   N/A

Optical Superoscillation for Label-free Subdiffraction Bioimaging
Guanghui Yuan   N/A

Implantable biochip for management of inflammation in lung cancer model
Guozhen Liu   N/A

In-Vitro Diagnostic Assays Enabled by Plasmon-Enhanced Photoacoustic Detection
Meng Lu   N/A

2D nanomaterials enhanced surface plasmon resonance for sensing applications
Shuwen Zeng   N/A

Hybrid metal/dielectric nanosystems and applications
Yali Sun   N/A

IS15: IS: Flexible and Wearable Microsystems for Sensing and Actuation

Trigger-Detachable Hydrogel Adhesives for Bioelectronic Interfaces
Ji LIU   N/A

Materials and Devices Designs for Bioelectronic Medicines
Jie ZHAO   N/A

Biodegradable materials for electronic medicine and biosensors
Lan YIN   N/A

Microscale Optoelectronic Devices for Biological Modulation and Sensing
Xing SHENG   N/A

Hybridized and Coupled Nanogenerators
Ya YANG   N/A

Deformable Structures for Flexible Smart Materials
Zunfeng LIU   N/A

IS16: IS: Advanced Scanning Probe Technology and Applications

Markerless fabrication of FinFET devices by Thermal Scanning Probe Lithography
Armin W. Knoll   N/A

Advanced scanning probes for nanofabrication and nanomeasurement
Huan Hu   N/A

Scanning Probe-Assisted Nanowire Circuitry
Pablo Ares   N/A

Advanced Lithography for Opto-Electronic Nanochips   N/A

Xiaorui Zheng

2D materials patterning by oxidation scanning probe lithography *Yu Kyoung Ryu*	N/A
Spatially Confined Surface Reactions and Nanopatterning by Cantilever-free Polymer Tip Arrays *Zhuang Xie*	N/A

IS17: IS: Advanced Microsystems for Biomedical Applications

Engineering and measuring systemic multi-organ interactions *Yi-Chin TOH*	N/A
Intelligent Drug Delivery System Based on Microneedle Technology *Jongho PARK*	N/A
Mechanosensing of biological samples with silicon-based microsystems *Mehmet Cagatay TARHAN*	N/A
Single-molecule analysis of bio-molecules and its application *Rikiya WATANABE*	N/A
Advanced biomedical microsystems for single-cell analysis *Soo Hyeon Kim*	N/A
Surface Enhanced Raman Scattering Sensors for Diseases Detection *Tianxun GONG*	N/A
Advanced microfluidics and PCR technology combined for point-of-care diagnostics *Sisi LI*	N/A

IS18: IS: Multidisciplinary Frontier: Advanced Micro/Nano Electronics Plus (Bio-, Photo-, Energy-)

Angstrom-Porous MOF Materials for Efficient Ion Adsorption and Sieving *Huacheng ZHANG*	N/A
Surface Free Energy Characterization of Soft Materials through Computational Experiments *Matteo FASANO*	N/A
Soliton microcombs: integrated photonics powering metrology *Qifan YANG*	N/A
Synthetic Embryology: Merging Stem Cells and Mechanical Microsystems to Forge an Embryo-Free Future for Human Embryology *Yue SHAO*	N/A
Controlled Synthesis and Devices Applications of 2D Crystal Arrays *Yu ZHOU*	N/A
Active Ionic Artificial Skin: Ion Transport Mechanisms and Prototype Design *Zicai ZHU*	N/A

S13: MEMS/NEMS III

Applications of Flexible Inorganic Thin-film Devices in Bioelectronics *Yuan LIN*	1248
An On-Chip Inductive-Capacitive Sensor for the Detection of Wear Debris and Air Bubbles in Hydraulic Oil *Haotian Shi, Hongpeng Zhang, Wei Li, Zhiwei Xu, Laihao Ma, Yucai Xie, Dian Huo*	1250
AlN Contour Mode Resonators with Half Circle Shaped Reflectors *Zhifang Luo, shuai shao, Tao Wu*	1255
Modeling of magnetic sensor based on BAW magnetoelectric coupling micro-heterostructure *Si Chen, Wanchun Ren, Junru Li, Chunrui Peng, Yang Gao*	1259
The 3D Capacitance Modeling of Non-parallel Plates Based on Conformal Mapping *Yue Feng, Zilong Zhou, Wenlong Wang, Zehong Rao, Yanhui Han*	1264
A Mode-localized Mass Sensor with Different Order Modes Coupling Induced by Asymmetric Structures *Jiahao Song, Jian Zhao, Ming Lyu, Pengbo Liu, Heng Zhong, Xianze Zheng, Yinghai Tang*	N/A

A Breathing Mode Dual-Ring Resonator with High Quality Factor Operating at Atmosphere Pressure
Wen Chen, Wenhan Jia, Guoqiang Wu

N/A

S14: Soft & Flexibe Devices and Applications II

Versatile E-Printing for Flexible Electronics
Yongan HUANG

1277

A High-Resolution Self-powered Flexible Pressure Sensor Matrix Based on ZnO Nanowires
Qifeng Lu, Fuqin Sun, Lili Li, Yuanyuan Bai, Zihao Wang, Ting Zhang

1279

Design and Preparation of a Microfluidic System for Stretching of Cells in Topographic Microstructures
Yingning He, Yuqian Yang, Yexin Gu, Tianjiao Mao, Yue Yu, Jiandong Ding

1283

Stochastic analysis of the elctrothermal microactuator with fabrication error
Haotian Liu, Hao Chen, Yun Cao, Weirong Nie, Zhanwen Xi, Shenghong Lei

1287

Dynamic Simulation of Nanogenerator Based on Finite Element Model Coupled with Lumped Parameter Elements
Yao Chu, Ruixing Han, Huiliang Liu, Xiongying Ye, Fei Tang

1291

Direct Ink Writing of Soft Microscale Structures Using Pure Polydimethylsiloxane
Huyue Chen, Qifan Ding, Wen-Ming Zhang, Lei Shao

N/A

All-printed Flexible Tactile Sensors with High Sensitivity and Large Detection Range
Qifan Ding, Huyue Chen, Wen-Ming Zhang, Lei Shao

1297

S15: Micro/Nano Sensors and Actuators II

Heterogeneous Integration for RF application
Jian ZHU

1301

Chitosan/graphene oxide/MoS2/AuNPs modified electrochemical sensor for trace mercury detection
Ri Wang, Chenyu Xiong, Yong Xie, Mingjie Han, Yuhao Xu, Chao Bian, Shanhong Xia

1303

Visual Chemical Detection Mechanism by a Liquid Gating System with Dipole-Induced Interfacial Molecular Reconfiguration
Yi Fan, Xu Hou

N/A

Quantitative Glucose Measurement on a Synthetic Paper Test Strip
Weijin Guo, Jonas Hansson, Wouter van der Wijngaart

1310

Organic Field-effect Transistors Based Gas Sensors for Volatile Toluene Detection
Tengfei Guo, Yuelin Wang, Jian Song, Tie Li

N/A

P-dopant Enhanced Organic Thin Film Transistors based Gas Sensors for Volatile Benzenes Detection
Meng Liu, Yuelin Wang, Jian Song, Tie Li

N/A

Coupling Mechanism of the Micro Mass-spring System and Electrothermal Actuator and its application in optical switches
Changlei Feng, Jin Xie, Kaiquan Li, Zhuang Xiong, Bin Tang, Jun Dai

N/A

Improved Design of Polymer Micromachined Transmission for Flapping Wing Nano Air Vehicle
RASHMI KANT, Daisuke Ishihara, Ryotaro Suetsugu, Sunao Murakami, Prakasha Ramegowda

1320

S16: Micro/Nano/Molecular Fabrication II

Atomic-scale Manufacturing based on TEM
Litao SUN

1326

Research on Melt Electrowriting TPU Hydrophobic Microfiber Mesh for Directional Water Transport
Zungui Shao, Jiaxin Jiang, Junyu Chen, Huatan Chen, Guoyi Kang, Gaofeng Zheng

1328

Object Manipulation with Freestanding Magnetic Microfibers Fabricated by FDM 3D Printing
Qing Lu, Ki-Young Song, Yue Feng

1332

Electrochemical Nanoimprint Lithography
Hantao Xu, Lianhuan Han, Bingqian Du, Qinghui Meng, Yang Wang, Zhen Ma, Zhong-Qun Tian, Zhao-Wu Tian, Dongping Zhan

1337

Adaptive Wavefront Shaping for Direct Laser Ablation Application — N/A
Chong Kuong Ng, Fan Zhang, Peng Tan, Yuanliu Chen

Simulation and Experimental Study on Single Pulse Ablation by a Femtosecond laser — N/A
Fan Zhang, Peng Tan, Chong Kuong Ng, Yuanliu Chen

Study on the Minimum Size of Molecule by Employing the Single-Molecule Plasmonic Optical Trapping Method — N/A
Yang Yang, Biao-Feng Zeng, Chun-An Huo, Jia Shi, Wenjing Hong, Zhong-Qun Tian

S17: M/NEMS IV

Soft Micro-robots and Micro-actuators: Materials, Design, Control and Applications — 1347
Lining SUN

Ultrasensitive Mass Sensing Utilizing High-Order Mode Localization of Clamped-Clamped Microbeams with Distributed Electrodes — N/A
Ming Lyu, Jian Zhao, Pengbo Liu, Jiahao Song, Heng Zhong, Xianze Zheng, Yinghai Tang

Influence of Mass Loading Effect on Radiation Quality Factor of BAW Magnetoelectric Antenna — 1353
Junru Li, Chunrui Peng, Yang Gao, Wanchun Ren, Xuefeng He

Optical Measurement of the Dynamic Response of an Electrothermal Microactuator — 1358
Chen Hao, Wang Xingjie, Xi Zhanwen, Wang Jiong, Nie Weirong

A Wearable Strain Sensor Based on Fiber-structured PU/MXene/CNT Composite with Ultra-high Sensitivity and Broad Sensing Range — 1362
Guoxi Luo, Qiankun Zhang, Yunyun Luo, Ke Chen, Wenke Zhou, Libo Zhao, Zhuangde Jiang

Nonlinear threshold mass sensor using the snap-through phenomenon of a clamped–clamped micromachined arch beam — N/A
Heng Zhong, Jian Zhao, Ming Lyu, Jiahao Song, Pengbo Liu, Xianze Zheng, Yu Huang

Feasibility Study of Wearable Muscle Disorder Diagnosing Based on Piezoelectric Micromachined Ultrasonic Transducer — 1370
Mengjiao Qu, Hong Ding, Xuying Chen, Dengfei Yang, Dongsheng Li, Ke Zhu, Jin Xie

Multifunctional Cardiomyocyte-Based Biosensor for Electrophysiology-Mechanical Beating-Growth Viability Monitoring — 1374
Dongxin Xu, Jiaru Fang, Ning Hu

S18: Soft & Flexibe Devices and Applications III

Developing biocompatible and implantable flexible pressure sensors for health monitoring applications — 1378
Guozhen SHEN

An Electrical Double Layer-based Iontronic Tactile sensor for Detection of Biological Ionic Liquid — N/A
Yulu Liu, Shuyi Huang, Menglu Li, Xiangyu Zeng, Wei Li, Xiaozhi Wang

Strain-Engineered Bistable Clamped Thin Films: From Macroscale to Nanoscale Fabrication — N/A
Guangchao Wan, Ziao Tian, Borui Xu, Yongfeng Mei, Zi Chen

Highly Sensitive and Flexible Tactile Sensor Based on the Fabrication of Porous Graphene/Silicone Rubber Composites — 1384
Zhijian Chen, Yancheng Wang, Deqing Mei, Jie Jin

Uniformly distributed self-filling micro-strips for high-performance pressure-sensitive sensor — 1390
Zhiping Chai, Xingxing Ke, Han Chen, Jiaqi Zhu, Chuanfei Guo, Zhigang Wu

A Sensitive Flexible Strain Sensor via Anisotropy Microstructured Sensitized Surface Resistive Change for Human Motion Monitoring — 1394
Wenjie Fei, Shuo Zhang, Zhigang Wu

A surface and interior material identification technology based on dual-mode sensor — 1398
Zhuhui Yin, Ning Li, Weifeng Chen, Yue Jiang, Kaiyang Lan, Jiabin Peng, Zhengchun Peng

S19: MEMS/NEMS V

MEMS-based platforms for characterization of advanced functional materials — 1403
Liviu NICU

Design and Fabrication of A latching Silicon-based MEMS Switch
Tongtong Cao, Tengjiang Hu, Yulong Zhao
1405

MULTI-MODAL RESONANCE MEASUREMENT CAPABLE OF DISCERNING AND VISUALIZING EFFECTS DUE TO EXCITATION SCHEMES IN MEMS RESONATOR
Bo Xu, Jiankai Zhu, Tongqiao Miao, Jing Li, Qingyang Deng, Song Wu, Ting Wen, Fei Wang, Xiaoping Hu, Xuezhong Wu, Dingbang Xiao, Zenghui Wang
1410

Nanostructure Vanadium-doped Zinc Oxide Film Sensor Endowed with Enhanced Piezoelectric Response
Wei Gao, Yu Zhang, Binghe Ma, Jian Luo, Jinjun Deng, Weizheng Yuan
1415

Humidity Sensor with High Resolution and Fast Response Based on AlN Cantilever with Two Groups of Segmented Electrodes
Dongsheng Li, Hanyong Dong, Zihao Xie, Mengjiao Qu, Qian Zhang, Jin Xie
1419

Design and Fabrication of Integrated Piezoelectric Micropump with Vortex enhancement Optimization
Boshen Liang, Louis Paquet, Yongbin Jeong, Paul Heremans, David Cheyns
1423

3D-Printed Sugar Scaffold for High-Precision and Highly Sensitive Active and Passive Wearable Sensors
Dong Hae Ho, Panuk Hong, Joong Tark Han, Sang - Youn Kim, S. Joon Kwon, Jeong Ho Cho
N/A

S20: Nanomaterials I

Structure-Controlled Synthesis of Single-Walled Carbon Nanotubes
Yan LI
1430

A Novel Low-Temperature Post-Curing Transfer Method Of Graphene Wrinkling Surface For Strain Engineering
Kai-Ming Hu
1432

Synthesis of Porous Co_3O_4-ZnO and its Performance for Sensitive Ethanol Detection
Xiao Zhang, Yaohua Xu, Feng Wei, Anjie Ming
1436

Simultaneous Sensing of Refractive Index and Temperature Using a Symmetry-breaking Silicon Metasurface with Multiple Fano Peaks
Luo Zhao, Guoguo Kang, Junyi Wang, Haiyang Li, Cheng Zhang
1441

Dynamic Curvature Nanochannel-Based Membrane with Anomalous Ionic Transport Behaviors
Miao Wang, Xu Hou
N/A

Switchable PDT by a selfassembled FRET quenching nanoparticle
Ziying Li, Yu Gao
N/A

Enhanced Ion Sensing Stability with Nanotextured Biosensors
Yanfang Wang, Yuanjing Lin
1451

PS3: Poster Session III

Photonic chip-based ultrafast Raman soliton source
Zhao Li, Qingyang Du, Chaopeng Wang, Jinhai Zou, Juejun Hu, Zhengqian Luo
N/A

Responsive DNA Hydrogel Facilitates Isolation and Retrieval of Circulating Tumor Cells
SHUGUANG XUAN, HONGTAO FENG, YULIN ZHOU, YUQING HUANG, Yan Chen
N/A

Electron-beam Grayscale Lithography Using Solid Anisole
Rui Zheng, Ding Zhao, Min Qiu
N/A

Velocity Random Walk Modelling of a Silicon MEMS Resonant Accelerometer based on Non-AGC Control Loop
Dong Li, QianCheng Zhao, Jian Cui
1461

A Piezoresistive Pressure Microsensor Based on Simplified Fabrication Processes
QingGang Meng, Yulan Lu, Junbo Wang, Deyong Chen, Jian Chen, Bo Xie
1465

Fabrication and Verification of a Novel Low-g MEMS Inertial Switch
Min Liu, Weidong Wang, Xiao Wang, D SY, Yingmin Zhu, Yijia Du
N/A

Sensitive Acetone Gas Sensors based on MOF-derived Carbon Nanoparticles-decorated Mesoporous Fe_2O_3 Nanorods on MEMS
Li-Yuan Zhu, Kai-Ping Yuan, Xue-Yan Wu, Tao-Tao Wu, Hong-Liang Lu
N/A

Hollow MXene Sphere-based Flexible E-skin for Multiplex Detection of Applied Force *Xue-Feng Zhao, Xiao-Hong Wen, Meng-Yang Liu, Hong-Liang Lu*	N/A
Triboelectric nanogenerator powered electrowetting-on-dielectric actuators *Dongyue Jiang, minyi xu, Jie Tan, Yutao Wang, Penghao Tian*	N/A
Mass fabrication of 4H-SiC high temperature pressure sensors by femtosecond laser etching *Lukang Wang, You Zhao, Yulong Zhao, Yu Yang, Bo Li, Taobo Gong*	1478
3D Hierarchical Nanoarchitecture AuNPs/MXene@PAMAM based Biosensor for cTnT Detection in Human Serum *Xin Liu, Yong Qiu, Hao Wan, Liujing Zhuang, Ping Wang*	1482
Insight of Volatile Benzenes Sensing Mechanisms for Conjugated Polymer Based Gas Sensors *Jian Song, Tengfei Guo, Meng Liu, Yuelin Wang, Tie Li*	N/A
Multidimensional characterization of optogenetically engineered cells based on an integrated platform *Jia Yang, Lipeng Zu, Wenxue Wang, Ning Xi, Lianqing Liu*	N/A
Design and Optimization of Microfluidic System Based on Energy Minimization *yike zhou, Daoheng Sun*	N/A
Effect of Scan Line Spacing on Laser-Reduced Graphene Oxide Based Temperature Sensing *Rui Chen, Wei Zhou, Chiqian Xiao, Tao Luo*	1493
High current density electron wind forces in metallic graphene nanoribbons *Ji Zhang, Tarek Ragab, Cemal Basaran*	N/A
A TENG-pressure combo-sensor for accurate material identification *Weifeng Chen, Ning Li, Zhuhui Yin, Yue Jiang, Jiabin Peng, Zhengchun Peng*	N/A
Investigation of Flexible Sweat Sensor for Sodium-Ion Concentration with a Combination of Two Sensing Mechanisms *Jiufu Zheng, Xuankai Xu, Yunzhi Hua, Xiaojin Zhao, Wei Xu*	1502
Long - line transmission design of ocean multi - parameter sensor data based on CR600 *Xuan Wang, Yongqiu Zheng, Juan Cui, Haoling Zhang*	N/A
Sensitive Wide-Frequency-Response Acoustic Sensor Based on Evanescent Field Excited in Semi-buried Optical Waveguide Ring Resonator *Yongqiu Zheng, chen chen, Jiamin Chen, Liyun Wu, Yuan Han, Chenyang Xue*	N/A
Fabrication of monodisperse magnetic nanorods for improving hyperthermia efficacy *Shan Zhao, Nanjing Hao, John X.J. Zhang, P.Jack Hoopes, Fridon Shubitidze, Zi Chen*	N/A
Pyro-Electrospinning for the Fabrication of Nanofiber Membrane-Embedded 3D Devices *Bin Qiu, feng xu, Shouju Yao, Wenchang Tu, yike zhou, Shanshan You, lingling liu, Tianhao Wu, Daoheng Sun*	N/A
Photothermal gel microvalve applied to precise flow control of microfluidic chip *Kehan Chen, Jingwen Pan, Minghao Xu, Shuai Wang, wenqiang zhang*	N/A
AlN Hybrid-Coupled Resonator With Phononic Crystal Reflector *Kangfu Liu, Yuxi Wang, Tao Wu*	1517
Multilayered Electret Generator for Self-powered Wireless Data Transmission *Zeyuan Cao, Shiwen Wang, Zibo Wu, Rong Ding, Xiongying Ye*	N/A
Optimization of S1 Lamb wave resonator with Al0.8Sc0.2N *shuai shao, Zhifang Luo, Tao Wu*	1523
Structure design and parameter exploration of a new type of plasma anemometer *xianlong Liu, Daoheng Sun*	N/A
Preparation of nanofiber membrane with controllable pore size and diameter based on auxiliary counter electrode *lingling liu, Dangheng Sun*	N/A
Optimizing Design, fabrication and calibration of thin film heat flux gauge on ITO/In2O3 thermopile *Xin Li*	N/A
Super-lightweight flexible wireless optogenetic stimulation device *Rui Luo, Ziqi Mei, Zheng You, Dahai Ren*	1535

Research on the Bouncing-Ball Based Triboelectric Nanogenerator for Self-powered Vibration Frequency Monitoring *Xusheng Zuo, Taili Du, Fangyang Dong, Shunqi Li, Yongjiu Zou, Junhao Zhao, Peng Zhang, Yuewen Zhang, Peiting Sun, Minyi Xu*	1539
Surface modification of 3D printed microfluidic chip *Shanshan You, Kunpeng Zhang, Shouju Yao, lingling liu, Tianhao Wu, yike zhou*	N/A
Design of Piezoelectric Micro-Actuators Design Based on LiNbO$_3$ Thin Film *Yushuai Liu, Zhiyuan Gao, Kangfu Liu, Tao Wu*	1545
Design of a Miniature Flat Plate-type Propellant actuator for MEMS Safe and Arm Devices *Yaoxiong Wang, Jinhong Huang, Bin Tang, Chongfei Zhang, Jun Dai*	N/A
Patterning multilayer alginate/PCL scaffolds for culturing myocardial tissue *feng xu, qiang gao, bin lin, daohen sun*	N/A
Electrospun Polyimide Nanofiber Separators for Lithium-ion Batteries *Wenwang Li, Bangzhou Che, Jinghua Lin, Sinan Fu, Jiaxin Jiang, Gaofeng Zheng, Xiang Wang*	1554
Design and Research on High Overload Resistance of a Micro-mirror Structure *Qiwei Wang, Jin Xie, Zhuang Xiong, Bing Tang, Jun Dai*	N/A
Simulation and Analysis of Nano Robotic Manipulators Inside SEM *Ziliang He, Lue ZHANG, Zhan Yang*	1560
PEDOT:PSS-based Nanopore Electrochemical Transistor Sensors for Particle Recognition *Lin Li, Feng Zhou, Qiannan Xue, Xuexin Duan*	N/A
Soft Robotic Manipulation System Capable of Stiffness Variation and Dexterous Operation for Safe Human-Machine Interactions *Changyong Cao*	N/A
Ultra-light Metamaterial for Sound Absorption Based on Miura-ori Sandwich Structure *Yixin Wang, Jingwen Guo, Xingru Chen, Yi Fang, Xin Zhang, Hongyu Yu*	N/A
High-temperature Thin Film Heat Flux Sensor Made of Polymer-derived Ceramics for Harsh Environment *Guochun Chen, Zaifu Cui, Xiaochuan Pan, Zhenyin Hai, Daoheng Sun*	N/A
CFD Analysis of Mixing Process in a Cross-shaped Micromixer *Shuai Yuan, Bingyan Jiang, Tao Peng, Mingyong Zhou*	N/A
High-detectivity Infrared Detector Based on Dual-layer Thermopile *zhaohui yang, Gaobin Xu, Shirong Chen, Jianguo Feng, Yuanming Ma, Xing Chen*	1574
A High-Precision Mode Matching Method for Rate-Integrating Honeycomb Disk Resonator Gyroscope *Sheng Yu, Xuezhong Wu, Xiang Xi, Jiangkun Sun, Qingsong Li, Dingbang Xiao, Yongmeng Zhang*	1579
Direct Electron-beam Patterning of Monolayer MoS2 with Water Ice *Guangnan Yao, Ding Zhao, Yu Hong, Min Qiu*	N/A
Acoustofluidic micromixer with bubble induced ultrasonic microstreaming *PENG Tao*	N/A
An Optical Microphone Based on Fabry-Perot Etalon Stability Structure *Liyun Wu, Yongqiu Zheng, Jiamin Chen, Chen Chen, Xiaoqiang Hua, Chenyang Xue*	N/A
Capacitive Stretchable Strain Sensors Based on Wavy-Structured Metal Electrodes *Siqi YU, Hongyu Yu*	1589
Fabrication of Soft Magnetic Microstructures for Modulation of the Magnetic Field Distribution on the Micrometer Scale *Fengshan Shen, Yan Yu, Yuexuan Li, Hongtao Feng, Yan Chen*	N/A
Soft Interface Design for Electrokinetic Energy Conversion *Jian Zhang, Xu Hou*	N/A
An Underwater Material Recognition Device Based On The Seebeck Effect *Nanxi Chen, Baichuan Shan, Jianhao Liu, Yuhang Fan, Kaiyuan Zhao, Mengze Li, Changxin Liu*	N/A
Patterned Diphenylalanine Nanotubes Regulate the Behavior of Hippocampal Neurons *Lipeng Zu, Huiyao Shi, Jia Yang, Yuanyuan Fu, Wenxue Wang, Ning Xi, Lianqing Liu*	N/A

Temperature Compensation for MEMS Mass Flow Sensors Based on Back Propagation Neural Network
Yan Wang, Shijin Xiao, Jifang Tao

1601

Pressure Induced Transition from Wrinkling to Period-Doubling Instability in Flexible Tactile Sensors
Xiuyuan Li, Kai-Ming Hu, Yi-hang Xin, Erqi Tu, Wenming Zhang

1605

Improvement on the Uniformity of Deep Reactive Ion Etch for Electrically Isolated Samples
Xiao Hu, Zhihan Zhen, Qiyu Huang

N/A

Design of a novel closed-loop electrostatic voltage sensor based on weakly coupled resonators
zilong wang, Zhengwei Wu, Xiangming Liu, Yahao Gao, Simin Peng, Ren Ren, Fengjie Zheng, Yao Lv, Jun Liu, Hucheng Lei, Zhouwei Zhang, Wei Zhang, Jiachen Li, Chunrong Peng

1611

Effect of the pitch on electroforming process of the microchannels
Yanzhuo Dong

N/A

The Tunable Deformation of Microfluidic Strain Sensor Based on Auxetic Metamaterial
Linna Mao, Dengji Guo, Sirong Huang, Taisong Pan

1617

Effect of the substrate size on accuracy of in-situ stress measurements
Jun Qiang, Bingyan Jiang

N/A

Measurement of Comb Frequency and Spacing Stability in Phononic Frequency Comb
Qiqi Yang, Liu Xu, Ronghua Huan, Zhuangde Jiang, Adarsh Ganesan, Xueyong Wei

1623

Inductive Effect of Nanofiber Deposition on Groove Structure on Cell Culture
Wenchang Tu, yike zhou, Bin Qiu, Shanshan You, lingling liu, Tianhao Wu

N/A

Patterning and immobilization of silver Nanowires for flexible electronics by using Microwave
sara khademi, Kiyumars Jalili, Hao Wang, MD ESHRAT E ALAHI, Tianzhun Wu

1629

Rapid Fabrication of Flexible Strain Sensor for Plants Growth Monitoring
Yicong Zhao, Jing Niu, Wenxing Huo, Xian Huang

N/A

Temperature and pressure dual-parameter sensing based on Fiber Bragg Grating
Na Zhao, Qijing Lin, Liangquan Zhu, Kun Yao, Bian Tian, Libo Zhao, Zhuangde Jiang

1635

Achieving High Energy-Storable Polythiourea Dielectric Materials in Molecular Design through Tuning H-Bonds
Yang Feng, Guanghao Qu, Liuqing Yang, Huan Niu, Shengtao Li

N/A

Optical analysis of perovskite light-emitting diodes with nanostructured emissive layer via electrical simulation-assisted dipole location
liyang chen, Zhuofei GAN, Dehu CUI, Jingxuan CAI, Wendi LI

N/A

Ultrathin elastic shape sensor used for endoscope shape reconstruction
Leixin Meng, Yuan Zhuang, Liqiang Wang, Qing Yang

N/A

Microfluidics for Single-cell Multi-omics Analysis
Xing Xu, Chaoyong Yang

N/A

A Biosensor Kit Based on Cu-MOF for Simultaneous Detection the Cortisol and Cyfra21-1 in Human Saliva
Xinyi Wang, Tao Zhang, Shuqi Zhou, Liubing Kong, Wencheng Lin, Hao Wan, Ping Wang

N/A

Laser Induced Graphene Patterns on a Thin Polyimide Film via a cooling plate
Qixiang Chen, Dezhi Wu, Zhiwen Chen, Hang Yu, Qinnan Chen, Daoheng Sun

1650

A Robotic Dynamic Tactile Sensing System based on Electronic Skin
Jishen Dai, Yu Xie, Dezhi Wu, Chen Songyue, Ting Fu, Wei Zhou

1655

Characterization of an Asymmetrical Capacitive MEMS Tilt Sensor
Yang Gao, Xiaopeng Zhang, Yonghong Qi, Maeda Ryutaro, Zhuangde Jiang, Xueyong Wei

1660

A Reflective Color Filter Based on ITO-Ba0.5Sr0.5TiO3-ITO Nanofilms Capitalizing on a Black Layer
Rui Wang, Jinying Zhang, Bingnan Wang, Xinye Wang, Defang Li, Jingyi Chen, Chenyu Guo, Xin Wang, Zhuo Li, Suhui Yang

1664

High-performance rubbery electronics enabled by elastomeric composite materials
Kyoseung Sim

N/A

A flexible rope-like structure sensor based on triboelectric nanognerator for multifunctional sensing

N/A

Cong Zhao, Anaeli Elibariki Mtui, Xiangyu Liu, Kun Jiang, Jianye Wang, Chuan Wang, Minyi Xu

Giant Optical Activity in Achiral Plasmonic Au Nanocones Embedded into Alumina Nanohole Arrays — N/A

Yuyi Feng

Ferroelectric Nanocrack-based Nanoelectromechanical (NEM) Switches for Memory and Complementary Logic — N/A
Yaodong Guan, Zhe Guo, Qiang Luo, Jeongmin Hong, Long You

Conical Helmholtz Resonator-Based Triboelectric Nanogenerator for Harvesting of Acoustic energy — 1676

HongYong Yu, Taili Du, Hongfa Zhao, Qiqi zhang, Ling Liu, yue Huang, Anaeli Elibariki Mtui, Xiu Xiao, Minyi Xu

3D Cell Electrical Impedance Biosensor for Real-Time Drug Permeability Gradient Effect Monitoring — N/A
Yong Qiu, Xin Liu, Hao Wan, Ping Wang, Liujing Zhuang

Triboelectric Nanogenerator for Wind Energy Harvesting and Speed Sensing — N/A
Yan Wang, Chuanqing Zhu, Mengwei Wu, Hao Wang, Xiu Xiao, Guochang Wang, Jianchun Mi, Minyi Xu

Flexible Triboelectric Nanogenerator for Flow Control — N/A
Chuanqing Zhu, Xiangyi Wang, Mengwei Wu, Jialin Zhang, Chenxing Jia, Guochang Wang, Jianchun Mi, Minyi Xu

Design and fabrication of a multimode optical fiber surface plasmon resonance urea biosensor — 1686
Jiayin Li, Mengdi Han, Ji Wan, Haixia Zhang

A Resonant Differential Pressure Microsensor With a Stress Isolation Layer — 1691
Yadong Li, Chao Cheng, Yulan Lu, Jian Chen, Deyong Chen, Junbo Wang

A Novel Dual-channel Helmholtz Resonance Acoustic Energy Converter Based on Friction Nanogenerator — 1695
Ling Liu, Hongfa Zhao, Zhenhui Lian, Hongyong Yu, Qiqi Zhang, Wenxiang Li, Minyi Xu, Xiu Xiao

Fabrication of Multi-oriented Composite Nanofibrous Membrane by Electrospinning — 1700
Xiang Wang, Yongfu Xu, Jinghua Lin, Sinan Fu, Jiaxin Jiang, Gaofeng Zheng, Wenwang Li

A Breathable and Flexible Epidermal Glucose Sensor — N/A
Hailong Chen, Zhihua Pu, Dachao Li

Design of Low Sidelobe Level Milimeter-wave Planar Slotted Array Antenna Fed by Ridge Waveguide — N/A
Yan Cao, Yiming Tang

Modularized Hydrogel-Gate Field-Effect-Transistor Biosensors — N/A
Xiaochuan Dai

Bandwidth Expansion in Acoustic Filters based on AlN-on-Silicon Resonators — N/A
Jiannan Chen, Cheng Tu, Feilong Li, XIAOSHENG ZHANG

An Electromagnetic Actuator with Nonlinear Planar Micro-Spring for extended output range — N/A
Xuhan Dai

A High DC Current Output Salt Battery Hybrid and Enhanced by TENG — 1714
shuangshuang liu, Jiayi Yang, Zihao Niu, Yan Meng, Wei Xu, Di Feng, Baolong Wang, Meiqi Wang, Xiuhan Li

The Influence of Electrode on Elastic Constant C33D Extraction of Scandium-doped Aluminum Nitride Thin Film by Thickness-extensional Mode FBAR — 1718
hu xia

One-step preparation of underwater superoleophobic aluminum-coated copper mesh by pulsed laser cladding for enhanced oil-water separation — N/A
Junjin Lai, Rui Zhou

A Resonant High Pressure Sensor Based on Dual Cavities Design — 1725
Jie Yu, Yulan Lu, Deyong Chen, Junbo Wang, Bo Xie

MICROFLUIDIC DEVICE FOR ISOLATING CIRCULATING FETAL CELLS — N/A
Huimin Zhang, Chaoyong Yang

Numerical analysis of laser ablated structural size effect on enhanced anti-icing property of TC4 surface — 1732
Zhekun Chen, Rui Zhou, Yuhang Lin, Yi Zhu, Huangping Yan

Facile access to solvent-dependent luminescent carbon nanoparticles by laser ablation of activated carbon powders *Zhibin Chen, Rui Zhou, Longfan Li, Jingqin Cui*	1737
Fabrication of Superhydrophobic Surface by Inkjet Printing of Nano Silver Seeds on Porous Paper *Xinghao Zhang, Jin Xie, YU LIU*	N/A
A MEMS Voice Coil Motor with a 3D solenoid coil *Jiamian Sun, Zhi Tao, Haiwang Li, Kaiyun Zhu, Donghui Wang, Tiantong Xu, Hanxiao Wu*	1745
Tribovoltaic and Tribo-thermoelectric coupling effect on metal-semiconductor interface *Zhi Zhang, Chi Zhang*	N/A
Full printed flexible pressure sensor based on microcapsule controllable structure and composite dielectrics *Xiangyou Meng, Jing Zhao, Yaqin Pan, ziyun Han, Lixin Mo*	N/A
Nanoscale Triboelectrification Gated Transistor *Tianzhao Bu, Chi Zhang*	N/A
Modulated flexible artificial synaptic transistor based on graphene and ionic gel *Di Feng*	N/A
Preparation and research of ITO thin film strain gauge *Tao Yang, daoheng sun*	N/A
Screen Printing-Based Wearable Multi-Sensing Double-Chain Thermoelectric Generator *Tao Feng, Danliang Wen, Haitao Deng, Xiaosheng Zhang*	N/A
Optical Fiber Waveguiding Soft Photoactuators Exhibiting Giant Reversible Shape Change *Yongcheng He, Haohua Liang, Jiulin Gan, Zhongmin Yang, Meihua Chen*	N/A
Study of strain sensing behavior of self-powered stretchable mechanoluminescent optical fiber *Haohua Liang, Yongcheng He, Meihua Chen, Jiulin Gan, Zhongmin Yang*	N/A
The large piezoelectricity and high-power density of a 3D-printed multilayer copolymer in a rugby the ball-structured mechanical energy harvester *Xiaoting Yuan, Zhanmiao Li, Zhonghui Yu, Shuxiang Dong*	N/A
Tuning and Visualizing Motional Signal Transduction Efficiency in Atomically-Thin Nanoelectromechanical Resonant Structures *Jiankai Zhu, Jing Li, Xu Bo, Song Wu, Fei Xiao, Yachun Liang, Ting Wen, Fei Wang, Zenghui Wang*	N/A
Particle-Filled PHPS Silazane-Based Insulation Coating with High-temperature Electrical Insulating Properties on Ni-based Superalloy Substrates *Ji'an Lin, Qinnan Chen, Zaifu Cui, Zhenyin Hai, Daoheng Sun*	N/A
Research on torque calculation model of MR fluid-based micro-brake in thermal environment *Yan Zhang, Ying Liu, Jun Dai*	N/A
A Low Power MEMS microheater for MOS Gas Sensor Applications *Ziwei Lian*	N/A
Electron-beam-induced 3D Direct-writing of a Nanoelectrode Array for Intracellular Electrophysiological Recording *Wenqi ZHANG, Lixin Dong*	N/A
Electrically-responsive colloidal particle assembly in ionic surfactant solutions *Minqi Yang, Lisha Luo, Haiyang Fu, Hongjie Yin, Zhibin Yan, Mingliang Jin, Huicheng Feng, Guofu Zhou, Lingling Shui*	N/A

CS3: 学术奥斯卡讲座

学术奥斯卡讲座：如何做好学术报告？ *Haixia ZHANG*	N/A

IS19: IS: Multidisciplinary Frontier: Micro/Nano-Engineered Materials for Sensing and Applications

Flexible Human-Machine Interacting Sensors *Yanchao MAO*	N/A

Soft Robotic Manipulation System Capable of Stiffness Variation and Dexterous Operation for Safe Human-Machine Interactions *Changyong CAO*	N/A
Borophene-Based Two-Dimensional Materials and Devices *Guoan TAI*	N/A
Functional photonic crystals with superwettability *Jingxia WANG*	N/A
Two-Dimensional Transition Metal Carbides and Nitrides for Terahertz Absorption Technology *Xu XIAO*	N/A
Room Temperature Gas Sensors Based on Transition Metal Dichalcogenides Heterostructures *Zhi YANG*	N/A

IS20: IS: Multidisciplinary Frontier: Functional Micro-nano Structures, Techniques and Applications

Nanoforests and nanoforest-based micro sensors *Haiyang MAO*	N/A
Bioinspired Catalytic Materials Based on Metal-Sulfur Clusters *Jian LIU*	N/A
Mass transport in atomic scale confinements *Sheng HU*	N/A
Quality Heterostructure from Two-Dimensional Materials *Yang CAO*	N/A
Fiber integrated multifunctional sensing and super-resolution imaging *Qing YANG*	N/A
The applications of vanadium dioxide thin films on multi-functional sensor *Min GAO*	N/A

S21: MEMS/NEMS VI

High Accuracy Resonant Accelerometers Based on QMEMS *Yulong ZHAO*	1800
INFRARED THERMAL IMAGING SENSOR USING TRANSFERRED RUPHEN-BASED TEMPERATURE SENSITIVE FILM *Jialiang LI, Jun Hu, Yufei Zhai, Song Li, Yiming Yuan, Min Wang*	N/A
Hybrid Frequency-time Domain Analysis of Bulk Acoustic Wave Circulator *yuan jing, zhang bingbin, Wu Chengfeng, gao yang*	1804
A Piezoelectric Micromachined Ultrasonic Array with Frequency-selectable Tubes *Wei Zhu, Lei Wang, Zhipeng Wu, Wenjuan Liu, Chengliang Sun*	N/A
A Micromachined Electrochemical Angular Accelerometer Based on Interdigital Electrodes *Tian Liang, Junbo Wang, Deyong Chen, Jian Chen, Bowen Liu, Chao Xu, Wenjie Qi, Xu She, Mingwei Chen, Anxiang Zhong, Yumo Duan*	1813
Research on the Stability Threshold and Subharmonic Oscillation in a Parametric Excitation MEMS Resonator *Kuo Lu, Kai Wu, Qingsong Li, Xin Zhou, Yongmeng Zhang, Ming Zhuo, Xuezhong Wu, Dingbang Xiao*	1817
Preparation and Detection of Micro-nano Electrode Array Based on Chipmunk Hypothalamus in Specific Brain Regions *Yiding Wang, Shengwei Xu, Chao Yang, Penghui Fan, Botao Lu, Gucheng Yang, Yinghui Li, Xinxia Cai*	N/A
Large-Swing Low-Distortion High Voltage Amplifier Design for MEMS Gyroscopes *Hua Chen, Ke Liu, Zhen Meng*	1825

S22: Bio Inspired Multiscale Interfaces

Bio-inspired multiscale adhesive interfacial materials *Shutao WANG*	1830
Nanokits for Single Cell Analysis	1832

Dechen JIANG

SWCNTs/PEDOT:PSS Modified Microelectrode Array For Detection Of The Neuronal Electrophysiological Activity Of Cortical Neurons Under Glutamate Stimulation *Shihong Xu, Yu Deng, Enhui He, Shenwei Xu, Longzhe Sha, Kui Zhang, Yiling Song, Jinping Luo, Qi Xu, Xinxia Cai*	N/A
MWCNT/PEDOT:PSS Nanocomposites-modified Microelectrode Array For Recording Neural Activity Of hiPSCs-derived Mature Neurons To Different Concentrations Of K+ And Glu *Enhui He, You Zhou, Shihong Xu, Kui Zhang, Shengwei Xu, Yilin Song, Jinping Luo, Wanwan Zhu, Qi Xu, Xinxia Cai*	N/A
Bioinspired Liquid Gating Membrane-based Catheter with Anticoagulation and Positionally Drug Release Properties *Chunyan Wang, Xu Hou*	N/A
Ultra-thin flexible neuro probe utilizing biodegradable collagen microneedle *Dong Huang, Junshu Li, Zhongyan Wang, Zhitong Zhang, Qining Wang, Zhihong Li*	1839
Fabrication of polymer/metal composite micro/nano array structures and their applications in biological interfaces and actuators *Hongxu Chen*	1843

S23: Nanomaterials II

Design of Sub-nanometre Pores for Precise Separation *Jian JIN*	1847
Self-powered Flexible Supercapacitor for Human Motion Monitoring *Zheng Fan, Hongmei Chen, Yanyun Fan, Danfeng Cui, Shubin Yan, Chenyang Xue*	N/A
Optoelectronic Synapse using MoS2 van der Waals Homojunction with Interfacial Passivation for Fast Potentiation *Yizhen Ke, Lin Lin, Tianxun Gong, XIAOSHENG ZHANG, Wen Huang*	N/A
Bionic Tactile Sensor based on Triboelectric Nanogenerator for Motion Perception *Siyuan Wang, Peng Xu, Xinyu Wang, Hao Wang, Changxin Liu, Liguo Song, Guangming Xie, minyi xu*	1853
Triboelectric Nanogenerator Based on Silver Nanowires *Fengru Fan*	N/A
Functionalization and application of inorganic porous nanofibers *Yunqian Dai*	N/A

S24: Nature-inspired Energy Conversion and Harvesting: from Fundamentals to Applicaitons

NEMS-enabled Innovations at Interfaces for Water-Energy Nexus *Zuankai WANG*	1861
Application and Prospect of MEMS Technology to Geophysics *Heting Hong, Lvchao Ni, Hongchun Sun*	1863
Design and Analysis of MEMS Biaxial Coupled Resonance Accelerometer *Huimin Zhang, Yating Zhang, Wei Zhang*	1867
Development of Annular-Shaped Bernoulli Gripper for Contactless Gripping of Large-Size Silicon Wafer *Shihang Wang, Yancheng Wang, Deqing Mei, Songqiao Dai*	1871
Ultrasound-mediated Tough Tissue Adhesives Using Nanoparticles *Shuaibing Jiang, Zhenwei Ma, Tony Jin, Edmond Lam, Audrey Moores, Jianyu Li*	N/A
A Self-Powered Angle Sensor at Nanoradian-Resolution for Robotic Arms and Personalized Medicare *Ziming Wang, Jie An, Jinhui Nie, Jianjun Luo, Jiajia Shao, Tao Jiang, Baodong Chen, Wei Tang, Zhong Lin Wang*	N/A
Dielectrophoretic microdevice with planar electrodes for concentrating microparticles *Salini Krishnan, Alia Mohammmed Shaker Alblooshi, Fadi Alnaimat, Bobby Mathew*	N/A

S25: MEMS/NEMS VII

Semiconductor Biosensors for Global Health in the Era of AIoT 1885
Yikuen LEE

Frequency Output, A Multi-modes Vacuum Gauge with Highly Output linearity Based on 1888
Electrostatic Nonlinearity
*wang chengxiang, Yunbing Kuang, Wu Yulie, Hou Zhanqiang, Zhang Yongmeng, Wu xuezhong,
Dingbang Xiao*

High-resolution noncontact electrostatic voltage meter based on microsensor chip 1893
Xiaolong Wen, Pengfei Yang, Bo Zhang, Chunrong Peng

Preparation and Characterization of Ni-AAO Composite Scaffold Getter with Induction Heating 1897
Lujiang Liu, Lingyun Wang, Zhenxiang Bu, Kuixi Wu, Wenlong Lv, Dandan Gu

S26: Soft & Flexibe Devices and Applications IV

Bio-Inspired Ionic Skin for Theranostics 1901
Peiyi WU

Ultra-flexible and highly transparent hydrogel-based triboelectric nanogenerator for physiological 1903
signal monitoring
*Jiahao Yu, Zhensheng Chen, Haozhe Zeng, Bowen Ji, Zixuan Wu, Jin Wu, Yunjia Li, Jianbing Xie,
Honglong Chang, Kai Tao*

A Flexible Tactile Sensor for Three-dimensional Force Detection Based on Piezoelectric Sensing 1908
*Yunyun Luo, Libo Zhao, Guoxi Luo, Zhikang Li, Ping Yang, Qiankun Zhang, Qijing Lin, Zhuangde Jiang,
Hongyan Wang, Yongshun Wu*

Spontaneous breath analysis and healthcare assessment enabled by triboelectric nanogenerator N/A
Yuanjie Su, Mingliang Yao, Guangzhong Xie, Huiling Tai, Yadong Jiang

S27: Soft & Flexibe Devices and Applications V

The Reinvention of silk 1914
Hu TAO

A Multifunctional Soft Robotic Finger based on Nano Temperature-Pressure Sensor for Material N/A
Recognition
Xiaoshuang Zhang, Wentuo Yang, Cheng Zhou, Xueyou Sun, Mengying Xie, Xuexin Duan

Readout of Neural Activity in Mammalian Peripheral Olfactory System via Microelectrode arrays N/A
Liujing Zhuang, Yan Duan, Suhao Wang, Qunchen Yuan, Jizhou Song, Ping Wang

A Three-electrode Multi-module Sensor for Accurate Bodily-kinesthetic Monitoring N/A
*Haobin Wang, Yu Song, Hang Guo, Ji Wan, Liming Miao, Chen Xu, Zhongyang Ren, Xuexian Chen,
Haixia Zhang*

S28: Micro & Nanotechnologies for Biomedicine

Biomolecular Needling System for Medicals 1921
Beomjoon KIM

Physical Forces Influence the Self-organization of the Formation of Leader Cells During Collective 1923
Cell Migration
Mengyun Pan, Yongliang Yang, Lianqing Liu

TENG Enhanced and Self-Powered Sensors for Biomedical Signal Detection N/A
Jieyu Dai, Zhou Li

Gap in pagination due to unavailable paper.

Pages 947-948

Proceedings of the 16th Annual IEEE International
Conference on Nano/Micro Engineered and Molecular Systems
April 25-29, 2021

A Method for Automatic Counting and Labeling of Cells Stained with Microporous Membrane

Jiangcheng Cao, Tingting Hun, Wenbo Zhou, Zheng Liu, Wei Wang and Yufeng Jin

Abstract—**Cells in the blood can be screened through a membrane. But the method of manually identifying the number of cells is time-consuming and laborious, and there may even be some errors. Therefore, an effective image recognition method was adopted in this paper to automatic count and label the stained cells on the membrane after filtration. We tested the images of cells without the influence of the membrane, and compared them with the actual number of cells with 90% accuracy. Under the influence of the filter membrane, the accuracy is about 80%. Therefore, this method can be used to calculate the number of cells in the images relatively accurately and efficiently.**

I. INTRODUCTION

Malignant tumor has become a serious threat to human health. In 2015, there were about 3.929 million cases of malignant tumors and about 2.338 million deaths. On average, more than 10,000 people are diagnosed with cancer every day, and 7.5 people are diagnosed with cancer every minute. In addition, the incidence and death of malignant tumors have been on the rise in recent decades, and the medical expenses caused by malignant tumors exceed 220 billion yuan every year. Therefore, the examination and treatment of malignant tumors is also a very important research hotspot. For example, the 5-year relative survival rate for patients with advanced lung cancer is only 5 percent, while the 5-year relative survival rate for patients with early stage lung cancer is significantly improved to 56 percent. Therefore, early diagnosis and timely treatment have been recognized as the key to improve the prognosis of patients [1]. In liquid biopsies, the use of membranes to screen for rare tumor cells in the blood shows great promise for the early detection of human cancers. In cell filtration experiments, it is necessary to screen and label the cells left on the membrane after filtration, so as to detect the metastasis of tumor cells [2]. Cell counting plays an important role in the research of medical biology. However, the traditional manual technique under the microscope is time consuming, low efficiency and high labor intensity. In addition, the subjective choice of the experimenter in the process of counting is easy to lead to non-repetition and large error in the process of counting. Therefore, the image processing method of cell automatic counting and labeling achieves the purpose of high efficiency and accuracy. It greatly reduces the subjective influence of laboratory

personnel and makes the counting results more objective, which is of great significance for biomedical experimental research. Therefore, an effective image recognition method was used in this paper to automatic count and label the stained cells on the membrane after filtration. Firstly, the cell image is processed with gray scale, followed by filtering operation, and finally the appropriate threshold is selected for recognition. Although an automatic counting method of staining cells was proposed in some papers, the influence of filtration membrane could not be ignored in practical experiments [3-7].

Therefore, this paper also proposes a method of automatic cell counting under the influence of filtration membrane. The influence of filter membrane can be maximized by using Fourier filtering operation to facilitate the selection of appropriate threshold value for identification. Through practical verification, this method has a high accuracy rate and can save a lot of time.

II. METHOD AND EXPERIMENTS

For cell counting, we use a Python program. Use Python specific OpenCV software library to carry out various operations of the cell images [8]. We divided the images into two categories for processing. As shown in Figure 1, one is the image of cells without filter membrane, and the other is the image of stained cells with filter membrane.

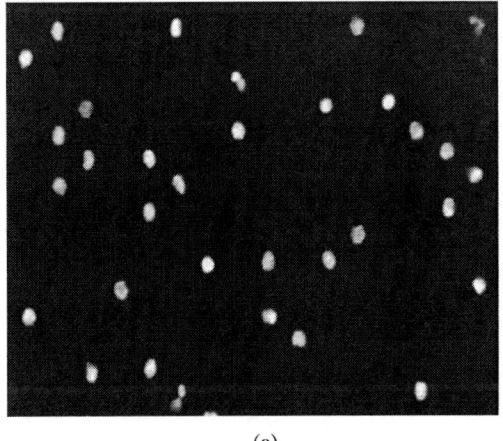

(a)

---

*Resrach supported by TSV 3D Integrated Micro/Nano system Lab and Shenzhen Science and Technology Innovation Committee. (Corresponding author: Wei Wang ang Yufeng Jin)

Jiangcheng Cao and Yufeng Jin are with Peking University ShenZhen Graduate School, Guangdong, 518055, China. (e-mail: 1801213274@pku.edu.cn, yfjin@pku.edu.cn).

Tingting Hun, Wenbo Zhou, Zheng Liu and Wei Wang are with Institute of Microelectronics, Peking University, Beijing, 100872, China. (e-mail: huntingting@pku.edu.cn, zhouwb1995@163.com, 756620476@qq.com, w.wang@pku.edu.cn).

(b)

Figure 1. Cell images. (a) is the image of cells affected by no filtration membrane, and (b) is the image of cells affected by filtration membrane.

The cell counting operation process is shown in Figure 2. After reading the cell image, the cell image is firstly gray processed, followed by filtering operation and selection of appropriate threshold value, and finally the cell recognition result is output.

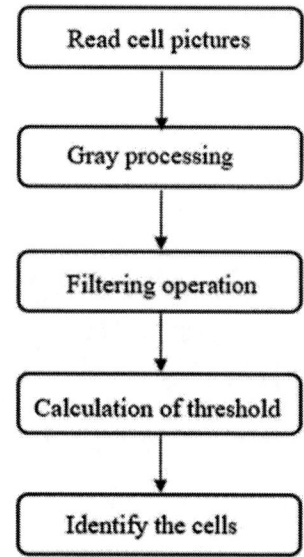

Figure 2. Flow chart of cell picture automatic counting. After grayscale, noise removal and threshold determination of cell images, the processed results are output.

The main steps in this operation are Gaussian filtering, selection threshold and Fourier transform. We will discuss each of these in the following sections.

A. Gaussian Filter

Filtering is a basic task in signal and image processing. The purpose is to selectively extract some important information from the image according to the different application environment. Filtering can remove noise from an image, extract visual features of interest, allow resampling of the image. Here we are using a Gaussian filter.

$$h(x, y) = e^{\frac{-(x^2+y^2)}{2\sigma^2}} \qquad (1)$$

Gaussian filter is a linear filter, it can use the two-dimensional Gaussian function distribution to smooth the image, will not change the original image edge direction, to ensure the characteristics of feature points and edges.

B. Threshold Selection

An image includes the target object, the background and the noise. In order to directly extract the target object from the multi-valued digital image, a common method is to set a threshold T, and use T to divide the image into two parts: the pixel group greater than T and the pixel group less than T. This is the most special method to study the gray transformation. The difference in gray characteristics is used to select a reasonable threshold value to determine whether each pixel in the image should belong to the target area or the background area, so as to produce the corresponding binary image. That is, the whole image is presented with only black and white visual effect.

$$\text{Pixel}(x, y) = \begin{cases} 1, & Pixel(x, y) > T \\ 0, & Pixel(x, y) < T \end{cases} \qquad (2)$$

C. Gaussian Filter

For an M×N image, the formula for its Fourier transform is as follows:

$$F(u, v) = \sum_{x=0}^{M-1} \sum_{y=0}^{N-1} f(x, y) e^{-j2\Pi(\frac{ux}{M} + \frac{vy}{N})} \qquad (3)$$

Fourier proposed that any periodic function could be expressed as a sum of sines function or cosines function of different frequencies. The frequency of cell image is an indicator of the intensity of gray scale change in an image, and it is the gradient of gray scale in plane space. The physical meaning of Fourier transform is to transform the gray distribution function of the image into the frequency distribution function of the image. And we can use the different frequency distribution of the image to select the desired target. The Fourier transform frequency diagram is shown in Figure 3.

Figure 3. Fourier transform of the cell image in the frequency domain. Different parts of the cell image correspond to different positions of the spectrum.

The bright spots that we see in the Fourier spectrum, that is, the magnitude of the gradient. Generally speaking, if the gradient is large, the brightness of the point is strong; otherwise, the brightness of the point is weak.

III. RESULT

Using the above operation, we experimented on two types of cell images respectively.

A. Cell Image Processing without Membrane Filtration

Because of the background noise has little influence, we can directly operate the images without the influence of filter membrane. The specific operation process is shown in Figure 1. First of all, we input the cell images into computer, and after graying, noise removing and thresholding, the final output is as shown in Figure 4. From the image, we can clearly identify each cell and calculate the specific number of cells.

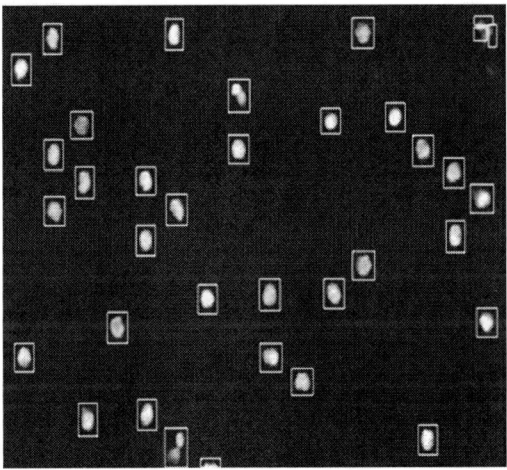

Figure 4. Automatic counting experiment without membrane influence.

TABLE I. Automatic Counting Results Of No Filter Membrane Influence

Cell Images Number	Experimental Values	Actual Values	Accuracy Rate
1	16	17	
2	20	24	
3	25	30	
4	20	23	
5	24	26	91.7%
6	33	34	
7	30	32	
8	23	23	
9	24	25	

After many experiments, the experimental results are shown in Table I. We found that the average accuracy of this method was over 90%.

B. Cell Images Processing with Filtration Membrane

The filter membrane used in this experiment is shown in Figure 5, with a diameter of 10 microns [9].

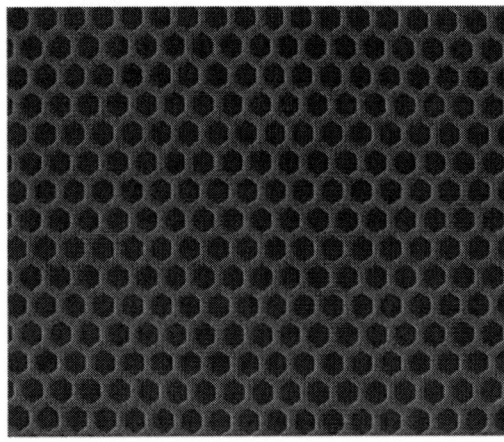

Figure 5. Filter membrane.

When we deal with the cell images under the influence of filter membrane, we directly operate the cell image and find that the error is very large. So here we do Fourier transform on the cell image. Fourier transform is used to transform the image in frequency domain. First of all, we input the cell image into computer. The filter membrane has a great influence on the screening of cells, but the background filter membrane and staining cells are different in the frequency domain of the image. The stained cells are more corresponding to the lower region of the frequency domain, while the filter membrane corresponds to the high frequency region. Therefore, Therefore, the cells to be detected can be screened out by removing the high frequency and retaining the low frequency information. The experimental results are shown in Figure 6.

Figure 6. Automatic Counting Experiment With Influence Membrane.

After many experiments, the experimental results are shown in Table II. We found that the average accuracy of this method was over 90%.

TABLE II. AUTOMATIC COUNTING RESULTS OF FILTER MEMBRANE INFLUENCE

Cell Images Number	Experimental Values	Actual Values	Accuracy Rate
1	62	72	
2	38	44	
3	58	61	
4	33	35	
5	58	65	
6	116	138	85.5%
7	59	76	
8	47	56	
9	15	15	
10	53	60	
11	15	27	
12	54	62	

IV. CONCLUSION

It takes a lot of time to manually identify the number of cells in an electron microscope image, but automatic computer recognition can save much more time and effort. We used Python program to count two types of cell images, and the accuracy of cell images without the influence of filtration membrane was as high as 90%. Under the influence of filter membrane, we use Fourier operation to reduce the influence of filter membrane on cells as much as possible, and the final screening accuracy is about 80%. Therefore, this method can be used to calculate the number of cells in the images relatively accurate and efficient.

ACKNOWLEDGMENT

The work was financially supported by TSV 3D Integrat ed Micro/Nano system Lab(ZDSYS201802061805105), th e Shenzhen Science and Technology Innovation Committe e(JCYJ20190808155007550).

REFERENCES

[1] Siegel RL, Miller KD, Jemal A. Canner statistics, 2019. CA Cancer J Clin. 2019; 69: 7-34.

[2] Tingting Hun, et al, "In-site electroporation on a micropore-array filter for cell labelling," 2020 IEEE 33rd International Conference on Micro Electro Mechanical Systems(MEMS), pp. 473-476, 2020.

[3] Yuefei Lin, et al, "Automatic cell counting and labeling for fluorescence microscope images," 2019 IEEE 13th International Conference on Anti-counterfeiting, Security, and Identification, pp. 197-201,2019.

[4] Mogeeb A. A. Mosleh, et al, "An automatic nuclei cells counting approach using effective image processing methods," 2019 IEEE 4th International Conference on Signal and Image Processing, pp. 865-869, 2019.

[5] Varun D Dvanesh, et al, "Blood Cell Count using Digitial Image Processing," 2018 International Conference on Current Trends towards Converging Technologies

[6] Daniel Riccio, "A new unsupervised approach for segmenting and counting cells in high-throughput microscopy image sets," IEEE Journal of Biomedical and Health Informatics, 2019.

[7] Tanapat Autaiem, "A novel iterative method for automatic avian red blood cell counting," 2019 12 th Biomedical Engineering International Conference.

[8] Raymond Joseph Meimban, "Blood cells counting using Python opencv," 2018 14th IEEE International Conference on Signal Processing, pp. 50-53, 2019.

[9] Li Tingyu and Liu Yaoping, et al., "A rapid liquid biopsy of lung cancer by separation and detection of exfoliated tumor cells from bronchoalveolar lavage fluid with a dual-layer 'PERFECT' filter system", THERANOSTICS, vol 10, pp.6517-6529, 2020.

Proceedings of the 16th Annual IEEE International
Conference on Nano/Micro Engineered and Molecular Systems
April 25-29, 2021

Design of a Controllable Push-triggered Microfluidic Chip for Vitrification Reagent Loading/unloading

Guangyi Cai[1], Boshi Jiang[1], Jiaxin Zhu[1], Fenglin Liu[1], Tianzhun Wu[1*]

1. Shenzhen Institute of Advanced Technology,Chinese Academy of Science
corresponding author e-mail: tz.wu@siat.ac.cn

Abstract— Microfluidic chips can be utilized to demonstrate the vitrification reagents loading/unloading for oocytes, reducing the physical damage resulting from the transfer of cells during manual operations. In this paper, we present a controllable press-triggered microfluidic chip, which realizes the automatic loading/unloading of vitrification reagents. Unlike the reagent exchangement in traditional manual operations that requires cell transfers, cells are fixed during the vitrification in order to facilitate the medium change operation. While simplifying the operation, it uses less reagents to complete the purpose of cell shrinkage, which provides a new solution for the automation and standardization of cell vitrification

I. INTRODUCTION

The cryopreservation of oocytes is of great significance in the field of human assisted reproduction and the preservation of rare and endangered species[1]-[3]. At present, the methods of cell cryopreservation mainly include vitrification and slow freezing. Vitrification refers to the process in which high-concentration cryoprotectants change directly from liquid to vitreous during ultra-fast freezing, which avoids the formation of intracellular and extracellular ice crystals, and the freezing effect is better than slow freezing[4]. However, the use of high concentrations of protective agents will bring many side effects to cells. During the loading and removal process of cryoprotectant, the rapid change of extracellular osmotic pressure will cause rapid contraction or expansion of cell volume, which will damage the cell membrane structure and cytoskeleton system[5]. Also, the chemical toxicity of cryoprotectants can cause toxic damage to cells[6]-[7]. In

order to reduce the osmotic effect and chemical toxicity of cryoprotectants on the oocytes, the method of stepwise loading and removal of cryoprotectants is usually used. Unfortunately, this method requires multiple transfers of cells to different reagents, which will cause permanent physical damage to the cells while complicating the operation. In addition, this method requires a large amount of reagents (280 μL equilibrium solution, 300 μL vitrification solution). To solve these problems, we designed a controllable push-type microfluidic chip, which can realize the loading/unloading for oocytes with a very small amount of reagents (30 μL equilibrium solution, 20 μL vitrification solution), and reducing the physical damage due to transfer of cells during manual operations.

II. EXPERIMENTAL SECTION

A. Fabrication Process of the microfluidic chip

SU8-3050 photo resist was spin-coated on the silicon wafers and patterned by photolithography technology (Figure1.B 1-2). Then, sample was etching using Inductively Coupled Plasma (ICP) (Figure 1.B 3). After that, silicon wafer was immersed into SU-8 developer to remove the photo-resist (Figure 1.B 4). Afterwards, PDMS prepolymer and curing agent were mixed at a ratio of 10:1, degassed in a vacuum desiccator, poured on the silicon wafer, and finally baked in a convection oven (30 min, at 80℃). PDMS were then peeled from the SU-8 master and reservoir holes were punched through (Figure 1.B 5-7). Finally, bonding to glass slides was done using a plasma cleaner (Figure 1.B 8).

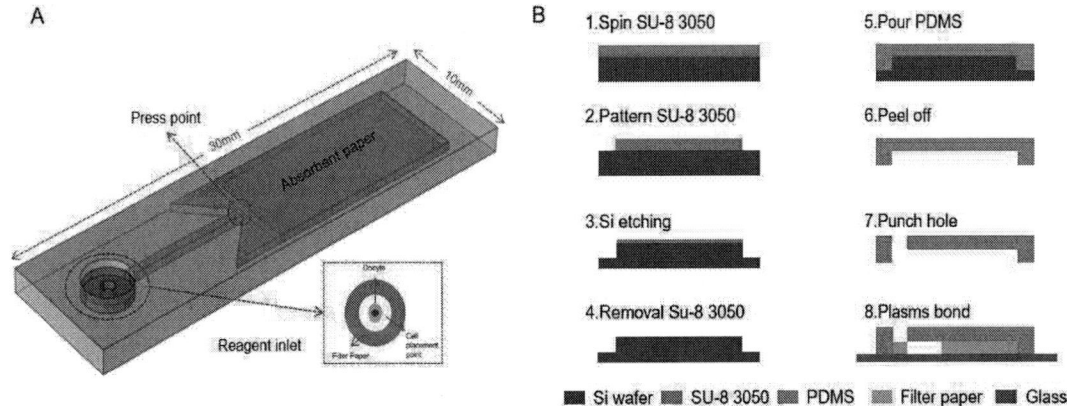

Figure 1. (A) Schematic illustration of the chip. (B) Fabrication process of the chip

978-1-6654-3008-1/21 $31.00 © 2021 IEEE

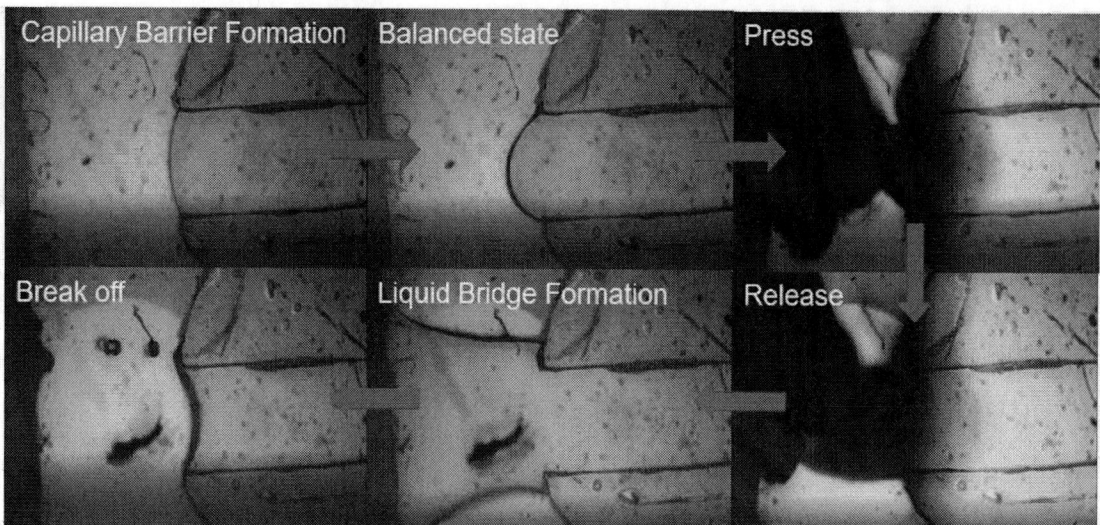

Figure 2. Formation of the capillary barrier and triggering of the press type valve

B. Reagent loading/unloading test

First, 30 μL equilibrium solution (ES) reagent was dropped on the chip, and let it stand for 10 min, then triggered the switch to unload ES. Afterwards, 20 μL vitrification solution (VS) was dropped, rested for 90 s, and then unloaded.

C. Shrinkage test of oocytes

The mouse's egg cells were placed in the hole in center of the filter paper, ES and VS were sequentially dropped, and immersed for 10 minutes and 2 minutes, respectively.

III. RESULTS AND DISCUSSION

A. Working principle of the chip

The microfluidic chip is composed of reagent inlet area, press deformation area and liquid absorption area as shown in Fig. 1A. The reagent inlet area is a circular groove with a diameter of about 5 mm. A piece of filter paper with a hole (diameter 2 mm) is placed inside. During the operation, the oocytes are placed inside the hole in the center of the filter paper. In this part, the filter paper acts as a fence, so that oocytes stay always inside the hole during the processes of medium changes. The press deformation area acts like a valve. The valve block the flow of liquid by creating a capillary barrier which can be suppressed by pressing lightly on the top. There is a micro-channel between the press point and the reagent inlet area. The Laplace pressure generated by the curved liquid surface formed by the cross section and the vertical section of the

	Rectangular channel with width, w and height, h
Driving force caused by hydrostatic pressure	$\rho g z_0 wh$
Inertial force of liquid in a tube or channel	$\rho wh(l+p)\dfrac{d^2 l}{dt^2}$
Inertial force at inlet of a tube or channel	$\rho wh\dfrac{q}{2}(\dfrac{dl}{dt})^2$
Capillary force	$2\sigma\cos(w+h)$
Viscous drag	$\dfrac{\pi^4\mu}{8\{1-\dfrac{2h}{\pi w}\tanh(\dfrac{\pi w}{2h})\}}\dfrac{w}{h}l\dfrac{dl}{dt}$
Pinning force	$\sigma\;(\cos\theta_a-\cos\theta_{pin\,max})$

Table 1. Comparison of terms of force governing interface motion in a rectangular tube

micro-channel and the feed pressure at the inlet are the driving forces to push the liquid forward (Figure 3, Table 1). The capillary barrier is achieved by abruptly expanding the cross section of the micro-channel to form a sharp edge, enhancing the pinning effect of the liquid. Pressing the top part of the press deformation area lowers the capillary barrier and draws the liquid into the absorption area by reducing the curvature radius of the curved liquid surface in the vertical direction, thereby enhancing the Laplace pressure and breaking the capillary barrier. Once the liquid entering the absorption area and contacts the absorbent paper, the absorbent paper begins to quickly absorb the liquid until the liquid breaks off at the capillary barrier point (Figure 2).

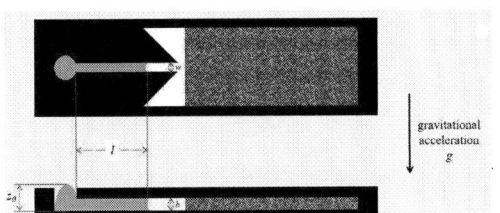

Figure 3.Analysis of forces on curved liquid surface in rectangular tube

B. Reagent loading/unloading test

Generally, oocytes need to be immersed in ES and VS to dehydrate the cells, and then put into liquid nitrogen for freezing. Two typical reagents, ES (yellow) and VS (blue), was applied to test the loading/unloading performance of the push-type microfluidic chip. As shown in Figure 4, ES was blocked by the capillary barrier at the end of the micro-channel until the top was pressed to trigger the switch. Once the switch was released, ES contacted the absorbent paper, and ES in the reagent inlet area was quickly pumped away until the flow broke off at the capillary barrier point. At this moment, a small amount of ES remained on the filter paper sheet in the reagent inlet area. Then VS was dropped, a new capillary barrier was formed. By pressing the top part of the chip and release, VS contacted the absorbent paper, the unloading process started. At last, a few VS remained on the filter paper sheet in the reagent inlet area.

C. Shrinkage test of oocytes

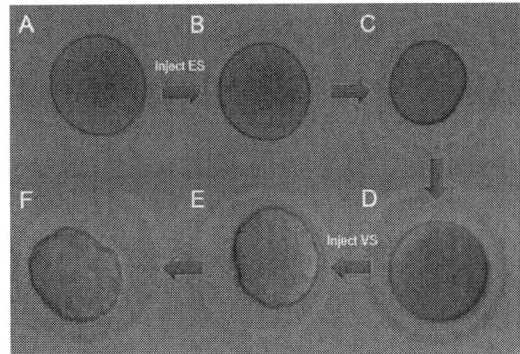

Figure 5. The shrinkage of oocyte during the reagent exchange

Cell shrinkage is an important indicator for judging the

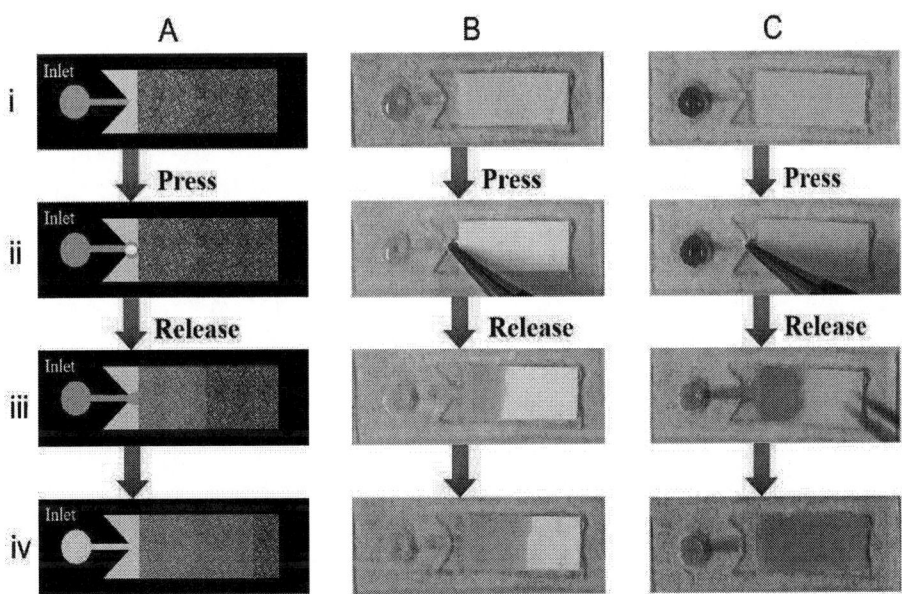

Figure 4. Working principle of the microfluidic chip.(A) Schematic diagram of liquid loading/unloading.(B) Loading/unloading of the equilibrium solution(ES). (C) Loading /unloading of the vitrification solution(VS).

success of vitrification. To observe the shrinkage of egg cells and whether the cells were lost during the operation, 30μL ES was dropped on the reagent inlet area. The cells remained position as shown in Figure 5.A and began to shrink, eventually shrinking to the size shown in Figure 5.C, and then slightly returning to the size shown in Figure 5.D. After that, 20 μL VS was dropped, and the cell quickly shrank to the size shown in Figure 5.F. During the whole process, the cells remained still (Figure 6.A-D), and only a very small amount of reagents were left around (Figure 6.D), which is very facile to rapid freezing in liquid nitrogen.

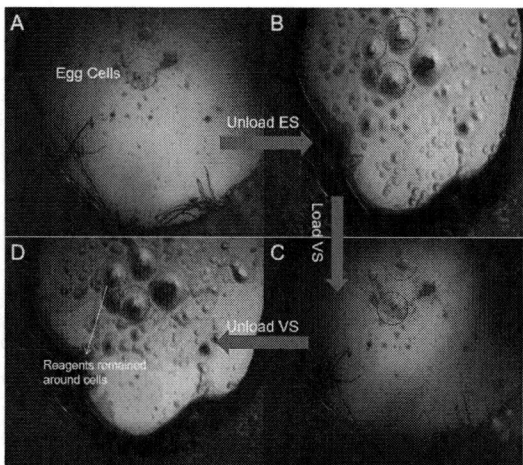

Figure 6.Cell movement during reagents loading and unloading

IV. CONCLUSION

In this paper, a controllable press-triggered microfluidic chip was proposed, which can realize the loading/unloading of reagents. Compared with the traditional manual operation, there is no need to transfer the cells for multiple times , and the amount of reagents is greatly reduced. While simplifying the operation process, it avoids physical damage to the cells. The filter paper is used to place the cells and acts as a fence to prevent cells from being lost during the operation. By using the simple but effective press-triggered switch, the time that the cells are immersed in various reagents can be controlled. By changing the flow channel size, material and other parameters, the chip can be adapted to any different brands of vitrification reagents. The fully automatic cell vitrification and freezing machine designed on the basis of this chip is expected to realize the standardization and automation of this process.

ACKNOWLEDGMENT

We thank Dr. Zhen Xu for the help and advice on egg cells test. This research was financially supported by Shenzhen Science and Technology Research Program (JCYJ20170818152810899, JCYJ20170818154035069, JCYJ20170818160050656), Guangdong Science and Technology Research Program (2015B020227002, 2016B020238003, 2019A050503008), CAS Key Laboratory on Health Bioinformatics (2011DP173015).

REFERENCES

1. Karlsson, J. O. M., Szurek, E. A., Higgins, A. Z., Lee, S. R. & Eroglu, A. Optimization of cryoprotectant loading into murine and human oocytes. *Cryobiology* **68**, 18–28 (2014).
2. Kuwayama, M., Vajta, G., Kato, O. & Leibo, S. P. Highly efficient vitrification method for cryopreservation of human oocytes. *Reproductive BioMedicine Online* **11**, 300–308 (2005).
3. Moussa, M., Shu, J., Zhang, X. & Zeng, F. Cryopreservation of mammalian oocytes and embryos: current problems and future perspectives. *Sci. China Life Sci.* **57**, 903–914 (2014).
4. Chen, S.-U. & Yang, Y.-S. Slow Freezing or Vitrification of Oocytes: Their Effects on Survival and Meiotic Spindles, and the Time Schedule for Clinical Practice. *Taiwanese Journal of Obstetrics and Gynecology* **48**, 15–22 (2009).
5. Dhali, A. *et al.* Gene expression and development of mouse zygotes following droplet vitrification. *Theriogenology* **68**, 1292–1298 (2007).
6. Vitrification of Oocytes: From Basic Science to Clinical Application | SpringerLink. https://link.springer.com/chapter/10.1007/978-1-4614-8214-7_6.
7. Mukaida, T. & Oka, C. Vitrification of oocytes, embryos and blastocysts. *Best Practice & Research Clinical Obstetrics & Gynaecology* **26**, 789–803 (2012).

Gap in pagination due to unavailable paper.

Pages 957-958

Proceedings of the 16th Annual IEEE International
Conference on Nano/Micro Engineered and Molecular Systems
April 25-29, 2021

Direct and All-dry Microfabrication of Ultramicroelectrode Based on Cold Atmospheric Microplasma Jet*

Y. Xi, L.C. Wang, Z.J. Guo, B. Yang and J. Q. Liu, *Member, IEEE*

Abstract—We develop a direct and all-dry microfabrication method of ultramicroelectrode based on cold atmospheric microplasma jet. The microfabrication process includes micropipette preparation, metal sputtering, chemical vapor deposition of parylene-C layer and microplasma jet micromachining (Fig.1). The generated microplasma jet can effectively remove the polymer layer on the tip. The exposed metal part can be used as ultramicroelectrode point or further modified with electrochemical deposition and other methods. The exposed area can be controlled by the treatment time and the average etching rate is approximately 1.1μm/s. Due to the properties of cold atmospheric microplasma jet, the microfabrication can be realized without high-precision machining and vacuum equipment.

I. INTRODUCTION

Ultramicroelectrode (UME) usually refers to an electrode that reaches the micron or nanometer scale in one dimension. Due to advantages in charging current [1], RC time constant, IR drop, mass transfer rate and sensitivity [2], ultramicroelectrodes are widely studied in many fields such as electron transfer kinetics, scanning electrochemical microscopy (SECM), scanning tunneling microscope (STM), atomic force microscope (AFM) high resolution imaging [3], chemical analysis and biological analysis [4]. Specific properties can be realized or optimized by modifying the electrode tip and the characteristic of the functional layer area on the tip also has a decisive influence on the ultramicroelectrode performance. For example, the geometrical morphology of functional area greatly affects the sensitivity and repeatability of the electrode.

Atmospheric pressure microplasma jets are operated under one atmosphere with geometric dimension confined to micrometers to millimeters [5-6]. Atmospheric pressure microplasma jets have drawn considerable attention in many applications such as surface patterning [7-9], material processing [10] and biomedical applications [11] because of advantages of highpower density, low temperature in localized area and no need of vacuum equipment. Besides, due to the small size, microplasma jets can be also used for maskless fabrication on planar and non-planar with no need of additional physical masks [12].

The gas plasma has been widely used to clean and activate surfaces by removing most organic films [13-14]. In

this paper, we report a novel and simple method for direct and all-dry microfabrication process for selective modification of UME based on cold atmospheric microplasma jet. UME is coated with parylene-C layer which is then selectively removed with microplasma jet. Selective modification can be achieved by modifying the exposed tip area. Due to advantages of low production cost, reactive chemical property and highly directional, this microfabrication method is greatly desired.

II. MATERIALS AND METHODS

A. Setup System of the Micromachining

The system for micromachining on the tip of UME based on cold atmospheric microplasma jet is shown in Figure 1. The system mainly includes three part: the microplasma jet generator (shown on the upper part of Figure 1), the gas supply controller (shown on the lower left part of Figure 1), the power supply controller (shown on the lower right part of Figure 1). The microplasma jet generator consists of the gas inlet, glass tube and copper electrode. The gas inlet was connected to the gas supply controller which controlled the flow rate of gas by the mass flow controller (MFC). Helium (He) and Oxygen (O_2) were used as working gases with total flow rate of 130 sccm (He: 125sccm, O_2: 5sccm). As the working electrode, a piece of 2 mm wide copper foil was wrapped around the other end of the glass tube. The copper electrode was connected with a ballast resistor of 150 kΩ and the power supply controller. A sinusoidal high voltage source (14kV, 18kHz) was applied to the copper electrode.

B. Setup System of the Micromachining

The microfabrication process for selective modification of ultramicroelectrode is shown in Figure 2. The borosilicate glass tube (B100-58-10, Sutter Instrument Company) was first prepared into micropipette by the laser-based micropipette puller system (P-2000, Sutter Instrument Company), as shown in Figure 2(a). Secondly, the micropipette was sputtered with Cr/Au on the outer surface as seen in Figure 2(b). Thirdly, Parylene-C layer was chemical vapor deposited in the coating system (PDS 2010, Specialty Coating Systems). Then the tip was polished to expose the metal ring as shown in Figure 2(c). With the help of micromaching system based on cold atmospheric microplasma jet (Figure 1), the parylene-C layer on the

---

* This work was partially funded by the National Key R&D Program of China under grant SQ2020YFB130047, the National Natural Science Foundation of China (No. 61728402), SJTU Trans-med Award (No.2019015), the Oceanic Interdisciplinary Program of Shanghai Jiao Tong University (No.SL2020ZD205), Scientific Research Fund of Second Institute of Oceanography, MNR (No.SL2020ZD205), the Program of Shanghai Academic/Technology Research Leader (18XD1401900). The

authors are also grateful to the Center for Advanced Electronic Materials and Devices (AEMD) of Shanghai Jiao Tong University.

Y. Xi, L.C. Wang, Z.J. Guo, B. Yang and J. Q. Liu are all with the National Key Laboratory of Science and Technology on Micro/Nano Fabrication, Shanghai Jiao Tong University, Shanghai, 200240, China, Department of Micro/Nano Electronics, Shanghai Jiao Tong University, Shanghai, 200240, China (corresponding auther to provide e-mail: jqliu@sjtu.edu.cn).

978-1-6654-3008-1/21 $31.00 © 2021 IEEE

UME tip was subsequently partial removed as seen in Figure 2(d). After plasma treatment and clean, part of metal layer on the UME tip was revealed as shown in Figure 2(e). Finally, the exposed part was selected modified with electrochemical deposition or other methods (Figure 2(f)).

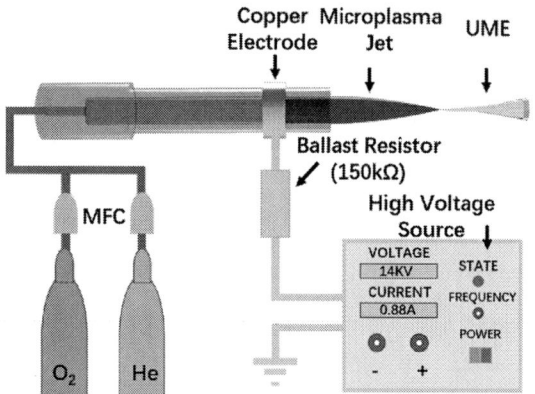

Figure 1: The schematic diagram of a system for micromachining on the tip of ultramicroelectrode based on cold atmospheric microplasma jet.

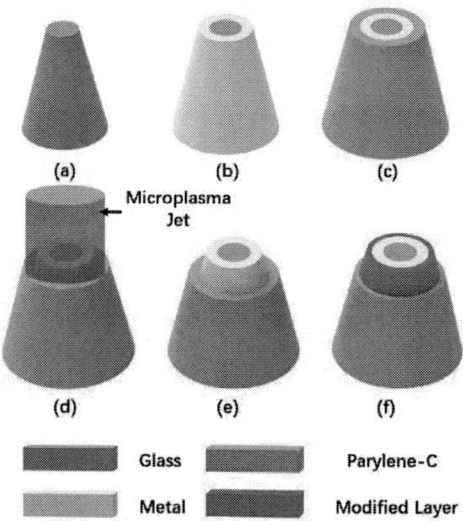

Figure 2: The schematic diagram of microfabrication process for selective modification of ultramicroelectrode based on cold atmospheric microplasma jet: (a) Glass micropipette fabrication, (b) deposit a layer of metal on micropipette surface, (c) deposit a layer of parylene-C, (d) microfabrication with cold atmospheric microplasma jet, (e) finish microfabrication with cold atmospheric microplasma jet and clean, (f) selective modification.

III. RESULTS AND DISCUSSION

A. Description of the Micromachining Process

Figure 3 shows the photograph of the cold atmospheric microplasma jet generator and UME. The microplasma jet generator and UME were fixed on a custom 3D-printed base which had several V-type grooves. The grooves worked as a guide rail to keep the microplasma and UME aligned (Figure 3(b)).

Figure 4(a-b) shows the cold atmospheric microplasma jet generator and UME before and after plasma initiation. The microplasma jet was emitted from the generator outlet and then acted on the electrode tip which coated with the parylene-C layer as shown in Figure 4(c). The UME tip showed a bright purple spot because of the reaction between the polymer and plasma. After a certain length of time, the parylene-C layer was partly removed from the electrode tip as shown in Figure 4(d).

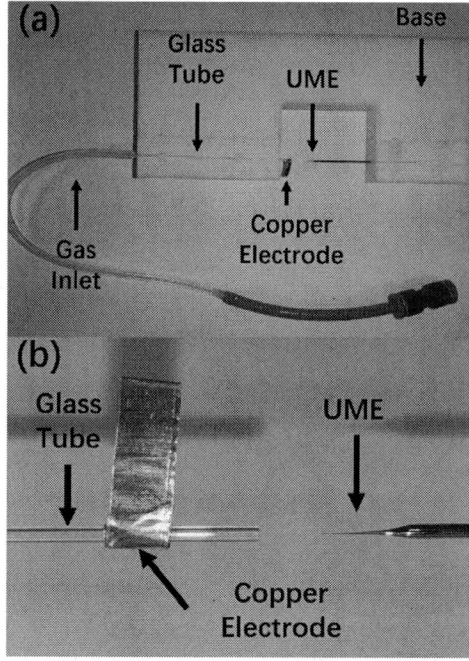

Figure 3: (a) Photograph of the cold atmospheric microplasma jet generator and UME; (b) The magnified image of the tip of microplasma generator and UME.

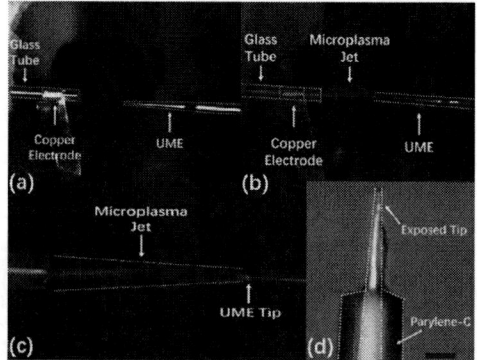

Figure 4: (a) Photograph of the cold atmospheric microplasma jet generator and UME before plasma initiation; (b) Photograph of the cold atmospheric microplasma jet generator and UME after plasma initiation; (c) Magnified image of the generated microplasma jet and UME tip; (d) Optical image of UME tip after microfabrication, Scale bar, 10μm.

B. Influence of Treatment Time

Figures 5 shows the etching length of parylene-C layer with different treatment time by the microplasma jet. The etching length was 33μm, 139μm and 312μm for treatment time of 30s, 120s and 300s, respectively. The average etching rate was approximately 1.1μm/s. The length of exposed tip area can be controlled by the treatment time to achieve the selective modification.

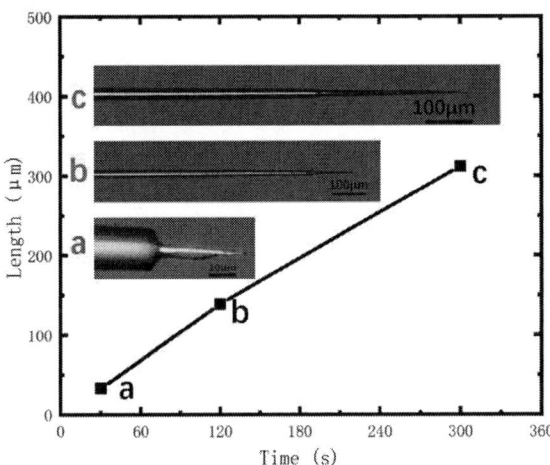

Figure 5: The etching length of parylene-C layer with different treatment time: (a) 30s; (b) 120s, (c) 300s.

C. Automatic Locking Effect

The preparation of UME usually requires high precision for operating and processing equipment. In practice, the cold atmospheric microplasma jet existed an automatic locking effect which could tolerate some alignment error between the microplasma jet generator and the electrode tip. As shown in Figure 6(a), there was no good alignment between the microplasma jet generator and the electrode tip and the microplasma could still act on the electrode tip well. The Figure 6(b) shows the magnified image of the generated microplasma jet which bent and eventually pointed to the electrode tip. This effect was influenced by the metal layer beneath the polymer film [15]. As a result, this effect can reduce the requirement of machining equipment precision and the processing cost.

IV. CONCLUSION

We develop a direct and all-dry microfabrication method for selective modification of ultramicroelectrode based on cold atmospheric microplasma jet. A system for micromachining on the tip of ultramicroelectrode based on cold atmospheric microplasma jet was successfully built. The microplasma jet was generated with a sinusoidal high voltage source (14kV, 18kHz). The generated microplasma jet effectively remove the polymer layer on the ultramicroelectrode tip. The exposed metal part can be selected modified with electrochemical deposition or other methods. The area available for decoration can be controlled by adjusting the treatment time and the average etching rate can reach approximately 1.1μm/s. Due to the special automatic locking effect of cold atmospheric microplasma jet, this machining method can be realized without high-precision machining equipment. All these results make this method suitable for selective modification of ultramicroelectrode.

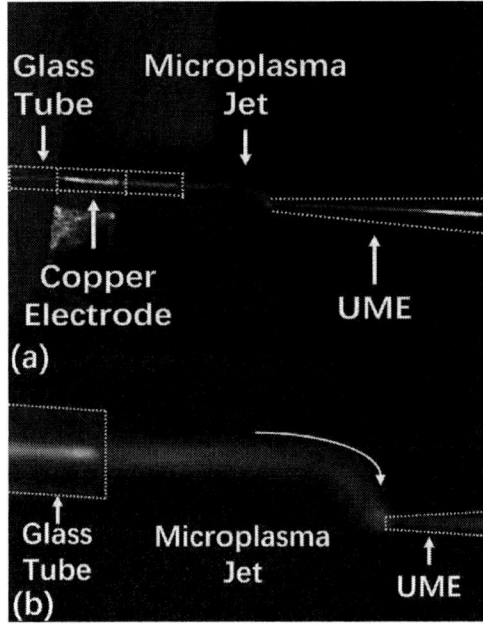

Figure 6: (a) Photograph of the cold atmospheric microplasma jet generator and UME after plasma initiation; (b) Magnified image of the generated microplasma jet with automatic lock.

ACKNOWLEDGMENT

This work was partially supported by the Strategic Priority Research Program of Chinese Academy of Sciences (Grant No. XDA 25040000), the National Key R&D Program of China under grant 2020YFB1313502, the National Natural Science Foundation of China (No. 61728402), SJTU Trans-med Award(No.2019015), the Oceanic Interdisciplinary Program of Shanghai Jiao Tong University (No.SL2020ZD205), Scientific Research Fund of Second Institute of Oceanography, MNR (No.SL2020ZD205), the Program of Shanghai Academic/Technology Research Leader (18XD1401900). The authors are also grateful to the Center for Advanced Electronic Materials and Devices (AEMD) of Shanghai Jiao Tong University.

REFERENCES

[1] S.M. Oja, M. Wood, B. Zhang, "Nanoscale Electrochemistry", *Anal. Chem.*, vol. 85, pp. 473-486, 2012.

[2] R. W. Murray, "Nanoelectrochemistry: Metal Nanoparticles, Nanoelectrodes, and Nanopores", *Chem Rev.*, vol. 108, pp. 2688-2720, 2008.

[3] C. Kranz, "Recent advancements in nanoelectrodes and nanopipettes used in combined scanning electrochemical microscopy techniques", *Analyst*, vol. 139, pp. 336-352, 2014.

[4] J. Lakbub, A. Pouliwe, A. Kamasah, C. Yang, P. Sun, "Electrochemical Behaviors of Single Gold Nanoparticles", *Electroanalysis,* vol. 23, pp. 2270-2274, 2011.

[5] F. Iza, G. J. Kim, S. M. Lee, J. K. Lee, J. L. Walsh, Y. T. Zhang, M. G. Kong, "Microplasmas: sources, particle kinetics, and biomedical applications", *Plasma Process. Polym.*, vol. 5, pp. 322-344, 2008.

[6] T. Wang, X. Wang, B. Yang, X. Chen, C. Yang, J. Liu, "Low temperature atmospheric microplasma jet array for uniform treatment of polymer surface for flexible electronics", *J. Micromech. Microeng.*, vol. 27, pp. 75005, 2017.

[7] L. Liu, D. Ye, Y. Yu, L. Liu, Y. Wu, "Carbon-based flexible micro-supercapacitor fabrication via mask-free ambient micro-plasma-jet etching", *Carbon*, vol. 111, pp. 121-127,2017.

[8] R.P. Gandhiraman, E. Singh, D.C. Diaz-Cartagena, D. Nordlund, J. Koehne, and M. Meyyappan, "Plasma jet printing for flexible substrates", *Appl Phys Lett*, vol. 108, pp.123103, 2016.

[9] M. Vandenbossche, L. Bernard, P. Rupper, K. Maniura-Weber, M. Heuberger, G. Faccio, D. Hegemann, "Micro-patterned plasma polymer films for bio-sensing", *Mater. Design*, vol. 114, pp. 123-128, 2017.

[10] S. Ghosh, E. Klek, C. A. Zorman, R.M. Sankaran, "Microplasma-Induced in Situ Formation of Patterned, Stretchable Electrical Conductors", *ACS Macro Lett.*, vol.6, pp. 194-199, 2017.

[11] E. J. Szili, S. Becker, R. D. Short, S. A. Al-Bataineh, "Microplasma jet treatment of bovine serum albumin coatings for controlling enzyme and cell attachment", *Eur. Phys. J. Spec. Top.*, vol. 226, pp. 2873-2885, 2017.

[12] E. Szili, S. Albataineh, P. Ruschitzka, G. Desmet, C. Priest, H. Griesser, N. Voelcker, F. Harding, D. Stell, R. Short, "Microplasma arrays: A new approach for maskless and localized patterning of materials surface", *RSC Adv.*, vol.2, pp. 12007-12010, 2012.

[13] Krügera, P.; Knesa, R.; Friedrich, J. Surf, "Surface cleaning by plasma-enhanced desorption of contaminants (PEDC)", *Coat. Technol*, Vol. 112, pp. 240-244, 1999.

[14] T. Sun, P. Blanchard, M.V. Mirkin, "Cleaning Nanoelectrodes with Air Plasma", Analytical Chemistry, vol. 87, pp. 4092-4095, 2015.

[15] T. Wang, B. Yang, X. Chen, X. Wang, C. Yang, J. Liu, "Distinct modes in the evolution of interaction between polymer film and atmospheric pressure plasma jet", *Plasma Process. Polym.*, vol. 14, pp. 1600067, 2017.

Proceedings of the 16th Annual IEEE International
Conference on Nano/Micro Engineered and Molecular Systems
April 25-29, 2021

Nanopore Surface Charge Sensing with Ion-Step Method

Jing Yang[1], Songyue Chen[1*]

Abstract— **We present a sensitive nanopore surface charge detection with ion-step as signal amplification method. An ion-step is applied to the inner and outer side of the glass nanopore, and ion rectification effect is recorded and analyzed. A comparison is performed for both bare glass nanopores that are negatively charged and silane modified nanopores, which are positively charged. We found that ion-step can greatly improve the current rectification during current-voltage scanning. The concentration difference and pore sizes have huge influence on the rectification ratio. With the ion-step, a nanopore with diameter of over 900 nm exhibits current rectification effect, and can be applied in surface charge detection. In addition, we herein also explore the influence of different voltage and solution pH to the rectification ratio. We conclude that the reason for the rectification ratio amplification is the incremental effect of electroosmotic flow and ionic diffusion, due to the surface charges present on the inner surface of the nanopore.**

I. INTRODUCTION

Nanopore detection technology has become a new and potential analysis and detection method in the chemical and biological fields. The conductivities of nanopores are assisted by ionic solution [1], which in turn affects the charge sensitivity due to the double layer capacitance formation. Nowadays, in order to ensure charge sensitivity for nanopore surface charge detection, a commonly used dimension is within 50 nm, and the electrolyte concentration is chosen below 10 mM [2]. Nanopore sensors based on surface charge detection have been used to detect a variety of metal ions [3, 4], gases [5], temperatures [6], and biomolecules [7, 8]. The stability and applicability of nanopore can be greatly promoted by improving the sensitivity of nanopores, especially by expanding the application of large aperture nanopores for surface charge detection. The surface charges presented inside the nanopore regulate the ion flood in and out, which can be characterized by the rectification ratio of the current-voltage (*I-V*) curve. The magnitude of ion current rectification is typically quantified by the rectification factor, R_f, which is defined by the absolute ratio of the current slope at positive voltage divided by the current slope at negative voltage [9].

Ion-step exists widely in the macro and micro of nature. In biological systems, the transmembrane ion-step is the basis of membrane potential, which regulates the function of ion channels and ion pumps on the cell membrane, thereby controlling the ion species in and out of the cell [10]. Inspired by nature, we applied ion-steps to the nanopores in this work, and explored the effect and regulation principle of the ion-step to the rectification factor, especially the rectification ratio improvement methods.

II. EXPERIMENTAL

A. Glass nanopore fabrication

The glass nanopores used in this paper are fabricated according to the production and characterization method reported by White [11] et al. The fabrication process of glass nanopore is shown in Fig. 1(a), which mainly consists of four steps. The first step was to electrochemically etch a sharp tip in 15% calcium chloride solution with a generator (64 Hz ac voltage, 5 V amplitude). Secondly, the platinum wire was encapsulated in the glass capillary with hydrogen-oxygen flame. After encapsulation, polishing of the glass capillary was finished with sandpaper (20,000 mesh) wetted in 0.02 M KCl solution until the Pt tip was exposed. In the end, the platinum was etched in boiled aqua regia for 12 hours to form a glass nanopore.

B. Characterization of glass nanopores

The nanopore radius was determined by measuring the steady-state diffusion limit current with a electrochemical workstation(CHI 660E) of the platinum wire electrode before etching with an empirical formula [12]. The size of the glass nanopores was measured by imaging with field emission scanning electron microscope (SEM, JSM-IT500A, Japan). 10 nm of gold was sputtered on the tip of the nanopores to enhance the conductivity. Fig. 1(b) shows the SEM of a glass nanopore with radius of 464 nm.

C. Experimental setup

Conductance is determined by scanning *I-V* curves, which were recorded with a picoammeter (Keithley 6487). The entire measurement system includes two Ag/AgCl electrodes, a glass nanopore, a shielding box, picoammeter and a PC, as shown in Fig. 1(c). All the measurements were performed with KCl solutions (pH 5.8, unless otherwise mentioned) in a Faraday cage at room temperature. Each *I-V* curve was repeated three times and average values were calculated as shown in the figures.

D. Surface functionalization

Surface functionalization of the glass nanopores was processed as follows. Firstly, the glass nanopores were cleaned with piranha solution (H_2SO_4:30%H_2O_2, 3:1) at 80°C for 2 h and dried in oven to obtain a clean glass surface with hydroxyl groups. Then, the nanopores were immersed in (3-aminopropyl)triethoxysilane (APTES) ethanol solution (5% APTES + 2% H_2O) for 2 h, cleaned with ethanol for three times, and dried at 120°C for 1 h to enhance the crosslinking of APTES on the oxide surface and form a stable layer.

---

1 the school of Aerospace Engineering, Xiamen University, Xiamen University, Xiamen 361005, China.

*corresponding author: phone: 13860487347; e-mail: s.chen@xmu.edu.cn

978-1-6654-3008-1/21 $31.00 © 2021 IEEE

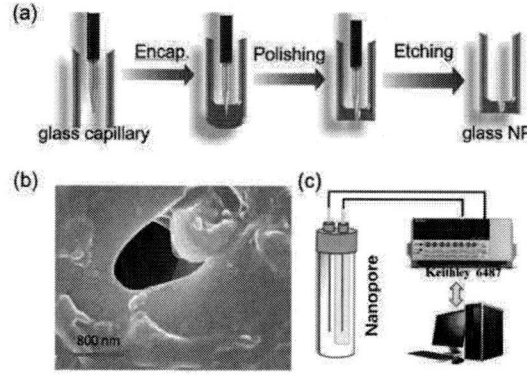

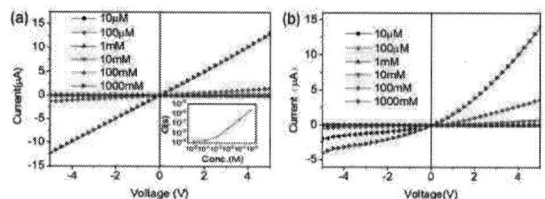

Figure 1. (a) Fabrication process of glass nanopore; (b) SEM image of the glass nanopore with a radiu of r = 464 nm; (d) schematic diagrams of *I-V* curve measurement.

Figure 2. *I-V* curves of glass nanopore (r = 464nm) (a) before and (b) after modification with APTES in different substrate solutions; The insert figure in (a) is the conductance of glass nanopore before modification.

III. RESULTS AND DISCUSSION

A. Performance of nanopores with ion-step

As shown in Fig. 2(a), the *I-V* curves of the bare glass nanopore with r = 464 nm show that the conductance increases with rising the electrolyte concentrations, and the curves are linear without rectification effect. Figure 2(a) inset shows the calculated conductance, which starts to level off for KCl concentrations below 1 mM. At lower salt concentrations, the negative surface charges present inside the glass nanopore affect the distribution and transmission of ions in the nanopore, where the ionic current carriers are mainly potassium ions. That is because the surface charge of nanopore plays a leading role when the ionic concentration is low, in which the thickness of the double layer is large and the electrostatic adsorption of the pore wall to the porous ions is stronger [13]. With the increase of ionic concentration, the thickness of the double layer decreases. As a result, the ionic concentration determines the conductance and the conductance increases linearly with increasing of ionic concentrations. Therefore, at lower ionic concentrations, the ionic conductance is related to the surface charge of the nanopore, which can be modulated by surface charge. The transition point from the two effects is mainly determined by the nanopore size and the surface charge density, in this case is 1 mM.

The functionalization of APTES on the glass, introduced amino terminal (-NH$_2$) to the nanopore surface, which has positive charges at the working pH. Ionic current carriers switched from potassium ion to chloride ion due to the polarity change of the surface charge. The presence of large amount of positive charge rendered rectification for the *I-V* curves, as shown in Fig. 2(b). The surface charge density of nanopore is determined by the surface properties of the materials and electrolytes.

In order to further study the rectifying characteristics, we applied ion-step on the nanopore: when the electrolyte concentration inside the nanopore is lower than the external, it is defined as forward ion-step, otherwise backward ion-step. It's shown in Figs. 3(a) and (b) that when forward ion-step was applied on the bare glass nanopore, the current under the positive voltage bias was smaller than that from the

negative bias; and in backward ion-step, the rectification direction reversed. We conclude that for the negatively charged surface, it is conducive for surface charge characterization when it is in backward ion-step, which brings signal amplification, as shown in Fig. 5(a). Figures 3(c)-(f) show the physical model we used for the ion-step method. For the negatively charged glass nanopore surface [14], positive ions are the main carriers, and play a key role in the ionic conductance of the nanopore. The direction of electroosmotic flow (EOF) is the same as that of applied potential. When the nanopore is in forward ion-step, the EOF and the diffusion direction have the same direction at negative bias voltage, so the current value is enhanced; the directions of the two effect is opposite to each other at positive bias, which reduces the ionic current. When apply a backward ion-step, the current in the nanopore at positive bias is always higher than that at negative bias. The rectification inversion is observed for inversed ion-step direction.

The *I-V* curves for APTES functionalized nanopore are shown in Fig. 4. Different from the bare glass nanopores, when the ion-step was applied to the APTES-modified nanopores, the rectification amplitude is either enhanced or reduced, but the rectification direction was less affected, as shown in Figs. 3(c) (d) and Fig. 5(b). Rectification direction reverse happens only when the concentration difference was enough large, 1M to 10 mM for backward ion-step. As the APTES modified glass surface has a larger positive charge density, the current is mainly affected by the surface charge, so the change of ion-step direction has little effect to the rectification ratio. We can also observe negative differential resistance (NDR) when the ion-step and bias voltage are large. NDR is a phenomenon that the current decreases when the voltage increases. It is widely used in traditional electronic devices such as multipliers, recent studies have found that NDR phenomenon also exists in asymmetric nanopores, which results from the superposition of EOF and diffusion direction change during voltage scanning [15-17]. Therefore, a reasonable ion-step value and direction should be chosen, since the rectification direction is different for positively and negatively charged nanopores. For positively charged surface, the forward ion-step is beneficial to increase its rectification ratio, which is shown in Fig. 5(b).

B. Influence of scanning voltage on rectification ratio

In order to explore the influence of bias voltage on the rectification ratio, we measured the *I-V* curves of different ion-step at ±5V and ±1V, as shown in Fig 6. It can be found that the rectification ratio of glass nanopore under ±5V is

978-1-6654-3008-1/21 $31.00 © 2021 IEEE

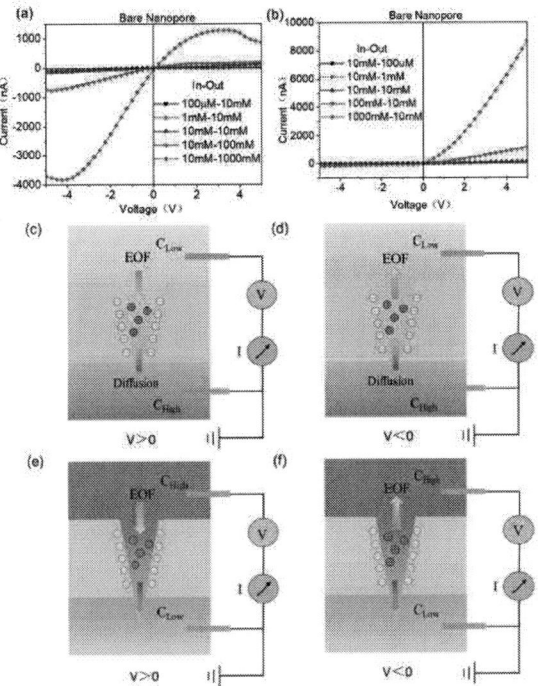

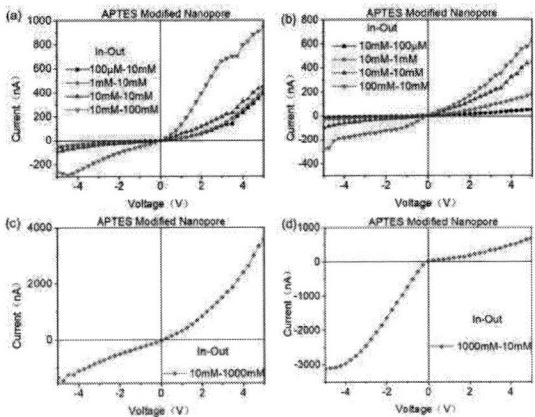

Figure 4. (a)-(d) *I-V* curves of glass nanopores (r = 464nm) after modification with APTES in different substrate solutions.

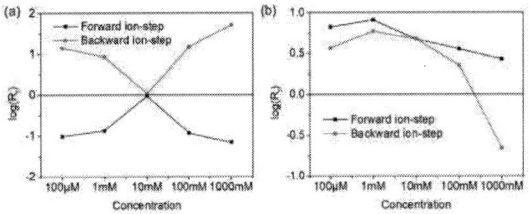

Figure 5. The retification factor of (a) bare nanopore and (b) APTES modified nanopore. When the negative bias conductance is larger, the rectification factor is positive; otherwise it is negative.

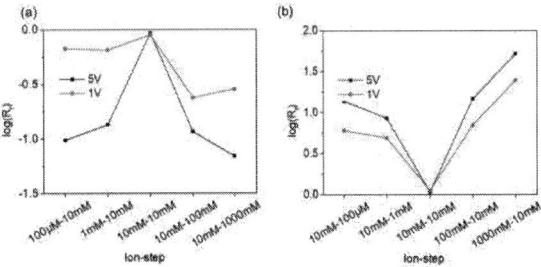

Figure 6. The rectification of (a) forward ion-step and (b) backward ion-step under 1V and 5V.

Figure 3. *I-V* curves of glass nanopores (r = 464nm) before modification in (a) forward ion-step and (b) backward ion-step. (c)-(f):Simplified physical models for ion-step of negatively charged glass nanopores (C_{Low} refers to low concentration, C_{High} refers to high concentration): (c) forward ion-step, positive bias; (d) backward ion-step, negative bias; (e) forward ion-step, positive bias; (f) backward ion-step, negative bias.

greater than that under ±1V. For negatively charged glass nanopore, the EOF under the positive voltage is from the inside of the nanopore to the outside, which is in the same direction as the ion diffusion under the backward ion-step, and the EOF increases with the increase of the applied bias voltage[18]. Therefore, under the positive bias, the number of potassium ions participating in the conduction increases and the rectification ratio increases significantly. Since the applied voltage in this study started from negative bias, more ions were accumulated or dissipated from - 5V to 0. But the voltage should not be too high as the electrodes used in this study are Ag/AgCl electrodes, which carried out redox reaction in the reaction process, and high bias voltage will result in unstable current due to the limited reactants on the electrodes.

C. Influence of pH on rectification ratio

As mentioned previously, at lower ionic concentration, the current rectification of nanopore is mainly influenced by the surface charge. Therefore, we also explored the influence of different pH on the *I-V* curve of bare glass nanopores, as pH affects the surface charge polarity and charge density. Figure. 7 (a) shows the forward ion-step experiment. There was obvious rectification phenomenon at pH5.8 when ion-step of 10 mM-100 mM is applied. The surface of glass nanopore is deprotonated when the pH is raised to 9, which increases negative surface charge and

the rectification ratio. The current under negative bias voltage increases as the EOF is in the same direction as the diffusion current. However, when the pH drops to 3, the glass nanopore is around the point-of-zero-charge, so the current rectification disappears [18]. For the backward ion-step situation, the *I-V* curve at pH3 did not show any rectification, and increase of pH to 9 significantly improved the rectification effect.

IV. CONCLUSION

In this study, the signal amplification for nanopore surface charge is realized with the help of ion-step, represented by the current rectification enhancement. By observing the change of rectification characteristics of glass

978-1-6654-3008-1/21 $31.00 © 2021 IEEE

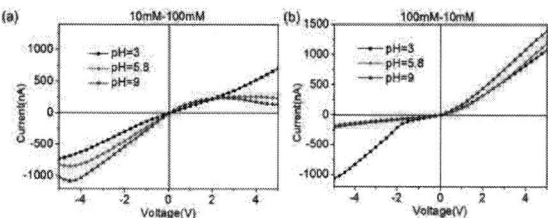

Figure 7. The *I-V* curves at different pH under the same ion-step : (a) forward ion-step with 10mM and 100mM; (b)backward ion-step with 10mM and 100mM.

nanopores after applying ion-step, it can be found that forward and backward ion-step had different effects on the rectification characteristics, and the influence is different at different charged surface: the negatively charged bare glass nanopore performed better under backward ion-step, while the positively charged nanopore after functionalized with silane, tended to have a better performance under forward ion-step. A larger bias voltage can greatly improve the rectification ratio. In addition, the pH can adjust the rectification ratio by affecting the surface charge polarity and charge density. If the surface is not charged, rectification will not occur even if there is ion-step. The detection signal can be amplified by applying ion-step especially when the surface charge density is low. In conclusion, an appropriate ion-step renders promotion of the electroosmotic flow and ionic diffusion effect. This technology is expected to be applicable for surface charge sensing of various sensors.

REFERENCES

[1] Chen, S.Y., et al. "Chemiresistive nanosensors with convex/concave structures", Nano Today,vol.20, pp. 84-100,2018.

[2] Tang, Y., et al. "Performance analysis of solid-state nanopore chemical sensor", Sensors and Actuators B: Chemical, vol.286, pp. 315-320, 2019.

[3] Cheng, L.J., and Guo, L.J. "Rectified ion transport through concentration gradient in homogeneous silica nanochannels", Nano Letters, vol.7, no.10, pp. 3165-3171, 2007.

[4] Zhao, X.P., et al. "Asymmetric nanochannel-ionchannel hybrid for ultrasensitive and label-free detection of copper ions in blood", Analytical Chemistry, vol.90, no.1, pp. 896-902, 2018.

[5] Shang, X.M., et al. "An artificial CO_2-driven ionic gate inspired by olfactory sensory neurons in mosquitoes", Advanced. Materials, vol.29, no.3, pp. 5, 2017.

[6] Wang, R., et al. "Temperature-sensitive artificial channels through pillar[5]arene-based host-guest interactions", Angew Chem Int Ed Engl, vol.56, no.19, pp. 5294-5298, 2017.

[7] Ali, M., et al. "Sequence-specific recognition of DNA oligomer using peptide nucleic acid (PNA)-modified synthetic ion channels: pna/dna hybridization in nanoconfined environment", ACS Nano, vol.4, no.12, pp. 7267-7274, 2010.

[8] Cai, S.L., et al. "Surface charge modulated aptasensor in a single glass conical nanopore", Biosens Bioelectron, vol.71, pp. 37-43, 2015.

[9] Yusko, E.C., et al. "Electroosmotic flow can generate ion current rectification in nano- and micropores", ACS Nano, vol.4, no.1, pp. 477-487, 2010.

[10] Cao, L., et al. "Concentration-gradient-dependent ion current rectification in charged conical nanopores", Langmuir, vol.28, no.4, pp. 2194-2199, 2012.

[11] Bo Zhang, et al. "Bench-top method for fabricating glass-sealed nanodisk electrodes, glass nanopore electrodes, and glass nanopore membranes of controlled size", Analytical Chemistry, vol.79, no.13, pp. 4778-4787, 2007.

[12] Zhang, B., et al. "Steady-state voltammetric response of the nanopore electrode", Analytical Chemistry, vol.78, no.2, pp. 477-483, 2006.

[13] Daiguji, et al. "Electrochemomechanical energy conversion in nanofluidic channels", Nano Letters, vol.4, no.12, pp. 2315-2321, 2004.

[14] Chen, S., et al. "A glass nanopore ionic sensor for surface charge analysis", RSC Advances, vol.10, no.36, pp. 21615-21620, 2020.

[15] Lin, C.Y., et al. "Electrodiffusioosmosis-induced negative differential resistance in ph-regulated mesopores containing purely monovalent solutions", ACS Appl Mater Interfaces, vol.12, no.2, pp. 3198-3204, 2020.

[16] Zuzanna S. Siwy, et al. "Calcium-induced voltage gating in single conical nanopores.", Nano Letters, vol.6, no.8, pp. 1729-1734, 2006.

[17] Long Luo, et al. "Negative differential electrolyte resistance in a solid-state nanopore resulting from electroosmotic flow bistability", ACS Nano, vol.8, no.3, pp. 3023-3030, 2014.

[18] Ai, Y., Zhang, et al. "Effects of electroosmotic flow on ionic current rectification in conical nanopores", Journal of Physical Chemistry C, vol.114, no.9, pp. 3883-3890, 2010.

Gap in pagination due to unavailable papers.

Pages 967-979

Proceedings of the 16th Annual IEEE International
Conference on Nano/Micro Engineered and Molecular Systems
April 25-29, 2021

Slippery coated Implantable flexible microelectrode array (fMEA) for High-Performance Neural Interface

Md Eshrat E Alahi[1], Zhou Tian[1], Sara Khademi[2], Hao Wang[1,*], and Tianzhun Wu[1,*]

Abstract— This study presents a novel approach of slippery liquid-infused porous surface (SLIPS) on the flexible microelectrode array (fMEA) for chronic neural interface with the advantage of reduced cell adhesion. In the demonstration, fMEA was microfabricated on the polyimide (PI) substrate, and Platinum (Pt) gray was used to create the porous nanocone structure for infusing the silicone oil. The combination of Pt gray and slippery oil layer has remained the low impedance level favorable for the neural stimulation and recording applications. The cytotoxicity study also shows that the coating does not have any cytotoxic potentiality; hence is biocompatible for human implantation.

I. INTRODUCTION

Neurological disorders or injuries of the central nervous system disturb any individuals' daily lives [1]. Therefore, engineers and neurophysiologists have shown increasing attention to restoring the neural activities. Neural implantation or neural interface is a technique that aims to restore the nervous system's function through communication with external prostheses devices such as limb prostheses. The objective of neural implantation is to record the neural signals from several neural cells and stimulating a designated area as per the activity requirements. Among all the challenges, high-performance neural interfaces [2] are the significant challenges for chronic recording/stimulation. One of the essential considerations for implantable electrodes is a low Young's modulus, which makes the implantable device mechanically similar to soft tissues [3, 4]. Soft materials with low Young's modulus can diminish the inequity in flexibility between the implantable electrodes and the neural tissue. Besides, they should be small enough on a specific region of the brain or neural location to minimize brain injury due to the implantation. A large-sized electrode might destroy a large number of neurons due to the insertion. Most of

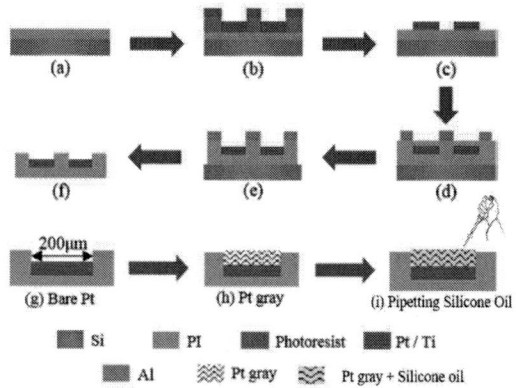

Figure 1: Step-by-step fabrication process of the fMEA, surface modification for SLIPS

the previously developed implantable electrodes were conductive metals, such as gold, platinum, iridium, stainless steel, and tungsten for clinical use or neurobiology research. Their surfaces are coated with a non-toxic, biocompatible insulating material on the electrode recording sites [5-7] to reduce inflammation,

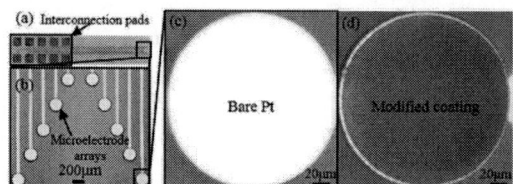

Figure 2: (a) Layout of the fabricated fMEA; (b) enlarged optical microscopic view of the recording sites; (c) Bare Pt surface before the surface modiciation; (d) modified surface with Pt Gray coating.

a common phenomenon in chronic recording.

When an electrode is implanted inside the brain, the blood-brain barrier (BBB) is disrupted and shows a series of inflammatory responses upon the electrodes' implantation into the intracortical tissue. The inflammatory responses are the cause of the degradation of the electrode-tissue interface location

---

* This work is supported by the National Science Foundation of China (NSFC) (61950410613), Chinese Academy of Sciences (2019PT0008), and Ministry of Science and Technology (MOST) (QNJ20200132001). . M.E.E.A., Z. T., H. W., and T.Z. are from (1). Institute of Biomedical and Health Engineering, Shenzhen Institute of Advanced Technology, Chinese

Academy of Sciences, Shenzhen 518055, China. S. K. is from (2). Sahand University of Technology, Tabriz, Iran. (corresponding authors: H.W.: hao.wang@siat.ac.cn; T.W. : tz.wu@siat.ac.cn)

978-1-6654-3008-1/21 $31.00 © 2021 IEEE

[8, 9]. The central nervous system (CNS) has a central defense mechanism, initiating the microglia around the implantation interface location to recover from the injury [10, 11]. The CNS also initiates the astrocytes within the first few days near the implantation location to heal the wound [12]. In the long-term implantation, micromotion between the implanted electrodes and the neural tissues can be seen due to the mechanical mismatch induced by vascular pulsations [8]. Due to this glial scar, the interface locations are damaged severely and lost the electrical coupling between the implanted electrode and the CNS's neural tissue. Because of these phenomena, both the SNR and the number of neural tissues recorded in the interface location decrease rapidly after the first several weeks

Figure 3: SEM of Pt gray and porous surface created for nanocone structure.

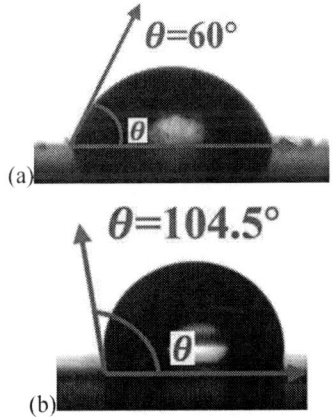

Figure 4: water droplet on the SLIPS surface (a) before the SLIPS coating, which is hydrophilic, and (b) after the SLIPS coating, which is hydrophobic.

of post-implantation.

Aizenberg et al. [6] have developed a new type of nonwetting surface named slippery liquid-infused porous surface (SLIPS) through a mimicry of nepenthes infusing the lubricant oil into a rough, porous, and lipophilic substrate. The overlying oil film layer is unmixable to water, which can be ensured as a protective layer for an extended period [13]. Beyond that, another exciting feature of SLIPS is that it can naturally recover the structural integrity through the inherent mobility of oil film when the surface is

damaged [14, 15]. It has successfully created highly "slippery," antifouling, and adhesion-free materials in various applications [16-18]. However, such ability has not yet been demonstrated for implantable devices to reduce cell adhesion of glial scar. Herein we propose the first attempt to combine SLIPS with fMEA based on Pt-nanocone structure for the anti-inflammatory applications.

II. EXPERIMENTAL PROCEDURES

A. Fabrication, and SLIPS coating procedures

A thin Pt seed layer was deposited and patterned on the PI substrate with an exposed diameter of 200 μm. Briefly, PI was spin-coated on a silicon (Si) wafer and

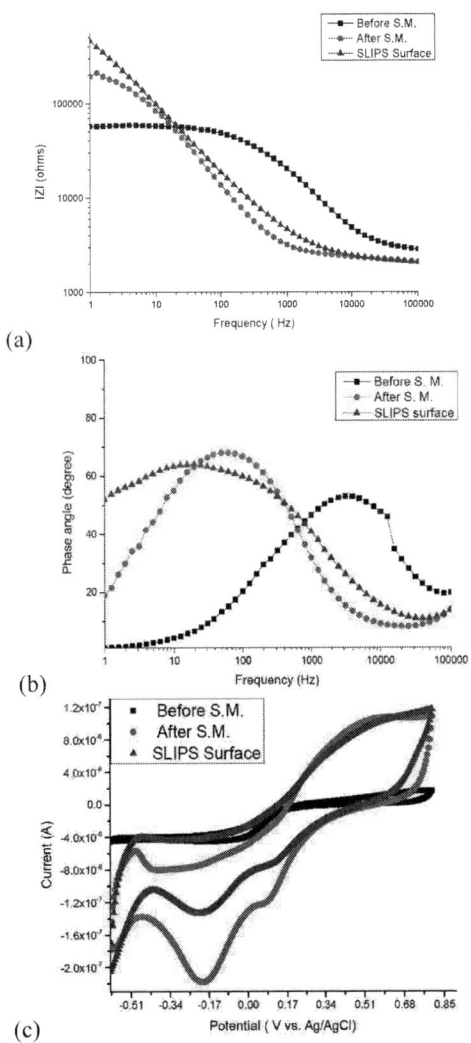

Figure 5: (a) and (b) Electrochemical impedance profile and phase angle for the surface modification and SLIPS for the coated fMEA. (c) Cyclic voltammograms of SLIPS coated microelectrodes at a sweep rate of 50 mV/s.

cured with the thickness of 5 μm (Fig. 1a), then the sputtering and lift-off process were used to deposit titanium (Ti) (20 nm-thick) and Pt (100 nm-thick) consecutively (Fig. 1b–c). After that, another layer of PI film was spin-coated and cured for the passivation layer. The reactive ion etching (RIE) was used to expose the Pt microelectrodes (Fig. 1d–e), and the fMEA were released from Si (Fig. 1f). A layer-by-layer electrodeposition method was used to deposit Pt gray on Pt microelectrodes as shown in Fig. 1h. to create the nanocone [19]. This nanocone enhances the surface area which reduces the impedance level further. This Pt-based nanocone is created for porous surfaces and infusing a small amount of viscous silicone oil (5 μL) by pipetting (Fig. 1(i)) to allow aqueous solution or cell in contact with electrodes to slide off readily. Cytotoxicity test conducts the biological evaluation and screening tests that use the tissue cells *in vitro* to observe the cell growth, reproduction, and morphological effects by the developed medical devices. The L-929 cell line was used to culture on the SLIPS coated extracted medium and counting the cell viability to identify the number of healthy cells for finding the cytotoxic effect due to the developed coating. The following formula is used for counting the cell viability-

$$\% \; viability = \frac{Test \; Sample}{Blank \; Sample} \; x \; 100 \quad (1)$$

Where, test samples are extracted medium from with SLIPS coating and without SLIPS coating fMEA. Blank sample is the medium without any extraction and can be used as blank control.

III. RESULTS AND DISCUSSIONS

Figure 2 showed the electrode layout (2x5 arrays) and the optical microscopic images of two different microelectrodes without and with modified SLIPS, respectively. Figure 3 shows the morphological study of the modified surface where the nanoporous surface is observed. Due to the slippery surface, the wettability has increased to 75%, and the SLIPS-coated surface becomes hydrophobic, which will be helpful to reduce the cell adhesion during the neural interface (Fig. 4). Electrochemical impedance and phase change are shown in figure 5. At 1 kHz, the impedance is slightly increased for a typical SLIPS coated electrode compared with the bare electrode. As shown in Table 1, Pt nanocone has increased the surface area, and therefore, up to 84.24 % impedance reduction is achieved. With the SLIPS coating, the impedance increased from 3.21 kΩ to 4.68 kΩ, yet the reduction of the impedance is almost 77%, which

remains excellent for chronic recording/stimulation of the neural cells. The characteristics curve of redox for the microelectrodes were further studied using cyclic voltammetry (CV), as shown in figure 5 (c). The Charge storage capacity (CSC) of a microelectrode is a crucial indicator for study the simulation performance. It shows that the charge stimulation capacity of the SLIPS coating has reduced compared to Pt gray surface. However, it is still excellent compared to the bare microelectrode.

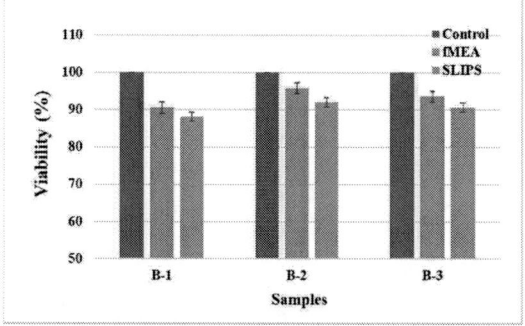

Figure 6: Comparison of Cell viability between the control, fMEA, and SLIPS coated electrode.

Figure 6 shows that more than 85% of cells are survived compared to the blank. It can be considered as excellent results and showed that the coating does not have any cytotoxic potentiality to the human body.

A. Conclusion

The procedure of the slippery coating on the electrode surface is explained. The developed SLIPS coated fMEA can record high-quality neural signals. The coating changes the impedance level but keeps it at a reasonable level (4.68 kΩ), a significant concern of chronic neural recording. It has excellent ability to provide neural interface which would be helpful for chronic neural recording. The cytotoxicity study also showed that the coating is biocompatible and not harmful for neural implantation.

REFERENCES

[1] R.L. Watts, W.C. Koller, Movement disorders: neurologic principles & practice: McGraw-Hill Professional; 2004.
[2] X. Navarro, T.B. Krueger, N. Lago, S. Micera, T. Stieglitz, P. Dario, A critical review of interfaces with the peripheral nervous system for the control of neuroprostheses and hybrid bionic systems, Journal of the Peripheral Nervous System, 10(2005) 229-58.

[3] E. Castagnola, A. Ansaldo, E. Maggiolini, T. Ius, M. Skrap, D. Ricci, et al., Smaller, softer, lower-impedance electrodes for human neuroprosthesis: a pragmatic approach, Frontiers in Neuroengineering, 7(2014) 8.

[4] J. Kim, J. Lee, D. Son, M.K. Choi, D.H. Kim, Deformable devices with integrated functional nanomaterials for wearable electronics, Nano Convergence, 3(2016) 4.

[5] V.S. Polikov, P.A. Tresco, W.M. Reichert, Response of brain tissue to chronically implanted neural electrodes, Journal of Neuroscience Methods, 148(2005) 1-18.

[6] P.K. Campbell, K.E. Jones, R.J. Huber, K.W. Horch, R.A. Normann, A silicon-based, three-dimensional neural interface: manufacturing processes for an intracortical electrode array, IEEE Transactions on Biomedical Engineering, 38(1991) 758-68.

[7] L. Berdondini, A. Bosca, T. Nieus, A. Maccione, Active Pixel Sensor Multielectrode Array for High Spatiotemporal Resolution: Springer New York; 2014.

[8] T.D.Y. Kozai, A.S. Jaquins-Gerstl, A.L. Vazquez, A.C. Michael, C. X Tracy, Brain tissue responses to neural implants impact signal sensitivity and intervention strategies, ACS Chemical Neuroscience, 6(2015) 48-67.

[9] P. Heiduschka, S. Thanos, Implantable bioelectronic interfaces for lost nerve functions, Progress in neurobiology, 55(1998) 433-61.

[10] T.D.Y. Kozai, A.L. Vazquez, C.L. Weaver, S.G. Kim, X.T. Cui, In vivo two-photon microscopy reveals immediate microglial reaction to implantation of microelectrode through extension of processes, Journal of Neural Engineering, 9(2012) 066001.

[11] T.D.Y. Kozai, N.B. Langhals, P.R. Patel, X. Deng, H. Zhang, K.L. Smith, et al., Ultrasmall implantable composite microelectrodes with bioactive surfaces for chronic neural interfaces, Nature materials, 11(2012) 1065-73.

[12] J.E. Burda, A.M. Bernstein, M.V. Sofroniew, Astrocyte roles in traumatic brain injury, Experimental Neurology, 275 Pt 3(2016) 305-15.

[13] Y. Tuo, H. Zhang, W. Chen, X. Liu, Corrosion protection application of slippery liquid-infused porous surface based on aluminum foil, Applied Surface Science, 423(2017) 365-74.

[14] Z. Shi, Y. Xiao, R. Qiu, S. Niu, P. Wang, A facile and mild route for fabricating slippery liquid-infused porous surface (SLIPS) on CuZn with corrosion resistance and self-healing properties, Surface and Coatings Technology, 330(2017) 102-12.

[15] T. Xiang, M. Zhang, H.R. Sadig, Z. Li, M. Zhang, C. Dong, et al., Slippery liquid-infused porous surface for corrosion protection with self-healing property, Chemical Engineering Journal, 345(2018) 147-55.

[16] A.C. Glavan, R.V. Martinez, A.B. Subramaniam, H.J. Yoon, R.M. Nunes, H. Lange, et al., Omniphobic "RF paper" produced by silanization of paper with fluoroalkyltrichlorosilanes, Advanced Functional Materials, 24(2014) 60-70.

[17] Q. Wei, C. Schlaich, S. Prévost, A. Schulz, C. Böttcher, M. Gradzielski, et al., Supramolecular polymers as surface coatings: Rapid fabrication of healable superhydrophobic and slippery surfaces, Advanced Materials, 26(2014) 7358-64.

[18] J. Zhang, A. Wang, S. Seeger, Nepenthes pitcher inspired anti-wetting silicone nanofilaments coatings: preparation, unique anti-wetting and self-cleaning behaviors, Advanced Functional Materials, 24(2014) 1074-80.

[19] Q. Zeng, K. Xia, B. Sun, Y. Yin, T. Wu, M.S. Humayun, Electrodeposited iridium oxide on platinum nanocones for improving neural stimulation microelectrodes, Electrochimica Acta, 237(2017) 152-9.

Gap in pagination due to unavailable paper.

Pages 984-985

Proceedings of the 16th Annual IEEE International
Conference on Nano/Micro Engineered and Molecular Systems
April 25-29, 2021

Equivalent Electrical Model and Experimental Analysis of a Novel Needle-ring Atmospheric Pressure Plasma Microjets Array*

Lingju Xia, Junfeng Yang, and Li Wen**

Abstract— We present a novel atmospheric pressure plasma microjets (APPμJ) array to achieve the localized mask-free parallel processing for material. Benefited from ease and low-cost microfabrication technology, the dimension and distance of adjacent jets can be designed at micro/nano scale. The equivalent electrical model of APPμJ is developed to investigate discharge electrical mechanism of plasma microjets. The simulation results show that the discharge is in Dielectric Barrier Discharge (DBD) mode, which has good consistency with experiment measurement. In addition, we use APPμJ array to realize maskless localized parallel etching of graphene film, and the results exhibit good surface quality and uniformity.

I. INTRODUCTION

In recent years, atmospheric pressure plasma jets (APPJ) array is widely used in etching, modification, deposition and biomedical applications for the advantages of simple structure, remote operation and three-dimensional parallel treatment [1-3]. However, the jet-to-jet distance of usual APPJ array, which is formed by integrating several independent plasma jets together, is at millimeter scale due to the limitation of tube size [4]. In this work, we propose a novel atmospheric pressure plasma microjets (APPμJ) array system, which combines APPJ with a microfabricated nozzle array. Thanks to Micro Electro Mechanical Systems (MEMS) fabrication technology, the dimension and jet-to-jet distance of nozzle array could be adjusted at the range of micron to nano scale, which has potential applications in MEMS and micro-biomedical field [5, 6].

The research on jet array discharge characteristic is mostly carried out by experimental methods such as ICCD image [7] and Schlieren photography [8]. Due to the limitation of the reactor structure, it is difficult to measure the internal parameters (e.g., dielectric barrier voltage, gas voltage, gas discharge current, and displacement current) of the reaction process. Therefore, the equivalent electrical modeling of Dielectric Barrier Discharge (DBD)has attracted growing interest to study the discharge characteristics and behaviors of DBD systems. Naudé et al. [9] researched the mechanism controlling the transition from a Townsend to a filamentary dielectric barrier discharge by electrical model. Guan et al. [10] studied the mechanism regarding the effect of plasma modified material on the DBD by an improved circuit model. Pal et al. [11] analyzed the discharge characteristic of a quartz coaxial dielectric-barrier discharge tube operated at different gas pressures and frequencies by multiswitch equivalent electrical model.

In this work, we modeled an equivalent electrical model of needle-ring structure, and simulated the electrical characteristic of discharge under different applied voltage. The transferred charge and discharge power were obtained by experiment and simulation. Finally, we used 1×2 plasma microjets array to realize micro-holes etching on graphene film, which provided an effective way for localized maskless parallel direct processing of material in micro/nano scale.

II. EXPERIMENTAL SETUP

Fig. 1 shows the schematic diagram of APPμJ array system based on the DBD principle. The system consists of power source, jet reactor and measuring unit. The power source is an AC sinusoidal resonant high voltage power source (Suman CTP-2000 K) with the frequency of 10 kHz. The jet reactor is made up of glass tube, high-voltage electrode and ground electrode. The glass tube with 4 mm inner diameter, 6 mm outer diameter and 90 mm length is used as the dielectric barrier layer. The high-voltage electrode with 60 mm length and 2 mm diameter locates the center of glass tube. An 8-mm-width copper collar is used as the ground electrode, which is covered on the outer wall of the glass tube. The distance between ground electrode and the end of glass tube is 2 cm.

The applied voltage, discharge current and discharge charge need to be measured in this study. The applied voltage is measured by a high voltage probe (Tektronix P6015A, 1000:1). The discharge current is obtained by a resistor with a value of 10 Ω in series in the discharge circuit. The transferred charge is gained from a capacitor of 330 pF in series in the discharge circuit. The measured voltage-current waveform and

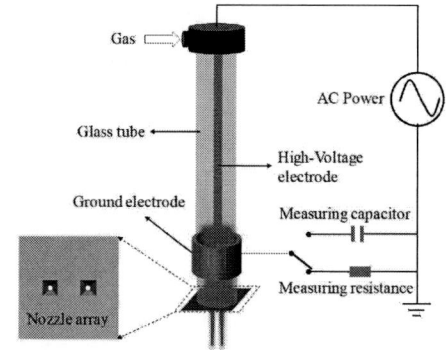

Figure 1. The schematic diagram of the atmospheric pressure plasma microjets array system.

---

*Resrach supported by National Natural Science Foundation of China (No. 51375469).

All authors are with the Department of Precision Machinery and Instrumentation, University of Science and Technology of China, Hefei, Anhui, 230027, P.R.China.

** Corresponding author to provide phone: +86-0551-6360-0214; e-mail: lilywen@ustc.edu.cn .

978-1-6654-3008-1/21 $31.00 © 2021 IEEE

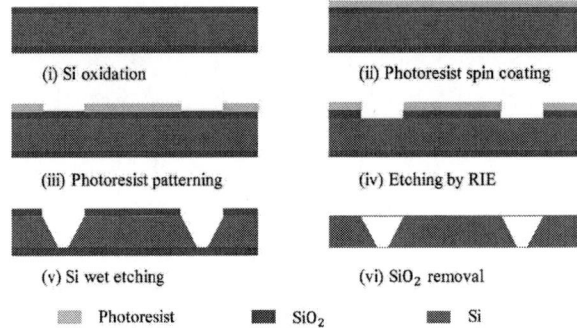

Figure 2. The schematic diagram of the silicon micro nozzle fabrication procedure.

Lissajous figures are recorded by a digital oscilloscope. Discharge image is captured by a digital camera with 1/60s exposure time.

As a key component of this system, the micro nozzle array of jet reactor is fabricated on a 150±10 μm thick silicon wafer by MEMS fabrication process as shown in Fig. 2. Firstly, the silicon wafer is oxidized on both side (i). Then, the photoresist is spin-coated (ii) and patterned (iii) on the upper side of the wafer. Next, SiO$_2$ layer is patterned by Reative Ion Etching (iv). Inverted pyramid microholes are fabricated by anisotropic KOH wet etching (v). Finally, SiO$_2$ is removed by hydrofluoric acid (vi).

III. ANALYSIS AND ELECTRICAL MODEL

A. Equivalent Electrical Model

In our work, the discharge unit is a needle-ring structure. Plasma discharge appears between needle electrode and ring electrode, and plasma ejects from the micro-nozzle array to form microjets array. Based on ionization channels, we establish an equivalent electrical model, as shown in Fig. 3. Comparing with Fig. 1, equivalent model is consistent with the schematic diagram of this system.

The red region in Fig. 3 represents high-voltage electrode, and gray rectangle refers to ground electrode. There are three capacitances between high voltage electrode and ground electrode, of which C_d and C_p are connected in parallel and then connected in series with C_g. The capacitance, C_d represents the equivalent capacitance of the glass tube. C_g represents the equivalent capacitance of the gas gap between two electrodes. C_p represents the equivalent capacitance of the plasma jet. The C_s, R_s and L_s refer to the stray capacitance, lead wire resistance and parasitic series inductance of this system, respectively. Taking these factors into consideration can reflect the actual discharge situation more accurately. In this structure, C_d and C_g can be calculate by the following equations:

$$C_d = \frac{2\pi\varepsilon_0\varepsilon_d l_d}{ln\,(R/r)} \quad (1)$$

$$C_g = \frac{\pi\varepsilon_0\varepsilon_d r^2}{l_g}. \quad (2)$$

Here, ε_0 is the permittivity of vacuum. ε_d is the relative permittivity of the dielectric barrier. l_d is the width of ground electrode. l_g is the gas gap distance. R is the outer diameter of the glass tube. r is the inner diameter of the glass tube.

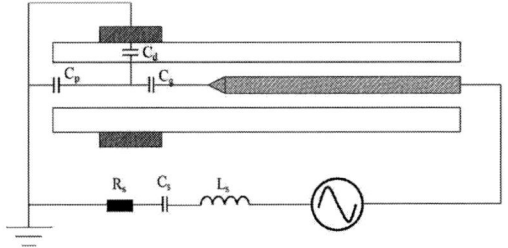

Figure 3. The equivalent electrical model of needle-ring electrode structure jet discharge.

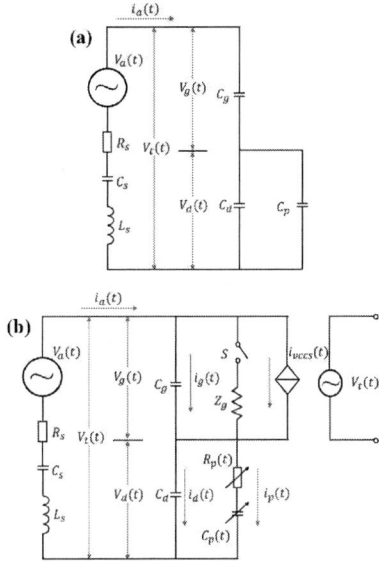

Figure 4. The equivalent electrical circuit of (a) pre-discharge and (b) discharge.

B. Equivalent Electrical Circuit

On the basis of the structure of Fig. 3 and correlative analysis, the equivalent electrical circuit is drawn in Fig. 4. The process of DBD discharge could be divided into two stages: the pre-discharge stage and discharge stage. As depicted in Fig. 4(a), pre-discharge stage takes place before the breakdown, and the equivalent circuit in this case can be simply modelled as an arrangement of capacitors in a serial and parallel connection. Here, $V_a(t)$ is applied voltage, $V_t(t)$ is reactor voltage. $V_g(t)$ is the voltage across the gas gap. $V_d(t)$ is the voltage across the dielectric barrier. $i_a(t)$ is the total circuit current. When the applied voltage exceeds the breakdown voltage, the DBD occurs between high-voltage electrode and ground electrode which is discharge stage. The discharge is generated in the tube and then ejects to the outside of the tube through the nozzle array to form an APPμJ. The equivalent electrical circuit of discharge stage is shown in Fig. 4(b). The discharge process can be equivalent to a voltage controlled current source (VCCS), microdischarge channel impedance Z_g and gas capacitance C_g. The microjets outside the tube is equivalent to variable resistance $R_p(t)$ and variable capacitance $C_p(t)$. The particles generated in the discharge tube moving with gas flow to the outside of the tube, then ionized the air to form an ionization channel and change its impedance. At this point, the capacitance of the plasma jet is no longer C_p. It can

978-1-6654-3008-1/21 $31.00 © 2021 IEEE

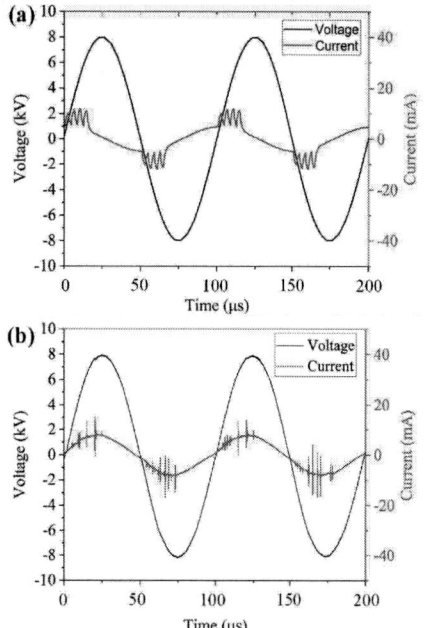

Figure 5. Applied voltage and current waveforms (peak applied voltage = 8 kV and frequency= 10 kHz): (a) simulated result; (b) measured result.

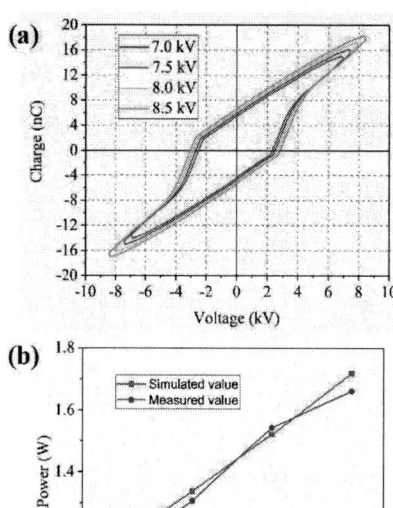

Figure 6. (a) Simulated Lissajous figures and (b) simulated and measured discharge powers under different applied voltage.

be replaced by a resistance-capacitive series circuit composed of variable $C_p(t)$ and $R_p(t)$.

Circuit will be difficult to analyze if considering nonlinear elements. Therefore, the microjet outside the tube is ignored provisionally, and the relationship of $i_{vccs}(t)$, $V_d(t)$ and $i_a(t)$ is obtained according to Kirchhoff's law, as shown in the following equations:

$$i_a(t) = C_d \frac{dV_d(t)}{dt} \qquad (3)$$

$$i_g(t) = i_a(t) - i_{vccs}(t) = C_g \frac{dV_g(t)}{dt} \qquad (4)$$

$$V_t(t) = V_d(t) + V_g(t). \qquad (5)$$

From above equations, we can conclude that the expression of the discharge current is:

$$i_{vccs}(t) = \left(1 + \frac{C_g}{C_d}\right) i_a(t) - C_g \frac{dV_t(t)}{dt}. \qquad (6)$$

We can see that $i_{vccs}(t)$ is controlled by the voltage $V_t(t)$ and strongly related to current $i_a(t)$. Then, we select the appropriate resistance $R_p(t)$ and capacitance $C_p(t)$ to represent the jet. After above discussion, a dynamic equivalent electrical circuit model of needle-ring atmospheric pressure plasma microjets array is established in Simulink software.

IV. RESULT AND DISCUSSIONS

A. Analysis of Discharge Voltage and Current

Fig. 5(a) shows the simulated voltage-current waveform under the condition that peak applied voltage is 8 kV and frequency is 10 kHz. Filamentary discharge occurs in both positive and negative half voltage period. The discharge current is about 15mA. The filamentary dielectric barrier discharge is due to charge accumulation between the power

electrode and the ground electrode [12]. The measured voltage-current waveform is shown in Fig. 5(b). By comparing simulated results and measured results, it can be seen that both of them are consistent in amplitude and duration of pulses. The simulated results verify the filamentary discharge characteristics of the DBD source, which illustrates the correctness of the electrical model and dynamic electrical circuit established in this article.

B. Analysis of Discharge Power

Discharge power, P and discharge charge, Q are two significant parameters to characterize discharge. The discharge power can get from the area of Lissajous figure, and discharge charge is obtained by measuring the voltage of measuring capacitor C in Fig. 1. The calculation method is shown in the following equations:

$$u_C(t) = \frac{1}{C} \int_{t_1}^{t_2} i(t) dt = \frac{Q(t)}{C} \qquad (7)$$

$$P = \frac{1}{T} \int_0^T u(t) i(t) dt = \frac{1}{T} \int_0^T u(t) dQ, \qquad (8)$$

where $u_C(t)$ is the voltage of measuring capacitor C and T is the period of applied voltage. According to (7), the simulated Lissajous figures under different applied voltage is shown in Fig. 6(a). It can be seen that the area of Lissajous figure increases with the increasing of applied voltage. According to (8), the discharge power is linearly related to the area of Lissajous figure, which is shown in Fig. 6(b). When the peak applied voltage is increased from 7.0 kV to 8.5 kV, the measured power increase from 1.1 W to 1.65 W, and simulated power increase from 1.15 W to 1.7 W. The simulated power value and curve trend are in great agreement with the measured value, which proves validity of equivalent electrical circuit.

978-1-6654-3008-1/21 $31.00 © 2021 IEEE

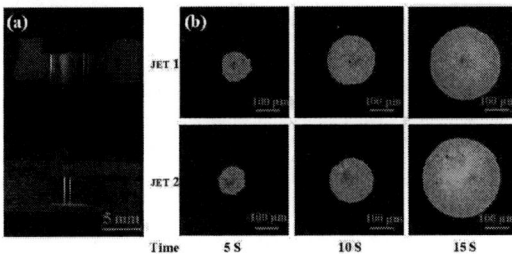

Figure 7. (a) The discharge photograph of 1×2 APPμJ array; (b) The micro holes array etched on the graphene film with the thickness of 20 μm by plasma jets array with the time of 5s, 10s, and 15s (peak applied voltage = 8 kV and helium flow rate = 0.8 L/min).

C. Experiment of Graphene Film Etching by APPμJ

In order to test the performance of needle-ring atmospheric pressure plasma microjets array, we perform experiments of graphene film etched by a 1×2 microjets array. Fig. 7(a) is an optical discharge photograph of the 1×2 plasma jets array. The dimension of micro nozzle is 90 μm and the jet-to-jet distance is 750 μm. The peak applied voltage is 8 kV and helium flow rate is 0.8 L/min. We can see that plasma is generated inside the glass tube and then ejects through micro nozzle to form microjets array. Fig. 7(b) shows the micro holes array etched on the graphene film with the thickness of 20 μm by plasma jets array. As time goes on, we can notice that the etching holes have good surface quality and uniformity. After 5 seconds of etching, the graphene film is etched through, and the etching aperture is about 110 μm, which is approximate to the microjet size. With the increase of time, the etching aperture gradually increases, which may attribute to the expansion effect of the plasma jet in ambient air conditions.

V. CONCLUSION

This article provides a novel needle-ring atmospheric pressure plasma microjets to achieve the localized mask-free parallel processing for material. An equivalent electrical model that can reflect the jet discharge process of needle-ring electrode structure is established. The discharge between the high-voltage electrode and ground electrode is equivalent to the circuit based on VCCS. Based on equivalent circuit, the voltage-current waveform and discharge power simulated by dynamic simulation model are in good agreement with the experimental results, which illustrates the correctness of the electrical model and provides guidance for further experiments. By using the system in this paper, we etched the graphene film which performs good surface quality and uniformity. The results in this article in-depth study of jet discharge characteristics, and lay a foundation for further applications in MEMS and biomedicine fields.

ACKNOWLEDGMENT

The authors acknowledge the Experimental Center of Engineering and Material Sciences at USTC for the fabrication and measuring of samples. The authors are also grateful to the colleagues for their essential contribution to this work.

REFERENCES

[1] I. Motrescu and M. Nagatsu, "Nanocapillary Atmospheric Pressure Plasma Jet: A Tool for Ultrafine Maskless Surface Modification at Atmospheric Pressure," *ACS Appl Mater Interfaces,* vol. 8, no. 19, 2016, pp. 12528-12533.

[2] T. Wang, X. L. Wang, B. Yang, X. Chen, C. S. Yang, and J. Q. Liu, "Low temperature atmospheric microplasma jet array for uniform treatment of polymer surface for flexible electronics," *Journal of Micromechanics and Microengineering,* vol. 27, no. 7, 2017, p. 075005.

[3] Q. Xie, H. Lin, S. Zhang, R. Wang, and F. Kong, "Deposition of SiCxHyOz thin film on epoxy resin by nanosecond pulsed APPJ for improving the surface insulating performance," *Plasma Science & Technology,* vol. 20, no. 2, 2018, p. 025504.

[4] D. Y. Kim, S. J. Kim, H. M. Joh, and T. H. Chung, "Characterization of an atmospheric pressure plasma jet array and its application to cancer cell treatment using plasma activated medium," *Physics of Plasmas,* vol. 25, no. 7, 2018, p. 073505.

[5] T. Abuzairi, M. Okada, Y. Mochizuki, N. R. Poespawati, R. W. Purnamaningsih, and M. Nagatsu, "Maskless functionalization of a carbon nanotube dot array biosensor using an ultrafine atmospheric pressure plasma jet," *Carbon,* vol. 89, 2015, pp. 208-216.

[6] L. Liu, D. Ye, Y. Yu, L. Liu, and Y. Wu, "Carbon-based flexible micro-supercapacitor fabrication via mask-free ambient micro-plasma-jet etching," *Carbon,* vol. 111, 2017, pp. 121-127.

[7] R. X. Wang, H. Sun, W. D. Zhu, C. Zhang, S. Zhang, and T. Shao, "Uniformity optimization and dynamic studies of plasma jet array interaction in argon," *Physics of Plasmas,* vol. 24, no. 9, 2017, p. 093507.

[8] M. Hasnain Qaisrani, C. Y. Li, P. Xuekai, M. Khalid, X. Yubin, and L. Xinpei, "Patterns of plasma jet arrays in the gas flow field of non-thermal atmospheric pressure plasma jets," *Physics of Plasmas,* vol. 26, no. 1, 2019, p. 013505.

[9] N. Naudé, J. P. Cambronne, N. Gherardi, and F. Massines, "Electrical model and analysis of the transition from an atmospheric pressure Townsend discharge to a filamentary discharge," *Journal of Physics D: Applied Physics,* vol. 38, no. 4, 2005, pp. 530-538.

[10] H. L. Guan, X. R. Chen, T. Jiang, H. Du, A. Paramane, and H. Zhou, "Electrical modeling of dielectric barrier discharge considering surface charge on the plasma modified material," *Chinese Physics B,* vol. 29, no. 7, 2020, p. 075204.

[11] U. N. Pal, P. Gulati, N. Kumar, R. Prakash, and V. Srivastava, "Multiswitch Equivalent Electrical Model to Characterize Coaxial DBD Tube," *IEEE Transactions on Plasma Science,* vol. 40, no. 5, 2012, pp. 1356-1361.

[12] T. Abuzairi, M. Okada, S. Bhattacharjee, and M. Nagatsu, "Surface conductivity dependent dynamic behaviour of an ultrafine atmospheric pressure plasma jet for microscale surface processing," *Applied Surface Science,* vol. 390, 2016, pp. 489-496.

Proceedings of the 16th Annual IEEE International
Conference on Nano/Micro Engineered and Molecular Systems
April 25-29, 2021

A Wearable Health Monitoring System Self-powered by Human-motion Energy Harvester

Chen Bao, Anxin Luo, Yi Zhuang, Junlong Huang and Fei Wang

Abstract— **In this paper, we have proposed a self-powered wireless wearable health data tracking system based on an inertial rotary electromagnetic energy harvester. The whole system can be divided into three parts. The first part is the charging circuit. The energy harvester, which generate 1.65 V Root Mean Square (RMS) voltage and 32.15 mW power at the frequency of one foot while running (1 Hz), could convert kinetic energy of human motion into electrical energy to charge the Li-ion battery. The second part contains an Arduino Microcontroller Unit (MCU) and two sensors. These two sensors could monitor the temperature and the blood oxygen saturation (which can be used in calculation of heart rate) information of the feet and toes, respectively. In the last part, A Bluetooth module is applied to communicate with the MCU and send the health data to the cloud. Meanwhile, this data could be displayed on the smartphone. We also demonstrated the workflow of the whole system. With the energy harvester, the energy from human motion is used to drive sensors and the function of health monitoring is realized.**

I. INTRODUCTION

Nowadays, with the increase pressure of people's life and work, many diseases occur in more younger people. Also, the worldwide outbreak situation of Covid-19 is still serious. The common symptoms of Covid-19 are fever, cough, fatigue [1], etc. Therefore, the monitoring of human health data is particularly important. The common method to get health information in daily life is using wearable device. With integrated biosensors, some wearable devices can obtain users' physiological indicators including heartrate, sleep situation and body temperature. However, the energy supply for these wearable devices is still a problem. Traditional methods of power supply need to recharge or change the battery frequently which may impede the adoption of wearable devices. Moreover, some chemical batteries are not environment friendly that may cause pollution. Recently, using the energy harvester to supply power for wearable device gives an appropriate solution [2][3][4][5]. According to the principle of operation, energy harvesters can be mainly divided into type of electromagnetic, piezoelectric [6][7], electrostatic [8][9] and triboelectric [10]. In the existing harvesters, electromagnetic vibration energy harvester makes use of the law of electromagnetic induction. When the magnet is shifted relative to the coil, an induced voltage will be generated in the coil, which supplies power to the external circuit. With the output voltage affected by magnetic field strength, coil turns and other parameters, this kind of energy harvester does not need external power supply and has better output characteristic [11][12]. Among different vibration sources, human movement is very abundant in our daily life, and it has the potential to

All the authors are with the School of Microelectronics, Southern University of Science and Technology (SUSTech), Shenzhen, China. (corresponding author e-mail: wangf@sustech.edu.cn).

generate enough energy for wearable devices. At present, vibration energy harvesters are generally based on resonant structures, which usually exhibit a natural frequency of tens of Hertz. However, the vibrations generated by human movement in daily activities (such as walking, running/jogging, etc.) are usually irregular vibrations with large amplitude and extremely low frequency [13][14]. If the optimal design is not carried out, the output power of the resonance energy harvester will drop sharply under the driving of human motion.

In this paper, we proposed a self-powered health monitoring system based on an inertial rotary electromagnetic energy harvester. The system is installed on the shoes which could collect the movement energy when the users run or walk (pressing the cover of the harvester). We use rectifier module to convert the AC output to DC output in order to charge the rechargeable lithium battery. The battery supplies power to subsequent circuits and sensor modules. There are two sensors acquiring the corresponding human body health data and transmitting it to the Arduino MCU (Micro Controller Unit) for processing. At the same time, the health data could also be sent to the cloud through Bluetooth transmission module and the real-time data can be directly viewed by APP on the mobile phone.

II. DESIGN OF THE SYSTEM

We designed the health monitoring system in three parts. As shown in the schematic of Fig. 1, the harvester collects the energy of human motion to charge the Li-ion battery firstly. Then, the battery supplies energy for the sensing circuit. Also, the sensors send the temperature and the blood oxygen saturation data to Arduino MCU to analysis. After that, the Bluetooth sends the data to cloud so that the user could get the

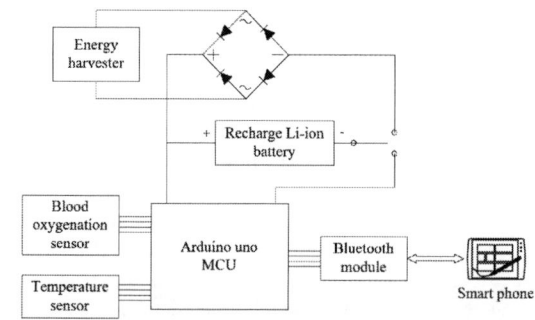

Fig 1. Schematic of the self-powered human health monitoring system.

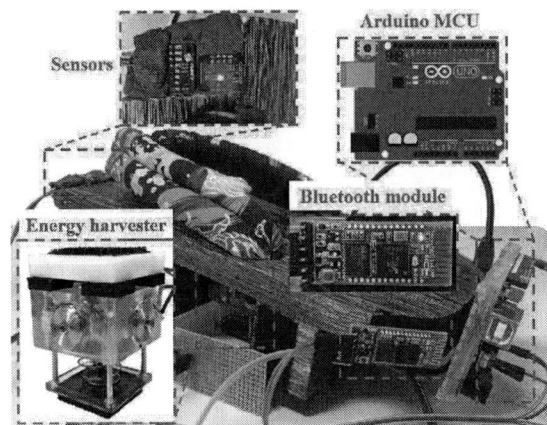

Fig 2. The assembling photo of the health sensor system.

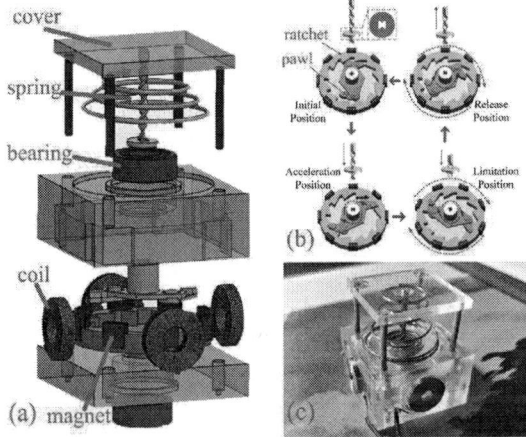

Fig 3. (a) 3D schematic (b) motion analysis (c) photo of device

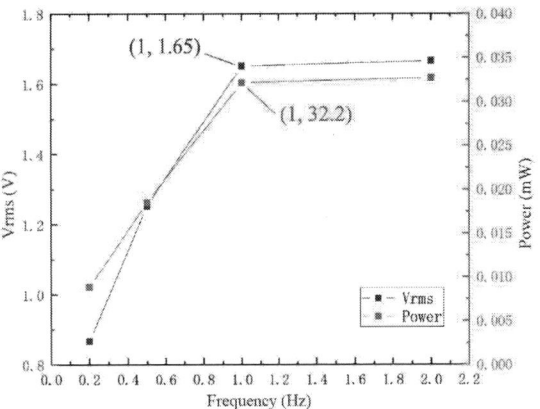

Fig 4. The output voltage and output power of the energy harvester device by collecting energy from human motion at different frequency.

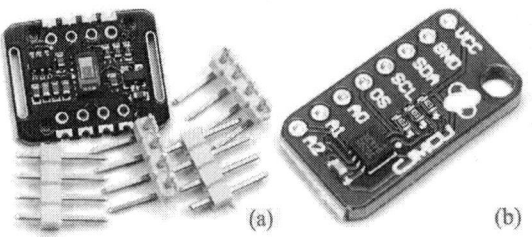

Fig 5. The schematic of (a) MAX30102 blood oxygen sensor (b) MAX30205 temperature sensor.

health information on the smart phone conveniently. Fig. 2 illuminates the assembling diagram of the system. We put the system on shoes so that the harvester can generate energy from walking or running. The sensors are fixed at the toe position which is close to skin to get health information.

A. Energy harvester

The core part of the self-powered system is the energy harvester. Fig. 3(a) and (b) shows the schematic and the motion analysis of the harvester. After the vertical compression process, the spiral torsion bar moves downward and drives the pawl to rotate, which interacts with the ratchet to accelerate the rotor. The spring can help the tablet to return to the initial position, and the tower spring structure can provide sufficient elastic force, while effectively saving volume. During each compression period, the rotor will remain inertially rotating due to the ratchet clutch system. Therefore, low frequency vibration is converted to high-speed rotation. Several magnets (Nd-Feb-N52) are symmetrically distributed on the side wall of the ratchet wheel, and four groups of 400 turns copper coils are assembled parallel to the magnets to form a magnetic coil system. According to Faraday's law of electromagnetic induction, the structure can successfully convert mechanical energy into electrical energy. The photo of

harvester is shown in Fig. 3. We used acrylic as the material of harvester body so it is light and small enough to be installed under the sole. In addition, the output characteristic of harvester was tested and the result is demonstrated in Fig. 4. It can generate 1.65 V voltage (RMS) and 32.16 mW power at 1 Hz, which satisfies the energy requirement of the sensing circuit.

B. Sensors

In the sensing circuit we used MAX30102 blood oxygen sensor and MAX30205 temperature sensor to collect body information. Fig. 5 shows the schematic of these two sensors. The MAX30102 blood oxygen sensor uses PPG (PhotoPlethysmoGraphy) technology [15] to measure pulse and oxygen saturation. The working principle of this technology is that the light intensity will attenuate as it passes through skin tissue and back to the photodetector. The absorption of light by muscles, bones, veins, and other connective tissues is essentially constant. As the blood moves through the arteries, the light absorption will be changed by the hemoglobin. When the sensor converts light signals into electrical signals, the result can be divided into AC and DC signals. By extracting the AC parts from these signals, we can get the characteristics of blood flow. The MAX30205 temperature sensor using the principle that the resistance of thermistor varies with temperature. The human body temperature can be measured from 25 °C to 40 °C, with 0.1 °C temperature measurement accuracy and 16-bit temperature resolution. These two sensors

Fig 6. Schematic of the HC-06 Bluetooth module.

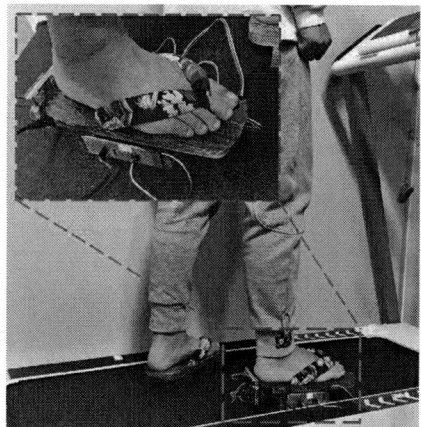

Fig 7. Photo of a volunteer wearing the system and walking on the treadmill.

both work under 2.7 V to 3.3 V input voltage condition and have very low working current.

C. MCU and Bluetooth

We use Arduino as the data processing unit of the whole system. Arduino is an open-source hardware product that is convenient, flexible and easy to use. It has a variety of interfaces, including digital I/O port, analog I/O port, and supports SPI, IIC and UART serial communication. The Bluetooth that we select is the HC-06, which could send data to Cloud. The Bluetooth module also make the system could be recognized and connected by smart phone and the users could get real-time health data of themselves. Fig.6 gives the schematic of bluetooth module.

III. TESTING RESULT

We demonstrated the working flow of the whole system. Fig. 7 presents a volunteer wearing the system and walking on the treadmill. After 30 minutes treading, the charging process was ended and the sensors started to work. The waveform of blood oxygen concentration changing is illuminated in Fig. 8. The light source of the sensor MAX30102 is a light-emitting electrode tube with a specific wavelength that is selective to the oxygenated hemoglobin (HbO_2) and hemoglobin (Hb) in the artery. Therefore, the heart rate information can be obtained by calculating the results of periodic changes in the signal. We could count the number of waves and calculate the average heartrate is about 82 BPM (BeatsPerMinute). The heartrate change in different respiratory states at the same time is also included in our research. The blood oxygen profile of a

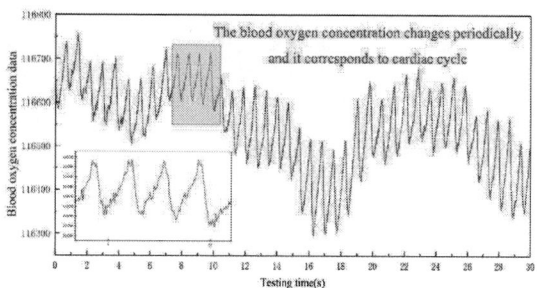

Fig 8. Waveform of blood oxygen concentration with normal breath in 30s.

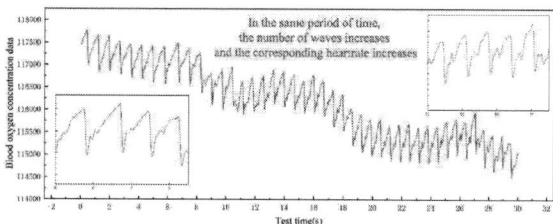

Fig 9. Waveform of blood oxygen concentration with holding breath in 30s.

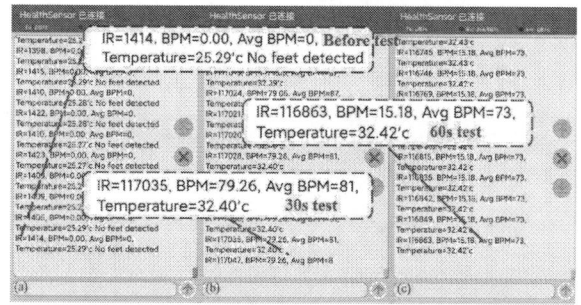

Fig 10. The data (a)before test (b)after 30s test and (c)after 60s test on the UI of smart phone.

volunteer holding his breath for 30 seconds is illustrated in Fig. 9. By intercepting the waveforms at the beginning and by the end of breath-holding for the same period of time, it can be easily informed from this figure that with the increasing of the time, the wave number increases (heartrate increases), and this matches scientific knowledge.

Moreover, we used smart phone to check the health information. Fig. 10 shows the UI scheme and it is convenient for users to check their health information.

IV. CONCLUSION

In summary, our team succeeded in designing and fabricating a self-powered health monitoring system. The system is based on a kind of electromagnetic energy harvester which could collect the energy of human motion. Two sensors are applied on the system to get temperature and blood oxygen data and after Arduino MCU processes the data, it will be sent to cloud by Bluetooth module. Users can check their health information by smart phones. The system finally realizes the function of self-powered health monitoring.

978-1-6654-3008-1/21 $31.00 © 2021 IEEE

ACKNOWLEDGMENT

This work was financially supported in part by the Shenzhen Science and Technology Innovation Committee under Grant JCYJ20200109105838951, in part by Guangdong Natural Science Funds under Grant 2018A050506001, and in part by "Climbing Program" Special Funds under Grant pdjh2021c0073.

REFERENCES

[1] J. Zeng et al, "Cardiac manifestations of COVID-19 in Shenzhen, China," in *2020 SPRINGER HEIDELBERG.*

[2] A. Luo, Y. Zhang, X. Dai, Y. Wang, W. Xu, Y. Lu, M. Wang, K. Fan, and F. Wang, "An inertial rotary energy harvester for vibrations at ultra-low frequency with high energy conversion efficiency," in *2020 Applied Energy*, 279, 115762.

[3] A. Luo, Y. Zhang, X. Guo, Y. Lu, C. Lee, and F. Wang, "Optimization of MEMS vibration energy harvester with perforated electrode," in *2021 Journal of Microelectromechanical Systems.*

[4] A. Luo, Y. Xu, Y. Zhang, M. Zhang, X. Zhang, Y. Lu, and F. Wang, "Spray-coated electret materials with enhanced stability in a harsh environment for an MEMS energy harvesting device," in *2021 Microsystems & Nanoengineering*, 7(1), 1-9.

[5] Y. Zhang, T. Wang, A. Luo, Y. Hu, X. Li, and F. Wang, "Micro electrostatic energy harvester with both broad bandwidth and high normalized power density," in *2018 Applied energy*, 212, 362-371.

[6] D. Shen, J. Park, J. Ajitsaria, S. Choe, H. Wikle, and D. Kim, "The design, fabrication and evaluation of a MEMS PZT cantilever with an integrated Si proof mass for vibration energy harvesting," in *2008 Journal of Micromechanics and Microengineering*, 18(5), 055017.

[7] H. Liu, C. Tay, C. Quan, T. Kobayashi, and C. Lee, "Piezoelectric MEMS energy harvester for low-frequency vibrations with wideband operation range and steadily increased output power," in *2011 Journal of Microelectromechanical Systems*, 20(5), 1131–1142.

[8] Y. Chiu, and Y. Lee, "Flat and robust out-of-plane vibrational electret energy harvester," in *2012 Journal of Micromechanics and Microengineering*, 23(1), 015012.

[9] H. Asanuma, H. Oguchi, M. Hara, R. Yoshida, and H. Kuwano, "Ferroelectric dipole electrets for output power enhancement in electrostatic vibration energy harvesters," in *2013 Applied Physics Letters*, 103(16), 162901.

[10] S. Wang, X. Mu, Y. Yang, C. Sun, A. Gu, and Z. Wang, L, "Flow-driven triboelectric generator for directly powering a wireless sensor node," in *2014 Advanced Materials*, 27(2), 240–248.

[11] S. Beeby, R. Torah, M. Tudor, P. Glynne-Jones, T. O'Donnell, C. Saha, and S. Roy, "A micro electromagnetic generator for vibration energy harvesting," in *2007 Journal of Micromechanics and Microengineering*, 17(7), 1257–1265.

[12] A. Foisa, C. Hong, and G. Chung, "Multi-frequency electromagnetic energy harvester using a magnetic spring cantilever," in *2012 Sensors and Actuators A: Physical*, 182, 106–113.

[13] K. Fan et al., "Achieving high electric outputs from low-frequency motions through a double-string-spun rotor," *Mechanical Systems and Signal Processing*, vol. 155, p. 107648, Jun. 2021.

[14] K. Fan, H. Qu, Y. Wu, T. Wen, and F. Wang, "Design and development of a rotational energy harvester for ultralow frequency vibrations and irregular human motions," *Renewable Energy*, vol. 156, pp. 1028–1039, Aug. 2020.

[15] H. Liu, F. Chen, V. Hartmann, S. Khalid, S. Hughes, and D. Zheng, "Comparison of different modulations of photoplethysmography in extracting respiratory rate: from a physiological perspective," *PHYSIOLOGICAL MEASUREMENT*

Proceedings of the 16th Annual IEEE International
Conference on Nano/Micro Engineered and Molecular Systems
April 25-29, 2021

Improved Reservoir Computing by Carbon Nanotube Network with Polyoxometalate Decoration

Shuo Wu[1], Wenli Zhou[1,2], *Member, IEEE*, Kaiqiang Wen[1], Chengzhu Li[1], and Qingfeng Gong[1]

[1]School of Optical and Electronic Information, Huazhong University of Science and Technology, Wuhan, China
[2]Wuhan National Laboratory of Optoelectronics, Wuhan 430074, China
Email: wlzhou@hust.edu.cn

Abstract—Physical reservoir computing (RC) is a recently introduced framework for information processing using the complex dynamics of physical systems. In this paper, a physical reservoir based on a molecular network of polyoxometalate (POM) decorated single-walled carbon nanotubes (SWCNT) with PBMA composite is fabricated. By the lab-built hardware platform, we experimentally demonstrate its excellent performance in a time series prediction benchmark with large short-term memory capacity (MC), indicating SWCNT/POM network as a promising substrate for reservoir computing because abundant inner charge and discharge in junctions leading to special electron transport characteristics that make rich dynamic and high-dimensional mapping properties appear in the POM-decorated SWCNT composite structure.

Keywords—Carbon nanotube, Physical reservoir computing, Molecular network, Polyoxometalate

I. INTRODUCTION

A reservoir computing (RC) system consists of a reservoir for mapping inputs into a high-dimensional space and a readout for pattern analysis from the reservoir. The reservoir is usually fixed and only the readout is trained with a simple method such as linear regression. Now it has become a competitive framework to deal with sequential processing problems because of its unique system and training procedure [1-2]. Recently, physical RC has been introduced using the complex dynamics of some physical systems, e.g., single-walled carbon nanotubes (SWCNT) /polymer reservoir to implement some simple calculation tasks like the simple logic gate, wave generator task, classification task, and so on [3-8]. However, its performance is not so desirable and it can be inferred to be due to its physical characteristics using the conductive network corresponding to its relatively low nonlinearity and simple dynamic behavior according to recent simulation research on RC behavior. Polyoxometalate (POM) molecule called 'electron sponge' has strong electron storage capabilities that can store up to 24 electrons by its molecular structural change [9-12]. The POM decorated SWNT junction shows many interesting electrical properties like negative differential resistance and inversion of current rectification direction [13-14]. The electrons transfer in SWCNT/POM can also induce electron cascading and generate spontaneous spikes and noise [15-16]. The simulation result of its reservoir computing capability using the cellular automata model is very exciting upon the assumption of the abundant high dimensional dynamics in SWCNT/POM network [15]. So we built a SWCNT/POM network with poly(butyl methacrylate)

(PBMA) composite on a microelectrode array as a physical reservoir and explored its computing capability in this paper. The good performance of the SWCNT/POM network in the NARMA10 (The 10th nonlinear auto-regressive moving average) task and its relative large memory capacity imply its promising potential in reservoir computing.

II. MATERIALS AND PLATFORM

A. Materials and Methods

The thin films of POM-decorated SWCNTs embedded within PBMA composites were prepared. First, SWCNT solutions contained 2.75 *mg*/ml pristine SWCNT and were prepared by ultrasonically dispersing CNT powder (1−10 *μm* long from Bucky, USA) in dimethylformamide (DMF) for 90 min. The POM-decorated CNT solutions were prepared by adding POM (400*μg* /mL) to the pristine SWCNT solutions with 15 min of stirring and 120 min of ultrasonication to absorb POM on the SWNT walls. Then, the superfluous POM was removed by filtering the POM-decorated CNT solutions using the anodic aluminum oxide membrane (pore diameter:100nm) and SWCNT/POM composites were obtained from that can remove. Then, the composites were added into 5ml anisole (Karma, analytical reagent grade) and 90 min of ultrasonication. The 1.375g PBMA powders (inherent viscosity: 0.47-0.56 dL/g from Karma, China) were added into the SWCNT/POM anisole solutions and 120 min of ultrasonication to mix them. The dispersions were visually uniform and stable over a day with no obvious precipitation of the SWCNTs from suspension.

Afterwards, 20*μL* SWCNT/POM mixtures with PBMA composites was drop-casted onto the microelectrode arrays on FR-4 circuit board substrates. Another 10×10 grid array was prepared in the middle of each board for reservoir output; the contact pads for the electrodes had a diameter of 80 *μm* (2*μm* thick gold/nickel on the surface) and a pitch of 300 *μm*.

The substrate were heated at 85 °C on a hot plate for 20 min and cool to room temperature for about 2h. The typical obtained SWCNT/POM with PBMA composites reservoir device is shown in Fig.1. The pristine SWCNT with PBMA composite reservoir devices were fabricated at the same time.

Electrical measurements on the devices in the air were performed using a semiconductor analyzer (Keysight, B1500A). The surface morphologies of SWCNT/POM/PBMA were characterized using a 3D laser confocal scanning microscope (Keyence, VK-X200K). The

978-1-6654-3008-1/21 $31.00 © 2021 IEEE

surface morphologies of SWNT/POM were characterized using atomic force microscopy (AFM, Bruker Dimension Edge) and scanning electron microscopy (SEM, Gemini SEM300, Carl Zeiss) with gold-coated samples.

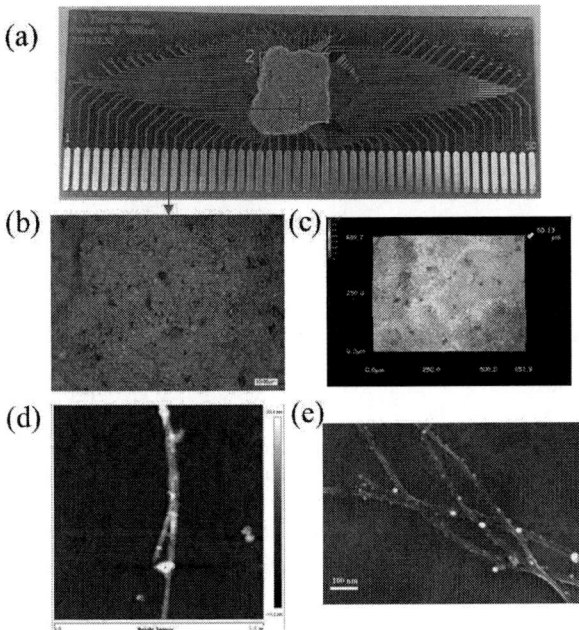

Fig.1: Optical (a) (b) and (c) 3-dimensional surface topography images of a fabricated SWCNT/POM reservoir device. AFM(d) and SEM(e) images of POM-decorated CNTs. The bright spots in (d,e) are POM molecules on SWCNT walls.

B. Experimental computing platform

It was necessary to apply and monitor many signals (up to 100 channels) on one sample for this work. An experimental platform was built to configure and evaluate the physical reservoir with its status information extracted. A schematic diagram of the platform is shown in Fig. 2. The MATLAB based user interface can configure, collect, store, and analyze data extracted from the reservoir. Each electrode in the fabricated device can be programmed as analog/digital input or output by the system interface compsed of three data acquisition cards (NI PCI-6255, 6723, and 6509).

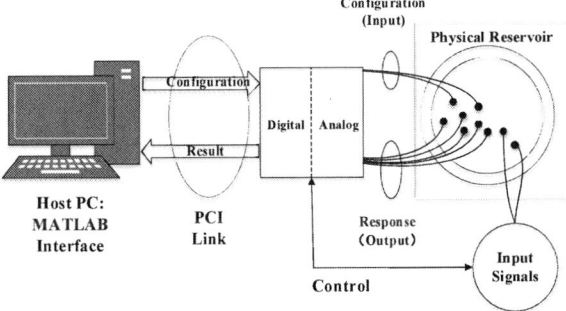

Fig.2: The diagram of the lab-built experimental systematic platform

III. RESULT AND DISCUSSION

A. The electrical properties

To investigate difference of the internal dynamics in the reservoir between the SWCNT/POM with PBMA

composites and SWCNT with PBMA composites, the characterization of current-voltage (I-V) and current-time (I-t) between every two adjacent electrodes were measured and analyzed.

These resistances in SWCNT/POM with PBMA composites spans between $2 \sim 10$ MΩ which were obviously lower without POM decorated (between $10 \sim 50$ MΩ). Several negative differential resistance (NDR) peaks in the I-V curve could be seen in SWCNT/POM reservoir devices as shown in Fig.3 while the I-V curve was linear in all pristine SWCNT reservoirs. In I-t measurements, many impulses (the max magnification: 600%) could be observed at 100mV with statistics of all the current amplitude variations in Fig. 3 (10 devices, about 10^6 values measured between two arbitary nodes). For the SWCNT/POM reservoir, 48% of them exhibited current fluctuation of >5% as contrast to its average current and 20% of them fluctuated over 17.5%, while it is in general within 2% for the pristine SWCNT reservoir. The ratio of current fluctuation between [0.15,1.3] and its percentage in the node-node measurements of SWCNT/POM reservior is shown in Fig.3. The impulses width was almost 2~6 ms and many current impulses may not be collected because of the instrument limitation (the maximum sampling rate: 500 Hz). When analyzing power spectral density (PSD) of spike signals, PSD of low-frequency noise could be fitted with flicker noise, GR noise, and thermal noise described using Eq.(1) [16]:

$$S(f) = \frac{A}{f^{\gamma}} + \frac{B}{1 + (f/f_c)^2} + C \qquad (1)$$

Where A, B, and C are respectively the amplitude of flicker noise, generation-recombination (GR) noise, and thermal noise; f_c is corner frequency of GR noise. it manifested a clear Lorentzian-shaped curve with corner = 48 Hz，indicating that the noise was mainly GR noise according to [16]. We also tested the response of 80 nodes (up to 80 acquisition channels in the platform) in the reservoirs to DC voltage. The voltage fluctuations in SWCNT/ POM reservoir were larger than the SWCNT reservoir which almost no change. The above phenomena indicated its frequent internal charge-discharge behavior in SWCNT/POM network and implied its abundant dynamics, but it could not be observed in the pristine SWCNT reservoir samples.

Using molecular dynamics simulation by first-principles software SIESTA, we investigated how the POM decoration significantly affected the electronic properties of CNTs [17]. When a POM molecule adsorbed on the zigzag-edged SWNT (8, 0) with an equilibrium distance of approximately 2.28 Å, the LUMO level of POM of approximately −6.46 eV was much lower than the Fermi level of the CNT, which was approximately −5.33 eV and induced a charge transfer of 0.11 e from the CNT to the POM molecule. And thus reduced the barrier heights between POMs in SWCNT and make electron transfer more easier in SWCNT/POM junction. Thus, the POM decorated CNTs network has more abundant inner charge and discharge than pristine CNTs network. So the SWCNT/POM reservoir could have richer dynamic properties and higher dimensional mapping capability to support computing more complex tasks.

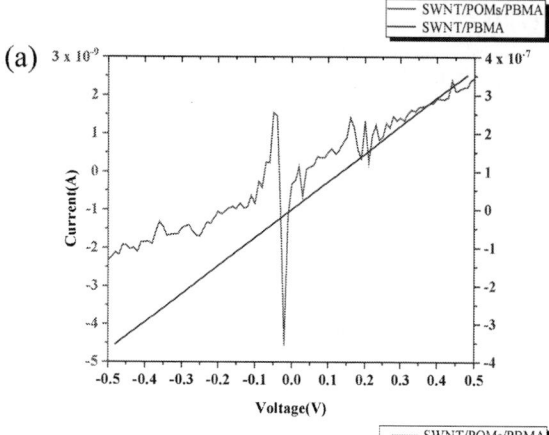

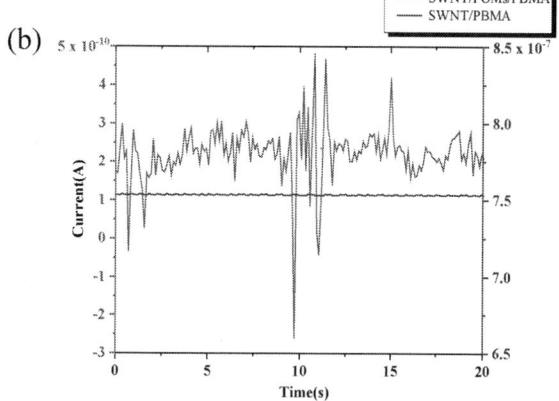

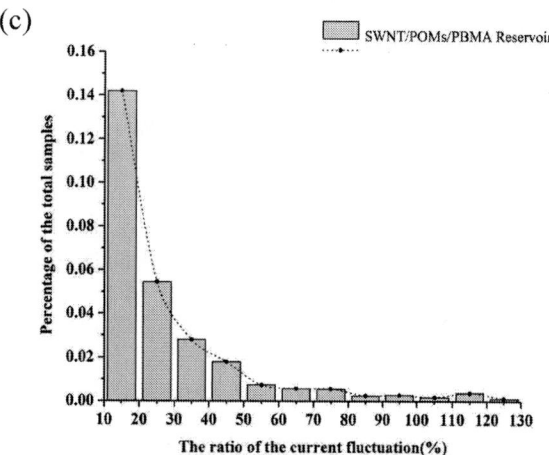

Fig.3: The electrical characteristics of the reservoir devices. (a) The I-V characteristic between one group of two adjacent electrodes (b) Time dependence of the current at 100mV of(a). (c) The ratio of current fluctuation and its percentage in the node-node measurements of SWCNT/POM reservoir.

B. Benchmark Tasks for Reservoir

To explore the performance and computational potential of the SWCNT/POM/PBMA reservoir, we used some benchmark tasks to test it. Nonlinear Auto-Regressive Moving Average (NARMA) task originates from training recurrent networks and it's often used to evaluates a reservoir's ability to model an n-th order highly non-linear

dynamical system [18]. NARMA10 sequence was generated by Eq.(2):

$$s(t+1) = 0.3s(t) + 0.05s(t)[\sum_{i=0}^{9} s(t-i)] + 1.5u(t)u(t-9) + 0.1 \quad (2)$$

where $u(t)$ consists of scalar random numbers with a uniform distribution in the intervals [0,0.5].

To implement the NARMA10 task, one electrode in the reservior was used as input, three electrodes connected to the ground, and other 80 electrodes were collected as readouts (Fig.4). The readouts' weights (W_{out}) were trained by the least square algorithm in real-time (online learning, 8×10^3 samples as training data, 10^3 samples as testing data). Fig. 5 showed the prediction result with the data length of 100 after weight training. It could be intuitively seen that the prediction of the SWCNT/POM reservoir was much better than that of the SWCNT reservoir. The average normalized root mean squared error (NRMSE) of SWCNT/ POM reservoir in all tests was 0.08, and the average NRMSE of SWCNT reservoir was 0.19. It was as high as 0.8 for the resistance reservoir consisting of 1- 2 MΩ ceramic resistors, due to the absence of any memory and information dimension enhancement ability. In the case of offline learning (length of one sample:100), the NRMSE of the SWCNT/POM reservoir was decreased to 0.0195, while the NRMSE of the SWCNT reservoir was lowered to 0.11.

At the same time, we used memory capacity (MC) as a quantitative measurement. Measuring the short-term memory capacity of a reservoir was first outlined in [19] as a quantitative measurement of the echo state property (fading memory). To determine the MC of a reservoir, we used a random number sequence as the input. By measuring how many delayed versions of the input $u(n-k)$ the outputs can recall or recover with precision using Eq.(3). The MC can be got by summing all delays which could be recovered. This is carried out by training individual output units according to the internal units' states to recall the input at time k.

$$MC = \sum_{k=1}^{\infty} MC_k = \sum_{k=1}^{\infty} \frac{\text{cov}^2(u(n-k), y(n))}{\sigma^2(u(n))\sigma^2(y(n))} \quad (3)$$

The calculated total memory capacity of the SWCNT/ POM reservoir was up to 20.5 while it was only 6.5 in the SWCNT reservoir (about 1 in the pure resistance reservoir which almost had no memory capability).

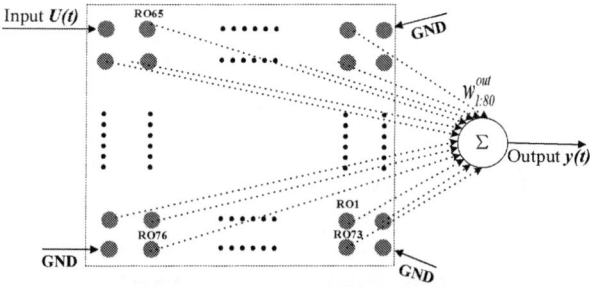

Fig.4: Schematic configuration for the physical reservoir to implement benchmark task. *U(t)* is the signal input, GND is the grounding of the corresponding electrode, and RO_n represents the state value of the reservoir. W_{out} matrix is the weight of output layer readouts.

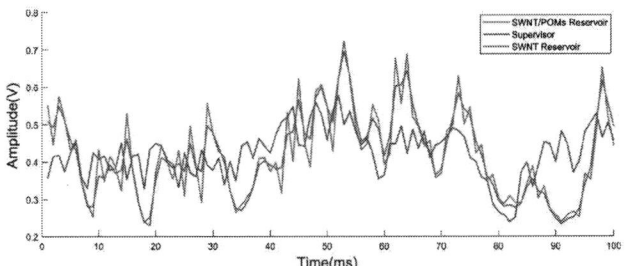

Fig.5: NARMA10 task performances of different reservoirs after training. The Emerald green is supervisor sequences, and the red and pale blue are the actual prediction of SWCNT/POM and pristine SWCNT reservoirs respectively.

IV. CONCLUSION

In this work. we proposed a physical reservoir consisting of the POM decorated SWCNT network with PBMA composites. Because of the stochastic and probabilistic electron transfer between POMs in SWCNT, the electrical properties of the SWCNT/POM device showed many unique phenomena like NDR peak in I-V, current impulse, and spontaneous noise. Those properties made SWCNT/POM reservoir have rich nonlinear physical dynamics than SWCNT reservoir. We used basic reservoir computing benchmark NARMA10 task to test the computing ability of SWCNT/POM reservoir and its good performance implied its excellent echo state property, non-linearity, and long-term dependencies due to its big short-time MC. These results indicated a big improvement computational ability in the carbon nanotube network by polyoxometalate decoration.

This work was a first step method in which the physical reservoir could be improved by changing the composite structure. In addition, the number of microelectrode array also affects the performance. By microelectronic processing technology, the SWCNT/POM network could be extremely more dense, although the size of the microelectrode array may be nanoscale limited. Moreover, by adding the feedback layer and drive signals in the reservoir, the performance would be better. These efforts might allow SWCNT/POM reservoir to undertake more complex tasks.

V. ACKNOWLEDGMENT

This work was supported by the National Natural Science Foundation of China (Grant No. 61774063) .

VI. REFERENCES

[1] K. Nakajima, "Physical reservoir computing—an introductory perspective," Jpn J Appl Phys, vol. 59, no. 6, p. 060501, 2020/05/15 2020.

[2] G. Tanaka et al., "Recent advances in physical reservoir computing: A review," Neural Networks, vol. 115, pp. 100-123, Jul 2019.

[3] M. Dale, J. F. Miller, S. Stepney, and M. A. Trefzer, "Evolving Carbon Nanotube Reservoir Computers," Lect Notes Comput Sc, vol. 9726, pp. 49-61, 2016.

[4] M. Dale, J. F. Miller, S. Stepney, and M. A. Trefzer, "A substrate-independent framework to characterize reservoir computers," Proceedings of the Royal Society A: Mathematical, Physical and Engineering Sciences, vol. 475, no. 2226, p. 20180723, 2019/06/28 2019.

[5] M. Mohid, J. F. Miller, S. L. Harding, G. Tufte, M. K. Massey, and M. C. Petty, "Evolution-in-materio: solving computational problems using carbon nanotube–polymer composites," Soft Computing, vol. 20, no. 8, pp. 3007-3022, 2016.

[6] M. Dale, S. Stepney, J. F. Miller, and M. Trefzer, "Reservoir computing in materio: An evaluation of configuration through evolution," in 2016 IEEE Symposium Series on Computational Intelligence (SSCI), 2016, pp. 1-8.

[7] M. Mohid et al., "Evolution-in-materio: Solving function optimization problems using materials," in 2014 14th UK Workshop on Computational Intelligence (UKCI), 2014: IEEE, pp. 1-8.

[8] M. Mohid et al., "Evolution-in-materio: A frequency classifier using materials," in 2014 IEEE International Conference on Evolvable Systems, 2014: IEEE, pp. 46-53.

[9] S. Z. Wen et al., "Theoretical insights into [PMo12O40](3-) grafted on single-walled carbon nanotubes," Phys Chem Chem Phys, vol. 15, no. 23, pp. 9177-9185, 2013.

[10] H. Wang et al., "In Operando X-ray Absorption Fine Structure Studies of Polyoxometalate Molecular Cluster Batteries: Polyoxometalates as Electron Sponges," J Am Chem Soc, vol. 134, no. 10, pp. 4918-4924, 2012/03/14 2012.

[11] Y.-F. Song and R. Tsunashima, "Recent advances on polyoxometalate-based molecular and composite materials," Chemical Society Reviews, 10.1039/C2CS35143A vol. 41, no. 22, pp. 7384-7402, 2012.

[12] X. López, J. J. Carbó, C. Bo, and J. M. Poblet, "Structure, properties and reactivity of polyoxometalates: a theoretical perspective," Chemical Society Reviews, 10.1039/C2CS35168D vol. 41, no. 22, pp. 7537-7571, 2012.

[13] L. Hong, H. Tanaka, and T. Ogawa, "Rectification direction inversion in a phosphododecamolybdic acid/single-walled carbon nanotube junction," Journal of Materials Chemistry C, 10.1039/C2TC00171C vol. 1, no. 6, pp. 1137-1143, 2013.

[14] A. Setiadi et al., "Room-temperature discrete-charge-fluctuation dynamics of a single molecule adsorbed on a carbon nanotube," Nanoscale, 10.1039/C7NR02534C vol. 9, no. 30, pp. 10674-10683, 2017.

[15] H. Tanaka et al., "A molecular neuromorphic network device consisting of single-walled carbon nanotubes complexed with polyoxometalate," Nat Commun, vol. 9, Jul 12 2018.

[16] K. L. Goh, H. Fujii, A. Setiadi, Y. Kuwahara, and M. Akai-Kasaya, "Spontaneous spike signals originated from redox-active molecules functionalised on carbon nanotubes," Jpn J Appl Phys, vol. 58, Aug 1 2019.

[17] Y.Zhu et al., "Phosphomolybdic Acid-Decorated Carbon Nanotubes for Low-Power Sensing of NH3 and NO2 at Room Temperature," ACS Applied Nano Materials, vol. 4, no. 2, pp. 1976-1984, 2021/02/26 2021.

[18] Atiya, A.F, Parlos, A.G: New results on recurrent network training: unifying the algorithms and accelerating convergence. IEEE Trans. Neural Netw. 11(3), 697–709 (2000).

[19] Jaeger, H : Short term memory in echo state networks. Tech. rep. no. GMD report 152. German National Research Center for Information Technology (2001)

Proceedings of the 16th Annual IEEE International
Conference on Nano/Micro Engineered and Molecular Systems
April 25-29, 2021

Hai-feng Dong, Chen Liu, Jun-jun Sang

2D Microscopy of Magnetic Field Using Atomic Vapor Cell

Abstract— **2D microscopy of magnetic field with high spatial resolution is realized using atomic vapor cell and digital micro mirror device. Cesium atom spins are polarized and measured using an elliptically polarized light simultaneously. The Allen variance of the field measurement is characterized to be from 18pT @ 109μm to 8pT @ 438μm.**

I. INTRODUCTION

Magnetic field is a key information in many research fields, such as oil detection, earthquake prediction and brain magnetic measurement. Magnetoencephalogram is widely used to psychiatric diagnosis, polygraph application, research on language, hearing induction, etc[1]. The millimeter size resolution of magnetic nanoparticle imaging does not meet application requirements[2]. So it is important for medical application to obtain magnetic field with high sensitivity and high spatial resolution. Magnetic field with high spatial resolution and field sensitivity will promote the analysis of pathology.

Different from giant magnetoresistive spin valve arrays[3], 2D field microscopy using atomic vapor spin have much longer relaxation time, thus high magnetic field sensitivity. In addition, as the spatial measurement is limited by optical diffraction and spin diffusion, a spatial resolution of several micron can be realized with high pressure buffer gas.

At present, there are four methods to realize 2D magnetic field measurement using atomic spin, including mobile cell scanning measurement[4], multiple light beam measurement[5], multiple cell measurement and optical magnetic resonance image[6]. We realize multiple light beams using digital micro mirrors, of which the element size is 13.7 μm. To overcome the background image caused by the light non-uniform distribution, we use one laser to realize spin pumping and probing at the same time. Furthermore, the light is filtered spatially and compensated for every element measurement.

II. THEORY

In this article, the cesium is polarized by the circular component of the elliptically light and the optical rotation generated by the interaction between magnetic field and atomic spin is measured by the linear component[7]. The Bloch equation given below can describe the dynamic of atomic spin[8, 9]

$$\frac{d}{dt}P = \gamma P \times B + R_p(v)(s\hat{z} - P) - R_{rel}P + D\nabla^2 P \quad (1)$$

where P is the spin polarization of cesium, B is the external magnetic field, γ is the gyromagnetic ratio of the cesium. R_{rel} is the relaxation rate, D is the diffusion constant. $R_p(v) = \frac{\sigma(v)}{A}\Phi(v)$ is the pumping rate as a function of the frequency of laser, where $\sigma(v)$ is the effective absorption cross section of cesium, $\Phi(v)$ is the photon flux. By solving the steady-state solution to Eq. (1) without considering diffusion for the case when B_x and B_v are equal to zero, we can get polarization in the x direction

$$P_x = \frac{R_p(v)(R_p(v) + R_{rel})}{(R_p(v) + R_{rel})^2 + (q\gamma B'_y)^2}\cos(\omega t)$$

$$- \frac{q\gamma R_p(v)B'_y}{(R_p(v) + R_{rel})^2 + (q\gamma B'_y)^2}\sin(\omega t) \quad (2)$$

where ω is the modulation frequency of the elliptically polarized light. $B'_y = B_y - \omega/\gamma$ is the equivalent magnetic field.

The rotation angle of the linear component of the elliptically polarized light is[10]

$$\theta' = -\frac{1}{2}lr_e cnf_{D1}P_x(v)D(v) \quad (3)$$

which can be detected by a balanced detector, where l is the length of the cell, r_e is the electron radius, c is the speed of light, n is the number of the cesium atom, f_{D1} is the oscillator strength corresponding to D1 line of cesium, $D(v) = \frac{v - v_{D1}}{(v - v_{D1})^2 + (\Delta v_1/2)^2}$, v_{D1} is the resonance frequency of the D1 transition, Δv_1 is the resonance line width of the D1 transition. Using a lock-in amplifier with a reference to $\sin(\omega t)$, the in-phase signal is proportional to B_y. The in-phase signal is given by

$$\theta = -\frac{1}{2}lr_e cnf_{D1}D(v)\frac{q\gamma R_p(v)B_y}{(R_p(v) + R_{rel})^2 + (q\gamma B_y)^2} \quad (4)$$

To generate the optimal rotation angle, the laser frequency is detuned from the resonance frequency. With small detuning, $D(v)$ is too small. With large detuning,

---

*This work was supported in part by the National Natural Science Foundation of China (61973021)

Hai-feng Dong is with the Beihang University, Xueyuan Lu 37#, Beijing, China, 100191; (e-mail: hfdong@buaa.edu.cn)

Chen Liu is with the Beihang University, Xueyuan Lu 37#, Beijing, China, 100191; (e-mail: zy1817227@buaa.edu.cn)

Jun-jun Sang is with the Beihang University, Xueyuan Lu 37#, Beijing, China, 100191; (e-mail: junjuns7@buaa.edu.cn)

978-1-6654-3008-1/21 $31.00 © 2021 IEEE

$R_p(\nu)$ is near to zero which makes the rotation signal very small. So there is an optimal detuning which makes the rotation angle maximum. The slope of the Eq. (4) about B_y as a function of the laser frequency detuning $\Delta\nu$ is given by

$$\text{Slope}(\Delta\nu) = \frac{lnqr_e cf_{D1}\Delta\nu}{4(\frac{\Delta\nu_1^2}{4}+(\Delta\nu)^2)(R_{rel}+\frac{cf_{D1}Pr_e\Delta\nu_1}{4Ah(\nu_{D1}+\Delta\nu)(\frac{\Delta\nu_1^2}{4}+(\Delta\nu)^2)})} \quad (5)$$

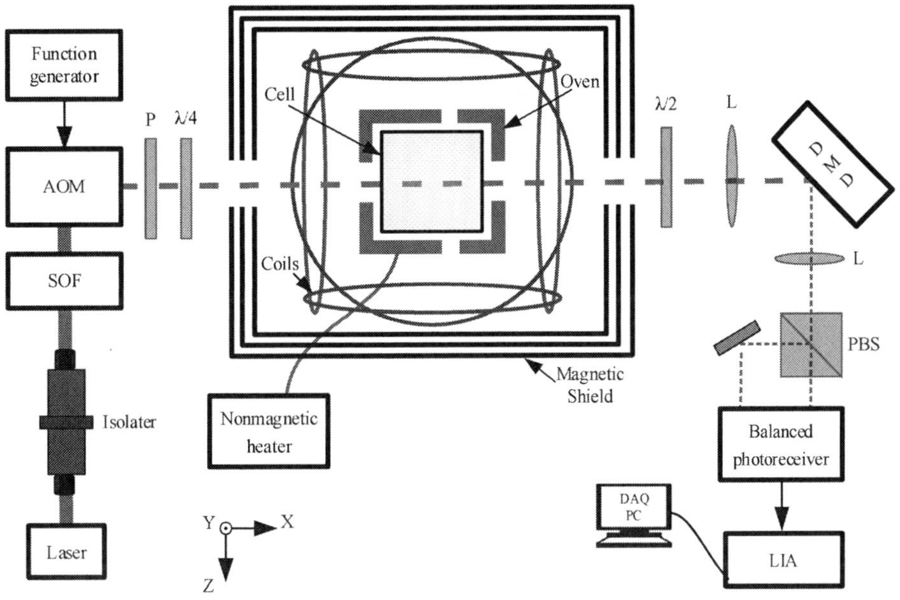

Fig. 1 Schematic of experimental setup (SOF: spatial optic filter; AOM: acoustic optical modulator; P: polarizer; λ/4: quart-wave plate; λ/2: half-wave plate; PBS: polarized beam splitter; LIA: lock-in amplifier)

III. EXPERIMENTAL SETUP AND RESULTS

A schematic of the experimental setup is shown in Fig. 1. A distributed Bragg reflector (DBR) diode laser with 895nm (D1 line of cesium) is used in this experiment. A spatial light filter is placed after opto-isolator. Light passed through an acoustic-optic modulator (AOM) is amplitude-modulated by a 6kHz, 50% duty cycle square signal[11]. Then the light is elliptically polarized by a polarizer oriented at $\theta=0$ and a $\lambda/4$ wave-plate oriented at $\theta=\pi/8$ [12]. The cesium atoms is put in a $2.5\times2.5\times2.5\text{cm}^3$ cubic cell which is placed in the center of three-layer magnetic shield. The cell filled with 600 Torr of ^{4}He buffer gas and 150 Torr of N_2 quenching gas is heated to 70 °C by the oven. Unlike the traditional beam splitting methods, the light is reflected from a Digital Micromirror Device (DMD) with a pixel size of $13.7\times13.7\mu\text{m}^2$ to realize micron beam detection[13]. The polarization rotation signal is detected by a polarized beam splitter (PBS) and balanced detectors.

The cesium is polarized by the circular component of the elliptically polarized light and the optical rotation generated by the atoms is measured by the linear component. We adjust the relative angle of the polarizer and $\lambda/4$ wave-plate close to $\pi/8$ for a maximum output signal[12].

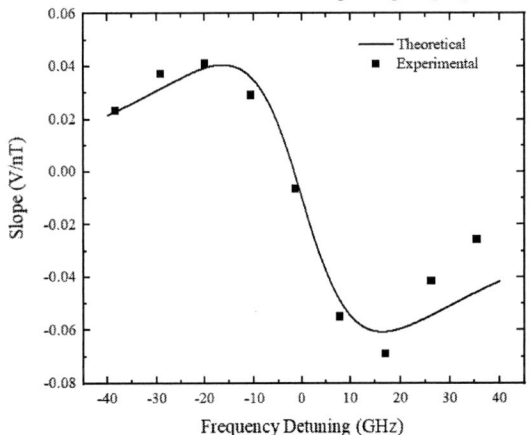

Fig. 2 The relationship between the slop of dispersion and the frequency detuning of laser. (Dots) experimental results. (Dash) predictions for the slope based on Eq. (5).

To get the maximum rotation signal, the laser frequency is also optimized experimentally. By only changing the frequency of the laser, we get the slope as a function of the frequency detuning of the laser. Fig. 2 shows the relationship between the detuning and the slope of the dispersion signal which agree with the theoretical analysis in Eq. (5) the theoretical analysis. The frequency detuning Δv=15GHz gives an optimal optical rotation.

The 2D magnetic field image is achieved by scanning the area using digital mirror device. Fig. 3 shows a field image of an area of about 0.5mm×0.6mm. The distance between adjacent measurement element is 109.6um. The maximum field difference in this area is about 11nT which is caused by the residual field of earth field after shielding and the field non-uniform of the coils.

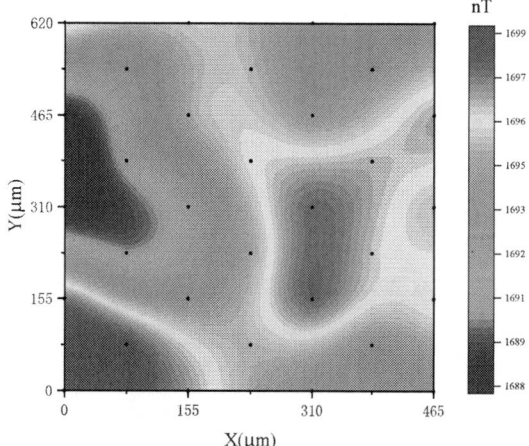

Fig. 3 Magnetic field image in the atomic vapor cell. The dots indicate the measured positions.

Fig. 4 shows continuous measurement of magnetic field at different positions. From Fig. 4, we can see that a minimum local gradient sensitivity of 0.16pT / μm can be measured.

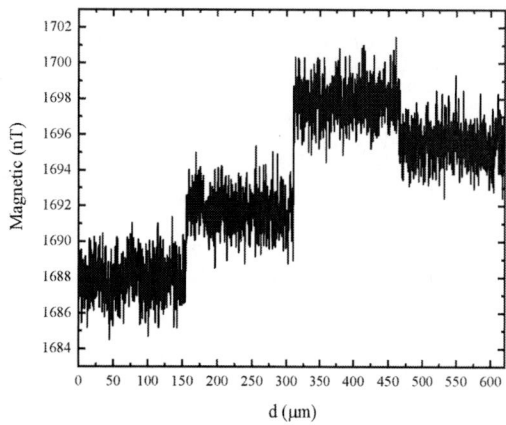

Fig. 4 The continuous measurement of magnetic field at different positions.

To explore the relationship between spatial resolution and magnetic field sensitivity of our method, we measure the Allan variance of the measurement for different element sizes[14]. By changing the size of the flipped mirror, we get the sensitivity of our method. Fig. 5 shows the relationship between measurement element size and the field sensitivity. According to the fitting result, the sensitivity and the distance between adjacent measurement element are

square-inversely proportional. The technical noise is independent to the size of the measurement element. So when we increase the size of the measurement element, the detection signal increase. Thus the equivalent magnetic field noise caused by the technical noise decrease square-inversely to the size.

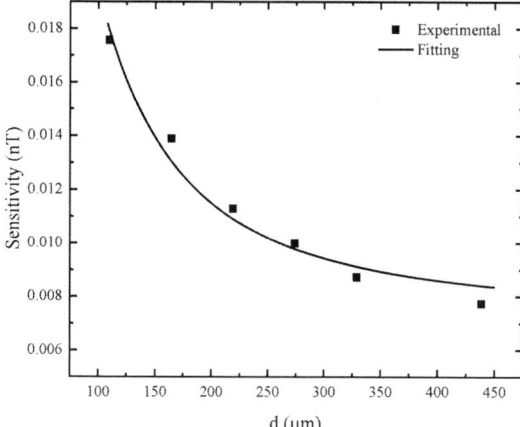

Fig. 5 The relationship between sensitivity of magnetometer and spatial resolution. d is the distance between adjacent measurement element..

IV. CONCLUSION

In conclusion, we have theoretically and experimentally examined the feasibility of long time stability of measurement of the magnetic field distribution. The optimal rotation signal of the magnetometer using elliptically polarized light is achieved by adjusting frequency detuning of the laser. In addition, the relationship between the size of measurement element and the sensitivity is explored which may help to research the diffusion of alkali-mental atoms in the cell. In addition, it may also help biomagnetic field detection to find the balance between sensitivity and the size of measurement element. The disadvantage of 2D field microscopy based on atomic vapor spin is the relative large setup at current stage.

REFERENCES

[1] E. Labyt et al., "Magnetoencephalography with optically pumped 4He magnetometers at ambient temperature," IEEE Transactions on Medical Imaging, vol. PP, no. 1, pp. 90-98, 2019.

[2] Y. W. Jun, J. H. Lee, and J. Cheon, "Chemical design of nanoparticle probes for high-performance magnetic resonance imaging," Cheminform, vol. 47, no. 28, pp. 5122-5135, 2010.

[3] Y. J. Kim, I. Savukov, J. H. Huang, and P. Nath, "Magnetic microscopic imaging with an optically pumped magnetometer and flux guides," Appl. Phys. Lett., vol. 110, no. 4, p. 043702, 2017.

[4] K. Kamada, Y. Ito, and T. Kobayashi, "Human MCG measurements with a high-sensitivity potassium atomic magnetometer," Physiological Measurement, vol. 33, no. 6, pp. 1063-1071, 2012.

[5] Johnson and N. Cort, "Magnetoencephalography with a two-color pump probe atomic magnetometer," in Proposed for Presentation at the IEEE International Frequency Control Symposium Held June, 2010.

[6]   A. Borna *et al.*, "A 20-Channel Magnetoencephalography System Based on Optically Pumped Magnetometers," *Physics in Medicine & Biology*, 2017.

[7]   V. Shah and M. V. Romalis, "Spin-exchange relaxation-free magnetometry using elliptically polarized light," *Phys. Rev. A*, vol. 80, p. 013416 2009.

[8]   S. J. Seltzer, "Developments in Alkali-Metal Atomic Magnetometry," Ph.D Ph.D thesis, Department of Physics, Princeton University, New Jersey, 2008.

[9]   D. Giel, G. Hinz, D. Nettels, and A. Weis, "Diffusion of Cs atoms in Ne buffer gas measured by optical magnetic resonance tomography," *Optics express*, vol. 6, no. 13, pp. 251-6, 2000.

[10] D. Budker and M. Romalis, "Optical magnetometry," *Nature Physics*.

[11] R. S. Grewal, G. Pati, and R. Tripathi, "Light-ellipticity and polarization-angle dependence of magnetic resonances in rubidium vapor using amplitude-modulated light: Theoretical and experimental investigations," 2020.

[12] H. F. Dong *et al.*, "Spin image of an atomic vapor cell with a resolution smaller than the diffusion crosstalk free distance," *Journal of Applied Physics*, vol. 125, no. 24, pp. 243904.1-243904.5, 2019.

[13] S. Taue, Y. Toyota, K. Fujimori, and H. Fukano, "AC magnetic field imaging by using digital micro-mirror device," in *2017 22nd Microoptics Conference (MOC)*, 2017.

[14] D. V. Land, A. P. Levick, and J. W. Hand, "The use of the Allan deviation for the measurement of the noise and drift performance of microwave radiometers," *Measurement Science & Technology*, vol. 18, no. 7, p. 1917, 2007.

Gap in pagination due to unavailable papers.

Pages 1002-1017

Mechanisms of branch tip fusion in meshwork patterns*

Shan Guo, Ming-zhu Sun, and Xin Zhao

Abstract—Meshwork pattern is a significant pattern for the development of biological tissues and organs. The formation of meshwork pattern is facilitated by branch tip fusion. However, the mechanism of branch tip fusion is still unclear. In this paper, based on a reaction-diffusion model, we found that the local high concentration substrate guided the fusion of branch tips and provided the quantitative ranges of environmental conditions for branch tip fusion. Then we explored the Turing wavelengths underlying branch tip fusions. We found that the width of local high concentration substrate region did not influence the Turing wavelength of branch tip fusion and the width less than two Turing wavelengths promoted the fusion of branches. We also found that the increasing Co (production rate of substrate) reduced the Turing wavelength and potential of branch tip fusion. We believe that the molecular mechanism and Turing mechanism of branch tip fusion in this paper will be beneficial to the in-depth understanding of network morphogenesis.

I. INTRODUCTION

Meshwork pattern is a significant pattern for the development of biological tissues and organs, such as alveolar microvascular network [1,2] and retinal vascular network [3,4], which are very important for nutrient transport or gas exchange. Our previous work has performed the formation of meshwork patterns in spherical shell domain and found that branch tip fusion promotes the formation of meshwork patterns [5]. However, it is still not clear to explain how the branch tip fusion occurs. Therefore, this paper aims to explore the mechanisms of branch tip fusion in meshwork patterns.

In this paper, a reaction-diffusion model (Meinhardt model) is used to investigate the mechanisms of branch tip fusion. The model is based on molecular mechanism and describes the interaction between morphogens [6]. It has been well used to explore the pattern formation and many works on the model are consistent with the biological experimental results, such as the pattern formation by vascular mesenchymal cells [7], the branching morphogenesis of pulmonary vascular development [8] and the branching patterns in lung development [9].

In nature, the formation of repetitive structures is often consistent with a reaction-diffusion mechanism, or Turing model, of self-organizing systems [10,11]. For example, by controlling the wavelength of a Turing-type mechanism, the Hox genes regulate the polydactyl in mouse [12]. As the Meinhardt model is a reaction-diffusion model based on the

Turing activator-inhibitor theory [6], in this paper we conduct the Turing instability analysis to explore the Turing mechanism underlying branch tip fusion.

In this paper, based on the Meinhardt model, we explored the molecular mechanism and Turing mechanism of branch tip fusion in meshwork patterns. First, we performed the local meshwork pattern formation including branch tip fusion behavior, and obtained the dynamic processes of branch tip fusion. Second, we verified the promotion effect of local high concentration substrate on branch tip fusion and provided the quantitative range of environmental conditions for branch tip fusion. Then, by calculating the Turing wavelength underlying branch tip fusion, we explored the Turing mechanism of local high concentration substrate affecting branch tip fusion.

II. MATHEMATICAL MODEL

The Meinhardt model is used in this paper. It is an activation-inhibition reaction-diffusion model with consumption of substrate, which are shown as (1)(2)(3)(4) [6]. The model describes the reaction and diffusion between activator A, inhibitor H and substrate S. And the cell differentiation state Y is marker for recording cell differentiation.

$$\frac{\partial A}{\partial t} = \frac{cA^2 S}{H} - \mu A + \rho_A Y + D_A \nabla^2 A \tag{1}$$

$$\frac{\partial H}{\partial t} = cA^2 S - vH + \rho_H Y + D_H \nabla^2 H \tag{2}$$

$$\frac{\partial S}{\partial t} = c_0 - \gamma S - \varepsilon YS + D_S \nabla^2 S \tag{3}$$

$$\frac{\partial Y}{\partial t} = dA - eY + \frac{Y^2}{1 + fY^2} \tag{4}$$

In the model, activator A is produced in autocatalytic effect and inhibited by inhibitor H, this reaction is supported by substrate S $((cA^2 S)/H)$. At the same time, activator A stimulates the production of inhibitor H under the participant of substrate S $(cA^2 S)$. Substrate S is produced at constant rate c_0. The differentiated cells Y secretes activator A at rate ρ_A $(\rho_A Y)$, secretes inhibitor H at rate ρ_H $(\rho_H Y)$, and consumes substrate S at rate ε $(-\varepsilon YS)$. Activator A, inhibitor H and substrate S diffuses with diffusion coefficient D_A $(D_A \nabla^2 A)$, D_H $(D_H \nabla^2 H)$ and D_S $(D_S \nabla^2 S)$ in turn. And they

---

*Resrach supported by the National Key R&D Program of China (grant 2018YFB1304905), the National Natural Science Foundation of China (grants U1813210, U1613220 and 61903201), and the Natural Science Foundation of Tianjin (grants 18JCYBJC19000 and 18JCZDJC39100).

Shan Guo is with the Institute of Robotics and Automatic Information Systems, Nankai University, Tianjin 300350, China (e-mail: guoshan@mail.nankai.edu.cn).

Ming-zhu Sun is with the Institute of Robotics and Automatic Information Systems, Nankai University, Tianjin 300350, China (e-mail: sunmz@nankai.edu.cn).

Xin Zhao is with the Institute of Robotics and Automatic Information Systems, Nankai University, Tianjin 300350, China (corresponding author, phone: 86-22-23505706-803; fax: 86-22-23500172; e-mail: zhaoxin@nankai.edu.cn).

degrades at rate μ ($-\mu A$), v ($-vH$) and γ ($-\gamma S$), respectively. Equation (4) indicates that the marker Y of cell differentiation is created by high concentrations of activator A (dA), degrades at rate e ($-eY$), has a positive feedback effect and saturates itself at high value of Y ($Y^2/(1+fY^2)$).

III. RESULTS

A. Dynamic Process of Branch Tip Fusion

The branch tip fusion behavior is very important for the formation of meshwork patterns. Our previous work have shown that the meshwork pattern is transformed form the tree-like pattern through branch tip fusion [5]. Here in order to explore the causes of branch tip fusion, a local region of branch tip fusion is extracted from a meshwork pattern in a two-dimensional (2D) square domain (shown in Fig. 1(A)) and the dynamic processes of branch tip fusion are shown in Fig. 1(B)(C)(D).

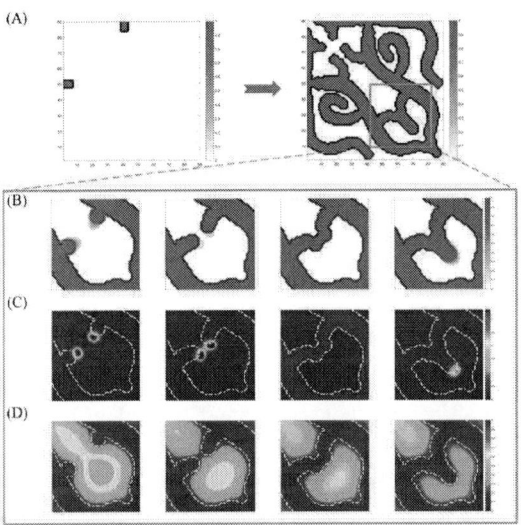

Figure 1. Dynamic processes of branch tip fusion. (A) A meshwork pattern including branch tip fusion formed in the 2D square domain. Grid size is 90 × 90. (B) The dynamic process of cell differentiation marker Y during the formation of branch tip fusion. (C) The dynamic process of activator A during the formation of branch tip fusion. The red indicates high concentration of activator while the blue indicates low concentration. The white dashed line indicates the boundary of branches. (D) The dynamic process of substrate S during the formation of branch tip fusion. The red indicates high concentration of substrate while the blue indicates low concentration. The white dashed line indicates the boundary of branches. (Parameters: $D_A = 0.02$, $D_H = 0.26$, $D_s = 0.06$, c = 0.002, $c_0 = 0.02$, $\mu = 0.16$, v = 0.04, $\gamma = 0.02$, ε = 0.4, d = 0.008, e = 0.1, f = 10, $\rho_A = 0.03$, $\rho_H = 0.00005$.)

Figure 1 shows the dynamic processes of branch tip fusion in a meshwork pattern. The meshwork structure including branch tip fusion is shown in Fig. 1(A), which is generated from two initial growth points located on the adjacent boundaries of the square domain. Figure 1(B) shows the branch behavior during the formation of tip fusion. Two branches approach to each other gradually and then the tips of branches meet and fuse to form one branch. Figure 1(C) shows the behavior of activator in the process of branch fusion. The activator peaks locate at the tips of branches. The migration of activator peaks lead to the

extension of branches. During the process of branch tip fusion, the activator peaks at the tips of two branches move closely to each other and then merge into one peak to continue migration. It suggests that the meeting and fusion of activator peaks lead to the fusion of two branches. Figure 1(D) shows the dynamic changes of substrate during branch fusion. Two branches extend towards the local high-concentration area of substrate along the gradient direction. And the branch tips merge at the area with local high-concentration substrate to form a new branch. Therefore, it can be seen that the high-concentration substrate drives the migration of activator peaks, which determines the extension of branches, and the local high-concentration substrate leads the meeting and fusion of activator peaks, which results in the fusion behavior of branch tips. We conclude that the branch tip fusion is guided by the local high concentration substrate.

B. Verification of Local High Concentration Substrate Guiding Branch Tip Fusion

Figure 1 shows the local high concentration substrate guides the formation of branch tip fusion. It seems like that the concentration and area size of local substrate paly significant roles in the formation of branch tip fusion. In order to verify this hypothesis, two groups of controlled experiments are performed in this section. The simulation environment setting of the simulation experiments is shown as Fig. 2. Two adjacent initial growth points are set on the top of square domain to generate two branches. A strip-shaped region with high concentration substrate is set in the middle of the two initial growth points to attract branch fusion. The substrate concentration in the strip region is set by the substrate generation rate Co and the initial value of substrate concentration $S0$. The size of the strip region is set by the width of the trip.

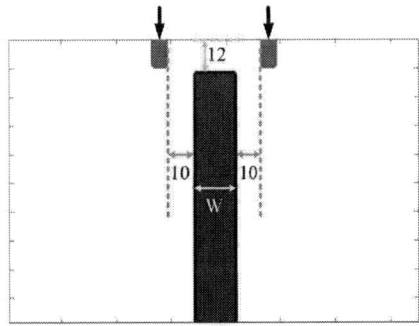

Figure 2. Simulation environment setting for branch tip fusion. The two small blue rectangular areas represent the two initial growth points (maked by black arrows). The red strip represents the region with different concentration of substrate. Grid size of the whole domain is 150×100. The horizontal distance between initial growth point and strip region is 10, and the distance between strip region and upper boundary of domain is 12.

For exploring the effects of substrate concentration on branch tip fusion, we performed the branch growth in the domain with local low concentration substrate, local high concentration substrate and uniformly distributed substrate, respectively. The initial state and simulation result of each simulation are shown in Fig. 3. The simulation with uniformly distributed substrate (Co=0.02, S0=1.0) is the control experiment (Fig. 3(B)). The local low concentration of substrate is set by Co=0.01 and S0=0.5 in the strip region (Fig. 3(A)), and the local high concentration of substrate is set by Co=0.05 and S0=2.5 in the strip region (Fig. 3(C)).

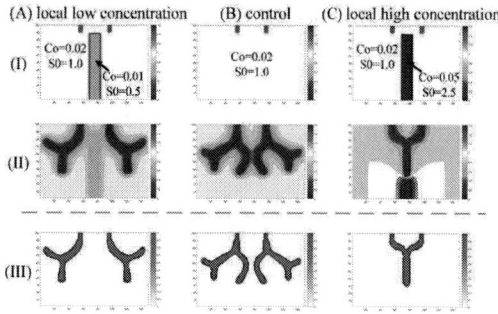

Figure 3. A group of controlled experiments about the effects of substrate concentration on branch tip fusion. (A) Co=0.01 and S0=0.5 in the strip region. (B) Co=0.02 and S0=1.0 in the whole region. (C) Co=0.05 and S0=2.5 in the strip region. (I) Initial state of simulation experiments. (II) Simulation results shown by substrate S. (III) Simulation results shown by cell differentiation marker Y. The width of the strip region is set W=15. In the rest region, Co=0.02 and S0=1.0. Other parameters: $D_A = 0.02$, $D_H = 0.26$, $D_s = 0.06$, c = 0.002, μ = 0.16, v = 0.04, γ = 0.02, ε = 0.7, d = 0.008, e = 0.1, f = 10, $\rho_A = 0.03$, $\rho_H = 0.0001$.

Figure 3 shows the effects of substrate concentration on branch tip fusion. In Fig. 3(A), the strip region with low concentration substrate provides repulse force to make the branches extend to the left and right sides, respectively. In the domain with uniformly distributed substrate (Fig. 3(B)), the branches grow independently, but when the branches are close enough, they will extend away from each other because of the excessive consumption of substrate between them. Figure 3(C) shows that the local high concentration substrate attracts the branches to approach and merge into a new branch. This group of controlled experiments in Fig. 3 confirm that the high concentration of local substrate facilitates the formation of branch tip fusion.

For exploring the effects of the region size of local high-concentration substrate on branch tip fusion, we performed the branch growth in the domains including local high-concentration substrate regions with different widths. The initial state and simulation result of each simulation are shown in Fig. 4. The widths of local high-concentration substrate regions are set to W=13 (Fig. 4(A)), W=19 (Fig. 4(B)) and W=25 (Fig. 4(C)) respectively for different branch behaviors.

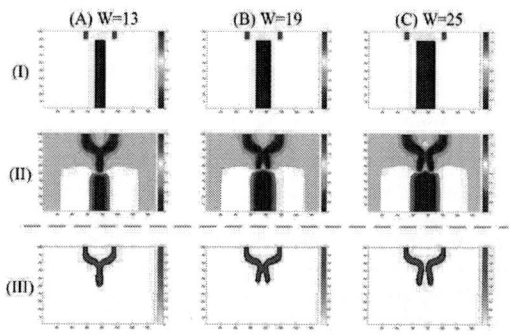

Figure 4. A group of controlled experiments about the effects of the region size of local high-concentration substrate on branch tip fusion. (A) W=13. (B) W=19. (C) W=25. (I) Initial state of simulation experiments. (II) Simulation results shown by substrate S. (III) Simulation results shown by cell differentiation marker Y. In the strip regions, Co=0.05 and S0=2.5; in the rest regions, Co=0.02 and S0=1.0. Other parameters: $D_A = 0.02$, $D_H = 0.26$, $D_s = 0.06$, c = 0.002, μ = 0.16, v = 0.04, γ = 0.02, ε = 0.7, d = 0.008, e = 0.1, f = 10, $\rho_A = 0.03$, $\rho_H = 0.0001$.

Figure 4 shows the effects of the region size of local high-concentration substrate on branch tip fusion. In Fig. 3(A), the two branches meet and merge into one branch in the local high-concentration substrate region. Figure 3(B) shows that the branches approach and adhere to each other but finally still extend independently in the larger region of local high-concentration substrate. In the much larger region of local high-concentration substrate (Fig. 4(C)), the adjacent branches repel each other and extend independently. This group of controlled experiments in Fig. 4 confirm that the small region of local high concentration substrate facilitates the formation of branch tip fusion.

C. Quantitative Range of Environmental Conditions for Branch Tip Fusion

Figure 3 and 4 have confirmed that the small region and high concentration of substrate are essential for branch tip fusion formation. In this section we aim to obtain the quantitative ranges of substrate concentration and region size, which are suitable for branch tip fusion. We performed the simulations of branch growth in the domain as Fig. 2, which included the local high concentration substrate with variable concentration and region size. The concentration of local high concentration substrate was varied by substrate generation rate Co. And the region size of local high concentration substrate was varied by the width of strip region. The simulation results are shown in Fig. 5.

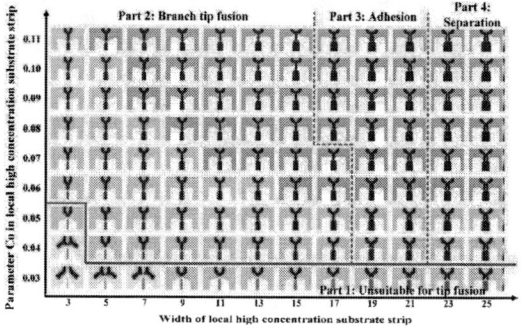

Figure 5. Simulation results of branch growth in the domain with local high concentration substrate strip under variable values of Co and strip width. Co ranges from 0.03 to 0.11 at interval 0.1 in the strip region of high concentration substrate. S0=Co/γ. In the rest region, Co=0.02 and S0=1.0. The width of the strip ranges from 3 to 25 at interval 2. The simulation results are divided into four parts based on the branch behaviors. Other parameters: $D_A = 0.02$, $D_H = 0.26$, $D_s = 0.06$, c = 0.002, μ = 0.16, v = 0.04, γ = 0.02, ε = 0.7, d = 0.008, e = 0.1, f = 10, $\rho_A = 0.03$, $\rho_H = 0.0001$.

Figure 5 shows the simulation results of branch growth influenced by Co and region width of local high concentration substrate strip. When Co is small (Co=0.03) or Co and strip width both are small (Co=0.04, 0.05 and W=3), the domains are not suitable for branch tip fusion. The fusion of branches are often failed in these domains. With larger Co ($Co\geq0.04$) and in the small region ($3\leq W\leq15$ or 17) of local high concentration substrate, the domains are suitable for branch tip fusion formation. As the region of local high concentration substrate increases further (17 or $19\leq W\leq21$), the adhesion behavior occurs between branches. When the region of local high concentration substrate is large enough ($W\geq23$), the growth of branches is separated in the region. Therefore, figure 5 suggests that the branch tip fusion is facilitated by the substrate with high concentration ($Co\geq0.04$) and small region ($3\leq W\leq15$ or 17).

D. Turing Mechanism underlying Branch Tip Fusion

In order to the further research of branch tip fusion, we explored the Turing mechanism underlying branch tip fusion in this section. First, we calculated the Turing wavelengths of branch tip fusions, which are formed in the local high concentration substrate strips with different widths, and plotted the curve of wave number and strip width (shown in Fig. 6). Then, we calculated the Turing wavelengths of branch tip fusions formed in local high concentration substrate strips with different concentrations and plotted the curve of wave number and parameter Co (shown in Fig. 7).

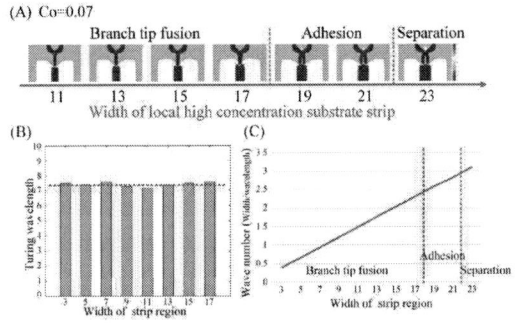

Figure 6. Turing mechanism of the width of local high-concentration substrate strip affecting branch tip fusion. (A) Part of simulation results of branch growth influenced by the width of substrate strip. In the strip region, Co=0.07 and S0=3.5; in the rest region, Co=0.02 and S0=1.0. Other parameters: $D_A = 0.02$, $D_H = 0.26$, $D_s = 0.06$, $c = 0.002$, $\mu = 0.16$, $v = 0.04$, $\gamma = 0.02$, $\varepsilon = 0.7$, $d = 0.008$, $e = 0.1$, $f = 10$, $\rho_A = 0.03$, $\rho_H = 0.0001$. (B) The Turing wavelengths underlying the branch tip fusions. (C) The wave number contained in the width of local high-concentration substrate strip.

Figure 6 shows the Turing mechanism of the width of local high-concentration substrate strip affecting branch tip fusion. Figure 6(A) shows that with the increase of strip width, the branch tip fusion behavior gradually changes into branch adhesion behavior, and finally becomes independently growing branches when the strip width is large enough. Figure 6(B) shows that the Turing wavelengths of branch tip fusions are consistent although the widths of high concentration substrate strips are different. In other words, it means that the width of high concentration substrate region does not affect Turing wavelength of branch tip fusion. Figure 6(C) shows that the wave number increases with the increase of strip width.

When the strip width is greater than two wavelengths, the branch tip fusion behavior changes into adhesion behavior or separated branches. Therefore, we conclude that the reason why small high concentration substrate region promotes branch fusion is that the width of small region is smaller than two Turing wavelengths so that only one branch is allowed to grow in the region.

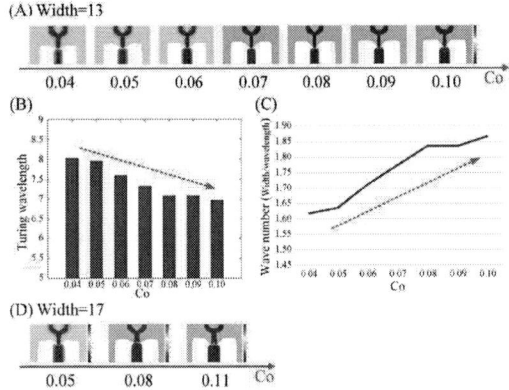

Figure 7. Turing mechanism of the concentration of local high-concentration substrate strip affecting branch tip fusion. (A) Part of simulation results of branch growth influenced by the concentration of substrate strip. The width of strip region is W=13. In the strip region, parameter Co ranges from 0.04 to 0.10 at interval 0.01, S0=Co/γ; in the rest region, Co=0.02 and S0=1.0. (B) The Turing wavelengths underlying the branch tip fusions with different Co. (C) The wave number contained in the width of local high-concentration substrate strip. (D) Part of simulation results of branch growth influenced by the concentration of substrate strip when W=17. In the strip region, Co ranges from 0.05 to 0.11 at interval 0.03, S0=Co/γ; in the rest region, Co=0.02 and S0=1.0. Other parameters: $D_A = 0.02$, $D_H = 0.26$, $D_s = 0.06$, $c = 0.002$, $\mu = 0.16$, $v = 0.04$, $\gamma = 0.02$, $\varepsilon = 0.7$, $d = 0.008$, $e = 0.1$, $f = 10$, $\rho_A = 0.03$, $\rho_H = 0.0001$.

Figure 7 shows the Turing mechanism of the concentration of local high-concentration substrate strip affecting branch tip fusion. Figure 7(A) shows some simulation results of branch tip fusion with different values of Co in local high concentration substrate when the width of strip region is W=13. Figure 7(B) shows that the Turing wavelength of branch tip fusion decrease with the increase of Co. So the wave number increases with the increase of Co (Fig. 7(C)). It is inferred that when Co is large enough in the strip region, the width of strip region will be larger than two Turing wavelengths, which will lead to the changes of branch tip fusion behavior. This hypothesis is verified in Fig. 7(D), which shows that the branch tip fusion behavior changes into branch adhesion behavior with the increase of Co ($Co=0.05\rightarrow0.08$) when the width of strip is W=17. Therefore, we conclude that with the increase of Co in local high-concentration substrate region, the Turing wavelength underlying branch tip fusion decreases and the potential of branch tip fusion decreases.

IV. CONCLUSION

In this paper, we explored the molecular mechanism and

Turing mechanism of branch tip fusion in meshwork patterns. We found that the formation of branch tip fusion is facilitated by the local high concentration substrate and provided the quantitative ranges of concentration (Co≥0.04) and region size (3≤W≤15 or 17) of local high concentration substrate for branch tip fusion. In addition, the Turing mechanism of branch tip fusion is that the width of small high concentration substrate region less than two Turing wavelengths promotes the fusion of branches. We believe that the molecular mechanism and Turing mechanism of branch tip fusion will be beneficial to the in-depth understanding of network morphogenesis.

APPENDIX

The model equations are dimensionless. The space and time units (L and T) of the model are L = 45 μm and T = 750 s, respectively. The parameter values in the paper are set to the same order of magnitudes in the previous work about Meinhardt model [8,9,13]. Our estimates agree with the experimentally determined values, for example, the value of the diffusion coefficient of activator A is $0.02\ L^2/T$, corresponding to $0.054 \times 10^{-8}\ cm^2 s^{-1}$, which is within the value of the effective diffusion coefficient of Dpp (a BMP homologue) in tissue to be $(0.1 \pm 0.05) \times 10^{-8}\ cm^2 s^{-1}$ [14].

REFERENCES

[1] J. H. Caduff, L. C. Fischer, and P. H. Burri, "Scanning electron microscope study of the developing microvasculature in the postnatal rat lung," *Anatomical Record*, vol. 216(2), pp. 154-164, 2010.

[2] M. Roth-Kleiner, T. M. Berger, M. R. Tarek, P. H. Burri, and J. C. Schittny, "Neonatal dexamethasone induces premature microvascular maturation of the alveolar capillary network," *Developmental Dynamics An Official Publication of the American Association of Anatomists*, vol. 233(4), pp. 1261-1271, 2010.

[3] R. F. Gariano, and T. W. Gardner, "Retinal angiogenesis in development and disease," *Nature*, vol. 438(7070), p. 960, 2005.

[4] F. Milde, S. Lauw, P. Koumoutsakos, and M. L. Iruela-Arispe, "The mouse retina in 3D: quantification of vascular growth and remodeling," *Integrative Biology*, vol. 5(12), pp. 1426-1438, 2013.

[5] S. Guo, C. Hong, M. Sun, and X. Zhao, "Meshwork pattern transformed from branching pattern in spherical shell domain," *Journal of Theoretical Biology*, vol. 455, pp. 293-302, 2018.

[6] H. Meinhardt, "Morphogenesis of lines and nets," *Differentiation*, vol. 6(2), pp. 117-123, 1976.

[7] A. Garfinkel, Y. Tintut, D. Petrasek, K. Bostrom, and L. L. Demer, "Pattern formation by vascular mesenchymal cells," *Proceedings of the National Academy of Sciences of the United States of America*, vol. 101(25), pp. 9247-9250, 2004.

[8] Y. Yao, S. Nowak, A. Yochelis, A. Garfinkel, and K. I. Boström, "Matrix GLA protein, an inhibitory morphogen in pulmonary vascular development," *Journal of Biological Chemistry*, vol. 282(41), pp. 30131-30142, 2007.

[9] Y. Guo, T. H. Chen, X. Zeng, D. Warburton, K. I. Boström, C. M. Ho, X. Zhao, and A. Garfinkel, "Branching patterns emerge in a mathematical model of the dynamics of lung development," *Journal of Physiology*, vol. 592(2), pp. 313-324, 2014.

[10] M. F. Bastida, and M. A. Ros, "How do we get a perfect complement of digits?" *Current Opinion in Genetics & Development*, vol. 18(4), pp. 374-380, 2008.

[11] R. Zeller, J. López-Ríos, and A. Zuniga, "Vertebrate limb bud development: moving towards integrative analysis of organogenesis," *Nature Reviews Genetics*, vol. 10(12), pp. 845-858, 2009.

[12] R. Sheth, L. Marcon, M. F. Bastida, M. Junco, L. Quintana, R. Dahn, M. Kmita, J. Sharpe, and M. A. Ros, "Hox genes regulate digit patterning by controlling the wavelength of a turing-type mechanism," *Science*, vol. 338(6113), pp. 1476-1480, 2012.

[13] M. Hagiwara, F. Peng, and C. M. Ho, "In vitro reconstruction of branched tubular structures from lung epithelial cells in high cell concentration gradient environment," *Scientific reports*, vol. 5, p. 8054, 2015.

[14] A. Kicheva, P. Pantazis, T. Bollenbach, Y. Kalaidzidis, T. Bittig, F. Jülicher, and M. González-Gaitán, "Kinetics of morphogen gradient formation," *Science*, vol. 315(5811), pp. 521-525, 2007.

Proceedings of the 16th Annual IEEE International
Conference on Nano/Micro Engineered and Molecular Systems
April 25-29, 2021

Leaf-like Self-assembled MXene/ZnOEP Hybrid Network for Highly-Sensitive Temperature Sensing in Electronic Skin

Shijie Wan[1], Weiguan Zhang[2], Chenyang Xing[1], and Zhengchun Peng[1]*

Abstract—Electronic skin (e-skin) has developed rapidly with promising applications in smart robots. Flexible temperature sensor plays an important role in robotic tactile sensing. In this work, we propose a hybrid temperature-sensitive material composed of $Ti_3C_2T_x$ (MXene) and zinc-octaethylphorphyrin (ZnOEP). Self-assembled ZnOEP fiber network can twist MXene sheets in a scaffold structure and form a leaf-like morphology on the substrate. Benefitting from the high temperature coefficient of MXene and low modulus of the ZnOEP network, the fabricated flexible temperature sensor achieves a high sensitivity of $2.34\%°C^{-1}$ and low hysteresis of 3.58%. These results suggest that the proposed temperature sensor can be used in e-skin to help robots accurately measure the temperature of their environment or the object they interact with

I. INTRODUCTION

E-skin which is proposed for mimicking the real functions of human skin has rising great research interest for application in wearable healthcare devices, smart robotics, prosthesis, and human-machine interactions. Among different tactile perceptions of e-skin, temperature is an important parameter that reflects human physiological status and the thermal conditions of touching objects and surrounding environments. Various flexible temperature sensors have been developed based on different sensing mechanisms, including thermosensitive resistors, thermocouples, infrared sensors, and optical sensors [1-5]. Owing to the simple fabrication and easy integration with other flexible electronics, thermo-resistive type of temperature sensor whose working principle mainly relies on the change of resistance with temperature is widely studied and adopted in e-skin applications. Sensitivity of the thermo-resistive temperature sensor is mainly determined by the thermal property of the thermosensitive material. Traditionally, bulky and rigid metals are mostly used for making temperature sensors, however, poor mechanical compliance and low temperature coefficient of resistance (TCR) limited their application flexible electronics. Thus, nanomaterials like silver nanowires or carbon nanotubes are then employed to incorporate with elastomer substrate for sensor fabrication [6,7]. However, due to the large modulus mismatch between organic and inorganic materials, cracking or delamination of these nanomaterials from flexible substrate still exist, and causing sensors'

stability issues for long-term usage. Apart from mechanical compliances, thermosensitive materials with larger TCR is demanded for higher sensitivity of the sensor. Recently, $Ti_3C_2T_x$ (MXene) as a novel two-dimensional (2D) material with superior electrical and thermal properties has attracted great attention in the field of flexible electronics [8-10]. However, its bonding with polydimethylsiloxan (PDMS) substrate is unstable due to over 10^4 times difference on Young's modulus [11]. On the other hand, Zinc-octaethylphorphyrin (ZnOEP) is a type of small organic monocular that can self-assemble into microscale fiber network with lower Young's modulus and non-polar bonding with PDMS [12]. A material that can merge the advantages of both MXene and ZnOEP is highly desired.

In this work, a flexible thermo-resistive temperature sensor is proposed by using MXene/ZnOEP hybridized inorganic/organic material. The hybrid material enables stable adhesion with the PDMS substrate that the relative resistance variation lies within 3% over 50 cycles of bending and recovering test. High sensitivity of 2.34% from $30 - 55$ °C and low hysteresis of 3.58% from $30 - 90$ °C are achieved benefitting from the combined properties of large TCR and low modulus of the hybrid material. With these advancements, the proposed flexible temperature sensor can be applied for temperature monitoring and detection in wearable e-skin of smart robotics.

II. MATERIALS AND METHODS

A. Preparation of MXene and ZnOEP solutions

The $Ti_3C_2T_x$ MXene was fabricated by etching Al layers of Ti_3AlC_2. First, hydrofluoric acid was prepared by adding 2 g LiF powder into 30 ml hydrochloric acid (9 mol/ml) and then magnetic stirring for 30 min. Then, 1g Ti_3AlC_2 was slowly added in the as-prepared HF solution. Continuous stirring for 24 hours with a speed of 600 rpm under room temperature is performed so that all the Al layers were etched, then, the bulk $Ti_3C_2T_x$ solution was successfully prepared. Deacidification is required by repeating the process of centrifuge until the solution is neutralized by adding DI water. Finally, thin MXene sheets in the mixed solution can be pipette out from the upper layer of the solution after centrifuging for 60 min with 3500 rpm. The detailed preparation process is shown in Figure 1(a).

ZnOEP solution was prepared by adding 2 mg ZnOEP powder into 4 ml mixed solution of dichloromethane, octane, and PGMEA with the volume mix ratio of 2:1:1, as depicted in Figure 1(b). To fully dissolve the ZnOEP powder, 30 min ultrasonic was performed.

This work was supported by the Shenzhen Science and Technology Program (JCYJ20190808142609414, KQTD20170810105439418), and the National Natural Science Foundation of China (61904112).
[1] Key Laboratory of Optoelectronic Devices and Systems of Ministry of Education and Guangdong Province, College of Physics and Optoelectronic Engineering, Shenzhen University, Shenzhen, China.
[2] Guangdong Provincial Key Laboratory of Micro/Nano Optomechatronic Engineering, College of Mechatronics and Control Engineering, Shenzhen University, Shenzhen, China.
Corresponding Author: Zhengchun Peng. (email: zcpeng@szu.edu.cn)

978-1-6654-3008-1/21 $31.00 © 2021 IEEE

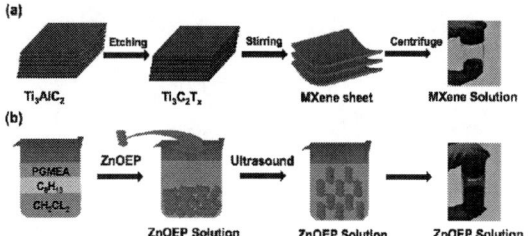

Figure 1. Preparation process of (a) MXene and (b) ZnOEP.

B. Fabrication of the flexible temperature sensor

The flexible PDMS substrate was first prepared by using 10:1 weight mix ratio of the base and curing agent. After heating at 80°C for 4 hours, the cured PDMS substrate was treated by O_2 plasma for a hydrophilic surface. Then, 300 μL ZnOEP and 120 μL MXene were dip-coated on the PDMS correspondingly with a defined dimension of 1 cm × 1 cm. Owing to π-π conjugation, ZnOEP self-assembled automatically during the solution evaporation process, and meanwhile, the assembled fiber networks will twist and combine the MXene sheets in-between the scaffold structures. For safety concern, the hybridization process was accomplished in a fume hood for over 2 hours. The resistance signal reading circuit was fabricated by flexible printed circuit board in an interdigital design so that more hybrid material can contact the electrodes to avoid noise. Then, the as-prepared sensor was mounted on the interdigital electrodes using PI tape. Fabrication process of the flexible temperature sensor is shown in Figure 2.

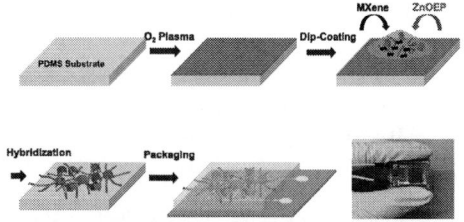

Figure 2. Fabrication process of the flexile temperature sensor.

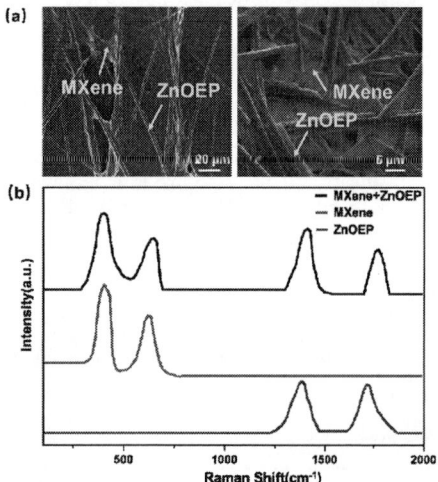

Figure 3. (a) SEM images of the leaf-like ZnOEP/MXene hybrid material. (b) Roman spectrum of the pure ZnOEP, MXene and ZnOEP/MXene hybrid.

Scanning electron microscope (SEM) images of the hybrid material are shown in Figure 3 (a). The MXene sheets were connected continuously with the support of ZnOEP network, and the hybrid material exhibits a leaf-like morphology. Enlarged view of the hybrid material shown in Figure 3 (b) shows the dense network of ZnOEP that bonds together the MXene. Roman spectrum of the pure MXene, ZnOEP and the ZnOEP/MXene hybrid material were characterized by using 532 nm laser, and the results are shown in Figure 3(b). Peaks of the hybrid material merged well with both pure MXene and ZnOEP, proving the success hybridization of MXene and ZnOEP.

C. Reliable test of the hybrid material on PDMS

Owing to the low young's modulus of ZnOEP and its non-polar bonding with PDMS, the hybrid material can adhere stably on the substrate. Also, the fiber network of ZnOEP hinges the MXene sheets together that under certain stretch or bending conditions of the sensor during working, no cracks or delamination of the hybrid material will happen, that guarantees the reliable performance of the thermoresistive temperature sensor. Figure 4 shows the resistance variation of the sensor over 50 cycles of bending-recovery test. After recovering from ~20° bending angle, resistance of the sensor was almost unchanged, and the maximum variation was within 3% for over 50 cycles. It proves the sensor composed by the hybrid material can work reliably.

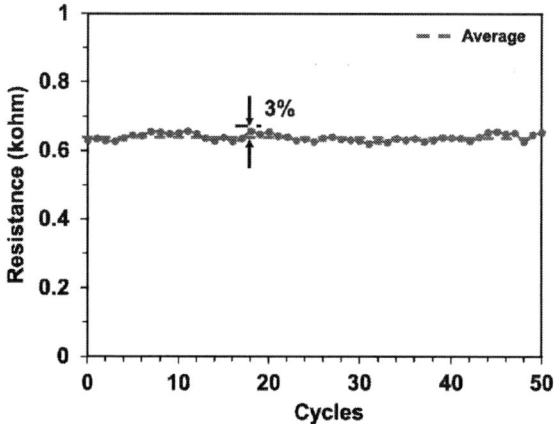

Figure 4. Resistance variation of the flexible temperature sensor with over 50 cycles of bending-recovery test. Bending angle for this test is 20°.

III. RESULTS AND DISCUSSION

Performance of the flexible temperature sensor was characterized in a calorstat to ensure temperature condition around the sensor is stable and constant. Resistance of the sensor was recorded by a digital multimeter. Sensitivity of the sensor which is defined by the relative resistance variation with temperature can be calculated by equation (1).

$$Sensitivity = \frac{(\Delta R / R) \times 100\%}{\Delta T} \qquad (1)$$

where $\Delta R/R$ indicates the relative resistance variation, and ΔT is the temperature change.

Figure 5 shows the measured response curve of the sensor with temperature change in every 5 °C. From 30 − 55 °C, high sensitivity of 2.34% °C^{-1} was achieved. After 55°C, sensitivity decreased to 0.066%°C^{-1} until temperature rose to 90 °C. High sensitivity is mainly attributed to the excellent thermal property of MXene, and such high TCR may only maintain in a limited temperature range.

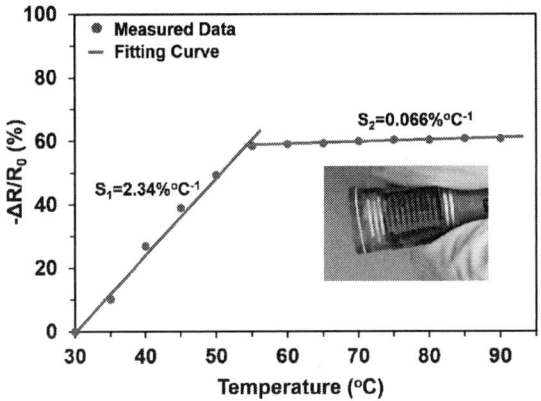

Figure 5. Measured resistance variation of the flexible temperature sensor as a function of temperature.

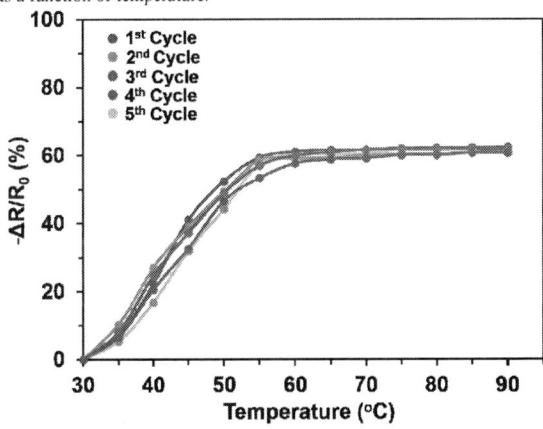

Figure 6. Repeating test of the flexible temperature sensor with 5 heating up cycles.

Repeatable response of the sensor is also an important feature, especially in real application. Within the full sensing range from 30 − 90 °C, multiple times measurement on the sensor response were performed to verify the repeatability of the sensor. Due to long waiting time for the sensor to cool down from 90°C to room temperature, we only repeated for 5 times as shown in Figure 6. It is noted from this figure that the five curves almost overlapped with each other, and no significant derivation was found between the 1st cycle and the 5th cycle.

Generally, temperature difference will introduce thermal stress to the interface between the hybrid material and PDMS substrate as they have different coefficient of thermal expansion, and cause resistance shift during the temperature change. However, benefitting to the soft ZnOEP fiber network that worked as a buffer layer, most of the stress will be released during the heating or cooling

process, that leads to a stable performance of the temperature sensor.

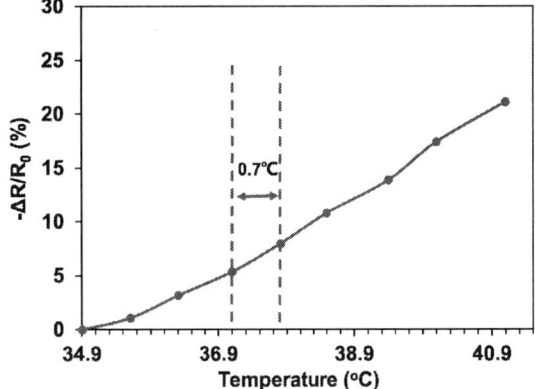

Figure 7. Resolution of the flexible temperature sensor.

Figure 7 shows the measured resolution of the flexible temperature sensor. 0.7 °C temperature resolution was obtained from 34 − 40 °C owning to high sensitivity of the sensor. At the current stage, the resolution may cannot be adaptable for some scenarios that requires high precise temperature control and detection (<0.5°C), however, better resolution can be achieved by further improving the sensitivity, like increasing TCR of the material through special treatment.

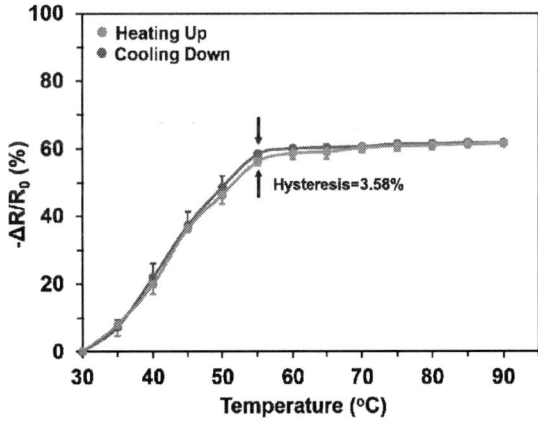

Figure 8. Heating and cooling response curves of the flexible temperature sensor.

Hysteresis which reflects the reluctance of the hybrid material to its initial condition after heating and cooling process determines the accuracy of the temperature sensor. Figure 8 shows the measured response curving under heating and cooling process from 30 − 90 °C. To enhance the reliability of the resistance under each temperature points, we carried out 5 times heating and cooling cycles, and took the average value of the recorded resistances. Standard error deviation was calculated and labelled on the data point in Figure 8. It shows the resistance at each temperature point for both heating and cooling process is reliable. After that, hysteresis of the flexible temperature sensor was studied. 3.58% hysteresis around 55°C is obtained, proving the hybridized material is stable to guarantee a high accurate temperature detection of the sensor.

Combining those above-mentioned advantages, the flexible temperature sensor fabricated by using the ZnOEP/MXene hybrid material exhibits great potential of integration with other tactile sensors on e-skin.

IV. CONCLUSION

In this work, a highly sensitive and low hysteresis flexible temperature sensor is proposed based on the ZnOEP/MXene hybrid material. High sensitivity of 2.34%°C^{-1} is achieved owing to excellent thermal property of MXene. Self-assembly of ZnOEP allows the continuous connection of MXene sheets and stable adhesion with flexible substrate with no cracking or delamination under certain bending or stretching. Resistance of the hybrid material almost maintained the same after 50 cycles of bending-recovering test. In addition, low hysteresis of 3.58% is achieved that ensuring the accurate temperature detection of the sensor. Through simple and fast fabrication technique of solution hybridization method, the high-performance flexible temperature sensor can be applied for smart robotics and human-machine interactions.

REFERENCES

[1] Q. Li, L.-N. Zhang, X.-M. Tao, and X. Ding, "Review of flexible temperature sensing network for wearable physiological monitoring", *Advanced Healthcare Materials*, vol. 6, pp. 1601371, May, 2017.

[2] F. Morais, C.-Dias. Pedro, Z. Yu, *et.al.*, "Low-Cost Control and Measurement Circuit for the Implementation of Single Element Heat Dissipation Soil Water Matric Potential Sensor Based on a SnSe2 Thermosensitive Resistor", *Sensors*, vol. 21, pp. 1490, May, 2020.

[3] Z. Zhang, B. Tian, Y. Liu, Z. Du, Q. Lin, and Z. Jiang, "Thermoelectric Characteristics of Silicon Carbide and Tungsten-Rhenium-Based Thin-Film Thermocouples Sensor with Protective Coating Layer by RF Magnetron Sputtering", *Materials*, vol. 12, pp. 1981, Jun, 2020.

[4] Q. Cai, Y. Ge, C. Sun, C. Chen, and H. Buni, "Immersive Interactive Virtual Fish Swarm Simulation Based on Infrared Sensors", *International Journal of Pattern Recognition and Artificial Intelligence*, vol. 35, pp. 2054027, Oct, 2020.

[5] B. Zhang, and M. Kahrizi, "High-temperature resistgance fiber bragg grating temperature sensor fabrication", *IEEE Sensors Journal*, vol. 7, pp. 586-591, April, 2007.

[6] L. Wu, *et.al.*, "Screen-printed flexible temperature sensor based on FG/CNT/PDMS composite with constant TCR", *Journal of Materials Science-Materials in Electronics*, vol. 30, pp. 9593-9601, May, 2019.

[7] M. Amjadi, A. Pichitpajongkit, S. Lee, S. Ryu, and I. Park, "Highly Stretchable and Sensitive Strain Sensor Based on Silver Nanowire-Elastomer Nanocomposite", *ACS Nano*, vol. 8, pp. 5154-5163, May, 2014.

[8] M. Naguib, V. N. Mochalin, M. W. Barsoum, Y. Gogotsi, "25th anniversary article: MXens: a new family of two-dimensional materials", *Advanced Materials*, vol 26, pp. 992-1005, Dec. 2013.

[9] Y. Ma, et. al., "A highly flexible and sensitive piezoresistive sensor based on MXene with greatly changed interlayer distances", *Nature Communications*, vol. 8, pp. 1207, Oct. 2017.

[10] Z. Cao, Y. Y. Yang, Y. Zheng, W. Wu, F. Xu, R. Wang, and J. Sun, "Highly flexible and sensitive temperature sensors based on $Ti_3C_2T_x$ (MXene) for electronic skin", *Journal of Materials Chemistry A*, vol. 7, pp. 25314-25323, Nov, 2019.

[11] K. Firestein, *et.al.*, "Review Young's Modulus and Tensile Strength of Ti_3C_2 MXene Nanosheets As Revealed by In Situ TEM Probing, AFM Nanomechanical Mapping, and Theoretical Calculations", *Advanced Nano Letters*, vol. 20, pp. 5900-5908, Aug, 2020.

[12] W. Zhang, *et.al.*, "A high-performance flexible pressure sensor realized by overhanging cobweb-like structure on a micropost array", *ACS Applied Materials & Interfaces*, vol. 12, pp. 48938-48947, Sep. 2020.

Gap in pagination due to unavailable papers.

Pages 1027-1034

Proceedings of the 16th Annual IEEE International
Conference on Nano/Micro Engineered and Molecular Systems
April 25-29, 2021

Preparation of Flame-retardant Lithium-ion Battery Separator by Coaxial Electrospinning

Fangqin Shao, Guoyi Kang, Huatan Chen, Xiang Wang, Zungui Shao, Wenwang Li, Gaofeng Zheng*

Abstract— Safety is an important part in the use of lithium-ion battery. This paper proposes a coaxial electrospinning method for preparing a flame-retardant lithium-ion battery of triphenyl phosphate (TPP) and polyvinylidene fluoride (PVDF) to prevent the battery from burning. For performance comparison, PVDF and TPP separators with different mass ratios were prepared by coaxial and uniaxial electrospinning. The separator prepared by coaxial electrospinning can wrap the flame retardant much better. The coaxial electrospinning was studied for the preparation of flame-retardant separators. The performances of TPP@PVDF fibrous separators were measured to optimize process parameter. The separator can be used in the best performances when the mass ratio of PVDF to TPP is 2:1. The TPP@PVDF fibrous separator has displayed a great application potential.

I. INTRODUCTION

As the lithium-ion batteries are prone to thermal runaway, many researchers use flame retardants to prevent thermal runaway of lithium-ion batteries. Flame retardants were originally used as electrolytes to prepare non-flammable electrolytes. Lithium-ion electrolyte is generally composed of lithium salt and organic solvents (ethylene carbonate: $C_3H_4O_3$; propylene carbonate: $C_4H_6O_3$; diethyl carbonate: $C_5H_{10}O_3$, etc.). These organic substances are highly flammable [1]. Jiang [2] et al added non-flammable solvents to the electrolyte, which reduced the flammability of the electrolyte and improved the safety of the lithium-ion battery. The additives are the mainly flame retardants, including halogen flame retardants, phosphorus flame retardants, nitrogen flame retardants, inorganic flame retardants. Among them, phosphorus flame retardants have been studied more frequently for their good flame retardancy and environmental friendliness, such as pentafluorocyclotriphosphazene (PFPN) [3], tris (2,2,2-trifluoroethyl) phosphorous (TTFP) [4], ethylene ethyl phosphate (EEP) [5], triethyl phosphate (TEP) [6]. Adding flame retardant additives to the electrolyte will increase the viscosity of the electrolyte and reduce not only the electrochemical performance but battery life of the lithium-ion battery.

In order to make the flame retardant achieve flame retardant without affecting the electrochemical performance of lithium ion, the method of coating the flame retardant in the separator is adopted [7]. However, excessive coating will affect the high energy density of lithium-ion batteries. The flame-retardant functional groups are also introduced to improve battery safety [8]. For example, Lou [9] et al added trifluoromethyl (-CF_3) groups to polyimide (PI) nanofibers in order to increase the flame retardancy of the separator, and at the same time increased the porosity, liquid absorption rate and lithium-ion migration speed of the separator. But the addition of functional groups will affect the thermal shutdown function of the separator destructively. Many people used hybrid spinning to prepare flame-retardant separators. Deng [10] et al mixed poly(m-phenylene isophthalamide) (PMIA) with flame retardants (with three elements of phosphorus, nitrogen, and sulfur) for electrospinning. Because it is difficult to completely encapsulate the flame retardant in mixed electrospinning, the flame retardant can float on the fiber surface easily.

In this paper, the coaxial electrospinning was used to prepare separators with a core-shell structure. A polymer material was used as the fiber shell layer, and the flame retardant was used as the fiber core layer. The flame retardant in the separator is prevent from penetrating into the electrolyte under normal conditions, which do not affect the electrochemical performance of the battery. When the battery is heated to a certain temperature, the casing melts and the flame retardant starts to act on the electrolyte to prevent further thermal runaway and battery burning. Thereby the safety performance of the separator is improved effectively.

II. EXPERIMENTAL DETAILS

Fig.1 shows the experimental device of coaxial and uniaxial electrospinning, including syringe pump (Harvard 11 Pico Plus, USA), coaxial nozzle (diameter of outer needle is 1.166 mm, diameter of inner needle is 0.424 mm), high-voltage power supply (DW-SA403-1ACE5, China) and collecting plate.

A. The basic steps of coaxial electrospinning

Polyvinylidene fluoride (PVDF) was dissolved in a mixed solvent of N, N-Dimethylformamide (DMF) and acetone ($v:v$ = 8:2) to prepare 10wt% PVDF solution. Triphenyl phosphate (TPP) was dissolved in the acetone to prepare 30 wt% TPP solution. The PVDF solution and the TPP solution was stirred for 8 hours until the solute was completely dissolved. Then, the solution was standing 2 hours to remove air bubbles.

---

* Research supported by the National Natural Science Foundation of China (No. 51805460), Science and Technology Planning Project of Fujian Province (No. 2019H0038, 2020H6003), Xiamen Municipal Science and Technology Project (No. 3502Z20193015).

Fangqin Shao, Guoyi Kang, Huatan Chen, Zungui Shao and Gaofeng Zheng are with the Department of Instrumental and Electrical Engineering,

Xiamen University, Xiamen 361102, Fujian, China (corresponding author: Gaofeng Zheng, phone: +86-592-2194957; fax: +86-592-2182221; e-mail: zheng_gf@xmu.edu.cn).

Xiang Wang and Wenwang Li are with the School of Mechanical and Automotive Engineering, Xiamen University of Technology, Xiamen 361024, Fujian, China.

978-1-6654-3008-1/21 $31.00 © 2021 IEEE

The coaxial electrospinning was performed with two solutions by coaxial, as shown in Fig. 1(a). The distance between the nozzle and the collecting plate is 15 cm. The supplied speed of the outer needle with a syringe pump is 1.2 mL/h and the supplied speed of the inner needle is 0.4 mL/h. After the solution flows out evenly, the high voltage is added to 13 kV slowly. The mixed solution of PVDF and TPP is stretched under the action of the high-voltage electric field to form a core-shell structure fiber. The coaxial electrospinning was working for 5 hours to obtain a separator with the mass ratio of PVDF to TPP of 1:1. Then, the supplied speed of the inner needle was changed to 0.2 mL/h and 0.13 mL/h successively, and the remaining experimental parameters and operations remained unchanged to prepare separators with a mass ratio of PVDF to TPP of 2:1 and 3:1.

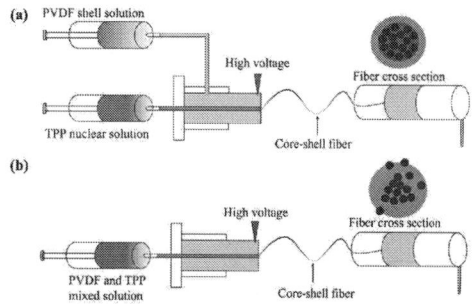

Figure 1. The schematic diagram of the preparation of flame-retardant Lithium-ion battery separator. (a) Coaxial electrospinning; (b) Uniaxial electrospinning.

B. The basic steps of Uniaxial electrospinning

PVDF and TPP were dissolved in the mixed solvent of DMF and acetone ($v:v = 8:2$) to obtain a solution with a mass ratio of PVDF to TPP of 1:1. The PVDF and TPP solution was stirred for 8 hours until the PVDF and TPP were completely dissolved. The solution was then allowed to stand for 2 hours to remove bubbles. The same method was used to prepare solutions with a mass ratio of PVDF to TPP of 2:1 and 3:1.

The distance between the nozzle and the collecting plate is 15 cm, and the supplied speed of the syringe pump is 1.2 mL/h. After the solution flows out evenly, the high voltage is increased to 13 kV slowly. The main reason for the formation of TPP@PVDF core-shell microstructure during electrospinning is the differences in solubility. The solubility of TPP is higher than that of PVDF. In the electrospinning process, as the solvent evaporates gradually, the precipitation time of PVDF is earlier than that of TPP, and the solid PVDF stays on the surface to form a core-shell structure, as shown in Fig. 1(b).

III. RESULTS AND DISCUSSION

A. Surface Morphology Characterization

The scanning electron microscope (SEM) image of the separator is shown in Fig. 2. The diameter distribution of the fiber prepared by uniaxial nozzle is more concentrated than that of the fiber prepared by coaxial nozzle. With the increase of TPP content, the fiber diameter distribution

becomes more concentrated. When the ratios between PVDF to TPP of separators prepared by coaxial electrospinning are 1:1, 2:1 and 3:1, the average fiber diameters are 739.55 nm, 741.52 nm and 683.17 nm respectively. While the ratios between PVDF and TPP of separators prepared by uniaxial electrospinning are 1:1, 2:1, and 3:1, the average fiber diameters are 375.14 nm, 48.38 nm and 352.46 nm, respectively. The diameter of the fibers prepared by coaxial nozzle is twice that of the fiber prepared by uniaxial nozzle. The uniaxial fiber has beaded structures, while the surface of coaxial fiber is smooth. As the TPP content in the separator prepared by uniaxial electrospinning increases, the beaded structure becomes larger and more numerous. This is mainly because TPP and PVDF are mixed to form a solution, and TPP prevents the formation of PVDF chain entanglement, so that the fiber forms a beaded structure.

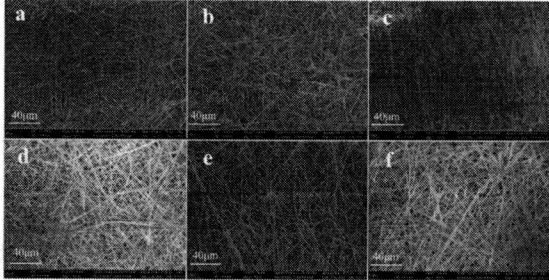

Figure 2. SEM image of separator. (a) Uniaxial PVDF：TPP=3:1; (b) Uniaxial PVDF：TPP=2:1; (c) Uniaxial PVDF：TPP=1:1; (d) Coaxial PVDF：TPP=3:1; (e) Coaxial PVDF：TPP=2:1; (f) Coaxial PVDF：TPP=1:1.

B. Flame Retardant Performance

In order to study the flame-retardant performance of TPP on the separator, the vertical burning time of the separator was tested. The combustion diagram is shown in Fig. 3. The self-extinguishing time is shown in Table I. After the PVDF separator was immersed in the electrolyte, the average burning time of the electrolyte was 8.75 s. The burning stopped when the electrolyte is exhausted. The burning time was reduced by adding TPP. The burning time of the separator with the mass ratio of PVDF to TPP of 3:1 is the longest. The burning time of the separator with the mass ratio of PVDF to TPP of 3:1 by coaxial nozzle is 2.07 s. The burning time of the separator with the mass ratio of PVDF to TPP of 3:1 by uniaxial nozzle is 1.93 s. It indicates that a small amount of TPP shows a good flame-retardant effect.

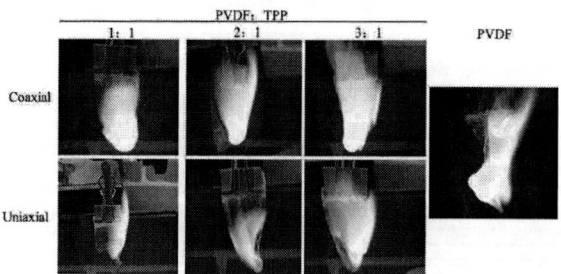

Figure 3. Separator burning vertically.

TABLE I. SEPARATOR SELF-EXTINGUISHING TIME

Sample	Coaxial self-extinguishing time (s)	Uniaxial self-extinguishing time (s)
PVDF: TPP=1:1	1.07	1.08
PVDF: TPP=2:1	0.43	1.13
PVDF: TPP=3:1	2.07	1.93
PVDF	8.75	

C. DSC Test Results

The encapsulation rate can be calculated by the latent heat of phase change. The DSC test of the separator with different ratio is shown in Fig. 4. The smaller the encapsulation rate is, the worse the effect of PVDF wrapping TPP will be. The greater encapsulation rate will cause, the better the effect of PVDF wrapping TPP. The formula for calculating the encapsulation rate is:

$$R = \frac{\Delta H_{mTPP@PVDF}}{\Delta H_{mTPP}} \tag{1}$$

ΔH_{mTPP} refers to the latent heat of phase change of the raw material TPP. $\Delta H_{mTPP@PVDF}$ refers to the latent heat of phase change of TPP in the separator.

The encapsulation rate of six separators is shown in Fig. 5. It can be seen that the separator prepared by coaxial electrospinning has a better encapsulation rate. In the coaxial electrospinning, the higher the proportion of TPP in the fiber of the separator is, the higher the encapsulation rate will be.

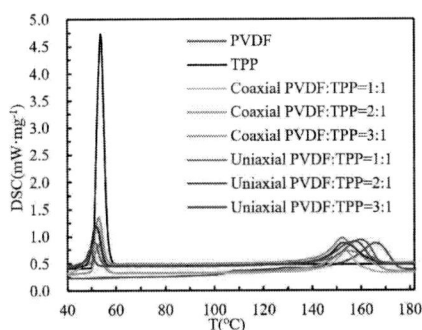

Figure 4. DSC curve of PVDF, TPP and separators in different ratio.

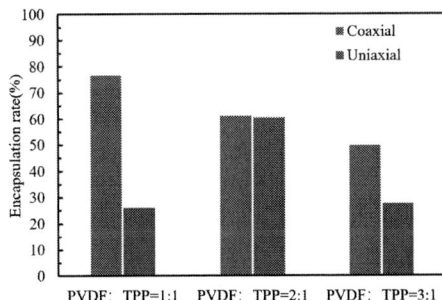

Figure 5. Separator encapsulation rate.

D. Ion conductivity

The AC impedance spectrum between commercial separators and separators prepared by electrospinning is shown in Fig. 6. The impedance formula of the AC impedance is $Z = Z' + jZ''$. The value of the disturbance signal j is 0. The impedance Z in the system is equal to Z'. The value of the intersection between the AC impedance spectrum and the real axis is Rb (Body resistance), which is the intrinsic resistance of the separator and electrolyte system.

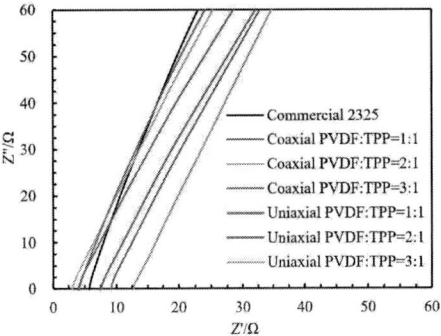

Figure 6. AC impedance spectrum of the separator.

The lithium-ion conductivity of commercial separators and electrospinning separators is shown in TABLE II. The lower the body resistance is, the higher the lithium-ion conductivity will be, which means the pass through of the lithium-ions will be more easily. The ion conductivity of separators prepared by electrospinning is higher than that of commercial separators. The less flame retardant TPP content in coaxial electrospinning will cause the higher ion conductivity. With the increase of the TPP content, the ionic conductivity of the uniaxial electrospinning separator increases. Because the uniaxial electrospinning separator has a beaded structure, making the TPP floats on the surface of the fiber. The surface of the separator prepared by the uniaxial nozzle is rougher than that prepared by the coaxial nozzle. The fiber prepared by uniaxial nozzle is thinner than the fiber prepared by coaxial nozzle. The rough surface of the fiber made the pore structure of the separator rough, which can increase the migration of the resistance to lithium-ion. The thinner fiber diameter will increase the number of separator layers with the same thickness and the lithium-ion migration channels, but the lithium-ion migration efficiency will reduce. The highest ionic conductivity is prepared by the uniaxial nozzle of PVDF and TPP with the mass ratio of 1:1 (4.779×10^{-4}S/cm). But overall, the ionic conductivity of the separator prepared by the coaxial nozzle is better than that by the uniaxial nozzle.

TABLE II. IONIC CONDUCTIVITY OF THE SEPARATOR

Sample	Body resistance (Ω)		Ion conductivity (10⁻⁴S/cm)	
	Coaxial	Uniaxial	Coaxial	Uniaxial
PVDF：TPP=1:1	9.2	3.9	1.829	4.779
PVDF：TPP=2:1	3.1	7.5	3.901	1.612
PVDF：TPP=3:1	4.1	12.7	4.347	1.583
Commercial 2325	5.7		1.548	

E. Electrochemical stability

The electrochemical stability window of the separator refers to the maximum voltage at which the electrolyte can stably exist in the battery. It determines the application potential of the separator in lithium-ion batteries. The larger the electrochemical window is, the greater the application potential of the separator will be. The curve in Fig. 7 refers to the linear sweep voltammetry test curves of commercial separators, uniaxial electrospinning separators and coaxial electrospinning separators after adding electrolyte. Table III shows the value of the electrochemical stability window of the separator. The electrochemical oxidation voltage of the commercial separator is 5.4 V. Among the separators prepared by electrospinning, only the electrochemical oxidation voltage of separator with the mass ratio of coaxial PVDF to TPP of 3:1 is slightly lower than that of commercial separators. The prepared separators are all satisfied with the voltage limit of the lithium-ion battery during the charging and discharging process.

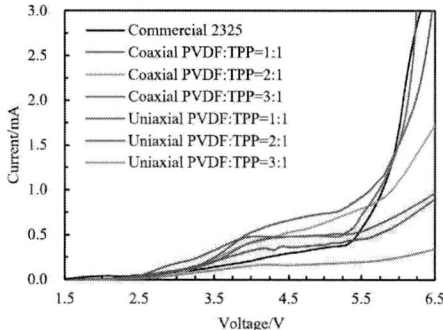

Figure 7. Separator electrochemical stability window.

TABLE III. THE ELECTROCHEMICAL STABILITY WINDOW OF THE SEPARATOR

Sample	Electrochemical stability window (V)	
	Coaxial	*Uniaxial*
PVDF：TPP=1:1	5.60	5.50
PVDF：TPP=2:1	5.85	5.70
PVDF：TPP=3:1	5.38	5.68
Commercial 2325	5.40	

IV. CONCLUSION

Nanofiber separators were prepared by coaxial electrospinning and uniaxial electrospinning. The surface morphology, flame retardant performance, DSC test and electrochemical performance of the separators were analyzed. The results show that the fiber diameter of the separator prepared by coaxial electrospinning is larger than that prepared by uniaxial electrospinning. The surface of the fiber prepared by the coaxial electrospinning is smoother. Coaxial and uniaxial electrospinning separators can achieve a good flame-retardant effect due to the addition of TPP. The self-extinguishing time of the separator with the mass ratio of PVDF to TPP of 1:1 and 2:1 are about 1 s. The self-extinguishing time of the separator with the mass ratio of PVDF to TPP of 3:1 is 2 s. The encapsulation efficiency of the separator prepared by coaxial electrospinning is better than that by uniaxial electrospinning. The electrochemical performance of the separator prepared by coaxial electrospinning is better than that of uniaxial electrospinning. According to the comparative analysis, the

separators prepared by the coaxial electrospinning with the mass ratio of PVDF to TPP of 2:1 have good flame retardancy and electrochemical properties, and their overall performance is the best.

ACKNOWLEDGMENT

This research was financially supported by the National Natural Science Foundation of China (No. 51805460), Science and Technology Planning Project of Fujian Province (No. 2019H0038, 2020H6003), Xiamen Municipal Science and Technology Project (No. 3502Z20193015).

REFERENCES

[1] J. Feng, L. Lu, "A novel bifunctional additive for safer lithium ion batteries," Journal of Power Sources, vol. 243, pp. 29-32, December 2013.

[2] L. Jiang, Q. Wang, K. Li, P. Ping, L. Jiang, J. Sun, "A self-cooling and flame-retardant electrolyte for safer lithium ion batteries," Sustainable Energy & Fuels, vol. 2, pp. 1323-1331, June 2018.

[3] L. Xia, Y. Xia, Z. Liu, "A novel fluorocyclophosphazene as bifunctional additive for safer lithium-ion batteries," Journal of Power Sources, vol. 278, pp. 190-196, March 2015.

[4] S. S. Zhang, K. Xu, T. R. Jow, "Tris(2,2,2-trifluoroethyl) phosphite as a co-solvent for nonflammable electrolytes in Li-ion batteries," Journal of Power Sources, vol. 113, no. 1, pp. 166-172, January 2003.

[5] H. Ota, A. Kominato, W. J. Chun, E. Yasukawa, S. Kasuya, "Effect of cyclic phosphate additive in non-flammable electrolyte," Journal of Power Sources, vol. 119, pp. 393-398, June 2003.

[6] S. Liu, J. Mao, Q. Zhang, Z. Guo, "An Intrinsically Non-flammable Electrolyte for High-Performance Potassium Batteries," Angewandte Chemie International Edition, vol. 59, no. 9, pp. 3638-3644, January 2020.

[7] X. L. Yao, S. Xie, C. H. Chen, Q. S. Wang, J. H. Sun, Y. L. Li, et al, "Comparative study of trimethyl phosphite and trimethyl phosphate as electrolyte additives in lithium ion batteries," Journal of Power Sources, vol. 144, no. 1, pp. 170-175, June 2005.

[8] J. J. Woo, S. H. Nam, S. J. Seo, S. H. Yun, W. B. Kim, T. Xu, et al, "A flame retarding separator with improved thermal stability for safe lithium-ion batteries," Electrochemistry Communications, vol. 35, pp. 68-71, October 2013.

[9] X. Luo, X. Lu, X. Chen, Y. Chen, C. Song, C. Yu, et al, "A robust flame retardant fluorinated polyimide nanofiber separator for high-temperature lithium–sulfur batteries," Journal of Materials Chemistry A, vol. 8, no. 29, pp. 14788-14798, August 2020.

[10] N. Deng, Y. Liu, L. Wang, Q. Yan, H. Yang, "Designing of a Phosphorus, Nitrogen, and Sulfur Three-Flame Retardant Applied in a Gel Poly-m-phenyleneisophthalamide Nanofiber Membrane for Advanced Safety Lithium-Sulfur Batteries," ACS applied materials & interfaces, vol. 11, no. 40, pp. 36705-36716, September 2019.

978-1-6654-3008-1/21 $31.00 © 2021 IEEE

Gap in pagination due to unavailable papers.

Pages 1039-1041

Proceedings of the 16th Annual IEEE International Conference on Nano/Micro Engineered and Molecular Systems
April 25-29, 2021

An In_2O_3 Nanotubes based Gas Sensor Array combined with Machine Learning Algorithms for Trimethylamine Detection

Wenjie Ren, Changhui Zhao, Yingming Liu and Fei Wang*, *Senior Member, IEEE*

Abstract— This work focuses on developing an electronic nose system with machine learning algorithm for detection of trimethylamine (TMA). Pure and Ga-doped In_2O_3 nanotubes are synthesized by a simple electrospinning method, and four kinds of gas sensors (pristine, 1% Ga, 10% Ga, and 20% Ga-doped In_2O_3) are fabricated to form a sensor array. Results show that the sensor array can classify TMA effectively from interference gases (xylene, ethanol, hydrogen sulfide) by a support vector machine (SVM) algorithm. Several algorithms, including radial basis function neural network (RBFNN), back propagation neural network (BPNN) and principal component analysis combined with linear regression (PCA-LR), are used to predict the concentration level of each gas. For TMA gas, the trained algorithms can predict its concentration with average relative errors of 1.22% for RBFNN, 2.5% for BPNN and 13.34% for PCA-LR. Furthermore, the binary mixtures of TMA and ethanol are measured and used to train the above algorithms, and the lowest average relative error of 1.74% is achieved in the case of RBFNN algorithm.

I. INTRODUCTION

Trimethylamine (TMA) is a stinking gas that smells like rotting fish. It is widely distributed in industrial and agricultural production. Long-time exposure to TMA gas often results in a greatly physical and psychological discomfort to humans [1-2]. Therefore, it is of practical significance to recognize TMA gas from environment and confirm its concentration. Among the existing methods that can monitor the stinking gas, electronic nose made of gas sensor array stands out for its rapid response and high selectivity [3-4]. There are two main parts in an efficient electronic nose system, which are gas sensor array and appropriate pattern recognition techniques.

To assemble the array of sensors, the gas sensitive materials for TMA need to be synthesized and selected. Among many gas sensing materials, metal oxide semiconductor (MOS) based materials are highly prized for their selectivity, fast response, low production cost and good stability [5]. To be specific, In_2O_3 is widely used and researched for its large amounts of oxygen vacancies at room temperature [6]. In_2O_3 materials doped with different elements are applied for various gas detections. Mesoporous $K_2O-In_2O_3$ nanowires are synthesized by template-calcined method in the work of Rehman et al. The as-obtained material shows improved sensing properties towards NO_x gas [7]. More than that, alkaline-earth metals (Ca, Sr and Ba) doped In_2O_3 nanotubes and Ag-modified In_2O_3 nanoparticles have been studied for the leakage monitoring of formaldehyde and ethanol [8-9]. Although a considerable number of researches have been performed on doped In_2O_3, little work has been done on Ga-doped In_2O_3. In this work, different percentages of Ga doped In_2O_3 nanotubes (NTs) are synthesized by electrospinning. This method is chosen for that it is suitable for preparing less-agglomerated nanotubes with high adjustability of doping levels and internal structures [10].

To complete the design of an efficient electronic nose, pattern recognition algorithms are needed apart from gas sensitive materials. As a result, SVM was used to identify TMA from other gases. SVM is widely used as a type of generalized linear classifier [12-13]. And radial basis function neural network (RBFNN), back propagation neural network (BPNN) and principal component analysis combined with linear regression (PCA-LR) algorithms are commonly seen in the system developed for quantitative analysis of gaseous mixtures. Many attempts based on RBFNN, BPNN and PCA have been reported [14-15]. Xiao et al. reported a binary classification and concentration prediction of combustible gas using BPNN and PCA, reaching a 100% accuracy for classification [16]. BPNN is one of the basic neural networks, which was firstly received attention in 1986 for the work of Rumelhart et al. [17]. BPNN actually realizes a mapping from input to output, making it particularly suitable for solving problems with complex internal mechanisms. The RBF network can approximate any nonlinear function with arbitrary accuracy and has the global approximation ability, which fundamentally solves the local optimal problem of some other networks. An active learning algorithm for RBFNN was proposed and applied to E-nose in 2017 by Jiang et al. [18]. PCA is mainly used to decrease the dimension of the features. It was combined with linear regression method to make a prediction in this work. For linear relationships between input and output, it can construct a mapping with high degree of fitting.

In this paper, Ga doped In_2O_3 sensitive materials are synthesized using electrospinning. Then gas sensors are fabricated to form an array. The gas sensing measurements are performed in different gases to train the algorithms. Finally, the best-performed algorithm will be selected and applied to the electronic nose system.

II. EXPERIMENTAL METHOD

A. Fabrication of the device

Electrospinning method was used to synthesize Ga-In_2O_3 NTs, which is detailly described in the paper of Zhao et al. [19]. Pristine In_2O_3 and In_2O_3 doped with 1%, 10%, 20% Ga (molar percentage) were obtained by similar procedures. As shown in Fig. 1, the sensors that are based on above materials were fabricated. A sensor array consisting of the four kinds of gas sensors was completed. The measurement was performed

---

W. Ren and C. Zhao are with the School of Microelectronics, Southern University of Science and Technology, Shenzhen 518055, China.

*Contacting Author: F. Wang is with the School of Microelectronics, Southern University of Science and Technology, Shenzhen 518055, China. (Phone: +86-755-88018509, email: wangf@sustech.edu.cn).

978-1-6654-3008-1/21 $31.00 © 2021 IEEE

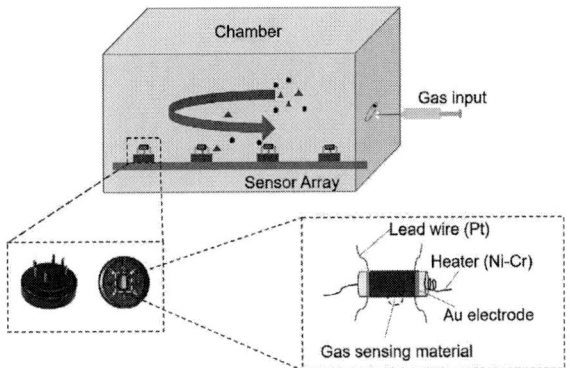

Fig. 1. The schematic of gas sensor array, single sensor and electrospinning device.

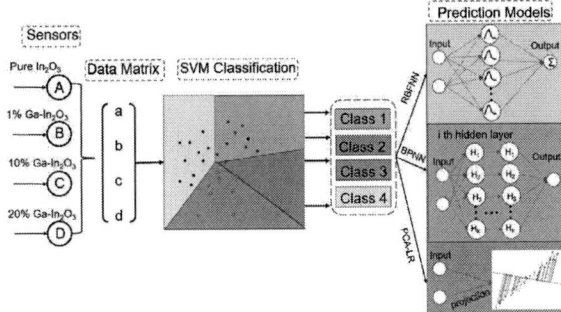

Fig. 2. The training flow of gas classification and the prediction of concentrations.

TABLE I.   GAS TYPES AND GAS CONCENTRAYIONS

Gas type	Concentration (ppm)						
TMA	5	10	20	40	60	80	100
Xylene	5	10	20	40	60	80	100
Ethanol	5	10	20	40	60	80	100
H$_2$S	0.1	0.2	0.4	0.6	0.8	1	2

in several different gases. The measurement process is similar to the work of Zhao et al. [19]. The gases used in this experiment and their corresponding concentrations are listed in Table 1. For binary mixture of TMA and ethanol, the concentration ratios of TMA to ethanol are: 10:10, 10:40, 10:100, 40:100, 100:100.

B. Algorithm training

After the gas sensing measurements, the responses (R_a/R_g) of the four sensors were collected and input into SVM algorithm for classification. The algorithm training flow is displayed in Fig. 2. The responses of the sensors will be divided into 5 classes, 1 for H$_2$S, 2 for TMA, 3 for xylene, 4 for ethanol and 5 for the mixture of TMA and ethanol. For a given sample set D, we have:

$$D = \{(x_1, y_1),(x_2, y_2),...,(x_n, y_n)\} \quad (1)$$

Where x is an m-dimensional input vector; y is the output; D is the data set including n samples. For the samples above, there is an optimal (most generalizing) hyperplane that separates the different classes of samples. If we assume that w is the normal vector and b is the displacement of the hyperplane, then it can be represented as follows:

$$w^T x + b = 0 \quad (2)$$

And for arbitrary x_i, the distance to the plane is:

$$\gamma = \frac{|w^T x_i + b|}{\|w\|} \quad (3)$$

If the sample is successfully classified by the hyperplane, then the following equations hold:

$$\begin{cases} w^T x_i + b \geq +1, y_i = +1 \\ w^T x_i + b \leq -1, y_i = -1 \end{cases} \quad (4)$$

$$\gamma = \frac{2}{\|w\|} \quad (5)$$

Supporting vectors are the vectors that make the two sides equal for above inequations. And the train target of SVM algorithm is to find a hyperplane which has the biggest distance γ to the samples. Finally, we construct the

classification problem as to minimize $\|w\|^2$ under the constraints in (4).

According to the hyperplane, the samples are divided into 5 groups in this experiment. Subsequently, the classification labels and the responses of the four sensors will be set as the input parameters of the prediction models. For each gas, there are 3 different prediction models, which are RBFNN, BPNN, PCA-LR. The results of the three models are compared to select the algorithm that performs the best. Actually, this system realizes the identification of TMA gas from different single gases and mixed gases. And also, it implements the prediction of its concentration.

III. RESULT AND DISCUSSION

A. Characterization of the materials

The morphologies of the as-obtained materials are characterized by the field emission scanning electron microscope (FE-SEM, Zeiss Gemini 300) in Fig. 3. All of them exhibit a porous tubular structure with average diameters of 40 to 70 nm. From the insets in the four figures, the detail information on the surface can be seen. As the concentration of Ga increased, the surfaces of the tubular structures become rougher. When the doping level exceeds 10%, small grains start to precipitate on the surface of the tube, which may due to the solid solubility of Ga in In$_2$O$_3$. The changes in morphologies and components result in different gas sensing properties for the four kinds of sensors, which can be selected as the classification features for the algorithms.

B. Gas sensing measurements

Firstly, the gas sensing measurements are performed in different levels of single gas atmospheres, where TMA, xylene, ethanol, H$_2$S are used. The relationship between the response and concentration of TMA is shown as an example for the gas sensing measurements in Fig. 4. The array is tested under the concentration range from 5 to 100 ppm for TMA, xylene, ethanol, H$_2$S separately (the range for H$_2$S is 0.1-2 ppm). As shown in Fig. 4, the responses of the four kinds of sensors are all increasing with the increasing concentrations. It

978-1-6654-3008-1/21 $31.00 © 2021 IEEE

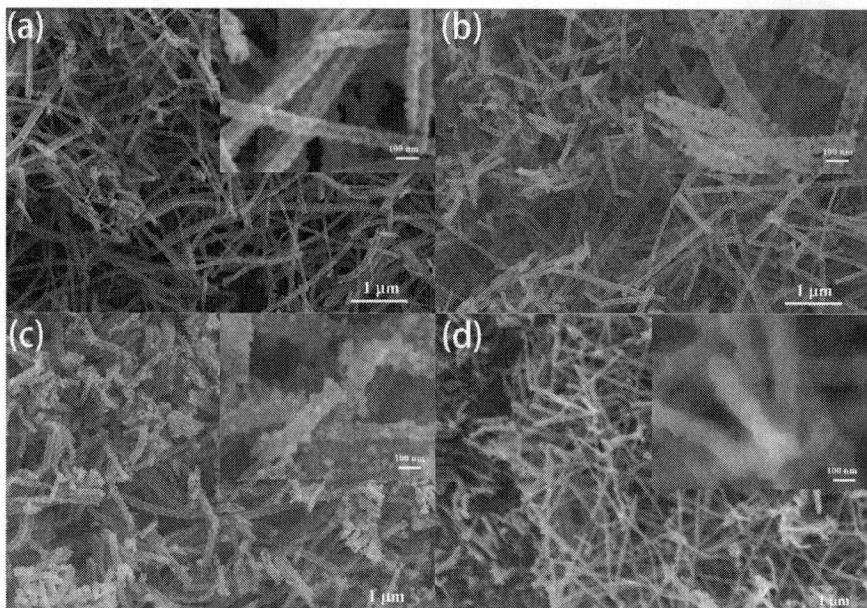

Fig. 3. SEM images of (a) pristine In_2O_3, (b) 1% Ga- In_2O_3, (c)10% Ga- In_2O_3, (d) 20% Ga- In_2O_3; the enlarged images are shown in the insets.

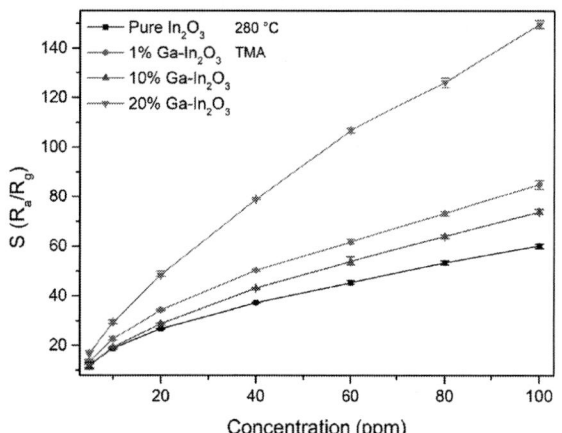

Fig. 4. The relationships between the device response and the concentration of TMA with error bars.

TABLE II. COMPARISON OF THE AVERAGE RELATIVE ERRORS

Gas type	ARE		
	RBFNN	BPNN	PCA-LR
TMA	1.22%	2.50%	13.34%
Xylene	6.48%	8.90%	2.64%
Ethanol	1.60%	2.56%	8.86%
H_2S	7.45%	6.81%	8.94%
TMA:Ethanol	1.74%	7.99%	53.74%

which reduces the classification difficulty. For the 66 test samples, the classification accuracy is 100%.

After classification, there is function leading to the corresponding prediction algorithms according to the class labels. The RBF networks are constructed by the MATLAB function *newrbe* with adjusted spread values. Similarly, the BP networks are constructed by function *newff* at a learning rate of 0.01 with adjusted hidden neurons. PCA is used to reduce the dimension of the input and 3, 2, 2, 3 main components are used for the regression of TMA, xylene, ethanol and H_2S. In MATLAB, the linear regression process is completed by the function *regress*.

The concentration predictions of four single gases are displayed in Fig. 5 for three prediction algorithms. 40% of the data set is used as testing data. It is indicated in the Fig. 5 that for each concentration there are 2 samples for testing. In general, the prediction accuracies for RBFNN and BPNN are better than PCA-LR. For all the prediction points, the closer they are to the diagonal line, the more accurate the results are. Afterwards, the prediction results for mixed TMA and ethanol are illustrated in Fig. 6. Z axis stands for the sum of the targeted TMA and ethanol concentrations. This time, PCA-LR still shows the worst performance with an accuracy less than 50%, which may be caused by the nonlinearity for mixed gases. To get the detailed evaluation of the prediction accuracy, the average relative (ARE) errors for all gases are plotted in Fig. 7. And ARE is calculated as below:

is found in Fig. 4 that there are higher responses for higher Ga doped sensors.

For the reason that the response values of ethanol and TMA are quite close. It can be more difficult to recognize TMA from the mixture of TMA and ethanol gases. As a result, TMA and ethanol are chosen to be mixed at different concentrations for algorithm trainings. The varying responses between the four devices makes it practical to recognize and predict the target gases. After the measurements, the response values will be selected as the inputs of the classification and concentration prediction algorithms.

C. Gas classification and concentration prediction

As shown in Table 1, for each gas and each concentration, there are 5 samples, 3 of which are used for classification and prediction trainings. The obtained responses for the five kinds of gases display significant variations for the four sensors,

978-1-6654-3008-1/21 $31.00 © 2021 IEEE

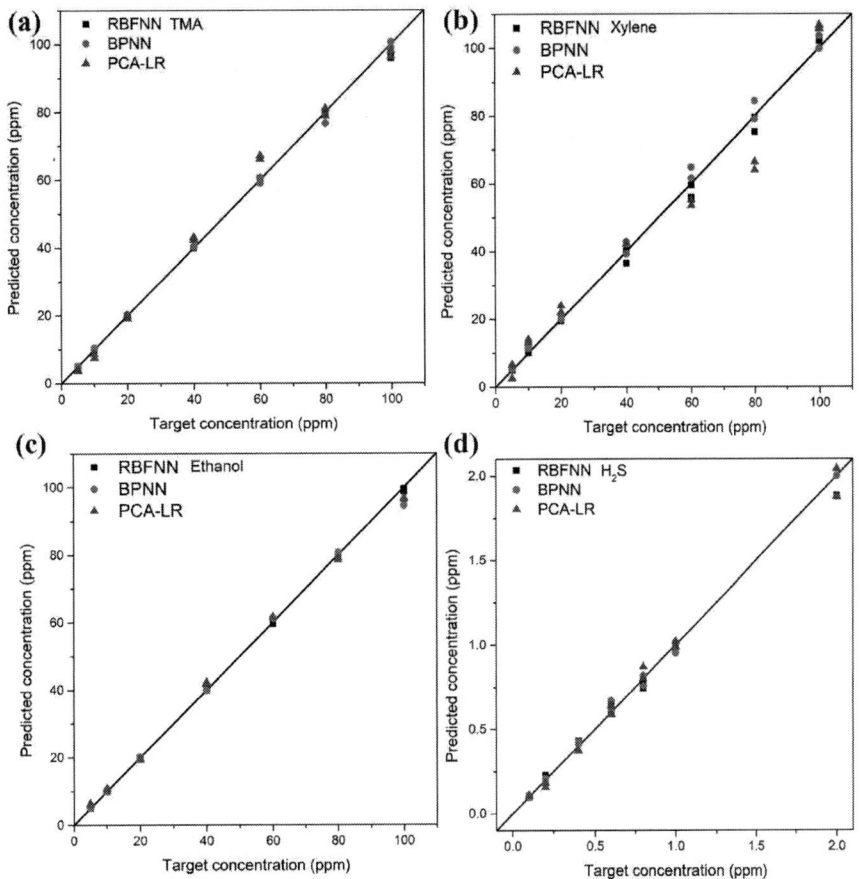

Fig. 5. The comparison between the actual values and predicted values by RBFNN, BPNN, PCA-LR algorithms for (a) TMA, (b) xylene, (c) ethanol, (d) H_2S.

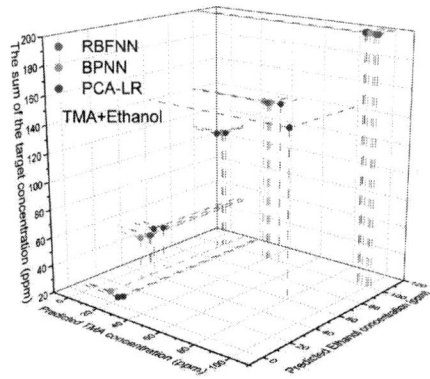

Fig. 6. The comparison between the actual values and predicted values by RBFNN, BPNN, PCA-LR algorithms for the mixture of TMA and ethanol.

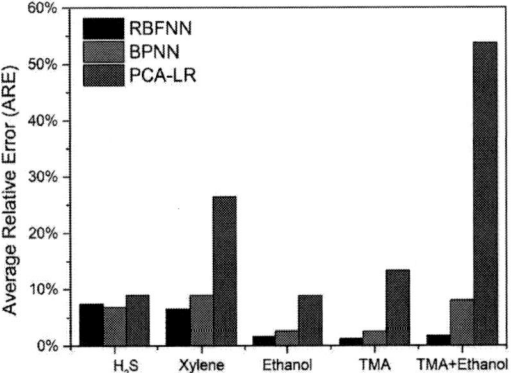

Fig. 7. The average relative error for five kinds of gases with RBFNN, BPNN, PCA-LR algorithms.

$$ARE = \frac{1}{n} \sum_{i=1}^{n} \frac{|y_i - y_i^{'}|}{|y_i|} \qquad (6)$$

Where y_i stands for the actual concentration and $y_i^{'}$ is the predicted value. From Fig. 7 we can find that the predictions of BP network have the lowest ARE for H_2S. However, for the other four kinds, RBFNN shows the lowest ARE. In particular,

the ARE for PCA-LR are relatively high with respect to the other two algorithms. It is observed that the average relative error for the mixture of TMA and ethanol has reached 53.74% for PCA-LR, which is due to the mismatch between the nonlinearity of the response and the linear regression algorithm. The lowest prediction errors for 5 kinds of gases among the 3 algorithms are 1.22%, 6.48%,1.6%, 6.81% and 1.74% for TMA, xylene, ethanol, H_2S and the mixed gas. The specific values of the errors are shown in the Table 2.

Therefore, RBFNN may be the most appropriate for the pattern recognition techniques in this experiment.

IV. CONCLUSION

In this work, Ga doped In_2O_3 sensitive materials were synthesized and used to fabricated gas sensors. Four kinds of gas sensors formed a gas sensor array, which were made from pure In_2O_3 and 1%, 10%, 20% Ga doped In_2O_3 NTs. The gas sensor array showed relatively high response to TMA, which is the basic for the detection of TMA and the prediction of its concentrations. Single gases and mixed gases are both used as the interference for the recognition of TMA. SVM is introduced as the classification algorithm, reaching a 100% recognition accuracy. Then, RBFNN, BPNN, PCA-LR are applied as the prediction methods for gas concentrations. Among them, the average relative error for TMA is the lowest (1.22%) with RBF network. The result reveals that RBFNN is the best for this electronic nose system. In conclusion, this study offers an attempt to implement an electronic nose system from the synthesis of the materials to the selection of appropriate pattern recognition algorithms. And it shows relatively high accuracy for the detection of TMA.

ACKNOWLEDGMENT

This work was supported in part by the The National Key Research and Development Program of China under Grant 2020YFB2008604, in part by the Shenzhen Science and Technology Innovation Committee under Grant JCYJ20170412154426330 and in part by "climbing Program" Special Funds under Grant pdjh2021c0075.

REFERENCES

[1] L. Vernetti, A. Gough, N. Baetz, S. Blutt, J. R. Broughman, J. A. Brown, J. Foulke-Abel, N. Hasan, J. In, E. Kelly, O. Kovbasnjuk, J. Repper, N. Senutovitch, J. Stabb, C. Yeung, N. C. Zachos, M. Donowitz, M. Estes, J. Himmelfarb, G. Truskey, J. P. Wikswo, and D. L. Taylor, "Functional coupling of human microphysiology systems: Intestine, liver, kidney proximal tubule, blood-brain barrier and skeletal muscle," *Sci. Rep.*, vol. 7, p. 42296, 2017.

[2] H. Greim, D. Bury, H. J. Klimisch, M. Oeben-Negele, and K. Ziegler-Skylakakis, "Toxicity of aliphatic amines: structure-activity relationship," *Chemosphere*, vol. 36, no. 2, pp. 271-295, 1998.

[3] A. Wilson and M. Baietto, "Applications and advances in electronic-nose technologies," *Sensors*, vol. 9, no. 7, pp. 5099-5148, 2009.

[4] Z. Chen, Y. Zheng, K. Chen, H. Li, and J. Jian, "Concentration estimator of mixed VOC gases using sensor array with neural networks and decision tree learning," *IEEE Sensors Journal*, vol. 17, no. 6, pp. 1884-1892, 2017.

[5] J. Zhang, Z. Qin, D. Zeng, and C. Xie, "Metal-oxide-semiconductor based gas sensors: screening, preparation, and integration," *Phys. Chem. Chem. Phys.*, vol. 19, no. 9, pp. 6313-6329, 2017.

[6] H. Cao, X. Qiu, Y. Liang, Q. Zhu, and M. Zhao, "Room-temperature ultraviolet-emitting In_2O_3 nanowires," *Applied Physics Letters*, vol. 83, no. 4, pp. 761-763, 2003.

[7] A. U. Rehman, J. Zhang, J. Zhou, K. Kan, L. Li, and K. Shi, "Synthesis of mesoporous K_2O-In_2O_3 nanowires and NO_x gas sensitive performance study in room temperature," *Microporous and Mesoporous Materials*, vol. 240, pp. 50-56, 2017.

[8] Q. Liang, X. Zou, H. Chen, M. Fan, and G.-D. Li, "High-performance formaldehyde sensing realized by alkaline-earth metals doped In_2O_3 nanotubes with optimized surface properties," *Sensors and Actuators B: Chemical*, vol. 304, p. 127241, 2020.

[9] J. Wang, Z. Xie, Y. Si, X. Liu, X. Zhou, J. Yang, P. Hu, N. Han, J. Yang, and Y. Chen, "Ag-modified In_2O_3 nanoparticles for highly sensitive and selective ethanol alarming," *Sensors*, vol. 17, no. 10, 2017.

[10] C.-S. Lee, I.-D. Kim, and J.-H. Lee, "Selective and sensitive detection of trimethylamine using ZnO–In_2O_3 composite nanofibers," *Sensors and Actuators B: Chemical*, vol. 181, pp. 463-470, 2013.

[11] D. G. Yu, M. Wang, X. Li, X. Liu, L. M. Zhu, and S. W. Annie Bligh, "Multifluid electrospinning for the generation of complex nanostructures," *Wiley Interdiscip Rev Nanomed Nanobiotechnol*, vol. 12, no. 3, p. e1601, 2020.

[12] Y. Yang, J. Li, and Y. Yang, "The research of the fast SVM classifier method," in *2015 12th International Computer Conference on Wavelet Active Media Technology and Information Processing (ICCWAMTIP)*. IEEE, 2015, pp. 121-124.

[13] A. J. Smola and B. Schölkopf, "A tutorial on support vector regression," *Statistics and Computing*, vol. 14, no. 3, pp. 199-222, 2004.

[14] H. Tai, G. Xie, and Y. Jiang, "An artificial olfactory system based on gas sensor array and back-propagation neural network," Springer Berlin Heidelberg, 2004, pp. 892-897.

[15] V. L. Skrobot, E. V. R. Castro, R. C. C. Pereira, V. M. D. Pasa, and I. C. P. Fortes, "Use of principal component analysis (PCA) and linear discriminant analysis (LDA) in gas chromatographic (GC) data in the investigation of gasoline adulteration," *Energy & Fuels*, vol. 21, no. 6, pp. 3394-3400, 2007.

[16] H. Xiao, J. Xu, B. Tang, and Z. Zhang, "Classification and concentration prediction of combustible gas based on BPNN and PCA," Atlantis Press, Conference Proceedings.

[17] D. Rumelhart, G. Hinton and R. Williams, "Learning representations by back-propagating errors", *Nature*, vol. 323, no. 6088, pp. 533-536, 1986.

[18] X. Jiang, P. Jia, R. Luo, B. Deng, S. Duan, and J. Yan, "A novel electronic nose learning technique based on active learning: EQBC-RBFNN," *Sensors and Actuators B: Chemical*, vol. 249, pp. 533-541, 2017.

[19] C. Zhao, B. Huang, E. Xie, J. Zhou, Z. Zhang, "Improving gas-sensing properties of electrospun In_2O_3 nanotubes by Mg acceptor doping," *Sensors and Actuators B: Chemical*, vol. 207, pp. 313–320, 2015.

Gap in pagination due to unavailable papers.

Pages 1047-1056

Proceedings of the 16th Annual IEEE International
Conference on Nano/Micro Engineered and Molecular Systems
April 25-29, 2021

The Fabrication of Three-layer Silicon Stacked Antenna

Cao Yang Lei, Zhu Jian, Member, IEEE, Hou Fang, Jiao Zong Lei and Huang Min

Abstract—**In this paper, a three-layer stacked structure patch antenna operating at W-band based on Three-Dimensional Silicon Integration Technology is proposed, designed and measured. The antenna is an 8*8 patch array antenna, made of three layers of high resistance silicon (HRSi) wafers. Through silicon via (TSV) technique and vertical interconnect technique are carried out to divide the feeding networks, signal interposer and radiation patches into different wafers, and then the multilayer wafer stacking technique is used to bond the wafers to form the three-layer structure antenna. The designed antenna is fabricated and measured, and the measured results basically match the simulation results. It is demonstrated that the fabrication process proposed in this paper is function well. The measured results show operating frequency band of antenna is 90.3-98.7GHz，and the peak gain is about 10dBi.**

I. INTRODUCTION

In recent years, with the further development of communication systems, higher requirements have been put forward for system capabilities, requiring the system to be smaller and cheaper in higher frequency bands, and have more functions. This requires the communication system to have higher integration capabilities [1]. Antenna is an important part of the modern radar communication system, and it occupies a large volume in the communication system. Therefore, how to miniaturize the antenna and integrate it with other devices is one of the keys to improving the integration of the communication system [2]. RF microsystem technology is a new concept that has been proposed in recent years. The main goal is to integrate various functional units including the RF front-end into a miniaturized system module as much as possible through three-dimensional integration technology. The three-dimensional silicon integration technology is based on the silicon-based MEMS process, and many key challenging processes are developed such as, TSV fabrication, wafer bonding, thinning, micro bumps, and passive device integration. Based these techniques, the three-dimensional silicon integration technology has high processing accuracy and can be integrated with other front-end devices on-chip. It provides an attractive and promising way for antenna design. Based on the three-dimensional silicon integration technology, the fabrication steps of a w-band 8*8 microstrip antenna array consisting of three layers are presented. And the antenna is tested to verify the feasibility of the process.

All authors are with Nanjing Institute of Electronic Devices, Nanjing, China (e-mail: cylcarlos@126.com)

II. FABRICATION PROCESS

The antenna generally includes a radiating unit and a feeding structure. When designing an antenna array, the feed network of the antenna occupies a large area, and at the same time it will greatly affect the performance of the antenna. In order to reduce the size of the antenna, the antenna has a three-layer structure, including: antenna radiation layer, interposer layer and divider layer. These three-layer structures are respectively located on three-layer high-resistance silicon substrates, and then stacked together. The Cross-sectional view of the antenna structure with TSV interconnection is shown in Figure 1.

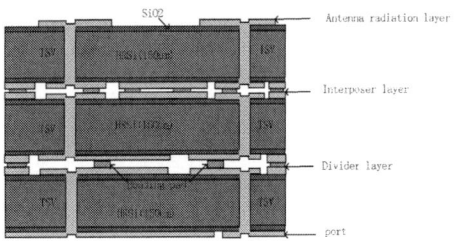

Figure 1 Cross-sectional view of the antenna structure with TSV interconnection

The three layers of the antenna needs to be processed separately and then bonded together. The complete details of fabrication steps are as follows:

1) Divider layer grows SiO_2 and SiN films with patterning;

2) Frontside TSV etch with $150\,\mu m$ depth and tapered via diameter of $20\mu m$;

3) Frontside TSV electroplating;

4) Frontside patterning;

5) Backside of wafer thinning;

6) Backside patterning;

7) Radiation layer and interposer layer repeat the step (1)-(6);

8) Bonding the frontside of divider layer with the backside of interposer layer;

9) Bonding the frontside of interposer layer with the backside of radiation layer.

Totally, eleven masks are used for fabrication of this cap wafer. The entire process flow is designed according to the antenna structure, among which TSV fabrication technology with a height of $150\mu m$ and a diameter of $20\,\mu m$, multi-layer wafer stacking technology and inter-layer interconnection technology are key technologies. We will discuss the issues of these

978-1-6654-3008-1/21 $31.00 © 2021 IEEE

technologies encountered during the preliminary development.

A. TSV fabrication technology

TSV fabrication technology is the key technology of three-dimensional silicon integration technology, which fundamentally solves the parasitic and reliability problems caused by traditional interconnection methods such as gold wire bonding. It helps to increase the density of devices on the substrate and achieve high-density interconnection. TSV with high aspect ratio and high diameter depth is the main technical difficulty in processing. The TSV processed in this paper has a size of $150*20\mu m$ and an aspect ratio of 7.5:1.

The specific flow diagram of TSV is shown in Figure 2, including the TSV deep hole etching, seed layer, barrier layer deposition, and conductive material electroplating filling. There are many etching methods for TSV, for example, wet etching, dry etching, electrochemical etching, laser etching, etc. Both laser etching and electrochemical etching are not suitable for TSV deep hole etching with high aspect ratio, and wet etching is prone to trapezoidal structure, and it is difficult to form vertical deep holes with oblique holes. This article mainly uses dry etching method for TSV deep hole etching. Dry etching is to etch the silicon through certain physical bombardment or chemical reaction methods, including plasma etching, reactive ion etching, chemically assisted ion etching, and gas phase chemical etching [3]. The etching process is the Bosch process, which uses an inductively coupled plasma (ICP) source, which is a typical time division multiplexing technology in which etching and protection are switched alternately.

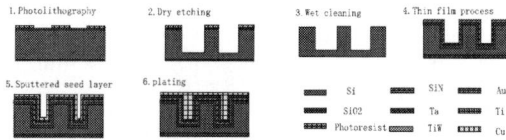

Figure 2 Process flow diagram of TSV

After the deep hole is etched, a copper electroplating technique is used to fill the deep hole with conductive material. Before copper electroplating, it is necessary to deposit the insulating layer film, the barrier layer metal and the seed layer metal. The purpose of the insulating layer is to isolate the metal layer and prevent short circuits. This article uses plasma-enhanced chemical vapor deposition (PECVD) to grow SiO_2 and SiN films. The main purpose of the barrier metal is to prevent the materials above and below the layer from mixing with each other. Because of its own characteristics, copper will quickly diffuse into silicon oxide and silicon, thereby affecting the performance of the device. This article uses metal Ta as the barrier material. A layer of Cu needs to be sputtered on the metal Ta as a seed layer for electroplating.

The filling effect of the TSV deep hole conductive material will directly affect the electrical performance of the TSV, thereby affecting the antenna performance. And compared to TSV deep hole etching, seed layer and barrier layer deposition and other technologies, the filling of TSV

deep hole conductive material is more difficult to control in the process, and there are not many reports on related results. Therefore, it can be used in high aspect ratio holes or the realization of defect-free conductive material filling in the through hole is one of the most important three-dimensional manufacturing technologies, and it is also the focus of this antenna processing.

Factors affecting the quality of TSV electroplating include: wafer design, the quality of the deposition of insulating layers and other films, the composition of the electroplating solution and the electroplating process. In the wafer design part, the TSV aspect ratio, aperture and hole shape, are determined. The larger the aspect ratio and the larger the aperture, the more difficult it is for copper electroplating, but it does not directly affect the quality of the electroplating [4]. At present, the process of film deposition is quite mature, and it has little effect on the quality of electroplating.

Among these factors, the composition of the electroplating solution and the electroplating process have the greatest impact on the electroplating quality. The basic composition of the electroplating solution is copper sulfate, sulfuric acid, copper chloride and water [5]. Most of the additives are organic ingredients, and their types include accelerators, levelers and inhibitors. During the electroplating process, various additives work together to ensure the electroplating performance. In the electroplating process, current density, current waveform, electroplating temperature and cycle time are the key factors.

In the process of $150*20\mu m$ TSV electroplating, a typical result of electroplating is shown in the figure below. From Figure 3 (a), it can be clearly seen that the upper part of the TSV is completely filled, but there is a void near the bottom, which is not filled. According to Figure 3（b）, the discontinuity in the TSV hole has a diameter of $11\mu m$, and the thickness of the copper layer on the hole wall is about $4\mu m$.

By adjusting the plating conditions, the TSV is basically completely filled, and there are still holes in the TSVs in a small number of positions. As shown in the Figure 5, the position of the holes is uncertain, and the diameter is about $6\mu m$.

By adjusting the current intensity, a fully filled TSV can be obtained, as shown in the Figure 6.

Figure 3 High aspect ratio TSV plating defects

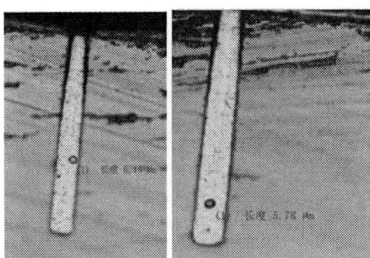

Figure4 the holes in TSV

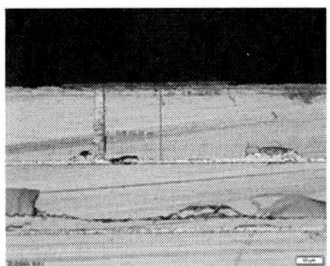

Figure 6 the image of bonding structure

Because the wafer is too thin, temporary bonding is required to prevent the wafer form being crushed during bonding. Temporary bonding is polymer bonding, in which an intermediate polymer is coated on the surface of one or two wafers and the bonded together by applying pressure. The suitable temperature for temporary bonding ranges from room temperature to 450 ℃, so the temporarily bonded wafers can be used in the Au-Sn bonding process.

Figure 5 the image of fully filling TSV

Fully filling TSV is a very challenging electroplating process, which requires extremely high requirements for the composition control of electroplating additives and electroplating solution. However, the current electroplating process lacks monitoring methods for the composition of the electroplating solution and effective additives. In this article, when the same electroplating solution is used for TSV electroplating, as the number of electroplating increases, the content of each component including additives in the solution changes, the changes of these factors have an important impact on the filling effect of TSV. Therefore, after electroplating is completed, it is necessary to use a certain inspection pattern to monitor the electroplating.

B. Multi-layer wafer stacking

Three-dimensional stacking technology is of great significance for reducing system size and developing richer device structures. Wafer-to-wafer bonding is a key process to realize three-dimensional stacking technology.

The bonding process is a low temperature bonding process——Au-Sn bonding. The eutectic temperature of Au-Sn is 280℃. At this temperature, the eutectic reaction takes place near the original Au-Sn interface, forming the alloy structure composed of different phase states. The Au-Sn interface gradually disappears, forming the bonding layer with good thermal and electrical properties and stable structure. Au-Sn alloy has good high temperature properties, excellent corrosion resistance, good electrical and thermal conductivity, high strength and other advantages. The excellent high temperature characteristics of Au-Sn alloy make its melting point much higher than the eutectic temperature of Au-Sn alloy, so that the second bonding of the already bonded wafer will not destroy the previous bonding.

The key conditions for Au-Sn bonding include bonding pressure, wetting temperature, alloy temperature and duration. In this paper, a three-layer silicon wafer with a size of 8 inches and a thickness of $150\mu m$ is bonded. The image of bonding structure is shown in Figure 6.

C. Interlayer via interconnection technology

Interlayer signal interconnection technology refers to the combination of TSV processing technology and wafer bonding other technologies, to achieve the purpose of microwave signal transmission between multilayer substrates. In the previous two sections, TSV processing and Au-Sn bonding have been introduced, but to achieve the purpose of signal interconnection between layers, the cooperation of other processes is still needed.

After the conductor electroplating of TSV is completed, a layer of copper with a thickness of several microns to more than ten microns will grow on the surface of the wafer. These coppers will completely cover and interconnect TSVs. Therefore, it is necessary to remove the copper on the surface of the wafer to expose the position of the TSVs. It should be noted that when removing copper, the insulation layer deposited before electroplating cannot be removed. The dielectric film thickness of the insulating layer is generally from thousands of angstroms to two microns. In order to ensure the accuracy, this paper adopts chemical mechanical polishing (CMP) to remove the copper.

After the copper removal, the front graphics are carried out to interconnect TSV and the transmission lines on the front of the wafer. For TSV fabrication, the wafer size was selected to be eight inches and 750microns thick. The final processing is to obtain a 150 μm wafer, with TSV visible from the back of the wafer. After the thinning is completed, the back graphics can be processed. The flow diagram is shown in Figure 7.

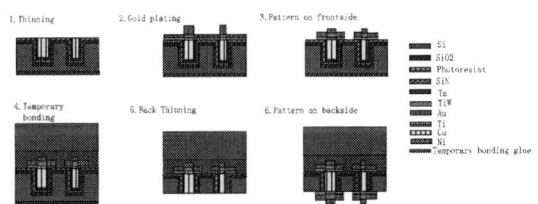

Figure 7 Process flow diagram before wafer bonding

D. Summaries

This chapter mainly introduces the key technologies in the antenna processing technology. First, it introduces the TSV technology, including its processing flow and related problems encountered in the processing. Then the details of wafer bonding technology and interlayer interconnection technology are introduced. The physical images of the three-layer structure of the antenna are shown in Figure 8. Figure 8 (a) is the ground plane and the input port of the antenna, Figure 8 (b) is the antenna power divider layer, Figure 8 (c) is the transfer layer, and Figure 8 (d) shows the radiating patch on the top of the antenna. Figure 9 is the cross view of the fabricated antenna.

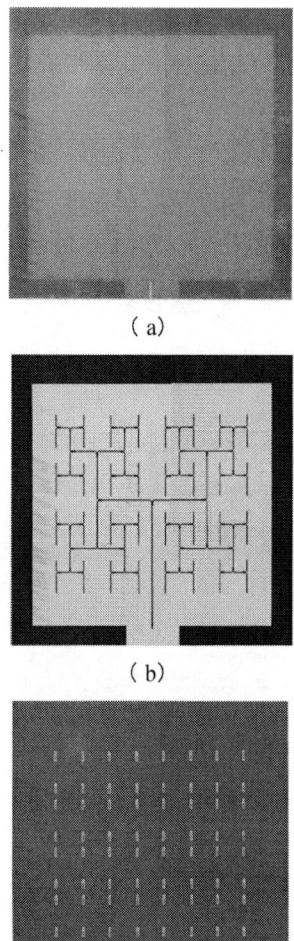

(a)

(b)

(c)

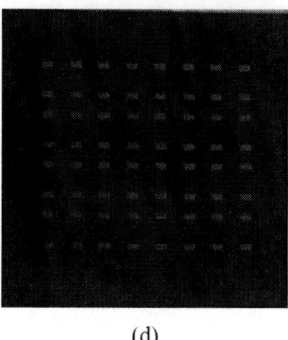

(d)

Figure 8 Prototype of the fabricated antenna. (a)The input port (b)The divider layer (c) The interposer layer (d)The patch layer

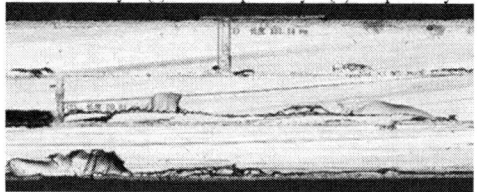

Figure 9 Cross view of the fabricated antenna

III. ANTENNA TEST RESULT

Figure 10 and Figure 11 show the simulated and measured results of the antenna. The measured results show the reflection coefficients bandwidth is 90.3-98.7GHz, and the gain in 95.8GHz is about 10dBi, indicating that the antenna is basically realized and the fabrication steps are feasible. Compared with antennas of other processes in w-band [6]-[9], antennas processed by three-dimensional silicon integration technology perform well in terms of size and gain, and have obvious advantages in subsequent system integration research. However, it can be seen from the results that the S11 exhibits broadband characteristics. The bandwidth is wider than the simulation result. Besides, it has a gap between the measured gain and the theoretical value, indicating that there is still room for improvement in the structure and process steps of the antenna.

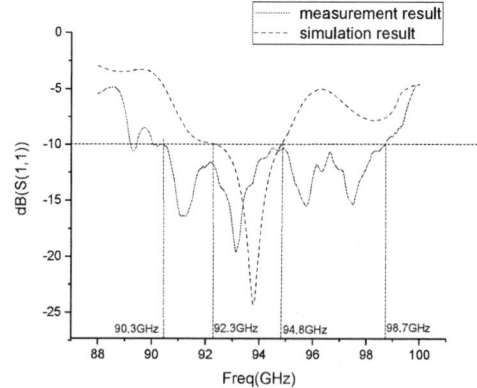

Figure 10 Measured frequency responses of the antenna

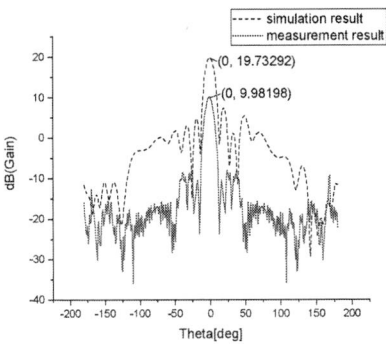

Figure 11 Simulation and test results of antenna E-plane pattern at95.8GHz

IV. CONCLUSION

In this paper, an antenna's fabrication flows based on the three-dimensional silicon integration technology was introduced. During the flows, many key challenging processes are developed such as, TSV fabrication technology with a height of $150\mu m$ and a diameter of $20\ \mu m$, multi-layer wafer stacking technology and inter-layer interconnection technology. The designed antenna was fabricated based the presented flows and measured. The measured results showed the radiation peek gain around 10dBi and the frequency band from 90.3 to 98.7GHz.

REFERENCES

[1] G. Y. Zhang and J. Mao, "An Overview of the Development of Antenna-in-Package Technology for Highly Integrated Wireless Devices," in Proceedings of the IEEE, vol. 107, no. 11, pp. 2265-2280, Nov. 2019, doi: 10.1109/JPROC.2019.2933267.

[2] C. Jin, V. N. Sekhar, X. Bao, B. Chen, B. Zheng and R. Li, "Antenna-in-Package Design Based on Wafer-Level Packaging With Through Silicon Via Technology," in IEEE Transactions on Components, Packaging and Manufacturing Technology, vol. 3, no. 9, pp. 1498-1505, Sept. 2013.

[3] Guo Qing Jiang,Lei Kuang,Jian Zhu. Research on TSV Dry Etching Technology. Key Engineering Materials,2015,3928(645).

[4] Elise Delbos,Laurent Omnès,Arnaud Etcheberry. Bottom-up filling optimization for efficient TSV metallization. Microelectronic Engineering,2009,87(3).

[5] Hongbin Xiao,Hu He,Xinyu Ren,Peng Zeng,Fuliang Wang. Numerical modeling and experimental verification of copper electrodeposition for through silicon via (TSV) with additives. Microelectronic Engineering,2017,170.

[6] J. S. Chieh, A. Pham, G. Kannell and A. Pidwerbetsky, "A W-Band 8 × 8 series fed patch array detector on Liquid Crystal Polymer," Proceedings of the 2012 IEEE International Symposium on Antennas and Propagation, Chicago, IL, 2012, pp. 1-2, doi: 10.1109/APS.2012.6348038.

[7] B. Cao, Y. Shi and W. Feng, "W-Band LTCC Circularly Polarized Antenna Array With Mixed U-Type Substrate Integrated Waveguide and Ridge Gap Waveguide Feeding Networks," in IEEE Antennas and Wireless Propagation Letters, vol. 18, no. 11, pp. 2399-2403, Nov. 2019, doi: 10.1109/LAWP.2019.2917774.

[8] S.Adamshick, A. Johnson, K. Moriarty, W. Tremblay and J. Burke, "Antenna on chip design utilizing 3D integration for mixed signal applications," 2017 IEEE 60th International Midwest Symposium on Circuits and Systems (MWSCAS), Boston, MA, 2017, pp. 209-212.

[9] V. Pano, I. Tekin, I. Yilmaz, Y. Liu, K. R. Dandekar and B. Taskin, "TSV Antennas for Multi-Band Wireless Communication," in

Gap in pagination due to unavailable papers.

Pages 1062-1064

Proceedings of the 16th Annual IEEE International
Conference on Nano/Micro Engineered and Molecular Systems
April 25-29, 2021

Influence of the Particle Size of Glass Powder on Sintering Characteristics in TGV Packaging

Yifang Liu[1], Hongbo Sang[1], Zhenxiang Bu[1], Yulong Zhang[2], Guangen Gao[3], and Lingyun Wang[1], *

Abstract— In the fabrication of TGV (through glass via) reflow process, nano-glass powder is used instead of glass substrate, and the structure is filled first and then heated to reflow, which can simplify the process steps and improve the filling effect. This paper represents the preparation of nano glass powder for TGV, and by studying the sintering characteristics of glass powders of different sizes at different temperatures, mixing particles before sintering can accelerate the melting of larger particles, reduce the requirements for grinding process, improve bonding strength and save costs. The silicon structure is filled with the mixed powder and thermally reflowed. The reflowed glass powder can completely fill the edges of the structure and is tightly combined with the silicon.

I. INTRODUCTION

In MEMS packaging, in order to ensure the stable interconnection and transmission of many complex signals, three-dimensional integration technology based on through silicon vias (TSV) has emerged [1]. However, the TSV process includes the fabrication of seed layer and insulating layer, and the complex via-refill process [2]. The free electrons around the via can cause parasitic effect by the electric or magnetic field, which will cause serious interference to nearby signals [3,4]. On the contrary, glass does not contain free electrons and has low dielectric loss. It has similar thermal expansion coefficient comparing with silicon [5]. Through glass vias (TGV) can be used instead of TSVs for packaging. Therefore, TGV technology has broad application prospects in three-dimensional packaging structures [6,7].

In the reflow method of manufacturing TGV, although the glass substrate reflow technology has made progress in stages, there are still many difficulties in the process that need to be overcome [8,9]. The glass substrate needs to be anodic bonded with the silicon wafer in high vacuum before reflowing [10,11], and the surface roughness of the bonding interface needs to be within 2nm. Besides, bubbles are easy to generate during the bonding, which is not good for the subsequent process. After reflowing, it will require a long time thinning and polishing to remove the excess glass, and the glass substrate is difficult to fill to the edge of the silicon structure during the reflow process, resulting in a cavity in the structure and affecting the sealing [12-14]. Anodic bonding, high-temperature reflowing, and mechanical thinning will all introduce residual stress, which will cause

excessive structural stress after reflowing, weaken the package strength, and even cause structural damage [15,16].

Here, we use nano-glass powder instead of glass sheet for filling and then reflow. Fig. 1 shows the process of glass powder reflow in TGV packaging. The feasibility is verified from the preparation and reflow of glass powder, and the method of sintering after mixing particles with different sizes is proposed. Thus, the bonding strength between glass and silicon wafer can be improved.

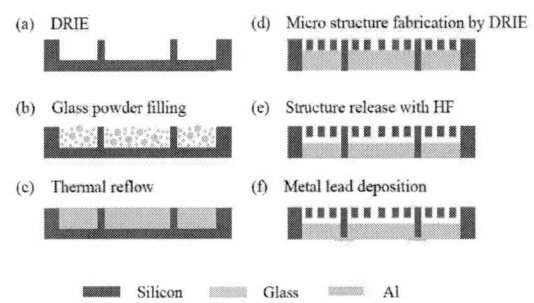

Figure 1. Fabrication of TGV packaging by glass powder reflow.

II. EXPERIMENTAL

In this experiment, borofloat 33 glass was selected as the raw material for the preparation of nano-glass powder. The high-speed ball grinding method was used to refine the glass powder. The factors that affect the particle size of glass powder in the grinding process include: the volume filling ratio of grinding tank, the weight ratio of grinding ball to the glass powder, grinding speed, and grinding time. Based on previous experience and the appropriate working conditions of the equipment (DECO-PBM-V-0.4L), it is determined that the best parameters suitable for the equipment are the volume filling rate of 15%, the weight ratio of 40:1, the grinding speed of 600r/min, and the grinding time of 18h[7]. The initial particles of glass powder are shown in Fig. 2a. Fig. 2b depicts the glass powder after grinding, which is a combination of small particles and large particles. Small particles agglomerate around large particles due to van der Waals and electrostatic forces. When filling the silicon

---

*Corresponding author: Lingyun Wang, Department of Mechanical and Electrical Engineering, Xiamen University, Xiamen, China, phone: 0592-218-5927; fax: 0592-218-5927; email: wangly@xmu.edu.cn.

1. Yifang Liu/ Hongbo Sang/ Zhenxiang Bu, Department of Mechanical and Electrical Engineering, Xiamen University, Xiamen, China.

2. Yulong Zhang, Pen-Tong Sah Institute of Micro-Nano Science and Technology, Xiamen University, Xiamen, China.

3. Guangen Gao, Aviation Key Laboratory of Science and Technology on Inertial Technology, FACRI, Xi'an, China.

978-1-6654-3008-1/21 $31.00 © 2021 IEEE

structure, smaller particles with a more even particle size distribution will get a better filling effect.

Figure 2. Particle graphs of (a) initial glass powder and (b) after grinding

However, it is not easy to prepare such small particles of uniform size. During the grinding process, large particles surrounded by small particles are more difficult to break due to the agglomeration effect. The breaking requires higher grinding speed and longer time until the equipment reached its working limit, which increases the cost of glass powder preparation.

III. SINTERING CHARACTERISTICS

From the sintering point of view, the free energy of the surface phase of the particles increases as the particle size decreases, so the powder melting point decreases as the particle size decreases. From this idea, we studied the sintering characteristics of powders of different particle sizes at different temperatures.

A. Single Size Particles

In order to obtain the glass powder with uniform particle size, the powder obtained by grinding in the previous step was dissolved in absolute ethanol. Using microporous membranes of different pore sizes for filtering, to obtain glass powder with four different size distributions: smaller than 1µm; 1-3µm; 3-5µm; larger than 5µm.

First, spread the nano-glass powder on a 10*10 mm silicon substrate and sinter in different temperature. Fig. 3 shows the SEM images of nano-glass powder at 600 °C, 800 °C, 1000 °C and 1100 °C. As the temperature rises, the glass powder gradually softens from the solid state and merges with other particles. The fusion phenomenon between the particles become significant, and the air between the particles gradually rises during the fusion, and finally escapes from the upper surface layer, leaving large traces of bubbles on the surface. At 1100 °C, the particles are all fused together, and the bubble traces on the surface become smaller. It indicates that the suitable temperature for sintering the glass powder is 1100 °C.

Second, spread large-particle powders of three different sizes on a 10*10 mm silicon substrate, and sinter according to the peak temperature of 1100 degrees and the holding time of 4 hours. The results are shown in Fig. 4. Under the same conditions, as the particle size increases, the sintering effect gradually becomes worse. Small particles have better fluidity after sintering. Although large particles are difficult to flow, the voids between the internal particles are smaller than those of small particles. To improve the bonding strength, it is necessary to minimize the voids between the powder particles. Therefore, compared with the use of

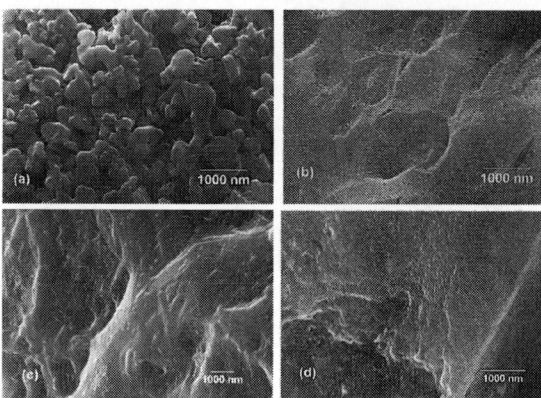

Figure 3. Particles smaller than 1µm at (a) 600°C, (b) 800°C, (c) 1000°C and (d) 1100°C.

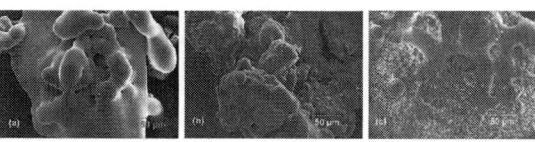

Figure 4. Larger particles at 1100°C; (a) Particles between 1µm and 3µm, (b) particles between 3µm and 5µm and (c) particles larger than 5µm.

single-size particles, the advantage of using mixed powder is that small particles with low melting point and good fluidity are used to make up for the defects of poor fluidity of larger ones, fill the voids between large particles, so as to accelerate the melting speed, reduce the requirements for the grinding process and save costs.

B. Mixed Particles of Different Sizes

In order to reduce the voids between particles, powders of different sizes are mixed. Theoretically, if powders of the same particle size are mixed, the void between the particles is the largest. It will become smaller when particles are composed of various sizes. Mixing the particles larger than 5um and particles smaller than 1um uniformly in proportion and then sintering. The specific parameters are given in TABLE I.

Fig. 5. shows the sintering state of particles with mass ratio of 2:1. Large particles are wrapped by molten small particles, forming a solid state similar to concrete as we supposed before. As the proportion of small particles increases, the distribution of large particles decreases in the solid formed after sintering.

TABLE I. SINTERING PARAMETERS

Mass ratio of large and small particles	Heating rate	Peak temperature	Holding time	Atmosphere
2:1	10°C/min	1100°C	4 hours	Air
1:1				
1:2				

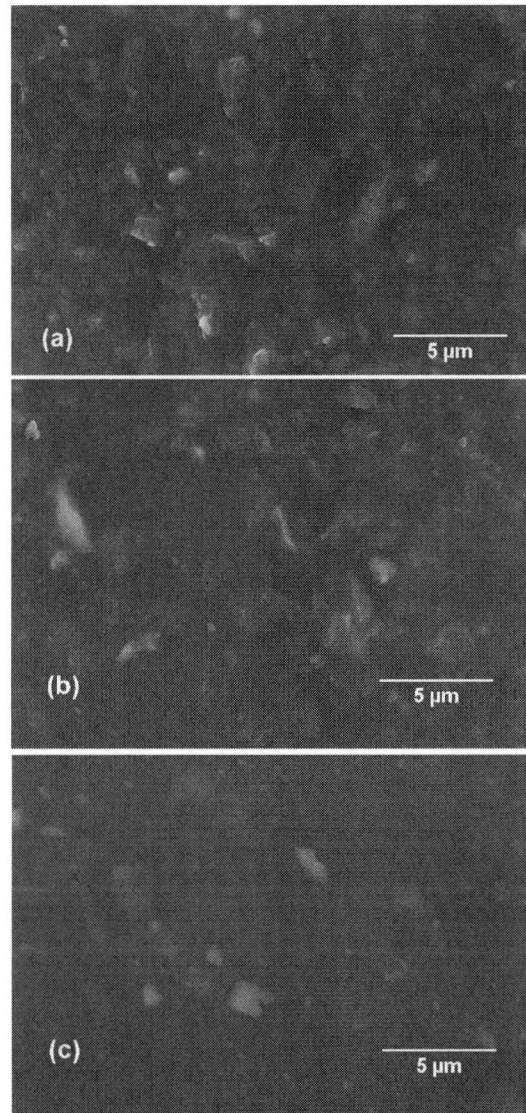

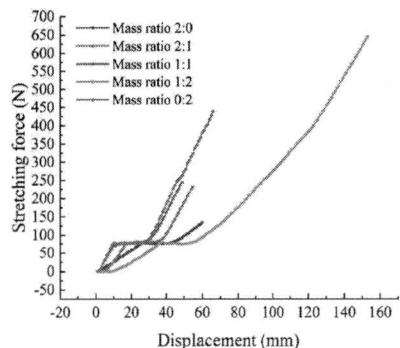

Figure 6. Stretching force vs. displacement curves of different mass ratios.

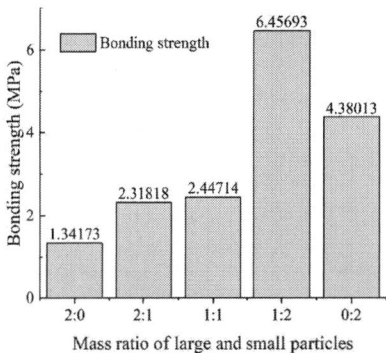

Figure 7. Bonding strenth results of different mass ratios.

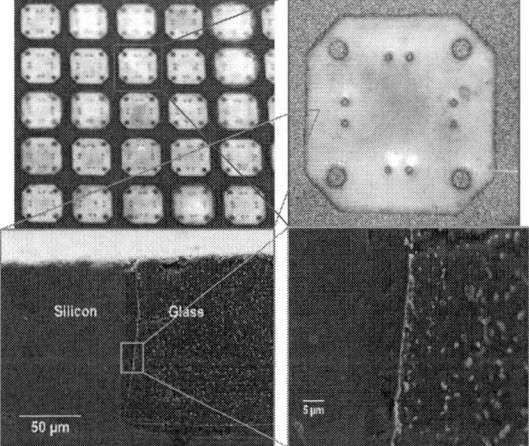

Figure 8. Silicon structure and its cross section after reflowing.

Figure 5. The sintering state of particles with mass ratio of (a)2:1, (b)1:1, and (c)1:2.

C. Bonding Strength Test and Filling

In order to explore the mixing ratio that can improve the bonding strength for packaging, several sets of bonding strength test were executed. Three groups of mixed particles with proportions set before are sandwiched between two 10*10mm silicon oxide samples and sintered. Add two sets of the largest and the smallest particles for contrast. The sintering parameters are the same as TABLE 1. Test results are shown in Fig. 6 and Fig. 7.

Theoretically, when 5um spherical particles are completely wrapped with 1um spherical particles, the mass ratio of large and small particles is about 1:1.838. Experiments have proved that when the mass ratio is 1:2, the glass powder can obtain greater bonding strength during sintering. This ratio is close to the value obtained by theoretical derivation, which proves the correctness of the viewpoint.

Fill the 150um deep silicon structure with the glass powder at a mass ratio of 1:2 for sintering. Fig. 8 graphs the silicon structure and cross section after reflowing. The mixed glass powder can completely fill the structure and be tightly combined with silicon.

IV. CONCLUSION

This article mainly presents the different sintering characteristics of glass powder. By sintering mixed particles of different sizes, the void between particles is reduced and the bonding strength is improved. With the 1:2 mass ratio of large and small particles mixing, the bonding strength can reach to 6.46MPa. The reflowed glass powder can completely fill the edges of the structure and is tightly combined with the silicon, which manifests that the glass powder reflow technology is a reliable option for TGV manufacture.

ACKNOWLEDGMENT

This work was financially supported by The National Natural Science Foundation of China (No.61674125), Aviation Science funds (Aviation Key Laboratory of Science and Technology on Inertia: 20180868001) and The Industry-University Research Collaboration of Xiamen (No. 3502Z20203001).

REFERENCES

[1] Jan Vardaman, E. 3D through- silicon via technology markets and applications. In Three Dimensional System Integration; Springer: Basel, Switzerland, 2011; pp. 237–242.

[2] M. J. Laakso et al., "Through-Glass Vias for Glass Interposers and MEMS Packaging Applications Fabricated Using Magnetic Assembly of Microscale Metal Wires," Ieee Access, vol. 6, pp. 44306-44317, 2018.

[3] W. Li, "Characterization of signal transfer performance of a through glass via (TGV) substrate with silicon vertical feedthroughs", Microelectronic Engineering, vol.165, pp. 52-56, 2016.

[4] R. M. Haque, K. D. Wise, "A Glass-in-Silicon Reflow Process for Three-Dimensional Microsystems", Journal of Microelectromechanical Systems, vol. 22, no. 6, pp. 1470-1477, 2013.

[5] U. Shah, J. Liljeholm, J. Campion, T. Ebefors, and J. Oberhammer, "Low-Loss, High-Linearity RF Interposers Enabled by Through Glass Vias," Ieee Microwave and Wireless Components Letters, vol. 28, no. 11, pp. 960-962, Nov 2018.

[6] K. Salah, TGV versus TSV: A Comparative Analysis (2016 3rd International Conference on Advances in Computational Tools for Engineering Applications). 2016, pp. 49-53.

[7] T. C. Lee et al., "Glass Based 3D-IPD Integrated RF ASIC in WLCSP," in 2017 ieee 67th Electronic Components and Technology Conference(Electronic Components and Technology Conference, 2017, pp. 631-636.

[8] S.Tanaka, "Wafer-level hermetic MEMS packaging by anodic bonding and its reliability issues", Microelectronics Reliability, 54(2014), pp. 875-881..

[9] A. Benali, M. Faqir, M. Bouya, A. Benabdellah, and M. Ghogho, "Analytical and finite element modeling of through glass via thermal stress," Microelectronic Engineering, vol. 151, pp. 12-18, Feb 2016.

[10] Y. Kuang D X, J. Zhou, M. Zhuo, W. Li, Z. Hou, H. Cui, and X. Wu, Theoretical model of glass reflow process for through glass via (TGV) wafer fabrication[J]. Journal of Micromechanics and Microengineering, 2018, 28(9):095004.

[11] Kuang Y, Xiao D, Zhou J, et al. Enhancing airtightness of TGV through regulating interface energy for wafer-level vacuum packaging[J]. Microsystem Technologies, 2018, 24(9):3645–3649.

[12] X. H. Du, S. Liu, M. J. Zhu, and Ieee, Research on Preparation Technology of Nano Glass-Particles for High Density TGV Application (Icept2019: The 2019 20th International Conference on Electronic Packaging Technology). 2019.

[13] W. Y. Li et al., "A new fabrication process of TGV substrate with silicon vertical feedthroughs using double sided glass in silicon reflow process," Journal of Materials Science-Materials in Electronics, vol. 28, no. 4, pp. 3917-3923, Feb 2017.

[14] Y. B. Kuang et al., "THEORETICAL ANALYSIS AND EXPERIMENTAL VERIFICATION FOR 3D TGV PACKAGING TECHNOLOGY," in 2018 Ieee Micro Electro Mechanical Systems(Proceedings IEEE Micro Electro Mechanical Systems, 2018, pp. 559-562.

[15] F. Yang et al., "Research on Wafer-Level MEMS Packaging with Through-Glass Vias," Micromachines, vol. 10, no. 1, Jan 2019, Art. no. 15.

[16] S. Yazdi, M. Garavaglia, A. Ghisi, A. Corigliano, and Ieee, Modelling and Simulation of Glass Frit Bonding of Silicon Wafers (2019 20th International Conference on Thermal, Mechanical and Multi-Physics Simulation and Experiments in Microelectronics and Microsystems). 2019.

Gap in pagination due to unavailable paper.

Page 1069

Proceedings of the 16th Annual IEEE International
Conference on Nano/Micro Engineered and Molecular Systems
April 25-29, 2021

An Electromagnetic-Piezoelectric-Triboelectric Hybridized Energy Harvester Towards Blue Energy*

Yunfei Li[1,2], Tianyi Tang[2], Manjuan Huang[2], Xin Ma[3], Zhaohui Chen[3], Peijuan Cui[4], Huicong Liu[2*], *Member IEEE*, and Lining Sun[1,2], *Member IEEE*

Abstract—This paper proposed a hybridized energy harvester combining electromagnetic, piezoelectric, and triboelectric transduction mechanisms for ocean wave energy harvesting. Two energy harvesting mechanisms of wave shaking and heaving are contained in this device. The shaking energy harvesting mechanism (S-EHM) consists of a compound pendulum mechanism and a gear mechanism. Together with an electromagnetic generator (EMG) and a piezoelectric generator (PEG) are used to harvest the shaking energy of ocean waves. The heaving energy harvesting mechanism (H-EHM) consists of a sliding sub, mass slider and springs. Meanwhile, the triboelectric nanogenerator (TENG) is used to harvest the undulations energy of ocean waves. They all supply energy to the electrical appliances inside the marine buoy, which can extend its working life. The diameter of the harvester is 140 mm and the height is 185 mm. The tested maximum output voltages of the electromagnetic, piezoelectric and triboelectric parts are 12V, 18V, and 55 V, respectively. Moreover, the device can light up at least 90 LEDs and can power the temperature and humidity sensor to display the ambient temperature.

Keywords—Ocean waves; Hybridized energy harvester; Self-powered wireless sensor system.

I. INTRODUCTION

The ocean covers 78% of the Earth which contains abundant renewable and clean energy. Blue energy, mainly in the form of ocean waves, has high potential to promote the sustainable development of resource [1,2]. Due to the low frequency and random nature of waves, the effective harvesting of wave energy and its conversion into electricity is a considerable challenge [3,4]. Ocean wave energy harvesting technology based on triboelectric [5,6], piezoelectric [7] and electromagnetic [8, 9, 10] mechanisms has received significant attention. In order to effectively harvest the low frequency motion of the waves, Tao et al. [11] designed an origami-inspired TENG that can store and convert the wave energy with relatively high power density. However, it only uses the triboelectric transduction mechanisms which does not harvest the energy of the ultra-low frequency waves very well. Liu et al. [12] and Chen et al. [13] have each designed an electromagnetic-triboelectric hybridized energy harvester using a pendulum mechanism. However, this design just can harvest wave energy in one single swing direction. Therefore, the output power is not high. Chen et al. [14] has developed a triboelectric

nanogenerator which harvests energy from the different directions of seawater movement. It needs to work in contact with seawater at all times, so it is not easy to power the buoy sensors. In this paper, we proposed an electromagnetic-piezoelectric-triboelectric hybridized energy harvester, which can be integrated into ocean buoys to power sensors. Owing to the high performance of this harvester in complex conditions, the service life of the marine buoy can be extended.

II. DESIGN OF THE GENERATOR

Figure 1(a) shows the structural schematic of the hybridized ocean-wave energy harvester, which is combined electromagnetic generator (EMG), piezoelectric generator (PEG) and triboelectric nanogenerator (TENG). With the diameter of 140 mm, and height of 185mm, the hybridized energy harvester has two energy harvesting mechanisms, shaking energy harvesting mechanism (S-EHM) and heaving energy harvesting mechanism (H-EHM). The photograph of the hybridized energy harvester is shown in Figure 1(b).

Figure 2(a) shows the EMG, which consists of three parts: coil, iron cores and magnets. The S-EHM with integrated EMG component is shown in Figure 2(b). The mechanism has a gear increaser consisting of a main gear and a secondary gear for increasing the output voltage of the EMG component. Three pendulum arms wtih different number of pendulums are used to adjust the natural frequency of the S-EHM to obtain more wave energy. When

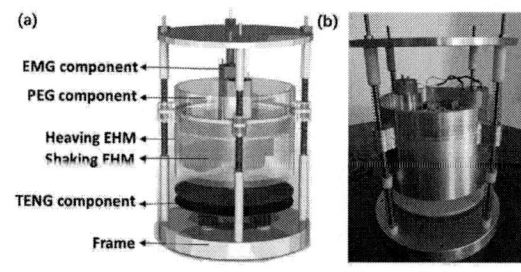

Figure 1 (a) Structural schematic and (b) photograph of a hybridized energy harvester.

---

*Resrach supported by ABC Foundation.

[1]State Key Laboratory of Robotics and System, Harbin Institute of Technology (HIT), Harbin 150001, China;

[2]School of Mechanical and Electric Engineering, Jiangsu Provincial Key Laboratory of Advanced Robotics, Soochow University, Suzhou, 215123, China;

[3]Physical Oceanography Laboratory, Ocean University of China, Qingdao 266100, China.

[4]Beijing Institute of Precision Mechatronic and Controls, Beijing 100076, China.

Contacting author: Huicong Liu (e-mail: hcliu078@suda.edu.cn).

978-1-6654-3008-1/21 $31.00 © 2021 IEEE

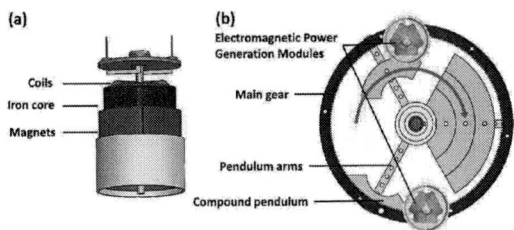

Figure 2 (a) Structure of the EMG component. (b) Structure of the S-EHM.

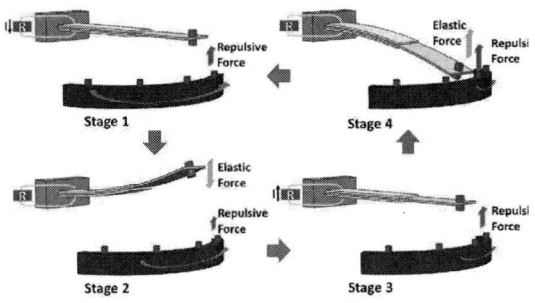

Figure 3 Working mechanism of the PEG component for different magnet positions.

the ocean waves swing the device at a low frequency, the natural frequency of the S-EHM is basically the same as the excitation frequency by adjusting the number of pendulums on three pendulum arms and the distance of pendulums to the rotation axis, resulting the pendulum swings or even rotates at high speed. The acquired energy is transferred to the EMG through the gear system, and the magnetic flux of the coil changes rapidly, which will make a large power output.

Figure 3 shows the working mechanism of the PEG component which consists of a copper-based piezoelectric thick film cantilever, a magnet on the end of the cantilever and several magnets fixed to the main gear. When the gear rotates, the change in the distance between the magnet on the gear and the magnet on the cantilever cause a periodic repulsive force, which drive the PEG to vibrate and generate voltage. When the gear speed is stable, the period of the repulsive force excited on the piezoelectric cantilever can be varied by changing the number of magnets evenly distributed on the gear. When the period of the repulsive force is the same as the natural frequency of the cantilever, the cantilever produces a large movement due to resonance, and the output performance of the piezoelectric is greatly increased. In addition, even if the wave cycle changes and the gear does not rotate, PEG can still generate power through the deformation caused by the heave of the magnet.

Figure 4 shows the working mechanism of TENG component. It consists of an upper electrode made of aluminum foil, a lower electrode made of copper foil, a friction layer made of polyimide and substrates made of

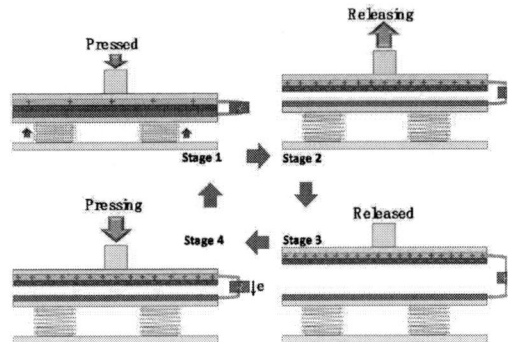

Figure 4 Working mechanism of the TENG component based on contact and separation.

polylactic acid (PLA). The lower electrode and the substrate are supported on the frame by springs. The upper electrode, the friction layer, and another substrate are fixed to the H-EHM. The H-EHM consists of four main parts: the springs, the guide track, the slider block and the frame. The slider block, which contains the mass of the S-EHM, is mounted on the guide tracks by using linear bearings. The H-EHM can move in a linear reciprocating motion under the heave of the ocean waves. When the waves are of a high frequency or high amplitude, the springs mounted on the top and bottom of the guide track can provide protection for the slider block. Also, the springs act as energy storage elements to make the slider move more regularly. With the reciprocating movement of the slider block on the H-EHM, the friction layer and the lower electrode produce periodic contact and separation movement. The directional reciprocating movement of electrons generates triboelectricity and provides energy to the load.

III. EXPERIMENTAL RESULTS

Figure 5 is the experimental set-up for testing the output voltage of the hybrid energy harvester. The 6-degree-of-freedom (6-DOF) motion platform is used to simulate different periods of ocean waves by adjusting the acceleration of the platform motion and excite the hybridized energy harvester. The frequency and amplitude of the 6-DOF motion platform are controlled by the upper computer. When the hybridized energy harvester is excited by the 6-DOF motion platform, two energy harvesting mechanisms couples together and drive the EMG, PEG and TENG to generate energy. The multi-channel oscilloscope is responsible for monitoring the output voltage of the harvester.

The output voltage waveforms of the EMG, PEG and TENG of the device at a period of 2s and an amplitude of 20cm are shown in Figure 6 (a-c). The peak open-circuit voltage of the EMG, PEG, and TENG components reach about 12V, 18V, and 55V, respectively. Since the different energy capture methods and frequency upscaling mechanism, the EMG has the widest voltage pulse width (Figure 6 (a)), the TENG has the highest output voltage (Figure 6 (b)), while the PEG component has the highest frequency (Figure 6 (c)).

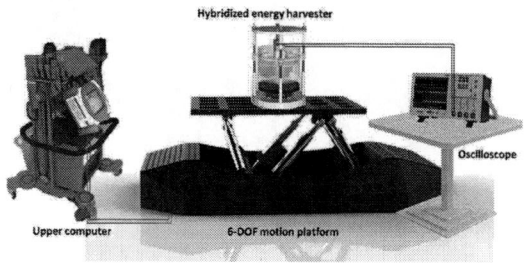

Figure 5 Schematic diagram of a 6-DOF motion platform driving a hybrid energy harvester.

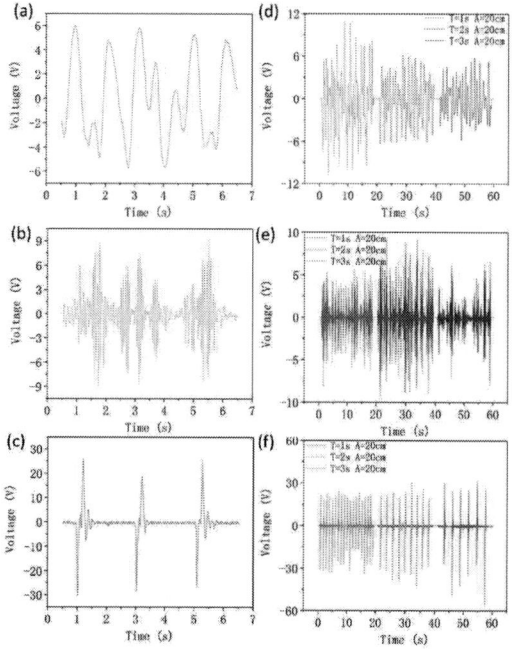

Figure 6 (a-c) Open circuit voltage output for EMG, PEG and TENG under a 20 cm amplitude wave within a period of 2 s. (d-f) Open circuit voltage output for EMG, PEG and TENG under a 20 cm amplitude wave within different periods.

The output voltage waveforms of the device for EMG, PEG and TENG components under different excitation periods of 1s, 2s, and 3s are shown in Figure 6 (d-f). It can be seen that the response of different power generation components to different frequency excitations is different. The EMG component has better response under high frequency excitation due to higher rotor speed (Figure 6 (d)). The TENG component produces higher voltage under low frequency excitation. The TENG produces higher voltages under low frequency excitation (Figure 6 (e)). Due to the motion range of H-EHM is greater and the instantaneous force of the friction layer colliding with the electrode is much greater. The PEG component has a better output performance with a wave period of 2 s (Figure 6 (f)). Science the gear speed is more beneficial to the resonance of the cantilever beam at this period.

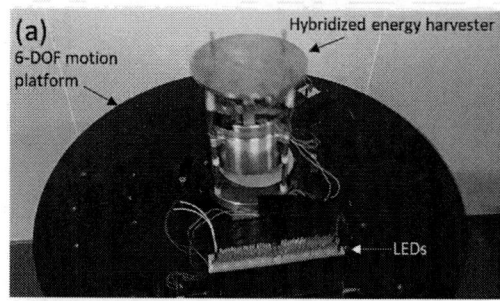

Figure 7 Hybridized energy harvester on the 6-DOF motion platform lights up 90 LEDs.

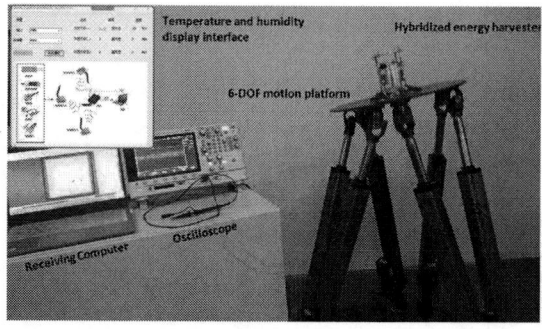

Figure 8 Hybridized energy harvester on the 6-DOF motion platform powers a temperature and humidity sensor.

Two application tests on the hybridized energy harvester have been completed using the 6-DOF motion platform. The first experiment was carried out with the device lighting up the LEDs. As shown in Figure 7(a), the energy harvester was fixed on the 6-DOF motion platform, and the generators of the device were connected to the LED array. The motion period of the platform was set to 2 s and the amplitude was 20 cm. In Figure 7(b), it can be evident observed that the device could power 90 LEDs and the light emitted from the LEDs is very bright during the experiment.

Figure 8 shows that the hybridized energy harvester is able to power the temperature and humidity sensor. The temperature and humidity sensor with the Bluetooth module is connected to the generators of the device via a capacitor and rectifier bridge chip. It sends information about the environmental temperature and humidity to the outside world when the supply voltage reaches a threshold. The receiving computer turns on its own Bluetooth module and starts the test software, ready to receive the signal from the temperature and humidity sensor and display it on the interface of the test software. In this experiment, after the harvester working for 6 minutes, the sensing signal of the environmental temperature and humidity is wirelessly transmitted and displayed on the computer.

IV. CONCLUSION

This paper proposed a hybridized energy harvester combining electromagnetic, piezoelectric and triboelectric transduction mechanisms for ocean wave energy harvesting. The device contains two energy harvesting mechanisms of wave shaking and heaving. The diameter of the harvester is 140 mm and the height is 185 mm. The tested maximum output voltages of the electromagnetic, piezoelectric and triboelectric parts are 12, 18 and 55 V, respectively. In experiments, the device can make 90 LED lights very bright. And it is able to power the temperature and humidity sensor. When the temperature and humidity sensor starts working, the values of temperature and humidity in the environment are displayed in the computer software via the Bluetooth module.

ACKNOWLEDGMENT

This work is funded by the National Science Foundation of China (Grant No. 51875377).

REFERENCES

[1] XL. Cheng, W. Tang, and Y. Song, "Power management and effective energy storage of pulsed output from triboelectric nanogenerator – ScienceDirect," *Nano Energy,* vol. 61, 2019, pp.517-532.

[2] X. Chen, L. Gao, and J. Chen, "A chaotic pendulum triboelectric-electromagnetic hybridized nanogenerator for wave energy scavenging and self-powered wireless sensing system, " *Nano Energy,* vol. 69, 2020, pp.104440.

[3] Z. L. Wang, T. Jiang, and L. Xu, "Toward the blue energy dream by triboelectric nanogenerator networks," *Nano Energy,* vol. 39, 2017, pp. 9-23.

[4] Z. Wen, H. Guo, Y. Zi, "Harvesting broad frequency band blue energy by a triboelectric–electromagnetic hybrid nanogenerator," *ACS nano,* vol. 10,2016, pp. 6526-6534.

[5] P. Cheng, H. Guo, Z. Wen, C. Zhang, X. Yin, X. Li, D. Liu, W. Song, X. Sun, J. Wang, Z.L. Wang, "Largely enhanced triboelectric nanogenerator for efficient harvesting of water wave energy by soft contacted structure," *Nano Energy,* vol. 57, 2019, pp. 432-9.

[6] Y. Wu, Q. Zeng, Q. Tang, W. Liu, G. Liu, Y. Zhang, "A teeterboard-like hybrid nanogenerator for efficient harvesting of low-frequency ocean wave energy," *Nano Energy,* vol. 67, 2020, pp. 104205.

[7] W.S. Hwang, J.H. Ahn, S.Y. Jeong, H.J. Jung, S.K. Hong, J.Y. Choi, J.Y. Cho, J.H. Kim, T.H. Sung, "Design of piezoelectric ocean-wave energy harvester using sway movement," *Sensors and Actuators A: Physical,* vol. 260, 2017, pp. 191-7.

[8] Y. Li, Q. Guo, M. Huang, X. Ma, Z. Chen, H. Liu, L. Sun, "Study of an electromagnetic ocean wave energy harvester driven by an efficient swing body toward the self-powered ocean buoy application," *IEEE Access,* vol. 7, 2019, pp. 129758-129769.

[9] H. Liu, T. Chen, L. Sun, and C. Lee, "An electromagnetic MEMS energy harvester array with multiple vibration modes," Micromachines, vol. 6, 2015, pp. 984-992.

[10] B. C. Boren, B. A. Batten, R. K. Paasch, "Active control of a vertical axis pendulum wave energy converter," IEEE, American Control Conference (ACC), 2014.

[11] K. Tao, H. Yi, Y. Yang, H. Chang, J. Wu, L. Tang, Z. Yang, N. Wang, L. Hu, Y. Fu, J. Miao, W. Yuan, "Origami-inspired electret-based triboelectric generator for biomechanical and ocean wave energy harvesting, " *Nano Energy,* vol. 67, 2020, pp. 104197.

[12] L. Liu, Q. Shi, C. Lee, "A novel hybridized blue energy harvester aiming at all-weather IoT applications, " *Nano Energy,* vol. 76, 2020, pp. 105052.

[13] X. Chen, L. Gao, J. Chen, S. Lu, H. Zhou, T. Wang, A. Wang, Z. Zhang, S. Guo, X. Mu, Z.L. Wang, Y. Yang, "A chaotic pendulum triboelectric-electromagnetic hybridized nanogenerator for wave energy scavenging and self-powered wireless sensing system, " *Nano Energy,* vol. 69, 2020, pp. 104440.

[14] B.D. Chen, W. Tang, C. He, C.R. Deng, L.J. Yang, L.P. Zhu, J. Chen, J.J. Shao, L. Liu, Z.L. Wang, "Water wave energy harvesting and self-powered liquid-surface fluctuation sensing based on bionic-jellyfish triboelectric nanogenerator, " *Materials Today,* vol. 21, 2018, pp.88-97.

Internal Resonant Oscillation in Coupled Resonators for High-resolution Mass Sensing with A Wider Coupling Range

Nanxing Li, Cao Xia, Guowen Zheng, Xu Du, Dong F. Wang, *Member, IEEE*

Abstract— **This paper, mainly reports a coupled cantilever structure applicable to mass or gas sensing applications with a higher resolution in a wider coupling range. Based on the coupled cantilevers with different rectangle geometry, we design a π-shaped coupled cantilever structure with triple frequency ratio, which provides mass or gas sensing with larger adsorption area. In the present work, internal resonance characteristics of the proposed structure are studied with different mass perturbations, and the coupling tongue characteristics with respect to the coupling range are further evaluated. Compared to that of the rectangular coupled cantilevers, on the one hand, a wider coupling range under the same applied voltage can be obtained, while on the other hand, a higher proportion of the third harmonic indicates a better resolution. The above preliminarily reveals the potential superiority of the internal resonant oscillation observed in the proposed the π-shaped coupled cantilever structure when applied to mass or gas sensing.**

Key words: internal resonant oscillation, π-shaped coupled cantilever structure, mass perturbation, coupling range, resolution, mass or gas sensing applications.

I. INTRODUCTION

In recent years, the nonlinear dynamic characteristics of cantilever-based oscillations have attracted much attention. The research of nonlinear oscillations mainly involves internal resonance, synchronous resonance and super harmonic resonance. Among them, because of the superior performance of internal resonance in detection performance and applied range, it has aroused great research enthusiasm.

If the resonance frequencies of two coupled vibration modes satisfy or approach an integer ratio relation, the driving force reaches the driving threshold, and the driving frequency approaches the resonance frequency of the low frequency vibration mode, nonlinear mode coupling and energy transfer between the two modes will become more intense, which is called internal resonance [1-6]. With the development of internal resonance research in recent years, the frequency ratio between the driven mode and internal resonance response mode can be 1:1 [7-8], 1:2 [7,9], 1:3 [2-5,10-11], 2:1 [12-17],3:1 [15,18], and even 23:1 [19].

Internal resonance is a kind of strong coupling, which is different from synchronous resonance with weak coupling. It can enhance the mode coupling and is accompanied by strong nonlinear energy transfer. Nonlinear mode coupling between different vibration modes is frequently observed in MEMS vibration structures, which is considered to have the potential to improve the performance of oscillations [20-22]. Therefore, researchers use the enhancement effect of internal resonance on mode coupling to study the effect of internal resonance on internal harmonic of vibration modes, and find that the response of higher harmonic of high frequency vibration mode in coupling modes can be enhanced through internal resonance [3,4,23]. This indicates that internal resonance is helpful to improve the sensing resolution when the resonator is used for mass or gas sensing.

In 2018, we study internal resonance in coupled resonator system with a frequency ratio of 1:3 for the first time. The occurrence of internal resonance is verified in the experiment using coupled copper cantilevers. The mass detecting mechanism is put forward to realize frequency shift multiplication and thus sensitivity enhancement in the coupled resonator system [2]. In 2019, we study internal resonance characteristics of the rectangular coupled cantilevers, and demonstrated that the rectangular coupled cantilevers can be used for mass sensing when internal resonance occurred. We defined internal resonance coupling range which is believed to be more suitable for mass sensing under high energy dissipation circumstance and studied the characteristics of internal resonance coupling tongues from the viewpoint of mass sensing [3], and proposed a mass sensitivity amplification scheme [4]. In 2020, we propose a new mass multi-warning scheme in a duffing nonlinear coupled system with 1:3 internal resonance, which can achieve the sensing, the first time warning and the second time warning with the increment of mass perturbations [5].

In recent years, we have a preliminary understanding of internal resonance, especially for the rectangular coupled cantilevers. And the mass sensing characteristics of rectangular coupled cantilevers are studied deeply. Although the sensitivity and resolution of mass sensing can be improved by internal resonance in the rectangular coupled

---

N. Li is with Micro Engineering and Micro Systems Laboratory (JML), School of Mechanical and Aerospace Engineering, Jilin University, Changchun 130022, China (e-mail: linx19@mails.jlu.edu.cn)

C. Xia is with Micro Engineering and Micro Systems Laboratory (JML), School of Mechanical and Aerospace Engineering, Jilin University, Changchun 130022, China (e-mail: xiacao19@mails.jlu.edu.cn)

G. Zheng is with Micro Engineering and Micro Systems Laboratory (JML), School of Mechanical and Aerospace Engineering, Jilin University, Changchun 130022, China (e-mail: zhenggw16@mails.jlu.edu.cn)

X. Du is with Micro Engineering and Micro Systems Laboratory (JML), School of Mechanical and Aerospace Engineering, Jilin University,

Changchun 130022, China. He is also with Department of Micro-Nano Mechanical Science and Engineering, Nagoya Univerisity, Furo-cho, Chikusa-ku, Nagoya, Aichi, 464-8603, Japan (e-mail: duxu15@mails.jlu.edu.cn).

D. F. Wang is with Micro Engineering and Micro Systems Laboratory (JML), School of Mechanical and Aerospace Engineering, Jilin University, Changchun, 130022 China. He is also with Research Center for Ubiquitous MEMS and Micro Engineering, National Institute of Advanced Industrial Science and Technology (AIST), Tsukuba, 305-8564 Japan (principal corresponding author: +86-(0)431-8509-4698; fax: +86-(0)431-8509-4698; e-mail: dongfwang@jlu.edu.cn).

cantilever system, the coupling range is narrower and the adsorption area is smaller when the rectangular coupled cantilevers are used for mass sensing. Therefore, the mass sensing performance needs to be further improved.

In this paper, internal resonance characteristics of the π-shaped coupled cantilevers with triple frequency ratio are studied, we prove the better performance of the π-shaped coupled cantilevers in mass sensing, and compared with the rectangular coupled cantilevers. Because this paper mainly studies the mass sensing characteristics of the π-shaped coupled cantilevers, and the comparison between the two coupled systems is limited to the coupling range and the proportion of the third harmonic in high frequency cantilever, its purpose is to highlight the better characteristics of π-shaped coupled cantilevers in the coupling range and sensing resolution. Therefore, due to the limitation of experimental conditions and the mass sensing characteristics of rectangular coupled cantilevers have been studied in detail in previous work, this paper does not do the experiment for the rectangular coupled cantilevers again.

II. Working Principle

A. Internal Resonance

Physically, we design a coupled cantilever system consists of low frequency cantilever (LFC) and high frequency cantilever (HFC), where the resonant frequency of LFC is w_1 and that of HFC is w_2, if w_1 and w_2 satisfy the relation: $w_2=nw_1$ (n is integer), internal resonance will be activated, when the driving force reaches a certain value and the driving frequency is close to w_1. And if a small perturbation causes that w_1 slightly shifts to $w_1^{'}=w_1+\Delta w$, w_2 will shift correspondingly to $w_2^{'}=nw_1^{'}=w_2+n\,\Delta w$. This means that the frequency shift can be multiplied by a factor of n if a mechanically-coupled system is constructed for two independent resonators. The characteristic provides theoretical support for improving sensitivity in mass or gas sensing applications.

B. Mass or Gas Detection

The working principle of mass or gas sensor based on cantilever is that the resonance frequency of cantilever will shift when the measured mass or gas is adsorbed on the cantilever. Mass or gas detection is realized indirectly by measuring frequency shift in the coupled cantilever system. The relationship between effective mass and resonance frequency is: $m=kw^{-2}$, k is linear coefficient. When the small mass perturbation Δm is applied to the cantilever, the resonant frequency w shifts to $w^{'}$. Consequently, we can get that: $\Delta m=kw^{'-2}-kw^{-2}$.

III. Experimental Results And Analysis

In this paper, we designed the π-shaped coupled cantilevers with triple frequency ratio by COMSOL, based on increasing adsorption area, as shown in Fig. 1. In the coupled system, the π-shaped cantilever with a low resonant frequency is for sensing, the rectangular cantilever with a high resonant frequency is for detecting. The coupled structure is made of brass, the real resonance frequency is different from the design, due to the manufacturing error. In

fact, the high resonant frequency is about 389 Hz and the low resonant frequency is about 130 Hz.

To study mass sensing characteristic of π-shaped coupled cantilevers, we use the instruments for the experiment, as shown in Fig. 2.

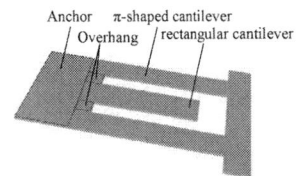

Figure 1. The π-shaped coupled cantilevers with triple frequency ratio.

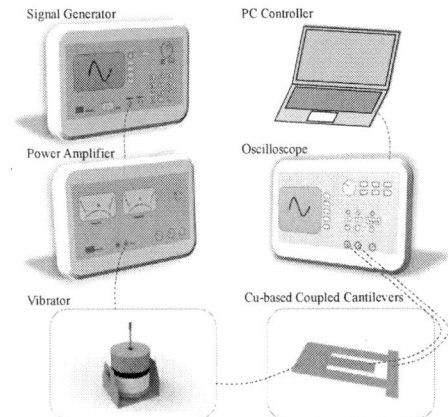

Figure 2. The diagram of experimental instruments.

Actually, we have known that there is more than one harmonic component in the output signal of a single cantilever in a coupled system, and in the Duffing nonlinear system, the main harmonic components are the first harmonic and the third harmonic. In the π-shaped coupled cantilever system, the mass detection is realized by detecting the third harmonic in the output signal of HFC. Normally, the amplitude of the third harmonic is lower than the first harmonic. However, when internal resonance occurs, the amplitude of the third harmonic in HFC will be enhanced because of the energy transfer between driving mode and response mode, so that it could be higher than the first harmonic, which is good for improving the sensing resolution when HFC is for detecting. Thus, we studied the harmonic components of the output signal in HFC.

In order to obtain a range where the third harmonic component is higher, we get the coupling range by increasing driving frequency under the specific applied voltage, the method for determining the boundary frequency of coupling range is the same as the rectangular coupled cantilevers, as shown in Fig. 3. In Fig. 3, Fig.3a and Fig.3e are in the uncoupling range, where the amplitude of the third harmonic is lower. From Fig.3b to Fig.3d are in the coupling range, where the amplitude of the third harmonic is higher than or equal to the first harmonic.

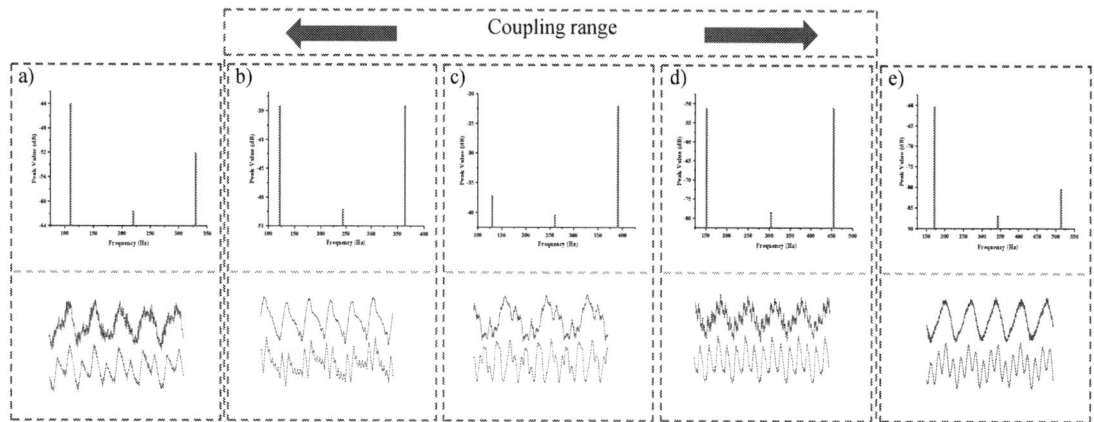

Figure 3. The coupling range in π-shaped coupled cantilevers. The above graphs are the FFT of output signal in HFC under gradually increased driving frequency. The below graphs are output waveforms of the coupled cantilevers corresponding to above graphs respectively. The LFC waveforms are in blue and the HFC waveforms are in red. They have the frequency ratio of 1:3 in the coupling range. The boundary frequency of coupling range could be acquired when the amplitude ratio is down to 1:1.

From Fig. 3, the amplitude of the second harmonic is much smaller than the first harmonic and the third harmonic. In order to express the coupling range more clearly, we only consider the relationship between the amplitude of the first harmonic and the third harmonic, as shown in Fig. 4. In Fig. 4, point A and point B represent the boundary frequency of coupling range, and the coupling range is between the two points.

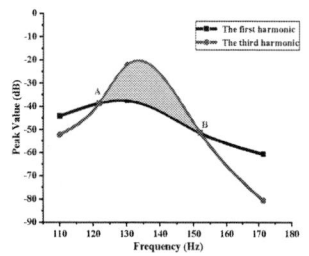

Figure 4. The coupling range in π-shaped coupled cantilevers.

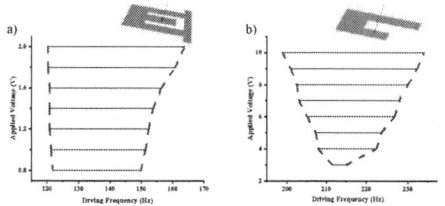

Figure 5. Variation of the coupling range with applied voltage. (a). π-shaped coupled cantilevers. (b). rectangle coupled cantilevers.

The coupling range affects the detectable range when the π-shaped coupled cantilever system is used for mass or gas sensing, and exploring the influence factors of coupling range is conducive to increasing the detectable range and improving the sensing performance. Consequently, using the same method, we get a coupling tongue via changing applied voltage, as shown in Fig. 5.

From Fig. 5, the change trend of coupling range in the π-shaped coupled cantilevers is the same as in the rectangular coupled cantilevers. It indicates that increasing the applied voltage could widen the coupling range. However, due to the limitations of the existing experimental equipment, when the applied voltage is lower than 0.8V or higher than 2V, the vibration of the exciter is intense, which results that vibration of the coupled cantilevers is unstable, so that the measurement of the coupling range boundary frequency cannot be continued. Therefore, only 0.8V to 2V complete experimental data are obtained. Although the range of applied voltage of the two coupled systems is different due to the limitation of experimental conditions, which makes the comparison a little more difficult. But from Fig. 5, in the sight of coupling range, the coupling range in π-shaped coupled cantilevers is significantly larger than rectangular coupled cantilevers, moreover, the coupling range increases with the increase of applied voltage, which lays a foundation for the comparison between the two coupled systems.

However, that is inaccurate, if ignoring applied voltage and eigen frequency. Therefore, we use relative coupling range, a dimensionless parameter, as a standard to compare, as shown in Table I.

TABLE I.   THE RELATIVE COUPLING RANGE

	Applied voltage (V)	Coupling range (Hz)	Eigen frequency (Hz)	Relative coupling range
π-shaped coupled cantilevers	2	120-163	130	$\frac{163-120}{130}=0.331$
rectangle coupled cantilevers	4	208-213	220	$\frac{213-208}{220}=0.023$

From Table I, when applied voltage is 2V, the relative coupling range of π-shaped coupled cantilevers is 0.331. According to the characteristics of coupling range, the relative coupling range will be larger if applied voltage is 4V, and the relative coupling range of rectangular coupled cantilevers is 0.023. The result indicates the relative coupling range of π-shaped coupled cantilevers is larger than rectangular coupled cantilevers.

We know when the coupled system is used for mass sensing, the mass measurement is achieved indirectly by measuring the frequency shift, so it is necessary to study the mass sensing characteristics of the coupling tongue. Therefore, we apply small mass ball onto the tip of LFC, and the variation trend of coupling tongue is obtained by applying different number of small mass balls as mass perturbations, as shown in Fig. 6.

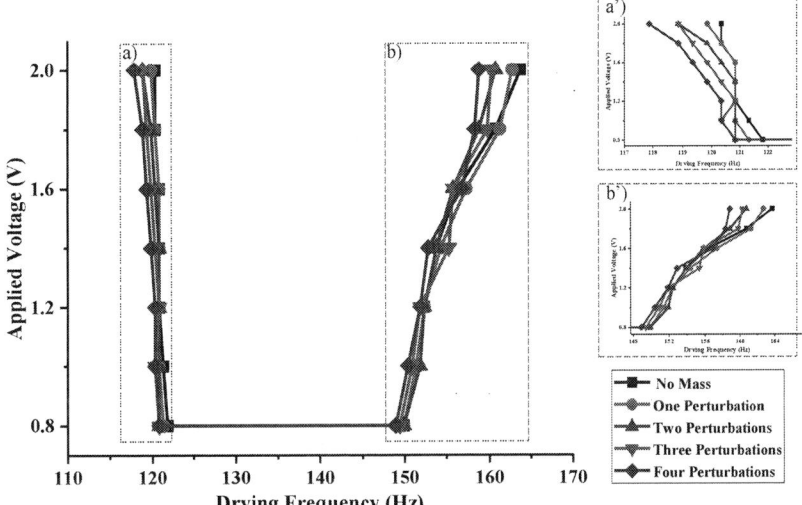

Figure 6. The coupling tongues were experimentally described via applied different number of small mass balls respectively. In the experiment, the variety trend of coupling tongue is not very obvious, so we enlarged the important part of the picture. Fig. 6a' is an enlarged view of part a and Fig. 6b' is an enlarged view of part b.

From Fig. 6, the coupling tongue will shift from high frequency to low frequency with the increase of mass perturbations, and the area of the coupling tongue did not change significantly. It indicates that the coupling tongue can be used for mass sensing and shift steadily in the π-shaped coupled cantilevers.

In previous work, we have known that the method of mass sensing utilizing internal resonance is to detect the third harmonic of high frequency cantilever. Nevertheless, there are three kinds of harmonic components in the output signal of HFC, and the proportion of third harmonic of HFC is an important parameter, which has a great influence on the sensing resolution. Thus, we study variation trend of the main harmonics' peak value with increasing applied voltage, as shown in Fig. 7.

From Fig. 7, the peak value is almost unchanged when applied voltage in certain extent. Thus, we think that when the applied voltage is not very large, the applied voltage has little effect on the peak value of the main harmonics, in other

word, the peak value hardly changes with the change of applied voltage.

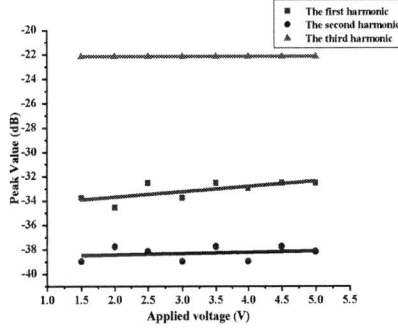

Figure 7. Peak value of the main harmonic with increasing applied voltage.

In order to better see whether the sensing resolution has been improved in the π-shaped coupled cantilevers, we take the rectangular coupled system as the comparison object, and compare the proportion of the first harmonic and the third harmonic, as shown in Fig. 8.

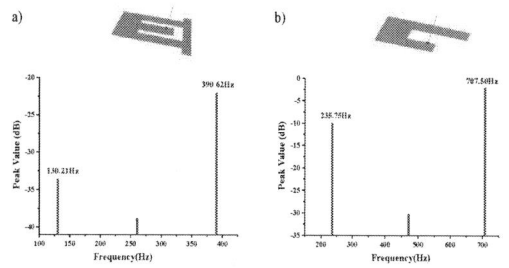

Figure 8. The proportion of the first harmonic and the third harmonic. (a). π-shaped coupled cantilevers. (b). rectangle coupled cantilevers.

From Fig. 8, we can observe the peak value of each harmonic intuitively, and we can clearly see that the relative component of the third harmonic and the first harmonic in the π-shaped coupled cantilevers is higher than that in the rectangular coupled cantilevers, but we can't directly obtain the proportion of each harmonic, so that we can't make an accurate comparison between the two coupled systems. Therefore, we compare the peak value ratio of the third harmonic and the first harmonic, as shown in Table II.

TABLE II.   THE PEAK VALUE RATIO

	Ratio (the third harmonic/the first harmonic)
π-shaped coupled cantilevers	$33.7485/22.1485 \text{ (dB)} = 10^{\frac{-22.1485+33.7485}{20}} \text{ (v)} = 3.802$
rectangle coupled cantilevers	$10.1727/2.2702 \text{ (dB)} = 10^{\frac{-2.2702+10.1727}{20}} \text{ (v)} = 2.484$

From Table II, we can clearly get that the component of the third harmonic in HFC is higher in π-shaped coupled cantilevers. It indicates the π-shaped coupled cantilevers own better sensing resolution than the rectangular coupled cantilevers.

IV. CONCLUSION

In this paper, the π-shaped coupled cantilevers with triple frequency ratio are designed with larger adsorption area, it separates sensing and detecting in two cantilevers respectively. we study internal resonance characteristics of the coupled cantilever structure, and prove that it can be used for mass or gas sensing. By comparing with the rectangular coupled cantilevers, the larger coupling range and sensing resolution of the π-shaped coupled cantilevers are verified, which shows that it is helpful to further improve the performance of oscillation.

ACKNOWLEDGMENT

This work was supported by National Natural Science Foundation of China (NSFC, Grant No. 51975250 & Grant No. 51675229). This work was also supported by Free Exploration Key Project of Jilin Natural Science Foundation (NSFJ, Grant No. 2020122366JC), Scientific Research Foundation for Leading Professor Program of Jilin University (Grant No. 419080500171 & No. 419080500246), as well as Graduate Innovation Fund of Jilin University (Grant No. 101832020CX101).

REFERENCES

[1] K. Asadi, J. Yu, and H. Cho, "Nonlinear couplings and energy transfers in micro- and nano-mechanical resonators: intermodal coupling, internal resonance and synchronization," Philosophical Transactions of the Royal Society Mathematical Physical and Engineering Sciences, vol. 376, 20170141, Jun. 2018.

[2] G. Zheng, X. Du, C. Xia, S. Wan, X. Wang, and D.F. Wang, "Oscillation in coupled resonator systems: Part IV - study on 1:3 internal resonance applicable to sensor devices of high sensitivity," in 20th Design, Test, Integration & Packaging of MEMS and MOEMS, Roma, Italy, 2018, pp. 141–144.

[3] X. Du, D. F. Wang, C. Xia, I. Shimoyama, and R. Maeda, "Internal resonance phenomena in coupled ductile cantilevers with triple frequency ratio–part I: experimental observations," IEEE Sensors Journal, vol. 19, pp. 5475–5483, Jul. 2019.

[4] X. Du, D. F. Wang, C. Xia, I. Shimoyama, and R. Maeda, "Internal resonance phenomena in coupled ductile cantilevers with triple frequency ratio-part II: a mass sensitivity amplification schemes," IEEE Sensors Journal, vol. 19, pp. 5484–5492, Jul. 2019.

[5] C. Xia, D. F. Wang, T. Ono, T. Itoh, and R. Maeda, "A mass multi-warning scheme based on one-to-three internal resonance," Mechanical Systems and Signal Processing, vol. 142, 106784, Mar. 2020.

[6] S. Shaw, "Internal resonances in tiny structures: New results and practical applications," in European Nonlinear Oscillations Conference, Budapest, Hungary, 2017, pp. 1–2.

[7] C. Samanta, P. R. Yasasvi Gangavarapu, and A. K. Naik, "Nonlinear mode coupling and internal resonances in MoS2 nanoelectromechanical system," Applied Physics Letters, vol. 107, 173110, Oct. 2015.

[8] X. Li, T. Ono, R. Lin, and M. Esashi, "Resonance enhancement of micromachined resonators with strong mechanical-coupling between two degrees of freedom," Microelectronic Engineering, vol. 68, pp. 1–12, Feb. 2002.

[9] R. Potekin, S. Dharmasena, H. Keum, X. Jiang, J. Lee, S. Kim, L. Bergman, A. F. Vakakis, and H. Cho, "Multi-frequency atomic force microscopy based on enhanced internal resonance of an inner-paddled cantilever," Sensors and Actuators, A: Physical, vol. 273, pp. 206–220, Apr. 2018.

[10] D. A. Czaplewski, D. H. Zanette, and D. López, "Frequency stabilization in nonlinear micromechanical oscillators," Nature Communications, vol. 3, 806, May. 2012.

[11] C. Chen, D. H. Zanette, D. A. Czaplewski, S. Shaw, and D. López, "Direct observation of coherent energy transfer in nonlinear micromechanical oscillators," Nature Communications, vol. 8, 15523, May. 2017.

[12] C. Van Der Avoort, R. Van Der Hout, J.J.M. Bontemps, P.G. Steeneken, K. Le Phan, R.H.B. Fey, J. Hulshof, and J.T.M. Van Beek, "Amplitude saturation of MEMS resonators explained by autoparametric resonance," Journal of Micromechanics and Microengineering, vol. 20, 105012, Sep. 2010.

[13] A. Sarrafan, B. Bahreyni, and F. Golnaraghi, "Design and characterization of microresonators simultaneously exhibiting 1/2 subharmonic and 2:1 internal resonances," in 19th International Conference on Solid-State Sensors, Actuators and Microsystems, 2017, pp. 102–105.

[14] A. Sarrafan, B. Bahreyni, and F. Golnaraghi, "Development and characterization of an h-shaped microresonator exhibiting 2: 1 internal resonance," Journal of Microelectromechanical Systems, vol. 26, pp. 993–1001, 2017.

[15] A. H. Ramini, A. Z. Hajjaj, and M. I. Younis, "Tunable resonators for nonlinear modal interactions," Scientific Reports, vol. 6, 34717, Oct. 2016.

[16] A. Vyas, D. Peroulis, and A. K. Bajaj, "A microresonator design based on nonlinear 1: 2 internal resonance in flexural structural modes," Journal of Microelectromechanical Systems, vol. 18, pp. 744–762, Jun. 2009.

[17] N. Noori, A. Sarrafan, F. Golnaraghi, and B. Bahreyni, "Utilization of 2:1 internal resonance in microsystems," Micromachines, vol. 9, pp. 448–448, Sep. 2018.

[18] H. M. Ouakad, H. M. Sedighi, and M. I. Younis, "One-to-one and three-to-one internal resonances in MEMS Shallow Arches," Journal of Computational and Nonlinear Dynamics, vol. 12, 051025, Jul. 2017.

[19] L. Lipiăinen, A. Jaakkola, K. Kokkonen, and M. Kaivola, "Nonlinear excitation of a rotational mode in a piezoelectrically excited square-extensional mode resonator," Applied Physics Letters, vol. 100, 153508, Apr. 2012.

[20] H. J. R. Westra, H. S. J. Van, and W. J. Venstra, "Modal interactions of flexural and torsional vibrations in a microcantilever," Ultramicroscopy, vol. 120, pp. 41–47, Sep. 2012.

[21] P. A. Truitt, J. B. Hertzberg, E. Altunkaya, and K. C. Schwab, "Linear and nonlinear coupling between transverse modes of a nanomechanical resonator," Journal of Applied Physics, vol. 144, 114307, Sep. 2012.

[22] Y. Yang, E. Ng, P. Polunin, Y. Chen, S. Strachan, V. Hong, C. H. Ahn, O. Shoshani, S. Shaw, M. Dykman, and T. Kenny, "Experimental investigation on mode coupling of bulk mode silicon MEMS resonators," in 28th IEEE International Conference on Micro Electro Mechanical Systems, Estoril, 2015, pp. 1008–1011.

[23] B. Jeong, C. Pettit, S. Dharmasena, H. Keum, J. Lee, J. Kim, D. Mcfarland, L. Bergman, and A. F. Vakakis. "Utilizing intentional internal resonance to achieve multiharmonic atomic force microscopy," Nanotechnology, vol. 27, 125501, Apr. 2016.

Gap in pagination due to unavailable papers.

Pages 1080-1083

Proceedings of the 16th Annual IEEE International
Conference on Nano/Micro Engineered and Molecular Systems
April 25-29, 2021

Multi-order Nonlinearities and Resulting Coherent Oscillations of the States in Quantum Dot-Mechanical Resonator Hybrid System

Lin Cong, Guang-Wei Deng, Xiaoyang Lin, *Senior Member, IEEE*, Wenjie Liang, Zaiqiao Bai, Weisheng Zhao, *Fellow, IEEE*, Kaili Jiang and Xinhe Wang, *Member, IEEE*

Abstract— The carbon nanotube (CNT) hybrid device composed of a gate-defined quantum dot embedded into a mechanical resonator is investigated under strong actuation condition. The Coulomb peaks are driven to the synchronous and large oscillation by the nonlinear mechanical vibration, and can be coherently modulated to the single-electron shuttle mode. Conversely, the mediated single-electron tunnelling current provides an in situ and sensitive characterization strategy for the multi-order mechanical nonlinear resonance. We show that, under the strong modulation applied via gate voltage, the swing and trampoline modes exhibit different nonlinearity behaviour, depending on both the quadratic and cubic coefficients. The nonlinearities in this hybrid system also give rise to coupling modes of quadratic and cubic resonances at combined frequencies and even fractional frequencies. Especially, the subharmonic resonance due to quadratic nonlinearity is predicted to enable the excitation of the coherent coupling state between two modes. Our theoretical mode of the intra- and intermodal nonlinearities shows a good agreement of our experimental observations quantitatively.

I. INTRODUCTION

Carbon nanotube (CNT) mechanical system has been a model system to approach the manifold physics of nonlinear dynamics[1]–[5] and the electromechanics that couples the individual electrons with nanometre-scale vibrations[6]–[12]. So far, based on the quantum-dot-embedded CNT mechanical resonator, the remarkable modulations of single-electron transport on the mechanical motion have got quite a bit of insight, including soften the restoring force[5], [7], [13], drive CNT into resonance spontaneously[6], and recently, induce a coherent oscillation[14]. These effects are exploited to control the CNT vibration in the quantum limit[15], [16], or detect the charge by the consequent mechanical clues[10]. In contrast, how the mechanical motion oppositely mediating single-electron transport hasn't been explored much yet[11]. Especially at large drive to the nonlinear regime, a lot of questions is still open, such as whether there is any higher order nonlinearity that beyond the ordinary material or mode[17], how to characterize the nonlinear amplitude (of the extremely small yet fast), and

what is the quantum mechanical signatures of nonlinear oscillators.

Here we construct a quantum dot-mechanical resonator hybrid system based on the doubly clamped CNT. Its flexural modes have an intrinsic nonlinear due to extension on bending[18]–[20]. When working in large amplitude regime, it can be exploited to realize the coherent control of single electron tunnelling, and enable an *in situ* and visual characterization for nonlinear dynamics of the system from the tunnelling current. it provides a platform for studying the nonlinear nanomechanics as well as hybrid quantum system.

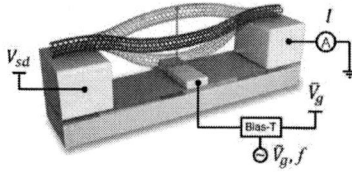

Figure 1. Schematic of the carbon nanotube mechanical resonator and the measurement circuit. The suspended length of the nanotube is 1.2 μm, and the height of the resonator from the gate electrode is 150 nm.

II. EXPERIMENTAL

Our samples are individual ultraclean CNT suspending over the gate electrode and connecting to the source and drain electrodes (Fig. 1). the pristine inner shell of the few-walled CNT with initial tension is assembled by our new deterministic transfer technique[21], [22]. A quantum dot can be defined on a small-bandgap CNT with controlling its charge states by the gate voltage. At the same time, the transverse mechanical vibration of the CNT can be actuated by the ac (microwave) gate electric field, while the dc gate electric field can modulate its tension thus tune the mechanical resonance. Therefore, our device serves as a hybrid quantum dot-mechanical resonator system. The details of the measurement can be found in Ref. [23]. Adopting the Coulomb ratification readout scheme, the displacement of vibration can be converted to the effective shift of the gate of the quantum dot, while the average tunnelling current at each detecting point is proportional to

---

Xinhe Wang, Xiaoyang Lin and Weisheng Zhao are with Fert Beijing Institute, MIIT Key Laboratory of Spintronics, School of Integrated Circuit Science and Engineering, Beihang University, Beijing 100191, China. Cong Lin and Kaili Jiang are with the State Key Laboratory of Low-Dimensional Quantum Physics, Department of Physics and Tsinghua-Foxconn Nanotechnology Research Center, Tsinghua University, Beijing 100084, China. Guang-Wei Deng is with Institute of Fundamental and Frontier

Sciences, University of Electronic Science and Technology of China, Chengdu 610054, China. Wenjie Liang is with Institute of Physics, Chinese Academy of Sciences, Beijing 100080, China, Zaiqiao Bai is with Department of Physics, Beijing Normal University, Beijing 100875, China.

Corresponding authors: Xinhe. Wang, e-mails: xinhe@buaa.edu.cn.

978-1-6654-3008-1/21 $31.00 © 2021 IEEE

the duration of sweeping the Coulomb peak in each cycle. The Coulomb peak serves as the probe (located by varying the DC gate voltage) to record the weight in the time domain of each transient position in a full cycle. Hence, just mapping the average tunnelling current can accurately learn the motion process of mechanical vibration. In particular, due to the weight getting the border and maximum at the maximum displacement of vibration, the broadening edge of the Coulomb peaks provides an accurate depiction of the nonlinear amplitude.

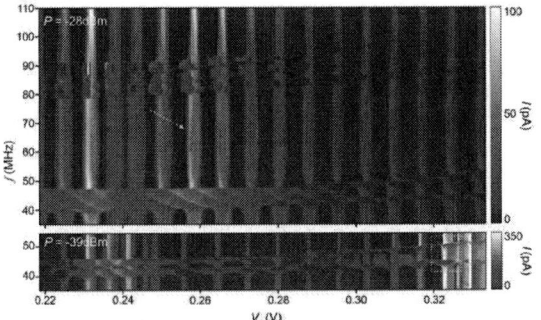

Figure 2. The map of resonance frequencies as a function of Vg (upper panel: -28 dBm, lower panel: -39 dBm). Two modes near 45 MHz are mode S and T (see the text). Three response lines around 90MHz correspond to 2S, S+T, and 2T, and four weaker signal lines (indicated by the arrow) are 3S/2, S+T/2, S/2+T and 3T/2. The dotted box highlights the different spread behaviors of the Coulomb peaks for S and T modes at a large gate voltage.

There are always two distinct mechanical modes near 40 MHz with a splitting of a few percent, which are the two orthogonal fundamental resonant modes. Based on the same static position, the "Swing" mode (labeled S) vibrates mainly parallel to the plane of the gate electrodes while the "Trampoline" mode (labeled T) perpendicular to it (z-direction). These two modes are originally degenerate at V_g = 0 V, and split gradually as the V_g increases because of the symmetry breaking induced by tension[23]. In a larger mapping range, the significant dependence of Coulomb peaks' spreads on the V_g and multi-order excitations of the

modes emerge (Fig.2) with striking dependency of gate voltage.

III. RESULTS AND DISCUSSION

A. The multi-order nonlinearities of coupled two modes

Although the eigenmodes with their tunability and Duffing nonlinearity in the CNT mechanical resonators have been widely investigated, the peculiar phenomenon in our experiments isn't reported yet, we will now give a description of the physics we observe by a multi-order nonlinear mode.

In the beginning, we write the static displacement under a certain DC gate voltage as $z_S(x)$. The two degrees of freedom vibration (as Fig.3b&c illustrated) of the doubly clamped beam can be described by the coupled equations of motion including the in-plane motion $Z(x,t)=z_S(x)+z(x,t)$ and the out-of-plane motion $Y(x,t)=y(x,t)$ as follows:

$$EI\frac{\partial^4 Z}{\partial x^4}+\rho S\frac{\partial^2 Z}{\partial t^2}+\eta_z\frac{\partial Z}{\partial t}-[T_0+T(\frac{\partial Z}{\partial x},\frac{\partial Y}{\partial x})]\frac{\partial^2 Z}{\partial x^2}=\frac{1}{2}\frac{\partial C[z(x,t)]}{\partial z(x,t)}V_g^2$$
(1)

$$EI\frac{\partial^4 Y}{\partial x^4}+\rho S\frac{\partial^2 Y}{\partial t^2}+\eta_y\frac{\partial Y}{\partial t}-[T_0+T(\frac{\partial Z}{\partial x},\frac{\partial Y}{\partial x})]\frac{\partial^2 Y}{\partial x^2}=\frac{1}{2}\frac{\partial C[y(x,t)]}{\partial y(x,t)}V_g^2=0$$
(2)

where E is Young's modulus, I is the moment of inertia about the longitudinal axis of the beam, ρ is the beam density, S is the cross-sectional area, η_z and η_y are the damping coefficient for the two directions respectively. The total tension is composed of the residual tension T_0 and the extra tension induced by beam bending $T(\frac{\partial Z}{\partial x},\frac{\partial Y}{\partial x})=\frac{ES}{2L}\int_0^L[(\frac{\partial z_s(x)}{\partial x}+\frac{\partial z(x,t)}{\partial x})^2+(\frac{\partial y(x,t)}{\partial x})^2]dx$, where L is the beam length. The capacitance per unit length $C[z(x,t)]$ is expanded to the fourth order of $z(x,t)$ as $C[z(x,t)]=K_0+K_1z(x,t)+K_2z(x,t)^2+K_3z(x,t)^3+K_4z(x,t)^4$. The partial differential of capacitance in the Y direction is zero since the gate electrode can be thought of as infinite with respect to the out-of-plane vibration of the beam.

TABLE I.   RESONANT FREQUENCY, QUADRATIC AND CUBIC NONLINEARITY COEFFICIENT OF S AND T MODE.

	S mode	T mode
Resonant frequency	$\sqrt{(\frac{EI}{3\rho S}+\frac{Ez_{dc}^2}{18\rho})(\frac{2\pi}{L})^4+\frac{T_0}{3\rho S}(\frac{2\pi}{L})^2}$	$\sqrt{(\frac{EI}{3\rho S}+\frac{Ez_{dc}^2}{6\rho})(\frac{2\pi}{L})^4+\frac{T_0}{3\rho S}(\frac{2\pi}{L})^2-\frac{K_2V_g^2}{\rho S}}$
Quadratic nonlinearity coefficient α_2	0	$\frac{Ez_{dc}}{6\rho}(\frac{2\pi}{L})^4-\frac{5}{2}\sqrt{\frac{2}{3}}\frac{K_3V_g^2}{\rho S}$
Cubic nonlinearity coefficient α_3	$\frac{E}{18\rho}(\frac{2\pi}{L})^4$	$\frac{E}{18\rho}(\frac{2\pi}{L})^4-\frac{35}{9}\frac{K_4V_g^2}{\rho S}$

Based on the Eq.(1)&(2) and using Galerkin procedure, we get the quadratic nonlinear coefficient (α_2) and cubic one (α_3) of S and T mode respectively, as shown in TABLE

I. the striking result is that the $\alpha_{2,3}$ for T mode are dependent on the gate voltage V_g, while those for S mode are constant, especially the α_2 for S mode is zero. On the other hand, from the forced-oscillation equation of the system with both quadratic and cubic nonlinearities, we get the Duffing-type response equation and extract the corresponding effective nonlinearity coefficient $\alpha_e = \alpha_3 - \frac{10}{9\omega_0^2}\alpha_2^2$. In fact, the effective nonlinearity coefficient can be quantitatively estimated by experimental parameters and the result of T mode is shown in Fig.4a. α_e decreases with gate voltage, and turns negative when the gate voltage exceeds about 0.28 V. A negative α_e means vibrating with larger amplitude is not stiffen but soften the resonator, thus there are no Duffing-type hysteresis, which is verified by Fig.2 (shown clearly under the driven of - 39dBm). In contrast, $\alpha_e = \alpha_3$ for S mode, which maintains a positive constant with gate voltage varying, so the umbrella-like spread always exists for the Coulomb peaks from all range of gate voltage in Fig.2.

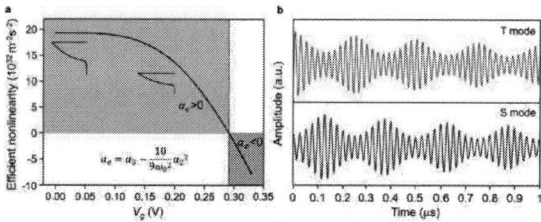

Figure 3 (a) The relationship between the effective nonlinearity of T mode and V_g calculated with the parameters set by the experiment. At lower V_g, the effective nonlinearity is positive and decreases as V_g increases, resulting the decrease of the Duffing-type hysteresis along the driving frequency. Eventually, the effective nonlinearity becomes negative at higher V_g, thereby the Duffing-type hysteresis disappears. The specific V_g at which the effective nonlinearity trend to zero is indicated by the dashed line. (b) The oscillations of the two distinct modes in time domain. The amplitudes of two modes oscillate with the same period but a phase different of $\pi/2$.

Next, our focus turns to the coupling between the distinct mechanical modes. Actually, Eq.(1)&(2) have shown that two freedom of degrees Z and Y can talk to each other via the tension induced by beam bending. The modal analysis equation and further forced-oscillation harmonic response equation indicates that the quadratic nonlinear coupled term yz dominates the coupling between the S and T modes. From the harmonic response equation, the subharmonic resonance[24] can occur when the driving frequency Ω satisfies: $\Omega/2 = \omega_{S,T}$, confirming the excitation lines of 2S (2T) in Fig.2.

What is more attractive is the subharmonic resonance of S+T, it means that the S and T mode is excited by the driving force with frequency $\Omega = \omega_S + \omega_T$. Within the framework of quadratic nonlinearities, the amplitudes of S mode (A_S) and T mode (A_T) are extracted as,

$$A_S \propto \exp\{-\mu_S t - \frac{\lambda F}{4\omega_S(\Omega^2 - \omega_T^2)(\Omega - 2\omega_S)}\cos[(\omega_T - \omega_S)t]\}$$

(3)

$$A_T \propto \exp\{-\mu_T t - \frac{\lambda F}{2\omega_T(\Omega^2 - \omega_T^2)(\Omega - 2\omega_T)}\sin[(\omega_T - \omega_S)t]\}$$

(4)

where λ is the coefficient of the quadratic nonlinearity (depend on the driving power), F is the driving force, and μ_S , μ_T is the damping coefficient of mode S and T, respectively. The amplitudes of two modes exhibit the decaying oscillations of exactly the same frequency, with a fixed phase difference of $\pi/2$. It indicates that the energy coherently transfers back and forth between the two coupled modes, and accompanied by the energy dissipation (Fig.4b). The amount of transferred energy can be tailored by adjusting driving power and gate voltage[25], proposing a coherent manipulation of phonon [26]–[29] by the pulsed excitation with frequency $\Omega = \omega_S + \omega_T$. As beneficial from the high performance of frequency, quality factor and tunability, the CNT resonator is promising for further coherently manipulating phonons in the quantum regime (f $Q > k_B T_{base}/h$, where f, Q_m, T_{base}, k_B, h is work frequency, quality factor of the system, base temperature, Boltzmann constant and Plank constant, respectively)[30], and pave an avenue for studying two-level mesoscopic quantum system, and generating superpositions of phonon states for further phononic quantum computing.

IV. CONCLUSION

In summary, the umbrella-like broadening of Coulomb peaks provides a rare insight into the multi-order nonlinearities in the nanomechanical system with two degrees of freedom. The quadratic and cubic nonlinearities not only determine the different gate-tuned nonlinear behaviours of two distinct modes, but also result in the plenty of intermodal nonlinearities, especially the excitation of the coherent coupling state between two modes. Our findings promote understanding and controlling the various nonlinearity and strong couplings in quantum dot-mechanical resonator hybrid system, and will benefit the mesoscopic or quantum coherent applications in both electronic and mechanical aspects.

ACKNOWLEDGMENT

This work was supported by the National Key Research and Development Program of China (2018YFA0306102), National Natural Science Foundation of China (No. 11904014, 91836102 and 61704164).

REFERENCES

[1] A. W. Barnard, V. Sazonova, A. M. van der Zande, and P. L. McEuen, "Fluctuation broadening in carbon nanotube resonators," *Proc. Natl. Acad. Sci.*, vol. 109, no. 47, pp. 19093–19096, Nov. 2012, doi: 10.1073/pnas.1216407109.

[2] A. Eichler, J. Moser, M. I. Dykman, and A. Bachtold, "Symmetry breaking in a mechanical resonator made from a carbon nanotube," *Nat. Commun.*, vol. 4, p. 2843, Nov. 2013, doi: 10.1038/ncomms3843.

[3] M. H. Matheny, L. G. Villanueva, R. B. Karabalin, J. E. Sader, and M. L. Roukes, "Nonlinear Mode-Coupling in

Nanomechanical Systems," *Nano Lett.*, vol. 13, no. 4, pp. 1622–1626, Apr. 2013, doi: 10.1021/nl400070e.

[4] O. Maillet *et al.*, "Nonlinear frequency transduction of nanomechanical Brownian motion," *Phys. Rev. B*, vol. 96, no. 16, p. 165434, Oct. 2017, doi: 10.1103/PhysRevB.96.165434.

[5] K. Willick, X. (Shirley) Tang, and J. Baugh, "Probing the non-linear transient response of a carbon nanotube mechanical oscillator," *Appl. Phys. Lett.*, vol. 111, no. 22, p. 223108, Nov. 2017, doi: 10.1063/1.4991412.

[6] G. A. Steele *et al.*, "Strong Coupling Between Single-Electron Tunneling and Nanomechanical Motion," *Science*, vol. 325, no. 5944, pp. 1103–1107, Aug. 2009, doi: 10.1126/science.1176076.

[7] B. Lassagne, Y. Tarakanov, J. Kinaret, D. Garcia-Sanchez, and A. Bachtold, "Coupling Mechanics to Charge Transport in Carbon Nanotube Mechanical Resonators," *Science*, vol. 325, no. 5944, pp. 1107–1110, Aug. 2009, doi: 10.1126/science.1174290.

[8] K. J. G. Götz, D. R. Schmid, F. J. Schupp, P. L. Stiller, Ch. Strunk, and A. K. Hüttel, "Nanomechanical Characterization of the Kondo Charge Dynamics in a Carbon Nanotube," *Phys. Rev. Lett.*, vol. 120, no. 24, p. 246802, Jun. 2018, doi: 10.1103/PhysRevLett.120.246802.

[9] Y. Wen, N. Ares, T. Pei, G. a. D. Briggs, and E. A. Laird, "Measuring carbon nanotube vibrations using a single-electron transistor as a fast linear amplifier," *Appl. Phys. Lett.*, vol. 113, no. 15, p. 153101, Oct. 2018, doi: 10.1063/1.5052185.

[10] P. Häkkinen, A. Isacsson, A. Savin, J. Sulkko, and P. Hakonen, "Charge Sensitivity Enhancement via Mechanical Oscillation in Suspended Carbon Nanotube Devices," *Nano Lett.*, vol. 15, no. 3, pp. 1667–1672, Mar. 2015, doi: 10.1021/nl504282s.

[11] G. Micchi, R. Avriller, and F. Pistolesi, "Mechanical Signatures of the Current Blockade Instability in Suspended Carbon Nanotubes," *Phys. Rev. Lett.*, vol. 115, p. 206802, 2015.

[12] A. Benyamini, A. Hamo, S. V. Kusminskiy, F. von Oppen, and S. Ilani, "Real-space tailoring of the electron-phonon coupling in ultraclean nanotube mechanical resonators," *Nat. Phys.*, vol. 10, no. 2, pp. 151–156, 2014, doi: 10.1038/nphys2842.

[13] A. Castellanos-Gomez, H. B. Meerwaldt, W. J. Venstra, H. S. J. van der Zant, and G. A. Steele, "Strong and tunable mode coupling in carbon nanotube resonators," *Phys. Rev. B*, vol. 86, no. 4, p. 041402, Jul. 2012, doi: 10.1103/PhysRevB.86.041402.

[14] Y. Wen, N. Ares, F. J. Schupp, T. Pei, G. a. D. Briggs, and E. A. Laird, "A coherent nanomechanical oscillator driven by single-electron tunnelling," *Nat. Phys.*, vol. 16, no. 1, pp. 75–82, Jan. 2020, doi: 10.1038/s41567-019-0683-5.

[15] S. Blien, P. Steger, N. Hüttner, R. Graaf, and A. K. Hüttel, "Quantum capacitance mediated carbon nanotube optomechanics," *Nat. Commun.*, vol. 11, no. 1, Art. no. 1, Apr. 2020, doi: 10.1038/s41467-020-15433-3.

[16] C. Urgell *et al.*, "Cooling and self-oscillation in a nanotube electromechanical resonator," *Nat. Phys.*, vol. 16, pp. 32–37, Oct. 2019, doi: 10.1038/s41567-019-0682-6.

[17] A. W. Barnard, M. Zhang, G. S. Wiederhecker, M. Lipson, and P. L. McEuen, "Real-time vibrations of a carbon nanotube," *Nature*, vol. 566, no. 7742, pp. 89–93, Feb. 2019, doi: 10.1038/s41586-018-0861-0.

[18] P. A. Greaney, G. Lani, G. Cicero, and J. C. Grossman, "Anomalous Dissipation in Single-Walled Carbon Nanotube Resonators," *Nano Lett.*, vol. 9, no. 11, pp. 3699–3703, Nov. 2009, doi: 10.1021/nl901706y.

[19] H. Cho, M.-F. Yu, A. F. Vakakis, L. A. Bergman, and D. M. McFarland, "Tunable, Broadband Nonlinear Nanomechanical Resonator," *Nano Lett.*, vol. 10, no. 5, pp. 1793–1798, May 2010, doi: 10.1021/nl100480y.

[20] R. Lifshitz and M. C. Cross, "Nonlinear Dynamics of Nanomechanical Resonators," in *Nonlinear Dynamics of Nanosystems*, G. Radons, B. Rumpf, and H. G. Schuster, Eds. Weinheim, Germany: Wiley-VCH Verlag GmbH & Co. KGaA, 2010, pp. 221–266.

[21] G.-W. Deng *et al.*, "Strongly Coupled Nanotube Electromechanical Resonators," *Nano Lett.*, vol. 16, no. 9, pp. 5456–5462, Sep. 2016, doi: 10.1021/acs.nanolett.6b01875.

[22] X. Wang *et al.*, "Stressed carbon nanotube devices for high tunability, high quality factor, single mode GHz resonators," *Nano Res.*, vol. 11, no. 11, p. 5812, May 2018, doi: 10.1007/s12274-018-2085-x.

[23] X. Wang *et al.*, "Visualizing nonlinear resonance in nanomechanical systems via single-electron tunneling," *Nano Res.*, no. 14, pp. 1156–1161, 2021, doi: 10.1007/s12274-020-3165-2.

[24] A. H. Nayfeh and D. T. Mook, *Nonlinear oscillations*. John Wiley & Sons, 2008.

[25] I. Kozinsky, H. W. Ch. Postma, I. Bargatin, and M. L. Roukes, "Tuning nonlinearity, dynamic range, and frequency of nanomechanical resonators," *Appl. Phys. Lett.*, vol. 88, no. 25, p. 253101, Jun. 2006, doi: 10.1063/1.2209211.

[26] H. Okamoto *et al.*, "Coherent phonon manipulation in coupled mechanical resonators," *Nat. Phys.*, vol. 9, no. 8, pp. 480–484, Aug. 2013, doi: 10.1038/nphys2665.

[27] T. Faust, J. Rieger, M. J. Seitner, J. P. Kotthaus, and E. M. Weig, "Coherent control of a classical nanomechanical two-level system," *Nat. Phys.*, vol. 9, no. 8, pp. 485–488, 2013, doi: 10.1038/nphys2666.

[28] D. Zhu *et al.*, "Coherent Phonon Rabi Oscillations with a High-Frequency Carbon Nanotube Phonon Cavity," *Nano Lett.*, vol. 17, no. 2, pp. 915–921, Feb. 2017, doi: 10.1021/acs.nanolett.6b04223.

[29] C. Chen, D. H. Zanette, D. A. Czaplewski, S. Shaw, and D. López, "Direct observation of coherent energy transfer in nonlinear micromechanical oscillators," *Nat. Commun.*, vol. 8, no. 1, p. 15523, Aug. 2017, doi: 10.1038/ncomms15523.

[30] R. A. Norte, J. P. Moura, and S. Gröblacher, "Mechanical Resonators for Quantum Optomechanics Experiments at Room Temperature," *Phys. Rev. Lett.*, vol. 116, no. 14, p. 147202, Apr. 2016, doi: 10.1103/PhysRevLett.116.147202.

Proceedings of the 16th Annual IEEE International
Conference on Nano/Micro Engineered and Molecular Systems
April 25-29, 2021

Investigation the Minimum Measurement Points for Calibration a High Precision NTC Thermistors in Cryogenic Field

Xun Gong, Xiaohe Tang, Yanjie Li, Minmin You Zude Lin and Jingquan Liu

Abstract—-For a negative temperature coefficient (NTC) thermistor, the thermistor characteristic equation should be fitting by Least Square Method (LSM). According to the characters of LSM, more measurement points can improve the precision of the thermistor characteristic equation, but those also increase the measurement cycle. As for our sensors, a temperature point will cost more than 9 hours at around 20K and 8 hours at around 100K. To decrease the sampling data quantity, in this paper, the optimal current of the NTC thermistors is discussed to reduce the system deviation (Class B Uncertainty) caused by self-heating. After optimizing the side effect and combining with the early work (a new calibration equations), the accuracy will be increased. In this way, the measurement temperature points can be reduced and the error, though it will increase, can still satisfy the demand that the deviation is below 1mK at around 20K in most cases. Finally, to balance the precision and the time overhead, the optimal number of measurement temperature points is 28 and it can save 21% of the time in a measurement cycle, at least.

I. INTRODUCTION

Thanks to the development of technology in aerospace and biomedical, such as radiofrequency ablation [1], the high precision cryogenic sensors have been proposed. And in cryogenic applications, one of the most important work is temperature measurement which is considered as a basic condition for those frontier technology. As for a commercial high precision temperature sensor, the deviation is one of the most significant indicators to evaluate the quality of a thermometer. In cryogenic temperature range (below 77K), the NTC thermistor is the most common thermometers. For a high accuracy NTC thermistor, the error range can be limited to less than 1mK at around some key points and 5mK at the entire temperature range. The measurement accuracy of temperature sensor is usually evaluated by measurement uncertainty, which is composed of the fitting standard deviation (class A Uncertainty) introduced by the temperature sensor in the calibration process and the system deviation (class B Uncertainty) introduced by the measuring instrument precision.

We can decrease the quantity of sample data to reduce the measurement cycle. But according to the principle of LSM, it may increase the fitting error and even exceed the allowable value of the deviation. To satisfy the requirements of the precision, the uncertainty of both class A and class B need to be reduced. In our early work, a new optimal piecewise

Chebyshev fitting method (OPCFM) was introduced [2], which reduce the class A Uncertainty successfully. In this paper, to reduce the class B Uncertainty, the optimal current was discussed to decrease the effect of self-heating based on the NTC thermistor. Combined with the OPCFM and the optimal current, the Uncertainty was reduced dramatically and the accuracy was increased significantly. Based on this situation, we try to reduce the sampling data quantity and discuss the optimal number of measurement points.

II. MATH

A. the Equipment and the Test Flow

During this test, the platinum resistance temperature detector (RTD) is a working standard showed in Fig. 1, and it has been calibrated according to the International Temperature Scale of 1990 [3]. And our commercial high precision NTC thermistor has been calibrated by the RTD.

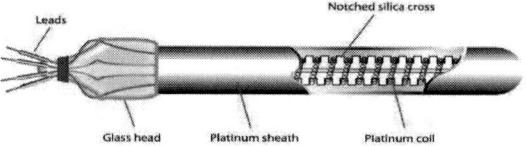

Fig. 1. Schematic diagram of a typical 25 Ω platinum RTD
(Illustration published on courtesy from FLUCK®)

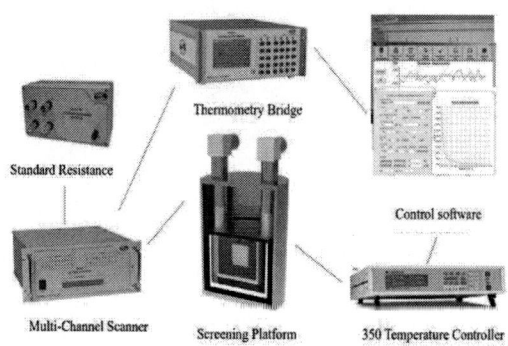

Fig. 2. Composition of Calibration System

The calibration system, as shown in Fig. 2, was composed with standard resistance, thermometry bridge, multi-channel scanner, screening platform, 350 temperature controller and

---

*Resrach supported by the Strategic Priority Research Program of Chinese Academy of Sciences under Grant XDA25040000, the National Key R&D Program of China under Grant 2020YFB1313502, the National Natural Science Foundation of China under Grant 61728402, SJTU Trans-med Award under Grant 2019015, the Oceanic Interdisciplinary Program of Shanghai Jiao Tong University under Grant SL2020ZD205, Scientific Research Fund of Second Institute of Oceanography , MNR under Grant SL2020ZD205, the Program of Shanghai Academic/Technology Research

Leader under Grant 18XD1401900, the Center for Advanced Electronic Materials and Devices (AEMD) of Shanghai Jiao Tong University.

Xun Gong, Xiaohe Tang, Yanjie Li, Minmin You, Zude Lin and Jingquan Liu are all with the National Key Laboratory of Science and Technology on Micro/Nano Fabrication, Shanghai Jiao Tong University, Shanghai, 200240, China, Department of Micro/Nano Electronics, Shanghai Jiao Tong University, Shanghai, 200240, China (corresponding auther to provide e-mail: jqliu@sjtu.edu.cn).

978-1-6654-3008-1/21 $31.00 © 2021 IEEE

control software. It can be divided into two groups: ambient temperature control system, where temperature in the vacuum chamber can be adjusted from 4K to 300K, and the measurement system where resistance of the NTC thermistor can be measurement with a high precision.

According to the standard of cryogenic temperature sensor aerospace applications in the United States [4], to reach the demand of accuracy and simplify the process, the test flow of this project extracts the corresponding necessary operations and is shown in Table. 1.

Procedure	Steps	Examination Content	Equipment
1	External inspection	Coating, lead damage, packing crack, sealing	OLYMPUS CX31 Stereo microscope and photographic system
2	Tightness inspection	Equivalent standard leakage rate L<1*10⁻⁸Pa•cm³/s	HELIOT 901 W1 Helium mass spectrometer leak detector
3	resistance screen	The resistance between200 and 400Ω at room temperature	Digital multimeter
4	1st temperature cycle	4.2-300K 20 times temperature cycle	Thermal shock platform
5	1st Particle impact noise detection	Sensors with excessive noise are screened by instruments	SD PIND 45LParticle impact noise detector
6	1st Stability test	10min Stability test in 19.6K	Screening platform
7	Calibration	A total of 30 points within 14-300K were calibrated	Screening platform
8	2nd temperature cycle	4.2-300K 20 times temperature cycle	Thermal shock platform
9	2nd Particle impact noise detection	Sensors with excessive noise are screened by instruments	SD PIND 45LParticle impact noise detector
10	2nd Stability test and Reproducibility test	10min Stability test in 19.6K and calculation the Reproducibility	Screening platform

Table. 1. The test flow

B. Self-heating Test

To meet those basic demand that the error range can be limited to less than 1mK at around some key points and 5mK at the entire temperature range, and self-heating was measured at our key point (around 20K).

The function of self-heating is shown in Eq. 1, and Eq. 2, where R_a and R_b are the resistance of the NTC thermistor in different working current, I_b is the working current, I_a is also the working current and the default value is $\sqrt{2}I_b$ or $2I_b$, $\frac{dR}{dt}$ is the sensitivity whose value only effects by the inherent characteristic and ambient temperature, ΔT is the self-heating of the sensors.

$$\Delta R = \frac{R_a - R_b}{I_a^2 - I_b^2} \tag{1}$$

$$\Delta T = \frac{\Delta R}{\frac{dR}{dt}} \tag{2}$$

The test of self-heating with NTC thermistor was carried out under 3 groups of different power, and five kinds of NTC thermistor were selected to avoid the contingency. The resistance of the sensors was measured by Thermometry Bridge (TB). When the current flows through the sensors, the self-heating increases the deviation, especially in cryogenic field. But if the current is too small, the measurement error increases instead. Hence, three groups were set up to find out the optimal power and the power was equal to 10^{-8} W, 10^{-7} W and 10^{-6} W. It notes that the self-heating is caused by power, and the current need to be adjust by the TB to change the power.

C. Fitting the Measurement Points

The Chebyshev polynomials are usually used for NTC thermistor fitting, as it has staggered positive and negative points, and can fit the thermistor characteristic equation better. The fitting function is shown in Eq. 3, where T is the absolute temperature, a_i is the fitting coefficient which is a multi-variate, x is the variable whose absolute value is less than 1, A and B are the normalization constants which are calculated by the Eq. 4 and Eq. 5, n is the fitting power and R is the resistance of the NTC thermistor. It should be paid attention that when the LSM is used, the fitting power must be set less than the quantity of the data set. Normally the quantity of sample data needs to be more than double of the fitting power.

As for the Chebyshev polynomials, the power of the equation can affect the accuracy of the curve significantly, which always shows the trend that the bigger the number is, the higher accuracy at most of the situations. On the other hand, higher fitting power may lead to overfitting, which decrease the fitting accuracy. To prevent this side effect, the fitting power of the Chebyshev polynomials is usually between 5 and 11.

$$T = \frac{a_0}{2} + \sum_{i=1}^{n} a_i \cos(i\cos^{-1} x) = \frac{a_0}{2} + \sum_{i=1}^{n} a_i \cos\left[i\cos^{-1}(A\ln R + B)\right] \tag{3}$$

$$A = \frac{2}{\ln(R_{max}) - \ln(R_{max})} \tag{4}$$

$$B = 1 - \frac{2\ln(R_{max})}{\ln(R_{max}) - \ln(R_{min})} \tag{5}$$

As for the fitting coefficient vector in the Chebyshev polynomials, $A = \{\frac{a_0}{2}, a_1, a_2, a_3, \dots, a_{0n-1}\}^T (6 \leq n \leq 12)$ can be calculated by LSM [5], showed in Eq. 6.The point are a series of R-T scatter plots. $C_{m \times n}$ denotes corresponding polynomial matrix, $R_{m \times 1}$ is a vector of corresponding

$$C_{m \times n} = (1_{n \times 1} \quad \cos[\cos^{-1}(A\ln R_{m \times 1} + B)] \quad \dots \quad \cos[(n-1)\cos^{-1}(A\ln R_{m \times 1} + B)]) \tag{6}$$

$$C_{m \times n} A_{n \times 1} = T_{m \times 1} \tag{7}$$

$$A = (C^T C)^{-1} C^T T \tag{8}$$

resistance, $T_{m \times 1}$ is a vector of measured temperature and m is the quantity of data set.

Moreover, the standard deviation DT_{std} was used to estimate the quality of the fitting curve. The function is expressed in Eq. 9. where m denotes the quantity of data set and T_{ci}, T_{oi} denote temperature after calculation and original temperature, respectively. m is usually set bigger than 2n to avoid overfitting. The goal of the method is to find an appropriate fitting power minimizing DT_{std} in a fixed segment.

$$DT_{std} = \sqrt{\sum_{i=1}^{m}(T_{ci} - T_{oi})^2/(m-n)} \qquad (9)$$

For this commercial high accuracy NTC· thermistor, the measurement range can cover 14K to 300K, which is hard to fitting with a curve at a high precision. Thus, we divide the temperature range into two parts and use the OPCFM to fitting [2], respectively. Calculated by the OPCFM algorithm, during the entire temperature range, in the worst case, the power of the Chebyshev polynomials is 8. Therefore, based on the principle of the LSM that sample data should more than double of the fitting power to avoid overfitting, the measurement points should be no less than 28. It is noticed that there are more than 4 overlap points in OPCFM algorithm for the piecewise Chebyshev polynomials fitting, so the minimum measurement points are 28 rather than 32.

D. the Minimum measurement points

The last step is to evaluate the minimum measurement points within the error limitation. The measurement uncertainty, as shown in Eq. 10, is composed of the class A Uncertainty (u_A) introduced by the temperature sensor in the calibration process and the class B Uncertainty (u_B) introduced by the measuring instrument precision.

$$u^2 = u_A^2 + u_B^2 \qquad (10)$$

In our previous work, the thermistor characteristic equation of NTC thermistors needs at least 34 points in the calibration fitting to meet the requirement of error limitation. After OPCFM function is used and the optimal power is applied, the accuracy has been increased dramatically. To avoid the contingency, five kinds of NTC thermistor were selected, showed in Fig. 3, and those 34 temperature points are measured in early work

The line segment is obtained by OPCFM algorithm to fitting the scatter with 34 points. It is shows that the Chebyshev polynomials get a great fitting with those points, and even has exceeded the precision requirements that the deviation is less than 1mK at 20K. As each of the temperature points needs a large amount of time, to reduce the measurement cycle, the feasibility of 28 measurement points, 30 measurement points and 32 measurement points are discussed within the error limitation.

As for the number of 28, it has been discussed in the part of *Fitting the Measurement Points*. The resistance-temperature (R-T) curve is fitted by LSM, which usually requires that the number of calibration points are more than double of equation parameters. The OPCFM function is an adaptive algorithm

which can calculate the optimal number of parameters to minimize the standard deviation. Although, the function requires only a few parameters, in some case, the number of points should be no less than 28, in the extreme situation,.

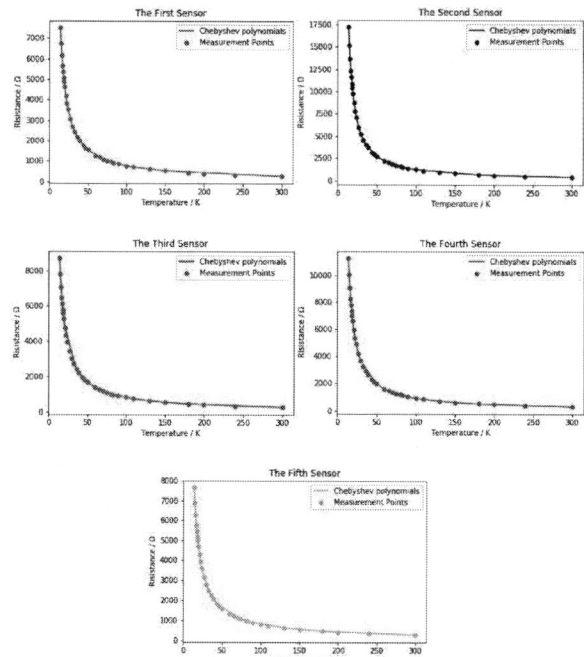

Fig. 3. the 34 measurement points from 14K to 300K and the fitting curve with Chebyshev polynomials about the five kinds of NTC Thermistors.

III. RESULTS

A. the Optimal Power

The self-heating was calculated in 3 different groups of the power, which is 10^{-8} W, 10^{-7} W and 10^{-6} W, as shown in Table. 2.

number	Self-heating (mK)		
	10^{-8}W	10^{-7}W	10^{-6}W
1	0.15	0.63	38.89
2	-0.01	0.03	1.75
3	0.23	0.67	40.06
4	0.05	0.10	2.79
5	0.12	0.51	38.87

Table. 2. the effect of self-heating with different power and batch number

Compared with those 5 kinds of sensors, although the effects of self-heating have a significant distinction on different sensors, the average trend shows that with the power increase, the error caused by self-heating grows to an unacceptable value. However, as the power is 10^{-8} W, the value of the self-heating of the 5th sensor become negative. Because any thermal and shot noise may cover the signal of the current and dramatically increase the uncertainty, in this situation, which is also unacceptable [6]. Therefore, the power of 10^{-7}W is a suitable value for most of sensors, but if the self-heating still has a significant effect, such as the 3rd sensors, the power can be adjusted to the value between 10^{-7}W and 10^{-8}W.

B. the Optimal Measurement Points

Finally, the 28 points, 30 points and 32 points to fitting the equation are compared, as shown in Table .3. The deviation of the 28 points, although is larger than others in most cases, was still limited at 1mK which is acceptable. So, to balance the accuracy and the time overhead, 28 points is the most suitable number of sample data.

number	Fitting deviation (mK) at 20K		
	32 points	30 points	28 points
1	0.11	0.23	0.25
2	-0.10	-0.58	-0.61
3	0.14	0.11	0.27
4	0.23	-0.11	0.34
5	0.08	-0.07	0.07

Table. 3 the deviation at around 20K with different number of measurement point and batch number

So, the final number of the points are 28 is the most suitable number which satisfy the error limitation and the minimum time cost at the same time and the specific temperature points are shown in Table. 4.

No.	Temperature(K)	No.	Temperature(K)
1	14	15	46
2	15	16	54.358
3	16	17	60
4	17.036	18	65
5	17.6	19	75
6	18.8	20	83.806
7	19.6	21	48
8	20.271	22	95
9	22.6	23	120
10	24.556	24	145
11	27	25	170
12	30	26	200
13	36	27	234.316
14	40	28	273.16

Table. 4. the 28 measurement temperature points

It is noticed that the triple point of water (273.16K) and other triple points need to be chosen to verify the accuracy of the NTC thermistor. Because the temperature of those triple points can be measured by other device with an extremely high precision.

IV. CONCLUSION

In this study, the commercial high precision NTC thermistor was calibrated and the deviation should be limited below 1mK at around 20K, and the thermistor characteristic equation fitting by LSM. In most situation, more sampling data leads to higher accuracy. So, to reduce the measurement point and also satisfy the deviation limit, the optimal power which can significantly decrease the class B Uncertainty was discussed and the OPCFM algorithm which can reduce the fitting error to optimized the class A Uncertainty was introduced.

To satisfy the deviation restriction and reduce the measurement point, 5 kinds of NTC thermistors were selected to evaluate the precision. After comparing three different power values (10^{-8} W, 10^{-7} W and 10^{-6} W), we found that, on the one hand, the self-heating effect, along with the power increasing, will significantly increase the class B Uncertainty, on the other hand, if the power is set too small, the amplification of ambient noise also leads to an increase in class B uncertainty. As evaluated, the power around 10^{-7} W is a suitable range to get a minor deviation.

After improving the uncertainty of Class A and Class B, the minimum sample data is discussed within the accuracy limitation. According to principle of LSM and OPCFM algorithm, the measurement points should be no less than 28, so we discuss the feasibility of 28 point, 30 point and 32 point. Finally, although the fitting curve with 28 points has the greatest uncertainty, it is still within the required range. Fortunately, it can save at least 21% of the time in one measurement cycle, at least.

ACKNOWLEDGMENT

This work was partially supported by the Strategic Priority Research Program of Chinese Academy of Sciences (Grant No. XDA 25040000), the National Key R&D Program of China under grant 2020YFB1313502, the National Natural Science Foundation of China (No. 61728402), SJTU Trans-med Award(No.2019015), the Oceanic Interdisciplinary Program of Shanghai Jiao Tong University (No.SL2020ZD205)，Scientific Research Fund of Second Institute of Oceanography, MNR (No.SL2020ZD205) , the Program of Shanghai Academic/Technology Research Leader (18XD1401900). The authors are also grateful to the Center for Advanced Electronic Materials and Devices (AEMD) of Shanghai Jiao Tong University.

REFERENCES

[1] K. Hong and C. Georgiades, "Radiofrequency Ablation: Mechanism of Action and Devices," J Vasc Interv Radiol, vol. 21, no. 8, pp. S179-S186, Aug 2010.
[2] X. Tang, "An optimal piecewise Chebyshev fitting method to calibrate cryogenic temperature sensors," Proceedings of SPIE, vol. 11617, 2020.
[3] Rusby R L, "The International Temperature Scale of 1990 at low temperatures, " Physica B Condensed Matter, 1990, 165-166:35-36.
[4] S. S. Courts, "A standardized diode cryogenic temperature sensor for aerospace applications," Cryogenics, vol. 74, pp. 172-179, Mar 2016
[5] Steinhart, J. S. and Hart S. R., "Calibration curves for thermistors," Deep-Sea Research and Oceanographic Abstracts, 15 (4),497-503 (1968).
[6] C. J. Yeager and S. S. Courts, "A Review of Cryogenic Thermometry and Common Temperature Sensors," IEEE Sens J, vol. 1, no. 4, pp. 352-360, Dec 2001.

Gap in pagination due to unavailable papers.

Pages 1092-1095

Proceedings of the 16th Annual IEEE International
Conference on Nano/Micro Engineered and Molecular Systems
April 25-29, 2021

A Facile Low-Cost Wireless Self-Powered Footwear System for Monitoring Plantar Pressure

Huayi Huang, Ziya Wang, Wenyu Zhao, Zhihao Zhu, Waner Lin and Zhengchun Peng*

Abstract—Feet is known as the second heart of human body. People stand, walk and run on their feet anytime and anywhere. These dynamic processes contain a variety of characteristic information and several biomechanical energy transitions. Previous studies have demonstrated the importance of collecting and analyzing pressure from all parts of the plantar foot in clinical medicine, industrial design, biomechanics, as well as in rehabilitation. This paper presents a lightweight self-powered large range plantar distribution mapping system, which uses a piezoresistive sensor with a porous structure and a low-power integrated data acquisition (DAQ) circuit board to process and transmit signals wirelessly to a host computer. With the program developed by LabVIEW software，the system can dynamically monitor the plantar pressure and display the changes of plantar pressure in real-time. More importantly, a triboelectric nanogenerator have been integrated into the insoles to collect the mechanical energy and continuously powers the whole system which provides a solution of sport/exercise biomechanics pressure monitoring.

I. INTRODUCTION

According to the Journal of the American Medical Association [1], one hundred thousand domestic adults were selected for the study, the results showed that the number of Chinese adults with diabetes reached 11.6%, the total number of people suffering from diabetes reached 113.9 million. What's more shocking is that the number of people with prediabetes was as high as 493 million. If it is not prevented and treated, there is always a risk of illness aggravation. Once it got worsens, all kinds of complications will come one after another.

Coupled with some other factors, such as foot deformity, joint activity is limited. About 1 / 6 of diabetic patients developed plantar ulcers [2]. At present, diabetes cannot be completely cured, amputation can only be the last helpless choice. Because of this kind of foot ulceration, the pressure distribution of the sole of the foot is different from ordinary people, the impulse value of some areas of the forefoot is 1.5 times 1.5 times that of the average value, and the internal and external pressure on the heels are also different from ordinary people obviously. With plantar pressure detection, it can be a good assessment of screening the disease [3], take targeted measures to make the plantar

pressure reasonably distributed, the probability of ulcerative foot can be reduced.

In order to measure the plantar pressure, intensive efforts have been devoted to developing a number of analysis systems such as platforms, in-shoe systems, and wearable sensors. TekScan's HR mat, the system is designed to accurately evaluate the foot pressure of children. The resistance sensing technology is used to measure the foot pressure when barefoot. The spatial resolution is 24 sensors per cm^2 and the maximum sampling frequency is 185hz, using USB transmission, and supports real-time data display in 2D or 3D mode, but the limit of the system is that it can only be used indoors. Germany company Medilogic has placed 64 FSR pressure sensing units in the insole, and has developed an insole-type dynamometer, which can be used to test the plantar pressure changes of the human body in different movement states and can send the data to the computer through wireless transmission for analysis and processing. The sampling frequency of pressure data is divided into 60Hz and 300Hz. This in-shoe system can be used to record the pressure distribution of the sole in the shoe outdoor, but the cost of the system remains high.

Recently, with the development of pressure sensors, triboelectric nanogenerators and wireless technology, some prototypes that have the potential to realize long-term plantar pressure monitoring and self-power system have been presented. This paper will show the development of a low-cost in-shoe device for measuring plantar pressures and high performance triboelectric nanogenerator for powering the wireless communication DAQ system.

II. DESIGN AND METHOD

A. Pressure Sensor

Flexible pressure sensors based on flexible materials can capture the pressure information in tactile perception, and are widely used in human-computer interface, health monitoring, electronic skin, Internet of things, robot tactile and other fields. There are four types of pressure sensors: piezoresistive, capacitive, piezoelectric, and triboelectric, among which piezoresistive sensor is the most widely used. According to the structure, piezoresistive sensors can be divided into three types, one-dimensional, two-dimensional,

*Research supported by Foundation KQTD20170810105439418, 2019.05－2024.04

Huayi Huang. College of Physics and Optoelectronic Engineering of Shenzhen University

Ziya Wang. Shenzhen Institute of Artificial Intelligence and robotics for society

Wenyu Zhao. Shenzhen Institute of Artificial Intelligence and robotics for society

Zhihao Zhu. College of Physics and Optoelectronic Engineering of Shenzhen University

Waner Lin. School of Electrical Information and Electrical Engineering of Shanghai Jiao Tong University

Prof. Zhengchun Peng. Center for Stretchable Electronics and Nano Sensors. College of Physics and Optoelectronic Engineering of Shenzhen University (*Corresponding author)

978-1-6654-3008-1/21 $31.00 © 2021 IEEE

and three-dimensional. These sensors are used in different fields due to their different structures [5]. One-dimensional piezoresistive sensor can detect bending, two-dimensional piezoresistive sensor can detect stress and strain, and three-dimensional piezoresistive sensor has a larger detection range due to its structural advantages, which can be used to detect large pressure.

Due to the more complex structure of the three-dimensional flexible pressure sensor, higher requirements are put forward for the fabrication. Researchers improve the performance of the sensor through the innovation of materials and the manufacturing process. At present new conductive materials such as MXene, CNT, silver nanowires and substrates materials like PDMS, eco flex, TPU are widely used, conductive polymer carbon black has the advantage of low price, while TPU has good adhesion with FPCB substrate of PI materials [4]. 3D printing, screen printing, template sacrifice, freeze-drying, lithography, or electroplating process are used to construct 3D structures, and the sacrificial template method has the advantage of simple operation.

The piezoresistive sensor with a porous structure has good pressure sensitivity and mechanical properties[6]. We can get a porous structure by sacrificing the template method. The following is the preparation process of the device. CB, sodium chloride and TPU dissolved in DMF are mixed, stirred, filled into the mold, cured, salted out, and finally dried to obtain the sensor. The sensor has a multi-stage pore structure. The first stage pore structure is the pore left after sodium chloride dissolves in the water bath, and the second stage pore is the pore formed after DMF volatilizes.

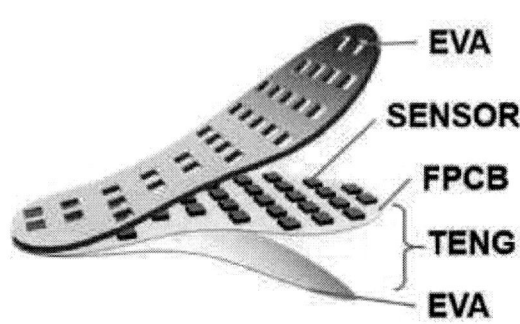

Figure 1. Construction of insole.

B. Assessment of sensor performance

A pressure testing platform was designed to specifically test the performance of the sensor, which consisted of a desktop multimeter (Keithley DMM6500, Tektronix company, USA), a material testing machine (Instron5943, Instron(shanghai) LTD, USA), and a desktop computer. The material testing machine could provide varying compressive force to measure the piezoresistive characteristics of the sensor. The sensor sample was placed on the cylindrical platform of the machine and connected to the desktop multimeter. The computer and the data acquisition module can synchronously record the pressure

value and the resistance value.

Sensitivity is a crucial parameter for the sensor. The purpose of this paper is to improve the sensitivity and working range of the sensor through the synergy of bionic structure design and material properties:

$$S = \frac{\Delta R}{\Delta P \times R_0} = \frac{\Delta I}{\Delta P \times I_0} \quad (1)$$

R_0 represent the initial resistance, I_0 represent the initial current of the pressure sensor ΔR and ΔI represent the resistance and current variation of the pressure sensor, and ΔP is the pressure applied on the pressure sensor.

We used the universal material testing machine to test the device, and used the multimeter to record the resistance change of the sensor.

C. The design of sensors array

The plantar pressure mapping system is designed as a 4×8 matrix, and each sensor size is 7.5mm×7.5mm, uniformly distributed on the FPCB (flexible printed circuit board) with a distance between sensors in row is 9mm and a distance among sensors in columns is 15mm to avoid a short circuit between sensors. And all the sensors are fixed on a custom-made FPCB design by Altium Designer software. The circuit is designed as a shape like an insole, in order to encapsulate to a real shoe-pad., as Fig1.

In order to continuously supply power to the acquisition system, a triboelectric nanogenerator (TENG) has been encapsulated into the shoe-pad. TENGs can convert low-frequency mechanical energy into electrical energy based on the coupling of tribo-electrification and electrostatic induction. As shown in fig.1, During the application of smart shoes based on our pressure matrix and friction electricity, we use ethylene-vinyl acetate copolymer (EVA) foam to wrap the whole pressure matrix. The EVA can not only protect the pressure sensor integrated into the sole from the damage caused by the complex pressure during walking or running, but also form a single motor friction nano generator with the induction electrode and friction layer on the bottom surface of FPCB. The nano generator can collect the mechanical energy from daily walking or joking.

D. Data Acquisition System

In consideration of the demand for the plantar pressure mapping system, the DAQ contains a signal amplification circuit, a Bluetooth module, and an anti-crosstalk analog processing circuit in a compact print circuit board. The low power-consumption integrated circuit is design as a single board with the size of 59.4×48.1 mm^2, as shown in Fig.2.

The main control chip of the system selected from STM32 series micro controller is STM32F103C8T6, which belongs to the Mainstream enhanced series. The chip contains 2 12bit A/D converter, the highest frequency of ADC is 14MHz, which can convert the analog signal into a digital signal in 1 us. With a 7-channel DMA controller, the chip can transport the pressure data to the Serial port directly without micro controller processing. The chip also includes 3 USART interfaces that can easily contact to the bluetooth module for transporting data to the plantar pressure mapping host system. In conclusion, the hardware configurations fully meet the requirements of fast acquisition of pressure matrix data and high-speed data communication.

Figure 2.   Hardware circuit system.

The sensing elements connect in 4×8 format arrays, the number of the connection electrodes would decrease from 64 to 12, However the crosstalk between adjacent elements will occur as long as the row–column electrode format is applied, which will affect the value of voltage corresponding to the resistance of a sensitive element reading by the A/D converter. Therefore, an anti-crosstalk analog processing circuit is necessary [7].

The design of the anti-crosstalk analog processing circuit is based on the zero-potential method as Fig.3 shows. The simplest circuit structures of the zero potential method is that use the concept of virtual short circuit the 0 voltage is fed back to non-current scanned sampling electrodes, and use the single pole double throw connect the 0 voltage to non-current scanned driving electrodes. (as shown in Fig.3, the row and column electrodes are defined as the driving and sampling electrodes, respectively).As for complexity, the Circuit needs a SPDT switches in each row and an OP-AMP in each column. Between the output of sensor matrix and the ADC, an inverse proportional amplifier is applying as sensors' amplification circuits. The MCU using the I/O control the 8 channel SPDT to select the row being driven, and 4 channel ADC sample the value of the sensor according to the time sequence column by column.

As Fig.5 shows, the resistance response of the flexible pressure sensor is non-linear. When the pressure sensor is under certain pressure, the sensor resistance decreases with the increase of the external pressure (The pressure is inversely proportional to the sensor resistance), and the amplify on the column outlet of the sensor array, so that the voltage of the amplifier input signal increase accordingly. The output volt will follow the formula:

$$V_{out} = -V_{in} \times (R_f/R_s) \qquad (2)$$

V_{in} is reference voltage, V_{out} is output voltage, R_f is the reference resistant, R_s is the sensor resistant.
In the inverse proportional circuit, the force electric function becomesliner, but the voltage is negative and the slope is

negative. In Fig.4, OP-AMP circuits were applicated after reading out the signal in order to reverse voltage and amplify the voltage signal of the pressure sensor, so that the output of the OP-AMP signal increase accordingly which could match the need of ADC of the MCU. Other sensors' amplification circuit, like single point amplification circuits, are no longer described detailly in the paper.

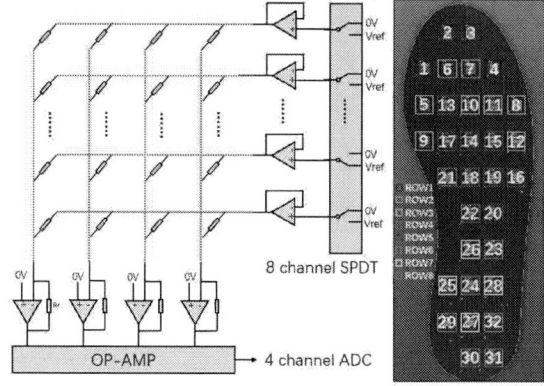

Figure 3.   Design of anti-crosstalk analog processing circuit and the arrangement of the sensors.

E. Data Processing

The OP-AMP circuits circuit was used to amplify the raw data. The output voltage values of 4 OP-AMP circuits 4 were separately connected to 4 ADC channels. The sampling frequency was 10Hz and the sampling resolution was 12-bit. Each data packet contained 64 bytes of data, which was the data of 32 sensors on the same insole. Each sensor data occupied 2 bytes. The sampling resistance value R sensor of the actual pressure sensor could be obtained through equation (2), where V_{in} was the reference voltage value set by voltage divider circuit, R_f was the divider resistance value with a value of 1kΩ, R_s is the resistance of the sensors. The LabVIEW software receives the data of the resistance and according to the electromechanical coefficient calculate the pressure and change in to pressure mapping.

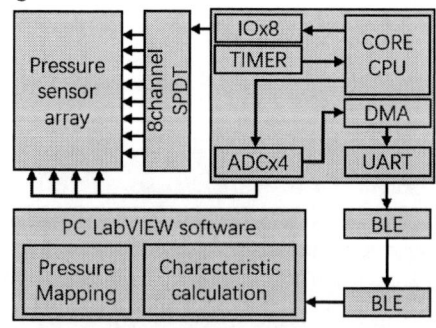

Figure 4.   Design of hardware circuit system.

F. Plantar Data Mapping

To achieve the host computer monitoring interface, the monitoring software of the host computer is developed by LabVIEW software. The MCU acquisition sensors data by reading the scanning the output of anti-crosstalk circuit and send to the Bluetooth Master module, software receive the data from the USB port which connect to the slave module. According to the voltage changes, the sensor resistance value is transformed, displayed, and stored. The man-machine interface of the data processing system is shown in Fig.7. As shown in. Combined with scan readout circuit, it can display the resistance distribution on the sensor surface in two or three dimensions at the rate of 20 frames per second. In pressure mapping, considering that 32 data points should be arranged in a certain area, because the correlation between the actual data is not high, there is an obvious transition between the adjacent data points, if the direct drawing will appear image discontinuity, it needs to interpolate the collected data.

III. RESULTS AND DISCUSSION

A. Performance of Pressure Sensor

1) Sensitivity Test Result

The resistance–pressure (R–P) curve of a sensor sample is shown in Fig.5. The resistance of the sensor decreases as the load increases. According to the equation (1), with the increase of pressure, t the sensitivity of the flexible pressure sensor will decrease. This is because, with the increase of pressure, the modulus of the flexible pressure sensor will increase, which leads to the smaller deformation and the lower sensitivity. But our flexible pressure sensor shows a large working range (20 pa-1.2 MPa) which can support the application of the plantar pressure mapping.

2) Response Time Test Result

As shown in Fig.5, the response time of the sensor is 100ms (orange area), which is much shorter than the stance phase and swing phase in the gait cycle. Even if a person's stride frequency during running can reach 5Hz, the response time of 100ms can also meet the measurement requirements, which shows that the sensor can be used for pressure monitoring in daily activities.

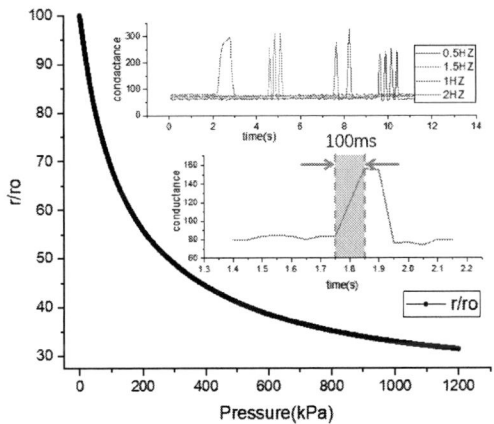

Figure 5. Pressure sensor performance

3) Frequency Response Test Result

Fig. 5 also shows the result of the frequency response test. It can be observed that in the frequency range of 0.5Hz-2Hz (the pink, green, red, and blue line), which includes person's normal walking frequency of 1.7Hz-2Hz, the

amplitude of the output signal of the pressure sensor can be stable during different compression cycles. This shows that the sensor can be used to collect the plantar pressure in different motion states in daily actual situations.

4) Energy Harvest Test Result

Fig.6 shows the results of energy harvest test, the triboelectric nanogenerators (TENGs) can charge the 4.4uF capacitance in about 500 seconds.

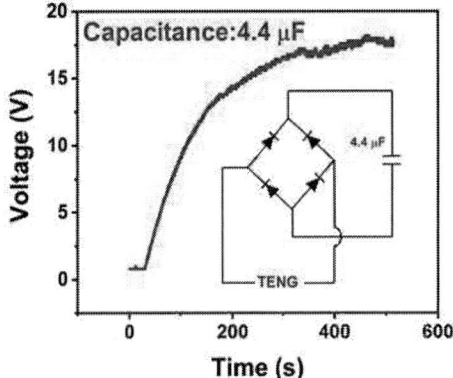

Figure 6. Energy harvest.

B. Performance of Single point

Through a simple experiment by pressing a single point of the sensor array, the data will be wirelessly transmitted to the pressure mapping software on the PC terminal. The pressure applied on the sensor causes the resistance reduction of the sensor and the change reading out by the DAQ system was shown as a pressure color nephogram. Three points of the senor array was pressed and the corresponding point the pressure color nephogram display as a color change. The date was recorded and demonstrate in Fig.8

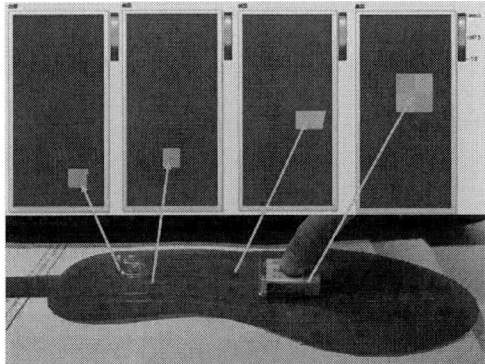

Figure 7. Experiment by pressing single point of the sensor array. The point under press change from purple to green and light blue.

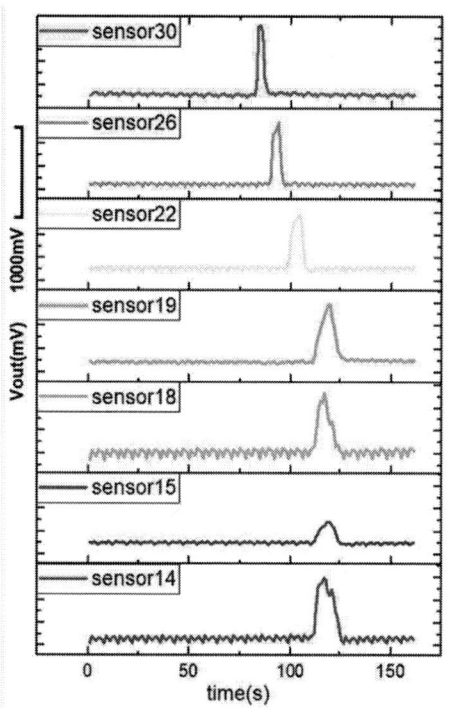

Figure 8. Data of the experiment by pressing single point of the sensor. array sequentially.

C. Performance of Walking Experiment

The normal gait process is divided into the supporting phase and swinging phase, in which the supporting phase accounts for 60% of the total time, and the swinging phase accounts for 40%. For normal people, the left and right feet are symmetrical gait processes, and the plantar pressure distribution also has symmetry. Therefore, in the analysis process, it can be characterized by the distraction of single foot pressure distribution. At the beginning of the gait, the right heel touches the ground, the second stage enters the middle stage of the right foot support phase, and then the right heel is off the ground and only the forefoot touches the ground. When the right foot only has the toe bone touching the ground, it is the end of the right foot support phase.

Figure 9. Walking experiment

The initial stage of the gait process is the stage of one foot following the ground, at this time, only the heel position contacts with the supporting surface, so the plantar pressure distribution is only in the heel position, while the arch and forefoot are suspended. So the sensor 31, 30, 29 were pressed and shown in Fig.10, the peak voltage of three sensors are clearly in front of the others, and the pressure mapping shown in Fig. 11(a). Fig. 11(b) is the stage of full foot landing, when the right foot is full foot landing and the left foot is in the stage of heel lifting, the center of gravity of the body is still in the rear direction, so the pressure distribution cloud is similar to that of standing. In Fig.10, all the sensors are at average pressure and all the sensors voltage shows a platform. When the right heel is lifted and only the toe touches the ground, as fig.11(c) shows, the left foot is full foot landing, the sensor 8, 9 and 10 reached voltage peaks in Fig10. Finally, the right foot left the floor and swings forward, and the center of gravity moves forward, and the voltage will no longer change until the next phase.

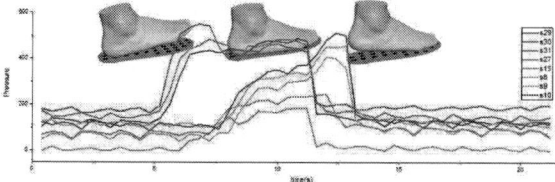

Figure 10. Walking data

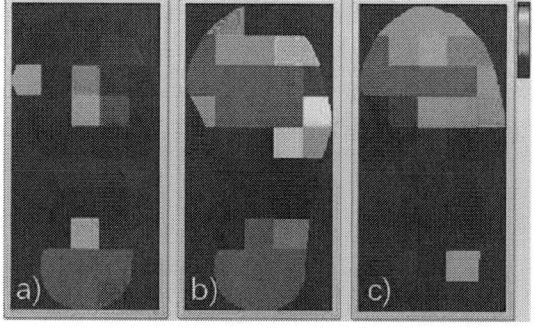

Figure 11. the plantar pressure distribution in real time

IV. CONCLUSION

In this paper, a wireless self-power smart insole system equipped with pressure-sensitive sensors has been developed, which is suitable for research laboratories, clinics, and daily outdoor and indoor activities. A user interface is provided to display and analyze the plantar pressure distribution in real-time. In biometric recognition, the stress center of the sole is calibrated in addition to the cloud map of the sole pressure. The recognition and analysis of the gait is the key to the cooperative control of biomechanics and exoskeleton robots. The results of the recognition of the gait cycle by using intelligent insole show that the intelligent insole can provide good feedback each time in the support phase. The foot pressure distribution in the stage can be analyzed by combining the two-dimensional cloud map of foot pressure distribution. Through a large number of data collection and comparative analysis to establish a biomechanical model, we can more accurately identify and diagnose the plantar pressure distribution.

V. ACKNOWLEDGMENT

The authors would like to thank Ms. Yihe Wang for her help in data collection.

VI. REFERENCEs

[1] Xu Y, Wang L, He J, et al. Prevalence and control of diabetes in Chinese adults[J]. JAMA, 2013,310(9): 948-959

[2] Awad E M, Tremaine D. Diabetic foot advisor[C] Mexico-USA Collaboration in Intelligent Systems Technologies. Proceedings. IEEE, 1996, 299-305

[3] Pham H, Armstrong D G,, et al. Screening techniques to identify people at high risk for diabetic foot ulceration: a prospective multicenter trial[J]. Diabetes care, 2000, 23(5): 606-611

[4] Guan X, Wang Z, et al. A flexible piezoresistive sensor with wide-range pressure measurement based on a graded nest-like architecture, ACS Appl. Mater. Interfaces 2020, 12, 26137–26144

[5] Lin W, He C, et al.Simultaneously Achieving Ultrahigh Sensitivity and Wide Detection Range for Stretchable Strain Sensors with an Interface-Locking Strategy, Adv. Mater. Technol. 2020, 2000008

[6] Wang Z, Guan X, et al. Full 3D Printing of Stretchable Piezoresistive Sensor with Hierarchical Porosity and Multimodulus Architecture. Adv. Funct. Mater. 2018, 1807569

[7] D'Alessio, T. Measurement errors in the scanning of piezoresistive sensors arrays. Sens. Actuators A 72, 71–76 (1999).

[8] Deng CR, Tang W. et al. Self -Powered Insole Plantar Pressure Mapping System. Adv. Funct. Mater. 2018, 1801606

[9] Pablo A, Rodrigo O, et al. Capacitive Sensors Array for Plantar Pressure Measurement Insole fabricated with Flexible PCB. 2018 IEEE

[10] Xinyao H, Fei Shen. et al. A Portable Insole for Foot Plantar Pressure Measurement Based on A Pressure Sensitive Etextile and Voltage Feedback Method.2018 IEEE

[11] Baitong W, et al. Free Walker: a smart insole for longitudinal gait analysis. IEEE 2015

[12] Guan L, Ligang C, et al. Design of Data Acquisition System for Surface Pressure Perception Matrix. IEEE 2017

[13] Shan shi G, et al.A metamaterial for wearable piezoelectric energy harvester. Smart Mater. Struct. 30 (2021) 015026

Proceedings of the 16th Annual IEEE International
Conference on Nano/Micro Engineered and Molecular Systems
April 25-29, 2021

Investigation of the Reliability of the Interconnection between Metal Electrode and Silicon Anchor in Silicon-on-Glass Process

Mengxia Liu, Xianshan Dong, Jian Cui and Qiancheng Zhao*

Abstract—The silicon-on-glass (SOG) process which relies on anodic bonding between the glass substrate and silicon device layer is widely used in the fabrication of a variety of MEMS devices. Electrodes are usually formed by sputtering some metals on glass and located between the silicon and the glass, so its reliability depends on the quality of contact between the silicon and the metal. In this paper, the problems occurred in the interconnection surface of our designed devices are analyzed from the perspective of experiment and simulation. Firstly, the samples are encapsulated with resin, and after the sample is cured, it is processed by longitudinal grinding. Later the slice surfaces are observed by FIB. From this, it can be observed that between the metal and the anchor point there are 13.14nm and 75.36nm wide cracks respectively. Then we carried out finite element simulation on the model by combining solid mechanics and solid heat transfer coupling physical field. The simulation results show that the maximum volume stress can reach 2.88×10^7Pa under 320℃ temperature difference of anodic bonding. We infer that the geometrical relationship between the metal and anchor points in the model is improper, due to the thermal stress of MPa generated during the bonding process, the silicon anchor point slips relative to the metal, and then cracks are generated. In this regard, we put forward the optimization and solution.

I. INTRODUCTION

Silicon-based MEMS sensors have a lot of significant advantages such as small size, low power consumption and low cost, and have been widely used in various fields [1-3]. As a mainstream packaging process, SOG process is easy to achieve and the production cost is low. In this process, the electrode is drawn by means of silicon anchor points and metal lapping. One of the key steps during SOG process -- the anodic bonding process is usually accompanied by a temperature difference of 320℃ [4-5]. Due to the different thermal expansion coefficient between glass and silicon, the bonding surface of silicon and glass will generate non-negligible thermal stress in this process. The stress radiates from the center of the anchor point, and the closer to the edge, the greater the stress. The metal is usually lapped at the edge of a silicon anchor point bonded to the glass to achieve electrical connection. The thermal stress caused by anodic bonding can affects the reliability of the interconnection between metal electrode and silicon anchor,

which degrades sensor performance. To study this problem, the gyroscope samples after preparing are first treated with resin encapsulation, curing and longitudinal sectioning, followed by FIB experiment. Meanwhile, simulation analysis based on solid mechanics and solid heat transfer coupling physical field are carried out with the help of COMSOL.

II. THE SAMPLE DESCRIPTION

The silicon micro-gyro samples in this study are prepared by the standard SOG process developed by the National Key Laboratory of Science and Technology on Micro/Nano Fabrication, Peking University. The specific process flow is shown in Fig.1. Firstly, we choose Pyrex 7740 glass, which has a similar coefficient of thermal expansion to silicon as the substrate. The metal electrode pattern is defined on the surface of the glass, then the glass is etched about 250nm deep and the metal is sputtered approximately 290 nm thick (that is the metal is about 40nm above the glass). Proceeding the lift-off process so that the metal electrode is formed on the glass substrate, shown in Fig.1(a)-(b). Secondly, high doping phosphorus doped silicon (grown by the Czochralski method) is selected. After the anchor area is defined on the back of the silicon, the 30 μm high anchor points are etched using DRIE (Deep Reaction Ion Etching), which described in shown in Fig.1(c)-(d). As shown in Fig.1(e), the silicon substrate is aligned with the glass and then carried out anodic bonding. In this step, the interconnection between electrode metal and the silicon anchor is achieved. After that, the wafer is immersed in KOH solution for thinning by wet etching process, and the remaining thickness of the silicon is 110 μm. Finally, the structure pattern is defined on the silicon surface, and through DRIE, the silicon wafer is anisotropic etched until the structure layer is penetrated to release the structure [6-7]. Fig.2 is a schematic diagram of the position relationship about the anchor points, the metal and the glass in our silicon micro-gyro samples. As illustrated in Fig.2, the part of the metal electrode connected to the silicon anchor has the same width as the anchor (wm=ws=20 μm), and the part of a silicon anchor that is anodic bonded to glass is separated from the part that interconnect with the metal. The

---

Mengxia Liu is with the Institute of Microelectronics, Peking University, Beijing 100871, China.

Xianshan Dong is with the Science and Technology on Reliability Physics and Application of Electronic Component Laboratory, No.5 Electronics Research institute of the Ministry of Industry and Information Technology, Guangzhou 510610, China.

Jian Cui is with the Institute of Microelectronics, Peking University, Beijing 100871, China

Qiancheng Zhao is with the Institute of Microelectronics, Peking University, Beijing 100871, China (phone: 8610-62745160; fax: 8610-62745160; e-mail: zqc@pku.edu.cn).

978-1-6654-3008-1/21 $31.00 © 2021 IEEE

material parameters of silicon and glass in this process are summarized in Table 1.

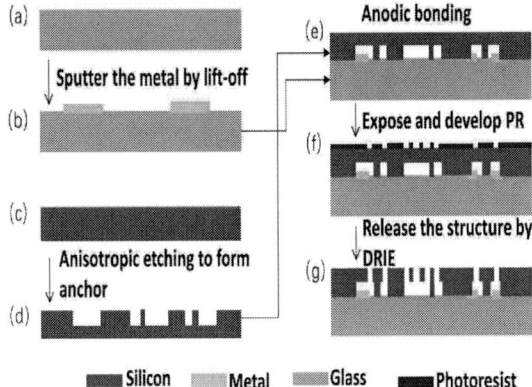

Figure 1.   SOG process flow.

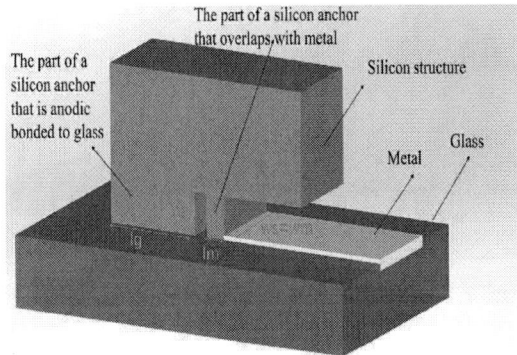

Figure 2.   Schematic diagram of the position relationship among the anchor points, metal and glass in the sample.

TABLE I.   MATERIAL PARAMETERS OF SILICON AND GLASS

Material characteristics	Silicon	Glass
Density(g/cm^3)	2.33	2.23
Young modulus (GPa)	130	64
Poisson's ratio	0.3	0.2
Coefficient of thermal expansion($10{-}6/\text{℃}$)	2.6	3.25

III.   THE EXPERIMENTAL ANALYSIS

After the samples are fabricated by SOG process, they are analyzed experimentally. Firstly, the samples are sealed with resin, and then carry longitudinal grinding after curing. The section surfaces are sliced by FIB for clear observation, after that the samples are observed and measured in detail with the help of SEM. The results are shown in Fig.3 and Fig.4.

Fig.3 shows the SEM observation of sample A. In Fig.3(a), the white part on the glass surface is a metal electrode, the gray part is the silicon anchor points and the silicon structure. It is easy to see that the left big anchor is bonding to glass and the right small anchor laps on the metal.

The contact part between anchor point and metal (the red circle part in Fig.(a)) is observed in detail as shown in Fig.3(b). It can be observed in SEM that there is a clear gap between the anchor point and the metal contact, and the measured height is 13.14nm.

Fig.4 shows the SEM observation of sample B. As illustrated in Fig.(a), the observation site is still where the metal and glass overlap. Observe the red circle in large multiples, as shown in Fig.4(b), we can also aware of the exist of cracks filled with glue. The sputtering thickness of the metal can be measured as 254.0nm, and the gap has a height of 75.36nm between the anchor point and the metal.

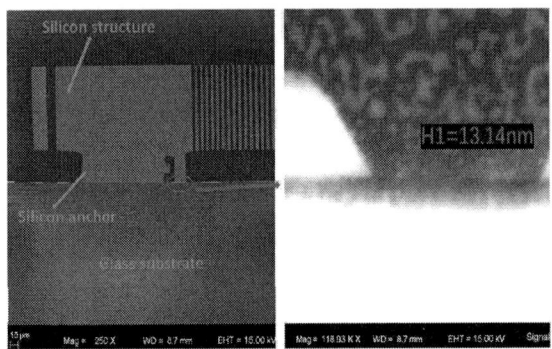

Figure 3.   Electron microscopy and local magnification picture of sample A. (a) The crack bewteen the anchor and the metal; (b) Details of (a).

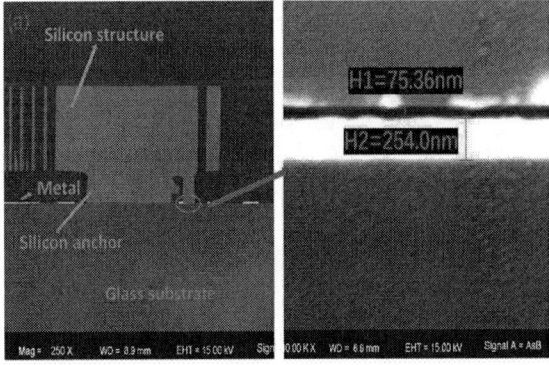

Figure 4.   Electron microscopy and local magnification picture of sample B. (a) The crack bewteen the anchor and the metal; (b) Details of (a).

IV.   THE FEM SIMULATIVE ANALYSIS

In order to study the distribution of thermal stress between silicon and glass due to the temperature difference during anodic bonding, simulation analysis based on solid mechanics and solid heat transfer coupling physical field are carried out with the help of COMSOL is carried out.

We firstly modeled in COMSOL according to the real geometry definition of gyro sample, whose model is called Model A, as shown in Fig.5(a). At the same time, we proposed an improved scheme for model geometry, called Model B, as shown in Fig.6(a). In the model A, the silicon anchor points are separated from the metal lapped parts and the glass-silica bonded parts, and there is a 10 μm wide gap between them. In the model B, the silicon anchor is a whole. The geometric dimension parameters of Model A and Model B are both shown in Table 2.

In order to reduce the amount of calculation, only the glass part in contact with the silicon is drawn in the model. The material parameters of silicon and anchor points are set according to Tab.1. Solid mechanics and solid heat transfer coupling physical field are used in simulation. The grid is divided by physical field control grid. The initial temperature is set at 293.15K and the target temperature is set at 613.15K. The simulation results are shown as Fig.5(b)-(d) and Fig.6(b)-(d), respectively.

TABLE II.   SIMULATION MODEL GEOMETRIC PARAMETERS

Geometric dimensioning (μ m)	Modal A	Modal B
hs	80	80
ha	30	30
lm	20	20
width of gap	10	no gap
lg	200	210
ws	20	20
The thickness of glass	500	500

Simulation results:

Model A -- Extract the stress distribution of silicon - glass contact surface. From Fig.5(b), we can see that the largest von mises stress is $2.88 \times 10^7 (N/m^2)$. Then extract the partial stress tensor of the yz component and the stress tensor of the z component, shown as Fig.5(c)-(d). The maximum value of the partial stress tensor of the yz component is $7.37 \times 10^6 (N/m^2)$, and the peak value of the stress tensor of the z component is $7.83 \times 10^6 (N/m^2)$.

Model B -- Similarly, the stress distribution of silicon - glass contact surface is extracted. The distribution of von mises stress is shown in the Fig.6(b), and we can see that the biggest value is $2.14 \times 10^7 (N/m^2)$. From Fig.6(c)-(d), we can see the partial stress tensor of the yz component and the stress tensor of the z component distribution situation on the surface. The biggest value of the partial stress tensor of the yz component is $7.17 \times 10^6 (N/m^2)$, and the biggest value of the stress tensor of the z component is $2.38 \times 10^6 (N/m^2)$.

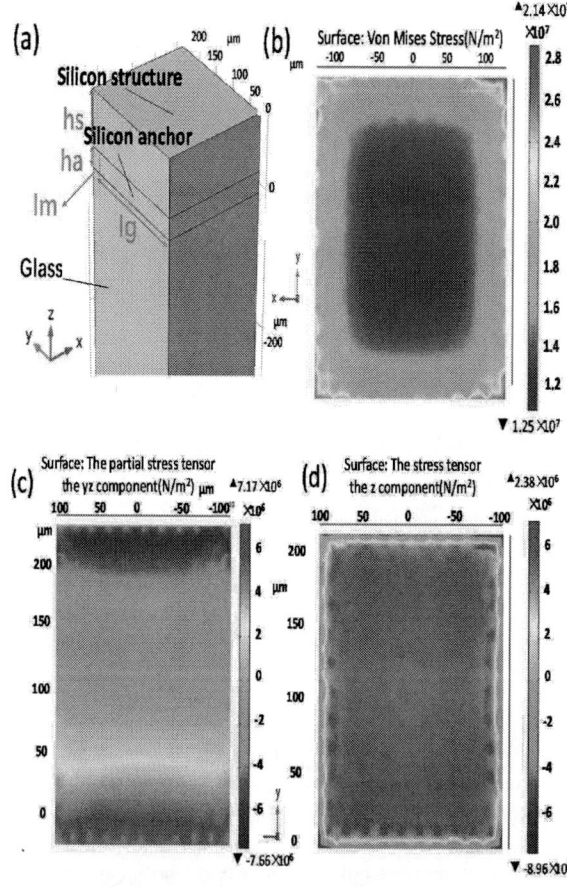

Figure 6.   Model B and its' simulation results. (a) Model B; (b) Von mises stress; (c) The partitial stress tnsor—the yz component; (d) The stress tensor—the z component.

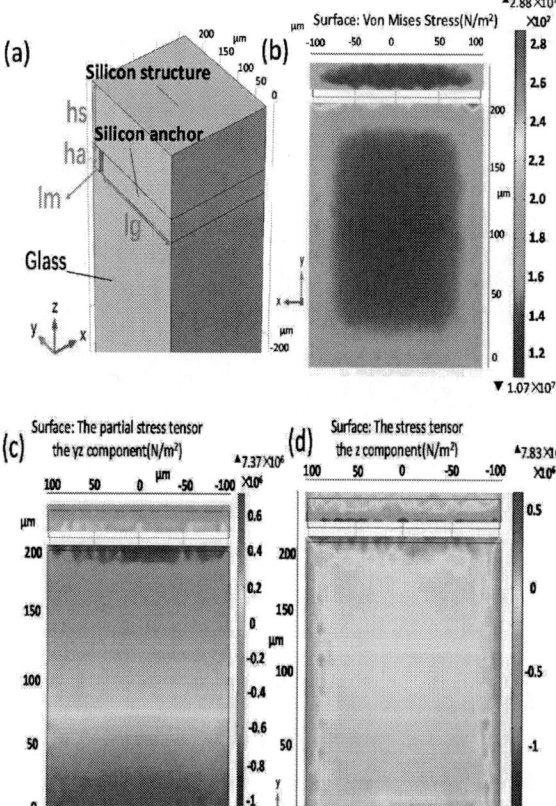

Figure 5.   Model A and its' simulation results. (a) Model A; (b) Von mises stress; (c) The partitial stress tnsor—the yz component; (d) The stress tensor—the z component.

In the process of anode bonding, there is a temperature difference of 320℃, and due to the inconsistency of thermal expansion coefficient between glass and silicon, thermal stress will be generated at the glass-silicon bonding interface. Through the simulation of Model A, we can see that, the maximum volume stress can reach 2.88×10^7Pa. On account of the generation of thermal stress, and one end of the anchor is completely lapped on the metal, this anchor in this section will produce transverse shear forces. Under the action of transverse shear forces, the anchor will slip relative to the metal driven by the thermal stress, resulting in the cracks which is observed in Fig.3 and Fig.4.

By comparing simulation results of Model A with model B, it is obvious that the von mises stress, the partial stress tensor of the yz component and the stress tensor of the z component of the model B are all smaller than model A. In particular, the maximum value of the stress tensor of the z component of the second model is much smaller than the model A, it's one third of the latter. It can be concluded that with a small geometric change, Model B can well alleviate the thermal stress effects of anodic bonding. The reduced stress means that cracks are less likely to develop between the silicon anchor and the metal, which increases the reliability of the silicon micro-gyro.

V. IMPROVEMENT MEASURE

According to the above simulation analysis of model A, it can be known that under the temperature difference of 320K, the interface of anchor point with a size of about 200 microns will have a stress greater than MPa; Because of the thermal stress and the fact that one end of the anchor is completely interconnected to the metal, this end will slides against the metal, creating a crack that can be observed in the experiment (see in section III). The occurrence of cracks leads to poor contact between the metal and silicon, which has a serious impact on the reliability of the device. In view of this, we propose an improvement scheme, as shown in the Fig.7.

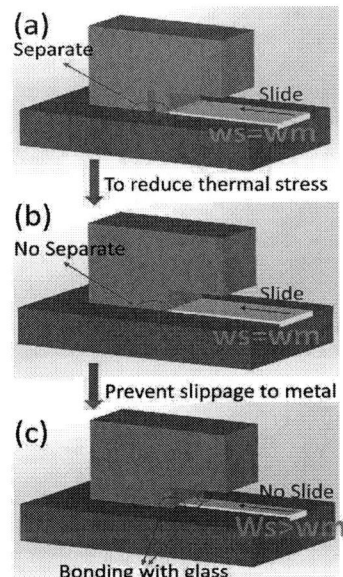

Figure 7. Diagram of improvement measures. (a) Schematic diagram of Model A. (b) Schematic diagram of Model B. (c) The model thoroughly solve the slippage.

The comparion of the above simulation results of Model A and Model B shows that Model B can relieve thermal stress better than Model A. Thus in the layout design, we should choose undivided anchor points to reduce the thermal stress effect as seen in Fig1. (b). In order to completely solve the sliding problem of silicon anchor point relative to metal, model C is proposed-- the width of the metal in model C is smaller than the width of the anchor point, so both sides of the anchor point are fixed to the glass, thus avoiding slippage relative to the metal, as described in Fig7. (c). In layout design, we should give priority to the geometric relationship of Model C, which can theoretically eliminate the relative slip between the silicon anchor and the metal from the source. However, when the anchor points are relatively small and the metal width cannot be smaller, we can choose Model B as the second-best choice to alleviate the thermal stress caused by bonding, thus alleviating the reliability problem of the interconnection between metal electrode and silicon anchor.

VI. CONCLUSION

In order to investigate the reliability of the interconnection between metal electrode and silicon anchor in SOG process, experimental observation and simulation analysis are carried out for the samples. After FIB treatment, it can be observed in SEM that there is a 13.14nm crack between the metal and the anchor point of sample A, and 75.36nm of sample B. Through simulation by combining solid mechanics and solid heat transfer coupling physical field method, it can be obtained that the stress at the anchor interface of our samples can reach the order of MPa magnitude under the temperature difference of 320K. In order to avoid the relative slip of silicon anchor under the action of thermal stress and improve the reliability of metal and silicon contact, structural optimization and improvement schemes are proposed. This method can be applied to physical sensors to improve the stability of electrical connection of devices.

REFERENCES

[1] M. X. Liu et al, "Characterization of thermal mismatch stress of micro-inertial devices based on silicon-glass bonding process," *Journal of Optical Precision Engineering*, vol.28 (2020), pp. 1715–1724.

[2] Z. Y. Song et al, "A Silicon Resonant Accelerometer with Vibrating Beam Integrated with Comb Fingers Sensing Structure," *IEEE NAMS*, Bangkok, Thailand, 2019, pp. 477-481.

[3] J. Cui et al, "A Silicon Resonant Accelerometer Embedded in An Isolation Frame with Stress Relief Anchor," *Journal of Micromachines*, vol.10 (2019), 10(9):571.

[4] T. L. Svetlana et al, "Bond-quality characterization of silicon-glass anodic bonding," *Journal of Sensors and Actuators A*, vol.60 (1997), pp.223-227.

[5] J. Wei et al, "Role of Bonding Temperature and Voltage in Silicon-to-Glass Anodic Bonding," *Electronics Packaging Technology Conference*, IEEE, 2002.

[6] Y. Y. Wang et al, "Study of Silicon-Based MEMS Technology and its Standard Process," *Journal of ACTA ELECTRONICA SINIC*, vol.30, (2002), pp.1577-1584.

[7] A. Lumbantobing et al, "Electrical contact resistance as a diagnostic tool for MEMS contact interfaces," *Journal of Microelectromechanical Systems*, vol.13, (2004), pp. 977-987.

Proceedings of the 16th Annual IEEE International
Conference on Nano/Micro Engineered and Molecular Systems
April 25-29, 2021

A bistable criterion for the V-beam mechanism

Min Liu, Junior Member, *IEEE*, Weidong Wang, Senior Member, *IEEE*, Xiao Wang, Junior Member, *IEEE*, Yingmin Zhu, Siyan Dong, Zimin Huo, Yijia Du

Abstract—**This study presents a method of dividing the beam into two elements beam constraint model (TEBCM) to analysis the bistable characteristic between the force and displacement. Moreover, a formula D_1 for defining the second order bending mode during the deformation of the V-beam is obtained. The finite elements analysis (FEA) is conducted in this study to validate the theoretical study of bistable feature. A comparison between the theoretical solutions and FEA results shows good agreement. Results further show that to design a bistable V-beam mechanism, the tilted angle should be larger for the critical beam length, simultaneously, the beam length should be longer for the critical tilted angle.**

I. INTRODUCTION

Bistable mechanisms have been extensively used in crash sensors [1], relays [2, 3], actuators [4] and switches [5] in micro electro mechanical system (MEMS) due to their unique mechanical behaviors of exhibiting stable states in two distinct positions. Bistable mechanisms demonstrate the ability to maintain two bistable positions without consuming power, as a result, effectively achieve a dramatic decrease in the energy [6]. For a compliant bistable mechanism, V-beam acts as a common structure. Hence, it is important to discuss the bistable phenomenon of the V-beam structure.

A double V-beams structure without prestress is shown in Fig. 1. To achieve more possible bistable characteristics, the symmetry and double beams is needed to provide a stiffer structure and restrict the lateral displacement [1]. The shuttle is regarded as a slider with only one degree of freedom. The bistable behavior of the V-beam structure can be described as the relationship between the load force and the displacement as shown in Fig. 2. The curve has two stable equilibrium positions and an unstable equilibrium position which occur at points where the reaction force is zero. The critical load $F_{\max}$ is the minimum force applied to convert the bistable mechanism from the first stable equilibrium position to the second position, while $F_{\min}$ is the minimum force used to transform the mechanism from the second into the first stable equilibrium position [7]. When a force F greater than the critical load $F_{\max}$ is applied to the shuttle, the state transition will happen from stable state I (solid line) into stable state II (dashed line in Fig. 1).

There are some methods are available for analyzing the performance of the bistable mechanism. Such as the pseudo rigid body model (PRBM) [8], the elliptic integral solutions [9] and the nonlinear finite element analysis (FEA) [7] and the beam constraint model (BCM) [10]. However, due to its low accuracy, PRBM is more suitable for a low-demand analysis and the initial design of the compliance mechanism. The finite element analysis method can be used to predict V-beam bistable characteristics. However, a three-hinge bistable structure does not conform to real boundary conditions. Nonlinear finite element models usually mis predicted a V-beam buckling mode unless a virtual constraint force is exerted on the shuttle. In the light of the geometrical nonlinear theory of large deflection elastic beams, Zhao [11] presented a governing differential equation for the clamped inclined beams. Holst [12] explored an axial bending deflection model of a fixed-guided beam. Different from existing methods of analyzing the large displacement post buckling characteristics of the fixed-guided beam. Based on the classical elastic solution, Kim [9] presented a novel curve decomposition method for the nonlinear analysis of fixed-guided beams. However, the methods proposed above all involve the strongly nonlinear equations formulated in terms of elliptic integrals.

II. CURRENT RESULTS

Considering the V-beam bistable mechanism and the applied load are centrally symmetric, a quarter analyzed beam is shown in Fig. 3(a). The parameters of the beam consist of the thickness T, the length L, the tilted angle θ and the width W. The V-beam subjected to transverse force F_o, moment M_o and axial force P_o at the beam guided end. The deflections of the guided end are denoted as axial X_o and radial Y_o, respectively. The beam is divided into two elements described in Fig. 3(b). For the whole fixed guide V-beam, the end deflection and load parameters are normalized with respect to the length of one element ($L/2$) as

$$x_o = \frac{2X_o}{L}, \ y_o = \frac{2Y_o}{L}, \ t = \frac{2T}{L},$$

$$f_o = \frac{F_o L^2}{4EI}, \ p_o = \frac{P_o L^2}{4EI}, \ m_o = \frac{M_o L}{2EI} \qquad (1)$$

where E is the elastic modulus of the material and $I=WT^3/12$ represents the moment of inertia of the beam.

For a given vertical displacement Δ_Y at the beam guided end, the corresponding deflections can be normalized as

---

* This work was supported by the Science and Technology on Vacuum Technology and Physics Laboratory (Grant No. HTKJ2019KL510007).

Weidong Wang is with the School of Mechano-Electronic Engineering, Xidian University, Xi'an 710071, China. (corresponding author to provide phone: +86-29-88201974; fax: +86-29-88201974; e-mail: wangwd@mail.xidian.edu.cn)

Min Liu, Xiao Wang, Siyan Dong and Yingmin Zhu are with the School of Mechano-Electronic Engineering, Xidian University, Xi'an 710071, China. (e-mail: minl381@163.com; wangxiao9626@outlook.com; ymzhu@xidian.edu.cn; sqy09042021@163.com; Huozimin@163.com).

Yijia Du is with the Microsystem and Terahertz Research Center, China Academy of Engineering Physics, Chengdu 610200, China (e-mail: duyijia@mtrc.ac.cn)

978-1-6654-3008-1/21 $31.00 © 2021 IEEE

$$x_o = -\frac{2\Delta_Y \sin\theta}{L}, \qquad y_o = -\frac{2\Delta_Y \cos\theta}{L} \qquad (2)$$

Owing to all solutions must be real numbers, when the beam thickness t is normalized, the terminal displacement x_o, y_o of the beam satisfy the inequality D_1 as

$$D_1 = -5.365t^2 - 1.163y_o^2 - 3.257x_o \geq 0 \qquad (3)$$

Solution **(a)**

$$\begin{cases} f_o = f_2 = f_1 = -4.8618 y_o \\ p_o = p_2 = p_1 = -9.8837 \\ m_o = m_2 + f_2(1+\delta_{x2}) - p_2\delta_{y2} \\ \quad = 9.8837\dfrac{y_o \mp \sqrt{D_1}}{2} \mp 1.2558\sqrt{D_1} - 4.8618y_o(1+\delta_{x2}) \end{cases} \qquad (4)$$

$$\begin{cases} \delta_{x1} = \delta_{y1}(0.1123y_o - 0.6141\delta_{y1}) - 0.8236t^2 - 0.08123y_o^2 \\ \delta_{y2} = \dfrac{y_o \mp \sqrt{D_1}}{2} \\ \delta_{x2} = x_o - \delta_{x1} \\ m_1 = -m_2 = \pm 1.2558\sqrt{D_1} \end{cases} \qquad (5)$$

There are two results for the bending moment M_o of Eq. (4), which correspond to two deformations paths may occur when the V-beam is deformed by second order bending. The problem of the bending moment M_o is positive or negative as described in [13].

Solution **(b)**

$$\delta_{y1} = \delta_{y2} = y_o/2, \qquad \delta_{x1} = \delta_{x2} = x_o/2 \qquad (6)$$

$$\begin{cases} k_1 = \dfrac{4t^2}{675} + \dfrac{y_o^2}{42000} \\ k_2 = \dfrac{16t^2}{45} - \dfrac{17y_o^2}{2100} - \dfrac{8x_o}{225} \\ k_3 = \dfrac{16t^2}{3} - \dfrac{96y_o^2}{175} - \dfrac{32x_o}{15} \\ k_4 = -\dfrac{48y_o^2}{5} - 32x_o \end{cases} \begin{cases} c_1 = -k_2^2 + 3k_1k_3 \\ c_2 = \Big[-2k_2^3 + 9k_1k_2k_3 - 27k_1^2k_4 + \\ \quad \sqrt{4c_1^3 + (-2k_2^3 + 9k_1k_2k_3 - 27k_1^2k_4)^2} \Big]^{1/3} \\ p_1 = -\dfrac{k_2}{3k_1} - \dfrac{\sqrt[3]{2}c_1}{3k_1c_2} + \dfrac{c_2}{3\sqrt[3]{2}k_1} \end{cases} \qquad (7)$$

$$f_o = \frac{y_o}{2}\left[(12 + \frac{6}{5}p_1 + \frac{1}{700}p_1^2) - \frac{(-6 - \frac{1}{10}p_1 + \frac{1}{1400}p_1^2)^2}{(4 + \frac{2}{15}p_1 - \frac{11}{6300}p_1^2)} \right]$$

$$p_o = p_1 = -\frac{k_2}{3k_1} - \frac{\sqrt[3]{2}c_1}{3k_1c_2} + \frac{c_2}{3\sqrt[3]{2}k_1} \qquad (8)$$

$$m_o = f_2(1 + \frac{x_o}{2}) - p_2\frac{y_o}{2}$$

Substituting Eq. (2) into Eq. (3) and combining the normalized thickness $t = 2T/L$, the following inequality is obtained

$$-4.652\frac{\cos^2\theta}{L^2}\Delta_Y + 6.514\frac{\sin\theta}{L}\Delta_Y - 21.46\frac{T^2}{L^2} \geq 0 \qquad (9)$$

It can be seen from the above derivation that the normalized end loads of the beam are related to the deformation x_o, y_o, and the thickness t of the end of the beam If the given vertical deformation Δ_Y of the beam satisfy inequality (9), then Eq. (4) should be used to solve the end loads of the V-beam, otherwise Eq. (8) should be used.

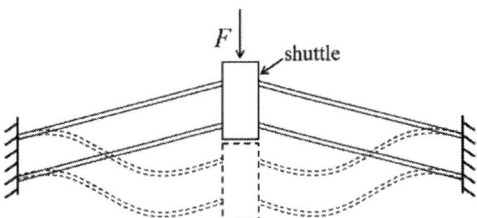

Figure 1. Structure of the V-beam bistable mechanism.

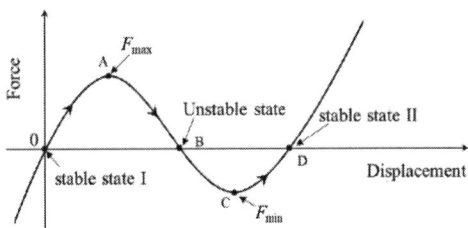

Figure 2. The bistable characteristic curve of the V-beam mechanism

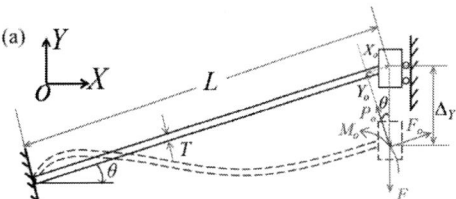

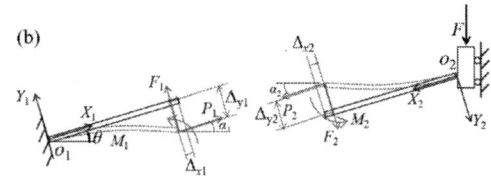

Figure 3. Diagrams of the two elements of the V-beam bistable mechanism. (a) a quarter of the V-beam, (b) two elements divided of the V-beam.

TABLE I.   THE GEOMETRIC PARAMETERS OF THE V-BEAM.

Parameters	L (µm)	T (µm)	W (µm)	θ (°)
Value	385	3	6	1.9

For the inequality (9), the range of displacement Δ_Y is the second order deformation of the beam can be obtained. When the deformation satisfies the inequality, the beam will exhibit the first order buckling. Based on the obtained equations in the above analysis, Table 1 gives the specific geometric parameters of the V-beam responding to Fig. 3(a) to verify the practicability of the above derivation method. The material of the bistable mechanism is polysilicon, its Young's modulus is 164 GPa, and Poisson's ratio is 0.23. Solving the inequality (9), the range of Δ_Y is

$$2.74\ \mu m \leq \Delta_Y \leq 15.15\ \mu m \qquad (10)$$

As shown in Fig. 3(a), for a given vertical displacement Δ_Y at the guided end, the corresponding deflections can be normalized as

$$x_o = -\frac{2\Delta_Y \sin\theta}{L}, \quad y_o = -\frac{2\Delta_Y \cos\theta}{L} \tag{11}$$

The force F required to displace the bistable beam can be expressed as

$$F = -\left(\frac{4EIf_o}{L^2}\cos\theta + \frac{4EIp_o}{L^2}\sin\theta\right) \tag{12}$$

The vertical displacement ΔY of the beam is 0 to 20 μm. When the displacement satisfies the range of Eq. (10), Eq. (4) is used to solve the end loads, otherwise, Eq. (8) is used to solve them. When the end loads (f_o, p_o and m_o) are obtained, the vertical load force F can be defined by using Eq. (12). The curve of the relationship between the load force F and the displacement Δ_Y is shown in Fig. 4. In addition, some FEA results are also given in Fig. 4 as comparing with the results of the TEBCM. It can be seen that the results of TEBCM agree well with FEA results. The V-beam is modeled by the ANSYS software using finite element analysis method, the model frequently predicts a third order mode or higher buckling modes as described in [12], which is obviously different from the bistable characteristics than the more realistic second mode and not usually occur in practice. ANSYS obtains the second buckling mode by using a small guiding force applied to the center of the beam. The guiding force improves the accuracy by inducing second mode buckling, but it slightly offsets the reaction force predictions because it is a nonrealistic loading condition.

Fig. 5 shows the comparison results of two guiding forces 0.5 μN and 0.05 μN with the TEBCM method (a partial enlarged view with a deformation displacement between 1.1 and 5 μm is given). It can be seen that the smaller the guiding force, the smaller offset to the load force F and closer to the formulation analytical solution. Therefore, the applied guiding force is as small as possible. In addition, the Eq. (12) indicates that the width W does not affect the force F value is positive or negative, so it also does not affect the existence of the bistable behavior of the V-beam. To investigate the effect of the beam thickness T, length L and the titled angle θ on the realization of bistable behavior, Figs. 6(a) and (b) show critical length and critical titled angle versus thickness. For a bistable mechanism, when the thickness T is given, each tilted angle has its corresponding critical length. If the beam length is larger than the critical value, the bistable phenomena will occur. Similarly, the length has its corresponding critical titled angle. If the tilted angle is larger than the critical value, bistable phenomena will occur.

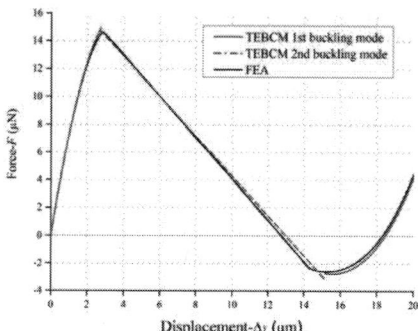

Figure 4. The relationship between the load force and displacement of the V-beam.

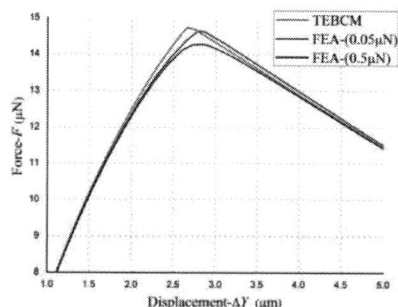

Figure 5. The comparison results of two guiding forces 0.5 and 0.05 (μN) with the TEBCM method.

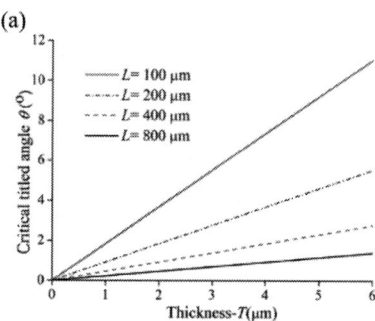

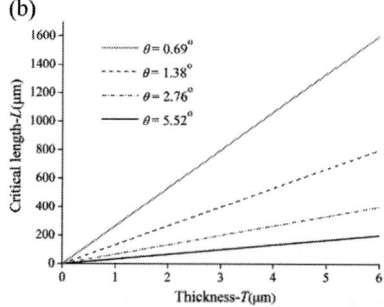

Figure 6. (a) Relationship between the critical tilted angle and thickness under different length.(b) Relationship between the critical length and thickness under different tilted angle.

III. CONCLUSION

In this paper, the bistable behavior of the V-beam is modeled by a new method of two elements beam constraint model (TEBCM). A formula D_1 is used to defining the second order bending mode of the V-beam. The FEA analysis results agree with TEBCM results well. Finally, for a designed thickness of the V-beam, each titled angle has a corresponding critical length and each length has a corresponding critical titled angle. If the beam length is larger than the critical length or the titled angle is larger than the critical titled angle, the bistable phenomenon would occur.

REFERENCES

[1] Sonmez U. Compliant MEMS Crash Sensor Designs: The Preliminary Simulation Results[C]// Intelligent Vehicles Symposium. IEEE, 2007: 303-308.

[2] Hyman D, Mehregany M. Contact physics of gold micro contacts for MEMS switches[J]. Components & Packaging Technologies IEEE Transactions on, 1999, 22(3): 357-364.

[3] Gomm T, Howell L L, Selfridge R H. In-plane linear displacement bistable microrelay[J]. Journal of Micro mechanics & Microengineering, 2002, 12(3): 257-264.

[4] Hwang I H, Shim Y S, Lee J H. Modeling and experimental characterization of the chevron-type bi-stable micro actuator[J]. Journal of Micro mechanics & Micro engineering, 2003, 13(13): 948.

[5] Joshitha C, Sreeja B S, Princy S S, et al. Fabrication and investigation of low actuation voltage curved beam bistable MEMS switch[J]. Micro system Technologies, 2016: 1-14.

[6] Qiu J, Lang J H, Slocum A H. "A curved-beam bistable mechanism," J. Microelectromechanical Systems Journal of, 2004, 13(2): 137-146.

[7] Cherry B B, Howell L L, Jensen B D. Evaluating three-dimensional effects on the behavior of compliant bistable micromechanisms[J]. Journal of Micromechanics & Microengineering, 2008, 18(9): 095001.

[8] Ümit Sönmez, Tutum C C. "A Compliant Bistable Mechanism Design Incorporating Elastica Buckling Beam Theory and Pseudo-Rigid-Body Model," Journal of Mechanical Design, 2008, 130(4): 137-139.

[9] Kim C, Ebenstein D. "Curve Decomposition for Large Deflection Analysis of Fixed-Guided Beams with Application to Statically Balanced Compliant Mechanisms," Journal of Mechanisms & Robotics, 2012, 4(4): 61-68.

[10] Awtar S, Slocum A H, Sevincer E. "Characteristics of Beam-Based Flexure Modules," Journal of Mechanical Design, 2007, 129(6): 625-639.

[11] Zhao J Jia, J, He X, and Wang H, "Post-Buckling and Snap-Through Behavior of Inclined Slender Beams," J. Appl. Mech., 2008, 75(4): 041020.

[12] Holst G L, Teichert G H, Jensen B D. Modeling and Experiments of Buckling Modes and Deflection of Fixed-Guided Beams in Compliant Mechanisms[J]. Journal of Mechanical Design, 2011, 133(5): 051002.

[13] Zhang A, Chen G. A Comprehensive Elliptic Integral Solution to the Large Deflection Problems of Thin Beams in Compliant Mechanisms[C]// ASME 2012 International Design Engineering Technical Conferences and Computers and Information in Engineering Conference. 2012: 420-431.

Proceedings of the 16th Annual IEEE International
Conference on Nano/Micro Engineered and Molecular Systems
April 25-29, 2021

Self-adaptive Microjet Array Cooling for RF High Power GaN Integration on Silicon

Miao Yu, Hao Zhang, Min Huang, Hongze Zhang, and Jian Zhu

Abstract—A novel Si interposer with microjet array cooling structure for RF high power GaN integration is presented in this work. The Si interposer is fabricated by stacking 3 wafers to form 2 layers of self-adaptive net microchannels and microjet array. The microfluids converge to form microjets in the bottom layer, then disperse in the top layer in 4 directions at net nodes. The cooling performance of the Si interposer is analyzed using finite element method (FEM) in COMSOL Multiphysics. The analysis results illustrate that the hotspot cooling capacity of the Si interposer depende largely on the flow rate of the coolant. The RF performance and temperature distribution of X-band GaN Monolithic Microwave Integrated Circuit (MMIC) power amplifier (PA) with static power density of ~ 480 W/cm^2 bonded on the Si interposer were measured in a test cube. The electrical measurement results validate that the cooling performance of the microjet impingement is able to provide the safety operating conditions of GaN MMIC PA. The thermal measurement results demonstrate that the hotspot cooling capacity of the Si interposer enabled GaN PA with hotspot heat density over 323.2 W/mm^2 to operate safely with low flow rate at 70 °C. The interposer is a promising solution for high power integration in silicon-based RF microsystem.

I. Introduction

The thermal management of microjet cooling has gained more attention recently due to the excellent heat transfer characteristics at the hotspots of the power devices in silicon-based microsystem. With the size of RF microsystem scaling down, Si interposer is a widely used in the stacked structure of RF microsystem, due to its low cost, high process precision, high dielectric constant and good thermal property [1]. As the integration density of the components and modules in RF microsystem improves rapidly, the increasing power density is the main bottleneck of the development of highly integrated RF microsystem. Especially for the RF microsystems integrated with GaN MMIC PAs [2,3], it suffers the deterioration of performance when the junction temperature of GaN PA climbs above the limited temperature. The high temperature concentrating at the junctions of transistors and non-uniform temperature distribution on GaN PA lead to an adverse impact on the performance and reliability of RF microsystem, thus the thermal management for the GaN PA integration is a key issue to develop the RF microsystem further. The microfluid structures have been widely applied in MEMS using Si material, and the cooling method evolves from microchannel fluid cooling to spray cooling [4] and microjet impingement cooling [5]. The latent heat of phase change is widely utilized for electronics cooling in many researches. The heat removal capability of the two-phase flow cooling can achieve 2-4 times increase by the benefits of boiling in the microchannels, conversely the application of two-phase flow is limited by the instability due to the flow topology designed inappropriately [6]. The convection flux of single-phase flow has been demonstrated at 1 kW/cm^2 in microchannels with nanoparticles [7], and that of two-phase flow has been demonstrated at 400 W/cm^2 in microchannels [8]. The repeatability of the spray pattern and the damage on the structure of microfluid by the flow are the primary concerns for the practical implementation of the spray cooling [9]. The microjet impingement cooling is the main scheme for hotspots, and it has been demonstrated that power density can be handled up to ~1 kW/cm^2 for high power density electronic cooling [10].

In this paper, it was demonstrated the thermal control capability of the Si interposer with the structure of microjet array is a good solution for GaN PA integration. The 3 Si wafers stacked interposer with 2 layers of net microchannels and microjet array is shown in Figure 1 (a), and the coolant flows along x-axis and y-axis in the top and bottom net microchannels and the microjets are formed along z-axis in the interposer. The structure of the Si interposer is illustrated in Figure 1 (b). The Si interposer is designed using COMSOL Multiphysics. A X-band GaN MMIC PA is bonded on the Si interposer to verify the cooling performance. The static power density of GaN PA is ~ 480 W/cm^2. The interposer is assembled in a test cube with two copper blocks. The circulating coolant tubes are sealed with rubber O-rings to the joint of the blocks. The microjet impingements in the Si interposer are designed beneath the transistors of the PA to achieve high heat transfer coefficiently and cool the transistor junctions efficiently. The microfluids disperse in 4 directions in the top net microchannel from microjets and transfer the heat from the hotspots to surroundings. The output power of assembled GaN PA was measured with 10 % / 30 % / 50 % pulse and continuous wave input RF signal and DC bias. It has been verified that the thermal control of the interposer is able to maintain the operation of the GaN PA instead of the failure with the traditional forced air cooling at continuous wave input. The thermal measurements of GaN PA were taken with DC bias. The microjet array cooled the chip with hotspot heat density of 323.2 W/mm^2 and controlled the junction temperature at 226.1 °C on a 70 °C hot stage when the flow rate was 59.7 mL/min. The simulation results were fitted to the curve of measurement results when the flow rate

---

[1]School of Electronic Science and Engineering, Nanjing University, Nanjing, Jiangsu Province, China.

[2]Nanjing Electronic Devices Institute, Nanjing, Jiangsu Province, China.

*Contacting Author: Miao Yu is with School of Electronic Science and Engineering, Nanjing University, Nanjing, Jiangsu Province, China (phone: +86-25-68005910; e-mail: yumiao1230@smail.edu.cn).

978-1-6654-3008-1/21 $31.00 © 2021 IEEE

was less than 1 mL/s. It was predicted that the thermal resistance of the assembled test cube would decrease to 1.15 K/W when the flow rate of coolant is 3 mL/s according to the simulation. It is equal to the thermal resistance of the GaN MMIC PA bonded on an 8 mm thick copper plate.

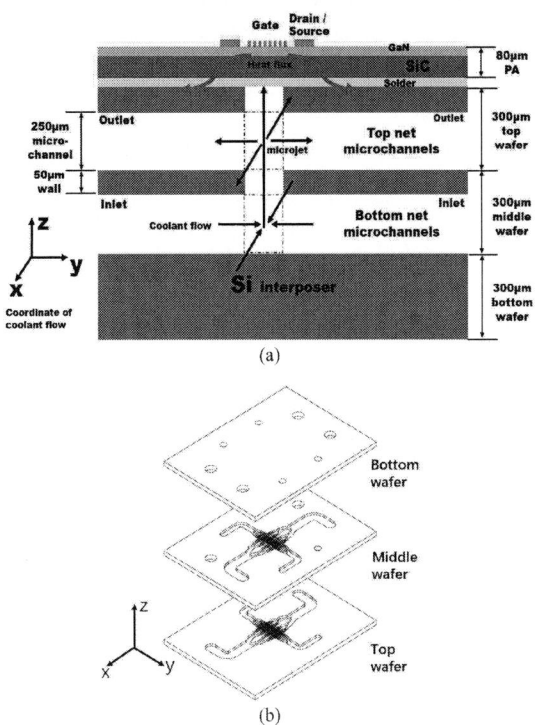

Figure 1. (a) Schematic diagram of the unit cooling structure with the cross-section view and (b) the Si interposer structure with the exploded view from bottom to top.

II. MODELING

The thermal characteristics of the Si interposer were studied in COMSOL Multiphysics. The Si interposer consists of three silicon wafers as Figure 1 (b) depicts: 1) the bottom wafer with 4 inlets and 4 outlets, 2) the middle wafer with inflow microchannels, the nozzles of microjets and 4 outlets from the top wafer and 3) the top wafer with outflow microchannels. 5/16 microchannels in the middle wafer and the top wafer are along x-axis / y-axis in horizontal plane, and the cross-over areas extend vertically as the nozzles of the microjet array beneath the transistors of GaN MMIC PA. The microfluids in the top wafer provide the microfluid convectional cooling and the microjet impingement with phase change cooling. The dryout phenomenon at some hotspots is probably to appear during the phase change. It results from high flow resistance if the microfluid structure is not designed appropriately. The liquid film at the hotspots is thinning when the temperature is above the boiling point at the local pressure, the liquid will dry out at the hotspots in the end if the inflow is insufficient. Then the thermal resistances of the hotspots increase sharply until the transistor is burnout. The net microfluids in the interposer are designed to compensate the inflow at the microjet impingement zone and avoid the dryout phenomenon when the heat density of GaN PA is high.

TABLE I.   STRUCTURE PARAMETERS IN SIMULATION

Component	Size parameters	Unit	Value
Si interposer Size: 6.3mm×4.5mm	Thickness of top wafer	μm	300
	Thickness of middle wafer	μm	300
	Thickness of bottom wafer	μm	300
	Number of x-channels	/	5
	Number of y-channels	/	16
	Pitch of x-channel	μm	375
	Pitch of y-channel	μm	285
	Gap between each 4 y-channels	μm	415
	Width of x-channel	μm	200
	Width of y-channel	μm	150
	Height of channel	μm	200
	Length of nozzle	μm	200
	Width of nozzle	μm	150
	Height of nozzle	μm	100
AuSn Solder	Solder thickness	μm	20
GaN MMIC PA Size: 5.3mm×3.5mm	SiC substrate thickness	μm	80
	Heat density of gates in 1st / 2nd / 3rd stage transistor	W/mm	3/3/7
	Pitch of gate	μm	20
	Width of gate	μm	90
	Number of gates in a HEMT	/	8

TABLE II.   MATERIAL PROPERTIES IN SIMULATION

Material	Density ρ (kg/m³)	Thermal conductivity λ (W/(m·K))	Specific heat C_p (J/kg·K)	Viscosity coefficient μ (Pa·s)
SiC	3300	350	550	-
AuSn	14700	57	128	-
Si	2330	130	700	-
Water	1000	0.6	4200	0.001

The structure size parameters and boundary conditions used in the simulations are listed in Table 1, and the material parameters are listed in Table 2. The model of the interposer is shown in Figure 2 (a) and (b). The temperature distribution of GaN MMIC PA and the interposer is simulated by energy equation for solid and liquid:

$$\rho C_p \frac{\partial T}{\partial t} = \nabla \cdot k \nabla T + Q + Q_p + Q_{vd},\qquad (1)$$

where ρ is the mass density, C_p is the heat capacitance, T is the temperature, k is the thermal conductivity, Q is the heat source, Q_p is the pressure work, and Q_{vd} is the viscous dissipation.

978-1-6654-3008-1/21 $31.00 © 2021 IEEE

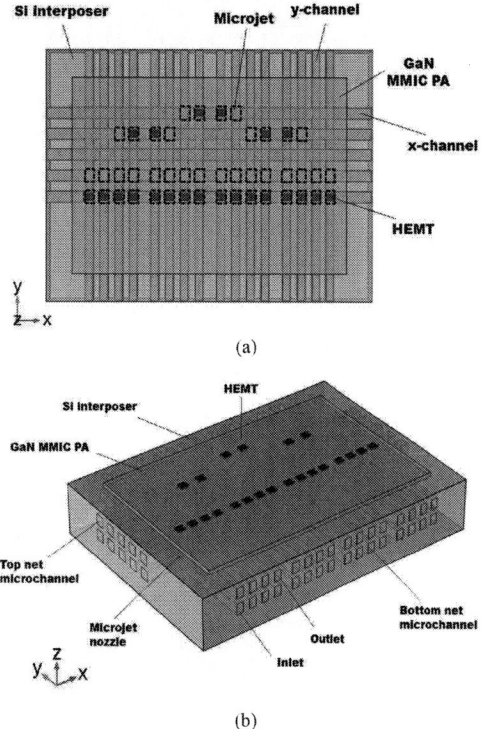

(a)

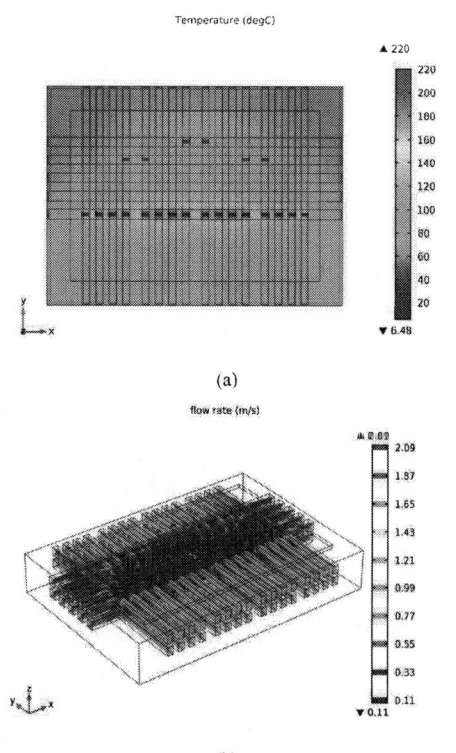

(b)

Figure 2. (a) The Top view and (b) the 3D view of the model of the GaN PA and the Si interposer.

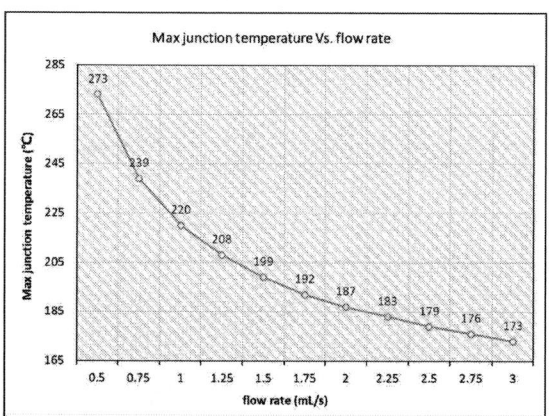

(a)

(b)

Figure 3. (a) The temperature distribution and (b) the flow rate distribution in the microchannels when the flow rate is 1 mL/s.

The Reynolds number of microfluids in the interposer is 100~300, the microfluid flow is laminar in the simulation. The GaN PA is attached on the Si interposer with AuSn solder with the lowest thermal conductivity in the model. The backside temperature of the interposer is set to 70 °C. The temperature distribution and the flow rate distribution are simulated in Figure 3 (a) and (b) on the top view when the flow rate of 1 mL/s. It is found that the flow rate of the coolant drops by ~ 60% through each node of the net microchannel. The results have verified that the design rule is to place the microjets at the peripheral nodes of the net. The maximum temperature of GaN MMIC PA is at the middle gates of the third stage. Therefore, the thermal management of the third stage gates is the major concern of GaN MMIC PA cooling. Keeping microjets beneath the third stage gates at the edge of the net structure is the first priority in the design rule.

The relation between the maximum junction temperature and the flow rate from 0.5 to 3 mL/s is graphed in Figure 4. The maximum junction temperature of the GaN PA decreases with the increase of flow rate. The junction temperature is 220 °C at the flow rate of 1 mL/s when the heat density of hotspots is 400 W/mm^2 in the simulation.

Figure 4. The relationship diagram of the maximum junction temperature of GaN PA vs. the flow rate in the simulation.

III. FABRICATION AND MICRO-ASSEMBLY

A. Fabrication of Si interposer

The structure of the Si interposer is designed based on the model in Figure 2 and the fabrication process flow of the Si interposer is schematically shown in Figure 5. The process contains three main steps: deep etching, metallization and wafer bonding. The deep etching is achieved by inductive coupled plasma (ICP) process. The microchannels in the top wafer and the inlets and outlets in the bottom wafer are etched from bonding interface. The microchannels, the nozzles and outlets in the middle wafer are etched side by side. The diameters of inlet and outlet are 1.5 mm and 2.4 mm respectively. Three wafers are metalized by evaporation and electroplating. The bonding interfaces of wafers are coated with 50 nm/ 100 nm/100 nm Ti/Ni/Au by electron beam evaporation, then electroplated with 4 μm/2 μm Au/Sn on the side without microchannel structure and 4 μm Au on the other side. Then wafer bonding is conducted by thermal compression bonding process. The wafers are clamped with compressive force of 9000 N for 30 minutes at 280 °C in the bonding. Three silicon wafers are bonded from top to bottom. The Si interposer is prepared after a 5 μm layer of Au electroplated on the topside for GaN PA soldering.

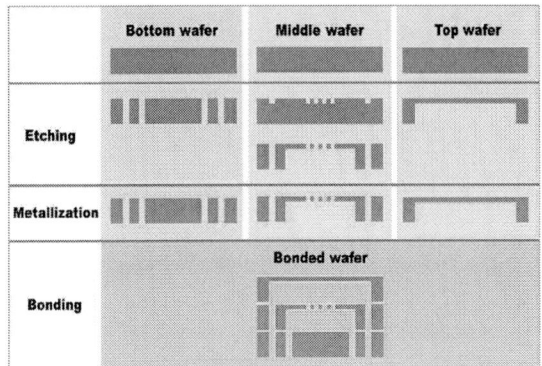

Figure 5. The process flow of the Si interposer.

B. Micro-assembly of test cube

The test cube consists of two copper functional parts as the sample presented in Figure 6. The top part is an assembled cube for device test, and the bottom part is composed of a cavity and two joint blocks for coolant circulation. The process of the micro-assembly is detailed as follows:

1) Installation for electrical measurement:

The micro-assembly of the top part for GaN MMIC PA test refers to the diagram plotted in Figure 7. The components for the micro-assembly contains 50 Ω transmission line substrates, four 50 Ω DC feed-thrus, two 50 Ω RF feed-thrus, two SubMiniature A (SMA) RF connectors, 150 pF and 1000 pF chip capacitors, 0.01 μF ceramic capacitors and 100 μF discrete capacitors. First, GaN MMIC PA and eight 150 pF capacitors are attached on the Si interposer with Au80Sn20 solder at 280 °C. Second, Rogers 5880 transmission lines are soldered with SAC305 at 217 °C on the interposer and top cube. Third, the DC and RF feed-thrus are sintered in the corresponding holes of top cube with SnPb solder at 183 °C. Fourth, six 1000 pF capacitors and two 0.01 μF capacitors are welded on the top cube with SnPbInAu solder at 145 °C. Fifth, Si interposer is clamped between the top cube and joint blocks by screws, then GaN MMIC PA is interconnected with capacitors and transmission lines by wire-bonding. Sixth, DC and RF feed-thrus are soldered to the transmission lines, and 100 μF capacitors and wires are soldered to the given feed-thrus in the diagram. Last, SMA connectors are fastened via the RF feed-thrus on the top cube.

2) Installation for coolant circulation:

The sealing structure of the coolant tubes and the test cube is presented in Figure 8. The O-rings are set in the holes of the joint blocks to seal the inlet and outlet joint, then the polypropylene coolant tubes with inner diameter of 0.8 mm are fastened by screw threads after the joint ends of tubes are inserted through taper joints tightly. Test cube is assembled after the bottom cavity installed with screws onto the top cube.

C. Measurement setup

RF performance and temperature distribution are measured in the microfluid cooling experiments, and the measurement setup is illustrated in Figure 9.

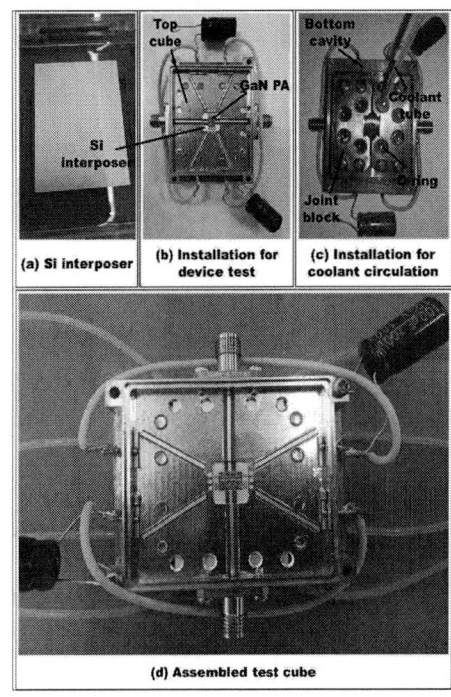

Figure 6. The assembly diagram of the test cube

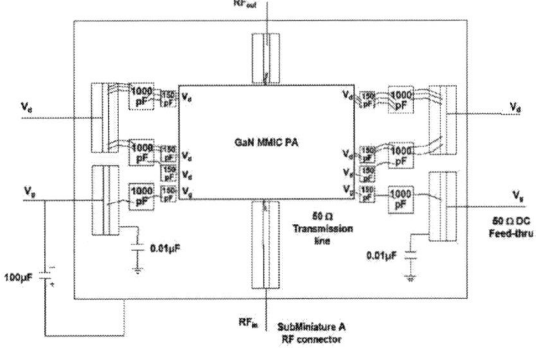

Figure 7. The micro-assembly diagram of GaN PA

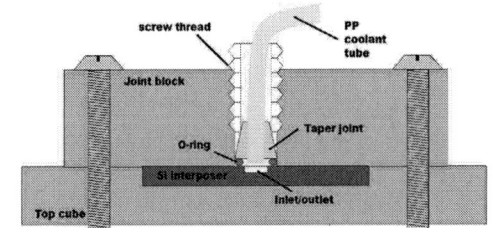

Figure 8. The cross-section of the coolant sealing structure.

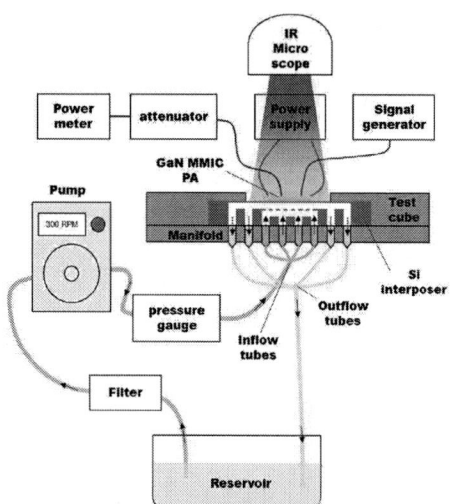

Figure 9. The measurement setup of the GaN PA cooling system

The coolant circulation system comprises a reservoir, filter, pressure gauge, a peristaltic pump and coolant tubes. The deionized water as coolant in the reservoir is driven by the peristaltic pump. It flows through a filter to remove the particle > 2 μm and back to the reservoir after cooling the test cube.

The equipment for RF performance measurement contains signal generator, power supply, and power meter connected with a 25 dB attenuator. The signal generator Agilent Technologies E8257D provides RF input signal for GaN MMIC PA. The power supply KEYSIGHT E3633A and E3634A provide the DC gate and drain bias of V_g and V_d, respectively. And the power meter KEYSIGHT N1912A measures the RF output power of the PA.

2D surface temperature distribution map of PA is obtained by using Quantum Focus instrument (QFI) MWIR-512 InSb infrared microscopy. The test cube is measured under DC bias and plated on a hot stage at 70 °C to acquire good accuracy of the thermal measurement.

IV. MEASUREMENT AND DISCUSSION

The flow rate of coolant is a critical parameter to cool the junction temperature according to the simulation. The flow rate is controlled by the rotation rate of the peristaltic pump, the highest flow rate is 59.7 mL/min (~ 1 mL/s) by the rotation rate of 300 RPM. The RF performance is measured with the highest flow rate, and the temperature distribution is obtained from 100 to 300 RPM.

The RF output power of GaN PA is measured with the following conditions: the drain voltage V_d = 28 V (the saturated drain current I_{DQ} = 3.6 A at the voltage), the gate voltage V_g = -1.6 V and RF input power P_{in} = 28 dBm. The input power is operated with duty cycle of 10 % / 30 %/ 50 % / 100 % and cycle time of 1 ms for comparison. The power density of the hotspots was 181 W/mm² in the GaN PA with a constant input of 28 dBm. The output power of GaN PA measured with 10 % / 30 % pulse input RF signal are drawn in Figure 10 (a) when the test cube was cooling with fan or not. And the GaN PA burned out with 50 % pulse input.

Compared with the condition of 10% pulse input, the output power with 30 % pulse input drops from 46 – 47 dBm to 41 - 43 dBm. The output power with 10 % pulse / 50 % pulse / continuous wave input when the test cube is cooled with microfluids were measured and plotted in Figure 10 (b). The latter two outputs dropped by 0.6 dBm and 1.6 dBm comparing with 10 % pulse input relatively. The Si interposer with microjet array cooling is able to provide the condition for GaN operating at continuous wave.

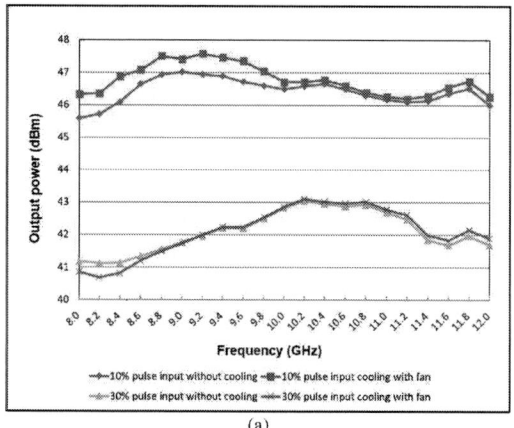

(a)

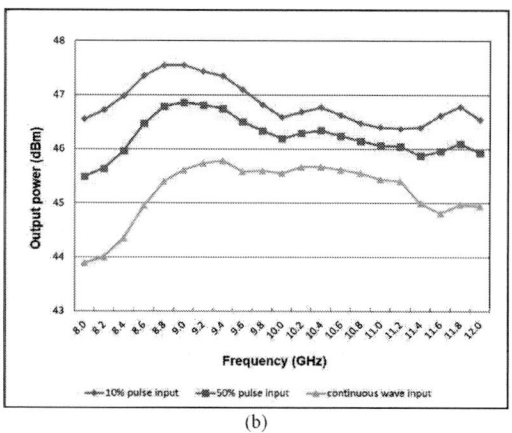

(b)

Figure 10. (a) The test copper cube measured with air cooling or not at 10 %/ 30 % input and (b) the test copper cube measured with microjet cooling at 10 %/ 50 %/ 100 % input

The temperature distribution map of GaN PA was measured after the temperature of test cube was heated up to 70 °C and the temperature amplitude remained in around 1 °C. The thermal measurement of GaN PA is under DC bias conditions: the drain voltage V_d = 28 V (the saturated drain current I_{DQ} = 3.2 A), the gate voltage V_g = -1.8 V without RF input power. The hotspot heat density is 323.2 W/mm² under the conditions. 2D surface temperature map with the flow rate of ~ 1 mL/s is presented in Figure 11, and the relations between the maximum junction temperature and the flow rate in the measurement and simulation are graphed in Figure 12. The highest temperature locates at the third stage transistors of GaN MMIC PA in the measurement. The junction temperature was 226.1 °C when the flow rate was 59.7 mL/min driven by 300 RPM approaching to the safety

978-1-6654-3008-1/21 $31.00 © 2021 IEEE

temperature 225 °C of GaN device. The thermal resistance R can be calculated by the following equation:

$$R = \frac{\Delta t}{\Phi} \qquad (2)$$

Where Δt is the difference between the junction temperature and the environment temperature, Φ is the heat transfer rate. The thermal resistance of the test cube was 1.74 K/W when it was cooled at the flow rate of ~1 mL/s.

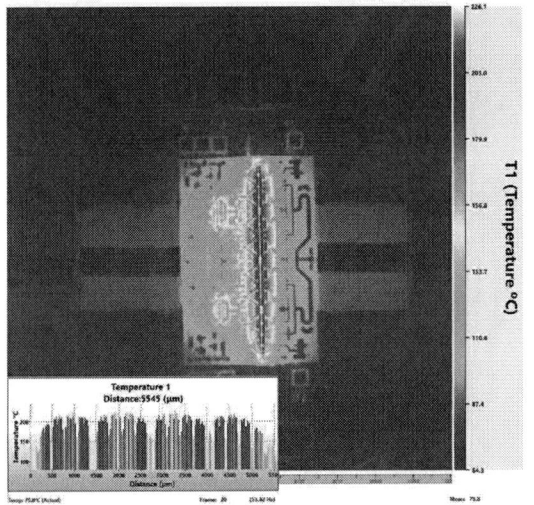

Figure 11. 2D surface temperature distribution map with the flow rate of ~ 1 mL/s.

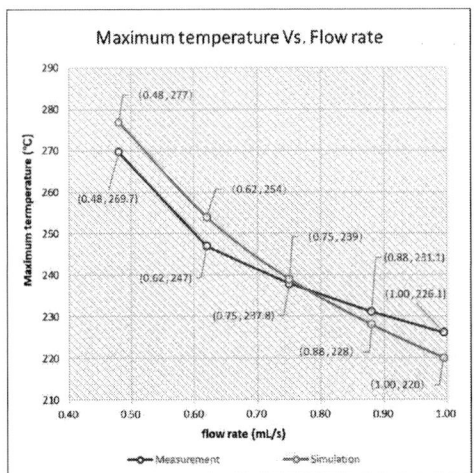

Figure 12. The relationship diagram of the maximum junction temperature of GaN PA vs. the flow rate in the measurement.

The maximum junction temperatures measured by 100 / 150 / 250 / 300 RPM pumping rate are presented in the relationship diagram. The relation acquired in the simulation is based on the model designed above. The measurement results are close to the simulation when the flow rate is less than 1 mL/s. The thermal resistance of the test cube is supposed to decrease to 1.15 K/W with a higher gain of GaN PA when the flow rate is 3 mL/s by comparing the simulation with the measurement. It is equivalent to the situation that the GaN PA is bonded on an 8 mm thick copper plate.

V. CONCLUSION

The test results have demonstrated that the interposer with microjet array cooling provided the operating condition for the high power GaN MMIC PA integrated on silicon. The junction temperature was 226.1 °C when GaN PA with hotspot heat density of 323.2 W/mm^2 was bonded on the interposer and cooled at low flow rate of 59.7 mL/min and high temperature of 70 °C. The microjet array cooling for the hotspots of Ga PA is an effective thermal control method, and it can be enhanced with the increasement of flow rate. The thermal resistance of the assembled test cube is predicted to be around 1.15 K/W by the simulation with the fitting curve to the thermal measurement results. The interposer with microjet array is a promising solution for high power integration in silicon-based RF microsystem.

REFERENCES

[1] Daniel S. Green, Carl L. Dohrman, Jeffrey Demmin, et al. "Heterogeneous integration for revolutionary microwave circuits at DARPA," Microwave Journal, 58 (6), pp. 22-38, 2015.

[2] James J. Komiak. "GaN HEMT: Dominant force in high-frequency solid-state PAs. IEEE Microwave Magazine," 16(3), pp. 97-105, 2015.

[3] Tyler J. Flack, Bejoy N. Pushpakaran, Stephen B. Bayne. "GaN technology for power electronic applications: a review," Journal of Electronic Materials, 45(6), pp. 2673-2682, 2016.

[4] Craig Green, Peter Kottke, Xuefei Han, et al. "A review of two-phase forced cooling in three-dimensional stacked electronics: technology integration," Journal of Electronic Packaging, 137(4): 040802, 2015.

[5] Yong Han, Boon Long Lau, Xiaowu Zhang, et al. "Thermal management of hotspots with a microjet-based hybrid heat sink for GaN-on-Si devices," IEEE Transactions on Components, Packaging and Manufacturing Technology, 4(9), pp. 1441-1450, 2014.

[6] David L. Saums, "Vaporizable dielectric fluid cooling of IGBT power semiconductors for vehicle powertrains," 5th IEEE Vehicle Power and Propulsion Conference, pp. 1-13, 2009.

[7] R. S. Prasher, J.-Y. Chang, I. Sauciuc, S. Narasimhan, D. Chau, G. Chrysler, A. Myers, S. Prstic, and C. Hu, "Nano and micro technology-based next-generation package-level cooling solutions," Intel Technology Journal, 9(4), pp. 285-296, 2015.

[8] A. Koşar, C.-J. Kuo, and Y. Peles, "Boiling heat transfer in rectangular microchannels with reentrant cavities," International Journal of Heat and Mass Transfer, 48(23), pp. 4867-4886, 2005.

[9] D. B. Tuckerman, R.F.W. Pease, "High performance heat sinking for VLSI," IEEE Electron Device Letters, 2 (5), pp. 126 – 129, 1981.

[10] A. Bhunia, Q. Cali; C. L. Chen, "Liquid impingement and phase change for high power density electronic cooling," in Proc. 41st AIAA Aerosp. Sci. Meeting Exhibit, January 2003.

Gap in pagination due to unavailable papers.

Pages 1116-1125

Proceedings of the 16th Annual IEEE International
Conference on Nano/Micro Engineered and Molecular Systems
April 25-29, 2021

A Contactless Switch for Cell Sorting by Area cooling*

Yigang Shen, Yaxiaer Yalikun, *Member, IEEE* , and Yo Tanaka *Member, IEEE*

Abstract— **We develop a flexible system that utilizes thermal convection for particle/cell sorting through a contactless switch in a continuous flow. In this platform, one microheater is applied to control the temperature gradient in the microchannel that induces thermal convection to change the particle lateral movement. An area cooling equipment is used to keep the safe temperature (under 37 °C) for biosamples. Different from the normal thermal manipulation system, the lateral displacement of the particle is changed by the cooling temperature which overcomes the problem of the thermal damage, and the impermanent bonding between microchannel and electrode offers low cost for disposable use, which has a high potential in point-of-care detection and commercialization.**

I. INTRODUCTION

Microfluidic platforms are increasingly attractive for biological applications due to their advantages, such as fast reaction rate, low reagent consumption, and high sensitivity [1]–[4]. Among these, non-contact approaches in microsystems for manipulating particles/cells have attracted extensive attention [5]–[8]. The core of contactless manipulation is to steer objects by harnessing an external force, such as an optical force, acoustic force, magnetic force, electrical force, and thermal force. For instance, optical tweezers have the precise manipulating ability, but could not steer large particles [9]–[11]. Acoustic methods can manipulate different sizes of objects include nanoparticles to microparticles, which need complex fabrication processes and operation equipment. Although electrical methods have simple equipment and can realize particle manipulation and

separation, these methods need to deal with some problems, like, the solution should be conductive and the electrical properties of particle will influence the steerability[12].

Recently, the thermal manipulation method has gotten consideration due for its advantages, such as optic-free mechanism, high integrating ability, no restriction for the conductivity of solution and simple equipment [13] However, since the thermal diffusion and the thermal force depends on the high temperature, most thermal manipulation approaches invariably have thermal damage and are only suitable for static liquid not for continuous flow [14].

Besides, the cost and time of electrode fabrication hamper the development and disposable use of most microsystems. Above all, the thermal method for manipulating particles, especially for biosamples, always suffers from thermal damage, which limits the application in biological research. Here, we develop a contactless switch system that combines with microheaters built on a glass sheet and a cooling system by Peltier element to reduce the thermal diffusion and cause thermal convection to realize particle sorting in continuous flow (Figure.1). The particle movement is altered by the area cooling temperature from 20 to 10 °C, which is different from the convection thermal manipulation by the heating temperature. In this case, the liquid temperature is kept under 37 °C, which makes a safe environment for bio samples. The results indicate that area cooling could flexibly tune the particle movement without thermal damage. Therefore, this simple and flexible

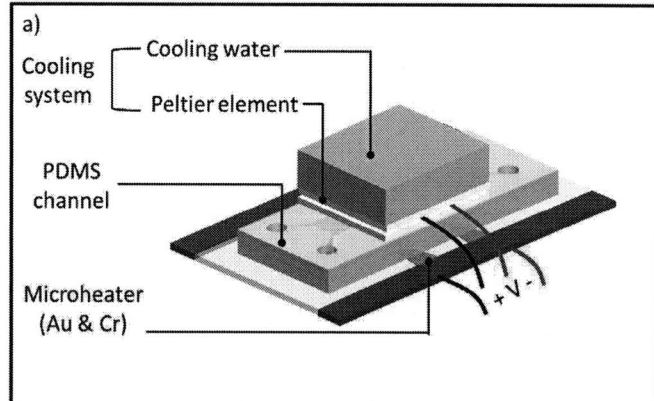
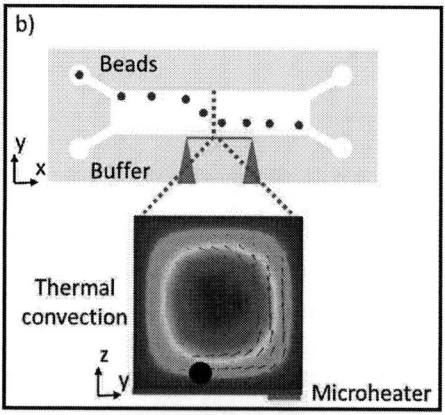

Figure. 1. Contactless switch system. (a) Schematics and (b) working principle of the thermal convection in continuous flow.

*Resrach supported by Foundation.
Yigang Shen is with the Graduate School of Frontier Biosciences, Osaka University, 565-0871 Japan (corresponding author to provide phone: 131-5790-0552; e-mail: yigang.shen@ riken.jp).

Yaxiaer Yalikun is now with the division of Materials Science, Nara Institute of Science and Technology, JAPAN (e-mail: yaxiaer@ms.naist.jp).
Yo Tanaka is with Laboratory for Integrated Biodevices, Center for Biosystems Dynamics Research (BDR), RIKEN, JAPAN (e-mail: yo.tanaka@riken.jp).

978-1-6654-3008-1/21 $31.00 © 2021 IEEE

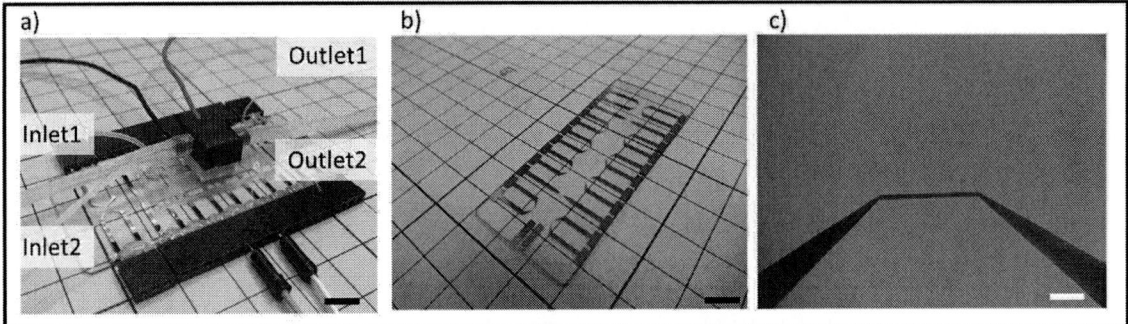

Figure. 2. The photograph of the fabricated particle sorting device. (a) Microfluidic system (b) Glass substrate with microheaters and (c) single microheater. (Scale bar: 100 µm white and 0.5 mm black)

platform will have high potential in biology research that relates to cell sorting.

II. METHODS

Device fabrication and experimental setup

This contactless switch system was consisted of glass sheet with microheaters, a polydimethylsiloxane (PDMS) microchannel bonded with a thin glass slide, and a cooling system combined Peltier and cooling water on the top (Figure. 2a). The microheaters were fabricated by the standard photolithography techniques, which were made of two metal layers (Cr and Au) (Figure. 2b-c). The microfluidic channel was built by a 3D printed model (Form 3, YOKOITO, Japan), and used the standard PDMS fabrication process. The microfluidic channel was connected to microheaters by gravity, which made the low cost for disposal use. By using this design concept, it only needs to fabricate microheaters once and replace the microfluidic channel every time for reducing contamination. In experiments, 10 µm polystyrene spheres (PS) (Polysciences. Inc, PA, USA) were suspended in deionized (DI) water. A microscope (MF-B1010C; Mitutoyo, Japan) was utilized to record particle movement, and the Fiji-ImageJ software was used to calculate the particle velocity. A DC power (PPS303; AS one, Japan) was utilized to supply power for the microheaters and the cooling system with the Peltier element. The temperature of microheaters was measured by the infrared camera (FlIRONE PRO, FLIR, USA).

Working principle

The solution in the microchannel can be heated when the microheater is applied by DC power, which leads to a change in the density of the solution. High temperature reduces the density of the local liquid, which causes the thermal buoyancy convection flow. Under this buoyancy convection, the particles close to the bottom layer are transferred to the other side with the microheater. In this paper, we tune the particle displacement by altering the area cooling temperature, which influences the strength of the buoyancy convection.

III. REULTS AND DISCUSSION

Temperature distribution

In the designed microfluidic system, the temperature of liquid in the microchannel is controlled by the microheater and area cooling system. Figure. 3a indicates the average temperature of microheater increases from 32 to 45 °C as the applied voltage rises from 2.5 to 4 V. For the area cooling system, the Peltier element is used to offer the cooling effect. The cooling temperature change is shown in Figure 3b, which indicates that the cooling temperature becomes stable in 60s, and 20,15 and 10°C cooling temperature are realized when the circuit applied to the Peltier is 0.6, 1.3 and 2.4A, respectively.

To understand the temperature distribution in the flow, a 3D physical model is built by using COMSOL software (COMSOL Multiphysics, 5.4). We mode a glass substrate

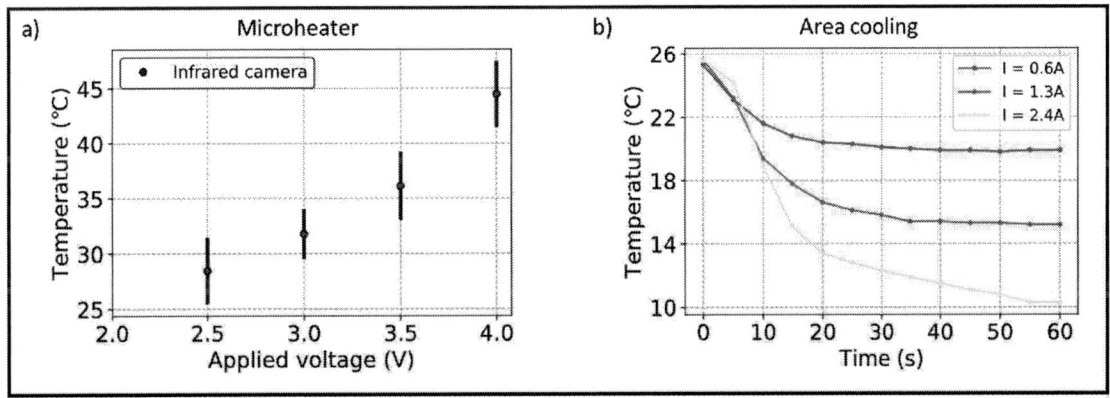

Figure. 3 Temperature distribution. (a) Temperature–voltage characteristics of electrode. (b) Temperature–current characteristics of area cooling system

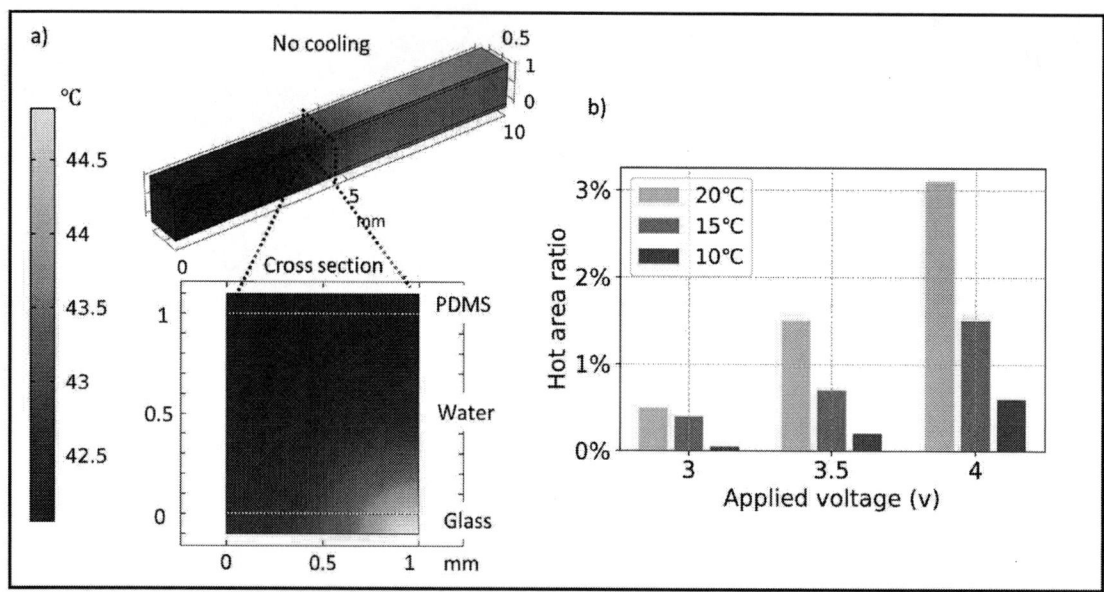

Figure. 4 (a) Temperature distribution in microchannel with on cooling (voltage amplitude, 4 Input velocity: 0.5 mm/s). (b) The area ratio between hot temperature area (> 37 ℃) and the whole area of cross-section in different cooling temperature.

(10 mm length, 1 mm width and 0.1 mm thick) with a channel (10 mm length, 1 mm width and 1 mm thick)), and a microheater (0.5 mm length and 0.1 mm width). Heat transfer in liquid and laminar flow models are used to calculate the temperature profile.

From the simulation of temperature distribution in the microchannel, the whole surface temperature of the middle cross-section is over 42 °C (Figure. 4a). In this situation, the biosamples are absolutely affected by thermal damage. When the cooling system is applied, the hot area (>37 °C) is hugely decreased even at the cooling temperature of 20 °C (Figure. 4b). When the cooling temperature decreases to 10°C, the hot area ratio is less than 1%. Generally, when the environmental temperature is over 37 ° C, live cells may have thermal damages. In contrast, the cold region (<37 °C) is significant for keeping cell viability. Besides, it is reported that live cells incubated at low temperature (<16 °C) keep a

good viability when they are recovered for a long time [15]. Thus, this platform can use the cooling temperature system to keeps a suitable temperature for these cells without thermal damage.

Particle sorting by area cooling

After observing the temperature profile in the microchannel, the performance of particle sorting is tested. Figure 5 shows the 10 µm particle migration in different cooling temperatures (particle velocity is around 0.5 mm/s). It could eliminate the thermal damage from the hot temperature by changing the cooling temperature instead of changing the temperature of microheaters and the results show that lowering the cooling temperature decreases the lateral displacement of particles (Figure 6), which means the thermal convection becomes weak when decreasing the cooling temperature.

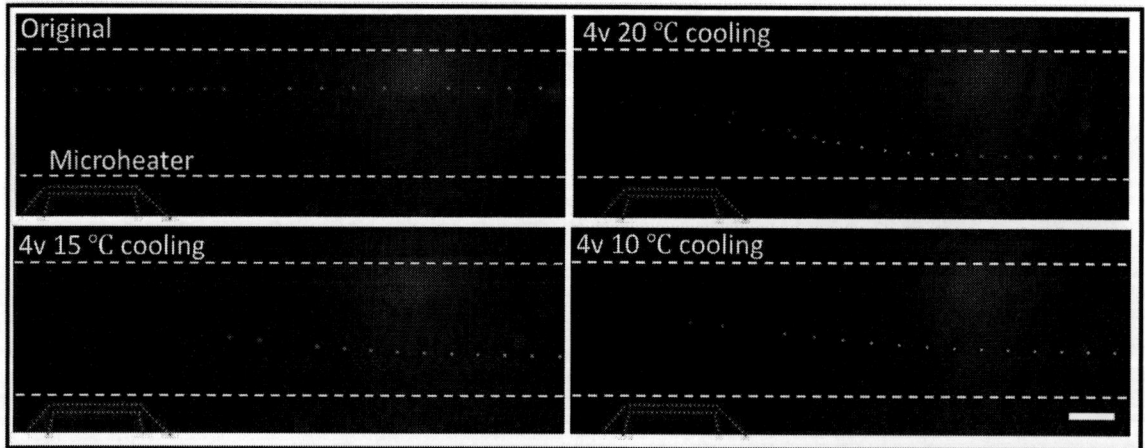

Figure. 5. Trajectory of a 10 µm PS particle migration in different cooling temperature. (Scale bar: 200 µm, time step, 0.2 s)

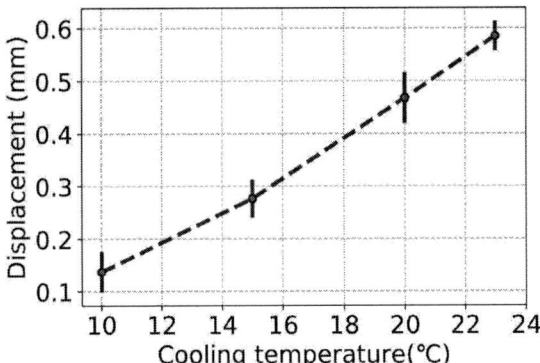

Figure. 6 The particle displacement in different cooling temperature

In most of the thermal manipulation systems, they controlled the hot temperature bellowed 37 °C to keep good biocompatibility. However, the thermal convection is not strong enough to steer the particle in continuous at this temperature. Although the particle has a large displacement by the thermal convection, the sorting ability is limited by the cooling performance. From the temperature experience, the cooling temperature will be stable after 60s, which means the sorting time is not fast enough. And we think this problem will be alleviated by optimizing the cooling system and using high thermal conductive material to build a microchannel.

IV. CONCLUSION

This work presented a contactless switch device that combined a microheater and an area cooling system for particle sorting by thermal convection. The ability of area cooling to reduce thermal diffusion in a continuous flow was characterized by using simulation. The ability to sort particles was demonstrated. This contactless thermal switch is able to steer particles in a large displacement (0~0.6 mm) and greatly limit thermal diffusion. This manipulation system offers three features. (1) Tunable temperature region: the system can effectively control temperature. (2) Biocompatibility: the system can keep a safe and low-temperature environment for biosamples. (3) Flexibility: the system with impermanent bonding between microchannel and electrode offers low cost for disposable use.

ACKNOWLEDGMENT

We acknowledge support received from Grant-in-Aid for Scientific Research (B) (20H02596) from the Japan Society for the Promotion of Science (JSPS), TEPCO memorial Fund, the Amada Foundation, Japan.

REFERENCES

[1] C. E. Majors, C. A. Smith, M. E. Natoli, K. A. Kundrod, and R. Richards-Kortum, "Point-of-care diagnostics to improve maternal and neonatal health in low-resource settings," *Lab Chip*, vol. 17, no. 20, pp. 3351–3387, 2017.

[2] Y. Shen, Y. Yalikun, and Y. Tanaka, "Recent advances in microfluidic cell sorting systems," *Sensors Actuators B Chem.*, vol. 282, no. August 2018, pp. 268–281, 2018.

[3] G. M. Whitesides, "The origins and the future of microfluidics," *Nature*. 2006.

[4] Y. Shen *et al.*, "Flow analysis on microcasting with degassed polydimethylsiloxane micro-channels for cell patterning with cross-linked albumin," *PLoS One*, vol. 15, no. 5, p. e0232518, 2020.

[5] L. Huang, P. Zhao, and W. Wang, "3D cell electrorotation and imaging for measuring multiple cellular biophysical properties," *Lab Chip*, vol. 18, no. 16, pp. 2359–2368, 2018.

[6] Q. Cao, Q. Fan, Q. Chen, C. Liu, X. Han, and L. Li, "Recent advances in manipulation of micro- and nano-objects with magnetic fields at small scales," *Mater. Horizons*, vol. 7, no. 3, pp. 638–666, 2020.

[7] Y. Shen, Y. Yalikun, Y. Aishan, N. Tanaka, A. Sato, and Y. Tanaka, "Area cooling enables thermal positioning and manipulation of single cells," *Lab Chip*, vol. 20, no. 20, pp. 3733–3743, 2020.

[8] Y. Yalikun, Y. Hosokawa, T. Iino, and Y. Tanaka, "An all-glass 12 μm ultra-thin and flexible micro-fluidic chip fabricated by femtosecond laser processing," *Lab Chip*, vol. 16, no. 13, pp. 2427–2433, 2016.

[9] J. Chen *et al.*, "Thermal gradient induced tweezers for the manipulation of particles and cells," *Sci. Rep.*, vol. 6, no. October, pp. 1–13, 2016.

[10] F. Falleroni, V. Torre, and D. Cojoc, "Cell mechanotransduction with piconewton forces applied by optical tweezers," *Front. Cell. Neurosci.*, vol. 12, no. May, pp. 1–11, 2018.

[11] Z. Zhu and C. J. Yang, "Hydrogel droplet microfluidics for high-throughput single molecule/cell analysis," *Acc. Chem. Res.*, vol. 50, no. 1, pp. 22–31, 2017.

[12] K. Zhao and D. Li, "Continuous separation of nanoparticles by type via localized DC-dielectrophoresis using asymmetric nano-orifice in pressure-driven flow," *Sensors Actuators, B Chem.*, vol. 250, pp. 274–284, 2017.

[13] H. Cong, J. Chen, and H. P. Ho, "Trapping, sorting and transferring of micro-particles and live cells using electric current-induced thermal tweezers," *Sensors Actuators, B Chem.*, vol. 264, pp. 224–233, 2018.

[14] K. Zhang *et al.*, "Efficient particle and droplet manipulation utilizing the combined thermal buoyancy convection and temperature-enhanced rotating induced-charge electroosmotic flow," *Anal. Chim. Acta*, vol. 1096, pp. 108–119, 2020.

[15] J. Wang *et al.*, "The analysis of viability for mammalian cells treated at different temperatures and its application in cell shipment," *PLoS One*, vol. 12, no. 4, pp. 1–16, 2017.

Gap in pagination due to unavailable papers.

Pages 1130-1135

Proceedings of the 16th Annual IEEE International
Conference on Nano/Micro Engineered and Molecular Systems
April 25-29, 2021

Real-time Monitoring and Analysis of Jet Behaviors in Electrohydrodynamic Direct-Writing

Yifang Liu, Junyu Chen, Jiaxin Jiang, Guoyi Kang, Huatan Chen, Jianyi Zheng, and Gaofeng Zheng*

Abstract— The jet behavior is one of the most important factors of the accurate control of electrohydrodynamic direct-writing (EDW). A real-time monitoring and analysis system of jet behaviors is built. The charge coupled device (CCD) images are processed by threshold segmentation and other methods to obtain the jet images. The experiment shows three jet modes of EDW. The area and center geometry position of droplet changes periodically with the jet modes, which are controllable. The correlation between the jet features and the jet modes is explored, which lays a foundation for the close-loop control of EDW system to improve the stability and uniformity of micro/nano fibers.

I. INTRODUCTION

Electrohydrodynamic direct-writing (EDW) [1] is a novel inkjet printing process, which has a good application prospect in flexible electronics [2-4], biological tissue engineering [5, 6], energy harvesting [7] and other fields with its low cost, easy integration, continuous, and accurate positioning [8-10]. The deposition of nanofiber is coordinated by the position and movement of the collector, and the spray pattern of the droplets [11]. By reducing the distance between the needle and the collector, EDW technology makes it easier to control the deposition position of the jet, which can achieve precise deposition of nanofibers [12].

The charged jet is unstable because of many factors, such as interference viscoelastic stress and charge repulsion [13]. There are several jet modes for EDW, including the micro-dripping mode, spindle mode, cone-jet mode, and multi-jet mode [14]. Different jet modes are directly related to the result of deposition, which has a great impact on the micro-nano structure [15]. The stable and uniform micro/nano fibers can only be obtained under the cone-jet mode, which is beneficial to control the precise deposition of fibers. Therefore, it is important to monitor and analyze the jet behaviors for accurate controlling the charged jet and realizing the closed-loop control of EDW technology.

Mounica et al [16] measured the images by ImageJ software to obtain the jet boundary and to analyze the variation of jet radius with the length of the jet. Li et al [17] used the high-speed camera to record the jet motions in three different melt electrospinning systems, which were obtained to study the effects of electric field distribution on the jet motion. Image recognition has been applied in the analysis of jet behaviors in EDW technology.

In order to explore the relationship between the jet modes and the jet behaviors, a system of jet behavior real-time monitoring and analysis is built. The jet modes are judged by the center geometry position and area of droplet. The applied voltage is regulated to verify the repeatability of the law and the controllability of the jet modes.

II. EXPERIMENTAL DETAILS

The schematic diagram of electrohydrodynamic direct-writing system is shown in Fig. 1, including a precision syringe pump (Harvard Apparatus Pump 11 Pico Plus Elite Series, Holliston, MA, USA), a high voltage power supply, a syringe (1 mL), a precise motion platform, a collector, a CCD camera (UI-2250-C, IDS Imaging Development Systems GmbH, Obersulm, German), and a host computer. Polyethylene oxide (PEO, molecular weight = 300,000 g/mol) aqueous solution with concentration of 8 wt% is used as the EDW liquid, which is dissolved in a solvent, of which the volume ratio of deionized water and ethanol is 3:1.

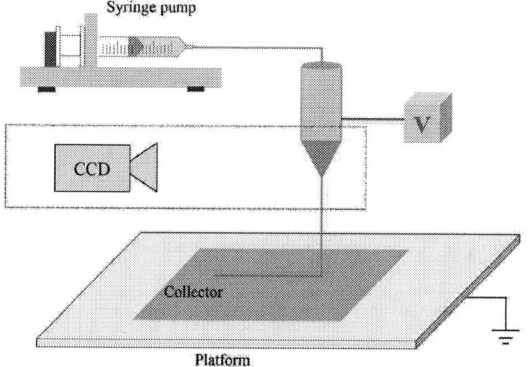

Figure 1. The schematic diagram of electrohydrodynamic direct-writing system

The host computer controls the precise motion platform and the output voltage of the power supply. The syringe pump supplies the solution to the nozzle at a constant rate. The process of the jet behavior monitoring and analysis program is divided into four steps. First, the jet images are collected by the CCD camera at the frequency of 50 Hz.

---

* Research supported by Science and Technology Planning Project of Fujian Province (No. 2020H6003, 2018J01082), Xiamen Municipal Science and Technology Project (No. 3502Z20193015), Aviation Key Laboratory of Science and Technology on Inertia (No. 20180868001), Natural Science Foundation of Guangdong Province (No. 2018A030313522), Science and Technology Planning Project of Shenzhen Municipality in China (No. JCYJ20180306173000073).

Yifang Liu, Junyu Chen, Jiaxin Jiang, Guoyi Kang, Huatan Chen, Jianyi Zheng, and Gaofeng Zheng are with the Department of Instrumental and Electrical Engineering, Xiamen University, Xiamen 361102, China and Shenzhen Research Institute of Xiamen University, Shenzhen 518000, China (corresponding author: Gaofeng Zheng, phone: +86-592-2194957; fax: +86-592-2182221; e-mail: zheng_gf@xmu.edu.cn).

978-1-6654-3008-1/21 $31.00 © 2021 IEEE

Second, the images are segmented by threshold value and perfected by screening the feature. Third, the needle area is calibrated to extract the droplet area. Last, the parameters of the jet behaviors are calculated and the data are saved in the host computer.

III. RESULTS AND DISCUSSIONS

The original images are obtained using a CCD camera as shown in Fig. 2(a). An appropriate threshold is selected to binarize the original images as shown in Fig. 2(b). The pixel points on the whole image whose gray level is within the selected threshold range are judged as objects, and the gray level is set to 255, which is black. The points outside the threshold range are judged as background, and the gray level is set to 0, which is white. There may be some noise in the binarization image due to the reflection of the droplet or environmental factors. As shown in Fig. 2(c), the perfect binarization images with needle and jet can only be obtained by screening the features such as area, number of columns and length. At this time, the image is consisted of the needle region, the droplet region and the jet region. After calibrating the needle area, a jet image is obtained. Through the open operation filter, the isolated small points, burrs and bridges are removed to obtain the droplet region, which is shown in Fig. 2(d). The area of the droplet is calculated by the area of the white region in the image. The operator "region features" is used to obtain the horizontal and vertical coordinates of the center geometry position of droplet, as indicated by the red dots in the Fig. 2(d).

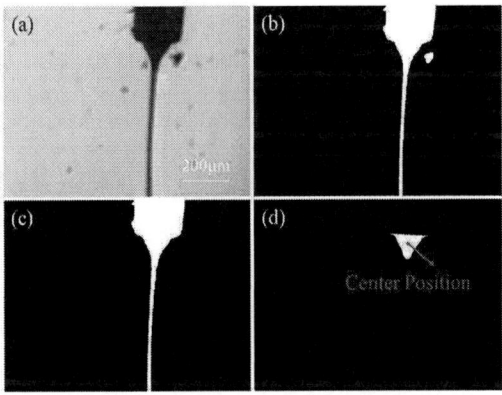

Figure 2.   The process of jet behavior analysis

The features of jet in the droplet ejection mode are shown in Fig. 3. The original images at the moment of 0 s, 0.5 s, 1.5 s and 2.5 s are shown in Fig. 3(a). After the image processing, the droplet images are obtained as shown in Fig. 3(b). The area of the droplet in the droplet ejection mode is between 0.029 mm² and 0.032 mm² and the center geometry position is between 0.928 mm and 0.935 mm, as shown in Fig. 3(c). The deviation of the center geometry position of droplet is related to the movement of the collector and the accumulation of droplets at the tip of the needle. In this state, the jet is unstable.

The features of jet in the Taylor cone ejection mode are shown in Fig. 4. It is a steady ejection process, which can obtain the stable fiber deposition. The original images at the moment of 0 s, 0.5 s,1 s and 2 s are shown in Fig. 4(a). After the image processing, the droplet images are obtained as shown in Fig. 4(b). The area of the droplet in the Taylor cone ejection mode is between 0.0135 mm² and 0.0152 mm² and the center geometry position is between 0.936 mm and 0.948 mm, as shown in Fig. 4(c). The deviation of the center geometry position of droplet is related to the movement of the collector. When the collection moves to the right, the jet is pulled, which causes the center geometry position of the droplet moving to the right.

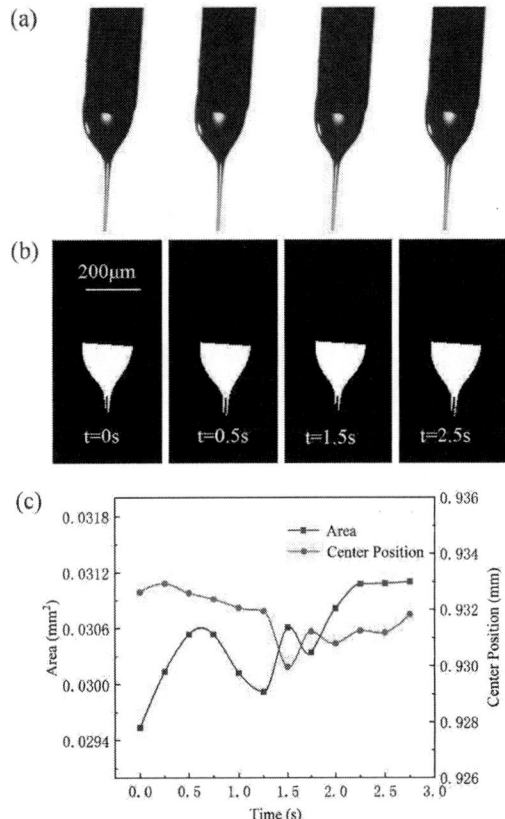

Figure 3.   Area and center geometry position of droplet in droplet ejection mode

The features of jet in the retractive ejection mode are shown in Fig. 5. The original images at the moment of 0.25 s, 1 s, 2 s and 2.75 s are shown in Fig. 5(a). After the image processing, the droplet images are obtained as shown in Fig. 5(b). The area of the droplet in the retractive ejection mode is between 0.006 mm² and 0.0068 mm² and the center geometry position is between 0.948 mm and 0.960 mm, as shown in Fig. 5(c). In this state, the center geometry position of the droplet moves greatly and was less related to the movement of the collector.

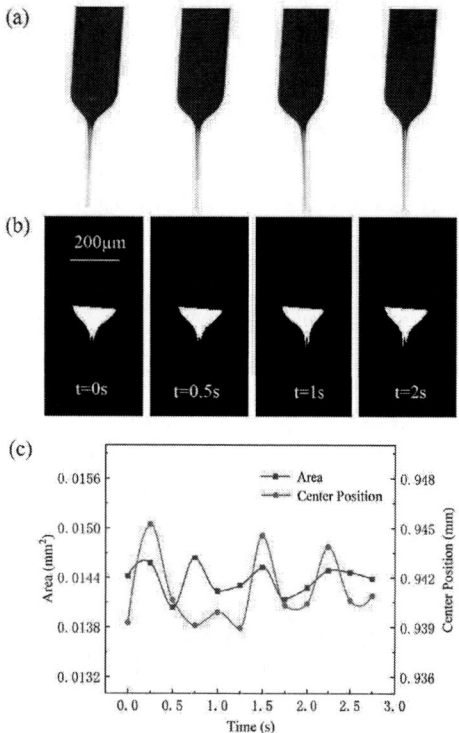

Figure 4. Area and center geometry position of droplet in Taylor cone ejection mode

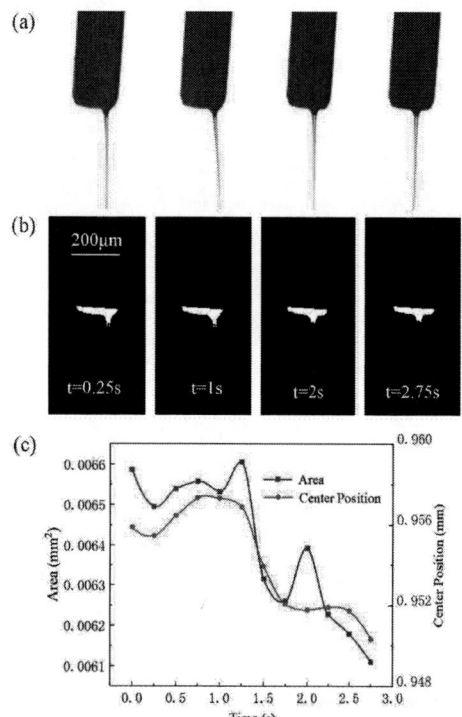

Figure 5. Area and center geometry position of droplet in retractive ejection mode

As shown in Fig. 6, changing the voltage applied to the needle, the area and the center geometry position of droplet changes periodically with the jet modes. The processes in 0 s to 6 s and 21 s to 43 s are in the droplet ejection mode with the voltage between 1.8 kV to 2.1 kV. The processes in 7 s to 20 s and 44 s to 63 s are in the Taylor cone ejection mode with the voltage between 2.1 kV to 2.2 kV. The process in 64 s to 72 s is in the Taylor cone ejection mode with the voltage between 2.2 kV to 2.4 kV. It can be found that when the jet mode changes, the features of droplet characteristics will change, which is a continuous and repeated changing process.

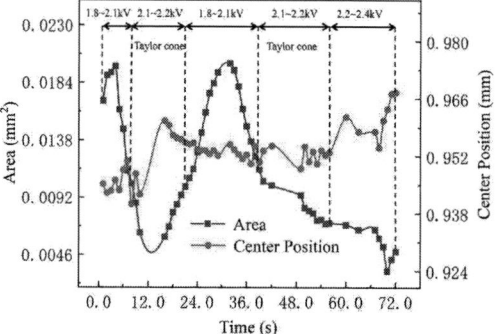

Figure 6. Change of area and center geometry position of droplet in different modes

IV. CONCLUSION

In this work, we build a real-time monitoring and analysis system of jet behaviors, which can extract the features of droplet and save the data in real time. The origin images are filtered with binarization, threshold segmentation and feature recognition. Especially, two open operations are used for the droplet images. The area and center geometry position of the droplet are analyzed to distinguish different jet modes. The features of droplet in three jet modes are studied. Moreover, a continuous process of the changing between three jet modes are explored, which proves the correlation between the features (area and center geometry position of droplet) and the jet modes. This work provides a way to identify different jet modes, and is expected to realize the close-loop control of EDW more stably and accurately.

ACKNOWLEDGMENT

This research was financially supported by Science and Technology Planning Project of Fujian Province (No. 2020H6003, 2018J01082), Xiamen Municipal Science and Technology Project (No. 3502Z20193015), Aviation Key Laboratory of Science and Technology on Inertia (No. 20180868001), Natural Science Foundation of Guangdong Province (No. 2018A030313522), Science and Technology Planning Project of Shenzhen Municipality in China (No. JCYJ20180306173000073).

REFERENCES

[1] Y. Huang, N. Bu, Y. Duan, Y. Pan, H. Liu, Z. Yin, et al., "Electrohydrodynamic direct-writing," Nanoscale, vol. 5, pp. 12007-12017, December 2013.

978-1-6654-3008-1/21 $31.00 © 2021 IEEE

[2] B. W. An, K. Kim, H. Lee, S. Y. Kim, Y. Shim, D. Y. Lee, et al., "High-Resolution Printing of 3D Structures Using an Electrohydrodynamic Inkjet with Multiple Functional Inks," Adv Mater, vol. 27, pp. 4322-4328, August 2015.

[3] Y. Liang, J. Yong, Y. Yu, A. Nirmalathas, K. Ganesan, R. Evans, et al., "Direct Electrohydrodynamic Patterning of High-Performance All Metal Oxide Thin-Film Electronics," ACS Nano, vol. 13, pp. 13957-13964, December 2019.

[4] G. Cai, P. Darmawan, M. Cui, J. Wang, J. Chen, S. Magdassi, et al., "Highly Stable Transparent Conductive Silver Grid/PEDOT:PSS Electrodes for Integrated Bifunctional Flexible Electrochromic Supercapacitors," Advanced Energy Materials, vol. 6, pp. 1501882, December 2016.

[5] S. Vijayavenkataraman, S. Thaharah, S. Zhang, W. F. Lu, and J. Y. H. Fuh, "Electrohydrodynamic jet 3D-printed PCL/PAA conductive scaffolds with tunable biodegradability as nerve guide conduits (NGCs) for peripheral nerve injury repair," Materials & Design, vol. 162, pp. 171-184, November 2019.

[6] S. H. Ahn, H. J. Lee, and G. H. Kim, "Polycaprolactone scaffolds fabricated with an advanced electrohydrodynamic direct-printing method for bone tissue regeneration," Biomacromolecules, vol. 12, pp. 4256-4263, December 2011.

[7] C. Chang, V. H. Tran, J. Wang, Y. K. Fuh, and L. Lin, "Direct-write piezoelectric polymeric nanogenerator with high energy conversion efficiency," Nano Lett, vol. 10, pp. 726-731, February 2010.

[8] W. Xu, S. Zhang, and W. Xu, "Recent progress on electrohydrodynamic nanowire printing," Science China Materials, vol. 62, pp. 1709-1726, September 2019.

[9] J. Chen, T. Wu, L. Zhang, X. Feng, P. Li, F. Huang, et al., "Fabrication of flexible organic electronic microcircuit pattern using near-field electrohydrodynamic direct-writing method," Journal of Materials Science: Materials in Electronics, vol. 30, pp. 17863-17871, September 2019.

[10] K. Zhao, D. Wang, K. Li, C. Jiang, Y. Wei, J. Qian, et al., "Drop-on-Demand Electrohydrodynamic Jet Printing of Graphene and Its Composite Microelectrode for High Performance Electrochemical Sensing," Journal of The Electrochemical Society, vol. 167, pp. 107508, March 2020.

[11] Z. Zhang, H. He, W. Fu, D. Ji, and S. Ramakrishna, "Electro-Hydrodynamic Direct-Writing Technology toward Patterned Ultra-Thin Fibers: Advances, Materials and Applications," Nano Today, vol. 35, pp. 100942, August 2020.

[12] Y. Liu and Y. Huang, "Theoretical and experimental studies of electrostatic focusing for electrohydrodynamic jet printing," Journal of Micromechanics and Microengineering, vol. 29, pp. 065002, April 2019.

[13] M. Rahmanpour and R. Ebrahimi, "Numerical simulation of electrohydrodynamic spray with stable Taylor cone–jet," Heat and Mass Transfer, vol. 52, pp. 1595-1603, September 2015.

[14] A. Jaworek and A. Krupa, "Classification of the Modes of EHD Spraying," Aerosol Science, vol. 30, pp. 873-893, December 1999.

[15] D. H. Reneker and A. L. Yarin, "Electrospinning jets and polymer nanofibers," Polymer, vol. 49, pp. 2387-2425, February 2008.

[16] M. J. Divvela and Y. L. Joo, "Design principles in continuous inkjet electrohydrodynamic printing from discretized modeling and image analysis," Journal of Manufacturing Processes, vol. 54, pp. 413-419, August 2020.

[17] X. Li, Y. Zheng, X. Mu, B. Xin, and L. Lin, "Investigation into Jet Motion and Fiber Properties Induced by Electric Fields in Melt Electrospinning," Industrial & Engineering Chemistry Research, vol. 59, pp. 2163-2170, January 2020.

Gap in pagination due to unavailable paper.

Pages 1140-1141

Proceedings of the 16th Annual IEEE International
Conference on Nano/Micro Engineered and Molecular Systems
April 25-29, 2021

Large-scale Uniformly Hybrid Micro-nano Structure Wetting Solid Substrate for Surface-enhanced Raman Spectroscopy

Fanhong Chen[1,2], Qi Qi[3], Yupeng Zhao[1,2,3], Shaoxun Zhang[1,2], Yongmin Zhao[1,2], Anjie Ming[1,2]*, and Changhui Mao[1,2]*

Abstract— Surface-enhanced Raman spectroscopy (SERS) is widely used in quantitative trace detection analysis fields benefit from its features of surface selectivity, fingerprint specificity, and rapidity without labeling. High sensitivity, reproducibility, analyte adsorption, and standardized preparation process are of critical importance to an ideal SERS solid substrate for detection applications in the actual circumstances. In this study, wafer-scale uniformly Si nanorods (SiNRs) arrays in conjunction with Au nanoparticles were fabricated as SERS solid substrate by polystyrene (PS) sphere self-assembly, nanosphere lithography and magnetron deposition techniques, which exhibited high sensitivity, reproducibility, and wettability. In detection of Rhodamine 6G (R6G) molecules, the SERS substrate reached an enhancement factor (EF) of 10^{10} and a limited detection of 5 ng/mL with a relative standard deviation (RSD) as low as 6.86% - 11.82%. The SERS substrates proposed in this work may serve as a promising detection scheme in chemical and biological fields.

Keywords: SERS; large-scale; sensitivity; reproducibility; R6G

INTRODUCTION

Surface-enhanced Raman spectroscopy (SERS) is a phenomenon associated with the amplification by several orders of magnitude of Raman signals of analytes located at or very close to metallic nanostructures [1]. Technique of SERS, which can realize an ultrahigh sensitivity down to the single-molecule level by means of Au, Ag and Cu metal, provide intrinsic chemical and fingerprint information of analytes molecule and possesses ability of label-free rapid on-site detection [2]. Therefore, SERS has been widely utilized in many fields such as medical diagnosis [3,4], environmental protection [5,6] and food safety [7,8] et al.

The SERS enhancement strongly relies on the optical resonance properties of substrate materials, which can significantly enhance the local electromagnetic field, largely owing to the excitation of surface plasmon resonance (SPR) [9,10]. To maximize the enhancement ability of SERS systems, strategies are to explore including novel electromagnetic materials with enhanced intrinsic activity and hybrid structure with exposed more hot spots. For this issues, tremendous efforts and progresses have been made in synthesizing SERS materials [11-14]. Detailed experimental and theoretical investigations also were conducted to understand SERS enhancement mechanism and develop highly sensitivity SERS substrates [15-18]. Besides, in order to realize the application of SERS technology in the actual environment, which requires the characteristics of strong enhancement, stability, uniformity, and reproducibility for SERS substrate. Therefore, to gain above substrate, it is important to fabricate highly ordered nanostructure SERS substrates with uniformly hot spots distribution. To this end, many methods have been adopted, such as electron beam lithography [19], nanoimprinting [20] and X-ray lithography [21]. Such as, Van Duyne's group in 2002 developed a metal film over a nanosphere electrode as SERS substrates [22]. And Wen-Jun Zhang' group [23] prepared Si nanowires array as SERS substrate by nanosphere lithography, metal-assisted chemical etching and Ag sputter deposition, which achieved high reproducibility with a relative standard deviation (RSD) of 14% and a low limited detection of 10^{-5} M for Rhodamine 6G (R6G) molecules. However, these methods are limited by the disadvantages of high cost and low throughput. Besides, surface wettability plays an important role in influencing the Raman signal intensity for detection of analyte. The effective wetting of analyte solution on SERS substrate can significantly improve the Raman signal [24,25]. Therefore, it is highly demanded to investigative modulated wettabilities of SERS substrate surface, while very few of these studies have been conducted.

Herein, large-scale ordered hexagonal-packed Si nanorods (SiNRs) arrays based on polystyrene (PS) sphere self-assembly and nanosphere lithography techniques were fabricated. Then, the SiNRs arrays with a diameter of ~250 nm, a gap of ~50 nm, and a length of ~700 nm was functionalized with homogeneous Au nanoparticles and clusters by depositing 20 nm Au films as SERS substrate. Owing to the ordered SiNRs arrays and wildly formed Au nanoparticles which providing symmetrical and substantial hot spots with extremely enhanced Raman signal, our SERS substrate has achieved highly reproducibility, sensitivity and hydrophilicity. The SERS system reached an enhancement factor (EF) of 10^{10}, a limited detection of 5 ng/mL, and an RSD 6.86% - 11.82% for detection of R6G. This study shall benefit the rational structure design and improved trace analysis methods of SERS substrate.

EXPERIMENTAL

The close-packed PS sphere self-assembled monolayer was prepared by following a process reported by Valeria Lotito et al [26] with a slight modification as sketched in Fig. 1. Typically, firstly the wafer substrate was placed on the sample table and filling the apparatus with water (Fig. 1a). Next, 6 mL Sodium dodecyl sulfate (SDS) in a concentration of 20 mM were added using the tilted glass slide in the area limited by the ring (Fig. 1b). Then, added the PS suspensions mixed with 66% ethanol to the air/water interface through the tilted glass slide in the area limited by the ring and let stand for an hour (Fig. 1c). Finally, self-assembled monolayer was transferred to the wafer substrate by letting water flow out of

---

1. State Key Laboratory of Advanced Materials for Smart Sensing, General Research Institute for Nonferrous Metals, Beijing 100088, P. R. China;
2. GRIMAT Engineering Institute Co., Ltd, Beijing 101407, P. R. China;
3. North China University of Technology, Beijing 100144, P. R. China.
*Contact Author: mingaj@grinm.com

978-1-6654-3008-1/21 $31.00 © 2021 IEEE

the apparatus (Fig. 1d). As demonstrated in Fig. 1d enlarged images that the PS self-assembly monolayer can transfer to the wafer with the thickness of 200 nm SiO_2 layer and exhibits the characterization of an excellent conservation of hexagonal order. This apparatus offers a smooth assembly and transfer process because using the ring and rods for confinement of the monolayer compare to other method relying on air/water interface self-assembly [27]. Overall, large-scale PS spheres monolayer on 4-inch silicon wafer with the thickness of 200 nm SiO_2 was fabricated successfully.

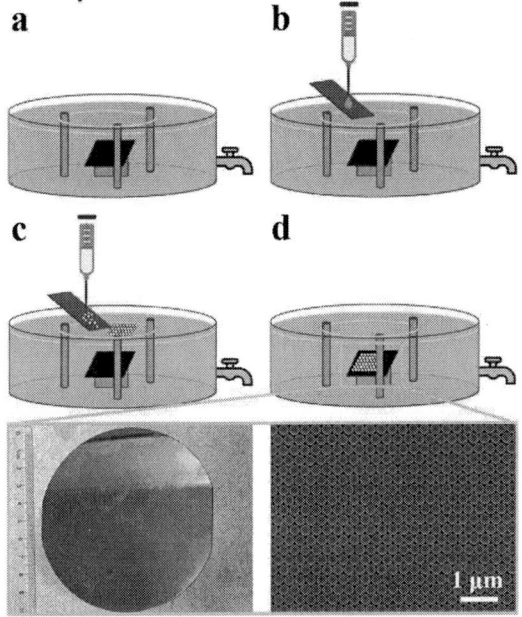

Figure. 1 PS sphere self-assembly at air/water interface: (a) apparatus installation and water filling; (b) adding amount of SDS; (c) adding PS suspension to the air-water interface; (d) PS monolayer transfer to the substrate by water discharge, the enlarged images are of large-scale PS monolayers on 4-inch wafer substrate and SEM.

SiNRs arrays were fabricated by nanosphere lithography and plasma etching (PE). As schematically shown in Fig. $2a_1$-c_1, firstly, the diameter of prepared ordered hexagonal-packed self-assembled PS spheres monolayer (Fig. $2a_2$) were reduced by reactive ion (RIE) with oxygen (O_2) gas by referring our previous report [28]. Here 200 nm SiO_2 was introduced as a dielectric layer to improve high aspect ratio of SiNRs arrays because PS spheres have poor corrosion resistance. Then SiO_2 nanorod arrays were formed by PE (Rainbow 4520), with CF_4 and CHF_3 gas, the gas flow rate was 10 sccm and 50 sccm respectively, the RF was 300 W, the press was 200 mT. Due to the energy emitted by etching caused PS spheres melting and connecting, in order to reduce the influence of temperature on the etching process, time-sharing etching was taken. The SiO_2 was etched for 150 s, divided into four times. By this method, high order and steepness SiO_2 arrays were fabricated as shown in Fig. $2b_2$. Finally, SiNRs arrays were prepared by PE (Rainbow 4420), with Cl_2 and HBr gas, the gas flow rate was 100 sccm and 20 sccm respectively, the RF was 300 W, the press was 300 mT, and the etching time was 60 s (Fig. $2c_2$).

Furthermore, SERS substrates were prepared by sputtering 20 nm Au film on the SiNRs arrays. As shown in Fig. 3a, abundant Au nanoparticles and clusters were coated on the surfaces of SiNRs arrays, which are favorable for the enhancement of the Raman signal [29]. It can be seen also from the element mapping result in Fig. 3a enlarged images that Au nanoparticles with size of about 20 nm uniformly were distributed on SiNRs, which resulted in uniformly hot spots distribution.

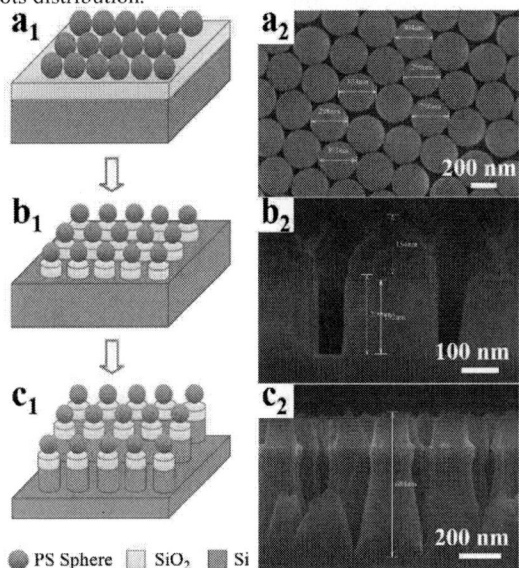

● PS Sphere ▢ SiO_2 ▨ Si

Figure. 2 (a_1-c_1) The fabrication schematic diagrams of SiNRs arrays; The SEM images of (a_2) ordered hexagonal-packed self-assembled PS spheres monolayer, (b_2) SiO_2 nanorods arrays, and (c_2) SiNRs arrays.

In addition, taking into account the analyte detection, SERS substrate surface wettability regulation is an essential step to enrich more analytes on surface. Therefore, the wetting behaviors of SERS substrate were examined. As demonstrated in Fig. 3b and c, the bubble exhibited a large contact angle (ca. 147.3°) and a small surface adhesion force (ca. 9 μN), and presented the droplet contact angle only 41.1° on SERS substrate. These results confirmed that SERS substrates exhibit super-hydrophilicity, which benefits to effective adsorption of analytes on the SERS substrate with improving Raman signal enhancement.

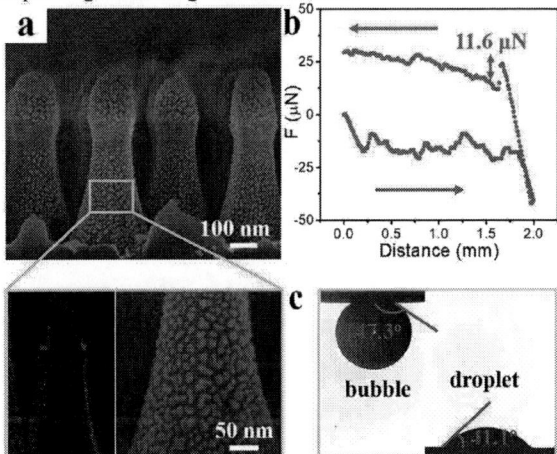

Figure. 3 (a) The SEM image of SiNRs arrays decorated with Au nanoparticle, and element mapping and zoomed micromorphology of Au. (b) The adhesion plot, and (c) contact angles of bubble and droplet on SERS substrate.

978-1-6654-3008-1/21 $31.00 © 2021 IEEE

RESULTS AND DISCUSSION

The SERS activity of the SERS substrates were evaluated by using R6G which is a typical model analyte as the probing adsorbate. Thus, the sensitivity of the SERS substrates was examined by using R6G molecules with concentration from 5 ng/mL to 5 μg/mL. Raman intensity increases with an increase in concentration of R6G with excitation wavelength at 633 nm as demonstrated in Fig. 4a. The high major peaks of R6G [30] and clear Raman signals of R6G at the concentration of 5 ng/mL (Fig. 4b) were observed on the SERS substrates, which indicates that the SERS substrates have high SERS activity.

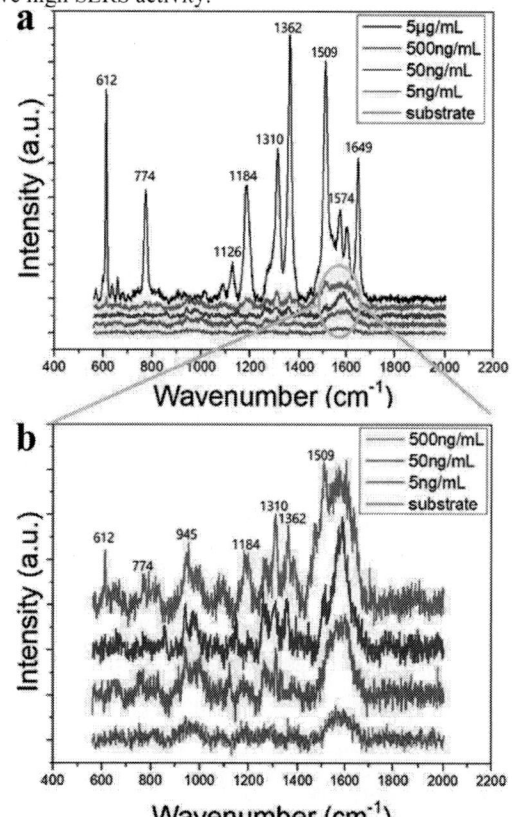

Figure. 4 (a) Raman spectra of R6G molecules from 5 ng/mL to 5 μg/mL under the 633 nm wavelength, and (b) zoomed Raman spectra of R6G molecules from 5 ng/mL to 500 ng/mL.

To confirm the enhancement effect of SERS substrate, EF value of substrate was further calculated by the following equation [31]:

$$EF = \frac{I_{SERS}}{I_{REF}} \cdot \frac{N_{REF}}{N_{SERS}} \qquad (1)$$

$$N_{SERS} = \frac{C_{SERS}V_{SERS}AN_A}{S} \qquad (2)$$

$$N_{REF} = C_{REF}V_{REF}N_A \qquad (3)$$

where I_{SERS} and I_{REF} are the Raman intensity of the adsorbed R6G molecules on SERS substrates and on planar Au films, respectively. N_{REF} is the number of R6G molecules excited by laser on the surface of planar Au film. N_{SERS} is the number of R6G molecules probed on SERS substrate, N_A is Avogadro constant, A is the area of laser spot (1 μm²), C_{SERS} and C_{REF}

are the molar concentrations of R6G solution on SERS substrates and planar Au films, and V_{SERS} and V_{REF} are the volumes of R6G solution added to the SERS substrate and planar Au films, respectively. Here, to obtain the Raman intensity of R6G on different substrates, 2 μL 5 μg/mL R6G molecules were uniformly dispersed on the surface of SERS substrate and the diffusion spot was measured to be about 4 mm in diameter, so the surface area (S) is S = π (2 mm)². 2 μL 10^{-2} M R6G molecules was dropped onto planar Au films. Substituting the above experimental results into (2) and (3), the values of N_{SERS} and N_{REF} are calculated as the following (4) and (5).

$$N_{SERS} = \frac{5\ \mu g/mL \times 2\ \mu L \times 1\ \mu m^2 \times N_A}{\pi(2mm)^2} \qquad (4)$$

$$N_{REF} = 10^{-2}\ mol/L \times 2\ \mu L \times N_A \qquad (5)$$

Thus, the N_{REF} / N_{SERS} is 1.1×10^{10}. Besides, the Raman intensity of R6G characteristic peaks can be obtained from Fig. 4. Then combining (1) and value of N_{REF} / N_{SERS}, the EF values of R6G on the SERS substrates are estimated to be 4.5×10^{10}, 3.19×10^{10}, 2.6×10^{10}, 1.69×10^{10}, 1.48×10^{10}, and 1.57×10^{10} at Raman shift of 1649, 1509, 1362, 1184, 744, and 612 cm⁻¹.

To further evaluate reproducibility of the SERS substrate, the experiments of spot-to-spot Raman scanning for the same substrate were conducted. As illustrated in Fig. 5, the 8 Raman spectra present highly reproducible and the RSD of R6G characteristic peaks from the 8 Raman spectra were also calculated (Table. 1). Table. 1 illustrates that the RSD of the SERS substrates reaches as low as 6.86% - 11.82%, which confirms the outstanding reproducibility of the SERS substrates.

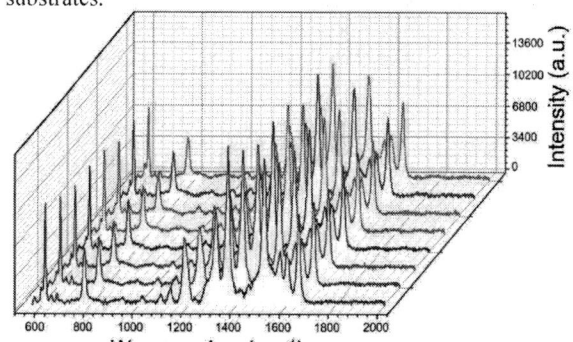

Figure. 5 Eight Raman spectra of R6G molecules on the SERS substrates.

Actually, the excellent reproducibility is attributed to the order distribution of Au nanoparticles on the SiNRs arrays, which provides homogeneous electromagnetic hot spots. Compared to previously reports [23] that similar ordered SiNW arrays deposited with Ag films (R6G 10^{-5}M, RSD 11% - 13%), the SERS substrates of this work show more excellent SERS activity i.e. outstanding sensitivity, EF and reproducibility.

Table. 1 Relative standard deviation of noticeable R6G Raman peaks from 8 Raman Spectra obtained on the same SERS substrate

Raman shift (cm⁻¹)	612	774	1126	1184
RSD	11.82	9.83	11.58	8.04
Raman shift (cm⁻¹)	1310	1362	1509	1649
RSD	8.61	11.68	11.41	6.86

CONCLUSIONS

In summary, low-cost and uniformly SERS substrate with high sensitivity, enhancement effect, and reproducibility was fabricated by coating Au nanoparticles onto SiNRs arrays that were prepared by PS spheres self-assembly and nanosphere lithography technology. The SERS substrate presents a highly EF of 10^{10}, an RSD of 6.86% - 11.82% with reasonable fluctuation, and the limited detection as low as 5 ng/mL in the detection of R6G. Based on the outstanding SERS activity that resulted from the spatial uniformity structure and abundant hot spots, the SERS substrate may serve as an excellent candidate for biomolecule detection.

ACKNOWLEDGMENT

This work was supported by the National Natural Science Foundation of China (No. 61874137), the National Key Research and Development Project (No. 2019YFB2005705), Shandong Provincial Key Research and Development Program (No. 2020CXGC010203).

REFERENCES

[1] S. Y. Ding, *et al.*, "Electromagnetic theories of surface-enhanced Raman spectroscopy", *Chem. Soc. Rev.*, 2017, 46 (13), 4042.

[2] S. Y. Ding, *et al.*, "Nanostructure-based plasmon-enhanced Raman spectroscopy for surface analysis of materials", *Nat. Rev. Mater.*, 2016, 1.

[3] Anne-Isabelle Henry, *et al.*, "Surface-Enhanced Raman Spectroscopy Biosensing: In Vivo Diagnostics and Multimodal Imaging", *Anal. Chem.*, 2016, 88 (13), 6638-6647.

[4] M. Nowaka, *et al.*, "Preparation and characterization of long-term stable SERS active materials as potential supports for medical diagnostic", *Appl. Surf. Sci.*, 2019, 472, 93-98.

[5] S. Y. Tang, *et al.*, "Efficient Enrichment and Self-Assembly of Hybrid Nanoparticles into Removable and Magnetic SERS Substrates for Sensitive Detection of Environmental Pollutants", *ACS Appl. Mater. Interfaces*, 2017, 9 (8), 7472-7480.

[6] Pandeeswar Makam, *et al.*, "SERS and fluorescence-based ultrasensitive detection of mercury in water", *Biosens. Bioelectron.*, 2018, 100, 556-564.

[7] Tehseen Yaseen, *et al.*, "SERS substrates and their applications in food safety evaluation: A review of recent research trends", *Trends Food Sci Technol*, 2018, 72, 162-174.

[8] N. Yang, *et al.*, "Fabrication of a Flexible Gold Nanorod Polymer Metafilm via a Phase Transfer Method as a SERS Substrate for Detecting Food Contaminants", *J. Agric. Food Chem.*, 2018, 66 (26), 6889-6896.

[9] M. Moskovits, *et al.*, "Surface roughness and the enhanced intensity of Raman scattering by molecules adsorbed on metals", *J. Chem. Phys.*, 1978, 69, 4159–4161.

[10] J. A. Creighton, *et al.*, "Plasma resonance enhancement of Raman scattering by pyridine adsorbed on silver or gold sol particles of size comparable to the excitation wavelength", *J. Chem. Soc. Faraday Trans.*, 1979, 2 (75), 790–798.

[11] J. L. Lu, *et al.*, "Silver Nanoparticle-Based Surface-Enhanced Raman Spectroscopy for the Rapid and Selective Detection of Trace Tropane Alkaloids in Food", *ACS Appl. Nano Mater.*, 2019.

[12] W. Zhu, *et al.*, "Rapid and low-cost quantitative detection of creatinine in human urine with a portable Raman spectrometer", *Biosens. Bioelectron.*, 2020, 154.

[13] S. Q. Xu, *et al.*, "Evaluation of cigarette flavoring quality via surface-enhanced Raman spectroscopy", *Chem Comm*, 2018, 54.

[14] J. F. Li, *et al.*, "Core-Shell Nanoparticle-Enhanced Raman Spectroscopy", *Chem. Rev.*, 2017, 6b00596.

[15] L. F. Xie, *et al.*, "Key Role of Direct Adsorption on SERS Sensitivity: Synergistic Effect among Target, Aggregating Agent, and Surface with Au or Ag Colloid as SERS Substrate", *J. Phys. Chem. Lett.*, 2020.

[16] X. R. Shen, *et al.*, "An Experimental and Theoretical Study of Surface-Enhanced Raman Spectra of Sulfadiazine Adsorbed on Nanoscale Gold Colloids", *J. Phys. Chem. A*, 2019.

[17] Z. W. Yang, *et al.*, "3D Hotspots Platform for Plasmon Enhanced Raman and Second Harmonic Generation Spectroscopies and Quantitative Analysis", *Adv. Optical Mater.*, 2019, 1901010.

[18] C. J. Zhang, *et al.*, "Observing the dynamic "hot spots" on two-dimensional Au nanoparticles monolayer film", *Chem Comm*, 2017.

[19] G. Das, *et al.*, "Plasmon based biosensor for distinguishing different peptides mutation states", *Sci. Rep.*, 2013, 3(1).

[20] S. J. Barcelo, *et al.*, "Fabrication of Deterministic Nanostructure Assemblies with Sub-nanometer Spacing Using a Nanoimprinting Transfer Technique", *ACS Nano*, 2012, 6(7), 6446–6452.

[21] D. Wouters, *et al.*, "Nanolithography and nanochemistry: Probe-related patterning techniques and chemical modification for nanometer-sized devices", *Angew. Chem. Int. Ed.*, 2004, 43, 2480.

[22] L. A. Dick, *et al.*, "Metal film over nanosphere (MFON) electrodes for surface-enhanced raman spectroscopy (SERS): Improvements in surface nanostructure stability and suppression of irreversible loss", *J. Phys. Chem. B*, 2002, 106, 853–860.

[23] J. A. Huang, *et al.*, "Averaging effect on improving signal reproducibility of gap-based and gap-free SERS substrates based on ordered Si nanowire arrays", *RSC Adv.*, 2017, 7, 5297-5305.

[24] B. B. Zhou, *et al.*, "Amphiphilic Functionalized Acupuncture Needle as SERS Sensor for In Situ Multiphase Detection", *Anal. Chem.*, 2018, 7b04348.

[25] H. Z. Li, *et al.*, "Bioinspired Micropatterned Superhydrophilic Au-Areoles for Surface-Enhanced Raman Scattering (SERS) Trace Detection", *Adv. Funct. Mater.*, 2018, 1800448.

[26] Valeria Lotito, *et al.*, "Self-Assembly of Single-Sized and Binary Colloidal Particles at Air/Water Interface by Surface Confinement and Water Discharge", *Langmuir*, 2016, 32(37), 9582-9590.

[27] Valeria Lotito, *et al.*, "Approaches to self-assembly of colloidal monolayers: A guide for nanotechnologists", *Adv. Colloid. Interface Sci*, 2017, 246, 217-274.

[28] Q. Q, *et al.*, "Fabrication, Characterization, and Application of Large-Scale Uniformly Hybrid Nanoparticle-Enhanced RamanSpectroscopy Substrates", *Micromachines*, 2019, 10, 282.

[29] Z. Yi, *et al.*, "Ordered array of Ag semishells on different diameter monolayer polystyrene colloidal crystals: An ultrasensitive and reproducible SERS substrate", *Sci. Rep.*, 2016, 6, 32314.

[30] D. D. Lin, *et al.*, "Large-Area Au-Nanoparticle-Functionalized Si Nanorod Arrays for Spatially Uniform Surface-Enhanced Raman Spectroscopy", *ACS Nano*, 2017, 11(2), 1478-1487.

[31] Z. Yi, *et al.*, "Mesoporous gold sponges: electric charge-assisted seed mediated synthesis and application as surface-enhanced Raman scattering substrates", *Sci. Rep.*, 2015, 5, 16137.

Gap in pagination due to unavailable papers.

Pages 1146-1156

Proceedings of the 16th Annual IEEE International
Conference on Nano/Micro Engineered and Molecular Systems
April 25-29, 2021

A Tunable Quasi-Zero Stiffness Mechanism for Thermal Compensation of a MEMS Gravimeter

Xiaopeng Zhang, Xueyong Wei, *Senior Member, IEEE*, Yang Gao, Minghui Zhao, Yonghong Qi, Libo Zhao, and Zhuangde Jiang

Abstract— **For the requirements of high sensitivity and long-term stability of a MEMS gravimeter. This paper presents a tunable quasi-zero stiffness mechanism by a complementary quasi-zero stiffness structure of the pre-shaped bistable beam and conventional folded springs. The V-beam actuators are used to drive a voltage controlled bulking compression for stiffness compensation. The system is theoretically modeled by *Galerkin* method based on the first buckling mode, and the behavior of nonlinear spring is discussed. The performance result shows that the value and interval of the negative stiffness are determined by the as-fabricated height-to-width and height-to-length ratios, and actively tuned by the effective axial load. The behavior of this mechanism shows a promising potential of realizing a highly stable gravimeter.**

I. INTRODUCTION

Recent years, precisely measuring the gravity field by a micro-electromechanical system (MEMS) accelerometer has a great significance of low-cost and mass manufacture for geophysical exploration, hazards monitoring, and gravity assisted navigation [1]. According to [2], the value of location-dependent gravitational field varies from 976 Gal to 982 Gal (1 Gal=1 cm/s²), and the time-dependent solid Earth tide changes quasi-statically with a period about one day and longer. These physical characteristics of a gravity field put forward two challenges for the design and fabrication of a MEMS gravimeter: instrumental sensitivity under g bias (1 g=980 Gal) and long-term stability. Firstly on the aspect of sensitivity, as the principle of mass-spring system goes that a gravity variation results in a relative displacement by multiplying mechanical gain $1/\omega_0^2$, where ω_0 is the resonant frequency. Researchers from University of Glasgow have progressively reported a nonlinear geometric anti-springs build-in MEMS gravimeter with a significant 2.3 Hz lower resonance and a 40 μGal/√Hz noise floor [3]. This quasi-zero stiffness (QZS) concept was then developed on structures, and obtained an improved sensitivity [4]. However with the performance of long-term stability, compared with commercial products, there still remain some gaps for a MEMS gravimeter. The main reason is that for a stable spring, the silicon based MEMS mechanical structure lies in a strong dependence on the temperature of the silicon's Young's modulus than of the fused silica based commercial gravimeters [5]. As the temperature varies, the stiffness of

MEMS spring changes, so that the related drifting of resonant frequency has thermal effects both on mechanical gain and mass displacement (output bias).

Here, a tunable QZS mechanism for thermal compensation is proposed in this paper. We subject a compressive force by the V-beam actuator (VBA) axially on the pre-shaped beam. And the stiffness can be tuned defectively by voltage adjustment. The system's mechanism is theoretically modeled, and the nonlinear-stiffness and QZS behavior of the suspension is discussed.

II. DESIGN AND THEORETICAL MODEL

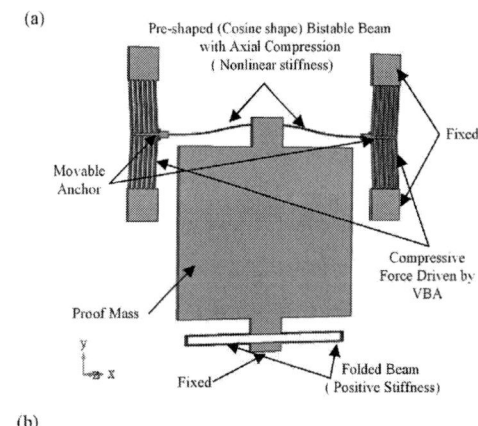

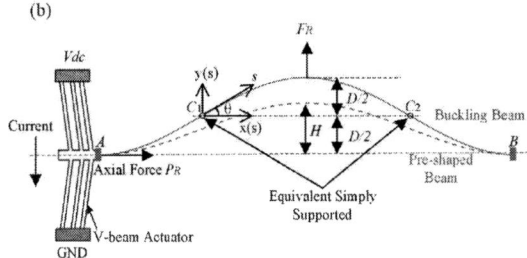

Figure 1. Mechanism of the MEMS gravimeter with tunable quasi-zero stiffness, (a) suspension structure, (b) schematic of the V-Beam actuator and coordinate of the pre-shaped beam

A. Suspension system

Figure 1(a) shows the proposed gravimeter with a QZS mechanism. The proof mass is suspended by the upside pre-shaped bistable beam with total arc length L, width W, and thickness T, and also is suspended by the downside conventional folded springs. Therefore these two kinds of springs would act as the mutual compensations, and obtains a lower resonance with a large dynamic range. Moreover, developed from [4], in which a clamped-clamped cosine beam

Corresponding author: Xueyong Wei (seanwei@mail.xjtu.edu.cn).

Xiaopeng Zhang is with the State Key Laboratory for Manufacturing Systems Engineering, Xi'an Jiaotong University, Xi'an 710049, China, and also with Northwest Institute of Nuclear Technology, Xi'an 710024, China.

Xueyong Wei, Yang Gao, Minghui Zhao, Yonghong Qi, Libo Zhao, and Zhuangde Jiang are with the State Key Laboratory for Manufacturing Systems Engineering, Xi'an Jiaotong University, Xi'an 710049, China

978-1-6654-3008-1/21 $31.00 © 2021 IEEE

with fixed anchors was used, we engage two VBAs on each side to deploy a compressive force on the as-fabricated beam. The coordinate is explained in Figure 1(b). Since the symmetry, it is only needed to analysis the upper half of equivalent simply supported C_1C_2, and there is no moment both at points C_1 and C_2. Another attention should be noted is that the transverse displacement y and the horizontal projection x along the beam from point C_1 are functions respected to arc length s of the pre-shaped beam. In the following, the theoretical model is established for a precise performance analysis of the suspension system.

B. Pre-shaped Bistable Beam Employed to an Axial Compression and a Transverse Force

With the relation between moment and curvature, we have the equilibrium equation along the left half of beam C_1C_2

$$-P_R y - \frac{F_R}{2} x = EI\left(\theta' - \theta'_0\right), \quad (1)$$

where $I = W^3 T/12$ is the second moment of the beam, E is the Young's modulus of the elasticity, the curvature denotes to $\theta' = d\theta / ds$, P_R, F_R is ether axial or transverse forces applied to the pre-shaped beam. To describe the bistable deformation on a relative large deflection, we use the more accurate approximated curvature $\theta' \approx y''(1 + y'^2 / 2)$ as a replacement of small deflection approximation y'', and all the primes are all derivatives respected to s. The as-fabricated shape of C_1C_2, with the middle of first mode of a clamped-clamped beam, can be written as:

$$\frac{y}{L} = \chi \sin\left(2\pi \frac{s}{L}\right), \quad (2)$$

where the variable χ denotes ether to normalized displacement $\delta = D/2L$ or pre-shaped height-to-length ratio $\delta_0 = H/2L$. For a long beam with a small height, the first mode approximation of the transverse deformation in the (2), would give a close and accurate with additional higher modes, and the error introduced by this approximation is within 2% [6], [7]. As the geometry shown in Figure 1(b), the horizontal projection x can be explained and approximated as

$$x = \int_0^s \sqrt{1 - y'^2} \, ds \approx \int_0^s \left(1 + y'^2 / 2\right) ds \quad (3)$$

With expressions of (2), (3), and boundary conditions $y(0)=0$, $y(L/4) = 0$. We apply the *Galerkin* method [8] to the residuals of deflection equation (1), and have the followed nondimentionalized equation that expresses the relation among axial compression, transverse force and displacement:

$$f_R\left(\frac{1}{\pi^2} - \frac{4}{3}\delta^2\right) = -\delta_0 - \frac{\pi^2}{2}\delta_0^3 - \left(p_R - 1\right)\delta + \frac{\pi^2}{2}\delta^3. \quad (4)$$

The normalized axial compression $f_R = F_R/P_{cr}$ and transverse force $p_R = P_R/P_{cr}$, where $P_{cr} = 4\pi^2 EI/L^2$ is the critical load of clamped-clamped buckling. As long as $p_R < 0$ (tensile), the transverse stiffness increases and meanwhile a larger f_R is needed, the stiffness decreases when $p_R > 0$ (compressive). In the case of compressive force lower than the second buckling load, it has been proved and verified in [6] and [8] that the first order mode approximation of (4) can describe the major behavior of the pre-shaped cosine beam, both in conditions of pre-buckling and post-buckling.

C. Coupling with the V-beam Actuator

As the scheme explained in Figure 1(b), an electro-thermal VBA is subjected on both ends of the pre-shaped bistable beam to subject a compressive force. The electric current passes through the beam and heats the actuator, thermal expansion of the structure drives a compressive force. Thus the VBA can be modeled as a voltage V_{dc} controlled spring with an actuating compression force

$$P_R = K_a\left(x_{max} - \Delta s_e - \Delta s_b\right). \quad (5)$$

where K_a is the VBA's stiffness, x_{max} denotes to the free displacement of the actuator's terminal. The value compressive force from VBA is $F_{VBA} = K_a x_{max}$ [9]. It also has been verified that x_{max} is in proportional to driving V_{dc}, and the detailed expressions can be found in [10]. According to Euler linear buckling theory, subscribing the expressions (refer to [11]) of pre-bulking (elasticity) and post-bulking induced length variations

$$\begin{aligned}\Delta s_e &= P_R / \left(AE / L\right), \\ \Delta s_b &= \pi^2 L \delta^2 - \pi^2 L \delta_0^2\end{aligned} \quad (6)$$

to (4), we have the force-and-displacement (F-D) relation of the pre-shaped bistable beam subjected to a VBA compression, by terms up to δ^3, as

$$f_R = -k_{b0} - k_{b1}\delta - k_{b2}\delta^2 + k_{b3}\delta^3. \quad (7)$$

The constant and square coefficients are given by

$$\begin{cases} k_{b0} = \delta_0\left(\pi^2 + \dfrac{\pi^4}{2}\delta_0^2\right) \\ k_{b2} = \delta_0\left(\dfrac{4}{3}\pi^4 + \dfrac{2}{3}\pi^6\delta_0^2\right) \end{cases}. \quad (8)$$

These two parameters indicate a bias force related to δ_0 and a square restoring force with terms δ^2. The linear and cubic coefficients in (7), which interpret the bistability, are

$$\begin{cases} k_{b1} = \left(f_{VBA}\gamma + \dfrac{3}{4}\eta^2 - 1\right)\pi^2 \\ k_{b3} = \left[\dfrac{4}{\pi^2}\lambda^2\left(1 - \gamma\right) - \dfrac{4}{3}f_{VBA}\gamma - \eta^2 + \dfrac{11}{6}\right]\pi^4 \end{cases}, \quad (9)$$

where the normalized VBA's force $f_{VBA} = F_{VBA}/P_{cr}$, actuator and axial stiffness fractional coefficient γ is

$$\gamma = \frac{AE / L}{K_a + AE / L}. \quad (10)$$

Constants λ and η are defined as the beam's length-to-width ratio $\lambda = L/W$ and height-to-width ratio

$$\eta = H / W. \quad (11)$$

According to [12] and also can be shown later, η is critical to the bistable behavior of the pre-shaped beam.

$$\begin{aligned}\frac{U}{2P_{cr}L} &= \left(-k_{b0} - k_{b1}\delta_0 - k_{b2}\delta_0^2 + k_{b3}\delta_0^3\right)\delta_\Delta \\ &\quad + \left(-\frac{k_{b1}}{2} - k_{b2}\delta_0 + \frac{3k_{b3}}{2}\delta_0^2\right)\delta_\Delta^2 \\ &\quad + \left(-\frac{k_{b2}}{3} + k_{b3}\delta_0\right)\delta_\Delta^3 \\ &\quad + \frac{k_{b3}}{4}\delta_\Delta^4, \end{aligned} \quad (12)$$

The potential energy U of the compressed pre-shaped beam, normalized by $2P_{cr}L$, is derived in (12) by integrating (7) respect to the relative displacement $\delta_\Delta = \delta - \delta_0$. The total energy E including the work of transverse force F_R, can be given by (13). Therefore, one can achieve the stable equilibrium state of the system by finding notches of E in the presence of F_R. This transverse force, in our proposed gravimeter shown in Figure 1(a), is regarded as a resultant force from mass gravity and folded springs.

$$\frac{E}{2P_{cr}L} = \frac{U}{2P_{cr}L} - \frac{F_R}{P_{cr}}\delta_\Delta. \tag{13}$$

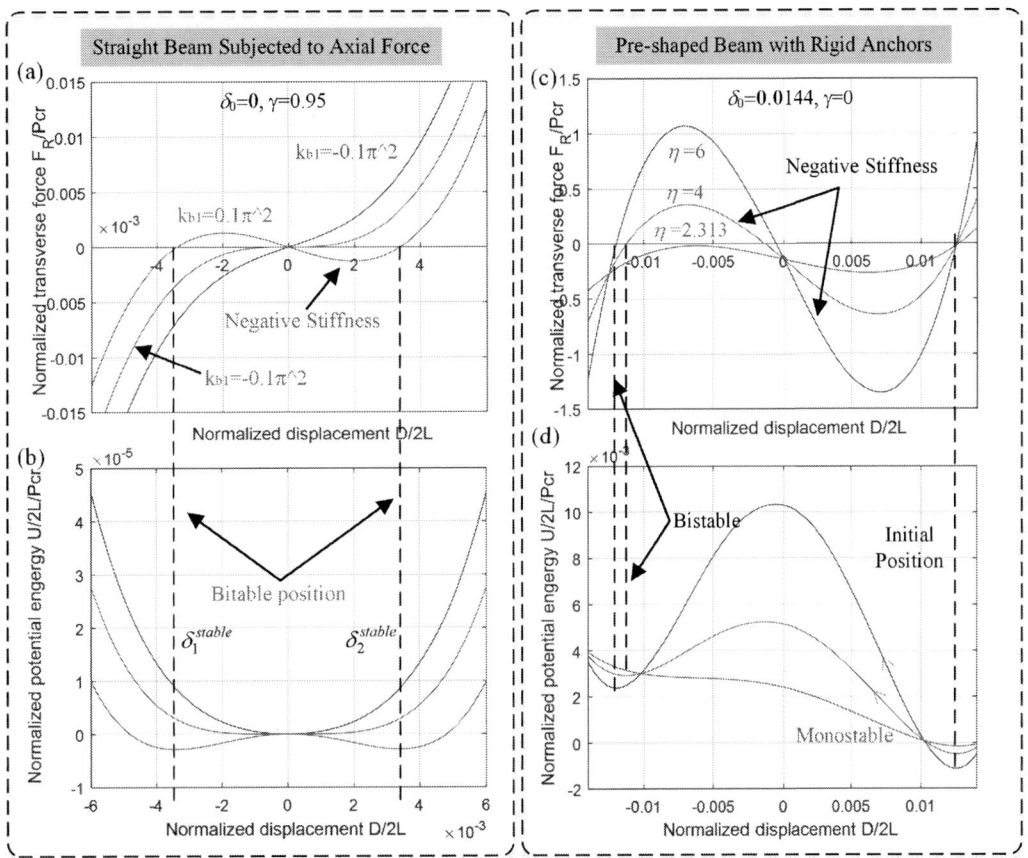

Figure 2. Normalized force-displacement and potential energy curves of the proposed non-linear spring. (a) and (b) are for a straight beam subjected to axial force, and there occurs a tunable decrease of the transverse stiffness when increasing the effective axial compression. (c) and (d) are for a pre-shaped beam with rigid anchors, and the value and interval of the negative stiffness of the bistable mechanism is determined by as-fabricated ratios λ and η.

III. PERFORMANCE ANALYSIS

To confirm the validation of the proposed theoretical model of the pressed bistable beam, we would carry out the F-D and energy predictions for two special cases that have been analyzed in [6] and [12]. Then, the influences of VBA's force and as-fabricated initial height to system mechanism are analyzed. The following would present the method for thermal compensation, and explain the principle both to balance the mass weight and to achieve a QZS by these complementary springs.

Special Cases

a) Straight Beam Subjected to Axial Force: $H=0$, then height-related ratios $\delta_0 = \eta = 0$, thus the constant and square coefficients $k_{b0} = k_{b1} = 0$. The F-D curve (7) and coefficients (9),

in the case of compressed straight beam, are reduced to a rather concise form:

$$f_R = -k_{b1}\delta + k_{b3}\delta^3 \tag{14}$$

and

$$\begin{cases} k_{b1} = \left(f_{VBA}\gamma - 1\right)\pi^2 \\ k_{b3} = \left[\dfrac{4}{\pi^2}\lambda^2\left(1-\gamma\right) - \dfrac{4}{3}f_{VBA}\gamma + \dfrac{11}{6}\right]\pi^4 \end{cases} \tag{15}$$

To analysis the physical interpretation of parameters in (14), the first object $f_{VBA}\gamma$ in linear stiffness k_{b1}, represents the effective fraction of VBA's force f_{VBA} shared by beam AB, and also resolves the linear stiffness around $\delta = 0$. For cubic stiffness that brings non-linear effect, there are two parts that influence the value of k_{b3}, the first object $4\lambda^2(1-\gamma)\pi^2$ is the influence due to the length-to-width ratio, and the other accounts for pre-stressed $f_{VBA}\gamma$ induced non-linearty. For a practical low resonant frequency oriented mass-spring system, especially with the proposed accelerometer build-in MEMS gravimeter, we usually have a higher length-to-width ratio λ to obtain a soft spring. If $\lambda > 100$, it can be calculated that λ dominates the cubic stiffness term k_{b3}.

Figure 2(a) and (b) present the either predicted normalized F-D curve or potential energy of the axial pressed straight beam. The coefficients $k_{b1} = -0.1\pi^2$, 0, $0.1\pi^2$, $\gamma = 0.95$, and $\lambda = 208.4$. It can be observed that as increasing the thermal actuated axial force, the linear stiffness decreases to zero as $F_{VBA}\gamma = P_{cr}$ or $k_{b1} = 1$, and becomes negative when $F_{VBA}\gamma > P_{cr}$ or $k_{b1} > 1$. And then, two breakeven points would take places in the predicted total energy. Thus, the system becomes bistable with two critical displacements

$$\delta_{1,2}^{stable} = \pm\sqrt{\frac{k_{b1}}{k_{b3}}}. \tag{16}$$

b) Pre-shaped Beam with Rigid Anchors: $K_a \to \infty$, $\gamma = 0$, which means that the fixed anchors share the entire force of F_{VBA}, and the curved beam bares nothing. With expressions in (9), we can derive the coefficients of linear and cubic terms in (7) as follow

$$\begin{cases} k_{b1} = \left(\dfrac{3}{4}\eta^2 - 1\right)\pi^2 \\ k_{b3} = \left[\dfrac{4}{\pi^2}\lambda^2 - \eta^2 + \dfrac{11}{6}\right]\pi^4 \end{cases}. \tag{17}$$

The predictions of F-D curve and also the potential energy of pre-shaped beam with rigid anchors, in normalized forms, are shown in Figure 2(c) and 2(d) with geometry parameters $\delta_0 = 0.0144$, and $\eta = 6, 4, 2.313$. It can be observed that the height-to-width ratio affects a lot to pre-shaped cosine beam, and these beams all obtain the nature of negative stiffness when $\eta > \sqrt{4/3}$, but only with bistability when $\eta > 2.313$. Compared with the F-D relation developed by energy variation method [12], the proposed model in (7) shares a similar critical value of η.

Tunable Quasi-Zero Stiffness

The technique to realize a highly sensitive and long-term stable MEMS gravimeter holds two aspects: 1) QZS for better mechanical sensitivity, and 2) tunable stiffness for thermal compensation. The QZS aspect has been demonstrated in [4] with rigid anchors of pre-shaped beam and folded spring to balance the mass and also obtain a large QZS section. Developing from [4], we employ axial force to pre-shaped beam in our design in Figure 1(a) for stiffness adjustment while Young's modulus varies with the temperature. Here comes the thermal compensation method for a long-term stable gravimeter:

1) Derive and predict the stiffness drift with the thermal effect on silicon's Young's modulus [5].

2) Subtract the stiffness drift to (7) to guarantee a fixed linear stiffness and come up with the VBA's axial force.

3) Calculate the driving voltage on VBA with the actuator force which is expressed in [10].

4) Slightly employ this temperature controlled voltage from a thermal-stable electric source on VBA and compensate the transverse stiffness.

IV. CONCLUSION

An analytical model of a pre-shaped and axial compressed bistable beam is proposed. The expression of transverse force and displacement is theoretically derived by the first mode based *Galerkin* method. The performance of the proposed model is analyzed proved and by two verified specific cases: straight beam with axial force and pre-shaped beam with rigid anchors. The key strength of this paper is a functional interpretation between the transverse displacement of a pre-shaped beam and the axial load from a spring-build-in actuator. The result shows that there is a bias force in the F-D curve due to the as-fabricated height-to-length ratio. This bias can be adjusted in the structure design, and it is essentially important for vertical gravimeter to balance with the proof mass. The results also indicate that the value and region of the linear stiffness can be defectively tuned by the VBA's Voltage load. Since the thermal performance of a commercial low dropout regulator (LDO) voltage source has reached as low as 3 ppm/°C, thus the active controlled mechanism shows a clear potential to realize a highly stable gravimeter.

ACKNOWLEDGMENT

This work is supported by the National Key Research and Development Program of China (2018YFB2002303) and the National Natural Science Foundation of China (52075432). The support from the International Joint Laboratory for Micro/Nano Manufacturing and Measurement Technologies is also appreciated.

REFERENCES

[1] S. Pearson-Grant, P. Franz and J. Clearwater, "Gravity measurements as a calibration tool for geothermal reservoir modelling," *Geothermics*, 2017.

[2] C. Hirt, S. Claessens, T. Fecher, M. Kuhn, R. Pail, and M. Rexer, "New ultra-high resolution picture of Earth's gravity field," *Geophysical Research Letters*, vol. 40, 2013.

[3] R. P. Middlemiss, A. Samarelli, D. J. Paul, J. Hough, S. Rowan, and G. D. Hammond, "Measurement of the Earth tides with a MEMS gravimeter," *Nature*, vol. 531, pp. 614-617, 2016.

[4] S. Tang, H. Liu, S. Yan, X. Xu, W. Wenjie, J. Fan, J. Liu, C. Hu, and L. Tu, "A high-sensitivity MEMS gravimeter with a large dynamic range," *Microsystems & Nanoengineering*, vol. 5, pp. 1-11, 2019.

[5] H. Liu, W. T. Pike, C. Charalambous, and A. E. Stott, "Passive Method for Reducing Temperature Sensitivity of a Microelectromechanical Seismic Accelerometer for Marsquake Monitoring Below 1 Nano- g," *Physical review applied*, vol. 12, 2019.

[6] M. T. A. Saif, "On a tunable bistable MEMS-theory and experiment," *Journal of Microelectromechanical Systems*, vol. 9, pp. 157-170, 2000.

[7] S. Park and D. Hah, "Pre-shaped buckled-beam actuators: Theory and experiments," *Sensors and Actuators A: Physical*, vol. 148, pp. 186-192, 2008.

[8] M. D. Greenberg, Foundations of Applied Mathematics. Englewood Cliffs, NJ: Prentice-Hall, 1978, p. 488.

[9] S. Huang, F. Lin and Y. Yang, "A novel single-actuator bistable microdevice with a moment-driven mechanism," *Sensors and Actuators A: Physical*, vol. 310, p. 111934, 2020.

[10] Y. Duan, X. Wei, H. Wang, M. Zhao, Z. Ren, H. Zhao, and J. Ren, "Design and numerical performance analysis of a microgravity accelerometer with quasi-zero stiffness," *Smart materials and structures*, vol. 29, p. 75018, 2020-01-01 2020.

[11] M. T. A. Saif and N. C. MacDonald, "A millinewton microloading device," *Sensors and Actuators A: Physical*, vol. 52, pp. 65-75, 1996.

[12] J. Qiu, J. H. Lang and A. H. Slocum, "A Curved-Beam Bistable Mechanism," *Journal of Microelectromechanical Systems*, vol. 13, pp. 137-146, 2004.

978-1-6654-3008-1/21 $31.00 © 2021 IEEE

Gap in pagination due to unavailable papers.

Pages 1161-1164

Proceedings of the 16th Annual IEEE International
Conference on Nano/Micro Engineered and Molecular Systems
April 25-29, 2021

The realization of ZrO_xN_y temperature sensors with good sensitivity and stability in the temperature range above 150K

Yanjie Li, Minmin You, Xun Gong, Xiuyan Li, Zude Lin, and Jingquan Liu, *Member, IEEE*

Abstract— In this work, the temperature range of ZrO_xN_y temperature sensors with high sensitivity is increased to 150 K~300 K for the first time. It was achieved by controlling the thickness of ZrO_xN_y thin films to 34 nm. In addition, we give an in-depth investigation of the stability of these temperature sensors utilizing three temperature shock methods. It was demonstrated that the method with the longest stabilization time could improve the stability to the greatest extent. We have achieved ZrO_xN_y thin films temperature sensors with good sensitivity and stability of 4.53 mK standard deviation under the measurement of water triple point. This work provides an instructive guide for developing temperature sensors with high performance.

I. INTRODUCTION

In materials science, transition metal nitrides of the IVb-Vb-VIb group can form cubic rocksalt-type crystals (B1-structure, Fm3m symmetry), which have excellent physical and chemical properties and constitute a very important class of technical materials [1]. Zirconium nitride, ZrN, is a material characterized by high electron conductivity and mobility, high melting point, fire resistance, chemical stability in harsh environments, biocompatibility, and excellent corrosion, wear resistance. Therefore, it can be used as potential alternative materials for plasmonics, ideal filler in fire resistance coating, protective and decorative coating, and implant materials [2-4]. Transition metal oxynitride is a new material that may have both oxide and nitride characteristics [5]. Doping oxygen into the ZrN-type growth films can obtain Zirconium oxynitride (ZrO_xN_y) thin films. By controlling the value of the gas flow to change the concentration of N, O, and Zr components during processing, the structure and performance of the ZrO_xN_y thin films can be both changed significantly, such as mechanical, electrical, and optical properties [6]. Therefore, the ZrO_xN_y thin films have recently received much attention. Studies have found that after long-term testing, the zirconia oxynitride coating had better corrosion resistance and oxidation resistance than ZrN in the application of conductive coatings [7]. In addition, ZrO_xN_y thin films have become a representative material for temperature sensing, especially for ultra-low temperature, due to their high sensitivity, low magnetoresistance, and excellent long-term stability [8].

Researches on high-resolution thermometers generally focus on sensors used at ultra-low temperatures, but sensors used at room temperature also have essential research values. With the emergence of intelligent sensing technology, wearable temperature sensors for continuous monitoring of human body temperature have attracted more attention [9]. In addition, changes in the internal temperature of the lithography system can cause the optical lens and metal parts to expand and contract, which will affect the final line width of the lithography machine. Therefore, temperature fluctuations are an important factor affecting the image quality of the lithography machine [10]. In order to ensure the long-term stability of the quality of the optical lens of the lithography machine, lithography systems have proposed extremely high requirements for the sensitivity and stability of the temperature sensor around 295K [11].

Negative temperature coefficient (NTC) thermistors based on ZrO_xN_y thin films have good sensitivity, response speed and stability, which shows potentials for temperature detection. Nevertheless, the currently developed zirconium oxynitride temperature sensors are proven to have high performance only at the cryogenic temperature range. According to the report of S. S. Courts and P. R. Swinehart, the zirconium oxynitride resistance temperature sensors could achieve a mean deviation of 0.788mK at 1.8K; however, it was only 28.267 mK at 270 K [12]. Thus, it has significant value to develop ZrO_xN_y thin film based temperature sensors with high sensitivity and stability applied in the ambient temperature range.

In this work, ZrO_xN_y thin films with a thickness of 34nm were deposited on a sapphire substrate using magnetron sputtering. Furthermore, temperature sensors based on ZrO_xN_y thin films with good sensitivity in the temperature range of 150 K to 300 K are achieved for the first time. It also exhibits excellent stability under the measurement of the water triple point.

II. EXPERIMENTS

A. Deposition of the films

In order to make the 2-inch sapphire substrate better heated by the infrared lamp during sputtering, graphite coatings were prepared by the spin coating method on the sapphire substrates. The homogeneous viscous liquid of non-photosensitive polyimide and graphite powder were intermixed evenly. Then, the mixture was spun on the back

---

*Resrach supported by the Strategic Priority Research Program of Chinese Academy of Sciences under Grant XDA25040000, the National Key R&D Program of China under Grant 2020YFB1313502, the National Natural Science Foundation of China under Grant 61728402, SJTU Transmed Award under Grant 2019015, the Oceanic Interdisciplinary Program of Shanghai Jiao Tong University under Grant SL2020ZD205, Scientific Research Fund of Second Institute of Oceanography , MNR under Grant SL2020ZD205, the Program of Shanghai Academic/Technology Research

Leader under Grant 18XD1401900, the Center for Advanced Electronic Materials and Devices (AEMD) of Shanghai Jiao Tong University.

Yanjie Li, Minmin You, Xun Gong, Xiuyan Li, Zude Lin and Jingquan Liu are all with the National Key Laboratory of Science and Technology on Micro/Nano Fabrication, Shanghai Jiao Tong University, Shanghai, 200240, China, Department of Micro/Nano Electronics, Shanghai Jiao Tong University, Shanghai, 200240, China (corresponding auther to provide e-mail: jqliu@sjtu.edu.cn).

of the sapphire substrate. After that, the coatings were cured through high-temperature baking. Magnetron sputtering was used to grow ZrO_xN_y thin films with a thickness of 34nm on the sapphire substrates. The deposition was carried out under 9 sccm (standard cubic centimeter per minute) nitrogen-oxygen mixed gas (containing 1% oxygen) at the temperature of 320 °C. The sputtering chamber's pressure was in the order of 10^{-5} Pa, and the sputtering power was 110W. Argon was used as the operating gas with a flow rate of 23 sccm.

B. Fabrication of the interdigital electrodes

Since the resistance of the thin films was tremendous, we need to adopt interdigital electrodes to obtain suitable resistance and reduce self-heating [13]. As shown in Fig. 1, 30nm Cr (adhesion layer) and 300nm Au interdigital electrodes were manufactured on the film surface by sputtered in Ar using an ion beam sputtering system.

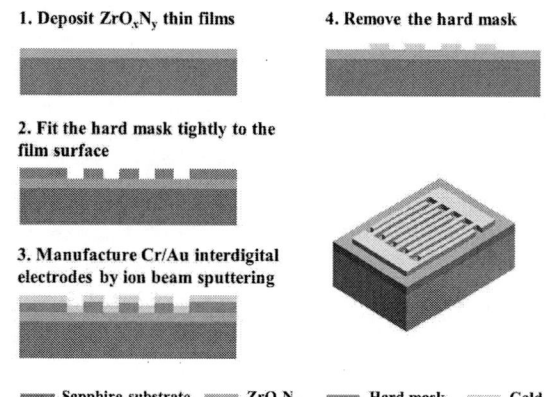

Figure 1. Manufacturing process for ZrO_xN_y thin film based temperature sensors.

C. Setup of the temperature shock methods

Fig. 2 is the schematic diagram of the temperature shock process. One cycle requires three steps: the first step is to soak the sensor in a liquid nitrogen environment (77K) for a certain period, then quickly remove it from the liquid nitrogen to the room temperature environment, finally place it in the room temperature also for a certain period. By setting different stabilization times and cycle times, we have studied the impact of three different shock methods on the stability of the sensor, as shown in Table 1. Method 1 and method 2 are characterized by short stabilization time and a large number of cycles. Method 2 increases the stabilization time slightly based on method 1. Method 3 significantly increased the stabilization time to sixty minutes and reduced the number of cycles from thirty to three, emphasizing the stabilization time.

TABLE I.   THE PARAMETERS OF TEMPERATURE SHOCK METHODS

Shock methods	Stabilization time (min)	Cycle number
1	1	30
2	10	30
3	60	3

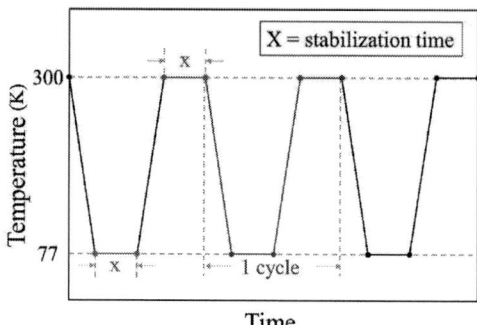

Figure 2. Schematic diagram of the temperature shock process.

D. Setup of the stability testing

In practical applications, stability is one of the most significant concerns of temperature sensors because they may cause false alarms [14]. In order to determine the stability of the ZrO_xN_y thin film based sensors, the degree of fluctuation of resistance over time in a stable temperature environment was tested. The water triple point was selected as the stable environment temperature in this work. The water triple point bottle was frozen by liquid nitrogen and stored in a constant temperature bath which pre-cooled in advance at 0.01°C.

Fig. 3 shows the experimental setup for stability testing, including a high-precision thermometry bridge, standard resistance, to-be-tested resistance sensor, and testing software. The sensor to be tested was placed in the water triple point bottle and connected to the test ports of the thermometry bridge by the four-wire method. The standard resistance was placed in a dry resistance box at room temperature. The standard resistance and the resistance to be measured were connected to the both ends of the bridge. The accurate resistance value of the resistance to be measured is determined by measuring the ratio of the two resistances. The resistance of the sensor was tested for ten consecutive hours under the water triple point. We extracted data of two hours after stabilization to obtain the resistance-time (R-t) curve, which can be converted into the temperature fluctuation curve using the typical sensitivity formula:

$$S = \delta R / \delta T \tag{1}$$

where S is the typical sensitivity calculated by the resistance-temperature (R-T) characteristic curve of the sensor.

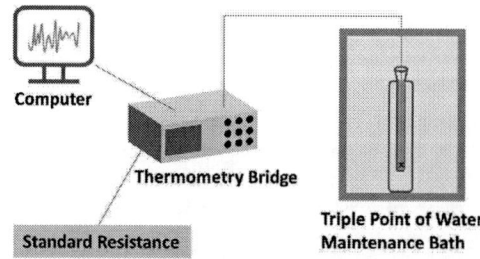

Figure 3. Experimental setup for the stability testing. The to-be-tested resistance sensor is placed in the water triple bottle, represented by the symbol "x" in this figure.

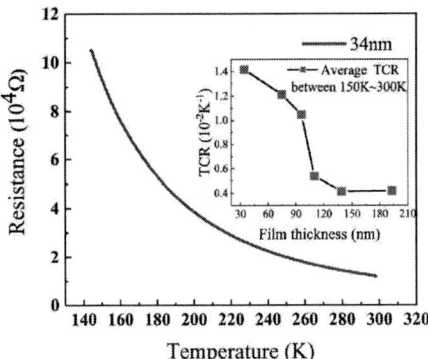

Figure 4. The R-T characteristic curve of 34nm sample. The inset figure presents the average TCR value of ZrO_xN_y temperature sensors with different film thicknesses in the temperature range of 150 K~300 K.

III. RESULTS AND DISCUSSIONS

In order to study the sensitive temperature zone of the ultra-thin ZrO_xN_y temperature sensor, we measured the R-T characteristic curve, as shown in Fig. 4. Unlike the traditional ultra-low temperature sensing ZrO_xN_y thin film based temperature sensors, when the thickness of the ZrO_xN_y thin film drops to 34nm, the sensor also has a certain sensitivity in the temperature range of 150K to 300K. Besides, the resistance value of the sensor increases monotonously as the temperature decreases, which is the characteristic of the negative temperature coefficient (NTC) temperature sensors [15]. In order to determine the relationship between the sensitivity and the thickness of ZrO_xN_y thin films in the temperature range of 150K to 300K, we compared the average temperature coefficient of resistance (TCR) value of sensors with different film thicknesses in this temperature zone. The average TCR of the sensors with film thickness more than 100nm is extremely low. When the thickness is less than 100nm, the average TCR increases significantly, which represents an increase in the sensitivity of the sensor. The results showed that the average TCR value of the 34nm sample was significantly higher than that of other samples.

Fig. 5 shows the results of the stability test of the samples. Here we use the temperature test standard deviation to measure the stability. Figure 5(a) shows the stability test result of the 34nm sample without temperature shock. Under this situation, the temperature test standard deviation of the sensor is 21.36mK. Figure 5(b) ~ (d) shows the stability test results of the 34nm sample after three different temperature shock methods. It can be seen that method 1 and method 2 significantly reduced the stability, and the standard deviation increased to 100mK. Unlike the previous two methods, method 3 improved the stability of the sensor and reduced the standard deviation to 5.51mK. The above results show that the long stabilization time method can significantly reduce the temperature test standard deviation. In contrast, the short stabilization time method has the opposite effect on the sensors. In addition, we implemented method 3 on another 34nm sample to further verify the effectiveness of this method. As shown in Fig. 6, the temperature test standard deviation of this sensor sample reached 4.53mK, which further verifies the effectiveness of method 3 in

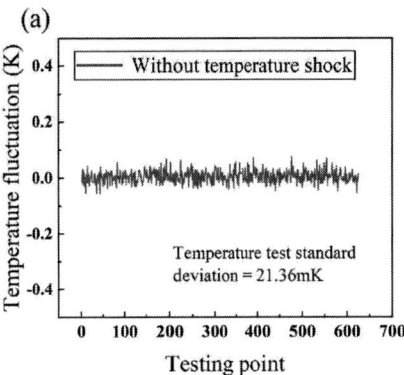

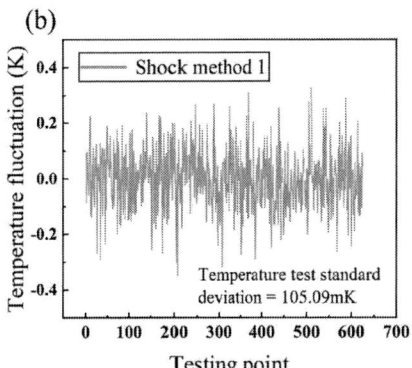

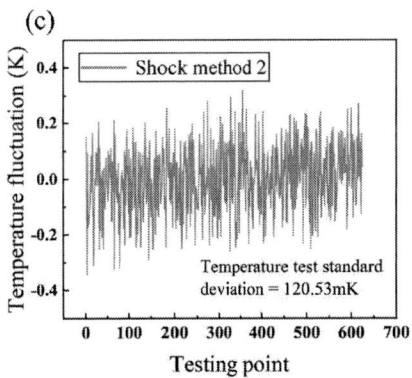

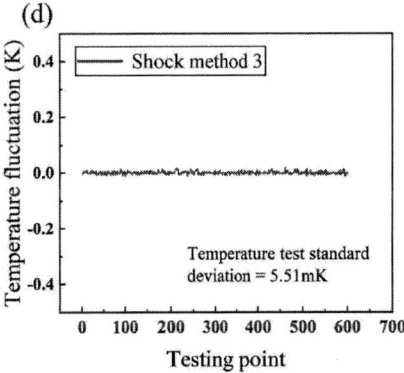

Figure 5. (a) The stability test result of the 34nm sample before temperature shock. (b) ~ (d) The stability test results of the same 34nm sample under three different shock methods.

978-1-6654-3008-1/21 $31.00 © 2021 IEEE

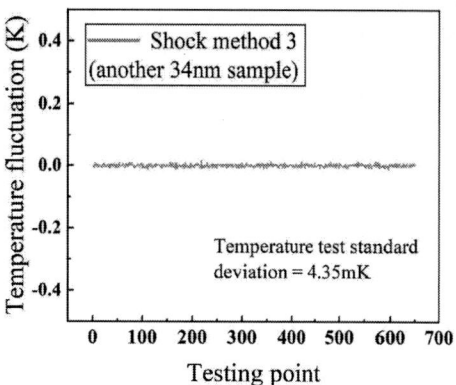

Figure 6. The stability test result of another 34nm sample under temperature shock method 3.

improving the stability of the ZrO_xN_y thin film based sensors.

IV. CONCLUSION

In this work, 34nm ultra-thin ZrO_xN_y films were deposited on the sapphire substrates using reactive magnetron sputtering system. The R-T characteristic curve and TCR values show that the temperature sensors based on the ultra-thin ZrO_xN_y films have good sensitivity in the temperature range of 150K to 300K. Temperature shock experiments and stability tests show that the shock method with a long stabilization time can significantly improve the stability of the sensors, and the temperature test standard deviation is finally improved to 4.53mK. These results indicate the broad prospects of temperature sensors based on ZrO_xN_y thin films for sensing in a wide temperature range.

ACKNOWLEDGMENT

This work was partially supported by the Strategic Priority Research Program of Chinese Academy of Sciences under Grant XDA25040000, the National Key R&D Program of China under Grant 2020YFB1313502, the National Natural Science Foundation of China under Grant 61728402, SJTU Trans-med Award under Grant 2019015, the Oceanic Interdisciplinary Program of Shanghai Jiao Tong University under Grant SL2020ZD205，Scientific Research Fund of Second Institute of Oceanography, MNR under Grant SL2020ZD205, the Program of Shanghai Academic/Technology Research Leader under Grant 18XD1401900. The authors are also grateful to the Center for Advanced Electronic Materials and Devices (AEMD) of Shanghai Jiao Tong University.

REFERENCES

[1] P. Patsalas, N. Kalfagiannis, S. Kassavitis, et al., "Conductive nitrides: growth principles, optical and electronic properties, and their perspectives in photonics and plasmonics," *Mater. Sci. Eng. R Rep.*, vol. 123, pp. 1–55, 2018.

[2] P. Patsalas, "Zirconium nitride: A viable candidate for photonics and plasmonics," *Thin Solid Films*, vol. 688, pp. 137438.1–137438.15, Oct. 2019.

[3] W. H. Feng, Y. F. Wang, B. Tang, "Suppression on heat and smoke diffusion by zirconium nitride (ZrN) in intumescent flame retardant epoxy coatings," *Prog. Org. Coat.*, vol. 146, pp. 105714.1–105714.7, 2020.

[4] T. Kuznetsova, V. Lapitskaya, A. Khabarava, et al., "The Influence of Nitrogen on the Morphology of ZrN Coatings Deposited by Magnetron Sputtering," *Appl. Surf. Sci.*, vol. 522, pp. 146508.1–146508.7, 2020.

[5] Z. D. Lin, G. H. Zhan, X. Y. Li, et al., "The conduction process of grain and grain boundary in the semiconductive zirconium oxynitride thin film," *Semicond. Sci. Technol.*, vol. 34, pp. 085008.1–085008.9, 2019.

[6] Y. Y. Yuan, R. Lan, C. Yan, et al., "Zirconium oxynitride films: Modulation of component as a function of the preparation parameters," *Mod. Phys. Lett. B*, vol. 32, pp. 1840066.1–1840066.8, 2018.

[7] X. Z. Wang, T. P. Muneshwar, H. Q. Fan, et al., "Achieving ultrahigh corrosion resistance and conductive zirconium oxynitride coating on metal bipolar plates by plasma enhanced atomic layer deposition," *J. Power Sources*, vol. 397, pp. 32–36, 2018.

[8] G. H. Zhan, Z. D. Lin, B. Xu, et al., "Study of temperature sensitivity and impedance spectroscopy of zirconium oxynitride thin film thermistors," *J. Mater. Sci. Mater. Electron*, vol. 28, pp. 9653–9657, 2017.

[9] I. Khan and Z. H. Lin, "Sputtered thermoelectric nanoparticles for an ultra-thin, flexible and cuttable self-powered temperature sensor," *ECS. Trans.*, vol. 97 (6), pp. 79–84, 2020.

[10] Y. Cui, J. Peng, M. Yu, P. Y. Li, "Assessment of the uncertainty in temperature measurements by industrial negative temperature coefficient thermistor sensors," *Electronic measurement technology*, vol. 8, pp. 104-106, 2014.

[11] L. L. Yu, "Research on high-precision temperature measurement technology of 100nm step-scan lithography machine," M.S. thesis, Dept. Control Eng., Tongji Univ., Shanghai, China, 2005.

[12] S. S. Courts and P. R. Swinehart, "Review of cernox (zirconium oxy-nitride) thin-film resistance temperature sensors," in *Proc. AIP.* 2003, pp. 393-398.

[13] M. Paeschke, U. Wollenberger, C. K ö hler, T. Lisec, U. Schnakenberg, R. Hinstsche, "Properties of interdigital electrode arrays with different geometries," *Anal. Chim. Acta.*, vol. 305, pp. 126-136, 1995.

[14] J. W. Gong, Q. F. Chen, M. R. Lian, N. C. Liu, C. Daoust, "Temperature feedback control for improving the stability of a semiconductor-metal-oxide (SMO) gas sensor," *IEEE Sens. J.*, vol. 6, pp. 139-145, Feb. 2006.

[15] Z. D. Lin, G. H. Zhan, M. M. You, et al., "NTC thin film temperature sensors for cryogenics region with high sensitivity and thermal stability," *Appl. Phys. Lett.*, vol. 113, pp. 133504.1-133504.5, 2018.

Proceedings of the 16th Annual IEEE International
Conference on Nano/Micro Engineered and Molecular Systems
April 25-29, 2021

NEMS Sensors Based on Suspended Graphene

Xuge Fan*, and Frank Niklaus, *Senior Member, IEEE*

Abstract—Graphene has exciting potential in nanoelectromechanical system (NEMS) applications due to its unique mechanical and electrical properties as well as its ultimate thinness. In this paper, we discuss the potential of using suspended graphene structures in NEMS sensors and provide an overview of our previous research results on piezoresistive graphene NEMS sensors, including pressure sensors and accelerometers.

INTRODUCTION

Graphene is an ultra-thin material with a single layer thickness of approximately 0.335 nm. Graphene is extremely strong and stiff, highly conductive, and chemically stable. Due to its atomically thinness, unique electrical and mechanical properties such as ultra-high carrier mobility and high Young's modulus and electromechanical properties [1], graphene is a very promising material for realizing ultra-thin suspended membranes, beams and ribbons. In these settings, the suspended graphene can be used both as structural material and as electromechanical transducers for nanoelectron-mechanical system (NEMS) sensors with potential for substantial NEMS device size reduction, while providing improved sensitivities. Various types of graphene-based NEMS resonators [2], [3] and pressure sensors [1], [4], [5] have been extensively studied, and graphene has been implemented in emerging NEMS sensors such as in NEMS microphones, bolometers and accelerometers, etc. [1], [6]. Here, we review our work on exploring the use of suspended graphene in NEMS pressure sensors and accelerometers. In particular we discuss suitable graphene NEMS device design, functionality and fabrication and the piezoresistive properties of graphene, along with the advantages and disadvantages of using graphene in NEMS sensors.

RESULTS

In our research we have explored the use of the piezoresistive effect in graphene to realize NEMS sensors in which the suspended graphene is at the same time the structural and the transducer element of the NEMS sensor. The piezoresistive effect of graphene is a result of mechanical strain in graphene that changes the electronic band structures of graphene, which can be commonly evaluated by gauge factors. For mechanically exfoliated graphene and chemical vapor deposited (CVD) graphene, gauge factors were reported to be on the order of between 2 and 6 [1], [7]. The piezoresistive behavior of graphene can

be ascribed to variations of both carrier density and carrier mobility, and the carrier mobility is the main factor in the piezoresistive effect of graphene [7]. The gauge factor of graphene is supposedly independent of crystallographic orientation and the doping concentration of graphene membranes [8]. Uniaxial strain and bi-axial strain in graphene (Fig. 1) results in comparable gauge factors [9]. As the strain is small, the intrinsic gauge factor of graphene was reported to be largely independent of the strain magnitude [8]. In addition, gauge factors can be significantly amplified in resonant structures consisting of very thin doubly-clamped transducers, including carbon nanotubes, silicon nanowires and graphene ribbons due to an asymmetric beam shape at rest [10].

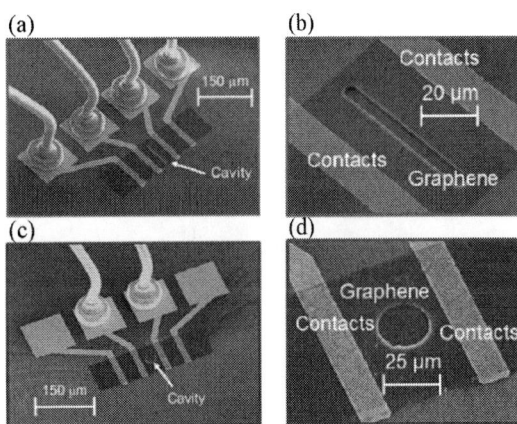

Figure 1. Uniaxial and biaxial strain in different types of suspended graphene. (a, b) SEM images of a pressure sensor device that is composed of a rectangular suspended graphene membrane featuring uniaxial strain in the graphene membrane. (c, d) SEM images of a pressure sensor device that is composed of a circular graphene membrane featuring biaxial strain in the graphene membrane [8].

In graphene NEMS sensors the suspended graphene is the essential functional and transducer element. As shown in Fig. 2, two typical ways to implement the suspended graphene structure in a NEMS sensor are 1) transferring graphene from the substrate where graphene is grown to a substrate that is fabricated with trenches, cavities or dielectric membranes, or 2) by transferring graphene from the substrate where graphene is grown to a substrate made of a flat SiO_2 or a polymer followed by the sacrificial etching of the material underneath the graphene [1], [11].

Research supported by the Swedish Research Council (GEMS, 2015-05112), the FLAG-ERA project (2DNEMS, VR 2019-03412), and the China Scholarship Council (CSC). All authors are with Division of Micro and Nanosystems, KTH Royal Institute of Technology, SE-10044 Stockholm, Sweden. (E-mail: xuge@eecs.kth.se).

978-1-6654-3008-1/21 $31.00 © 2021 IEEE

The PMMA-based wet transfer is a commonly used method to transfer graphene from the growth substrate to the target substrate [12]–[16], however alternative methods that are potentially more mass manufacturable are also being explored [17].

We have realized piezoresistive graphene NEMS pressure sensors by suspending CVD graphene membranes over cavities etched into the SiO_2 layer on a silicon substrate with metal electrodes [7], [8], [18], in which graphene was used as the membrane and the electromechanical piezoresistive transducer simultaneously (Fig. 3). To realize electrical signal readout, the devices were wire bonded and the chips were placed in a package. These piezoresistive graphene NEMS pressure sensors could substantially reduce the device sizes and their sensitivities per unit area were better than that of conventional silicon and carbon nanotube-based piezoresistive pressure sensors [7]. This is because for a given membrane area, the membrane deflection increases with decreasing membrane thickness. Thus, the ultimate thinness of graphene enables increased membrane deflection and enhanced sensitivity of the NEMS sensor [7].

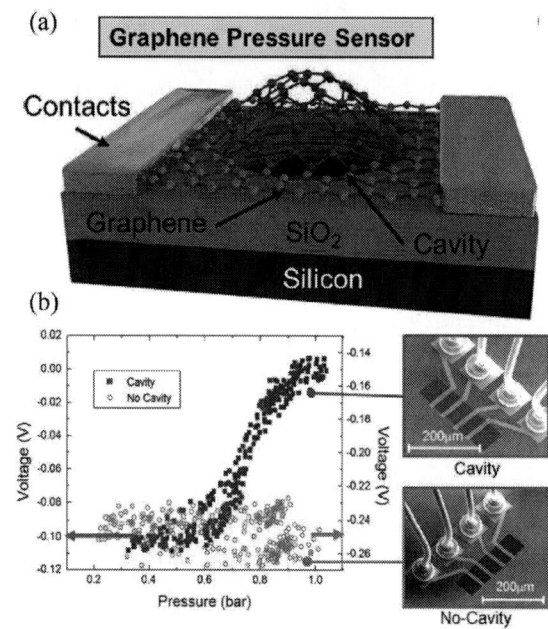

Figure 3. Piezoresistive graphene pressure sensor. (a) Schematic of the graphene pressure sensor [8]. (b) Voltage measurement of a graphene pressure sensor (blue squares) and a reference device without a cavity (red hollow circles) at different applied pressures [7].

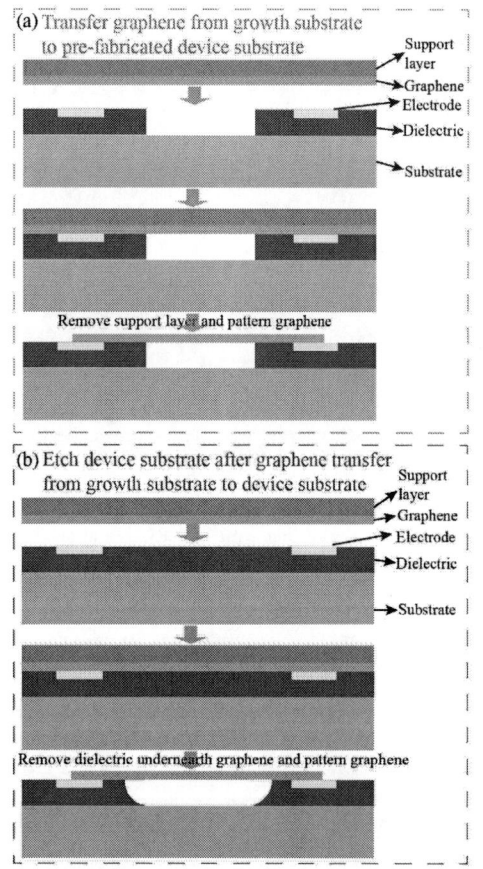

Figure 2. Two typical ways to realize suspended graphene. (a) The graphene is transferred from the growth substrate to the device substrate with pre-processed device structures. (b) The graphene is transferred from growth substrate to a flat dielectric and then the dielectric underneath the desired area of suspended graphene is sacrificially etched.

We have also explored the advantages of thin suspended graphene structures for realizing piezoresistive graphene NEMS accelerometers by attaching large Si proof masses (from 5 μm × 5 μm × 16.4 μm to 100 μm × 100 μm × 16.4 μm) onto the double-layer graphene ribbons or membranes [11], [19], [20]. In these sensors, the graphene ribbons or membranes with attached Si proof mass have the function of both the spring-mass system and the transducer elements using the piezoresistivity of graphene (Fig. 4). The graphene accelerometers were fabricated using silicon-on-insulator (SOI) substrate as starting material. Specifically, they were fabricated by using the evaporation of metal electrodes, etching trenches, etching backside openings in the SOI substrate, transferring and patterning graphene, and releasing the proof mass by dry etching and vapor HF etching (Fig. 5). All steps in the fabrication process are compatible with large-scale semiconductor manufacturing technologies. The suspended graphene ribbons and membranes with an attached Si proof mass feature excellent robustness. For instance, they can withstand large AFM tip indentation forces (thousands of nN) and yield relatively high strain in the graphene without device failure [11], [19], [20]. During sensor operation, an acceleration that is acting on the Si proof mass causes a deflection and related strain in the suspended graphene ribbons or membranes. The strain ultimately results in a resistance change in the graphene ribbons or membranes that can be electrically measured and that correlates with the acceleration (Fig. 6). The die areas of our piezoresistive graphene accelerometers are of the order of two magnitudes smaller than the areas of conventional silicon accelerometers, while achieving reasonable sensor performance [19], [20]. For instance, the normalized relative resistance change per proof mass volume in our graphene ribbon accelerometers was significantly larger compared to that of typical silicon accelerometers [19].

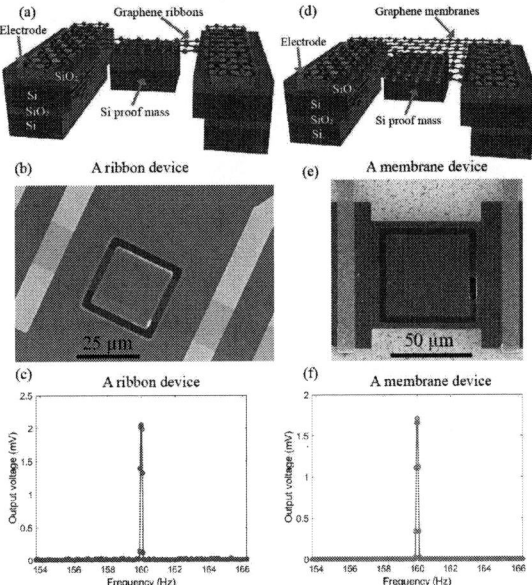

Figure 4. Graphene NEMS accelerometers. (a) Schematic of a graphene ribbon accelerometer, (b) SEM image of a graphene device, (c) the output signal of a graphene ribbon accelerometer [14]. (d-f) Corresponding schematic, SEM image and output signal of a graphene membrane accelerometer structure [15].

Compared to the graphene ribbon accelerometers, the graphene membrane accelerometers were more robust, have longer lifetime and fabrication yield, are able to withstand higher measurement currents and can sustain more heavy proof masses, but feature lower sensing signals [20]. The resonant frequency of our graphene ribbon and membrane accelerometer ranged from tens of kHz to hundreds of kHz (Fig. 7) [11], [19]. We believe that there is further potential to improve the sensitivity of graphene NEMS sensors by optimizing the graphene transducer elements, or by using the heterolayers of graphene with other 2D materials with higher piezoresistive gauge factors such as MoS_2 [21].

The built-in stress in the suspended graphene is a critical characteristic of graphene NEMS such as resonators, accelerometers and pressure sensors because the built-in stress substantially impacts the resonance frequency and the force-deflection properties of the suspended graphene structures. Therefore, it is important to characterize the built-in stress in suspended graphene and explore ways to reduce the built-in stress. Using graphene ribbons with an attached Si mass, both the built-in stress and the Young's modulus of graphene can be conveniently measured by using AFM indentation. With this approach we measured the built-in stress in our graphene ribbons with attached Si masses and found that they were typically on the order of hundreds of MPa, which are non-negligible stress levels [19]. We also measured the Young's modulus of our double-layer graphene to be about 0.21-0.25 TPa, which is

lower than that of typical single layer graphene. Possible approaches to reduce the built-in stress in the suspended graphene structures may include the optimization of the process of transferring graphene, of the final graphene substrate surface and of the graphene source material [19].

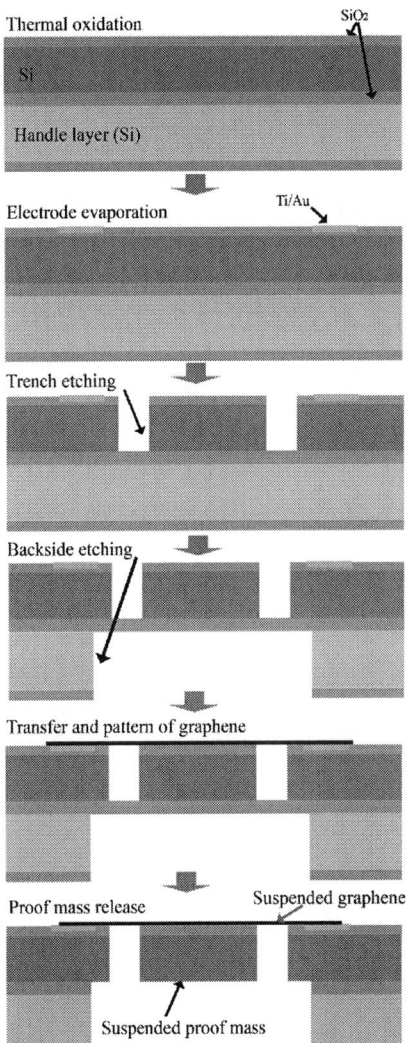

Figure 5. Schematic of fabrication process of graphene accelerometers.

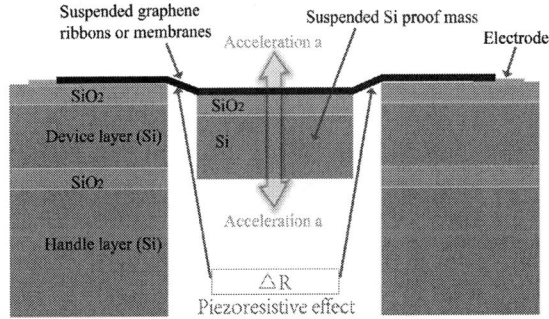

Figure 6. Schematic diagram of the working principle of the piezoresistive NEMS accelerometer based on suspended graphene ribbons or membranes with an attached Si proof mass.

978-1-6654-3008-1/21 $31.00 © 2021 IEEE

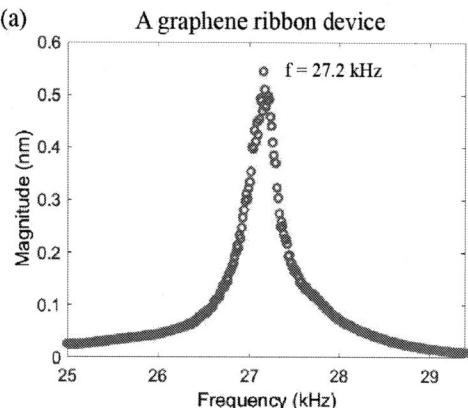

(a)

A graphene ribbon device

f = 27.2 kHz

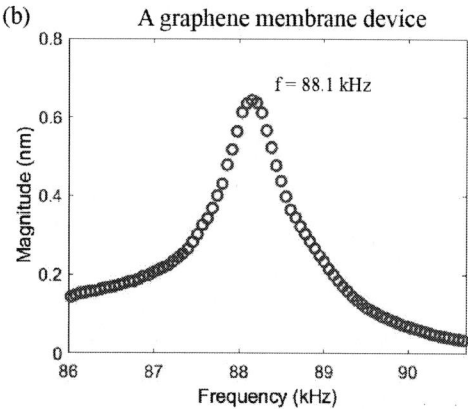

(b)

A graphene membrane device

f = 88.1 kHz

Figure 7. Resonance frequencies of graphene accelerometers. (a) Measured resonance frequency of an accelerometer based on graphene ribbons with an attached Si proof mass [19]. (b) Measured resonance frequency of a accelerometer based on graphene membranes with an attached Si proof mass [23].

In summary, we have demonstrated that graphene has exciting potential for use as structural and transducer material in ultra-miniaturized NEMS sensors.

REFERENCES

[1] M. C. Lemme *et al.*, "Nanoelectromechanical Sensors Based on Suspended 2D Materials," *Research*, vol. 2020, pp. 1–25, Jul. 2020, doi: 10.34133/2020/8748602.

[2] Y. Oshidari, T. Hatakeyama, R. Kometani, S. Warisawa, and S. Ishihara, "High Quality Factor Graphene Resonator Fabrication Using Resist Shrinkage-Induced Strain," *Appl. Phys. Express*, vol. 5, no. 11, p. 117201, Oct. 2012, doi: 10.1143/APEX.5.117201.

[3] S. Afyouni Akbari, V. Ghafarinia, T. Larsen, M. M. Parmar, and L. G. Villanueva, "Large Suspended Monolayer and Bilayer Graphene Membranes with Diameter up to 750 μm," *Sci. Rep.*, vol. 10, no. 1, Dec. 2020, doi: 10.1038/s41598-020-63562-y.

[4] R. J. Dolleman, D. Davidovikj, S. J. Cartamil-Bueno, H. S. J. van der Zant, and P. G. Steeneken, "Graphene Squeeze-Film Pressure Sensors," *Nano Lett.*, vol. 16, no. 1, pp. 568–571, Jan. 2016, doi: 10.1021/acs.nanolett.5b04251.

[5] M. Lee *et al.*, "Sealing Graphene Nanodrums," *Nano Lett.*, vol. 19, no. 8, pp. 5313–5318, Aug. 2019, doi: 10.1021/acs.nanolett.9b01770.

[6] S. Wittmann, C. Glacer, S. Wagner, S. Pindl, and M. C. Lemme, "Graphene Membranes for Hall Sensors and Microphones Integrated with CMOS-Compatible Processes," *ACS Appl. Nano*

Mater., vol. 2, no. 8, pp. 5079–5085, Aug. 2019, doi: 10.1021/acsanm.9b00998.

[7] A. D. Smith *et al.*, "Electromechanical Piezoresistive Sensing in Suspended Graphene Membranes," *Nano Lett.*, vol. 13, no. 7, pp. 3237–3242, Jul. 2013, doi: 10.1021/nl401352k.

[8] A. D. Smith *et al.*, "Piezoresistive Properties of Suspended Graphene Membranes under Uniaxial and Biaxial Strain in Nanoelectromechanical Pressure Sensors," *ACS Nano*, vol. 10, no. 11, pp. 9879–9886, Nov. 2016, doi: 10.1021/acsnano.6b02533.

[9] A. D. Smith *et al.*, "Biaxial strain in suspended graphene membranes for piezoresistive sensing," in *2014 IEEE 27th International Conference on Micro Electro Mechanical Systems (MEMS)*, Jan. 2014, pp. 1055–1058, doi: 10.1109/MEMSYS.2014.6765826.

[10] M. Sansa, M. Fernández-Regúlez, J. Llobet, Á. San Paulo, and F. Pérez-Murano, "High-sensitivity linear piezoresistive transduction for nanomechanical beam resonators," *Nat. Commun.*, vol. 5, no. 1, Art. no. 1, Jul. 2014, doi: 10.1038/ncomms5313.

[11] X. Fan *et al.*, "Manufacture and characterization of graphene membranes with suspended silicon proof masses for MEMS and NEMS applications," *Microsyst. Nanoeng.*, vol. 6, no. 1, p. 17, Dec. 2020, doi: 10.1038/s41378-019-0128-4.

[12] X. Fan *et al.*, "Humidity and CO2 gas sensing properties of double-layer graphene," *Carbon*, vol. 127, pp. 576–587, Feb. 2018, doi: 10.1016/j.carbon.2017.11.038.

[13] A. Quellmalz *et al.*, "Influence of Humidity on Contact Resistance in Graphene Devices," *ACS Appl. Mater. Interfaces*, vol. 10, no. 48, pp. 41738–41746, Dec. 2018, doi: 10.1021/acsami.8b10033.

[14] A. D. Smith *et al.*, "Toward effective passivation of graphene to humidity sensing effects," in *2016 46th European Solid-State Device Research Conference (ESSDERC)*, Sep. 2016, pp. 299–302, doi: 10.1109/ESSDERC.2016.7599645.

[15] A. D. Smith *et al.*, "Graphene-based CO2 sensing and its cross-sensitivity with humidity," *RSC Adv.*, vol. 7, no. 36, pp. 22329–22339, Apr. 2017, doi: 10.1039/C7RA02821K.

[16] X. Fan *et al.*, "Direct observation of grain boundaries in graphene through vapor hydrofluoric acid (VHF) exposure," *Sci. Adv.*, vol. 4, no. 5, p. eaar5170, May 2018, doi: 10.1126/sciadv.aar5170.

[17] A. Quellmalz *et al.*, "Large-area integration of two-dimensional materials and their heterostructures by wafer bonding," *Nat. Commun.*, vol. 12, no. 1, Art. no. 1, Feb. 2021, doi: 10.1038/s41467-021-21136-0.

[18] Anderson. D. Smith *et al.*, "Pressure sensors based on suspended graphene membranes," *Solid-State Electron.*, vol. 88, pp. 89–94, Oct. 2013, doi: 10.1016/j.sse.2013.04.019.

[19] X. Fan *et al.*, "Graphene ribbons with suspended masses as transducers in ultra-small nanoelectromechanical accelerometers," *Nat. Electron.*, vol. 2, no. 9, pp. 394–404, Sep. 2019, doi: 10.1038/s41928-019-0287-1.

[20] X. Fan *et al.*, "Suspended Graphene Membranes with Attached Silicon Proof Masses as Piezoresistive Nanoelectromechanical Systems Accelerometers," *Nano Lett.*, vol. 19, no. 10, pp. 6788–6799, Oct. 2019, doi: 10.1021/acs.nanolett.9b01759.

[21] S. Manzeli, A. Allain, A. Ghadimi, and A. Kis, "Piezoresistivity and Strain-induced Band Gap Tuning in Atomically Thin MoS2," *Nano Lett.*, vol. 15, no. 8, pp. 5330–5335, Aug. 2015, doi: 10.1021/acs.nanolett.5b01689.

Gap in pagination due to unavailable paper.

Pages 1173-1174

Proceedings of the 16th Annual IEEE International
Conference on Nano/Micro Engineered and Molecular Systems
April 25-29, 2021

A Miniature Infrared Emitter with Ultra-high Emissivity

Dongsheng Shu[1], Jifang Tao[1†], *Member, IEEE*, Yan Li[1] and Xinye Fan[2]

Abstract— **A miniature infrared emitter has been developed based on MEMS technology. The platinum-black material is used as an emission layer with ultra-high emissivity covering wavelength range from 2 μm to 14 μm, in which the average emissivity is up to 99.5% that is much higher than conventional infrared emitter designs and infrared lamps. The modulation depth is 100% at 10 Hz and 42% at 100 Hz. This emitter has potential to be adopted in high-end infrared gas sensors for medical, industry, environmental applications, etc.**

I. INTRODUCTION

With the merits of high reliable, long lifetime and high accuracy, infrared gas sensors are widely used for industry safety, gas analyzers and breath monitoring[1-4]. Infrared emitter as one key component to provide middle infrared light covering finger-print absorption wavelength of gas molecules. Compared conventional miniature infrared lamps, MEMS emitters have broadband infrared emission spectrum, low power consumption, small size, and fast modulation speed. The emitter generates infrared waves following the principle of the black body radiation. Recently, some researchers developed new emitters with narrow-band radiation spectrum which aligns with a target gas' absorption wavelength. For example, by use of metamaterial structures, Xu obtain a emissivity of 0.9 in the band of 5.5-7.6 um, while below 0.06 in other wavelength band[5]. Gerald used a non-periodic multilayer stack of dielectric layers (Si and SiO2) to generate a certain narrow wavelength , and the center wavelength can be tuned by changing the layer thickness and number of layers[6]. However, the fabrication processing technology causes the deviation of the center wavelength of the designed emitter radiation, and the temperature drift will also cause the deviation of the center wavelength after a long time of operation. Compared narrow band infrared emitter, the infrared emitter with broadband radiation spectrum can be as a universal components in single gas detection or multi-gas detection [7]. For broadband spectrum emitter, it requires the radiation layer with high emissivity in the middle infrared region. According to Kirchhoff's law, when the transmittance is 0, the emissivity (E) of the materials is related to its absorption coefficient (A) and reflectivity (R)[8]

$$E = A = 1\text{-}R \qquad (1)$$

Therefore, increasing the absorption coefficient of the radiation layer can increase its emissivity. Li achieve absorption reach to 75% in 2-14 um by using a graphene oxide layer[9]. Syed chose carbon nanotubes and achieved an emissivity approximately 1 within 2-14 μm[10].

In this article, we develop the emitter with an emissivity up to 99.5% covering 2-14 μm infrared region by electroplating a platinum black layer on a MEMS hotplate. With the merits of low heat capacity of the MEMS membrane, the emitter can be rapidly heated-up to a target working temperature (400~500°), thus increasing the modulation rate of the emitter. In addition, platinum has stable chemical properties and temperature coefficient of resistance, which can suppress the temperature drift caused by the emitter after a long-time operation.[11]

II. DESIGN , SIMULATION AND FABRICATION

A. Design of the MEMS Emitter

Fig. 1 shows the design of the MEMS emitter, which consists of a micro-heater, a platinum black acting as an

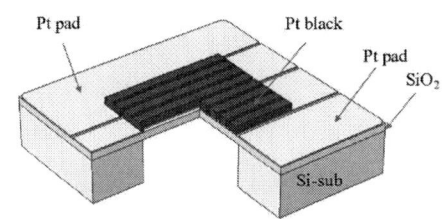

Fig. 1. The 3D structure of emitter

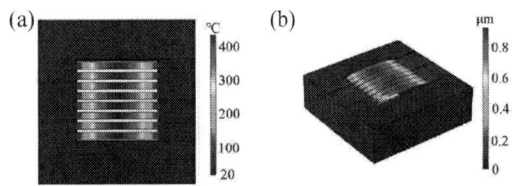

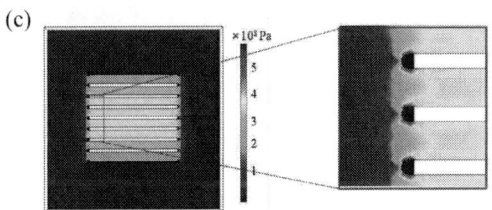

Fig. 2. Simulation of the MEMS emitter driven by 3V. (a) Thermal distribution. (b) displacement. (c) 1st principle stress of the film at high working temperature.

---

[1]School of Information Science and Engineering, Shandong University, No. 72 Binhai Road, Qingdao City, Shandong Province, 266000, China (Corresponding author: J. Tao, phone: +86 17806243369; email: taojf@sdu.edu.cn)

[2]Pacific Optoelectronic Technology Co., Ltd., No.259 Huanghe Rd, Liaocheng City, Shandong Province, 252022, China

978-1-6654-3008-1/21 $31.00 © 2021 IEEE

infrared radiation layer, a silicon dioxide layer and a silicon substrate. The emitter is 1.8mm × 1.8mm × 0.38mm (Width × Length × Height) with a radiation area of 0.8mm × 0.8mm. In the MEMS emitter design, the main influence on the temperature of the film is the heat conduction between the film and the silicon substrate. The back surface of the silicon substrate can be etched into a cavity structure to reduce the contact area between the film and the substrate, thereby reducing the heat conduction between the heating layer and the silicon substrate. Meantime, the oxide layer acts as the thermal insulator which isolates the heating element from the bulk silicon, thereby reducing power consumption.

B. Simulation

The heat transfer loss (P) of the heat radiation emitter is divided into three parts, the heat conduction P_{cond} between the film and the substrate, the heat convection P_{conv} between the film and the air, and the heat radiation P_{rad} of the film.[12]

$$P = P_{cond} + P_{conv} + P_{rad}$$
$$= \kappa A_c \frac{dT}{dL} + h A_s (T_{hot} - T_{amb}) + \varepsilon \sigma A_s (T_{hot}^4 - T_{amb}^4) \quad (2)$$

In the formula, T is the temperature, L is the length between the hot and cold ends, κ is the thermal conductivity of the material, A_c is the contact cross-sectional area of the film and the substrate, and T_{hot} is the temperature of the film. T_{amb} is the temperature of atmosphere. ε is the emissivity of the material, and σ is the Stefan-Boltzmann constant.

Import the 3D model of the emitter into the finite element simulation software. Select the above three heat loss methods and physical field of solid mechanics to simulate the temperature distribution and stress distribution. Fig. 2 shows the simulation results of the infrared emitter. Fig. 2(a) shows

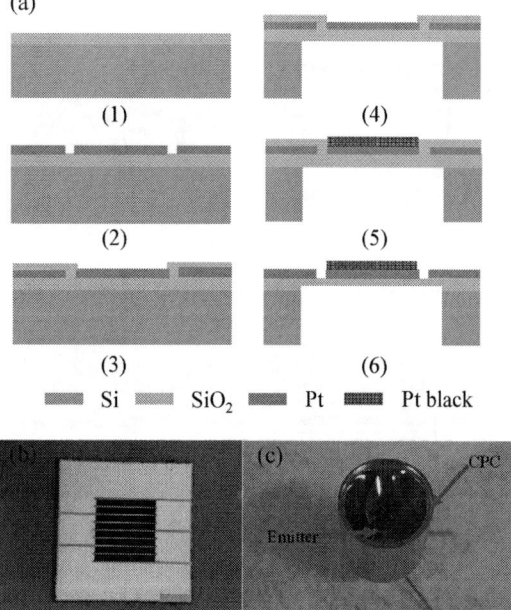

Fig. 3. (a) Fabrication processes of the MEMS emitter. (b) optical image of the MEMS emitter. (c) the MEMS emitter with a TO package.

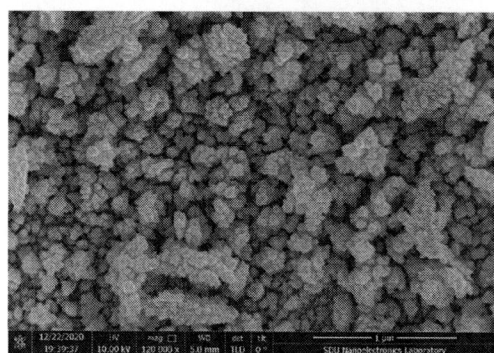

Fig.4 SEM of platinum black formed by electroplating.

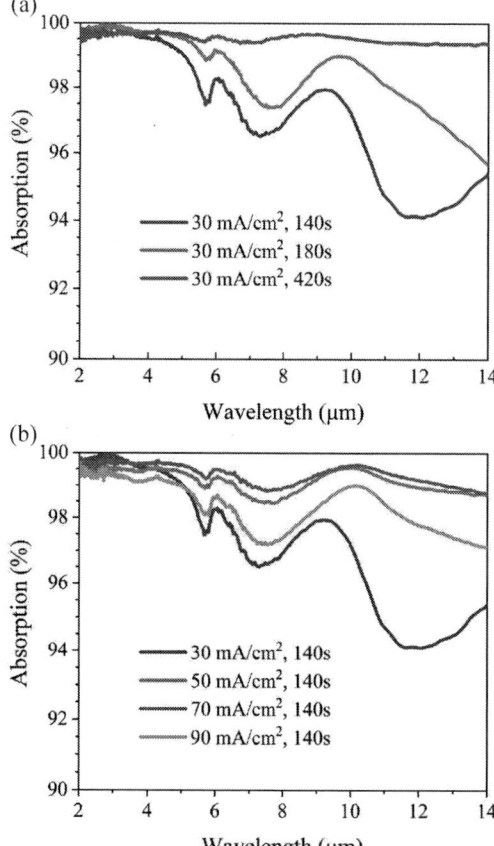

Fig. 5. The absorption spectrum of platinum black with different plating conditions. (a) same electroplating current density but different time (b) same electroplating time but different current density.

the temperature distribution of the emitter with 3 V driven voltage. The maximum temperature is around 400°C at the center of the infrared radiation layer. Fig. 2(c) shows the thermal stress of the infrared emitter in which the stress is mainly concentrated at the corners of membrane. Therefore, round-shaped chamfers are performed in these areas to disperse the stress that results of the maximum thermal stress are reduced from 1.1 GPa to 578 MPa.

C. Emitter Fabrication Process

The processing flow starts with a 4-inch double-side polished silicon test wafer. The sequence of fabrication is shown in Fig. 3(a) and summarized as follows:

(1) deposite 2 μm SiO2 by PECVD process; (2) sputter 15 nm Ti/200 nm Pt metal film and lift -off; (3) SiO2 deposition which is used to pattern electroplating region; (4) Etching the backside cavity; (5) Electroplating the Pt-black; (6) Remove

the top SiO2. Fig. 3(b) shows the optical image of a structure of MEMS emitter on Si wafer. Fig. 3(c) is the infrared emitter with a Compound Parabolic Concentrator (CPC) package to collimate infrared light for test.

III. MEASUREMENT RESULTS AND DISCUSSION

Fig. 4 shows the SEM of Pt-black. The surface of the electroplated Pt-black consists of nano-porous structure that absorb the incident infrared light efficiently. To obtain high absorptivity nano platinum black plating conditions. A simple electroplating platform was built to electroplate nano-platinum black under different conditions, and then the absorption rate of the nano-platinum black layer in the range of 2-14 μm in the spectrum was measured by using (FTIR). Fig. 5(a) and Fig. 5(b) shows the absorption spectrum of Pt-black with equates to emissivity at zero transmittance. It shows that the absorption (or emissivity) of the Pt-black within the spectrum region from 2 μm to 14 μm is higher than 99.5% when plating current density is 30 mA/cm^2 and 420 seconds. Fig. 5(b) shows that the emissivity of Pt-black reaches the maximum value when the plating current density is 70 mA/cm^2.

Use a DC power supply to drive the emitter to test the power consumption of the emitter under different driving voltages. Observe the temperature distribution on the surface

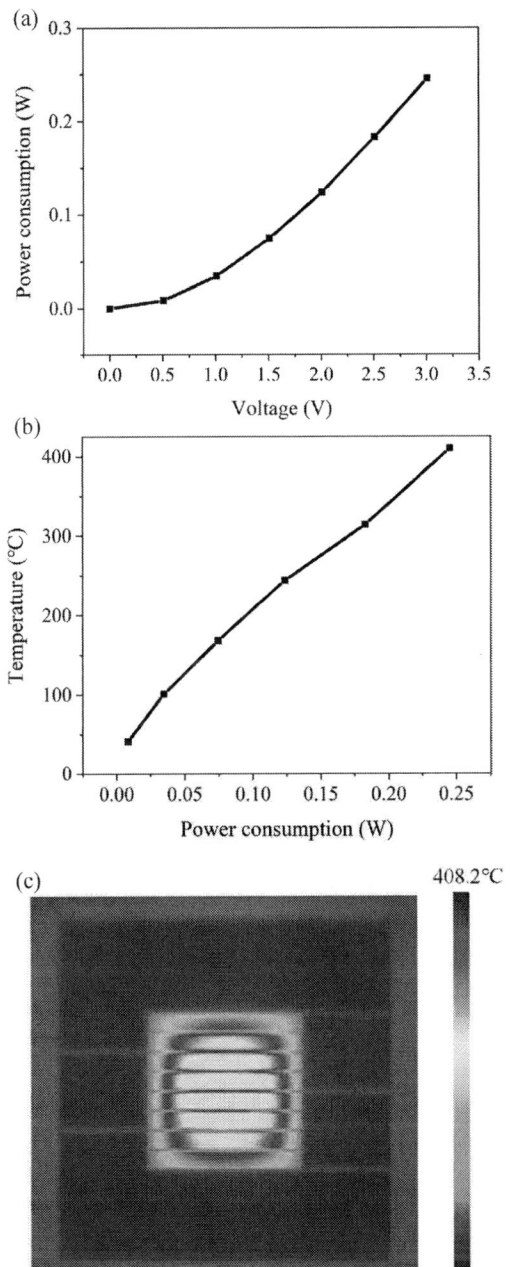

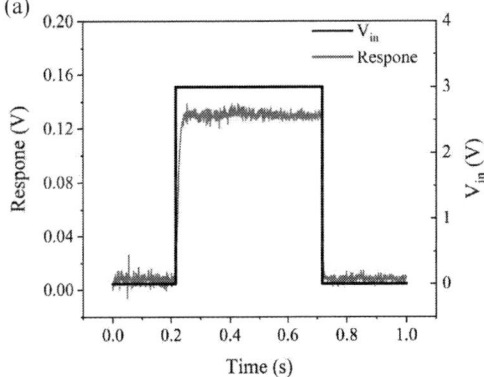

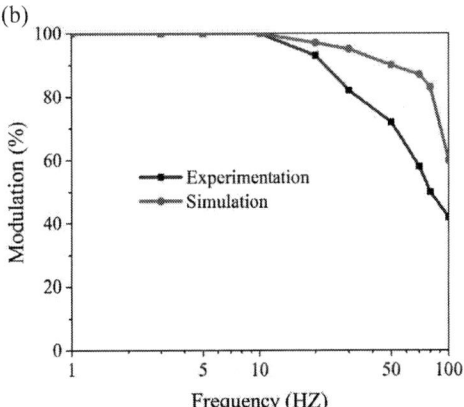

Fig. 6. (a) Power consumption vs. Driven voltage. (b) The maximum working temperature vs. power consumption. (c) the infrared image of emitter under 3V driven voltage.

Fig. 7. (a) The time response of the MEMS emitter, rise time is 16ms and fall time is 5ms. (b) the modulation depth of the MEMS emitter.

978-1-6654-3008-1/21 $31.00 © 2021 IEEE

of the emitter with an infrared thermal imager and record the highest temperature when the emitter is working. Fig. 6(a) shows the power consumption rising with the driving voltage of the infrared emitter. Fig. 6(b) shows the relationship between the power consumption and the operating temperature. Fig.6(c) shows that the maximum temperature of the emitter film is 408.2℃ under the voltage of 3 V.

Drive the emitter at different frequencies and use the mid-infrared photodetector (Thorlabs PDA07P2) to detect the voltage response of the emitter. The time required for the emitter to rise from 0 to 90% of the maximum value is the rise time, and the time required for the emitter to fall from the maximum value to 10% is the fall time. According to the peak-to-peak value of the response voltage, the modulation depth of the IR emitter can be calculated by the following equation[13]

$$m(f) = \frac{V(f)_{p-p}}{V(1HZ)_{p-p}} \times 100\% \qquad (3)$$

Where $V(f)_{p-p}$ is the peak-to-peak value of the detector output voltage when the frequency of the emitter. $V(1HZ)_{p-p}$ is the peak-to-peak value of the detector output voltage when the emitter is driven by 1 HZ.

Fig. 7(a) shows the dynamic response of the infrared emitter at 1 HZ. The rise time of the emitter is 16ms and the fall time is 5ms. Fig. 7(b) shows the infrared emitter working performance at different modulation frequency. The infrared emitter keeps 100% modulation depth with 10 Hz, after that the modulation depths decreases to 42% at 100 Hz gradually. The difference between the test results and the simulation results at high frequencies is because the platinum black radiation layer is not considered in the simulation.

IV. CONCLUSION

In summary, a MEMS infrared emitter with an ultra-high emissivity and a fast modulation speed is designed, fabricated and experimental demonstrated. The emissivity is up to 99.5% in 2-14um, in which there are more than 30 gas species' fingerprint wavelengths the experimental results show that the rise time is 16 ms and fall time is 5 ms. The modulation depth of the emitter can reach to 42% at 100 Hz modulation frequency. The MEMS emitter can be adopted by high performance infrared gas sensors and portable FTIR for medical, industry, environmental applications, etc.

V. REFERENCES

[1] T. A. Vincent and J. W. Gardner, "A low cost MEMS based NDIR system for the monitoring of carbon dioxide in breath analysis at ppm levels," *SENSORS AND ACTUATORS B*, 2016.

[2] C. Calaza, M. Salleras, N. Sabaté, J. Santander, C. Cané, and L. Fonseca, "A MEMS-based thermal infrared emitter for an integrated NDIR spectrometer," *Microsystem Technologies*, vol. 18, pp. 1147-1154, 2012.

[3] A. R. Berlanga, A. D. Castro, F. C. Martiniez, C. López-Ongil, and F. L. Martinez, "A light compact and rugged IR sensor for space applications," in *Infrared Sensors, Devices, and Applications IX*, 2019.

[4] M. F. Chowdhury, R. Hopper, S. Z. Ali, J. W. Gardner,

and F. Udrea, "MEMS Infrared Emitter and Detector for Capnography Applications," *Procedia Engineering*, vol. 168, pp. 1204-1207, 2016.

[5] Cuilian, Xu, Shaobo, Qu, Yongqiang, Pang, Jiafu, Wang, Mingbao, and Yan, "Metamaterial absorber for frequency selective thermal radiation," *Infrared Physics & Technology*, vol. 88, pp. 133-138, 2018.

[6] Z. Cheng and H. Toshiyoshi, "Design of CMOS-MEMS broadband infrared emitter arrays integrated with metamaterial absorbers based on CMOS back-end-of-line," *Iet Micro & Nano Letters*, vol. 11, pp. 602-605, 2016.

[7] C. Calaza, M. Salleras, N. Sabaté, J. Santander, C. Cané, and L. Fonseca, "A MEMS-based thermal infrared emitter for an integrated NDIR spectrometer," *Microsystem Technologies*, vol. 18, pp. 1147-1154, 2012.

[8] L. R. R. Langoju, M. K. Singh, K. M. Subramaniam, N. Jampana, and S. Asokan, "Molybdenum Microheaters for MEMS-Based Gas Sensor Applications: Fabrication, Electro-Thermo-Mechanical and Response Characterization," *IEEE Sensors Journal*, p. 1-1, 2017.

[9] N. Li, H. Yuan, L. Xu, J. Tao, and D. Zhao, "Radiation Enhancement by Graphene Oxide on MEMS Emitters for Highly Selective Gas Sensing," *Acs Sensors*, vol. 4, pp. 2746-2753, 2019.

[10] S. Z. Ali, A. De Luca, R. Hopper, S. Boual, J. Gardner, and F. Udrea, "A Low-Power, Low-Cost Infra-Red Emitter in CMOS Technology," *IEEE Sensors Journal*, vol. 15, pp. 6775-6782, 2015.

[11] S. Akasaka, E. Boku, Y. Amamoto, H. Yuji, and I. Kanno, "Ultrahigh temperature platinum microheater encapsulated by reduced-TiO2 barrier layer," *Sensors & Actuators A Physical*, vol. 296, 2019.

[12] G. Saxena and R. Paily, "Simulation study of power loss components in a microheater," in *International Conference on Power & Energy in Nerist*, 2013.

[13] P. Zhou, R. Chen, N. Wang, S. Haisheng, and X. Chen, "Reliability Design and Electro-Thermal-Optical Simulation of Bridge-Style Infrared Thermal Emitters," *Micromachines*, vol. 7, p. 166, 2016.

Gap in pagination due to unavailable papers.

Pages 1179-1247

April 25-29 , 2021   Xiamen, China

Keynote Speaker

Applications of Flexible Inorganic Thin-film Devices in Bioelectronics

Yuan Lin

School of Materials and Energy, University of Electronic Science and Technology of China, Chengdu 610054, Sichuan, P.R. China

ABSTRACT

Flexible bioelectronics have greatly improved the way of human-machine interaction due to the fact that they can provide seamless interactions with humans. Due to the advantages of mature processibility and rich physicochemical properties, flexible and stretchable inorganic thin-film electronics play an increasingly important role in the emerging and exciting flexible electronic field. Serving as sensor elements or biomimetic actuators, flexible inorganic thin film bioelectronics could dynamically sense physiological signals and provide timely stimulations or treatments. Thus, these types of electronics may change the future of healthcare. This talk summarizes our recent work in the development of flexible inorganic thin film devices focusing on their biomedical applications, including biosensing and non-pharmacological stimulation treatments. A future perspective into the challenges and opportunities for the next-generation flexible bioelectronics will also be discussed.

BIOGRAPHY

Yuan Lin is currently a Professor at School of Materials and Energy, University of Electronic Science and Technology of China. Dr. Lin received her Ph.D. degree in Condensed Matter Physics from University of Science and Technology of China in 1999. After that, she had worked in the University of Houston and Los Alamos National Lab as a postdoc, and in Intel Corp as a senior engineer. In 2008, she joined the faculty of University of Electronic Science and Technology of China as a Yangtze River Scholars Distinguished Professor. Dr. Lin is active in the field of electronic thin films and devices. Her main research interests are focused in the development of various thin films (such as ferroelectric oxide, vanadium oxide and other oxides) for applications in electronic devices, especially in stretchable and flexible electronic devices. She has co-authored more than 150 papers in peer-reviewed journals and her publications have been cited for more than 2000 times. She also has more than 20 Chinese patents and 4 US patents awarded.

978-1-6654-3008-1/21 $31.00 © 2021 IEEE   1248

This page intentionally left blank.

Proceedings of the 16th Annual IEEE International
Conference on Nano/Micro Engineered and Molecular Systems
April 25-29, 2021

An On-Chip Inductive-Capacitive Sensor for the Detection of Wear Debris and Air Bubbles in Hydraulic Oil *

Haotian Shi, Hongpeng Zhang, Wei Li, Zhiwei Xu, Laihao Ma, Yucai Xie and Dian Huo

Abstract— An on-chip inductive-capacitive sensor is proposed to detect metal debris and air bubbles in hydraulic oil. The sensor includes inductance and capacitance detection units. The inductance detection unit detects and distinguishes the properties of metal debris by monitoring the inductance change of the solenoid coil. The capacitance detection unit detects air bubbles by monitoring the capacitance change of the cylindrical capacitor. Inductance tests are conducted using iron and copper debris with different sizes. The test results showed that the inductance pulse amplitude increases nonlinearly with debris size. Capacitance tests are conducted using air bubbles with different sizes. The test results showed that the cylindrical capacitor can detect and count air bubbles with high sensitivity. The design concept presented in this paper can be extended to oil conditioning sensors for multi-contaminants detection and differentiation.

I. INTRODUCTION

Hydraulic equipment is widely used in marine engineering fields such as ships, working platforms and wind turbines. Hydraulic oil is not only the power transmission medium of the hydraulic system, but also has the functions of reducing wear, inhibiting oxidation, and reducing system temperature rise. The running status of the system can be displayed by oil condition monitoring. For example, the metal debris characterize the wear status of the parts, the moisture content is related to the corrosion and oxidation of the parts, and the air will affect the response speed and accuracy of the hydraulic system.

Presently, the measurement of oil contamination mainly adopts ferrograph analysis method [1], spectrum detection method [2], microscope comparison method [3], etc. These methods can accurately and qualitatively analyze the various components and physicochemical states in the oil, but the detection cycle is long and it is difficult to achieve portable and rapid detection. Especially for the marine hydraulic equipment, the timeliness of detection is insufficient. With the development of microelectronics and micromachining technology [4, 5], various laboratory-on-chip devices for different purposes continue to emerge. These sensors can realize high sensitivity detection and counting of oil contaminants, and make the equipment miniaturized, automated and highly integrated. The particle counting methods used in oil condition monitoring mainly include optical method [6], resistance method [7], capacitance method [8], inductance method [9], etc. Among the above

methods, the optical method is greatly affected by the light transmittance of the oil, the resistance and capacitance methods cannot distinguish the solid contaminants, while the inductance method can distinguish the properties of metal debris, and has good robustness.

In the last decade, a series of debris micro-sensors have been designed. Jiang et al. [10, 11] introduced a particle counter by winding coil on a capillary glass tube, then the team used LC resonance method to obtain higher accuracy [12]. But due to the limitation of detection channel, the sensitivity still has a large room for improvement. Zhang et al. [13] innovatively proposed a mold casting method to fabricate the microfluidic oil detection chip, so that the distance between the coil inner hole and the detection channel is close to 0. In addition, our group has further improved the detection sensitivity by adding paramagnetic materials [14, 15].

However, the inductance method cannot detect the non-metallic contaminants, but the air bubbles can be detected by capacitance method. In this paper, the sensor integrated two detection units of inductance and capacitance on the microfluidic chip is presented, which greatly improves the detection sensitivity and obtains more accurate information. Using the inductive-capacitive sensor, the ferromagnetic metal debris, nonferromagnetic metal debris and air bubbles can be effectively identified and counted.

II. SENSOR DESIGN

As shown in Figure 1, the sensor is mainly composed of PDMS substrate, channel, inductance detection unit and capacitance detection unit.

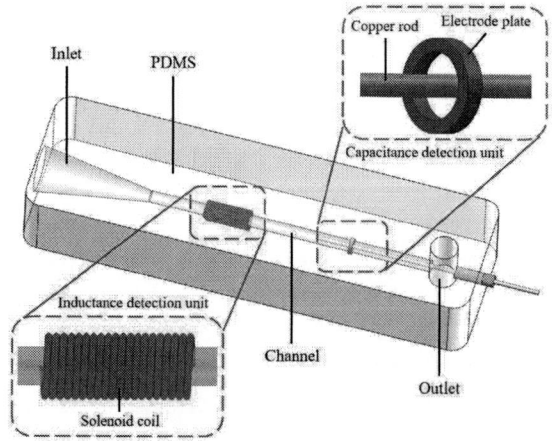

Figure 1.   The structure of inductance-capacitive sensor.

---

* This work was supported by the Natural Science Foundation of China (Grant No. 51679022), the Dalian Science and Technology Innovation Fund (Grant No. 2019J12GX023).

H. Shi, H. Zhang, W. Li, L. Ma, Z. Xu, Y. Xie, D. Huo, are with the Marine Engineering College, Dalian Maritime University, Dalian, 116026, China (e-mail: dmu6hao@163.com; zhppeter@dlmu.edu.cn).

The channel is 900 μm in diameter, the inductance detection unit is a solenoid coil (40 turns) with 900 μm inner diameter, the capacitance detection unit is a cylindrical capacitor. The inner electrode plate of the cylindrical capacitor is an aluminum rod with 300 μm diameter, and the outer electrode plate is a copper ring with 900 μm inner diameter and 60 μm height. The inductance detection unit can distinguish metal debris (ferromagnetic and nonferromagnetic), the capacitance detection unit can detect air bubbles.

III. DETECTION PRINCIPLE

A. Inductance detection

As shown in Figure 2, the metal debris in the coil will be magnetized and generate eddy current. According to Lenz's law, the magnetic field produced by eddy current weakens the original magnetic flux, while the magnetic field produced by magnetization enhances the original magnetic flux.

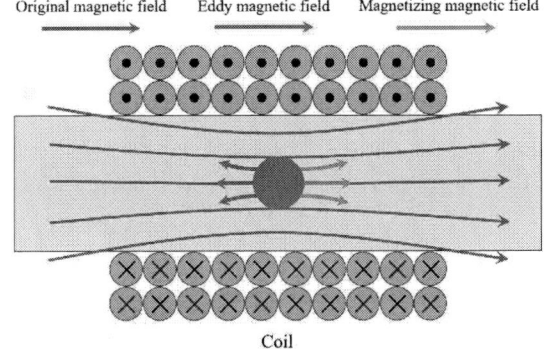

Figure 2.   Inductance detection principle.

According to prior research [16], the inductance of the coil can be calculated

$$L = \sum_{n=1}^{N} \Phi_d \qquad (1)$$

Here, N is the number of coil turns. Φ_d is the magnetic flux change of the coil, which is given by

$$\Phi_d = \iint_{S_n} B_d ds \qquad (2)$$

S is the cross-sectional area of each turn in the coil, B_d is the magnetic induction vector produced by metal debris.

The inductance change generated by the metal debris is

$$L_d = \sum_{n=1}^{N} \Phi_d = \sum_{n=1}^{N} \iint_{S_n} B_d ds \qquad (3)$$

For ferromagnetic debris, the magnetization effect is more severe, B_d is positive, which will increase the coil inductance. For nonferromagnetic debris, the eddy current effect is more severe, B_d is negative, which will decrease

the coil inductance. According to the direction of inductance pulse, the properties of metal debris can be judged.

B. Capacitance detection

Figure 3 is the schematic diagram of the capacitor in the capacitance detection unit. The outer plate height h is 70 μm, the inner plate radius r is 150 μm, and the outer plate radius R is 900 μm.

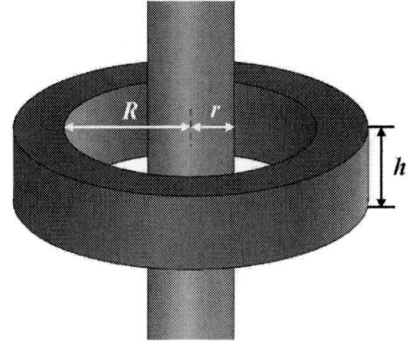

Figure 3.   The schematic diagram of the cylindrical capacitor.

Ignoring the edge effect, the capacitance calculation formula of cylindrical capacitor is

$$C = \frac{2\pi\varepsilon h}{\ln\dfrac{R}{r}} \qquad (4)$$

Here, ε is the permittivity of the medium. Equation (4) shows that the capacitance value depends on the geometric parameters and the medium.

The contaminants will change the medium between the two electrodes. According to Maxwell's equations [17], the complex permittivity of the mixed medium in the capacitance detection region is as follows.

$$\tilde{\varepsilon}_{mix} = \tilde{\varepsilon}_o \frac{V_s\left(\tilde{\varepsilon}_p + 2\tilde{\varepsilon}_o\right) + V_p\left(\tilde{\varepsilon}_p - \tilde{\varepsilon}_o\right)}{V_s\left(\tilde{\varepsilon}_p + 2\tilde{\varepsilon}_o\right) - V_p\left(\tilde{\varepsilon}_p - \tilde{\varepsilon}_o\right)} \qquad (5)$$

Here, V_p is the particle volume, V_s is the detection volume, $\tilde{\varepsilon}_o$ is the complex permittivity of hydraulic oil, $\tilde{\varepsilon}_p$ is the complex permittivity of contaminant.

The capacitance change of cylindrical capacitor caused by spherical contaminant is

$$\Delta C = \frac{2\pi\left[real\left(\tilde{\varepsilon}_{mix}\right) - \varepsilon_o\right]rh}{\ln\dfrac{R}{r}} \qquad (6)$$

When the capacitor structure and excitation are determined, the complex permittivity of the mixed medium increases with the contaminant volume, so the capacitance change represents the size of contaminants.

978-1-6654-3008-1/21 $31.00 © 2021 IEEE

IV. EXPERIMENTS

As shown in Figure 4, a micro-injection pump, an inductive-capacitive sensor, a microscope, an LCR meter and a computer installed with signal acquisition program form the detection system.

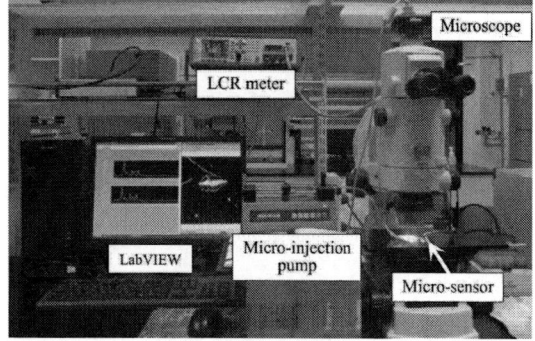

Figure 4.   The detection system.

A. Inductance experiment

The spherical metal debris with different diameters were mixed into hydraulic oil to prepare the oil sample. The micro-injection pump injected the oil sample into the channel at a constant speed (250 µL/min), and making the debris can pass through the coil repeatedly. The debris size was measured by microscope. Connect the coil to the LCR meter, and the excitation parameters of LCR meter were set to 2.0 V, 2.0 MHz.

In the laboratory, the corresponding relationship between the inductance amplitude and the debris size was sorted out, as shown in Figure 5.

It can be found from the Figure 5 that the inductance pulse amplitude increases nonlinearly with the debris size. However, the inductance change produced by iron debris is much larger than that of copper debris, and its nonlinear curve is steeper. This is because that the magnetization effect in ferromagnetic debris is more significant. Therefore, the inductance detection unit has stronger detection ability for ferromagnetic debris, which can detect 42 µm iron particles, as shown in the Figure 6. And the detection ability for nonferromagnetic debris is relatively weak, which only detects 130 µm copper particles, as shown in the Figure 7.

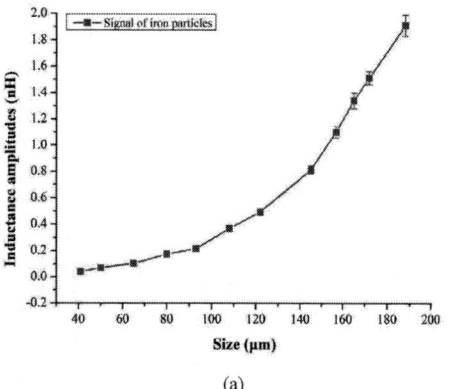

(a)

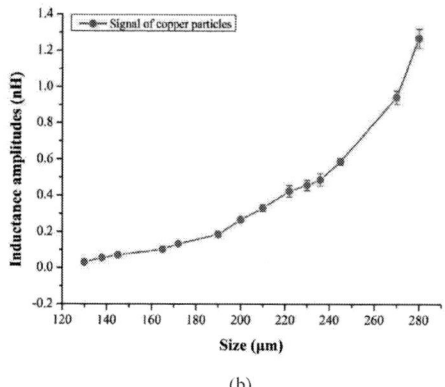

(b)

Figure 5.   The inductance pulse amplitudes of metal particles with different diameters. (a) detection results of iron particles; (b) detection results of copper particles.

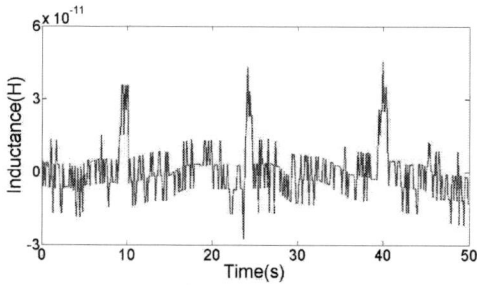

Figure 6.   The detection results of 42 µm iron particles.

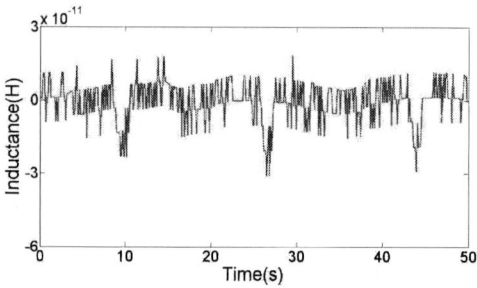

Figure 7.   The detection results of 130 µm copper particles.

This sensor can measure the size of metal debris with high sensitivity. At the same time, it can distinguish the properties of the metal debris according to the pulse direction, and can count the particles according to the number of pulses.

B. Capacitance experiment

The oil sample containing air bubbles was prepared by shaking the mixture of hydraulic oil and air with different proportions. The cylindrical capacitor was connected to the LCR meter, and the excitation parameters of LCR meter were set to 2.0 V and 0.5 MHz.

In the laboratory, the corresponding relationship between the capacitance amplitude and the air bubble size was sorted out, as shown in Figure 8.

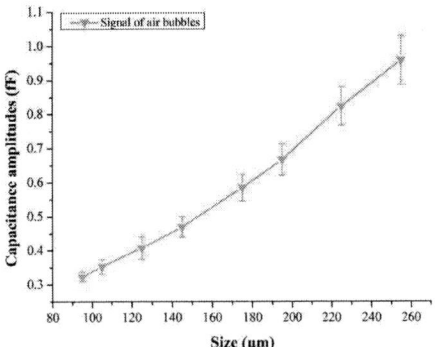

Figure 8. The capacitance pulse amplitudes of air bubbles with different diameters.

Air bubbles will decrease the complex permittivity of the mixture in capacitance detection region. The negative capacitance pulse signal will be generated by air bubbles. Capacitance detection unit can detect 95 μm air bubbles, as shown in the Figure 9.

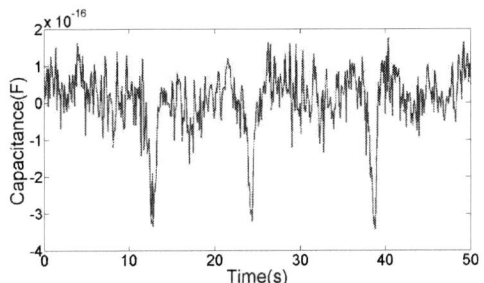

Figure 9. The detection results of 95 μm air bubbles.

The metal debris and water droplets will increase the complex permittivity of the mixture in capacitance detection region, which will generate positive capacitance pulse. Therefore, the capacitance detection unit cannot distinguish between metal debris and water droplets. In the next work, our group will study to distinguish multi-contaminants (metal debris, air and water) in hydraulic oil by combining multiple parameters.

V. CONCLUSION

An on-chip inductive-capacitive sensor is introduced to detect metal debris and air bubbles in hydraulic oil. The sensor uses a solenoid coil as an inductance detection unit, and a cylindrical capacitor as a capacitance detection unit. In the inductance detection experiment, 42 μm iron debris and 130 μm copper debris were successfully detected, and the debris properties can be distinguished according to the inductance pulse direction. The inductance amplitude increases nonlinearly with debris size. The inductance parameter has stronger ability to detect ferromagnetic debris, and its nonlinear curve is steeper. In the capacitance experiment, 95 μm air bubbles were successfully detected, and high-sensitivity counting was achieved. The inductive-capacitance sensor proposed in this paper

provides a technical solution for distinguishing and detecting multi-contaminants in hydraulic oil, and can provide key information for the health monitoring and condition warning of offshore hydraulic equipment.

ACKNOWLEDGMENT

The authors would like to thank the Microfluidic Chip Laboratory in Dalian Maritime University for the device support.

REFERENCES

[1] Biswas, R. K. , M. C. Majumdar , and S. K. Basu . "Vibration and Oil Analysis by Ferrography for Condition Monitoring." Journal of the Institution of Engineers 94.3(2013):267-274.

[2] Raposo, Jorge Luiz , et al. "Determination of Silicon in Lubricant Oil by High-Resolution Continuum Source Flame Atomic Absorption Spectrometry Using Least-Square Background Correction and Internal Standardization." Analytical Letters 44.12(2011):2150-2161.

[3] Mabe, Jon , J. Zubia , and E. Gorritxategi . "Photonic Low Cost Micro-Sensor for in-Line Wear Particle Detection in Flowing Lube Oils." Sensors 17.3(2017):586.

[4] Zhang H., Chon C., Pan X., et al. "Methods for counting particles in microfluidic applications." Microfluidics & Nanofluidics 7.6(2009):739.

[5] Zhou, T., Ji, X., Shi, L., Zhang, X., Deng, Y. and Joo, S.W. "Dielectrophoretic choking phenomenon in a converging - diverging microchannel for Janus particles." ELECTROPHORESIS, 40(2019): 993-999.

[6] Hamilton A, Cleary A, Quail F. Development of a novel wear detection system for wind turbine gearboxes. IEEE Sens J 2014;14:465–73.

[7] Jagtiani A V , Carletta J , Zhe J . A microfluidic multichannel resistive pulse sensor using frequency division multiplexing for high throughput counting of micro particles. Journal of Micromechanics & Microengineering, 2011, 21(6):065004.

[8] Shi H , Zhang H , Ma L , et al. A multi-function sensor for online detection of contaminants in hydraulic oil[J]. Tribology International, 2019, 138:196-203.

[9] Zhang, Y. , Cao, Y. , Si, E. , & Zheng, L. . (2019). Study on a high sensitive online sensor for wear particles in lubricant. Journal of Physics Conference Series, 1237, 042082.

[10] Du L , Zhe J . A high throughput inductive pulse sensor for online oil debris monitoring[J]. Tribology International, 2011, 44(2):175-179.

[11] Zhu, Xiaoliang , L. Du , and J. Zhe . "A 3×3 wear debris sensor array for real time lubricant oil conditioning monitoring using synchronized sampling." Mechanical Systems and Signal Processing (2017).

[12] Du L , Zhu X , Han Y , et al. Improving sensitivity of an inductive pulse sensor for detection of metallic wear debris in lubricants using parallel LC resonance method[J]. Measurementence & Technology, 2013, 24(7):660-664.

[13] Ma, Laihao , et al. "High-sensitivity distinguishing and detection method for wear debris in oil of marine machinery." Ocean Engineering 215(2020):107452.

[14] Shi, Haotian , et al. "An Impedance Debris Sensor Based on a High-Gradient Magnetic Field for High Sensitivity and High Throughput." IEEE Transactions on Industrial ElectronicsPP.99(2020):1-1.

[15] Zeng, Lin , et al. "A High Sensitivity Micro Impedance Sensor Based on Magnetic Focusing for Oil Condition Monitoring." IEEE Sensors Journal 20.7(2020):3813-3821.

[16] Liu, D. , H. P. Zhang , and X. M. Zhang . "Metal particle magnetization and resistance pulse detection method in harmonic field." Journal of Dalian Maritime University 42.2(2016):96-101.

[17] Morgan, Hywel , et al. "Single cell dielectric spectroscopy." Journal of Physics D Applied Physics 40(2007):61-70.

978-1-6654-3008-1/21 $31.00 © 2021 IEEE

Proceedings of the 16th Annual IEEE International
Conference on Nano/Micro Engineered and Molecular Systems
April 25-29, 2021

AlN Contour Mode Resonators with Half Circle Shaped Reflectors

Zhifang Luo[1,2,3,*], Shuai Shao[1,2,3], Tao Wu[1,*]

Abstract— AlN contour mode resonators (CMRs) with half circle shaped reflectors are designed to reduce the anchor loss. In this work, we use finite element analysis (FEA) and demonstrate that the half circle shaped reflector can effectively reduce the energy dissipation through the anchor to the plate, and then boost the CMR quality factor Q. Furthermore, the measured experimental data of AlN CMR with half circle shaped reflectors design yields a Q of 1605 operating at the resonance of around 400 MHz, which provides over 80% improvement compared to an AlN CMR with a normal plate configuration.

I. INTRODUCTION

Piezoelectric microelectromechanical resonators have shown many advantages in radio frequency (RF) filters, and miniature transducers [1–4]. Among various piezoelectric materials, such as Zinc Oxide (ZnO), piezoelectric ceramics (PZT), Lithium Niobate (LN) and Aluminum Nitride (AlN) [5–10], AlN-based resonators play a more important role in 5G communication, due to the high phase velocity of the AlN thin film and compatibility with CMOS process [9, 11, 12]. In the last decades, film bulk acoustic resonators (FBAR), surface acoustic wave (SAW) and contour mode resonators (CMRs) have been widely investigated [13–15]. The CMRs have the advantages of photolithographic defined high operating frequencies compared to other techniques. However, there are several loss mechanisms in CMRs, such as interface loss from the connection of electrodes and AlN thin film, damping loss from the resonators and anchor loss, which limit the applications of AlN-based CMRs [16–20]. A large amount of energies leak through the anchor to the piezoelectric plate in the AlN resonators, i.e. the anchor loss [21–25]. Zou and Chih-Ming *et al.* [26] have designed the beveled and rounded butterfly-

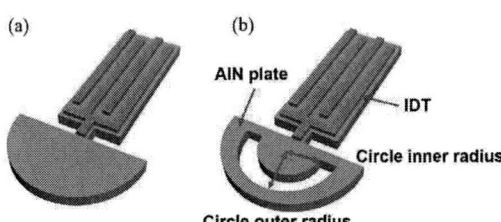

Fig. 1 Illustrations of AlN CMRs with (a) normal plate, (b) half circle shaped reflector.

[1]School of Information Science and Technology, ShanghaiTech University
[2]Shanghai Institute of Microsystem and Information Technology, Chinese Academy of Sciences
[3]University of Chinese Academy of Sciences

shaped AlN CMRs. Acoustic wave reflectors are formed at the transition areas between the anchor and vibration region, such design enables to push waves away from the anchors, so the energy is suppressed. Yung-Yu Chen *et al.* [16] have investigated the AlN CMRs convex free edges to reduce the anchor loss and then improve Q factor. In our work, another efficient way to reduce the anchor loss of the AlN-based CMRs is demonstrated utilizing half circle shaped reflectors to push acoustic waves back to the resonate region.

To decrease the energy leakage from anchor loss, we demonstrate an AlN CMR with half circle shaped reflectors. Fig. 1(a) illustrates a normal AlN CMR employing the normal plate. As shown in Fig. 1(b), The AlN CMR with half circle shaped reflector is utilized to reduce the anchor loss in AlN CMRs. Unlike the AlN CMRs with normal plate, the CMRs with half circle shaped reflectors can limit the propagation of acoustic wave to the substrate plate. We utilize the COMSOL Multiphysics® to show the improvement on the Q factor of the CMRs. In addition, AlN CMRs with half circle shaped reflectors are experimentally fabricated, and present a higher Q factor compared to the CMRs with a normal plate configuration.

II. DESIGN AND SIMULATION

The main energy dissipation mechanisms for piezoelectric resonators can be expressed as [16],

$$\frac{1}{Q} = \frac{1}{Q_{interface}} + \frac{1}{Q_{anchor\ loss}} + \frac{1}{Q_{other}} \qquad (1)$$

In order to investigate the influence of anchor loss in the AlN CMRs, we assume that the main vibration energy dissipation in the CMRs is anchor loss, and $Q_{anchor\ loss}$ is mainly contributes to the Q of AlN CMRs. Anchor loss is determined by the displacement fields in the anchor and plate. Our design aims to decrease the displacement fields in the plate to reduce anchor loss, and then increase the total Q factor of the AlN CMRs, the geometric dimensions of AlN CMRs are listed in the Table 1. In the AlN CMRs with

Table 1 Geometric dimensions of AlN CMRs

Parameters	Normal	Half circle shaped
IDT numbers	4	4
IDT aperture	180 μm	180 μm
IDT coverage	0.5	0.5
Anchor length	4.9 μm	4.9 μm
Anchor width	8.2 μm	8.2 μm
Wavelength	20 μm	20 μm
Circle inner radius	NA	17.5 μm
Circle outer radius	NA	27.5 μm

*Corresponding Authors:
Zhifang, Luo. (e-mail: luozhf@shanghaitech.edu.cn),
Tao Wu. (e-mail: wutao@shanghaitech.edu.cn),

978-1-6654-3008-1/21 $31.00 © 2021 IEEE

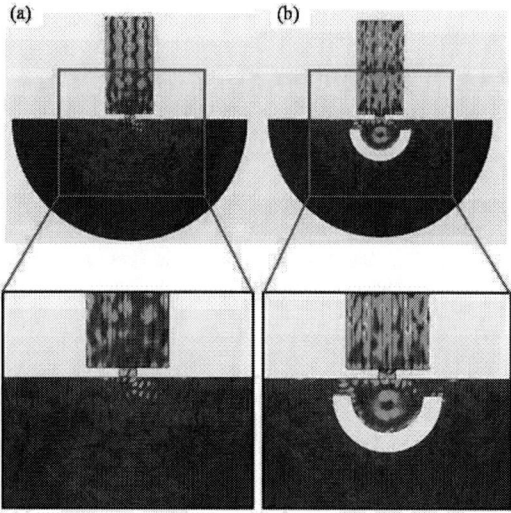

Fig. 2 Simulated displacement fields of AlN CMRs with (a) normal plate, (b) half circle shaped reflector.

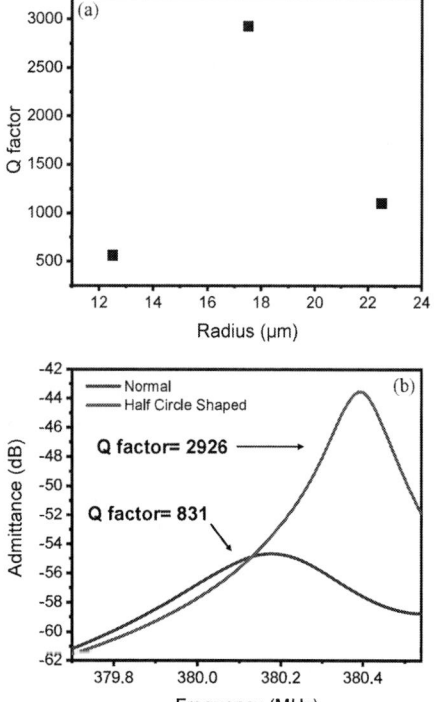

Fig. 3 (a) Simulated Q factor of AlN CMRs with different inner circle radius. (b) Simulated admittance response of AlN CMRs with normal plate and half circle shaped reflector.

normal plate, the acoustic wave in the vibration region would propagate through anchor to the substrate plate, resulting large anchor loss. By using the half circle shaped reflectors, the acoustic wave reflector can reflect acoustic wave effectively to the active region.

Perfectly matched layer (PML)-based finite element analysis from COMSOL Multiphysics® software are used to investigate the displacement fields on the vibration region and plate [16, 17, 26]. The AlN CMRs with anchors are attached to the semi-cylinder plates and then covered by the PMLs, the radius of semi-cylinder plates is set as three times wavelength, and the radius of PMLs are set as five times wavelength, which aims to absorb the acoustic wave efficiently. The design of anchor width and length is also essential to reduce anchor loss. Yung-Yu Chen *et al.* [16], B P Harrington and R Abdolvand [17] have investigated the influence of different designs, so our anchor length and width are designed based on their findings. Fig. 2 presents the displacement fields of AlN CMRs with the normal plate and half circle shaped reflectors at S0 mode utilizing FEA simulation. Brighter color means larger displacement fields. $V_{RF} = 1$ V is applied to the IDT to induce S0 mode of AlN CMRs. To reduce memory of computation, only half of the CMR and plate are simulated. As illustrated in Fig. 2(a), there is large displacement in the piezoelectric thin film of the normal design, which means a lot of energies leaked from piezoelectric layer, resulting in the big anchor loss of the AlN CMRs. In addition, as a lot of energy loss in plate via anchor, the displacement fields near the anchor in AlN CMRs is reduced dramatically by energy dissipation. Fig. 2(b) shows that the displacement fields away from the anchor are similar with those of normal design, but the displacement fields in the piezoelectric plate are obliviously less than those in the normal plate. Since a large amount of energies are reflected back to the CMRs by half circle shaped reflectors, only small part is absorbed by PMLs. It makes much larger displacement fields in the vibration region near the anchor. As seen in Fig. 3(a), the best performance for simulated resonators is found for inner circle radius of 17.5 μm. Fig. 3(b) illustrates the admittance response of AlN CMRs with and without half circle shaped reflectors. The admittance response is simulated in the 3D frequency domain analysis by a frequency spacing of 10 kHz. The simulated Q of S0 mode in the AlN CMR with half circle shaped reflectors is 2926, while the Q of the AlN CMR with normal plate is only 831. The simulated Q is increased by 252 %.

III. FABRICATION AND CHARACTERIZATION

In order to validate the results of the simulation results, AlN CMRs with normal plates and half circle shaped reflectors are fabricated on the same wafer.

For the sample preparation, 1μm (0002) polar AlN is deposited on high-resistive silicon wafer at 300°C by reactive magnetron sputtering system with a 12 inch Al target. Fig. 4(a) illustrates the surface morphology of AlN thin film by scanning electron microscopy Carl Zeiss Gemini300. AlN grains are uniformly and densely aligned, which indicates the great crystal quality of the sputtered thin films. It can be seen from Fig. 4(b) that the AlN thin film is well deposited on the silicon wafer with great c-axis oriented crystal columns. As shown in Fig. 4(c), the roughness of AlN surface is measured by Atomic Force Microscopy from Asylum Research, and the value of RMS is only 970 pm, which shows great surface uniformity of AlN piezoelectric thin film. Such uniform and great z-axis oriented AlN thin film is necessary to fabricate devices based on it, which enables less energy dissipation in the piezoelectric thin film,

978-1-6654-3008-1/21 $31.00 © 2021 IEEE

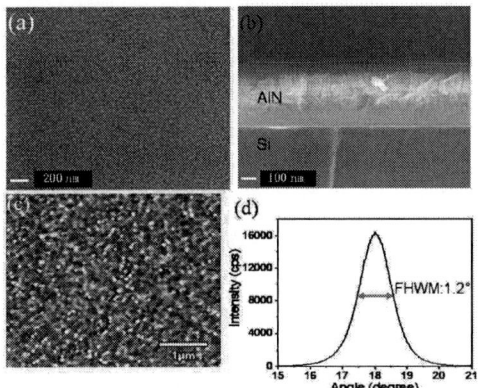

Fig. 4 SEM images of (a) (0002) polar AlN surface, (b) (0002) polar AlN crosssection view. (c) AFM height image of (0002) polar AlN surface. (d) XRD rocking curve of AlN thin film.

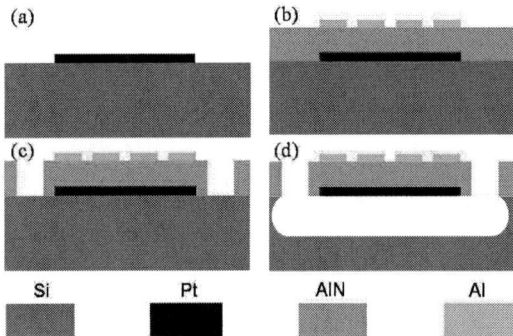

Fig. 5 (a) Lift-off Ti/Pt layer as bottom electrode, (b) deposit AlN thin film and pattern Al as top electrode, (c) ICP dry etching, (d) release AlN CMRs via XeF$_2$.

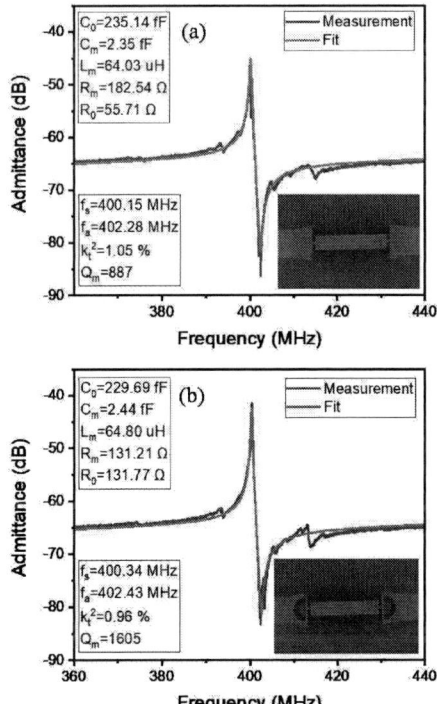

Fig. 6 The admittance response of AlN CMRs with (a) normal plate, and (b) half circle shaped reflector.

resulting in higher performance devices. The crystal orientation of AlN is measured by X-ray diffraction (XRD) by PANalytical® Empyrean system, the full width at half maximum (FWHM) of AlN thin film is 1.2°, as illustrated in Fig. 4(d).

The fabrication of AlN CMRs is a traditional 4-mask process on high-resistive (100) silicon wafer, as seen in Fig. 5 (a)-(d). First, the physical vapor deposition (PVD) is utilized to deposit and pattern10nm Ti/ 100 nm Pt via a lift-off process as bottom electrode. Then, 1 μm AlN piezoelectric thin film is deposited utilizing EVATEC CLN200 reactive magnetron sputtering system. Then, a 200 nm Al layer is deposited and patterned to define the top interdigitated electrodes (IDT). Next, the ICP etching is used to define the active region of AlN CMRs as well as the half circle shaped reflectors. Finally, the CMRs are released by XeF$_2$. To reduce the fabrication process variations, all the AlN CMRs we discussed here are fabricated on the same wafer and placed in the vicinity.

According to our FEA simulation and analysis, the quality factor Q of AlN CMR with half circle shaped reflectors is significantly higher than the AlN CMR with normal plate. The fabricated AlN CMRs are characterized by Keysight® N5234B PNA-L Network Analyzer. The measured Q of CMRs are extracted from the admittance response by dividing the resonance frequencies by the -3 dB bandwidth. Fig. 6(a) and Fig. 6(b) present the one-port admittance response for the AlN CMRs based on normal plate and half circle shaped reflectors, respectively. The AlN CMR with half circle shaped reflectors yields a Q factor of 1605, upwards 80 % over a normal AlN CMR which has a Q of 887, which shows higher Q improvement compared to the previous works. Zou and Chih-Ming et al. [26] present 67 % improvement of measured Q, Chih-Ming et al. [18] indicate 50 % measured Q improvement to the normal design. In the FEA simulation, loss of the resonator is set larger than the actual situation, so that the measured Q is a little larger than the simulated Q in the CMRs with normal plate. Due to the loss of the manufacturing process, the measured Q of CMRs with half circle shaped reflectors is lower than the simulated value. The electrical response of piezoelectric resonators is usually represented with the modified Butterworth-Van Dyke (mBVD) model [27]. After fitting into the mBVD model, the C_0, C_m, L_m, R_m, R_0 are all extracted, where C_0 and R_0 are the static capacitance and static resistance of resonators, respectively. C_m, L_m, R_m are the motional capacitance, motional inductance, motional resistance of resonators, respectively.

IV. CONCLUSION

A design of AlN CMR with half circle shaped reflectors is presented in this work. PML-based FEA simulation confirms the influence of half circle shaped reflectors on the reduction of displacement fields in the piezoelectric plate and resonate region near the anchor and improvement of Q. Such reflector can efficiently reflect the acoustic wave back to the resonator, significantly reduce the energy loss. The AlN CMR with half circle shaped reflectors yields a Q of 1605, showing 80% increase in Q over AlN CMR with normal plate.

V. ACKNOWLEDGMENTS

The authors appreciate the support from the ShanghaiTech Quantum Device Lab (SQDL), and Analytical Instrumentation Center (SPSTAIC10112914) XRD Lab, Natural Science Foundation of Shanghai (19ZR1477000) and National Natural Science Foundation of China (61874073).

REFERENCES

[1] A. Gao, K. Liu, J. Liang, T. Wu, "AlN MEMS filters with extremely high bandwidth widening capability," *Microsyst. Nanoeng.*, vol. 6, no. 1, p. 74, Dec. 2020.

[2] C. Caliendo, P. Imperatori, "High-frequency, high-sensitivity acoustic sensor implemented on ALN/Si substrate," *Appl. Phys. Lett.*, vol. 83, no. 8, pp. 1641–1643, Aug. 2003.

[3] V. Yantchev, I. Katardjiev, "Thin film Lamb wave resonators in frequency control and sensing applications: a review," *J. Micromechanics Microengineering*, vol. 23, no. 4, p. 043001, 2013.

[4] L. Wang, B. Wu, J. Chen, H. Liu, P. Hu, Y. Liu, "Monolayer hexagonal boron nitride films with large domain size and clean interface for enhancing the mobility of graphene-based field-effect transistors," *Adv. Mater.*, vol. 26, no. 10, pp. 1559–1564, 2014.

[5] Q.-X. Su, P. Kirby, E. Komuro, M. Imura, Q. Zhang, R. Whatmore, "Thin-film bulk acoustic resonators and filters using ZnO and lead-zirconium-titanate thin films," *IEEE Trans. Microw. Theory Tech.*, vol. 49, no. 4, pp. 769–778, Apr. 2001.

[6] J. L. Hockel, T. Wu, G. P. Carman, "Voltage bias influence on the converse magnetoelectric effect of PZT/terfenol-D/PZT laminates," *J. Appl. Phys.*, vol. 109, no. 6, Art. no. 6, 2011.

[7] M.-H. Li, C.-Y. Chen, R. Lu, Y. Yang, T. Wu, S. Gong, "Power-Efficient Ovenized Lithium Niobate SH0 Resonator Arrays with Passive Temperature Compensation," in *2019 IEEE 32nd International Conference on Micro Electro Mechanical Systems (MEMS)*, Seoul, Korea (South), Jan. 2019, pp. 911–914.

[8] M.-A. Dubois, P. Muralt, "Properties of aluminum nitride thin films for piezoelectric transducers and microwave filter applications," *Appl. Phys. Lett.*, vol. 74, no. 20, Art. no. 20, 1999.

[9] T. Wu, G. Chen, C. Cassella, W. Z. Zhu, M. Assylbekova, M. Rinaldi, N. McGruer, "Design and fabrication of AlN RF MEMS switch for near-zero power RF wake-up receivers," in *2017 IEEE SENSORS*, Glasgow, Oct. 2017, pp. 1–3.

[10] Z. Luo, S. Shao, T. Wu, "Characterization of ALN and ALSCN film ICP etching for micro/nano fabrication," *Microelectron. Eng.*, p. 111530, Feb. 2021.

[11] T. Wu, Z. Qian, M. Rinaldi, "Low cost thin film encapsulation for AlN resonators," in *2018 IEEE Micro Electro Mechanical Systems (MEMS)*, Belfast, Jan. 2018, pp. 1024–1027.

[12] J. Zou, A. Gao, A. P. Pisano, "Spectrum-clean S$_1$ AlN Lamb wave resonator with damped edge reflectors," *Appl. Phys. Lett.*, vol. 116, no. 2, p. 023505, Jan. 2020.

[13] G. Wingqvist, "AlN-based sputter-deposited shear mode thin film bulk acoustic resonator (FBAR) for biosensor applications — A review," *Surf. Coat. Technol.*, vol. 205, no. 5, pp. 1279–1286, Nov. 2010.

[14] J. G. Rodriguez-Madrid, G. F. Iriarte, J. Pedros, O. A. Williams, D. Brink, F. Calle, "Super-High-Frequency SAW Resonators on AlN/Diamond," *IEEE Electron Device Lett.*, vol. 33, no. 4, pp. 495–497, Apr. 2012.

[15] M. Giovannini, S. Yazici, N.-K. Kuo, G. Piazza, "Apodization technique for spurious mode suppression in AlN contour-mode resonators," *Sens. Actuators Phys.*, vol. 206, pp. 42–50, Feb. 2014.

[16] Y.-Y. Chen, Y.-T. Lai, C.-M. Lin, "Finite element analysis of anchor loss in AlN Lamb wave resonators," in *2014 IEEE International Frequency Control Symposium (FCS)*, Taipei, Taiwan, May 2014, pp. 1–5.

[17] B. P. Harrington, R. Abdolvand, "In-plane acoustic reflectors for reducing effective anchor loss in lateral–extensional MEMS resonators," *J. Micromechanics Microengineering*, vol. 21, no. 8, p. 085021, Jul. 2011.

[18] C. Lin, J. Hsu, D. G. Senesky, A. P. Pisano, "Anchor loss reduction in ALN Lamb wave resonators using phononic crystal strip tethers," in *2014 IEEE International Frequency Control Symposium (FCS)*, May 2014, pp. 1–5.

[19] J. Segovia-Fernandez, M. Cremonesi, C. Cassella, A. Frangi, G. Piazza, "Anchor Losses in AlN Contour Mode Resonators," *J. Microelectromechanical Syst.*, vol. 24, no. 2, pp. 265–275, Apr. 2015.

[20] J. Zou, C.-M. Lin, A. P. Pisano, "Anchor loss suppression using butterfly-shaped plates for AlN Lamb wave resonators," in *2015 Joint Conference of the IEEE International Frequency Control Symposium & the European Frequency and Time Forum*, 2015, pp. 432–435.

[21] C.-M. Lin, Y.-J. Lai, J.-C. Hsu, Y.-Y. Chen, D. G. Senesky, A. P. Pisano, "High-Q aluminum nitride Lamb wave resonators with biconvex edges," *Appl. Phys. Lett.*, vol. 99, no. 14, p. 143501, 2011.

[22] C. Cassella, N. Singh, B. W. Soon, G. Piazza, "Quality factor dependence on the inactive regions in AlN contour-mode resonators," *J. Microelectromechanical Syst.*, vol. 24, no. 5, pp. 1575–1582, 2015.

[23] B. Gibson, K. Qalandar, C. Cassella, G. Piazza, K. L. Foster, "A study on the effects of release area on the quality factor of contour-mode resonators by laser doppler vibrometry," *IEEE Trans. Ultrason. Ferroelectr. Freq. Control*, vol. 64, no. 5, pp. 898–904, 2017.

[24] C. Tu, J. E.-Y. Lee, "VHF-band biconvex AlN-on-silicon micromechanical resonators with enhanced quality factor and suppressed spurious modes," *J. Micromechanics Microengineering*, vol. 26, no. 6, p. 065012, Jun. 2016.

[25] C. Tu, J. E.-Y. Lee, "A semi-analytical modeling approach for laterally-vibrating thin-film piezoelectric-on-silicon micromechanical resonators," *J. Micromechanics Microengineering*, vol. 25, no. 11, p. 115020, Nov. 2015.

[26] J. Zou, C. Lin, G. Tang, A. P. Pisano, "High-Q Butterfly-Shaped AlN Lamb Wave Resonators," *IEEE Electron Device Lett.*, vol. 38, no. 12, pp. 1739–1742, Dec. 2017.

[27] J. D. Larson, P. D. Bradley, S. Wartenberg, R. C. Ruby, "Modified Butterworth-Van Dyke circuit for FBAR resonators and automated measurement system," in *2000 IEEE Ultrasonics Symposium. Proceedings. An International Symposium*, 2000, vol. 1, pp. 863–868.

Proceedings of the 16th Annual IEEE International
Conference on Nano/Micro Engineered and Molecular Systems
April 25-29, 2021

Modeling of magnetic sensor based on BAW magnetoelectric coupling micro-heterostructure

Si Chen, Wan-chun Ren*, Junru Li, *Student Member, IEEE*, Chun-rui Peng, Yang Gao*

Abstract — Sensors actuated by bulk acoustic wave (BAW) have attracted lots of attention due to their high sensitivity, GHz-level high frequency, and small size. Different from the previous studies, the performance optimization of magnetoelectric (ME) micro-heterostructure has been adequately considered in this work. We constructed a micro-level model of the magnetic sensor based on BAW ME coupling 2~5 layers micro-heterostructure. Its properties working under both direct current bias and high-frequency alternating magnetic field were systemically analyzed. The results show that the sensitivity and linearity of the sensor can be optimized by adjusting the layer number of the device and the bias magnetic field applied to the ME heterostructure. Eventually, the highest sensitivity of 3.24 V/Oe • cm has been achieved on the magnetic sensor with three-layer structure at the first resonant frequency of 2.2 GHz, and linearity can be strictly controlled within 2% with measuring range of 75~150 Oe. These results has demonstrated that the micro-model of the BAW ME coupling micro-heterostructure, and realized a method to improve its sensitivity and linearity by optimizing the device structure and test conditions. This achievement will further guide the structural design and performance optimization of ME coupling devices.

Abstract — ME heterostructure, bulk acoustic wave, magnetic sensor, FEA, ME coefficient.

I. INTRODUCTION

Magnetoelectric (ME) heterostructure is composed of the ferromagnetic and ferroelectric material, its ME coupling effect is derived from the piezoelectric effect of ferro-electric phase and magnetostrictive effect of ferromagnetic phase. There are many advantages for ME heterostructure, such as free conversion of energy between

* Supported by SciTech. Foundation of Southwest University of Science and Technology (Grant No. 20ZX7114), Robot Technology Used for Special Environment Key Laboratory of Sichuan Province (Grant No. 20KFKT02), Science and Technology on Electronic Information Control Laboratory (Grant No. 6142105200203), Science and Technology on Reliability Physics and Postgraduate Innovation Fund Project by Southwest University of Science and Technology (Grant No. 20YCX0062). (*Corresponding authors:* Wan-chun Ren, *Yang Gao*.)

Si Chen and Wan-chun Ren are with the School of Information Engineering, Southwest University of Science and Technology, Mianyang 621010, China (742871054@qq.com; rwch_qw@163.com).

Jun-ru Li is with the College of Optoelectronic Engineering, Chongqing University, Chongqing 400044, China (e-mail: li_junru@foxmail.com).

Chun-rui Peng is with the School of Electronic Science and Engineering, University of Electronic Science and Technology of China, Chengdu 611731, China (e-mail: 17865916198@163.com).

Yang Gao is with the Robot Technology Used for Special Environment Key Laboratory of Sichuan Province, Mianyang 621010, China (e-mail: gaoy@swust.edu.cn).

magnetic and electric field, and large ME conversion coefficient [1-2]. Many investigations about developing and preparing of the magnetic sensor with ME heterostructure has been reported, however, its device size was cm-level or bigger and difficult to be shrinked due to bulk materials using in ME heterostructure previously [3-5]. On the contrary, the micro-magnetic sensor based on ME thin film became a hot topic for its advantages of

miniaturization, low cost, and easy integration with conventional CMOS process [6-9]. There are two types of magnetic sensors actuated by acoustic waves: surface acoustic wave (SAW) and bulk acoustic wave (BAW). The sensor based on SAW type is limited to working in the low-mid frequency band of kHz or measuring static/quasi-static magnetic field signals despites its high-static sensitivity [10-14]. Whereas the sensor based on BAW excitation has attracted tremendous attention in recent years because of its high frequency characteristics, high power capacity, and high energy conversion efficiency. At present, many researches on BAW sensors through experimental methods have been reported in the literature. Hui et al reported a MEMS resonant magnetic field sensor based on AlN/FeGaB bilayer nano-plate resonator [15], and Nan et al enhanced ME coupling by depositing composite ME heterostructure of monolayer AlN/10 layers FeGaB/Al$_2$O$_3$ on AlN CMR on this basis [16]. A ME structure based on nano-plate resonators was reported by Nan et al, it had good resolution [17]. Simultaneously, some work on BAW sensors through modeling and simulation methods are also reported. Wu et al reported a flexible magnetic sensor based on thin film bulk acoustic resonator, the equivalent circuit of Mason model of the sensor was established, and its sensitivity was improved by selecting the electrode of giant magnetostrictive material with large frequency offset [18]. Martos et al proposed a circuit simulation model of a novel miniaturized magnetoelectric antenna with applications in low-power sensing [19]. However, there is few systematic investigation about the structure simulation and performance optimization of the micro-magnetic sensor based on BAW actuation yet.

In this work, we constructed a micro-magnetic sensor model of 2-5 layers AlN/FeGaB alternating heterostructures based on BAW actuation, its operating frequency can reach several GHz. Through finite element analysis (FEA) simulation, the effect of DC bias and high frequency alternating magnetic field is used to simulate the magnetic and electric coupling behavior, furthermore, the sensitivity and linearity of the magnetic sensor are analyzed and calculated. Eventually, a method to improve the sensitivity and linearity of the BAW excitation micro magnetic sensor is obtained.

978-1-6654-3008-1/21 $31.00 © 2021 IEEE

II. Sensor Coupling Model

A. FEA Model

The sensitivity of the magnetic sensor based on BAW excitation is characterized by the ME coefficient of the ME heterostructure. In order to clarify the coupling principle of the device, the force, magnetic and electric field must be coupled. A typical configuration of a ME heterostructure includes a magnetostrictive/piezoelectric layer.

The magnetic sensor based on 2-5 layers ME heterostructure was simulated by the software of Comsol Multiphysics. In figure 1(a), the FEA models of ME heterostructure, including magnetostrictive layer, piezoelectric layer, and air domain, were constructed by coupling the magnetic field, solid mechanics module, and electrostatic module in the 3D geometric model. Take the three-layer cantilever structure as an example in Figure 1(b), a bias magnetic field is applied along Y-direction in the external air domain, and the lower surface of piezoelectric layer is grounded, the left and right side in this model is the fixed and free end respectively. Furthermore, average voltage on the upper surface of piezoelectric layer under the conditions of DC bias and high-frequency alternating magnetic field is measured by the steady-state and small signal frequency solution respectively. Finally, the performance of 2-5 layers magnetic sensor with cantilever is analyzed in details.

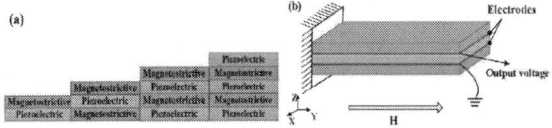

Figure 1.   (a) 2-5 layers ME heterostructure, and (b) the FEA model of three-layer magnetic sensor with device size of 100 μm×50 μm, and thickness of each layer was 1 μm. Material parameters of AlN and FeGaB films were used in reference [17].

B. Fundamental Equation

1) Piezoelectric Layer

The constitutive relationship between strain and charge is established in the piezoelectric layer [17], because the change of the piezoelectric layer is driven by the strain of the magnetostrictive layer. The external magnetic field deforms the magnetostrictive layer and then couples it to the piezoelectric layer. The relationship between the strain, stress, electric field and electric displacement of the piezoelectric layer is given by formulas (1) and (2):

$$S_E = S_{E_0} + C_E \cdot T_E - d_{33,p} \cdot E$$
$$D = D_r + d_{33,p} \cdot T_E + \varepsilon \cdot E \tag{1}$$

The electrostatics modules provide the following constraint conditions:

$$E = -\nabla \varphi$$
$$\nabla \cdot D = \rho_\upsilon \tag{2}$$

E and D is electric field intensity and electric flux density, S_E and T_E, T_{E_0} are strain and stress tensors, initial

stress of the piezoelectric layer respectively, C_E and $d_{33,p}$ is elasticity matrix and piezoelectric coefficients of the piezoelectric layer respectively, D_r is the remanent electric displacement, ε, ρ_υ and φ is relative permittivity matrix, electric charge density, and electric potential respectively.

2) Magnetostrictive Layer

The relationship between stress and strain of the magnetostrictive layer is established by setting the properties of magnetoelastic with strong magnetic field as nonlinear isotropy:

$$T_H = T_{H_0} + C_H \cdot \left(S_H - S_{H_0} \right) \tag{3}$$

$$S_H = \frac{1}{2} \left[(\nabla u)^T + \nabla u \right] \tag{4}$$

S_H, T_H, T_{H_0}, C_H and u are strain and stress tensors, initial stress, elasticity matrix, and dis-placement of the magnetostrictive layer respectively, Equation 4 shows that strain of the magnetostrictive layer is related to its displacement gradient.

The relationship between magnetic flux density (B) and magnetic field intensity (H) can be separated as two cases: Firstly, the piezoelectric layer and air domain is non-magnetic phases, which follows constitutive equation: $B = \mu_0 \mu_r H$. μ_0 and μ_r is vacuum permeability and relative permeability respectively; Secondly, the magnetostrictive layer adopts nonlinear magnetization relation of $B = \mu_0 (H + M)$. M is the magnetization current density.

C. Performance Parameter

The ME coefficient   is used to characterize the sensitivity of the magnetic sensor, and it is the most important parameter to evaluate the performance of the sensor. Its calculation formula is as follows [20]:

$$\alpha_{ME} = \frac{\partial E_z}{\partial H_{bias}} = \frac{\partial E_z}{\partial H_y} \tag{5}$$

In the formula $E_z = V/t_p$, V and t_p is the average voltage on the upper surface and thickness of the piezoelectric layer respectively. The DC bias H_{dc} is the magnetic field added in the Y direction H_y.

Linearity (δ) is another important index describing the static characteristics of the sensor. The maximum deviation (ΔY_{max}) between the calibration curve of the sensor and its fitting straight line divided by full scale output (Y) is called linearity (Equation 6). The linearity is better as the value gets smaller [20].

$$\delta = \frac{\Delta Y_{max}}{Y} \times 100\% \tag{6}$$

III. RESULTS AND DISCUSSION

A. DC Bias Magnetic Field

The properties of ME heterostructure was simulated by the established model in the FEA software. DC bias magnetic field H_{dc}=0~5000 Oe is applied to the structure in the external air domain along Y-direction, then its influence to the stress, strain, and output voltage of 2-5 layers magnetic sensor are analyzed respectively.

1) Influences of Basic Properties

As can be seen in Figure 2(a), as the DC magnetic field gradually increases, the stress of the piezoelectric layer first increases rapidly, and then reaches saturation. And the three-layer ME heterostructure can obtain the maximum stress, which is about twice that of two-layer structure. The inseted image shows the stress value of piezoelectric layer is significantly greater than that of the magnetostrictive layer. This is because under the same bias magnetic field, the upper and lower magnetostrictive layers squeeze the middle piezoelectric layer in the Z direction. The stress distribution nephogram of the piezoelectric layer in the three-layer magnetic sensor shows that due to the edge effect of the magnetic field, in the Y direction of the ME heterostructure, the stress distribution in the middle is more uniform than on both sides. In addition, because the fixed end is constrained, the stress on the left side is significantly greater than the free end on the right side, and the maximum stress value is at the edge of the fixed end.

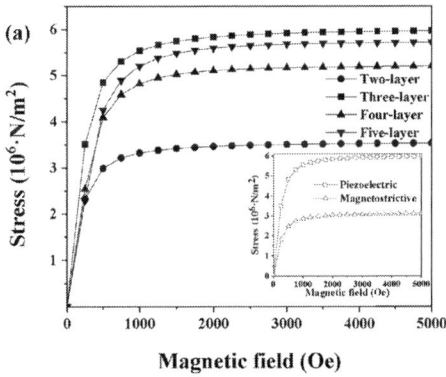

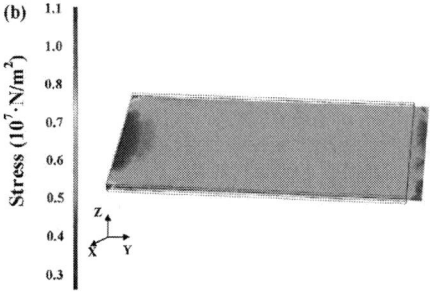

Figure 2. (a) With the increase of the bias DC magnetic field, the change trend of the stress in ME heterostructure with different layers, and (b) the stress distribution nephogram of the piezoelectric layer in three-layer magnetic sensor

As shown in Figure 3(a), the strain in the magnetostrictive layer is significantly greater than that in the piezoelectric layer, and the strain in the two-layer structure is the largest. This is because the strain of the magnetostrictive layer is affected by the displacement gradient (Equation 4), and its strain change law is consistent with that of the displacement gradient. It can be seen from the displacement distribution cloud diagram in Figure 3(b) that the displacement deflection of the two-layer structure is the largest, about 10^2 times that of the three-layer structure.

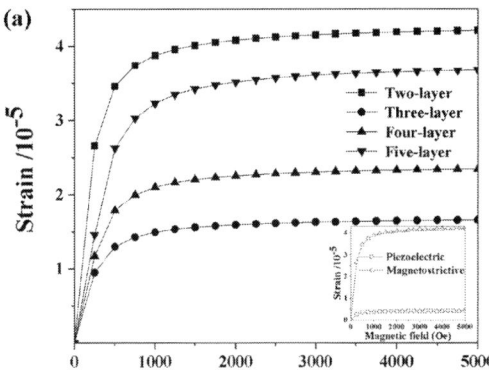

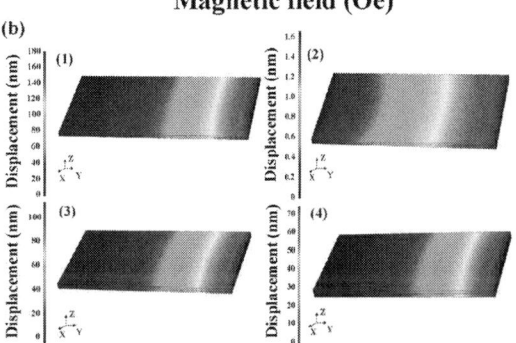

Figure 3. (a) With the increase of the bias DC magnetic field, the variation curve of the strain with different layers, and (b) the displacement distribution nephogram of magnetostrictive layer, (1)-(4): 2-5 layers structure.

As shown in Figure 4(a), the voltage value obtained on the piezoelectric layer increases with the increase of the magnetic field, and finally tends to be saturated. As can be seen from the voltage distribution nephogram of the three-layer magnetic sensor in Figure 4(b), the voltage value of the three-layer structure is the largest, the middle voltage is evenly distributed, and the voltage at the left fixed end is greater than the right free end.

978-1-6654-3008-1/21 $31.00 © 2021 IEEE 1261

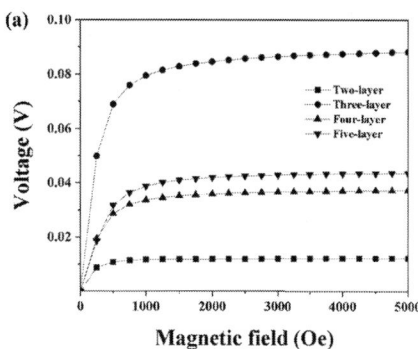

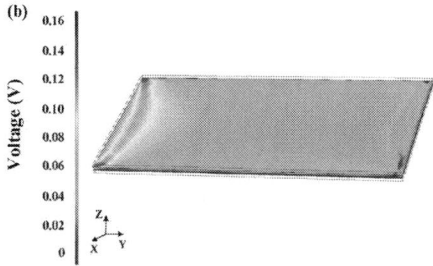

Figure 4. (a) With the increase of the bias DC magnetic field, the variation curve of the output voltage with different layers, and (b) the voltage distribution nephogram of magnetostrictive layer.

2) Performance Analysis

Under different bias magnetic fields, the performance of the BAW magnetic sensor is analyzed through the magnetic-electromechanical coupling of the ME heterostructure. Its coupling generates an induced charge on the surface of the piezoelectric layer, thereby generating an induced voltage. As shown in Figure 5(a), the variation law of ME coefficient under the bias magnetic field (0~500 Oe) is analyzed by equation 5. The ME coefficient firstly increases and then decreases with the addition of the bias magnetic field. The sensor with the highest ME coefficient has the highest sensitivity, which also means the best magnetoelectric conversion efficiency and the largest output voltage. Near the bias magnetic field of 125 Oe, the ME coefficients of the two-layer and three-layer ME heterostructures reach the lowest of 0.52 V/Oe·cm and highest of 2.81 V/Oe·cm, respectively. Therefore, the sensitivity of the magnetic sensor based on the three-layer ME heterostructure with bias maganetic field of 125 Oe is the largest. In summary, the result demonstrates a method to improve the sensitivity of the magnetic sensor by optimizing the number of ME heterostructure layers and the bias magnetic field.

Calculated from Equation 6, under a bias magnetic field of 0-500 Oe, the linearity of the 2-5 layer ME heterostructure is greater than 13%, which cannot meet the requirements of sensor performance. As shown in Figure 5(b), narrowing the measurement range of the magnetic field to 75-150 Oe can control the linearity of the 3-5 layer structure within 2%, which are 1.68%, 1.39%, and 1.86% respectively. However, the worse linearity of the two-layer structure is 6.65%. In order to make its linearity less than 2%, its measurement range have to be further limited within 75-100 Oe.

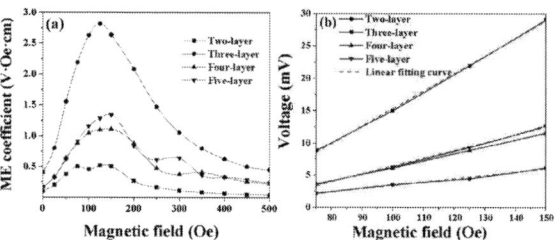

Figure 5. Figure.5 (a) With the increase of the bias DC magnetic field, the variation curve of ME coefficient with different layers, and (b) the linear fitting curve of the output voltage with different DC magnetic field.

B. High-frequency alternating magnetic field

1) Output voltage at resonant frequency

In order to analyze the output voltage changes of the magnetic sensor in practical applications, a small signal frequency domain analysis method is used to simulate the high frequency alternating magnetic field. Take the three-layer magnetic sensor with the highest sensitivity as an example, an alternating current (AC) disturbance signal of H_{ac}=1 Oe is applied in the FEA model. Therefore, the external magnetic field is divided into: DC bias and alternating excitation magnetic field, and $|H_{ac}| \ll |H_{dc}|$. Figure 6(a) shows the admittance curve obtained by applying an AC voltage to the piezoelectric layer of the ME heterostructure without a bias magnetic field. The first and second-order resonance frequency is approximately 2.1 GHz and 2.7 GHz, respectively.

When the DC bias magnetic field is less than 300 Oe, the output voltage at the resonance frequency changes irregularly and is insignificant. Therefore, as shown in Figure 6(b), the small signal frequency domain analysis method is used to investigate the voltage output performance with DC bias magnetic field of 300-500 Oe. It can be found that both the first and second-order resonance frequencies are slightly shifted to the right, increasing to approximately 2.2 GHz and 2.9 GHz, respectively. In addition, the out-put voltage at the resonance frequency is significantly higher than others, and it increases with the addition of the bias magnetic field. When the bias magnetic field is 500 Oe, the maximum output voltage at the first-order resonance frequency is 0.18 V, which is much larger than the pure DC magnetic field bias mode under the same conditions. Therefore, the sensitivity of the magnetic sensor is higher in the small signal frequency domain analysis mode.

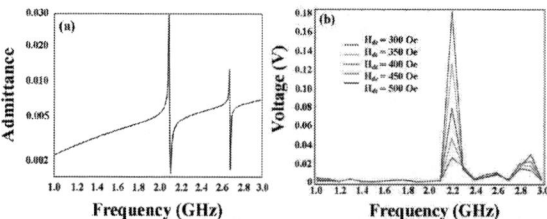

Figure 6. Taking the three-layer heterostructure magnetic sensor as an example, (a) the admittance curve obtained by applying an AC voltage to the piezoelectric layer of the ME heterostructure without a bias magnetic field, and (b) the output voltage variation curve with frequency under different DC bias magnetic field.

978-1-6654-3008-1/21 $31.00 © 2021 IEEE

2) ME Coefficient

Figure 7 shows the change rule of the ME coefficient with the DC bias magnetic field at the first-order resonance frequency of 2.2GHz in the high-frequency alternating magnetic field mode. When the bias magnetic field is about 90 Oe, the maximum value of the ME coefficient is 3.24 V/Oe·cm, which is significantly greater than 2.81 V/Oe·cm in the pure DC bias mode. Therefore, the sensitivity of the magnetic sensor can be further improved by adjusting the optimal bias magnetic field at the resonance frequency.

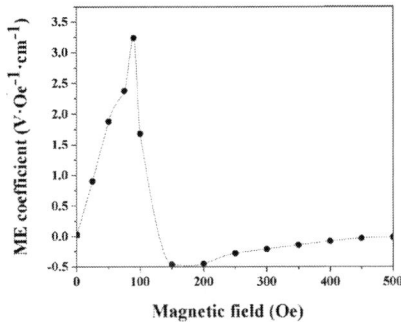

Figure 7. In the high-frequency alternating magnetic field mode, the variation curve of ME coefficient with DC bias magnetic field.

IV. CONCLUSIONS

A simulation model of the micro-magnetic sensor based on BAW actuated ME coupling heterostructure has been established, and the performance parameters of the sensor in the DC bias and high-frequency alternating magnetic field mode have been calculated and analyzed, respectively. Compared the stress, strain and output voltage performance of the 2-5 layer ME heterostructure, and analyzed the sensitivity and linearity of the sensor, it can be concluded that in the DC mode, the three-layer magnetic sensor has the highest sensitivity of 2.81 V/Oe·cm when the bias magnetic field is 125 Oe. In addition, the linearity of the sensor can be controlled within 2% by reducing the range of the bias magnetic field to 75-150 Oe. In the high-frequency alternating magnetic field mode, the output voltage at the resonance frequency is higher, and the maximum value of the ME coefficient of 3.24 V/Oe·cm at the first-order resonance frequency. Therefore, by adjusting the layer number of the sensor and optimizing the bias magnetic field to improve the output voltage at the resonance frequency, and a highly sensitive magnetic sensor can be achieved. The investigation results are of great significance for guiding the structural design and performance optimization of high-frequency magnetic sensors.

REFERENCES

[1] J. Jin, S. G. Lu, C. Chanthad, Q. Zhang, M. A. Haque, Q. Wang, "Multiferroic Polymer Composites with Greatly Enhanced Magnetoelectric Effect under a Low Magnetic Bias," Advanced Materials, vol. 23, no. 33, pp. 3853-3858, Sept. 2011.

[2] X. Zhuang, S. Saez, M. Lam Chok Sing, C. Cordier, C. Dolabdjian, J. Li, D. Viehland, S. K. Mandal, G. Sreenivasulu, G. Srinivasan, "Investigation on the magnetic noise of stacked Magnetostricitive-Piezoelectric laminated composites," Sens. Lett. vol. 10, no. 3-4, pp. 961–965, Mar. 2012.

[3] S. Dong, J. Zhai, J. F. Li, D. Viehland, "Small dc magnetic field response of magnetoelectric laminate composites," Applied Physics Letters, vol. 88, no. 8, Feb. 2006.

[4] J. Das, J. Gao, Z. Xing, J. F. Li, D. Viehland. "Enhancement in the field sensitivity of magnetoelectric laminate heterostructures," Applied Physics Letters, vol. 95, no. 9, Aug. 2009.

[5] J. Gao, L. Shen, Y. Wang, D. Gray, J. Li, D. Viehland. "Enhanced sensitivity to direct current magnetic field changes in Metglas/Pb (Mg$_{1/3}$Nb$_{2/3}$)O$_3$-PbTiO$_3$ laminates," Journal of Applied Physics, vol. 109, no. 7, Apr. 2011.

[6] E. Lage, C. Kirchhof, V. Hrkac, L. Kienle, J. Robert, R. Knöchel, E. Quandt, D. Meyners. "Exchange Biasing of Magnetoelectric Composites," Nature Materials, vol. 11, no. 6, pp. 523-529, Jun. 2012.

[7] P. Hayes, S. Salzer, J. Reermann, E. Yarar, V. Röbisch, A. Piorra, D. Meyners, M. Höft, R. Knöchel, G. Schmidt, E. Quandt, "Electrically modulated magnetoelectric sensors," Applied Physics Letters, vol. 108, no. 18, May. 2016.

[8] P. Hayes, V. Schell, S. Salzer, D. Burdin, E. Yarar, A. Piorra, R. Knöchel, Y. Fetisov, E. Quandt. "Electrically modulated magnetoelectric AlN/FeCoSiB film composites for DC magnetic field sensing," Journal of Physics D Applied Physics, vol. 51, no. 35, Sept. 2018.

[9] X. Liang, C. Dong, H. Chen, J. Wang, Y. Wei, M. Zaeimbashi, A. He, A. Matyushov, C. Sun, N. Sun. "A Review of Thin-Film Magnetoelastic Materials for Magnetoelectric Applications," Sensors, vol. 20, no. 5, Mar. 2020.

[10] H. Zhou, A. Talbi, N. Tiercelin, O. Bou Matar. "Multilayer magnetostrictive structure based surface acoustic wave devices," Applied Physics Letters, vol. 104, no. 11. Mar. 2014.

[11] L. Huang; Q. Lyu, D. Wen, Z. Zhong, H. Zhang, F. Bai. "Theoretical investigation of magnetoelectric surface acoustic wave characteristics of ZnO/Metglas layered composite," AIP Advances, vol. 6, no. 1, Jan. 2016.

[12] V. Polewczyk, K. Dumesnil, D. Lacour, M. Moutaouekkil, H. Mjahed, N. Tiercelin, S. P. Watelot, H. Mishra, Y. Dusch, S. Hage-Ali, O. Elmazria, F. Montaigne, A. Talbi, O.B. Matar, M. Hehn. "Unipolar and Bipolar High-Magnetic-Field Sensors Based on Surface Acoustic Wave Resonators," Physical Review Applied, vol. 8, no. 2, Aug. 2017.

[13] A. Kittmann, P. Durdaut, S. Zabel, J. Reermann, J. Schmalz, B. Spetzler, D. Meyners, N. X. Sun, J. McCord, M. Gerken, G. Schmidt, M. Höft, R. Knöchel, F. Faupel, E. Quandt. "Wide Band Low Noise Love Wave Magnetic Field Sensor System," Scientific Reports, vol. 8, no. 1, pp. 1-10, Jan. 2018.

[14] X. Liu, J. Ou-Yang, B. Tong, S. Chen, Y. Zhang, B. Zhu, X. Yang. "Influence of the delta-E effect on a surface acoustic wave resonator," Applied Physics Letters, vol. 114, no. 6, pp. 1-5, Feb. 2019.

[15] Y. Hui, T. X. Nan, N. X. Sun, M. Rinaldi. "MEMS resonant magnetic field sensor based on an AlN/FeGaB bilayer nano-plate resonator," Proc. 26th IEEE MEMS Conf., pp. 721-724, Jan. 2013.

[16] T. Nan, Y. Hui, M. Rinaldi, N. X. Sun, "Self-biased 215 MHz magnetoelectric NEMS resonator for ultra-sensitive DC magnetic field detection," Sci. Rep. 3, pp. 1-6, Jun. 2013.

[17] T. Nan, H. Lin, Y. Gao, A. Matyushov, G. Yu, H. Chen, N. Sun, S. Wei, Z. Wang, M. Li, X. Wang, A. Belkessam, R. Guo, B. Chen, J. Zhou, Z. Qian, Y. Hui, M. Rinaldi, M. E. McConney, B. W. Howe, Z. Hu, J. G. Jones, G. J. Brown, N. X. Sun. "Acoustically actuated ultra-compact NEMS magnetoelectric antennas," Nature Communication, vol. 8, pp. 1-8, Aug. 2017.

[18] Y. Wu, S. Dong, H. Jin, X. Wang, G. Chen. "Flexible Magnetic Sensor Based on FBAR," Proc. 2016 IEEE International Nanoelectronics Conference, pp. 1-2, 2016.

[19] I. Martos-Repath, A. Mittal, M. Zaeimbashi, D. Das, N. X. Sun, A. Shrivastava, M. Onabajo. "Modeling of Magnetoelectric Antennas for Circuit Simulations in Magnetic Sensing Applications," 2020 IEEE 63rd International Midwest Symposium on Circuits and Systems (MWSCAS). IEEE, pp. 49-52, Sept. 2020.

[20] S. Reis, M. P. Silva, N. Castro, V. Correia, P. Martins, A. Lasheras, J. Gutierrez, J. M. Barandiarán, J. G. Rocha, S. Lanceros-Mendez, "Characterization of Metglas/poly(vinylidene fluoride)/Metglas magnetoelectric laminates for AC/DC magnetic sensor applications," Materials & Design, vol. 92, pp. 906-910, Dec. 2016.

Proceedings of the 16th Annual IEEE International
Conference on Nano/Micro Engineered and Molecular Systems
April 25-29, 2021

The 3D Capacitance Modeling of Non-parallel Plates Based on Conformal Mapping

Yue Feng, *Member, IEEE*, Zilong Zhou, *Student Member, IEEE*, Wenlong Wang, Zehong Rao, and Yanhui Han*

Abstract—A highly accurate three-dimensional (3D) capacitance analytical model of parallel plates is important for the optimal design of electrostatic MEMS devices. However, due to constraints of fabrication processes, two parallel conducting plates are ineluctably tilted at an angle. Basically, finite element analysis (FEA) seems the preferred access to evaluate fringing effects across the non-parallel plates, while the analytical solution is not available. In this work, we present an analytical model for the 3D capacitance of a non-parallel-plate structure using the conformal mapping method. Taking the fringing effect into account, the proposed model shows a high accuracy of 95%, compared with other models.

Keywords—Fringing effect, 3D capacitance model, non-parallel-plate capacitor, conformal mapping

I. INTRODUCTION

Electrostatic Microelectromechanical Systems (MEMS) devices, such as capacitive MEMS accelerometers [1], electrostatic MEMS actuators [2], electrostatic MEMS energy harvester [3] has been widely used, owing to high sensitivity, broad bandwidth, and low-temperature dependence [4-6]. The micro-scale effect in electrostatic MEMS devices introduces the non-negligible fringe capacitances [7]. For example, R. Chen and Y. Suzuki measured the fringing capacitance of 45pF in the vibration energy harvester [8]. Feng *et al.* evaluated that the fringing capacitance accounted for 28.7% of the total effective capacitance in the rotational energy harvester [9]. Limited by the precision of fabrication processes, a slight incline between the movable electrodes and fixed electrode causes a huge variation in capacitance, which is mostly predicted by the finite element analysis (FEA) [9, 10]. However, a valid capacitance analytical model is required for the optimal design and precious performance evaluation of electrostatic MEMS devices. Therefore, this paper presents a high-accurate three-dimensional (3D) capacitance analytical model for the non-parallel plates using the conformal mapping method. Based on the previous works [11, 12], a low-error two-dimensional (2D) capacitance per unit width formula of the non-parallel plates has been obtained *via* conformal mapping. Considering the fringing effect, the 3D capacitance analytical model for finite-width plates is established from the 2D capacitance analytical model superposition. Furthermore, the accuracy of the 3D

model as a function of three structure parameters is discussed in detail.

II. FIELD REGION MAPPING

The capacitive structure considered in the literatures consists of two non-parallel conducting plates with infinity length, whose cross-section is shown in Fig. 1. The extended lines of the two plates intersect at a point with the angle θ. The distances between the intersection point and endpoints of a plate are l and R, respectively. Thus, the width of a plate is R-l. The electric potentials φ=+V and φ=-V are severally applied on the upper conducting plate and the lower conducting plate, where the zero potential line φ=0 passes between the centers of the plates. Ignoring the fringe capacitance, the ideal capacitance per unit length can be calculated as follows:

$$C_i = \frac{\varepsilon_0}{\theta}\ln\frac{R}{l} \qquad (1)$$

where ε_0 is the vacuum permittivity. To calculate the fringe capacitance of the non-parallel plates, Xiang divided the electrostatic field into two parts: one is confined by two electrode plates to the interior of the angle and another exists outside the angle [11]. However, this model neglects the fringe electric field across extended lines of two plates, leading to the underestimation of capacitance. Besides, the 2D model for the non-parallel plates given by Xiang involves elliptic integrals, which evaluate to infinity at a small angle (<5°) even with ultra-high numerical precision [13]. Taking the fringing effect and singularity into consideration, the logarithm transformation is adopted rather than the exponential transformation proposed by Xiang [11].

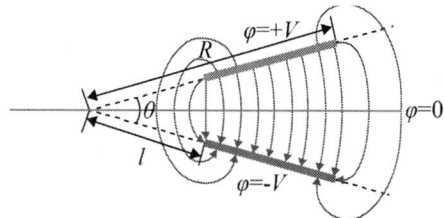

Figure 1. Cross-section of a non-parallel-plate capacitor

*Research supported by the National Natural Science Foundation of China (NSFC) (No. 520770055 and No. 51607007).

Y. Han is with Beijing Orient Insititute of Measurement and Test, Haidian District 100081 Beijing, 100080 China (corresponding author to provide e-mail: hui920718@163.com).

Y. Feng is with Beijing Institute of Technology, Beijing, Haidian District 100081 China (e-mail: fengyue@bit.edu.cn).

Z. Zhou is with Beijing Institute of Technology, Beijing, Haidian District 100081 China (e-mail: 3120190281@bit.edu.cn).

W. Wang is with Beijing Institute of Technology, Beijing, Haidian District 100081 China (e-mail: 15385895890@163.com).

Z. Rao is with Beijing Institute of Technology, Beijing, Haidian District 100081 China (e-mail: 1461829213@qq.com).

978-1-6654-3008-1/21 $31.00 © 2021 IEEE

A. 2D field region mapping

As shown in Fig. 2(a), the boundary conditions of the non-parallel plates are plotted in the complex Z plane ($z=x+iy$). The longitudinal dimension of the two plates is infinite, reducing the 3D problem to two-dimensional. In the Z plane, the coordinates A, B, C, and D, are $z_A=le^{i\theta/2}$, $z_B=Re^{i\theta/2}$, $z_C=le^{-i\theta/2}$, and $z_D=Re^{i\theta/2}$, respectively. Taking account of all of the whole electrostatic field, the conformal mapping is as follows:

$$w = \ln z \tag{2}$$

where the whole Z plane is mapped into the stripe region in the W plane. As shown in Fig. 2(b), the non-parallel-plate capacitive structure is transformed into the parallel-plate capacitive structure, And the zero potential line in the Z plane is transformed into three parallel lines $w=i\pi$, $w=0$, and $w=-i\pi$, respectively. In the W plane, the coordinate A, B, C, and D, are $w_A=\ln l+i\theta/2$, $w_B=\ln R+i\theta/2$, $w_C=\ln l-i\theta/2$, $w_D=\ln R-i\theta/2$, respectively.

Owing to the good symmetrical structure of the parallel-plates in the W plane, it can be regarded as the series and parallel connections of the capacitor in region A. After calculation, all of the capacitance in the W plane, which is the whole capacitance value of the non-parallel plates, is equal to the capacitance value in region A.

Using Schwarz–Christoffel mapping, while the inclined angle θ is small ($\theta<30°$), the 2D analytical capacitance per unit length C_a can be approximately expressed as[14]:

$$C_a = \frac{\varepsilon_0}{\theta}\ln\frac{R}{l} + 2\varepsilon_0\left[\frac{\ln 2\pi/\theta}{2\pi-\theta} + \frac{\ln 2\pi/2\pi-\theta}{\theta}\right] \tag{3}$$

In the expression of the capacitance C_a, the first item is the ideal capacitance; the last item is the fringe capacitance.

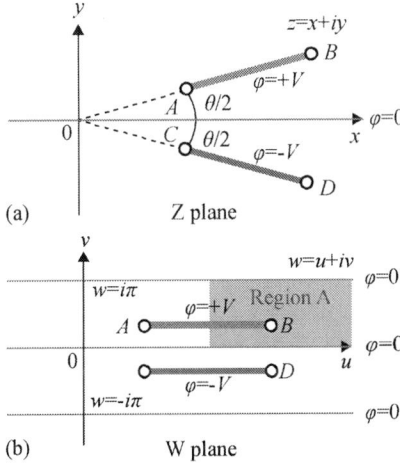

(a)   Z plane

(b)   W plane

Figure 2. 2D conformal mapping from (a) the Z plane to (b) the W plane

B. 3D field region mapping

Figure 3(a) presents the 3D capacitor structure considered in this paper consists of the width of a plate is h. Using the logarithm mapping in the y-direction, the non-parallel plates are transformed into the parallel plates with the length $\ln R/l$, the width h, and the gap θ [Fig. 3(b)]. According to Matick's research [15], the 3D capacitance of parallel plates can be approximated as the summation of three components: capacitance of parallel plates assuming no fringe fields, fringe capacitance in y-direction caused by finite width h, and fringe capacitance in z-direction caused by finite length $\ln R/l$.

Therefore the 3D analytical capacitance C_a of a non-parallel-plate capacitor is described as

$$C_a = C_{xz} + C_{yz} + \frac{h\varepsilon_0}{\theta}\ln\frac{R}{l} \tag{4}$$

where C_{xz} and C_{yz} are the fringe capacitance in the y-direction and the fringe capacitance in the x-direction, respectively [Fig. 3(a)]. In 3D space, C_{xz} and C_{yz} is the are expressed as[16]

$$\begin{cases} C_{xz} = 2h\varepsilon_0\left[\dfrac{\ln 2\pi/\theta}{2\pi-\theta} + \dfrac{\ln 2\pi/2\pi-\theta}{\theta}\right] \\ C_{yz} = \displaystyle\int_l^R \frac{\varepsilon_0}{\pi}\ln\frac{R}{l}\left\{1+\ln\left[1+\frac{2\pi h}{\theta x}+\ln\left(1+\frac{2\pi h}{\theta x}\right)\right]\right\}dx \end{cases} \tag{5}$$

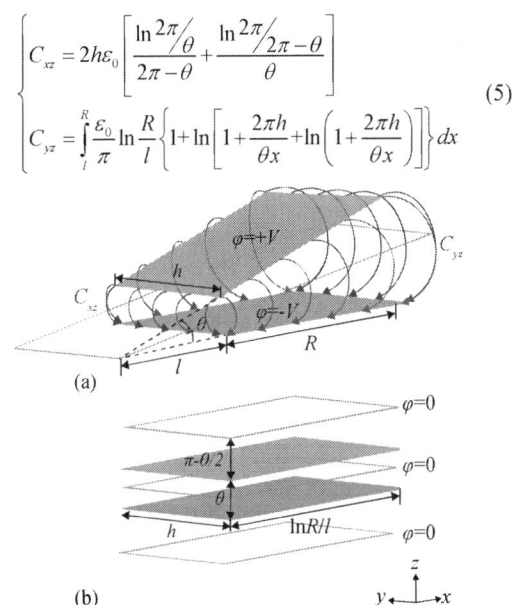

(a)

(b)

Figure 3. (a) Schematic diagram and (b) 3D conformal mapping of a non-parallel-plate capacitor

III. ELECTROSTATIC CAPACITANCE

To confirm the accuracy of the proposed analytical capacitance model, the FEA method was implemented by the commercial software COMSOL Multiphysics®, which has a specific module to solve electrostatic problems.

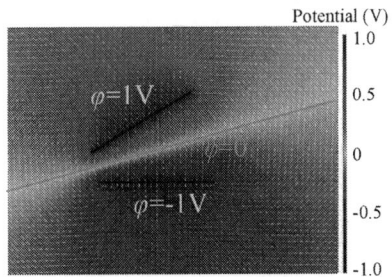

Figure 4. Electric field and potential distributions on the non-parallel plates capacitor with $\theta = 30°$, $l = 1$mm, and $R = 3$mm

978-1-6654-3008-1/21 $31.00 © 2021 IEEE

A. 2D capacitance

The simulated 2D electric field and potential distributions of the non-parallel plates are plotted, as shown in Fig. 4. Considering the fringe electric field, an air domain circle is set as ten times the length of a plate [9]. The zero potential line $\varphi=0$ passes between the centers of two plates, while the electric potentials of two conducting plates are opposite.

To eliminate the influence of structure scale, a dimensionless parameter is introduced: the length ratio r_l, defined as $r_l = (R-l)/l$. Fig. 5 shows the comparison of the ideal capacitance C_i, the analytical capacitance C_a, and the simulated capacitance C_c, between two inclined conducting plates with respect to different inclined angle θ. The analytical capacitance values is closely similar to the simulated capacitance values, and both of them are higher than the ideal capacitance values. The relative errors E between the analytical capacitance C_a and the simulated capacitance C_c is defined as follows

$$E = \frac{C_a - C_c}{C_c} \times 100\% \quad (6)$$

Figure 6 presents the relative errors E as a function of the length ratio r_l and the inclined angle θ. With the increase of the inclined angle, the accuracy of Eqn. (3) degenerates, and the derivation between the analytical values and the simulated values increase slightly. Therefore, while the length of the plate is too short ($r_l < 1$) and the tilt angle between the two plates is too large ($\theta > 20°$), the accuracy of the 2D approximate model is relatively low ($E > 3\%$). However, while the length ratio $r_l > 3$ and the inclined angle $\theta < 15°$, a high accuracy ($E < 1\%$) can be obtained.

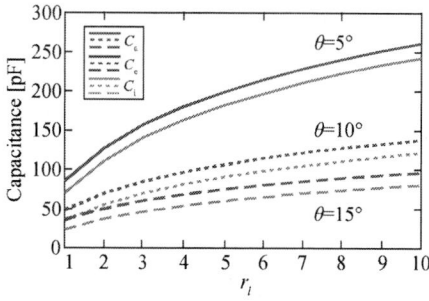

Figure 5. Comparison of the whole capacitance concerning to varied inclined angles

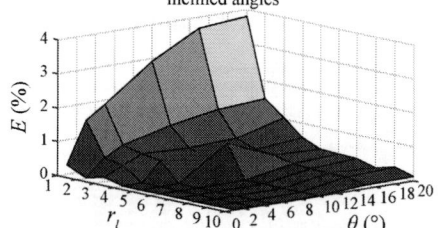

Figure 6. 2D relative errors E as a function of r_l and θ

Table I shows the comparison between 2D analytical capacitance values from the proposed model and the previous models with 2D simulated capacitance values. Obviously, compared to previous models, the accuracy of

the proposed 2D analytical capacitance model has greatly improved. In addition, the analytical model remarkably reduces the calculated amount, dispensing with complicated grids. The calculation cost (memory and time) of the analytical model is one thousandth of that of the finite element model.

TABLE I. COMPARISON BETWEEN DIFFERENT 2D ANALYTICAL MODELS

Parameters		Capacitance Values (pF)				Relative errors (%)		
r_l	θ (°)	C_c	C_a	$C_a{}^a$	$C_a{}^b$	E	E^c	E^d
1	1	370.7	372.0	NaN	338.7	0.3	NaN	-8.6
	5	84.8	86.3	39.3	68.3	1.7	-53.6	-19.4
	10	48.0	49.4	21.7	34.8	2.9	-54.8	-27.5
3	1	724.4	724.6	NaN	609.4	0.03	NaN	-15.8
	5	157.0	157.6	NaN	122.8	0.3	NaN	-21.7
	10	85.0	85.5	39.9	62.2	0.6	-53.0	-26.8
5	1	930.9	931.0	NaN	725.7	0.01	NaN	-22.0
	5	199.0	199.4	NaN	146.2	0.2	NaN	-26.5
	10	106.4	106.7	50.6	74.0	0.3	-52.4	-30.4

aEstimated capacitance values from Ref. 9.

bEstimated capacitance values from Ref. 11.

cEstimated relative errors from Ref. 9.

dEstimated relative errors from Ref. 11.

B. 3D capacitance

The simulated 3D electric field and potential distributions of the non-parallel plates are plotted, as shown in Fig. 7. In consideration of the computational complexity, an air domain sphere with an infinite element domain layer is set. In the 3D model, the zero potential line is transformed into the zero potential plane (Fig. 7).

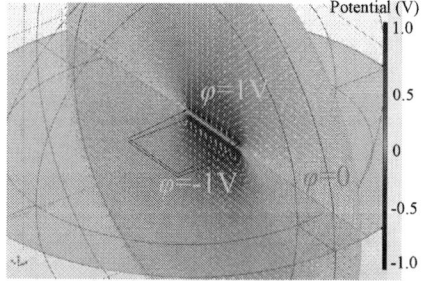

Figure 7. Electric field and potential distributions on the non-parallel-plate capacitor with $\theta = 30°$, $l = 1$mm, $h = 2$mm, and $R = 3$mm

Also, a dimensionless parameter is introduced: the width ratio r_w, defined as $r_w=h/l$. Fig. 8 gives the relative errors E as a function of the length ratio r_l and the inclined angle θ with the different width ratio r_w. Obviously, the analytical 3D capacitances are in good agreement with simulated ones ($E < 5\%$). At the larger width ratio ($r_w = 3, 5$), while the inclined angle increase, the accuracy of C_{xz} degenerates, and the fringe capacitance values in the x direction C_{xz} deviate from the simulated values, leading to the increase of the relative error values. In addition, the fringe capacitance values in the y-direction C_{yz} is the integration of the approximated capacitance of the parallel plates over the length. While the width ratio is small ($r_w = 1$), the formula of C_{xz} would underestimate capacitance, resulting in negative errors shown in Fig. 8(a).

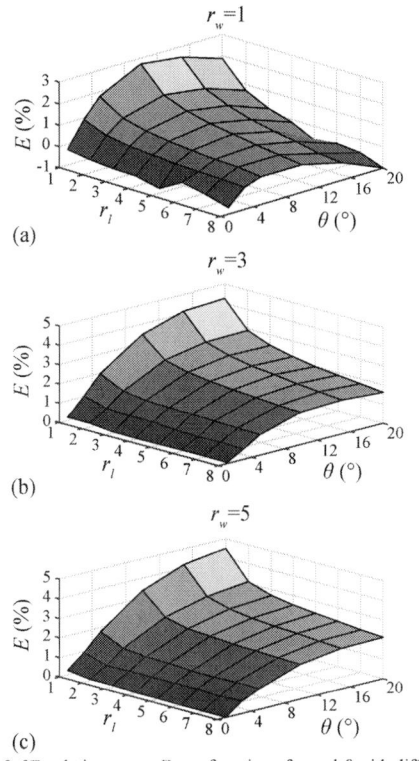

Figure 8. 3D relative errors E as a function of r_l and θ with different r_w

IV. CONCLUSION

In this paper, high-accuracy analytical models for both 2D and 3D capacitances of the non-parallel plates based on the conformal mapping method are proposed. Using the logarithm transformation, all of the fringe capacitances are taken into consideration. Therefore, the accuracies of both proposed 2D and 3D capacitance models have significantly improved, compared to previous models. The relative deviation of the proposed 3D theoretical model was within 5% compared with simulation data for $\theta \le 20°$, $1 \le r_l \le 5$, and $1 \le r_l \le 8$. The proposed model featuring the simple analytical form, wide applicable dimension range, and high accuracy, is appropriate for the structure design and optimization of electrostatic MEMS devices.

REFERENCES

[1] M. Daeichin, M. Ozdogan, S. Towfighian, and R. Miles, "Dynamic response of a tunable MEMS accelerometer based on repulsive force," Sens. Actuators A Phys., Vol. 289, pp. 34-43, April 2019.

[2] M. Pallay, R. Miles, and S. Towfighian, "Merging parallel-plate and levitation actuators to enable linearity and tunability in electrostatic MEMS," J. Appl. Phys., Vol. 126, pp. 014501, June 2019.

[3] K. Murotani and Y. Suzuki, "MEMS electret energy harvester with embedded bistable electrostatic spring for broadband response," J. Micromech. Microeng., Vol. 28, pp. 104001, June 2018.

[4] Y. Suzuki, "Recent Progress in MEMS Electret Generator for Energy Harvesting," IEEJ Trans. Electr. Electron. Eng., Vol 6, pp. 101-111, January 2011.

[5] W. de Groot, J. Webster, D. Felnhofer, and E. Gusev, "Review of Device and Reliability Physics of Dielectrics in Electrostatically Driven MEMS Devices," IEEE Trans. Device Mater. Reliab., Vol. 9, pp.190-202, 2009.

[6] W. Chuang, H. Lee, P. Chang, and Y. Hu, "Review on the Modeling of Electrostatic MEMS," Sensors, Vol. 10, pp. 6149-6171, April 2010.

[7] G. Nielson and G. Barbastathis, "Dynamic pull-in of parallel-plate and torsional electrostatic MEMS actuators," J. Microelectromech. Syst., Vol. 15, pp. 811-821, August 2006.

[8] R. Chen and Y. Suzuki, "Suspended electrodes for reducing parasitic capacitance in electret energy harvesters." J. Micromech. Microeng. Vol. 23, pp.125025, November 2013.

[9] Y. Feng, B. Shao, X. Tang, Y. Han, T. Wu, and Y, Suzuki, "Improved Capacitance Model Involving Fringing Effects for Electret-Based Rotational Energy Harvesting Devices," IEEE Trans. Electron Devices. Vol. 65, pp. 1597-1603, April 2018.

[10] F. Tay, X. Jun, Y. Liang, V. Logeeswaran, and Y. Yufeng, "The effects of non-parallel plates in a differential capacitive microaccelerometer," J. Micromech. Microeng., Vol. 9, pp. 283-293, August 1999.

[11] Y. Xiang, "Further study on electrostatic capacitance of an inclined plate capacitor," J. Electrostat., Vol. 66, pp. 366-368, July 2008.

[12] R. Das, S. Ghoshb, and R. Chakrabortya, "Analysis of electric field for inclined electrodes and use of such configuration for generating tunable differential polarization phase," Eur. Phys. J. Appl. Phys., Vol. 72, pp. 30501, November 2015.

[13] B. Patla, "Small angle approximation for non parallel plate capacitors with applications in experimental gravitation," ArXiv:1208.2984v3, August 2013.

[14] W. Zhou, H. Lan, H. Yu, L. Lai, B. Peng, and X. He, "Consideration of the fringe effects of capacitors in micro accelerometer design," IEEE Trans Inst. Meas. Contr., Vol. 40, pp. 1-6, June 2017.

[15] R. Matick and A. Ruehli, "Accurate 3-D Capacitance of Parallel Plates From 2-D Analytical Superposition," IEEE Trans. Compon. Packaging Manuf. Technol., Vol. 3, pp. 299-305, February 2013.

[16] V. Leus and D. Elata, "Fringing field effect in electrostatic actuators", January 2004.

Gap in pagination due to unavailable papers.

Pages 1268-1276

The 16th IEEE International Conference on Nano/Micro Engineered & Molecular Systems

Versatile E-Printing for Flexible Electronics

YongAn Huang

State Key Laboratory of Digital Manufacture Equipment and Technology, Huazhong University of Science and Technology, Wuhan 430074, China.

Website : http://faculty.hust.edu.cn/huang

ABSTRACT

Electrohydrodynamic (EHD) printing (E-Printing), which adopts electric field force to pull fluid flows from microcapillary nozzles, is becoming a promising technology for flexible electronics manufacture due to its unique advantages of high-resolution (<1μm) and compatibility with wide range of inks (1-10000 cps). Three types of printing mode that are E-jet printing, EHD direct-writing, and electrospray, can be switched through process adjustment to deposit dots, fibers/lines and thin films directly. Specially, the E-Printing technology enables large-scale deposition of highly aligned nanofibers in an additive, noncontact, real-time adjustment, and individual control manner on rigid or flexible substrates, making it rather attractive in the fabrication of flexible electronics.

This talk will briefly introduce our recent researches on E-Printing techniques, including the basic principles, alignment strategies and their applications. Several attractive applications in flexible electronics are also presented herein: (1) a hyper-stretchable (>300%) self-powered sensor that can be used to measure different physical quantities simultaneously; (2) organic thin-film transistor (OTFT) arrays that are fabricated in a large-area, high-efficiency and photolithography-free manner; (3) Au metal-network electrodes that are fabricated to obtain a flexible and transparent touch panel with transmittance of 94.3% and sheet resistance of 36.7 $\Omega\square^{-1}$; (4) full color perovskite display and photodetector, including full-color pixel and high-resolution dot matrix of 1 μm are fabricated. This technology overcomes the limitations on the resolution of fabrication and viscosity of ink of conventional inkjet printing, and represents major advances in manufacturing of flexible electronics.

BIOGRAPHY

YongAn Huang, professor of Huazhong University of Science and Technology (HUST), vice director of State Key Laboratory of Digital Manufacturing Equipment and Technology. Dr. Zhang received her Bachelor, Master and Ph.D. degrees from the Northwestern Polytechnical University. After finishing his postdoctoral research at HUST, he joined the faculty of HUST in 2009. In 2012, he was a visiting professor at the Northwestern University. He is a winner of The National Science Fund for Distinguished Young Scholars. He focuses on the research of design and manufacturing for flexible electronics, including 1) EHD printing and laser lift-off, and 2) flexible printed display, smart skin of aircrafts, electronic skin of robots and epidermal electronics of human. He has published more than 100 papers in journals including Science Advances, Advanced Materials, Advanced Functional Materials, Materials Horizon, Nano Energy, Small, Nature Communication, etc, and also published 3 monographs. He has won first prize of Natural Science in Hubei Province, Gold/Special Gold Award on International Exhibition of Inventions in Geneva, Switzerland.

978-1-6654-3008-1/21 $31.00 © 2021 IEEE   1278

Proceedings of the 16th Annual IEEE International
Conference on Nano/Micro Engineered and Molecular Systems
April 25-29, 2021

A High-Resolution Self-powered Flexible Pressure Sensor Matrix Based on ZnO Nanowires

Qifeng Lu, Fuqin Sun, Lili Li, Zihao Wang, and Ting Zhang*

Abstract—Flexible pressure sensor matrix, which can perceive the external mechanical stimuli with high spatial resolution, is the essential component for the electronic skin (E-skin) and is also able to be applied in various areas such as health monitoring and disease diagnosis. However, the design and fabrication of high-resolution pressure sensors with a high linearity in medium pressure range (10 kPa-100 kPa) at no power consumption is seldom reported. Here, we fabricate a self-powered pressure sensor matrix based on wurtzite structured ZnO nanowires (NWs) grown by hydrothermal method. The geometry and density of ZnO NWs were analyzed using a finite element method (FEM) to optimize the sensing performance. The matrix with an area of 25 mm² consists of 25 sensors with the dimension of each unit being 500 µm * 500 µm (density of 100/cm²). Each sensor can be accessed independently and shows a linear response for the pressure ranging from 15 kPa to 50 kPa. The proposed self-powered pressure sensor matrix in this study shows a great potential in the application of wearable devices.

I. INTRODUCTION

With the great advances in the areas of healthcare and human-machine, flexible and wearable electronics have attracted significant attentions from both academic and industrial societies due to their potential applications in physiological activities monitoring, chemical detection and disease diagnosis [1-5]. Among these flexible devices, pressure sensor, which can perceive the environmental stimuli, is considered as a promising candidate to electronic skin (E-skin) applications and has become one of the most fascinating research of interest. Generally, capacitance, piezoelectricity, and piezoresistivity are the three main transduction mechanisms to convert external mechanical stimuli into electrical signals for pressure sensors [6]. A piezoresistive-type pressure sensor, consisting of a stacking structure and microstructured sensing materials, usually shows high sensitivity and fast response time. However, the uniformity in each pixel is a challenge to be addressed in the sensor matrix. For the capacitive type sensor, the susceptibility to neighboring interference is the main limitation despite its simple device structure. With regard to the piezoelectricity-type device, it shows advantages in overcoming the above mentioned challenges and no external power source is required during operation. However, the sensitivity and reliability should be considered in the design and fabrication of piezoelectric pressure sensors matrix.

Recently, various piezoresistive-type and capacitance-type pressure sensors have been fabricated due to the simple structure and attractive device performance [7, 8]. For example, pressure sensors with the carbon nanotubes or fabric as the conductive materials and porous elastic materials as the substrate or dielectric layer exhibit high sensitivity in the pressure of several Pa and fast response time in the order of milliseconds [9-11]. However, most of the works are demonstrated based on discrete device with flexible substrate or pressure sensor matrix with rigid substrate, which can hardly meet the requirements in spatial resolution and flexibility for the applications in healthcare and human-machine. Besides, continuous power supply was required for these devices, which is not desired in the wearable electronics, especially in the case of pressure sensor matrix with large areas. In addition, in contrast to detection limit of several Pascal, the sensitivity of human skin is on the order to kilo Pascal, typical of about 20 kPa [12]. Therefore, the pressure sensor with low power consumption and linear response in medium pressure range (10 kPa-100 kPa) is of great interest in object manipulation. However, the flexible sensor matrix, which is able to overcome the above limitations and satisfy the mentioned demands simultaneously, is seldom reported.

Herein, a high-resolution self-powered flexible pressure sensor matrix is fabricated on polyimide (PI) substrate based on ZnO nanowires (NWs) grown by hydrothermal method and patterned using photolithograph. The pressure sensor matrix with an area of 5 mm * 5 mm contains 25 sensors with the dimension of each unit being 500 µm * 500 µm and each sensor can be accessed independently. Benefitting from the piezoelectric effect of ZnO NWs, no external voltage source is required during operation. In addition, a finite element method (FEM) was employed to optimize the geometry and density of ZnO NWs in the sensing unit. The fabricated pressure sensor shows a linear response in the pressure raning from 15 kPa to 50 kPa. The results indicate that the proposed flexible pressure sensors may open new opportunities for applications in human-machine interface, and wearable health monitoring devices.

II. EXPERIMENT

Fig. 1 (a)-(h) illustrate the schematic diagrams for the fabrication process of the pressure sensor matrix and the optical image of the devices. In detail, the fabrication of the

All authors are with i-lab, Key Laboratory of Multifunctional Nanomaterials and Smart Systems, Suzhou Institute of Nano-Tech and Nano-Bionics, Chinese Academy of Sciences, 398 Ruoshui Road, Suzhou, 215123, P. R. China. *Corresponding Author: Ting Zhang, email: tzhang2009@sinano.ac.cn. Phone: +86 512 62872706

The authors acknowledge the funding support from the National Key R&D Program of China (2018YFB1304700), the National Natural Science Foundation of China (61574163, 61801473).

978-1-6654-3008-1/21 $31.00 © 2021 IEEE

pressure matrix starts with a 2 inch glass wafer. After the wafer was cleaned in acetone, isopropanol, and DI water and dried in nitrogen flow, a diluted PI solution was spinning-coated on the wafer followed by the annealing in atmospheric ambient at 300 °C for one hour and the ramping rate should be controlled carefully in case of delamination of the PI film from the substrate. The thickness of the PI film is adjusted to be about 10 μm to ensure the easy peel-off of the film from the substrate after the fabrication of sensor matrix. Then, the bottom electrodes (BE, 10 nm Ti/100 nm Au) and 50 nm ZnO seed were deposited by sputtering and patterned followed by the growth of ZnO NWs using hydrothermal method as described in our previous research output [13]. Afterwards, a polymethyl methacrylate (PMMA) layer was spin-coated to encapsulate the ZnO NWs and cured at 90 °C for one minute. Next, 10 nm Ti/100 nm Au was deposited as the top electrodes (TE). Finally, the PI film was peeled-off from the glass wafer. The optical image in Fig. 1 (i) illustrates the flexibility of the pressure sensor matrix.

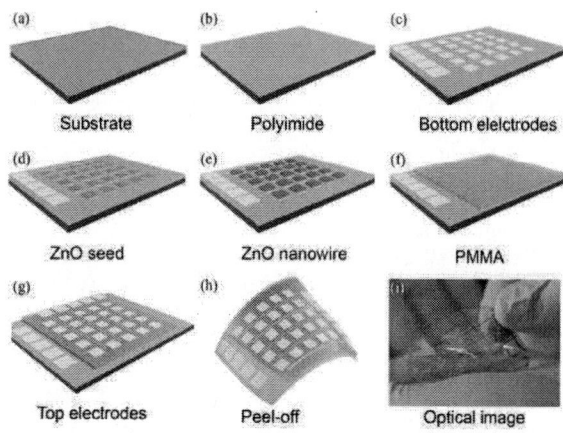

Fig. 1 The schematic diagrams for the fabrication process and optical image of the self-powered flexible pressure sensor matrix. (a) 2 inch glass substrate. (b) PI film on glass substrate. (c) Deposition of patterned bottom electrodes by sputtering. (d) Deposition of patterned ZnO seed. (e) Selective growth of ZnO NWs by hydrothermal method. (f) Spin-coating of PMMA to encapsulate the ZnO NWs. (g) Deposition of patterned top electrodes to form the sensor matrix. (h) Delamination of the flexible sensor matrix on PI film from the glass substrate. (i) Optical image of the pressure sensor matrix on PI substrate.

III. RESULTS AND DISCUSSIONS

Fig. 2 (a) and (b) shows the X-ray diffractions (XRD) patterns for the ZnO seed and ZnO NWs on silicon substrates, respectively. Except for the diffraction peaks from the silicon substrate, an obvious diffraction peak centered at around 34.3°, indexed to (002) plane of wurtzite structured ZnO, is observed for both samples, which indicates that both ZnO NWs and ZnO seed grown with preferential c-axis orientation. Besides, the diffraction peak for ZnO NWs is more intensive than that of ZnO seed, which is also an evidence for the well-oriented ZnO NWs in c-axis. In addition, as indicated by the scanning electron microscope (SEM) images in Fig. 2 (c)-(f), uniform and dense ZnO NWs only grew on the patterned ZnO seed and all the unit cells in the matrix are isolated from each other.

Furthermore, from the high-magnification SEM image, most of the ZnO NWs are aligned vertically and exhibit hexagonal cross-sections with a diameter ranging from 200 nm to 300 nm. Therefore, from XRD and SEM results, a high-resolution ZnO NWs matrix can be constructed using this method and the obtained NWs are single-crystal wurtzite structured ZnO NWs with a preferential growth in [002] direction. As a result, a self-powered pressure sensor matrix can be possibly realized by virtue of the piezoelectric effect of the wurtzite structured ZnO NWs, whose working mechanism can be described by the transient flow of inductive charges driven by the piezopotential [14]. In brief, when ZnO NWs are exposed to a compressive stress, a piezopotential field is formed along the direction of NWs. Due to the electrostatic force, inductive charges are established on the top and bottom electrodes. Once the external stress is removed, piezopotential field will disappear. Therefore, a voltage pulse can be measured responding to a mechanical deformation [15].

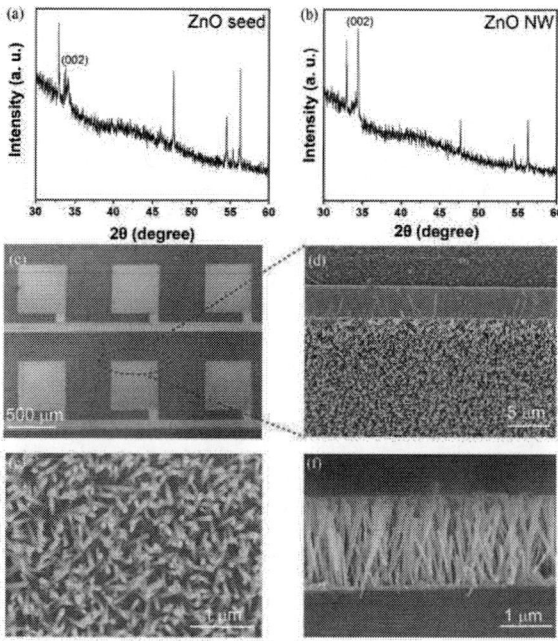

Fig. 2 Crystallinity and morphology of the ZnO NWs. XRD diffraction patterns for (a) ZnO seed and (b) ZnO NWs on silicon substrates. (c) SEM image of one segment of pressure sensor matrix. The sensing unit was well-defined by photolithography and patterned growth of ZnO NWs was realized. (d) Clear boundary between the NWs and bare PI was observed. (e) High-magnification SEM image shows the vertically oriented ZnO NWs with a diameter of 200 nm to 300 nm. The NWs grows densely but isolated from each other. (f) The cross-sectional view indicates that the thickness of the ZnO seed is about 50 nm and the length of the ZnO NWs is around 2 μm.

In order to optimize the sensing performance, it is crucial to study the geometry and density of ZnO NWs using a finite element method (FEM). As illustrated by the SEM images in Fig. 3 (a), Fig. 2 (c), and Fig. 3 (b) for ZnO NWs grown by hydrothermal method for one hour, four hours, and seven hours, respectively, ZnO NWs with a larger dimension are obtained with the increase of the growth time. However, the shape remains to be hexagonal structure and the ratio of length to diameter for the NWs keeps to be about 10 for the

978-1-6654-3008-1/21 $31.00 © 2021 IEEE

all the NWs. With regard to the NW density, it is controlled by the crystallinity of the ZnO seed. Therefore, the sensing performance can be investigated theoretically by adjusting the diameter, length, and space of the ZnO NWs as defined in Fig. 3 (c), which corresponds to the geometry and density of the NWs. In the FEM simulation, the piezoelectric effect is described by the following equations and vanishing normal stress at the boundary is considered.

$$T = C_E S - eE$$
$$D = e^T S + \epsilon_s E$$

Where T is the stress vector, D is the electric flux density vector, S is the strain vector, E is the electric field vector, C_E is the elasticity matrix, e is the piezoelectric stress vector, and ϵ_s is the dielectric matrix.

The related parameters of the materials used in FEM are presented in Table 1. The coefficients in elasticity matrix, piezoelectric stress vector and dielectric matrix of typical ZnO NWs were used in the simulation [16].

Table 1: Material parameters used for numerical FEM.

Materials	Density (kg/m³)	Dielectric Constant	Poisson's Ratio	Young's Modulus (GPa)
PI	1300	/	0.3	3.1
ZnO NW	5680	10.204	/	/
PMMA	1190		0.3	4
Au	19300	1	0.44	9

Fig. 3 (d) shows the dependence of the output voltage on NW density, which is determined by the space between the neighboring NWs in the simulation. It is clear that an increase in the output voltage is obtained with the space between the NWs and the increasing rate becomes smaller when the space is larger than 700 nm regardless of the NW diameter. Thus, in the study of the influence of the NW geometry, length and diameter of the NW, on the output voltage, a fixed space of 700 nm between the neighborng ZnO NWs was used as shown in Fig. 3 (e). This process corresponds to the growth time of the NWs, which has a significant influence on the sensing performance according to the theoretical analysis. In detail, a dramatic increase in the output voltage was observed before the maximum output voltage of about 100 mV was achieved with the NW diameter being 220 nm for a given pressure of 50 kPa. With the further increase of the diameter, a decrease in the output voltage was obtained and dropped to 40 mV when the diameter reaching to 800 nm. The theoretical response of the sensor to various pressure is investigated and summarized in Fig. 3 (f) with the diameter of the ZnO NW of 200 nm and space between neighboring ZnO NW of 700 nm. From the simulation result, an excellent linearity can be obtained in the pressure ranging from 10 kPa to 200 kPa, which is preferable for a pressure sensor.

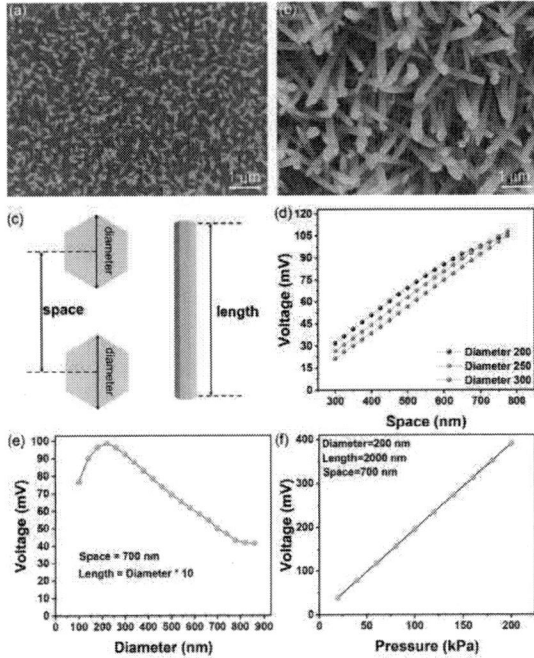

Fig. 3 The geometry and density of NWs with various growth time and the optimization of the sensing response. (a) and (b) show the SEM images for the ZnO NWs grown by hydrothermal method for one hours and seven hours respectively. (c) Definition of the parameters used in the FEM simulation. (d) The influence of space on the output voltage with various diameters of the NWs for a pressure of 50 kPa. (e) The dependence of the output voltage on the NW diameter. A maximum voltage of about 100mV was obtained with a NW diameter of 220 nm. (f) The sensing performance of the self-powered pressure sensor.

In addition, in order to confirm the reliability of the simulation results, the dependence of FEM results on the number of mesh was investigated as shown in Fig. 4. It is clear that when the number of mesh is larger than 80,000, the output voltage is almost independent on the mesh number.

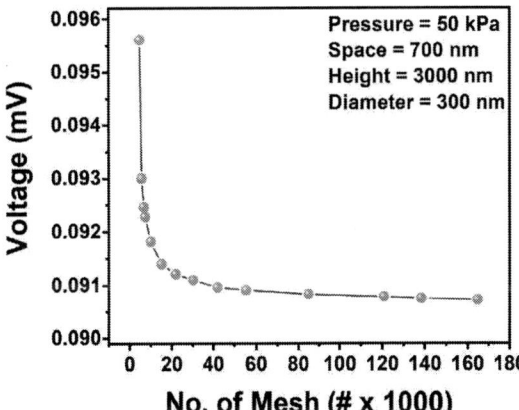

Fig. 4 Dependence of the output voltage on mesh number with an applied pressure of 50 kPa. The diameter and space of the ZnO NW are 300 nm and 700 nm respectively.

After the optimization of the geometry and density of the ZnO NWs, a pressure sensor matrix was fabricated and the sensing performance of the device is to be studied. As indicated in the inset of Fig. 5 (f), a weight was gently dropped on the tip of the cantilever placed close to a unit of the sensor to exert the dynamic force during the measurement and the corresponding output voltage from the sensor was monitored. From Fig. 5 (a) and (b), there is negligible change when the applied pressure increase from 9.6 kPa to 15.9 kPa. However, when the pressure increases from 15.9 kPa to 47.8 kPa as shown in Fig. 5 (c), (d), and (e), the output voltage increase steadily. The results indicate that the sensor shows a good linearity in the pressure ranging from 15 kPa to 50 kPa as confirmed by the linear fitting curve in Fig. 5 (f), which is consistent with the simulation analysis presented in Fig. 3 (f). In addition, a fast response time of about 50 ms was observed as representatively demonstrated by the response of the sensor to the pressure of 47.8 kPa in the inset of Fig. 5 (f). These results imply that the flexible pressure sensor matrix has a potential to be used to monitor the physiological activities in the areas of human-machine interface and health care with high spatial resolution.

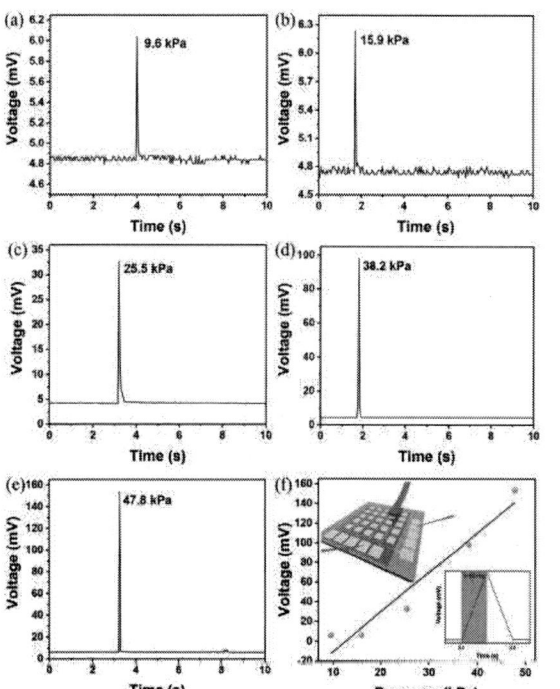

Fig. 5 Sensing performance of the pressure sensor matrix. (a)-(e) Response of the ZnO NWs based sensor to various pressure exerted by different weights. (f) Relationship between the output voltage and applied pressure. A good linearity was observed in the pressure ranging from 15 kPa to 50 kPa.

IV. CONCLUSION

In this research, a flexible pressure sensor matrix with a density of 100/cm^2 was designed and fabricated on PI substrate. An FEM was employed to optimize the sensing performance of the devices. The maximum output voltage can be obtained with the NW diameter of 220 nm and space between the neighbors of 700 nm in theory, which corresponds to the growth time of four hours. The fabricated

sensor exhibits a linear response in the pressure range between 15 kPa and 50 kPa. Benefitting from the piezoelectric effect of ZnO NWs, the device is able to be used to monitor dynamic pressure change with a high spatial resolution without any power supply. Therefore, the proposed pressure sensor matrix shows a potential to be applied in the areas of healthcare and human-machine with no power consumption.

REFERENCES

[1] W. Gao *et al.*, "Fully integrated wearable sensor arrays for multiplexed in situ perspiration analysis," *Nature,* vol. 529, no. 7587, pp. 509-514, 2016.

[2] N. Matsuhisa *et al.*, "Printable elastic conductors with a high conductivity for electronic textile applications," *Nature communications,* vol. 6, p. 7461, 2015.

[3] M. Park, Y. J. Park, X. Chen, Y. K. Park, M. S. Kim, and J. H. Ahn, "MoS$_2$-based tactile sensor for electronic skin applications," *Advanced Materials,* vol. 28, no. 13, pp. 2556-2562, 2016.

[4] H. Kudo, T. Sawada, E. Kazawa, H. Yoshida, Y. Iwasaki, and K. Mitsubayashi, "A flexible and wearable glucose sensor based on functional polymers with Soft-MEMS techniques," *Biosensors and Bioelectronics,* vol. 22, no. 4, pp. 558-562, 2006.

[5] S. Gong *et al.*, "A wearable and highly sensitive pressure sensor with ultrathin gold nanowires," *Nature Communications,* vol. 5, no. 1, pp. 1-8, 2014.

[6] L. Pan *et al.*, "An ultra-sensitive resistive pressure sensor based on hollow-sphere microstructure induced elasticity in conducting polymer film," *Nature Communications,* vol. 5, no. 1, pp. 1-8, 2014.

[7] J. J. Park, W. J. Hyun, S. C. Mun, Y. T. Park, and O. O. Park, "Highly stretchable and wearable graphene strain sensors with controllable sensitivity for human motion monitoring," *ACS Applied Materials & Interfaces,* vol. 7, no. 11, pp. 6317-6324, 2015.

[8] Y. R. Jeong, H. Park, S. W. Jin, S. Y. Hong, S. S. Lee, and J. S. Ha, "Highly stretchable and sensitive strain sensors using fragmentized graphene foam," *Advanced Functional Materials,* vol. 25, no. 27, pp. 4228-4236, 2015.

[9] M. Liu *et al.*, "Large-area all-textile pressure sensors for monitoring human motion and physiological signals," *Advanced Materials,* vol. 29, no. 41, p. 1703700, 2017.

[10] X. Wang, Y. Gu, Z. Xiong, Z. Cui, and T. Zhang, "Silk-molded flexible, ultrasensitive, and highly stable electronic skin for monitoring human physiological signals," *Advanced Materials,* vol. 26, no. 9, pp. 1336-1342, 2014.

[11] A. V. Tran, X. Zhang, and B. Zhu, "The development of a new piezoresistive pressure sensor for low pressures," *IEEE Transactions on Industrial Electronics,* vol. 65, no. 8, pp. 6487-6496, 2017.

[12] A. Kaneko, N. Asai, and T. Kanda, "The influence of age on pressure perception of static and moving two-point discrimination in normal subjects," *Journal of Hand Therapy,* vol. 18, no. 4, pp. 421-425, 2005.

[13] F. Sun *et al.*, "Bioinspired Flexible, Dual-Modulation Synaptic Transistors toward Artificial Visual Memory Systems," *Advanced Materials Technologies,* vol. 5, no. 1, p. 1900888, 2020.

[14] G. Zhu, A. C. Wang, Y. Liu, Y. Zhou, and Z. L. Wang, "Functional electrical stimulation by nanogenerator with 58 V output voltage," *Nano Letters,* vol. 12, no. 6, pp. 3086-3090, 2012.

[15] Y. Hu, Y. Zhang, C. Xu, G. Zhu, and Z. L. Wang, "High-output nanogenerator by rational unipolar assembly of conical nanowires and its application for driving a small liquid crystal display," *Nano Letters,* vol. 10, no. 12, pp. 5025-5031, 2010.

[16] I. Kobiakov, "Elastic, piezoelectric and dielectric properties of ZnO and CdS single crystals in a wide range of temperatures," *Solid State Communications,* vol. 35, no. 3, pp. 305-310, 1980.

Proceedings of the 16th Annual IEEE International
Conference on Nano/Micro Engineered and Molecular Systems
April 25-29, 2021

Design and Preparation of a Microfluidic System
for Stretching of Cells in Topographic Microstructures

Yingning He, Yuqian Yang, Yexin Gu, Tianjiao Mao, Yue Yu, and Jiandong Ding*

Abstract— **In this work, a new method was proposed to construct an elastic membrane with specially designed micropillar arrays and integrate the membrane into a microfluidic chip. RFP-transfected human mesenchymal stem cells were seeded in the micropillar arrays. Cyclic stretching of cells in the topographic microstructures resulted in different cell activities (cell spreading, orientation, and migration). The cell activities were enhanced with a proper level of spatial constraint on cells. This study explored a new approach to spatially influence cell behaviors.**

I. INTRODUCTION

Interaction between cells and their microenvironment is a fundamental topic in tissue engineering and biomaterials [1, 2]. Among these interactions, mechanotransduction is the process that cells sense and respond to mechanical stimuli [3, 4]. It is well known that mechanical cues can affect cell proliferation, differentiation, and apoptosis, thereby regulating various physiological activities, such as cardiomyocyte orientation, osteogenesis, and angiogenesis [5, 6]. In addition, many human tissues and organs (such as blood vessels, muscles, lungs, heart, and intestine) are affected by periodic stretching stimulation. Therefore, studies of cells on or inside a material under mechanical stimulation is especially important for biomedical and biomaterial research.

Studies have shown that cell behaviors, such as morphology, proliferation, and differentiation [7-10] can be greatly modulated by cyclic cell stretching. Among the related research, cyclic stretching can be generally categorized into two kinds: two-dimensional (2D) stretching of cells on an elastic polymer material and three-dimensional (3D) stretching of cells in a biocompatible matrix [11, 12]. 2D and 3D stretching lead to cell responses in different ways. In the 2D cell stretching, cells orient perpendicularly to the stretching direction [13-18], which is the direction with the least physical disturbance. In the 3D cell stretching, cells tend to orient to the direction that is "parallelly" to the stretching direction [19-22]. Besides, in a 3D matrix, the cell behaviors such as differentiation is normally faster than cells on a 2D substrate [23-25]. In general, the 2D cell stretching is simpler and relatively easy to be investigated as a model system, but the 3D cell stretching mimics in vivo cell

behavior that may have a better influence on biomaterial research.

Is it possible to own both advantages by using a 2D elastic membrane and, meanwhile, mimicking the 3D cell adhesion as it was in a matrix? We present a novel approach to create topographic micropillars on an elastic membrane so that a cell can be constrained by a micropillar array, forming a quasi-3D cell adhesion. By stretching the elastic membrane, the quasi-3D stimuli could be delivered from the micropillars to the cell.

II. EXPERIMENTAL

A. Fabrication of micropillars on an elastic membrane

A positive photoresist (AZ ECI 3012, MicroChem) was spin-coated on a silicon wafer at a speed of 2000 rpm, and then the photoresist was baked at 90 °C for 90 s. A photomask was used to selectively expose the baked photoresist on a mask aligner (MA6, Karl Suss) equipped with a UV light source. The process was followed by a post-bake at 90 °C for 90 s. The wafer with photoresist was then immersed in a TMAH-based developer for 60 s. The photoresist pattern with a thickness of ~2 μm was then used as the mask of dry etching.

The parameters of dry etching were personalized according to the etching area and depth. The dry etching resulted in an array of cylindrical microholes with a diameter of ~5 μm and a depth of ~10 μm in the silicon wafer. After removing the photoresist with oxygen plasma (800 W, 5 min), the silicon wafer was immediately vapor-deposited with a release agent (1H, 1H, 2H, 2H-perfluorooctyltrichlorosilane, Alfa Aesar). The silicon templet with microhole array was thus ready to use.

The poly(dimethyl siloxane) (PDMS, Dow Corning Sylgard 184) base elastomer and curing agent were mixed at a ratio of 10 : 1, degassed under vacuum for 2 h, and spin-coated (3000 rpm, 60 s) on the silicon template. After curing at 70 °C for 6 h, the micropillar layer was prepared on silicon for further process.

B. Integration of micropillars into a quasi-3D microfluidic device

For photolithography, SU-8 (Microchem 2150) was used as a negative photoresist. SU-8 prepolymer was spin-coated

This work was financially supported by National Natural Science Foundation of China (grant No. 51803032, 2191101058), National Key R&D Program of China (grant No. 2016YFC1100300), and Science and Technology Commission of Shanghai Municipality (grant No. 17JC1400200).

State Key Laboratory of Molecular Engineering of Polymers, Department of Macromolecular Science, Fudan University, Shanghai, 200433, China.

*Contacting Author: Jiandong Ding is a Distinguished Professor and the Director of State Key Laboratory of Molecular Engineering of Polymers. (phone: +86-21-31243506; e-mail: jdding1@fudan.edu.cn).

978-1-6654-3008-1/21 $31.00 © 2021 IEEE

on a silicon wafer at a low speed of 500 rpm for 10 s and then a high speed of 1500 rpm for 30 s. The wafer was baked on a hot plate at 65 °C for 7 min, and 95 °C for 80 min. A photomask was used to selectively expose the baked SU-8 for 6 min (wavelength: 300-436 nm, energy: 1050 μW cm-2). The process was followed by a post-bake at 95 °C for 25 min. The wafer with SU-8 was then immersed in propylene glycol monomethyl ether acetate for 25 min. The SU-8 templet was derived after rinsed with isopropanol and deionized water, and dried with nitrogen, each triplicate. The SU-8 templet was treated with oxygen plasma (100 W, 90 s) and immediately vapor-deposited with the release agent, 1H, 1H, 2H, 2H-perfluorooctyltrichlorosilane (Alfa Aesar). The SU-8 template with a thickness of about 300 μm was ready to use.

PDMS (Dow Corning Sylgard 184) components were mixed at a 1:10 ratio and poured on the SU-8 template in a 4-inch crystallizing dish (33 g and 8.8 g, for a thick and a thin PDMS microchannel layer, respectively). After proper degassing, the mixture was placed on a 70 °C hot plate for 12 h. The PDMS rubber was demolded from the templates and finely cut. The thick PDMS microchannel layer was punched holes for the creation of microfluidic ports.

The microfluidic chip was encapsulated by plasma bonding of the four layers, which are a thick microchannel layer, a membrane with micropillars, a thin microchannel layer, and a glass layer. All plasma bonding processes used the parameters of 100 W for 90 s and cured at 70 °C for 15 min. The thin microchannel layer was aligned and bonded to the membrane with micropillars. The bonded product was then aligned and bonded with the thick microchannel layer. The chip was fixed on a glass slide by plasma bonding. A PDMS etchant (tetrabutylammonium fluoride: 1-Methyl-2-pyrrolidinone = 1 : 3) was injected into the side-channels for 1 min to dissolve the unwanted membrane.

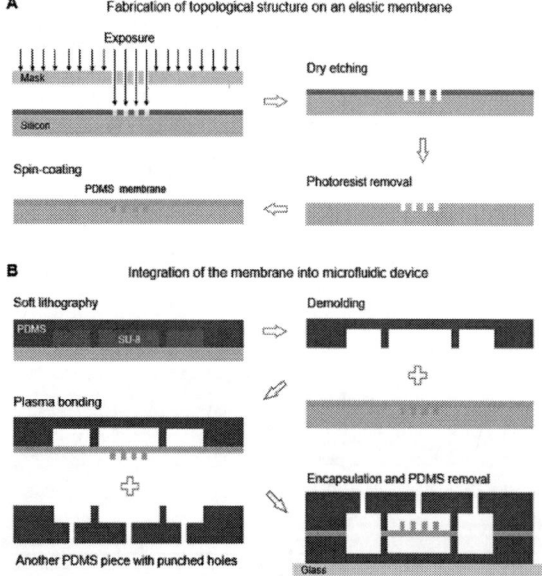

Figure 1. (A) Fabrication of micropillars on a PDMS membrane. (B) Integration of the membrane with micropillars into a microfluidic chip.

C. Personalized stretching system and cell stretching experiments

The stretching platform included a computer-controlled pressure controller, a syringe pump, and a live-cell imaging system (Figure 2). The pressure controller (OB1, Elveflow) was connected to a vacuum pump with an ultimate vacuum of -100 kPa as the source of negative pressure and a supply of high purity nitrogen at a pressure of 200 kPa as the source of positive pressure. The syringe pump (NE-1800, New era) supplied culture medium and PBS to the chip's upper and lower culture chambers. The live-cell imaging system consisted of a live-cell culture system equipped with a CO_2 generator and a heating chamber, and an inverted fluorescence microscope (Axiovert 200, Zeiss).

For the cell stretching experiments, microfluidic chips were sterilized under ultraviolet light for 30 min, rinsed with 75% alcohol for 15 min, and washed with phosphate-buffered saline. The RFP-tranfected hMSCs were digested by 0.25% trypsin / ethylenediaminetetraacetic acid. The cells were resuspended at a density of 2×10^5 cells ml^{-1}, and then injected into the chip. After incubation at 37 °C for 4 h, the cells were well adhered to in the micropillars. The microfluidic chip was connected to stretching system and it was placed on the stage of the inverted fluorescence microscope.

During the cell stretching, the culture medium was supplied by the syringe pump with a flow speed of 1 μL min^{-1}. Cyclic stretching was motivated by negative pressure controlled by the pressure controller. Snapshots of cells were automatically captured every 10 min with phase-contrast imaging by Zeiss's free Zen software.

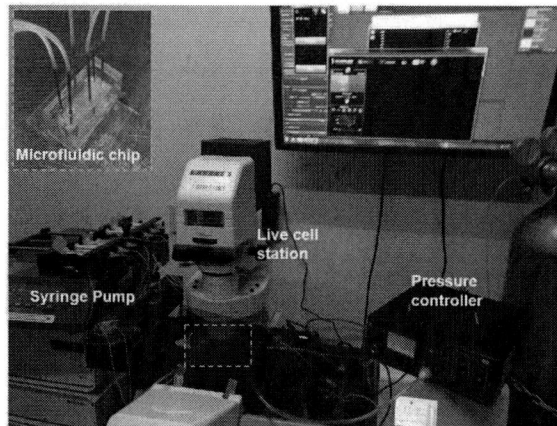

Figure 2. Microfluidic stretching system.

III. RESULTS AND DISCUSSIONS

A. As-fabricated microfluidic chip

Figure 3A shows a bonded PDMS chip and, on the right side, a zoom-in image of micropillars located between two microchannel walls. A small slice (~1 mm thick) was cut from the chip with a blade, and the cross-section of the microfluidic chip was exposed as shown in Figure 3B. The elastic membrane with micropillars was sandwiched between the two microchannel layers. The zoomed-in image of the elastic membrane shows that the micropillars were formed on the elastic membrane (~20 μm thick). Each micropillar had a diameter of ~5 μm and a height of ~10 μm. Three kinds of micropillars were designed to form three kinds of microwells with diameters of 10 μm (each with 6 micropillars), 25 μm (each with 12 micropillars), and 40 μm (each with 18 micropillars).

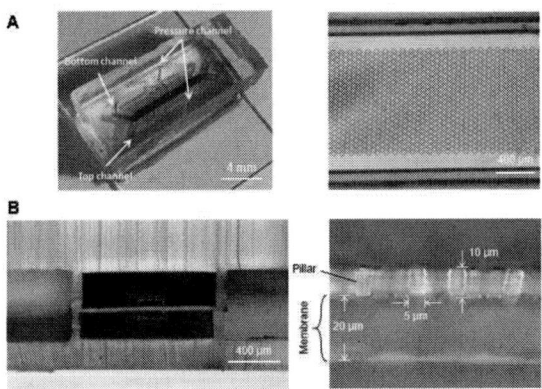

Figure 3. (A) A photo of the as-fabricated microfluidic chip. The right figure shows a microscopic image of the micropillars. (B) Cross-section of the microfluidic chip. The left figure shows a membrane located between two microchannel layers. The right figure shows a zoom-in image of the micropillars and the membrane.

3.6. Cell activities during cyclic stretching

We investigated the cell responses (cell spreading, orientation, and migration) to dynamic topographic surfaces, which are cyclic stretching of cells in small-well array (S), middle-well array (M), large-well array (L), and flat surface (F). The cell behaviors were significantly altered during 10 h cyclic stretching (1 Hz, 10%), as the experimental results are shown in Figure 4.

The apparent spreading of cells was obtained by counting the number of spreading cells ($N_{\text{spreading}}$) and round cells (N_{round}) through

$$f = \frac{N_{\text{spreading}}}{N_{\text{spreading}} + N_{\text{round}}} \tag{1}$$

In the three micropillar arrays (S, M, and L), the apparent spreading decreased within 0-2 h of stretching (5-8%), while the decrease for F is not obvious (~1%). During the stretching of 2-10 h, the apparent spreading reached steady states with the values of 38% (S), 59% (M), 52% (L), and 75% (F).

The cell orientation was characterized with order parameter, which was calculated by

$$R = \sqrt{\langle \sin 2\theta \rangle^2 + \langle \cos 2\theta \rangle^2} \tag{2}$$

where θ is the angle of a cell determined by marking the long axis of a cell with respect to the stretching direction.

Before stretching, the cells were oriented randomly, while during stretching, the order parameters raised and

reached steady states with the values of 0.41 (S), 0.71 (M), 0.63 (L), and 0.86 (F).

Cell migration was significantly influenced by the micropillar arrays during cyclic stretching. To quantify the cell migration, we calculated migration velocity through

$$V(\text{contour}) = \sum_{i=0}^{n} \frac{l_i}{t} \tag{3}$$

where l is the contour length and t is the record time.

The results of migration velocity are $V = 10$ μm h^{-1} (S), 29 μm h^{-1} (M), 24 μm h^{-1} (L), 76 μm h^{-1} (F).

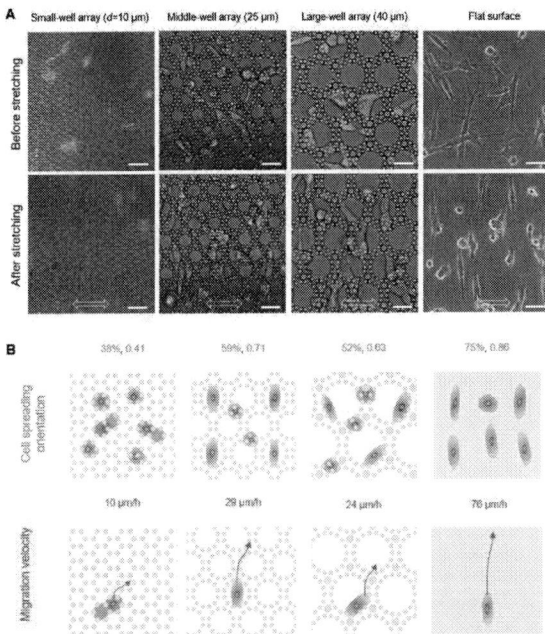

Figure 4. (A) Cell stretching experiments of RFP-hMSCs in three kinds of micropillar arrays and on flat surface. The double-sided arrows in the bottom indicated the stretching directions. Scale bars: 30 μm. (B) Schematic illustration of cell behaviors under cyclic stretching.

IV. CONCLUSION

We present a novel microfluidic chip with a stretchable topographic microstructure to construct the quasi-3D adhesion state of the cells and performed cyclic stretching stimulation. An elastic PDMS membrane with micropillars was fabricated and integrated into a microfluidic chip. A microfluidic system was built to support the chip with precise control of air pressure and fluid supply. Cyclic stretching of the cells in different micropillar arrays led to different levels of cell spreading, orientation, migration velocity. Moderate-size microwells surrounded by polymer micropillars triggered the strongest cell responses. This study has provided a useful tool for stretching cells in a quasi-3D microenvironment and revealed complicated effects of topological features of biomaterials on cells.

REFERENCES

[1] Y. Sun, C. S. Chen and J. Fu, Forcing stem cells to behave: a biophysical perspective of the cellular microenvironment *Annu Rev Biophys* 41 519-42, 2012

[2] J. J. Green and J. H. Elisseeff, Mimicking biological functionality with polymers for biomedical applications *Nature* **540** 386-94, 2016

[3] X. Yao, R. Peng and J. Ding, Cell–material interactions revealed via material techniques of surface patterning *Adv Mater* **25** 5257-86, 2013

[4] B. Ladoux and R. -M. Mege, Mechanobiology of collective cell behaviours *Nat Rev Mol Cell Biol* **18** 743-57, 2017

[5] T. Mammoto, A. Mammoto and D. E. Ingber *Annual Review of Cell and Developmental Biology, Vol 29*, ed R Schekman pp 27-61, 2013

[6] K. A. Jansen, D. M. Donato, H. E. Balcioglu, T. Schmidt, E. H. J. Danen and G. H. Koenderink, A guide to mechanobiology: Where biology and physics meet *Biochim Biophys Acta* **1853** 3043-52, 2015

[7] S. -H. Kook, H. -J. Lee, W.-T. Chung, I.-H. Hwang, S.-A. Lee, B.-S. Kim and J.-C. Lee, Cyclic mechanical stretch stimulates the proliferation of C2C12 myoblasts and inhibits their differentiation via prolonged activation of p38 MAPK *Mol Cells* **25** 479-86, 2008

[8] A. Livne, E. Bouchbinder and B. Geiger, Cell reorientation under cyclic stretching *Nat Commun* **5** 3938, 2014

[9] Y. Cui, F. M. Hameed, B. Yang, K. Lee, C. Q. Pan, S. Park and M. Sheetz, Cyclic stretching of soft substrates induces spreading and growth *Nat Commun* **6** 6333, 2015

[10] N. F. Jufri, A. Mohamedali, A. Avolio and M. S. Baker, Mechanical stretch: physiological and pathological implications for human vascular endothelial cells *Vasc Cell* **7** 8, 2015

[11] B. D. Riehl, J. H. Park, I. K. Kwon and J. Y. Lim, Mechanical stretching for tissue engineering: two-dimensional and three-dimensional constructs *Tissue Eng Part B Rev* **18** 288-300, 2012

[12] C. Sears and R. Kaunas, The many ways adherent cells respond to applied stretch *J Biomech* **49** 1347-54, 2016

[13] A. Tondon, H.-J. Hsu and R. Kaunas, Dependence of cyclic stretch-induced stress fiber reorientation on stretch waveform *J Biomech* **45** 728-35, 2012

[14] S. Jungbauer, H. Gao, J. P. Spatz and R. Kemkemer, Two characteristic regimes in frequency-dependent dynamic reorientation of fibroblasts on cyclically stretched substrates *Biophys J* **95** 3470-8, 2008

[15] T. Mao, Y. He, Y. Gu, Y. Yang, Y. Yu, X. Wang and J. Ding, Critical frequency and critical stretching rate for reorientation of cells on a cyclically stretched polymer in a microfluidic chip, *ACS Appl Mater Interfaces*, accepted for publication

[16] M. Rabbani, M. Tafazzoli-Shadpour, M. A. Shokrgozar, M. Janmaleki and M. Teymoori, Cyclic Stretch Effects on Adipose-Derived Stem Cell Stiffness, Morphology and Smooth Muscle Cell Gene Expression *Tissue Eng Regen Med* **14** 279-86, 2017

[17] Y. He, T. Mao, Y. Gu, Y. Yang, and J. Ding, A simplified yet enhanced and versatile microfluidic platform for cyclic cell stretching on an elastic polymer Biofabrication 12 045032, 2020

[18] N. Y. Hui, B. Pingguan-Murphy, A. A. Abbas, A. M. Merican and T. Kamarul, Uniaxial Cyclic Tensile Stretching at 8% Strain Exclusively Promotes Tenogenic Differentiation of Human Bone Marrow-Derived Mesenchymal Stromal Cells *Stem Cells Int* **2019** 9723025, 2019

[19] P. C. Dartsch and E. Betz, RESPONSE OF CULTURED ENDOTHELIAL-CELLS TO MECHANICAL STIMULATION *Basic Res Cardiol* **84** 268-81, 1989

[20] A. Nieponice, T. M. Maul, J. M. Cumer, L. Soletti and D. A. Vorp, Mechanical stimulation induces morphological and phenotypic changes in bone marrow-derived progenitor cells within a three-dimensional fibrin matrix *J Biomed Mater Res A* **81A** 523-30, 2007

[21] L. Krishnan, C. J. Underwood, S. Maas, B. J Ellis, T. C. Kode, J. B. Hoying and J. A. Weiss, Effect of mechanical boundary conditions on orientation of angiogenic microvessels *Cardiovasc Res* **78** 324-32, 2008

[22] W. Zhang, C. W Kong, M. H. Tong, W. H. Chooi, N. Huang, R. A. Li and B. P. Chan, Maturation of human embryonic stem cell-derived cardiomyocytes (hESC-CMs) in 3D collagen matrix: Effects of niche cell supplementation and mechanical stimulation *Acta Biomater* **49** 204-17, 2017

[23] H. Baharvand, S. M. Hashemi, S. Kazemi Ashtian and A. Farrokhi, Differentiation of human embryonic stem cells into hepatocytes in 2D and 3D culture systems in vitro *Int J Dev Biol* **50** 645-52, 2006

[24] C. Hildebrandt, H. Bueth, S. Cho, Impidjati and H. Thielecke, Detection of the osteogenic differentiation of mesenchymal stem cells in 2D and 3D cultures by electrochemical impedance spectroscopy *J Biotechnol* **148** 83-90, 2010

[25] R. Edmondson, J. J. Broglie, A. F. Adcock and L. Yang, Three-Dimensional Cell Culture Systems and Their Applications in Drug Discovery and Cell-Based Biosensors *Assay Drug Dev Technol* **12** 207-18, 2014

Stochastic analysis of the elctrothermal microactuator with fabrication error

Haotian Liu, Hao Chen, Yun Cao, Weirong Nie, Zhanwen Xi, and Shenghong Lei

Abstract— **In order to study the characteristic of the electrothermal microactuator with machining error, the actuator is first fabricated by deep reactive-ion etch (DRIE) technology. Key sizes of the fabricated actuators are measured by microscope. According to the distribution feature of the sizes, normal distribution is chosen and fitted to depict the fabrication error. Based on the joint normal distribution in the sizes, 1000 sample points are decided by Monte Carlo method. Meanwhile, FEM model about the thermal actuator is established so that the static displacement and maximum temperature of the sample points are obtained. With the data processing, the displacement and temperature also meet normal distribution. The displacement of the thermal actuator is more sensitive to the machining tolerance than temperature. The machining error has a great effect on the stiffness of the actuator.**

I. INTRODUCTION

The U-shaped electrothermal actuator, as shown in Fig 1, is composed of the hot beam, cold beam, flexure beam and anchor [1]. The uncertainty in its fabrication usually results in the degradation of the performance such as displacement. Even more the yield of the actuator also decreases. Therefore, stochastic analysis of the thermal actuator from machining error becomes important.

Much research work [2-4] on the machining error in the microelectromechanical system (MEMS) has been done. In [2], because of the manufacture tolerance existing in the sizes, such as width, thickness, sidewall angle of the spring and the gap between two electrodes, the relationship between the sizes and the threshold of the universal inertia switch was studied. Given the fabrication error in the sizes of the electrostatic actuator, the stochastic sample was done in [3]. The displacement of the actuator in each sample point was calculated by FEM analysis. Finally, the statistical feature of the displacement was obtained. An intrusive polynomial chaos expansions (IPCE) model was established [4], where the relationship between elastic modulus, residual stress and machining tolerance can be depicted precisely. With the fabrication errors, the probability density distribution of the elastic modulus and residual was achieved and validated by Monte Carlo method.

Here is the sketch of the paper: the fabrication process is presented in section 2; In section 3, the tolerance of sizes in the actuator is obtained and the normal probability density curve is fitted; the stochastic analysis based on the FEM model and Monte Carlo sampling method is depicted in section 4; in the section 5 is the conclusion.

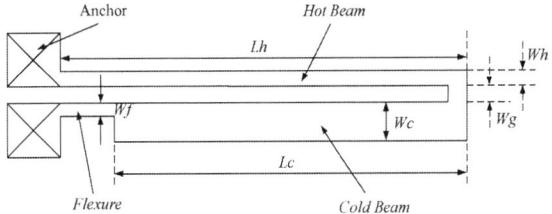

Figure 1. Sketch of the U-shaped thermal actuator

II. FABRICATION PROCESS

The U-shaped thermal actuator is made of polysilicon, where the deep reactive-ion etch (DRIE) technology [5] is applied. To begin with, the electrical resistivity $\rho 0$ of the sample (Si), heavily doped p-type single-crystal silicon wafer, is about 2e-4 $\Omega \cdot$m (Fig 2.a). Lithography is conducted in the backside of the sample (Fig 2.b). Then a trench (about 10 μm) is etched with KOH (Fig 2.c). The anodic bonding is applied to bond the silicon wafer and glass together (Fig 2.d). The sample is etched again to ensure its thickness 100 μm (Fig 2.e). A thin Ti/Au film is shaped on the silicon wafer (Fig 2.f). Lithography is done for the anchor of the actuator (Fig 2.g). The Ti/Au film, not covered with photoresist, is removed (Fig 2.h). The photoresist is also removed (Fig 2.i). Lithography is done for the actuator (Fig 2.j). The silicon wafer is etched with KOH and the actuator is shaped (Fig 2.k). The photoresist is removed with acetone solution (Fig 2.m).

---

*Resrach supported by the National Natural Science Foundation of China (No.51805268), the Fundamental Research Funds for the Central Universities (No.30920021101). (Corresponding author: Yun Cao.)

Haotian Liu is with School of Mechanical Engineering, Nanjing University of Science and Technology, Nanjing, China (e-mail: liuht1111@ 163.com).

Hao Chen, Yun Cao, Weirong Nie, Zhanwen Xi, and Shenghong Lei are with School of Mechanical Engineering, Nanjing University of Science and Technology, Nanjing, China.

978-1-6654-3008-1/21 $31.00 © 2021 IEEE

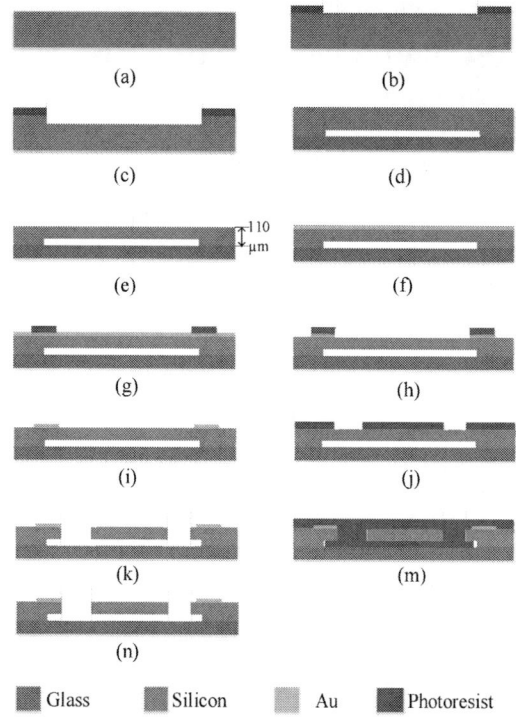

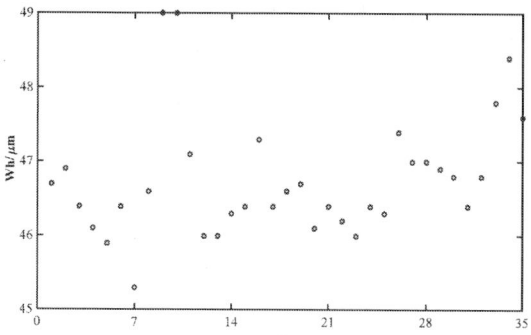

Figure 3. Width of the hot beam Wh in different samples

Figure 2. Process of the deep reactive-ion etch (DRIE) technology for thermal actuator

III. FABRICATION ERROR

In Fig 1, the width of the hot Wh and the gap Wg are smaller than the length of the beams in the U-shaped actuator. Hence, the relative machining errors in these two sizes have larger effect on the actuator's performance than that in the sizes of length. After the fabrication, 35 samples of the actuator are obtained. Those two sizes in these samples are measured by microscope, where 1 µm represents about 40 pixels. The measuring results of Wh and Wg in every sample are presented in Fig 3 and 4, respectively. From the figures, the sizes are mainly concentrated in some region. The number of the sizes will gradually decrease when the sizes are away from the region. Therefore, an assumption that Wh and Wg both satisfy normal distribution is made here. This is the characteristic feature of the normal distribution, as expressed in (1).

$$f(x) = \frac{1}{\sqrt{2}\sigma} e^{-2(x-u)^2/\sigma^2} \quad (1)$$

where u is the mean value and σ is the standard deviation. According to the measuring results, their first, second moment and the parameters of the fitting normal distribution are given in Table 1. The cumulative density curves of the fitting normal distribution of Wh and Wg are presented in Fig 5 and 6, respectively. From the figures, considering the few sample size, the cumulative density from the sample still has a good agreement with that from the fitting distribution. What is more, the dispersion degree in the distribution of Wh is larger than that of Wg.

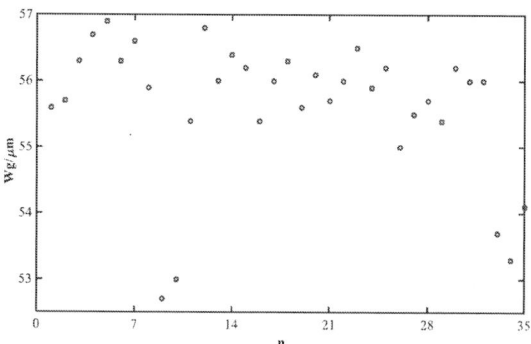

Figure 4. Width of the gap Wg in different samples

TABLE I. FIRST AND SECOND ORDER STATISTICAL MOMENT AND THE PARAMETERS OF THE FITTING NORMAL DISTRIBUTION

Parameter	Wh/µm		Wg/µm	
	Statistics	*Fitting*	*Statistics*	*Fitting*
mean	46.7	46.6	55.6	56.6
variance	0.66	0.4624	1.099	0.64

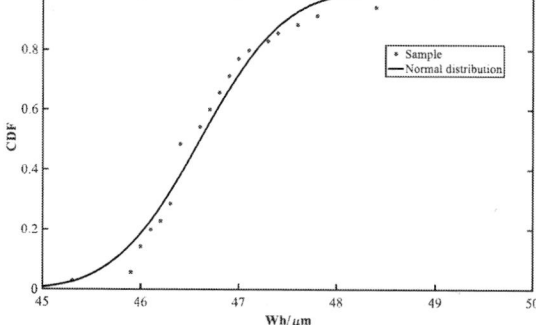

Figure 5. Cumulative density function (CDF) curve of the width of the hot beam

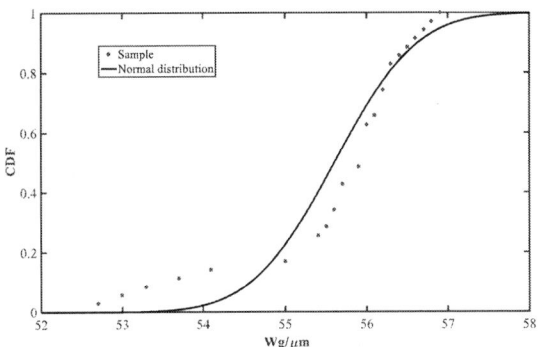

Figure 6. Cumulative density function (CDF) curve of the width of the gap

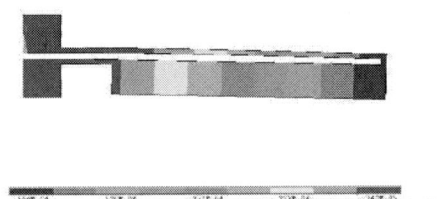

Figure 7. Displacement of the thermal actuator from FEM analysis

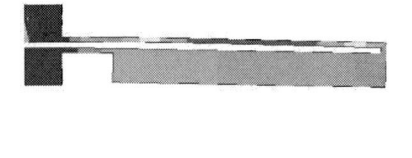

Figure 8. Temperature distribution of the thermal actuator from FEM analysis

IV. STOCHASTIC ANALYSIS

The performance of the thermal actuator, such as temperature and displacement, is studied by FEM analysis here. In the analysis, the geometric model of the actuator is established first while the material model is also built. Then a SLID226 20-node hexahedral coupled-field element is picked for the meshing. More, the constant temperature, voltage, displacement constraint and thermal convection condition are applied on the actuator. Finally, the FEM model is solved. The material parameters at room temperature is given in Table 2. The material property with different temperature can be obtained by [6]. The sizes of the actuator used in the analysis are given in Table 3. In the table, b is the thickness of the actuator and bg is the gap between the actuator and substrate. The static temperature and displacement of the thermal actuator from the FEM analysis are shown in Fig 7 and 8, respectively. From the figures, the maximum displacement lies in the end of the beam and the maximum temperature exists in the middle of the hot beam. When 18 V volt is applied on the actuator, its maximum displacement and temperature are about 66.8 μm and 861.1 K.

To further study the characteristic of the actuator with machining error, the sample points are picked first, where each point represents an actuator. Because of the linear independence between Wh and Wg, the covariance between them should be 0. The joint normal distribution can be expressed in (2).

$$f(x, y) = \frac{1}{\sqrt{2}\sigma_1\sigma_2} e^{-2(x-u_1)^2/\sigma_1^2 - 2(x-u_2)^2/\sigma_2^2} \qquad (2)$$

Where u_1 and u_2 are the mean value of Wh and Wg in the fitted normal distribution, the standard deviations σ_1 and σ_2 have similar definition. According to the joint normal distribution, 1000 sample points are decided by Monte Carlo method, as shown in Fig 9. The FEM analysis on the actuators with different Wh and Wg is done. According to the result from FEM analysis, the statistical results of temperature and displacement are presented in Fig 10 and 11, respectively. From the pictures, temperature and displacement almost meet the Gaussian distribution. A minor range of fluctuation (about 4 K) exists in the maximum temperature of the thermal actuators under fabrication error in Fig 10. In another word, the manufacturing tolerance has less influence on the temperature. Considering the larger span of the displacement in Fig 11, the machining error mainly changes the stiffness of the actuator, which affects the displacement of the actuator.

TABLE II. MATERIAL PARAMETERS OF POLYSILICON AT ROOM TEMPERATURE

Parameter	Symbol	Value	Unit
density	ρ_d	2300	kg·m^{-3}
thermal conductivity	k_0	146	W·m^{-1}·K^{-1}
electrical resistivity	ρ_0	2e-4	Ω·m
thermal convection	k_v	70	W·m^{-2}·K^{-1}
thermal expansion	α_0	2.56e-6	K^{-1}
young's modulus	E	1.1e11	GPa
poisson ratio	v	0.28	

TABLE III. VALUE OF THE SIZES IN THE THERMAL ACTUATOR FOR FEM ANALYSIS

Size	Wh	Wg	Wc	Wf
value/μm	46.9	55.7	325.2	46.9
Size	Lc	Lh	b	bg
value/μm	2698	503	100	10

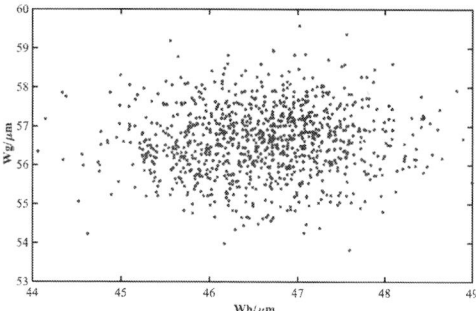

Figure 9. Sample points of Wh and Wg by Monte Carlo method

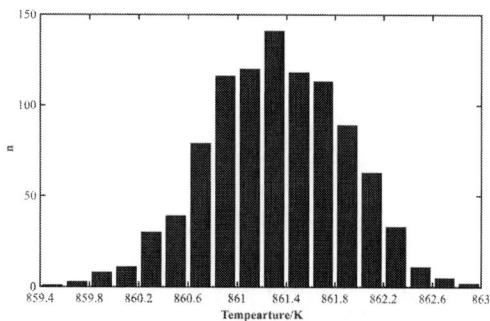

Figure 10. Maximum temperature distribution of the thermal actuator with fabrication error

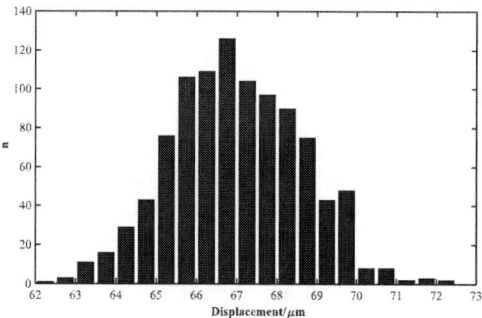

Figure 11. Displacement distribution of the thermal actuator with fabrication error

V. CONCLUSION

In this paper, the U-shaped actuator is fabricated by deep reactive-ion etch (DRIE) technology. By measuring the key sizes in the actuator, the characteristics of the normal distribution exists in them. Fitting normal distribution curves of Wh and Wg of the actuator are also obtained. Based on the Monte Carlo sampling method, the displacement and temperature of different actuators are acquired by FEM analysis. Under manufacturing tolearnce, the displacement and temperature of the actuators still meet normal distribution. What is more, the fabrication error exerts more influence on the displacement than on the temperature. That is to say the variation in the key sizes of the thermal actuator leads to the fluctuation of its stiffness, which further affects the static displacement of the actuator.

On the one hand, the fabrication error analysis gives access to quantifying tolerance effect on the performance of the actuator. On the other hand, the performance can be also improved when the fabrication error is taken into account in design. However, the assessment in this paper still needs more refinement to improve its accuracy and efficiency in the future.

ACKNOWLEDGMENT

The authors gratefully acknowledge the funding support from the National Natural Science Foundation of China (No.51805268), the Fundamental Research Funds for the Central Universities (No.30920021101).

REFERENCES

[1] Hussein, et al. "Dynamic electro-thermo-mechanical modelling of a U-shaped electro-thermal actuator." Journal of Micromechanics & Microengineering (2016).

[2] Yun Cao, Zhan-wen Xi, Jiong Wang, et al. "Influence of UV-LIGA fabrication error on performance of a MEMS ominidirectional inertial switch." Journal of Zhejiang University (Engineering Science) (2017).

[3] Kong, J. S., et al. "A methodology for analyzing the variability in the performance of a MEMS actuator made from a novel ceramic." Sensors & Actuators A 116.2(2004):336-344.

[4] Gao, Lili, Z. F. Zhou, and Q. A. Huang. "A Generalized Polynomial Chaos-Based Approach to Analyze the Impacts of Process Deviations on MEMS Beams." Sensors 17.11(2017):2561-.

[5] Bouhadda, I., et al. "Dynamic characterization of an electrothermal actuator devoted to discrete MEMS positioning." 2017:1-4.

[6] Pu, Xuan, W. Li, and Z. Zhou. "Electrothermal-driven gap adjustable MEMS comb structure: modeling and simulation of the equivalent circuit macromodel." Microsystem Technologies 20.6(2014):1205-1212.

Proceedings of the 16th Annual IEEE International
Conference on Nano/Micro Engineered and Molecular Systems
April 25-29, 2021

Dynamic Simulation of Nanogenerator Based on Finite Element Model Coupled with Lumped Parameter Elements*

Yao Chu, Ruixing Han, Huiliang Liu, Xiongying Ye, and Fei Tang

Abstract — Nanogenerator has demonstrated its capability as motion, displacement and pressure sensor in various sensing scenarios. This work presents the first time-dependent finite element modeling of nanogenerator with resistive load. Compared with previous research on the theoretical study and modeling of nanogenerator, three distinctive advances have been achieved: (1) the finite element modeling of the nanogenerator is established and coupled with the lumped-parameter elements for resistive load; (2) the time-dependent study is conducted rather than the quasi-static study based on parametric sweep to reveal the transient process; (3) the results of finite element simulations are compared with the derived results based on the simplified theoretical model to reveal the influence of fringe effect.

I. INTRODUCTION

Triboelectric nanogenerators (TENG), as a promising candidate in IoT applications, can act as not only energy harvesters but also self-powered sensors. There are several configurations for TENG to convert relative motion to electric signal, and meanwhile, to transform the mechanical energy to electric energy, including the vertical contact-separation mode, single-electrode mode, lateral sliding mode and freestanding triboelectric-layer mode. Among them, the contact-separation mode is the most widely adopted one for motion and pressure sensing, of which the capability has been verified and demonstrated in previous work [1, 2]. For a fabricated TENG device, there still exists different setups for measurement and the feasible measurands include the open-circuit voltage, the short-circuit current, the transferred charge and the voltage across the resistive load. To study the output performance of TENG, a simplified model based on lumped-parameter elements is proposed [3-5] and the optimization of TENG as an effective power source can be conducted accordingly [6, 7]. For the sensing characteristics of TENG, such as sensitivity, linearity and fidelity, systematic studies are carried out for both the transient process [8] and the steady state [9] based on the simplified lump-parameter model (LPM). Besides the analytical or numerical study on the derived differential equations from the LPM, finite element model (FEM) is another way to study the performance of TENG. Compared with LPM, FEM is more accurate and closer to actual

situations. For example, the equivalent capacitance of TENG in LPM are simplified as infinite parallel plates without considering fringe effect, which can be, however, modelled and calculated in FEM. Thus, FEM has been used as an effective tool to verify the results of theoretical analysis [10]. Nevertheless, most of previous work solves the static or quasi-static problem rather than the time-dependent problem.

In this paper, the FEM of TENG is coupled with the lumped-parameter resistor in the external circuit to study the output voltage across load. The time-dependent output is solved to obtain the transient and steady-state responses under a sinusoidal input motion. The results of the coupled model are compared with that of LPM to not only verify the validity of the simplified theoretical model but also reveal the effect of fringe effect.

II. MODEL

The schematic of TENG in contact-separation mode for motion sensing is shown in **Figure 1**. The upper and lower plates both contain an electret layer attached on a metal layer. The two electret layers are placed facing to each other and a resistor is connected between the two electrodes in the external circuit. In this work, it is supposed that there are positive and negative charges of same amount on the upper and lower electret layers respectively. We also assume that the amount of charges on the electret layers keep constant without decay or increase during the whole working process. Then, when the upper and lower plates get close to or move far away from each other, the induced charges on the two electrodes will be redistributed and there will be current flowing back and forth in the external circuit, namely across the resistor. Thus, the relative displacement between the upper and lower plates can be detected accordingly by measuring the voltage across resistor R_L as the output. In this way, the TENG in contact-separation mode works as a motion/displacement sensor. In the following simulation and calculation, the input motion $x(t)$ is supposed to change sinusoidally with time as expressed in **Equation 1**.

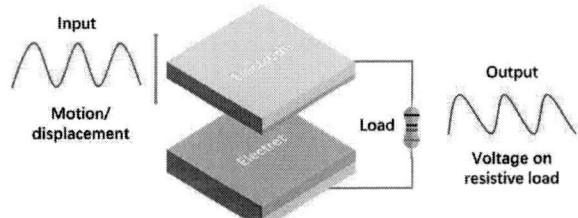

Figure 1: Schematic of the nanogenerator in contact-separation mode with resistive load as a motion sensor. The input is the relative motion between the upper and lower plates and the output is the voltage across the load resistor.

---

*Research partially funded by the National Natural Science Foundation of China (81873912, 51675305), China Postdoctoral Science Foundation (2020M670361), Natural Science Foundation of Beijing (3202011) and Shuimu Tsinghua Scholar Program.

Y. Chu, R. Han, X. Ye, and F. Tang are with State Key Laboratory of Precision Measurement Technology and Instruments, Department of Precision Instrument, Tsinghua University, Beijing, 100084, CHINA (corresponding author: F. Tang, e-mail: tangf@mail.tsinghua.edu.cn)

H. Liu is with the Institute of Telecommunication and Navigation Satellite, China Academy of Space Technology, Beijing, 100094, CHINA

978-1-6654-3008-1/21 $31.00 © 2021 IEEE

As illustrated in **Figure 2**, the LPM simplifies TENG as a variable capacitor C_{TENG} in series with a voltage source V_{oc}. The expression of C_{TENG} and V_{oc} as a function of input $x(t)$ is given in **Equations 2** and **3**, in which the representations of symbols are listed in **Table I**. In LPM, the output voltage is calculated by solving the differential equations (**Equations 4** and **5**) numerically and the computation is conducted with MATLAB.

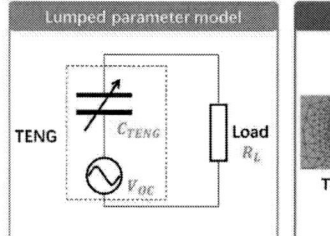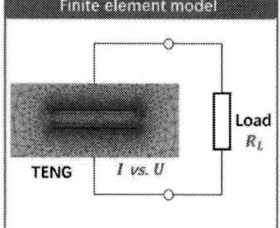

Figure 2: Comparison between the lumped parameter model and finite element model for nanogenerator with resistive load.

$$x(t) = A(1 + \sin(2\pi f t)) \tag{1}$$

$$C_{TENG}(t) = \frac{S\varepsilon_0\varepsilon_r}{2d_r + \varepsilon_r(d_0 + x(t))} \tag{2}$$

$$V_{oc}(t) = \frac{\sigma}{\varepsilon_0}(d_0 + x(t)) \tag{3}$$

$$V_{out}(t) = R_L \frac{dQ(t)}{dt} \tag{4}$$

$$R_L \frac{dQ(t)}{dt} = V_{oc}(t) - \frac{Q(t)}{C_{NG}(t)} \tag{5}$$

TABLE I. PARAMETER SETTINGS FOR MODELING.

Parameters	Symbols	Values
Surface area	S	$5 \times 5\ cm^2$
Thickness of electret	d_r	$300\ \mu m$
Thickness of electrode	d_m	$100\ \mu m$
Air gap	d_0	$2.7\ mm$
Amplitude of movement	A	$1\ mm$
Relative permittivity	ε_r	3.4
Charge density	σ	$8\ \mu C/m^2$
Load resistance	R_L	$10\ G\Omega$
Frequency of movement	f	$100\ Hz$

For the FEM method, the two-dimensional geometric model of TENG is constructed, and boundary conditions are set properly. The lower electrode of TENG is grounded and the upper one is left as a terminal to connect with a lumped-parameter resistor to implement the model of TENG in resistive-load mode. In the coupled model, the voltage between the two electrodes of TENG is regarded as a voltage source connecting with an ideal resistor. In this way, the FEM of TENG is combined with the circuit model of a resistor. The transient output voltage is computed based on the coupled model, which is established and solved in COMSOL Multiphysics. The default geometric parameters and material properties for modeling are listed in Table I.

III. RESULTS AND DISCUSSIONS

Assuming that the transferred charge through the resistor is zero at the starting point ($Q(0) = 0$), then the simulated results based on the coupled model of FEM and LPM, along with the calculated LPM results corresponding to $100\ Hz$ stimuli are shown in **Figure 3**. In general, the solution for both LPM and FEM roughly coincides with each other. The time-dependent output can be regarded as a superposition of an exponential decay and a periodical signal. The decay process is similar to a typical resistor-capacitor (RC) discharging curve. For the RC circuit, the resistor is the load R_L in the external circuit and the TENG acts as the capacitor. For the periodical signal, it changes synchronously with the sinusoidal input $x(t)$. These results are consistent with our previous work on the theoretical study of TENG of contact-separation mode with resistive load [8].

Comparing the two curves in Figure 3a, it can be noticed that the results of LPM is larger than that of FEM at the initial stage. From the enlarged details shown in Figure 1b, the steady-state output of LPM and FEM are in phase with each other but there is a deviation in amplitude. The amplitudes of periodical output simulated based on FEM is larger than that of LPM. It can be explained as the followings. In LPM, the equivalent capacitance of TENG is calculated as Equation 2 with the assumption of infinite parallel plates. The fringing effect is neglected in LPM and the charges are supposed to be distributed uniformly on the electrodes. In FEM, however, the distribution of charge on electrodes is no longer even. As illustrated in **Figure 4**, the large gradient of electric potential at the fringe indicates the increase and the nonuniformity of the electric field. With the neglecting of fringe effect, the equivalent capacitance of TENG in LPM is an underestimation of the actual one. Thus, the amplitude of output voltage on load resistor calculated by LPM is lower than that of FEM.

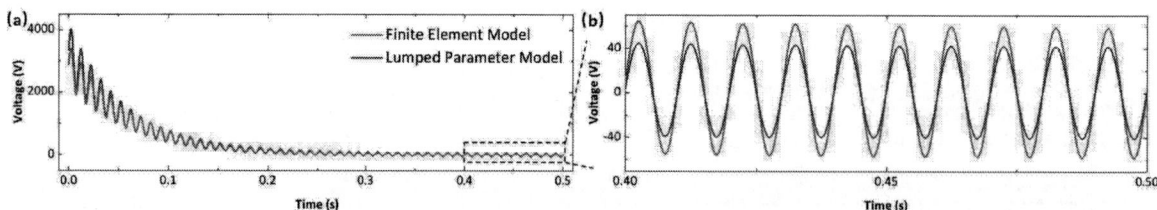

Figure 3: Simulated and calculated results of nanogenerator under sinusoidal input motion based on the coupled model containing FEM (red) and LPM (black), respectively, (a) for the overall process and (b) at the steady state. The parameters for modeling are set as the default values as listed in Table I.

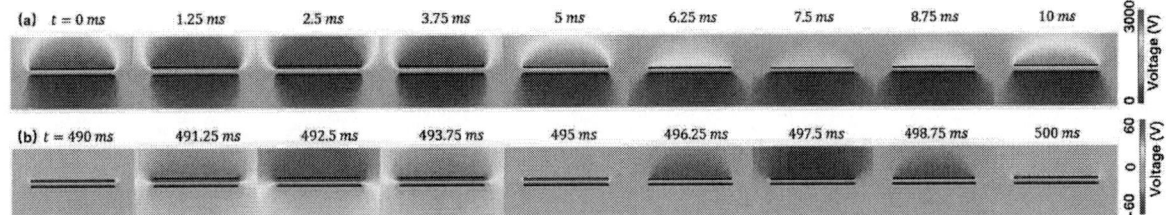

Figure 4: The potential profiles of the nanogenerator during one cycle of motion (100 Hz) at (a) the initial transient stage and (b) the steady state.

Figure 4 shows the potential profiles of the nanogenerator during one cycle of motion ($100\ Hz$) at the initial stage ($0 - 10\ ms$) and at the steady state ($490 - 500\ ms$). The colormap in Figure 4a and 4b is set with different scale to reveal the potential distribution more clearly. For the nine snapshots in one cycle, the upper plate starts from the central position (the first snapshot, $t = 0$ and $490\ ms$). Then it moves upwards until the farthest separation (the third snapshot, $t = 2.5$ and $492.5\ ms$). Afterwards it moves downwards through the central position (the fifth snapshot, $t = 5$ and $495\ ms$) and reaches the closest approach (the seventh snapshot, $t = 7.5$ and $497.5\ ms$). Finally, it moves upwards and return to the central position (the ninth snapshot, $t = 10$ and $500\ ms$). Comparing the first, fifth and ninth snapshots for both Figure 4a and 4b in which the upper plates are all at the same relative position to the lower plates, it can be observed that the potential profiles become weaker for the initial stage while those for the steady state are almost the same. The reason is that the output at the transient process is dominated by the fast decay while the output at steady state is periodic and changing synchronously with the input motion. In Figure 4b, it is illustrated that the potential of the upper plate changes from positive (red) to negative (blue) as the upper plate moves from higher than to lower than the central position. Such symmetry in output also follows the harmonic input motion.

Figure 5 presents the responses under different input frequencies and device sizes. The insets are the corresponding enlarged views at the steady state ($0.4 - 0.5\ s$). From Figure 5a, it can be observed that the decay processes are similar for all frequencies with the same device. The attenuation in the transient process is governed by the device parameters rather than the inputs. Such phenomenon can be explained as the followings. Referring to the expression of time constant for RC circuit ($\tau = RC$), τ is a characteristic parameter of the system itself, which is only related to the equivalent resistance and capacitance in the circuit, and is independent of the input frequency. From the inset of Figure 5, it can be noticed that the amplitude of outputs under different frequency is almost the same. According to our previous work [8], as the input frequency increases, the amplitudes of output will increase first and then saturates at a plateau. The cut-off frequency for entering the saturation can be calculated as **Equation 6**. For the device in Figure 5a, the cut-off frequency is about $2\ Hz$. Since the three frequencies in simulation ($10, 50$ and $100\ Hz$) are all much larger than the cut-off frequency, all of them are within the saturation zone and thus the output amplitudes are almost the same.

$$f_{cutoff} = \frac{d_0 + d_r/\varepsilon_r}{2\pi\varepsilon_0 R_L S} \tag{6}$$

For devices with different dimensions, as illustrated in Figure 5b, the larger one experiences slower attenuation. It is reasonable since the time constant for the decay process is proportional to the capacitance, namely to the surface area of the device. From the enlarged view in Figure 5b, the outputs from the four devices with various sizes are very close to each other. As the size of device increases 5.25 times (from $4\ cm^2$

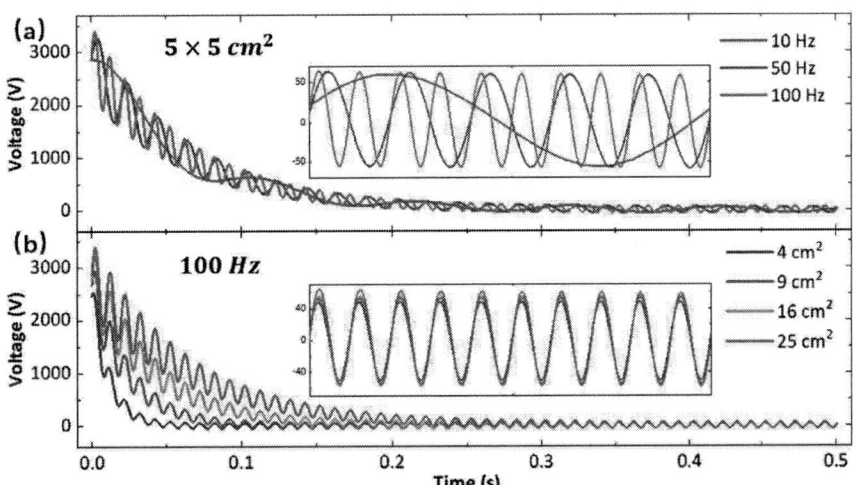

Figure 5: Simulated results based on the coupled model for different (a) input frequencies and (b) device sizes. Insets show steady-state outputs.

978-1-6654-3008-1/21 $31.00 © 2021 IEEE   1293

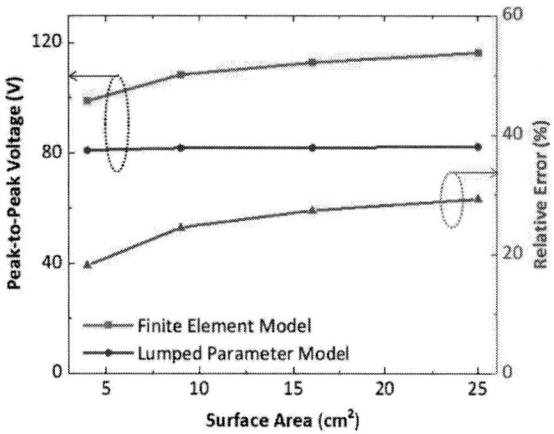

Figure 6: Comparison of output amplitudes at steady-state with various sizes.

to $25\ cm^2$), the peak-to-peak voltage at the steady state only increases about 17% (from $98.9\ V$ to $116.4\ V$). The reason is described as the followings. As the surface area increases, the equivalent capacitance of TENG rises as well. So the impedance of the capacitive component decreases, which results in an increase of the voltage on the resistive load. However, such improvement of output amplitude by increasing the device size is limited and will get saturated when the capacitive reactance is relatively small enough compared to the resistance.

Figure 6 presents the peak-to-peak amplitude of output at steady state with various device sizes. In general, the results of FEM are larger than that of LPM. As the surface area of the device increases, the peak-to-peak amplitudes of LPM saturates around the plateau of $82.4\ V$. For that of FEM, the amplitude grows in a slow-down manner as the surface area increases. As explained above, it can be inferred that the amplitude of output voltage solved by FEM will saturate with larger surface areas in a similar way as LPM does. The relative error between FEM and LPM increases from 18.0% to 29.2% as the surface area changes from $4\ cm^2$ to $25\ cm^2$. It is reasonable to infer that the relative error between the two methods will also become stable when the amplitude enters the saturation stage.

The relative error between FEM and LPM can be ascribed partially to fringe effect as well as the chamfering in the geometric model of FEM. It is worth noting that the relative error increases as the surface area gets larger, which seems contradictory with our intuitive thought that the assumption of infinite parallel plates gets closer to the real situations with the increasing of surface area and the FEM will approach to LPM. To look for the cause of relative error between FEM and LPM, further study will be conducted, including comparing the results of FEM in three dimensions with that in two dimensions, the settings of the interface between electrode and the electret layer, the influence of the thickness of metal layer on the result, and so on.

IV. CONCLUSION

In summary, a coupled model consisting of the FEM for TENG and the lumped-parameter resistor is established in this work and the time-dependent output is solved to obtain the transient and steady-state responses under a sinusoidal input motion. The consistency between the result of FEM and LPM verifies the validity of the simplified model and the difference reveals the fringe effect. This work provides a more accurate way for analyzing the sensing performance of TENG in complementation of LPM.

On the one hand, the coupled model of FEM and lump-parameter element proposed in this work is a successful attempt and an initial demonstration on solving time-dependent problem of TENG with external circuit using COMSOL. It is possible to conduct such analysis when more complicated components are added into the circuit, for example in the study of power management circuit for TENG. On the other hand, considering that the time and computing resource that the FEM consumes is much larger than that of LPM, it is meaningful to improve and optimize LPM for a quick and accurate calculation in the future. For example, the expression of equivalent capacitance considering fringing effect can be substituted into the LPM to improve its accuracy and make the result more comparable with that of FEM.

REFERENCES

[1] Ouyang, H., Tian, J., Sun, G., Zou, Y., Liu, Z., Li, H., Zhao, L., Shi, B., Fan, Y., Fan, Y., Wang, Z.L., and Li, Z.: 'Self-Powered Pulse Sensor for Antidiastole of Cardiovascular Disease', Adv. Mater. Processes, 2017, 29, (40)

[2] Chu, Y., Zhong, J., Liu, H., Ma, Y., Liu, N., Song, Y., Liang, J., Shao, Z., Sun, Y., Dong, Y., Wang, X., and Lin, L.: 'Human Pulse Diagnosis for Medical Assessments Using a Wearable Piezoelectret Sensing System', Adv. Funct. Mater., 2018, 28, (40), pp. 1803413

[3] Niu, S., Wang, S., Lin, L., Liu, Y., Zhou, Y.S., Hu, Y., and Wang, Z.L.: 'Theoretical study of contact-mode triboelectric nanogenerators as an effective power source', Energy & Environmental Science, 2013, 6, (12), pp. 3576

[4] Niu, S., Liu, Y., Chen, X., Wang, S., Zhou, Y.S., Lin, L., Xie, Y., and Wang, Z.L.: 'Theory of freestanding triboelectric-layer-based nanogenerators', Nano Energy, 2015, 12, pp. 760-774

[5] Niu, S., and Wang, Z.L.: 'Theoretical systems of triboelectric nanogenerators', Nano Energy, 2015, 14, pp. 161-192

[6] Zi, Y., Wu, C., Ding, W., and Wang, Z.L.: 'Maximized Effective Energy Output of Contact-Separation-Triggered Triboelectric Nanogenerators as Limited by Air Breakdown', Adv. Funct. Mater., 2017, 27, (24), pp. 1700049

[7] Zi, Y., Niu, S., Wang, J., Wen, Z., Tang, W., and Wang, Z.L.: 'Standards and figure-of-merits for quantifying the performance of triboelectric nanogenerators', Nat. Commun., 2015, 6, pp. 8376

[8] Chu, Y., Cao, Z., Xu, J., Zhou, J., Wang, S., Han, R., Feng, R., Ye, X., and Tang, F.: 'Theoretical study of nanogenerator with resistive load and its sensing performance as a motion sensor', Nano Energy, 2021, 81, pp. 105628

[9] Cao, Z., Chu, Y., Wang, S., and Ye, X.: 'Theoretical analysis of sensor properties of contact-separation mode nanogenerator-based sensors', Nano Energy, 2021, 79, pp. 105450

[10] Niu, S., Liu, Y., Wang, S., Lin, L., Zhou, Y.S., Hu, Y., and Wang, Z.L.: 'Theoretical Investigation and Structural Optimization of Single-Electrode Triboelectric Nanogenerators', Adv. Funct. Mater., 2014, 24, (22), pp. 3332-3340

Gap in pagination due to unavailable paper.

Pages 1295-1296

Proceedings of the 16th Annual IEEE International
Conference on Nano/Micro Engineered and Molecular Systems
April 25-29, 2021

All-Printed Flexible Tactile Sensors with High Sensitivity and Large Detection Range*

Qifan Ding, Huyue Chen, Wen-Ming Zhang, and Lei Shao

Abstract— This paper presents an all-printed flexible capacitive tactile sensor fabricated by a single desktop printer. The main components of the sensor are a soft silica gel dielectric layer sandwiched between two layers of silver electrodes. The fabrication process includes inkjet printing, used to deposit silver layer as the electrodes on polyethylene terephthalate (PET) substrates, and direct ink writing (DIW), used to print silica gel elastic microstructures as the dielectric layer. Cross-sectional micro-dome shaped silica gel lines are printed to improve the sensitivity for low pressure range and increase the detection range for high pressure range, similar to the responsivity behavior as human tactile sensation. As a result, the sensor has a high sensitivity of 1 kPa^{-1} below 0.4 kPa and a large detection range up to 2 MPa but with a low sensitivity of 0.0006 kPa^{-1}. Finally, loading cycles applied on the printed sensor shows good stability and small hysteresis.

I. INTRODUCTION

Flexible and stretchable tactile sensors have attracted a lot of research attention due to their applications in cooperative robots, wearable devices and biological sensors. To imitate functions of human skin, tactile sensors with high sensitivity and large detection range are highly desired to detect very small pressure level (similar as the force exerted by a bug resting on one's arm), and to measure large pressure (similar as human holding heavy objects) as well. Prior work has proved that dielectric layers with microstructures like pyramids [1-3], pores [4,5], domes [6] and other microscale patterned structures [7,8] can significantly improve the sensitivity of a capacitive sensor. Their fabrication process mostly involves micromachining such as lithography, etching and deposition. Although these conventional approaches are mature and has been widely used in micro-scaled devices, they also result in high costs, complex process, and long fabrication cycles.

To overcome these challenges, printing techniques like fused filament fabrication, inkjet printing, aerosol jet printing, direct ink writing (DIW) and electrohydrodynamic printing (E-jet) [9] have been developed to produce wearable and stretchable electronic products with low cost and high efficiency. For instances, Guo et al. developed a method to continuously print four layers of a tactile sensor with sinter-free ink but required a complex process using four independent nozzles and supporting structures [10]. Cai et al. printed silver nanowires as electrodes for stretchable capacitive sensors [11] and Peng et al. used inkjet printing to print Polydimethylsiloxane (PDMS) droplets as a dielectric layer on ITO/glass surface for a capacitive pressure sensor [12]. However, most of these achievements either show a low sensitivity or fabricated on stiff substrates, or still require fabrication processes other than printing.

In this work, we present a flexible capacitive tactile sensor which is fully printed using a single desktop printer, combining inkjet printed silver electrodes on polyethylene terephthalate (PET) substrate and DIW printed silica gel dielectric microstructures on electrodes. To improve sensitivity and at the same time simplify the fabrication process, the microstructure of the dielectric layer is designed to be cross-sectional microdome shaped lines. As a result, this sensor has a biomimetic nonlinear capacitance-pressure relationship with a high sensitivity up to 1 kPa^{-1} at a low-pressure range and a smaller sensitivity for a large linear detection range up to 2 MPa as well. In the repeated loading/unloading test, the sensor features small hysteresis and good cycling stability. With a performance comparable to or even better than those sensors fabricated by micromachining, the full printing production strategy shows promising results for soft and flexible devices.

II. MATERIALS AND FABRICATION

The fabrication process flow of the all-printed flexible capacitive tactile sensor is shown in Fig. 1. First, 10×10 mm thin-film electrodes are inkjet printed on a 50 μm-thick PET substrate by squeezing a reactive silver ink solution out of 20 μm diameter piezoelectric nozzles with a voltage of 25 V, followed by baking at 150 °C for 20 minutes. This solution (BASE-CP12, Prtronic) is obtained as the clear supernatant by mixing silver acetate, aqueous ammonium hydroxide and formic acid, followed by mixing with 2,3-butanediol (10% by volume) as a humectant and viscosifying aid. This preparation process can yield a long-term stable ink and can also prevent agglomeration of the silver nanoparticles during inkjet printing using an optimized custom ejection waveform applied to the piezoelectric. Its silver content is 10 ± 3 wt.%. After heat treatment, the organic solvent

---

*Research supported by the Shanghai Sailing Program of Shanghai Science and Technology Committee, China (Grant No. 19YF1425000), and the National Science Fund for Young Scientists of China (Grant No. 12002201).

Qifan Ding is with the University of Michigan–Shanghai Jiao Tong University Joint Institute, Shanghai Jiao Tong University, Shanghai, China 200240 (e-mail: ash_541@sjtu.edu.cn).

Huyue Chen is with the University of Michigan–Shanghai Jiao Tong University Joint Institute, Shanghai Jiao Tong University, Shanghai, China 200240 (e-mail: huyue_chen@sjtu.edu.cn).

Wen-Ming Zhang is with the School of Mechanical Engineering and the State Key Laboratory of Mechanical System and Vibration, Shanghai Jiao Tong University, Shanghai, China 200240 (e-mail: wenmingz@sjtu.edu.cn).

Lei Shao is with the University of Michigan–Shanghai Jiao Tong University Joint Institute, Shanghai Jiao Tong University, Shanghai, China 200240 (corresponding author, phone: +86-21-34206567-5421; fax: +86-21-34206545, e-mail: lei.shao@sjtu.edu.cn).

978-1-6654-3008-1/21 $31.00 © 2021 IEEE

evaporates and silver particles rapidly form, resulting in a highly conductive layer [13] with a volume resistivity around 3-10 μΩ · cm.

Second, elastic silica gel is patterned into lines with a certain periodicity on the bottom electrode using DIW printing using a 0.16 mm diameter syringe needle with a pressure of 100 kPa and at a speed of 1 mm/s, followed by baking at 80 °C for more than two hours. The silica gel we used for dielectric layer is a transparent viscous fluid with a viscosity of 15 Pa·s. After curing, the dielectric constant was measured to be 2.96 at 60 Hz. During DIW printing, a constant air pressure is applied to extrude the viscous ink while the syringe needle is controlled to move at a set speed along a designed pathway. We have explored several printing parameters and found that a larger syringe needle diameter, or a higher air pressure, or a lower speed, results in a larger linewidth of the printed silica gel line. We have also investigated the resulted sensitivity of the capacitive sensor as a function of the coverage percentage of silica gel, and concluded that the smaller this coverage is, the higher the sensitivity as long as enough silica gel is printed to prevent capacitor shortage. It is also worth noting that to improve the shape fidelity and increase the height of printed silica gel, the printer platform used to hold the PET substrate could also be heated during the ink dispensing process.

Third, another electrode, serving as the top electrode, is inkjet printed on a blank PET substrate as in the first step, and then aligned and bonded with the silica gel dielectric layer. Finally, wires are bonded to both the top and bottom electrodes for capacitance measurement. Fig. 2 presents photographs of the printed sensor, the silica gel dielectric layer, and a bent sensor by two fingers showing its good flexibility. The structural profile of the DIW printed silica gel layer with a periodicity of 800 μm is measured using a scanning surface profiler and it shows a dome-like cross-sectional geometry with a height of about 20 μm and a footprint of about 500 μm (Fig. 3).

III. RESULTS AND DISCUSSIONS

To study the capacitance-pressure relationship, an experimental setup is built, including a 3-axis manual stage, a linear translational stage (MTS25-Z8, Thorlabs), a 20 N force gauge with a resolution of 1 mN (HP-20, Handpi) and an LCR meter with 20,000 counts resolution and 0.2% basic accuracy (U1733C, Keysight), as shown in Fig. 4a. The force gauge is fixed on the translational stage whose motion can be controlled by a computer. The sensor is glued to the manual stage, and the LCR meter is used to measure the capacitance with a frequency of 100 kHz. The measured capacitance-pressure relationship is shown in Fig. 4b, where C_0 is the initial capacitance, and ΔC is the change in capacitance. A high sensitivity up to 1 kPa^{-1} for the initial 0.4 kPa range is exhibited. There exists a significant decrease in sensitivity at 2 kPa, and higher pressure leads to lower sensitivity. The sensitivity is 0.0006 kPa^{-1} in the range larger than 50 kPa. In addition, the sensor works well up to a pressure range of 2 MPa (data not shown here).

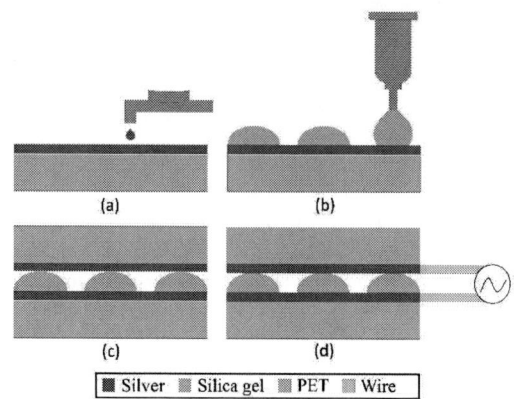

Figure 1. Fabrication process for all-printed flexible tactile sensor. (a) Inkjet printing of the electrode layer on PET with conductive silver solution ink. (b) DIW printing of elastic silica gel microstructures as the dielectric layer on the silver electrode. (c) Aligning and bonding of the structure in (b) with another top electrode as in (a). (d) Wire bonding.

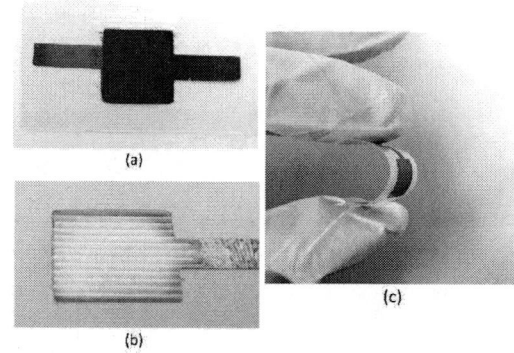

Figure 2. Photographs of (a) a printed tactile sensor, (b) dielectric layer dispensed on a bottom electrode by DIW printing and (c) a bent sensor.

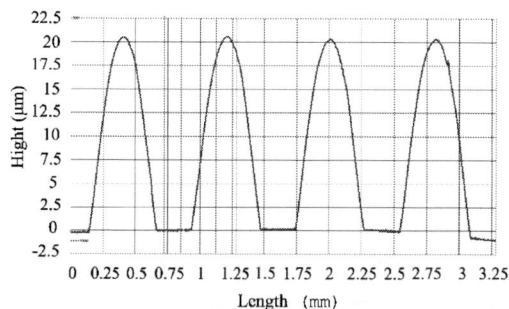

Figure 3. Height profile of the DIW printed dielectric layer measured by a stylus profiler.

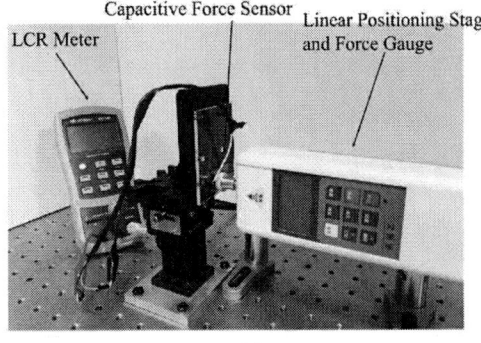

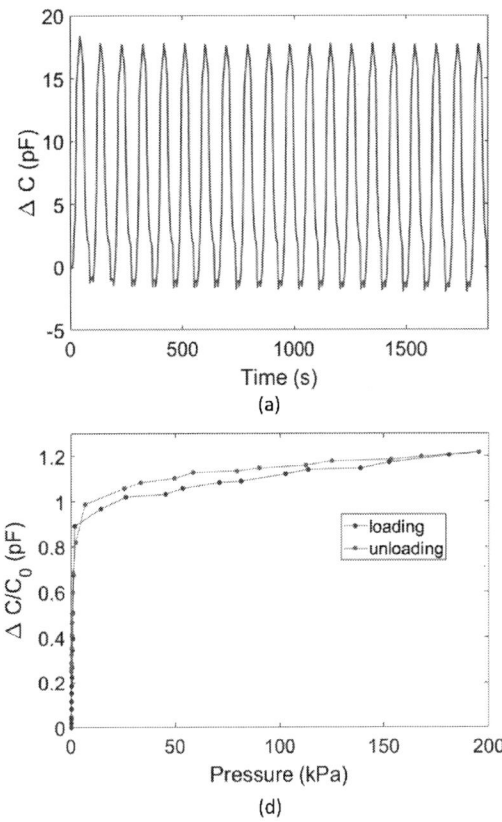

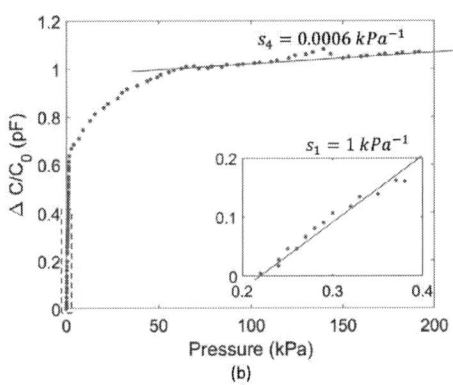

Figure 4. (a) Experimental setup for measuring capacitance and force/pressure. (b) Relative change of capacitance under normal pressure in the range of 0-200 kPa with the inset showing the low-pressure detection range for 0-0.4 kPa.

Figure 5. (a) Capacitance change for 20 cycles of repeated loading and unloading of 200 kPa. (b) Relative capacitance changes for one period of loading and unloading of 200 kPa.

Then, twenty cycles of repeated loading/unloading pressure between 0 to 200 kPa is carried out, and the relationship between capacitance and time is shown in Fig. 5a, demonstrating good stability and repeatability. Fig. 5b is one cycle of the sensor's capacitance-pressure record for loading and unloading of 200 kPa. The nearly symmetric response demonstrates small hysteresis.

The sensor shows a high sensitivity for a low pressure and a large pressure detection range with a much lower sensitivity. To demonstrate the benefit of this highly nonlinear capacitance response over pressure, weights of 2 g, 100 g, 500 g and 1000 g are applied on the sensor separately and the capacitance change is recorded in Fig. 6a. The result indicates that this sensor, like human skins, can on one side detect a gentle touch represented by a 2 g weight and, on the other side, identify heavy presses represented by a 1000 g weight. Clearly, small fluctuations in a large pressure level cannot be easily felt by human skins, which corresponds to the low sensitivity of the sensor for a large pressure. Furthermore, to explain the rapid decrease in sensitivity, Figs. 6b-6d depict the cross-sectional profile of a sensor under different pressure levels. Without pressure, the top surface of the silica gel is curved, which can be easily pressed into a flat surface under a low pressure, resulting in a fast increase in capacitance. However, further deformation

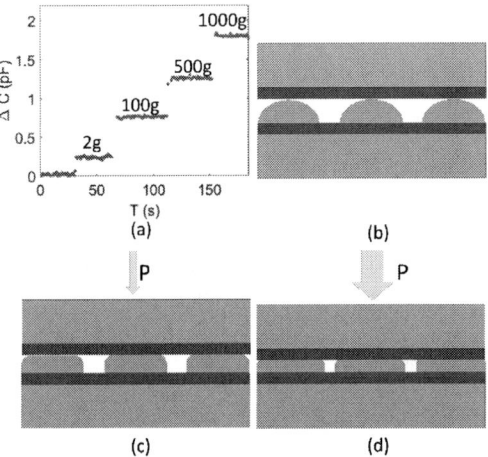

Figure 6. (a) Capacitance change of the sensor for an applied weight ranging from 2 g to 1000 g. Schematic deformation of the dielectric layer under (b) zero pressure, (b) a small pressure and (d) a large pressure.

978-1-6654-3008-1/21 $31.00 © 2021 IEEE

of the silica gel layer becomes precipitously harder as the dielectric structure is already squeezed to a shape with a flat top surface under high pressure, resulting in a much lower sensitivity, which in turn yields a large linear detection range.

IV. CONCLUSION

We have developed an all-printed flexible capacitive tactile sensor consisting of micro-dome cross-sectional dielectric lines sandwiched between two silver electrodes. Electrodes are fabricated by inkjet printing of ionic silver solution on PET substrates, while the dielectric layer is fabricated by DIW printing of silica gel directly on one of the silver electrodes. The sensor shows good stability and the curved top surface of the dielectric layer results in a high sensitivity at low pressure and also a large detection range as well. Such sensors can be printed by a single desktop printer with less costs and a significantly faster turnaround period, demonstrating its promising potential applications in artificial somatosensation and health monitoring.

REFERENCES

[1] G. Liang, Y. Wang, D. Mei, K. Xi, and Z. Chen, "Flexible Capacitive Tactile Sensor Array With Truncated Pyramids as Dielectric Layer for Three-Axis Force Measurement," Journal of Microelectromechanical Systems, vol. 24, no. 5, pp. 1510–1519, 2015.

[2] S. C. Mannsfeld, B. C. Tee, R. M. Stoltenberg, C. V. H. Chen, S. Barman, B. V. Muir, A. N. Sokolov, C. Reese, and Z. Bao, "Highly sensitive flexible pressure sensors with microstructured rubber dielectric layers," Nature materials, vol. 9, no. 10, pp. 859, 2010

[3] C. M. Boutry, M. Negre, M. Jorda, O. Vardoulis, A. Chortos, O. Khatib, and Z. Bao, "A hierarchically patterned, bioinspired e-skin able to detect the direction of applied pressure for robotics," Science Robotics, vol. 3, no. 24, 2018.

[4] B. Nie, J. Geng, T. Yao, Y. Miao, Y. Zhang, X. Chen, and J. Liu, "Sensing arbitrary contact forces with a flexible porous dielectric elastomer," Materials Horizons, 2021.

[5] D. Kwon, T. Lee, J. Shim, S. Ryu, M. Kim, S. Kim, T. Kim, and I. Park, "Highly sensitive, flexible, and wearable pressure sensor based on a giant piezocapacitive effect of three-dimensional microporous elastomeric dielectric layer," ACS Appl. Mater. Interfaces, vol. 8, no. 26, pp. 16922–16931, 2016.

[6] S. Wang, K. Huang and Y. Yang, "A Highly Sensitive Capacitive Pressure Sensor with Microdome Structure for Robot Tactile Detection," 2019 20th International Conference on Solid-State Sensors, Actuators and Microsystems & Eurosensors XXXIII (TRANSDUCERS & EUROSENSORS XXXIII), 2019, pp. 458-461

[7] Y. Wan, Z. Qiu, Y. Hong, Y. Wang, J. Zhang, Q. Liu, Z. Wu, and C. F. Guo, "A Highly Sensitive Flexible Capacitive Tactile Sensor with Sparse and High-Aspect-Ratio Microstructures," Advanced Electronic Materials, vol. 4, no. 4, 2018.

[8] J. Wang, R. Suzuki, M. Shao, F. Gillot, and S. Shiratori, "Capacitive Pressure Sensor with Wide-Range, Bendable, and High Sensitivity Based on the Bionic Komochi Konbu Structure and Cu/Ni Nanofiber Network," ACS Applied Materials & Interfaces, vol. 11, no. 12, pp. 11928–11935, 2019.

[9] K. Senthil Kumar, P.-Y. Chen, and H. Ren, "A review of printable flexible and Stretchable tactile sensors," Research, vol. 2019, pp. 1–32, 2019.

[10] S.-Z. Guo, K. Qiu, F. Meng, S. H. Park, and M. C. McAlpine, "3D Printed Stretchable Tactile Sensors," Advanced Materials, vol. 29, no. 27, p. 1701218, 2017.

[11] L. Cai, S. Zhang, Y. Zhang, J. Li, J. Miao, Q. Wang, Z. Yu, and C. Wang, "Direct Printing for Additive Patterning of Silver Nanowires for Stretchable Sensor and Display Applications," Advanced Materials Technologies, vol. 3, no. 2, p. 1700232, 2017.

[12] Y. Peng, S. Xiao, J. Yang, J. Lin, W. Yuan, W. Gu, X. Wu, and Z. Cui, "The elastic microstructures of inkjet printed polydimethylsiloxane as the patterned dielectric layer for pressure sensors," Applied Physics Letters, vol. 110, no. 26, 2017.

[13] H. W. Tan, J. An, C. K. Chua, and T. Tran, "Metallic Nanoparticle Inks for 3D Printing of Electronics," Advanced Electronic Materials, vol. 5, no. 5, 2019.

Heterogeneous Integration for RF application

Jian Zhu

Nanjing Electronic Devices Institute, Nanjing, 210096, P.R. China

ABSTRACT

Complementary metal oxide semiconductor (CMOS) offers ultimate advantages in terms of maturity, complexity and integration density among semiconductor technologies, especially in digital function applications. III-V compound semiconductor (CS) technology is dominating radio frequency (RF) applications due to their superior transport properties. CS devices offer better RF performance in power, efficiency, bandwidth, dynamic range and frequency, comparing to their Si based competitors. Therefore, integration of high performance CS circuits with CMOS is essential for a RF module with both optimum performance and compact size. Conventional printed circuit board (PCB) or low temperature co-fired ceramic (LTCC) based assemblies are cost-effective and reliable solutions for CS and CMOS integration. However, drawbacks in pattern resolution, stacking accuracy, and interconnect size restrict these techniques in higher frequency and more compact applications. Heterogeneous Integration technology is emerging as an attractive approach.

This talk will introduce our latest research activities on heterogeneous integration technology for RF application. A 38 GHz T/R module will be used as an example to illustrate the idea. A silicon based IPD structure combined with Cu TSV interposer is suitable for RF applications. Compound material based MMICs are integration through heterogeneous integration scheme. The processing techniques are compatible with the standard CMOS technology.

BIOGRAPHY

Jian Zhu is currently a Chief scientist at CETC. Dr. Zhu is a joint professor at Nanjing University. Dr. Zhu received her Ph.D. degree from Southeast University. Dr. Zhu is active in the field of RF MEMS and microsystem. She is fellow of Chinese Society of Micro-Nano Technology （CSMNT）, vice president of Chinese semiconductor Society MEMS board, board member of DTIP OF MEMS/MOEMS Technical Program Committee, and expert in national science foundation committee. She publishes over 130 papers and holds over 40 patents.

978-1-6654-3008-1/21 $31.00 © 2021 IEEE

Proceedings of the 16th Annual IEEE International
Conference on Nano/Micro Engineered and Molecular Systems
April 25-29, 2021

Chitosan/graphene oxide/MoS₂/AuNPs modified electrochemical sensor for trace mercury detection[*]

Ri Wang[1,2], Chenyu Xiong[1,2], Yong Xie [1,2], Mingjie Han [1,2], Yuhao Xu [1,2], Chao Bian [1,2], and Shanhong Xia [1,2]

Abstract——This study reports a novel electrochemical sensor to detect trace mercury based on an electrode coated by chitosan/graphene oxide/MoS₂/AuNPs and thymine-Hg^{2+}-thymine base pair. This method effectively utilizes the biocompatibility of molybdenum disulfide to DNA and constructs a detection platform for mercury ions. The electrocatalytic activity of MoS₂ and the specific binding of thymine-Hg^{2+}-thymine base pair significantly reduce the detection limit of mercury and improve the sensitivity of the electrode. When there exists Hg^{2+}, the voltammetric current signal of redox peak decreases after the interaction between mercury and DNA. The sensor showed good linearity from 0.01μg/L to 4μg/L with a correlation coefficient of 0.991 and the limit of detection was found to be 5.8ng/L. This simple, sensitive, low cost and easy preparation electrochemical sensor has the potential application in trace mercury detection.

Keywords——molybdenum disulfide, thymine-Hg^{2+}-thymine, electrochemical sensor, trace mercury detection

I. INTRODUCTION

Mercury ions (Hg^{2+}) in water environment have the characteristics of trace, high biological toxicity and serious pollution[1]. It is of great significance to detect Hg^{2+} to ensure water quality and water health. Cold vapor atomic absorption spectrometry (CV-AAS)[2], cold vapor atomic fluorescence spectrometry (CV-AFS)[3] and inductively coupled plasma mass spectrometry (ICP-MS)[4] are traditional detection methods, which have high sensitivity and low detection limit. However, these methods rely on large equipments, which are expensive, time-consuming and not portable. Several new methods have been studied to realize the miniaturization and improve the sensitivity of the Hg^{2+} sensors, such as fluorimetry[5], colorimetry[6], electrochemistry[7], surface plasmon resonance (SPR) [8] and surface-enhanced raman scattering (SERS)[9]. Among these methods, electrochemical sensors have been widely used in heavy metal ions detection due to its portability, low cost and easy operation. In order to realize trace mercury detection, some new materials such as graphene, MOFs, and nanoparticles have been studied to modify the surface of the sensing electrode to improve the performance of the sensors[10]-[12].

Molybdenum disulfide (MoS₂) is a kind of 2D transition metal dichalcogenides (TMDs) and has been widely studied due to its tunable electric properties, fluorescence quenching ability, electrocatalytic activity, and adsorption performance[13]-[16]. 2D MoS₂ nanosheets show a good biocompatibility with DNA, which makes it being able to connect with biological molecules[15]. In MoS₂-based electrochemical sensors, the MoS₂ nanosheets can be used not only as a substrate to immobilize aptamers, but also as an electrochemical catalyst to amplify signals[13].

Herein, we report a sensitive electrochemical biosensor to detect trace Hg^{2+} in water. The whole sensing system includes chitosan/graphene oxide (GO) layer, a monolayer

---

[*]Research supported by the National Basic Research Program of China (Grant No. 2015CB352100).

1. State Key Laboratory of Transducer Technology, Aerospace Information Research Institute, Chinese Academy of Sciences, Beijing 100190, China;

2. School of Electronic, Electrical and Communication

Engineering, University of Chinese Academy of Sciences, Beijing 100190, China.

Corresponding author: Shanhong Xia (phone: +86-10-5888-7180; e-mail: shxia@mail.ie.ac.cn)

978-1-6654-3008-1/21 $31.00 © 2021 IEEE

MoS$_2$, gold nanoparticles (AuNPs) and thiolated DNA probe (DNA$_P$). The chitosan/GO layer was used to increase the adhesion of a glassy carbon electrode (GCE). MoS$_2$ nanosheets can improve the electronic conductivity and electrochemical activity of the electrode. DNA$_P$ were connected to AuNPs on the surface of MoS$_2$ by Au-S bond. When Hg^{2+} exists, target DNA (DNA$_T$) in the test solution could connect with DNA$_P$ and form a double-stranded T-Hg^{2+}-T structure, which changed the oxidation current of K$_3$[Fe(CN)$_6$]. This method can enhance the electrocatalytic activity of the sensor and the chitosan/GO/MoS$_2$/AuNPs modified electrode shows an obvious response to Hg^{2+}. We successfully detected trace Hg^{2+} and reduced the detection limit effectively.

II. EXPERIMENTAL SECTION

A. Apparatus

In this study, the Gamry Reference 600 electrochemical workstation was used to experiment the electrochemical detection. Glassy carbon electrode, platinum electrode and Ag/AgCl (3M KCl, aq) electrode were used as working electrode (WE), counter electrode and reference electrode respectively.

B. Preparation of chitosan/GO/MoS$_2$/AuNPs modified electrode

The modified GCE was made as follows. Firstly, the mixture of chitosan and graphene oxide was dripped onto the GCE surface and dried. Then, MoS$_2$ dispersion was prepared by ultrasonication assisted exfoliated process[17]. Briefly, 30mg MoS$_2$ powder was dispersed in 30ml formamide solvent, and treated with ultrasonic water bath for 12h first. Then the dispersion was treated by ultrasonic probe for 30 min with a power of 125 W, MoS$_2$ nanosheets were obtained after centrifugation and then dispersed in ultrapure water. The clean electrode surface was modified by adding 10μL suspension and dried for 24h. AuNPs were synthesized by sodium citrate reduction method.

C. DNA modification

DNA oligonucleotides were obtained from Shanghai Sangon Biotechnology Co. Ltd. The DNA base sequences were shown as follow:

DNA$_P$:5'- CACTGCTTTTTTGGTCACAAAAAA - (CH$_2$)$_6$-SH-3'

DNA$_T$:5'- GTGACCATTTTTGCAGTGAAAAAA -3'

The DNA was centrifuged at 1000 rpm for 5 min and the dispersed in 50mM Tris-HCl buffer (pH = 7.4). TCEP was added to DNA$_P$ to activate sulfhydryl group and prevent the formation of disulfide bonds. Then, 10μL DNA$_P$ was modified on the surface of the GCE and dried at 57°C for 1h.

D. Electrochemical Measurements

In the work, electrochemical impedance spectroscopy (EIS) was used to study the conductivity of the sensor. The applied potential and ac voltage amplitude were set as 0.26 V and 5 mV respectively. The voltage frequencies was 10^6 Hz to 0.5 Hz. Differential pulse stripping voltammetry (DPSV) method was used for Hg^{2+} detection. The test solution included 0.1M KCl, 10mM K$_3$[Fe(CN)$_6$] and different concentrations of Hg^{2+}. The precondition was held at -0.7V for 300s. In order to clean the electrode surface, the WE was held at 0.6V for 60s before each scanning.

III. RESULTS AND DISCUSSION

A. EIS response of different electrodes

In order to characterize electrical performance of the electrode fabricated. The bare GCE and MoS$_2$ modified GCE were scanned by EIS in 50mM K$_3$[Fe(CN)$_6$]. As illustrated in Fig.1, the Nyquist diagram demonstrated that MoS$_2$ nanosheets enhanced the electrical conductivity of the GCE.

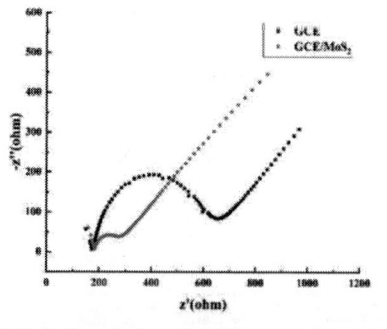

Fig.1. Nyquist diagrams of bare GCE and MoS2 modified GCE

B. Mechanism of mercury detection

Fig.2 indicates the mechanism of the electrochemical sensor to detect Hg^{2+} based on T-Hg^{2+}-T base pair. MoS_2 nanosheets show a good biocompatibility with DNA and the AuNPs above helps to fix DNA_P. AuNPs also played a synergistic catalytic effect. When there exists Hg^{2+} in the test solution, DNA_T will connect with DNA_P through Hg^{2+}, forming a double-stranded structure of T-Hg^{2+}-T on the electrode surface. The formation of double chain changed the sensing environment of the electrode surface, which affected the redox ability of the electrode to $K_3[Fe(CN)_6]$. The concentration of Hg^{2+} was detected by monitoring the peak current of $K_3[Fe(CN)_6]$ in the solution.

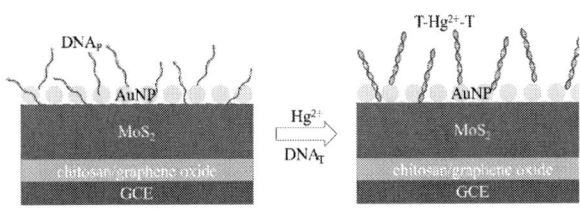

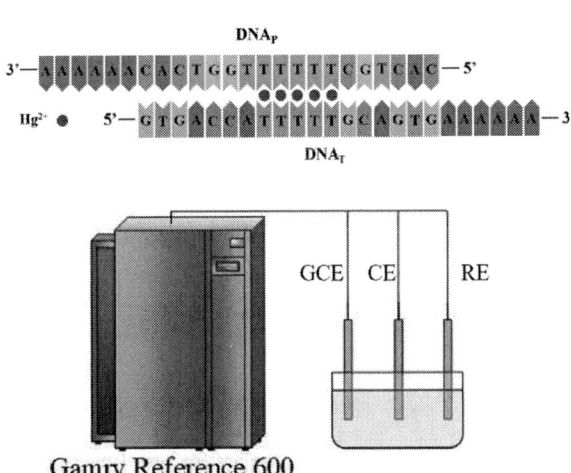

Fig.2 Schematic diagram of mercury ion detection mechanism

C. Mercury determination by DPSV

In this experiment, DPSV was used to characterize the current change of $K_3[Fe(CN)_6]$. As can be seen from Fig.3, in the absence of Hg^{2+}, an obvious oxidation peak can be observed at around 0.25V. When Hg^{2+} was added, the peak value decreases, which indicated that the DNA_T in the solution was successfully connected to the DNA_P on the

electrode surface through Hg^{2+} and formed a double-stranded structure. With the increase of Hg^{2+} concentration, the peak value further decreased. Fig.4 is the correlation curve between Hg^{2+} concentration and $K_3[Fe(CN)_6]$ peak current. The linear response range was from 0.01μg/L to 4μg/L with a correlation coefficient of 0.991 and the LOD was 5.8ng/L (S/N=3).

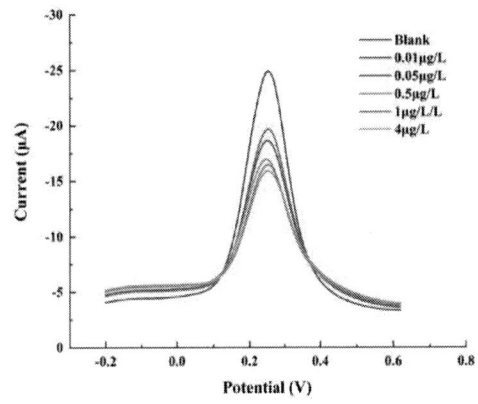

Fig.3 DPV plots of 10mM $K_3[Fe(CN)_6]$ for different concentration of Hg^{2+} in the range of 0-4μg/L

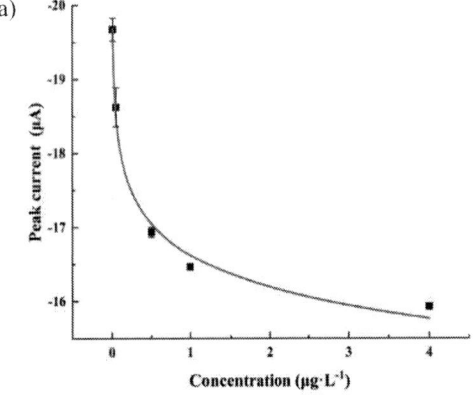

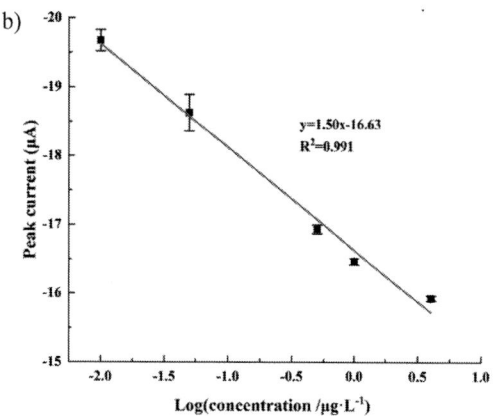

b)

$y=1.50x-16.63$
$R^2=0.991$

Fig.4 a) The curve of relationship between Hg^{2+} concentration and $K_3[Fe(CN)_6]$ current peak; b) the linear range of peak current vs. the logarithm of Hg^{2+} concentration

IV. CONCLUSION

In this study, we have fabricated a chitosan/GO/MoS$_2$/AuNPs modified GCE to detect trace mercury ions in water. The modified MoS$_2$ is well compatible with DNA and forms a Hg^{2+} sensing platform. MoS$_2$ nanosheets improved electrochemical activity of the sensor and amplified the electrochemical signal. By forming the T-Hg^{2+}-T structure on the surface of the electrode, trace Hg^{2+} detection was realized. The sensor showed a good linearity from 0.01μg/L to 4μg/L with a correlation coefficient of 0.991. The limit of detection was found to be 5.8ng/L. This electrochemical sensor is sensitive, low cost, simple, easy to operate and has the potential application in trace mercury detection.

REFERENCE

[1]. Li P , Feng X , Qiu G , et al. Mercury exposures and symptoms in smelting workers of artisanal mercury mines in Wuchuan, Guizhou, China[J]. Environmental Research, 2008, 107(1):108-114.

[2]. Hight, S. C, Cheng J . Determination of total mercury in seafood by cold vapor-atomic absorption spectroscopy (CVAAS) after microwave decomposition[J]. Food Chemistry 2005, 91, (3):557-570.

[3]. Zhang R, Peng M, Zheng C, et al. Application of flow injection-green chemical vapor generation-atomic fluorescence spectrometry

to ultrasensitive mercury speciation analysis of water and biological samples[J]. Microchemical Journal 2016,127:62-67.

[4]. Mcshane W J , Pappas R S , Wilson-Mcelprang V , et al. A rugged and transferable method for determining blood cadmium, mercury, and lead with inductively coupled plasma-mass spectrometry[J]. Spectrochimica Acta Part B Atomic Spectroscopy, 2008, 63(6):463-475.

[5]. Helena M.R. Gonçalves, Duarte A J, Silva J C G E D. Optical fiber sensor for Hg(II) based on carbon dots[J]. Biosensors & Bioelectronics, 2010, 26(4):1302-1306.

[6]. Sun X, Liu R, Liu Q, et al. Colorimetric sensing of mercury (II) ion based on anti-aggregation of gold nanoparticles in the presence of hexadecyl trimethyl ammonium bromide[J]. Sensors and Actuators B: Chemical, 2018:S0925400518300832.

[7]. Zou Y, Zhang Y, Xie Z, et al. Improved sensitivity and reproducibility in electrochemical detection of trace mercury (II) by bromide ion & electrochemical oxidation[J]. Talanta, 2019, 203:186-193.

[8]. Rithesh Raj D, Prasanth S, Vineeshkumar T V, et al. Surface Plasmon Resonance based fiber optic sensor for mercury detection using gold nanoparticles PVA hybrid[J]. Optics Communications, 2016, 367:102-107.

[9]. Xu L, Yin H, Ma W, et al. Ultrasensitive SERS detection of mercury based on the assembled gold nanochains[J]. Biosensors and Bioelectronics, 2015, 67:472-476.

[10]. P. Butmee, J. Mala, C. Damphathik, K. Kunpatee, G. Tumcharern, M. Kerr, E. Mehmeti, G. Raber, K. Kalcher, A. Samphao, A portable selective electrochemical sensor amplified with Fe3O4@Au-cysteamine-thymine acetic acid as conductive mediator for determination of mercuric ion, Talanta 221, 2021, 12.

[11]. Xiong C, Xu Y, Bian C, Wang R, Xie Y, Han M, Xia S. Synthesis and Characterization of Ru-MOFs on Microelectrode for Trace Mercury Detection. Sensors. 2020; 20(22):6686.

[12]. L. Fu, K. Xie, A. Wang, F. Lyu, J. Ge, L. Zhang, H. Zhang, W. Su, Y.-L. Hou, C. Zhou, C. Wang, S. Ruan, High selective detection of mercury (II) ions by thioether side groups on metal-organic frameworks, Anal. Chim. Acta 1081 (2019) 51-58.

[13]. Kukkar M , Mohanta G C , Tuteja S K , et al. A Comprehensive Review on Nano-Molybdenum Disulfide/DNA Interfaces as Emerging Biosensing Platforms[J]. Biosensors & Bioelectronics, 2018, 107.

[14]. Zhu C , Zeng Z , Li H , et al. Single-Layer MoS2-Based Nanoprobes for Homogeneous Detection of Biomolecules[J]. Journal of the American Chemical Society, 2013, 135(16):5998.

[15]. Chu Y , Cai B , Ma Y , et al. Highly sensitive electrochemical detection of circulating tumor DNA based on thin-layer MoS2/graphene composites[J]. RSC Advances, 2016, 6(27):22673-22678.

[16]. Wang X , Nan F , Zhao J , et al. A label-free ultrasensitive electrochemical DNA sensor based on thin-layer MoS2 nanosheets with high electrochemical activity[J]. Biosensors & Bioelectronics, 2015, 64C:386-391.

[17]. B, Siddharth Kaushik A, et al. Rapid detection of Escherichia coli using fiber optic surface plasmon resonance immunosensor based on biofunctionalized Molybdenum disulfide (MoS$_2$) nanosheets[J]. Biosensors and Bioelectronics, 2019, 126:501-509.

Gap in pagination due to unavailable paper.

Pages 1308-1309

Proceedings of the 16th Annual IEEE International
Conference on Nano/Micro Engineered and Molecular Systems
April 25-29, 2021

Quantitative Glucose Measurement on a Synthetic Paper Test Strip

Weijin Guo, Jonas Hansson, and Wouter van der Wijngaart*, *Member, IEEE*

Abstract— **To measure the glucose concentration in blood is an important target for point-of-care diagnostics. We integrated a colorimetric assay on synthetic paper, a novel lateral flow test substrate, for quantitative glucose measurement. At first, we dropped a colorimetric enzyme assay in silk solution on one end of the synthetic paper test strip and dried it in a vacuum chamber for overnight. After that, we flowed glucose solution on the test strip, imaged the test strip by smartphone, and analyzed the images with ImageJ. Exponential fitting of the experimental data revealed a low limit of detection of the assay to be 0.21 mM, and the R^2 is 0.99 for the concentration (0 – 15 mM), which covers the clinically relevant blood glucose concentration (2 – 12 mM). This test strip can be further developed into a test for glucose quantification in whole blood.**

I. INTRODUCTION

Glucose concentration in blood is a very important marker of human health condition, and it is of high interest for point-of-care diagnostics. Now there are more than 400 million people with diabetes in the world, who need to monitor the glucose concentration in blood regularly. Diabetes can lead to very serious syndromes, and the medical cost on diabetes reaches to more than 500 billion dollars every year world widely. Concerning market share, glucose measurement test strip is one of the most popular products in the market of point-of-care diagnostics [1]. There are many scientific reports on glucose measurement, as well as many commercial products in the market. The most common method for glucose measurement is by electrochemical methods [2-11]. Glucose concentration in other body liquids including urine, sweat and tear is also related to the glucose concentration in blood [12-16]. Nowadays some portable detection platforms based on commercial electrochemical readers or smartphones are developed for glucose detection in human body [17-22]. Due to the fast development of biosensing and bioelectronics technology, continuous or wireless glucose monitoring can be achieved by smart wearables, such as contact lens [23-26].

Glucose measurement can also be performed on a nitrocellulose paper test strip by a colorimetric assay [27,28]. Nitrocellulose paper is a suitable substrate for such colorimetric assays because it is a porous media with high specific surface area. However, the nonuniform microstructures and short optical path limit further applications of nitrocellulose. Off-stoichiometry thiol–ene (OSTE) synthetic paper is a novel lateral flow test substrate with uniform microstructures and good optical properties [29-31]. Synthetic paper is transparent, and with

low autofluorescence [30,31]. Moreover, the pore size of synthetic paper is tunable and its specific surface area is big enough to ensure strong intensity for colorimetric assays. Here, we report a glucose measurement test strip by combining synthetic paper and a colorimetric enzyme assay. The glucose measurement on synthetic paper is run in a lateral flow test format. The enzyme assay reagents are prepared in silk solution, which can help to increase the sensitivity of the enzyme assay and the shelf-life of the test strips [32,33]. In addition, we use a smartphone to capture the experimental images to get the colorimetric signal, which makes it possible to build a portable glucose detection system based on synthetic paper and smartphones in the near future.

II. EXPERIMENTS

For the preparation of synthetic paper test strips, at first we fabricated a big piece of synthetic paper (10 cm × 10 cm, shown in Figure 1(a)) by OSTE lithography. OSTE (Ostemer 220) was provide by Mercene Labs, Sweden. In detail, at first we prepared the OSTE mixture by mixing component A and component B in a weight ratio of 1.86:1. The mixing was conducted by a highspeed mixer (Synergy Devices Ltd, UK). Then we degassed the mixture by putting it in a vacuum chamber. After that, we did the multi-directional lithography of OSTE using a customized mirror setup [30]. Lastly, we developed the structures in acetone. We cut the big piece into small pieces with dimension 5 mm × 15 mm by a cutting potter (Cutting Plotter CE5000-60, Graphtech, Japan), and did the hydrophilic treatment of the synthetic paper test strips by dipping them inside Tween 20 solution (0.05% in DI water; VWR, Sweden), then took them out and let them dry at room temperature. The microstructure dimension of synthetic paper (shown in Figure 1(b) and 1(c)) is: the diameter t of the micropillars is 50 μm, the pitch distance p between the micropillars is 100 μm, and the thickness t of the micropillars is 100 μm. The synthetic paper is fabricated on a plastic film (Write-On Transparency Film, Office Depot, USA) with a thickness of 100 μm as the rigid support on the bottom.

The following experimental procedures are shown in Figure 2. For the preparation of the enzyme assay, at first, we prepared the enzyme assay solution by mixing silk solution (5.3 mL, concentration of silk fibroin: 7%, the detailed preparation of silk solution can be found in Ref. [32]), glucose oxidase (GOx) (4 mg, G7141, Sigma Aldrich, USA), horseradish peroxidase (HRP) (6 mg, P8375, Sigma Aldrich, USA), and 2,2'-azino-bis(3-ethylbenzothiazoline-6-sulphonic acid) diammonium salt (ABTS) (24 mg, A1888,

---

Weijin Guo is with Shantou University, Department of Biomedical Engineering, 243 Daxue Road, 515063 Shantou, China, and KTH Royal Institute of Technology, Division of Micro and Nanosystems, Malvinas väg 10, 100 44 Stockholm, Sweden

Jonas Hansson and Wouter van der Wijngaart are with KTH Royal Institute of Technology, Division of Micro and Nanosystems, Malvinas väg 10, 100 44 Stockholm, Sweden
*Corresponding Author: Wouter van der Wijngaart. Email: wouter@kth.se

978-1-6654-3008-1/21 $31.00 © 2021 IEEE

Sigma Aldrich, USA). Then we degassed the mixture in a vacuum chamber shortly to remove the bubbles. The mixture after degassing looked slightly yellowish. Then we dropped the enzyme assay solution (1 µL) on one end of the test strip, and let it dry at room temperature. The assay solution of 1 µL can fill half area of the synthetic paper test strip by capillary action. After that, we put the test strips in a vacuum chamber (with a cup of water beside) for overnight to let it fully dry. After drying of the assay solution, the test strips still looked clear. The immobilization of enzyme assay on synthetic paper relies on the physical absorption. The enzyme assay in the fully dried silk can keep the bioactivity for at least 7 months [32].

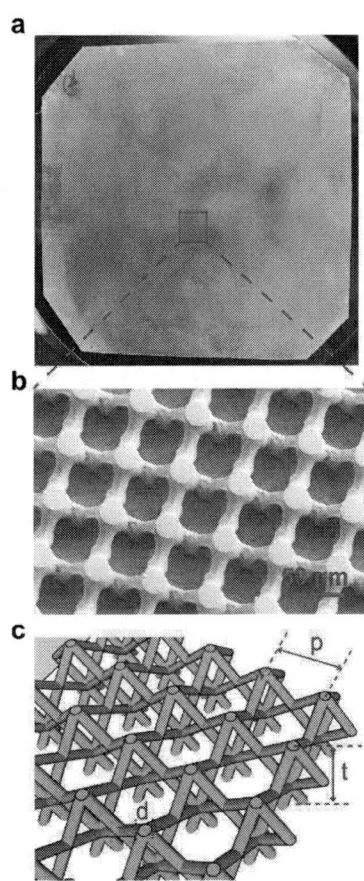

Figure 1. A synthetic paper sheet of 10 cm x 10 cm (a), SEM picture of synthetic paper (b), and 3D schematic picture of synthetic paper (c). For the parameters in (c), d indicates the diameter of the pillars, p indicates the pitch distance of pillars, and t indicates the thickness of the pillars. (c) is reprinted (adapted) from [31], Copyright (2020), with permission from Elsevier.

For the preparation of glucose solution, we prepared glucose solution of 0, 2, 4, 8, 15 µM by dissolving glucose in PBS. Then we flowed glucose solution (8 µL) from the other end of the test strips. When the glucose solution reached the region coated with enzyme assay, the enzyme reaction will happen and show some color changes. At first, the color will change from clear to green, then to purple and become stable after ~10 min. For each concentration, we repeated the experiments for four times. After the experiments, we used a smartphone (iPhone SE, Apple,

USA) to take the pictures of the test strips under the same light environment, and used ImageJ (http://rsb.info.nih.gov/ij/) to analyze the signal intensity of reaction region on the test strips. In detail, we split the color channels of the pictures and measured the intensity of reaction region on the green channel. We chose a square of same area for each test strip, and did an exponential fitting between color intensity and glucose concentration.

1. fabricate synthetic paper

2. drop silk solution with enzyme (1 uL) on half of the synthetic paper surface

3. let silk solution dry in room temperature

4. cure silk in vacuum chamber for overnight

5. flow glucose solution (8 uL)

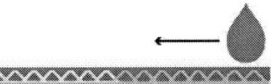

6. let it dry and take pictures

Figure 2. The experimental procedures of glucose detection on the glucose test strip: 1. we fabricate synthetic paper and prepare the test strip, 2. we drop 1 µL silk solution with enzyme on half of the test strip, 3. we let the silk dry at room temperature, 4. the test strip is put into a vacuum chamber for overnight, 5. we flow the glucose solution from the other end of the test strip, 6. we take pictures of the test strip after the color of reaction region becomes stable.

III. RESULTS AND DISCUSSIONS

As shown in Figure 3, after flowing glucose solution on the test strips, we can see a color change of the reaction region from clear to purple. The chemical reaction during this process is described below [32,33]:

$$\beta - D - glucose + O_2 \rightarrow D - glucono - 1,5 - lactone + H_2O_2 \quad (1)$$

$$H_2O_2 + ABTS \xrightarrow{Horseradish\ Peroxidase} H_2O + ABTS^{+\cdot} \quad (2)$$

$$ABTS^{+\cdot} + Tyrosine \rightarrow ABTS - Tyrosine\ bond \quad (3)$$

. With the increase of glucose concentration, c_g, we can see a change of the intensity, I, of the purple color of the reaction region on the test strips. The exponential fitting (the blue line in Figure 3) can be described by Equation (4):

$$I = 50 + 139 * \exp\left(-0.32 * c_g\right) \qquad (4)$$

. The low limit of detection is 0.21 mM, and the fitting is very good with a $R^2 = 0.99$ for the glucose concentration $0 \sim 15$ mM, which covers the clinically relevant glucose concentration in blood (2 - 12 mM).

Although the glucose test strip shows good performance in the detection range and low limit of detection, there is still room for further improvement. As we can notice from Figure 3, the error bar at 15 mM is larger than that at other concentrations and the slope of the curve becomes smaller with the increase of glucose concentration, which means the sensitivity of glucose detection near 15 mM is lower compared to that near 0 mM. We can further optimize the concentrations of assay reagents in the silk solution to achieve a higher sensitivity at high glucose concentrations. In addition, the areas of reaction region on the test strips show some variations, which may be caused by manual liquid dropping. To make the areas of reaction region consistent from strip to strip, we can look for another way to load assay solution, either by a mechanical arm or by a spotter machine. Other than the smartphone, we can also use a commercial optical detector or build a customized optical reader to read the color intensity or light absorbance with a higher accuracy.

Silk has been proved to be with good optical properties, and can be used as photonics components [34,35]. A recent report introduced a glucose sensor based on a thin silk film for glucose detection in whole blood sample, and used a customized optical setup to detect the light absorbance as the signal [33]. Synthetic paper is a very good substrate for integration of such enzyme assays dissolved in silk solution since synthetic paper also has very good optical properties. In comparison with other nontransparent porous substrates (nitrocellulose paper or silicon pillar forest), synthetic paper will facilitate the absorbance measurement while it is not convenient to measure the absorbance on other nontransparent substrates.

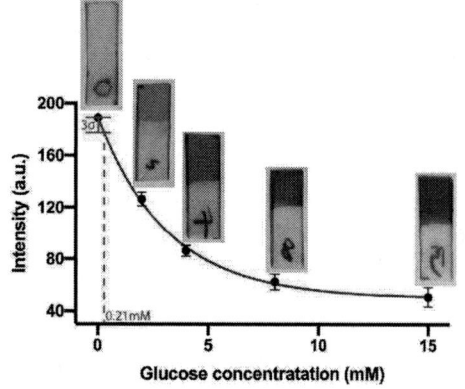

Figure 3. The intensity of the reaction region on test strips versus glucose concentration. The relation between intensity and glucose concentration follows an exponential fitting with $R^2 = 0.99$.

For the outlook of this work, we can develop this glucose test strip into a test strip for glucose measurement in whole blood sample. By integrating a blood filtration membrane in a similar way as Ref. [36], synthetic paper can be used to filter the blood cells and run the colorimetric enzyme assay on blood plasma. Alternatively, we can precoat the synthetic paper with a layer of agglutination antibody, which can be used to cause local agglutination of red blood cells and achieve blood plasma separation [37]. With the function of blood filtration, the test strip can directly use whole blood sample for glucose detection, which fits better for point-of-care diagnostics since usually there will be few suitable tools at home for blood plasma separation. By avoiding the usage of centrifuge for blood plasma separation, we can also reduce the human intervention and increase the accuracy of detection.

By immobilizing an enzyme assay on synthetic paper, it can be developed into a point-of-care diagnostic tool for glucose detection. We think this platform can be easily adapted for other targets in point-of-care diagnostics, such as the alcohol and lactate in human body since enzyme assays for such analytes are already commercially available. Because it is easy to manufacture synthetic paper by cutting plotter or laser cutter, we can come up with a new design which can integrate multiple enzyme assays on a single synthetic paper test strip for multiple detections. Furthermore, the size of reaction region in this work can be further reduced to millimeter or submillimeter scale, similar to the spot size of microarray by spotter machine. By spotting the enzyme assay solution on synthetic paper, it can be used for biomarker detection in a high-throughput way. It can increase the efficiency of detection, and save the cost on enzyme assay at the same time. Besides the colorimetric enzyme assay, it is also very promising to apply fluorescent enzyme-linked immunoassay on synthetic paper [38,39], since synthetic paper has been proved to be compatible with fluorescent assays [31,40].

IV. CONCLUSION

We integrated a colorimetric glucose measurement assay on the novel lateral flow test substrate – OSTE synthetic paper. The enzyme assay was prepared in silk solution, with a high sensitivity and a long shelf-life. The experimental pictures were captured by a smartphone, and analyzed by ImageJ. The intensity of purple color on the reaction region of this test strip shows an exponential relation with the glucose concentration, with a low limit of detection of 0.21 mM, and a $R^2 = 0.99$ for the glucose concentration from 0 mM to 15 mM, which includes the concentration of clinical interest (2 ~ 12 mM). Considering that synthetic paper can be used for plasma separation of whole blood samples, we believe this test strip can be further developed for quantitative glucose measurement of whole blood samples.

ACKNOWLEDGMENT

We acknowledge financial support from the European Union's Horizon 2020 research and innovation programme under the Marie Sklodowska-Curie grant 675412, and Shantou University at Shantou, China (STU Scientific Research Foundation for Talents: NTF20034). We thank Augusto Márquez and Xavier Muñoz-Berbel from Instituto de Microelectrónica de Barcelona (IMB-CNM, CSIC) at Barcelona, Spain for their help in the experiments by providing the experimental chemicals and assisting the preparation of enzyme assay solution.

REFERENCES

[1] S. Vashist, P. Luppa, L. Yeo, A. Ozcan and J. Luong, "Emerging Technologies for Next-Generation Point-of-Care Testing", *Trends in Biotechnology*, vol. 33, no. 11, pp. 692-705, 2015.

[2] J. Wang, "Electrochemical Glucose Biosensors", *Chemical Reviews*, vol. 108, no. 2, pp. 814-825, 2008.

[3] Y. Chen et al., "Skin-like biosensor system via electrochemical channels for noninvasive blood glucose monitoring", *Science Advances*, vol. 3, no. 12, p. e1701629, 2017.

[4] D. Hwang, S. Lee, M. Seo and T. Chung, "Recent advances in electrochemical non-enzymatic glucose sensors – A review", *Analytica Chimica Acta*, vol. 1033, pp. 1-34, 2018.

[5] A. Saei, J. Dolatabadi, P. Najafi-Marandi, A. Abhari and M. de la Guardia, "Electrochemical biosensors for glucose based on metal nanoparticles", *TrAC Trends in Analytical Chemistry*, vol. 42, pp. 216-227, 2013.

[6] C. Sun et al., "Applications of antibiofouling PEG-coating in electrochemical biosensors for determination of glucose in whole blood", *Electrochimica Acta*, vol. 89, pp. 549-554, 2013.

[7] F. Xie, X. Cao, F. Qu, A. Asiri and X. Sun, "Cobalt nitride nanowire array as an efficient electrochemical sensor for glucose and H_2O_2 detection", *Sensors and Actuators B: Chemical*, vol. 255, pp. 1254-1261, 2018.

[8] K. Dhara and D. Mahapatra, "Electrochemical nonenzymatic sensing of glucose using advanced nanomaterials", *Microchimica Acta*, vol. 185, no. 1, 2017.

[9] S. Zaidi and J. Shin, "Recent developments in nanostructure based electrochemical glucose sensors", *Talanta*, vol. 149, pp. 30-42, 2016.

[10] M. Picher et al., "Nanobiotechnology advanced antifouling surfaces for the continuous electrochemical monitoring of glucose in whole blood using a lab-on-a-chip", *Lab on a Chip*, vol. 13, no. 9, p. 1780, 2013.

[11] J. Noiphung, T. Songjaroen, W. Dungchai, C. Henry, O. Chailapakul and W. Laiwattanapaisal, "Electrochemical detection of glucose from whole blood using paper-based microfluidic devices", *Analytica Chimica Acta*, vol. 788, pp. 39-45, 2013.

[12] J. Lankelma, Z. Nie, E. Carrilho and G. Whitesides, "Paper-Based Analytical Device for Electrochemical Flow-Injection Analysis of Glucose in Urine", *Analytical Chemistry*, vol. 84, no. 9, pp. 4147-4152, 2012.

[13] X. Zhu, Y. Ju, J. Chen, D. Liu and H. Liu, "Nonenzymatic Wearable Sensor for Electrochemical Analysis of Perspiration Glucose", *ACS Sensors*, vol. 3, no. 6, pp. 1135-1141, 2018.

[14] S. Oh et al., "Skin-Attachable, Stretchable Electrochemical Sweat Sensor for Glucose and pH Detection", *ACS Applied Materials & Interfaces*, vol. 10, no. 16, pp. 13729-13740, 2018.

[15] A. Abellán-Llobregat et al., "A stretchable and screen-printed electrochemical sensor for glucose determination in human perspiration", *Biosensors and Bioelectronics*, vol. 91, pp. 885-891, 2017.

[16] B. Peng, J. Lu, A. Balijepalli, T. Major, B. Cohan and M. Meyerhoff, "Evaluation of enzyme-based tear glucose electrochemical sensors over a wide range of blood glucose concentrations", *Biosensors and Bioelectronics*, vol. 49, pp. 204-209, 2013.

[17] Z. Nie, F. Deiss, X. Liu, O. Akbulut and G. Whitesides, "Integration of paper-based microfluidic devices with commercial electrochemical readers", *Lab on a Chip*, vol. 10, no. 22, p. 3163, 2010.

[18] A. Bandodkar et al., "Re-usable electrochemical glucose sensors integrated into a smartphone platform", *Biosensors and Bioelectronics*, vol. 101, pp. 181-187, 2018.

[19] S. Zhao et al., "ISFET and Dex-AgNPs based portable sensor for reusable and real-time determinations of concanavalin A and glucose on smartphone", *Biosensors and Bioelectronics*, vol. 151, p. 111962, 2020.

[20] R. Bandi et al., "Cellulose nanofibrils/carbon dots composite nanopapers for the smartphone-based colorimetric detection of hydrogen peroxide and glucose", *Sensors and Actuators B: Chemical*, vol. 330, p. 129330, 2021.

[21] N. Alizadeh, A. Salimi and R. Hallaj, "Mimicking peroxidase-like activity of Co3O4-CeO2 nanosheets integrated paper-based analytical devices for detection of glucose with smartphone", *Sensors and Actuators B: Chemical*, vol. 288, pp. 44-52, 2019.

[22] D. Ji et al., "Smartphone-based cyclic voltammetry system with graphene modified screen printed electrodes for glucose detection", *Biosensors and Bioelectronics*, vol. 98, pp. 449-456, 2017.

[23] J. Park et al., "Soft, smart contact lenses with integrations of wireless circuits, glucose sensors, and displays", *Science Advances*, vol. 4, no. 1, p. eaap9841, 2018.

[24] M. Elsherif, M. Hassan, A. Yetisen and H. Butt, "Wearable Contact Lens Biosensors for Continuous Glucose Monitoring Using Smartphones", *ACS Nano*, vol. 12, no. 6, pp. 5452-5462, 2018.

[25] D. Keum et al., "Wireless smart contact lens for diabetic diagnosis and therapy", *Science Advances*, vol. 6, no. 17, p. eaba3252, 2020.

[26] J. Kim et al., "Wearable smart sensor systems integrated on soft contact lenses for wireless ocular diagnostics", *Nature Communications*, vol. 8, no. 1, 2017.

[27] D. Sechi, B. Greer, J. Johnson and N. Hashemi, "Three-Dimensional Paper-Based Microfluidic Device for Assays of Protein and Glucose in Urine", *Analytical Chemistry*, vol. 85, no. 22, pp. 10733-10737, 2013.

[28] W. Zhu et al., "Bienzyme colorimetric detection of glucose with self-calibration based on tree-shaped paper strip", *Sensors and Actuators B: Chemical*, vol. 190, pp. 414-418, 2014.

[29] C. Carlborg, T. Haraldsson, K. Öberg, M. Malkoch and W. van der Wijngaart, "Beyond PDMS: off-stoichiometry thiol–ene (OSTE) based soft lithography for rapid prototyping of microfluidic devices", *Lab on a Chip*, vol. 11, no. 18, p. 3136, 2011.

[30] J. Hansson, H. Yasuga, T. Haraldsson and W. van der Wijngaart, "Synthetic microfluidic paper: high surface area and high porosity polymer micropillar arrays", *Lab on a Chip*, vol. 16, no. 2, pp. 298-304, 2016.

[31] W. Guo, L. Vilaplana, J. Hansson, M. Marco and W. van der Wijngaart, "Immunoassays on thiol-ene synthetic paper generate a superior fluorescence signal", *Biosensors and Bioelectronics*, vol. 163, p. 112279, 2020.

[32] A. Márquez, S. Aznar-Cervantes, J. Cenis, C. Dominguez and X. Muñoz-Berbel, "Silk Fibroin Pads for Whole Blood Glucose Determination", *Proceedings*, vol. 2, no. 13, p. 886, 2018.

[33] A. Márquez et al., "Nanoporous silk films with capillary action and size-exclusion capacity for sensitive glucose determination in whole blood", *Lab on a Chip*, vol. 21, no. 3, pp. 608-615, 2021.

[34] S. Kujala, A. Mannila, L. Karvonen, K. Kieu and Z. Sun, "Natural Silk as a Photonics Component: a Study on Its Light Guiding and Nonlinear Optical Properties", *Scientific Reports*, vol. 6, no. 1, 2016.

[35] F. Omenetto and D. Kaplan, "A new route for silk", *Nature Photonics*, vol. 2, no. 11, pp. 641-643, 2008.

[36] A. Homsy et al., "Development and validation of a low cost blood filtration element separating plasma from undiluted whole blood", *Biomicrofluidics*, vol. 6, no. 1, p. 012804, 2012.

[37] W. Guo, J. Hansson and W. van der Wijngaart, "Synthetic Paper Separates Plasma from Whole Blood with Low Protein Loss", *Analytical Chemistry*, vol. 92, no. 9, pp. 6194-6199, 2020.

[38] L. Jiao, L. Zhang, W. Du, H. Li, D. Yang and C. Zhu, "Hierarchical manganese dioxide nanoflowers enable accurate ratiometric fluorescence enzyme-linked immunosorbent assay", *Nanoscale*, vol. 10, no. 46, pp. 21893-21897, 2018.

[39] S. Piletsky, E. Piletska, A. Bossi, K. Karim, P. Lowe and A. Turner, "Substitution of antibodies and receptors with molecularly imprinted polymers in enzyme-linked and fluorescent assays", *Biosensors and Bioelectronics*, vol. 16, no. 9-12, pp. 701-707, 2001.

[40] W. Guo, J. Hansson and W. van der Wijngaart, "Synthetic microfluidic paper with superior fluorescent signal readout", in *The 23rd International Conference on Miniaturized Systems for Chemistry and Life Sciences (µTAS 2019)*, Basel, Switzerland, 2019, pp. 1056-1057.

Gap in pagination due to unavailable papers.

Pages 1314-1319

Proceedings of the 16th Annual IEEE International
Conference on Nano/Micro Engineered and Molecular Systems
April 25-29, 2021

Improved Design of Polymer Micromachined Transmission for Flapping Wing Nano Air Vehicle*

Rashmikant, Daisuke Ishihara**, Ryotaro Suetsugu, Sunao Murakami, *Member, IEEE*, Prakasha C. Ramegowda

Abstract— This paper presents an improved design of polymer micromachined transmission for flapping-wing Nano Air Vehicles (FWNAVs) in comparison with our last design. Design Improvement includes (1) reduction of the crack and fracture during microfabrication, and (2) performance improvement of the transmission. The novelty of our polymer micromachined transmission includes (1) the transmission mechanism based on the geometrical nonlinear bending deformation, and (2) the complete 2.5-dimensional structure that can be fabricated using standard microfabrication techniques including the etching, the photolithography, the deposition, and the curing process. The complete 2.5-dimensional structure requires no post-assembly and leads to further miniaturization without so much difficulty. Each wing of FWNAVs will be driven by the proposed transmission and actuator separately. The present transmission can produce around 40° stroke angle, which is about 50% higher than that of our previous work without considering the mass effect.

I. INTRODUCTION

In recent years, insect-inspired flapping-wing nano air vehicles (FWNAVs) have received much attention because of their wide application in areas including surveillance, entertainment, agriculture, media, and especially in a dangerous environment where humans cannot survive. Various groups [1-4] are working on the FWNAVs whose sizes are ranging from approximately 1cm to 10cm. One of the FWNAVs is fabricated which can produce enough lift [1]. However, further miniaturization is difficult because of the use of complicated mechanical transmission system. One of the engineering concerns on this topic is to minimize their size as a scale of minimum insect size (about 1mm). The problem in reducing the size results in a reduction of lift to drag ratio which makes Nano air vehicles difficult to fly. To produce enough lift to drag ratio, FWNAV should have a large stroke angle during the flapping motion.

Our main concern here is the production of large flapping motion in the development of the FWNAV since other motions such as the feathering motion can be produced by the fluid-structure interaction [5-11]. Other groups [2&4] mostly use complex mechanical transmission, that requires post-assembly of components, which might lead to difficulty in reducing the size. Some research groups have fabricated FWNAV [2,3,12] using microfabrication techniques but in these groups, stroke angle is obtained through resonance,

which is used to maximize flight dynamics parameters [13]. However, it will increase the difficulty of the fluid-structure interaction design of FWNAVs [5].

In our previous study [14], we have proposed and developed the novel polymer micromachined transmission for FWNAVs. This transmission has a complete 2.5-dimensional structure that can be fabricated using standard microfabrication techniques including the etching, the photolithography, the deposition, and the curing process. This novel transmission is based on the geometrical nonlinear bending deformation, which requires flexible material to fabricate the transmission. Different from the other studies, it can produce enough stroke angle reaches about 40° without any resonance.

In this study, we present an improved design of polymer micromachined transmission for FWNAVs in comparison with our last design. The transmission is changed from the two-wing type to the one-wing type to increase the yield ratio for the polymer micromachined transmission. Different from the previous design, the piezoelectric bimorph actuator with the metal shim is considered instead of the piezoelectric bimorph actuator without metal shim to evaluate the forced displacement applied by the actuator to the transmission more accurately. The advantages of our transmission over the other research groups are as follows: (1) no post-assembly. (2) enough stroke angle generation without resonance assist. Therefore, our transmission is suitable for further miniaturization.

II. METHODOLOGY

A. An Abstract Idea of The Polymer Micromachined FWNAV

The polymer micromachined FWNAV consists of the micro transmission, the supporting frame, the piezoelectric bimorph actuator, and the wings as shown in Fig. 1. These are the main components of the FWNAV [15]. The main difference from the previous studies is that this has a complete 2.5-dimensional structure that can be fabricated using standard microfabrication techniques [14]. The production of a large stroke angle of the wing during flapping motion is the main consideration in the design and fabrication of the FWNAV. The other fundamental motion of the wing to produce the lift force is the feathering motion which can be evaluated by the fluid-structure interaction

---

*Resrach supported by JSPS KAKENHI Grant Numbers 20H04199 and 19F19379.

All authors are with the Department of Computer Science and System Engineering, Kyushu Institute of Technology, 680-4 Kawazu, Iizuka, Fukuoka 820-8502 Japan.

**Corresponding Author: D. Ishihara is with the Faculty of Computer Science and System Engineering, Kyushu Institute of Technology, 680-4 Kawazu, Iizuka, Fukuoka 820-8502 Japan (e-mail: ishihara@mse.kyutech.ac.jp).

978-1-6654-3008-1/21 $31.00 © 2021 IEEE

analysis [5-11]. The FSI design of FWNAV wings reduces the electromechanical complexity [6].

B. Fundamental Mechanism of The Transmission

The transmission is based on geometrically nonlinear bending deformation [14], where small translational displacement produces large rotational displacement. In Fig. 2, a small translational displacement (u_x) is applied on the cantilever beam to produce a large deflection angle (ϕ) at the tip of the cantilever beam.

C. Detailed Design of the Transmission

The basic design of the transmission [14] includes (1) the parallel elastic hinge, (2) the hinge supporting beam, (3) the wing attachment part, and (4) the actuator attachment part as shown in Fig. 3. Fig. 4 (A) shows the reference of transmission with detailed dimension in plain view whereas Fig. 4 (B) shows the deformed transmission where the small translation displacement (u_x) given at point A in the actuator attachment part to produce large deflection (θ) in the wing attachment parts.

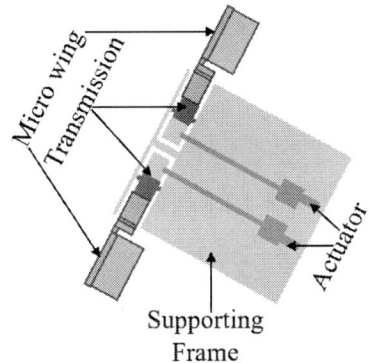

Figure 1. A Schematic diagram of polymer micromachined insect inspired FWNAV proposed in this study.

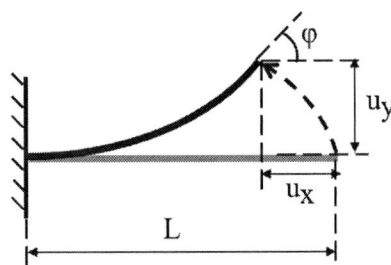

Figure 2. Transmission using the geometric nonlinear bending of the cantilever beam.

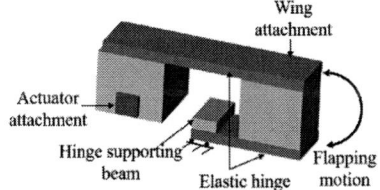

Figure 3. Basic design of the transmission.

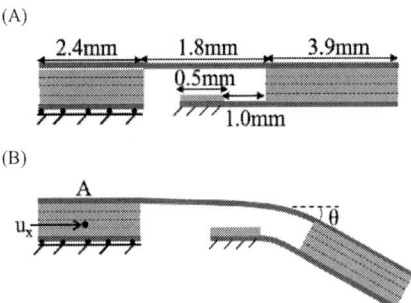

Figure 4. (A) Reference transmission with detailed dimension, and (B) deformed transmission.

D. Static Structural Finite Element Analysis of the Transmission and the Supporting Frame

This part of the paper provides the finite element analysis of the transmission and the supporting frame without considering the mass effect. Fig. 5 represents the problem setup for the geometrically nonlinear static analysis of the transmission, the forced displacement ($u_x = 83\mu m$) is applied in the transmission area where the upper part of the actuator is attached whereas the fixed boundary condition is applied to the supporting frame at the area where the fixed end of the actuator is attached. The forced displacement ($u_x = 83\mu m$) in boundary condition is evaluated by the equilibrium point between transmission reaction force (dot line) and actuator reaction force (dash line) as shown in Fig. 6. The transmission reaction force is obtained using geometrically nonlinear finite element analysis with the hexahedral secondary element in the marc solver. The actuator reaction force (F) is obtained theoretically [16,17] using the below equations. Fig. 7 and 8 represent the finite element analysis results, Fig. 7 shows that the stroke angle is around 40° which is comparable with the actual insect during flapping motion [18]. Fig. 8 shows the von Mises stress distribution during flapping motion. The maximum von Mises stress is around 16 MPa which is about 15% of PI sheet breaking strength (109MPa). Therefore, the chance of break or fail of the transmission during actual working will be very less.

$$F = \frac{3E_p w^* \delta}{2L^3}\left[\frac{t_p^3}{3} + t_p\left(t_m + t_p\right)^2 + \frac{E_m t_m^3}{6E_p}\right], \quad (1)$$

$$\delta = \frac{6E_p d_{31} V\left(t_m t_p + t_p^2\right)L^2}{2E_p t_p\left(3t_m^2 t_p + 6t_m t_p^2 + 4t_p^3\right) + E_m t_m^3}, \quad (2)$$

where d_{31} is the piezoelectric constant (m/V), L is the length of the actuator (20mm), E_p is the young modulus of piezoelectric (61GPa), E_m is the young modulus of the metal (140GPa), t_p is the thickness of piezoelectric (0.2mm), t_m is the thickness of metal (0.1mm), w is the width of micro actuator (0.5mm) and V is the applied voltage (200V).

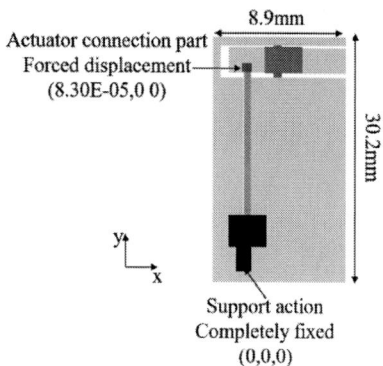

Figure 5. Problem setup: load and boundary condition for the finite element analysis of the transmission

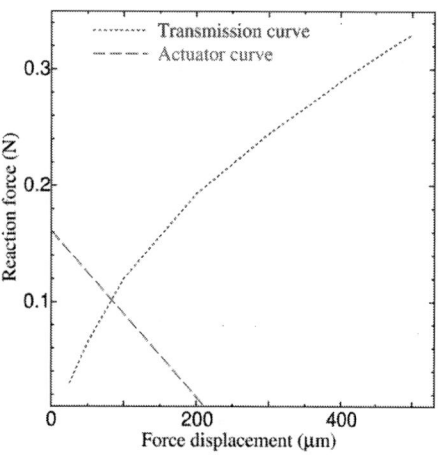

Figure 6. Relationship between forced displacement and reaction force. Dot line is transmission curve which is obtained using geometrical nonlinear finite element analysis. Dash line is actuator curve which is obtained through theoretical formula.

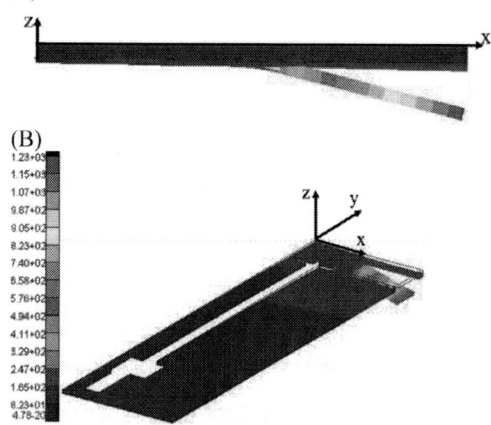

Figure 7. Geometrically nonlinear finite element analysis of transmission. The color contour indicates the magnitude of the deformation. (A: x-z plane view, B: 3-D view).

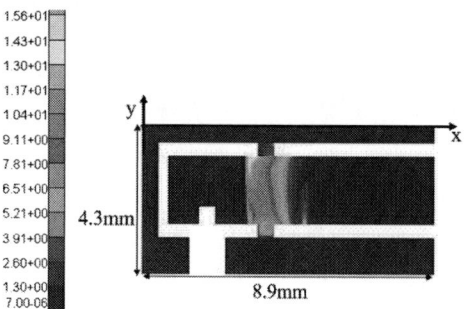

Figure 8. Von-misses stress distribution of the upper part of the transmission compared to figure 5 during flapping motion. Color contour shows the magnitude of the miss's stress.

E. Fabrication Process of Transmission and supporting frame using the Polymer Micromachining

The transmission and the supporting frame are fabricated together using standard microfabrication steps and the polyimide adhesive sheets with a thickness of 40μm are used as a material [14]. The details of the steps are given in Fig. 9 where the lamination, the exposure, and the development process of the polyimide sheet over the silicon wafer are repeated. After the required repetition of the exposure and development process, the transmission and the supporting frame are cured and finally, the transmission is released from the substrate.

III. RESULTS AND DISCUSSION

A. The Precision of the Fabricated Transmission

The in-plane dimension of the fabricated transmission is measured using the microscope and the micrometer, and it is compared with the design values. Fig. 10 shows the precision of the transmission, and it is observed that the precision level of the fabricated transmission is around 95% as shown in Table 1.

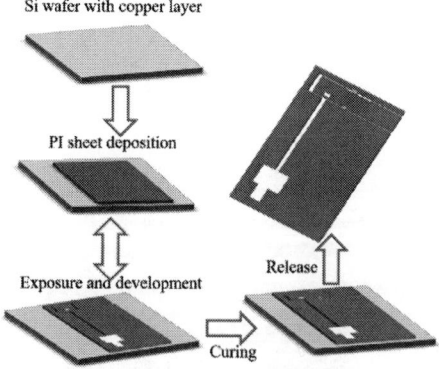

Figure 9. Standard microfabrication steps which include (a) the deposition (b) the etching (c) the photolithography (d) the curing process.

978-1-6654-3008-1/21 $31.00 © 2021 IEEE

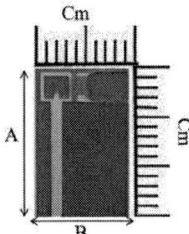

Figure 10. Observation of precision of fabricated transmission

Table 1. Observation of precision of fabricated transmission

Measured Dimension (see Figure 10)	Observed Value (mm)	Design Value (mm)	Precision
A	14.865	15.0	99.1
B	8.621	8.90	96.8

B. Evaluation of Performance of the Transmission

The static performance of the transmission is evaluated and compared between finite element analysis (numerical) and experimental results. Fig. 11 provides the detailed experimental setup for the static driving test [14]. Taking the account of the experimental setup, numerical problem setup for upward and downward flapping motion is shown in Fig. 12 (A) and (B), respectively. Numerical problem setup is used for finite element analysis with the hexahedral secondary element in the marc solver. Fig. 13 describes the upward flapping motion of the transmission. Fig. 13 (A) shows the finite element analysis of the transmission with forced displacement ($u_x = 83\mu m$). Fig. 13 (B) shows the deformation of the transmission during the experiment with a forced displacement ($u_x = 80\mu m$). The forced displacement is applied statically during the experiment. Fig. 13 (C) indicates the comparison between the experimental results (two samples) and the numerical result for the upward flapping motion. Similarly, Fig. 14 is for the downward motion. The stroke angle of the FWNAV is calculated by adding the flapping angular displacement in the upward and downward directions. The stroke angle of the transmission is about 40° without considering the mass effect. There exists a difference of around 20% between experimental and numerical results due to deviation between design and fabricated structural shapes as shown in Fig. 15.

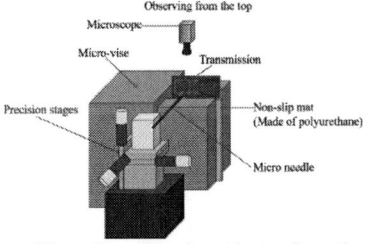

Figure 11. Experimental setup for static analysis

(A)

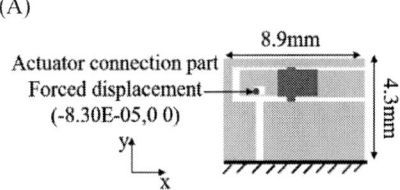

(B)

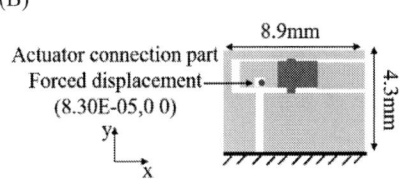

Figure 12. (A) Problem setup for upward flapping motion (B) Problem setup for downward flapping motion.

(A)

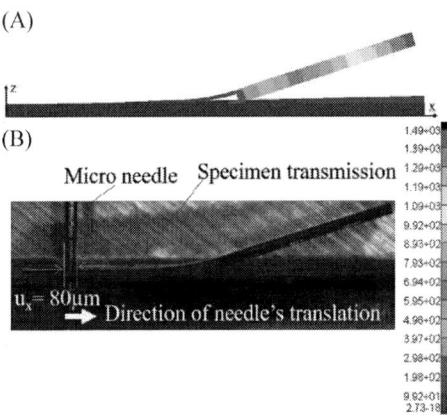

(B)

(C)

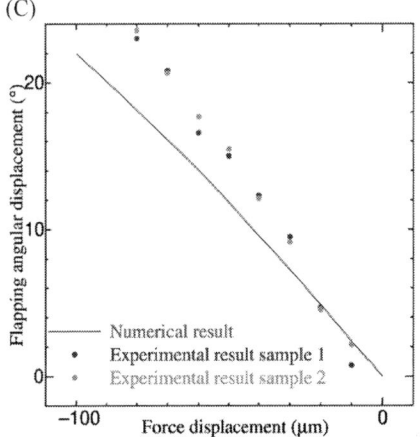

Figure 13. (A) Static deformation of transmission in finite element analysis for upward flapping motion. (B) Static deformation of transmission during experiment for upward flapping motion. (C) Performance comparison between numerical result and experimental result for upward flapping motion.

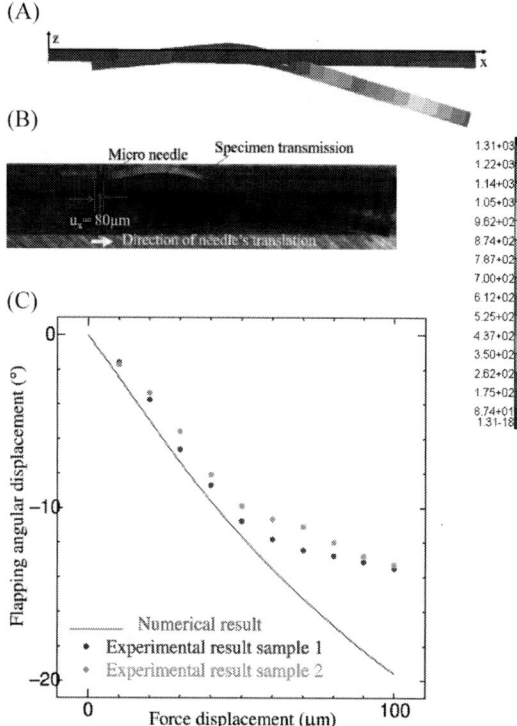

(A)

(B) Micro needle   Specimen transmission

$u_z = 80\mu m$ Direction of needle's translation

(C)

Figure 14. (A) Static deformation of transmission in finite element analysis for downward flapping motion. (B) Static deformation of transmission during experiment for downward flapping motion. (C) Performance comparison between numerical result and experimental result for downward flapping motion.

Figure 15. Fabricated structural shape (the cross section view in the xz-plane). As shown in this figure, the actuator and wing attachment parts are not in the same xy-plane of the frame exactly, that is, these parts are slightly distorted around the hinge supporting beam (y-axis).

IV. CONCLUSION

The paper presented here shows an improved design of the micro transmission using polymer micromachining, which is based on geometrically nonlinear bending, and it can give further miniaturization for FWNAVs. The proposed transmission is different from our previous fabrication in two way: (1) The change from the two-wing type to the one-wing type to increase the yield ratio, and (2) the consideration of the piezoelectric bimorph actuator with the metal shim to evaluate the displacement applied by the actuator to the transmission more accurately. The improvement from the previous design includes the performance increase of the transmission and no fracture during microfabrication. In this design, the maximum von Mises stress during the flapping motion only 15% of the breaking strength of the PI sheet. Therefore, there is little

chance of failure during the working of the transmission. The stroke angle of the proposed transmission is about 40° without considering the mass effect which is comparable with the natural insect. In this design, there exists about a 20% difference between the finite element analysis and experimental results in the static test due to the slight distortion of the fabricated structural shape, although the precision level of the fabricated transmission is about 95% agreement in the representative dimensions. It follows from these results that the proposed transmission can give the further miniaturization of the FWNAVs because of no post assembly.

ACKNOWLEDGMENT

This work was supported by JSPS KAKENHI Grant Numbers 19F19379 and 20H04199. We appreciate the support from Toray Industries, Inc.

REFERENCES

[1] R.J. Wood, "The First Takeoff of a Biologically Inspired at-Scale Robotic Insect", *IEEE Transactions on Robotics*, vol. 24, pp. 341-347,2008.

[2] A. Bontemps, T. Vanneste, J. B. Paquet, T. Dietsch, S. Grondel, E. Cattan, "Design and Performance of an Insect-Inspired Nano Air Vehicle", *Smart Material and Structures*, vol. 22, 014008, 2013.

[3] T. Dargent, X. Q. Bao, S. Grondel, G. Le Brun, J. B. Paquet, C. Soyer E. Cattan, "Micromachining of an SU-8 Flapping-Wing Flying Micro-Electro-Mechanical System", *Journal of Micromechanics and Microengineering*, vol. 19,085028,2009.

[4] Y. Chen, H. Zhao, J. Mao, P. Chirarattananon, E.F. Helbling, N.S.P. Hyun, D.R. Clarke, R. J. Wood, "Controlled Flight of a Microrobot Powered by Soft Artificial Muscles", *Nature*, vol. 575, pp. 324-329, 2019.

[5] D. Ishihara, N. Ohira, M. Takagi, S. Murakami, T. Horie, "Fluid-Structure Interaction Design of Insect-like Micro Flapping Wing", *Proceedings of VII International Conference on Computational Methods for coupled problems in science and Engineering*, June 12-14, 2017, pp. 870-875.

[6] D. Ishihara and T. Horie, "Fluid Structural Interaction Modeling of Insect Flight", *Transactions of the Japan Society of Mechanical Engineers B*, vol. 72 (718), pp. 1410-1417,2006.

[7] D. Ishihara, T. Horie, M. Denda, "A Two-Dimensional Computational Study on the Fluid-Structure Interaction Cause of Wing Pitch Changes in Dipteran Flapping Flight", *The Journal of Experimental Biology*, vol. 212, pp. 1-10,2009.

[8] D. Ishihara, Y.Yamashita, T.Horie, S. Yoshida, T. Niho, "Passive Maintenance of High Angle of Attack and its Lift Generation During Flapping Translation in Crane Fly Wing", *The Journal of Experimental Biology*, vol. 212, pp. 3882-3891,2009.

[9] D. Ishihara, T. Horie, T. Niho, An Experimental and Three-Dimensional Computational Study on the Aerodynamic Contribution to the Passive Pitching Motion of Flapping Wing in Hovering Flies", *Bioinspiration and Biomimetics*, vol. 9, 046009, 2014.

[10] D. Ishihara and T. Horie, "Passive Mechanism of Pitch Recoil in Flapping Insect Wings", *Bioinspiration and Biomimetics*, vol. 12, 016008,2017.

[11] D. Ishihara, "Role of Fluid-Structure Interaction in Generating the Characteristic Tip Path of a Flapping Flexible Wing", *Physical Review E*, vol. 98(3), 032411, 2018.

[12] T. Ozaki, and K. Hamaguchi, "Electro-Aero-Mechanical Model of Piezoelectric Direct-Driven Flapping Wing Actuator", *Applied Sciences*, vol. 8(9), 1699, 2018.

[13] J. Zhang and X. Deng, "Resonance Principle for the Design of Flapping Wing Micro Air Vehicles", *IEEE transactions on Robotics*, vol. (99), pp. 1-15, 2017.

[14] D. Ishihara, S. Murakami, Member IEEE, N. Ohira, J. Ueo, M. Takagi, "Polymer Micromachined Transmission for Insect- Inspired Flapping Wing Nano Air Vehicle", *Proceedings of the 15th Annual IEEE International Conference on Nano/Micro Engineered and Molecular System*, September 27-30, 2020, pp. 176-179.

[15] C. Zhang and C. Rossi, "A Review of Compliant Transmission Mechanisms for Bio-Inspired Flapping-Wing Micro Air Vehicles", *Bioinspiration and Biomimetics*, 12,025005,2017.

[16] Q.M. Wang, L.E. Cross, "Performance Analysis of Piezoelectric Cantilever Bending Actuator", *Ferroelectrics*, vol.215 no. 1, pp. 187-213, 1998.

[17] P.C. Ramegowda, D. Ishihara, T. Niho, T. Horie, "Performance evaluation of numerical finite element coupled algorithms for structure- electric interaction analysis of MEMS piezoelectric actuator", *International Journal of Computational Methods*, vol. 16, 1850106, 2019.

[18] X. Cheng, M.Sun, "Wing-kinematics measurement and aerodynamics in a small insect in hovering flight", *Sci. Rep. 6*, 25706, 2016.

The 16th IEEE International Conference on Nano/Micro Engineered & Molecular Systems

Atomic-scale Manufacturing based on TEM

Litao Sun

SEU-FEI Nano-Pico Center, Key Lab of MEMS of Ministry of Education,
School of Electronic Science and Engineering, Southeast University, Nanjing, 210096, P.R. China

Website : http://www.seu-npc.com

ABSTRACT

With the rapid development of semiconductor technology, the 5 nm feature size of fabrication is already achieved. It is thus quite essential to explore more precise manufacturing and characterization method to evaluate the shape/structure stability and possible new properties of sub-5nm material components, especially under external stimuli such as strain, electric, or thermal fields. Along with the reduction of dimensions, the surface-to-volume ratio of the materials increases. Surface atoms are preferably reconstructed for adapting their geometrical and electronic structure to the environment. Thus, the surface structure begins to dominate material properties ranging from electronic and structural aspects when the characteristic dimension is reduced to sub-5 nm. Here we review our recent progress in atomic resolution manufacturing and dynamic characterization of individual nanostructures and nanodevices based on the idea of "setting up a nanolab inside a transmission electron microscope". The electron beam can be used as a tool to induce manufacturing on the atomic scale. Additional probes from a special-designed holder provide the possibility to further manipulate and measure the electric/mechanical properties of the nanostructures in the small specimen chamber of a TEM. Recently, the optical signal also was introduced into the electron microscope to enrich the coverage of investigation inside the "multifunctional nanolab". All phenomena from the in-situ experiments can be recorded in real time with atomic resolution.

References:
[1] L. Sun, F. Banhart, et. al., *Science* 312, 1199 (2006)
[2] J. R-Manzo, M. Terrones, et.al., *Nature Nanotechnology* 2, 307 (2007)
[3] X. Liu, T. Xu, et al., *Nature Communications* 4, 1776 (2013)
[4] X. Guo, G. Fang, et al., *Science* 344, 616 (2014)
[5] J. Sun, L. He, et al., *Nature Materials* 13, 1007 (2014)
[6] Q. Zhang, K. Yin, et al., *Nature Communications* 8, 14889 (2017)
[7] C. Zhu, S. Liang, et al., *Nature Communications* 9, 421 (2018)
[8] H. Dong, F. Xu, et al., *Nature Nanotechnology* 14, 950 (2019)

BIOGRAPHY

Prof. Litao Sun received his PhD from the Shanghai Institute of Applied Physics, Chinese Academy of Sciences in 2005. He worked as a research fellow at University of Mainz, Germany from 2005 to 2008, and a visiting professor at University of Strasbourg, France from 2009 to 2010. Since 2008, he joined SEU and honored as a Distinguished Professor. He currently serves as the head of School of Electronic

Science and Engineering, Southeast University (SEU), the director of SEU-FEI Nano-Pico center, the director of Key Lab of MEMS of Ministry of Education, and the director of Center for Advanced Materials and Manufacture, Joint Research Institute of Southeast University and Monash University. He is the founding chairman of IEEE Nanotechnology Council Nanjing Chapter. He is the author and co-author of around 200 papers on international journals including 2 in Science, 15 in Nature and Nature series journals, etc. He was selected the Global Highly Cited Researchers list from the Web of Science Group. He holds around 80 patents and has given more than 160 invited presentations. He is the Review Panel member of Graphene Flagship, European Union and Member of European Science Foundation College of Expert Reviewers. He has obtained National Science Fund for Distinguished Young Scholars of China, New Century Excellent Talents in University, Young Leading Talent in Science and Technology Innovation, Cheung Kong Scholar Chair Professor from Ministry of Education etc.

978-1-6654-3008-1/21 $31.00 © 2021 IEEE

This page intentionally left blank.

Proceedings of the 16th Annual IEEE International
Conference on Nano/Micro Engineered and Molecular Systems
April 25-29, 2021

Research on Melt Electrowriting TPU Hydrophobic Microfiber Mesh for Directional Water Transport

Zungui Shao, Jiaxin Jiang, Junyu Chen, Huatan Chen, Guoyi Kang, Xiang Wang, Wenwang Li, and Gaofeng Zheng*

Abstract— Self-pumping textiles have been widely researched for the function of directional perspiration, which can realize comfortable wearing. However, there are still many problems to be solved in the study of hydrophobic layer. A good match between the wettability of hydrophobic layer and the breakthrough pressure is the key to realize the directional water transfer. In addition, it is also very important to ensure its mechanical strength and to improve its durability, which is the key to ensure its continuous and multiple use. Herein, a flexible and elastic thermoplastic polyurethanes (TPU) hydrophobic microfiber mesh (TPU-HMM) produced by melt electrowriting (MEW) is reported. TPU-HMMs with different wettability were fabricated by orderly and controllable stacking. By controlling the pore size and stacking thickness, the relationship between the wettability and the breakthrough pressure of TPU-HMM was explored. This research has a guiding role for the realization of on-demand water directional transport, and the ordered TPU-HMM has great potential in the development of wearable devices, anti-adhesion textiles and wound dressings.

I. INTRODUCTION

The flexible fabric with asymmetric wettability surface has the function of moisture directional transmission, which is an important means to achieve moisture absorption and perspiration [1-3]. According to the difference of wettability gradient, the construction of asymmetric wettability surface can be divided into Janus wettability, wettability gradient and the structure combining the characteristics of these two methods [4-6]. These methods all achieve the good matching between the wettability and the breakthrough pressure of the hydrophobic layer, which is the key to realize the directional water transport. In this case, improving the durability of hydrophobic layer to ensure its long-term use is of great significance to the production of moisture absorbing and sweat wicking textiles.

The orderly and controlled piling of the fiber is an important method to achieve the stable use of the fabric [7, 8]. For disorderly arranged fiber, it was easy to make the fabric deform under the effect of external force, which would not only affect its mechanical properties, but also reduce its stability of use, and thus have a negative impact on the direct transport of water [9]. In addition, the controlled piling of the fiber could achieve the uniform appearance and the controlled wettability, which could guarantee the stable and directed transmission of water as required. Therefore, the controllable construction of orderly stacked hydrophobic fiber layer can achieve stable and on-demand water transmission, which is conducive to the exploration and development of functional flexible materials.

Melt electrowriting (MEW) is a simple and solvent-free method for fiber manufacturing, which is widely used in many fields such as filtration, textiles, sensors and so on [10-12]. Under the effect of the electric field, the melt stretched out and cooled down into a fiber, which was then attached to the collection plate. With a suitable movement speed and collection distance, the straight fiber could be collected and orderly deposit. The fiber strength produced by melt electrospinning is generally higher than that of solution electrospinning, which has a broad application prospect [13-15].

Thermoplastic polyurethane (TPU) is a popular elastomeric material, which can be easily processed by melt technique [16, 17]. For the excellent stretchability as well as biocompatibility, TPU is widely used in biomedicine and flexible electronics, becoming an excellent candidate for the development of flexible functional fabrics [18-20].

Herein, a MEW TPU hydrophobic microfiber mesh (TPU-HMM) was fabricated and attached to the hydrophilic textile to form a self-pumping textile, realizing the unidirectional water transport. Moreover, the TPU-HMM had good elasticity and shape-preserving ability so that it can maintain a good shape after multiple uses and maintain a stable unidirectional water transport performance, showing a promising application for the functional moisture wicking textiles and dressing.

II. EXPERIMENTAL DETAILS

The schematic diagram of MEW setup is illustrated in Fig. 1. A purchased MEW special device (Foshan Qingzi precision measurement and Control Technology Co., Ltd, Foshan, China) was used to prepare the MEW fibers. TPU particles (90A) were melted in a heating chamber at 220℃ for 2 hours. Then, Molten TPU was pushed into the metal nozzle (25G, Shenzhen dalicheng Plastic Hardware Products Co., Ltd, Shenzhen, China) by the air at 0.4 MPa. The distance between the nozzle and the collector was 3 mm. The voltage of +2.0 kV was applied to the collector and the

---

*Research supported by the National Natural Science Foundation of China (No. 51805460), Science and Technology Planning Project of Fujian Province (No. 2019H0038, 2020H6003), Xiamen Municipal Science and Technology Project (No. 3502Z20193015).

Zungui Shao, Jiaxin Jiang, Junyu Chen, Huatan Chen, Guoyi Kang and Gaofeng Zheng are with the Department of Instrumental and Electrical

Engineering, Xiamen University, Xiamen 361102, Fujian, China (corresponding author: Gaofeng Zheng, phone: +86-592-2194957; fax: +86-592-2182221; e-mail: zheng_gf@xmu.edu.cn).

Xiang Wang and Wenwang Li are with the School of Mechanical and Automotive Engineering, Xiamen University of Technology, Xiamen 361024, Fujian, China.

978-1-6654-3008-1/21 $31.00 © 2021 IEEE

nozzle was grounded. The collector was mounted on an X-Y-Z three-axis linear motion platform with the resolution of 1 μm, which was controlled by the host computer. The TPU microfiber was deposited alternately layer by layer on the movement path of the plexiglass collector. In order to obtain a certain degree of orientation and a certain density of TPU-HMM, the grid was set to a square with a side length of 0.15 mm. Many square grids of the same size formed a TPU-HMM with a side length of 50 mm.

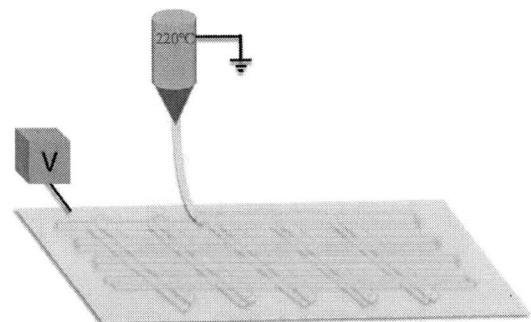

Figure 1.   The schematic diagram of the process of MEW.

The thickness of TPU-HMM was measured by a hand-held digital thickness gauge (Shenzhen Yuanhengtong Technology Co., Ltd, Shenzhen, China). The average value was taken from five different points of each sample.

A charge coupled device (CCD) camera (UI-2250-C, IDS Imaging Development Systems GmbH, German) was used to capture the image of water droplets, and ImageJ was used to measure the water contact angle (WCA).

The TPU-HMM was attached to the hydrophilic fabric (WCA = 0) to form a self-pumping textile. The TPU-HMM / self-pumping textile was placed at the bottom of the tubular container and then slowly dripped into it. When the water penetrated into the TPU-HMM / self-pumping textile, the height was recorded as the breakthrough pressure.

III.   RESULTS AND DISCUSSIONS

As shown in Fig. 2 (a), the TPU microfibers were straight and stacked orderly. The TPU-HMM was very soft and elastic so that it can be easily folded and restored, which shows that it can be used as a candidate material for flexible electronic device substrate, as shown in Fig. 2 (b) and Fig. 2 (c).

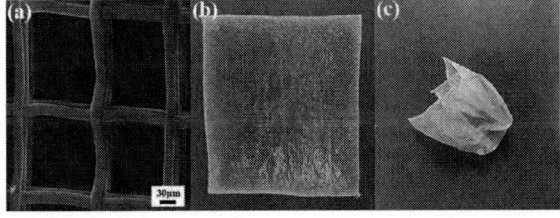

Figure 2.   (a) SEM image of 4-layers TPU-HMM. (b) Image of unfolded TPU-HMM (50 mm * 50 mm). (c) Image of fold TPU-HMM.

The thickness of TPU-HMM increased linearly with the number of stacking layers from 35 μm to 137 μm when the number of layers of TPU-HMM increased from 2 to 6, indicating an ordered stacking, as shown in Fig. 3. What's more, the WCA increased from 90 ° to 133 ° with the increase of stacking TPU-HMM layers, showing a decreased wettability, as shown in Fig. 4. It can be seen that the hydrophobicity increased with the increase of the thickness of TPU-HMM with the same pore size. This may be due to the fact that with the increasing thickness of TPU-HMM, the liquid tended to form the Cassie state, resulting in higher hydrophobicity [21, 22]. Therefore, different degrees of hydrophobicity can be achieved by controlling the thickness of the TPU-HMM.

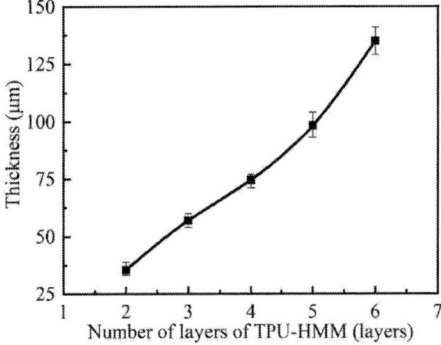

Figure 3.   The thickness of TPU-HMM with different deposited layers.

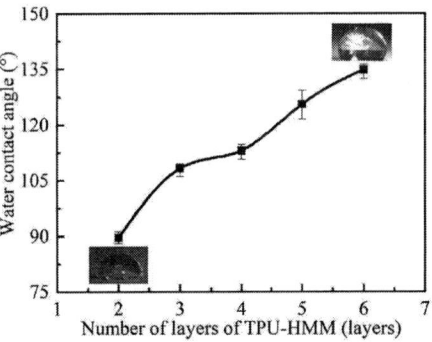

Figure 4.   The water contact angle of TPU-HMM with different deposited layers.

In order to further explore the directional water transport behavior of the TPU-HMM, the breakthrough pressure of TPU-HMM and self-pumping textile were measured. As shown in Fig. 5, the breakthrough pressure of TPU-HMM increased with the increase of the number of layers. With the increase of the number of layers, the breakthrough pressure of TPU-HMM increased from 16 mm H_2O to 47 mm H_2O. It showed that the decrease of surface wettability was helpful to the improvement of breakthrough pressure. Furthermore, TPU-HMM was compounded with different hydrophilic substrates to form a self-pumping textile. The breakthrough pressure of self-pumping textiles obviously reduced and were much lower than that of the pure TPU-HMM, showing the water transmission in positive direction and water blocking in reverse direction. It was apparent that the breakthrough pressure of self-pumping textile all increased slowly with the increasing layer of TPU-HMM. These results proved that the TPU-HMM had the effect of water blocking and can be attached to many hydrophilic textiles to form a self-pumping textile, realizing the directional water transmission. Moreover, it suggested that the hydrophilic textiles seemed to play a pump role in the phenomenon of self-pumping, offering the traction so as to decrease the breakthrough pressure. In addition, by controlling the thickness of TPU-HMM, different breakthrough pressure gradients were obtained, so as to realize the on-demand liquid transportation.

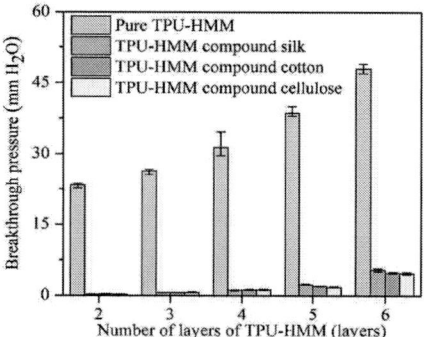

Figure 5. Breakthrough pressure of pure TPU-HMM and TPU-HMM with different hydrophilic textiles.

The schematic diagram of directional water transport process is illustrated in Fig. 6. In the positive direction (hydrophobic layer to hydrophilic layer), the hydrophobic force provided by the TPU-HMM made the water droplet unable to penetrate the hydrophobic layer. When the water droplet increased to have the opportunity to contact the hydrophilic fiber, the capillary force provided by the hydrophilic layer quickly dragged the water droplets to the hydrophilic layer and then the water droplets spread in the hydrophilic layer. However, in the reverse direction (hydrophilic layer to hydrophobic layer), the water droplets spread directly in the hydrophilic layer and cannot penetrate the hydrophobic layer for it did not have enough hydrostatic pressure to break it.

At present, Young's equation is widely used to explain this process with the following equation:

$$F_{Laplace} = \frac{4\gamma cos\theta}{D} \quad (1)$$

Where, γ is the surface tension of the liquid, θ is the local WCA, D is the pore size of the fibrous material, which can be approximately considered as the inscribed circle diameter of square mesh of TPU-HMM [23-25].

Hence, it was precisely because of this pressure difference that the water transmission had opposite characteristics in two directions.

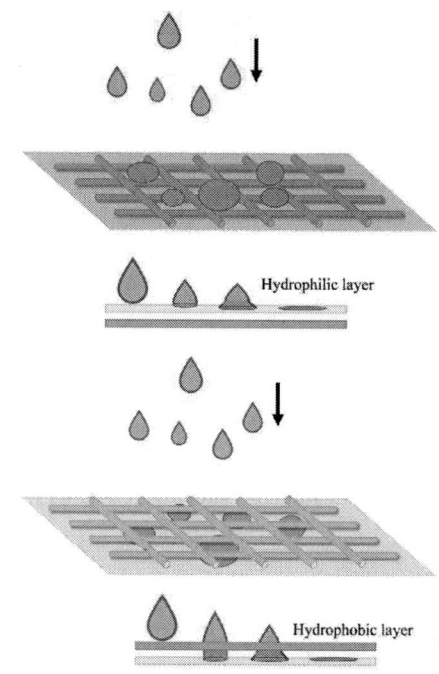

Figure 6. The schematic diagram of directional water transport process.

IV. CONCLUSION

In summary, a flexible and elastic ordered TPU-HMM was successfully produced via MEW, which had certain strength and anti-deformation ability so that it can be used stably for many times. The thickness of TPU-HMM can be controlled by orderly stacking to achieve different hydrophobic effects. Moreover, TPU-HMM can be attached to the hydrophilic textile directly to prepare the self-pumping textiles, which had a low breakthrough pressure in the positive direction (i.e. water transport from TPU-HMM to hydrophilic textile layer) and an excellent resistance in the reverse direction (i.e. water transport from hydrophilic textile layer to TPU-HMM). For achieving the on-demand water directional transport, the preparation of ordered TPU-HMM had great potential in improving the comfort of wearable devices and developing anti-adhesion textiles and wound dressings.

ACKNOWLEDGMENT

This research was financially supported by the National Natural Science Foundation of China (No. 51805460), Science and Technology Planning Project of Fujian Province (No. 2019H0038, 2020H6003), Xiamen Municipal Science and Technology Project (No. 3502Z20193015).

REFERENCES

[1] J. Tabor, K. Chatterjee, and T. K. Ghosh, "Smart Textile-Based Personal Thermal Comfort Systems: Current Status and Potential Solutions," Advanced Materials Technologies, vol. 5, pp. 1901155, March 2020.

[2] Y. Dong, J. Kong, C. Mu, C. Zhao, N. L. Thomas, and X. Lu, "Materials design towards sport textiles with low-friction and

moisture-wicking dual functions," Materials & Design, vol. 88, 82-87, August 2015.

[3] Y. Dong, J. Kong, S. L. Phua, C. Zhao, N. L. Thomas, and X. Lu, "Tailoring surface hydrophilicity of porous electrospun nanofibers to enhance capillary and push-pull effects for moisture wicking," ACS Appl Mater Interfaces, vol. 6, pp. 14087-14095, Aug 2014.

[4] D. Miao, Z. Huang, X. Wang, J. Yu, and B. Ding, "Continuous, Spontaneous, and Directional Water Transport in the Trilayered Fibrous Membranes for Functional Moisture Wicking Textiles," Small, vol. 14, pp. 1801527, Aug 2018.

[5] Q. Zhang, Y. Li, Y. Yan, X. Zhang, D. Tian, and L. Jiang, "Highly Flexible Monolayered Porous Membrane with Superhydrophilicity-Hydrophilicity for Unidirectional Liquid Penetration," ACS Nano, vol. 14, pp. 7287-7296, June 2020.

[6] W. Yan et al., "Multi-scaled interconnected inter- and intra-fiber porous janus membranes for enhanced directional moisture transport," J Colloid Interface Sci, vol. 565, pp. 426-435, January 2020.

[7] Y. Wang et al., "Fabrication of high-performance wearable strain sensors by using CNTs-coated electrospun polyurethane nanofibers," Journal of Materials Science, vol. 55, pp. 12592-12606, May 2020.

[8] M. Gong, L. Zhang, and P. Wan, "Polymer nanocomposite meshes for flexible electronic devices," Progress in Polymer Science, vol. 107, pp. 101279, June 2020.

[9] T. D. Brown, P. D. Dalton, and D. W. Hutmacher, "Melt electrospinning today: An opportune time for an emerging polymer process," Progress in Polymer Science, vol. 56, pp. 116-166, January 2016.

[10] J. C. Kade and P. D. Dalton, "Polymers for Melt Electrowriting," Adv Healthc Mater, vol. 10, pp. 2001232, August 2021.

[11] S. Florczak et al., "Melt electrowriting of electroactive poly(vinylidene difluoride) fibers," Polymer International, vol. 68, pp. 735-745, January 2019.

[12] T. D. Brown, P. D. Dalton, and D. W. Hutmacher, "Direct writing by way of melt electrospinning," Adv Mater, vol. 23, pp. 5651-5657, November 2011.

[13] P. D. Dalton, "Melt electrowriting with additive manufacturing principles," Current Opinion in Biomedical Engineering, vol. 2, pp. 49-57, May 2017.

[14] G. Hochleitner et al., "Additive manufacturing of scaffolds with sub-micron filaments via melt electrospinning writing," Biofabrication, vol. 7, pp. 035002, June 2015.

[15] B. L. Farrugia, T. D. Brown, Z. Upton, D. W. Hutmacher, P. D. Dalton, and T. R. Dargaville, "Dermal fibroblast infiltration of poly(epsilon-caprolactone) scaffolds fabricated by melt electrospinning in a direct writing mode," Biofabrication, vol. 5, pp. 025001, February 2013.

[16] M. Enayati et al., "Assessment of a long-term in vitro model to characterize the mechanical behavior and macrophage-mediated degradation of a novel, degradable, electrospun poly-urethane vascular graft," J Mech Behav Biomed Mater, vol. 112, pp. 104077, August 2020.

[17] Y. Jiang et al., "Stretchable, Washable, and Ultrathin Triboelectric Nanogenerators as Skin-Like Highly Sensitive Self-Powered Haptic Sensors," Advanced Functional Materials, vol. 31, pp. 2005584, August 2020.

[18] P. Mistry et al., "Fabrication and characterization of starch-TPU based nanofibers for wound healing applications," Mater Sci Eng C Mater Biol Appl, vol. 119, pp. 111316, August 2021.

[19] H. Li et al., "A highly stretchable strain sensor with both an ultralow detection limit and an ultrawide sensing range," Journal of Materials Chemistry A, vol. 9, pp. 1795-1802, December 2021.

[20] J. Lin et al., "Modification of thermoplastic polyurethane nanofiber membranes by in situ polydopamine coating for tissue engineering," Journal of Applied Polymer Science, vol. 137, pp. 49252, March 2020.

[21] D. A. L. E, "Hydrodynamic friction of fakir-like super-hydrophobic surfaces," Journal of Fluid Mechanics, vol. 661, pp. 402-411, 2010.

[22] C. Ybert, C. Barentin, C. Cottin-Bizonne, P. Joseph, and L. Bocquet, "Achieving large slip with superhydrophobic surfaces: Scaling laws for generic geometries," Physics of Fluids, vol. 19, pp. 123601, October 2007.

[23] Z. Zhang et al., "Design of a biofluid-absorbing bioactive sandwich-structured Zn-Si bioceramic composite wound dressing for hair follicle regeneration and skin burn wound healing," Bioact Mater, vol. 6, pp. 1910-1920, December 2021.

[24] J. Wu, N. Wang, L. Wang, H. Dong, Y. Zhao, and L. Jiang, "Unidirectional water-penetration composite fibrous film via electrospinning," Soft Matter, vol. 8, pp. 5996, April 2012.

[25] Z. J. Krysiak, J. Knapczyk-Korczak, G. Maniak, and U. Stachewicz, "Moisturizing effect of skin patches with hydrophobic and hydrophilic electrospun fibers for atopic dermatitis," Colloids Surf B Biointerfaces, vol. 199, pp. 111554, December 2020.

Proceedings of the 16th Annual IEEE International
Conference on Nano/Micro Engineered and Molecular Systems
April 25-29, 2021

Object Manipulation with Freestanding Magnetic Microfibers Fabricated by FDM 3D Printing

Qing Lu, *Student Member, IEEE*, Ki-Young Song*, *Member, IEEE*, and Yue Feng, *Member, IEEE*

Abstract— In this research, a new object manipulation method is proposed via the functional surface composed of the freestanding magnetic microfibers to realize the movement of objects under the magnetic fields. An improved multi-fragment extrusion (MFE) manufacturing method by fused deposition modeling (FDM) 3D printing has been developed to fabricate uniform magnetic microfibers in the range of 200 μm ~ 400 μm in thickness and 2 ~ 4 cm in length. The geometric parameters of magnetic microfibers have been studied to optimize the performance of object manipulation. As a result, it is found that microfiber with thickness of 200 μm and length of 4 cm has a better mechanical response, such as less energy consumption, small hysteresis error, and larger bending angle. In addition, different pitches of the magnetic functional surfaces are investigated, and the result indicates that the surface with 600 μm pitch performs better manipulation to convey an object due to higher driving force supported by more dense microfibers.

I. Introduction

Accurate and fast manipulation and separation of objects are critical activities in conveyor systems in macro-/micro-sizes [1]. Current object manipulation technologies include hydrodynamic [2], acoustic [3], optical [4], electrical [5], and magnetic methods [6]. The hydrodynamic method relies on the shear-gradient lift force originated from the curvature of the shear flow profile and on the repulsion force generated at microchannel walls, and the particles are pushed forward along the central axis. The acoustic method manipulates particles to move to the pressure node by standing waves formed by the interference between a reflective layer and a matching layer [7]. Optical tweezers move an object forward to the center of beam when a laser beam with Gaussian intensity hit the object [8]. A typical method of electrical particle manipulation method is Dielectrophoresis (DEP) which employs non-uniform (often Alternating Current) electric fields to drive a non-charged dielectric particle by dielectric force [9]. The magnetic method applies higher magnetic fields to attract magnetic particles. The attraction force depends on the strength of the magnetic field and the volume of the magnetic particles [6, 10].

However, each of the abovementioned methods still presents limitations. The hydrodynamic method faces the difficulty to realize remote control. The design and fabrication of acoustic and optical systems are tedious, complex, and high cost. Although the electrical and the magnetic method can achieve non-contact control, they are

limited with the special object material, for instance, low-conductivity media and magnetic particles, respectively [11, 12].

Recently, functional surfaces consisted of micro-/nano-pillars or particles have received great attentions by the manipulation of small objects, such as tiny mechanical parts and liquid droplets [13, 14]. In particular, the unique micro-nano surface of lotus leaves is highlighted as an emerging strategy for the manipulation on superhydrophobic surfaces [15].

Inspired by the abovementioned functional micro-structured surfaces for object manipulation, we propose a functional surface with freestanding magnetic microfibers fabricated by FDM 3D printing. Through this method, non-contact remote manipulation can be realized, and there is no requirement for the material property of the manipulated object. Moreover, the functional surface with magnetic microfibers are adjustable for various masses and sizes of objects by changing the geometric manufacturing parameters.

Microfibers are fine filaments made of polymers with diameters (or thicknesses) in a micrometer range, which are widely used in actuator and sensor applications [16, 17]. Currently, various microfiber fabrication methods have been studied, including electrospinning (far field and near field) [18, 19] and direct writing method [20]. However, the current fabrication methods have limitations to realize freestanding structures due to complex preparation, and the high cost is also to be concerned.

In our previous research work [21], we introduced a novel and simple method to fabricate suspending polymer microfibers by a conventional fused deposition modeling (FDM) 3D printer. The viscoelastic filament melts on the nozzle formed liquid bridges in the range of 25 ~ 180 μm in thickness and a centimeter range in length. On the basis of the previous study, here we improve the liquid bridge-based microfiber by multi-fragment extrusion (MFE) which significantly enhances the uniformity of microfibers. Adapting MFE, uniform magnetic microfibers are achievable in the range of 200 ~ 400 μm in thickness and several centimeters in length by a commercial FDM 3D printer.

Consequently, the magnetic microfibers are aligned in various pitches to form magnetic functional surfaces to manipulate objects. Under applied magnetic fields, the

All authors are with the School of Mechatronical Engineering, Beijing Institute of Technology, Beijing, China
* Contacting Author: kiyoungsong@bit.edu.cn.
Acknowledgment: This work was supported by grants from Beijing Institute of Technology Research Fund Program for Young Scholars (grant

number 3020012222017) and National Natural Science Foundation of China (grant number 52077005).

978-1-6654-3008-1/21 $31.00 © 2021 IEEE

magnetic surface (i.e, microfibers) bends with an orientation by the attracting magnetic fields, yielding the object on top of the surface to move, thereby realizing a personalized object manipulation.

II. MATERIALS AND METHOD

Magnetic microfibers were fabricated by a commercial fused deposition modeling (FDM) 3D printer (CREASEE CS20, China) with 0.4 mm of nozzle diameter, and a commercial iron metal-filled polylactic acid (magnetic PLA) filament (Proto-Pasta, USA, 1.75mm) with a density of 1.85 g/cc was purchased for the experiment [22]. During the printing process, the recommended temperatures of the nozzle and the printing bed of the 3D printer were kept at 210 °C and 50 °C, respectively.

Since the mechanical property of the magnetic PLA is not available in the literature as far as we know, we conducted a static tensile test by an MTS system (SUNS, CMT4104, China). Three specimens were printed and then melted by fusion casting to avoid the structure effect by 3D printing on the mechanical property. The sample size and the test process were followed by ASTM D638 standard which determines the tensile properties of plastics throughout the experiment [23]. Three samples showed similar stress strain relationship as shown in Fig. 1. Below the strain of 1.5 %, the material exhibited a linear elastic relationship during stretching. The average Young's modulus of this magnetic PLA was 3124 MPa, which indicated magnetic PLA has higher stiffness than conventional PLA (~2900 MPa). This higher stiffness was conducive to the recovery of the deformed microfiber and to the resistance to severe distortion during object manipulation.

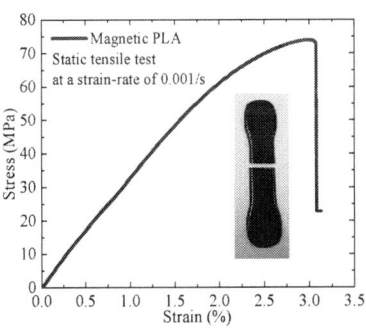

Figure 1.   Stress strain relationship of the specimen made of the magetic PLA

We made our own G-codes to print the magnetic microfibers, and the fabrication process of freestanding microfiber is shown in Fig. 2. Initially, on both sides of the microfiber, frames were printed with the conventional 3D printing method to support the liquid bridge-based microfibers (Fig. 2(a)). For the frames, the extrusion parameter was set as E0.08 / mm with the printing speed of F500. On top of the frame layer, the microfiber started as an anchor in 4 mm length by the same conventional printing method to fix the microfiber during the fabrication. After the anchor, multi-fragment extrusion (MFE) was applied to print multiple liquid bridges to form a microfiber. A liquid bridge was formed by an extrusion of filament melts. From the

experiment, ten sub-liquid bridges were proper to produce a magnetic microfiber by CS20 3D printer. The improved liquid bridges by MFE achieved more uniform microfibers. In this process, the extrusion was set as E0.01 / mm, E0.03 / mm, and E0.05 / mm to fabricate 200 μm, 300 μm, and 400 μm microfibers and the printing speed was set as F1000 (Fig. 2(b)). The printed microfiber ended with another anchor at the other end to form a suspending microfiber (Fig. 2(c)). Completing a layer of magnetic microfibers, we printed multiple frame layers to control layer gaps for specific pitch sizes between microfibers, resulting in three-dimension structure (Fig. 2(d)). Finally, we removed one side of the frame to form a 3D freestanding microfiber structure.

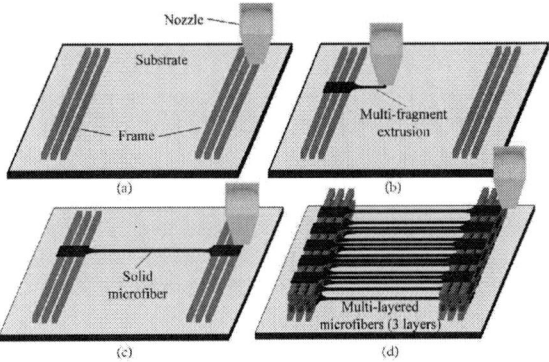

Figure 2.   Multi-fragment extrusion (MFE) microfiber fabrication process for multi-layered (3D) microfibers

III. CURRENT RESULT

The optimized MFE method produced highly uniform microfibers as shown in Fig. 3(a). In the figure, the grains on the microfibers are the magnetic particles and cause rough surfaces of the microfiber. After printing multi-layered microfibers in three thicknesses (200 μm, 300 μm, and 400 μm) and in 3 cm length, three microfiber samples from each size were randomly taken, and the thickness was measured under an optical microscope. Coefficient of variation (CV = standard deviation / mean of diameter of microfiber) was applied to evaluate the uniformity from the measurement (uniformity if CV < 5%), and we observed that CVs of the fabricated microfibers were less than 7% (very close to the uniformity), which indicated high repeatability and controllability of our method for microfiber fabrication. In addition, we adopted pitch to define the density of microfibers per unit area to form a freestanding magnetic functional surface as illustrated in Fig. 3(b). For the experiment, we controlled pitch (the distance between two microfibers) to have the same distance between neighboring microfibers (except for diagonal ones).

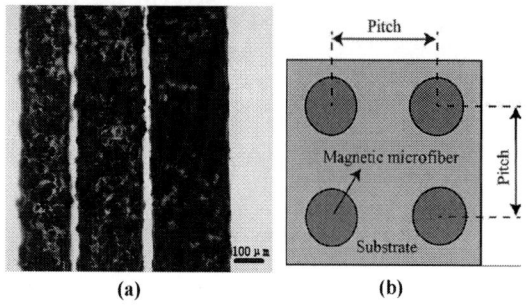

(a)   **(b)**

Figure 3. (a) Optical image of microfibers in different thickness, and (b) Schematics of the arrangement of magnetic microfibers to form a freestanding magnetic functional surface.

Initially, we investigated the mechanical responses of the microfiber with different geometric parameters in DC magnetic fields. The mechanical responses were eventually the index for us to select optimal dimension for better object manipulation. By controlling the printing parameters, three thicknesses (200 μm, 300 μm, and 400 μm) of microfibers in 3 cm length were fabricated, and the deformation angles were measured under various magnetic fields as seen in Fig. 4(a).

It was observed that the different thicknesses affected the bending angle under the magnetic attracting force. Among three thicknesses, 200 μm thick microfibers presented the maximum recoverable bending angle in weaker magnetic fields, which indicates comparing with thicker microfibers 200 μm thick microfibers require less energy (magnetic energy) for bending and thus manipulating object on the surface. Therefore, 200 μm thick microfibers were selected for the further experiment. Next, different lengths (2 cm, 3 cm, and 4 cm) of microfiber were tested for bending under various magnetic fields as shown in Fig. 4(b). The optical images of experiment are displayed in Fig. 5. As expected, a longer (4 cm) microfiber presented a better bending performance. It should be noted that the microfiber reached the maximum recoverable bending angle within 30 mT (the bent microfiber was not recoverable under > 30 mT). Shorter (2 cm and 3 cm) microfibers exhibited relatively higher stiffness, which required stronger magnetic fields to obtain maximum bending angles.

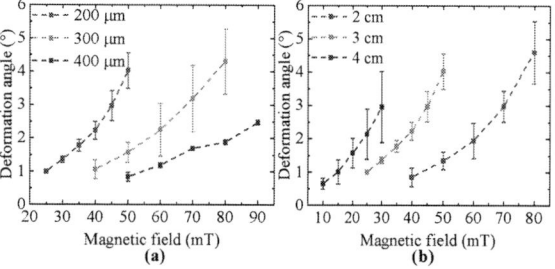

Figure 4. Single microfiber deformation by magnetic fields in different conditions: (a) different thickness and (b) different length

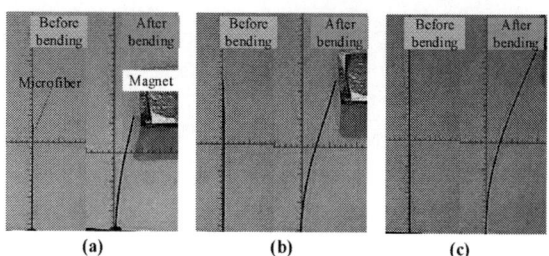

(a)   **(b)**   **(c)**

Figure 5. Deformation of magnetic microfibers: before and after bending of (a) 2 cm, (b) 3 cm, and (c) 4cm

Next, we conducted a deformation hysteresis of 200 μm thick microfibers in the three lengths (2 cm, 3 cm and 4 cm) to estimate the recovery process of the microfiber after the deformation. It should be noted that hysteresis represents the dependence of the state of a system on its history. Thus, hysteresis evaluated the recoverability and repeatability of microfiber deformation. As shown in Fig. 6(a), 4 cm long microfiber showed smaller hysteresis error during the bending and the recovery processes.

In addition, we measured tension force of the microfiber by increasing the bending angles. Since the freestanding microfiber was fixed at one end, which was a typical cantilever beam, tension force on a microfiber should be changing by varying bending angle and inversely proportional to the square of the length as shown in Fig. 6(b). Compared with 2 cm and 3 cm long microfibers, the tension force on a 4 cm long microfiber was lower, which was not prone to unrecoverable elastic deformation. This result supported the fact that 4 cm length presented smaller hysteresis error, indicating a better performance for actuation and manipulation.

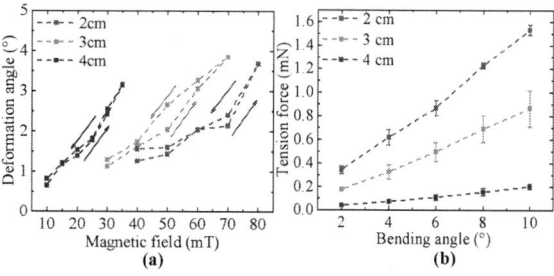

(a)   **(b)**

Figure 6. (a) Deformation hysteresis of 200 μm microfibers (arrows represent increasing and decreasing. (b) Surface stress on the microfiber under the magnetic fields.

After confirming the thickness and length of microfibers from the analysis, we fabricated freestanding magnetic functional surfaces in 4 different pitches, 600 μm, 800 μm, 1000 μm and 1200 μm as shown in Fig. 7(a), and the surfaces were tested for object manipulation as the experimental setup shown in Fig. 7(b). We applied 3 plastic objects in different mass and size as listed in Table 1. A permanent magnet (~200 mT) was applied to form the magnetic fields on top of the functional surfaces and to induce the bending of the microfiber. At 5 mm above the surface, the measured magnetic field strength was 60 mT, and the strength was sufficient to attract the magnetic microfiber to the maximum bending angle. During the manipulation process, the object

978-1-6654-3008-1/21 $31.00 © 2021 IEEE   1334

on top of the magnetic functional surface was completely isolated from the magnet by a transparent acrylic plate in order to avoid any contact by the magnet as shown in Fig. 7(b). At the initial state, the magnetic microfibers stood upright, supporting the object as shown in Fig. 8(a). While the magnet moved to one direction above the object, the magnetic microfibers were induced by the magnetic fields and bent to carry the object to the same direction as the magnet moved as shown in Fig. 8(b). It should be noted that the object was successfully manipulated under certain velocity (~0.1 m/s) of the magnet because slow motion (bending and recovery) of the microfibers by slower magnet movement was not able to produce sufficient stick-slip phenomenon to advance the object. The effect of the moving velocity and acceleration of the magnet (i.e., change of magnetic fields) for object manipulation on the surface of the magnetic microfibers will be studied in the further research.

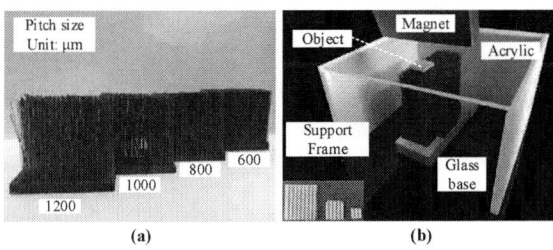

(a)   **(b)**

Figure 7. (a) Freestanding magnetic functional surfaces with microfibers in different pitch sizes, and (b) object manipulation setup

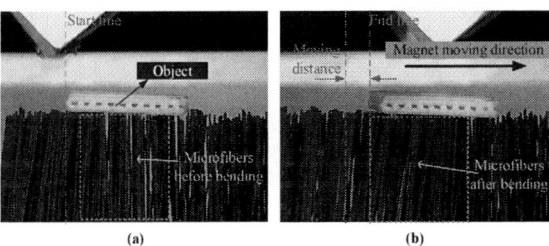

(a)   **(b)**

Figure 8. Object manipulation process under the magnetic fields: (a) without the magnetic fields and (b) under the magnetic fields.

From the manipulation experiment, we evaluated the manipulation performance of the surfaces in different pitches by counting the number of driving magnet movement to advance the object 1 cm forward, and the result is listed in Table 1. It was observed that smaller and lighter objects required a smaller number of driving magnet movement. Furthermore, it was also observed the lightest object (#1) presented similar movement with four different pitches. However, heavier objects (#2 and #3) exhibited different movement with different pitches, resulting in smaller pitch required a smaller number of driving magnet movement. The result explained that denser microfibers declared better manipulation performance because more microfibers per unit area with a smaller pitch provided higher driving force to convey the object.

TABLE I.   OBJECT MANIPULATION PERFORMANCE IN DIFFERENT PITCH SIZES.

Pitch (μm)	Number of driving magnet movement		
	Object #1 0.5 x 0.5 x 0.2 (cm) 0.05 (g)	Object #2 1 x 1 x 0.2 (cm) 0.16 (g)	Object #3 2 x 2 x 0.2 (cm) 0.54 (g)
600	6	9	20
800	10	13	22
1000	8	15	34
1200	7	19	56

IV. CONCLUSION

In this study, a new magnetic functional surface was proposed for object manipulation. With a commercial FDM 3D printer, magnetic microfibers with thickness of 200 μm ~ 400 μm and length of 2 ~ 4 cm were successfully fabricated by multi-fragment extrusion (MFE). Single magnetic microfibers with different geometric parameters were analyzed under the magnetic fields, and optimal parameters for freestanding magnetic functional surfaces were determined. It was found that a microfiber in thickness of 200 μm and length of 4 cm required less energy consumption for deformation and presented smaller hysteresis error due to its smaller surface tension force during bending. Furthermore, magnetic functional surfaces consisted of the microfiber were studied by four pitch sizes (600 μm, 800 μm, 1000 μm, and 1200 μm) to analyze the effect on object manipulation. The objects on top of the surface in different pitches were manipulated and conveyed by the applied magnetic fields. The result showed that smaller pitch (600 μm) performed better manipulation to convey an object because more dense microfibers supported higher driving force.

Although this study is preliminary, it contributes a new path of magnetic functional surfaces for object manipulation, which will promote the development of manipulation in various engineering applications.

REFERENCES

[1] M. Li, W. Li, J. Zhang, G. Alici, and W. Wen, "A review of microfabrication techniques and dielectrophoretic microdevices for particle manipulation and separation," *Journal of Physics D: Applied Physics*, vol. 47, no. 6, p. 063001, 2014.

[2] A. Karimi, S. Yazdi, and A. Ardekani, "Hydrodynamic mechanisms of cell and particle trapping in microfluidics," *Biomicrofluidics*, vol. 7, no. 2, p. 021501, 2013.

[3] A. Ozcelik, J. Rufo, F. Guo, Y. Gu, P. Li, J. Lata, and T. J. Huang, "Acoustic tweezers for the life sciences," *Nature methods*, vol. 15, no. 12, pp. 1021-1028, 2018.

[4] A. Lenshof and T. Laurell, "Continuous separation of cells and particles in microfluidic systems," *Chemical Society Reviews*, vol. 39, no. 3, pp. 1203-1217, 2010.

[5] X. Xuan, "Recent advances in direct current electrokinetic manipulation of particles for microfluidic applications," *Electrophoresis*, vol. 40, no. 18-19, pp. 2484-2513, 2019.

[6] I. K. Puri and R. Ganguly, "Particle transport in therapeutic magnetic fields," *Annual review of fluid mechanics*, vol. 46, pp. 407-440, 2014.

[7] T. Laurell, F. Petersson, and A. Nilsson, "Chip integrated strategies for acoustic separation and manipulation of cells and particles," *Chemical Society Reviews*, vol. 36, no. 3, pp. 492-506, 2007.

[8] J. E. Molloy and M. J. Padgett, "Lights, action: optical tweezers," *Contemporary physics*, vol. 43, no. 4, pp. 241-258, 2002.

[9] H. A. Pohl, "The motion and precipitation of suspensoids in divergent electric fields," *Journal of Applied Physics*, vol. 22, no. 7, pp. 869-871, 1951.

[10] A. Rida and M. Gijs, "Manipulation of self-assembled structures of magnetic beads for microfluidic mixing and assaying," *Analytical chemistry,* vol. 76, no. 21, pp. 6239-6246, 2004.

[11] S. Zhang, Y. Wang, P. Onck, and J. den Toonder, "A concise review of microfluidic particle manipulation methods," *Microfluidics and Nanofluidics,* vol. 24, no. 4, pp. 1-20, 2020.

[12] R, Tornay., T, Braschler., N, Demierre., B, Steitz., A, Finka., H, Hofmann., and P, Renaud., "Dielectrophoresis-based particle exchanger for the manipulation and surface functionalization of particles," *Lab on a Chip,* vol. 8, no. 2, pp. 267-273, 2008.

[13] C, H. Seo, H. Jeong, Y. Feng, K. Montagne, T. Ushida, Y. Suzuki, and K. S. Furukawa , "Micropit surfaces designed for accelerating osteogenic differentiation of murine mesenchymal stem cells via enhancing focal adhesion and actin polymerization," *Biomaterials,* vol. 35, no. 7, pp. 2245-2252, 2014.

[14] K.-Y. Song, K. Morimoto, Y.-C. Chen, and Y. Suzuki, "Electrostatic instability of liquid droplets on MEMS-based pillared surfaces," *Sensors and Actuators B: Chemical,* vol. 225, pp. 492-497, 2016.

[15] W. Kim, D. Kim, S. Park, D. Lee, H. Hyun, and J. Kim, "Engineering lotus leaf-inspired micro-and nanostructures for the manipulation of functional engineering platforms," *Journal of industrial and engineering chemistry,* vol. 61, pp. 39-52, 2018.

[16] F. Li, W. Liu, C. Stefanini, X. Fu, and P. Dario, "A novel bioinspired PVDF micro/nano hair receptor for a robot sensing system," *Sensors,* vol. 10, no. 1, pp. 994-1011, 2010.

[17] J. Pu, X. Yan, Y. Jiang, C. Chang, and L. Lin, "Piezoelectric actuation of direct-write electrospun fibers," *Sensors and Actuators A: Physical,* vol. 164, no. 1-2, pp. 131-136, 2010.

[18] Z. Liu, C. Pan, L. Lin, and H. Lai, "Piezoelectric properties of PVDF/MWCNT nanofiber using near-field electrospinning," *Sensors and Actuators A: Physical,* vol. 193, pp. 13-24, 2013.

[19] A. S. Motamedi, H. Mirzadeh, F. Hajiesmaeilbaigi, S. Bagheri-Khoulenjani, and M. Shokrgozar, "Effect of electrospinning parameters on morphological properties of PVDF nanofibrous scaffolds," *Progress in biomaterials,* vol. 6, no. 3, pp. 113-123, 2017.

[20] Z. Liu, C. Pan, L. Lin, J. Huang, and Z. Ou, "Direct-write PVDF nonwoven fiber fabric energy harvesters via the hollow cylindrical near-field electrospinning process," *Smart materials and structures,* vol. 23, no. 2, p. 025003, 2013.

[21] Q. Lu, K.-Y. Song, Y. Feng, J. Xie, "Fabrication of suspended uniform polymer microfibers by FDM 3D printing," *CIRP Journal of Manufacturing Science and Technology,* vol. 32, pp. 179-187, 2021.

[22] Proto-Pasta, "Proto-Pasta Magnetic Iron PLA Technical Data Sheet," *available at https://www.proto-pasta.com/pages/technical-data-sheets,* 23 December 2020

[23] S. A. Kumar and Y. S. Narayan, "Tensile testing and evaluation of 3D-printed PLA specimens as per ASTM D638 type IV standard," in *Innovative Design, Analysis and Development Practices in Aerospace and Automotive Engineering (I-DAD 2018):* Springer, 2019, pp. 79-95.

Electrochemical Nanoimprint Lithography

Hantao Xu, Lianhuan Han*, Bingqian Du, Yang Wang, Zhen Ma, Zhong-Qun Tian, Zhao-Wu Tian, and Dongping Zhan*

Abstract: **Electrochemical Nanoimprint Lithography (ECNL) works directly on semiconductor wafer, free of thermoplastic and photocuring resists, and without any auxiliary process. The principle of ECNL is the spatially confined electrochemical corrosion caused by the contact potential across the metal/semiconductor phase boundaries exposed to electrolyte solution. ECNL has been proved successful in fabricating various functional micro/nano-structures directly on gallium arsenide and silicon wafer. By virtue of photoelectric effect of semiconductors, the corrosion process can be well accelerated, and the ECNL efficiency is improved. The machining accuracy is determined by the diffusion coefficient of the holes and the corrosion rate of the semiconductor ($\mu = (D_{hole} / k_{corr})^{1/2}$), which is usually smaller than the Debye length of the space charge layer at the metal/semiconductor phase boundary. Here we would like to report our recent progresses in ECNL.**

I. INTRODUCTION

Corrosion is a spontaneously electrochemical phenomenon in the natural world. It is reported that one third of the world's steel products are corroded each year. Therefore, significant efforts have been made for anti-corrosion. On the contrary, as an electrochemical approach of material removal, corrosion can be employed for micro/nano-machining if the corrosion process is confined spatially at a micrometer or nanometer-scale [1]. We demonstrated that the contact potential between metal and semiconductor can induce the corrosion of the latter strictly along the metal/semiconductor/electrolyte 3-phase boundaries, and develop the termed ECNL technique for the template forming of 3D micro/nano-structures (3D-MNSs).

Indeed, metal-assisted chemical etching (MacEtch) is based on the same principle [2]. When patterned metal film or nanoparticles were deposited on the wafer surface, the electrons in the semiconductor phase will spontaneously flow across their boundaries to the metal phase because the Fermi energy level of the semiconductor is higher than that of the metal. Therefore, a contact electric field is established across the metal/semiconductor boundaries where the metal side is negatively charged and the semiconductor side is positively charged. If the metal/semiconductor boundaries were immersed in an electrolyte solution containing favorable electron acceptors, upright 3D-MNSs will be obtained on the surface of semiconductor wafer. However, MacEtch has to employ photolithography to fabricate the metallic micro/nano-patterns, which can't be repetitively

used. Furthermore, the collapse of the metallic micro/nano-patterns makes MacEtch unqualified for the fabrication of 3D-MNSs with large feature size (> 10 μm).

Nanoimprint Lithography (NIL) has been paid great attention because of its simple process, high throughput and low cost. Directly imprint for semiconductor wafers was also proposed by Chou's group in 2002, termed as laser-assisted direct imprinting (LADI) [3]. A high powerful density laser was employed to melt the silicon surface. Meanwhile, a silica mold was imprinted into the melted silicon by extremely pressure (> 1 MPa). However, LADI requires extreme experimental conditions that may damage the workpiece, so it has been seldom reported since then.

We proposed the principle of Electrochemical Nanoimprint Lithography (ECNL), registered the patent in 2013 and published the research paper in 2016 [4, 5]. As depicted in Fig. 1, by employing a metalized template, ECNL is capable of fabricating the large-area 3D-MNSs directly on semiconductor wafer, and the size scale can be nanometer though micrometer to millimeter. More importantly, ECNL is free of thermoplastic and photocuring resists used in conventional nanoimprint lithography, and the photoresists used photolithography. The machining accuracy is determined by the diffusion coefficient of holes and the corrosion rate of the semiconductor ($\mu = (D_{hole} / k_{corr})^{1/2}$). ECNL is believed as a prospective technique for the mass production of functional 3D-MNSs.

Compared with LADI process, the experimental conditions of ECNL are milder, the cost is much lower, and the equipment is simpler, which is more competitive for semiconductor industry. Distinct from MachEtch, ECNL employs a metalized mold instead of patterned metal film, which avoids the metal contaminations and can be reused repetitively. In the last five years, many research papers have reported on the application of ECNL to fabricate round holes, stripes, cylinders, inverted pyramids, and continuous curved surfaces on porous silicon (Si), single crystalline Si, and gallium arsenide (GaAs) wafers [6-11].

II. ELECTROCHEMICAL NANOIMPRINT LITHOGRAPHY

A. The principle of ECNL

The basic principle of ECNL is to confine the anodic corrosion of semiconductors wafer by holes strictly at the metal/semiconductor/electrolyte solution 3-phase interface. Fig. 1 shows the platinum (Pt)/GaAs/electrolyte solution 3-

---

State Key Laboratory of Physical Chemistry of Solid Surfaces, Department of Chemistry, College of Chemistry and Chemical Engineering, and Department of Mechanical and Electrical Engineering, School of Aerospace Engineering, Xiamen University, Xiamen 361005, China

*Contacting Author: Prof. LianhuanHan (hanlianhuan@xmu.edu.cn); Prof. Dongping Zhan (dpzhan@xmu.edu.cn).
*This study is supported by the National Natural Science Foundation of China (21827802, 22021001).

978-1-6654-3008-1/21 $31.00 © 2021 IEEE

phase interface as an example [7-8]. The processes of ECNL include three steps as followed:

1) A compact contact between the Pt metallized mold and GaAs wafer to form the interior electron pathway by the applied pressure. The electrons flow spontaneously from GaAs to the Pt and leave the same amounts of holes at the GaAs boundaries, due to the difference of electron work functions between them.

$$h^+e^-(GaAs) \rightarrow h^+(GaAs) + e^-(Pt) \quad (1)$$

2）At the Pt/solution interface, permanganate anions (MnO_4^-) capture the electrons on the Pt surface, and are reduced wherein.

$$8H^+ + MnO_4^- + 6e^- \rightarrow Mn^{2+} + 4H_2O \quad (3)$$

3）GaAs is anodized or corroded by the accumulated holes to form the aimed 3D-MNSs at the contact area.

$$GaAs + 3H_2O + 6h^+ \rightarrow Ga^{3+} + AsO_3^{3-} + 6H^+ \quad (2)$$

It should be noted that the corrosion of GaAs is induced by the contact potential across the Pt/GaAs boundaries. If the local GaAs is corroded and separated from Pt metalized mold, the corrosion process will be ceased automatically, which ensures the machining accuracy of ECNL. The machining accuracy μ is determined by the electron-hole pair recombination rate, the corrosion rate k_{corr}, the hole diffusion coefficient D_{hole}. Generally, the diffusion rate of holes in GaAs is much slower comparing to the electron-hole recombination rate, and the machining accuracy can be expressed as:

$$\mu = (D_{hole}/k_{corr})^{1/2} \quad (4)$$

If $k_{corr} = 10^{-6}$ s^{-1}, $D_{hole} = 10^{-8}$ cm$^2 \cdot$ s^{-1}, μ will reach 1 nm. It predicts that the machining accuracy of ECNL can be no more than the theoretical Debye length of the space charge layer at the GaAs side of the contacted Pt/GaAs boundaries.

B. Fabrication of Sub-micron structure by ECNL

An imprint mold with simple two-dimensional structure was selected to verify the feasibility of ECNL [7]. A polymethyl methacrylate (PMMA) mold coated with 100-nm-thickness Pt film was pressed on the n-GaAs wafer at about 0.5 atm pressure. 40 mM KMnO$_4$ is used as an oxidant, and 1.84 M H$_2$SO$_4$ provides an acidic electrolyte environment to ensure the dissolution of etching products. Atomic force microscopy (AFM) proved that nano-trenches with a width of 350 nm (Fig. 2a) were transferred to the n-GaAs wafer to form nanowires with a width of 430 nm (Fig. 2b). Nanowires with a width of 510 nm (Fig. 2c) were also imprinted onto the n-GaAs wafer to form nano-trenches with a width of 700 nm (Fig. 2d). The plane machining accuracy is maintained at the sub-micron level.

In the ECNL experiment, good contact is required. The working pressure is as low as 0.5 atm, which means that the mold can be reused without tool wear and thermal effects. The Pt film can also be renewed by electroplating or magnetron sputtering. Compared with NIL, ECNL is free of mold filling and demolding. Different from LADI, ECNL can work at low pressure and ambient temperature without

any expensive equipment, and without surface damage caused by high-energy laser beam and wafer fragmentation by high pressure.

C. Fabrication of multilevel 3D microstructure by ECNL

The capability of ECNL to fabricate multilevel 3D-MNSs is further demonstrated [8]. A diffractive microlens array with eight-phase level was fabricated on GaAs wafer by using a 100-nm-thickness Pt film metallized PMMA mold. The SEM image in Fig. 3a and Fig. 3b proves that the eight steps are completely transferred onto the semiconductor surface. The profile curve statistics in Fig. 3c and Fig. 3d show that the total height of the steps on the template is 1.14 μm, and the total height of the GaAs surface is 1.03 μm. The square in the center of GaAs is the lowest point, and gradually increases outward with the development of the steps, presenting a complementary state with the imprint mold.

The ability to fabricate continuously curved structure was also demonstrated, which is difficult to be completed by traditional mechanical machining [8]. Fig. 4a shows a Pt metalized PMMA mold contained convex lens array with 105.4 μm in diameter and 4.7 μm in height. The concave microlens array on the n-GaAs wafer is shown in Fig. 4b, where the depth is 1.50 μm and the diameter is 58.43 μm (Fig. 4c). When the height of the convex microlens on the mold is 1.5 μm from the top point, the diameter of the spherical crown is 57.69 μm (Fig. 4c). The machining tolerance is only 0.74 microns. The results show that ECNL is more suitable for continuous curved surfaces or special-shaped structures comparing to the traditional precision mechanical machining.

"Buckling effect" occurs in the ECNL experiment when mold is undergoing an elastic deformation, which is termed as electrochemical buckling microfabrication (ECBM) [4]. As shown in Fig. 5a, regular secondary 3D nanostructures generated on the mold electrode surface by a constant force because of the elastic-modulus difference between the PMMA mold and Pt film. Then, bromine (Br$_2$) is electrogenerated on-site and confined close to the surface of Pt metalized PMMA mold as the etchant. Fig. 5b shows the 23-level concentric-circle nanostructure fabricated on the Ga$_x$In$_{1-x}$P wafer surface when a contact force between the Pt metalized PMMA mold and the Ga$_x$In$_{1-x}$P wafer as held at 20 mN. It breaks the limitation that such structures can only be fabricated by high-energy direct writing technology, and the machining accuracy was also improved by the electrochemical micromachining method.

D. Kinetics of corrosion

Electrochemical kinetics of the semiconductor corrosion process is very complicated. The traditional "hole injection" theory in MacEtch has not been studied on the point of reaction kinetics. Our group believes that contact potential across the metal/semiconductor boundaries is the driving force of semiconductor corrosion or etching. In order to study the cathode and anode reaction kinetics, we used a three-electrode system to measure the open circuit potentials (OCP) of the Pt-mold electrode/solution interface and the n-

GaAs wafer/solution interface, and to study the corrosion kinetics by polarization curve method [12].

As shown in Fig. 6a, the n-GaAs wafer/solution interface and the Pt-mold electrode/solution interface are in a thermodynamic equilibrium state when they separated from each other, and the OCPs were measured at -0.63 V and 0.84 V vs. an Hg/Hg$_2$SO$_4$ reference electrode, respectively. When n-GaAs wafer and Pt-mold electrode come into contact at 0.5 atm, the interfaces were polarized to 0.13 V and 0.67 V respectively, the corrosion current will be generated in this case. The potential of the two interfaces returned to their initial OCPs when they separated again from each other. The polarization curve of n-GaAs wafer/solution interface (Fig. 6c) shows that the corrosion current density of the anodic dissolution process of n-GaAs is 6.7 μA/cm^2, showing the rate-determined step in ECNL processes. Furthermore, the contact potential across the Pt/n-GaAs boundary can be enhanced by the photoelectric effect of the semiconductor (Fig. 6b). The OCP of n-GaAs wafer/solution interface under illumination was increased to 0.38 V, and the corrosion current was increased by an order of magnitude to 77.9 μA/cm^2 (Fig. 6d). The photoelectric effect enhanced semiconductor corrosion kinetics provides a new technical guidance for improving ECNL's machining accuracy and efficiency by external physical-field modulations.

E. The advantages of ECNL

In brief, ECNL is a one-step electrochemical corrosion process, but not a real "imprint forming" process. It has been proven to be a mature technology in the micromachining of semiconductor surface through chemical removal. Based on the recent progresses in this emerging field, the advantages of ECNL are summarized as followed:

1) ECNL breaks the limitation of the two-electrode and three-electrode system used in conventional electrochemical machining, and overcomes the technical difficulties such as the uneven potential distribution and the alignment between the working electrode and workpieces.

2) ECNL is a controllable template-forming technique for the mass production of various complex 3D-MNSs with high efficiency and accuracy. The machining processes can be self-ceased when the mold electrode is separated from the semiconductor during the corrosion process.

3) Compared with photolithography and nanoimprint lithography, ECNL is one-step forming technique free of photoresist, thermoplastic and photocuring resists, and without any auxiliary process.

4) Compared with precision mechanical machining, ECNL works in a gently chemical removal mode at low contact pressure. Thus, it avoids tool wear, thermal effect residual stress and mechanical damage to the fabricated 3D-MNSs.

6) Compared with energy-beam direct-writing techniques, ECNL can avoid the surface damages in the surface and subsurface of the workpiece.

III. PROSPECTS OF FUTURE RESEARCH

ECNL has shown broad prospects for the micro/nano-fabrications on semiconductor wafer. Future investigations

should be focused on solving the following scientific and technical issues:

1) The reaction kinetics of the contact potential induced semiconductor corrosion. It is essential to clarify the reaction mechanism and kinetics, to reveal the coupling effect of various physical field, to find the modulation strategies of the corrosion process and, finally, to promote the technological level of ECNL to the practical industrial applications.

2) The mass transfer and mass balance during the ECNL machining process. There is very limited electrolyte solution between the mold electrode and semiconductor wafer. The depletion of the electron acceptors will slow down or even cease the corrosion reaction, and the precipitation of corrosion product will make the surface of semiconductor passivated. These problems limited the material removal amount. Consequently, ECNL is only applicable to fabricate bas-relief 3D MNSs. It is essential to solve the mass transfer and mass balance problem in order to fabricate 3D MNSs with high-aspect ratio.

3) To promote ECNL from the feasibility to the level of functionalized microdevices needs the multidiscipline cooperation of the researchers in various domains.

IV. REFERENCE

[1] D. Zhan, L. Han, J. Zhang, Q. He, Z. W. Tian, Z. Q. Tian, "Electrochemical micro/nano-machining: Principles and practices", Chem. Soc. Rev., Vol. 46, pp. 1526-1544, 2017.

[2] H. Hana, Z. Huang, W. Leea, "Metal-assisted chemical etching of silicon and nanotechnology applications", Nano Today Vol. 9, 271-304, 2014.

[3] S. Y. Chou, C. Keimel, J. Gu, "Ultrafast and direct imprint of nanostructures in silicon", Nature, Vol. 417, pp. 835-837, 2002.

[4] D. Zhan, J. Zhang, Z.-Q. Tian, J. Jia, L. Han, Y. Yuan, Z.-W. Tian, "A confined etching layer method for fabricate complex 3D multistage micro-nano structures", CN 103325674 B, 2013.

[5] J. Zhang, B. Y. Dong, J. Jia, L. Han, F. Wang, C. Liu, Z. Q. Tian, Z. W. Tian, D. Wang, D. Zhan, "Electrochemical buckling microfabrication", Chem Sci, Vol. 7, pp. 697-701, 2016.

[6] B. P. Azeredo, Y.-W. Lin, A. Avagyan, M. Sivaguru, K. Hsu, P. Ferreira, "Direct imprinting of porous silicon via metal-assisted chemical etching", Adv. Funct. Mater., Vol. 26, pp. 2929-2939, 2016.

[7] J. Zhang, L. Zhang, W. Wang, L. Han, J. C. Jia, Z. W. Tian, Z. Q. Tian, D. Zhan, "Contact electrification induced interfacial reactions and direct electrochemical nanoimprint lithography in n-type gallium arsenate wafer", Chem Sci, Vol. 8, pp. 2407-2412, 2017.

[8] J. Zhang, L. Zhang, L. Han, Z. W. Tian, Z. Q. Tian, D. Zhan, "Electrochemical nanoimprint lithography: When nanoimprint lithography meets metal assisted chemical etching", Nanoscale, Vol. 9, pp. 7476-7482, 2017.

[9] K. Kim, B. Ki, K. Choi, S. Lee, J. Oh, "Resist-free direct stamp imprinting of GaAs via metal-assisted chemical etching", ACS Appl. Mater. Interfaces, Vol. 11, pp. 13574-13580, 2019.

[10] H. Li, J. Niu, G. Wang, E. Wang, and C. Xie, "Direct production of silicon nanostructures with electrochemical nanoimprinting", Vol. 1, pp. 1070–1075, 2019.

[11] S. Bastide, E. Torralba, M. Halbwax, S. Le Gall, E. Mpogui, C. Cachet-Vivier, V. Magnin, J. Harari, D. Yarekha, J. P. Vilcot, "3D patterning of Si by contact etching with nanoporous metals", Front in Chem., Vol. 7, pp. 256, 2019.

[12] C. Guo, L. Zhang, M. M. Sartin, L. Han, Z. W. Tian, Z. Q. Tian, D. Zhan, "Photoelectric effect accelerated electrochemical corrosion and nanoimprint processes on gallium arsenide wafers", Chem Sci, Vol. 10, pp. 5893-5897, 2019.

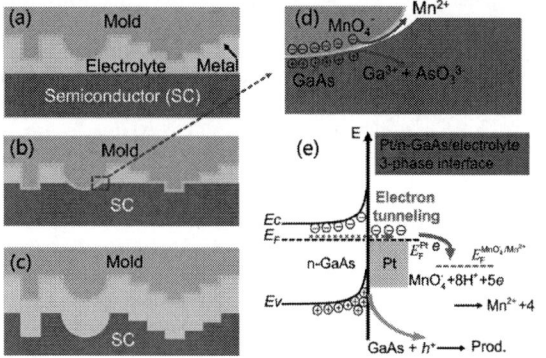

Fig. 1. Schematic diagram of Electrochemical Nanoimprint Lithography.

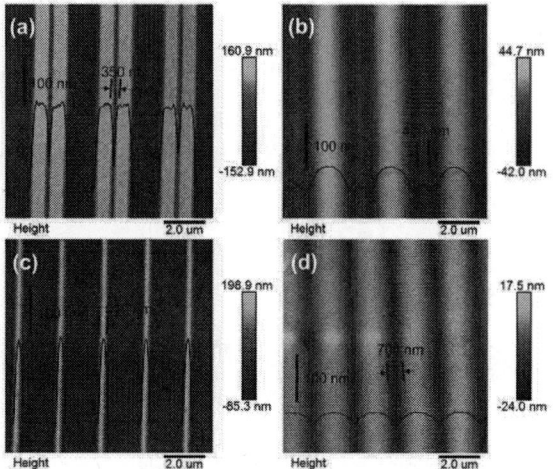

Fig. 2. (a) AFM image of nano-grooves on the mold and (b) nano-grooves transferred onto n-GaAs. (c) AFM image of nanowires on the mold and (d) nanowires transferred onto n-GaAs.

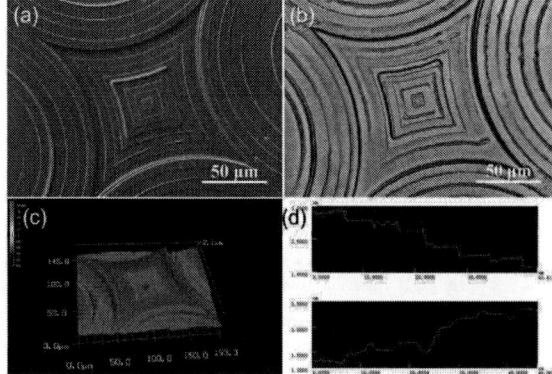

Fig. 3. The multilevel 3D-MNSs fabricated on GaAs by ECNL. (a) the SEM image of diffractive microlens with eight-phase level on mold and (b) diffractive microlens transferred onto GaAs. (c) The 3D height image of (b). (d) The cross-sectional profiles of mold and GaAs wafer.

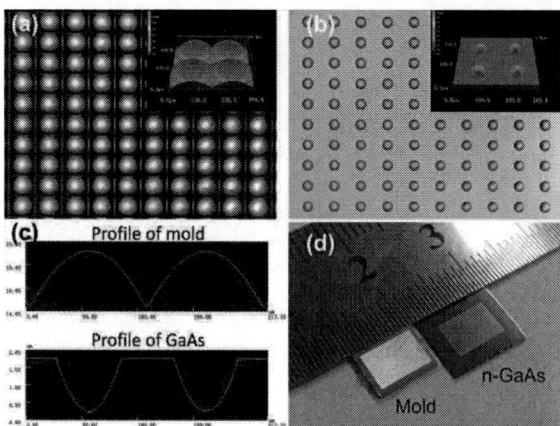

Fig. 4. The continuously curved 3D-MNSs fabricated on n-GaAs wafer. (a) The confocal laser microscopy image of mold and (b) the concave microlens transferred onto GaAs. (c) the cross profiles of on mold and GaAs. (d) The large area optical image of microlens array fabricated on n-GaAs.

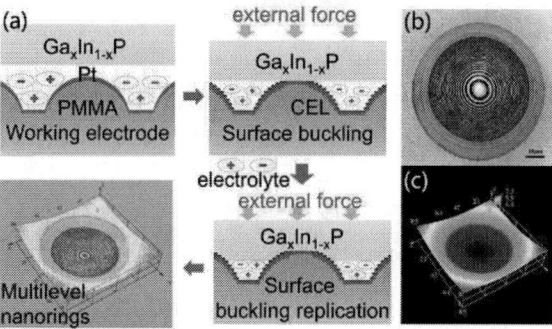

Fig. 5. (a) Schematic illustration of the ECBM process. (b) and (c) Hierarchical Fresnel nanostructures fabricated on the $Ga_xIn_{1-x}P$ workpiece by 20 mM contact force and 20 mM KBr as precursor.

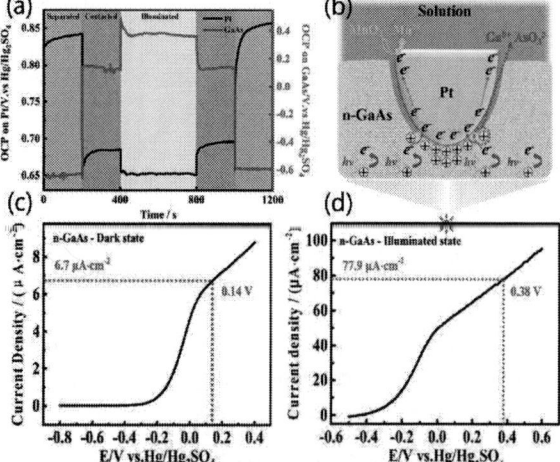

Fig. 6. (a) The potential measurements of the Pt/solution interface (black line) and the n-GaAs/solution interface (red line). (b) Schematic diagram of the photoelectric effect accelerated ECNL processes. (c) and (d) Linear scan voltammograms of the n-GaAs electrode at a scan rate of 100 mV·s^{-1} at dark state and light state.

978-1-6654-3008-1/21 $31.00 © 2021 IEEE

Gap in pagination due to unavailable papers.

Pages 1341-1346

April 25-29 , 2021   Xiamen, China

Soft Micro-robots and Micro-actuators: Materials, Design, Control and Applications

Lining Sun

Soochow University, Suzhou 215000 P.R. China

Website : group or personal website if applicable

ABSTRACT

Microrobots have drawn extensive attentions due to their great potential for biomedical and MEMS applications. In this talk, I would like to share my experiences of creating soft micro-robots and micro-actuators with stimulus-responsive materials. Then, I will report our field based teleoperation actuation systems to accurately control these micro-robots. Finally, I will show our attempts of applying these micro-robots and micro-actuators to execute targeted drug delivery and therapy trial and cooperatively micro-manipulation.

BIOGRAPHY

Prof. Lining Sun is currently the director of Robotics and Microsystems Center of Soochow University, and is the dean of the College of Mechatronic Engineering of Soochow University. He is also the vice-director of the State Key Laboratory for Robotics and System in Harbin Institute of Technology. He is the Chang-Jiang Professor and the obtainer of The National Natural Science Fund for Distinguished Young Scholars. Prof. Sun has directed more than 30 projects from "863" Program, "973" Program and NSFC, and obtained many academic honors and awards, including National Science and Technology Progress Awards and Provincial Science and Technology Prizes, the Ho Leung Ho Lee Foundation Science and Technology Innovation Award, to name a few.

978-1-6654-3008-1/21 $31.00 © 2021 IEEE

Gap in pagination due to unavailable paper.

Pages 1348-1352

Proceedings of the 16th Annual IEEE International
Conference on Nano/Micro Engineered and Molecular Systems
April 25-29, 2021

Influence of Mass Loading Effect on Radiation Quality Factor of BAW Magnetoelectric Antenna

Junru Li, *Student Member, IEEE*, Chunrui Peng, Yang Gao, Wanchun Ren, and Xuefeng He

Abstract—Stacked piezomagnetic/piezoelectric heterostructure have attracted wide attention in antenna community for their potential application in a subminiature device, and the number of layers has a significant influence on the performance of antenna especially when considering the mass loading effect. However, its physical essence and device structure optimization still keep unidentified. Therefore, the analytical models of energy and radiated power for bulk acoustic wave magnetoelectric (BAW ME) antenna were constructed respectively to derive the corresponding analytic results of stacked structures with 2 to 6-layer under the conditions of fixed thickness of unit layer and resonant frequency respectively. The calculation data shows that the 2-layer and 3-layer structures enjoy the optimal radiation quality factor respectively when the thickness of unit layer or resonant frequency is fixed. The finite element analysis is exploited to simulate the stress field distribution of the stacked layers, which reveals the physical essence of BAW ME antenna with different radiation quality factor. Furthermore, the analytical models are validated by the numerical results.

I. INTRODUCTION

As the antenna technology develops, mechanical antenna has gradually come into people's view, which decouples antenna size from free space wavelength and releases the application of wireless communication systems and radars in mobile platforms [1, 2]. Bulk acoustic wave (BAW) magnetoelectric (ME) antenna is a typical strain powered (SP) mechanical antenna, whose basic elements include piezoelectric phase for excitation and piezomagnetic phase for radiation. The operating principle of magnetic oscillation actuated by acoustics rather than current determines that BAW ME antenna is 1 to 2 orders of magnitude smaller than the existing smallest electrically small antenna [3]. Some numerical analysis have considered multiferroic heterostructures as an avenue to reduce antenna size over a decade ago [4, 5]. Other efforts have been taken on the radiated efficiency modeling which shows that the SP antenna enjoys higher radiated power and efficiency compared to the electrically small antenna with the same size [6-8]. However, the relationship between internal energy and external radiation for multilayer heterostructures is still unclear. So it's necessary to dig deeply into the mechanism of energy storage and radiation of stacked structure for the physical design and performance optimization.

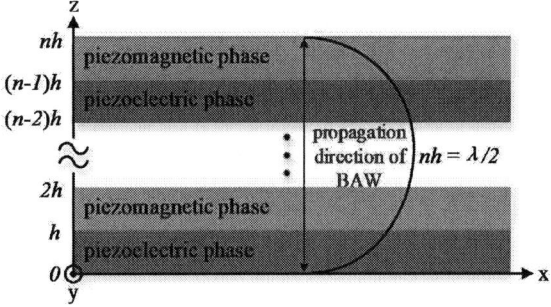

Fig. 1. Schematic of the stacked layers. The total thickness of layered piezoelectric/piezomagnetic heterostructures meets the acoustic resonance condition.

The energy in BAW ME antenna is transferred through two phases coupling effect, as shown in Fig. 1. It has been shown that the conversion between mechanical and magnetic energy can be effectively realized by coupling strong strains in the two phases [9-12]. Recently, Yao et al. calculated the radiation performance of BAW ME antenna by a finite-difference time-domain (FDTD) method and found out its time-dependent character of stress and electric field [13, 14]. The study shows that the 3-layer structure can produce higher strain locating on the piezomagnetic phase, which is more conducive to electromagnetic (EM) radiation. However, other researchers fabricated BAW ME antenna with 2-layer structure which achieved effective coupling between the two phases in 2017 [15]. In fact, if the thickness of unit layer keeps constant, the resonant frequency of the device will shift with the addition of stacked layers because of the mass loading effect [16]. Accordingly, it is a crucial task to understand the principle of BAW ME antenna and optimize its stacked structure for the design of device.

In this work, we present a calculation method of energy and average radiated power of multi-layer BAW ME antenna based on the stress field distribution. Considering the mass loading effect, the relationship between the number of stacked layers and radiation quality factor is demonstrated. By establishing the finite element analysis (FEA) model of stress in ultra-high frequency (UHF) band, the physical essence of different stacked structures with different radiation quality factor is further revealed, and the analytical models are verified.

---

*Research supported jointly by the Science and Technology on Electronic Information Control Laboratory under Grant No. 6142105200203, Fund of Robot Technology Used for Special Environment Key Laboratory of Sichuan Province under Grant No. 20kfkt02, and Science and Technology Foundation of Southwest University of Science and Technology under Grant No. 20zx7114. (*Corresponding authors: Wanchun Ren, and Xuefeng He.*)

Junru Li and Xuefeng He are with the College of Optoelectronic Engineering, Chongqing University, Chongqing 400044, China (e-mail: li_junru@foxmail.com; hexuefeng@cqu.edu.cn).

Chunrui Peng is with the School of Electronic Science and Engineering, University of Electronic Science and Technology of China, Chengdu 611731, China (e-mail: 17865916198@163.com).

Yang Gao and Wanchun Ren are with the School of Information Engineering, Southwest University of Science and Technology, Mianyang 621010, China (e-mail: 29636791@qq.com; rwch_qw@163.com).

978-1-6654-3008-1/21 $31.00 © 2021 IEEE

II. ANALYTICAL MODEL

In this section, the analytical models of potential energy, kinetic energy and average radiated power, under the weak magnetic field condition, are developed to compare the energy storage and the EM radiation of the antenna with different number of stacked layers. Considering BAW as a longitudinal wave propagating along the z-axis, the constitutive relations of piezoelectric and piezomagnetic phase are defined as 1D form by (1) and (2) respectively:

$$D = \varepsilon_T E + d_E T \tag{1}$$

$$B = \mu_T H + d_H T \tag{2}$$

where T is stress tensor of the film, E and H are the electric and magnetic field intensity vector respectively, D and B are the electric and magnetic flux density vector respectively, ε_T and μ_T are the dielectric constant and permeability of the two phases respectively under stress-free condition, d_E and d_H are the longitudinal piezoelectric and piezomagnetic coefficients respectively.

According to the constitutive relation, taking the mechanical quantities as a medium is a theoretical foundation of coupling the piezoelectric and piezomagnetic phases. To simulate the operation of device in lossless condition and simplify the calculation process, the constitutive relation of piezoelectric phase is simplified under an open-circuit excitation condition (i.e. the value of electric flux density is zero), and the piezomagnetic phase, under a weak magnetic field condition (i.e. the magnetic field intensity is zero), yield:

$$E = -\left(d_E / \varepsilon_T\right) T \tag{3}$$

$$B = d_H T \tag{4}$$

The relationships between the stress and the electric/magnetic fields are constructed by (3) and (4) respectively. Furthermore, the total potential energy can be solved, which includes the piezoelectric and piezomagnetic phases. The exchange of electric energy and magnetic energy with the outside world is negligible compared to the stored energy. As a result, the piezoelectric potential energy W_{PE} can be described as the sum of mechanical and electric energy, as shown in (5). And the piezomagnetic potential energy W_{PM} can be described as the sum of mechanical and magnetic energy, as shown in (6).

$$W_{PE} = \frac{1}{2}\iiint s_D |T|^2 \, dv + \frac{1}{2}\iiint \frac{|D|^2}{\varepsilon_T} \, dv \tag{5}$$

$$W_{PM} = \frac{1}{2}\iiint s_B |T|^2 \, dv + \frac{1}{2}\iiint \frac{|B|^2}{\mu_T} \, dv \tag{6}$$

where s_D and s_B are the compliance constants of the two phases respectively. The subscripts indicate that the corresponding parameters measured under the constant electric and magnetic flux density condition respectively. Combining open-circuit excitation condition with (5), the simplified (7) is yielded. Based on the definition of magneto-mechanical coupling coefficient $k_H^2 = d_H^2 / \left(s_H \mu_T\right)$ and $s_B = s_H (1 - k_H^2)$, and combining (4)

with (6), a simplified equation can be presented as (8).

$$W_{PE} = \frac{1}{2}\iiint s_D |T|^2 \, dv \tag{7}$$

$$W_{PM} = \frac{1}{2}\iiint s_H |T|^2 \, dv \tag{8}$$

where s_H is compliance constant measured under a constant magnetic field condition. The potential energy W_P in BAW ME antenna can be expressed as the total mechanical energy of the system stored in the form of mechanical stress, thus yield:

$$\begin{aligned} W_P &= \frac{1}{2}\iiint_{v_1} s_D |T|^2 \, dv + \frac{1}{2}\iiint_{v_2} s_H |T|^2 \, dv \\ &= \frac{1}{2} s_D A \int_{z_1} |T|^2 \, dz + \frac{1}{2} s_H A \int_{z_2} |T|^2 \, dz \end{aligned} \tag{9}$$

where A is the area of the radiation surface (i.e. normal to the z-axis), v_1 and v_2 are the volume range of the two phases respectively, z_1 and z_2 are the thickness range of the two phases respectively. Similarly, the kinetic energy W_V can be yielded based on the theorem of the kinetic energy of particles.

$$W_V = \iiint_{v_1} |T| \, dz + \iiint_{v_2} |T| \, dz = A\int_{z_1} |T| \, dz + A\int_{z_2} |T| \, dz \tag{10}$$

It is worth noting that due to the difference of the compliance constants between the piezoelectric and piezomagnetic phases, $s_H = n s_D$, the stress corresponding in the two phases is subject to their constitutive relations.

$$\begin{aligned} & s_D T_1 + \frac{d_E}{\varepsilon_T} D = s_B T_2 + \frac{d_H}{\mu_T} B \\ & \because D = 0, \quad B = d_H T_2, \\ & \quad k_H^2 = d_H / s_H \mu_T, \quad s_B = s_H \left(1 - k_H^2\right) \\ & \therefore s_D T_1 = s_H \left(1 - k_H^2\right) T_2 + s_H k_H^2 T_2 \\ & and \therefore T_1 / T_2 = s_H / s_D = n \end{aligned} \tag{11}$$

where T_1 and T_2 are the stress of piezoelectric and piezomagnetic phases respectively.

BAW ME antenna is excited by a sinusoidal signal. If the total thickness, d, of the two phases meets the BAW resonant condition (i.e. $d = \lambda_{ac}/2$), the stress field distribution can be expressed as:

$$|T| = \begin{cases} nT_0 \sin(\dfrac{2\pi}{\lambda_{ac}} z) & z \in z_1 \\[2mm] T_0 \sin(\dfrac{2\pi}{\lambda_{ac}} z) & z \in z_2 \end{cases} \tag{12}$$

where λ_{ac} is the wavelength of the BAW, T_0 is the stress amplitude, n is the ratio of stress amplitude between the piezoelectric and piezomagnetic phases, which is from (11). Since the strain is transmitted continuously between the two phases, the potential and kinetic energy of BAW ME antenna with 2 to 6-layer can be calculated by integrating (9) and (10) in the thickness direction. The analytic results are shown in Table I.

Antenna has better radiation performance when the

radiation quality factor is smaller. To calculate the quality factor, an analytic model of average radiated power $\overline{P}_{rad}$ needs to be established firstly. The definition of $\overline{P}_{rad}$ is as follows:

$$\overline{P}_{rad} = \frac{1}{2\eta_0} \iint_s |\boldsymbol{E}_0|^2 \, ds \tag{13}$$

Where η_0 is the wave impedance of free space, $\boldsymbol{E}_0$ is the induced electric field. For BAW ME antenna, $\boldsymbol{E}_0$ indicates the aperture electric field derived from the surface of the piezomagnetic phase. Combining (4) with the Faraday's law of electromagnetic induction in time-harmonic domain, $|\boldsymbol{E}_0| = \omega h |\boldsymbol{B}|$, the average radiated power can be expressed in the form of mechanical stress, as shown in (14).

$$
\begin{aligned}
\overline{P}_{rad} &= \frac{\omega^2 h^2 d_H^2}{2\eta_0} \iint_s |\boldsymbol{T}|^2 \, ds = \frac{T_0^2 \omega^2 h^2 d_H^2 A}{2\eta_0} \sin^2\left(\frac{2\pi}{\lambda_{ac}} z\right) \\
&= \frac{T_0^2 \omega^2 h^2 d_H^2 A}{2\eta_0} \cdot \frac{1}{h_2 - h_1} \int_{h_1}^{h_2} \sin^2\left(\frac{2\pi}{\lambda_{ac}} z\right) dz
\end{aligned} \tag{14}
$$

where ω is the angular frequency in operation, h is the thickness of each phase, h_1 and h_2 are the beginning and ending coordinates of the piezomagnetic phase respectively. Using this equation, the average radiated power of BAW ME antenna with 2 to 6-layer can be calculated and the analytic results is as shown in Table I.

Table 1. Analytic results of potential energy, kinetic energy and average radiated power

Layers	Potential Energy	Kinetic Energy	Average Radiated Power
2(d=2h)	$0.25\left(n^2 s_D + s_H\right) A h T_0^2$	$2(n+1) A h T_0 / \pi$	$0.25 T_0^2 \omega^2 h^2 d_H^2 A / \eta_0$
3(d=3h)	$\left(0.293 n^2 s_D + 0.457 s_H\right) A h T_0^2$	$3(n+1) A h T_0 / \pi$	$0.457 T_0^2 \omega^2 h^2 d_H^2 A / \eta_0$
4(d=4h)	$0.5\left(n^2 s_D + s_H\right) A h T_0^2$	$4(n+1) A h T_0 / \pi$	$0.5 T_0^2 \omega^2 h^2 d_H^2 A / \eta_0$
5(d=5h)	$\left(0.605 n^2 s_D + 0.645 s_H\right) A h T_0^2$	$5(n+1) A h T_0 / \pi$	$0.645 T_0^2 \omega^2 h^2 d_H^2 A / \eta_0$
6(d=6h)	$0.75\left(n^2 s_D + s_H\right) A h T_0^2$	$6(n+1) A h T_0 / \pi$	$0.75 T_0^2 \omega^2 h^2 d_H^2 A / \eta_0$

It is clear to see from the analytic results that the average radiated power depends on the operating frequency and the thickness of unit layer. However, there is a trade-off between the frequency and the thickness for BAW resonator because of mass loading effect. So it is necessary that fix one quantity to discuss the change of another quantity.

III. NUMERICAL CALCULATION

To obtain the optimal stacked structure of BAW ME antenna, it is necessary to compare the radiation quality factor Q of antenna with different number of stacked layers. The Q can be defined as:

$$Q = \omega \frac{W_{total}}{\overline{P}_{rad}} = \omega \frac{W_P + W_V}{\overline{P}_{rad}} \tag{15}$$

where W_{total} is the sum of potential and kinetic energy. To simplify the calculation, the Q of the potential and kinetic energy is considered separately. As shown in Table I and (15), the Q is related to the resonant frequency and the thickness of unit layer. So the key to calculate the Q is to determine the resonant frequency corresponding to different number of stacked layers. Taking the existing thin film parameters of AlN and FeGaB as an example[15, 17], the FEA

model is constructed to investigate the relationship among resonant frequency, thickness of unit layer and stacked structure.

A. Fix Thickness of Unit Layer: Stacking Effect

When the thickness of each piezoelectric and piezomagnetic phases is fixed, the resonant frequency correspondingly moves downward with the addition of stacked layers. The relationship between stacked layers and resonant frequency is obtained by FEA simulation, and then the normalized average radiated power with different number of stacked layers is calculated by using the data of Table I, as shown in Fig. 2.

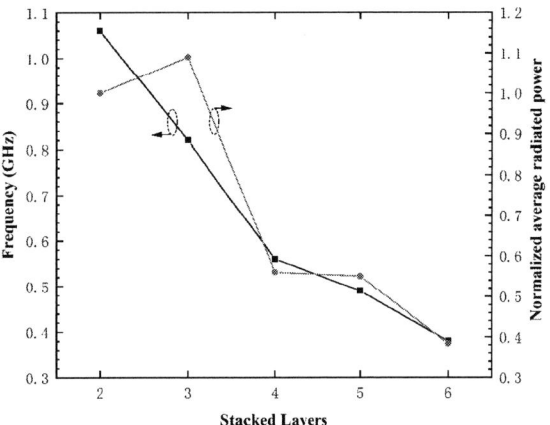

Fig. 2. Correlation between the frequency (left) & normalized average radiated power (right) with the stacked layers. The value of 2-layer is taken as a baseline.

It can be clearly seen from Fig. 2 that the resonant frequency decreases significantly with the addition of stacked layers, which is a dominant factor affecting the average radiated power. Similarly, the data of Table I can also be used to obtain the relationship between the energy and the number of layers, as shown in Fig.3. Then, the Q and radiation energy can be calculated.

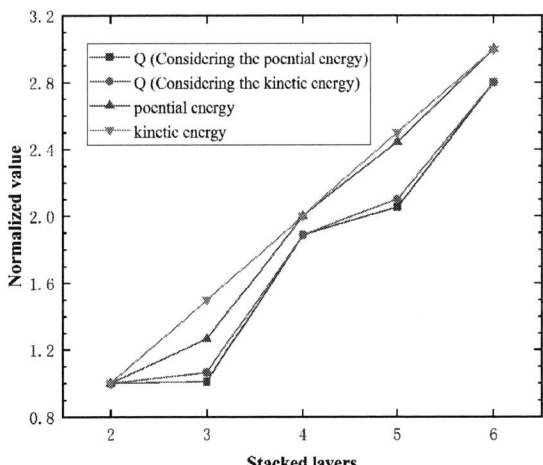

Fig. 3. Influence of stacked layers with fixed thickness of unit layer on the energy and the Q. The value of 2-layer is taken as a baseline. The compliance constant of FeGaB is about six times that of AlN, which means $s_H \approx 6 s_D$.

Energy calculation shows that the energy storage increases with the addition of stacked layers, which is consistent with the physical law. Compared to the average radiated power, the potential energy increases more rapidly from the 2-layer to the 3-layer. It tells that the 2-layer structure prefer radiating energy rather than storing energy. The interesting fact is that the 3-layer structure has the higher average radiated power and the higher Q. From the physical point of view, the reasons are that the frequency shift is caused by the mass loading effect and the stress attenuation is caused by the influence of stacking. To verify this analysis, we use COMSOL to establish the stress simulation model of different number of stacked layers in UHF band, study the stress field distribution of stacked layers at resonant frequency, and calculate the average stress of piezomagnetic phase.

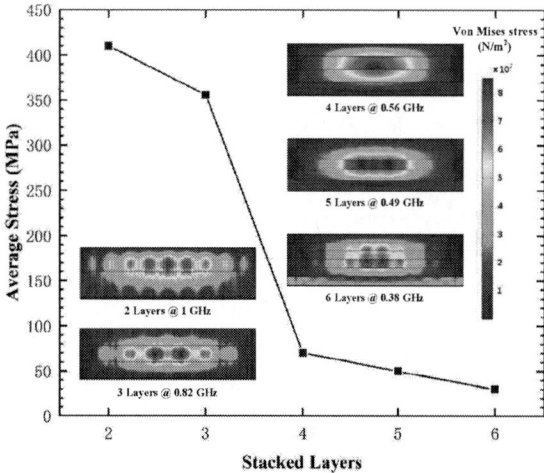

Fig. 4. Average stress of 2 to 6-layer stacked structure with fixed thickness of unit layer. The inset is the stress field distribution of 2 to 6-layer stacked structure. In the FEA model, the piezoelectric material AlN and ferromagnetic material FeGaB are used respectively with identical thickness of 1 μm.

Perfect matched layers are set on both the left and right boundary of the FEA model to suppress the lateral stray of BAW. A sinusoidal RF signal source is applied on the upper and lower electrodes of the bottom piezoelectric phase. Then the stress field distribution of continuous phases is excited in the stacked layers. During the simulation, the resonant frequency of the antenna with different number of stacked layers is identified by frequency scanning. Then as shown in Fig. 4, the value of the average stress in the piezomagnetic phase is calculated by the recorded distribution of stress field.

As shown in the inset of Fig. 4, the stress field shows the sinusoidal distribution along the thickness direction. The phenomenon complies with the assumed stress distribution condition in the analytical model. From the simulation, the average stress value of the 2-layer structure is the highest. The results are consistent with the analytical calculation of the Q, which validates the analytical model. Therefore, for the stacked structure with fixed thickness of unit layer, the more stacked layers, and the worse radiation performance of BAW ME antenna.

B. Fix Frequency: Stacking Effect

By adjusting the thickness of unit layer, BAW ME antenna with different number of stacked layers operates at a fixed resonant frequency (~ 1 GHz). The relationship between the number of stacked layers and the thickness of unit layer is obtained by FEA, then the normalized average radiated power with different number of stacked layers is calculated by using the data of Table I, as shown in Fig. 5.

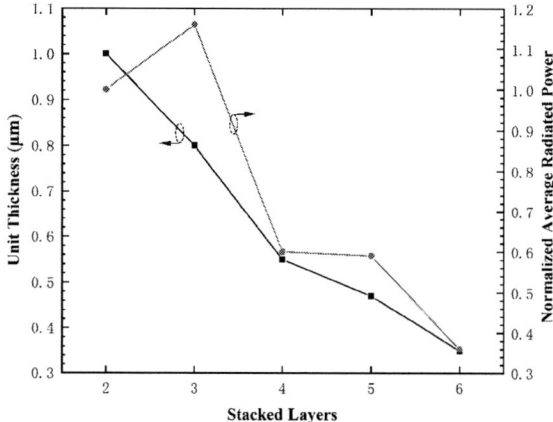

Fig. 5. Correlation between the unit thickness (left) & the normalized average radiated power (right) with the stacked layers. The value of 2-layer is taken as a baseline.

Obviously, in order to ensure that BAW ME antenna with different number of stacked layers operates at the same frequency, it is necessary to reduce the thickness of unit layer. Furthermore, the trend of normalized average radiated power is basically the same as that in Fig. 2, which indicates that the thickness of unit layer also plays a dominant role in the average radiated power. To comprehensively evaluate and predict the radiation performance of BAW ME antenna with different number of stacked layers at a fixed frequency, the data in Table I should be used to calculate the energy and the Q. The results are shown in Fig. 6.

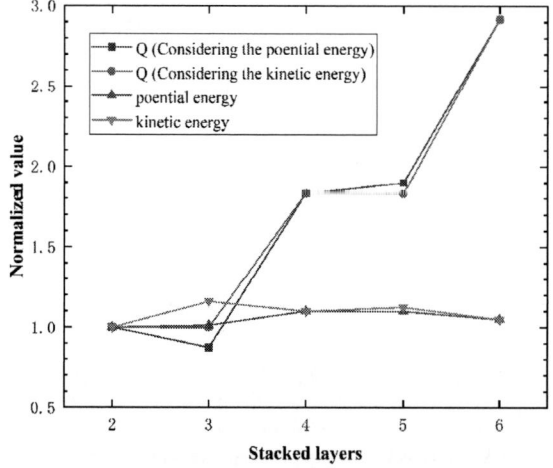

Fig. 6. Influence of stacked layers with a fixed frequency on the energy and the Q. The value of 2-layer is taken as a baseline. The compliance constant of FeGaB is about six times that of AlN, which means $s_H \approx 6 s_D$.

When the resonant frequency is fixed, it can be found that the thickness of unit layer has a great influence on the stored energy. And the radiation performance of the 3-layer structure is slightly better than that of the 2-layer structure in terms of the Q. This result is completely different from that of fixed thickness of unit layer (see Fig. 3). From the point of physics view, the reason for the decrease of the Q is that the mass loading effect is weakened with the decrease of the thickness of unit layer. And the highest density region of the stress field distribution is concentrated on the piezomagnetic phase. The stress of the piezomagnetic phase determines its ability to radiate EM waves when the resonant frequency is fixed. The root cause of this phenomenon is further revealed through the FEA, as shown in Fig. 7.

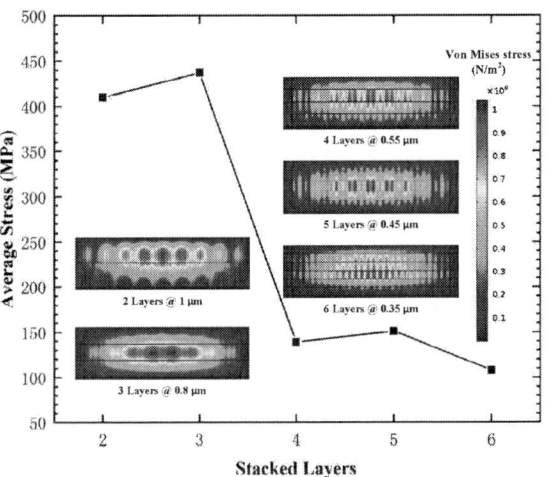

Fig. 7. Average stress of the 2 to 6-layer stacked structure with fixed frequency. The inset shows the distribution of stress field in 2 to 6-layer stacked structure. In the FEA model, the operating frequency is kept at 1 GHz.

It can be seen from the simulation results that the average stress of the 3-layer structure is the largest, the average stress decreased significantly from 3-layer to 4-layer, and the average stress of the 5-layer structure is also better than that of the 4-layer structure, which is highly consistent with the analytical results of the Q in Fig. 6. The correctness of the radiation models is further indicated by the mutual verification between the analytical solution and numerical solution. Therefore, the 3-layer structure has the best radiation performance at a fixed frequency.

IV. CONCLUSION

In this paper, a method is proposed for analyzing the radiation quality factor of BAW ME antenna with different number of stacked layers. The relationship among the different stacked layers, the different thickness of unit layer and the shift of resonant frequency is discussed. Based on the distribution function of stress field at UHF band, the influence of stacked layers on energy, average radiated power and the Q is evaluated under considering the mass loading effect. The result shows that the Q of the stacked structure deteriorates with the increase of the number of layers in the case of a certain thickness of unit layer, and the 2-layer structure is the best. However, the Q of the 3-layer

structure is the best when the operating frequency is fixed. Finally, the physical essence of stacked structure with different radiation performance is further revealed by using the stress simulation model. The numerical results are entirely consistent with the analytical results, which validates the both. This research can provide strong support for understanding the physical mechanism and give a solution for designing BAW ME antenna.

REFERENCES

[1] J. P. Gianvittorio, and Y. Rahmat-Samii, "Fractal antenna: a novel antenna miniaturization technique, and applications," *IEEE Antenna Propagation Magazine*, vol. 44(1), pp. 20-36, 2002.

[2] H. Lin, M. R. Page, M. McConney, J. Jones, B. Howe and N. X. Sun, "Integrated magnetoelectric devices: Filters, pico-Tesla magnetometers, and ultracompact acoustic antenna," *Mrs Bulletin*, vol. 43(11), pp. 841-847, 2018.

[3] C. Tu, Z. Chu, B. Spetzler, P. Hayes, C. Dong, and X. Lin, et. al., "Mechanical - Resonance - Enhanced Thin - Film Magnetoelectric Heterostructures for Magnetometers, Mechanical Antenna, Tunable RF Inductors, and Filters," *Materials*, vol. 12, No. 2259, 2019.

[4] R. Chang, S. Li, M. V. Lubarda, B. Livshitz and V. Lomakin, "FastMag: Fast micromagnetic simulator for complex magnetic structures (invited)," *Journal of Applied Physics*, vol. 109, No. 07D358, 2011.

[5] S. M. Keller, A. E. Sepulveda, and G. P. Carman, "Effective MagnetoElectric Properties of MagnetoElectroElastic (Multiferroic) Materials and Effects on Plane Wave Dynamics," *Progress in Electromagnetics Research*, vol. 154, pp. 115-126, 2015.

[6] John P. Domann and Greg P. Carman, "Strain powered antenna," *Journal of Applied Physics*, vol. 121(4), No. 044905, 2017.

[7] J. Xu, C. M. Leung, X. Zhuang, J. Li, S. Bhardwaj, J. Volakis and D. Viehland, "A Low Frequency Mechanical Transmitter Based on Magnetoelectric Heterostructures Operated at Their Resonance Frequency," *Sensors*, vol. 19(4), No. 853, 2019.

[8] C. Dong, Y. He, M. Li, C. Tu, Z. Chu and X. Liang, et al., "A Portable Very Low Frequency (VLF) Communication System Based on Acoustically Actuated Magnetoelectric Antenna," *IEEE Antenna and Wireless Propagation Letters*, vol. 19, No. 3, 2020.

[9] M. Bibes, "Multiferroics: towards a magnetoelectric memory," *Nature Materials*, vol. 7(6), pp. 425-426, 2008.

[10] S. Dong, J. Zhai, Z. Xing, J. F. Li, and D. Viehland, "Extremely low frequency response of magnetoelectric multilayer composites," *Apply Physics Letter*, vol. 86(10), No. 102901, 2005.

[11] J. Das, Y. Song, N. Mo, P. Krivosik, and C. E. Patton, "Electric - Field - Tunable Low Loss Multiferroic Ferrimagnetic - Ferroelectric Heterostructures," *Advanced Materials*, vol. 21(20), pp. 2045-2049, 2009.

[12] N. X. Sun, and G. Srinivasan, "Voltage control of magnetism in multiferroic heterostructures and devices," *Spin*, vol. 2, No. 3-1240004, 2012.

[13] Z. Yao, Y. E. Wang, S. Keller and G. P. Carman, "Bulk acoustic wave-mediated multiferroic antenna: architecture and performance bound," *IEEE Transactions on Antenna and Propagation*, vol. 63(8), pp. 3335-3344, 2015.

[14] Z. Yao, S. Tiwari, T. Lu, J. Rivera, K. Q. T. Luong, and R. N. Candler, et al., "Modeling of Multiple Dynamics in the Radiation of Bulk Acoustic Wave Antennas," *IEEE Journal on Multiscale and Multiphysics Computational Techniques*, vol. 5, pp. 5-18, 2020.

[15] T. Nan, H. Lin, Y. Gao, A. Matyushov, G. Yu and H. Chen, et al., "Acoustically actuated ultra-compact NEMS magnetoelectric antenna," *Nature Communications*, vol. 8, No. 296, 2017.

[16] A. Ruimi, Y. Liang, R. M. McMeeking, "Improved Prediction of Electrodes' Mass Loading Effect on MEMS FBAR Structure in Longitudinal Resonance," *ASME Conference on Smart Materials, Adaptive Structures and Intelligent Systems*, Maryland, Oct. 2008, pp. 397-402.

[17] C. Dong, M. Li, X. Liang, H. Chen, H. Zhou and X. Wang, et al., "Characterization of magnetomechanical properties in FeGaB thin films," *Applied Physics Letters*, vol. 113(26), No. 262401, 2018.

Proceedings of the 16th Annual IEEE International Conference on Nano/Micro Engineered and Molecular Systems
April 25-29, 2021

Optical Measurement of the Dynamic Response of an Electrothermal Microactuator

Hao Chen
School of Mechanical Engineering
Nanjing University of Science and Technology
Nanjing, China
17766106120@163.com

Xinjie Wang*
School of Mechanical Engineering
Nanjing University of Science and Technology
Nanjing, China
xjwang@njust.edu.cn

Zhanwen Xi
School of Mechanical Engineering
Nanjing University of Science and Technology
Nanjing, China
4317045xi@sina.com

Jiong Wang
School of Mechanical Engineering
Nanjing University of Science and Technology
Nanjing, China
wjiongz@njust.edu.cn

Weirong Nie
School of Mechanical Engineering
Nanjing University of Science and Technology
Nanjing, China
niewrhappy@163.com

Abstract— In order to study the dynamic characteristic of an electrothermal actuator, a high-speed camera-based optical measurement system is first established. Then, a combination of the background differencing and canny edge detection algorithm is utilized to deal with the images from the measurement system above. Hence, the picture that represents the starting point of the motion of the actuator is distinguished between these pictures and the transient displacement can also be calculated. The time that the actuator reach steady state is no more than 100 ms when a constant voltage is applied on it. Finally, the dynamic behavior of the actuator is further studied under changeable voltage. For the sine voltage, an obvious hysteresis appears at the beginning and gradually vanishes with the time growing. When applied to a periodic voltage such as square or sine, the motion of the actuator is also gradually cyclical and has the same time period.

Keywords—electrothermal actuator, dynamic, optical measurement, picture process, periodic

I. INTRODUCTION

Compared with electrostatic, piezoelectric and electromagnetic actuator, the electrothermal one owns larger displacement, great output force and compatibility[1]. These advantages have been arousing many researchers' interest on the actuator.

The study on the dynamic performance of the thermal actuator gives the access to decreasing its response time, which is very important in the application such as optical switch[2] and micro-positioning[3]. However, because of its little scale and fast response, it is always difficult to measure the transient displacement. A complicated non-contacting laser reflectance system is used for time and frequency response of the actuator in [4]. By this means, the received light intensity, which can be sampled 200 kHz, is dependent on the displacement of the actuator. But the precise relationship between them needs to be obtained by calibration, before the measurement. In [5], a 3D digital image correlation system is utilized for the measurement. By image correlation algorithm, the

same objective in the images at different moments can be calculated. In this paper, an optical measurement system including high-speed camera is built. A novel hybrid algorithm combining background differencing and canny edge detection algorithm together is applied to calculating the displacement of the actuator.

II. FABRICATION AND EXPERIMENTAL SETUP

The V-shaped actuator, made of polysilicon, is fabricated by the deep reactive-ion etch (DRIE) technology[6] in this paper. The acrylic plate is employed for its packaging, as shown in Figure 1. The optical measurement system including high-speed camera, computer and microscope is presented in Figure 2. In the system, the arbitrary function generation (Tektronix AFG3022C) can supply the voltage for the actuator. The voltage signal on the actuator is monitored by oscilloscope (Tektronix MDO3024). What is more, the motion image of the actuator in the microscope is captured by high-speed camera. The fill-in light, under the packaged chip, is driven by the power source and used to improve the quality of the image above.

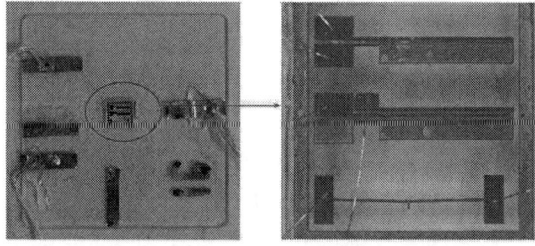

Figure 1. Structure of the packaging chip

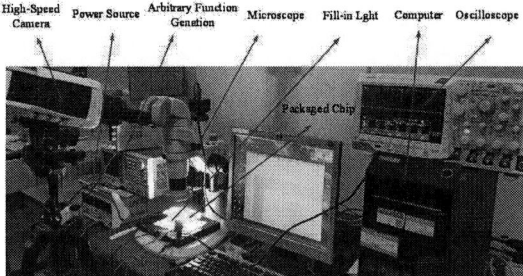

Figure 2. High-speed camera-based optical measurement system

III. IMAGE PROCESSING ALGORITHM

Because of the fast response of the actuator, external environment brings about little effect on the image during the filming. Thus, the background differencing method is able to be used for detecting the movement of the actuator swiftly. According to the difference between the background picture and the current picture, this method obtains the region that the objective moved through. Its calculation is expressed as

$$D_i(x, y) = \left| I_i(x, y) - B(x, y) \right| \tag{1}$$

and

$$M_i(x, y) = \begin{cases} 1 & D_i > T \\ 0 & D_i \leq T \end{cases} \tag{2}$$

where i is the index of the gray picture, (x, y) is the location of the pixels in the picture, I is the picture of the objective, B is the picture before the movement of the objective, T is the threshold value and M is binary image. As shown in Figure 3, the middle of the actuator is chosen to be the objective. The rectangle represents the region that the middle of the actuator passed over. More, the length of the rectangle is about the displacement of the V-shaped actuator. Hence, the picture, at the appearance of the rectangle with this method, denotes the beginning of the motion. The results of the 19th and 20th picture with background differencing algorithm are presented in Figure 4 and 5 when 12 V voltage is loaded on the actuator. Compared with Figure 4, a rectangle-shape region lies in the middle of Figure 5. Therefore, the 19th picture should be the initial moment of the actuator's motion. Based on the frequency of 500 frames a second in high-speed camera, the moment of the ones after the 19th picture is achieved.

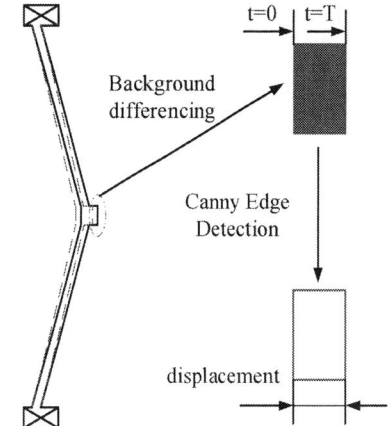

Figure 3. Sketch of the picture processing for the V-shaped actuator

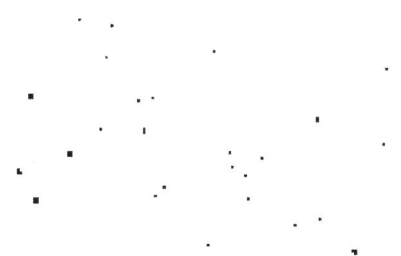

Figure 4. The 19th picture with background differencing algorithm

Figure 5. The 20th picture with background differencing algorithm

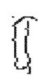

Figure 6. Result of the middle in Figure 5 with canny edge detection algorithm

The rising temperature in the actuator generates noise in the image, which will lead to the larger error in the calculation of the displacement. To eliminate this, the canny edge detection algorithm is used for the process after the differencing of equation (1). In the algorithm, the 2-D gaussian function, which can reduce the effect of the noise, is given in (3).

$$G(x, y) = \frac{1}{2\pi\sigma^2} e^{\frac{x^2 + y^2}{2\sigma}} \qquad (3)$$

The differenced picture $D(x, y)$ is convolved with the Gaussian function, as

$$F(x, y) = D(x, y) * G(x, y) \qquad (4)$$

where $*$ is convolution. With the picture $F(x, y)$, the gradient $M(x, y)$ and gradient direction $\theta(x, y)$ can be separately calculated as

$$M(x, y) = \sqrt{F_x^2(x, y) + F_y^2(x, y)} \qquad (5)$$

and

$$\theta(x, y) = \arctan(F_y(x, y) / F_x(x, y)) \qquad (6)$$

On the basis of gradient and its direction, the initial and current edge of the middle of the V-shaped actuator in the middle of the 20th picture is presented in Figure 6. When the V-shaped actuator is applied on 12 V voltage, its middle location at different moment in the processed images are shown in Figure 7. Combining the real width of the beam of the actuator with its number of pixels, one pixel is about 0.8 μm in the picture, which can further be used for calculating the displacement of the actuator.

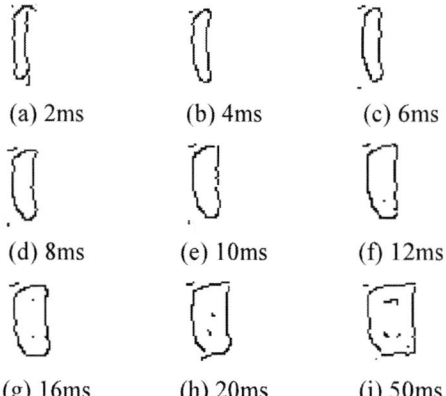

(a) 2ms   (b) 4ms   (c) 6ms

(d) 8ms   (e) 10ms   (f) 12ms

(g) 16ms   (h) 20ms   (j) 50ms

Figure 7. Middle location of the V-shaped actuator at different moment with 12V voltage

IV. DYNAMIC RESPONSE

The transient displacement of the actuator is shown in Figure 8 when 12 V and 16 V voltage are applied to it. From the figure, the displacement of the actuator increases with the voltage growing. The actuator moves fast at the initial time and tends to be stable. After the time of 100 ms, little change can be observed in the actuator, which is also called steady state. With

12 V and 16 V voltage, its steady-state displacement is 28.8 μm and 44.6 μm, respectively.

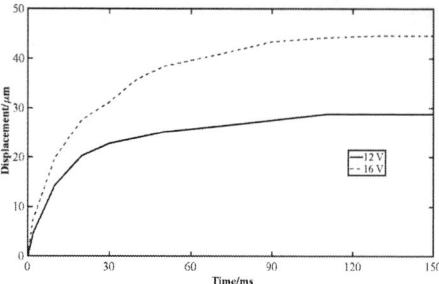

Figure 8. Transient displacement of the actuator under 12V and 16V voltage

When 10Hz sin and square voltages are applied on the actuator, its transient displacement in the first cycle is given in Figure 9. For the sine voltage, the maximum displacement is not at the time (25 ms) when the voltage is the largest. This is because although the voltage after 25 ms is small than 12 V, the actuator still does not reach steady state. In other words, the generated heat is more than the lost heat in the actuator. This will lead to a slight rise in its displacement. Similarly, its displacement decreases to the minimum slower than its sine voltage. That the energy from the applied voltage after 75 ms is less than that dissipated contributes to this phenomenon above. Therefore, a distinct hysteresis exists in the actuator in the response of displacement to sine voltage. For the square voltage, the displacement of the actuator at 50 ms is very close to that in its steady state. Thus, its maximum value lies in 50 ms. What is more, its displacement diminishes sharply at the beginning when the actuator is rid of the voltage. Yet more time it is also required to move back to its initial location.

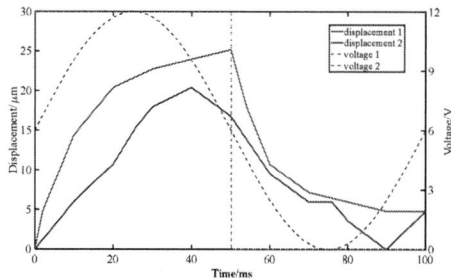

Figure 9. Transient displacement of the actuator in first cycle with sine and square voltage of 10Hz

Figure 10 is the dynamic displacement of the actuator under the phase angle of 72° in sine and square voltage, respectively. The frequency of these two voltages is 100 Hz. Although the phase angle, frequency and equivalent voltage between them is the same, the displacement of square voltage is always larger than that of sine voltage. With the time growing, the displacement at the same phase angle eventually evens out. The displacement of the actuator in two cycles under the sine and square voltage is presented in Figure 11. For the sine voltage, the hysteresis of displacement at initial moment vanishes. The change of displacement is correspondence with that of the sine voltage. In addition, dynamic stability finally appears in the V-shaped actuator when the sine or square voltage is applied on it. The displacement tends to be periodic and has same time period with the voltage.

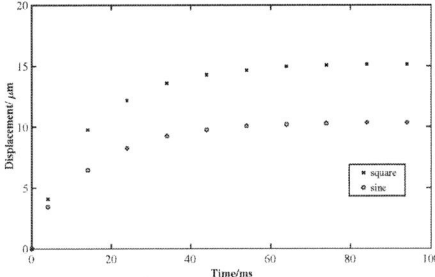

Figure 10. Transient displacement of the actuator under the phase angle of 72 ° in sine and square voltage of 100 Hz

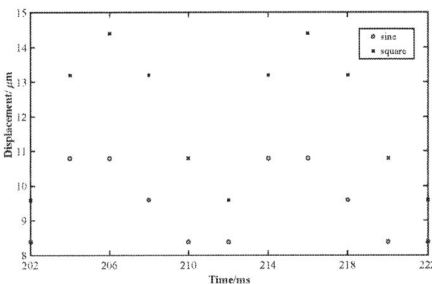

Figure 11. Steady-state displacement of the actuator under sine and square voltage of 100 Hz

V. CONCLUSION

In this paper, an optical measurement system based on high-speed camera has been built to capture the motion of the V-shaped actuator. To calculate the displacement in the images, an improved canny edge detection algorithm, incorporating with the background differencing method, is put forward. According the results from image processing, the steady-state time of the V-shaped actuator under constant voltage is about 100 ms. When sine voltage is applied on the actuator, a hysteresis in the response of displacement to the voltage gradually disappears with the time increasing. What is more, the motion of the V-shaped actuator tends to be cyclical with periodic voltage such as sine and square voltage. The

periodic time of the displacement is equal to that of voltage.

On the one hand, the measurement method contributes to the study on the dynamic response of the electrothermal micro actuator. On the other hand, the dynamic behavior of the actuator with different kinds of voltage also paves the way for the control on its displacement. However, the image processing algorithm in this paper still needs more refinement to improve its accuracy in the future.

REFERENCE

[1] Hao. C, Xin-jie. W, and Jiong, W. "Analysis of the dynamic behavior of a V-shaped electrothermal microactuator." Journal of Micromechanics & Microengineering 30.8(2020):5.

[2] Lee, Chengkuo, and C. Y. Wu. "Study of electrothermal V-beam actuators and latched mechanism for optical switch." Journal of Micromechanics & Microengineering 15.1(2004):11.

[3] Henneken, V. A., Tichem, M., & Sarro, P. M. "In-package mems-based thermal actuators for micro-assembly." Journal of Micromechanics & Microengineering 16(6), S107-S115.

[4] Ryan, et al. "Time and frequency response of two-arm micromachined thermal actuators." Journal of Micromechanics and Microengineering (2003).

[5] Pustan, Marius, et al. "Reliability design of thermally actuated MEMS switches based on V-shape beams." Microsystem Technologies (2017).

[6] Kolahdoozan, M., Esfahani, A. R., & Hassani, M. "Experimental and Numerical Investigation of the Arms Displacement in a New Electrothermal MEMS Actuator." International Journal of Advanced Design and Manufacturing Technology 10.2(2017).

Proceedings of the 16th Annual IEEE International
Conference on Nano/Micro Engineered and Molecular Systems
April 25-29, 2021

A Wearable Strain Sensor Based on Fiber-structured PU/MXene/CNT Composite with Ultra-high Sensitivity and Broad Sensing Range

Guoxi Luo[1,2,*], Qiankun Zhang[1,2], Yunyun Luo[1,2], Ke Chen[1,2], Wenke Zhou[1,2], Libo Zhao[1,2], Zhuangde Jiang[1,2]

Abstract— As the critical branch of wearable sensors, strain sensors with lightweight and high stretchability have been widely demanded in human gesture recognition and robotics engineering. However, most strain sensors cannot simultaneously satisfy the requirements of high sensitivity and a broad detection range, attributed to the poor stretchability and fixed connection of commonly used sensing elements. Here, a single-fiber structure composed of polyurethane (PU) as the stretchable core and MXenes/Carbon nanotubes (CNTs) as the conductive shell was fabricated by wet spinning followed with repeated dip-coating. The obtained highly stretchable fiber exhibited superior strain sensing performance, including ultra-high sensitivity (gauge factor up to 2504.1), a large detection range (up to 250%), and good stability (over 1000 cycles).Furthermore, the PU/MXene/CNT single-fiber strain sensor (SFSS) can be utilized to detect human motion signals, such as folding of finger, bending of wrist, arm and knee. Benefiting from the integrated features of high sensitivity and stretchability, this SFSS promises a great application potential in the field of human gesture recognition and smart robotics.

I. INTRODUCTION

With the development of flexible electronic devices, an increasing amount of wearable devices enter individuals' life. As a considerable part of wearable electronic devices, there is wide use of flexible strain sensors in smart robotics and physiological state monitoring[1-4]. For practical applications, strain gauge sensors need to satisfy the essentials of small size, high sensitivity, and wide strain range. Unfortunately, traditional metal and semiconductor-based strain sensors are hard to satisfy the above requirements simultaneously owing to their poor stretchability and the fixed connection of the sensitive unit[5]. One effective strategy to solve this problem is that the resistance of sensitive materials can vary dramatically under small deformation and maintain the structural connection under large deformation[6].

Nowadays, CNTs have become the preferred choice for strain sensor sensitive elements due to their extremely high conductivity, outstanding flexibility, and huge aspect ratio[7, 8]. The high aspect ratio and super conductivity of the CNTs can ensure structural connection and obvious resistance variation during the stretching process. However, the strain sensor made up of only elastic matrix and CNTs sensitivity elements is difficult to ensure high sensitivity and large measurement range at the same time. Li et al. firstly

prepared PU fibers by electrospinning, then developed PU-CNTs fiber strain sensors by ultrasonic adsorption, the maximum strain of the composite sensor can reach 100%, while the gauge factor(GF) was only 1.67[9]. Pan et al. braided the elastic substrate and 8 PET fibers to implement the core-shell composite structure and then coated the CNTs to produce the strain sensor, the sensitivity (GF) of the sensor reached up to 980, while the maximum stretchable range was only 50%[10].

In this work, a highly stretchable strain sensor based on PU fiber decorated with MXenes and CNTs was developed. As illustrated in Fig. 1, firstly, the PU micro fiber was fabricated by the technique of wet spinning, and then the functional MXenes and CNTs were decorated coaxially onto the fiber with repeated dip-coating process. MXenes and CNTs nanomaterials have many surface functional groups, which enable the assembly onto the surface of PU fiber firmly through hydrogen bonding. Furthermore, attributed to high intrinsic conductivity, excellent mechanical properties, and large surface area of MXenes and CNTs, the SFSS exhibited ultra-high sensitivity (gauge factor up to 2504.1), large detection range (up to 250 %), and good stability (over 1000 cycles). Based on its superior performance, the SFSS can successfully detect human motions, demonstrating its great potential for practical applications.

II. EXPERIMENT SECTION

A. Preparation of PU fibers

The high-performance stretchable PU fiber was fabricated by wet-spinning technology. Firstly, 4 g PU (Elastollan 1185A, BASF Co. Ltd) in powder form was dispersed with 5g of acetone using a magnetic stirrer. In the next step, 5g of N,N-Dimethylformamide (DMF) was added into the PU/acetone suspension by magnetic stirring at least 5 hours to reach a good homogeneity. Thirdly, the prepared PU precursor was transferred into a 5ml syringe and then injected into the coagulation bath of deionized water for spinning. Finally, the wet-spun fiber was immersed in deionized water for further 2h for curing, subsequently dried in air to obtain a PU fiber.

B. Fabrication of PU/MXene/CNT SFSS

CNTs ink was prepared in the following two steps. Firstly 50 mg SWCNTs was dispersed into 100ml deionized water with 50 mg surfactant SDBS(Sodium dodecyl sulfate), next, the dispersion was ultrasonicated for 30 mins with an Ultrasonic Homogenizer (Scientz IID, Ningbo Scientz Biotechnology Co., China) under the power of 300 W.

A PU/MXenes fiber was fabricated by dip-coating MXenes ink on the surface of a wet-spun PU fiber followed with dried in an oven at 80 degrees for 5 minutes. And then

---

1 State Key Laboratory for Manufacturing Systems Engineering, International Joint Laboratory for Micro/Nano Manufacturing and Measurement Technologies, Xi'an Jiaotong University, Xi'an, Shannxi 710049, PR China.

2 School of Mechanical Engineering, Xi'an Jiaotong University, Xi'an, Shannxi 710049, PR China.

*E-mail: luoguoxi@mail.xjtu.edu.cn

978-1-6654-3008-1/21 $31.00 © 2021 IEEE

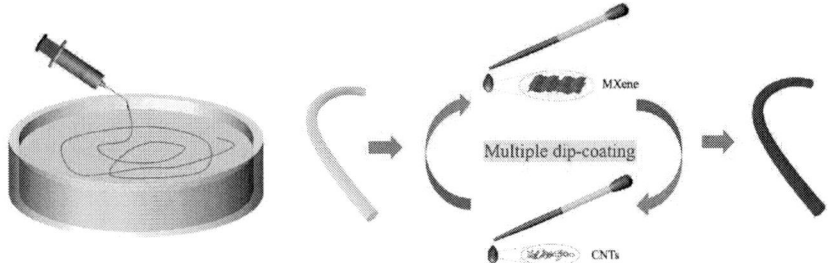

Fig 1.   Schematic illustration of the fabrication process for the PU/MXene/CNT SFSS

drop SWNTs ink on the prepared PU/MXene fiber with the same strategy to obtain a PU/MXenes/CNTs fiber. After that, repeat the above steps 3-5 times to get PU/MXene/CNT SSFS. Based on the hydrogen bonding, PU, MXenes, CNTs can be tightly assembled together.

C. Characterization

The morphology and structure of the as-obtained samples were observed with a field-emission scanning electron microscope (FESEM; SU-8010, HITACHI). The crystal structure was characterized by X-ray diffraction (XRD; D8 Advanced, Bruker) using Cu Kα radiation over a 2θ range from 5 to 70°. And the resistance change of the prepared sensor was measured by a home-made linear motor and a multimeter (KESIGHT 34556A).

III. RESULT AND DISCUSSION

A. Morphology and Materials characterization

Fig. 2a and 2c present the pristine PU fiber with a uniform diameter of ca. 200 μm and a coarser surface. Correspondingly, The microstructure of the as-prepared PU/MXene/CNT fiber is exhibited in Fig. 2b and 2d, it can be clearly seen that the MXenes and CNTs are uniformly cross-linked and intertwined onto the surface of PU fiber to form a conductive network.

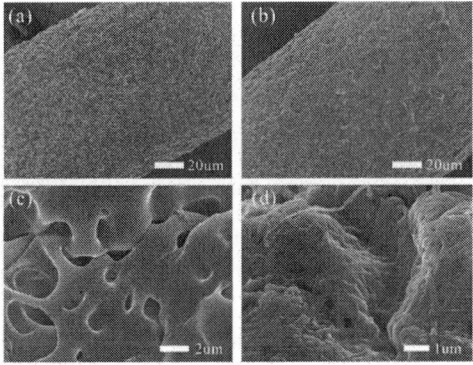

Fig 2.   SEM images of (a) wet-spun PU fiber and (b) PU/MXene/CNT fiber. The magnified views of (c) pristine PU fiber and (d) PU/MXene/CNT fiber.

The X-ray diffraction(XRD) patterns of the PU/MXene/CNT composite is shown in Fig. 3, revealing the principal diffraction peak of MXenes at 6.5° and the additional peak of CNTs at 20.5°, further indicating the successful assembly of MXenes and CNTs[11].

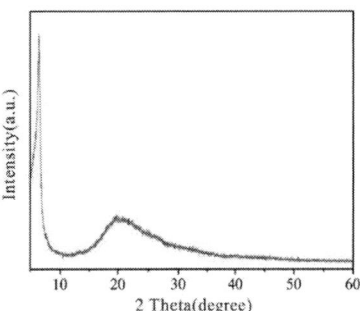

Fig. 3.   XRD patterns of the as-prepared fiber.

B. Sensing performance analysis

Fig. 4a highlights that the PU/MXene/CNT fiber can be stretched up to around 800 % tensile rate without any visible breakage due to excellent stretchability of PU fiber. To further characterize the excellent tensile properties of PU/MXene/CNT fiber. Fig. 4b shows the results of tensile

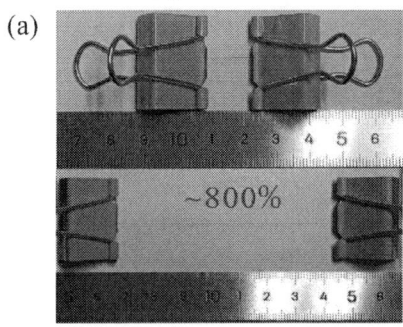

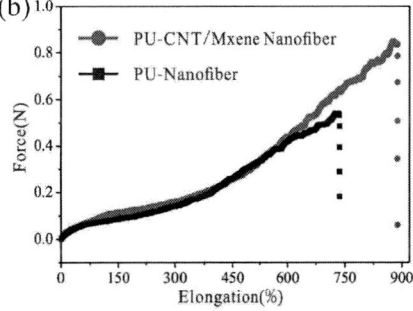

Fig 4.   (a) Optical image shows the excellent stretchability of this PU/MXene/CNT SFSS (around 800% tensile strain). (b) Force-elongation curves of the as-prepared fiber.

978-1-6654-3008-1/21 $31.00 © 2021 IEEE

tests on pristine PU fiber and PU/MXene/CNT fiber. Experimentally, pristine PU fiber possesses outstanding extensibility intrinsically, which can stretch up to 750% tensile rate. After dip-coating with MXenes and CNTs, the elongation rate of the composite fiber was further increased to 900%, indicating a substantial broadening of sensing range.

The change of relative resistance of the as-prepared fiber under 0~250 % strain with strain rate of 60 mm/min was recorded in Fig. 5. It shows the typical piezoresistive properties of PU/MXenes/CNTsthe SSFS. To characterize the sensitivity of the device, the gauge factor (GF), is defined as $GF = (R-R_0)/ R_0\varepsilon$, Where ε is the strain, R_0 is the resistance under no strain and R represents the resistance under strains. The curve could be divided into at least three linear regions and the GF can be calculated as: 52.9 (0~100 %), 533.5 (100-160%), and 2504.1 (160-250%). The different GF originate from the point that the different extent of damage with the increasing of strain. It could be concluded that the performance of the SSFS is superior to most MXene/CNT strain sensors in terms of sensitivity and response range.

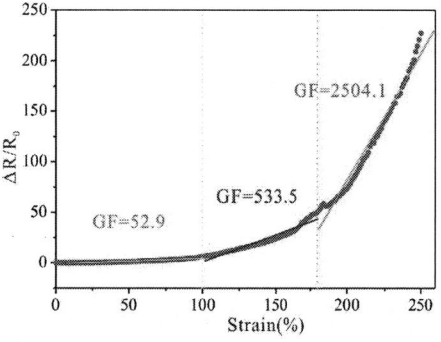

Fig 5. The relationship between relative resistance changes and strain.

To further analyze the sensing performance, the response curves at different tensile rates and strains were measured. Fig. 6 exhibits the change of relative resistance of the SFSS at a strain of 50% under different strain rates (from 60 to 200 mm/min), it can be clearly seen that the strain rates have almost no influence on the change of relative resistance, this is a very critical feature for developing reliable strain sensor.

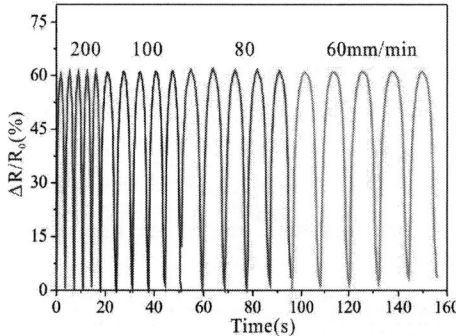

Fig 6. The changes of relative resistance at different frequencies under 50% strain.

The resistance variations under different strains (50%, 100%, 150%, 200%) were measured in Fig. 7a, and the SFSS shows a stable response under various strain conditions. Besides, the test also shows that SSFS can maintain a superb response even under 200% strain conditions, revealing the broad measurement range of the SSFS. To evaluate the cyclic stability of the PU/MXenes/CNTs SFSS, the performance under cyclic loading of 15% strain with a tensile rate of 100mm/min for 1000 cycles was recorded in Fig. 7b, the insets of 200th and 800th cycles show that the change of relative resistance almost stays the same, demonstrating the excellent durability of this PU/MXenes/CNTs SFSS.

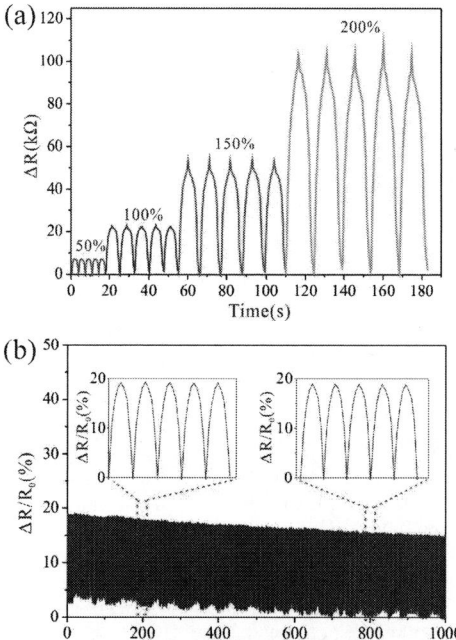

Fig 7. (a) The resistance changes with various strains at a strain rate of 200mm min-1. (b) The performance stability of the SFSS with applied strain of 15% during 1000 stretching-releasing cycles, the inset shows the detailed curves of the selected areas.

Due to the microstructure, fiber-based architecture, and excellent stretchability, the PU/MXenes/CNTs SFSS can be integrated with any part of human body to monitor physical motion. Fig. 8(a-d) shows the physical activity signals measured by the sensors attached to the fingers, wrists, elbows, and knees.

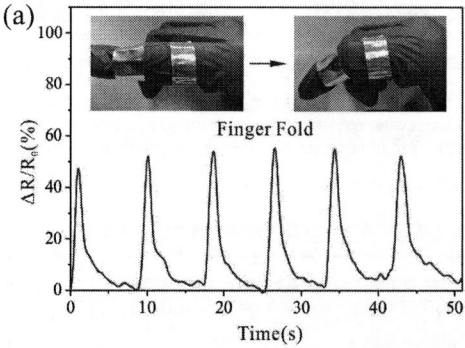

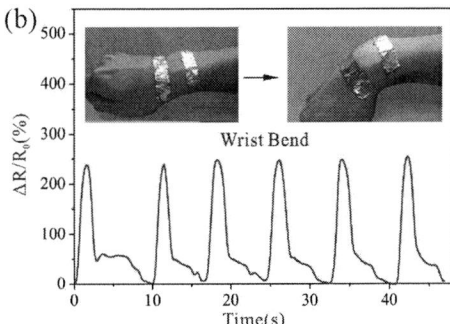

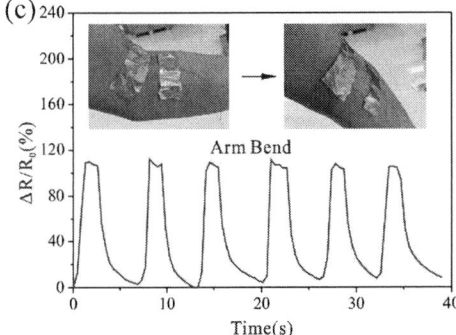

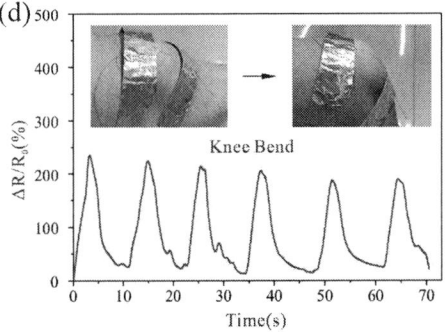

Fig 8. The monitoring of human motions using the PU/MXene/CNT SFSS: (a) finger fold, (b) wrist bend, (c) arm bend and (d) knee bend.

The measured results exhibit good response to each motion with decent stability and repeatability, illustrating the great potential of this PU/MXenes/CNTs SFSS for advancement in the field of human gesture recognition and smart robotics.

IV. CONCLUSION

In this work, a highly stretchable strain sensor based on wet-spun PU fiber decorated with dip-coated MXenes and CNTs was developed. Benefit from the superb elasticity and stretchability of PU and the mechanical toughness of CNT and Mxene materials, the fabricated strain sensor promises both features of high sensitivity and a broad sensing range simultaneously, the maximum response range can reach 250%, and the sensitivity can reach 2504.1. In addition, the sensor can still maintain excellent performance stability even after 1000 cycles under 20% strain. Based on the characteristic structure of a PU/MXene/CNT, it can be attached to any part of the human body to monitor physical signals and have a stabilized response. This research provides expansive design ideas for manufacturing flexible, high-sensitivity, broad response range, and high-elasticity electronic devices.

ACKNOWLEDGMENT

The authors thank Instrument Analysis Center of Xi'an Jiaotong University for the SEM and XRD facilities. This work is supported by the National Key Research & Development (R&D) Program of China (grant number 2018YFB2002402), the National Natural Science Foundation of China (grant numbers 51705409, 91748207), and the the Fundamental Research Funds for the Central Universities (grant number xpt012020004).

REFERENCES

[1] Cheng, Y., et al., *A Stretchable and Highly Sensitive Graphene-Based Fiber for Sensing Tensile Strain, Bending, and Torsion.* Adv Mater, 2015. **27**(45): p. 7365-71.

[2] Liao, X., et al., *Flexible and Highly Sensitive Strain Sensors Fabricated by Pencil Drawn for Wearable Monitor.* Advanced Functional Materials, 2015. **25**(16): p. 2395-2401.

[3] Shi, J., et al., *Smart Textile-Integrated Microelectronic Systems for Wearable Applications.* Adv Mater, 2020. **32**(5): p. e1901958.

[4] Xu, K., Y. Lu, and K. Takei, *Multifunctional Skin-Inspired Flexible Sensor Systems for Wearable Electronics.* Advanced Materials Technologies, 2019. **4**(3).

[5] Paul, O., J. Gaspar, and P. Ruther, *Advanced silicon microstructures, sensors, and systems.* IEEJ Transactions on Electrical and Electronic Engineering, 2007. **2**(3): p. 199-215.

[6] Yang, K., et al., *A highly flexible and multifunctional strain sensor based on a network-structured MXene/polyurethane mat with ultra-high sensitivity and a broad sensing range.* Nanoscale, 2019. **11**(20): p. 9949-9957.

[7] Yang, K., et al., *A highly flexible and multifunctional strain sensor based on a network-structured MXene/polyurethane mat with ultra-high sensitivity and a broad sensing range.* Nanoscale, 2019. **11**(20): p. 9949-9957.

[8] Hu, N., et al., *Investigation on sensitivity of a polymer/carbon nanotube composite strain sensor.* Carbon, 2010. **48**(3): p. 680-687.

[9] Park, S., M. Vosguerichian, and Z. Bao, *A review of fabrication and applications of carbon nanotube film-based flexible electronics.* Nanoscale, 2013. **5**(5): p. 1727-52.

[10] Li, Y., et al., *Continuously prepared highly conductive and stretchable SWNT/MWNT synergistically composited electrospun thermoplastic polyurethane yarns for wearable sensing.* Journal of Materials Chemistry C, 2018. **6**(9): p. 2258-2269.

[11] Pan, J., et al., *Highly sensitive and durable wearable strain sensors from a core-sheath nanocomposite yarn.* Composites Part B: Engineering, 2020. **183**. 570–578, July 1993.

Gap in pagination due to unavailable paper.

Pages 1366-1369

Proceedings of the 16th Annual IEEE International
Conference on Nano/Micro Engineered and Molecular Systems
April 25-29, 2021

Feasibility Study of Wearable Muscle Disorder Diagnosing Based on Piezoelectric Micromachined Ultrasonic Transducer

Mengjiao Qu, Hong Ding, Xuying Chen, Dengfei Yang, Dongsheng Li, Ke Zhu and Jin Xi*e*,
Member, IEEE

Abstract— This work demonstrates a high-performance imager of muscle-like phantoms based on a piezoelectric micromachined ultrasonic transducer (pMUT), which aims at wearable muscle disorder diagnostics. A 23×26 pMUT array is fabricated for both transmit ultrasonic pulses and receive echoes, and its resonant frequency is 5 MHz in mineral oil with a -6 dB bandwidth of 40%. Ecoflex and polydimethylsiloxane (PDMS) are used to mimic different tissues including muscles and fat. The PMUT array is excited by 2 cycles 5 MHz sine wave pulse signal with amplitude of 10 Vpp, and mechanically scanned two muscle-like phantoms that standing for normal and disordered muscles respectively. B-mode images are generated using the received pulse-echo signals, and the theoretical axial resolution of 2-pulse excitation is calculated to be 0.75mm. These preliminary results demonstrate the potential of pMUT in medical ultrasonic imaging, especially muscle imaging.

I. INTRODUCTION

Muscle diseases such as inclusion body myositis (IBM) and Duchenne muscular dystrophy (DMD) make many people suffer from weakness and impaired cardiorespiratory function. Fatigue and pain are also common symptoms, which lead to a decline in the quality of life [1], [2]. Heavy manual labor and improper exercise contribute to disorders of joint and muscle injuries including hematoma, atrophy and laceration [3]. Congenital muscle disease or muscle impairment have a serious impact on human health. It is necessary to adopt appropriate diagnosis and timely treatment. Diagnostic imaging clarification is essential to assess the exact location of the affected muscles, to evaluate the extent of the lesion or injury as well as to define possible concomitant complications [1],[4].

Compared to other modalities such as MRI and CT, sonography is quickly available, inexpensive and easily accessible [4],[5]. The role of ultrasound is to assess the longitudinal extent of the lesion, calculate the volume of the hematoma and detect possible compression of the adjacent structures. This non-invasive technique could also be useful in assessing the extent of pathological change in dystrophic patients and could prove a valuable diagnostic aid. However, imaging based on traditional ultrasonic transducers (UTs) are bulky and complex thus not suitable for portable devices. Benefit from the development of microelectromechanical

systems (MEMS), micromachined ultrasonic transducer (MUT) has the potential for ultrasonic imaging in wearable mode due to their good acoustic impedance matching, broad bandwidth, low cost and easy integration with supporting electronics compared with traditional bulky ultrasound transducer [4]. Although capacitive MUT (cMUT) have been developed for decades and demonstrated for pulse-echo imaging, it requires high DC bias voltage (normally tens to hundreds of volts) to achieve satisfying sensitivity[6], [7]. Instead, piezoelectric MUT (pMUT) shows advantages including no requirement of additional DC bias voltage, no constraints on the gap design of the capacitor, and better CMOS-compatibility [8]. In this work, we constructed muscle-like phantoms mimicking normal and disordered muscle and fabricated a 23×26 pMUT array for ultrasonic imaging. The high performance of B-mode scans demonstrates the pMUT has great potential in application of wearable muscle disorder diagnostic.

II. DEVICE DESIGN AND CHARACTERIZATION

The device used in this work is a 23×26 pMUT array that has a resonant frequency of 6 MHz (in air), and the optical microscope image is shown in Fig. 1a. The area of the array

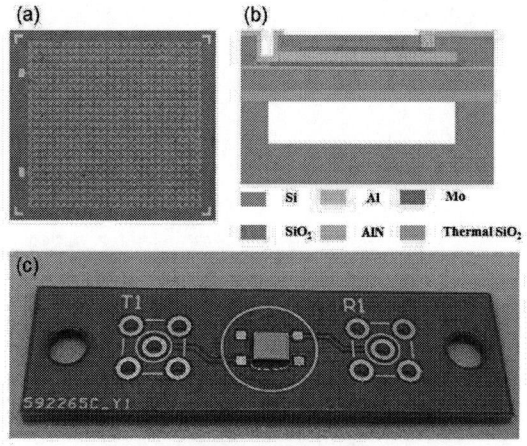

Figure. 1. (a) Optical microscope image of the 23×26 pMUT array; (b) the cross-sectional view of a single element; (c) the tested pMUT array mounted on a print circuit board.

This work is supported by the "Zhejiang Provincial Natural Science Foundation of China (LZ19E050002)", the "National Natural Science Foundation of China (51875521)", and the "Science Fund for Creative Research Groups of National Natural Science Foundation of China (51821093)".

Mengjiao Qu, Xuying Chen, Dengfei Yang, Dongsheng Li, Ke Zhu and Jin Xie are with the State Key Laboratory of Fluid Power and Mechatronic Systems, Zhejiang University, Hangzhou, 310027, China. Corresponding author: Jin Xie, tel: +86-571-87951271; e-mail: xiejin@zju.edu.cn).

Hong Ding is with University of California, San Diego, USA (e-mail: hding@eng.ucsd.edu)

978-1-6654-3008-1/21 $31.00 © 2021 IEEE

(a)

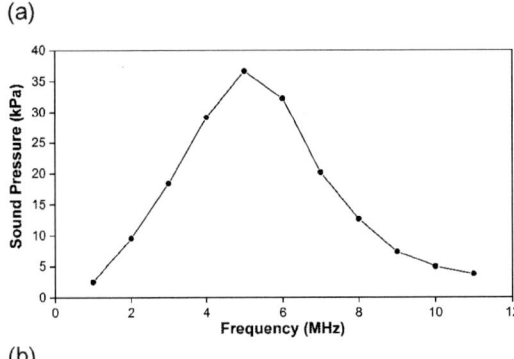

(b)

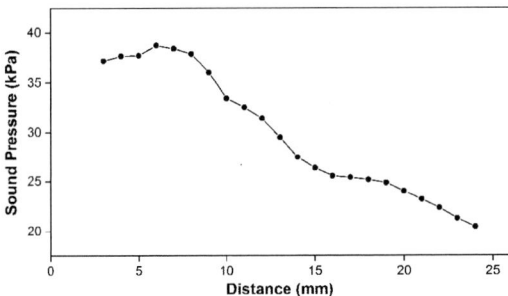

Figure. 2. Transmitted sound pressure of the 23×26 pMUT array measured by hydrophone (a) at a distance of 10 mm with applied excitation signals from 1 MHz to 11 MHz; (b) with applied excitation signals at 5MHz and spacing from 3 mm to 24mm.

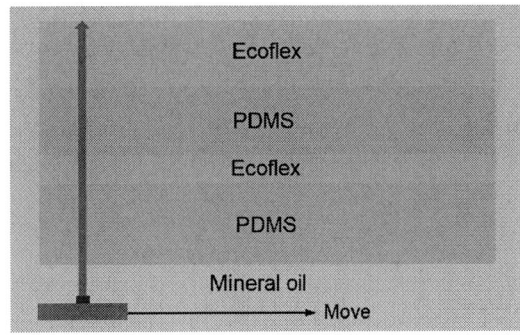

Figure. 3. Working principle of the muscle-like phantom imaging using pMUT array.

hydrophone and record the sound pressure, we found that the sound pressure remained unchanged when the distance increases from 3mm to 8mm, thus distance between pMUT array and hydrophone is 8 mm for imaging experiments.

III. RESULTS AND DISCUSSION

A. Results

As for muscle-like phantoms imaging, the working principle of the B-mode imaging using pMUT array is shown in Fig. 3. Ecoflex and PDMS blocks with various thicknesses are fabricated to mimic muscle and fat tissues respectively, and the disordered muscle phantom has multi-layers of mixed PDMS and Ecoflex that emulate muscle atrophy [11]. To get satisfactory axial resolution, the pMUT array is excited by 2 cycles 5 MHz sine wave pulse signal with an amplitude of 10 Vpp and moves mechanically with a step of 1.5 mm in the lateral direction, while the phantoms are placed in a fixed position at approximately 8 mm distance from the surface of the pMUT array.

The fundamental principle of ultrasonic diagnostic imaging is to reflect sound from the tissue interface. Fig. 4 shows the transmission and receiving signals of the pMUT array operating at 5MHz in mineral oil. The ultrasonic echoes are obviously obtained when reflected at the boundaries as shown in Fig. 4b. The first to fifth echoes locate the interfaces between PDMS and Ecoflex, and the last echo is caused by the large impedance mismatch between mineral oil with air. After subsequent signal processing include band-pass filtering and enveloping, the normalized amplitude of the received voltage signals is utilized to create a gray-scale B- mode image.

Fig. 5 shows the B-scan image constructed from lateral scan of the phantom mimicking normal muscle tissues. Especially, the 0.6mm Ecoflex layer between two PDMS layers stands for fascia lata, a kind of connective tissue. As we can see, both the interface between PDMS and Ecoflex layers and the upside boundary are clear to recognize. Besides, the thickness obtained from the image is consistent with the designed dimensions, demonstrating that the proposed pMUT array can achieve great accuracy in B-mode imaging. In the meantime, we also fabricate another phantom emulating disordered muscle. The pathological change in atrophied muscle probably increases the number of reflecting surfaces within the muscle, thus we use multi-

is 3.2 ×3.2 mm². All the 498 elements are connected in parallel and each single element is a clamped circle pMUT with a diameter of 100 μm. Fig.1b shows the cross-sectional view of a single element, which consists of 1 μm AlN piezoelectric layer sandwiched by 0.2 μm top Mo electrode and 0.2 μm bottom Mo electrode. The PMUT array is fabricated on a cavity-SOI wafer. The detailed process flow has been introduced in our previous works [9], [10]. The fabricated pMUT array is wire-bonded on a print circuit board (PCB), as shown in Fig. 1c, and is utilized to both transmit ultrasound and receive echoes with the help of MD0100 T/R switch.

In order to reduce the impact to acoustic wave transmission caused by impedance mismatch and simulate a body-coupled condition, the imaging experiments are conducted with the pMUT array and muscle-like phantoms immersed in mineral oil, an insulating fluid which has a similar acoustic impedance with that of human tissues. Before imaging experiments, the transmitting performance of the pMUT array is characterized using a needle hydrophone with 1 mm effective diameter (Precision Acoustics Ltd, Dorchester, United Kingdom). The pMUTs are driven by 10 Vpp continuous waves from 1 MHz to 11 MHz with a step of 1 MHz using a waveform generator (33500B Series, Agilent Technologies) and the oscilloscope (DSO-X 4052A, Agilent Technologies) is used to display the signal received by hydrophone in a fixed position. According to the measured sound pressure as shown in Fig. 2a, we choose 5 MHz as imaging operating frequency. Besides, we change the distance between pMUT array and

978-1-6654-3008-1/21 $31.00 © 2021 IEEE

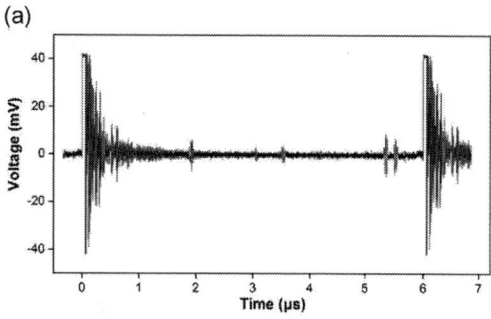

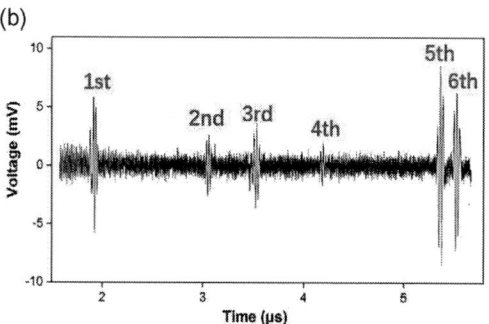

Figure. 4. (a) Pulse-echo response by the pMUT array at 10 mm from one of the phantoms. (b) Ultrasonic echoes received when reflected at the boundaries. The 6th echo represent the interface between mineral oil with air.

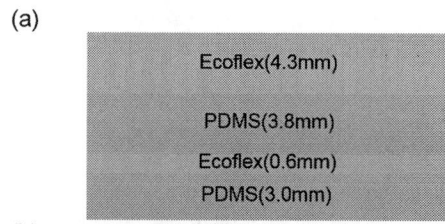

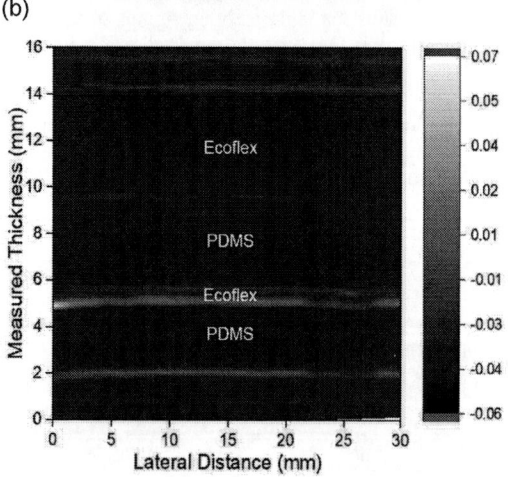

Figure. 5. (a) The cross-sectional view of the phantom representing normal muscle. (b) B-mode imaging of the artificial phantom emulating normal muscle tissue.

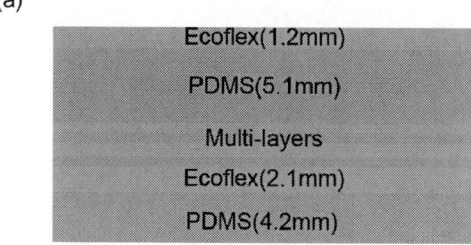

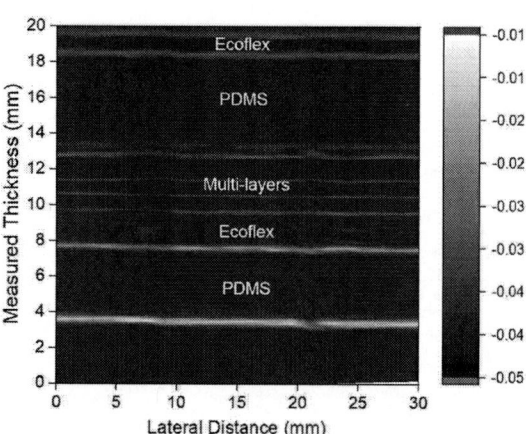

Figure. 6. (a) The cross-sectional view of the phantom mimicking muscle atrophy. (b) B-mode imaging of the artificial phantom emulating muscle atrophy.

layers of thin PDMS and Ecoflex stacked alternately on top of each other to mimic muscle atrophy[12], [13]. The obtained image is shown in Fig. 6b. The interfaces of multi-layers are clearly identified, proving that the fabricated pMUT array has the potential for muscle atrophy diagnosing.

B. Discussion

Benefit from the advantages of small size, low cost, easy fabrication and low power consumption, pMUT is suitable for wearable consumer electronics, such as fingerprint sensing and blood vessel motion tracking [14], [15]. In this work, we demonstrate the feasibility of pMUT for ultrasound diagnostic imaging using two muscle-like phantoms. We fabricate a 23×26 pMUT array for both ultrasound transmission and reception, the circuit is simple and the overall size is very small, therefore can be integrated into a probe or in a wearable mode. The resonant frequency is found by measuring the generated sound pressure while the pMUT array is simulated by excitation signals with different frequency. However, the receiving performance which has a significant impact on the quality of imaging has not been characterized yet. Besides, the noise is obvious in the pulse-echo response, echoes with low amplitude can be easily submerged by noise, such as the forth echo as marked in Fig 4b. Therefore, a filter is needed to increase the SNR (signal to noise ratio) and thereby achieve high-quality imaging. Moreover, reducing the step size of mechanical scanning will also improve the image quality.

As for the imaging experiments, we only fabricated two muscle-like phantoms to mimic normal and disordered muscles. PDMS and Ecoflex are chosen to emulate muscle tissue and adipose tissue because they have similar acoustic impedance with that of body tissues [16]. But in fact, the composition of human tissues is complex and not homogenous, thus echoes generated within human body are more difficult to collect. In addition, the achieved imaging depth currently is only a few centimeters. In our future work, we aim to optimize the performance of the pMUT array to achieve high-resolution and large imaging depth simultaneously and be used on human body.

IV. CONCLUSION

This work preliminarily demonstrates that the pMUT can be used in ultrasound diagnostic imaging, especially for muscle disease or impairment. A 23×26 pMUT array is constructed and used for both transmitting and receiving ultrasonic signals. B-mode images of two fabricated muscle-like phantoms are created by using pulse-echo responses and the interfaces are clearly obtained. These results imply that our device has the potential for muscle diagnostic imaging. When combine with wireless transmission, advanced imaging algorithm, it can be used in wearable mode and will open widely practical applications in medical diagnosis.

REFERENCES

[1] N. B. M. Voet, E. L. van der Kooi, B. G. M. van Engelen, and A. C. H. Geurts, "Strength training and aerobic exercise training for muscle disease," *Cochrane Database Syst. Rev.*, vol. 2019, no. 12, 2019, doi: 10.1002/14651858.CD003907.pub5.

[2] J. Witherick and S. Brady, "Update on muscle disease," *J. Neurol.*, vol. 265, no. 7, pp. 1717–1725, 2018, doi: 10.1007/s00415-018-8856-1.

[3] F. Draghi, M. Zacchino, M. Canepari, P. Nucci, and F. Alessandrino, "Muscle injuries: Ultrasound evaluation in the acute phase," *J. Ultrasound*, vol. 16, no. 4, pp. 209–214, 2013, doi: 10.1007/s40477-013-0019-8.

[4] A. Loizides, H. Gruber, S. Peer, and M. Plaikner, "Muskelverletzungen des Sportlers: Stellenwert des Ultraschalls," *Radiologe*, vol. 57, no. 12, pp. 1019–1028, 2017, doi: 10.1007/s00117-017-0292-1.

[5] C. R. Heier *et al.*, "Non-invasive MRI and spectroscopy of mdx mice reveal temporal changes in dystrophic muscle imaging and in energy deficits," *PLoS One*, vol. 9, no. 11, 2014, doi: 10.1371/journal.pone.0112477.

[6] D. T. Yeh, Ö. Oralkan, I. O. Wygant, M. O'Donnell, and B. T. Khuri-Yakub, "3-D Ultrasound imaging using a forward-looking CMUT ring array for intravascular/intracardiac applications," *IEEE Trans. Ultrason. Ferroelectr. Freq. Control*, vol. 53, no. 6, pp. 1202–1210, 2006, doi: 10.1109/TUFFC.2006.1642519.

[7] B. T. Khuri-Yakub and Ö. Oralkan, "Capacitive micromachined ultrasonic transducers for medical imaging and therapy," *J. Micromechanics Microengineering*, vol. 21, no. 5, 2011, doi: 10.1088/0960-1317/21/5/054004.

[8] Y. Lu, Q. Wang, and D. A. Horsley, "Piezoelectric micromachined ultrasonic transducers with increased coupling coefficient via series transduction," *2015 IEEE Int. Ultrason. Symp. IUS 2015*, pp. 3–6, 2015, doi: 10.1109/ULTSYM.2015.0093.

[9] X. Chen, J. Xu, H. Chen, H. Ding, and J. Xie, "High-Accuracy Ultrasonic Rangefinders via pMUTs Arrays Using Multi-Frequency Continuous Waves," *J. Microelectromechanical Syst.*, vol. 28, no. 4, pp. 634–642, 2019, doi: 10.1109/JMEMS.2019.2912869.

[10] X. Chen, C. Liu, D. Yang, X. Liu, L. Hu, and J. Xie, "Highly Accurate Airflow Volumetric Flowmeters via pMUTs Arrays Based on Transit Time," *J. Microelectromechanical Syst.*, vol. 28, no. 4, pp. 707–716, 2019, doi: 10.1109/JMEMS.2019.2916987.

[11] H. Ding, S. Akhbari, B. E. Eovino, Y. Wu, J. Xie, and L. Lin, "Ultrasonic imaging of muscle-like phantoms using bimorph pmuts toward wearable muscle disorder diagnost ICS," in *Proc. IEEE 31st Int. Conf. MEMS*, Belfast, Northern Ireland, Jan. 2018, pp. 396–399.

[12] S. Schiaffino, K. A. Dyar, S. Ciciliot, B. Blaauw, and M. Sandri, "Mechanisms regulating skeletal muscle growth and atrophy," *FEBS J.*, vol. 280, no. 17, pp. 4294–4314, 2013, doi: 10.1111/febs.12253.

[13] S. Pillen and N. van Alfen, "Skeletal muscle ultrasound," *Neurol. Res.*, vol. 33, no. 10, pp. 1016–1024, 2011, doi: 10.1179/1743132811Y.0000000010.

[14] X. J. Jiang *et al.*, "Piezoelectric Micromachined Ultrasonic Transducers for Blood Vessel Motion Tracking," no. January, pp. 423–425, 2021.

[15] Y. Lu *et al.*, "Ultrasonic fingerprint sensor using a piezoelectric micromachined ultrasonic transducer array integrated with complementary metal oxide semiconductor electronics," *Appl. Phys. Lett.*, vol. 106, no. 26, 2015; doi: 10.1063/1.4922915.

[16] A. Cafarelli, P. Miloro, A. Verbeni, M. Carbone, and A. Menciassi, "Speed of sound in rubber-based materials for ultrasonic phantoms," *J. Ultrasound*, vol. 19, no. 4, pp. 251–256, 2016, doi: 10.1007/s40477-016-0204-7.

Proceedings of the 16th Annual IEEE International
Conference on Nano/Micro Engineered and Molecular Systems
April 25-29, 2021

Multifunctional Cardiomyocyte-Based Biosensor for Electrophysiology-Mechanical Beating-Growth Viability Monitoring

Dongxin Xu, Jiaru Fang, and Ning Hu*

Abstract— **Establishing a long-term and multi-parameter cell physiological assessment platform is of great significance for drug screening and cardiology research. Invasive and label-based recording platform usually destruct cell viability, hinder the long-term chronic monitoring of cardiomyocytes. Here, we develop a multifunctional cardiomyocyte-based biosensor, which integrates microelectrodes and interdigitated electrodes for noninvasive simultaneous monitoring of cardiomyocyte electrophysiology, mechanical beating and growth viability. Through further analysis of the typical and chronic physiological signals, it is demonstrated that the cardiomyocytes possess a stable and mature state after 192 h culture, which is applicable to the pharmaceutical analysis. It is believed that the multifunctional biosensor will provide a practical platform for the study of cardiac physiology/pathology, and be widely used in the biomedical field.**

I. INTRODUCTION

Cardiotoxicity induced by drugs is the major cause of drug development cessation, restricted use and withdrawal, which not only greatly endangers the life safety of patients, but also leads to a large waste of resources [1-3]. Therefore, it is urgent to develop an effective and stable platform for drug cardiotoxicity assessment. Most of cardiotoxic drugs cause arrhythmias, which are mainly related to the inhibition of the human ether-a-go-go-related gene (hERG) channel [4, 5]. In the early stage of drug development, hERG receptor protein and hERG inhibitor for drug cardiotoxicity assessment are high-throughput, but limited by low clinical relevance and is difficult to directly reflect the physiological characteristics of cardiomyocytes [6-8]. In the middle and late stages of drug development, cardiotoxicity assessment models based on isolated heart tissue and *in-vivo* animals have high clinical relevance, but they have disadvantages such as low throughput, complex operation and high cost [9, 10]. Human induced pluripotent stem cell-derived cardiomyocytes are an effective candidate for assessing cardiac safety of drugs, which possess unique gene and protein channels of human, and has an accurate response to drug action [11-13].

Drugs can cause changes in physiological parameters of cardiomyocytes, many research techniques have been

applied to record the electrical signals, mechanical beating signals and growth activity of cardiomyocytes in real time. Patch clamp is considered the gold standard technique for action potential recording and ion channel detection, but its irreversible damage to cells and complex operation process are not conducive to the high-throughput long-term electrophysiological monitoring [14-16]. Voltage-sensitive dyes and Ca^{2+} fluorescent dyes can record the electrophysiological and mechanical properties of cardiomyocytes simultaneously, but they are limited by drug side effects and phototoxicity [17-19]. At present, non-invasive, label-free microelectrodes can realize long-term cell electrophysiological monitoring [20]; cellular impedance detection technique can achieve quantitative detection of cardiomyocyte mechanical beating and growth viability [21]. These extracellular physiological detection techniques are non-invasive and high-throughput, but their parameters are relatively single, which is not conducive to the simultaneous multi-parameter assessment of cardiomyocytes.

In this study, a multifunctional cardiomyocyte-based biosensor is designed integrated with microelectrodes and interdigitated electrodes. Typical chronic electrophysiological signals, mechanical beating signals, and growth viability are simultaneously monitored by this multifunctional biosensor, providing a powerful tool for multi-parameter drug screening and cardiology research.

II. EXPERIMENTAL METHODS

A. Principle

The multifunctional cardiomyocyte-based biosensor integrates microelectrodes and interdigitated electrodes for simultaneous monitoring of electrophysiology, mechanical beating and growth viability in a non-invasive, label-free manner (Fig. 1a). Electrophysiological detection is achieved by extracellular recording with microelectrodes, which has no effect on cell viability, and is conducive to long-term monitoring of electrical signals. Two microelectrodes on the chip can improve the detection efficiency of the extracellular potential, and the combination with large reference electrodes is beneficial to achieve a stable reference potential. Meanwhile, mechanical beating and growth

---

*Resrach supported by the National Natural Science Foundation of China (Grant Nos. 82061148011), Guangdong Basic and Applied Basic Research Foundation (Grant No. 2020A1515010665), Department of Science and Technology of Guangdong Province Project (Grant No. 2020B1212060030), Foundation of Sun Yat-sen University (Grant Nos. 76120-18821104, 20lgpy47), and Open Project of Chinese Academy of Sciences (Grant No. SKT2006).

All authors are with State Key Laboratory of Optoelectronic Materials and Technologies, Guangdong Province Key Laboratory of Display

Material and Technology, School of Electronics and Information Technology, Sun Yat-sen University, Guangzhou 510006, China.

*Contacting Author: Ning Hu is with State Key Laboratory of Optoelectronic Materials and Technologies, Guangdong Province Key Laboratory of Display Material and Technology, School of Electronics and Information Technology, Sun Yat-sen University, Guangzhou 510006, China; State Key Laboratory of Transducer Technology, Chinese Academy of Sciences, Shanghai 200050, China (corresponding author to provide phone: +86-15381353839; fax: 303-555-5555; e-mail: huning3@mail.sysu.edu.cn).

978-1-6654-3008-1/21 $31.00 © 2021 IEEE

viability are detected by interdigitated electrodes. Cardiomyocytes cultured in vitro can produce rhythmic pulsation (contraction and relaxation), which changes the coupling with interdigitated electrodes and the ion current generated by the excitation signal of the impedance sensing. Thus, the mechanical beating of cardiomyocytes can be recorded in real time by interdigitated electrodes. Based on the impedance changes caused by cell morphology and attachment, the growth viability of cells can be estimated by long-term impedance monitoring. In contrast to slow growth signals, the mechanical beating signals require high temporal resolution recording with high-speed ECIS technology, and the size and arrangement of interdigitated electrodes are determined by the signal detection sensitivity. In addition, compared with the large excitation signal of impedance detection, the weak extracellular electrical signal needs to be separated with low noise based on frequency differences. Therefore, electrophysiology, mechanical beating and growth viability of cardiomyocytes can be monitored simultaneously by the multifunctional biosensor.

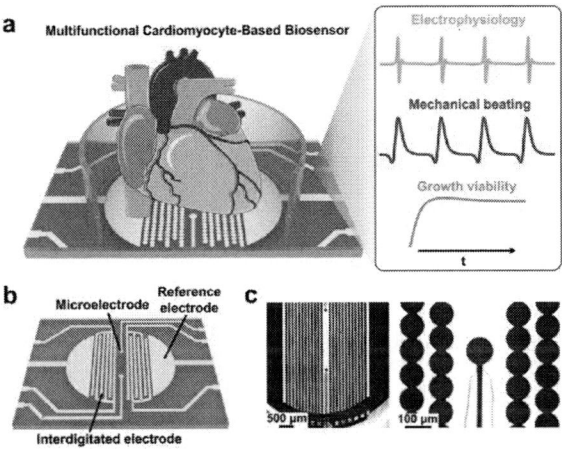

Figure. 1 Multifunctional cardiomyocyte-based biosensor for electrophysiology-mechanical beating- growth viability monitoring. (a) Schematic illustration of multifunctional cardiomyocyte-based biosensor for electrophysiological signals, mechanical beating and growth viability signals detection. (b, c) Schematic and optical images of multifunctional cardiomyocyte-based biosensor.

B. Device Fabrication and Characterization

The multifunctional biosensor chip is fabricated by a standard photolithography process. A rectangular borosilicate glass substrate is spin-coated with positive photoresist s1813 at 3000 rpm. After a soft bake at 100°C for 60 s, the electrically conductive pattern is defined with constant intensity UV exposure source at a dose of 20 mW/cm^2 and developed for 40 s using the photoresist developer CD-26. Thereafter, the patterned substrate is cleaned by oxygen plasma, the titanium (Ti)- gold (Au) (20/100 nm) is sputtered on the substrate and the photoresist is stripped by acetone. Finally, a 2-μm-think SU-8 2002 insulating layer is spin-coated at 3000 rpm and soft baked at 95°C for 1 min. After exposure, it is postexposure baked at 95°C for 1 min, developed in propylene glycol methyl ether acetate (PGMEA) for 1 min, rinsed in isopropanol, dried, and hard baked at 150°C for 30 min. For cell culture and insulation, the biocompatible polystyrene (PS) chambers are

fixed on the chip. The multifunctional cardiomyocyte-based biosensor consists of 100 μm microelectrodes located in the central region, 90 μm circle-on-line interdigitated electrodes with a 120 μm branch center spacing, and two reference electrodes located outside the interdigitated electrodes (Fig. 1b), which is characterized by optical microscope (Fig. 1c). To obtain the basic performance of microelectrodes and the characteristic frequency range of interdigitated electrodes, the multifunctional cardiomyocyte-based biosensor is characterized by electrochemical impedance spectroscopy in the phosphate-buffered saline (PBS).

C. Cell Culture

Each device is rinsed by 75% ethanol and sterilized under ultraviolet exposure for 2 h in biosafety cabinet. Prior to cell seeding, the device is coated with 10 mg/ml fibronectin to form the surface coverage of ~ 2.5 μg/cm^2 for improving the cell adhesion, and then placed in the 37°C, 5.0% CO_2 incubator for at least 2 hours. Human induced pluripotent stem cell-derived cardiomyocytes (iPSC-CMs) are thawed, resuspended, and plated inside the chamber at a density of 15,000 cells/well. After seeding, the devices are maintained in the 37°C and 5% CO_2 humidified incubator, and the cell culture medium is refreshed every 48 h for long-term recording.

III. RESULTS AND DISCUSSION

A. Electrophysiological Recording and Assessing

Electrophysiological detection of cardiomyocytes is performed by microelectrodes on the multifunctional biosensor. The extracellular rhythmic firing signals are recorded from the cardiomyocytes with mature status, which present biphasic pulses with amplitude of ~4 mV (Fig. 2a).

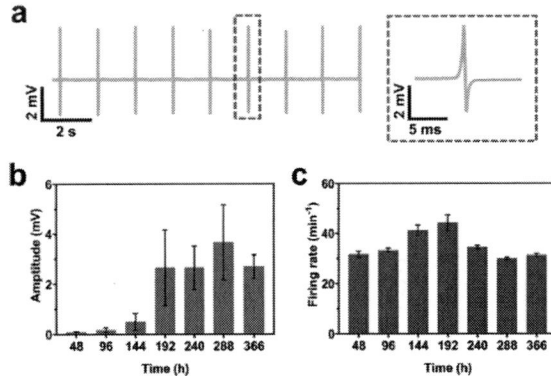

Figure. 2 Electrophysiological recording and assessing. (a) The typical electrophysiological signals of cardiomyocytes. (b, c) Statistical results of mean amplitudes and firing rate (n=4).

To further quantitatively assess the electrophysiological characteristics of cardiomyocytes, continuous monitoring for 366 h are performed, the amplitude and firing rate of recorded signals are statistically analyzed (Fig. 2b and 2c). The amplitude of the extracellular potential increases from 97.40 ± 11.19 μV (48 h) to 2.650 ± 1.503 mV (192 h), and the firing rate increases from to 31.66 ± 1.280 min^{-1} (48 h) to 44.18 ± 3.270 min^{-1} (192 h). The increasing amplitude and firing rate may be attributed to the deposition, adhesion and diffusion of cardiomyocytes after seeding, which form a close coupling with the electrode and greatly improve the signal quality. After 192 h, the firing rate of cardiomyocytes decreases to 34.44 ± 0.7359 min^{-1} (240 h), which may reflect the gradual decline in cell viability. Therefore, based on the comprehensive analysis of the statistical results, the state of cardiomyocytes is relatively stable after 192 h culture, which is suitable for drug screening and disease modeling.

B. Mechanical Beating Recording and Assessing

Mechanical beating signals of cardiomyocytes is recorded by interdigitated electrodes on the multifunctional biosensor. The rhythmic beating signals are recorded from well-conditioned cardiomyocytes using high-temporal resolution ECIS techniques (Fig. 3a).

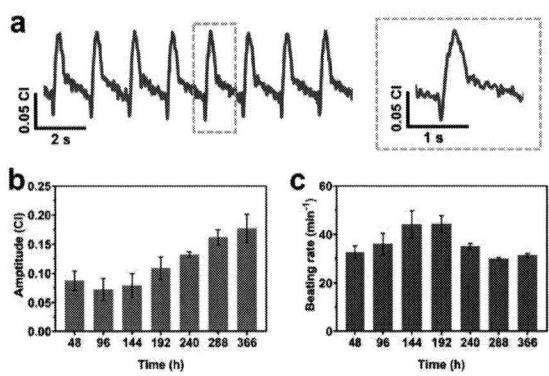

Figure. 3 Mechanical beating recording and assessing. (a) The typical mechanical beating signals of cardiomyocytes. (b, c) Statistical results of mean amplitudes and beating rate (n=4).

To quantitatively assess the mechanical beating characteristics of cardiomyocytes, the cell index (CI) is defined as the ratio of the impedance change to the background impedance for calculating the amplitude of mechanical beating. Through continuous monitoring for 366 h, the amplitude of mechanical beating increases from 0.08713 ± 0.01652 CI (48 h) to 0.1088 ± 0.01902 CI (192 h) (Fig. 3b), and the firing rate increased from 32.62 ± 2.550 min^{-1} (48 h) to 44.37 ± 3.370 min^{-1} (192 h) (Fig. 3c). The change of impedance signal is influenced by confounding factors such as cell morphology, proliferation and attachment, which reflects the vitality, number, morphology and attachment of cells. The increasing amplitude and beating rate indicate the gradual maturation and good viability of cardiomyocytes. After 192 h, the beating rate of cardiomyocytes gradually decreases to 35.00 ± 1.182 min^{-1} (192 h). Based on the comprehensive analysis of the statistical results, the beating state of cardiomyocytes is relatively stable after 192 h culture, which is suitable for drug experiments. In addition, the simultaneous recording of electrophysiology and mechanical beating is beneficial to explore the cardiac excitation-contraction coupling. By comparing the statistical data of electric signal and beating signal, it is indicated that the frequency of the two signals is basically similar and remains stable at ~44 min^{-1}, which provides a new idea for exploring the physiological and pathological mechanisms of the heart.

C. Growth Viability Recording and Assessing

Growth viability of cells is assessed by interdigitated electrodes fabricated on the multifunctional biosensor. Based on the statistical analysis of cell index, growth viability of cardiomyocytes increases from 17.68 ± 2.262 CI (48 h) to 25.58 ± 3.659 CI (96 h), and remains stable within 366 h (Fig. 4). After being seeded onto the device, the cardiomyocytes deposit on the electrodes in the shape of spheres, gradually adhere and diffuse, forming tight coupling with the electrodes, which results in the increase of the impedance. After that, a stable state is formed between cardiomyocytes and electrodes, with the impedance maintained at ~20 CI. By the simultaneous detection of the electrophysiology, mechanical beating and growth viability, it can be concluded that the stable and mature state of cardiomyocytes after 192 h culture is conducive to drug analysis.

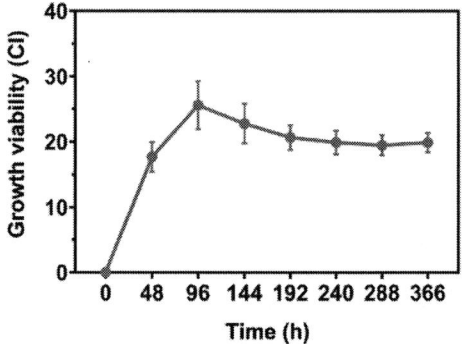

Figure. 4 Growth viability chronic recording and statistical analysis of cardiomyocytes for 366 h (n=4).

IV. CONCLUSION

In the study, multifunctional cardiomyocyte-based biosensor is developed integrated with microelectrodes and interdigitated electrodes for electrophysiology, mechanical beating and growth viability monitoring. Typical extracellular electrical signals and mechanical beating signals are recorded by microelectrodes and interdigitated electrodes, respectively. Through long-term recording for 366 h, the growth trend of cells is visualized by the statistical data of electrophysiological signals, beating signals and growth viability. Based on further comprehensive analysis, it can be concluded that the cardiomyocytes cultured in vitro reach a stable and mature state after 192 h culture, and are suitable for cardiology research and drug screening. Simultaneous detection of electrophysiological signals, beating signals and growth viability is important for exploring the mechanism of cardiac excitation-coupling, as well as assessing the short-term and long-term effects of the drugs.

Further work will focus on improving the performance of multifunctional cardiomyocyte-based biosensors: (i) Single cell detection. Large electrodes are limited by low throughput and signal overlap interference from multiple cells. Reducing the size of the electrode is beneficial to improve the throughput of the biosensor and achieve accurate single-cell monitoring. (ii) Signal quality optimization. Reducing the electrode size may lead to a worse signal-to-noise ratio. Combined with three-dimensional nanostructure is a potential strategy to increase the effective surface area of the electrode and form a tight coupling with the cell for high-quality signal detection. In addition, compared with extracellular potential, intracellular potential is closer to transmembrane potential, which contains key information about the type, status and density of various ion channels. Intracellular recording has outstanding advantages in screening of ion channel drugs and assessment of cardiomyocyte type and maturation stage. Three-dimensional nanoelectrodes combined with spontaneous/assisted membrane penetration have been demonstrated to be an effective intracellular recording platform for cardiology studies [22-26]. (iii) Multifunctional expansion. Combined with electrical/optical stimulation or microfluidic delivery is conducive to precise regulation and synchronous monitoring of cells. Integrated with high content analysis technology is beneficial to realize automatic visual cell analysis. Further development of multifunctional biosensors promises major breakthroughs in cardiology and the broader biomedical field.

REFERENCES

[1] J. Bowes et al., "Reducing safety-related drug attrition: the use of in vitro pharmacological profiling," Nat Rev Drug Discov, vol. 11, no. 12, pp. 909-22, Dec 2012.

[2] C. L. Lawrence, C. E. Pollard, T. G. Hammond, and J. P. Valentin, "In vitro models of proarrhythmia," Br J Pharmacol, vol. 154, no. 7, pp. 1516-22, Aug 2008.

[3] B. Fermini and A. A. Fossa, "The impact of drug-induced QT interval prolongation on drug discovery and development," Nat Rev Drug Discov, vol. 2, no. 6, pp. 439-47, Jun 2003.

[4] D. Rampe and A. M. Brown, "A history of the role of the hERG channel in cardiac risk assessment," Journal of pharmacological toxicological methods, vol. 68, no. 1, pp. 13-22, 2013.

[5] M. C. Sanguinetti and M. Tristani-Firouzi, "hERG potassium channels and cardiac arrhythmia," Nature, vol. 440, no. 7083, pp. 463-469, 2006.

[6] I. Cavero and W. Crumb, "Native and cloned ion channels from human heart: laboratory models for evaluating the cardiac safety of new drugs," European heart journal supplements, vol. 3, no. suppl_K, pp. K53-K63, 2001.

[7] A. Lacerda, J. Kramer, K.-Z. Shen, D. Thomas, and A. Brown, "Comparison of block among cloned cardiac potassium channels by non-antiarrhythmic drugs," European heart journal supplements, vol. 3, no. suppl_K, pp. K23-K30, 2001.

[8] S. Zhang, Z. Zhou, Q. Gong, J. C. Makielski, and C. T. January, "Mechanism of block and identification of the verapamil binding domain to HERG potassium channels," Circulation research, vol. 84, no. 9, pp. 989-998, 1999.

[9] P. Kannankeril, D. M. Roden, and D. Darbar, "Drug-induced long QT syndrome," Pharmacological reviews, vol. 62, no. 4, pp. 760-781, 2010.

[10] T. Meyer, P. Sartipy, F. Blind, C. Leisgen, and E. Guenther, "New cell models and assays in cardiac safety profiling," Expert opinion on drug metabolism toxicology, vol. 3, no. 4, pp. 507-517, 2007.

[11] I. Kehat et al., "Human embryonic stem cells can differentiate into myocytes with structural and functional properties of cardiomyocytes," The Journal of clinical investigation, vol. 108, no. 3, pp. 407-414, 2001.

[12] P. Liang et al., "Drug screening using a library of human induced pluripotent stem cell–derived cardiomyocytes reveals disease-specific patterns of cardiotoxicity," Circulation, vol. 127, no. 16, pp. 1677-1691, 2013.

[13] E. G. Navarrete et al., "Screening drug-induced arrhythmia using human induced pluripotent stem cell–derived cardiomyocytes and low-impedance microelectrode arrays," Circulation, vol. 128, no. 11_suppl_1, pp. S3-S13, 2013.

[14] B. Sakmann and E. Neher, "Patch clamp techniques for studying ionic channels in excitable membranes," Annual review of physiology, vol. 46, no. 1, pp. 455-472, 1984.

[15] Y. Zhao, S. Inayat, D. A. Dikin, J. H. Singer, R. S. Ruoff, and J. B. Troy, "Patch clamp technique: Review of the current state of the art and potential contributions from nanoengineering," Proceedings of the Institution of Mechanical Engineers, Part N: Journal of Nanoengineering and Nanosystems, vol. 222, no. 1, pp. 1-11, 2009.

[16] M. Bebarova, "Advances in patch clamp technique: towards higher quality and quantity," General physiology and biophysics, vol. 31, no. 2, pp. 131-40, Jun 2012.

[17] P. Lee et al., "Simultaneous voltage and calcium mapping of genetically purified human induced pluripotent stem cell-derived cardiac myocyte monolayers," Circ Res, vol. 110, no. 12, pp. 1556-63, Jun 8 2012.

[18] M. Scanziani and M. Häusser, "Electrophysiology in the age of light," Nature, vol. 461, no. 7266, pp. 930-939, 2009.

[19] S. Awasthi et al., "Multimodal SHG-2PF Imaging of Microdomain Ca2+-Contraction Coupling in Live Cardiac Myocytes," Circ Res, vol. 118, no. 2, pp. e19-28, Jan 22 2016.

[20] K. Asakura et al., "Improvement of acquisition and analysis methods in multi-electrode array experiments with iPS cell-derived cardiomyocytes," Journal of Pharmacological and Toxicological Methods, vol. 75, pp. 17-26, 2015/09/01/ 2015.

[21] N. Hu et al., "High-performance beating pattern function of human induced pluripotent stem cell-derived cardiomyocyte-based biosensors for hERG inhibition recognition," Biosensors and Bioelectronics, vol. 67, pp. 146-153, 2015/05/15/ 2015.

[22] C. Xie, Z. Lin, L. Hanson, Y. Cui, and B. Cui, "Intracellular recording of action potentials by nanopillar electroporation," Nature nanotechnology, vol. 7, no. 3, pp. 185-90, Feb 12 2012.

[23] Z. C. Lin, C. Xie, Y. Osakada, Y. Cui, and B. Cui, "Iridium oxide nanotube electrodes for sensitive and prolonged intracellular measurement of action potentials," Nature communications, vol. 5, p. 3206, 2014.

[24] M. Dipalo et al., "Intracellular and Extracellular Recording of Spontaneous Action Potentials in Mammalian Neurons and Cardiac Cells with 3D Plasmonic Nanoelectrodes," Nano Letters, vol. 17, no. 6, pp. 3932-3939, 2017.

[25] B. X. E. Desbiolles, E. de Coulon, A. Bertsch, S. Rohr, and P. Renaud, "Intracellular Recording of Cardiomyocyte Action Potentials with Nanopatterned Volcano-Shaped Microelectrode Arrays," Nano Letters, vol. 19, no. 9, pp. 6173-6181, Sep 11 2019.

[26] N. Hu et al., "Intracellular recording of cardiomyocyte action potentials by nanobranched microelectrode array," Biosens Bioelectron, vol. 169, p. 112588, Sep 12 2020.

The 16th IEEE International Conference on Nano/Micro Engineered & Molecular Systems

Developing biocompatible and implantable flexible pressure sensors for health monitoring applications

Guozhen Shen

State Key Laboratory for Superlattices and Microstructures, Institute of Semiconductors, Chinese Academy of Sciences, Beijing 100083, P.R. China

Website : https://www.x-mol.com/groups/flextronics

ABSTRACT

Advances in digital health care have driven innovations in high-performance wearable and smart sensors. One requirement in this field is establishing healthy, secure and reliable medical devices for precisely monitoring vital signs of the human body or the surrounding environment through flexible sensors with not only high sensing performance, but also excellent biofunctionality. Smart wearable sensors with excellent biofunctionality furnish medical devices with various smart functions such as biocompatibility, biodegradability, and self-healing, which have attracted widespread interest from device engineers and materials scientists. Here, we focus on the various types of materials used in biomedical and implantable devices and their bio-multifunctional (biocompatible, biodegradable, and self-healing) designs. The subsequent content highlights the most advanced medical applications of bio-multifunctional wearable sensors, classified into three main subfields (Figure 1): 1) biophysical monitoring (heart rate/pulse, human motion, and temperature), 2) biochemical tracking (biomolecule, blood glucose, and Na+/K+), and 3) real-time environmental information detection (gas molecules and humidity). Finally, this work concludes with an overview of key challenges and a summary of opportunities, ultimately determining that advances in smart wearable sensors are critical to their continuing progress.

BIOGRAPHY

Guozhen Shen is currently a Professor and Group Leader at Institute of Semiconductors, Chinese Academy of Sciences. Dr. Shen received his Ph.D degree in Chemistry from University of Science and Technology of China. From 2004 to 2013, he did his research in Hanyang University (Korea), National Institute for Materials Science (Japan), University of Southern California (US) and Huazhong University of Science and Technology (China). Dr. Shen's research focuses on design of low-dimensional nanostructures for flexible electronic applications, including flexible sensors, flexible energy storage devices and flexible multifunctional integrated electronic systems. He is the Editor/ Editorial board member of Nanoscale Research Letters, Advanced Materials Technologies, Journal of Semiconductors, etc. Dr. Shen has published more than 300 peer-reviewed articles and edited 2 books with an H-factor of 76.

Dr. Shen won several important awards, including the National Science Fund for Distinguished Young Scholars, the 2nd prize of Science and Technology Award of Beijing, and 1st prize of Science and Technology of C-MRS.

Gap in pagination due to unavailable papers.

Pages 1379-1383

Proceedings of the 16th Annual IEEE International
Conference on Nano/Micro Engineered and Molecular Systems
April 25-29, 2021

Highly Sensitive and Flexible Tactile Sensor Based on the Fabrication of Porous Graphene/Silicone Rubber Composites

Zhijian Chen, Yancheng Wang*, *Member, IEEE*, Deqing Mei, and Jie Jin

Abstract—This paper presents a novel flexible tactile sensor by the fabrication of porous graphene nanoplate (GNP)/silicone rubber (SR) composites as sensing material. The designed tactile sensor has 3×3 (=9) sensing units, each unit has three layers: upper electrode and cover layer, middle porous composites, and bottom electrode layer. The processes of preparing the sensing material and fabricating the tactile sensor are presented. The porous GNP/SR composite possess an extremely high sensitivities of 8.45 kPa^{-1} at 0~55 kPa and 195.02 kPa^{-1} at 55~80 kPa for pressure sensing, its Young's modulus is low (~85.72 kPa) and can endure large strain over 60%. After fabrication of the tactile sensor, its sensing performances are characterized. Experimental testing results showed that the developed tactile sensor has high sensitivity, good dynamic response performance and repeatability under different applied pressures. Results indicated that the developed tactile sensor has potential and could be used to detect human body motion and health monitoring.

Index Terms–Flexible tactile sensor; Porous graphene composite; Wearable electronics; Distributed detection.

I. INTRODUCTION

With the rapid development of flexible electronic sensors research, wearable and flexible electronic devices have been widely applied in prosthetic hand, intelligent robots, wearable devices, et al [1-3]. As one of the important parts of them, tactile sensors with high sensitivity and flexibility are able to provide signals such as contact, pressure, vibration and so on for joint movement detection, biomedical usage, and other health consultations [4, 5]. In the past decade, several types of sensing principles have been proposed to develop flexible tactile sensors, including piezoelectric [6], capacitive [7], piezoresistive [8], etc. Among them, the piezoresistive tactile sensors have been widely concerned for their generally good flexibility, simple structure, broad measuring range, et al [9]. However, the sensitivity of piezoresistive sensors is mainly limited due to preparation of piezoresistive sensing material and its electrical conductivity. Therefore, the sensing material's preparation to develop the tactile sensor still needs to be conducted to improve its sensing performances, such as mechanical properties, low hysteresis and high sensitivity.

To obtain high performance tactile sensors, many sensitive materials such as nano-silver sheets [10], carbon

Y.C. Wang and D.Q. Mei are with the State Key Laboratory of Fluid Power & Mechatronic Systems, School of Mechanical Engineering, Zhejiang University, Hangzhou, China (*Corresponding author: Yancheng Wang, phone: 86-13675828104, e-mail: yanchwang@zju.edu.cn).
Z.J. Chen and J. Jin are with the Key Laboratory of Advanced Manufacturing Technology of Zhejiang Province, School of Mechanical Engineering, Zhejiang University, Hangzhou, China.

nanotubes [11] and graphene [12-14] are used as sensitive materials. Graphene features high mechanical strength, large specific surface area and excellent electrical, and is considered to be an ideal candidate material for high sensitivity sensors [15]. As for the graphene, it can be divided into 2D graphene and 3D graphene according to its morphology. Thus, the structural of graphene-based tactile sensor can be classified into two groups [12-14]. The graphene-based thin film tactile sensor relies on the deformation of the flexible matrix to change the conductive network in the graphene material. Generally, this type of sensor can only undergo tensile deformation instead of compression. Moreover, the thin film graphene layer is weak to connect each other and difficult to maintain stable electrical conductivity under large strains. On the contrary, graphene with 3D geometry is able to address these issues. The tactile sensors based on 3D graphene are scalable in both horizontal and vertical directions when subjected to forces, so they can simultaneously withstand large pressures and strains while maintaining stability of structure and performance.

As a typical three-dimensional structure of graphene, graphene aerogel, has been reported to be used to fabricate tactile sensors [16, 17]. The prepared graphene aerogels have large surface area, low density and excellent electrical conductivity [18], while this material has generally poor mechanical properties. Thus, using the graphene with flexible and thin substrate to enhance its mechanical properties could be a potential approach to designing tactile sensors with high performance. Zhang *et al.* [19] using the dip-coating process to coat the graphene oxide (GO) on the polyurethane (PU) foam, and reduced the GO by heating to obtain the graphene foam coated with a thin-layer of graphene. Compared with original graphene aerogel, the PU foam was used as a skeleton, it can enhance the performances such as dynamic stability and repeatability of the graphene foam. Hu *et al.* [15] developed a method by using freeze-drying to fabricate the graphene foam, and coated it with polydimethylsiloxane (PDMS) to improve its mechanical property. However, the complicated and relatively high-cost fabrication processes limit the wide application of it to fabricate sensing foam. Therefore, a simple and low-cost fabrication method to develop 3D graphene porous composite with high sensitivity and flexibility still needs to be studied and be an objective of this research.

Herein, we report a highly sensitive and flexible tactile sensor based on the preparation of porous graphene nanoplate (GNP)/silicone rubber (SR) composite. The proposed tactile sensor has 9 sensing units arranged in 3 rows and 3 columns, thus able to measure the distributed tactile signals. The porous GNP/SR composites are fabricated through a simple and low-cost fabrication method.

978-1-6654-3008-1/21 $31.00 © 2021 IEEE

In this paper, the structural design and processes to fabricate flexible tactile sensor are presented, and the performance testing of the prepared porous composite sensing material and fabricated tactile sensor are performed. Results show that the porous composites possess good mechanical properties, high sensitivity as well as good repeatability, which indicated that the developed tactile sensor has potential to be used in human health monitoring.

II. TACTILE SENSOR DESIGN AND FABRICATION

A. Tactile Sensor's Design

The structural design schematic diagram of the tactile sensor is illustrated in Fig. 1(a). The proposed tactile sensor array consists of 9 sensing units distributed in 3 rows and 3 columns, and each sensing unit has the truncated pyramid-shaped porous GNP/SR composite as the sensing material, which is sandwiched between the top and bottom electrode layers. The sensing unit of the sensor has a dimensions of top side length of 1.5 mm and bottom length of 3.0 mm, the shape factor of the sensing unit can be calculated as 0.256 that is half of the rectangular-shaped geometry [20]. The pattern of the top electrode and bottom electrode is designed according to the contact area with the sensing unit, and is coated on the surface of the thin-film polyethylene terephthalate (PET) substrates. To facilitate the signals acquisition and design of scanning circuit, the upper electrode and the lower electrodes are arranged vertically. The overall dimensions of the proposed tactile sensor array are designed as 10 mm in length, 10 mm in width, and 3 mm in thickness.

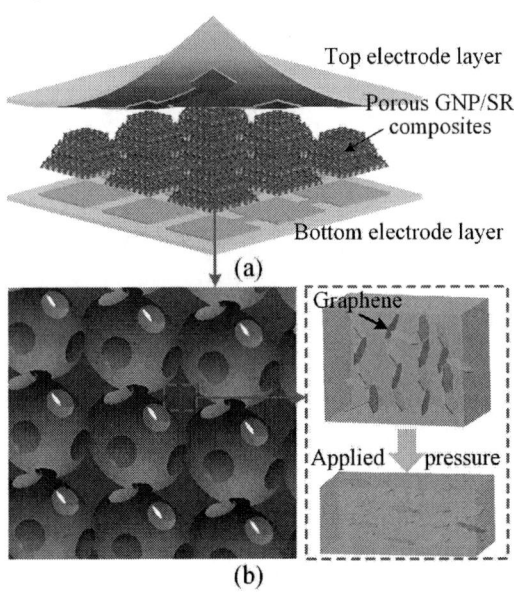

Fig. 1 (a) Structural design of flexible tactile sensor, (b) porous GNP/SR composite as sensing material and its pressure sensing principle during compression.

The porous GNP/SR composites are made of graphene as conductive filler and silicone rubber as flexible matrix. The graphene nanoplates can be uniformly distributed inside the silicone rubber matrix, thus the position and orientation of the graphene nanoplates would be more stable. As shown in Fig. 1(b), the conductive network in the porous GNP/SR composite will be easily destroyed by external compression due to the graphene nanoplates having a high aspect ratio [21], which leads to the increasing resistance of the porous composite and contributes to high sensitivity of the tactile sensor for pressure or tactile force sensing.

B. Tactile Sensor's Fabrication

The prepared porous GNP/SR composites will affect the sensing performance of the tactile sensor. Here, a simple method to fabricate the porous GNP/SR composite into a truncated pyramid shape is developed. The process mainly has four steps and is shown in Fig. 2.

In **Step 1**, the silicone rubber (GD401, Zhonghao Chenguang Research Institute of Chemical Industry, China), Azodicarbonamide (AC foaming agent, LK-8000, Jia Shi chemical industry Co., Ltd, China), and graphene nanoplates (diameter <10 μm, The Sixth Element Materials Technology Co., Ltd, China) were added into a test tube with mass ratios of 44.8 wt%, 6.7 wt%, and 2.2 wt%, respectively. Then, ZnO powder with mass ratio of 1.3 wt% as the catalyst for AC foaming and Polyphenylmethylsiloxane (PPMS, Weng Jiang Reagent Co., Ltd, China) with mass ratio of 44.8 wt% as the solvent were added into the mixture. These materials were mixed together by using a planetary mixer at 2000 rpm for 10 mins to obtain a uniform pasty mixture.

In **Step 2**, the GNPs and AC foaming in the mixture were further dispersed by using ultrasonic stirring (FS-300N, Shanghai Sonxi Ultrasonic Ins. China) for 10 mins.

In **Step 3**, the mixed material was poured into a pre-designed mold made by aluminum with nine inverted pyramid shape patterned structure, and then put it into a vacuum drying oven for 30 mins to remove air bubbles. Then heat the mixture at 110 °C for 2 hours, the AC foaming would break down to create gases and meanwhile silicone rubber would gradually solidify, so the diffused pores would be created in the composite material. Thus, the porous GNP/SR composites with diffuse tiny pores inside can be fabricated, and its volume is larger than the mixture before curing.

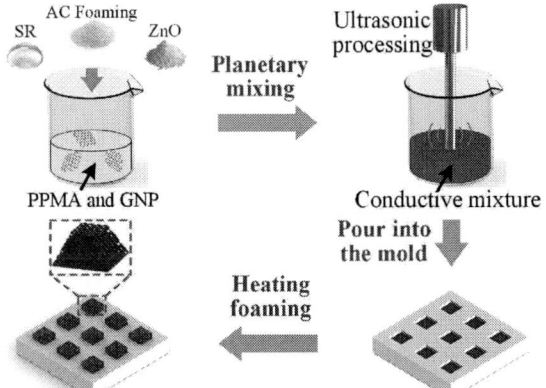

Fig. 2 Diagram of the fabrication process to prepare the porous GNP/SR composites.

In **Step 4**, the porous composites were taken out from the mold and immersed in anhydrous ethanol, then ultrasonic cleaned for 5 minutes to remove the residual PPMS on the surface of the porous composites. Finally, the porous GNP/SR composites with truncated pyramid shape could be obtained.

After preparation of the porous GNP/SR composite sensing material, it was used to fabricate the tactile sensor. Figs. 3(a-g) exhibit detailed processes to fabricate the tactile sensor. The fabrication procedure and process are described as below: 1) The shape of the top and bottom copper electrodes was determined by photoetching firstly, and then 250 nm Cu were put on the surface of the PET substrates through magnetron sputtering, as shown in Figs. 3(a)-(d); 2) A conductive paste with thickness of 100 μm was coated on the copper electrodes by screen-printing, which was used to connect the patterned porous GNP/SR composite and electrode layers, as in Fig. 3(e); 3) The prepared porous composites with truncated pyramid shape were placed on the corresponding positions of the conductive paste (Fig. 3(e)), and gently pressed them to fully contact with the corresponding electrodes. Then, the whole device was heated up to 90 °C on a heating platform for 2 hours so that the conductive adhesive could be fully solidified to ensure a solid adhesion between porous composites and electrodes; 4) Repeating the above steps to make a top electrode layer, then the top electrode layer was covered on the top surface and heated up to 90 °C for another 2 hours.

The final finished tactile sensor is shown in Fig. 3(h), the overall dimensions of the tactile sensor are measured about 10 mm × 10 mm × 3 mm. The sensor has nine sensing units, and it is 3.5 mm between adjacent units.

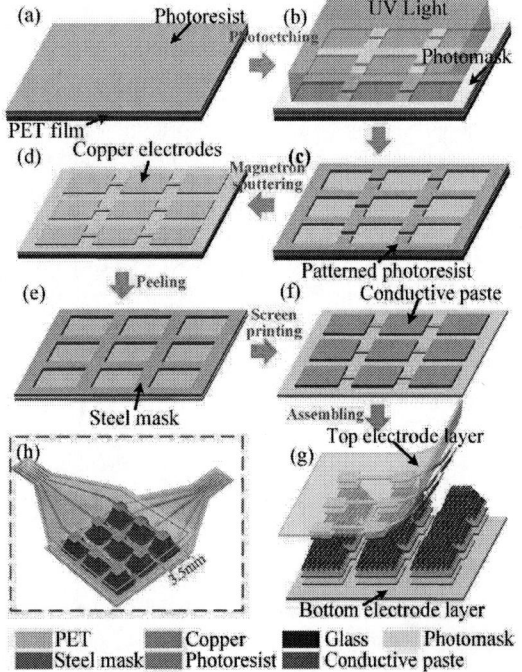

Fig. 1 (a-g) Fabrication procedure of the developed flexible tactile sensor, (h) the final fabricated tactile sensor with nine sensing units.

III. RESULTS AND DISSCUSION

A. Performance Testing of Porous Composite

For the fabricated tactile sensor, the mechanical and electrical properties of porous GNP/SR composite will greatly affect the sensing performances. To explore the mechanical and electrical properties of the prepared material, a specimen of the porous GNP/SR composite with dimensions of 3 mm × 3 mm × 3 mm is prepared for material testing. The cyclic loading/unloading tests were carried on by a uniaxial universal experimental machine (UTM2203, Shenzhen Suns Technology Stock Co., Ltd).

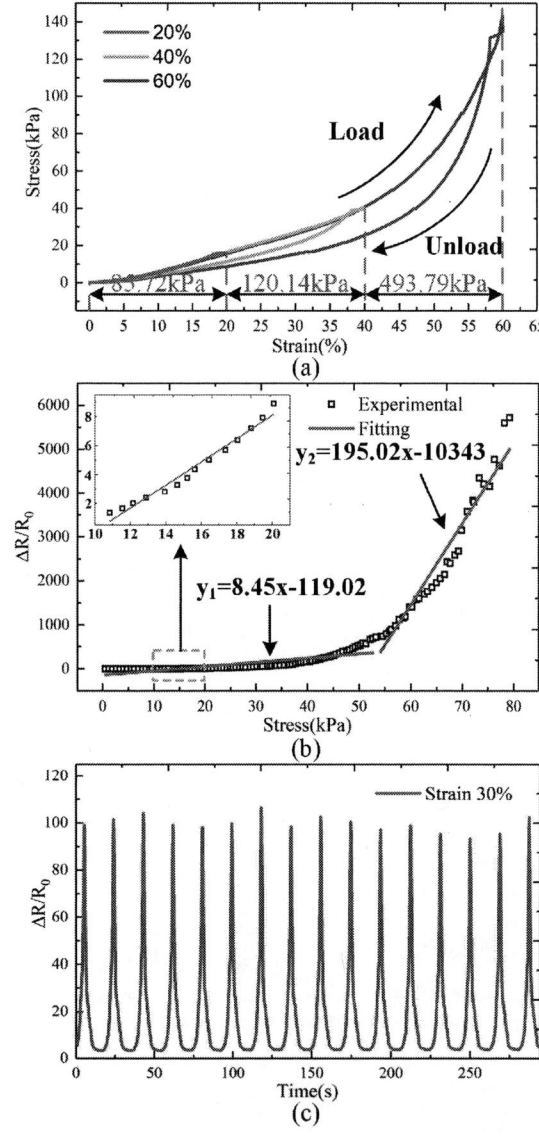

Fig. 4 The measured properties of porous GNP/SR composites: (a) mechanical property at different strains, (b) resistance changes of porous composite under compression, (c) Repeated cyclic loading and unloading tests under 30% strain.

TABLE 1 COMPARSION OF THE SENSING PERFORMANCE OF OUR POROUS GNP/SR COMPOSITE WITH OTHER WORKS

Materials	Sensitivity(kPa^{-1})	Gauge factor	Measuring range	Foaming agent	Refs
GNP/Silicone rubber	8.45(0~55kPa) 195.02(55~80kPa)	15614	0~80kPa	AC foamimg	Our work
SWCNTs/G801 rubber	4.3×10^{-3}	--	0~200kPa	DPT	[22]
GO/PDMS	--	1.6(<10%) 60(30~50%)	ε<50%	F127	[15]
CNT/PDMS	--	6.4	ε<80%	Brown sugar	[23]
LSG	0.96	--	50~113kPa	--	[24]
CNT/PDMS	0.03(0~15kPa) <0.008(>15kPa)	~1.5	0~150kPa	Cube sugar	[25]
GO/PI foam	0.36(0~4kPa) 0.01(4~14kPa)	<6	0~14kPa	--	[26]
MWNTs/rGO/PU foam	0.022(0~2.7kPa) <0.088(>2.7kPa)	0.05(<50%) -2.13(50~86%) -2.3(86~100%)	0~50kPa	--	[27]
Graphene/PU foam	7.62(0~50kPa) 0.14(50~200kPa)	16.6	0~200kPa	DMSO	[28]

Firstly, the induced stress under different strains (ε = 20%, 40%, 60%) were measured. During the tests, the loading and unloading velocities were set as 1.0 mm/min, and results are shown in Fig. 4(a). We can see that the porous composite has generally good consistency even at different applied strains. When ε < 20%, The loading curve and unloading curve of porous composites coincide nearly, showing that the material has almost no hysteresis and the Young's modulus can be calculated as about 85.72 kPa; When 20% < ε < 40%, the Young's modulus of the material becomes greater with the value of 120.14 kPa and the hysteresis of the porous composite becomes larger; When 40% < ε < 60%, as the densification of porous composites is increasing, the Young's modulus is further increasing to 493.79 kPa, which is much larger than that under small strains. And a distinct hysteresis can be observed due to the elasticity of silicone rubber rather than the porous skeleton playing a major role in its mechanical properties. This phenomenon also indicates that the porosity of the produced porous composite is around 60%. When the applied pressure increases to about 60%, there will be no pores inside the composites.

Secondly, the resistance changes of porous GNP/SR composite under different applied stress were also performed. During the experiments, a loading bar with a diameter of 3.0 mm was mounted on a biaxial motion platform which is constructed on Newport's UTC linear stages (Newport Corporation, USA) and underneath is an ATI triaxial force sensor (ATI Industrial Automation, Inc., USA). The force sensor has an accuracy of 0.01 N, it was used to measure the applied normal force and pressure. The real-time resistance of porous composites was recorded by using a digital multimeter (34465A, Keysight Technologies Co., Ltd, Beijing, China). And the relative resistance change can be acquired as $(R_C - R_0)/R_0$, where R_0 is the initial resistance, R_C is the measured specimen's resistance in compression process. The calculated results of resistance change versus stress are plotted in Fig. 4(b). As the exerted pressure and induced stress increased, the resistance changes also increased gradually. Typically, in Fig. 4(b),

we can see that the porous composites have two sensitivities under pressure: a relative low sensitivity of 8.45 kPa^{-1} at 0~55 kPa, and a high sensitivity of 195.02 kPa^{-1} at 55~80 kPa. Although the porous composite shows relatively low sensitivity under 0~55 kPa, it still exhibits significantly different resistance responses under applied pressures, as in Fig. 4(b). Therefore, the prepared porous composites have the characteristics of maintaining high sensitivity under a broad range of pressures.

The repeated cyclic tests under a strain of 30% were conducted on the porous composites and results are shown in Fig. 4(c). We can see that the $\Delta R/R_0$ of the specimen changed regularly from about zero at the strain of 0 to about 100 at the strain of 30% with negligible fluctuations. This demonstrated that the fabricated porous composites have both good repeatability and recovery.

The gauge factor (GF = $\Delta R/ (R_0 \varepsilon)$) is a key parameter to characterize the sensing performance of the prepared material. Here, we used the measured results to calculate the GF and found the value of GF can reach up to 15614 when the strain is 37%. Thus, the prepared porous composite features an extremely high sensitivity for external force and pressure sensing. Furthermore, a comparison was made on the sensing performances such as sensitivity and GF of our fabricated porous composites and some other sensing materials from recent studies, as shown in Table 1. Through the comparison of various material performance in the table, we can draw a conclusion that our porous composites have extremely high sensitivity and GF. Such a high sensitivity indicates that the developed porous composites have great prospects to be used in tactile sensors for tiny force or weak signals sensing, such as pulse detection, muscle vibration, etc.

B. Performance Characterization of Tactile Sensor

Connecting the tactile sensor with a circuit, the resistance change of the sensing units is able to be converted to voltage output. In this study, the scanning circuit designed by our previous work [29] was utilized for real-time voltage acquisition of the fabricated tactile sensor, and real-time voltage signals of the tactile sensor can be collected by scanning row by row and column simultaneously.

The experimental results when increasingly applied normal force is applied to the sensing units are plotted in Fig. 5(a). With the increasing of applied normal force, the resistance of the sensing units will be changed accordingly and leads to changes of the generated voltages. Fig. 5(a) exhibits that with the applied force increasing from 0.02 N to 0.56 N, the output voltage showed monotonically increasing results. When the exerted normal force is between 0.02 N and 0.2 N, the sensing units have a relatively low sensitivity of 1.54 V/N, while the sensing units show a high sensitivity of 4.81 V/N when the force increased from 0.2 N to 0.56 N. After that, as the applied force increased, the instability of the resistance change made the voltage output unstable. In consequence, the fabricated flexible tactile sensor possess a good measuring accuracy when the applied force is below 0.6 N that the corresponding pressure is about 66.7 kPa.

Cyclic loading/unloading test at different magnitude force were exerted to a sensing unit by a 3.0 mm loading bar, and the measured response is shown in Fig. 5(b). During the experiments, the applied force was changed from 0.05 N to 0.5 N gradually, and applied frequency was kept constant as 0.5 Hz. We can see that in 5(b), the measured voltage signals of the sensing unit changed evidently from about 0 V to 0.5 V at 0.3 N, 1.05 V at 0.4 N, and 1.3 V at 0.5 N, respectively. This indicated that the sensing unit has generally good voltage responses under relative high pressure. The results in the inset of Fig. 5(b) showed the voltage responsiveness of the sensing unit under small normal force, and the voltage of the sensing unit changes from about 0 V to 0.03 V at 0.05 N, 0.11 V at 0.1 N and 0.22 V at 0.2 N, so that the sensing units also possess an obvious voltage response even though there is a slight disturbance exerted on the sensing units.

The effects of the force loading frequency on the output voltage response of the sensing unit were also conducted. With the increasing of loading force frequency, the responses of the output voltage changed correspondingly, while the magnitude of the peak voltages maintains almost the same value, demonstrating that each sensing unit has good dynamic response characteristics as well as stability. In addition, to further explore the recovery, repeatability and durability of the sensing units, 500 loading/unloading cyclic test was carried on, during which the applied normal force is 0.5 N and the frequency is 0.5 Hz. The output voltage of the measured unit is plotted in Fig. 5(c). We can see that in the first few cycles, the output voltage signals decreased a little significantly due to the internal graphene nanoplates of the porous GNP/SR composites being relatively unstable after being fabricated, the almost same phenomenon has been reported in [27]. After tens of times of loading and unloading, the whole graphene nanoplates inside the porous composites gradually stabilized, making the sensing unit's response tends to be steady. From the illustration in Fig. 5(c), the results of ten cyclic tests between 1250 seconds and 1300 seconds can be seen clearly. When the applied force quickly rose up to 0.5 N, the output voltage of the sensing unit increased to about 1.2

V accordingly, and it is consistent with the results shown in Figs. 5(a) and (b). Therefore, the experimental results indicated that the fabricate tactile sensor possesses good repeatability, recovery and durability for cyclic external force sensing. In consequence, the developed flexible tactile sensor has the advantages of high sensitivity, good dynamic response stability, durability and repeatability for external force sensing.

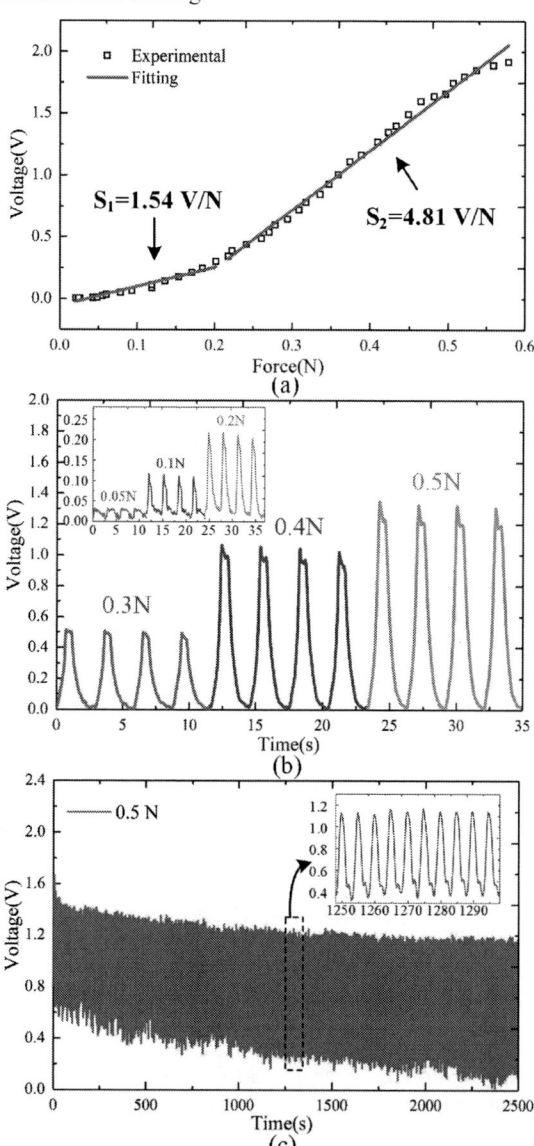

Fig. 5 Sensing performance of the tactile sensor: (a) output voltage during calibration test, (b) output voltage of the sensing unit during the cyclic loading/unloading test with incremental forces, (c) 500 cyclic loading/unloading tests when the force of 0.5 N at frequency of 0.5 Hz.

IV. CONCLUSIONS

In this work, we developed a tactile sensor with high sensitivity and good flexibility based on truncated pyramid-shaped porous GNP/SR composites as the sensing units. A simple and low-cost process to fabricate the porous GNP/SR composite and tactile sensor is presented. The truncated pyramid-shaped sensing units have a low shape factor of 0.256, thus the sensing units would be more prone to deformation under applied pressures, and resulting in greater resistance change for highly sensitive tactile force sensing. Material testing experiments showed that the fabricated truncated pyramid-shaped porous GNP/SR composite has low Young's modulus of 85.72 kPa. In addition, the porous composite features two sensitivities for pressure sensing: a low sensitivity of 8.45 kPa^{-1} at 0~55 kPa and a high sensitivity of 195.02 kPa^{-1} at 55~80 kPa. Calibration tests exhibited that the fabricated tactile sensor has an extremely high voltage sensitivity of 4.81 V/N for external force sensing. Further, dynamic force response, cyclic loading and unloading tests demonstrated that the tactile sensor possesses good voltage response capability, recovery and repeatability. Therefore, the developed tactile sensor would have great potential for use in neuroprosthetics, wearable electronics, and other biomedical applications. For further work, the sensing mechanism behind the porous GNP/SR composite for highly sensitive tactile sensing will be investigated, and the applications of the developed tactile sensor in human motion and health monitoring will be conducted.

ACKNOWLEDGMENT

This research is supported by the Zhejiang Provincial Funds for Distinguished Young Scientists of China (LR19E050001), Open Fund Project of Zhijiang Laboratory (2019MC0AB02), and Creative Research Groups of National Natural Science Foundation of China (51821093).

REFERENCES

[1] Yin, Y., J. Wang, S. Zhao, W. Fan, et al., "Stretchable and Tailorable Triboelectric Nanogenerator Constructed by Nanofibrous Membrane for Energy Harvesting and Self-Powered Biomechanical Monitoring," *Adv. Mater. Technol.*, vol. 3, pp. 1700370-7, 2018.

[2] Wang, Y., Y. Shi, D. Mei, and Z. Chen, "Wearable thermoelectric generator to harvest body heat for powering a miniaturized accelerometer," *Appl. Energ.*, vol. 215, pp. 690-698, 2018.

[3] Wu, C., T.W. Kim, J.H. Park, B. Koo, S. Sung, et al., "Self-Powered Tactile Sensor with Learning and Memory," *ACS Nano*, vol. 14, pp. 1390-1398, 2020.

[4] Ren, Y. and J. Feng, "Skin-Inspired Multifunctional Luminescent Hydrogel Containing Layered Rare-Earth Hydroxide with 3D Printability for Human Motion Sensing," *ACS Appl. Mater. Inter.*, vol. 12, pp. 6797-6805, 2020.

[5] Yamada, T., Y. Hayamizu, Y. Yamamoto, A. Izadi-Najafabadi, et al., "A stretchable carbon nanotube strain sensor for human-motion detection," *Nat. Nanotechnol.*, vol. 6, pp. 296-301, 2011.

[6] Lin, W., B. Wang, G. Peng, Y. Shan, and Z. Yang, "Skin-Inspired Piezoelectric Tactile Sensor Array with Crosstalk-Free Row+Column Electrodes for Spatiotemporally Distinguishing Diverse Stimuli," *Adv. Sci.*, vol. 8, pp. 2002817-11, 2021.

[7] Liu, Y., H. Wo, S. Huang, et al., "A Flexible Capacitive 3D Tactile Sensor With Cross-Shaped Capacitor Plate Pair and Composite Structure Dielectric," *IEEE Sens. J.*, vol. 21, pp. 1378-1385, 2021.

[8] Wang, Y., L. Zhu, D. Mei, and W. Zhu, "A highly flexible tactile sensor with an interlocked truncated sawtooth structure based on stretchable graphene/silver/silicone rubber composites," *J. Mater. Chem. C*, vol. 7, pp. 8669-8679, 2019.

[9] Stassi, S., V. Cauda, G. Canavese, and C.F. Pirri, "Flexible tactile sensing based on piezoresistive composites: a review," *Sensors-Basel*, vol. 14, pp. 5296-5332, 2014.

[10] Kim, K.H., N.S. Jang, S.H. Ha, J.H. Cho, and J.M. Kim, "Highly Sensitive and Stretchable Resistive Strain Sensors Based on Microstructured Metal Nanowire/Elastomer Composite Films," *Small*, vol. 14, pp. e1704232-10, 2018.

[11] Kim, I., K. Woo, Z. Zhong, P. Ko, Y. Jang, et al., "A photonic sintering derived Ag flake/nanoparticle-based highly sensitive stretchable strain sensor for human motion monitoring," *Nanoscale*, vol. 10, pp. 7890-7897, 2018.

[12] Wu, Z., C. Xu, C. Ma, Z. Liu, H.M. Cheng, and W. Ren, "Synergistic Effect of Aligned Graphene Nanosheets in Graphene Foam for High-Performance Thermally Conductive Composites," *Adv. Mater.*, vol. 31, pp. e1900199-8, 2019.

[13] Yang, Z., D.Y. Wang, Y. Pang, Y.X. Li, Q. Wang, et al., "Simultaneously Detecting Subtle and Intensive Human Motions Based on a Silver Nanoparticles Bridged Graphene Strain Sensor," *ACS Appl. Mater. Inter.*, vol. 10, pp. 3948-3954, 2018.

[14] Luo, N., Y. Huang, J. Liu, S.C. Chen, C.P. Wong, and N. Zhao, "Hollow-Structured Graphene-Silicone-Composite-Based Piezoresistive Sensors: Decoupled Property Tuning and Bending Reliability," *Adv. Mater.*, vol. 29, pp. 1702675-9, 2017.

[15] Zhang, R., R. Hu, X. Li, Z. Zhen, Z. Xu, et al., "A Bubble-Derived Strategy to Prepare Multiple Graphene-Based Porous Materials," *Adv. Funct. Mater.*, vol. 28, pp. 1705879-10, 2018.

[16] Pengfei Liu, Xiaofeng Li, Xiyuan Chang, Peng Min, et al., " Highly anisotropic graphene aerogels fabricated by calcium ion-assisted unidirectional freezing for highly sensitive sensors and efficient cleanup of crude oil spills ," *Carbon*, vol. 178, pp. 301-309, 2021.

[17] Li, C., L. Qiu, B. Zhang, and C.Y. Liu, "Robust Vacuum-/Air-Dried Graphene Aerogels and Fast Recoverable Shape-Memory Hybrid Foams," *Adv. Mater.*, vol. 28, pp. 1510-1516, 2016.

[18] Shehzad, K., Y. Xu, C. Gao, and X. Duan, "Three-dimensional macro-structures of two-dimensional nanomaterials," *Chem. Soc. Rev.*, vol. 45, pp. 5541-5588, 2016.

[19] Zhang, B.-X., Z.-L. Hou, W. Yan, Q.-L. Zhao, and K.-T. Zhan, "Multi-dimensional flexible reduced graphene oxide/polymer sponges for multiple forms of strain sensors," *Carbon*, vol. 125, pp. 199-206, 2017.

[20] Choong, C.L., M.B. Shim, B.S. Lee, S. Jeon, D.S. Ko, et al., "Highly stretchable resistive pressure sensors using a conductive elastomeric composite on a micropyramid array," *Adv. Mater.*, vol. 26, pp. 3451-8, 2014.

[21] Chen, L., G.H. Chen, and L. Lu, "Piezoresistive Behavior Study on Finger-Sensing Silicone Rubber/Graphite Nanosheet Nanocomposites," *Adv. Funct. Mater.*, vol. 17, pp. 898-904, 2007.

[22] Hsiao, F.-R., I.F. Wu, and Y.-C. Liao, "Porous CNT/rubber composite for resistive pressure sensor," *J. Taiwan Inst. Chem. E*, vol. 102, pp. 387-393, 2019.

[23] Wu, S., J. Zhang, R.B. Ladani, A.R. Ravindran, A.P. Mouritz, et al., "Novel Electrically Conductive Porous PDMS/Carbon Nanofiber Composites for Deformable Strain Sensors and Conductors," *ACS Appl. Mater. Inter.*, vol. 9, pp. 14207-14215, 2017.

[24] Tian, H., Y. Shu, X.F. Wang, M.A. Mohammad, Z. Bie, et al., "A graphene-based resistive pressure sensor with record-high sensitivity in a wide pressure range," *Sci. Rep.*, vol. 5, pp. 8603-8609, 2015.

[25] Song, Y., H. Chen, Z. Su, X. Chen, L. Miao, et al., "Highly Compressible Integrated Supercapacitor-Piezoresistance-Sensor System with CNT-PDMS Sponge for Health Monitoring," *Small*, vol. 13, pp. 1702091-10, 2017.

[26] Yang, J., Y. Ye, and R. Chen, "Flexible, conductive, and highly pressure-sensitive graphene-polyimide foam for pressure sensor application," *Compos. Sci. Technol.*, vol. 164, pp. 187-194, 2018.

[27] Tewari, A., S. Gandla, S. Bohm, C.R. McNeill, and D. Gupta, "Highly Exfoliated MWNT-rGO Ink-Wrapped Polyurethane Foam for Piezoresistive Pressure Sensor Applications," *ACS Appl. Mater. Inter.*, vol. 10, pp. 5185-5195, 2018.

[28] Feng, C., Z. Yi, X. Jin, S.M. Seraji, Y. Dong, et al., "Solvent crystallization-induced porous polyurethane/graphene composite foams for pressure sensing," *Compos. Part B-Eng.*, vol. 194, pp. 108065-108075, 2020.

[29] Zhu, L., Y. Wang, D. Mei, and X. Wu, "Highly Sensitive and Flexible Tactile Sensor Based on Porous Graphene Sponges for Distributed Tactile Sensing in Monitoring Human Motions," *J. Microelectromech. S.*, vol. 28, pp. 154-163, 2019.

Proceedings of the 16th Annual IEEE International
Conference on Nano/Micro Engineered and Molecular Systems
April 25-29, 2021

Uniformly distributed self-filling micro-strips for high-performance pressure-sensitive sensor

Zhiping Chai
*State Key Laboratory of Digital
Manufacturing Equipment and
Technology
Huazhong University of Science and
Technology*
Wuhan, China
zhipingchai@hust.edu.cn

Xingxing Ke
*State Key Laboratory of Digital
Manufacturing Equipment and
Technology
Huazhong University of Science and
Technology*
Wuhan, China
xxke@hust.edu.cn

Han Chen
*State Key Laboratory of Digital
Manufacturing Equipment and
Technology
Huazhong University of Science and
Technology*
Wuhan, China
chenhan9861@hust.edu.cn

Jiaqi Zhu
*State Key Laboratory of Digital
Manufacturing Equipment and
Technology
Huazhong University of Science and
Technology*
Wuhan, China
jqzhu@hust.edu.cn

Chuan Fei Guo
*Department of Materials Science and
Engineering
Southern University of Science and
Technology*
Shenzhen, China
guocf@sustech.edu.cn

Zhigang Wu
*State Key Laboratory of Digital
Manufacturing Equipment and
Technology
Huazhong University of Science and
Technology*
Wuhan, China
zgwu@hust.edu.cn

Abstract—Porous structure is one of widely utilized structures for flexible sensors. Researchers have endeavored for years to improve the performance of porous sensors. Apart from materials, the sensitivity of tactile sensors is heavily dependent on the microstructures of dielectric materials between the electrodes. Consequently, the study on structure of micropores becomes important. To obtain better sensitivity, we re-design the microstructure of porous elastomer with uniformly distributed self-filling micro-strips. And this design shows lower equivalent modulus, which is further verified by Finite Element Analysis (FEA). The fabricated sensor possesses a high sensitivity of 0.849 kPa^{-1}, and a low detection limit of 3 Pa. What's more, it also maintains a good stability after 1000 cyclic compression test. This sensor was demonstrated to be pragmatic enough in diverse practical scenarios, such as airflow detection and wrist pulse monitoring.

Keywords—self-filling distribution, Finite Element Analysis, high sensitivity, practical demonstrations.

I. INTRODUCTION

One of the key method to significantly improve the performance of flexible sensors is to microengineer the active layer [1]. By engineering microstructures, sensor performance can be greatly improved. As a widely used microstructure, micropores have been intensively studied. Micropores contribute to not only lower modulus, but also higher permittivity when under pressure [2,3]. These two features couple when the low permittivity almost no modulus air is squeezed out. One of the most important way to do both effectively is to optimize the pattern of micropores. Highly-ordered spherical voids[4,5] or randomly distributed micropores[6-9] have been discussed earlier. However, the most effective way to deploy micropores still remains unclear.

Compared to other micropore patterns, self-filling structure allows active layer to decrease the modulus or deform easily under compression. It replaces materials with air, and thus less resistance will be overcome in the path of deformation. By introducing uniformly distributed self-filling micro-strips into our capacitive sensors, better sensitivity as well as a good response consistency and stability can be obtained. Additionally, The low detection limit and high sensitivity of the sensor make the perception of subtle stimulus possible, e.g. detection of gentle touches and wrist pulses.

II. SENSOR DESIGN

The cross-sectional views of two typical architectures of porous elastomers are illustrated in Fig.1. Uniformly stacked micropores (Fig.1(a)) will prevent further deformation of elastomers. Instead of filling into nearby voids, the column between two electrodes are compressed severely (Fig.1(c)), and consequently would result in a lower sensitivity of sensor. By contrast, when the micropores are aligned in a staggered manner, the upper elastomer fills into the voids

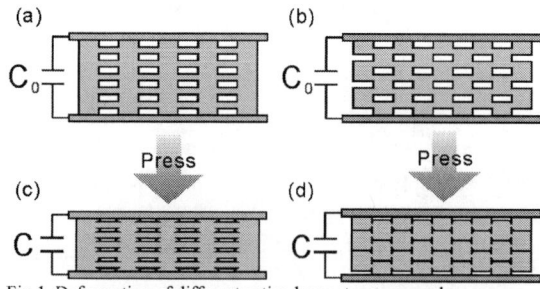

Fig.1. Deformation of different active layer structures under pressure.

beneath easily, or so-called self-filling, and thus will lead to a higher sensitivity (Fig. 1b and 1d).

Deformation of these two typical structures are further proved by Finite Element Analysis. As shown in Fig.2, a

978-1-6654-3008-1/21 $31.00 © 2021 IEEE

same pressure is exerted on the two structures (electrodes of these two sensors are unseen). During the simulation, the porosity of structures was kept at 0.5. In Fig. 2a and c,it's the columns between two electrodes that withstand compression. Compared to the pattern in Fig2(a), pattern in Fig.2(b) shows its advantage in compressibility. It will maintain a relatively

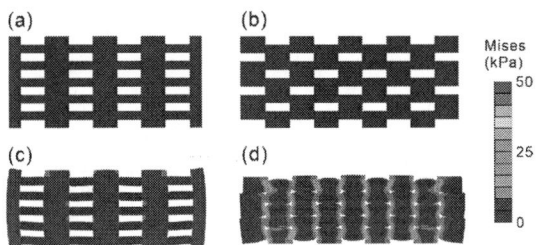

Fig.2. Simulation of two typical microporous structures under compression. (a, b) represent initial state, (c, d) is under compression.

low equivalent modulus until the voids in structure are finally replaced by elastomers.

Compressive strain of the structures is also extracted from the simulation, which is plotted in Fig.3. It's clear that compressive strain of uniformly stacked porous pattern grows almost linearly with the increment of pressure. By comparison, porous elastomer designed in a staggered pattern undergoes a two stage compression, which have a smooth transition. As seen in Fig.3, the slope of red curve is about four times as large as that of blue curve, when pressure is low. However, after the voids have been filled with elastomers, elastomers become to contact with each other. With nowhere to deform, they are squeezed severely, and the slope of red curve gradually declines to the same level of blue curve.

Stress in Fig.2(c) concentrates in the compression region, and stress in Fig.2(d) concentrates mainly in shape deformation region. This phenomenon may suggests that instead of compressing, deforming can be a more effective way to improve compressibility of the active layers. Therefore, it may provide a guideline for later porous active

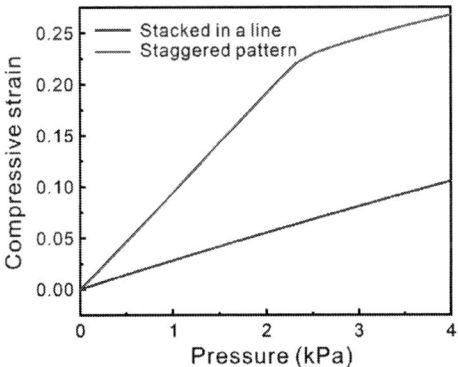

Fig.3. Compressive strain of typical structures under pressure.

layer design.

The structure of porous active layer is thus designed according to the simulation results. As illustrated in Fig.4,

this active layer is constructed by layers of microstrips. Actually, from two perpendicular angles, the pattern of

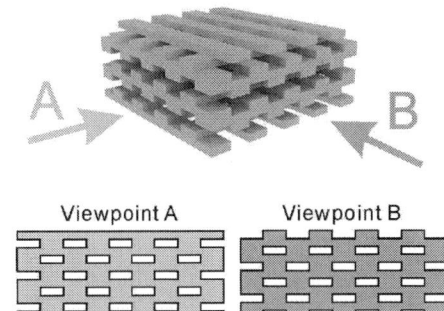

Fig.4. structure of active layer of pressure-sensitive sensors.

active layer keeps the same, which is a staggered micropores pattern. This 3D self-filling structure provides more voids to accommodate deformed elastomer, and thus will improve the sensitivity to an even larger extent.

To realize this self-filling microstrips structure, a water-soluble PVA based molding was used to make this structure. The mold was fabricated by a commercial 3D printer. As conductive carbon nanotubes have positive impact on the sensitivity of sensors [8], the active layer is made of silicone mixed with multi-wall carbon nanotube (Ecoflex0030/0.3wt%MWCNT). When the active layer material was ready, PVA molds were immersed into the composites. After the composite was cured, PVA molds were then dissolved by hot water.

After that, porous elastomer was trimmed into cubes for later fabrication. The SEM images of this elastomer was shown in Fig.5. It's obvious that microstrips follow a layer-

Fig.5. Cross-sectional SEM images of (a) microstrips array and (b) a single microstrip.

by-layer fabrication process. And each strip have a uniform hourglass-like sectional view. Width of each microstrips is about 200 μm. Therefore, they are narrow enough to be filled into the voids below.

Finally, two sheets of conductive fabric were used for electrodes. They were placed on a thin layer of PDMS, and glued together with the active layer by Sil-Poxy. For later characterization and demonstrations of this sensor, copper wires were attached to electrodes with Ag paste.

III. CHARACTERIZATION OF SENSOR

Electrical responses of this sensor to external stimulus were investigated systematically.

A. Sensitivity

A sensor with good sensitivity will also have a decent Signal-to-Noise Ratio (SNR), which secures reliability of signals. Our sensor is tested to be possess a high sensitivity of 0.849 kPa^{-1} in low pressure region and 0.134 kPa^{-1} in high pressure region (Fig. 6a).

This result can be interpreted by its physical behaviors, as the equivalent modulus in low pressure region is much higher than that in high pressure region. Capacitance change of the sensor is in accordance with this behavior, though it has a smoother transition.

B. Response and relaxation time

A force was exerted on the sensor with a commercial force gauge. This sensor exhibit a response time of 100 ms and a relaxation time of 180 ms (Fig. 6b), which is comparable with other capacitive or resistive sensors [8,10].

C. Response to multi-frequency force

A multi-frequency force was produced during testing. Frequency of force ranges from 0.024 Hz to 0.0118 Hz. The sensor showed decent consistency to different frequencies (Fig. 6c).

D. Repeatability

An eligible tactile sensor also response consistently to the same force. Inconsistency will results in inaccurate measurement and thus invalidity of obtained signals. Hence, a multi-level force was applied on the sensor. Each level of force was repeated for five times. The sensor responded consistently under each level of force (Fig. 6d).

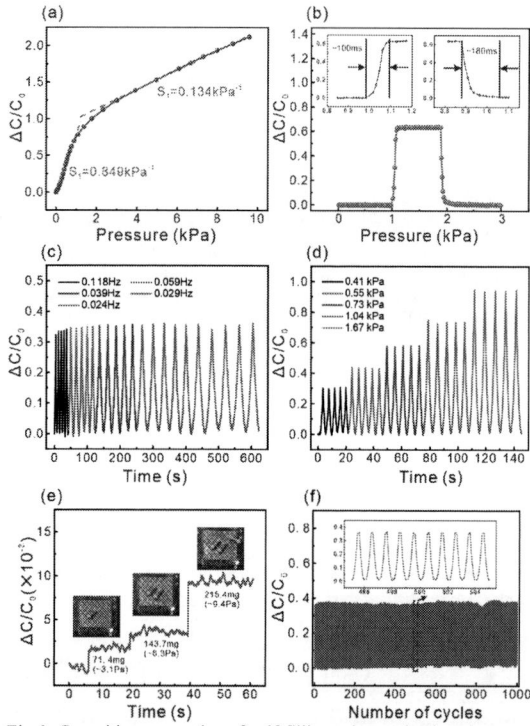

Fig.6. Capacitive properties of self-filling micro-strips-based porous sensor. (a) Capacitive change under pressure. (b) Response time of sensor under rapid exerted force. (c) The consistency of sensor under reciprocating force with different frequency. (d) Response of sensor under a multi-level force. (e) Sequential ultra-light weight detection. (f) Stability under a force cycle of over 1,000 times.

E. Detection threshold

Detection threshold represents the ability of a sensor to detect subtle changes of the environments. As a tactile sensor, low detection threshold endow it with various potential applications and wider practical uses, such as the motion of a water [11] and the weight of a fly [12].

To quantify the detection threshold of our sensor, resistances were used for ultralight weight exertion. The weight of a resistance (wires of the resistance were cut off) is about 70 mg, which corresponds to 3.1 Pa. Three resistances were placed on the sensor sequentially, and the sensor responded well to each of the resistances (Fig. 6e), which proves that the sensor possess a detection threshold of 3 Pa and a resolution of 3 Pa.

F. Stability under cyclic force

Validity of obtained signals are also determined by the stability of sensors. Drift of signals may exists, because of the instability. A well fabricated and stable sensor thus needs less frequent calibrations and have higher precision.

Our sensor were tested under a cyclic force of 0.5 kPa for over 1000 times (Fig. 6f). No apparent signal drift or response inconsistency can be observed.

IV. DEMONSTRATIONS

Our sensor was demonstrated to be used for various applications, including proximity and press, gentle touches of tissue, airflow sensing and wrist pulse monitoring.

A. Proximity and press

Proximity of objects changes ambient electric field. They interfere signals of the sensor, thus they can be detected.

When a finger is approaching the sensor, the electrodes of sensor form capacitance between the finger and themselves, which consequently lower the measured capacitance. As seen in Fig. 7a, two presses were accomplished. Each time the finger approaches sensor, capacitance reduced. And when the finger pressed down, capacitance grows.

B. Gentle touches of tissue

A sheet of tissue was used to touch the fabricated sensor. There is an about 6 seconds pause between two touch periods. The ups and downs in figure represent frequent touches of tissue. Gentle touch of the tissue could lead to a 1% capacitive change of the sensor (Fig. 7b).

C. Airflow sensing

An air blower was used to produce airflows with different intensity. Each peak represents a successful detection of airflow. The relative change of capacitance represents the intensity of the flow. As air was blown from gently to fiercely, amplitude of capacitance change grows gradually (Fig. 7c).

D. Wrist pulses monitoring

Based on its performance, this sensor has also been investigated for its potential applications in biomedical field. A small pressure-sensitive sensor was place on the artery of wrist and wrist pulses can be precisely monitored (Fig. 7d). Three individual peripheral artery pressure waves are clearly displayed in the graph, of which the peaks are P_1, P_2, P_3, respectively. Radial artery augmentation index (AI_r) is defined as P_2/P_1. And ΔT_{DVP} is defined as the interval of first and second peaks, it has been used as a measurement of arterial stiffness. These two parameters are clinically measured in a noninvasive evaluation of cardiovascular system. In this figure, AI_r and ΔT_{DVP} are calculated to be 0.54 and 268 ms, respectively, which are normal values of a healthy man in his early twenties [13].

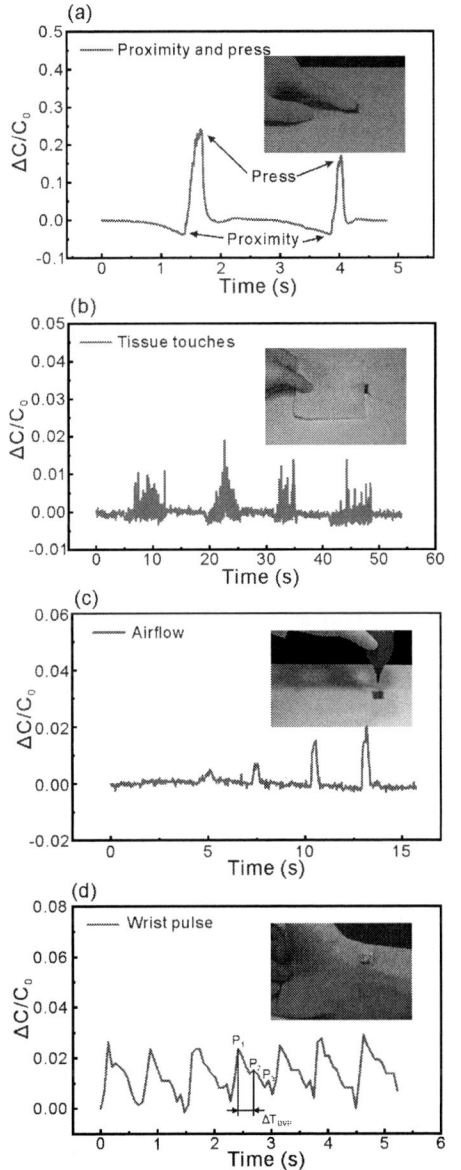

Fig.7. Demonstrations of porous capacitive sensors. (a) Proximity and press of finger. (b) A sheet of tissue touch and slide on the sensor. (c) Detection of gentle airflow. (d) Monitoring of wrist pulses.

V. SUMMARY

This paper proposes a self-filling distribution of micropores for porous sensors. The self-filling distribution greatly lessen equivalent modulus of active layer, and thus leads a way to fabricate high sensitivity sensors. Leveraging this design principle, we managed to fabricate a high-performance capacitive sensor, which demonstrated to be sensitive enough to detect both contact and non-contact mechanical stimulus.

ACKNOWLEGEMENT

This work is supported by the National Natural Science Foundation of China (Grant No. U1613204). The authors would also like to acknowledge the support of the Flexible Electronics Research Center of the HUST.

REFERENCES

[1] S. R. A. Ruth, V. R. Feig, H. Tran, and Z. Bao, "Microengineering Pressure Sensor Active Layers for Improved Performance," *Adv. Funct. Mater.*, vol. 30, no. 39, p. 2003491, 2020.

[2] D. Kwon *et al.*, "Highly Sensitive, Flexible, and Wearable Pressure Sensor Based on a Giant Piezocapacitive Effect of Three-Dimensional Microporous Elastomeric Dielectric Layer," *ACS Appl. Mater. Interfaces*, vol. 8, no. 26, pp. 16922–16931, 2016.

[3] P. Wei, X. Guo, X. Qiu, and D. Yu, "Flexible capacitive pressure sensor with sensitivity and linear measuring range enhanced based on porous composite of carbon conductive paste and polydimethylsiloxane," *Nanotechnology*, vol. 30, no. 45, p. 455501, 2019.

[4] S. Kang *et al.*, "Highly Sensitive Pressure Sensor Based on Bioinspired Porous Structure for Real-Time Tactile Sensing," *Adv. Electron. Mater.*, vol. 2, p. 1600356, 2016.

[5] J. O. Kim *et al.*, "Highly Ordered 3D Microstructure-Based Electronic Skin Capable of Differentiating Pressure, Temperature, and Proximity," *ACS Appl. Mater. Interfaces*, vol. 11, no. 1, pp. 1503–1511, 2019.

[6] B. Y. Lee, J. Kim, H. Kim, C. Kim, and S. D. Lee, "Low-cost flexible pressure sensor based on dielectric elastomer film with micro-pores," *Sensors Actuators, A Phys.*, vol. 240, pp. 103–109, 2016.

[7] J. Qiu *et al.*, "Rapid-Response, Low Detection Limit, and High-Sensitivity Capacitive Flexible Tactile Sensor Based on Three-Dimensional Porous Dielectric Layer for Wearable Electronic Skin," *ACS Appl. Mater. Interfaces*, vol. 11, no. 43, pp. 40716–40725, 2019.

[8] J. Choi *et al.*, "Synergetic Effect of Porous Elastomer and Percolation of Carbon Nanotube Filler toward High Performance Capacitive Pressure Sensors.pdf," *Appl. Mater. interfaces*, vol. 12, pp. 1698–1706, 2020.

[9] C. W. Visser, D. N. Amato, J. Mueller, and J. A. Lewis, "Architected Polymer Foams via Direct Bubble Writing," *Adv. Mater.*, vol. 31, p. 1904668, 2019.

[10] H. Tian *et al.*, "A Graphene-Based Resistive Pressure Wide Pressure Range," *Sci. Rep.*, vol. 5, p. 8603, 2015.

[11] C. Pang *et al.*, "A flexible and highly sensitive strain-gauge sensor using reversible interlocking of nanofibres," *Nature Materials*, vol. 11, no. 9. pp. 795–801, 2012.

[12] S. C. B. Mannsfeld *et al.*, "Highly sensitive flexible pressure sensors with microstructured rubber dielectric layers," *Nat. Mater.*, vol. 9, no. 10, pp. 859–864, 2010.

[13] W. W. Nichols, "Clinical measurement of arterial stiffness obtained from noninvasive pressure waveforms," *Am. J. Hypertens.*, vol. 18, pp. 3S-10S, 2005.

Proceedings of the 16th Annual IEEE International
Conference on Nano/Micro Engineered and Molecular Systems
April 25-29, 2021

A Sensitive Flexible Strain Sensor via Anisotropy Microstructured Sensitized Surface Resistive Change for Human Motion Monitoring

Wenjie Fei
State Key Laboratory of Digital Manufacturing Equipment and Technology
Huazhong University of Science and Technology
Wuhan, China
wenjie_fei@hust.edu.cn

Shuo Zhang,
State Key Laboratory of Digital Manufacturing Equipment and Technology
Huazhong University of Science and Technology
Wuhan, China
shuo_zhang@hust.edu.cn

Zhigang Wu
State Key Laboratory of Digital Manufacturing Equipment and Technology
Huazhong University of Science and Technology
Wuhan, China
zgwu@hust.edu.cn

Abstract—**We reported here a sensitive flexible strain sensor via anisotropy microstructured sensitized surface resistance change. The strain sensor consists of a flexible elastomer and a thin Ag film coated on the microstructures from laser-treated elastomeric surface. The laser scanned microstructures on the surface and led to an anisotropy crack propagation of the metallic conducting layer under strain as sensing mechanism. The developed strain sensor possessed GFs as high as 49 ($\varepsilon < 2\%$) and 3,580 ($4\% < \varepsilon < 7\%$) with a good linear relationship. Sensing performances of the sensor suggested that laser-assisted fabrication is an effective way to improve sensitivity of strain sensors. The anisotropy microstructured strain sensor mounted on the wrist and the throat successfully detected the motion in real time indicating the potential in human motion monitoring.**

Keywords—*sensitive strain sensor; anisotropy microstructure; human motion monitoring*

I. BACKGROUNDN

Resistive-type or capacitive-type strain sensors can detect the mechanical deformations by the change of resistance or capacitance [1,2]. Recently, strain sensors have been widely used in the field of wearable electronic devices for their excellent performance including stretchability, sensitivity, and robustness [3]. When the strain sensor was attached on the human skin, the electrical signals (resistance or capacitance) collected by strain sensors can be transferred to the deformation of the human skin, which indicates human motion. Hence, the sensitivity of flexible strain sensor has been regarded as an important performance index for precise monitoring. The gauge factor (GF) is defined to describe the sensitivity of strain sensors. The GF is defined as GF=$(\Delta R/R0)/\varepsilon$, where ΔR, $R0$ and ε are the relative resistance, initial resistance and strain, respectively.

In order to achieve accurate detection of human motion, numerous researches have taken efforts to improve the gauge factor. These solutions mainly focused on embedment of various materials, e.g. nanoparticles [4,5], nanowires [6], nanofibers [7,8] and nanofilms [9,10] through bulky resistive change. Except consideration of materials, wrinkled structure [11] was also proposed as possible solutions. However, complex fabrication processes of materials or structures decreased productivity and increased the cost. Despite impressive advances of science and technology, the demand for high performance strain sensor still remains a challenge. Therefore, a convenient fabrication to combinate materials and designed structures is expected to be an effective way to overcome such limitations [12].

Laser direct writing is an emerging micro-nano fabrication technique. The laser engraves designed pattern on the substrate by etching materials in an ordinary laboratory environment, and has been proved a simple, fast and low-cost process. The size of engraved groove on the surface of substrate can be tuned by the operational parameters of the laser marker such as wavelength, power, pulse width and scanning speed. In addition, anisotropy microstructures especially generate on the laser-scanned surface of deformed elastomeric polymer substrate [13].

The vacuum evaporation technology is a common method in conductive film fabrication. In a high-vacuum chamber, target of source made of metals is heated until evaporates and adheres to the surface of the substrate. By controlling the thickness of vapor coating film, the surface of the substrate can turn into conductive.

In this work, we demonstrate a sensitive flexible strain sensor via sensitized surface resistive change, by depositing conductive nanofilms on an anisotropy microstructured surface after laser ablation. The experiment results revealed high sensitivity and durability of the sensor, which could be potential used in human motion monitoring.

II. EXPERIMENTAL AND METHODS

A. Materials

Polydimethylsiloxane (PDMS) was served as the material for the substrate of the strain sensor, which was prepared by mixing a prepolymer (Sylgard 184, Dow Corning Corporation) and its curing agent in a weight ratio of 10:1. It is hard for transparent materials to absorb ultraviolet effectively, which makes UV-laser difficult to pattern. In order to enhance the light absorbability of PDMS, 0.2 g of carbon black (XC72R, CABOT) was added into 11 g of standard PDMS. After stirring and vacuuming, the carbon added PDMS (cPDMS) mixture was prepared. A flat PET film was used to support the cPDMS substrate. Isopropanol (IPA) was used to clean the surface of the PET film as well as the cPDMS substrate. Silver was selected as target of source.

B. Sensor Fabrication

Fig. 1 shows the fabrication of the strain sensor. The mixed cPDMS solution was poured on a flat PET film for about 300 μm thickness and heated at 90°C for 10 minutes. A UV-laser marker (HGL-LSU3/5EI, Huagong Laser, Wuhan, China) was employed for laser direct writing on the surface of the cured cPDMS substrate at the scanning speed of 200 mm/s. The size for the sensor was designed as 15 mm×50 mm. The middle

978-1-6654-3008-1/21 $31.00 © 2021 IEEE

part (15 mm×30 mm) was used as sensing region and the side parts were used as electrode. The pattern was designed as lines and the hatch spacing of adjacent laser scanning lines was 0.05 mm. After cleaned with IPA, the laser-scanned cPDMS substrate was loaded into thermal evaporator (BOX-RH400, SKY Technology Development Co., Ltd, CAS, China) and coated with a thin Ag film of about 300 nm thickness. After connected with copper wires by silver paste and peeled off from the PET film, a strain sensor was obtained.

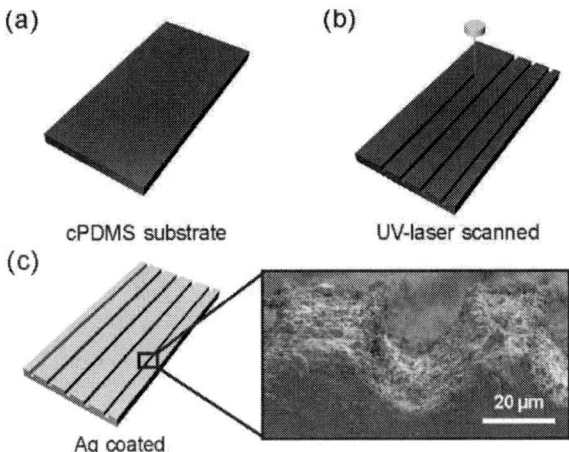

Fig. 1. Microfabrication process of the anisotropy microstructured strain sensor.

C. Analysis

1) Field Scanning Electron Microscope: The micro surface morphology was observed by a field scanning electron microscope (FSEM, GeminiSEM300, Carl Zeiss, Jena, Germany).

2) Three-dimensional Finite Element Analysis: Three-dimensional finite element analysis was used to study the distribution of the stress for a flexible elastomer substrate with anisotropy microstructured surface upon stretching by using COMSOL Multiphysics. A simplified linear elastic model (Young's modulus, 750 kPa and Poisson's ratio, 0.49) was used to describe the elastomer's behaviour upon stretching.

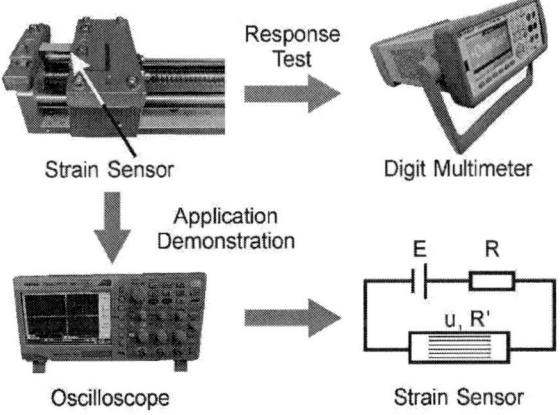

Fig. 2. Experiment setup

3) Electrical characterization: The experiment setup has been shown in Fig 2. A digital multimeter (34461A, KeySight Technologies) was employed to measure the sensor's resistance during the bending test, strain test and cyclic test. For real time monitoring, the strain sensor was connected into a circuit in series with DC power source, E, and a resistance, R, during application demonstration. The voltage of the strain sensor was measures by an oscilloscope (TBS 1202B-EDU, Tektronix, USA) as u. The resistance of the sensor, R', can be calculated through the formula, $R'=Ru/(E-u)$.

III. RESULTS AND DISCUSSION

A. Sensing Mechanism

The microview of the sensor was observed by an FSEM as shown in Fig. 3. When the UV-laser scanned the cPDMS substrate following designed pattern, undesired parts were removed from the surface. Meanwhile, uneven microstructures appeared on the scanning path. During the process of evaporation, silver vapor surrounded these anisotropy microstructures, which made Ag thin film rough.

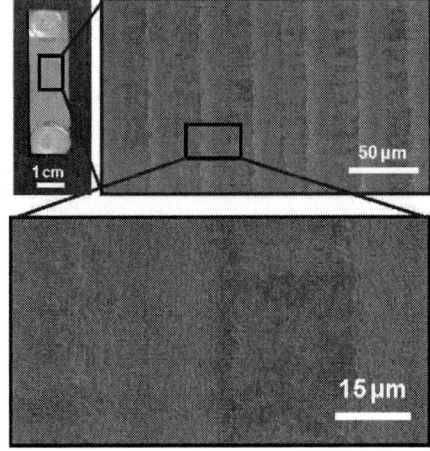

Fig. 3. Top view of the anisotropy microstructured strain sensor observed by an FSEM.

To understand the effect of anisotropy microstructures on the strain distribution, a flexible elastomer substrate with anisotropy microstructured surface was simplified as a cone-shaped elastomer in the simulation. A three-dimensional finite element analysis was carried out by COMSOL Multiphysics. As shown in Fig. 4a, when the elastomer was stretched, the stress concentration generated around the cone along the stretching direction. The simulation indicated that the anisotropy microstructure increased stress concentrations.

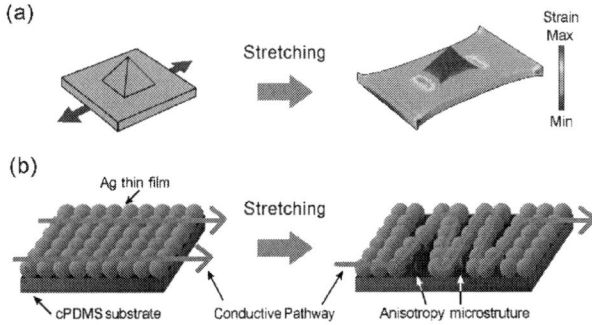

Fig. 4. Sensing mechanism investigation. (a) Finite element analysis of the strain distribution for a flexible elastomer substrate with anisotropy microstructured surface. (b) Schematic diagram of sensing mechanism based on crack propagation.

In this work, the resistive-type sensor responds to the strain based on crack propagation. Fig. 4b shows the schematic diagram of the sensing mechanism. When the fabricated sensor was stretched, anisotropy microstructures led to coated Ag thin film periodically zigzag crack upon stretching as the result of stress concentrations. With the increase of cracks, the number of conductive paths decreased while the length of conductive paths increased, consequently it increased the resistance of metal conductive layer. Such a relation between crake during stretching and resistance change can be explained as sensing mechanism of the strain sensor.

B. Strain Sensor Response

A bending and strain test were carried out to measure the strain sensor's electrical performance, Fig. 5. Cylinders with different radius were employed for the bending test. When a flat strain sensor was attached to the cylinder and conformed the curved surface, the strain sensor performed as stretched and the resistance of the sensor increased. Through the experiment result, Fig. 5a, it can be found that, with the decrease of bending radius from 100 mm to 10 mm, the sensor was further stretched, which caused the increase of relative resistance change from 7% to 96%.

During the strain test, when the sensor was stretched, the stress concentration was server than that in the bending state. Cracks generated around anisotropy microstructures and propagated with increasing axial tension load. As a result, the sensor's resistance increased under the strain more obviously than under bending. The developed strain sensor possessed GFs as high as 49 ($\varepsilon < 2\%$) and 3,580 ($4\% < \varepsilon < 7\%$) and the coefficient of determination was 0.93 and 0.91, respectively, which indicated good linearity, Fig. 5b.

When the strain sensor was stretched by an excessive load, the conductive path will be totally blocked by cracks leading to failures of the strain sensor. Hence, the strain sensors have limited reliable operation range. In order to ensure the strain sensor's sensing characteristics without fatigue failure, the strain sensor has to be tested before application. To investigate the long-term stability and repeatability of the strain sensor, a cyclic test (100 cycles at 3% strain) was applied to the strain sensor. As seen in Fig. 6, the sensor provided stable performance and showed good stability and repeatability.

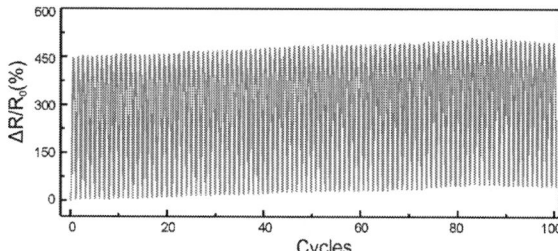

Fig. 6. Cyclic test of the sensitive flexible strain sensor at 3% strain

C. Demonstrations

To tap the sensor's potential for wearable electronic devices, the sensor was mounted on the wrist of a volunteer, Fig. 7a. As revealed in Fig. 8, the resistance of the sensor changed following the motions of wrist. When the volunteer bent the wrist, the strain sensor was stretched with the skin. Consequently, the resistance of the sensor increased and the relative resistance change could reach up to 130%. When the wrist became flat, the strain released by the skin and the resistance of the sensor turned to its initial value.

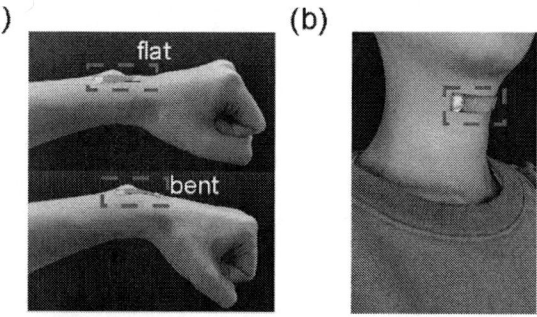

Fig. 7. Photographs of the anisotropy microstructured strain sensor mounted on (a) the wrist and (b) the throat.

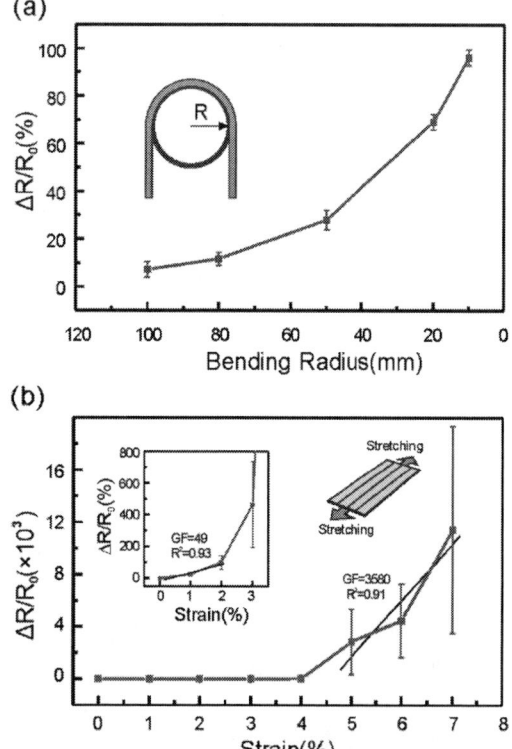

Fig. 5. Relative resistance changes under (a) bending and (b) straining.

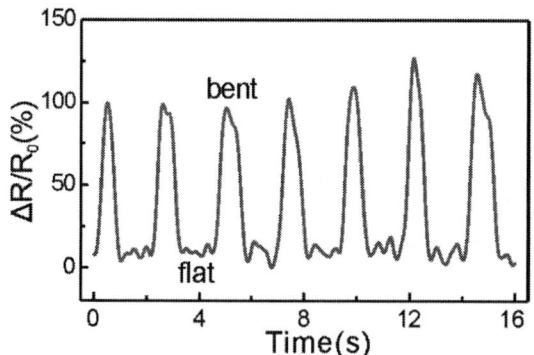

Fig. 8. The anisotropy microstructured strain sensor for detection of the bending of the wrist.

Further, the sensor could be mounted on the throat to detect its motions, Fig. 7b. When the volunteer spoke words "hello" for many times, the sensor could recognize motions of the throat with similar data, Fig. 9. What's more, the sensor detected the vibration sensitively with maximal relative resistance change for about 500%. When the volunteer spoke different words such as "nihao" (hello in Chinese), the vibrations of the throat changed as well as the change of data caught by the sensor, which can be easily distinguished from before. These results showed good stability and repeatability of the sensor to catch the motions of the throat with distinguishable data, which indicated the potential in speech recognition.

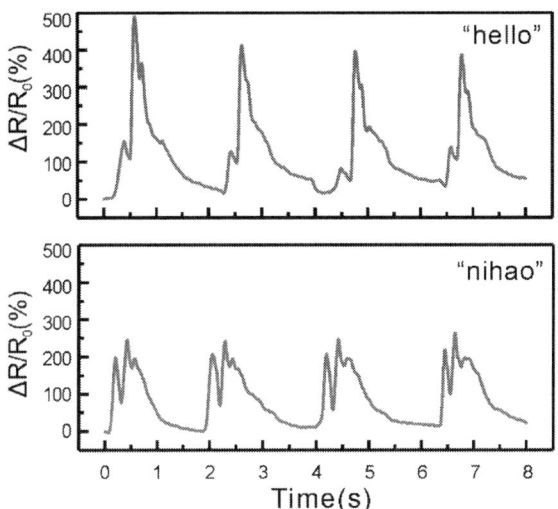

Fig. 9. The anisotropy microstructured strain sensor for detection of the motions of the throat.

IV. CONCLUSION

In this work, an anisotropy microstructured strain sensor was fabricated by laser assisted patterning and vacuum evaporation. The anisotropy microstructured surface of cPDMS patterned by a UV-laser marker was coated with a Ag thin film, which was sensitized to the strain in the form of resistive change. The fabricated strain sensor possessed GFs as high as 49 (ε <2%) and 3,580 (4% < ε < 7%) with good stability and repeatability. These results demonstrated excellent sensitivity and highly potential in human motion monitoring.

ACKNOWLEDGMENT

This work is supported by the National Natural Science Foundation of China (No. U1613204). The authors would also like to acknowledge the support of the Analytic Testing Center and the Flexible Electronics Research Center of the HUST.

REFERENCES

[1] J. Park, S. Park, S. Ahn, Y. Cho, J. J. Park, and H. Shin, "Wearable strain sensor using conductive yarn sewed on clothing for human respiratory monitoring," IEEE Sens. J., vol. 20, no. 21, pp. 12628–12636, 2020.

[2] C. Deng et al., "High-performance capacitive strain sensors with highly stretchable vertical graphene electrodes," J. Mater. Chem. C, vol. 8, no. 16, pp. 5541–5546, 2020.

[3] Y. Lu, M. C. Biswas, Z. Guo, J. W. Jeon, and E. K. Wujcik, "Recent developments in bio-monitoring via advanced polymer nanocomposite-based wearable strain sensors," Biosens. Bioelectron., vol. 123, no. June 2018, pp. 167–177, 2019.

[4] G. Y. Lee et al., "Highly Sensitive Solvent-free Silver Nanoparticle Strain Sensors with Tunable Sensitivity Created Using an Aerodynamically Focused Nanoparticle Printer," ACS Appl. Mater. Interfaces, vol. 11, no. 29, pp. 26421–26432, 2019.

[5] W. Zhang, Q. Liu, and P. Chen, "Flexible strain sensor based on carbon black/silver nanoparticles composite for humanmotion detection," Materials (Basel)., vol. 11, no. 10, pp. 1–13, 2018.

[6] N. Tang et al., "A Highly Aligned Nanowire-Based Strain Sensor for Ultrasensitive Monitoring of Subtle Human Motion," Small, vol. 16, no. 24, pp. 1–9, 2020.

[7] S. Yang et al., "Facile Fabrication of High-Performance Pen Ink-Decorated Textile Strain Sensors for Human Motion Detection," ACS Appl. Mater. Interfaces, vol. 12, no. 17, pp. 19874–19881, 2020.

[8] Z. Yang et al., "Graphene Textile Strain Sensor with Negative Resistance Variation for Human Motion Detection," ACS Nano, vol. 12, no. 9, pp. 9134–9141, 2018.

[9] S. Lu et al., "Wearable graphene film strain sensors encapsulated with nylon fabric for human motion monitoring," Sensors Actuators, A Phys., vol. 295, pp. 200–209, 2019.

[10] F. Chen et al., "Low-cost highly sensitive strain sensors for wearable electronics," J. Mater. Chem. C, vol. 5, no. 40, pp. 10571–10577, 2017.

[11] J. Jung, K. M. Lee, S. H. Baeck, and S. E. Shim, "Piezoresistive behavior of a stretchable carbon nanotube-interlayered poly(dimethylsiloxane) sheet with a wrinkled structure," RSC Adv., vol. 5, no. 89, pp. 73162–73168, 2015.

[12] X. Liao, Z. Zhang, Z. Kang, F. Gao, Q. Liao, and Y. Zhang, "Ultrasensitive and stretchable resistive strain sensors designed for wearable electronics," Mater. Horizons, vol. 4, no. 3, pp. 502–510, 2017.

[13] S. Zhang, Q. Jiang, Y. Xu, C. F. Guo, and Z. Wu, "Facile fabrication of self-similar hierarchical micro-nano structures for multifunctional surfaces via solvent-assisted UV-lasering," Micromachines, vol. 11, no. 7, 2020.

Proceedings of the 16th Annual IEEE International
Conference on Nano/Micro Engineered and Molecular Systems
April 25-29, 2021

A surface and interior material identification technology based on dual-mode sensor

Zhuhui Yin, Ning Li, Weifeng Chen, Yue Jiang, Jiabin Peng, and Zhengchun Peng*

Abstract— In this paper, a novel hybrid-sensor integrating TENG and inductive sensor is proposed to acquire the triboelectric and inductive signals from different objects. A matching CNN algorithm was used to process the above signals in order to achieving highly accurate recognition. The identification accuracy of different solid surfaces reached 98.89% and the identification accuracy of different solution reached 95%. As a demonstration, we designed a real-time identification system for effective sorting of different items.

I. INTRODUCTION

Robot-oriented artificial intelligence (AI) and sensor technology enable service robots and cobots to execute complex operations. Service robots, in particular, often encounter complex work tasks and operate with various irregular objects. However, the position, appearance, and mechanical properties of these objects are usually undetermined. Thus, service robots require kinds of multidimensional sensing capability to perceive the external environment. To acquire such a capability sufficiently, scientists not only further study machine vision[1] but also robot-oriented sensor technology[2]. However, most research works about electronic skin focus on pressure detection[3], information distribution, and drawing force [4]. Perceiving the material information of an object is difficult for traditional pressure.

Moreover, such an information is unavailable for visual technology, which identifies the object by analyzing its shape and color. Therefore, material information is the "blind spot" for the present intelligent recognition system. Based on the above analysis, developing a sensor on the basis of a new mechanism for surface and interior material recognition is urgently necessary.

Triboelectric nanogenerator (TENG) can collect transferred electron between its friction layer and the measured object. When materials contact with TENG, electron transfer appears due to triboelectrification[5]. Given that the output signal is related to the property of the material, the features of the output signal correspond with the material information. Therefore, identifying the surface material information of the object by analyzing the TENG signal[6] [7] is a feasible idea.

However, triboelectrification only occurs on the surface of the object, and its internal material information remains unattainable. Obtaining the internal material information of the object is necessary for service robots. For example, when service robots grasp an opaque cup, they must know what kind of beverage is inside the cup. Owing to magnetic induction can penetrate an insulating material, and an inductive sensor provides a way to perceive the interior material information of an object[8]. Different materials have different electromagnetic induction abilities due to their varying electrical conductivity. Thus, the inductance signal varies when measuring different materials. Furthermore, identifying the material inside a container by an inductance sensor[9]. It is a feasible method.

AI techniques, which provide reliable solutions to analyze sensory information, can highly amplify the intelligence of wearable electronics. Lee et al. presented a smart floor monitoring system through advanced deep learning-based data analytics[10]. Chen et al. successfully combined machine vision and convolutional neural network (CNN) to run human gesture recognition tasks[11]. Therefore, decoupling and analyzing an output signal by using a neural network algorithm is an effective method.

Herein, we present a hybrid sensor coplanarly integrating TENG and an inductor sensor. The fabricated hybrid sensor is loaded on the fingertips of the robot hand, which is set to grasp an object with a fixed posture and velocity. The fluctuant signal of triboelectric and inductance can be gathered and decoupled. The material information of the surface and interior of the object can be identified on the basis of a 1D-CNN neural network processing with high accuracy. Given these advantages, this kind of hybrid sensor exhibits great potential in the application of item classification robots and intelligent robots.

II. DEIGN AND METHOD

The schematic of the hybrid sensor is shown in in Fig. 1a. Structurally, the hybrid sensor consists of two parts. In the center part is a TENG, combining an electrode and a Polydimethylsiloxane(PDMS) layer. To enhance the triboelectric effect, we fabricate a special PDMS layer with a rough surface structure as a charge-generating layer. It is a simple fabrication process in which the PDMS solution spreads on a sandpaper to be cured, followed by demolding. Furthermore, the peripheral part is a copper coil fabricated as an inductance sensor, which can create a magnetic field interaction with the internal material of the object. The whole

---

*Resrach supported by the Science and Technology Innovation Commission of Shenzhen (JCYJ20180305124942832, KQTD20170810105439418), the National Nature Science Foundation of China (61903259).

Zhuhui Yin is with the College of Physics and Optoelectronic Engineering of Shenzhen University, Shenzhen 518060, China

Ning Li is with the College of Physics and Optoelectronic Engineering of Shenzhen University, Shenzhen 518060, China

Weifeng Chen is with the College of Physics and Optoelectronic Engineering of Shenzhen University, Shenzhen 518060, China

Yue Jiang is with the College of Physics and Optoelectronic Engineering of Shenzhen University, Shenzhen 518060, China

Jiabin Peng is with the College of Physics and Optoelectronic Engineering of Shenzhen University, Shenzhen 518060, China

Zhengchun Peng is with the Key Laboratory of Optoelectronic Devices and Systems of Ministry of Education and Guangdong Province, College of Physics and Optoelectronic Engineering, Shenzhen University, Shenzhen 518060, China (corresponding author to provide e-mail: zcpeng@szu.edu.cn).

978-1-6654-3008-1/21 $31.00 © 2021 IEEE

size of the hybrid sensor is a circle with a radius of 10 mm and a thickness of 100 μm, as illustrated in Fig. 1b. Benefiting from the small size and thinness, our hybrid sensor can be easily assembled to various devices and curved surfaces, such as fingertip, making it adaptable for intelligent electronic skin.

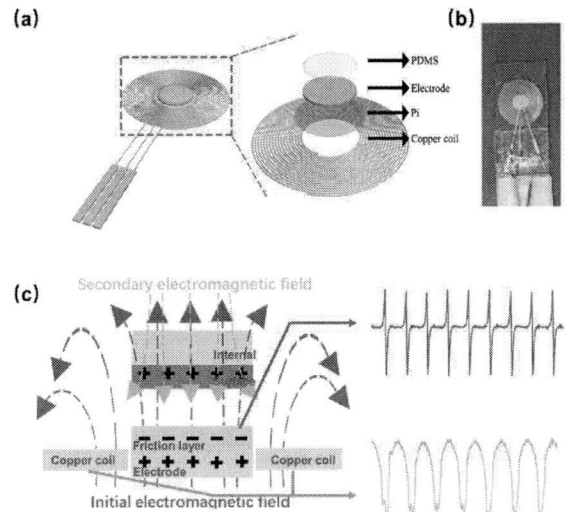

Figure 1. (a) The schematic diagram and (b)The photograph of the hybrid sensor. (c) The detailed working mechanism.

The working mechanism of the hybrid sensor is illustrated in Fig. 1c. When a contact-separation touching occurs on the TENG sensor, a transfer of surface charges appears due to different abilities to electron affinities, driving electrons to flow to electrodes, thus generating an alternating voltage related to the triboelectric ability of a touching object. Moreover, triboelectric signals reflect the ability of the object to gain and lose electrons. Meanwhile, the magnetic field created by the copper coil due to magnetic induction induces currents in an interior material. Then, the interior material generates a secondary magnetic field, opposing the initial one. As a result, a magnetic coupling, which continuously interacts with the internal material, appears between the sensor and the target. Therefore, the characteristics of the signal produced by an inductive sensor relate to the internal material, as a basis for interior material identification.

III. FUNDAMENTAL CHARACTERISTICS OF TENG

When contact-separation touching occurs on the TENG sensor due to triboelectrification, the electric potential between electrodes and the surface material of the object fluctuates. As displayed in Fig. 2a, the movement of contacting can produce a pulse signal, and the separation generates a reversed pulse signal. Comparing Fig. 2a with Fig. 2b, the characteristics of the TENG signals generated by different materials always vary, as previously mentioned. Additional signals are shown in Fig. 2d. However, the material of the touching object is not the only element that influences the output TENG signal but also velocity. To investigate the influence of the speed, comparable voltage responses to the varied velocities are observed when the velocity is retained from 55 mm/s to 440 mm/s. As illustrated in Fig. 2c, the amplitude of the output signal increases with the velocity and implies a linear relationship

between them. The TENG signal exhibits maximal voltage when operating at the fastest velocity of 440 mm/s. Meanwhile, the period of waveform is inversely proportional to the velocity. To improve the accuracy of the intelligent identification system, we should set the speed of contact separation into a constant. To exhibit the toughness and stability of our device under external forces, we perform the voltage measurement under contact-separation touching, which is repeated 1,000 times. The output signal is stable throughout the whole process (Fig. 2e). This result indicates that our TENG sensor can stably produce a signal throughout the repeated grasping process and provides a stable database for the subsequent algorithm.

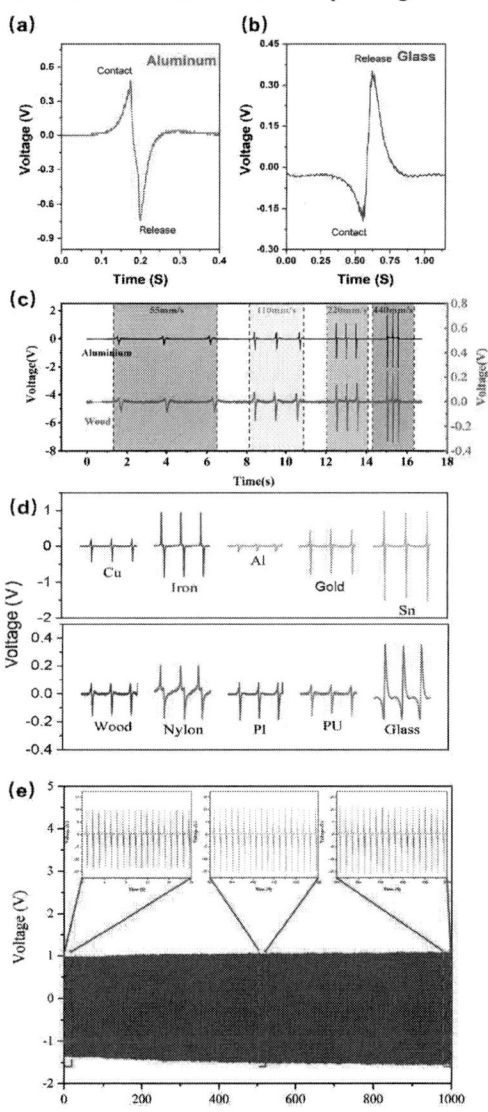

Figure 2. Enlarged curve of the signal for (a) aluminum and (b) glsss. (c) The voltage output generated from the contact speed testing sequence.(d) Output voltage signals of TENG for ten kinds of material. (e)Triboelectric output generated by fatigue testing program during 1000s.

IV. INDUCTANCE PROPERTY OF THE DEVICE

We load our hybrid sensor onto the fingertip of a robotic hand and set the robotic hand to grasp various objects with a fixed velocity. To prove that the hybrid sensor responds to the whole body of different objects, different liquids in the same kinds of glass cups are selected as the test sample. Moreover, five kinds of metals are tested because they are equipotential and difficult to distinguish by TENG signals. Comparing Fig. 3a with Fig. 3b, the inductance signals generated by different materials vary. The output signal generated by iron is in the range from 1.8 μH to 4.8 μH, which means its amplitude is greater than the amplitude of the signal from beer. Meanwhile, the waveform of the signal from iron is sharper than the other one. Other differential characteristics are presented in Fig. 3c. Benefiting from the contactless sensing, the inductance signals enable sensitive detection for the interior material.

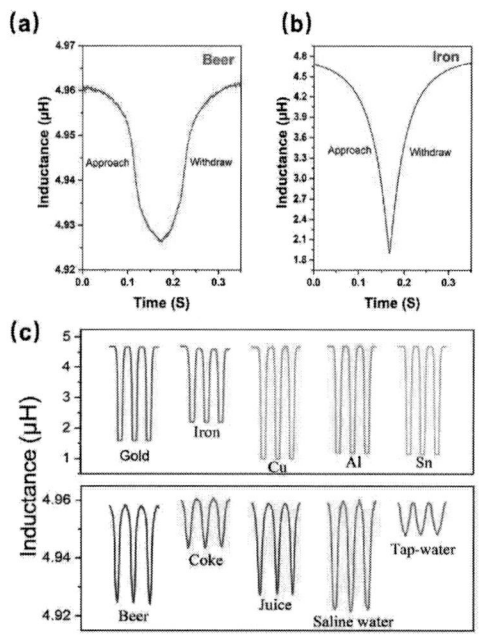

Figure 3. Enlarged curve of the inductance signal for (a) beer and (b) iron. (c) Various inductance signal waveform generated by ten kinds different material.

When the hybrid sensor contacts different objects with a fixed velocity, the time sequence and amplitude of the output voltage signals of the TENG sensor are different due to their varying electronic abilities. Meanwhile, the process of approach and departing can generate an inductance signal on the basis of the conductivity of the object (Figs. 4a and 4b). Moreover, this kind of bimodal sensing can take effect with the surface and interior material of an object, such as a bottle and the conductive liquid inside of it. Hence, both kinds of signals complement each other to perform a surface and an interior material identification, which judges an uncertain object not only on the basis of exterior factors but also on interior ones.

V. CONVOLUTIONAL NEURAL NETWORK FOR DATA ANALYSIS

Deep learning technologies, due to their excellent computing performance, have been showing a great potential for data processing and data analysis in recent years. For example, Lee et al. successfully used CNN to realize the gait recognition by triboelectric intelligent socks[12]. In this work, the features of signals and the unobvious relationships among them are the key basis to identify a material. 1D-CNN, which can automatically extract features from a signal, is proven to be suitable for the analysis of time sequences of sensor data. Thus, realizing surface and interior material identification by using 1D-CNN to analyze complex triboelectric and inductance signals is an efficient method.

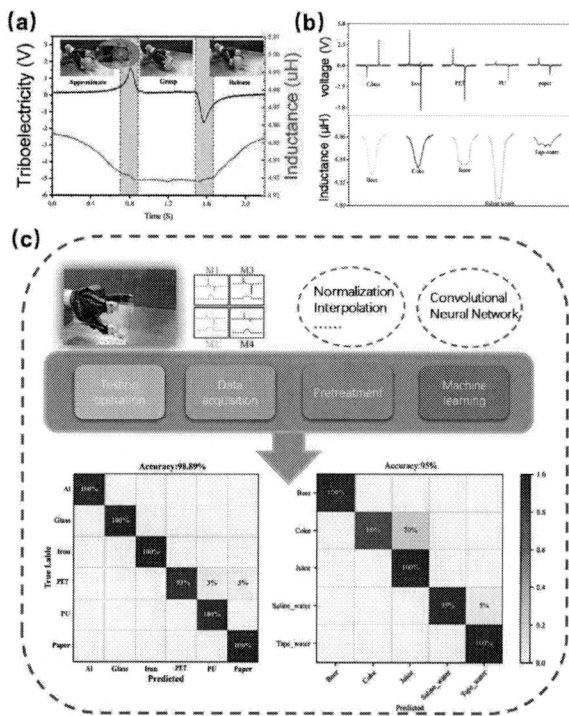

Figure 4. Recognition algorithm process based on 1D CNN. (a) The data acquisition method of material identification process. (b) TENG and inductance signals of different objects (c) Recognition algorithm process based on 1D CNN (insert figure: the identification accuracy based on inductance and TENG signals data set).

By using the 1D-CNN algorithm to analyze the output signal of the hybrid sensor, the convolution layer can automatically extract features of the signal, including the shape of waveform, amplitude, frequency, period, time sequence, and other subtle features. The obtained information of features flows to the fully connected layer as a basis for material identification. In return, the full connection layer implements deep learning on the relationship between the material and the features and makes a judgment. To build the dataset of the TENG and inductance signals, we build a data acquisition system consisting of an oscilloscope, an LCR bridge, a computer-controlled robotic hand, and a hybrid sensor (Fig. 4a). The analog voltage signal and inductance signal generated by the TENG sensor and inductance coil are collected by oscilloscope and LCR bridge.

The collected signals for diverse objects are visualized in Fig. 4b, and the data length for each channel is 1000. In this process, the tested objects, such as aluminum, glass, iron, PET, PU, and paper as surface materials and beer, coke, juice, saline

978-1-6654-3008-1/21 $31.00 © 2021 IEEE

water, and tap water as interior materials, can be tested by repeating touching and releasing motions. As illustrated in Figs. 4a and 4b, differences in certain features exist between the triboelectric signals of different objects, and inductance signals exhibit a similar phenomenon. Here, the above time series information contains various features, suggesting that the data include the information of the contact force, speed, sequences, contact positions, latency, and contact durations. Such multidimensional information can be learned and decoupled in the neural network. Each object is grasped for 100 times to ensure the reliability of the dataset in the data collection process. The 100 samples of each category are randomly split into two groups at the ratio of 8:2 (training: 80 samples, testing: 20 samples). As displayed in Fig. 2c, after the data collection and the dataset establishment, the signals generated during the grasping process are normalized, interpolated into a fixed length, and then imported into 1D-CNN for training. The 1D-CNN algorithm used here comprises three convolutional layers, three max-pooling layers, and two full-connection layers. After 30 cycles of neural network training, the accuracy of the training set is higher than 95%. As shown in Fig. 2c, the verification results indicate that the trained model has a high positive predictive value for object recognition. The identification accuracy of different surface material information reaches 98.89% based on the triboelectric signal dataset. The recognition rate of different solutions in cups of different materials reaches 95% based on the inductance signal dataset.

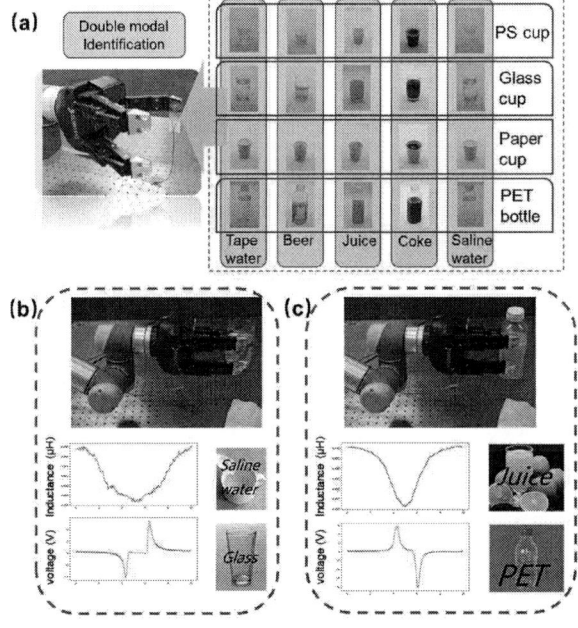

Figure 5. A functionalized demo. (a) The real-time target recognition system was used to identify combinations of five drinks and four cups. (b) The recognition system accurately identified a glass cup filled with saline water and (c) a plastic cup filled with juice.

As a demonstration demo, we design a real-time target recognition system by integrating a customized 1D-CNN-based algorithm, the hybrid sensor, and a robotic hand, as illustrated in Fig. 3a. The hybrid sensor is initially assembled on the fingertip of the robotic hand. The robotic hand and the hybrid sensor are connected to a computer, which controls the robotic arm to perform the grasping operation. When the robotic hand grasps a cup, the tactile signal is collected by the data acquisition instrument and transmitted to the computer. The computer uses a trained CNN model to analyze the signal and recognizes the material of a cup and its internal solution.

In this demonstration, we select five different cups as surface materials and four different solution as interior materials, and fill a cup with solution for recognition system to test. And then, the robotic hand was set to grasp the test object. The signals acquired by hybrid sensor during gripping process transmit to the computer immediately. After trained CNN model processing the signals, the real-time target recognition system successfully identifies both the cup and the solution, as shown in Fig. 5b and 5c. Both the obtained signal and the corresponding recognized result in the gripping process are shown on the bottom of the figure. It is a feasible method to identify the material and the solution inside the cup by the real-time target recognition system, which can be further applied to logistic factory for sorting and detecting.

VI. CONCLUSION

We developed a novel hybrid sensor to dectect the information of surface and internal material, and employed 1D-CNN algorithm to identify different surface materials and interior materials. As a demonstration demo, we designed a real-time target recognition system to identify different kinds of cups filled with different solution with satisfied recognition accuracy of the system. It proved that such a hybrid sensor have great potential in service robot applications.

REFERENCES

[1] I. Sharmin, N.F. Islam, I. Jahan, T.A. Joye, and M.T. Habib, "Machine vision based local fish recognition," *SN Appl. Sci.*, vol. 1, pp.1529, Nov. 2019.

[2] N. Bai, L. Wang, Q. Wang, J. Deng, Y. Wang, and P. Lu, "Graded intrafillable architecture-based iontronic pressure sensor with ultra-broad-range high sensitivity," *Nat. Commun.*, vol. 11, pp.209, Jan. 2020.

[3] X. Sun, J. Sun, T. Li, S. Zheng, and N. Xue, "Flexible Tactile Electronic Skin Sensor with 3D Force Detection Based on Porous CNTs/PDMS Nanocomposites," *Nano-Micro Lett.*, vol. 11, pp.57, Jul. 2019.

[4] J. Zhou, Y. Gu, P. Fei, W. Mai, Y. Gao, and R. Yang, "Flexible Piezotronic Strain Sensor," *Nano Lett.*, vol.8, pp. 3035-3040, Sep. 2008.

[5] B. Yang, W. Zeng, Z. H. Peng, S. R. Liu, K. Chen, and X. M. Tao, "A fully verified theoretical analysis of contact-mode Triboelectric nanogenerators as a wearable power source," *Adv. Energy Mater.*, vol. 6, pp. 1600505, Aug. 2016.

[6] X. Rong, J. Zhao, H. Guo, G. Zhen, and G. Dong, "Material Recognition Sensor Array by Electrostatic Induction and Triboelectric Effects," *Adv. Mater. Technol.*, vol. 5, pp. 2000641, Sep. 2020.

[7] T. Jin, Z. Sun, L. Li, Q. Zhang, and C. Lee, "Triboelectric nanogenerator sensors for soft robotics aiming at digital twin applications," *Nat. Commun.*, vol. 1, pp. 5381, Oct. 2020.

[8] S. Tumanski, "Induction coil sensors—a review," *MEAS SCI TECHNOL*, vol. 18, pp. R31-R46, Mar. 2007.

[9] M.R. Nabavi, and S.N. Nihtianov, "Design Strategies for Eddy-Current Displacement Sensor Systems: Review and Recommendations," *IEEE SENS J*, vol. 12, pp. 3346-3355, Dec. 2012.

[10] Q. Shi, Z. Zhang, T. He, Z. Sun, B. Wang, Y. Feng, ... and Lee, "Deep learning enabled smart mats as a scalable floor monitoring system," *Nat. Commun.*, vol. 11, pp. 4609, Sep 2020.

[11] S. Crandall, S. Cruikshank, and B. Connors, "A corticothalamic switch: controlling the thalamus with dynamic synapses," *Neuron*, vol. 86, pp. 1-15, May 2015.

[12] Z.X. Zhang, T. He, M. Zhu, Z. Sun, and C. Lee, "Deep learning-enabled triboelectric smart socks for IoT-based gait analysis and VR applications," *npj Flexible Electronics*, vol. 4, pp. 29, Oct 2020.

April 25-29 , 2021   Xiamen, China

MEMS-based platforms for characterization of advanced functional materials

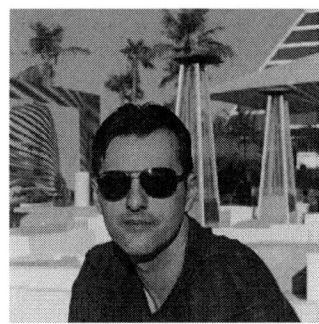

Liviu Nicu[1], Susan Trolier-McKinstry[2], Gabor Molnar[3], Azzedine Bousseksou[3], Maria Dolores Manrique Juarez[1], Fabrice Mathieu[1], Daisuke Saya[1], Karsten Haupt[4], Denis Dezest[5], Thierry Leïchlé[1,6]

[1]LAAS-CNRS, Université de Toulouse, France
[2]Department of Materials Science and Engineering and Materials Research Institute, The Pennsylvania State University, PA, USA
[3]LCC-CNRS, Université de Toulouse, France
[4]Université de Technologie de Compiègne, CNRS Laboratory of Enzyme and Cell Engineering, Compiègne, France
[5]CEA-Leti, Grenoble, France
[6]Georgia Tech-CNRS Joint International Laboratory, GA, USA

Website: www.laas.fr

Liviu NICU is currently senior scientist at the French National Scientific Research Center and is General Director of the CNRS Systems Analysis and Architecture Laboratory in Toulouse, France. After obtaining his PhD degree in Electrical Engineering from Paul Sabatier University in Toulouse, he joined Thales Avionics Valence (France) where he took charge of the company's inertial silicon microsensor projects from 2001 to 2003. He joined the Systems Analysis and Architecture Laboratory since 2003 where he conducted research in the fields of bioMEMS and MEMS for the physical and chemical characterization of advanced functional materials. Liviu NICU is co-author of over 100 peer-reviewed articles, 2 books / book chapters and has co-invented 3 patents. The laboratory he is heading since January 2016 brings together more than 700 people working in the fields of automation, robotics, computing and micro-nanotechnologies.

Abstract

The integration of new advanced functional materials in the production chain of microelectromechanical systems (MEMS) requires a fundamental in-depth knowledge of their physical and chemical properties. In recent years, there has been a significant increase in publications reporting on performant transduction properties of classes of materials partially or completely unknown by the production chain mentioned above [1], [2]. These materials are generally synthesized and characterized in a form (powders, single crystals, etc.) that is not suitable for MEMS manufacturing processes, rather requiring homogeneous thin films. Thus, despite the announced performances, which are very convincing and promising, there is no guarantee that they can be easily integrated into MEMS devices, nor can there be any guarantee that their performances will be preserved once the integration phase is successful.

Our work over the past few years was to assess the performance of materials, which either by their nature or by the way they are structured, still remain exotic for conventional micro / nanofabrication pathways. This assessment goes through the same stages, which are: the deposition and patterning of these materials at the micro-scale (by conventional or alternative techniques), their integration into basic micrometric structures (passive or active levers) and their characterization.

In this talk we will present our contribution to the characterization of three classes of materials: molecularly imprinted polymers (MIP), spin crossover materials (SCO) and piezoelectric materials (sol-gel lead zirconate titanate or PZT). In each of the cases presented, the same process flow and characterization methodology are followed, namely:

- in case of SCO materials (Figure 1.a): evaporation onto microfabricated cantilevers with actuation and sensing capabilities, assessment of the post-process spin cross-over capabilities;

- in case of MIPs (Figure 1.b): deposition and patterning of the MIP (by spin-coating and photolithography) fully integrated within the process flow of the microcantilevers' fabrication, assessment of the post-process mechanical integrity of the MEMS and chemical functionality preservation;

978-1-6654-3008-1/21 $31.00 © 2021 IEEE   1403

- in case of sol-gel PZT (Figure 1.c): integration of the active material into the process flow of microcantilevers' fabrication by means of micro-contact printing in order to assess the actuation and sensing capabilities.

As a general conclusion, we will demonstrate that by appropriately integrating thin films of these advanced materials and using the MEMS transduction capability, MEMS platforms are powerful tools for the physical and chemical characterization of these materials. As exemplified by the works presented here, these materials can be organic or inorganic, deposited by evaporation, micro-contact printing, spray coating or other types of techniques adaptable to MEMS manufacturing chain. The anticipation of the compatibility with MEMS process flow should allow, when a new material with high transduction potential is identified, its integration into the whole device architecture starting with the design phase.

References

[1] M. D. Manrique-Juárez *et al.*, « Switchable molecule-based materials for micro- and nanoscale actuating applications: Achievements and prospects », *Coord. Chem. Rev.*, vol. 308, p. 395 - 408, févr. 2016, doi: 10.1016/j.ccr.2015.04.005.

[2] E. Ahmed, D. P. Karothu, et P. Naumov, « Crystal Adaptronics: Mechanically Reconfigurable Elastic and Superelastic Molecular Crystals », *Angew. Chem. Int. Ed.*, vol. 57, nº 29, p. 8837 - 8846, juill. 2018, doi: 10.1002/anie.201800137.

Biography

Liviu NICU is currently senior scientist at the French National Scientific Research Center and is General Director of the CNRS Systems Analysis and Architecture Laboratory in Toulouse, France. After obtaining his PhD degree in Electrical Engineering from Paul Sabatier University in Toulouse, he joined Thales Avionics Valence (France) where he took charge of the company's inertial silicon microsensor projects from 2001 to 2003. He joined the Systems Analysis and Architecture Laboratory since 2003 where he conducted research in the fields of bioMEMS and MEMS for the physical and chemical characterization of advanced functional materials. Liviu NICU is co-author of over 100 peer-reviewed articles, 2 books / book chapters and has co-invented 3 patents. The laboratory he is heading since January 2016 brings together more than 700 people working in the fields of automation, robotics, computing and micro-nanotechnologies.

Proceedings of the 16th Annual IEEE International
Conference on Nano/Micro Engineered and Molecular Systems
April 25-29, 2021

Design and Fabrication of A latching Silicon-based MEMS Switch

Tongtong Cao, Tengjiang Hu* and Yulong Zhao

Abstract— In this paper, a new electro-thermal MEMS switch is presented. The application of silicon-based resistive switch is mainly restricted by the two factors of high contact resistance and high power consumption. In the proposed switch, ion implantation technology is adopted to solve the problems of high contact resistance and electrical signal isolation simultaneously. In addition, the proposed switch can realize state self-locking through the motion design, that is, it can maintain OFF/ON state without any power supply. The power consumption of this switch is limited to state switching process. The fabrication of this device is compatible with traditional semiconductor technology and easy to be integrated with other devices. The test results show that the switch can achieve a displacement of more than 55 microns at 15 volts driving voltage, which meets the requirement of circuit closure and can realize state locking.

Index Terms—MEMS switch, ion implantation, physical latching mechanism

I. INTRODUCTION

The switch is mainly used to control the ON-OFF state of the circuit. MEMS switch is an important part of MEMS system, and its rise provides strong technical support for the development of signal control system. Compared with traditional mechanical switches, MEMS switches have the advantages of low power consumption, high integration, small size, and mass production. At present, the demand for MEMS switches mainly comes from military security systems [1], medical devices [2, 3], automotive industry [2, 4], wireless communications [5] and other fields. Leveraging on the development of MEMS switch technology, products in these fields have reached or are about to reach unprecedented scale and performance levels.

The MEMS switch is mainly composed of two parts: MEMS actuators and contacts. The characteristics of the drive principle and contacts have a decisive influence on the performance of the switch. Compared with other drive methods, the electrothermal switch overcomes the dependence of distance between contacts, so it has a larger stroke (lateral displacement> 30 um) [1] and output force (>500 μN)[6]. The driving voltage of the electrothermal switch is not high (<20 V), and the fabrication process is simple. However, there are also some urgent problems that hinder the introduction of electrothermal switches into the market, mainly in the following aspects. (a) High power consumption. The excess power consumption will not only increase the cost of the switch, but also cause local heating of the device, which leads to a decrease in reliability, and ultimately a failure or safety of the switch. The latching

mechanisms are often used to reduce power consumption [7]. (b) High contact resistance. Silicon-based switches have excessively high contact resistance. In order to reduce the contact resistance, a layer of low-resistivity metal is often coated on the contact surface of the electrode [8], which cannot completely solve the problem. Furthermore, the substitution of metal for silicon structure is often used to reduce contact resistance [9, 10]. However, due to the immature process, the preparation of metal structure based on micro-electroplating falls into the dilemma of low shape accuracy and low yield. (c) Signal interference. For an active switch, signal interference should be avoided between the drive signal and the switch-on signal. For MEMS switches made on SOI wafers, the insulating layer cannot be prepared by surface micromachining under the movable structure layer, so signal isolation is a problem.

In this paper, a silicon-based electro-thermal MEMS switch is proposed. For one thing, a mechanical self-locking structure is designed, which means that the closed state of the switch is maintained by its own structural characteristics without continuous input of energy. The use of the latching mechanism can greatly reduce power loss and help ensure the stability of the switching state during long-term storage. For another, the problems of high contact resistance and insulating layer are solved simultaneously by ion implantation.

II. MODEL

A. The overall structure of the MEMS switch

The switch is prepared on a SOI (Silicon-on-Insulator) wafer, which is vertically composed of a handle layer, a silicon dioxide sacrifice layer and a device layer. The overall structure of the switch is shown in Fig. 1(a), which is mainly composed of the actuation structure, the latch structure, contacts, and pads. The actuation part uses three V-shaped electrothermal actuators, which are respectively used to drive the sliding terminal and two sets of limit rods. Actuator 1 is used to drive the sliding terminal to output linear displacement forward along the shuttle beam, and actuator 2 and actuator 3 are used to drive the limit rods to produce linear displacement outward. As shown in Fig. 1(b), the bridge contact designed on the sliding terminal has a certain gap *a* from the fixed contacts. The fixed contacts are connected to the signal lines by the cantilever beams. The sliding terminal is also designed with two pairs of slots, so that the limit rods can restrain or release the sliding terminal (maximum displacement *>b*). In addition, the stoppers are designed on both sides of the limit rods to restrict the radial position of the limit rods. To improve heat utilization, back cavities are designed on the substrate under the three sets of actuators.

[1]All author are with State Key Laboratory for Manufacturing System Engineering, Xi'an Jiaotong University, Xi'an 710049 China.
Contacting Author: T. J. Hu is with State Key Laboratory for Manufacturing System Engineering, Xi'an Jiaotong University, Xi'an 710049 China (email: htj047@xjtu.edu.cn)

978-1-6654-3008-1/21 $31.00 © 2021 IEEE

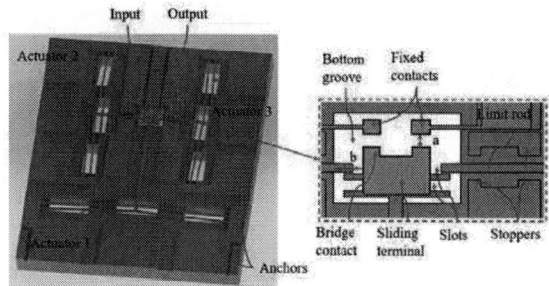

Figure 1. Schematic of the proposed switch:(a) The overall structure of the switch;(b).The close-up of the contact area.

B. The working principle

The realization of the switch's self-locking and unlocking functions requires two pulse voltages. The phase relationship between the two pulse voltages is shown in Fig. 2(A). The voltage U1 is applied to the actuator 1, and the voltage U2 is applied to the actuator 2 and the actuator 3. Fig. 2(B) is the working principle of the designed MEMS switch. When the switch is in the initial state, neither the limit rod nor the contact is stressed. To close and lock the switch, first the voltage U2 is apply to the actuator 2 and the actuator 3 to make the limit rods move away until it leaves the front slots on the sliding terminal. Keep this voltage at high level. Second, the voltage U1 is applied to the actuator 1, and the sliding terminal is driven to move toward the fixed contacts. Notably, the displacement of the sliding contact will be slightly larger than the gap *a* between the contacts, so that the restoring force generated by the cantilever beams will further enhance the contact force between the contacts. Next, keep the voltage U1 at a high level and remove the voltage U2, the limit rods return to the initial position under the action of the restoring force. Since the sliding terminal has produced a certain displacement, the limit rods will enter the latter slots on the sliding terminal. Finally, the voltage U1 is removed. The sliding terminal tends to move backwards under the action of the restoring force, but it is stuck by the limit rods. In this way, the closed state is latched.

Similarly, the voltage U2 is apply again to move the limit rods outward. The sliding terminal will automatically return to the initial position to unlock the switch. At this time, remove the voltage U2. The switch is switched to the normally open state.

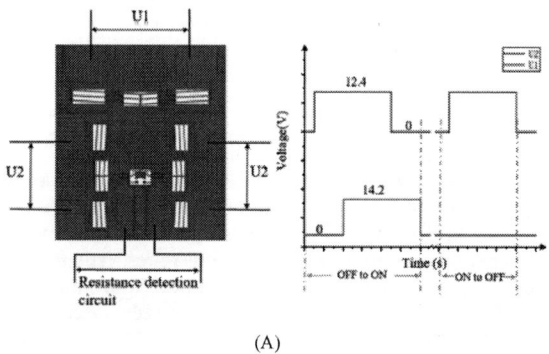

(A)

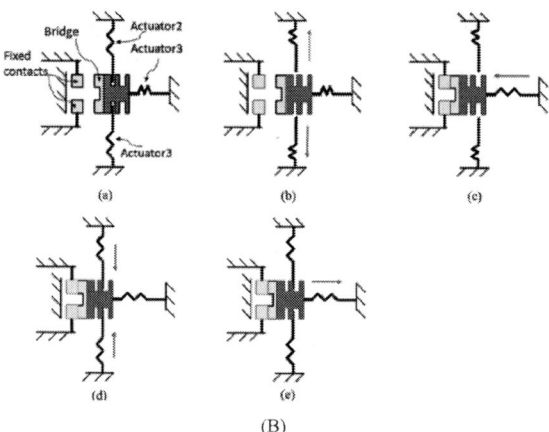

(B)

Figure 2.(A) The phase diagram of pulse voltages. (B) The diagram of working principle: (a) Original state; (b) Driving actuator 2, 3 to move in opposite directions; (c) Driving actuator 1 forward; (d) Releasing actuator 2, 3; (e) Releasing actuator 1.

C. The geometric parameter design of actuators

The displacement of the electrothermal actuator determines whether the switch can be closed and realize self-locking. Therefore, it is necessary to analyze the influencing factors of displacement. The working process of the electrothermal actuator is actually a multi-physics coupling process of electro-thermal and mechanical fields. When a current input to the V-shaped actuator, the V-shaped beams are equivalent to a resistance that generates heat in the beams. The heat loss is mainly transferred to the substrate and the surrounding environment by heat conduction. The temperature distribution along the length of the V-shaped beam can be solved according to the following equation [11, 12]:

$$k_\mathrm{p} \frac{d^2 T(x)}{dx^2} + J^2 \rho = \frac{S}{h} \frac{T(x) - T_\mathrm{s}}{R_\mathrm{T}} \qquad (1)$$

where k_p is the thermal conductivity of silicon; J is the current density related to the voltage applied and the geometry of the V-beam; ρ is the electrical resistivity of silicon; S is the shape factor of beam heat conduction; R_T is the thermal resistance between the beam and the substrate; T_s is the ambient temperature; h is the thick of the beam. After obtaining the temperature distribution function from (1), the average temperature of the V beam can be obtained, which can be written as (2):

$$\overline{T} = \frac{1}{L} \int_0^L T(x) dx \qquad (2)$$

Where L is the span of the V beam. The two ends of the V beam have been fixed. When it is heated and expanded, the displacement is along the direction of the symmetry axis of the V beam, as shown in Fig. 3(a). According to the geometrical relation of V beam, the displacement *d* can be expressed as (3):

$$d = \frac{L}{2}\left(\sqrt{\frac{[1+\alpha(\overline{T}\text{-}T_{\text{s}})]^2}{\cos^2\theta}} - 1 - \tan\theta\right) \qquad (3)$$

where α is the thermal expansion coefficient of silicon.

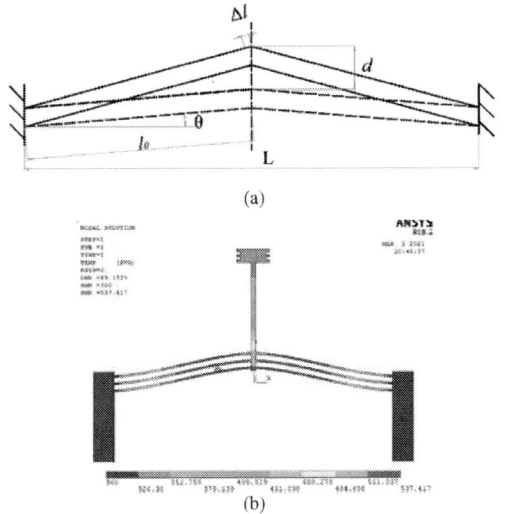

Figure 3. (a) The thermal expansion model of V-shaped beam; (b) Simulation results of the output displacement and temperature distribution of the V-shaped actuator.

From the above equations, it can be seen that the temperature rise and the geometric parameters of the V-shaped actuator, such as the length and the angle of the beam have a direct influence on its output displacement. To clarify the influence of these factors on the output displacement, the finite element simulation analysis of the actuator was carried out, as shown in Fig. 3(b). According to the simulation results, the output displacement is positively correlated with the length of the beam and the driving voltage; inversely related to the width and the angle of the beam. Moreover, the effect of changing angle on output displacement is the most obvious. Through theoretical calculation and simulation analysis of the v-shaped actuator, the main geometric parameters of the actuators are determined (see Table 1).

TABLE I.   THE MAIN GEOMETRICAL PARAMETERS OF THE SWITCH

	Actuator 1	Actuator 2/3
Length of Beam(μm)	2000	1300
Width of Beam (μm)	30	30
Thick of Beam (μm)	50	50
Angle of Beam (°)	2	2
Gap of a (μm)	50	
Gap of b(μm)	28	

III. FABRICATION

The fabrication of the MEMS switch chip mainly includes six steps, Fig. 4 explains the fabrication process of the switch. In short, (a). Masking layer preparation. Deposit a 6μm SiO$_2$ layer on the upper surface of the device layer by PECVD, then lithography and pattern the SiO$_2$ layer with BOE solution. (b). Ion implantation and annealing. Implant boron in the pattern window and anneal at 1050 °C. (c). Alignment mark preparation. Etch the SiO2 layer a second time, leaving only alignment marks. (d). Sputter and pattern Cr and Au layers to form metal pads and conductive lines. (e). Etch the bottom grooves on the handle layer, and then etch the structure on the device layer with the ICP system. (f). Releasing. Etch away the buried silicon dioxide with HF solution to release the movable structure.

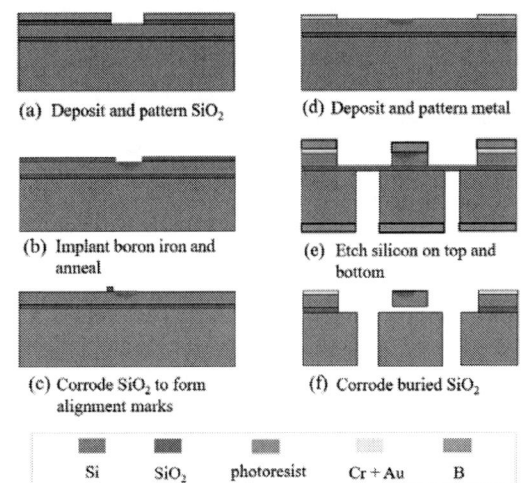

Figure 4. The fabrication process flow of the proposed switch chip.

Ion implantation and structure etching of SOI wafer are the key process steps. For one thing, the ion implantation process simultaneously achieves both signal isolation and reducing contact resistance. The device layer of SOI wafer itself is n⁻ type silicon. By injecting boron atom on the upper surface of the contact, p⁺ type silicon is formed on its upper layer, and PN junction is formed at the junction, which will isolate the electrical signals on both sides of the boundary layer. Meanwhile, the resistivity of the contacts after heavy doping is reduced. For another, the bottom groove has two functions: to improve heat utilization and to facilitate the release of movable structures. As for the use of intermittent grooves instead of through grooves, it is found that if the bottom groove too large, the heat conduction during the etching process of the device layer will be affected, causing local overheating and affecting the front etching effect.

After a round of processing, the structure of the switch has been slightly adjusted in combination with the process conditions. The comparison before and after structural optimization is shown in Fig. 5. Firstly, the mechanical self-locking structure was optimized to reduce the limit of the size on the output displacement of the actuator. Second, the original distance between the fixed contact and the movable contact was shortened to increase the contact force. Third, the bottom grooves were enlarged to reduce heat loss to the substrate, thereby increasing the displacement of the electrothermal actuators. The minimum gap of the structure is between the stoppers and the limit bars, which is 3μm (subject to fabrication design rules).

978-1-6654-3008-1/21 $31.00 © 2021 IEEE   1407

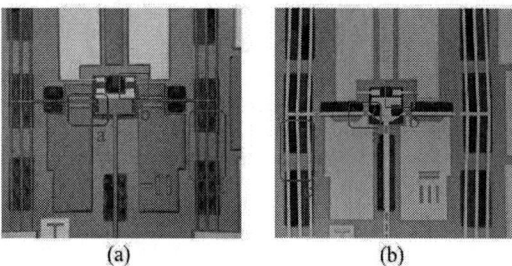

Figure 5. The comparison before and after structural optimization: (a) the switch before optimization; (b) the switch after optimization.

After the switch chip is fabricated, bond the chip with PCB through gold wires (see Fig. 6(a)). A partial close-up of the manufactured switch is shown in Figure 6(b). It can be seen that the proposed switch has good structural verticality and precise shape even at the minimum line width of the device.

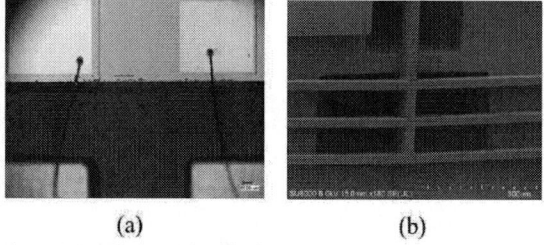

Figure 6. (a) The bonded gold wires; (b) SEM picture of the V-shaped beams of the proposed switch.

IV. TEST AND DISSCUSSION

The performance of the fabricated switch was tested. Fig. 7(a) is the test platform built for the proposed switch. The test results show that the current and the displacement of the actuators increase with the increase of voltage. The output displacements used to drive the sliding terminal and to drive the limit rods of the switch can reach more than 55 μm and 32 μm respectively at 15 V (see Fig. 7(b)). The experimental results verify the correctness of theoretical estimation and simulation results. Burn tests were carried out on the switches, and actuators 2 and 3 burned out at 17 V and actuator 1 burned out at 19 V. Therefore, to ensure safe operation and extend the service life of the device, the voltage applied should not exceed 15 V.

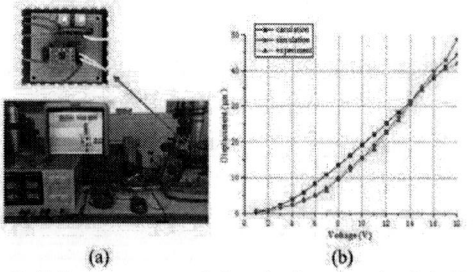

Figure 7. (a) Performance test platform for the proposed switch; (b) The test results and theoretical values of displacement vs voltage.

The motion test of the switch was investigated. The actuators acted in sequence under two pulse voltages with phase difference, as shown in Fig. 8. The voltage applied to actuator 2 and actuator 3, and the voltage applied to actuator 1 are 12.4 V and 14.2 V, respectively. The fabricated switch could realize locking and unlocking. After the contacts were in contact, the switch circuit had an obvious current signal. The mechanical self-locking structure can ensure that the switch has power loss only during the state switching process.

In order to compare the resistance changes before and after doping, the loop resistance the switch was measured with a multimeter. The measurement results show that the loop resistance before doping is about 3-5 MΩ, and the loop resistance is about 8 KΩ, indicating that the idea of using ion implantation to reduce contact resistance is feasible. Moreover, the insulating layer formed by the PN junction prevents the signals from interfering with each other, which is also one of the reasons for the decrease of loop resistance. This research is devoted to studying the effect of ion implantation, so gold has not been sputtered on the contact sidewalls to reduce the contact resistance.

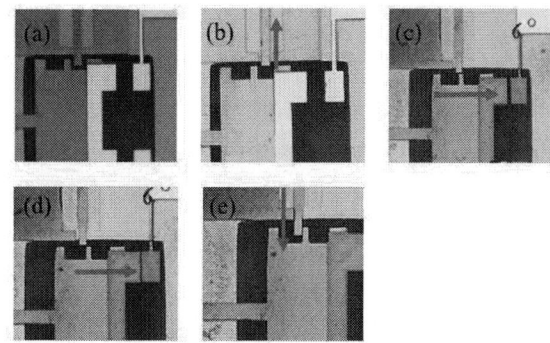

Figure 8. The sequential action of the movable structure of the switch.(a) Initial position;(b) the limit rod moves outward;(c) the slider moves in the direction of the fixed contact;(d) the sliding contact and the fixed contacts contact;(e) the limit rod returns to the position to achieve locking.

V. CONCLUSION AND PERSPECTIVE

In this paper, a MEMS switch based on SOI wafers is proposed. A mechanical latching structure is designed to reduce power consumption. The switch is fabricated using mature semiconductor technology, so it has an accurate morphology and line width. Ion implantation is used to prepare signal isolation layers and reduce contact resistance. The measured loop resistance is nearly 5 times lower than before doping. The fabricated chip measures 4.5 mm × 4.7 mm.

Although the proposed switch can achieve the expected functions, its reliability has not been verified, and the contact resistance can be further reduced. The follow-up work will include these two aspects, namely, to reduce the contact resistance through the research of silicon-based switches with metal contacts; and to improve the long-term reliability of the switches.

ACKNOWLEDGMENT

This work was supported by the National Key R&D Program of China (2017YFB1102900) and the China Postdoctoral Science Foundation (2018M640977).

REFERENCES

[1] T.J. Hu, K. Fang, Y. Zhao, et al. , "Design and research on large displacement bidirectional MEMS stage with interlock mechanism," *Sensors & Actuators A Physical*, vol.283, pp. 26–33,2018.

[2] J.I. Lee, Y. Song, H. Jung, et al., "Deformable Carbon Nanotube-Contact Pads for Inertial Microswitch to Extend Contact Time," *IEEE Transactions on Industrial Electronics*, vol. 59: pp. 4914–4920, 2011.

[3] A. Ongkodjojo, F. E. H. Tay., "Optimized design of a micromachined G-switch based on contactless configuration for health care applications," *Journal of Physics: Conference Series*, vol. 34, pp.1044-1052, 2006.

[4] L. Du, W. Wang, C. Du, et al., "A novel contact-enhanced low-g inertial switch with low-stiffness fixed electrode," *Microsystem Technologies*, vol. 26, pp. 395–404, 2009.

[5] Y. Xu, Y. Tian, B. Zhang, et al., "A novel RF MEMS switch on frequency reconfigurable antenna application," *Microsystem Technologies*, vol. 24, pp. 3833–3841, 2018.

[6] J. C. Hsieh, D. T. W. Lin, M. S. Suen, "The design of high strength electro-thermal micro-actuator based on the genetic algorithm," *Microsystem Technologies*, vol. 26, pp. 1113–1119, 2019.

[7] T. T. Cao, T. J. Hu and Y. L. Zhao, "Research Status and Development Trend of MEMS Switches: A Review," *Micromachines*, vol. 11, pp. 694, July 2020.

[8] C. Joshitha, B. S. Sreeja, S. S. Princy, et al., "Fabrication and investigation of low actuation voltage curved beam bistable MEMS switch," *Microsystem Technologies*, vol. 23, no. 10, pp. 4553-4566, 2017.

[9] J. Doutreloigne, D. Dellaert, "Compact thermally actuated latching MEMS switch with large contact force," *Electronics Letters*, vol. 51, pp. 80-81, Jan. 2015.

[10] M. Bakri-Kassem and R. R. Mansour, "High Power Latching RF MEMS Switches," in *IEEE Transactions on Microwave Theory and Techniques*, vol. 63, no. 1, pp. 222-232, Jan. 2015.

[11] E.T. Enikov, S.S. Kedar, K.V. "Lazarov, Analytical model for analysis and designof V-shaped thermal microactuators," *J. Microelectromech. Syst.*, vol. 14, pp. 788–798, 2005.

[12] M. Mayyas, H. Stephanou. "Electrothermoelastic modeling of MEMS gripper," *Microsystem Technologies*, vol.15, pp. 637-646, 2009.

Proceedings of the 16th Annual IEEE International
Conference on Nano/Micro Engineered and Molecular Systems
April 25-29, 2021

MULTI-MODAL RESONANCE MEASUREMENT CAPABLE OF DISCERNING AND VISUALIZING EFFECTS DUE TO EXCITATION SCHEMES IN MEMS RESONATOR

Bo Xu, Jiankai Zhu, Tongqiao Miao, Jing Li, Qingyang Deng, Song Wu, Ting Wen, Fei Wang,
Xiaoping Hu, Xuezhong Wu, Dingbang Xiao, and Zenghui Wang

Abstract— **The capability of characterizing MEMS resonant sensors using different driving mechanisms and measuring their resonant response in multiple physical domains can offer insights into the resonant response of the devices. Here, we present our work using a multi-modal resonance measurement system that enables driving the same MEMS resonator using electrical, optothermal and mechanical excitation schemes, and provides information of the resonant response in both spectral and spatial domains. We show that for the exact same device under the same measurement environment, different driving mechanisms can lead to important differences in resonance frequency, quality factor, and mode shape. Our findings can lead to new insights into the dynamics of such resonant sensors, and important implications for characterization protocols.**

I. INTRODUCTION

MEMS resonant sensors have become increasingly popular due to a number of advantages, such as small size, low cost, and high performance, and have demonstrated great potential towards applications such as inertial sensing [1], biomedical sensing [2], and environmental sensing [3]. In particular, MEMS resonators leverage the amplification effect of mechanical motion on resonance, as well as their capability of signal sensing and transduction in both displacement and frequency domains. Therefore, accurate measurements of MEMS resonant sensors are important for determining how precise the physical quantities can be sensed and transduced by these devices.

In resonant sensors that involves more than one modes, such as mode-coupling resonators [4,5] or mode-localization sensors [6,7], the details of each resonance mode, such as the frequency, quality factor (Q), and actual mode shape, can have strong effects on (and sometimes are important indicators of) the coupling strength, which are critical for the operation of the devices. Therefore, it is important to be able to precisely characterize these key attributes of MEMS resonant sensors from all the necessary physical domains.

However, in many cases it is challenging to experimentally measure all the key device attributes, and

B. X., J. Z., T. M., J. L., Q. D., S. W., T. W., F. W. and Z. W. are with the University of Electronic Science and Technology of China, Chengdu, China (e-mail: zenghui.wang@uestc.edu.cn).
T. M., X. H., X. W. and D. X. are with National University of Defense Technology, Changsha, China (e-mail: dingbangxiao@nudt.edu.cn).
X. W. and D. X. are with Hunan MEMS Research Center, Changsha, China.

numerical simulation must be used to infer some of the critical information. For example, using electrical readout scheme one can measure resonance characteristics in the frequency domain, but it is difficult to obtain spatial resolution and thus the mode shape. In other cases, the driving mechanism of a device could be different between during device characterization (to obtain more comprehensive information of the device in order to understand and analyze its performance) and actual application (in which only one signal domain is used for sensing), which could lead to unexpected discrepancies that are difficult to characterize.

In this paper, we demonstrate a custom-build MEMS resonant sensor characterization system for multi-modal resonance measurement, which simultaneously enables multiple excitation schemes, including electrical, optothermal, and mechanical excitation, all for the same MEMS resonator under the same test environment. Such greatly enhanced measurement capability facilitates meaningful comparison of the device response across different driving schemes. Using a MEMS resonator, we show that interesting and important differences in resonant characteristics can arise under different driving schemes, in all facets of the device response, ranging from frequency and quality factor to mode shape. The findings offer new insights into the dynamics in resonator based sensors, and can have important implication for future testing protocols for MEMS resonant sensors.

II. DEVICE STRUCTURE

The MEMS device used in this paper is a ~ 3 mm × 3 mm suspended silicon resonator structure (40 μm thick) with silicon electrodes (2μm thick) underneath (Fig. 1). The electrode layer resides on a 2μm thick SiO_2 box layer over the handle layer. The fabrication process is analogous to that of similar devices found in the literature [8]. The device exhibits a number of resonant modes, including two torsional modes that differ in the relative phase between different parts of the structure, which are referred to the in-phase mode (IP mode) and anti-phase mode (AP mode). Both modes involve out-of-plane vibration and have clear symmetry, offering great opportunities for the studies in this work.

978-1-6654-3008-1/21 $31.00 © 2021 IEEE

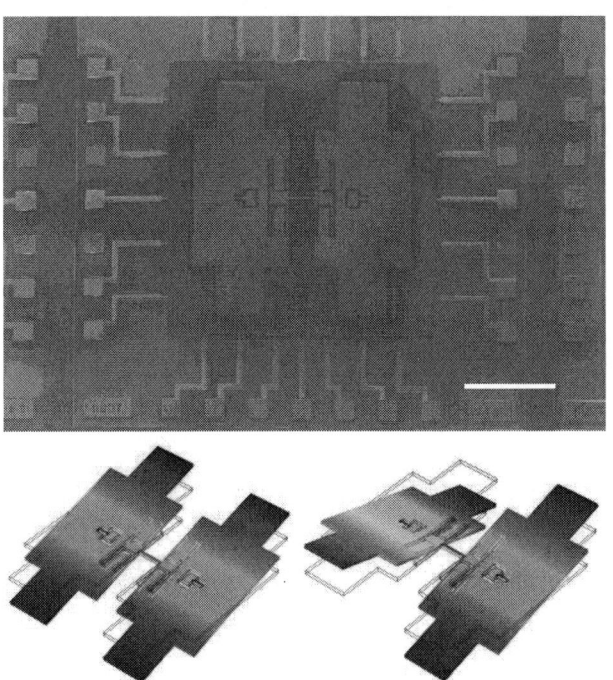

Figure 1: MEMS resonator in this work. Top: SEM image; scale bar: 1mm. Bottom: Mode shapes of the two resonance modes of interest.

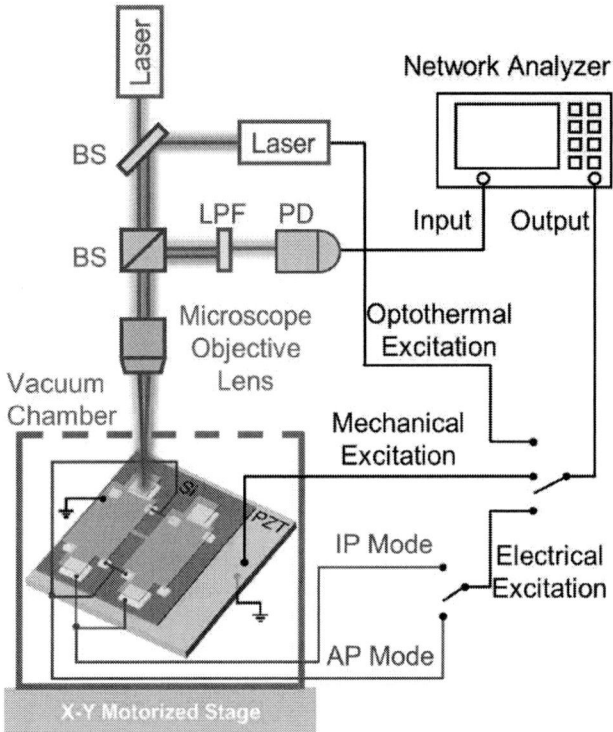

Figure 2: Schematic of the measurement system. The driving signal can be routed through three different excitation paths, illustrated using the 1-3 selection switch. The two different resonance modes can be selectively excited (with better efficiency) by connecting to different pairs of electrodes, as indicated by the 1-2 switch.

III. RESONANCE MEASUREMENT

A. Overview of the measurement system

The measurements setup used in this study is a custom-built system that is capable of exciting a MEMS/NEMS resonator using a number of different schemes and detecting its resonance in multiple signal domains. As shows in Fig. 2, the sample is mounted in a vacuum chamber with electrical feedthrough and optical access. All measurements are performed under vacuum $\sim 1 \times 10^{-6}$ torr. The sample position can be precisely controlled and scanned using an x-y translational stage. The driving signal is generated by the output of a network analyzer, and can be routed through a number of circuit connections to enable the different excitation schemes.

B. Electrical Excitation

The resonant motion can be excited electrically. The electrical excitation scheme is realized by connecting the driving signal from the network analyzer to the bottom electrodes underneath the resonator body while grounding the resonator. Different electrodes can be selected to improve the efficiency in exciting different resonant modes, based on the simulated mode shape. For example, connecting the driving signal to electrode configuration A is expected to excite the IP mode, while configuration B is designated to effectively drive the AP mode. The electrical excitation scheme is typically used then the resonant device is used for sensing applications.

C. Optothermal Excitation

The resonant motion can also be excited optically using optothermal effect [9]. This excitation scheme is realized by connecting the driving signal from the network analyzer to the modulation signal input of a 405-nm modulated diode laser. The driving signal controls the modulation depth and frequency of the laser power. The modulated light intensity, when incident on the device structure, produces a periodic optothermal force, which drives the device motion. The resonator body and electrodes are grounded in this excitation scheme. The all-optical measurement scheme allows efficient characterization of resonant responses without the requirement of electrical connection to the device.

D. Mechanical Excitation

The resonant motion can further be excited mechanically using a piezo shaker on which the resonator device is mounted. The mechanical excitation scheme is realized by applying the driving signal from the network analyzer across the piezo shaker, while grounding all the electrodes on the device wafer as well as the resonator body. Under the driving voltage, the piezo shaker deforms periodically, and thus drives the device into vibration through base excitation. The mechanical driving scheme enables measurements without effects typically associated with the other schemes such as capacitive softening and laser heating.

E. Resonance Measurement

The resonance is measured optically [10]. A collimated 633-nm laser beam is focused on the device through a long-working-distance objective. As the device vibrates, the reflected light intensity changes, which is measured by a photodetector and then collected by the network analyzer. Using the optical measurement allows us to compare the vibration amplitude across different driving schemes. In addition, the optical probe provides local information about the resonator, allowing spatial resolution of the resonance mode shape.

F. Resonance Mode Shape Mapping

The resonance mode shape is resolved using the "spectromicroscopy" technique [11]. During the measurement, a resonance spectrum is taken for each location, and the laser spot scans over the entire device surface. The amplitude of on-resonance mechanical vibration at each location is then extracted through fitting the measured frequency response curve to a damped harmonic resonator model. The extracted vibration amplitude is then plotted as a function of laser position in a 2D color plot (Fig. 4), with red color representing larger amplitude. The actual resonance mode shape can be then visualized using this technique, and thus facilitating comparison across different driving schemes.

IV. RESULTS AND DISCUSSIONS

We perform both spectral and spatial measurements for the two resonant modes illustrated in Fig. 1 using the different driving schemes. The frequency response curves for electrical, optothermal, and mechanical excitations are shown in Fig. 3 for both resonant modes. The frequency scan range is adjusted for each resonance feature, while the plotting range is kept same for easy comparison.

A. Effect of Driving Scheme on Frequency Response

A number of interesting and important results are revealed by the measurement data. First, the measured resonance frequency f_{res} and quality factor (Q) of the same resonator, even under the temperature and pressure, can vary notably across the different driving schemes. In particular, for both modes the electrical excitation scheme produces the highest Q and lowest frequency, while under mechanical excitation the measured Q values are significantly lower (by one order of magnitude) for both modes, with clearly higher frequency. The optothermal excitation gives intermediate Q values.

Note that for all the different driving schemes, we adjust the excitation power to maintain similar measured motional signal level to ensure that the device vibrate with comparable amplitude. Further, the driving intensity is varied within each driving scheme to ensure absence of nonlinearity in the transduction, and for electrical driving, also the capacitive softening effect. Therefore, the observed variation in f_{res} and Q are attributed to the different driving mechanisms. Detailed

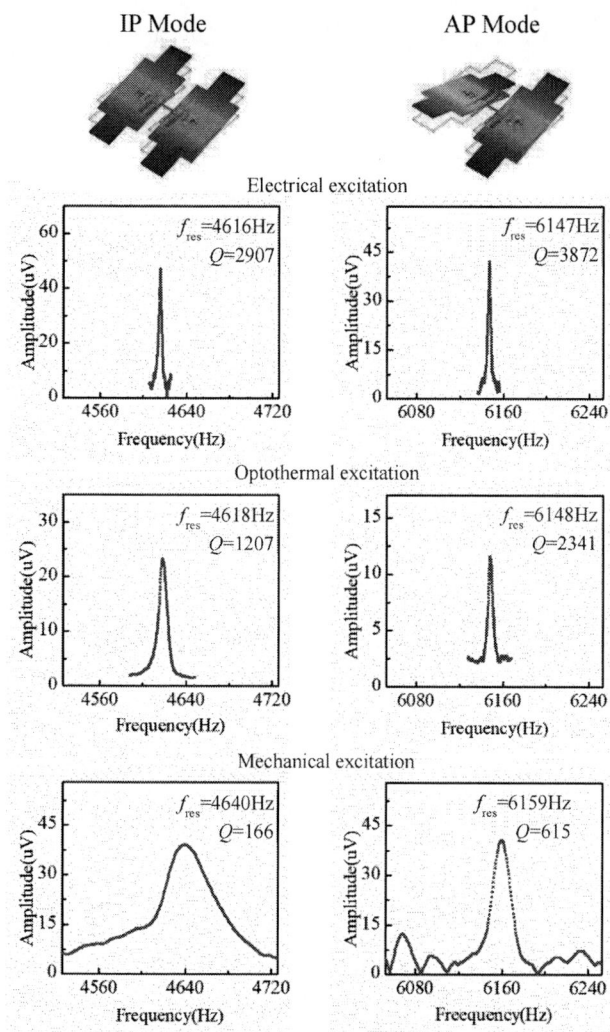

Figure 3: Frequency response of the two resonant modes measured under different excitation schemes. Left column: the IP mode. Right column: the AP Mode. From top to bottom are data measured with electrical, optothermal, and mechanical excitation schemes, showing clear contrast in both frequency and Q.

analysis of the observed effects, including analytical models capturing the key features, will be discussed during the conference presentation.

B. Dependence of Mode Shape on Driving Scheme

Further, the spatially-resolved measurement reveals that the actual mode shape can also be dependent on the driving scheme. As shown in Fig. 4, the measured mode shapes exhibit clear asymmetry for electrical excitation, but much less so for mechanical excitation. This suggest that the simulated mode shape using finite element model may not always precisely represent the actual behavior of the device, and the choice of excitation scheme can have unexpected effects on the actual mode shape. More detailed analysis of the specific example will be presented during the actual talk.

V. Conclusion

A multi-model resonance measurement technique is developed and tested for its capability using a MEMS resonator. By exciting the same device using different driving schemes, i.e., electrical excitation, optothermal excitation and mechanical excitation, we find that both the resonant response and the actual mode shape of the device can strongly depend on the excitation scheme, and such conclusion holds for multiple resonance modes. This suggests that caution must be exerted when extracting certain attributes of a resonator, such as quality factor, from a single-modal measurement, and the actual mode shape can differ significantly from the simulated ones, depending on the specific driving scheme. The results can have important implications for characterizing and analyzing the performance of resonant sensors, which strongly depends on the intrinsic performance and actual mode shape of the devices.

Acknowledgment

The authors thank supports from Ministry of Science and Technology (grants 2018YFE0115500 and 2019YFE0120300), National Natural Science Foundation of China (grant 61774029), Sichuan Provincial Science and Technology Department (grants 21CXTD0088, 21YYJC3079, 2019JDTD0006 and 2019YFSY0007), the Excellent Youth Project of Natural Science Foundation of Hunan Province under Grant No. 2020JJ2033, and the National Natural Science Foundation of China (again, if one notices) under Grant No. 51935013.

References

[1] D. K. Shaeffer, "MEMS inertial sensors: A tutorial overview", *IEEE Communication Magazine*, vol. 51, pp. 100-109, 2013.

[2] P. Steglich, M. Hülsemann, B. Dietzel and A. Mai, "Optical biosensors based on silicon-on-insulator ring resonators: A review", *Molecules*, vol. 24, 519, 2019.

[3] F. Ejeian, S. Azadi, A. Razmjou, Y. Orooji, A. Kottapalli, M. E. Warkiani and M. Asadnia, "Design and applications of MEMS flow sensors: A review", *Sensors and Actuator A: Physical*, vol. 295, pp. 483-502, 2019.

[4] X. Zhou, C. Zhao, D. Xiao, J. Sun, G. Sobreviela, D. D. Gerrard, Y. Chen, L. Flader, T. W. Kenny, X. Wu and A. A. Seshia, "Dynamic modulation of modal coupling in microelectromechanical gyroscopic ring resonators", *Nature Communications*, vol. 10, 4980, 2019.

[5] L. Xu, S. Wang, Z. Jiang and X. Wei, "Programmable synchronization enhanced MEMS resonant accelerometer", *Microsystems & Nanoengineering*, vol. 6, 63, 2020.

[6] H. Zhang, G. Sobreviela, D. Chen, M. Pandit, J. Sun, C. Zhao and A. A. Seshia, "A High-Performance Mode-Localized Accelerometer Employing a Quasi-Rigid Coupler" *IEEE Electron Device Letters*, vol. 41, pp. 1560-1563, 2020.

[7] H. Zhang, J. Huang, W. Yuan and H. Chang, "A high-sensitivity micromechanical electrometer based on mode localization of two degree-of-freedom weakly coupled resonators", *Journal of Microelectromechanical Systems*, vol. 25, pp. 937-946, 2016.

[8] T. Miao, X. Zhou, Z. Hou, X. Hu, X. Wu and D. Xiao, "A Million-order Effective Quality Factor MEMS Resonator by Mechanical Pumping", in *2020 IEEE International Symposium on Inertial Sensors and Systems*, Hiroshima, Japan, March 23-26, 2020.

[9] J. Lee, Z. Wang, K. He, R. Yang, J. Shan and P. X.-L. Feng, "Electrically tunable single-and few-layer MoS_2 nanoelectromechanical systems with broad dynamic range", *Science Advances*, vol. 4, 6653, 2018.

[10] J. Lee, Z. Wang, K. He, J. Shan and P. X.-L. Feng, "High frequency MoS_2 nanomechanical resonators", *ACS Nano*, vol. 7, pp. 6086-6091, 2013.

[11] Z. Wang, J. Lee and P. X. L. Feng, "Spatial mapping of multimode Brownian motions in high-frequency silicon carbide microdisk resonators", *Nature Communications*, vol. 5, 5158, 2014.

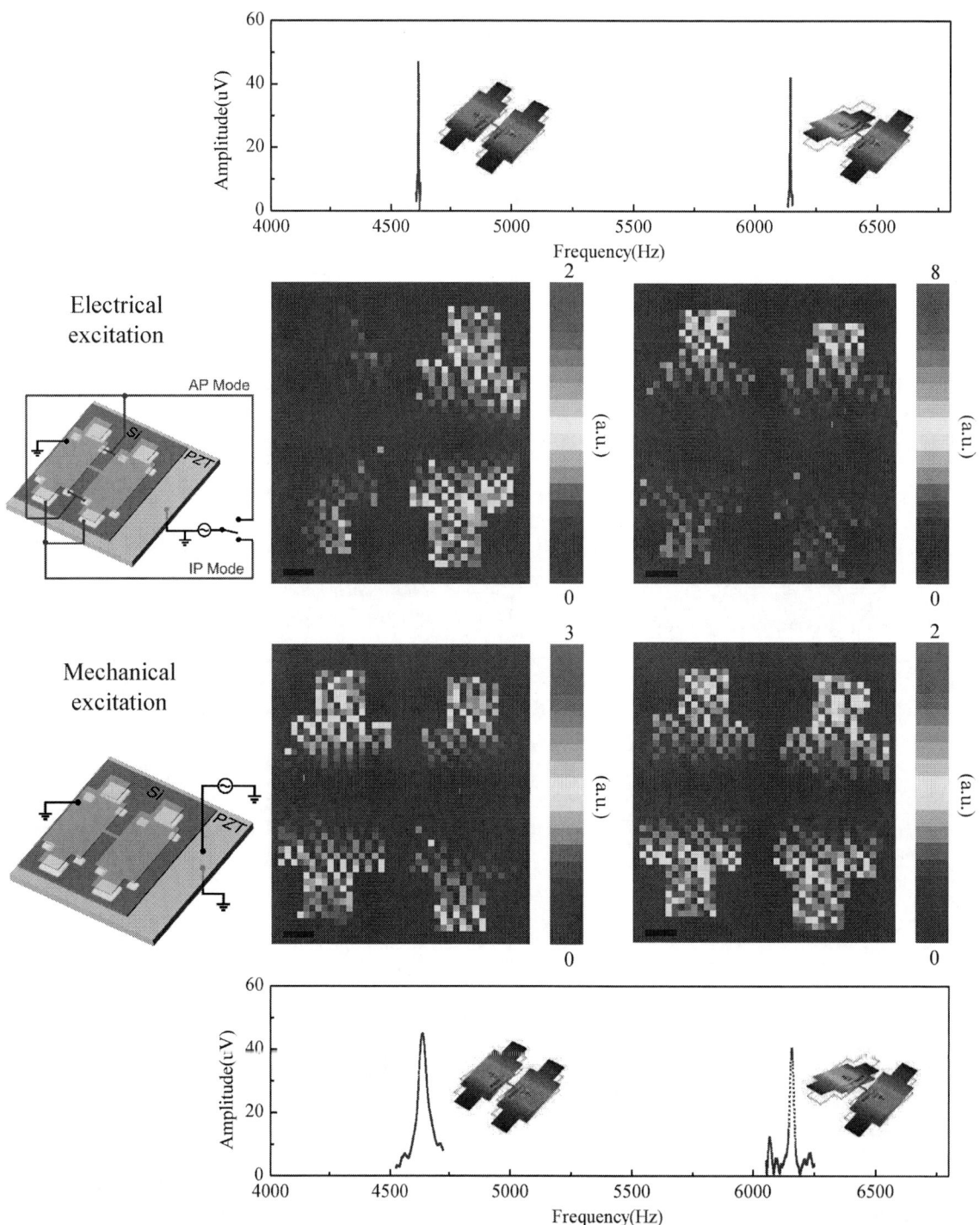

Figure 4: Measured mode shapes of the two resonance modes using electrical and mechanical excitation schemes. The frequency response curves (top and bottom) are plotted for both modes, showing the frequency separation in the resonance peaks. The left column shows the lower frequency IP mode and the right column shows the higher frequency AP mode. Clear contrasts can be visualized in the measured mode shapes.

Proceedings of the 16th Annual IEEE International Conference on Nano/Micro Engineered and Molecular Systems
April 25-29, 2021

Nanostructure Vanadium-doped Zinc Oxide Film Sensor Endowed with Enhanced Piezoelectric Response *

Wei Gao, Yu Zhang, Binghe Ma*, Jian Luo, Jinjun Deng, and Weizheng Yuan

Abstract— We present a performance-enhanced piezoelectric sensor utilizing nanostructure Vanadium-doped zinc oxide (ZnO) thin film aiming at sensitively measuring underwater sound signals. The sensor is fabricated by MEMS technology using a 6-mask ZnO-on-SOI process platform. Individual sensor with a low foot-print of 2 mm×2mm could be easily assembled working as a hydrophone. Sensing element based on the V-doped ZnO film is endowed improved piezoelectricity with the piezoelectric constant of 110.3 pm/V. Demonstrations in a plane-wave tube test system confirm that the output voltage of the V-doped ZnO film sensor obviously increases by an order of magnitude compared with the pure ZnO film sensor with same dimensions, showing great potential in the field of underwater acoustic detection.

I. INTRODUCTION

The accurate measurement of underwater sound signals is critical for various applications, such as noise monitoring, underwater communication, and sound navigation etc.[1]

Over the past few decades, a variety of sensors based on piezo-resistive type, optical method, and piezoelectric type were developed for underwater acoustic measurement [2][3][4]. Among all these methods, the hydrophone with a piezoelectric thin film based on the crystalline deformation, has been proven an effective method due to its merits of lower power consumption, high sensitivity and simple readout circuits etc. The sensitivity of the piezoelectric hydrophone is determined by the ratio between its piezoelectric coefficient (d_{33}) and dielectric constant [5]. Many piezoelectric materials such as lead zirconate titanate (PZT), polyvinylidene fluoride (PVDF), Aluminum nitride (AlN) and ZnO have been utilized for the development of micromachined pressure sensor [6]. Since being discovered, PZT dominates the piezoelectric applications due to its excellent piezoelectric coefficient, but the high dielectric constant also restricts the capacitance characteristic of the device. Moreover, the PZT ceramic and PVDF become piezoelectric requiring an electric-thermal poling at high temperature conditions, restricting the processing compatibility. Thus, AlN and ZnO based devices have attracted more attention due to the unique semiconducting properties and outstanding batch fabrication process compatibility. However, the d_{33} for an oriented AlN and ZnO film is about 8 pC/N and 12.4 pC/N, respectively, limiting the piezoelectric response of the devices [7][8]. Therefore, a high piezoelectric coefficient coupled with a lower dielectric constant is desired for a thin film of hydrophones.

Contributions

● Design and fabrication of a micromachining pressure sensor utilizing nanostructure V-doped ZnO film.
● Investigation of the influence of doping V in the ZnO thin film on the piezoelectric properties.
● Calibrations of the ZnO film sensor in a plane-wave tube system.
● Evaluation of the sensitivity of the presented V-doped ZnO film sensor.

Figure 1: Schematic view of the presented V-doped ZnO film sensor.

II. DESIGNING AND FABRICATION

Figure 2 shows the schematic view of the mainly fabrication processes of the V-doped ZnO film sensors. SOI-based V-doped ZnO thin film sensors are fabricated by MEMS technology. Firstly, about 300nm of silicon oxides (SiO_2) is grown by thermal oxidation (Figure 2a), and 500nm of silicon nitride (Si_3N_4) is deposited by chemical vapor deposition (Figure 2b) as the insulation layer. Secondly, 50nm/250nm Ti/Pt films are sputtered and patterned through lift-off process as bottom electrodes. Thirdly, about 1μm V-doped ZnO film is deposited by RF magnetron sputtering from a sintered stoichiometric circular $Zn_{0.98}V_{0.02}O$ target. The detailed sputtering parameters used for the films deposition are shown in Table 1. Photolithography and wet-etched are used to pattern the piezoelectric layer. Then, 50nm/250nm Ti/Au films are sputtered as top electrodes. Lastly, the diaphragm is released by inductively coupled plasma process. Vacuum thermal annealing is carried out at 300°C for 4 hours at the base pressure lower than 10^{-3}Pa to relieve device residual stress and improve film defects in the film sensors.

---

* Resrach supported by the National Natural Science Foundation of China (Grant No. 51775446).

W Gao, Y Zhang, B Ma, J Luo, J Deng and W Yuan are with the Key Laboratory of Micro and Nano systems for Aerospace, Ministry of Education, Northwestern Polytechnical University, Xi'an, 710072, China. (e-mail: mabh@ nwpu.edu.cn).

978-1-6654-3008-1/21 $31.00 © 2021 IEEE

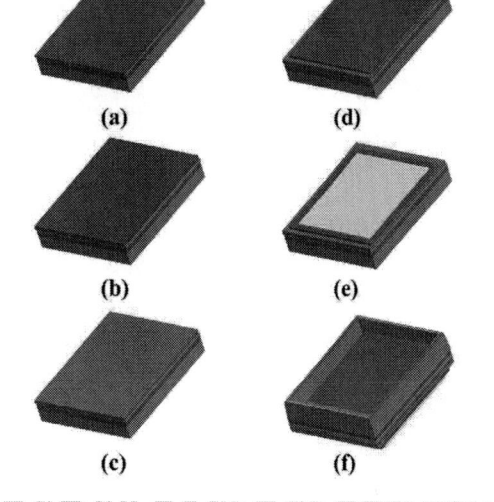

■ Si ■ Si₃N₄ ■ ZnVO ■ SiO₂ ■ Ti/Pt ▨ Ti/Au

Figure 2: Microfabrication process for V-dope ZnO film sensor. (a) Growing of SiO_2 (300nm) by thermal oxidation, (b) deposition of Si_3N_4 (500nm) by chemical vapor deposition, (c) sputtering of Ti/Pt (50nm/250nm) films as bottom electrodes, (d) magnetron sputtering of ZnVO film (1μm), (e) sputtering of Ti/Au (50nm/250nm) films as top electrodes, (f) releasing the diaphragm by inductively coupled plasma process.

TABLE 1 SPUTTERING PARAMETERS USED FOR ZNO FILM DEPOSITION

Ultimate pressure	8×10^{-4} mtorr
RF power	100W
Ar ratio	40 sccm
Working pressure	4.3 mtorr
Deposition rate	2 ± 0.2 nm/min

A sensor with foot-print of 2mm×2mm is bonding on the test printed circuit board (PCB) by the gold wires, as shown in Figure 3a. The geometry is clearly depicted in the scanning electron micrograph in Figure 3b, showing a typical sandwich structure.

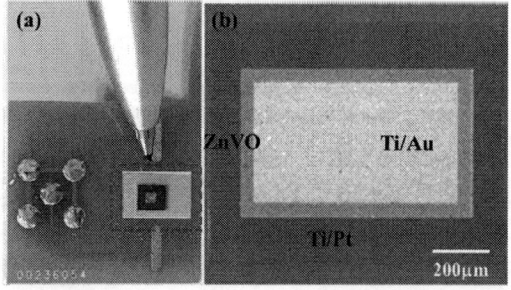

Figure 3: (a) A sensor with a foot-print of $2 \times 2mm^2$ bonding on a PCB by cold wire, (b) SEM micrograph of the sensing element.

III. EXPERIMENTAL RESULTS

Morphologies

Detailed morphologies of the surface and cross-sectional microstructures of doped and undoped ZnO films characterized by SEM are shown in Figure 4. Films show a columnar structure perpendicular to the surface of the substrate, confirming the oriented growth. However, the surface morphology of the V-doped ZnO film has a relative improvement and the unit cells are denser. The possible reason for the improvement in the structural properties is attributed to the doped v ions which enhances the crystal growth.

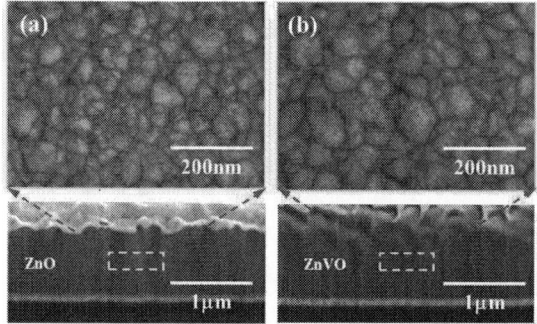

Figure 4: (a) (b) SEM images of the pure ZnO film and V-doped ZnO film, respectively.

Structural Evolution

In order to further investigate the influence of doping on the piezoelectric properties of the thin film, X-ray diffraction (XRD, SHIMADZU XRD-7000) is used for measure the crystal structure, as shown in Figure 5. The results show that all the peaks in the patterns of the doped and undoped films are attributed to the wurtzite phase of ZnO. The reflections at about 34.36°,47.50° , 62.80° and 67.86° for undoped ZnO film represent the (002), (102), (103) and (112) planes of hexagonal wurtzite structure, which are in good agreement with the JCPDS card (No.36-1451). The average grain size D for the samples could be determined from the XRD peaks by the Scherer's formula [9][10],

$$D = k\lambda/\beta\cos\theta, \qquad (1)$$

where k is the Scherer constant (k=0.89) and βis the full width at half maximum (FWHM), as shown in Table 2.

TABLE 2 AVERAGE GRAIN SIZE FOR THE PURE AND DOPED ZNO FILM

	Samples	
	Pure ZnO	Doped ZnO
2-theta (°) (002)	34.279	34.421
FWHM (°)	0.561	0.613
Average grain size (nm)	14.58	13.42

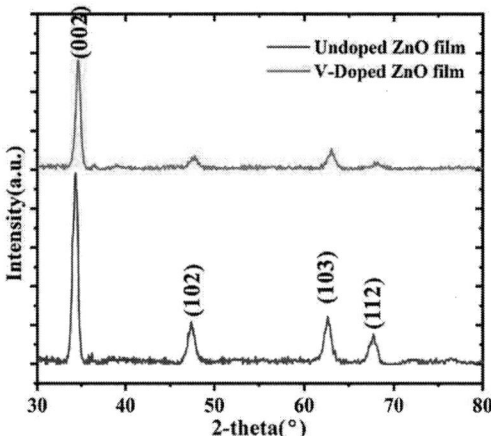

Figure 5: XRD patterns of undoped and V-doped ZnO film samples.

The average crystallite size for undoped ZnO films are estimated to be about 15nm, while that of the V-doped ZnO films is about 13nm. The crystal orientation of the ZnO film is improved after appropriate doping. The grain size is decreased after the V doping in ZnO lattice, which is mainly attributed to the substitution of V^{5+} ions in Zn^{2+} ions [11][12].

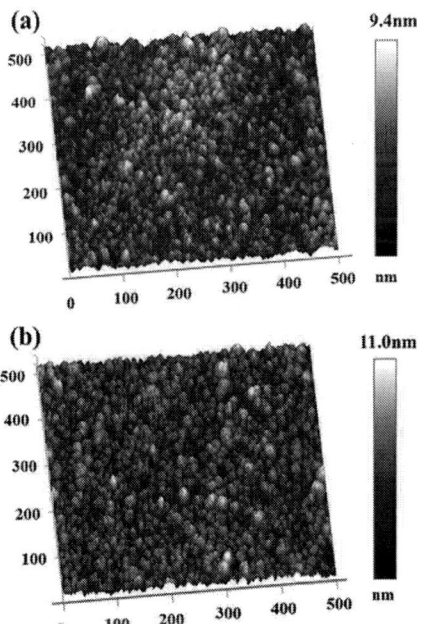

Figure 6: (a) AFM images of the Pure ZnO films and (b) V-doped ZnO films.

Piezoelectric Properties Analysis

In situ piezo-response signal and piezoelectricity are evaluated by piezoelectric force microscopy (PFM, Bruker DFDI, Germany). The film surface morphologies could also be investigated, as shown in Figure 6. Compared with undoped ZnO films, V-doped ZnO films possess more

dense grains and uniform dimensional crystal pillars. Moreover, the values of the mean roughness (Ra) and the root mean square roughness (Rq) of the films are measured, for the pure films Ra=1.10nm, Rq=1.51nm and the doped films Ra=0.973nm, Rq=1.22nm. The results show that the surface roughness of the doped films decreases, which is sufficient to deposit as piezoelectric thin film for device development.

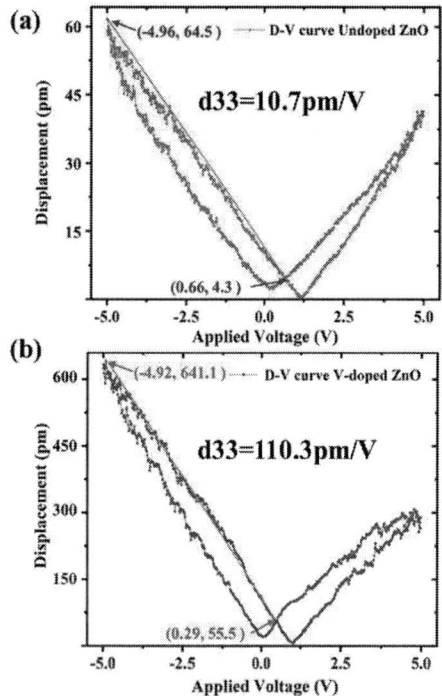

Figure 7: (a) (b) Displacement-voltage butterfly loops for the pure and doped ZnO films, respectively.

Furthermore, the piezo-response dependence of the applied voltage for the pure ZnO and V-doped ZnO films show typical electrical displacement-voltage butterfly loops in Figure 7. For the V-doped ZnO films, the d_{33} value of 110.3 pm/V is obtained, while the d_{33} value of pure ZnO film is only 10.57 pm/V.

Piezoelectric Response

Demonstrations in a plane-wave tube test system, as shown in Figure 8, are conducted under the sinusoidal excitation of 500 Hz at the applied voltage of 5V to validate the piezoelectric response characteristic of the presented sensor.

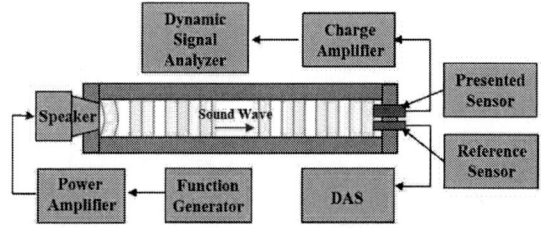

Figure 8: Schematic diagram of the plane-wave tube test system.

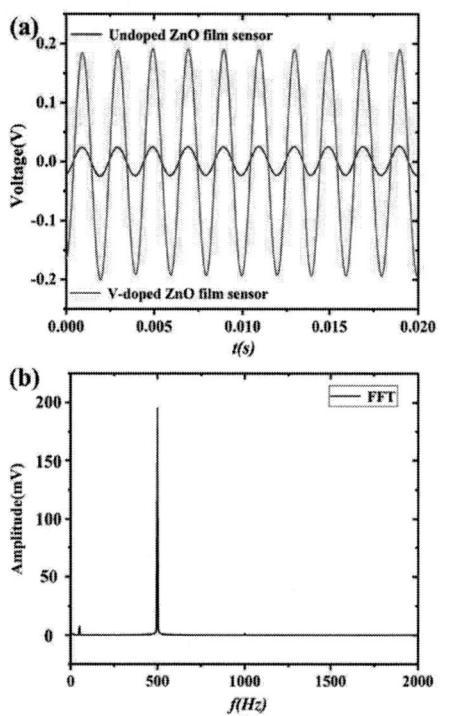

Figure 9: (a) Piezoelectric response output wave of the undoped and V-doped ZnO film sensor under same sinusoidal excitation of 500 Hz, (b) FFT of the doped ZnO film sensor.

Compared to the pure ZnO film sensor with same dimensions, the output voltage of the V-doped ZnO film sensor increases by an order of magnitude as shown in Figure 9a, confirming that the presented nanostructure V-doped ZnO film sensor has improved output sensitivity. Moreover, the results in Figure 9b also show good frequency response characteristics for the presented sensor, indicating a better performance in dynamic measurements.

IV. CONCLUSION

In summary, a nanostructure V-doped ZnO film sensor was designed, fabricated and characterized. The presented nanostructure V-doped ZnO films exhibit enhanced piezoelectric response with the piezoelectric constant of 110.3 pm/V. Moreover, a calibration in a plane-wave tube was conducted. An improved reliability and sensitivity has been achieved compared with the pure ZnO film sensor with same dimensions. On the basis of the results, the proposed nanostructure V-doped ZnO film sensor endowed improved piezoelectricity, shows a great application potential in the field of underwater acoustic detection.

ACKNOWLEDGMENT

This work is supported by the National Natural Science Foundation of China (Grant No. 51775446). The author would like to thank Analytical & Testing Center of Northwestern Polytechnical University.

REFERENCES

[1] M. Asadnia et al. "Flexible, zero powered, piezoelectric MEMS pressure sensor arrays for fish-like passive underwater sensing in marine vehicles." 2013 IEEE 26th International Conference on Micro Electro Mechanical Systems (MEMS). IEEE, 2013.

[2] X. Ni, Y. Zhao, and J. Yang. "Research of a novel fiber Bragg grating underwater acoustic sensor." Sensors and Actuators A: Physical 138.1 (2007): 76-80.

[3] S. Choi, H. Lee, and W. Moon. "A micro-machined piezoelectric hydrophone with hydrostatically balanced air backing." Sensors and Actuators A: Physical 158.1 (2010): 60-71.

[4] J. Song et al. "A bionic micro-electromechanical system piezo-resistive vector hydrophone that suppresses vibration noise." Journal of Micromechanics and Microengineering 29.11 (2019): 115007.

[5] J. Li, C. Wang, W. Ren, and J. Ma. "ZnO thin film piezoelectric micromachined microphone with symmetric composite vibrating diaphragm." Smart Materials and Structures 26.5 (2017): 055033.

[6] H. Li, Z. D. Deng, and T. J. Carlson. "Piezoelectric materials used in underwater acoustic transducers." Sensor Letters 10.3-4 (2012): 679-697.

[7] J. Xu, X. Zhang, S. N. Fernando, K. T. Chai, and Y. Gu, "AlN-on-SOI platform-based micro-machined hydrophone," Applied Physics Letters, vol. 109, p. 032902, 2016.

[8] MN, Suma, et al. "Study on the suitability of ZnO thin film for dynamic pressure sensing application." International Journal on Smart Sensing & Intelligent Systems 13.1 (2020).

[9] W. Gao, et al. "Efficient carbon nanotube/polyimide composites exhibiting tunable temperature coefficient of resistance for multi-role thermal films." Composites Science and Technology 199 (2020): 108333.

[10] B. Sun , et al. "Effects of annealing on the temperature coefficient of resistance of nickel film deposited on polyimide substrate." Vacuum 160 (2019): 18-24.

[11] J. T. Luo , et al. "Filtering performance improvement in V-doped ZnO/diamond surface acoustic wave filters." Applied surface science 256.10 (2010): 3081-3085.

[12] F. Pan, et al. "Giant piezoresponse and promising application of environmental friendly small-ion-doped ZnO." Science China Technological Sciences 55.2 (2012): 421-436.

Proceedings of the 16th Annual IEEE International
Conference on Nano/Micro Engineered and Molecular Systems
April 25-29, 2021

Humidity Sensor with High Resolution and Fast Response Based on AlN Cantilever with Two Groups of Segmented Electrodes

Dongsheng Li, Hanyong Dong, Zihao Xie, Mengjiao Qu, Qian Zhang and Jin Xie*, *Member, IEEE*

Abstract— Piezoelectric resonant cantilever is one of the most promising real-time humidity sensing platforms. This paper presents a humidity sensor based on an aluminum nitride (AlN) piezoelectric cantilever operated at a high-order mode and a sensing layer of MoS₂. The top electrode of the cantilever is divided into two groups of segmented electrodes in order to obtain a high intensity of the resonance peak of the cantilever resonator operated at a high-order mode. Compared with the humidity sensor based on a standard cantilever with the same dimension, the sensitivity of the proposed humidity sensor is increased from 5.99 to 778 Hz/%RH when the humidity is about 80%RH. The resolution is increased to 0.025%RH because of the improvement of the ratio of sensitivity to noise. The sensor shows a low hysteresis in a wide humidity sensing range from 10%RH to 90%RH. Moreover, the proposed humidity sensor has good short-term repeatability, fast response (0.6 s) and recovery (8 s) to humidity changes, indicating its great potential for fast-response detection.

I. INTRODUCTION

Piezoelectric resonant cantilever is one of the most promising real-time sensing platforms in physical, chemical and biological sensing due to its small size, low requirement of analysis, quick response, and ability of self-exciting and self-sensing [1], [2]. Humidity is usually monitored through detecting the shift of resonant frequency of piezoelectric resonant sensors [3]. The sensitivity of piezoelectric resonant sensor can be improved by using resonators with high resonant frequency [4]. High-order modes are used to increase the resonant frequency of the piezoelectric cantilever [5]. However, modes of too high order are often not available for sensing due to the low intensities of resonance peaks of the cantilever resonator [6].

Typically, a sensing material which can absorb water molecules is deposited on the piezoelectric cantilever and properties of this sensing layer greatly affect the responses of the humidity sensor [7]. Two-dimensional (2D) materials are ones of the most promising materials for humidity sensing applications, because of their large surface areas, excellent mechanical properties, and capabilities of precise detection at room temperature [8], [9]. Recently, humidity sensors based on molybdenum disulfide (MoS₂) have shown good performance [10]. In the graphite-type layered structure of MoS₂, an atomic layer of molybdenum is sandwiched between two atomic layers of sulfur. MoS₂-

based humidity sensors were reported to have high sensitivity, fast response and recovery [11].

Aluminum nitride (AlN) film has received extensive attention owing to its good thermal and chemical stabilities and CMOS process compatibility [12]. In this study, we fabricated an AlN/silicon cantilever resonator. We divided the top electrode of the cantilever into two groups of segmented electrodes to obtain high intensity of the resonance peak of the cantilever resonator operated at a high-order mode. Then we deposited a thin film of MoS₂ on the surface of an AlN cantilever to form a humidity sensor, which was then exposed to a range of humidity conditions. The sensitivity, resolution, hysteresis, short term repeatability, response and recovery time of the proposed humidity sensor were investigated.

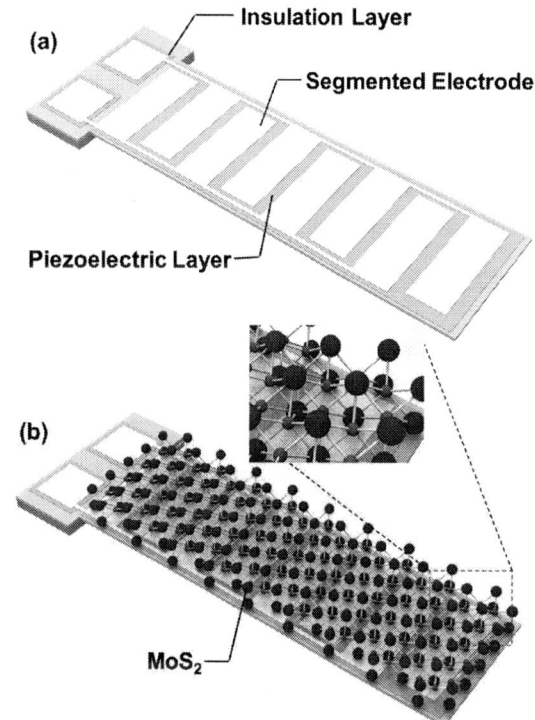

Figure 1. The schematic of proposed piezoelectric cantilever before (a) and after (b) coating with MoS₂.

Resrach supported by the "National Natural Science Foundation of China (51875521)", the "Zhejiang Provincial Natural Science Foundation of China (LZ19E050002)" and the "Science Fund for Creative Research Groups of National Natural Science Foundation of China (51821093)".

Dongsheng Li, Hanyong Dong, Zihao Xie, Mengjiao Qu, Qian Zhang and Jin Xie are with the State Key Laboratory of Fluid Power and Mechatronic Systems, Zhejiang University, Hangzhou, Zhejiang 310027, China.
*Corresponding Author: Jin Xie (e-mail: xiejin@zju.edu.cn).

978-1-6654-3008-1/21 $31.00 © 2021 IEEE   1419

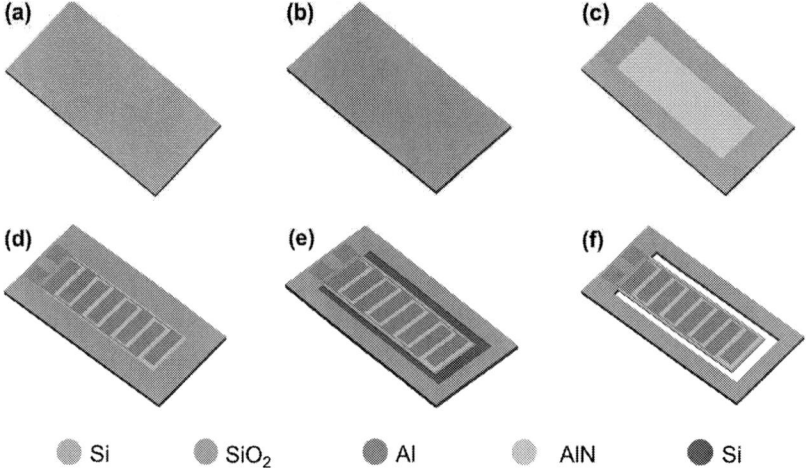

Figure 2. The fabrication process of the AlN cantilever.

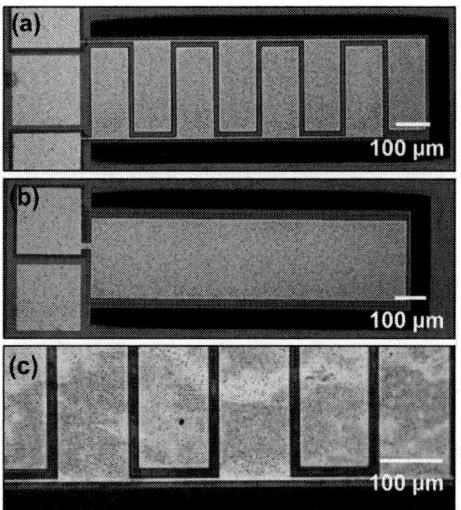

Figure 3. Photograph of the AlN cantilever with two groups of segmented electrodes before (a) and after (c) coating with MoS₂. (b) Photograph of the standard AlN cantilever.

II. DESIGN AND FABRICATION

The proposed humidity sensor is composed of a piezoelectric cantilever based on AlN and a sensing layer of MoS₂, as shown in Fig. 1. The area of the cantilever is 300x1000 μm². The top electrode is divided into two groups of segmented electrodes to improve the intensity of the resonance peak of the cantilever resonator operated at a high-order mode, in which one group is used as the activating electrode and the other group is used as the receiver for signals.

Fig. 2 shows the detailed fabrication process of the AlN-based cantilever. The piezoelectric cantilever is made up of layers of 0.5 μm-thick AlN and 10 μm-thick Si. A multilayer metal including 20 nm chrome and 1 μm aluminum is deposited and patterned on the surface of AlN layer by evaporation and lift-off process to form electrodes. The cantilever was released from backside by etching silicon substrate through a standard deep reactive ion etching

process. The fabricated piezoelectric cantilever is shown in Figs. 3a and 3b. Then the dispersed (0.4 μL) MoS₂ in deionized water with a concentration of 0.1 mg/ml was dropped on the cantilever and dried in a vacuum chamber, thus forming a sensing layer, as shown in Fig. 3c.

III. RESULTS AND DISCUSSION

Fig. 4a shows the experimentally obtained T/R transmission curve from 10 kHz to 500 kHz of the normal piezoelectric cantilever shown in Fig. 3c. There is no obvious resonance peak when the frequency exceeds 500 kHz due to the low intensity of the resonance peak of the cantilever. Fig. 4c shows the T/R transmission curve of the proposed piezoelectric cantilever within the frequency range from 2.78 MHz to 2.83 MHz. There are remarkable resonance peaks in the transmission curve of the proposed cantilever. The insets in Figs. 4a and 4c present the corresponding vibration mode from finite element analysis. Figs. 4b and 4d show typical transmission curve of the normal cantilever and the proposed cantilever after coating with MoS₂ film. The resonant frequency is slightly reduced after coating.

Fig. 5a shows resonant frequency shifts of the conventional cantilever-based humidity sensor operated at different modes when the humidity level is changed from 10%RH to 90%RH with an interval of 10%RH. The sensitivity of the sensor is quite high when the humidity level is relatively high. The sensitivities of the sensor based on the first three modes of the standard cantilever are 5.99, 27.4, 63.7 Hz/%RH, respectively, when the humidity is about 80%RH. Frequency shifts of the newly proposed humidity sensor with the changes of humidity are shown in Fig. 5b. The sensitivity is 778 Hz/RH% when the humidity is about 80%RH, which is much higher than that of the standard sensor. Mass sensitivity of the humidity sensor based on piezoelectric cantilever can be calculated by [13], [14]:

$$\Delta f / \Delta M = (-v_n^2 / 8\sqrt{3}\pi L^3 \omega\rho)\sqrt{E/\rho} \qquad (1)$$

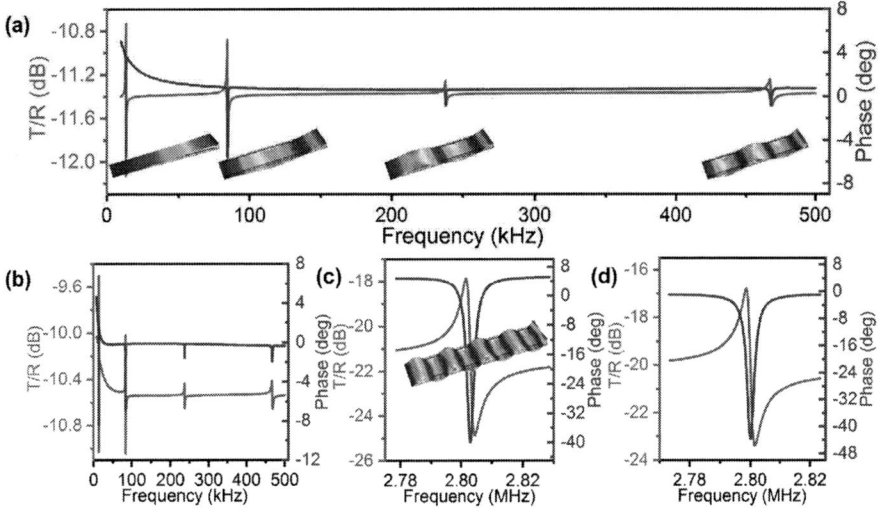

Figure 4. (a) T/R transmission curve of the normal piezoelectric cantilever. Transmission curve of the proposed AlN cantilever before (b) and after (c) coating.

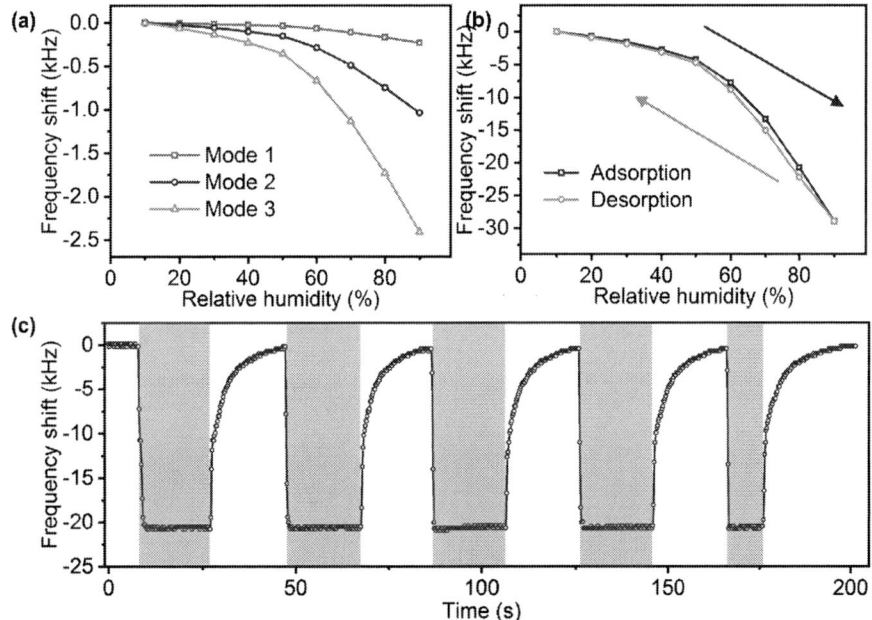

Figure 5. Frequency responses of the normal cantilever-based humidity sensor working at different modes (a) and the newly proposed humidity sensor (b) when relative humidity changes from 10%RH to 90%RH. (c) Dynamic response of the proposed humidity sensor.

where f, M, v_n, L, w, ρ, and E are resonant frequency, mass, dimensionless eigenvalue, length, width, density and Young's modulus of the humidity sensor, respectively. The sensitivity of the newly proposed humidity sensor is improved with the increase of the frequency of the cantilever resonator owing to the change of v_n at different resonant modes.

The resolution of the humidity sensor, i.e. the minimum humidity change that can be distinguished, is calculated based on the signal-to-noise ratio (S/N > 3) [15]. After calculation based on the frequency fluctuation, the resolutions of our newly proposed sensor are up to 0.3%RH and 0.025%RH at 20%RH and 80%RH, respectively. The

improvement of resolution is mainly attributed to the suppression of noise level because of the improved quality factor. The proposed sensor also shows a very low hysteresis in a wide humidity range, as shown in Fig. 5b.

The frequency shifts of the proposed humidity sensor were further recorded with a time interval of 0.2 s when the humidity was changes between 10%RH and 80%RH for five cycles. An excellent repeatability is observed over these five cycles of tests, as shown in Fig. 5c. The proposed sensor has both fast response (0.6 s) and recovery (8 s) to humidity change, which demonstrates its great potential for high-speed humidity detections.

978-1-6654-3008-1/21 $31.00 © 2021 IEEE

IV. CONCLUSION

We propose a humidity sensor based on an AlN piezoelectric cantilever operated at a high-order mode and a sensing layer of MoS$_2$. The top electrode of the cantilever is divided into two groups of segmented electrodes in order to obtain a high intensity of the resonance peak of the cantilever resonator operated at a high-order mode. The sensitivity of the proposed humidity sensor is up to 778 Hz/%RH, and the resolution is up to 0.025%RH because of the improvement of the ratio of sensitivity to noise, when the humidity is ~80%RH. Moreover, the proposed sensor has low hysteresis, good short-term repeatability, fast response (0.6 s) and recovery (8 s) to humidity change.

REFERENCES

[1] J. Zhao, Y. Zhang, R. Gao, and S. Liu, "A new sensitivity improving approach for mass sensors through integrated optimization of both cantilever surface profile and cross-section," *Sens. Actuators, B,* vol. 206, pp. 343-350, Jan. 2015.

[2] W. Pang, H. Zhao, E. S. Kim, H. Zhang, H. Yu, and X. Hu, "Piezoelectric microelectromechanical resonant sensors for chemical and biological detection," *Lab on a Chip,* vol. 12, no. 1, pp. 29-44, 2012.

[3] Y. Guan, X. Le, M. Hu, W. Liu, and J. Xie, "A noninvasive method for monitoring respiratory rate of rats based on a microcantilever resonant humidity sensor," *J. Micromech. Microeng.,* vol. 29, no. 12, p. 125001, Oct. 2019.

[4] F. Lochon, I. Dufour, and D. Rebière, "An alternative solution to improve sensitivity of resonant microcantilever chemical sensors: comparison between using high-order modes and reducing dimensions," *Sens. Actuators, B,* vol. 108, no. 1, pp. 979-985, Jul. 2005.

[5] R. Gao, Y. Huang, X. Wen, J. Zhao, and S. Liu, "Method to Further Improve Sensitivity for High-Order Vibration Mode Mass Sensors with Stepped Cantilevers," *IEEE Sens. J.,* vol. 17, no. 14, pp. 4405-4411, 2017.

[6] X. Le, L. Peng, J. Pang, Z. Xu, C. Gao, and J. Xie, "Humidity sensors based on AlN microcantilevers excited at high-order resonant modes and sensing layers of uniform graphene oxide," *Sens. Actuators, B,* vol. 283, pp. 198-206, Mar. 2019.

[7] Y. Pang, J. Jian, T. Tu, Z. Yang, J. Ling, Y. Li, X. Wang, Y. Qiao, H. Tian, Y. Yang, and T.-L. Ren, "Wearable humidity sensor based on porous graphene network for respiration monitoring," *Biosens. Bioelectron.,* vol. 116, pp. 123-129, Sep. 2018.

[8] J. Cai, C. Lv, E. Aoyagi, S. Ogawa, and A. Watanabe, "Laser Direct Writing of a High-Performance All-Graphene Humidity Sensor Working in a Novel Sensing Mode for Portable Electronics," *ACS Appl. Mater. Interfaces,* vol. 10, no. 28, pp. 23987-23996, Jul. 2018.

[9] B. Du, D. Yang, X. She, Y. Yuan, D. Mao, Y. Jiang, and F. Lu, "MoS2-based all-fiber humidity sensor for monitoring human breath with fast response and recovery," *Sens. Actuators, B,* vol. 251, pp. 180-184, Nov. 2017.

[10] J. Zhao, N. Li, H. Yu, Z. Wei, M. Liao, P. Chen, S. Wang, D. Shi, Q. Sun, and G. Zhang, "Highly Sensitive MoS2 Humidity Sensors Array for Noncontact Sensation," *Adv. Mater.,* vol. 29, no. 34, p. 1702076, Sep. 2017.

[11] S. Yadav, P. Chaudhary, K. N. Uttam, A. Varma, M. Vashistha, and B. C. Yadav, "Facile synthesis of molybdenum disulfide (MoS2) quantum dots and its application in humidity sensing," *Nanotechnol.,* vol. 30, no. 29, p. 295501, Apr. 2019.

[12] D. Li, X. Le, J. Pang, L. Peng, Z. Xu, C. Gao, and J. Xie, "A SAW hydrogen sensor based on decoration of graphene oxide by palladium nanoparticles on AlN/Si layered structure," *J. Micromech. Microeng.,* vol. 29, no. 4, p. 045007, Feb. 2019.

[13] K. Eom, H. S. Park, D. S. Yoon, and T. Kwon, "Nanomechanical resonators and their applications in biological/chemical detection: Nanomechanics principles," *Phys. Rep.,* vol. 503, no. 4, pp. 115-163, Jun. 2011.

[14] A. Boisen, S. Dohn, S. S. Keller, S. Schmid, and M. Tenje, "Cantilever-like micromechanical sensors," *Rep. Prog. Phys.,* vol. 74, no. 3, p. 036101, Feb. 2011.

[15] W. Xu, W.-B. Huang, X.-G. Huang, and C.-y. Yu, "A simple fiber-optic humidity sensor based on extrinsic Fabry–Perot cavity constructed by cellulose acetate butyrate film," *Optl. Fiber Technol.,* vol. 19, no. 6, pp. 583-586, Dec. 2013.

Proceedings of the 16th Annual IEEE International
Conference on Nano/Micro Engineered and Molecular Systems
April 25-29, 2021

DESIGN AND FABRICATION OF INTEGRATED PIEZOELECTRIC MICROPUMP WITH VORTEX ENHANCEMENT OPTIMIZATION

Boshen Liang[12], Louis Paquet[1], Yongbin Jeong[12], Paul Heremans[12], and David Cheyns[1]
[1]IMEC, Leuven, BELGIUM
[2]KU Leuven, Leuven, BELGIUM

ABSTRACT

In this study, a 3D acoustic-solid interaction model is combined with a time-dependent vorticity enhancement method for design optimization of nozzle/diffuser micropumps. Compared with the conventional method where static pressure loss ratio is used as the performance indicator of individual nozzle/diffuser element, the vortex method provides a direct and good prediction of output characteristics of the micropumps.

An integrated piezoelectric micropump prototype is fabricated with flat-panel display compatible processing techniques. The frequency characterization shows good agreement of the micropump working frequency with that predicted by finite element simulation where additional damping effects are considered. A high-speed particle image velocimetry shows successfully the movement of fluid elements driven by the piezoelectric actuator.

KEYWORDS

Micropump, Nozzle/Diffuser, Lab-on-Chip, Vorticity, Optimization

INTRODUCTION

Micropumps are one of the indispensable components for integrated Lab-on-Chip (LOC) devices [1]. Among different concepts, nozzle/diffuser micropumps attract scientific attention since its flow regulation function is achieved with non-moving geometrical structures, which decrease dramatically the risk of issues such as channel clogging, fluid leakage and thus device failure [2].

Similar to diodes for electronics, the effectiveness of a diffuser for flow directing can be defined as Diodicity [3]. However, following the conventional method, the nozzle/diffuser elements with larger diodicity calculated based on static fluidic simulation fail to achieve higher flowrate for micropumps. This may be caused by the pulsatile nature of the flow when micropumps are driven by periodic forces, which existing models do not take into account.

In this study, a 3D acoustic-solid interaction model is used to set the working frequency range, and then the proposed time-dependent vortex-based enhancement is used for design optimization. A prototype device is fabricated following the optimization result, and frequency and flowrate characterization are conducted.

MODELLING

A schematic of the micropump is shown in Fig.1. In our design, a 30μm deep microfluidic channel is covered with 15μm polyimide (PI) membrane, which is driven by a micromachined 500nm thick piezoelectric (polyvinylidene difluoride, PVDF) transducer unit [4] to provide periodic force into the fluid system. The working fluid used for this work is water, which is featured with much higher inertial effect compared to the case when air is used as working fluid. Two nozzle/diffuser valve units are added to the left and right side of the central chamber for modulation of the net flow direction.

In this study, the central circular chamber radius is constant (450μm), while several geometrical diffuser parameters shown in Fig.1(b) are optimized to get the best micropump performance, i.e. the highest flowrate. In the model, D_{in} and D_{out} are the inlet and outlet width, α is the opening angle, L is the length, H_{ch} is the connecting fluid channel height, and R is the corner fillet radius.

Frequency response

One of the major features of micropumps driven by periodic forces is that the flow inside the fluid system is pulsatile. Analytically, for a microfluidic channel with circular cross-sections, providing there is no flow rotation and the flow is fully developed, the Navier-Stoke equation in polar coordinate system can be simplified to

(a)

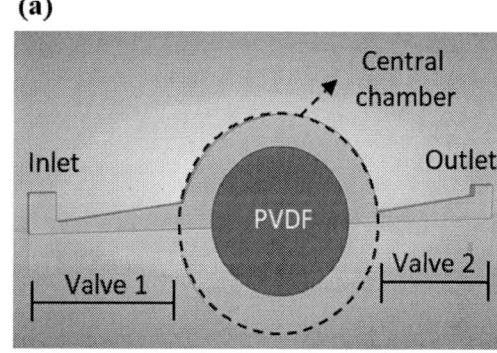

(b)

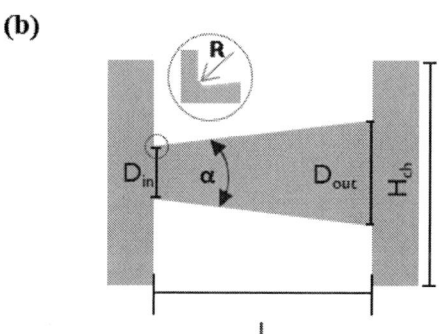

Figure 1: Schematics of diffuser/nozzle micropump driven by a piezoelectric actuator (a) Configuration of the micropump with a central chamber connected by two diffuser/nozzle elements as flow rectifying diodes. (b) Design parameters of a diffuser/nozzle valve unit.

978-1-6654-3008-1/21 $31.00 © 2021 IEEE

$$\frac{\partial P(x,t)}{\partial x} = \mu\left(\frac{\partial^2 u(r,t)}{\partial r^2} + \frac{1}{r}\frac{\partial u(r,t)}{\partial r}\right) - \rho\frac{\partial u(r,t)}{\partial t} \qquad (1)$$

where P is the periodic input force expressed as pressure term, μ and ρ are the viscosity and density of water, x and r are the longitudinal and polar axis, u is the velocity of fluid, and t is the time.

Following a similar approach introduced in previous research [5], by assuming the pressure term to the left side of eqn (1) is periodic with the form $P_s e^{i\omega t}$, the fluid velocity $u(r,t)$ in the channel can be expressed to have the form

$$u_\emptyset(r,t) = \frac{iP_s R^2}{\mu W^2} \times \left(1 - \frac{J_0(i^{3/2}W\frac{r}{R})}{J_0(i^{3/2}W)}\right) \times e^{i\omega t} \qquad (2)$$

where P_s is the pressure amplitude, R is the radius of the channel, J_0 is the order 0 Bessel function of first kind, and W is the Womersley number with the form $W = R\sqrt{\rho\omega/\mu}$. Here ω is the angular frequency of the external periodic driving force. The three terms in eqn (2) represents the amplitude factor, shape factor and periodic term of the pulsatile flow velocity profiles, respectively.

The solution of eqn (2) is quantitatively shown in Fig.2. It shows that with the increase of driving frequency and thus the Womersley number α, the fully developed flow profile changes from a parabolic shape to a flatter shape, while the velocity amplitude of the fluid decreases dramatically. The nozzle/diffuser elements will eventually lose its fluid directing ability at extremely high frequency range since the inertial effect of the working fluid stop itself from moving at any meaningful velocity.

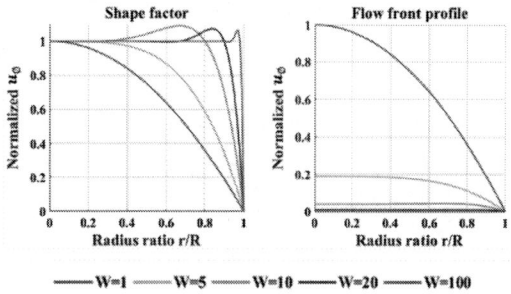

Figure 2: Solution of simplified Navier-Stroke equation. The shape factor and flow front profile are graphically plotted for different Womersley numbers.

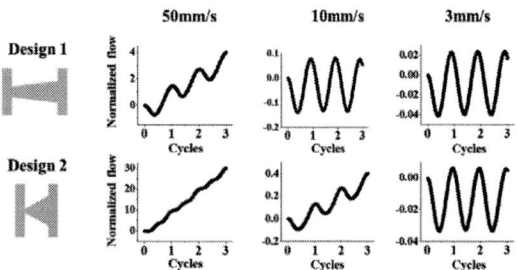

Figure 3: Comparison of full micropump flowrate results for two distinct nozzle/diffuser designs with same D_i values, which fails to predict the apparent higher performance of Design 2 compared to Design 1.

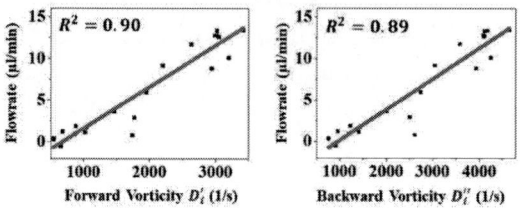

Figure 4: Relationship of micropump flowrate with forward/reverse vorticity of diffuser/nozzle element. A good linear regression indicates the vorticity can be used as a good performance indicator of micropump output performance.

A 3D acoustic-solid interaction model is built up in COMSOL to calculate the resonant frequency of the vibrating membrane. Compared with the analytical Rayleigh-Ritz method [11-12] which is only capable of calculating the resonance frequency of a circular membrane in contact with fluid half space, the COMSOL model takes into account extra damping effects imposed by shallow microfluidic channel and diffuser/nozzle elements. Based on the fixed chamber radius and chamber depth used in this study, the simulated 1st mode resonant frequency of the PI membrane with water as working fluid is in the range of 3 kHz – 5 kHz. Eqn (2) reveals that this frequency range wouldn't degrade the nozzle/diffuser performance towards non-working zones while it can ensure the maximum vibrating amplitude.

Vortex-based optimization method

A better diffuser design is figured with higher ability of allowing forward flow while inhibiting reverse flow. Conventionally, the ratio of the pressure drops in backward direction to that in forward direction is named as Diodicity:

$$D_i = \frac{\Delta P_{backward}}{\Delta P_{forward}} \qquad (3)$$

However, this definition fails to take into account the transient of the pulsatile flow inside micropumps driven by periodic forces. As an example shown in Fig.3, two distinct diffuser designs with same D_i as 1.35 are plugged into full 3D micropump model to calculate flowrates at three different membrane vibrating peak velocities. Apparently, the pressure drop ratio from static simulations cannot predict well the micropump performance difference.

To overcome this discrepancy between static diffuser performance indicator and micropump performance, two time-dependent parameters are proposed as

$$D_i' = \frac{\int_0^{T/2} V_{forward}\,dt}{T/2} \qquad (4)$$

$$D_i'' = \frac{\int_{T/2}^{T} V_{backward}\,dt}{T/2} \qquad (5)$$

where V is the vorticity with the unit $1/second$, and T is the period of a full pumping cycle obtained from previous frequency response analysis. Despite the same D_i values for the two diffuser designs in Fig.3, the D_i' and D_i'' values are different and correspond well to the different micropump performance. Furthermore, 22 extra diffuser designs with a combination of different diffuser/nozzle parameters are simulated. As shown in Fig.4, a good linear

978-1-6654-3008-1/21 $31.00 © 2021 IEEE

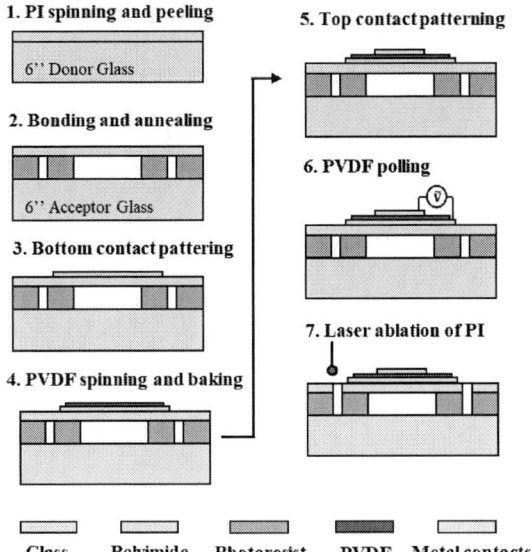

Figure 5: A flat-panel display compatible fabrication process of the micropump.

relationship can be established, which indicates that the two parameters from time-dependent simulation can be used as a good performance indicator for the micropump.

FABRICATION

The fabrication process is schematically shown in Fig.5. Firstly, a polyimide resin was spin-coated on 6-inch a donor glass wafer, and then fully cured in N_2 atmosphere at 300 °C for 3 hours. The 15 μm PI membrane was then peeled off from the donor glass substrate and applied to the acceptor wafer where a 30 μm photoresist was spin-coated and patterned on top by standard lithography step to form the microfluidic channel. A bonding and annealing in vacuum were followed to achieve permanent bonding between PI and microfluidic channel. Then, 60 nm aluminum was deposited and patterned on top of PI as bottom contact. P(VDF-TrFE) powder was dissolved in Methyl Ethyl Ketone and the solution was spin-coated on top of the wafer to form a 0.5 μm piezoelectric layer. The PVDF layer was cured in N_2 atmosphere at 140 °C for 1 hour. Another 60 nm aluminum layer was deposited and patterned as top metal contact. In this study, the metal coverage ratio is set to 67% based on a previous study [8] to have the optimal driving efficiency.

The PVDF layer was polarized by applying 80 V peak-to-peak triangular waves at 100 Hz for 200 pulses. Finally, the inlet and outlet of the microfluidic chip were opened by laser ablation of the PI.

EXPERIMENTS & RESULTS

A prototype device is successfully manufactured based on the above optimization and fabrication methods. Each micropump unit occupies less than 1.5 mm x 1 mm area on glass wafer. A function generator and a voltage amplifier were used to apply square waveform signal to drive the PVDF membrane. The applied peak-to-peak voltage (V_{pp}) amplitude was 1 V for vibration

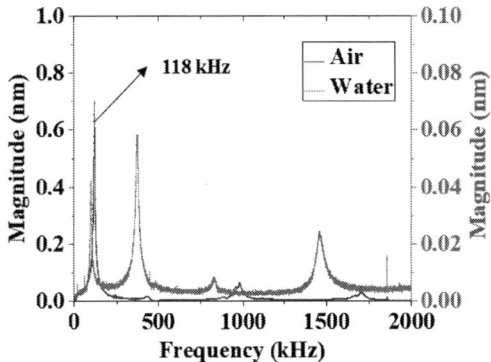

Figure 6: Frequency response of the fabricated micropump with or without fluid channel filled with water. This result is measured by LDV high frequency unit. Magnitude is measured at $V_{pp}=1V$.

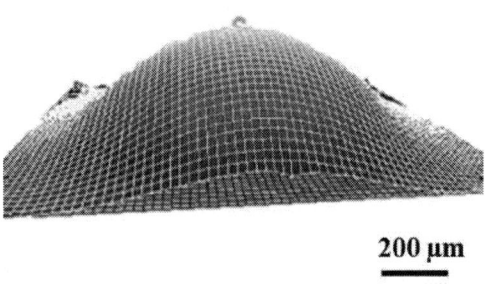

Figure 7: 3D reconstructed shape of the vibrating membrane when the fluid channel is filled with water. This result is measured by LDV low frequency unit at unit voltage.

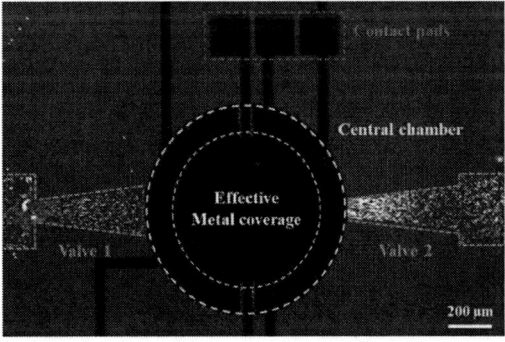

Figure 8: MicroPIV test image. The fluorescent microspheres aggregate to the right side of the micropump, which illustrates a net flow from left to right side.

characterization and 80 V for flowrate characterization.

Laser Doppler vibrometer (LDV, Polytec MSA500) was used for membrane vibration characterization. Firstly, a high frequency unit was used to measure the vibration of the PI membrane with and without the channel filled with water. As seen in Fig.6, as a consequence of additional damping imposed by introducing water into the microfluidic channel, both the resonant frequency peaks

978-1-6654-3008-1/21 $31.00 © 2021 IEEE

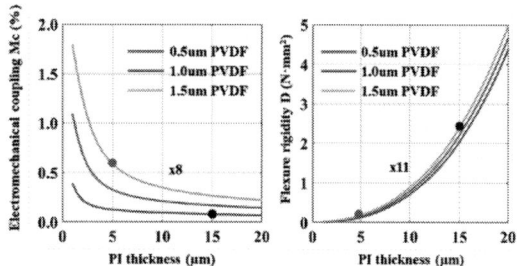

Figure 9: Impact of different PVDF and PI thickness on the micropump vibrating amplitude. Black dots are the current thickness combination, and red dots are proposed new combination for higher flowrate.

and amplitudes changed after the membrane is contacted with water. For air, the resonance peak at 118 kHz corresponds to the 1st mode vibration. In order to capture the 1st mode in water, the LDV was switched to low frequency measurements. In Fig. 7, the 1st mode vibration of the membrane at 4.4 kHz is reconstructed in 3D form. The measured frequency matches well with the frequency range predicted by 3D acoustic-solid interaction model.

Micron resolution particle image velocimetry (Micro-PIV) was used for micropump flowrate characterization. Carboxylate-modified microspheres (FluoSpheres™, 2.0 µm diameter, 515nm fluorescence peak) was diluted in water and then injected in the fluid channel, and an inverted microscope setup with a high-speed camera was used to capture the movement of the microspheres. As shown in Fig.8, after turning on the driving voltage, the florescent microspheres start aggregating towards the right side of the micropump, which illustrates a net flow from left to right side. Driven by periodic square voltage input at 80 V_{pp}, the PI membrane vibrated at the velocity of 3 mm/s, and the overall net flowrate was estimated to be around 20 nl/min from particle tracking result.

The achievable flowrate with current technology can be easily increased to µl/min range. From the 1st mode frequency drop from 118 kHz in air to 4.4 kHz in water, it can be deduced that the vibrating membrane is in near over-damping status. Based on the simplified 1D spring-mass theory, the vibrating amplitude of the piezo-driven membrane w can be modeled as

$$w = \frac{F_0}{k} \sim \frac{E \times D_{31} \times M_c}{D} \quad (6)$$

where F_0 is external force amplitude, k the equivalent spring constant of the multi-layer membrane system, E is the electric field applied to PVDF, D_{31} is the PVDF piezoelectricity constant, M_c is the electromechanical coupling efficiency [9], and D is the flexural rigidity of the multi-layer membrane system. For a larger membrane vibrating amplitude and hence a higher micropump flowrate, apart from increasing electric field E or replacing PVDF with other piezoelectric materials for higher D_{31}, changing the thickness of PVDF and PI can be two other effective choices. As illustrated in Fig. 9, a combination of thinner PI and thicker PVDF layer could boost the vibrating amplitude and thus the flowrate by 10-100 times via enhancing the electromechanical coupling efficiency and decreasing the equivalent system spring constant.

CONCLUSION

We have developed an efficient and effective method for nozzle/diffuser micropump optimization by combining a 3D acoustic-solid interaction model with a time-dependent vorticity enhancement method. It overcomes the drawback of the conventional static optimization method where the transient feature of pulsatile flow for micropumps driven by periodic force is omitted. The fabrication method of our prototype device is featured with superior compatibility to the flat-panel display manufacturing process, and it can be easily extended to cover a huge flowrate range from nl/min range to µl/min range. This guarantees the proposed device is not restricted in labs, but has the potential for mass production in fabs in the mature phase.

ACKNOWLEDGEMENTS

We greatly appreciate the contributions and meaningful discussion from several members of the imec Large Area Electronics team.

REFERENCES

[1] P. Woias, "Micropumps—past, progress and future prospects," *Sensors Actuators B Chem.*, vol. 105, no. 1, pp. 28–38, Mar. 2005.

[2] A. Olsson, P. Enoksson, G. Stemme, and E. Stemme, "Micromachined flat-walled valveless diffuser pumps," *J. Microelectromechanical Syst.*, vol. 6, no. 2, pp. 161–166, Jun. 1997.

[3] A. R. Gamboa, C. J. Morris, and F. K. Forster, "Improvements in fixed-valve micropump performance through shape optimization of valves," *J. Fluids Eng. Trans. ASME*, vol. 127, no. 2, pp. 339–346, Mar. 2005.

[4] Y. Jeong, C. Huang, D. Cheyns, and G. B. Torri, "pMUT device compatible with large-area display technology," *16th Int. Symp. Electrets*, p. 2004, 2016.

[5] M. Zamir, "Basic Elements of Pulsatile Flow," Springer, Cham, 2016, pp. 81–122.

[6] M. Kwak, "Vibration of circular plates in contact with water," 1991.

[7] M. Kwak, K. K.-J. of S. and Vibration, and undefined 1991, "Axisymmetric vibration of circular plates in contact with fluid," *Elsevier*.

[8] B. Shieh, K. G. Sabra, and F. Levent Degertekin, "A Hybrid Boundary Element Model for Simulation and Optimization of Large Piezoelectric Micromachined Ultrasonic Transducer Arrays," *IEEE Trans. Ultrason. Ferroelectr. Freq. Control*, vol. 65, no. 1, pp. 50–59, Jan. 2018.

[9] A. Ben Amar, H. Cao, and A. B. Kouki, "Modeling and process design optimization of a piezoelectric micromachined ultrasonic transducers (PMUT) using lumped elements parameters," *Microsyst. Technol.*, vol. 23, no. 10, pp. 4659–4669, Oct. 2017.

CONTACT

*Boshen Liang, phone: +32-016-283-402;
Email: boshen.liang@imec.be

Gap in pagination due to unavailable paper.

Pages 1427-1429

April 25-29 , 2021   Xiamen, China

Structure-Controlled Synthesis of Single-Walled Carbon Nanotubes

Yan Li

College of Chemistry and Molecular Engineering and Academy for Advanced Interdisciplinary Studies, Peking University, Beijing National Laboratory for Molecular Sciences, Beijing, China

ABSTRACT

Single-walled carbon nanotubes (SWCNTs) can be either semiconducting and metallic. Semiconducting SWCNTs (s-SWCNTs) present direct bandgap and balanced high mobility toward electrons and holes, which is desired in electronic applications. However, the bandgap is determined by their structure which is described by their chiral index (n,m). Therefore, the structure-controlled preparation of SWCNTs is a crucial issue for carbon-based nanoelectronics and has been a great challenge since 1990s.

We developed a strategy to synthesize SWCNTs with a specific chirality by using intermetallic compound Co_7W_6 nanocrystals. The catalyst nanocrystals act as the structural templates for the formation of SWNTs with designed chiralities. SWCNTs of high chirality purity are obtained under optimized kinetic growth conditions. This strategy has shown success in the synthesis of SWCNTs of several different chiralities. Further in situ studies on the catalysts under reactive environment reveals the mechanism at atomic scale.

BIOGRAPHY

Yan Li joined the faculty of Chemistry, Peking University in 1995 and has been the full professor of chemistry at Peking University since 2002. She was a visiting associate professor in Department of Chemistry, Duke University from 1999 to 2001. From 2016 to 2019, she was appointed as Distinguished Visiting Professor at School of Engineering, University of Tokyo and was awarded as Fellow of the School in 2020. Her research is focused on carbon nanomaterials, especially the preparation, modification, characterization and application of carbon nanotubes. She has published over 150 papers in premier journals such as Nature, Science, J. Am. Chem. Soc., Adv. Mater., Acc. Chem. Res, Chem. Rev. etc. Currently, she is serving as the associated editor of ACS Nano. She is also on the editorial or advisory board of several journals including Chemical Society Reviews, Materials Horizons, Nano Research, and Carbon.

Dr. Li has received many awards and honors, such as National Science Fund for Distinguished Young Scholars by NSFC (2011), Chang Jiang Scholar by Chinese Ministry of Education (2013), Fellow of Royal Society of Chemistry (2014), First Class Science and Technology Award by China Association for Instrumental Analysis (2015), Outstanding Researcher by China Association for Science and Technology (2016), First Prize in Natural Science by Ministry of Education (2017), Chemical Innovation Award Distinguished Scientist by Chinese Chemical Society (2018). She was also awarded as Top Ten Best Teacher (for undergraduate education) in Peking University (2008), Famous Teacher of Beijing City (2013), and Top Ten Best Supervisor (for graduate education) in Peking University (2015).

978-1-6654-3008-1/21 $31.00 © 2021 IEEE

This page intentionally left blank.

A Novel Low-Temperature Post-Curing Transfer Method Of Graphene Wrinkling Surface For Strain Engineering

Kai-Ming Hu*, Xiu-Yuan Li, Yi-Hang Xin, Er-Qi Tu and Wen-Ming Zhang

Abstract—Strain engineering has attracted great interest for being regarded as a straightforward and universal approach to tune the intrinsic properties of two-dimensional (2D) materials and improve the performances of nanodevices from 2D materials. We proposed a novel low-temperature post-curing transfer method to fabricate mono-scale conformal wrinkles to apply strains on graphene membranes. Because the low-temperature post-curing process not only ensures sufficient interface adhesion strength, but also avoids diffusion-induced gradient interface, mono-scale conformal graphene wrinkle patterns are triggered. Mechanical bending-induced cracks are introduced to make wrinkles align in one-dimensional (1D) order. The mono-scale graphene wrinkles can be used for strain engineering of 2D materials and ultrathin semiconductor applications.

I. INTRODUCTION

Atomically thin 2D materials are perfectly suitable for strain engineering because they can bear large nonhomogeneous stresses before fracture [1-5]. Mechanical strains in 2D materials can be easily altered due to their atomic thickness and extremely low bending rigidity compared to bulk three-dimensional (3D) materials [6-8]. Strain engineering can shed light on modifying the band structure of 2D materials, further improving the mechanical, optical, catalytic, electrical and thermal performances of 2D-material-based devices [9-14].

Surface corrugations including ripples, wrinkles, folds, crumples and buckle-induced delamination can introduce a local topological disorder to modify the charge transport, the appearance of electron-hole puddles, band-gap opening and carrier scattering [15, 16]. However, the thermal-fluctuation-induced ripples dynamically changed with time as observed via scanning tunneling microscopy, which are disordered and the wavelength and amplitude of ripples are difficult to control [17]. Crumples containing fold and wrinkle surface instability behaviors have dense out-of-plane deformations in 2D and 3D, and are usually multidirectional [18, 19]. Therefore, the ripples and crumples are not suitable for the modulation of local strains in 2D materials. Surface wrinkles can be an effective way to regulate the localized strains in 2D nanomaterials induced by out-of-plane deformations [11, 20].

However, the wrinkle patterns of 2D materials are mostly random and disordered, which is difficult to accurately control the amplitude and direction of strains in 2D material-substrate systems. Many methods, such as pre-stretched substrate [1, 19], compressive stresses [21], prepattern template [22], thermal treatment [23-25], stretch-induced compression [26], have been proposed to trigger surface instability and corresponding surface morphology. However, the directions and scale diversity of the wrinkles triggered by the above methods is hard to regulate. Therefore, the effective fabrication method of graphene wrinkles with well controlled wavelength, amplitude and orientation is essential for strain engineering of 2D materials.

II. RESULTS AND DISCUSSION

A. A low-temperature post-curing transfer method

The graphene is transferred onto the soft substrate in liquid environment and the interfacial liquid is trapped during the wet transfer [27], which greatly reduces interfacial adhesion energy between graphene and substrate surface. To improve the interfacial adhesion energy of interfaces between surface film and substrates, a high-temperature post-curing transfer method of graphene is proposed [28]. However, the multiscale conformal graphene wrinkles are triggered due to a gradient interface layer caused by high temperature thermal diffusivity, which makes it difficult to control wrinkle features including wavelength, amplitude, orientation and location.

To ensure sufficient interfacial adhesion energy and avoid gradient interface simultaneously, a novel low-temperature post-curing transfer method is proposed to fabricate mono-scale conformal graphene wrinkles in graphene/PMMA-PDMS systems (Figure 1). First, single layer graphene films are grown on germanium (Ge) surface by Chemical Vapor Deposition (CVD) (Figure 1(a)), and PMMA solution with 4% concentration is spin-coated onto the graphene/Ge composite with 3000 rpm (Figure 1(b)). And then the mixed PDMS is slowly poured onto the surface of PMMA/graphene/Ge laminate (Figure 1(c)). Second, the mixed PDMS is cured at room temperature (25 °C) for 48 h (Figure 1(d)) to reduce PDMS polymer diffusion caused by Brownian motion. The Ge layer is etched with HF: H_2O_2 : H_2O solution for 4 h (Figure 1(e-f)). The samples are rinsed with deionized water and dry on the heating plate to fully flatten PMMA/graphene film. The mono-scale graphene wrinkles are triggered by the heating

All authors are with State Key Laboratory of Mechanical System and Vibration, School of Mechanical Engineering, Shanghai Jiao Tong University - Shanghai 200240, China (corresponding author to provide phone: -86-18321842038; e-mail: hukaiming@sjtu.edu.cn).

treatment (110 °C) of the graphene/PMMA-PDMS system (Figure 1(g-h)).

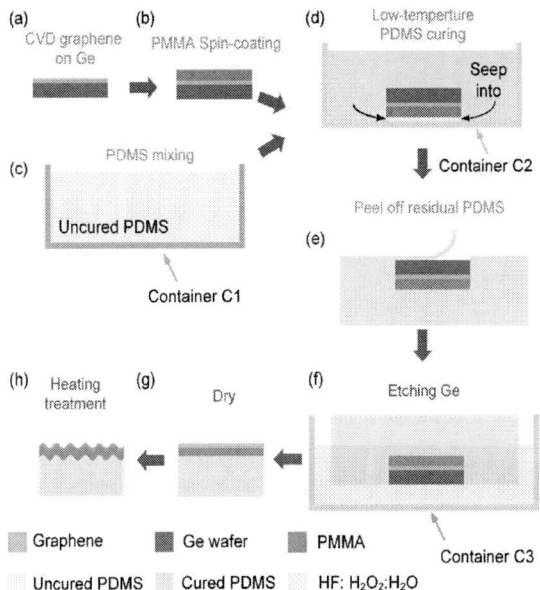

Figure 1: Fabrication flow of mono-scale conformal graphene wrinkles in graphene/PMMA-PDMS systems: (a) single layer graphene films are grown on Ge(110) surface by chemical vapor deposition (CVD); (b) thin PMMA polymer is spin-coated onto the graphene/Ge composite; (c) the silicone base and curing agents are mixed at a 10:1 mass ratio in Container C1 to obtain uncured PDMS; (d) the PMMA/graphene/Ge film is put on the bottom of Container C2 and the mixed PDMS is slowly poured into Container C2 and cured at 25 °C for 48 h; (e) the residual PDMS on back side on Ge wafer is peeled off; (f) the Ge layer is etched in Container C3 with HF:H2O2:H2O solution for 4 h; (g-h) the samples are dried and micro-wrinkles are triggered by the heating treatment of the graphene/PMMA-PDMS system.

B. Experimental Results

Raman D, G and 2D characteristic peaks of graphene indicate that the graphene/PMMA is well transferred onto PDMS substrates (Figure 2). According to 2D Raman mapping of the graphene/PMMA-PDMS, Raman ratio $I_{2D}/I_G \approx 2.5$ (Figure 2(b)), which demonstrated that the transferred graphene membranes are high quality and low defect.

Compare with the multiscale conformal wrinkles in Ref. [28], we observed the mono-scale conformal graphene wrinkle patterns in Figure 3(a) by optical microscope. The 3D topological morphologies of disordered wrinkle patterns are characterized by 3D laser scanning confocal microscope (LSCM) images (Figure 3(b)). The topological curves of the mono-scale wrinkles for two different positions marked in

Figure 3(b) indicated that the wavelength and amplitude of conformal graphene wrinkles are 15 μm and 700 nm, respectively.

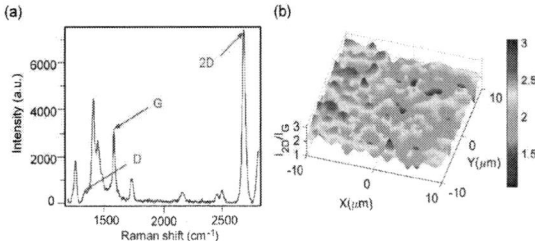

Figure 2: Raman characterization of graphene membranes onto the PMMA-PDMS substrates transferred by the low-temperature post-curing transfer method: (a) Raman spectra of graphene/PMMA-PDMS, where Raman D, G and 2D characteristic peaks for graphene and the other Raman peaks due to the PDMS; (b) 2D Raman mapping of the graphene/PMMA-PDMS, where $I_{2D}/I_G \approx 2.5$.

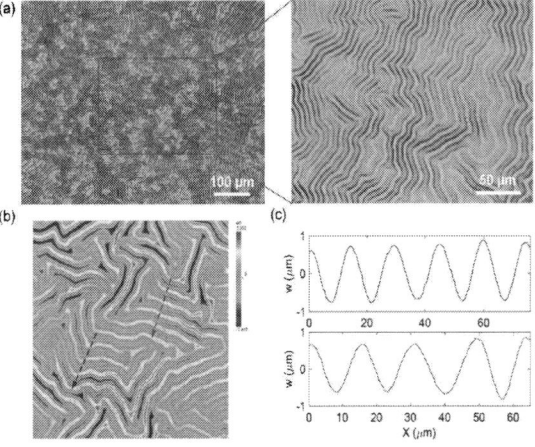

Figure 3: (a) Optical micro-images of mono-scale conformal graphene wrinkles in graphene/PMMA- PDMS systems prepared by the low-temperature post-curing transfer method; (b) 3D LSCM images of 2D disordered wrinkle patterns; (c) the topological curves of the mono-scale wrinkles for two different positions marked in Figure 3(b).

To make disordered graphene wrinkle patterns in Figure 3 into ordered patterns, mechanical bending-induced cracks are introduced to defined boundaries. As shown in Figure 4, large scale (centimeter scale) 1D ordered graphene wrinkle pattern is observed due to parallel cracks. The reason is that the stress can relax at the crack boundaries and the inhomogeneous stress distributions give rise to the onset of unidirectional wrinkle morphology (Figure 4(c)). The tunable wrinkled graphene functional surface can create opportunities for the study of new fundamental physics of 2D materials and functional surface nano-devices from strained 2D materials.

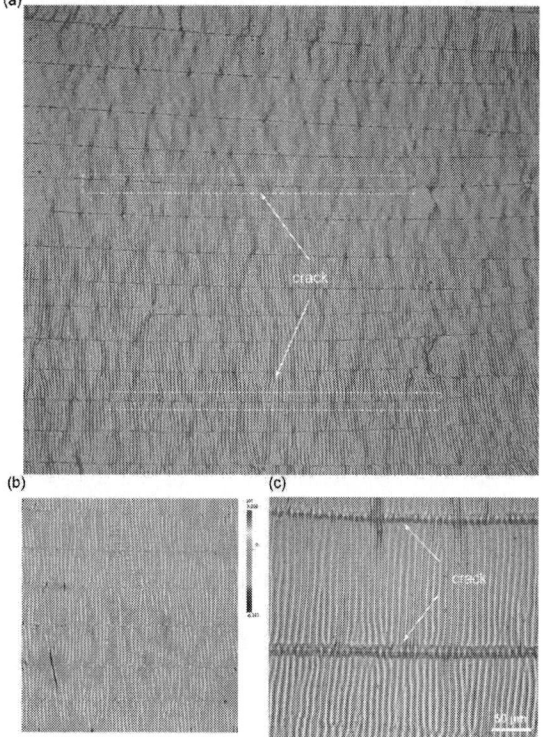

Figure 4: Crack-induced 1D ordered wrinkles in graphene/PMMA-PDMS tri-layer systems: (a) optical micrograph of large-scale 1D ordered wrinkling pattern; (b)-(c) 3D laser scanning confocal microscope (LSCM) images; (c) enlarged view of crack-induced 1D ordered wrinkle pattern.

III. CONCLUSION

In this paper, we proposed a novel low-temperature post-curing transfer method to fabricate mono-scale conformal wrinkles of graphene membranes. The low-temperature post-curing process not only avoids diffusion-induced gradient interface strength, but also produces sufficient interface adhesion, which can guarantee the onset of surface instability-induced graphene wrinkles. Raman characterization of graphene indicates that the graphene/ PMMA is well transferred onto PDMS substrates. The 3D topological morphology characterization shows that the wavelength and amplitude of conformal graphene wrinkles are 15 μm and 700 nm. Mechanical bending-induced cracks are introduced to make wrinkles align in 1D order. The mono-scale graphene wrinkles can be used for strain engineering of 2D materials and ultrathin semiconductor applications.

ACKNOWLEDGMENT

The authors gratefully acknowledge the supports by the National Science Foundation for Distinguished Young Scholars (11625208), Major Program (12032015) and Young Scientists of China (11802173), the Program of

Shanghai Academic/Technology Research Leader (19XD 1421600).

REFERENCES

[1] A. Castellanos-Gomez *et al.*, "Local strain engineering in atomically thin MoS2," *Nano Letters*, vol. 13, no. 11, pp. 5361-5366, 2013.

[2] C. Lee, X. Wei, J. W. Kysar, and J. Hone, "Measurement of the elastic properties and intrinsic strength of monolayer graphene," *Science*, vol. 321, no. 5887, pp. 385-388, 2008.

[3] S. Bertolazzi, J. Brivio, and A. Kis, "Stretching and breaking of ultrathin MoS2," *Acs Nano*, vol. 5, no. 12, pp. 9703-9709, 2011.

[4] A. Castellanos‐Gomez, M. Poot, G. A. Steele, H. S. Van Der Zant, N. Agraït, and G. Rubio‐Bollinger, "Elastic properties of freely suspended MoS2 nanosheets," *Advanced Materials*, vol. 24, no. 6, pp. 772-775, 2012.

[5] C. Androulidakis, E. N. Koukaras, G. Paterakis, G. Trakakis, and C. Galiotis, "Tunable macroscale structural superlubricity in two-layer graphene via strain engineering," *Nature Communications*, vol. 11, no. 1, pp. 1-11, 2020.

[6] Z. H. Dai, L. Q. Liu, and Z. Zhang, "Strain Engineering of 2D Materials: Issues and Opportunities at the Interface," (in English), *Advanced Materials.*, Article vol. 31, no. 45, p. 1805417, Nov 2019, Art no. 1805417, doi: 10.1002/adma.201805417.

[7] K.-M. Hu *et al.*, "Resonant nano-electromechanical systems from 2D materials," *EPL (Europhys Letter)*, vol. 131, no. 5, p. 58001, 2020.

[8] K.-M. Hu *et al.*, "Tension-Induced Raman Spectrum Enhanced Phenomena of Graphene Membrane," in *International Design Engineering Technical Conferences and Computers and Information in Engineering Conference*, 2018, vol. 51791: American Society of Mechanical Engineers, p. V004T08A016.

[9] P. Xiong *et al.*, "Strain engineering of two-dimensional multilayered heterostructures for beyond-lithium-based rechargeable batteries," *Nature Communications*, vol. 11, no. 1, pp. 1-12, Jul 3 2020, Art no. 3297, doi: 10.1038/s41467-020-17014-w.

[10] T. P. Darlington *et al.*, "Imaging strain-localized excitons in nanoscale bubbles of monolayer WSe(2)at room temperature," *Nature Nanotechnology*, vol. 15, no. 10, pp. 854–860, Oct 2020, doi: 10.1038/s41565-020-0730-5.

[11] B. Liu *et al.*, "Strain-Engineered van der Waals Interfaces of Mixed-Dimensional Heterostructure Arrays," *Acs Nano*, vol. 13, no. 8, pp. 9057-9066, Aug 2019, doi: 10.1021/acsnano.9b03239.

[12] K.-M. Hu, W.-M. Zhang, H. Yan, Z.-K. Peng, and G. Meng, "Nonlinear pull-in instability of suspended graphene-based sensors," *EPL (Europhys Letter)*, vol. 125, no. 2, p. 20011, 2019.

[13] K.-M. Hu, K.-C. Bai, H. Yan, B. Peng, and W.-M. Zhang, "Effect of Built-in Stresses on Defects of Graphene Based Gas Sensors," in *2019 20th International Conference on Solid-State Sensors, Actuators and Microsystems & Eurosensors XXXIII (TRANSDUCERS & EUROSENSORS XXXIII)*, 2019: IEEE, pp. 2388-2391.

[14] R. J. Nicholl *et al.*, "The effect of intrinsic crumpling on the mechanics of free-standing graphene," *Nature Communications*, vol. 6, no. 1, pp. 1-7, 2015.

[15] W. Zhu *et al.*, "Structure and electronic transport in graphene wrinkles," *Nano Letters*, vol. 12, no. 7, pp. 3431-3436, 2012.

[16] V. M. Pereira, A. C. Neto, H. Liang, and L. Mahadevan, "Geometry, mechanics, and electronics of singular structures and wrinkles in graphene," *Physical Review Letters*, vol. 105, no. 15, p. 156603, 2010.

[17] P. Xu *et al.*, "Unusual ultra-low-frequency fluctuations in freestanding graphene," *Nature Communications*, vol. 5, no. 1, pp. 1-7, 2014.

[18] J. Luo *et al.*, "Compression and aggregation-resistant particles of crumpled soft sheets," *Acs Nano*, vol. 5, no. 11, pp. 8943-8949, 2011.

[19] J. Zang *et al.*, "Multifunctionality and control of the crumpling and unfolding of large-area graphene," *Nature Materials*, vol. 12, no. 4, pp. 321-325, 2013.

[20] M. G. P. Carbone, A. C. Manikas, I. Souli, C. Pavlou, and C. Galiotis, "Mosaic pattern formation in exfoliated graphene by mechanical deformation," *Nature Communications*, vol. 10, no. 1, p. 1572, 2019.

[21] P. Y. Chen *et al.*, "Multiscale graphene topographies programmed by sequential mechanical deformation," *Advanced Materials,* vol. 28, no. 18, pp. 3564-3571, 2016.

[22] B. Pacakova *et al.*, "Mastering the wrinkling of self-supported graphene," *Scientific Reports,* vol. 7, no. 1, pp. 1-11, 2017.

[23] K. Sampathkumar *et al.*, "Sculpturing graphene wrinkle patterns into compliant substrates," *Carbon,* vol. 146, pp. 772-778, 2019.

[24] W. Bao *et al.*, "Controlled ripple texturing of suspended graphene and ultrathin graphite membranes," *Nature Nanotechnology,* vol. 4, no. 9, pp. 562-566, 2009.

[25] L. Meng *et al.*, "Wrinkle networks in exfoliated multilayer graphene and other layered materials," *Carbon,* vol. 156, pp. 24-30, 2020.

[26] W. H. Duan, K. Gong, and Q. Wang, "Controlling the formation of wrinkles in a single layer graphene sheet subjected to in-plane shear," *Carbon,* vol. 49, no. 9, pp. 3107-3112, 2011.

[27] K. M. Hu *et al.*, "Tension ‐ Induced Raman Enhancement of Graphene Membranes in the Stretched State," *Small,* vol. 15, no. 2, p. 1804337, 2019.

[28] K.-M. Hu *et al.*, "Delamination-Free Functional Graphene Surface by Multiscale, Conformal Wrinkling," *Advanced Functional Materials,* vol. 30, no. 34, p. 2003273, Aug 2020, Art no. 2003273, doi: 10.1002/adfm.202003273.

Proceedings of the 16th Annual IEEE International
Conference on Nano/Micro Engineered and Molecular Systems
April 25-29, 2021

Synthesis of Porous Co_3O_4-ZnO and its Performance for Sensitive Ethanol Detection

Xiao Zhang*, Yaohua Xu, Feng Wei, and Anjie Ming

Abstract—**Ethanol detection has urgent application needs in traffic safety, biomedical and food safety fields. Porous Co_3O_4-ZnO with nitrogen-doped carbon (Co-Zn-O/NC) were successfully fabricated by pyrolysis bimetallic precursor. The as-prepared Co-Zn-O/NC exhibited high response, reliable reversibility, and good selectivity towards ethanol. Specifically, the response value of the Co50Zn50-O/NC is as high as 49.2 towards 50 ppm ethanol at 400 °C which is better than pure Co_3O_4 and ZnO. The results were attributed to the porous structure and the p-n heterojunction formed between Co_3O_4 and ZnO. In particular, the continuous Co_3O_4, which acts as the conductive channel, leads to a relatively lower resistance while ZnO greatly enhances gas sensing properties. This simple synthetic route and design procedure will provide ideas for the development of metal oxide-based semiconductor gas sensors.**

I. INTRODUCTION

Ethanol is a common utilized chemical reagent and widely used in industry, food, health, and transportation departments. However, ethanol is also a kind of flammable explosive reagent. Furthermore, excessive consumption ethanol may lead to serious health problems such as headache and irritation of eyes [1]. Thus, effective detection of ethanol is urgently necessary.

Metal oxide-based semiconductor gas sensors (MOS gas sensor), which have the advantages of low cost, high sensitively, and fast response, are widely used in ethanol detection. However, MOS gas sensor suffers from high energy consumption and low selectivity [2]. The effective method to enhance the sensing performance is the improving of sensing materials. In past studies, many semiconductor oxides have been used in gas detection. Among these oxides, ZnO is widely investigated. ZnO is an n-type semiconductor with a band gap of 3.37 eV and it shows high electronic mobility, good photo response, and excellent chemical, and thermal mechanical stability [3]. ZnO is also widely investigated as gas detection material. For example, Chitra et al. synthesized ZnO via so-gel method by using rice husk as template [4]. The obtained ZnO nanoparticles were uniform distributed over the interwoven fibrous network and its response to 300 ppm ethanol was about 1.41 at Room Temperature. Zhu et al. fabricated ZnO with different microstructures via hydrothermal route. The results showed that the responses of ZnO nanoparticles, ZnO nanoplates and ZnO nanoflowers to 400 ppm ethanol at 350 °C were about 20, 23 and 30, respectively [5]. However, n-type semiconductors suffer from

[1]State Key Laboratory of Advanced Materials for Smart Sensing, General Research Institute for Nonferrous Metals, Beijing 100088, China.

[2]GRIMAT Engineering Institute Co., Ltd, Beijing 101407, China.

*Corresponding author: Xiao Zhang is with the State Key Laboratory of Advanced Materials for Smart Sensing, General Research Institute for Nonferrous Metals; Beijing 100088, China. (e-mail: zhangxiao@grinm.com).

unstable baseline and humidity sensitive in gas detection. Thus, it is necessary to modify ZnO. Co_3O_4 is p-type semiconductor material with band gap of 1.5 eV, and it shows excellent gas sensing, electrochemical, catalytic, and electromagnetic performance. Co_3O_4 is applied in gas detection, supercapacitors, air cells, catalysts, and electromagnetic material fields. For example, Sun et al. prepared Co_3O_4 nanocubes with lateral size about 20 nm by microwave-assisted solvothermal method [6]. The gas sensing of the Co_3O_4 nanocubes was 5 to 100 ppm ethanol at 200 °C. Wen et al. synthesized rhombus-shaped Co_3O_4 nanorod arrays and the Co_3O_4 nanorod showed a response of 71.0 to 500 ppm ethanol at 160 °C [7]. Zhang et al. fabricated 3D microporous Co_3O_4-carbon hybrids (Co_3O_4@C) biotemplated from butterfly wings [8]. The Co_3O_4@C is sensitivity to VOCs and its response value to 100 ppm ethanol at 170 °C is 14.7. What's more, the baseline of Co_3O_4 is stable. Thus, the combine of ZnO and Co_3O_4 might be an effective way to enhance material sensing properties.

In this work, we synthesized Co-Zn-O/NC bimetallic oxides by thermal decomposition Co/Zn bimetallic organic precursors. The precursors were prepared by a simple co-precipitation method. The ethanol gas sensing performance of the Co-Zn-O/NC samples were investigated, and the Co50Zn50-O/NC showed highest ethanol response value among the Co-Zn-O/NC. Further, the sensing mechanism of the sensing process was discussed and the influence of the p-n heterostructure was analyzed in detail. The results will provide theoretical guidance for developing gas sensitive polymetallic oxides.

II. MATERIALS AND METHODS

A. Material preparation

The Co-Zn-O/NC samples were synthesized by a simple co-precipitation method and followed by an annealing process according to our previous work [9]. In a typical synthesis process, 3 mmol $Co(NO_3)_2 \cdot 6H_2O$ (99%, Aladdin Bio-Chem Technology Co., Ltd) and 3 mmol $ZnSO_4 \cdot 7H_2O$ (99.5%, Aladdin Bio-Chem Technology Co., Ltd) were dissolved into a mixed solvent with 20 mL deionized water and 20 mL ethanol. The obtained solution marked as solution A. Then, 24 mmol 2-Methylimidazole (2-MI, 98%, Aladdin Bio-Chem Technology Co., Ltd) was dissolved in another mixed solvent with the same volume ratio and forming solution B. Later, the solution A was dropwise into the solution B with stirring. The purple precipitation was obtained after allowed to stand at room temperature for 24 h. The precipitation was filtered and washed by ethanol. The precipitation was then dried at 60 °C overnight as precursor Co50Zn50-2MI. The obtained precursor was collected and annealed at 400 °C with a heating rate of 2 °C min^{-1} under atmosphere. The final bimetallic oxide

978-1-6654-3008-1/21 $31.00 © 2021 IEEE

was named as Co50Zn50-O/NC. Other Co_3O_4-ZnO bimetallic oxides were synthesized by changing Co/Zn input molar ratio from 2.0 to 0.5, and the final samples were named as Co66Zn34-O/NC and Co34Zn66-O/NC. By the way, the Co100Zn0-O/NC and Co0Zn100-O/NC were synthesized as reference material.

B. Physicochemical characterization and gas detection

The X-ray diffraction (XRD) patterns were performed on Rigaku D/max-2550 X-ray diffraction with the 2θ ranging from 15° to 90°. The scanning electron microscopy (SEM) images were performed on Hitachi S-4800 to observe the morphologies of the samples. The Energy dispersive X-ray spectrum (EDS) was tested on a HORIBA EMAX-350. XPS spectrum of the sample was characterized by X-ray photoelectron spectrometer (ESCALAB 250Xi, Thermo Scientific, USA), while the binding energy was referenced to 284.8 eV (C 1s).

Sensing performance of the Co-Zn-O/NC samples was performed on JF02F sensing analysis system (Guiyan Jinfeng Technology Co., Kunming, China). In a typical testing process, the material was loaded on ceramic substrate with a pair of Ag electrodes. The substrate was then transferred to the test chamber and heated to the operating temperature. The operating temperature ranged from 200 to 450 °C, because the active oxygen ions (O^-, O^{2-}) will form on the surface of the material under high temperature conditions. The optimal operating temperature was confirmed by comparing response values of samples to 100 ppm ethanol. Moreover, the sample with the highest response values was tested on different ethanol concentration and various target gas. The gas response value of samples was defined as the $R=R_g/R_a$, where R_g and R_a were the resistance of the material exposed to target gas and air, respectively.

III. RESULTS AND DISCUSSIONS

A. The structure of the materials

The morphology and composition of Co-Zn-O/NC were investigated by SEM and EDS (Fig. 1). Fig. 1(a) is the SEM image of the Co100Zn0-O/NC and it owns interconnected structure formed by nanorods (the size is about 20-50nm). Fig. 1(b) is the SEM image of the Co0Zn100-O/NC and it formed blocky structure. Fig. 1(c) shows typical SEM of the Co66Zn34-O/NC. The Co66Zn34-O/NC also owns interconnected structure with agglomeration and the diameter of the basic constitution is about 150-200 nm. With the decrease of the Co content, obvious changes are shown in structure and morphology of the Co-Zn-O/NC. Fig. 1(d) shows the SEM of the Co50Zn50-O/NC. The diameter of the Co50Zn50-O/NC nanoparticles are about 20-40 nm and they are apparently agglomerated. Fig. 1(e) is the SEM image of the Co34Zn66-O/NC. Unlike the Co66Zn34-O/NC and Co50Zn50-O/NC, the porosity performance of the Co34Zn66-O/NC is well improved. It also owns interconnected structure while the basic constitution is the nano-flake with a thickness about 20-40 nm. Fig. 1(f) and Fig. 1(g) shows the EDS spectrum and element mapping of Co50Zn50-O/NC, suggesting the existence and well dispersed of Co, Zn and O elements.

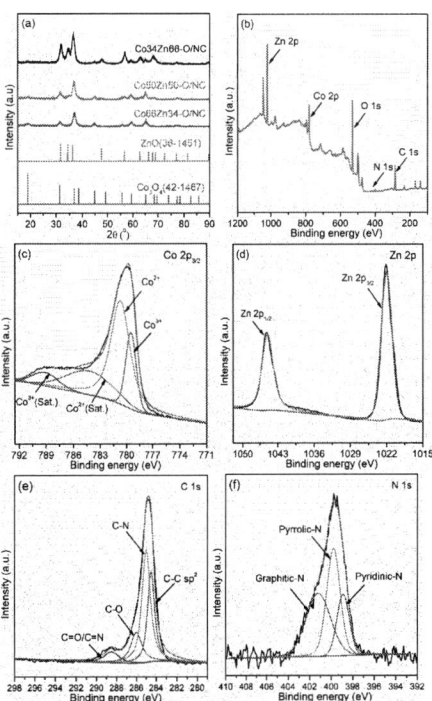

Fig. 1 SEM images of (a) Co100Zn0-O/NC, (b) Co0Zn100-O/NC, (c) Co66Zn34-O/NC, (d) Co50Zn50-O/NC, (e) Co34Zn66-O/NC, and (f) EDS spectrum, (g) element mapping of Co50Zn50-O/NC.

Fig. 2 (a) XRD patterns of the Co34Zn66-O/NC, Co50Zn50-O/NC, Co66Zn34-O/NC, and (b) survey, (c) Co 2p$_{3/2}$, (d) Zn 2p, (e) C 1s, (f) N 1s XPS spectrum of Co50Zn50-O/NC.

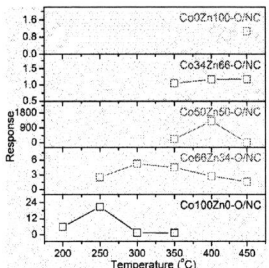

Fig. 3 Response values to 100 ppm ethanol of different materials as a function of operating temperature.

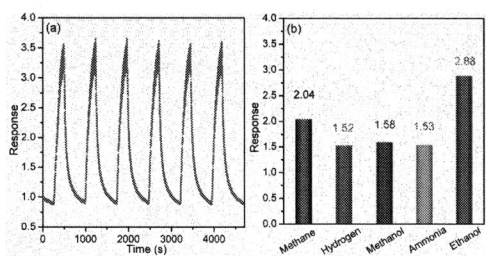

Fig. 5 (a) Dynamical responses of the Co50Zn50-O/NC to 10 ppm ethanol at 400 °C and (b) the selectivity of the Co50Zn50-O/NC to 10 ppm various gases at 400 °C.

The XRD patterns of samples are shown in Fig. 2(a). The XRD patterns of the Co-Zn-O/NC own similar characteristic peaks and the peaks around ($2\theta=$) 31.5°, 37.1°, 59.2° and 65.4° are indexed to the (220), (311), (511) and (440) planes of cubic Co_3O_4 (JCPDS No. 42-1467). With the increase of ZnO content, the peaks of the Co-Zn-O/NC samples are shifted to the left. The peaks at 36.3°, 47.5, 56.8° and 62.3° of the Co34Zn66-O/NC correspond to the (101), (102), (110) and (103) planes of hexagonal ZnO (JCPDS No. 36-1451). The peak at 31.6° and 56.6°of the Co50Zn50-O/NC correspond to the (100) and (110) planes of ZnO (JCPDS No. 36-1451), indicating the formation of Co_3O_4-ZnO bimetallic oxide. In addition, the diffraction peak of the Co-Zn-O/NC gets narrower with the increase of ZnO content, suggesting well crystallizing of the samples.

XPS analysis of the Co50Zn50-O/NC was performed to confirm the valence state of the elements. Fig. 2(b) shows the XPS survey spectrum, which proved the existence of Co, Zn, O, N elements. Fig. 2(c) displays the Co $2p_{3/2}$ XPS spectrum. The peaks at 779.5 eV and 780.7 eV can be assigned to Co^{3+} and Co^{2+}, while the peaks at 789.1 eV and 784.2 eV corresponding to the satellite peak of Co^{3+} and Co^{2+} [10, 11]. The binding energies of Zn 2p is shown in Fig. 2(d). The peaks centered at 1022.0 eV and 1045.0 eV are corresponded to Zn $2p_{3/2}$ and Zn $2p_{1/2}$, indicating the presence of Zn^{2+} [12, 13]. In Fig. 2(e), the C 1s XPS spectrum of Co50Zn50-O/NC can be deconvoluted into four peaks. Specifically, the peak at 284.5 eV, 285 eV, 286 eV and 288.5 eV are assigned to C-C sp^2, C-N, C-O and C=O/C=N, respectively [14]. The N 1s XPS spectrum of Co50Zn50-O/NC is shown in Fig. 2(f) and the peaks at 398.9 eV, 399.8 eV and 401.2 eV are corresponding to pyridinic-N, pyrrolic-N, and graphitic-N [15, 16].

B. Gas sensing properties

The operating temperature dependence of the material was evaluated by changing temperature from 200 to 450 °C (Fig. 3) [9]. As shown in Fig. 3, the Co100Zn0-O/NC shows good sensing performance at relatively low temperature and optimum operating temperature of the Co100Zn0-O/NC is about 250 °C. The Co0Zn100-O/NC shows sensing properties at relatively high temperature. Other Co-Zn-O/NC samples works at temperature range from 300 to 450 °C. Among these Co-Zn-O/NC samples, Co50Zn50-O/NC shows the best response at 400 °C, and its response value reaches at least 1300, while the response of Co0Zn100O/NC, Co34Zn66-O/NC, Co66Zn34-O/NC, and Co100Zn0-O/NC are 1.1, 1.2, 5.2 and 20.9, respectively. This is might be related to the enough ZnO distributing on the surface of Co_3O_4, while suitable amounts of ZnO can avoid cluster. As for Co_3O_4, its good sensing response might be related to its extraordinary structure inherited from ZIF-67. Considering the limitation of the sensing analysis system, further tests of the Co50Zn50-O/NC are carried on a relatively low concentration at 400 °C. Furthermore, the resistances of Co-Zn-O/NC in air at optimum operating temperature are shown in Table 1. It is obviously that the Co0Zn100-O/NC shows the highest resistance while the Co100Zn0-O/NC owns the lowest resistance in air. As for Co-Zn-O/NC, the resistance of the samples decreased with the increase of Co contents, and the bimetallic oxides show the properties of Co_3O_4 and ZnO.

The dynamic sensing characteristics and its liner relation between response and concentration of the Co50Zn50-O/NC are shown in Fig. 4. The ethanol concentration ranges from 1 to 50 ppm. As shown in Fig. 4(a), the detection limitation of the Co50Zn50-O/NC is as low as 1 ppm. Furthermore, the Co50Zn50-O/NC exhibits a remarkable liner relation between response and concentration at 1 to 10 ppm (Fig. 4b).

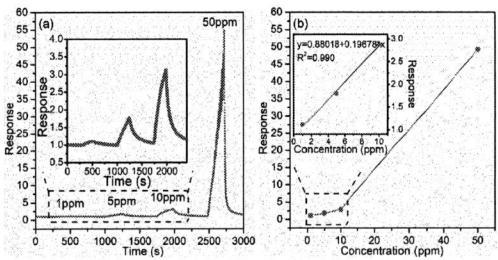

Fig. 4 (a) Dynamic sensing characteristics and (b) liner relation between response and concentration of the Co50Zn50-O/NC to different ethanol concentration at 400 °C.

TABLE I.   THE OPTIMUM OPERATING TEMPERATURE OF Co-Zn-O/NC AND ITS RESISTANCES IN AIR.

Sample	Optimal Operating Temperature (°C)	Resistances in Air (kΩ)
Co0Zn100-O/NC	450	1.49×10^6
Co34Zn66-O/NC	400	1.30×10^6
Co50Zn50-O/NC	400	7.67×10^4
Co66Zn34-O/NC	300	2.50×10^3
Co100Zn0-O/NC	250	1.51×10^2

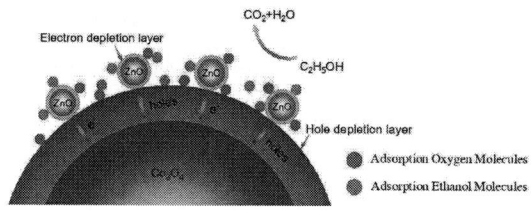

Fig. 6 Schematic illustration for gas sensing mechanism.

The reversibility of the Co50Zn50-O/NC towards 10 ppm ethanol at 400 °C is demonstrated in Fig. 5(a). It is obviously that the response can be repeated 6 cycles without marked decay, suggesting the remarkable reversibility of Co50Zn50-O/NC. Fig. 5(b) shows the response values of the Co50Zn50-O/NC to different target gases. The material exhibits the highest response towards ethanol and the response value is about 1.41, 1.89, 1.82 and 1.88 times higher than target gases as methane, hydrogen, methanol, and ammonia, which indicate the Co_3O_4-ZnO bimetallic oxides can improve the selectivity towards ethanol.

C. Discussion

The response value of gas sensing materials relates to its resistance in target gas and air and it is inextricably bound up with the material surface adsorption state. The chemical adsorption and desorption of the gas molecules is the key factor for enhancing material performance. As is known to all, the oxygen molecules are adsorbed on material surface when material exposes in air and the states of adsorbed oxygen depends on the temperature. When the temperature is below 150 °C, the oxygen molecules were adsorbed as O_2^- while it changes to atomic oxygen ions (O^-, O^{2-}) with temperature between 150 and 400 °C [17]. Once exposed to ethanol vapor, the ethanol molecules will react with adsorbed oxygen atoms and forming CO_2 and H_2O (Fig. 6). In this process, the electrons release back to the material and the resistance will change [18, 19].

According to dynamical responses of material, Co50Zn50-O/NC shows typical p-type oxides characteristic which means Co_3O_4 is the continuous phase and ZnO is distributed on Co_3O_4 surface. Because the difference Fermi level between Co_3O_4 and ZnO, the electron migrated on the interface until an equalization value and thus p-n heterojunction is formed. The enhanced sensing performance of Co-Zn-O/NC may be due to the p-n heterojunction formed on the surface between Co_3O_4 and ZnO. Upon exposure to air, the adsorbed oxygen leads to a decreasing electron density on ZnO surface. As a result, the number of electrons that transferred from ZnO to Co_3O_4 is reduced and resistance decrease. When ethanol vapor is introduced, the ethanol molecules react with adsorption oxygen and electrons are released to Co_3O_4 and ZnO. In this process, the number of electrons on the ZnO surface increase. It is easier for electrons transferring from ZnO to Co_3O_4 [20, 21]. In other words, the transferred electrons from ZnO to Co_3O_4 lead to higher resistances of Co_3O_4-ZnO than pure Co_3O_4. In addition, density of the p-n heterojunction depended on the amount of ZnO. However,

excessive ZnO will increase cluster and reduce sensitive of the sensing material. Therefore, Co50Zn50-O/NC bimetallic oxide demonstrates better sensing performance owing the greater resistance variation and suitable Co/Zn ratio. Notably, the response to ethanol might be related to the presence of graphitic-N and pyridinic-N species in the metal oxide (Figure 2f). These N species may increase the electron-donating ability of the metal oxide by doping carbon which is benefited for enhancing conductivity, thereby promoting the reductive adsorption of oxygen [22]. In addition to the reasons discussed above, the collision between the gas molecules and material as well as specific surface area are related to the porous structure of the sensing material. Thus, porous material with appropriate pore size will be conducive to improving material sensing performance [23].

IV. CONCLUSIONS

In summary, porous Co_3O_4-ZnO were fabricated via co-precipitation and thermal decomposition route. The response curves indicate the p-type characteristic of the bimetallic oxide and the Co50Zn50-O/NC demonstrates the best response properties towards ethanol. Co-Zn-O/NC exhibits high response value, excellent reversibility, and remarkable selectivity. The reasons to these sensing properties are linked to the porous structures and p-n heterojunction formed on the interface between Co_3O_4 and ZnO. In particular, the results show that ZnO particles are distributed on the surface of continuous Co_3O_4. The combine of sensing sensitive ZnO and low resistance Co_3O_4 greatly enhances the gas sensing performance. More significantly, the facile fabrication strategy can be extended to expand various metallic oxides which can be used in high-performance gas sensors.

REFERENCES

[1] R. A. Kadir, R. A. Rani, A. S. Zoolfakar, J. Z. Ou, M. Shafiei, W. Wlodarski and K. Kalantar-zadeh, "Nb2O5 Schottky based ethanol vapour sensors: Effect of metallic catalysts," *Sens. Actuator. B: Chem.*, vol. 202, pp. 74-82, 2014.

[2] B. T. Raut, P. R. Godse, S. G. Pawar, M. A. Chougule, D. K. Bandgar and V. B. Patil, "Novel method for fabrication of polyaniline-CdS sensor for H2S gas detection," *Measurement*, vol. 45, pp. 94-100, 2012.

[3] R. Sankar Ganesh, E. Durgadevi, M. Navaneethan, V. L. Patil, S. Ponnusamy, C. Muthamizhchelvan, S. Kawasaki, P. S. Patil and Y. J. Hayakawa, "Low temperature ammonia gas sensor based on Mn-doped ZnO nanoparticle decorated microspheres," *J. Alloys. Compd.*, vol. 721, pp. 182-190, 2017.

[4] M. Chitra, K. Uthayarani, N. Rajasekaran, N. Neelakandeswari, E. K. Girija and D. P. Padiyan, "Rice Husk Templated Mesoporous ZnO Nanostructures for Ethanol Sensing at Room Temperature," *Chinese Phys. Lett.*, vol. 32, pp. 078101, 2015.

[5] L. Zhu, Y. Li and W. Zeng, "Hydrothermal synthesis of hierarchical flower-like ZnO nanostructure and its enhanced ethanol gas-sensing properties," *Appl. Surf. Sci.*, vol. 427, pp. 281-287, 2018.

[6] C. Sun, X. Su, F. Xiao, C. Niu and J. Wang, "Synthesis of nearly monodisperse Co3O4 nanocubes via a microwave-assisted solvothermal process and their gas sensing properties," *Sens. Actuator. B: Chem.* vol. 157, pp. 681-685, 2011.

[7] Z. Wen, L. Zhu, W. Mei, L. Hu, Y. Li, L. Sun, H. Cai, Z. Ye, "Rhombus-shaped Co3O4 nanorod arrays for high-performance gas sensor," *Sens. Actuator. B: Chem.*, vol. 186, pp. 172-179, 2013.

[8] J. Zhang, Y. Liang, J. Mao, X. Yang, Z. Cui, S. Zhu and Z. Li, "3D microporous Co3O4-carbon hybrids biotemplated from butterfly

wings as high performance VOCs gas sensor," *Sens. Actuator. B: Chem.*, vol. 235, pp. 420-431, 2016.

[9] X. Zhang, Y. Xu, H. Liu, W. Zhao, A. Ming and F. Wei, "Preparation of homogeneous porous Zn-CoO$_x$ and its response to alcohols under relative low operating temperature," *Chinese Chem. Lett.*, vol. 31, pp. 2059-2062, 2020.

[10] X. Zhang, Y. Xu, H. Liu, W. Zhao, A. Ming and F. Wei, "Preparation of porous Co$_3$O$_4$ and its response to ethanol with low energy consumption," *RSC Adv.*, vol. 10, pp. 2191-2197, 2020.

[11] T. Hu, Y. Wang, L. Zhang, T. Tang, H. Xiao, W. Chen, M. Zhao, J. Jia and H. Zhu, "Facile synthesis of PdO-doped Co$_3$O$_4$ nanoparticles as an efficient bifunctional oxygen electrocatalyst," *Appl. Catal. B-Environ.*, vol. 243, pp. 175-182, 2019.

[12] Y. Xiong, W. Xu, Z. Zhu, Q. Xue, W. Lu, D. Ding and L. Zhu, "ZIF-derived porous ZnO-Co$_3$O$_4$ hollow polyhedrons heterostructure with highly enhanced ethanol detection performance," *Sens. Actuator. B: Chem.*, vol. 253, pp. 523-532, 2017.

[13] B. Li, J. Liu, Q. Liu, R. Chen, H. Zhang, J. Yu, D. Song, J. Li, M. Zhang and J. Wang, "Core-shell structure of ZnO/Co$_3$O$_4$ composites derived from bimetallic-organic frame-works with superior sensing performance for ethanol gas," *Appl. Surf. Sci.*, vol. 475, pp. 700-709, 2019.

[14] Z. Liu, H. Xing, L. Lin, X. Ji and Z. Shen, "Facial Synthesized Co-doped SnO$_2$@Multi-Walled Carbon Nanotubes as an Efficient Microwave Absorber in High Frequency Range," *Nano*, vol. 12, pp. 1750118, 2017.

[15] X. Li, Y. Fang, X. Lin, M. Tian, X. An, Y. Fu, R. Li, J. Jin and J. Ma, "MOF derived Co$_3$O$_4$ nanoparticles embedded in N-doped mesoporous carbon layer/MWCNT hybrids: extraordinary bi-functional electrocatalysts for OER and ORR," *J. Mater. Chem. A*, vol. 3, pp. 17392-17402, 2015.

[16] T. Y. Ma, S. Dai, M. Jaroniec and S. Z. Qiao, "Graphitic carbon nitride nanosheet-carbon nanotube three-dimensional porous composites as high-performance oxygen evolution electrocatalysts," *Angew. Chem.*, *vol.* 53, pp. 7409-7413, 2014.

[17] M. Karmaoui, S. G. Leonardi, M. Latino, D. M. Tobaldi, N. Donato, R. C. Pullar, M. P. Seabra, J. A. Labrincha and G. Neri, "Pt-decorated In$_2$O$_3$ nanoparticles and their ability as a highly sensitive (< 10 ppb) acetone sensor for biomedical applications," *Sens. Actuator. B: Chem.*, vol. 230, pp. 697-705, 2016.

[18] C. Jin, S. Park, H. Kim and C. Lee, "Ultrasensitive multiple networked Ga$_2$O$_3$-core/ZnO-shell nanorod gas sensors," *Sens. Actuator. B: Chem.*, vol. 161, pp. 223-228, 2012.

[19] A. Mirzaei, K. Janghorban, B. Hashemi, M. Bonyani, S. G. Leonardi and G. Neri, "Highly stable and selective ethanol sensor based on α-Fe$_2$O$_3$ nanoparticles prepared by Pechini sol-gel method," *Ceram. Int.*, vol. 42, pp. 6136-6144, 2016.

[20] L. L. Xing, S. Yuan, Z. H. Chen, Y. J. Chen and X. Y. Xue, "Enhanced gas sensing performance of SnO$_2$/α-MoO$_3$ heterostructure nanobelts," *Nanotechnology*, vol. 22, pp. 225502, 2011.

[21] S. Park, S. Kim, G.-J. Sun, S. Choi, S. Lee and C. Lee, "Ethanol sensing properties of networked In$_2$O$_3$ nanorods decorated with Cr$_2$O$_3$-nanoparticles," *Ceram. Int.*, vol. 41, pp. 9823-9827, 2015.

[22] X. Yi, X. He, F. Yin, B. Chen, G. Li and H. Yin, "Co-CoO-Co$_3$O$_4$/N-doped carbon derived from metal-organic framework: The addition of carbon black for boosting oxygen electrocatalysis and Zn-Air battery," *Electrochim. Acta*, vol. 295, pp. 966-977, 2019.

[23] X. Zhang, Y. Xu, H. Liu, W. Zhao, A. Ming and F. Wei, "Facile fabrication of cobalt-doped SnO$_2$ for gaseous ethanol detection and the catalytic mechanism of cobalt," *CrystEngComm*, vol. 21, pp. 7528-7534, 2019.

Simultaneous Sensing of Refractive Index and Temperature Using a Symmetry-breaking Silicon Metasurface with Multiple Fano Peaks

Luo Zhao
Beijing Institute of Technology
Beijing, China
zl18401670335@sina.com

Junyi Wang
Beijing Institute of Technology
Beijing, China
3220190429@bit.edu.cn

Haiyang Li
Beijing Institute of Technology
Beijing, China
1204625528@qq.com

Cheng Zhang
Beijing Institute of Technology
Beijing, China
3120205336@bit.edu.cn

Guoguo Kang
Beijing Institute of Technology
Beijing, China
kgg@bit.edu.cn

Abstract—**In this paper, an asymmetric all-dielectric metasurface is used to measure the temperature and refractive index (RI) simultaneously. We use FDTD Solutions to obtain reflection spectrum, and use the two Fano peaks to realize two parameter sensing. The Fano resonance of narrow line width increases the figure of merit (FoM, ratio of sensitivity to full width at half maximum) by two orders of magnitude compared to previous proposed dual parameter sensors. Simulation reveals the sensor possessing RI sensitivities of 444.52 nm/RIU and 100.06 nm/RIU, FoMs of 1778.08 and 227.41, temperature sensitivities of 34.9 pm/°C and 83.8 pm/°C. The device has the advantages of simple structure, easy to manufacture, small volume and low loss. It has a good prospect in the application of biological and chemical sensing, especially in the measurement of substance concentration.**

Keywords—*Metasurface, Refractive index sensing, Temperature sensing, Simultaneous measurement, Fano resonance.*

I. Introduction

As a very important parameter, the refractive index (RI) can provide us with a lot of information, such as the type, concentration, biochemical reaction, and optical properties of substances. Therefore, RI, as a detection target, has been widely used in environmental monitoring [1], clinical medicine [2], food industry [3] and other fields [4,5]. RI usually changes with temperature. For accurate measurement of RI, it is necessary to carry out the measurement under constant temperature, however, it's an impossible mission in practice. Consequently, the temperature measurement needs to be well considered.

To measure the temperature and RI simultaneously, several requirements on the peaks appearing in the spectrum need to be fulfilled, namely, more than two can be found, stable under various parameters, responsive to the changes of temperature and refractive index, and as sharp as possible. Certainly, easy manufacturing and convenient experiment should also be our goal to pursue.

Recently, researches on sensors that simultaneously measure RI and temperature sprang up like mushrooms. Metal waveguide sensors capable of RI and temperature simultaneous measurement have been proposed [6-8]. Free electrons in the metal are excited by incident light to cause collective oscillations, that is, surface plasmon resonance (SPR). The shape of metal structure determines the direction of surface current[9]. It is possible to realize selective absorption and scattering of light, and form resonance peak in different wavebands, but complex structure is needed. Due to the ohmic loss of metal, the resonance peak would be broadened, which leads to low quality factor (Q factor, reciprocal of peak width) [10]. Moreover, the absorption characteristics of metals leads to local heating, which makes applications for biological sensing greatly restricted. More extensive research is carried out on fiber optic sensors [11-15]. Fiber optic sensors have the characteristics of complex manufacturing process and high manufacturing cost. The coupling of light should also be considered in practical application.

Compared with devices above, all-dielectric devices are more advantageous for temperature and RI sensing [16-18]. Lately, all-dielectric metasurface has become a cutting edge of research [19-22]. Dielectric materials possess smaller imaginary part of dielectric constant and low absorption, which results in little non-radiation loss (such as ohmic loss). In all-dielectric metasurface, exists the displacement current, which is different from the transmission current of metal, and enable the simple structure of dialectic to produce sharp resonance peaks. To sum up, the advantages of simple structure, low processing cost, undemanding machining accuracy and high Q factor, promise it good prospects in sensing applications [23-25].

There have been few reports on the simultaneous measurement of RI and temperature using all-dielectric metasurfaces. The FoM available is small, less than twenty, denoting poor sensing performance [26]. In this paper, an all-dielectric periodic structure made of a single unit is used to measure the ambient RI and temperature at one time. The sensor shows sharper peaks, higher sensitivities, higher FoMs and better sensing performance than existed all-dielectric sensors. The sensing simulation is obtained using a commercial finite difference time domain (FDTD) solver (from Lumerical Inc.). Compared with the device made of composite units, Fano rasonance induced by the symmetry breaking of single unit structure is more robust against processing error. To sum up, the structure we proposed is simple-structured, small-sized and easy to manufacture, and it has small fabrication error, non-radiation loss and scattering loss, as well as high FoMs close to the ideal value.

II. PRINCIPLE

Temperature or RI of the ambient medium can be obtained by measuring frequency shift of resonance peaks. This is the basic principle of temperature or RI sensing. In actual sensing applications, the change in RI is not only related to the type and concentration of ambient medium, but also to the temperature [27]. Considering the response of sensor to temperature is due to the following two effects: the change of RI of the ambient media and sensor material, both induced by the temperature. We can use two resonance peaks to perform dual-parameter measurement to work out the corresponding change of RI. This process can be formulated as [28]:

$$\frac{d\lambda_i}{dn} \times \Delta n + \frac{d\lambda_i}{dn} \times \frac{dn}{dT} \times \Delta T + \frac{d\lambda_i}{dT} \times \Delta T = \Delta\lambda_i \ (i = 1,2) \quad (1)$$

where n is refractive index of the ambient medium, T is temperature, $\lambda_i (i=1,2)$ refers to the two resonance peaks. The first part refers to the independent RI variation of ambient medium, the second part refers to the thermo-optic effect of ambient medium (dn/dT is the thermo-optical coefficient of the ambient medium obtained by checking relative literature or performing a calibration experiment which aims to calculate dn/dT under the assumption $\Delta n = 0$), and the third part refers to independent temperature variation. The formula can be further expressed as follows:

$$\begin{bmatrix} \Delta n + \frac{dn}{dT} \times \Delta T \\ \Delta T \end{bmatrix} = \begin{bmatrix} K_{n,1} & K_{T,1} \\ K_{n,2} & K_{T,2} \end{bmatrix}^{-1} \begin{bmatrix} \Delta\lambda_1 \\ \Delta\lambda_2 \end{bmatrix} \quad (2)$$

Where $K_{n,i}$ is the RI sensitivity coefficient, $K_{T,i}$ is the temperature sensitivity coefficient. The change of RI and environmental temperature can be derived from Eq. (2).

A figure of merit (FoM) is used to evaluate the performance of a sensor, and it can be formulated as [29]:

$$FoM = \frac{K_{i,j}}{FWHM} \quad (3)$$

where K refers to the sensitivity, i refers to the peaks, and j is n or T, which represents the RI of the ambient environment or the temperature. FWHM represents full width at half maximum. For the asymmetric Fano line type, we use peak width (the wavelength difference between resonance peak and valley) in the formula.

There are two types of resonance in all-dielectric devices with simple structure. One is basic magnetic dipole or electric dipole resonance [26,30,31] based on the Mie resonance theory [32]. The other is Fano resonance [33]. Fano peak is formed by the coupling of two resonance modes: bright mode, which has a large radiation bandwidth and can be directly coupled with the incident light, and dark mode, which cannot be directly excited and has a small radiation bandwidth. Compared to the former generation method, Fano resonance displays much narrower bandwidth which in turn creates higher FoM [34-37]. FoM is an important evaluation function that comprehensively evaluates the performance of the sensor. High FoM brings low detection limit and high signal-noise ratio (SNR). In the following chapter, sharp Fano resonance will be introduced to achieve a large FoM value.

III. SIMULATION AND DISCUSSIONS.

A. Resonance analysis

As discussed above, multiple resonance peaks is required for further sensing of the RI and temperature change. The unit structure diagram of multi-resonance generation ability is shown in Fig. 1(a)&(b). The unit is a silicon disk with an air hole, and the substrate is Silica. The simulation is carried out by FDTD Solutions based on the finite difference time domain method. The period Px and Py in x and y directions are both 0.8 μm. The radius R and height H of the disk are 0.6 μm and 0.22 μm respectively. The air hole of which radius is 50nm, locates 50 nm from the center. The reflection spectrum is shown in Fig. 1(c), there are multiple stable peaks in the spectrum, and peak 1 and peak 2 are Fano resonances.

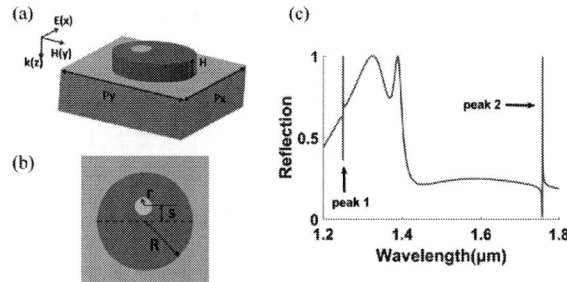

Fig. 1. (a) Diagram of the metasurface structure, the x-polarized incident plane wave propagates along z axis. (b) A cross-sectional view of the unit cell (r: air hole radius, s: offset between hole cylinder axis and disk cylinder axis). (c) Simulated reflection spectrum with two resonance peaks at $\lambda \approx 1.25\mu m, 1.75\mu m$ (denoted with peak 1,2 respectively).

To further investigate the origin of peak 1 and peak 2, the cross-sectional distributions of the electric and magnetic fields in a unit cell are depicted in Fig. 2. We combine the simulated electromagnetic field distribution and the multipole decomposition method [38] (calculate scattering power of various multipole moments) to distinguish the nature of the two Fano peaks. The Fano resonance here is mainly due to the existence of air holes in the structure that breaks symmetry of the structure, and excites the dark state which is coupled with the bright state. Fig. 2(a)-(d) are field intensity and vector distribution of electric and magnetic field at peaks 1 and 2, respectively. The black arrow in Fig. 2(c) indicates the existence of a circular displacement current, generating a magnetic moment along the z-axis. It is mainly induced by the uneven charge distribution caused by asymmetry of the structure. Peak 2 is formed by the coupling of electric dipole (dark state) and magnetic dipole (bright state) in the z-axis, that is trapped mode [39]. The displacement current in Fig. 2(a) has four loops, the upper and lower are counterclockwise, and the left and right are clockwise, resulting in four moments along the z-axis. So peak 1 is formed by the coupling of an electric quadrupole (dark state) and a magnetic dipole (bright state) along the z-axis. Fig. 2(e)-(f) show the calculated scattering power of the electric dipole (ED), magnetic dipole (MD), toroidal dipole (TD), electric quadrupole (EQ), and magnetic quadrupole (MQ) near resonance wavelengths, which derived by the multipole decomposition method. MD and ED dominate in Fig. 2(e) and MD and EQ dominate in Fig. 2(f), which corresponds to different coupling modes.

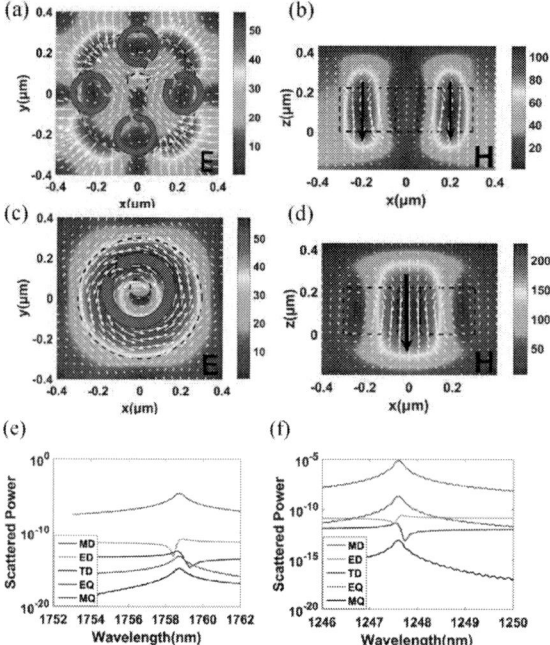

Fig. 2. (a) Electric field and vector distribution at peak 1 ($\lambda \approx 1.25\mu m$) in x-y plane. (b) Magnetic field and vector distribution at peak 1($\lambda \approx 1.25\mu m$) in x-z plane. (c) Electric field and vector distribution at peak 2 ($\lambda \approx 1.75\mu m$) in x-y plane. (d) Magnetic field and vector distribution at peak 2 ($\lambda \approx 1.75\mu m$) in x-z plane. (e) Scattering power of various multipole moments at peak 2, the y-axis is logarithmic. (f) Scattering power of various multipole moments at peak 1, the y-axis is logarithmic. The blue arrows represent the direction of the displacement current. The black arrows represent the direction of the polar moment.

According to the analysis above, peak 1 and peak 2 display larger field enhancement, wider field distribution and shaper shape which directly contributes to larger FoM and higher sensitivity.

B. Sensing Performance

After analyzing the principle of peaks, the sensing characteristics need to be studied by adjusting the structural parameters. We targeted to obtain a larger modulation depth (RI difference between the peak and the valley) and a sharper

resonance peak, that is to say, smaller peak width. Peak 1 and peak 2 are formed by introducing asymmetry excitation. The way to control the asymmetry is to change the radius of the hole and its location on the disk. The sweep results (sensitivity, modulation depth) are shown in Fig. 3. And data analysis of peak 1 and peak 2 are included in Table 1&2. The RI sensitivity can be obtained by fixing the temperature in 300 K, ranging the background RI from 1.32 to 1.52. Likewise, the temperature sensitivity can be obtained by fixing RI to 1, ranging the temperature to range from 0°C to 100°C. In order to improve work efficiency, we use two sets of data (0°C&100°C, 1.32RIU&1.52RIU) to obtain the sensitivity coefficient.

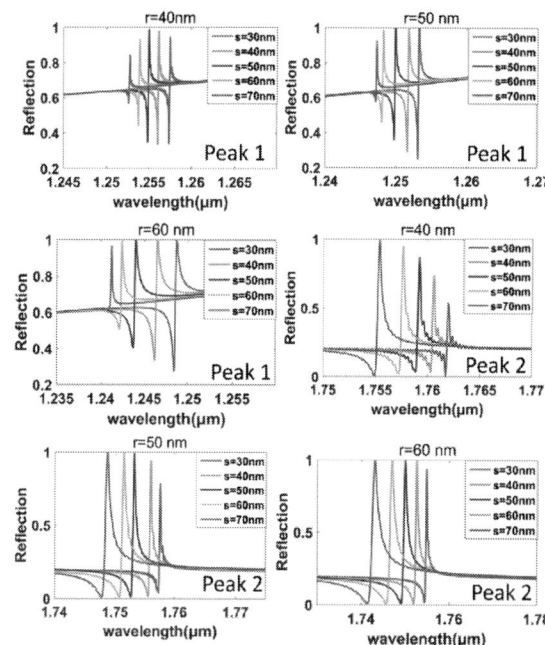

Fig. 3. (a) Simulated reflection spectrum with different parameters r and s.

TABLE I. SIMULATION RESULTS OF PEAK 1 AFTER ADJUSTING THE PARAMETERS R AND S

r (nm)	s (nm)	modulation depth	peak width	sensitivity coefficient	
				RI(nm/RIU)	T(pm/°C)
40	30	0.32-0.98	0.19	443.4	35.9
	40	0.32-0.99	0.3	444.3	35.4
	50	0.35-0.98	0.29	444.45	35.3
	60	0.44-0.94	0.25	444.4	34.8
	70	0.56-0.84	0.24	444.7	34.8
50	30	0.25-1	0.73	446.8	35.2
	40	0.29-1	0.77	446.95	34.7
	50	0.36-1	0.82	446.6	34.7
	60	0.46-0.98	0.73	446.1	34.5
	70	0.57-0.93	0.49	445.3	34.6
60	30	0.27-1	2.04	454.35	34.6
	40	0.34-1	2.12	452.6	34.1
	50	0.41-1	1.88	449.35	33.9
	60	0.51-1	1.34	446	33.5
	70	none	none	none	none

TABLE II. SIMULATION RESULTS OF PEAK 2 AFTER ADJUSTING THE PARAMETERS R AND S

r (nm)	s (nm)	modulation depth	peak width	sensitivity coefficient RI(nm/RIU)	sensitivity coefficient T(pm/°C)
40	30	0.0005-0.53	0.16	99.30	84.20
	40	0.026-0.74	0.30	99.60	84.30
	50	0.01-0.84	0.34	100.00	84.10
	60	0.003-0.93	0.44	100.10	83.50
	70	0.0003-0.98	0.52	100.75	83.50
50	30	0.015-0.76	0.34	99.85	84.10
	40	0.002-0.94	0.44	100.15	83.50
	50	0.0008-0.99	0.59	100.50	83.40
	60	0.001-0.99	0.84	101.40	83.30
	70	0.001-1	1.07	101.85	83.30
60	30	0.003-0.92	0.40	100.40	83.70
	40	0.001-0.99	0.72	100.85	83.40
	50	0.001-1	1.04	101.55	83.20
	60	0.001-1	1.43	102.40	83.00
	70	0.001-1	1.87	103.45	82.60

As can be seen from the Fig.3 and Table 1&2, when adjusting the offset s under different r, the modulation depth of peak 1 gradually decreases, which is on the contrary for that of peak 2. It can be seen from section 3.1 that the field distributions of peak 1 and peak 2 are completely different. From table 1 we know that when r exceeds 60 nm, peak 1 is unstable. There is such a rule for r of 40-60 nm: the peak widths of the two peaks increase together with r. When adjusting the distance s under a constant r, the modulation depth of peak 1 gradually decreases, which is on the contrary for that of peak 2. And in addition, the peak width of peak 1 first becomes larger and then smaller, while that of peak 2 gradually increases.

Peak width is proportional to the light energy radiated to the external field through the air hole. When the air hole is bigger (bigger r) or located at a position (different s) with a greater field strength, the energy radiated to the external field is greater and the peak width is smaller. As the offset s increases, along the moving trajectory of the air hole field strength grows from larger to smaller for peak 1, and always larger for peak 2 (illustrated in Fig.2). According to the mechanism discussed above, further analysis of s exceeding 70nm is no longer necessary for the degradation of sensing performance of both peaks.

In addition, when changing the parameters, the RI sensitivity and temperature sensitivity change only a few nanometers or picometers. That's because the mode of each peak is certain, and the field enhanced area is also unchanged. As is shown in Fig. 2, the field enhanced area is equivalent to a probe of a sensor, and the closeness of the probe to the measured regions leads to a strong response. The field enhanced area at peak 1 is mainly distributed around the structure, and contact area between the probe and ambient medium is larger, so the RI sensitivity of peak 1 is higher. In the same way, the probe for peak 2 is mainly inside the structure because the field enhanced area is mainly distributed inside the structure. The contact area between

peak 2 and dielectric materials is larger, so the temperature sensitivity of peak 2 is higher.

Based on the simulation data in Table 1&2, we choose r = 40 nm, s = 60 nm as the ultimate parameters. First, we discuss the the simulation of RI sensitivity coefficient. Simulated temperature is set at 300K, and background RI changes from 1.32 to 1.52. As the RI increases, both peak 1 and peak 2 are red-shifted, as shown in Fig. 4(a) -(b). Due to the high sensitivity, peak 1 gradually overlaps with peak 2 and 3 (see Fig. 4(a)) producing electromagnetic induced transparency (EIT) [40,41]. This EIT phenomenon still validate the sensing performance of peak 1 for the valley position still maintains a good linear relationship with RI (see Fig. 4(a)). According to Fig. 4(c)-(d), $K_{n,1}$ and $K_{n,2}$ are linear fitted to be 444.52 nm/RIU and 100.06 nm/RIU.

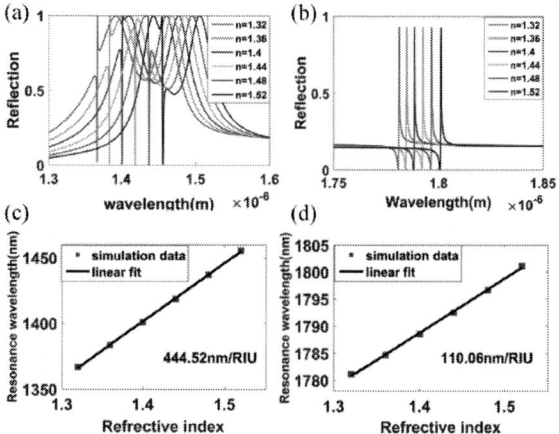

Fig. 4. (a) The reflection curve of peak 1 under different background RI. (b)The reflection curve of peak 2 under different background RI. (c) The distribution of the resonance wavelength of peak 1 under different background RI and the linear fitting curve of RI sensitivity coefficient. (d) The distribution of the resonance wavelength of peak 2 under different background RI and the linear fitting curve of RI sensitivity coefficient.

Next is the simulation of temperature sensitivity coefficient. Background RI is set to 1, and temperature is changed from 0°C to 100°C. Due to the small size of the structure, the coefficient of thermal expansion is negligible. Using the thermo-optical coefficients of Si and Silica in the simulation, where the thermo-optical coefficient of Silica is 8.6×10^{-6}/K, and the thermo-optical coefficient of Si is 1.84×10^{-4}. From fig. 5(a)-(d) we can see, as the temperature increases, both peaks are red-shifted, and the frequency shift is smaller than which caused by RI. $K_{T,1}$ and $K_{T,2}$ are 34.9 pm/°C and 83.8 pm/°C by linear fitting. After obtaining these four parameters, Eq. (4) can be used to calculate the change in RI and temperature:

$$\begin{bmatrix} \Delta n + \dfrac{dn}{dT} \times \Delta T \\ \Delta T \end{bmatrix} = \begin{bmatrix} 444.52\,nm/RIU & 34.9\,pm/°C \\ 100.06\,nm/RIU & 83.8\,pm/°C \end{bmatrix}^{-1} \begin{bmatrix} \Delta\lambda_1 \\ \Delta\lambda_2 \end{bmatrix}$$

(4)

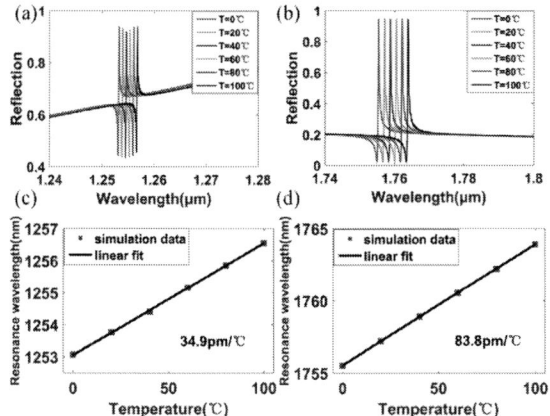

Fig. 5. (a) The reflection curve of peak 1 under different temperature. (b) The reflection curve of peak 2 under different temperature. (c) The distribution of the resonance wavelength of peak 1 under different temperature and the linear fitting curve of temperature sensitivity coefficient. (d) The distribution of the resonance wavelength of peak 2 under different temperature and the linear fitting curve of temperature sensitivity coefficient.

According to the inset of Figure 5(a)-(b), the widths of peak 1 and peak 2 are 0.25 nm and 0.44 nm. $FOM_{1,n}$ and $FOM_{2,n}$ are 1778.08 and 227.41 calculated by formula (3). For temperature, since the frequency shift is at pm level, FOM is small, but it is still a relatively large value compared to the existing all-dielectric devices.

IV. CONCLUSION

We use the asymmetry of the all-dielectric metasurface to obtain two Fano peaks to achieve simultaneous temperature and RI measurement. Through simulation, RI sensitivity coefficients of the two peaks are 444.52 nm/RIU and 100.06 nm/RIU, and temperature sensitivity coefficients are 34.9 pm/°C and 83.8 pm/°C. The FoMs are 1778.08 and 227.41, respectively, which proves that the sensor has good sensing properties.

Compared with other temperature and refractive index dual parameter sensors, the structure we proposed has following advantages. Firstly, there is no resonance broadening caused by non-radiation loss. Secondly, the precision of manufacturing process is low because of simple construction. Thirdly, it's a simple experimental without considering the coupling of light as the optical fiber sensor. Most importantly, the FoM value is two orders of magnitude higher than the existing sensors of the same material [26].

Since the index relates to both concentration and temperature, when measuring with the refractometer, the temperature must be kept the same with the standard solution. This is difficult in practical operation, especially in trace analysis, where changes in temperature have a great influence on RI. Our structure can be used in this application by measuring temperature and refractive index simultaneously to exclude the interference of temperature on the index, and obtain more accurate concentration information. At the same time, higher FOM is conducive to the detection for subtle concentration variation. That means the structure is advantageous to in trace detection where the

value of detection limit is expected to be as small as possible. That means the structure is advantageous to in trace detection where the value of detection limit is expected to be as small as possible.

ACKNOWLEDGMENT

This work was supported by National Natural Science Foundation of China (NSFC) (No. 61675020)

REFERENCES

[1] S. Wang S , and L. Zu , "Simultaneous measurement of the BODconcentration and temperature based on atapered microfiber for water pollution monitoring," J. Appl. Opt., vol. 59, pp. 396831, 2020.

[2] J. Tang , L. Jin, and S. Li, "Density,Refractive Index and Volume Property of Imidazolium Chloride Ionic Liquid Aqueous Solution," J. Chemistry & Bioengineering, vol. 34, pp. 37-43, 2017.

[3] J. Hee, S. Choi, and O. Gyeongsik, "Qualitative identification of food materials by complex refractive index mapping in the terahertz range," J. Food chemistry, vol. 245, pp. 282-288, 2018.

[4] S. Pavel, and T. Brunger, "Refractive index-based determination of detergent concentration and its application to the study of membrane proteins," J. Protein Science, vol. 14, pp. 051543805, 2005.

[5] A. Jian, M. Jiao, and Y. Zhang, "Enhancement of the volume refractive index sensing by ROTE and its application on cancer and normal cells discrimination," J. Sensors and Actuators A Physical, vol. 313, pp. 112177, 2020.

[6] Y. Kong, P. Qiu, and Q. Wei, "Refractive index and temperature nanosensor with plasmonic waveguide system," J. Opt. Commun., vol. 371, pp. 132-137, 2016.

[7] F. Chen, and J. Li, "Refractive index and temperature sensing based on defect resonator coupled with a MIM waveguide," J. Modern Phys. Lett. B, vol. 33, pp. 9, 2019.

[8] W. Luo, R. Wang, H. Li, "Simultaneous measurement of refractive index and temperature for prism-based surface plasmon resonance sensors, "J. Opt. Express, vol. 27, pp. 576-589, 2019.

[9] Boris, Luk'yanchuk, I. Nikolay, and Zheludev, "The Fano resonance in plasmonic nanostructures and metamaterials," J. Nature materials, vol. 9, pp. 707-715, 2010.

[10] Pile, and F. P. David, "Gaining with loss," J. Nature Photonics, vol. 11, pp. 742-743, 2017.

[11] T. Liu, Y.Chen, and Q Han, "Sensor based on macrobent fiber Bragg grating structure for simultaneous measurement of refractive index and temperature," J. Appl. Opt., vol. 55, pp. 791-795, 2016.

[12] W. Zhang, Y. Liu, and T. Zhang, "Integrated Fiber-optic Fabry-Pé rotInterferometerSensor for Simultaneous Measurement of Liquid Refractive Index and Temperature," J. IEEE Sens. Journal, vol. 99, pp. 1, 2019.

[13] W. Wang, X. Dong, and D. Chu D, "Refractive index and temperature-sensing characteristics of a cladding-etched thin core fiber interferometer," J. Aip Advances, vol. 8, pp. 055104, 2018.

[14] Y. Liu, X. Liu, and T. Zhang, "Integrated FPI-FBG composite all-fiber sensor for simultaneous measurement of liquid refractive index and temperature," J. Optics and Lasers in Engineering, vol. 111, pp. 167-171, 2018.

[15] C. Zhang, S. Xu, and J. Zhao, "Multipoint refractive index and temperature fiber optic sensor based on cascaded no core fiber-fiber Bragg grating structures," J. Opt. Engineering, vol. 56, pp. 027102, 2017.

[16] D. Urbonas, and A. Balcytis, "Air and dielectric bands photonic crystal microringresonator for refractive index sensing," J. Opt. Lett., vol. 41, pp. 3655-3658, 2016.

[17] Y. Borwen, J. Lu, and C. Yu, "Terahertz refractive index sensors using dielectric pipe waveguides," J. Opt. Express, vol. 20, pp. 5858-5866, 2012.

[18] C. Zhang, G. Kang, and Y. Xiong, " Photonic thermometer with sub-millikelvin resolution and broad temperature range by waveguide-microring Fano resonance," J. Opt. Express, vol. 28, pp. 12599-12608 , 2020.

[19] A. Rahimzadegan, D. Arslan, and R. N. S. Suryadharma, "Disorder-Induced Phase Transitions in the Transmission of Dielectric Metasurfaces," J. Phys. Rev. Lett., vol. 122, pp. 015702.1-015702.6, 2019.

[20] S. Jahani S, and Z. Jacob, "All-dielectric metamaterials," J. Nature Nanotechnology, vol. 11, pp. 23-36, 2016.

[21] A. I. Kuznetsov, A. E. Miroshnichenko, and M. L. Brongersma, "Optically resonant dielectric nanostructures," J. Science, vol. 354, pp. 2472, 2016.

[22] G. Zhang, C. Lan, and R. Gao, "Trapped-Mode-Induced Giant Magnetic Field Enhancement in All-Dielectric Metasurfaces," J. The Journal of Phys. Chem. C, vol. 123, pp. 28887-28892, 2019.

[23] J. Hu, T. Lang, and M. Wu, "Refractive index sensing using all-dielectric metasurface with analogue of electromagnetically induced transparency," C. 2017 16th International Conference on Optical Communications and Networks (ICOCN). IEEE, 2017.

[24] K. Shih, P. Pitchappa, and L. Jin, "Nanofluidic terahertz metasensor for sensing in aqueous environment," J. Appl. Phys. Lett., vol. 113, pp. 071105, 2018.

[25] S. Xiao, T. Wang, and X. Jiang, "Strong interaction between graphene layer and Fano resonance in terahertz metamaterials," J. Journal of Phys. D Appl. Phys., vol. 50, pp. 195101, 2017.

[26] J. Hu, T. Lang, and G. Shi, "Simultaneous measurement of refractive index and temperature based on all-dielectric metasurface," J. Opt. Express, vol. 25, pp. 15241-15251, 2017.

[27] C. Tan, and Y. Huang, "Dependence of Refractive Index on Concentration and Temperature in Electrolyte Solution, Polar Solution, Nonpolar Solution, and Protein Solution," J. Journal of Chemical & Engineering Data, vol. 60, pp. 150903131520003, 2015.

[28] H. Liu, F. Pang, and H. Guo, "In-series double cladding fibers for simultaneous refractive index and temperature measurement," J. Opt. Express, vol. 18, pp. 13072-13082, 2010.

[29] G. Liu, X. Zhai, and L. Wang, "Actively Tunable Fano Resonance Based on a T-Shaped Graphene Nanodimer," J. Plasmonics, vol. 11, pp. 381-387, 2016.

[30] I. Staude, A. E. Miroshnichenko, and M. Decker, "Tailoring directional scattering through magnetic and electric resonances in subwavelength silicon nanodisks," J. Acs Nano, vol. 7, pp. 7824-7832, 2013.

[31] Bontempi, Nicolò, K. E. Chong, and H. W. Orton, "Highly sensitive biosensors based on all-dielectric nanoresonators," J. Nanoscale, vol. 9, pp. 4972-4980, 2017.

[32] G. Mie G, "Beitrge zur Optik trüber Medien, speziell kolloidaler Metallsungen," J. Annalen der Physik, vol. 330, pp. 377-445, 1908.

[33] U. Fano, "Effects of Configuration Interaction on Intensities and Phase Shifts,"J. Phys. Rev., vol. 124, pp. 1866-1878, 1961.

[34] G. Liu, X. Zhai, and L. Wang X, "A High-Performance Refractive Index Sensor Based on Fano Resonance in Si Split-Ring Metasurface," J. Plasmonics, vol. 13, pp. 15-19, 2017.

[35] Z. Liu, G. Fu, and X.Liu, "High-quality multispectral bio-sensing with asymmetric all-dielectric meta-materials," J. Journal of Phys. D: Appl. Phys., vol. 50,pp. 6384, 2017.

[36] K. S. Modi, J. Kaur, and S. P. Singh, "Extremely high figure of merit in all-dielectric split asymmetric arc metasurface for refractive index sensing," J. Opt. Commun., vol. 462, pp. 12532, 2020.

[37] Y. Zhang, W. Liu, and Z. Li, "High-quality-factor multiple Fano resonances for refractive index sensing,"J. Opt. Lett., vol. 43, pp. 1842, 2018.

[38] V. Savinov, V. A. Fedotov and N. I. Zheludev, "Toroidal dipolar excitation and macroscopic electromagnetic properties of metamaterials," J. Phys. Rev. B, vol. 89, pp. 205112, 2014.

[39] V. A. Fedotov, M. Rose, and S. L. Prosvirnin, "Sharp trapped-mode resonances in planar metamaterials with a broken structural symmetry," J. Phys. Rev. Lett., vol. 99, pp. 147401, 2007.

[40] M. F. Limonov, M. V. Rybin, and A. N. Poddubny, "Fano resonances in photonics," J. Nature Photonics, vol. 11, pp. 543-554, 2017.

[41] J. Gu, R. Singh and X. Liu, " Active control of electromagnetically induced transparency analogue in terahertz metamaterials,"J. Nature Communications, vol. 3, pp. 1151, 2012.

Gap in pagination due to unavailable papers.

Pages 1447-1450

Enhanced Ion Sensing Stability with Nanotextured Biosensors

Yanfang Wang, and Yuanjing Lin[*], *Member, IEEE*

Abstract— **The rapidly growing market on healthcare wearable electronics has been witnessed and raised enormous research interest in biosensors for noninvasive health monitoring. Electrolytes are present as an essential in the human body, while properly maintaining the balance of the electrolytes in our bodies is critical to ensure the normal function of our body. Herein, in order to achieve reliable detection of the sodium (Na^+) concentrations in the body fluids, nanotextured biosensors were designed and fabricated. The ionic detection is realized by a Na^+ ion selective membrane assembled onto a nanotextured Au @ PEDOT: PSS layer. The sensor prepared on flexible substrate allows monitoring of Na^+ concentration levels through a highly selective and sensitive electrochemical process, and is capable to achieve long-term stability with significantly reduced signal drift, which has been rarely studied. The results proved that nanotextured biosensors can be a practical and effective method to enhance ion sensing stability.**

Keywords— nanotextured biosensor; ion sensing; noninvasive health monitoring; long-term stability

I. INTRODUCTION

The noninvasive health monitoring technique has gained increasing interest in clinical diagnosis and personalized health care in recent years because it is more convenient and more friendly compared with the traditional clinical procedures. Normally, it relies on sensors as the key component to monitor critical analytes through body fluids, and refer to an integrated product in clothing or accessories fashion, such as watches [1], sweatbands [2], shirts [3], gloves [4], glasses [5], patch [6, 7] and contact lenses [8], to fulfill the intelligent functions. Body fluids like sweat, saliva, tears and urine that can be acquired in noninvasive approaches contain a huge library of biomarkers as those can be invasively obtained by blood tests [4]. These biomarkers can serve as human health indicators. Human body normally has a well maintained and proportioned mixture of varies electrolyte ions and other components, such as proteins, glucose and lactate. Studies have showed that a disproportionate amount of specific ion could interfere with the ion equilibrium status inside human body and cause a corresponding variation in the sweat secretion composition. Sodium ion, for instance, is a key biomarker to indicate the balance state of the body fluids and reflect our health state. A massive loss of sodium ions in body fluid is a note of alarm for patients with cystic fibrosis, because it can cause hyponatremia, while a long term continues increasing on the concertation of sodium ion could be considered as a sign which relate to high blood pressure[9].

With the increasing demand for clinical diagnosis and heightened awareness of personalized healthcare, wearable biosensors are believed to provide ideal approaches to implement noninvasive health monitoring in real time, which is necessary for daily and personalized healthcare. Biosensors fabricated on flexible platforms can closely adhere to human skin and monitor the physiological signals of the human body without disturbing our daily life. In the real application of the wearable potentiometric ion sensors, selectivity, sensitivity and stability are crucial factors. While a large amount of research effort has been invested to achieve high selectivity and sensitivity, the long-term stability of noninvasive biosensing still requires much improvement.

In this work, we developed a nanotextured biosensor for selective Na^+ sensing. The sensor electrode is modified with Au nanodendritic structure and well controlled active materials mass loading. The largely increased electrode surface area could facilitate the ion diffusion and charge transfer. Thus, the potentiometric response of the as-fabricated sensor delivers significantly enhanced sensing stability and largely suppressed signal drift over 30 hours continuous ion sensing. Besides, its sensitivity is desirable for a relatively low concentration range, especially for noninvasive monitoring via sweat. The as-developed nanotextured biosensors fabrication strategy can be applied to a large library of biomarkers, including various electrolyte ions, for their promising applications in health monitoring.

II. RESULTS AND DISCUSSIONS

For Na^+ sensing, a two-electrode system that contains a Ag/AgCl reference electrode and a functionalized working electrode was designed and fabricated on a flexible polyethylene terephthalate (PET) substrate to achieve noninvasive potentiometric detection in biofluids, as shown in Figure 1. The conductive electrode patterns were made through evaporation of Au on a thin and flexible PET. The working electrode was then nanotextured with Au dendritic nanostructures to enhance biosensing response and better adhesion of active materials. The dendritic nanostructure grown on the Au pattern by over-potential deposition via applying a periodic voltage wave with amplitude of -2 V, frequency of 50 Hz and duty cycle of 50% in gold planting solution (50 mM $HAuCl_4$ and 50 mM HCl) for 3000 cycles following previous reports [10]. Following that, a layer of conducting polymer Poly(3,4-ethylenedioxythiophene) modified with Poly(sodium-p-styrene-sulfonate) (PEDOT-PSS) was electrochemically deposited on the Au dendrites. PEDOT has the characteristics of high conductivity, simple molecular structure, and small energy gap. And it is widely used in the research of thin film materials, especially in the field of electronic devices due to its almost transparent thin layer. PEDOT is doped with a water-soluble polymer electrolyte PSS to obtain a PEDOT/PSS thin layer with high conductivity, high mechanical strength and good stability. This layer serves as an ion-to-electron transducer, and ensures stable electroactivity due to its low sensitivity to O_2 and pH to ensure reliable biosensing [11, 12]. The as-fabricated dendritic Au @ PEDOT: PSS nanostructures is shown in the SEM image in Figure 1. The selective potentiometric sensing of Na$^+$ is achieved by assembling an ion selective membrane consisted of Na ionophore, lipophilic ion exchanger Na-TFPB, PVC as supporting polymer and DOS as plasticizer.

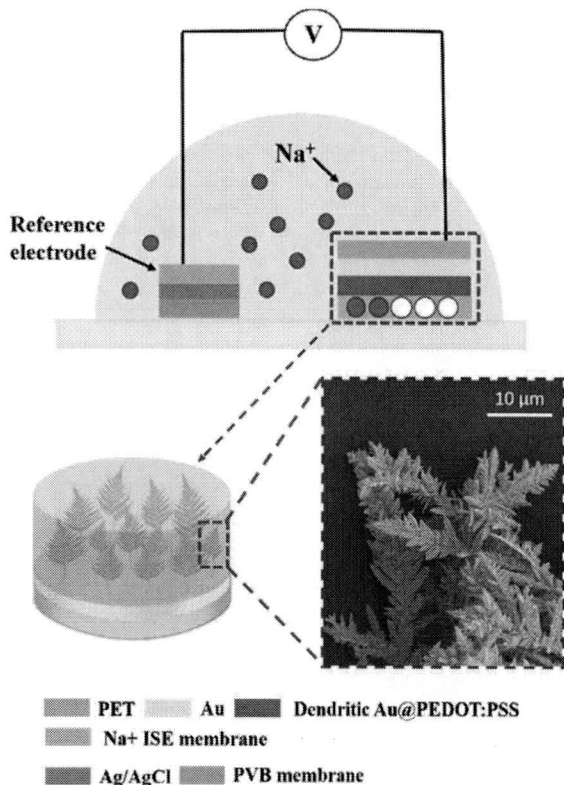

Figure 1. Schematic illustration of the Na$^+$ biosensor and SEM image of dendritic nanostructures.

The ion sensing performance of the as-fabricated sensor was tested by electrochemical methods, including instance open circuit voltage-time and electrochemical impedance. The electrochemical behavior of the as-fabricated Na$^+$ sensors were characterized in standard solutions with Na$^+$ concentrations from 15 mM to 120 mM, which is in the range as that in sweat. The corresponding results were studied systematically. As shown in Figure 2a, both the as-prepared sensors with thin film or nanotextured electrodes deliver potentiometric response to the Na$^+$ concentration variation, while the nanotextured one contributes to higher response signals. This is mainly due to the reduced intrinsic resistance with dendritic nanostructures, which can be proved with the electrochemical impedance spectrum (EIS) in Figure 2b. As demonstrated in Figure 2b, it is very clear that a smaller arc at higher frequencies was observed at nanotextured electrodes than thin film sensor, which indicates a reduced charge-transfer resistance at the nanotextured sensor interface. Besides, compared with the thin film sensor electrode, the nanotextured sensor delivers a superior sensing recovery with negligeable hysteresis in repeated response to the same ion concentration, as indicated by the dash line in Figure 2a. It is believed that the large surface area of dendritic Au@ PEDOT: PSS layer facilitates ion absorption and reduces the ion charge transport impedance, which contributes to the suppressed hysteresis.

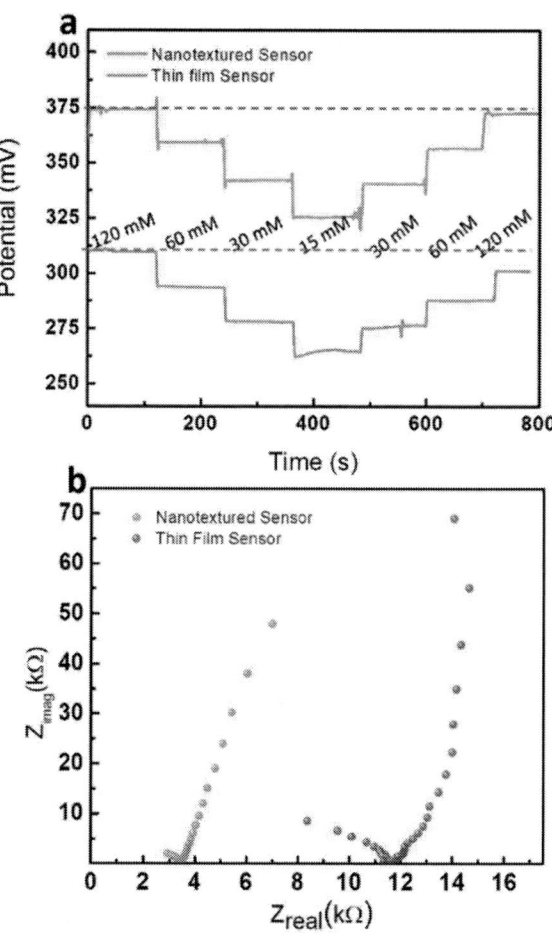

Figure 2. Electrochemical performance comparison of nanotextured and thin film sensors. (a) Potentiometric response in standard Na$^+$ solutions and (b) EIS curves.

Serving as the ion-to-electron layer, the optimization of PEDOT: PSS has a critical influence on the ion sensing performance without doubt. Compared with constant potential deposition, pulse deposition is easier to obtain a conformal coating layer. By using pulse deposition, the resulted mass loading and morphology of the active materials can be controlled by parameters such as waveform, frequency, duty cycle and average current density. By optimizing the parameters in this electrodeposition process, it is possible to obtain a uniform active material coating on the nanotextured electrode. During pulse deposition process, after applying a pulse potential, the electrolyte ions consumed at the electrode-electrolyte interface can be replenished within the pulse interval. Thus, the electrolyte concentration polarization at the interface is effectively reduced. Besides, it hinders the growth of large particles, so that it is not easy to form coarse particles.

To achieve conformal coating of PEDOT: PSS onto the nanodendritic Au electrode, pulse deposition under a periodic voltage wave with amplitude of 0.865 V, frequency of 1 Hz, duty cycle of 25 % was utilized [13]. Figure 3a shows the performance comparison between the sensor with conformal PEDOT: PSS layer deposited under pulse sensor and the one using a constant potential of 0.865 V for PEDOT: PSS deposition. It can be seen that the pulse deposition strategy contributes to a larger response signal. This result proves that the pulse deposition realizes successful PEDOT: PSS mass loading, and indicates that a conformal layer coating can reduce the sensor resistance, and thus facilitate electron transfer.

Apart from providing a conformal material coating on complicated nanostructures, the layer thickness can be well controlled under pulse deposition method by adjusting the deposition time represented by the deposition cycle numbers. The ion sensing performance of sensors with different PEDOT: PSS thickness is displayed in Figure 3b and the corresponding sensitivities are plotted in Figure 3c. In this work, PEDOT: PSS was deposited with a cycle number of 630, 840 and 1050, respectively. The results indicate that the optimized performance was achieved with 840 cycles of pulse deposition, generating a high sensitivity of 55.5 mV/decade and eliminated hysteresis in varying ion concentrations. It can be concluded that inadequate mass loading of ion-to-electron transfer materials could result in relatively poor sensing recovery, while the sensitivity could be sacrificed when the layer thickness keeps increasing.

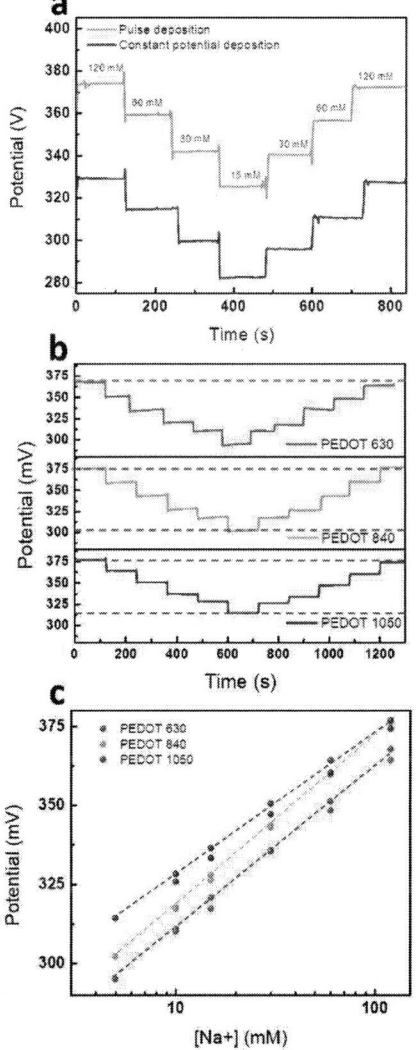

Figure 3. Optimization of deposition on PEDOT:PSS. Potentiometric response comparison between (a) different deposition methods and (b) different deposition cycle numbers and (c) the corresponding sentivities.

The as-fabricated nanotextured electrodes with optimized active material coating deliver remarkable enhancement on long-term ion sensing stability. It is worth mentioning that signal drift is one of the critical form factors for sensing performance evaluation, especially for wearable healthcare applications. Sensor drifting refers to the phenomenon that the output of the sensor changes with time when the input is unchanged. Reliable sensing is unlikely to be demonstrated without significant suppressed drift, simply because large errors will show up when the drift keeps accumulating over long-term using. There are two reasons for sensor drifting, one is the structural parameters of the sensor itself and another is the surrounding environment (such as temperature, humidity, etc). As shown in Figure 4a and b, the nanotextured sensor shows a largely reduced potential drift than thin film sensor in different concentrations over a long-term test of 35 hours. For non-invasive sensors, long-term stability is still a huge challenge, such long-term continuous monitoring of up to 35 hours has rarely been seen before. The nanotextured sensor has a potential drift of less than 1 mV hr^{-1}, while the thin film one shows a signal drift of over two times larger, as shown in Figure 4c. The results indicate that the nanotextured sensor with conformal active material coating can significantly enhance the long term sensing stability. This desirable form factor can be attributed to the suppressed capacitance on the nanotextured electrode with reduced charge accumulation. Besides, a robust adhesion between the ion selective membrane and the electrode can be achieved with the nanotextured morphology, so as to eliminate possible mechanical interference, such a introduced undesirable capacitance due to the formation of hydration layer between layers of materials.

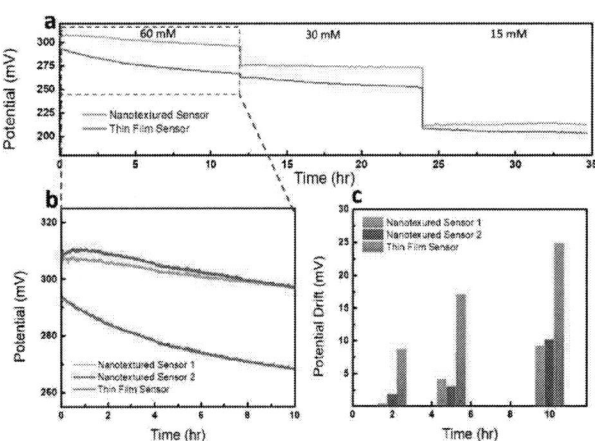

Figure 4. Ion sensing stability with nanotextured sensors. (a) 35 hours long-term testing and (b) repeatability with two nanotextured sensors compared with thin film sensor. (c) Potential drift of the sensors.

III. CONCLUSION

In this work, nanotextured sensors for Na$^+$ ion sensing were fabricated and systematically characterized. Optimized fabrication approach was explored and the remarkable electrochemical performance enhancement was demonstrated. The as-fabricated nanotextured sensors achieve a superior sensing stability with eliminated hysteresis and largely suppressed signal drift without sacrificing selectivity and sensitivity. The as-demonstrated enhanced ion sensing stability is desirable for long-term and real time noninvasive health monitoring via body fluids like sweat. The biosensor fabrication approach developed in this work can also be applied for sensitive and reliable monitoring of a variety of ions and other health relevant biomarkers, which is critical for wearable biosensors innovation.

ACKNOWLEDGMENT

This work was supported by Engineering Research Center of Integrated Circuits for Next-Generation Communications Grant(Y01796303) and Southern University of Science and Technology Grant (Y01796108, Y01796208).

REFERENCES

[1] J.Q. Zhao, Y.J. Lin, J.B. Wu, H.Y.Y. Nyein, M. Bariya, Li-Chai Tai, M.H. Chao, W.B. Ji. G. Zhang, Z.Y. Fan, A. Javey, "A Fully Integrated and Self-Powered Smartwatch for Continuous Sweat Glucose Monitoring," *ACS Sensors*, vol. 4, pp. 1925-1933, July 2019.

[2] S.Q. Wang, Y.J. Gu, T. Li, H. Luo, L. H. Li, Y.Y. Bai, L.L. Li, L. Liu, Y.D. Cao, H.Y. Hai, T, Zhang, "Wearable Sweatband Sensor Platform Based on Gold Nanodendrite Array as Efficient Solid Contact of Ion-Selective Electrode," *Anal. Chem.*, vol. 89, pp. 10224-10231, October 2017.

[3] L. Wang, L.Y. Wang, Y. Zhang, J. Pan, S.Y. Li, X.M. Sun, B. Zhang, H.S. Peng, "Weaving Sensing Fibers into Electrochemical Fabric for Real-Time Health Monitoring," *Adv. Funct. Mater.*, vol. 28, p. 1804456, October 2018.

[4] M. Bariya, L. Li, R. Ghattamaneni, C.H. Ahn, H.Y.Y. Nyein, L.C. Tai and A. Javey, "Glove-based sensors for multimodal monitoring of natural sweat," *Sci. Adv.*, vol. 6, p. eabb8308, 2020.

[5] J.R. Sempionatto, T. Nakagawa, A. Pavinatto, S.T. Mensah, S. Imani, P. Mercier and J. Wang, "Eyeglasses based wireless electrolyte and metabolite sensor platform," *Lab Chip*, vol: 17, pp:1834-1842, April 2017.

[6] X. Gang et al., "Smartphone-based battery-free and flexible electrochemical patch for calcium and chloride ions detections in biofluids," *Sens. Actuators B Chem.* vol: 297, p.126743, 2019.

[7] V. Mazzaracchio, L.Fiore, S. Nappi, G. Marrocco and F. Arduini, "Medium-distance affordable, flexible and wireless epidermal sensor for pH monitoring in sweat," *Talanta*, vol:222, p. 121502, 2021.

[8] R. Badugu, H. Szmacinski, E.A. Reece, B.H. Jeng and J.R. Lakowicz, "Sodium-sensitive contact lens for diagnostics of ocular pathologies," Sens. Actuators B Chem., vol: 331, p. 129434, 2021.

[9] Y. Lin, M. Bariya and A. Javey, "Wearable biosensors for body computing," *Adv. Funct. Mater.*, 2008087, 2020.

[10] Y. Lin et al., "Porous Enzymatic Membrane for Nanotextured Glucose Sweat Sensors with High Stability toward Reliable Noninvasive Health Monitoring," *Adv. Funct. Mater.* , vol. 29, no. 33, 2019.

[11] A. Elschner, S. Kirchmeyer, W. Lovenich, U. Merker, K. Reuter, " PEDOT: Principles and Applications of an Intrinsically Conductive Polymer," *Taylor & Francis*, London, UK, 2010.

[12] J. Bobacka, "Conducting polymer-based solid-state ion-selective electrodes," *Electroanalysis*, vol. 18, pp: 7-18, 2006.

[13] Y.J. Lin, Y. Gao, F. Fang and Z.Y. Fan, "Printable Fabrication of Nanocoral-Structured Electrodes for High-Performance Flexible and Planar Supercapacitor with Artistic Design," *Adv. Mater.*, vol: 29, November 2017.

Gap in pagination due to unavailable papers.

Pages 1455-1460

Proceedings of the 16th Annual IEEE International
Conference on Nano/Micro Engineered and Molecular Systems
April 25-29, 2021

Velocity Random Walk Modelling of a Silicon MEMS Resonant Accelerometer based on Non-AGC Control Loop

Dong Li, Qiancheng Zhao and Jian Cui*

***Abstract*—Automatic gain control (AGC) is widely used in the closed-loop drive circuit of silicon resonant accelerometer, but the non-AGC scheme is rarely reported. In this paper, a simplified velocity random walk model of a non-AGC scheme which uses comparator instead of AGC module is established. Firstly，for linear devices, we construct the phase noise model of the according to the Leeson model. Then, we get the transmission characteristics of the comparator noise by measurement. Finally, we predict that the white frequency noise (2.706 μg/√Hz) at output is introduced by the amplifier (1.754 μg/√Hz) and the comparator (2.060μg/√Hz). The predict value is in good agreement with the measurement result of 2.818 μg/√Hz. This model provides guidance for further optimization of resonant accelerometer performance.**

I. INTRODUCTION

MEMS accelerometers are widely used in industry, military, consumer and other fields because of its low power consumption and compact size.[1] Among them, MEMS silicon resonant accelerometer(SRA) has been widely concerned because of its high precision, quasi digital output and strong anti-interference.[2-5] Generally, the driving circuit of resonant accelerometer includes two types: automatic gain control(AGC) scheme[4,6] and non-AGC scheme[7,8]. AGC module can control the resonance displacement constant, so as to reduce the influence of environment such as temperature on SRA. However, AGC has complex structure, high power consumption and high 1/f noise. Therefore, more researchers focus on the non AGC scheme. The comparator is the most commonly used to replace AGC modular in non-AGC schemes. Using comparator to drive MEMS resonant accelerometer can simplify the circuit structure, reduce the noise introduced by AGC module, and reduce power consumption. Many researchers use comparator instead of AGC module to drive resonant accelerometer and obtain good results [7, 8]. However, as a nonlinear device, it is difficult to analyze the noise characteristics of the comparator systematically, so the influence of the comparator on the performance of the accelerometer is hard to get.

In this paper, the velocity random walk noise model of the closed-loop drive circuit including the comparator is constructed by measuring the noise transmission characteristics of the comparator. By analyzing the phase noise model, the white frequency noise of the resonant

accelerometer is predicted, which is in good agreement with the measurement results.

II. NOISE MODELING OF RESONANT LOOP

A. Phase Noise Modelling for Liner Modules

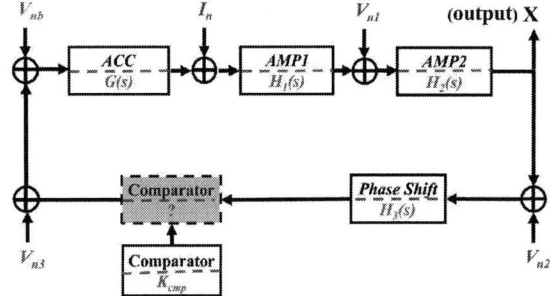

Figure 1.   Example of a figure caption. (figure caption)

The phase noise model of linear modules such as amplifiers can be constructed according to the Leeson model[9]. The resonant system is shown in Fig 1. Among them, "*AMP1*", "*AMP2*" and "*Phase Shift*" correspond to preamplifier, differential amplifier and phase shifter respectively, which can be considered as linear module. "*Comparator*" is nonlinear module which is ignored temporarily and only linear devices are considered. *G(s)*, *H1(s)*, *H2(s)* and *H3(s)* are the transfer function of the corresponding module, while V_{nb}, V_{n1}, V_{n2} and I_n are the noise sources.

First, we analyze the effect of the resonator noise V_{nb} on the output. The transfer function from resonator noise V_{nb} to the output noise X is:

$$\frac{X(s)}{V_{vb}(s)} = \frac{G(s)H_1(s)H_2(s)}{1 - H_1(s)H_2(s)H_3(s)G(s)} \quad (1)$$

G (s) is the transfer function of the resonator:

$$G(s) = \frac{(K/M)s}{s^2 + \frac{\omega_n}{Q}s + \omega_n^2} \quad (2)$$

K is a constant coefficient, which is determined by the structure and size of the resonator, M is the mass of the

---

*Resrach supported by ABC Foundation.

Dong. Li is with the National Key Laboratory of Science and Technology on Micro/Nano Fabrication, Institute of Microelectronics, Peking University, Beijing 100871, P. R. China (corresponding author to e-mail: a67870007a@ 163.com).

Jian Cui is with the National Key Laboratory of Science and Technology on Micro/Nano Fabrication, Institute of Microelectronics,

Peking University, Beijing 100871, P. R. China. (corresponding author to e-mail: eric.cuijian@pku.edu.cn).

Qiancheng. Zhao is with the National Key Laboratory of Science and Technology on Micro/Nano Fabrication, Institute of Microelectronics, Peking University, Beijing 100871, P. R. China. (corresponding author to e-mail: zqc@pku.edu.cn).

978-1-6654-3008-1/21 $31.00 © 2021 IEEE

resonator, ω_n is resonant frequency and Q is the quality factor. In the resonant state, by taking equation (2) into equation (1), the transfer function can be obtained as follows:

$$\frac{X(s)}{V_{nb}(s)} = \frac{\omega_n / Q}{H_3(s)} \times \frac{s}{s^2 + \omega_n^2} \tag{3}$$

The noise power spectral density of V_{nb} at the output X can be obtained by squaring equation (3):

$$S_{x\_Vnb}(\omega) = \frac{\overline{V_{nb}^2}}{|H_3(j\omega_n)|^2 Q^2} \left| \frac{j\omega\omega_n}{\omega_n^2 - \omega^2} \right| \tag{4}$$

Let $\omega = \Delta\omega + \omega_n$, equation (4) can be written as follows:

$$S_{x\_Vnb}(\Delta\omega) = \frac{\overline{V_{nb}^2}}{|H_3(j\omega_n)|^2 Q^2} \left(\frac{\omega_n}{2\Delta\omega} \right)^2 \tag{5}$$

The power spectrum of phase noise at the output X introduced by resonator noise V_{nb} is:

$$S_{\varphi\_Vnb}(\Delta\omega) = \frac{\overline{V_{nb}^2}}{2C|H_3(j\omega_n)|^2} \left(\frac{\omega_n}{2Q\Delta\omega} \right)^2 \tag{6}$$

C is the average power of carrier. The influence of other noise sources on the phase noise at the output can also be calculated by the above method. Finally, the total phase noise at the output is as follows:

$$S_{\varphi}(\Delta\omega) = b_0 + \frac{b_{-2}}{\Delta\omega^2}$$
$$= \frac{1}{2C} \left\{ \begin{array}{l} (H_1 H_2)^2 \overline{I_n^2} + H_2^2 \overline{V_{n1}^2} + [(H_1 H_2)^2 \overline{I_n^2} \\ + H_2^2 \overline{V_{n1}^2} + \overline{V_{n2}^2} + \left(\frac{1}{H_3} \right)^2 \overline{V_{nb}^2}] (\frac{\omega_n}{2Q \cdot \Delta\omega})^2 \end{array} \right\} \tag{7}$$

B. Measurement of Noise Transmission Characteristics of Comparator

The noise characteristics of the comparator are very complex, and are related to the threshold voltage and the amplitude of the input signal. We have measured the noise characteristics of the comparator at the point of work in closed-loop circuit, which can be seen as linear, as shown in Fig 2. According to the transmission characteristics of comparator noise, the output noise of the comparator can be simplified as the sum of the input noise multiplied by a fixed gain (K_{cmp}) and a fixed noise(V_{nc}) introduced by the comparator, as shown in Fig 1. The noise transfer characteristics of the comparator can be simplified as:

$$Y = K_{cmp} X + V_{nc} \tag{8}$$

The modified spectral density equation of phase noise is as follows:

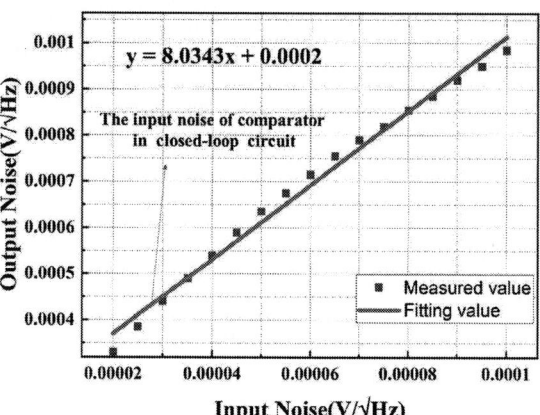

Figure 2. Noise transfer characteristic curve of comparator

$$S_{\varphi}(\Delta\omega) = b_0 + \frac{b_{-2}}{\Delta\omega^2}$$
$$= \frac{1}{2C} \left\{ \begin{array}{l} (H_1 H_2)^2 \overline{I_n^2} + H_2^2 \overline{V_{n1}^2} + [(H_1 H_2)^2 \overline{I_n^2} \\ + H_2^2 \overline{V_{n1}^2} + \overline{V_{n2}^2} + \left(\frac{1}{K_{cmp} H_3} \right)^2 (\overline{V_{nc}^2} + \overline{V_{nb}^2})] (\frac{\omega_n}{2Q \cdot \Delta\omega})^2 \end{array} \right\} \tag{9}$$

From equation (9), we can see that the white phase noise is mainly affected by the noise of the amplifiers: Amp1 and Amp2, while the white frequency noise is affected by all modules in the resonant loop.

III. Experiment

In order to verify the proposed noise model, we test the noise performance of the SRA which is driven by comparator. Figure 3 shows the structure of SRA that mainly consists of a proof mass, a pair of double-ended tuning forks (DETF)[10], two micro-leverages, four folded beams, and a supporting structure.

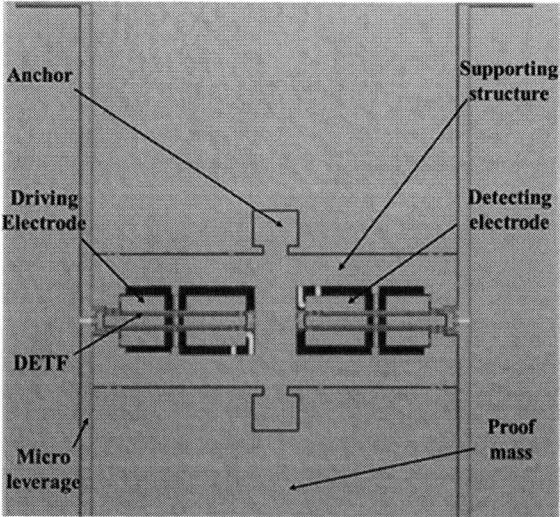

Figure 3. Optical photos of resonant accelerometer structure

Figure 4. Closed loop driving circuit of resonant accelerometer based on comparator

Figure 4 is the circuit used in the experiment, which uses a comparator to drive the SRA. The circuit consists of four modules: preamplifier, differential amplifier, phase shifter and comparator.

First, we will measure the power spectral density of the voltage noise at the output and calculate the white frequency noise according to it. Then, we will measure the SRA output for one hour and calculate its Allan variance. Theoretically, the velocity random walk obtained by Allan variance should be consistent with the white noise frequency calculated previously.

A. Power Spectrum Noise Measurement and Frequency White Noise Calculation

The voltage power spectral density at output X is shown in Figure 5, where the voltage white noise is 15.9 μV/√Hz. The white phase noise can be calculated by voltage white noise. According to equation (9), if the white phase noise is b_0, the white frequency noise b_{-2} can be calculated by b_0 and comparator noise.

TABLE I. NOISE CALCULATION

Frequency noise spectrum	
h_0 (White frequency noise)	
Amplifier contribution	Comparator contribution
1.754 μg/√Hz	2.060 μg/√Hz

According to the noise model, the white noise contributed by the comparator is 2.060 μg/√Hz, and the noise of the amplifier is 1.754 μg/√Hz.

B. Time Domain Test and Allan Variance Analysis

We use the comparator-driven circuit to drive SRA and measure the output of SRA for one hour. The sampling rate is 1 Hz. The time domain diagram of SRA output is shown in Figure 6. It can be seen in Figure 6 that the output acceleration of SRA has a significant drift due to the influence of temperature. Because the output noise of SRA is obviously less than the drift, the frequency white noise of SRA can not be obtained from the time domain diagram. We use Allan variance to analyze the short-term stability of SRA[11]. Generally, the lowest point of Allan variance represents the flicker noise corresponding to its bias instability, and when $\tau = 1$, it represents the velocity random walk, which is related to frequency white noise.

Figure 7 shows the Allan deviation of the SRA output one hour acceleration. It can be seen that the bias instability is 1.93 μg and the corresponding white noise is 2.82 μg.

According to the relationship between Allan variance and frequency spectral density:

$$\sigma_y^2(\tau) = \frac{h_0}{2}\tau^{-1} \tag{10}$$

We can get that the white frequency noise is 2.818 μg/√Hz, which is very consistent with the result predicted by our noise model.

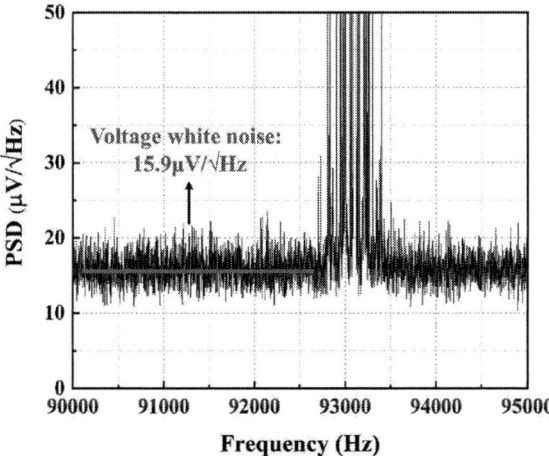

Figure 5. Spectral density of voltage noise at output

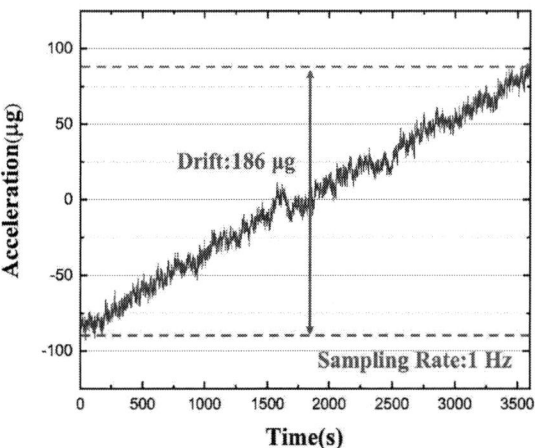

Figure 6. Time domain diagram of SRA output; measurement time: 1 hour; smapling rate:1 Hz.

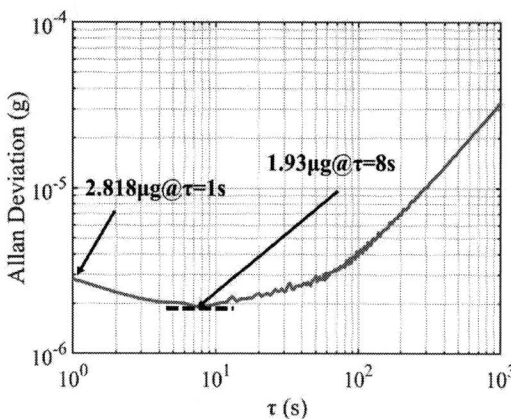

Figure 7. Allan deviation of resonant accelerometer driven by comparator, measurement time: 1 hour; smapling rate:1 Hz.

IV. CONCLUSION

In this paper, the velocity random walk noise model of SRA resonant loop driven by comparator is established. The comparator is regarded as a linear device in the proposed model. We predict the output white frequency noise of SRA by measuring the voltage power spectral density, and the result is 2.598 µg/ $\sqrt{}$ Hz, which is in good agreement with Allan variance test results(2.818 µg/$\sqrt{}$Hz). In the future, we can use this noise model to further optimize the driving circuit of SRA, so as to improve the performance of SRA.

REFERENCES

[1] Priyanka Aggarwal; Naser El-Sheimy; Aboelmagd Noureldin; Zainab Syed, MEMS-Based Integrated Navigation , Artech, 2010.

[2] J. Juillard, P. Prache, P. M. Ferreira and N. Barniol, "Resolution of phase difference and frequency measurements of mutually injection-locked oscillators for resonant sensing applications," 2018 Symposium on Design, Test, Integration & Packaging of MEMS and MOEMS (DTIP), Rome, Italy, 2018, pp. 1-4

[3] M. Pandit et al., "An Ultra-High Resolution Resonant MEMS Accelerometer," 2019 IEEE 32nd International Conference on Micro Electro Mechanical Systems (MEMS), Seoul, Korea (South), 2019, pp. 664-667

[4] J. Zhao et al., " A 0.23- µ g bias instability and 1- µ g/ √ Hz acceleration noise density silicon oscillating accelerometer with embedded frequency-to-digital converter in PLL," in IEEE Journal of Solid-State Circuits, vol. 52, no. 4, pp. 1053-1065, April 2017

[5] Y. Yin et al., "Design and test of a micromachined resonant accelerometer with high scale factor and low noise", Sensors and Actuators A: Physical, Volume 268, 2017, pp 52-60.

[6] Suk-Chang Yun, Sangkyung Sung, Taesam Kang and Young Jae Lee, "A study on the automatic gain control loop design for the resonant accelerometer," 2007 International Conference on Control, Automation and Systems, Seoul, Korea (South), 2007, pp. 1378-1382

[7] C. Comi, A. Corigliano, G. Langfelder, A. Longoni, A. Tocchio, B. Simoni, "A resonant microaccelerometer with high sensitivity operating in an oscillating circuit," J Microelectromech Syst, vol. 19, pp. 1140-1152, May, 2010.

[8] J. Zhang, A. Qui, Q. Shi, G. Xia, Y. Zhao, " A compact low-power oscillation circuit for the high performance silicon oscillating accelerometer," in Proc. MSREE 2017

[9] D. B. Leeson, "A simple model of feedback oscillator noise spectrum," in Proc. IEEE, vol. 54, no. 2, pp. 329-330, Feb. 1966

[10] J. Cui, H. Yang, D. Li, Z. Song, and Q. Zhao, "A Silicon Resonant Accelerometer Embedded in An Isolation Frame with Stress Relief Anchor," Micromachines, vol. 10, no. 9, p. 571, Aug. 2019.

[11] Allan D W and Levine J 2016 A Historical Perspective on the Development of the Allan Variances and Their Strengths and Weaknesses. IEEE transactions on ultrasonics, ferroelectrics, and frequency control 63 513-519.

Proceedings of the 16th Annual IEEE International
Conference on Nano/Micro Engineered and Molecular Systems
April 25-29, 2021

A Piezoresistive Pressure Microsensor Based on Simplified Fabrication Processes

Qinggang Meng, Yulan Lu, Junbo Wang*, *Member, IEEE,* Deyong Chen, *Member, IEEE,* Jian Chen
and Bo Xie

Abstract— This paper presents a piezoresistive pressure microsensor composed of a Silicon on Insulator (SOI) layer with sensing elements (a pressure-sensitive diaphragm and four piezoresistors) and a glass cap for hermetic package. The pressure under measurement bends the pressure-sensitive diaphragm, producing resistance changes of underlining piezoresistors. Numerical simulations were conducted, where key structure parameters of piezoresistors were optimized, producing a sensitivity of 18.61 mV/(V·MPa) and a linearity of 0.74%FS. The proposed microsensor was fabricated based on simplified fabrications, which included only two etching processes and one anodic bonding. Compared to other microfabrications, the fabrication of the developed microsensor does not need ion implantation and thin-film deposition, leading to high uniformity with low residual stress. Fabricated microsensors were characterized, obtaining the sensitivity of 17.278 mV/(V·MPa) and the linearity of 0.613 %FS (0-2.5 MPa).

I. INTRODUCTION

The piezoresistive pressure microsensors are widely used in the field of medical, petrochemical, aerospace, and power machinery [1-4]. Compared with other types of pressure microsensors such as resonant pressure microsensors, capacitive pressure microsensors, and piezoelectric pressure microsensors, the piezoresistive pressure microsensor is featured with low cost, high linearity and large output single [1-2]. The key structure of the piezoresistive pressure microsensor is the pressure-sensitive diaphragm above the cavity and the piezoresistors on the surface of the diaphragm. For absolute pressure microsensors, the cavity should have a high degree of vacuum. In addition, the residual stress must be low enough to ensure the accuracy of the measurement and the long-term stability of the microsensor. However, due to the influence of some fabricating processes such as thin film deposition, thermal annealing, and ion implantation in the traditional manufacturing process [5], the manufactured microsensors always have a lot of residual stress and production errors inevitably, which make the performance of the microsensor not optimal. For example, T.A. Vang proposed a stress concentration structure based on a cross-beam membrane and peninsula (CBMP) diaphragm [5], the fabrication process contains once thermal oxidation, twice ion implantation, three-step thin film deposition, and five-step etching, which is so complex that it brings a lot of residual

stress to the device, reduce the reliability of the microsensor. M. Manjunath use once fusion bonding, several-step sputtering and ion implantation in the fabrication process of his bossed diaphragm coupled fixed guided beam structure [6], which has relatively higher cost and longer production cycle because of the complex process. Therefore, it is necessary to find a simple microsensor manufacturing method without complicated process steps, to reduce the cost, shorten the production cycle and reduce the fabrication error.

In this work, a simplified fabrication process is proposed. The core part of the microsensor can be fabricated by only two etching processes: by etching the Silicon on Insulator (SOI) handle layer, the vacuum cavity is formed, and the structure of the Wheatstone bridge is formed by etching the SOI device layer. All the structures of the Wheatstone bridge, including the piezoresistors, the connections, and the electrode are all made of bulk silicon, which has a thickness of 2 μm and a resistivity of 0.02 Ω·cm. In this way, many processes are eliminated, including thin film deposition, ion implantation, thermal annealing, etc. Thus, the fabrication error generated in the microsensor manufacturing process is controlled to a minimum extent.

The Finite Element Method (FEM) simulation is used to optimize the structure parameters of the microsensor to maximize the sensitivity. The characteristics of the microsensor are also shown in this paper.

II. MICRO-STRUCTURE OF MICROSENSOR

Two key structures of piezoresistive microsensors are the press-sensitive diaphragm and the piezoresistors located on the diaphragm. The diaphragm can neither be too thick to reduce the sensitivity of the microsensor nor too thin to increase the non-linearity. On the other hand, for the consideration of heat, the resistance of piezoresistors should not be too small, otherwise, the microsensor will generate a large current and affect the output performance. Therefore, an SOI wafer with 2 μm thickness device layer, 1 μm thickness oxide layer, and 400 μm thickness handle layer was selected for microsensor fabrication. The resistivity of device layer and handle layer is 0.02 Ω·cm.

Fig. 1 shows the microstructure of the piezoresistive pressure microsensor. The dark yellow region is the device

---

This work was financially supported by the National Key R&D Program of China (2018YFB2002302), the National Natural Science Foundation of China (Grant No. U1930206), and the National Science Fund for Distinguished Young Scholars (Grant No. 61825107).

Q.G. Meng, Y.L. Lu, J.B. Wang, D.Y. Chen and J. Chen are with the State Key Laboratory of Transducer Technology, Aerospace Information Research Institute, Chinese Academy of Sciences, Beijing 100190, China, and also with the School of Electronic, Electrical and Communication Engineering,

University of Chinese Academy of Sciences, Beijing 100049, China (e-mail: mengqinggang19@mails.ucas.ac.cn, corresponding author to provide phone: 135-2008-9501; e-mail: jbwang@mail.ie.ac.cn).

B. Xie is with the State Key Laboratory of Transducer Technology, Aerospace Information Research Institute, Chinese Academy of Sciences, Beijing 100190, China.

978-1-6654-3008-1/21 $31.00 © 2021 IEEE   1465

layer of SOI, the light yellow region is the oxide layer, the gray region is the handle layer, and the blue region is the glass cap for the hermetic package. As can be seen from Fig. 1(a), four piezoresistors are located at the center of four edges to obtain maximum stress. The connections have the same material and thickness as the piezoresistors, but the width is much wider than them, so the connections have a much smaller resistance than piezoresistors. The square structures on the four corners of microsensor are electrodes, which are used to deposit aluminum to transmit electrical signal. Fig. 1(b) shows the details of piezoresistor. The oxide layer (light yellow region in Fig. 1(b)) under device layer can isolate device layer and handle layer.

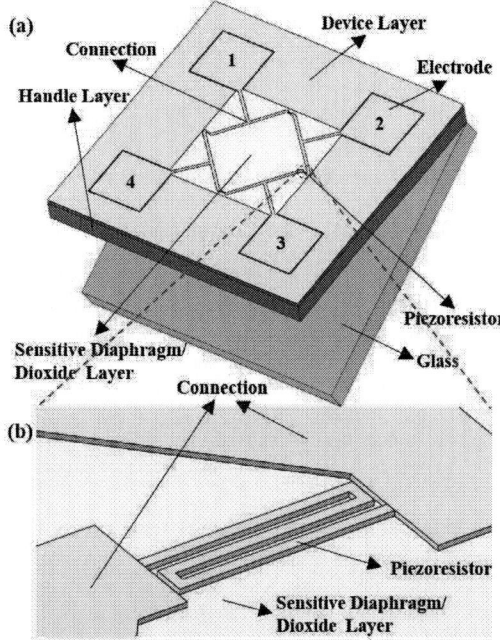

Fig. 1: Schematic diaphragm of the proposed piezoresistive microsensor, which comprises two sections, an SOI wafer - composed of a device layer, an oxide layer and a handle layer, and a glass cap for hermetic package (a) and details of piezoresistor, the dark yellow area is the etched device layer, the light yellow area is oxide layer which act as part of the sensitive diaphragm and achieve electrical insulation (b).

When the microsensor is under pressure, the pressure-sensitive diaphragm is deformed, and the four piezoresistors at the edge of diaphragm are subjected to different stress, resulting in different resistance change. When 5 V DC voltage was applied to Wheatstone bridge, a voltage output proportional to the relative change of resistance is generated between electrode 2 and 4. The designed pressure scale of the microsensor is 0-2.5 MPa.

III. DESIGN AND FABRICATION

A. Design and Analysis

With the increase of doping concentration, the ohmic contact between silicon and metal will be easier to make, but its piezoresistive coefficient will gradually decrease [8], resulting in a decrease in sensitivity of device output. In order to solve this contradiction, researchers use a low doping concentration in piezoresistor area to obtain a high piezoresistance coefficient and a high doping concentration in electrode area to obtain a good ohmic contact [9]. However, because the dose of ion implantation is difficult to control accurately, the concentration of doping is easy to deviation, resulting in inconsistent parameters of different batches [10]. Hence, the same silicon material with a certain doping concentration is selected to make the piezoresistor, the connection and the electrode in this work, so that there is no need to worry about the inconsistency of doping. Considering piezoresistive coefficient and ohmic contact of the microsensor, 5×10^{18} cm^{-3} doping concentration is chosen (corresponding to the resistivity of 0.02 $\Omega\cdot$cm). Besides, the resistance of piezoresistors should also be several thousand ohm, ensure that the working current of microsensor is several milliamps when the working voltage is 5 V. Therefore, the thickness of the device layer of SOI is selected to be 2 µm, in which case, the square resistance of device layer is 100 Ω. As long as the aspect ratio of the piezoresistors is greater than 10, the resistance of the piezoresistors can meet the requirements.

For pressure microsensors, thinner thickness of diaphragm means higher sensitivity. However, as the thickness becomes thinner, the non-linearity of the pressure-sensitive diaphragm will increase. Hence, researchers limit the maximum displacement thickness ratio of the diaphragm to less than 20%, which ensures that the nonlinearity of diaphragm deformation is less than 1% [7]. To shrink down the dimensions of microsensor, a square-shaped pressure-sensitive diaphragm with a side length of 2000 µm was used in this paper. Then, using FEM simulation to optimize diaphragm thickness. It can be concluded that the maximum displacement thickness ratio will decrease as the diaphragm thickness increases. The optimal diaphragm thickness is set to 70 µm, at which the maximum displacement thickness ratio of the diaphragm is 18.8%, as shown in Fig. 2(a).

Then, set the aspect ratio of the piezoresistor to 20, change the length of piezoresistor for simulation. Apply 5V DC voltage between electrodes 1 and 3, at the same time, measure the output voltage between electrodes 2 and 4, the simulation result is shown in Fig. 2(b). It can be seen from the figure that when the length of piezoresistor is 80 µm, the output has the highest sensitivity of 84.22 mV/MPa. Hence, the length of 80 µm is selected as the optimal parameter, and the width of piezoresistor is 4 µm. Then, in order to make more use of the maximum stress area of diaphragm and increase the resistance, the piezoresistor is designed as a snake shape. Considering the difficulty of fabrication, the piezoresistor is finally designed as the shape shown in Fig. 1(b).

Finally, the complete microsensor is simulated. The thickness of each layer of simulation model is set to 2 µm~1 µm~400 µm; a 330 µm deep hole is dug in handle layer to form a pressure-sensitive diaphragm with an area of 2×2 mm² and a thickness of 70 µm. The piezoresistor has a length of 80 µm, a width of 4 µm and a snake shape. The width of the connection is set to 50 µm. The experimental results show that the full-scale output voltage of microsensor is 232.601 mV, the sensitivity is 18.61 mV/(V·MPa), the R² value (correlation coefficient) of the output curve is 0.9998, and the linearity is 0.74%FS.

978-1-6654-3008-1/21 $31.00 © 2021 IEEE

(a)

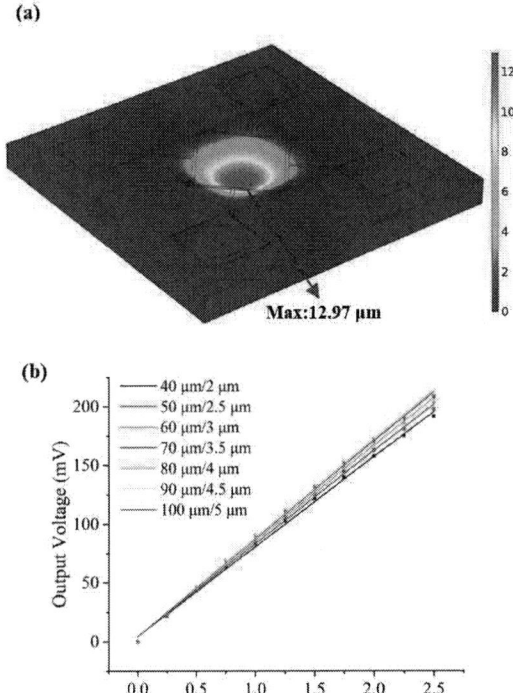

(b)

Fig. 2: the deformation simulation of the diaphragm when the thickness is 70 μm (a) the output voltage of Wheatstone bridge under different resistance sizes (b).

B. Fabrication and Packing

Fig. 3(a) shows the fabrication process of the microsensor in details. An SOI wafer with 2 μm thick device layer, 1 μm thick oxide layer, and 400 μm thick handle layer was selected for microsensor fabrication. The resistivity of device layer and handle layer is 0.02 Ω·cm. The selected SOI wafer has two (1 0 0) p-type silicon layer and a diameter of 4 inch.

(I) The AZ4903 photoresist was spin-coated on the backside of the SOI wafer. After photolithography and development, some certain areas of the photoresist are removed (because the selection ratio of photoresist is relatively low, a more viscus photoresist AZ4903 was chosen to obtain a thicker thickness of 12 μm). Then, the handle layer of SOI was partially removed by Deep Reactive Ion Etching (DRIE) to form the vacuum cavity.

(II) A dilute photoresist (AZ1500) is used in the patterning of the device layer. After patterning, DRIE was used to etch the device layer. Through etching, all of the key structures are fabricated at one time, including piezoresistors, connections and electrodes.

(III) The SOI wafer and BF33 glass were bonded together by anodic bonding. The bonding voltage is 800 V, the applied pressure is 1000 mbar, and the bonding temperature is 350°C.

(IV) A hard mask is patterned for Al electrodes. Through the holes in the hard mask, the round aluminum electrodes are evaporated onto the four pins of the Wheatstone bridge, whose thickness is about 1 μm.

The fabricated microsensor chip is shown in Fig. 3(b), which has the size of 5 mm×5 mm×0.94 mm.

The microsensor chip is mounted on the Kovar base with epoxy resin. The pressure-sensitive diaphragm of the microsensor is placed outwards to sense the pressure, the four electrodes of microsensor and the pins of Kovar base are connected by wire bonding. Finally, weld the seat and the cylindrical shell together by soldering.

(a)

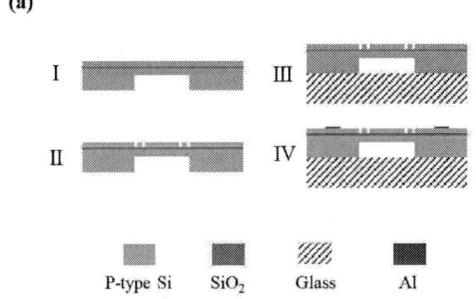

(b)

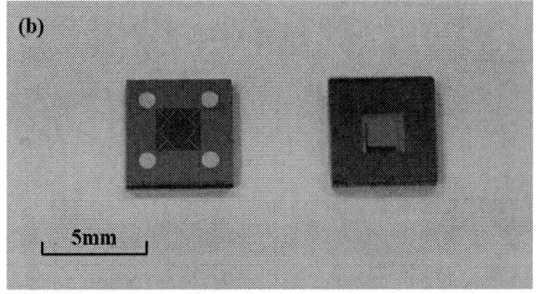

Fig. 3: Fabrication processes of the piezoresistive microsensor. (I) SOI handle layer etching to form the sensitive diaphragm (II) SOI device layer etching to form the Wheatstone bridge structure, including the piezoresistors, the connections and the electrodes (III) anodic bonding of BF33 glass and SOI wafer (IV) Al electrode evaporation (a), and the fabricated microsensor chip with 5mm side length (b).

IV. EXPERIMENTS AND RESULTS

The microsensor test system includes a thermostatic oven (SH 241, ESPEC, JPN), a digit multimeter (KEITHLEY 2100, Tektronix, USA), a DC power supply (KEITHLEY 2231, Tektronix, USA), a pressure controller (ATM/A 700K, DH Instruments, USA) and a computer used to record data.

978-1-6654-3008-1/21 $31.00 © 2021 IEEE

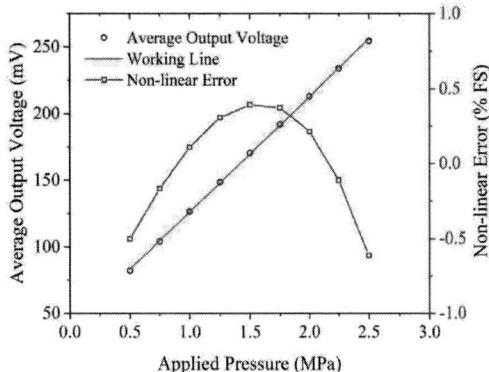

Fig. 4: The static characteristic of the microsensor, which comprise three groups of data - the average output voltage in three cycles of measurement, the working curve obtained by fitting the average output voltage, and the non-linearity error obtained by comparing the average output voltage with the working curve.

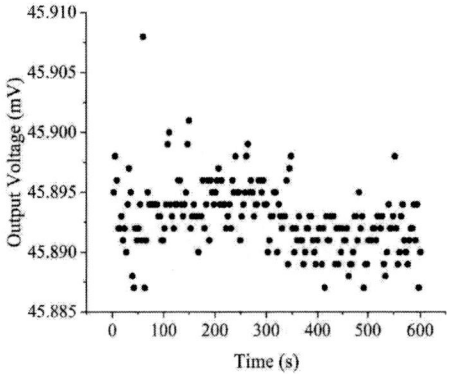

Fig. 5: The zero drift of the microsensor at room temperature and atmospheric pressure.

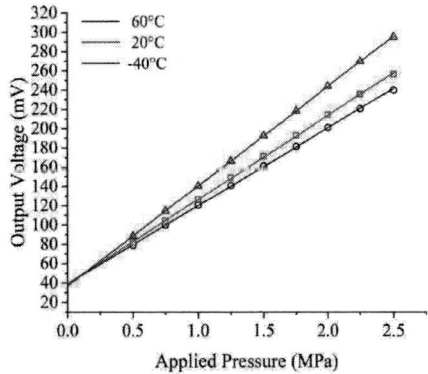

Fig. 6: The sensitivity drift of the microsensor between three different temperatures

In order to evaluate the static characteristics of the microsensor, use thermostatic oven to set the temperature to 20°C, then apply an increasing pressure from 0.5 to 2.5 MPa in steps of 0.25 MPa and record the output voltage of microsensor for 120 seconds at each pressure point. After that,

gradually reduce the pressure and record the voltage output to complete the first cycle. The whole test consists of three cycles.

The static characteristic test result of the microsensor is shown in Fig. 4. According to the extrapolation of the working line, it can be calculated that the zero output of the microsensor is 39.921 mV, the full-scale output is 215.974 mV, and the pressure sensitivity is 17.278 mV/(V·MPa). By comparing the average output voltage and working line of the microsensor, the maximum nonlinear error of the microsensor is obtained as 0.613 %FS. In addition, the hysteresis, repeatability, and accuracy of the microsensor are 0.051 %FS, 0.116 %FS, and 0.730 %FS respectively.

In order to characterize the stability of the microsensor, the zero drift and thermal sensitivity drift is tested. Connect the test system and apply 5V DC voltage to the microsensor, preheat it for 30 min. After that, continuously record 10 min output voltage data of the microsensor in steps of 3s. The result is shown in Fig. 5. It can be concluded from the test results that the zero drift of the microsensor is 0.0074%FS. Then, put the microsensor in a thermostat oven, keep it at room temperature (20°C), upper limit working temperature (60°C), and lower limit working temperature (-40°C) for 1h respectively, record the output voltage of each temperature point, and get the thermal sensitivity drift of 0.31 %/°C, as shown in Fig. 6.

V. CONCLUSION

A piezoresistive pressure microsensor based on simplified fabrication processes is proposed to simplify the manufacturing process and improve the performance of the microsensor. Different from traditional complicated piezoresistive microsensor manufacturing method, the manufacturing process of this microsensor is very simple. All the key structures can be fabricated by twice etching, many processes such as ion implantation and thin film deposition are omitted to obtain high uniformity with low residual stress.

The microsensor chip was bonded to the Kovar base with epoxy resin, and the metal shell was welded for subsequent testing.

The characteristic experiments of the microsensor show that the sensitivity and linearity of the microsensor are 17.278 mV/(V·MPa) and 0.613 %FS respectively when the external pressure changes from 0 to 2.5 MPa, which is consistent with the simulation results.

ACKNOWLEDGMENT

The author acknowledges the contributions of the research group members who have collaboratively contributed to this work, in particular, to Prof. Deyong Chen, Prof. Junbo Wang, Prof. Jian Chen and A.P. Bo Xie.

REFERENCES

[1] A.S. Fiorillo, C.D. Critello, S.A. Pullano, "Theory, technology and applications of piezoresistive sensors: A review," Sensors and Actuators A: Physical, vol. 281, 2018, pp. 156-175.

[2] S.C. Fan, *Sensing technology and Application.* Beijing, CN: Beijing University of Aeronautics and Astronautics Press, 2010.

[3] Pramanik C., Saha H., "Low pressure piezoresistive sensors for medical electronics applications," Materials and Manufacturing Processes, 21(3), 2006, pp. 233-238.

[4] Fiorillo A.S., "A piezoresistive tactile sensor," IEEE Transactions on Instrumentation & Measurement, 46(1), 1997, pp. 15-17.

[5] T.A. Vang, "Research on a Novel Structure MEMS Piezoresistive Pressure Sensor," Guangzhou, CN: South China University of Technology, 2018.

[6] M. Manuvinakurake, U. Gandhi, M. Umapathy, et al. "Bossed diaphragm coupled fixed guided beam structure for MEMS based piezoresistive pressure sensor," Sensor Review, 39(4), 2019, pp. 586-597.

[7] X.G. Yuan, *Sensor Technology Handbook*. Beijing, CN: National Defense Industry Press, 1986, pp. 107-108.

[8] O.N. Tufte, E.L. Stelzer. "Piezoresistive Properties of Silicon Diffused Layers," Journal of Applied Physics, 34(2), 1963, pp. 313-318.

[9] P. Song, C. Si, M. Zhang, et al. "A Novel Piezoresistive MEMS Pressure Sensors Based on Temporary Bonding Technology," Sensors (Basel, Switzerland), 20(2), 2020.

[10] L.N. Large, R.W. Bicknell, "Ion-implantation doping of semiconductors," Journal of Materials Science, 2(6), 1967, pp. 589-609.

Gap in pagination due to unavailable papers.

Pages 1470-1477

Proceedings of the 16th Annual IEEE International
Conference on Nano/Micro Engineered and Molecular Systems
April 25-29, 2021

Mass fabrication of 4H-SiC high temperature pressure sensors by femtosecond laser etching*

Lukang Wang
the State Key Laboratory for Mechanical Manufacturing Systems,
Xi'an Jiaotong University
Xi'an, China
wanglukang@stu.xjtu.edu.cn

You Zhao*
the State Key Laboratory for Mechanical Manufacturing Systems,
Xi'an Jiaotong University
Xi'an, China
zhaoyou628@xjtu.edu.cn

Yulong Zhao
the State Key Laboratory for Mechanical Manufacturing Systems,
Xi'an Jiaotong University
Xi'an, China
zhaoyulong@xjtu.edu.cn

Yu Yang
the State Key Laboratory for Mechanical Manufacturing Systems,
Xi'an Jiaotong University
Xi'an, China
yangyuu@stu.xjtu.edu.cn

Bo Li
the State Key Laboratory for Mechanical Manufacturing Systems,
Xi'an Jiaotong University
Xi'an, China
li.bo.123.666@stu.xjtu.edu.cn

Taobo Gong
the State Key Laboratory for Mechanical Manufacturing Systems,
Xi'an Jiaotong University
Xi'an, China
gongtaobo@stu.xjtu.edu.cn

Abstract—**4H Silicon carbide (4H-SiC) is one of the most promising materials for pressure sensing in harsh environments. Due to the low efficiency and high energy consumption of dry etching process, mass manufacturing of SiC pressure sensors has always been a challenging problem. A Yb:KGW femtosecond laser was used to perform the circular diaphragms of bulk 4H-SiC by parallel straight line scanning. Small dimension errors and low surface roughness in rapid etching process were measured. This indicates the potential of utilising the femtosecond laser scanning approach to mass-fabricate bulk SiC sensor chips for pressure sensing.**

Keywords—4H-SiC, pressure sensor, femtosecond laser, mass fabrication

I. Introduction

Silicon carbide (SiC) has the potential to be an alternative to silicon (Si) for pressure sensing under harsh environments, especially at severe chemical or heat corrosion [1,2]. This is due to its superior mechanical strength, attractive electrical properties, high thermal conductivity, and excellent corrosion resistance [3]. The experimental testing proved that bulk 4H-SiC has a promising future for pressure sensing at extremely high temperature [4-6]. The extreme physical and chemical inertness of single crystal SiC leads to difficulties in bulk micromachining of deep structures. Traditional dry etching methods suffer from low etch rates, poor selectivity and the complicated masking procedures. Femtosecond lasers have proved to be a fast alternative with high material removal rate and little thermal impact [7-10].

In this paper, we described mass fabrication process for bulk 4H-SiC pressure chips in detail. Femtosecond laser used for rapid etching of sensor chip back cavity structure was proved to be feasible.

II. Manufacturing and Testing

A. Manufacture of sensor Wheatstone bridge

The Wheatstone full bridge configuration on the front surface of the 4H-SiC pressure sensor chip is designed in embossment pattern, in order to avoid the metal fracture caused by the mismatch of thermal expansion coefficient between piezoresistive element and metal layer. The piezoresistors are defined by the short circuit of the interconnect leads.

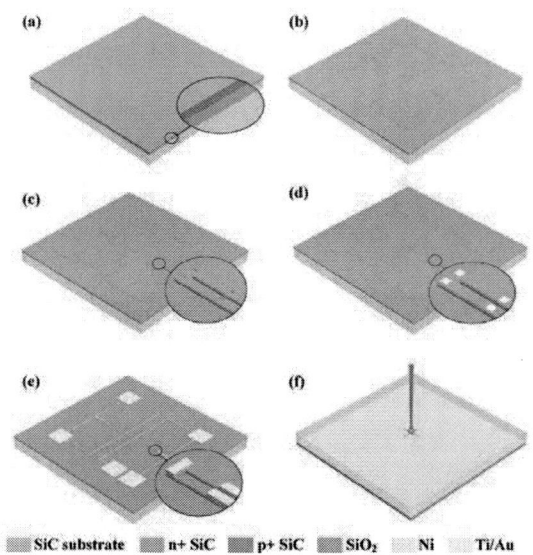

Fig. 1. Overview of the fabrication process.

Fig. 1 shows the fabrication process of the 4H-SiC pressure sensor chip. (a) Here a 350 μm thick n-type 4H silicon carbide wafer (SK siltron Co., Ltd., South Korea) was used as the starting material. The p and n type homoepitaxial layers grown on the 4° off-axis Si-face of the 4H-SiC wafer in succession were manufactured at EpiWorld International Co., Ltd., China. (b) The embossed Wheatstone bridge configuration is structured by ICP etching technology on the n type homoepitaxial 4H-SiC layers. The piezoresistors had the height of 2.0 μm, which ensured that the unmasked n-type functioning layer was thoroughly etched and the robust p-n junction isolation was formed. (c) As a passivation layer, 400 nm thick silicon dioxide (SiO_2) was deposited by plasma enhanced chemical vapor deposition process (PECVD). Contact holes were opened by BOE etching in the oxide to expose the ends of the piezoresistors that acted as lead terminals. (d) 100 nm thick Ni was sputtered in the contact holes as ohmic contact layer metal. The ohmic metallization

978-1-6654-3008-1/21 $31.00 © 2021 IEEE

was achieved through a rapid thermal annealing process using UNITEMP RTP-100 in nitrogen at 500°C for 5 minutes and 1000°C for 3 minutes. (e) The multiple layers of metal Ti(50nm)/Au(300nm) were sputtered sequentially on the embossed 4H-SiC Wheatstone bridge configuration to form interconnection bridges. (f) Metal alignment marks for laser drilling and cutting were deposited on the backside surface of the 4H-SiC wafer.

B. Femtosecond laser fast etching for back cavity

A three-axis ultrafast laser micromachining system GCC A4060 (Xi'an Micromach Technology Co., Ltd., China) equipped with a femtosecond laser was used for the mass production of the 4H-SiC pressure sensor diaphragms. The laser micromachining system comprises a lasing chamber, a machining platform, a laser scanning module and a control panel. The femtosecond laser technical parameters are shown in Table 1.

TABLE I.   TECHNICAL PARAMETERS OF FEMTOSECOND LASER

Laser parameter	Values
Centre Wavelength	1026 nm
Pulse duration	290 fs
Repetition rate	100 kHz
Average power	4 W
Scanline spacing	10 μm
Scanline deflection	30°
Scan times	310
Laser on delay	220 μs
The turning delay	10 μs
The end delay	10 μs

Fig. 2.   The 4H-SiC sensor chip was mass-fabricated by drilling and cutting from half a wafer through a three-axis ultrafast laser micromachining system.

As shown in Fig. 2, a silicon carbide sample being etched in the lasing chamber. Before the laser is turned on, adjust the displacement platform so that the focal plane of the laser beam is slightly lower than the surface of the material. The circular sensor back cavities are processed by directly writing a series of parallel straight-line passes, which are preset in the CAXA CAD. Based on the cross-shaped alignment mark on the surface of the silicon carbide wafer, the same pattern is copied in advance in this software. Scan times dominates the depth of the sensor back cavity. The thickness of the material layer removed by each laser scan is controlled by other energy parameters. Although back cavity manufacturing is a serial process, the rapid movement of laser scanning module makes wafer-scale processing possible. 128 back cavities of the pressure sensor chips were manufactured by laser etching in 2 hours using the technical parameters in table 1. Since the jump time of the laser beam is almost negligible, the average processing time for a single back cavity is about 1 minute. Table 2 shows the depth of 40 back cavities measured with a digital dial indicator altimeter. The average machining error of less than 7% demonstrates the feasibility of using femtosecond lasers for mass production of sensor back cavities.

TABLE II.   BACK CAVITY DEPTHS AND MACHINING ERRORS

Depth (mm)	error	Depth (mm)	error	Depth (mm)	error	Depth (mm)	error
0.102	2.0%	0.105	5.0%	0.106	6.0%	0.11	10.0%
0.103	3.0%	0.111	11.0%	0.105	5.0%	0.106	6.0%
0.105	5.0%	0.114	14.0%	0.104	4.0%	0.104	4.0%
0.112	12.0%	0.108	8.0%	0.105	5.0%	0.106	6.0%
0.104	4.0%	0.106	6.0%	0.104	4.0%	0.109	9.0%
0.103	3.0%	0.106	6.0%	0.104	4.0%	0.11	10.0%
0.104	4.0%	0.105	5.0%	0.106	6.0%	0.108	8.0%
0.105	5.0%	0.109	9.0%	0.102	2.0%	0.108	8.0%
0.105	5.0%	0.113	13.0%	0.107	7.0%	0.111	11.0%
0.109	9.0%	0.118	18.0%	0.108	8.0%	0.108	8.0%

Fig. 3(a) shows a cross section of the pressure sensor back cavity. The problem of over-etched grooves on the edge of the diaphragm mentioned in [7] are overcome, thanks to the optimized laser delay parameters. The smooth sidewall and flat underside of the back cavity can be seen in Fig. 3(b). The microscopic morphology of the bottom surface of the back cavity is shown in Fig. 3(c). The undulation of the rough stripe structure on the bottom surface is below 200 nm. Laser scanning confocal microscopy and atomic force microscope were used to further illustrate the flatness of the sensor diaphragm and the smoothness of the bottom surface, as shown in Fig. 4. The fluctuation of the whole cavity bottom is less than 1 μm. In the range of 5×5 μm, the surface roughness of the bottom of the cavity is 101 nm. The silicon carbide wafer and chip manufactured through the integration of MEMS technology and femtosecond laser processing are shown in Fig. 5. The pressure sensor chip has a size of 4 × 4 mm. In the next work, the mass-produced sensors will be calibrated and pressure tested.

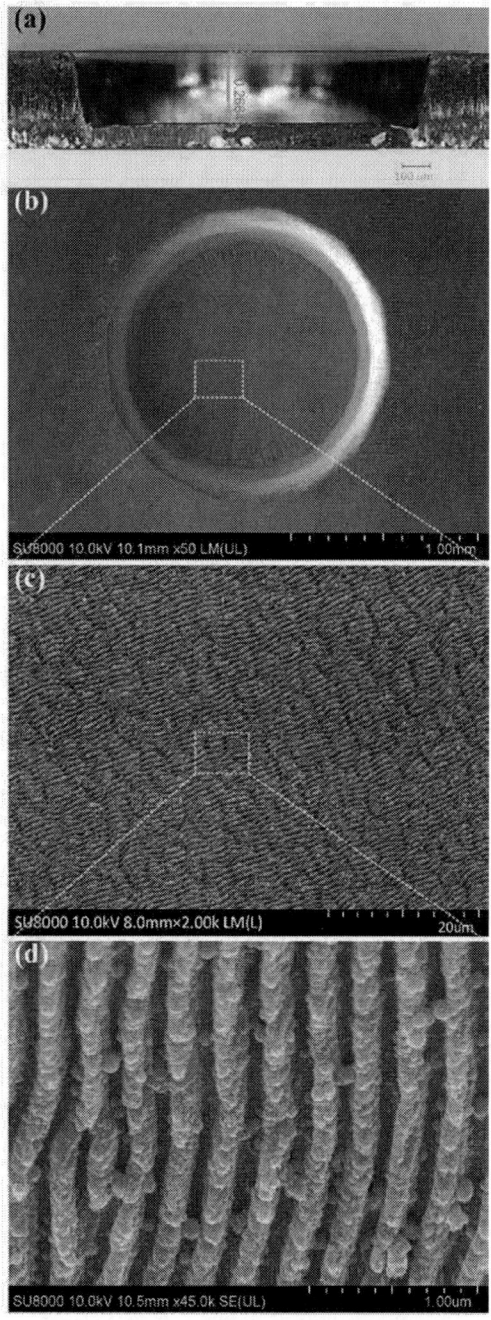

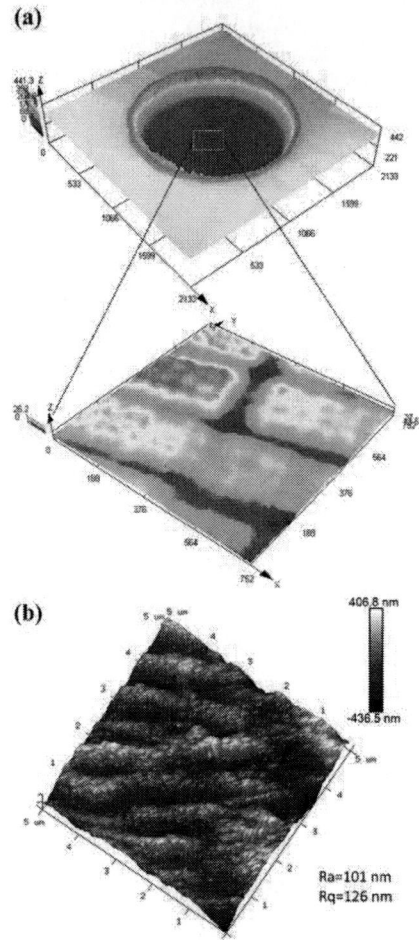

Fig. 4. The 4H-SiC pressure sensor wafer manufactured by MEMS process and the sensor chip manufactured by laser processing.

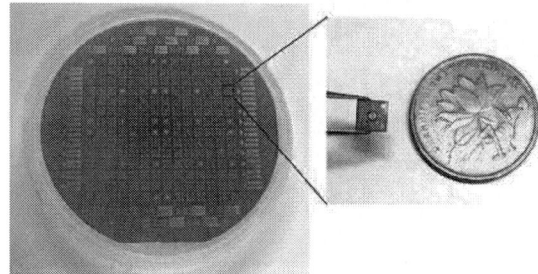

Fig. 5. The 4H-SiC pressure sensor wafer manufactured by MEMS process and the sensor chip manufactured by laser processing.

Fig. 3. (a) Cross-sectional photo of the sensor diaphragm. (b) SEM image of the back cavity in a 4H-SiC pressure sensor. (c) SEM image of the bottom surface of the back cavity. (d) Enlarged view of (c).

III. SUMMARY

This paper investigated a mass fabrication process for bulk 4H-SiC pressure sensor chips. 128 sensor back cavities are quickly manufactured, and the single-cavity etching time is about 1 minute. The small dimension error (<7%) and low surface roughness (Ra=101 nm) show that the femtosecond laser technology has the significance in replacing dry etching in the integrated manufacturing of silicon carbide pressure sensors.

ACKNOWLEDGMENT

This research was funded by National Key R&D Program of China (2017YFB1102900), Key Research and Development Projects of Shaanxi Province (Grant No. 2020ZDLGY14-10), and Postdoctoral Foundation of Shaanxi Province (Grant No. 2018BSHYDZZ06). Special Fund for Technology Innovation Guidance of Shaanxi Province (Grant No. 2019CGXNG-016/Grant No. 2018XNCG-G-19), National Natural Science Foundation of China (Grant No. 51705408).

REFERENCES

[1] H. P. Phan, D. V. Dao, K. Nakamura, S. Dimitrijev, and N. T. Nguyen, "The Piezoresistive Effect of SiC for MEMS Sensors at High Temperatures: A Review," *Journal of Microelectromechanical Systems,* vol. 24, no. 6, pp. 1663-1677, 2015.

[2] D. G. Senesky, B. Jamshidi, K. B. Cheng, and A. P. Pisano, "Harsh Environment Silicon Carbide Sensors for Health and Performance Monitoring of Aerospace Systems: A Review," *IEEE Sensors Journal,* vol. 9, no. 11, pp. 1472-1478, 2009.

[3] C. M. Zetterling, *Process Technology for Silicon Carbide Devices.* London, U.K.: Insitution of Electrical Engineers, 2002.

[4] R. S. Okojie, D. Lukco, V. Nguyen, and E. Savrun, "4H-SiC Piezoresistive Pressure Sensors at 800 °C With Observed Sensitivity Recovery," *IEEE Electron Device Letters,* vol. 36, no. 2, pp. 174-176, 2015.

[5] T. Akiyama, D. Briand, and N. De Rooij, "Piezoresistive n-type 4H-SiC pressure sensor with membrane formed by mechanical milling," *Proceedings of IEEE Sensors,* vol. 1, no. 5, pp. 222-225, 2011.

[6] T. K. Nguyen, H. P. Phan, T. Dinh, A. R. M. Foisal, and D. V. Dao, "High-temperature tolerance of piezoresistive effect in p-4H-SiC for harsh environment sensing," *Journal of Materials Chemistry C,* vol. 6, no. 32, 2018.

[7] Y. Zhao, Y.-L. Zhao, and L.-K. Wang, "Application of femtosecond laser micromachining in silicon carbide deep etching for fabricating sensitive diaphragm of high temperature pressure sensor," *Sensors and Actuators A: Physical,* vol. 309, p. 112017, 2020.

[8] L. Wang *et al.*, "Femtosecond laser micromachining in combination with ICP etching for 4H–SiC pressure sensor membranes," *Ceramics International,* vol. 47, no. 5, pp. 6397-6408, 2021.

[9] L. Wang *et al.*, "Design and Fabrication of Bulk Micromachined 4H-SiC Piezoresistive Pressure Chips Based on Femtosecond Laser Technology," *Micromachines,* vol. 12, no. 1, p. 56, 2021.

[10] E. H. Ransom, K. M. Dowling, D. Rocca-Bejar, J. W. Palko, and D. G. Senesky, "High-throughput pulsed laser manufacturing etch process for complex and released structures from bulk 4H-SiC," in *2017 IEEE 30th International Conference on Micro Electro Mechanical Systems (MEMS),* 2017: IEEE, pp. 671-674.

Proceedings of the 16th Annual IEEE International
Conference on Nano/Micro Engineered and Molecular Systems
April 25-29, 2021

3D Hierarchical Nanoarchitecture AuNPs/MXene@PAMAM based Biosensor for cTnT Detection in Human Serum*

Xin Liu, Yong Qiu, Hao Wan, Liujing Zhuang, and Ping Wang, *Member, IEEE*

Abstract— 2D MXene-$Ti_3C_2T_\chi$ has demonstrated promising application prospect in various fields, however, failing to function properly in biosensor setup due to restacking and anodic oxidation problem. Herein, we reported a synthesis study of covalently grafting MXene by first-generation poly(amidoamine) (PAMAM) dendrimers with AuNPs self-adsorption (AuNPs/MXene@PAMAM) to consider it as a potential functional nanoplatform for electrochemical sensing. In this hybrid system, MXene provided a highly conductive substrate and specific 2D architecture; PAMAM not only acted as an efficient stabilizer simultaneously suppressing serious restacking and oxidation of MXene under anodic potential and consequently improving the electrochemical performance but also as a signal amplifier offering large specific surface areas and massive amino terminals to adsorb Au nanoparticles (AuNPs) for the final formation of 3D hierarchical nanoarchitecture. This functional platform containing abundant active sites was allowed to directly and covalently immobilize thiol-linked anti-cardiac troponin T (cTnT) antibody as the biorecognition element to fabricate the cTnT immunosensor. Preliminary results indicated that this immunosensor was validated with a fast and sensitive response toward cTnT in presence of $[Fe(CN)_6]^{3-/4-}$ redox marker and displayed a wide detection range from 0.1 to 1000 ng/mL and a limit of detection (LOD) of 0.069 ng/mL. This high-performance MXene-based nanobiosensing platform has the potential applicability in bioanalysis of cTnT or other biomarkers and opens a new pathway for expanding its application in the electrochemical biosensing field.

I. INTRODUCTION

MXene, a newly emerging class of postgraphene two-dimensional (2D) transition metal carbides and nitrides materials with a general formula of $M_{n+1}X_nT_\chi$ (where n = 1 to 3, M represents a transition metal, X is C and/or N, and T denotes surface functional groups such as F, O and OH), has drawn growing interest since early 2011 [1, 2]. Unlike most other 2D materials such as graphene and graphitic carbon nitride, MXene nanoflakes possess a unique combination of outstanding electric conductivity, excellent mechanical stability, hydrophilicity, distinguished accordion-like multiplayer (e.g. MXene $Ti_3C_2T_\chi$) and large redox-active surface area apt to high-density incorporation of designed functional groups or nanomaterials, making it a superior candidate for electrochemical sensing [3, 4]. So far, MXene has been applied to develop diverse biosensors for target detection. Despite good behavior in biocompatibility and charge-carrier mobility, it suffered from stacking and

anodic oxidation and eventually led to the reduction of specific surface area for biomolecules loading and narrow electrochemical window, thus seriously affecting biosensing performance [5]. Manipulations to solve these limits of MXene are rarely reported in the fields of electrochemical sensing. More comprehensive explorations into MXene are therefore highly expected to broaden and strengthen its application in electrochemical sensing and biosensor-related fields.

II. MATERIALS AND METHODS

A. Synthesis of $Ti_3C_2T_\chi$-MXene

Layered $Ti_3C_2T_\chi$-MXene was prepared via the aqueous acid etching method. 5 g of Ti_3AlC_2 powders were immersed in 100 mL of a 40% concentrated HF solution and slightly stirred at 30°C for 24 h to remove the Al layer. Subsequently, the obtained sediments were washed repeatedly by deionized water addition, ultrasonication for 15 min, centrifugation at 3500 rpm for 5 min, and decanting until the pH of the supernatant was closed to 6.0, followed by a further wash with ethanol. To prevent dehydration of the hydroxyls on the surface of $Ti_3C_2T_\chi$, the resulting precipitate was dried under vacuum and low temperature to obtain $Ti_3C_2T_\chi$-MXene with abundant hydroxyl groups.

B. Growth of first-generation PAMAM dendrimers on $Ti_3C_2T_\chi$-MXene

The active carboxyl group was introduced into $Ti_3C_2T_\chi$-MXene before its functionalization of the first-generation poly(amidoamine) (PAMAM) dendrimers. First, 1 g of the prepared $Ti_3C_2T_\chi$-MXene and 10 g of succinic anhydride were dispersed in 100 mL of ethanol and held at 25°C for 24 h under a stirring rate of 200 rpm. After the esterification, the products were rinsed using deionized water and then collected centrifuging at 1500 rpm for 15 min, which was repeated seven times to entirely remove extra succinic anhydride. Finally, the sample underwent vacuum drying at 25°C for 12 h to obtain $Ti_3C_2T_\chi$-MXene functionalized carboxyl groups (MXene-COOH).

1g of the as-prepared MXene-COOH was dissolved in 50 mL of $SOCl_2$ with continuous stirring at 70°C for 24 h. The reaction mixture was filtered and washed with anhydrous tetrahydrofuran twice, followed by vacuum drying at 25°C for 24 h. Then, the resultant acylated MXene powder was further dispersed in 100 mL of ethylenediamine under an ultrasonic condition of 5 h at 60°C and stirred for

---

*Resrach is supported by National Natural Science Foundation of China (Grant No. 31627801, 61901412) and the Science and Technology Project of Zhejiang Province (Grant No. 2019C03066, LGF19H180022).

All authors are with Biosensor National Special Laboratory, Key Laboratory for Biomedical Engineering of Education Ministry,

Department of Biomedical Engineering, Zhejiang University, China. Contacting author: Liujing Zhuang; phone: +0571-87952832; email: thisiszlj@163.com. Ping Wang; phone: +0571-87952832; email: cnpwang@zju.edu.cn.

978-1-6654-3008-1/21 $31.00 © 2021 IEEE

another 24 h. After suction filtration, washing with anhydrous methanol thrice and vacuum drying, the dendrimer initiator MXene-CONH₂ was obtained. 0.1 g of dendrimer initiator dispersed in 20 mL of anhydrous methanol was carefully dropped into a solution containing methyl acrylate (20 mL) and anhydrous methanol (50 mL) in 20 min with continuous stirring. The reaction mixture then was held at 40 kHz ultrasonic condition for 7 h at 50°C and stirred for another 24 h. Thereafter, to remove non-grafted polymer and unreacted reactants, the resulting precipitate was rinsed using anhydrous methanol three times and then dried at 25°C overnight to obtain half-generation dendrimer grafted MXene. This product was further dispersed in 40 ml of 1:1 methanol/ethylenediamine solution, and reacted for 5 h at 50°C with ultrasonic treatment, followed by stirring for 24 h, filtered and washed thrice with anhydrous methanol to obtain the first-generation dendrimer grafted MXene (MXene@PAMAM). After dried at 25°C for two days under vacuum, the resulting MXene@PAMAM was then dissolved in ultrapure water (5 mg/mL) sonicating for 20 min under N₂ atmosphere at 25°C and centrifuged for 5 min at 3500 rpm. After centrifugation, the obtained supernatant of MXene@PAMAM nanocomposite was stored at 4°C for further study.

C. Fabrication of AuNPs/MXene@PAMAM biosensing nanoplatform

Before this experiment, the Au nanoparticles (AuNPs) were synthesized as our previous report [6], and thiol-linked anti-cTnT monoclonal antibody (SH-mAb) was prepared according to the reported method. [7]. To fabricate this nanoplatform, the screen-printed carbon electrode (SPCE) was firstly activated in 0.1 M deaerated NaOH with cyclic voltammetry (CV) scanning from -0.6 V to 1.3 V at a scan rate of 100 mV/s for twenty-four cycles and then thoroughly rinsed with ultrapure water and dried with nitrogen. After that, 12 μL of MXene@PAMAM suspension was dropped onto the surface of activated SPCE and dried under ambient conditions. Subsequently, MXene@PAMAM modified SPCE (MXene@PAMAM/SPCE) was exposed in the fresh AuNPs solution incubating for 2 h at 4°C. Thereafter, the modified electrode was washed gently with ultrapure water to remove excess material and dried under nitrogen. Next, 10 μL of thiol-linked mAb against cTnT was dropped onto the as-prepared AuNPs/MXene@PAMAM/SPCE and incubated for 90 min at 25°C, followed by blocking unspecific active sites with 0.5% BSA and incubating for 40 min. After rinsing with PBS, the modified SPCE was incubated in 12 μL of cTnT target at 25°C for 40 min.

D. Electrochemical measurements

Cyclic voltammetry (CV) at the potential window of -0.2 to 0.6 V with a scan rate of 100 mV/s was used to characterize the electrochemical properties of the prepared nanomaterial. Differential pulse voltammetry (DPV) was recorded in the potential ranging from -0.4 to 0.4 V at a pulse amplitude of 0.05 V and a pulse width scan of 0.05 s. All the above electrochemical measurements were performed in a deaerated PBS solution containing 5 mM $K_3[Fe(CN)_6]/K_4[Fe(CN)_6]$ as a redox probe and 0.1 M KCl as supporting electrolyte and carried out in triplicate at room temperature.

III. CURRENT RESULTS

A. Construct of AuNPs/MXene@PAMAM platform

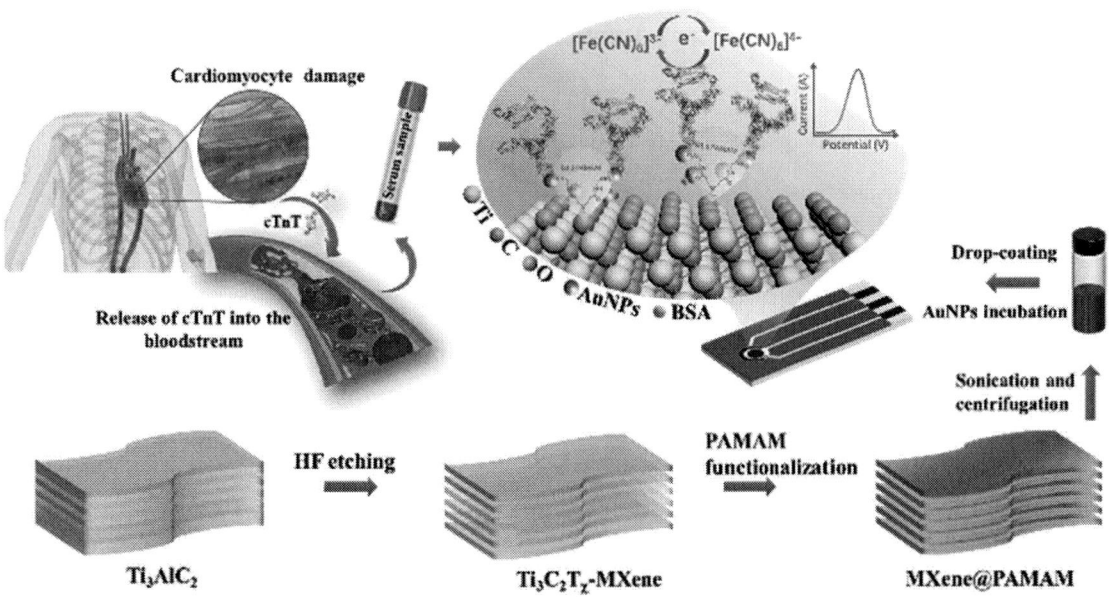

Fig. 1. Schematic illustration of the fabrication process of the AuNPs/MXene@PAMAM platform and its electrochemical application for cTnT detection.

Fig. 1 illustrated the fabrication process of the AuNPs/MXene@PAMAM platform and its electrochemical application for cTnT detection. 2D $Ti_3C_2T_\chi$ nanosheets, the typical type of MXene that we prepared by acid etching aluminum layers from titanium aluminum carbide (Ti_3AlC_2), were covalently functionalized with the first-generation PAMAM dendrimers by in-situ growth method using $Ti_3C_2T_\chi$ as the core bringing about simultaneous improvement in the stability of $Ti_3C_2T_\chi$-MXene and the conductivity of PAMAM dendrimers. The resulting MXene@PAMAM not only inherited the structural integrity and superior conductivity like MXene but also provided favorable fractal architecture with large specific surface areas, abundant active sites, and improved stability for versatile applications. On the other hand, due to the rich surface functionality, sufficient nitrogen functional groups of PAMAM's branch end offer multiple interaction sites for anchoring other metal nanomaterials such as AuNPs. The obtained MXene@PAMAM suspension was dropped onto the surface of activated SPCE, which was then exposed in the fresh AuNPs solution to construct the AuNPs/MXene@PAMAM platform. Subsequently, through the thiol linkers of antibody that were introduced with antibody bioconjugation chemistry, a single-step modification could be realized to further enhance coverage onto the surface of AuNPs/MXene@PAMAM via Au-S bond omitting cumbersome surface functionalization steps. Next, BSA was used to block the unspecific active sites of the sensor. This work aimed at establishing a high-performance MXene-based nanoplatform to develop a simple and reliable electrochemical immunosesnsor for rapidly and sensitively monitoring cTnT in serum.

B. Structural and morphological analysis

This preparation process was also confirmed by X-ray diffraction (XRD) observation. As shown in Fig. 2, the elimination of the (104) peak containing Al (~39°) after HF treatment revealed the removal of Al from the Ti_3AlC_2-MAX phase and a successful etching. A series of (00l) peaks for $Ti_3C_2T_\chi$-MXene broadened and downshifted toward lower angles compared to the original MAX phase, indicating an increase in interlayer spacing of MXene and a significant loss of crystallinity. Such diffractogram results were well in accordance with previous reports [8]. Additionally, a similar decrease in the peak position was observed after the interfacial chemical grafting PAMAM. The (002) peak at 9.04° 2Θ (MXene) sharply shifted to 6.52° 2Θ (MXene@PAMAM), which corresponded to a clear increase in the d-spacing between multilayers from 9.8 Å to 13.5 Å suggesting that the firm anchor of PAMAM on MXene nanosheets tended to expand the interlayer spacing and prevent these nanosheets from restacking. On account of the amorphous nature of PAMAM polymer and their close attachment onto layers, MXene@PAMAM hybrid exhibited a distinct reduction in crystallinity with the absence of (004) and (006) peaks compared to that of MXene.

To investigate the variation of the chemical structure of MXene before and after PAMAM grafting, the Fourier transform infrared (FTIR) spectrum was performed. As seen from the FTIR spectrum (Fig. 3), the stretching vibration at 3434 cm^{-1} and bending vibration in the plane at 1400 cm^{-1}

within the pristine $Ti_3C_2T_\chi$-MXene represented strong hydrogen-bond resulting from the -OH terminal groups on the MXene surface [9]. These highly active -OH groups easily underwent a ring-opening esterification reaction with succinic anhydride in an anhydrous environment, thereby through which the carboxylic acid functional groups were covalently introduced into the surface of the Ti layer to obtain MXene-COOH and for further growth of PAMAM. In the FTIR spectrum of MXene-COOH, the band at 2933 cm^{-1} corresponded to the C-H antisymmetrical stretching vibration. The absorption bands at 1739, 1658, 1452, and 1236 cm^{-1} were associated with the formation of the carboxyl group, and the two at 1739 and 1658 cm^{-1} were assigned to the vibrational-coupling effects of the two carbonyls. As for MXene@PAMAM, the C-H peak split into two enhanced absorption peaks at 2921 cm^{-1} and 2851 cm^{-1}, and the typical peaks for O=C-N-H stretching vibration at 1629 cm^{-1} and N-H in-plane bending vibration at 1565 cm^{-1} could be seen. The results exactly indicated the successful grafting and the growth of dendrimers on MXene. Also, the band at ~3400 cm^{-1} showed a small shift compared with carboxylated MXene, which might be attributed to the overlapping of O-H and N-H stretching vibration absorption. The result of the XRD and FTIR spectrum confirmed that the MXene was synthesized successfully.

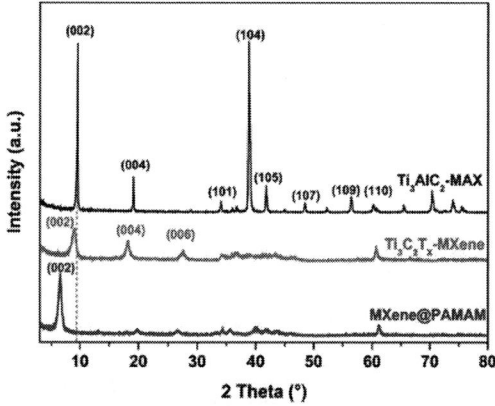

Fig. 2. XRD patterns of Ti_3AlC_2-MAX, $Ti_3C_2T_\chi$-MXene, and MXene@PAMAM.

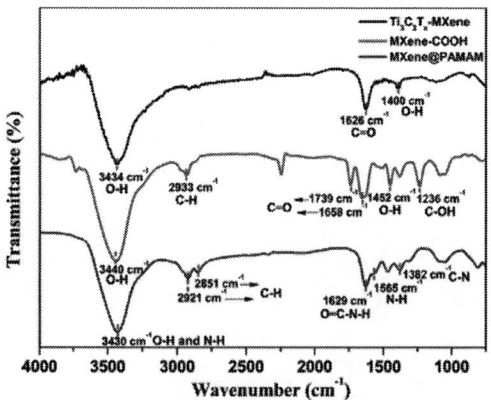

Fig. 3. FTIR spectrum of $Ti_3C_2T_\chi$-MXene, carboxylic MXene (MXene-COOH), and MXene@PAMAM.

Further, we characterize the morphological structure of the AuNPs/MXene@PAMAM hybrid using a transmission electron microscope (TEM) and scanning microscope (SEM). As shown in Fig. 4a. a number of well-shaped AuNPs were successfully attached onto the rough surface of MXene@PAMAM nanosheets and the SEM images (the inset of Fig. 3a) indicated that the AuNPs distributed relatively uniform. Moreover, the X-ray energy spectrum (EDS) elemental mapping (Fig. 4b) presented a uniform distribution of Au elements, suggesting the relatively homogeneous dispersion of AuNPs on the functionalization nanosheets. Those results verified the efficient and favorable immobilization of AuNPs on MXene@PAMAM nanosheets.

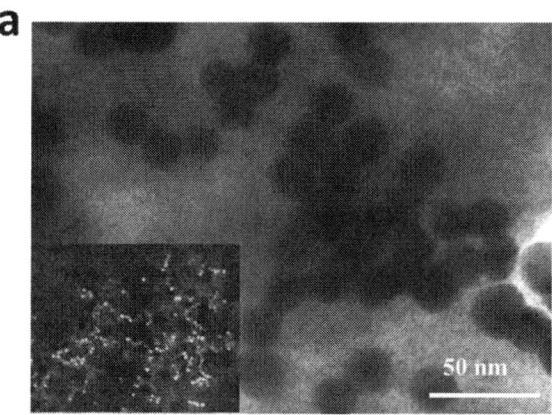

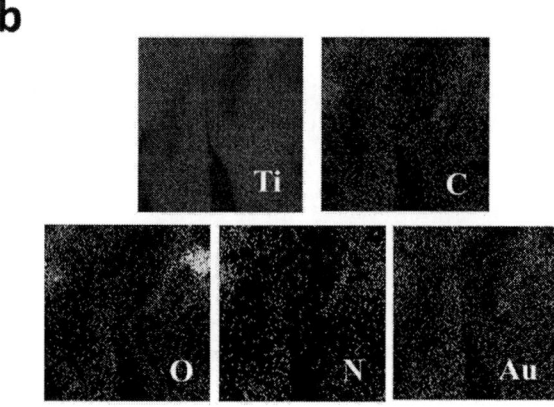

Fig. 4. (a) TEM images of AuNPs/MXene@PAMAM (inset: the corresponding SEM images). (b) EDS elemental map of AuNPs/MXene@PAMAM.

C. Electrochemical properties of AuNPs/ MXene@PAMAM nanohybrid for biosensing

The irreversible oxidation of $Ti_3C_2T_\chi$-MXene upon exposure to anodic potentials over 0.2 V resulting in the reduction of electrochemical reactive activity and the narrow electrochemical window is a striking problem that limits its extended application in the field of electrochemical sensing. Herein, in order to investigate the effect of PAMAM modification on the electrochemical stability of MXene under an anodic potential, we simultaneously fabricated two types of electrodes by modifying pre-activated SPCE with MXene and MXene@PAMAM nanocomposites

(designated as MXene/SPCE and MXene@PAMAM/SPCE respectively). The involved pristine MXene was pre-prepared following the same dispersed condition as MXene@PAMAM. The CV analysis results of the two types of modified SPCEs in a range of -0.2 to 0.6 V were depicted in Fig. 5. Significantly, a pair of well-defined redox peaks could be observed at MXene@PAMAM/SPCE in the presence of a fairly reversible redox couple $[Fe(CN)_6]^{3-/4-}$. However, SPCE after MXene modification changed into the atypical oxidation-reduction peak with a remarkable current decrease and peak-to-peak separation increase and even worse emerging a sharp rise in anodic oxidation current density at more positive potentials (marked by the gray dotted boxes). Such an anomalous electrochemical behavior of the MXene at higher positive potentials reflected the oxidation of $Ti_3C_2T_\chi$-MXene upon the poor electrochemical stability, leading to much less electrochemical activity and the impediment of the quasi-reversible redox behavior of $[Fe(CN)_6]^{3-/4-}$. By comparison, MXene@PAMAM/SPCE behaved improved electrochemical performance that was attributed to the rapid electron transfer rate and fast ion transport toward the redox-active center in the heterostructure of MXene. Also, the stability has been enhanced due to the robust stability of the nanohybrid endowed by PAMAM polymer, making the novel and unique MXene@PAMAM a suitable candidate in electrochemical sensing. Moreover, the next self-adsorption of AuNPs onto the MXene@PAMAM/SPCE showed obvious enhancement in redox current, which stemmed from the excellent conductivity of AuNPs to accelerate electron transport.

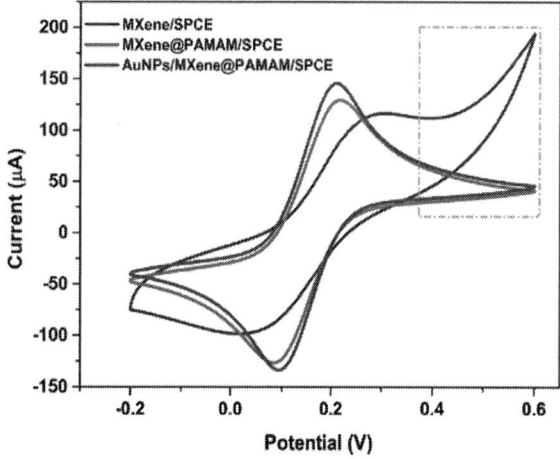

Fig. 5. CVs of MXene, MXene@PAMAM, AuNPs/MXene@PAMAM modification on SPCE.

D. Electrochemical characterization of the immunosensor

DPV was employed to monitor the assembly performance of the proposed immunosensor. As shown in Fig. 6, a well-performed peak was observed after the bare SPCE (curve a) was activated by NaOH (curve b). Then, with the stepwise modification of MXene@PAMAM (curve c) and AuNPs (curve d), the redox peak current dramatically increased and reached a peak as a result of the strong electrical conductivity of the two nanomaterials. Noticeably, the AuNs/MXene@PAMAM/SPCE exhibited a remarkably

978-1-6654-3008-1/21 $31.00 © 2021 IEEE

enhanced response signal achieving a peak current magnification of 5.11 times compared with that of bare SPCE, manifesting the favorable electrochemical properties of this functional nanoplatform and the potential availability for biosensing. After coating the thiol-linked anti-cTnT mAb onto the electrode surface, an insulating blocking layer formed by the immobilized mAb hindered the diffusion of $[Fe(CN)_6]^{3-/4-}$ and the electron transfer across the electrode/electrolyte interface, leading to a great decrease in the peak current (curve e). The assembly of BSA molecule also yielded a decrease in current density (curve f), for which a blockage of BSA to electronic transportation might be accountable. In the subsequent incubation of cTnT, the modified SPCE showed a similar decline in the current peak, which could be ascribed to the specific binding between anti-cTnT antibody and cTnT that retarded the electron transfer. The above characterization manifested the successful fabrication of the cTnT immunosensor. The result of DPV showed that the 3D hierarchical nanoarchitecture AuNPs/MXene@PAMAM distinctly enhanced the response signal in electrochemical transduction.

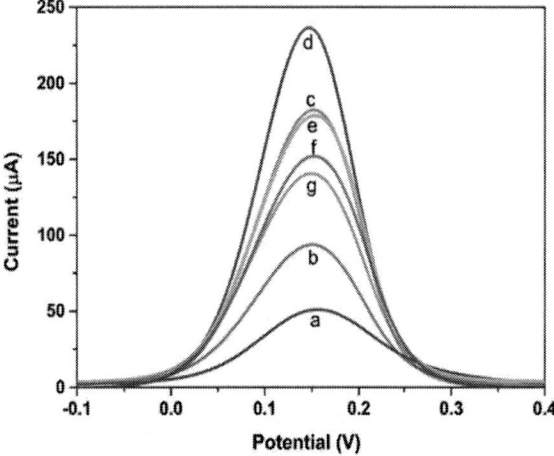

Fig. 6. DPV of a bare SPCE, b NaOH activation/SPCE, c MXene@PAMAM/SPCE, d AuNPs/MXene@PAMAM/SPCE, e SH-mAb/AuNPs/MXene@PAMAM/SPCE, f BSA/SH-mAb/AuNPs/MXene@PAMAM/SPCE, g cTnT/BSA/SH-mAb/AuNPs/MXene@PAMAM/SPCE.

E. Analytical performance of the immunosensor

The analytical performance of the immunosensor was investigated by testing a series of different target concentrations. As shown in Fig 7, the DPV peak current gradually declined with an increase in the concentration of cTnT owing to the hindrance of immune-complex formed by antigen-antibody-specific binding at the interface. A good linear positive correlation between the current responses and the logarithmic values of target cTnT levels ranging from 0.1 to 1000 ng/mL could be obtained. Its linear regression equation was $I (\mu A) = 141.874 - 8.200 \log c$ (ng/mL) with a determination coefficient (R^2) of 0.989 and a detection limit of 0.069 ng/mL (detection limit defined as Mb + 3 SD, where Mb and SD were the mean value and standard deviation of the blank, respectively).

IV. CONCLUSIONS

In summary, a novel and facile electrochemical biosensing strategy based on the 3D AuNPs/MXene@PAMAM nanoplatform was proposed for biosensor fabrication, and we hereby developed a label-free immunosensor for the rapid and sensitive detection of cTnT. The proposed sensor for cTnT detection achieved desirable results. In this work, the combination of thiol-linked antibody and AuNPs/MXene@PAMAM nanobiosensing platform provided an efficient and reliable strategy for electrochemical quantification of antigen and broadened the application of MXene in biosensor-related fields.

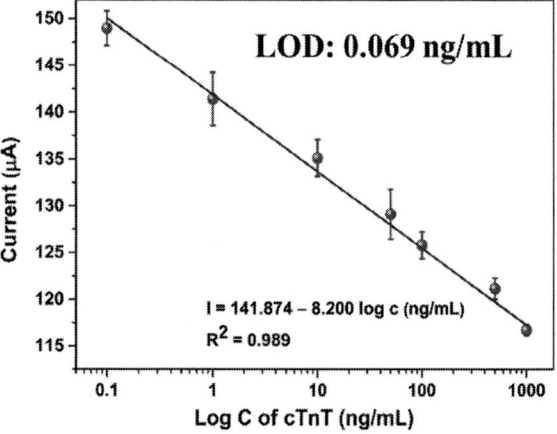

Fig. 7. Calibration curve of the DPV peak currents for various cTnT concentration and obtained a LOD of 0.069 ng/mL.

REFERENCES

[1] M. Ghidiu, M. R. Lukatskaya, M. Zhao, Y. Gogotsi, M. W. Barsoum. Conductive Two-Dimensional Titanium Carbide 'Clay' with High Volumetric Capacitance. *Nature* 2014, 516, 78-81.

[2] M. Naguib, M. Kurtoglu, V. Presser, J. Lu, J. Niu, M. Heon, L. Hultman, Y. Gogotsi, M. W. Barsoum. Two-Dimensional Nanocrystals Produced by Exfoliation of Ti$_3$AlC$_2$. *Adv. Mater.* 2011, 23, 4248-4253.

[3] M. Boota, B. Anasori, C. Voigt, M. Zhao, M. W Barsoum, Y. Gogotsi. Pseudocapacitive Electrodes Produced by Oxidant-Free Polymerization of Pyrrole between the Layers of 2D Titanium Carbide (MXene). *Adv. Mater.* 2016, 28, 1517-1522.

[4] L. Liu, Y. Wei, S. Jiao, S. Zhu, X. Liu. A Novel Label-Free Strategy for the Ultrasensitive miRNA-182 Detection Based on MoS$_2$/Ti$_3$C$_2$ Nanohybrids. *Biosens. Bioelectron.* 2019, 137, 45-51.

[5] H. Xu, A. Ren, J. Wu, Z. Wang. Recent Advances in 2D MXenes for Photodetection. *Adv. Funct. Mater.* 2020, 30, 2000907.

[6] S. Zhou, Y. Gan, L. Kong, J. Sun, T. Liang, X. Wang, H. Wan, P. Wang, A Novel Portable Biosensor Based on Aptamer Functionalized Gold Nanoparticles for Adenosine Detection. *Anal. Chim. Acta* 2020, 1120, 43-49.

[7] Y. Dai, C. Wang, L. Chiu, K. Abbasi, B. S. Tolbert, G. Sauvé, Y. Yen, C. Liu. Application of Bioconjugation Chemistry on Biosensor Fabrication for Detection of TAR-DNA Binding Protein 43. *Biosens. Bioelectrons.* 2018, 117, 60-67.

[8] D. Kim, T. Y. Ko, H. Kim, G. H. Lee, S. Cho, C. M. Koo, Nonpolar Organic Dispersion of 2D Ti$_3$C$_2$T$_x$ MXene Flakes via Simultaneous Interfacial Chemical Grafting and Phase Transfer Method. *ACS Nano* 2019, 13, 13818-13828.

[9] Q. Peng, J. Guo, Q. Zhang, J. Xiang, B. Liu, A. Zhou, R. Liu, Y. Tian. Unique Lead Adsorption Behavior of Activated Hydroxyl Group in Two-Dimensional Titanium Carbide. *J. Am. Chem. Soc.* 2014, 136, 4113-4116.

Gap in pagination due to unavailable papers.

Pages 1487-1492

Proceedings of the 16th Annual IEEE International
Conference on Nano/Micro Engineered and Molecular Systems
April 25-29, 2021

Effect of Scan Line Spacing on Laser-Reduced Graphene Oxide Based Temperature Sensing

Rui Chen, Wei Zhou, Chiqian Xiao, Tao Luo*

Abstract—This paper systematically studies the effect of scan line spacing on the sensitivity of laser-reduced graphene oxide based temperature sensor. As for the fabrication, we firstly sputter and pattern a pair of 80 nm thick gold electrodes on PET film, and then coat a layer of graphene oxide solution in the middle of the electrode. Second, graphene oxide is reduced by UV laser scanning to bridge the electrodes with the laser-reduced graphene oxide, which is a temperature sensitive material with temperature-dependent resistivity. A group of rGO temperature sensors are fabricated in this way with different scan line spacings ranging from 0.02 to 0.12 mm with a step of 0.02 mm. Raman spectroscopy analysis shows that the defects of graphene first decrease and then increase with increasing scanning spacing. The temperature calibration results show that defects of graphene generally enhance the sensitivity of rGO temperature sensor, and the larger sensitivity is achieved with a scan line spacing of 0.02 mm and 0.12 mm.

I. INTRODUCTION

Human body temperature is one of the important physiological information, which reflects the health status of people to a certain extent. COVID-19 swept the world and brought great harm to all human beings. The infection of COVID-19 leads to the rise of patients' body temperature [1]. Therefore, the development of wearable flexible temperature sensors with high sensitivity can enable the 24-hours online monitoring of the body temperature, which could greatly help containing the spreading virus [2]. As a high thermal-conductive material, graphene allows rapid conduction of temperature, making it a very good material for realizing temperature sensor with fast response and high sensitivity [3]. Moreover, the technique of laser reduction of graphene oxide not only allows facile production of graphene rapidly but also allows patterning graphene on various substrates easily[4]. In this work, we use laser processing method to prepare temperature sensor by photoreduction and patterning of graphene oxide, and adjust the scan line spacing of the laser to study the effect of scan line spacing on the sensitivity of temperature sensor. Gaining this understanding will greatly facilitate the optimization of laser processing parameter for fabricating flexible temperature sensor with high sensitivity for various applications.

II. EXPERIMENTAL

A. Design and fabrication of temperature sensors

Resistive type temperature sensor fabrication requires a pair of planar electrodes bridged by a conductive film. The temperature sensor is composed of rGO sensitive material,

gold electrode, 125μm-thick PET substrate, conductive silver glue and wire. The fabrication of rGO temperature sensors in this paper is shown in Fig.1, which mainly includes four steps: sputtering of a gold layer on PET, UV laser processing of gold layer to form a pair of patterned electrodes with the distance of 4 mm, drip coating of the graphene oxide (GO) solution. The samples were then air dried to obtain electrical contact, and UV laser reduction of graphene oxide (rGO) with an area of 4×4 mm^2. The rGO film between the electrodes acts as the semiconducting sensing element, which has been fabricated sequentially as illustrated from Fig.1(a) to Fig.1(d). Finally, the sensor is packaged with a layer of polyimide tape.

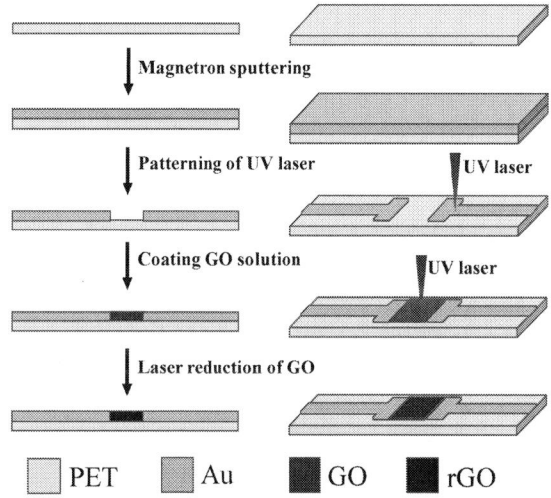

Figure 1. (a) Magnetron sputtering gold film. (b) Laser processing electrode pattern. (c) Drop coating of GO solution. (d) Laser reduction of graphene oxide.

The photo of the rGO temperature sensor is shown in Figure 2. The GO solution used in this paper was prepared by Hummers method [5]. The concentration of GO solution was 2 mg/ml. The pulsed fiber laser (YLP-1-100-20-20-CN IPG, Germany) can generate 100 ns pulses with repetition rate of 20kHz and emission wavelength center of 1064 nm. The laser beam was controlled to focus on the GO surface by the linear movement of the Z axis. The scan line spacings are set to be 0.02 mm, 0.04 mm, 0.06 mm, 0.08mm, 0.1mm and 0.12 mm.

Department of Mechanical & Electrical Engineering, Xiamen University,Xiamen 361005, China.

*corresponding author: Tao Luo, phone: 18216231870; e-mail: luotao@xmu.edu.cn.

978-1-6654-3008-1/21 $31.00 © 2021 IEEE

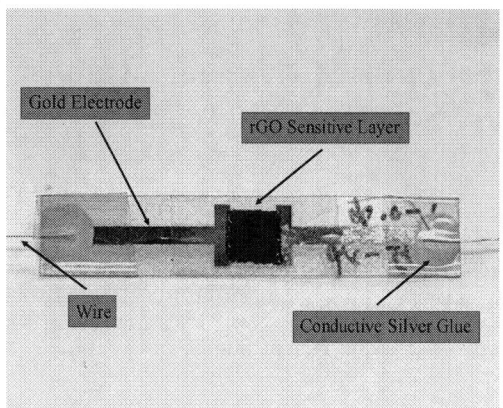

Figure 2. Photo of temperature sensor.

B. Temperature sensor measurement

The entire measurement system is shown in Fig. 3, which includes a temperature inspection instrument, a Kithley 2400 digital source meter, a refrigeration chip and a power supply. In the temperature measurement process, the rGO sensitive layer was placed toward the surface of the refrigeration chip, and a thermocouple connected to the temperature inspection instrument was placed beside the rGO sensitive layer as the reference temperature sensor for the calibration of the proposed sensor. Then the temperature of the refrigeration chip is controlled, and the resistance of the rGO temperature sensor and the temperature measured by the thermocouple are recorded simultaneously.

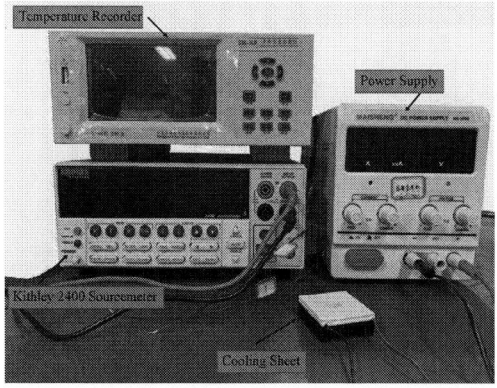

Figure 3. Calibration platform for the temperature response of fabricated sensors.

III. RESULTS AND DISCUSSION

A. Characterization of reduced graphene oxide

The surface morphology of the rGO film was imaged using scanning electron microscopy. Fig. 4 shows the SEM image of the rGO film with a scanning distance of 0.08mm. It is evident from Fig.4 (a) that the rGO is uniformly dispersed on the sample surface, the bulge and layered structure on the surface of rGO can be clearly observed, and Fig.4 (b) shows the microstructure of the rGO film, which obviously has layered and porous micro/nano-structures. IDSpec ARCTIC Micro Confocal Raman Spectrometer was used to characterize rGO films processed with six sets of parameters.

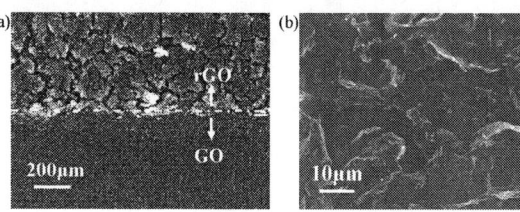

Figure 4. (a) SEM image of GO film before and after laser processing. (b) SEM image of rGO film at 2000 magnification.

Using 523nm laser for testing, Fig.5 (a) shows the Raman spectra for rGO film at different scan line spacing. The ratio of D-peak and G-peak intensities (I_D/I_G) is used to determine the purity of the laser-derived graphene material structure [6,7]. The line shape of 2D-peak and its intensity relative to G-peak (I_{2D}/I_G) can be utilized to characterize the number of graphene layers present[8]. Monolayer graphene usually has a very sharp and symmetric Lorentzian 2D-peak with intensity greater than twice G-peak's intensity. As the number of layers increases, 2D-peak broadens, becomes less symmetric, and the also intensity decreases[8].

It can be seen from the figure that the scanning distances of 0.06mm and 0.08mm have the maximum D - peak, G - peak and 2D - peak intensity and the scanning distances of 0.02mm and 0.12mm have the minimum intensity. As shown in Fig.5(b), shows the I_D/I_G and I_{2D}/I_G curves of Raman spectra with different scan line spacings. The larger the I_D/I_G, the more graphene defects. The smaller I_{2D}/I_G, the more graphene layers.

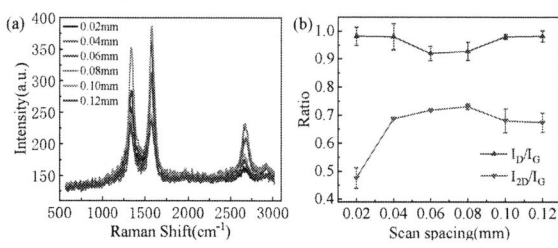

Figure 5. (a) Raman spectra for rGO film at different scan line spacing. (b) Raman Spectrum Characterization: the change curve of I_D/I_G and I_{2D}/I_G ratio of rGO film processed with different laser scanning spacing parameters with scanning spacing.

It can be seen from the figure that as the scan line spacing increases, I_D/I_G first decreases and then increases, and the change trend of I_{2D}/I_G is opposite. There is a non-monotonous trend of resistance and I_{2D}/I_G with the changing scan spacing, because the 0.02mm scan line spacing is too close, which causes the GO layer to be over reduced by laser, resulting in poor reduction effect. As the scan line spacing increases, the reduction effect of GO was enhanced, resulting in an increase of I_{2D}/I_G. When the scan line spacing reaches 0.1mm, the reduction effect is reduced due to too wide line spacing. The rGO fabricated with the scan line spacing of 0.06 mm and 0.08 mm has the smallest defects and the smallest number of layers.

B. Temperature sensor performance

Figure 6 shows the initial resistance of the sensor with different scan line spacings before and after packaging. It can be seen from the figure that the resistance value after packaging is smaller than the resistance before packaging due to that the rGO film is pressed during packaging to become compact, which reduces the resistance. The sensor with scanning spacing of 0.06 mm has the lowest resistance.

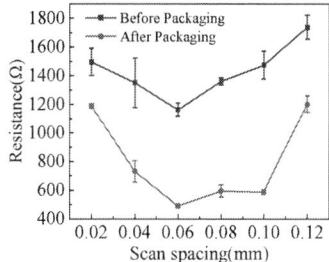

Figure 6. The change curve of the resistance value of the unpacked sensor processed with different laser scanning spacing parameters.

In order to test the performance of the sensor, Kithley 2400 digital source meter is used to measure the resistance change of the sensor, and the temperature of the refrigeration chip is controlled from 30°C to 100°C. At the same time, the resistance change was recorded with each measurement step of 5°C. The temperature inspection instrument records the temperature of the refrigeration chip as a reference temperature for data calibration. The temperature response of two rGO temperature sensors is shown in Fig.7, in which Fig.7 (a) shows the temperature response curve with the scan line spacing of 0.08 mm, and Fig.7 (b) shows the temperature resistance curve with the scan line spacing of 0.12 mm.

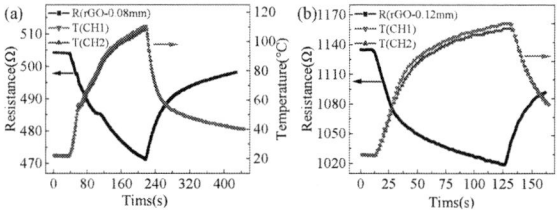

Figure 7. The experimental results of rGO temperature sensor are as follows: (a) the resistance variation curve with temperature when the scanning spacing is 0.08mm. (a) the resistance variation curve with temperature when the scanning spacing is 0.12mm.

It is clear that the resistance decreases as the temperature increases. The change trend of resistance is opposite to that of temperature, which has good consistency with the data acquired by two reference thermocouples. This change in the resistance of the rGO temperature sensor is due to that graphene exhibits semiconducting properties, the dependence of the resistance on the temperature is determined by its thermally activated charge carriers. As the temperature increases, the mobility of the charge carriers increases, and thus, the resistance decreases[9,10]. The rGO used in this paper was made by chemical method, and the

performance of our rGO temperature sensor is consistent with the above theory, which shows semiconducting properties.

The temperature coefficient of the resistance (TCR) is often used to describe the temperature-sensitive properties popularly known as sensitivity[11].

$$TCR = (R - R_0)/(R_0 \times \Delta T) \qquad (1)$$

where R, R_0 are the measured resistance, and ΔT are the measured temperature.

The change in resistance with temperature is illustrated in Fig. 8. As can be seen from the figures, the variation in resistance with temperature of the different scan line spacings decreases linearly with the increase of the temperature. In addition, the extracted corresponding sensitivities were found to be 0.219%, 0.1149%, 0.1198%, 0.1553%, 0.1212% and 0.2043% per degree Celsius for the scan line spacings of 0.02 mm, 0.04 mm, 0.06 mm, 0.08 mm, 0.1 mm and 0.12 mm.

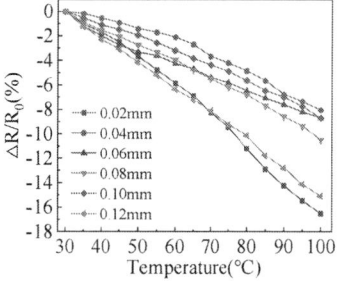

Figure 8. The resistance change rate of the sensors with different scan line spacings at temperature from 30 °C to 100 °C.

Figure 9 shows the sensitivity of the rGO temperature sensor processed with different scan line spacings within 100°C resistance change. The sensitivity of sensors processed with scan line spacing of 0.02mm and 0.12mm is higher than that of sensors processed with other parameters, which are 21.87% and 20.43% per 100°C.

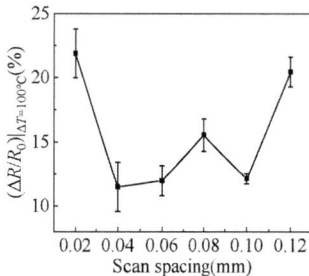

Figure 9. The resistance change rate of the sensor with different scanning distance when the temperature changes 100°C.

The sensitivity of the sensor increases with the scan line spacing form 0.04mm to 0.08mm, the change trend of sensitivity is similar to that of I_D/I_G in Fig.5 (b). indicating that the sensitivity of rGO temperature sensor is related to the defects of graphene, and the defects generally can enhance the sensitivity of rGO temperature sensor.

Finally, we conducted respiration monitoring tests using the fabricated sensors. The rGO temperature sensor is attached to the tester's nostrils for tests. The result of tests is shown in Fig. 10. It can be seen from the figure that the exhalation process and the inhalation process are accompanied by heating and cooling, the sensor resistance changes accordingly.

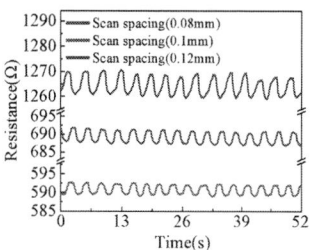

Figure 10. The resistance change of the sensor with 0.08mm,0.1mm and 0.12mm scanning distance when the subject breathes.

IV. CONCLUSION

The influence of different laser scanning line spacing parameters on the temperature sensitive characteristics of temperature sensor were demonstrated in this study, with sensitive materials made of rGO. In order to determine the relationship between the scan line spacing and the temperature sensor performance, six different scan line spacings ranging from 0.02 to 0.12 mm with a step of 0.02 mm was used We have compared the sensitivities of sensors fabricated using six different scan line spacings. The results show that the sensitivity of 0.02mm and 0.12mm is the highest, exceeding 0.2% per degree Celsius. The rGO temperature sensor we designed is simple in structure, light in weight, easy to manufacture and low in cost. All the above factors show that the rGO temperature sensor is suitable for robot skins and electronic skins and it can be widely used for various applications.

REFERENCES

[1] Tharakan, Serena, et al. "Body temperature correlates with mortality in COVID-19 patients." *Critical Care.* vol.24, no.1,2020.

[2] Guanyu, Liu, et al. "A Flexible Temperature Sensor Based on Reduced Graphene Oxide for Robot Skin Used in Internet of Things." *Sensors.*vol.18, no.5, 2 May 2018.

[3] A, Pei Huang, et al. "Graphene film for thermal management: A review." *Nano Materials Science*, 23 Sep 2020.

[4] Mensing, Johannes Ph., et al. "Advances in research on 2D and 3D graphene-based supercapacitors." *Advances in Natural Sciences Nanoscience & Nanotechnology.*vol.8, no.3, 17 July 2017.

[5] Marcano D C, Kosynkin D V, Berlin J M, et al. "Improved synthesis of graphene oxide." *ACS Nano*, vol.4, no.8, pp.4806-4814, July 22 2010.

[6] Ni Z, Wang Y, Yu T, et al. "Raman spectroscopy and imaging of graphene" *Nano Research*, vol.1, no.4, pp.273-291, 28 August 2008.

[7] Pimenta, M. A., et al. "Studying disorder in graphite-based systems by Raman spectroscopy." *Physical Chemistry Chemical Physics.* Vol.9, no.11, pp.1276-1290, 7 December 2006.

[8] Wang, Ying Ying, et al. "Raman studies of monolayer graphene: the substrate effect." *The Journal of Physical Chemistry C* vol.112, no.29, pp.10637-10640, 23 April 2008.

[9] Davaji, Benyamin, et al. "A patterned single layer graphene resistance temperature sensor." *Scientific Reports* vol.7, no.1, pp.1-10, 18 August 2017.

[10] Shao, Q., et al. "High-temperature quenching of electrical resistance in graphene interconnects." *Applied Physics Letters.* Vol.92, no.20, 19 May 2008.

[11] Yi, Xiao-Su, Lie Shen, and Yi Pan. "Thermal volume expansion in polymeric PTC composites: a theoretical approach." *Composites Science and Technology.* Vol.61, no.7, pp.949-956, 18 August 2000.

Gap in pagination due to unavailable papers.

Pages 1497-1501

Proceedings of the 16th Annual IEEE International
Conference on Nano/Micro Engineered and Molecular Systems
April 25-29, 2021

Investigation of Flexible Sweat Sensor for Sodium-Ion Concentration with a Combination of Two Sensing Mechanisms

Jiufu Zheng, Xuankai Xu, Yunzhi Hua, Xiaojin Zhao, and Wei Xu

Abstract—We present a flexible microelectrode array for analyzing the NaCl electrolyte of the sweat of human beings based on two sensing mechanisms. The sensor is fabricated on polyimide (PI) substrate which is low-cost, flexible, and biocompatible as the wearable application for healthcare monitoring. Based on the electrolyte impedance measurement transduction, the two adjacent microelectrodes show a linear response with a sensitivity of 8.78 mS/M to the electrolyte concentration of human sweat range from 10 to 120 mM. In comparison with the impedance method, the open-circuit potential (OCP) measurements are carried out with a modified ion-selective electrode versus a commercial reference electrode. The OCP results show the rational Nernst response (66.2 mV/decade) in a range of $10^{-5} \sim 10^{-1}$ M and with the limit of detection (LOD) of 8.2×10^{-6} M.

I. INTRODUCTION

Sweat secretions contain a variety of chemical analytes including ions, glucose, lactate, hormones, and so on, and the concentrations of these analytes are very useful for the indication of one's health state. Sweat analysis is an important detection method in-situ to noninvasively monitor health status [1]. The ions Na^+ and Cl^- constitute the most abundant components of sweat electrolytes which is usually 10-120 mM respectively. To date, many wearable sensors have been reported for sweat component analysis. Most of the devices use the ion-selective electrodes (ISEs) as the transducing medium, which is fast and highly accurate for sweat secretion detection.

The ISE utilizes the membrane potential to measure the activity or concentration of ions in the electrolyte solution. When the ISE is combined with a reference electrode that can provide a fixed potential to form a basic electrochemical cell, the potential of ISE can be obtained by measuring the open-circuit potential, which conforms to the Nernst equation. Besides, the NaCl solution is a strong electrolyte where the sodium and chloride ions can dissociate entirely to form free-moving ions that can conduct electricity. The activity of the free-moving ions in the electrolyte can also be characterized by the impedance measurement of the electrolyte.

Previous articles have reported the sensor for sodium-ion [2-5], pH [6], ammonium ion [7], glucose [8], ethanol [9], etc. Gao et al. [4] have presented a fully integrated sensor array for multiplexed sweat secretion analysis. Most of these sensors for ion detection are usually based on potentiometric, and few have detailed the difference in sensing performance between the potentiometric and impedimetric. In this paper, we propose a new

microelectrode array sensor on a polyimide substrate for the detection of sodium-ion based on impedimetric and potentiometric, simultaneously. The schematic of our proposed sweat sensor is shown in Fig.1, with the microelectrode array for impedimetric and the modified ISE for potentiometric. The performances of the sensor between these two methods are discussed and the discrepancies are experimentally investigated, especially when targeting the limit of detection (LOD).

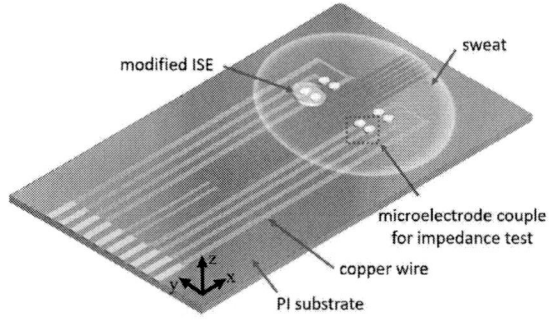

Figure 1. Schematic of the microelectrode array-based flexible sweat sensor with two sensing mechanisms.

II. EXPERIMENTS

A. Fabrication

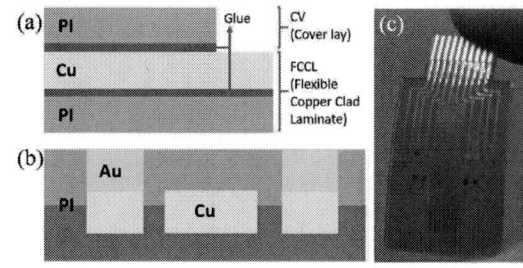

Figure 2. (a) Key fabrication process for the microelectrode array sensor, (b) the sectional view of the fabricated sensor, (c) photograph of the fabricated sensor.

The fabrication process of the polyimide (PI) based microelectrode array sensor is shown in Fig. 2(a). A 100 µm thick Flexible Copper Clad Laminate (FCCL) on the PI structure is adopted. The photolithography and copper etching are performed for the sensor pattern, where a passivation cover lay (PI) is attached to FCCL by a hot stamping process. The copper (Cu) is used as the wire and the exposed microelectrodes are coated with gold (Au) that

---

J. Zheng, X. Xu, X. Zhao, and W. Xu are with College of Electronics and Information Engineering, Shenzhen University, 518060, Shenzhen, China (e-mail: weixu@szu.edu.cn).

Y. Hua is with the College of Electronics and Information Engineering, Shenzhen Institute of Information Technology, 518060, Shenzhen, China.

978-1-6654-3008-1/21 $31.00 © 2021 IEEE

is highly stable and anti-corrosion. The sectional view of the fabricated flexible sensor is shown in Fig. 2(b). Each gold microelectrode has an equal square size of 760 μm × 760 μm with a 1.5 mm spacing from the counter-electrode. A pair of adjacent microelectrodes are used for the impedance measurement. A photograph of the fabricated sensor is shown in Fig. 2(c), which has excellent flexibility.

B. Preparation of the Ion Selective Electrode

For the potential test, the ion selective electrode (ISE) is prepared for the specific ion detection which converts the concentration to the Nernst-response potential. The ion selective membrane (ISM) tightly adheres to the conducting contact, forming the solid ISE. The membrane potential generates at the phase interface between the sensitive membrane and the electrolyte. For this work, the sodium ionophore cocktail membrane is prepared as the ISM targeting sodium ions, which is a mixture of Selectophore™ grade high molecular weight polyvinyl chloride (PVC), sodium ionophore X, sodium tetrakis [3,5-bis(trifluoromethyl) phenyl] borate (Na-TFPB), bis(2-ethylhexyl) sebacate (DOS) and tetrahydrofuran (THF) [2]. The cocktail membrane is drop-casting on the surface of the working electrode and the modified sensor is preserved overnight till THF evaporation entirely at room temperature before experiments.

C. Experimental Setup

The experimental setup for the sensor characterization is shown in Fig. 3. For the electrolyte impedance measurement, one pair of microelectrodes connect with the LCR meter Keysight E4980A, as shown in Fig. 3(a). A small excitation voltage of 20mV is applied on the microelectrodes under sweep frequency from 1kHz to 300kHz. The NaCl electrolyte samples are prepared shortly before the measurements, with the ion concentration range from 0.01 to 0.12 M, simulating the concentration range of NaCl electrolyte in sweat.

For the potential test, the modified membrane electrode is taken as the working electrode. Combined with a commercial Ag/AgCl/KCl (Saturated) reference electrode, the open-circuit potential can be measured by the potentiometer, which reflects the concentration of the electrolyte. The electrochemical interface PalmSens4 is taken as the potentiometer, as shown in Fig. 3(b). The potential test and the impedance test are going simultaneously in a chamber. Both the instruments are carefully calibrated before experiments.

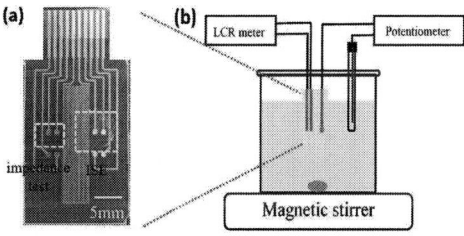

Figure 3. (a) The microelectrode array sensor with a microelectrode modified by the ionophore cocktail as the ion-selective electrode (ISE) and one couple of microelectrodes for the electrolyte impedance test, (b) The experimental setup for the sensor characterization in NaCl electrolyte of various concentrations.

III. RESULTS AND DISCUSSION

A. Impedance Test

For the electrolyte impedance analysis, the frequency sweeping impedance result of the sensor tested in NaCl electrolyte under different concentrations is shown in Fig. 4. At low frequencies, the impedance decreases as frequency increases, while in high frequencies (>50kHz), the value of impedance gets stable. In the electrode-electrolyte-electrode system, the charged ions gather at the electrode-electrolyte phase interface, forming the double layer with the capacitance. At low frequencies, the electrical double layer in the phase interface dominates the total impedance of the system. At high frequencies (>50kHz), the impedance of the electrical double layer can be ignored, and the electrolyte impedance dominates, which is the resistance of the electrolyte solution. For the strong electrolyte, sodium ions and chloride ions are completely dissociated in the NaCl aqueous solution. The charged ions in the electrolyte solution move freely thus forming the resistance of solution (R_s) at the voltage excitation on the microelectrodes. At high frequencies, the total impedance value (Z) of the system can be expressed by the equation of (1):

$$Z \approx R_s = \sigma^{-1} \cdot l / A \qquad (1)$$

where σ is the conductivity, l is the distance of the adjacent microelectrodes, and A is the cross-sectional area. When other factors are controlled, conductivity is the key factor that affects the value of the total impedance. For the strong electrolyte, the number of the dissociated charged particles determines the conductivity of the electrolyte when the temperature is constant.

We measure the impedance values at the high-frequency of 100kHz and analyze the change with the concentration. Based on (1), 1/Z which is the conductance of the electrolyte at the specified frequency of 100 kHz shows a linear response to the concentration ranging from 0.01 to 0.12 M, as shown in Fig. 5(a). This is due to the migration of free-moving ions in the electrolyte solution accelerates when the concentration increases.

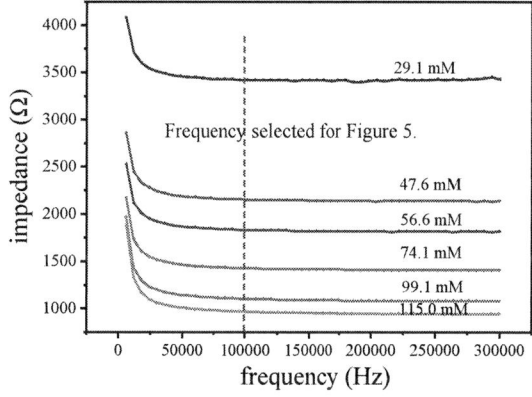

Figure 4. The electrical impedance of NaCl electrolyte under sweep frequency from 1kHz to 300kHz, and the impedance of electrolyte solution at a frequency of 100kHz is selected for sensing calibration.

978-1-6654-3008-1/21 $31.00 © 2021 IEEE   1503

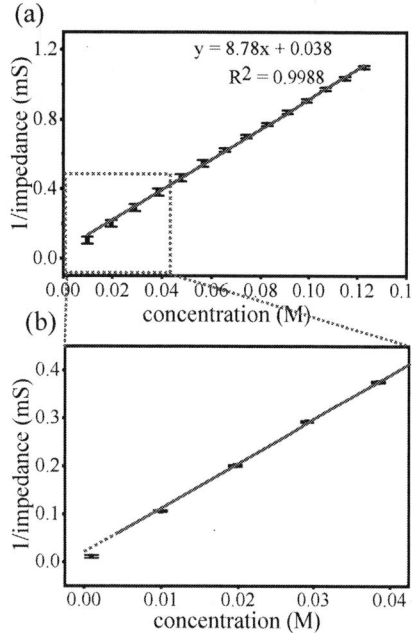

Figure 5. (a) The determined conductivity of NaCl electrolyte has a linear response to the ion concentration range from 0.01 to 0.12 M; (b) The experimental conductivity of the extended concentration of 0.001M deviates from the linear trend.

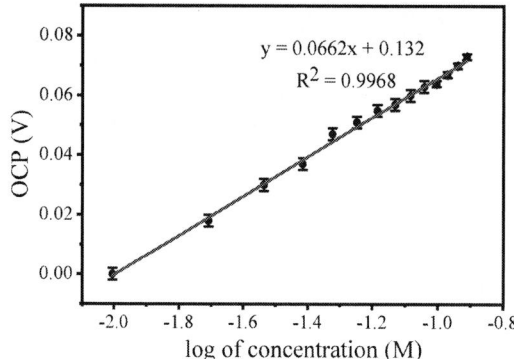

Figure 6. Calibration curve of the modified ion-selective electrode (ISE). The concentration of the test solution was changed from 0.01 to 0.12 M. Open-circuit potentials (OCPs) in NaCl solutions of different concentrations are measured *vs* an std. RE (Ag/AgCl/ KCl (Saturated)).

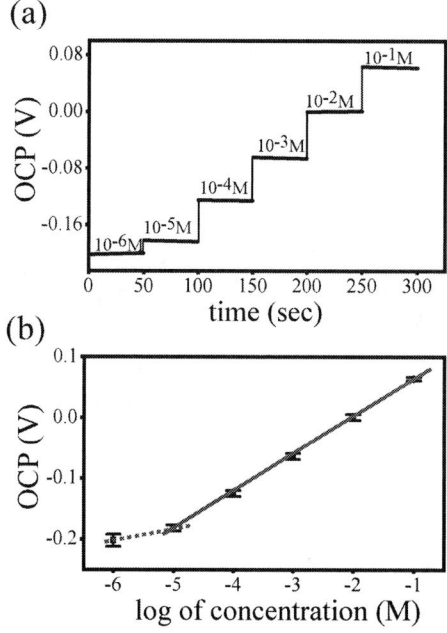

Figure 7. (a) The potential response of the ISE sensor measured in NaCl electrolyte of 10^{-6}, 10^{-5}, 10^{-4}, 10^{-3}, 10^{-2}, and 10^{-1} M, respectively; (b) Sensitivity of the ISE in NaCl electrolyte of concentration as low as 10^{-5} M based on the OCP measurement.

The impedance measurement is characterized with high sensitivity and stability in the range of sweat concentration. For the detection limits of the electrolyte impedance method, the sensor is tested with the extended concentration range to 10^{-6} M. As shown in Fig. 5(b), the conductivity of 10^{-3} M ion solution deviates from the linear trend, indicated a LOD of larger than 10^{-3} M.

B. Potential Test

By modifying the microelectrode with the ion selective membrane, the open-circuit potentials (OCPs) can be measured for various concentrations. For the OCP analysis, the modified microelectrode by the sodium ionophore cocktail as the ion selective electrode (ISE) is linked with a commercial Ag/AgCl/KCl (Saturated) reference electrode and tested under the same conditions as the impedance method. As shown in Fig. 6, the modified ISE shows the rational Nernst response (66.2 mV/decade) in the concentration level of 0.01 ~ 0.12 M. It indicates that the potential of the modified membrane can respond sensitively to small changes of the electrolyte concentration.

For the detection limits of the ISE based on the OCP method, the sensor is tested with the extended concentration range of 10^{-6}, 10^{-5}, 10^{-4}, 10^{-3}, 10^{-2}, and 10^{-1} M. As shown in Fig. 7(a), the OCP almost keeps the constant value under the respective concentration. As shown in Fig. 7(b), the OCP shows a larger range of $10^{-5} \sim 10^{-1}$ M and the lowest LOD of 8.2×10^{-6} M, according to the method for the determination of LOD, which is the concentration of the target ion where the error of the analysis is 100% reported in [5]. As compared to the impedance measurement method, the LOD of the ISE sensor is greatly improved.

For the electrochemical impedance spectra (EIS) analysis of the above two tests, the result of the Nyquist plot is shown in Fig. 8. The bare microelectrode couple for the impedance test shows a slope line which is related to the diffusion process on the electrode surface. For the modified ISE, at low frequencies, the slope indicates the diffusion process in the membrane on the electrode surface, and at high frequencies, the slight semi-circle indicates the electrochemical reaction on the modified membrane with the charge transfer process. The fitting parameters of the Nyquist plot with the equivalent circuit model is shown in Table 1. The charge transfer resistance R_{ct} is consistent with the electrochemical reaction on the modified membrane.

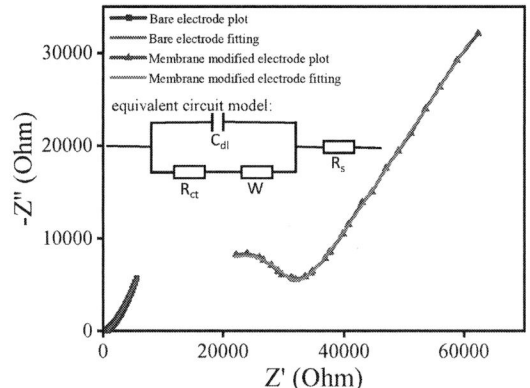

Figure 8. Nyquist plot of the bare microelectrodes and the ISE by the modified membrane in 0.1 M NaCl solution. (AC: 10mV, frequency: 1 Hz - 1 MHz).

TABLE I. PARAMETERS OF CIRCUIT FITTING FOR NYQUIST PLOT

Parameters	R_{ct} / Ω	C_{dl} / nF	$W / k\sigma$	R_s / Ω
Bare electrode	928.5	2.412	186.8	657.5
Modified electrode	1.5×10^4	0.109	871.7	1.5×10^4

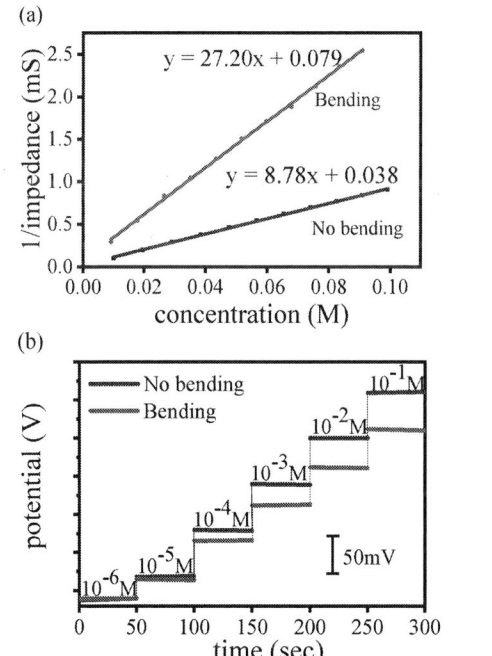

Figure 9. Flexibility performance test with the sensor bending with a curvature of 0.3 cm^{-1}.

To investigate the flexibility performance of the sensor, the impedance test and the potential test are carried out with the sensor bending along the y axis (shown in Fig. 1) with a curvature of 0.3 cm^{-1}, and the result is shown in Fig. 9. The $1/Z$ which is the conductance of the electrolyte shows a greater slope for the bending sensor. This is due to the

distance of the adjacent microelectrodes and the cross-sectional area have changed when the sensor is bending, while the value of the conductivity remains unchanged, which is related to the properties of the electrolyte. For the potential measurement, the potential output of the bent ISE sensor is decreased as compared to the unbent counterpart. The possible reason is the attachment between the modified membrane and the microelectrode has changed in the bent condition, such effect will be investigated in the future.

The OCP response indicates the modified ISE performs well for sweat detection with a wide linear range and high stability. Good reversibility of this proposed ISE sensor by testing for hysteresis is also observed in Fig. 10, which is a crucial factor in wearable applications. The test is carried out in the same NaCl solution from 10^{-5} M to 10^{-1} M for three cycles. Table 2 summarizes the performance achieved by our microelectrode sensor and compares it with the other sweat sensors for Na$^+$ detection [2-5]. Our low-cost sweat sensor array shows the high sensitivity, the competitive linear range, and LOD as compared with other works. After the modification of counter-electrode with Ag/AgCl, future work will focus on in-situ human sweat testing.

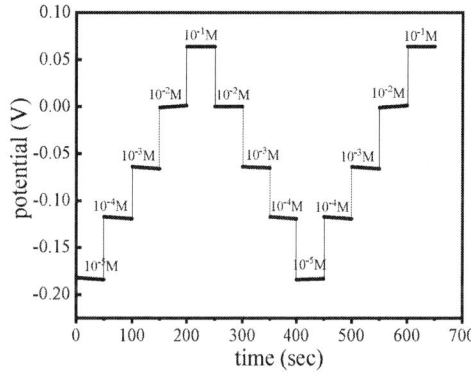

Figure 10. Reversibility test in potential response for the ISE.

TABLE II. COMPARISON OF PERFORMANCE WITH OTHERS' WORK

Reference	Mechanism	Sensitivity	Range (M)	LOD(M)
Bandodkar [2]	potentiometry	63.8 mV/decade	1×10^{-4} to 1×10^{-1}	N/A
Rose [3]	potentiometry	25 mV/decade	1×10^{-2} to 9×10^{-2}	N/A
Gao [4]	potentiometry	64.2 mV/decade	0.01 to 0.16	N/A
Roy [5]	potentiometry	58 mV/decade	7.08×10^{-7} to 1	3.16×10^{-6} #
This work	impediametry	8.78 mS/M	1×10^{-2} to 1×10^{-1}	$> 10^{-3}$
	potentiometry	66.2 mV/decade	1×10^{-5} to 1×10^{-1}	8.2×10^{-6} #

LOD is defined as the target ion concentration where the measured error is 100%.

IV. CONCLUSION

A flexible microelectrode array fabricated on PI substrate for sweat analysis is proposed in this work. Different from the single transduction mechanism reported in previous sweat sensors, the sweat sensor based on two different transduction mechanisms is introduced. For the sweat analysis, the impedimetric method is concerned with the conductivity of the strong electrolyte of the sweat, which is accurate and efficient for the NaCl electrolyte detection within the range of human sweat concentration. The potentiometric method shows the distinct advantages on the detection range of the ion concentration and the limit of detection (LOD). By using the two sensing methods simultaneously for the sweat sensor, the impedance analysis can improve the accuracy of the detection on basis of the potential results, while the potential analysis performs in the larger concentration range and the lower LOD.

ACKNOWLEDGMENT

This research was partially supported by a grant from the Natural Science Foundation of Guangdong Province (2020A1515011555), Natural Science Foundation of SZU (860-2110309 and 85304-211), and HighTalent Research Funding (827-000451).

REFERENCES

[1] M. Bariya, H. Y. Y. Nyein, and A. Javey, "Wearable sweat sensors," *Nature Electronics*, vol. 1, pp. 160-171, 2018.

[2] A. J. Bandodkar, et al. "Epidermal tattoo potentiometric sodium sensors with wireless signal transduction for continuous non-invasive sweat monitoring," *Biosensors and Bioelectronics*, vol. 54, pp. 603-609, 2014.

[3] D. P. Rose, et al. "Adhesive RFID Sensor Patch for Monitoring of Sweat Electrolytes," *IEEE Transactions on Biomedical Engineering*, vol. 62, pp. 1457-1465, 2015.

[4] W. Gao, et al. "Fully integrated wearable sensor arrays for multiplexed in situ perspiration analysis," *Nature*, vol. 529, pp. 509-514, 2016.

[5] S. Roy, M. David-Pur, and Y. Hanein, "Carbon Nanotube-Based Ion Selective Sensors for Wearable Applications," *ACS Applied Materials & Interfaces*, vol. 9, pp. 35169-35177, 2017.

[6] S. Nakata, M. Shiomi, Y. Fujita, T. Arie, S. Akita, and K. Takei, "A wearable pH sensor with high sensitivity based on a flexible charge-coupled device," *Nature Electronics*, vol. 1, pp. 596-603, 2018.

[7] Y. Hua, et al. "Flexible Sweat Monitoring based on All-Solid-State Metal-Organic Frameworks/graphene Composite Sensors," in *2019 IEEE SENSORS*, 2019, pp. 1-4.

[8] J. Monge, et al. "Glucose Detection in Sweat using Biosensors," in *2019 E-Health and Bioengineering Conference (EHB)*, 2019, pp. 1-5.

[9] J. Kim *et al.*, "Noninvasive Alcohol Monitoring Using a Wearable Tattoo-Based Iontophoretic-Biosensing System," *ACS Sensors*, vol. 1, pp. 1011-1019, 2016.

Gap in pagination due to unavailable papers.

Pages 1507-1516

AlN Hybrid-Coupled Resonator With Phononic Crystal Reflector

Kangfu Liu, Yuxi Wang, Tao Wu*

Abstract—This work proposes a method to design phononic crystal reflectors efficiently. For a prove of concept, a kind of quasi-surface acoustic wave and bulk acoustic wave (SAW/BAW) hybrid-coupled resonator with phononic crystal as reflectors is studied. The proposed design is implemented on the aluminum nitride (AlN) thin film with a high acoustic velocity layer. The resonator achieves an electromechanical coupling coefficient of 5.4%, which is comparable to AlN thin film bulk acoustic wave resonator (FBAR). With the help of phononic crystal reflector, the proposed resonator achieves a quality factor (Q) up to 3146 in simulation, which is 1.7 times of the resonator with traditional Bragg reflectors, thus achieving high coupling and quality factor simultaneously. Considering the possibility of monolithic chip with multiple frequency, this technique might be a favorable candidate of FBAR technology.

I. INTRODUCTION

Filters are the keys to define these frequency bands in the communication system, which select the desired frequency and suppress irrelevant signals. The radio frequency (RF) piezoelectric micro-electro-mechanical-system (MEMS) resonators/filters have been playing an important role in filtering technology, due to the excellent performance, compact size, and low cost. The bandwidth (BW) and insertion loss (IL) of filter based on piezo-MEMS is fundamentally determined by the electromechanical coupling coefficient (k_t^2) and quality factor (Q) of the constructed resonator [1]. Therefore, the figure-of-merit (FoM) are defined by $k_t^2 \cdot Q$, which has always been one of the goals pursued by researchers.

Conventional SAW filters based on Lithium Niobate (LN) and that has a high k_t^2 are commercially successful. However, due to the low phase velocity of SAW in LN substrate, the frequency of LN SAW is limited below 2.5 GHz, and LN substrate is not compatible with the complementary metal-oxide-semiconductor (CMOS) process. The maturation of AlN bulk acoustic wave (BAW) resonator (including FBAR), has allowed the replacement of off-chip SAW products. However, its resonance frequency (f_s) is defined by the thickness of the AlN plate (h), thus losing the ability to be tuned by lithography. Meanwhile, traditional AlN based SAW has limited coupling factor ($k_t^2 \approx 0.3\%$).

In our previous research, a SAW/BAW hybrid-coupled resonator implemented on AlN/6H-SiC layer is proposed, which achieves a high k_t^2 up to 6% [2], which is comparable with

All authors are 1 School of Information Science and Technology, ShanghaiTech University, 2 Shanghai Institute of Micro-system and Information Technology, Chinese Academy of Sciences, 3 University of Chinese Academy of Sciences, China (E-mail: liukf, wangyx3, wutao@shanghaitech.edu.cn)

FBAR. Only the unit cell in a resonator with Floquet periodic boundaries was analyzed in the previous study. However, without a proper reflector design, the energy will escape from the effective region, reducing the Q.

One of the resonator quality factor enhanced techniques is phononic crystal (PnC) [3–5]. Phononic crystals are defined as artificial materials made of periodic arrangement of scatters embedded in a medium. Under certain conditions, the band structure of PnC may present absolute band gaps: they display frequency ranges in which waves cannot propagate. The results of finite element analysis (FEA) software (COMSOL) show that, with the help of PnC reflector, the proposed hybrid-coupled AlN resonator might achieve high k_t^2 and Q simultaneously.

This article will discuss in detail how to synthesize a complete hybrid-coupled resonator and improve its quality factor with the PnC reflectors. The design method of PnC will greatly improve design efficiency with only using of Eigenfrequency study, especially in the case of numerous grids or 3D model. Meanwhile, the band structure of the PnC provides the guidance of optimization, avoiding the blind parameter searching.

II. REFLECTOR DESIGN

A. Resonator Structure

The cross-section view of the proposed resonator structure shown in Fig. 1. The proposed resonator consists of inter-digital electrode transducer (IDT) and the reflectors on both sides. Positive and negative potentials are applied alternately on the IDT, while the metal electrode at the reflector is floating. Different from traditional SAW, there are many grooves in AlN layer (piezoelectric layer). The etched grooves can be obtained by using Ni IDT as an etching mask, and naturally form periodic structures.

The resonator contains multiple layers, which consist of Ni/AlN/Mo and a high acoustic speed layer (HASL). The Ni layer is used not only as IDT electrodes, but also as the mask for etching grooves. The selection of Ni electrode is because of the high selectivity of Ni to AlN (up to 30) in the inductively coupled plasma (ICP) etching [6]. The Mo layer is utilized to provide the floating potential to enhance the electric field between the upper and lower electrodes. Thereby utilizing e_{33} of AlN efficiently. Due to the large different in the acoustic impedance between the HASL and AlN layer, the HASL effectively prevents the acoustic energy from leaking downward.

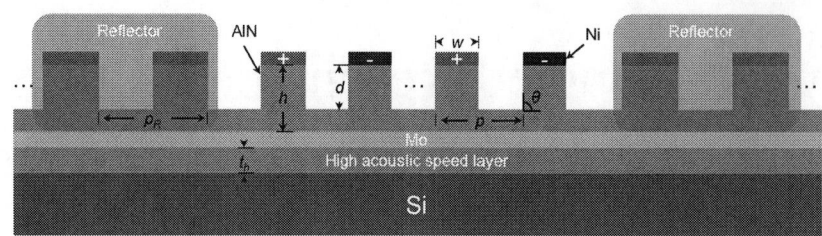

Fig. 1. Cross-section view of the proposed structure of the resonator, which consist of IDTs and reflectors.

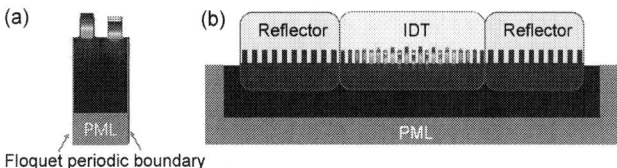

Fig. 2. The 2D simulation setup for (a) a unit cell of reflector (b) a complete resonator. The color and deformation represent the displacement when the quasi-SAW is generated.

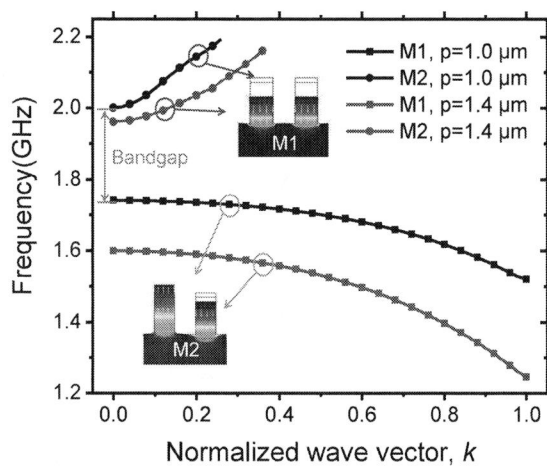

Fig. 3. The band structure of Bragg reflector and optimized PnC reflectors, when p_R is 1 and 1.4 μm respectively. The vignettes show two different SAW mode shapes, M1 and M2.

Previous study [2] shows that the selection of material of HASL is critical, which will greatly affect the k_t^2 of resonator. By using diamond or SiC as HASL, the resonator exhibits the best k_t^2 performance. However, the cost of diamond substrate or thin film is extremely high. For practical purposes, the HASL is substituted by 6H-SiC in this work.

In Fig. 1, h, t_h, and d represent the thickness of AlN, 6H-SiC, and depth of grooves, respectively; w, p and p_R represent the width and pitch of IDT and pitch of reflector, respectively; θ is the angle of the etched AlN sidewall. The period of IDT is $\lambda = 2p$.

Higher d/h, h/λ, and θ, indicates higher k_t^2 of resonator [2]. For the consideration of high k_t^2, the default geometry parameters adopted in this work are $h = 1$μm, $d/h = 0.9$, $t_h = 4$μm, $\theta = 90°$, $\lambda = 2$μm. The thickness of Ni and Mo layer is both 100 nm.

B. Band Structure of Reflector

Periodic grating reflectors are commonly used in SAW devices to efficiently reflect and confine the acoustic energy within the resonator. In the implementation of SAW device, periodic structures of metal on the top of the piezoelectric film are used as reflectors. The acoustic speed in the metal-covered and the non-metal-covered area are different, resulting in the acoustic wave is partially reflected when it propagates through one period of grating structure. When the number of period is higher enough, most of the energy is reflected. The width and spacing of metal grating reflector are usually set to one quarter of the wavelength of the medium, thus it is named as $\lambda/4$ reflector or Bragg reflector. Essentially, the grating reflectors can be considered as a one-dimensional PnC [7]. Therefore, the analysis method of PnC can be used to analyze the operation frequency and stop bandwidth of grating reflectors.

The 2D simulation setup for the band structure of the reflector unit cell is shown in Fig. 2 (a). The floquet periodic boundary are employed to both side of the unit cell. The perfectly matched layer (PML) at the bottom is used to absorb the leaked energy.

By changing the wave vector of the periodic boundary, the Eigenfrequency variation of different modes can be scanned. However, there are many bulk acoustic mode cannot be exited. By removing these modes, the band structure of SAW modes can be obtained. There are two SAW modes in this design, denoted as M1 and M2, whose mode shapes are shown in the vignette in Fig. 3. M1 is an unwanted SAW mode, while M2 represents the quasi-SAW mode need to be reflected by the reflector. The band gap between the two surface wave modes indicates the bands where the SAW can not pass through.

In this work, the unit cell of the resonator and the reflector are in the same structure. When a traditional Bragg reflector design is adopted, the resonant frequency of IDT is located right on the edge of the band gap, thus reflecting the wave energy. However, the number of grating reflector in reality is finite, the reflectivity is limited. When the IDT operating frequency is placed within the center of band gap, the highest reflection efficiency can be obtained [8]. The p_R of traditional Bragg reflector is $\lambda/4$, the corresponding $p_R = 1$

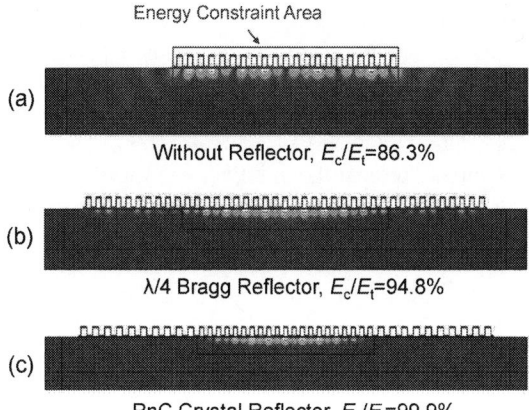

Fig. 4. The displacement and energy constraint efficiency ($\eta = E_c/E_t$) of different designs. (a) without reflector (b) $\lambda/4$ Bragg reflector (c) Optimized PnC reflector. To highlight the propagation range of surface waves, the displacement of IDT and reflector is hidden.

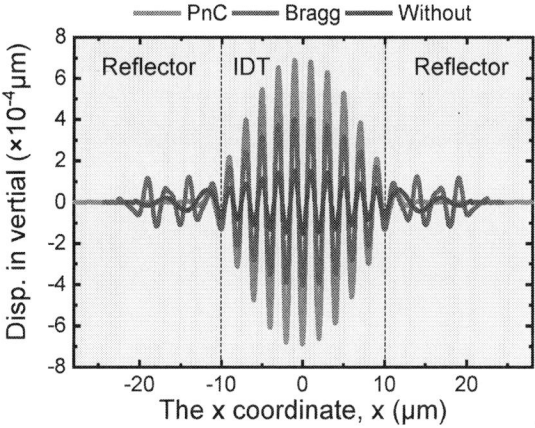

Fig. 5. The displacement in vertical direction at the surface of HASL with different reflector designs.

μm. Optimized the p_R by adjusting it to make the operation frequency of IDT around the center of the band gap of PnC. An optimized value of p_R is 1.4 μm. The SAW mode band structure of Bragg reflector and optimized PnC are shown in Fig. 3, where the wave vector (k) is normalized to $2\pi/\lambda$. The stopband of traditional Brag reflector is 1741-2001 MHz, while the optimized PnC is 1599-1962 MHz. The band gap is expanded by 39.6%. The optimized PnC will be used as reflector for the performance analysis in the following section.

III. SIMULATION OF RESONATOR

A. Energy and Displacement Distribution

A complete resonator is obtained by combining IDT with the reflectors. To verify the superiority of the optimized PnC reflector, a series of simulation are conducted. The resonators with three different grating reflector design are studied, which include the cases of without reflector, traditional Bragg reflector, and optimized PnC reflector. Fig. 2 (b) shows the simula-

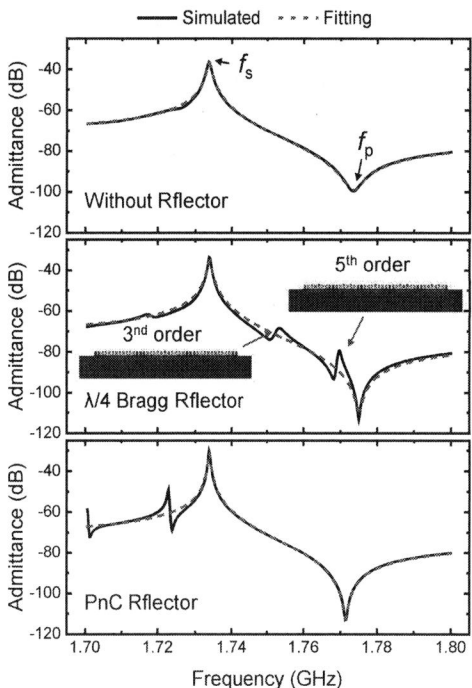

Fig. 6. The simulated admittance performance with different reflector.

TABLE I
THE PERFORMANCE OF RESONATOR WITH DIFFERENT REFLECTOR

Reflector structure	k_t^2	Q_s	Q_p
Without	5.69	1436	448
Bragg reflector	5.91	1780	1310
PnC reflector	5.41	3146	2166

tion setup of resonators. The surrounding of the substrate is set to be PML and highlighted in light blue color. The number of IDT and grating reflector are set to be 20 and 10, respectively. The mechanical quality factor of AlN is setting as 2000 to simulate the loss of piezoelectric materials.

For qualitative analysis the reflective effect of different design, we define an energy constraint efficiency ($\eta = E_c/E_t$). E_c represents the mechanical elastic strain energy in the constraint area, which includes the IDT and the region of one wave length depth below it; E_t is the total mechanical elastic strain energy within the whole device. E_t and E_c can be obtained by monitoring with the domain probe in COMSOL. The energy constraint efficiency (η) and displacement distribution of the resonator in the resonant state can be obtained by Eigenfrequency study.

Fig. 4 shows the displacement of the 2D model and η with different reflector designs. To highlight the propagation range of surface waves, the displacement of IDT and reflector is hidden. Obviously, without reflector, the acoustic wave propagates laterally and part of the energy is leaked to the substrate. Its energy confinement effect is the worst, and has the lowest η. Although the Bragg reflector restrict part of the energy leakage, there still is 5.2% energy pass through the reflector. The PnC reflector confine almost all of the energy within the IDT region, and has the highest η.

To visualize the acoustic propagation at resonance, the frequency domain analysis is conducted. The surface displacement at the top of HASL with different reflectors is shown in Fig. 5. They are all operating at the resonant frequency and exited by the same voltage. Under the same excitation conditions, the displacement of IDT region without reflector is obviously weaker than the other two designs. The Bragg reflector enhances the vibration of the IDT region, but still leaks part of energy. The amplitude of the acoustic wave decreases dramatically to zero after a distance of about twice of wavelength, when it through the optimized PnC reflector. These simulations have shown the effectiveness of the PnC reflector.

B. The Performance of Resonator with Different Reflector

To specify the influence of different reflectors on the performance of the resonator, the frequency domain study is adopted. Fig. 6 shows the simulated and fitting admittance responses of the resonators with three different reflector designs that is shown in Fig. 4. The Modified Butterworth-Van Dyke model [9] is used for fitting and parameters extraction. Their key performances are listed in Table I. The electromechanical coupling coefficient (k_t^2) is defined as:

$$k_t^2 = \frac{\pi^2}{8} \cdot \frac{(f_p^2 - f_s^2)}{f_p^2} \tag{1}$$

where f_s and f_p represent the resonant and anti-resonant frequency, respectively. And Q_s and Q_p represent the quality factor at f_s and f_p, respectively.

As expected, a resonator without reflectors has the worst performance, i.e. the lowest both Q_s and Q_p. The resonator with optimized PnC reflectors, has the highest Q_s and Q_p of 3146 and 2166, respectively . It is approximately 70% larger than the resonator with traditional $\lambda/4$ Bragg reflectors.

It is worth noting that the presence of lateral spurious modes of 3rd- and 5th- order in the resonator with Bragg reflectors, resulting a slightly higher k_t^2. The cause of this phenomenon is that the resonant frequency of the Bragg reflectors and the IDT are close to each other, thus generating regional resonance. The PnC reflectors and IDT have significant difference in resonant frequency, which avoids the lateral high order modes in the pass band.

IV. CONCLUSION

In this work, a convenient method of designing reflector is proposed. By simulating the unit cell model with Floquet periodic boundaries in FEA software, the band structure of the reflector and the energy constraint efficiency (η) of the resonator can be obtained, with only using of the Eigenfrequency study in COMSOL. The analysis and design method in this article is theoretically applicable to any periodic grating reflector.

By synthesizing the reflector and IDT to a complete resonator and analyzing the frequency response, the superiority of the optimized PnC reflector is verified. Compared to the Bragg reflectors in the same conditions, PnC reflector significantly improves the Q of the resonators. The proposed Hybrid-coupled resonator with PnC reflectors achieves the electromechanical coupling coefficient (k_t^2) of 5.4% and the quality factor (Q) up to 3146. Considering the possibility of monolithic chip with multiple frequency, this design might be a favorable candidate of FBAR technology.

ACKNOWLEDGMENT

The authors appreciate the support from National Natural Science Foundation of China (61874073), and Natural Science Foundation of Shanghai (19ZR1477000).

REFERENCES

[1] A. Gao, K. Liu, J. Liang, and T. Wu, "AlN MEMS filters with extremely high bandwidth widening capability," *Microsystems & Nanoengineering*, vol. 6, no. 76, Sep. 2020.

[2] K. Liu, J. Liang, and T. Wu, "Aln Hybrid-Coupled Resonators With High Acoustic Velocity Layer," in *2020 IEEE International Ultrasonics Symposium (IUS)*, Sep. 2020, pp. 1–4.

[3] J. Liu, T. B. Workie, T. Wu, Z. Wu, K. Gong, J. Bao, and K.-y. Hashimoto, "Q-Factor Enhancement of Thin-Film Piezoelectric-on-Silicon MEMS Resonator by Phononic Crystal-Reflector Composite Structure," *Micromachines*, vol. 11, no. 12, p. 1130, Dec. 2020.

[4] L. Binci, C. Tu, H. Zhu, and J. E.-Y. Lee, "Planar ring-shaped phononic crystal anchoring boundaries for enhancing the quality factor of Lamb mode resonators," *Applied Physics Letters*, vol. 109, no. 20, p. 203501, Nov. 2016.

[5] S.-Y. Yu, J.-Q. Wang, X.-C. Sun, F.-K. Liu, C. He, H.-H. Xu, M.-H. Lu, J. Christensen, X.-P. Liu, and Y.-F. Chen, "Slow surface acoustic waves via lattice optimization of a phononic crystal on a chip," *Physical Review Applied*, vol. 14, 12 2020.

[6] Z. Luo, S. Shao, and T. Wu, "Chrarcterization of AlN and AlScN film ICP etching for micro/nano fabrication," *Microelectronic Engineering*, p. 111530, Feb. 2021.

[7] N. Y. Kozlovski and D. C. Malocha, "SAW Phononic Reflector Structures," in *2007 IEEE International Frequency Control Symposium Joint with the 21st European Frequency and Time Forum*, May 2007, pp. 1229–1234.

[8] A. Kochhar, A. Mahmoud, Y. Shen, N. Turumella, and G. Piazza, "X-cut lithium niobate-based shear horizontal resonators for radio frequency applications," *Journal of Microelectromechanical Systems*, vol. 29, no. 6, pp. 1464–1472, 2020.

[9] J. D. Larson, P. D. Bradley, S. Wartenberg, and R. C. Ruby, "Modified Butterworth-Van Dyke circuit for FBAR resonators and automated measurement system," in *2000 IEEE Ultrasonics Symposium. Proceedings. An International Symposium*, vol. 1, Oct. 2000, pp. 863–868 vol.1.

978-1-6654-3008-1/21 $31.00 © 2021 IEEE

Gap in pagination due to unavailable paper.

Pages 1521-1522

Proceedings of the 16th Annual IEEE International
Conference on Nano/Micro Engineered and Molecular Systems
April 25-29, 2021

Optimization of S_1 Lamb wave resonators with $Al_{0.8}Sc_{0.2}N$

Shuai Shao[1,2,3,*], Zhifang Luo[1,2,3], and Tao Wu[1,*]

Abstract—We demonstrate the optimized design of $Al_{0.8}Sc_{0.2}N$-based S_1 Lamb wave resonators. The S1 mode shows a phase velocity of over 20,000 m/s, and an electromechanical coupling factor of over 7% can be achieved using the enhancement of the AlN piezoelectric characteristics by Sc doping. To exploit the high phase velocity region with thickness/lambda (h/λ) less than 0.3, a clean spectral response can be obtained by utilizing the double-IDT electrode configuration. The electrode coverage can significantly suppress the spurious modes, and high order modes free resonance has been achieved by utilizing 20% electrode coverage design. Although Sc-doping in AlScN film causes a slight decrease in phase velocity, significant enhancement of the electromechanical coupling coefficient provides potential for high frequency and large bandwidth applications.

I. BACKGROUND

Because of the high phase velocity, strong polarity, and the ability to be integrated with CMOS processes, aluminum nitride based resonators are now widely used in sensors and RF filters[1]–[5]. The demand for high-frequency and wideband filters in 5G communication is difficult to be satisfied by the conventional AlN resonators[6]. Thin film bulk resonator (FBAR) has become a commercial success case due to high quality factor and electromechanical coupling coefficient[7], [8]. However, the lack of wide range frequency tunability ability of FBAR limits the application for small footprint and multi-band RF front-end module.

Lamb wave modal resonators have become a research hot topic due to the combination of lithographically defined frequencies and electromechanical coupling coefficients[9]–[11]. The dispersion characteristics, electrode design and material selection of S_0 mode resonators have been discussed in detail. The S_0 mode resonator coupled d_{31} with d_{33} piezoelectric constants can obtain electromechanical coupling coefficients close to FBAR. However, the phase velocity of S_0 mode is low, and the frequency is only about 2 GHz when the wavelength is around 5 μm[12]. It is not sufficient to meet the demand for high frequency and wide passband filters for 5G communications. Further the wavelength reduction will face serious power capacity and resistance loss problems. High-order modes with higher phase velocities urgently need to be developed.

AlN Lamb wave resonators with S_1 mode have been shown to be promising in achieving the above objectives. The Lamb wave resonator consists of the piezoelectric film and double-side interdigital transducer (IDT), as shown in Fig. 1 (a). The S_1 mode displacement of the finite element analysis (FEA) simulation of is shown in Fig. 1 (b). In contrast to the S_0 mode, the S_1 mode has a period of standing waves in the film thickness direction. Displacement is

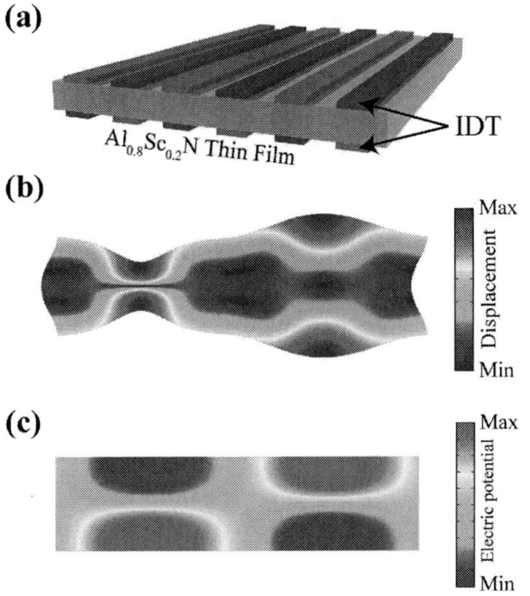

Fig. 1. (a) Illustrations of the lamb wave resonator with IDT. FEA simulation of the S_1 mode lamb wave resonator. (b) total displacement, (c) electric potential.

mainly concentrated in the thickness direction, and high coupling coefficients can be achieved by d_{33}. The electric potential is also alternately distributed on both sides of the film to fit the IDT arrangement as shown in Fig. 1 (c). The S_1 mode has a 20,000 m/s higher phase velocity than the S_0 mode. The coupling coefficient is comparable to the S_0 mode with a reasonable electrode design. The S_1 mode was reported that the group velocity is 0 at h_{AlScN}/λ=0.3[13]. Therefore, this property is exploited to suppress other modes using damping boundaries. The resonance frequency of 3 GHz can be achieved in a design with the wavelength of 20 μm[14]. However, use of higher phase velocity region ($h_{AlScN}/\lambda < 0.3$) and further improvement of electromechanical coupling coefficients become problems to be solved. Compared with other piezoelectric materials such as $LiNbO_3$ and PZT, the d_{33} of aluminum nitride is only around 6 pC/N. The low piezoelectric constant of aluminum nitride limits the maximum electromechanical coupling coefficient to 7%. It has been reported that the piezoelectric constant and performance of resonators can be significantly enhanced by doping scandium(Sc) into aluminum nitride[15]–[19].

Here, we discuss the use of AlScN thin film resonators to enhance device performance and the design optimization

---

[1]School of Information Science and Technology, ShanghaiTech University, Shanghai, China
[2]Shanghai Institute of Microsystem and Information Technology, Chinese Academy of Sciences

[3]University of Chinese Academy of Sciences, China
*Corresponding Authors:
Shuai Shao. (e-mail: shaoshuai@shanghaitech.edu.cn)
Tao Wu. (e-mail: wutao@shanghaitech.edu.cn)

978-1-6654-3008-1/21 $31.00 © 2021 IEEE

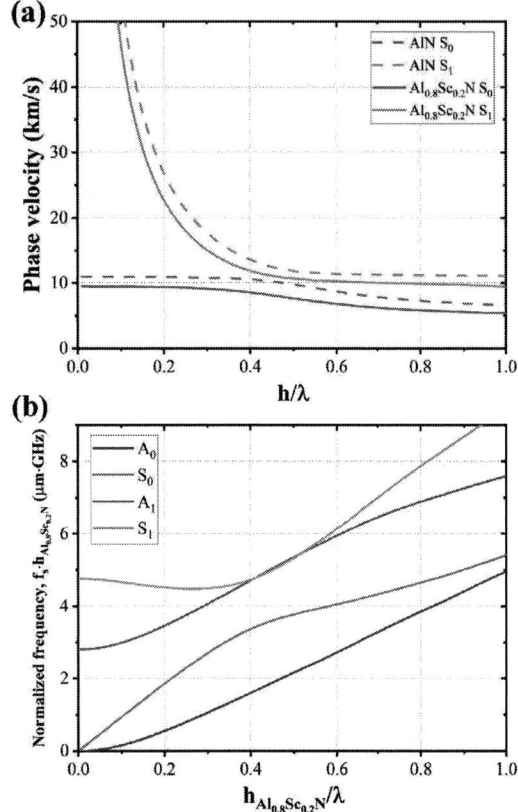

Fig. 2. (a) Phase velocity dispersion curves with AlN and $Al_{0.8}Sc_{0.2}N$, (b) Normalized frequency dispersion curves.

TABLE I. MATERIAL PARAMETERS OF $AL_{0.2}SC_{0.8}N$

Parameters	Value	Parameters	Value
c_{11}^{E} (GPa)	353.57	e_{15} (C/m²)	-0.34
c_{12}^{E} (GPa)	143.74	e_{31} (C/m²)	-0.70
c_{13}^{E} (GPa)	127.80	e_{33} (C/m²)	1.86
c_{33}^{E} (GPa)	287.02	$\varepsilon_{33}^{eff}/\varepsilon_0$	15.08
c_{44}^{E} (GPa)	110.25	ρ (kg/m³)	3286

of AlScN-based S_1 mode resonators for achieving ultra-high frequency and electromechanical coupling coefficients.

II. CHARACTERISTIC OF S_1 MODE IN $AL_{0.2}SC_{0.8}N$ FILM

Fig. 2 (a) shows the phase velocities of AlN and AlScN in the S_0 and S_1 modes, respectively. The phase velocity with the normalized thickness (h_{AlScN}/λ) is analyzed by using the Floquet periodicity boundary condition method in COMSOL Finite Element Analysis (FEA). Scandium doping causes a slight decrease in the S_1 mode phase velocity, but still achieves the phase velocity of over 20,000 m/s at h_{AlScN}/λ less than 0.2. This allows resonant frequencies in excess of 2 GHz to be obtained using a wavelength with 10 μm pitch. Moreover, the phase velocity increases sharply as h_{AlScN}/λ decreases, which means less resistive loss. Therefore, a design needs to be developed for a clean spectral response without spurious modes at h_{AlScN}/λ less than 0.2.

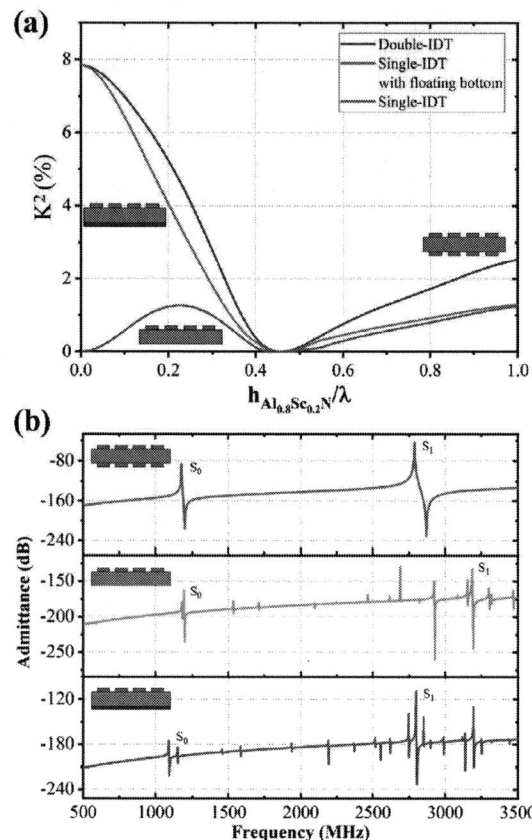

Fig. 3. (a) S_1 mode intrinsic coupling coefficient of the three electrode configurations, (b) Simulated admittance responses of the three electrode configurations

Fig. 2 (b) shows the dispersion characteristics of the first four modes of Lamb waves in $Al_{0.8}Sc_{0.2}N$ films by FEA simulations. To normalize the thickness of the film to the wavelength, the angular frequency is transformed to $f*h$. As shown in Table I, the material parameters of $Al_{0.8}Sc_{0.2}N$ based on the ab-initio equations are reported in [20], [21]. When h_{AlScN}/λ is between 0.4 and 0.6, the normalized frequencies of S_1 and A_1 modes overlap and are close to the S_0 mode, which is more easily to form spurious modes. In order to obtain a clean frequency response, the region with h_{AlScN}/λ less than 0.4 needs to be chosen to avoid the other modes interference.

III. DESIGN FOR S_1 MODE IN $AL_{0.2}SC_{0.8}N$ FILM

The difference in phase velocity between the free surface v_0 and the metalized surface v_m can be used to estimate the intrinsic coupling coefficient (K^2)[12], [22].

$$K^2 = \frac{v_0^2 - v_m^2}{v_0^2} \qquad (1)$$

Fig. 3 (a) shows the intrinsic coupling coefficient of three electrode configurations in the conventional Lamb wave resonator using AlScN films. As expected, the single-IDT configuration has the simplest processing difficulty, but the lowest coupling coefficient due to the absence of the bottom electrode. The single-IDT with floating electrode and

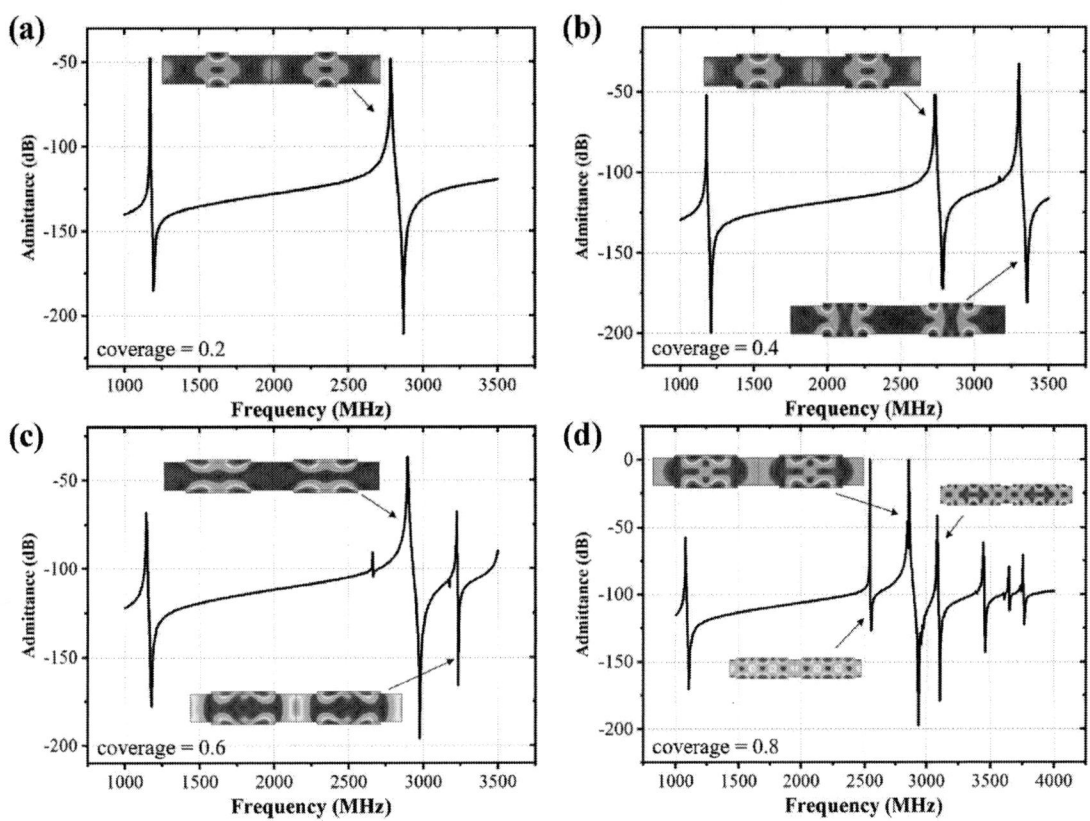

Fig. 4. Simulated admittance responses of the S_1 Lamb mode resonators with different electrode coverage, (a) 0.2, (b) 0.4, (c) 0.6, (d) 0.8.

double-IDT configurations obtain approximate coupling coefficients due to a better vertical electric field distribution. Although they require an additional layer of metal deposition and lithography. Because of the $Al_{0.8}Sc_{0.2}N$ films, the S_1 lamb wave resonator can obtain an electromechanical coupling coefficient of up to nearly 8%, which is higher than the 7% for the AlN-based FBAR. In addition to the phase velocity and electromechanical coupling coefficient, a spurious-free resonance is also critical for RF filtering component implementation. Although the process difficulty of single-IDT with floating electrode is simpler than double-IDT, only double-IDT can obtain a clean spectrum response, as shown in Fig. 3 (b). Unlike the S_0 mode where a clean spectrum can be obtained in all three settings, the S_1 mode requires a symmetrical electrodes arrangement and vertical electric field distribution. For single-IDT and single-IDT with floating bottom, the asymmetric mass loading and the absence of a periodic vertical electric field distribution lead to the formation of a large number of spurious modes. To obtain high phase velocities and electromechanical coupling

coefficients when h_{AlScN}/λ is less than 0.3, only the double-IDT electrode configuration can acquire clean spectrum. Corresponding to the phase velocity in Fig. 2 (a), the double-IDT electrode setup can obtain both high phase velocity and electromechanical coupling coefficient.

To optimize the design of double-IDT configuration, the electrode coverage is also investigated. Since different mechanical loads and electric field distributions directly affect the generation of spurious modes. The electrode coverage is defined as the ratio of the width of the IDT to the pitch. Using 2D FEA simulations, four coverage ratios of 0.2, 0.4, 0.6, and 0.8 were analyzed. The specific design parameters are summarized in Table II. The frequency response with the different electrode coverages at double-IDT configuration and h_{AlScN}/λ of 0.125 is shown in Fig. 4. Mo was selected as the electrode material. When the coverage is 0.4 and 0.6 (as shown in Fig. 4 (b) and (c)), spurious modes appear near the expected S_1 modes. This mode is considered to be excited by the vertical electric field. The stresses and deformations occur mainly at the electrode edges. Since the stress is concentrated at the edge of the electrode and not distributed in the center of the electrode, it is not beneficial to the charge collection. Fig. 4 (d) shows that when the coverage reaches 0.8, the overtone mode of S_0 also appear in addition to the spurious modes. As the coverage decreases, the spurious modes gradually disappear and finally a clean spectrum response is obtained at a coverage of 0.2, as indicated in Fig.4 (a). At a small coverage, the vertical electric field distribution is more concentrated, suppressing the appearance of spurious modes, and the displacement occurs at the center of the electrode

TABLE II. DESIGN PARAMETERS OF $AL_{0.2}SC_{0.8}N$ S_1 LAMB WAVE RESONAOTR

Parameters	Value
Thickness of AlScN (μm)	1
Thickness of electrodes (μm)	0.1
Pitch (μm)	4
Coverage	0.2, 0.4, 0.6 and 0.8
Number of IDTs	8

IV. CONCLUSION

This work demonstrates the optimized design of an $Al_{0.8}Sc_{0.2}N$-based S_1 Lamb wave resonator. The S_1 mode shows a phase velocity of over 20,000 m/s, and an electromechanical coupling factor of over 7% is obtained using the enhancement of the AlN piezoelectric characteristics by Sc doping. AlScN causes a slight decrease in phase velocity, but significantly improves the electromechanical coupling coefficient. Three electrode configurations are compared for higher electromechanical coupling coefficients and the clean spectrum response. Symmetrical structure of double-IDT allows for symmetrical mechanical loading and nice vertical electric field distribution is more suitable for S_1 mode resonator designs. In addition, the electrode coverage can significantly suppress the spurious modes and high order modes free resonance has been achieved utilizing 20% electrode coverage.

ACKNOWLEDGMENT

The authors appreciate the support from the National Natural Science Foundation of China (61874073) and Natural Science Foundation of Shanghai (19ZR1477000).

REFERENCES

[1] T. Wu, R. Lu, A. Gao, C. Tu, T. Manzaneque, and S. Gong, "A Chip-Scale RF MEMS Gyrator via Hybridizing Lorentz-Force and Piezoelectric Transductions," in *2019 IEEE 32nd International Conference on Micro Electro Mechanical Systems (MEMS)*, Jan. 2019, pp. 887–890.

[2] T. Wu *et al.*, "Design and fabrication of AlN RF MEMS switch for near-zero power RF wake-up receivers," in *2017 IEEE SENSORS*, Oct. 2017, pp. 1–3.

[3] Y. Qiu *et al.*, "Piezoelectric Micromachined Ultrasound Transducer (PMUT) Arrays for Integrated Sensing, Actuation and Imaging," *Sensors*, vol. 15, no. 4, pp. 8020–8041, Apr. 2015.

[4] M. D. Williams, B. A. Griffin, T. N. Reagan, J. R. Underbrink, and M. Sheplak, "An AlN MEMS Piezoelectric Microphone for Aeroacoustic Applications," *J. Microelectromechanical Syst.*, vol. 21, no. 2, pp. 270–283, 2012.

[5] Z. Luo, S. Shao, and T. Wu, "Characterization of AlN and AlScN film ICP etching for micro/nano fabrication," *Microelectron. Eng.*, p. 111530, Feb. 2021.

[6] S. Mahon, "The 5G Effect on RF Filter Technologies," *IEEE Trans. Semicond. Manuf.*, vol. 30, no. 4, pp. 494–499, Nov. 2017.

[7] R. Ruby, P. Bradley, J. Larson, Y. Oshmyansky, and D. Figueredo, "Ultra-miniature high-Q filters and duplexers using FBAR technology," in *2001 IEEE International Solid-State Circuits Conference. Digest of Technical Papers. ISSCC*, Feb. 2001, pp. 120–121.

[8] R. C. Ruby, P. Bradley, Y. Oshmyansky, A. Chien, and J. D. Larson, "Thin film bulk wave acoustic resonators (FBAR) for wireless applications," in *2001 IEEE Ultrasonics Symposium. Proceedings. An International Symposium*, Oct. 2001, vol. 1, pp. 813–821 vol.1.

[9] G. Piazza, P. J. Stephanou, and A. P. Pisano, "Single-chip multiple-frequency ALN MEMS filters based on contour-mode piezoelectric resonators," *J. Microelectromechanical Syst.*, vol. 16, no. 2, pp. 319–328, 2007.

[10] C. Lin, V. Yantchev, Y. Chen, V. V. Felmetsger, and A. P. Pisano, "Characteristics of AlN Lamb wave resonators with various bottom electrode configurations," in *2011 Joint Conference of the IEEE International Frequency Control and the European Frequency and Time Forum (FCS) Proceedings*, May 2011, pp. 1–5.

[11] J. Zou, C.-M. Lin, C. S. Lam, and A. P. Pisano, "Transducer design for AlN Lamb wave resonators," *J. Appl. Phys.*, vol. 121, no. 15, p. 154502, Apr. 2017.

[12] J. Zou, C. Lin, A. Gao, and A. P. Pisano, "The Multi-Mode Resonance in AlN Lamb Wave Resonators," *J. Microelectromechanical Syst.*, vol. 27, no. 6, pp. 973–984, Dec. 2018.

[13] J. Zou, A. Gao, and A. P. Pisano, "Spectrum-clean S1 AlN Lamb wave resonator with damped edge reflectors," *Appl. Phys. Lett.*, vol. 116, no. 2, p. 023505, Jan. 2020.

[14] A. Gao, J. Zou, and S. Gong, "A 3.5 GHz AlN S1 lamb mode resonator," in *2017 IEEE International Ultrasonics Symposium (IUS)*, Sep. 2017, pp. 1–4.

[15] M. Akiyama, T. Kamohara, K. Kano, A. Teshigahara, Y. Takeuchi, and N. Kawahara, "Enhancement of Piezoelectric Response in Scandium Aluminum Nitride Alloy Thin Films Prepared by Dual Reactive Cosputtering," *Adv. Mater.*, vol. 21, no. 5, pp. 593–596, 2008.

[16] L. Colombo, A. Kochhar, C. Xu, G. Piazza, S. Mishin, and Y. Oshmyansky, "Investigation of 20% scandium-doped aluminum nitride films for MEMS laterally vibrating resonators," in *2017 IEEE International Ultrasonics Symposium (IUS)*, Sep. 2017, pp. 1–4.

[17] A. Lozzi, E. T.-T. Yen, P. Muralt, and L. G. Villanueva, "Al0.83Sc0.17N Contour-Mode Resonators With Electromechanical Coupling in Excess of 4.5%," *IEEE Trans. Ultrason. Ferroelectr. Freq. CONTROL*, vol. 66, no. 1, pp. 146–153, 2019.

[18] Y. Lu *et al.*, "Elastic modulus and coefficient of thermal expansion of piezoelectric Al $_{1-x}$ Sc $_x$ N (up to x = 0.41) thin films," *APL Mater.*, vol. 6, no. 7, p. 076105, Jul. 2018.

[19] F. Parsapour *et al.*, "Free standing and solidly mounted Lamb wave resonators based on Al0.85Sc0.15N thin film," *Appl. Phys. Lett.*, p. 6, 2019.

[20] M. A. Caro *et al.*, "Piezoelectric coefficients and spontaneous polarization of ScAlN," *J. Phys. Condens. Matter*, vol. 27, no. 24, p. 245901, Jun. 2015.

[21] N. Kurz *et al.*, "Experimental determination of the electro-acoustic properties of thin film AlScN using surface acoustic wave resonators," *J. Appl. Phys.*, vol. 126, no. 7, p. 075106, Aug. 2019.

[22] J. H. Kuypers and A. P. Pisano, "Green's function analysis of Lamb wave resonators," in *2008 IEEE Ultrasonics Symposium*, Nov. 2008, pp. 1548–1551.

Gap in pagination due to unavailable papers.

Pages 1527-1534

Proceedings of the 16th Annual IEEE International
Conference on Nano/Micro Engineered and Molecular Systems
April 25-29, 2021

Super-lightweight Flexible Wireless Optogenetic Stimulation Device

Rui Luo, *Student Member, IEEE*, Ziqi Mei, *Student Member, IEEE*, Zheng You, *Member, IEEE*, Dahai Ren*, *Member, IEEE*

Abstract— We present a super-lightweight (<0.05 g) flexible device for wireless optogenetic stimulation that utilizes photolithography microfabrication methods and manual flip-chip bonding for flexible substrates. Over 5 mW/mm² of adjustable optical power can be obtained, exceeding the expectation of optogenetic stimulation (>1 mW/mm²[1]). Deformation curvature of the receiving device and biological tissue block thickness between the transmitting device and the receiving device have been evaluated to have negligible influence on the output electrical power. Temperature increase of the device is less than 2°C after 10 minutes of continuous working, which guarantees low phototoxicity.

I. INTRODUCTION

Optogenetics has been a remarkable method in research on excitable cells, which are related to several important biomedical research fields such as brain science [2, 3], cardiac modulation [4, 5], ophthalmic therapeutics [6, 7], etc. It realizes cell excitation or prohibition by the aid of genetically expressed light-gated ion channel, which can be specifically gated by the illumination of certain wavelength of light. Compared with conventional neuroscience methodologies like electrical or mechanical stimulation, optogenetic stimulation shows great superiority in contactless control, cell specific targeting, simpler circuitry [2], etc. Flexible devices for wireless optogenetic stimulation have been conspicuously attractive due to their advantage of building an untethered optogenetic experiment system, which is especially crucial for chronic behavioral experiments and guarantees more reliable results. Existing wireless optogenetic stimulation devices [8] are often too heavy with batteries or complicated circuit boards and chips, being burdens on experimental animals and may lead to inaccurate results. Here we designed and fabricated a super-light-weight flexible wireless optogenetic stimulation device, which is electrically, optically and thermally characterized.

II. METHODS

A. Principle, design and simulation

Wireless optical stimulation is realized based on magnetic resonant coupling theory [9, 10], relying mainly on a receiving coil. As shown in Fig. 1a, a transmitting coil, which is tuned to the same resonant frequency as the receiving coil, is powered by the signal generator. Alternating magnetic field induces alternating current on the receiving coil, which lights up the µ-LED.

*Research supported by Major Projects of Beijing Municipal Education Commission.

Rui Luo, Ziqi Mei, Zheng You, Dahai Ren are with Department of Precision Instrument, Tsinghua University, Beijing, 100084, China (corresponding author to provide phone: +86-10-62776000; e-mail: rendh@tsinghua.edu.cn).

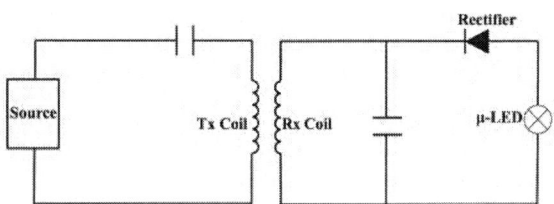

Figure 1. Schematic diagram of system principle.

The device is designed as Fig. 2a, in which the receiving coil is on one side and the connecting pads are on the other. Connecting wires on the back side are designed into serpentine pattern to ensure high flexibility. Different metals are chosen for the circuit in different fabrication methods as described in the fabrication part. Polyimide (PI) is chosen as the substrate polymer of the device for the thermostability and its good adhesiveness to metal. Structural parameters of the flexible system are shown in Fig. 2c, in which the device is 14mm in diameter, and consists of a copper (18 µm) - polyimide (50 µm) - copper (18 µm) vertical structure. The line width of the coil is 50 µm and the interval between each turn is 150 µm. The weight of the device is comprised of a capacitor (01005 package, 0.4 mm × 0.2 mm × 0.2 mm) for impedance matching and µ-LED (0.22 mm × 0.27 mm × 0.1 mm). The weight of the two chips is about 0.005 mg. Together with the relatively thin PI substrate and metal wires, the total weight of the device is much lighter than 0.05 g. The design can be adjusted simply into different application scenarios, as is shown in Fig. 2d.

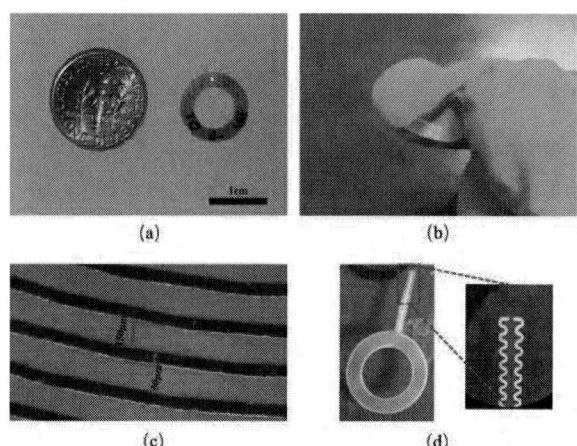

Figure 2. (a) Prototype of the device photo; (b) bending of the device; (c) detailed structure of the device under the microscope and (d) a different design for application on cortex optogenetic stimulation.

Resonant frequency of the coil is determined by the following equation:

$$f_0 = \frac{1}{2\pi\sqrt{LC}}, \qquad (1)$$

in which inductance (L) of the coil can be simulated by COMSOL (Fig. 3a) and matching capacitor (C) can be calculated afterwards. Here, 1nF capacitor is selected to do impedance matching. Fig. 3b shows a lowest reflection coefficient (S11) at the frequency around 3.5 MHz, indicating the resonant frequency.

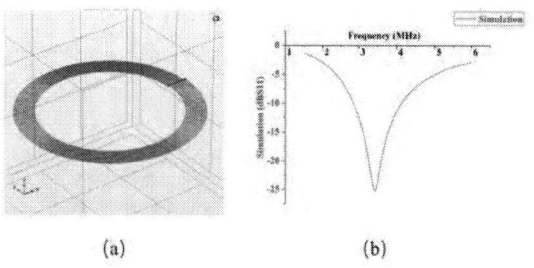

(a)   (b)

Figure 3. (a) Simulation model in COMSOL and (b) simulated resonant frequency of the coil.

B. Fabrication

Two different methods have been taken for device fabrication.

The first fabrication method is shown in Fig. 4a. Firstly, a layer of polyvinyl alcohol (PVA), which is water-soluble at 110°C, was spun coated onto a 2-inch silicon wafer at 3000 rpm for 30 seconds and baked at 70°C for 15 minutes. Then, the commercially available Kapton Cu-PI-Cu foil was cut into 2-inch wafer size by a circular blade, and taped onto the wafer via PVA adhesive layer. Next, a layer of positive photoresist AZ4620 was spun onto the foil at 3000 rpm followed by 110°C baking for 3 minutes. After photolithography and development, wet etching was conducted using FeCl₃ solution for about 5 minutes to get the copper coil pattern. During wet etching, the related container should be wobbled slightly to ensure uniform spread of the etchant and the wafer needs to be taken to the stereomicroscope every 30 seconds to monitor its etching status. To release the foil from the wafer, it is emerged into 110°C water for 5 minutes to solve the PVA sacrificial layer. Then the other side of copper was fabricated into solder and testing pads by the same procedure above. The via holes connecting the two copper layers are drilled by laser processing equipment, and the chips are soldered by manual flip-chip bonding using low-temperature solder under the stereomicroscope.

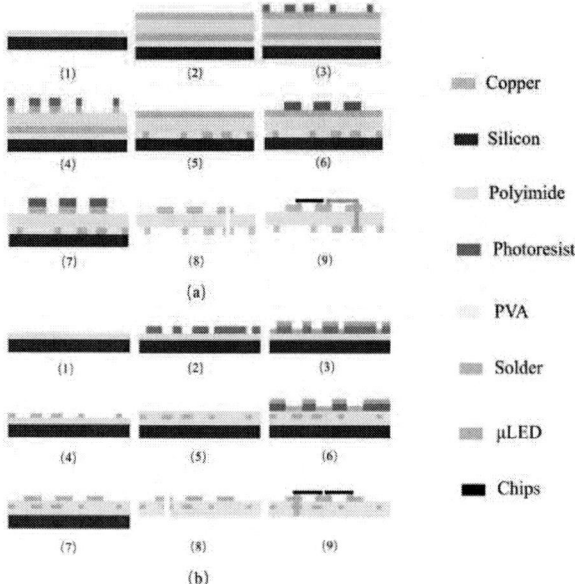

Figure 4. (a) Wet-etching method of fabrication and (b) lift-ff method of fabrication.

Lift-off technology [11, 12] was also employed to fabricate the device, as shown in Fig. 4b. First of all, a layer of PI was spun coated onto a 2-inch silicon wafer at 5000 rpm for 90 seconds followed by a baking procedure in Table 1. Then, a layer of lift-off negative photoresist NR9-3000PY was spun coated above the PI layer at 3000 rpm for 40 seconds. After photolithography and development, 30nm-thick titanium and 300-nm-thick platinum layers were deposited onto the sample. For metal lift-off, the sample was emerged into acetone in ultrasound cleaner for about 30 minutes. Also, in order to monitor its lift-off status, the sample needs to be taken to the stereomicroscope every 5 minutes. Another layer of PI and metal was fabricated in the same way. The via holes with laser drilling and manual flip-chip bonding, which are same with the method above.

TABLE I.   BAKING PROCEDURE OF PI FILM

Temperature (°C)	Time (minutes)
80	10
150	20
200	10
250	10
260	10

The wet-etching method in Fig. 4a requires less microfabrication steps, especially is free of metal deposition, making it simpler and more economical. On the other hand, lift-off method guarantees more uniform and precise line widths and spacing distances. Another benefit of the lift-off method is that thinner polymer and metal films can be fabricated. As is shown in Fig. 5a, the total thickness is about 4.4 μm (measured by a white-light interferometer, the white and black arrows in the colorful inlet indicate the two measuring points), by which a much lighter device can be achieved, as shown in Fig. 5b. For the characterization part below, we utilized devices fabricated in the wet-etching method in Fig. 4a. For further application in biological environment, Parylene-C deposition should be employed to ensure better biocompatibility and device insulation.

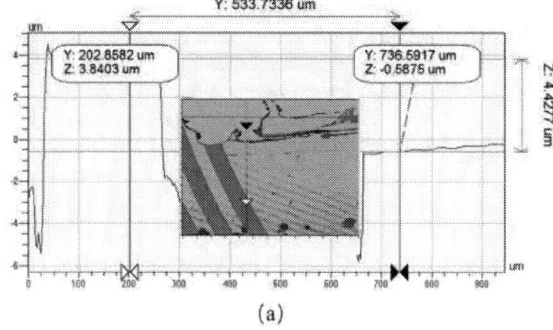

(a)

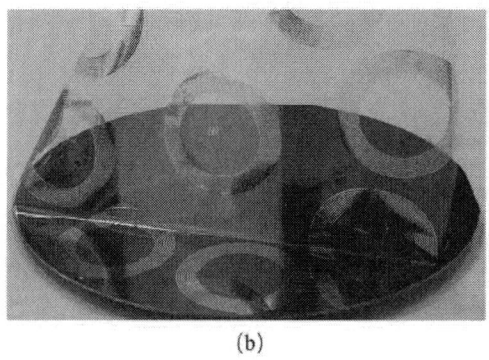

(b)

Figure 5. (a) PI film measured to be ~4 μm by white-light interferometer and (b) thinner device fabricated by lift-off process.

III. RESULTS AND DISCUSSIONS

A. Electrical Characterization

Experimental system was set up in Fig. 6a to complete electrical characterization, the frames and holders in which were 3D-printed. The transmitting coil (TX coil) is powered by a signal generator (Rigol DG1062Z, sine wave, 3.5 MHz, 10 Vpp).

Fig. 6b shows that the deformation curvature (generated by 3D-printed cylinders with different curvature, Fig. 6c) within 30% impacts little on electrical behavior of the device. By filling the separation distance between the RX and TX coil with pork slices (~3 cm × 3 cm × 3 mm in size for each slice), We found that biological tissue medium also causes little impact to device behavior, revealing the fact that biological tissue absorbs little energy from the transmitting path, because the typical absorbing frequency of biological tissue is around gigahertz, which is far from the functional frequency of our device (Fig. 6d). As calculated by experiment data, 15 mW of electrical power can be obtained at a typical operation distance (10 mm) for biomedical applications.

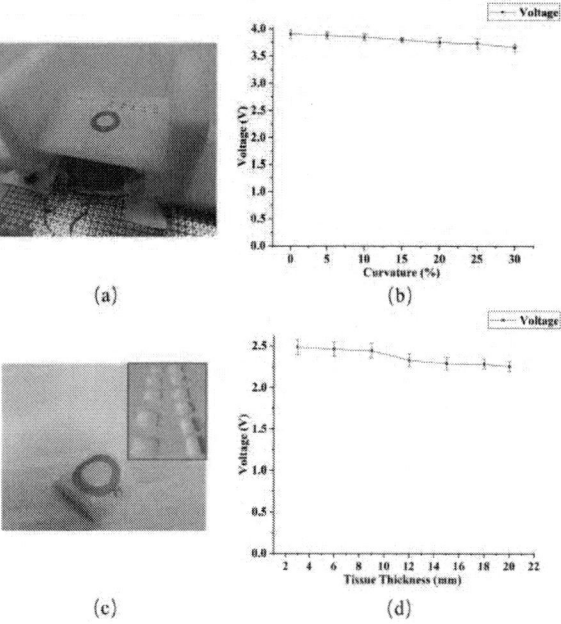

(a)   (b)

(c)   (d)

Figure 6. (a) 3D-printed frames and holders for electrical characterization experimental setup; (b) output voltage V.S. curvature (separation distance =15 mm); (c) 3D-printed cylinders with different curvature and (d) output voltage V.S. biological tissue medium thickness (separation distance = 21 mm).

B. Optical Characterization

Optical characterization was completed with the setup in Fig.7a, in which optical power is measured by optical power meter. Fig.7b displays good consistency between the experiments and simulation results, in which both resonant frequencies are about 3.5 MHz. Slight difference may arise from fabrication uneven artifacts and several approximating treatments in simulation, in which metal thickness was set to 0 (2D model), the round Archimedean spiral was approximated by 24-gon, and connecting wires on the back side was set to be straight instead of serpentine. Dividing optical power by μ-LED chip area (220 μm × 270 μm), the irradiance is calculated and about 1.5 mW/mm² can be obtained at the typical operation distance of 10mm. More than 5 mW/mm² can be obtained at a closer distance or by adjusting the signal generator output.

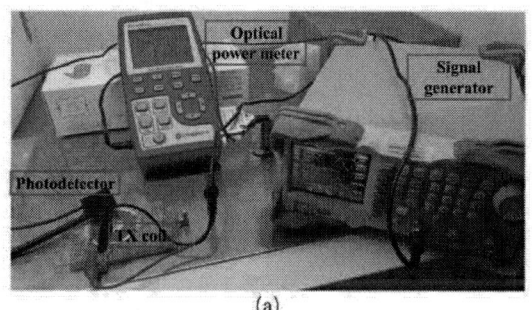

(a)

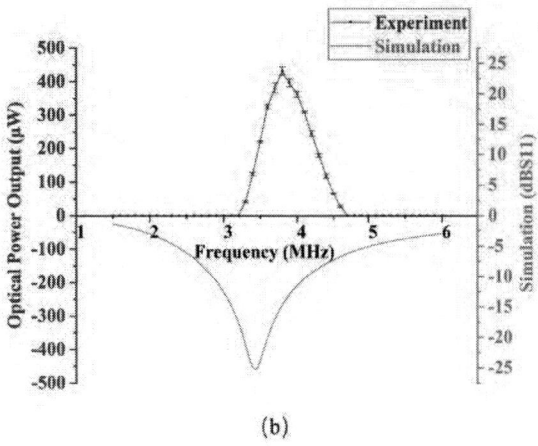

(b)

Fig. 7. (a) Optical characterization setup and (b) optical power output V.S. frequency, compared with simulation results.

C. Thermal Characterization

Thermal characterization (Fotric 346) was done by monitoring temperature change of the device continuously for 10 minutes. Fig. 8 (a - c) shows that temperature increase of the device is less than 2°C after working continuously for 10min, which guarantees little phototoxicity to human tissue. Fig. 8d shows the infrared thermography of a working device with a mouse eyeball directly placed on the electrodes of the device for 30 minutes, displaying negligible temperature elevation and no visible harm to human tissues.

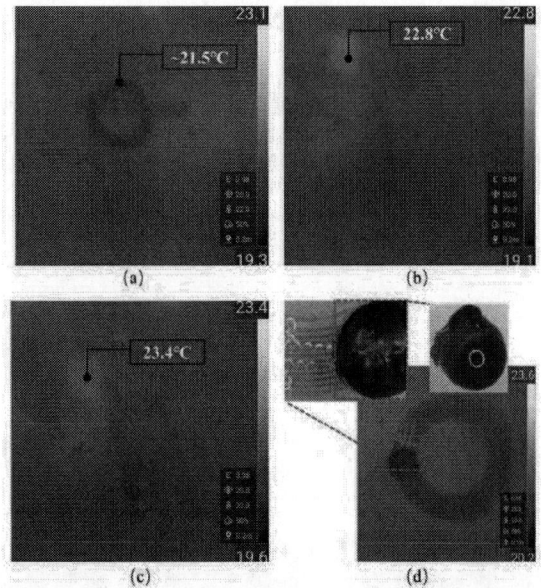

Fig. 8. (a) Infrared thermography before LED illumination; (b) Infrared thermography after 3 minutes LED illumination and (c) Infrared thermography after 10 minutes LED illumination; (d) Infrared thermography of device electrodes directly function on a mouse eyeball for 30 minutes.

IV. Summary

This paper reports a super-lightweight wireless optogenetic stimulator. Simulations were conducted for impedance matching, and the RX coil is tuned to resonant at around 3.5 MHz. Wet-etching method and lift-off method were employed for device fabrication. Several influencing factors on electrical and optical behavior of the device were investigated and compared, low phototoxicity was verified as well. In- vitro and in-vivo experiments on B6 mouse is still underway. Due to the simplified design and the optimized circuitry, a super-lightweight device was achieved, which largely broaden its application possibilities in untethered optogenetic stimulation and related biomedical research scenarios.

V. Acknowledgment

The authors would like to thank Tsinghua Nanofabrication Center for fabrication support. This work was supported by the Key Project of Beijng Municipal Education Commission and the Project of Beijing Laboratory of Biomedical Detection Technology and Instrument. The authors are grateful for the support of the Beijing Innovation Center for Further Chips.

References

[1] T. Xie, X. Bi, R. Luo, F. Bin, Z. Wang, and W. Li, "Optical Propagation of Blue LED Light in Brain Tissue and Parylene-C," in *Proc. IEEE NEMS 2017*, pp. 599-602.

[2] K. Deisseroth, "Optogenetics," *Nature Methods*, vol. 8(1), pp. 26-29, 2011.

[3] X. Xu, T. Mee, and X. Jia, "New era of optogenetics: from the central to peripheral nervous system," *Critical Reviews in Biochemistry and Molecular Biology*, vol. 55(1), pp. 1-16, 2020.

[4] J. Joshi, M. Rubart, and W. Zhu, "Optogenetics: background, methodological advances and potential applications for cardiovascular research and medicine," *Frontiers in bioengineering and biotechnology*, vol. 7, 466, 2020.

[5] J. C. Williams, and E. Entcheva, "Optogenetic versus electrical stimulation of human cardiomyocytes: modeling insights," *Biophysical Journal*, vol. 108(8), pp. 1934-1945, 2015.

[6] T. Cronin, and J. Bennett, "Switching on the lights: the use of optogenetics to advance retinal gene therapy", *Molecular Therapy*, vol. 19(7), pp. 1190-1192, 2011.

[7] L. Montazeri, N. E. Zarif, S. Trenholm, and M. Sawan, "Optogenetic stimulation for restoring vision to patients suffering from retinal degenerative diseases: current strategies and future directions", *IEEE transactions on biomedical circuits and systems*, vol. 13(6), pp. 1792-1807, 2019.

[8] M. H. M. Kouhani, R. Luo, F. Madi, A. J. Weber, and W. Li, "A wireless, smartphone controlled, battery powered, head mounted light delivery system for optogenetic stimulation," in *Proc. EMBC 2018*, pp. 3366-3369.

[9] A. Kurs, A. Karalis, R. Moffatt, J. D. Joannopoulos, P. Fisher, and M. Soljačić, "Wireless power transfer via strongly coupled magnetic resonances," *Science*, vol. 317, pp. 83-86, 2007.

[10] A. Torrisi, and D. Brunelli, "Magnetic Resonant Coupling Wireless Power Transfer for Lightweight Batteryless UAVs," in *Proc. IEEE SPEEDAM 2020*, pp. 751-756.

[11] D. Kim, H. Kang, and Y. Nam, "Compact 256-channel multi-well microelectrode array system for in vitro neuropharmacology test," *Lab on a Chip*, vol. 20(18), pp. 3410-3422, 2020.

[12] E. Mafi, N. Calvano, J. Patel, M. S. Islam, M. S. H. Khan, and M. Rana, "Electro-Optical Properties of Sputtered Calcium Lead Titanate Thin Films for Pyroelectric Detection", Micromachines, vol. 11(12), 1073, 2020.

Proceedings of the 16th Annual IEEE International
Conference on Nano/Micro Engineered and Molecular Systems
April 25-29, 2021

Research on the Bouncing-Ball Based Triboelectric Nanogenerator for Self-powered Vibration Frequency Monitoring*

Xusheng Zuo, Taili Du, Fangyang Dong, Shunqi Li, Yongjiu Zou, Junhao Zhao, Peng Zhang, Yuewen Zhang*, Peiting Sun* and Minyi Xu*

Abstract—Traditional vibration sensor of ship machinery using cable for power supply cannot meet the requirements of a large number of distributed sensors of the intelligent ship. A bouncing-ball based triboelectric nanogenerator (BB-TENG) to be used as a self-powered vibration frequency monitoring sensor is proposed. In the sensing test, the vibration frequency of 15 ~ 50 Hz under different amplitudes can be measured accurately. The frequency calculation of the voltage signal can accurately reflect external vibration frequency, while the change of short-circuit current has linear relationship with the frequency, and the correlation coefficient is about 0.99. In addition, electrical signal test results and photos taken by high-speed cameras show that the limit of the available detection frequency of BB-TENG increases with the increase of amplitude and support height. The electrical output of BB-TENG under different working conditions are tested, which provides the basis for selecting suitable BB-TENG according to different vibration characteristics.

I. INTRODUCTION

To realize intelligent ship [1], multiple-point monitoring is the first step have to be taken, which needs a huge distributed sensor network. Due to the vibration frequency can reflect the working condition of the ship machinery, which is most reciprocating or rotary machinery working at constant frequency in most instances, accurately and timely, researchers have carried out a series of studies on fault diagnosis and optimal control of the ship machinery based on vibration signal [2,3]. Nonetheless, the research of vibration sensor itself should be strengthened. In addition, it can be imagined that the cables arrangement for so large number of sensors will lead to high cost and difficulty of the ship, let alone the low reliability resulting from damage of the cables. Hence, it is urgent to develop the self-powered vibration sensor for the intelligent ships.

Thanks to the theory of triboelectric nanogenerator (TENG) which is firstly proposed by Wang [4], which provides a new way for the development of self-powered sensors. Because of its low cost, simple and flexible structure, easy production and manufacturing, high integration level and high efficiency, it has attracted the attention of many scholars in the world. Based on this theory, this technology has been applied to energy harvesting in different fields, such as mechanical energy [5-8], wind energy [9], wave energy [10-11], acoustic energy

[12] and human motion energy [13]. In addition, it is also used as a variety of self-powered sensors, such as vibration [14], wind speed [15], flow rate [16], displacement [17], pressure [18], tilt [19] and wave sensors [20]. For the vibration energy harvesting and sensing, a great number of excellent achievements had been accomplished, such as TENGs based on resonant structure [21] and non-resonant structure [22], and hybrid nanogenerators [23]. However, such TENGs are unsuitable for sensors on ship due to complex structure, large size and damage of spring, which is mostly installed in narrow space and needs easy to dismounting for machinery maintenance because of the complex structure, large size, or difficulties in fabrication. Therefore, a new compact designed bouncing-ball based TENG (BB-TENG) is proposed to solve the actual needs for the distributed and self-powered vibration monitoring of the intelligent ship.

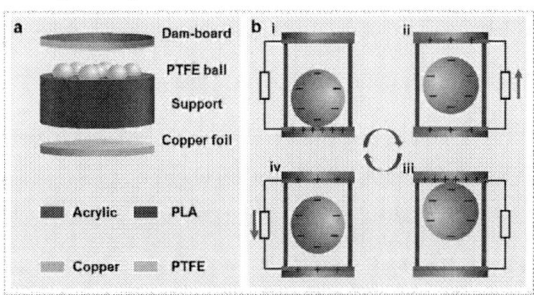

Figure1. Composition and working principle of the BB-TENG. (a) Structure of BB-TENG; (b) Working principle of BB-TENG

II. EXPERIMENT SECTION

A. Working principle of the BB-TENG

The BB-TENG consists of 7 bouncing balls made of Polytetrafluoroethylene (PTFE), circular support with 7 hollows, two acrylic dam-boards and copper foil pasted on the dam-board, as shown in Fig. 1a. Due to the external vibration, the PTFE balls bounce in the support, so the working process of the BB-TENG can be divided into four stages. Fig. 1b depicts the working principle of the BB-TENG: (i) after several contacts between the ball and copper, they will be negatively or positively charged; (ii) when the ball bounces up, it will lead to potential difference between the upper and lower copper, so the electron will

*Research supported by National Science Foundation of China (Grant Nos. 51879022, 51979045), the Fundamental Research Funds for the Central Universities, China (Grants Nos. 3132019330)

All authors are with Marine Engineering College, Dalian Maritime University, No.1 Linghai Road, Dalian, 116026, China

*Contacting Authors: Yuewen Zhang is with Marine Engineering College, Dalian Maritime University, No.1 Linghai Road, Dalian, 116026,

China (email: zhangyuewen@dlmu.edu.cn). Peiting Sun is with Marine Engineering College, Dalian Maritime University, No.1 Linghai Road, Dalian, 116026, China (sunptg@dlmu.edu.cn). Minyi Xu is with Marine Engineering College, Dalian Maritime University, No.1 Linghai Road, Dalian, 116026, China (phone: +86 18941134769; email: xuminyi@dlmu.edu.cn).

978-1-6654-3008-1/21 $31.00 © 2021 IEEE

flow from upper copper to the lower one in external circuit to produce current; (iii) a new equilibrium is obtained when the ball contacts with the upper copper; (iv) falling down of the ball results in the adverse current in the external circuit.

B. Experiment setup

In order to test the sensing performance of the BB-TENG, a test system as shown in Fig. 2 is set up. The signal generator is used to generate vibration signal, the vibration signal will be amplified by the power amplifier, and then used to drive the shaker. The electrical signal including open-circuit voltage, short-circuit current and transferred charge, and acceleration signal of BB-TENG are measured by the electrometer and accelerometer respectively and illustrated on the computer.

Totally, 25 BB-TENGs with different ball diameters are fabricated, and the electrical signals of the BB-TENGs working at different vibration frequencies from 15 Hz to 50 Hz, and amplitudes from 0.5 mm to 3 mm are tested.

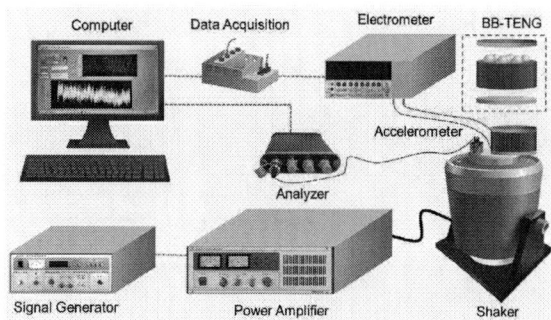

Figure 2. Test system for BB-TENG

III. RESULTS AND DISCUSSION

The machinery arranged on ships are mostly reciprocating or rotating type. The working performance of these machinery can be reflected by the main working frequency easily and accurately. For example, the air compressor, which used to provide compressed air for the starting of main engine and generator set, and for miscellaneous use, has 2 or more cylinders. It can be imagined that every cylinder produces the compressed air one time in every revolution, and the main working frequency of each cylinder can be stabilized at a certain value. However, in the case of mechanical or electrical malfunction, the main working frequency changes and gives the signal that the air compressor needs to be stopped, checked and repaired. Therefore, the vibration frequency sensing performance of BB-TENG sensor is tested. According to the test results, the main frequency monitoring characteristics of BB-TENG sensor with ball diameter of 5 mm and support height of 7 mm under different amplitudes and frequencies are shown in Fig. 3. As shown in Fig. 3a, the open-circuit voltage of BB-TENG increases at first and then decreases, and finally stabilizes around 6 V. When the external vibration frequency is less than 15 Hz, the voltage value is almost the noise of the electrometer, which can be ignored. This is because the vibration energy of the shaker is not enough to bounce the ball up and generate effective voltage signal. In the range of 15-25 Hz, open-circuit voltage

of BB-TENG gradually increases from 4.73 V to 8.06 V, and then gradually decreases and stabilizes at about 6 V from 25 Hz to 50 Hz. Fig. 3b shows the short-circuit current of BB-TENG, which is similar to the open-circuit voltage signal with almost no current generation below 15 Hz. From 15 Hz, the short-circuit current shows almost linear increase with the frequency increase, from 0.143 uA at 15 Hz to 0.57 uA at 50 Hz.

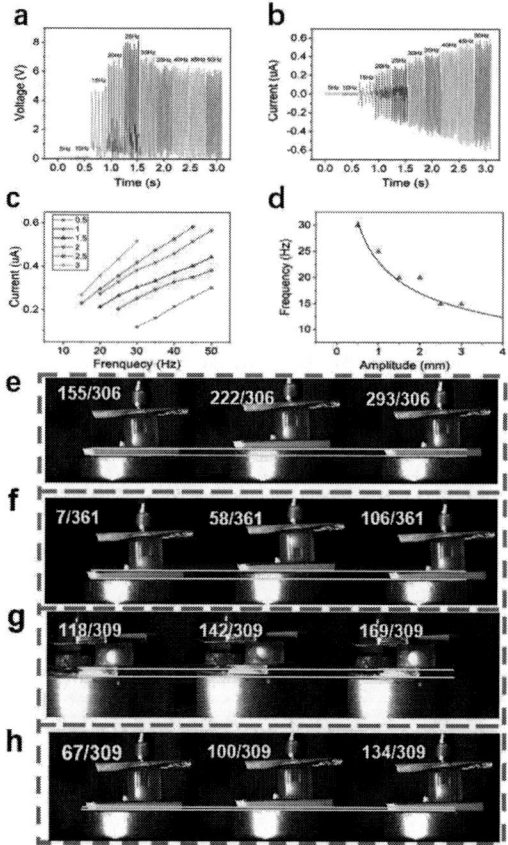

Figure 3(a) Open-circuit voltage, (b) Short-circuit current under different frequency with amplitude of 2.5 mm. (c) Relationships among the vibration frequency, vibration amplitude, and output short-circuit current of the BB-TENG sensor. (d) The lowest stable frequency at different amplitude. Bouncing status of BB-TENG with ball diameter of 5mm, support height of 10 mm, frequency of (e) 15 Hz, (f) 20 Hz and amplitude of 1.5 mm. Bouncing status of BB-TENG with ball diameter of 5 mm, frequency of 30 Hz, amplitude of 0.5 mm and support height of (g) 6 mm and (h)10 mm.

In order to verify the relationship between the short-circuit current generated by BB-TENG and the frequency furtherly, the short-circuit current under different frequencies and amplitudes are tested, which can be seen in Fig. 3c. It can be noted that the short-circuit current shows almost linear relationship with the increase of the frequency at all amplitudes.

In addition, it can be found that the effective electrical signals are generated at different frequencies under different amplitudes, which means there exists the lowest stable frequency under different amplitude as shown in Fig. 3d.

The lowest stable frequency is 30 Hz under amplitude of 0.5 mm, while it is 15 Hz under amplitude of 3 mm. In order to know the working process of the BB-TENG under different amplitudes, the dynamic analysis of the bouncing status of the PTFE ball in the support was demonstrated much more clearly by using high-speed camera. The vibration excitation energy is not enough so that the ball moves with the bottom copper all the time, which is illustrated in Fig. 3e. However, the ball will be bounced up when the frequency is larger than 20 Hz, as shown in Fig. 3f. Fig. 3g shows that the bouncing status of ball with diameter of 5 mm is accordance with the vibration status of the shaker. When the support height is changed to 10 mm with all else being equal, as shown in Fig. 3h, the bouncing status of the ball would be chaos unless the frequency increased to 35 Hz or more. Therefore, the support height should be short to enlarge the frequency range of application. This is owing to the external vibration energy, the Poisson's ratio, and Young's modulus of the PTFE ball, acrylic plug, and copper electrode, and the bouncing status of the PTFE ball is also influenced by the coefficient of restitution and the collision duration, which will be discussed in our future work.

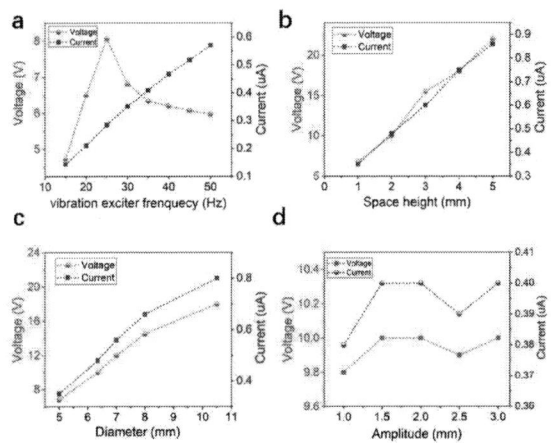

Figure 4. The relationships among open-circuit voltage, short-circuit current of BB-TENG and (a) Shaker frequency, (b) Space height, (c) Ball diameter and (d) Amplitude respectively.

In order to study the performance of BB-TENG furtherly, a series of factors that may affect the output electrical signal are tested as shown in Fig. 4. The relationship among the open-circuit voltage, short-circuit current of BB-TENG and the vibration frequency of the exciter with ball diameter of 5 mm, ball bouncing space height of 2 mm, and the amplitude of 2.5 mm, which is got according to the results shown in Fig. 3a and 3b, is shown in Fig. 4a. It can be got that the linear correlation coefficient of current-frequency curve reaches about 0.99, which means the short-circuit current signal can be used to represent the main vibration frequency of the ship machinery. The variation tendency of the voltage, current signals of BB-TENG in the wake of the increase of bouncing space height of the PTFE ball with ball diameter of 6.35 mm, amplitude of 2.5 mm, and vibration frequency of 30 Hz is shown in Fig. 4b. The results show that the voltage and current increase linearly with the increase of the height of the bouncing space. When the PTFE ball has enough space to accelerate in the support, the impact

velocity between it and the copper foil will be greater and the contact will be more sufficient, resulting in a gradually increase of the voltage and current. It can be found that the voltage and current gradually increase with the increase of PTFE ball diameter with the vibration frequency of 30 Hz, amplitude of 2.5 mm and space height of 2 mm as shown in Fig. 4c. The reason for this phenomenon is that when the diameter of the PTFE ball increases, the contact area between it and the copper foil increases, resulting in the increase of both voltage and current. Fig. 4d demonstrates the relationship among the voltage, current and amplitude with the frequency of 30 Hz, ball diameter of 6 mm, space height of 2 mm. It can be seen that the voltage and current are both stable in certain scope, so it can be inferred that the amplitude of the external environment rarely affect the electrical performance after the vibration of BB-TENG reaches a stable state, which is another advantage of the BB-TENG sensor.

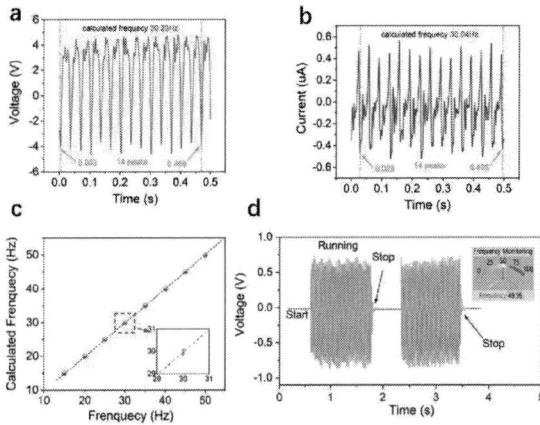

Figure 5. (a) 14 open-circuit voltage peaks were randomly selected with frequency of 30 Hz, and the calculation frequency was 30.23 Hz. (b) 14 short-circuit current peaks were randomly selected and the calculation frequency was 30.04 Hz. (c) The relationship between the set frequency and the calculated frequency, illustrated with the Y error at 30 Hz. (d) BB-TENG is used to monitor the working condition and the main working frequency of an air compressor.

The open-circuit voltage and short-circuit current with the working frequency of 30Hz is shown in Fig. 5a and 5b. Here, 14 voltage and current peaks were randomly selected and the corresponding variation frequencies are calculated by Eq. (1):

$$f = \lambda / (T_2 - T_1) \tag{1}$$

Where T_2 and T_1 is the end and start time randomly selected, and λ is the corresponding number of signal peaks between T_2 and T_1. Based on Eq. (1), the result calculated from the voltage signal in Fig. 5a is 30.23 Hz, and the result calculated from the current signal in Fig. 5b is 30.04 Hz, both of which are in good agreement with the set vibration frequency of 30 Hz. In addition, various frequencies ranging from 15 Hz to 50 Hz were tested and shown in Fig. 5c indicating that the calculated vibration frequency was in good agreement with the set frequency. The inset in Fig. 5c shows a zoomed in view of the 30Hz point with an error of just 0.1%. Moreover, the noise of the voltage is about 0.15

978-1-6654-3008-1/21 $31.00 ©2021 IEEE   1541

V as shown in Fig. 3a and the open-circuit voltage is about 8 V as shown in Fig. 5a, so the voltage signal-noise-ratio (SNR) is equal to $20\log*(8/0.15) = 34.54$ dB, which is high enough to be used to monitor the main working frequency of the machinery. Therefore, the open-circuit voltage should be taken as the original signal so as to enlarge the signal-noise ratio on account of its high signal output compared to the short-circuit current.

In addition, as shown in Fig. 5d, the BB-TENG sensor can monitor the starting, running and stopping states of an air compressor, which proves it's potential to become an accurate vibration frequency sensor. The working speed of this double-piston double-cylinder air compressor is about 1500 revolutions per minute, and each cylinder works once in every revolution. Therefore, its working frequency is $1500 \times 2/60 = 50$ Hz. The vibration frequency detected by the BB-TENG is 49.95 Hz, which shows good performance in the monitoring of mechanical vibration frequency.

IV. CONCLUSION

In this study, a self-powered BB-TENG sensor with high precision for vibration frequency monitoring is proposed and developed. It is found that BB-TENG can measure vibration frequency at different amplitudes from 15 Hz to 50 Hz, and the lowest stable frequency of BB-TENG is lower along with the increase of the amplitude. The frequency of the voltage signal can accurately show the vibration frequency of machinery, and the change of short-circuit current has a linear relationship with the frequency, and the correlation coefficient is about 0.99. Moreover, the influence of vibration frequency, PTFE ball diameter, space height and vibration amplitude on the electrical performance of BB-TENG is systematically studied, which is helpful to select the appropriate size of BB-TENG for different vibration conditions. The vibration frequency of the air compressor is also monitored by the BB-TENG accurately. The results show that the BB-TENG has high accuracy and shows its potential as a sustainable power source due to its self-powered characteristics and has great application potential in intelligent monitoring of ship mechanical condition.

REFERENCES

[1] R. Tang, Q. An, F. Xu, X. Zhang, X. Li, J. Lai, and Z. Dong, "Optimal operation of hybrid energy system for intelligent ship: An ultrahigh-dimensional model and control method," *ENERGY*, vol. 211, 2020.

[2] S. Yang, G. Lu, A. Wang, J. Liu, and P. Yan, "Change detection in rotational speed of industrial machinery using Bag-of-Words based feature extraction from vibration signals," *MEASUREMENT*, vol. 146, pp. 467-478.

[3] N. Shimizu, H. Tsukui, and T. Fujikawa, "Bending vibration analysis of rotating machineries using multibody dynamics technology — development of computer program RotB," *THEOR APPL*, vol. 2, no. 6, pp. 063008.

[4] F.-R. Fan, Z.-Q. Tian, and Z. Lin Wang, "Flexible triboelectric generator," *NANO ENERGY*, vol. 1, no. 2, pp. 328-334.

[5] A. Yar, "High performance of multi-layered triboelectric nanogenerators for mechanical energy harvesting," *ENERGY*, vol. 222, pp. 119949.

[6] S. A. Graham, S. C. Chandrarathna, H. Patnam, P. Manchi, J.-W. Lee, and J. S. Yu, "Harsh environment–tolerant and robust triboelectric nanogenerators for mechanical-energy harvesting, sensing, and energy storage in a smart home," *NANO ENERGY*, vol. 80, pp. 105547.

[7] Y. Liu, Y. Zheng, Z. Wu, L. Zhang, W. Sun, T. Li, D. Wang, and F. Zhou, "Conductive elastic sponge-based triboelectric nanogenerator (TENG) for effective random mechanical energy harvesting and ammonia sensing," *NANO ENERGY*, vol. 79, pp. 105422.

[8] J. Sintusiri, V. Harnchana, V. Amornkitbamrung, A. Wongsa, and P. Chindaprasirt, "Portland Cement-TiO2 triboelectric nanogenerator for robust large-scale mechanical energy harvesting and instantaneous motion sensor applications," *NANO ENERGY*, vol. 74, pp. 104802.

[9] X. Ren, H. Fan, C. Wang, J. Ma, H. Li, M. Zhang, S. Lei, and W. Wang, "Wind energy harvester based on coaxial rotatory freestanding triboelectric nanogenerators for self-powered water splitting," *NANO ENERGY*, vol. 50, pp. 562-570.

[10] C. Rodrigues, M. Ramos, R. Esteves, J. Correia, D. Clemente, F. Gonçalves, N. Mathias, M. Gomes, J. Silva, C. Duarte, T. Morais, P. Rosa-Santos, F. Taveira-Pinto, A. Pereira, and J. Ventura, "Integrated study of triboelectric nanogenerator for ocean wave energy harvesting: Performance assessment in realistic sea conditions," *NANO ENERGY*, vol. 84, pp. 105890.

[11] U. T. Jurado, S. H. Pu, and N. M. White, "Grid of hybrid nanogenerators for improving ocean wave impact energy harvesting self-powered applications," *NANO ENERGY*, vol. 72, pp. 104701.

[12] W. Qiu, Y. Feng, N. Luo, S. Chen, and D. Wang, "Sandwich-like sound-driven triboelectric nanogenerator for energy harvesting and electrochromic based on Cu foam," *NANO ENERGY*, vol. 70, pp. 104543.

[13] J. Fu, K. Xia, and Z. Xu, "A triboelectric nanogenerator based on human fingernail to harvest and sense body energy," *MICRO ENGN*, vol. 232, pp. 111408.

[14] S.-N. Lai, C.-K. Chang, C.-S. Yang, C.-W. Su, C.-M. Leu, Y.-H. Chu, P.-W. Sha, and J. M. Wu, "Ultrasensitivity of self-powered wireless triboelectric vibration sensor for operating in underwater environment based on surface functionalization of rice husks," *NANO ENERGY*, vol. 60, pp. 715-723.

[15] D. Kim, I.-W. Tcho, and Y.-K. Choi, "Triboelectric nanogenerator based on rolling motion of beads for harvesting wind energy as active wind speed sensor," *NANO ENERGY*, vol. 52, pp. 256-263.

[16] Phan, W. Song, Xiao, Pan, M. Xu, and Mi, "A Self-Powered and Low Pressure Loss Gas Flowmeter Based on Fluid-Elastic Flutter Driven Triboelectric Nanogenerator," *SENSORS*, vol. 20, pp. 729.

[17] H. Liu, H. Wang, Y. Lyu, C. He, and Z. Liu, "A novel triboelectric nanogenerator based on carbon fiber reinforced composite lamina and as a self-powered displacement sensor," *MICRO ENGN*, vol. 224, pp. 111231.

[18] Z. Zhao, Q. Huang, C. Yan, Y. Liu, X. Zeng, X. Wei, Y. Hu, and Z. Zheng, "Machine-washable and breathable pressure sensors based on triboelectric nanogenerators enabled by textile technologies," *NANO ENERGY*, vol. 70, pp. 104528.

[19] S. Wang, Y. Wang, D. Liu, Z. Zhang, W. Li, C. Liu, T. Du, X. Xiao, L. Song, H. Pang, and M. Xu, "A robust and self-powered tilt sensor based on annular liquid-solid interfacing triboelectric nanogenerator for ship attitude sensing," *SENSOR ACTUAT A-PHYS*, vol. 317, pp. 112459.

[20] B. D. Chen, W. Tang, C. He, C. R. Deng, L. J. Yang, L. P. Zhu, J. Chen, J. J. Shao, L. Liu, and Z. L. Wang, "Water wave energy harvesting and self-powered liquid-surface fluctuation sensing based on bionic-jellyfish triboelectric nanogenerator," *MATER TODAY*, vol. 21, no. 1, pp. 88-97.

[21] Y. F. Hu, J. Yang, Q. S. Jing, S. M. Niu, W. Z. Wu, and Z. L. Wang, "Triboelectric Nanogenerator Built on Suspended 3D Spiral Structure as Vibration and Positioning Sensor and Wave Energy Harvester," *ACS NANO*, vol. 7, no. 11, pp. 10424-10432, Nov. 2013.

[22] B. B. Zhang, L. Zhang, W. L. Deng, L. Jin, F. J. Chun, H. Pan, B. N. Gu, H. T. Zhang, Z. K. Lv, W. Q. Yang, and Z. L. Wang, "Self-Powered Acceleration Sensor Based on Liquid Metal Triboelectric Nanogenerator for Vibration Monitoring," *ACS NANO*, vol. 11, no. 7, pp. 7440-7446, Jul. 2017.

[23] Y. Pang, Y. Cao, M. Derakhshani, Y. Fang, Z. L. Wang, and C. Cao, "Hybrid Energy-Harvesting Systems Based on Triboelectric Nanogenerators," *MATTER*, vol. 4, no. 1, pp. 116-143

Gap in pagination due to unavailable paper.

Pages 1543-1544

Proceedings of the 16th Annual IEEE International
Conference on Nano/Micro Engineered and Molecular Systems
April 25-29, 2021

Design of Piezoelectric Micro-Actuators Based on LiNbO₃ Thin Film

Yushuai Liu[1, 2, 3], Zhiyuan Gao[1], Kangfu Liu[1, 2, 3] and Tao Wu[1]*, *Member, IEEE*

Abstract— **The paper reports on the micro-actuators based on 36Y-cut LiNbO₃ thin film. Two different designs of SiO₂ layer below (design 1) and above (design 2) LiNbO₃ thin film have been proposed and analyzed. The two micro-actuators both have a mirror size at 15×80 µm².Through the analysis of different design parameters of micro-actuator structures, the designs of large displacement have been offered. The design 1 has a driving efficiency of 21.02 °/V at 236 kHz and design 2 has a driving efficiency of 4.11 °/V at 193 kHz. LiNbO₃ thin film based micro-actuators have been proved to have lithographically defined operating frequency and rotating displacement. Our designs show great potential for single axis rotating micro-actuators applications where high scanning angle and scalable frequency are needed.**

I. INTRODUCTION

Microelectromechanical systems (MEMS) micro-actuators have been widely used for optical scanning applications [1]. The preferred microfabrication technology as well as the performance is closely linked to the size of the micro-actuator. Different applications may have different actuator size and rotating angle requirements for the scanning plate. The size between 1 mm and 1 cm can be used for scanning display imaging; the size between 1 mm and 100 µm is primarily used for 2D-MEMS switch and optical sensing, while the size below 100 µm is widely used for projection display and optical attenuators [2]. Except for the dimension of micro-actuator, the performance of micro-actuator is also determined by the optical beam deflection angle θ_{opt}, and the scanning frequency f [3]. Among the various applications of light detection and ranging (LiDAR), fingerprint sensing and raster actuator, electrostatic, electromagnetic and piezoelectric actuations can all be used as the driving mechanisms of the scanning mirror [4, 5]. Among them, piezoelectric actuations don't rely on large driving voltage, and have better integration capability and lower cost. MEMS micro-actuators based on piezoelectric material can excite resonant frequency of the torsional mode, which can achieve the scanning devices of both high speeds and wide angles for display. Therefore, it has become one of the most advantageous driving mechanisms [4].

Among the state-of-the-art piezoelectric micro-actuators, lead zirconate titanate (PZT) and zinc oxide (ZnO) have been widely demonstrated. PZT micro-actuator has a higher piezoelectric constant and better performance compared with ZnO. The piezoelectric coefficients d_{31} of PZT ceramics and ZnO single crystal are -93.5 pC/N and-5.4 pC/N, which means that ZnO micro-actuators need larger driving voltage than PZT to achieve the same driving efficiency [3]. In addition, ZnO film has good piezoelectric properties when its crystal structure is oriented primarily along the c-axis, which also adds complexity of the fabricated process. Considering these reasons, the majority

of piezoelectric micro-actuators uses PZT. For example, thick PZT with stainless steel substrate was used to resonate at 28 kHz with 1 mm mirror aperture and achieved a 41° optical angle [6]. A bulk PZT stack micro-actuator was used to resonate at 38 kHz with 1 mm mirror aperture and achieved a 54° optical angle [7]. Other published approaches use thin/thick PZT on different types of substrates. Pyrochlore-free PZT films with perovskite structure can be grown on Pt-coated substrates or other substrates such as Si or MgO [1, 8]. Although the PZT has superior performance of micro-actuators, the piezoelectric coefficients d_{31} determines that PZT micro-actuators is mainly excited by 3-direction electric field and the upper and bottom electrode structure must be constructed. In addition, the PZT synthesis typically yields polycrystalline quality thin films and thus low mechanical quality factors at resonant frequency [4].

As a new piezoelectric material, single crystal lithium niobate (LiNbO₃) thin films has become available as a piezoelectric material via film transfer techniques [9]. In contrast to PZT, it has demonstrated the large piezoelectric coefficients of d_{16} in the Y36 cut, which means that it only relys on the upper electrodes which can achieve the micro-actuators. However, micro-actuators based on LiNbO₃ thin film have not been extensively explored. In this paper, we have proposed MEMS micro-actuator composed of piezoelectric LiNbO₃ thin film and silicon oxide (SiO₂) layer on the Si substrate and firstly analyzed the performance of micro-actuator under two different designs and different design parameters. Finally, the designs of micro-actuator with simple fabrication process, high angular drive efficiency and low power consumption (1-5V) have been demonstrated.

II. DESIGN AND ANALYSIS

A. Micro-Actuator Design and Structure

Different from PZT or ZnO, LiNbO₃ thin films can be available via film transfer techniques, which brings great challenge for depositing bottom electrodes. We designed a torsional MEMS scanner mirror suspended by one pairs of torsion bars that are driven by piezoelectric cantilevers of only applying the X-direction electric field based on LiNbO₃ thin film, as shown in Fig. 1 [10]. When sinusoidal voltages with opposite phases are applied to the right and left sides of the actuators, they deflect in opposite directions to each other because of the piezoelectric effect [4]. The torsional bars generate rotating torques, which allow the mirror to be rotated from A to A'. As shown in Fig. 1, the micro-actuator consists of a LiNbO₃ layer and a silicon oxide (SiO₂) layer below LiNbO₃ layer. The design 1 has a SiO₂ layer below LiNbO₃ layer (shown Fig. 1 (a)) and design 2 has a SiO₂ layer above LiNbO₃ layer (shown Fig. 1 (b)). The shear

---

Research supported by the National Natural Science Foundation of China (61874073), and Natural Science Foundation of Shanghai (19ZR1477000).

[1]School of Information Science and Technology, ShanghaiTech University

[2]Shanghai Institute of Microsystem and Information Technology, Chinese Academy of Sciences

[3]University of Chinese Academy of Sciences

*Correspondence: liuysh2@shanghaitech.edu.cn;
wutao@shanghaitech.edu.cn

978-1-6654-3008-1/21 $31.00 © 2021 IEEE

stress in combination with the SiO$_2$ layer acts as the bending layer.

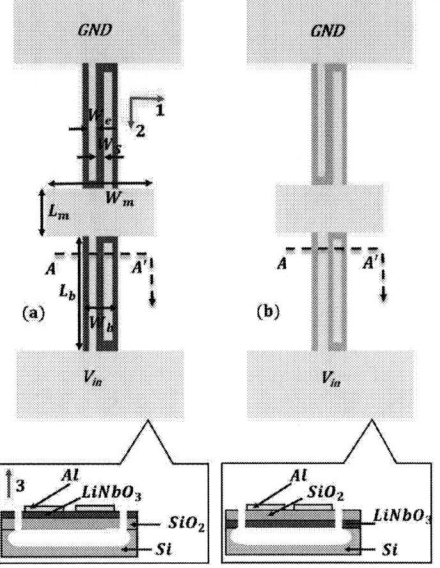

Figure 1 The structures of (a) design 1: SiO$_2$ layer is below LiNbO$_3$ and (b) design 2: SiO$_2$ layer is between LiNbO$_3$ and top Al electrodes

B. LiNbO$_3$ Orientation Selection

Since LiNbO$_3$ is a highly anisotropic crystal which has different material properties in different cut orientations, the orientation (cut) and the direction of the applied electric field have to be carefully designed for the desired mode of operation [11] in order to maximize the actuation efficiency of the LiNbO$_3$ micro-actuator. Based on our torsional beam structure design, a large piezoelectric coupling coefficient e16 is preferred to actuate the center mirror with a pair of top electrodes. Therefore, we have examined different orientations of LiNbO$_3$ crystal and 36Y-cut LiNbO$_3$ rotated e-matrix is shown as below:

$$e = \begin{bmatrix} 0 & 0 & 0 & 0 & 0.1 & -4.5 \\ -1.7 & -2.3 & 2.6 & 0.5 & 0 & 0 \\ -1.94 & -1.6 & 4.5 & -0.3 & 0 & 0 \end{bmatrix} C/m^2 \quad (1)$$

As shown in Equ. (1), the 36Y-cut LiNbO$_3$ has a large piezoelectric stress constant component of -4.5 (C/m^2) in e_{16} [4], which can excite the torsional mode vibration with top-only interdigitated transducers (IDT). The orientation is identified as the X-axis of LiNbO$_3$ crystal, along which the electrical field will also be applied to fully harness the maximized e_{16} for maximum actuation efficiency. Here, (φ, β, γ) were used to represent the Euler rotation angle, for 36Y-cut is (φ, 54, 0). The electrode arrangement direction is along the x-axis direction after Euler rotation, then φ represents the direction of wave propagation.

C. Structure Parameters Analysis

Different from the single MEMS resonator, micro-actuator consists of a rotated mirror plate and two actuators. As is shown in Fig. 2, we analyzed the influence of different parameters on the displacement of micro-actuator through COMSOL simulation. The displacement represents the distance of rotated mirror plate along the 3-direction. Fig. 2 (a) shows the relation between maximum static displacement and oxide-to-LiNbO$_3$ thickness ratio α. In

general, the displacement of both design 1 and design 2 have similar trend that they firstly increase and then decrease as the thickness ratio α increases. The design geometry is set to L_m = 22 μm, W_m = 40 μm, L_b = 55 μm and W_b = 10 μm and electrodes coverage is 0.5. Electrodes coverage can be defined as the ratio of W_e and W_b (W_e = W_s). As indicated in Fig.2 (a), design 1 has a maximum displacement at α = 0.7 while design 2 has a maximum displacement at α = 0.3. Fig. 2 (b) shows the displacement as a function of electrodes coverage. Electrodes coverage almost has no effect for the displacement of design 1. The displacement of design 2 increases rapidly when electrodes coverage goes from 0.2 to 0.6 and then almost saturate as electrodes coverage further increases to 0.8. Note that the α of design 1 is set as 0.7 and the α of design 2 is set as 0.3 for the electrodes coverage analysis in Fig. 2 (b). It can be concluded that different LiNbO$_3$/SiO$_2$ stack structure could generate different effects on the displacement of micro-actuators. Optimal oxide-to-LiNbO$_3$ thickness ratio and electrodes coverage should be designed to achieve maximum displacement.

For other four design parameters, design 1 and design 2 have the same features, therefore we only present and analyze the design space for design 2. W_m and L_b are positively correlated with the displacement. The displacement of micro-actuators has a rapid decreasing as W_b goes from 8 μm to 20 μm. However, L_m has little impact on the scanner torsional rotation and displacement. It means that the displacement response of the micro-actuator can be adjusted by changing the size of the micro-actuator structure.

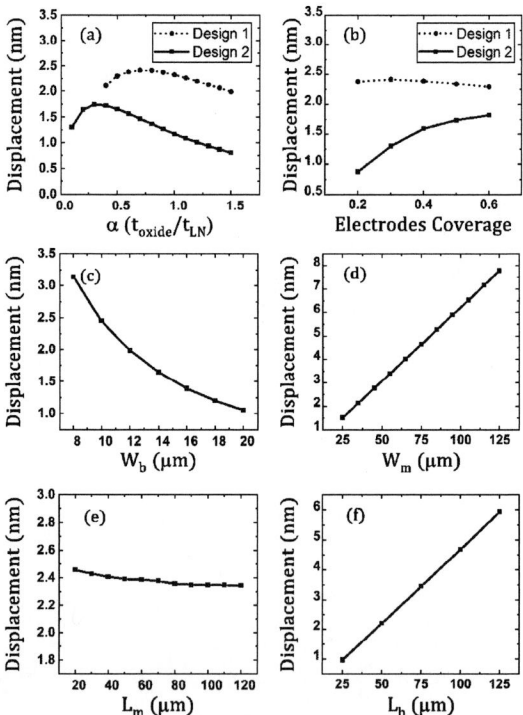

Figure 2. The relations between structure parameters and rotated displacement

Besides the driving displacement, the resonant frequency is another important parameter of micro-actuator. It can dictate the maximum scan speed of micro-actuator which is determined by the mechanical properties of the micro-actuator, specifically the mass and spring constant. The natural frequency for torsional displacement is given by [2]:

$$f = (1/2\pi)\sqrt{k/I} \tag{2}$$

where k is the spring constant, I is the mirror moment of inertia, k and I can be expressed as follows:

$$k = 2J_bG/L_b \tag{3}$$

$$I = \rho TL_mW_m{}^3/12 \tag{4}$$

where J_b is the torsional constant, G is the shear modulus. T is the total thickness of LiNbO$_3$ thin film (T_{LN}) and SiO$_2$ (T_o). J_b can be expressed as follows:

$$J_b = W_bT^3(\frac{1}{3} - 0.21\frac{T}{W_b}(1 - \frac{T^4}{12W_b{}^4}) \tag{5}$$

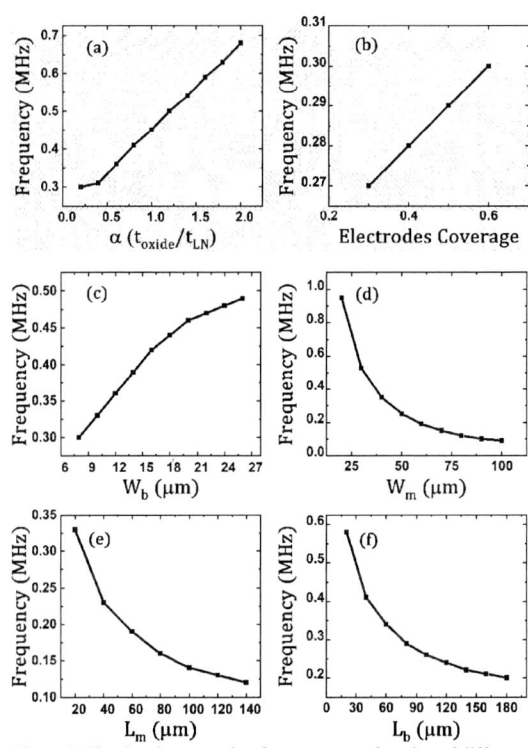

Figure 3. Simulated resonanting frequency as a function of different structure parameters

Equ. (2) shows the mathematical relationship between the torsional resonating frequency and the structure parameters. Fig. 3 illustrates the influence of different parameters on the resonance frequency based on COMSOL simulation. For the relations of the frequency and structure parameters, design 1 and design 2 have the same features, so we only demonstrated the trends of design 2. W_b, α and electrodes coverage are positively correlated with the resonant frequency, as expected from Equs. (2-5). The frequency of micro-actuators has a trend of decreasing in the structure parameters of W_m, L_m and L_b. Except for the electrodes cover and L_m, other four parameters exist trading off corresponding to parameter itself in the displacement and frequency as shown in Fig. 2 and 3. According to this feature, we can achieve high frequency or high deflected displacement LiNbO$_3$ micro-actuator. Micro-actuator can achieve lithography defined frequency, which provides

guidance to implement multi-frequency and multi-displacement micro-actuators on the same sample.

D. Fabrication and Parameters Analysis

The fabrication process of design 1 is shown in Fig. 4 (a). (1) Design 1 starts with transfer-bonding a 36Y-cut LiNbO$_3$ thin film to SiO$_2$ layer on a silicon carrier). (2) The electrodes are then defined by lifting off sputtered aluminum (Al). (3) SiO$_2$ layer is deposited using Plasma-enhanced chemical vapor deposition (PECVD) and patterned using reactive ion etching (RIE) to serve as the hard mask. (4) Inductively coupled plasma (ICP) with Cl$_2$-based is used to etch the LiNbO$_3$. (5) the micro-actuator is released by using XeF$_2$ isotropic etching to remove the silicon underneath.

The fabrication process of design 1 is shown in Fig. 4 (b). (1) Design 2 starts with transfer-bonding a 36Y-cut LiNbO$_3$ thin film to a silicon carrier. The SiO$_2$ layer is deposited using PECVD. (2) Before the etching process, a hard baking is performed for AZ5214 to harden the photoresist (PR) to serve the mask for etching SiO$_2$ layer and LiNbO$_3$ layer. (3) RIE is used to etch SiO$_2$ and ion beam etching (IBE) to etch LiNbO$_3$. (4) Al metal is subsequently defined on top of the LiNbO$_3$ thin film as the IDT electrodes using a lift-off process. (5) The micro-actuator is released by using XeF$_2$ isotropic etching to remove the silicon underneath.

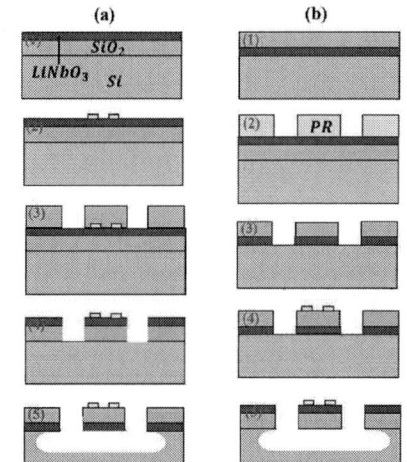

Figure 4 The fabrication process of (a) design 1 and (b) design 2

E. The designs of large displacement

The angular drive efficiency can be defined as θ_{mech} per applied drive voltage. θ_{mech} is the zero-to-peak mechanical angle (marked in Fig. 5 (a)). Large rotated displacement can help achieve the large rotated angle. It can help better deflect for light. Therefore, the designs of large displacement micro-actuators are of great significance to actual optical applications. The design 1 and design 2 of large displacement have been demonstrated in Fig. 3 (a) and (b). The simulated angular drive efficiency is analyzed in COMSOL (assuming Q = 1500). The structure parameters of the two designs are shown in Table 1. The design 1 has a driving efficiency of 21.02 °/V at 236 kHz and design 2 has a driving efficiency 4.11 °/V at 193 kHz. The design 1 has a larger rotated angle than design 2, which can be explained by the different shear stress distribution.

The simulated driving efficiency of design 1 and design

978-1-6654-3008-1/21 $31.00 © 2021 IEEE

2 in Fig. 5 are based on Q value of 1500. However, the real Q may not be ideal due to the limitation of the fabricated process. The effect of different Q values on the driving efficiency of micro-actuator has been shown in Fig. 6 (a). Q is positively correlated with θ_{mech}. The low quality may come from the large size of micro-actuator and the rough fabrication that can bring low θ_{mech}. And, high Q can help achieve larger θ_{mech}.

TABLE I.   PARAMETERS USED FOR MICRO-ACTUATOR

Parameter	Description	design 1 [μm]	design 2 [μm]
L_b	Beam length	100	100
W_b	Beam width	10	10
L_m	Mirror length	15	15
W_m	Mirror width	70	70
W_e	Electrode width	1.5	2.5
W_s	Electrode spacing	1.5	2.5
T_o	SiO_2 thickness	0.49	0.21
T_{LN}	$LiNbO_3$ thickness	0.7	0.7
T_A	Al thickness	0.2	0.2

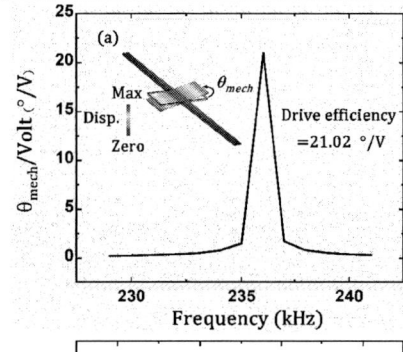

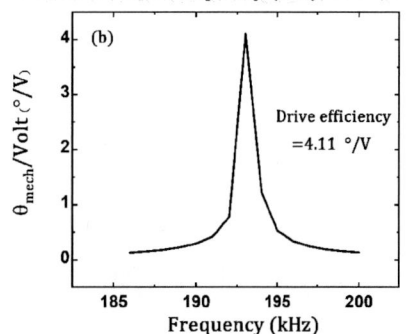

Figure 5 Simulated angular drive efficiency of (a) design 1 and (b) design 2 of large displacement (drive voltage set to 1 V and Q set to 1500)

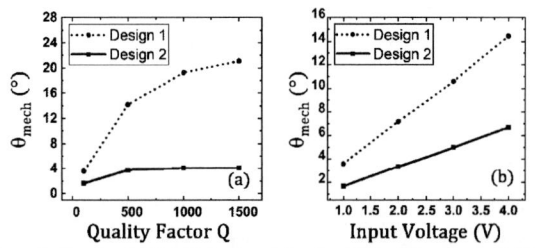

Figure 6. Simulated of (a) design 1 and 2 under different Q setting and (b) design 1 and 2 under different driving voltage (Q = 100).

The simulated mechanical angle (θ_{mech}) of design 1 and design 2 based on different input voltage level using COMSOL (Q = 100) are plotted in Fig. 6 (b). The θ_{mec} is positively correlated with input voltage. So, high input voltage can help achieve larger θ_{mech}. The driving efficiency of design 1 and 2 are 14.45 °/V and 6.7 °/V at Q = 100 and

input voltage is 4 V, respectively, which means it can improve the low driving efficiency caused by micro-actuator of low Q by increasing the input voltage.

Both design 1 and 2 have their advantages. Design 2 provides a lot of convenience for the fabrication process of micro-actuator by depositing SiO_2 above the $LiNbO_3$ thin film, which offers an idea of adjusting the working frequency and displacement of the micro-actuator by adjusting the thickness of SiO_2 layer. But The driving efficiency of design 1 is much larger than design 2, which can bring larger optical rotated angle.

III.   CONCLUSION

Two different micro-actuator designs of SiO_2 layer below (design 1) and above (design 2) 36 Y-cut $LiNbO_3$ thin film have been demonstrated. The driving efficiency of micro-actuators based on $LiNbO_3$ thin films are analyzed under different design parameters. Through the analysis of different design parameters of micro-actuator structure, the designs of large displacement have been offered. The design 1 has a driving efficiency of 21.02 °/V at 236 kHz and design 2 has a driving efficiency of 4.11 °/V at 193 kHz, which have been shown from COMSOL simulation when the quality factor (Q) is set 1500. The value of Q and input voltage are positively correlated with θ_{mech}. So, one method of increasing the input voltage can improve the low driving efficiency caused by micro-actuator of low Q. It is crucial that $LiNbO_3$ micro-actuators have lithographically defined frequency and rotated displacement. Our designs show great potential of scanning micro-actuators applications where it needs high scanning angle.

REFERENCES

[1] T. Iseki, M. Okumura, T. J. S. Sugawara, and A. A. Physical, "Shrinking design of a MEMS optical scanner having four torsion beams and arms," vol. 164, no. 1-2, pp. 95-106, 2010.

[2] P. R. Patterson, D. Hah, M. Fujino, W. Piyawattanametha, and M. C. Wu, "Scanning micromirrors: An overview," in Optomechatronic Micro/Nano Components, Devices, and Systems, 2004, vol. 5604: International Society for Optics and Photonics, pp. 195-207.

[3] T. Iseki, M. Okumura, and T. Sugawara, "High - Speed and Wide - Angle Deflection Optical MEMS Scanner Using Piezoelectric Actuation," IEEJ Transactions on Electrical and Electronic Engineering, vol. 5, no. 3, pp. 361-368, 2010.

[4] A. Emad, R. Lu, M.-H. Li, Y. Yang, T. Wu, and S. Gong, "Resonant Torsional Micro-Actuators Using Thin-Film Lithium Niobate," in 2019 IEEE 32nd International Conference on Micro Electro Mechanical Systems (MEMS), 2019: IEEE, pp. 282-285.

[5] D. Wang, C. Watkins, and H. Xie, "MEMS mirrors for LiDAR: a review," Micromachines, vol. 11, no. 5, p. 456, 2020.

[6] J.-H. Park, J. Akedo, H. J. S. Sato, and A. A. Physical, "High-speed metal-based optical microscanners using stainless-steel substrate and piezoelectric thick films prepared by aerosol deposition method," vol. 135, no. 1, pp. 86-91, 2007.

[7] T. Iseki, M. Okumura, T. J. I. T. o. E. Sugawara, and E. Engineering, "High - Speed and Wide - Angle Deflection Optical MEMS Scanner Using Piezoelectric Actuation," vol. 5, no. 3, pp. 361-368, 2010.

[8] S. Matsushita, I. Kanno, K. Adachi, R. Yokokawa, and H. J. M. t. Kotera, "Metal-based piezoelectric microelectromechanical systems scanner composed of Pb (Zr, Ti) O 3 thin film on titanium substrate," vol. 18, no. 6, pp. 765-771, 2012.

[9] M. Levy et al., "Fabrication of single-crystal lithium niobate films by crystal ion slicing," vol. 73, no. 16, pp. 2293-2295, 1998.

[10] S. T. Holmström, U. Baran, and H. Urey, "MEMS laser scanners: a review," Journal of Microelectromechanical Systems, vol. 23, no. 2, pp. 259-275, 2014.

[11] S. Gong, G. J. I. T. o. M. T. Piazza, and Techniques, "Design and analysis of lithium–niobate-based high electromechanical coupling RF-MEMS resonators for wideband filtering," vol. 61, no. 1, pp. 403-414, 2012.

Gap in pagination due to unavailable papers.

Pages 1550-1553

Electrospun Polyimide Nanofiber Separators for Lithium-ion Batteries

Wenwang Li, Bangzhou Che, Jinghua Lin, Sinan Fu, Jiaxin Jiang, Gaofeng Zheng, and Xiang Wang[*]

Abstract—The electrospun polyimide (PI) nanofiber membrane was prepared and used as lithium-ion battery separator. The electrochemical properties of the PI membrane were test. The experimental results showed that the ionic conductivity of the PI membrane was 1.310e-3 S/cm. The electrochemical stability window is 5.5 V. The PI membrane had a specific discharge capacity of 110.5 mAh/g and an efficiency of 97.3% after 111 charging and discharging cycles. The physical properties such as surface morphology, thermal stability, and contact angle of PI film was also measured. Our results show that the electrospun nanofiber membrane has potential application in lithium-ion battery separators.

I. INTRODUCTION

Lithium-ion (Li-ion) battery has become the most popular energy for a wide variety of electronic devices in modern society because of its high energy density, high working voltage and long cycle life [1]. With the development of Li-ion battery in the field of power and energy storage, its performance has gradually unable to meet the needs of emerging fields [2]. Typical Li-ion battery consists of anode, cathode, electrolyte, and separator. The charge and discharge function of Li-ion battery is achieved by lithium ions pass through the electrolyte between two electrodes. Although the separator is not directly involved in the internal reaction of Li-ion battery, it still plays an important role in the system[3]. The main function of the separator is to soak in the electrolyte to avoid physical contact between anode and cathode. The lithium ions in the electrolyte can migrate freely between the two electrodes while the passage of free electrons is prevented. The structure and composition of separator directly determines the interface structure, internal resistance, stability, and other indicators of the battery, which greatly affects the electrochemical performance of the battery, including the energy and power density, cycle life and safety [4]. Therefore, the research of separator has great significance for the development of battery technology. [5, 6].

Commercial separators for Li-ion battery are usually made of polyethylene (PE) membranes generated by the dry or wet stretching process. Though being used for decades, the traditional separator suffers from poor thermal stability and limited porosity and wettability [7, 8]. In recent years, nanofibers have been widely used in many fields due to their unique physical and chemical properties [9, 10]. The kind of nanofibrous membrane is highly porous with a large surface area to volume ratio, flexible surface, and mechanical properties, makes it suitable for separator in Li-ion battery applications [11, 12].

Electrospinning [13-15] is a simple and versatile technique to fabricate nanofibers and nanofibrous membranes. The principle is to stretch the solution by electric field force. Nanofibers can be obtained after stretching, whipping and solidification of solution. The morphology of electrospun membrane can be adjusted by changing the process parameters. Due to the outstanding properties such as high porosity, high specific surface area and controllable fiber diameters, electrospun nanofibers have attracted extensive attention in many fields [16, 17]. Rapid progress has been made to the electrospinning of battery separators, offered appealing features to separator and rendered performance enhancement for Li-ion applications [18-20].

In this work, the application of electrospun PI nanofiber membrane in lithium-ion battery separator was studied, and the performance of the separator was tested.

II. EXPERIMENTAL DETAILS

The polyimide (PI) solution was purchased from Hangzhou Surmount Technology Co., Ltd., and was used without any treatment. The PI nanofiber membrane was prepared by a traditional electrospinning equipment. The applied voltage was 16 kV, and the nozzle-to-substrate was 12 cm. The thickness of electrospun PI nanofiber membrane was about 20 μm (Fig. 1(a)). A commercial PE membrane (Fig. 1(b)) was also employed for performance comparison. The morphography of membrane was examined by a scanning electron microscope (SEM, Zeiss Sigma 500). The contact angle was captured by a high-speed camera (Phantom Miro 110).

The separator to be tested was assembled into the CR2016 button battery case (Fig. 1(c)) according to different battery systems. The electrolyte used was DLC301 purchased from Guotai Huarong New Material Co., Ltd. The assembled battery was then stand for 24 hours before testing.

Resrach supported by National Natural Science Foundation of China (No. 51805460), Science and Technology Planning Project of Fujian Province (No. 2019H0038, 2020H6003), Xiamen Municipal Science and Technology Project (No. 3502Z20193015), and Xiamen University of Technology (No. XPDKT20022).

W. Li, B. Che, S. Fu, X. Wang are with the School of Mechanical and Automotive Engineering, Xiamen University of Technology, Xiamen,

Fujian 361024, China (*corresponding author: X. Wang; e-mail: wx@xmut.edu.cn).

J. Lin is with the Institute of Manufacturing Engineering, Huaqiao University, Xiamen, Fujian 361021, China.

J. Jiang, G. Zheng are with the Department of Instrumental and Electrical Engineering, Xiamen University, Xiamen, Fujian 361102, China.

978-1-6654-3008-1/21 $31.00 © 2021 IEEE

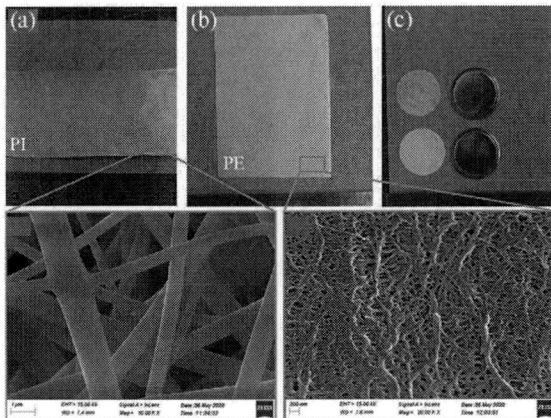

Fig. 1. (a) Electrospun PI nanofiber membrane. (b) Commercial PE membrane. (c) The assembled CR2016 battery with PI and PE membranes.

The AC impedance was tested by a CHI-660E electrochemical workstation. The scanning frequency was in the range of 0.1 Hz to 100 kHz, and the disturbance voltage was 10 mV. The assembly system of the battery was stainless steel/membrane/stainless steel.

The electrochemical stability window was also examined by CHI-660E. The assembly system of the battery was stainless steel/membrane/Li sheet. The test method was linear sweep voltammetry (LSV) with a scanning range of 0-6 V and a scanning rate of 10 mV/s.

A BTS-5V charging and discharging test system was used to test the performance of Li-ion battery. The assembly system of the battery was LiFePO$_4$/membrane/Li sheet. The constant current charging and discharging test was carried out in the range of 2.8-3.7 V under a charging and discharging rate of 1C.

III. RESULTS AND DISCUSSION

Fig. 1(a) shows the SEM image of the electrospun PI membrane, of which the collected nanofibers had typical randomly arrangement. The PI membrane prepared by electrospinning was made of nanofibers overlapping each other, and the average diameter of the nanofibers was less than 1 μm. Numerous through-holes formed by those overlapping nanofibers, and the average pore diameter of these through-holes was at the same order of magnitude as the diameter of the nanofibers. The good insulation properties of PI and the pore size of electrospun nanofiber membrane were sufficient to prevent direct contact between the positive and negative electrodes of the Li-ion battery and prevent the self-discharge. In addition, due to the large number of nanofibers and the random and scattered position of the overlapping nanofibers, many through-holes formed under the layer-by-layer stacking, and these through-holes can act as channels for lithium-ion shuttling. Therefore, the membrane made by electrospinning technology had highly porosity. Fig. 1(b) shows the SEM image of the commercial PE membrane. The traditional commercial PE Li-ion battery separator was made by wet drawing process. The pores were produced through solid-liquid or liquid-liquid

phase separation, and the porosity was usually lower than that of membrane prepared from electrospinning method. As a result, the electrospun PI membrane had a lower impedance and higher ion conductivity than the traditional PE membrane.

The contact angles between electrolyte and PI/PE membranes at different times were investigated, as shown in Fig. 2. The electrolyte formed a conical droplet on the surface of the membrane due to the effect of surface tension, and then gradually absorbed as time goes on. It can be observed that the absorption rate of PI membrane was significantly higher than that of PE membrane, and the contact angle of PI membrane was always smaller than that of PE membrane. On one hand, the more polar groups of PI molecule led to better affinity for liquid with strong polarity. On the other hand, the larger specific surface area of electrospun membrane made it more fully contact with electrolyte. These characteristics makes the Li-ion battery separator made of electrospun PI membrane have strong electrolyte absorption and retention ability. The separator can be more easily immersed in the electrolyte and increase the contact area with the electrolyte, thus improving the overall performance of the battery.

Fig. 3 shows the AC impedance spectrum of Li-ion batteries with PI and PE separator. It can be obtained that the bulk resistance of PI and PE membrane were 0.77 Ω and 3.3 Ω, respectively. The ionic conductivity of PI membrane can be calculated as 1.310e-3 S/cm, and that of PE membrane was 0.201e-3 S/cm. The ionic conductivity of PI membrane was one order of magnitude higher than that of PE membrane.

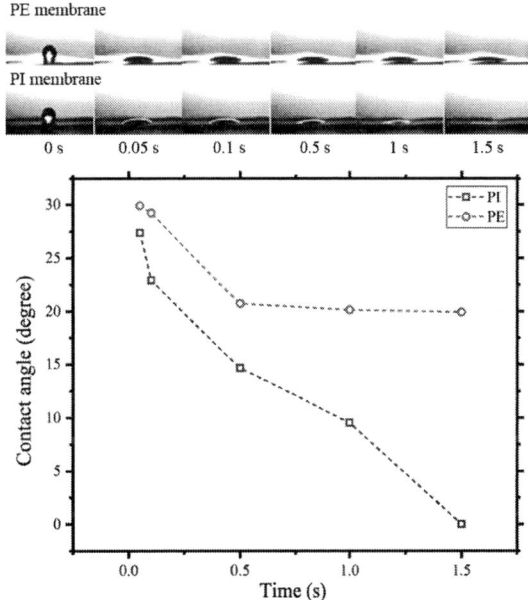

Fig. 2. The contact angle of PI and PE membranes.

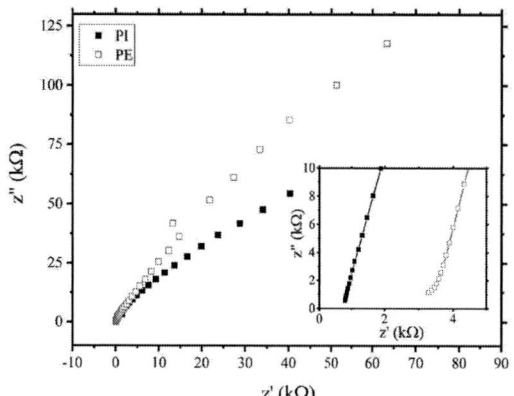

Fig. 3. The electrochemical impedance spectrum of batteries with PI and PE separators.

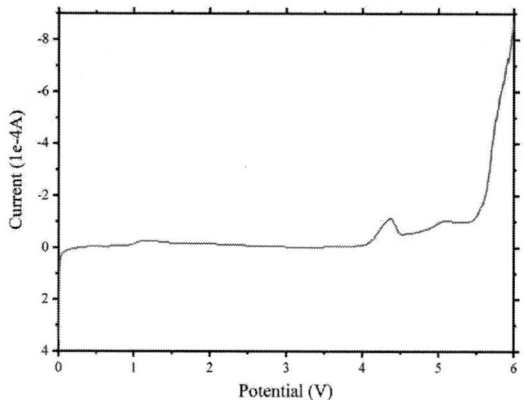

Fig. 4. The linear sweep voltammetry curve of battery loaded with PI membrane.

The linear sweep voltametric curve of the battery loaded with PI membrane is shown in Fig. 4. When the voltage was higher than 5.5 V, there is an inflection point of sharp rise in current, thus that 5.5 V should be treated as the electrochemical stability window of PI membrane. It indicates that the PI membrane can be applied to a wider range of electrode materials to meet the requirements of Li-ion batteries.

Fig. 5 shows the first charge and discharge performance of Li-ion batteries with PI and PE membrane. The PI membrane have charge and discharge specific capacities of 146.2 mAh/g and 130.4 mAh/g, respectively, with a coulomb efficiency of 89.2%. The first charge and discharge specific capacities of PE film are 147.7 mAh/g and 126.0 mAh/g respectively, and the coulomb efficiency of charge and discharge is 85.3%. In contrast, the first charge and discharge specific capacities of the PE membrane was 147.7 mAh/g and 126.0 mAh/g, respectively, and the coulombic efficiency was 85.3%.

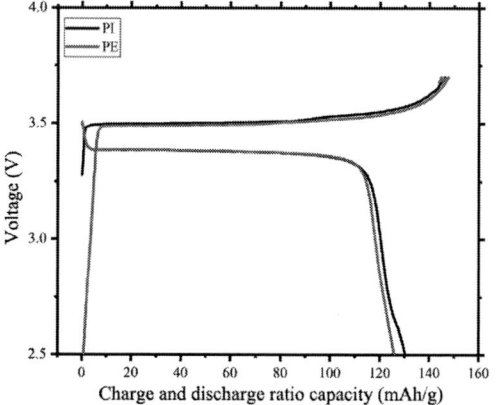

Fig. 5. First charge and discharge performances of PI and PE membranes.

The cycle performance of Li-ion batteries with PI membrane and PE membrane are illustrated in Fig. 6. The specific discharge capacity of the battery gradually decreased as the number of cycles increased. After 111 cycles, the specific discharge capacity of battery equipped with PI membrane decreased from 130.4 mAh/g to 110.5 mAh/g, which was 15.3% lower than that of the first cycle; while the battery equipped with PE membrane decreased from 126.0 mAh/g to 103.0 mAh/g and was a 18.3% decrease compared to the first cycle. The charging and discharging efficiency of PI membrane (excluding the first cycle) dropped to 97.3%, and that of PE membrane was 96.1%. From the above results, the cycling performance of Li-ion battery equipped with PI membrane is better than that of equipped with PE membrane.

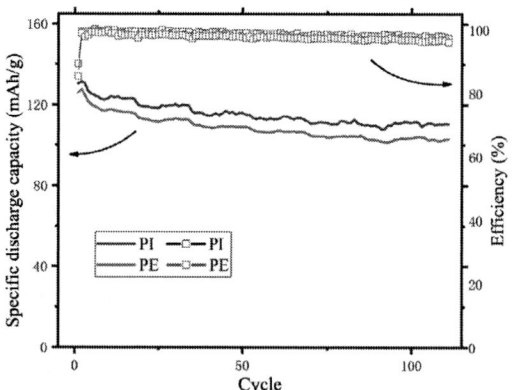

Fig. 6. Cycling performance of batteries.

978-1-6654-3008-1/21 $31.00 © 2021 IEEE

The thermal stability of separator is one of the important aspects related to the safety performance of Li-ion batteries. The temperature of Li-ion battery may rise due to incorrect connection methods or changes in the external environment. If the separator cannot withstand high temperature and deforms or even ruptures, the positive and negative electrodes of the battery will be in direct contact, causing serious consequences such as combustion and explosion. Therefore, the thermal stability of PI membrane was also tested, as shown in Fig. 7. The PI and PE membranes were heated in the oven for 30 minutes, the temperatures were set to be 50 °C, 100 °C, 150 °C, and 200 °C, respectively. It can be observed that there were no significant changes in the morphology of PI membrane during these heating processes. As a contrast, the PE membrane began to warp when the temperature was 100 °C, and shrunk rapidly at 150 °C. When the temperature was 200 °C, the PE membrane was completely melted. Therefore, the Li-ion battery separator prepared by electrospun PI membrane may has higher safety than the traditional PE membrane.

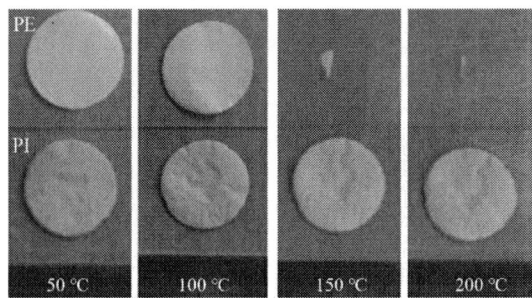

Fig. 7. Thermal stability of PI and PE membranes heating at various tempture for 30 minutes.

IV. CONCLUSION

In this study, the electrospun PI nanofiber membrane was served as separator and assembled into Li-ion battery for electrochemical tests. The electrospun separator had highly porosity due to the large number and the random and dispersed overlapping deposition of nanofibers. It provided more channels for lithium ions to pass through in the Li-ion battery system. The experimental results indicated that the electrospun PI separator obtained lower impedance and higher ionic conductivity than that of traditional commercial PE separator. The ionic conductivity, impedance, and electrochemical stability window of PI separator was 1.310e-3 S/cm, 0.77 Ω, and 5.5 V, respectively. The PI membrane had a specific discharge capacity of 110.5 mAh/g and an efficiency of 97.3% after 111 charging and discharging cycles. In addition, the electrospun PI membrane had better hydrophilicity and thermal stability than that of traditional PE membrane. These results support that the electrospun membrane has the feasibility and advantages for next generation of Li-ion battery separator.

REFERENCES

[1] B. Diouf, and R. Pode, "Potential of lithium-ion batteries in renewable energy," *Renewable Energy*, vol. 76, pp. 375-380, 2015.

[2] F. Wu, J. Maier, and Y. Yu, "Guidelines and trends for next-generation rechargeable lithium and lithium-ion batteries," *Chemical Society Reviews*, vol. 49, no. 5, pp. 1569-1614, 2020.

[3] C. F. J. Francis, I. L. Kyratzis, and A. S. Best, "Lithium-Ion Battery Separators for Ionic-Liquid Electrolytes: A Review," *Advanced Materials*, vol. 32, no. 18, pp. 1904205, 2020.

[4] C. M. Costa, Y.-H. Lee, J.-H. Kim, S.-Y. Lee, and S. Lanceros-Méndez, "Recent advances on separator membranes for lithium-ion battery applications: From porous membranes to solid electrolytes," *Energy Storage Materials*, vol. 22, pp. 346-375, 2019.

[5] M. F. Lagadec, R. Zahn, and V. Wood, "Characterization and performance evaluation of lithium-ion battery separators," *Nature Energy*, vol. 4, no. 1, pp. 16-25, 2019.

[6] C. M. Costa, E. Lizundia, and S. Lanceros-Méndez, "Polymers for advanced lithium-ion batteries: State of the art and future needs on polymers for the different battery components," *Progress in Energy and Combustion Science*, vol. 79, pp. 100846, 2020.

[7] P. Arora, and Z. Zhang, "Battery Separators," *Chemical Reviews*, vol. 104, no. 10, pp. 4419-4462, 2004.

[8] J. Nunes-Pereira, C. M. Costa, and S. Lanceros-Méndez, "Polymer composites and blends for battery separators: State of the art, challenges and future trends," *Journal of Power Sources*, vol. 281, pp. 378-398, 2015.

[9] E. P. S. Tan, and C. T. Lim, "Mechanical characterization of nanofibers – A review," *Composites Science and Technology*, vol. 66, no. 9, pp. 1102-1111, 2006.

[10] B.-S. Kim, and I.-S. Kim, "Recent Nanofiber Technologies," *Polymer Reviews*, vol. 51, no. 3, pp. 235-238, 2011.

[11] J. Liu, K. Yang, Y. Mo, S. Wang, D. Han, M. Xiao, and Y. Meng, "Highly safe lithium-ion batteries: High strength separator from polyformaldehyde/cellulose nanofibers blend," *Journal of Power Sources*, vol. 400, pp. 502-510, 2018.

[12] K. Yuriar-Arredondo, M. R. Armstrong, B. Shan, W. Zeng, W. Xu, H. Jiang, and B. Mu, "Nanofiber-based Matrimid organogel membranes for battery separator," *Journal of Membrane Science*, vol. 546, pp. 158-164, 2018.

[13] M. S. Islam, B. C. Ang, A. Andriyana, and A. M. Afifi, "A review on fabrication of nanofibers via electrospinning and their applications," *SN Applied Sciences*, vol. 1, no. 10, pp. 1248, 2019.

[14] J. Xue, J. Xie, W. Liu, and Y. Xia, "Electrospun Nanofibers: New Concepts, Materials, and Applications," *Accounts of chemical research*, vol. 50, no. 8, pp. 1976-1987, 2017.

[15] D. Li, and Y. Xia, "Electrospinning of Nanofibers: Reinventing the Wheel?," *Advanced Materials*, vol. 16, no. 14, pp. 1151-1170, 2004.

[16] T. Subbiah, G. S. Bhat, R. W. Tock, S. Parameswaran, and S. S. Ramkumar, "Electrospinning of nanofibers," *Journal of Applied Polymer Science*, vol. 96, no. 2, pp. 557-569, 2005.

[17] S. Thenmozhi, N. Dharmaraj, K. Kadirvelu, and H. Y. Kim, "Electrospun nanofibers: New generation materials for advanced applications," *Materials Science and Engineering: B*, vol. 217, pp. 36-48, 2017.

[18] Y. Li, Q. Li, and Z. Tan, "A review of electrospun nanofiber-based separators for rechargeable lithium-ion batteries," *Journal of Power Sources*, vol. 443, pp. 227262, 2019.

[19] N. Angulakshmi, and A. M. Stephan, "Electrospun Trilayer Polymeric Membranes as Separator for Lithium–ion Batteries," *Electrochimica Acta*, vol. 127, pp. 167-172, 2014.

[20] V. Aravindan, J. Sundaramurthy, P. Suresh Kumar, Y.-S. Lee, S. Ramakrishna, and S. Madhavi, "Electrospun nanofibers: A prospective electro-active material for constructing high performance Li-ion batteries," *Chemical Communications*, vol. 51, no. 12, pp. 2225-2234, 2015.

Gap in pagination due to unavailable paper.

Pages 1558-1559

Simulation and Analysis of Nano Robotic Manipulators Inside SEM

Ziliang He, Lue Zhang*, Zhan Yang, Senior Member, *IEEE*

Abstract—The nano robotic manipulators inside scanning electron microscopy (SEM) were developed to perform nano assemblies. However, it was challenging to design path planners and controllers with real manipulators because it was difficult to see the motion inside the specimen chamber, what's more, the crash caused by mismatching path planners and controllers will result in great losses. To resolve the issue of a low-cost inspection on path planner, controller and motion features, the simulation and analysis of Nano Robotic Manipulator inside SEM chamber were developed. The kinematics were described by D-H formulation. The structural of the manipulators were built on Solidworks. The visible simulate model was created on Matlab/Simulink. Finally, a case study was conducted to verify the effectiveness of the model.

I. INTRODUCTION

Nano Robotic Manipulation has shown great prospects in the manufacturing of nano devices, such as MEMS devices; carbon-based chips and CNT-FETs. Prof. Hatamura *et al.* purposed a nano robotic manipulator built in SEM specimen chamber [1]. The manipulating experiment of 5μm diameter iron particles were conducted on that manipulator. Fukuda *et al.* purposed a 16-degrees of freedom(16-DoFs) nano robotic manipulator system that enables nano-processing, nano-measurement, and nano-assembly[2,3]. Fatikow *et al.* built a nanorobotic system that combines electrothermal microgrippers and mobile microrobots inside a scanning electron microscope with 6-DoFs, which can achieve the spatial motion [4,5], achieved real-time tracking and positioning of carbon nanotubes (CNTs). A Nano Robotic Manipulation platform based on nano positioning systems and SEM was demonstrated in our previous work[6,7].

In addition to control manipulators manually, many studies on automated manipulating have been conducted. Prof. Sun *et al.* achieved automated picking and measuring of nanowires[8]. However, it was challenging to design path planning algorithms and controllers without viewing motion of manipulators, which played an important role in auto manipulating.

Before applying path planning algorithms and controllers on actual robotic manipulators, testing on a simulating environment is always beneficial. Mohammed Abu *et al.* presented the development of a visual software package for kinematic simulating and path planning problems. The package was created by utilizing Matlab/Simulink and AutoCAD[9]. Jamali *et al.* presented virtual prototype

modeling, simulation and optimization of a 3 DOF SCARA robot as an example of robot manipulators. The software utilized in the work were Solidworks and Matlab. Based on the model, Multi-Variable control process was performed with PID controller for controlling the robot[10].

Though simulations of industrial robots has been studied widely, however, few works were conducted to simulate the moving properties and kinematics of Nano Robotic Manipulators. Since Nano Robotic Manipulators were assembled inside the specimen chamber of SEM, it is vital to know the location of the equipment in the SEM and to avoid any part of the manipulators colliding with the equipment during manipulation. While designing path planners and controllers, tests on real nano robotic manipulators may cause great loss if collide happened. So a simulation and analysis of manipulators and the environments inside SEM will play a great role in the design of controllers and development of path planning algorithms for Nano Robotic Manipulators.

To address the issue discussed above, geometry model of nano robotic manipulators were analyzed in this paper, as well as the environments inside SEM. The kinematic was discussed by D-H formulation subsequently. Then, the kinematic simulation of manipulators were performed on Matlab/Simulink software.

II. INTRODUCTION OF NANO ROBOTIC MANIPULATORS

Each nano robotic manipulator consists of a nano linear positioning system and a Picomotor actuator. The linear motion in three dimensional Cartesian space was achieved by positioning system. Total degrees of freedom(DOF) were 21. The manufacturers, types, sizes, of the nano linear positioners system and the rotary motor are as the as the follow table shows:

TABLE I.   COMPOSITION OF THE MANIPULATOR

	nano positioning system	Picomotor actuator
Manufactures	Smart Act	New Focus
Types	SLC-1720-s	8301-UHV
Resolution	2 nm	30 nm
Motion ranges	x, y, z: ±6 mm	-360 °- 360 °

The nano robotic manipulators inside the SEM specimen chamber were mounted as shown in Figure 1. There were two manipulators inside SEM named 'Unit 1' and 'Unit 2'.

---

*Resrach supported by National Key Research and Development Program of China, grant number 2018YFB1304901 and National Natural Science Funds of China (Grant No. 61773275).

Ziliang He, Lue Zhang, Yang Zhan are with School of Mechanical and Electric Engineering, Soochow University, Suzhou, China.

Figure 1. Nano Robotic Manipulators inside SEM.

Each manipulator in this research had three prismatic joints and a rotary joint. The translations in X, Y and Z axis were achieved by nano positioning system, while rotation was obtained by Picomotor actuator. The nano positioning system was driven by MCS controller box. Vacuum compatible cables for power and signal transporting were configured to pass through the vacuum chamber through a vacuum flange. The gold-coated copper grippers were fastened on the shafts of the two Picomotor actuators by screws. Grounding wires were attached to grippers to prevent electrical charge under electron beam. Nano picking, nano transporting and nano assembling of nano materials could be achieved with nano robotic manipulators.

III. KINEMATIC ANALYSIS

The kinematic model of nano robotic manipulators was analyzed by diagram shown in Fig. 2.

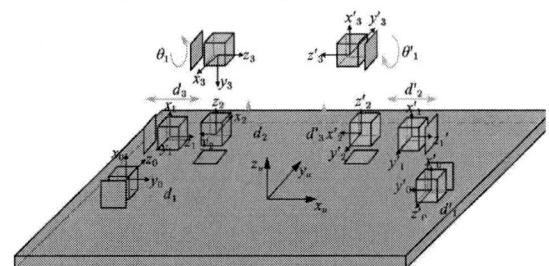

Figure 2. kinematic model of Nano Robotic Manipulators.

Z axis of the world coordinate system was denoted as z_u. Then, z_0, z_1, and z_2 were the motion axis of three prismatic joints of unit1, and z_3 was the rotate axis of the rotary joint of Unit 1, Unit 2 had a similar structural. Those two-way arrows represented the directions of motion, d_x represent displacements of prismatic joints and θ represent the rotary

degrees of rotary joints. The dynamic transporting process was also analyzed, as shown in Fig. 3.

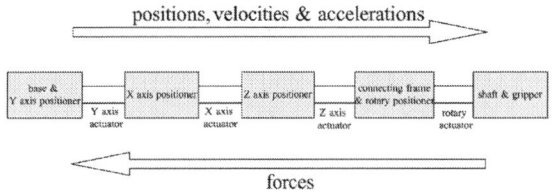

Figure 3. dynamic model of Nano Robotic Manipulators.

The base and the Y axis positioner were seen as a solid part, then the Y axis actuator connected it with X axis positioner. The models of other parts were shown in the figure. Position, velocity, and acceleration were transmitted from the base to the shaft and gripper, whereas forces were transmitted from the shaft and gripper to the base.

A. Forward and Inverse Kinematics

D-H formulation was adopted to describe the forward kinematics of nano robotic manipulators. The homogeneous transformation matrix between tow links were obtained firstly, then all matrix of joints were multiplied together. The homogeneous transformation matrix between end effector and the world coordinate were achieved finally.

The kinematic model of nano robotic manipulators was shown above. The D-H parameters of Unit 1 are as follows:

TABLE II.   D-H PARAMETERS OF THE NANO ROBOTIC MANIPULATOR

Link No.	a_i	α_i	d_i	θ_i
1	-90 °	a_1	d_1	0 °
2	-90 °	0	d_2	-90 °
3	-90 °	0	d_3	180 °
4	0 °	0	d_4	-90 °

The homogeneous transformation matrices are:
For the first link:

$$T_1^0 = \begin{bmatrix} 1 & 0 & 0 & a_1 \\ 0 & 0 & 1 & 0 \\ 0 & -1 & 0 & d_1+y \\ 0 & 0 & 0 & 1 \end{bmatrix} \qquad (1)$$

For the second link:

$$T_2^1 = \begin{bmatrix} 0 & 0 & 1 & 0 \\ -1 & 0 & 0 & 0 \\ 0 & -1 & 0 & d_2+x \\ 0 & 0 & 0 & 1 \end{bmatrix} \qquad (2)$$

For the third link:

$$T_3^2 = \begin{bmatrix} -1 & 0 & 0 & 0 \\ 0 & 0 & -1 & 0 \\ 0 & -1 & 0 & d_3+z \\ 0 & 0 & 0 & 1 \end{bmatrix} \qquad (3)$$

For the fourth link:

$$T_4^3 = \begin{bmatrix} \sin\theta_4 & \cos\theta_4 & 0 & 0 \\ -\cos\theta_4 & \sin\theta_4 & 0 & 0 \\ 0 & 0 & 1 & d_4 \\ 0 & 0 & 0 & 1 \end{bmatrix} \quad (4)$$

Then the overall manipulator transformation matrix is as follow:

$$T_4^0 = T_1^0 T_2^1 T_3^2 T_4^3 = \begin{bmatrix} \cos\theta_4 & -\sin\theta_4 & 0 & a_1+d_3+z \\ 0 & 0 & 1 & d_2+d_4+x \\ -\sin\theta_4 & -\cos\theta_4 & 0 & d_1+y \\ 0 & 0 & 0 & 1 \end{bmatrix} (5)$$

In which: a_1 - Vertical distance between link1 and link2
d_1 - Horizon distance between link1 and link2
d_2 - Horizon distance between link2 and link3
d_3 - Vertical distance between link3 and link4
d_4 - Horizon distance between link4 and link5
x - displacement of joint 1
y - displacement of joint 2
z - displacement of joint 3
θ_4 - rotation angle of Picomotor actuator

By analysis of structural and experimental measures, those constant parameters are as follows: $a_1 = 8.5$ mm, $d_1 = 0$ mm, $d_2 = 36$ mm, $d_3 = 33.53$ mm, $d_4 = 37$ mm.

The overall transformation matrix of Unit 2 could be solved by a rotation of 180 °:

$$T''^0_4 = Rot(x, 180°)T_4^0$$

$$= \begin{bmatrix} \cos\theta_4' & -\sin\theta_4' & 0 & a_1'+d_3'+z' \\ 0 & 0 & -1 & -d_2-d_4-x \\ \sin\theta_4' & \cos\theta_4' & 0 & -d_1-y \\ 0 & 0 & 0 & 1 \end{bmatrix} (6)$$

In which $Rot(x, 180°)$ represents rotate 180 degrees along the X axis.

Finally, the inverse kinematic expression of Unit 1 is as follow:

$$\begin{aligned} x &= y_d - d_2 - d_4 \\ y &= z_d - d_1 \\ z &= x_d - a_1 - d_3 \\ \theta &= \theta_d \end{aligned} \quad (7)$$

In which: x_d - Desired displacement in X direction
y_d - Desired displacement in Y direction
z_d - Desired displacement in Z direction
θ_d - Desired rotation angle of the gripper

The inverse kinematic expression of Unit 2 is:

$$\begin{aligned} x' &= y_d' - d_2' - d_4' \\ y' &= z_d' - d_1' \\ z' &= x_d' - a_1' - d_3' \\ \theta' &= \theta_d' \end{aligned} \quad (8)$$

When the desired trajectory of the end point was generated, the trajectories of joints could be calculated by the inverse kinematic expression.

IV. CONSTRUCTION OF SIMULATING MODEL

A. Drawing of manipulators and environments

In order to realize the visible simulating, it is necessary to build a three-dimensional geometry model of manipulators and the environment inside SEM. A geometry model of manipulators and environment was created by drawing on Simulink, as shown in Figure 4. The side view, top view, front view, and the three-dimensional structure were shown in four images respectively in this figure. The position of the pole piece of SEM was shown directly, the position and shape of pole piece is important to avoid colliding.

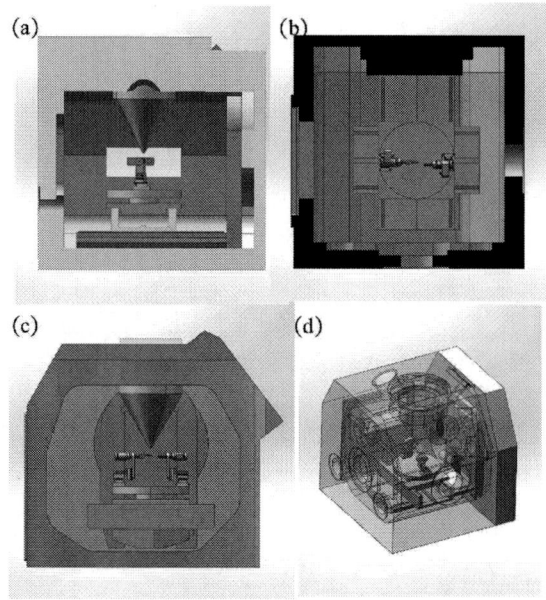

Figure 4. CAD model of Nano Robotic Manipulators inside SEM: (a) side view (b) top view (c) front view (d) three-dimensional structure.

After established the geometry model on Simulink, the model was imported into Simulink as a robotic model.

B. Importing of robotic model

The geometry model was imported to Simulink, as shown in Figure 5.

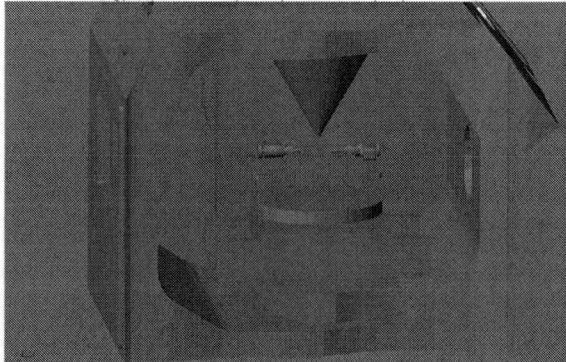

Figure 5. Simulating platform created on Simulink.

Nano linear positioners were represented by prismatic joints. Similarly, the rotary joint represents the Picomotor actuator. The Unified Robot Description Format (URDF) was created to describe the kinematic relationships and geometry models of manipulators and environment. After importing URDF into Matlab, the simulating model was created automatically. Through the visual simulation function of Simulink, the simulation results of the Nano Robotic Manipulators were presented at the interface. Then the movements could be observed easily.

A case of simulation results was shown in Fig. 6. The initial state and final state of motion that the simulating platform performed were shown on the left, similarly, the motion initial state and final state of the real platform were shown on the left. Blue arrows and the green arrow represents the moving direction of Unit 1 and Unit 2 respectively.

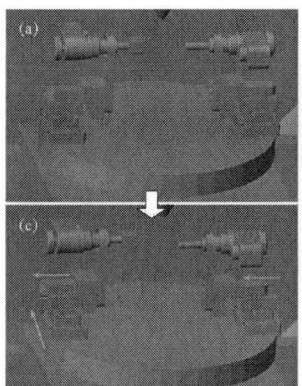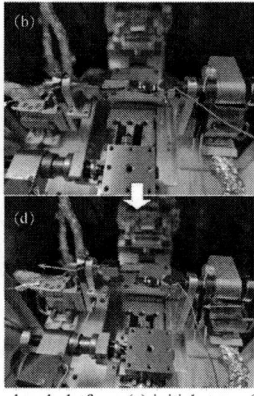

Fig. 6. Motion of simulating platform and real platform (a) initial state of simulating platform (b) initial state of real platform (c) final state of simulating platform (d) final state of real platform.

By testing controller and path planning algorithms on the simulating platform, the performance of the control algorithm and the effectiveness of the collide avoidance path planning algorithm can be verified, without repeating experiments and debugging on high-cost instruments.

V. CONCLUSION

In this paper, the visual simulate model was created by utilization of both Solidwork and Matlab/Simulink software. The kinematics was constructed by D-H formulation.

Through the visual simulate model, the path planning algorithms and controllers can be tested and simulated. The results of this research will reduce the time and costs of designing path planners and controllers. Such a method can be applied to other types of nano robotic manipulators.

REFERENCES

[1] Y. Hatamura and H. Morishita, "Direct Coupling System between Nanometer World and Human world," 1990.

[2] T. Fukuda, F. Arai, and L. Dong, "Assembly of nanodevices with carbon nanotubes through nanorobotic manipulations," *Proceedings of the IEEE,* vol. 91, no. 11, pp. 1803-1818.

[3] L. Dong, F. Arai, and T. Fukuda, "Destructive constructions of nanostructures with carbon nanotubes through nanorobotic manipulation," *IEEE/ASME Transactions on Mechatronics,* vol. 9, no. 2, pp. 350-357, 2004.

[4] Eichhorn V, Fatikow S, Wortmann T, et al. NanoLab:A nanorobotic system for automated pick-and-place handling and characterization of CNTs.Proceedings of IEEE International Conference on Robotics & Automation, IEEE, 2009: 1826-1831.

[5] Eichhorn V, Carlson K, Andersen K N, et al. Nanorobotic manipulation setup for pick-and-place handling and nondestructive characterization of carbon nanotubes. International Conference on Intelligent Robots and Systems, IEEE, 2007: 291-296.

[6] Z. Yang et al., "Mechatronic Development and Vision Feedback Control of a Nanorobotics Manipulation System inside SEM for Nanodevice Assembly," (in English), Sensors, vol. 16, no. 9, Sep 2016.

[7] Q. Shi et al., "A Vision-Based Automated Manipulation System for the Pick-Up of Carbon Nanotubes," IEEE/ASME Transactions on Mechatronics, vol. 22, no. 2, pp. 845-854, 2017.

[8] Changhai Ru *et al.,* "Automated Four-Point Probe Measurement of Nanowires Inside a Scanning Electron Microscope," pp. 674-681.

[9] Q. Mohammed Abu, I. Abuhadrous, and H. Elaydi, "Modeling and Simulation of 5 DOF educational robot arm," in 2nd International Conference on Advanced Computer Control, pp. 2010, 569-574.

[10] P. Jamali and K. Heidari Shirazi, "Robot Manipulators: Modeling, Simulation and Optimal Multi-Variable Control," Applied Mechanics and Materials, vol. 232, pp. 383-387, 2012.

Gap in pagination due to unavailable papers.

Pages 1564-1573

Proceedings of the 16th Annual IEEE International
Conference on Nano/Micro Engineered and Molecular Systems
April 25-29, 2021

High-detectivity Infrared Detector Based on Dual-layer Thermopile

Zhaohui Yang, Gaobin Xu*, Shirong Chen, Jianguo Feng, Yuanming Ma and Xing Chen

Abstract— **This paper presents a high-detectivity micromachined infrared detector based on dual-layer thermopile. The dual-layer thermopile was designed by using n/p-polysilicon on beams radially distributed on the surface of the support layer. The absorber made by black silicon was released by dry etching of polysilicon using XeF$_2$. The etch-stop structure was designed to prevent the damage of the detector caused by isotropic etching. We simulated the thermal conduction of the detector in steady state and in transient state, respectively. The results indicated that the presented thermopile detector performed a responsivity of 179.3 V/W, a detectivity of 2.99×10^8 cm·Hz$^{1/2}$/W and a time constant of 9.6 ms, respectively at room temperature.**

I. INTRODUCTION

Infrared (IR) detectors have been widely used in remote temperature sensing, gas analysis, and thermal imaging [1-3], which can be classified into photon detectors and heat detectors, based on infrared heat effect and photoelectric effect. The thermopile IR detector is one of the heat detectors [4, 5], which has become a hot topic due to their advantages in room temperature measurements without extra cooling equipment and wide range of spectrum. The thermopile consists of a sequence of thermocouples connected in series. The thermocouples are made of two materials with different Seebeck coefficients [6]. According to the Seebeck effect, a voltage will be generated when there is a temperature difference between the cold and the hot junction of the thermocouple (Fig. 1).

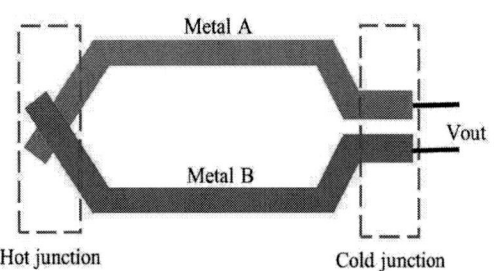

Fig. 1 Seebeck effect of the thermocouple.

In recent years, great progress has been made in the development of high-responsivity and high-detectivity thermopile IR detector based on the thermoelectric materials

* This work was supported by following foundations: National Key Research and Development Program (2020YFB2008900), Key Research and Development Program of Anhui (1804a09020018) and University Collaboration and Innovation Project of Anhui (GXXT-2019-030).
  Zhaohui Yang, Gaobin Xu*, Shirong Cheng, Jianguo Feng, Yuanming Ma and Xing Cheng are with the School of Electronic Science and Applied Physics, Hefei University of Technology, Hefei, Anhui 230009, People's Republic of China. (phone: +86-13505518762; e-mail: gbxu@hfut.edu.cn).

[7]. For example, Ryan Shea et al. published a cave-coupled thermoelectric infrared detector using Bi$_2$Te$_3$/Sb$_2$Te$_3$ material with a peak detectivity of 3×10^9 cm·Hz$^{1/2}$/W [8]. However, these materials cannot be compatible with complementary metal-oxide-semiconductor (CMOS) processes, which makes it difficult for mass production. Alternative materials including aluminum and polysilicon are also widely used [9]. Kaiqun Wang et al. achieved thermal isolation by etching the back of bulk silicon. The thermocouple arrays were made by p-polysilicon /Al, and a layer of Au/Ti was evaporated on the surface of the cold junction to block incident radiation. However, the detector performed a responsivity of 31.65 V/W, and a detectivity of 1.16×10^8 cm·Hz$^{1/2}$/W, due to the low Seebeck coefficient of Al limits the performance of the thermopile IR detector [10, 11].

Moreover, the traditional four-end beam structure detector is also widely used [12, 13]. Wenjian Ke et al. designed a thermopile infrared detector with this structure and responsivity and detectivity of the detector were calculated as 160.03 V/W and 9.75×10^7 cm·Hz$^{1/2}$/W, respectively. They added a Pt heater to the absorption area of the detector to test its performance. Compared with other thermopile infrared detectors relying on complex equipment, this method is more convenient and simpler. However, four corner areas of the detector are not fully utilized, and the cold junction distribution was relatively concentrated, which decreased the responsivity and detectivity of the detector.

In this paper, a high-responsivity and high-detectivity micromachined thermopile IR detector based on dual-layer thermopile was proposed. The thermocouple arrays were made by n/p-polysilicon and radially distributed on the surface of the support layer. Black silicon material was used for the absorber to improve the infrared absorption rate of the detector. Moreover, we optimized the structure of thermopile IR detector by performing the theoretical calculation and simulation analysis. Finally, we designed the fabrication processes of thermopile IR detector.

II. DESIGN AND OPTIMIZATION

The thermopile IR detector consists of silicon substrate, supporting layers, thermopile and absorber as shown in Fig. 2(a). The thermocouple arrays were radially distributed on a supporting layer made by SiO$_2$. They made use of four corner areas of the detector, resulting in the increasing of absorber area. The SiO$_2$ film can support the thermopile without breaking, and at the same time, it was used as a thermal insulation layer to increase the temperature difference between the hot and the cold junctions of the thermopile. The thermopile materials have a significant impact on detector performance. It was great importance to choose materials with low thermal conductivity, low resistivity and large Seebeck coefficient.[14] Polysilicon

978-1-6654-3008-1/21 $31.00 © 2021 IEEE

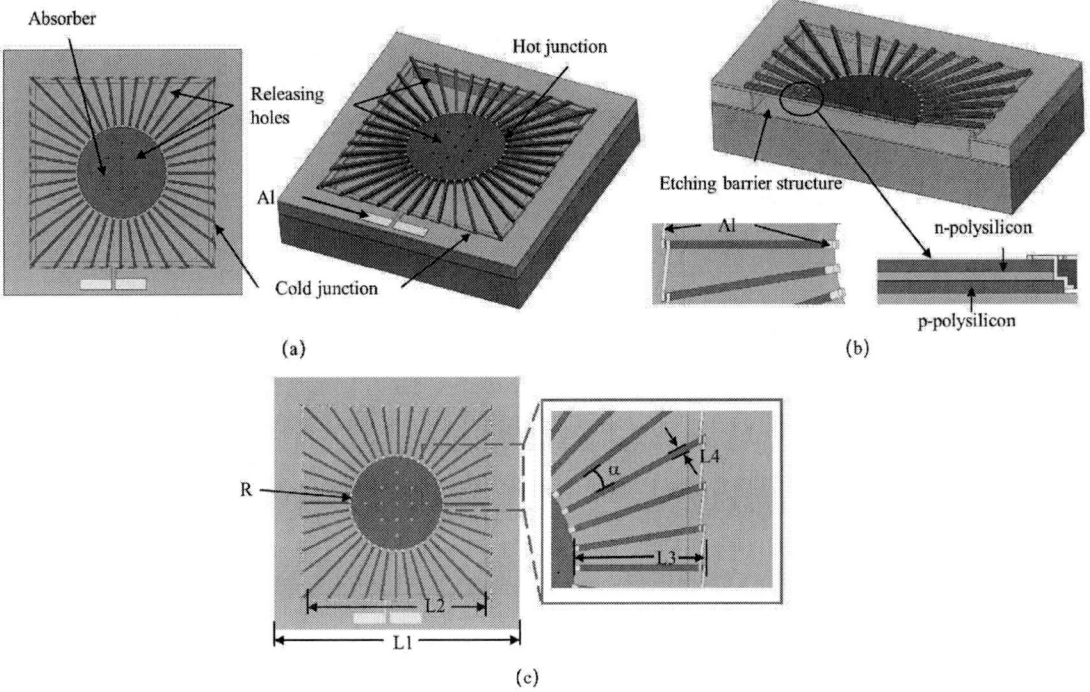

Fig. 2 The design of the thermopile IR detector. (a) The overall design of the detector, consisting of silicon substrate, supporting layers, thermopile and absorber. (b) The design of the double-layer thermocouple structure. (c) The geometric parameters of the thermopile IR detector.

compatible with CMOS process satisfies these requirements. Therefore, a double-layer thermocouple structure was formed by stacking n-type and p-type polysilicon. The hot junctions of the thermopile contact with the infrared absorption zone, while the cold junctions locate on the top of substrate to keep consistent temperature with ambient as shown in Fig. 2(b). The absorber made by black silicon converts incident IR radiation energy into heat energy, increasing the temperature of hot junctions. Both the infrared absorption zone and the hot junctions were suspended to reduce the heat loss. According to the Seebeck effect, the thermopile detector outputs the thermal response voltage, due to the energy conversion of light-heat-electricity.

The responsivity, the detectivity, and the response time are important parameters for evaluating the performance of thermopile IR detectors. Generally, increasing the thermal resistance between the absorber and the cold junction can improve the responsivity and detectivity of the detector, but at the same time it will reduce the response speed. When designing the structural parameters of the detector, it is of great importance to find a balance in this contradiction to achieve better performance. Based on the above analysis, we optimized the structural parameters of the thermopile infrared detector (Fig. 2(c)). The lengths of the square detector and cavities are $L_1 = 1100$ μm and $L_2 = 800$ μm, respectively. The radius of the circular infrared absorber is R = 200 μ m. The number of thermocouple pairs is 40. Thermocouple structural parameters are $L_3 = 230\mu$m, $L_4 = 10\mu$m and $\alpha = 9°$, respectively.

We simulated the heat transfer of the presented IR detector by using the thermal finite element method with ANSYS Workbench 15. The thermal energy converted by the infrared absorption zone was transferred to the substrate through solid heat conduction, thermal radiation, and thermal convection. Since the thermopile infrared detector was vacuum-encapsulated and the working temperature was relatively low, the heat convection and thermal radiation can be ignored. We analyzed the solid heat conduction of the detector. Firstly, the model of the detector was established by SolidWorks, and then the model was meshed after imported to ANSYS. After that, the detector was firstly analyzed in a steady state. We set the boundary conditions of heat transfer on the detector. The temperature of the silicon substrate was 22 °C, while the radiant power density applied on infrared absorption was set to 100W/m². The simulation results are shown in (Fig. 3(a)(b)). The center of the infrared absorber had the highest temperature, and the heat was transmitted to the cold junction along with the thermocouple. The temperature decreases linearly along the direction from the hot junction to the cold junction of the thermocouple. The average temperature difference between the hot junctions and the cold junctions was about 0.352 °C. The responsivity and detectivity of thermopile IR detector were calculated as 179.3 V/W and 2.99×10^8 cm·Hz$^{1/2}$, respectively. We also analyzed the response time of the detector by performing the transient thermal simulation as shown in Fig. 3(c). The response time represents the time when the temperature reaches 63.2% of the saturation level, which was evaluated as 9.6 ms.

978-1-6654-3008-1/21 $31.00 © 2021 IEEE

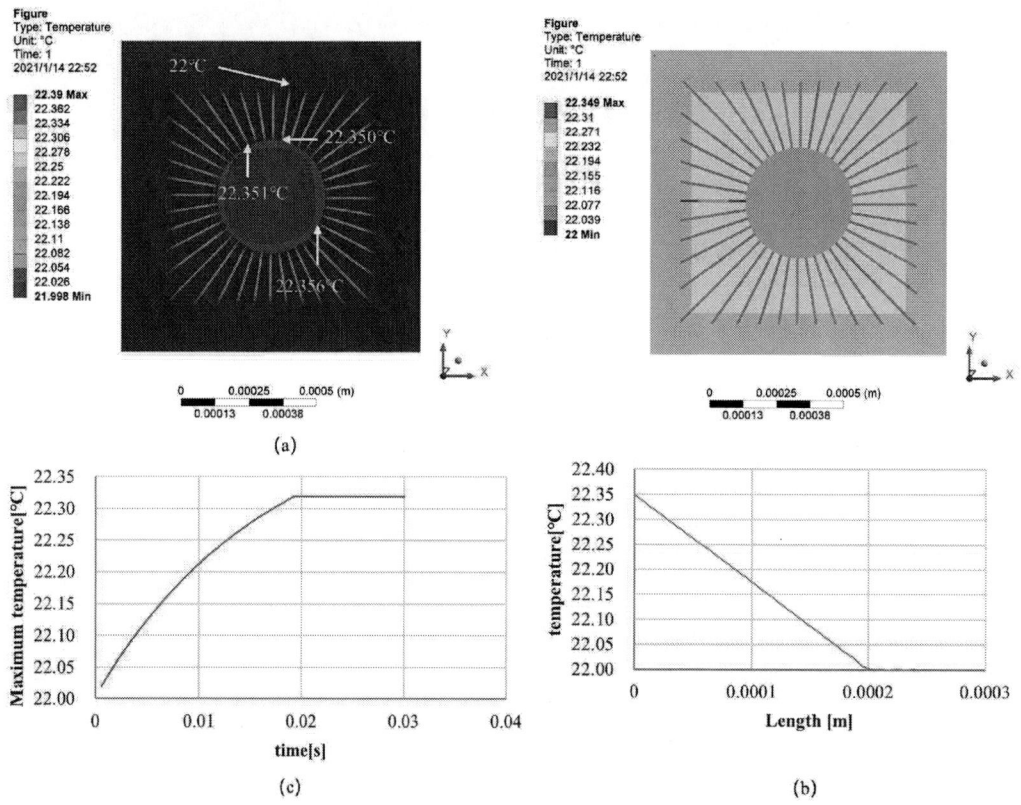

Fig. 3 The thermal ANSYS simulation of thermopile IR detector (a) Temperature distribution of thermopile IR detector. (b) Temperature distribution of the thermocouple, (c) The transient simulation results of response time.

Table 1 Figure of merits of the thermopile IR detector

Figure of merits	Theoretical results	Simulation results	Relative error
Responsivity, R_v(V/W)	173.5	179.3	3.2%
Detectivity, D^*(cm·Hz$^{1/2}$)	2.90×10^8	2.99×10^8	3.0%
Noise equivalent power, NEP (nW/Hz$^{1/2}$)	1.22	1.18	3.3%

We also calculated the responsivity, detectivity, and noise equivalent power theoretically (Table 1), and compared to the simulation results. It can be seen that the relative errors between simulation results and theoretical results are less than 5%.

III. DETECTOR FABRICATION

The CMOS compatible fabrication process of the thermopile IR detector is shown in Fig. 4. First, A layer of SiO₂ film was grown on the silicon wafer by thermal oxidation, and then polysilicon layer was deposited on it by low pressure chemical vapor deposition (LPCVD). The thickness of the SiO₂ and polysilicon were 0.2µm and 1.5µm, respectively (Fig. 4 (a)). Second, in order to build a etch-stop structure, the deep trench structure was formed by reactive ion etching (RIE). Then, the SiO₂ was deposited onto the polysilicon surface by LPCVD to fill the deep trench. The SiO₂ film was used as a dielectric support layer (Fig. 4 (b)). After that, the first layer of the thermocouple was fabricated by depositing a 0.3 µm p-polysilicon using LPCVD, which was then phosphorus doping and patterned (Fig. 4(c)). The 0.2 µm SiO₂ film was deposited by the LPCVD process, which served as an isolation layer between the two polysilicon films (Fig. 4(d)). The second layer of thermocouple with a thickness of 0.3 µm was also deposited using n-polysilicon (Fig. 4(n)). Then the SiO₂ isolation layer at the absorption zone and both ends of the thermocouple were removed (Fig. 4(e)). The Ti/Al film was sputtered and

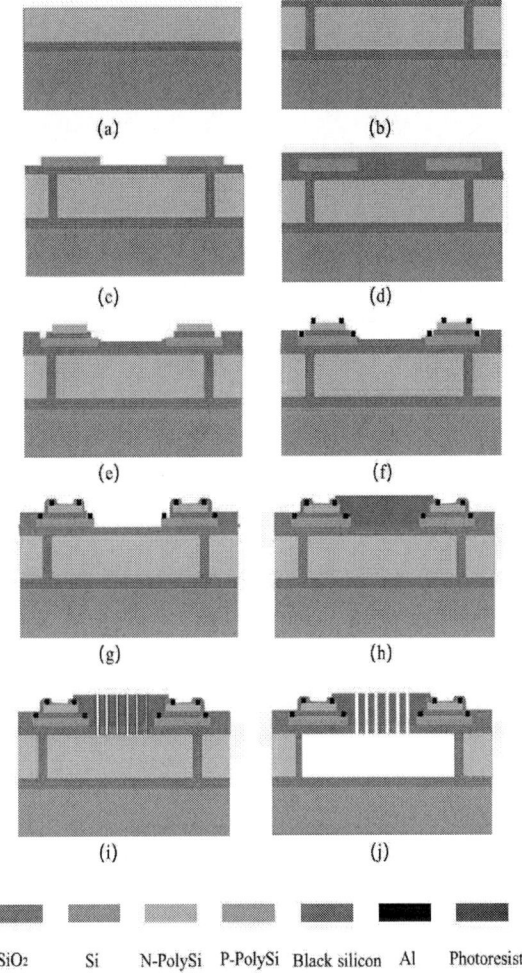

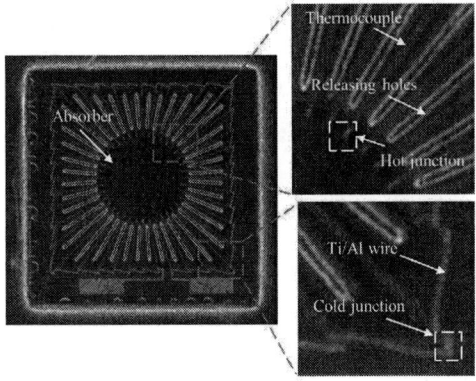

protection photoresist was removed. The fabricated thermopile IR detector chip is shown in Fig. 5. Two close-ups of the junction show the details of the device. We will then perform the experiments to test the performance by measuring its resistance and response voltage under a black body radiation.

Fig. 5 Microscopic pictures of the fabricated device.

IV. CONCLUSION

In summary, we present a high-detectivity IR detector based on dual-layer thermopile. The thermopile IR detector consists of silicon substrate, supporting layers, thermopile and absorber. The thermopile structure and absorber were suspended by XeF_2 dry etching polysilicon. In order to prevent the damage of the detector caused by isotropic etching of XeF_2, the etch-stop structure was utilized. Simulation and theoretical results indicated that the presented detector achieved an excellent detectivity of 2.99×10^8 cm·Hz$^{1/2}$. This thermopile IR detector can be used in environment detection, such as gas and temperature analysis.

SiO₂   Si   N-PolySi   P-PolySi   Black silicon   Al   Photoresist

Fig. 4 The fabrication process of the infrared detector. (a) Si wafer thermal oxidation and deposition of polysilicon by LPCVD. (b) Formation of the deep trench structure. (c) polysilicon deposition, phosphorus doping and patterned. (d) SiO_2 isolation layer deposition by LPCVD. (e) Polysilicon deposition, boron doping and pattern. (f) forming Ti-Al connection. (g) Deposition of SiO_2 passivation layer by PECVD. (h) The formation of black silicon. (i) Releasing hole etching and photoresist protection. (j) XeF_2 dry etching.

patterned, forming a mental connection between the two layers of polysilicon (Fig. 4 (f)). Then a SiO_2 layer with a thickness of 0.4 µm was deposited on the wafer surface by plasma enhanced chemical vapor deposition (PECVD) as a passivation layer to protect thermopile and Ti/Al wire. The passivation layer was etched to expose the hot junctions of the thermocouple and absorption area (Fig. 4(g)). We deposited a layer of polysilicon with a thickness of 0.8 µm using LPCVD, and it was then dry-etched by hydrogen bromide to form black silicon as an absorber (Fig. 4(h)). The releasing holes was etched by RIE. The photoresist was spin-coated and baked to prevent the black silicon from being damaged during the XeF_2 etching process (Fig. 4(i)). Finally, the absorber and thermopile were released by dry etching of polysilicon using XeF_2 (Fig. 4(j)), and the

REFERENCES

[1] A. Van Herwaarden, "Low-cost satellite attitude control sensors based on integrated infrared detector arrays," *IEEE Transactions on Instrumentation and Measurement*, vol. 50, no. 6, pp. 1524-1529, 2001.

[2] T. A. Vincent and J. Gardner, "A low cost MEMS based NDIR system for the monitoring of carbon dioxide in breath analysis at ppm levels," *Sensors and Actuators B: Chemical*, vol. 236, pp. 954-964, 2016.

[3] C. Calaza *et al.*, "An uncooled infrared focal plane array for low-cost applications fabricated with standard CMOS technology," *Sensors and Actuators A: Physical*, vol. 132, no. 1, pp. 129-138, 2006.

[4] D. Xu, Y. Wang, B. Xiong, and T. Li, "MEMS-based thermoelectric infrared sensors: A review," *Frontiers of Mechanical Engineering*, vol. 12, no. 4, pp. 557-566, 2017.

[5] A. Graf, M. Arndt, M. Sauer, and G. Gerlach, "Review of micromachined thermopiles for infrared detection," *Measurement Science and Technology*, vol. 18, no. 7, p. R59, 2007.

[6] V. Leonov, T. Torfs, P. Fiorini, and C. V. Hoof, "Thermoelectric Converters of Human Warmth for Self-Powered Wireless Sensor Nodes," *IEEE Sensors Journal*, vol. 7, no. 5, pp. 650-657, 2007.

[7] L. Goncalves, C. Couto, P. Alpuim, D. Rowe, and J. Correia, "Thermoelectric microstructures of Bi2Te3/Sb2Te3 for a self-calibrated micro-pyrometer," *Sensors and Actuators A: Physical,* vol. 130, pp. 346-351, 2006.

[8] R. Shea, A. Gawarikar, and J. Talghader, "Process integration of co-sputtered bismuth telluride/antimony telluride thermoelectric junctions," *Journal of microelectromechanical systems,* vol. 23, no. 3, pp. 681-688, 2013.

[9] H. Zhou, P. Kropelnicki, and C. Lee, "CMOS compatible midinfrared wavelength-selective thermopile for high temperature applications," *Journal of Microelectromechanical Systems,* vol. 24, no. 1, pp. 144-154, 2014.

[10] K. Wang *et al.,* "Thermopile Infrared Detector with Detectivity Greater Than 10 8 cmHz (1/2)/W," *Journal of Infrared, Millimeter, and Terahertz Waves,* vol. 31, no. 7, pp. 810-820, 2010.

[11] G. Li, S. Liu, Y. Piao, B. Jia, Y. Yuan, and Q. Wang, "Joint improvement of conductivity and Seebeck coefficient in the ZnO: Al thermoelectric films by tuning the diffusion of Au layer," *Materials & Design,* vol. 154, pp. 41-50, 2018.

[12] C. Lian-Min, Z. Ya-Zhu, S. Shi-Jiao, G. Hui, and Z. Zhen, "Structure design and test of MEMS thermocouple infrared detector," *Microsystem Technologies,* vol. 24, no. 5, pp. 2463-2471, 2018.

[13] W. Ke, Y. Wang, H. Zhou, T. Li, and Y. Wang, "Research on Self-Test Method Based on Thermopile Infrared Sensor," in *2018 IEEE SENSORS,* 2018: IEEE, pp. 1-4.

[14] A. D. McConnell, S. Uma, and K. E. Goodson, "Thermal conductivity of doped polysilicon layers," *Journal of Microelectromechanical Systems,* vol. 10, no. 3, pp. 360-369, 2001.

Proceedings of the 16th Annual IEEE International
Conference on Nano/Micro Engineered and Molecular Systems
April 25-29, 2021

A High-Precision Mode Matching Method for Rate-Integrating Honeycomb Disk Resonator Gyroscope*

Sheng Yu, Xuezhong Wu, Xiang Xi, Jiangkun Sun, Qingsong Li, Dingbang Xiao and Yongmeng Zhang*

Abstract— This paper presents a precise mode matching method based on the quadrature control forces for the honeycomb disk resonator gyroscope working on the rate-integrating mode. Rate-integrating gyroscope (RIG) allows the pattern angle to precess freely to observe the output forces of quadrature control loop in specific directions which contains the full information of frequency mismatch for mode matching process. This method is proved by experiments to be highly effective and precise, reducing the frequency mismatch from 4Hz to about 5mHz. And the gyroscope achieves an angle-dependent bias drift (ADB) of 0.05°/s and a threshold value of less than ±0.005°/s after mode matching. Such tuning voltages are also suitable for the MEMS rate gyroscope (RG).

I. INTRODUCTION

MEMS vibratory gyroscope has wide market in compact and low-cost inertial applications because of its highly-integration design and batch-processing technic [1]. Its working principle is based on the Coriolis coupling between two perfectly matched degenerated modes of an ideal MEMS resonator. However, mass and stiffness imperfection caused by fabrication tolerance limits, assembly errors and material inhomogeneity will lead to frequency mismatch of two modes. This greatly degrades the performance of RG by affecting its sensitivity and resolution [2], as well as RIG by causing quadrature error and rate threshold [3]. Compared with the mass tuning method of selective polysilicon deposition [4] or laser trimming [5], a stiffness tuning method relying on the effect of electrostatic spring softening is more adjustable, flexible and allows real-time or closed-loop implementation [6]. The key to electrostatic stiffness tuning is to apply accurate DC tuning voltages, which can be determined in RG by comparing Bode plots of two modal frequency responses [6], using phase relationship between quadrature and drive signals [7] or monitoring amplitude of residual quadrature signal [8]. Nevertheless, RIG allows the pattern angle to stay at arbitrary directions, which inspires more convenient and accuracy mode matching approaches, such as evaluating the stiffness imperfections from the recorded resonance frequencies or minor-axis amplitudes in different pattern angles with the help of a rate table [9].

In this paper, a control scheme of RIG based on Lynch's model [10] is established and a new mode matching method based on self-precession and quadrature-force observation is presented for the honeycomb disk resonator gyroscope. A high mode matching accuracy of about 5mHz is achieved by this method, which greatly improve the performance of RIG by reducing the bias drift and improving the rate threshold.

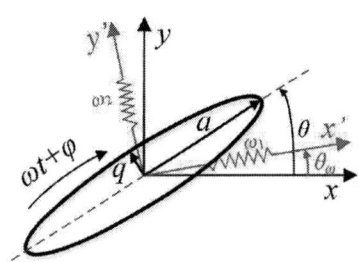

Fig. 1. The elliptical orbit of the motion of RIG and frequency mismatch with the principal-stiffness-axis angle θ_ω and frequency split ($\omega_1 \neq \omega_2$).

II. THEORY OF MODE MATCHING METHOD

A. Theory of Stiffness Tuning

When the resonator has the frequency mismatch with the principal-stiffness-axis angle θ_ω and frequency split ($\omega_1 \neq \omega_2$) as shown in Fig. 1, the stiffness matrix can be described as

$$\begin{bmatrix} \omega^2 - \omega\Delta\omega\cos2\theta_\omega & -\omega\Delta\omega\sin2\theta_\omega \\ -\omega\Delta\omega\sin2\theta_\omega & \omega^2 + \omega\Delta\omega\cos2\theta_\omega \end{bmatrix} \quad (1)$$

where

$$\omega^2 = \frac{\omega_1^2 + \omega_2^2}{2}, \omega\Delta\omega = \frac{\omega_1^2 - \omega_2^2}{2}$$

The electrostatic force between a pair of parallel-plate electrodes under a constant DC voltage V_{dc} can be given as

$$f \approx f_{dc} + kx, f_{dc} = \frac{C_0}{2d_0}V_{dc}^2, k = \frac{C_0}{d_0^2}V_{dc}^2 \quad (2)$$

where C_0 and d_0 are the initial capacitance and gap of the parallel-plate electrodes, x is the displacement of the moveable electrode and k represents the magnitude of the spring softening.

Therefore, it is possible to create certain spring softening effects by applying suitable DC voltages on corresponding electrodes for modifying the stiffness matrix to be perfect. For example, a pair of spring softening with the magnitudes of k_1 and k_2 in the pattern angle of 45° and 0° respectively will transform the stiffness matrix into

$$\begin{bmatrix} \omega^2 - \omega\Delta\omega\cos2\theta_\omega - k_2/m & -\omega\Delta\omega\sin2\theta_\omega - k_1/m \\ -\omega\Delta\omega\sin2\theta_\omega - k_1/m & \omega^2 + \omega\Delta\omega\cos2\theta_\omega \end{bmatrix} \quad (3)$$

where m represents the equivalent mass.

---

*Resrach supported by the National Key R&D Program of China (2018YFB2002300), National Natural Science Foundation of China (51905538) and Natural Science Foundation of Hunan Province (2020JJ2033). (Corresponding author: Dingbang Xiao, Qingsong Li)
All authors are with the College of Intelligence Science and Technology,

National University of Defense Technology, Changsha, 410073, China.
*Contacting Author: Yongmeng Zhang is with College of Intelligence Science and Technology, National University of Defense Technology, Changsha, 410073, China. (phone: +86 15974159741; email: zhangym@nudt.edu.cn).

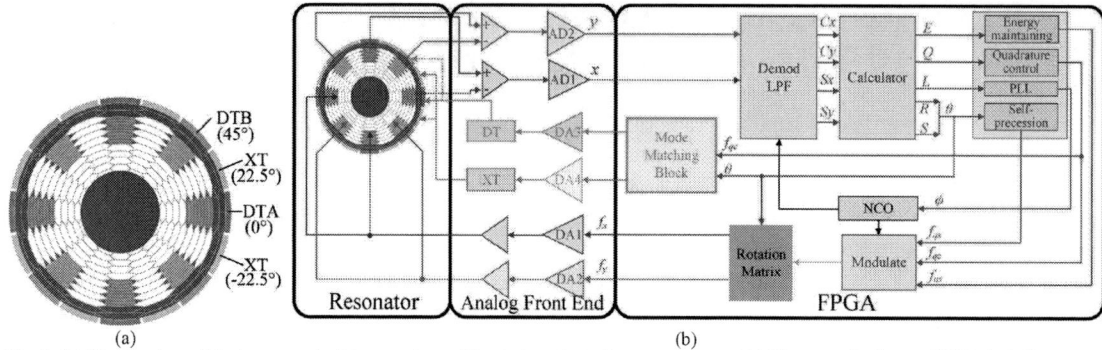

Fig. 2. (a) The structure of the honeycomb disk resonator and the tuning electrodes arrangements. (b) The control scheme of RIG, including energy maintaining loop, quadrature control loop, phase locked loop (PLL), self-precession loop and mode matching block.

According to (3), in the case of $\Delta\omega<0$ and $\theta_\omega>0$, when the magnitude of k_1 satisfy

$$k_1 = -m\omega\Delta\omega\sin 2\theta_\omega \qquad (4)$$

the alignment of the principal-stiffness axis can be realized on no dependency of the magnitude of k_2, namely

$$\theta'_\omega = \frac{1}{2}\arctan\left(\frac{\omega\Delta\omega\sin 2\theta_\omega + k_1/m}{\omega\Delta\omega\cos 2\theta_\omega + k_1/2m}\right) = 0 \qquad (5)$$

After that, adjusting the magnitude of k_2 has no influence on the position of the principal-stiffness axis and when the magnitude of k_2 satisfy

$$k_2 = -2m\omega\Delta\omega\cos 2\theta_\omega \qquad (6)$$

the elimination of frequency split can be realized, namely

$$\frac{\omega_1'^2 - \omega_2'^2}{2} = -\omega\Delta\omega\cos 2\theta_\omega - \frac{k_2}{2m} = 0 \qquad (7)$$

B. Theory of Frequency-Mismatch Identification

The motion of the gyroscope working on rate-integrating mode is an elliptical orbit with the major-axis a, minor-axis q, pattern angle θ and vibration phase $\omega t+\varphi$, as shown in Fig. 1. According to Lynch's RIG model [10], the dynamic equation of quadrature error can be expressed as

$$\dot{Q} = -\frac{2}{\tau}Q - \Delta\omega\sin 2\left(\theta - \theta_\omega\right)E + \frac{\sqrt{E}}{\omega}f_{qc} \qquad (8)$$

where

$$Q = 2aq, E = a^2 + q^2$$

Q represents the quadrature error, E represents energy, τ is the decay-time and f_{qc} is the in-phase component of the force applied along the minor-axis which can be used to control the quadrature error.

Under the control strategies of rate-integrating mode, the energy is kept constant as E_0 and the quadrature error is suppressed at zero. Therefore, the stable output forces of the quadrature control loop can be expressed as

$$f_{qc} = \omega\Delta\omega\sin 2\left(\theta - \theta_\omega\right)\sqrt{E_0} \qquad (9)$$

According to (9), the information of frequency mismatch with the principal-stiffness-axis angle and frequency split can be obtain through observations of f_{qc} in different pattern angles. More specially, in the pattern angle of 0°, the magnitude of f_{qc} reflects the misalignments of the principal stiffness axis, and in the pattern angle of $45°+\theta_\omega$, the magnitude of f_{qc} reflects the level of the frequency split, which are shown in (10).

$$\begin{cases} f_{qc}\big|_{\theta=0°} = -\omega\Delta\omega\sin 2\theta_\omega\sqrt{E_0} \propto \sin 2\theta_\omega \\ f_{qc}\big|_{\theta=45°+\theta_\omega} = \omega\Delta\omega\sqrt{E_0} \propto \omega\Delta\omega \end{cases} \qquad (10)$$

III. IMPLEMENTATION OF MODE MATCHING METHOD

The mode matching method is implemented on a honeycomb disk resonator gyroscope, whose structure and tuning electrodes arrangements are shown in Fig. 2a. For n=2 working modes of the gyroscope, the cross-axis stiffness tuning (XT) electrodes are placed at ±22.5° which correspond to the spring softening with the magnitudes of k_1 in the pattern angle of ±45°, and the direct (frequency) tuning (DT) electrodes of two modes are placed at 0° (DTA) or 45° (DTB) which correspond to the spring softening with the magnitudes of k_2 in the pattern angle of 0° or 90°.

The control scheme of RIG is implemented in a Xilinx FPGA, including energy maintaining loop, quadrature control loop, phase locked loop (PLL), self-precession loop and mode matching block, as shown in Fig. 2b. The self-precession loop can be used to rotate or keep the pattern angle at a required position without the turntable. The mode matching block has two inputs of pattern angle θ and control force f_{qc} of quadrature loop to identify the frequency mismatch by observation f_{qc} in special pattern angle, as well as two outputs of DA3 and DA4 to tune the stiffness matrix by adjusting the DT and XT voltages.

According to the theory of mode matching method, the practical operation contains two steps. In the first step, by using the self-precession loop to make the pattern angle stay at 0° where the magnitude of f_{qc} reflects the misalignments of the principal stiffness axis and then iteratively adjusting XT voltage to make f_{qc} converge to zero, the cross components in stiffness matrix can be canceled to realize the alignment of principal-stiffness axis ($\theta'_\omega=0$).

978-1-6654-3008-1/21 $31.00 © 2021 IEEE

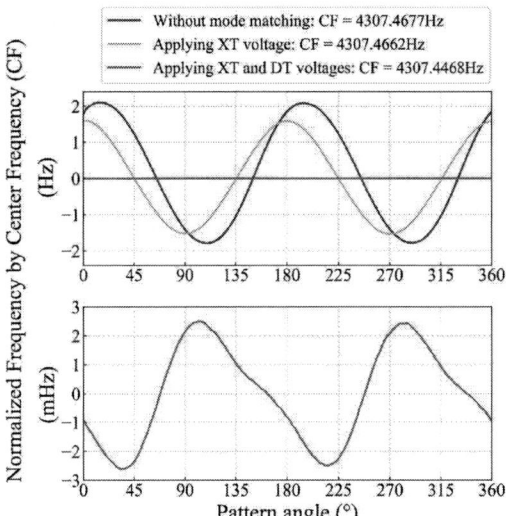

Fig. 3. Frequency mismatch in three cases: without mode matching, applying XT voltage and applying both XT and DT voltages. A final matching accuracy of 5mHz is achieved, with a reduction of approximately 99.8% from 4Hz.

Subsequently in the second step, by using the self-precession loop to make the pattern angle stay at 45° where the magnitude of f_{qc} reflects the level of the frequency split and then iteratively adjusting DT voltage to make f_{qc} converge to zero, the diagonal components in stiffness matrix can be balanced to realize the elimination of frequency split ($\omega'_1 = \omega'_2$).

IV. EXPERIMENTAL RESULTS

Following the operation steps of the mode matching method illustrated in the last chapter, a XT voltage of 5.135V and a DT voltage of 23.768V are obtained to apply in the corresponding electrodes for a honeycomb disk resonator gyroscope. The mode matching result is shown in Fig. 3, by recording the frequency output of PLL in different pattern angles. The initial frequency mismatch of the resonator is about 4Hz. After applying XT voltage, the maximum and minimum value of frequency occurs at 0° and 90° respectively, proving that the alignment of the principal-stiffness axis has been realized and indicating that the DT voltage should be applied in the DTA electrode. After applying both XT and DT voltages, a final mode matching accuracy of 5mHz is achieved, with a reduction of approximately 99.8% from 4Hz.

Enormous frequency mismatch leads to large quadrature error which cannot be controlled by f_{qc} and brings extra drifts. As shown in Fig. 4, the bias drifts in different pattern angle are measured. After mode matching, the angle-dependent bias drifts (ADB) of RIG reduces about 97.5% from 2°/s to 0.05°/s and almost become a sinusoidal shape mainly from anisodamping [11].

Fig. 5 demonstrates the threshold value test for RIG under the sequential rate inputs of 0°/s and ±0.005°/s in a time interval of 30s. The rate outputs of RIG are obtained by the derivation operation on the output angles. Without mode matching, the rate output of RIG has a jump and then decay to zero gradually under small rate input. The pattern angle

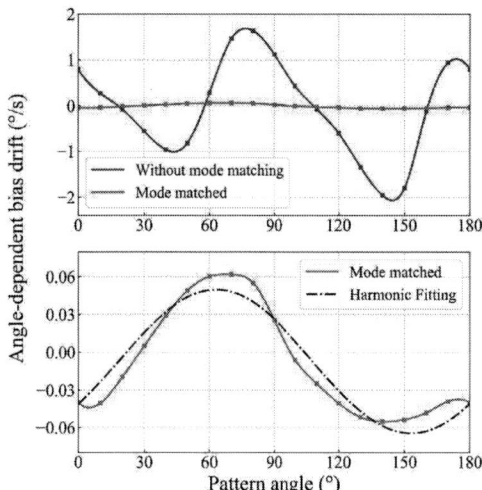

Fig. 4. Angle-dependent bias drifts (ADB) of RIG without mode matching and under matched modes. ADB reduces about 97.5% from 2°/s to 0.05°/s after mode matching.

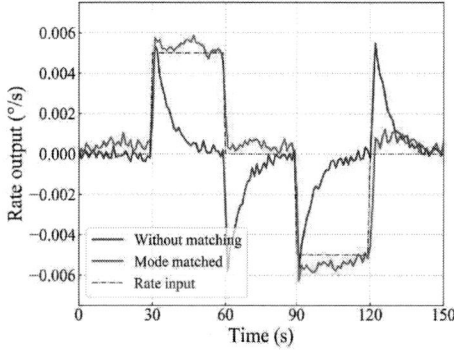

Fig. 5. The threshold value test for RIG under rate inputs of ±0.005°/s without mode matching and with matched modes. Without mode matching, RIG tends to stall at certain pattern angle under small rate input. Matched modes can improve the performance of RIG with a threshold value of less than ±0.005°/s.

of RIG tends to stall at certain position and stop precession. While after mode matching, RIG has obvious responses to these rate inputs, which confirms a rate threshold value of less than ±0.005°/s.

Mode matching operations are performed under different vibrating amplitudes which can be represented by the peak-to-peak voltages of AD1 signals of RIG vibrating in the pattern angle of 0°. Fig.6 shows the recorded frequency outputs of PLL and force outputs of quadrature control loop in different pattern angles after mode matching. As the vibrating amplitudes of RIG rise, the residual frequency mismatch significantly increase from a few mHz to about 20mHz with increasing residual quadrature forces from a few mV to about 90mV. Moreover, the waveform of residual frequency mismatch consists of both 2θ-harmonic and 4θ-harmonic, and the waveform of residual quadrature force is mainly made up of 4θ-harmonic. Such phenomenon may be caused by the nonlinearity of electrostatic actuation [12]. Therefore, future work will focus on the elimination of the nonlinearity to further promote the matching accuracy.

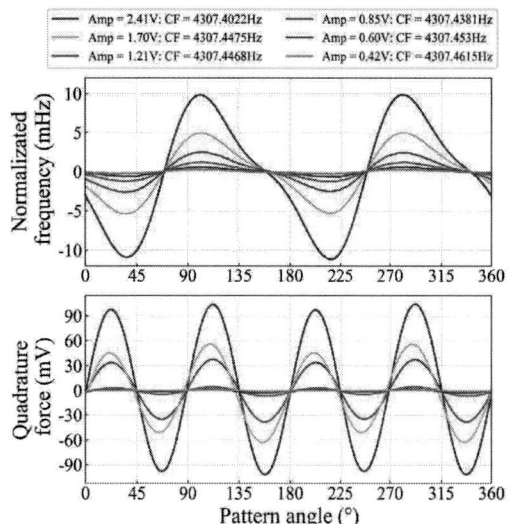

Fig. 6. Residual frequency mismatch and quadrature output force after mode matching under different vibration amplitudes of RIG.

V. CONCLUSION

This paper presents a high-precision mode matching method for rate-integrating honeycomb disk resonator gyroscope. With the help of self-precession of rate-integrating control scheme, the output forces of quadrature control loop in different pattern angles can be observed to identify the frequency mismatch, which are the indicators for mode matching operations. A final matching accuracy of 5mHz is achieved, with the reduction of bias drift and the improvement of rate threshold of RIG. The tuning voltages obtained from this method are also suitable for the gyroscopes working on the force-rebalance mode, which supplies an accurate, efficient and convenient alternative of mode matching approaches.

REFERENCES

[1] G. Zhanshe, C. Fucheng, L. Boyu, C. Le, L. Chao, and S. Ke, "Research development of silicon MEMS gyroscopes: a review," *Microsystem Technologies*, vol. 21, no. 10, pp. 2053-2066, 2015.

[2] Z. X. Hu, B. J. Gallacher, J. S. Burdess, S. R. Bowles, and H. T. D. Grigg, "A systematic approach for precision electrostatic mode tuning of a MEMS gyroscope," *Journal of Micromechanics & Microengineering*, vol. 24, no. 12, p. 125003, 2014.

[3] D. D. Lynch, "MRIG frequency mismatch and quadrature control," in *2014 International Symposium on Inertial Sensors and Systems (ISISS)*, 2014.

[4] D. Joachim and L. Lin, "Characterization of selective polysilicon deposition for MEMS resonator tuning," *Microelectromechanical Systems Journal of,* vol. 12, no. 2, pp. 193-200, 2003.

[5] M. A. Abdelmoneum, M. M. Demirci, Y. W. Lin, and T. C. Nguyen, "Location-dependent frequency tuning of vibrating micromechanical resonators via laser trimming," in *IEEE International Frequency Control Symposium & Exposition*, 2010.

[6] B. J. Gallacher, J. Hedley, J. S. Burdess, A. J. Harris, A. Rickard, and D. O. King, "Electrostatic Correction of Structural Imperfections Present in a Microring Gyroscope," *Journal of Microelectromechanical Systems,* vol. 14, no. 2, pp. p.221-234, 2005.

[7] S. Sonmezoglu, S. E. Alper, and T. Akin, "An Automatically Mode-Matched MEMS Gyroscope With Wide and Tunable Bandwidth," *Journal of Microelectromechanical Systems,* vol. 23, no. 2, pp. 284-297, 2014.

[8] A. Sharma, M. F. Zaman, M. Zucher, and F. Ayazi, "A 0.1°/HR bias drift electronically matched tuning fork microgyroscope," in *2008 IEEE 21st International Conference on Micro Electro Mechanical Systems*, 2008.

[9] Z. Hu and B. J. Gallacher, "Precision mode tuning towards a low angle drift MEMS rate integrating gyroscope," *Mechatronics,* pp. 306-317, 2017.

[10] D. D. Lynch, "Vibratory Gyro Analysis by the Method of Averaging," in *Proc. 2nd Saint Petersburg Int. Conf. on Gyroscopic Technology and Navigation*, pp. 26-34, May 1995.

[11] Z. Hu and B. Gallacher, "Control and damping imperfection compensation for a rate integrating MEMS gyroscope," in *Inertial Sensors & Systems Symposium*, 2015.

[12] S. H. Nitzan, P. Taheri-Tehrani, M. Defoort, S. Sonmezoglu, and D. A. Horsley, "Countering the Effects of Nonlinearity in Rate-Integrating Gyroscopes," *IEEE Sensors Journal,* vol. 16, no. 10, pp. 3556-3563, 2016.

Gap in pagination due to unavailable papers.

Pages 1583-1588

Capacitive Stretchable Strain Sensors Based on Wavy-Structured Metal Electrodes

Siqi Yu[1], Hongyu Yu[1*]

Abstract—**Stretchable strain sensors are of great importance in developing intelligent robotics, virtual reality, and e-skins. Here, we propose a capacitive stretchable strain sensor based on wavy structured interdigital metal electrodes encapsulated in elastomers, where wavy design not only endows metal electrodes with stretchability, ensuring long-term stability, but also provides sidewalls to form 3D nonplanar capacitor. Distances between electrodes change with the elongation when stretching, resulting in capacitance variation with high sensitivity. The devices are fabricated, and the sensing properties are characterized. The simulation on device deformation, stress distribution, and sensitivity is also performed for validation and comparison. The presented stretchable strain sensor is reliable, suitable with high-volume manufacturing, and promising in various applications.**

I. INTRODUCTION

Strain sensors have attracted significant interest because of their broad applications, such as health monitoring, wearable electronics, and soft robotics[1, 2]. High sensitivity, stretchability, and stability are essential [3]. Conventional strain sensors, such as strain gauges made of metal foil or ceramics, exhibit high sensitivity. But these sensors are rigid with little stretchability, resulting in low dynamic range and limited applications. In recent years, stretchable strain sensors have gathered momentum in development due to their extended detection range and compatibility with the human body [4, 5] for wearable application. Carbon black, carbon nanotube, graphene, nano silver wires, and other conductive nanoparticles, wires, and fibers are commonly used together with silicone elastomer to achieve capacitance or resistance readout of stretchable strain sensors [6-10]. However, there are uniformity and stability issues for the mixture of nanoparticles and elastomers [11, 12]. This type of combination has demonstrated significant signal drift in long-term operation; nanoparticles may dislocate and lose their bonds within the elastomer, and random cracks may also occur during stretching, especially under cycling operation, causing non-ideal changes of electrical conductivity or dielectric coefficient [13-15]. For instance, Kim et al. designed wearable strain sensors using stretchable graphene PDMS composite film, utilizing crack junction mechanisms to realize sensing property[16]. In their design, cracks are essential in their sensing mechanism but will also cause instability in the long run. Therefore, a long-term stable sensing mechanism needs to be developed.

PVD sputtering deposited platinum is a mature material in MEMS fabrication process and is stable in various environments. But its non-stretchability limits its application in stretchable research. Thus, a particular structure is needed to apply this rigid metal material in stretchable devices [17-19]. Parylene C is a kind of bio-compatible thin film material which can be chemical vapor deposited uniformly on any complex surface at a low cost. It is usually applied as dielectric and protection film against moisture and chemical attack. A specifically designed stretchable structure and Parylene C are utilized in this paper to realize platinum-based electrodes in practical stretchable strain sensors.

Here, a capacitive strain sensor is designed composed of a stretchable interdigital wavy-structure platinum electrode array as the sensing readout mechanism, and silicone elastomer as the packaging material. The metal electrodes sit on the sidewalls of wavy structures and interdigitally face each other. Distances between electrodes can be changed under strain, resulting in the variation of the device's capacitance. In the as-designed device, 3D wavy structure endows interdigital metal electrodes with stretchability. At the same time, interdigital metal electrodes are controllable, uniform, and stable for a long time.

II. EXPERIMENTAL

A. Wavy structure

As shown in Fig.1, a 0.5 μm silicon dioxide layer was thermally grown on a bare silicon wafer, patterned using photolithography afterward. Anisotropic Oxide Etching was used to etch SiO_2, and TMAH solution was then utilized to etch 27 μm deep grooves to form wavy structures on the wafer, which corresponded to 50 μm opening of grooves. A 2 μm photoresist layer was spray-coated onto the wavy structure to smooth the sharp corners, which also worked as a sacrificial layer to help the subsequent peeling-off process.

B. Interdigital electrode

Then, a 3 μm Parylene C layer was coated onto the surface. Another photoresist layer was spray-coated and then patterned, followed by sputtering a 50 nm/100 nm TiW/Pt layer on top and using a lifting-off process to obtain metal electrodes. Another 2 μm Parylene C layer was coated onto the whole device, leaving the end of the electrodes exposed.

C. Encapsulation

An 80-μm PDMS layer was spin-coated onto the device after surface modification via 5 min O_2 plasma etching, which changes the Parylene-C surface to hydrophilic and thins down Parylene C layer to 0.6 μm.

---

[1]Department of Mechanical and Aerospace Engineering, Hong Kong University of Science and Technology, Hong Kong, China

*Contacting Author: Hongyu Yu. is with the Department of Mechanical and Aerospace Engineering, Hong Kong University of Science and Technology, Kow Loon, Hong Kong, China (e-mail:hongyuyu@ust.hk).

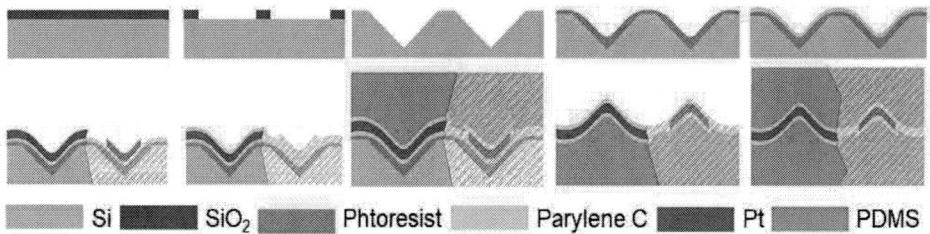

Figure 1. Fabrication process flow of capacitive stretchable strain sensor.

The sample was then immersed into acetone to remove the photoresist sacrificial layer and released the device from the silicon substrate. After flipping, it was tiled on a new bare wafer and exposed to another 10 min O_2 plasma for the same purpose as above. Finally, the second 80 μm PDMS layer was spin-coated on another side of the device.

III. RESULTS AND DISCUSSION

A. Device structure and its working principle

Fig. 2(a) describes the schematic of the proposed capacitive strain sensor. It is one of 200 units of the whole device. Interdigital electrodes are in the middle of the sample, embedded by 0.6 μm parylene and 80 μm PDMS on both sides, encapsulating a nearly symmetric device. PDMS provides stretchability for the device. Parylene C bonds to the metal electrodes, protecting metal from the damage while stretching. 3D wavy-structure interdigital metal electrodes are the functional elements, sensing strain and transferring it to capacitance signal. There are five electrodes shown in fig. 2 (a) as positive, negative, positive, negative, positive, respectively or in contrast. Every two adjacent electrodes form one unit of the capacitor as explained in fig. 2 (a). Upon stretching, the angle between every two electrodes increases. According to the capacitance formula,

$$ C = \varepsilon \frac{A}{d} \qquad (1) $$

$$ C = \int dC = \frac{\varepsilon_0 \varepsilon_r w}{\alpha} \int_{R_1}^{R_2} \frac{dr}{r} = \frac{\varepsilon_0 \varepsilon_r w}{\alpha} \log \frac{R_2}{R_1} \qquad (2) $$

Where ε_0 is vacuum dielectric permittivity, ε_r is relative dielectric constant, w is the width of the capacitor, α is the angle between two electrodes, R_2 and R_1 refers to the radius of the outer circle and inner circle of the capacitor when the point of the two electrodes' intersection is presumed to be the center of the circle. Thus, capacitance decreases as the angle between two electrodes increases.

Fig. 2(b) shows the photograph of a 20 mm long 2 mm wide stretchable strain sensor. Its detailed electrode structure is shown in Fig. (c, d), including the device morphology after the lift-off process and interdigitated electrodes on the wavy surface's top side. The shortest distance between positive electrode and the negative electrode is 14 μm. The depth of the wavy structure is around 30 μm.

B. One Cycle Performance of The Device

Fig. 3(a) describes the capacitance change when stretching and releasing the sample. The output signal is $\Delta C/C_0$, C_0 is initial capacitance value, and ΔC is capacitance change before and after stretching. Capacitance is recorded every 2.5% strain. For the sample with 50 μm unit width, as

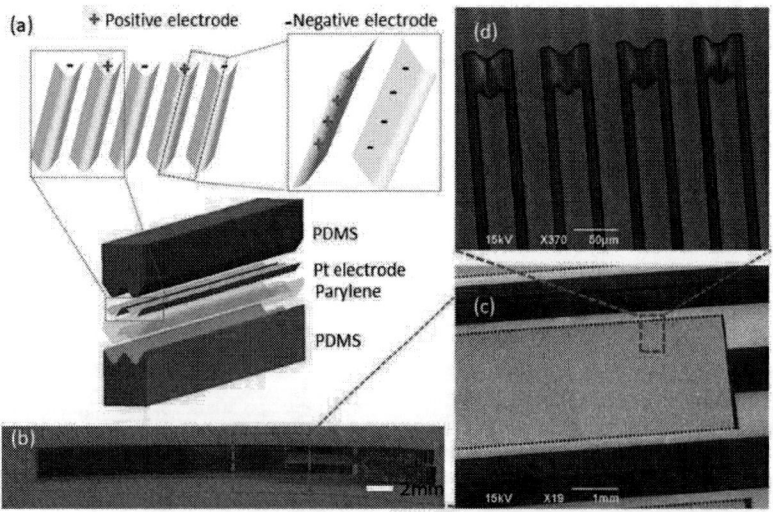

Figure 2. (a) Schematic of one unit of the strain sensor; (b) Fabricated strain sensors right after releasign from the substrate; (c) Scanning electronic microscope picture of the specified part; (d) Zoom-in view of the area framed in (c).

978-1-6654-3008-1/21 $31.00 © 2021 IEEE

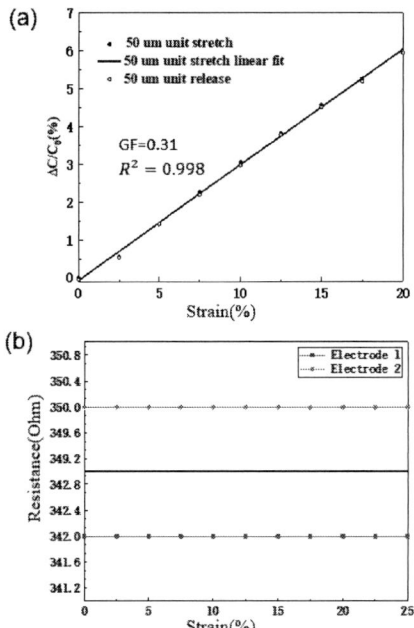

Figure 3. (a) Relationship between strain and capacitance change and its corresponding linear fit result. (b) Relationship between the resistance of the two connecting electrodes for interdigital circuit

strain changes from 0 to 20.0%, capacitance change varies from 0 to 6%. The capacitance change responds to strain linearly with the linear function R^2=0.998. Gauge factor (GF) is the slope of the linear fit curve, representing its sensitivity.

$$GF = \frac{C - C_0}{C_0 \varepsilon} \qquad (3)$$

The measured GF equals 0.31. The sample almost recovers totally in one stretching and releasing cycle with 1.49% hysteresis value.

Fig. 3(b) shows the stability of the interdigital electrodes while stretching. It depicts resistance of two connecting

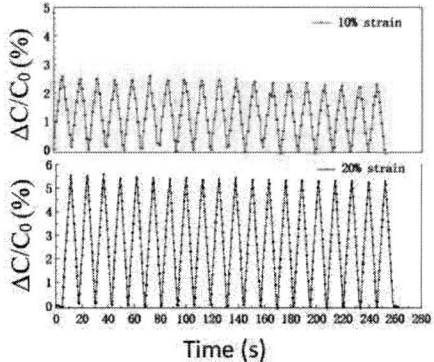

Figure 4. Capacitance change under specific strain range in multiple cycles. (c) Capacitance change under 20% peak strain at different frequencies. (d) Durability test for 20 cycles with 10% and 20% strain.

electrodes in the interdigital circuit. As strain rises from 0 to 25%, the resistance of positive connecting electrode of the interdigital circuit remains at 342.0 Ohm, and the resistance of negative connecting electrode of the interdigital circuit remains at 350.0 Ohm. The stability of the electrode resistance reflects the stability of the 3D wavy-structure electrodes, which further demonstrates the stability of the capacitive strain sensor to some degree.

C. Dynamic characteristics

Fig. 4 represents the sample with its capacitance change over 20 cycles under 10% and 20% strain. No significant difference could be found over 20 cycles of testing, indicating the stability of the proposed sensor. The excellent stability of the sensor results from its sensing principle. PDMS encounters the stress from outside first, and experiences strain along with the stress, driving the 3D wavy-structure interdigital electrodes stretching itself to become more planar. Subsequently, while releasing stress from outside, PDMS recovers to its initial state. This drives the stretched 3D wavy-structure interdigital electrodes to return to their initial state. Therefore, strain response only depends on the structure change of the 3D wavy-structure electrodes, and electrode thin films do not experience much tensile strain, leading to the long-term integrity and stability of the sensor.

D. Simulation Results

The finite element analysis simulation is conducted to study the relationship between capacitance change and strain theoretically. The two ends of the eight wavy structure units are pulled to a specific location, and then its stress

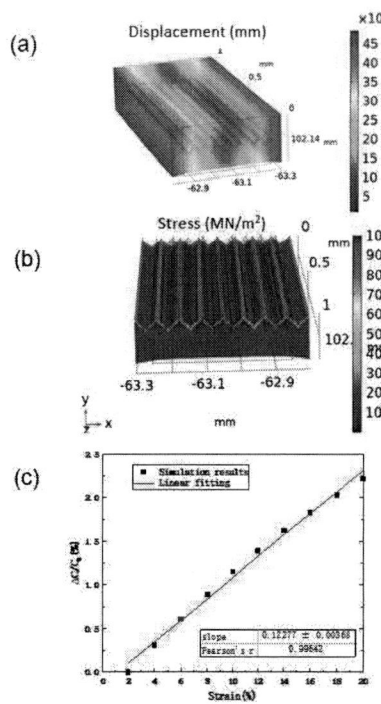

Figure 5. Simulation results of the displacement(a), stress distribution(b) and capacitance change(c).

distribution and capacitance change are studied. Simulation results in Fig. 5(a,b) illustrate the device's displacement and stress distribution. The stress is mainly distributed on the metal electrodes, but well below the metal's tensile strength. The relationship between the capacitance variation and strain is also simulated and is plotted in Fig. 5(c). The trend is consistent with the experiment, but the amplitude is different because only eight wavy structure units are included in the simulation, while the whole set of 200 units exceeds the simulation power.

IV. CONCLUSION

In summary, a capacitive strain sensor with novel 3D interdigital electrodes on the wavy-structured side walls is designed and fabricated successfully. PDMS endows the device with stretchability. Parylene C protects the metal electrodes from damage while stretching. Wavy-structured interdigital metal electrodes serve as the functional element to sense strain change based on the angle variation between wavy structure sidewalls. A low hysteresis (1.49%), high linearity (R^2=0.998), and high stability, are observed from the capacitive strain sensor. The high stability is attributed to the sensing principle, where structure change results in capacitance change.

ACKNOWLEDGMENTS

We thank Nano Fabrication Facility at HKUST for the support with the fabrication process. This work was supported by grants from the Innovation and Technology Commission (project ITS/051/18) of HKSAR, Foshan HKUST Projects (Grant No.: FSUST19-FYTRI05), and Zhongshan Municipal City Introducing Innovation Projects from High-end Scientific Research Institutions (Grant No.: ZSST21EG07).

REFERENCES

[1] A. Qiu, P. Li, Z. Yang, Y. Yao, I. Lee, and J. Ma, "A Path Beyond Metal and Silicon:Polymer/Nanomaterial Composites for Stretchable Strain Sensors," *Advanced Functional Materials*, vol. 29, no. 17, p. 1806306, 2019.

[2] M. Amjadi, K.-U. Kyung, I. Park, and M. Sitti, "Stretchable, Skin-Mountable, and Wearable Strain Sensors and Their Potential Applications: A Review," *Advanced Functional Materials*, vol. 26, no. 11, pp. 1678-1698, 2016.

[3] G. Ge, W. Huang, J. Shao, and X. Dong, "Recent progress of flexible and wearable strain sensors for human-motion monitoring," *Journal of Semiconductors*, vol. 39, no. 1, p. 011012, 2018.

[4] A. Frutiger *et al.*, "Capacitive soft strain sensors via multicore-shell fiber printing," *Adv Mater*, vol. 27, no. 15, pp. 2440-6, Apr 17 2015.

[5] Y. Li *et al.*, "Continuously prepared highly conductive and stretchable SWNT/MWNT synergistically composited electrospun thermoplastic polyurethane yarns for wearable sensing," *Journal of Materials Chemistry C*, vol. 6, no. 9, pp. 2258-2269, 2018.

[6] A. Qiu *et al.*, "Stretchable and calibratable graphene sensors for accurate strain measurement," *Materials Advances*, vol. 1, no. 2, pp. 235-243, 2020.

[7] S. B. Choi, C. J. Han, C. R. Lee, and J. W. Kim, "Interfaceless Strain and Pressure - Sensitive Stretchable Capacitor Based on Self - Bonding and Surface Morphology Control of a

[8] Reversibly Crosslinkable Silicone Elastomer," *Advanced Materials Technologies*, vol. 5, no. 2, p. 1900757, 2020.
F. Xu and Y. Zhu, "Highly conductive and stretchable silver nanowire conductors," *Adv Mater*, vol. 24, no. 37, pp. 5117-22, Sep 25 2012.

[9] E. Roh, B. U. Hwang, D. Kim, B. Y. Kim, and N. E. Lee, "Stretchable, Transparent,Ultrasensitive, and Patchable Strain Sensor for Human Machine Interfaces Comprising a Nanohybrid of Carbon Nanotubes and Conductive Elastomers," *ACS NANO* vol. 9, no. 6, 2015.

[10] J. T. Muth *et al.*, "Embedded 3D printing of strain sensors within highly stretchable elastomers," *Adv Mater*, vol. 26, no. 36, pp. 6307-12, Sep 2014.

[11] G. Cai, J. Wang, K. Qian, J. Chen, S. Li, and P. S. Lee, "Extremely Stretchable Strain Sensors Based on Conductive Self-Healing Dynamic Cross-Links Hydrogels for Human-Motion Detection," *Adv Sci (Weinh)*, vol. 4, no. 2, p. 1600190, Feb 2017.

[12] J. Gu, D. Kwon, J. Ahn, and I. Park, "Wearable Strain Sensors Using Light Transmittance Change of Carbon Nanotube-Embedded Elastomers with Microcracks," *ACS Appl Mater Interfaces*, vol. 12, no. 9, pp. 10908-10917, Mar 4 2020.

[13] C. Shao *et al.*, "Mussel-Inspired Cellulose Nanocomposite Tough Hydrogels with Synergistic Self-Healing, Adhesive, and Strain-Sensitive Properties," *Chemistry of Materials*, vol. 30, no. 9, pp. 3110-3121, 2018.

[14] Y. Cai *et al.*, "Stretchable Ti3C2Tx MXene/Carbon Nanotube Composite Based Strain Sensor with Ultrahigh Sensitivity and Tunable Sensing Range," *ACS Nano*, vol. 12, no. 1, pp. 56-62, Jan 23 2018.

[15] S. Choi, S. Kim, H. Kim, B. Lee, T. Kim, and Y. Hong, "2-D Strain Sensors Implemented on Asymmetrically Bi-Axially Pre-Strained PDMS for Selectively Switching Stretchable Light-Emitting Device Arrays," *IEEE Sensors Journal*, vol. 20, no. 24, pp. 14655-14661, 2020.

[16] Q. Zheng *et al.*, "Sliced graphene foam films for dual-functional wearable strain sensors and switches," *Nanoscale Horiz*, vol. 3, no. 1, pp. 35-44, Jan 1 2018.

[17] Q. Hua *et al.*, "Skin-inspired highly stretchable and conformable matrix networks for multifunctional sensing," *Nat Commun*, vol. 9, no. 1, p. 244, Jan 16 2018.

[18] X. Huang *et al.*, "Materials and Designs for Wireless Epidermal Sensors of Hydration and Strain," *Advanced Functional Materials*, vol. 24, no. 25, pp. 3846-3854, 2014.

[19] J. Park, I. You, S. Shin, and U. Jeong, "Material approaches to stretchable strain sensors," *Chemphyschem*, vol. 16, no. 6, pp. 1155-63, Apr 27 2015.

Gap in pagination due to unavailable papers.

Pages 1593-1600

Proceedings of the 16th Annual IEEE International
Conference on Nano/Micro Engineered and Molecular Systems
April 25-29, 2021

Temperature Compensation for MEMS Mass Flow Sensors Based on Back Propagation Neural Network

Yan Wang, Shijin Xiao, and Jifang Tao*, *Member, IEEE*

Abstract— **A temperature compensation method based on Back Propagation (BP) Neural Network is designed by taking advantage of the characteristics of neural network, whose performance is demonstrated in a mass flow sensor. The mass flow sensor is tested under different temperatures to obtain the sample data. The algorithm compensates the temperature effects by establishing a non-linear mapping relationship between temperature and mass flow rate. To settle the problem of accuracy degradation induced by ambient temperature variation, both BP neural network and polynomial fitting method are developed to compensate the drift of the mass flow sensor. The result shows that the method based on the BP Neural Network has high compensation accuracy and fast convergence speed, which can effectively compensate the influence of temperature on MEMS mass flow sensor and improve the sensor output accuracy.**

I. INTRODUCTION

MEMS mass flow sensors play an important role in automotive electronics, industrial gas control, biological medicine, semiconductor manufacturing, scientific research experiments and so on by virtue of their excellent performance in integration, detection accuracy and sensitivity [1]. Due to the working principle of heat exchange between gas flow and mass flow sensors, it is inevitable that the measurement is affected by the ambient temperature, which reduces the measurement accuracy, resulting in large error of the measurement results and even wrong results. Therefore, the influence of ambient temperature must be considered in the design of thermal mass flow sensor system, e.g., temperature compensation via hardware compensation or software compensation. Hardware compensation in the circuit is affected by the drift of electronic components and the precision of component welding, which leads to the low reliability of the whole measurement circuit, and it is costly. Software compensation methods mainly include least square method, interpolation method and polynomial fitting method [2]. The most frequently used method is that the compensation coefficient is calculated by selecting several temperature points, the compensation coefficients of other temperature points are obtained by linear interpolation. The linear interpolation is a non-linear method, which is using piecewise linearization method to fit the compensation coefficient. It is difficult to achieve high compensation accuracy with this method [3], and its calculation is very complex. BP Neural Networks are widely applied in function approximation, pattern recognition, classification and data compression because of their strong non-linear mapping ability and learning ability [4,5]. In recent years, they have been applied to the measurement error compensation and shown their robustness

and effectiveness [6,7]. Here, we apply BP Neural Network to establish the non-linear mapping relationship between temperature and mass flow rate, to achieve the purpose of temperature compensation for mass flow sensor.

II. TEMPERATURE DRIFT ANALYSIS OF THE MASS FLOW SENSOR

The mass flow sensor mainly consists of a micro-heater and a pair of thermopiles symmetrically placed upstream and downstream of the micro-heater. Based on the measuring principle of heat exchange between gas flow and mass flow sensors, the measurement of mass flow is related to some physical parameters of the fluid, such as heat conduction, density, viscosity, etc., and these physical parameters change with the fluid temperature increasing or decreasing, thus causing errors in mass flow measuring [2]. At the same time, after a mass flow sensor chip is packaged into a device, the electronic components in its signal amplification circuit and the power supply system used for supplying power to the micro-heater and the circuit are also affected by the change of ambient temperature, which makes the output voltage of the device can not remain constant even if there is no gas flowing through. Therefore, the phenomenon of temperature drift occurs very commonly when mass flow sensors are used for gas flow measuring, and which is often non-liner.

III. TEMPERATURE COMPENSATION PRINCIPLE BASED ON BP NEURAL NETWORK

A. BP Neural Network

BP Neural Network is a widely used multi-layer feedforward neural network, typically including an input layer, one or more hidden layers and an output layer, its structure is shown in Fig. 1. Each layer consists of processing units, called neurons (mathematically, activation or transfer functions), which receive their input from units of the previous layer and send their output to units in the successive layer. Neurons are connected by weight values $w_{i,j}$, and each neuron has a corresponding threshold value $b_{i,j}$ except the input layer, it simply distributes the training samples, without modification, to the first hidden layer. The training samples are trained by the hidden layers and the output layer. The Sigmoid activation functions (tan-Sigmoid, log-Sigmoid and linear-Sigmoid) are usually adopted as transfer functions for hidden layers to introduce non-linearity in the model, which enable BP Neural Network to have strong adaptive capability to deal with complex high-no-linear issues, and this adaptive capability increases with the number of processing nodes with a typical learning curve trend, too many nodes are meaningless, which

---

All authors are with the School of Information Science and Engineering, Shandong University, Qingdao, 266000, China
*(phone: +86 17806243369; email: taojf@sdu.edu.cn)

978-1-6654-3008-1/21 $31.00 © 2021 IEEE

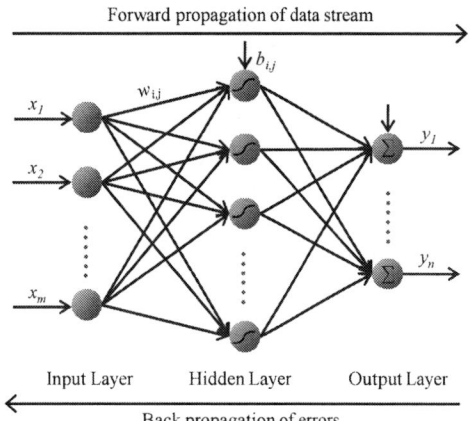

Fig. 1. Structure of three-layer BP Neural Network.

will not only lead to a linear increase of the computational load, but also cause over fitting [8]. In theory, a BP Neural Network with a single hidden layer can approximate any nonlinear mapping provided that the sample data are sufficient and the parameter settings of the network are reasonable [9], and the number of nodes in the single hidden layer is usually determined by an empirical formula:

$$h = m + n + a \qquad (1)$$

where h, m, n, are the number of notes of the hidden layer, the input layer and the output layer respectively, and a is an adjustment constant between 1 and 10 [10]. The linear function is regularly used as the transfer function for the output layer to weight and sum the output data of the last hidden layer [8].

BP Algorithm includes two processes: forward calculation of data stream (forward propagation) and back propagation of error signals. In the forward propagation, the input samples, transmitted from the input layer, are processed by the hidden layers layer by layer, and then passed to the output layer to get the predicted outputs. Then it turns to the back propagation, according to the errors between the actual values and the predicted values, the inverse feedback is performed to update the weight values and the threshold values in the neural network, Gradient Descent Algorithm is usually used to update these parameters. The two processes are alternately carried out until the errors reach the requirements, the training is finished [11] and the parameters obtained from the training process are stored in the weight values between neurons and the threshold values of each neuron.

B. Temperature compensation principle

The block diagram of temperature compensation for mass flow sensor based on BP Neural Network is shown in Fig. 2. The input signal is composed of the output voltage signal of the mass flow sensor and the ambient temperature, the target output signal is the actual flow rate. Through the above training process, the internal relationship between input and output can be simulated without establishing the specific model of sensor output with temperature changing. After temperature compensation by the trained BP Neural Network, the output can better approach the actual value, which almost eliminates the effects caused by the ambient temperature

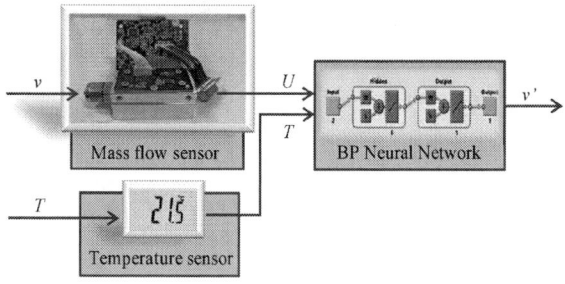

Fig. 2. Temperature compensation principle block diagram of mass flow sensor.

variations and improve the measurement accuracy and stability of the mass flow sensor.

IV. TEMPERATURE COMPENSATION BASED ON BP NEURAL NETWORK AND POLYNOMIAL FITTING

A. Sample data acquisition and analysis

A mass flow controller is used to control the mass flow rate of the gas (dry pure nitrogen in this experiment), and then the gas flow of some certain flow rates, through a section of heat exchange tube, arrives at the mass flow sensor. The mass flow sensor and the gas heat exchange tube are placed in a programmable constant temperature chamber, which is used to control the experiment temperature. The experiment is carried out at 7 different temperatures (0, 10, 20, 30, 40, 50, 60 ℃), and 11 different mass flow rates (0, 20, 40, 60, 80, 100, 120, 140, 160, 180, 200 sccm) is measured at each temperature. The power supply system of the mass flow sensor consists of two parts: the dual power supply of 3.3 V and - 3.3 V is used to supply power to the circuit, mainly for the amplification circuit, and the heating voltage of the micro-heater is given as 1.8 V. The output voltages of the mass flow sensor are collected and recorded by the signal acquisition system.

Fig. 3 reports the relationship between output voltage and flow rate of the mass flow sensor under different ambient temperatures, Fig. 3(b) is the partial enlargement of Fig. 3(a) at the flow of 0 sccm. With a same mass flow rate, the output of the mass flow sensor has obvious temperature drift with the change of temperature, including the condition that there is no gas flowing through. It can be seen that the change of temperature has a great influence on the gas flow, the circuit and the whole mass flow sensor system, and appears a high non-linearity.

B. Construction and training of the network

Take the 11 groups of data at the flow rate of 140 sccm as testing samples, the other 66 groups of data as training samples. In order to speed up the convergence of network training to achieve a better compensation effect, normalize these samples before training. The BP Neural Network is built through MATLABR2019b, and its self-contained neural network toolbox can be used for the network training and simulation. To establish the network, two nodes are selected in the input layer, corresponding to the output voltage U of the mass flow sensor and ambient temperature T. According to the empirical

978-1-6654-3008-1/21 $31.00 © 2021 IEEE

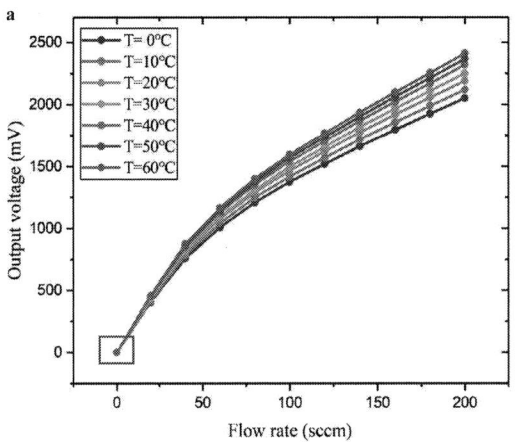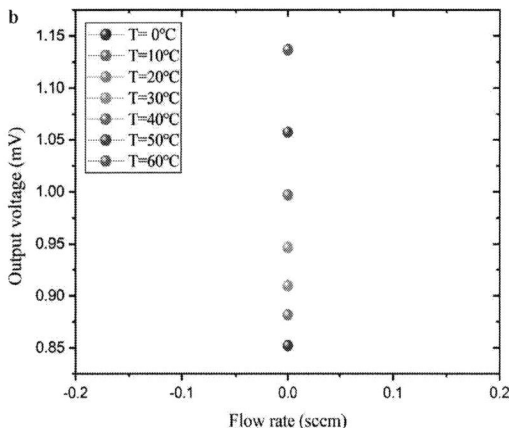

Fig. 3. Curve of relationship between output voltage and flow rate of the mass flow sensor under different temperatures.
a, Flow rate range from 0 to 200 sccm. **b**, Flow rate at 0 sccm.

formula, and after debugging for many times, single hidden layer with 6 nodes is selected. The output layer only needs one node to correspond to the actual mass flow rate v. Levenberg-Marquart Algorithm is used for the network training, tansig function is selected as the transfer function of the hidden layer and purelin function as the transfer functions of the output layer. After the network structure is established, some key training parameters needs to be set: the performance goal (the minimum error expected, mean square error is selected as the error function) is 10^{-6}; the maximum number of epochs to train

is 3000 epochs, means that the training will automatically end if the network fails to meet the error requirement after having been trained over 3000 epochs; the maximum validation failures is 16 epochs, which means that the system judges whether the error of verification set has not decreased after 16 consecutive tests, if the error does not decrease or even increases, meaning that the error of training can not decrease any more, further training can not achieve better results and is meaningless, stop training at this time, otherwise it will fall into the state of over fitting. Fig. 4 shows the flow chart of the BP Neural Network training process, and Fig. 5 shows the learning curve of the BP Neural Network with the hidden layer of 6 nodes, it can be seen that after 50 epochs, the mse is less than 10^{-5}. Less number of nodes can't reach the purpose of high precision compensation, while more nodes may cause the mse of testing set much higher than the training set, because the sample points may be not enough for a more complex network.

C. Polynomial fitting

The same as the BP Neural Network, the 11 groups of data at the flow rate of 140 sccm are selected as testing samples, the other 66 groups of data are used for polynomial fitting. This article [12] introduces a temperature drift correction algorithm for sensors. Referring to this method, the sample data are fitted by mainly using the polyfit function, which is also realized by MATLAB R2019b software in this paper.

V. DISCUSSION AND CONCLUSIONS

To compare the performance, we substitute testing samples into the trained Neural Network and the fitted polynomial. Fig. 6 shows the absolute values of the relative error after temperature compensation by BP Neural Network (blue line) and polynomial fitting (red line). The maximum relative errors of BP Neural Network and polynomial fitting are 0.2558% and 1.1016% respectively. Both them can effectively compensate the error caused by temperature drift, while the BP Neural Network has higher accuracy, and the output with PB Neural Network compensation keeps constant as temperature increasing from 0 to 60℃.

Fig. 4. Flow chart of the BP Neural Network training process.

978-1-6654-3008-1/21 $31.00 © 2021 IEEE

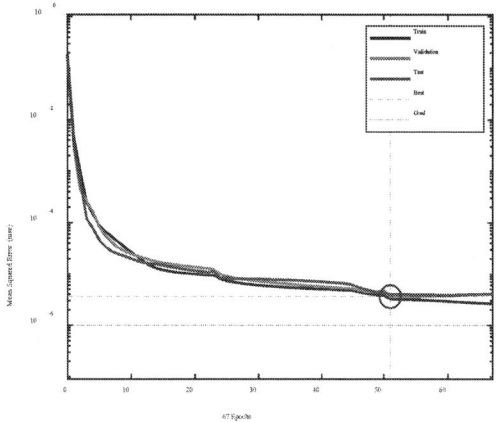

Fig. 5. Machine learning curve.

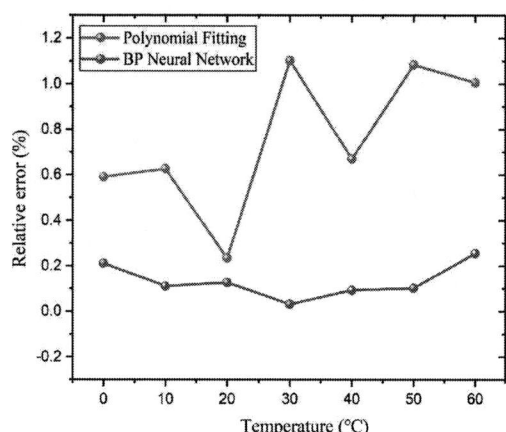

Fig. 6. Relative error of sensor at different operation temperature after compensation.

In this paper, BP Neural Network is used for the temperature compensation which applied to a mass flow sensor. The maximum relative errors is only 0.2558% over the temperature range from 0 to 60°C, which is 4 times smaller than conventional polynomial fitting. This compensation model can be used as a universal approach for temperature drift compensation, which also can be used for humidity sensors, gas sensors, pressure sensors, etc.

REFERENCES

[1] V. Balakrishnan, H. P. Phan, T. Dinh, D. V. Dao, and N. T. Nguyen, "Thermal Flow Sensors for Harsh Environments," *Sensors (Basel)*, vol. 17, 2017-09-08 2017.

[2] WANG Chuan, YIN Wen-qing, YANG Zhi-jun, FAN Li, CAO Peng, "Temperature compensation of thermal gas flowmeter based on RBF neural networks," *Transducer and Microsystem Technologies*, vol. 35, pp. 99-102, 2016-12-06 2016.

[3] Qian Zhang and Qiang Wu, "Approaches to realize temperature compensation based on BP neural network," *Electronic Design Engineering*, vol. 19, pp. 152-154, 2011.

[4] Y. Jia, F. Chen, P. Wu, Q. Liu, Y. Chen and Y. Liu, "A Study of Online Function Approximation System Based on BP Neural Network," *2019 International Conference on Cyber-Enabled Distributed Computing and Knowledge Discovery (CyberC)*, Guilin, China, 2019, pp. 147-150.

[5] S. Li, L. He, S. Liu, Y. Zhang and H. Yu, "Recognition of locomotion patterns based on BP neural network during different walking speeds," *2017 Chinese Automation Congress (CAC)*, Jinan, China, 2017, pp. 5215-5218.

[6] R. Zhang, Y. Duan, Y. Zhao, and X. He, "Temperature compensation of elasto-magneto-electric (EME) sensors in cable force monitoring using BP neural network," *Sensors (Switzerland)*, vol. 18, no. 7, 2018, doi: 10.3390/s18072176.

[7] W. He, J. Yi, B. Zhou, B. Chen, and L. He, "Nonlinear Correction and Temperature Compensation Method of Turbine Flowmeter Based on Neural Network," *IOP Conference Series Materials Science and Engineering*, vol. 782, p. 052049, 2020.

[8] R. Fontanella, D. Accardo, R. S. Lo Moriello, L. Angrisani, and D. De Simone, "MEMS Gyros Temperature Calibration through Artificial Neural Networks," *Sensors & Actuators A Physical*, p. S0924424717312116, 2018.

[9] L. Bingying, L. Yongxin, W. Haitao, M. Yuming, H. Qiang, and G. Fangli, "Compensation of automatic weighing error of belt weigher based on BP neural network," *Measurement*, vol. 129, pp. 625-632, 2018.

[10] J. Li, J. H. Cheng, J. Y. Shi, and F. Huang, *Brief Introduction of Back Propagation (BP) Neural Network Algorithm and Its Improvement*: Springer Berlin Heidelberg, 2012.

[11] Y. Li, J. Li, J. Huang, and H. Zhou, "Fitting analysis and research of measured data of SAW micro-pressure sensor based on BP neural network," *Measurement*, vol. 155, p. 107533-, 2020.

[12] https://blog.csdn.net/liyuanbhu/article/details/7725824

Proceedings of the 16th Annual IEEE International
Conference on Nano/Micro Engineered and Molecular Systems
April 25-29, 2021

Pressure Induced Transition from Wrinkling to Period-Doubling Instability in Flexible Tactile Sensors

Xiu-Yuan Li, Kai-Ming Hu*, Yi-Hang Xin, Er-Qi Tu and Wen-Ming Zhang*

Abstract— **Wrinkle structures are widely used in flexible tactile sensors due to ease of fabrication and significant improvement of sensitivity. However, period-doubling instability induced by pressure hinders the improvement of its measure range and accuracy. We conduct experiments and finite element analysis to elucidate the mechanisms and characteristics of period-doubling instability. Influence of pre-stretch and friction coefficient on critical load is discussed. A pre-shaping method is proposed to restrain period-doubling and extend the limit of pressure measurement. The influence of defects and average effect in actual flexible tactile sensors are discussed. The profound comprehension of period-doubling instability can instruct the design of flexible tactile sensors with high stability and accuracy, and give inspires to new concepts of flexible tactile sensors with versatile functions and unique capabilities.**

I. INTRODUCTION

Flexible tactile sensors have wide applications in robotics and healthcare, such as touch sensing and pulse detection [1, 2], which require conformal contact, high sensitivity and stability. Surface microstructures can reduce compressive stiffness and introduce contact area change that bring higher sensitivity to resistive and capacitive tactile sensors. The pressure sensing capabilities of microdome, micropyramid and micropillar structured flexible tactile sensors are studied with finite-element-methods by Park et al. [3]. Bae et al. [2, 4] introduced statistical models of profile height distribution to characterize the sensitivity and linearity of flexible tactile sensors. Wrinkles are common microstructures that can be easily obtained by pre-stretching, swelling, and heating process of film-substrate systems [5-8]. It is facile and tunable for large area fabrication, and has great potentials for flexible tactile sensors. A representative wrinkle structure commonly used in flexible tactile sensors is composed of a stiff conductive PEDOT: PSS film and a soft PDMS substrate [9]. By pre-stretching the substrate before coating the conductive film, wrinkles generate after releasing the substrate. Huang et al. and Song et al. [10, 11] have derived the relation of wrinkling wavelength and amplitude with pre-stretch analytically by energy method, which can guide the design of wrinkle profile. Joseph [12] has solved the contact problem of wrinkles with flat elastic half-space, that can describe the infinitesimal contact deformation of wrinkle based flexible tactile sensors.

However, under large tactile pressure, surface wrinkling could transit to period-doubling that the surface profile suddenly changes, with saltation of reaction force and contact area. The period-doubling instability can vary the

sensing properties, even deteriorate the stability and accuracy of flexible tactile sensors. Brau et al. and Cao et al. [13-15] have studied the spontaneous period-doubling instability induced by substrate pre-stretch theoretically. The critical pre-stretch of substrate to trigger period-doubling instability is about 20%, and his morphology bifurcation is attributed to the intrinsic non-linearity of materials and elasticity equations. For flexible tactile sensors under pressure, partial contact of wrinkle surface will generate non-linear reaction force that could lead to period-doubling instability. But the critical pressure and mechanical properties are not clear. Therefore, it is necessary to reveal the mechanical characteristics and the affecting factors of period-doubling instability. And its influences to flexible tactile sensors should be elucidated. Base on this phenomenon, new concept of flexible tactile sensor is also hopeful.

II. RESEARCH METHOD

A meso-scale wrinkle structure is fabricated to achieve facile observing and measuring of wrinkling to period-doubling transition. In Fig. 1a, a pre-stretched elastomer substrate (ECO-FLEX 0030, original length 100 mm, width 15 mm, thickness 5 mm, E=30 kPa) and a stress-free PU film (length 80 mm, width 10 mm, thickness 0.1 mm, E=20 MPa) are treated with ultraviolet ozone for 15 min to obtain active bounding surfaces. The counter surfaces are bonded by a thin layer of uncured elastomer, and cured at 80 °C for 2 h to enhance the adhesion of the film and the substrate. After releasing the pre-stretch of the substrate, millimeter-scale surface wrinkles are obtained. In Fig. 1b, the wrinkle structure is sandwiched by two glass slides and pressed by a force gauge in controlled displacement of indentation. The evolution of surface profile and contact area can be observed on top and side positions.

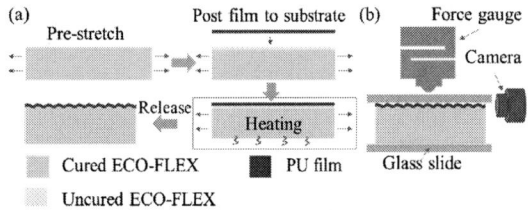

Figure 1. *Specimen fabrication and experiment setup: (a) Fabrication of PU/ECO-FLEX meso-scale wrinkle specimen; (b) experiment setup of loading and observing devices.*

All authors are with State Key Laboratory of Mechanical System and Vibration, School of Mechanical Engineering, Shanghai Jiaotong University, 200240, Shanghai, China (corresponding author

Kai-Ming Hu, Tel.: +86-18321842038, e-mail: hukaiming@sjtu.edu.cn; corresponding author Wen-Ming Zhang, Tel.: +86-2154744990, e-mail: wenmingz@sjtu.edu.cn).

978-1-6654-3008-1/21 $31.00 © 2021 IEEE   1605

Finite element method is also used to investigate the affecting parameters of period-doubling instability. The wrinkle structure is assumed to be frictionless supported at both sides and bottom, and pressed by a rigid flat indenter on the top. Incompressible Neo-Hookean hyperelastic materials and CPE4RH elements are used for the film and the substrate in plane strain condition. The displacement of indentation, load and contact area are traced in the whole loading process.

III. RESULTS AND DISCUSSION

A. Pressure induced period-doubling instability

The profile evolution of wrinkles under pressure without and with period-doubling instability is showed in Fig. 2a-b. The FEM and experiment results agree well in both situations. Fig. 2c shows the strain energy evolution of film and substrate during loading process in FEM. Without period-doubling instability, wrinkle crests are continuously flatted under pressure, and troughs become dense creases which have the same period with wrinkles. The total strain energy is dominated by the film bending energy due to its dense and severely curved profile, and increase rapidly with indentation. When period-doubling instability happens, one wrinkle trough rises and form a large flat area with neighboring crests, while neighboring troughs descend and form creases. The final profile alternates by crease and flat area in double period length. The film bending energy declines due to larger flat areas and sparser creases, and part of the film bending energy transforms to the localized strain energy around the crease in the substrate. Dot lines in Fig. 2c indicates that the total energy decreases by this transition. Therefore, when the pressure exceeds a critical value, the period-doubling mode is preferred.

B. Characteristics of period-doubling instability

In Fig. 3a-c, the mechanical behavior of wrinkles with different pre-stretch is discussed. In the initial stage, load and stiffness steadily increase, where most flexible tactile sensors can work stably in this situation. When period-doubling instability happens, load decreases and exhibits negative contact stiffness. Contact area sharply increase due to the large flat contact area in period-doubling profile. Finally, contact area is saturated and contact stiffness grows into high level that approximate the bulk material of substrate. The system is stable again but the sensitivity is low in this situation.

Fig. 3d shows the relation of critical load and pre-stretch. Higher pre-stretch can induce wrinkles with higher aspect ratio that can improve sensitivity, but the residual stress will lower the critical load that reduce stability. For small pre-stretch, the critical load decreases slowly, but for pre-stretch higher than 15%, the critical load decreases sharply. When the pre-stretch exceeds 20%, period-doubling instability happens spontaneously without any external pressure which corresponds to the previous research [13-15]. This indicates that flexible tactile sensors should adopt intermediate pre-stretch to take balance of sensitivity and stability.

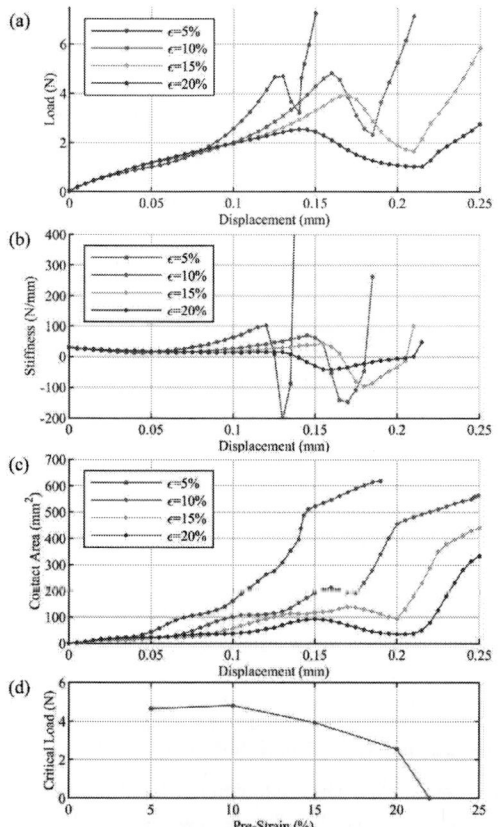

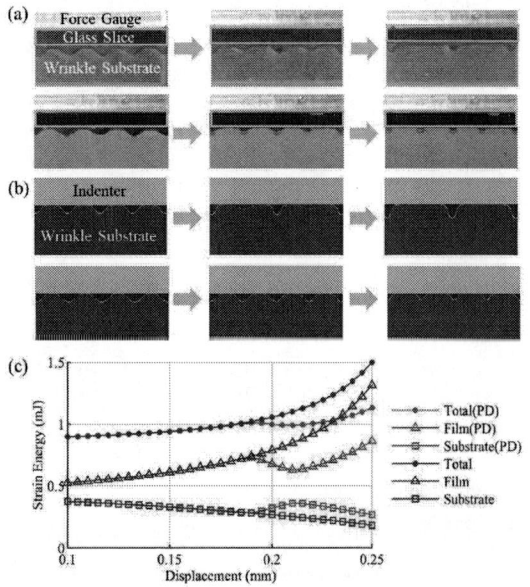

Figure 2. Wrinkle profile and energy evolution during loading: (a) Side-view profile of wrinkle specimen, upper: with period-doubling, under: without period-doubling; (b) corresponding results of FEM; (c) evolution of total energy, film energy and substrate energy in FEM, red line: with period-doubling, blue line: without period-doubling.

Figure3. Mechanical behaviors of wrinkles under pressure with various pre-stretch in FEM: (a) Load-displacement relation during loading process; (b)contact stiffness evolution during loading process; (c) contact area evolution during loading process; (d) relation of critical load and pre-stretch.

For fixed pre-stretch, surface friction coefficient (denoted by μ) also influences the critical load. In Fig. 4, when friction coefficient increases, the period-doubling instability is delayed and the critical load is greatly enhanced. For $\mu \geq 0.2$, period-doubling instability can be eliminated. Period-doubling instability requires surface sliding between film and indenter. Friction hinders this tendency and creates extra energy dissipation for the transition, so period-doubling instability is delayed or eliminated. Flexible tactile sensors will have better stability if high friction coefficient is attained between wrinkle microstructures and electrodes.

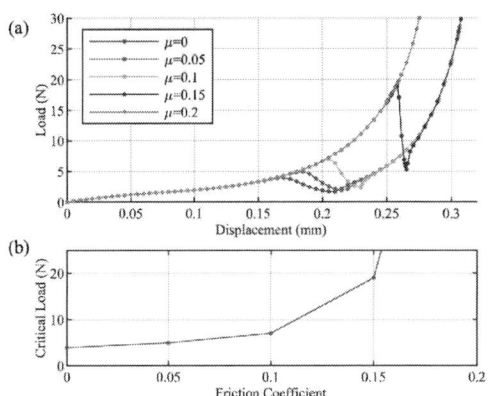

Figure 4. Influence of friction coefficient to period-doubling instability in FEM: (a) Load-displacement relation of specimens with various interface friction coefficient; (b) influence of friction coefficient to the critical load

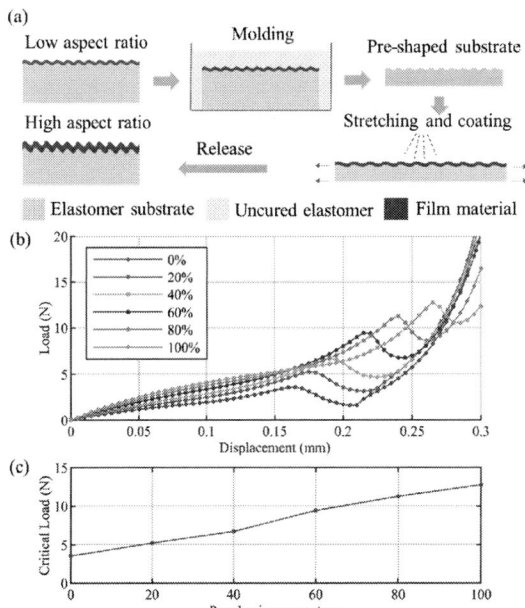

Figure 5. Pre-shaping method to enhance critical load: (a) Illustration of pre-shaping method; (b) load-displacement relation of specimens with various pre-shaping percentage; (c) influence of pre-shaping percentage to the critical load.

If the sensitivity and stability is required simultaneously, period-doubling instability can also be avoided by pre-shaping method. In Fig. 5a, by molding the low aspect ratio wrinkle surface, sinusoidal shape substrate without initial stress can be obtained. Then lower pre-stretch is needed to form the final profile, which means higher critical load. Fig. 5b-c shows that the critical load increases with higher pre-shaping percentage. Wrinkles made by completely molding method (100% pre-shaping) can have best stability.

In experiments, specimens with different lubrication conditions are compared. The friction coefficient of clean PU film and glass slide is above 0.5, and period-doubling instability does not happen under this condition (Upper pictures in Fig. 1a). When the surface is coated with lubrication grease, the friction coefficient decreases to approximate 0.1, then the period-doubling instability is observed (under pictures in Fig. 1a). The load-displacement relation is showed in Fig. 6a. In the first stage, two situations coincide with each other due to the same deformation mode. In the intermediate stage, specimen with lubrication shows lower load, due to the period-doubling instability that reduce contact stiffness. At last, two load-displacement curves are almost straight and parallel, which mimic the buck material property of the substrate.

Different from the sharp transition from wrinkling to period-doubling in FEM, the period-doubling instability happens gradually in experiments, and saltation of load and negative stiffness is not notably displayed. This variation can be explained by several reasons. First, the film and substrate have initial defects, such as local variation of film thickness, substrate modulus, and adhesion strength. The residual plastic deformation of previous loading process can also become defects. Period-doubling instability is easily triggered by these defects of surface profile.

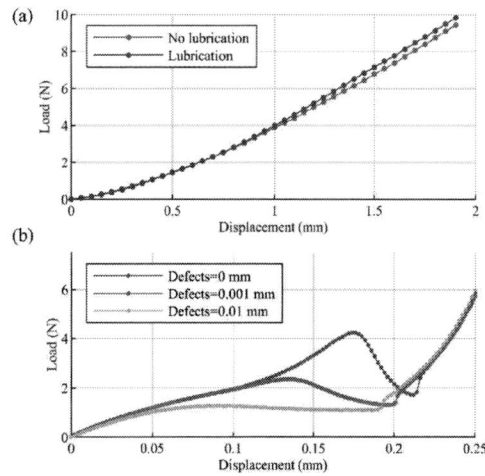

Figure 6. Experiment results of load-displacement relation and FEM results on sensitivity of defects: (a) load-displacement relation of specimens with and without surface lubrication; (b) load-displacement relation of wrinkles with different amplitudes of initial defects.

The sensitivity to defects of period-doubling instability is investigated by FEM. A small amplitude of double period sinusoidal profile is added to the initial profile as perturbation. The profile evolution with defects amplitude 0 mm, 0.001 mm, and 0.01 mm is showed in Fig. 6b. The result indicates that very small defect will lower the period-doubling critical load, and elongate the span of period-doubling instability. Figure 6c shows the residual plastic deformation of film when the substrate recovers to initial length. Other factors will also influence the period-doubling instability, such as the ununiform deformation along the width direction, imperfectly confined boundaries, and curved wrinkle surface due to the poison effect of substrate contraction. Period-doubling instability could happen gradually in wrinkle length direction, and contact process is stabilized by similar effects of bevel gear. For actual flexible tactile sensors, period-doubling instability of numerous wrinkles happens in different times, further homogenize the sensing properties.

In another perspective, instantaneous period-doubling instability can be utilized for special sensing abilities. When the pressure exceeds a certain value, the sudden change in contact resistance can act as a switch, that can be used as alertor. For sensors that require high sensitivity in certain load range, traditional sensors only have fixed sensitivity in the whole sensing range. But for sensors adopting period-doubling, the sensitivity can be extremely high at period-doubling instability period. These new concepts may be achieved by using materials with better elasticity, and methods to fabricate stricter confined boundaries. Micro silicon ribbon wrinkles on PDMS substrate may be an alternative choice. Wrinkles of smaller size are also easier for integration and array, which fits the trend of flexible tactile sensor development.

IV. CONCLUSION

We investigated the mechanical and morphological behavior of wrinkling to period-doubling transition under pressure. Period-doubling instability exhibits sharp decrease of stiffness and increase of contact area. The critical load of period-doubling instability is affected by pre-stretch and friction coefficient. By pre-shaping method, wrinkles with high aspect ratio and high stability can be simultaneously achieved. Experiments shows period-doubling instability is sensitive to initial defects, and the sensing properties can be stabilized by average effects of numerous wrinkles and defects. We also propose new types of sensor using period-doubling instability to achieve versatile sensing functions and higher sensitivity. This work takes advances to the wrinkling instability mechanisms, and extend the comprehension of flexible tactile sensors.

V. ACKNOWLEDGEMENTS

The authors gratefully acknowledge the supports by the National Science Foundation for Distinguished Young Scholars (11625208), Major Program (12032015) and Young Scientists of China (11802173), the Program of Shanghai Academic/Technology Research Leader (19XD 1421600).

REFERENCES

[1] Z. Y. Huang, W. Hong, and Z. Suo, "Nonlinear analyses of wrinkles in a film bonded to a compliant substrate," Journal of the Mechanics and Physics of Solids, vol. 53, no. 9, pp. 2101-2118, 2005, doi: 10.1016/j.jmps.2005.03.007.

[2] S.-J. Park, J. Kim, M. Chu, and M. Khine, "Flexible Piezoresistive Pressure Sensor Using Wrinkled Carbon Nanotube Thin Films for Human Physiological Signals," Advanced Materials Technologies, vol. 3, no. 1, 2018, doi: 10.1002/admt.201700158.

[3] J. Park et al., "Tailoring force sensitivity and selectivity by microstructure engineering of multidirectional electronic skins," NPG Asia Materials, vol. 10, no. 4, pp. 163-176, 2018, doi: 10.1038/s41427-018-0031-8.

[4] G. Y. Bae et al., "Linearly and Highly Pressure-Sensitive Electronic Skin Based on a Bioinspired Hierarchical Structural Array," Adv Mater, vol. 28, no. 26, pp. 5300-6, Jul 2016, doi: 10.1002/adma.201600408.

[5] S. Yang, K. Khare, and P.-C. Lin, "Harnessing Surface Wrinkle Patterns in Soft Matter," Advanced Functional Materials, vol. 20, no. 16, pp. 2550-2564, 2010, doi: 10.1002/adfm.201000034.

[6] H. Hou et al., "Reversible Surface Patterning by Dynamic Crosslink Gradients: Controlling Buckling in 2D," Adv Mater, p. e1803463, Jul 31 2018, doi: 10.1002/adma.201803463.

[7] L. Zhou, K. Hu, W. Zhang, G. Meng, J. Yin, and X. Jiang, "Regulating surface wrinkles using light," National Science Review, vol. 7, no. 7, pp. 1247-1257, 2020, doi: 10.1093/nsr/nwaa052.

[8] K. M. Hu et al., "Delamination‐Free Functional Graphene Surface by Multiscale, Conformal Wrinkling," Advanced Functional Materials, vol. 30, no. 34, 2020, doi: 10.1002/adfm.202003273.

[9] Z. Wen et al., "A Wrinkled PEDOT:PSS Film Based Stretchable and Transparent Triboelectric Nanogenerator for Wearable Energy Harvesters and Active Motion Sensors," Advanced Functional Materials, vol. 28, no. 37, 2018, doi: 10.1002/adfm.201803684.Z. Y. Huang, W. Hong, and Z. Suo, "Nonlinear analyses of wrinkles in a film bonded to a compliant substrate," Journal of the Mechanics and Physics of Solids, vol. 53, no. 9, pp. 2101-2118, 2005, doi: 10.1016/j.jmps.2005.03.007.

[10] Z. Y. Huang, W. Hong, and Z. Suo, "Nonlinear analyses of wrinkles in a film bonded to a compliant substrate," Journal of the Mechanics and Physics of Solids, vol. 53, no. 9, pp. 2101-2118, 2005, doi: 10.1016/j.jmps.2005.03.007.

[11] J. Song et al., "Buckling of a stiff thin film on a compliant substrate in large deformation," International Journal of Solids and Structures, vol. 45, no. 10, pp. 3107-3121, 2008, doi: 10.1016/j.ijsolstr.2008.01.023.

[12] J. M. Block, and L. M. Keer, "Periodic contact problems in plane elasticity," Journal of Mechanics of Materials and Structures, Vol. 3, No. 7, pp. 1207–1237, 2008, doi: 10.2140/jomms.2008.3.1207.

[13] F. Brau, H. Vandeparre, A. Sabbah, C. Poulard, A. Boudaoud, and P. Damman, "Multiple-length-scale elastic instability mimics parametric resonance of nonlinear oscillators," Nature Physics, vol. 7, no. 1, pp. 56-60, 2010, doi: 10.1038/nphys1806.

[14] Y. Cao and J. W. Hutchinson, "From wrinkles to creases in elastomers: the instability and imperfection-sensitivity of wrinkling," Proceedings of the Royal Society A. Mathematical, Physical and Engineering Sciences, vol. 468, no. 2137, pp. 94-115, 2011, doi: 10.1098/rspa.2011.0384.

[15] Y. Zhao, Y. Cao, W. Hong, M. K. Wadee, and X. Q. Feng, "Towards a quantitative understanding of period-doubling wrinkling patterns occurring in film/substrate bilayer systems," Proc Math Phys Eng Sci, vol. 471, no. 2173, p. 20140695, Jan 8 2015, doi: 10.1098/rspa.2014.0695.

Gap in pagination due to unavailable paper.

Pages 1609-1610

Proceedings of the 16th Annual IEEE International
Conference on Nano/Micro Engineered and Molecular Systems
April 25-29, 2021

Design of a novel closed-loop electrostatic voltage sensor based on weakly coupled resonators

Zilong Wang[1,2], Zhengwei Wu[1], Xiangming Liu[1,2], Yahao Gao[1,2], Simin Peng[1,2], Ren Ren[1]，Fengjie Zheng[1], Yao Lv[1], Jun Liu[1,2], Hucheng Lei[1,2], Zhouwei Zhang[1,2], Wei Zhang[1,2], Jiachen Li[1,2] and Chunrong Peng[1,2*]

Abstract—A novel high-performance closed-loop electrostatic voltage sensor based on electrically weakly coupled resonator was designed. The MEMS static voltage sensor is designed based on mode localization effect. A three-degree-of-freedom micromechanical sensor which is weakly coupled by electrostatic voltage was fabricated to achieve high-sensitivity voltage measurement. Currently under the open-loop test, the sensor can achieve voltage measurement with a sensitivity of 0.0967/V and a linearity of 6.11%. The experimental results show that the voltage sensitivity of the sensor based on the amplitude ratio output is 87.9 times higher than the sensitivity based on the frequency output.

I. INTRODUCTION

Electrostatic voltage sensors are widely used in space detection [1], mass spectrometry[2], electrostatic monitoring and so on. At present, commonly used static voltage sensors mainly include contact type and non-contact type. Among them, the contact type requires a very high input impedance. The non-contact type generally includes three working principles: induction probe type, field mill type and feedback type. the inductive probe type is based on the principle of current integration [3]. Therefore, calibration must be performed before each reading, so it is not suitable for continuous reading of surface voltage. The principle of the field mill is a mechanical shutter alternately shields and then exposes a measuring plate. These meters have high accuracy but tend to be expensive and vulnerable to mechanical damage. The principle of feedback type is that, when the electrostatic voltage between the measured surface and the sensor is equal to zero, the detected conductor voltage can be detected by adjusting the feedback loop. Although higher accuracy can be achieved with this method, there are problems such as high static voltage contacts due to the presence of boost equipment. Therefore, it is of great significance to develop a high-performance, low-cost static voltage sensor. Accordingly, based on the mode localization effect of the multi-degree-of-freedom weakly coupled resonant structure, this paper proposes a high-performance micro static voltage sensor.

II. PRINCIPLE AND DESIGN

The schematic of closed-loop electrostatic voltage sensor based on mode localization effect is shown in Figure 1. With the stiffness perturbation, the mode shape will

drastic change [4]. Based on this principle, the electrostatic perturbation which is generated by the electrostatic voltage causes the phenomenon of mode localization. The coupling method in this paper is electrostatic coupling. Every two neighboring resonators are weakly coupled by the voltage difference between the resonators [5]. The advantage of using electrostatic coupling is that the sensitivity can be adjusted by change the coupling coefficient. With the application of stiffness perturbation, the phase difference between the input and output signals of resonator 1 and resonator 3 are minus 90 degrees in the in-phase mode. Thus, we can shift the phase of output signal by 90 degrees, which can be fed back to the input to lock it in in-phase mode. In addition, the output amplitude is constant by AGC to ensure the output will not be influenced by the change of environment. With a closed-loop feedback control for mode localized sensors, its vibration modes can be selectively locked and the amplitude noises are greatly reduced. Ultimately, a high-performance electrostatic voltage sensor could be achieved by measuring the amplitude ratio under the closed-loop.

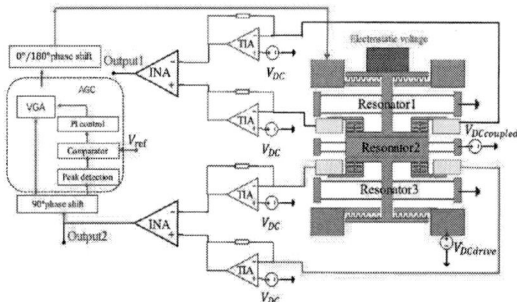

Figure 1. The schematic of closed-loop electrostatic voltage sensor based on mode localization effect

III. CHARACTERIZATION ANALYSIS

For the static voltage sensor designed in this paper, the model can be equated to a three-degree-of-freedom system model as shown in Figure 2 [6]. Since the device is measured in a vacuum, the damping can be ignored.

Where x_1, x_2, x_3 represent the displacements of resonators 1, 2, and 3, m_1, m_2, and m_3 represent the equivalent ideal

---

1. State Key Laboratory of Transducer Technology, Aerospace Information Research Institute, Chinese Academy of Sciences, Beijing 100190;

2. School of electronic, electrical and communication engineering, University of Chinese Academy of Sciences, Beijing 100049;

*Contacting Author is with the State Key Laboratory of Transducer Technology, Aerospace Information Research Institute, Chinese Academy of Sciences, Beijing 100190 (corresponding author to provide phone: +86-10-58887590; e-mail: crpeng@mail.ie.ac.cn).

978-1-6654-3008-1/21 $31.00 © 2021 IEEE

mass of the resonator, and $m_1=m_2=m_3=m$, k_1, k_2, and k_3 represent the mechanical stiffness of the resonator, where $k_1=k_3=k$, k_c represents the electrostatic coupling stiffness between the resonators, Δk represents the perturbation of the electrostatic stiffness of the resonator 1 caused by the electric field input electrode.

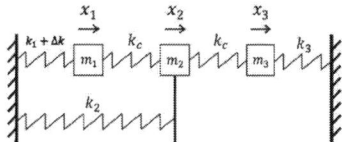

Figure 2 Mass-spring system (neglecting damping)

Solve the differential equations for this model. The relationship between the resonant frequency and the stiffness perturbation could be obtained, and the relationship between the amplitude ratio and the stiffness perturbation also could be obtained.

$$\omega_{ip} \approx \sqrt{\frac{1}{m}\left[k+k_c+\frac{1}{2}\left(\Delta k-\frac{2k}{\beta}-\sqrt{\Delta k^2+\left(\frac{2k}{\beta}\right)^2}\right)\right]} \quad (1)$$

$$\omega_{op} \approx \sqrt{\frac{1}{m}\left[k+k_c+\frac{1}{2}\left(\Delta k-\frac{2k}{\beta}+\sqrt{\Delta k^2+\left(\frac{2k}{\beta}\right)^2}\right)\right]} \quad (2)$$

$$\left|\frac{x_3}{x_1}\right|_{ip} \approx \left|\frac{\sqrt{\beta^2(\Delta k/k)^2+4}+\beta(\Delta k/k)}{2}\right| \quad (3)$$

$$\left|\frac{x_3}{x_1}\right|_{op} \approx \left|\frac{\sqrt{\beta^2(\Delta k/k)^2+4}-\beta(\Delta k/k)}{2}\right| \quad (4)$$

where ω_{ip} and ω_{op} denote the frequencies of the in-phase and out-of-phase modes of the system, respectively, $|x_3/x_1|_{ip}$ and $|x_3/x_1|_{op}$ denote the amplitude ratios of the resonators 3 and 1 of the in-phase and out-of-phase modes of the system, respectively, and β is defined as.

$$\beta = \frac{k(k_2-k+k_c)}{k_c^2} \quad (5)$$

When $\beta\Delta k/k > 10$, variation of amplitude ratio and stiffness are approximately linearly related as follows.

$$\left|\frac{x_3}{x_1}\right| = \beta\left|\frac{\Delta k}{k}\right| \quad (6)$$

For the sensor structure in this paper, the stiffness perturbation generated by the static voltage is [7]

$$\Delta k = -\frac{\varepsilon A V^2}{d^3} \quad (7)$$

where d is the distance between the perturbation electrode and the resonator, and A is the square area between the perturbation electrode and the resonator. ε is the dielectric constant.

Thus, the relationship between the amplitude ratio and the input static voltage is obtained as:

$$\left|\frac{x_3}{x_1}\right| = \frac{\varepsilon A(k_2-k+k_c)}{k_c^2 d^3} V^2 \quad (8)$$

So, the RMS value of amplitude ratio is linear to the applied static voltage as:

$$\sqrt{\left|\frac{x_3}{x_1}\right|} = \sqrt{\frac{\varepsilon A(k_2-k+k_c)}{k_c^2 d^3}} V \quad (9)$$

IV. DEVICE FABRICATION

The device was fabricated in a commercial SOIMUMPS process [8]. The SOIMUMPs process is a simple 4-mask level SOI patterning and etching process. The microscopic photo of the sensor is shown in Figure 3.

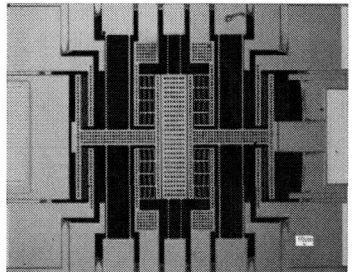

Figure 3. The microscopic photo of the sensor

V. EXPERIMENTAL RESULTS

In this paper, the open-loop test system is composed of POWER part, SIGNAL part, and etc. The sensor and the interface circuit are placed in a vacuum chamber at 2Pa, and the coupling voltage is 40V, the driving DC signal is 4Vdc. As shown in Figure 4, the POWER section is to provide power to the system. The output of the resonators is transmitted to the locking amplifier. The vacuum control equipment is used to control the vacuum level in the vacuum chamber. PC is used to control lock-in amplifier and process the data.

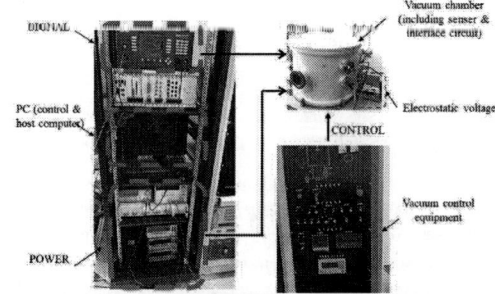

Figure 4. Schematic diagram of the experimental setup

Currently, we performed an open-loop test on this sensor. The amplitude output signals of resonators 1 and 3 are connected to the lock-in amplifier through a transimpedance amplifier and a differential amplifier. Then we perform a division operation on the output signal through the PC terminal. Based on the data measured by the lock-in amplifier, a curve of the amplitude ratio versus the input voltage is plotted.

VI. RESULT

A. Magnitude-frequency responses

The magnitude-frequency responses of resonator 1 and resonator 3 outputs are shown in Figure 5.

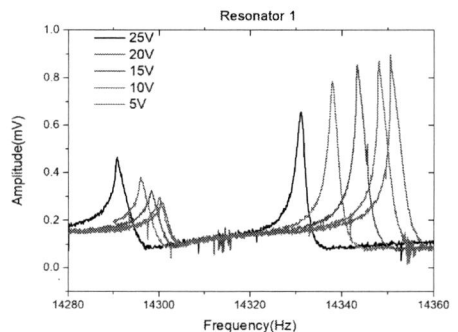

(a) Magnitude-frequency responses of Resonator 1

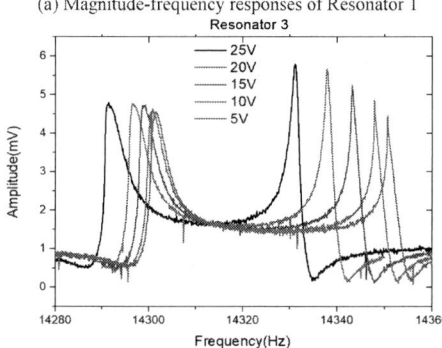

(b) Magnitude-frequency responses of Resonator 3
Figure 5. Magnitude-frequency responses of Resonator 1(a) and Resonator 3(b) versus input voltage

As shown in figure 5, the peaks with smaller frequencies are in-phase modes, and the peaks with lager frequencies are out-of-phase modes. The variation of the measured voltage leads to a frequency change where the frequency change of the in-phase mode is smaller with the out-of-phase mode. Thus, the sensor is asymmetrical with a positive stiffness [9]. The reason for the asymmetry of the system structure is due to the process error, which results in a higher stiffness of resonator 1 than resonator 3. Therefore, the operating mode selected in this paper is the in-phase mode.

B. Sensitivity Characterization

In order to achieve static voltage measurements, the following three cases were considered for this purpose, as the sensor operates in the positive stiffness region.

1) Frequency output in out-of-phase mode

The test results are shown in Figure 6. The sensitivity based on the frequency output in out-of-phase mode is calculated and fitted as follows:

$$s_{f\text{-}outphase} = \frac{\partial(\Delta f/f_0)}{\partial V} = 0.0011V^{-1} \quad . \tag{10}$$

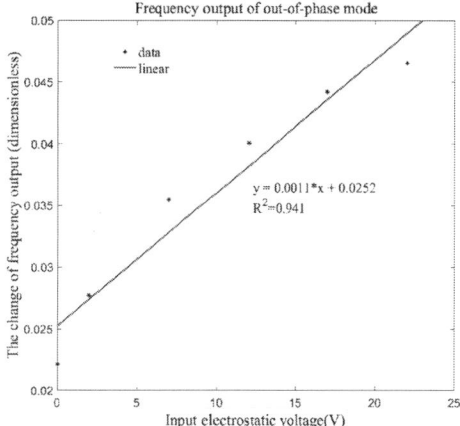

Figure 6. Frequency output in out-of-phase mode

2) Amplitude ratio output in in-phase mode

The amplitudes ratio of resonator 1 and resonator 3 in in-phase mode was shown in Figure 7. which has a good linearity in this range with a linear correlation coefficient of 0.9841 and a linearity of 6.11%.

Sensitivity of in-phase mode amplitude ratio output as follows:

$$s_{AR\text{-}Inphase} = 0.0967V^{-1} \tag{11}$$

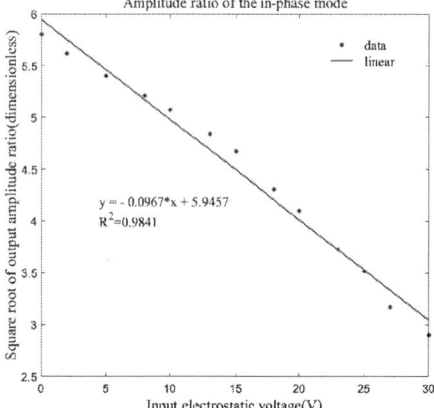

Figure 7. Amplitude ratio output of in-phase mode

3) Amplitude ratio output in out-of-phase mode

The amplitudes ratio of resonator 1 and resonator 3 in out-of-phase mode was shown in Figure 8S. The sensitivity can be obtained as 0.0507 according to the figure, and the linear correlation coefficient is 0.8996. The linearity is poor.

According to equations (9) and (10), The sensitivity of the amplitude ratio (0.0967/V) is 87.9 times higher than that of the frequency (0.0011/V). And the selected in-phase mode output has a good linearity.

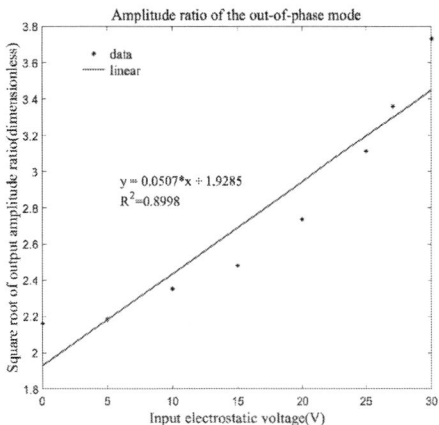

Figure 8. Amplitude ratio output of out-of-phase mode

VII. CONCLUSION

Based on SOI process, a high-performance MEMS static voltage sensor based on the mode localization effect has been developed. And closed-loop test systems are proposed. The sensor has a linearity of 6.11% and can achieve high sensitivity static voltage measurement.

ACKNOWLEDGMENT

The research was funded by the National Natural Science Foundation of China (Grant No. 62031025,61971398,41775024), National Key R&D Program of China (Grant No. 2018YFF01010800) and the Scientific Instrument Developing Project of the Chinese Academy of Sciences, Grant No. YJKYYQ20200026 , GJJSTD20210004.

REFERENCES

[1] C. Calle, J. Mantovani, C. Buhler, E. Groop, MBuehler. A. Nowicki. Embedded electrostatic sensors for Mars exploration missions", *J. Electrostat*, vol 61, pp. 245-257, 2004.

[2] E. R. Badman and R. G. Cooks, "Miniature mass analyzers," *J. Mass Spectrom*, vol. 35, pp. 659–671, 2000.

[3] A D G, A G H, A W M, et al. "An electrostatic charge meter using a microcontroller offers advanced features and easier ATEX certification", *Journal of Electrostatics*, vol. 67, pp. 473-476, 2009.

[4] C. Pierre, "Mode localization and eigenvalue loci veering phenomena in disordered structures", *Journal of Sound and Vibration*, vol. 126, pp. 485-502,1988.

[5] Zhao C, Wood G, Xie J, et al. "A sensor for stiffness change sensing based on three coupled resonators with enhanced sensitivity", in *2015 28th IEEE International Conference on Micro Electro Mechanical Systems. IEEE, 2015.*

[6] ZHAO C, WOOD G S, XIE J, et al. "A force sensor based on three weakly coupled resonators with ultrahigh sensitivity". *Sensors & Actuators A Physical*, vol. 232, pp. 151-162, 2015.

[7] P. Thiruvenkatanathan, J. Woodhouse, J. Yan and A. A. Seshia, "Manipulating Vibration Energy Confinement in Electrically Coupled Microelectromechanical Resonator Arrays," *Journal of Microelectromechanical Systems*, vol. 20, pp. 157-164, 2011.

[8] Yang, Pengfei , et al. "Design, fabrication and application of an SOI-based resonant electric field microsensor with coplanar comb-shaped electrodes." *Journal of Micromechanics & Microengineering* vol. 23, 2013.

[9] H. Zhang, J. Huang, W. Yuan and H. Chang, "A High-Sensitivity Micromechanical Electrometer Based on Mode Localization of Two Degree-of-Freedom Weakly Coupled Resonators," in *Journal of Microelectromechanical Systems*, vol. 25, pp. 937-946, 2016,

Gap in pagination due to unavailable paper.

Pages 1615-1616

Proceedings of the 16th Annual IEEE International
Conference on Nano/Micro Engineered and Molecular Systems
April 25-29, 2021

The Tunable Deformation of Microfluidic Strain Sensor Based on Auxetic Metamaterial*

Linna Mao, Dengji Guo, Sirong Huang, Taisong Pan

Abstract— **Microfluidic system has played a key role in realizing wearable biosensing and biochemical analysis. With the intrinsic flexibility and stretchability, the microfluidics is also a promising candidate for assembling the flexible sensors for the force signals. As the force detection with microfluidics highly relies on the structural deformation of the microfluidic channel, the mechanical property of the microfluidic device is closely related to its performance in a wide range. In this study, the feasibility to regulate performance of the microfluidic strain sensor by modulating its mechanical properties with the auxetic metamaterial is discussed with the finite element analysis. By analyzing the strain distributions in the sensors with/without the embedded auxetic frame, the abilities of the auxetic frame to override the device's intrinsic mechanical property and regulate the device's performance are presented. The influences of thickness, modulus, and location of the auxetic frame on the regulation effect are also discussed with the simulation results.**

I. INTRODUCTION

As an important element of the flexible electronics, the flexible strain sensor has been intensively studied in recent years[1-3]. The microfluidic strain sensor, with the intrinsic flexibility and good biocompatibility, is a promising candidate for the wearable force sensing[4, 5]. The sensing of external force with the microfluidic strain sensor relies on the deformation of the microfluidic channel with the applied strain. For example, when the sensor is stretched, the deformation of microfluidic channel leads to the changes of channel's length and cross-sectional area, which contribute to the change of resistance[6]. The deformation-dependent sensing mechanism provides the opportunity to regulate the sensor's performance by modulating the mechanical properties of the sensor. However, as the microfluidic devices are commonly constructed by the silicone elastomers, such as polydimethylsiloxane (PDMS) and Ecoflex, the limited modulation ranges of the mechanical properties for these materials make it difficult to modulate

*Resrach supported by the Natural Science Foundation of China under Grant 61825102 and Grant 61901085 and in part by the "111" Project under Grant B13042 and Grant B18011.

Linna.Mao. is with School of Materials and Energy, University of Electronic Science and Technology of China, Chengdu 610054, P.R.China. (e-mail: Linna_M00@outlook.com).

Dengji.Guo. is with School of Materials and Energy, University of Electronic Science and Technology of China, Chengdu 610054, P.R.China. (e-mail: 991917072@qq.com).

Sirong.Huang. is with School of Materials and Energy, University of Electronic Science and Technology of China, Chengdu 610054, P.R.China. (e-mail: 913815186@qq.com).

Taisong.Pan is with School of Materials and Energy, University of Electronic Science and Technology of China, Chengdu 610054, P.R.China.(Corresponding author, email: tspan@uestc.edu.cn).

the sensor's performance by regulating the structural deformation.

The development of mechanical metamaterial provides the opportunity to override the intrinsic mechanical property of the material by adopting different structure designs[7, 8]. Mechanical metamaterials as an important branch of metamaterials refer to a group of artificial structures with unique mechanical properties, which is indicated by four elastic constants: elastic modulus E, shear modulus G, bulk modulus K, and Poisson's ratio v[9-12]. Accordingly, auxetic metamaterials[13] with negative Poisson's ratio is one of the most important subfield in mechanical metamaterials[14, 15]. By forming the metamaterial with specific unit cells, transverse expansion of the metamaterial under uniaxial elongation can be realized. Moreover, the Poisson's ratio of the metamaterial can be significantly adjusted with the different design of unit cell. The metamaterial provides an alternative way to regulate the deformation of the microfluidic strain sensor without the limitation of the intrinsic mechanical properties of the device.

In this study, a microfluidic strain sensor embedded with auxetic metamaterial is proposed. The modulation effect of the auxetic metamaterial on the device's deformation is discussed with the finite element analysis (FEA). By theoretically analyzing the influence of the auxetic metamaterial on the strain distribution in the device, the auxetic frame embedded in the device is demonstrated to be able to regulate the device's deformation. Moreover, the roles of the metamaterial's structural parameter in determining the amplitude and uniformity of the regulation of deformation are discussed with the corresponding simulation results.

II. CURRENT RESULTS

The structure of the microfluidic strain sensor is shown in Fig. 1a. The sensor consists of three components, which are the microfluidic channel prepared by PDMS, the conductive liquid filled in the channel, and the auxetic frame. The Poisson's ratio (v), which represents the deformation of a material in a direction perpendicular to the direction of the applied force, is defined as the negative of the ratio of transverse strain to the axial strain. The principle of the auxetic metamaterial is illustrated in Fig. 1b. Compared with the conventional materials, the auxetic metamaterial will expand in the direction vertical to applied tensile strain, behaving with the negative Poisson's ratio. In this study, we mainly discussed the situation that the microfluidic strain sensor is stretched in the radial direction of microfluidic channel, as illustrated in Fig. 2a. The

978-1-6654-3008-1/21 $31.00 © 2021 IEEE

cross-sectional view of the device is shown in Fig. 2b. The thickness of the auxetic frame layer and the distance between the frame and microfluidic channel are defined as h (μm) and d (μm) respectively.

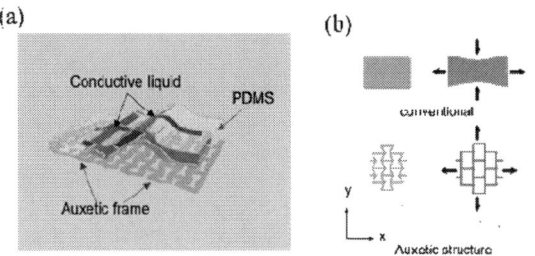

Figure 1.(a) Schematic diagram of the microfluidic strain sensor embedded with auxetic frame; (b) Mechanical deformation of conventional structure and auxetic structure.

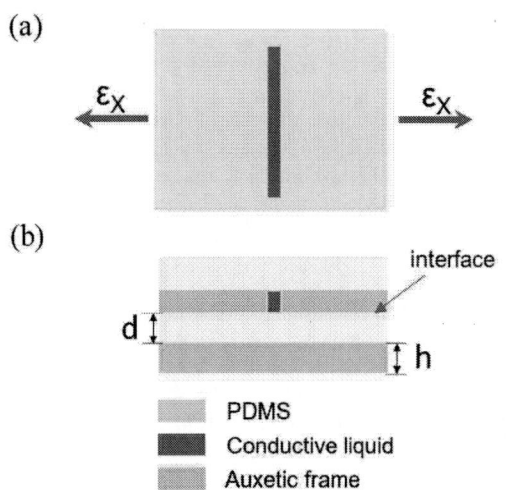

Figure 2. Simulation diagram: (a) Plane view; (b) Cross-sectional view.

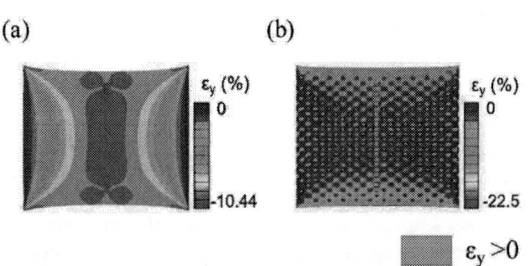

Figure 3. The strain distribution in the interface layer with ε_x =20%: (a) the microfluidic strain sensor without the auxetic frame; (b) the microfluidic strain sensor with the auxetic frame. The structural parameters of the auxetic frame are as follows: h=100 μm, d=200 μm, Elastic modulus (E) =100 MPa.

As displayed in Fig. 3, the strain distribution in the interface layer (labeled in Fig. 2b) can be derived from the FEA results of strain in the microfluidic strain sensor with/without the auxetic frame after applying 20% tensile

strain. The thickness (h) and location (d) of auxetic frame of the microfluidic strain sensor in Fig .3(b) was respectively set to 100 μm and 200 μm, and the elastic modulus of frame is 100 MPa. When the tensile strain is applied in the radial direction (ε_x), the auxetic frame effectively regulates the device's deformation in axial direction (ε_y). While the device is compressed without the auxetic frame (ε_y <0) due to the positive Poisson's ratio of PDMS (~0.5), the existence of auxetic frame leads to the appearance of expanded region in the interface layer (ε_y >0). As the applied tensile strain in the radical direction solely depends on the external loading, the change of the deformation in the axial direction can significantly modulate the dependence of device's resistance on applied strain (the gauge factor). However, the strain distribution shown in Fig. 3b also indicates a non-uniform regulation effect. As the auxetic frame is a mesh-like structure, the deformation of PDMS around the "hollow" of the frame may not be effectively regulated when the auxetic frame is not "stiff" enough. As shown in Fig. 4, the positively maximum ε_y and negatively maximum ε_y in the interface layer are derived from the FEA results under 20% tensile strain in radial direction, when considering the change of frame thickness (h), frame location (d) and frame modulus (E). With different frame thicknesses, it can be noted that the values of positively maximum ε_y all increase with the larger modulus of frame material, indicting the amplitude of the regulation can be improved by adopting high modulus material to form the auxetic frame. Meanwhile, merely increasing the modulus leads to the more obvious non-uniform distribution of the strain, since the value of negatively maximum ε_y also significantly increases with the increasing of the modulus. The uniformity of strain distribution can be modified by increasing the frame thickness, as indicated by the decrease of negatively maximum ε_y with the thicker frame. The amplitude of the regulation also demonstrates obvious dependence on the location of the auxetic frame, as indicated by Fig. 4d. Consequently, the elastic modulus, the thickness and location of auxetic frame played an important role in redistributing the applied strain and regulating the deformation of the microfluidic strain sensor.

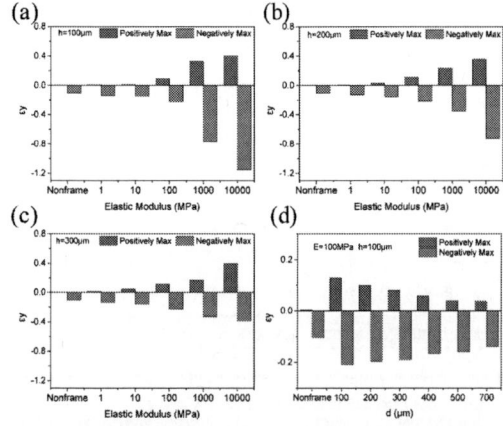

Figure 4. The distributions of positively maximum ε_y and negatively maximum ε_y in the interface layer with different thicknesses (h), elastic modulus (E) and location (d) of the auxetic frame.

Furthermore, the strain distribution along the microfluidic channel with/without the auxetic frame are also derived from the simulation results to investigate the dependence of the uniformity of regulation effect on the structural parameters of the metamaterial, as in Fig.5. The frame thickness of microfluidic strain sensor in Fig.5(a-c) is respectively set to 100μm, 200μm, 300μm. Fig.5(a-c) respectively demonstrated the trend of the strain distribution along microfluidic channel the length of the microfluidic channel when the elastic modulus increases with the fixed thickness and location of auxetic frame layer (d=200 μm). It can be noted that the strain in the microfluidic channel without auxetic frame are all negative and uniformly distributed along the microfluidic channel length, which indicates the device is transversely compressed ($\varepsilon_y<0$) with the positive Poisson's ratio of PDMS and axially tensile strain. When the auxetic frame is in the device, positively maximum strain appears, indicating the microfluidic channel deforms with the negative Poisson's ratio in some regions of the channel. Meanwhile, the amplitude of the positively maximum strain increases with the larger elastic modulus of the embedded auxetic frame. However, as the PDMS is a non-compressive material, the increased maximum strain also leads to more significant non-uniform distribution of strain in microfluidic channel. Larger positively maximum strain corresponds to larger negative maximum strain in other regions of the microfluidic channel. Fig.5(d) shows the influence of location on the uniformity of auxetic frame effect. the strain distribution becomes more uniform as the distance between the auxetic frame layer and microfluidic channel increases. Being similar with the modulus dependence of the strain distribution, the improvement of the uniformity by increasing the distance is also accompanied with the decrease of the positively maximum strain. Besides, when the elastic modulus of the auxetic frame is large enough to introduce the negative Poisson's ratio in the microfluidic channel, the uniformity of strain distribution along the microfluidic channel can be further improved by increasing the thickness of the auxetic frame, as shown in Fig.5(e) and (f). In general, it can be concluded from the simulation results that, the increase of the amplitude of the negative maximum Poisson's effect can be realized by increasing the elastic modulus of auxetic frame, reducing the placement distance, or reducing the thickness of the frame. However, it also brings a non-uniform strain distribution in the microfluidic channel. Therefore, the two consequences of modifying the structure parameters of the auxetic frame, the amplitude of positively maximum strain and the uniformity of the strain distribution, are mutually restricted. Thus, a balance in the design of the auxetic frame should be considered when adopting the auxetic frame.

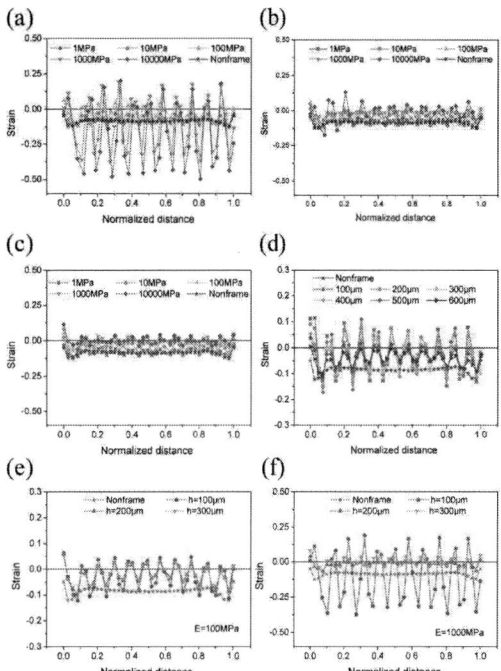

Figure 5. strain distribution along the channel length for different situation. The auxetic frame thickness(h) and the frame location(d) were respectively set to fixed values with different elastic modulus (E):a) h=100 μm,d=200 μm;b) h=200 μm,d=200 μm;c) h=300 μm,d=200 μm;d) The auxetic frame thickness(h=100 μm) and the elastic modulus (E=100 MPa)were set to fixed values with different location(d). The strain distribution along the channel length with different frame thickness and the elastic modulus of frame was set to fixed values: e) E=100 MPa; f) E=1000 MPa.

III. CONCLUSION

In conclusion, the introduction of mechanical metamaterial in the microfluidic strain sensor can effectivity regulate the deformation of the device by overriding the intrinsic mechanical properties of microfluidic strain sensor. By comparing the strain distribution at the interface layer, it is verified that the deformation trend of the microfluidic channel can be modulated by the auxetic frame. By modifying the thickness, modulus and location of the auxetic frame, the amplitude and uniformity of the regulation effect can be effectively modulated. This work demonstrates the feasibility of using the mechanical metamaterial to regulate the electro-mechanical coupling in the stretchable electronics, and also sheds a light on the application of mechanical metamaterial in the stretchable electronics.

REFERENCES

[1] Y. Yang and W. Gao, "Wearable and flexible electronics for continuous molecular monitoring," *Chem Soc Rev*, vol. 48, no. 6, pp. 1465-1491, Mar 18 2019.

[2]   J. Chen *et al.*, "Superelastic, Sensitive, and Low Hysteresis Flexible Strain Sensor Based on Wave-Patterned Liquid Metal for Human Activity Monitoring," *ACS Appl Mater Interfaces,* vol. 12, no. 19, pp. 22200-22211, May 13 2020.

[3]   T. Gong *et al.*, "Highly responsive flexible strain sensor using polystyrene nanoparticle doped reduced graphene oxide for human health monitoring," *Carbon,* vol. 140, pp. 286-295, 2018.

[4]   S. Li, Z. Ma, Z. Cao, L. Pan, and Y. Shi, "Advanced Wearable Microfluidic Sensors for Healthcare Monitoring," *Small,* vol. 16, no. 9, p. e1903822, Mar 2020.

[5]   M. Bechthold and J. C. Weaver, "Materials science and architecture," *Nature Reviews Materials,* vol. 2, no. 12, 2017.

[6]   D. Y. Choi *et al.*, "Highly Stretchable, Hysteresis-Free Ionic Liquid-Based Strain Sensor for Precise Human Motion Monitoring," *ACS Appl Mater Interfaces,* vol. 9, no. 2, pp. 1770-1780, Jan 18 2017.

[7]   K. E. Evans and A. Alderson, "Auxetic Materials: Functional Materials and Structures from Lateral Thinking!," *Advanced Materials,* vol. 12, no. 9, pp. 617-628, 2000.

[8]   Y. Jiang *et al.*, "Auxetic Mechanical Metamaterials to Enhance Sensitivity of Stretchable Strain Sensors," *Adv Mater,* vol. 30, no. 12, p. e1706589, Mar 2018.

[9]   K. Bertoldi, V. Vitelli, J. Christensen, and M. van Hecke, "Flexible mechanical metamaterials," *Nature Reviews Materials,* vol. 2, no. 11, 2017.

[10]   J. H. Lee, J. P. Singer, and E. L. Thomas, "Micro-/nanostructured mechanical metamaterials," vol. 24, no. 36, pp. 4782-810, Sep 18 2012.

[11]   Y. Liu, K. He, G. Chen, W. R. Leow, and X. Chen, "Nature-Inspired Structural Materials for Flexible Electronic Devices," *Chem Rev,* vol. 117, no. 20, pp. 12893-12941, Oct 25 2017.

[12]   J. L. Silverberg *et al.*, "Using origami design principles to fold reprogrammable mechanical metamaterials," (in English), *Science,* vol. 345, no. 6197, pp. 647-650, Aug 8 2014.

[13]   Y.-J. Lee *et al.*, "Auxetic elastomers: Mechanically programmable meta-elastomers with an unusual Poisson's ratio overcome the gauge limit of a capacitive type strain sensor," *Extreme Mechanics Letters,* vol. 31, p. 100516, 2019.

[14]   Y. Shi *et al.*, "An analytic model of two-level compressive buckling with applications in the assembly of free-standing 3D mesostructures," *Soft Matter,* vol. 14, no. 43, pp. 8828-8837, Nov 7 2018.

[15]   H. Chen *et al.*, "The equivalent medium of cellular substrate under large stretching, with applications to stretchable electronics," *J Mech Phys Solids,* vol. 120, pp. 199-207, Nov 2018.

Gap in pagination due to unavailable paper.

Pages 1621-1622

Proceedings of the 16th Annual IEEE International
Conference on Nano/Micro Engineered and Molecular Systems
April 25-29, 2021

Measurement of Comb Finger and Comb Spacing Stability in Phononic Frequency Comb

Q. Yang, L. Xu, R. Huan, Z. Jiang, A Ganesan and X. Wei, *Senior Member, IEEE*

Abstract—In this paper, we present a method for measuring the frequency stability of each comb finger and comb spacing in a phononic frequency comb. Based on this method, we present the experimental results of the stability of the comb fingers and comb spacing in a piezoelectrically actuated free-free beam microstructure for the first time. The theoretical and experimental results show that the Allan deviation of the comb finger increases approximately linearly with the increase of the comb order, while the Allan deviation of comb spacing is independent of comb order. These results provide a reference for the application of frequency comb and are expected to accelerate its application.

I. INTRODUCTION

An optical frequency comb is a comb like spectrum composed of discrete and equidistant frequencies, which is a powerful tool for spectroscopy and optical frequency measurement [1-2]. The optical frequency comb can be generated by using a mode-locked laser and high Q toroidal micro-resonators. In Kerr micro-ring optical frequency combs, the fluctuation of optical frequency is reflected by linewidth and frequency stability, which are measured by delayed self-heterodyne method. Increasing the stability of the optical frequency comb will greatly improve the performance of the GPS, and do coherent Raman transitions [3]. Therefore, the generation of stable optical frequency comb has always been an active area of research.

As the similarities between phonons and photons, theoretical work has demonstrated that the frequency combs can be generated in a coupled-mode system represented by Fermi-Pasta-Ulam α (FPU−α) chains [4]. And the phononic comb was experimentally observed in internal resonance system [5-8], parametric pumping system [9], modal coupling system [10], and so on. Despite there are more and more experimental observations of phononic frequency combs in mechanical resonators, the relationship between comb finger frequency stability and comb order is still unknown. Due to the multi-frequency characteristics of the frequency comb, band-pass filtering is required before measuring the comb fingers stability. However, a narrow bandwidth bandpass filter is very difficult to implement, and when measuring the stability of different comb fingers, the filter needs to be replaced. Therefore, the measurement of the stability of the frequency comb is challenging. Here in this paper, using a recently established phononic frequency comb pathway [5], we present a method for measuring the frequency stability of each comb finger in a phononic frequency comb. This method can measure the stability of different comb fingers

only by adjusting the reference frequency, and it is easier to implement narrowband filtering. Using this method, we measure the comb fingers stability and comb spacing stability.

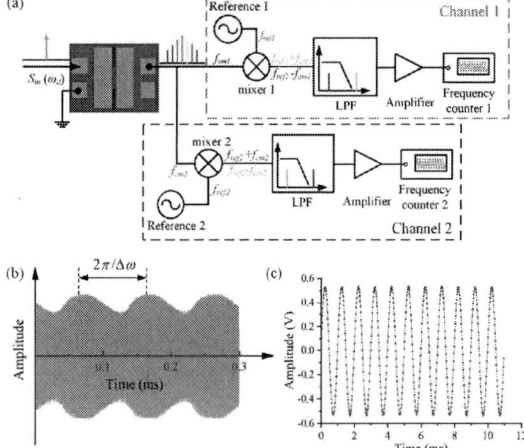

Figure 1: (a) The setup of experiment circuit. The resonator is a flat plate structure on which AlN thin film is deposited as a piezoelectric transducer. (b) Temporal Signature of frequency combs. The pulse train with a pulse period of $2\pi/\Delta\omega$. (c) Temporal waveform corresponding to the beat signal.

II. THEORETICAL EXPLANATION

Based on the understanding of the recently established frequency comb generation via 'Two-Mode Three-Wave Mixing', we now know that the emergence of frequency comb requires parametric resonance and a suitable drive frequency [5]. The resonator is driven by an electrical signal ω_d, when the resonant mode is driven outside the dispersion band, the frequency comb can be generated at ω_1 and $\omega_1/2$. This behavior is described by two coupled phonon modes with quadratic coupling nonlinearities and a 2:1 parametric resonance [11]. And the coupled equations of motion can be written as

$$\ddot{x}_1 + 2\gamma_1\dot{x}_1 + \omega_1^2 x_1 + \alpha x_2^2 = f\cos(\omega_d t) \qquad (1.a)$$

$$\ddot{x}_2 + 2\gamma_2\dot{x}_2 + \omega_2^2 x_2 + \beta x_1 x_2 = 0 \qquad (1.b)$$

Q. Yang, L. Xu, Z. Jiang and X. Wei are with the State Key Laboratory for Manufacturing Systems Engineering, Xi'an Jiaotong University, Xi'an, 710049, China. (phone: 86-29-82668839; fax: 86-29- 83234716; e-mail: senawei@mail.xjtu.edu.cn).

R. Huan is with the Zhejiang University, Hang Zhou, 310027, China.

978-1-6654-3008-1/21 $31.00 © 2021 IEEE   1623

where x_1 and x_2 denote the displacements of the modes at ω_1 and ω_2 respectively, γ_1 and γ_2 denote damping ratios, α and β denote the nonlinear coupling terms, f denotes the amplitude of the external driving force, and ω_d denotes the driving frequency. When the driving frequency matches the resonant mode frequency and the driving force is large enough, the resonant mode ω_d and the sub-harmonic mode $\omega_d/2$ triggered by the automatic parameter are excited. For the larger drive power f, when ω_d is beyond the dispersion band, the spectral lines of $\omega_1+n(\omega_1-\omega_d)$ and $\omega_1/2+n(\omega_1-\omega_d)$ are generated by the coupling term defined in (1). We interpret the appearance of the frequency as the nonlinear process of 'Two-Mode Three-Wave Mixing' since the comb fingers can be expressed as mixing products of the drive and resonant frequencies [5]. We use Allan deviation to evaluate the frequency stability of frequency comb [12]. Assuming that the nominal value of the central comb finger at ω_1 is f_1 and the relative frequency deviation is $x_1(t)$; the nominal value of ω_d is f_d and the relative frequency deviation is $x_d(t)$, then their actual frequencies are

$$f_1(t) = f_1(1 + x_1(t)) \tag{2}$$

$$f_d(t) = f_d(1 + x_d(t)) \tag{3}$$

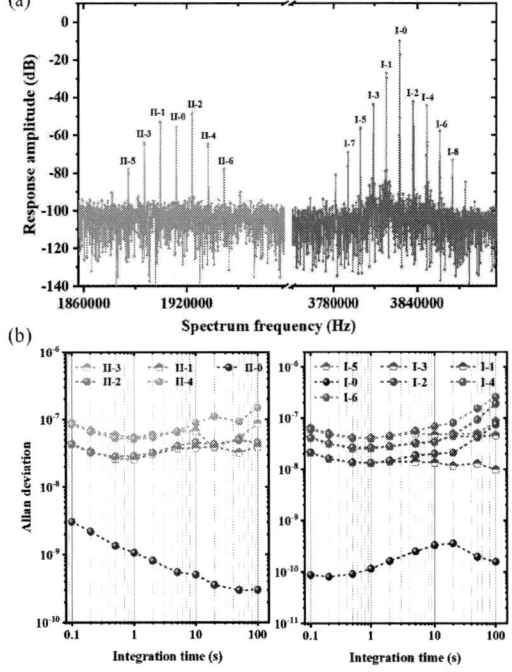

Figure 2: (a) Phononic frequency combs generated at ω_1 (red line) and $\omega_1/2$ (yellow line). (b) The Allan deviation of the comb fingers at ω_1 (right) and $\omega_1/2$ (left).

The Allan deviation $\sigma_{y_{1,n}}(\tau)$ of the n-order comb and the Allan deviation $\sigma_{y_{cs}}(\tau)$ of the comb spacing are given by (see appendix for detail)

$$\sigma_{y_{1,n}}(\tau) = \frac{f_1 \cdot \sigma_{y_{1,0}}(\tau)}{f_1 + nf_1 - nf_d} + \frac{nf_1 \cdot \sigma_{y_{1,0}}(\tau)}{f_1 + nf_1 - nf_d} \tag{4}$$

$$\sigma_{y_{cs}}(\tau) = \frac{f_1}{f_1 - f_d}\sigma_{y_{1,0}}(\tau) \tag{5}$$

For low-order combs (i.e., $n \leq 5$), since f_1 is much larger than $n(f_1 - f_d)$, it can be considered that the Allan deviation of the comb fingers increases approximately linearly with the increase of the comb order. Eq. (5) shows that the Allan deviation of comb spacing is a constant value independent of the comb order.

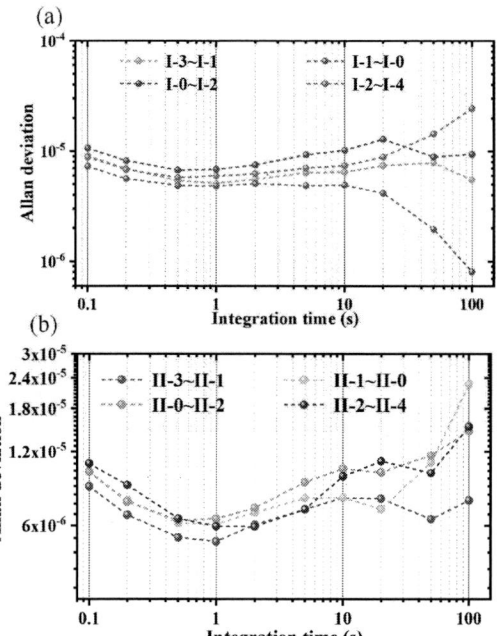

Figure 3: The Allan deviation of the comb spacing at (a). ω_1 and (b). $\omega_1/2$

III. EXPERIMENT SETUP

Now, to experimentally validate this, we consider a specific micromechanical device in Fig. 1(a). The employed micromechanical resonator consists of two-coupled Si-based micromechanical free-free beam structures of dimensions $1100 \times 350 \times 11$ μm. The device also consists of the 0.5 μm thick AlN and 1 μm thick Al layers deposited on the Si surface for piezoelectric actuation. The micro-mechanical resonator is fabricated using a standard AlN-on-Si foundry process. Fig.1(a) shows the test method of the stability of the comb fingers and comb spacing based on 'beat difference method'. Channel 1 and Channel 2 are used to measure the frequency stability of different comb fingers, and the two channels are identical. In Channel 1, the comb response is mixed with the reference signal f_{ref1} through a frequency mixer, where f_{ref1} is different from the frequency of the targeted comb f_{cf1}. The process of signal mixing is equivalent to the multiplication of two input signals, after mixing, a high frequency signal $(f_{cf1} + f_{ref1})$ and a low frequency signal $(f_{cf1} - f_{ref1})$ will appear in the spectrum. Through the adjustable low-pass filter (LPF) to obtain the low-frequency signal, and the frequency counter to measure the low frequency signal, the frequency stability of the targeted comb is calculated. The key of the 'beat difference method' is to mix the targeted comb and the reference signal, and then select the beat signal through the LPF. Therefore, it is important to set the 3dB bandwidth of the LPF reasonably. For the measurement of the stability of comb spacing, it is necessary to use two channels to simultaneously demodulate the frequencies of adjacent comb fingers.

IV. RESULTS AND DISCUSSION

To begin with, drive the resonator with driving strength V_{ac} = 15 dBm and driving frequency ω_d = 3.828 MHz, a comb-like response with few equidistant fingers in the frequency domain is observed, and the comb spacing is 9.33 kHz. Fig. 1(b) shows the response of the resonator as a function of time when a frequency comb appears. It can be seen that in the time domain, the frequency comb is in the form of 'pulse train', and the period of the 'pulse train' corresponds to the comb spacing. The frequency of I-1 is 3818540 Hz, and its frequency stability is measured by Channel 1. The reference 1 is set to 3817540 Hz, and the 3dB bandwidth of LPF is set to 500Hz. Fig. 1(c) shows the time domain waveform of the beat signal. After mixing, low-pass filtering and amplification, the output signal of the resonator is no longer a 'pulse train' in the time domain but a sine wave with a frequency of $f_{cf1} - f_{ref1}$. This signal is conveyed to the frequency counter to count the high-amplitude tone at gate time of 100ms. Generally, the relative frequency fluctuation of the reference signal can be ignored, and the frequency fluctuation of the target comb finger can be represented by the relative fluctuation of the beat signal frequency.

Fig 2(a) shows the frequency spectrum of the resonator, in which we can see the comb-like spectrum at ω_1 (marked as I-0, I-1, I-2, I-3, I-4, I-5 and I-6) and $\omega_1/2$ (marked as II-0, II-1, II-2, II-3, II-4, II-5 and II-6). Fig. 2(b) shows the measured frequency stability of each comb finger. From the experimental results, it can be seen that as the comb order increases, the Allan deviation also increases. When the integration time is 1s, the Allan deviation of the same order comb fingers on the left and right sides of the center comb are roughly the same, which indicates that their frequency stability is roughly similar. In addition, we can also find that the amplitudes of I-2 and I-4 are equivalent, but the Allan deviation of I-2 is smaller, which shows that the comb fingers stability are independent of with their energy. The measured signal of the central comb contains the feed-through signal, so the Allan deviation value is extremely small.

The stability of the comb spacing measured experimentally is shown in Fig.3. When measuring the spacing stability of I-1 (3818540 Hz) and I-3 (3809210 Hz), the reference 1 of Channel 1 is set to 3817540 Hz to demodulate I-1, and the reference 2 of Channel 2 is set to 3808210 Hz to

demodulate I-3. The beat signals of two channels are measured simultaneously by two identical frequency counters. The experimental results show that there is no definite relationship between the spacing stability and the comb order. Unlike the stability of the comb fingers, when the integration time is 1s, the stability of the comb spacing will not deteriorate with the increase of the comb order, but remains at 5-7ppm, which is consistent with the theoretical prediction.

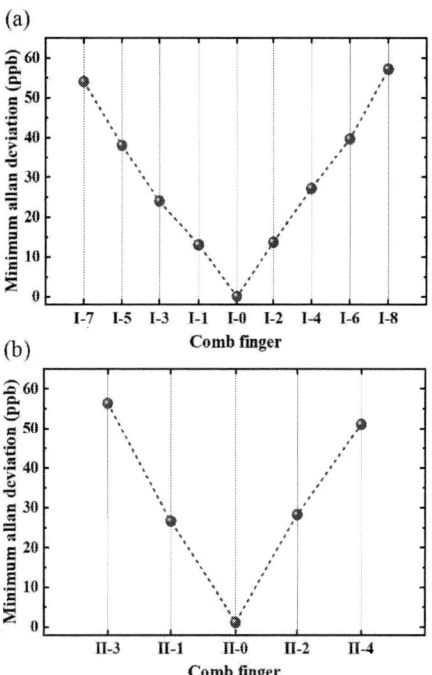

Figure 4: The minimum Allan deviation of comb fingers at (a). ω_1 and (b). $\omega_1/2$

Fig.4 depicts the frequency stability of the comb at ω_1 and $\omega_1/2$ when the integration time is 1s. Obviously, the center comb has the best frequency stability. The higher the comb order, the worse the frequency stability of the comb, and the Allan deviation increases linearly with the increase of comb order, which validates the result of Eq. (4).

CONCLUSION

In summary, this paper presented a method for measuring the frequency stability of phononic frequency comb, which is used to measure the comb fingers stability and comb spacing stability of phononic frequency comb in a piezoelectric resonator. Experimental results show that the Allan deviation of the comb fingers increase approximately linearly with the increase of the comb order; the Allan deviation of comb spacing is a constant value independent of the comb order. The Experimental results are in good agreement with the theoretical ones. This work will be beneficial to the application of frequency combs as frequency reference, signal processing and sensing [13-14].

Appendix

Suppose the nominal value of the central comb finger at ω_1 is f_1 and the relative frequency deviation is $x_1(t)$, then the actual frequency is

$$f_1(t) = f_1(1 + x_1(t)) \tag{A1}$$

Similarly, the driving frequency $f_d(t)$ can be expressed as

$$f_d(t) = f_d(1 + x_d(t)) \tag{A2}$$

Then the actual frequency of n-order comb is

$$f_{1,n}(t) = f_1(t) + n\left(f_1(t) - f_d(t)\right) \tag{A3}$$

The relative fluctuation of the driving frequency $x_d(t)$ is small and can be ignored. The relative frequency deviation $Y_{1,n}(t)$ of n-order comb is given by

$$
\begin{aligned}
Y_{1,n}(t) &= \frac{f_1(t) + n\left(f_1(t) - f_d(t)\right) - (f_1 + nf_1 - nf_d)}{f_1 + nf_1 - nf_d} \\
&= \frac{f_1 x_1(t) + nf_1 x_1(t) - nf_d x_d(t)}{f_1 + nf_1 - nf_d}
\end{aligned}
\tag{A4}
$$

Allan deviation $\sigma_y(\tau)$, is a "two-sample" variance of a time series measurement [12]. It is computed as follows

$$\sigma_y(\tau) = \sqrt{\frac{1}{2(M-1)} \sum_{i=1}^{M-1} \left[y_{i+1}(\tau) - y_i(\tau)\right]^2} \tag{A5}$$

where the y_n is the average value of the relative frequency deviation within a certain sampling time τ, and M is the number of measurements. Averaging (A4) and inserting in (A5), the Allan deviation $\sigma_{y_{1,n}}(\tau)$ of n-order comb can be obtained

$$
\begin{aligned}
\sigma_{y_{1,n}}(\tau) &= \frac{f_1}{f_1 + nf_1 - nf_d} \sigma_{y_{1,0}}(\tau) \\
&+ \frac{nf_1}{f_1 + nf_1 - nf_d} \sigma_{y_{1,0}}(\tau)
\end{aligned}
\tag{A6}
$$

For low-order combs (i.e., $n \leq 5$), since f_1 is much larger than $n(f_1 - f_d)$, $n(f_1 - f_d)$ can be ignored. Eq. (A6) can be written in the form

$$\sigma_{y_{1,n}}(\tau) = (n+1)\sigma_{y_{1,0}}(\tau) \tag{A7}$$

Similarly, the actual frequency of comb spacing f_{cs} is

$$f_{cs}(t) = f_1(1 + x_1(t)) - f_d(1 + x_d(t)) \tag{A8}$$

The relative frequency deviation $Y_{cs}(t)$ of comb spacing is given by

$$Y_{cs}(t) = \frac{f_1 x_1(t) - f_d x_d(t)}{f_1 - f_d} \tag{A9}$$

Averaging (A9) and inserting in (A5), the Allan deviation $\sigma_{y_{cs}}(\tau)$ of comb spacing can be obtained

$$\sigma_{y_{cs}}(\tau) = \frac{f_1}{f_1 - f_d} \sigma_{y_{1,0}}(\tau) \tag{A10}$$

Theoretical results show that the Allan deviation of the comb finger increases approximately linearly with the increase of the comb order, and the Allan deviation of comb spacing is a constant value independent of the comb order. The derivation process of frequency stability at $\omega_1/2$ is similar to the above.

Acknowledgment

This work is supported by the National Key Research and Development Program of China (2018YFB2002303), the National Natural Science Foundation of China (11772293, 52075432) and Key Research and Development Program of Shaanxi Province (2018ZDCXL-GY-02-03). The support from the International Joint Laboratory for Micro/Nano Manufacturing and Measurement Technologies is also appreciated.

References

[1] P. Del'Haye, A. Schliesser, O. Arcizet, T. Wilken, R. Holzwarth, and T. J. Kippenberg, "Optical frequency comb generation from a monolithic microresonator," *Nature*, vol. 450, no. 7173, pp. 1214-7, Dec 20 2007.

[2] T. Udem, R. Holzwarth, and T. W. Hänsch, "Optical frequency metrology," *Encyclopedia of Materials Science & Technology*, vol. 416, no. 6877, pp. 1-5, 2005.

[3] N. R. Newbury, "Searching for applications with a fine-tooth comb," *Nature Photonics*, vol. 5, no. 4, pp. 186-188, 2011.

[4] L. S. Cao, D. X. Qi, R. W. Peng, M. Wang, and P. Schmelcher, "Phononic frequency combs through nonlinear resonances," *Phys Rev Lett*, vol. 112, no. 7, p. 075505, Feb 21 2014.

[5] A. Ganesan, C. Do, and A. Seshia, "Phononic Frequency Comb via Intrinsic Three-Wave Mixing," *Phys Rev Lett*, vol. 118, no. 3, p. 033903, 2017.

[6] A. Ganesan, C. Do, and A. Seshia, "Excitation of coupled phononic frequency combs via two-mode parametric three-wave mixing," *Physical Review B*, vol. 97, no. 1, 2018.

[7] D. A. Czaplewski et al., "Bifurcation Generated Mechanical Frequency Comb," *Phys Rev Lett*, vol. 121, no. 24, p. 244302, 2018.

[8] D. A. Czaplewski, S. W. Shaw, O. Shoshani, M. I. Dykman, and D. Lopez, "Frequency Comb Generation in a Nonlinear Resonator through Mode Coupling Using a Single Tone Driving Signal," presented at the 2018 Solid-State, Actuators, and Microsystems Workshop Technical Digest, 2018.

[9] M. Park and A. Ansari, "Phononic Frequency Combs in Stand-Alone Piezoelectric Resonators," in *2018 IEEE International Frequency Control Symposium (IFCS)*, 2018.

[10] R. L. Kubena, W. S. Wall, J. Koehl, and R. J. Joyce, "Phononic comb generation in high-Q quartz resonators," *Applied Physics Letters*, vol. 116, no. 5, 2020.

[11] Z. Qi, C. R. Menyuk, J. J. Gorman, and A. Ganesan, "Existence conditions for phononic frequency combs," *Applied Physics Letters*, vol. 117, no. 18, 2020.

[12] D. W. Allan, "Statistics of Atomic Frequency Standards," *Proceedings of the IEEE*, vol. 54, no. 2, pp. 221-230, 1966.

[13] O. O. Soykal, R. Ruskov, and C. Tahan, "Sound-based analogue of cavity quantum electrodynamics in silicon," *Phys Rev Lett*, vol. 107, no. 23, p. 235502, Dec 2 2011.

[14] A. Ganesan and A. Seshia, "Resonance tracking in a micromechanical device using phononic frequency combs," *Sci Rep*, vol. 9, no. 1, p. 9452, Jul 1 2019.

Gap in pagination due to unavailable paper.

Pages 1627-1628

Proceedings of the 16th Annual IEEE International
Conference on Nano/Micro Engineered and Molecular Systems
April 25-29, 2021

Patterning and Immobilization of Silver Nanowires for Flexible Electronics by Using Microwave

Sara Khademi[1,2], Kiyumars Jalili[1*], Hao Wang[2], Md Eshrat E Alahi[2], Tianzhun Wu[2*]

Abstract— In most electronic devices such as flexible transparent conductive electrodes, silver nanowires (AgNWs) need to be patterned with time-consuming processing. Herein we propose a versatile technique to pattern silver nanowires with enhanced adhesion to polyimide substrate. By grafting cysteamine on the substrate via microwave irradiation and then by spin-coating, AgNWs can be grafted to the polymeric substrate. After scotch tape test on the robustness of AgNWs on PI, the results exhibited excellent stability. Hence this method has the potential to be a simple, fast, and straightforward technique to graft cysteamine and AgNWs on the polymeric surfaces.

Keywords
AgNWs, cysteamine, grafting, microwave, adhesion, patterning

I. INTRODUCTION

Due to the fact that ITO is brittle and expensive for flexible electronics, silver nanowires (AgNWs), owing to their outstanding optoelectronic properties and superior mechanical flexibility, are widely utilized in flexible transparent electronics such as optoelectronic devices, light-emitting displays, wearable electronics, and flexible solar cells. Adhesion to the substrates and precise patterning of AgNWs are crucial in a flexible electronic device. In order to commercialize AgNWs devices, suitable patterning methods should be developed. For high-resolution flexible electronics, desired patterns at an accurate position are required. To date, a lot of research groups have studied AgNWs patterning, such as photolithography, laser ablation, and spray or drop coating. [1-3] These methods are limited by their disadvantageous. For instance, the polymeric substrate tends to be damaged by the acid wet etching and high laser power during the photolithography and laser ablation process, respectively.

So far, improvement adhesion among AgNWs and substrate by chemical and physical methods such as photonic sintering technique [1] have been reported. Jiu et al. showed that AgNWs adhesion on the PET could be improved by using a high-intensity pulsed light technique [1]. Although AgNWs adhesion to the substrate could be improved, to prohibit NWs damage, the light power, and the pulse length need to be controlled. Several research groups widely investigate contact transfer methods by using PDMS, but it is not an appropriate method for large scale fabrication [2]. The other group attempted to solve the AgNWs adhesion problem by spray coating of AgNWs composite solution on the PET substrate. To made a composite solution, dopamine hydrochloride and alginic acid were added to the AgNWs ethanol solution. They reported that the electrical conductivity was reduced three times after one hundred tape tests. Therefore, the adhesion problem was partially solved by this method [3].

Herein, we proposed a facile and accurate AgNWs patterning by grafting them to the polymeric substrate. In this paper, cysteamine grafting on the substrate via MW irradiation is more discussed. In order to immobilize AgNWs to the substrate, cysteamine was grafted on the polyimide (PI) and polyethylene terephthalate (PET) substrate via microwave (MW) irradiation. Cysteamine is the simplest stable aminothiol which its -SH functional group prone to covalently bonded to the inorganic material such as Ag. Therefore, this interaction can be used to bind AgNWs on the polymeric substrate.

Microwave as electromagnetic radiation can significantly reduce the reaction times from several hours to a few minutes due to speed up the reaction rate. Besides, enhancement of reaction yield, elimination of side product formation, and reproducibility improvement are MW's main advantages. In microwave irradiation, only polar compounds able to adsorb microwave efficiently and can transform electromagnetic into heat [4].

II. EXPERIMENTAL METHODS

A. Plasma treatment and Grafting cysteamine on the substrate

Since MW can be adsorbed by polar component, thus, the cysteamine can be selectively grafted on the polymeric surface by surface treatment. For the aim of precise cysteamine grafting, photoresist was firstly deposited and patterned using the photo-lithography process. Free radicals and functional groups, especially the -COO-, -C=O, -COOH, -OH groups were created on the polyimide (PI) surface by plasma treatment. After the plasma treatment, substrates were inserted into a cysteamine solution and put them to microwave. Then samples rinsed in ethanol and water, following dried by nitrogen gas. The schematic of this process is represented in Figure 1 (a-c). In the aim to obtain optimal conditions, different reaction conditions were investigated. Whereby the tested reaction conditions were limited to two different MW irradiation time and two different MW power. Reaction conditions of grafting cysteamine on the substrate are summarized in Table 1.

TABLE 1. DIFFERENT REACTION CONDITIONS FOR THE MWs ASSISTED CYSTEAMINE GRAFTING.

sample	Reaction time (min)	Mw Power
1	1	Low power
2	2	Low power
3	1	High power

This work was supported by the grant from Guangdong Science and Technology Research Program (2019A1515110843, 2019A050503007), Shenzhen Science and Technology Research Program (JCYJ20170818152810899, JCYJ20170818154035069; JSGG20170824170930929) and UNSW-CAS Collaborative Research Seed Program (172644KYSB20190077).
[1]Sahand University of Technology, Tabriz, Iran (Kiyumars Jalili, Tel: +98-41-33459087, Fax: +98-41-33444313, email: k_jalili@sut.ac.ir)
[2]Shenzhen Institute of Advanced Technology, Chinese Academy of Sciences, Shenzhen, P.R. China (Tianzhun Wu, email:tz.wu@siat.ac.cn)

978-1-6654-3008-1/21 $31.00 © 2021 IEEE

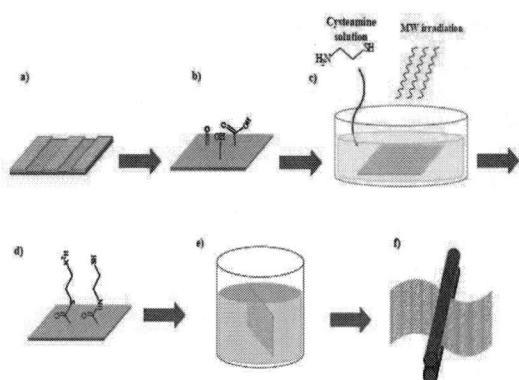

Figure 1. Schematic representation of PI surfaces grafted by cysteamine (a) surface treatment by O_2 plasma to create polar functional groups; (b) Insert substrate into cysteamine solution and cysteamine grafted on the surfaces by MW irradiation; (c) rising substrate which grafted with cysteamine in ethanol and DI water following dried by N_2 gas; (d) AgNWs spin-coated on the substrate (e) modified substrate insert to the AgNWs (0.3 %wt) dispersed in DI water and (f) substrate passed through an office laminator in order to continue the reaction between AgNWs and cysteamine, and more effective contact between AgNWs and substrate.

B. Grafting silver NWs on the PI surface

After the cysteamine had been grafted on the substrate, silver NWs (0.3 % wt dispersed in water, purchased from Guangdong Nanhai ETEB Technology Co.) was deposited on the substrate by using the dip-coating technique. In order to water evaporation and react silver NWs with -SH functional groups, put the substrate on a hot plate at 100°C for 5min. Subsequently, the substrate passed through a commercial hot-roll laminator at 100°C as shown in Figure 1-e). By using laminator in addition to improve silver NWs junctions, reaction between surface functional groups and NWs can be completed and enhanced interfacial surface among them. Finally, samples sonicate and rinsed in DI water and dried with nitrogen gas.

III. RESULT AND DISCUSSION

A. Contact Angle Analysis

In order to uniformly graft cysteamine on the PI surfaces, uniform and high-density radicals and functional groups such as -COO-, -C=O, -COOH, -OH must be created on the surfaces. Plasma treatment is one of the best methods to create radicals and chemical groups on the surfaces. Contact angle measurements is an effective and feasible method to can provide good information about surface energy. To characterize surface energy and polarity of pristine and treated substrates, two types of liquids (water and glycerol) are used. In order to obtain the best surface free energy, PI substrate was treated by O_2 plasma at different exposure time (35, 55, 75, and 90 sec). The measurements are performed with the sessile drop technique in which a 5μl drop was placed on the surface. An image was taken from a drop and was analyzed by drop analysis LB-ADSA in image j software. Five contact angle measurements for each sample are made and average. Then, the surface free energy

substrates were calculated using the Owens–Wendt method [5] via equation (1). Which θ is contact angle (°), γ_l, γ_l^p and γ_l^d are total surface tension of solution, polar component and dispersive component of surface tension, respectively; γ_s, γ_s^p and γ_s^d are total surface tension of sample, polar component and dispersive component of sample, respectively.

$$(\gamma_s^p \, \gamma_l^p)^{0.5} + (\gamma_s^d \, \gamma_l^d)^{0.5} = 0.5 \, \gamma_l(1+\cos(\theta)) \quad (1)$$

$$\gamma_l = \gamma_l^p + \gamma_l^d \quad (2)$$

$$\gamma_s = \gamma_s^p + \gamma_s^d \quad (3)$$

The variation of contact angle with exposure time with O_2 plasma to understand the surface affinity is shown in Figure 2a. Contact angle remarkably reduced for both water and glycerol. In the case of water, the contact angle reduced from 59.65° for pristine PI to 9.52° only after 35 sec exposure time. Obviously, the PI surfaces after plasma exposure become completely hydrophilic.

The surface free energy, polar, and dispersion components of the surface free energy are calculated and plotted as depicted in Figure 2-b to show the effect of exposure time. The polar component significantly enhanced with a negligible dispersion component for all exposure time. Consequently, a dichotomy is shown between high-polar and low-dispersion characteristics on the PI surfaces because of increasing amount of polar groups such as -COO-, -C=O, -COOH, -OH on the PI surfaces, which can significantly affect on the cysteamine grafting.

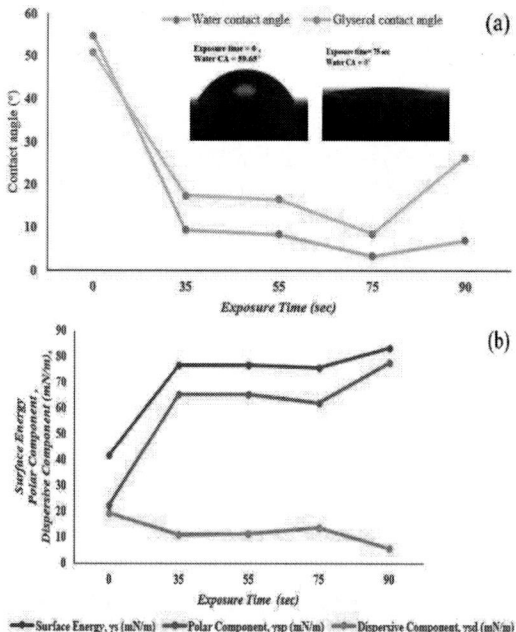

Figure 2. (a) surface contact angle measurement of PI with water and glycerol after treatment with O_2 plasma at different exposure time, water contact angle for the samples before and after O_2 plasma treatment are represented on the chart and (b) variation of surface free energy (γ_s), polar component (γ_s^p) and dispersive component (γ_s^d) vs. different exposure time.

B. AFM Characterization

Atomic force microscopy (AFM) can provide high-resolution images from material surfaces because of the force interaction between the probe and the sample. The surface morphology of PI was grafted with cysteamine at different MW conditions were characterized by AFM analysis. Surface topography has been obtained with uncoated silicon cantilever tips of about 10 nm apex diameters over scan areas of 6×6 μm^2. AFM images have been processed with Nanoscope analysis software.

Figure 3 illustrates the AFM images for a PI substrate treated with O$_2$ plasma and grafted with cysteamine at low and high MW power for different irradiation time. It is apparent from phase images that cysteamine grafted on PI surfaces for all samples at low and high MW power. In comparison between phase images, it is clear that cysteamine is uniformly grafted on PI surfaces at high MW power rather than the lower power. The probability of recombination of a radical pair may be increased by a strong magnetic field that can be controlled by MW irradiation [6]. This acceleration is due to a collective vibration of electrons induced by MW irradiation [7]. The effect of cysteamine grafting time investigated as well. It is obvious from phase images that at the same MW power, cysteamine grafted more uniform in 2 min irradiation than 1 min irradiation. The surface roughness of Ra and root-mean-square roughness (RMS) values were calculated from AFM height images and are listed in Table 2 for all samples. The smooth surfaces were roughed at low power for both irradiation times, which indicates ununiform grafting of cysteamine on the surfaces. The surface roughness for high MW power is about 0.585 nm indicate uniform cysteamine grafting on the surface.

C. XPS

Successful grafting of thiols has been confirmed by X-ray photoelectron spectroscopy (XPS), which clearly indicated the oxygen concentration created on PI surfaces after plasma treatment decreased after thiol grafting. On the other hand, the sulphur concentration increased (Table 3).

D. SEM Characterization

The scanning electron microscopy (SEM) images of AgNWs graft on the PI and PET substrate is shown in Figure 4. The density of AgNWs patterns could be readily tuned by changing AgNWs concentration. The density of AgNWs strongly effects on optical and electrical properties. The line width in PI substrate is 6μm and It is obvious that AgNWs can be accurately patterned in small size which rely on photolithographic step. The SEM image demonstrates that AgNWs uniformly have been patterned and grafted on both PI and PET substrates.

E. Adhesion tests

Scotch tapes were used to evaluate the adhesion between AgNWs and PI grafted with cysteamine. To make

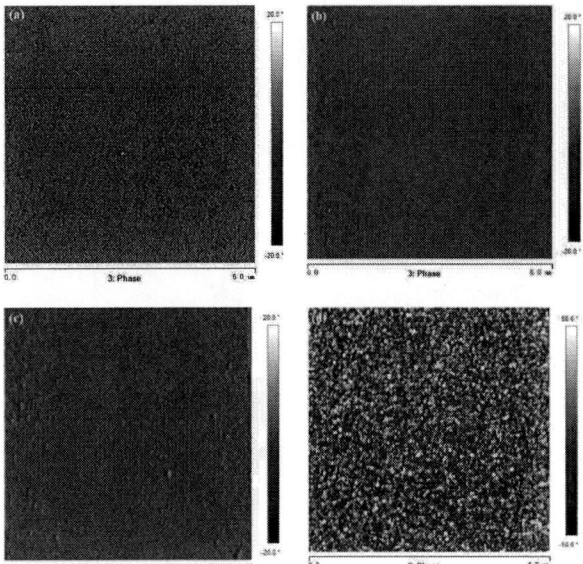

Figure 3. Tapping mode of AFM for (a) PI treated by O2 plasma, treated PI was grafted with cysteamine (b) at low MW power for 1 min MW irradiation; (c) at low MW power for 2 min MW irradiation; (d) at high MW power for 1 min MW irradiation.

TABLE 2: SURFACE ROUGHNESS OF DIFFERENT SAMPLES.

Samples	Ra (nm)	RMS (nm)
Surface treatment by O₂ plasma	0.659	0.852
Low MW power/ irradiation for 1 min	1.041	1.412
Low MW power/ irradiation for 2 min	1.072	1.559
High MW power/ irradiation for 1 min	0.585	0.831

TABLE 3: ATOMIC CONCENTRATION OF C(1s), O(1s), N(1s) AND S(2p)

Samples	C (%)	N (%)	O (%)	S (%)
Pristine PI	49.82	1.06	24.67	-
PI treated with O2 Plasma	45.66	7.95	25.8	-
PI grafted with Cysteamine	60.42	8.82	17.62	3.99

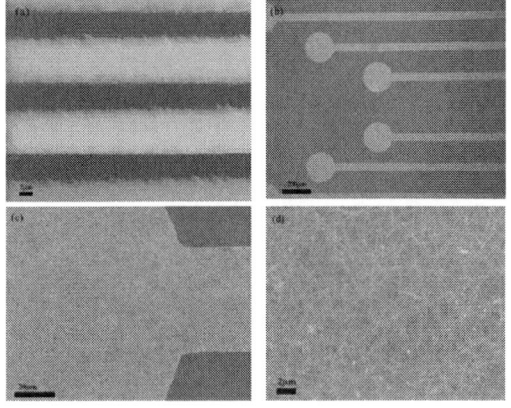

Figure 4. SEM images of (a) AgNWs patterned and grafted on (a) PI substrate with 6μm line width and (b,c and d) PET substrate with 200 and 50μm.

sure that the tape is adequately attached to the sample, the scotch tape was firmly pressed on the AgNWs coated surface by fingers. Then the scotch tape was peeled off from the PI surface. This Scotch tape test was repeated 10 times and at 5 different points. Scotch tape test image from on AgNWs grafted on PI substrate is shown in *Figure 5*. The sample showed the same conductivity before and after the test. This result confirms that the adhesion between AgNWs and the PI film grafted with cysteamine is very strong.

Resistances of AgNWs Microelectrodes with Different Width from 50 to 500 μm and the same length (2.3cm) are obtained with 2-point prob, which varies between 34.6 to 11.2 Ω. The resistance results are shown in Fig. 5a. Electrical conductivity enhanced by line width decrease.

In order to characterize this method performance an LED array was designed. When the bias was applied to the 7 LED array, all LEDs of the array emitted light brightly at flat and folding state with exhibiting a clear background image (Figure 6- b and c). when the LED was connected with the electrodes and then the bending was performed, the LED still turn on with no significant change of brightness at the same applied voltage which indicate appropriate adhesion between substrate and AgNWs.

I. CONCLUSION

In this paper, AgNWs were covalently bonded to the PI and PET substrate and enhanced NWs adhesion to the substrate. Since cysteamine can graft on the polymeric substrate such as PET, PI, PS, PTFE, PVDF and so on, we anticipate that this technique can be widely used for immobilization inorganic nanoparticles such as gold and silver on different polymeric substrates in various applications. Besides, by using MW to graft cysteamine on the PI surfaces, reaction time is remarkably reduced from 24 hours to 1 min. We believe this method has great potential in transparent and stretchable devices with long-lasting performance.

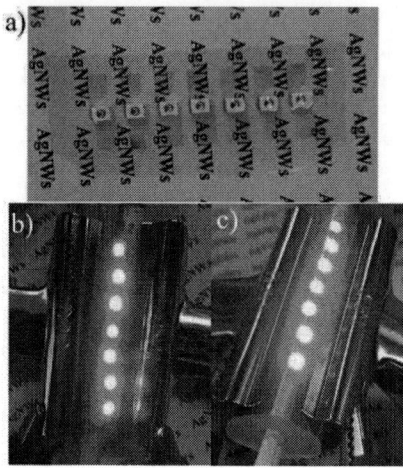

Figure 6. Photographs of operation of 7 LED array at a) without applied bias, b) flat and c) bending state and with applied bias.

REFERENCES

[1] J. Jiu, M. Nogi, T. Sugahara, T. Tokuno, T. Araki, N. Komoda, K. Suganuma, H. Uchida and K. Shinozaki, "Strongly adhesive and flexible transparent silver nanowire conductive films fabricated with a high intensity pulsed light technique", *J. Mater, Chem.*, 22, 23561 (2012)

[2] A. R. Madaria, A. Kumar and C. Zhou, "Large scale, highly conductive and patterned transparent films of silver nanowires on arbitrary substrates and their application in touch screens", *Nanotechnol.*, 22, 245201 (2011)

[3] Y. Jin, D. Deng, Y. Cheng, L. Kong and F. Xiao, "Annealing-free and strongly adhesive silver nanowire networks with long-term reliability by introduction of a nonconductive and biocompatible polymer binder", *Nanoscale*, 6, pp 4812-4818 (2014).

[4] Aʹ. Dıʹaz-Ortiz, P. Prieto, and A. de la Hoz, "A Critical Overview on the Effect of Microwave Irradiation in Organic Synthesis", *Chem. Rec.* 2019, 19, 85 –97

[5] Jia Wang, Kan Zhang, Fuguo Wang and Weitao Zheng, "Improving frictional properties of DLC films by surface energy manipulation", *RSC Adv.*, 2018, 8, 11388–11394

[6] P. Klaʹn, V. Cıʹrkva, Microwaves in Photochemistry in Microwaves in Organic Synthesis 2nd ed. A. Loupy Ed., Wiley, 2006.

[7] F. Kishimoto, T. Imai, S. Fuji, D. Mochizuki, M. M. Maitani, E. Suzuki, Y. Wada, Sci. Rep. 2015, 5, 11308.

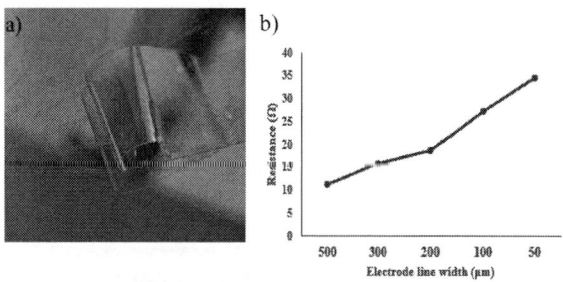

Figure 5. a) Scotch tape test performed on AgNWs grafted on PI substrate, b) Resistances of AgNW-Based Microelectrodes with Different line width and 2.3cm in length on PET.

Gap in pagination due to unavailable paper.

Pages 1633-1634

Proceedings of the 16th Annual IEEE International
Conference on Nano/Micro Engineered and Molecular Systems
April 25-29, 2021

Temperature and pressure dual-parameter sensing based on Fiber Bragg Grating

Na Zhao, Qijing Lin*, Liangquan Zhu, Kun Yao, Bian Tian, Libo Zhao, Zhuangde Jiang

Abstract— This paper experimentally studies the temperature and pressure response of fiber Bragg grating (FBG) and gold-plated FBG. The incident light is transmitted through the sensor and is reflected after encountering FBG. As the temperature and pressure changes, the reflection peak will change. The changes in the reflection spectrum are studied through experiments to get the temperature and pressure changes at the measured point. The experimental results show that the temperature response sensitivity of the FBG is 0.011 nm/ °C, and the response sensitivity of gold-plated FBG is 0.031 nm/ °C . The pressure response sensitivities of the sensors are 0.07nm/Mpa and 0.09nm/Mpa, respectively. Through the comparison of FBG and gold-plated FBG, temperature and pressure dual-parameter sensing is realized. The probe structure is simplified, the measurement accuracy of the sensor is improved, and the temperature and pressure dual-parameter sensor measurement in the medical, chemical, engine and other research fields are realized.

Keywords—optical fiber sensor; multimode fiber Bragg grating (MMFBG); multimode fiber; temperature sensing

I. INTRODUCTION

Due to fiber optic sensors have the advantages of stable chemical performance, anti-electromagnetic interference, compact structure, low cost, sensing in harsh environments, good insulation performance, light weight, etc., in aerospace, petrochemical and other fields, pressure and temperature sensing measurement has important applications. The work of optical fiber sensor researchers mainly focuses on Michelson interferometer [1-5], Mach-Zehnder interferometers [6-11], Fabry-Perot interferometers [12-16], long period gratings [17-19], fiber Bragg grating grating [20-30] and other extensive research. As mentioned above, various sensors have many shortcomings. For example, Mach-Zehnder interferometer and Fabry-Perot interferometer will weaken the mechanical strength during

the fusion process, and there will be a complex combination of two optical fiber components. Fiber grating has attracted great attention due to its small size, low cost, easy manufacturing, and high maturity. It has become an important sensor in many fields. In 1989, G. Meltz of the East Hartford Joint Technology Center in the United States opened up a new way to prepare gratings flexibly and efficiently [31]. However, one method has higher requirements on the light source, so the grating can also be used for greater practical applications. In 2001, NASA installed a fiber grating sensing network on the X-38 aircraft to monitor temperature and strain in real time. The high-sensitivity fiber Fabry-Perot cavity with nano-silver film has an extremely high pressure sensitivity of 70.5nm/kpa in the pressure range of 0-50kpa [32]. CiDRA was used in an oil well with a depth of 670 meters in Cohen River, California, USA. And they installed a commercial temperature and pressure dual-parameter fiber grating sensor in an oil well with a depth of 4,500 meters in the Gulf of Mexico. American Halliburton Petroleum Technology Service Company [33] provides industry-leading oil downhole optical fiber sensors. It not only has optical fiber dispersion temperature sensors, but also developed optical fiber sensors that measure temperature and pressure at the same time. Qiao Xueguang et al. [34]proposed a dual-parameter measurement between 0-7Mpa and 22.6-112.6 ℃ based on a double-FBG combined sensor with a thin-walled tubular shell. The accuracy can reach ± 0.5 ℃ , and the pressure accuracy can reach ±0.02Mpa. ; Wu Haifeng et al. [35] also chose to design and use dual FBGs to conduct experimental tests in the temperature range of 30℃ -100℃ and pressure 1Mpa-40Mpa, and conducted field tests on the dual parameters of oil well temperature and pressure, which proved that the sensor has better performance.

In this paper, FBG and gold-plated FBG are studied and used for temperature and pressure measurement. The working principle of the sensor is discussed. Based on the

---

*This work is supported by the Postdoctoral Innovative Talent Support Program (Grant No. BX20200274) and the National Natural Science Foundation of China (Grant Nos. 51805421, 51720105016, and 91748207) for financial support.

Na Zhao, Liangquan Zhu, Kun Yao, Bian Tian, Libo Zhao, Zhuangde Jiang are all with State Key Laboratory for Manufacturing Systems Engineering, Xi'an Jiaotong University, Xi'an 710049, China (e-mail: zn2015@stu.xjtu.edu.cn; 395918884@qq.com vinsent@stu.xjtu.edu.cn;

t.b12@mail.xjtu.edu.cn; libozhao@mail.xjtu.edu.cn; zdjiang@mail.xjtu.edu.cn).

Qijing Lin is with State Key Laboratory for Manufacturing Systems Engineering, Xi'an Jiaotong University, Xi'an 710049, China and Collaborative Innovation Center of High-End Manufacturing Equipment, Xi' an Jiaotong University, Xi'an 710054, China (corresponding author to provide phone: +86-029-82668616; e-mail: xjjingmi@163.com).

different temperature and pressure response sensitivities, the purpose of constructing a temperature and pressure dual-parameter sensor monitoring on a sensing structure is realized.

II. PRINCIPLE AND DESIGN

A. Principle

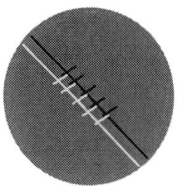

Figure 1.   Schematic diagram of layout of FBG and gold-plated FBG

A sensor test calibration system is established to calibrate the sensor. The schematic diagram of the measure system is shown in figure 1. The experiment designed to lay FBG and gold-plated FBG on a pressure diaphragm. The number of lines of the grating is about 1000 lines per centimeter, and the FBG is written along the longitudinal direction. After the light emitted from the light source enters the FBG through the input fiber, the light of each order is reflected back, and the temperature of the external environment is monitored by monitoring the light of each order. The temperature and pressure response sensitivity of sensors of FBG and gold-plated FBG are different. Finally, the demodulation device is used to monitor each order sensing peak. The core diameter of the multimode fiber is 65 μ m (1.4723) and the cladding diameter is 125 μ m (1.4573). When the temperature and pressure change, the corresponding reflection peak changes.

For the FBG, thermo-optical effect and thermal expansion effect affect the change of optical path difference. Therefore, when the ambient temperature changes, the length and effective index of the FBG will change. The optical path difference can be expressed as the following formula

$$\lambda_B^m = 2n_{eff}^m \Lambda \tag{1}$$

Among them, Λ is the pitch of the grating, and neff is the effective refractive index of the fiber core. When the ambient temperature acts on the FBG, the reflection peak will drift. In the formula, the thermal expansion coefficient and the optical path are constant. The reflection peak will change as the external environment changes. We can monitor the spectrum to get the ambient temperature.

B. Experiment and Discussions

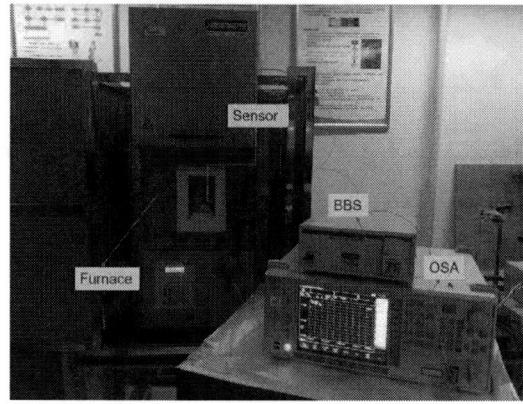

Figure 2.   The actual photograph for the temperature measurement system.

Figure 2 is the photo of the measurement system which consists of a broadband light source(BBS) with a spectral range from 1500 to 1600 nm and an optical spectrum analyzer (OSA) with a resolution of 0.02 nm. The measurement accuracy of the high temperature furnace is 1 ℃ . As the temperature increases, the spectra of FBG and gold-plated FBG both shift to the long wave direction. The temperature of the point to be measured can be deduced by detecting the movement of the spectrum.

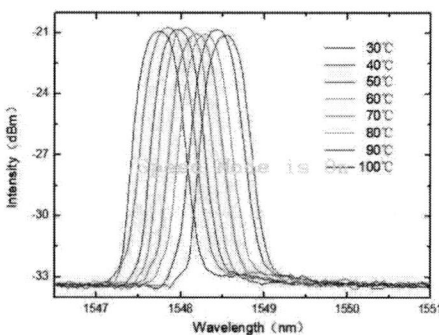

Figure 3.   FBG reflection spectrum change with temperature.

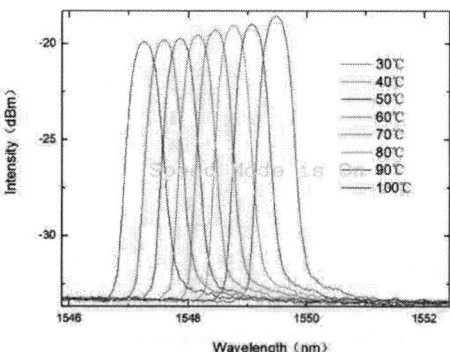

Figure 4.   Gold-plated FBG reflection spectrum change with temperature.

During the entire experiment, the temperature was heated to 100℃ to ensure the coating was removed. Then, fix the MMFBG in a thermostat with a temperature range from 30℃ to 100℃. The transmission spectra are shown in figure 3 and figure 4.

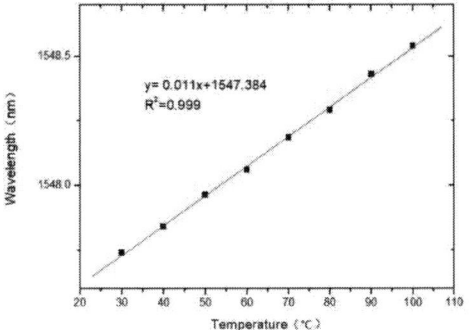

Figure 5.　Temperature response of gold-plated FBG

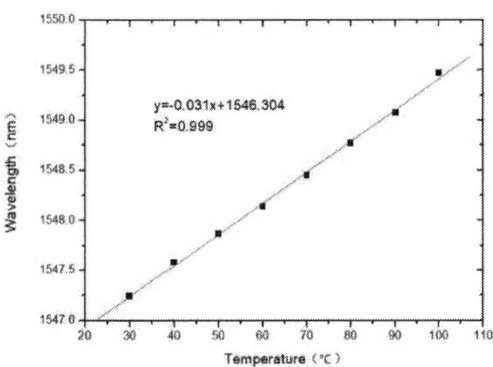

Figure 6.　Temperature response of FBG.

The experiment monitored the peak at 1547.256 nm from 30°C to 100°C. These points in figures are measured data. The measured temperature response sensitivity of FBG and gold-plated FBG are 0.011 nm/℃ and 0.031 nm/℃, respectively.

Figure 7.　Hydraulic measuring equipment.

In the same way, we probed the pressure response of the sensors. In the range of 1MPa and 36MPa, the pressure response sensitivity are 0.07nm/Mpa and 0.09nm/Mpa respectively. As shown in figure 8 to figure 11.

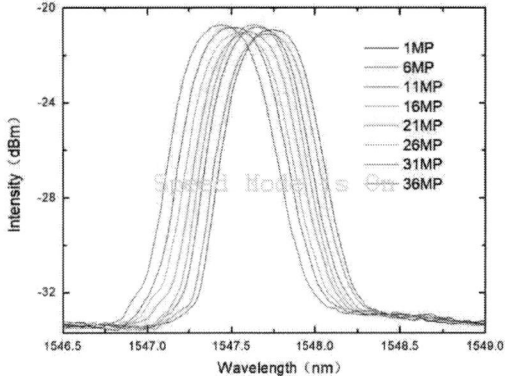

Figure 8.　FBG reflection spectrum change with pressure.

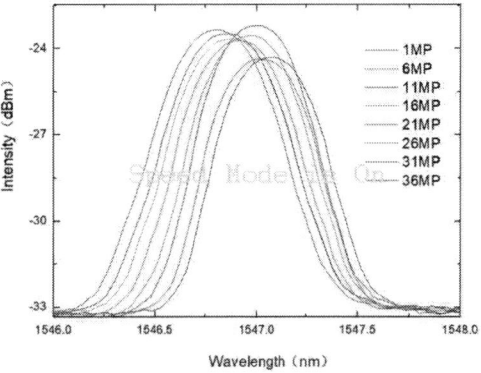

Figure 9.　Gold-plated FBG reflection spectrum change with pressure.

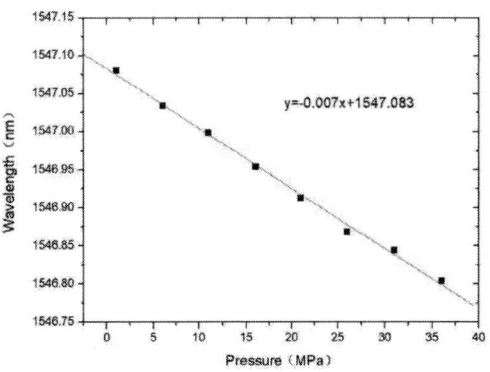

Figure 10. Pressure response of FBG

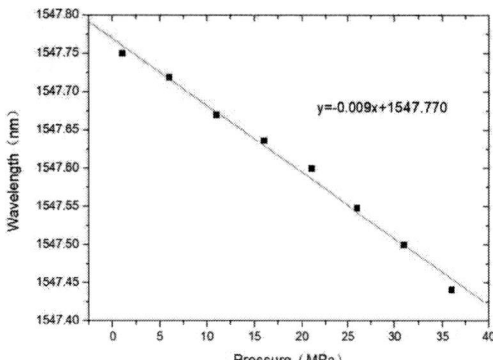

Figure 11. Pressure response of gold-plated FBG

By comparing the temperature and pressure response sensitivity of different types of FBG, it is found that their sensitivity coefficients are different. In this way, using FBG and gold-plated FBG with known temperature and pressure sensitivity coefficients, temperature and pressure dual- parameter sensing can be realized.

III. CONCLUSION

An intelligent sensor configuration for temperature and pressure measurement based on FBG is proposed. The temperature response sensitivity of FBG is 0.011 nm/℃ in the temperature range of 30 ℃ to 100 ℃, and the pressure response sensitivity of FBG is 0.07nm/Mpa in the pressure range of 1Mpa to 36Mpa. Gold-plated FBG temperature response sensitivity is 0.031 nm/℃ in the temperature range of 30 ℃ to 100 ℃, and pressure response sensitivity is 0.09nm/Mpa in the pressure range of 1 Mpa to 36 Mpa. Combination of gold-plated FBG and FBG, the detection system has great potential for temperature and pressure measurement, which shows that it can provide effective parameters for diagnosis system.

REFERENCES

[1] Becker A , K. Hler W , B. Müller. A Scanning Michelson Interferometer for the Measurement of the Concentration and Temperature Derivative of the Refractive Index of Liquids[J]. Ztschrift fr Elektrochemie, Berichte der Bunsengesellschaft fr physikalische Chemie, 2010, 99(4):600-608.

[2] Xu L , Jiang L , Wang S , et al. High-temperature sensor based on an abrupt-taper Michelson interferometer in single-mode fiber[J]. Applied Optics, 2013, 52(10):2038-2041.

[3] Jiangtao, Zhou, Kaiming, et al. Temperature-insensitive refractive index sensor based on in-fiber Michelson interferometer[J]. Sensors and Actuators, B. Chemical, 2014, 199:31-35.

[4] Lu P , Men L , Sooley K , et al. Tapered fiber Mach–Zehnder interferometer for simultaneous measurement of refractive index and temperature[J]. Applied Physics Letters, 2009, 94(13):5267.

[5] Liao C R , Wang Y , Wang D N , et al. Fiber In-Line Mach–Zehnder Interferometer Embedded in FBG for Simultaneous Refractive Index and Temperature Measurement[J]. IEEE Photonics Technology Letters, 2010, 22(22):1686-1688.

[6] Na Z , Qijing L , Zhuangde J , et al. High Temperature High Sensitivity Multipoint Sensing System Based on Three Cascade Mach–Zehnder Interferometers[J]. Sensors, 2018, 18(8):2688.

[7] Cao Y , Liu H , Tong Z , et al. Simultaneous measurement of temperature and refractive index based on a Mach–Zehnder interferometer cascaded with a fiber Bragg grating[J]. Optics Communications, 2015, 342:180-183.

[8] Cao Y , Zhao C , Tong Z R . All fiber sensor based on Mach-Zehnder interferometer for simultaneous measurement of temperature and refractive index[J]. Optoelectronics Letters, 2015, 11(6):438-443.

[9] Zhao N , Lin Q , Jing W , et al. High temperature high sensitivity Mach-Zehnder interferometer based on waist-enlarged fiber bitapers[J]. Sensors and Actuators A Physical, 2017, 267:491-495.

[10] Furuya K , Nemoto T , Kato K , et al. Athermal Operation of a Waveguide Optical Isolator Based on Canceling Phase Deviations in a Mach-Zehnder Interferometer[J]. Journal of Lightwave Technology, 2016, 34(8):1699-1705.

[11] Jin X , Huanhuan L , Fufei P , et al. Cascaded Mach-Zehnder interferometers in crystallized sapphire-derived fiber for temperature-insensitive filters[J]. Optical Materials Express, 2017, 7(4):1406.

[12] Njegovec, Matej, onlagic, et al. Temperature Measurement Using all Fiber Fabry-Perot Interferometers Based on Phase Measurement Between Reference and Sensing Interferometer Spectral Characteristic.[J]. AIP Conference Proceedings, 2010, 1236(1):309-313.

[13] Xu B , Yang Y , Jia Z , et al. Hybrid Fabry-Perot interferometer for simultaneous liquid refractive index and temperature measurement[J]. Optics Express, 2017, 25(13):14483.

[14] Islam M , Ali M , Lai M H , et al. Chronology of Fabry-Perot Interferometer Fiber-Optic Sensors and Their Applications: A Review[J]. Sensors, 2014, 14(4):7451-7488.

[15] Islam M , Ali M , Lai M H , et al. Chronology of Fabry-Perot Interferometer Fiber-Optic Sensors and Their Applications: A Review[J]. Sensors, 2014, 14(4):7451-7488.

[16] Ben, Xu, Yi, et al. Hybrid Fabry-Perot interferometer for simultaneous liquid refractive index and temperature measurement.[J]. Optics Express, 2017.

[17] Grochowski J , Mysliwiec M , Mikulic P , et al. Temperature Cross-Sensitivity for Highly Refractive Index Sensitive Nanocoated Long-Period Gratings[J]. Acta Physica Polonica, 2013, 124(3):421-424.

[18] Tripathi S M , Bock W J , Kumar A , et al. Temperature insensitive high-precision refractive-index sensor using two concatenated dual-resonance long-period gratings[J]. Optics Letters, 2013, 38(10):1666-1668.

[19] Jmietana Mateusz, Magdalena D , Predrag M , et al. Temperature and refractive index sensing with Al2O3 nanocoated long-period gratings working at dispersion turning point[J]. Optics & Laser Technology, 2018, 107:268-273.

[20] Yan B , Sun L , Luo Y , et al. Temperature Self-Compensated Refractive Index Sensor Based on Fiber Bragg Grating and the Ellipsoid Structure[J]. Sensors, 2019, 19(23):5211.

[21] Jiang B , Hao Z , Feng D , et al. Hybrid Grating in Reduced-Diameter Fiber for Temperature-Calibrated High-Sensitivity Refractive Index Sensing[J]. Applied Sciences, 2019, 9(9):1923.

[22] Liao C , Yang K , Wang J , et al. Helical Microfiber Bragg Grating Printed by Femtosecond Laser for Refractive Index Sensing[J]. IEEE Photonics Technology Letters, 2019, 31(12):971-974.

[23] Tian Y , Xu B , Chen Y , et al. Liquid Surface Tension and Refractive Index Sensor Based on a Side-Hole Fiber Bragg Grating[J]. IEEE Photonics Technology Letters, 2019, 31(12):947-950.

[24] Zhi Y , Li X , Li Y , et al. Superstructure microfiber grating characterized by temperature, strain, and refractive index sensing[J]. Optics Express, 2020, 28(6).

[25] Ran Y , Long J , Xu Z , et al. Temperature monitorable refractometer of microfiber Bragg grating using a duet of harmonic resonances[J]. Optics Letters, 2019, 44(13):3186.

[26] Korganbayev S , Ayupova T , Sypabekova M , et al. Partially etched chirped fiber Bragg grating (pECFBG) for joint temperature, thermal profile, and refractive index detection[J]. Optics Express, 2018, 26(14):18708.

[27] Shuo Y , Daniel H , Gary P , et al. Fiber Bragg grating fabricated in micro-single-crystal sapphire fiber[J]. Optics Letters, 2018, 43(1):62-65.

[28] Yang Y , Wang M , Shen Y , et al. Refractive Index and Temperature Sensing Based on an Optoelectronic Oscillator Incorporating a Fabry–Perot Fiber Bragg Grating[J]. IEEE Photonics Journal, 2018, 10(1):1-9.

[29] Xiaohe, Li, Sheng, et al. Novel refractive index sensor based on fiber bragg grating in nano-bore optical fiber[J]. Optical and Quantum Electronics, 2019, 51(4):1-23.

[30] Bandyopadhyay S , Shao L , Smietana M , et al. Employing Higher Order Cladding Modes of Fiber Bragg Grating for Analysis of Refractive Index Change in Volume and at the Surface[J]. IEEE Photonics Journal, 2020, 12(1):1-13.

[31] [31] G. Meltz, W. W. Morey, and W. H. Glenn, Formation of Bragg Gratings in Optical Fibers by a Transverse Holographic Method [J], Optics Letters, vol. 14, pp. 823- 825, Aug 1989.

[32] F. Xu, D. Ren, X. Shi, C. Li, W. Lu, L. Lu, et al., High- sensitivity Fabry-Perot interferometric pressure sensor based on a nanothick silver diaphragm [J], Optics Letters, vol. 37, pp. 133-135

[33] D. L. Gysling, Changing paradigms in oil and gas reservoir monitoring the introduction and commercialization of in-well optical sensing systems. New York: IEEE [C], 2002.

[34] Qiao Xueguang, Chen Yi, Jia Zhenan, "Research on the simultaneous differential measurement of temperature and pressure based on dual fiber gratings[J], Optoelectronics. Laser, pp. 12-14, 2010.

[35] Wu Haifeng, "Research on theory and field test of downhole high temperature and high pressure fiber grating sensor[D], Master, Xi'an Shiyou University, 2009.

Gap in pagination due to unavailable papers.

Pages 1640-1649

Proceedings of the 16th Annual IEEE International
Conference on Nano/Micro Engineered and Molecular Systems
April 25-29, 2021

Laser Induced Graphene Patterns on a Thin Polyimide Film via a cooling plate

Qixiang Chen, Dezhi Wu*, Zhiwen Chen, Hang Yu, Qinnan Chen and Daoheng Sun

Abstract—Laser induced graphene technology on a polyimide (PI) film substrate for supercapacitors, flexible sensors and heaters, has been a hot topic due to its fast fabrication speed and 3D controllable microstructure formation. However, the thin PI film will suffer burnthrough of the film under the required high energy density to form graphene. Herein, we introduced an auxiliary cooling semiconductor plate beneath the PI substrate to decrease the temperature of the substrate to reduce the burnt thickness. Experiment results showed that the burnt thickness will decrease to about 33.39 μm for -2°C on the thin film with thickness about 12 μm and the value of resistance can still be kept at 8.33 kΩ. The effects of the scanning speed and power on the size of the wire also discussed. Therefore, such method may open a way to fabricate graphene patterns on thin films with thickness less than a few micrometers, which will play an important role in applications including electronic skins and multi-layer flexible plastic circuits.

Keywords—Direct laser writing, Polyimide film, Graphene

I. INTRODUCTION

Laser induced graphene (LIG) have been widely used in the fabrication of flexible devices [1]. The applied fields include supercapacitors [2], multi-functional sensors [3], thermal actuators [4] and nanogenerators [5]. Direct laser writing (DLW) can induce pattern tracks in one step, which is characterized of high processing speed, aptness to Roll-to-Roll manufacturing. Polyimide (PI) film is a mostly used substrate in flexible electronics because it owns superb chemical durability, good mechanical properties and easiness of machining. Nowadays, pulsed laser technology has become an important method for reprocessing of PI film, and research work to induce graphene material by pulsed laser has been extensively carried out. Luo et al. [6] applied a CO_2 laser with wavelength of 10.6 μm and diameter of ~100 μm to irradiate PI film with thickness of 127 μm to generate flexible and conductive graphitic porous patterns or arrays. With the pulse per inch or pixel per inch (PPI) fixed at 400, the laser power (1.5-7.5W) and/or rasterring/scanning speed (25.4-88.9 mm s⁻¹) were varied to examine their effects on the morphologies. The minimum height of cross-section morphologies of the graphitic line was 40 μm. Wang et al. [7] investigated the effect of substrate on LIG generated on polyimide films and then fabricate the patterned LIG structures including gratings and Fresnel zone plates for terahertz (THz)-wave modulations. A continuous laser with a wavelength of 450 nm was used to fabricate LIG samples on commercial PI film with a thickness of 80 μm, and the beam size of the laser spot was 120 μm. The cross section of LIG generated on the PET substrate under laser power values of 1.5W was about 9.45 μm in thickness.

Compared with traditional processes, which has problems of high cost and environmental pollution, LIG has the advantages of fast processing and high spatial resolution [8]. However, most researchers use thicker films with thickness of 24-127 μm to design flexible products and machine microstructure on thin film by nano wavelength laser. In fact, flexible devices are developing in trend of thinner and smaller size, so thin PI film (thickness less than 12 μm) is more suitable for fabrication of flexible electronic devices, which has smaller size and is easier for conformation, and can be used to manufacture multilayer flexible circuit and electronic skin. But, during processing of LIG, it is easy to burn through the thin PI film and uncapable to achieve LIG with structure stability because of thin thickness and overtop laser power.

In this work, we firstly brought forward a method that an auxiliary cooling semiconductor plate is placed beneath the film to sharply decrease temperature of the film during process of LIG. Experiment results show that such method can work to decrease the burn thickness to about 33.39 μm for -2°C. The relationship between the temperature of refrigeration plate and the height of graphene wire was explored and the effects of the scanning speed and power on the size and resistance of the wire were also discussed.

II. MATERIALS AND METHODS

As seen in Fig. 1b, the experimental setup of the direct laser writing process includes a CO_2 laser system (Wuhan Jinhuo, JHCV-30W), a focal flat top beam shaper, focusing optics, a scanning head, PI film and a cooling system. The wavelength of CO_2 laser was 10.6 μm and the focused beam spot size was approximately 200 μm at a working distance of 185 mm. The collimated Gaussian input beam is shaped into flat topped beam with uniform intensity distribution by focal flat top beam shaper, then focusing optics can focus the laser to make the spot smaller. The laser can scan processing in a certain range through two galvanometers in the scanning head in order to fabricate graphene patterns on thin films. The PI film with thickness of 12μm was supplied by McMaster-Carr (Kapton®) and used as receiver without any further treatment. Cooling system is composed of cooling semiconductor plate, radiator, temperature sensor, MCU and power supply, which can maintain the rated temperature of the substrate.

978-1-6654-3008-1/21 $31.00 © 2021 IEEE

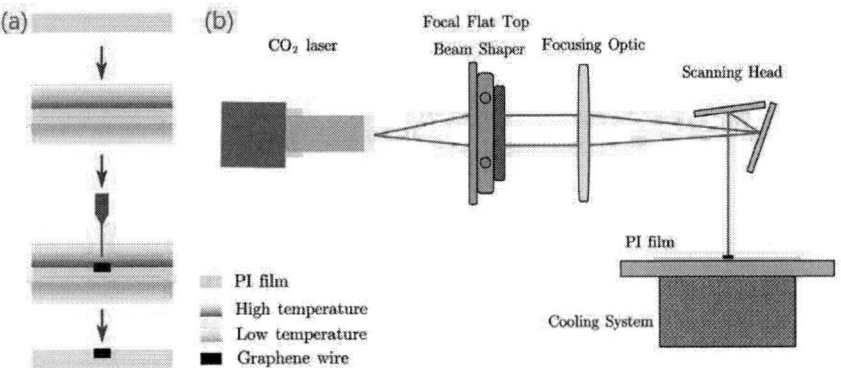

Fig. 1. Technological process based on cooling system (a) and the experimental setup of LIG process (b)

Owing to the excellent laser effect, laser is often applied in the fields of precision machining [9], includes thermal effect, photochemical effect, mechanical effect, electromagnetic field effect and biological stimulation effect. There are thermal effect and photochemical effect in laser processing. Thermal effect is both collision heat generation and absorption heat generation and exists in all laser irradiation. The extent of this influence is related to the power density, irradiation area and irradiation time of the laser, as well as the absorptivity, specific heat and thermal conductivity of the material. Photochemical effect refers to the biochemical reaction under the action of light, occurring at low power laser irradiation. The extra energy is consumed in the breaking or formation of its own chemical bond, and a chemical reaction takes place. Mechanical effect will be generated if the energy density exceeds a certain threshold when biological tissue absorbs laser energy. Electromagnetic field effect is produced because of the vibration of charged particles. Biological stimulation effect is reflected by the response of the organism that excitation or inhibition when a weak laser is used to irradiate an organism.

In direct laser writing, the environment of high temperature and high pressure is formed in the process of instantaneous laser irradiation, so that PI film is converted into graphene. We take advantage of thermal effect of laser, as different thermal effects occur on PI films with changing the temperature and atmosphere. We firstly spray deionized water on the substrate, then make film stick to the substrate evenly with the adsorption capacity of water. After 6 hours of natural air drying, the water evaporates and the film adheres tightly to the substrate. In this way, the effect of laser on the film is well-distributed. At present, cooling semiconductor plate and black reflective glass (Fig. 2) are main application substrates to research on laser induced graphene process.

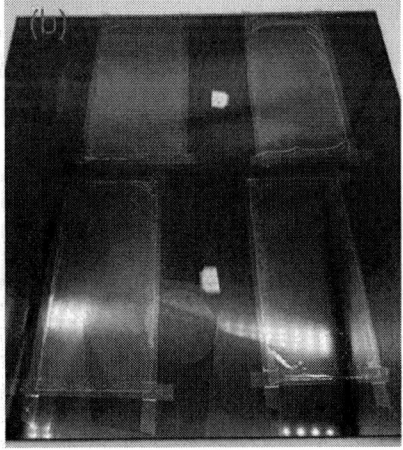

Fig. 2. The cooling semiconductor plate (a) and black reflective glass (b) to assist LDW

III. RESULT AND DISCUSSION

A. Thick PI film experimental results

At first, graphene wire was written on PI film at a scanning speed 100 mm s^{-1} and repetition rate 10 kHz through a complete processing. After applying different power to film, observe the morphology of the wire by scanning electron microscope and measure the resistance. From Fig. 3a, we can find that the resistance of graphene wire steadily decreased with the increase of power and the value of resistance was about 5kΩ to 20kΩ.

According to different scanning speeds, the power range was set reasonably to satisfy stable power density of CO_2 laser. However, when the power of 28% was selected at the same time, the resistance of graphene wire went up with the increase of scanning speed. From the SEM image of the graphene wire in Fig. 3d, it can be seen that there is a half width groove in the middle of the graphene wire and it's a porous.

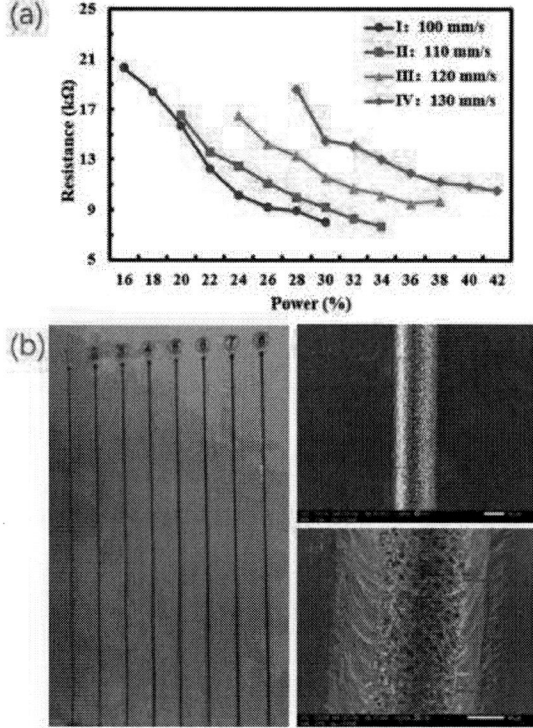

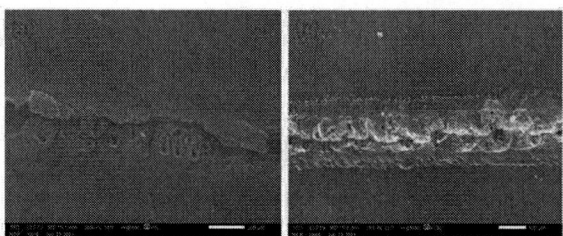

Fig. 3. The curve of power vs. resistance at scanning speeds of 100mm s⁻¹, 110mm s⁻¹, 120mm s⁻¹ and 130 mm s⁻¹ (a), single graphene wire length 80 mm (b), top-view SEM iamge of the graphene wire with 410 μm in width (c) and SEM of the groove of graphene wire with 205 μm in width (d)

B. Thin PI film experimental results

In order to avoid the phenomenon of overtop power density in thin film processing in Fig. 4, we have two solution. First one is adopting plate with cooling semiconductor and it can make the surface temperature of film back lower 15°C than room temperature, which is approximately equal to the surface temperature of the front of the film.

Fig. 4. SEM of the graphene wire without cooling system (a) and with cooling system (b)

We controlled the burnt thickness by adjusting the temperature of the substrate, and observed the formation thickness with SEM. We discovered that the burnt thickness decreases with the decrease of temperature and the minimum burnt thickness will decrease to about 33.39 μm for -2 °C on the thin film with thickness about 12 μm. As shown in the cross-sectional SEM view of the graphene wire in Fig. 5b, the wire was even and complete.

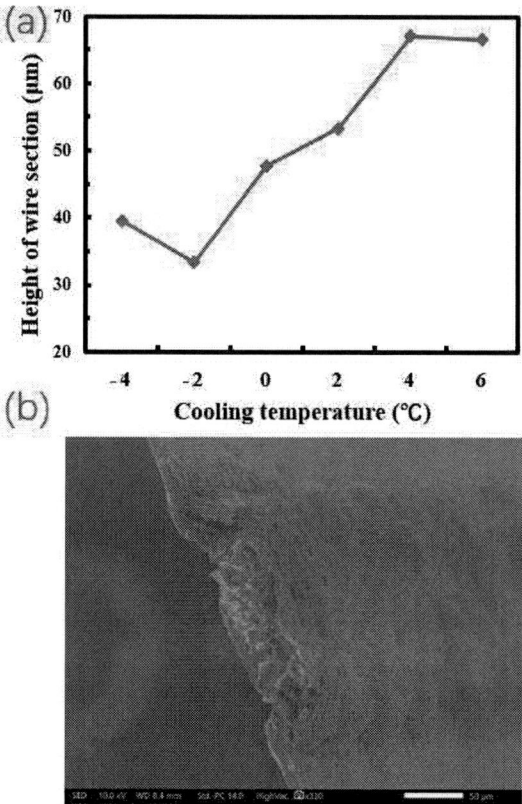

Fig. 5. The curve of cooling temperature vs. height of wire section (a) and cross-sectional SEM view of the graphene wire (b)

By setting power at 3 W and repetition rate at 10 kHz through two complete processing of 50 mm length, we can find that the width (Fig. 6a, b) and resistance (Fig. 6c, d) change with the scanning speed with the substrate temperature -4 °C and 5 °C, respectively. When the temperature of substrate was high, the resistance and width changed slowly. On the contrary, the width of wire dropped dramatically and resistance of wire fell substantially on the substrate with low temperature.

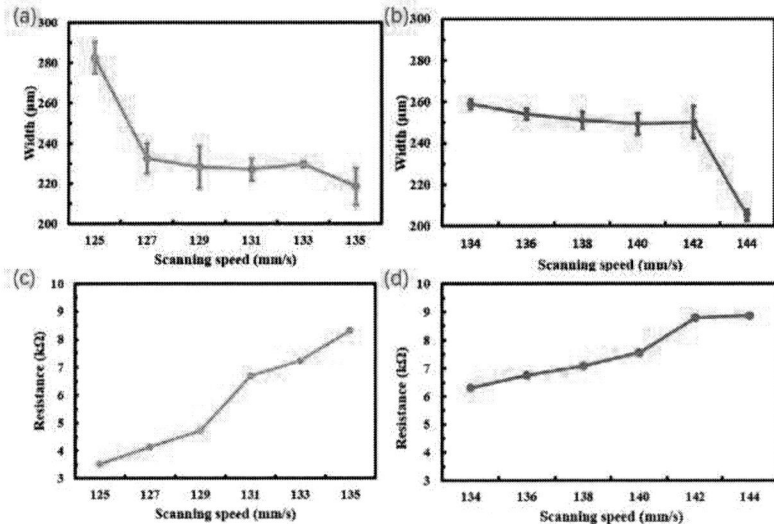

Fig. 6. The curve of scanning speed vs. width (a) and scanning speed vs. resistance (b) at -4 °C substrate temperature, the curve of scanning speed vs. width (c) and scanning speed vs. resistance (d) at 5 °C substrate temperature

The second one is choosing black reflective glass and we make use of the principle of black absorption. After passing through the film, the laser irradiated on the glass and the bottom thermal effect becomes weak because of light reflection and absorption. Graphene wire with length of 80 mm was written on PI film at a scan speed of 180 mm s⁻¹ and repetition rate of 20 kHz through a complete processing (Fig. 7).

Among the obtained graphene wires, the value of resistance is entirely higher, more than 20kΩ, but we acquired the minimum width wire of 178.3μm when power of 25%. With the increase of laser power, the area acting on the PI film grew due to the increase of irradiation energy and the resistance of wire was reduced because of the graphitization of the insulating film

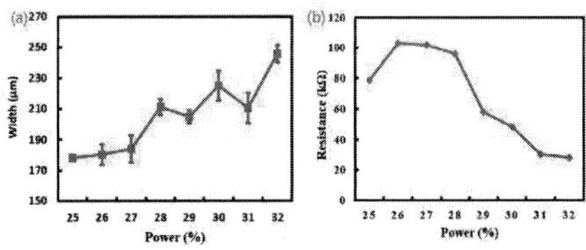

Fig. 7. The curve of width vs. power (a) and resistance vs. power (b) on black reflective glass substrate

IV. CONCLUSION

In summary, an auxiliary cooling semiconductor plate was put beneath the PI film to decrease the PI film temperature to successfully fabricate graphene patterns on thinner PI film via laser direct writing. The process parameters of laser irradiated PI film to generate graphene wires, and realize controllable machining on 12 μm thick film on refrigeration substrate and black reflective glass substrate. When the cooling substrate is -4 °C, the resistance value of graphene wire with length of 60 mm can be as low as 3.52kΩ, and width of graphene wire machining on the black reflective glass can be as low as 178.3 μm. Through

the application of cooling system, we can effectively reduce the burnt thickness of PI film and the minimum burnt thickness was about 33.39 μm. Therefore, we can design physiological sensors that are more attached to human skin, and make flexible circuit boards to reduce the complexity of traditional processes by this improved technology.

ACKNOWLEDGMENTS

This work was in part supported by National Natural Science Foundation of China (52075464, 91648114) and Science and Technology Program of Shenzhen City (JCYJ20180306172700388).

REFERENCES

[1] J. Lin, Z. Peng, Y. Liu, F. Ruiz-Zepeda, R. Ye, E. L. G. Samuel, M. J. Yacaman, B. I. Yakobson, and J. M. Tour, "Laser-induced porous graphene films from commercial polymers", *Nature Communications*, vol. 5, no. 1, p. 5714, 2014.

[2] L. Li, J. Zhang, Z. Peng, Y. Li, C. Gao, Y. Ji, R. Ye, N. D. Kim, Q. Zhong, Y. Yang, H. Fei, G. Ruan, and J. M. Tour, "High-Performance Pseudocapacitive Microsupercapacitors from Laser-Induced Graphene", *Advanced Materials*, vol. 28, no. 5, pp. 838–845, 2016.

[3] B. Sun, R. N. Mccay, S. Goswami, Y. Xu, C. Zhang, Y. Ling, J. Lin, and Z. Yan, "Gas-Permeable, Multifunctional On-Skin Electronics Based on Laser-Induced Porous Graphene and Sugar-Templated Elastomer Sponges", *Advanced Materials*, vol. 30, no. 50, p. 1804327, 2018.

[4] D. Wu, L. Deng, X. Mei, K. S. Teh, W. Cai, Q. Tan, Y. Zhao, L. Wang, L. Zhao, G. Luo, D. Sun, and L. Lin, "Direct-write graphene resistors on aromatic polyimide for transparent heating glass", *Sensors and Actuators A: Physical*, vol. 267, pp. 327–333, 2017.

[5] P. Zhao, G. Bhattacharya, S. J. Fishlock, J. G. M. Guy, A. Kumar, C. Tsonos, Z. Yu, S. Raj, J. A. Mclaughlin, J. Luo, and N. Soin, "Replacing the metal electrodes in triboelectric nanogenerators: High-performance laser-induced graphene electrodes", *Nano Energy*, vol. 75, p. 104958, 2020.

[6] S. Luo, P. T. Hoang, and T. Liu, "Direct laser writing for creating porous graphitic structures and their use for flexible and highly sensitive sensor and sensor arrays", *Carbon*, vol. 96, pp. 522–531, 2016.

[7] Wang, Zongyuan , et al. "Patterned laser induced graphene for terahertz wave modulation." *Journal of the Optical Society of America B*, vol. 37, no. 2, pp. 546–551, 2020.

[8] L. Huang, J. Su, Y. Song, and R. Ye, "Laser-Induced Graphene: En Route to Smart Sensing", *Nano-Micro Letters*, vol. 12, no. 1, 2020.

[9] R. K. Biswas, N. Farid, G. O'Connor, and P. Scully, "Improved conductivity of carbonized polyimide by CO_2 laser graphitization", *Journal of Materials Chemistry C*, vol. 8, no. 13, pp. 4493–4501, 2020.

Proceedings of the 16th Annual IEEE International
Conference on Nano/Micro Engineered and Molecular Systems
April 25-29, 2021

A Robotic Dynamic Tactile Sensing System based on Electronic Skin

Jishen Dai[1], Yu Xie[1*], Dezhi Wu[1], Songyue Chen[1], Ting Fu[2], Wei Zhou[1]

Abstract— **Robot tactile is getting more and more practical with the rapid development of electronic skin (e-skin). In this work, we propose a dynamic tactile sensing system based on e-skin, a high-density array tactile sensor. The system is comprised of three algorithms to track motion and detect slip. With real-time slip detecting, an adaptive controller that can rapidly tune applied grasping force without having a priori knowledge of the objects was designed and experimentally validated.**

I. INTRODUCTION

In the field of robotic control, vision servo has been applied frequently [1]. However, vision servo has made a rigid request to the working environment. When the object contacts and interacts with the end effectors, the vision servo cannot work efficiently. Electronic skin (e-skin) can function as a flexible, high-density and compact array sensor, and its development can help robots solve the above issue [2]. The e-skin plays an essential role in dynamic tactile sensing for robots to measure the force distribution on the contact surface and dig much hidden information. Therefore, dynamic tactile sensing system with the e-skin is attracting more and more attention [3].

Tactile sensing can increase the application of robots by taking the advantage of rich tactile features. Heever et al. [4] enhanced the resolution of tactile image and implemented the fusion of images to analyze the pressure distribution. With hidden tactile information being extracted, several studies have explored object recognition [5-7]. Besides, in terms of robotic manipulation, dynamic tactile sensing provides solid feedback information of grasp condition. Sriram et al. [8] proposed a fuzzy logic controller to avoid slippage. And other robotic systems with capabilities of tactile sensing have also been reported for manipulation tasks [9-12]. However, in the complex manipulation task, it is a dynamic condition between effector and grasped object. So, a dynamic tactile sensing system is necessary for robot to learn that dynamic condition.

In this paper, based on the e-skin, a dynamic tactile sensing system that can track the motion of object and detect slip is developed. This system is comprised of three sensing algorithms including two motion tracking algorithms and a slip detection algorithm. After that, the slip detection algorithm is incorporated into a grasping force controller. Experiments were conducted to test the performance of

system. The results showed that the applied grasping force can be rapidly tuned in grasping tasks.

II. MOTION TRACKING ALGORITHM

A. Optimization Method

In order to acquire the parameters of translation and rotation, an optimization method was proposed by processing two adjacent tactile images. Firstly, tactile images were preprocessed by binarization, dilation and erosion to reflect the contact surface of the object more accurately. The transformation matrix M is used to build the map between the two adjacent images as follow.

$$CoordinatesB = M \times CoordinatesA$$
$$M = \begin{bmatrix} cos\theta & -sin\theta & -C_x cos\theta + C_y sin\theta + C_x + t_x \\ sin\theta & cos\theta & -C_x sin\theta - C_y cos\theta + C_y + t_y \\ 0 & 0 & 1 \end{bmatrix} (1)$$

Where $CoordinatesA$ and $CoordinatesB$ are the corresponding pixel coordinates in the first and second image. (C_x, C_y) is the centroid coordinates of the first image. θ is the rotation angle between the images, t_x and t_y are the distance of translation along the x-axis and y-axis. The transformation matrix M actually processes the tactile images to rotate the first image θ degrees counterclockwise along its centroid and translate the first image t_x and t_y pixel-distance along the x-axis and y-axis.

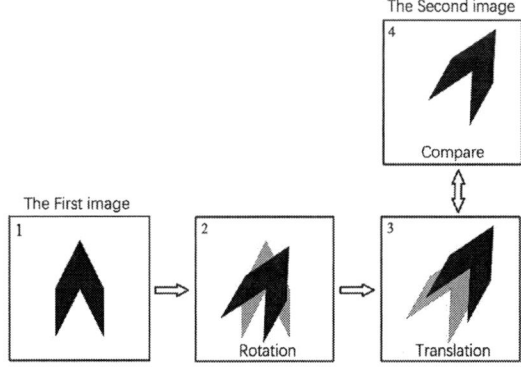

Figure 1. the transformation of the optimization method.

After that, modified Powell method is used to search the optimal parameters which can make the first image, after transformation, have the slightest difference with the second image. The difference χ between two adjacent images is given by

---

*Resrach supported in part by the Open Fund of Hubei Key Laboratory of Mechanical Transmission and Manufacturing Engineering at Wuhan University of Science and Technology under Grant MTMEOF2019A01, and in part by the Science and Technology on Space Intelligent Control Laboratory for National Defense under Grant KJGZDSYS-2018-07.

1 the Department of Mechanical and Electrical Engineering, Xiamen University, Xiamen, 361005, China.
2 the School of Machinery and Automation, Wuhan University of Science and Technology, Wuhan, 430081, China.
*corresponding author: phone: 15160061679; e-mail: xieyu@xmu.edu.cn

$$\chi = \sum_{i=1}^{m}\sum_{j=1}^{n}\left(I_1(i,j) - I_2(i,j)\right)^2 \qquad (2)$$

where m and n refer to the number of rows and columns. I_1, I_2 are the coordinate systems of two adjacent tactile image and (i,j) is the coordinate of each pixel. This method uses conjugate direction to find the minimum solution instead of using the gradient. So, to avoid the local minimum due to the presence of trigonometric functions, three tiny random values are added to a set of parameters. Then this step is repeated ten times to reduce the difference χ and the optimized parameters can be set in the matrix M to transform the first image to the second image. The optimized parameters are available for motion tracking.

B. Corner Matching Method

To improve the accuracy of detection, only dilating and eroding are firstly employed in this method. There is no binarization here since the symmetrical contact surface will cause a large error. Then the centroid coordinates of two images are calculated. By comparing that, the distance t_x and t_y can be obtained to translate the first image.

In order to detect the rotation angle in two tactile images, Harris corners of images are utilized in this method. By means of a build-in function in OpenCV, Harris corners are extracted from images. With further adjusting the parameters of the function, the appropriate corners are select to match between two images. A 7×7 window around the corners of the first image is chosen and compared with those windows in the second images. Two corners whose difference of window is smallest are selected to generate a relative angle ($\angle\alpha$, $\angle\beta$) with their centroids as shown in Fig. 2. The difference between $\angle\alpha$ and $\angle\beta$ is the rotation angle θ of two images. At last, the distance between centroids t_x and t_y, the rotation angle θ are obtained for the motion tracking.

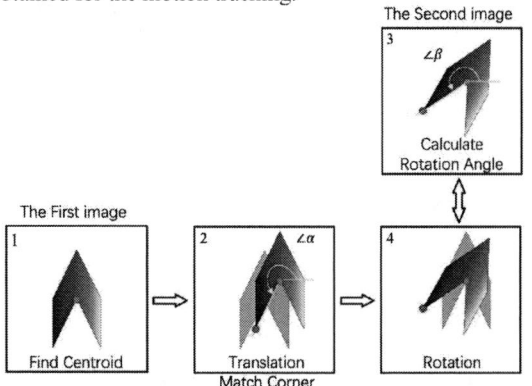

Figure 2. The transformation of the corner matching method. (green point and red point present the centroid and the matching corners)

C. Experiment And Result Evaluation

To experimentally validate the proposed algorithms, the high-density e-skin (Pressure Mapping Sensor 5101, Tekscan, USA) was applied to collect tactile images. The e-skin whose rows and columns are made of piezoresistive material has 44 rows and 44 columns with a gap of 2.5 mm.

The sensor was mounted on a six-axis parallel positioner to achieve relative translation and rotation with the object. A 5 mm thick foamed silicone was stuck under the e-skin to act as muscle tissue.

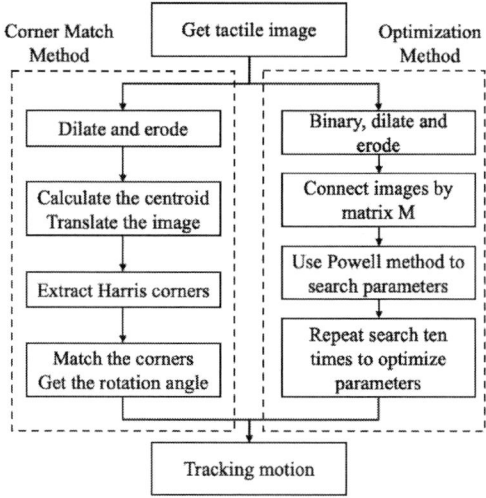

Figure 3. The process of two algorithms.

The first experiment was set to mainly track the orientation since the translation distance is easy to acquire. The images set was collected under the pressure of 4.5 kg with tiny random differences in position. Each image is separated by $45°$. The preprocess is shown in Fig. 4. And the steps of experiments are presented in Fig. 5. The difference χ, the pixel-distance d_c of centroid and the pixel-distance d_{cop} of center of pressure between each image and the first image were calculated to be the evaluation indexes. The results were presented in TABLE I and TABLE II.

TABLE I. THE RESULTS OF OPTIMIZATION METHOD

Optimization Method							
θ	45°	90°	145°	180°	225°	270°	315°
χ	69	56	72	136	89	54	105
d_c	1	1.41	2	0	2.24	1.41	0
d_{cop}	3.16	3.16	4.12	4.12	3.61	2	1.41

TABLE II. THE RESULTS OF CORNER MATCHING METHOD

Corner Matching Method							
θ	45°	90°	145°	180°	225°	270°	315°
χ	279	336	345	176	313	394	237
d_c	0	1	1	0	1	1	1
d_{cop}	1.41	1	2	3.16	3	3	2.24

Images in the set had 780 pixels averagely, the result of the optimization method had an average difference χ of 83, whose relative error was 10.64%. Due to the presence of surface roughness, this result was acceptable. However, the corner matching method got a terrible result in difference χ. The average was 290 and the relative error reached to 37.16%. As for translation distance, corner matching method had less error, but performances of two methods were both acceptable. Consequently, Optimization method has better robustness in orientation tracking than corner matching method, and both of them have good property in position tracking.

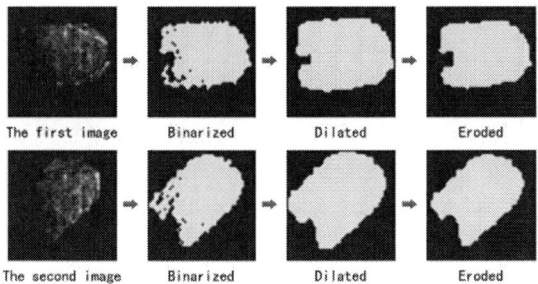

Figure 4. The preprocess of the optimization method.

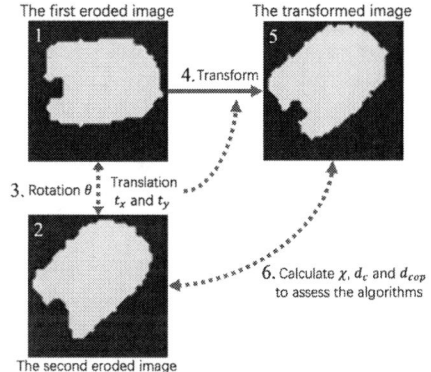

Figure 5. The steps of experiments.

Actually, it is difficult to keep a constant pressure when tracking the object. So, the second experiment was designed to examine the robustness to the pressure variation. The first images set was sampled under the pressure of 4.5kg. And the other images were sampled under 1kg, 2kg, 4kg and 4.5kg with 45° counterclockwise differences in orientation and tiny random differences in position with the first one. The change range of χ, d_c and d_{cop} were concerned to evaluate the influence of pressure variation. The experiment results are shown in TABLE III and TABLE IV.

TABLE III. THE RESULTS OF OPTIMIZATION METHOD IN DIFFERENT PRESSURE

Optimization Method					
Pressure	1 kg	2 kg	4 kg	4.5 kg	Range
χ	103	63	62	69	41
d_c	1	1	1	1	0
d_{cop}	1	2.24	2.83	3.16	2.16

TABLE IV. THE RESULTS OF CORNER MATCH METHOD IN DIFFERENT PRESSURE

Corner Matching Method					
Pressure	1 kg	2 kg	4 kg	4.5 kg	Range
χ	385	273	269	279	116
d_c	1	0	0	0	1
d_{cop}	5	1	1	1.41	4

The tables revealed that the change of χ, d_c and d_{cop} increased while the pressure varied greatly. When the pressure is 1 kg, the relative error of χ is obvious since the pressure is slight to conduct a well-distributed interaction. Consequently, it is validated that there was great robustness in both methods when the object is in good contact with the e-skin.

III. SLIP DETECTION ALGORITHM

The tactile image collected from the e-skin is a matrix that can be defined as follow.

$$\mathcal{F}(t_k) = \begin{bmatrix} f_{0,0}(t_k) & \cdots & f_{0,M}(t_k) \\ \vdots & \ddots & \vdots \\ f_{N,0}(t_k) & \cdots & f_{N,M}(t_k) \end{bmatrix} \quad (3)$$

where $\mathcal{F}(t_k)$ is the contact force of the e-skin at time t_k; $f_{i,j}(t_k)$ is the force in the normal direction on row i and column j; N and M are the number of rows and columns in the e-skin array; and $\mathcal{F}(t_k)$ is defined as the flame of the tactile image. To clearly represent that change of $\mathcal{F}(t_k)$, the difference between two adjacent tactile frames is given by

$$\Delta\mathcal{F}(t_k) = \mathcal{F}(t_k) - \mathcal{F}(t_{k-1})$$
$$= \begin{bmatrix} \Delta f_{0,0}(t_k) & \cdots & \Delta f_{0,M}(t_k) \\ \vdots & \ddots & \vdots \\ \Delta f_{N,0}(t_k) & \cdots & \Delta f_{N,M}(t_k) \end{bmatrix} \quad (4)$$

and $\Delta f_{i,j}(t_k) = f_{i,j}(t_k) - f_{i,j}(t_{k-1})$. When the gripper applies a constant force or regulates the grasping force to the object, the value of every element in $\mathcal{F}(t_k)$ is concentrated near their mean value. When the object is slipping, the change generated by the random vibration is also random. This means the value of each element is more spread out. So, the proposed method calculates the standard deviation of $\Delta\mathcal{F}(t_k)$ to observe how far the element value of $\Delta\mathcal{F}(t_k)$ is spread out from the mean value .

$$STD(\Delta\mathcal{F}(t_k)) = \sqrt{\frac{1}{NM}\sum_{i=0}^{N}\sum_{j=0}^{M}\left(\Delta f_{i,j}(t_k) - \overline{\Delta f_{i,j}}\right)^2} \quad (5)$$

where $\overline{\Delta f_{i,j}}$ is the mean value of all elements in the matrix $\Delta\mathcal{F}(t_k)$. Then the SLIP signal can be defined by

$$SLIP = \begin{cases} 1, STD(\Delta\mathcal{F}(t_k)) \geq \text{THRESHOLD} \\ 0, STD(\Delta\mathcal{F}(t_k)) < \text{THRESHOLD} \end{cases} \quad (6)$$

where THRESHOLD is a preset experimental value. When $STD(\Delta\mathcal{F}(t_k))$ exceeds that threshold, the SLIP signal is generated and fed back to the robot system for grasping force regulation. The structure of the slip detection algorithm is shown in Fig. 6.

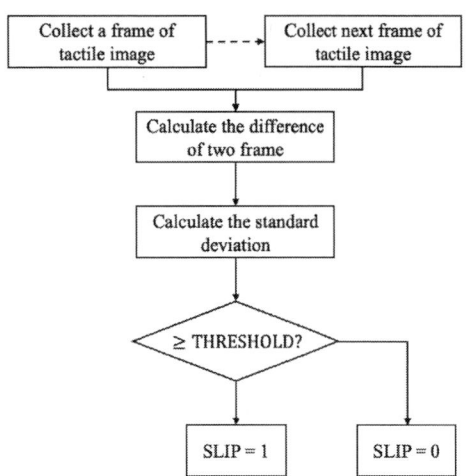

Figure 6. The structure of slip detection algorithm.

IV. ADAPTIVE GRASPING FORCE CONTROLLER

A. The structure of controller

The adaptive grasping force controller consists of two parts: closing and grasping. Closing is the first step, aiming to contact the object gently. To detect whether the object and the gripper are in contact, a small pre-grasping force needs to be set as the input of the controller. Once the grasping force reaches that value, the controller switches to grasping. Based on the slip detection algorithm, the controller regulates the grasping force to conduct a grasping attempt until the object is no longer slipping. Generally, the grasping force that just holds the object without slipping is the minimum. The block diagram of controller is shown in Fig. 7.

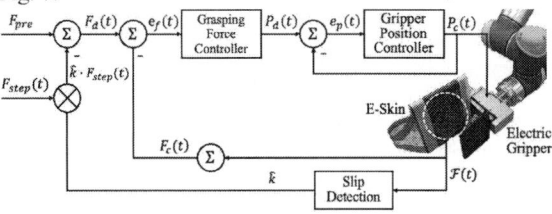

Figure 7. Block diagram of the adaptive grasping force controller.

The kernel of the grasp controller utilizes the double closed-loop proportional–integral–derivative (PID) control strategy. The inner loop provides the position control for the gripper, and the outer loop provides the grasping force control. First, the desire grasping force $F_d(t)$ and the force error $e_f(t)$ are given by

$$F_d(t) = F_{pre} - \hat{k} \cdot F_{step}(t) \qquad (7)$$

and

$$e_f(t) = F_d(t) - F_c(t) \qquad (8)$$

where the current contact force $F_c(t)$ is the sum of the tactile frames $\mathcal{F}(t_k)$. The step of increased grasping force, F_{step}, is set according to the experimental environment. The desired position of the gripper $P_d(t)$ is given by

$$P_d(t) = K_p e_f(t) + K_i \int_o^t e_f(t)dt + K_d \frac{de_f(t)}{dt} \qquad (9)$$

When slip is detected, the algorithm assigns a Boolean value to the signal SLIP based on Eq. (6). The slip detection algorithm updates $\hat{k}$ as follows:

$$\hat{k} = \begin{cases} k, \text{SLIP} = 1 \\ 0, \text{SLIP} = 0 \end{cases} \qquad (10)$$

Based on the current position $P_c(t)$ recalled from the electric gripper, the desired position is given by

$$e_p(t) = P_d(t) - P_c(t) \qquad (11)$$

The updated gripper position applies a new grasping force on the object. Eventually, the grasp is conducted with the minimum grasping force when the object stops slipping.

B. Grasping tasks for unknown objects

To test the performance of the controller and detection algorithm, the grasping task was set to grasp an egg as shown in Fig. 8. The e-skin was mounted on an electric gripper (XEG-32, HIWIN, China) and a UR5 (Universal Robots, Denmark) robotic arm. The grasping task was repeated ten times and conducted as follow steps. First, the gripper closed until the contact force reached the pre-grasping force, which was set as 0.5 N. Then the robotic arm moved upward to generate slip. This step mimics a human's behavior when grasping an unknown object. During the upward moving process, the controller was repeatedly updating the grasping force with the result of slip detection. When slip was no longer detected, the egg was steadily held under the final grasping force. The curves of grasping force and $STD(\Delta\mathcal{F}(t_k))$ versus time are shown in Fig. 9. As can be seen, the grasping force stabilized at 0.5 N in phase B. Then the robotic arm was moved up in phase C attempting to grasp the egg. With the slip being detected as shown in Fig. 9(b), the grasping force increased. In phase D, a stable grasp was achieved without dropping and damaging the egg.

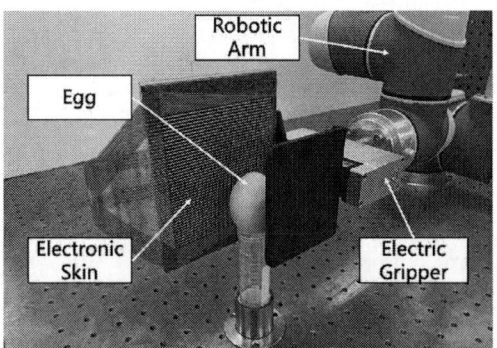

Figure 8. Grasping an egg to test the controller and detection algorithms.

V. CONCLUSION

In this paper, a dynamic tactile sensing system that includes the motion tracking algorithm and the slip detection algorithm was developed. In motion tracking, the optimization method has a much better performance in orientation tracking, and both of them show good robustness to pressure variation. Based on the standard deviation of frame difference, the slip detection algorithm was proposed. And an adaptive grasping force controller was designed with real-time slip detecting. Experiments were conducted to validate that the detection methods and the controller could achieve the expected performance.

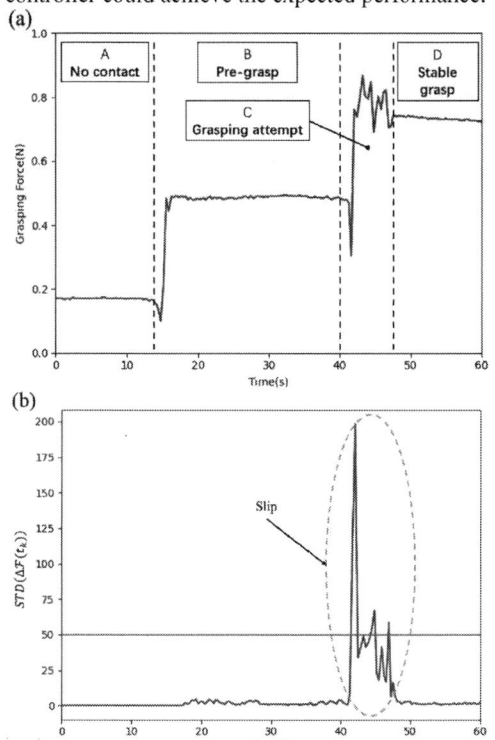

Figure 9. The grasping force and $STD(\Delta\mathcal{F}(t_k))$ in grasping an egg.

REFERENCES

[1] N. Papanikolopoulos, "Vision and control techniques for robotic visual tracking," *IEEE Int.conf.robotics & Automation Sacramento Ca*, vol. 1, pp. 857-864 vol.1, 1991.

[2] M. L. Hammock, A. Chortos, C. K. Tee, B. H. Tok, and Z. Bao, "25th anniversary article: The evolution of electronic skin (e-skin): a brief history, design considerations, and recent progress," *Advanced Materials*, vol. 25, no. 42, pp. 5997-6038, 2013.

[3] P. S. Girão, P. M. P. Ramos, O. Postolache, and J. M. D. Pereira, "Tactile sensors for robotic applications," *Measurement*, vol. 46, no. 3, pp. 1257-1271, 2013.

[4] D. J. van den Heever, K. Schreve, and C. Scheffer, "Tactile sensing using force sensing resistors and a super-resolution algorithm," *IEEE Sensors Journal*, vol. 9, no. 1, pp. 29-35, 2008.

[5] U. Martinez-Hernandez, T. J. Dodd, L. Natale, G. Metta, T. J. Prescott, and N. F. Lepora, "Active contour following to explore object shape with robot touch," in *IEEE World Haptics*, 2013.

[6] A. Drimus, M. B. Petersen, and A. Bilberg, "Object texture recognition by dynamic tactile sensing using active exploration," in *2012 IEEE RO-MAN: The 21st IEEE International Symposium on Robot and Human Interactive Communication*, 2012.

[7] Y. Xie, C. Chen, D. Wu, W. Xi, and H. Liu, "Human-Touch-Inspired Material Recognition for Robotic Tactile Sensing," *Applied Sciences*, vol. 9, no. 12, p. 2537, 2019.

[8] G. Sriram, A. N. Jensen, and S. C. Chiu, *Slippage control for a smart prosthetic hand prototype via modified tactile sensory feedback*. 2014.

[9] A. Delgado, J. A. Corrales, Y. Mezouar, L. Lequievre, C. Jara, and F. Torres, "Tactile control based on Gaussian images and its application in bi-manual manipulation of deformable objects," *Robotics and Autonomous Systems*, p. S092188901630478X, 2017.

[10] Z. Kappassov, J. A. Corrales, and V. Perdereau, "ZMP features for Touch Driven Robot Control via Tactile Servo," in *Experimental Robotics*, 2016.

[11] Y. Chebotar, O. Kroemer, and J. Peters, "Learning robot tactile sensing for object manipulation," in *2014 IEEE/RSJ International Conference on Intelligent Robots and Systems*, 2014: IEEE, pp. 3368-3375.

[12] A. Drimus, G. Kootstra, A. Bilberg, and D. Kragic, "Design of a flexible tactile sensor for classification of rigid and deformable objects," *Robotics & Autonomous Systems*, vol. 62, no. 1, pp. 3-15, 2014.

Proceedings of the 16th Annual IEEE International
Conference on Nano/Micro Engineered and Molecular Systems
April 25-29, 2021

Characterization of an Asymmetrical Capacitive MEMS Tilt Sensor

Y. Gao, X. Zhang, Y. Qi, H. Zhang, M. Ryutaro, Z. Jiang and X. Wei, *Senior Member, IEEE*

Abstract—This paper presents a uniaxial tilt sensor with the asymmetric differential capacitors for the unique tilt measurement, which has the advantages of low cross-sensitivity and low fabrication cost. The MEMS capacitive tilt sensor was designed, characterized and verified for high accuracy and reliability. The experiment results indicated the average sensitivity and bias instability of 5 fF/° and 0.0028° (at 0° position). The temperature drift of bias is 0.034 °/°C (above room temperature) and 0.146 °/°C (below room temperature). After calibration compensation, the temperature sensitivity drift can be stabilized to be only 15.8 ppm/°C. In addition, the practical performance has been validated by the prototype of the test board, which indicates the feasible application potential of the asymmetrical capacitive MEMS tilt sensor.

Keywords: MEMS, tilt sensing, asymmetrical capacitors, bias instability, temperature effect

I. INTRODUCTION

In recent years, the importance of tilt measurement has been widely concerned in strategic applications, such as robot attitude control, spacecraft docking, antenna alignment, navigation [1-4], and also in some unexpected applications like portable digital devices [5]. The traditional tilt sensor can be uniaxial, biaxial or tri-axial. In some published multiple axis sensors, the sensors offer the advantages of high accuracy and application convenience. However, these sensors usually need expensive, complex manufacturing technology [6] and complex signal processing system [7-8], which increase the system complexity and chip occupancy. Moreover, in some industrial applications such as multi-point measurement and smart wearable devices application, the miniaturization, fabrication cost and response time of tilt sensors should also be taken into consideration. Besides, the calibration error of the uniaxial or multiaxial sensor will have a great impact on the accuracy of measurement [9]. In some special applications, the sensors need to be calibrated, which is a challenge for users without self-calibration equipment [10]. Therefore, the uniaxial tilt sensor with asymmetric differential capacitors and self-calibrate, offers the advantages of quick response to the specific angle without complex signal processing circuit.

In this paper, an asymmetric capacitive tilt sensor with serpentine beams is proposed, which can quickly measure the single-axis tilt between 10° and 15° from horizontal position. The experimental results including sensitivity, nonlinear, temperature effect, bias stability and the

prototype verification, have verified the high precision and stability of the asymmetric tilt sensor.

II. PRINCIPLE AND DESIGN

In some unique applications, in order to quickly and precisely measure the tilt change from 10° to 15°, the asymmetric differential capacitors was designed in the tilt sensor. The measurement schematic is shown in Fig. 1. The asymmetric differential capacitors (C_A and C_B) can be realized by changing the parameters in the following ways: the different design of the comb fingers' number, gap distance, effective overlap area and dielectric material between the capacitors on the two sides of the proof-mass [10]. In this paper, the numbers of comb fingers are different in the asymmetric differential tilt sensor. The structural parameters of the MEMS tilt sensor is shown in TABLE I.

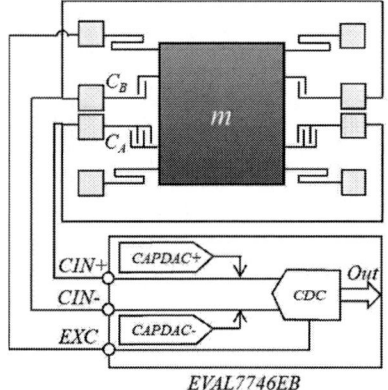

Figure 1. The measurement schematic of the tilt sensor.

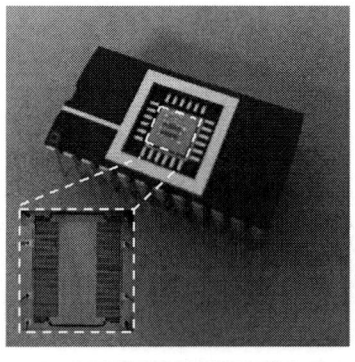

Y. Gao, X. Zhang, Y. Qi, H. Zhang, M. Ryutaro, Z. Jiang and X. Wei are with the State Key Laboratory for Manufacturing Systems Engineering, Xi'an Jiaotong University, Xi'an, 710049, China. X. Zhang is also with Northwest Institude of Nuclear Technology, Xi'an 710024,

China. (Phone: 86-29-82668839; Fax: 86-29-83234716; E-mail: senawei@mail.xjtu.edu.cn).

978-1-6654-3008-1/21 $31.00 © 2021 IEEE

Figure 2. The tilt sensor in the chip carrier. Inset is the micrograph of the sensor.

The two differential capacitors of the tilt sensor are not equal. In the horizontal position, the capacitor C_A is larger than of the capacitor C_B. When the angle is turned in the range from 10° to 15°, the capacitor C_A is smaller than C_B. The specific angle variation (between 10° and 15°) will result in the quick response of the output voltage polarity. Meanwhile, the tilt angle in other measurement range can be also precisely calculated from output analog voltage. Fig. 2 presents the micrograph of the device packaged in a DIP-24 chip carrier.

TABLE I. THE STRUCTURAL PARAMETERS OF THE ASYMMETRICAL CAPACITIVE MEMS TILT SENSOR

Parameter	Value	Unit
Die area	3×3	mm²
Thickness	25	μm
Wieght	260.4	μg
Stiffness of the spring	1.56	N/m
Capacitor plate gap	3	μm
Capacitor C_A	3.98158	pF
Capacitor C_B	3.85968	pF

III. EXPERIMENT VERIFICATION

A. Frequency response

The vector network analyzer (VNA) was used to measure the frequency response of the MEMS tilt sensor to verify the fabrication accuracy and the dynamic performance in the vacuum environment (20 Pa) at room temperature (24.5 °C). The measurement schematic of the open-loop frequency response is shown in Fig. 3. The excitation signal from VNA will result in the motion of the sensor at the first-order resonance, and is collected by the input port after amplification. With the scan scale of 400 Hz, the first-order resonant frequency is 626 Hz with the quality factor of 1042.6, as shown in Fig. 4. Considering the inevitable fabrication tolerances and influence of the experiment environment including vacuum pressure and temperature, the measurement result proves a good agreement with the simulation results by COMSOL.

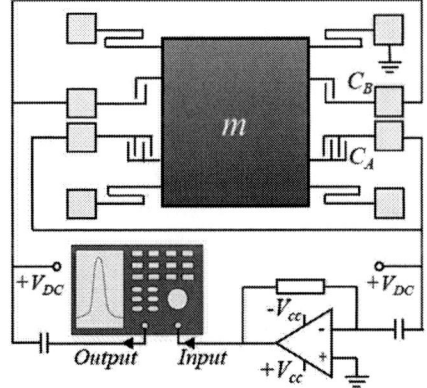

Figure 3. The schematic of the open loop frequency response test.

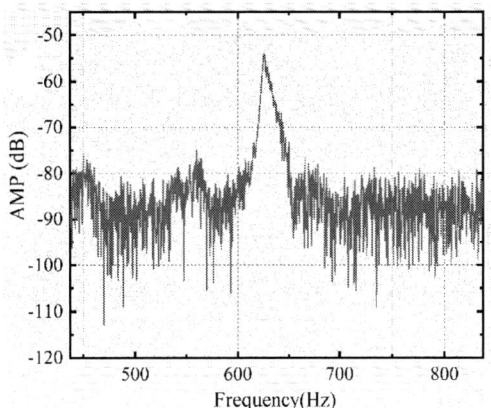

Figure 4. The open loop resonance amplitude curve with the scan scale of 400 Hz by VNA.

B. Sensitivity and nonlinearity

With a high-precision rotary table, the output of the tilt sensor in the tilt range from -180° to +180° is measured by a second-order sigma-delta capacitance readout circuit (AD7745, as shown in Fig. 1). The sigma-delta modulator has an inherent high resolution (up to 21-bit effective ADC), high linearity (0.01%), and high accuracy (±4 aF calibrated in factory). As illustrated in Fig. 5, it indicates that the average sensitivity of the tilt sensor is 5 fF/°. Within its linear measurement range from -30° to +30°, the sensor has a good sensitivity of 7.7 fF/g and nonlinearity of 0.63%. Additionally, the output of the MEMS tilt sensor has a perfect sine-shape curve. Converting the output angle to the acceleration, the sensor has the nonlinearity of 0.54% between ±1 g.

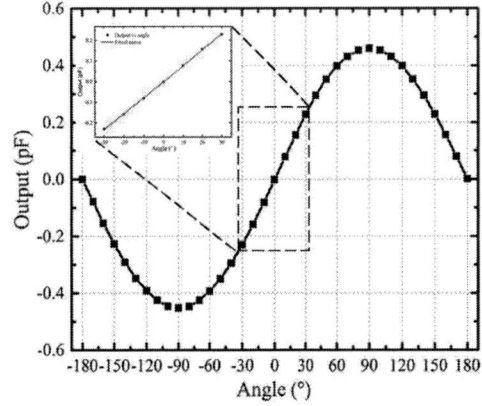

Figure 5. Sensitivity test result of the tilt sensor. Inset is the relation between the acceleration and outpu in the linear range.

C. Bias instability

Bias instability is generated by electronic or other components, which is susceptible to random flickering. It usually occurs as bias fluctuation in the data due to its low-frequency character. Bias instability of the sensor can be calculated by Allan deviation as shown in Fig.6. The static data was collected for half an hour in the quiet laboratory environment at room temperature (24.5 ℃) to ensure the reliability of test results. The sampling data rate is set to be 9.1 Hz to reduce the quantization noise. The lowest part of the Allan deviation curve with zero slope represents the bias instability. At 0° and 90° position, the bias instability is 0.0028° with an integral time of 5.71s and 0.0041° with an integral time of 0.33s, respectively, which implies the ultimate resolution.

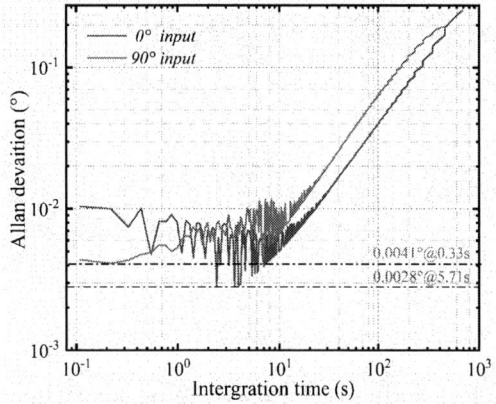

Figure 6.   Allan deviation of the MEMS tilt sensor's output with 0° and 90° input.

D. Temperature effect

The temperature effect of silicon-based sensors will significantly affects the accuracy of output and should be considered. The temperature bias and sensitivity drift of the sensor were measured, corresponding to an input temperature range from -10 ℃ to 100 ℃. The temperature bias and sensitivity drift increase with the temperature as shown in Fig. 7 and Fig. 8. As a result, the temperature bias below room temperature is 0.146 °/℃, which is larger than that above room temperature (0.034 °/℃). As shown in Fig. 8, without compensation, the temperature drift of sensitivity is 66.9 ppm/℃ in the temperature range from 10 ℃ to 100 ℃; after calibration compensation, it can be stabilized to be only 15.8 ppm/℃, which is far less than that of common commercial tilt sensors. Both results indicate high temperature stability of the tilt sensor.

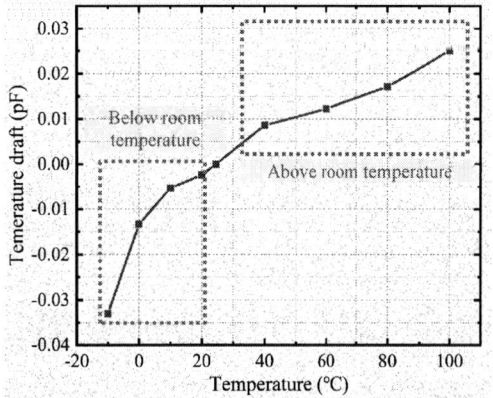

Figure 7.   Temperature bias of the tilt sensor from -10 ℃ to 100 ℃.

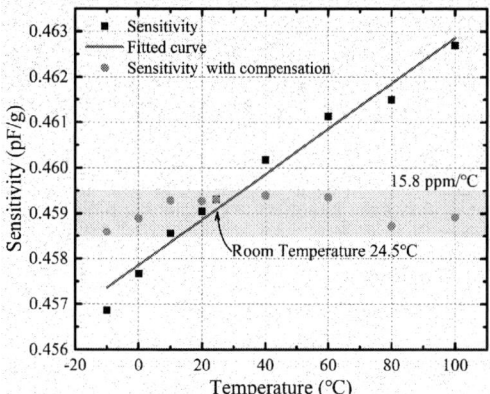

Figure 8.   The sensitivity drift with the temperature from -10 ℃ to 100 ℃.

E. Practical test

In order to validate the quick and precise measurement of tilt around 10° to 15° from horizontal position, practical test was carried out to measure the angle of rotation around the y-axis of the test board. The test board can be powered by a 3.3V button battery or 5V micro USB interface. The capacitance bridge on the test board detects the relationship between two differential capacitors (C_A and C_B). The test results are shown in Fig. 10 (a) and (b). The red and blue light indicate the power on and position signal. When the board is powered on, the red light is on. When the tilt is turned 15° from the horizontal position, the signal of green light can be used as switch signal of the actuators or used for self-calibration in practical applications. For this unique application, the design of asymmetric differential capacitors can simplify the circuit and reduce power consumption.

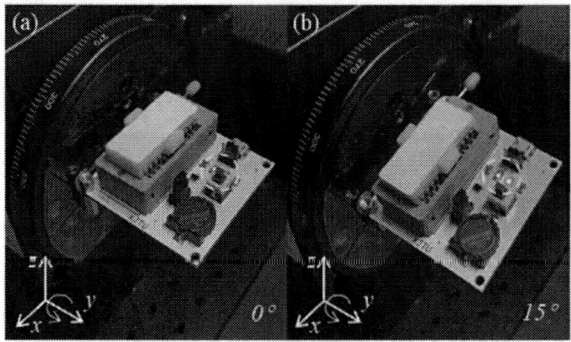

Figure 9.   Practical verification of the MEMS tilt sensor. (a) and (b) are the response of the sensor separately in horizontal position and 15° tilt position. The green light indicates that it has turned 15° from the horizontal position.

Compared with commercial MEMS tilt sensors shown in TABLE II, the tilt sensor designed in the paper has a good resolution, small temperature drift and small temperature sensitivity drift. However, the commercial tilt sensors like JMI-100 has larger linear measurement range and smaller nonlinearity. In order to extend linear measurement range, a group of tilt sensors assembled at special angles is an appropriate choice.

TABLE II. COMPARISON OF THE PROPOSED MEMS TILT SENSOR WITH COMMERCIAL MEMS TILT SENSORS

MEMS Sensor	SCA100T-D02[11]	JMI-100 [12]	CXTA-02 [13]	This work
Full Range	±90°	±90°	±75°	±90°
Linear Range	±90°	±90°	±20°	±30°
Sensitivity	35 mV/°	28 mV/°	35 mV/°	5 fF/°
Non-linearity	0.57%	0.05%	<0.06% (small angle)	0.63%
Resolution	0.0035°	0.004°	0.05°	0.0028°
Temperature drift	0.008 °/°C	0.14 °/°C	0.03 °/°C	0.034 °/°C
Temperature sensitivity drift	140 ppm/°C	150 ppm/°C	100 ppm/°C	15.8 ppm/°C

The characterization results of the MEMS tilt sensor indicate small bias instability and stable temperature effect. In the future, to improve the better accuracy and stability in the practical application, the calibration of the temperature effect and integrations of the tilt sensor are still necessary to study.

IV. CONCLUSION

This paper presents the principle design, characterization and practical validation of the asymmetrical capacitive MEMS tilt sensor. The first modal resonant frequency of the MEMS tilt sensor is 626Hz. The micro structure designed for sensing tilt has a full measurement range of ±90°, a linear measurement range of ±30° with a scale factor of 7.7 fF/g with non-linearity of 0.63%. The ultimate resolution of the MEMS tilt sensor is evaluated by the Allan deviation and indicates an angle resolution of 0.0028° with integral time of 5.71 s. The tilt sensor has good temperature stability of 0.146 °/°C below room temperature and 0.034 °/°C above room temperature, low sensitivity drift of 15.8 ppm/°C after calibration compensation. The practical verification with a prototype proves that the MEMS tilt sensor has a quick response at the specific tilt angle of 15° from the horizontal position.

ACKNOWLEDGMENT

This work is supported by the National Key Research and Development Program of China (2018YFB2002303), the National Natural Science Foundation of China (52075432) and Key Research and Development Program of Shaanxi Province (2018ZDCXL-GY-02-03). The support from the International Joint Laboratory for Micro/Nano Manufacturing and Measurement Technologies is also appreciated.

REFERENCES

[1] N. Neda, Jr. Landry René, C.J. Hua, and G. Denis,. "A new technique for integrating mems-based low-cost IMU and GPS in vehicular navigation." Journal of Sensors,2016,(2016-6-8), 2016, 1-16.

[2] A. Abdelrahman, and E.S. Naser, "Low-cost mems-based pedestrian navigation technique for GPS-denied areas." Journal of Sensors,2013,(2013-8-22), 2013, 572-575.

[3] B. Arnold, D. Wohlrab, C. Meinecke, et al. "Design, fabrication and characterisation of high-precision MEMS tilt sensor for surgical robot navigation" Thomas. (2017).

[4] S. Z. Guo, K. Qiu, F. Meng, S. H. Park, and M. C. McAlpine, "3D printed stretchable tactile sensors," Adv. Mater., vol. 29, no. 27, p. 1701218, 2017.

[5] C. Chaovalit, et al. "Applications of Smartphone-Based Sensors in Agriculture: A Systematic Review of Research." Journal of Sensors (2015).

[6] V. Kaajakari, Practical MEMS, Small Gear Publishing: Las Vegas, NV, USA, 2009; ISBN 978-0-9822991-0-4

[7] L. Łuczak, S. Sergiusz, R. Grepl and M. Bodnicki. "Selection of mems accelerometers for tilt measurements." Journal of Sensors,2017,(2017-3-30), 2017, ID 9796146.

[8] S. Łuczak, "Guidelines for tilt measurements realized by MEMS accelerometers, " International Journal of Precision Engineering and Manufacturing, Vol. 15, no. 3, pp. 489 – 496, 2014.

[9] L. Sergiusz. "Effects of Misalignments of MEMS Accelerometers in Tilt Measurements." Mechatronics 2013. Springer International Publishing, 2014.

[10] S. Łuczak, "Dual-axis test rig for MEMS tilt sensors" Metrology and Measurement Systems, Vol. 21, No. 2, pp. 351 – 362, 2014.

[11] Murata, SCA100T-D01, Datasheet. Available online: https://www.murata.com/-/media/webrenewal/products/sensor/inclinometer/sca100t/sca100t_inclinometer_datasheet_8261800b2_0.ashx?la=en-us (accessed on 21 January 2021)

[12] Jewell, JMI-100/200-S, Datasheet. Available online: http://www.jewellinstruments.com/wp-content/uploads/2018/05/JMI-S-Rev-C.pdf

[13] MEMSIC, CXTA-02, Datasheet. Available online: http://www.memsic.com/userfiles/files/Datasheets/Tilt-Sensors-Datasheets/ CXTA_Series_Datasheet.pdf (accessed on 21 January 2021).

Proceedings of the 16th Annual IEEE International
Conference on Nano/Micro Engineered and Molecular Systems
April 25-29, 2021

A Reflective Color Filter Based on ITO-Ba$_{0.5}$Sr$_{0.5}$TiO$_3$-ITO Nanofilms Capitalizing on a Black Layer

Rui Wang[1, #], Jinying Zhang[1, #, *], Bingnan Wang[1], Xinye Wang[1], Defang Li[1], Jingyi Chen[2], Chenyu Guo[3], Xin Wang[1], Zhuo Li[1] and Suhui Yang[1]

Abstract—**A reflective color filter was proposed based on ITO (Indium Tin Oxide)-Ba$_{0.5}$Sr$_{0.5}$TiO$_3$-ITO multiple nanofilms. A black layer was capitalized on to absorb the incoherent scattered light and to enhance the contrast ratio of the reflected light. While the thickness of BST (Ba$_{0.5}$Sr$_{0.5}$TiO$_3$) thin film changed from 100 nm to 140 nm, the peak wavelength of the reflected light shifted from 380 nm to 500 nm as the refractive index of BST thin film was set as 2.4. Meanwhile, the visualized color of the multiple nanofilms was changed from purple to turquoise. The measured reflected spectra fitted well with the simulation results based on transfer matrix method. The fabricated samples proved that the black layer played an important role in the contrast ratio increase. When the BTO thin film was 170 nm thick, the reflected peak wavelength shifted from 595 nm to 510 nm as its refractive index was tuned from 2.4 to 2.0 under a driven DC voltage. The unique characteristics show great potential in selectively reflective chromatic systems.**

I. INTRODUCTION

Reflective color filters, with the ability to selectively reflect a portion of the incident light, are key elements in various applications. [1-3] The potential for applications in dynamic optical display, electronic papers, stealth and camouflage has stimulated efforts to improve the reconfigurability and color saturation (reflection efficiency), e.g., via optimizing the materials and structure design of the filters. Nanostructured photonic crystals, metasurfaces, and nanofilms [4-6] have been extensively studied to acquire dynamic colors through electrical modulation [7], thermal response [8-9], self-assemble technology [10], etc. However, it is quite challenging to fabricate nanostructured photonic crystals and metasurfaces [11]. Nanofilms have great potential to be applied because of its feasibility in mass production. Researchers studied the color changing performance based on nanofilms using thermal response [8]. Suffering from slow rate of heating and cooling, the color changing speed is constrained significantly.

BST (Ba$_{0.5}$Sr$_{0.5}$TiO$_3$) is a ferroelectric material, and its dielectric constant will change with the driven DC (direct current) voltage in the frequency range from DC to terahertz band [12]. In addition, the reported results show that BST film possesses a high electro-optic coefficient (~100 pm/V) [13], making it a promising candidate in fast color changing

application. Based on this active film, we proposed a novel ITO-BST-ITO multiple nanofilms to realize dynamic colors. A black layer was introduced to enhance the reflection efficiency and to provide a high contrast ratio.

II. STRUCTURE DESIGN

The schematic of multiple nanofilms was shown in Fig. 1. It was composed of a black layer, a glass substrate, and a BST thin film sandwiched by two layers of ITO thin films. The black layer was capitalized on to absorb the incoherent scattered light and to enhance the contrast ratio of the reflected light. ITO thin films were used to provide voltage. BST thin film was used as the active material to tune the refractive index. When a voltage was applied to the two ITO films, the refractive index of BST will change. Consequently, the device could produce dynamic vivid colors due to the interference of light. The thickness of the ITO and BST films was t_m and d_c, respectively.

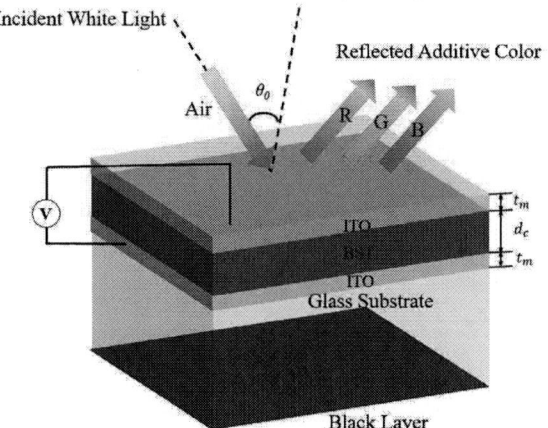

Figure 1 Schematic of ITO-BST-ITO nanofilms

III. METHODS

3.1 Simulation

The reflectivity of the nanofilms can be calculated by transfer matrix method [14]. When the light was incident to the nanofilms, the electric field and magnetic field of the

---

[1]Beijing Key Laboratory for Precision Optoelectronic Measurement Instrument and Technology, School of Optics and Photonics, Beijing Institute of Technology, Beijing 100081, China

[2]School of Information and Electronics, Beijing Institute of Technology, Beijing, 100081, China

[3]College of Sciences, Xi'an University of Science and Technology, Xi`an, 710054, China

[#]Co-first author

[*]Corresponding author: Jinying Zhang; e-mail: jyzhang@bit.edu.cn

Research supported by National Key Research and Development Program of China (2018YFF01010304, 2018AAA0100301), NSFC of China (No. 61704166, 61875011) and Beijing Institute of Technology Research Fund Program for Young Scholars.

incident plane and the outflow plane in each layer should satisfy a certain relationship, which can be described by the transfer matrix. Transform matrix for the k-th layer can be expressed as:

$$M_k = \begin{bmatrix} cos\,\delta_k & -\frac{i}{\eta_k}sin\,\delta_k \\ -\eta_k i\,sin\,\delta_k & cos\,\delta_k \end{bmatrix} \quad (1)$$

In Eq. (1), η_k is the impedance of k-th layer. δ_k is the phase shift. As shown Eq. (2):

$$\begin{cases} \delta_k = \frac{2\pi}{\lambda} \cdot \frac{n_k d_k}{cos\,\theta_k} \\ \eta_k = \sqrt{\frac{\varepsilon_k}{\mu_k}} / cos\,\theta_k\ (TM) \\ \eta_k = \sqrt{\frac{\varepsilon_k}{\mu_k}}\,cos\,\theta_k\ (TE) \end{cases} \quad (2)$$

In k-th layer, n_k is the refractive index, d_k is the physical thickness; ε_k and μ_k are the permittivity and permeability, respectively. Refraction angle θ_k can be calculated according to Snell theorem:

$$n_k\,sin\,\theta_k = n_{k-1}\,sin\,\theta_{k-1} = \ldots n_0\,sin\,\theta_0 \quad (3)$$

For the nanofilms composed of m-layer dielectrics, it can be regarded as a cascade of m transmission matrices. The transmission equation can be expressed as:

$$M = \prod_{m=1}^{m} M_m = \begin{bmatrix} m_{11} & m_{12} \\ m_{21} & m_{22} \end{bmatrix} \quad (4)$$

Then the reflection coefficient of incident electromagnetic wave is derived from the relation:

$$r' = \frac{(m_{11}+m_{12}p_m)p_1 - (m_{21}+m_{22}p_m)}{(m_{11}+m_{12}p_m)p_1 + (m_{21}+m_{22}p_m)} \quad (5)$$

$$p_1 = \sqrt{\frac{\varepsilon_1}{\mu_1}}\,cos\,\theta_1 \quad (6)$$

$$p_m = \sqrt{\frac{\varepsilon_m}{\mu_m}}\,cos\,\theta_m \quad (7)$$

Then The reflectance r can be calculated by the equation:

$$r = |r'|^2 \quad (8)$$

3.2 Spectra to RGB transformation

Colors are observed when light interacts with objects and the interacted light changes its spectrum through wavelength-dependent absorption, reflection, or refraction. Different spectra will produce different colors. In the spectrum, the peak wavelengths of 564–580 nm, 534–545 nm and 420–440 nm respond to red, green, and blue, respectively. The color is also related to the type of light source. Many methods of converting spectra to RGB values have been reported [15]. In this work, the D65 light source was selected to calculate the structural color. RGB color matching function, as illustrated in Fig. 2, is used to calculate the stimulus values (X, Y, Z) from the spectral data based on Eqs. (8)-(11). In Eq. (11), k is a normalized factor. Then the normalized values (x, y, z) are obtained based on Eq. (12)-(14). From the coordinates (x, y) in CIE diagram, the color can be obtained corresponding to the spectrum.

$$X = \frac{1}{k}\int_\lambda \bar{x}(\lambda)I(\lambda)R(\lambda)d\lambda \quad (8)$$

$$Y = \frac{1}{k}\int_\lambda \bar{y}(\lambda)I(\lambda)R(\lambda)d\lambda \quad (9)$$

$$Z = \frac{1}{k}\int_\lambda \bar{z}(\lambda)I(\lambda)R(\lambda)d\lambda \quad (10)$$

$$k = \int_\lambda \bar{y}(\lambda)I(\lambda)d\lambda \quad (11)$$

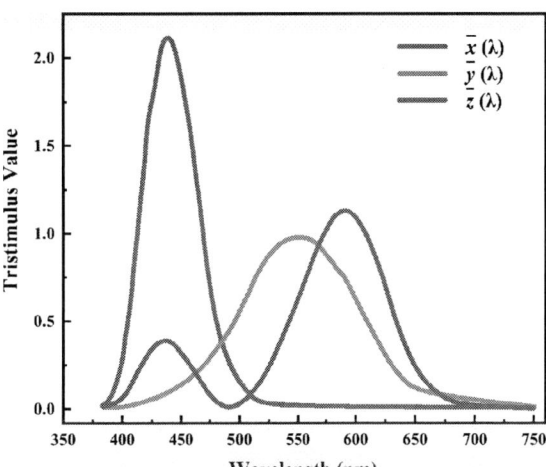

Figure 2 RGB color matching function

$$x = \frac{X}{X+Y+Z} \quad (12)$$

$$y = \frac{Y}{X+Y+Z} \quad (13)$$

$$z = \frac{Z}{X+Y+Z} \quad (14)$$

IV. RESULTS AND DISCUSSION

Based on transfer matrix method, we calculated the contour maps for the reflection spectra in response to refractive index of BST with a thickness of d_c =120 nm. Fig. 3 presented the results for ITO films with different thickness of t_m = 200 nm, 70 nm, and 30 nm, respectively. Evidently, when the thickness of ITO was 30 nm, the device performed most desired tuning effect: (1) the reflection peak was single rather than multiple in the wavelength range of 360 nm to 800 nm, and the color tuning was simpler to be controlled. (2) with the decrease of the refractive index of BST, the reflection peak moved widely from 585 nm to 537 nm.

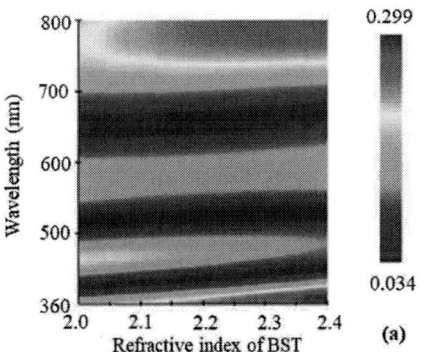

978-1-6654-3008-1/21 $31.00 © 2021 IEEE

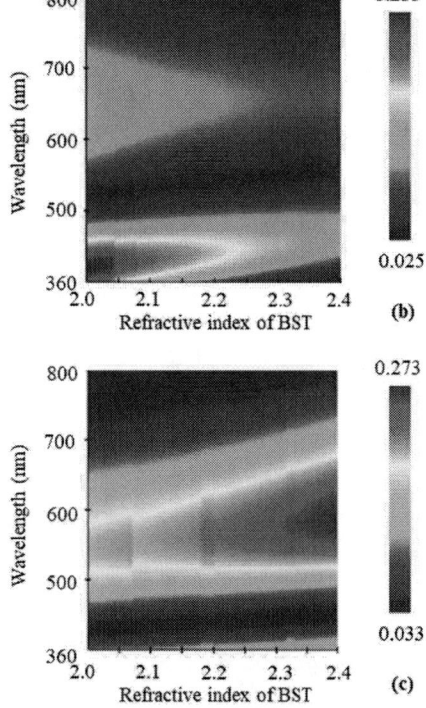

Figure 3 Contour maps for the reflection spectra in response to refractive index of BST with different ITO thickness: (a) $t_m = 200$ nm; (b) $t_m = 70$ nm; (c) $t_m = 30$ nm.

Therefore, a 30-nm-thick ITO film was magnetron sputtered on quartz with DC power of 150 W, argon and oxygen flow of 50 sccm and 2 sccm, respectively. After annealing for 2 hours at 280 °C, the ITO film became more transparent and the resistance of the sample at a distance of 1cm decreased from 8.4 k Ohm to 2.36 k Ohm, which indicated its conductivity was increased by 72%. Then a BST film was magnetron sputtered on the ITO film with RF power of 120 W, argon and oxygen flow of 50 sccm and 10 sccm, respectively. Fig. 4 presented the samples with BST films of different thickness. In Fig. 4 (a)-(c), the thickness of BST film was 100 nm. In Fig. 4 (d)-(f), the thickness of BST film was 140 nm. Moreover, in Fig. 4 (a) and (d), the fabricated samples had a white layer beneath the quartz substrate. While in Fig. 4 (c) and (f), the fabricated samples had a black layer. Fig. 4 (b) and (e) displayed a direct comparison of the visualized colors on white and black layers. Several interesting phenomena could be observed: (1) on white layer, the samples seemed quite transparent and exhibited the same color as the white layer even if the samples had different thickness of BST films. (2) the same sample showed different colors on different layers. For example, the one with BST film of 100 nm thick, was white in white layer but purple on black layer. (3) the thickness of BST film played an important role in color visualization for the samples on black layer. For instance, the one with BST of 100 nm thick was purple while the one of 140 nm thick

was turquoise. These phenomena could be explained by simulated and measured results as below.

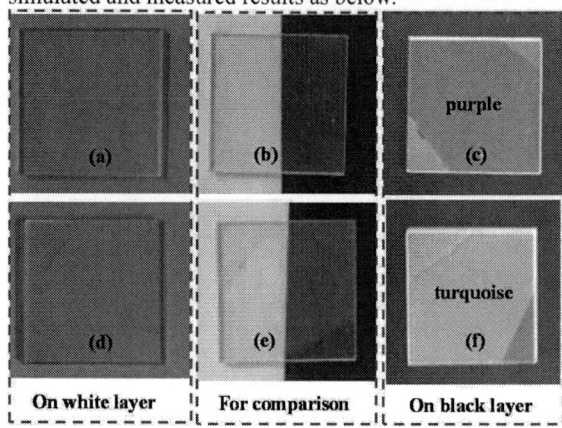

Figure 4 Photos of the fabricated samples with ITO-BST-ITO nanofilms on white or black layers

The simulated electric field distribution for 140-nm-thick BST film with a black layer at the wavelength of 500 nm was shown in Fig. 5. The black layer absorbed the incident light and had no contribution to the reflected spectrum which influence the visualized color. The light in the BST film interference and the electric field distribution was influenced significantly by the thickness of BST film. Fig. 6 presented the measured and simulated reflective spectrum in the wavelength range of 360 nm to 760 nm. First, the simulation results agreed well with the measured results. Second, the thickness of BST film determined the reflected peak. For example, the peak with 100-nm-thick BST film located at 390 nm and that with 140-nm-thick BST film was at 500 nm. Third, when the refractive index of BST film was set as a constant (here it was 2.40), the reflected peak was red-shifted with the thickness increase of BST film.

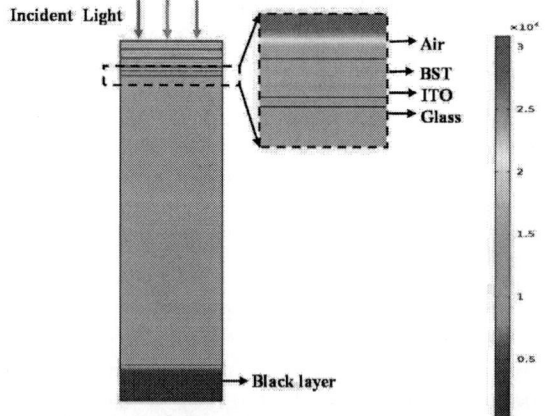

Figure 5 Simulated electric field distribution of the multiple nanofilms at the wavelength of 500 nm

As stated in Section 3.2, the RGB color matching function was used to calculate the stimulus values (X, Y, Z) from the spectral data in Fig. 6. The corresponding values of (x, y) was derived and determined the position in CIE diagram, as illustrated in Fig. 7. The calculated colors were consistent with those of the fabricated samples.

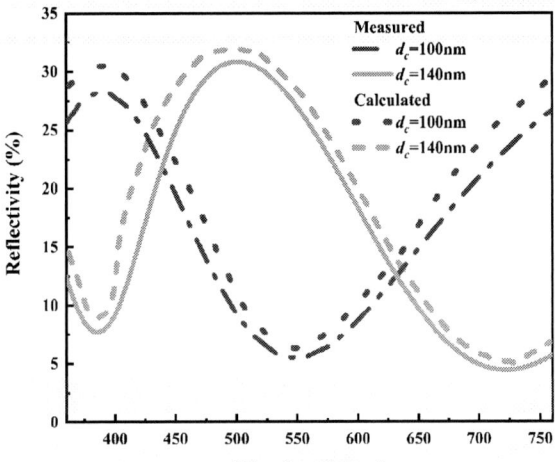

Figure 6 The measured and simulated reflected spectrum

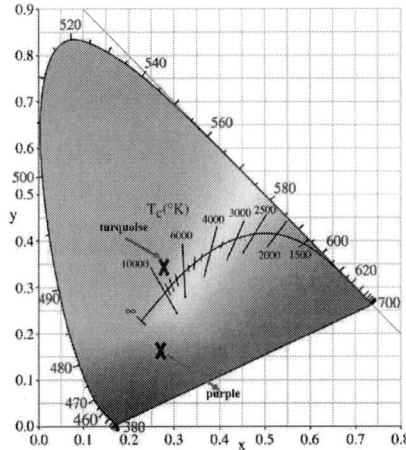

Figure 7 The position of the samples' colors in the CIE standard colorimetric system

Based on reported literatures[13], the BST film possessed a high electro-optic coefficient around 100 pm/V. When the BST film was 170 nm thick and was driven by a DC voltage of 170 V, its refractive index would change from 2.40 to 2.16. Furthermore, barium titanate (BTO) as a ferroelectric material, possessed much larger electro-optic coefficient. For example, its r_{15} in BTO crystal was around 1300 pm/V [16]. It indicated the refractive index of BTO could be tuned in a wider range. When a BTO film was 170 nm thick and was driven by a DC voltage of 23 V, its refractive index would change from 2.4 to 2.0. Fig. 8 presented the reflected spectra for BTO films with different refractive indices. The reflected peak shifted from 595 nm to 510 nm as its refractive index was tuned from 2.4 to 2.0. The unique characteristics show great potential in selectively reflective chromatic systems.

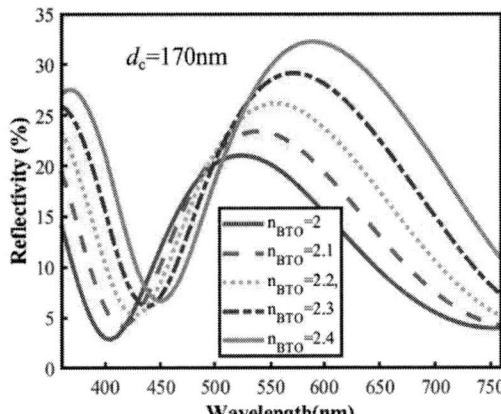

Figure 8 Simulated reflective spectrum with BTO thin film of different refractive index and 170 nm thick

REFERENCES

[1] S. Bang, J. Kim, G. Yoon et al., "Recent Advances in Tunable and Reconfigurable Metamaterials," Micromachines, vol. 9, pp. 560, 10/31, 2018.

[2] W.-J. Joo, J. Kyoung, M. Esfandyarpour et al., "Metasurface-driven OLED displays beyond 10,000 pixels per inch," Science, vol. 370, no. 6515, pp. 459, 2020.

[3] C.-S. Park, V. R. Shrestha, S.-S. Lee et al., "Trans-Reflective Color Filters Based on a Phase Compensated Etalon Enabling Adjustable Color Saturation," Scientific Reports, vol. 6, no. 1, pp. 25496, 2016/05/06, 2016.

[4] I. Kim, J. Yun, T. Badloe et al., "Structural color switching with a doped indium-gallium-zinc-oxide semiconductor," Photonics Research, vol. 8, 06/22, 2020.

[5] R. Meade, S. Johnson, and J. Winn, Photonic Crystals: Molding the Flow of Light - Second Edition, 2008.

[6] B. Yang, H. Cheng, S. Chen et al., "Structural colors in metasurfaces: principle, design and applications," Materials Chemistry Frontiers, vol. 3, 02/26, 2019.

[7] G. Yoon, S. So, M. Kim et al., "Electrically tunable metasurface perfect absorber for infrared frequencies," Nano Convergence, vol. 4, no. 1, pp. 36, 2017/12/21, 2017.

[8] N. Raeis-Hosseini, S. Lim, H. Hwang et al., "Reliable Ge2Sb2Te5-Integrated High-Density Nanoscale Conductive Bridge Random Access Memory using Facile Nitrogen-Doping Strategy," Advanced Electronic Materials, vol. 4, no. 11, pp. 1800360, 2018/11/01, 2018.

[9] S. Yoo, T. Gwon, T. Eom et al., "Multi-color changeable optical coating by adopting multiple layer of the ultrathin phase change material film," ACS Photonics, vol. 3, 06/13, 2016.

[10] G. Isapour, and M. Lattuada, "Bioinspired Stimuli-Responsive Color-Changing Systems," Advanced Materials, vol. 30, pp. 1707069, 04/26, 2018.

[11] E. Højlund-Nielsen, J. Clausen, T. Mäkelä et al., "Plasmonic Colors: Toward Mass Production of Metasurfaces," Advanced Materials Technologies, vol. 1, pp. 1600054, 08/01, 2016.

[12] N. Xi, S. Chen, J. Zhang et al., "Dielectric microwave properties of Si-integrated pulsed laser deposited (Ba, Sr)TiO3 thin films up to 110 GHz," Applied Physics Letters, vol. 107, pp. 052905, 08/03, 2015.

[13] D. Y. Wang, J. Wang, H. L. W. Chan et al., "Structural and Electro-Optic Properties of Ba0.7Sr0.3TiO3 Thin Films Grown on Various Substrates Using Pulsed Laser Deposition," Journal of Applied Physics, vol. 101, pp. 043515-043515, 02/15, 2007.

[14] F. Wang, Y. Cheng, X. Wang et al., "Effective modulation of the photonic band gap based on Ge/ZnS one-dimensional photonic crystal at the infrared band," Optical Materials, vol. 75, pp. 373-378, 01/01, 2018.

[15] Brainard, and H. David, "The Science of Color ‖ Color Appearance and Color Difference Specification," pp. 191-216, 2003.

[16] S. M. Liu, R. Guo and J. J. Xu. Photorefractive nonlinear optics and Its Applications. Beijing: Science Press, 2004, pp. 222-231.

Gap in pagination due to unavailable papers.

Pages 1668-1675

Proceedings of the 16th Annual IEEE International
Conference on Nano/Micro Engineered and Molecular Systems
April 25-29, 2021

Conical Helmholtz Resonator-Based Triboelectric Nanogenerator for Harvesting of Acoustic energy *

Hongyong Yu, Taili Du, Hongfa Zhao, Qiqi zhang, Ling Liu, Anaeli Elibariki Mtui, Yue Huang, Xiu Xiao, Minyi Xu*

Abstract-This paper present the conical Helmholtz resonator-based triboelectric nanogenerator (CH-TENG) designed for highly efficient harvesting of acoustic energy in various occasions. This CH-TENG consists of an acoustic collecting tube, a conical Helmholtz resonant cavity and an aluminum film with uniform distributed pinholes, and a dielectric soft film with one side ink-printed for electrode. Furthermore, the effects of resonant cavity structure, sound wave frequency, sound pressure level and sound collection tube on the performance of CH-TENG are systematically studied. Compared with previous research, the proposed CH-TENG can generates higher acoustic sensitivity per unit area and power density per unit sound pressure, because the acoustic collecting tube structure gathers more sound waves and sound wave amplification capability of the Conical Helmholtz Resonator stronger. Meanwhile, this new technique to some extent can serve as a sensor.

I. INTRODUCTION

With the development of the times, modern society pays more and more attention to the development and utilization of new energy sources (such as solar energy, wind energy, mechanical movement, geothermal energy, etc.). As a kind of energy generated by vibration and propagated in the multiple medium, acoustic energy belong to new energy sources. acoustic energy is widely distributed in nature. For example: ship engine room, factory machinery premises, traffic roads and other places have a large amount of acoustic energy and are sustainable energy sources [1-3]. Although the acoustic energy is widely distributed and rich in energy, it is almost wasted due to the low energy density of sound waves and the lack of effective collection technology. Most of the current researches are based on electromagnetic induction or piezoelectric effect to convert acoustic energy, which is vibration energy [4-7], into electrical energy. These two conversion mechanisms are difficult to effectively convert acoustic energy into electrical energy, the main reason is that the energy density of acoustic energy becomes very low due to the diffusion of acoustic energy in the process of propagation in the medium, and acoustic energy is a high-frequency vibration energy, which causes the above two conversion mechanisms [8-10] to be unable to efficiently collect acoustic energy. Therefore, it is very important to propose an efficient acoustic energy collection device. In recent years, with the rise of the triboelectric

nanogenerator (TENG) technology, new ideas have been provided for efficient collection of acoustic energy. Since the triboelectric nanogenerator can use flexible materials as electrodes, TENG will be very sensitive response for external disturbance, this feature is very suitable for collecting sound wave energy. There has been work to collect acoustic energy using TENG for a long time [11]. Yang et al [12] reported an acoustic energy collection device based on a Helmholtz resonance cavity and organic thin film TENG. The acoustic collection device they reported can generate a voltage of 60.5V and a short circuit current of 15.1μA at acoustic pressure level of 110dB. Fan et al [13] reported a result of making copper and paper into a micro-hole array to enhance the collection of acoustic energy, and designed a thin and flexible acoustic energy collection TENG. Under the action of 117dB acoustic pressure, the device can generate a maximum instantaneous power density of 121mW/m2. The TENG uses paper to capture acoustic energy. Although the structure is innovative, the electrical output is still limited. Zhao et al [14]. reported an acoustic energy collection device based on a double-tube Helmholtz resonant cavity. Due to the unique double-tube design, the Helmholtz resonant cavity's amplification effect on acoustic energy is improved, making it the most efficient currently reported. Acoustic energy TENG, the above report is enough to show that TENG is very suitable for acoustic energy collection [15-18].

In this work, we propose a triboelectric nanogenerator based on a conical Helmholtz cavity (CH-TENG). The CH-TENG consists of a conical Helmholtz resonance cavity and a power generation unit. The power generation unit is composed of an aluminum film with pinholes and an FEP film. The sound waves are first collected and amplified by the conical Helmholtz cavity, and then the FEP film and the aluminum film are excited to make alternate contact and separation. The two dielectric materials undergo electron transfer to convert acoustic energy into electrical energy output. This research systematically analyzes the output performance of CH-TENG in theory and experiment. Compared with the traditional Helmholtz resonant cavity TENG, the Through a large number of experiments and simulations, we found that CH-TENG has better output performance, because the conical Helmholtz resonant cavity has better sound wave amplification capabilities than other

* *Research supported by National Science Foundation of China (Grant Nos. 51879022, 51979045), the Fundamental Research Funds for the Central Universities, China (Grants Nos. 3132019330)

All authors are with Marine Engineering College, Dalian Maritime University, No.1 Linghai Road, Dalian, 116026, China

*Contacting Authors: Minyi Xu is with Marine Engineering College, Dalian Maritime University, No.1 Linghai Road, Dalian, 116026, China (phone: +86 18941134769; email: xuminyi@dlmu.edu.cn).

978-1-6654-3008-1/21 $31.00 © 2021 IEEE

forms of resonant Helmholtz cavity. Then we verified the influencing factors of CH-TENG's output performance through experiments and simulations. The results of experiments and simulations showed that sound wave frequency, sound pressure level, and conical cavity structure all have a significant impact on the output performance of CH-TENG.

II. RESULTS AND DISCUSSION

A. CH-TENG structure and working principle

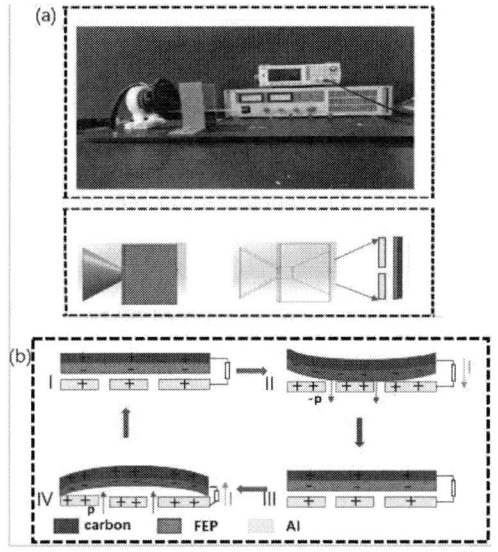

Figure1. (a) Acoustic test bench and Structure scheme of the CH-TENG. (b) Working mechanism of the CH-TENG.

As shown in Figure1.1b, the CH-TENG consists of an acoustic collecting tube, conical Helmholtz cavity, an aluminum film with uniform distributed pinholes, and a FEP film. Among them: acoustic collecting tube can gather sound waves to improve the output performance of CH-TENG. The conical Helmholtz resonant cavity is a kind of resonant cavity with powerful sound wave amplification function. The aluminum film and FEP film are important components of the power generation unit. The conical Helmholtz resonant cavity used in this paper is a special form of resonant cavity. Compared with the traditional Helmholtz resonant cavity, it has a better sound wave energy amplification performance, and in order to improve the collection of sound waves by the cavity. Ability to add a cone-shaped collection tube in front of the conical resonant cavity. In order to improve the output performance of TENG, we have performed surface micro-treatment on FEP film. Fig. 1b shows the working principle of CH-TENG intuitively. First, there is a cone-shaped sound wave collecting tube at the front to collect the sound waves into the conical Helmholtz resonant cavity, the conical Helmholtz resonant cavity expands the amplitude of the sound wave, and finally the amplified sound wave acts on the FEP through the uniform small holes in the aluminum film. On the membrane, because the sound wave is a form of energy in the form of vibration, the pressure between the FEP film and the aluminum film changes periodically, and the periodic

change in pressure causes the FEP film and the aluminum film to repeatedly contact and separate. When the two films are in contact, the electrons of the FEP film are negatively charged due to the different electronegativity of the two films (FEP is higher than that of aluminum), and the aluminum film loses electrons with a positive point (Fig1.bi). When the pressure in the cavity changes due to the effect of sound waves, The FEP film is far away from the aluminum film under the action of acoustic pressure, and free electrons flow from the conductive ink electrode to the aluminum electrode through an external circuit to balance the local electric field, thereby generating a positive charge on the conductive ink electrode (Fig1.1bii). Push to the aluminum film. At this stage, the voltage difference weakens, and the free electrons in the aluminum flow back through the external circuit to the conductive ink electrode (Fig1.iv). The two surfaces of the FEP film and the aluminum film contact again, and the charge distribution returns to its original state (Fig1.i). At this point, the entire power generation cycle is completed. Therefore, CH-TENG produces an AC pulse electrical output under the action of sound waves.

B. Influence of Helmholtz resonant cavity geometry

Currently reported devices that use electromagnetic induction, piezoelectric effect or TENG mechanism to convert acoustic energy into electrical energy, almost all of them use Helmholtz resonant cavity to collect and increase the amplitude of incident sound waves. The traditional Helmholtz resonant cavity is composed of the sealed cavity is composed of a short tube. The volume of the short tube is much smaller than the volume of the resonant cavity. However, there are many types of resonant cavity. After analyzing the performance of various forms of resonant cavity, we finally determined the conical Helmholtz cavity has the best acoustic amplification performance. The Fig2. shows the geometry of the double-tube Helmholtz resonant cavity and the Conical Helmholtz resonant cavity. We have built a systematic acoustic test bench to study the performance of CH-TENG's acoustic energy collection. As shown in the Fig1.(a), the sine sound wave of the test bench is provided by a speaker with a rated power of 100w, and a

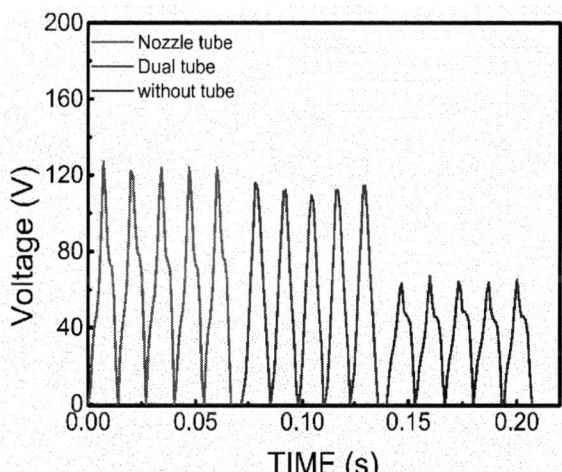

Figure2. Open-circuit voltage of the different tube of HR-TENGs under the same sound wave condition.

978-1-6654-3008-1/21 $31.00 © 2021 IEEE

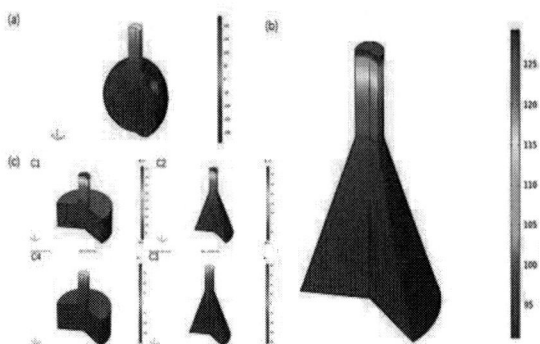

Figure3. The simulation results of different acoustic pressure level based on multiple shapes of resonant cavities.

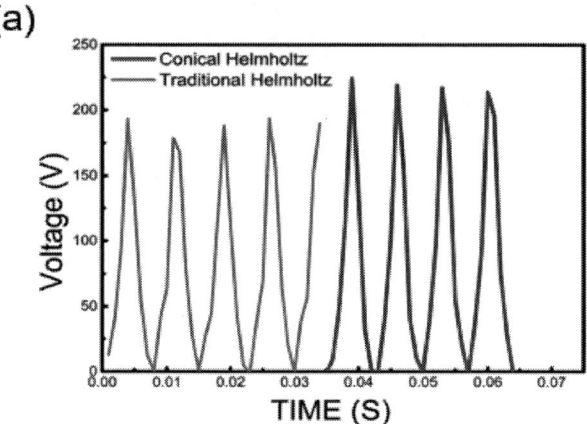

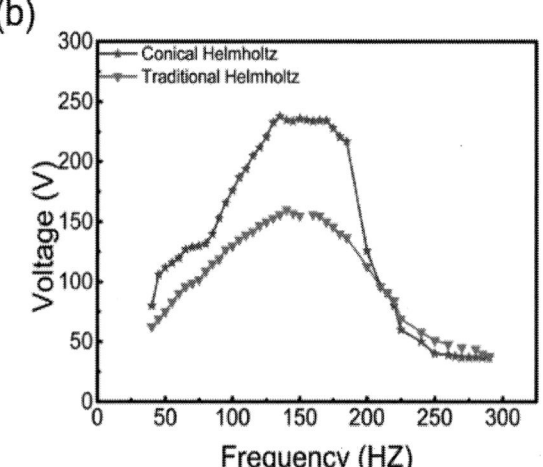

Figure4. (a)Open-circuit voltage of the CH-TENGs and TH-TENG under the same sound wave condition. (b)Open-circuit voltage of the Conical Helmholtz and Traditional Helmholtz at different acoustic frequencies.

function signal generator is used to provide a sine wave signal. Adjust the frequency of the acoustic source, the output signal of CH-TENG is collected by the data acquisition card and displayed on the computer programmed with LabVIEW software. In addition, a decibel meter with an accuracy of 1.5dB and 0.1 resolution is placed near the FEP membrane to measure the acoustic pressure level. In order to ensure the accuracy of the experiment, all the experiments in this article were done under the condition of changing different types of Helmholtz while keeping the TENG power generation unit unchanged, so the influence of different TENG performance on power generation performance was excluded. Fig2. shows the open-circuit voltage of the different tube of TENGs under the same sound wave condition. As expected, the nozzle tube can increase the power output of the TENG. The peak voltage of the Nozzle tube TENG is 130V, compared to that with a dual tube, the Helmholtz resonator with nozzle tubes can improve the output performance of the TENG by 18.1%. Fig3. shows that the simulation results of different acoustic pressure level based on multiple shapes of resonant cavities, this result indicates that compared to with the traditional Helmholtz resonator with can improve the sound waves amplification capability. Fig4.(a) displays the open-circuit voltage of the CH-TENG and traditional Helmholtz resonator-based triboelectric nanogenerator (TH-TENG) under the same sound wave condition, in order to compare the acoustic pressure amplification performance of the two resonators. The maximum open-circuit voltage of CH-TENG can reach 185 V, and the maximum open-circuit voltage of CH-TENG can reach 225V, this result indicates that, 21.6% higher than the maximum voltage of the CH-TENG with the conventional resonator. Fig4(b) shows their peak open-circuit voltages at different acoustic frequencies. With the application of the conical Helmholtz resonator, the electrical output of the TENG is improved. The maximum open-circuit voltage can reach 230 V at the optimal frequency of 130Hz. Therefore, it is proved that CH-TENG has better performance for harvesting of acoustic energy.

We further studied the influence of the taper of the conical cavity on the output performance of TENG. The taper was changed by changing the size of the front-end circular hole D1, and other conditions remained unchanged.

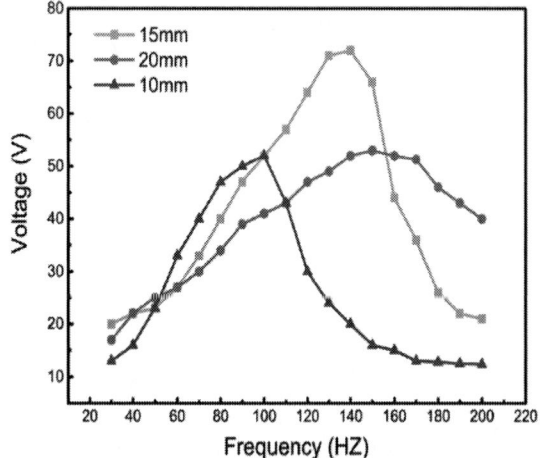

Figure5. The influence of the size of the front hole on CH-TENG

The Fig5. shows three different front-end circular holes D1 CH-TENG in Peak open circuit voltage at different acoustic frequencies. As the diameter of the hole changes, the output performance of CH-TENG has made a significant change. The TENG output performance has changed significantly. The acoustic source frequency range is 30-

200Hz and the acoustic pressure level range is 70.1-92.3dB. When the acoustic source frequency is 50-100Hz, the CH-TENG with a hole diameter of 10mm has the best output performance. The voltage continues to increase with the increase of the acoustic source frequency and reaches a peak voltage of 52.6V when it reaches 100Hz. However, when the frequency is higher than 80Hz, the output performance begins to decrease. When the frequency range is 100-160Hz, the output performance of the 15mm hole TENG is the best, reaching a peak voltage of 75.1V at a frequency of 140Hz. When the frequency is greater than 160Hz, the 20mm hole TENG has the best output performance. What is more interesting is that as the hole diameter increases, the optimal response frequency of CH-TENG gradually increases, and at the same time the frequency bandwidth gradually increases, but the output performance first increases and then decreases.

ACKNOWLEDGMENT

The authors are grateful for the joint support from the National Natural Science Foundation of China (Grant Nos. 51879022, 51822901, 51979045, 51906029), the Fundamental Research Funds for the Central Universities, China (Grant No. 3132019330).

III. CONCLUSION

In summary: the conical Helmholtz resonant cavity has better acoustic pressure amplification performance, and has a wider response bandwidth, so it is very suitable for making an acoustic energy collection device. This device has a wide range of application scenarios. It can be used as an acoustic energy sensor and placed in mechanical equipment. The acoustic information of mechanical operation is collected nearby, and the acoustic information collected by CH-TENG can be analyzed for failure detection of mechanical equipment. This is a new type of self-powered acoustic sensor. You can also integrate CH-TENG and place it on the side of the highway to make it into a highway noise barrier that can collect acoustic energy and reduce vehicle noise.

REFERENCES

[1] H. Zhao, X. Xiao, P. Xu, T. Zhao, L. Song, X. Pan, J. Mi, M. Xu, and Z. L. Wang, "Dual‐Tube Helmholtz Resonator‐Based Triboelectric Nanogenerator for Highly Efficient Harvesting of Acoustic Energy," *Advanced Energy Materials*, vol. 9, no. 46, 2019.

[2] Y. Wang, X. Zhu, T. Zhang, S. Bano, H. Pan, L. Qi, Z. Zhang, and Y. Yuan, "A renewable low-frequency acoustic energy harvesting noise barrier for high-speed railways using a Helmholtz resonator and a PVDF film," *Applied Energy*, vol. 230, pp. 52-61, 2018.

[3] J. Tian, X. Chen, and Z. L. Wang, "Environmental energy harvesting based on triboelectric nanogenerators" Nanotechnology, vol. 31, no. 24, pp. 242001, Mar 27, 2020.

[4] Y. F. Hu, J. Yang, Q. S. Jing, S. M. Niu, W. Z. Wu, and Z. L. Wang, "Triboelectric Nanogenerator Built on Suspended 3D Spiral Structure as Vibration and Positioning Sensor and Wave Energy Harvester," *Acs Nano*, vol. 7, no. 11, pp. 10424-10432, Nov, 2013.

[5] Self-Powered Acceleration Sensor Based on Liquid Metal Triboelectric Nanogenerator for Vibration Monitoring," *ACS Nano*, vol. 11, no. 7, pp. 7440-7446, Jul 25, 2017.

[6] S. A. Graham, S. C. Chandrarathna, H. Patnam, P. Manchi, J.-W. Lee, and J. S. Yu, "Harsh environment‐tolerant and robust triboelectric nanogenerators for mechanical-energy harvesting, sensing, and energy storage in a smart home," *Nano Energy*, vol. 80, 2021.

[7] S. H. Yang, G. L. Lu, A. Q. Wang, J. Liu, and P. Yan, "Change detection in rotational speed of industrial machinery using Bag-of-Words based feature extraction from vibration signals," *Measurement*, vol. 146, pp. 467-478, Nov, 2019.

[8] Padi, and Gyula, "Electrodynamic loudspeaker with electromagnetic impedance sensor coil," The Journal of the Acoustical Society of America, vol. 94, no. 3, pp. 1755-1755, 1993.

[9] F. U. Khan, and Izhar, "Electromagnetic-Based Acoustic Energy Harvester," 2013 16th International Multi Topic Conference (Inmic), pp. 125-130, 2013.

[10] T. H. Lai, C. H. Huang, and C. F. Tsou, "Design and Fabrication of Acoustic Wave Actuated Microgenerator for Portable Electronic Devices," Dtip 2008: Symposium on Design, Test, Integration and Packaging of Mems/Moems, pp. 28-33, 2008.

[11] F. U. Khan, and Izhar, "Hybrid acoustic energy harvesting using combined electromagnetic and piezoelectric conversion," Review of Scientific Instruments, vol. 87, no. 2, Feb, 2016.

[12] F. Q. Chen, Y. H. Wu, Z. Y. Ding, X. Xia, S. H. Li, H. W. Zheng, C. L. Diao, G. T. Yue, and Y. L. Zi, "A novel triboelectric nanogenerator based on electrospun polyvinylidene fluoride nanofibers for effective acoustic energy harvesting and self-powered multifunctional sensing," Nano Energy, vol. 56, pp. 241-251, Feb, 2019.

[13] Yang J, Chen J, Liu Y, et al. Triboelectrification-based organic film nanogenerator for acoustic energy harvesting and self-powered active acoustic sensing.[J]. Acs Nano, 2014, 8(3):2649-57.

[14] Ultrathin, Rollable, Paper-Based Triboelectric Nanogenerator for Acoustic Energy Harvesting and Self-Powered Sound Recording[J]. Acs Nano, 2015, 9(4):4236-43.

[15] J. Yang, J. Chen, Y. Liu, W. Q. Yang, Y. J. Su, and Z. L. Wang, "Triboelectrification-Based Organic Film Nanogenerator for Acoustic Energy Harvesting and Self-Powered Active Acoustic Sensing," Acs Nano, vol. 8, no. 3, pp. 2649-2657, Mar, 2014.

[16] J. Liu, N. Cui, G. Long, X. Chen, B. Suo, Y. Zheng, C. Hu, and Q. Yong, "A three-dimensional integrated nanogenerator for effectively harvesting sound energy from the environment," Nanoscale, vol. 8, 2016.

[17] X. Fan, J. Chen, J. Yang, P. Bai, Z. L. Li, and Z. L. Wang, "Ultrathin, Rollable, Paper-Based Triboelectric Nanogenerator for Acoustic Energy Harvesting and Self-Powered Sound Recording," *Acs Nano*, vol. 9, no. 4, pp. 4236-4243, Apr, 2015

[18] Y. Pdf, Y. Pdf, Y. Pdf, O. Gif, P. Gif, and P. Gif, "Simulation of Acoustic Energy Harvesting Using Piezoelectric Plates in a Quarter-Wavelength Straight-Tube Resonator," 2012.

Gap in pagination due to unavailable papers.

Pages 1680-1685

Proceedings of the 16th Annual IEEE International
Conference on Nano/Micro Engineered and Molecular Systems
April 25-29, 2021

Design and fabrication of a multimode optical fiber surface plasmon resonance urea biosensor†

Jiayin Li[1,2], Ji Wan[2], Mengdi Han[3] and Haixia Zhang[1,2,*], *Senior Member, IEEE*

Abstract— Optical fiber surface plasmon resonance (SPR) sensors are widely used in concentration measurement. Conventional approaches usually exploit wet electroless plating to deposit silver film on the surface of the core of optical fiber to produce surface plasmon resonance effect. The inability to accurately control and measure the thickness of metal growth is a drawback of chemical methods to prepare optical fiber SPR sensors. Here, we propose an indirect method to measure the thickness of metal film and design an experiment to control the thickness of metal film by using gradient concentration solution. We also discuss the design and fabrication of a multimode optical fiber SPR urea biosensor. In addition, we build a reflective fiber optic sensor system. We test the performance of different thicknesses of the sensor probe in solution. The best sensitivity is about 31.7 nm/M from 0 to 1 M.

Keywords—optical fiber biosensor; surface plasmon resonance; urea sensor; electroless plating

I. INTRODUCTION

Surface plasmon resonance sensors are widely used in the study of concentration measurement, due to their advantages in high sensitivity. SPR effect is a kind of physical optical phenomenon caused by the resonance of evanescent wave and metal surface plasmon wave. The first optical fiber SPR sensor was proposed by R.C. Jorgenson in 1992 [1]. Since then, many different structures of optical fiber SPR sensors for chemical and biochemical sensing have been reported. Compared to traditional prism SPR sensors, optical fiber SPR sensor has the advantage of high sensitivity, miniaturization, ease of integration and fast response [2]. The methods of preparing optical fiber SPR sensors mainly include vacuum sputtering [3], chemical synthesis [4], and photolithography [5].

Vacuum sputtering is the most common way to prepare optical fiber SPR sensors, but it has some disadvantages, such as expensive equipment and complex experiment. Compared to vacuum sputtering, the chemical method is more convenient and inexpensive. The inability to accurately control and measure the thickness of metal growth is a drawback of chemical method.

Here, we propose an indirect method to measure the thickness of metal film and design an experiment to control

the thickness of metal film by using gradient concentration solution in chemical method. We also discuss the design and fabrication of a multimode optical fiber SPR urea biosensor. Finally, we test the sensor in reflective sensing system.

II. EXPERIMENTAL METHODS

In order to measure the liquid in the beaker more conveniently, the completely sensing system is designed as reflective measurement and the optical fiber SPR sensor is designed as probe. The SPR phenomenon in optical fiber is shown as the production of resonance peak. The sensing system itself should have flat transmission in order to observe the resonance peak.

Metal silver is deposited on the surface of the core of optical fiber by wet electroless plating (ELP). The basic principle of this ELP method is the silver mirror reaction between Tollens' reagent and Glucose solution. The SPR effect occurs on the interface between the core of optical fiber and the silver film. The thickness of silver film is an important factor affecting the SPR phenomenon. Generally, the thickness of the metal film should be 50 to 100 nm [6]. The thickness of the metal film depends on the concentration of the Tollens' reagent in wet electroless plating.

The preparation of silver film in solution is different from vacuum sputtering, because growth thickness of silver film depends on the concentration of silver mirror reaction. We prepare the Tollens' reagent in different concentration for different metal thicknesses. In wet electroless plating, we dilute the original Tollens' reagent to different concentration in order to acquire different thicknesses of the silver film.

The preparation method of the original Tollens' reagent is as follows:

1) Prepare silver nitrate solution with molecular concentration: 0.1 mol/L

2) Prepare potassium hydroxide solution with molecular concentration: 0.8 mol/L

3) Place 2.8 mL potassium hydroxide solution in 6 mL silver nitrate solution

4) Add the ammonia solution drop by drop until the precipitate formed in step 3) dissolves and the solution becomes clear again

---

†This work was supported by the National Natural Science Foundation of China (Grant No. 61674004).

[1]Academy for Advanced Interdisciplinary Studies, Peking University, China.

[2]National Key Laboratory of Science and Technology on Micro/Nano Fabrication, Institute of Microelectronics, Peking University, China.

[3]Department of Biomedical Engineering, College of Future Technology, Peking University, China.

*Corresponding author. E-mail address: zhang-alice@pku.edu.cn (H. Zhang).

978-1-6654-3008-1/21 $31.00 © 2021 IEEE

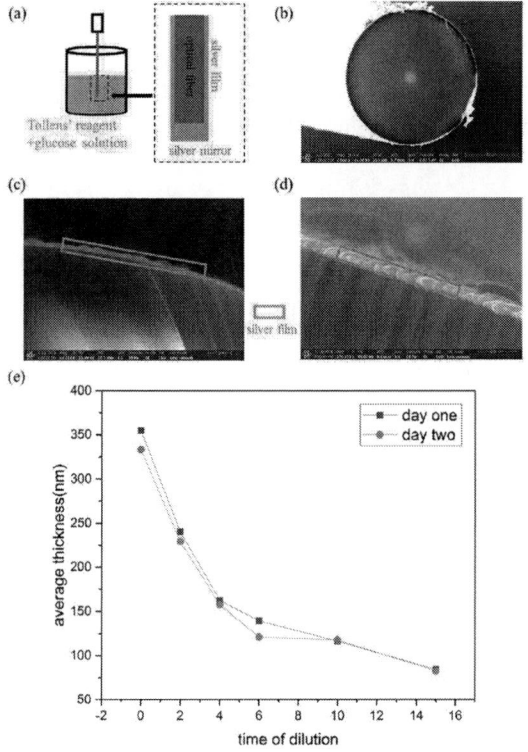

Figure 1. (a) the principle of the silver mirror reaction; (b) Scanning electron microscope image of optical fiber interface in a large space; (c) Scanning electron microscope image of optical fiber interface in a middle space; (d) Scanning electron microscope image of optical fiber interface in a small space; (e) the relationship between the average thickness of the silver film and the concentration of the reaction solution.

Then we get the Tollens' reagent. As shown in Fig. 1a, the length of sensing area is Equal to the length of the stripped cladding layer of the fiber, which is 10mm in this paper. The preparation of the silver film is as follows:

1) Prepare 0.2% SnCl2 solution

2) Prepare glucose solution with molecular concentration: 0.05 mol/L

3) Place the sample that need to be silvered (such as optical fiber or oxidized silicon wafer) in SnCl2 solution for 20 min

4) Place 1 mL original Tollens' reagent in a beaker and dilute it to specific multiples

5) Rinse the sample with distilled water and place it in the diluent Tollens' reagent

6) Pour 1.14 mL glucose solution into the diluent Tollens' reagent (the silver mirror reaction begin)

7) After 10 min, pick out the sample and rinse with distilled water.

8) Dry the sample

We use three samples to prepare the silver film. First sample is SMF-28 single-mode optical fiber (Cladding

diameter is 125 μm and Core diameter is 8 μm) from YOFC (Changfei company), and the size of its cladding is more versatile. Its core and the cladding layer of it are made of SiO2 principally. We can use a common cutter to cut this optical fiber and observe the end face after preparing the silver film.

Second sample is 15 mm*15 mm oxidized silicon wafer (the thickness of oxide layer is 300 nm) from ShunSheng Electronic Science And Technology Company. Its surface is flat enough that we can measure the thickness of the silver film with a step profiler after preparing the silver film.

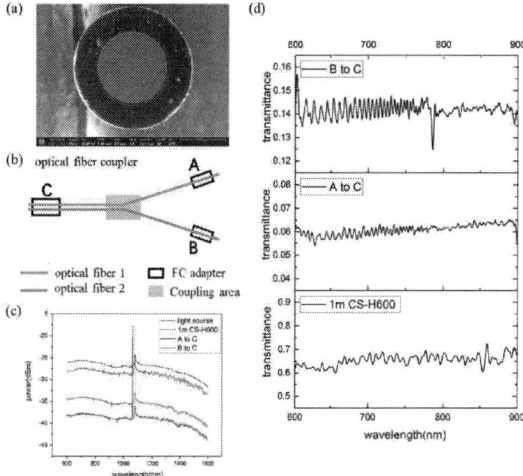

Figure 2. (a) optical fiber interface of CS-H600 multimode optical fiber by SEM; (b) the schematic plot of 1*2 optical fiber coupler; (c) the spectrum of the light source and the light source through basic components; (d) transmittance curves of basic components

Third sample is 10 cm CS-H600 multimode optical fiber (Cladding diameter is 630 μm and Core diameter is 600 μm) from Nanjing Chunhui Company as shown in Fig. 2a. Its core is made of SiO2 while the cladding layer of it is made of hard resin. Hard resin can be corroded by concentrated sulfuric acid or be scraped off by a scalpel, so we can remove the cladding layer of this multimode optical fiber to expose the core of optical fiber. Without the cladding layer, the evanescent wave on the optical fiber surface contacts with environment directly. In order to make its structure more convenient for solution concentration measurement, we use it as a reflective optical fiber probe. As shown in Fig. 3a (B), one end of it is made into FC adapter and the other end is worn down and stripped 10 mm (only the core is left). The end has a little reflection because of wearing down. We hope move light can reflect on the worn end which can enhance the light receive to the spectrometer. Therefore, we use the original Tollens' reagent to silvering once again on the worn end after silvering on the side of the reflective optical fiber probe by diluent Tollens' reagent. The schematic diagram of final reflective optical fiber probe shows in Fig. 1a.

CS-H600 multimode optical fiber is also made into a 1*2 optical fiber coupler as shown in Fig. 2b in order to ensure the operation of the reflective sensing system. Two CS-H600 multimode optical fibers constitute this coupler. This coupler has three interfaces A, B, C and light can through A to C as well as B to C.

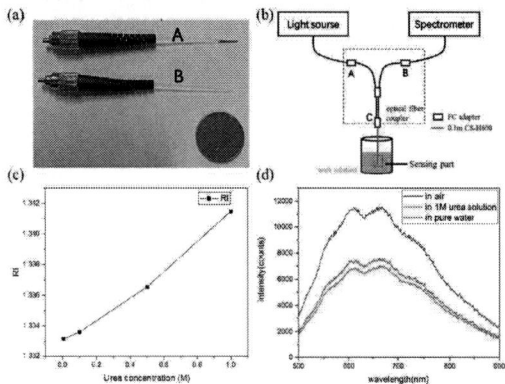

Figure 3. (a) material object of 10 cm reflective optical fiber probe (A:after silvering, B:before silvering); (b) schematic diagram of the measurement process of urea solution; (c) the refractive indexes of urea solution of different concentration from 0.005M to 1M; (d) the spectrum of the no silvering optical fiber probe in different environment

We measure the refractive indexes of urea solution with different concentrations, because the optical fiber SPR sensor measures the difference of environmental refractive index in principle. We also prepare a batch of optical fiber SPR urea biosensor of different metal thicknesses and compare their SPR phenomenon.

The light source is SUPERK COMPACT SUPERCONTINUUM LASERS from NKT Photonics Company and the spectrometers are from YOKOGAWA Company and Ocean Insight Company.

III. RESULTS AND DISCUSSION

A. The morphology and the thickness of the silver film

We use the above silver mirror reaction method to coat silver on the side of the SMF-28 single-mode optical fiber. Then we cut the optical fiber vertically along the axis of the optical fiber and scan the cross section by scanning electron microscope (SEM).

Fig. 1b, Fig. 1c and Fig. 1d show the image of the cross section of silvered SMF-28 single-mode optical fiber in a large space, middle space and small space. The conductivity of Ag is better than that of SiO2, so the edge of the section and the side of the cylinder are brighter than other area. It can be seen from the figures that the thickness of the silver film prepared by this method is uniform.

To investigate the relationship between the thickness of the silver film and the concentration of the reaction solution, we silvering oxidized silicon wafer by diluent Tollens' reagent with different concentration. Then we measure the thickness of the silver film with a step profiler. The results are shown in Fig. 1e, the times of dilution is from 0 (means no dilution) to 15, while the results of thickness is from about 350nm to less than 100nm. It gives us guidance that if the silver film thickness should be between 50 to 100 nm, the time of dilution must be 15 approximately. Considering the oxidation of silver in the air, we put the samples in the

air for one day to measure again. By comparing the data on day one and day two, we found differences than the data on day two is larger than day one. Metal oxide layer will greatly affect the performance of SPR sensor. All the sensors in this paper are tested within one hour after the silver mirror reaction.

B. The reflective optical fiber sensing system

Firstly, we test the Light transmission characteristics of basic components (such as original CS-H600 multimode optical fiber and 1*2 optical fiber coupler).

As shown in Fig. 2c, the spectrum of light source is black line. The spectrum of light source through 1 m CS-H600 multimode optical fiber is red line. The spectrums of light source through optical fiber coupler are green and blue lines.

For optical fiber surface plasmon resonance sensor with the silver film, the wavelength of resonance peak is mainly less than 900nm. Fig. 2d shows the transmittance when light source through 1 m CS-H600 multimode optical fiber and optical fiber coupler (A to C and B to C) from 600 nm to 900 nm. We can get the transmittance of 1 m CS-H600 multimode optical fiber is about 0.65, A to C is about 0.06, and B to C is about 0.14. These three transmittance curves are both quite flat, which is instrumental in searching resonance peak. We can also get than the loss increases, when CS-H600 multimode optical fiber made into optical fiber coupler.

Secondly, we should test the reflective optical fiber sensing system. Fig. 3b shows the schematic diagram of the reflective optical fiber sensing system for measurement of urea solution. Fig. 3a shows the material object of 10 cm reflective optical fiber probe (A: after silvering, B: before silvering).

The working principle of reflective optical fiber sensing system is as follows approximately:

The light E(x, y, z, t) from light source to FC adapter C, then a part of light E1(x, y, z, t) go directly to adapter B because of end reflection in adapter C, while the rest part light E2(x, y, z, t) go into optical fiber probe. The light E2(x, y, z, t) return to adapter C after SPR effect and end reflection in optical fiber probe, and turn into E3(x, y, z, t). The major differences between E2 an E3 is the resonance peak. Then light E3 go into both A and B of optical fiber coupler. If the part that light E3 go into B is E4(x, y, z, t), the light E1 and E4 occur optical interference before go into spectrometer.

In order to get the background information of this system, we test one no silvering optical fiber probe in different environment as shown in Fig. 3d. We can find no different in Spectral shape between air (RI=1) and pure water (RI= 1.333) because of no SPR effect. At the same time, because of different reflection on the worn end, the spectral intensity of air is larger than the spectral intensity of pure water.

At last, we measure the refractive indexes of urea solution with different concentrations because in principle, SPR effect is related to the refractive index of the surrounding environment rather than the concentration of the surrounding solution. As shown in Fig. 3c, we measured

978-1-6654-3008-1/21 $31.00 © 2021 IEEE

the refractive indexes of urea solution of different concentration (molecular concentration: 1 M, 0.5 M, 0.1 M and 0.005 M) by automatic refractometer (JIAHANG Instruments). The refractive index of urea solution rises when the concentration of urea solution rises. The concentration of urea in human sweat is about from 0.0018 M to 0.046 M [7]. The refractive indexes of these urea solutions is between 1.333 and 1.334.

C. The comparison of sensors with different thickness of silver film

According to the above work, using the silver mirror reaction concentration corresponding to the silver film thickness of 50 to 100 nm for preparing some reflective optical fiber probe. Four probes are prepared in diluent Tollens' reagent with 6 times dilution, 12 times dilution, 18 times dilution and 24 times dilution. Their normalized reflection curves from 500 nm to 900 nm in pure water (RI= 1.333) show in Fig. 4a, while their normalized reflection curves from 500 nm to 900 nm in 1M urea solution (RI= 1.341) show in Fig. 4b.

The results show that the probe prepared in diluent Tollens' reagent with 18 times dilution has sharper resonance peak than others in both Fig. 4a and Fig. 4b. The black curve (18 times dilution) has larger fluctuation than other curves in both Fig. 4a and Fig. 4b. This conclusion according to the suitable thickness of silver film in theory (50 to 100 nm) and relationship between the thickness of the silver film and the concentration of the reaction solution during Fig. 1e. Too thick a silver film make the evanescent wave unable to contact with the environment, while too thin silver film leads to insufficient resonance of SPR effect.

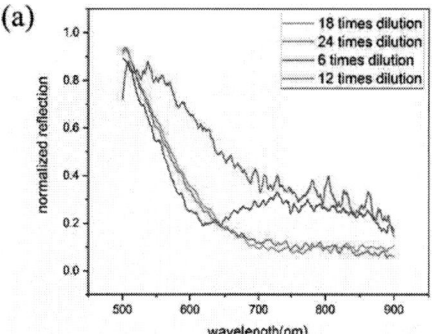

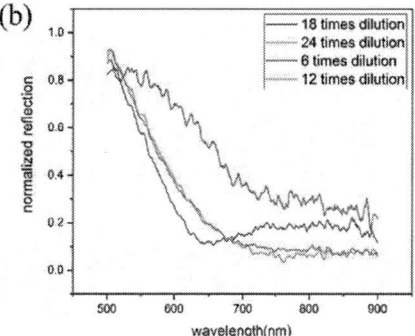

Figure 4. (a) normalized reflection curves from 500 nm to 900 nm in pure water; (b) normalized reflection curves from 500 nm to 900 nm in 1M urea solution.

After filtering and smoothing for more normalized reflections of the probe prepared in diluent Tollens' reagent with 18 times dilution, we find the different resonance peak wavelength in different urea solutions as shown in Fig. 5. The slope of the fitted curve of Fig.5 is about 31.7 nm/M from 0 to 1 M.

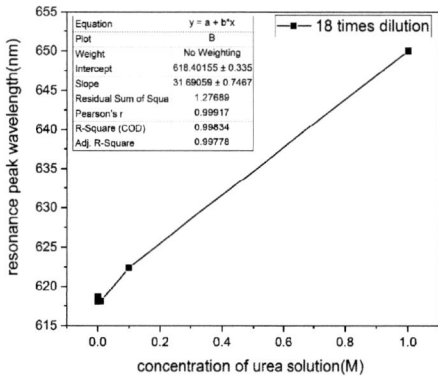

Figure 5. the relationship between the concentration of urea solution with the resonance peak wavelength

We plan to do more concentration tests about different solutions in the future using this special silver film thickness

978-1-6654-3008-1/21 $31.00 © 2021 IEEE

probe. We also plan to study other parameters that affect the SPR effect in optical fibers, such as the thickness of optical fiber, the length of sensing area, the type of metal and the modification of metal surface.

IV. CONCLUSION

In summary, we use oxidized silicon wafer to propose an indirect method to measure the thickness of metal film and the time of dilution must be 15 approximately for silver film thickness of 50 to 100 nm. We observed the silvering single-mode optical fiber by SEM to prove the uniformity of this wet electroless plating. In addition, we build a reflective fiber optic sensor system, and the components of this system have quite flat transmittance curves. We test the background information of this system to obtain the reference curve of SPR resonance peak.

We determined the preparation process of a multimode optical fiber surface plasmon resonance urea biosensor. We control the thickness of metal film by using gradient concentration solution in chemical methods. We test the performance of different thickness of the sensor probe in solution, and found the preparation condition of the suitable thickness is 18 times dilution. The sensitivity of this sensor probe is about 31.7nm/M from 0 to 1 M. In a word, we study the relationship between the concentration of silver mirror reaction solution and the thickness of silver plating.

REFERENCES

[1] R. C. Jorgenson and S. S. Yee, "A FIBEROPTIC CHEMICAL SENSOR-BASED ON SURFACE-PLASMON RESONANCE," (in English), Sensors and Actuators B-Chemical, Article vol. 12, no. 3, pp. 213-220, Apr 1993.

[2] J. Zeng, D. Liang, Z. Zeng, and Y. Du, "Reflective optical fiber surface plasma wave resonance sensor," Acta Optica Sinica, vol. 27, no. 3, pp. 404-9, March 2007.

[3] H. Suzuki, M. Sugimoto, Y. Matsui, and J. Kondoh, "Effects of gold film thickness on spectrum profile and sensitivity of a multimode-optical-fiber SPR sensor," Sensors and Actuators B-Chemical, vol. 132, no. 1, pp. 26-33, May 28 2008.

[4] Y. zhao, Z.-Q. Deng, and Q. Wang, "Fiber optic SPR sensor for liquid concentration measurement," Sensors and Actuators B-Chemical, vol. 192, pp. 229-233, Mar 2014.

[5] Y. Shen et al., "Plasmonic gold mushroom arrays with refractive index sensing figures of merit approaching the theoretical limit," (in English), Nature Communications, Article vol. 4, p. 9, Aug 2013, Art. no. 2381.

[6] M. Iga, A. Seki, and K. Watanabe, "Gold thickness dependence of SPR-based hetero-core structured optical fiber sensor," (in English), Sensors and Actuators B-Chemical, Article vol. 106, no. 1, pp. 363-368, Apr 2005.

[7] C. J. Harvey, R. F. LeBouf, and A. B. Stefaniak, "Formulation and stability of a novel artificial human sweat under conditions of storage and use," (in English), Toxicology in Vitro, Article vol. 24, no. 6, pp. 1790-1796, Sep 2010.

Proceedings of the 16th Annual IEEE International
Conference on Nano/Micro Engineered and Molecular Systems
April 25-29, 2021

A Resonant Differential Pressure Microsensor With a Stress Isolation Layer

Yadong Li, Chao Cheng, Yulan Lu, Jian Chen, Junbo Wang, *Member, IEEE* and Deyong Chen*,
Member, IEEE

Abstract— This article presents a resonant differential pressure microsensor with a stress isolation layer, which is mainly composed of three parts, an SOI wafer, including a handle layer, an oxide layer and a device layer, a GOS wafer, including a glass layer and a silicon layer and a stress isolation structure, including a two-layer glass. The SOI device layer including the central beam in the central area and the side beam in the side area is bonded to GOS glass wafer for vacuum packaging of the beams. The diaphragm including SOI device layer and the GOS wafer is coupled to the two beams through anchors. In order to realize stress isolation, the glass base is designed as a convex structure, which is bonded to sensor chip by anodic bonding technology and fixed to Kovar pedestal by epoxy glue, respectively. Experimental characterizations were carried out, indicating differential pressure sensitivity of – 144.85 Hz/kPa (~2079 ppm/kPa) and static pressure sensitivity of –0.98 Hz/kPa (~14 ppm/kPa) and temperature sensitivity of -2.19 Hz/°C (~31 ppm/°C). Compared with existing research, the low temperature sensitivity was realized, improving the temperature long-term stability of the sensor.

I. INTRODUCTION

Differential pressure sensors are widely used in biomedicine, aerospace, industrial control, electronic information and other fields [1-3]. Compared to other types of differential pressure sensors, resonant differential pressure sensors are featured with high resolution, high accuracy, good reliability and strong anti-jamming ability [3], which are widely used in the precision differential pressure measurement field.

Resonant differential pressure sensors measure the differential pressure by detecting the intrinsic frequency drift of the silicon resonant beam [4-7]. In order to isolate the two pressure sources, the sensor chips need to be fixed to a plastic [8], ceramic [9-11] or metal [12-14] pedestal. However, packaging stresses caused by the mismatch of thermal expansion coefficients between the package materials and chips of the sensor will affect the accuracy and stability of the sensor. Therefore, it is necessary to isolate the assembly stress.

Honeywell bonded the sensor chip to a glass tube by silicon-glass anodic bonding technology to achieve a low-stress assembly [15]. Thermal expansion coefficient between glass tube and silicon chip is similar, which reduces the thermal stress of assembly to the maximum extent. In addition, the glass tube also plays the role of ventilation.

However, the fabrication and assembling processes of these sensors are complex, leading to compromised assembly efficiency and high-cost.

Druck bonded the sensor chip and a long glass tube together by glass frit bonding, and then fixed the glass tube to the TO8 pedestal by epoxy glue to achieve low-stress assembly of the sensor [16]. When the temperature changes, the pedestal deforms more, and the sensor chip deforms less. The long glass tube can relieve the large deformation of the pedestal and reduce the thermal stress of the sensor chip, thereby effectively isolating the thermal stress caused by the inconsistent thermal expansion coefficient between the pedestal and sensor chip. However, the sensor chip and the glass tube can only be bonded together by single-chip bonding technology. In addition, the alignment and bonding processes in the bonding process are complicated, resulting in high manufacturing costs and low assembly efficiency.

Yokogawa designed a stress isolation structure, which was bonded to sensor chip by silicon-silicon bonding technology and bonded to Kovar pedestal by Au-Si eutectic bonding technology, respectively [17]. The stress isolation structure released the thermal stress of the Kovar pedestal and improved the temperature stability of the sensor. Nevertheless, these bonding processes have high requirements on the roughness and flatness of the wafer surface, resulting in low yield.

The Fraunhofer Institute in Germany proposed a stress isolation structure based on V-shaped groove [9]. The stress generated by the assembly of the sensor is easily released through the deformation of the V-shaped isolation groove, thereby realizing the stress isolation of the sensor chip. This fabrication process is simple, and the bonding process between the sensor chip and the V-shaped groove stress isolation structure can realize wafer-level bonding. The simulation results showed that the assembly stress can be reduced by more than 90%. However, the V-shaped groove isolation structure is thin and fragile, so it can only be fixed to the glass tube by adhesion, and the stress will be introduced during the bonding process.

This article presents a resonant differential pressure microsensor with a two-layer glass stress isolation layer. The two resonant beams are designed in an SOI wafer, which is bonded to a glass-on-silicon (GOS) wafer for vacuum packaging of the beams. The glass base is designed as a convex structure, which is bonded to sensor chip by anodic

---

This study was supported by the National Key Research and Development Program of China (Grant No. 2018YFB2002302).

Y. Li, C. Cheng, Y. Lu, J. Chen, J. Wang and D. Chen are with the State Key Laboratory of Transducer Technology, Aerospace Information

Research Institute, Chinese Academy of Sciences, Beijing 100190, China, and also with the School of Electronic, Electrical and Communication Engineering, University of Chinese Academy of Sciences, Beijing 100049, China (e-mail: liyadong16@mails.ucas.ac.cn; dychen@mail.ie.ac.cn)

978-1-6654-3008-1/21 $31.00 © 2021 IEEE

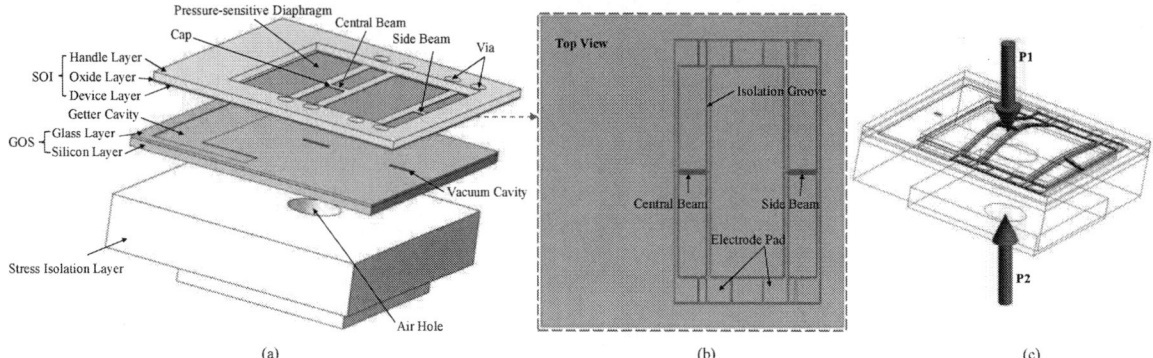

Fig. 1. Schematic diagram of the sensor. (a) The sensor consists of three parts, an SOI wafer, including a handle layer, an oxide layer and a device layer, a GOS wafer, including a glass layer and a silicon layer and a stress isolation structure, including two-layer glass. (b) The SOI device layer, including the central beam in the central area and the side beam in the side area. (c) When the sensor works, two pressures (P1 and P2) are applied to the top and bottom of the diaphragm sensitive to pressure, respectively. The diaphragm deforms under differential pressure (P1-P2), resulting in changes in the stiffness of the resonant beams. Furthermore, the axial stresses of the beams changes, leading to the resonant frequencies shift of the beams.

bonding technology and fixed to Kovar pedestal by epoxy glue, respectively. FEA simulations were carried out, showing that thermal stress of the two beams is significantly reduced. Experimental characterizations were carried out, indicating that the sensor possessed low temperature sensitivity, further confirming the effectively of the stress isolation layer.

II. DESIGN

A. Sensor Design and Working Principle

Fig. 1(a) and (b) show the stacked structure of the sensor which is composed of three parts, an SOI wafer, including a handle layer, an oxide layer and a device layer, a GOS wafer, including a glass layer and a silicon layer and a stress isolation structure, including a two-layer glass. The SOI device layer including the central beam in the central area and the side beam in the side area is bonded to GOS glass wafer for vacuum packaging of the beams. The isolation groove in the SOI device layer insulates the resonant beams from the frame. The electrode pads of the resonant beams can be led out through the vias in the SOI handle layer. The

vacuum cavities in the GOS glass layer provides space for the resonant beams to vibrate. The diaphragm including SOI device layer and the GOS wafer is coupled to the two beams through anchors. In order to realize stress isolation, a 3 mm-thick convex stress isolation structure of two-layer glass with a 2 mm diameter air hole is bonded to the GOS silicon layer by anodic bonding technology and fixed to Kovar pedestal by epoxy glue, respectively. The air hole serves to introduce pressure to the diaphragm. In addition, the getter sputtered into the getter cavity can absorb the gas generated during the bonding process, which improves the vacuum degree of the vacuum cavities and thereby improves the quality factors of the resonant beams.

Fig. 1(c) shows working principle of the microsensor. When the sensor works, two pressures (P1 and P2) are applied to the top and bottom of the diaphragm sensitive to pressure, respectively. The diaphragm deforms under differential pressure (P1-P2), resulting in changes in the stiffness of the beams. Furthermore, the central and side beams are respectively subjected to tensile stress and compressive stress, leading to an increase and a decrease in

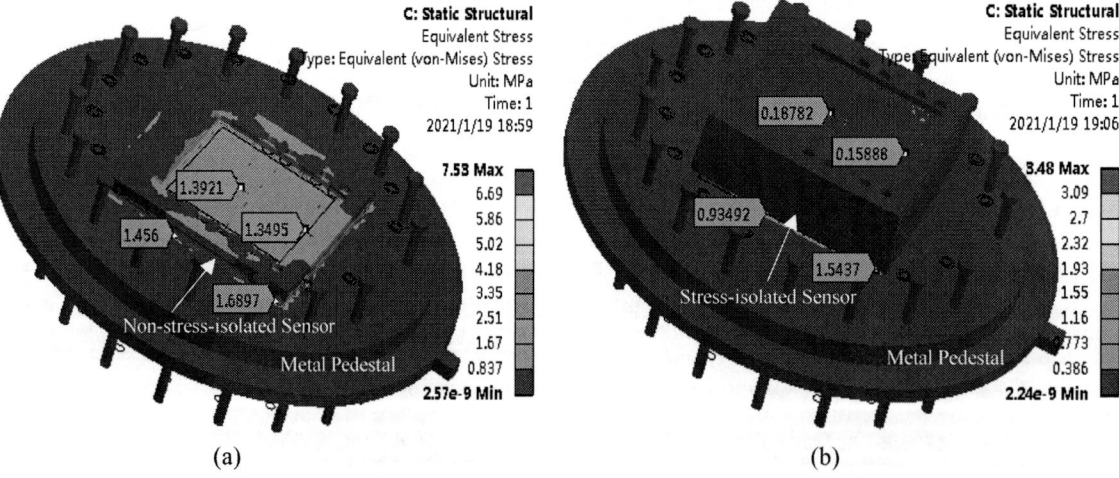

Fig. 2. Simulated stress distribution caused by temperature changes of non-stress-isolated (a) and stress-isolated (b) sensor.

978-1-6654-3008-1/21 $31.00 © 2021 IEEE

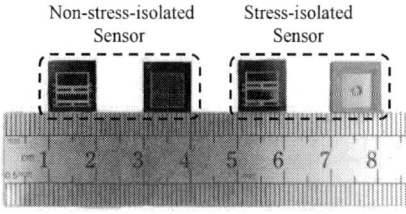

(a)

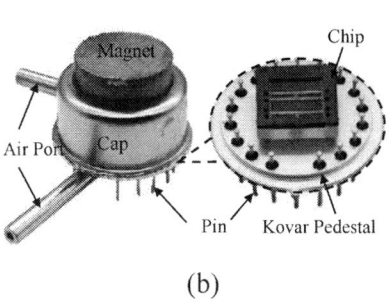

(b)

Fig. 3. The fabricated and encapsulated sensor. (a) The front and back views of chip with no stress isolation and a stress isolation. (b) Prototype of the encapsulated sensor.

the resonant frequency of the central and side beams, respectively. The differential output of the frequencies of the two beams can improve the sensitivity of the sensor [18].

B. Finite Element Analysis

In order to optimize the structure of the sensor, FEA simulation was used to observe thermal stress in response to temperature variances. Fig. 2 shows the simulation model of the chip and Kovar pedestal fixed together and the stress contour of the beams. As Fig. 2(a) and (b) show, positive stress (tensile) of 1.39 and 1.35 MPa were observed in the central and side beams of the non-stress-isolated sensor, while positive stress (tensile) of 0.19 and 0.16 MPa were observed in the central and side beams of the stress-isolated sensor when the temperature changes by 300°C, which indicate that the stress isolation structure greatly reduced the thermal stress of temperature. After parametric simulation, the optimal size of the two-layer glass stress isolation structure can be determined.

III. FABRICATION AND CHARACTERIZATION

In order to verify the validity of the design and the correctness of the simulation for the two-layer glass stress isolation structure, the sensor chip was fabricated and the sensor performance was characterized. The front and back views of the fabricated chip with no stress isolation and a stress isolation are as shown in Fig. 3(a). For the sensor with stress isolation, the glass base with a 2 mm diameter air hole was anodic bonded with the fabricated chip. In order to extract the electrodes and test pressure performance of the sensor, an encapsulation structure was designed and the encapsulation process was as follows. Firstly, the fabricated chip was fixed on the Kovar pedestal by epoxy glue. Then, the pins of the Kovar pedestal and the metal electrode pads of the chip were connected by wire bonding. Gold wire was selected as the conductor to lead out the beam signal, because of its low resistivity. Finally, the Kovar pedestal was welded together with two caps with air ports to complete the assembly of the sensor. Moreover, the magnet was fixed together with the cap by epoxy glue to provide a magnetic field for subsequent vibration of the beams. The encapsulated sensor is shown in Fig. 3(b).

To obtain the difference of differential pressure sensitivity between the resonant beams, the differential pressure applied to the sensor varies from 0 kPa to 100 kPa [see Fig. 4(a)]. To obtain the difference of static pressure sensitivity between the resonant beams, the static pressure applied to the sensor varies from 110 kPa to 210 kPa. Besides, the temperature applied to the sensor is 25 °C [see Fig. 4(b)]. To obtain temperature sensitivity of the sensor, the temperature applied to the sensor varies from -10 °C to 40 °C. Besides, the pressures applied to the bottom and top of the diaphragm are 110 kPa, which means the differential pressure applied to the sensor is 0 kPa [see Fig. 4(c)]. As a result, the characterization results reveal that the non-stress-isolated sensor demonstrated differential pressure sensitivity of -142.76 Hz/kPa (~2078 ppm/kPa) [see Fig. 4(a)], static pressure sensitivity of 1.60 Hz/kPa (~23 ppm/kPa) [see Fig. 4(b)] and temperature sensitivity of -10.72 Hz/°C (~156 ppm/°C) [see Fig. 4(c)] and the stress-isolated sensor demonstrated differential pressure sensitivity of -144.85 Hz/kPa (~2079 ppm/kPa) [see Fig. 4(a)], static pressure sensitivity of -0.98 Hz/kPa (~14 ppm/kPa) [see Fig. 4(b)] and temperature sensitivity of -2.19 Hz/°C (~31 ppm/°C) [see Fig. 4(c)]. To sum up, under the premise that the

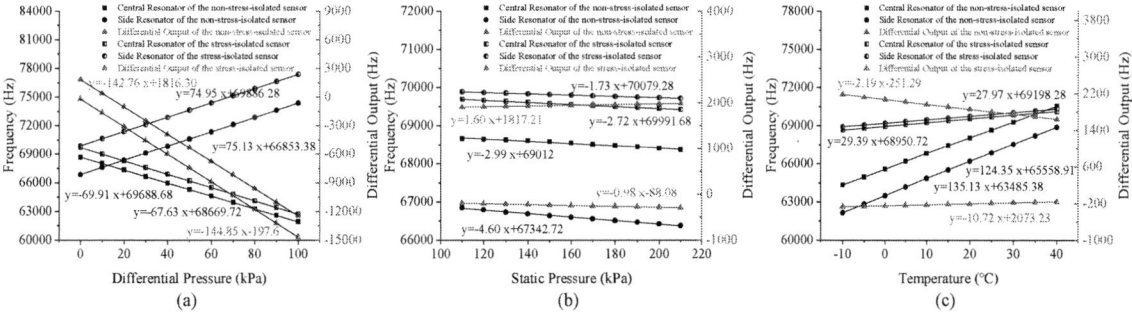

Fig. 4. Key performances of the proposed sensor. (a) The resonant frequency drifts with the differential pressure. (b) The resonant frequency drifts with the static pressure. (c) The resonant frequency drifts with the environmental temperature.

differential pressure sensitivity was basically not affected, the temperature sensitivity of the sensor with a two-layer glass stress isolation layer was reduced by 80% (from -10.72 Hz/°C to -2.19 Hz/°C) in comparison to the sensor without isolation layer. Furthermore, the ratio of differential pressure sensitivity to temperature sensitivity of the stress-isolated sensor were approximately 66 times, which confirmed that the sensor realized a low temperature sensitivity.

IV. CONCLUSION

A resonant differential pressure microsensor with a two-layer glass stress isolation layer was simulated, fabricated, and tested. The sensor was featured by anodic bonding of SOI wafer and GOS wafer to form pressure-sensitive diaphragms and realize vacuum packaging. Compared with the non-stress-isolated sensor, the sensor with the stress isolation structure achieves high differential pressure sensitivity and at the same time low temperature sensitivity, which verifies the effectiveness of the low-stress assembly. In the future, the calibration method and compensation algorithm of the sensor will be paid attention to.

REFERENCES

[1]   H. Hu, L. ZHONG, and Q. ZHOU, "Actuality and development of the differential pressure sensor technology," Machine Tool & Hydraulics, vol. 41, no. 11, pp. 187-190. Nov. 2013.

[2]   K. Harada, K. Ikeda, H. Kuwayama, and H. Murayama, "Various applications of resonant pressure sensor chip based on 3-D micromachining," SeAcA, vol. 73, no. 3, pp. 261-266. Mar. 1999.

[3]   T. Saigusa, and H. Kawayama, "Intelligent differential pressure transmitter using micro-resonators," in Proceedings of the 1992 International Conference on Industrial Electronics, Control, Instrumentation, and Automation, Dec. 1992, pp. 1634-1639.

[4]   C. J. Welham, J. Greenwood, and M. M. Bertioli, "A high accuracy resonant pressure sensor by fusion bonding and trench etching," Sensors and Actuators A Physical, vol. 76, no. 1-3, pp. 298-304. 1999.

[5]   Z. Luo, D. Chen, J. Wang, Y. Li, and J. Chen, "A High-Q Resonant Pressure Microsensor with Through-Glass Electrical Interconnections Based on Wafer-Level MEMS Vacuum Packaging," Sensors, vol. 14, no. 12, pp. 24244-24257. 2014.

[6]   S. Xiaodong, Y. Weizheng, Q. Dayong, S. Ming, and R. Sen, "Design and Analysis of a New Tuning Fork Structure for Resonant Pressure Sensor," Micromachines, vol. 7, no. 9, pp. 148. 2016.

[7]   W. Liying, D. Xiaohui, W. Lingyun, X. Zhanhao, Z. Chenying, and G. Dandan, "High-Q Wafer Level Package Based on Modified Tri-Layer Anodic Bonding and High Performance Getter and Its Evaluation for Micro Resonant Pressure Sensor," Sensors, vol. 17, no. 3, pp. 599. 2017.

[8]   Chang, Y.-Y., Chung, Lwo, B.-J., Tseng, and K.-F., "In Situ Stress and Reliability Monitoring on Plastic Packaging Through Piezoresistive Stress Sensor," IEEE Transactions on Components, Packaging and Manufacturing Technology, vol. 3, no. 8, pp. 1358-1363. 2013.

[9]   H. L. Offereins, H. Sandmaier, B. Folkmer, U. Steger, and W. Lang, "Stress free assembly technique for a silicon based pressure sensor," in International Conference on Solid-state Sensors & Actuators, 2002, pp.

[10]   R. N. Dean, S. Surgnier, J. Pack, N. Sanders, P. Reiner, C. W. Long, R. Fenner, and W. P. Fenner, "Porous Ceramic Packaging for a MEMS Humidity Sensor Requiring Environmental Access," IEEE Transactions on Components Packaging & Manufacturing Technology, vol. 1, no. 3, pp. 428-435. 2011.

[11]   S. A. Wright, H. Z. Harvey, and Y. B. Gianchandani, "A Microdischarge-Based Deflecting-Cathode Pressure Sensor in a Ceramic Package," JMemS, vol. 22, no. 1, pp. 80-86. 2013.

[12]   Y. Hao, W. Yuan, J. Xie, Q. Shen, and H. Chang, "Design and Verification of a Structure for Isolating Packaging Stress in SOI MEMS Devices," IEEE Sens. J., vol. PP, no. 99, pp. 1-1. 2016.

[13]   L. I. Yuxin, D. Chen, and J. Wang, "Vacuum Adhesive Bonding and Stress Isolation for MEMS Resonant Pressure Sensor Package," MSF, vol. 694, pp. 896-900. 2011.

[14]   T. Yun, Y. Lin, Z. Shao-Ping, and T. Shan-Tung, "An Improved Metal-Packaged Strain Sensor Based on A Regenerated Fiber Bragg Grating in Hydrogen-Loaded Boron–Germanium Co-Doped Photosensitive Fiber for High-Temperature Applications," Sensors (Basel, Switzerland), vol. 17, no. 3. 2017.

[15]   D. Burns, J. Zook, R. Horning, W. Herb, and H. Guckel, "Sealed-cavity resonant microbeam pressure sensor," SeAcA, vol. 48, no. 3, pp. 179-186. 1995.

[16]   J. C. greenwood, "Miniature silicon resonant pressure sensor." 1988.

[17]   K. Ikeda, H. Kuwayama, T. Kobayashi, T. Watanabe, T. Nishikawa, T. Yoshida, and K. Harada, "Silicon pressure sensor integrates resonant strain gauge on diaphragm," SeAcA, vol. 21, no. 1-3, pp. 146-150. Feb. 1990.

[18]   C. F. Chiang, A. B. Graham, B. J. Lee, C. H. Ahn, and T. W. Kenny, "Resonant pressure sensor with on-chip temperature and strain sensors for error correction," in 2013 IEEE 26th International Conference on Micro Electro Mechanical Systems (MEMS), 2013, pp.

Proceedings of the 16th Annual IEEE International
Conference on Nano/Micro Engineered and Molecular Systems
April 25-29, 2021

An Ultra-high Power Density Helmholtz Resonance Acoustic Energy Converter Based on Triboelectric Nanogenerator *

Ling Liu [1], Hongfa Zhao[1], Zhenhui Lian[1], Hongyong Yu [1], Qiqi Zhang[1], Wenxiang Li[1], Minyi Xu[1], and Xiu Xiao[1]*

Abstract—**Sound energy is a ubiquitous renewable energy source. How to comprehensively harvest and utilize this energy has become a hot research topic. Triboelectric Nanogenerator(TENG) can convert different forms of mechanical energy into kinetic energy, and has significant advantages in micro-nano power generation, self-driving sensing, and device performance control. This work designs an ultra-high power density Helmholtz resonant cavity based on triboelectric nanogenerator with a printed electrode(H-TENG) for harvesting acoustic energy efficiently. We proposed an optimized scheme for harvesting sound energy effectively by the special treatment of the electrode flexible film. Then we researched the structure of the resonant cavity, acoustic conditions, and the tension of flexible film, they can affect the output characteristics of the H-TENG acoustic energy converter. We find H-TENG follows the best output frequency deviation law, that is, introducing appropriate tension according to a specific sound wave range can improve the best frequency response. Compared with the best results in the early literature, the H-TENG has a high power density of 1.47 W/m² sound pressure. The study showed that the high power output of H-TENG can light up 457 LEDs at the same time. Finally, this research also proposes the application potential of H-TENG in energy conversion and acoustic sensing.**

I. INTRODUCTION

Acoustic energy as a clean, abundant, and sustainable energy source exists in large quantities in our surrounding environment. Energy harvesting is the process of capturing environmental energy and converting it into electrical energy. Due to the low energy density of sound waves and the lack of effective energy harvesting technology, most sound wave energy is wasted[1]. Acoustic energy collection is usually harvested by means of an acoustic resonance device. Specifically, when the acoustic resonator is excited by the incident wave at its resonance frequency, acoustic energy in the form of a standing wave is harvested inside the resonator [2-8]. Since 2012, the triboelectric nanogenerator (TENG) has been considered the most promising way to realize distributed energy harvesting and self-powered sensing [9-15]. TENG is usually composed of dielectric materials and metal electrodes. TENG's materials are more environmental and reliable. In recent years, triboelectric nanogenerators (TENG) have made it possible to harvest sound energy with flexible electrode materials. Some previous work tried to use TENG to convert sound energy into electrical energy. For example, Cui et al. developed a new mesh TENG acoustic wave collector [16]. Yang et al. used a Helmholtz resonant cavity with an adjustable narrow neck back and a circular friction nanogenerator to make an

acoustic energy harvester [17], which reached the maximum resonance frequency of 240 Hz. With a voltage of 60.5 V, a maximum current of 15.1 μA, and a maximum power density of 60.2 mW/m², it can light up 17 commercial LEDs. Xing et al. developed an ultra-thin, rollable paper-based sonic energy harvester [18], with a maximum power density of 121 mW/m² at 117 dB sound pressure level. Cui et al. developed a new type of mesh TENG acoustic wave harvester [19], with a maximum power density of mW/m². But these TENG-based acoustic energy collectors have low output power and low collection efficiency. Therefore, advanced sound energy harvesters with high output performance and strong practicability have become a prerequisite for sound energy utilization. Our TENG-based dual-necks Helmholtz resonance acoustic energy converter (H-TENG) is easy to design and manufacture, and the generated electric energy can well be matched with the sensor system.

II. WORKING MECHANISM OF THE H-TENG

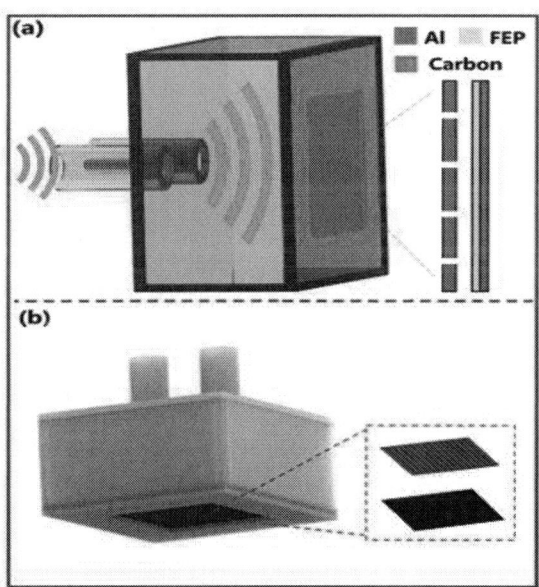

Figure 1. (a) The three-dimensional structure diagram of the dual-necks Helmholtz resonator(b) The actual detail diagram of the H-TENG.

We designed a high-power density triboelectric nanogenerator made by dual-necks Helmholtz resonator triboelectric nanogenerator. The structure of H-TENG is shown in Figure 1(a). It consists of an improved dual-necks

*Resrach Support from the Fundamental Research Funds for the Central Universities, China .

[1]Marine Engineering College, Dalian Maritime University, Dalian, 116026, China.

Corresponding author Dr. Xiao Xiu graduated from Tsinghua University in 2017. From 2014 to 2016, she was a visiting scholar at Purdue University. Now, she is an associate ptrofessor in the Marine Engineering College of Dalian Maritime University.
(Telephone:13610865112;Email: xiaoxiu@dlmu.edu.cn)

978-1-6654-3008-1/21 $31.00 © 2021 IEEE   1695

Helmholtz resonator, an aluminum film with uniform acoustic hole distribution, and a fluorinated ethylene-propylene copolymer (FEP) with conductive ink printed electrodes It's photograph structure is shown in Figure 1(b), the size of the resonant cavity is 70mm × 70mm × 20 mm, the working area of H-TENG is 40 mm × 40mm × 20 mm.

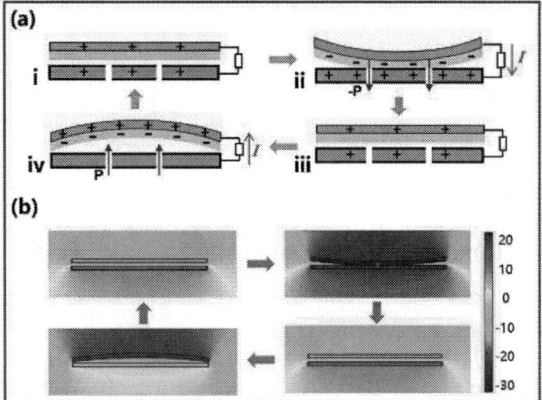

Figure 2(a) The working principle of acoustic driven H-TENG(b) COMSOL simulation of the potential distribution between the two electrodes of the H-TENG.

To describe the electric field changes in the working process of the H-TENG. The working principle of H-TENG is shown in Figure 2(a), The relative contact separation movement between FEP film and aluminum film regularly generates AC pulse output. Through the software, COMSOL MULTIPHYSICS to use the finite element method to construct a two-dimensional plane model of H-TENG. It performs related simulation calculations and analyses the electric field distribution between the dielectric film and the aluminum film. As can be seen from Figure 2(b), the relative contact and separation movement between the two films causes the electric field distribution to change significantly.

III. OPTIMAL DESIGN

In the preparation process of the triboelectric nanogenerator, the surface of the power generation material is usually with special treatment to improve the output performance. Pyramid, hemispherical and other microstructures usually are made on the surface material to increase the contact area between the dielectric material and the metal material for increasing the number of overlapping electron clouds between atoms of different materials. With the surface contact stress and the number of atoms increasing, the depth of overlap between electron clouds enhances the charge transfer between materials. Another method is to increase the negative charge on the surface of the dielectric material by ion spraying. With a stronger built-in electric field during the contact and separation process of dielectric material and metal material, so as to increase the output of the triboelectric nanogenerator. In this experiment, the FEP material was polished with 10,000-mesh fine sandpaper to destroy its smooth surface and make the surface more uneven, causing more surface structural defects. At the same time, the FEP material was

treated with a high-voltage electric field of 20,000 volts for adding more charge to its surface.

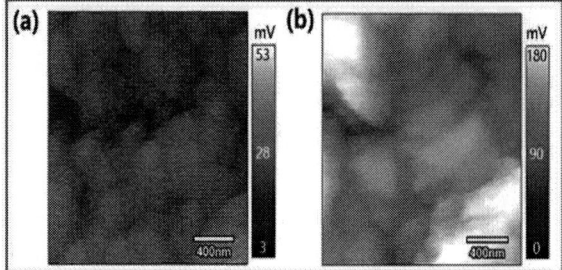

Figure 3. (a) The FEP film not treated with the electric field and sandpaper (b) The FEP film not treated with electric field and sandpaper.

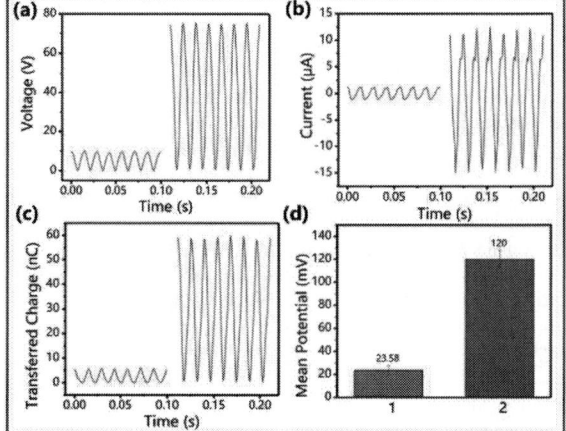

Figure 4 (a~c) Comparison of output voltage V_{oc}, short-circuit current I_{sc}, transferred charge Q_{sc} of H-TENG constructed by two different FEP films, (d) Comparing average surface potential of two FEP films.

Figure 3 shows the FEP film treated in two different ways. The experimental data is shown in Figure 4. The open-circuit voltage of the constructed H-TENG is shown in Figure 4(a). The output voltage of the H-TENG constructed with the FEP film without polishing treatment is about 10 V, and the output voltage of H-TENG constructed with polished FEP film is about 75 V, which is about 7.5 times that of the former. As the output short-circuit current shown in Figure 4(b), we can see that the unpolished FEP film is about 2.5 μA, while the polished one reaches 15 μA, which is about 5 times higher than that. In Figure 4(c), the transferred charge of the H-TENG has similar obvious results. The output performance results obtained are consistent with the theoretical derivation and prediction, which proves that the surface film treatment of the triboelectric nanogenerator has a significant effect on the improvement of electric energy output and can improve the efficiency of sound energy harvesting.

IV. EXPERIMENTAL RESULTS

In order to research the sound energy collection performance of H-TENG, a loudspeaker was used to generate a sine wave electric signal in the experiment, and a signal generator was used to adjust the frequency and sound pressure of the sound wave. As the experimental setup is shown in Figure 5.

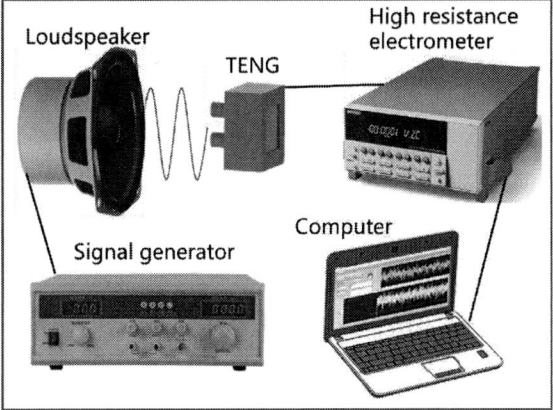

Figure 5. H-TENG's sound energy harvesting experimental equipment diagram.

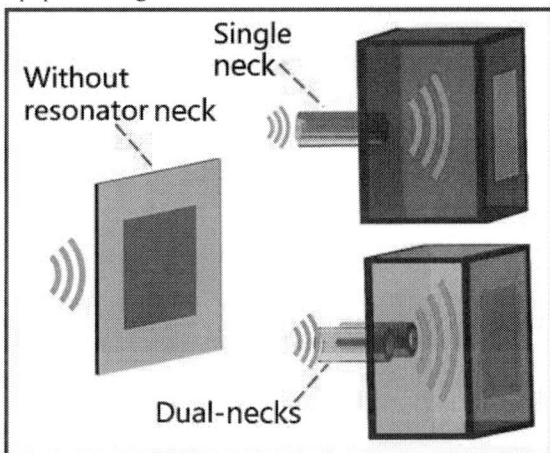

Figure 6. The structure of the TENG without neck, single-neck, and dual -necks resonant cavity.

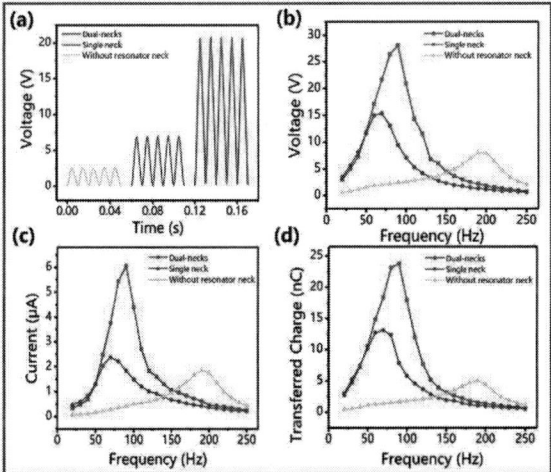

Figure 7. (a) The peak of the voltage of H-TENG with the same sound frequency and sound pressure (b~d) The output performance of H-TENG under different sound frequencies.

Experiment with three Helmhertz resonators with different structures in Figure 6. As Figure 7(a) shows the output voltages of the three resonant cavity TENGs under the 100 Hz and 80dB sound waves, their respective the peak of voltage is about 2.8V, 7.3V, and 21.7V. Compared with traditional single-neck, the output of dual-necks is improved by 197%. Figure 7(b) tests the output voltages of three TENGs with different structures at different audio frequencies. It can be clearly seen that the output voltage of the Helmholtz resonator of different necks is quite different. Among them, the maximum open-circuit voltage of the dual-necks TENG can reach 28V when it stays at the optimal frequency of 90Hz. The peak open-circuit voltage of the TENG without a resonant cavity at the optimal frequency of 190Hz is 8V, which is 269% lower than the corresponding output of the dual-necks Helmholtz cavity. The maximum open-circuit voltage of the optimal frequency of the traditional single-neck resonant cavity TENG is 15.2V, which is 87% lower than the corresponding output of the dual-necks Helmholtz cavity, and when the frequency exceeds 190 Hz, this performance difference gradually disappears. Figures 7(c) and 7(d) show the short-circuit current Isc and the transferred charge Qsc curves, we can find similar results. Compared with the TENG without cavity neck, with the single-neck Helmholtz cavity using, the maximum output power is significantly increased, and the best output frequency is significantly reduced. After using the dual-necks Helmholtz resonant cavity, the output performance of the H-TENG is further improved. So the experiment shows that the dual-necks Helmholtz resonant cavity not only broadens the response frequency band but also dramatically enhances the output power.

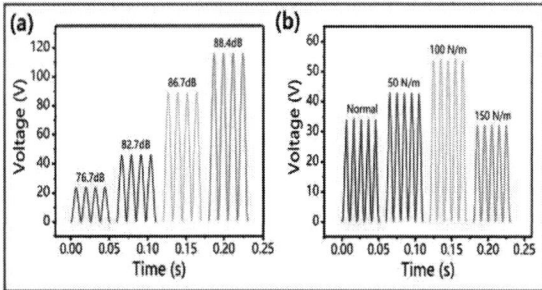

Figure 8. (a)Comparison of open-circuit voltage of H-TENG with four types of sound pressure levels. (b) Comparison of the output voltage of H-TENG with different FEP film tension at 100 Hz frequency.

The experiment researched the effect of sound pressure level on the H-TENG. It is mainly reflected in the effect of sound pressure level on the vibration amplitude of H-TENG's FEP film generates different output voltages. Figure 8(a) depicts the output characteristics at 100 Hz corresponding to four groups of different sound pressure levels. At 77dB sound pressure level, the peak voltage of H-TENG is 24.8 V. At 88dB sound pressure level, its peak voltage is 118.6 V, which is about 4.8 times than the former. It also can be seen that the open-circuit voltage obviously enhanced with the increasing sound pressure.

To further study the influencing factors of the output performance of the Helmholtz cavity. We use the fixed sound wave frequency at 100 Hz and the sound pressure level at 85 dB, Figure 8(b) shows four types of tension at an interval of 50N/m to compare the open circuit of the H-TENG by different film tensions. It can be clearly seen that under tension conditions from none to 150N/m, the output voltage of H-TENG does not simply increase with the increase in tension. And when the film tension reaches 100 N/m, H-TENG has the best output voltage. It Shows that proper film tension is beneficial to AC output.

V. APPLICATION OF THE H-TENG

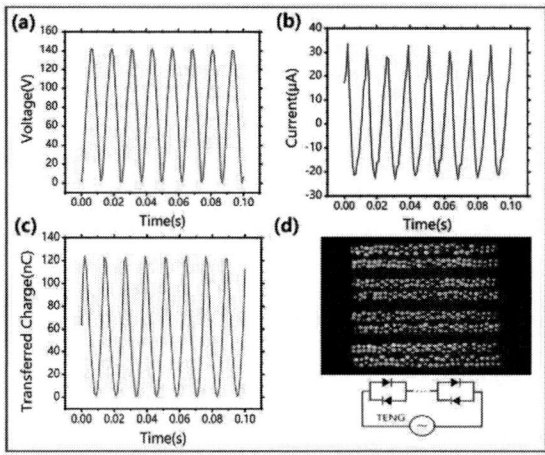

Figure 9. (a~c)The output performance of H-TENG at 85 frequency （d） H-TENG lights up the LED lights.

As a good output performance sound energy harvester, H-TENG can be used to be a power source . When the H-TENG harvesting acoustic wave circuit is connected in series with an internal resistance of 2MΩ, we can see the output performance of H-TENG from Figure 9(a~c), the short circuit current reaches 34.3 μA at 85 frequency. As shown in Figure 9(d), the output power can be directly supplied to the LED lights, making 457 LEDs light up at the same time. It can be calculated that H-TENG has a high power density of 1.47 W/m^2.

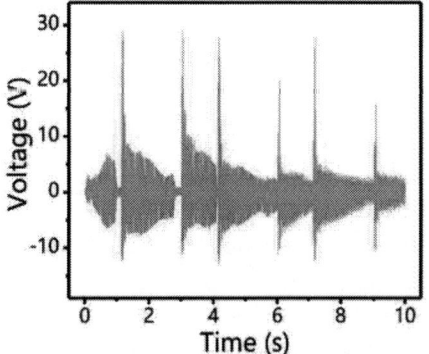

Figure 10. The voltage waveform of H-TENG used as a self-driven acoustic sensor.

H-TENG can be used as a self-driven sensor. In the experiment, H-TENG recorded 10s of audio. Figure 10

shows the voltage waveform of the sensor's response. When the audio plays, the H-TENG output voltage changes with the continuous changes of the audio sound wave characteristics. In addition, we can also use MATLAB's neural network algorithm to restore the response voltage waveform to the original audio. The neural network is shown in Figure 11 shows the high degree of restoration. At the same time, a certain amount of power output will be generated during the recording process, which can achieve the effect of self-driving. So the H-TENG can be optimized for applications such as microphones, recorders, mobile phones, and other devices.

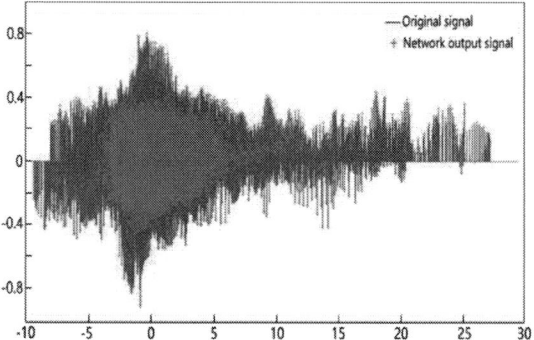

Figure 11. MATLAB's neuron network when H-TENG restores the music waveform.

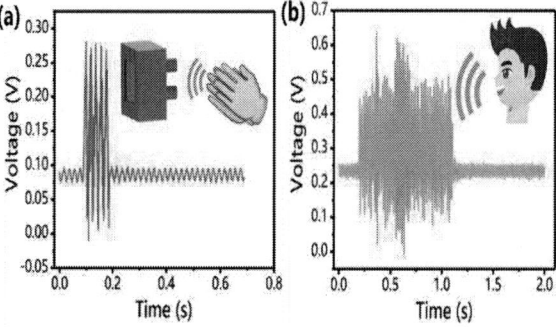

Figure 12. HR-TENG is used to perceive people's (a) applause and (b) voice.

H-TENG can also be used to sense people's applause and voice, and generate corresponding voltage and current output. As shown in Figure 12(a), a person is 5 meters away from the H-TENG. When a person claps his palms, the sound produced by the H-TENG is sensed by the H-TENG 5 meters away, and a set of signals is output. At the same distance of 5 meters, Figure 12(b) shows when a person makes a voice, H-TENG senses the sound wave of the voice, and the power generating film vibrates with the voice and generates a voltage output. It can be seen that the two waveforms are not the same. It can be seen that the characteristics of the two signals are quite different, which can be used for speech analysis and recognition.

VI. CONCLUSION AND FUTURE WORK

This paper proposes a novel ultra-high power density Helmholtz resonance acoustic energy converter based on a triboelectric nanogenerator. And its maximum output voltage is increased by 87%. Then studied the influence factors of the output characteristics of the H-TENG. The result shows that optimal treatment of the film is beneficial to H-TENG's output. The number of necks of the Helmholtz resonator affects the resonance frequency and transmission loss, so has a significant effect on the output voltage among them, the dual-necks H-TENG has the best output performance. In addition, the acoustic conditions and the tension of flexible film will also affect the output characteristics of the H-TENG acoustic energy converter. H-TENG follows the best output frequency deviation law, and with appropriate tension according to a specific sound wave range can improve the best frequency response, thereby improving the acoustic-electric conversion performance. Finally, through experiments, it is found that H-TENG has a high power density of 1.47 W/m^2.Obviously, H-TENG can be used for acoustic energy harvesting in the environment. It can also be used as a stable power supply with high power output, It also can sense people's applause and voice as an acoustic sensing, music recording for special detection. Therefore, H-TENG has good output performance in acoustic energy harvesting and self-driving sensing and has great application potential in the sensing and energy supply of wireless sensor networks and the Internet of things.

ACKNOWLEDGMENT

Support from the Fundamental Research Funds for the Central Universities, China (Grant No. 3132019331) is appreciated.

REFERENCES

[1] Zhao, H. Xiao, X. , Xu, P. , Zhao, T. , & Wang, Z. L. . (2019). Dual tube helmholtz resonator-based triboelectric nanogenerator for highly efficient harvesting of acoustic energy. Advanced Energy Materials, 9(46W.-K. Chen, Linear Networks and Systems (Book style). Belmont, CA: Wadsworth, 1993, pp. 3-9.

[2] D. T. Blackstock, A. A. Atchley, Fundamentals of Physical Acoustics. The Journal of the Acoustical Society of America[J]. 2001, 109 (4), 1274-1276.

[3] C. H. Sohn, J. H. J. A. S. Park, Technology, A comparative study on acoustic damping induced by half-wave, quarter-wave, and Helmholtz resonators. 2011, 15 (8), 606-614.

[4] F. U. Khan, Izhar, Electromagnetic energy harvester for harvesting acoustic energy. Sādhanā[J]. 2016, 41 (4), 397-405.

[5] Izhar, U. Khan Farid, Electromagnetic based acoustic energy harvester for low power wireless autonomous sensor applications. Sensor Review[J]. 2018, 38 (3), 298-310.

[6] S. B. Horowitz, M. Sheplak, L. N. Cattafesta, et al., A MEMS acoustic energy harvester. Journal of Micromechanics and Microengineering[J]. 2006, 16 (9), S174-S181.

[7] S. Tomioka, S. Kimura, K. Tsujimoto, et al., Lead–Zirconate–Titanate Acoustic Energy Harvesters with Dual Top Electrodes. Japanese Journal of Applied Physics[J]. 2011, 50 (9), 09ND16.

[8] A. Yang, P. Li, Y. Wen, et al., Enhanced Acoustic Energy Harvesting Using Coupled Resonance Structure of Sonic Crystal and Helmholtz Resonator. Applied Physics Express[J]. 2013, 6 (12), 127101.

[9] Z. L. Wang, Triboelectric Nanogenerator (TENG)—Sparking an Energy and Sensor Revolution. Advanced Energy Materials[J]. 2020, 10 (17).

[10] H. Zou, L. Guo, H. Xue, et al., Quantifying and understanding the triboelectric series of inorganic non-metallic materials. Nat Commun[J]. 2020, 11 (1), 2093.

[11] K. Zhao, B. Ouyang, C. R. Bowen, et al., One-Structure-Based Multi-Effects Coupled Nanogenerators for Flexible and Self-Powered Multi-Functional Coupled Sensor Systems. Nano Energy[J]. 2020, 71, 104632.

[12] J. Shao, D. Liu, M. Willatzen, et al., Three-dimensional modeling of alternating current triboelectric nanogenerator in the linear sliding mode. Applied Physics Reviews[J]. 2020, 7 (1), 011405.

[13] S. F. Leung, H. C. Fu, M. Zhang, et al., Blue energy fuels: converting ocean wave energy to carbon-based liquid fuels via CO2 reduction. Energy & Environmental Science[J]. 2020, 13.

[14] W. Harmon, D. Bamgboje, H. Guo, et al., Self-driven Power Management System for Triboelectric Nanogenerators. Nano Energy[J]. 2020, 71, 104642.

[15] M. Bouza, Y. Li, C. Wu, et al., Large-Area Triboelectric Nanogenerator Mass Spectrometry: Expanded Coverage, Double-Bond Pinpointing, and Supercharging. Journal of the American Society for Mass Spectrometry[J]. 2020, 31 (3).

[16] N. Cui, L. Gu, J. Liu, et al., High performance sound driven triboelectric nanogenerator for harvesting noise energy. Nano Energy[J]. 2015, 15, 321-328.

[17] J. Yang, J. Chen, Y. Liu, et al., Triboelectrification-Based Organic Film Nanogenerator for Acoustic Energy Harvesting and Self-Powered Active Acoustic Sensing. ACS Nano[J]. 2014, 8 (3), 2649-2657.

[18] X. Fan, J. Chen, J. Yang, et al., Ultrathin, Rollable, Paper-Based Triboelectric Nanogenerator for Acoustic Energy Harvesting and Self-Powered Sound Recording. ACS Nano[J]. 2015, 9 (4), 4236-4243.

[19] N. Cui, L. Gu, J. Liu, et al., High performance sound driven triboelectric nanogenerator for harvesting noise energy. Nano Energy[J]. 2015, 15, 321-328.

Proceedings of the 16th Annual IEEE International
Conference on Nano/Micro Engineered and Molecular Systems
April 25-29, 2021

Fabrication of Multi-oriented Composite Nanofibrous Membrane by Electrospinning

Xiang Wang, Yongfu Xu, Jinghua Lin, Sinan Fu, Jiaxin Jiang, Gaofeng Zheng, and Wenwang Li[*]

Abstract—In order to enhance the tensile strength of electrospun nanofibrous membrane, a method of preparing composite nanofiber was proposed. A rotation drum was introduced into electrospinning, the rotation speed could be changed to adjust the orientation of nanofibers. The diameter and orientation of nanofiber collected at various rotation speed were examined, and the tensile stress as well as average porosity were investigated. The tensile stress of nanofibrous membranes increased while the strain and porosity decreased with the increasing of the orientation arrangement. To achieve both high porosity and large tensile strength, multi-oriented composite nanofibrous membrane with different fiber orientations were prepared. The maximum tensile strength was 28.8 MPa and the porosity was 81.7%.

I. INTRODUCTION

Electrospinning [1-3] is a typical technology to prepare nanofibers at a controllable process. Due to the small diameter, high porosity and large specific surface area, electrospinning nanofibrous membranes have shown great advantages in many fields [4-6]. While the nanofibers inside the membrane were combined with each other based on the electrostatic force and the frictional force, the tensile strength of the membrane was usually worse than that of traditional film prepared by dry or wet process [7, 8], limiting its further industrial applications.

In the recent years, many progresses have been made to enhance the mechanical strength of electrospun nanofibrous membrane. One is to enhance the tensile strength of the fiber on a single orientation by changing the arrangement of the fibers. One of the typical methods is to increase the tensile strength in one direction by changing the arrangement of nanofibers. Wang [9] et al. utilized oriented fibers to enhance the tensile strength of fiber mat in wet-electrospinning, the tensile strength of fiber mat increased with fiber alignment, and the effects of operational parameters were evaluated. Gong [10] et al. reported that the electrospun oriented poly(phthalazinone ether sulfone ketone) fibrous membranes hot pressed with two pieces overlaid perpendicularly shown high tensile strength of 22.8 MPa at both horizontal and vertical directions and high porosity of 70%. Another method to improve the tensile strength is to make the fiber crosslinked. Zhang [11] et al. put the electrospun gelatin fiber in saturated glutaraldehyde

vapor and the mechanical properties of nanofiber was enhanced after crosslinking. Arifeen [12] et al. investigated the effects of oven-drying and thermal-pressing conditions on the porosity and tensile strength of polyacrylonitrile nanofibrous membranes. Moreover, modification of nanofiber can also be used. Liu [13] et al. improved the mechanical properties and porosity of polyimide nanofiber membrane by adding SiO2 nanoparticles into the solution.

In this work, nanofibers with different orientations were compounded to enhance the tensile strength of electrospun membrane. The orientation of nanofibers was adjusted by controlling the rotation speed of drum that served as collector. Multi-oriented composite nanofibrous membrane with both high porosity and large tensile strength can be generated by this simple method.

II. EXPERIMENTAL DETAILS

An electrospinning system including a high-voltage power source, a stainless-steel capillary nozzle, a steel drum, a pump, and a speed controller was built up to generate nanofibrous membrane. The drum was served as collector, and the linear speed of the outer surface can be controlled by the speed controller. The high-voltage power source (DW-SA403-1ACE5) was used to generate high electrical field between the nozzle and the drum, as the anode connected to the nozzle and the cathode connected to the drum. The distance between nozzle and drum was 12 cm. The applied voltage was 15 kV. The precise syringe pump (Harvard 11 Pico Plus) continuously supplied the solution out of the nozzle at a controllable feed rate. Here, the polyimide (PI, purchased from Hangzhou Surmount Technology Co., Ltd.) solution was used without any further purification.

Preparation of the nanofibrous membrane was carried out by controlling the rotation speed of collecting drum. The randomly whipped nanofibers would be twining with the drum when they touch the rotating drum during solidification and deposition, and the drum would give the deposited nanofibers a tangential force to make the nanofibers aligned. The orientation of the nanofibers will be enhanced with the increase of the rotation speed of the drum.

Fig. 1 shows the preparation process of the multi-oriented composite nanofibrous membrane. The nanofiber

---

Resrach supported by National Natural Science Foundation of China (No. 51805460), Science and Technology Planning Project of Fujian Province (No. 2019H0038, 2020H6003), Xiamen Municipal Science and Technology Project (No. 3502Z20193015), and Xiamen University of Technology (No. XPDKT20022).

X. Wang, Y. Xu, S. Fu, W. Li are with the School of Mechanical and Automotive Engineering, Xiamen University of Technology, Xiamen,

Fujian 361024, China (*corresponding author: W. Li; e-mail: xmlww@xmut.edu.cn).

J. Lin is with the Institute of Manufacturing Engineering, Huaqiao University, Xiamen, Fujian 361021, China.

J. Jiang, G. Zheng are with the Department of Instrumental and Electrical Engineering, Xiamen University, Xiamen, Fujian 361102, China.

978-1-6654-3008-1/21 $31.00 © 2021 IEEE

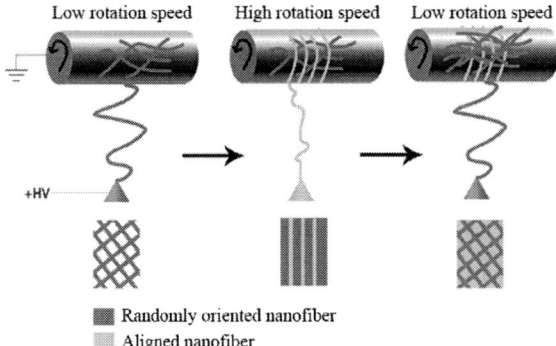

Fig. 1. Preparation process of the multi-oriented composite nanofibrous membrane.

collected with high rotation speed (518 m/min) was used as the core layer to guarantee the tensile strength of the composite membrane, and the outer layer collected with low rotation speed (35 m/min) plays the advantages of high porosity and high specific surface area for applicable functionality.

The morphography of electrospun PI nanofibrous membrane was examined by a scanning electron microscope (SEM, Zeiss Sigma 500). The tensile strength of the membrane was tested by a computerized electronic universal testing system (CTM8010).

The porosity of electrospun membrane was determined by using the n-Butanol absorption method. The membrane was cut into cycle with a diameter of 10 cm, dried in an oven at 60 °C for 6 hours, and then immersed in n-Butanol for 12 hours. After taking the membrane from n-Butanol, the residual n-Butanol on the surface of the membrane was absorbed with filter paper. The weights of the membrane before and after immersion were recorded, and the porosity of the membrane, P, can be calculated as:

$$P = (M_2/\rho_2)/[(M_2/\rho_2) + (M_1/\rho_1)] \tag{1}$$

Where M_1 was the dry weight of membrane, M_2 was the weight of n-Butanol, ρ_1 was the true density of membrane, and ρ_2 was the density of n-Butanol.

III. RESULTS AND DISCUSSION

The SEM images of nanofibrous membrane collected at various rotation speed are shown in Fig. 2. Nanofibrous membrane generated at the rotation speed of 35 m/min was made up of many interlaced fibers. The deposition position of the nanofiber was uniform with many interlaced holes distributed. With the increase of the rotation speed of the drum, the orientation of the fibers changed obviously. The alignment degree increased with the increases of rotation speed of drum, and the holes on the nanofibrous membrane decreased as well as the fiber entanglement gradually appeared. The entanglement of the fibers collected at the rotation speed of 518 m/min was most serious, indicated a lowest porosity.

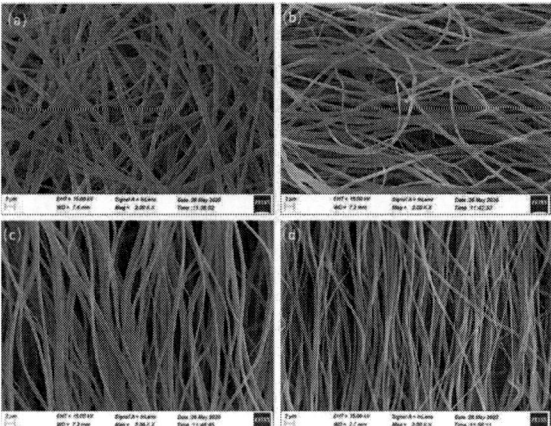

Fig. 2. The SEM images of nanofibrous membrane collected at various rotation speed of drum: (a) 35 m/min, (b) 172 m/min, (c) 345 m/min and (d) 518 m/min.

The distributions of nanofiber diameter generated at different rotation speeds are shown in Fig. 3, and the averaged diameter and coefficient of variation (CV) are calculated as shown in Fig. 4. The average diameter of nanofiber decreased, and the uniformity of nanofiber increased with the increasing of rotation speed. The average diameters were 0.668, 0.603, 575, and 0.483 μm under the rotation speeds of 35, 172, 345, and 518 m/min respectively, with the CV of 41.2%, 35.4%, 29.7% and 34.0%. At low drum rotating speed, the stretching effect of the electric field force on the nanofiber was the main factor affecting the fiber diameter. The whipping of charged solution during deposition resulting in uneven diameter distribution. As the drum rotating speed increased, the stretching effect of the drum on the nanofibers gradually increased, the diameters of nanofibers began to shrink further and the difference in nanofiber diameter was partially diminished, resulting in a uniform fiber distribution. Noted that the CV enlarged when the rotation speed increased from 345 m/min and 518 m/min, the reason was that faster drum speed caused faster surrounding airflow, which would affect the movement of nanofiber and resulted in an uneven stretching effect as well as non-uniform diameter distribution.

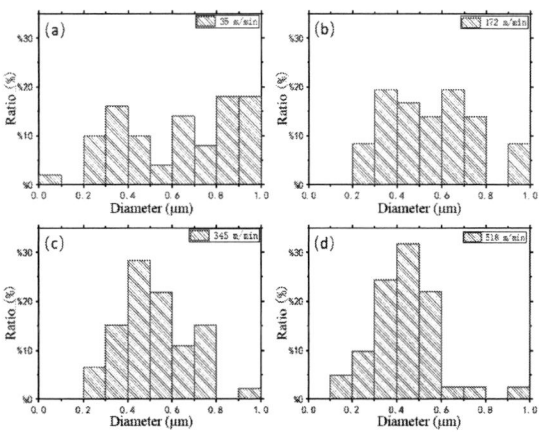

Fig. 3. The distribution of nanofiber diameter at different rotation speeds of drum.

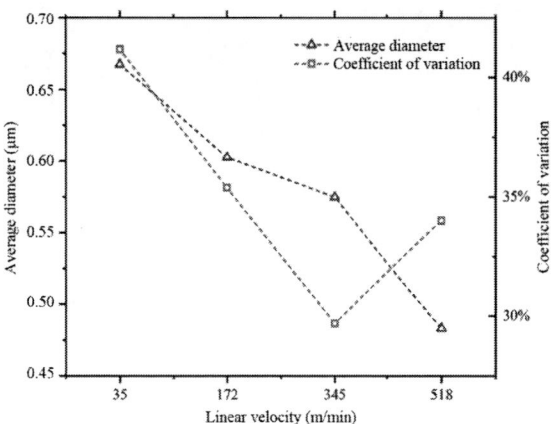

Fig. 4. The average diameter of nanofiber and coefficient of variation at different rotation speeds of drum.

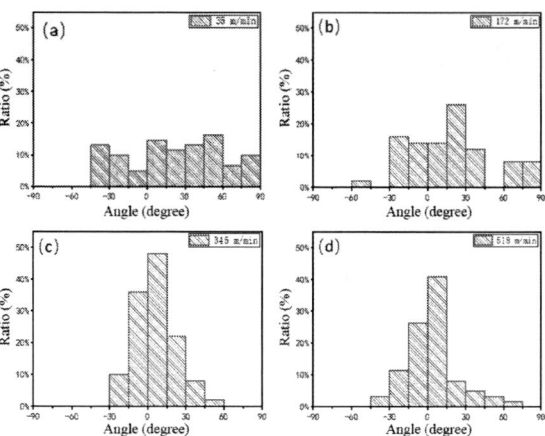

Fig. 5. The distribution of nanofiber orientation at different rotation speeds of drum.

Fig. 5 shows the distribution of nanofiber orientation at different rotation speeds. The orientation degree of the nanofiber increased with the increase of rotation speed. The nanofibers received at 345 m/min had the most concentrated deposition orientation, as 71% of the nanofibers were in the range of -15 ° to +15 °. The airflow caused by the high rotation drum speed influenced the deposition speed and position of the nanofiber, thus that the orientation generated at 518 m/min was worse than that at 345 m/min.

Fig. 6 shows the tensile stress-strain curves of different nanofibrous membranes. The maximum tensile stress of nanofibrous membranes increased and the strain of nanofibrous membranes decreased with the increasing of the orientation arrangement. At low rotation speed, the orientation of the nanofibers was disordered, only part of the nanofibers was stressed according to their own material strength, so the maximum tensile stress was small, and the maximum strain was large. When the nanofiber was in good orientation arrangement, many fibers were stressed at the same time and would increase the mechanical property, thus that the maximum tensile stress was large, and the maximum strain was low.

The tensile stress of the multi-oriented composite nanofibrous membrane was between the surface layer (generated from 35 m/min) and the core layer (generated from 518 m/min). The tensile stress of the composite nanofibrous membrane was 28.7 MPa, and was 142.9% higher than that of membrane collected at 35 m/min.

Fig. 7 shows the relationship between the average porosity of nanofibrous membrane and the rotation speed. It can be indicated that the porosity of nanofibrous membrane decreased with the increase of orientation degree. This is because the pores of nanofibrous membrane are formed by overlapping fibers. When the fibers were oriented, the probability of overlapping fibers would be reduced, resulting in a decreasing of porosity. The porosity of the multi-oriented composite nanofibrous membrane was 81.7%.

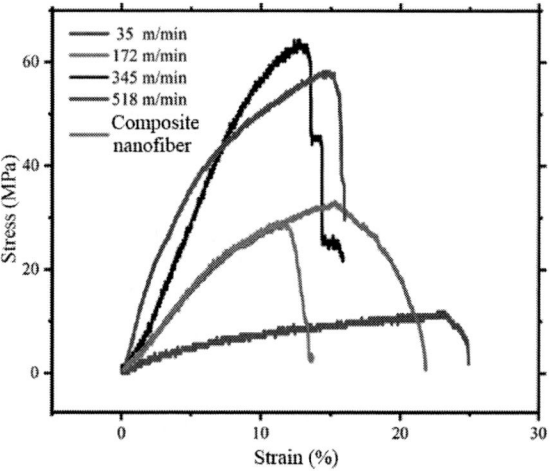

Fig. 6. The tensile stress-strain curves of different nanofibrous membranes.

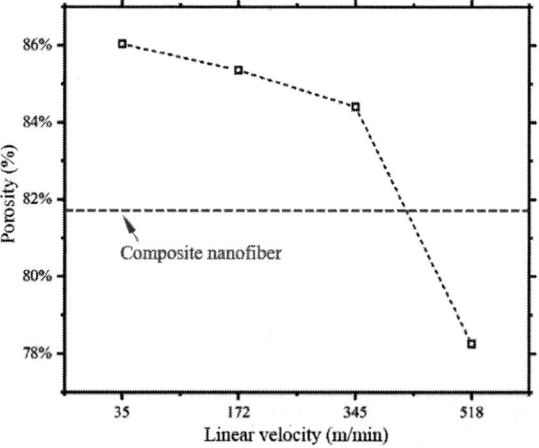

Fig. 7. The relationship between the average porosity of nanofibrous membrane and the rotation speed of drum.

IV. CONCLUSION

In this work, multi-oriented composite nanofibrous membrane with both high porosity and large tensile strength was realized by electrospinning method. A drum was introduced as collector to adjust the orientation of nanofibers by controlling the rotation speed. The diameter and orientation of nanofiber collected at various rotation speed were examined, and the tensile stress as well as average porosity were investigated. The experimental results show that the increasing of orientation arrangement of nanofiber would increase the maximum tensile stress and decrease the porosity of membrane. For the multi-oriented composite nanofibrous membrane, nanofiber collected with high rotation speed (518 m/min) was used as core layer to increase the tensile strength, and outer layer collected with low rotation speed (35 m/min) played the advantages of high porosity and high specific surface area for applicable functionality. The maximum tensile strength and porosity were tested to be 28.8 MPa and 81.7%, respectively. Our work provides a sample method for regulating the balance between porosity and mechanical properties of fiber membranes.

REFERENCES

[1] D. Li, and Y. Xia, "Electrospinning of Nanofibers: Reinventing the Wheel?," *Advanced Materials,* vol. 16, no. 14, pp. 1151-1170, 2004.

[2] T. Subbiah, G. S. Bhat, R. W. Tock, S. Parameswaran, and S. S. Ramkumar, "Electrospinning of nanofibers," *Journal of Applied Polymer Science,* vol. 96, no. 2, pp. 557-569, 2005.

[3] B.-S. Kim, and I.-S. Kim, "Recent Nanofiber Technologies," *Polymer Reviews,* vol. 51, no. 3, pp. 235-238, 2011.

[4] M. S. Islam, B. C. Ang, A. Andriyana, and A. M. Afifi, "A review on fabrication of nanofibers via electrospinning and their applications," *SN Applied Sciences,* vol. 1, no. 10, pp. 1248, 2019.

[5] S. Thenmozhi, N. Dharmaraj, K. Kadirvelu, and H. Y. Kim, "Electrospun nanofibers: New generation materials for advanced applications," *Materials Science and Engineering: B,* vol. 217, pp. 36-48, 2017.

[6] J. Xue, J. Xie, W. Liu, and Y. Xia, "Electrospun Nanofibers: New Concepts, Materials, and Applications," *Accounts of chemical research,* vol. 50, no. 8, pp. 1976-1987, 2017.

[7] E. P. S. Tan, and C. T. Lim, "Mechanical characterization of nanofibers – A review," *Composites Science and Technology,* vol. 66, no. 9, pp. 1102-1111, 2006.

[8] S. Nauman, G. Lubineau, and H. F. Alharbi, "Post Processing Strategies for the Enhancement of Mechanical Properties of ENMs (Electrospun Nanofibrous Membranes): A Review," *Membranes,* vol. 11, no. 1, pp. 39, 2021.

[9] H. Wang, L. Kong, and G. R. Ziegler, "Aligned wet-electrospun starch fiber mats," *Food Hydrocolloids,* vol. 90, pp. 113-117, 2019.

[10] W. Gong, J. Gu, S. Ruan, and C. Shen, "A high-strength electrospun PPESK fibrous membrane for lithium-ion battery separator," *Polymer Bulletin,* vol. 76, no. 10, pp. 5451-5462, 2019.

[11] Y. Z. Zhang, J. Venugopal, Z. M. Huang, C. T. Lim, and S. Ramakrishna, "Crosslinking of the electrospun gelatin nanofibers," *Polymer,* vol. 47, no. 8, pp. 2911-2917, 2006.

[12] W. U. Arifeen, M. Kim, J. Choi, K. Yoo, R. Kurniawan, and T. J. Ko, "Optimization of porosity and tensile strength of electrospun polyacrylonitrile nanofibrous membranes," *Materials Chemistry and Physics,* vol. 229, pp. 310-318, 2019.

[13] J. Liu, Y. Liu, W. Yang, Q. Ren, F. Li, and Z. Huang, "Lithium ion battery separator with high performance and high safety enabled by tri-layered SiO2@PI/m-PE/SiO2@PI nanofiber composite membrane," *Journal of Power Sources,* vol. 396, pp. 265-275, 2018.

Gap in pagination due to unavailable papers.

Pages 1704-1713

Proceedings of the 16th Annual IEEE International
Conference on Nano/Micro Engineered and Molecular Systems
April 25-29, 2021

A High DC Current Output Salt Battery Hybrid and Enhanced by TENG

Shuangshuang Liu, Jiayi Yang, Zihao Niu, Yan Meng, Wei Xu, Di Feng, Baolong Wang,

Meiqi Wang, Xiuhan Li*

Abstract—The conversion of water energy into sustainable electricity has attracted more and more attention. Conventional water energy harvesting technologies are limited by manufacturing materials and output performance. The triboelectric nanogenerator (TENG) shows good performance in mechanical energy harvesting, while it has a relatively low short-circuit current and large inherent impedance. A novel high DC current output salt battery (SB) hybrid and enhanced by TENG (DC-SBHET) is proposed in this paper. The high open-circuit voltage generated by TENG can accelerate the velocity of mobile ions in the SB. The combination of TENG and SB can successfully increase the DC short- circuit current up to 18.1mA. Nano-foam electrodes were also chosen to further increase the battery performance.

Keywords: Salt Battery; Triboelectric Nanogenerator; Water Energy; High DC Current; Droplet Energy Generator

I. INTRODUCTION

With the fast development of internet of things, a large amount of electronic devices are scattered everywhere. Hence there is an urgent demand for distributed power supply. In order to meet the demand, more and more energy harvesting technologies have emerged (e.g., solar, mechanical, chemical and water energy) [1-5]. Water occupies 10% of the surface of the earth. Due to its abundant and renewable advantages, water energy has attracted more and more attention. At present, hydroelectric power generation mainly relies on electromagnetic generators to convert the dynamic flow of water into electrical energy, but this method is restricted by the large-scale water source [6]. Therefore, more research efforts are devoted to collect energy from raindrops, rivers, and waves through methods such as contact electrification effect [7-8] and bulk effect [9-10]. The triboelectric nanogenerator (TENG) proposed by Wang in 2012 [11] provides a new idea for collecting water energy [12-15]. Through triboelectrification and the electrostatic induction between water and triboelectric materials electrical energy can be collected effectively. Although TENG has many advantages, the low short-circuit current (Isc) and high inherent impedance of TENG limit its range of applications. Therefore, it is still a big challenge to improve the output performance to drive the load.

Salt batteries (SB) which can be easily acquired can generate DC current through chemical reaction. The Isc of SB is much higher than that of the TENG. More, the open-circuit voltages (Voc) of TENG can greatly accelerate the velocity of mobile ions in SB, hence we can combine the TENG and the SB to build a novel battery structure with a higher output current.

In this work, first, a small salt battery based on triboelectrification (SBBT) was fabricated to verify that the triboelectrification can accelerate the velocity of mobile ions of the SB and increase the output performance. Then, a salt battery hybrid and enhanced by TENG with a high DC current output （DC-SBHET） was made. The whole article is divided into three parts: experimental process, analysis and testing. Firstly, the structure and manufacturing process of the SBBT and DC-SBHET are introduced. Secondly, the working principle of the SBBT and DC-SBHET is analyzed. Finally, the electrical output of SBBT and DC-SBHET are tested, and some parameters that affect the electrical output characteristics are analyzed, such as friction distance, frequency, pressure, etc. When 36 μL sodium chloride (NaCl) solution droplet slips on the triboelectric material for 10 cm, the Isc reaches 40 μA in the experiment. The combination of TENG and SB can successfully increase the DC output current up to 18.1 mA.

II. EXPERIMENTS PROCEDURES

Fabrication of SBBT: A PTFE (polytetrafluoroethylene, Hongfu Insulating Material Factory) tape is attached to a PMMA (polymethyl methacrylate) board, and Cu (copper) and Al (aluminium) tapes are attached on the PTFE tape as electrodes. The structure diagram is shown in Fig. 1a. The injection pump (LSP01-3A) is used to push the NaCl solution at a constant speed to produce NaCl solution droplets of the same volume.

Fabrication of DC-SBHET: The basic structure of DC-SBHET consists of a TENG, a rectifier bridge and a SB. TENG is composed of PMMA, Cu electrode, PTFE tape, Cu triboelectric layer. SB is formed by Cu nano-foam, Al nano-foam (Guangjiayuan Electronic Materials), and NaCl solution. The structure diagram is shown in Fig. 1b.

---

*Xiuhan Li is corresponding author.
*Xiuhan Li works in Beijing Jiaotong University (corresponding Dr. Xiuhan Li author to provide phone: +86-10-5168-3981; fax: +86-10-5168-3682; e-mail: lixiuhan@bjtu.edu.cn).

Shuangshuang Liu, Jiayi Yang, Zihao Niu, Yan Meng, Wei Xu, Di Feng, Baolong Wang, Meiqi Wang are with Beijing Jiaotong University (e-mail: 19120012@bjtu.edu.cn, 20111006@bjtu.edu.cn).

978-1-6654-3008-1/21 $31.00 © 2021 IEEE

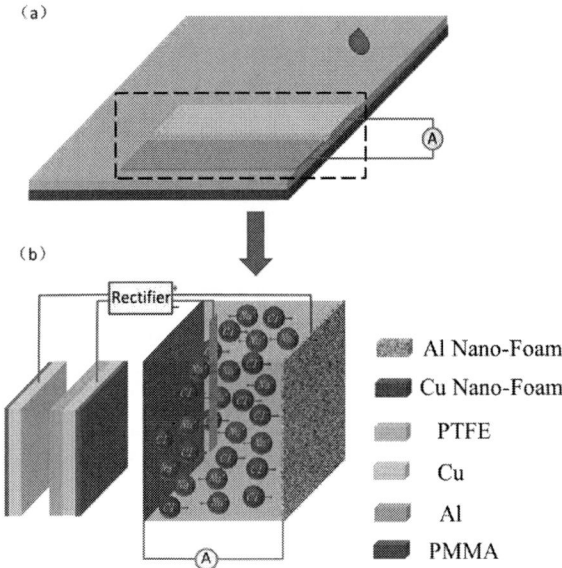

(a)

(b)

Rectifier

- Al Nano-Foam
- Cu Nano-Foam
- PTFE
- Cu
- Al
- PMMA

Fig. 1. The structure of SBBT and DC-SBHET (a) The structure of SBBT (b) The structure of DC-SBHET.

III. RESULTS AND DISCUSSION

When the NaCl solution drops down on PTFE, the droplets will be positively charged according to the triboelectric series. Three different electrode pairs: Al-Al, Cu- Cu and Al-Cu were used to construct SBBT, and it was found that the Isc reached the maximum value of 17.5 μA when using Al-Cu electrode, as shown in Fig. 2a. Using Al-Cu electrodes, when the droplet falls into the gap between two electrodes, the SB is formed. The droplet and PTFE form a TENG to generate a voltage which can accelerate the movement rate of sodium ions and chloride ions in SB and a larger Isc can be obtained. Changing the sliding distance of the droplet, it is found when a droplet (36 μL) slides for a distance of 10cm on the PTFE, the Isc reaches about 40 μA, as shown in Fig. 2b. According to this phenomenon, TENG and SB can be combined to construct DC-SBHET with an enhanced DC current.

(a)

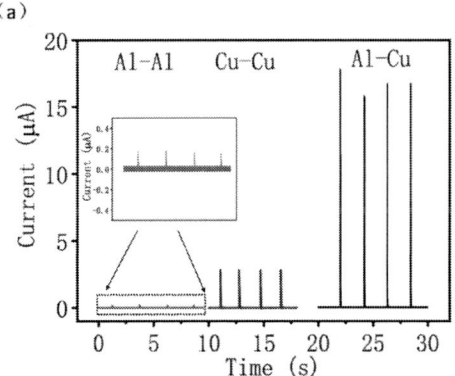

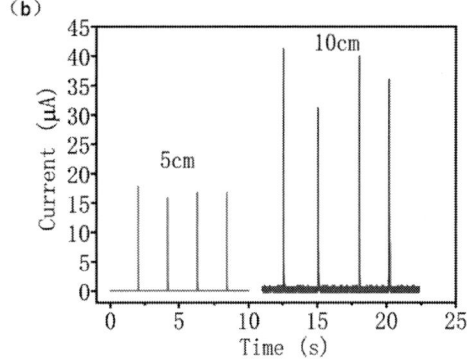

Fig. 2. The electricity generation of SBBT. (a) The short-circuit current of SBBT with three different electrode pairs: Al-Al, Cu-Cu and Al-Cu (b) The short-circuit current of SBBT under different sliding distances.

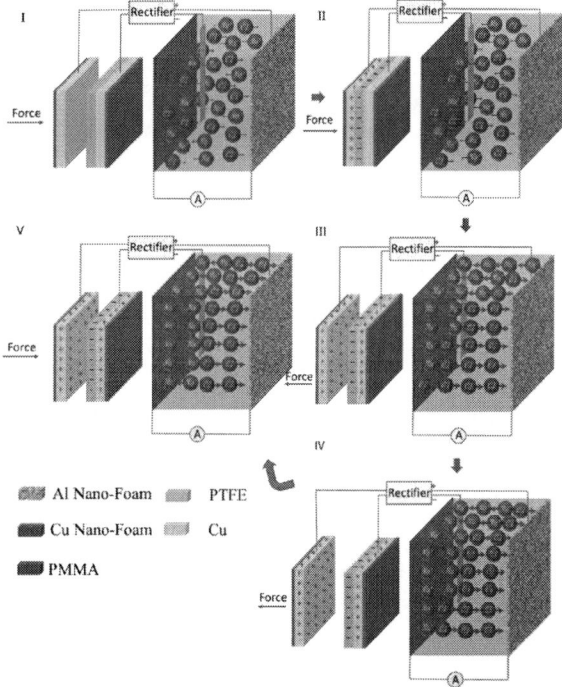

- Al Nano-Foam
- PTFE
- Cu Nano-Foam
- Cu
- PMMA

Fig. 3. Working principle of the DC-SBHET. (I) In the equilibrium state, the short-circuit current consists only of the current of the SB. (II) The contact between the two triboelectric materials will generate an equal amount of charge of opposite polarity on the surface of PTFE and Cu. (III) During the separation process, the two Cu electrodes generate a potential difference to accelerate the movement of ions in the solution. (IV) When the separation reaches the maximum, the potential difference and ion concentration difference are at the maximum. (V) When force is applied again, the charges of the two plates are neutralized and another cycle is entered.

The working principle of DC-SBHET is shown in Fig. 3. The working principle of SB is based on the working principle of the primary cell. It uses the difference in potential between Al and Cu electrodes to generate a potential difference, so that electrons flow and current is generated. When TENG is in an equilibrium state, all layers are not charged, and the output current of DC-SBHET is only composed of the current of the SB (Fig. 3I). When

978-1-6654-3008-1/21 $31.00 © 2021 IEEE   1715

PTFE and Cu are in complete contact, an equal number of charges of opposite polarity are generated on the surface of Cu and PTFE (Fig. 3II). When the force is released, PTFE and Cu separate from each other, which leads to a potential difference, that drives electrons from the right electrode to the left electrode to balance the potential difference. The output of TENG is connected to the electrode of the SB through a rectifier bridge. The high voltage generated by TENG accelerates the movement of sodium ions and chloride ions in the NaCl solution. Therefore, more sodium ions are transported to the Cu nano-foam electrode, and more chloride ions are transported to the Al nano-foam electrode (Fig. 3III). When PTFE and Cu reach the maximum separation distance, the Voc of the TENG reach the maximum (Fig. 3IV). When force is applied again, the charge on Cu and PTFE decreases, and electrons flow back to the right electrode. The sodium ions at the Cu foam and the chloride ions at the Al foam continue to increase (Fig. 3V).

In order to obtain the highest Isc of the SB, the concentration of NaCl solutions were optimized. We prepared different concentrations of 100 mL NaCl solution. Based on our experiment results, the maximum SB output current of 10 mA was obtained at the concentration of 10%, as shown in Fig. 4. The electrodes with nano structures can greatly improve the output current of SB. Hence the SB with Cu nano-foam and Al nano-foam was measured and compared with the SB which uses normal electrodes, as shown in Fig. 5. It is indicated that the Isc of the SB with nano-foam electrode is 10mA which has been increased by 10 times with that of normal electrodes (0.8 mA).

TENG with the area of 2×2 cm^2, 8×8 cm^2 and 10×10 cm^2 were designed and fabricated for comparison. The Voc are respectively 30 V, 66 V and 123 V (Keithley 6514), as shown in Fig. 6(a). TENG with an area of 8×8 cm^2 was selected to build the DC-SBHET. In order to obtain a higher Voc, PTFE was polarized under a strong electric field (ET 2673A), and the Voc after polarization is 183 V, as shown in Fig. 6(b). The output current of the DC-SBHET combining TENG and SB is shown in Fig. 7.

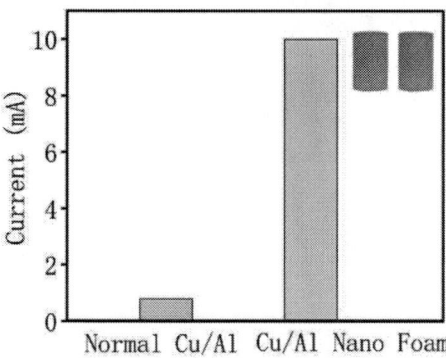

Fig. 5. The short-circuit current of SB with different electrodes, the illustration shows a schematic diagram of the nano-foam Al electrode.

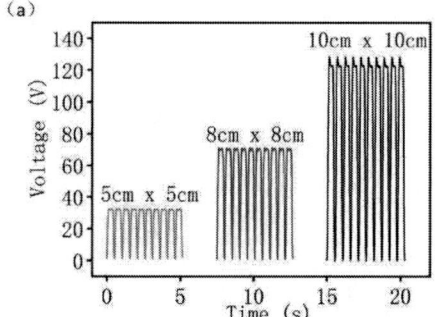

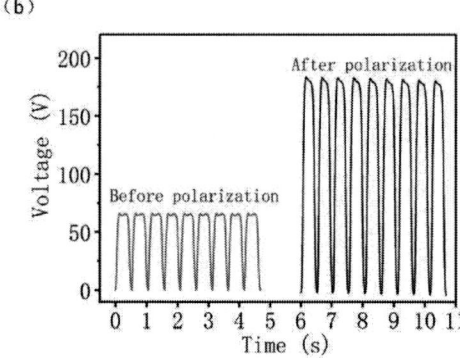

Fig. 6. Open-circuit voltages of different TENGs (a) The output voltages of TENGs with different areas of 2×2 cm^2, 8×8 cm^2 and 10×10 cm^2. (b) The output voltages before and after polarization of 8x8 cm^2 TENG.

Finally, the influence of TENG under different frequencies and pressures on the Isc of DC-SBHET is explored. As shown in Fig. 8 and Fig. 9, the baseline of the short-circuit current is 10 mA. The Isc increased with the frequency and contact force. When the frequency reaches 5 Hz, the Isc reaches the maximum value of 18.1 mA. Under the frequency of 5 Hz, the Isc changes with contact force was measured. As the force increases, the Isc gradually increases and finally reach the maximum value of 18.1 mA.

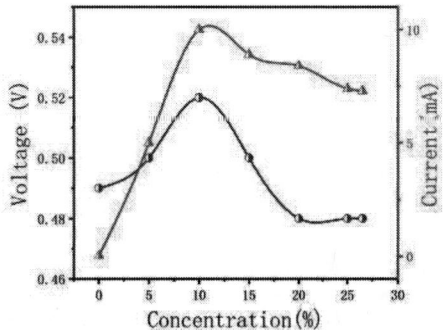

Fig. 4. The short-circuit current of SB with different NaCl concentrations.

978-1-6654-3008-1/21 $31.00 © 2021 IEEE

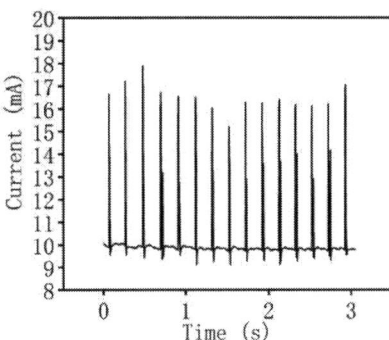

Fig. 7. DC-SBHET short-circuit current with TENG working.

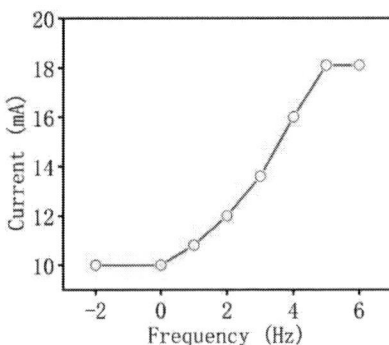

Fig. 8. Short-circuit current under different working frequencies of DC-SBHET from 1 Hz to 6 Hz.

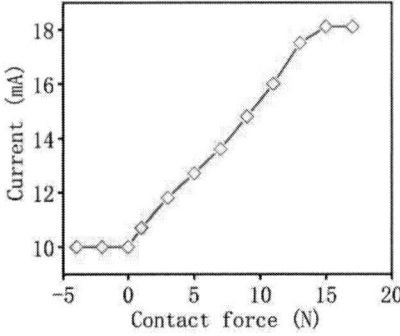

Fig. 9. Short-circuit current under different contact force of DC-SBHET from 1 N to 17 N.

IV. CONCLUSION

In this work, a high DC current output salt battery hybrid and enhanced by TENG was made. It is verified that the triboelectrification can accelerate velocity of mobile ions in SB, so a DC-SBHET with high short-circuit current is manufactured by combining TENG and SB. Its short-circuit current can reach 18.1 mA. DC-SBHET solves the problem of low short-circuit current of TENG and provides a new idea for collecting the energy of droplets and the ocean.

ACKNOWLEDGMENT

This work is supported by the Fundamental Research Funds for the Central Universities+2020JBZD011, National Natural Science Foundation of China (60706031 and 61574015), Beijing; National Science Foundation (4122058), National Key R&D Program of China (2016YFB1200203).

REFERENCES

[1] P.A. Owusu, S. Asumadu-Sarkodie, "A review of renewable energy sources, sustainability issues and climate change mitigation" *Cogent Eng*, 3 (2016) 1167990.

[2] M. Salauddin, R. M. Toyabur, P. Maharjan, M. S. Rasel, H. Cho, J. Y. Park, "Design and experimental analysis of a low-frequency resonant hybridized nanogenerator with a wide bandwidth and high output power density" *Nano Energy*, 66 (2019), 104122.

[3] D. Jiang, M. Xu, M. Dong, F. Guo, X. Liu, G. Chen, Z.L. Wang, "Water-solid triboelectric nanogenerators: An alternative means for harvesting hydropower" *Renew. Sustain. Energy Rev.*,115 (2019), 109366.

[4] W. Tang, B. D. Chen, Z. L. Wang, "Recent Progress in Power Generation from Water/Liquid Droplet Interaction with Solid Surfaces" *Adv. Funct. Mater.*, 29 (2019), 1901069.

[5] J. Yin, J. Zhou, S. Fang, W. Guo, "Hydrovoltaic energy on the way" *Joule*, 4(2020) 1852-1855.

[6] N. Zhang, H. J. Gu, X. F. Zhou, et al. "A universal single electrode droplet-based electricity generator (SE-DEG) for water kinetic energy harvesting" *Nano Energy*, 82(2021), 105735.

[7] S.-B. Jeon, D. Kim, G.-W. Yoon, J.-B. Yoon, Y.-K. Choi, "Self-cleaning hybrid energy harvester to generate power from raindrop and sunlight" *Nano Energy*, 12 (2015) 636-645.

[8] L.E. Helseth, "A water droplet-powered sensor based on charge transfer to a flow through front surface electrode" *Nano Energy*, 73 (2020), 104809.

[9] J. Yu, T. Ma, "Triboelectricity-based self-charging droplet capacitor for harvesting low-level ambient energy" *Nano Energy*, 74 (2020), 104795.

[10] H. Wu, S G. Zhou, F. Mugele, et al. "Charge Trapping-Based Electricity Generator (CTEG): an ultrarobust and high efficiency nanogenerator for energy harvesting from water droplets" *Adv. Mater.*, 32 (2020), 2001699.

[11] F. R. Fan, Z. Q Tian, Z. L. Wang, "Flexible triboelectric generator" *Nano Energy*, 1(2012) 328-334.

[12] N. Zhang, H. J. Gu, X. F. Zhou, et. al. "Boosting the output performance of volume effect electricity generator (VEEG) with water column" *Nano Energy*, 73 (2020), 104748.

[13] X. Li, L. Zhang, Y. Feng, X. Zhang, D. Wang, F. Zhou, "Solid–liquid triboelectrification control and antistatic materials design based on interface wettability control" *Adv. Funct. Mater.*, 29 (2019), 1903587.

[14] J. Nie, Z. Wang, Z. Ren, S. Li, X. Chen, Z. L. Wang, "Power generation from the interaction of a liquid droplet and a liquid membrane" *Nat. Commun.*, 10 (2019) 2264–2273.

[15] Y. Liu, Y. Zheng, T. Li, D. Wang, F. Zhou, "Water-solid triboelectrification with self-repairable surfaces for water-flow energy harvesting" *Nano Energy*, 61 (2019) 454–461.

Proceedings of the 16th Annual IEEE International
Conference on Nano/Micro Engineered and Molecular Systems
April 25-29, 2021

The Influence of Electrode on Elastic Constant C_{33}^{D} Extraction of Scandium-doped Aluminum Nitride Thin Film by Thickness-extensional Mode FBAR

Hu Xia[1,2], Su Ouyang[1,2] and Lifeng Qin[1,2, *)]

Abstract— Recently, scandium-doped aluminum nitride ($Sc_xAl_{1-x}N$) has drawn a lot of attention as a promising material for RF-MEMS due to its large piezoelectric constants. In this paper we present the study of electrode effect on material constant extraction of $Sc_xAl_{1-x}N$ by measuring parallel resonant frequency of thickness-extensional mode thin film bulk acoustic resonator. Analytical method based on Mason model was adopted to investigate the effect of electrode material and thickness on the impedance spectrum of resonator. The accuracy of elastic constant C_{33}^{D} extraction for ignoring the electrode was evaluated. The results show that electrode has a great effect on the impedance of resonator, a correction to electrode effect should be made for accurate extraction of C_{33}^{D}.

I. INTRODUCTION

As a well-established piezoelectric material, aluminum nitride thin film (AlN) has been widely used for filters, oscillators, sensors and other fields due to its high acoustic velocity, high thermal conductivity, good thermal stability and compatibility with CMOS technology [1-7]. Moreover, it has been shown that the doping of scandium (Sc) into AlN to form $Al_{1-x}Sc_xN$ can resulting in 3 times improvement in the piezoelectric coefficient (d_{33}= 27.6pC/N for x = 0.43, and 6pC/N for pure AlN) and 2 times increase in the electromechanical coupling coefficient (k_t^2=15% for x=0.3, and 7% for pure AlN) [8]. Thus, scandium-doped aluminum nitride ($Sc_xAl_{1-x}N$) has attracted a lot of attention as a promising material for RF-MEMS because of its large piezoelectric constants compared with AlN [9]. For performance evaluation and design of $Sc_xAl_{1-x}N$ based device, the material properties of thin film such as elastic constants, piezoelectric coefficient and electromechanical coupling coefficient are essential. Ab initio calculations based on density functional theories (DFT) can be used to theoretically predict the complete sets of material properties with good precision [10]. While the actual properties of thin film are greatly dependent on the deposition processing. And in fact, there exists a lot of discrepancy on the material constant of $Sc_xAl_{1-x}N$ from the reported literatures. Hence, to achieve accurate value of material constant, it needs to experimentally characterize the thin film. One general way to determine bulk piezoelectric material is to produce

different shape resonators, and full material constants can be achieved by the analysis of impedance spectrum of resonators, where the electrode effect on the extraction of material constant usually could be ignored considering the thickness of electrode is much less than that of the piezoelectric bulk material layer [11]. However, in contrast to resonator based on bulk or thick film piezoelectric material, the electrode thickness in piezoelectric thin film-based MEMS resonator is comparable to that of the piezoelectric layer, and large errors can occur if their effect is ignored for determine electrical impedance spectrum. Hence, in this paper we study the influence of electrode on C_{33}^{D} extraction of $Sc_xAl_{1-x}N$ by measuring the impedance spectrum of thickness-extensional mode thin film bulk acoustic resonator (FBAR). Mason model was used to calculate the parallel resonant frequency shift for different electrode by MATLAB programming. The error of C_{33}^{D} extraction for different electrode thickness was systematically evaluated. The results show that a correction to mass loading of electrode should be performed for accurate C_{33}^{D} extraction of $Sc_xAl_{1-x}N$.

II. METHOD

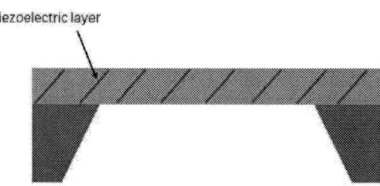

Fig. 1. Schematic view of FBAR with electrode ignored

For a thickness-extensional mode FBAR with electrode ignored (Fig. 1), the electric impedance can be calculated by:

$$z = \frac{1}{j\omega C_0}\left(1 - k_t^2 \frac{tan(kh)}{kh}\right) \qquad (1)$$

---

* This work was supported in part by the National Natural Science Foundation of China (Grant No. 51775465) and in part by Basic and Applied Basic Research Foundation (Grant No.2020A1515011486) and in part by the Knowledge Innovation Program of Shenzhen City (Fundamental Research, Free Exploration) under Grant JCY 20190809162001746.

1. Department of Mechanical and Electrical Engineering, Xiamen University, Xiamen 361005, China

2. Shenzhen Research Institute of Xiamen University, Shenzhen 518000, China

*Contacting Author: Lifeng, Qin (Email: liq@xmu.edu.cn)

978-1-6654-3008-1/21 $31.00 © 2021 IEEE

where $C_0 = \varepsilon^S S/d$ is the clamped capacitor of the resonator with area S, $\omega = 2\pi f$ is the angular frequency, $k = \omega/v$, v is the longitudinal acoustic wave velocity in piezoelectric layer along the direction normal to the resonator surface, $h = e/\varepsilon^S$ where e is piezoelectric coefficient and ε^S is the permittivity of piezoelectric materials, $\varepsilon^S = \varepsilon_r \varepsilon_0$, ε_r is the relative permittivity and ε_0 is the permittivity of vacuum.

By measuring the parallel resonant frequency fp (the frequency corresponding to the maximum impedance) of thickness-extensional mode FBAR (Fig. 1), the elastic constant $C_{33}{}^D$ can be determined by

$$f_p = \frac{1}{2t}\sqrt{\frac{c_{33}^D}{\rho}} \tag{2}$$

where t is the thickness of piezoelectrical layer and ρ is density of piezoelectrical layer.

However, the electrode effect may be substantial when thickness of the piezoelectric layer is small or the electrode thickness is increased enough, the parallel resonant frequency measured by experiment will shift due to the influence of electrode layer, thus the extraction error of elastic constant $C_{33}{}^D$ is caused.

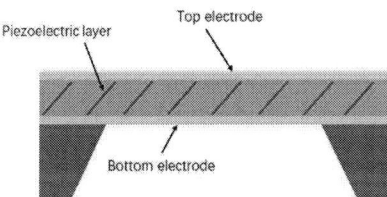

Fig. 2. Schematic view of FBAR with upper and lower electrodes

According to the Mason model, the electric input impedance (z_{in}) of a three-layer thickness-extensional mode FBAR (Fig. 2) could be achieved as

$$z_{in} = \frac{1}{j\omega C_0}\left[1 - \frac{k_t^2}{\gamma}\frac{(z_1 + z_2)\sin\gamma + j2(1-\cos\gamma)}{(z_1 + z_2)\cos\gamma + j(1 + z_1 z_2 \cos\gamma)}\right] \tag{4}$$

where $z_1 = Z_1/Z_0$ and $z_2 = Z_2/Z_0$ are the normalized acoustic impedance of the upper electrode layer and the lower electrode, $\gamma = vd/\omega$ is the phase delay of the acoustic wave in the piezoelectric film; d is the thickness of the piezoelectric layer, $Z_0 = S\rho v$ is the acoustic impedance of the piezoelectric layer, Z_1 is the acoustic load of top electrode layer, and Z_2 is the acoustic load of the bottom electrode layer and liquid loading.

Based on equation (4), the error caused by the electrode layer when extracting the material parameters can be obtained by simulation. The error of $C_{33}{}^D$ is calculated by

$$Err_{C_{33}^D} = \left|\frac{C_{33}^{D^0} - C_{33}^{D'}}{C_{33}^{D^0}}\right| * 100\% \tag{5}$$

where $C_{33}^{D^0}$ is the elastic constant when the thickness of the electrode layer is zero, $C_{33}^{D'}$ is calculated by equation (2) when the thickness of electrode layer is not zero.

In the simulation, Sc_0Al_1N, $Sc_{0.14}Al_{0.86}N$ and $Sc_{0.26}Al_{0.74}N$ thin film are chosen as the piezoelectric layer with thickness $1\mu m$ or $2\ \mu m$. Parameters of the electrode materials used in simulation are shown in Table I based on reference [12], and material properties of Sc_0Al_1N, $Sc_{0.14}Al_{0.86}N$ and $Sc_{0.26}Al_{0.77}N$ thin film in Table II was specifically assumed for simulation, which is based on reference [13].

TABLE I. Parameters of electrode material used in the simulation

Material	Density (kg/m³)	Acoustic Velocity (m/s)	Mechanical quality factor	Area (μm²)
Au	19320	1963	480	300 × 300
Al	2700	6418	2000	300 × 300

TABLE II. Parameters of piezoelectrical material used in the simulation

Material	Density (kg/m³)	$C_{33}{}^D$ (GPa)	k_t^2 (%)	ε_r	e_{33} (C/m²)
Sc_0Al_1N	3512	379.9	6	9.37	1.46
$Sc_{0.14}Al_{0.86}N$	3530	340	9	10.68	1.81
$Sc_{0.26}Al_{0.74}N$	3560	300	13.9	13.06	2.333

III. RESULT AND DISCUSSION

Based on equation (4), the effect of electrode on parallel resonant frequency can be studied. The simulation of impedance spectrum of a three-layer thickness-extensional mode FBAR with different electrode and different piezoelectric layer is shown Fig. 3 and Fig. 4. From Fig. 3 and Fig. 4 it can be clearly seen that for $Sc_xAl_{1-x}N$ with different doping rate and thickness, the resonant peaks all continue to shift from right to left with the increasing of electrode layer thickness, which means parallel resonant frequency decreases with the thickness of electrode layer increases. Comparing Fig. 3(a) with Fig. 4(a), it can be found that for a thicker piezoelectric layer, the influence of the thickness of the electrode layer on the parallel resonant frequency shift will be reduced. Fig. 3(a) and Fig. 3(b) show the effect of electrode materials on parallel resonant frequency, and it can be seen that the electrode material with higher density has greater influence on the parallel resonant frequency.

978-1-6654-3008-1/21 $31.00 © 2021 IEEE

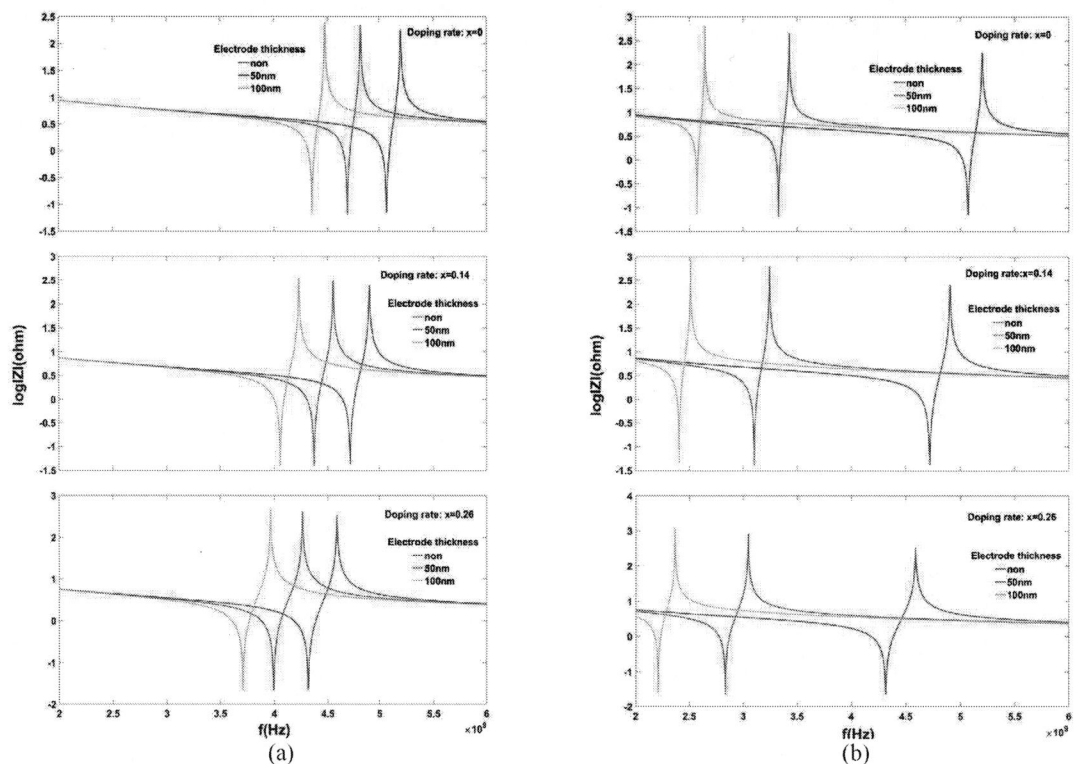

Fig. 3. Simulation of impedance spectrum of $1\mu m$ $Sc_xAl_{1-x}N$ with different electrode thickness (0nm, 50nm, 100nm) and material: (a)Al (b)Au

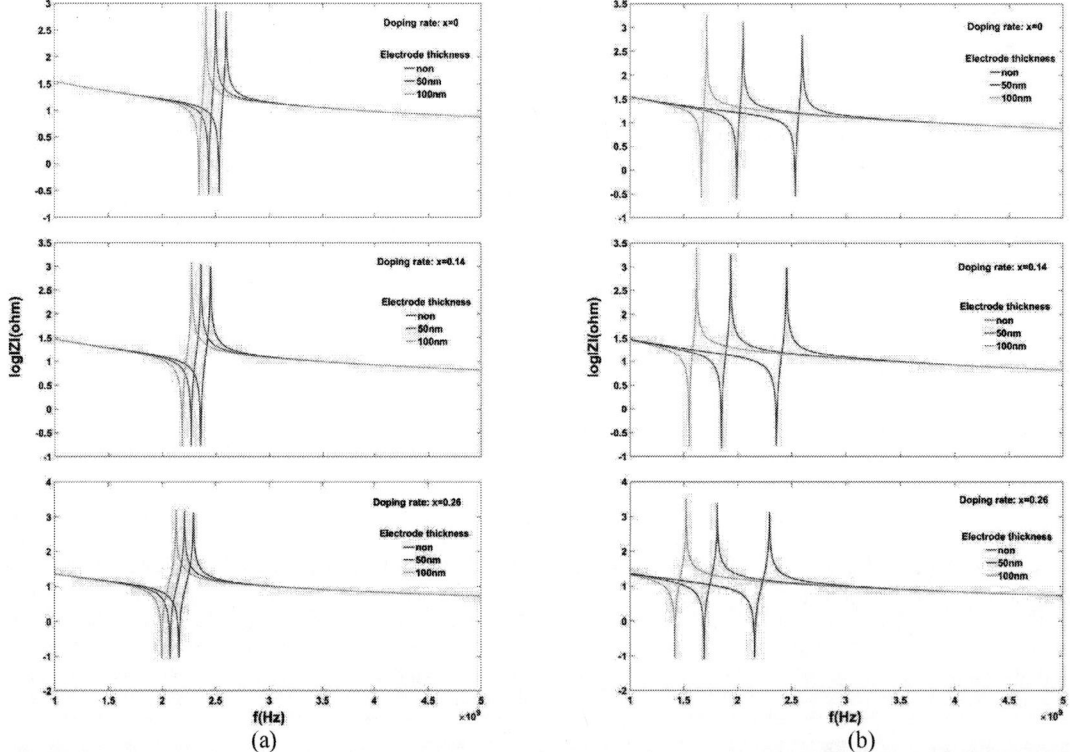

Fig. 4. Simulation of impedance spectrum of $2\mu m$ $Sc_xAl_{1-x}N$ with electrode thickness (0nm, 50nm, 100nm) and material: (a)Al (b)Au

978-1-6654-3008-1/21 $31.00 © 2021 IEEE

In order to have an overall understanding of electrode effect on the extraction of C_{33}^D, we gradually increase the thickness of electrode layer from 0nm to 300nm. Fig. 5 and Fig. 6 show the $Err_{C_{33}^D}$ change with thickness of electrode layer for FBAR with different Sc doping and thickness ($1\mu m$, $2\mu m$). As shown in Fig. 5 and Fig. 6, if the thickness of the electrode increases, $Err_{C_{33}^D}$ will continue to increase. From Fig. 5(a) and Fig. 6(a), it can be seen that the $Err_{C_{33}^D}$ of a $1\mu m$ $Sc_{0.14}Al_{0.86}N$ reaches 25.7% and a $2\mu m$ $Sc_{0.14}Al_{0.86}N$ reaches 13.9% when the thickness of aluminum electrode is 100nm, which means that the smaller the thickness of the piezoelectric layer is, the higher $Error_{C_{33}^D}$ is. Comparing Fig. 5(a) with Fig. 5(b), it can be found that the $Err_{C_{33}^D}$ of a $1\mu m$ $Sc_{0.14}Al_{0.86}N$ with aluminum electrode reaches 25.7% and reaches 76.4% with gold electrode when the thickness of electrode is 100nm, which indicates that the electrode material with higher density has greater influence on the resonant frequency, thus causing greater $Err_{C_{33}^D}$ when the thickness of piezoelectric layer is the same. All in all, the electrode layer will cause the resonance frequency shift from right to left, which leads to $Err_{C_{33}^D}$ when using equation (2).

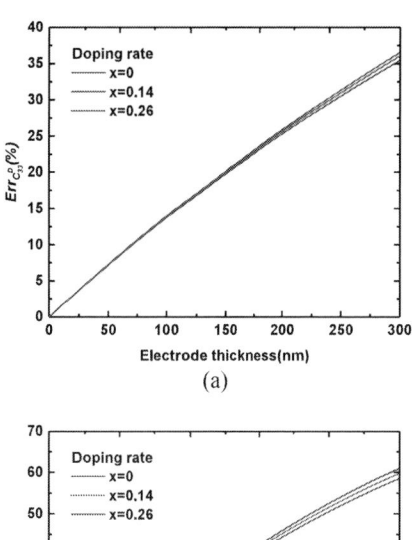

(a)

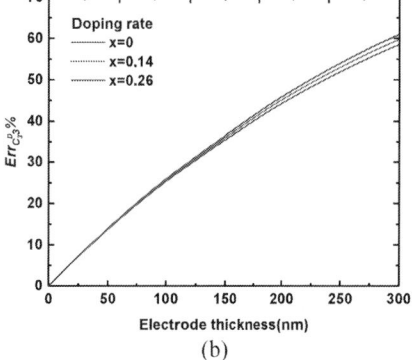

(b)

Fig. 6. $Err_{C_{33}^D}$ of $2\mu m$ $Sc_xAl_{01-x}N$ changes with electrode thickness (0~300nm) and material: (a)Al (b)Au

In fact, according to sauerbrey equation, the effect of electrode layer can be regarded as mass loading effect and Δf_p and Δm is approximately equal when the thickness of the electrode layer is very small, Δf_p and Δm are defined as

$$\Delta f_p = \frac{f_p^0 - f_p'}{f_p'} \tag{6}$$

$$\Delta m = \frac{\rho_e h_e}{\rho h} \tag{7}$$

where f_p^0 is parallel resonant frequency without mass loading effect, f_p' is parallel resonant frequency with mass loading effect. ρ_e, ρ, h_e and h represent electrode layer density, piezoelectric layer density, electrode layer thickness and piezoelectric layer thickness. Based on equation (6) and (7), the correction formula of parallel resonant frequency can be put forward as equation (8)

$$f_P^0 = \frac{f_P'}{1 - \Delta m} \tag{8}$$

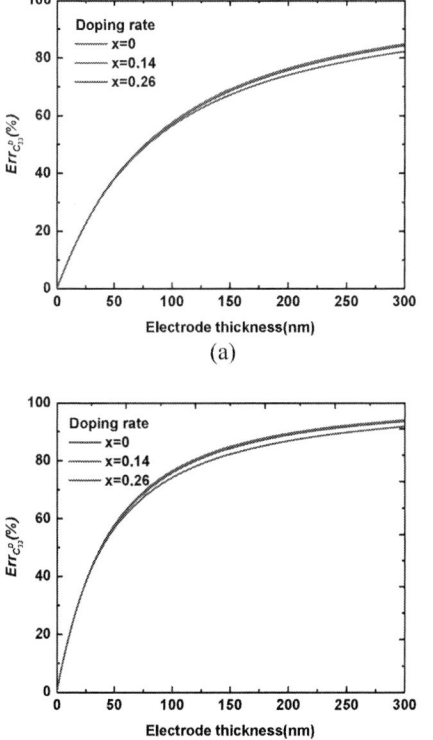

(a)

(b)

Fig. 5. $Err_{C_{33}^D}$ of $1\mu m$ $Sc_xAl_{01-x}N$ change with electrode thickness (0~300nm) and material: (a)Al (b)Au

equation (8) can be used to calculate the resonant frequency without mass loading effect, thus accurate C_{33}^D can be calculated.

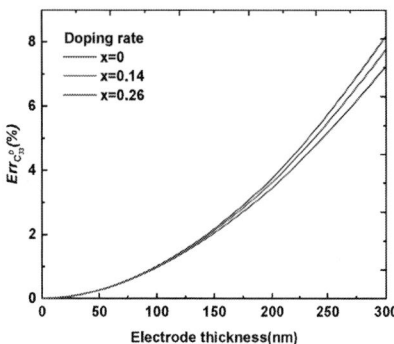

Fig. 7. $Err_{C_{33}^D}$ of 2μm $Sc_xAl_{1-x}N$ with different doping rate change with Al electrode thickness.

Fig. 7 shows the $Err_{C_{33}^D}$ calculated by the resonant frequency corrected by the equation (8). Compared with Fig.6(a), it can be found that the $Err_{C_{33}^D}$ significantly reduced, the $Err_{C_{33}^D}$ reduces from 7.3% to 0.27% when the thickness of electrode layer is 50nm. When the mass loading is large, such as the electrode thickness is greater than 250nm in Fig. 7, the $Err_{C_{33}^D}$ will be more than 5%. For large mass loading, for precision of C_{33}^D finite element analysis or Data fitting based on Mason model need to be done.

IV. CONCLUSION

In summary, the effects of electrode on the extraction error of elastic constant C_{33}^D based on was systematically evaluated by simulation. The Mason model of a three-layer thickness-extensional mode FBAR was used to calculate the offset of parallel resonant frequency due to electrode, and the $Err_{C_{33}^D}$ was investigated based on impedance spectrum. The result shows that the $Err_{C_{33}^D}$ will increase with the mass loading, and the correction formula of parallel resonant frequency based on sauerbrey equation is proposed, it shows that the $Err_{C_{33}^D}$ significantly reduced for small mass loading after the parallel resonant frequency is corrected for small mass loading. As for large mass loading, for precision of C_{33}^D finite element analysis or Data fitting by Mason model need to be done.

REFERENCES

[1] A. Ding, L. Kirste, Y. Lu, R. Driad, N. Kurz, V. Lebedev, T. Christoph, N. M. Feil, R. Lozar, T. Metzger, O. Ambacher, and A. Zukauskait, "Enhanced electromechanical coupling in SAW resonators based on sputtered nonpolar Al0.77Sc0.23N thin films," J. Appl. Phys., vol. 116, no, 101903, Mar. 2020.

[2] C. C. W. Ruppel, "Acoustic Wave Filter Technology–A Review, " IEEE Transactions on Ultrasonics, Ferroelectrics, and Frequency Control, vol. 64, no. 9, pp. 1390-1400, 2017.

[3] Y. Liu, Y. Cai, Y. Zhang, A. Tovstopyat, S. Liu, and C. Sun, "Materials, Design, and Characteristics of Bulk Acoustic Wave Resonator: A Review," Micromachines, vol. 11, no. 7, p. 630, Jun. 2020.

[4] G. Chen and M. Rinaldi, "Aluminum Nitride Combined Overtone Resonators for the 5G High Frequency Bands," Microelectromechanical Systems, vol. 29, no. 2, April. 2020.

[5] F. Bartoli et al., "Theoretical and experimental study of ScAlN/Sapphire structure based SAW sensor," 2017 IEEE SENSORS, Glasgow, UK, pp. 1-3 2017.

[6] B. Jo, M. Ghatge and R. Tabrizian, "d15-Enhanced shear-extensional aluminum nitride resonators with k2t > 4.4% for wide-band filters," 2017 19th International Conference on Solid-State Sensors, Actuators and Microsystems (TRANSDUCERS), Kaohsiung, pp. 94-97 ,2017

[7] D. K. T. Ng, T. Zhang, L. Y. Siow, L. Xu, C. P. Ho, H. Cai, L. Y. T. Lee, Q. Zhang, and N. Singh, "A functional CMOS compatible MEMS pyroelectric detector using 12%-doped scandium aluminum nitride," Appl. Phys. Lett. 117, 183506 .2020.

[8] Y. Lu et al., "Elastic modulus and coefficient of thermal expansion of piezoelectric Al1−xScxN (up to x = 0.41) thin films," APL Mater. 6, 076105, 2018.

[9] H. Ichihashi, T. Yanagitani, M. Suzuki, S. Takayanagi and M. Matsukawa, "Effect of Sc concentration on shear wave velocities in ScAlN films measured by micro-Brillouin scattering technique," 2014 IEEE International Ultrasonics Symposium, Chicago, IL, USA, pp. 2521-2524 2014

[10] L. N. McCartney, L. Wright, M. G. Cain, J. Crain, G. J. Martyna, and D. M. Newns, "Methods for determining piezoelectric properties of thin epitaxial films: Theoretical foundations," J. Appl. Phys., vol. 116, no. 1, art. no. 014104, Jul. 2014.

[11] L. Qin, Y. Sun, Q. M. Wang, Y. Zhong, M. Ou, Z. Jiang, and W. Tian, "Fabrication and characterization of thick-film piezoelectric lead zirconate titanate ceramic resonators by tape-casting," IEEE Trans. Ultrason. Ferroelectr. Freq. Control, Vol. 59, pp. 2803-2812, 2012

[12] Q. Chen and Q.M. Wang, "The effective electromechanical coupling coefficient of piezoelectric thin film resonators," Appl. Phys. Lett. vol. 86, 022904, 2005.

[13] M.A. Caro et al., "Erratum: Piezoelectric coefficients and spontaneous polarization of ScAlN," J. Phys. Matter vol.27 245901, 2015

Gap in pagination due to unavailable paper.

Pages 1723-1724

A Resonant High Pressure Sensor Based on Dual Cavities Design

Jie Yu, Yulan Lu, Deyong Chen*, *Member, IEEE*, Junbo Wang, *Member, IEEE*, and Bo Xie

Abstract— **This paper presents a new type of resonant high-pressure sensor with measuring range of 100 MPa, which composed of SOI layer and a glass cap. Frequency shifts of resonators caused by bending pressure sensitive diaphragm under high pressure. The sensor structure was optimized by using finite element analysis to meet the applications in the fields of the ocean science and the petrochemical industry, especially in the packaging strength and sensitivities. In addition, the sensor chip was fabricated by simplified SOI-MEMS fabrication technology, including only two photolithography steps. The experimental results show that 1) the sensor structure can withstand a pressure of 100 MPa; 2) the pressure sensitivity of the dual resonators is -0.0677 kHz/MPa (resonator I) and 0.0649 kHz/MPa (resonator II) respectively, which meets the requirements of the ocean science and the petrochemical industry.**

I. Introduction

With the development of the ocean science and the petrochemical industry, the urgent demands for high-precision and high-pressure sensors are increasing, especially in ocean depth measurement, surge measurement, and accurate measurement of fluid pressure [1-3]. Based on a variety of sensing principles [4-10], silicon-based high-pressure sensors have been reported. Compared to other types of high-pressure sensors, resonant sensors behave better performances on the accuracies, stabilities, and reliabilities, and thus resonant pressure sensor of the higher range is under the intensive studies [11-13]. For the resonant sensor, when the sensor is placed in high pressure environment, the strain is biggest, which may damage the sensor structure and exceed the linearity limit. Thus, the main problems of resonant high-pressure sensors are to find a vacuum packaging method which can withstand high pressure, and how to design and optimize the structure to make it have appropriate sensitivity and obtain good linearity in the pressure measurement range.

In order to overcome these defects, a resonant high-pressure sensor with measuring range of 100 MPa with plate electrostatic excitation and piezoresistive detection was proposed. The sensor adopts dual cavities design to minimize the strain during the measurement process of high pressure, so that this design can be used to expand the scope of the pressure measurement. Besides, the vacuum cavities for vibrations of double resonators were formed by using the anodic bonding process.

II. Design and Simulation

A. Design

Figure 1 shows the schematic of the high-pressure sensor, which is mainly composed of an SOI wafer and a glass cap. The SOI wafer (including a device layer, an oxide layer and a handle layer) is used to form sensing elements. The glass cap is used to form the vibrating space of the dual resonators and vacuum cavities. Meanwhile, Ti getters are used to maintain the vacuum environment. The dual cavities are used to form vibration cavities for dual resonators, which is deployed on the glass cap. The dual small cavities can minimize the stresses induced by high pressure applied, such as 100 MPa. The silicon vias on the handle layer (as shown in the backside view) are used to draw out the electric signals of the dual resonators.

The dual resonators on device layer are driven by electrostatic force and detected through piezoresistors changes (as shown in the resonator view and resonator details). The dual resonators are used to form a differential structure, which has the same dimensions and are placed in the different positions of the dual small cavities through anchor, respectively. When high pressure applied on the sensor, compressive stresses and tensile stresses lead the intrinsic resonant frequency of resonator II to increase and the frequency of resonator I to decrease, respectively. Thus, the differential output results in a doubling of the sensor sensitivity when the pressure sensitivity of the dual resonators are same. Besides, differential outputs could improve the linear of each resonator. Furthermore, with the change of surroundings temperature, the effect of temperature on the resonant frequencies of dual resonators is consistent so that differential outputs can realize the temperature compensation of the sensor and get a better performance.

B. Finite Element Analysis

Static structural stress and sensitivities match of dual resonators is the key factors of the design and optimization of resonant high-pressure sensors [14]. In this paper, the maximum equivalent stress of the sensor was calculated by using the finite element analysis (FEA). The sensor was optimized to enhance the strength of the sensor by using static structural stress when high pressure applied, such as 100MPa. Besides, the position of the dual resonators was optimized by modal simulations, so that the dual pressure sensitivities are matched well.

Jie Yu, Yulan Lu, Deyong Chen, and Junbo Wang, the State Key Laboratory of Transducer Technology, Aerospace Information Research Institute, Chinese Academy of Sciences, Beijing 100190, China, and University of Chinese Academy of Sciences, Beijing 100049, China (e-mail: yujie18@mails.ucas.ac.cn, corresponding author phone: 010-58887182; e-mail: dychen@mail.ie.ac.cn).

Bo Xie, the State Key Laboratory of Transducer Technology, Aerospace Information Research Institute, Chinese Academy of Sciences, Beijing 100190, China.

This work was financially supported by the National Key Research and Development Program (2018YFB2002302), the National Natural Science Foundation of China (Grant No. U1930206), and the National Science Fund for Distinguished Young Scholars (Grant No. 61825107).

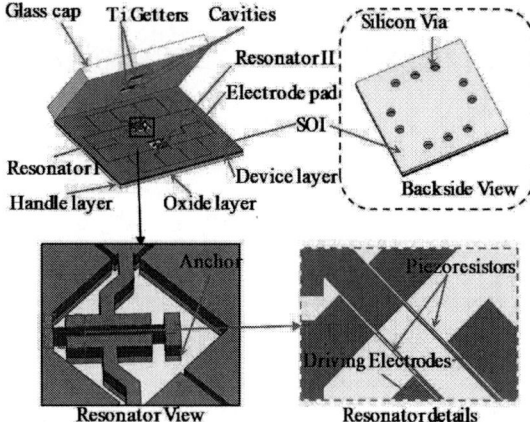

Figure 1. Schematic of the high-pressure sensor. The high-pressure sensor is composed of an SOI wafer and a glass cap. Two cavities with sides of 1 mm× 1 mm, which are used to form vibration cavities for two resonators, are deployed on the glass cap. Silicon vias are used to form electrode connections of inner signals. The sensing element consists of a pressure diaphragm and two resonators, which are clamped on the diaphragm based on electrostatic excitation and piezoresistive detection.

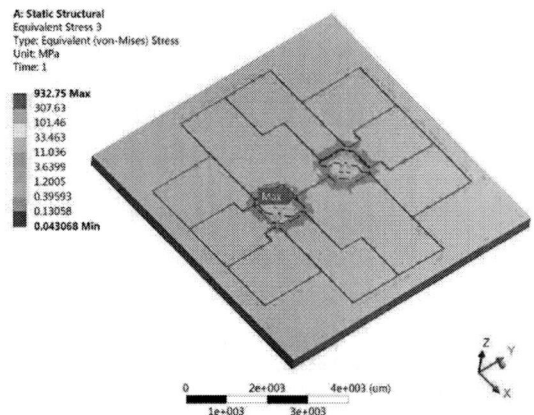

Figure 2. Stress analysis of the sensor chip. The maximum equivalent stress distributed in the high-pressure sensor chip is 933 MPa with the dual cavity sides of 1 mm × 1 mm, which is within the yield strength of silicon of 7 GPa.

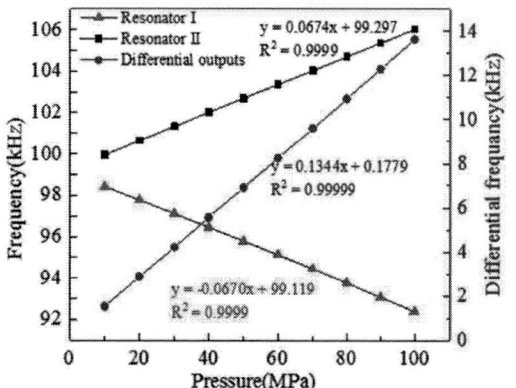

Figure 3. The designed sensitivities of dual resonators are -0.0670 kHz/MPa (resonator I) and 0.0674 kHz/MPa (resonator II), which match the applications of a high-pressure sensor to keep a linear sensitive characteristic for resonant sensor.

In this paper, the tetrahedron was selected as the model mesh method, the sensor was meshed by using an element size of 200 μm and the sixteen sides of dual resonators was meshed by using an element size of 4 μm. The boundary condition is that apply distributed pressure to all external surfaces of the sensor chip. A series of pressures with values from 10 MPa to 100 MPa were chosen to apply on the external surfaces of the sensor chip. The static stress of the sensor structure obtained through finite element analysis is shown in Figure 2. By optimizing the size of the dual cavities of the sensor chip, the maximum equivalent stress was reduced when high pressure applied, such as 100 MPa. And the maximum equivalent stress is 933 MPa with the dual cavity sides of 1 mm × 1 mm, which is within the yield strength of silicon of 7 GPa [15].

Based on the results of the static stress simulations, the positions of the dual resonators were adjusted so that the designed sensitivities of the dual resonators can be matched well. The simulation results of designed sensitivities are shown in the Figure 3. The designed sensitivities of dual resonators are -0.0670 kHz/MPa (resonator I) and 0.0674 kHz/MPa (resonator II). The differential outputs of dual resonators are 0.1344 kHz/MPa within the full range of 100 MPa, which the linear correlation is 0.99999. These simulation results show that the range of the sensor can be extended and the sensitivity of the sensor can be doubled with the dual cavities design, which meets the applications of a high-pressure sensor to keep a linear sensitive characteristic for resonant sensor.

III. FABRICATION

The fabrication processes of the resonant high-pressure sensor based on SOI-MEMS technology is shown in Figure 4. There are mainly two parts in the whole fabrication process, which are an SOI wafer and a glass wafer processing.

The total thickness of the sensor chip is 1342μm, which is composed of SOI and glass. The thickness of SOI wafer is 342 μm and the thickness of glass wafer is 1mm. For the SOI wafer, the first step of high-pressure sensor fabrication was clean the SOI wafer to remove organic matters and particulate matters on both surfaces. Next, AZ4620 photoresist was used as a barrier layer on the handle layer, when the handle layer was etched to form the silicon vias, as shown in Figure 4 (a). Then, the dual resonators of the sensor were fabricated by pattering and DRIE technology, shown in Figure 4 (b). Afterwards, the oxide layer under the resonators and in silicon vias was removed by using the gaseous HF and isopropanol as shown in Figure 4 (c). For the BF33 glass wafer, the dual cavities of BF33 wafer were etched through the gaseous HF. Then, Ti getters were deposited in the dual cavities by sputtering, which maintained a high vacuum situation, shown in Figure 4 (d). In order to form the vibration cavities for the dual resonators, the dual pattern wafers were bonded by anodic bonding in this paper, shown in Figure 4 (e). In the end, the Al electrodes were deposited in silicon vias by evaporation technology, shown in Figure 4 (f).

The fabricated sensors are shown in Figure 5. The picture of anodic bonding of dual pattern wafers at the wafer-level is shown in Figure 5 (a). Figure 5 (b) shows the

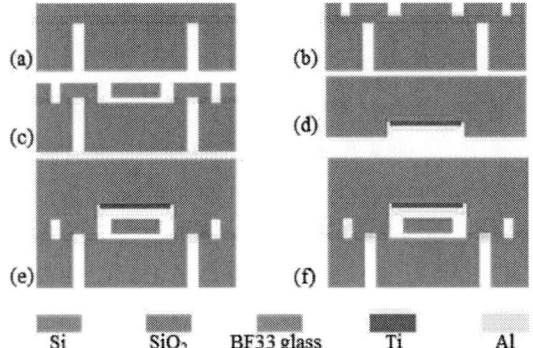

Figure 4. Fabrication processes of the resonant high-pressure sensor: (a) DRIE to form silicon vias; (b) Pattering and DRIE for resonators; (c) Removing the silicon dioxide by HF buffer solution in the silicon vias and beneath the resonators; (d) Etching glass to form cavities and sputtering Ti metal to form getters; (e)Anodic bonding to form vacuum packaging for resonators; (f) Metallization of electrodes.

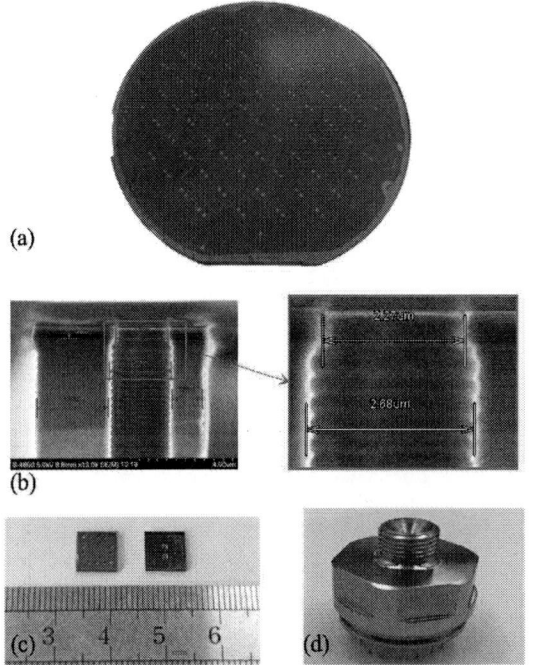

Figure 5. Fabrication results: (a) Wafer after anodic bonding; (b) Undercuts of the DRIE, which are with 200 nm; (c) Sensor chips after dicing; and (d) Sensor after assembly.

DRIE results, which undercuts of the DRIE are with 200 nm. Figure 5 (c) shows that the both views of the fabricated high-pressure sensor units after diced, and the sides of the sensor are 8 mm× 8 mm. The picture of high-pressure sensor after assembly is shown in Figure 5 (d). The fabricated sensor chip was packaged in the Kovar base and stainless cap. Then, the fabricated electrode pads on the sensor chip were attached to the Kovar pins, which forms wire connections.

IV. EXPERIMENTAL CHARACTERIZATIONS

Figure 6 shows the tested sensitivity of the fabricated high-pressure sensor. The sensitivities of the dual resonators are -0.0677 kHz/MPa (resonator I) and 0.0649 kHz/MPa (resonator II) within pressure range of 0.11~ 2.5 MPa under room temperature, which matches the FEA simulations.

Furthermore, the value of the sensitivity of the fabricated sensor is doubled to 0.1326 kHz/MPa within pressure range of 0.11~ 2.5 MPa under room temperature, which the high linear correlation coefficient is 0.9992 by using the different frequency output.

V. CONCLUSION

A new type of resonant high-pressure sensor with measurement range of 100 MPa is proposed in this paper. The sensor minimized the strain during the measurement process of high pressure based on dual cavities design by using finite element analysis. The presented sensors are fabricated by SOI-MEMS fabrication technology. Compared to the previous resonant high-pressure sensors, the sensitivity is doubled to 0.1326 kHz/MPa with the high linear correlation coefficient is 0.9992 by using the different frequency outputs. The experimental results confirmed the design of the proposed high-pressure sensor, which meets the applications in the fields of the ocean science and the petrochemical industry.

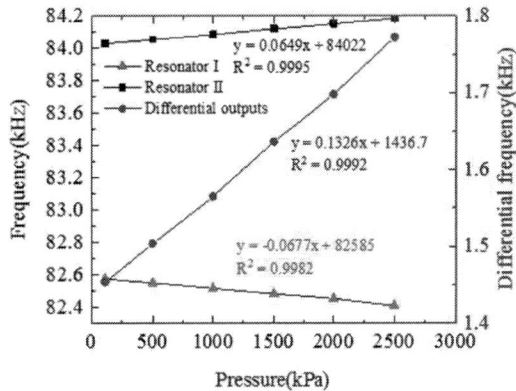

Figure 6. The tested pressure sensitivities of one sensor, which are -0.0677 kHz/MPa (resonator I) and 0.0649 kHz/MPa (resonator II) within pressure range of 0.11~ 2.5 MPa under room temperature, which matches the FEA simulations.

ACKNOWLEDGMENT

Thank my four teachers for their guidance. Thank you for my elder brother Yulan Lu for his help in this work. Thank my team members for their help in this work. Besides, this work was financially supported by the National Key R&D Program of China (2018YFB2002302), the National Natural Science Foundation of China (Grant No. U1930206), and the National Science Fund for Distinguished Young Scholars (Grant No. 61825107).

REFERENCES

[1] Z. Niu, Y. Zhao, and B. Tian, "Design optimization of high pressure and high temperature piezoresistive pressure sensor for high sensitivity," *Review of Scientific Instruments*, vol. 85, pp. 2107-2109, Jan. 2014.

[2] H. Terabe, H. Arashima, N. Ura, and K. Suzuki, "A silicon pressure sensor with stainless diaphragm for high temperature and chemical application," in *Proc. IEEE. Int Conf. Solid-State Sensors and Actuators*, Chicago, 1997, pp. 1481-1484.

[3] X. Guo and R. Hebibul, "A novel piezoresistive sensitive structure for micromachined high-pressure sensors," in *Proc. 12th Intl Conf.*

Nano/Micro Engineered and Molecular, Los Angeles, 2017, pp. 728-731.

[4] L.B. Zhao, X D. Fang, and Y. L. Zhao, "A high pressure sensor with circular diaphragm based on MEMS technology," *Key Engineering Materials*, vol. 483, pp. 206-211, June. 2011.

[5] J.K. Reynolds, D. Catling, and R.C. Blue, "Packaging a piezoresistive pressure sensor to measure low absolute pressures over a wide sub-zero temperature range," *Sensors and Actuators A: Physical*, vol. 83, pp. 142-149, May. 2000.

[6] L. Chen, and M. Mehregany, "A silicon carbide capacitive pressure sensor for high temperature and harsh environment applications," in *Proc. 2007 IEEE Int Solid-State Sensors, Actuators and Microsystems Conference,* Lyon, 2007, pp. 2597-2600.

[7] W.J. Bock, T. Eftimov, and G.F. Molinar, "Free active element bulk-modulus high-pressure transducer based on fiber-optic displacement sensor," *IEEE Trans. Instrumentation and Measurement*, vol. 47, pp. 179-182, Feb. 1998.

[8] M. Akiyama, Y. Morofuji, and T. Kamohara, "Flexible piezoelectric pressure sensors using oriented aluminum nitride thin films prepared on polyethylene terephthalate films," *J. applied physics*, vol. 100, pp. 114318-114318-5, Dec. 2006.

[9] J.C. Greenwood and T. Wray, "High accuracy pressure measurement with a silicon resonant sensor," *Sensors and Actuators A: Physical*, vol. 37, pp. 82-85, June. 1993.

[10] P.K. Kinnell and R. Craddock, "Advances in silicon resonant pressure transducers," *Procedia Chemistry*, vol. 1, pp. 104-107, Sep. 2009.

[11] X. Du, L. Wang, A. Li, L. Wang and D. Sun, "High accuracy resonant pressure sensor with balanced-mass DETF resonator and twinborn diaphragms," *J. Microelectromechanical Systems.* vol. 26, pp. 235-245, Feb. 2017.

[12] Y. Jiang, "Design and simulation of fully-symmetrical resonant pressure sensor," *7th IEEE Int Conf. Nano/Micro Engineered and Molecular Systems (NEMS)*, Kyoto, 2012, pp. 702-707.

[13] C. Chiang, "Resonant pressure sensor with on-chip temperature and strain sensors for error correction," in *Proc. 26th Int Conf. Micro Electro Mechanical Systems (MEMS)*, Taiwan, 2013, pp. 45-48.

[14] H. Jiao, B. Xie, and J.B. Wang, "Electrostatically driven and capacitively detected differential lateral resonant pressure microsensor," *Micro and Nano Letters*, vol. 8, pp. 650-653, Sep. 2013.

[15] Y. Lu, P. Yan, and X. Chao, "A resonant pressure microsensor with the measuremet range of 1 MPa based on sensitivities balanced dual resonators," *Sensors*, vol. 19, pp. 2272-2280. May. 2019.

Gap in pagination due to unavailable paper.

Pages 1729-1731

Proceedings of the 16th Annual IEEE International
Conference on Nano/Micro Engineered and Molecular Systems
April 25-29, 2021

Numerical analysis of laser ablated structural size effect on enhanced anti-icing property of TC4 surface*

Zhekun Chen, Rui Zhou†, Yuhang Lin, Yi Zhu and Huangping Yan

Abstract— Superhydrophobic surface with micro/nano structures has attracted extensive attention because of its excellent performance in passive anti-icing strategy. In this study, a 2D Computational Fluid Dynamics model was proposed to simulate the icing process of droplet on surface at low temperature by coupling the Volume of Fluid model with solidification/melting model. Then, the icing process of droplet on surfaces with different micro structures were simulated and compared. Furthermore, the samples with different micro-nano structures were obtained by laser ablation and their anti-icing performance were measured. Results showed that the icing time of bouncing droplet on the surface with micro/nano structures is significantly delayed by seven times, where the nucleation rate is related to the contact area of solid-liquid phases. It is also found that the larger size of the air cavity brought by the structure is more conducive to obtain enhanced anti-icing performance. Furthermore, the anti-icing mechanism of superhydrophobic surface was revealed by investigating variations in solidification/melting contour and solid volume fraction.

I. INTRODUCTION

The icing phenomenon of cold metal surface is widely involved in important industries and daily life such as aerospace, transportation, energy and so on. The common icing phenomenon will not only affect the social production status and efficiency, but even threaten people's life and property safety in serious cases[1].In view of the serious hazards caused by icing on metal, researchers have proposed and developed a variety of anti-icing/deicing systems and related technologies. However, the existing active/passive anti-icing systems have problems such as high energy consumption, low efficiency, environmental pollution and limited by time and space. In recent years, superhydrophobic surface with micro/nano structures has attracted extensive attention because of its excellent performance in passive anti-icing strategy[2, 3]. Up to now, many experimental demonstrations have supported the anti-icing characteristics of superhydrophobic surface. However, due to droplet icing on structural surface is a complex heat and mass transfer phenomenon, the dynamic mechanism of anti-icing performance is still not fully understood. Aiming at the status of research in related fields, the main purpose of this study was to analyze the dynamic anti-icing mechanism of

different textured surfaces and experimentally verify it. Firstly, the proposed 2-D calculation model was used to simulate the icing process of a bouncing droplet on the cold surface. By comparing and analyzing the effect of the surface with different micro-structures on the droplet freezing process, revealing the anti-icing mechanism of the superhydrophobic surface[4] and studying the special phenomena of the icing process, such as nucleation process and solid-liquid interface, etc. Subsequently, characteristic micro-nano surface structures with anti-icing performance are obtained by laser ablation, which was environmentally friendly and high-efficient[5]. Finally, the dynamic anti-icing performance of such surfaces are measured by recording the impinging and icing process of the bouncing droplet.

II. EXPERIMENT SETUP

A. Simulation method

In this study, a 2D Computational Fluid Dynamics model was proposed to simulate the icing process of droplet on surface at low temperature by coupling the Volume of Fluid model[6] with solidification/melting model by employing Ansys Fluent software. As is shown in Fig. 1(a), The uniform grid of quadrilateral is used to divide the quadrilateral grid into a total of 16,000 to 16,371 which depend on the geometric morphology. In the simulation, the liquid phase is set as water by patch a circle with 1 mm diameter, and its specific physical parameters are: density ρ = 1,000 kg / m^3, viscosity μ = 1 × 10^{-3} Pa · s, surface tension σ = 7. 3 × 10^{-2} N / m, phase transition temperature = 0 ℃ and pure solvent melting heat = 3.34 × 10^5 J / kg. Set the acceleration of gravity g = 9.8 m^2 / s, and its direction is vertical down, the initial shape of the droplet spherical diameter is 0.1 mm. The solver used in the numerical model was pressure based implicit algorithm. The wall of the lower boundary of the simulation area is set as Dirichlet boundary condition with temperature = -50 ℃ in order to reduce the time required for the phase transition process to save computational cost. According to the actual situation, the upper surface of the solid is set as the superhydrophobic surface, the contact angle of the wall is set to be 150 ° and the rest of the boundaries are set as the pressure outlet

---

*Research supported by Fujian Provincial Science and Technology Programme (Industry Oriented Key Project, No. 2020H0006), Guangdong Basic and Applied Basic Research Foundation (2020A1515010519).

Zhekun Chen is with the Department of Mechanical & Electrical Engineering, School of Aerospace Engineering, Xiamen University, Fujian Province, China, 361102 (e-mail: chenzk@stu.xmu.edu.cn).

Rui Zhou is with the Department of Mechanical & Electrical Engineering, School of Aerospace Engineering, Xiamen University, Fujian Province, China, 361102 (†Corresponding author, phone:86-592-2186970; e-mail: rzhou2@xmu.edu.cn).

Yuhang Lin is with the Department of Mechanical & Electrical Engineering, School of Aerospace Engineering, Xiamen University, Fujian Province, China, 361102 (e-mail: lin_yuhang@qq.com).

Yi Zhu is with the Department of Mechanical & Electrical Engineering, School of Aerospace Engineering, Xiamen University, Fujian Province, China, 361102 (e-mail: zhuyi2y@stu.xmu.edu.cn).

Huangping Yan is with the School of Aerospace Engineering, Xiamen University, Fujian Province, China, 361102. She is also with Shenzhen Research Institute of Xiamen University, Shenzhen, China, 518000 (e-mail: hpyan@xmu.edu.cn)

978-1-6654-3008-1/21 $31.00 © 2021 IEEE

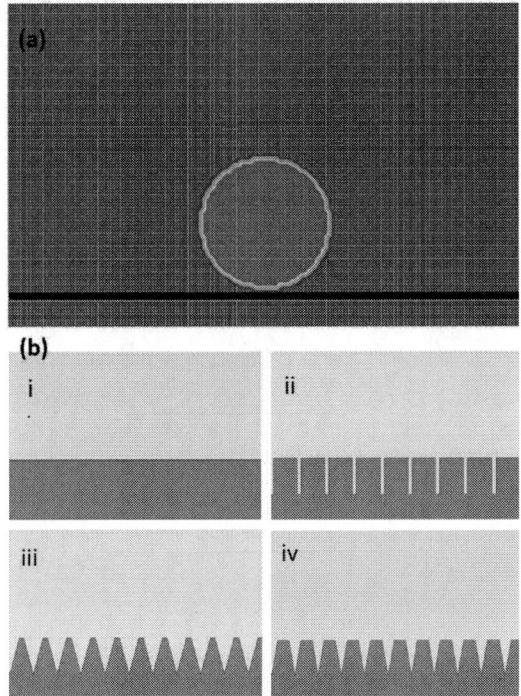

Figure 1. (a)Meshing and model setup of simulation by coupling VOF model with S&M model.(b) Surfaces with different micron-scale structures. (i) original sample with smooth surface; (ii) square column with 300 μm depth and 200 μm width; (iii) cone with 300 μm depth and 40 μm upper boundary width; (iv) frustum of a cone with 300 μm depth and 100 μm upper boundary width.

boundary and the pressure is zero. The size of time step is set to 5×10^{-5} s. Fig. 1(b) shows the geometric modeling of different micron-scale structures that will be calculated to study the differences in the freezing behavior of droplets on different textured surfaces. The specific geometrical parameters are described in the title of the Fig. 1.

B. Sample preparation and morphology characterization

On the basis of simulation, expected structural surface is obtained on TC4 by Nd:YAG pulsed laser ablation with a wavelength is 355 nm. Then the samples were modified with

FAS-17/ethanol (Qufu Jiaye Chemical New Material Co. Ltd, China) for 6 hours to realize the superhydrophobicity[7].

The morphology of the textured surfaces was observed by super depth of field microscope (VHX-5000, KEYENCE, Japan) and SEM (Zeiss SUPRA 55, Carl Zeiss AG, German).

C. Dynamic anti-icing performance test

The dynamic anti-icing performance of the above four samples is evaluated through a dynamic droplet icing test. Droplet of 7 μL is freely dropped at a height of 4.5 mm from the substrate surface. The cooling of the sample is realized by a semiconductor refrigerating board and a water-cooling device. The surface temperature is controlled by a temperature sensor and a computer, and the temperature was set to -15 °C in this experiment. The impact and icing process were recorded by using a 1080p at 240 fps slow motion camera (iPhone Xs, Apple, America).

III. RESULT AND DISCUSSION

The icing process of droplet on surfaces with different structural parameters was simulated and compared. The time step was set as 5×10^{-5} s to ensure the calculation accuracy, which bring a huge amount of computation to reproduce the complete icing process of the droplets on the cold surface. As shown in Equations (1-3), the icing process of droplets on the cold surfaces needed to release a certain amount of heat (Q), which mainly came from the decrease of the internal temperature of the droplets (Q_1) and the latent heat of solidification (Q_2). The former was related to the specific heat capacity of droplets (C_P), the temperature of droplets (T_2) and the temperature of the surfaces (T_1), while the latter was related to the specific latent heat (β) of droplets[8].

$$Q_1 + Q_2 = Q \qquad (1)$$

$$Q_1 = mC_P (T_1 + T_2) \qquad (2)$$

$$Q_2 = \beta m \qquad (3)$$

The problem of the large time scales of droplet freezing process could be overcome by modifying the surface temperature. In this case, the boundary temperature is set to -50 °C and the initial room temperature is set to 5 °C. Fig. 2(a) shows the temperature field distribution of a droplet after 10,000 time steps (corresponding to 0.5 s), it is obvious that the temperature inside the droplet is much higher than

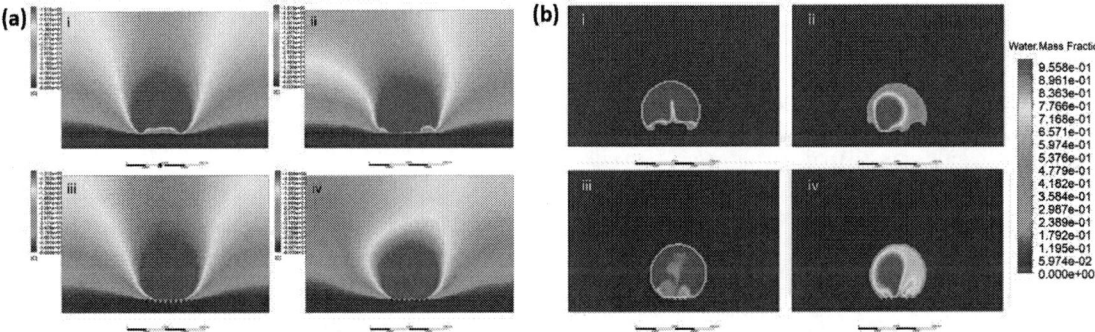

Figure 2. (a) Contour of temperature field. (b) Contour of mass fraction of liquid water.

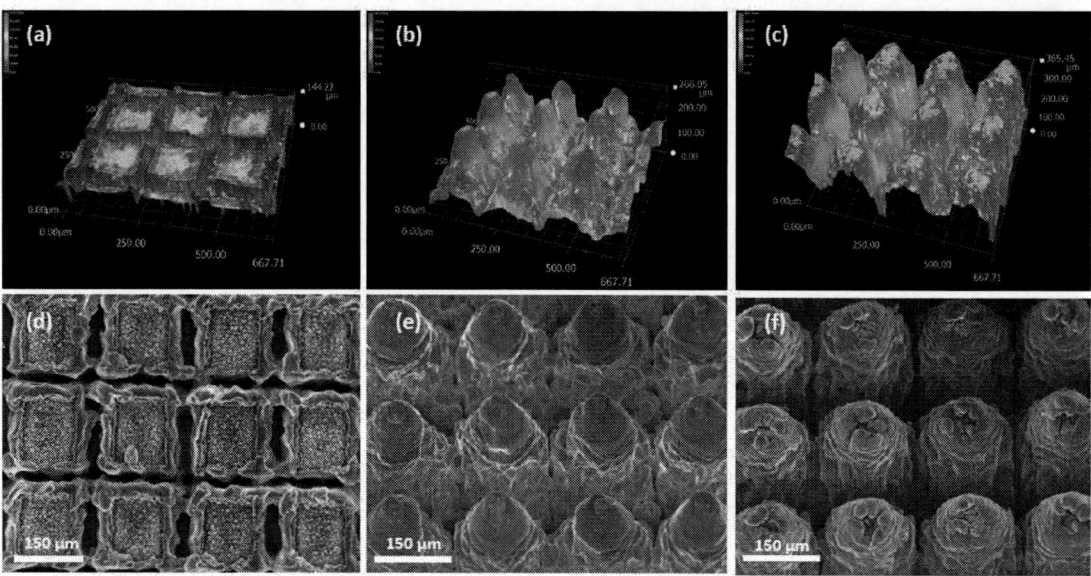

Figure 3. Super depth of field microscope and scanning electron microscopy (SEM) images of different as-prepared surface.

the surroundings due to the high specific heat coefficient and the flow inside the droplet. In Fig. 2 (b), the red color in contour represents the volume fraction at one, which means that none of the components have started a phase transition. The blue areas inside the droplet showed areas that had been frozen. Ice appears at the bottom of droplets on the surface of sample (i) and (ii), which can be attributed to the obvious inhomogeneity of the temperature distribution inside the droplet shown as Fig. 2(a). Further, this inhomogeneity is caused by the difference in surface structure that sample (i) and (ii) provided a larger solid-liquid contact area compared with (iii) and (iv) at Cassie-Baxter state. The rate of heat transfer inside the metal is much greater than the loss of heat from droplets on the surface and from the air above. Therefore, the upper surface of the substrate can also be assumed to have a constant temperature of -50 °C by setting Dirichlet boundary condition. As the droplets fall and touch the surface, large temperature differences and turbulence create strong thermal convection so that the bottom of the droplet cools rapidly and crystallizes. The nucleation area inside the droplet was located at the place of the contact region between the droplet and cold surface. However, the temperature of the storage air between the textures will not drop in a very short time. There's enough time inside the droplet for the heat to be distributed more evenly through the flow as shown in Fig. 2(iii) and (iv). Results showed that the icing time of droplet on the surface with micro/nano structures is significantly delayed, and the nucleation rate is related to the contact area of solid-liquid[9]. It is also found that the larger size of the air layer brought by the structure is more conducive to obtain enhanced dynamic anti-icing performance because the thermal resistance coefficient of air is much greater than the substrate metal.

The general outline of the characteristic microstructure is presented in Fig. 3 (a-c). By regulating the path and times of laser scanning to control the position and depth of material to be removed. The size of the top area of a single

unit which is not completely flat due to thermal effect of laser ablation is controlled by changing the area unscanned. The material at the edges of the scan path will melt and pile up outward as shown obviously in Fig. 3 (d) and (f). This phenomenon is used to make the materials on the upper edge of the cylinder pile up with each other to form the morphology of an approximate cone[10]. In order to avoid excessive thermal effect resulting in the structure of the square column cannot be produced, the ultraviolet laser with the wavelength of 355 nm was selected. The side of the cylinder is not smooth due to the sputtering of particles caused by the laser interaction with the material. However, the surface droplets will not directly contact this part and the influence of roughness can be ignored according to Cassie-Baxter state. Micro-nano structures was fabricated on titanium alloy with 1 cm × 1 cm ablation area, that is, there are at least 2,500 structural units on the surface of each sample. The static contact angles of the above samples are all greater than 150° after surface chemical modification and will be ready for dynamic anti-icing performance test.

Fig. 4 shows the process by which a droplet impinging superhydrophobic surfaces and completely freezes. The zero point of the timing is the frame before the droplet impinge on the surface of different sample. As shown in Frame 1, the height of the snapshot may vary slightly but the error is small due to the frame rate limit. The Frame 2 shows the maximum height at which the droplet bounces off the surface with different micro-nano structure. The supercooled surface that has not been processed by the laser ablation does little to bounce the droplet while the droplet bounces completely into the air on the surface (iii) that the micro structure of approximate frustum of a cone array provide a moderate scale of impinging area. The low temperature affected the wettability of sample that tests at room temperature all showed a stronger ability to rebound droplets. It was because that a 7 μL water drop cannot avoid gravity deformation [31]. The Frame 3 shows the time it takes for a droplet to reach

	Frame 1	Frame 2	Frame 3	Frame 4	Frame 5
Surface (i)	00'00.00	00'00.03	00'01.02	00'11.26	00'41.02
Surface (ii)	00'00.00	00'00.05	00'09.06	00'14.26	03'25.24
Surface (iii)	00'00.00	00'00.05	00'06.24	02'40.10	03'36.16
Surface (iv)	00'00.00	00'00.08	00'12.07	00'54.10	04'51.26

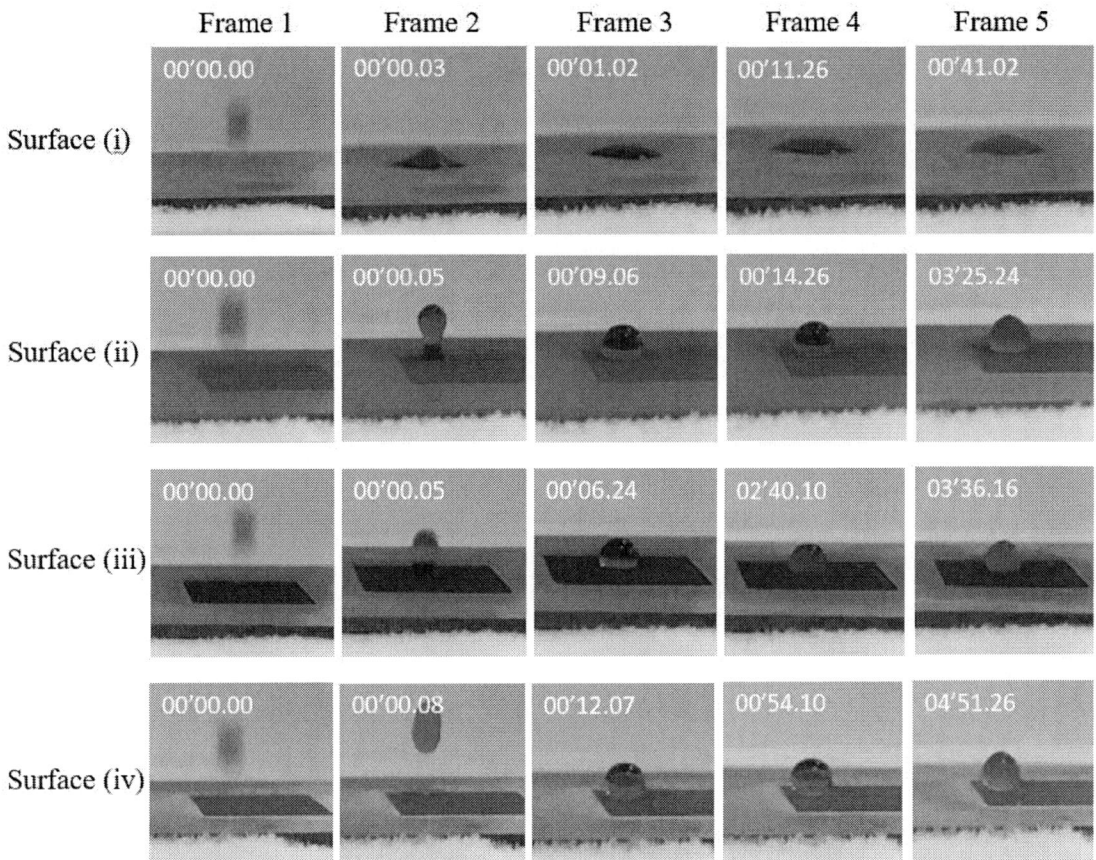

Figure 4. Droplet impinging on supercooled surfaces at -15℃ with different textured micro-nano structure.

steady state while it is at a liquid state. This time is determined by both the rate of heat transfer to the droplet and the bouncing property caused by the sample. Under the temperature of -15°C on the surface of the sample, the droplets do not crystallize before the end of the bouncing process but after the steady state. The Frame 4 shows the time when crystallization began. The lowest part of droplet on surface (i) forms a layer of crystals very early because the square column array structure provides a large contact area with the water droplets so that the heat of water in contact quickly dissipates. The same situation presents on surface (iii), but the time is much later indicating that the air cavity between the droplet and the metal acts as a thermal barrier. Just as the surface (ii) with the most air cavity volume, the low heat transfer efficiency is even less than the heat transfer velocity inside the droplet. The droplet stays transparent for a long time until the whole droplet is supercooled and at some time thin layer crystals form suddenly on the entire surface of the droplet shown as Frame 4 of surface (ii). The Frame 5 represents the time when the overall crystallization of the droplet is completed. The experimental results show that weather the rebound effect of the droplet on the supercooled surface or the delay of the freezing time are considered, the approximate frustum of a cone array micro structure has certain advantages, while the approximate

square column structure has the least obvious advantages although it performs much better than the original surface.

IV. CONCLUSION

In this study, by coupling the Volume of Fluid (VOF) with solidification/melting (S&M) model, a two-dimensional Computational Fluid Dynamics (CFD) model was used to reveal the anti-icing mechanism of superhydrophobic surface. Moreover, the process of the droplets icing on surfaces with different micro structures was compared by discussing variations in solidification fraction and temperature contour. Furthermore, by regulating the parameters of laser process, the surfaces of different patterns were obtained to be tested in the process that droplet impinging on supercooled surfaces and icing. Some important observations were noted and summarized below: (1) The trends of simulation and experiment are consistent in delay time of droplet icing with different structural surface. For the surface structures with large air layer, the delay time of droplet icing is significantly increased; (2) The difference of solid-liquid contact area leads to the difference of icing process that large one tends to freeze the bottom of the droplet first, rather than forming ice shells on the surface first; (3) The height at which the droplet bounces off the surface does not correlate linearly with the solid-liquid contact area that the moderate-scaled sample had the strongest capability to rebound droplets.

REFERENCES

[1] D. Wang *et al.*, "Design of robust superhydrophobic surfaces," *Nature,* vol. 582, no. 7810, pp. 55-59, 2020/06/01 2020.

[2] M. J. Kreder, J. Alvarenga, P. Kim, and J. Aizenberg, "Design of anti-icing surfaces: smooth, textured or slippery?," *Nature Reviews Materials,* vol. 1, no. 1, Jan 2016, Art. no. Unsp 15003.

[3] R. Ramachandran, M. Kozhukhova, K. Sobolev, and M. Nosonovsky, "Anti-Icing Superhydrophobic Surfaces: Controlling Entropic Molecular Interactions to Design Novel Icephobic Concrete," *Entropy,* vol. 18, no. 4, Apr 2016, Art. no. 132.

[4] A. A. Yancheshme, G. Momen, and R. J. Aminabadi, "Mechanisms of ice formation and propagation on superhydrophobic surfaces: A review," *Advances in Colloid and Interface Science,* vol. 279, no. 0, May 2020, Art. no. 102155.

[5] Z. Liu, M. Zhang, H. Tao, and J. Lin, "The Anti-icing characteristics of micro/nano surface of stainless steel prepared by femtosecond laser," in *Advanced Laser Processing and Manufacturing Iii,* vol. 11183, R. Xiao, M. Hong, J. Liu, J. Yao, and Y. Sano, Eds. (Proceedings of SPIE, 2019.

[6] Z. Yang, C. Zhu, J. Zhou, N. Zheng, D. Le, and Iop, "Numerical Study on Size Effects in the Wettability of Textured Surfaces," in *2018 4th International Conference on Environmental Science and Material Application,* vol. 252(IOP Conference Series-Earth and Environmental Science, 2019.

[7] M. Liu *et al.*, "Tunable Hierarchical Nanostructures on Micro-Conical Arrays of Laser Textured TC4 Substrate by Hydrothermal Treatment for Enhanced Anti-Icing Property," *Coatings,* vol. 10, no. 5, May 2020, Art. no. 450.

[8] C. Liu, Q. Liu, R. Jin, Z. Lin, H. Qiu, and Y. Xu, "Mechanism analysis and durability evaluation of anti-icing property of superhydrophobic surface," *International Journal of Heat and Mass Transfer,* vol. 156, Aug 2020, Art. no. 119768.

[9] G. Momen, R. Jafari, and M. Farzanehnserc, "Ice repellency behaviour of superhydrophobic surfaces: Effects of atmospheric icing conditions and surface roughness," *Applied Surface Science,* vol. 349, pp. 211-218, Sep 15 2015.

[10] J. Zhang, B. Song, Q. Wei, D. Bourell, and Y. Shi, "A review of selective laser melting of aluminum alloys: Processing, microstructure, property and developing trends," *Journal of Materials Science & Technology,* vol. 35, no. 2, pp. 270-284, Feb 2019.

Proceedings of the 16th Annual IEEE International Conference on Nano/Micro Engineered and Molecular Systems April 25-29, 2021

Facile access to solvent-dependent luminescent carbon nanoparticles by laser ablation of activated carbon powders*

Zhibin Chen, Rui Zhou[†], Longfan Li and Jingqin Cui

Abstract—Carbon nanoparticles modified with functional groups have been widely studied for photoluminescence (PL) features. Laser ablation as a promising *in-situ* method can be applied to synthesize carbon nanoparticles (CNPs) with sizes below 10 nm by restricting the particle aggregation. In this work, the as-prepared CNPs demonstrate solvent-dependent luminescence by laser ablation of activated carbon powders in various organic solvents, which could be attributed to excitation-dependent emission. It showed that spherical nanoparticles were obtained in the diameter range of 2–10 nm. The existing defect centers and oxygen-related functional groups on the CNPs fabricated in different solvents were responsible for their PL properties.

I. INTRODUCTION

Carbon-based nanomaterials include different classes of nanostructures with various shapes, sizes and dimensions, such as fullerenes (0D), nanotubes (1D) and nanosheets (2D). Researchers happened to discover luminescent fragments, while separating single-walled carbon nanotubes (SWCNTs) using gel electrophoresis techniques [1]. Afterwards, the luminescent carbon materials were extensively developed, such as carbon nanoparticles (CNPs), graphene oxide quantum dots (GOQDs) and graphene quantum dots (GQDs), according to selection of the structure and physicochemical properties of the carbon precursors.

Recently, a variety of interest in luminescent CNPs is claimed, attributed to their desirable properties, such as high quantum yields, water solubility, non-toxic and etc. [2]. These remarkable properties contribute to the profound impact on multiple applications, for instance, bio-imaging, catalysis, cytotoxicity, surface enhanced Raman scattering (SERS) sensing and etc. Luminescent CNPs have attracted much attention due to the property of excitation-dependent photoluminescence (PL) emission, resulting in broad emission spectrum in the visible region. However, the actual mechanism of PL is still under active discussion among researchers in this field [3]. Most scholars assume that PL mechanism mainly consists of the bandgap (or HOMO-LUMO) transitions corresponding to conjugated π-domains, which depends on the size of the sp^2 clusters as well as the surface traps [4]. It is found that modifying the surface by creating defect centers and introducing oxygen-related functional groups is the common route to synthesize luminescent CNPs.

Various methods have been employed for synthesizing CNPs, which can be divided into two routes, bottom-up and top-down. The bottom-up method is the synthesis of CNPs from specific organic molecules by microwave-assisted methods, hydrothermal methods, ultrasonic synthesis, and carbohydrates. Top-down methods are used to create CNPs

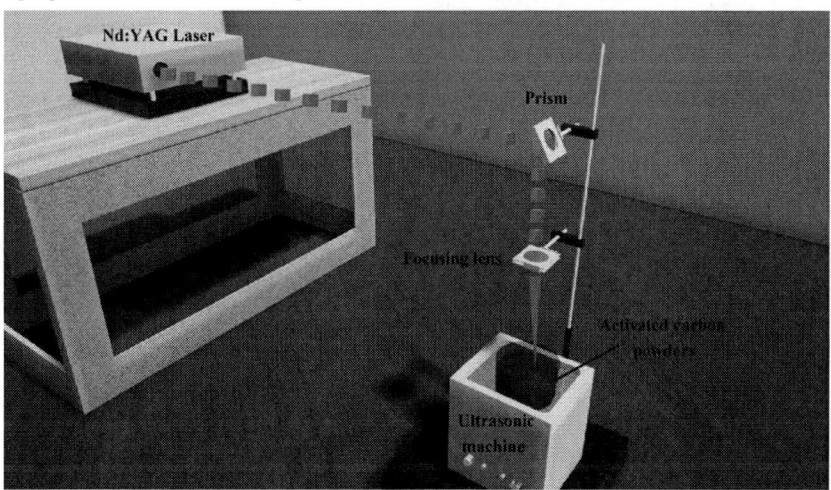

Figure 1. Schematic diagram of fabricating CNPs by laser ablation.

*Research supported by Fujian Provincial Science and Technology Programme (Industry Oriented Key Project, No. 2020H0006), National Natural Science Foundation of China (No. 61605162).

Zhibin Chen is with the Department of Mechanical & Electrical Engineering, School of Aerospace Engineering, Xiamen University, Fujian Province, China, 361102 (e-mail: chenzhibin@stu.xmu.edu.cn).

Rui Zhou is with the Department of Mechanical & Electrical Engineering, School of Aerospace Engineering, Xiamen University,

Fujian Province, China, 361102 ([†]Corresponding author, phone:86-592-2186970; e-mail: rzhou2@xmu.edu.cn).

Longfan Li is with the Department of Mechanical & Electrical Engineering, School of Aerospace Engineering, Xiamen University, Fujian Province, China, 361102 (e-mail: 34520182201541@stu.xmu.edu.cn).

Jingqin Cui is with the Pen-Tung Sah Institute of Micro-Nano Science and Technology, Xiamen University, Fujian Province, China, 361005 (e-mail: jqcui@xmu.edu.cn).

978-1-6654-3008-1/21 $31.00 © 2021 IEEE

by breaking up larger carbon structures, such as exfoliation, explosive techniques, arc discharge, laser ablation. For example, Thomas Nesakumar Jebakumar Immanuel Edison et al. [2] successfully synthesized nitrogen-doped CNPs by microwave assisted method, which applied in the cytotoxicity analysis. S. Sarkar et al. [5] reported the preparation of water soluble CNPs from sucrose in a Teflon lined autoclave reactor. Haitao Li et al. [6] prepared the CNPs from glucose by ultrasonic treatment, which have desirable up-conversion PL properties. Compared to other techniques, pulsed laser ablation has been widely reported as a novel, facile and green approach to produce nanomaterials with different precursor materials. In addition, well-dispersed nanoparticles can be fabricated in a confining solvent via laser ablation by precisely tuning the processing parameters such as laser power, wavelength, ambient solvent and etc.

In this paper, an effective approach is employed to synthesize luminescent CNPs by laser ablation of activated carbon powders in various organic solvents. The synthesis and surface modification of luminescent CNPs are considered to be carried out simultaneously, resulting in solvent-dependent luminescent carbon nanoparticles, whose surface state could be easily modified to realize tunable PL properties.

II. Experimental method

A. Fabrication of CNPs

Activated carbon powders with an average size of 2 μm (Shunhang Metal Material Co., Ltd). About 20 mg of activated carbon powders were added to ethanol and acetone. Ultrasonic treatment was utilized during the laser irradiation to expedite the movement of carbon powders. A pulsed laser of the wavelength 1064 nm and pulse width ~100 ns was employed to irradiate the solvent at a repetition frequency 20 Hz and constant power 22.5 W. The suspension changed to black after 40 mins and was centrifuged into precipitate and supernatant for further analysis. The schematic of laser ablation on activated carbon powders in ethanol and acetone solvents is shown in Fig. 1.

B. Characterizations of CNPs

The morphology and size of CNPs was recorded by transmission electron microscopy (JEM 1400). The absorption spectra were investigated by the UV-visible spectroscopy (UV 2600 220V CH). Raman measurements were characterized by microscopic confocal Raman spectrometer (IDSPeC ARCTIC). The functional groups on the surface of CNPs were analyzed by Fourier transform infrared spectroscopy (MICOET iS10). HORIBA fluorescence spectrophotometer Fluorolog was employed to record the PL emission.

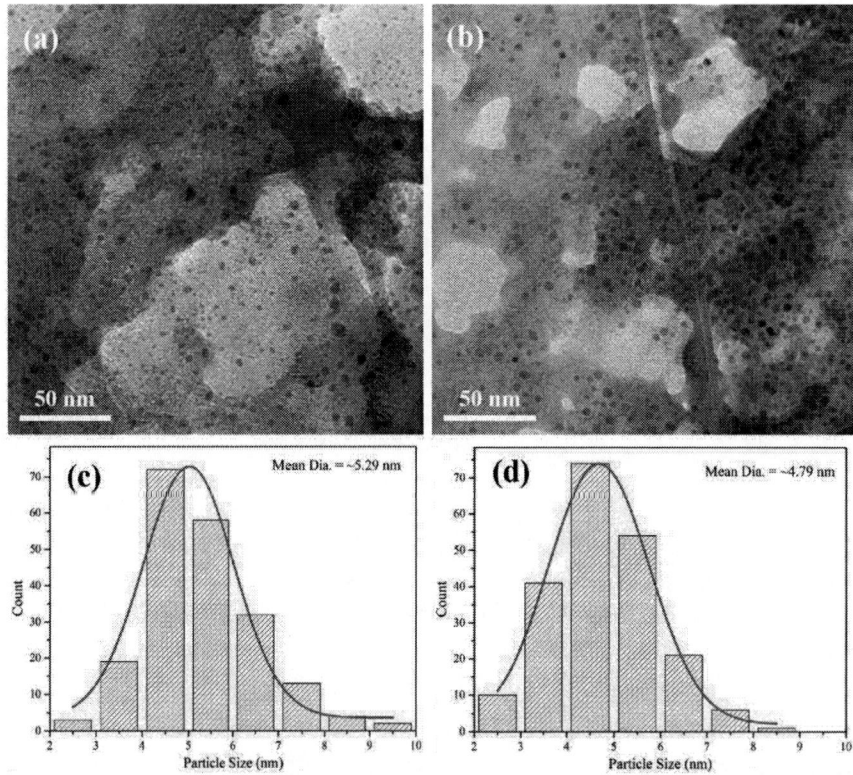

Figure 2.   TEM images and size distribution of the CNPs synthesized in solvents (a,c) ethanol and (b,d) acetone.

III. RESULTS AND DISCUSSION

A. Morphological analysis

Fig. 2 (a-b) illustrate the TEM micrographs of CNPs synthesized by laser ablation in ethanol and acetone. As can be seen, the quasi-spherical CNPs synthesized in organic solvents were well-dispersed. ImageJ software was performed to estimate the average particle size by statistical evaluation of various particle sizes. The analysis report shows a narrow distribution of particles size and the average particle is found to be around 5.29 nm and 4.79 nm corresponding to ethanol and acetone, respectively. The formation of CNPs strongly depends on solvent. Organic solvent as a highly viscous liquid can slow down the expansion of the plasma, which forms on the surface of the target and increases the shockwave intensity of the plasma bubbles, leading to the destruction of the carbon powders. On the other hand, intra-interaction of as-prepared nanoparticles during the laser ablation process significantly contributed to the resulted morphologies, for example, surface charge of fabricated nanoparticle plays an important role in the formation of final products. Generally, solvent is very important to the surface charge of nanoparticles. Alcohol and acetone as highly polar solvents provide an electrical double layer on the surface of the CNPs, preventing aggregation and precipitation of the final products [7]. Thus, during laser ablation, the plasma species generated are quenched by the heated vapor at very high pressure. The kinetic energy of carbon atoms is decreased by expansion through multiple collisions with surrounding molecules and atoms. The confinement effects of organic solvent lead to rapid nucleation and then rapid growth of the nanoparticles.

B. UV-Vis and FTIR characterizations

In the laser ablation process, once the laser irradiation is projected onto the activated carbon powders, a dense plasma plume is formed and then expands into the solvent, which is followed by cooling process of the surrounding solvent. During the cooling process, it nucleates and interacts with the solvent, which results in various functional groups on the surface of the CNPs. As can be seen in Fig. 3, typical absorption peaks of carbon nanostructures were observed

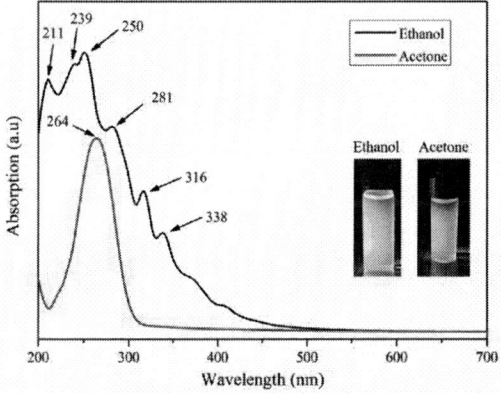

Figure 3.   UV spectra of the CNPs synthesized in (a) ethanol and (b) acetone. (Inset are the direct photographs of the samples under a 365 nm UV lamp.)

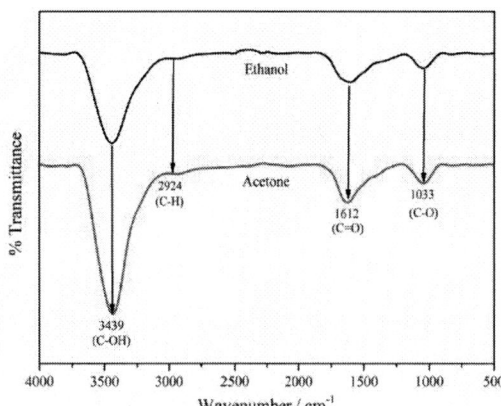

Figure 4.   FTIR spectra of the CNPs synthesized in ethanol and acetone.

around 250-281 nm, which is attributed to π-π* transition of C=C of aromatic sp^2 domains. The absorption peak of the nanoparticles shifting to longer wavelengths (red shift) suggests an increment in particle size, and vice versa. Compare to the acetone, the absorption peak of ethanol shifts from 264 nm to 281 nm has confirmed that the CNPs produced in this solvent are larger than acetone in average, which is in accordance with the size distribution report. There are two minor absorption peaks located at 316 nm and 338 nm corresponding to the ethanol due to n-π* transition of C=O. The absorbance peaks observed below 250 nm are due to the polyynes carbons. In ethanol solvent, the absorbance peak features at 211 nm and 239 nm belong to the polyynes carbon structure C_8H_2. During the laser ablation process, the high-energy photochemical process can overcome the activation barrier of the polyynes moiety in a short time, thus generating short-sized polyynes moiety. The snapshot images of Fig. 3 illustrate that samples show green and yellow color corresponding to ethanol and acetone under irradiation with a 365 nm UV lamp.

As shown in Fig. 4, FTIR spectra were used to identify the functional groups of as-prepared CNPs. Two samples possess -OH stretching vibration at 3439 cm⁻¹, indicating the existence of large numbers of residual hydroxyl groups on the CNPs [8]. Minor absorption band can be observed at 2924 cm⁻¹, 1612 cm⁻¹ and 1033 cm⁻¹, which are ascribed to C–H, C=O and C-O, respectively [9].

It is shown that the CNPs have structural features similar to most commonly reported CNPs, namely a carbon core of graphitic nature and a surface with functional groups. The abundant functional groups lead to surface traps on the surface of CNPs, acting as excitation energy traps and leading to the tunable PL properties. The presence of these functional groups equip the synthesized CNPs with desired water solubility for potential applications [10].

C. Raman measurements

Raman spectroscopy plays an important part in the efficient, non-contact and non-destructive identification of various carbon species. A laser was operated with wavelength of 532 nm as incident beam, which provided a spot size of the order of 0.1 μm. The spectra were recorded

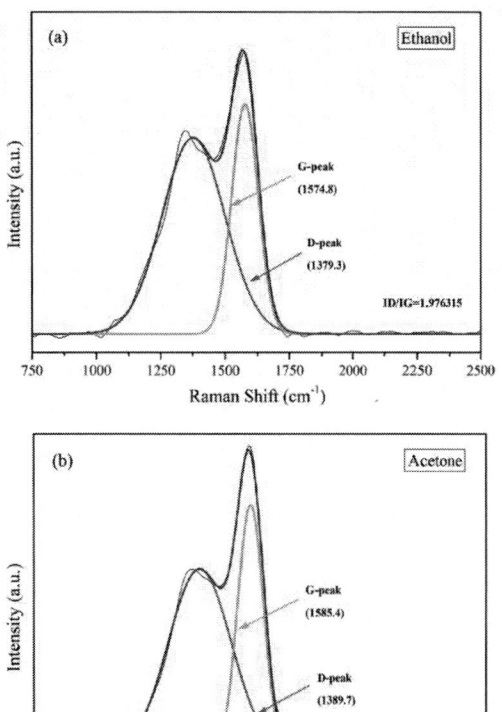

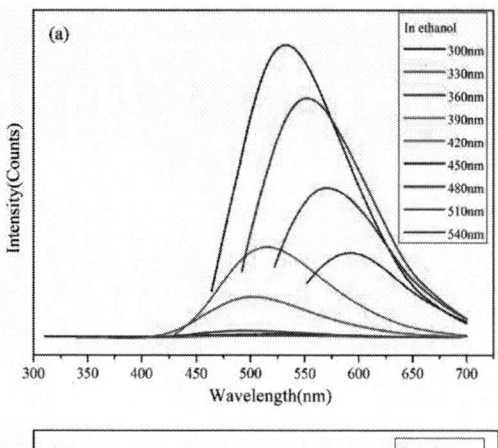

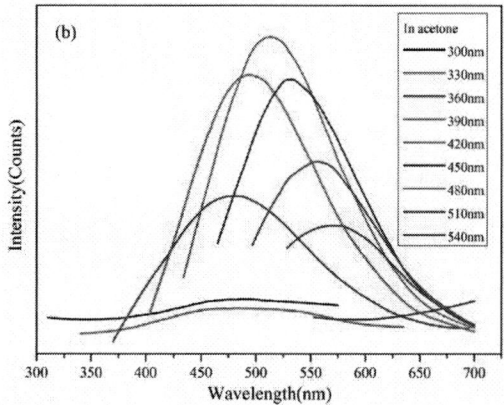

Figure 5. Raman spectra of the CNPs synthesized in (a) ethanol and (b) acetone.

Figure 6. PL spectra of the CNPs synthesized in (a) ethanol and (b) acetone.

in the range from 750 to 2500 cm^{-1}. These are the so-called 'G-band' (for 'graphite'), at 1574.8 cm^{-1} in Fig. 5 (a) and 1585.4 cm^{-1} in Fig. 5 (b), and 'D-band' (for 'disorder'), at 1379.3 cm^{-1} in Fig. 5 (a) and 1389.7 cm^{-1} in Fig. 5 (b). The band peak characteristics strongly rely on the graphitized structure and the Raman spectra are accounted for by the model and theory [11]. The G-band is ascribed to an in-plane transverse optical and longitudinal optical (iTO and LO) double degenerate phonon mode (E$_{2g}$ symmetry) in the center of the Brillouin zone with Raman activity for the sp^2 carbon network. The D-band attributed to the A$_{1g}$ phonon breathing pattern, which can be noticed when disorder and defects originate in the transverse optical (TO) phonons near the K point coinciding with the first Brillouin zone appearing in the pure graphite sample.

The ratios of the intensities to D-peak (I$_D$) of G-peak (I$_G$) are 1.976315 and 1.809287 for as-prepared CNPs fabricated by laser ablation in the ethanol and acetone, respectively. Since the intensity ratios I$_D$/I$_G$ corresponding to the D and G bands are in inverse proportion to the sp^2 cluster size, the higher I$_D$/I$_G$ is reported to be the result of grain size reduction, where the D mode becomes the dominant active mode, implying a higher degree of disorder. In addition, the HOMO-LUMO gap gradually increases by decreasing the grain size. From the above results, it can be inferred that the CNPs synthesized in organic solvents possess defective

structures and the PL property is owing to defect states and surface vacancies entrapping excited state energy.

D. Photoluminescence analysis

Fig. 6 illustrates the steady-state PL emission spectra of CNPs. The PL emission profile of CNPs displays excitation-dependent PL emission. When the excitation wavelength is varied from 300 nm to 540 nm by an increment of 30 nm, the emission peak shows red shift by monotonically from 450 nm to 592 nm. The excitation-dependent mechanism is attributed to local energy levels (π-π* states of C=C), stemming from the size effect of non-uniform carbon nanodomains, which are an isolated aromatic structure in CNPs. Aromatic domains of a certain size dominate PL emission once such aromatic domains are excited at a certain wavelength.

The PL emission is the result of the π-π* transition of the sp^2 carbon cluster localized in the carbon-oxygen sp^3 carbon matrix and facilitates the geminate recombination of the electron-hole (e-h) pair, which appears to be a chromophoric luminescent region. Electrons in π orbitals can be excited to different vibrational levels of π* orbitals by excitation wavelengths, and then some of the electrons in π* orbitals could directly return to π orbitals, recombining with holes to emit PL emission, which is regarded as intrinsic emission. Numerous studies have shown that the PL behavior of CNPs hinges on the size of sp^2 carbon clusters. The electrons can

978-1-6654-3008-1/21 $31.00 © 2021 IEEE

be excited to local energy levels induced by the size effect of sp^2 clusters. It is suggested that PL emission may arise from surface traps, that is, functional groups and defect states can act as energy traps. Some electrons from π^* orbitals are trapped by surface traps before returning to π orbitals. It is worth noting that a large number of electrons can also be excited an directed into the surface trap, and eventually, the electrons in the surface trap recombine with the holes in the π orbitals to emit PL emission.

IV. CONCLUSION

Laser ablation of activated carbon in organic solvents is a promising strategy to synthesize luminescent CNPs by simultaneous modification of surface functional groups, leading to excitation-dependent emission. It shows that the spherical CNPs are well-dispersed with an average particle of 5.29 nm and 4.79 nm corresponding to ethanol and acetone, respectively. FTIR and Raman results demonstrate that the surface of CNPs are modified by introducing oxygen-related functional groups and defect centers. Detailed analysis of structures and chemical compositions for CNPs indicates that the intense PL emission arises from the sp^2 carbon clusters, defect centers and surface functional groups. This approach has great potential for fabrication of novel luminescent carbon materials used in a wide range of fields, such as biology and imaging.

REFERENCES

[1] X. Xu *et al.*, "Electrophoretic Analysis and Purification of Fluorescent Single-Walled Carbon Nanotube Fragments," *Journal of the American Chemical Society,* vol. 126, pp. 12736-12737, Oct. 2004.

[2] T. N. J. I. Edison, R. Atchudan, M. G. Sethuraman, J.-J. Shim, and Y. R. Lee, "Microwave assisted green synthesis of fluorescent N-doped carbon dots: Cytotoxicity and bio-imaging applications," *Journal of Photochemistry and Photobiology B: Biology,* vol. 161, pp. 154-161, Aug. 2016.

[3] I. Y. Goryacheva, A. V. Sapelkin, and G. B. Sukhorukov, "Carbon nanodots: Mechanisms of photoluminescence and principles of application," *TrAC Trends in Analytical Chemistry,* vol. 90, pp. 27-37, May. 2017.

[4] G. K. Yogesh, E. P. Shuaib, and D. Sastikumar, "Photoluminescence properties of carbon nanoparticles synthesized from activated carbon powder (4% ash) by laser ablation in solution," *Materials Research Bulletin,* vol. 91, pp. 220-226, Jul. 2017.

[5] S. Sarkar, D. Banerjee, U. K. Ghorai, N. S. Das, and K. K. Chattopadhyay, "Size dependent photoluminescence property of hydrothermally synthesized crystalline carbon quantum dots," *Journal of Luminescence,* vol. 178, pp. 314-323, Oct. 2016.

[6] H. Li *et al.*, "One-step ultrasonic synthesis of water-soluble carbon nanoparticles with excellent photoluminescent properties," *Carbon,* vol. 49, pp. Jan. 2011.

[7] H. Sadeghi, E. Solati, and D. Dorranian, "Producing graphene nanosheets by pulsed laser ablation: Effects of liquid environment," *Journal of Laser Applications,* vol. 31, pp. 042003, Nov. 2019.

[8] C. Chen, W. Chen, and Y. Zhang, "Synthesis of carbon nano-tubes by pulsed laser ablation at normal pressure in metal nano-sol," *Physica E: Low-dimensional Systems and Nanostructures,* vol. 28, pp. 121-127, Jul. 2015.

[9] B. De and N. Karak, "A green and facile approach for the synthesis of water soluble fluorescent carbon dots from banana juice," *RSC Advances,* vol. 3, pp. 8286-8290, Mar. 2013.

[10] L. Shen, L. Zhang, M. Chen, X. Chen, and J. Wang, "The production of pH-sensitive photoluminescent carbon nanoparticles by the carbonization of polyethylenimine and their use for bioimaging," *Carbon,* vol. 55, pp. 343-349, Apr. 2013.

[11] A. C. Ferrari, "Interpretation of Raman Spectra of Disordered and Amorphous Carbon," *Phys. Rev. B,* vol. 61, pp. 20, May. 2000.

Gap in pagination due to unavailable paper.

Pages 1742-1744

Proceedings of the 16th Annual IEEE International
Conference on Nano/Micro Engineered and Molecular Systems
April 25-29, 2021

A MEMS Voice Coil Motor with a 3D Solenoid Coil*

Jiamian Sun, Zhi Tao, Haiwang Li, Kaiyun Zhu, Donghui Wang, Hanxiao Wu and Tiantong Xu*

Abstract—Previously-reported electromagnetic (EM) linear micromotors typically applied meander coils or planar spiral coils because there are many difficulties in solenoid coil fabrication using integrated micro electro-mechanical system (MEMS) technologies. However, solenoid microcoils with iron cores have a higher inductance density and a lower leakage magnetic flux than planar coils. In this study, a MEMS voice coil motor (VCM) with a 3D solenoid voice coil was designed, fabricated and tested. The VCM was comprised by a voice coil fabricated using MEMS processes as the traveler and a combination of two permanent magnets (PMs) and an E-shape silicon-steel core which was inserted into the voice coil as the stator. The copper voice coil in silicon substrate was a solenoid-shape coil of high aspect ratio (coil height was 1.4 mm and line width was 0.1 mm). The coil was embedded tightly in the silicon substrate, resulting in a good heat dissipation performance and thus a high current-carrying capacity. The VCM weighed 0.07 g with a size of approximately 4 mm × 2.5 mm × 4 mm before circuit connection. The testing results showed that the effective travelling distance of the traveler was 1 mm with stable thrust force of approximately 3.9 mN/A and the motor vibrated stably when a 50 Hz square-wave current of 0.5 A was applied. This MEMS VCM has low weight and high thrust force, which has bright application prospects in micro actuators.

I. INTRODUCTION

Micro linear motors play a key role in micro robotic systems, microfluidic driving systems and micro manipulators. There are two major types of micro linear motors: electrostatic and electromagnetic (EM). Compared to electrostatic motors, EM motors typically have higher power density but more complex structures. However, there have not been many studies on micro EM linear motors due to the difficulty in fabricating solenoid coils, the crucial structures of EM motors, by micro electro-mechanical system (MEMS) processes.

Most reported integrated coils compatible with complementary metal–oxide–semiconductor (CMOS) processes are 2D planar coils, such as meander and spiral coils. As a result, previously established micro linear motors commonly had planar coils. Gatzen *et al.* [1] reported a batch fabricated micro linear synchronous motor with planar spiral coils. Feldmann *et al.* [2] reported a linear variable reluctance micromotor with a 3D meander coil. Fujiwara [3] reported a micro linear permanent magnet synchronous motor with a planar meander coil.

However, planar coils have a lower inductance per unit area compared to solenoid coils and cannot incorporate magnetic cores. As a result, integrated solenoid coils are generally considered to be more promising because of their high inductance per area and low magnetic resistivity. Some researchers have reported integrated solenoid coils [4–7] with higher inductance than planar coils but they focused on its high-frequency performance. Few researchers paid attention to their low-frequency application, such as linear motors working at relatively low frequency.

To improve the power density of Power MEMS, we focus on improving coil design by developing iron-core 3D solenoid coils of high aspect ratio. In previous studies, we reported the electroplating process of high aspect ratio through silicon vias (TSVs) and then the MEMS fabrication of iron-core 3D solenoid coils of high inductance per area and high current-carrying capacity [8-9]. Then we applied this kind of coils to a rotary micromotor [10] and a micro energy harvester [11]. According to the test results, the coils were proved to be efficient in improving power density of EM MEMS.

Based on our research above, we design and fabricate a MEMS voice coil motor (VCM) with a 3D solenoid voice coil in this paper. VCM is a type of linear motors which have been widely used in driving cellphone camera lenses, hard disk recording heads and precise displacement platforms. A VCM are comprised by a voice coil as the traveler to carry currents and a stator to provide magnetic field. Due to the low-density substrate, high current-carrying capacity and inserted iron core of the solenoid coil in this paper, this MEMS VCM has low weight and high thrust force, which has bright application prospects in micro actuators.

II. DESIGH AND STRUCTURE

Based on the principles of VCMs and the structures of our high-performance voice coil, a MEMS VCM with 3D solenoid coils was designed.

The overall size of this VCM is approximately 4 mm × 4 mm × 2.5 mm before packaging. It contains a traveler and a stator, as shown in Fig. 1. The traveler is a voice coil comprised by silicon (Si) substrate and copper (Cu) conductors. The stator contains two permanent magnets and an E-shape iron core inserted into the central cavity of the voice coil. The magnetic field is generated by the two permanent magnets and conducted by the iron core. When the coil is energized，the magnetic field will apply ampere force on the traveler and the traveler will move along the core center beam.

---

*Research funded by the National Natural Science Foundation of China (Nos. 51906008 and 51822602) and the High-speed moving component dynamic tester (2017YFF0107601 and 2017YFF0107604).

J. Sun, Z. Tao T. and H. Wu are with the School of Energy and Power Engineering, Beihang University, Beijing 100191, China (e-mail:

sunjiamian@buaa.edu.cn; tao_zhi@buaa.edu.cn; wuhanxiao7652@buaa.edu.cn).

H. Li, K. Zhu and T. Xu are with the Research Institute of Aero Engine, Beihang University, Beijing 100191, China (e-mail: 09620@buaa.edu.cn; 506279150@qq.com; gele_baizi@163.com).

978-1-6654-3008-1/21 $31.00 © 2021 IEEE

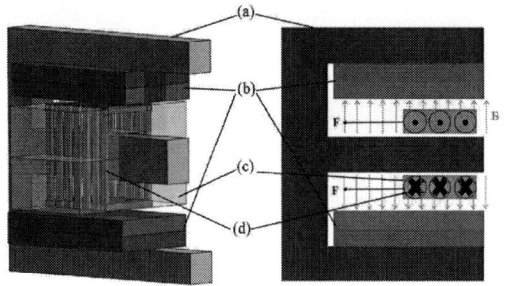

Fig. 1. Structure stereograph and schematic diagram of the motor: (a) E-shape iron core. (b) PMs. (c) Voice coil as the traveler. (d) Cu conductors to form the solenoid coil. The magnetic flux density, marked as B, is shown around the voice coil. The ampere force, marked as F is shown on the voice coil.

The 7-turn voice coil is core structure of this motor (shown in Fig. 2). The 3D solenoid coil is comprised of horizontal conductors on the bottom and top sides of silicon substrates and vertical conductors in TSVs, which means it is tightly embedded in silicon. Due to the low density (2.33 g/cm³) and high thermal conductivity (150 W/(m·K)) of single-crystal silicon, the coil has low weight and good heat dissipation performance. The overall size of the coil is 1.5 mm in moving direction, 1.6 mm in height and 2.5 mm in width including a 1.1-mm-wide Cu winding. There is a cuboid-shape cavity for iron core insertion in the middle of this solenoid coil of high aspect ratio (approximately 14). The rectangular cross-section of the conductors is 0.1 mm × 0.1 mm, which can tolerate large currents.

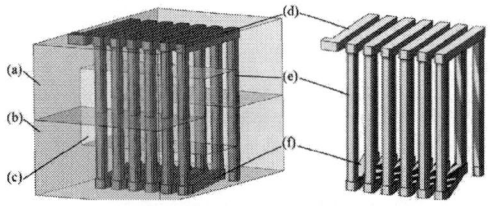

Fig. 2. Structure stereographs of the 7-turn voice coil before iron core insertion (the substrates are hidden in the right graph): (a) (b) Top and bottom Si substrates respectively which are boneded. (c) Center cavity for iron core insertion. (d) Top, vertical and bottom Cu conductors respectively to form the coil.

The stator is comprised of two 0.5-mm-thick Nd-Fe-B magnets (N40H) and a 0.7-mm-thick E-shape silicon-steel core as the stator which was inserted into the voice coil. The iron core establishes a low-resistivity magnetic circuit and conducts the magnetic flux to be almost perpendicular to the horizontal conductors of the coil, resulting in lower leakage magnetic flux and thus higher magnetic flux density perpendicular to the currents.

To summarize, the voice coil has high current-carrying capacity due to the excellent heat dissipation performance and large conductor cross-sectional area. The magnetic flux density perpendicular to the currents are high due to the rational structure design of the inserted iron core.

The Ampere force on a current element is

$$dF=Idl\times B \qquad (1)$$

where I represents the current, dl represents length of the current element and B represent the magnetic flux density through the current element.According to Formula (1), for this VCM design, high thrust Ampere force can be obtained by high current and high magnetic flux density perpendicular to the currents.

III. Fabrication and Assembly

To establish a VCM above, a fabrication and assembly scheme is designed and verified to be feasible involving two stages: voice coil fabrication by MEMS technolgies, and motor assembly.

The substrate and in-chip copper conductors of the voice coil were fabricated by MEMS technologies, which are compatible with complementary metal oxide semiconductor (CMOS) processes. The process flow showing simplified cross-sectional drawings shows how key structural features changes during the fabrication (shown in Fig. 3).

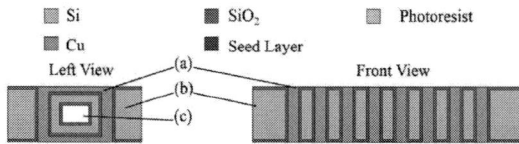

Fig. 3. Simplified cross-sectional drawings of the voice coil: (a) Cu windings. (b) Si substrates (c) Trench for iron core insertion.

In Part 1, two wafers to be bonded in the next part are processed by following steps (shown in Fig. 4): **(1.1)** Prepare a oxidized double-side-polished wafer (4 inch, 800-μm-thick, resistivity 1000 Ω·cm, with 2-μm-thick oxide on both sides). **(1.2)** Pattern horizontal trenches and through holes on the top side photoresist (S1813 from MicroChem Corp.) and iron-core trenches and through holes on the bottom side photoresist by lithography. **(1.3)** Wet etch the exposed oxide in the buffered oxide etchant (BOE). Remove the PR in a piranha solution (98%H₂SO₄:30%H₂O₂ = 3:1). **(1.4) (1.5)** Pattern the through holes on the top side photoresist (AZ4620) and DRIE (deep reactive ion etching) etch the patterns in an inductive coupled plasma (ICP) etcher (SPTS Co.). **(1.6) (1.7)** Remove the PR. DRIE Etch from the top side and then from the bottom side until the holes are through.

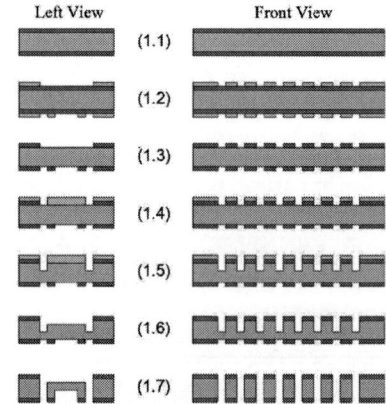

Fig. 4. Process flow of Part 1: fabrication of two wafers to be bonded.

978-1-6654-3008-1/21 $31.00 © 2021 IEEE

In Stage 2, two wafers are bonded and sawed into dies by following steps (shown in Fig. 5): **(2.1)** Wet etch the SiO_2 layers of the two wafers in DHF (diluted HF) etchant. **(2.2)** RCA clean the wafers and pre-bond them by 3000-N pressure in an aligned bonder (AML Co.). **(2.3)** Anneal and oxidize the wafer stack in a wet-oxidation diffusion furnace to create 0.4-μm-thick insulating oxide layers. Saw the stack into dies using a die saw machine.

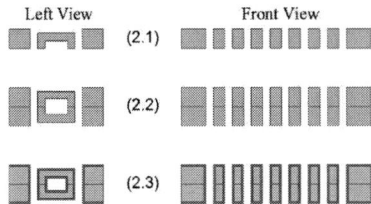

Fig. 5. Process flow of Part 2: wafer bonding and die saw.

In Part 3, a 26 mm × 24 mm die with coil arrays is fabricated by following die-level steps (shown in Fig. 6): **(3.1)** Sputter a Cu film on the bottom side as the plating seed layer. **(3.2)** Electroplate from bottom to top until Cu fills the bottom-side horizontal trenches and seals the bottom end of the through holes. Electroplate from the bottom to the top to almost fill the through holes. **(3.3)** Sputter another plating seed layer. **(3.4)** Electroplate until Cu fills the top-side horizontal trenches. **(3.5)** Remove the Cu beyond the die surfaces using a lapping machine (Logitech Co.). Dice the die into voice coils using a die saw machine.

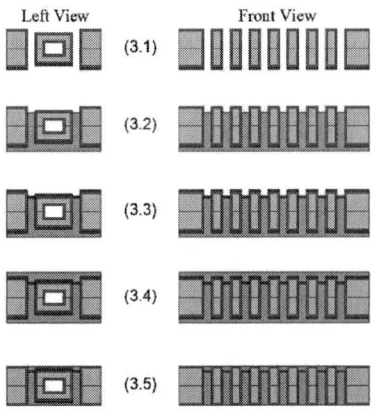

Fig. 6. Process flow of Part 3: die-level fabrication.

This MEMS process scheme combines some key process steps of bulk silicon fabrication including DRIE, Si-Si direct bonding, and Cu electroplating in high-aspect-ratio TSV to establish a complex 3D structure, which is proved to be qualified for batch silicon-based production.

After the MEMS fabrication, the voice coil is assembled with the iron core and magnets. Firstly, the magnets are attached to the E-shape iron core. Then the iron core is inserted into the coil. At last, the two terminal pads are connected to a power supply, which means the whole motor prototype is completed, as shown in Fig. 7.

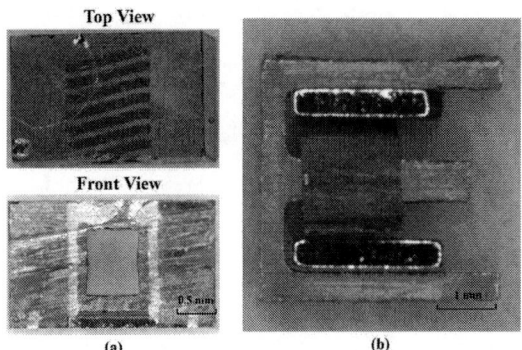

Fig. 7. VCM prototypes. (a) Top view and front view of the coil. (b) Front view of the assembled motor before circuit connecting.

IV. SIMULATION AND TESTING RESULTS

The performance of the motor was modeled using Ansoft Maxwell EM. In simulation, the motor carried a 1 A direct current and its traveling range was 1 mm. The simulation results confirmed the high vertical magnetic flux density perpendicular to the horizontal conductors (more than 0.45 T all over the 1-mm traveling range) and thus high thrust ampere force (more than 6 mN in the whole range) (shown in Fig. 8).

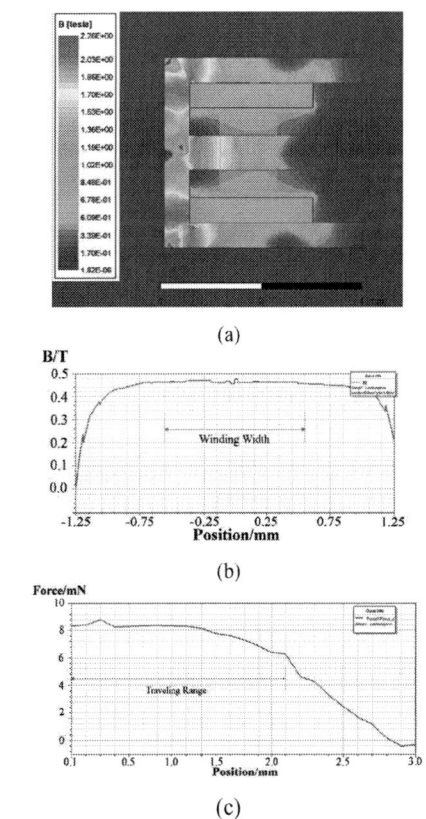

Fig. 7. Modeling results. (a) Magnetic flux density distribution. (b) Magnetic flux density (B) vs Position curve in winding-width direction. (c) Force vs Position curve in traveling direction.

978-1-6654-3008-1/21 $31.00 © 2021 IEEE

According to testing results, the weight of the VCM before circuit connection was 0.07 g.

In power-on test, the coil carried a solenoid current with a peak of 1 A at 50 Hz, and the maximum temperature was maintained under 45 ℃, which proved its high current-carrying capacity.

In static thrust force test (shown in Fig. 8), the effective travel distance is about 1 mm with stable thrust force of approximately 3.9 mN/A.

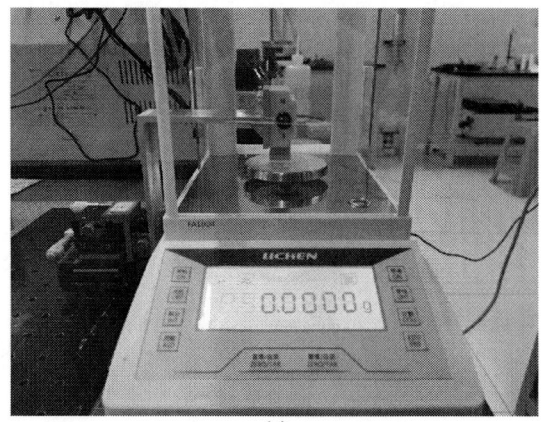

(a)

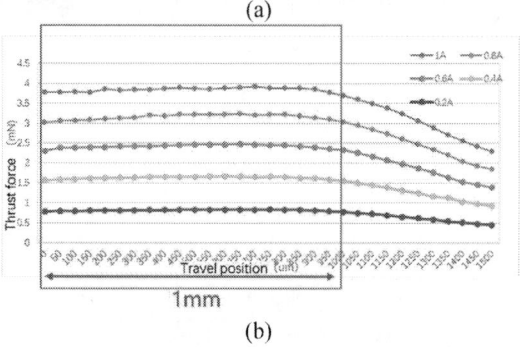

(b)

Fig. 8. Images of static thrust force test. (a) images of the test rig comprised of a precision electronic weighing scale, a motor clamp. (b) thrust force vs travel position curves for different currents.

In operating testing, the motor vibrated stably when a 50 Hz square-wave current of 0.5 A was applied.

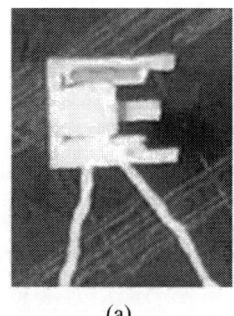 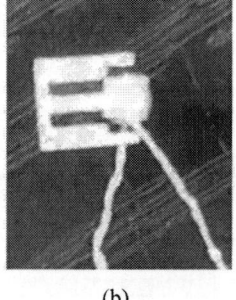

(a)   (b)

Fig. 9. Images of motor testing. (a) Coil at the inner end; (b) Coil at the outer end.

V. CONCLUSION

The VCM in this paper weighed approximately 0.07 g with a size of approximately 4 mm × 4 mm × 2.5 mm before packaging. Its effective travel distance is about 1 mm with stable thrust force of approximately 3.9 mN/A. It vibrates stably when a 50 Hz square-wave current of 0.5 A was applied.

The VCM contains a high-performance MEMS-fabricated solenoid voice coil and a high-efficiency stator magnetic circuit. As is proved by the modeling and testing results, it has a very low weight, a high current-carrying capacity and a high magnetic flux density around the coil. Therefore, the motor performed high thrust force despite the low weight, which proved the high power density of micro linear motors and thus bright application prospects in micro actuators.

We are working on its applications in flapping-wing aircrafts and crawling robots and will report the results in the future.

REFERENCES

[1] M. Föhse, D.- Hanover, J. Edler, and D.- Hanover, "IMECE2003-41388," pp. 1–8, 2016.

[2] M. Feldmann and S. Büttgenbach, "Linear Variable Reluctance (VR) micro motors with compensated attraction force: Concept, simulation, fabrication and test," *IEEE Trans. Magn.*, vol. 43, no. 6, pp. 2567–2569, 2007.

[3] R. Fujiwara, T. Shinshi, and M. Uehara, "Positioning Characteristics of a MEMS Linear Motor Utilizing a Thin Film Permanent Magnet and DLC Coating," *International Journal of Automation Technology*, vol. 7, no. 2. pp. 148–155, 2016.

[4] J. B. Yoon, B. K. Kim, C. H. Han, E. Yoon, and C. K. Kim, "Surface micromachined solenoid on-Si and on-glass inductors for RF applications," *IEEE Electron Device Lett.*, vol. 20, no. 9, pp. 487–489, 1999.

[5] D. W. Lee, K. P. Hwang, and S. X. Wang, "Fabrication and analysis of high-performance integrated solenoid inductor with magnetic core," *IEEE Trans. Magn.*, vol. 44, no. 11 PART 2, pp. 4089–4095, 2008.

[6] F. Jensen *et al.*, "Fabrication of 3D air-core MEMS inductors for very-high-frequency power conversions," *Microsystems Nanoeng.*, vol. 4, no. March 2017, p. 17082, 2018.

[7] S. Zhou, L. Xu, J. Lu, and Y. Yang, "Simulation and Optimization of High Performance On-Chip Solenoid MEMS Inductor," *Proc. - 2018 19th Int. Conf. Electron. Packag. Technol. ICEPT 2018*, pp. 710–715, 2018.

[8] H. Li, J. Liu, T. Xu, J. Xia, X. Tan, and Z. Tao, "Fabrication and Optimization of High Aspect Ratio Through-Silicon-Vias Electroplating for 3D Inductor," *Micromachines*, vol. 9, no. 10, p. 528, 2018.

[9] T. Xu, J. Sun, H. Wu, H. Li, H. Li, and Z. Tao, "3D MEMS In-Chip Solenoid Inductor with High Inductance Density for Power MEMS Device," *IEEE Electron Device Lett.*, vol. 40, no. 11, pp. 1816–1819, 2019.

[10] Z. Tao *et al.*, "A Radial-flux Permanent Magnet Micromotor with 3D Solenoid Iron-core MEMS In-chip Coils of High Aspect Ratio," *IEEE Electron Device Lett.*, vol. 41, no. 7, pp. 1090–1093, 2020.

[11] Z. Tao *et al.*, "Theoretical Model and Analysis of an Electromagnetic Vibration Energy Harvester with Nonlinear Damping and Stiffness Based on 3D MEMS Coils," *J. Phys. D. Appl. Phys.*, 2020.

Gap in pagination due to unavailable papers.

Pages 1749-1799

Keynote Speaker

High Accuracy Resonant Accelerometers Based on QMEMS

Yulong Zhao

State Key Laboratory for Manufacturing System Engineering, Xi'an Jiaotong University, Xi'an 710049, China;

ABSTRACT

It is well known that there are many kinds of micro accelerometers, such as capacitive sensors, piezoresistive sensors, resonant sensors and so on. Compared with capacitive or piezoresistive sensors, resonant accelerometers based on QMEMS have the ability to achieve better performances such as high accuracy, high resolution and low power consumption. Benefiting from digital output signals, resonant accelerometers can be easily interfaced with digital electronics, highly compatible with digital instrumentation and data-processing units on account of the frequency output.

This talk will introduce some new research activities on high accuracy resonant accelerometers based on QMEMS such as working principle, some novel optimization of structural design and fabrication. We also focus on integrated packaging and temperature compensation technology, which makes significant impact on the performance of resonant accelerometers.

By improvements of structures and packaging processes, tests show that the QMEMS accelerometers obtain the characteristic of higher stability and reliability. The resolution of the designed sensor is less than 1μg and bias stability is improved to 5μg. this sensor can be used in the fields of high accuracy micro inertial measurement unit (MIMU) system.

BIOGRAPHY

Yulong Zhao is currently a professor of Xi'an Jiaotong University, Xi'an, China. Dr. Zhao received his Ph.D. degree on high temperature MEMS pressure sensor in Xi'an Jiaotong University. His main research fields include MEMS sensors, biosensors, precise instrument and micro/nano manufacturing technology. and he has published about 300 scientific articles including 160 SCI indexed papers. He also owns more than 100 China invention patents. Those papers and patents are mainly about MEMS sensors and biosensors on piezoresistive and resonance theory, or based on new materials. It also includes MEMS actuators, PCR technology and femtosecond micro machining.

Dr. Zhao is a Doctoral Advisor & the Vice Director of State Key Laboratory for Manufacturing Systems Engineering. The rewards that Prof. Zhao received include: National Nature Science Foundation for Distinguished Young Scholars, National Talents Project, National High-level Personnel of Special Support Program (Scientific and Technological Innovation Leader), Tengfei Scholars Program of Xi'an Jiaotong University, Program for New Century Excellent Talents in University.

Gap in pagination due to unavailable paper.

Pages 1801-1803

Proceedings of the 16th Annual IEEE International
Conference on Nano/Micro Engineered and Molecular Systems
April 25-29, 2021

Hybrid Frequency-time Domain Analysis of Bulk Acoustic Wave Circulator

Jing Yuan, Bing-bin Zhang, Cheng-feng Wu, Yang Gao*

Abstract—This paper introduces the architecture and working principle of the magnet-less bulk acoustic wave (BAW) circulator, which consists of three identical MEMS bulk acoustic wave resonators (BAWRs) and spatiotemporal modulation (STM) circuit based on varactors connected in wye configuration. In order to be able to predict the performance of the BAW circulator accurately, we present a hybrid frequency-time domain simulation analysis method. First, the MBVD model of BAWR and circuit simulation model of BAW circulator are respectively established. And then, the harmonic balance method is used to simulate scattering properties of BAW circulator. Meanwhile, the influence of modulation signal amplitude on performance of circulator is analyzed and the optimal modulation condition is determined. Finally, BAW circulator case at 1.805 GHz is designed by this method. The simulation results show that insertion loss is less than 3.6 dB, and isolation is greater than 14 dB.

I. Introduction

Circulators are nonreciprocal devices with three or several ports, which enable one-way signal transmission between ports. In view of this characteristic, the circulator is used to isolate the transmitted signal from the received signal, which is commonly used in radar, antenna, satellite communication and telecommunications [1,2]. At present, most of the commercial circulators are achieved by applying a strong magnetic field to ferromagnetic materials. Although, they are mature and broadly available, it can hardly be used for the miniaturization design of 5G communication transceiver system due to their bulky dimensions, high cost and most importantly the incompatibility of magnetic materials with IC technologies. Therefore, the research of magnet-less circulator has attracted extensive attention in recent years [3-5]. Especially a new class of magnet-less circulators with higher integration and sufficient nonreciprocity is achieved by the application of linear-periodically-time-variant (LPTV) circuits [6-14]. Spatiotemporal modulation is introduced into the LPTV-based circulators to lift degeneracy, and thus break reciprocity. Among, N. A. Estep et al proposed the first magnet-less circulator by applying angular-momentum biasing to three coupled resonators, which achieved a large isolation in a compact size compared to ferrite circulators [9-11]. However, due to the lower quality factor of LC circuit, high modulation signal is needed to achieve strong nonreciprocity, which increases the power consumption. Owing to MEMS bulk acoustic wave resonator (BAWR) much higher quality factor and better integrability than LC components, BAW circulator enable greatly reduce the modulation frequency, as well as improve the circulator's insertion loss (IL) and isolation (IX) potentially. Based on this condition, there been many attempts realizing BAW circulators [12-14]. However, these demonstrations almost only give the simulation and experimental results of BAW circulator, the specific design procedure and simulation method are not over mentioned.

In this work, we discuss physical mechanism of magnet-less BAW circulator based on spatiotemporal modulation (STM) angular-momentum biasing, and the specific simulation method of the bulk acoustic circulator is introduced. We used the ADS software to set up a BAWR MBVD (Modified Butterworth-Van Dyke) equivalent circuit model and circuit simulation model of BAW circulator. The scattering properties of BAW circulator are calculated by harmonic balance simulation, and the influence of modulation signal amplitude on isolation of circulator is also analyzed. Finally, BAW circulator at 1.805 GHz was designed by the above method.

II. Theory

A. Architecture

The BAW circulator is composed of three identical MEMS BAWRs and STM circuit based on varactors connected in a wye topology. The structure diagram is shown in Fig. 1.

---

*Supported by SciTech. Foundation of Southwest University of Science and Technology (Grant No. 20ZX7114), Science and Technology on Electronic Information Control Laboratory (Grant No. 6142105200203) and Science and Technology on Reliability Physics and Application of Electronic Component Laboratory (Grant No. 61428060303) (*Corresponding authors: Yang Gao.*)

Jing Yuan, Bing-bin Zhang, Cheng-feng Wu, and Yang Gao are with the School of Information Engineering, Southwest University of Science and Technology, Mianyang 621010, China (e-mail: gaoy@swust.edu.cn).

978-1-6654-3008-1/21 $31.00 © 2021 IEEE

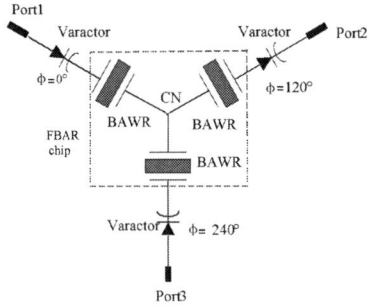

Fig. 1. Structure diagram of BAW circulator.

Modulation circuit is used to provide signals of equal frequency and amplitude and phases of 0°, 120°, and 240° to varactor. Schematic diagram of varactor modulation circuit is shown in Fig. 2, AC and DC are signal source, LC band-pass filter has linear bandwidth, which is used to select specific modulation signal frequency and isolate RF input signal. R_b is used for circuit protection as bleed resistor.

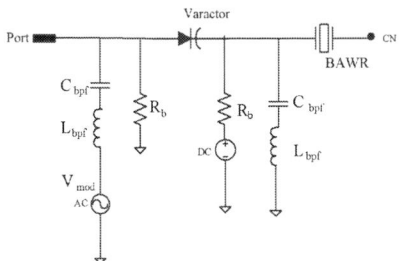

Fig. 2. Schematic diagram of varactor modulation circuit [12].

B. Angular Momentum Biasing

Basic physical principle of STM circulator is that using effective synthesized angular momentum by modulating the resonator in space and time to break the degeneracy of its counter-rotating states and induce non-reciprocity [15]. The angular momentum biasing diagram of BAW circulator is shown in the figure. 3, where the resonance frequency of the nth resonant circuit is given as:

$$f_n = f_0 + kV_m \cos(\omega_n t + \varphi_n), \quad n=1, 2, 3 \quad (1)$$

$$\varphi_n = (n-1)120° \quad (2)$$

Where, n is the coefficient of the resonant circuit, f_0 is the resonant frequency of BAWR without modulation signal, ω_m is the angular frequency of the modulation signal, V_m is the voltage amplitude of the modulation signal, and k is a constant, which is used to quantify the modulation depth of the modulation voltage on the resonant frequency of the resonant circuit; φ_n is the phase of the modulated signal, it can be seen from equation (2) that the phase difference of the modulated signals of the three resonant circuits is 120 °.

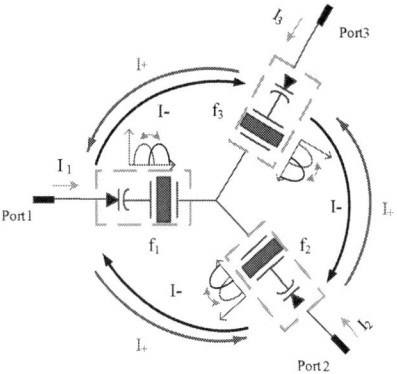

Fig. 3. Schematic of angular momentum biasing of BAW circulator through modulation.

The modulation scheme of equation (1) is called the STM angular momentum biasing, because it involves the phase change in space (φ-direction) and time, and effectively synthesized angular momentum to the circuit. The direction of angular momentum biasing is the direction of phase increase of modulated signal. The current through three resonant circuits can be regarded as the superposition of the two counter-rotating modes I+ and I-, which can be expressed as follows

$$i_n = i_+ e^{j(n-1)\alpha} + i_- e^{-j(n-1)\alpha} \quad (3)$$

Where, $i_+ e^{j(n-1)\alpha}$ and $i_- e^{-j(n-1)\alpha}$ are the aforementioned counter-rotating modes.

Without any modulation, the resonant frequencies of the three resonant circuits are the same exactly, and these modes are degenerate. When a signal is excited from one port, and the transmission to the other two ports is equal, so the network is reciprocal. However, when the modulation signal is applied to the system in equation (1), the STM bias can lift the degeneracy of two counter-rotating modes, because one of the rotation modes has the same direction as the STM bias and the other is opposite to the STM bias. STM bias makes the resonance frequency of two counter-rotating modes separate, which plays the role of magnetic bias in ferrite circulators. Therefore, the operation of a circulator can be realized by choosing appropriate modulation parameters.

III. DESIGN

A. BAWR equivalent circuit model

The equivalent circuit model of an BAWR is shown in Fig. 4, comprising 6 lumped parameter components [16]. L_m , C_m , R_m are motional inductance, motional capacitance, motional resistance respectively; C_0 is static capacitance; R_0 is the dielectric loss of piezoelectric film;

978-1-6654-3008-1/21 $31.00 © 2021 IEEE

R_s is the ohmic loss of the electrode and the resistance loss of the lead.

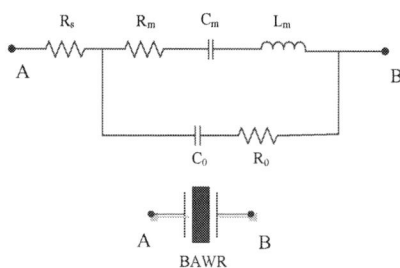

Fig. 4. MBVD model of a BAWR.

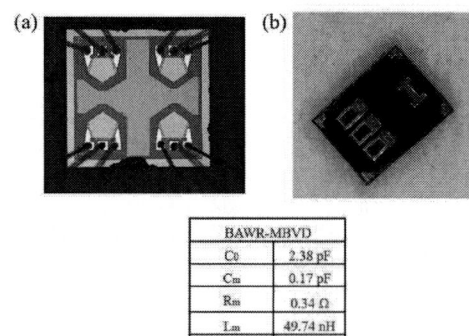

BAWR-MBVD	
C0	2.38 pF
Cm	0.17 pF
Rm	0.34 Ω
Lm	49.74 nH
R0	0.03 Ω
Rs	0.35 Ω

(c)

Fig. 5. (a) BAWR chip photograph, (b) BAWR picture after packaging, (C) MBVD model parameters of the BAWR resonator.

Bulk acoustic resonator is fabricated by MEMS processing technology. Fig. 5(a) shows the BAWR chip photograph. Fig. 5(b) shows BAWR picture after packaging, the size of ceramic package is 0.85mm × 0.65mm × 0.22mm. The MBVD model parameters of the BAWR resonator can be accurately extracted by fitting the measured S parameters, the parameters after fitting as shown in Figure. 5(c).

The measurement and fitted frequency impedance curve of BAWR are shown in Fig. 6. The MBVD model fits well with measured data, and the BAWR is a center frequency of 2.5 GHz, kt² of 7.9% and Q of 791.

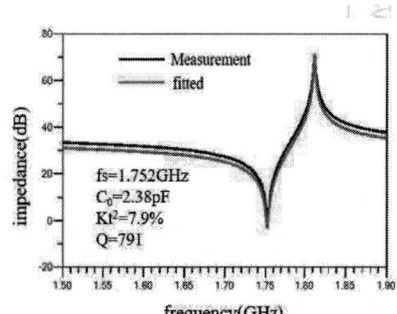

Fig. 6. Measured and fitted frequency impedance curve of BAWR.

B. BAW circulator circuit model

It is difficult to describe nonlinear devices such as varactor in frequency domain for BAW circulator simulation. However, it can easily get the nonlinear model of nonlinear components in time domain. Therefore, a hybrid frequency domain and time domain analysis method are needed. The Harmonic Balance (HB) simulator in Keysight ADS is a fitting and forceful tool to analyze the response of stable state of nonlinear circuit in frequency domain. So, this simulator was used to predict the performance of the circulator. Fig. 7(a) shows the simulation model of BAW circulator circuit. Fig. 7(b) shows the simulation model corresponds to the actual PCB-assembled BAW circulator. BAWR chip is wire bonded to the PCB, which occupy an area of 8×8 mm². The size of all the BAW circulator components is approximately 40×25 mm². It is obvious that most of the area of the BAW circulator is occupied by the varactor modulation circuit.

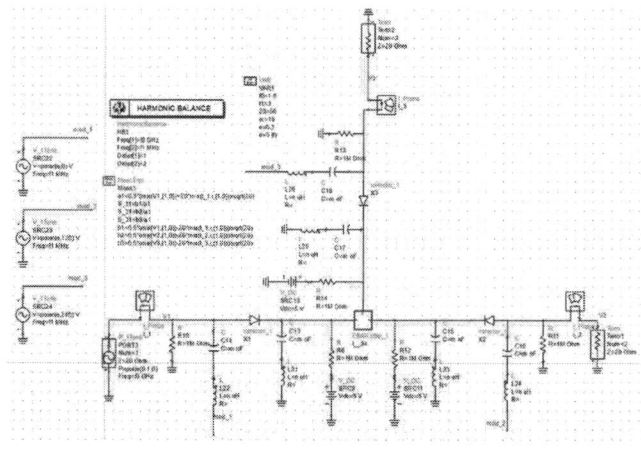

(a)

(b)

Fig. 7. (a) Simulation model of BAW circulator circuit, (b) PCB-assembled BAW circulator.

As shown in Fig. 7a, f0 and f1 are carrier and modulation frequency, respectively. a is amplitude of modulated signal(mod_1-3), The mix function is used for index carrier frequency, and spectral S-parameters (S_11,

978-1-6654-3008-1/21 $31.00 © 2021 IEEE   1806

S_21, and S_31 in this figure) is obtained by sweep the carrier frequency. The S-parameters are calculated by the ratio of reflected wave(b1) or forwarded waves (b2, b3) to the incident wave(a1).

The simulation result shows that: without the modulation signal amplitude (a=0 V), the transmission parameters S_21 and S_31 of the BAW circulator are exactly the same, the system is fully reciprocal, and the input power is equally split to the output ports as shown in Figure 8.

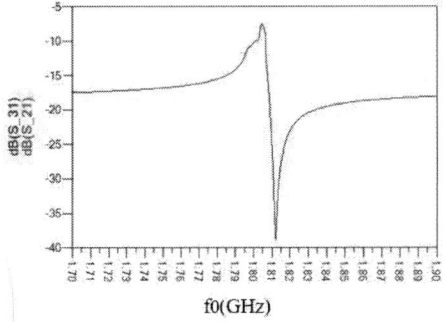

Fig. 8. simulated transmission from port 1 to ports 2 and 3 without modulation (a=0 V).

When the modulation signal amplitude a=5 V, simulation result shows that the forward insertion loss of the BAW circulator is less than 3.6 dB, the isolation between S_21 and S_31 is greater than 14 dB, as shown in Figure 9.

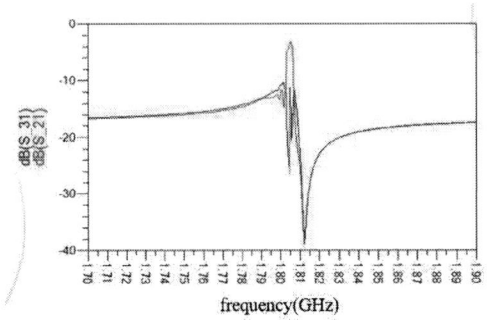

Fig. 9. simulated transmission from port 1 to ports 2 and 3 with modulation (a=5 V).

To further understand the influence of modulation signal amplitude on operation of circulator, Fig. 10 shows the isolation between ports 2 and 3 versus V_m. For V_m =0 V, the isolation is zero, as expected. The isolation increases with the increase of V_m until V_m =8 V, where the maximum isolation is 27 dB. Past this point, isolation decreases with the increase of V_m, because excessive modulation makes the counter-rotating states far away from the resonant frequency, resulting in signal is greatly attenuated.

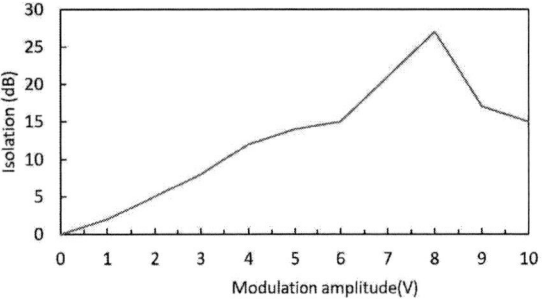

Fig. 10. Influence of modulation amplitude on the isolation of circulator.

Although it has the greatest isolation for V_m =8 V, but the insertion loss of the circulator is greater than 5 dB, as shown in Figure. 11. Modulation amplitude V_m =5 V is a better choice compared to V_m =8 V. Therefore, the optimal modulation condition can be obtained through this simulation method.

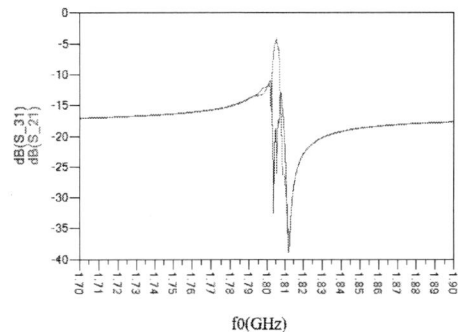

Fig. 11. Simulated transmission from port 1 to ports 2 and 3 with modulation (a=8 V).

IV. CONCLUSION

In this paper, a hybrid frequency-time domain simulation method is proposed for the bulk acoustic wave circulator. Using this method, a BAW circulator case at 1.805 GHz is designed. The simulation result shows that insertion loss is less than 3.6 dB, isolation is greater than 14 dB. The influence of modulation amplitude on the isolation of the BAW circulator is also analyzed, the optimal modulation condition can be obtained by this simulation method. A PCB for BAW circulator is designed and fabricated, Further device performance test will be carried out, and the test results will be compared with the simulation results. This research provides strong support for device design and optimization.

REFERENCES

[1] Jin Zhou, Negar Reiskarimian, "Jelena Diakonikolas, et al. Integrated full duplex radios," *IEEE Communications Magazine*, vol. 55(4), pp. 142-151, 2017.

[2] J. Zhou, T. H. Chuang, T. Dinc, et al. "Integrated Wideband Self-Interference Cancellation in the RF Domain for FDD and Full-Duplex Wireless," *IEEE Journal of Solid-State Circuits*, vol. 50(12), pp. 3015-3031, 2015.

[3] J. Kenneth. "Microwave ferrites and ferrimagnetics," *New York: MCGRAW-HILL*, 1962.

[4] C. E. Fay, R. L. Comstock. "Operation of the Ferrite Junction Circulator," *IEEE Transactions on Microwave Theory and Techniques*, vol. 13(1), pp. 15-27, 2003.

[5] A. Kord, D. L Sounas, A. Alu. "Pseudo-Linear Time-Invariant Magnetless Circulators Based on Differential Spatiotemporal Modulation of Resonant Junctions," *IEEE Transactions on Microwave Theory & Techniques*, vol. 66(6), pp. 2731-2745, 2018.

[6] Qin S, Xu Q, Wang Y E. "Nonreciprocal Components with Distributedly Modulated Capacitors," *IEEE Transactions on Microwave Theory & Techniques*, vol. 62(10), pp. 2260-2272, 2014.

[7] R. Lu, T. Manzaneque, Y. Yang, et al. "A Radio Frequency Nonreciprocal Network Based on Switched Acoustic Delay Lines," *IEEE Transactions on Microwave Theory and Techniques*, vol. 67(4), pp. 1516-1530, 2019.

[8] R. Fleury, D. L. Sounas, C. F. Sieck, et al. "Sound isolation and giant linear nonreciprocity in a compact acoustic circulator," *Science*, vol. 343(6170), pp. 516-519, 2014.

[9] N. A. Estep, D. L. Sounas, A. Alu. "On-chip non-reciprocal components based on angular-momentum biasing," *IEEE MTT-S International Microwave Symposium.* May 2015.

[10] N. A. Estep, D. L. Sounas, J. Soric, et al. "Magnetic-free non-reciprocity and isolation based on parametrically modulated coupled-resonator loops," *Nature Physics*, vol. 10(12), pp. 923-927, 2014.

[11] N. A. Estep, D. L. Sounas, A. Alu. "Magnet-less microwave circulators based on spatiotemporally modulated rings of coupled resonators," *IEEE Transactions on Microwave Theory and Techniques*, vol. 64(2), pp. 502-518, 2016.

[12] M. M. Torunbalci, T. J. Odelberg, S. Sridaran, et al. "An FBAR circulator," *IEEE Microwave and Wireless Components Letters*, vol. 28(5), pp. 395-397, 2018.

[13] Yao. Yu, G. Michetti, A. Kord, et al. "Magnetic-free radio frequency circulator based on spatiotemporal commutation of MEMS resonators," *Northern Ireland: IEEE Micro Electro Mechanical Systems (MEMS) Belfast*, Jan. 2018.

[14] M. M. Torunbalci, S. Sridaran, R. C. Ruby, et al. "Mechanically modulated microwave circulator," *20th International Conference on Solid-State Sensors, Actuators and Microsystems & Eurosensors*, Jun. 2019.

[15] Y. Yu, G. Michetti, A. Kord, et al. "Highly-Linear Magnet-Free Microelectromechanical Circulators," *Journal of Microelectromechanical Systems*, vol. 28(6), pp. 933-940, 2019.

[16] J. D. L. Iii, P. D. Bradley, S. Wartenberg, et al. "Modified Butterworth-Van Dyke circuit for FBAR resonators and automated measurement system," *Ultrasonics Symposium*, IEEE. Oct. 2000.

Gap in pagination due to unavailable paper.

Pages 1809-1812

Proceedings of the 16th Annual IEEE International
Conference on Nano/Micro Engineered and Molecular Systems
April 25-29, 2021

A Micromachined Electrochemical Angular Accelerometer Based on Interdigital Electrodes

Tian Liang, Junbo Wang*, Deyong Chen*, Jian Chen, Bowen Liu, Chao Xu, Wenjie Qi, Xu She,
Mingwei Chen, Anxiang Zhong, Yumo Duan

Abstract— This study proposes a micromachined electrochemical angular accelerometer based on interdigital electrodes, with the ability to measure low-frequency angular acceleration. In this paper, the design, fabrication, and characterization of the angular accelerometer were included, with a sensitivity of 100 V/(rad/s²) at 0.01Hz. Compared with commercial sensors, the micro sensor presented in this study has the characteristics of a simple electrode structure and manufacturing process, which may provide a new perspective for mass fabrication and further used in seismic monitoring.

I. INTRODUCTION

Angular vibration is one of the most basic forms of motion in nature, and it exists widely in the objective world [1]. The vibration of any object can be divided into linear motion and angular motion. The accurate measurement of angular motion is of great significance in the applications such as resource exploration and seismic monitoring [2-3].

Many angular accelerometers with different principles were developed, and they can be divided into piezoelectric type [4-5], piezoresistive type [6], electromagnetic type [7-8], capacitive type [9], optical type [10], and electrochemical type [11-18]. Different from these sensors that use solid as inertial mass, electrochemical angular accelerometers which adopt liquid as inertial mass is having these advantages of simple fabrication process and strong shock resistance due to there is no mechanical components.

The first electrochemical angular accelerometer [12] was proposed by V.A. Kozlov by changing the method of the electrochemical translational vibration sensor component in 2006, however, its performance was limited. Subsequently, Robert Leugoud put forward a three-axis device of a rotating electrochemical seismograph [16] which adopts magnetohydrodynamic technology for performance optimization. However, the electrodes of the previous reported devices were manufactured from platinum mesh by ceramic technology, leading to complicated manufacturing processes and poor consistency.

With the development of microelectromechanical systems, sensors based on MEMS technology have low cost, high consistency, and mass production [19-20]. To deal with the aforementioned problems, a micromachined electrochemical angular accelerometer based on interdigital electrodes was put forward with a simple and reliable structure and a high sensitivity.

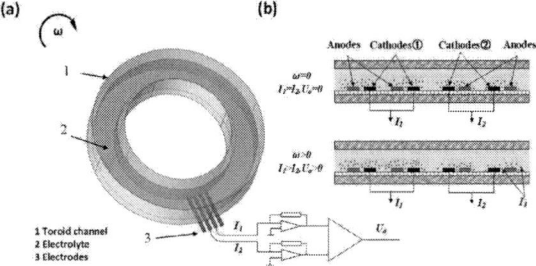

Figure 1. (a) The working principle of this micromachined electrochemical angular accelerometer based on interdigital electrodes, which mainly includes a toroid channel, an electrolyte, and interdigital micro electrodes. (b) When there is an inputting angular acceleration, the active ions will move relative to the sensitive electrodes, resulting in the two cathode currents to be unequal ($I_1>I_2$), which can be translated into differential current output.

II. WORKING PRINCIPLE

Fig 1(a) is the working principle of this presented micromachined electrochemical angular accelerometer based on interdigital electrodes, which mainly includes a toroid channel, interdigital sensitive electrodes, and filled with the electrolyte solution. The electrolyte is functioned as inertial mass to sense external angular acceleration. The micro electrodes were set up as ACAC-CACA (where A represent anode and C represent cathode), forming an interdigital structure (see Fig 1(b)). The electrode pair spacing should be much larger than the cathode and anode spacing to achieve differential output.

Fig 1(b) indicates how the micromachined electrochemical angular accelerometer works. The electrochemical redox reactions will appear on the electrode surface when a positive potential is applied on anode. Furthermore, the redox reactions on anode and cathode are $3I^- - 2e \to I_3^-$ and $I_3^- + 2e \to 3I^-$, respectively. When there is no input angular acceleration, there is a stable active ion (I_3^-, red dots as shown in Fig 1(b)) gradient distribution between anode and cathode. Due to the symmetry, the concentration of I_3^- near the two cathodes is equal, therefore the currents of the two cathodes are equal ($I_1 = I_2$), and the output is zero. When there is an inputting angular acceleration, the active ions will move relative to the sensitive electrodes, resulting in the two cathode currents to be unequal ($I_1>I_2$), which can be translated into a voltage output.

T. Liang, J. Wang, D. Chen, J. Chen, B. Liu, C. Xu, W. Qi, X. She, M. Chen, A. Zhong, Y. Duan are with the State Key Laboratory of Transducer Technology, Aerospace Information Research Institute, Chinese Academy of Sciences, Beijing, China, and also with School of Electronic, Electrical

and Communication Engineering, University of Chinese Academy of Sciences, Beijing, China.
*Corresponding authors:
Junbo Wang, E-mail: jbwang@mail.ie.ac.cn

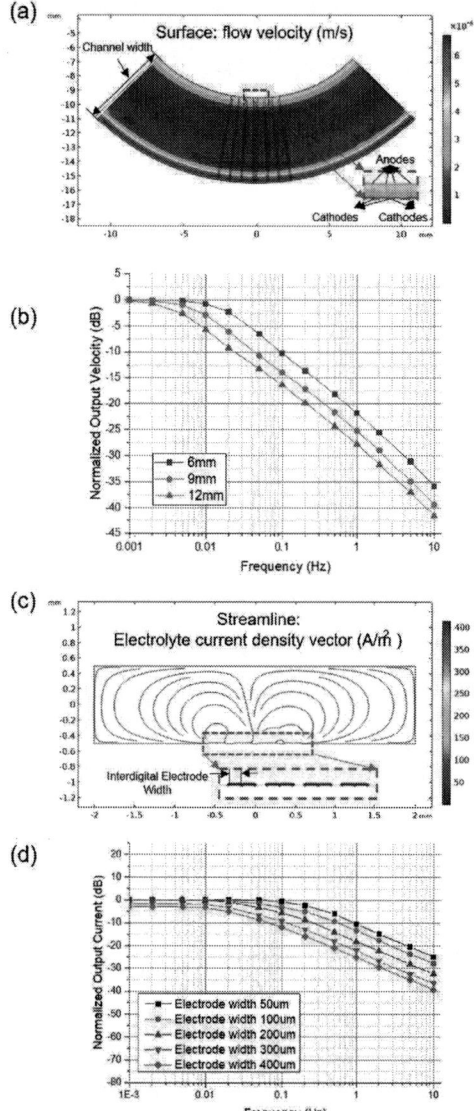

Figure 2. Simulation results of this micromachined electrochemical angular accelerometer based on interdigital electrodes. (a) Simulation results of mechanical link, and relationship between input angular acceleration frequency and the outputting flow velocity of electrolyte was obtained and shown in (b). (c)Simulation results of electrochemical link, and relationship between the frequency of inputting flow rate of electrolyte and the outputting current of the interdigital electrodes was obtained and shown in (d).

III. THEORY AND SIMULATION

The transduction process of this micromachined electrochemical angular accelerometer based on interdigital electrodes can be divided into two links including a mechanical link and an electrochemical link. The mechanical link converts the external angular acceleration signal into a relative movement between the electrolyte and the sensitive electrode, while the electrochemical module converts the relative movement between the sensitive

electrode and the electrolyte into an electrical signal for output. The influence of the key parameters of each module on the performance of the sensor is studied through the combination of theoretical derivation and finite element simulation analysis.

Since the transduction process of this micromachined electrochemical angular accelerometer is complex and difficult to calculate, this study adopts the finite element simulation method for numerical analysis, and establishes a two-dimensional numerical simulation model to analyze these key parameters that affect the output.

In mechanical module, according to the transfer function, the flow resistance has an important influence on the output. The larger the flow resistance, the better the device performance. Since the electrode is a planar structure, the flow resistance is mainly determined by the geometric size of this electrolyte flow channel. In the simulation of this mechanical module, changing the width of the flow channel can change the flow resistance. Fig 2(a) is the simulation model of the electrolyte flow velocity around the sensitive electrodes where the relationship between the electrolyte flow rate and the input angular acceleration frequency under different flow channel widths was show in Fig 2(b). The increase in the frequency of the input angular acceleration indicates that the output speed of the electrolyte is reduced, due to the low-pass model of this mechanical module. Meanwhile, a decrease of the width of toroid channel (6-12 mm) was noticed to increase the outputting velocity of electrolyte which can produce a higher sensitivity.

In the electrochemical module, current research points out that reducing the anode and cathode spacing can improve the performance of the device, but there is no research on the important parameter of electrode width. In the simulation of this electrochemical module, the influence of the electrode width on the output was simulated, keeping the total area of the electrode constant, reducing the electrode width and increasing the amount of electrode groups. Fig 2(c) is the simulation model of the electrolyte current density vector and the relationship between the output current of the electrodes and the input electrolyte velocity with different electrodes width was shown in Fig 2(d). The increase in the frequency of the electrolyte input rate indicates that the output current of the electrode is reduced, due to the low-frequency features of the electrochemical module. Meanwhile, the decrease of the width of electrodes was noticed to improve the device performance because of corresponding increase in the effective electrode area for chemical reactions.

According to the simulation results, the key parameters of the micromachined electrochemical angular accelerometer are determined as follows: the toroid channel width is set as 6 mm, the electrode width is set as 50 μm, and the spacing between cathode and anode is set as 10 μm. Furthermore, the spacing between the interdigital electrodes was set as 300μm to ensure that there is no crosstalk between the electrodes.

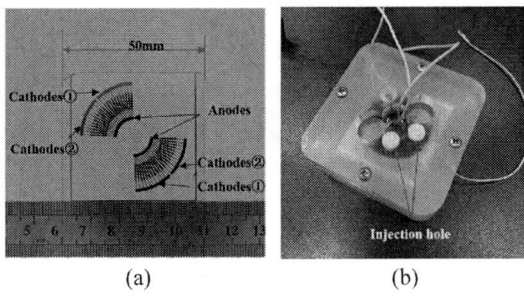

(a)　　　　　(b)

Figure 3. Prototypes of the fabricated micromachined electrochemical angular accelerometer based on interdigital electrodes, including (a) the sensing chip with sensitive microelectrodes, and (b) a testable angular accelerometer.

IV. FABRICATION

The prototype of the fabricated micromachined electrochemical angular accelerometer based on interdigital electrodes is shown in Fig 3. Fig 3(a) is the image of fabricated sensing chip with interdigital electrodes where two sets of interdigital electrodes are placed symmetrically and the blank space is reserved for the magnetohydrodynamic cell. The micro electrodes were fabricated by microstructuring technology called lift-off. Fig 3(b) is a prototype of the assembled micromachined electrochemical angular accelerometer which adopts a sandwich structure where the sensing chip was caught in the middle by two PMMA layers. The PMMA layers were functioned as toroid channel due to its good processability. Furthermore, a magnetohydrodynamic unit was contained in the prototype for characterizing in the laboratory.

V. DEVICE CHARACTERIZATION

Fig 4 shows the measurement results of this micromachined electrochemical angular accelerometer based on interdigital electrodes, including the relationship between input angular acceleration frequency and the (a) angular acceleration sensitivity, and (b) phase, which were obtained by magnetohydrodynamic excitation method in the laboratory and calibrated by the national-standard angular acceleration turntable at the Beijing Precision Engineering Institute for Aircraft Industry (BPEI). As shown in Fig 4, the micromachined electrochemical angular accelerometer based on interdigital electrodes produced the angular acceleration sensitivity as high as 100 V/(rad/s^2) at low frequency domain and acted like a low-pass link. More specifically, the sensitivity curve of the angular accelerometer tested in laboratory was coincided with the curve calibrated at BPEI which proved the test method in the laboratory was reliability. And the low pass feature of this presented micromachined electrochemical angular accelerometer was confirmed because of the consistency between the simulation results and test results.

VI. CONCLUSION

In this study, design, fabrication and characterization of the micromachined electrochemical angular accelerometer based on interdigital electrodes were included. The key parameters of the angular accelerometer were determined

and the frequency response of this angular accelerometer were characterized. The optimized parameters of the angular accelerometer were determined as 6 mm for the width of toroid channel and 50 μm for the width of electrodes, respectively, and the angular acceleration sensitivity were characterized as 100 V/(rad/s^2) at 0.01Hz.

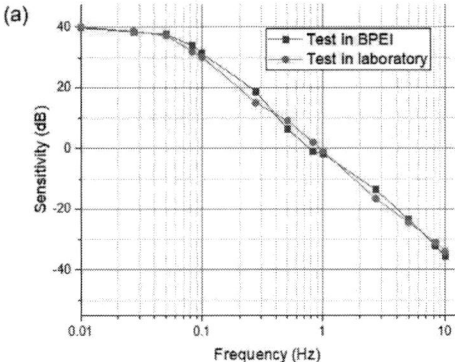

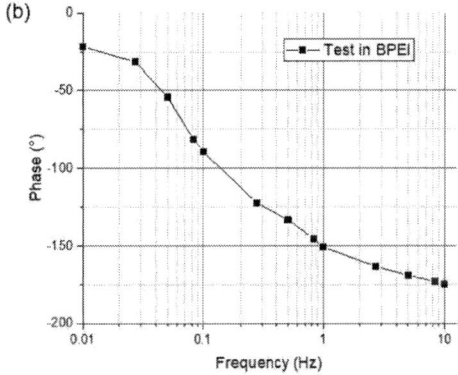

Figure 4. Measuring results of this micromachined electrochemical angular accelerometer based on interdigital electrodes, including the relationship between input angular acceleration frequency and the (a) angular acceleration sensitivity, and (b) phase.

ACKNOWLEDGEMENT

We are grateful to the support of the Chinese Academy of Sciences (Grant No. XDA22020302), the National Natural Science Foundation of China for Distinguished Young Scholars (Grant No. 61825107) and the National Natural Science Foundation of China (U1930206, 62071454, 62061136012).

REFERENCES

[1] F. Yuan, G. Xinping, and M. Junfa, "Measurement and Demonstration System Design for Molecular Circular Angular Accelerometer," (in Chinese), *Navigation and Control*, vol. 14, no. 02, pp. 76-82, 2015.

[2] L. Jaroszewicz *et al.*, "Review of the usefulness of various rotational seismometers with laboratory results of fibre-optic ones tested for engineering applications," *Sensors*, vol. 16, no. 12, p. 2161, 2016.

[3] D. L. Zaitsev, V. M. Agafonov, E. V. Egorov, A. N. Antonov, and V. G. Krishtop, "Precession azimuth sensing with low-noise molecular electronics angular sensors," *J. Sensors*, vol. 2016, pp. 1-8, May 2016.

[4] R. Marat-Mendes, C. J. Dias, J. N. Marat-Mendes, "Development of a piezoelectric sensor to measure angular acceleration," presented at the *10th International Symposium on Electrets*, Greece, Athens, Sept. 22-24, pp. 759-762, 1999.

[5] Y. Tomikawa, S. Okada, "Piezoelectric angular acceleration sensor," in *IEEE Symposium on Ultrasonics*, Honolulu, HI, USA, Oct. 5-8, pp. 1346-1349, 2003.

[6] T. Hidetoshi, K. Tetsuo, N. Akihito, et al., "Highly sensitive and low-crosstalk angular acceleration sensor using mirror-symmetric liquid ring channels and MEMS piezoresistive cantilevers," *Sensors and Actuators*, vol. 287, pp. 39-47, 2019.

[7] G. Feng and Y. Xueshan, "Study on active servo ultra-low frequency rotational accelerometer," (in Chinese), *Earthquake Eng. Eng. Dyn.*, vol. 38, no. 5, pp. 171-178, 2018.

[8] Q. Mingzhe et al., "Research on passive servo seismic rotational acceleration sensor with large damping," (in Chinese), *Seismol. Geol.*, vol. 40, no. 5, pp. 1170-1178, 2018.

[9] X. Yang, F. Gao, Q. Chi, T. She, L. Yang, and N. Wang, "Study of strong earthquake rotational accelerometer based on a spoke-type mass-string system," (in Chinese), *J. Natural Disasters*, vol. 24, no. 3, pp. 37-45, 2015.

[10] Z. Xin, Z. Bowen, D. Zhiguang, Z. Zheng, "Research on Response Angle Acceleration of High-precision FOG," (in Chinese), *Command Control and Simulation*, vol. 41, no. 4, pp. 130-134, 2019.

[11] V. A. Kozlov, V. M. Agafonov, J. Bindler, and A. V. Vishnyakov, "Small, low-power, low-cost sensors for personal navigation and stabilization systems," MET Tech. Inc., Raleigh, NC, USA, Tech. Rep., Jan. 2006. [Online]. Available: http://www.mettechnology.com, doi: 10.1142/9789812701626_0034.

[12] D. Zaitsev, V. M. Agafonov, V. E. Egorov, A. Antonov, and A. Shabalina, "Molecular electronic angular motion transducer broad band self-noise," *Sensors*, vol. 15, no. 11, pp. 29378-29392, 2015.

[13] D. Zaitsev, V. M. Agafonov, E. V. Egorov, A. N. Antonov, and V. G. Krishtop, "Precession azimuth sensing with low-noise molecular electronics angular sensors," *J. Sensors*, vol. 2016, pp. 1-8, May 2016.

[14] V. E. Egorov, V. I. Egorov, V. M. Agafonov, "Self-Noise of the MET Angular Motion Seismic Sensors," *J. Sensors*, vol. 2015, pp.1-5, 2015.

[15] V. A. Kozlov, M. V. Safonov, "Dynamic Characteristic of an Electrochemical Cell with Gauze Electrodes in Convective Diffusion Conditions," *Russian Journal of Electrochemistry*, vol. 40, no.4, pp. 518-520, 2004.

[16] R. Leugoud, A. Kharlamov, "Second generation of a rotational electrochemical seismometer using magnetohydrodynamic technology," *Journal of Seismology*, vol. 16, no. 4, pp.587-593, 2012.

[17] E. Egorov, V. Agafonov, S. Avdyukhina, and S. Borisov, "Angular molecular–electronic sensor with negative magnetohydrodynamic feedback," *Sensors*, vol. 18, no. 1, p. 245, 2018.

[18] B. Liu et al., "A MEMS Based Electrochemical Angular Accelerometer with Integrated Plane Electrodes for Seismic Motion Monitoring." *IEEE Sensors Journal*, vol. 20, no. 18, pp. 10469-10475, 2020.

[19] H. Huang et al., "A micro seismometer based on molecular electronictransducer technology for planetary exploration," *Appl. Phys. Lett.*,vol. 102, no. 19, May 2013, Art. no. 193512.

[20] W. T. He, D. Y. Chen, G. B. Li, and J. B. Wang, "Low frequencyelectrochemical accelerometer with low noise based on MEMS," *KeyEng. Mater.*, vol. 503, pp. 75-80, Feb. 2012.

Proceedings of the 16th Annual IEEE International
Conference on Nano/Micro Engineered and Molecular Systems
April 25-29, 2021

Research on the Stability Threshold and Subharmonic Oscillation in a Parametric Excitation MEMS Resonator

Kuo Lu[1], *Student Member, IEEE*, Kai Wu[1], *Student Member, IEEE*, Qingsong Li[*,1,3], *Member, IEEE*, Xin Zhou[1,3], *Member, IEEE*, Yongmeng Zhang[1,3], *Member, IEEE*, Ming Zhuo[1,3], *Member, IEEE*, Xuezhong Wu[1,2,3], *Member, IEEE*, and Dingbang Xiao[*,1,2,3], *Member, IEEE*.

Abstract—**The parametric excitation is an effective method to improve the quality factor of nano-micro resonators, which is achieved by superimposing a twice frequency pump signal on the fundamental frequency driving signal. This paper presents a research on the stability threshold and subharmonic oscillation in a parametric excitation MEMS resonator. It is demonstrated that there is a stability boundary of the parametric excitation, which can be defined by the threshold of parametric pump signal. When the pump signal's amplitude is lower than the threshold, the resonator's response will remain a stable vibration and with the increasing of the parametric voltage the fundamental frequency driving signal will decrease. Especially, when the amplitude of the parametric pump signal reaches its stability threshold, the fundamental frequency driving voltage will drop to zero. In this case, the resonator operates in a generalized parametric resonance condition, which is called subharmonic oscillation. Compared with the stable parametric excitation, the subharmonic response proves to have an additional time-varying response in addition to a steady-state response, indicating that it is a combination of different signals and depends on the oscillation time. The work presented in this paper can help better understand the principle of stability in parametric excitation MEMS resonators and the impact of subharmonic oscillation.**

I. INTRODUCTION

Recently, nano-micro resonators have attracted the attention of many research teams for their typical applications in advanced sensors, actuators and other fields [1-3]. With advantages of small size, low cost and high performance, nano-micro resonators have become one of the main development trends of the new generation of sensors, exhibiting great scientific research value and market potential [4-5].

There are many factors that can affect the performance of micro-nano resonators, among which the quality factor is one of the core influencing factors. The quality factor reflects the energy dissipation during the operation process of the resonator, which is induced by the air damping, the surface loss, the thermoelastic damping, and so on [6]. In the process of the structural design and device processing, the intrinsic mechanical quality factor of the resonator has been determined. In this case, the energy dissipation caused by the

inherent damping can be compensated by supplementing energy from the outside. As a result, the resonator's effective quality factor can be modulated by pumping external energy into the vibration mode, thereby improving its performance [7-8].

Parametric excitation is a typical method to enhance the effective quality factor by pumping energy into the oscillation system, which is usually realized by modulating the device's stiffness with a twice resonance frequency signal. The efficiency of parametric amplification is related to the pump signal's amplitude and phase [9-10]. Moreover, there is a stability boundary in the parametric excitation process, when the parametric excitation voltage is larger than its threshold, the resonator steps into the sustained oscillation state [11]. In this case, as long as the amplitude of the twice frequency pump signal is maintained at its stability threshold (or even larger), the resonator will keep vibrating even if the fundamental frequency driving signal is removed. The resonator operates in a generalized parametric resonance condition, which is called the subharmonic excitation. In general, the appearance of subharmonic excitation is caused by the quadratic nonlinear term of the system [12].

In this paper, the stability threshold and subharmonic oscillation in a parametric excitation MEMS resonator are theoretically analyzed associated with experiment results. A vacuum-packaged disk resonator is employed as the experimental testbed in this work. Firstly, the parametric excitation and its stability boundary is theoretically analyzed. Then, based on the multi-scale method, the dynamic model of the resonator with subharmonic excitation is considered. Furthermore, the related experiments are carried out to verify the correctness of the stability boundary and subharmonic excitation. Finally, the basic principles of the stability threshold and subharmonic oscillation in a parametric excitation MEMS resonator are concluded.

II. THEORETICAL ANALYSIS AND SIMULATIONS

A. Structural Description

The experimental device is a disk resonator which is a kind of structure composed by several concentric rings and connecting beams. This multi-ring nested structure provides

[1]College of Intelligence Science, National University of Defense Technology, Changsha, Hunan, China.
[2]Laboratory of Science and Technology on Integrated Logistics Support, National University of Defense Technology, Changsha, Hunan, China.
[3]Hunan MEMS Research Center, Changsha, Huna, China.
[*]Contacting Author: Qingsong Li and Dingbang Xiao are with the National University of Defense Technology; 109th DeYa Road, KaiFu

District, Changsha, Hunan, China (e-mail: liqingsong12@nudt.edu.cn & dingbangxiao@nudt.edu.cn).
Research supported by National Key R&D Program of China (Grant No. 2018YFB2002304), National Natural Science Foundation of China (Grant No.51705527) and Excellent Youth Foundation of Hu'nan Scientific Committee (Grant No. 2020JJ2033).

978-1-6654-3008-1/21 $31.00 © 2021 IEEE

convenience for arranging electrodes with different functions, effectively increasing the number of electrodes. Since the quality factor is one of the key parameters determining the performance of the resonator, in order to reduce the thermoelastic damping of the resonator, multiple masses are suspended on the outer ring [13]. Besides, in order to maintain the resonator operating in a vacuum environment (0.1Pa) to reduce the air damping, the experimental device uses a vacuum packaging process. Figure 1 shows the schematic diagram of the resonator.

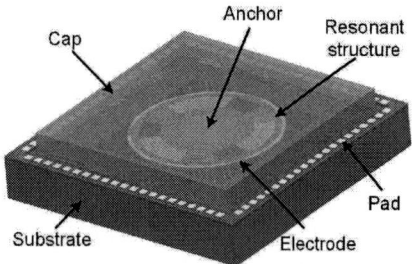

Figure 1. The schematic of the vacuum-packaged multiring disk resonator, containing the multiring resonant structure, the substrate, electrodes and the cap.

B. Parametric Excitation

Compared with the traditional resonant excitation methods where generally only fundamental frequency excitation signals are applied to the system, as for the parametric excitation, fundamental frequency driving signals are applied to the resonator associated with twice frequency pump signals simultaneously. In the case of parametric excitation, the resonator's response can be amplified by the parametric pump signal, which appears as a time-dependent dynamic stiffness parameter in the resonator's governing equation [14].

In the actual parametric excitation resonator system as shown in Figure 2, on the basis of a DC voltage bias V_{dc}, a fundamental frequency driving voltage $V_d(t) = V_d\sin\omega_d t$ superposed with a twice frequency pump signal $V_p(t) = V_p\sin(\omega_p t + \phi)$ is applied on the drive electrode, where $\omega_p = 2\omega_d$. When the resonator operates under the parametric excitation, its dynamical motion can be described by the Mathieu-Hill equation [12].

$$\ddot{x} + \frac{\omega_0}{Q_0}\dot{x} + \left[\omega_0^2 - K_1 - K_2\sin(\omega_p t + \phi)\right]x = F_3\sin\omega_d t \quad (1)$$

Where:

$$\begin{cases} \beta_1 = \dfrac{\varepsilon_r\varepsilon_0 A_{eff}}{2md_0^3}, \beta_2 = \dfrac{\varepsilon_r\varepsilon_0 A_{eff}}{md_0^2} \\ K_1 = (2V_{dc}^2 + V_d^2 + V_p^2)\beta_1 \\ K_2 = 4V_{dc}V_p\beta_1 \\ F_3 = V_{dc}V_d\beta_2 \end{cases} \quad (2)$$

Here, ω_0 is the natural frequency, Q_0 is the intrinsic mechanical quality factor of the resonator, ε_r is the relative permittivity, ε_0 is the vacuum dielectric constant, d_0 is the

initial capacitive gap, and A_{eff} is the equivalent capacitance area between the drive electrodes and the resonant structure.

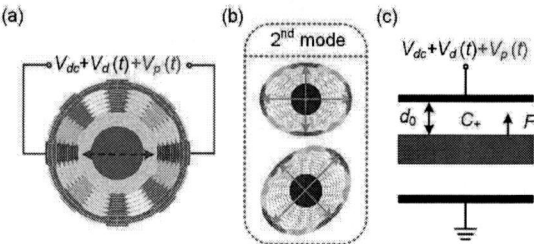

Figure 2. (a) The diagram for the resonator's parametric excitation; (b) The 3rd operation modes of the disk resonator; (c) The equivalent dynamic model of the resonantor under the parametric excitation.

Based on the harmonic balance method [12], when the vibration of resonator is an approximately periodic oscillation, the steady-state solution of Equation (1) can be expressed as:

$$x(t) = A_0\sin(\omega_d t + \psi) \quad (3)$$

Where:

$$A_0 = \frac{2F_3\sqrt{4\hat{\omega}_1^4 + 4\zeta^2 + K_2^2 - 4K_2(\hat{\omega}_1^2\sin\phi + \zeta\cos\phi)}}{4\hat{\omega}_1^4 + 4\zeta^2 - K_2^2}$$

$$\psi = \tan^{-1}\left(-\frac{2\zeta - K_2\cos\phi}{2\hat{\omega}_1^2 - K_2\sin\phi}\right) \quad (4)$$

$$\omega_1^2 = \omega_0^2 - \omega_d^2 - K_1$$

$$\zeta = \frac{\omega_0\omega_d}{Q_0}$$

It is obvious that the parametrically amplified amplitude A_0 is related to the frequency mistuning ω_1, the parametric voltage V_p, and the phase advance ϕ. The maximum amplitude of A_0 appears at $\phi = \pm\pi$, while the amplitude will be suppressed when $\phi = 0$, indicating that parametric excitation is phase-sensitive. As a result, the resonator's effective quality factor can be tuned by modifying the phase ϕ.

To evaluate the effect of the parametric amplification, its gain factor G is defined as:

$$G = \frac{A_0\mid_{V_p\neq0}}{A_0\mid_{V_p=0}} = \frac{2\zeta}{2\zeta - K_2} \quad (5)$$

It represents the ratio of the resonator's steady-state vibration amplitude with and without parametric excitation under the condition of the maximum oscillation amplitude ($\omega_1 = 0$ and $\phi = \pm\pi$). Obviously, it is a function about the parametric excitation voltage V_p. Substituting Equation (5) into Equation (4), it can be obtained that there is a linear relationship between the reciprocal of G and the parametric excitation voltage V_p:

$$\frac{1}{G} = 1 - \frac{K_2}{2\zeta} = 1 - \left(\frac{2\beta_1 V_{dc}Q_0}{\omega_0\omega_d}\right)V_p \quad (6)$$

978-1-6654-3008-1/21 $31.00 © 2021 IEEE

Noticeably, it is demonstrated that the following condition needs to be satisfied for the stable oscillation of the resonator when parametrically excited:

$$B = 4\omega_1^4 + 4\zeta_1^2 - K_2^2 > 0 \qquad (7)$$

As the value of B gradually approaches 0, the resonator's vibration will rapidly increase and approach infinity. Therefore, the stability boundary of the system is determined by the amplitude of the parametric excitation voltage V_p. It can be observed from Equation (4) that the sustained oscillation occurs when the vibration system satisfies:

$$B = 4\omega_1^4 + 4\zeta_1^2 - K_2^2 = 0 \qquad (8)$$

Theoretically, when the above-mentioned equation is satisfied, the vibration amplitude of the resonator will become infinite, and it will no longer be a steady-state response. In this case, the resonator operates in a parametric resonance condition, which determines the stability boundary of the parametrically excited system. The stability threshold between the steady-state and the unsteady-state can be obtained by solving Equation (8):

$$V_t = \frac{\omega_0 \omega_d}{2\beta_1 Q_0 V_{dc}} \qquad (9)$$

When $V_p = V_t$, the resonator will maintain its vibration even when the driving signal $V_d(t)$ is removed. Besides, the oscillation amplitude of the resonator will increase without bound theoretically when $V_p > V_t$. In this case, the effective quality factor will grow infinitely with the input of parametric energy. However, it should be noted that the oscillation amplitude cannot grow infinitely due to the pull-in effect in the realistic system.

C. Subharmonic Oscillation

When the fundamental frequency driving signal is removed while the twice frequency parametric excitation signal remains at the stability threshold (or even larger), the resonator will step into a special parametric resonance condition, which is named as the subharmonic oscillation. Generally, the appearance of subharmonic oscillation is caused by the resonator's nonlinearities, and in this paper, it is generated by the system's quadratic nonlinearity. It has been proved in previous researches that the electrostatic nonlinearity is the main nonlinearity in capacitive resonators, which is caused by electrostatic driving forces [13].

In the case of subharmonic oscillation, the fundamental frequency driving signal is removed, and only maintained the twice frequency parametric pump signal. Therefore, the resonator's nonlinear dynamic equation can be expressed as:

$$\ddot{x} + 2\gamma\dot{x} + (\omega_0^2 + \alpha_1)x + \alpha_2 x^2 + \alpha_3 x^3 = F_0 \sin(\omega_p t) \qquad (10)$$

where,

$$\begin{cases} \alpha_1 = \dfrac{2\varepsilon_0 \varepsilon_r A_{eff} V_{dc}^2}{md_0^3}, \quad \alpha_2 = \dfrac{3\varepsilon_0 \varepsilon_r A_{eff} V_{dc}^2}{md_0^4} \\[3mm] \alpha_3 = \dfrac{4\varepsilon_0 \varepsilon_r A_{eff} V_{dc}^2}{md_0^5}, \quad F_0 = \dfrac{2\varepsilon_0 \varepsilon_r A_{eff} V_{dc} V_d}{md_0^2} \end{cases} \qquad (11)$$

In equation (10), γ is the resonator's damping coefficient. And the resonator's nonlinear elastic coefficients are represented as α_1, α_2, and α_3, respectively.

In order to study the influence of high-order nonlinear terms on resonator's dynamic characteristics, the multi-scale method is used to analyze this resonant system. Firstly, a small parameter ε is introduced to unify the dimensions in Equation (10). In this case, the resonator's vibration amplitude x, the amplitude of the driving force F_0, and the resonator's damping coefficient γ can be expressed as $x = \varepsilon u$, $F_0 = \varepsilon f_0$, and $\gamma = \varepsilon\mu$, respectively. Therefore, the dimension of nonlinear terms, damping terms and driving terms in Equation (10) are unified so that it can be rewritten as:

$$\ddot{u} + \omega_0^2 u = -2\varepsilon\mu\dot{u} - \varepsilon\alpha_2 u^2 - \varepsilon^2 \alpha_3 u^3 + f_0 \sin\omega_p t \quad (12)$$

Then, based on the principle of the multi-scale method, the traditional time scale is extended to the multiple time scales, where $T_0 = t$, $T_1 = \varepsilon t$. The dynamic response of the resonator depends on the relationship between the frequency of the excitation force and the natural frequency of the eigenmode. Especially, in the case of the subharmonic excitation, the frequency of the excitation signal is $\omega_p = 2\omega_0 + \varepsilon\sigma$, where σ is the frequency detuning. The dynamic oscillation in equation (12) can be expressed as:

$$u = a \exp(\lambda\varepsilon t)\sin\left(\frac{1}{2}\omega_p t \pm \theta\right) + \frac{f_0}{\omega_0^2 - \omega_p^2}\sin(\omega_p t) \qquad (13)$$

where:

$$\lambda = -\mu \pm \sqrt{\frac{\alpha_2^2 f_0^2}{4\omega_0^2\left(\omega_0^2 - \omega_p^2\right)^2} - \frac{\sigma^2}{4}} \qquad (14)$$

According to the equation (13), it is obvious that the resonator's oscillation is composed by a fundamental frequency response and a twice frequency response. Besides, it is noticeable that the resonator's fundamental frequency response is a time-varying response which is determined by λ as shown in equation (14), while its twice frequency response is a steady state response. The value category and range of λ determine the type of the resonator response as shown in Table I. In the first case, λ is a complex number, the resonator's fundamental frequency response is a kind of continuous oscillation attenuation oscillation which determines the response of the device. When it comes to the real number case, the resonator's response will show different evolutionary laws for different λ. As for the second case, $\lambda < 0$, based on the equation (13) and (14), the resonator exhibits a typical attenuation response without oscillation. Finally, when $\lambda < 0$, the resonator exhibits a typical rising response without oscillation.

TABLE I. THE RELATIONSHIP BETWEEN THE RESONATOR'S RESPONSE AND λ

No.	Value category and range of λ	Response type
1	$\lambda = a + bi$ (a and b are real)	continuous oscillation attenuation
2	$\lambda < 0$	attenuation without oscillation
3	$\lambda > 0$	rising without oscillation

In general, the resonant frequency of the resonator is locked by the phase-locked loop in the actual experimental system, ensuring that the frequency detuning σ is limited within a small range. As a result, according to the Equation (14), λ can be guaranteed to be a real number and if $\lambda = 0$ is established, the amplitude of excitation force needs to meet:

$$f_{th} = \frac{6\mu\omega_0^3}{\alpha_2} \quad (15)$$

The simulation results of the resonator responses with different λ ($\lambda < 0$ and $\lambda > 0$) are displayed in Figure 3.

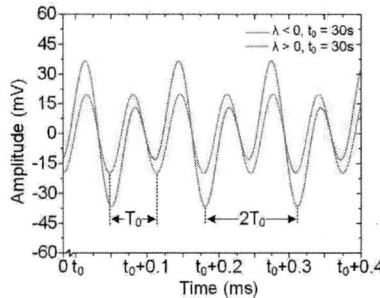

Figure 3. Simulation results of the disk MEMS resonator's subharmonic responses when it has quadratic nonlinearity with different λ.

It is obvious that at the time of 30s, for different values of λ, the resonator's time domain response exhibits different manifestations. Since $\lambda < 0$, the fundamental frequency response signal of the resonator quickly decays to infinitesimal, and the overall time domain response of the resonator appears as a typical twice frequency response signal. In contrast, when $\lambda > 0$, at $t_0 = 30$s, the time-varying term of the resonator is gradually increasing. At this time, the overall performance is a mixed signal of the fundamental frequency response signal and the twice frequency signal. Then as time continues, the fundamental frequency signal in the resonator's time domain response continues to increase to play a dominant role (before the pull-in effect occurs), and then the twice frequency response signal can be ignored.

III. EXPERIMENT AND DISCUSSION

In parametric excitation experiments, a lock-in amplifier is applied to produce the excitation signals. Moreover, the resonator's response enters this instrument for analysis after the demodulation. The schematic diagram of the parametric excitation experiment control loop is demonstrated in Figure 4. The details about the resonator used in the experiments are shown in Table II.

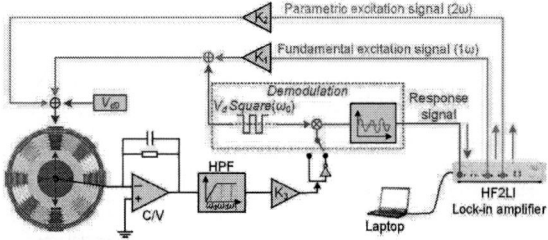

Figure 4. Schematic diagram of the control loop for the parametric excitation experiment.

TABLE II. THE STRUCTURAL PARAMETERS OF THE DISK RESONATOR

Structural parameter	Value
Radius of the outermost ring r	4mm ± 1μm
Height of the ring h	150μm ± 1μm
Gap between ring and outer electrodes d_0	12μm ± 0.5μm
Effective (2nd mode) mass m	1.8mg ± 0.1mg
Natural frequency f_0	4323.1Hz
Intrinsic mechanical quality factor Q_0	292,995

A. Threshold Voltage and Stability Boundary

In order to evaluate the parametric gain and find the stability boundary experimentally, different parametric pump voltages are applied on the driven electrodes while keeping the fundamental frequency driving voltage at a constant value (10mV). Meanwhile, to obtain a better parametric amplification effect, the phase difference between the parametric pump signal and the fundamental frequency driving signal is set as π. The resonator's response amplitudes at different parametric pump voltages are recorded and compared with the theoretical results as shown in Figure 5.

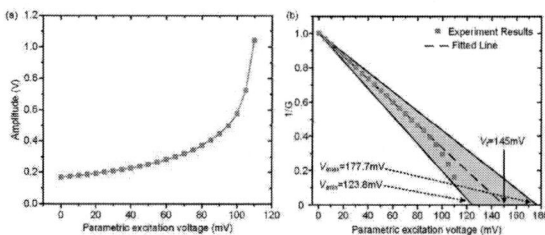

Figure 5. The resonator's response and gain factors at different parametric pump voltages. The 1ω driving voltage is maintained at 10mV, and the phase of $V_p(t)$ is ahead of $V_d(t)$ by π. The minimum and maximum theoretical value of the threshold voltage are determined by the errors of the resonator's parameters as shown in Table II.

It is apparent that the vibration amplitude of the resonator increases rapidly with the increment of the parametric pump voltage V_p, and the reciprocal of the parametric gain factor $1/G$ has an approximately linear relationship with the parametric voltage V_p, which is consistent with the theorical analysis. Noticeably, the parametric gain increases sharply especially when the parametric voltage exceeds 90mV. This is because the high order nonlinear term plays an important role in the electrostatic driving force when the resonator is working on large displacement, which causes the sharp change in the gain factor. Obviously, with the increasing of the parametric excitation voltage, the resonator's response grows rapidly, especially after the parametric excitation voltage exceeds 90mV, it shows a growing trend close to infinity. In this case, the steady-state response of the resonator is broken and gradually enters the parametric resonance state.

B. Subharmonic Oscillation

Based on the subharmonic oscillation theoretical analysis, it is obvious that the time domain response of the resonator will evolve over time, and the way it evolves and the final form of expression are determined by the magnitude of the excitation force. Therefore, the different subharmonic oscillations of the resonator can be studied and analyzed by changing the amplitude of the external excitation force. According to the basic parameters of the resonator, in subharmonic excitation experiments, the subharmonic excitation voltage V_p was set as 0.2V and 0.5V, to satisfy the condition of $\lambda < 0$ and $\lambda > 0$, respectively.

When $V_p = 0.2$V, the subharmonic excitation force does not reach its threshold condition ($f_p < f_{th}$), where the fundamental frequency response signal in the time domain is an attenuation signal without oscillation, as shown in Figure 6a. In this case, the fundamental frequency response signal of the resonator gradually decays over time. In the I stage as shown in Figure 6b, the fundamental frequency response signal of the resonator is stronger than the twice frequency response signal, and it is dominant in the overall time domain response signal. At this time, the response of the resonator appears as a typical mixed wave of two frequency signals. Then in the II stage as shown in Figure 6c, the fundamental frequency response signal gradually weakens, and the proportion of the twice frequency response signal gradually increases. In the last III stage as shown in Figure 6d, the fundamental frequency response signal is almost attenuated to negligible, at this time the resonator exhibits an approximate twice frequency response signal.

(b) The resonator's response in the I stage; (c) The resonator's response in the II stage; (d) The resonator's response in the III stage.

On the contrary, when $V_p = 0.5$V, the subharmonic excitation force has reached its threshold condition ($f_p > f_{th}$), where the fundamental frequency response signal in the time domain is a rising signal without oscillation, as shown in Figure 7a. Compared with the first case, the resonator's response displays an opposite rising trend. In the I stage as shown in Figure 7b, the fundamental frequency response signal is equivalent to the twice frequency response signal, so that the response of the resonator appears as a typical mixed wave of two frequency signals. Then in the II stage as shown in Figure 7c, the fundamental frequency response signal grows, and its proportion gradually increases. In the last III stage as shown in Figure 7d, the fundamental frequency response signal occupies an absolute dominant position, at this time the resonator exhibits an approximate fundamental frequency response signal.

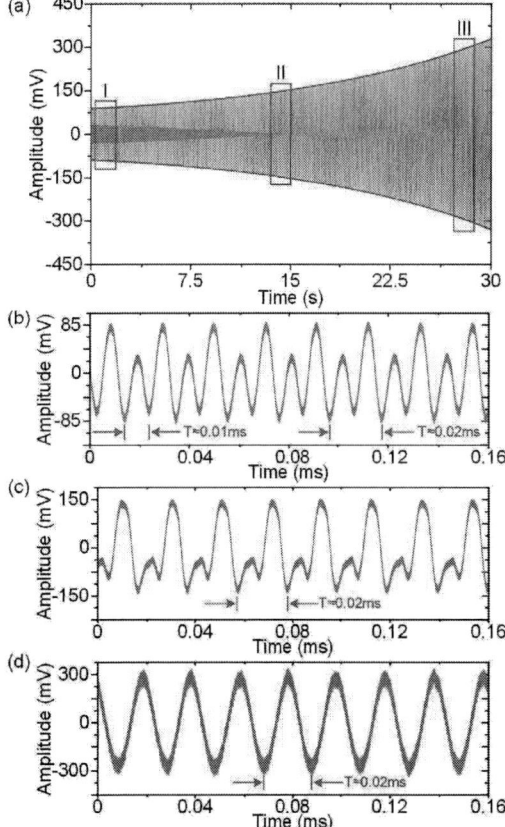

Figure 7. Responses of the disk resonator under subharmonic excitation when $\lambda > 0$. (a) The resonator's response in the time domain when $\lambda > 0$; (b) The resonator's response in the I stage; (c) The resonator's response in the II stage; (d) The resonator's response in the III stage.

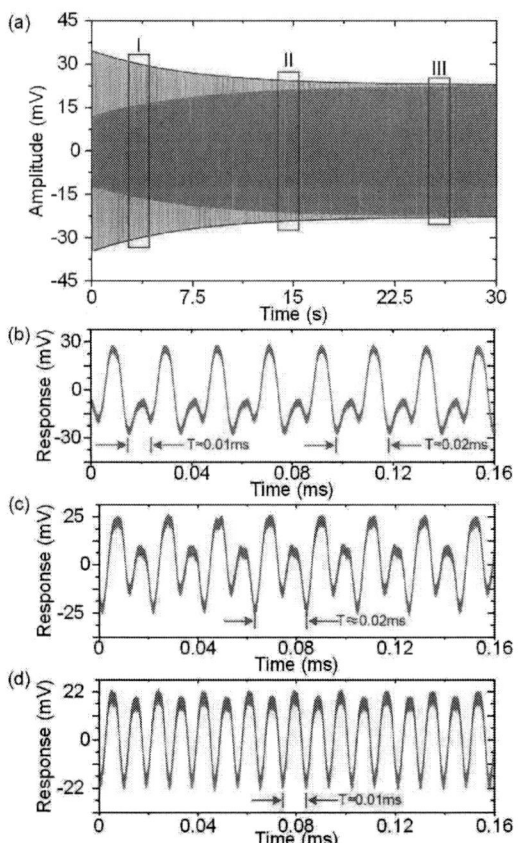

Figure 6. Responses of the disk resonator under subharmonic excitation when $\lambda < 0$. (a) The resonator's response in the time domain when $\lambda < 0$;

Obviously, due to the existence of the quadratic nonlinearity which is mainly induced by the large electrostatic force, when subharmonic (twice frequency) excitation signals are applied on electrodes, the resonator's response converts to a mixed signal of the fundamental frequency and twice frequency. Compared with the primary excitation, the subharmonic response proves to have an additional time-varying response in addition to a steady-state response, indicating that it is a combination of different signals and depends on the oscillation time. The evolution of the periodic solution is closely related to the strength of subharmonic excitation signal, which also determines the final manifestation of the resonator's response.

IV. CONCLUSION

In this paper, the stability threshold and subharmonic oscillation in a parametric excitation MEMS resonator is researched.

Based on the harmonic balance method, the capacitive MEMS resonator's dynamic parametric excitation model is analyzed and simulated. It is demonstrated that the resonator's response will increase with the increasing of the parametric pump signal and it will grow to infinity theoretically. However, it is impossible in a real experiment system due to the existence of pull-in effect and other limiting factors. As a result, when the response amplitude of the resonator is maintained at a constant value, the amplitude of the twice frequency pump signal is continuously increased, so that the amplitude of the fundamental driving signal is reduced to 0, where the parametric voltage reaches its stability threshold. In this case, the steady-state of the resonator is broken from and it steps into the subharmonic oscillation condition. The resonator will keep vibration under the sole action of parametric pump signals even if the fundamental frequency driven signal has been removed.

In this capacitive MEMS resonator, it has been proved that the quadratic nonlinearity induced by the electrostatic force causes the resonator's subharmonic oscillation. In the subharmonic excitation condition, the time domain response signal of the resonator is a mixed signal composed of a fundamental frequency response signal and a twice frequency response signal. Besides, the fundamental frequency response signal of the resonator is a time-varying signal, which will evolve over time according to the amplitude of the sub-harmonic excitation force, so that the resonator exhibits different mixed wave responses.

The parametric excitation is an effective method to enhance the performance of nano-micro resonators, especially its core parameter: quality factor. Based on this work, the stability threshold and subharmonic oscillation in a parametric excitation resonator has been deeply analyzed, which does help to complete the theoretical system of parametric excitation. It has been proved that theory and experiment are consistent, which is meaningful for enhancing the performance of nano-micro resonators.

ACKNOWLEDGMENT

The authors would like to thank the Laboratory of Microsystem, National University of Defense Technology, China, for equipment access and technical support. This work was supported by National Key R&D Program of China (Grant No. 2018YFB2002304), National Natural Science Foundation of China (Grant No. 51705527) and Excellent Youth Foundation of Hu'nan Scientific Committee (Grant No. 2020JJ2033).

REFERENCES

[1] J. Pirkkalainen, M. Brandt, and F. Massel, "Squeezing of quantum noise of motion in a micromechanical resonator," *Phys. Rev. Lett.*, vol. 115, pp. 243601, July 2015.

[2] Q. Li, D. Xiao and Y. Xu, "Nonlinearity reduction in disk resonator gyroscopes based on the vibration amplification effect," *IEEE T. Ind. Electron.*, vol. 67, pp. 6946-6954, September 2019.

[3] K. Lu, Q. Li, X. Zhou and D. Xiao, "Modal coupling effect in a novel nonlinear micromechanical resonator," *Micromachines*, vol.11, no. 5, pp. 472, April 2020.

[4] Y. Tsaturyan, A. Barg, and E. Polzik. "Ultra-coherent nanomechanical resonators via soft clamping and dissipation dilution," *Nat. Nanotechnol.*, vol. 12, pp. 776-783, June 2017.

[5] K. Lu, X. Zhou, Q. Li, and D. Xiao, "Coherent phonon manipulation in a disk resonator gyroscope with inertial resonance," in *7th IEEE International Symposium on Inertial Sensors and Systems (INERTIAL)*, Hiroshima, March 2020, pp. 9090063.

[6] K. Wu, K. Lu, Q. Li and D. Xiao, "Analysis of parametric and subharmonic excitation in push-pull driven disk resonator gyroscopes," *Micromachines*, vol. 12, no.1, pp. 61, January 2021.

[7] K. L. Turner, S. A. Miller, P. G. Hartwell, and S. G. Adams, "Five parametric resonances in a microelectromechanical system," *Nature*, vol. 396, pp. 149–152, November 1998.

[8] R. B. Karabalin, R. Lifshitz, M. C. Cross, and M. L. Roukes, "Signal amplification by sensitive control of bifurcation topology," *Phys. Rev. Lett.*, vol. 106, no.9, pp. 094–102, March 2011.

[9] I. Mahboob, H. Yamaguchi, "Piezoelectrically pumped parametric amplification and Q enhancement in an electromechanical oscillator," *Appl. Phys. Lett.*, vol. 92, no.17, pp. 173109, April 2008.

[10] K. M. Harish, B. J. Gallacher, J. S. Burdess, and J. A. Neasham, "Experimental investigation of parametric and externally forced motion in resonant MEMS sensors," *J. Micromechanics Microengineering.*, vol. 19, no. 1, pp. 15–21, December 2008.

[11] Z. X. Hu, B. J. Gallacher, J. S. Burdess, and K. A. Townsend, "Parametrically amplified MEMS rate gyroscope," *Sens. Actuators Phys.*, vol. 167, pp. 249–260, February 2011.

[12] A. Nayfeh, and D. Mook, Nonlinear Oscillations, 1st ed., vol. 4, Willey: New York, 1979, pp. 195–201.

[13] X. Zhou, C. Zhao, D. Xiao, and A. Seshia, "Dynamic modulation of modal coupling in microelectromechanical gyroscopic ring resonators," *Nat. Commun.*, vol. 10, no. 4980, October 2019.

[14] G. Sobreviela, C. Zhao, M. Pandit, and A. Seshia, "Parametric noise reduction in a high-order nonlinear MEMS resonator utilizing its bifurcation points," *J. Microelectromechanical Syst.*, vol. 26, no. 6, pp. 1189–1195, December 2017.

Gap in pagination due to unavailable paper.

Pages 1823-1824

Proceedings of the 16th Annual IEEE International
Conference on Nano/Micro Engineered and Molecular Systems
April 25-29, 2021

Large-Swing Low-Distortion High Voltage Amplifier Design for MEMS Gyroscopes

Hua Chen*, *Member, IEEE*, Ke Liu, and Zhen Meng*

Abstract— This paper presents a high-performance high-voltage (HV) amplifier with an on-chip charge pump for MEMS gyroscopes. The HV amplifier is realized by an HV operational amplifier (op-amp) and a resistance negative feedback. The HV op-amp is a single-stage differential scheme with an HV NMOS input pair and an HV PMOS current mirror. Two integrated polysilicon resistors detect the output voltage and feed it back to the negative input of the op-amp. The on-chip charge pump is the Dickson type, works at skipping mode, and provides as high as 300 µA load current for the HV amplifier. Based on the HHGrace 0.35 µm Bipolar-CMOS-DMOS process, the proposed HV amplifier and the high-efficiency charge pump were designed. Post-simulation results show that the charge pump can output a stable voltage of 20V with a ripple of ± 320 mV_{p-p} for the HV amplifier. The gain, bandwidth, harmonic distortion, and output swing of the HV amplifier are 6, 2.2 MHz, -40 dBc, and 12 V_{p-p}, respectively. The total power consumption is 7.4 mW under a 5 V supply.

I. INTRODUCTION

Microelectromechanical system (MEMS) sensors have many advantages such as low cost, small size, light weight, low power, etc. They are widely used in many applications, such as consumer electronics, automotive electronics, medical health, and industrial control. As a core component of the inertial measurement unit, the MEMS gyroscope determines the measurement accuracy. To achieve high resolution and low bias-instability, modulation and filtering techniques are used in the readout circuit to eliminate the low-frequency noise and the direct-current (DC) offset [1]. The gyro requires a high-voltage (HV) signal to be applied to the mass. The DC component of the HV signal provides a polarization voltage, and the alternating-current (AC) one provides a modulation carrier. Therefore, how to design an HV amplifier with a large swing, low distortion, and low power has become an important topic. Thanks to the pure capacitive characteristic exhibited by the gyro, the HV amplifier does not need to be designed with strong driving ability. This feature undoubtedly reduces the complexity and power consumption of the circuit.

Generally, the HV amplifier is a kind of closed-loop amplifier with negative voltage feedback. Hence the internal HV operational amplifier (op-amp) has become the design focus because it determines the open-loop characteristic. Literature [2] reported a fully differential HV op-amp with adjustable output common-mode level for MEMS scanning micromirrors. The op-amp consists of three stages: low voltage (LV) amplifier, HV amplifier, and HV Class-AB

push-pull output, with a load capacitance of 100 pF. The work in [3] proposed a two-stage architecture: an HV amplifier cascading an HV Class-AB output. The load is 300 pF. The circuit exhibits low harmonic distortion of -56 dBc, and a large output swing of 180 V. The work in [4] presented a three-stage op-amp which consists of an LV folded-cascode amplifier, an HV amplifier, and an HV Class-B output. It innovatively uses three poles and two zeros within the unity-gain bandwidth to stabilize the whole circuit. The load is 100 pF~10 nF, and the differential output swing is 290 V_{p-p}. For less than 20 pF load capacitance applications, the works in [5][6][7] removed the Class-AB or Class-B push-pull output, simplified the op-amp into two-stage, and used Miller-capacitor compensation to stabilize the circuit. Although the gain provided by the two-stage circuit is sufficient, there are two disadvantages in the circuit: 1) need a compensation circuit that works in the HV environment; 2) use the current source as active load in the second stage, which leads to partial cancellation between the pull-down current and the pull-up one when the output voltage changes from high to low. This results in a long tail and small swing.

Based on the above reviews, in this work, we referred to the literature [5][6][7] due to the 10 pF capacitance exhibited by the gyro and adopted a one-stage scheme. We utilized the NMOS-type input pair instead of the PMOS-type due to the low input common-mode level of the HV amplifier. Because the highest frequency and the largest amplitude of the input signal are 500 kHz and 1 V, respectively, the bias current was set to 300 µA to obtain high bandwidth and fast slew rate (SR). Moreover, to supply the proposed HV amplifier, an on-chip high-efficiency charge pump was also designed. The charge pump could output a 20 V voltage with a 300 µA load current capability. Based on the HHGrace 0.35 µm Bipolar-CMOS-DMOS (BCD) process [8][9], the HV amplifier and the charge pump were implemented. Post-simulation results show that the HV amplifier achieves a large swing and low distortion characteristic.

II. CIRCUIT DESIGN

The block diagram of the proposed HV amplifier is shown in Fig. 1. The power supply is 5V and the load capacitance exhibited by the mass is 10 pF. The input signal comes from the previous Digital-to-Analog (DAC) module, with a common-mode level of 1.65 V. The maximum input frequency is 500 kHz, and the largest signal amplitude is 1V. The target of the HV amplifier is generating a rail-to-rail HV signal for the gyro within a 5 mW power dissipation. The

---

All authors are with Smart Sensing Center, Institute of Microelectronics Chinese Academy of Sciences (CAS), 3 Beitucheng West Road, Beijing, 100029, China.

*Contacting Author: Zhen Meng and Hua Chen are both with Institute of Microelectronics CAS; (phone: +86-010-82995709; e-mail: mengzhen@ime.ac.cn; chenhua111@mails.ucas.ac.cn).

978-1-6654-3008-1/21 $31.00 © 2021 IEEE

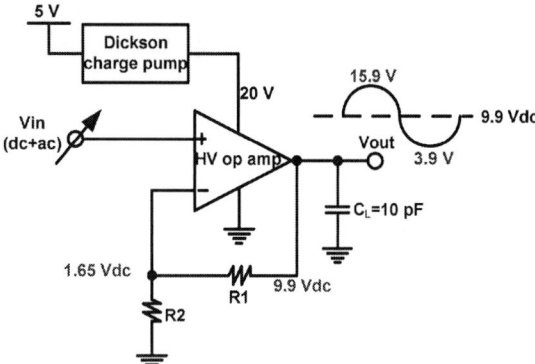

Figure 1. Block diagram of the proposed HV amplifier.

second-order harmonic distortion must be less than -30 dBc.

To maximize the swing and minimize the distortion, the common-mode level of the output signal was set to about 10 V. To achieve a good resistor-matching, the resistance ratio of R1 and R2 was specified to 5. Therefore, for the DC part, a 1.65V input DC voltage leads to a 9.9V DC output; for the AC part, a 1V input AC amplitude brings about a 6V AC output. Hence, the maximum output voltage is 15.9 V and the minimum one is 3.9 V as illustrated in Fig.1. Besides, to supply the HV op-amp, a high-efficiency charge pump that generates a 20 V source was accordingly designed.

The internal HV op-amp is illustrated in Fig. 2. To obtain unconditional stability and large swing, the op-amp is single-stage. To completely switch the bias current when the input signal is at the highest frequency and the largest amplitude, the op-amp utilizes current-mirror load instead of current-source which is commonly used in [5][6][7]. Since the input DC voltage is 1.65 V, the input pair was realized by the NMOS-type HV transistor. Specifically, it is the NLDMOS transistor with a threshold voltage of 0.66 V, a safety Vgs of 6 V, and a maximum Vds of 25 V. For the current-mirror load, M3 and M4 are both the PHDMOS transistor with a threshold voltage of 1.71 V and a largest Vgs and Vds of 25 V. For the tail-current circuit, M5, M6, and M7 are all the ordinary NMOS transistor with a

threshold voltage of 0.45 V. Note that the M7 was used to filter the bias-current noise.

The bias current I_B of the circuit depends on the circuit requirement when the input signal is at the maximum frequency and the largest amplitude, that is, 500 kHz and 1 V. At this time, the circuit works at a large-signal state which calls for high SR. Based on the equation in the literature [10], we got $SR = V_{out\_peak} / (T / 4)$, wherein T is the signal period. Combined with the formula of $SR = I_B / C_L$, we could get the value of I_B. That is, 120 µA. To reduce the nonlinearity and distortion of the output voltage, I_B was designated as 300 µA. To achieve a large output swing, the overdrive voltages of M2, M4, and M5 were allocated as 0.3 V, 0.15 V, and 0.15 V, respectively. Based on the bias current and overdrive voltage, the size of each component was calculated as shown in Table I.

The on-chip charge pump is the inductor-less Dickson-type [11], as depicted in Fig. 3. The function of the charge pump is to boost the 5V power supply to 20V to provide a high-quality on-chip supply for the HV op-amp. As shown in the figure, as a load, the HV op-amp is equivalent to a 300 µA constant current source. Overall, the charge pump works at skipping mode, that is, when the HV output is near the final value, the logic control circuit periodically turns on or off the ring oscillator to stabilize the HV output. In terms of power supply, to achieve high performance, the bandgap reference and comparator use independent power and ground, and other modules share another set of power and ground. The two power supplies are both 5V and shorted together outside the chip.

The key submodule design determines the overall charge pump performance. For the oscillator, a current-hungry five-stage inverter structure with an inter-stage resistor was used to lower power consumption [12]. The oscillation frequency is 100 MHz with phase noise of -23 dBc/Hz at 10 kHz offset and -83 dBc/Hz at 1 MHz offset. For the charge pump core, a high-performance n-type diode instead of an NMOS

TABLE I.   COMPONENT SIZING OF THE PROPOSED HV OP-AMP.

Components	Size (width/length, number)
M1~M2	18 µm/1.8 µm, m=20
M3~M4	55 µm/5.5 µm, m=30
M5	10 µm/1 µm, m=15
M6	10 µm/1 µm, m=1
M7	10 µm/10 µm, m=8
R1=750KΩ	1 µm/10 µm, seg=30
R2=150KΩ	1 µm/10 µm, seg=6

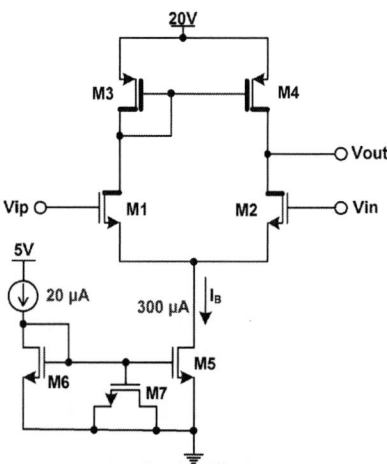

Figure 2.   The proposed HV op-amp.

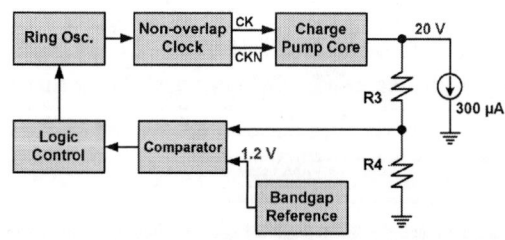

Figure 3.   Block diagram of the presented Dickson charge pump.

978-1-6654-3008-1/21 $31.00 © 2021 IEEE

transistor was adopted [13]. The pump stage is 6, the pump capacitance is 4 pF, and the storage capacitance in the output stage is 35 pF. For the bandgap reference, the popular Banba style was employed [14] and a low noise reference voltage was generated for the comparator.

III. EXPERIMENT RESULT

Based on the HHGrace 0.35 μm BCD process, the tape-out layout of the proposed HV amplifier was drawn, as displayed in Fig. 4. The silicon area without IO parts is 270 μm * 940 μm. It can be seen from the figure that the charge pump occupies most of the chip area. Specifically, the six pump capacitors (each is 4 pF) and one storage capacitor (35 pF) take up a lot of space. Such large capacitors are because the charge pump must be able to continuously provide at least 300 μA current in addition to providing a 20V source for the HV op-amp. It can be seen that high-performance HV op-amps need charge pumps with large driving capacity as support, and this is at the expense of a large chip area. The gray part in the picture is other circuits, which were inserted next to the HV amplifier to save the layout area.

The parasitic capacitance and resistance of the tape-out layout (including IO parts) were extracted, and complete post-simulations were done. As the working basis of the HV amplifier, the operation status of the charge pump plays a key role. The output voltage of the charge pump is shown in Fig. 5. It can be seen that although the operation of the HV op-amp directly causes ripples on the output voltage of the charge pump, the charge pump can still provide a stable 20

V supply for the HV op-amp. As displayed in Fig. 5, the charge pump output is about 20 V, with a ripple voltage of 640 mV$_{p-p}$. That is, the fluctuation range of the charge pump output is 20 V ±1.6%, which is much smaller than the off-chip supply (±15%). Besides, the charge pump dissipated 1.48 mA current under a 5 V supply.

After the charge pump works stably, the first thing that needs to be observed is the open-loop characteristics of the proposed HV amplifier, that is, the frequency response of the HV op-amp. Because only when the open-loop gain and phase margin are large enough, the closed-loop amplifier can work stably and the closed-loop gain error is small. The post-simulated frequency response of the HV op-amp is illustrated in Fig. 6. The low-frequency gain is 56.01 dB, that is, the gain is 631, which will lead to a 0.95% gain error at the closed-loop gain of 6. Generally, a gain error of less than 1% is acceptable. Hence the open-loop gain is sufficient. On the other hand, the unity-gain bandwidth is 20.57 MHz. If there is only one pole in the unity-gain bandwidth, it will bring about a 3.33 MHz closed-loop bandwidth at the closed-loop gain of 6. However, from the phase response curve, it can be seen that there are 2 poles within the unity-gain bandwidth. Especially the non-dominant pole is near the unity-gain bandwidth, which will reduce the closed-loop bandwidth a little. Considering that the maximum frequency of the input signal is 500 kHz, this bandwidth is ample. Finally, the phase margin is 64.7°, which is also acceptable.

The closed-loop frequency response of the proposed HV amplifier is shown in Fig. 7. The closed-loop gain is 15.5 dB, that is, a gain of 6. The -3 dB bandwidth is 2.152 MHz,

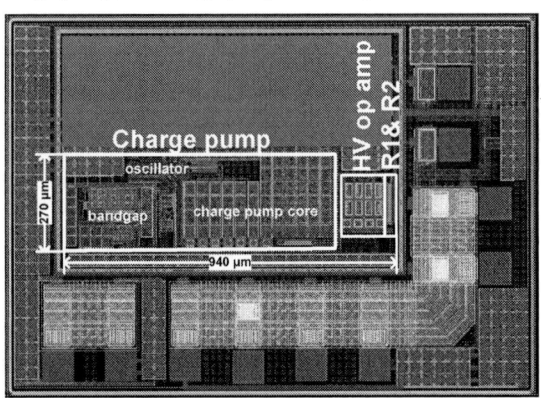

Figure 4. The tape-out layout of the proposed HV amplifier.

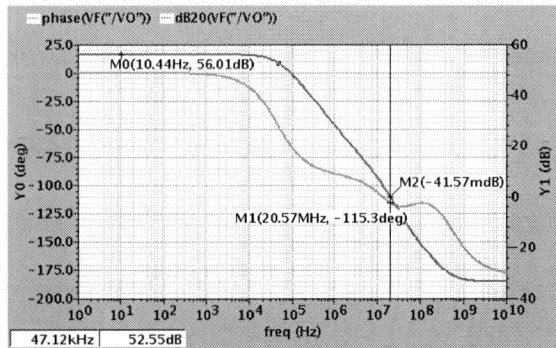

Figure 6. Gain response (red line) and phase response (green line) of the HV op-amp.

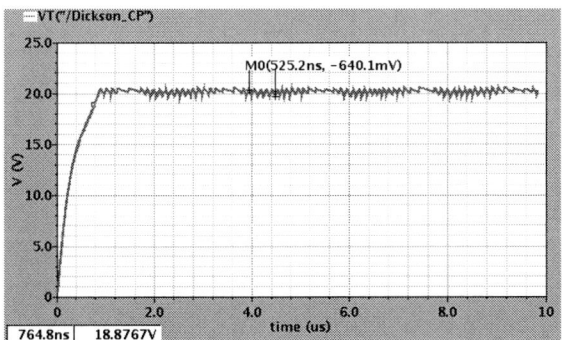

Figure 5. Charge pump output when worked with the HV op-amp.

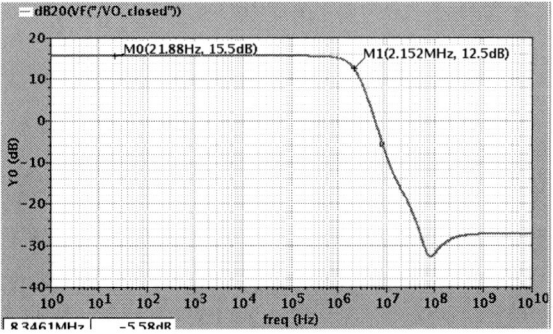

Figure 7. The closed-loop gain response of the proposed HV amplifier.

which is a little smaller than expected because of the non-dominant pole within the unity-gain bandwidth. However, the circuit can still handle a 500 kHz input signal. From the perspective of the frequency domain, the amplifier works normally. Next, we need to observe the time-domain behavior of the circuit, especially when the input signal amplitude is 1 V and the frequency is 500 kHz.

The transient response of the presented HV amplifier is displayed in Fig. 8. The lower green waveform is the input signal, with a DC level of 1.65 V, AC amplitude of 1 V, and frequency of 500 kHz. The upper red is the output voltage, with a DC level of 9.9 V, and a voltage swing of 11.91 V_{p-p}. For the output signal, the highest voltage is 15.83 V and the lowest one is 3.929 V. Moreover, from the waveform of the output signal, almost no distortion can be seen with the naked eye. Next, we will perform DFT (Discrete Fourier Transform) analysis on this time-domain signal and observe the harmonic distortion in the frequency domain.

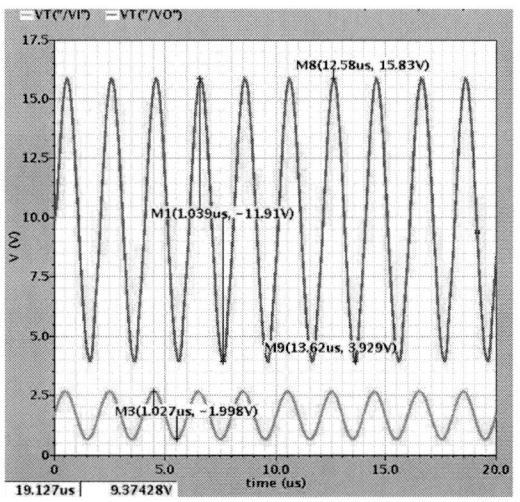

Figure 8. Transient response of the proposed HV amplifier when the input signal amplitude is 1 V and the frequency is 500 kHz.

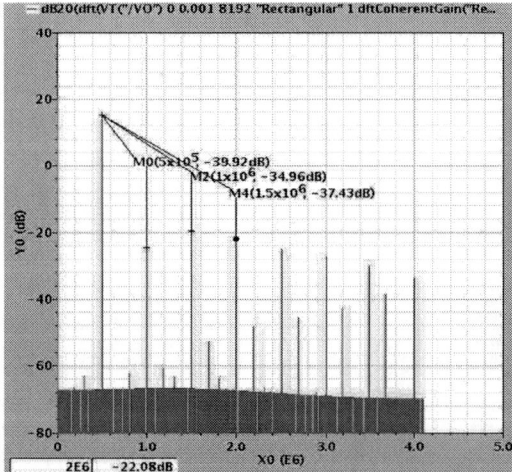

Figure 9. Harmonic distortion of the proposed HV amplifier when the input signal amplitude is 1 V and the frequency is 500 kHz.

The harmonic distortion of the proposed HV amplifier is depicted in Fig. 9. From the picture, it can be seen that the second-order harmonic distortion is -39.92 dBc, the third-order one is -34.96 dBc, and the fourth-order one is -37.43 dBc. All of these are below -30 dBc and meet the design requirement of the gyro system.

The post-simulation results of the proposed HV amplifier and the comparison with other related works are shown in Table II. Compared with other works, the proposed HV amplifier with an integrated charge pump exhibits a lower distortion, larger bandwidth, and less power consumption characteristics. Noted that the output swing of the presented HV amplifier is relatively small, which is determined by the requirement of the gyro system.

IV. CONCLUSION

Based on the driving demand of MEMS gyroscopes, this paper presents a high-performance HV amplifier with a gain of 6, a bandwidth of 2.2 MHz, second-order harmonic distortion of -40 dBc, and output swing of 12 V. To provide a 20 V supply power to the HV amplifier, a Dickson charge pump was also integrated on the chip, with a load current of 300 μA, output voltage of 20 V, and ripple of 640 mV_{p-p}. The overall supply is 5 V and the power consumption is 7.4 mW. The proposed HV amplifier can also be used in other MEMS sensors with similar drive requirements.

ACKNOWLEDGMENT

Dr. Hua Chen thanks Zhi Li and Dr. Jianzhong Zhao for valuable discussions about the selection of the high-voltage MOS device and its layout problems.

REFERENCES

[1] S. D. Senturia, *Microsystem Design.* Dordrecht, The Netherlands: Kluwer Academic Publishers, 2001.

[2] S. Saponara, T. Baldetti, L. Fanucci, E. Volpi, and F. D'Ascoli, "Design of an Integrated Scanning Micromirror Driver in BCD Technology," *J. Circuits Syst. Comput.*, vol. 20, no. 4, pp. 780-799, 2011.

[3] K. Sun, Z. Gao, P. Gui, R. Wang, I. Oguzman, X. Xu, K. Vasanth, Q. Zhou, and K. K. Shung, "A 180-Vpp Integrated Linear Amplifier for Ultrasonic Imaging Applications in a High-Voltage CMOS SOI Technology," *IEEE Trans. Circuits Syst. II-Express Briefs*, vol. 62, no. 2, pp. 149-153, 2015.

[4] S. Dai, R. W. Knepper, and M. N. Horenstein, "A 300-V LDMOS Analog-Multiplexed Driver for MEMS Devices," *IEEE Trans. Circuits Syst. I-Regul. Pap.*, vol. 62, no. 11, pp. 2806-2816, 2015.

[5] S. Yang and C. Wang, "Domestic Indirect Feedback Compensation of multiple-stage amplifiers for multiple-voltage level-converting amplification," in *Proc. ICICDT 2011*, pp. 1-4.

[6] R. Fang, W. Lu, G. Wang, T. Tao, Y. Zhang, Z. Chen, and D. Yu, "A Low-noise High-voltage Interface Circuit for Capacitive MEMS Gyroscope," *J. Circuits Syst. Comput.*, vol. 22, no. 9, 2013.

[7] J. Ning and K. Hofmann, "A 120V high voltage DAC array for a tunable antenna in communication system," in *Proc. DDECS 2014*, pp. 65-70.

[8] M. Rose and H. J. Bergveld, "Integration Trends in Monolithic Power ICs: Application and Technology Challenges," *IEEE J. Solid-State Circuit*, vol. 51, no. 9, pp. 1965-1974, 2016.

[9] W. Sun, B. Zhang, S. Xiao, W. Su, and J. Cheng. "Development and trend of power semiconductor devices and power integrated technology," *Sci. China-Inf. Sci.*, vol. 42, no. 12, pp. 1616-1630, 2012.

[10] W. M. C. Sansen, *Analog Design Essentials.* Dordrecht, The Netherlands: Springer, 2006.

[11] F. Pan and T. Samaddar, *Charge Pump Circuit Design.* McGraw-Hill Education, 2006.

[12] H. Hwang, B. Jo, S. Park, S. Kim, C. Jeong and J. Moon, "A 13.56 MHz CMOS ring oscillator for wireless power transfer receiver system," in Proc. TENCON 2014, pp. 1-4.

[13] J. Wu and K. Chang, "MOS charge pumps for low-voltage operation," *IEEE J. Solid-State Circuit,* vol. 33, no. 4, pp. 592-597, 1998.

[14] H. Banba, H. Shiga, A. Umezawa, T. Miyaba, T. Tanzawa, S. Atsumi, and K. Sakui, "A CMOS bandgap reference circuit with sub-1-V operation," *IEEE J. Solid-State Circuit,* vol. 34, no. 5, pp. 670-674, 1999.

TABLE II. POST-SIMULATION RESULTS OF THE PROPOSED HV AMPLIFIER AND COMPARISON WITH OTHER RELATED WORKS.

References		[2]	[4]	[5]	This work
HV amplifier	Gain	15.8	100	12	6
	Bandwidth, MHz	0.5	NA	NA	2.2
	Second-order harmonic distortion, dBc	NA	NA	NA	-40
	Output swing, V_{p-p}	20	145	50	12
Internal op-amp	Gain, dB	150	>120	133/ 128	56
	GBW, MHz	10	3	2.6/ 1.3	20
	Phase margin	45°	35°	85°	65°
	Load, pF	<100	100	10	10
	Supply, V	25	300	60	20
	Current dissipation, µA	550	300	2128	300
On-chip charge pump		No	No	No	Yes
Power supply, V		1.8	5	5	5
Power dissipation (w/ charge pump), mW		NA	NA	NA	7.4

The 16th IEEE International Conference on Nano/Micro Engineered & Molecular Systems

Bio-inspired multiscale adhesive interfacial materials

Shutao Wang

CAS Key Laboratory of Bio-inspired Materials and Interfacial Science, Technical Institute of Physics and Chemistry, Chinese Academy of Sciences, Beijing 100190, P. R. China

ABSTRACT

Bio-interfacial adhesion has become a frontier hot in interfacial chemistry. It is not only helpful for us to understand the mystery of living systems, but also important for the development of new functional interfacial materials and related technologies. Leaning from nature, our group has recently investigated several special adhesion phenomena on biointerfaces and developed a series of bio-inspired adhesive interfacial materials. 1) We discovered the superdurability of bird feathers against tears originated from their cascaded slide-lock system, not from the "hook–groove system" proposed centuries ago; Inspired by the arrester system of dragonfly, we developed a new mechanical interlocker with a nylon pestle instead of the traditional hook, which breaks the limitation of traditional Velcro with undesirable deformation, breaking and noise. 2) Inspired by immune system, we proposed the concept of synergistic effect of biointerface adhesion based on structural matching and molecular recognition for detecting circulating tumor cell (CTC). We have developed a series of CTC detecting biochips by chemical etching, vapor deposition, electrochemical deposition, template replication, electrospinning and others; We also developed an emulsion interfacial polymerization strategy to fabricate bio-inspired immunomagnetic bead (spanning from Janus to porous) with controllable topology and surface chemistry. 3) We disclosed the microstructure of wound blood scab and developed a series of wound dressing, greatly promoting wound healing.

BIOGRAPHY

Shutao WANG, doctoral supervisor, project leader, was selected into the National Outstanding Youth, youth top talents. He mainly engaged in the research of bionic multi-scale adhesion controllable interface materials, such as anti-adhesive interface materials, efficient bio-identification adhesion interfaces and devices, and early diagnosis of diseases. He firstly proposed the biomarker adhesion effect of "structural matching and molecular recognition" which is used for cancer detection. It is 1000 times more sensitive than the traditional cell separation method, so it was nominated for the 2010 World Science and Technology Award. He has won major scientific advances in China in 2006, outstanding doctoral thesis of the Chinese Academy of Sciences in 2008 (50 papers), National Outstanding Doctoral Dissertation Nomination Award in 2009, and China Chemical Society Youth Chemistry Award in 2013. He has published over 140 SCI papers with H-index of 39, including Angew. Chem.(9), Adv. Mater.(20), JACS (2), Nat. Protocol.(1), which have been cited over 4200times. He has 8 patents authorized by China and 2 by USA. He has published Invitation reviews and 3 English articles on well-known international journals such as Chemical Review, Chemical Society Review, Annual Review Nano Research, Account of Chemical Research, Journal of Photochemistry and Photobiology C: Chemistry Review, MRS Bulletin. His researching results have been reported in highlight in Nature, Nature medicine, Science Daily, Chem&Eng News, Materials Views China, Material Today, Lab Chipfor many times. He has accepted interview of "The Sceptical Chemist" of Nature Chemistry, "People Watch" of Asia-Pacific Biothech News and China Science Daily.

This page intentionally left blank.

Nanokits for Single Cell Analysis

Dechen Jiang

State Key Laboratory of Analytical Chemistry for Life Science, School of Chemistry and Chemical Engineering, Nanjing University, Nanjing, 210092, China

Website : group or personal website if applicable

ABSTRACT

Single cell analysis can obtain more accurate and comprehensive information reflecting the physiological state and process of cells, which is of great significance for life science research and early diagnosis of major diseases. At present, the main analysis strategy is to design specific recognition molecules (probes) to realize the detection of intracellular biomolecules. However, due to the differences in the uptake of probes by different cells, it is difficult to accurately regulate the content and distribution of these recognition probes in cells. Consequently, this approach could only provide relatively quantitative results

In the past six years, our laboratory has developed a single-cell nanokit strategy that addresses the challenging as mentioned above. This talk will focus on the introduction of nanokit, and the application to determine the activity of biomolecules in a single cell, and even in a single lysosome. Compared with the current methods, the nanokit has adapted features of the well-established biological kits, and thus, could provide a specific device to characterize the individual cell activity.

BIOGRAPHY

Dr. Dechen Jiang is currently a professor in State Key Laboratory of Analytical Chemistry for Life, School of Chemistry and Chemical Engineering at Nanjing University. He obtained his B.S. M.S and Ph.D. degrees from Nanjing University (2000), Fudan University (2003) and Case Western Reserve University (2008). After three years postdoctoral training at UNC-Chapel Hill, Dr. Jiang joined Nanjing University to start his independent research. Dr. Jiang has published over 50 research publications as the corresponding author, such as PNAS, JACS, Angew Chem and Anal. Chem. His research focuses on the development of cutting-edge technology and instrumental for the characterization of cellular activity at single cells. He has received multiple awards, such as the NSFC Award for Distinguished Young Scientist, Changjiang Young Professor of the Ministry of Education, Young Innovation Award for Analytical Instrument (China Instrument and Control Society).

Gap in pagination due to unavailable papers.

Pages 1833-1838

Proceedings of the 16th Annual IEEE International
Conference on Nano/Micro Engineered and Molecular Systems
April 25-29, 2021

Ultra-thin Flexible Neuro Probe Utilizing Biodegradable Collagen Microneedle*

Dong Huang, Junshi Li, Zhongyan Wang, Zhitong Zhang, Qining Wang, *Member, IEEE* and Zhihong Li, *Member, IEEE*

Abstract— **In this paper, we proposed a flexible neuro probe with the biodegradable microneedle as a carrier for reliable and long-term neuro recording. The ultra-thin Parylene microelectrode mesh was assembled on a needle-shaped collagen suture with assistance of surface tension. After penetration into brain, the collagen suture can be degraded and absorbed gradually, resulting in the recovery of the glial scar and leaving ultra-thin Parylene mesh embedded in tissue. Hence, such a strategy can significantly reduce the potential safety hazard and conduct satisfactory contact between microelectrodes and nerve cells in the long-term implantation. This device is expected to offer a general tool for experimental neuroscience study and clinic application.**

I. INTRODUCTION

The implantable neuro probe is a crucial tool for brain science, which can conduct observation and measurement of the neural activities at cellular level to realize long-term brain-machine interface. [1,2] However, conventional neural implants have large size and considerable surgical footprints inducing substantial damage and disruption to local cellular and vascular network. [3,4] Moreover, The chronic stability and unstable electrical contact are main barrier for the conventional neuro probe. [2,5-6] Benefitting from microfabrication technologies, researchers found that optimizing the mechanical properties and shrinking size to obtain more flexible and smaller devices can help overcome these drawbacks in long-term implantation. [7-8]

The flexible microelectrode array prepared by microfabrication has been utilized to achieve safer and low-damage implantation and reduce deleterious tissue response, enabling electrical recording and neuro modulation. [9] Parylene [10] and SU-8 [9] are common materials of implantable flexible neuro probes, due to excellent biocompatibility, good mechanical properties, and excellent flexibility. [11] However, the pliable probe still requires a carrier with enough strength to penetrate into brain tissue. Particular fabrication strategies have been explored for smooth implantation of flexible microelectrode, including surface tension [12] and assisted shuttle [13]. The optical fiber or the carbon fiber are commonly chosen as the carriers, remaining potential safety risk in the long-term implantation. However, there are few of reports on flexible

microelectrode array on biodegradable carries to form reliable and long-term neural integration.

This paper proposed a new neuro probe integrated with biodegradable collagen suture and ultra-thin flexible microelectrode array with mesh structure. After implantation into the brain tissue, the collagen suture can be degraded and absorbed gradually. Thus, the recovery of the glial scar can keep ultra-thin Parylene mesh embedded in tissue, promoting interpenetration and close integration with the neural networks. Based on this operation, such a strategy can significantly reduce the potential safety hazard and conduct satisfactory contact between microelectrodes and nerve cells in the long-term implantation.

II. DESIGN AND FABRICATION

A. Design and Materials

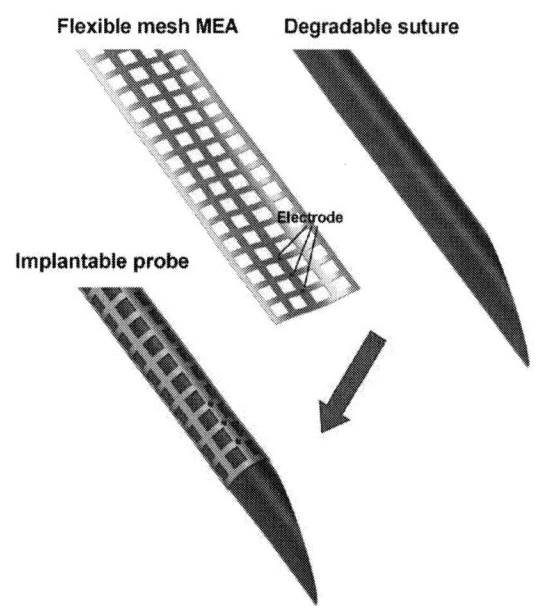

Figure 1: Schematic diagram of the biodegradable flexible neuro probe, wchich contains the flexible mesh microelectrode array (MEA) and a degradable suture with sharp tip.

---

*Resrach supported by National Natural Science Foundation of China (62004007), National Key Research and Development Program (2016YFA0200802), China Postdoctoral Science Foundation Grant (2019M660309), and the China Capital Health Research and Development of Special (2018-1-4111).

D. Huang, J. Li, Z. Wang, Z. Zhang and Z. Li are with the National Key Laboratory of Nano/Micro Fabrication Technology, Institute of

Microelectronics, Peking University, Beijing 100871, China (corresponding e-mail: zhhli@pku.edu.cn).

D. Huang and Q. Wang are with the College of Engineering, Peking University, Beijing 100871, China (corresponding e-mail: qiningwang@pku.edu.cn)

978-1-6654-3008-1/21 $31.00 © 2021 IEEE

In order to achieve higher biocompatibility and reliability, we have utilized biodegradable suture as the base of the probe and mesh electrode structure to form fixation after implantation. As shown in Figure 1, the proposed neuro probe contains two components, a flexible microelectrode array based on an ultra-thin Parylene film mesh and a surgical needle-shape collagen suture with diameter of 500 µm that is biodegradable and absorbable. The collagen suture is rigid enough to penetrate into the brain, and the microelectrode array is able to contact the nerve cells and record the electrical signal. After degradation of the suture, the injury caused by puncture on the brain tissue may recover and the ultra-thin Parylene film mesh is covered tightly by the neurons to promote interpenetration and close integration with neural tissue, leading to a long-term and stable contact for signal recording.

The collagen suture is general tool in clinic surgeries due to its good biocompatibility. And this kind implant may degrade in the aqueous environment and be absorbed by tissue. So we choose this tool as the base substrate of our device. Parylene is widely used in implantable equipment and biomedical applications for its excellent biocompatibility and high resistance to harsh physical and chemical environments. This material is also approved for implantation by U. S. Food and Drug Administration (FDA). Moreover, this kind of polymer is compatible with microfabrication for its superior conformality in vacuum deposition. Hence, Parylene has been selected as the base material of the microelectrode array. The geometry size of mesh microelectrode is shown in Figure 2. The witdh of the mesh structure is 540 µm, and the line is 40 µm. The microelectrode and gold conductive trace are embedded in two Parylene film. The radius of the exposed site of the microelectrode is 25 µm, which can provide a good conduct with tissue at cell level. The toal thickness of the mesh microelectrode is 1 µm that offers high flexibility. The exceptional geometrical morphology yields mechanical interactions at tens of micrometers (cellular level).

The surface tension-assisted mechanism has been utilized to assembly the film on the suture as shown in Figure 3. This assembly process is achieved by aligning the microelectrode film with the suture in DI water and slowly lift both devices in the vertical direction. Because the thickness t of the film was ultra small, surface tension forced the film to conformally adhere on the curved suture after crossing the water−air interface. This strategy is achieved bu reducing the surface energy at the liquid-air interface. Zhao et. al. has analytically studied this by considering an elastic film with thickness t in loose contact with a rigid cylinder with radius R. The film and cylinder are both wetted with the same liquid. If R is large enough, the deformation of the film packaging the cylinder is small. If R is very small, the deformation of the film is relatively large. So there is the critical radius Rc, lower than which the film with thickness of t can't conduct self-package on the cylinder by the surface tension force. In the ideal case, taking no account of surface roughness and geometric dimensioning, Zhao et. al. has obtained [12]:

$$R_c = \sqrt{\frac{Et^3}{24\gamma(1-v^2)}} \qquad (1)$$

Where E is the Young's modulus of the film, γ is the surface tension of the liquid, and v is the Poisson ratio of the film. For Parylene film with thickness of 1 µm. $E_{parylene}$ is 400 psi, $v_{parylene}$ is 0.4,[13] and γ of DI water is 71.97 mN/m [12]. The critical radius is roughly 45 µm. However, taking account of surface roughness of the cylinder and geometric dimensioning of the film, the critical parameter should be remarkably greater than theoretical value. Thus the suture with radius of 250 µm is selected as the base microneedle, much larger than the theoretical radius. So in our design, theoretically, the surface tension is large enough for the mesh microelectrode array to realize self-package.

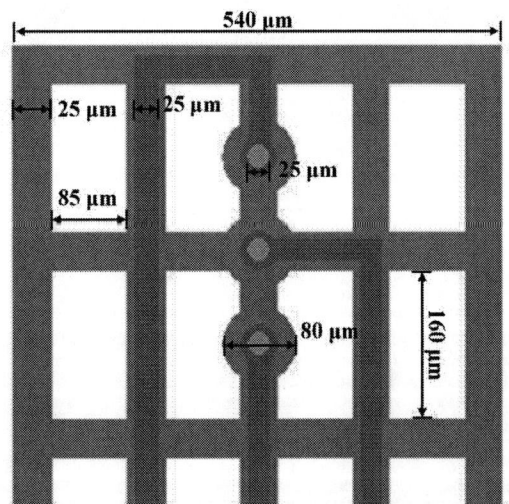

Figure 2: Geometrical morphology of the mesh microelectrode. Red, ultra thin Parylene C film. Blue, gold conductive layer. Gray, exposed site of microelectrode.

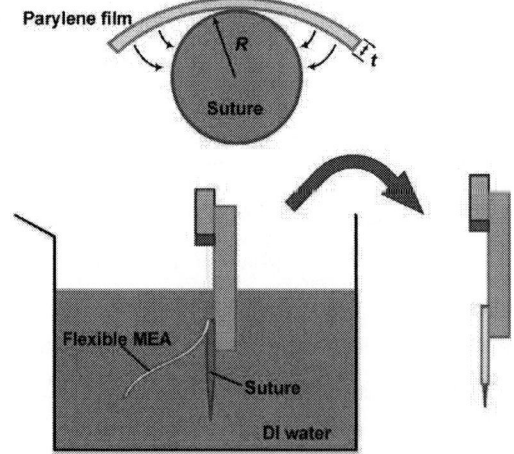

Figure 3: Schematic illustration of self-packaging of the ultra-thin Parylene film on the suture by surface tension. t,thickness of the mesh microelectrode array. R, the radius of the suture.

B. Fabrication Process

The fabrication method of ultra-thin Parylene microelectrode array is illustrated in Figure 3. First, a 0.5-μm-thinckness Parylene film was deposited on a silicon wafer as the substrate by chemical vapor deposition (PDS 2010, Special Coating Systems, USA). The excellent conformality of vapor Parylene coating ensure flatness of the whole film, and the biocompatibility can be further improved. Then, 300 Å chromium and 1000 Å gold were sputtered on the substrate Parylene layer as the adhesion layer and the electrode layer, respectively. After that, the lithography patterning and wet etching were conducted on the mental layer to pattern the conductive trace and microelectrodes. Another Parylene film with thickness of 0.5 μm was deposited to form isolation layer. Then, pattern the microelectrode site by the second lithography. Expose the microelectrode site and form mesh structure by the oxygen dry etching on Parylene film. After the dry etching, remove the photo resist and release the flexible microelectrode array from the wafer. The flexible mesh microelectrode array was finally completed.

After the microfabrication process of the mesh microelectrode array, the surface tension- assisted assebly method was performed to finally fabricate the neuro probe. As illustrated in Figure 3, the connection pad and aligning mark were pre-designed in a printed circuit board. The biodegradable suture with diameter of 500 μm was first stuck on the preset position on the board by the medical epoxy resin. The connection points on the microelectrode array were adhered to pads on the PCB by the conductive silver paste in 90 °C for 30 minutes. After the fixation of microelectrode array and suture on the PCB, the aligning process was finished. Finally, we immersed the whole device into the dilute water and slowly lift the device across the liquid-air interface, during which the surface tension was able to attach the ultra-thin Parylene film onto the wall of biodegradable suture.

III. RESULTS

A. Prototype

Figure 5 representes pictures of the proposed biodegradable flexible neuro probe in different scales. The overall view of the probe was shown in Figure 5(a). The probe was fixed on a printed circuit board, and the microelectrodes connected to the connector through traces on the board. The detail of the probe tip by the stereomicroscope was shown in Figure 5(b). To form a sharp tip at the end of the suture, we cut the tip at a certain angle with a surgical scissor. Due to good mechanical property of the collagen, the tip was sharp and tough enough to penetrate into the soft tissue like brain.

The microscope image in Figure 5(c) represents the detail of the flexible microelectrode array attached on the suture. The diameter of the collagen suture was 500 μm. The radius of the microelectrode was 15 μm, and there were 3 recording points on the Parylene mesh. The mesh structure was conformally attached on the surface of the suture. The hydrolyzing time of the suture was roughly 2 weeks.

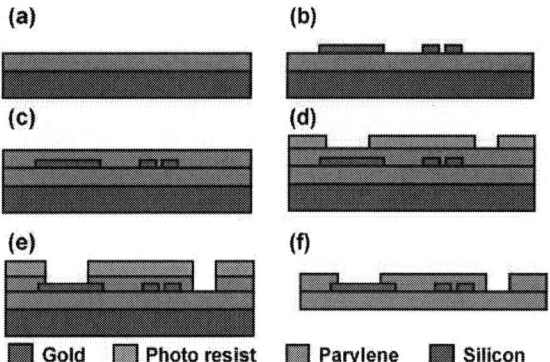

Figure 4: Fabrication process of flexible Parylene MEA. (a) First Parylene depostion. (b) Metalisation and patterning. (c) Second Parylene deposition. (d) Patterning photo resist. (e) Oxygen dry etching. (f) Release from subtrate wafer.

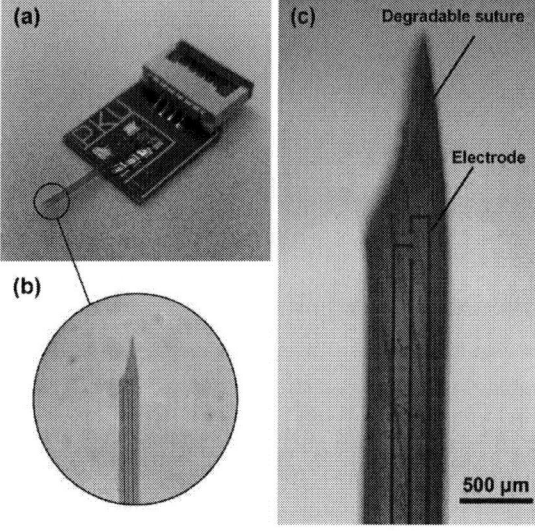

Figure 5: Photos of the prototypes. (a) The neuro probe soldered on a PCB. (b) The detail of the probe tip by the stereomicroscope. (c) Microphotograph of the probe tip.

B. AC impedance

To measure the impedance characterization of the proposed neuro probe, we conducted AC impedance tests on our devices in PBS solution. The neuro probe, Pt mesh electrode, and general Ag/AgCl reference electrode were immersed into PBS solution as working electrode, counter electrode, and reference electrode, respectively. The AC test was performed from 1 Hz to 10 kHz by electrochemical workstation (CHI 660, CH Instruments Ins, USA).

Figure 5 and 6 showed the impedances characteristics of the biodegradable flexible neuro probe in the PBS. At the typical frequency of 100 Hz, the average impedance magnitude of the microelectrode was 50 kΩ. The frequency-dependent impedance magnitude demonstrated a large variation in the low frequency band (1~100 Hz), whereas it had a smooth and steady decrease from 100 Hz to 10 kHz. By comparison, the impedance phase curve was relatively

978-1-6654-3008-1/21 $31.00 © 2021 IEEE

more stable in terms of frequency, gently increase from -15° to -10° in the band between 1 Hz and 10 kHz.

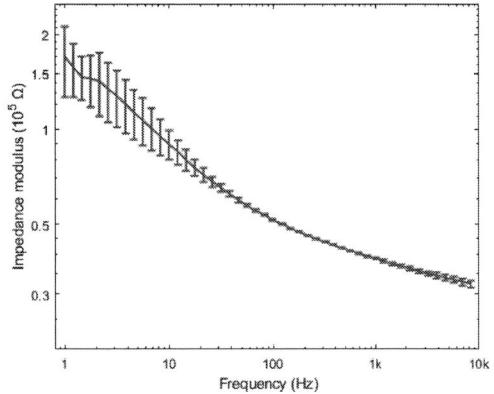

Figure 6: Frequency-dependent characteristics of impedance magnitude of the microelectrode in PBS. Error bar is standard deviation with n = 5.

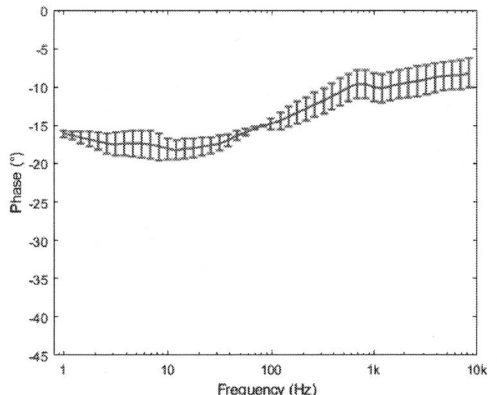

Figure 7: Frequency-dependent characteristics of impedance phase of the microelectrode in PBS. Error bar is standard deviation with n = 5.

IV. CONCLUSION

We proposed a novel ultra-thin flexible neuro probe that utilized biodegradable suture as base substrate and the fabrication strategy of this device. The sharp tip of the biodegradable suture can penetrate into soft brain tissue, and the tissue environment can degrade and absorb the suture gradually. After fully biodegradation of the suture, the mesh structure would promote interpenetration and close integration with neural tissue during self-recovery process, leading to a long-term and stable contact for signal recording. Owing to theoretical calculation and physical design, the surface tension-assisted assembly method was introduced into the fabrication process. This technique remarkably reduced the difficulty of integration with flexible film and curved substrate. The results of the AC impedance tests demonstrated great and stable electrical properties, showing high potential in implantation.

REFERENCES

[1] Alivisatos, A. Paul, Miyoung Chun, George M. Church, Karl Deisseroth, John P. Donoghue, Ralph J. Greenspan, Paul L. McEuen et al. "The brain activity map." *Science* 339, no. 6125 (2013): 1284-1285.

[2] Polikov, Vadim S., Patrick A. Tresco, and William M. Reichert. "Response of brain tissue to chronically implanted neural electrodes." *Journal of neuroscience methods* 148, no. 1 (2005): 1-18.

[3] Yu, Huaiqiang, Nenggan Zheng, Wei Wang, Shuo Wang, Xiaoxiang Zheng, and Zhihong Li. "Electroplated nickel multielectrode microprobes with flexible parylene cable for neural recording and stimulation." *Journal of microelectromechanical systems* 22, no. 5 (2013): 1199-1206.

[4] Yu, Huaiqiang, Wenjie Xiong, Hongze Zhang, Wei Wang, and Zhihong Li. "A parylene self-locking cuff electrode for peripheral nerve stimulation and recording." *Journal of Microelectromechanical Systems* 23, no. 5 (2014): 1025-1035.

[5] Li, Junshi, Dong Huang, Yufeng Chen, and Zhihong Li. "Low-Cost, Metal-Based Micro-Needle Electrode Array (M-MNEA): A Three-Dimensional Intracortical Neural Interface." *20th International Conference on Solid-State Sensors, Actuators and Microsystems & Eurosensors XXXIII (TRANSDUCERS & EUROSENSORS XXXIII)*, pp. 1635-1638. IEEE, 2019.

[6] HajjHassan, Mohamad, Vamsy Chodavarapu, and Sam Musallam. "NeuroMEMS: neural probe microtechnologies." *Sensors* 8, no. 10 (2008): 6704-6726.

[7] Sohal, Harbaljit S., Andrew Jackson, Richard Jackson, Gavin J. Clowry, Konstantin Vassilevski, Anthony O'Neill, and Stuart N. Baker. "The sinusoidal probe: a new approach to improve electrode longevity." *Frontiers in neuroeng*ineering 7 (2014): 10.

[8] Kozai, Takashi D. Yoshida, Nicholas B. Langhals, Paras R. Patel, Xiaopei Deng, Huanan Zhang, Karen L. Smith, Joerg Lahann, Nicholas A. Kotov, and Daryl R. Kipke. "Ultrasmall implantable composite microelectrodes with bioactive surfaces for chronic neural interfaces." *Nature materials* 11, no. 12 (2012): 1065-1073.

[9] Xie, Chong, Jia Liu, Tian-Ming Fu, Xiaochuan Dai, Wei Zhou, and Charles M. Lieber. "Three-dimensional macroporous nanoelectronic networks as minimally invasive brain probes." *Nature materials* 14, no. 12 (2015): 1286-1292.

[10] Wang, Renxin, Xianju Huang, Guangfeng Liu, Wei Wang, Fangtian Dong, and Zhihong Li. "Fabrication and characterization of a parylene-based three-dimensional microelectrode array for use in retinal prosthesis." *Journal of Microelectromechanical Systems* 19, no. 2 (2010): 367-374.

[11] Rodger, Damien C., Andy J. Fong, Wen Li, Hossein Ameri, Ashish K. Ahuja, Christian Gutierrez, Igor Lavrov et al. "Flexible parylene-based multielectrode array technology for high-density neural stimulation and recording." *Sensors and Actuators B: chemical* 132, no. 2 (2008): 449-460.

[12] Zhao, Zhengtuo, Lan Luan, Xiaoling Wei, Hanlin Zhu, Xue Li, Shengqing Lin, Jennifer J. Siegel, Raymond A. Chitwood, and Chong Xie. "Nanoelectronic coating enabled versatile multifunctional neural probes." *Nano letters* 17, no. 8 (2017): 4588-4595.

[13] Luan, Lan, Xiaoling Wei, Zhengtuo Zhao, Jennifer J. Siegel, Ojas Potnis, Catherine A. Tuppen, Shengqing Lin et al. "Ultraflexible nanoelectronic probes form reliable, glial scar–free neural integration." *Science advances* 3, no. 2 (2017): e1601966.

[14] http://www.mit.edu/~6.777/matprops/parylene.htm

Proceedings of the 16th Annual IEEE International
Conference on Nano/Micro Engineered and Molecular Systems
April 25-29, 2021

Fabrication of Polymer/Metal Composite Micro/Nano Array Structures and the Applications in Biological Interfaces and Actuators

Hongxu Chen*

Key Laboratory of Yarn Materials Forming and Composite Processing Technology of Zhejiang Province, Jiaxing University, Jiaxing 314001, China

Abstract—The chameleon change color when the environment changes, the gecko's feet have super adhesion ability, and the surface of lotus leaf is hydrophobic. All of these are due to their micro/nano structures. So the biomimetic micro/nano structures have attracted much attention due to their broad applications in self-cleaning, microelectronic devices, information storage, plasma optics, biomedical interfaces, micro/nanofluidic and sensors. My research is about the fabrication and application of biomimetic micro/nano structures based on the colloidal lithography. The colloidal lithography is a technique for fabricating ordered micro/nano structures using 2D or 3D colloidal crystal as template or mask. This method is simple, low cost and suitable for large area preparation. Then the metal nanoparticles or metal layer were combined with the polymer micro/nano structure by adsorption or deposition. The composite micro/nano structures can be used as barcode nanorods, coaxial gold nanorings, biological interfaces and integrated micro/nanomotors.

INTRODUCTION

Periodic micro/nanostructures are abundant in nature, and have provided great inspirations for scientists to simulate them in many important applications, such as super-wetting surfaces, antireflection membranes, plasma optics, micro/nanofluidic devices and sensors [1]. Traditional methods including chemical vapor deposition, laser ablation, electron beam lithography, and interference lithography have been used to prepare ordered micro/nano arrays on many substrates. However, these top-down lithographic technologies require complex equipments, which require a great deal of time and cost in practical applications. Recently, colloidal lithography technique has been widely used in the fabrication of micro/nano arrays due to the characteristics of inexpensive, high efficiency, repeatable and high yield. In this paper, several ordered micro/nano structures (nanorods, microcones) have been successfully fabricated using micro/nanospheres as templates, and their properties and applications have also

*Resrach supported by the National Natural Science Foundation of China (Grant No. 51905526) and Jiaxing Science and Technology Project (2020AY10018), Open Project Program of Key Laboratory of Yarn Materials Forming and Composite Processing Technology of Zhejiang Province (MTC 2020-04).

H. Chen is with College of Material and Textile Engineering, Jiaxing University, Jiaxing 314001, Zhejiang, China. (phone: +86-573-83640063; e-mail:hx.chen@zjxu.edu.cn).

been explored. In the following, we will introduce the micro/nano arrays in the view of applications [2-6]. Firstly, we generated barcode nanorods with a controlled coding capacity by combining colloidal lithography, selective ion-exchange, and the in situ reduction of Ag ions (Ag^+). Secondly, three-dimensional (3D) coaxial metallic nanorings arrays were prepared by colloidal lithography and polymer-assisted self-assembly of gold nanoparticles. Thirdly, a living biological interface based on microcone array that matches the cellular characteristic size of neurons has been developed, which was used to study the behavior of cells. Furthermore, an integrated micro/nanomotor composed of gold hollow microcone array was fabricated by colloidal lithography for the first time, which can realize various controllable motions under near-infrared light illumination.

RESULTS AND DISCUSSION

1. Ag nanoparticle/polymer composite nanorods with coding capabilities

Nano-coded materials with large amount of pattern discrimination information have important applications from product tracking to biological detection, and thus have attracted widespread attention. Unlike the black-and-white product identification codes that are visible to the naked eye in supermarkets, nano-coded materials are invisible to the naked eye, which make them useful as codes for valuable items or important materials. The wide application of nano-coded materials requires high coding capacity, low energy consumption, high particle yield and suitable and accurate detection system. In face of above requirements, many coding material technologies have been developed rapidly. Among them, the barcode nanorod materials, due to the advantages of low cost, high yield, high length-diameter ratio, easy preparation and surface modification, have received widespread attention [7, 8]. The length of nanorods is advantageous to increase the encoding capacity and facilitate surface modification, which make nanorods a very important structure. However, the current construction of barcode nanorods mainly relies on the alumina template method, which can only deposit specific metals, greatly limiting their coding capacity. Therefore, it is of great significance to study a method to increase the encoding capacity of barcode nanorods.

978-1-6654-3008-1/21 $31.00 © 2021 IEEE

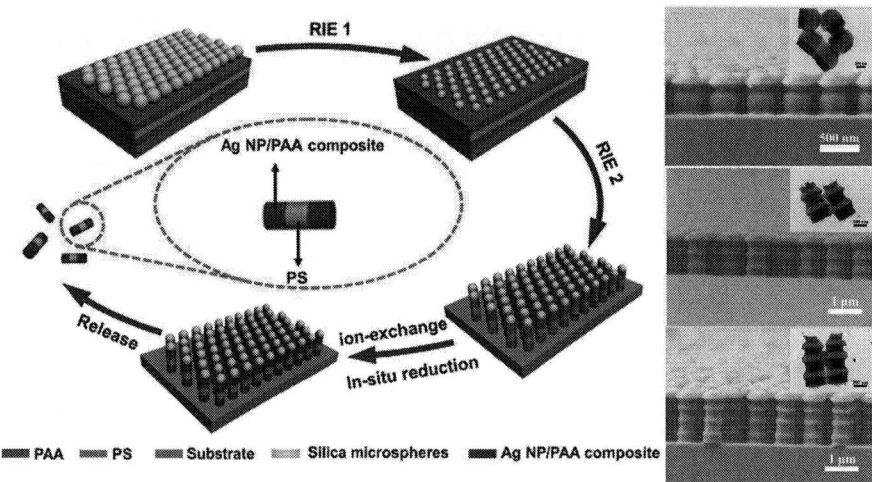

Fig. 1 Fabrication procedure of the Ag NP/polymer composite nanorods based on colloidal lithograph.

On the basis of a variety of highly ordered micro/nano structures previously prepared by colloidal lithography techniques [9-11]. The multi-segmented barcode nanorods with a controlled coding capacity were fabricated using this method, as shown in the Fig. 1 [2]. The multi-segmented polymer nanorods were firstly fabricated by colloidal lithography, then Ag nanoparticles (Ag NPs) were incorporated into the polyacrylic acid (PAA) segments by an ion exchange and the in situ reduction of the Ag^+. The figures show the morphology of the polymer nanrods and demonstrate the Ag NPs were selectively incorporated into the PAA segments of the nanorods. The encoding information of these novel barcode nanorods can be achieved by adjusting the number and length of nanorods and the sequence of PAA layer. Furthermore, the density of Ag NPs in different layers of the same nanorod can be adjusted by changing the composition of the PAA layers. And the intensity decreased as the volume fraction of PAA decreased, which greatly improve the coding capacity of barcode nanorods. In addition, we believe that many coding materials can be obtained by introducing other functional nanoparticles into the nanorod structure, which will provide a new way to prepare nano-coded materials.

2. Polymer-assisted fabrication of gold nanoring arrays

Plasma optical properties of noble metal nanostructures have been widely used in subwavelength optics and surface-enhanced Raman scattering. In the past few decades, many metal nanostructures have been synthesized and studied, such as triangular nanocrystals, bipyramids, nanorods, nanocrystals and nanocrystals. In these structures, nanoring has been widely used in biological sensors, medical diagnosis, food safety, environmental monitoring, metamaterials and optics since it has large specific surface area, strong coupled plasma from the surface of inside and outside ring, highly tunable plasmonic resonance from the visible to near infrared wavelengths [12, 13]. Early studies on gold nanorings

showed that the plasma properties of the rings could be regulated by changing their diameter, thickness and geometric arrangement. Current methods for constructing metal nanorings include colloidal etching, electron beam etching, anodic alumina template-assisted, on-wire lithography and molecular template-assisted methods [14-17]. But the resulting nanorings are not self-supporting and the size and spacing of the nanorings cannot be well regulated. In addition, it is still difficult to construct three-dimensional coaxial gold nanorings. Therefore, it is essential to develop a simple and low-cost method for large-scale construction of self-supporting and monodisperse coaxial metal nanorings structures with controllable ring height, diameter and ring spacing.

We constructed a self-supporting three-dimensional coaxial metal nanoring array structure with controllable nanoring parameters by combining colloidal lithography with polymer-assisted self-assembly of gold nanoparticles [3], as shown in the Fig. 2. The multi-segmented polymer nanorods were also fabricated by colloidal lithography. Then the gold nanoparticles were absorbed on poly(4-vinyl pyridine) (P4VP) segments of the polymer nanorods and the gold nanoparticles rings can grow into gold nanaorings. In the process, the P4VP segments were quaternized and positively charged. The gold nanoparticles are coated by citrate, which are negatively charged. Then negatively charged gold nanoparticles were site-selectively adsorbed on the surface of positively charged P4VP segment by electrostatic interactions. This method can accurately adjust the diameter and spacing of the nanorings. By this method, a three-dimensional coaxial gold nanoring structure with one, two, and three gold nanorings on a polymer nanorod was constructed for the first time. Coaxial Au nanoparticle (NP) rings with different diameters were also formed on the same polymer nanorod by simply modification of the etching process, and the diameters of the nanorings increase gradually from the top to the bottom. In addition, Au NP rings can be grown into coaxial gold nanorings by electroless deposition of gold. The morphology of the nanoring is clear, and the inner diameter of the nanoring is

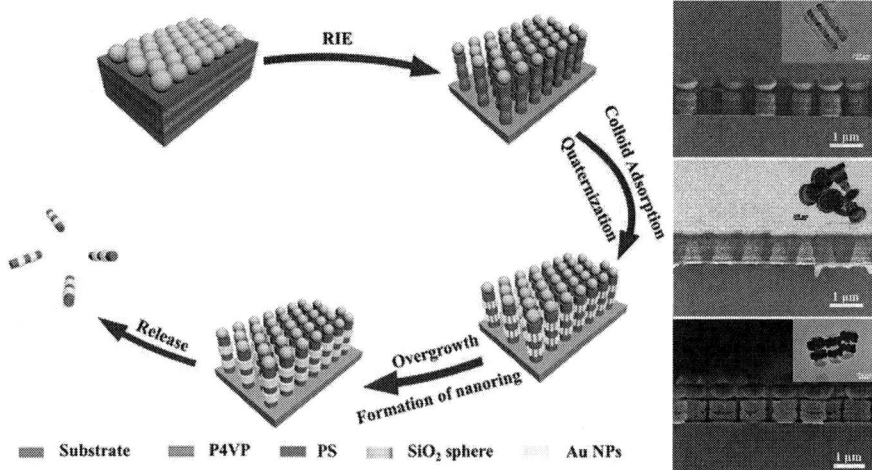

Fig. 2 Schematic diagram of polymer-assisted self-assembly of Au NPs and the fabrication of gold nanoring arrays.

the same as that of the polymer nanorods. Futhermore, this method can also be extended to the fabrication of palladium (Pd) nanorings, and Pd nanorings with good morphology and distribution can be also obtained. We believe that our method opens up a new way for the construction of a variety of metal nanorings, and the resulting metal nanorings have potential applications in plasma antennas, plasma semiconductors, etc.

3. Bioinspired microcone-array-based living biointerfaces: enhancing the anti-inflammatory effect and neuronal network formation

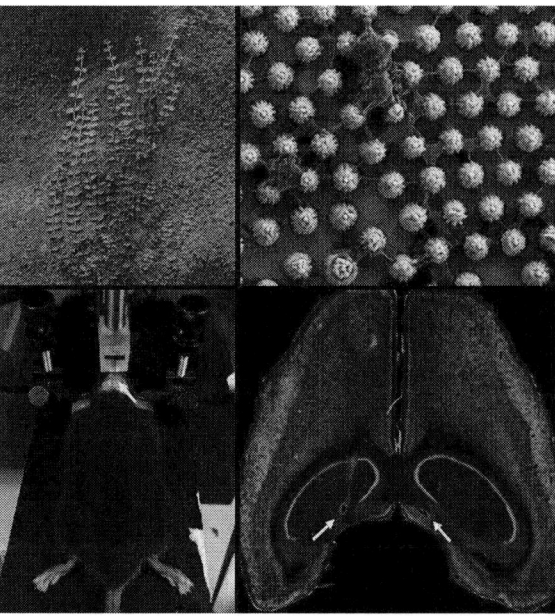

Fig. 3 Bioinspired living biological interface based on microcone array represents an unprecedented attempt for enhancing anti-inflammatory effect and creeper-like neuron network formation, which were experimentally clarified via in vitro and in vivo studies.

Implantable neural interfaces and systems have attracted attention due to their extensive application in the treatment of a variety of neurological diseases such as Parkinson's disease, epilepsy, blindness, essential tremor, anxiety and depression [18-20]. Traditional implantable devices are usually made of polymers, organic matter, metals, or composites, which are quickly recognized by the central nervous system as foreign bodies and thus trigger an inflammatory response [21, 22]. The continuation of this process can lead to scar tissue encapsulation, leading to a significant reduction in the performance or even loss of functionality of the implantable device. Therefore, enhancing anti-inflammatory effects to ensure long-term reliability is a key factor for these implantable devices to reach their full potential.

Recently, the design of implanted devices has benefited from the advantages and opportunities presented by bionic design strategies [23, 24]. These methods include designing stimulus-responsive materials with dynamic stiffness to facilitate insertion of the device, intelligent devices with shape conversion capabilities, and neuron-like nerve probes [25]. Given these successful applications of biomimetic concepts in neural implants, it is noteworthy that neural interfaces simulating micro/nanostructures on the surface of rose petals have not yet been explored.

In the above research context, we developed a living biological interface [6], as shown in the Fig. 3. Patterned SU-8 probes were firstly obtained by photolithography. Then SU-8 microcone-array-based probes (MAPs) were fabricated via colloidal lithography. Platinum (Pt) layer was deposited onto the SU-8 microcone array by magnetron sputtering. Compared with other advanced neural interfaces, the microcone array matches the characteristic size of neuron cells, thus ensuring a denser expansion of neuronal processes and forming a highly interconnected network of neurons that behave like reptiles. Interestingly, the characteristic size and roughness of the microcone array promoted neurite adhesion and differentiation but inhibited astrocyte adhesion, indicating a significantly enhanced anti-inflammatory effect compared to previous studies. Notably, the microcone array probes are the basis for neuronal migration, suggesting that it has the potential to direct specific cells to damaged sites where they can help repair or regenerate a variety of damaged tissues. In addition,

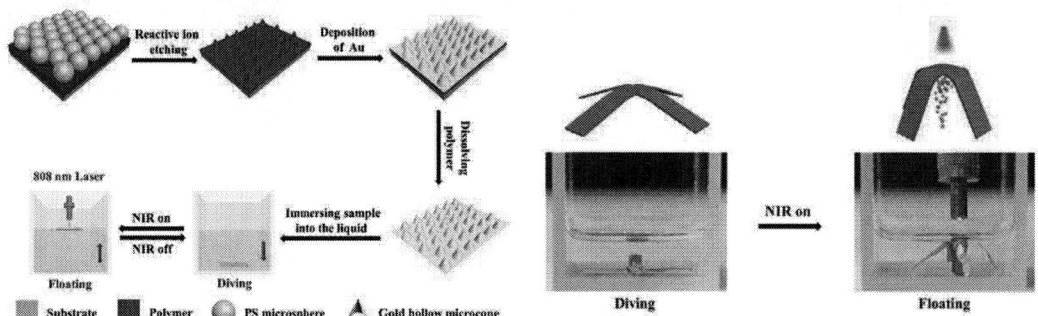

Fig. 4 The integrated micro/nanomotors composed of gold hollow microcone array (AuHMA), are fabricated by colloidal lithography. The vapor bubbles induced by plasmonic heating effect of gold hollow microcones under NIR irradiation can drive various motion behaviors of AuHMA film.

the patterned design of the microcone array can contribute to the formation of patterned neural networks, suggesting that it has the potential to stimulate or record individual neurons. In combination with other functional materials, our bioinspired structural approach with enhanced anti-inflammatory effects and neuronal network formation may be a platform technology that not only opens the way to a new generation of artificial neural networks and brain-computer interfaces, but also provides a universal biomedical treatment approach.

4. Near-Infrared Light-Driven Controllable Motions of Gold-Hollow-Microcone Array

The continuous development of fuel-free micro/nano motors provides a new method for the construction of new driving components and the controlled movement of soft robots. A micro/nanomotor is a micro/nano particle or device that converts other forms of energy into its own kinetic energy. Micro/nanomotor shows excellent movement performance under the driving of magnetic, optical, ultrasonic, thermal and other external fields [26-28]. The sensitivity and repeatable control behavior of these external physical stimulation is widely used in biomedical and drug transport [29, 30]. However, the integration of micro/nanomotors into one monolithic device remains a challenge.

As a noncontact and harmless stimulus, near-infrared light provides an effective way for the movement of micro/nanomotors. Moreover, near-infrared light has many advantages such as fast response, remote control, repeated use and strong penetration ability. Among the previously reported individual micro/nanomotors, the microcone structure with rocket morphology showed superfast propulsion ability under near-infrared light irradiation, and the asymmetrical morphology of the microcone was more conducive to the directional movement of the micromotor with rocket morphology. In view of the characteristics of the monodisperse individual micro/nanomotor to convert the external field energy into its own kinetic energy, especially the advantages of remote control of near-infrared light. The microcone array structure with near-infrared light response was constructed by using the microcone with rocket morphology as driving unit, and the microcone array

is integrated into the bionic soft robot as the driving part to drive its controllable motion.

As shown in the Fig. 4, the integrated micro/nanomotors, composed of gold hollow microcone array (AuHMA), are fabricated by colloidal lithography. The SEM images show the morphologies of the fabrication process of the AuHMA film. The optical properties of the AuHMA film were also studied, there is a peak in near-infrared region. The finite difference time domain simulation results show that the electric field is mainly distributed in the inner tip region of the gold hollow microcone. Under NIR light illumination, the temperature of gold hollow microcone array can reach about 80°C within 8 seconds. Due to the excellent plasma-heating properties of the AuHMA, the integrated micro/nanomotors can generate vapor bubbles in the liquid under near-infrared irradiation, which induces floating-diving motion through near-infrared on/off irradiation. The AuHMA film has the ability to transfer macroscopic objects directionally at high near-infrared radiation intensity. In addition, the precise manipulation of a variety of motion behaviors and jellyfish-like floating motions can also be achieved. Moreover, the AuHMA film can be used as a motor to drive the foam boat as exposure to NIR irradiation. This method will open up a new way to drive bionic devices and is expected to solve the driving and motion control problems faced by bionic soft robots at present.

REFERENCES

[1] A. Valsesia et al., "Selective Immobilization of Protein Clusters on Polymeric Nanocraters," *Adv. Funct. Mater.*, Vol. 16, pp. 1242-1246, 2006.

[2] H. Chen, J. Zhang et al., "Ag Nanoparticle/Polymer Composite Barcode Nanorods," *Nano Research*, Vol. 8, pp. 2871-2880, 2015.

[3] H. Chen, J. Zhang et al., "Polymer-Assisted Fabrication of Gold Nanoring Arrays," *Nano Research*, Vol. 10, pp. 3346-3357, 2017.

[4] H. Chen, X. Du et al., "Light-Powered Micro/Nanomotors," *Micromachines*, Vol. 9, pp. 41, 2018.

[5] H. Chen, X. Du et al., "Near-Infrared Light-Driven Controllable Motions of Gold-Hollow Microcone Array," *ACS Appl. Mater. Interfaces*, Vol. 11, pp. 15927–15935, 2019.

[6] H. Chen, L. Wang, Y. Lu, X. Du, "Bioinspired microcone array-based living bio-interfaces: enhancing anti-inflammatory effect and neuronal network formation," *Microsystems & Nanoengineering*, Vol. 6, pp. 58, 2020.

[7] D. Seo et al., "Ag-Au-Ag heterometallic nanorods formed through directed anisotropic growth," *J. Am. Chem. Soc.*, Vol. 130, pp. 2940–2941, 2008.

[8] J. B.-H. Tok et al., "Metallic striped nanowires as multiplexed immunoassay platforms for pathogen detection," *Angew. Chem., Int. Ed.*, Vol. 45, pp. 6900–6904, 2006.

[9] J. Zhang et al., "Colloidal self-assembly meets nanofabrication: From two-dimensionalcolloidal crystals to nanostructure arrays," *Adv. Mater.*, Vol. 22, pp. 4249–4269, 2010.

[10] J. Zhang et al., "Patterning colloidal crystals and nanostructure arrays by soft lithography," *Adv. Funct. Mater.*, Vol. 20, pp. 3411–3424, 2010.

[11] X. Li, J. Zhang et al., "Controlled fabrication of fluorescent barcode nanorods," *ACS Nano*, Vol. 4, pp. 4350–4360, 2010.

[12] J. Aizpurua et al., "Optical properties of gold nanorings," *Phys. Rev. Lett.*, Vol. 90, pp.057401, 2003.

[13] F. Hao et al., "Shedding light on dark plasmons in gold nanorings," *Chem. Phys. Lett.*, Vol. 458, pp. 262−266, 2008.

[14] M. G. Banaee et al., "Gold nanorings as substrates for surface-enhanced raman scattering," *Opt. Lett.*, Vol. 35, pp. 760–762, 2010.

[15] R. Near et al., "Pronounced effects of anisotropy on plasmonic properties of nanorings fabricated by electron beam lithography," *Nano Lett.*, Vol. 12, pp. 2158–2164, 2012.

[16] C. Y. Tsai et al., "High sensitivity plasmonic index sensor using slablike gold nanoring arrays," *Appl. Phys. Lett.*, Vol. 98, pp.153108, 2011.

[17] C. Liusman et al., "Free-standing bimetallic nanorings and nanoring arrays made by on-wire lithography," *ACS Nano*, Vol. 4, pp.7676–7682, 2010.

[18] Y. Maet al., "Mammalian near-infrared image vision through injectable and self-powered retinal nanoantennae," *Cell*, Vol. 177, pp. 243–255, 2019.

[19] J. Tang et al., "Nanowire arrays restore vision in blind mice," *Nat. Commun.*, Vol. 9, pp. 786, 2018.

[20] S. M. Wellman et al., "A materials roadmap to functional neural interface design," *Adv. Funct. Mater.*, Vol. 28, pp.1701269, 2018.

[21] J. Rivnay et al., "Next-generation probes, particles, and proteins for neural interfacing," *Sci. Adv.*, Vol. 3, pp. e1601649, 2017.

[22] G. Hong et al., "A method for single-neuron chronic recording from the retina in awake mice," *Science*, Vol. 360, pp. 1447–1451, 2018.

[23] M. Montgomery et al., "Flexible shape-memory scaffold for minimally invasive delivery of functional tissues," *Nat. Mater.*, Vol. 16, pp. 1038–1048, 2017.

[24] Q. Zhao et al., "Programmed shape-morphing scaffolds enabling facile 3D endothelialization," *Adv. Funct. Mater.*, Vol. 28, pp. 1801027, 2018.

[25] Y. Zhang et al., "Climbing-inspired twining electrodes using shape memory for peripheral nerve stimulation and recording," *Sci. Adv.*, Vol. 5, pp. eaaw1066, 2019.

[26] M. Medina-Sánchez et al., "Swimming Microrobots: Soft, Reconfigurable, and Smart," *Adv. Funct. Mater.*, Vol. 28, pp. 1707228, 2018.

[27] T. Li et al., "Janus Microdimer Surface Walkers Propelled by Oscillating Magnetic Fields," *Adv. Funct. Mater.*, Vol. 28, pp. 1706066, 2018.

[28] C. Chen et al., "Light-Steered Isotropic Semiconductor Micromotors," *Adv. Mater.*, Vol. 29, pp. 1603374, 2017.

[29] J. Li et al., "Micro/Nanorobots for Biomedicine: Delivery, Surgery, Sensing, and Detoxification," *Sci. Robot.*, Vol. 2, pp. eaam6431, 2017.

[30] M. Uygun et al., "Micromotor-Based Biomimetic Carbon Dioxide Sequestration: Towards Mobile Microscrubbers," *Angew. Chem., Int. Ed.*, Vol. 54, pp.12900−12904, 2015.

978-1-6654-3008-1/21 $31.00 © 2021 IEEE

978-1-6654-3008-1/21 $31.00 © 2021 IEEE   1846c

The 16th IEEE International Conference on Nano/Micro Engineered & Molecular Systems

Design of Sub-nanometre Pores for Precise Separation

Jian Jin

Chemical Engineering and Materials Science, Soochow University, Suzhou, 215123, P. R. China

Website: http://www.film-sinano.com/

ABSTRACT

With the rapid development of modern industry, the development of materials and technologies for energy-efficient precise separation, and the realization of precise separation of ions and molecules will have a revolutionary impact on energy, water, chemical, pharmaceutical and other fields, which is our urgent need and goal. Membrane separation has the advantages of low energy consumption, mild separation conditions and easy operation. It has been widely used in various fields of industrial production and daily life. However, the preparation of nanoporous membranes with highly uniform pore size and accurate separation of ions or small molecules are still facing great challenges.

Nanofiltration membrane prepared by interfacial polymerization is the main form of commercial nanoporous membrane, which has great potential in the application of water treatment, food, medicine, metallurgy and so on. This talk will introduce our recent progress on developing nanofiltration membranes with high flux and high rejection for precise separation of ions and small molecules. Our strategies include: (1) design of new porous support membrane based on one dimensional nanowires network; (2) increasing the effective filtration area of the selective layer by using nanoparticles and salt crystals as sacrificing template; (3) regulation of the reaction process of interfacial polymerization via a dynamic, self-assembled network of surfactants monolyer. As a result, we achieved polyamide nanofiltration membranes with highly uniform sub-nanometre pores for sub-1 Å precision separation.

BIOGRAPHY

Jian Jin is currently a Professor at Chemical Engineering and Materials Science, Soochow University. Dr. Jin received her Ph.D. degree in Institute of Chemistry from Jilin University. She was a postdoctoral fellow in Tokyo University from 2001 to 2003. From 2004 to 2009, she worked as a senior researcher in National Institute for Materials Science, Japan. In 2009, she joined Suzhou Institute of Nano-Tech and Nano-Bionics, Chinese Academy of Sciences as a group leader. She moved to Soochow University as a distinguished professor in 2017. Her research interests are designing porous materials and membranes for ion/molecule separation, wastewater purification, gas separation, and the applications in other fields. She has published more than 130 peer-reviewed scientific publications with over 9600 citations and h-index 52. She applied 40 invention patents and received 22 licensed patents (include 2 US patents).

Dr. Jin was awarded "Distinguished Yong Scholar Award" from National Natural Science Foundation of China and "Hundred Talents Program" of Chinese Academy of Sciences. She was honored as the Leading Innovative Talent Scholars of Young and Middle Age in the Ministry of Science and Technology of People's Republic of China and elected as a member in the Ten Thousand-Talent Program in China. She won the "Zhu Li Yue Hua" Excellent Supervisor Award. As an outstanding woman scholar, she was awarded as "National Woman Pacesetter". In 2018, she won the first prize of Jiangsu Science and Technology Award.

978-1-6654-3008-1/21 $31.00 © 2021 IEEE

Gap in pagination due to unavailable papers.

Pages 1848-1852

Proceedings of the 16th Annual IEEE International
Conference on Nano/Micro Engineered and Molecular Systems
April 25-29, 2021

Bionic Tactile Sensor based on Triboelectric Nanogenerator for Motion Perception

Siyuan Wang[1], Peng Xu[1], Xinyu Wang[1], Hao Wang[1], Changxin Liu[1], Liguo Song[1], Guangming Xie[2], Minyi Xu[1*]

Abstract— Several species of terrestrial and marine organisms like seals, use their vibrissae (whiskers) as important tactile sensors to find and follow underwater wakes. Here, we propose a bionic-tactile sensor (BTS) based on triboelectric nanogenerator to capture complex stimuli by organisms moving through water, and utilize received signal to track the hydrodynamic trails. One of the BTS's essential features is that triboelectric nanogenerator coupled with inspired by the structure characteristics of seal whiskers, which leads to the highly-sensitive detection of stimuli. Another is that our design is self-powered without additional power supply. Experiments show that the designed BTS has the ability to perceive the hydrodynamic trails left by underwater moving objects. Moreover, the relationship between the output signal and the movement parameters is linear, and we use this principle to perceive the relative motion state of the moving object. These new insights into the biological basis of tactile perception using whiskers provides new design guidelines to develop efficient underwater robotic sensors.

I. INTRODUCTION

Remote operated vehicles (ROVs) use a large number of sensors aiming at emerging applications like environmental monitoring [1] and pipe manipulation tasks [2]. Generally, the traditional strategies reports the use of ROVs equipped with acoustic and visual perception [1,2]. The drawbacks of these strategies are multiple including the limitations and high operation costs. Specially, most of the sonar devices have a relatively larger weight and volume, and higher cost. Visual perception devices have poor perception of harsh underwater environments. However, the vast majority of ROVs are space-constrained, with finite internal volume and extremely limited payload capacity [3]. In addition, the neutral buoyancy requirements must be considered.

In order to break the dilemma, some research groups got inspiration from marine organisms, which involves a strong ability to perceive their surroundings through detecting hydrodynamic signals independently of acoustic, visual, or chemical cues. In particular, blind fish with their sideline system can perceive waterborne disturbances [4] dolphins and whales can achieve the navigation and localization tasks through echolocation. Seals lack specialized sideline and echolocation, but possess the ability to perceive flow changes and water tracking using the tiny vibrations of their whiskers [5]. Moreover, the unique geometric shape of the seal whiskers also has better performance in resisting vortex-induced vibration (VIVs). Some researchers have started to

develop new underwater sensors to enhance the environmental perception of underwater devices by learning from the way marine organisms perceive environment independently of sight and hearing, such as robotic fish sensing systems based on bionic sidelines system [6,7], piezoelectric material-based whisker sensors [8] and graphene-based whisker sensors [9], but the difficulty of single-channel acquisition, complex manufacturing processes and high material costs limit the further development of bionic underwater sensors.

Recently, triboelectric nanogenerator (TENG) based on the coupling technology of contact electrification and electrostatic induction has been developed as a new electromechanical conversion technology, which has ultra-high sensitivity to mechanical stimuli due to its unique working mechanism [10]. TENG can detect mechanical motion within a certain range, and has wide applications in both energy collection and self-powered sensing. In this paper, a bionic-tactile sensor (BTS) based on triboelectric nanogenerator, which appearance of the seal whiskers is bionic, is developed to sense changes of the underwater environment. The sensor is composed of silica gel, conductive ink, fluorinated ethylene propylene copolymer (FEP), polyethylene terephthalate (PET) and electrostatic shielding material (A-PET/CPP), which has the advantages of low cost, high accuracy and self-powered. This bionic-tactile sensor can be used as a supplement to traditional sensing systems to enhance the ability of ROVs to perceive their working environment.

II. STRUCTURE AND WORKING PRINCIPLE OF BTS

As shown in Fig.1(a), the seal whiskers show unique wavy-like appearance, and it contributes to the ability to reduce the influences of the vortex-induced vibrations (VIVs). With the specific structure schematically shown in Fig.1(b), the design of bionic-tactile sensor refers to the wavy-like appearance of the seal whiskers, consists of an internal triboelectric nanogenerator unit and an external flexible shell. The internal generator unit is composed of a fluorinated ethylene propylene fluoride copolymer (FEP) film and polyethylene terephthalate (PET) film printed on conductive inks, and a plastic gasket made of PET with a pore in the middle to make up a contact-separated triboelectric nanogenerator section, which is externally encapsulated with an electrostatic shielding material (A-PET/CPP) to reduce the influences of generation performance in underwater environment.

---

1 Marine Engineering College, Dalian Maritime University, Dalian, 116026, China
2 College of Engineering, Peking University, Beijing, 100871,
China
*Address correspondence to xuminyi@dlmu.edu.cn

978-1-6654-3008-1/21 $31.00 © 2021 IEEE

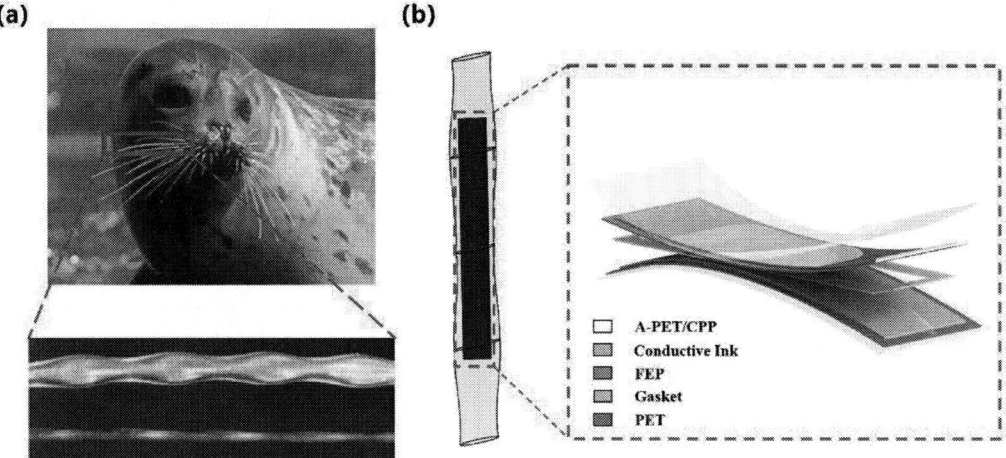

Figure 1. (a) The unique geometric structure of the seal whiskers; (b) The external structure and internal structure of the bionic tactile sensor.

To verify the resistance of seal whiskers to vortex-induced vibrations, in this paper, we used ICEM + ANSYS CFX to simulate the vortex generation situation of a cylindrical whisker model and a seal whisker model of the same diameter under different conditions, which is shown in Fig.2(a) and 2(b). It can be seen from Fig.2(a) that under incoming flow conditions, the vortices formed around the cylindrical whisker but vortices formed far from the seal whisker. And these vortices formed near the whisker cause obviously the VIVs when contact with the whisker surface. Due to seal whisker's wavy-like appearance, the VIVs will not produce violently. However, it can be found in Fig.2(b) that both the cylindrical whisker and the seal whisker respond significantly to the proximity of the vortices under turbulence conditions, where a cylinder is set up as a turbulence column. Both the cylindrical whisker and the seal

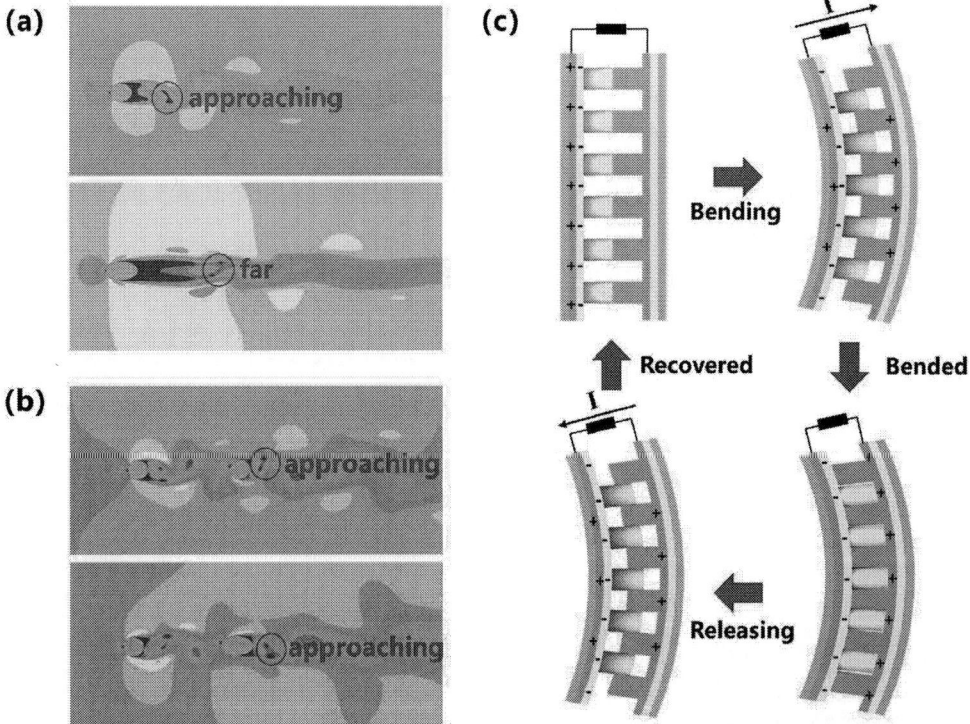

Figure 2. (a) Schematic diagram of cylindrical whisker and seal whisker vortex simulation under inflow conditions; (b) Simulated diagram of cylindrical whisker and seal whisker vortex simulation under turbulence conditions; (c) Schematic diagram of bionic tactile sensor power generation.

978-1-6654-3008-1/21 $31.00 © 2021 IEEE   1854

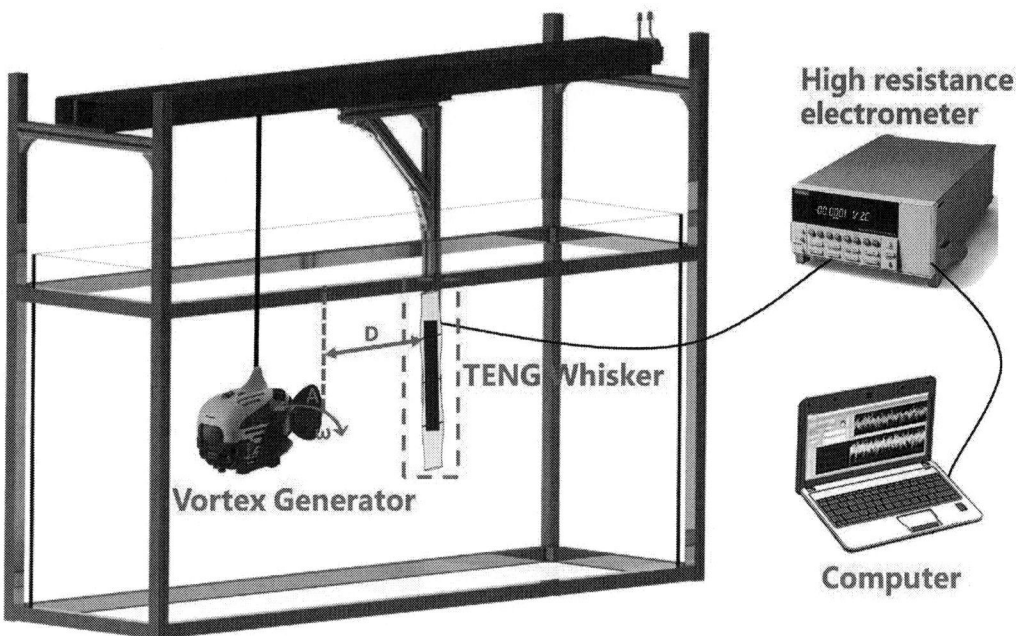

Figure 3. Schematic diagram of a performance test platform for the BTS

whisker generate reciprocating oscillations due to the VIVs, and the amplitude and frequency of the oscillations are affected by the closed vortex. So the BTS we designed, which not generate oscillation when it moving, but when the nearby flow changes it can sense the variety by output a corresponding electrical signal.

The specific principle of the BTS is shown in Fig.2 (c). In the initial state, the charge on the FEP film and the charge on the conductive ink are in balance of the PET film. When the whisker structure swings, the micro-nano structure on the surface of the FEP film and the conductive ink squeeze each other, because the coupling of contact electrification and electrostatic induction effect, the surface charge is transferred on the contact area, a net negative charge is generated on the surface of the FEP film, and a net positive charge is formed on the surface of the conductive ink of the PET film. Thus, an electrical signal output can be collected, and the surface charge is completely transferred when it is in full-contact. When the BTS begins to return to its original state, in order to balance the charge difference between the FEP film and the conductive ink of the PET film, the electrons on the conductive ink of the PET film, which is as the sensing electrode, will be driven by the electric field force to transfer to the conductive ink of the FEP film, the electrical signal output in the opposite direction is formed.

III. BIONIC-TACTILE SENSOR MEASURES THE DISTANCE AND MOTION STATUS OF THE UNDERWATER MOVING OBJECT

As shown in Fig.3, a BTS performance test platform had built to test the sensor's ability to perceive underwater

moving objects. The vortex is generated by the swing of a caudal fin, so we can control the amplitude A and the angular velocity ω of the caudal fin to imitate the motion of underwater objects. The Keithley 6514 high resistance electrometer is used to measure the electrical signal output of bionic-tactile sensor and the data input into the computer for storage via the National Instruments acquisition card. The distance between the vortex-generator and the bionic-tactile sensor can be adjusted via the upper slideway, to imitate the distance D between the moving object and the BTS.

We systematically study the effect on the output performance of the sensor from the distance D, the amplitude A and the angular velocity ω. As shown in Fig. 4 (a), when D = 30mm and A = 0.628rad, the BTS output signal has a strong positive correlation with the change of ω. As ω increases from 0.74 rad / s to 2.62 rad / s, the output signal of the sensor increased from 0.52V to 0.88V. But it is worth noting that the change of ω will also cause the change of the output period in the same time. When ω = 0.74 rad / s, the BTS only outputs a half-cycle signal within 4s, and when ω = 2.62 rad / s, the BTS outputs four cycles signal within 4s. When A = 0.628 rad, the change trend of the BTS output with the swing angular velocity ω and distance D is shown in Fig.4 (b). It can be found that the change of the BTS output has an obvious linear law overall, and the non-linearity at D = 10mm could be explained as the incomplete formation of the vortex because the excessively short distance, which effects the incomplete swing of the BTS. At the same time, as the distance D increases, the difference of

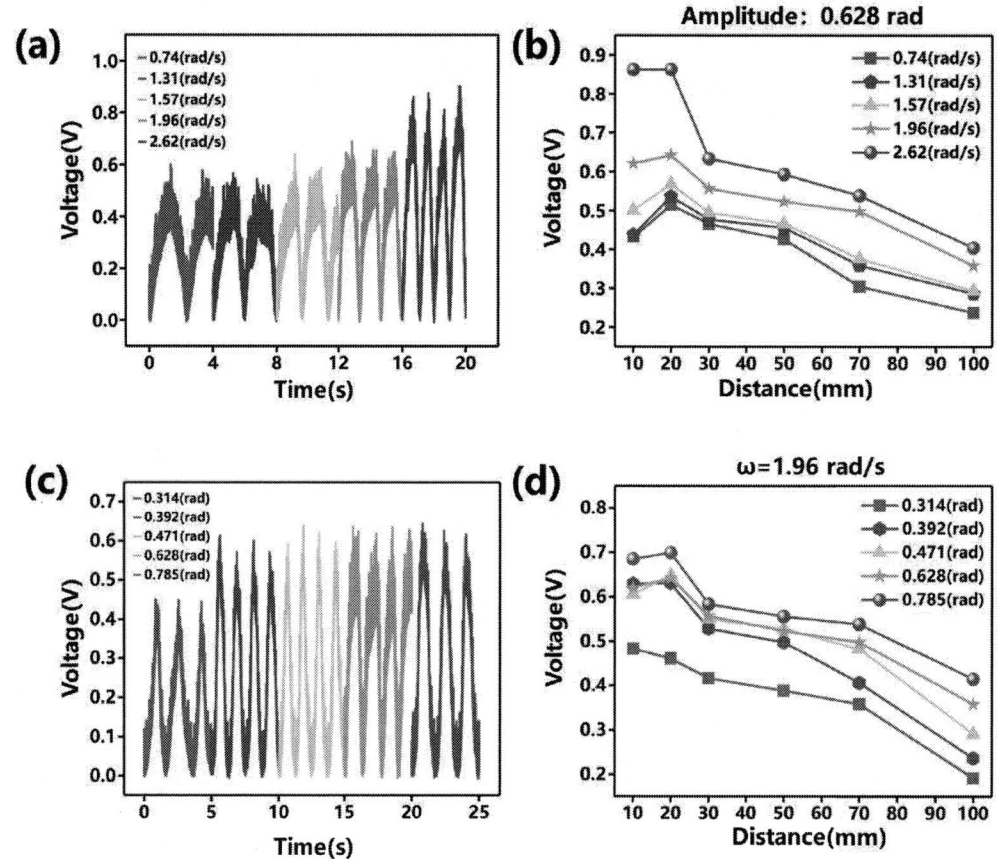

Figure 4. (b) Schematic diagram of sensor output when D=30mm, A=0.628rad, ω=0.74rad/s, 1.31rad/s, 1.57rad/s, 1.96rad/s, 2.62rad/s (c) Schematic diagram of sensor output changing with distance D and swing angular velocity ω when A = 0.628rad (d) Schematic diagram of sensor output signal when D=30mm,ω=1.96rad/s, A=0.314rad, 0.392rad, 0.471rad, 0.628rad, 0.785rad (e) Schematic diagram of sensor output changing with distance D and swing amplitude A when ω=1.96rad/s.

the output under different ω is more obvious, which is conducive to accurate measurement.

As shown in Fig.4 (c), when ω is constant, the BTS output and A also show a positive correlation. As A increases from 0.314 rad to 0.785 rad, the BTS output increases from 0.44V to 0.61V. The abrupt change of the output period at A = 0.314 rad could be explained as the swing amplitude is too small to generate the vortex had enough strength to drive the BTS swing completely. When ω = 1.96 rad / s, the change trend of output with A and D is shown in Fig.4 (d).

IV. BIONIC-TACTILE SENSOR MEASURES THE DEFLECTION OF UNDERWATER MOVING OBJECT

We use the BTS to measure the deflection of moving object. As shown in Fig.5 (a), the turning process of a moving object can be imitated, by changing the unilateral deflection angle of the caudal fin. In order to facilitate the experiment, the total swing angle of the caudal fin is limited to ± 36 °, and only the unilateral deflection angles α + and

α- during the swing process are changed, ensure the sum of α+ and α- is always 72 °.

Fig.5 (b) shows the variation of the output signal of the BTS under different deflection angles. When α + = α -, the BTS output is bilaterally symmetric, and the peak value of the output voltage can reach the maximum value. When α + ≠ α -, the BTS output will have a sudden sag at the peak of each output cycle compared with the condition when α + = α-, and the peaks on the left and right sides of the sag show differences. As the difference between α + and α- increases, the sag exhibits a significantly increasing trend. When α +> α-, the waveform of the output at the right side of the sag is higher than the left side, when α + <α-, the waveform of the output at the left side of the sag is higher than the right side, and the peak attenuation of the output voltage under two conditions is 1/3 - 1/2 compared with the condition when α + = α-. Fig.5 (c) shows the variation of the output signal of the BTS under different vortex rotation conditions. The variation trend is completely opposite. By monitoring the variation trend of the BTS output, the initial motion state of the underwater moving object can be predicted, and the

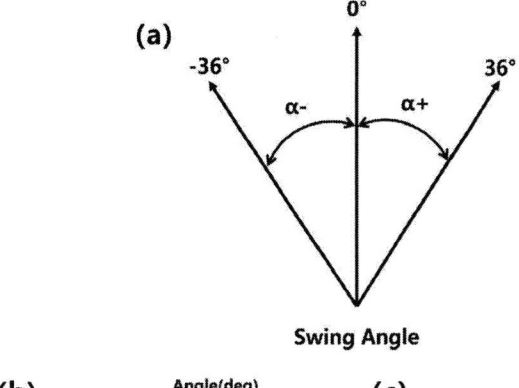

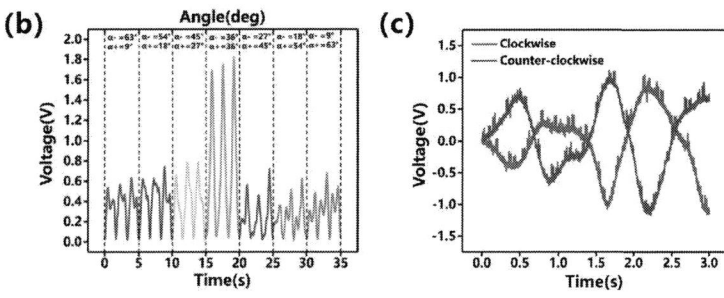

Figure 5. (a) Schematic diagram of the deflection angle of the caudal fin (b) Schematic diagram of the change of the sensor output signal under different deflection angles (c) Schematic diagram of the effect of the sensor output signal with different vortex rotation

motion change of moving object can be more accurately sensed.

V. CONCLUSION

In this paper, we have designed a bionic-tactile sensor (BTS) based on a triboelectric nanogenerator, which can be used as a supplement to the perception system of the ROVs. On the basis of demonstrating the structure, bionic principle and generator principle of the BTS, this work provides the variation trend of the BTS output sensing the flow field under the disturbance of the underwater moving object. In the case of different motion parameters, the output peak shows a good linear trend supporting the accurate measurement of motion information. In addition, the BTS possesses an ability to dynamically capture the effect. The relative motion state of the moving object can be determined by the variation trend of its output signal. The design of the BTS herein advance our ability to replicate the seal's remarkable mechanoreceptive sensory abilities.

ACKNOWLEDGMENT

The authors are grateful for the joint support from the National Natural Science Foundation of China (Grant Nos. 51879022, 51979045), the Fundamental Research Funds for the Central Universities, China (Grant No. 3132019330). Minyi Xu are Corresponding authors.

REFERENCES

[1] Nicosevici T, Garcia R, Carreras M, et al. A review of sensor fusion techniques for underwater vehicle navigation[C]//Oceans' 04

MTS/IEEE Techno-Ocean'04 (IEEE Cat. No. 04CH37600). IEEE, 2004, 3: 1600-1605.

[2] Agus Budiyono. Advances in unmanned underwater vehicles technologies: Modeling, control and guidance perspectives[J]. Indian Journal of Marine Sciences, 2009, 38(3): 282-295.

[3] Wang W H, Chen X Q, Marburg A, et al. Design of low-cost unmanned underwater vehicle for shallow waters[J]. International Journal of Advanced Mechatronic Systems, 2009, 1(3): 194-202.

[4] Hassan E S. Hydrodynamic imaging of the surroundings by the lateral line of the blind cave fish Anoptichthys jordani[M]//The mechanosensory lateral line. Springer, New York, NY, 1989: 217-227.

[5] Krüger Y, Hanke W, Miersch L, et al. Detection and direction discrimination of single vortex rings by harbour seals (Phoca vitulina)[J]. Journal of Experimental Biology, 2018, 221(8): jeb170753.

[6] Zheng X, Wang C, Fan R, et al. Artificial lateral line based local sensing between two adjacent robotic fish[J]. Bioinspiration & biomimetics, 2017, 13(1): 016002.

[7] Zheng X, Wang M, Zheng J, et al. Artificial lateral line based longitudinal separation sensing for two swimming robotic fish with leader-follower formation[C]//2019 IEEE/RSJ International Conference on Intelligent Robots and Systems (IROS). IEEE, 2019: 2539-2544.

[8] Beem H R. Passive wake detection using seal whisker-inspired sensing[R]. MASSACHUSETTS INST OF TECH WOODS HOLE MA DEPT OF APPLIED OCEAN PHYSICS AND ENGINEERING, 2015.

[9] Gul J Z, Su K Y, Choi K H. Fully 3D printed multi-material soft bio-inspired whisker sensor for underwater-induced vortex detection[J]. Soft robotics, 2018, 5(2): 122-132.

[10] Wang Z L, Wang A C. On the origin of contact-electrification[J]. Materials Today, 2019.Ed. New York: McGraw-Hill, 1964, pp. 15–64.

Gap in pagination due to unavailable papers.

Pages 1858-1860

April 25-29 , 2021   Xiamen, China

NEMS-enabled Innovations at Interfaces for Water-Energy Nexus

Zuankai Wang

Department of Mechanical Engineering, City University of Hong Kong, Hong Kong, 999077, P.R. China

Website : https://wangzuankai.wixsite.com/wanglab

ABSTRACT

Water is the origin of life and energy. In spite of its ubiquity and seemingly simplicity, the water is probably the least understood matter in the world. The phase transition, transport, and manipulation of water, normally spanning different time and length scales, constitute the basic paradigm of numerous biological systems and industrial processes such as thermal management, energy, agriculture, and healthcare. Over the past decade, the advances in NEMS manufacturing and visualization provide new dimensions in our fundamental and controlling of interfacial and transport phenomena of water, especially on textured surfaces and under complicated working environments involving varying temperatures.

In this talk, I will discuss recent innovations enabled by NEMS at the interfaces to address one of the most important challenges facing us today, i.e., water-energy nexus. In particular, I will highlight how the rational design and control of topological structures enables us to fundamentally change the triple-phase interaction and achieve the preferred functionalities. Examples include how to efficiently collect water from air, how to use one droplet to cool down hot surfaces by several hundreds of degrees, and how to use one droplet to light up 100 LEDs.

BIOGRAPHY

Dr. Zuankai Wang is a full professor in the Department of Mechanical Engineering and Associate Dean in the College of Engineering at the City University of Hong Kong. He earned his B.S. degree in Mechanical Engineering from Jilin University in 2000 and Master degree in Microelectronics from Shanghai Institute of Microsystem and Information Technology, Chinese Academy of Sciences, in 2003, and Ph. D. degree in Mechanical Engineering at Rensselaer Polytechnic Institute in 2008. After one-year postdoc training in Biomedical Engineering at Columbia University, he joined in the City University of Hong Kong in September 2009 as an assistant professor.

Dr. Wang is the founding member of Young Academy of Science of Hong Kong, fellow of the International Society of Bionic Engineering and Changjiang Chair Professor awarded by Ministry of Education of China. He has won many awards including the 2020 Xplorer Prize, World Cultural Council Special Recognition Award, Outstanding Youth Award from International Society of Bionic Engineering, Outstanding Research Award and President's Lectureship at the City University of Hong Kong. The Ph.D. students he supervised have won a number of prestigious awards including MRS Graduate Student Gold Award, MRS Graduate Student Silver Award, Hiwin Doctoral Dissertation Silver Award, Hong Kong Young Scientist Award (2015, 2019).

978-1-6654-3008-1/21 $31.00 © 2021 IEEE

978-1-6654-3008-1/21 $31.00 © 2021 IEEE   1862

Application and Prospect of MEMS Technology to Geophysics

Heting Hong, *Student Member, IEEE*, Lvchao Ni, *Student Member, IEEE*, Hongchun Sun[*]

Abstract — **Earthquake early warning is one of the most effective means to reduce earthquake disasters. At present, many countries and regions in the world have established their own earthquake early warning systems. Compared with traditional geophone, the MEMS seismic sensor system has some advantages, including high sensitivity and compact structure, so it has been widely used in earthquake early warning systems. This paper mainly describes some projects of MEMS technology in earthquake early warning, and uses artificial excitation compares the MEMS seismic sensor system with traditional geophone. It is found that the sensitivity of the MEMS seismic sensor system is about 1.7 times that of the traditional geophone. Finally, some prospects are put forward for the application of MEMS technology in the field of earthquake early warning.**

I. INTRODUCTION

With the continuous deepening of human research on geophysics, the role of seismographs in exploring the pulsation of the earth has become more and more important. American scientist Cooper first proposed the concept of earthquake early warning in 1868. Earthquake early warning systems and seismic sensors play an irreplaceable role in the field of seismology. They are key elements for detecting earthquakes and generating electrical signals[1-4]. The earthquake early warning system can automatically shut down the nuclear power plants and other large projects before the destructive seismic waves reach them, minimizing casualties and property losses. High-density seismic network, such as Hi-Net, F-Net and V-Net seismic network in Japan[5], can quickly estimate the source parameters (seismic center, earthquake level and time of earthquake), generate strong ground motion maps and earthquake warning signal, Post earthquake disaster assessment, these parameters are very important for earthquake rescue, but the cost and maintenance cost of traditional seismic stations are very high, and increasing the density of the network requires a lot of investment. However, nowadays, low-cost MEMS sensor technology and devices based on MEMS technology have become very common in our lives[6]. It has the advantages of compact size, low energy consumption, high sensitivity, and low price. After continuous experiments in various countries, series of research found that its performance is grater than traditional force feedback sensors. With the continuous development of science and technology, a large number of application examples of MEMS accelerometers show that the use of high-density MEMS sensors can provide strong

All author are with the School of Mechanical Engineering and Automation, Northeastern University, Shenyang, China.

*Contacting Author: Hongchun Sun is with the School of Mechanical Engineering and Automation, Northeastern University; 11 Lane 3, Wenhua Rd, Shenyang, Liaoning, 110004, China (phone: +86 13190082316; e-mail: Hchsun@mail.neu.edu.cn).

ground motion maps within a few seconds after an earthquake, also compared with the traditional geophone, it has higher sensitivity and S/N ratio which provides critical seismic information and reduces[7]. There are great application prospects in earthquake early warning research such as early warning blind zone and shortening early warning time. According to Figure 1, the amount of seismic research related to MEMS technology is also increasing all over the world[8].

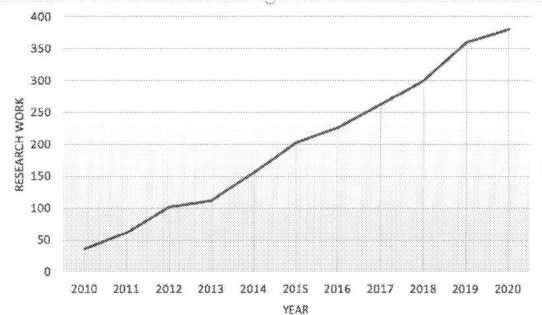

Figure 1.   Amount of seismic studies related to MEMS technology

II. MEMS IN SEISMIC FIELD

A. CSN Project

The CSN project was established in 2009 as a summer research project by Michael Olson, a PhD student, working with four undergraduates at Caltech, supervised by Mani Chandy and Ben Krause in Computer Science and Robert Clayton and Thomas Heaton in Geophysics. Shortly thereafter, they were joined by Monica Kohler in Civil Engineering. The team built a working system with an accelerometer connected to a desktop computer that sent messages to the cloud, implemented on the Google App Engine. Since Los Angeles is located near the San Andreas fault and other faults, the US Geological Survey pointed out that when an earthquake of magnitude 7.8 or greater occurs in the area, it will cause huge casualties and property losses. Therefore, the project was originally established in Los Angeles. In 2011, the Community Seismic Network (CSN) began taking data from small, inexpensive accelerometers in the greater Pasadena area. Able to measure both weak and strong ground movement along three axes, these accelerometers promise to provide very high-resolution data of shaking produced by seismic activity in the region. The 2014 survey results show that there are more than 400 sensors distributed in volunteers' homes, as well as some schools and high-rise buildings. The primary mission of CSN are: 1. Provide maps of maximum shaking immediately following a major earthquake to help direct first responders. 2. Monitor health and safety of structures. 3. Create zonation maps of populated areas[9]. At present the CSN sensor package consists of a 3-axis, class-C MEMS accelerometer (currently a Phidget 1043) in Figure 2 and Linux micro-computer (Raspberry Pi 3b). The sensor has a native sampling rate of 250 samples per second (sps), which is decimated in the microcomputer to 50 sps to reduce the data volume and to ensure the device does not function in the audible range.

Figure 2.   Picture of Phidget 1043

B. QCN Project

The QCN project was established in 2007. By 2014, more than 2,000 volunteers have joined the project, mainly in California, Mexico, etc[10]. QCN links volunteer hosted computers into a real-time motion-sensing network. The volunteer computers monitor vibrational sensors called micro electro-mechanical systems (MEMS) accelerometers and digitally transmit "triggers" to QCN's servers whenever strong motions are observed. QCN's servers sift through these signals, and determine which ones represent earthquakes, and which ones represent cultural noise. At present, the most used Joy-Warrior 12-bit to 16-bit sensors in Figure 3, it uses a MEMS solid state 3 axis acceleration sensor for acceleration or inclination measurement. The effective recording range is $\pm 2g$, and the sensitivity is $4 \times 10^{-2}g$&$.2.4 \times 10^{-8}g$, It takes about $30 to $150. The user connects the MEMS to the computer and provides network connection and maintenance work, thus greatly reducing the system cost. QCN instruments are mainly installed in volunteers' homes, schools, offices, and other highly noisy environments. BOINC (Berkeley Open Infrastructure for Network Computing) monitors and records data from sensors, and transmits the data to QCN servers.he QCN project was established in 2007. By 2014, more than 2,000 volunteers have joined the project, mainly in California, Mexico, etc. At present, the most used (Joy-Warrior and O-NAVI)12-bit to 16-bit sensors can detect seismic waves in the range of 0.1 Hz to 20 Hz. The effective recording range of acceleration is $\perp 2g$, and the sensitivity is $4 \times 10^{-2}g$&$2.4 \times 10^{-8}g$. It takes about US$30 to US$150. The user connects the MEMS to the computer and provides network connection and maintenance work, thus greatly reducing the system cost. QCN instruments are mainly installed in volunteers' homes, schools, offices, and other highly noisy environments. Through BOINC (Berkeley Open Infrastructure for Network Computing) monitors and records data from sensors, and transmits the data to QCN servers.

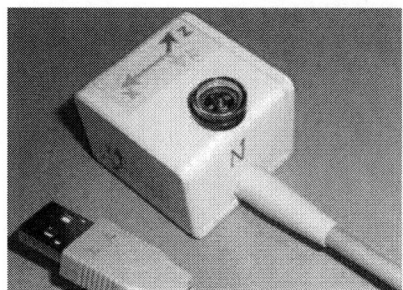

Figure 3.   Picture of   Joy-Warrior sensor

C. My shake Project

Based on the communication and networking functions of smart phones and the diversified characteristics of built-in MEMS, researchers have proposed a new type of earthquake early warning system --My Shake, which was originally established at the University of California (UC) Berkeley, uses P waves to trigger alarms. In February 2016, My Shake was publicly released for download on Google. As of August 2016, My Shake has recorded 237 earthquake events, including earthquakes in Chile, Argentina, Mexico, New Zealand, Taiwan, and Japan[11-12]. This is the first world seismic network to use personal smart phones to provide acceleration wave forms. Its advantage is that all hardware is encapsulated in the phone, but the disadvantage is that the location of the phone is constantly changing. The mission of My shake project is to build a worldwide earthquake early warning network so that communities can reduce the impact of earthquakes. Since My Shake uses smart phones as earthquake sensors, it can be used everywhere - even in countries without access to traditional seismic technology.

By placing the mobile phone in the basement to record data, the background noise of the mobile phone accelerometer is determined, which includes the internal noise of the mobile phone and the environmental noise. The magnitude of the earthquake can be estimated if the ground amplitude exceeds the noise level. The research report pointed out that the seismic data recorded by the personal smart phone MEMS can be used for earthquake research, earthquake disaster information evaluation, earthquake early warning and earthquake rupture process inversion. My Shake is sensitive to earthquakes of magnitude 2.5 or greater with a frequency range of 1 Hz to 10 Hz that occur within 10 km, and has the ability to recognize earthquakes from daily vibrations.

The My Shake project illustrates the feasibility of a smart phone based earthquake network. The project provides ground motion records when a destructive earthquake occurs, but the key is that the system is designed and tested on private smart phones, and there are billions of mobile phones worldwide. In order to make full use of the potential of "crowd-sourced"[11], researchers must develop software based on commonly used mobile phone sensors, rather than specific mobile phones. The software needs to have the least impact on users and provide real benefits to participants. Screen shots of its operation is shown in Figure 4.

978-1-6654-3008-1/21 $31.00 © 2021 IEEE

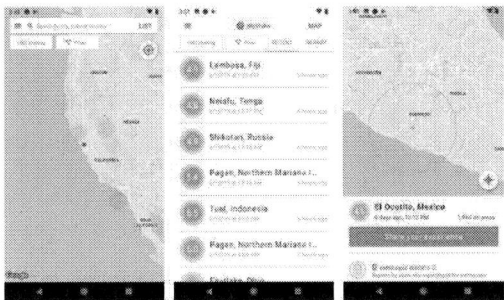

Figure 4. Screenshot of My shake app

III. COMPARISON BETWEEN MEMS SENSOR AND TRADITIONAL GEOPHONE

A. Traditional Geophone

According to the structural diagram shown in Figure 5, it is currently the most widely used moving coil geophone in seismic exploration[13]. The permanent magnet, the magnetic boots and the shell are fixed together, the coil mounted on the coil holder and the spring plate form an inertial body, and the spring plate plays the role of supporting the coil. Its working principle is also relatively simple: the tail cone at the lower end of the shell is coupled with the ground surface, and the permanent magnet fixed to the shell vibrates with the earth, and the coil remains inertially stationary due to the elastic support of the spring plate. As a result, the permanent magnet and the coil produce relative motion, the induced current is generated in the coil, and the detector will have an electrical signal output. Because of its low price and easy installation, it is widely used at present, but its disadvantage is that the sensitivity and dynamic range are small. Its operating frequency is generally 15Hz to 60Hz , so it's difficult to meet the demand of high-precision seismic exploration.

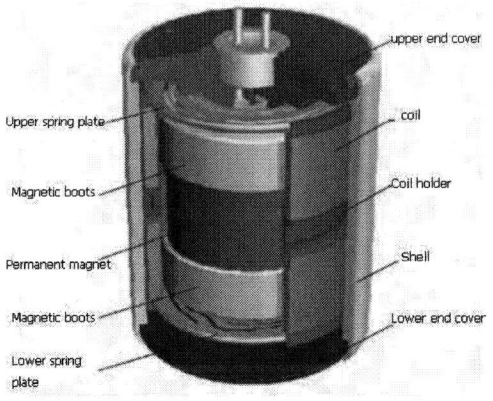

Figure 5. Traditional moving coil geophone

B. MEMS Geophone

According the assembly diagram of the new generation MEMS seismic sensor system in Figure 6, which consists of a fixed housing, a stainless steel base and a probe. The MEMS seismic sensor and circuit board are assembled into a stainless steel housing and base. The probe can be buried in the ground. The base uses a fixed plate to fix the MEMS seismic sensor so that the z-axis of the sensor is parallel to the plane in the base, and the wireless module and the GPS module is assembled into an external plastic box to facilitate the carrying of the entire system. The peripheral module composed of the wireless module and the GPS module is connected to the wiring terminal of the seismic acquisition unit through a 12-core cable, and the wireless module is added with an antenna to increase the communication distance. The picture is shown in Figure 7.

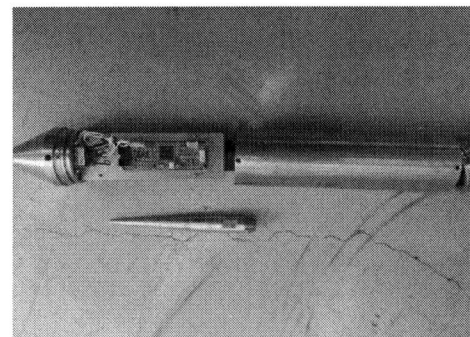

Figure 6. Assembly diagram of the new generation MEMS seismic sensor system

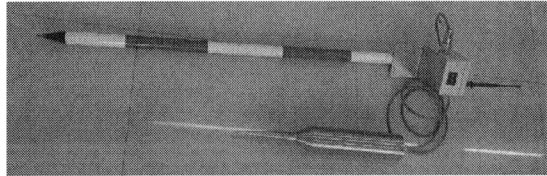

Figure 7. picture of the new MEMS seismic sensor system

The MEMS acceleration seismic sensor in the new MEMS seismic sensor system includes an upper capacitor, a lower capacitor plate and a movable middle capacitor plate. When the acceleration generated during an earthquake reaches a certain value, the middle capacitor plate will move, and the upper and lower capacitor plate's distance will change, then the upper and lower capacitance will change. It can be seen from Eq. (1) that the capacitance change is proportional to the acceleration. Therefore, after digital processing of the voltage, a digital signal is output to determine the level and intensity of the earthquake. It has the characteristics of strong impact resistance, small temperature drift and high reliability.

$$a = \frac{kd*\left(\frac{Cs1-Cs2}{Cs1+Cs2}\right)*Vm}{mV_m} \tag{1}$$

Where kd is damping coefficient, $Cs1$ and $Cs2$ are two opposite capacitors, Vm is input voltage, m is block mass.

C. Comparison

In the comparison, we use the traditional moving coil geophone and the new MEMS seismic sensor system with the same setting conditions, fix it on the marble floor with plaster, make artificial hammering at a distance of 1.5

meters, and use the two instruments to collect acceleration data for about 10 seconds at the same condition. The two sets of data wave forms in the vertical direction are shown in Figure 8 and Figure 9.

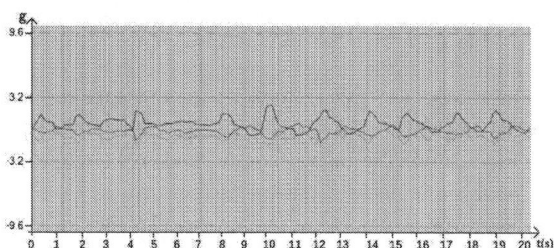

Figure 8.   Vertical acceleration of traditional moving coil geophone

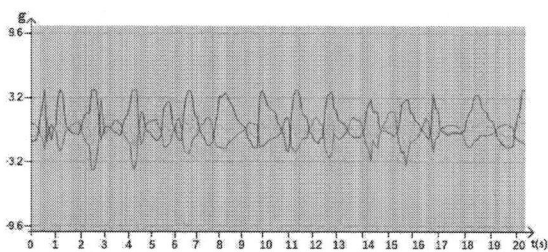

Figure 9.   Vertical acceleration of   new MEMS seismic sensor system

In the vertical record, all records are very similar. In order to compare the sensitivity, record the signal component and the noise component, and compare the sensitivity of the S/N ratios obtained by dividing the signal component by the noise component, as shown in Table 1.

TABLE I.   RESULT OF THE SENSITIVITY COMPARISON

Sensor	Signal	Noise	Sensitivity relative to traditional geophone
Traditional	184346	533	1.000
MEMS	567133	931	1.712

In Table 1, we can see that the MEMS seismic sensor system has the higher sensitivity than that of traditional geophone.

IV.   CONCLUSION

In this paper, we first introduced the advantages of MEMS technology in seismic applications. Secondly, through the CSN project, QCN project and My shake project, the application of MEMS technology in the field of seismic reconnaissance and earthquake early warning in the current era was elaborated. Finally, we compared the advantages and disadvantages of the two technologies by describing the structure and principle of traditional geophone and the structure and principle of the new MEMS seismic sensor system, using the response of traditional geophone and MEMS seismic sensor system to artificial excitation under the same conditions. Compared with traditional geophone, MEMS seismic sensor system

has obvious advantages in sensitivity. In the continuous investigation, it is found that the number of seismic research related to MEMS technology in the world is gradually increasing. We believe that with the application and promotion of MEMS technology, more and more countries will adopt MEMS technology for seismic monitoring and early warning. Work, MEMS seismic sensor system will also have broader application prospects.

REFERENCES

[1]   Havskov J and Alguacil G, "Instrumentation in Earthquake Seismology," *Springer Netherlands*, 2010.

[2]   Chistyakov, V. A. . "Portable seismic sensor," *Seismic Instruments* 47.1(2011):8-14.

[3]   Trifonov, N. V. . "The digital seismometer, a modern instrument." *Seismic Instruments* 45.1(2009):83-85.

[4]   Bashilov, I. P. , S. G. VolosoV, YN Zubko, and SA Korolyov. "Portable digital seismometer." *Seismic Instruments* 47.1(2011):80-88.

[5]   AIZAWA Takao, et al."Application of MEMS accelerometer to geophysics." *International Journal of the JCRM*, 2009, 4(2):33-36.

[6]   Huang, Xin , et al. "CrowdQuake: A Networked System of Low-Cost Sensors for Earthquake Detection via Deep Learning." *The 26th ACM SIGKDD Conference on Knowledge Discovery and Data Mining ACM*, 2020.

[7]   Ramos, J. . "Sensitivity enhancement in lateral capacitive accelerometers by structure width optimsation." *Electronics Letters* 33.5(2002):384-386.

[8]   Kalita, Sanjib . "USES OF MEMS ACCELEROMETER IN SEISMOLOGY." *International Journal of Advanced Research in Engineering & Technology* (2013).

[9]   Clayton, R., Heaton, T., Chandy, et al.. Community seismic network. *Annals of Geophysics*, 54.6 (2011), 738-747.

[10] Cochran, Elizabeth, et al. "A novel strong-motion seismic network for community participation in earthquake monitoring." *IEEE Instrumentation & Measurement Magazine* 12.6 (2009): 8-15.

[11] MLA Finazzi, Francesco, and Fassò, Alessandro. "A statistical approach to crowd sourced smart phone based earthquake early warning systems." *Stochastic Environmental Research and Risk Assessment* (2015):1-10.

[12] Kong, Qingkai, Richard M. Allen, Louis Schreier and Young-Woo Kwon. "MyShake: A smartphone seismic network for earthquake early warning and beyond." *Science advances* 2.2 (2016): e1501055.

[13] Bertolini, A. , Desalvo R, Fidecaro F, and A Takamori. "Monolithic folded pendulum accelerometers for seismic monitoring and active isolation systems." *IEEE Transactions on Geoscience and Remote Sensing* 44.2(2006):p.273-276.

Proceedings of the 16th Annual IEEE International
Conference on Nano/Micro Engineered and Molecular Systems
April 25-29, 2021

Design and Analysis of MEMS Biaxial Coupled Resonance Accelerometer*

Huimin Zhang, Yating Zhang, Wei Zhang

Abstract— In this paper, a new type of biaxial coupled resonant accelerometer is designed, which mainly includes proof mass, four double-ended tuning forks, micro-lever, excitation comb and detection comb. The biaxial coupled resonant accelerometer can measure acceleration along the X and Y axes with coupled structure. In addition, the manufacturing process of the resonant accelerometer is described. The modal analysis, harmonic response analysis and thermodynamic analysis of the tuning fork are carried out based on the relative parameters of silicon material.

I. INTRODUCTION

With the development of MEMS technology, the development of inertial sensors is the most successful. As one of the typical representatives of inertial devices, the MEMS accelerometer has received extensive attention and research. MEMS accelerometers are characterized by small size, high level of integration, low power consumption, low cost and wide application range, so they have a huge application market in national defense, medical, industrial and electronic products. MEMS accelerometer is a miniature device to measure the acceleration of an object, so as to measure the force acting on the object. MEMS accelerometers can be divided into piezoresistance, capacitance, piezoelectricity, resonance, tunnel, optical, and thermal convection according to sensing principle[1].

Compared with capacitive MEMS accelerometers, MEMS resonant accelerometer is a typical inertial device which measures the frequency shift of the vibrating beams. The variation of resonant frequency is proportional to the acceleration. Resonant sensing, with respect to other sensing principles, has the advantage of direct frequency output, high potential sensitivity and large dynamic range[2].

Draper Laboratory is one of the first units to successfully develop the silicon microresonant accelerometer, and the reported prototype performance has reached the level of strategic accuracy. Its scale stability is up to , and its spiritual sensitivity stability is up to 190ng[3]. UC Berkeley took the lead in adopting surface silicon technology and manufacturing the structure of double-ended fixed tuning fork resonator with micro-lever structure. This multi-stage micro-lever structure can amplify the force by about 80

times and improve the sensitivity by 30Hz/g[4]. In 2010, Comi et al. of the Polytechnic Institute of Milan, Italy, announced a new type of highly sensitive silicon micro-resonant accelerometer. The structure of the accelerometer is made by surface silicon micromachining technology, and its thickness is 5μm. the sensitive mass area is 400 μm^2, and the overall device area is 0.25mm^{2}[5].

In this paper, a new type of biaxial coupled resonant accelerometer is introduced, and its structural design, modal analysis, harmonic response analysis, thermodynamic analysis and manufacturing process are described in detail.

II. PRINCIPLE AND DESIGN

The biaxial coupled resonance accelerometer proposed in this paper is shown in Fig 1. As shown in Fig.1(a), the biaxial coupled resonance accelerometer is composed of proof mass, the DETF(double-ended turning fork), micro-lever structure, detection comb, excitation comb, support structures and so on. Micro-lever can have the amplification effect of inertia force. The proof mass produces inertia force under the effect of acceleration, which is amplified several times by the micro-lever mechanism and transferred to the resonant beam so that the resonant frequency of the beam changes. The measured acceleration value can be obtained by detecting the variation of resonant beam frequency. In order to reduce interference and improve measurement accuracy, the resonator structure adopts two symmetrical distributed double-ended tuning fork structure. When there is an input acceleration, one is subjected to tension and the other is subjected to compression, and their resonant frequency increases and decreases correspondingly. After differential calculation, the differential frequencies of the signals are obtained within a certain range of input acceleration, and the difference value is proportional to the input acceleration.

At present, most of the research is the resonant accelerometer measuring single-direction acceleration. The resonant accelerometer is composed of a resonator with four microcavities and four resonant structures as shown on Fig.1(b). When a X-axial force is generated and coupled onto the proof mass, the resonators changes along the sensing X-axis. The resonant frequency of each DETF changes under the applied load. At the same time, The resonator along the Y-axis continues to remain stationary due to the nature of the material and its structure. The structure is designed so that the accelerations measured in both directions do not affect each other[6].

Huimin Zhang is with College of Enigeering, Peking University, Beijing 100871, China(1901213621@pku.edu.cn)

Yating Zhang is with Beijing zhixin sensing technology limited company, Beijing 100871, China(zhangyt@ime.pku.edu.cn)

Wei Zhang is with the National Key Laboratory of Nano/Micro Fabrication Technology, Peking University, Beijing 100871, China(weizhang@pku.edu.cn)

978-1-6654-3008-1/21 $31.00 © 2021 IEEE

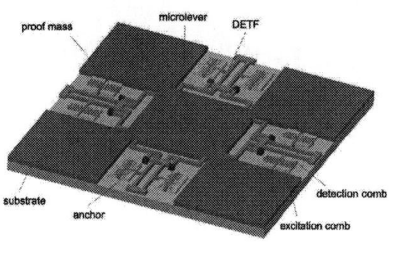

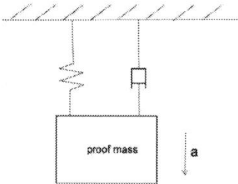

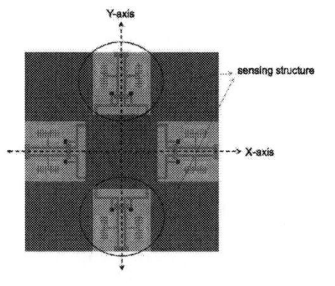

Figure.1.(a) The structure of the biaxial coupled resonance accelerometer (b) The sensing structure of two directions(X-axis and Y-axis)

This structure can be equivalent to a simple spring-mass-damped second-order physical model in Fig.2 Among of this, the equation of Newton's second law can be obtained by force analysis of proof mass.

$$m\frac{d^2x(t)}{dt^2}+c\frac{dx(t)}{dt}+kx(t)=ma(t) \qquad (1)$$

Where x is the displacement of the proof mass, m is mass, c is the damping coefficient, k is the spring constant, a is acceleration. The Laplace transform of equation (1) can be obtained:

$$\frac{1}{s^2+2\xi s\omega+\omega^2}=\frac{x(s)}{a(s)} \qquad (2)$$

Where s is Laplace operator, ω is the operating frequency of tuning fork in accelerometer; ξ is damping ratio, $\xi=\dfrac{c}{2\sqrt{km}}$.

Therefore, the input acceleration can change the resonant frequency of the tuning fork by changing the equivalent stiffness K, and the value of the acceleration can be obtained by measuring the resonant frequency of the tuning fork[7][8].

Micro-lever structure mainly includes: fulcrum, input, output and lever. Depending on the distribution of fulcrum, input and output, microlever mechanisms can be designed in different ways. In order to make the output displacement offset greater than the input displacement offset and play a role of amplifying lever, the three design methods proposed in this paper all satisfy that the input lever arm is smaller than the output lever arm in Fig.3.

Figure.2. Spring-mass-damped second-order physical model

Considering the structure design and position distribution of the proof mass and tuning fork, plan (a) is adopted finally[9].

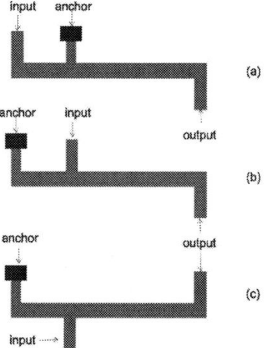

Figure.3.Three kinds of micro-lever structure design methods

The vibration of the DETF can be regarded as the result of the superposition of single degree of freedom vibration, assuming that the two resonant beams with symmetrical distribution of the DETF resonator are independent of each other. The following parameters are used in the calculation and simulation as shown in Table 1:

TABLE I. THE PARAMETERS USED IN SIMULATIN

Modulus of elasticity	E=190GPa
Poisson's ratio	υ=0.278
Density of silicon	ρ=2.33g/cm³

When the deformations in the X and Y directions are the same, the external forces corresponding to them are not equal. The deformation of DETF in X direction is shown in (3).

$$\Delta x=\frac{F_x l}{EA} \qquad (3)$$

Where F_x is the axial force on the DETF in the X direction, l is the length of the DETF, and A is the cross-sectional area.

The deformation of DETF in Y direction is shown in (4), and the angle variation is shown in (5).

$$\Delta x_{max}=\frac{F_y l^3}{3EI} \qquad (4)$$

$$\alpha_{max}=\frac{F_y l^2}{2EI} \qquad (5)$$

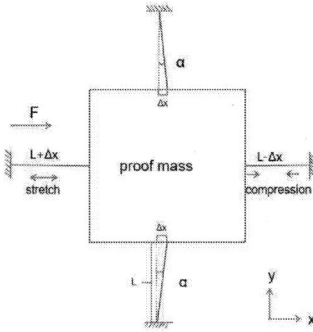

Figure.4.Force diagram in the direction of X axis

Where Δx_{max} is the maximum displacement change obtained by the DETF in the Y direction, F_y is the external bending force received by DETF, I is the moment of inertia.

When the mass is subjected to a force along the X direction, the two DETFs symmetrically distributed in the X direction will be subjected to tension and compression, and the two DETFs symmetrically distributed in the Y direction will have a displacement Δx perpendicular to the axial direction. From a mechanical point of view, when the displacements in the X direction and the Y direction are equal, their forces differ greatly, which is shown in (6). From the frequency analysis, it shows that when the DETF has the same bending deformation and tensile deformation, the bending deformation caused by frequency variation, compared to the frequency of tension deformation caused by the change is almost negligible, thus, the coupled structure can be achieved when the accelerometer is subjected to a force from one direction, the output of another direction is almost zero.

$$\frac{F_x}{F_y} = \frac{4l^2}{b^2} \qquad (6)$$

l=560μm, b=12μm.

In addition, one DETF in the X direction is subjected to tension and the other to compression. The final output is to make difference calculation of the frequency change of the two DETFs, and then obtain the acceleration in this direction. However, the direction of the bending stress received by the two DETFs in the Y direction is the same, and the value is still negligible after the difference calculation.

Meanwhile, considering the influence of processing error, assuming that the tuning fork beam width is 12 μm and the process error is estimated to be about 0.1 μm, it is impossible to achieve the exact same two DETFs with symmetrical distribution, which will also lead to the phenomenon that the output in the Y direction is not zero when the acceleration is only in the X direction, namely the so-called frequency drift.

The biaxial coupled accelerometer can also measure the acceleration along the Z axis in theory. When the force is exerted along the Z direction, the deformation directions of the four DETF are consistent. The acceleration values along the Z direction can be obtained by adding up the frequency changes of the four DETF and sorting out the data.

III. FABRICATION

As shown in Fig.5, the fabrication process of the biaxial coupled resonance accelerometer is proposed. These methods are completely based on the existing MEMS processing technology to design, so there is a high feasibility. First, prepare glass as substrate(Fig.5(a)). Then, the metal is sputtered on a glass substrate as an electrode(Fig.5(b)). Carve cavities into the back of silicon(Fig.5(c)).Silicon is bonded to glass(Fig.5(d)). Next, the front of the silicon is coated with photoresist(Fig.5(e)), and finally the structure of the accelerometer is etched on the front of the silicon.(Fig.5(f)).

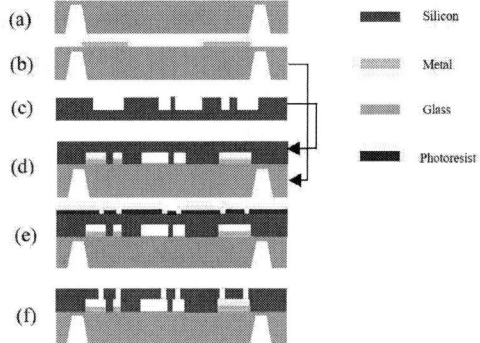

Figure.5. The process of the accelerometer fabrication.

IV. SIMULATION AND RESULTS

A. Modal analysis

The resonant frequency distribution and vibration mode of the tuning fork model can be obtained by using ANSYS software. The boundary condition is that one end of the tuning fork is fixed and the other end is freely telescopic. The resonant frequency of the turning fork is shown in Table 2.

TABLE II. THE RESONANT FREQUENCY OF THE TURNING FORK

Mode	Frequency(MHz)
1	6.197×10^{-2}
2	9.067×10^{-2}
3	0.3949
4	0.4147
5	0.4204
6	0.4909

B. Harmonic response analysis

Harmonic response analysis is a special time domain analysis method used to determine the steady-state response of a structure under various loads with simple harmonic variations[10][11]. In harmonic response analysis, the harmonic force is applied to simulate the external force of the MEMS resonant accelerometer in the process of application. According to the frequency results obtained from the modal analysis, we input the simulation frequency range of the resonant accelerometer with 200 output values of 0-0.5MHz. It can be obtained that near the second natural frequencies, the MEMS resonant accelerometer can resonate and has obvious displacement deviation in Fig.6.

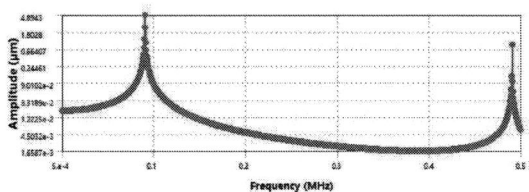

Figure.6. The relationship between amplitude and frequency of a resonator

C. Thermodynamic analysis

The key technical problem of high precision silicon MEMS resonant accelerometer technology lies in its adaptability to thermal environment. Compared with quartz vibration beam accelerometer, the thermal expansion coefficient of silicon material is significantly higher than that of quartz material. Meanwhile, due to the thermal mismatch between different materials, the temperature performance of silicon MEMS resonant accelerometer has become a recognized technical problem.

Temperature influences the performance of MEMS resonant accelerometer mainly include three points[12]:

a. The crystalline structure of silicon makes it a preferred material for MEMS resonant accelerometers, but silicon is highly sensitive to temperature, and silicon's temperature coefficient is -25ppm/ $^\circ$C ~-75ppm/ $^\circ$C. The relationship between thermal expansion coefficient and temperature is shown in the expression (7).

$$\alpha_{Si}(t) = \left(3.725\left\{I - \exp\left[-5.88 \times 10^{-3}(t-124)\right]\right\} + 5.548 \times 10^{-4} t\right) \times 10^{-6}$$
(7)

b. Ideally, the structure and size of the MEMS resonant accelerometer are consistent, and the effects of temperature changes on the resonator are symmetric and consistent, offsetting each other. The two resonators are not completely symmetrical due to the errors in the machining process, and the offset temperature effect can be retained.

c. Wafer bonding is the key technology of device integration in MEMS processing. Anchor point bonding is widely used in MEMS resonant accelerometer and machining process because of its high bonding quality. When the anchor point is bonded, the base material is heated to 300-400 $^\circ$C. Due to the difference of thermal expansion coefficient and thermal conductivity coefficient between glass and silicon, when the bonding chip is cooled to room temperature, a large residual stress will occur. Therefore, the residual stress affects the stability of the resonant frequency by affecting the mechanical properties of the resonator.

V. CONCLUSION

In this paper, a new type of biaxial coupled resonance accelerometer is designed, which mainly includes proof mass, four double-ended tuning forks, micro-lever, excitation comb and detection comb. The biaxial resonant accelerometer measures acceleration along the X and Y axes. In addition, the manufacturing process of the resonant accelerometer is described. The modal analysis, harmonic response analysis and thermal analysis of the tuning fork are carried out based on the relative parameters of silicon material. The modes and frequencies of a tuning fork in different modes can be obtained by modal analysis. In harmonic response analysis, it can be found that a tuning fork produces a large amplitude near the second natural frequencies. And the influence factors of temperature on accelerometer are analyzed in thermodynamic analysis.

ACKNOWLEDGMENT

I would like to thank professor Zhang for his guidance on my academic and scientific research, and professor Zhao for his opinions on my paper.

REFERENCES

[1] Zhang Yangxi, Research of Some Key Technologies in Typical MEMS Accelerometer[D], Beijing. Peking University.2015.

[2] M. Aikele, K. Bauer, W. Ficker, F. Neubauer, U. Prechtel, J. Schalk and H. Seidel, Resonant accelerometer with self-test. Sensors and Actuators A, vol. 92, pp. 161-167, 2001.

[3] R Hopkins, J Miola, W Sawyer, The silicon oscillating accelerometer: A high-performance MEMS accelerometer for precision navigation and strategic guidance application, Proceedings of the 61st Annual Meeting of The Institute of Navigation, 2005: 1043-1052.

[4] A A Seshia, M Palaniapan, T A Roessig et al., A vacuum packaged surface micromachined resonant accelerometer, Journal of Microelectro mechanical Systems,2002,11 (6):784-793.

[5] Claudia Comi et al., A High Sensitivity Uniaxial Resonant Accelerometer, in Proc.2010 IEEE MEMS.

[6] Bo Yang, Hui Zhao et al., A new silicon biaxial decoupled resonant micro-accelerometer, Microsystem Technologies volume 21, pages 109－115(2015).

[7] HU H, DONG J X, LIU Y F et al., Calculation and improvement on amplification effect of MEMS leverage mechanism, Journal of Chinese Inertial Technology, 2011, 19 (1):91-94 (in Chinese).

[8] Duan Xiaomin. Han Ziq et al., Design of the High Sensitivity Silicon Micro Resonant Accelerometer Structure Based on Two-Wing Distribution Method, Semiconductor Devices,2018.03.003:171-176.

[9] Xiao-Ping S. Su. Henry S. Yang, Design of compliant microleverage mechanisms, Sensors and Actuators, A 87 (2001) 146±156.

[10] ZHAO L, DAI B, YANG B, et al., Design and simulations of a new biaxial silicon resonant microaccclerometer, Microsystem Technologies, 2016, 22 (12):2829-2834.

[11] YAN L, GUO Z S, QU Y et al., Design and simulations of a resonant Accelerometer, Microsystem Technologies, 2017 (5):1-11.

[12] Dong Jinhu. Study on temperature characteristics of silicon resonant accelerometer[D].Nanjing. Nanjing University of Science and Technology. 2012.

[13] Aung Thura, B. M. Simonov, S. P. Timoshenkov et al. Studying the Influence of Temperature on the Operation of a Resonator of a Frequency Micromechanical Accelerometer. Russian Microelectronics, 2019, Vol. 48, No. 7, pp. 485－489.

Proceedings of the 16th Annual IEEE International
Conference on Nano/Micro Engineered and Molecular Systems
April 25-29, 2021

Development of an Annular-Shaped Bernoulli Gripper for Contactless Gripping of Large-Size Silicon Wafer

Shihang Wang, Yancheng Wang*, *Member, IEEE*, Deqing Mei, and Songqiao Dai

Abstract—Silicone wafer gripping and transportation have been widely utilized in wafer fabrication process, such as chemical mechanical polishing, grinding, epitaxial growth, etc. This paper develops a novel annular-shaped gripper based on Bernoulli effects with the aim of contactless gripping of large size and thin silicon wafer. The structural design and working principle of this annular-shaped Bernoulli gripper are presented. A computational fluid dynamic (CFD) model is established to study the adsorption performance of the gripper when gripping an 8-inch silicon wafer. Numerical simulation results showed that the gripper can generate an annular-shaped pressure distribution to adsorb the silicon wafer which can reduce the wafer's deformation. Then, the Bernoulli gripper was fabricated, and the 8-inch silicon wafer's gripping experiments demonstrated that the gripper has generally low suction curve gradient. Thus, the developed annular-shaped Bernoulli gripper could be utilized for contactless gripping and transportation of large-size and thin silicon wafers in potential industrial applications.

Index Terms – Bernoulli principle; Gripper; Silicon wafer; Adsorption force; Contactless gripping.

I. INTRODUCTION

Silicon wafer can be utilized as the substrate for the fabrication of microelectronic sensors and devices [1-2]. With the development of integrated circuit (IC) technology, the silicon wafer's size has been greatly increased while its thickness decreased. The industrial used silicon wafer's thickness will be reduced to less than 120 μm and its diameter will be enlarged to 12 and 16-inches in the next decade [3]. As for the fabrication of electronic sensors and integrated circuits, the gripping and transportation of silicon wafers are utilized, such as in wafer polishing, grinding, epitaxial growth process, etc. Due to the fragile properties, the large size and thin silicon wafers are prone to fracture, deflect and wear in contact gripping and transporting by using mechanical manipulators [4]. Thus, the defects will be increased and affect the productivity and yield in automatic production lines for silicon wafers [5]. So the gripping and transportation approaches for large-size and thin silicon wafers still need to be investigated.

In industrial applications, the commonly used end effectors for gripping and transporting silicon wafer can be divided into two groups [6]: manipulator for contact

Y.C. Wang and D.Q. Mei are with the State Key Laboratory of Fluid Power & Mechatronic Systems, School of Mechanical Engineering, Zhejiang University, Hangzhou, China (*Corresponding author: Yancheng Wang, phone: 86-13675828104, e-mail: yanchwang@zju.edu.cn).

S.H. Wang and S.Q. Dai are with Key Laboratory of Advanced Manufacturing Technology of Zhejiang Province, School of Mechanical Engineering, Zhejiang University, Hangzhou, China.

gripping and vacuum sucker for contactless gripping. For contact gripping using manipulators, the damage induced by the contacts between manipulators and wafers is usually unavoidable. As for contactless gripping, both vortex and Bernoulli principles have been used to design the end effectors to reduce the mechanical damages to the fragile silicon wafers [7]. Figs. 1(a)-(b) show the schematic view of working principles for the vortex and Bernoulli grippers, respectively. In vortex grippers, the centrifugal force generated by the rotating of fluid flows can make the pressure in the central region lower than that on the periphery area, as shown in Fig. 1(a). Then, negative pressure will be generated in the central area of the gripper to adsorb the objects [8]. As for Bernoulli grippers, the induced pressure on the inner perimeter is lower than that on the periphery area based on Bernoulli principle, the pressure will be increased as the decreasing of flow velocity, as in Fig. 1(b). Thus, a negative pressure zone can be generated in the region of the deflector for contactless gripping of objects [9]. Compared to Bernoulli grippers, the vortex gripper generally has lower energy consumption while needs greater upstream pressure when gripping the same parts and/or workpiece. Meanwhile, the induced deformation of the gripped objects in vortex gripping would be greater due to generated stress distribution and bending moments on the objects [10-14]. Thus, this paper adopted the Bernoulli effects to design a novel gripper for contactless gripping and transporting of large-size and thin silicon wafers with the purpose to reduce the deformation of the wafers.

There are two types of Bernoulli grippers using different structural designs, such as single Bernoulli gripper [15] and distributed Bernoulli gripper [16]. For the single Bernoulli gripper, the gas flows in the gripper more stably and uniformly at the working condition. It can change the structure of the deflector to adjust the size of the adsorption area, while the adsorption area remains at the central region of the gripper [17]. With the increasing of gripper's diameter, the adsorption force will be reduced and the distribution of adsorption force may lead to a large deflection when gripping large-sized parts [18]. For the distributed Bernoulli gripper, it can grip an 8-inch silicon wafer softly by adjusting the number and layout of Bernoulli heads to satisfy the size of wafers [16]. The used multiple Bernoulli heads may create a positive pressure at the center of the gripper, thus may introduce large deformation to the gripped objects. Therefore, both types of grippers may generate large deformation and induce damage to large-sized and thin disc-shaped parts during the gripping process. In this study, we proposed a novel structural design of annular-shaped Bernoulli gripper by using different sized deflectors to adjust the position of adsorption area for gripping the silicon wafer with different

978-1-6654-3008-1/21 $31.00 © 2021 IEEE

sizes. We assumed that the annular-shaped adsorption area could reduce the deformation of silicon wafers by decreasing the pressure distribution at the central area of the designed gripper.

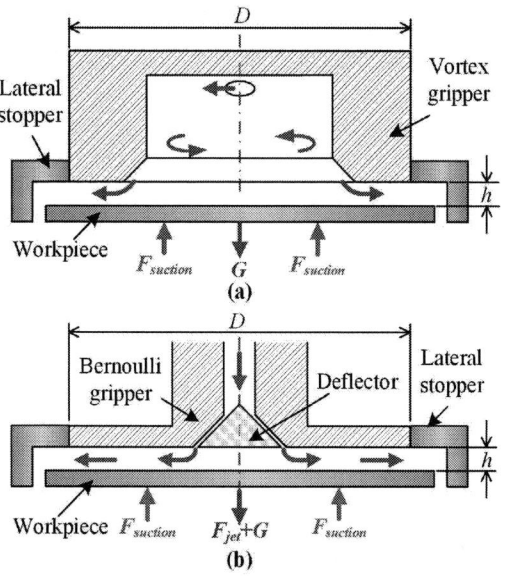

Fig. 1 Schematic view of contactless grippers: (a) vortex gripper and (b) Bernoulli gripper.

In this study, we developed a novel annular-shaped Bernoulli gripper for contactless gripping of 8-inch silicon wafers. Firstly, the structural design and working principle of the annular-shaped Bernoulli gripper are described. A computational fluid dynamic (CFD) model is developed to investigate the adsorption force and pressure distributions generated on the silicon wafer under different gas flow rates and adsorption heights. This is followed by the fabrication of annular-shaped Bernoulli gripper and experimental setup for wafer gripping. Finally, results and discussions are conducted. The contributions of our work can be described as:

1) A novel annular-shaped Bernoulli gripper for contactless gripping of large-size and thin silicon wafer is developed;

2) A CFD simulation model is established to study the adsorption performances of the Bernoulli gripper for wafer gripping, optimal inputs of gas flow rate and gap height are obtained;

3) Experimental setup of the gripper for 8-inch silicon wafer gripping is constructed, the adsorption performances of the developed gripper are validated by both numerical prediction and experimental tests.

II. DESIGN OF ANNULAR-SHAPED BERNOULLI GRIPPER

A. Structural Design and Working Principle

Structural design: The structural design of the annular-shaped Bernoulli gripper is illustrated as shown in Fig. 2(a), it mainly consists of several layers: upper-end layer, flow channel layer, four lateral stoppers, and adsorption layer. The upper-end layer is used to contact and fix the gripper to external motion platform, four gas

inlets are designed on the upper-end layer to enable the airflow into the gripper. The structural design of the flow channel is critical to affecting the gas flow velocity and pressure distribution for wafer gripping. Four lateral stoppers are used to constraint the movement of silicon wafers parallel to the lower surface of the gripper. In the adsorption layer, the adsorption force will be generated by the jet stream. Each layer of the gripper has inner and outer sealing grooves to ensure the gas flows inside the gripper without leakage.

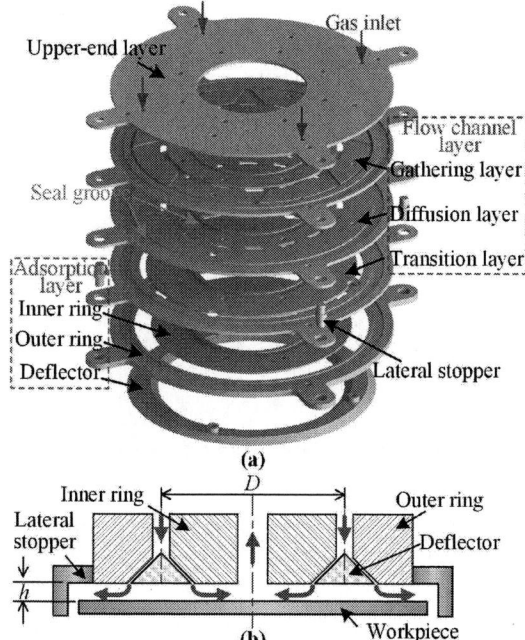

Fig. 2 (a) Schematic view of the designed annular-shaped Bernoulli gripper and (b) Working principle of the annular Bernoulli gripper.

The key component of the gripper is the flow channel layer. As in Fig. 3(a), it consists of a gathering layer, a diffusion layer, and a transition layer. The gathering layer has two annular-shaped grooves and eight triangular-shaped grooves to converge all the gas from four inlets, which can balance the gas pressure. The diffusion layer is based on a large-width and low-depth annular groove to decrease the height of flow channel, which can make the gas spread uniformly. As for the transition layer, an annular-shaped channel is designed on the deflector to let the gas flow into the adsorption layer, and the flow path of the gas is shown in Fig. 3(a).

The two-dimensional (2D) structure of the adsorption layer is shown in Fig. 3(b). It mainly has the inner ring, outer ring, and annular deflector. The structural parameters of the annular-shaped deflector are the location dimension of section B (D), bottom edge width (w), and inclination angle of the cone (α). The D, w, and α are set as 67 mm, 20 mm, and 30°, respectively. Both inner ring and outer ring have cones parallel to the cones of the annular-shaped deflector, which generates a small gap distance (g) of 0.4 mm. And they connect and fix all the layers with the upper-end layer, as in Fig. 3(a). The structure of lateral stopper is shown in Fig. 3(b). It can slide vertically to avoid

978-1-6654-3008-1/21 $31.00 © 2021 IEEE

rigid impact with the objective table while gripping silicon wafers and provide thrust for lateral transportation.

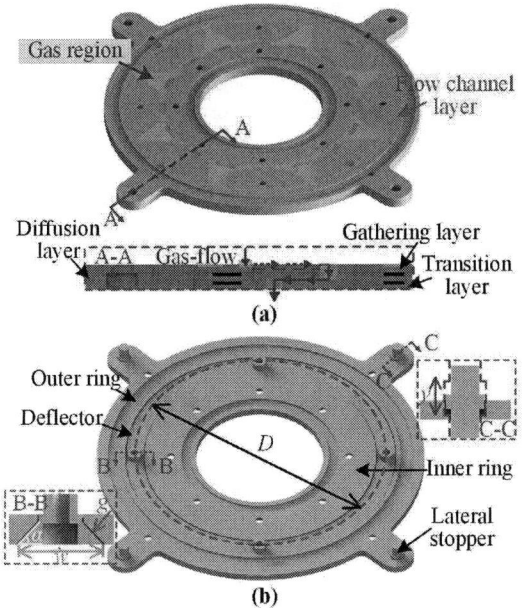

Fig. 3 Structure of (a) flow channel layer and (b) adsorption layer and lateral stopper.

Working principle: The working principle of this annular-shaped Bernoulli gripper is illustrated as shown in Fig. 2(b). The air first flows into the adsorption layer through the above flow channel, and flows out from the small gap formed by the cones of inner ring, outer ring, and annular-shaped deflector. Then, the air gas towards the inner side goes through the central hole and spreads to the atmosphere due to the below solid silicon wafer. While the gas towards the outer side goes through the interval to the atmosphere. The diameter of the central hole is set as 40 mm and the height of interval can be adjusted by moving the gripper. So, negative pressure can be generated in the area under the annular-shaped deflector based on Bernoulli principle. Typically, with the aim of gripping different-sized silicon wafers, we can adjust the structural parameters (D and w) of the annular-shaped deflector to meet the requirement. In this study, the designed gripper can be used to grip the silicon wafer from the diameters of 183 mm to 206 mm. Therefore, the proposed gripper has a greater adsorption area than that of other grippers, thus has the potential to grip different-sized workpieces in real applications.

B. CFD Modeling

To study the adsorption performances of the gripper to grip the silicon wafer, a CFD simulation model is developed by using the software of ABAQUS v.16. The developed CFD model can be seen as shown in Fig. 4(a), the main geometry is the gas area surrounded by inner ring, outer ring, annular-shaped deflector, and a silicon wafer. The gas flow first enters the area above the annular-shaped deflector and then flows from the edge of the silicon wafer and the central hole of the inner ring. The key parameters of the geometry model are the gap distance of the cones (g) and gap height (h). Here, the value of g is set as 0.4 mm, and h is larger than 0.1 mm in general.

For element meshing, the FC3D4 grid unit is selected and utilized to mesh the geometry model. It can avoid the complicated bottom-up hexahedral meshes and guarantee the quality of the meshes. Further, due to the dis-shaped geometry model, the global seeds with an approximate size of 1 mm, maximum deviation factor (height/length of mesh body) of 0.1, and minimal size of 0.1 mm are adopted to ensure one layer of the grip in the bottom. Totally, the mesh number of the CFD model is about 466,945.

For boundary conditions, the gas flow rate (Q_v) at the inlet can be adjusted by changing the gas velocity in the direction of the Z-axis. In this study, the gas flow rate value changes from 0 to 300 L/min. At the central outlet and periphery outlet, the ambient pressure and temperature are set as 0 Pa and 293 K, respectively. Apart from the inlet and outlets, the rest of the model surfaces are the wall, and they are set to be non-slip, smooth, and adiabatic. The convergence criterion for each variable-normalized residual is set as $< 10^{-6}$.

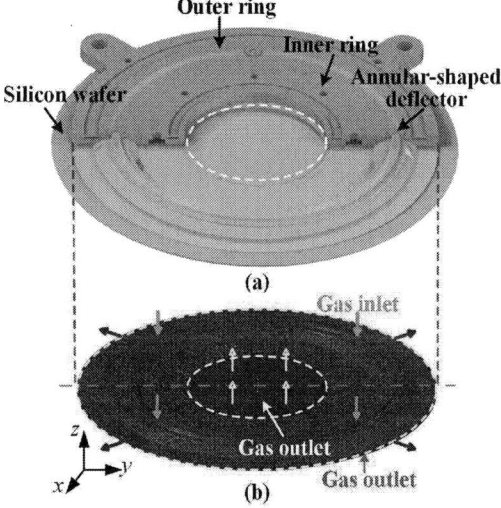

Fig. 4 (a) CFD model of annular-shaped Bernoulli gripper and (b) element meshing.

The pressure distribution and adsorption force are critical to affect the performance of the gripper to adsorb the silicon wafer, thus the pressure distribution and adsorption force need to be analyzed. Since the gap height (h) ranges from 0.1 mm to 5 mm, the Reynolds numbers in the flow channel and gripper can be calculated and the value ranges from 27 to 1300. Because these values are smaller than 2300, the gas flow in the model can be assumed as steady-state, laminar, and incompressible. The governing equations for the calculation of the gas flow and pressure distribution, including continuity conservation, momentum conservation, and energy conservation equations, are utilized.

Continuity equation:

$$\nabla \cdot \rho \vec{v} = 0 \qquad (1)$$

Momentum conservation equation:

$$\int_V \rho \vec{v} \otimes \vec{v} \cdot \vec{n} dS = -\int_V \nabla \cdot p dV + \int_S \tau \cdot n dS + \int_V f dV \quad (2)$$

Energy conservation equation:

$$\int_V \rho c_p T_v \cdot n dS = \int_V r dV - \int_S q \cdot n dS \quad (3)$$

where ρ is the density of air-fluid; $\vec{v}$ is the velocity vector of fluid; V is any grid cell with surface area S; $\vec{n}$ is the outer normal vector of S; p is the pressure; f is the body force; τ is the viscous shear stress; c_p is the specific heat at constant pressure; T_v is the temperature; q is the heat flux generated by the Fourier's law at the defined condition, and r is the heat provided by entering the outer of the grid cell. Here, $\rho = 1.16$ kg/m^3, $c_p = 1004$ J/(kg·K), $T_v = 293$ K, and $\mu = 1.76e{-}05$.

By changing the gas velocity in the Z-axis and the values of gap height formed by the silicon wafer, inner ring, and outer ring, the gas flow characteristics, adsorption force, and pressure distribution on silicon wafer can be simulated and analyzed by the developed CFD model.

III. EXPERIMENTAL SETUP AND PROCEDURE

The system setup of Bernoulli gripper to grip silicon wafers is established. The system was constructed with linear motion stages (331, Shenzhen Zhida Automation Equipment Co., Ltd) and a gas controller (MF5712-200, Siargo, Ltd), the positional accuracy of linear motion stages is 0.1 mm. The gas flow controller can accurately control the gas flow rate, whose block diagram is shown in Fig. 5(a). The adsorption force was measured by using an electronic balance (LQ-A20002, Ruian Ante Weighting Equipment Co. Ltd), the minimum measured force is about 0.001 N. The final experimental setup is constructed as shown in Fig. 5(b), the fabricated Bernoulli gripper and gas pipes are assembled and then fixed onto the motion stage.

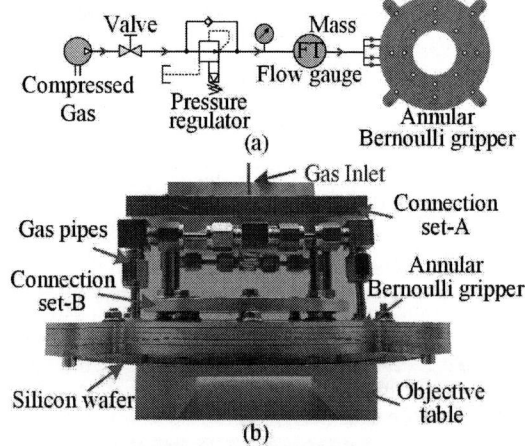

Fig. 5 (a) Block diagram of the pneumatic system and (b) experimental setup of the gripper for gripping of a silicon wafer.

For experimental procedures, we adjusted firstly the parallelism between the object table and the lower surface of gripper using a leveling instrument. Then an 8-inch silicon wafer with a thickness of 700 μm was placed on the objective table and reset the reading of electronic balance, as shown in Fig. 5(b). Afterwards, the gripper was brought into contact with the silicon wafer by controlling the linear motion stage where the electronic balance suddenly had a nonzero reading. This position of gripper was regarded as the original location where the gap height is zero. Subsequently, the gap height was controlled by linear motion stages and the gas flow rate was adjusted by the gas controller.

$$F_{suction} = \left| m_{reading} \right| \cdot g \quad (4)$$

Thus, as air flowed into the gripper, the adsorption force at various working conditions can be calculated by recording the weight values ($m_{reading}$) through electronic balance.

IV. RESULTS AND DISCUSSION

A. CFD Simulation Results

Pressure distribution. During the simulation, the initial gap height was set as 1 mm. The generated pressure distribution on the surface of a silicon wafer at different gas flow rates is calculated by using the developed CFD model, and the results are shown in Fig. 6. When the gas flow rate is 150 L/min, we can see that the pressure distribution curve can be divided into seven regions, both regions II and II' have generally large negative pressure values (-15 Pa) thus can be regarded as the adsorption area. Because regions II and II' are located below the annular-shaped deflector, the size and position of regions II and II' can be adjusted by changing the width (w) and location (D) of the deflector. Thus, the developed gripper can also be applied to grip a large-sized dis-shaped workpiece. The pressure distributions in regions I, I' and III, III' are consistent with the Bernoulli equation. The pressure along the gas flow direction increases gradually from the position of 10 mm to 0 mm and 170 mm to 180 mm in regions I and I', while decreases from the position of 30 mm to 55 mm and 150 mm to 125 mm in regions III and III'. The highest pressure (~ 4 Pa) occurs at the sudden decrease of the gas volume, as shown in Fig. 6. The pressure in region IV almost equals zero due to its direct connection with the air atmosphere.

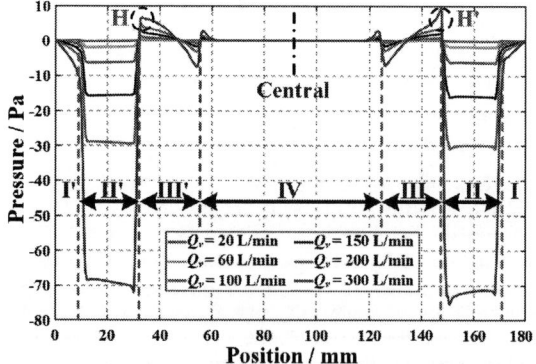

Fig. 6 The simulation results of pressure distribution on the top surface of the silicon wafer at gap height of 1 mm.

With gas flow rate increasing, the trend of pressure distribution curve almost has no change. The pressure values have slight variations in regions I, I', III, III', and IV while having significant changes in regions II and II' (the highest-pressure value increases from 4 Pa to 5 Pa but the pressure value in adsorption regions II and II' decrease from -15 Pa to -30 Pa with gas flow rate increasing from 150 L/min to 200 L/min). Thus, the developed gripper can be applied to grip heavy workpieces by increasing the gas flow rate.

Adsorption force. When the gas flow rate was set as 200, 250, and 300 L/min, the adsorption force curves at different gap heights are shown in Fig. 7(a). We can see that the generated adsorption force firstly can be increased with the increase of gap heights from 0 mm to 0.5 mm. As the gap height keeps on increasing, the generated adsorption force will be slowly decreased and flattens to maintain almost constant values, as in Fig. 7(a). The reasons can be explained as that the gas flow paths will be changed when turning from the slant jet to horizontal interval.

When the gap height is respectively set to 0.5, 1.0, and 1.5 mm, the adsorption force curves at different gas flow rates are shown in Fig. 7(b). We can see that there is no adsorption force generated from gas flow rate of 0 to 40 L/min. As gas flow rate increases and is greater than 40 L/min, adsorption force increases faster and faster. From these adsorption force curves, gripping a silicon wafer weighted 0.14 N at gap height of 1 mm requires the gas flow rate of 135 L/min.

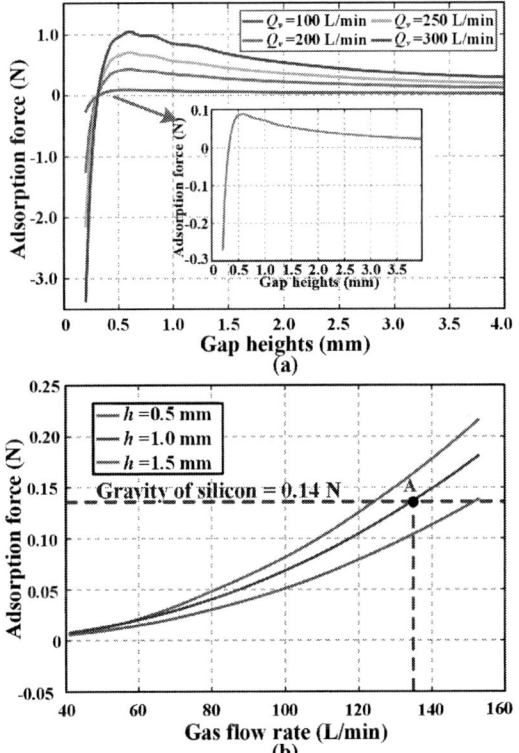

Fig. 7 (a) The simulated adsorption force versus gap height and (b) adsorption force versus input gas flow rate.

B. Experimental Results

To verify the simulation results, experimental tests were conducted.

Wafer adsorption under different gap heights. During the experiment, the initial gas flow rate was set to 100 L/min, the measured adsorption force values were plotted and shown in Fig. 8(a). We can see that the generated adsorption force firstly can be increased as the increase of gap heights from 0 to 1.2 mm. As the gap height keeps on increasing, the generated adsorption force will be slowly decreased and flattens to maintain an almost constant value. The maximum adsorption force is 0.063 N at the gap height of 1.2 mm lower than the maximum adsorption force of 0.08 N at the gap height of 0.6 mm from simulation results. Meanwhile, the adsorption curve is very smooth from gap heights of 0.8 mm to 1.6 mm, where the maximum difference value is lower than 0.005 N.

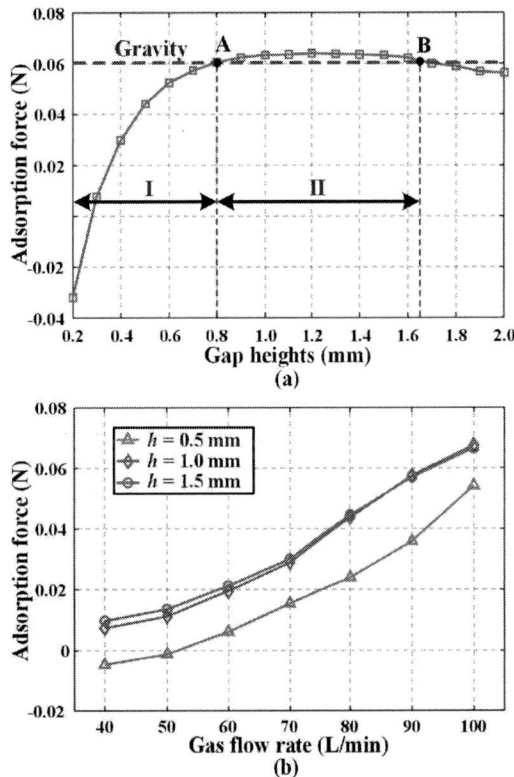

Fig. 8 (a) The measured adsorption force versus different gap heights at gas flow rate of 100 L/min and (b) the measured adsorption force under different gas flow rates.

Wafer adsorption with different gas flow rates. When the gap height is respectively set to 0.5, 1.0, and 1.5 mm, the measured adsorption force values were plotted and shown in Fig. 8(b). We can see that adsorption force almost equals zero at the gas flow rate of 40 L/min and increases faster and faster with the increase of gas flow rate. The adsorption force reaches 0.068 N at the gap height of 1.0 mm and gas flow rate of 100 L/min, which equals the simulation results at the same working conditions.

Comparing the simulation results with experimental results, we can find that the trend of adsorption force curve versus gap height matches the simulation results (increase firstly, then decrease, and flatten to a constant value) while the amplitude of adsorption force and key points' coordinates of adsorption curves exist slight differences (the adsorption force is lower and the turning point is located at higher gap height in experimental tests). The reasons can be explained as that the gap (g) formed by the annular-shaped deflector and adsorption layer becomes bigger for invisible deformation of gripper's parts. So, the differences between simulation results and experimental results are reasonable and acceptable. Meanwhile, the trends of adsorption force curves versus the gas flow rates are similar to the simulation results. Thus, the correctness of the CFD simulation model gets validated.

V. CONCLUSIONS

In this study, we developed a novel annular-shaped Bernoulli gripper for gripping 8-inch sized silicon wafers. The design of annular-shaped deflector helps the gripper grip silicon wafers without contact based on Bernoulli effect and reduce the deformation damage. The CFD model of this gripper was established, and the simulation results showed that the adsorption region was annular-shaped under the deflector and the gripper could grip the large-size silicon wafer by adjusting the parameters of the deflector (this gripper can grip the silicon wafer from the diameters of 183 mm to 206 mm). The adsorption force could reach 1.0 N at the gap height of 0.6 mm and the gas flow rate of 300 L/min. With the gas flow rate increasing, the adsorption force would become larger for gripping heavy workpieces. Meanwhile, the annular-shaped Bernoulli gripper was fabricated and experiment tests were conducted verifying the correctness of the CFD simulation model.

The preliminary study of the Bernoulli gripper in this paper opens up the opportunity to design the end-effector for contactless gripping and transportation of disc-shaped workpiece and/or silicon wafers. In future work, optimal structural design of the gripper with the goal to increase the adsorption force and improve the pressure distribution when gripping large-sized silicon wafer will be studied. Also, more experiments will be performed to demonstrate the capability of the developed Bernoulli gripper for contactless gripping of large size and fragile workpieces.

ACKNOWLEDGMENTS

This research is supported by the Zhejiang Provincial Funds for Distinguished Young Scientists of China (No. LR19E050001), Key Research and Development Program of Zhejiang Province (Grant No. 2020C01034), and Creative Research Groups of National Natural Science Foundation of China (No. 51821093).

REFERENCES

[1] "International Technology Roadmap for Semiconductors Reports (ITRS-2015)," Semiconductor Industry Association, Washington, DC, USA, Version 2.0, Accessed: July 17, 2020, [online] Available: http://www.itrs2.net/.

[2] "International Roadmap for Devices and Systems (IRDS™-2017)," Institute of Electrical and Electronics Engineers, Piscataway, New Jersey, USA, Version 1.0, Accessed: June 15, 2020, [online] Available: http://irds.ieee.org/.

[3] T. Giesen, R. Wertz, C. Fischmann, G. Kreck, J. Govaerts, J. Vaes, M. Debucquoy and A. Verl, "Advanced production challenges for automated ultra-thin wafer handling," in *Proceeding 27th European Photovoltaic Solar Energy Conf. and Exhibition*, Germany, pp. 1165-1170, 2012.

[4] X. F. Brun and S. N. Melkote, "Analysis of stresses and breakage of crystalline silicon wafers during handling and transport," *Sol. Energ. Mat. Sol. C.*, vol. 93, pp. 1238-1247, 2008.

[5] H. T. Cheng, H. P. Chen, and B. W. Mooring, "Accuracy analysis of dynamic-wafer-handling robotic system in semiconductor manufacturing," *IEEE T. Ind. Electron.*, vol. 61, pp. 1402-1410, 2014.

[6] P. Bryan, S. Kumar and F. Sahin, "Design of a Soft Robotic Gripper for Improved Grasping with Suction Cups", *IEEE International Conf. on Systems, Man and Cybernetics (SMC)*, pp. 2405-2410, 2019.

[7] B. Ozcelik, F. Erzincanli and F. Findik, "Evaluation of handling results of various materials using a non-contact end-effector," *Ind. Robot*, vol. 30, pp. 363-369, 2003.

[8] S. Iio, M. Umebachi, X. Li, T. Kagawa and W. Ikeda, "Performance of a non-contact handling device using swirling flow with various gap height," *J. Visual.*, vol. 13, pp. 319-326, 2010.

[9] X. Lin, W. Zhong, T. Kagawa, H. Liu and G. L. Tao, "Development of a Pneumatic Sucker for Gripping Workpieces with Rough Surface," *IEEE T. Autom. Sci. Eng.*, vol. 13, pp. 639-646, 2014.

[10] K. Shi and X. Li, "Experimental and Theoretical Study of Dynamic Characteristics of Bernoulli Gripper," *Precis. Eng.*, vol. 52, pp. 323-331, 2018.

[11] T. Giesen, E. Bürk, C. Fischmann, W. Gauchel, M. Zindl and A. Verl, "Advanced gripper development and tests for automated photovoltaic wafer handling," *Assembly Autom.*, vol. 33, pp. 334-344, 2013.

[12] D. Liu, T. C. Sing, W. Y. Liang and K. K. Tan, "Soft-Acting, Noncontact Gripping Method for Ultrathin Wafers Using Distributed Bernoulli Principle," *IEEE T. Autom. Sci. Eng.*, vol. 16, pp. 654-667, 2019.

[13] X. F. Brun and S. N. Melkote, "Modeling and Prediction of the Flow, Pressure, and Holding Force Generated by a Bernoulli Handling Device," *J. Manuf. Sci. Eng.*, vol. 131, pp. 031018-7, 2009.

[14] X. Li, N. Li, G. L. Tao, H. Liu and T. Kagawa, "Experimental Comparison of Bernoulli Gripper and Vortex Gripper," *Int. J. Precis. Eng. Man.*, vol. 16, pp. 2081-2090, 2015.

[15] G. Dini, G. Fantoni and F. Failli, "Grasping leather plies by Bernoulli grippers," *CIRP Ann. Manuf. Techn.*, vol. 58, pp. 21-24, 2009.

[16] D. Liu, W. Y. Liang, H. Zhu, C. S. Teo and K. K. Tan, "Development of a Distributed Bernoulli Gripper for Ultra-thin Wafer Handling," *IEEE International Conf. on Advanced Intelligent Mechatronics (AIM)*, pp. 265-270, 2017.

[17] X. Brun and S. N. Melkote, "Effect of Substrate Flexibility on the Pressure Distribution and Lifting Force Generated by a Bernoulli Gripper," *J. Manuf. Sci. Eng.*, vol. 134, pp. 051010-8, 2012.

[18] K. Shi and X. Lin, "Optimization of outer diameter of Bernoulli gripper," *Exp. Therm. Fluid Sci.*, vol. 77, pp. 284-294, 2016.

Gap in pagination due to unavailable papers.

Pages 1877-1884

The 16th IEEE International Conference on Nano/Micro Engineered & Molecular Systems

Semiconductor Biosensors for Global Health in the Era of AIoT

Prof. Yi-Kuen Lee
Department of Mechanical and Aerospace Engineering
Department of Chemical and Biomolecular Engineering
Nanosystem Fabrication Facility
Hong Kong Center for Construction Robotics
Nanosystem Fabrication Facility
Hong Kong University of Science and Technology
Clear Water Bay, Kowloon, Hong Kong SAR
http://meyklee.people.ust.hk/

ABSTRACT

Semiconductor technologies, especially silicon-based micro/nano CMOS processes, have been making a tremendous impact on human society in the past 60 years. In recent years, Field Effect Transistors (FET) biosensors, e.g., Ion Sensitive FET (ISFET), have been developed for the detection of various ions and biomolecules for biomedical applications. One of the most prominent applications is CMOS ISFET-based DNA sequencing, such as the Ion Torrent Ion Proton II chip with 660 million ISFET sensors. With the synergy of a fundamental principle of electrochemistry, generalized dimensional analysis from mechanics, analog IC design technique, and fabricated devices' experimental data, we propose a generalized gm/Id theory for the design of FET-based biosensors for detection of ions, DNA, and cells using commercial CMOS/CMOS MEMS foundry processes (TSMC 0.18 μm CMOS process and InvenSense 0.18 μm CMOS MEMS process). The nonlinear coupling behavior of electrochemical-FET biosensors can be described as the sensitivity as a function of critical normalized parameters such as gm/Id, the normalized drain current (Inversion Coefficient), the electrochemical reference voltage, and various noises. A series of CMOS biosensors for detection of ions, DNA, and bacteria with excellent performance. In addition, we are working on developing a new CMOS Bio-MEMS platform for integration with micro PCR/RT-PCR, the on-chip microelectronic circuit, microfluidics and edge-AI using heterogeneous integration of commercial semiconductor foundry processes, in-house-designed fabrication processes at HKUST NFF. This will be promising in the era of AIoT for detecting various diseases, such as infectious diseases (COVID-19 and emerging infectious diseases), cancer diagnosis, etc.

BIOGRAPHY

Yi-Kuen Lee receive his BS degree with honor at National Taiwan University (NTU) in 1992. He received his MS degree under the supervision of late Prof Yih-Hsing Pao (member of US NAE and Academia Sinica) in the Institute of Applied Mechanics, NTU in 1995. He went to US and obtained Ph.D. degree in Mechanical Engineering with Major in MEMS under the guidance of Prof Chih-Ming Ho (member of US NAE and Academia Sinica) at UCLA in 2001. He was an Assistant Professor from 2001 to 2007 at HKUST. He received substantiation and was promoted to Associate Professor in 2007. He was a Visiting Associate at Caltech in 2011. He has published two book chapters, more than 120 refereed journal and international conference papers. His current research topics include microfluidics for enumeration of Circulation Tumor Cells (CTCs) for cancer diagnostics, microchips for DNA transfection, micro/nano heat transfer, micro/nano electrophoresis for large DNA molecules, MEMS sensors for environmental monitoring and energy-efficiency building. He is the Associate Editor, HKIE Transaction. He is the President of Hong Kong Society of Theoretical and Applied Mechanics since 2014. He is the co-founder of the annual Nano/Micro Engineered and Molecular Systems (IEEE NEMS) conferences since 2006; the co-founder of International Contest of Application in Nano-micro Technology (ICAN) Association; Technical Program Committee (TPC) member of IEEE MEMS 2007, Kobe, Japan; TPC member of IEEE Nano 2007; TPC member of APCOT 2008, 2010, 2012, 2014; TPC member of IEEE NEMS 2009, 2010, 2011 & 2012, IEEE Transducers 2009 & 2011, 2013, 2015. He is also the Chair for the 5th International Workshop on Innovation and Commercialization of Micro & Nano Technologies (ICMAN 2011), Shenzhen, 5-8 Nov 2011. He is the Associate Editor of Microfluidics and Nanofluidics, Nature Springer. He chaired APCOT 2018 (Asia-Pacific Conference of Transducers and Micro-Nano Technology), Hong Kong in 2018. He also serves the Editorial Board of Micro/Nanotechnology Journal in China, and the Bio-Design and Manufacturing (BDM), Springer.

978-1-6654-3008-1/21 $31.00 © 2021 IEEE

Gap in pagination due to formatting issues.

Pages 1886-1887

Proceedings of the 16th Annual IEEE International Conference on Nano/Micro Engineered and Molecular Systems
April 25-29, 2021

Frequency Output, A multi-modes Vacuum Gauge with Highly Output linearity Based on Electrostatic Nonlinearity

Chengxiang wang, Yunbin Kuang, Yulie Wu, Zhanqiang Hou*,
Yongmeng Zhang, Xuezhong Wu and Dingbang Xiao*

Abstract— Based on electrostatic nonlinearity, this paper mainly introduces a MEMS resonant vacuum gauge (RVG) for low vacuum measurement. With the merits of small size, low power consumption and wonderful output linearity, the RVG could be widely used in various condition, especially for embedded measurement. The gauge has multi-working modes due to different driving methods. There are three modes has been measured under different pressure, named the second-order resonant mode (the second mode in short), the third mode and the fourth mode, respectively. Measurement results illustrates that the second mode has nearly no relationship with pressure variation, the third-order mode measures 1Pa~500Pa, linear fitting shows a wonderful output linearity (R^2=0.995) and resolution (superior than 0.5Pa); measurement range of the fourth mode covers 1Pa~5000Pa, fitting results shows a highly linearity (R^2=0.997) and about 1Pa in resolution.

INTRODUCTION

As an advanced manufacturing technology platform, MEMS has been rapidly developed in recent decades. Through the MEMS process, micro and nano-scale processing operations can be realized on a silicon-based wafer and micron-level structures can be obtained for various sensors' utility. Due to mature technology, a large number of MEMS sensors such as gyroscopes, acceleration sensors and pressure sensors have emerged [1-5]. Those sensors with many merits such as small size, low power consumption and low price that meets the development trend of instrument miniaturization.

As a classic MEMS product, MEMS type vacuum gauges have been widely studied. Such as MEMS capacitance diaphragm vacuum gauge, MEMS resistance gauge, MEMS Pirani vacuum gauge and MEMS resonant gauge has appeared. Those gauges have broad market prospects in space detection and industrial control [6-9].

Vacuum gauge based on gas damping have some unique merits than others, for example, measurement result has no relationship with gas composition, highly resolution. Such as the vacuum gauge in [10], that has been used as measurement standard for medium and high vacuum measurement. However, the gauge has some shortages of long reactive time (2-3mins), because of low driving frequency (200Hz, approximately), and complex driving

---

National University of Defense Technology, Changsha, China;
Laboratory of Science and Technology on Integrated Logistic Support, National University of Defense Technology, Changsha, China and Hunan MEMS Research Center, Changsha, China;
Corresponding author: houzhanqiang@nudt.edu.cn
Corresponding author: dingbangxiao@nudt.edu.cn

circuits.

In previous research, a butterfly-like gyroscope has been developed [11-12]. The resonant frequency of the gyroscope will drift with the variation of sealing pressure has been observed during air damping measurement process. So, a resonant vacuum gauge based on previous gyroscope has been developed for low vacuum measurement. The particularity of gyro structure makes the RVG has different working modes, named the second mode, the third mode and the fourth mode. In this paper, the measurement principle and measurement steps has been introduced in detail, the relationship between frequency, phase and gas pressure of each mode has been measured.

GAUGE FABRICATION, MODEL INTRODUCTION AND MEASUREMENT

A. fabrication of the RVG

Air damping of torsional block has complex mechanisms, which has been detailed analyzed by Ref [13]. In this paper, working modes, fabricated process and measurement results were mainly introduced.

The gauge has three layers, named bottom electrode layer, sensing layer and cover layer, respectively. Fabrication process as show in figure.1.

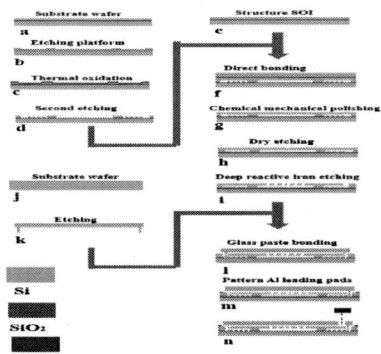

Fig 1 illustration of fabrication process

A 500μm thick (100) SOI silicon wafer with 10μm thick structure layer was used as the bottom electrode. After two times of etching and oxidation, the anchor points and bottom electrodes were released (a to d). Another SOI wafer was used to construct the sensing layer (single-crystal silicon 100 crystal orientation) (e) and directly bonded to anchor points through silicon-silicon bonding (f). The support layer of the SOI was removed by chemical mechanical polishing process(g). To adjust the azimuth of the sensing axis, dry etching was adapted to etch a micro-groove to generates the second order mode of the gauge (h), the width and depth of micro-groove is 40μm and 24μm, respectively. After that, deep-reactive

978-1-6654-3008-1/21 $31.00 © 2021 IEEE

etching method was used to release the mass block (i). Then, in order to protects the sensing layer from large particle impurities of test environment, a 500μm thick single-crystal silicon wafer (100) (j) was etched as the upper cover plate (k), and bonded by glass paste bonding(l). Magnetron sputtering was used to plate aluminum pads(m). Finally, some holes through the cover layer were drilled by femtosecond laser to ensure a connection of inside and outside (n) that the dynamic part can be directly connected with external environment. Fabricated gauge shows in Fig.2. Detailed bottom electrodes and driving steps can be read in reference [14][12], the sensing layer as show in Fig.2(b) with two sensing blocks, anchor points, sensing axes and a coupling axis is used to connect the blocks, shows in Fig.2(c).

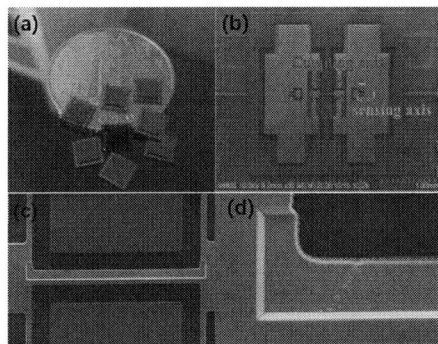

Fig.2 (a) samples of fabricated gauge; (b) SEM picture of sensing structure, with two sensing blocks, anchor points, sensing axes and a coupling axis; (c) SEM photograph of amplified coupling axis; (c) SEM picture of etched groove on the sensing axis.

B. mode introduction of the gauge

Multi-modes can be driven by changing the position and frequency of the electrostatic force. Near to resonance frequency, a drastic change, both amplitude and phase, can be observed. So, resonant frequency can be investigated by frequency sweeping, the block movement could be demonstrated by simulated result. There are three modes can be driven and has been demonstrated by COMSOL simulation. Those modes, named the second mode, the third mode, and the fourth mode.

The second mode, sensitive axis bent in horizontal direction, two mass blocks take the anchor point as a vibration center. The sensing block, shows in Fig.2(b), was driven by electrostatic force and slight vibrates in out-of-plane direction and higher movement component in in-plane direction can be realized by azimuth angle of the sensing axis. Micro grove was etched, as shows in Fig.1(step h) and Fig.2(d), to adjust azimuth angle [14]. Figure.3(a) shows sweeping and simulation results of the second mode, the red line represents phase change and the black line is amplitude variation. A significant phase and amplitude changes can be observed near to resonance frequency the resonant frequency is 5883.62 Hz (34.2 Pa, 296K), near to the simulation (6130 Hz). This little variation could be owing to structure deviation and air damping.

The third mode is a torsional movement of the sensing axis and the sensing block vibrates out of plane direction,

sensitive axis as the vibrating center, squeeze film damping as the main damping source. Measurement result of the frequency sweep and simulation result, shows in Fig.3(b), is 6209.85Hz (12.2Pa, 296K) and 6417Hz, respectively. The amplitude changes drastically, shows a peak-like shape and the phase changes negative at first and reverses into positive increase at the behind.

The fourth mode, shows in Fig.3(c), is a vertical bending motion of the sensitive axis. Center line of the block width, through anchor point, as the vibration center, air squeeze film contributes as the main damping component. The simulation result shows that the resonant frequency is 2035Hz and the sweep frequency measurement is 18695.42Hz (0.4Pa, 296K).

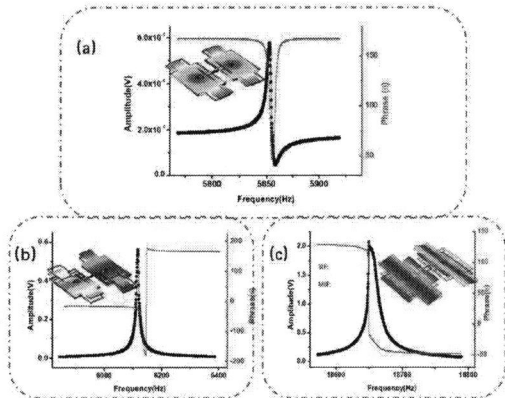

Fig.3 (a) Frequency sweeping of the second mode, the resonant frequency is 5883.62 Hz (34.2 Pa, 296K) and simulated frequency is 6130Hz;(b) Frequency sweeping of the third mode, resonance frequency is 6209.85Hz (12.2Pa, 296K) that near to simulated frequency 6417H;(c) Frequency sweeping of the fourth mode, measurement frequency is 18695.42Hz (0.4Pa, 296K) and simulation result is 2035Hz

C. RVG measurement

Driving signal contains a dc bias voltage and an ac bias voltage [3]. The ac frequency is controlled by a phase controller, (Zurich instrument, MFLI).

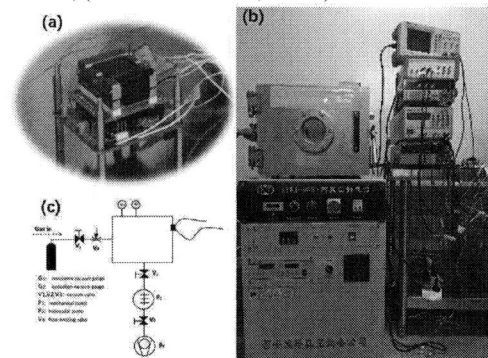

Fig.4 (a) jig and fixture of the gauge with signal circuit; (b) schematic diagram of measurement instrument;(c) illustration of measurement instrument;

The gauge, fixed in a jig with signal circuit (Fig.4(a)), sets in the vacuum chamber (Fig.4(b)). Illustration of the vacuum system as show in Fig.4(c), a capacitance diaphragm vacuum gauge was used as standard instrument for low vacuum measurement (ReBorn, ZDMVO1, measurement range 1.33Pa~13300Pa) due to especially high and stable measurement capability in low vacuum

978-1-6654-3008-1/21 $31.00 © 2021 IEEE

condition.

After pumping the vacuum system into ultimate pressure (0.2Pa, approximately), the vacuum valve between pump and vacuum chamber was closed, a little gas was inflated into the vacuum chamber through another valve to adjust gas pressure. When a stable reading of the capacitor membrane vacuum gauge has obtained, sweeping the resonance frequency. Repeat the test steps as aforementioned to get the relationship between resonance frequency and gas pressure.

RESULTS AND DISSCUSSION

A. measurement results

All measurements were implemented under room temperature(296K). Three modes were measured independently, the relationship of pressure and resonance frequency under different pressure were dedicated studied.

Fig.5(a) shows the measured results of the second mode. The pressure increases from 3.4Pa to 5360Pa, resonant frequency has a slight change, drops from 5877.69Hz to 5864.72 Hz and the amplitude decrease from 1.44mV to 0.48mV. According to former research, resonance frequency of the second mode is especially sensitive to temperature variation [12], this measured deviation may come from temperature variation of the chamber during inflation, or heat dissipation from the vacuum system. Sweeping results under 3.4Pa and 4987Pa shows in Fig.5(b) and Fig.5(c), respectively. Measurement results of resonance frequency under a lower pressure are clear but turn into more mess under higher pressure, this was resulted by smaller and smaller movement amplitude of the sensing block in higher pressure that result in poor signal-to-noise ratio.

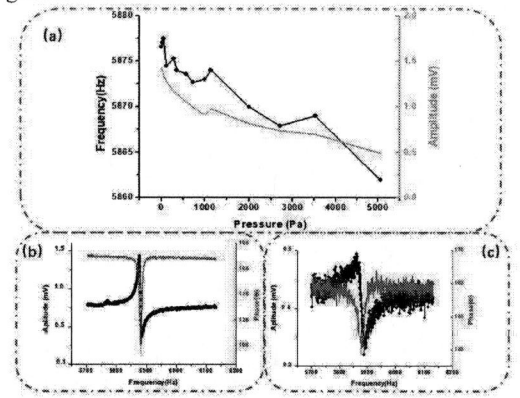

Fig.5 (a) measurement result between frequency and amplitude of the second mode under different pressure; the resonance frequency drops from 5877.69Hz to 5864.72 Hz and the amplitude decrease from 1.44mV to 0.48mV (b) scanning result of the second mode under 3.4Pa, (c) scanning result of the second mode under 4987Pa

Fig.6(a) shows measurement result of the third mode, amplitude decreases sharply with the increases of gas pressure but resonance frequency has a nearly linear relation with pressure. The result illustrates that measurement range of this mode covers 1Pa~500Pa with an excellent output linearity and a high resolution (superior than 0.5Pa), the relationship between pressure and resonance frequency can be concluded as

$$y = 6160 + 2.86x, \quad R^2 = 0.995 \qquad (1)$$

Frequency sweeping under 0.1Pa and 345Pa as show in Fig.6(b) and Fig.6(c), the amplitude is 0.223V under 0.1Pa but decrease dramatically to 1.507mV under 345Pa. The resonance frequency of 0.1Pa is 6160.27Hz but increase to 7146.42Hz under 345Pa. By comparing, it can be seen that driving amplitude decreases and resonance frequency increases with pressure ascending, half width of the resonance peak becomes wider and the phase turns into more and more messy at the same time. This phenomenon can be explained through degrade of quality factor(Q-factor) by air damping. But it is still doubt, whether exist some physical phenomenon or mode coupling effects in vibrating and results in this phenomenon.

Fig.7(a) shows experimental results of the fourth mode. Measurement range of this mode covers 1Pa~4000Pa with a highly output linearity (R^2=0.997) and resolution (probably 1Pa). The relationship between resonance frequency and pressure of this mode can be concluded as

$$y = 18692 + 0.81x, \quad R^2 = 0.997 \qquad (2)$$

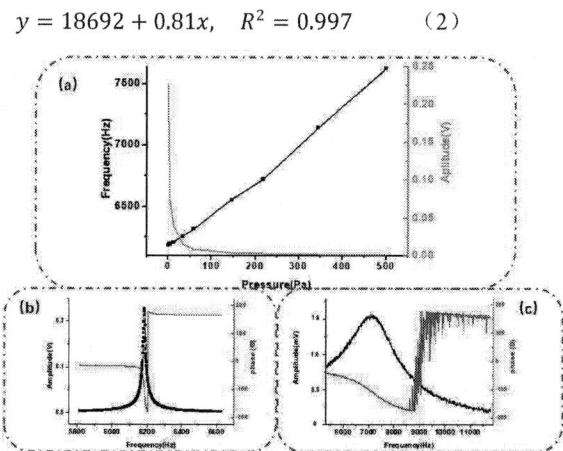

Fig.6 (a) measurement results between frequency and amplitude to pressure of the third mode, the frequency increase from 6160.27Hz under 0.1Pa to 7146.42Hz of 345Pa (b) scanning result under 0.1Pa, with resonance frequency 6160.27Hz and amplitude 0.223V (c) scanning results under 345Pa, resonance frequency and amplitude is 7146.9Hz and 1.596mV.

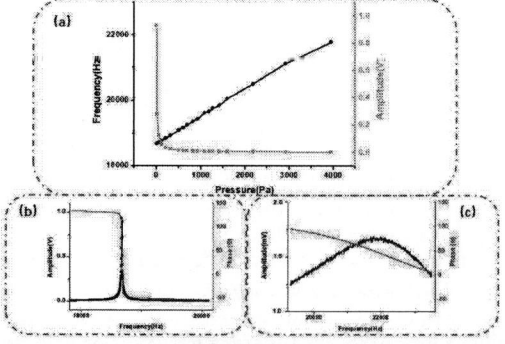

Fig.7 (a) measurement relashionship between frequency and amplitude to pressure of the fourth mode, measurement range covers 1Pa~4000Pa, with highly output linearity; (b) sweeping result under 1.1Pa, resonance frequency is 18653Hz and the amplitude is 0.96V; (c) measurement result under 3953Pa, resonance frequency is 21778Hz and the amplitude is 1.69mV;

978-1-6654-3008-1/21 $31.00 © 2021 IEEE

Sweeping results of 1.1Pa and 3953Pa are respectively shown in Fig.7(b) and Fig.7(c). Resonance frequency under 1.1Pa is 18653Hz and sharply increase to 21778Hz of 1953Pa, the amplitude decreases from 0.96V to 1.69mV.

The phase changes obviously under a low pressure but gradual decreases at the higher pressure, this comes from degrade of Q-factor. For the third mode and the fourth mode, movement of the sensing axis are torsion and bending, respectively. Comparing with the third mode, phase change in a mess has not appear, this distinguish maybe caused by some physical phenomenon or mode coupling that effects vibrating phase.

B. discussion

Total Q-factor of the gauge can be calculated from contribution of different dissipation mechanism, air damping loss, anchor loss, surface loss, thermoelastic damping loss and additional loss mechanism. The Q-factor can be calculated as Eq. 3. It can be found the total Q-factor is mainly determined by the loss mechanism with the lowest Q-factor [2].

$$\frac{1}{Q} = \frac{1}{Q_{gas}} + \frac{1}{Q_{anchor}} + \frac{1}{Q_{surf}} + \frac{1}{Q_{TED}} + \frac{1}{Q_{etc.}} \quad (3)$$

Q_{gas}, is the most dominant factors when the resonator working at atmospheric conditions. It can be easily eliminated by operating the resonator in high vacuum condition. Q_{surf}, are mainly caused by the defects and surface roughness. Anchor loss, Q_{anchor}, is mainly caused by the energy loss from the resonator through its supporting anchor structure. Thermoelastic damping loss, Q_{TED}, is caused by the interaction between elastic strain and thermal effects. Despite the same sensing axis, the third mode and the fourth mode have different Q_{TED} due to different movement.

The gauge works on a low vacuum condition, Q_{gas} contributes the main component of lowest Q-factor. Peak width of the amplitude at the half height increasing due to lowing Q-factor caused by air damping, shows by measurement results of three modes.

Resonance frequency of the second mode is not sensitive to pressure variation while the third mode and the fourth mode has a linear relationship with gas pressure. Measurement range of the third mode covers 1Pa~500Pa with a highly resolution of more than 0.5Pa and the fourth mode measurement range covers 1Pa~4000Pa with the resolution of 1Pa. For the second mode, a bent movement of the sensing axis and the synovial damping as the main damping source, the block vibrates in the in horizontal direction and vertical movement could be ignored. Higher pressure means higher synovial damping that reduce the amplitude in horizon direction. A same degrade of Q-factor can be observed by half peak width due to higher air synovial damping. The electrostatic driving force can be seemed as a constant due to negligible movement in vertical direction. So, different from the third mode or the fourth mode, obvious frequency change, comparing to pressure change, has not be observed, the negligible frequency drift can be owing to temperature disturbance.

For the third mode, a torsional movement of the sensing axis, the sensing block vibrates in vertical direction, the amplitude reduces drastically by the influence of increasing squeeze-film damping. The driving force by electrostatic has reduced, follows with amplitude decreasing, that results in frequency increase at final. This explanation can be illustrated by the measurement result of the second and the fourth mode. Comparing with the third mode, the measurement range has been extended nearly for an order of magnitude but the resolution has decreased about 70%. This difference can be explained by air damping which is influenced by the shape of the sensing block, inherent vibrating frequency, movement velocity of the sensing block, electrodes distance and gas pressure.

This gauge based on air damping to measure gas pressure, measurement frequency of the second mode independent with gas pressure, but it effected by environment temperature. So, this gauge could precisely measure pressure by the third mode or the fourth mode before temperature standardization process that implemented by the second mode. Measurement range of the fourth mode covers the third mode but with a lower resolution. If the third mode and the fourth mode is used to measure a same pressure, a self-calibration process can be realized for a higher accuracy measurement. Damping torque can be adjust by reshape the size of the sensing block or some small hole [15] can be etched in the block for a wider measurement range.

CONCLUSION

In this paper, a multi-modes resonance vacuum gauge has been proposed. This gauge has the advantages of small size, light weight, low price, excellent output linearity and multiple working modes. Measurement ranges of the two modes has overlap part so that a same gas pressure can be measured through two modes under a same pressure to realize an independent calibration of a single gauge and improve measurement accuracy. By structural optimizing or etching small hole on the sensing block, damping torque of the sensing block can be optimized, measurement range of the fourth mode can be extended and lower pressure sensitivity of the third mode can be enhanced. Subsequently work will focus on the characteristics of full temperature range and gauge optimization.

ACKNOWLEDGEMENTS

The author would like to thank the laboratory of Microsystem, National University of Defense Technology, China, for fabrication and technical support. This work was supported by National Nature Science Foundation of China (Grant No.51905538).

REFERENCES

[1] Sorenson L, Ayazi F. Effect of structural anisotropy on anchor loss mismatch and predicted case drift in future micro-Hemispherical Resonator Gyros[C]. Position, Location and Navigation Symposium - PLANS 2014, 2014 IEEE/ION, 2014: 493-498.

[2] Qingsong Li, Dingbang Xiao, Xin Zhou, et al. 0.04 degree-per-hour MEMS disk resonator gyroscope with high-quality factor (510 k)

and long decaying time constant (74.9s) [J]. Microsystems & Nanoengineering, 2018, 4(1): 32.

[3] Xin Zhou, Dingbang Xiao, Zhanqiang Hou, et al. Influence of the Structure Parameters on Sensitivity and Brownian Noise of the Disk Resonator Gyroscope[J]. IEEE Journal of Microeletromechanical Systems, 2017, 26(3): 519-527.

[4] Roy Anindya Lal, Sarkar Hrishikesh, Dutta Anupam. A high precision SOI MEMS–CMOS ±4g piezoresistive accelerometer[J]. Sensors and Actuators A, 2014,210: 77-85

[5] John K. Coultat, Colin H.J. Fox, Stewart McWilliama, et al. Application of optimal and robust design methods to a MEMS accelerometer[J]. Sensors and Actuators A, 2008.142 (1):88-96

[6] https://www.canon-anelva.co.jp

[7] https://www.sensorsportal.com

[8] https://www.sens4.com

[9] David C. Catling. High-Sensitivity Silicon Capacitive Sensors for Measuring Medium-Vacuum Gas Pressure[J]. Sensors and Actuators, 1998,A64:157-164

[10] https://www.mksinst.com/

[11] Fenlan Ou, Zhanqiang Hou, Xuezhong Wu, et al. Analysis and Design of a Polygonal Obique Beam with Improved Robustness to Fabrication Imperfections[J]. Micromachines, Vol.9, No.198.

[12] Fenlan Ou, Zhanqiang Hou, Xuezhong Wu, et al. A New Stress-released Structure to Improve the Temperature Stability of the Butterfly Vibratory Gyroscope[J]. Vol.10, No.82.

[13] Minhang Bao, Yuanchen Sun, Jia Zhou, Yiping Huang. Squeeze-film Air Damping of a Torsion mirror at a Finite Tilting Angle[J]. J. Micromech. Microeng. 2006, 16: 2330-2335

[14] K L, Qingsong Li, Xin Zhou. Modal coupling effect in a novel nonlinear micromechanical resonator[J]. Micromachines, 2020, 11,472;

[15] Bao Minghang. Analysis and Design Principles of MEMS Devices[M]. Elsevier, 2005

Proceedings of the 16th Annual IEEE International
Conference on Nano/Micro Engineered and Molecular Systems
April 25-29, 2021

High-resolution noncontact electrostatic voltage meter based on microsensor chip*

Xiaolong Wen, Pengfei Yang+, Bo zhang, Chunrong Peng

Abstract—Electrostatic voltage is a vital parameter in industrial production lines for reducing electrostatic discharge harms and improving yields. Due to such drawbacks as package shielding and low resolution, previously reported electric field microsensors are still not applicable in industrial lines. In this paper, we introduce a new designed microsensor package structure which enhances the field strength inside the package cavity remarkably. This magnification effect is studied and optimized by both theoretical calculation and ANSYS simulation. By means of digital synthesizer and digital coherent demodulation method, the compact signal processing circuit for the packaged microsensor is developed. The meter protype is calibrated above a charged metal plate, and the measuring error is less than 1V from -1kV to 1kV in 2.5 centimeters distance. The meter is also installed into a production line and shows good consistency with and better resolution than a traditional vibratory capacitance sensor.

I. INTRODUCTION

Electrostatic voltage is drawing much attention recently as causing damages in Organic Light Emitting Diode (OLED) manufacturing processes. Long-term noncontact electrostatic voltage meters, besides those ionizing air blowers, are promised to be the key for the electrostatic discharge (ESD) control and evaluating the effectiveness of ESD protection [1].

Fill mill and the vibrating capacitor are the most common used instruments for electrostatic voltage measurement. With a rotating or vibrating component, the induced charge is modulated into an alternating current, whose amplitude is proportional to the outer electric field. Despite they have good performance in use, the complicated actuating mechanical structure makes them difficult in assembling, leading to an unwelcome cost and vulnerable entity. Based on Micro-electro-mechanical Systems (MEMS) technology, the resonant electric field microsensor (also known as the micro field mill) employs the similar working principle and has been studied since 1990s. Benefitting from the advantages such as batch manufacturing and wear-free, it is promising that the microsensors will be used in the future. Most related articles focus on optimizing the chips' microelectrodes [2,3]. Nevertheless, the resolution of electric field is still insufficient for such OLED's low static voltage sensing applications due to their small sensing area and weak induced current. The packaging of the electric

field microsensor is also important but is not fully considered yet. Several previously reported packaging protypes were prone to shield the outer electric field by the package cap and sidewalls of the package cavities, therefore causing an attenuation of field resolution [4].

In this paper, we bring out an innovative high-resolution microsensor package structure which enhances the field strength inside the package cavity to gain an ultra-high resolution. Also, by means of digital driving and signal demodulating circuit, a compact meter protype is developed, calibrated, and then applied in an OLED manufacturing line.

II. METHOD

Historically, the electric field microchips were mostly exposed into the external environment for test. This paper proposes a new package structure for the electric field microsensor, and the resolution is enhanced remarkably. We also developed a compact circuit for the microchip, and ran a series of tests on the meter protype.

A. Working principle of the electric field microsensor

Figure 1 shows the sketch and the photo of the electric field sensor (EFS) based on MEMS technology, consisting of two sense electrodes, an earthed resonant shield electrode, a driven electrode, and a folded beam. From the vibration of shield electrode, time-varying electrical charge signal is induced on the sense electrodes whereas the signal amplitude is proportional to the measured field. Two groups of differential sensing electrodes are designed to eliminate the cross-interference from the surrounding circuits. As reported by our group [5], the EFS is fabricated from a Silicon-on-Insulator (SOI) wafer, where all the components shown in Figure 1 distribute in the same plane.

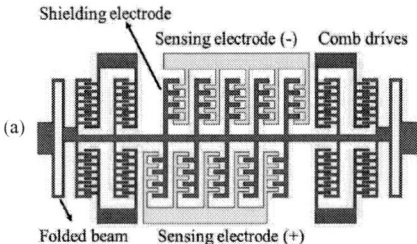

---

*Resrach supported by the Key Projects of National Natural Science Foundation of China (62031025), the National Key Research and Development Plan of China (2018YFF01010800), and the Fundamental Research Funds for the Central Universities (FRF-TP-19-045A2, FRF-BD-19-017A, and FRF-BD-20-12A).

X. Wen and B. Zhang is with the Department of Physics, University of Science and Technology Beijing, Beijing Engineering Research Center of

Detection and Application for Weak Magnetic Field, Beijing 100083, China (e-mail: xiaolongwen@ustb.edu.cn).

+Contacting Author: P. Yang was with the School of Applied Science, Beijing Information Science and Technology University, Beijing 100192, China. (e-mail: pfy@bistu.edu.cn).

C. Peng is with the Aerospace Information Research Institute, Chinese Academy of Sciences, State Key Laboratory of Transducer Technology, Beijing 100190, China (e-mail: crpeng@mail.ie.ac.cn).

978-1-6654-3008-1/21 $31.00 © 2021 IEEE

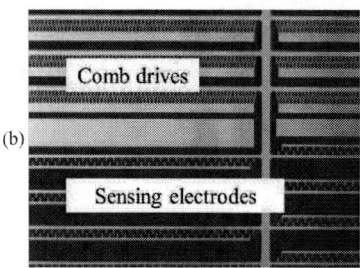

Figure 1. (a) Structure and (b) picture of the electric field micro chip.

The reported resolution of this EFS is approximately 40 V/m under the parallel plate calibration. However, it is still insufficient for the microchip to be used in the low static voltage measurement, because the microchip might be damaged without protection, the signal processing system is too bulky, and the target static voltage error should be under about 10 V.

B. Innovative package structure design and simulation

Previously reported sealed packages mainly contained a flat package cap on an insulative support. The flat cap is either metallic or insulative, either of which formed a float-potential body for the electric field to pass through. Although the vulnerable microchip was well protected, the electric field penetrating the package cap is attenuated too, therefore causing the resolution to be worse.

The relationship between the electric field inside and outside the package cavity is given by (1):

$$E_{in} = c \cdot E_{out} \tag{1}$$

where the E_{in} and E_{out} are the electric field in and out of the package cavity respectively, and c is the amplification coefficient which is usually less than 1 in existent reports.

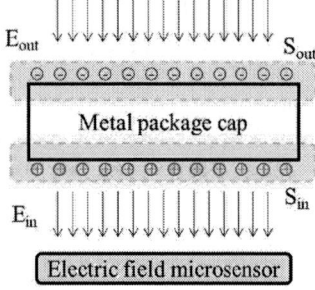

Figure 2. Mathematical analysis model of the microsensor package.

Here we take the metal package cap for calculation, as shown in Figure 2. The fringe effect of charge distribution and the support for the metal package cap are neglected in this model. Since the cap is float potential, the induced charge on the top and bottom of the metal package cap in Figure 2 are equal in quantity but opposite in sign. Based on the electric field Gauss theorem, the integrals around the two surfaces are written as:

$$\oint_{S_{out}} E_{out}\, ds = \frac{Q_{out}}{\varepsilon_0} \tag{2}$$

$$\oint_{S_{in}} E_{in}\, ds = \frac{Q_{in}}{\varepsilon_0} \tag{3}$$

where the Q_{out}, and S_{out} represent the charge and area on the top respectively. Q_{in}, and S_{in} are defined in the same way. Since $Q_{out} = -Q_{in}$, and the electric field inside the metal package cap is zero, from (2) and (3) we can get:

$$E_{in} = \frac{S_{out}}{S_{in}} \cdot E_{out} \tag{4}$$

Equation (4) shows that the amplification coefficient c is decided by the area ratio of the two surfaces. For the previous flat cap structure, $c = 1$ in this model. But if the support for the cap is further considered and calculated, the amplification coefficient $c < 1$.

In order to gain a high-sensitive packaged microsensor, a new package structure is proposed by using (4). Different from the simple flat structure, the new concept is designing a taper shape package cap, in which the out surface is large enough to collect the induced charge, and the inside surface is small to concentrate the opposite induced charge. Therefore, the electric field can be enhanced inside the package cavity.

To make a field-enhancing structure, the outer electrode which connects the package cap with a metal wire is proposed. The outer electrode is exposed to the measured electric field outside, and can be supported by the shell of the total meter with insulative material. Therefore, its area is to the magnitude of the voltage meter. Figure 3(a) shows the finite element simulation result of electric field distribution by ANSYS. The electric field around the proposed package structure is significantly distorted, while the electric field strength inside the package cavity is enhanced by one order of magnitude. As shown in Figure 3(b), an air tight package made of highly reliable ceram is fabricated. The cap is made of covar for further connecting to the outer electrode with wire. Although the induced charge on the tiny sense area is weak theoretically, benefitting from the optimization on the shape of package and the outer electrode connected to the package cap, local electric field near the sense chip inside the sealed space is not attenuated but remarkably strengthened.

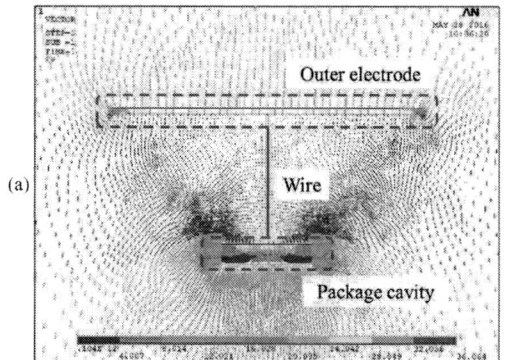

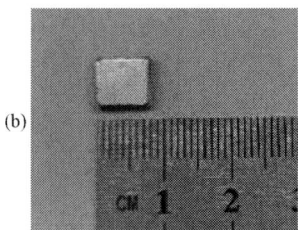

Figure 3. Package of the electric field microchip (a) Electric field simulation on the enhancement effect of package, (b) Picture of the sealed cavity using ceram.

C. Signal processing circuit

The signal processing circuit includes the amplifier, A/D sampling chip, etc. shown in Figure 4. Regarding the detecting part, the weak output alternative current of the microsensor is first magnified by a transimpedance amplifier, and then magnified by an instrument amplifier. High speed and high accuracy A/D sampling chip AD7660 is used to digitize the magnified microsensor output and the driving voltage. The resonant driving electrostatic force is generated by the Direct Digital Synthesizer (DDS), which is controlled by the Microcontroller Unit (MCU). A unique feature of the signal processing circuit is that the amplitude of the alternative signal is demodulated by the digital coherent demodulation method, which mainly contains a multiplication and a filter algorithm. It is anti-interference from the environment, and fast enough given the high-speed MCU. The meter's measuring results are transmitted in accordance with RS485 protocol. Total power of the meter is 600mW, and the size of the sensor protype is 99 mm × 23 mm × 19mm.

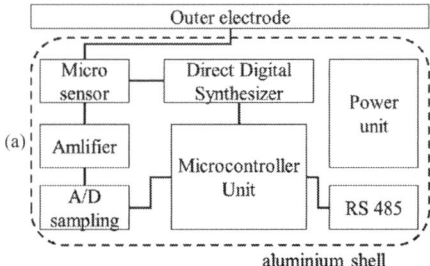

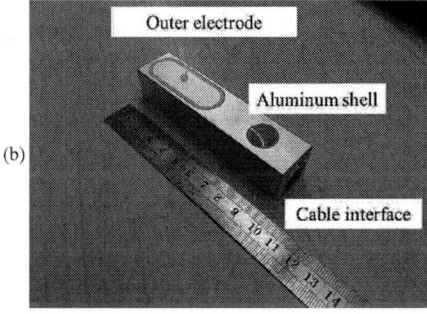

Figure 4. Signal processing circuit of the sensor. (a) Diagram (b) Picture of the sensor.

D. Calibration and accuracy test

The calibration system consisted of a high voltage power supply and a metal plate is shown in Figure 5. The meter was placed perpendicularly to the plate, while the distance between them was adjustable. The high voltage power was Keithley 2410, and the output voltage range was from -1.1 kV to +1.1 kV.

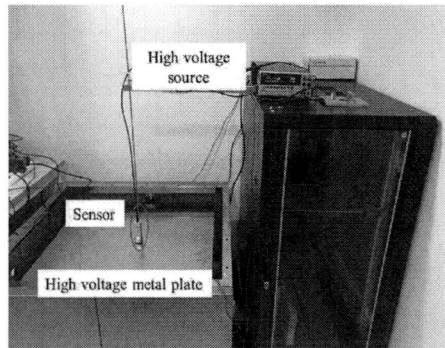

Figure 5. The electrostaic voltage calibration system of the sensor.

In the meter calibration, the gap distance between the meter and the plate was set as 1 centimeter. By applying step voltages from -1 kV to 1 kV, the output of the meter was recorded synchronously. The slope of the fitting straight line was calculated as the sensitivity.

In the accuracy test, the gap distance was set to be 2.5, 5, and 10 centimeters. Before the test, the cite coefficient needs to be calibrated because the distance was changed. This could be accomplished easily by applying a known voltage to the plate, and calculating the multiple between the input and the output direct voltage.

E. Real test in an OLD manufacturing line

The meter was also installed into an OLED manufacturing line to monitor the static voltage on the glass along with a traditional vibrating capacitor sensor, as shown in Figure 6.

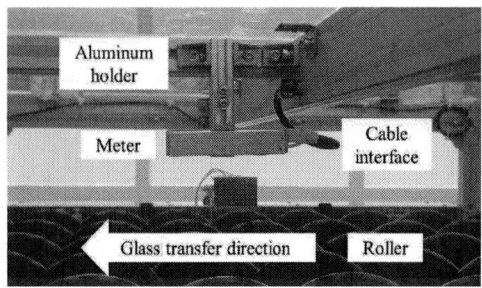

Figure 6. Picture of test in the OLED product line.

III. RESULTS

A. Laboratory test result

Figure 7 shows the calibration results and the fitting line when applying voltages ranging from -1kV to 1kV. The output of the meter before the calibration was in mV, which represented the amplitude of the alternative voltage getting from the amplifier.

978-1-6654-3008-1/21 $31.00 © 2021 IEEE

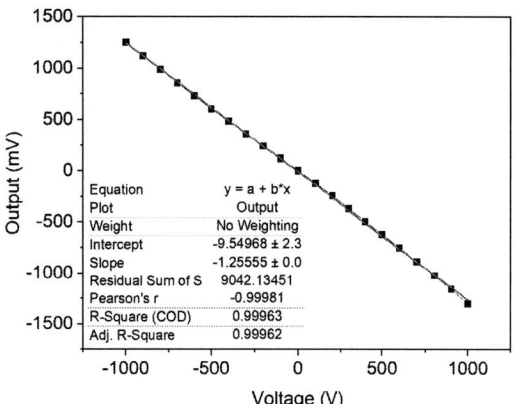

Figure 7.   Calibration results of the meter protype on a charged metal plate

Figure 8 shows the long-term accuracy test curve. The gap distance was 2.5 centimeters, and the cite coefficient had already been calibrated before. When the applied voltage changed from -1 kV to 1kV, the measured result by our meter agreed well with an absolute error less than 1 V.

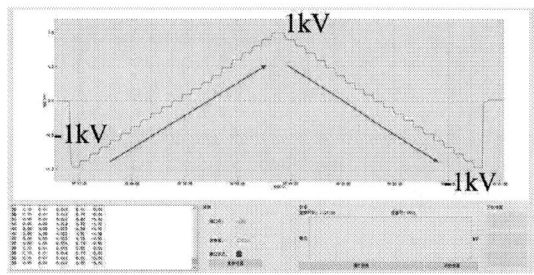

Figure 8.   Long-term accuracy measurement result.

E. Real test result in an OLD manufacturing line

The MEMS-based meter was installed next to a commercial vibrating capacitor type meter. The resolution of the vibrating capacitor type was 10 V. As shown in Figure 6, their measuring results agree well while the MEMS sensor curve is more elegant because of the better resolution.

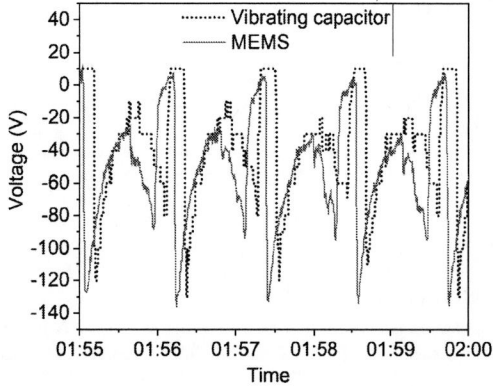

Figure 9.   Data compared with the vibratory capacitance meter in the OLED manufacturing process.

IV.   DISCUSSION

In this paper, a new MEMS-based electrostatic voltage meter is proposed, designed, and tested. Significantly different from the existent microsensor chip designs, our design focuses on the package of the microsensor, and realizes a distinct improvement in the electric field resolution. The resolution is both verified both by the charged plate test and the real test.

Even though, several drawbacks exist in this design and still need further consideration. First, since the sensitivity is magnified, the measuring range is narrowed. Therefore, if larger measurement range is required, the meter should be placed farther away from the object, meanwhile the sensitivity is lower then. Second, the package cap structure is also sensitive to the accumulated electrostatic charge which might be transferred in the installation process. Although in this paper we neglect the accumulated charge and consider the package cap is an electroneutral conductor, it is not easy to keep neutral in actual test. It takes hours or even days for the accumulated charge to dissipate.

V.   CONCLUSION

Electrostatic voltage measurement is a key factor in the ESD control area. Based on the electrostatic microsensor, a high-resolution contactless voltage meter is proposed. Rather than designing the sensor's micro structure, here we bring out a new package structure with a large area outer electrode and small area package cap. By inducing more charge under the electric field, and concentrating the charge to a narrow space, the electric field inside the package cavity is enhanced significantly. Both theoretical analysis and finite element simulation have proved this conception. Furthermore, we designed the signal processing circuit and made a meter protype. After calibration, lab test and real environment test, it was proven that our protype had a resolution better than 1 V in 2.5 centimeters distance, and it performs better than the traditional vibrating capacitor in the OLED production line. With the various advantages of the MEMS technology, the proposed electrostatic voltage meter protype is promised to be widely used in future.

REFERENCES

[1]   J. Smallwood, N. G. Green and K. Robinson, "Understanding Electrostatic Field Meter Field and Voltage Measurements from Conductors and Insulators," *2020 42nd Annual EOS/ESD Symposium (EOS/ESD)*, Reno, NV, USA, 2020, pp. 1-10.

[2]   Yan Z, Liang J, Hao Y, et al. A Micro Resonant DC Electric Field Sensor Based on Mode Localization Phenomenon. *32th International Conference on Micro Electro Mechanical Systems (MEMS) IEEE*, Seoul, Korea (South), 2019: 849-852. doi: 10.1109/MEMSYS.2019.8870880

[3]   Zhaozhi C, Chunrong P, Ren R, et al. A High Sensitivity Electric Field Microsensor Based on Torsional Resonance. *Sensors*, 2018, 18 (1): 286.

[4]   Mou Y, Yu Z, Huang K, et al. Research on a Novel MEMS Sensor for Spatial DC Electric Field Measurements in an Ion Flows Field. *Sensors*, 2018, 18 (6): 1740.

[5]   Yang P, Peng C, Fang D, et al. Design, fabrication and application of an SOI-based resonant electric field microsensor with coplanar comb-shaped electrodes. *Journal of Micromechanics & Microengineering*, 2013, 23(23): 055002.

Proceedings of the 16th Annual IEEE International
Conference on Nano/Micro Engineered and Molecular Systems
April 25-29, 2021

Preparation and Characterization of Ni-AAO Composite Scaffold Getter with Induction Heating

Lujiang Liu[1], Zhenxiang Bu[1], Kuixi Wu[1], Wenlong Lv[2], Dandan Gu[2], Lingyun Wang[1,*]

Abstract—Non-evapotranspiration getter (NEG) is widely used in the field of MEMS vacuum packaging, the development of vacuum packaging devices is restricted by the low adsorption capacity and long activation time of 2D planar thin film getter. In view of this article, a composite porous scaffold getter which can be activated rapidly by induction heating was proposed. Through experimental investigation, the Si-Ni-Al composite scaffold with a pore distance of 650 nm and a pore diameter of over 700 nm was finally prepared. The test results showed that the prepared composite porous scaffold base getter could be heated to over 700 ℃ by micro-flexion induction within 15s to realize the activation and reactivation of the Ti film getter. The composite porous getter can adsorb gas molecules with a mass of 6.01% with its maximum adsorption rate of $89.4 \times 10^{-4} mg/(s.g)$.

Index Terms—Ni-AAO composite scaffold getter, SSA, TGA, adsorption rate, induction heating.

I. INTRODUCTION

Micro-electro-mechanical system (MEMS) makes traditional sensors more miniaturization, and it has significant advantages such as small size, low power consumption, high sensitivity, good dynamic range and low production cost[1-3].After the vacuum encapsulation of MEMS sensor, the measurement of vacuum degree in the vacuum chamber and the maintenance of vacuum have great influence on the quality and life of the device [4-6].

Getter is one of the most commonly used methods to maintain the vacuum degree of MEMS devices. It can absorb active gases by physical and chemical methods in the vacuum or inert gas environment during the packaging process of MEMS devices. [7-10]However, the activation of the existing getter requires a long time of high-temperature vacuum annealing, which in many cases is in contradiction with the packaging temperature of the device, so that the getter cannot be fully activated and the adsorption capacity cannot be fully exerted, and the cost is high to active getter[11-13]. Therefore, many researchers have explored the activation of getter, including the selection of lower activation temperature of the material, such as the getter material doping and so on. However, these methods are complex and cannot achieve getter reactivation. [11-13]

In this article, we puts forward electromagnetic induction heating way to activate getter, sputtering, electroplating, secondary anodic oxidation and so on

preparation technology was prepared on the silicon substrate success Si-Ni-Al multilayer composite porous anodic aluminum oxide (AAO) getter, the getter by composite induction heating layer structure of the building of the getter rapid activation and local heating function, using the electromagnetic induction heating needed to build a getter activation micro zone temperature field, which can effectively solve the problem of activation temperature and packaging of contradictions, and it has large specific surface area, can obtain high pumping speed and high absorption capacity, solve the contradiction between small space and large suction capacity in MEMS packaging. [14]

II. EXPERIMENTAL

A. Preparation of induction heating layer

The Ni-AAO composite scaffold getter structure is composed of seed layer、micro-zone induction heating layer、AAO Scaffold layer and getter, as shown in Figure 1. The Ti/Cu layer is used as the conductive seed layer for the preparation of the micro-zone induction heating layer, the micro-zone induction heating [15, 16] layer is prepared by electroplating, and AAO is used to deposit getter to increase its getter capacity. Micro-zone induction heating layer preparation is made by using electroplating solution (Table I) and MP-01BR-CU wafer electroplating mechanism. The current density is 1A/dm², and the electroplating effect is good after 30 minutes. The surface roughness can be measured by DEKTAK-XT, between 10 and 20nm，which lays a good foundation for the next production process.

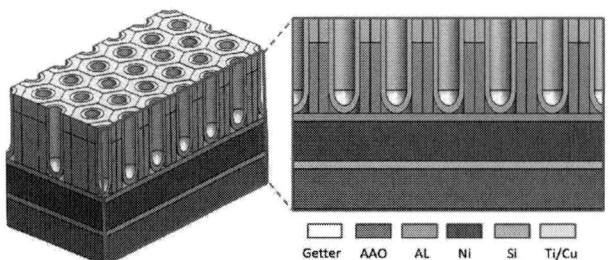

Figure 1. Schematic diagram of Ni-AAO Composite Scaffold Getter

TABLE I. COMPOSITION OF SOLUTION

Element	Concentration
$NiSO_4 \cdot 6H_2O$	300g/L
$NiCl_2 \cdot 6H_2O$	45g/L
H_3BO_3	35g/L
$C_6H_5COSO_2NH$	2g/L
$C_{12}H_{25}SO_4Na$	0.2g/L
$C_6H_5O_2SNa$	0.2g/L

*Corresponding author: Lingyun Wang, Department of Mechanical and Electrical Engineering, Xiamen University, Xiamen, China, phone: 0592-218-5927; fax: 0592-218-5927; e-mail: wangly@xmu.edu.cn.

1.Lujiang Liu/ Zhenxiang Bu/ Kuixi Wu, Department of Mechanical and Electrical Engineering, Xiamen University, Xiamen, China.

2.Wenlong Lv/ Dandan Gu, Pen-Tong Sah Institute of Micro-Nano Science and Technology, Xiamen University, Xiamen, China.

B. Preparation of Ni-AAO composite scaffold getter

In order to apply induction heating to NEG, it is necessary to prepare AAO on the induction heating layer. The preparation process is shown as follows: firstly, an Al film with a thickness of 2μm was sputtered on the surface of the nickel layer by magnetron sputtering system EXPLORER 14. Secondly, it was anodized in the 0.15M citric acid system at 4℃ with an oxidation voltage of 80V. The porous morphology was obtained initially, but the pore size was small and relatively disordered at this time (Fig.2 (a)). The oxidation film of the first anodic oxidation was removed by soaking in the mixture of 6wt% phosphoric acid and 1.8wt% chromic acid at 60℃ for 15min. On this basis, the second anodic oxidation experiment was carried out. The technological parameters of the second anodic oxidation were completely consistent with those of the first anodic oxidation. AAO was prepared by secondary oxidation on active substrates, and the homogeneous porous structure controlled at 600-700nm can be obtained without any other treatment (Fig.2 (b)).

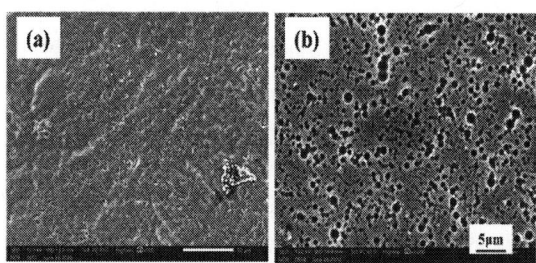

Figure 2.　SEM image of AAO on Ni

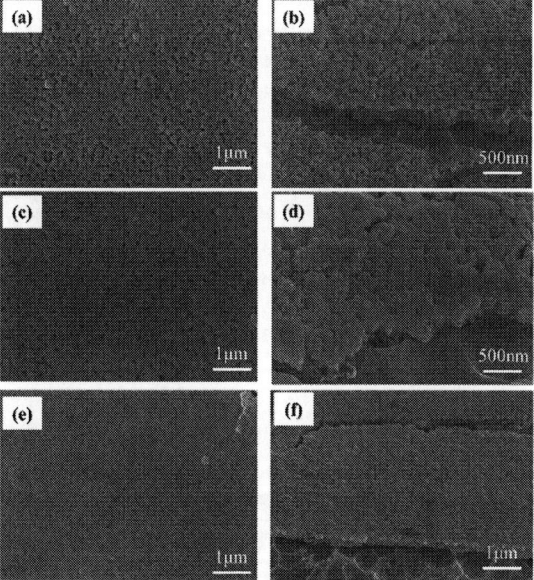

Figure 3.　SEM images of Ti film getter with different thickness deposited on Ni-AAO composite:(a)、(b)350nm;(c)、(d)450nm;(e)、(f)550nm

After that, we sputtered Ti film getter with thickness of 350nm, 450nm and 550nm on the composite scaffold. According to SEM observation, with the increase of sputtering thickness, the pore structure becomes less and less obvious. It can be seen from the figure that the pore diameter of Ti film getter with a thickness of 350nm is still relatively large. The measured pore diameter is over 200nm and the maximum pore diameter can reach 500nm. It has not been fully utilized and can be further deposited. When the 450nm thick Ti film getter was deposited on the AAO scaffold, the getter holes were uniform and orderly, with a pore diameter of about 200nm, of which the maximum pore diameter was up to 400nm and the deposition depth was up to 603nm; At the sputtering parameter of 550nm, the pore structure is basically covered by Ti film without obvious morphological characteristics. Therefore, it can be concluded that the porous scaffold structure is conducive to the deposition of 450nm thickness Ti getter.

III. RESULTS AND DISSCUSSIONS

A. Specific surface area test

In this paper, Tristar-3020 automatic nitrogen adsorption apparatus was used to measure the specific surface area of different 3D nano-scaffolds. The specific surface measurement range was nitrogen adsorption 0.01m2/g to no upper limit, the pore size analysis range was 3.5 Å ~ 5000 Å, the repeatability was better than 1%, and the test sample mass was about 0.1g. Fig.4 shows the N_2 adsorption-desorption[17] curves of AAO nano-scaffold and corresponding getter. We can see from the diagram, at smaller relative pressure, the adsorption capacity of N_2 increases slowly with the increase of relative pressure. When the relative pressure is greater than 0.8, the adsorption[18, 19] amount of nitrogen increases sharply, and the adsorption isotherm shows a sudden jump, indicating that the sample has a macro-porous structure, and most of its hysteresis rings show a vertical state, indicating that the sample structure is straight cylinder shape, which is consistent with the actual situation of pore morphology of porous alumina. Moreover, the N_2 adsorption capacity of porous alumina scaffold was higher than that of the adsorbent deposited. This was because the deposition of getter made the pores of porous alumina smaller. In the middle and low pressure section, the desorption curve was in good agreement with the adsorption curve, indicating that the pore size was uniform.

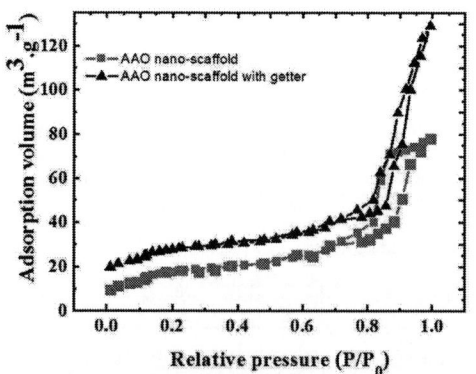

Figure 4.　Nitrogen adsorption-desorption curve of AAO getter

Table II shows that the specific surface area of porous alumina scaffolds can be reached 35.632m²/g. After the deposition of 450nm Ti film getter, the specific surface area decreases, but still has 26.685 m²/g, which indicates that porous alumina scaffold getter has larger specific surface area than the Non-scaffold sample(0.1476m²/g), and also has a strong adsorption performance.

TABLE II. THE SPECIFIC SURFACE AREA OF DIFFERENT AAO AND AAO WITH GETTER

Sample	BET(m²/g)
Non-scaffold	0.1476
AAO	35.632
AAO with getter	26.685

B. Thermogravimetric Analysis(TGA)

As can be seen from Figure 5, the adsorption trend of AAO nano-scaffold-based getter is basically the same, and there is almost no change of the adsorption mass at 550℃ due to amount of water vapor and the existence of residual gas. With temperature increasing, the sample mass begins to increase, which is due to the chemisorption and diffusion effect of the sample. The mass was measured by STA449F3, and its accuracy can be reached 1e⁻⁸g.

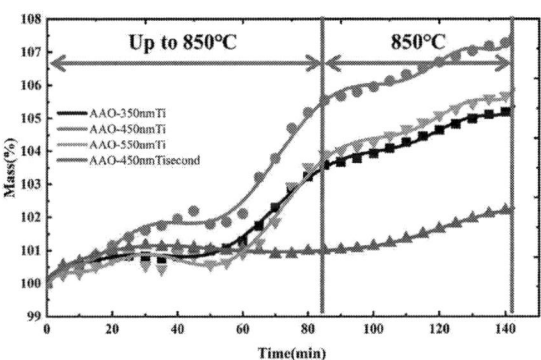

Figure 5. Thermogravimetric analysis of different samples

From the table III, the adsorption capacity [20] of 350nm Ti getter (4.85%) and 550nm Ti getter (5.22%) were less than 450nmTi (7.51%) getter due to their limited diffusion depth and inability to effectively utilize porous nano-scaffolds, which is also consistent with the SEM. In addition, when the 450nm Ti getter was reactivated under the same conditions, it still had little adsorption capacity. However, since the first activation had basically reached the saturation state, the adsorption capacity of getter decreased significantly.

TABLE III. ADSORPTION CAPACITY AND ADSORPTION RATE OF DIFFERENT SAMPLES

ample	Adsorption capacity (%)	Adsorption rate (10⁻⁴mg/(s*g))
350nm Ti	4.85	57.7
450nm Ti	7.51	89.4
550nm Ti	5.22	62.14
reactivation	1.14	13.57

C. The composite scaffold getter induction heating test

Fig. 6 shows the temperature variation of induction heating and reheating of porous AAO on silicon base and on nickel base. The samples were heated by DDCGP series high frequency induction heating electric equipment, and the heating current was 20A. On the same time, the temperature changes during the induction heating process are monitored in real time by the infrared thermal imager FLIR T400.

As can be seen from Fig. 6, the getter film on silicon substrate can be rapidly heated to 250℃ in the first 15s, and the sample can be reheated to 270℃, but it does not reach the optimal temperature for activating the getter on titanium film. As can be seen from the composite scaffold getter, the temperature of the porous AAO film on the electroplated nickel base can reach 600℃ quickly within 10s of heating，and its highest temperature can reach 745℃ (Fig.7), which corresponds to the activation and reactivation of the titanium film getter. By comparing these two curves, it can be seen that the cooling time of porous AAO on the electroplated nickel base lasts for more than 50 seconds, which is conducive to the subsequent activation and reactivation of the titanium film getter.

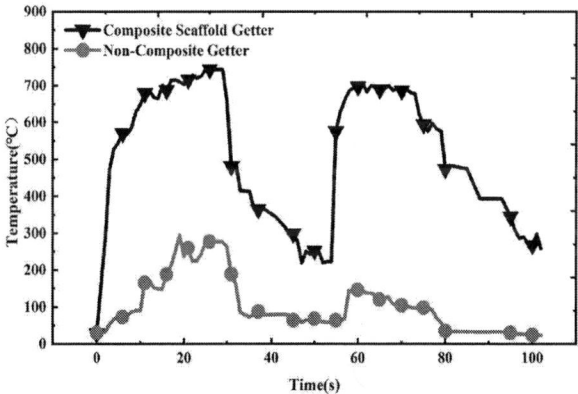

Figure 6. Induction heating and reheating of different sample structures

Figure 7. Maximum temperature for induction heating test by LFA467

IV. CONCLUSION

This paper mainly solves two problems: the first is the contradiction between small MEMS package volume and large getter capacity; the second is the contradiction between MEMS package temperature and getter activation temperature. To solve the first problem, porous AAO nano-scaffold was used to deposit the getter, so that the specific surface area of the getter reached $26.685 m^2/g$ and the getter capacity reached 7.51%, which effectively solved the problem of small getter capacity. To solve the second problem, we used The Ni substrate as the induction heating layer, and the local temperature measured by the induction heating temperature could reach more than 700℃ within 15S, effectively realizing the rapid activation and reactivation of the getter without damaging the structure of the device. In this paper, the induction heating of large aperture getter can effectively increase the adsorption rate and adsorption capacity of getter, and thus reduce the air damping of MEMS devices, improve their service life and quality factor.

ACKNOWLEDGMENT

This work was financially supported by The National Natural Science Foundation of China (No.61674125) and Aeronautical Science Funds (No.20180868001) and The Industry-University Research Collaboration of Xiamen (No. 3502Z20203001)

REFERENCES:

[1] C. Zhou, D. Li, H. Zhou, K. Zhang, and S. Cao, "Non-evaporable Getter Films for Vacuum Packaging of MEMS Devices: an Overview," Materials Reports, 2019.

[2] W. D. Sawyer, M. S. Prince and G. J. Brown, "SOI bonded wafer process for high precision MEMS inertial sensors," Journal of Micromechanics & Microengineering, vol. 15, p. 1588, 2005.

[3] R. Naveh and E. Zussman, "Vacuum Packaging of MEMS by Self-Assembly," Journal of Electronic Packaging, Transactions of the ASME, vol. 134, p. 011009, 2012.

[4] M. F. Monteiro, J. S. D. Aguila, C. D. O. Pessoa, and R. A. Kluge, "VACUUM PACKAGING IS EFFICIENT TO REMOVE ASTRINGENCY AND TO MAINTAIN THE FIRMNESS OF 'GIOMBO' PERSIMMON," Rev.bras.frutic, vol. 39, 2017.

[5] Z. Gan, D. Huang, X. Wang, D. Lin, and S. Liu, "Getter free vacuum packaging for MEMS," Sensors & Actuators A Physical, vol. 149, pp. 159-164, 2009.

[6] S. Askari, M. H. Asadian, K. Kakavand, and A. M. Shkel, "Vacuum sealed and getter activated MEMS Quad Mass Gyroscope demonstrating better than 1.2 million quality factor," in 2016 IEEE International Symposium on Inertial Sensors and Systems, 2016.

[7] R. Mehalso, "MEMS packaging and microassembly challenges," 1999.

[8] Y. Xu, J. Cui, H. Cui, H. Zhou, Z. Yang, and J. Du, "Influence of deposition pressure, substrate temperature and substrate outgassing on sorption properties of Zr‐Co‐Ce getter films," Journal of Alloys and Compounds, vol. 661, pp. 396-401, 2016.

[9] O. B. Malyshev, K. J. Middleman, J. S. Colligon, and R. Valizadeh, "Activation and measurement of nonevaporable getter films," Journal of Vacuum ence & Technology A Vacuum Surfaces & Films, vol. 27, pp. 321-327, 2009.

[10] R. K. Sharma, A. K. Sinha, Jagannath, D. C. Basak, and S. K. Gupta, "Surface studies and measurement of pumping characteristic of NEG coating (Ti-V-Zr)," Proceedings International Symposium on Discharges & Electrical Insulation in Vacuum Isdeiv, pp. 529-532, 2014.

[11] K. Celebi, "Chemical vapor deposition of graphene on copper," Journal of the Electrochemical Society, vol. 140, pp. 1793-1801, 2013.

[12] A. Conte, M. Moraja, G. Longoni, and A. Fourrier, "High and stable Q-factor in resonant MEMS with Getter film," in MOEMS-MEMS 2006 Micro and Nanofabrication, 2006.

[13] C. Li, J. L. Huang, R. J. Lin, H. K. Chang, and J. M. Ting, "Fabrication and characterization of non-evaporable porous getter films," Surface and Coatings Technology, 2005.

[14] S. Lin, X. Huang, D. Gu, W. Lv, and L. Wang, "Investigation on Nickel-Based Nano-Scaffold Getter With Induction Heating and Rapid Activation," IEEE Transactions on Nanotechnology, vol. PP, p. 1-1, 2019.

[15] M. J. A, R. I. A, L. M. P. B, and J. P. B. A, "Effect of surface functionalization on the heating efficiency of magnetite nanoclusters for hyperthermia application," Journal of Alloys and Compounds, 2020.

[16] A. Kucukkomurler, "Environmental, low cost, energy efficient, electromagnetic indoor induction space heating system design," in Eurocon, Eurocon 09 IEEE, 2009.

[17] S. J. Rothenberg, D. K. Flynn, A. F. Eidson, J. A. Mewhinney, and G. J. Newton, "Determination of specific surface area by krypton adsorption, comparison of three different methods of determining surface area, and evaluation of different specific surface area standards," Journal of Colloid & Interface Science, vol. 116, pp. 541-554, 1987.

[18] Z. C. Feng, C. Wang, D. Zhou, and D. Zhao, "Variation law of adsorption heat of methane and coal with inhomogeneous potential well," Adsorption Science and Technology, vol. 36, p. 026361741878894, 2018.

[19] A. A. Pribylov and K. A. Murdmaa, "Gas Adsorption onto an MN-270 Polymer Adsorbent within a Supercritical Range of Temperatures and Pressures," Protection of Metals and Physical Chemistry of Surfaces, vol. 56, pp. 245-251, 2020.

[20] L. Liu, C. Jin, L. Li, C. Xu, and L. An, "Coalbed methane adsorption capacity related to maceral compositions," Energy Exploration & Exploitation, p. 014459871987032, 2019.

[21] J. D. Cui, H. T. Wu, Y. Zhang, Y. H. Xu, and Z. M. Yang, "Structure and properties of ZrCoCe getter film with Pd protection layer," Rare Metals, 2020.

[22] L. N. Dinh, H. N. Sharma, S. M. Matt, W. Mclean, and R. S. Maxwell, "Phase Change Driven Negative Activation Energies in Pd/Carbon-Based/Organic Getter Hydrogenation Reactions," The Journal of Physical Chemistry A, vol. 124, 2020.

April 25-29 , 2021   Xiamen, China

Bio - Inspired Ionic Skin for Theranostics

Peiyi Wu

State Key Laboratory for Modification of Chemical Fibers and Polymer Materials, College of Chemistry, Chemical Engineering and Biotechnology, Donghua University

Website : www.peiyiwu.cn

ABSTRACT

The first - generation ionic skins demonstrate great advantages in the tunable mechanical properties, high transparency, ionic conductivities, and multiple sensory capacities. However, little attention is paid to the interfacial interactions among the ambient environment, natural organisms, and the artificial skins. In particularly, current ionic skins based on traditional synthetic hydrogels suffer from dehydration in vitro and lack of substance communication channels with biological tissues. Herein, this work develops a bio - inspired hydrogel to address these key challenges. The hydrogel is designed with natural moisturizing factors to lock water, biomineral ions to transmit signals, and biomimetic gradient channels to transport substances from non - living to living interfaces. It is stable in ambient condition, adhesive and hydrated on mammal skins, and capable of non - invasive point - to - point theranostics. This theranostic ionic skin realizes sensitive detection, enhanced treatment efficacy, and reduced side effects toward major diseases in vitro. It will shed light on the hydrogel bioelectronics with excellent biocompatibility, bio - protection, and bio - integration for human–machine interfaces and intelligent theranostics.

BIOGRAPHY

Dr. Peiyi Wu is currently a Professor and Dean at College of Chemistry, Chemical Engineering and Biotechnology, Donghua University. Dr. Wu received his Ph.D. degree in Chemistry from University Essen, Germany. He was a Professor(2000-2017) and dean(2005-2012) of the Department of Macromolecular Science, Fudan University. Prof. Wu is active in the fields of two dimensional correlation spectroscopy and smart materials. He is corresponding-author of 300 peer-reviewed scientific publications including Science Advances, Nature Communications and Advanced Materials, *etc.*.

Professor Wu is a distinguished young scholar of the national natural science foundation of China(2004) and fellow of the royal society of chemistry

978-1-6654-3008-1/21 $31.00 © 2021 IEEE

978-1-6654-3008-1/21 $31.00 © 2021 IEEE

Proceedings of the 16th Annual IEEE International
Conference on Nano/Micro Engineered and Molecular Systems
April 25-29, 2021

Ultra-flexible and highly transparent hydrogel-based triboelectric nanogenerator for physiological signal monitoring*

Jiahao Yu, Zhensheng Chen, Haozhe Zeng, Bowen Ji, Zixuan Wu, Jin Wu, Yunjia Li, Jianbing Xie, Honglong Chang, and Kai Tao

Abstract— Wearable electronics and electric skin attract increasing attention, which demand high transparency, extensibility, and biocompatibility. In this work, a transparent and highly stretchable pyramidal hydrogel-based triboelectric nanogenerator (HTG) for tactile sensation and neck movement detection is reported. Hydrogel can be easily stretched to 800% of the initial length. The output property of HTG could reach up to 360V and 1.27mW. Meanwhile, a slight signal like blowing could also be detected by the proposed device. Besides, the charging time of the capacitor reveals the dependence of output property and frequency. This work provides a brilliant method for self-powered physiological signal monitoring.

I. INTRODUCTION

With the development of Internet of things (IoT), increasing amounts of sensors would be utilized to render human beings' life more convenient [1-4]. Due to the challenges in reducing the volume, maintaining, and recycling, the traditional power might be unreliable for such a tremendous sensors array, especially for wearable electronic devices. Besides, the pollution problems of the batteries are nonnegligible [5-8]. Wang et al. invented triboelectric nanogenerators (TENG) in 2012 [9], which could harvest energy from ambient environment and human motions, such as rain drops, winds, ocean wave, oscillation and walk [10-11]. Compared with other renewable energy, the TENG possesses the merits of low cost, stability, and a wide choice of materials [12-14]. In other words, nanogenerators are seldom influenced by the factors like weather and season. They can work efficiently both in the daytime and nighttime.

As for the previous materials of TENG, carbon nanotube, graphene and AgNWs were widely used in conductive electrode, but the applications of these materials are severely limited in the field of wearable devices and flexible sensors due to the transparency, stretchability, durability, and biocompatibility [15-18]. The ionic conductive hydrogel could be an alternative material for electrode in contact-separation TENG (CS-TENG). The double network (DN) ionic hydrogel is one of the research

focuses in wearable devices, which possesses ultra-high transparency, excellent stretchability, wonderful electrical conductivity, and admirable biocompatibility [19-20]. After proper treatment, hydrogel could become anti-drying and anti-freezing, expanding the operating temperature range of hydrogel-based devices [21-22].

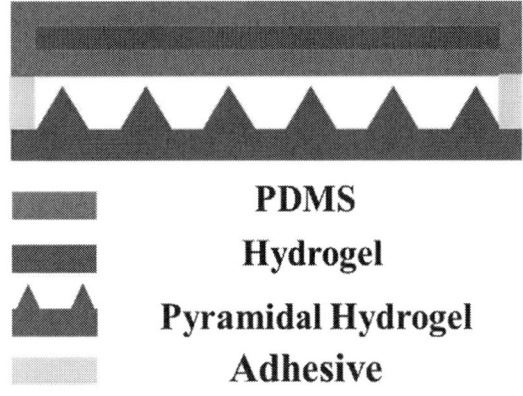

Figure 1. The structure of hydrogel-based TENG.

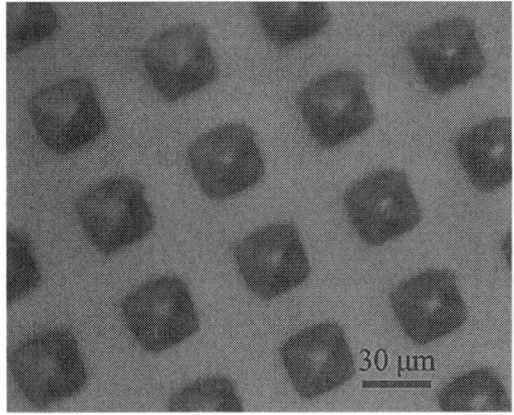

Figure 2. Optical microscope image of pyramidal hydrogel.

---

*Resrach supported by National Natural Science Foundation of China Grant (No. 51705429).

Jiahao Yu, Zhensheng Chen, Haozhe Zeng, Bowen Ji, Jianbing Xie, Honglong Chang, and Kai Tao are with Ministry of Education Key Laboratory of Micro and Nano Systems for Aerospace, School of Mechanical Engineering, Northwestern Polytechnical University, Xi'an 710072, China (corresponding author to provide phone: +86-29-88460434; fax: +86-29-88460434; e-mail: taokai@ nwpu.edu.cn).

Bowen Ji is also with Unmanned System Research Institute, Northwestern Polytechnical University, Xi'an 710072, China (e-mail: bwji@nwpu.edu.cn).

Zixuan Wu and Jin wu are with School of Electronics and Information Technology, Sun Yat-sen University, Guangzhou 510275, China (e-mail: wujin8@mail.sysu.edu.cn).

Yunjia Li is with School of Electrical Engineering, Xi'an Jiaotong University, Xi'an 710049, China (e-mail: liyunjia@xjtu.edu.cn).

Jiahao Yu and Zhensheng Chen contributed equally.

978-1-6654-3008-1/21 $31.00 © 2021 IEEE

In this paper, the hydrogel electrodes accompanied by polydimethylsiloxane (PDMS) dielectric are introduced to the nanogenerator. Figure 1 illustrates the structure of the hydrogel-based triboelectric nanogenerator (HTG) vividly. As for the top sandwiched structure, hydrogel is embedded in PDMS as the top electrode, where PMDS could serve not only as the dielectric layer but also as a protective layer of the hydrogel electrode. The bottom electrode of HTG is the pyramidal hydrogel and it plays the role of both electrode and triboelectric layer. Surface microstructures were widely used in dielectric layer to reduce the elastic resistance of the material, generate more surface charges and improve the sensitivity of the sensor [23-25].

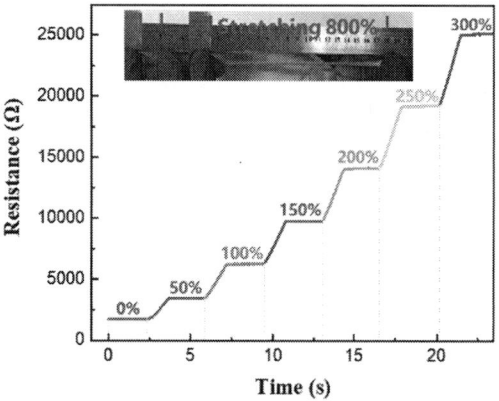

Figure 3. Resistance variation of the hydrogel sensor under different tensile strains.

Figure 4. Image shows the transparency of the device.

Here, surface modified electrode instead of the dielectric is introduced to CS-TENG; the optical microscope image of the pyramid-like structured hydrogel is shown in figure 2. Moreover, the resistance of the hydrogel varies with the stretched length, and figure 3 demonstrates the dependence of them. The characteristic accords hydrogel the function of strain sensing. The inset image of figure 3 shows the prominent extensibility of hydrogel, which could be stretched to 800% of its initial

length easily. The transparency of the whole device could be seen clearly in figure 4. Both the excellent stretchability and high transparency render the hydrogel a promising material in the field of flexible sensors and wearable electronic devices. Besides, excellent biocompatibility makes it an attractive choice.

II. DEVICE FABRICATION

A. Fabrication of the sandwiched top electrode

Figure 5 illustrates the fabrication processes of the sandwiched top electrode. Firstly, PDMS mold I and mold II were prepared by three-dimensional printing, followed by hydrophobic treatment in order to take out PDMS easily. And then, pour the previous PDMS prepolymer solution into the mold I. After curing at 70℃ for 2 hours, the PDMS was transferred to mold II and then pour hydrogel into mold II. Finally, pour the PDMS solution and cure once more.

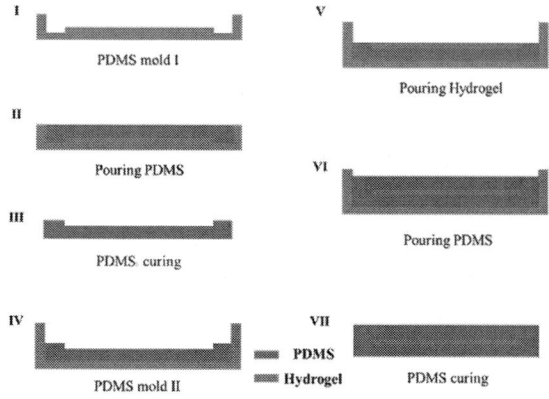

Figure 5. Fabrication processes of sandwiched top electrode.

B. Fabrication of the pyramidal bottom electrode and the device

The silicon wafers with pyramid grooves were prepared by etching initially. Then the bottom pyramidal hydrogel was spin-coated on Si wafer. After forming hydrogel, the top sandwiched electrode and bottom hydrogel electrode were bonded, which brought to an end the fabrication of HTG.

III. OPERATING PRINCIPLES

A. General principle of the device

The working principle of the hydrogel-based triboelectric nanogenerator can be summarized by the coupling of triboelectric and electrostatic induction. First, the presence of adhesive provides the gap between the top PDMS and the bottom pyramid-like structured hydrogel at the initial stage. When the external force is applied to HTG, the two friction layers contact each other. Meanwhile, the triboelectric effect occurs owing to the different electron affinity between PDMS and hydrogel. Since PDMS is easier to gain electrons, electrons would be transferred from top pyramidal hydrogel to PDMS. Once released, sandwiched hydrogel will generate positive charges due to electrostatic induction. Meanwhile, the electrons will move to the bottom pyramidal hydrogel via an external circuit because of the electric potential difference between the two electrodes. When the force is applied once more, the electrons will flow back to the top sandwiched electrode. Periodic output is generated by cyclical of external forces.

The working principle of HTG was verified by using finite element software COMSOL. Here, two working modes were simulated, including pressing and stretching mode, which are exactly the working patterns of HTG. For the sake of the beauty of the image, the top plate refers to the bottom pyramid-like structured hydrogel and the bottom plate represents the sandwiched electrode. In the view of the micro-scale pyramid, the plate is divided into 187656 units to render the pyramid comprehensively studied. The primary hypothesis is that the total charges of the two electrodes are constant during pressing and stretching.

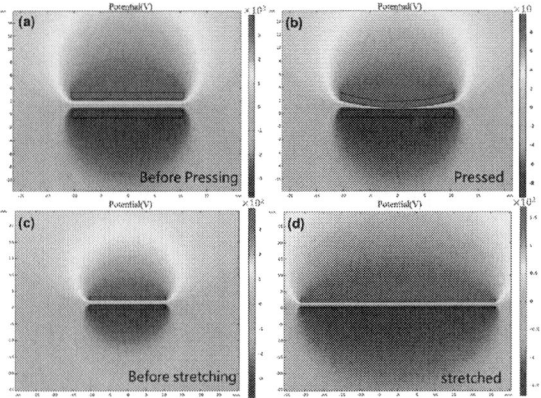

Figure 6. Simulation results show the potential distribution of HTG: (a)The initial state before pressing; (b)Pressed state; (c)The initial state before stretching; (d)Stretched state.

B. The simulation of pressing mode

As for the pressing pattern, Figure 6 (a) illustrates the device's initial stage before pressing. The red color indicates a positive electric potential, while blue means the negative potential. Figure 6 (b) shows the pressed state of HTG. Once pressed, two electrodes approach each other and the potential decreases with the distance between two electrodes declining. In other words, there is a positive correlation between potential and the distance. When the distance reaches its highest value, HTG outputs the maximum voltage accordingly, which corresponds to the electrical theory of CS-TENG.

C. The simulation of stretching mode

Another working pattern of the HTG is stretching mode, figure 6 (c) and (d) demonstrate the state before and after stretching, respectively. It can be seen clearly that the potential decreases with the extended length increasing. That could be attributed to the change of charge density σ and the distance h between two electrodes with the continuous increase of plate area. As shown in (1), the voltage V_{OC} has a close relation to h and σ. And $\mathcal{E}$ stands for the dielectric constant.

$$V_{OC} = h\sigma / \mathcal{E} \qquad (1)$$

During stretching, charge density σ decreases due to the increase of area of electrode on the basis of constant quantity of charges. Meanwhile, the distance h decreases owing to the presence of adhesive during stretching. In other words, the two electrodes are connected in the vertical direction. Once stretched, two electrodes would approach each other. In short, both the distance h and charge density σ descend during stretching and V_{OC} decreases accordingly. As shown in figure 6 (c) and (d), the simulation results are in agreement with the theoretical analysis.

IV. CHARACTERIZATION

The corresponding output of the proposed hydrogel-based triboelectric nanogenerator is also characterized. Figure 7 demonstrates the output voltage of the fabricated device by continuous finger tapping. Owing to the inhomogeneity of the tapping frequency and pressure, the output varies slightly. The maximum output voltage could reach up to 360V approximately and meanwhile the output power attains 1.27 mW. The inset image in figure 7 shows that the proposed HTG device can light up 34 LEDs easily. Moreover, the brightness of LED is also considerable. High output voltage could be attributed to the microstructure of bottom hydrogel electrode. Pyramid-like structure improves the contact area between hydrogel and PDMS effectively and then the amounts of generated charges augment. Besides, a rougher friction surface could induce the triboelectric effect more efficiently.

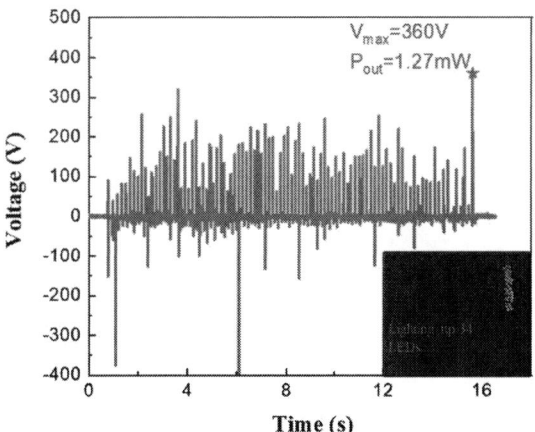

Figure 7. The output voltage of the proposed HTG device.

The proposed hydrogel-based TENG is also used to charge capacitor via a rectifier circuit. To minimize the influence of other factors, a vibration platform is employed to drive the HTG at different frequencies. As shown in figure 8, the charging time varies from the input frequency of HTG. Generally, the charging time is shorted with the increase of the external force frequency, which could be attributed to the current increase.

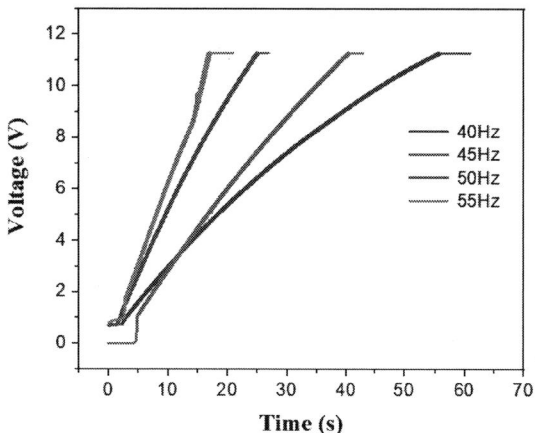

Figure 8. Charging the capacitor by using HTG under different frequencies.

V. Testing Results

Owing to the high transparency, excellent flexibility, and favorable biocompatibility, hydrogel-based triboelectric nanogenerators could be an alternative for wearable devices. The feasibility of HTG as the physiological sensor is further studied by detecting signals including tactile sensation, neck motion and even slight blowing. Figure 9 illustrates the constant and stable output of the HTG device under finger touching. The inset image is the testing picture of tactile sensation. To demonstrate the sensitivity of HTG, the force of touching is carefully controlled and the frequency reaches up to 5 Hz approximately.

The proposed device could also be attached to human body parts directly, such as the neck in this paper. As shown in figure 10, the stable electrical output is detected with the movement of the neck. That endows the device with the function of drivers' behavior monitoring and feedback, which means the HTG could save lives during driving to some extent.

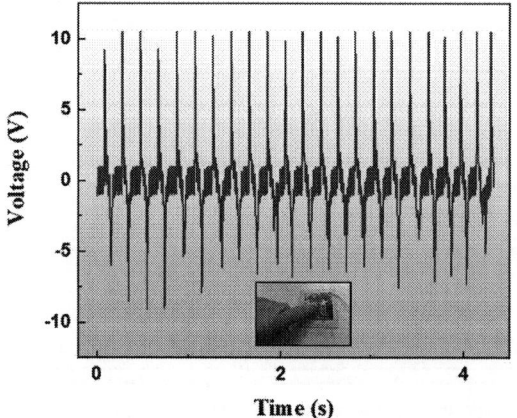

Figure 9. HTG serves as tactile sensor.

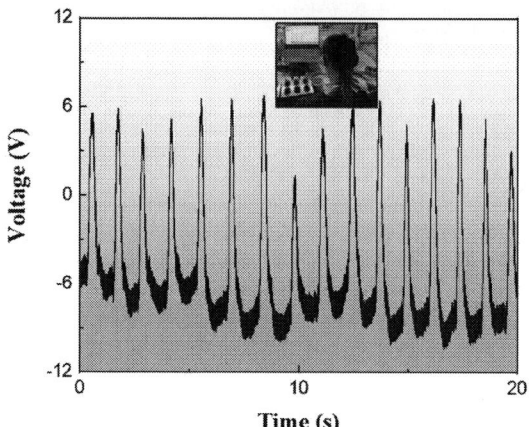

Figure 10. HTG serves as sensor for neck motion detection.

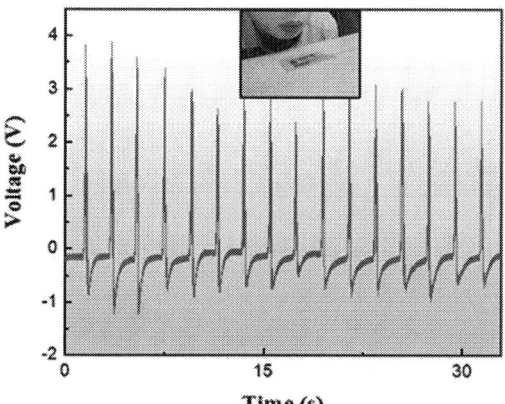

Figure 11. HTG serves as sensor for blowing detection.

Figure 11 demonstrates the testing result of HTG for blowing detection. It can be seen clearly that HTG could generate a stable output as a response to human blowing, which means that the proposed device could not only output an admirable voltage but also detect slight motion. That could be attributed to the presence of pyramid-like structured hydrogel electrode. Once blown, the edge of pyramid could contact and then separate from PDMS dielectric. Therefore，HTG could generate constant output with the blowing.

VI. Conclusion

In this paper, a novel pyramidal hydrogel-based triboelectric nanogenerator for physiological signal monitoring has been successfully developed. The proposed device consists of top sandwiched PDMS-hydrogel-PDMS structure and a pyramidal bottom ionic hydrogel electrode, which could output considerable voltage and detect the slight motion like blowing at the same time. This work demonstrates that HTG has excellent potential in wearable sensors owing to its high transparency, excellent stretchability, admirable biocompatibility，and favorable sensitivity.

978-1-6654-3008-1/21 $31.00 © 2021 IEEE

ACKNOWLEDGMENT

This research is supported by National Natural Science Foundation of China Grant (No. 51705429), the Fundamental Research Funds for the Central Universities, Guangdong Natural Science Funds Grant (2018A030313400), Science, Technology and Innovation Commission of Shenzhen Municipality.

REFERENCES

[1] Y. Li, K. Tao, B. George and Z. Tan, "Harvesting Vibration Energy: Technologies and Challenges," *IEEE Industrial Electronics Magazine*, doi: 10.1109/MIE.2020.2978219.

[2] X. Pu, M. Liu, X. Chen, J. Sun, C. Du, Y. Zhang, J. Zhai, W. Hu, and Z. L. Wang, "Ultrastretchable, transparent triboelectric nanogenerator as electronic skin for biomechanical energy harvesting and tactile sensing," *Sci Adv*, vol. 3, no. 5, p. e1700015, May 2017.

[3] K. Tao, J. Wu, L. Tang, X. Xia, S. W. Lye, J. Miao, and X. Hu, "A novel two-degree-of-freedom MEMS electromagnetic vibration energy harvester," *Journal of Micromechanics and Microengineering*, vol. 26, no. 3, 2016.

[4] K. Tao, H. Yi, L. Tang, J. Wu, P. Wang, N. Wang, L. Hu, Y. Fu, J. Miao, and H. Chang, "Piezoelectric ZnO thin films for 2DOF MEMS vibrational energy harvesting," *Surface and Coatings Technology*, vol. 359, pp. 289-295, 2019.

[5] H. Wang, M. Han, Y. Song, and H. Zhang, "Design, manufacturing and applications of wearable triboelectric nanogenerators," *Nano Energy*, vol. 81, 2021.

[6] K. Tao, J. Wu, L. Tang, L. Hu, S. W. Lye, and J. Miao, "Enhanced electrostatic vibrational energy harvesting using integrated opposite-charged electrets," *Journal of Micromechanics and Microengineering*, vol. 27, no. 4, 2017.

[7] J. B. Yu, X. J. Hou, M. Cui, S. N. Zhang, J. He, W. P. Geng, J. L. Mu, and X. J. Chou, "Highly skin-conformal wearable tactile sensor based on piezoelectric-enhanced triboelectric nanogenerator," *Nano Energy*, vol. 64, Oct 2019.

[8] K. Tao, H. P. Yi, Y. Yang, L. H. Tang, Z. S. Yang, J. Wu, H. L. Chang, and W. Z. Yuan, "Miura-origami-inspired electret/triboelectric power generator for wearable energy harvesting with water-proof capability," *Microsystems & Nanoengineering*, vol. 6, no. 1, Aug 10 2020.

[9] F.-R. Fan, Z.-Q. Tian, and Z. Lin Wang, "Flexible triboelectric generator," *Nano Energy*, vol. 1, no. 2, pp. 328-334, 2012.

[10] X.-S. Zhang, M.-D. Han, B. Meng, and H.-X. Zhang, "High performance triboelectric nanogenerators based on large-scale mass-fabrication technologies," *Nano Energy*, vol. 11, pp. 304-322, 2015.

[11] T. Jing, B. Xu, Y. Yang, M. Li, and Y. Gao, "Organogel electrode enables highly transparent and stretchable triboelectric nanogenerators of high power density for robust and reliable energy harvesting," *Nano Energy*, vol. 78, 2020.

[12] T. Tat, A. Libanori, C. Au, A. Yau, and J. Chen, "Advances in triboelectric nanogenerators for biomedical sensing," *Biosens Bioelectron*, vol. 171, p. 112714, Jan 1 2021.

[13] X. Cheng, L. Miao, Y. Song, Z. Su, H. Chen, X. Chen, J. Zhang, and H. Zhang, "High efficiency power management and charge boosting strategy for a triboelectric nanogenerator," *Nano Energy*, vol. 38, pp. 438-446, 2017.

[14] K. Tao, L. Tang, J. Wu, S. W. Lye, H. Chang, and J. Miao, "Investigation of Multimodal Electret-Based MEMS Energy Harvester With Impact-Induced Nonlinearity," *Journal of Microelectromechanical Systems*, vol. 27, no. 2, pp. 276-288, 2018.

[15] S. Han, C. Liu, X. Lin, J. Zheng, J. Wu, and C. Liu, "Dual Conductive Network Hydrogel for a Highly Conductive, Self-Healing, Anti-Freezing, and Non-Drying Strain Sensor," *ACS Applied Polymer Materials*, vol. 2, no. 2, pp. 996-1005, 2020.

[16] K. Tao, S. W. Lye, J. Miao, L. Tang, and X. Hu, "Out-of-plane electret-based MEMS energy harvester with the combined nonlinear effect from electrostatic force and a mechanical elastic stopper," *Journal of Micromechanics and Microengineering*, vol. 25, no. 10, 2015.

[17] W. Fan, Q. He, K. Meng, X. Tan, Z. Zhou, G. Zhang, J. Yang, and Z. L. Wang, "Machine-knitted washable sensor array textile for precise epidermal physiological signal monitoring," *Sci Adv*, vol. 6, no. 11, p. eaay2840, Mar 2020.

[18] K. Tao, J. Miao, S. W. Lye, and X. Hu, "Sandwich-structured two-dimensional MEMS electret power generator for low-level ambient vibrational energy harvesting," *Sensors and Actuators A: Physical*, vol. 228, pp. 95-103, 2015.

[19] J. Wu, Z. Wu, X. Lu, S. Han, B. R. Yang, X. Gui, K. Tao, J. Miao, and C. Liu, "Ultrastretchable and Stable Strain Sensors Based on Antifreezing and Self-Healing Ionic Organohydrogels for Human Motion Monitoring," *ACS Appl Mater Interfaces*, vol. 11, no. 9, pp. 9405-9414, Mar 6 2019.

[20] Y. Wang, L. N. Zhang, and A. Lu, "Highly stretchable, transparent cellulose/PVA composite hydrogel for multiple sensing and triboelectric nanogenerators," *J Mater Chem A*, vol. 8, no. 28, pp. 13935-13941, Jul 28 2020.

[21] J. Wu, Z. Wu, S. Han, B. R. Yang, X. Gui, K. Tao, C. Liu, J. Miao, and L. K. Norford, "Extremely Deformable, Transparent, and High-Performance Gas Sensor Based on Ionic Conductive Hydrogel," *ACS Appl Mater Interfaces*, vol. 11, no. 2, pp. 2364-2373, Jan 16 2019.

[22] J. Wu, Z. Wu, Y. Wei, H. Ding, W. Huang, X. Gui, W. Shi, Y. Shen, K. Tao, and X. Xie, "Ultrasensitive and Stretchable Temperature Sensors Based on Thermally Stable and Self-Healing Organohydrogels," *ACS Appl Mater Interfaces*, vol. 12, no. 16, pp. 19069-19079, Apr 22 2020.

[23] B. Meng, W. Tang, Z. H. Too, X. S. Zhang, M. D. Han, W. Liu, and H. X. Zhang, "A transparent single-friction-surface triboelectric generator and self-powered touch sensor," *Energy & Environmental Science*, vol. 6, no. 11, pp. 3235-3240, Nov 2013.

[24] M. Muthu, R. Pandey, X. Wang, A. Chandrasekhar, I. A. Palani, and V. Singh, "Enhancement of triboelectric nanogenerator output performance by laser 3D-Surface pattern method for energy harvesting application," *Nano Energy*, vol. 78, 2020.

[25] J. Pignanelli, K. Schlingman, T. B. Carmichael, S. Rondeau-Gagné, and M. J. Ahamed, "A comparative analysis of capacitive-based flexible PDMS pressure sensors," *Sensors and Actuators A: Physical*, vol. 285, pp. 427-436, 2019.

Proceedings of the 16th Annual IEEE International
Conference on Nano/Micro Engineered and Molecular Systems
April 25-29, 2021

A Flexible Tactile Sensor for Three-dimensional Force Detection Based on Piezoelectric Sensing

Yunyun Luo[1,2], Libo Zhao[1,2]*, Guoxi Luo[1,2], Zhikang Li[1,2], Ping Yang[1,2], Qiankun Zhang[1,2], Qijing Lin[1,2], Hongyan Wang[3], Yongshun Wu[3], Zhuangde Jiang[1,2]

Abstract—**A flexible tactile sensor was proposed to measure the three-dimensional forces. The sensor consisted of a bump structure, top and bottom electrodes, a piezoelectric sensitive layer and a substrate layer. The top and the bottom electrodes were prepared by lift-off process. When a three-dimensional force was applied, the bottom four electrodes under the bump structure would collect different charge signals, which can be detected by the different electric current variations. In the experiments, when the forces range from 0 to 10N with different angles were applied to the bump structure, the sensor can measure forces with a best non-linearity of 1.89%. This proposed tactile sensor could be used in the intelligent robotics to measure multi-dimensional force.**

BACKGROUD

Touch sensing has attracted much attention from researchers due to the various application requirements of robotic dexterous manipulation and artificial hand [1-4]. Typically, most of tactile sensors can only detect normal force or pressure. However, the tangential force widely exists in the actual grasping motion. To detect both the normal and shear forces, silicon based triaxial tactile sensors using micro-electromechanical systems (MEMS) technology were proposed. Even though some silicon based triaxial sensors possessed high sensitivity and good linearity, but the rigid silicon restricted their application in some situations with big mechanical deformation [5]. Therefore, several flexible polymer based three axial tactile sensors were developed in recent years. Sun [6] et al proposed and fabricated a three-dimensional (3D) force sensor using carbon nanotubes/polydimethylsiloxane (CNTs/PDMS) nanocomposite as sensitive materials, it exhibited a high sensitivity and good repeatability, while its output was nonlinear. Theoretically, the force sensor based on the piezoelectric effect can work with a good linearity under both normal and shear load [7]. Yu [8] et al presented a flexible piezoelectric tactile sensor for dynamic response in both normal and shear forces, but the range of force

Resrach supported by the National Natural Science Foundation of China (Grant Nos. U1909221, 91748207), the Key Research and Development Project of Shaanxi Province (Grant Nos. 2018ZDXM-GM-103), the Shaanxi Province Natural Science Basic Research Project (2019JC-06).

[1]State Key Laboratory for Manufacturing Systems Engineering, the International Joint Laboratory for Micro/Nano Manufacturing and Measurement Technologies, Overseas Expertise Introduction Center for Micro/Nano Manufacturing and Nano Measurement Technologies Discipline Innovation, and Xi'an Jiaotong University (Yantai) Research Institute for Intelligent Sensing Technology and System, Xi'an Jiaotong University, Xi'an 710049, China;

[2]School of Mechanical Engineering, Xi'an Jiaotong University, Xi'an 710049, China;

[3]Shaanxi Institute of Metrology Science, Xi'an, 710065, China.

*Contacting Author is with State Key Laboratory for Manufacturing Systems Engineering, School of Mechanical Engineering, Xi'an Jiaotong University, Xi'an 710049, China (phone: +86-29-82668616; email: libozhao@mail.xjtu.edu.cn)

detection was only 0-1.5 N. Here, a flexible three axial tactile sensor based on piezoelectric polyvinylidene fluoride (PVDF) film was proposed and fabricated by patterning electrodes, casting the bump structure and assembling the sensor layer by layer. The sensor exhibited a good linearity under the measured forces at different angles from 0-10 N, which was attributed to the rational design of the bottom four electrodes and the bump structure.

EXPERIMENT

A. Materials

The piezoelectric film was a commercial PVDF film (Kureha, Japan) with a thickness of 40 μm. The PDMS (sylgard184) was purchased from Dow corning Corporation, USA. The photoresist (EPG 535, Everlight Chemical) was used, and acetone and sodium hydroxide were purchased from Sinopharm Group, China. The 50 μm PET film was obtained by Lucky group, China.

B. Fabrication of the sensor

The schematic of proposed flexible tactile sensor was showed in Fig.1, where the bump structure, patterned electrodes and a PVDF film were assembled layer by layer. The electrodes of the tactile sensor were prepared by patterning Cr/Au on the PET film as showed in Fig. 2(a). The PET films were cleaned in the ethanol for cleaning the surface by ultrasonic treatment. Then, PET films were baked at 90 ℃ for 5 min and then were coated with EPG 535 by spin coater (500 rpm, 15 s/1000rpm, 30 s). Before being exposed by ABM6 (ABM, USA), the coated films were baked at 90 ℃ for 10 min. After exposure with ABM6, the coated PET films were developed by NaOH (0.5 wt%) and the pattern of electrodes were transferred from the mask. Then, 20 nm Cr and 100 nm Au were deposited on the patterned PET films by electron beam evaporation (HHV TF500, UK) and the PET films deposited with Cr/Au were placed into acetone solution to lift-off the unnecessary parts. The bottom electrode was four square-shaped electrodes and the top electrode was one common square-shaped electrode. The PVDF film was sandwiched between the top electrode and the bottom electrode by PDMS, thus four piezoelectric sensing units were formed. The size of each piezoelectric unit was 2.5×2.5 mm^2. The fabricated electrodes were observed by a confocal laser microscope (LEXT OLS4000, OLYMPUS) as shown in Fig.3.

The bump structure of the tactile sensor was fabricated by casting PDMS. The PDMS prepolymer and the curing agent were mixed with the weight ratio of 10:1, then stirred for 5 min and degassed in a vacuum chamber for 10 min. The mixture was then poured into a stainless mold and cured in a 70 ℃ oven for 3 hours. Then, the bump PDMS structure was peeled off from the mold. The liquid PDMS

978-1-6654-3008-1/21 $31.00 © 2021 IEEE

was used to bond between layers. The structure parameters of the flexible tactile sensor were shown in Fig. 4. When the flexible tactile sensor was assembled, the signal output of the sensor was obtained from wires connected to pads by conductive adhesive. The photo of the fabricated flexible tactile sensor was shown in Fig. 5.

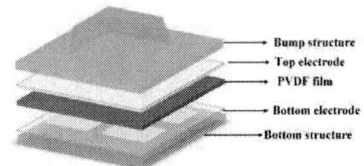

Figure1. Schematic of flexible piezoelectric three axial tactile sensor

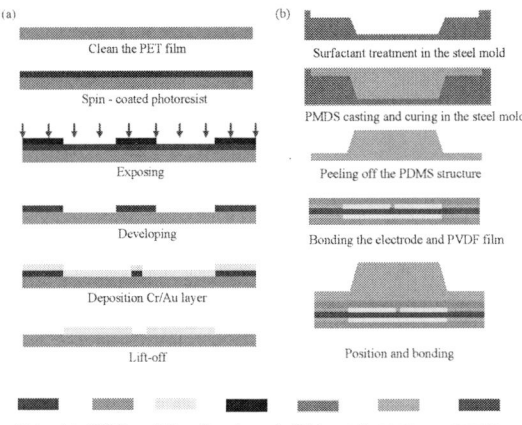

Figure 2(a) Schematic of electrode fabrication process, (b) Schematic of sensor fabrication process

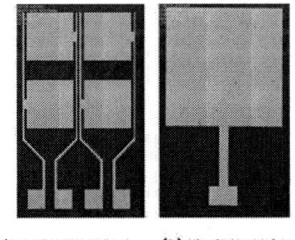

(a) the top electrode   (b) the bottom electrode

Figure 3. The confocal laser microscope image of fabricated electrodes

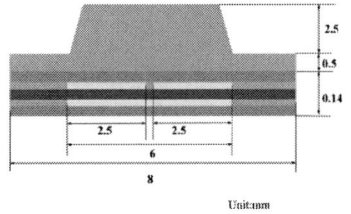

Figure 4. The cross-sectional view of the tactile sensor

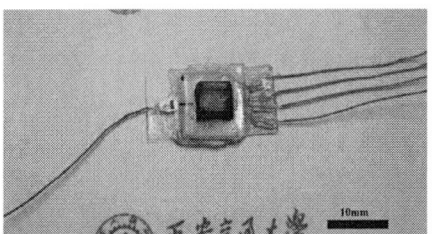

Figure 5. The fabricated tactile sensor

C. MEASUREMENT

In order to test the performance of the flexible tactile sensor, a test platform with different tilt angles was designed to apply three-dimensional forces in different directions. In Fig. 6, a force was generated by a mechanical shaker (MS100, YMC Piezotronics Inc, China) that was controlled by a signal generator (YMC9200) and a power amplifier (LA-200, YMC Piezotronics Inc, China). The signal generator was used to control the mechanical shaker to generate forces with different frequency or magnitudes. There was a calibrated force sensor (LH-S09A, China) to calibrate the applied force on the top rod of the shaker. The current output of the flexible tactile sensor was collected by an electrochemical workstation (Reference 600+, Gamry Instruments, USA). Due to the mechanical shaker was hard to move, we designed some movable platform with different tilt angles to place the flexible tactile sensor. It could be easily to measure the three-dimension forces in different directions by changing the sensor fixed platform.

D. sensing principle

The flexible PVDF film as a piezoelectric material generates charges when it is deformed [9]. According to the linear theory of piezoelectricity, the tensor relation between mechanical stress, mechanical strain, electric field, and electric displacement is [10]

$$s_p = s_{pq}^E T_q + d_{kp} E_k \qquad (1)$$

$$D_i = d_{iq} T_q + \varepsilon_{ik}^T E_k \qquad (2)$$

where s_p is the mechanical strain in the p direction, D_i is electric displacement in the i direction, T_q is mechanical stress in the q direction, E_k is the electric field in the k direction, s_{pq}^E is elastic compliance at a constant electric field, ε_{ik}^T is the dielectric constant under constant stress, and d_{iq} is the piezoelectric constant.

In the tactile sensor, the PVDF film was used in the thickness mode, the Eq. (2) can be written as follows with no external electric field.

$$D_3 = d_{33} T_3 \qquad (3)$$

When a three-dimension force is applied to the bump structure, the bump structure is compressed and deformed, then four sensing piezoelectric units generate charges under the mechanical stress. The force could be decomposed into a normal force component and two shear force components.

978-1-6654-3008-1/21 $31.00 © 2021 IEEE

The normal component force can cause the same stress in the four bottom regions and the shear forces will produce torques at fixed end. The charges of the four sensing units would be different under the normal component force and the two shear forces.

CURRENT RESULT

The structure of the flexible tactile sensor was assembled layer by layer and the patterned electrodes were fabricated by lift-off process. Compared with depositing and patterning electrodes on PVDF films, preparing electrodes on PET films and bonding them to a PVDF film are much simpler and convenient because patterning electrodes on a PVDF film needs to be aligned on both sides of the film, which is a time consuming and expensive work because of the complex process. The electrodes deposited on the PET film could be accurately and easily positioned during the assembly of the electrodes, the bump structure and a PVDF film. When the flexible tactile sensor was fabricated, the output of the sensor was tested by pressing the bump structure with a finger. As shown in Fig.7, when pressing the bump, a positive current generated and when the force was released, there would be a negative current. Due to the force applied to the top of the bump contained only a normal component, the current outputs of four electrodes were almost equal. This flexible piezoelectric sensor was suitable for dynamic tactile force sensing.

To further test the tactile sensor, the normal forces with different amplitudes were applied to the top surface of the sensor. The output signals of four bottom electrodes were showed in Fig. 8, which were approximately equal due to the same stress in the four sensing regions produced by the normal force. According to the relationship between the current outputs and applied force, the flexible tactile sensor had a measuring sensitivity of 110pA/N. Furthermore, the sensor was placed on the inclination angles of 60°, 45°, 30°, respectively. The current outputs of the flexible tactile sensor were shown in Fig. 9, 10 and 11. The output signals of the four electrodes appeared to be different increasing trends at different inclination angles when the force was applied from 0 N to 10 N. As the inclination angle decreased, the output signals of electrode 1 and electrode 2 increased and the output signals of electrode 3 and electrode 4 decreased because of the tangential component increased. As the testing results demonstrated, the direction of applied force could be obtained by the different increasing trend. When the inclination angle of the moveable platform was 60°, the average slope of the output signals from electrode 1 and 2 divided by the average slope of the output signals from electrode 3 and 4 was 1.2. When the inclination angles were 45° and 30°, the ratios of the average slope were 1.46 and 7.4. The output results showed that there may be a relatively coupling effect between the normal component and the tangential component of the force applied to the sensor under the loading mode.

Because of the bump structure, tangential components can be transmitted by the bump structure and applied to the PVDF sensitive layer, which resulted in positive pressure on one side and negative pressure on the other. Due to the change of the force position, the deformation of the PDMS bump structure would lead to a part of the coupling effect between the normal component and the tangential component. Besides, all the output signals of this piezoelectric three-dimensional tactile force sensor shown good linearity. According to the output results under different inclination angles, the output currents of the four electrodes can be used to calculate three-dimensional forces in different directions. Furthermore, the flexibility of this tactile sensor is helpful for the application in different situations. This three-dimensional force sensor can be used to assist robotic operation in the intelligent robot.

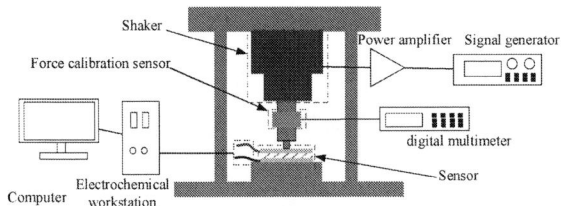

Figure. 6. The schematic of sensor testing platform

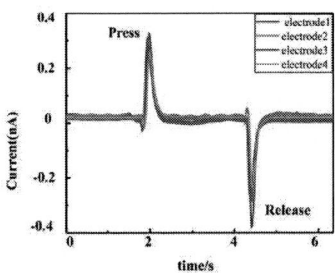

Figure.7. The current outputs of the flexible tactile sensor under the finger pressing

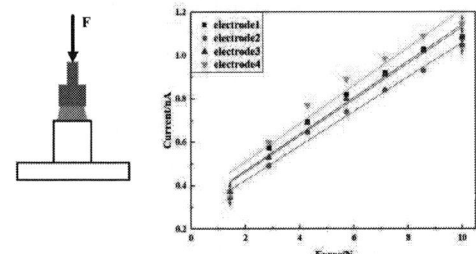

Figure. 8. The current outputs of the tactile sensor under normal force

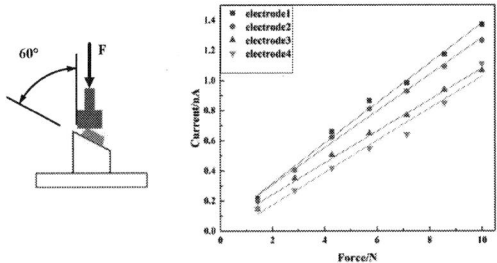

Figure. 9. The current outputs of the tactile sensor under the force with the inclination angel of 60°

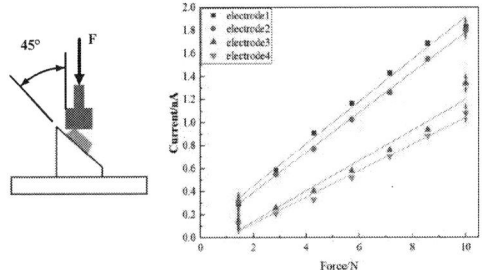

Figure. 10. The current outputs of the tactile sensor o under the force with the inclination angel of 45°

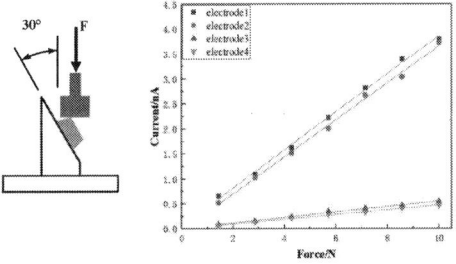

Figure. 11. The current output of the tactile sensor under the force with the inclination angel of 30°

CONCLUSION

This research work reports a flexible tactile sensor for the detection of three-dimension force. A PVDF film was used as the sensing layer to measure the dynamic force. The flexible tactile sensor demonstrated good linearity when the forces range from 0 to 10N with different angles were applied on, which implies that the flexible tactile sensor can work in a wide variety of applications such as intelligent robot, human-computer interaction application. However, the coupling effect of the sensor due to its own deformation needs to be further studied.

REFERENCES

[1] H. Yousef, M. Boukallel, and K. Althoefer, "Tactile sensing for dexterous in-hand manipulation in robotics—a review," Sensors and Actuator A Physical, vol. 167, pp. 171–187, 2007.

[2] K. Xi, Y. Wang, D. Mei, G. Liang and Z. Chen, "A flexible tactile sensor array based on pressure conductive rubber for three-axis force and slip detection," 2015 IEEE International Conference on Advanced Intelligent Mechatronics (AIM), Busan, Korea (South), pp. 476-481, 2015

[3] Kappassov, Zhanat , J. A. Corrales , and Perdereau, Véronique. "Tactile sensing in dexterous robot hands — Review." Robotics & Autonomous Systems, vol.74, pp. 195-220,2015.

[4] Yang, Tingting , et al. "Recent advances in wearable tactile sensors: Materials, sensing mechanisms, and device performance." Materials Science & Engineering R Reports ,vol.115, pp.1-37, 2017.

[5] Takahashi, H, Nakai, A. , Thanh-Vinh, N. , Matsumoto, K , and Shimoyama, I, "A triaxial tactile sensor without crosstalk using pairs of piezoresistive beams with sidewall doping," Sensors and Actuators A Physical, vol.199, pp.43–48, 2013.

[6] Sun, X, Sun, J, Li, T, Zheng, S , and Xue, N, "Flexible tactile electronic skin sensor with 3d force detection based on porous CNTs/PDMS nanocomposites," Nano Micro Letters,vol. 4, pp. 57, 2019

[7] Dagdeviren, C, Joe, P, Tuzman, O.L, Park, K. I, and Rogers, J. A., "Recent progress in flexible and stretchable piezoelectric devices for mechanical energy harvesting, sensing and actuation," Extreme Mechanics Letters, vol. 9, pp. 269-281,2016.

[8] P. Yu, W. Liu, C. Gu, X. Cheng, and X. Fu, "Flexible Piezoelectric Tactile Sensor Array for Dynamic Three-Axis Force Measurement," Sensors, vol. 16, no. 6, pp. 819, 2016

[9] Rajala S , Tuukkanen S , Halttunen J . "Characteristics of Piezoelectric Polymer Film Sensors With Solution-Processable Graphene-Based Electrode Materials," Sensors Journal IEEE, vol.15, pp.3102-3109, 2015

[10] R. S. Dahiya, M. Valle and L. Lorenzelli, "SPICE model for lossy piezoelectric polymers," in IEEE Transactions on Ultrasonics, Ferroelectrics, and Frequency Control, vol. 56, no. 2, pp. 387-395, February 2009

Gap in pagination due to unavailable paper.

Pages 1912-1913

The 16th IEEE International Conference on Nano/Micro Engineered & Molecular Systems

The Reinvention of silk

Hu TAO

Shanghai Institute of Microsystem and Information Technology, Chinese
Academy of Sciences, P.R. China

ABSTRACT

A paradigm shift for implantable medical devices lies at the confluence between regenerative medicine, where materials remodel and integrate in the biological milieu, and technology, through the use of recently developed material platforms based on biomaterials and bioresorbable technologies such as optics and electronics. The union of materials and technology in this context enables a class of biomedical devices that can be optically or electronically functional and yet harmlessly degrade once their use is complete. The talk will discuss the use of silk protein as a sustainable material in transient optics and photonics, electronics and optoelectronic applications. The favorable properties of the material certainly make a favorable case for the use of silk, yet serve as a broad inspiration to further develop biological foundries for both the synthesis and processing of Nature's materials for high technological applications.

BIOGRAPHY

Hu TAO received his Ph.D. in Mechanical Engineering with the Best Dissertation Award from Boston University, in 2010. His research interests have mainly focused on terahertz metamaterials using MEMS technology. He is currently a Professor at Shanghai Institute of Microsystem and Information Technology, CAS and his research interests focus on green nanotechnology, micro/nano- technology enhanced novel electronic and photonic devices for biomedical applications. Dr. Tao has published over 70 papers in peer-reviewed scientific journals including Science, Nature, Nature Photonics, Nature Nanotechnology, Nature Communications, PNAS, Advanced Materials and Physical Review Letters.

Gap in pagination due to unavailable papers.

Pages 1915-1920

Biomolecular Needling System for Medicals

Beomjoon Kim

Institute of Industrial Science, The University of Tokyo, Japan

Website : http://www.kimlab.iis.u-tokyo.ac.jp/

ABSTRACT

Based on nano/micro components systems for the fabrication of novel nano devices, we have very successfully investigated to develop various micro sensors for biological applications, health care as well as environmental monitoring. Here, new transdermal drug delivery system by using dissoluble micro needle patch will be introduced. Recently, in the transdermal drug delivery methods, the microneedle-mediated drug delivery system (DDS) has been developed to replace the hypodermic injection-mediated DDS, to provide painless self-administration of biological drug with patient friendly manner. Especially, dissolving microneedles, which deliver the target drugs as the drug-loaded microneedle dissolves into the skin, have been spotlighted recently.

We investigate a novel fabrication method to achieve the user-friendliest, low-cost, and safest way for dissoluble microneedle patches with vaccine delivery, several medical treatments, and even glucose sensing. We have developed a porous microneedle patch for monitoring glucose levels using a paper sensor. The device painlessly monitors fluid in the skin within seconds. Anyone can use the disposable patch without training, making it highly practical. Additionally, fabrication is easy, low cost, and the glucose sensor can be swapped for other paper-based sensors that monitor other important biomarkers.

Furthermore, to monitor the biological information through the skin continuously, a novel microneedles' structure is proposed and investigated by combining flexible "sponge-like" porous PDMS matrix and the coating by dissoluble hyaluronic acid (HA).

BIOGRAPHY

Beomjoon Kim is currently a full professor of Institute of Industrial Science, the University of Tokyo, Japan (Dept. of Precision Engineering, The Univ. of Tokyo) and a director of LIMMS/CNRS-IIS UMI 2820. He received the B.S. degree from Seoul National University, Dept. of Mechanical Design and Production Engineering, Seoul, Korea, in 1993, and received M.Eng., and Ph.D. in Department of Precision Engineering, the University of Tokyo, Tokyo, Japan, in 1995 and 1998, respectively. Recently he is also in charge as a chair of corporate sponsored research division of "Virological Medicine", at I.I.S., the University of Tokyo, and active as executive advisor of BNS Medicals, Co., Ltd., Japan.

He was a CNRS Associate Researcher for Microsensors, Nano-instruments for Nanotechnology in Centre National de la Recherché Scientifique at Besancon, France at 1998. He worked also in research orientation NanoLink, MESA+ Research Institute, University of Twente in the Netherlands, to September 2000. Since September 2000, he has been an Associate Professor of Institute of Industrial Science, the University of Tokyo until April 2014. He visited Institute of Micro-engineering, EPFL (group of Prof. Juergen Brugger), Swiss, Dept. Chemical Eng. (group of Prof. D. Schwartz), The University of Washington, USA in 2005, Dept. Chemistry, (group of Prof. Y. Chen) École Normale Supérieure (ENS) Paris, in 2010, and Dept. of Medicine, Harvard Medical School, Harvard-MIT HST, Cambridge, USA in 2013 as well for collaboration about micro/nano bio-engineering.

He investigates several aspects of bio-sensors components to accomplish portable Point-of-Care diagnostic devices, which are disposal, user-friendly, low-cost, and highly sensitive. Moreover, he is interested to develop self-powered, energy harvesting micro sensors as well as smart monitoring system with network like "internet of things". Recently, main research topic is focused to study on new transdermal drug delivery system by using dissoluble micro needle patch. He has published 103 peer reviewed Journal papers, 200 international conference papers, 189 domestic conference papers as well as several patents, books publications so far.

978-1-6654-3008-1/21 $31.00 © 2021 IEEE

Proceedings of the 16th Annual IEEE International
Conference on Nano/Micro Engineered and Molecular Systems
April 25-29, 2021

Physical Forces Influence the Self-organization of the Leader Cell Formation During Collective Cell Migration

Mengyun Pan, Yongliang Yang, and Lianqing Liu, *Senior Member, IEEE*

Abstract—Many dynamic physiological processes, including embryo development and wound healing, involve collective cell migration. During this process, leader cells emerge in self-organization manner to guide other cells migration. The mechanisms of their formation, however, are not clear. We hypothesized that the physical forces among cells, and between cell and substrate influence the formation of leader cells. Here, we presented a computational model of cell monolayer migration based on the Glazier-Graner-Hogeweg (GGH) model. Our results indicated that the adhesion forces among cells and the cell-substrate adhesion forces regulate the formation of leader cells. Increasing cellular mobility also affected the characters of leader cells. In summary, our results provided evidence that physical forces among cells and between cell and substrate, along with mobility of individual cells, affect the self-organization process of leader cell formation during collective cell migration.

I. INTRODUCTION

Collective cell migration is a basic and complicated biological phenomenon, which occurs in morphological development, tissue repair, and wound healing processes. Previous research suggested that collective cell migration also has an important impact in pathological procedures such as cancer invasion and metastasis. Comparing with individual cell migration, cells in collective migration form cohesive groups with cell-cell adhesion to coordinate their movement [1]. This complex migration process is guided by chemical and mechanical cues during migration, which has been well studied. For example, the physical forces arising from the tensions in actomyosin rings transmitted to the matrix through focal adhesion. Actin stress fibers in cells at the migration frontiers protrude towards the migration direction to drive epithelial monolayer collective migration in *in vitro* wound healing [2].

In this process, the leader cells emerge to explore the tissue environment and interact mechanically with the follower cells to guide the other cells during collective cell migration [3-6]. Therefore, it is generally believed that leader cells make vital impacts in the migration process. Definitive evidence established that leader cells at the leading edge of cell monolayer would exert traction forces on their surroundings to drive collective cell migration [7].

Current studies have demonstrated that a variety of factors can regulate and control the emergence of leader cells. Nevertheless, most reports describing regulation and

control of the formation of leader cells focused on biological and chemical signals, such as promoting the transmission of Notch1-DLL4 signal and diminishing activity of RhoA [8-11]. While mechanical forces have a significant impact on regulating the formation of leader cells. Reducing wound stress promoted leader cell formation during the migration of smooth muscle cells [12]. The organization of the actin cytoskeleton and its collective polarization also induced the formation of leader cells [13]. The mechanical interactions between the follower cells could effectively select the leader cell of the wound edge [14]. Although these findings reveal leader cell formation during collective cell migration, the role of adhesion forces in leader cell formation remains poorly understood. Remarkably, recent evidence has shown that the interactions of cell-cell and cell-substrate adhesion could affect collective cell migration [15]. However, if and how the adhesion forces might control the formation of leader cells at the margin is not clear.

We presented a computational model of cell monolayer migration using the Glazier-Graner-Hogeweg (GGH), whose results confirm that adhesion forces affected the formation of leader cells. We proved that without extra stimulation, varying adhesion energies by varying either one of E-cadherin–E-cadherin adhesion between cells, integrin–collagen adhesion between cells and substrate, density of adhesion molecules, and cell migration mobilities or all of them together will cause cells to migrate heterogeneously to emerge leader cells. In the simulation, we also observed that increasing cellular mobilities will make the movement of cell clusters more intense and easier to spread and transfer. In summary, our data provided evidence that the leader cell formation is affected by cell-cell adhesion, cell-substrate adhesion and cellular mobilities during collective cell migration.

II. METHODS

The simulated cell behavior is based on the Glazier-Graner-Hogeweg (GGH) model, which could depict cellular activities and interplays, such as its shape, motility, adhesion, and response to extracellular signal by using an effective-energy formulation [16-17]. The GGH in this paper uses cell-substrate adhesion, cell-cell adhesion, volume restriction and surface area restriction to build the effective energy function governing cellular mobility. The boundary energy is defined to represent the variations of energy, which is due to adhesion between different types of cells.

---

*Resrach supported by the Key Research Program of Frontier Sciences, CAS (Grant No. QYZDB-SSW-JSC008).

Mengyun Pan, Yongliang Yang and Lianqing Liu are with the State Key Laboratory of Robotics, Shenyang Institute of Automation, Chinese Academy of Sciences, Shenyang 110016, China, and also with the Institutes for Robotics and Intelligent Manufacturing, Chinese Academy

of Sciences, Shenyang 110169, China. (e-mail: yishuiaoi@live.com; lqliu@sia.cn).

Mengyun Pan is also with the University of Chinese Academy of Sciences, Beijing 100049, China. (e-mail: panmengyun@sia.cn).

978-1-6654-3008-1/21 $31.00 © 2021 IEEE

Considering the boundary energy, volume restriction, and surface area restriction together, the basic GGH effective energy can be expressed as:

$$H_{GGH} = \sum_{i,j} J(\sigma(i),\sigma(j))(1-\delta(\sigma(i),\sigma(j)))$$
$$+ \sum_{\sigma} \lambda_v (v_\sigma - V)^2 + \sum_{\sigma} \lambda_s (s_\sigma - S)^2 \quad (1)$$

Where, v_σ and s_σ denote volume and area. V and S denote the target of volume and surface area. λ_v and λ_s is defined separately as the elasticity parameter of volume and area. $\sigma(i)$ and $\sigma(j)$ are two different cells. $J(\sigma(i),\sigma(j))$ and $\delta(\sigma(i),\sigma(j))$ is defined separately as the adhesion energy and the usual Kronecker delta function.

In our model, we take into consideration two different cell types: cell and medium. Taking into account that contacts occur between cell-medium and cell-cell, the adhesion energy terms in (1) can be presented by cell-substrate and cell-cell adhesion energy terms respectively. In this model, a variety of adhesion molecules, which determine the adhesion energy at cell-cell and cell-substrate, are considered as the principal reasons [16-17]. Furthermore, adhesion energy is regulated by adhesion molecule densities and molecule binding affinities. So we can get the formulation of adhesion energy

$$W(\sigma(i),\sigma(j)) = -\sum_{m,n} L_{m,n} \cdot \min(X^m_{\sigma(i)}, X^n_{\sigma(j)}) \quad (2)$$

where, m and n denote different adhesion molecules. The binding affinity is defined by $L_{m,n}$, which is between a pair of adhesion molecules of cells. $X^m_{\sigma(i)}$ and $X^n_{\sigma(j)}$ denote the density of different adhesion molecules.

In this paper, the adhesion between cells is mediated by E-cadherins $X^{Ecad}_{\sigma(i)}$, and the adhesion between cell and substrate is mediated by integrins $X^{Int}_{\sigma(i)}$ and collagen $X^{Coll}_{\sigma(j)}$. The binding affinity of E-cadherins is defined by $L_{Ecad,Ecad}$. The binding affinity of integrins and collagen is defined by $L_{Int,Coll}$.

When we investigate that varying density of integrin and integrin-collagen binding affinity also affects simulation results, so we use the following formulation,

$$X^{Int}_\sigma(t+1) = \alpha \cdot x + X^{Int}_\sigma(t) \quad (3)$$

where, $x = 0.01$, α is linear coefficient. $X^{Int}_\sigma(t)$ denotes initial density of integrins. So we can get integrin expression of all cells increasing linearly.

Finally, combining all the above equations, the GGH effective energy can be expressed as,

$$H_{GGH} =$$
$$-\sum_{i,j}[L_{Ecad,Ecad} \cdot \min(X^{Ecad}_{\sigma(i)}, X^{Ecad}_{\sigma(j)})$$
$$+ L_{Int,Coll} \cdot \min(X^{Int}_{\sigma(i)}, X^{Coll}_{\sigma(j)})] \times (1-\delta(\sigma(i),\sigma(j))) \quad (4)$$
$$+ \sum_{\sigma} \lambda_v (v_\sigma - V)^2 + \sum_{\sigma} \lambda_s (s_\sigma - S)^2$$

In our final formalism, six factors determining the effective energy are as following: E-cadherin–E-cadherin binding affinity, E-cadherin density, integrin–collagen binding affinity, integrin density, volume restriction, and surface restriction.

III. RESULTS

We used CompuCell3D to implement the formulation of GGH described in the methods section and get the following results. During cell monolayer migration, leader cells and follower cells emerged at the wound frontier (Fig. 1a). We hypothesized that the intercellular adhesions and cell-substrate adhesions affect the leader cell formation. In our model, the intercellular adhesions were mediated by the expression and binding affinities of E-cadherin molecules. The cell-substrate adhesions were mediated by the expression of integrin molecules and its binding affinity with collagen molecules (Fig. 1b). We tested the effects of these factors under conditions with low and high cellular mobilities (Fig. 1c-d). But we found that compared with low cellular mobility, cell clusters are easier to spread and transfer under high cellular mobility (Fig. 1c-d).

First, we tested that how intercellular and cell-substrate binding forces regulate leader cell characters with constant integrin and E-cadherin expressions. In both low and high cellular mobilities, the density of the leader cells increased when the affinity between E-cadherin molecules reduced or the binding affinity between integrin–collagen increased (Fig. 2 a-b). Meanwhile, the density of leader cells was higher with low cellular mobility (Fig. 2 a-b). The size of leader cells formed a contour with specific patterns of E-cadherin and integrin-collagen binding affinities. Increasing the cellular mobility increased the size of leader cells (Fig. 2 c-d).

The expression of integrin may increase during wound healing process. We further tested how this affects the

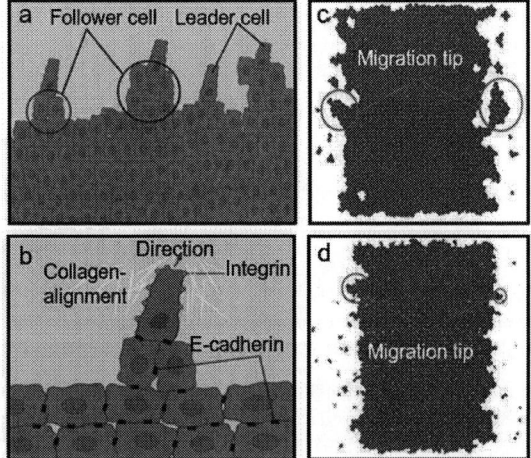

Figure 1. Leader cells in collective cell migration. (a) The formation of leader cells and follower cells during collective cell migration; (b) E-cadherin–E-cadherin and integrin–collagen binding forces in leader cell during collective cell migration; (c) simulation result of collective cell migration in low cellular mobility; (d) simulation result of collective cell migration in high cellular mobility.

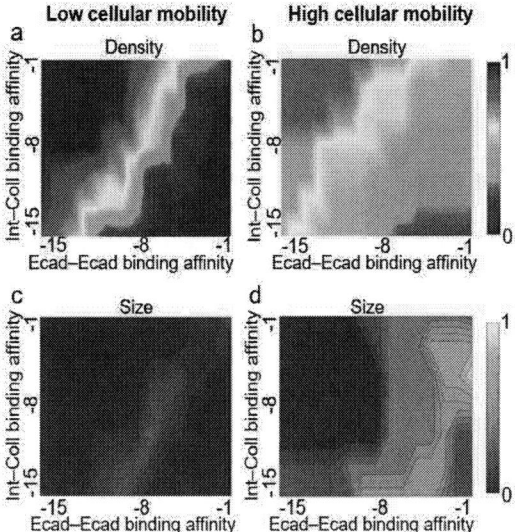

Figure 2. Intercellular and cell-substrate binding forces regulate leader cell characters with constant integrin and E-cadherin expressions. The density of leader cells with low cellular mobility (a) and with high cellular mobility (b); the size of leader cells with low cellular mobility (c) and with high cellular mobility (d). Ecad- Ecad: E-cadherin- E-cadherin. Int-Coll: integrin–collagen.

formation of leader cells. The binding affinities between E-cadherin molecules and between integrin and collagen molecules affected the density and size of leader cells in a similar manner as the case of constant E-cadherin expression during the wound healing process (Fig. 3 a-b). The leader cells were larger with the cellular expression of integrin linearly increasing compared with the constant integrin

expression case (Fig. 3 c-d). But when the binding affinity between integrin and collagen molecules with linearly increasing integrin expression was bigger and the binding affinity between E-cadherin molecules was smaller, it formed larger leader cells (Fig. 3 c-d).

We evaluated the effect of the rate of integrin expression leader cell characters. Under low cellular mobility, the faster the integrin expression increases, the smaller the average density and size of leader cells (Fig. 4). The average density and size of leader cells had the maximum value at a medium rate of integrin expression with high cellular mobility (Fig. 4).

Adhesion forces not only correlate with the binding affinity between molecules, but also influence by the density of adhesion molecules. So we further evaluated the effect of the expression density of E-cadherin and integrin on the leader cell characters with constant integrin–collagen binding affinity and E-cadherin–E-cadherin binding affinity. In both low and high cellular mobility cases, increasing E-cadherin expression density and reducing integrin expression density increased the density of the leader cells (Fig. 5 a-b). But increasing the cellular mobility reduced the density of leader cells (Fig. 5 a-b). Increasing cellular mobility enlarged the size of the leader cells (Fig. 5 c-d). But when the density integrin and E-cadherin were smaller, the leader cells were larger compared with low cellular mobility (Fig. 5 c-d).

IV. CONCLUSION

In our research, we have built a simulation model of collective cell migration to evaluate the roles of cell-cell adhesion, cell-substrate adhesion, and cellular mobility in

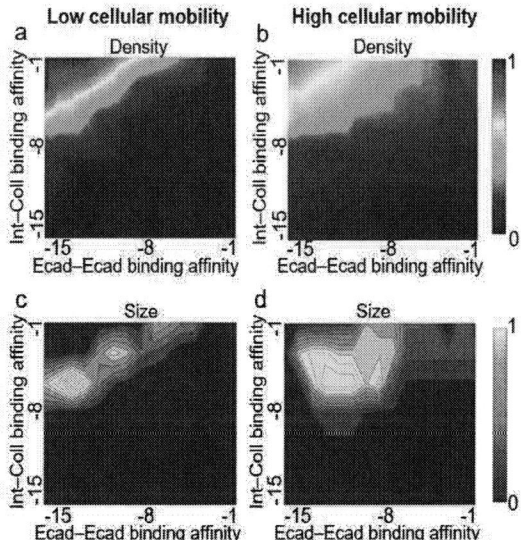

Figure 3. Intercellular and cell-substrate binding forces regulate leader cell characters with linearly increasing integrin expression and constant E-cadherin expression. The density of leader cells with low cellular mobility (a) and with high cellular mobility (b); the size of leader cells with low cellular mobility (c) and with high cellular mobility (d).

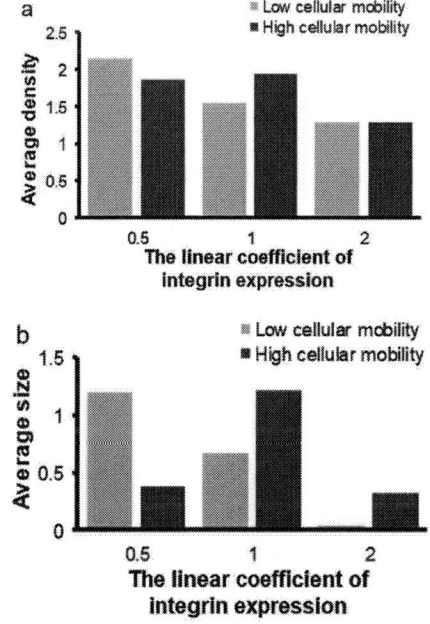

Figure 4. The linear coefficient of integrin expression affects the density (a) and size (b) of leader cells.

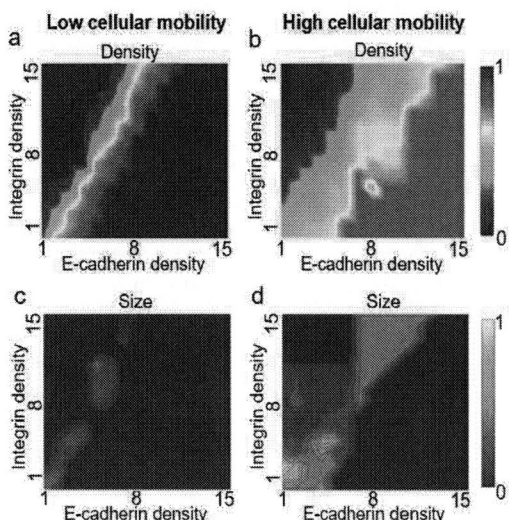

Figure 5. Intercellular and cell-substrate binding forces regulate leader cell characters with constant integrin–collagen binding affinity and E-cadherin–E-cadherin binding affinity. The density of leader cells with low cellular mobility (a) and with high cellular mobility (b); the size of leader cells with low cellular mobility (c) and with high cellular mobility (d).

the characteristic of leader cells. The density of leader cells increased as the binding forces among cells decreased and the cell-substrate binding forces increased. However, when the adhesion forces between cells were too small and cell-substrate adhesion forces were too large, cell clusters would tend to spread and transfer. The size of leader cells formed a contour with specific patterns of cell-cell adhesion and cell-substrate adhesion. The cellular mobility affected the role of these two types of binding forces in leader cell formation. These results indicated that the adhesion forces among cells, and between cell and substrate regulate the formation of leader cells. They thus affected the direction and speed of collective cell migration.

Via changing the adhesion forces of cell-cell and the cell-substrate adhesion force, it is possible to control the emergence of leader cells, which is beneficial to investigate the physiological and pathological processes involving collective cell migration, such as accelerating wound healing and reducing cancer metastasis. Furthermore, the work on the leader cells proposed here not only demonstrates the significance of physical forces in leader cell formation, but also provides a new mechanism framework for future research on the collective migration of cells in physiological and pathological aspects.

REFERENCES

[1] X. Trepat, Z. Chen, and K. Jacobson, "Cell Migration," *Comprehensive Physiology*, vol. 2, pp. 2369-2392, Oct. 2012.

[2] A. Brugues, E. Anon, V. Conte, J. H. Veldhuis, M. Gupta, J. Colombelli, J. J. Muñoz, G. W. Brodland, B. Ladoux, and X. Trepat "Forces driving epithelial wound healing," *Nature Physics,* vol. 10, pp. 684-691, Sep. 2014.

[3] A. A. Khalil and P. Friedl, "Determinants of leader cells in collective cell migration," *Integrative Biology*, vol. 2, pp. 568-574, 2010.

[4] M. Reffay, M. C. Parrini, O. Cochet-Escartin, B. Ladoux, A. Buguin, S. Coscoy, F. Amblard, J. Camonis, and P. Silberzan, "Interplay of

RhoA and mechanical forces in collective cell migration driven by leader cells," *Nature Cell Biology*, vol. 16, pp. 382-382, Apr. 2014.

[5] D. A. Chapnick and X. Liu, "Leader cell positioning drives wound-directed collective migration in TGF beta-stimulated epithelial sheets," *Molecular Biology of the Cell*, vol. 25, pp. 1586-1593, May 2014.

[6] N. Yamaguchi, T. Mizutani, K. Kawabata, and H. Haga, "Leader cells regulate collective cell migration via Rac activation in the downstream signaling of integrin beta 1 and PI3K," *Scientific Reports*, vol. 5, Jan. 2015.

[7] X. Trepat, M. R. Wasserman, T. E. Angelini, E. Millet, D. A. Weit, J. P. Butler, and J. J. Fredberg, "Physical forces during collective cell migration," *Nature Physics,* vol. 5, pp. 426-430, Jun. 2009.

[8] R. Riahi, J. Sun, S. Wang, M. Long, D. D. Zhang, and P. K. Wong, "Notch1-Dll4 signalling and mechanical force regulate leader cell formation during collective cell migration," *Nature Communications*, vol. 6, Mar. 2015.

[9] T. Omelchenko, J. M. Vasiliev, I. M. Gelfand, H. H. Feder, and E. M. Bonder, "Rho-dependent formation of epithelial "leader" cells during wound healing," *Proceedings of the National Academy of Sciences of the United States of America*, vol. 100, pp. 10788-10793, Sep. 2003.

[10] S. Suchting, C. Freitas, F. I. Noble, R. Benedito, C. Breánt, A. Duarte, and A. Eichmann, "The Notch ligand Delta-like 4 negatively regulates endothelial tip cell formation and vessel branching," *Proceedings of the National Academy of Sciences of the United States of America*, vol. 104, pp. 3225-3230, Feb. 2007.

[11] C. Yang, M. Cao, Y. Liu, Y. He, Y. Du, G. Zhang, and F. Gao "Inducible formation of leader cells driven by CD44 switching gives rise to collective invasion and metastases in luminal breast carcinomas," *Oncogene*, vol. 38, pp. 7113-7132, Nov. 2019.

[12] Z. S. Dean, N. Jamilpour, M. J. Slepian, and P. K. Wong, "Decreasing Wound Edge Stress Enhances Leader Cell Formation during Collective Smooth Muscle Cell Migration," *Acs Biomaterials Science & Engineering*, vol. 5, pp. 3864-3875, Aug. 2019.

[13] S. Rausch, T. Das, J. RD. Soiné, T. W. Hofmann, C. HJ. Boehm, U. S. Schwarz, H. Boehm, and J. P. Spatz, "Polarizing cytoskeletal tension to induce leader cell formation during collective cell migration," *Biointerphases*, vol. 8, p. 32, Nov. 2013.

[14] M. Vishwakarma, J. Di Russo, D. Probst, U. S. Schwarz, T. Das, and J. P. Spatz, "Mechanical interactions among followers determine the emergence of leaders in migrating epithelial cell collectives," *Nature Communications*, vol. 9, p.3469, Aug. 2018.

[15] C. Wang, S. Chowdhury, M. Driscoll, C. A. Parent, S. K. Gupta, and W. Losert, "The interplay of cell-cell and cell-substrate adhesion in collective cell migration," *Journal of the Royal Society Interface*, vol. 11, Oct. 2014.

[16] V. Andasari, D. Lu, M. Swat, S. Feng, F. Spill, L. Chen, X. Luo, M. Zaman, and M. Long, "Computational model of wound healing: EGF secreted by fibroblasts promotes delayed re-epithelialization of epithelial keratinocytes," *Integrative Biology*, vol. 10, pp. 605-634, Oct. 2018.

[17] M. Swat, Gilberto L. Thomas, Julio M. Belmonte, A. Shirinifard, D.Hmeljak, and J. A. Glazier, "Multi-Scale Modeling of Tissues Using CompuCell3D," *Computational Methods in Cell Biology, Methods in Cell Biology*, vol. 110, pp. 325-366, 2012.

[18] Y. Xiao, R. Riahi, P. Torab, D. D. Zhang, and P. K. Wong, "Collective Cell Migration in 3D Epithelial Wound Healing," *American Chemical Society Biomaterials Science & Engineering*, vol. 13, pp. 1204-1212, Feb. 2019.

AUTHOR INDEX

Aishan, Yusufu 203
Alahi, Md Eshrat E 980, 1629
Alcheikh, N. 425
Bai, Yu 282, 299, 513
Bai, Zaiqiao 1084
Bao, Chen 184, 990
Bao, Weijie 904
Bian, Chao 1303
Bian, Tian 826
Bian, Xiongheng 909
Bingfei, Zhang 826
Bu, Zhenxiang 1065, 1897
Cai, Guangyi 953
Cai, Xiaoyu 255
Cai, Yao 228, 931
Cai, Yaxing 271
Cai, Yu 26
Cao, Jiangcheng 949
Cao, Liang 158
Cao, Qingpeng 762
Cao, Tongtong 1405
Cao, Wenping 770
Cao, Yongqi 528
Cao, Yun 699, 1287
Cao, Zhen 180, 459
Chai, Zhiping 1390
Chang, Honglong 55, 1903
Chang, Xiaocong 798
Che, Bangzhou 1554
Chen, Daner 840
Chen, Deyong 193, 1465, 1691, 1725, 1813
Chen, Fangshuai 731
Chen, Fanhong 1142
Chen, Guobin 941
Chen, Haifeng 319
Chen, Han 1390
Chen, Hao 1287, 1358
Chen, Hongxu 1843
Chen, Huatan 1035, 1136, 1328
Chen, Hua 197, 1825
Chen, Hui-Jiuan 503, 509
Chen, Huyue 1297
Chen, Jiang 399
Chen, Jian 193, 1465, 1691, 1813
Chen, Jinbo 919
Chen, Jingyi 1664
Chen, Jing 647
Chen, Junyu 491, 1136, 1328

Chen, Ke 1362
Chen, Liguo 909
Chen, Mingwei 1813
Chen, Qianhuang 845
Chen, Qinnan 1650
Chen, Qixiang 1650
Chen, Rui 1493
Chen, Shirong 1574
Chen, Shixing 30, 676
Chen, Si 1259
Chen, Songyue 963, 1655
Chen, Tao 60
Chen, Weifeng 1398
Chen, Wu 826
Chen, Xiaojun 319
Chen, Xiaoli 474
Chen, Xing 144, 1574
Chen, Xin 158
Chen, Xi 715
Chen, Xuying 1370
Chen, Yanan 731
Chen, Yao 282
Chen, Ye 882
Chen, Yunfei 539
Chen, Yusa 889
Chen, Zeji 455
Chen, Zhaohui 1070
Chen, Zhekun 1732
Chen, Zhensheng 1903
Chen, Zhibin 1737
Chen, Zhijian 1384
Chen, Zhiwen 1650
Cheng, Chao 193, 1691
Cheng, Zhenxing 30
Cheyns, David 1423
Chu, Jinkui 372
Chu, Yao 271, 1291
Cong, Bo 21
Cong, Lin 1084
Cui, Jian 1102, 1461
Cui, Jingqin 1737
Cui, Peijuan 1070
Dai, Jishen 1655
Dai, Songqiao 1871
Deng, Guang-Wei 1084
Deng, Jinjun 872, 1415
Deng, Qingyang 1410
Deng, Tao 68

Ding, Guanghui	872
Ding, Haojun	303
Ding, Hong	1370
Ding, Jiandong	1283
Ding, Qifan	1297
Ding, Zan	345
Djoulde, Aristide	919
Dong, Fangyang	1539
Dong, Hai-Feng	998
Dong, Hanyong	463, 1419
Dong, Hao	144
Dong, Jiale	660
Dong, Siyan	1106
Dong, Xianshan	1102
Du, Bingqian	1337
Du, Taili	660, 1539, 1676
Du, Xu	1074
Du, Yijia	1106
Duan, Yongqing	652
Duan, Yumo	1813
Dun, Guanhua	680
Fan, Xinye	1175
Fan, Xuge	1169
Fan, Yuanyi	372
Fan, Zhongqi	660
Fan, Zixiao	378
Fang, Hou	1057
Fang, Jiaru	1374
Fang, Lu	26
Fang, Xudong	334
Fei, Wenjie	1394
Feng, Di	941, 1714
Feng, Heng-Zhen	935
Feng, Jianguo	55, 1574
Feng, Linhao	737
Feng, Xue	731
Feng, Yanlu	687, 915
Feng, Yue	1264, 1332
Feng, Zhihong	334
Fu, Gang	491
Fu, Sinan	1554, 1700
Fu, Ting	1655
Fu, Yunxia	255
Ganesan, A.	1623
Gao, Chao	931
Gao, Guangen	1065
Gao, Guowei	392
Gao, Jiaming	378
Gao, Lingxiao	158
Gao, Wei	1415
Gao, Y.	1660
Gao, Yahao	1611
Gao, Yang	1157, 1259, 1353, 1804
Gao, Zhiyuan	1545
Geng, Jiangjun	60
Gong, Cheng	826
Gong, Qingfeng	994
Gong, Taobo	1478
Gong, Xun	1088, 1165
Gong, Zhuhao	695
Gu, Chaoming	180, 459
Gu, Dandan	1897
Gu, Guoqiang	711, 715
Gu, Xiyu	931
Gu, Yexin	1283
Guan, Chuanlong	372
Guo, Chenyu	1664
Guo, Chuan Fei	1390
Guo, Dengji	1617
Guo, Hang	148
Guo, Shan	1018
Guo, Shishang	931
Guo, Weijin	1310
Guo, Xinyan	882
Guo, Yixuan	691
Guo, Z. J.	959
Hai, Zhenyin	804
Han, Feng	47
Han, Lianhuan	1337
Han, Mengdi	148, 1686
Han, Mingjie	1303
Han, Ruixing	1291
Han, Weilong	443
Han, Yanhui	1264
Hansson, Jonas	1310
He, Bo	935
He, Ronggana	528
He, Xuefeng	386, 1353
He, Yinghao	898
He, Yingning	1283
He, Yunqian	51
He, Ziliang	1560
Heremans, Paul	1423
Hong, Hao	174
Hong, Heting	1863
Hosokawa, Yoshihiro	251
Hou, Cheng	60
Hou, Weiguo	271
Hou, Zhanqiang	1888
Hu, Cuiwen	319
Hu, Jie	407
Hu, Jingfang	392
Hu, Kai-Ming	433, 1432, 1605
Hu, Ning	1374

Hu, Tengjiang 1405
Hu, Tengjing 215
Hu, Xiaoping 1410
Hua, Yunzhi 1502
Huan, R. 1623
Huang, Dong 1839
Huang, Huayi 1096
Huang, Hui 386
Huang, Junlong 184, 990
Huang, Kai 737
Huang, Manjuan 1070
Huang, Min 1110
Huang, Shuang 503, 509
Huang, Sirong 1617
Huang, Xinshuo 503, 509
Huang, Yao 680
Huang, Yongan 652
Huang, Youchao 665
Huang, Yue 1676
Huang, Yunyi 228
Huang, Yun 546, 889
Huang, Zhiping 30
Hun, Tingting 949
Huo, Dian 1250
Huo, Zimin 1106
Ishihara, Daisuke 1320
Izhar 813
Jalili, Kiyumars 1629
Jeong, Yongbin 1423
Ji, Bowen 528, 1903
Jia, Lingjie 941
Jia, Qianqian 455
Jian, Zhu 1057
Jiang, A. Feng 412
Jiang, Boshi 953
Jiang, Fan 524
Jiang, Hongjie 64
Jiang, Jiaxin 140, 1136, 1328, 1554, 1700
Jiang, Kaili 1084
Jiang, Shulan 339
Jiang, Yong 474
Jiang, Yue 1398
Jiang, Z. 1623, 1660
Jiang, Zhuangde 47, 282, 299, 334, 513, 1157, 1362,
.... 1635, 1908
Jiangjiang, Liu 826
Jiao, Binbin 21
Jin, Chuanghong 459
Jin, Guoqing 518
Jin, Jie 1384
Jin, Shengxiao 889
Jin, Yiming 691

Jin, Yufeng 898, 949
Jin, Zhonghe 691
Jing, Weixuan 47
Kang, Guoguo 1441
Kang, Guoyi 140, 1035, 1136, 1328
Kang, Qiang 334
Kang, Tian 889
Ke, Xingxing 1390
Khademi, Sara 980, 1629
Kong, Lingli 919
Kong, Yanmei 21
Kuang, Yunbin 1888
Lee, Sang-Seok 251
Lee, Yi-Kuen 813
Lei, Cao Yang 1057
Lei, Hucheng 1611
Lei, Jiao Zong 1057
Lei, Shenghong 699, 1287
Lei, Yongjun 437
Li, Bolun 704
Li, Bo 1478
Li, Chengzhu 994
Li, Chunlong 386
Li, Defang 1664
Li, Dongsheng 463, 1370, 1419
Li, Dong 1461
Li, Haiwang 1745
Li, Haiyang 1441
Li, Huaan 749
Li, Hui 647
Li, Jiachen 518, 1611
Li, Jiayin 1686
Li, Jie 299
Li, Jing 1410
Li, Junru 1259, 1353
Li, Junshi 1839
Li, Keer 509
Li, Ke 174
Li, Lei 47
Li, Lili 1279
Li, Lin 647
Li, Liye 889
Li, Longfan 1737
Li, Longqiu 798
Li, Nanxing 1074
Li, Ning 1398
Li, Qingsong 1579, 1817
Li, Qi 845
Li, Shunqi 1539
Li, Tianyu 762
Li, Tie 30, 51, 676
Li, Wei 1250

Li, Wenwang 1035, 1328, 1554, 1700
Li, Wenxiang 1695
Li, Xiang 513
Li, Xiaojie 180, 459
Li, Xinghui 43
Li, Xiu-Yuan 433, 1432, 1605
Li, Xiuhan 941, 1714
Li, Xiuyan 1165
Li, Yadong 193, 1691
Li, Yang 474
Li, Yanjie 1088, 1165
Li, Yanqing 60
Li, Yansheng 392
Li, Yan 1175
Li, Yinqiao 271
Li, Yuan 255
Li, Yuhua 680
Li, Yunfei 1070
Li, Yuning 68
Li, Yunjia 1903
Li, Z T 311
Li, Zhihong 1839
Li, Zhikang 299, 1908
Li, Zhuo 1664
Li, Zixuan 299
Lian, Haishan 319
Lian, Zhenhui 1695
Liang, Boshen 1423
Liang, Bo 26, 762
Liang, Renrong 680
Liang, Tian 1813
Liang, Wenjie 1084
Liao, Jianhui 915
Lin, Guanzhou 546, 889
Lin, Jinghua 1554, 1700
Lin, Qijing 299, 513, 826, 1635, 1908
Lin, Waner 1096
Lin, Xiaoyang 1084
Lin, Yuanjing 1451
Lin, Yuhang 1732
Lin, Zude 1088, 1165
Liu, Bowen 1813
Liu, Changxin 749, 1853
Liu, Chenglong 845
Liu, Chen 998
Liu, Fenglin 953
Liu, Guochang 770
Liu, Haotian 1287
Liu, Huicong 60, 518, 1070
Liu, Huiliang 271, 695, 1291
Liu, J. Q. 959
Liu, Jianhao 749

Liu, Jieyu 931
Liu, Jingquan 1088, 1165
Liu, Juan 140
Liu, Jun 1611
Liu, Kangfu 1517, 1545
Liu, Ke 197, 1825
Liu, Lianqing 832, 1923
Liu, Lianqin 849
Liu, Lingling 236
Liu, Ling 1676, 1695
Liu, Lujiang 1897
Liu, Mei 919
Liu, Mengxia 1102
Liu, Mingwei 437
Liu, Mingxin 737
Liu, Ming 47
Liu, Min 498, 1106
Liu, Ruimin 491
Liu, Ruiwen 21
Liu, Shuangshuang 941, 1714
Liu, Wenjuan 656
Liu, Wenli 455
Liu, Wenxin 144
Liu, Xiangming 1611
Liu, Xingqi 30
Liu, Xin 1482
Liu, Yang 180, 459
Liu, Yan 931
Liu, Yaoping 898
Liu, Yifang 140, 491, 840, 1065, 1136
Liu, Yingming 1042
Liu, Yi 242
Liu, Yue 498
Liu, Yufei 386, 704
Liu, Yushuai 1545
Liu, Zewen 68, 174, 695
Liu, Ze 372
Liu, Zheng 949
Liu, Zhijian 378
Liu, Zichen 299
Liu, Ziqi 503, 509
Lou, Wen-Zhong 935
Lu, Chennan 469
Lu, Kuo 1817
Lu, Peimin 546, 889
Lu, Qifeng 1279
Lu, Qing 1332
Lu, Yulan 193, 1465, 1691, 1725
Luo, Anxin 184, 990
Luo, Guoxi 282, 513, 1362, 1908
Luo, Jian 872, 1415
Luo, Ningning 563, 571, 864

Luo, Rui	1535
Luo, Tao	1493
Luo, Yunyun	1362, 1908
Luo, Zebang	647
Luo, Zhifang	1255, 1523
Lv, Wenlong	1897
Lv, Yao	1611
Lv, Yuanjie	334
Ma, Binghe	872, 1415
Ma, Laihao	1250
Ma, Long	563
Ma, Shenhui	665
Ma, Shuang	832
Ma, Xin	1070
Ma, Yinji	731
Ma, Yintao	282
Ma, Yuanming	55, 1574
Ma, Zhen	1337
Ma, Zhipeng	691
Mao, Changhui	1142
Mao, Linna	1617
Mao, Tianjiao	1283
Mao, Xiyu	26, 762
Mao, Zhuofan	749
Matsunaga, Tadao	251
Mbarek, S. Ben	425
Mei, Deqing	1384, 1871
Mei, Ziqi	1535
Meng, Qinggang	1465
Meng, Qingwang	563, 571, 864
Meng, Yanfang	731
Meng, Yan	941, 1714
Meng, Zhen	197, 1825
Miao, Jiahao	242
Miao, Liming	148
Miao, Tongqiao	1410
Min, Huang	1057
Ming, Anjie	1142, 1436
Mizutani, Fumikazu	251
Mo, Deyun	319
Mori, Akinori	251
Mtui, Anaeli Elibariki	1676
Mu, Xiaojing	158
Murakami, Sunao	1320
Ni, Lvchao	1863
Nie, Dezhi	528
Nie, Weirong	699, 1287, 1358
Niklaus, Frank	1169
Niu, Bo	443
Niu, Gaoqiang	820
Niu, Zihao	1714
Ochoa, Manuel	64

Ono, Takahito	43, 670
Ouyang, Su	1718
Pan, Hongzhi	158
Pan, Mengyun	1923
Pan, Taisong	1617
Pan, Wenhao	762
Pan, Xinxiang	378
Pang, Jintao	463
Paquet, Louis	1423
Peng, Chun-Rui	1259
Peng, Chunrong	1611, 1893
Peng, Chunrui	1353
Peng, Hao	140
Peng, Jiabin	1398
Peng, Shi	826
Peng, Simin	1611
Peng, Zhengchun	1023, 1096, 1398
Pu, Jiangtao	47
Qi, Qi	1142
Qi, Wenjie	1813
Qi, Y	1660
Qi, Yonghong	1157
Qi, Yun	358
Qian, Jin	882
Qian, Lei	407
Qin, Guangyu	228
Qin, Lei	392
Qin, Lifeng	1718
Qiu, Weixiang	699
Qiu, Yong	1482
Qu, Guanghao	749
Qu, Mengjiao	1370, 1419
Ramegowda, Prakasha C.	1320
Rao, Jinjun	919
Rao, Wei	469
Rao, Zehong	1264
Rasadujjaman, Md.	417
Rashmikant	1320
Ren, Dahai	1535
Ren, Hangxu	26
Ren, Hongling	60
Ren, Ren	1611
Ren, Tian-Ling	680
Ren, Wan-Chun	1259
Ren, Wanchun	1353
Ren, Wei	47, 215
Ren, Wenjie	1042
Riaud, Antoine	809
Rong, Mingzhe	443
Ryutaro, M.	1660
Ryutaro, Maeda	513
San, Haisheng	232, 345, 399

Sang, Hongbo 1065
Sang, Jun-Jun 998
Sarro, Pasqualina M. 174
Shang, Zhengguo 386
Shao, Fangqin 1035
Shao, Lei 1297
Shao, Shuai 1255, 1523
Shao, Zungui 140, 1035, 1328
She, Xu 1813
Shen, Wenjiang 265, 358, 407
Shen, Yajing 293
Shen, Yigang 1126
Shi, Haotian 780, 1250
Shi, Huiyao 849
Shi, Jialin 849
Shimomura, Michio 251
Shu, Dongsheng 1175
Song, Ki-Young 1332
Song, Liguo 1853
Song, Yongxin 378
Song, Yu 392
Su, Weilin 919
Su, Wenqu 704
Su, Yi 647
Suetsugu, Ryotaro 1320
Sun, Changhe 704
Sun, Chengliang 656, 931
Sun, Daoheng 236, 1650
Sun, Fuqin 1279
Sun, Hongchun 1863
Sun, Jiamian 1745
Sun, Jiangkun 1579
Sun, Lining 60, 1070
Sun, Ming-Zhu 1018
Sun, Peiting 1539
Sun, Shengnan 904
Sun, Tiancheng 509
Sun, Yi 935
Sun, Yuyang 60
Tan, Haoyu 770
Tan, Yong 339
Tanaka, Yo 203, 1126
Tang, Fei 1291
Tang, Tianyi 1070
Tang, Xiaohe 1088
Tao, Jifang 1175, 1601
Tao, Kai 303, 528, 1903
Tao, Zhi 1745
Tavakkoli, Hadi 813
Tian, Bian 1635
Tian, He 680
Tian, Yuan 242

Tian, Yun 539
Tian, Zhao-Wu 1337
Tian, Zhong-Qun 1337
Tian, Zhou 980
Toda, Masaya 43
Tong, Daqiao 158
Tu, Er-Qi 433, 1432, 1605
Tu, Wenchang 236
Van Der Wijngaart, Wouter 1310
Wan, Hao 1482
Wan, Ji 148, 1686
Wan, Shijie 1023
Wang, Baolong 941, 1714
Wang, Bingnan 1664
Wang, Chengxiang 1888
Wang, Chenying 47
Wang, Dawei 469
Wang, Di 144
Wang, Dong F. 670, 1074
Wang, Donghui 1745
Wang, Fayang 158
Wang, Fei 184, 820, 990, 1042, 1410
Wang, Fengxia 60
Wang, Haobin 148
Wang, Hao 660, 980, 1629, 1853
Wang, Hongyan 1908
Wang, Jiaqiang 909
Wang, Jiong 1358
Wang, Jiyang 392
Wang, Junbo 193, 1465, 1691, 1725, 1813
Wang, Junyi 1441
Wang, K F 311
Wang, Kexin 215
Wang, L. C. 959
Wang, Lei 647, 656
Wang, Lingyun 840, 1065, 1897
Wang, Lukang 1478
Wang, Luming 563, 571
Wang, Lu 513
Wang, Meiqi 1714
Wang, Mengjie 498
Wang, Nan 524
Wang, Na 232, 345, 399
Wang, Ping 1482
Wang, Qilu 652
Wang, Qining 1839
Wang, Qi 55
Wang, Renxin 770
Wang, Ri 1303
Wang, Rong 670
Wang, Rui 1664
Wang, Shihang 1871

Wang, Siyuan..1853
Wang, Tianlu...832
Wang, Weidong................................498, 1106
Wang, Wei..................................809, 898, 949
Wang, Wenlong..1264
Wang, Wenxue...832
Wang, Xiang....................1035, 1328, 1554, 1700
Wang, Xiaohong...469
Wang, Xiao...498, 1106
Wang, Xinhe...1084
Wang, Xinjie...1358
Wang, Xinye...1664
Wang, Xinyu..1853
Wang, Xin...1664
Wang, Yancheng........................1384, 1871
Wang, Yanfang...1451
Wang, Yang..1337
Wang, Yanrong...417
Wang, Yan..1601
Wang, Yaohong..904
Wang, Yiwei...528
Wang, Yongkang..539
Wang, Yuechao...832
Wang, Yuelin.......................................30, 51, 676
Wang, Yuxi...1517
Wang, Zaichen...909
Wang, Zengbo..715
Wang, Zenghui...1410
Wang, Zhen..............................232, 345, 399
Wang, Zhihai...904
Wang, Zhiming...919
Wang, Zhongbao...804
Wang, Zhongyan..1839
Wang, Zihao...1279
Wang, Zilong..1611
Wang, Ziya...1096
Wang, Z...311
Wei, C. Yu..412
Wei, Feng..1436
Wei, Jiasi...255
Wei, Shuhua...417
Wei, X...1623, 1660
Wei, Xueyong..1157
Wei, Yaoming...303
Wen, Jian...443
Wen, Kaiqiang..994
Wen, Li...986
Wen, Ting...1410
Wen, Xiaolong...................................474, 1893
Wen, Zhiwei..931
Wu, Cheng-Feng...1804
Wu, Chen...334

Wu, Dezhi..........................804, 1650, 1655
Wu, Dong...437
Wu, Hanxiao..1745
Wu, Jin..303, 528, 1903
Wu, Junjie...255
Wu, Kai...1817
Wu, Kuixi..1897
Wu, Sen...378
Wu, Shuo...994
Wu, Song..1410
Wu, Tao...........................1255, 1517, 1523, 1545
Wu, Tianze..378
Wu, Tianzhun.....................................953, 980, 1629
Wu, Wengang...546, 889
Wu, Xuezhong............1410, 1579, 1817, 1888
Wu, Yaming...704
Wu, Yifei..443
Wu, Yigen...804
Wu, Yi...443
Wu, Yongshun...1908
Wu, Yulie...1888
Wu, Zhengwei...1611
Wu, Zhigang.....................................1390, 1394
Wu, Zhipeng...656
Wu, Zixuan..303, 1903
Xi, Xiang...1579
Xi, Y...959
Xi, Zhanwen......................699, 1287, 1358
Xia, Cao...670, 1074
Xia, Hu...1718
Xia, Lingju..986
Xia, Shanhong...1303
Xiang, Chao...193
Xiang, Zehua...148
Xiao, Chiqian...1493
Xiao, Dingbang 1410, 1579, 1817, 1888
Xiao, Haifeng...563
Xiao, Shijin..1601
Xiao, Shuyu..392
Xiao, Xiu...........................660, 1676, 1695
Xie, Bo...1465, 1725
Xie, Chao...665
Xie, Dan..680
Xie, Guangming...1853
Xie, Jianbing.....................................528, 1903
Xie, Jin.............................463, 1370, 1419
Xie, Yanbo..809
Xie, Yong..1303
Xie, Yucai...780, 1250
Xie, Yu...1655
Xie, Zihao...1419
Xin, Yi-Hang....................433, 1432, 1605

Xing, Chenyang	1023
Xing, Fei	382
Xing, Xiaoxing	228
Xing, Yan	845, 882
Xiong, Chenyu	1303
Xiong, Heng	840
Xiong, Menghan	571, 864
Xiong, Xy	311
Xu, Bo	1410
Xu, Chao	1813
Xu, Chen	148
Xu, Dongxin	1374
Xu, Gaobin	55, 1574
Xu, Guangyuan	731
Xu, Hantao	1337
Xu, Jia	242
Xu, L.	1623
Xu, Minyi	660, 1539, 1676, 1695, 1853
Xu, Na	699
Xu, Peng	1853
Xu, Shiyi	26, 762
Xu, Tiantong	1745
Xu, Wei	941, 1502, 1714
Xu, Xuankai	1502
Xu, Yaohua	1436
Xu, Yongfu	1700
Xu, Yuhao	1303
Xu, Zhiwei	1250
Xue, Gaopeng	43
Yalikun, Yaxiaer	203, 1126
Yan, Bing	715
Yan, Huangping	1732
Yan, Jiang	417
Yan, Xin	491
Yan, Yuchao	872
Yan, Yuefei	334
Yang, B.	959
Yang, Cheng	503
Yang, Dengfei	1370
Yang, Fuhua	455
Yang, Hui	711, 715
Yang, Jiayi	941, 1714
Yang, Jing	963
Yang, Jinling	455
Yang, Junfeng	986
Yang, Pengfei	1893
Yang, Ping	299, 513, 1908
Yang, Q.	1623
Yang, Quan	919
Yang, Suhui	1664
Yang, W H	311
Yang, Xiaokang	386
Yang, Xuanlin	30
Yang, Yinhuan	319
Yang, Yi	676, 680
Yang, Yongliang	1923
Yang, Yuanyuan	293, 358
Yang, Yuqian	1283
Yang, Yu	1478
Yang, Zhan	1560
Yang, Zhaohui	1574
Yao, Kun	1635
Yao, Yuan	271
Ye, Li	174
Ye, Xiongying	1291
Ye, Xuesong	26, 762
Ye, Yuxin	21
Ye, Zhi	180, 459
Yin, Zhuhui	1398
Yoshida, Yoshikazu	251
You, Minmin	1088, 1165
You, Shanshan	236
You, Zheng	1535
Younis, M. I.	425
Yu, Duli	228
Yu, Hang	1650
Yu, Hongyong	1676, 1695
Yu, Hongyu	1589
Yu, Huijun	265, 407
Yu, Jiahao	1903
Yu, Jie	1725
Yu, Lihang	21
Yu, Lingke	236
Yu, Miao	1110
Yu, Mingzhi	282
Yu, Peng	849
Yu, Sheng	1579
Yu, Siqi	1589
Yu, Xiaomei	242
Yu, Yue	1283
Yu, Zhoubin	459
Yu, Zitong	715
Yuan, Jiawei	299
Yuan, Jing	1804
Yuan, Quan	455
Yuan, Weizheng	872, 1415
Yuan, Ziyi	378
Zeng, Haozhe	1903
Zhan, Dongping	1337
Zhang, B. Wen	412
Zhang, Bing-Bin	1804
Zhang, Bo	474, 1893
Zhang, Cheng	1441
Zhang, Dapeng	437

Zhang, Guannan 652
Zhang, Guoqi 174
Zhang, H. 1660
Zhang, Hainan 680
Zhang, Haixia 148, 1686
Zhang, Hanyuan 652
Zhang, Hao 1110
Zhang, Hongpeng 30, 780, 1250
Zhang, Hongze 1110
Zhang, Huimin 687, 1867
Zhang, Jian 158, 528
Zhang, Jiaona 665
Zhang, Jing 417
Zhang, Jinming 417
Zhang, Jinying 1664
Zhang, Kenan 546, 889
Zhang, Li 382
Zhang, Lue 1560
Zhang, Min 665
Zhang, Nan 21
Zhang, Pengcheng 711, 715
Zhang, Peng 1539
Zhang, Qiankun 1362, 1908
Zhang, Qian 463, 1419
Zhang, Qiqi 1676, 1695
Zhang, Ran 372
Zhang, Shanshan 26
Zhang, Shaoxun 1142
Zhang, Shuo 1394
Zhang, Tengfei 691
Zhang, Ting 1279
Zhang, Weiguan 1023
Zhang, Wei 518, 687, 915, 1611, 1867
Zhang, Wen-Ming 433, 1297, 1432, 1605
Zhang, Wendong 770
Zhang, X. 1660
Zhang, Xiaopeng 1157
Zhang, Xiaosheng 737
Zhang, Xiao 1436
Zhang, Xingcheng 197
Zhang, Yang 68
Zhang, Yating 1867
Zhang, Yaxin 47
Zhang, Yijun 47
Zhang, Yongmeng 1579, 1817, 1888
Zhang, Yuewen 1539
Zhang, Yulong 695, 1065
Zhang, Yuwei 780
Zhang, Yu 1415
Zhang, Zhaohao 68
Zhang, Zhimin 563, 571, 864
Zhang, Zhitong 1839

Zhang, Zhouwei 1611
Zhao, Changhui 820, 1042
Zhao, Cong 749
Zhao, Hongfa 1676, 1695
Zhao, Junhao 1539
Zhao, Libo 282, 299, 513, 1157, 1362, 1635, 1908
Zhao, Luo 1441
Zhao, Minghui 1157
Zhao, Na 1635
Zhao, Qiancheng 1102, 1461
Zhao, Weisheng 1084
Zhao, Wenyu 1096
Zhao, Xiaodong 915
Zhao, Xiaojin 1502
Zhao, Xin 1018
Zhao, Xu 813
Zhao, Yifan 524
Zhao, Yihe 299
Zhao, Yongmin 1142
Zhao, You 1478
Zhao, Yue-Cen 935
Zhao, Yulong 215, 1405, 1478
Zhao, Yunpeng 660
Zhao, Yupeng 1142
Zhao, Zhe 528
Zhaojun, Liu 826
Zheng, Fengjie 1611
Zheng, Gaofeng 140, 491, 1035, 1136, 1328, 1554, 1700
Zheng, Guowen 1074
Zheng, Jianyi 1136
Zheng, Jiufu 1502
Zheng, Renrong 232, 345, 399
Zheng, Xudong 691
Zhong, Anxiang 1813
Zhong, Hongsheng 737
Zhongkai, Zhang 826
Zhou, Dekai 798
Zhou, Lingfei 509
Zhou, Peng 265
Zhou, Rui 1732, 1737
Zhou, Wei 1493, 1655
Zhou, Wenbo 949
Zhou, Wenke 1362
Zhou, Wenli 994
Zhou, Xin 1817
Zhou, Zhenghui 749
Zhou, Zilong 1264
Zhu, Jiankai 1410
Zhu, Jian 1110
Zhu, Jiaqi 1390
Zhu, Jiaxin 953
Zhu, Jia 546

Zhu, Kaiyun 1745
Zhu, Ke 1370
Zhu, Liangquan 1635
Zhu, Wei 656
Zhu, Xin 180, 459
Zhu, Yinfang 455
Zhu, Yingmin 1106
Zhu, Yi 1732
Zhu, Zhengfang 647
Zhu, Zhihao 1096
Zhuang, Liujing 1482
Zhuang, Rencheng 798
Zhuang, Yi 990
Zhuangde, Jiang 826
Zhuo, Ming 1817
Zou, X D 311
Zou, Yongjiu 1539
Zuo, Xiuli 60
Zuo, Xusheng 1539

IEEE
445 Hoes Lane
Piscataway, NJ 08854-4141

ISBN 978-1-6654-3008-1